Problems and Solutions of
Mathematics
For
Engineers
With Essential Theory

Problems and Solutions of
Mathematics
for
Engineers

With Essential Theory
(Includes Over 1050 Problems with Solutions)

1st & 2nd Semesters of B.E., B.Tech., B.Arch., B.Sc.(I.T.)
and other examinations of various universities
and engineering colleges of India and abroad

I.J.S. Sarna
B.E. Tech. (Mech. Engg.)
Delhi College of Engineering
Delhi University, Delhi

C B S

CBS PUBLISHERS & DISTRIBUTORS PVT. LTD.
New Delhi • Bengaluru • Pune • Kochi • Chennai

ISBN : 978-81-239-1968-3

First Edition : 2011

Published by Satish Kumar Jain and produced by V.K. Jain for
CBS Publishers & Distributors Pvt. Ltd.,
CBS Plaza, 4819/XI Prahlad Street, 24 Ansari Road, Daryaganj,
New Delhi - 110002, India. • Website: www.cbspd.com
e-mail: delhi@cbspd.com, cbspubs@vsnl.com, cbspubs@airtelmail.in
Ph.: 23289259, 23266861, 23266867 • Fax: 011-23243014

Branches:
• *Bengaluru:* Seema House, 2975, 17th Cross, K.R. Road,
  Bansankari 2nd Stage, Bangalore - 560070 Ph.: 26771678/79
  Fax: 080-26771680   • e-mail: bangalore@cbspd.com
• *Pune:* Bhuruk Prestige, Sr. No. 52/12/2+1+3/2,
  Narhe, Haveli (Near Katraj-Dehu Road by Pass), Pune-411051
  Ph.: +91-20-32404169 • Fax: 020-24464059
  e-mail: pune@cbspd.com
• *Kochi:* 36/14, Kalluvilakam, Lissie Hospital Road,
  Cochin - 682018, Kerala • e-mail: cochin@cbspd.com
  Ph.: 0484-4059061-65 • Fax: 0484-4059065
• *Chennai:* 20, West Park Road, Shenoy Nagar, Chennai - 600030
  e-mail: chennai@cbspd.com Ph.: 044-26260666-26202620
  Fax: 044-45530020

Printed at :
J.S. Offset Printers, Delhi

Dedicated
to the memory of

My Respected Parents
Late Saran Kaur and
Late Rawel Singh Sarna
Who are responsible for what I am today

and

My Esteemed Professor
Late G.S. Sharma
**Who taught Mathematics to Engineering Students at
Delhi College of Engineering (Delhi University), Delhi
for over four decades**

Preface

I have the pleasure in presenting to my readers this comprehensive book incorporating therein a large number of typical solved problems and illustrations. It also contains number of supplementary problems with answers which are specifically designed for the aspirants intending to appear in Engineering, Technology and Applied Sciences examinations of various Universities and Engineering Colleges.

While writing this book I have made constant endeavour to give large number of typical solved problems with essential theory part as needed. A remarkable thing about this book is that it has solved and unsolved problems collected from the past years examination papers of various universities and engineering colleges. So far as unsolved problems are concerned vital hints have been provided wherever necessary so as to enable the readers to arrive at correct answer.

Deliberate elementary approach to subject matter with many examples, solutions and diagrams should appeal to my readers and this is particularly suited as effective self-study guide with minimum guidance from others.

Most sincere gratitude to various esteemed authors whose books I have consulted during accomplishing the task of writing this book.

I am grateful to my publishers for their hard work and enthusiasm displayed in making this book in a short span of time.

I, however, always welcome suggestions and criticisms, if any, from my esteemed readers for further improvement of this book.

New Delhi   **Author**

Contents

– Divergence and Curl of a vector Field – Physi cal Interpretation of Divergence – Physical Interpretation of C url – Vector Identities – Illustrative Examples – Line Integrals – Applications of Line Integrals (work done by a forc e, Circulation) – Line Integrals Independent of the (*i*) path i in plane, (*ii*) path in space – Green's Theorem in the plane – Illustrative Examples – Surface Integral – Illustrative Example s – Divergence Theorem of Gauss – Illustrative Examples – Stokes's Theorem – Illustrative Examples.

Introduction – Definition – Some types of Matrices – Matrix Algebra – Operations on Matrices – Transpose of a Matrix – Symmetric and Skew-symmetric Matrices – Determinants – Product of two determinants – Rank of a Matrix – Adjoint of a square Matrix – Singular and Non-singular Matrices – Inverse of a square matrix – Systems of $n \times n$ linear equations and their solution – Vector spaces – Subspace – Linear independence of vectors – Dimension and Basis – Linear Transformations – Elementary Transformations (or operations) on a matrix – Echelon form of a Matrix – Normal form of a Matrix – Solution of General linear system of equations – Gauss Elimination method – Gauss Jordon Method – Homogeneous system $m \times n$ linear equations – Gauss-Jordon method to find the inverse of a Matrix – Inverse transformation.

Eigenvalues and Eigen vectors – Cayley – Hamilton Theorem – Similar Matrices – Diagonalizable Matrices – Inner Product (dot product) of vectors – Norm (Length) of a vector – unit vector – Orthogonal Vectors – Orthogonal Matrix.

Introduction – Formation of differential equations – Solution of a differential equation – Analytical solution of a differential equation of the first order and first degree – Variable Separable form – Equations reducible to separable form – Homogeneous equations – Equations of the form: $dy/dx = (a_1x + b_1y + c_1)/(a_2x + b_2y + c_2)$ – Linear First order differential equations – Equations reducible to Linear form – Exact differential equations – Integrating factor found by Inspection – Rules for finding

Integrating factors – Differential equations of the first order and higher degree – Equations solvable for p-Equations solvable for y-Equations solvable for x – Clairaut's equations – Equations reducible to Clairaut's form.

– Unit step function or Heaviside Function – Second shifting theorem or Translation on the t-axis – Impulse function – Important Deductions.

Chapter 1

Successive Differentiation

1.1 Definition and Notation

If $y = f(x)$ be a differentiable function of x, then dy/dx is called the *first differential coefficient* of y w.r.t. x or *first derivative*. If this derivative is again a differentiable function, then its derivative that is, $\dfrac{d}{dx}\left(\dfrac{dy}{dx}\right)$ is called the *second differential coefficient* of y w.r.t. x or *second derivative* and is denoted by d^2y/dx^2. If this also admits the differentiation, then its derivative is called the *third differential coefficient* or *third derivative* and is denoted by d^3y/dx^3. In general, the nth differential coefficient of y w.r.t. x is denoted by d^ny/dx^n and is given by

$$\frac{d^ny}{dx^x} = \frac{d}{dx}\left(\frac{d^{n-1}y}{dx^{n-1}}\right).$$

If $y = f(x)$, then the successive differential coefficients are denoted by $y_1, y_2, ..., y_n, ..$ or by $y', y'', y''', ..., y^n, ...$ The nth differential coefficient of y w.r.t. x is denoted by any of the following ways:

$$\frac{d^ny}{dx^n}, D^ny, y_n, f^{(n)}(x), D^nf(x), \frac{d^n}{dx^n}f(x), \text{ etc.}$$

The second differential coefficient is of special importance to engineering students owing to its constant recurrence in dynamical problems. For example linear acceleration is denoted by $\dfrac{d^2s}{dt^2}$ and angular acceleration by $\dfrac{d^2\theta}{dt^2}$, the letters s, θ, and t having their usual meanings.

1.2 Illustrative Examples

Example 1.1 (a) Find $\dfrac{d^2y}{dx^2}$ for the cycloid $x = a(\theta - \sin\theta), y = a(1 - \cos\theta)$.

(b) If $x = a \cos t$, $y = a \sin t$. Show that $\dfrac{d^3y}{dx^3} = -\dfrac{3\cos t}{a^2 \sin^5 t}$.

(c) If $y = 2 \cos t - \cos 2t$, $x = 2 \sin t - \sin 2t$, find the value of $\dfrac{d^2y}{dx^2}$ at $t = \dfrac{\pi}{2}$.

(d) Find $\dfrac{d^2y}{dx^2}$ when $x = a\left(\cot t + \log \tan \dfrac{t}{2}\right)$, $y = a \sin t$.

Solution. (*a*) We have $\dfrac{dx}{d\theta} = a(1 - \cos\theta)$ and $\dfrac{dy}{d\theta} = a \sin\theta$.

$\therefore \dfrac{dy}{dx} = \dfrac{dy/d\theta}{dx/d\theta} = \dfrac{a \sin\theta}{a(1 - \cos\theta)} = \dfrac{\sin\theta}{1 - \cos\theta}$. Differentiating again w.r.t. x, we obtain

$$\dfrac{d^2y}{dx^2} = \dfrac{\cos\theta(1 - \cos\theta) - \sin^2\theta}{(1 - \cos\theta)^2} \cdot \dfrac{d\theta}{dx} = \dfrac{\cos\theta - 1}{(1 - \cos\theta)^2} \cdot \dfrac{1}{a(1 - \cos\theta)}$$

$$= -\dfrac{1}{a(1 - \cos\theta)^2} = -\dfrac{1}{4a\sin^4(\theta/2)}.$$

(*b*) We have $\dfrac{dx}{dt} = -a \sin t$ and $\dfrac{dy}{dt} = a \cos t$. Then,

$\dfrac{dy}{dx} = \dfrac{dy/dt}{dx/dt} = -\cot t$. Differentiating again w.r.t. x.

$$\dfrac{d^2y}{dx^2} = \left(\operatorname{cosec}^2 t\right) \cdot \dfrac{dt}{dx} = -\dfrac{1}{a} \operatorname{cosec}^3 t. \qquad \qquad ...(i)$$

Differentiating (i) w.r. to x, we obtain

$$\dfrac{d^3y}{dx^3} = \dfrac{3}{a} \operatorname{cosec}^2 t \cdot \operatorname{cosec} t \cot t \cdot \dfrac{dt}{dx}$$

$$= \dfrac{3}{a} \dfrac{\cos t}{\sin^4 t} \cdot \left(\dfrac{1}{-a \sin t}\right) = -\dfrac{3 \cos t}{a^2 \sin^5 t}, \text{ which is same as desired.}$$

(*c*) We have

$$\frac{dy}{dt} = 2\left(\sin 2t - \sin t\right) \text{ and } \frac{dx}{dt} = 2\left(\cos t - \cos 2t\right).$$

$$\therefore \quad \frac{dy}{dx} = \frac{dy/dt}{dx/dt} = \frac{\sin 2t - \sin t}{\cos t - \cos 2t} = \frac{2\cos(3t/2)\sin(t/2)}{2\sin(3t/2)\sin(t/2)} = \cot\frac{3t}{2}.$$

Differentiating this w.r. to *x*, we obtain

$$\frac{d^2y}{dx^2} = -\left(\frac{3}{2}\right)\mathrm{cosec}^2\left(\frac{3t}{2}\right)\cdot\frac{dt}{dx} = -\frac{3}{2}\,\mathrm{cosec}^2\frac{3t}{2}\cdot\frac{1}{2(\cos t - \cos 2t)}$$

$$= -\frac{3}{2}\cdot\frac{1}{\sin^2 3t/2}\cdot\frac{1}{4\sin(3t/2)\sin(t/2)} = -\frac{3}{8}\cdot\frac{1}{(\sin 3t/2)^3}\cdot\frac{1}{\sin(t/2)}.$$

$$\therefore \quad \left(\frac{d^2y}{dx^2}\right)_{t=\pi/2} = -\frac{3}{8}\cdot\frac{1}{(\sin \pi/4)^4} = -\frac{3}{2}.$$

(*d*) We have

$$\frac{dx}{dt} = a\left[-\sin t + \frac{1}{2\tan t/2}\cdot\sec^2 t/2\right] = a\left[-\sin t + \frac{\cos t/2}{2\sin t/2\,\cos^2 t/2}\right]$$

$$= a\left(-\sin t + \frac{1}{\sin t}\right) = a\frac{\cos^2 t}{\sin t}, \text{ and } \frac{dy}{dt} = a\cos t.$$

$$\therefore \quad \frac{dy}{dx} = \frac{dy/dt}{dx/dt} = \frac{a\cos t}{(a\cos^2 t)/\sin t} = \tan t. \qquad \qquad ...(i)$$

Differentiating (*i*) w.r.t. *x*, we obtain

$$\frac{d^2y}{dx^2} = \sec^2 t\cdot\frac{dt}{dx} = \sec^2 t\cdot\frac{1}{(a\cos^2 t)/\sin t} = \frac{\tan t\sec^3 t}{a}.$$

Example 1.2 (a) Given $x = \left(\frac{t^2}{2}\right) + t,\ y = \left(\frac{t^2}{2}\right) - t$; find $\dfrac{d^2y}{dx^2}$ and $\dfrac{d^2x}{dy^2}$.

(b) Find $f'''\left(\dfrac{\pi}{3}\right)$, **given** $f(x) = \sin x \cos 3x$.

(c) Find d^2y/dx^2, **given** $y = e^{-x}\log x$. **(d) If** $y = e^{-2x}\sin 3x$, **find** y''.

Solution. (a) We have $\dfrac{dx}{dt} = t+1$, $\dfrac{dy}{dt} = t-1$. Then

$$\frac{dy}{dx} = \frac{dy/dt}{dx/dt} = \frac{t-1}{t+1}.$$ Differentiating w.r.t. x, we obtain

$$\frac{d^2y}{dx^2} = \frac{(t+1)-(t-1)}{(t+1)^2} \cdot \frac{dt}{dx} = \frac{2}{(t+1)^2} \cdot \frac{1}{t+1} = \frac{2}{(t+1)^3}.$$

Also, $\dfrac{dx}{dy} = \dfrac{dx/dt}{dy/dt} = \dfrac{t+1}{t-1}$. Differentiating this w.r.t. y, we obtain

$$\frac{d^2x}{dy^2} = \left\{ \frac{(t-1)-(t+1)}{(t-1)^2} \right\} \cdot \frac{dt}{dy} = -\frac{2}{(t-1)^2} \cdot \frac{1}{t-1} = -\frac{2}{(t-1)^3}.$$

(b) We have $f(x) = \sin x \cos 3x = (\sin 4x - \sin 2x)/2$

$$\therefore f'(x) = [4\cos 4x - 2\cos 2x]/2, \quad f''(x) = -[4^2 \sin 4x - 2^2 \sin 2x]/2,$$

$$f'''(x) = -[4^3 \cos 4x - 2^3 \cos 2x]/2$$

Hence, $\qquad f'''\left(\dfrac{\pi}{3}\right) = -\dfrac{1}{2}\left[64\left(-\dfrac{1}{2}\right) - 8\left(-\dfrac{1}{2}\right)\right] = \dfrac{1}{2}(32-4) = 14.$

(c) We have $y = e^{-x} \log x$.

$$\therefore y' = e^{-x}\frac{d}{dx}(\log x) + \log x \frac{d}{dx}(e^{-x}) = e^{-x}\cdot\frac{1}{x} - e^{-x}\log x = \frac{e^{-x}}{x} - y.$$

$$y'' = \frac{x\dfrac{d}{dx}(e^{-x}) - e^{-x}\dfrac{d}{dx}(x)}{x^2} - y' = \frac{(-xe^{-x} - e^{-x})}{x^2} - \frac{e^{-x}}{x} + e^{-x}\log x$$

$$= -e^{-x}\left(\frac{2}{x} + \frac{1}{x^2} - \log x\right), \text{ the required result.}$$

(d) $y = e^{-2x} \sin 3x$. Therefore, $y' = e^{-2x}\dfrac{d}{dx}(\sin 3x) + \sin 3x \dfrac{d}{dx}(e^{-2x})$

$$= 3e^{-2x}\cos 3x - 2e^{-2x}\sin 3x = 3e^{-2x}\cos 3x - 2y,$$

and $y'' = 3e^{-2x}\dfrac{d}{dx}(\cos 3x) + 3\cos 3x \dfrac{d}{dx}(e^{-2x}) - 2y'$

$$= -9e^{-2x}\sin 3x - 6e^{-2x}\cos 3x - 2(3e^{-2x}\cos 3x - 2e^{-2x}\sin 3x)$$

$$= -e^{-2x}(5\sin 3x + 12\cos 3x).$$

Example 1.3 (a) If $\sqrt{y+x} + \sqrt{y-x} = c$**, show that**

$$\frac{dy}{dx} = \frac{y}{x} - \sqrt{\frac{y^2}{x^2} - 1}, \text{ and find the value of } \frac{d^2y}{dx^2}.$$

(b) If $y = a\cos(\log x) + b\sin(\log x)$**, show that** $x^2 y_2 + xy_1 + y = 0$.

Solution. (*a*) On squaring both sides, we obtain

$$y + x + y - x + 2\sqrt{y^2 - x^2} = c^2 \text{ or } 2\sqrt{y^2 - x^2} = c^2 - 2y.$$

Squaring again, $4y^2 - 4x^2 = c^4 - 4c^2 y + 4y^2$ or $4c^2 y = c^4 + 4x^2$.

Differentiating w.r.t. x, $4c^2\dfrac{dy}{dx} = 8x$ so that $\dfrac{dy}{dx} = \dfrac{2x}{c^2} = \dfrac{x}{y + \sqrt{y^2 - x^2}}$

$$= \frac{x}{y + \sqrt{y^2 - x^2}} \cdot \frac{y - \sqrt{y^2 - x^2}}{y - \sqrt{y^2 - x^2}} = \frac{x\left(y - \sqrt{y^2 - x^2}\right)}{x^2} = \frac{y}{x} - \sqrt{\frac{y^2}{x^2} - 1}.$$

We have $\dfrac{dy}{dx} = \dfrac{2x}{c^2}$. Differentiating it w.r.t. x, we obtain $\dfrac{d^2y}{dx^2} = \dfrac{2}{c^2}$.

(*b*) We have, $y = a\cos(\log x) + b\sin(\log x)$.   ...(i)

Therefore, $y_1 = -a(1/x)\sin(\log x) + b(1/x)\cos(\log x) \cdot$

or $x \cdot y_1 = -a\sin(\log x) + b\cos(\log x)$.

Differentiating again, we obtain

$$xy_2 + y_1 \cdot 1 = -a\cos(\log x) - b\sin(\log x) \cdot$$

$$= -\left[a\cos(\log x) + b\sin(\log x)\right]/x = -y/x, \text{ from (i)}.$$

or $x^2 y_2 + xy_1 = -y$ or $x^2 y_2 + xy_1 + y = 0$.

Example 1.4 (a) If $y = e^{ax}\cos bx$**, prove that** $\dfrac{d^2y}{dx^2} - 2a\dfrac{dy}{dx} + \left(a^2 + b^2\right)y = 0$.

(b) If $x = \dfrac{\left(1 - \sqrt{y}\right)}{\left(1 + \sqrt{y}\right)}$**, show that** $\dfrac{d^3y}{dx^3} = \dfrac{24(x-3)}{(x+1)^5}$.

(c) If $y^3 - 3ax^2 + x^3 = 0$, then show that $\dfrac{d^2y}{dx^2} + \dfrac{2a^2x^2}{y^5} = 0$. (D.C.E., 2004)

Solution. (a) We have $y = e^{ax} \cos bx$.

Therefore, $\dfrac{dy}{dx} = e^{ax} \dfrac{d}{dx}(\cos bx) + \cos bx \dfrac{d}{dx}\left(e^{ax}\right)$

$$= -be^{ax} \sin bx + ae^{ax} \cos bx = -be^{ax} \sin bx + ay. \qquad \ldots(i)$$

$$\dfrac{d^2y}{dx^2} = \dfrac{d}{dx}\left(\dfrac{dy}{dx}\right) = -be^{ax}\dfrac{d}{dx}(\sin bx) - b\sin bx \dfrac{d}{dx}\left(e^{ax}\right) + a\dfrac{dy}{dx}$$

$$= -b^2\left(e^{ax}\cos bx\right) + a\left(-be^{ax}\sin bx\right) + a\dfrac{dy}{dx}$$

$$= -b^2y + a\left[\left(\dfrac{dy}{dx}\right) - ay\right] + a\cdot\left(\dfrac{dy}{dx}\right), \text{ from } (i)$$

or $\dfrac{d^2y}{dx^2} - 2a\left(\dfrac{dy}{dx}\right) + \left(a^2 + b^2\right)y = 0.$

(b) We may write $x = \dfrac{1-\sqrt{y}}{1+\sqrt{y}}$ in an equivalent form as $\sqrt{y} = \dfrac{-x+1}{x+1}$.

Squaring, we obtain $y = \dfrac{(x-1)^2}{(x+1)^2}$. Differentiating, w.r.t. x, we obtain

$$\dfrac{dy}{dx} = \dfrac{(x+1)^2\cdot 2(x-1) - (x-1)^2\cdot 2(x+1)}{(x+1)^4} = \dfrac{4(x-1)}{(x+1)^3}.$$

Again, differentiating, w.r.t. x, we get

$$\dfrac{d^2y}{dx^2} = \dfrac{(x+1)^3\cdot 4 - 4(x-1)\cdot 3(x+1)^2}{(x+1)^6} = \dfrac{-8x+16}{(x+1)^4}.$$

Differentiating once more, we obtain

$$\dfrac{d^3y}{dx^3} = \dfrac{-8(x+1)^4 + 32(x-2)(x+1)^3}{(x+1)^8} = \dfrac{24(x-3)}{(x+1)^5}.$$

(c) Differentiating the given equation w.r. to x, we obtain

$$3y^2 \frac{dy}{dx} = 6ax - 3x^2, \text{ or } \frac{dy}{dx} = \frac{2ax - x^2}{y^2}. \qquad \text{...(i)}$$

Differentiating (i) w.r.t. x, we get

$$\frac{d^2y}{dx^2} = \frac{y^2(2a - 2x) - 2y\frac{dy}{dx}(2ax - x^2)}{y^4}$$

$$= \frac{y^2(2a - 2x) - \left\{2y(2ax - x^2)(2ax - x^2)/y^2\right\}}{y^4}$$

$$= \frac{2(a-x)y^3 - 2(2ax - x^2)^2}{y^5} = \frac{2(a-x)(3ax^2 - x^3) - 2(2ax - x^2)^2}{y^5} = -\frac{2a^2x^2}{y^5}.$$

Thus
$$\frac{d^2y}{dx^2} + \frac{2a^2x^2}{y^5} = 0.$$

Example 1.5 (a) Find A and B such that y = A sin 5x + B cos 5x, satisfies the

equation $\dfrac{d^2y}{dx^2} + \dfrac{1}{5}\dfrac{dy}{dx} + 15y = 101\sin 5x.$

(b) If $p^2 = a^2\cos^2\theta + b^2\sin^2\theta,$ **prove that** $\dfrac{d^2p}{d\theta^2} + p = \dfrac{a^2b^2}{p^3}.$

Solution. (a) We have $y = A\sin 5x + B\cos 5x.$

Differentiating w.r.t. x, we obtain $\dfrac{dy}{dx} = 5A\cos 5x - 5B\sin 5x.$

Again differentiating, we get $\dfrac{d^2y}{dx^2} = -25A\sin 5x - 25B\cos 5x.$

Substituting these values in differential equation $\dfrac{d^2y}{dx^2} + \dfrac{1}{5}\cdot\dfrac{dy}{dx} + 15y = 101\sin 5x$

and collecting like terms, we get $\sin 5x(-10A - B) + \cos 5x(A - 10B) = 101\sin 5x.$

Now comparing the coefficients of $\sin 5x$ and $\cos 5x$, we obtain

$-10A - B = 101$ and $A - 10B = 0.$

Solving simultaneously these two equations, we get $A = -10,\ B = -1.$

(b) We have $p^2 = a^2 \cos^2\theta + b^2 \sin^2\theta$. ...(i)

Differentiating (i) w.r.t. θ, we obtain

$$2p \cdot \frac{dp}{d\theta} = -2a^2 \cos\theta\sin\theta + 2b^2 \sin\theta\cos\theta = \left(b^2 - a^2\right)\sin 2\theta.$$ ...(ii)

Again differentiating (ii), we obtain

$$2p \cdot \frac{d^2 p}{d\theta^2} + 2\frac{dp}{d\theta} \cdot \frac{dp}{d\theta} = 2\left(b^2 - a^2\right)\cos 2\theta.$$

Therefore, $p \cdot \left(d^2 p/d\theta^2\right) = \left(b^2 - a^2\right)\cos 2\theta - \left(dp/d\theta\right)^2$

$$= \left(b^2 - a^2\right)\cos 2\theta - \left\{\left(b^2 - a^2\right)^2 \sin^2 2\theta\right\}/4p^2 \text{ , from (ii).}$$

Adding p^2 to both sides, we have

$$p \cdot \frac{d^2 p}{d\theta^2} + p^2 = \left(b^2 - a^2\right)\cos 2\theta - \frac{\left(b^2 - a^2\right)^2 \sin^2 2\theta}{4p^2} + p^2$$

or $p\left(\dfrac{d^2 p}{d\theta^2} + p\right) = \dfrac{4p^2\left(b^2 - a^2\right)\cos 2\theta - \left(b^2 - a^2\right)^2 \sin^2 2\theta + 4p^4}{4p^2}$.

Substituting the value of p^2 in R.H.S., we obtain

$$p\left(\frac{d^2 p}{d\theta^2} + p\right) = \frac{\begin{array}{c}4\left(b^2 - a^2\right)\left(\cos^2\theta - \sin^2\theta\right)\left(a^2 \cos^2\theta + b^2 \sin^2\theta\right)\\ -4\left(b^2 - a^2\right)^2 \sin^2\theta\cos^2\theta + 4\left(a^2\cos^2\theta + b^2\sin^2\theta\right)^2\end{array}}{4p^2}.$$

Right hand side on simplification reduces to

$$\left(a^2 b^2 \cos^4\theta + 2a^2 b^2 \sin^2\theta\cos^2\theta + a^2 b^2 \sin^4\theta\right)/p^2$$

i.e., $a^2 b^2 \left(\cos^2\theta + \sin^2\theta\right)^2/p^2$

Thus $\left(d^2 p/d\theta^2\right) + p = a^2 b^2\left(\cos^2\theta + \sin^2\theta\right)^2/p^3 = a^2 b^2/p^3$.

Exercise 1.1

1. If $ax^2 + 2hxy + by^2 = 1$, prove that $\dfrac{d^2 y}{dx^2} = \dfrac{h^2 - ab}{\left(hx + by\right)^3}$.   (A.M.I.E., W-1996)

2. If $y = \dfrac{(1+x)^n}{(1-x)^n}$, show that $\dfrac{dy}{dx} = \dfrac{2ny}{1-x^2}$ and $\dfrac{d^2 y}{dx^2} = \dfrac{2(n+x)}{1-x^2} \cdot \dfrac{dy}{dx}$.

(Rewa, 1998)

3. If $y = x^3 \log(1/x)$, prove that $y_2 - (2/x)y_1 + 3x = 0$.

4. If $y = ax + b\sqrt{x}$, prove that $2x^2 y_2 - xy_1 + y = 0$.

5. If $x = a(\cos t + t \sin t)$, $y = a(\sin t - t \cos t)$, find d^2y/dx^2.

   Ans. $(\sec^3 t)/at$.   (*Pondicherry, 1998S*)

6. If $x = at^2$, $y = 2at$, prove that $4ax^3 y_2^2 + 2xy\ y_1^3 - 5a^2 = 0$. (*A.M.I.E., W-2004*)

7. If $x = f(t)$, $y = g(t)$, find d^2y/dx^2 **Ans.** $\dfrac{g''(t) \cdot f'(t) - g'(t) \cdot f''(t)}{\{f'(t)\}^3}$.

8. If $y = a \sin mx + b \cos mx$, prove that $y_2 + m^2 y = 0$.

9. If $y = e^{ax} \sin bx$, prove that $y_2 - 2ay_1 + (a^2 + b^2)y = 0$.

10. If $y = \log(\log x)$, prove that $xy'' + y' + xy'^2 = 0$.   (*A.M.I.E. W-2004*)

11. If $y = (\log x)/x$, show that $d^2y/dx^2 = (2\log x - 3)/x^3$.

12. If $y = (1/x)^x$, show that $y_2(1) = 0$.

13. Find (*i*) the fifth derivative of $1/(ax + b)$ **Ans.** $-120a^5(ax + b)^{-6}$
    (*ii*) the second derivative of $\log(3 - x)$.   **Ans.** $-(3 - x)^{-2}$

14. If $y = ae^{mx} + be^{nx}$, prove that $(d^2y/dx^2) - (m + n)(dy/dx) + mny = 0$.

15. If $y = (ax + b)/(cx + d)$, prove that $2y_1 y_3 = 3y_2^2$.

16. If $x = \sin t$, $y = \sin pt$, prove that $(1 - x^2)\dfrac{d^2y}{dx^2} - x\dfrac{dy}{dx} + p^2 y = 0$. (*Rewa, 1998*)

17. If $y = \left(x + \sqrt{x^2 + 1}\right)^n$, show that $(x^2 + 1)\ y_2 + xy_1 = n^2 y$.

18. If $y = \sin(\sin x)$, prove that $\dfrac{d^2y}{dx^2} + \tan x\ \dfrac{dy}{dx} + y\cos^2 x = 0$.

19. Find d^2y/dx^2, when $x = a\cos^3 \theta$, $y = b\sin^3 \theta$.

    Ans. $b(\operatorname{cosec} \theta \cdot \sec^4\theta)/3a^2$.

20. Show that $y = A \sin(pt + \alpha) + B \sin(2pt + \beta)$ satisfies the differential equation

    $$\left(\dfrac{d^4 y}{dt^4}\right) + 5p^2\left(\dfrac{d^2 y}{dt^2}\right) + 4p^4 y = 0,$$ for all values of the constants A, B,

    α and β.

21. Find the values of the constants A and B so that
$y = e^x(A \sin 2x + B \cos 2x)$ satisfies the differential equation
$\left(d^4y/dx^4\right) + 2y = e^x \sin 2x$.   **Ans.** A = −5/601, B = 24/601

22. Find A and B such that $y = A \sin 3x + B \cos 3x$ satisfies the differential equation $y_2 + 4y_1 + 3y = 10 \cos 3x$.   **Ans.** A = 2/3, B = −1/3.

23. If $x\sqrt{1-y^2} + y\sqrt{1-x^2} = a$, show that $d^2y/dx^2 = -a\big/\left(1-x^2\right)^{3/2}$.

[**Hint.** On putting $x = \sin\alpha, y = \sin\beta$, we get $\sin(\alpha+\beta) = a$ or

$\alpha + \beta = \sin^{-1} a$ or $\sin^{-1} x + \sin^{-1} y = \sin^{-1} a$. Differentiating

w.r.t. x, we get $dy/dx = -\sqrt{1-y^2}\big/\sqrt{1-x^2}$].

24. If $y = \left(\tan^{-1} x\right)^2$, prove that

$$\left(1+x^2\right)^2 \frac{d^2y}{dx^2} + 2x\left(1+x^2\right)\frac{dy}{dx} = 2.$$ (*AMIE, W-2008*)

25. If $x = \cosh(\log y/m)$, prove that $(x^2 - 1)y_2 + xy_1 = m^2 y$.   (*AMIE, S-2010*)

1.3 Calculation of the nth Derivative

Some Standard Results

(1) **To find the nth differential coefficient of $(ax+b)^m$, where $m > n$.**
Let $y = (ax + b)^m$.

∴ $y_1 = ma\,(ax + b)^{m-1}$, $y_2 = m\,(m-1)\,a^2(ax + b)^{m-2}$
$y_3 = m\,(m-1)(m-2)\,a^3\,(ax + b)^{m-3}$
....
....
$y_n = m\,(m-1)(m-2)\,...\,(m-n+1)\,a^n(ax + b)^{m-n}$,
which can be written as

$$y_n = \frac{m!}{(m-n)!}\,a^n\left(ax+b\right)^{m-n}.$$ ...(1.1)

Cor. 1. If $m = n$, then we shall have

$$y_n = \frac{d^n}{dx^n}(ax+b)^n = n!a^n$$ ...(1.2)

Cor. 2. If $m = -1$, we obtain $y_n = (-1)(-2)(-3)\cdots\cdots(-n)a^n\left(ax+b\right)^{-1-n}$

$$\frac{d^n}{dx^n}\left(\frac{1}{ax+b}\right) = \frac{(-1)^n\, n!a^n}{(ax+b)^{n+1}}.$$ ...(1.3)

(2) To find the nth differential coefficient of log $(ax+b)$.

Let $y = \log(ax+b)$. We have $y_1 = a/(ax+b)$.

Hence, $y_n = \dfrac{(-1)^{n-1}(n-1)!\,a^n}{(ax+b)^n}$.   ...(1.4)

(3) To find the nth differential coefficient of a^{mx}.

Let $y = a^{mx}$.

Therefore, $y_1 = ma^{mx}\log a, \quad y_2 = m^2 a^{mx}(\log a)^2$

$$y_3 = m^3 a^{mx}(\log a)^3$$

....

....

$y_n = m^n a^{mx}(\log a)^n$.   ...(1.5)

Cor. 3. $\dfrac{d^n}{dx^n}(e^{mx}) = m^n e^{mx}$.   ...(1.6)

(4) To find the nth differential coefficient of $\sin(ax+b)$ and $\cos(ax+b)$.

Let $y = \sin(ax+b)$.

Therefore, $y_1 = a\cos(ax+b) = a\sin\left[ax+b+(\pi/2)\right]$

$$y_2 = a^2\cos\left[ax+b+(\pi/2)\right] = a^2\sin\left[ax+b+(\pi/2)+(\pi/2)\right]$$

$$= a^2\sin\left[ax+b+2(\pi/2)\right]$$

$$y_3 = a^3\cos\left[ax+b+2(\pi/2)\right] = a^3\sin\left[ax+b+2(\pi/2)+(\pi/2)\right]$$

$$= a^3\sin\left[ax+b+3(\pi/2)\right]$$

....

....

$y_n = a^n\sin\left[ax+b+\dfrac{n\pi}{2}\right]$.   ...(1.7)

Similarly, $\dfrac{d^n}{dx^n}[\cos(ax+b)] = a^n\cos\left(ax+b+\dfrac{n\pi}{2}\right)$ ...(1.8)

(5) To find the nth differential coefficient of $e^{ax}\sin(bx+c)$.

Let $y = e^{ax}\sin(bx+c)$.

Therefore, $y_1 = ae^{ax}\sin(bx+c) + e^{ax}\cdot b\cos(bx+c)$

$$= e^{ax}[a\sin(bx+c) + b\cos(bx+c)].$$

Put $a = r\cos\theta$ and $b = r\sin\theta$, so that $r = \sqrt{a^2+b^2}$ and $\theta = \tan^{-1}(b/a)$.

Then, $y_1 = e^{ax}r[\sin(bx+c)\cdot\cos\theta + \cos(bx+c)\cdot\sin\theta] = re^{ax}\sin(bx+c+\theta)$.

Similarly, $y_2 = r^2 e^{ax}\sin(bx+c+2\theta)$.

$$y_3 = r^3 e^{ax}\sin(bx+c+3\theta).$$

... ...

... ...

$$y_n = r^n e^{ax}\sin(bx+c+n\theta).$$

Putting the values of r and θ, we get

$$y_n = (a^2+b^2)^{n/2}\,e^{ax}\sin\left(bx+c+n\tan^{-1}(b/a)\right). \qquad ...(1.9)$$

Similarly,

$$\frac{d^n}{dx^n}\left[e^{ax}\cos(bx+c)\right] = \left(a^2+b^2\right)^{n/2}e^{ax}\cos\left(bx+c+n\tan^{-1}(b/a)\right). \quad ...(1.10)$$

Remarks.

1. In order to find the nth derivative of any algebraic rational functions, we should resolve them into partial fractions.

2. When we cannot break up the denominator of a given algebraic fraction into real linear factors, we have to use **De Moivre's Theorem** after resolving the denominator into its linear factors, real or imaginary. **De Moivre's Theorem** states that if n be an integer, positive or negative
   $(\cos\theta + i\sin\theta)^n = \cos n\theta + i\sin n\theta$, where $i = \sqrt{-1}$.

3. Sometimes functions which are the product of the powers of sines and cosines can be expressed as the sum of the sines and cosines of multiple angles by proper trigonometrical substitutions.

1.4 Illustrative Examples

Example 1.6 Find the nth derivative of following expressions

(i) $\dfrac{2x-3}{x^2-3x+2}$. $\quad$ (ii) $\dfrac{3}{(x+1)(2x-1)}$.

(iii) $\dfrac{x-2}{6x^2-7x+2}$.   (iv) $\dfrac{x^2+x-1}{x^3+x^2-6x}$.   (A.M.I.E., W-2000)

Solution. (i) We have $\dfrac{2x-3}{x^2-3x+2}=\dfrac{2x-3}{(x-1)(x-2)}$.

Resolving into partial fractions $\dfrac{2x-3}{(x-1)(x-2)}=\dfrac{1}{x-1}+\dfrac{1}{x-2}$.

$$\therefore \quad \frac{d^n}{dx^n}\left[\frac{2x-3}{(x-1)(x-2)}\right]=\frac{d^n}{dx^n}\left[\frac{1}{x-1}+\frac{1}{x-2}\right]=\frac{(-1)^n\,n!}{(x-1)^{n+1}}+\frac{(-1)^n\,n!}{(x-2)^{n+1}}$$

$$=(-1)^n\,n!\left\{\frac{1}{(x-1)^{n+1}}+\frac{1}{(x-2)^{n+1}}\right\}.$$

(ii) Resolving into partial fractions, we obtain $\dfrac{3}{(x+1)(2x-1)}=\dfrac{2}{2x-1}-\dfrac{1}{x+1}$.

The nth derivative of $\dfrac{1}{2x-1}$ is $\dfrac{(-1)^n\,n!\,2^n}{(2x-1)^{n+1}}$, and that of $\dfrac{1}{x+1}$ is $\dfrac{(-1)^n\,n!}{(x+1)^{n+1}}$.

Therefore, $\dfrac{d^n}{dx^n}\left[\dfrac{3}{(x+1)(2x-1)}\right]=2\cdot\dfrac{(-1)^n\,n!\,2^n}{(2x-1)^{n+1}}-\dfrac{(-1)^n\,n!}{(x+1)^{n+1}}$

$$=(-1)^n\,n!\left[\frac{2^{n+1}}{(2x-1)^{n+1}}-\frac{1}{(x+1)^{n+1}}\right].$$

(iii) We have $\dfrac{x-2}{6x^2-7x+2}=\dfrac{x-2}{(2x-1)(3x-2)}$.

Resolving into partial fractions, we have $\dfrac{x-2}{6x^2-7x+2}=\dfrac{3}{2x-1}-\dfrac{4}{3x-2}$.

Hence, $\dfrac{d^n}{dx^n}\left[\dfrac{x-2}{6x^2-7x+2}\right]=3\cdot\dfrac{(-1)^n\,n!\,2^n}{(2x-1)^{n+1}}-4\cdot\dfrac{(-1)^n\,n!\,3^n}{(3x-2)^{n+1}}$

$$=(-1)^n\,n!\left[\frac{3\cdot 2^n}{(2x-1)^{n+1}}-\frac{4\cdot 3^n}{(3x-2)^{n+1}}\right].$$

(iv) We have $\dfrac{x^2+x-1}{x^3+x^2-6x} = \dfrac{x^2+x-1}{x(x+3)(x-2)} = \dfrac{1}{6x}+\dfrac{1}{3(x+3)}+\dfrac{1}{2(x-2)}.$

(Resolving into partial fractions)

Hence, $\dfrac{d^n}{dx^n}\left(\dfrac{x^2+x-1}{x^3+x^2-6x}\right) = \dfrac{1}{6}\cdot\dfrac{(-1)^n n!}{x^{n+1}} + \dfrac{1}{3}\dfrac{(-1)^n n!}{(x+3)^{n+1}} + \dfrac{1}{2}\cdot\dfrac{(-1)^n n!}{(x-2)^{n+1}}$

$$= \dfrac{(-1)^n n!}{6}\left[\dfrac{1}{x^{n+1}}+\dfrac{2}{(x+3)^{n+1}}+\dfrac{3}{(x-2)^{n+1}}\right].$$

Example 1.7 (a) Find the nth differential coefficient of $\log\sqrt{\dfrac{x-1}{x+1}}$.

(b) Find the fifth and the nth differential co-efficients of $\dfrac{x^4}{(x-1)(x-2)}$.

(c) Find the nth derivative of $1/(x^2+a^2)$.

(d) Find the *n*th differential coefficient of $y = \tan^{-1}(x/a)$.

Solution. *(a)* We have $\log\left(\dfrac{x-1}{x+1}\right)^{1/2} = \dfrac{1}{2}\left[\log(x-1)-\log(x+1)\right].$

Therefore, $\dfrac{d^n}{dx^n}\left(\log\sqrt{\dfrac{x-1}{x+1}}\right) = \dfrac{1}{2}\left[\dfrac{(-1)^{n-1}(n-1)!}{(x-1)^n}-\dfrac{(-1)^{n-1}(n-1)!}{(x+1)^n}\right]$

$$= \dfrac{(-1)^{n-1}(n-1)!}{2}\left[\dfrac{1}{(x-1)^n}-\dfrac{1}{(x+1)^n}\right].$$

(b) Dividing the numerator by its denominator, we get

$$\dfrac{x^4}{(x-1)(x-2)} = x^2+3x+7+\dfrac{15x-14}{(x-1)(x-2)}.$$

Resolving $\dfrac{15x-14}{(x-1)(x-2)}$ into partial fractions, we obtain

$$\dfrac{x^4}{(x-1)(x-2)} = \left(x^2+3x+7\right)+\dfrac{16}{x-2}-\dfrac{1}{x-1}.$$

Hence, $\dfrac{d^n}{dx^n}\left[\dfrac{x^4}{(x-1)(x-2)}\right]=(-1)^n\,n!\left[\dfrac{16}{(x-2)^{n+1}}-\dfrac{1}{(x-1)^{n+1}}\right],\;n>2.$

Now putting $n=5$, the required fifth derivative $=-120\left[\dfrac{16}{(x-2)^6}-\dfrac{1}{(x-1)^6}\right]$

(c) Let $y=\dfrac{1}{x^2+a^2}=\dfrac{1}{(x+ia)(x-ia)}=\dfrac{1}{2ia}\left[\dfrac{1}{x-ia}-\dfrac{1}{x+ia}\right],$

(resolving into partial fractions.)

Differentiating n times with respect to x, we get

$$y_n=\dfrac{1}{2ia}\left[D^n\left(\dfrac{1}{x-ia}\right)-D^n\left(\dfrac{1}{x+ia}\right)\right]$$

$$=\dfrac{1}{2ia}\left[\dfrac{(-1)^n\,n!}{(x-ia)^{n+1}}-\dfrac{(-1)^n\,n!}{(x+ia)^{n+1}}\right]$$

$$=\dfrac{(-1)^n\,n!}{2ia}\left[(x-ia)^{-n-1}-(x+ia)^{-n-1}\right]$$

$$=\dfrac{(-1)^n\,n!}{2ia}\left[r^{-n-1}(\cos\theta-i\sin\theta)^{-n-1}-r^{-n-1}(\cos\theta+i\sin\theta)^{-n-1}\right],$$

where $x=r\cos\theta$, and $a=r\sin\theta$.

$$=\dfrac{(-1)^n\,n!}{2iar^{n+1}}\left[\{\cos(n+1)\theta+i\sin(n+1)\theta\}-\{\cos(n+1)\theta-i\sin(n+1)\theta\}\right]$$

[*By De-Moivre's Theorem*]

$$=\dfrac{(-1)^n\,n!}{ar^{n+1}}\sin(n+1)\theta,\text{ where }r=a\,\operatorname{cosec}\theta$$

$$=\dfrac{(-1)^n\,n!}{a^{n+2}}\sin^{n+1}\theta\sin(n+1)\theta,\text{ where }\theta=\tan^{-1}\left(\dfrac{a}{x}\right).$$

(d) We have $y=\tan^{-1}\left(\dfrac{x}{a}\right).$ ...(i)

Therefore, $y_1 = \dfrac{a}{a^2 + x^2} = \dfrac{a}{(x + ia)(x - ia)}$, where $i = \sqrt{-1}$,

$$= \frac{1}{2i}\left[\frac{1}{x - ia} - \frac{1}{x + ia}\right], \text{ (by resolving into partial fractions.)} \quad ...(ii)$$

Differentiating (ii), $(n - 1)$ times w.r.t. x, we obtain

$$y_n = \frac{1}{2i}\left[\frac{(-1)^{n-1}(n-1)!}{(x - ia)^n} - \frac{(-1)^{n-1}(n-1)!}{(x + ia)^n}\right]$$

$$= \frac{(-1)^{n-1}(n-1)!}{2i}\left[(x - ia)^{-n} - (x + ia)^{-n}\right]. \quad ...(iii)$$

To render the result free from 'i' and express the same in real form, we determine two numbers r and θ such that $x = r\cos\theta$ and $a = r\sin\theta$.

From the above two relations, we obtain $r = \sqrt{x^2 + a^2}$ and $\theta = \tan^{-1}(a/x)$.

Now $(x - ia)^{-n} = (r)^{-n}(\cos\theta - i\sin\theta)^{-n} = (r)^{-n}\left[\cos(-\theta) + i\sin(-\theta)\right]^{-n}$

$$= (r)^{-n}[\cos n\theta + i\sin n\theta], \text{ by De Moivre's Theorem .}$$

Similarly $(x + ia)^{-n} = (r)^{-n}(\cos n\theta - i\sin n\theta)$. Substituting in (iii), we obtain

$$\frac{d^n y}{dx^n} = \frac{(-1)^{n-1}(n-1)!(r)^{-n}}{2i}[\cos n\theta + i\sin n\theta - \cos n\theta + i\sin n\theta]$$

$$= (-1)^{n-1}\cdot(n-1)!(r)^{-n}\sin n\theta = \frac{(-1)^{n-1}\cdot(n-1)!\sin n\theta}{(a/\sin\theta)^n} \qquad (\because a = r\sin\theta)$$

$$= (-1)^{n-1}(n-1)!a^{-n}\sin^n\theta\sin n\theta, \text{ where } \theta = \tan^{-1}(a/x).$$

Example 1.8 Find the nth differential coefficients of the following:
(i) $\sin 4x \cos 7x$. (ii) $\sin^3 x$. (iii) $\cos^4 x$. (iv) $\sin^5 x \cos^3 x$. (v) $\sin 2x \sin 3x \cos 4x$.

Solution. (i) We have $\sin 4x \cos 7x = (\sin 11x - \sin 3x)/2..$

Therefore, $\dfrac{d^n}{dx^n}(\sin 4x \cos 7x) = \dfrac{1}{2}\dfrac{d^n}{dx^n}(\sin 11x) - \dfrac{1}{2}\dfrac{d^n}{dx^n}(\sin 3x)$

$$= \frac{1}{2}\left[11^n \sin\left(11x + \frac{n\pi}{2}\right) - 3^n \sin\left(3x + \frac{n\pi}{2}\right)\right].$$

(*ii*) We know that $\sin 3x = 3\sin x - 4\sin^3 x$. Therefore, we have

$$\sin^3 x = (3\sin x - \sin 3x)/4 = \{(3\sin x)/4\} - \{(\sin 3x)/4\}.$$

Hence, $\qquad \dfrac{d^n}{dx^n}\left(\sin^3 x\right) = \dfrac{3}{4}\sin\left(x + \dfrac{n\pi}{2}\right) - \dfrac{3^n}{4}\sin\left(3x + \dfrac{n\pi}{2}\right).$

(*iii*) We know that $\cos^2 x = (1 + \cos 2x)/2$.

Therefore, $\quad \cos^4 x = \left(\dfrac{1 + \cos 2x}{2}\right)^2 = \dfrac{1}{4} + \dfrac{\cos 2x}{2} + \dfrac{\cos^2 2x}{4}$

$$= \dfrac{1}{4} + \dfrac{1}{2}\cos 2x + \dfrac{1}{8}(1 + \cos 4x) = \dfrac{3}{8} + \dfrac{1}{2}\cos 2x + \dfrac{1}{8}\cos 4x.$$

Hence, $\qquad \dfrac{d^n}{dx^n}\left(\cos^4 x\right) = 2^{n-1}\cos\left(2x + \dfrac{n\pi}{2}\right) + 2^{2n-3}\cos\left(4x + \dfrac{n\pi}{2}\right).$

(*iv*) We have $\sin^5 x \cos^3 x = \sin^2 x(\sin x \cos x)^3 = (1 - \cos 2x)(\sin^3 2x)/16$

$$= \dfrac{1}{16}(1 - \cos 2x)\left(\dfrac{3\sin 2x - \sin 6x}{4}\right)$$

$$= \dfrac{1}{64}\left[3\sin 2x - 3\sin 2x \cos 2x - \sin 6x + \sin 6x \cos 2x\right]$$

$$= \dfrac{1}{64}\left[3\sin 2x - \dfrac{3}{2}\sin 4x - \sin 6x + \dfrac{1}{2}(\sin 8x + \sin 4x)\right]$$

$$= \dfrac{1}{128}(6\sin 2x - 2\sin 4x - 2\sin 6x + \sin 8x).$$

$\therefore \quad \dfrac{d^n}{dx^n}\left(\sin^5 x \cos^3 x\right) = \dfrac{1}{2^7}\left\{6\cdot 2^n \sin\left(2x + \dfrac{n\pi}{2}\right) - 2\cdot 4^n \sin\left(4x + \dfrac{n\pi}{2}\right)\right.$

$$\left. -2\cdot 6^n \sin\left(6x + \dfrac{n\pi}{2}\right) + 8^n \sin\left(8x + \dfrac{n\pi}{2}\right)\right\}.$$

(*v*) Let $y = \sin 2x \sin 3x \cos 4x = \sin 2x(2\sin 3x \cos 4x)/2$

$$= \sin 2x(\sin 7x - \sin x)/2 = (2\sin 7x \sin 2x)/4 - (2\sin 2x \sin x)/4$$

$$= (\cos 5x - \cos 9x)/4 - (\cos x - \cos 3x)/4 = (\cos 5x - \cos 9x - \cos x + \cos 3x)/4.$$

Therefore, $y_n = \left[D^n (\cos 5x) - D^n (\cos 9x) - D^n (\cos x) + D^n (\cos 3x) \right]/4$

$$= \frac{1}{4}\left[5^n \cos\left(5x + \frac{1}{2}n\pi \right) - 9^n \cos\left(9x + \frac{1}{2}n\pi \right) \right.$$

$$\left. - \cos\left(x + \frac{1}{2}n\pi \right) + 3^n \cos\left(3x + \frac{1}{2}n\pi \right) \right].$$

Example 1.9 Find the nth differential coefficients of

(*i*) $e^{3x}\sin 4x$. (*ii*) $e^x \sin^4 x$. (*iii*) $e^x \cos x \cos 2x$. (*iv*) $e^{3x} \cos^2 x \sin x$.

Solution. (*i*) $\dfrac{d^n}{dx^n}\left(e^{3x} \sin 4x \right) = \left(3^2 + 4^2 \right)^{n/2} \cdot e^{3x} \sin\left(4x + n\tan^{-1}\frac{4}{3} \right)$

$$= 5^n e^{3x} \sin\left(4x + n\tan^{-1}\frac{4}{3} \right).$$

(*ii*) We know that $\sin^2 x = \dfrac{1 - \cos 2x}{2}$. Therefore, $\sin^4 x = \left(\dfrac{1 - \cos 2x}{2} \right)^2$

$$= \frac{1}{4} - \frac{\cos 2x}{2} + \frac{\cos^2 2x}{4} = \frac{1}{4} - \frac{1}{2}\cos 2x + \frac{1}{8}(1 + \cos 4x)$$

$$= \frac{3}{8} - \frac{1}{2}\cos 2x + \frac{1}{8}\cos 4x.$$

Hence $\dfrac{d^n}{dx^n}\left(e^x \sin^4 x \right) = \dfrac{3}{8}\dfrac{d^n}{dx^n}e^x - \dfrac{1}{2}\dfrac{d^n}{dx^n}\left(e^x \cos 2x \right) + \dfrac{1}{8}\dfrac{d^n}{dx^n}\left(e^x \cos 4x \right)$

$$= \frac{3}{8}e^x - \frac{1}{2}\left(1 + 2^2 \right)^{n/2} e^x \cos\left(2x + n\tan^{-1} 2 \right)$$

$$+ \frac{1}{8}\left(1 + 4^2 \right)^{n/2} e^x \cos\left(4x + n\tan^{-1} 4 \right)$$

$$= \frac{e^x}{8}\left[3 - 4 \cdot 5^{n/2} \cos\left(2x + n\tan^{-1} 2 \right) + 17^{n/2} \cos\left(4x + n\tan^{-1} 4 \right) \right].$$

(*iii*) We know that $\cos x \cos 2x = (\cos 3x + \cos x)/2$.

Hence $\dfrac{d^n}{dx^n}\left(e^x \cos x \cos 2x \right) = \dfrac{1}{2}\dfrac{d^n}{dx^n}\left(e^x \cos 3x \right) + \dfrac{1}{2}\dfrac{d^n}{dx^n}\left(e^x \cos x \right)$

$$= \frac{1}{2}\left(1+3^2\right)^{n/2} e^x \cos\left(3x + n\tan^{-1}3\right) + \frac{1}{2}\left(1+1\right)^{n/2} e^x \cos\left(x + n\tan^{-1}1\right)$$

$$= \frac{1}{2} \cdot e^x \left[10^{n/2} \cos\left(3x + n\tan^{-1}3\right) + 2^{n/2}\cos\left(x + \frac{n\pi}{4}\right)\right].$$

(iv) $\cos^2 x \sin x = (1+\cos 2x)(\sin x)/2 = (\sin x)/2 + (2\cos 2x \sin x)/4$

$$= (\sin x)/2 + (\sin 3x - \sin x)/4 = (\sin x + \sin 3x)/4 .$$

Hence $\dfrac{d^n}{dx^n}\left(e^{3x}\cos^2 x \sin x\right) = \dfrac{1}{4}\dfrac{d^n}{dx^n}\left(e^{3x}\sin x\right) + \dfrac{1}{4}\dfrac{d^n}{dx^n}\left(e^{3x}\sin 3x\right)$

$$= \frac{1}{4}\left(3^2+1\right)^{n/2} e^{3x} \sin\left(x + n\tan^{-1}\frac{1}{3}\right) + \frac{1}{4}\left(3^2+3^2\right)^{n/2} e^{3x}\sin\left(3x + n\tan^{-1}1\right)$$

$$= \frac{1}{4} e^{3x}\left[10^{n/2}\sin\left(x + n\tan^{-1}\frac{1}{3}\right) + 18^{n/2}\sin\left(3x + \frac{n\pi}{4}\right)\right].$$

1.5 Leibnitz's Theorem. The nth derivative of the product of two functions

This theorem is useful for finding the nth differential coefficient of a product.
Statement. If u, v be any two functions of x possessing derivatives of nth order,

$$(uv)_n = {}^nc_0 u_n v + {}^nc_1 u_{n-1}v_1 + {}^nc_2 u_{n-2}v_2 + \ldots + {}^nc_r u_{n-r}v_r + \ldots + {}^nc_n uv_n,$$

Proof We shall prove this theorem by *Mathematical induction.*
We know that

$$(uv)_1 = u_1 v + uv_1 ,$$
$$(uv)_2 = u_2 v + u_1 v_1 + u_1 v_1 + uv_2$$
$$= {}^2c_0 u_2 v + {}^2c_1 u_1 v_1 + {}^2c_2 uv_2. \qquad \ldots(1.11)$$

Thus, the theorem is true for $n = 1, 2$.
We assume that the theorem is true for particular value of $n = m$ (say) so that we have

$$(uv)_m = {}^mc_0 u_m v + {}^mc_1 u_{m-1}v_1 + {}^mc_2 u_{m-2}v_2 + \ldots + {}^mc_r u_{m-r}v_r + \ldots + {}^mc_m uv_m.$$

Differentiating both sides w.r.t. x, we obtain

$$(uv)_{m+1} = {}^mc_0\left(u_{m+1}v + u_m v_1\right) + {}^mc_1\left(u_m v_1 + u_{m-1}v_2\right) + \ldots$$

$$+ {}^mc_r\left(u_{m-r+1}v_r + u_{m-r}v_{r+1}\right) + \ldots + {}^mc_m\left(uv_{m+1}\right)$$

$$= {}^mc_0 u_{m+1}v + \left({}^mc_0 + {}^mc_1\right)u_m v_1 + \left({}^mc_1 + {}^mc_2\right)u_{m-1}v_2 + \ldots$$

$$\ldots + \left({}^mc_{r-1} + {}^mc_r\right)u_{m-r+1}v_r + \ldots + {}^mc_m uv_{m+1}.$$

But we know that $^m c_{r-1} + {}^m c_r = {}^{m+1} c_r$.

Putting $r = 1, 2, \ldots$

$$^m c_0 + {}^m c_1 = {}^{m+1} c_1$$

$$^m c_1 + {}^m c_2 = {}^{m+1} c_2$$

$$\ldots \quad \ldots \quad \ldots \quad \text{and so on.}$$

Also $^m c_0 = 1 = {}^{m+1} c_0$ and $^m c_m = 1 = {}^{m+1} c_{m+1}$.

Hence $(uv)_{m+1} = {}^{m+1} c_0 u_{m+1} v + {}^{m+1} c_1 u_m v_1 + {}^{m+1} c_2 u_{m-1} v_2 + \ldots$

$$\ldots + {}^{m+1} c_r u_{m-r+1} v_r + \ldots + {}^{m+1} c_{m+1} u v_{m+1}.$$

Therefore, if the theorem is true for $n = m$, it is true for $n = m + 1$.

Conclusion

By (1.11) we have seen that the theorem is true for $n = 2$. Hence it is true for $n = 2 + 1$ that is, 3 and therefore, for $n = 3 + 1$ that is, 4 and so on. Thus it is true for every positive integral value of n.

Remark. The Leibnitz's theorem can also be stated as follows:

$$D^n (uv) = (D^n u) \cdot v + {}^n c_1 D^{n-1} u \cdot Dv + {}^n c_2 D^{n-2} u \cdot D^2 v + \cdots$$

$$+ {}^n c_r D^{n-r} u \cdot D^r v + \cdots + u \cdot D^n v.$$

Example 1.10 Find the nth derivative of

(*i*) $x^2 \log x$. (*ii*) $e^{ax} x^2$. (*iii*) $e^x \log x$. (*iv*) $x^2 e^x \cos x$. (*v*) $(\log x)/x$.

Solution. (i) Let $u = \log x$, and $v = x^2$.

Therefore, $u_n = \dfrac{(-1)^{n-1} (n-1)!}{x^n}, \quad v_1 = 2x, v_2 = 2, v_3 = 0.$

Employing Leibnitz's theorem

$$(uv)_n = {}^n c_0 u_n v + {}^n c_1 u_{n-1} v_1 + {}^n c_2 u_{n-2} v_2 + \ldots + {}^n c_n u v_n, \text{ we obtain}$$

$$\frac{d^n}{dx^n}\left(x^2 \log x\right) = \frac{(-1)^{n-1} (n-1)!}{x^n} x^2 + n \cdot \frac{(-1)^{n-2} (n-2)!}{x^{n-1}} \cdot 2x$$

$$+ \frac{n(n-1)}{2!} \cdot \frac{(-1)^{n-3} (n-3)!}{x^{n-2}} \cdot 2$$

$$= \frac{(-1)^{n-1} (n-3)!}{x^{n-2}} \left[(n-1)(n-2) - 2n(n-2) + n(n-1)\right] = \frac{(-1)^{n-1} \cdot 2(n-3)!}{x^{n-2}}.$$

(*ii*) Let $\qquad\qquad u = e^{ax}$ and $\qquad v = x^2$

Therefore, $\qquad\qquad u_1 = ae^{ax} \qquad\qquad v_1 = 2x$

$$u_2 = a^2 e^{ax} \qquad\qquad v_2 = 2$$

$$\cdots \quad \cdots \qquad\qquad\qquad v_3 = 0$$

$$\cdots \quad \cdots$$

$$u_n = a^n e^{ax}$$

Hence, $\qquad \dfrac{d^n}{dx^n}\left(e^{ax} x^2\right) = a^n e^{ax} x^2 + na^{n-1} e^{ax} \cdot 2x + \dfrac{n(n-1)}{2!} a^{n-2} e^{ax} \cdot 2$

$$= e^{ax}\left[a^n x^2 + 2na^{n-1}x + n(n-1)a^{n-2}\right].$$

(*iii*) Let $u = e^x$, $\qquad\qquad$ and $v = \log x$

Therefore, $\qquad u_n = e^x, u_{n-1} = u_{n-2} = \cdots\cdots u_2 = u_1 = e^x.$

$$v_1 = \frac{1}{x}, v_2 = -\frac{1}{x^2}, v_3 = 2/x^3 \cdots\cdots v_n = \frac{(-1)^{n-1}(n-1)!}{x^n}.$$

By Leibnitz's theorem, we have

$$\frac{d^n}{dx^n}\left(e^x \log x\right) = e^x \log x + ne^x \cdot \frac{1}{x} + \frac{n(n-1)}{2!}e^x \cdot\left(-\frac{1}{x^2}\right) + \dots$$

$$\dots + e^x \frac{(-1)^{n-1}(n-1)!}{x^n}$$

$$= e^x\left[\log x + \frac{n}{x} - \frac{n(n-1)}{2x^2} + \dots + \frac{(-1)^{n-1} n!}{nx^n}\right].$$

(*iv*) Let $u = e^x \cos x$ $\qquad\qquad\qquad\qquad\qquad\qquad$ and $v = x^2$

$$\therefore \quad u_n = 2^{n/2} e^x \cos(x + n\tan^{-1} 1) = 2^{n/2} e^x \cos\left(x + n\frac{\pi}{4}\right)$$

$$u_{n-1} = 2^{(n-1)/2} e^x \cos\left(x + \overline{n-1}\frac{\pi}{4}\right) \qquad\qquad v_1 = 2x$$

$$u_{n-2} = 2^{(n-2)/2} e^x \cos\left(x + \overline{n-2}\frac{\pi}{4}\right). \qquad\qquad v_2 = 2, v_3 = 0.$$

....

....

Now by Leibnitz's theorem, we have

$$\frac{d^n}{dx^n}\left(x^2 e^x \cos x\right) = 2^{n/2} e^x \cos\left(x + \frac{n\pi}{4}\right) \cdot x^2 + n \cdot 2^{(n-1)/2} e^x \cos\left\{x + (n-1)\frac{\pi}{4}\right\} \cdot 2x$$

$$+ \frac{n(n-1)}{2} \cdot 2^{(n-2)/2} e^x \cos\left\{x + (n-2)\frac{\pi}{4}\right\} \cdot 2$$

$$= 2^{(n-2)/2} e^x \left[2x^2 \cos\left(x + n\frac{\pi}{4}\right) + 2^{3/2} nx\cos\left(x + (n-1)\frac{\pi}{4}\right)\right.$$

$$\left. + n(n-1)\cos\left\{x + (n-2)\frac{\pi}{4}\right\}\right].$$

(v) Let $u = 1/x$ and $v = \log x$. Then,

$$D^n(u) = D^n\left(\frac{1}{x}\right) = \frac{(-1)^n n!}{x^{n+1}}, \quad D^{n-1}\left(\frac{1}{x}\right) = \frac{(-1)^{n-1}(n-1)!}{x^n}$$

$$D^{n-2}\left(\frac{1}{x}\right) = \frac{(-1)^{n-2}(n-2)!}{x^{n-1}}, \text{ etc.}$$

and $D(\log x) = 1/x, \; D^2(\log x) = -1/x^2, D^3(\log x) = 2/x^3 ,$

$$D^n(\log x) = \frac{(-1)^{n-1}(n-1)!}{x^n}.$$

Therefore, By Leibnitz's theorem

$$D^n\left(\frac{\log x}{x}\right) = D^n\left(\frac{1}{x}\right) \cdot \log x + nD^{n-1}\left(\frac{1}{x}\right).D(\log x)$$

$$+ \frac{n(n-1)}{2!} D^{n-2}\left(\frac{1}{x}\right) D^2(\log x) + \frac{n(n-1)(n-2)}{3!} D^{n-3}\left(\frac{1}{x}\right) D^3(\log x)$$

$$+ \cdots + \frac{1}{x} D^n(\log x)$$

$$= \frac{(-1)^n n!}{x^{n+1}} \log x + \frac{n(-1)^{n-1}(n-1)!}{x^n}\frac{1}{x} + \frac{n(n-1)}{2!} \cdot \frac{(-1)^{n-2}(n-2)!}{x^{n-1}} \cdot \left(-\frac{1}{x^2}\right)$$

$$+ \frac{n(n-1)(n-2)}{3!} \cdot \frac{(-1)^{n-3}(n-3)!}{x^{n-2}} \cdot \left(\frac{2}{x^3}\right) + \cdots + \frac{1}{x}\frac{(-1)^{n-1}(n-1)!}{x^n}$$

$$= \frac{n!}{x^{n+1}}\left[(-1)^n \log x + (-1)^{n-1} - \frac{(-1)^{n-2}}{2} + \frac{(-1)^{n-3}}{3} + \cdots + \frac{(-1)^{n-1}}{n}\right]$$

$$= \frac{(-1)^n n!}{x^{n+1}}\left[\log x - \left(1 + \frac{1}{2} + \frac{1}{3} + \cdots + \frac{1}{n}\right)\right].$$

1.6 Illustrative Examples

Example 1.11 (a) Find the third differential coefficient of $x^2 \sin 3x + 5x^2$, using Leibnitz's theorem.

(b) Find the fourth order derivative of $e^{ax} \sin bx$ at the point $x = 0$.

(c)If $f(\theta) = \left(\dfrac{\sin\theta}{\theta}\right)^2$, show that $\theta^2 f''(\theta) + 4\theta f'(\theta) = 2\left[1 - f(\theta) - 2\theta^2 \cdot f(\theta)\right]$.

(d) If $y(x) = \tan x$, Show that

$$y_n(0) - {}^nc_2 y_{n-2}(0) + {}^nc_4 y_{n-4}(0) - {}^nc_6 y_{n-6}(0) + \cdots = \sin\frac{n\pi}{2}.$$

Solution. *(a)* We have $D^3(x^2 \sin 3x + 5x^2) = D^3(x^2 \sin 3x) + D^3(5x^2)$

$$= D^3(x^2 \sin 3x) \qquad\qquad \{\because D^3(5x^2) = 0\}$$

$$= {}^3c_0\,(x^2)''' \sin 3x + {}^3c_1\,(x^2)''(\sin 3x)' + {}^3c_2\,(x^2)'(\sin 3x)'' + {}^3c_3\,x^2(\sin 3x)'''$$

$$= 0 + 3(2)(3\cos 3x) + 3(2x)(-9\sin 3x) + x^2(-27\cos 3x)$$

$$= \left(18 - 27x^2\right)\cos 3x - 54x \sin 3x.$$

(b) Let $u = e^{ax}$, $v = \sin bx$. Using the Leibnitz's theorem, we obtain

$$\frac{d^4}{dx^4}\left(e^{ax} \sin bx\right) = {}^4c_0\left(e^{ax}\right)^{(4)} \sin bx + {}^4c_1\left(e^{ax}\right)^{(3)}(\sin bx)'$$

$$+ {}^4c_2\,(e^{ax})^{(2)}(\sin bx)'' + {}^4c_3\,(e^{ax})^{(1)}(\sin bx)''' + {}^4c_4 e^{ax}(\sin bx)''''$$

$$= e^{ax}\left[a^4 \sin bx + 4a^3 b \cos bx - 6a^2 b^2 \sin bx - 4ab^3 \cos bx + b^4 \sin bx\right]$$

Hence, $\qquad \dfrac{d^4}{dx^4}\left(e^{ax} \sin bx\right)_{at\,x=0} = 4a^3 b - 4ab^3 = 4ab\left(a^2 - b^2\right).$

(c) We have $\sin^2\theta = \theta^2 \cdot f(\theta).$

By Leibnitz's theorem, $\dfrac{d^2}{d\theta^2}(\sin^2\theta) = f''(\theta)\cdot\theta^2 + 2f'(\theta)\cdot 2\theta + f(\theta)\cdot 2$

We know that $\dfrac{d}{d\theta}(\sin^2\theta) = 2\sin\theta\cdot\cos\theta = \sin 2\theta$.

Therefore, $\dfrac{d^2}{d\theta^2}(\sin^2\theta) = \dfrac{d}{d\theta}\left[\dfrac{d}{d\theta}(\sin^2\theta)\right]$

$$= \dfrac{d}{d\theta}(\sin 2\theta) = 2\cos 2\theta = 2(1 - 2\sin^2\theta) = 2\left[1 - 2\theta^2\cdot f(\theta)\right].$$

Hence, $2 - 4\theta^2 f(\theta) = \theta^2\cdot f''(\theta) + 4\theta\cdot f'(\theta) + 2f(\theta)$

that is, $\theta^2 f''(\theta) + 4\theta\cdot f'(\theta) = 2\left[1 - f(\theta) - 2\theta^2\cdot f(\theta)\right]$.

(*d*) We have $y(x) = \tan x = \sin x/\cos x$ or, $(\cos x)\cdot y(x) = \sin x$.

Differentiating this *n* times by Leibnitz's theorem, we obtain

$$(\cos x)\cdot y_n(x) + {}^nc_1(-\sin x)\cdot y_{n-1}(x) + {}^nc_2(-\cos x)\cdot y_{n-2}(x)$$

$$+ {}^nc_3(\sin x)\cdot y_{n-3}(x) + {}^nc_4(\cos x)\cdot y_{n-4}(x) + \cdots = \sin\left(x + \dfrac{n\pi}{2}\right). \qquad \ldots(i)$$

Putting $x = 0$ in (*i*), we get

$$y_n(0) - {}^nc_2 y_{n-2}(0) + {}^nc_4 y_{n-4}(0) - {}^nc_6 y_{n-6}(0) + \cdots = \sin\dfrac{n\pi}{2}.$$

Example 1.12 (*a*) **If** $y = x^2 e^x$, **then prove that**

$$y_n = \dfrac{1}{2}n(n-1)\dfrac{d^2y}{dx^2} - n(n-2)\dfrac{dy}{dx} + \dfrac{1}{2}(n-1)(n-2)y.$$

(b) **If** $y = x\log\left(\dfrac{x-1}{x+1}\right)$, **show that** $y_n = (-1)^n(n-2)!\left[\dfrac{x-n}{(x-1)^n} - \dfrac{x+n}{(x+1)^n}\right]$

(*UPTU 2003; D.U. 1999*)

Solution. (*a*) We have $y = x^2 e^x$.

∴refore, by Leibnitz's theorem

$$\dfrac{d^ny}{n} = D^n(x^2 e^x) = D^n(e^x)\cdot x^2 + nD^{n-1}(e^x)D(x^2)$$

$$+ \dfrac{n(n-1)}{2!}D^{n-2}(e^x)D^2(x^2)$$

$$e^x\cdot 2x + \dfrac{n(n-1)}{2}\cdot e^x\cdot 2 = e^x\left[x^2 + 2nx + n(n-1)\right] \qquad \ldots(i)$$

Putting $n = 1, 2$ in (i), we have

$$y_1 = e^x(x^2 + 2x), \quad y_2 = e^x(x^2 + 4x + 2).$$

We have to express y_n in terms of y_2, y_1 and y.

Let $y_n \equiv Ay_2 + By_1 + Cy.$ ...(ii)

Then $x^2 + 2nx + n(n-1) \equiv A(x^2 + 4x + 2) + B(x^2 + 2x) + Cx^2.$

Equating coefficients of x^2, x and constant terms, we obtain

$$1 = A + B + C, \quad 2n = 4A + 2B, \quad \text{and} \quad n(n-1) = 2A.$$

On solving, these equations, we get

$$A = n(n-1)/2, \ B = -n(n-2); \text{and } C = (n-1)(n-2)/2.$$

Putting these values in (ii), we get the required result.

(b) We have $y = x \log\left(\dfrac{x-1}{x+1}\right) = x\left[\log(x-1) - \log(x+1)\right].$

Differentiating, we obtain $y_1 = x\left(\dfrac{1}{x-1} - \dfrac{1}{x+1}\right) + \log(x-1) - \log(x+1)$

$$= \dfrac{1}{x-1} + \dfrac{1}{x+1} + \log(x-1) - \log(x+1)$$ ...(i)

Differentiating (i), $(n-1)$ times w.r.t. x, we obtain

$$y_n = \frac{(-1)^{n-1} \cdot (n-1)!}{(x-1)^n} + \frac{(-1)^{n-1} \cdot (n-1)!}{(x+1)^n} + \frac{(-1)^{n-2} \cdot (n-2)!}{(x-1)^{n-1}} - \frac{(-1)^{n-2} \cdot (n-2)!}{(x+1)^{n-1}}$$

$$= (-1)^{n-2}(n-2)!\left[\frac{-(n-1)+(x-1)}{(x-1)^n}\right] + (-1)^{n-2}(n-2)!\left[\frac{-(n-1)-(x+1)}{(x+1)^n}\right]$$

$$= (-1)^{n-2}(n-2)!\left[\frac{x-n}{(x-1)^n} - \frac{x+n}{(x+1)^n}\right], \ = (-1)^n(n-2)!\left[\frac{x-n}{(x-1)^n} - \frac{x+n}{(x+1)^n}\right].$$

Example 1.13 If $y = A\left(x + \sqrt{x^2 - 1}\right)^n + B\left(x - \sqrt{x^2 - 1}\right)^n,$

prove that $(x^2 - 1)y_2 + xy_1 = n^2 y,$ and $(x^2 - 1)y_{n+2} + (2n+1)xy_{n+1} = 0.$

Solution. We have $y = A\left(x + \sqrt{x^2 - 1}\right)^n + B\left(x - \sqrt{x^2 - 1}\right)^n.$

Differentiating w.r.t. x, we obtain

$$y_1 = nA\left(x + \sqrt{x^2 - 1}\right)^{n-1} \cdot \left(1 + \frac{x}{\sqrt{x^2 - 1}}\right) + nB\left(x - \sqrt{x^2 - 1}\right)^{n-1}\left(1 - \frac{x}{\sqrt{x^2 - 1}}\right)$$

$$= \frac{nA\left(x + \sqrt{x^2 - 1}\right)^n}{\sqrt{x^2 - 1}} - \frac{nB\left(x - \sqrt{x^2 - 1}\right)^n}{\sqrt{x^2 - 1}}.$$

Squaring and simplifying, we obtain $\left(x^2 - 1\right)y_1^2 = n^2 y^2 - 4n^2 AB.$

Differentiating again, we have $2\left(x^2 - 1\right)y_1 y_2 + 2xy_1^2 = 2n^2 y\, y_1.$

Dividing throughout by $2y_1$, we get $\left(x^2 - 1\right)y_2 + xy_1 - n^2 y = 0.$

Differentiating this n times by Leibnitz's theorem, we obtain

$$\left[\left(x^2 - 1\right)y_{n+2} + n(2x)\,y_{n+1} + \frac{n(n-1)}{2!} \cdot 2 \cdot y_n\right] + \left[xy_{n+1} + n \cdot 1 \cdot y_n\right] - n^2 y_n = 0,$$

or $\left(x^2 - 1\right)y_{n+2} + (2n + 1)xy_{n+1} = 0,$ the required result.

Example 1.14 (a) If $\cos^{-1}(y/b) = \log(x/n)^n$, prove that

$$x^2 y_{n+2} + (2n+1)xy_{n+1} + 2n^2 y_n = 0.$$

(Mangalore, 1999; Marathwada, 1998; GGSIPU-2006)

(b) If $y = \left(x^2 - 1\right)^n$, then prove that $\left(x^2 - 1\right)y_{n+2} + 2xy_{n+1} - n(n+1)y_n = 0$

(UPTU 2003, V.T.U., 2000S)

(c) If $P_n = \dfrac{d^n}{dx^n}\left(x^2 - 1\right)^n$, show that $\dfrac{d}{dx}\left\{\left(1 - x^2\right)\dfrac{dP_n}{dx}\right\} + n(n+1)P_n = 0.$

(A.M.I.E. S-2006)

Solution. (*a*) We have $\cos^{-1}(y/b) = \log(x/n)^n$

or $\quad y/b = \cos\{n\log(x/n)\} \quad$ or $\quad y = b\cos\{n\log(x/n)\}.$ ...(*i*)

Differentiating (*i*) w.r.t. x, we get

$$y_1 = -b\sin\left[n\log(x/n)\right]n \cdot \frac{1}{x/n} \cdot \frac{1}{n} \quad \text{or} \quad xy_1 = -bn\sin\left[n\log(x/n)\right].$$

Differentiating again, we have

$$xy_2 + y_1 = -bn^2\cos\left[n\log(x/n)\right]/x = -n^2(y/x), \text{ from (}i\text{)}$$

or
$$x^2 y_2 + xy_1 + n^2 y = 0. \qquad \text{...(ii)}$$

Now differentiating (ii) n times by Leibnitz's theorem, we obtain

$$\left[x^2 y_{n+2} + n(2x)\, y_{n+1} + \frac{n(n-1)}{2!} \cdot (2)\, y_n \right] + \left[xy_{n+1} + n \cdot 1 \cdot y_n \right] + n^2 y_n = 0$$

or
$$x^2 y_{n+2} + (2n+1) xy_{n+1} + 2n^2 y_n = 0.$$

(b) Taking logarithms, we obtain $\log y = n \log (x^2 - 1)$.
Differentiating w.r.t. x, we get

$$y_1 / y = n \cdot 2x / (x^2 - 1) \quad \text{or} \quad (x^2 - 1) y_1 - 2nxy = 0.$$

Differentiating again, we obtain $(x^2 - 1) y_2 + 2xy_1 - 2n(xy_1 + y) = 0$

or,
$$(x^2 - 1) y_2 + 2x(1-n) y_1 - 2ny = 0.$$

Differentiating it n times by Leibnitz's theorem, we obtain

$$\left[(x^2 - 1) y_{n+2} + n(2x)\, y_{n+1} + \frac{n(n-1)}{2!} (2)\, y_n \right] + 2(1-n)\left[xy_{n+1} + n \cdot 1 \cdot y_n \right] - 2ny_n = 0,$$

or $(x^2 - 1) y_{n+2} + 2xy_{n+1} - n(n+1) y_n = 0.$

(c) Let $y = (x^2 - 1)^n$. $\qquad \text{...(i)}$

Then, $P_n = \dfrac{d^n}{dx^n}(x^2 - 1)^n = \dfrac{d^n}{dx^n}(y) = y_n.$

Therefore, $\dfrac{d}{dx}\left\{ (1 - x^2) \dfrac{dP_n}{dx} \right\} = \dfrac{d}{dx}\left\{ (1 - x^2) y_{n+1} \right\}, \qquad (\because P_n = y_n)$

$$= (1 - x^2) y_{n+2} - 2xy_{n+1} = -\left\{ (x^2 - 1) y_{n+2} + 2xy_{n+1} \right\} \qquad \text{...(ii)}$$

Differentiating (i), we obtain $y_1 = n(x^2 - 1)^{n-1} \cdot 2x$

or $y_1 (x^2 - 1) = 2nxy.$ Differentiating again, we get

$$(x^2 - 1) y_2 + 2x(1 - n) y_1 - 2ny = 0.$$

Differentiating it n times by Leibnitz's theorem, we obtain

$$\left[(x^2 - 1) y_{n+2} + n(2x)\, y_{n+1} + \frac{n(n-1)}{2!} \cdot 2 \cdot y_n \right] + 2(1-n)\left[xy_{n+1} + ny_n \right] - 2ny_n = 0$$

or $\left(x^2 -1\right)y_{n+2} +2xy_{n+1} -n(n+1)y_n =0$

or $\left(x^2 -1\right)y_{n+2} +2xy_{n+1} = n(n+1)y_n$ ...(iii)

Equ. (ii) may be rewritten as

$$\frac{d}{dx}\left\{\left(1-x^2\right)\frac{dP_n}{dx}\right\} = -n(n+1)y_n = -n(n+1)P_n, \text{ from ...(iii)}$$

or $\frac{d}{dx}\left\{\left(1-x^2\right)\frac{dP_n}{dx}\right\}+n(n+1)P_n =0$, which is same, as required.

Example 1.15 (a) If $y = \left(x+\sqrt{1+x^2}\right)^m$, **show that**

$$\left(1+x^2\right)y_{n+2} +(2n+1)xy_{n+1} +\left(n^2 - m^2\right)y_n = 0.$$

(A.M.I.E. S-2007, Madras 2000).

(b) **If** $y = \left[\log_e\left(x+\sqrt{a^2 +x^2}\right)\right]^2$, **prove that**

$$\left(a^2 +x^2\right)y_{n+2} +(2n+1)xy_{n+1} +n^2 y_n = 0.$$

(c) **If** $y^{1/m} +y^{-1/m} = 2x$, **prove that**

$$\left(x^2 -1\right)y_{n+2} +(2n+1)xy_{n+1} +\left(n^2 - m^2\right)y_n = 0.$$ (UPTU 2001)

Solution. (a) We have, $\log y = m\log\left(x+\sqrt{1+x^2}\right)$.

Differentiating, $\frac{1}{y}\cdot y_1 = m\cdot\frac{1}{x+\sqrt{1+x^2}}\cdot\left[1+\frac{x}{\sqrt{1+x^2}}\right] = \frac{m}{\sqrt{1+x^2}}$

On squaring both sides, we obtain $\left(1+x^2\right)y_1^2 = m^2 y^2$.

Differentiating again, w.r.t. x, we get

$$2\left(1+x^2\right)y_1 y_2 +2xy_1^2 = 2m^2 yy_1 \text{ or } \left(1+x^2\right)y_2 +xy_1 - m^2 y = 0.$$

Differentiating this n times by Leibnitz's theorem, we obtain

$$\left[\left(1+x^2\right)y_{n+2} +n(2x)y_{n+1} +\frac{n(n-1)}{2!}(2)y_n\right] +[xy_{n+1} +n\cdot1\cdot y_n] -m^2 y_n = 0$$

or $\left(1+x^2\right)y_{n+2} +(2n+1)xy_{n+1} +\left(n^2 - m^2\right)y_n = 0.$

(b) We have $y = \left[\log_e \left(x + \sqrt{a^2 + x^2} \right) \right]^2$.

$$\therefore \quad y_1 = 2\log_e\left(x + \sqrt{a^2 + x^2}\right) \cdot \frac{1}{x + \sqrt{a^2 + x^2}} \cdot \left(1 + \frac{x}{\sqrt{a^2 + x^2}}\right)$$

$$= \frac{2}{\sqrt{a^2 + x^2}} \log_e\left(x + \sqrt{a^2 + x^2}\right) = \frac{2}{\sqrt{a^2 + x^2}} \cdot \sqrt{y}.$$

Squaring and transposing, we get $y_1^2\left(a^2 + x^2\right) = 4y$.

Differentiating again w.r.t. x, we obtain

$$2\left(a^2 + x^2\right)y_1y_2 + 2xy_1^2 = 4y_1, \text{ or } \left(a^2 + x^2\right)y_2 + xy_1 - 2 = 0.$$

Differentiating this n times by Leibnitz's theorem, we have

$$\left[\left(a^2 + x^2\right)y_{n+2} + n(2x)y_{n+1} + \frac{n(n-1)}{2!}(2)y_n\right] + \left[xy_{n+1} + n \cdot 1 \cdot y_n\right] = 0,$$

or $\left(a^2 + x^2\right)y_{n+2} + (2n+1)xy_{n+1} + n^2y_n = 0$, which is same as desired.

(c) We have $y^{1/m} + y^{-1/m} = 2x$, that is, $y^{2/m} - 2xy^{1/m} + 1 = 0$. This equation is a quadratic equation. Solving for $y^{1/m}$, we have

$$y^{1/m} = \frac{2x \pm \sqrt{4x^2 - 4}}{2} = x \pm \sqrt{x^2 - 1}, \text{ or } y = \left(x \pm \sqrt{x^2 - 1}\right)^m. \qquad \ldots(i)$$

Therefore, $\quad y_1 = m\left(x \pm \sqrt{x^2 - 1}\right)^{m-1}\left(1 \pm \frac{x}{\sqrt{x^2 - 1}}\right)$

$$= m\left(x \pm \sqrt{x^2 - 1}\right)^{m-1}\left(\frac{\sqrt{x^2 - 1} \pm x}{\sqrt{x^2 - 1}}\right)$$

$$= \pm \frac{m}{\sqrt{x^2 - 1}}\left(x \pm \sqrt{x^2 - 1}\right)^m \text{ or, } y_1 = \pm \frac{m}{\sqrt{x^2 - 1}} \cdot y, \text{ from } (i).$$

Squaring both sides, we obtain $y_1^2\left(x^2 - 1\right) = m^2 y^2$.

Differentiating again w.r.t. x, we get $2y_1y_2\left(x^2 - 1\right) + 2xy_1^2 = 2m^2 yy_1$.

Dividing throughout by $2y_1$, we obtain $\left(x^2 - 1\right)y_2 + xy_1 - m^2 y = 0$.

Differentiating it n times by Leibnitz's theorem, we obtain

$$\left[\left(x^2 - 1\right)y_{n+2} + n(2x)\,y_{n+1} + \frac{n(n-1)}{2!}\cdot(2)\cdot y_n\right] + \left(xy_{n+1} + n\cdot 1\cdot y_n\right) - m^2 y_n = 0$$

or, $\left(x^2 - 1\right)y_{n+2} + (2n+1)\,xy_{n+1} + \left(n^2 - m^2\right)y_n = 0$.

Exercise 1.2

1. Find the nth differential coefficient of the following

(i) $\dfrac{x+1}{x^2 - 4}$, (ii) $\dfrac{4x}{(x-1)^2(x+1)}$, (iii) $\dfrac{1}{1 - 5x + 6x^2}$, (UPTU-2005; A.M.I.E., S-2004)

(iv) $\dfrac{ax+b}{cx+d}$, (v) $\dfrac{x^2 + 4x + 1}{x^3 + 2x^2 - x - 2}$, also find the tenth differential coefficient.

(vi) $\dfrac{x^2}{\left(2x^2 + 7x + 6\right)}$ (Mangalore, 1999), (vii) $\dfrac{ax+b}{x^2 - c^2}$, (viii) $\dfrac{x^2}{(x-1)^2(x+2)}$.

Ans. (i) $\dfrac{(-1)^n\, n!}{4}\left[\dfrac{3}{(x-2)^{n+1}} + \dfrac{1}{(x+2)^{n+1}}\right]$,

(ii) $(-1)^n\, n!\left[\dfrac{1}{(x-1)^{n+1}} - \dfrac{1}{(x+1)^{n+1}} + \dfrac{2(n+1)}{(x-1)^{n+2}}\right]$,

(iii) $n!\left[3^{n+1}(1-3x)^{-n-1} - 2^{n+1}(1-2x)^{-n-1}\right]$,

(iv) $(-1)^n\, n!\, c^{n-1}(bc - ad)(cx+d)^{-n-1}$,

(v) $(-1)^n\cdot n!\left[\dfrac{1}{(x-1)^{n+1}} + \dfrac{1}{(x+1)^{n+1}} - \dfrac{1}{(x+2)^{n+1}}\right]$;

$$10!\left[\dfrac{1}{(x-1)^{11}} + \dfrac{1}{(x+1)^{11}} - \dfrac{1}{(x+2)^{11}}\right],$$

(vi) $(-1)^n n! \left[\dfrac{9(2)^{n-1}}{(2x+3)^{n+1}} - \dfrac{4}{(x+2)^{n+1}} \right]$,

(vii) $\dfrac{(-1)^n n!}{2c} \left[\dfrac{ac-b}{(x+c)^{n+1}} + \dfrac{ac+b}{(x-c)^{n+1}} \right]$,

(viii) $\dfrac{(-1)^n \cdot (n+1)!}{3(x-1)^{n+2}} + \dfrac{5(-1)^n \cdot n!}{9(x-1)^{n+1}} + \dfrac{4(-1)^n \cdot n!}{9(x+2)^{n+1}}$.

2. Find the *n*th derivative of
   (i) $\sin x \sin 2x$, (ii) $e^x \cos^2 x$, (iii) $\cos x \cos 2x \cos 3x$.

Ans. (i) $\dfrac{1}{2} \cos\left(x + \dfrac{1}{2} n\pi \right) - \dfrac{1}{2} 3^n \cos\left(3x + \dfrac{1}{2} n\pi \right)$.

(ii) $\dfrac{1}{2} e^x \left\{ 1 + \left(\sqrt{5}\right)^n \cos\left(2x + n \tan^{-1} 2 \right) \right\}$.

(iii) $y_n = \dfrac{1}{4} \left\{ 2^n \cos\left(2x + \dfrac{n\pi}{2} \right) + 4^n \cos\left(4x + \dfrac{n\pi}{2} \right) + 6^n \cos\left(6x + \dfrac{n\pi}{2} \right) \right\}$.

3. (i) If $y(x) = \tan^{-1}\left(\dfrac{2x}{1-x^2} \right)$, then show that nth derivative of $y(x)$ is

$$\dfrac{d^n y}{dx^n} = 2(-1)^{n-1} (n-1)! \left(1+x^2\right)^{-n/2} \sin\left[n \tan^{-1}\left(\dfrac{1}{x} \right) \right].$$

(UPTU-2001, AMIE, W-1997)

(ii) Find $\dfrac{d^n}{dx^n} \left[\tan^{-1}\left(\dfrac{1+x}{1-x} \right) \right]$ **Ans.** $(-1)^{n-1} \cdot (n-1)! \sin^n \theta \cdot \sin n\theta$; $\theta = \cot^{-1} x$.

(iii) Let $y(x) = (2-3x)^{10}$, then evaluate $d^9 y / dx^9$.   **Ans.** $3^9 (10!)(2-3x)$

(A.M.I.E., W-2009)

4. If $y = x\left(a^2 + x^2\right)^{-1}$, prove that $y_n = (-1)^n n! a^{-n-1} \sin^{n+1} \phi \cos(n+1)\phi$,

   where $\phi = \tan^{-1}(a/x)$.

5. If $u = \sin px + \cos px$, show that $u_n = p^n \left[1 + (-1)^n \sin 2px \right]^{1/2}$, where u_n
   denotes the nth derivative of u w.r.t. x.

6. If $y = \tan^{-1} x$, show that $y_n = (-1)^{n-1}(n-1)! \sin n\theta \sin^n \theta$; $\theta = \cot^{-1} x$.

7. Find the nth derivative of (i) $x^3 \cos x$, (*D.U. 1999*) (ii) $x^3 \log x$, (iii) $x^2 e^{ax}$.

Ans.   (i) $x^3 \cos\left(x + \dfrac{1}{2}n\pi\right) + 3nx^2 \cos\left\{x + \dfrac{1}{2}(n-1)\pi\right\}$

$$+3n(n-1)x\cos\left\{x + \dfrac{1}{2}(n-2)\pi\right\} + n(n-1)(n-2)\cos\left\{x + \dfrac{1}{2}(n-3)\pi\right\},$$

(ii) $(-1)^n (n-4)! 6x^{-n+3}$,   (iii) $a^{n-2}e^{ax}\left\{a^2 x^2 + 2nax + n(n-1)\right\}$.

8. If $I_n = \dfrac{d^n}{dx^n}\left(x^n \log x\right)$, prove that $I_n = nI_{n-1} + (n-1)!$ and hence show that

$$I_n = n!\left(\log x + 1 + \dfrac{1}{2} + \dfrac{1}{3} + ... + \dfrac{1}{n}\right).$$

(DCE-2005; AMIE S-2003; VTU-2001; D.U. 1998).

[Hint: $I_n = D^n\left(x^n \log x\right) = D^{n-1} \cdot D\left(x^n \log x\right)$

$$= D^{n-1}\left(nx^{n-1}\log x + x^{n-1}\right) = nI_{n-1} + (n-1)!$$

Now $\dfrac{I_n}{n!} = \dfrac{I_{n-1}}{(n-1)!} + \dfrac{1}{n}$. Therefore, $\dfrac{I_{n-1}}{(n-1)!} = \dfrac{I_{n-2}}{(n-2)!} + \dfrac{1}{n-1}$

and so on till $\dfrac{I_2}{2!} = \dfrac{I_1}{1!} + \dfrac{1}{2}$. Adding, we get the result].

9. If $y = \sin^{-1} x$, then show that $\left(1 - x^2\right)y_{n+2} - (2n+1)xy_{n+1} - n^2 y_n = 0$.

10. Differentiate the differential equation

$$\left(1 - x^2\right)y_2 - xy_1 - a^2 y = 0, \ n \text{ times with respect to } x.$$

Ans. $\left(1 - x^2\right)y_{n+2} - (2n+1)xy_{n+1} - \left(n^2 + a^2\right)y_n = 0.$

11. If $y = a\cos(\log x) + b\sin(\log x)$, show that

$$x^2 y_2 + xy_1 + y = 0 \text{ and } x^2 y_{n+2} + (2n+1)xy_{n+1} + \left(n^2 + 1\right)y_n = 0.$$

(UPTU 2004; Madras, 2000 PT; AMIE., S-2000)

12. If $y = e^{\tan^{-1} x}$, show that $\left(1 + x^2\right)y_{n+2} + \left[2(n+1)x - 1\right]y_{n+1} + n(n+1)y_n = 0.$

(D.C.E., 2005)

13. If $x = \tan(\log y)$, prove that

(i) $(1+x^2)y_{n+1} + (2nx-1)y_n + n(n-1)y_{n-1} = 0$. [**Hint.** We have $y = e^{\tan^{-1}x}$].

(ii) $(1+x^2)y_n = \{1 - 2(n-1)x\}y_{n-1} - (n-1)(n-2)y_{n-2}$ *(AMIE, S-2009)*
[**Hint.** In result (i), replace n by $(n-1)$]

14. If $y = \cos(m \log x)$, prove that $x^2 y_{n+2} + (2n+1)xy_{n+1} + (m^2 + n^2)y_n = 0$.

15. If $y = \tan^{-1} x$, then prove that $(1+x^2)y_{n+1} + 2nxy_n + n(n-1)y_n = 0$.

16. If $y = x^{n-1} \log x \, (n \geq 1)$, show that $y_n = (n-1)!/x$, y_n being the nth derivative of y with respect to x.   *(A.M.I.E., W-2005)*

[**Hint:** $y_1 = (n-1)x^{n-2} \log x + x^{n-2}$.

Therefore, $xy_1 = (n-1)x^{n-1} \log x + x^{n-1} = (n-1)y + x^{n-1}$.
Differentiating $(n-1)$ times by Leibnitz theorem,

$xy_n + (n-1)y_{n-1} = (n-1)y_{n-1} + (n-1)!$ etc.]

17. If $y = \dfrac{x^n}{1+x}$, show that $y_n = n!/(1+x)^{n+1}$.   *(A.M.I.E., W-2005, W-2004)*

[**Hint:** $y_n = n! \dfrac{1}{1+x} + n \cdot (n!x)\left[-\dfrac{1}{(1+x)^2} \right] + ... + \dfrac{(-1)^n n!}{(1+x)^{n+1}} \cdot x^n$

$$= \frac{n!}{1+x}\left[1 - \frac{nx}{1+x} + \frac{n(n-1)}{2!}\left(\frac{x}{1+x}\right)^2 + ... + (-1)^n\left(\frac{x}{1+x}\right)^n \right]$$

$$= \frac{n!}{1+x}\left(1 - \frac{x}{1+x}\right)^n = n!/(1+x)^{n+1} \,\Bigg].$$

18. If $y = \dfrac{\log(1+x)}{1+x}$, show that $(1+x^2)y_{n+2} + (2n+3)(1+x)y_{n+1}$

$+ (n+1)^2 y_n = 0$.   *(A.M.I.E., S-2004)*

19. Given $y = \dfrac{1}{\sqrt{1+2x}}$, prove that $(1+2x)y_{n+1} + (2n+1)y_n = 0$.

    (A.M.I.E., S-2005)

20. If $y = x^n \log x$, prove that (i) $y_{n+1} = n!/x$, (ii) $y_n = ny_{n-1} + (n-1)!$

(iii) Show that $D^n \left(x^{n-1} \log x \right) = (n-1)!/x$ (UPTU 2001)

21. If $x + y = 1$, prove that

$$\frac{d^n}{dx^n} \left(x^n y^n \right) = n! \left[y^n - \left({}^n c_1 \right)^2 y^{n-1} x + \left({}^n c_2 \right)^2 y^{n-2} x^2 + \cdots\cdots + (-1)^n x^n \right].$$

1.7 Determination of the value of the nth derivative of a function for x = 0

In some cases it is possible to find the value of nth derivative of a function for $x = 0$ directly without finding the general expression for the nth derivative.

The following examples will make the procedure clear.

Example 1.16 If $y = \left(\sin^{-1} x \right)^2$, show that $\left(1 - x^2 \right) y_2 - xy_1 - 2 = 0$ and hence, find the value of y_n at x = 0. (A.M.I.E., W-2006; Nagpur, 1997)

Solution: We have $y = \left(\sin^{-1} x \right)^2$ ...(i)

Therefore, $y_1 = 2 \sin^{-1} x \cdot \dfrac{1}{\sqrt{1 - x^2}}$.

Squaring, we have $\left(1 - x^2 \right) y_1^2 = 4 \left(\sin^{-1} x \right)^2 = 4y.$ ...(ii)

Differentiating w.r.t. x, we obtain $2 \left(1 - x^2 \right) y_1 y_2 - 2xy_1^2 = 4y_1.$

Dividing throughout by $2y_1$, we get $\left(1 - x^2 \right) y_2 - xy_1 - 2 = 0.$ ...(iii)

Differentiating this equation n times by Leibnitz's theorem, we get

$$\left[\left(1 - x^2 \right) y_{n+2} + {}^n c_1 (-2x) y_{n+1} + {}^n c_2 (-2) y_n \right] - \left[xy_{n+1} + {}^n c_1 (1) y_n \right] = 0$$

or, $\left(1 - x^2 \right) y_{n+2} - 2nxy_{n+1} - n(n-1) y_n - xy_{n+1} - ny_n = 0$

or, $\left(1 - x^2 \right) y_{n+2} - (2n+1) xy_{n+1} - n^2 y_n = 0$ ...(iv)

Putting $x = 0$, (i), (ii), (iii) and (iv), and denoting the value of y_n for $x = 0$ by $y_n(0)$, we have

$$y(0) = 0, \; y_1 (0) = 0, \; y_2 (0) = 0, \; y_{n+2} (0) = n^2 y_n (0).$$ ...(v)

Putting $n = 1, 2, 3, 4, \ldots$ in (v), we obtain

$$y_3(0) = 1^2 \cdot y_1(0) = 0, \qquad\qquad y_4(0) = 2^2 \cdot y_2(0) = 2 \cdot 2^2$$

$$y_5(0) = 3^2 \, y_3(0) = 0, \qquad\qquad y_6(0) = 4^2 \cdot y_4(0) = 2 \cdot 2^2 \cdot 4^2$$

$$y_7(0) = 5^2 \, y_5(0) = 0, \qquad\qquad y_8(0) = 6^2 \cdot y_6(0) = 2 \cdot 2^2 \cdot 4^2 \cdot 6^2$$

....

....

In general, $(y_n)_0 = \begin{cases} 0 & \text{if } n \text{ is odd,} \\ 2 \cdot 2^2 \cdot 4^2 \cdot 6^2 \cdots (n-2)^2, n \neq 2, & \text{if } n \text{ is even.} \end{cases}$

Example 1.17 If $y = \sin(m \sin^{-1} x)$, show that

$(1 - x^2) y_2 - xy_1 + m^2 y = 0$ and

$(1 - x^2) y_{n+2} - (2n+1) xy_{n+1} - (n^2 - m^2) y_n = 0$. Hence, find the value of nth derivative of y for $x = 0$. $\qquad$ (A.M.I.E., W-2009, S-2002; UPTU 2005).

Solution. We have $y = \sin(m \sin^{-1} x)$ $\qquad\qquad\qquad\qquad$ ...(i)

$$\therefore \quad y_1 = \cos(m \sin^{-1} x) \cdot \frac{m}{\sqrt{1 - x^2}} \qquad\qquad\qquad\qquad \text{...(ii)}$$

Squaring and multiplying by $(1 - x^2)$, we obtain

$$(1 - x^2) y_1^2 = m^2 \cos^2(m \sin^{-1} x) = m^2 \left[1 - \sin^2(m \sin^{-1} x)\right] = m^2(1 - y^2)$$

or $\quad (1 - x^2) y_1^2 + m^2 y^2 - m^2 = 0.$

Differentiating w.r.t. x, we obtain $2(1 - x^2) y_1 y_2 - 2xy_1^2 + 2m^2 yy_1 = 0.$

Dividing throughout by $2y_1$ we get $(1 - x^2) y_2 - xy_1 + m^2 y = 0$ $\qquad$ ...(iii)

Differentiating this n times by Leibnitz's theorem, we have

$$\left[(1 - x^2) y_{n+2} + {}^n c_1(-2x) y_{n+1} + {}^n c_2(-2) y_n\right] - \left[xy_{n+1} + {}^n c_1(1) y_n\right] + m^2 y_n = 0$$

or $\quad (1 - x^2) y_{n+2} - 2nxy_{n+1} - n(n-1) y_n - xy_{n+1} - ny_n + m^2 y_n = 0$

or $\quad (1 - x^2) y_{n+2} - (2n+1) xy_{n+1} - (n^2 - m^2) y_n = 0.$ $\qquad$ ...(iv)

By Putting $x = 0$ in (i), (ii), (iii) and (iv), we get

$$y(0) = \sin(m \sin^{-1} 0) = 0, \qquad y_1(0) = (\cos 0) \cdot m = m,$$

$$y_2(0) = -m^2 \cdot y(0) = 0, \quad y_{n+2}(0) = \left(n^2 - m^2\right)y_n(0). \qquad ...(v)$$

Putting $n = 1, 2, 3, 4, ...,$ in (v), we obtain

$$y_3(0) = \left(1^2 - m^2\right)y_1(0) = \left(1^2 - m^2\right)m, \quad y_4(0) = \left(2^2 - m^2\right)y_2(0) = 0$$

$$y_5(0) = \left(3^2 - m^2\right)y_3(0) = \left(3^2 - m^2\right)\left(1^2 - m^2\right)m, \quad y_6(0) = \left(4^2 - m^2\right)y_4(0) = 0$$

....

....

In general, $y_n(0) = \begin{cases} m\left(1^2 - m^2\right)\left(3^2 - m^2\right)\cdots\left[(n-2)^2 - m^2\right], & \text{when } n \text{ is odd} \\ 0, & \text{when } n \text{ is even.} \end{cases}$

Example 1.18 If $y = e^{m\cos^{-1}x}$, **show that**

$$(1-x^2)y_{n+2} - (2n+1)xy_{n+1} - (n^2 + m^2)y_n = 0.$$ **Hence, determine** $y_n(0)$.

Solution. We have $y = e^{m\cos^{-1}x}$ $\qquad ...(i)$

Therefore, $y_1 = e^{m\cos^{-1}x} \cdot (-m)/\sqrt{1-x^2}$ $\qquad ...(ii)$

Squaring, we obtain $(1-x^2)y_1^2 = m^2 y^2$. Differentiating w.r.t. x, we have

$$2(1-x^2)y_1 y_2 - 2xy_1^2 = 2m^2 yy_1.$$

Dividing throughout by $2y_1$, we get $(1-x^2)y_2 - xy_1 = m^2 y$ $\qquad ...(iii)$

Differentiating (iii) n times by Leibnitz theorem, we obtain

$$\left[(1-x^2)y_{n+2} - 2nxy_{n+1} - n(n-1)y_n\right] - \left[xy_{n+1} + n\cdot 1\cdot y_n\right] = m^2 y_n$$

or $(1-x^2)y_{n+2} - (2n+1)xy_{n+1} - (n^2 + m^2)y_n = 0.$ $\qquad ...(iv)$

By putting $x = 0$ in (i), (ii), (iii) and (iv), we obtain

$$y(0) = e^{m\cdot\pi/2}, \quad y_1(0) = -me^{m\cdot\pi/2},$$

$$y_2(0) = m^2 \cdot y(0) = m^2 \cdot e^{m\cdot\pi/2}, \quad y_{n+2}(0) = \left(n^2 + m^2\right)y_n(0). \qquad ...(v)$$

Putting $n = 1, 2, 3, 4, ...$ in (v), we get

$$y_3(0) = \left(1^2 + m^2\right)y_1(0) = -m\left(1^2 + m^2\right)e^{m\cdot\pi/2}$$

$$y_4(0) = \left(2^2 + m^2\right)y_2(0) = m^2\left(2^2 + m^2\right)e^{m\cdot\pi/2}$$

$$y_5(0) = \left(3^2 + m^2\right)y_3(0) = -m\left(1^2 + m^2\right)\left(3^2 + m^2\right)e^{m\cdot\pi/2}$$

$$y_6(0) = (4^2 + m^2) y_4(0) = m^2 (2^2 + m^2)(4^2 + m^2) e^{m \cdot \pi/2}$$

....

....

In general,

$$y_n(0) = \begin{cases} -m \cdot e^{m \cdot \pi/2} (1^2 + m^2)(3^2 + m^2) \cdots \left[(n-2)^2 + m^2 \right], \text{ when } n \text{ is odd,} \\ m^2 \cdot e^{m \cdot \pi/2} (2^2 + m^2)(4^2 + m^2) \cdots \left[(n-2)^2 + m^2 \right], \text{ when } n \text{ is even.} \end{cases}$$

Exercise 1.3

1. If $y = \sin^{-1} x$, show that $(1 - x^2) y_{n+2} - (2n+1) xy_{n+1} - n^2 y_n = 0$. Hence, find $y_n(0)$.

   Ans. $y_n(0) = \begin{cases} 1^2 \cdot 3^2 \cdot 5^2 \cdots \cdots (n-2)^2, & \text{when } n \text{ is odd,} \\ 0, & \text{when } n \text{ is even.} \end{cases}$

2. If $y = e^{m \sin^{-1} x}$ or $x = \sin\left(\dfrac{1}{m} \log y \right)$, show that $(1 - x^2) y_2 - xy_1 = m^2 y$, and $(1 - x^2) y_{n+2} - (2n+1) xy_{n+1} - (n^2 + m^2) y_n = 0$. Hence find the value of y_n when $x = 0$. 　　　*(GGSIPU-2006; A.M.I.E., S-2005) (Gorakhpur, 1999)*

   Ans. $y_n(0) = \begin{cases} m(1^2 + m^2)(3^2 + m^2) \cdots \left[(n-2)^2 + m^2 \right], \text{ when } n \text{ is odd,} \\ m^2 (2^2 + m^2)(4^2 + m^2) \cdots \left[(n-2)^2 + m^2 \right], \text{ when } n \text{ is even.} \end{cases}$

3. If $y = \tan^{-1} x$, show that $(1 + x^2) y_2 + 2xy_1 = 0$ and $(1 + x^2) y_{n+2} + 2(n+1) xy_{n+1} + n(n+1) y_n = 0$. Hence find the $y_n(0)$. 　　　　　　　　　　　　　　　*(AMIE S-2008; UPTU-2003)*

   Ans. $y_n(0) = \begin{cases} (-1)^n \cdot (2n)!, \text{ when } n \text{ is odd,} \\ 0, \text{ when } n \text{ is even.} \end{cases}$

4. If $y = \left[\log\left(x + \sqrt{a^2 + x^2} \right) \right]^2$, show that $a^2 y_{n+2}(0) = -n^2 y_n(0)$.

5. If $y = \sin^{-1} x / \sqrt{1 - x^2}$, prove that $(1 - x^2) y_{n+1} - (2n+1) xy_n - n^2 y_{n-1} = 0$.

   Also show that $y_n = (n-1)^2 y_{n-2}$ for $x = 0$.

6. Given $y = 1 / \sqrt{1 + 2x}$, prove that $(1 + 2x) y_{n+1} + (2n+1) y_n = 0$. Also show that $y_n(0) = (1 - 2n) y_{n-1}(0)$. 　　　　　　　　　　　*(A.M.I.E., S-2005)*

7. If $y = (3x+2)/(x^2 - 2x + 5)$, prove that for $n \geq 2$,

   $5y_n(0) = 2ny_{n-1}(0) - n(n-1)y_{n-2}(0)$. Hence compute $y_2(0)$.

   Ans. $56/125$.

8. If $y = \log\left(x + \sqrt{1+x^2}\right)$, find $y_n(0)$.

   Ans. $y_{2n}(0) = 0; y_{2n+1}(0) = (-1)^n \cdot 1^2 \cdot 3^2 \cdot 5^2 \dots\dots (2n-1)^2$.

9. If $y = \left[x + \sqrt{(x^2+1)}\right]^m$, show that $(1+x^2)y_{n+2} + (2n+1)xy_{n+1} + (n^2 - m^2)y_n = 0$. Hence find $y_n(0)$.

   Ans.

   $$y_n(0) = \begin{cases} m(m^2 - 1^2)(m^2 - 3^2)(m^2 - 5^2)\cdots\left[m^2 - (n-2)^2\right]; \text{when } n \text{ is odd,} \\ m^2(m^2 - 2^2)(m^2 - 4^2)\cdots\left[m^2 - (n-2)^2\right]; \text{when } n \text{ is even.} \end{cases}$$

Mean Value Theorems, Taylor's Theorem with Remainder and Taylor's Series, Expansions of Functions

2.1 Rolle's Theorem

Let a real valued function f(x) be continuous at every point on a closed interval [a, b] that is, on a $\leq x \leq b$, differentiable at every point in the open interval (a, b) that is, on a < x < b, and f(a) = f(b), then there exists at least one real number c between a and b (a < c < b) such that f'(c) = 0.

Proof. Since the function $f(x)$ is continuous on the closed interal [a, b], it is bounded and attains its maximum value M and minimum value m at some points on [a, b]. Let the function $f(x)$ attain respectively its minimum and maximum values at the points c and $d \in [a, b]$, that is

$$f(c) = m \quad and \quad f(d) = M.$$ There are two possibilities

Case I. If $m = M$, then the function $f(x)$ is constant over [a, b], which means that for all values of x it has a constant value $f(x) = m$ and therefore, its derivative $f'(x)$ is zero for all $x \in [a, b]$.

Case II. If $m \neq M$, then both of these cannot be equal to the same quantity $f(a)$ or $f(b)$. We note that $f(a) = f(b)$. Thus, at least one of these numbers say m, is different from $f(a)$ and $f(b)$. Hence,

$$f(c) = m \neq f(a), \quad \text{implies} \quad c \neq a,$$

$$f(c) = m \neq f(b), \quad \text{implies} \quad c \neq b.$$

This means that c lies in the open interval (a, b). We shall now show that c is the point where $f'(c) = 0$.

If $f'(c) < 0$, then for every x in the interval $(c, c + \delta_1)$, $\delta_1 > 0$, $f(x) < f(c) = m$, which contradicts the assumption that m is the minimum value of $f(x)$. If $f'(c) > 0$, then for every x in the interval $(c - \delta_2, c)$, $\delta_2 > 0$, $f(x) < f(c) = m$, which is again a contradiction. Hence, the only possibility. $f'(c) = 0$.

The truth of the above theorem is obvious from the following geometrical interpretation of the same.

Geometrical interpretation of Rolle's Theorem

If a continuous curve is drawn in [a, b] such that the ordinates at a and b are equal, then there exists at least one point c lying between a and b such that the tangent line at (c, f(c)) is parallel to the x-axis. [Refer Fig. 2.1(a)]

Remark. There can be more than one point $c \in (a, b)$ such that $f'(c) = 0$ [Refer Fig. 2.1(b)].

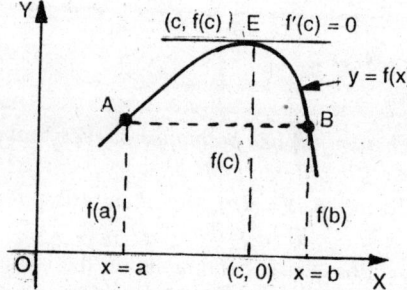

Fig. 2.1(a) Rolle's theorem

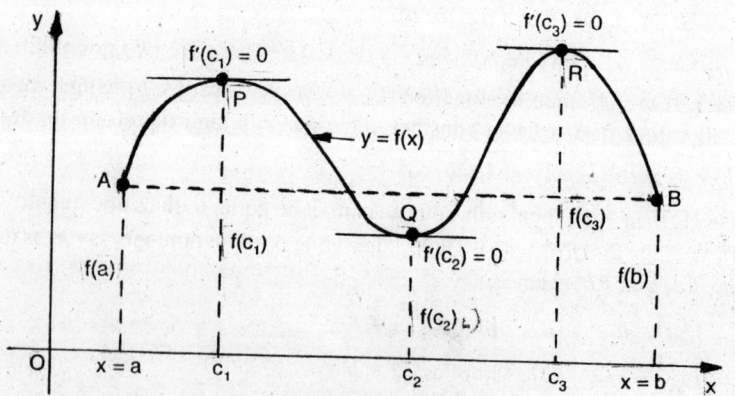

Fig. 2.1(b) Rolle's theorem

2.2 Solved Problems

Example 2.1 Does Rolle's theorem apply to the functions:

(a) $f(x) = (x^2 - 4x)/(x-2)$, and (b) $f(x) = (x^2 - 4x)/(x+2)$ on [0, 4]?

Solution. (a) $f(x) = 0$ when the values of x are 0 and 4. Since $f(x)$ has a discontinuity at $x = 2$, a point on [0, 4], the theorem does not apply.

(b) $f(x) = 0$ when $x = 0, 4$. $f(x)$ has a discontinuity at $x = -2$, a point not on [0,4]. In addition,

$$f'(x) = \frac{(2x-4)(x+2)-(x^2-4x)}{(x+2)^2} = \frac{(x^2+4x-8)}{(x+2)^2}$$

exists everywhere except at $x = -2$.

So, the theorem applies and $c = \left(-4 \pm \sqrt{16+32}\right)/2 = 2\left(\sqrt{3}-1\right) = 1.464$, the positive root of $x^2 + 4x - 8 = 0$.

Example 2.2 Verify Rolle's theorem for each of the following functions:

(i) $f(x) = x^3 + 5x^2 - 6x, \ x \in [0,1]$.

(ii) $f(x) = (x-1)(x-2)(x-3), \ x \in [1,3]$. *(GGSIPU 2006)*

(iii) $f(x) = x(x+2) \ e^{-x/2}$, **on** [-2, 0]. *(A.M.I.E. S. 1996)*

Solution. (i) $f(x) = x^3 + 5x^2 - 6x$, being a polynomial function is continuous in the closed interval [0, 1].

$f'(x) = 3x^2 + 10x - 6$, which exists in the open interval (0, 1). Therefore, $f(x)$ is derivable in the open interval (0, 1). $f(0) = 0 = f(1)$. Thus, $f(x)$ satisfies all the conditions of Rolle's theorem. Hence there must be atleast one point $x = c$ in the open interval $0 < x < 1$ such that $f'(c) = 0 \Rightarrow 3c^2 + 10c - 6 = 0$ or

$$c = \left(-10 \pm \sqrt{100+72}\right)/6 = (-5 \pm 6.56)/3 = -3.85 \text{ or } 0.52.$$

Of these two values of c, for which $f'(c) = 0$, 0.52, belongs to the interval (0, 1) under consideration. As $0 < 0.52 < 1$. Hence, Rolle's theorem is verified.

(ii) $f(x)$ being a polynomial function is continuous on [1, 3] and derivable in (1, 3). Also $f(1) = 0 = f(3)$. Thus, all the conditions of Rolle's theorem are satisfied by $f(x)$. Now, we have to show that there exists at least one, point $x = c$ in the open interval $1 < x < 3$, such that $f'(c) = 0$.

we have $f(x) = (x-1)(x-2)(x-3) = x^3 - 6x^2 + 11x - 6$.

$f'(x) = 3x^2 - 12x + 11. \ f'(c) = 0 \Rightarrow 3c^2 - 12x + 11 = 0$.

Therefore, $c = \left(12 \pm \sqrt{144-132}\right)/6 = 1.42 \text{ or } 2.57$.

Both these values of c, for which $f'(c) = 0$, belongs to the interval (1, 3) under consideration. Hence Rolle's theorem is verified.

(*iii*) We have $f(x) = x(x+2)e^{-x/2}$. $f(x)$ is continuous in the closed interval $(-2, 0)$ derivable in the open interval $(-2, 0)$ and also $f(-2) = 0 = f(0)$. Thus, $f(x)$ satisfies all the conditions of Rolle's theorem.

Hence there is atleast one point $x = c$ in the open interval $-2 < x < 0$ such that $f'(c) = 0$.

$$\text{Now } f'(x) = (2x+2)e^{-x/2} + x(x+2)e^{-x/2}(-1/2) = \frac{\left(-x^2+2x+4\right)}{2}e^{-x/2}.$$

$f'(c) = 0 \Rightarrow c^2 - 2c - 4 = 0$ or $c = 1 \pm \sqrt{5}$. Of these two values of c for which $f'(c)$ is zero, $1 - \sqrt{5}$ that is, -1.24 belongs to the interval $(-2, 0)$. Hence, Rolle's theorem is verified.

Example 2.3. It is given that the Rolle's theorem holds for the function $f(x) = x^3 + qx^2 + px, 1 \le x \le c$ **at the point x = 4/3. Find the values of p and q.**

Solution: The conclusion of Rolle's theorem gives $f'(c) = 0$. $f'(c) = 3c^2 + 2qc + p$. Therefore $3c^2 + 2qc + p = 0$. But $c = 4/3$ (given). Therefore

$$3p + 8q + 16 = 0. \qquad \qquad ...(i)$$

Also, Rolle's theorem holds for $f(x)$ on $[1, 2]$.

$$\therefore f(1) = f(2) \Rightarrow p+q+1 = 2p+4q+8 \quad \text{or} \quad p+3q+7 = 0 \qquad ...(ii)$$

Solving these two equations for p and q, we obtain $p = 8$ and $q = -5$.

Example 2.4 Verify Rolle's theorem for the following functions

(i) $f(x) = \begin{cases} x^2 + 1, & \text{when } 0 \le x \le 1 \\ 3 - x. & \text{when } 1 < x \le 2 \end{cases}$

(ii) $f(x) = e^x (\sin x - \cos x)$ on $[\pi/4, 5\pi/4]$.

Solution. (*i*) Since a polynomial function is ever where continuous and differential, therefore $f(x)$ is continuous and differentiable at all points except possibly at $x = 1$.

Now, we consider the differentiability of $f(x)$ at $x = 1$.

We have $Rf'(1) = \lim\limits_{x \to 1^+} \dfrac{f(x)-f(1)}{x-1} = \lim\limits_{x \to 1} \dfrac{(3-x)-(1+1)}{x-1} = \lim\limits_{x \to 1} \dfrac{-(x-1)}{x-1} = -1.$

Also $Lf'(1) = \lim\limits_{x \to 1^-} \dfrac{f(x)-f(1)}{x-1} = \lim\limits_{x \to 1} \dfrac{(x^2+1)-(1+1)}{x-1} = \lim\limits_{x \to 1} \dfrac{x^2-1}{x-1}$

$= \lim\limits_{x \to 1} (x+1) = 2.$

Since $Rf'(1) \neq Lf'(1)$, it follows that $f'(1)$ does not exist.

Thus, the condition of differentiability at each point of the given interval is not satisfied.

Hence, Rolle's theorem is not applicable to $f(x)$ on the given interval.

(ii) $f(x) = e^x (\sin x - \cos x)$.

Since an exponential function and sine and cosine functions are everywhere continuous and differentiable, therefore $f(x)$ is continuous on $[\pi/4, 5\pi/4]$ and differentiable on $(\pi/4, 5\pi/4)$.

Now, $\qquad f'(x) = e^x (\cos x + \sin x) + e^x (\sin x - \cos x) = 2e^x \cdot \sin x$,

Also, $f(\pi/4) = e^{\pi/4} (\sin \pi/4 - \cos \pi/4) = e^{\pi/4} (1/\sqrt{2} - 1/\sqrt{2}) = 0$,

$f(5\pi/4) = e^{5\pi/4} (\sin 5\pi/4 - \cos 5\pi/4) = e^{5\pi/4} (-1/\sqrt{2} + 1/\sqrt{2}) = 0$.

$\therefore f(\pi/4) = f(5\pi/4)$.

Thus, all the three conditions required for the applicability of Rolle's theorem are satisfied by $f(x)$. Hence there exists $c \in (\pi/4, 5\pi/4)$ such that $f'(c) = 0$.

$\Rightarrow 2e^c \sin c = 0 \Rightarrow \sin c = 0 (\because e^c \neq 0 \text{ for real } c) \Rightarrow c = \pi \in (\pi/4, 5\pi/4)$.

Hence, Rolle's theorem is verified.

Example 2.5 Prove that if $a_0, a_1, \ldots\ldots, a_n$ are real numbers such that

$$\frac{a_0}{n+1} + \frac{a_1}{n} + \ldots\ldots + \frac{a_{n-1}}{2} + a_n = 0, \qquad \ldots(i)$$

then there exists at least one real number x between 0 and 1 such that

$a_0 x^n + a_1 x^{n-1} + \ldots\ldots + a_{n-1} x + a_n = 0$.

Solution. Define a function f as follows:

$$f(x) = \frac{a_0}{n+1} x^{n+1} + \frac{a_1}{n} x^n + \ldots + \frac{a_{n-1}}{2} x^2 + a_n x, \ x \in [0,1] \qquad \ldots(ii)$$

Since f is a polynomial in x, therefore

(i) f is continuous on $[0, 1]$, (ii) f is derivable on $(0, 1)$,

(iii) $f(0) = f(1)$, since $f(0) = 0$ and $f(1) = 0$, by (i) and (ii). Thus there exists some $x \in (0, 1)$ such that $f'(x) = 0$.

Now $f'(x) = \dfrac{a_0}{(n+1)} \cdot (n+1) x^n + \dfrac{a_1}{n} \cdot n x^{n-1} + \ldots + \dfrac{a_{n-1}}{2} \cdot 2x + a_n \cdot 1 = 0$

Hence $a_0 x^n + a_1 x^{n-1} + \ldots + a_{n-1} x + a_n = 0$.

Example 2.6 Show that there is no real number k for which the equation $x^2 - 3x + k = 0$ has two distinct roots in [0, 1].

Solution. Suppose, on the contrary, that there is a real number k for which the given equation has two distinct roots α and β in [0, 1], where $\alpha < \beta$. Obviously,

$[\alpha, \beta] \subset [0,1]$. ...(i) Consider $f(x) = x^2 - 3x + k$ in $[\alpha, \beta]$.

Since $f(\alpha) = f(\beta) = 0$, the conditions of Rolle's theorem are satisfied. Hence there exists some $c \in (\alpha, \beta)$ such that

$$f'(c) = 0 \Rightarrow 2c - 3 = 0 \Rightarrow c = 3/2 \notin (0, 1) \supset (\alpha, \beta).$$

This is a contradiction to the condition that $c \in (\alpha, \beta)$. Hence our assumption is wrong and therefore the result follows.

Example 2.7 Show that between any two roots of $e^x \cos x = 1$, there exists atleast one root of $e^x \sin x - 1 = 0$.

Solution. Let α and β be any two distinct roots of $e^x \cos x = 1$.

∴ $\qquad\qquad e^\alpha \cos \alpha = 1$ and $e^\beta \cos \beta = 1$. ...(i)

Define a function f as follows:

$$f(x) = e^{-x} - \cos x \ \forall x \in [\alpha, \beta]. \qquad\qquad ...(ii)$$

Obviously f is continuous in $[\alpha, \beta]$ and f is derivable in (α, β).

Now $f'(x) = -e^{-x} + \sin x \ \forall x \in (\alpha, \beta)$.

From (ii), $f(\alpha) = e^{-\alpha} - \cos \alpha = \dfrac{1 - e^\alpha \cos \alpha}{e^\alpha} = 0$, by (i).

Similarly, $f(\beta) = 0$ and so $f(\alpha) = f(\beta)$.

Thus f satisfies all the conditions of Rolle's theorem in $[\alpha, \beta]$ and so there exists some $\gamma \in (\alpha, \beta)$ such that $f'(\gamma) = 0 \Rightarrow \sin \gamma - e^{-\gamma} = 0 \Rightarrow e^\gamma \sin \gamma - 1 = 0$, $\alpha < \gamma < \beta$.

Hence γ is a root of $e^x \sin x - 1 = 0$, $\alpha < \gamma < \beta$.

2.3 Lagrange's mean value theorem (First mean value theorem of Differential Calculus)

Theorem: *Let f(x) be a real valued function which is continuous on a closed interval [a, b] and differentiable in the open interval (a, b), then there exists atleast one real number c between a and b that is, a < c < b, such that*

$$f'(c) = \frac{f(b) - f(a)}{b - a}.$$

Proof. Consider the function

$$F(x) = f(x) + Ax, \qquad a \le x \le b$$

where A is a constant to be determined such that F(a) = F(b). We have

$F(a) = f(a) + Aa$ and $F(b) = f(b) + Ab$. Using $F(a) = F(b)$, we obtain

$$A = -\frac{f(b) - f(a)}{b - a}. \text{ Therefore, } F(x) = f(x) - \left[\frac{f(b) - f(a)}{b - a}\right]x.$$

Now, $F(x)$ is continuous on the closed interval $[a, b]$ and differentiable in the open interval (a, b) and F(a) = F(b). Since, the function F(x) satisfies all the conditions of Rolle's Theorem, there exists a point c, $a < c < b$ such that

$$\mathbf{F'(c) = 0, \text{ or } f'(c) = \frac{f(b) - f(a)}{b - a}.} \qquad \qquad ...(2.1)$$

or $$\mathbf{f(b) = f(a) + (b - a)f'(c)} \qquad \qquad ...(2.2)$$

2.3.1 Another form of the Lagrange's mean value theorem (L.M.V.Th.)

If in the Formula (2.2), b is replaced by $a + h$, then the number c between a and b may be written as $a + \theta h$, where $0 < \theta < 1$. Thus

$$f(a + h) - f(a) = hf'(a + \theta h)$$

or $\mathbf{f(a + h) = f(a) + hf'(a + \theta h)}$, where $0 < \theta < 1$. ...(2.3)

2.3.2 Geometrical Interpretation

The theorem states that somewhere between two points A and B of the graph of f there exists at least one point where the tangent to the curve at this point is parallel to the chord AB.

Refer Fig. 2.2. The slope of the tangent to f at $P(c, f(c))$ is given by $f'(c)$. Also the slope of the chord joining the points $A(a, f(a))$ and $B(b, f(b))$ is $\dfrac{f(b) - f(a)}{b - a}$. Since the two lines are parallel, we have

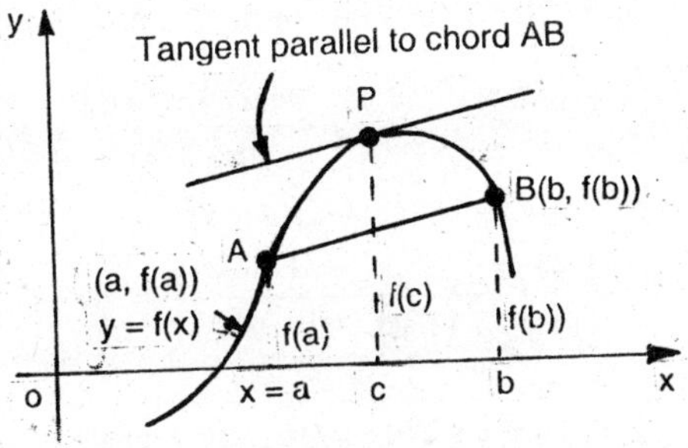

$$f'(c) = \frac{f(b) - f(a)}{b - a}.$$

Fig. 2.2

Geometrically, the Lagrange's mean value theorem states that there exists a point $(c, f(c))$, $a < c < b$, on the curve $y = f(x)$, $a \le x \le b$, such that the tangent to the curve at this point is parallel to the chord joining the points $(a, f(a))$ and $(b, f(b))$ on the curve.

Remark If $f(a) = f(b)$, then Lagrange's mean value theorem reduces to the Rolle's theorem.

Example 2.8 **Verify Lagrange's mean value theorem for the following functions:**

(*i*) $f(x) = x(x^2 - 6)$ on $[1, 3]$.

(*ii*) $f(x) = x(x-1)(x-2)$ on $[0, 1/2]$.   (*A.M.I.E., W-2006*)

(*iii*) $f(x) = |x|$ on $[-1, 1]$.

Solution. (*i*) $f(x) = x^3 - 6x$ is a polynomial function. Since polynomial functions are continuous everywhere on R, $f(x)$ is continuous on $1 \le x \le 3$.

Also, $f'(x) = 3x^2 - 6$, exists for all $x \in (1, 3)$. So $f(x)$ is differentiable on $(1, 3)$.

Thus, $f(x)$ satisfies both the conditions required for the applicability of L.M.V.T. and so there must be at least one point $x = c$ on $1 < x < 3$ such that

$$\frac{f(3) - f(1)}{3 - 1} = f'(c) \Rightarrow \frac{(3^3 - 6 \times 3) - (1 - 6)}{3 - 1} = 3c^2 - 6$$

$$\Rightarrow \frac{9 + 5}{2} = 3c^2 - 6 \Rightarrow 3c^2 = 13 \Rightarrow c = \pm\sqrt{13/3} = \pm 2.08.$$

Of these two values, $c = 2.08$ belongs to $(1, 3)$. Hence the Lagranges's Mean Value theorem is verified.

(*ii*) $f(x) = x(x-1)(x-2) = x^3 - 3x^2 + 2x$, which is a polynomial function and so continuous on $[0, 1/2]$.

Also $f'(x) = 3x^2 - 6x + 2$ is also a polynomial function and exists on $(0, 1/2)$. The L.M.V.T. can be applied to $f(x)$. Hence, there is at least one point $x = c$ on $0 < x < 1/2$ such that $\dfrac{f(1/2) - f(0)}{(1/2) - 0} = f'(c)$

$$\Rightarrow \frac{(1/2)^3 - 3(1/2)^2 + 2(1/2) - 0}{(1/2) - 0} = 3c^2 - 6c + 2 \Rightarrow 12c^2 - 24c + 5 = 0$$

$$\Rightarrow c = \frac{24 \pm \sqrt{576 - 240}}{24} = \frac{24 \pm 4\sqrt{21}}{24} = 1 \pm \frac{1}{6}\sqrt{21}.$$

Taking the minus sign, $c = 0.236$ which lies between 0 and 1/2. Hence L.M.V.T. is verified for $f(x)$ on $[0, 1/2]$.

(*iii*) $f(x) = |x| = \begin{cases} x, & x \ge 0 \\ -x & x < 0 \end{cases}$

Since absolute value functions are continuous every where on R, $f(x)$ is continuous in the closed interval $[-1, 1]$.

Let us find the differential coefficient of $f(x) = |x|$ at $x = 0$ which is a point on $-1 < x < 1$.

We have $Rf'(0) = \lim\limits_{x \to 0^+} \dfrac{f(x) - f(0)}{x - 0} = \lim\limits_{h \to 0} \dfrac{f(0+h) - f(0)}{h}$

$$= \lim\limits_{h \to 0} \dfrac{h - 0}{h} = 1 \quad \left[\text{Since } |h| = h\right]$$

Also $Lf'(0) = \lim\limits_{x \to 0^-} \dfrac{f(x) - f(0)}{x - 0} = \lim\limits_{h \to 0} - \dfrac{f(0-h) - f(0)}{-h}$

$$= \lim\limits_{h \to 0} \dfrac{h - 0}{-h} = -1 \quad \left[\text{Since } |-h| = h\right]$$

Since $Rf'(0) \neq Lf'(0)$, it follows that $f'(0)$ does not exist.

Thus L.M.V.T. is not applicable to $f(x)$.

Example 2.9 At what point the tangent to the curve $y = \log x$ is parallel to the chord joining the points A (1, 0) and B (e, 1)? *(A.M.I.E., W-2003)*

Solution. Slope of the tangent to the given curve at any point $(x, f(x))$ is given by $f'(x) = 1/x$. Also the slope of the chord joining $(1, 0)$ and $(e, 1)$ is $\dfrac{1-0}{e-1}$, that is,

$1/(e-1)$. By L.M.V.Th, the chord is parallel to the tangent at some point c.

Therefore $f'(c) = 1/(e-1) \Rightarrow 1/c = 1/(e-1) \Rightarrow c = (e-1)$.

$\left[\text{Note that } (e-1) \in (1, e)\right] \Rightarrow f(c) = \log c = \log(e-1)$.

Hence, the point where the tangent to the curve is parallel to the given chord is $(e-1, \log(e-1))$.

Example 2.10 Let f be defined and continuous on $[a-h, a+h]$ and derivable on $(a-h, a+h)$. Prove that there is a real number θ between 0 and 1 such that

$$f(a+h) - f(a-h) = h[f'(a+\theta h) + f'(a-\theta h)]. \qquad \textit{(D.U., 1995)}$$

Solution. Define a function $\phi(x)$ on $[0, 1]$ as follows:

$$\phi(x) = f(a+hx) - f(a-hx). \qquad \qquad ...(i)$$

$\therefore \qquad \phi'(x) = hf'(a+hx) + hf'(a-hx). \qquad \qquad ...(ii)$

As x varies over $[0, 1]$, $a + hx$ varies over $[a, a+h]$ and $a - hx$ varies over $[a-h, a]$. Thus ϕ is continuous on $[0, 1]$, ϕ is derivable on $(0, 1)$. Hence, by Lagrange's mean value theorem, there exists θ, $0 < \theta < 1$, satisfying

$$\frac{\phi(1) - \phi(0)}{1 - 0} = \phi'(\theta) \text{ or } \phi(1) - \phi(0) = \phi'(\theta).$$

Hence $f(a+h) - f(a-h) = h[f'(a+\theta h) + f'(a-\theta h)]$, by (i) and (ii).

Example 2.11 Let f be defined and continuous on $[a-h, a+h]$ and derivable on $(a-h, a+h)$. Prove that there is a real number θ between 0 and 1 for which

$$\mathbf{f(a+h) - 2f(a) + f(a-h) = h[f'(a+\theta h) - f'(a-\theta h)].} \quad (D.U., 1996)$$

Solution. Consider $\phi(x) = f(a+hx) + f(a-hx)$ on $[0, 1]$. \quad ...(i)

ϕ is continuous in $[0, 1]$, ϕ is derivable on $(0, 1)$. Thus by Lagrange's mean value theorem,

$$\frac{\phi(1) - \phi(0)}{1 - 0} = \phi'(\theta), \text{ for some } \theta \in (0, 1).$$

$$\text{or } \phi(1) - \phi(0) = \phi'(\theta).$$

or $[f(a+h) + f(a-h)] - [f(a) + f(a)] = h[f'(a+\theta h) - f'(a-\theta h)]$

Hence $\quad f(a+h) - 2f(a) + f(a-h) = h[f'(a+\theta h) - f'(a-\theta h)]$.

Example 2.12 Using the Lagrange's mean value theorem, show that

(i) $|\tan^{-1}a - \tan^{-1}b| \le |a - b| \, \forall a, b \in R$,

(ii) $1 + x < e^x < 1 + xe^x \, \forall \, x > 0$,

(iii) $\dfrac{x}{1+x^2} < \tan^{-1}x < x$, if $x > 0$.

Solution. (i) Let $f(x) = \tan^{-1} x$ in $[a, b]$, where $a < b$.

By Lagrange's mean value theorem, there exists some $c \in (a, b)$ such that

$$f(b) - f(a) = (b-a) f'(c).$$

or $$f(b) - f(a) = (b-a) \cdot \frac{1}{1+c^2} \qquad \left[\because f'(x) = \frac{1}{1+x^2} \right]$$

$$\therefore \qquad f(b) - f(a) \le (b - a). \qquad\qquad \left[\because 1 + c^2 \ge 1\right]$$

Similarly $f(a) - f(b) \le (a - b)$, when $b < a$

$$\therefore \qquad \left|f(a) - f(b)\right| \le \left|a - b\right|.$$

Hence $\qquad \left|\tan^{-1} a - \tan^{-1} b\right| \le \left|a - b\right| \forall a, b \in R.$

(*ii*) Let $f(x) = e^x$ on $\left[0, x\right]$. The function $f(x) = e^x$ is derivable on $\left(0, x\right)$ and there exists some c, $0 < c < x$.

Applying Lagrange's mean value theorem to the function $f(x) = e^x$ on $\left[0, x\right]$,

we obtain $\dfrac{e^x - e^0}{x - 0} = f'(c) = e^c$

or $\dfrac{1}{x}\left(e^x - 1\right) = e^c$ ...(*i*). Since $0 < c < x$, so $1 = e^0 < e^c < e^x$ ...(*ii*)

From (*i*) and (*ii*), $1 < \dfrac{1}{x}\left(e^x - 1\right) < e^x \Rightarrow x < e^x - 1 < xe^x$ $(\because x > 0)$

Hence $1 + x < e^x < 1 + xe^x$ $\forall$ $x > 0$.

(*iii*) Let $f(x) = \tan^{-1} x$ in $\left[0, x\right]$. f satisfies the conditions of Lagrange's mean value theorem in $\left[0, x\right]$, there exists some θ satisfying $0 < \theta < 1$ such that

$$\frac{f(x) - f(0)}{x - 0} = f'(\theta x) \qquad \left[f'(x) = \frac{1}{1 + x^2} \text{ and } f(0) = 0\right]$$

or $\qquad \tan^{-1} x = x / \left(1 + \theta^2 x^2\right).$ ...(*i*)

Now $0 < \theta < 1$ and $x > 0 \Rightarrow \theta^2 x^2 < x^2 \Rightarrow 1 + \theta^2 x^2 < 1 + x^2$

$$\Rightarrow \frac{x}{1 + \theta^2 x^2} > \frac{x}{1 + x^2} \Rightarrow \frac{x}{1 + x^2} < \frac{x}{1 + \theta^2 x^2} \qquad \text{...(}ii\text{)}$$

Again $0 < \theta < 1$ and $x > 0 \Rightarrow 1 < 1 + \theta^2 x^2$

$$\Rightarrow \frac{1}{1 + \theta^2 x^2} < 1 \Rightarrow \frac{x}{1 + \theta^2 x^2} < x \qquad \text{...(}iii\text{)}$$

From (*ii*) and (*iii*), we obtain

$$\frac{x}{1 + x^2} < \frac{x}{1 + \theta^2 x^2} < x \qquad \text{...(}iv\text{)}$$

From (*i*) and (*iv*), we obtain

$$\frac{x}{1+x^2} < \tan^{-1} x < x, \quad x > 0.$$

Example 2.13 Applying Lagrange's mean value theorem to the function defined by $f(x) = \log(1+x)$ for all $x > 0$, show that

(i) $\dfrac{x}{1+x} < \log(1+x) < x, \qquad x > 0$

(ii) $0 < \left[\log(1+x)\right]^{-1} - x^{-1} < 1, \quad$ **whenever $x > 0$.** *(A.M.I.E., W-2004)*

Solution. (*i*) Let $f(x) = \log(1+x)$ in $[0, x]$, so that $f'(x) = \dfrac{1}{1+x}$. Since f is

continuous in $[0, x]$ and derivable in $(0, x)$, so by Lagrange's mean value theorem, there exists some $\theta, 0 < \theta < 1$ such that

$$\frac{f(x) - f(0)}{x - 0} = f'(\theta x) \qquad (\because (0, x) = (0x, 1x), c = \theta x)$$

or $\log(1+x) = x/(1+\theta x)$ ...(i) $\qquad (\because f(0) = 0)$

Now $0 < \theta < 1$ and $x > 0 \Rightarrow \theta x < x$

$\Rightarrow 1 + \theta x < 1 + x \Rightarrow \dfrac{1}{1+\theta x} > \dfrac{1}{1+x} \Rightarrow \dfrac{x}{1+\theta x} > \dfrac{x}{1+x} \Rightarrow \dfrac{x}{1+x} < \dfrac{x}{1+\theta x},$ ...(ii)

Again $0 < \theta < 1$ and $x > 0 \Rightarrow 1 < 1 + \theta x \Rightarrow \dfrac{1}{1+\theta x} < 1 \Rightarrow \dfrac{x}{1+\theta x} < x$ ...(iii)

From (*ii*) and (*iii*), we obtain

$$\frac{x}{1+x} < \frac{x}{1+\theta x} < x \qquad \qquad ...(iv)$$

From (*i*) and (*iv*), we obtain $\dfrac{x}{1+x} < \log(1+x) < x$, the required result.

(*ii*) In part (*i*), we have prove $\dfrac{x}{1+x} < \log(1+x) < x$

$\Rightarrow \dfrac{1+x}{x} > \dfrac{1}{\log(1+x)} > \dfrac{1}{x} \Rightarrow 1 + \dfrac{1}{x} > \dfrac{1}{\log(1+x)} > \dfrac{1}{x}$

$\Rightarrow 1 > \dfrac{1}{\log(1+x)} - \dfrac{1}{x} > 0.$ Hence $0 < \left[\log(1+x)\right]^{-1} - x^{-1} < 1,$ $\qquad x > 0.$

Example 2.14 (i) Prove that

$$\frac{b-a}{1+b^2} < \tan^{-1}b - \tan^{-1}a < \frac{b-a}{1+a^2}, \text{ if } a < b.$$

(ii) Show that $\dfrac{\pi}{4} + \dfrac{3}{25} < \tan^{-1}\dfrac{4}{3} < \dfrac{\pi}{4} + \dfrac{1}{6}.$ (*A.M.I.E., W-2008; D.U. 1995*)

Solution. (*i*) Let $f(x) = \tan^{-1}x$. Since $f'(x) = \dfrac{1}{1+x^2}$ and $f'(c) = \dfrac{1}{1+c^2}$, we

have by the mean value theorem

$$\frac{\tan^{-1}b - \tan^{-1}a}{b-a} = \frac{1}{1+c^2}, a < c < b$$

Since $c > a \Rightarrow 1+c^2 > 1+a^2 \Rightarrow \dfrac{1}{\left(1+c^2\right)} < \dfrac{1}{\left(1+a^2\right)}.$

Since $c < b \Rightarrow 1+c^2 < 1+b^2 \Rightarrow \dfrac{1}{\left(1+c^2\right)} > \dfrac{1}{\left(1+b^2\right)}.$ Then

$$\frac{1}{1+b^2} < \frac{\tan^{-1}b - \tan^{-1}a}{b-a} < \frac{1}{1+a^2}$$

Hence $\dfrac{b-a}{1+b^2} < \tan^{-1}b - \tan^{-1}a < \dfrac{b-a}{1+a^2}.$

(*ii*) Taking $a = 1$ and $b = 4/3$ in the result of part (*i*). Then since $\tan^{-1}1 = \pi/4$,

we have

$$\frac{3}{25} < \tan^{-1}\frac{4}{3} - \tan^{-1}1 < \frac{1}{6}, \text{ or } \frac{3}{25} < \tan^{-1}\frac{4}{3} - \frac{\pi}{4} < \frac{1}{6}.$$

Hence $\qquad \dfrac{\pi}{4} + \dfrac{3}{25} < \tan^{-1}\dfrac{4}{3} < \dfrac{\pi}{4} + \dfrac{1}{6}.$

Example 2.15 A twice differentiable function f is such that $f(a) = f(b) = 0$ and $f(c) > 0$ for $a < c < b$. Prove that there is at least one value ξ between a and b for which $f''(\xi) < 0$.   (*A.M.I.E., S-2006*)

Solution. Consider the function $f(x)$ defined on $[a, b]$. Since $f''(x)$ exists, both f and f' exist and are continuous on $[a, b]$. Since c is a point on (a, b). Applying Lagrange mean value theorem to $f(x)$ on $[a, c]$ and $[c, b]$ separately, we obtain

$$\frac{f(c)-f(a)}{c-a}=f'(\xi_1), a<\xi_1<c, \text{ and } \frac{f(b)-f(c)}{b-c}=f'(\xi_2), c<\xi_2<b.$$

Using $f(a)=f(b)=0$, we obtain from the above equations

$$f'(\xi_1)=\frac{f(c)}{c-a} \quad \text{and} \quad f'(\xi_2)=-\frac{f(c)}{b-c}.$$

Now, $f'(x)$ is continuous and differentiable on $\left[\xi_1,\xi_2\right]$. Using the Lagrange mean value theorem again, we obtain

$$\frac{f'(\xi_2)-f'(\xi_1)}{\xi_2-\xi_1}=f''(\xi), \qquad\qquad \xi_1<\xi<\xi_2.$$

Substituting the values of $f'(\xi_1)$ and $f'(\xi_2)$, we get

$$f''(\xi)=-\frac{f(c)}{\xi_2-\xi_1}\left[\frac{1}{b-c}+\frac{1}{c-a}\right]=-\frac{(b-a)f(c)}{(b-c)(c-a)(\xi_2-\xi_1)}<0.$$

2.4 Cauchy's mean value theorem (Second mean value theorem)

Let $f(x)$ and $\phi(x)$ be two real valued functions defined on a closed interval $\left[a,b\right]$ such that (*i*) they are continuous on $\left[a,b\right]$, (*ii*) they are differentiable in (a, b) and (*iii*) $\phi'(x)\neq 0$ for every value of x in (a, b), then, there exists a point $c\in(a,b)$ such that

$$\frac{f(b)-f(a)}{\phi(b)-\phi(a)}=\frac{f'(c)}{\phi'(c)}, \quad a<c<b.$$

Proof. Consider a function $F(x)=f(x)+A\phi(x), a\leq x\leq b$

where A is a constant to be determined such that $F(a)=F(b)$. We have

$$F(a)=f(a)+A\phi(a), \quad F(b)=f(b)+A\phi(b).$$

Using $F(a)=F(b)$, we get $A=-\dfrac{f(b)-f(a)}{\phi(b)-\phi(a)}.$

Therefore, $F(x)=f(x)-\left[\dfrac{f(b)-f(a)}{\phi(b)-\phi(a)}\right]\phi(x).$

Now, $F(x)$ is continuous on $a\leq x\leq b$ and differentiable in $a<x<b$ and $F(a)=F(b)$. Since, the function $F(x)$ satisfies all the conditions of the Rolle's Theorem, there exists a point c, $a<c<b$, such that $F'(c)=0$.

or $\quad f'(c) - \left[\dfrac{f(b)-f(a)}{\phi(b)-\phi(a)}\right]\phi'(c) = 0$, or $\quad \dfrac{f(b)-f(a)}{\phi(b)-\phi(a)} = \dfrac{f'(c)}{\phi'(c)}$, $a < c < b$.

...(2.4)

2.4.1 Another form of Cauchy's mean value theorem

If two functions f, ϕ defined on $[a, a+h]$ are continuous on $[a, a+h]$, derivable on $(a, a+h)$ and $\phi'(x) \neq 0$ for any $x \in (a, a+h)$, then there exists at least one real number θ between 0 and 1 such that

$$\frac{f(a+h)-f(a)}{\phi(a+h)-\phi(a)} = \frac{f'(a+\theta h)}{\phi'(a+\theta h)}, \quad 0 < \theta < 1. \qquad ...(2.5)$$

Remarks. (a) For $\phi(x) = x$, Cauchy's mean value theorem reduces to Lagrange's mean value theorem.

Proof. Take $\phi(x) = x \; \forall x \in (a, b)$ in Cauchy's mean value theorem, we obtain

$$\frac{f(b)-f(a)}{b-a} = f'(c) \quad (\because \phi'(c) = 1)$$

which is Lagrange's mean value theorem.

(b) Let a curve be represented parametrically as $x = f(t), y = \phi(t), a \le t \le b$. Then, Cauchy's mean value theorem states that there exists a point $(f(c), \phi(c)), c \in (a, b)$ on the curve such that the slope $\phi'(c)/f'(c)$ of the tangent to curve at this point is equal to the slope of the chord joining the end points of the curve. Hence, Cauchy's mean value theorem has the same geometrical interpretation as the Lagrange's mean value theorem.

Example 2.16 Verify the Cauchy's mean value theorem for

(i) $f(x) = x^2 + 2$, $\phi(x) = x^3 - 1$ in [1, 2]. (ii) $f(x) = e^x$, $\phi(x) = e^{-x}$ in [0, 1].

(iii) $f(x) = \sin x$, $\phi(x) = \cos x$ in $[-\pi/2, 0]$.

Solution. Clearly $f(x) = x^2 + 2$, $\phi(x) = x^3 - 1$ are continuous in [1, 2] and derivable in (1, 2). Further $\phi'(x) = 3x^2 \neq 0 \; \forall x \in (1, 2)$. Thus the conditions of the Cauchy's mean value theorem are satisfied and so there exists some point $c \in (1, 2)$ such that

$$\frac{f(2)-f(1)}{\phi(2)-\phi(1)} = \frac{f'(c)}{\phi'(c)}, \quad 1 < c < 2$$

or $\quad\dfrac{6-3}{7-0}=\dfrac{2c}{3c^2}$ or $\dfrac{3}{7}=\dfrac{2}{3c}$ or $c=\dfrac{14}{9}=1.55\in(1,2).$

Hence, Cauchy's mean value theorem is verified.

(ii) $f(x)=e^x$ and $\phi(x)=e^{-x}$ satisfy the conditions of the Cauchy's mean value theorem and so there exists some $c\in(0,1)$ such that

$$\frac{f(1)-f(0)}{\phi(1)-\phi(0)}=\frac{f'(c)}{\phi'(c)}, \quad 0<c<1$$

or $\quad\dfrac{e-1}{e^{-1}-1}=\dfrac{e^c}{-e^{-c}}\Rightarrow\dfrac{e(e-1)}{1-e}=-e^{2c}$ or $e^{2c}=e$

or $\quad 2c=1$ or $c=1/2\in(0,1).$ Hence Cauchy's mean value theorem is verified.

(iii) Clearly $f(x)=\sin x,\phi(x)=\cos x$ are continuous in $[-\pi/2,0]$ and derivable in $(-\pi/2,0).$ Further $\phi'(x)=-\sin x\neq0$ for all $x\in(-\pi/2,0).$ Thus the conditions of the Cauchy's mean value theorem are satisfied and so there exists some point $c\in(-\pi/2,0)$ such that

$$\frac{f(0)-f(-\pi/2)}{\phi(0)-\phi(-\pi/2)}=\frac{f'(c)}{\phi'(c)}, \text{ or } \frac{0+1}{1-0}=\frac{\cos c}{-\sin c}$$

or $\tan c=-1,$ which gives $c=-\pi/4\in(-\pi/2,0).$

Hence Cauchy's mean value theorem is verified.

Example 2.17 Show that

$$\frac{\sin\alpha-\sin\beta}{\cos\beta-\cos\alpha}=\cot\theta, \text{ where } 0<\alpha<\theta<\beta<\pi/2.$$

Solution. Let $f(x)=\sin x,$ and $\phi(x)=\cos x\ \forall x\in[\alpha,\beta],0<\alpha<\theta<\beta<\pi/2.$

$f'(x)=\cos x$ and $\phi'(x)=-\sin x.$

Functions f and ϕ are both continuous on $[\alpha,\beta]$ and differentiable on $(\alpha,\beta).$ Therefore by Cauchy's mean value theorem,

$$\frac{f(\beta)-f(\alpha)}{\phi(\beta)-\phi(\alpha)}=\frac{f'(\theta)}{\phi'(\theta)}, \text{ for some } \theta\in(\alpha,\beta),$$

that is, $\quad\dfrac{\sin\beta-\sin\alpha}{\cos\beta-\cos\alpha}=\dfrac{\cos\theta}{-\sin\theta},\ \alpha<\theta<\beta.$

Hence $\dfrac{\sin\alpha-\sin\beta}{\cos\beta-\cos\alpha}=\cot\theta,$ where $0<\alpha<\theta<\beta<\pi/2.$

2.5 Increasing and Decreasing Functions

Let $y = f(x)$ be a function defined on an interval I. Let x_1 and x_2 be any two points in I, where x_1, x_2 are not the end points of the interval. On the interval I, the function $f(x)$ is said to be

(i) an **increasing** function, if $f(x_1) \le f(x_2)$ whenever $x_1 < x_2$.

(ii) a **strictly increasing** function, if $f(x_1) < f(x_2)$ whenever $x_1 < x_2$.

(iii) a **decreasing** function, if $f(x_1) \ge f(x_2)$ whenever $x_1 < x_2$.

(iv) a **strictly decreasing** function, if $f(x_1) > f(x_2)$ whenever $x_1 < x_2$.

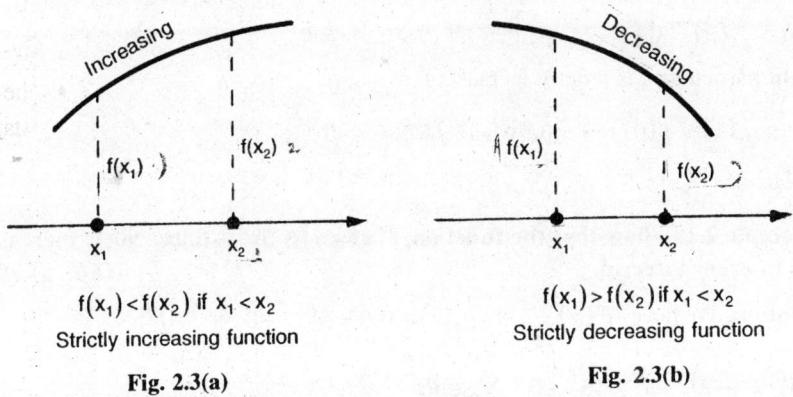

$f(x_1) < f(x_2)$ if $x_1 < x_2$

Strictly increasing function

Fig. 2.3(a)

$f(x_1) > f(x_2)$ if $x_1 < x_2$

Strictly decreasing function

Fig. 2.3(b)

A function is said to be *monotonic* function in an interval I if it is either increasing or decreasing in the entire interval I.

The First Derivative Test for Increasing and Decreasing

Let f be a function that is continuous on a closed interval $[a, b]$ and differentiable on the open interval (a, b).

(a) f is increasing on $[a, b]$ if $f'(x) > 0$ for every value of x in (a, b).

(b) f is decreasing on $[a, b]$, if $f'(x) < 0$ for every value of x in (a, b).

Proof. Let x_1 and x_2 be two points in $[a, b]$ with $x_1 < x_2$.

Let $f'(x) > 0 \forall x \in (a, b)$. By Lagrange's mean value theorem, there exists a point c between x_1 and x_2 such that

$$\frac{f(x_2) - f(x_1)}{x_2 - x_1} = f'(c)$$

$\Rightarrow f(x_2) - f(x_1) = (x_2 - x_1) f'(c) > 0$, Since $f'(c) > 0$ and $x_2 - x_1 > 0$

$\Rightarrow f(x_2) > f(x_1)$. Thus, we have

$x_2 > x_1 \Rightarrow f(x_2) > f(x_1)$, for all x_1, x_2 in $[a, b]$.

Hence f is an increasing function in $[a, b]$. Similarly, we can show that if $f'(x) < 0 \; \forall \; x \in (a, b)$, then f is decreasing on $[a, b]$.

Now we shall prove some inequalities with the help of above Theorem.

Remark A differential function increases when its graph has positive slopes and decreasing when its graph has negative slopes.

Example 2.18 Show that $x^3 - 6x^2 + 15x + 3 > 0, \; \forall \; x > 0$.

Solution. Let $f(x) = x^3 - 6x^2 + 15x + 3$.

Therefore, $f'(x) = 3x^2 - 12x + 15 = 3(x^2 - 4x + 5)$.

$\Rightarrow f'(x) = 3\left[(x-2)^2 + 1\right] \Rightarrow f'(x) > 0 \; \forall x \in R$.

In particular, f is strictly increasing $\forall \; x > 0$.

$\Rightarrow f(x) > f(0) \; \forall \; x > 0$, where $f(0) = 3 > 0$.

Hence $f(x) > 0 \qquad \forall \; x > 0$.

Example 2.19 Show that the function $f(x) = x^5 - 5x^3 + 100x - 90$ is increasing in every interval. (D.U., 1998)

Solution. We have $f'(x) = 5x^4 - 15x^2 + 100 = 5(x^4 - 3x^2 + 20)$

$= 5\left[(x^2 - 2)^2 + x^2 + 16\right] > 0 \qquad \forall x \in R$.

Hence f is increasing for all $x \in R$ and so f is increasing in every subinterval.

Example 2.20 Show that

(i) $x - \dfrac{x^2}{2} < \log(1+x) < x - \dfrac{x^2}{2(1+x)}$, $\qquad x > 0$ $\qquad$ (D.U.. 1996)

(ii) $\dfrac{x^2}{2(1+x)} < x - \log(1+x) < \dfrac{x^2}{2}$, $\qquad x > 0$ $\qquad$ (D.U., 1999)

(iii) $1 - x < e^{-x} < 1 - x + \dfrac{x^2}{2}$ $\qquad\qquad \forall \; x, x > 0$

Solution. Let $f(x) = \log(1+x) - \left(x - \dfrac{x^2}{2}\right)$, so that $f(0) = 0$.

$\therefore f'(x) = \dfrac{1}{1+x} - 1 + x = \dfrac{x^2}{1+x} > 0$. $\qquad\qquad\qquad (\because x > 0)$

$\Rightarrow f(x)$ is an increasing function of x, for $x > 0$

$\Rightarrow f(x) > f(0)$, for $x > 0 \Rightarrow f(x) > 0$ $\qquad \left(\because f(0) = 0 \right)$

$\Rightarrow \log(1+x) - \left(x - \dfrac{x^2}{2} \right) > 0$, for $x > 0$

$\therefore \quad x - \dfrac{x^2}{2} < \log(1+x), x > 0.$ $\qquad\qquad ...(i)$

Let $\phi(x) = x - \dfrac{x^2}{2(1+x)} - \log(1+x)$, so that $\phi(0) = 0$.

$\therefore \phi'(x) = 1 - \dfrac{1}{2}\left\{ \dfrac{2x+x^2}{(1+x)^2} \right\} - \dfrac{1}{1+x} = \dfrac{x^2}{2(1+x)^2} > 0$ $\qquad (\because x > 0)$

$\Rightarrow \phi(x)$ is an increasing function, for $x > 0$.

$\Rightarrow \phi(x) > \phi(0)$, for $x > 0 \Rightarrow \phi(x) > 0$ $\qquad \left(\because \phi(0) = 0 \right)$

$\Rightarrow x - \dfrac{x^2}{2(1+x)} - \log(1+x) > 0$, for $x > 0$.

$\therefore \log(1+x) < x - \dfrac{x^2}{2(1+x)}$, for $x > 0$ $\qquad\qquad ...(ii)$

Combining the above two results, we obtain

$$x - \dfrac{x^2}{2} < \log(1+x) < x - \dfrac{x^2}{2(1+x)}, \text{ for } x > 0.$$

(ii) By part (i), we have

$$-x + \dfrac{x^2}{2} > -\log(1+x) > \dfrac{x^2}{2(1+x)} - x, \quad x > 0$$

or $\quad \dfrac{x^2}{2} > x - \log(1+x) > \dfrac{x^2}{2(1+x)}, \quad x > 0$

Hence $\qquad\qquad \dfrac{x^2}{2(1+x)} < x - \log(1+x) < \dfrac{x^2}{2}, x > 0.$

(iii) Let $f(x) = e^{-x} + x - 1$, so that $f(0) = 0$.

Now, $f'(x) = 1 - e^{-x} > 0$ for all $x > 0$. Hence, $f(x)$ is an increasing function for all $x > 0$. Therefore, $f(x) > f(0) = 0$, or $e^{-x} + x - 1 > 0$ or $e^{-x} > 1 - x$. $\qquad\qquad ...(i)$

Now, consider $\phi(x) = e^{-x} - 1 + x - \dfrac{x^2}{2}$. Therefore $\phi'(x) = 1 - x - e^{-x} < 0$ for all $x > 0$.

Hence, $\phi(x)$ is a decreasing function for all $x > 0$.

$\therefore \ \phi(x) < \phi(0) = 0$, or $e^{-x} < 1 - x + \dfrac{x^2}{2}$. ...(ii)

Combining the above two results, we obtain

$$1 - x < e^{-x} < 1 - x + \frac{x^2}{2}, \quad x > 0.$$

Example 2.21 Establish the following inequalities

(i) $x < \sin^{-1} x < \dfrac{x}{\sqrt{1 - x^2}}$, if $0 < x < 1$. (ii) $\dfrac{\tan x}{x} > \dfrac{x}{\sin x}$, whenever $0 < x < \dfrac{\pi}{2}$.

(D.U. 1999)

Solution. (i) Let $f(x) = \sin^{-1} x - x$, so that $f(0) = 0$.

Now $f'(x) = \dfrac{1}{\sqrt{1 - x^2}} - 1 = \dfrac{1 - \sqrt{1 - x^2}}{\sqrt{1 - x^2}} > 0$, for $0 < x < 1$

$\Rightarrow f(x)$ is strictly increasing in $0 < x < 1$.

$\Rightarrow f(x) > f(0)$, for $0 < x < 1$ $\qquad\qquad (\because x > 0)$

$\Rightarrow f(x) > 0$, for $0 < x < 1$.

$\Rightarrow \sin^{-1} x - x > 0$, for $0 < x < 1 \Rightarrow \sin^{-1} x > x$, for $0 < x < 1$

$\Rightarrow x < \sin^{-1} x$, for $0 < x < 1$. ...(i)

Let $\phi(x) = \dfrac{x}{\sqrt{1 - x^2}} - \sin^{-1} x$, so that $\phi(x) = 0$

$\therefore \ \phi'(x) = \dfrac{\sqrt{1 - x^2} - \dfrac{x(-2x)}{2\sqrt{1 - x^2}}}{(1 - x^2)} - \dfrac{1}{\sqrt{1 - x^2}}$

$= \dfrac{1}{(1 - x^2)\sqrt{1 - x^2}} - \dfrac{1}{\sqrt{1 - x^2}} = \dfrac{x^2}{(1 - x^2)^{3/2}}.$

$\Rightarrow \phi'(x) > 0,$ for $0 < x < 1 \Rightarrow \phi(x)$ is strictly increasing, for $0 < x < 1$.

$\Rightarrow \phi(x) > \phi(0),$ for $0 < x < 1$ $\hspace{3cm} (\because x > 0)$

$\Rightarrow \phi(x) > 0,$ for $0 < x < 1 \Rightarrow \dfrac{x}{\sqrt{1-x^2}} > \sin^{-1} x, \hspace{0.5cm}$ for $0 < x < 1$

$\Rightarrow \sin^{-1} x < \dfrac{x}{\sqrt{1-x^2}}, \hspace{3cm}$ for $0 < x < 1$ $\hspace{1cm}$ ...(ii)

Combining the above two results, we obtain

$x < \sin^{-1} x < \dfrac{x}{\sqrt{1-x^2}}, \hspace{3cm}$ for $0 < x < 1.$

(*ii*) We have to show that

$$\frac{\tan x}{x} - \frac{x}{\sin x} > 0, \hspace{0.3cm} \text{or} \hspace{0.3cm} \frac{\sin x \tan x - x^2}{x \sin x} > 0, \hspace{1cm} \text{for } 0 < x < \pi/2.$$

Since $x \sin x > 0$, for all x in $(0, \pi/2)$, therefore, we need to show that $\sin x \tan x - x^2 > 0 \forall x \in (0, \pi/2)$. Let c be any real number in $(0, \pi/2)$.

Let $f(x) = \sin x \tan x - x^2 \; \forall \; x \in [0, c].$

Then f is continuous as well as derivable in $[0, c]$.

Now $f'(x) = \cos x \tan x + \sin x \sec^2 x - 2x$

$= \sin x + \sin x \sec^2 x - 2x = \sin x (\sec^2 x + 1) - 2x.$

The form of $f'(x)$ is such that we cannot decide about its sign because of the presence of the $2x$ term.

Let $\phi(x) = f'(x)$ for all x in $[0, c]$.

The function ϕ is continuous as well as derivable on $[0, c]$.

Now $\phi'(x) = \cos x + \cos x \sec^2 x + 2 \sin x \sec^2 x \tan x - 2$

$= \left(\sqrt{\sec x} - \sqrt{\cos x} \right)^2 + 2 \sin^2 x \sec^3 x.$

Since $\phi'(x) > 0$ for all $x \in (0, c)$. This implies that ϕ is strictly increasing in $[0, c] \Rightarrow \phi(x) > \phi(0)$, whenever $0 < x < c$. Since $\phi(0) = 0$, this means that $\phi(x) > 0$, whenever $0 < x < c \Rightarrow f'(x) > 0$, whenever $0 < x < c \Rightarrow f$ is strictly increasing on $[0, c] \Rightarrow f(c) > f(0) = 0 \Rightarrow f(c) > 0$

$$\Rightarrow \sin c \tan c - c^2 > 0 \Rightarrow \frac{\tan c}{c} - \frac{c}{\sin c} > 0.$$

Since c is any point on $(0, \pi/2)$, it follows that $\dfrac{\tan x}{x} > \dfrac{x}{\sin x}$,

whenever $0 < x < \dfrac{\pi}{2}$.

Remark. The above inequality can be put in the form

$$\cos x < \left(\frac{\sin x}{x}\right)^2, \qquad 0 < x < \pi/2.$$

Exercise 2.1

1. Verify Rolle's theorem for the following functions:

   (*i*) $f(x) = 8x - x^2$ on [0, 8], (*ii*) $f(x) = 3x^4 - 4x^2 + 5$ on [−1, 1].

   (*iii*) $f(x) = \sin x + \cos x$ on [0, $\pi/2$]. (*iv*) $f(x) = \log\left(\dfrac{x^2 + 6}{x}\right)$ on [2, 3].

   (*v*) $f(x) = (\sin x)/e^x$ on [0, π]. (*vi*) $f(x) = (x - 1)(x - 2)^2$ on [1, 2].

   (A.M.I.E. W-2002)

   (*vii*) $f(x) = x^2,\ x \in [-1, 1]$ (*D.U. 1999*). (*viii*) $f(x) = x^2 - 5x + 6$ on [2, 3].

   (A.M.I.E., S-1998)

   (*ix*) $f(x) = (x + 1)^m(x - 1)^n,\ x \in [-1, 1]$, where m and n are positive integers.

   (A.M.I.E., S-1999)

   (*x*) $f(x) = x(x-3)^2,\ 0 \le x \le 3$.

2. Discuss the applicability of Rolle's theorem to the following functions

   (*a*) $f(x) = 1 - \sqrt[3]{x^2}$ on [−1, 1]. Illustrate your answer by a rough sketch.

   (*b*) $f(x) = 2 + (x - 1)^{2/3}$ on [0, 2]. **Ans.** Not applicable

   (*c*) $f(x) = \begin{cases} x^2 - 4, & x \le 1 \\ 5x - 8, & x > 1 \end{cases}$ on [-2, 1.6]. **Ans.** Not applicable as $f(x)$ is not derivable at $x = 1 \in (-2, 1.6)$

   (*d*) $f(x) = |x|$ on $[-1, 1]$. **Ans.** Not applicable

   [**Hint.**

   (*a*) Since $f'(x) = -2/3\sqrt[3]{x}$ does not exist at $x = 0$, a point of the open interval $(-1, 1)$, Rolle's theorem is not applicable. It is clear that $f'(x)$ does not vanish for any point in the interval.

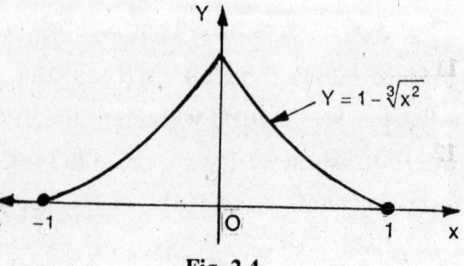

Fig. 2.4

From the Fig. 2.4 it is clear that at no point of the curve for $-1 < x < 1$, the tangent is parallel to x-axis.]

3. It is given that for the function f given by $f(x) = x^3 + qx^2 + px$, $x \in [1, 3]$ Rolle's theorem holds with $c = 2 + \left(1/\sqrt{3}\right)$. Find the values of p and q.

Ans. $p = 11, q = -6$.

4. Find the point on the curve $y = 12(x+1)(x-2)$, in the interval $[-1, 2]$, where the tangent is parallel to x-axis. **Ans.** $(1/2, -27)$

5. Show that between any two real roots of $e^x \sin x = x$, there is atleast one root of $\cos x + \sin x = e^{-x}$. *(D.U. 1999)*

Verify Lagrange's mean value theorem for the following functions:

6.   (i) $f(x) = \sqrt{x^2 - 4}$ on $[2, 4]$. (ii) $f(x) = \log x$ on $[1/2, 2]$.

    (iii) $f(x) = x^3 + x$ on $[0, 1]$. *(A.M.I.E. S-2002)*

    (iv) $f(x) = \sqrt{x-1}$ on $[1, 3]$.

    (v) $f(x) = x^3 + 3x^2 - 5x$ on $(1, 2)$. *(A.M.I.E. S-1998)*

[**Hint.** (ii) Here $a = 1/2$, $b = 2$, $f(x) = \log x$, $f'(x) = 1/x$.

$f(b) - f(a) = (b - a) f'(c) \Rightarrow \log 2 - \log(1/2) = 3/2c$

or $\log 2 - \log 1 + \log 2 = 3/2c \Rightarrow c = 3/(4 \log 2) = 3/(4 \times 0.693) = 1.08$

which lies between 0.5 and 2.0. Hence L.M.V.T. is verified.]

7. Find c of Lagrange's mean value theorem for the following functions
(i) $f(x) = 3x^2 + 4x - 3$ on $[1, 3]$. (ii) $f(x) = (x - 1)(x - 2)(x - 3)$ on $[0, 4]$.

(iii) $f(x) = x - 2 \sin x$ on $[-\pi, \pi]$. **Ans.** (i) 2, (ii) 0.845 and 3.155, (iii) $\pi/2$.

8. Applying mean value theorem, prove that $f(x) = e^x - 1 - x$, cannot vanish for any value of x other than $x = 0$. *(A.M.I.E., S. 2003)*

9. If a and b are distinct real numbers, show that there exists a real number c between a and b such that $a^2 + ab + b^2 = 3c^2$.

$\left[\textbf{Hint.} \text{ Applying Lagrange's mean value theorem to } f(x) = x^3 \text{ in } [a, b]\right]$

10. Using the Lagrange's mean value theorem, show that $|\sin b - \sin a| \le |b - a|$ for all real a and b. *(A.M.I.E., W-2005)*

11. Find a point on the curve $f(x) = (x-3)^2$, where the tangent is parallel to the chord joining the points $(3, 0)$ and $(4, 1)$. **Ans.** $(7/2, 1/4)$.

12. (i) Prove that

$$\frac{b-a}{\sqrt{1-a^2}} < \sin^{-1} b - \sin^{-1} a < \frac{b-a}{\sqrt{1-b^2}}, \text{ if } b > a.$$

    (ii) Show that $\dfrac{\pi}{6}-\dfrac{1}{2\sqrt{3}}<\sin^{-1}\dfrac{1}{4}<\dfrac{\pi}{6}-\dfrac{1}{\sqrt{15}}$.

13. Use mean value theorem to show that there is a point on the graph of $y = x^2 - 3$ between the points $A(2,1)$ and $B(-1,-2)$, where the tangent is parallel to the chord AB. Also find the point. *(A.M.I.E., S-1999)*   **Ans.** $(1/2, -11/4)$.

14. Find the coordinates of the point at which the tangent to the curve given by $f(x) = x^2 - 6x + 1$ is parallel to the chord joining the points $(1, -4)$ and $(3, -8)$.   **Ans.** $(2, -7)$.

15. Find the points on the curve $y = (x^3/3) - (3x^2/2) + 2x$, where the tangents will be parallel to the line joining the origin and the point $x = 3$ on the curve.   *(A.M.I.E., S-2005)* **Ans.** $(0.634, 0.75)$; $(2.366, 0.75)$.

16. With the help of the mean value theorem, show that if $x > 0$, $\log_{10}(x+1) = (x\log_{10}e)/(1+\theta x)$, where $0 < \theta < 1$.   *(A.M.I.E., S-2000)*

17. Verify Cauchy's mean value theorem for the following functions:

    (i) $f(x) = x^2 + 2x - 3,\ \phi(x) = x^2 - 4x + 6$ on $[0, 1]$.

    (ii) $f(x) = x^2,\ \phi(x) = x^3$ on $[1, 2]$.

18. If in the Cauchy's mean value theorem, $f(x) = e^x$ and $g(x) = e^{-x}$, show that c is arithmetic mean between a and b.

19. Using mean value theorem for the function $f(x) = x^{1/3}$, prove that

$$3 + \frac{1}{28} < \sqrt[3]{28} < 3 + \frac{1}{27}. \qquad\qquad\qquad \textit{(A.M.I.E., S-2005)}$$

20. If the functions $f(x), \phi(x)$ and $\Psi(x)$ satisfy the conditions of mean value theorem in $[a, b]$, show that there exist at least one point ζ such that

$$\begin{vmatrix} f(a) & \phi(a) & \Psi(a) \\ f(b) & \phi(b) & \Psi(b) \\ f'(\zeta) & \phi'(\zeta) & \Psi'(\zeta) \end{vmatrix} = 0. \quad (a < \zeta < b) \qquad\qquad \textit{(A.M.I.E. W-2007)}$$

21. Prove the following inequalities:

    (i) $\dfrac{x}{1+x} < \log(1+x) < x,\ x > 0$

    (ii) $x - x^2 + \dfrac{x^3}{3(1+x)} < \log(1+x) < x - \dfrac{x^2}{2} + \dfrac{x^3}{3}$, if $x > 0$.

(iii) $x > \log(1+x) > x - \dfrac{x^2}{2}, x > 0.$ *(A.M.I.E., W-2003)*

(iv) $\dfrac{2}{\pi} < \dfrac{\sin x}{x} < 1,$ whenever $0 < x < \pi/2.$

(v) $\dfrac{1}{2}x^2 < x - \log(1+x) < \dfrac{1}{2}\dfrac{x^2}{(1+x)},$ if $-1 < x < 0.$

(vi) $\dfrac{x}{1+x} < \log(1+x) < x,$ for $x > -1, x \ne 0.$

(vii) $x < -\log(1-x) < x(1-x)^{-1},$ if $0 < x < 1.$

(viii) $\tan x > x > \sin x,$ for $0 < x < \pi/2.$

2.6 Taylor's Theorem

Taylor's theorem can be regarded as an extension of the mean value theorems to higher order derivatives. Mean value theorems relate the value of the function and its first order derivative, whereas the Taylor's theorem (Taylor's formula) relates the value of the function and its higher order derivatives.

2.6.1 Taylor's theorem with Lagrange's form of remainder

If $f(x)$ and all its derivatives upto $(n-1)$th order be continuous on $[a, a+h]$ and if $f^n(x)$ exists for every value of x in $(a, a+h)$, then there exists some real number θ between 0 and 1 $(0 < \theta < 1)$, such that

$$f(a+h) = f(a) + \frac{h}{1!}f'(a) + \frac{h^2}{2!}f''(a) + \frac{h^3}{3!}f'''(a) + \ldots\ldots$$

$$\ldots\ldots + \frac{h^{n-1}}{(n-1)!}f^{n-1}(a) + R_n(x)$$

where $\qquad R_n(x) = \dfrac{h^n}{n!}f^n(a+\theta h), 0 < \theta < 1$

is the remainder or the error term of the expansion.

Proof: Consider a function $\phi(x)$ defined on $[a, a+h]$ as

$$\phi(x) = f(x) + \frac{(a+h-x)}{1!}f'(x) + \frac{(a+h-x)^2}{2!}f''(x) + \ldots$$

$$\ldots + \frac{(a+h-x)^{n-1}}{(n-1)!}f^{n-1}(x) + \frac{(a+h-x)^n}{n!} \cdot A \qquad\qquad \ldots(2.6)$$

where A is a constant to be determined such that $\phi(a) = \phi(a+h)$ ...(2.7)

From (2.6) and (2.7), we obtain

$$f(a) + hf'(a) + \frac{h^2}{2!}f''(a) + ... + \frac{h^{n-1}}{(n-1)!}f^{n-1}(a) + A\frac{h^n}{n!} = f(a+h). \qquad ...(2.8)$$

By the given hypothesis, the function $f(x), f'(x), f''(x), ..., f^{n-1}(x)$ being all continuous on $[a, a+h]$, and $f^n(x)$ exists on $(a, a+h)$, $\phi(x)$ is continuous on $[a, a+h]$ and $\phi'(x)$ exists on $(a, a+h)$.

Also $\phi(a) = \phi(a+h)$.

Thus, the function $\phi(x)$ satisfies all the conditions of Rolle's theorem and so there exists atleast one real number θ, between 0 and 1 such that $\phi'(a+\theta h) = 0$.

Differentiating (2.6) and simplifying, we obtain

$$\phi'(x) = \frac{(a+h-x)^{n-1}}{(n-1)!}\left[f^n(x) - A\right],$$

$$\therefore \quad \phi'(a+\theta h) = \frac{[a+h-(a+\theta h)]^{n-1}}{(n-1)!}\left[f^n(a+\theta h) - A\right] = 0 \quad \left[\because \phi'(a+\theta h) = 0\right]$$

$$\Rightarrow \frac{(1-\theta)^{n-1}}{(n-1)!} \cdot h^{n-1}\left[f^n(a+\theta h) - A\right] = 0$$

$$\therefore \qquad\qquad A = f^n(a+\theta h) \qquad\qquad [\because (1-\theta) \neq 0].$$

Substituting this value of A in (2.8), we get

$$\mathbf{f(a+h) = f(a) + hf'(a) + \frac{h^2}{2!}f''(a) + ... + \frac{h^{n-1}}{(n-1)!}f^{n-1}(a) + \frac{h^n}{n!}f^n(a+\theta h).}$$

$$...(2.9)$$

The $(n+1)th$ term is $\mathbf{R_n(x) = \frac{h^n}{n!}f^n(a+\theta h), \quad 0 < \theta < 1}$

is called the *Lagrange's form of the remainder.*

Remark. Writing $a+h = x$ in equation (2.9), we obtain

$$\mathbf{f(x) = f(a) + \frac{(x-a)}{1!}f'(a) + \frac{(x-a)^2}{2!}f''(a) + ... + \frac{(x-a)^{n-1}}{(n-1)!}f^{n-1}(a)}$$

$$\mathbf{+ \frac{(x-a)^n}{n!}f^n[a+\theta(x-a)], \quad 0 < \theta < 1} \qquad ...(2.10)$$

which is the second form of Taylor's theorem with Lagrange's form of the remainder. The *remainder* or *error* term can also be written as

$$R_n(x) = \frac{(x-a)^n}{n!} f^n(c), \text{ where } c \text{ lies between } a \text{ and } x. \qquad ...(2.11)$$

2.6.2 Taylor's theorem with Cauchy's form of remainder

Consider a function $\phi(x)$ defined on $[a, a+h]$ as

$$\phi(x) = f(x) + (a+h-x)f'(x) + \frac{(a+h-x)^2}{2!} f''(x) + ...$$

$$...... + \frac{(a+h-x)^{n-1}}{(n-1)!} f^{(n-1)}(x) + (a+h-x)A,$$

where A is a constant to be determined such that $\phi(a+h) = \phi(a)$.

The function $\phi(x)$ satisfies all the conditions of the Rolle's theorem. Therefore, $\phi'(a+\theta h) = 0, \quad 0 < \theta < 1$.

Now $\phi'(x) = \frac{1}{(n-1)!}(a+h-x)^{n-1} f^n(x) - A$

and $\phi'(a+\theta h) = \frac{h^{n-1}}{(n-1)!}(1-\theta)^{n-1} f^n(a+\theta h) - A = 0$.

Hence, $A = \frac{h^{n-1}}{(n-1)!}(1-\theta)^{n-1} f^n(a+\theta h)$.

From $\phi(a+h) = \phi(a)$, we obtain

$$f(a+h) = f(a) + hf'(a) + \frac{h^2}{2!}f''(a) + ... + \frac{h^{n-1}}{(n-1)!}f^{n-1}(a) + hA.$$

Substituting the value of A, we get

$$f(a+h) = f(a) + hf'(a) + \frac{h^2}{2!}f''(a) + ... + \frac{h^{n-1}}{(n-1)!}f^{n-1}(a)$$

$$+ \frac{h^n}{(n-1)!}(1-\theta)^{n-1} f^n(a+\theta h), \quad 0 < \theta < 1 \qquad ...(2.12)$$

The $(n+1)^{th}$ term $R_n(x) = \frac{h^n}{(n-1)!} \cdot (1-\theta)^{n-1} \cdot f^n(a+\theta h), \quad 0 < \theta < 1$

is called the Cauchy's form of the remainder after n terms in the expansion of $f(a+h)$ in ascending powers of h.

Remark. Setting $a+h=x$, we obtain second form of Taylor's theorem with Cauchy's form of the remainder

$$f(x)=f(a)+\frac{(x-a)}{1!}f'(a)+\frac{(x-a)^2}{2!}f''(a)+......$$

$$+\frac{(x-a)^{n-1}}{(n-1)!}f^{n-1}(a)+\frac{(x-a)^n}{(n-1)!}(1-\theta)^{n-1}f^n(a+\theta h),\quad 0<\theta<1,\qquad ...(2.13)$$

2.7 Maclaurin's theorem

On taking $a=0$, in Taylor theorem (2.10), we get

$$f(x)=f(0)+\frac{x}{1!}f'(0)+\frac{x^2}{2!}f''(0)+......+\frac{x^{n-1}}{(n-1)!}f^{n-1}(0)$$

$$+\frac{x^n}{n!}f^n(\theta x),\quad 0<\theta<1 \qquad ...(2.14)$$

which is called the *Maclaurin's theorem with Lagrange's form of the remainder.*

The error term in equation (2.13) is $R_n(x)=\frac{x^n}{n!}f^n(\theta x),\quad 0<\theta<1$.

This can also be written as $R_n(x)=\frac{x^n}{n!}f^n(c),\quad 0<c<x$.

Further, the *Maclaurin's theorem* of $f(x)$ in the interval $[0,x]$ with *Cauchy's form of the remainder* after n terms is given below:

$$f(x)=f(0)+\frac{x}{1!}f'(0)+\frac{x^2}{2!}f''(0)+......+\frac{x^{n-1}}{(n-1)!}f^{n-1}(0)$$

$$+\frac{x^n}{(n-1)!}(1-\theta)^{n-1}f^n(\theta x),0<\theta<1. \qquad ...(2.15)$$

2.8 Taylor's Series

In the Taylor's formula with remainder (Eqs. (2.10), (2.11)), if the remainder $R_n(x)\to 0$ *as* $n\to\infty$, then we obtain

$$f(x)=f(a)+\frac{(x-a)}{1!}f'(a)+\frac{(x-a)^2}{2!}f''(a)+......+\frac{(x-a)^{n-1}}{(n-1)!}f^{n-1}(a)+...$$

$$...(2.16)$$

which is *called the Taylor's series* in powers of $(x-a)$.
When $a = 0$, we obtain the *Maclaurin's series.*

$$f(x) = f(0) + \frac{x}{1!} f'(0) + \frac{x^2}{2!} f''(0) + \ldots\ldots + \frac{x^{n-1}}{(n-1)!} f^{n-1}(0) + \ldots\ldots \quad \ldots(2.17)$$

2.9 Illustrative Examples

Example 2.22 Expand $e^{\sin x}$ by Maclaurin's series up to the term containing x^4.
                                                                (A.M.I.E., W-2003)
Solution. Let $f(x) = e^{\sin x}$ $f(0) = 1,$
Taking derivatives, we obtain

$$f'(x) = e^{\sin x} \cdot (\cos x) = f(x) \cdot \cos x, \qquad\qquad f'(0) = 1,$$

$$f''(x) = f'(x) \cdot (\cos x) - f(x) \cdot \sin x, \qquad\qquad f''(0) = 1,$$

$$f'''(x) = f''(x) \cdot (\cos x) + f'(x) \cdot (-\sin x)$$

$$-f'(x) \sin x - f(x) \cos x$$

$$= f''(x) \cdot (\cos x) - 2f'(x) \cdot (\sin x) - f(x) \cdot \cos x, \qquad f'''(0) = 0,$$

$$f^{iv}(x) = f'''(x) \cdot (\cos x) - f''(x) \cdot (\sin x) - 2f''(x) \sin x$$

$$-2f'(x) \cdot (\cos x) - f'(x) \cos x + f(x) \sin x$$

$$= f'''(x) \cdot \cos x - 3f''(x) \sin x - 3f'(x) \cos x + f(x) \sin x, \qquad f^{iv}(0) = -3$$

....

....

Substituting the values of $f(0), f'(0), f''(0)$ etc. in the Maclaurin's series, we obtain

$$e^{\sin x} = 1 + x(1) + \frac{x^2}{2!}(1) + \frac{x^3}{3!}(0) + \frac{x^4}{4!}(-3) + \ldots\ldots$$

$$= 1 + x + \frac{x^2}{2} - \frac{x^4}{8} - \ldots\ldots\ldots, \text{ the required result.}$$

Example 2.23 Assuming the validity of expansion, show that

$$\sin^{-1}\left(\frac{2x}{1+x^2}\right) = 2\left(x - \frac{x^3}{3} + \frac{x^5}{5} - \ldots\right)$$

Solution. Let $f(x) = \sin^{-1} \frac{2x}{1+x^2}$. Putting $x = \tan\theta$.

$$\therefore f(x) = \sin^{-1}\left(\frac{2\tan\theta}{1+\tan^2\theta}\right) = \sin^{-1}(\sin 2\theta) = 2\theta = 2\tan^{-1}x, \qquad f(0) = 0$$

Taking derivatives, we obtain

$$f'(x) = \frac{2}{1+x^2} = 2(1+x^2)^{-1} = 2(1-x^2+x^4-x^6+\ldots), \qquad f'(0) = 2,$$

$$f''(x) = 2(-2x+4x^3-6x^5+\ldots), \qquad f''(0) = 0,$$

$$f'''(x) = 2(-2+12x^2-30x^4+\ldots), \qquad f'''(0) = -4,$$

$$f^{iv}(x) = 2(24x-120x^3+\ldots), \qquad f^{iv}(x) = 0,$$

$$f^{v}(x) = 2(24-360x^2+\ldots), \qquad f^{v}(x) = 48,$$

and so on.

Substituting these values of $f(0),\ f'(0),\ f''(0),\ f'''(0).\ \ldots$ in Maclaurin's series, we obtain

$$\sin^{-1}\left(\frac{2x}{1+x^2}\right) = 2x - 4\frac{x^3}{3!} + 48\frac{x^5}{5!} - \ldots = 2\left(x - \frac{x^3}{3} + \frac{x^5}{5} - \ldots\right).$$

Example 2.24 Apply Maclaurin's series to prove

$$\log(1+\sin x) = x - \frac{x^2}{2} + \frac{x^3}{6} - \frac{x^4}{12} + \frac{x^5}{24} - \ldots\ldots \qquad \textit{(A.M.I.E. S-2002)}$$

Solution. Let $f(x) = \log(1+\sin x)$. $\qquad\qquad f(0) = 0,$

Taking derivatives, we obtain

$$f'(x) = \frac{\cos x}{1+\sin x} = \frac{\cos^2 x/2 - \sin^2 x/2}{(\cos x/2 + \sin x/2)^2} = \frac{\cos x/2 - \sin x/2}{\cos x/2 + \sin x/2}$$

$$= \frac{1 - \tan x/2}{1 + \tan x/2} = \tan\left(\frac{\pi}{4} - \frac{x}{2}\right), \qquad f'(0) = 1,$$

$$f''(x) = -\frac{1}{2}\sec^2\left(\frac{\pi}{4} - \frac{x}{2}\right) = -\frac{1}{2}\left[1+\tan^2\left(\frac{\pi}{4} - \frac{x}{2}\right)\right] = -\frac{1}{2}\left[1+\{f'(x)\}^2\right],$$

$$f''(0) = -1,$$

$$f'''(x) = -f'(x)f''(x), \qquad f'''(0) = 1,$$

$$f^{iv}(x) = -\left\{f''(x)\right\}^2 - f'(x)f'''(x), \qquad\qquad f^{iv}(0) = -2,$$

$$f^{v}(x) = -3f''(x)f'''(x) - f'(x)f^{iv}(x), \qquad\qquad f^{v}(0) = 5,$$

and so on.

Putting these values in

$$f(x) = f(0) + xf'(0) + \frac{x^2}{2!}f''(0) + \frac{x^3}{3!}f'''(0) + \frac{x^4}{4!}f^{iv}(0) + \dots :$$

$$\therefore \ \log(1 + \sin x) = 0 + x(1) + \frac{x^2}{2!}(-1) + \frac{x^3}{3!}(1) + \frac{x^4}{4!}(-2) + \frac{x^5}{5!}(5) + \dots$$

$$= x - \frac{x^2}{2} + \frac{x^3}{6} - \frac{x^4}{12} + \frac{x^5}{24} + \dots\dots$$

Example 2.25 Using Maclaurin's series, expand tan x upto the term containing x^5. *(A.M.I.E., W-2001)*

Solution. Let $f(x) = \tan x$, $\qquad\qquad\qquad\qquad\qquad\qquad f(0) = 0,$

$$\therefore \ f'(x) = \sec^2 x = 1 + \tan^2 x, \qquad\qquad\qquad\qquad f'(0) = 1,$$

$$f''(x) = 2\tan x \sec^2 x = 2\tan x\left(1 + \tan^2 x\right) = 2\tan x + 2\tan^3 x, \qquad f''(0) = 0,$$

$$f''' \ x \ \ne 2\sec^2 x + 6\tan^2 x \ \sec^2 x = 2\left(1 + \tan^2 x\right) + 6\tan^2 x\left(1 + \tan^2 x\right)$$

$$= 2 + 8\tan^2 x + 6\tan^4 x, \qquad\qquad\qquad f'''(0) = 2,$$

$$f^{iv}(x) = 16\tan x \cdot \left(\sec^2 x\right) + 24\tan^3 x \sec^2 x$$

$$= 16\tan x\left(1 + \tan^2 x\right) + 24\tan^3 x\left(1 + \tan^2 x\right)$$

$$= 16\tan x + 40\tan^3 x + 24\tan^5 x, \qquad\qquad f^{iv}(0) = 0,$$

$$f^{v}(x) = 16\sec^2 x + 120\tan^2 x \sec^2 x + 120\tan^4 x \sec^2 x$$

$$= 16\left(1 + \tan^2 x\right) + 120\tan^2 x\left(1 + \tan^2 x\right) + 120\tan^4 x\left(1 + \tan^2 x\right)$$

$$= 16 + 136\tan^2 x + 240\tan^4 x + 120\tan^6 x, \qquad f^{v}(0) = 16,$$

$$f^{vi}(x) = 272\tan x \sec^2 x + 960\tan^3 x \sec^2 x + 720\tan^5 x \sec^2 x$$

$$= 272\tan x\left(1 + \tan^2 x\right) + 960\tan^3 x\left(1 + \tan^2 x\right) + 720\tan^5 x\left(1 + \tan^2 x\right)$$

$$= 272\tan x + 1232\tan^3 x + \dots \text{ terms containing higher power of } \tan x, \ f^{vi}(0) = 0$$

$$f^{vii}(x) = 272\sec^2 x + 3696\tan^2 x \sec^2 x + \ldots$$

$$= 272\left(1 + \tan^2 x\right) + \text{terms containing higher power of } \tan x, \ f^{vii}(0) = 272,$$

....

....

Substituting the values of $f(0), f'(0)$, etc. in the Maclaurin's series, we obtain

$$\tan x = 0 + x \cdot 1 + \frac{x^2}{2!}(0) + \frac{x^3}{3!}\cdot(2) + \frac{x^4}{4!}\cdot(0) + \frac{x^5}{5!}\cdot(16) + \frac{x^6}{6!}\cdot(0) + \frac{x^7}{7!}\cdot(272) + \ldots$$

$$= x + \frac{x^3}{3} + \frac{2x^5}{15} + \frac{17}{315}x^7 + \ldots\ldots$$

Example 2.26 Assuming the validity of expansion, prove that

$$e^x(\cos x) = 1 + x - \frac{2x^3}{3!} - \frac{2^2 x^4}{4!} - \frac{2^2 x^5}{5!} + \frac{2^3 x^7}{7!} + \ldots \qquad \textit{(MDU 2001)}$$

Solution. We have $f(x) = e^x (\cos x)$, $\qquad\qquad\qquad\qquad\qquad f(0) = 1,$

$$f^n(x) = (1+1)^{n/2} e^x \cos\left(x + n\tan^{-1} 1\right) = 2^{n/2} e^x \cos\left(x + \frac{n\pi}{4}\right).$$

$$\therefore \quad f'(0) = 1, \ f''(0) = 0, \ f'''(0) = -2, \ f^{iv}(0) = -4, \ f^v(0) = -4,$$

$$f^{vi}(0) = 0, \ f^{vii}(0) = 8 \text{ etc.}$$

Substituting these values of $f(0), f'(0), f''(0),\ldots$ in Maclaurin's series, we obtain

$$e^x \cos x = 1 + x - \frac{2x^3}{3!} - \frac{2^2 x^4}{4!} - \frac{2^2 x^5}{5!} + \frac{2^3 x^7}{7!} + \ldots\ldots$$

Example 2.27 If $y = \left[\log\left(x + \sqrt{a^2 + x^2}\right)\right]^2$, prove that

$\left(x^2 + a^2\right)y_{n+2} + (2n+1)xy_{n+1} + n^2 y_n = 0.$ **Hence write down the first four terms of the series of y, when expanding in ascending powers of x.**

Solution. We have $y = \left[\log\left(x + \sqrt{a^2 + x^2}\right)\right]^2.$ $\qquad\qquad\qquad$ ...(i)

$$\therefore \ y' = \frac{2}{\sqrt{a^2 + x^2}}\log\left(x + \sqrt{a^2 + x^2}\right) = \frac{2\sqrt{y}}{\sqrt{a^2 + x^2}}, \qquad\qquad \text{...(ii)}$$

and $y''\left(a^2 + x^2\right) + xy' - 2 = 0.$ $\qquad\qquad\qquad\qquad\qquad\qquad\qquad$ ...(iii)

Differentiating this n times by Leibnitz's theorem, we obtain

$$\left(x^2 + a^2\right)y_{n+2} + (2n+1)xy_{n+1} + n^2 y_n = 0 \qquad \text{...(iv)}$$

Putting $x = 0$ in (i), (ii) and (iii), we obtain $y(0) = (\log a)^2$, $y'(0) = (2/a)\log a$

and $y''(0) = 2/a^2$.

Now Putting $n = 1$ in (iv), we obtain $(a^2 + x^2)y''' + 3xy'' + y' = 0$

Therefore, $y'''(0) = -\dfrac{1}{a^2}y'(0) = \dfrac{-2}{a^3}\log a$.

Substituting these values in Maclaurin's series

$$y(x) = y(0) + xy'(0) + \frac{x^2}{2!}y''(0) + \frac{x^3}{3!}y'''(0) + ...; \text{ we obtain}$$

$$\left[\log\left(x + \sqrt{a^2 + x^2}\right)\right]^2 = (\log a)^2 + \left(\frac{2}{a}\log a\right)x + \frac{x^2}{a^2} - \left(\frac{1}{3a^3}\log a\right)x^3 - \ldots\ldots$$

Example 2.28 If $y = \sin(m \sin^{-1}x)$, show that $(1-x^2)y_2 - xy_1 + m^2 y = 0$,

and $(1-x^2)y_{n+2} - (2n+1)xy_{n+1} + (m^2 - n^2)y_n = 0$.
Hence expand $\sin m\theta$ in powers of $\sin\theta$.

Solution. Refer Ex. 1.17, we have

$$y(0) = 0,\ y'(0) = m,\ y''(0) = 0,\quad y'''(0) = (1^2 - m^2)m,$$

$$y^{iv}(0) = 0,\ y^{v}(0) = (3^2 - m^2)(1^2 - m^2)m,\quad y^{vi}(0) = 0\ \text{etc.}$$

Substituting these values in Maclaurin's series, we obtain

$$\sin(m\sin^{-1}x) = 0 + x\cdot m + \frac{x^2}{2!}(0) + \frac{x^3}{3!}m(1^2 - m^2)$$

$$+\frac{x^4}{4!}(0) + \frac{x^5}{5!}m(1^2 - m^2)(3^2 - m^2) + \ldots\ldots$$

$$= mx + \frac{m(1^2 - m^2)}{3!}x^3 + \frac{m(1^2 - m^2)(3^2 - m^2)}{5!}x^5 + \ldots \qquad \text{...(i)}$$

Putting $\sin^{-1}x = \theta$, that is, $x = \sin\theta$ in (i), we obtain

$$\sin m\theta = m\sin\theta\left[1 - \frac{m^2 - 1}{3!}\sin^2\theta + \frac{(m^2 - 1)(m^2 - 3^2)}{5!}\sin^4\theta - \ldots\ldots\right].$$

Example 2.29 Expand log sin (x + h) in powers of h by Taylor's series up to terms in h³. *(A.M.I.E., W-2002)*

Solution. Let $f(x+h) = \log \sin(x+h)$ so that $f(x) = \log \sin x$.

Taking derivatives, we obtain

$$f'(x) = \cos x / \sin x = \cot x, \quad f''(x) = -\operatorname{cosec}^2 x, \quad f'''(x) = 2\operatorname{cosec}^2 x \cot x, \text{ etc.}$$

Substituting these values in the Taylor's series

$$f(x+h) = f(x) + hf'(x) + \frac{h^2}{2!} f''(x) + \frac{h^3}{3!} f'''(x) + \dots\dots, \text{ we obtain}$$

$$\log \sin(x+h) = \log \sin x + h \cot x - \frac{h^2}{2}\operatorname{cosec}^2 x + \frac{h^3}{3}\operatorname{cosec}^2 x \cot x + \dots\dots$$

Example 2.30 Obtain the Taylor's series of following functions

(i) f (x) = log x, in powers of (x–2) and hence find the value of log 2.2,

(ii) $f(x) = x^3 - 2x^2 + 5x - 7$ in powers of (x–1),

(iii) $f(x) = \cos x$ at $x = \pi/4$, (iv) $f(x) = e^{x/2}$ at x = 2.

Solution. *(i)* We have $f(x) = \log x$ and $a = 2$. We need to find $f(2), f'(2)$, $f''(2), f'''(2), \dots\dots$. Taking derivatives, we obtain

$$
\begin{aligned}
& f(x) = \log x, && f(2) = \log 2, \\
& f'(x) = 1/x, && f'(2) = 1/2, \\
& f''(x) = -1/x^2, && f''(2) = -1/4, \\
& f'''(x) = 2/x^3, && f'''(2) = 1/4, \\
& f^{iv}(x) = -6/x^4, && f^{iv}(2) = -3/8,
\end{aligned}
$$

....
....

Hence, the Taylor's series of $f(x) = \log x$ about $x = 2$ is given by

$$f(x) = \log x = f(2) + (x-2)f'(2) + \frac{(x-2)^2}{2!} f''(2) + \frac{(x-2)^3}{3!} f'''(2)$$

$$+ \frac{(x-2)^4}{4!} f^{iv}(2) + \dots..$$

$$= \log 2 + \frac{(x-2)}{2} - \frac{(x-2)^2}{4.2!} + \frac{(x-2)^3}{4.3!} - \frac{3(x-2)^4}{8.4!} + \dots\dots$$

$$= \log 2 + \frac{(x-2)}{2} - \frac{(x-2)^2}{8} + \frac{(x-2)^3}{24} - \frac{(x-2)^4}{64} + \dots\dots$$

Putting $x = 2.2$, we obtain

$$\log 2.2 = \log 2 + \frac{0.2}{2} - \frac{(0.2)^2}{8} + \frac{(0.2)^3}{24} - \frac{(0.2)^4}{64} + \dots$$

$$= 0.69315 + 0.1 - 0.005 + 0.00033 - 0.00003 + \dots \approx 0.78845$$

(ii) Here $f(x) = x^3 - 2x^2 + 5x - 7$ and $a = 1$. We need to find

$f(1), f'(1), f''(1), f'''(1), \dots$. Taking derivatives, we obtain

$$f(x) = x^3 - 2x^2 + 5x - 7, \qquad\qquad f(1) = -3,$$
$$f'(x) = 3x^2 - 4x + 5, \qquad\qquad f'(1) = 4,$$
$$f''(x) = 6x - 4, \qquad\qquad f''(1) = 2,$$
$$f'''(x) = 6, \qquad\qquad f'''(1) = 6,$$

The Taylor expansion of $f(x) = x^3 - 2x^2 + 5x - 7$ about $x = 1$ is given by

$$f(x) = f(1) + (x-1)f'(1) + \frac{(x-1)^2}{2!}f''(1) + \frac{(x-1)^3}{3!}f'''(1)$$

$$= -3 + (x-1)4 + \frac{(x-1)^2}{2!} \cdot 2 + \frac{(x-1)^3}{2!} \cdot 6 = -3 + 4(x-1) + (x-1)^2 + (x-1)^3.$$

(iii) The Taylor series generated by f at $x = a$ is

$$f(x) = f(a) + (x-a)f'(a) + \frac{(x-a)^2}{2!}f''(a) + \frac{(x-a)^3}{3!}f'''(a) + \dots$$

Here $f(x) = \cos x$ and $a = \pi/4$. We need to find

$f(\pi/4), f'(\pi/4), f''(\pi/4), f'''(\pi/4), \dots$. Taking derivatives, we get

$$f(x) = \cos x, \qquad\qquad f(\pi/4) = 1/\sqrt{2},$$
$$f'(x) = -\sin x, \qquad\qquad f'(\pi/4) = -1/\sqrt{2},$$
$$f''(x) = -\cos x, \qquad\qquad f''(\pi/4) = -1/\sqrt{2},$$
$$f'''(x) = \sin x, \qquad\qquad f'''(\pi/4) = 1/\sqrt{2}.$$

Hence Taylor's series of $\cos x$ about $x = \pi/4$ is given by

$$\cos x = \frac{1}{\sqrt{2}} - \frac{1}{\sqrt{2}}\left(x - \frac{\pi}{4}\right) - \frac{1}{2\sqrt{2}}\left(x - \frac{\pi}{4}\right)^2 + \frac{1}{6\sqrt{2}}\left(x - \frac{\pi}{4}\right)^3 + \dots$$

$$= \frac{1}{\sqrt{2}}\left[1 - \left(x - \frac{\pi}{4}\right) - \frac{1}{2}\left(x - \frac{\pi}{4}\right)^2 + \frac{1}{6}\left(x - \frac{\pi}{4}\right)^3 + \dots\right].$$

(iv) Here $f(x) = e^{x/2}$ and $a = 2$. We need to find $f(2), f'(2), f''(2) f'''(2)$....

Taking derivatives, we get

$$f(x) = e^{x/2}, \qquad\qquad\qquad f(2) = e,$$

$$f'(x) = e^{x/2}/2, \qquad\qquad\qquad f'(2) = e/2,$$

$$f''(x) = e^{x/2}/4, \qquad\qquad\qquad f''(2) = e/4,$$

$$f'''(x) = e^{x/2}/8, \qquad\qquad\qquad f'''(2) = e/8.$$

The Taylor's expansion of $f(x) = e^{x/2}$ about $x = 2$ is given by

$$f(x) = f(2) + (x-2)f'(2) + \frac{(x-2)^2}{2!}f''(2) + \frac{(x-2)^3}{3!}f'''(2) + \ldots\ldots$$

Therefore, $\quad e^{x/2} = e + \frac{(x-2)e}{2} + \frac{(x-2)^2}{2!}\cdot\frac{e}{4} + \frac{(x-2)^3}{3!}\cdot\frac{e}{8} + \ldots\ldots$

$$= e\left[1 + \frac{(x-2)}{2} + \frac{(x-2)^2}{8} + \frac{(x-2)^3}{48} + \ldots\ldots\right].$$

Example 2.31 Calculate the approximate value of $\sqrt{10}$ to four decimal places by taking the first four terms of an appropriate Taylor's expansion.

(GGSIPU 2006)

Solution. Let $f(x+h) = (x+h)^{1/2}$ so that $f(x) = x^{1/2}$.

Taking derivatives, we obtain

$$f'(x) = 1/2\sqrt{x}, \quad f''(x) = x^{-3/2}/4, \quad f'''(x) = 3x^{-5/2}/8, \text{ etc.}$$

By Taylor series, we have

$$f(x+h) = f(x) + hf'(x) + \frac{h^2}{2!}f''(x) + \frac{h^3}{3!}f'''(x) + \ldots$$

$$\therefore \quad (x+h)^{1/2} = \sqrt{x} + \frac{h}{2}\cdot\frac{1}{\sqrt{x}} - \frac{h^2}{8}\cdot\frac{1}{x\sqrt{x}} + \frac{h^2}{16}\cdot\frac{1}{x^2\sqrt{x}} + \ldots\ldots$$

Put $x = 9$ and $h = 1$, in the above series, we obtain

$$\sqrt{10} = 3 + \frac{1}{2}\cdot\frac{1}{3} - \frac{1}{8}\cdot\frac{1}{27} + \frac{1}{16}\cdot\frac{1}{243} + \ldots\ldots$$

$$= 3 + 0.16667 - 0.00463 + 0.00025 + \ldots = 3.1623.$$

Example 2.32 Use Taylor's series, to prove that

$$\tan^{-1}(x+h) = \tan^{-1}x + (h\sin\theta)\cdot\frac{\sin\theta}{1} - (h\sin\theta)^2\cdot\frac{\sin 2\theta}{2}$$

$+(h \sin \theta)^3 \cdot \dfrac{\sin 3\theta}{3} - \ldots\ldots,$ where $\theta = \cot^{-1}x.$ $\hfill$ *(D.C.E., 2001)*

Solution. We have $\cot \theta = x$ $\hfill$ ...(i)

Therefore $\qquad -\csc^2\theta \cdot \dfrac{d\theta}{dx} = 1$ or $\dfrac{d\theta}{dx} = -\sin^2\theta.$ $\hfill$ ...(ii)

Now Let $f(x+h) = \tan^{-1}(x+h)$, so that $f(x) = \tan^{-1} x$.

Taking derivatives, we obtain

$$f'(x) = \frac{1}{1+x^2} = \frac{1}{1+\cot^2\theta} = \sin^2\theta, \hspace{2cm} \text{[By (i)]}$$

$$f''(x) = 2\sin\theta\cos\theta \cdot \frac{d\theta}{dx} = \sin 2\theta \cdot (-\sin^2\theta), \hspace{1cm} \text{[By (ii)]}$$

$$f'''(x) = -\left[2\cos 2\theta \cdot \left(\sin^2\theta\right) + (\sin 2\theta)\cdot 2\sin\theta\cos\theta\right]\frac{d\theta}{dx}$$

$$= -2\sin\theta\left[\sin\theta\cos 2\theta + \sin 2\theta\cos\theta\right]\left(-\sin^2\theta\right)$$

$$= 2\left(\sin^3\theta\right)\cdot\sin(\theta+2\theta) = 2\sin^3\theta\sin 3\theta, \text{ and so on.}$$

Substituting these values in the Taylor's series

$$f(x+h) = f(x) + hf'(x) + \frac{h^2}{2!}f''(x) + \frac{h^3}{3!}f'''(x)+\ldots, \text{ we get the required result.}$$

2.10 Expansions of functions in power Series

The function of x to be expanded in series of powers of x is generally a complicated function and so the direct use of Taylor's or Maclaurin's series, which
involves successive differentiation is not always convenient. The following are the
methods used in problems on series expansions of functions.

(A) Method of using standard series. The standard series are given below for
reference:

$$\sin x = x - \frac{x^3}{3!} + \frac{x^5}{5!} - \frac{x^7}{7!} + \frac{x^9}{9!} - \ldots \qquad \cos x = 1 - \frac{x^2}{2!} + \frac{x^4}{4!} - \frac{x^6}{6!} + \frac{x^8}{8!} - \ldots$$

$$\tan x = x + \frac{x^3}{3} + \frac{2}{15}x^5 + \ldots\ldots \qquad \sinh x = x + \frac{x^3}{3!} + \frac{x^5}{5!} + \frac{x^7}{7!} + \frac{x^9}{9!} + \ldots\ldots$$

$$\cosh x = 1 + \frac{x^2}{2!} + \frac{x^4}{4!} + \frac{x^6}{6!} + \frac{x^8}{8!} + \ldots\ldots \qquad \tanh x = x - \frac{x^3}{3} + \frac{2}{15}x^5 - \ldots\ldots$$

$$e^x = 1 + x + \frac{x^2}{2!} + \frac{x^3}{3!} + \frac{x^4}{4!} + \ldots \ldots \qquad \log_e(1+x) = x - \frac{x^2}{2} + \frac{x^3}{3} - \frac{x^4}{4} + \frac{x^5}{5} - \ldots \ldots$$

$$\log(1-x) = -\left(x + \frac{x^2}{2} + \frac{x^3}{3} + \frac{x^4}{4} + \ldots\right) \qquad \tanh^{-1} x = x + \frac{x^3}{3} + \frac{x^5}{5} + \ldots \ldots$$

$$\tan^{-1} x = x - \frac{x^3}{3} + \frac{x^5}{5} - \frac{x^7}{7} + \ldots \ldots, \ |x| \le 1$$

$$(1+x)^n = 1 + nx + \frac{n(n-1)}{2!}x^2 + \frac{n(n-1)(n-2)}{3!}x^3 + \ldots \ldots$$

Example 2.33 (*a*) **prove that**

$$e^{x \sin x} = 1 + x^2 + x^4/3 + x^6/120 + \ldots \ldots \qquad \qquad \textit{(Marathwada, 1998)}$$

(*b*) **Find the first three non zero terms in the Maclaurin's series for the function** $f(x) = e^{-x^2} \cdot \tan^{-1} x.$

Solution. (*a*) Here if we use the Maclaurin's series we need successive derivatives of $e^{x \sin x}$ which are inconvenient to obtain.

We therefore make use of the two standard series *viz.*,

$$e^\theta = 1 + \theta + \frac{\theta^2}{2!} + \frac{\theta^3}{3!} + \frac{\theta^4}{4!} + \ldots \ldots ; \text{ and } \sin\theta = \theta - \frac{\theta^3}{3!} + \frac{\theta^5}{5!} - \frac{\theta^7}{7!} + \ldots \ldots$$

Let $y = x \sin x$, then we have

$$e^{x \sin x} = e^y = 1 + y + \frac{y^2}{2!} + \frac{y^3}{3!} + \frac{y^4}{4!} + \ldots \ldots$$

$$= 1 + (x \sin x) + \frac{(x \sin x)^2}{2!} + \frac{(x \sin x)^3}{3!} + \frac{(x \sin x)^4}{4!} + \ldots \ldots$$

$$= 1 + x(\sin x) + \frac{x^2}{2!}(\sin x)^2 + \frac{x^3}{3!}(\sin x)^3 + \frac{x^4}{4!}(\sin x)^4 + \ldots \ldots \qquad \ldots(i)$$

Substituting in (*i*) the series for sin *x*, we obtain

$$e^{x \sin x} = 1 + x\left\{x - \frac{x^3}{3!} + \frac{x^5}{5!} - \ldots \ldots\right\} + \frac{x^2}{2!}\left\{x - \frac{x^3}{3!} + \frac{x^5}{5!} - \ldots \ldots\right\}^2$$

$$+ \frac{x^3}{3!}\left\{x - \frac{x^3}{3!} + \frac{x^5}{5!} - \ldots \ldots\right\}^3 + \frac{x^4}{4!}\left[x - \frac{x^3}{3!} + \frac{x^5}{5!} - \ldots \ldots\right]^4 + \ldots \ldots$$

$$= 1 + x^2\left(1 - \frac{x^2}{6} + \frac{x^4}{120} - \ldots\ldots\right) + \frac{x^4}{2}\left[1 - \left(\frac{x^2}{6} - \frac{x^4}{120} + \ldots\ldots\right)\right]^2$$

$$+ \frac{x^6}{6}\left[1 - \left(\frac{x^2}{6} - \frac{x^4}{120} + \ldots\ldots\right)\right]^3 + \frac{x^8}{24}\left[1 - \left(\frac{x^2}{6} - \frac{x^4}{120} + \ldots\ldots\right)\right]^4 + \ldots\ldots$$

Expanding the second and onward bracket in the above expression by the Binomial theorem, we have

$$e^{x \sin x} = 1 + x^2\left(1 - \frac{x^2}{6} + \frac{x^4}{120} - \ldots\ldots\right) + \frac{x^4}{2}\left[1 - 2\left(\frac{x^2}{6} - \frac{x^4}{120} + \ldots\ldots\right) + \ldots\ldots\right]$$

$$+ \frac{x^6}{6}\left[1 - 3\left(\frac{x^2}{6} - \ldots\ldots\right)\right] + \ldots\ldots$$

$$= 1 + x^2 + \frac{x^4}{3} + \frac{x^6}{120} + \ldots\ldots$$

(b) We can find the Maclaurin's series for $e^{-x^2} \cdot \tan^{-1} x$, by multiplying the Maclaurin series for e^{-x^2} by Maclaurin series for $\tan^{-1} x$.

The simplest way to obtain the Maclaurin's series for e^{-x^2} is to replace x by $-x^2$ in the Maclaurin's series of $e^x = 1 + x + \frac{x^2}{2!} + \frac{x^3}{3!} + \frac{x^4}{4!} + \ldots.$ to obtain

$$e^{-x^2} = 1 - x^2 + \frac{x^4}{2!} - \frac{x^6}{3!} + \ldots.$$

$$\therefore \quad e^{-x^2} \cdot \tan^{-1} x = \left(1 - x^2 + \frac{x^4}{2} - \frac{x^6}{6} + \ldots\right)\left(x - \frac{x^3}{3} + \frac{x^5}{5} - \ldots\ldots\right)$$

$$= x - \frac{4}{3}x^3 + \frac{31}{30}x^5 - \ldots\ldots$$

Example 2.34 Expand $e^{\cos x}$ by Maclaurin's series or otherwise upto the term containing x^6.

Solution. We have $e^{\cos x} = e \cdot e^{(\cos x - 1)}$.

Using $e^{\theta} = 1 + \theta + \dfrac{\theta^2}{2!} + \dfrac{\theta^3}{3!} + \ldots$ and $\theta = \cos x - 1 = \left(-\dfrac{x^2}{2!} + \dfrac{x^4}{4!} - \dfrac{x^6}{6!} + \ldots \right)$,

we find $\quad e^{\cos x} = e \left\{ 1 + \left(-\dfrac{x^2}{2!} + \dfrac{x^4}{4!} - \dfrac{x^6}{6!} + \ldots \right) + \dfrac{1}{2!} \left(\dfrac{x^4}{(2!)^2} - \dfrac{2x^6}{2!4!} + \ldots \right) \right.$

$$\left. + \dfrac{1}{3!} \left(-\dfrac{x^6}{(2!)^3} + \ldots \right) + \ldots \right\} = e \left\{ 1 - \dfrac{x^2}{2} + \dfrac{x^4}{6} - \dfrac{31}{720} x^6 + \ldots \right\}.$$

Example 2.35 Show that $(1+x)^x = 1 + x^2 - x^3/2 + 5x^4/6 - \ldots$

Solution. We may write $(1+x)^x = e^{x \log(1+x)}$

$$= 1 + x \log(1+x) + \dfrac{x^2}{2!} \left[\log(1+x) \right]^2 + \dfrac{x^3}{3!} \left[\log(1+x) \right]^3 + \ldots$$

Next using the known series for $\log(1+x)$. Then expanding by the Binomial Theorem and collecting the coefficients of various powers of x, we obtain the required result.

Example 2.36 Expand $\dfrac{\log(1+x)}{1+x}$ in powers of x for $|x| < 1$. (*A.M.I.E; 2000*)

Solution. If $|x| < 1$, then $\log(1+x) = x - \dfrac{x^2}{2} + \dfrac{x^3}{3} - \dfrac{x^4}{4} + \ldots$

Since $\quad \dfrac{1}{1+x} = (1+x)^{-1} = 1 - x + x^2 - x^3 + \ldots,$

$\therefore \quad \dfrac{\log(1+x)}{1+x} = \left(x - \dfrac{x^2}{2} + \dfrac{x^3}{3} - \dfrac{x^4}{4} + \ldots \right) \left(1 - x + x^2 - x^3 + \ldots \right)$

$$= x - \left(1 + \dfrac{1}{2} \right) x^2 + \left(1 + \dfrac{1}{2} + \dfrac{1}{3} \right) x^3 - \left(1 + \dfrac{1}{2} + \dfrac{1}{3} + \dfrac{1}{4} \right) x^4 + \ldots$$

$$= x - \dfrac{3}{2} x^2 + \dfrac{11}{6} x^3 - \dfrac{25}{12} x^4 + \ldots$$

Example 2.37 Expand $\log(1 + \sin^2 x)$ in powers of x as far as the term in x^6.
Solution. As it is very cumbersome to determine the successive derivatives of $\log(1 + \sin^2 x)$, therefore the method of using standard series expansion is preferable to Maclaurin's series method.

We have $\quad \sin^2 x = \left(x - \dfrac{x^3}{3!} + \dfrac{x^5}{5!} - \ldots\ldots\right)^2 = \left[x - \left(\dfrac{x^3}{6} - \dfrac{x^5}{120} + \ldots\ldots\right)\right]^2$

$= x^2 - 2x\left(\dfrac{x^3}{6} - \dfrac{x^5}{120} + \ldots\ldots\right) + \left(\dfrac{x^3}{6} - \dfrac{x^5}{120} + \ldots\right)^2 = x^2 - \dfrac{x^4}{3} + \dfrac{2x^6}{45} + \ldots = z$ (say).

Now $\log(1 + \sin^2 x) = \log(1 + z) = z - \dfrac{z^2}{2} + \dfrac{z^3}{3} - \dfrac{z^4}{4} + \ldots \qquad\qquad …(i)$

Substituting the value of z in (i) we obtain

$$\log(1 + \sin^2 x) = \left(x^2 - \dfrac{x^4}{3} + \dfrac{2x^6}{45} + \ldots\right) - \dfrac{1}{2}\left(x^2 - \dfrac{x^4}{3} + \ldots\right)^2 + \dfrac{1}{3}\left(x^2 - \ldots\right)^3 - \ldots$$

$$= x^2 - \dfrac{x^4}{3} + \dfrac{2x^6}{45} - \dfrac{1}{2}\left(x^4 - \dfrac{2x^6}{3} + \ldots\right) + \dfrac{1}{3}\left(x^6 + \ldots\right) + \ldots$$

$$= x^2 - \dfrac{5}{6}x^4 + \dfrac{32}{45}x^6 + \ldots\ldots$$

(B) The Method of Substitution. Sometimes the complicated function of x, to be expanded in series of powers of x, can be reduced to the standard form by a convenient substitution. The following problems illustrate this method.

Example 2.38 Expand $\cos^{-1}\left(\dfrac{x - x^{-1}}{x + x^{-1}}\right)$ **in the ascending powers of x.**

Solution. Put $x = \cot(\theta/2)$, we obtain

$$\cos^{-1}\left(\dfrac{x - x^{-1}}{x + x^{-1}}\right) = \cos^{-1}\left(\dfrac{\cot(\theta/2) - \tan(\theta/2)}{\cot(\theta/2) + \tan(\theta/2)}\right) = \cos^{-1}\left(\dfrac{\cos^2(\theta/2) - \sin^2(\theta/2)}{\cos^2(\theta/2) + \sin^2(\theta/2)}\right)$$

$$= \cos^{-1}(\cos\theta) = \theta = 2\cot^{-1} x = \pi - 2\tan^{-1} x.$$

∴ The expansion for the given function

$$\cos^{-1}\left(\dfrac{x - x^{-1}}{x + x^{-1}}\right) = \pi - 2\tan^{-1} x = \pi - 2\left(x - \dfrac{x^3}{3} + \dfrac{x^5}{5} - \dfrac{x^7}{7} + \ldots\ldots\right).$$

Example 2.39 Prove that

$$\sec^{-1}\dfrac{1}{1 - 2x^2} = 2\left(x + \dfrac{1}{2}\dfrac{x^3}{3} + \dfrac{1}{2}\dfrac{3}{4}\dfrac{x^5}{5} + \ldots\right)$$

Solution. The substitution $x = \sin\theta$, gives

$$\sec^{-1}\frac{1}{1-2x^2} = \sec^{-1}\left[\frac{1}{1-2\sin^2\theta}\right] = \sec^{-1}\left[\frac{1}{\cos^2\theta - \sin^2\theta}\right] = \sec^{-1}\left(\frac{1}{\cos 2\theta}\right)$$

$= \sec^{-1}(\sec 2\theta) = 2\theta = 2\sin^{-1}x$. Using the series of $\sin^{-1}x$, we obtain

$$\sec^{-1}\frac{1}{1-2x^2} = 2\sin^{-1}x = 2\left(x + \frac{1}{2}\cdot\frac{x^3}{3} + \frac{1}{2}\cdot\frac{3}{4}\cdot\frac{x^5}{5} + \ldots\ldots\right)$$

(C) Method of differentiation or integration of known series

In this method the given function on differentiation or integration gives a function of which the series is known or can be found out very easily.

Example 2.40 Obtain the Maclaurin's series for $\sqrt{1+\sin x}$.

Solution. We have $\sqrt{1+\sin x} = \sqrt{\left(\sin\frac{x}{2} + \cos\frac{x}{2}\right)^2} = \sin\frac{x}{2} + \cos\frac{x}{2}$.

Replacing x by $x/2$ in the series for $\sin x$, we obtain

$$\sin\frac{x}{2} = \frac{x}{2} - \frac{x^3}{2^3\cdot 3!} + \frac{x^5}{2^5\cdot 5!} - \frac{x^7}{2^7\cdot 7!} + \ldots\ldots \qquad \ldots(i)$$

Differentiating the series (i), we obtain

$$\cos\frac{x}{2} = 2\left\{\frac{1}{2} - \frac{x^2}{2^3\, 2!} + \frac{x^4}{2^5\cdot 4!} - \frac{x^6}{2^7\cdot 6!} + \ldots\ldots\right\}$$

$$= 1 - \frac{x^2}{2^2\cdot 2!} + \frac{x^4}{2^4\cdot 4!} - \frac{x^6}{2^6\cdot 6!} + \ldots\ldots \qquad \ldots(ii)$$

Adding (i) and (ii), we get

$$\sqrt{1+\sin x} = \sin\frac{x}{2} + \cos\frac{x}{2}$$

$$= 1 + \frac{x}{2} - \frac{x^2}{2^2\cdot 2!} - \frac{x^3}{2^3\, 3!} + \frac{x^4}{2^4\cdot 4!} + \frac{x^5}{2^5\cdot 5!} - \ldots\ldots, \text{ for all values of } x.$$

Example 2.41 Find the Maclaurin's series for $\tan^{-1}x$.

Solution. It would be tedious to find the maclaurin's series directly. A better approach is to start with the formula

$$\int\frac{1}{1+x^2}dx = \tan^{-1}x + c. \text{ We have}$$

$$\frac{1}{1+x^2} = \left(1+x^2\right)^{-1} = 1-x^2+x^4-x^6+x^8-\ldots\ldots(-1<x<1) \qquad \ldots(i)$$

Now integrate the Maclaurin's series (i) term by term. This yields

$$\tan^{-1}x + c = \int \frac{1}{1+x^2}dx = \int\left(1-x^2+x^4-x^6+x^8-\ldots\ldots\right)dx$$

or $$\tan^{-1}x = \left(x-\frac{x^3}{3}+\frac{x^5}{5}-\frac{x^7}{7}+\frac{x^9}{9}-\ldots\ldots\right)-c.$$

The constant of integration can be evaluated by substituting $x = 0$ and using the condition $\tan^{-1}0 = 0$. This gives $c = 0$, and we obtain

$$\tan^{-1}x = x-\frac{x^3}{3}+\frac{x^5}{5}-\frac{x^7}{7}+\frac{x^9}{9}-\ldots\ldots(-1<x<1)$$

Exercise 2.2

Obtain the Maclaurin's series expansion of each of the following functions:

1.   (i) $\cos x$, (ii) $e^x/(e^x+1)$ *(MDU 2005)*, (iii) $e^x \sec x$, (iv) $\cos(6\sin^{-1}x)$,

   (v) $\log(1+e^x)$ *(MDU 2000)*, (vi) $\sin^{-1}x$.   *(GGSIPU 2006; D.C.E. 2001)*

Ans.: (i) $1-\dfrac{x^2}{2!}+\dfrac{x^4}{4!}-\dfrac{x^6}{6!}+\ldots\ldots$,   (ii) $\dfrac{1}{2}+\dfrac{x}{4}-\dfrac{x^3}{48}+\ldots\ldots$

   (iii) $1+x+x^2+\dfrac{2}{3}x^3+\ldots\ldots$,   (iv) $1-18x^2+48x^4-32x^6+\ldots\ldots$

   (v) $\log 2+\dfrac{x}{2}+\dfrac{x^2}{8}-\dfrac{x^4}{192}+\ldots\ldots$,   (vi) $x+\dfrac{1}{2}\cdot\dfrac{x^3}{3}+\dfrac{1\cdot3}{2\cdot4}\cdot\dfrac{x^5}{5}+\dfrac{1\cdot3\cdot5}{2\cdot4\cdot6}\cdot\dfrac{x^7}{7}+\ldots\ldots$

2. Verify the following expansions of functions by Maclaurin's series.

   (i) $\log\cos x = -\dfrac{x^2}{2}-\dfrac{x^4}{12}-\dfrac{x^6}{45}-\ldots\ldots$ *(A.M.I.E., S-2000)*

   (ii) $\log\sec x = \dfrac{x^2}{2}+\dfrac{x^4}{12}+\dfrac{x^6}{45}+\ldots\ldots$

   (iii) $\tan^{-1}x = x-\dfrac{x^3}{3}+\dfrac{x^5}{5}-\dfrac{x^7}{7}+\ldots\ldots$

(*iv*) $\log(1+\cos x) = \log 2 - \dfrac{1}{4}x^2 - \dfrac{1}{96}x^4 - \ldots\ldots$

(*v*) $\sec x = 1 + \dfrac{1}{2}x^2 + \dfrac{5}{24}x^4 + \ldots\ldots$

(*vi*) $\dfrac{e^x}{e^x+1} = \dfrac{1}{2} + \dfrac{1}{4}x - \dfrac{1}{48}x^3 + \ldots\ldots$

(*vii*) $\log(1+\tan x) = x - \dfrac{x^2}{2} + \dfrac{2}{3}x^3 - \ldots\ldots$

(*viii*) $e^{x\cos x} = 1 + x + \dfrac{x^2}{2} - \dfrac{x^3}{3} - \dfrac{11}{24}x^4 - \ldots\ldots$ (*A.M.I.E., S-2009, S-2007*)

(*ix*) $\dfrac{\sin^{-1} x}{\sqrt{1-x^2}} = x + \dfrac{2}{3}x^3 + \dfrac{8}{15}x^5 + \ldots\ldots$

[**Hint.** Let $y(x) = \dfrac{\sin^{-1} x}{\sqrt{1-x^2}}$ so that $(1-x^2)y_1 = 1 + xy$,

and $(1-x^2)y_2 - 3xy_1 - y = 0$.

Differentiating this n times by Leibnitz's theorem, we get

$$(1-x^2)y_{n+2} - (2n+3)xy_{n+1} - (n+1)^2 y_n = 0.$$

From these results, we obtain $y(0) = 0$, $y_1(0) = 1$, $y_2(0) = y_4(0) = \ldots = 0$,

$y_3(0) = 4$, $y_5(0) = 64$ etc. Substituting these values in Maclaurin's series

$$y(x) = y(0) + xy_1(0) + \dfrac{x^2}{2!}y_2(0) + \dfrac{x^3}{3!}y_3(0) + \ldots\ldots.$$ We obtain the required

series].

(*x*) $e^{ax}\cos bx = 1 + ax + \dfrac{a^2-b^2}{2!}x^2 + \dfrac{a(a^2-3b^2)}{3!}x^3 + \ldots\ldots$

3. Show that $\log\left(x + \sqrt{x^2+1}\right) = x - \dfrac{1}{2}\cdot\dfrac{x^3}{3} + \dfrac{1\cdot3\cdot x^5}{2\cdot4\cdot5} - \dfrac{1\cdot3\cdot5}{2\cdot4\cdot6}\cdot\dfrac{x^7}{7} + \ldots\ldots$

    Hence find the value of log 2 by putting $x = 0.75$ to four places of decimals, and find the percentage error.   **Ans.** 0.69152; 0.24%

    [**Hint.** Actual value of log 2 = 0.69315...

$$\therefore \%Error = \dfrac{0.69315 - 0.69152}{0.69315} \times 100 = 0.24]$$

4. Obtain the Maclaurin's series expansion of $\tan\left(\dfrac{\pi}{4}+x\right)$, and hence find the value of $\tan 46°\ 30'$ to four decimal places. *(DCE 2005)*

Ans. $1+2x+2x^2+8x^3/3+10x^4/3+\dots\dots$

[**Hint.** $\tan\left(\dfrac{\pi}{4}+x\right)=1+2x+2x^2+8x^3/3+\dots\dots$ Put $x=\pi/120$ in both sides.

We obtain $\tan 46°\ 30'=1+0.05236+0.00137+0.00005+\dots\dots=1.0538$ correct to four decimal places].

5. Find the Taylor's series generated by f at $x=a$.

(i) $f(x)=\sin x,\quad a=\pi/2$ *(MDU 2003)*, (ii) $f(x)=\log\sin x,\quad a=2,$

(iii) $f(x)=\tan x,\quad a=\pi/4,$ *(A.M.I.E., W-2005)*

(iv) $f(x)=\log x,\quad a=1.$ Hence find the value of $\log 1.1.$ *(GGSIPU 2006)*

Ans.: (i) $1-\dfrac{1}{2!}\left(x-\dfrac{\pi}{2}\right)^2+\dfrac{1}{4!}\left(x-\dfrac{\pi}{2}\right)^4-\dots\dots$

(ii) $\log\sin 2+(x-2)\cot 2-\dfrac{(x-2)^2}{2}\operatorname{cosec}^2 2+\dfrac{(x-2)^3}{3}\cot^2 2\operatorname{cosec}^2 2+\dots\dots$

(iii) $1+2\left(x-\dfrac{\pi}{4}\right)+2\left(x-\dfrac{\pi}{4}\right)^2+\dfrac{8}{3}\left(x-\dfrac{\pi}{4}\right)^3+\dots\dots$

(iv) $(x-1)-\dfrac{1}{2}(x-1)^2+\dfrac{1}{3}(x-1)^3-\dfrac{1}{4}(x-1)^4+\dots\dots;\ 0.095308.$

6. Use Taylor's series expansions to show that

(i) $\log\cos(x+h)=\log\cos x-h\tan x-\dfrac{h^2}{2}\sec^2 x-\dfrac{h^3}{3}\sec^2 x\tan x+\dots\dots$

(ii) $\tan^{-1}(x+h)=\tan^{-1}x+\dfrac{h}{1+x^2}-h^2\cdot\dfrac{x}{\left(1+x^2\right)^2}+\dfrac{h^3}{3!}\cdot\dfrac{6x^2-2}{\left(1+x^2\right)^3}+\dots\dots$

(D.C.E, 2004)

(iii) $\sin(x+h)=\sin x+h\cos x-\dfrac{h^2}{2!}\sin x-\dots\dots$

7. Find the Taylor's series expansions of each of the following functions:

(i) $f(x)=x^3$ in powers of $(x-1),$

(ii) $f(x) = 2x^3 + 7x^2 + x - 1$ in powers of $(x-2)$,

(iii) $f(x) = x^4 - 3x^3 + 4x^2 - 2x + 5$ in powers of $(x-1)$,   *(A.M.I.E., W-97)*

(iv) $f(x) = x^4 + x^2 + 1$ in powers of $(x+2)$.

Ans.: (i) $1 + 3(x-1) + 3(x-1)^2 + (x-1)^3$,

   (ii) $45 + 53(x-2) + 19(x-2)^2 + 2(x-2)^3$,

   (iii) $5 + (x-1) + (x-1)^2 + (x-1)^3 + (x-1)^4$,

   (iv) $21 - 36(x+2) + 25(x+2)^2 - 8(x+2)^3 + (x+2)^4$.

8. Use the Taylor's series to compute to five-decimal places accuracy the following:

(i) $\sin 31°$, $(\text{Take } \cos 30° = 0.8660254)$, (ii) $\log 0.97$,

(iii) $\tan^{-1} 1.003$ (Take $\pi = 3.141593$), (iv) $\tan 44°$ (Take $1° = 0.01745$ radian).

Ans.: (i) 0.51504, (ii) −0.03046, (iii) 0.78690, (iv) 0.96569.

9. By Maclaurin's series or otherwise find the series expansion of following functions

(i) $f(x) = \sin(e^x - 1)$,   (ii) $f(x) = e^x(\sin x)$,   *(GGSIPU 2006)*

(iii) $f(x) = e^x \cdot (\sin^2 x)$,   (iv) $\dfrac{x}{\sin x}$ *(UPTU 2000)*

Ans.: (i) $x + \dfrac{x^2}{2} - \dfrac{5x^4}{24} + \ldots\ldots$, (ii) $x + x^2 + \dfrac{x^3}{3} - \dfrac{x^5}{30} + \ldots\ldots$,

   (iii) $x^2 + x^3 + \dfrac{1}{6}x^4 + \ldots\ldots$, (iv) $1 + \dfrac{x^2}{6} + \dfrac{7x^4}{360} + \ldots\ldots$

10. Arrange $f(x) = 7 + (x+2) + 3(x+2)^3 + (x+2)^4 - (x+2)^5$. in powers of x, by using Taylor's series.   *(A.M.I.E., W-2004)*

[Hint. Let $f(x) = 7 + (x+2) + 3(x+2)^3 + (x+2)^4 - (x+2)^5$. Here $a = 0$. Taking derivatives, we get

$f'(x) = 1 + 9(x+2)^2 + 4(x+2)^3 - 5(x+2)^4$,

$f''(x) = 18(x+2) + 12(x+2)^2 - 20(x+2)^3$,

$f'''(x) = 18 + 24(x+2) - 60(x+2)^2$, $f^{iv}(x) = 24 - 120(x+2)$,

$f^{v}(x) = -120$, $f^{vi}(x) = 0$.

$\therefore \quad f(0) = 17, f'(0) = -11, f''(0) = -76, f'''(0) = -174,$

$f^{iv}(0) = 216, f^{v}(0) = -120, f^{vi}(0) = 0.$

Substituting these values in the Taylor's series

$$f(x) = f(0) + x f'(0) + \frac{x^2}{2!} f''(0) + \frac{x^3}{3!} f'''(0) + \frac{x^4}{4!} f^{iv}(0) + \frac{x^5 f^{v}}{5!}(0), \text{ we}$$

obtain the result $17 - 11x - 38x^2 - 29x^3 - 9x^4 - x^5$].

11. If $f(x) = x^3 + 8x^2 + 15x - 24$, calculate the value of $f(11/10)$ by the application of Taylor's series. *(MDU 2004)* **Ans. 3.511**

12. If $f(x) = f(0) + xf'(0) + \frac{x^2}{2!} f''(\theta x), 0 < \theta < 1,$ find the value of θ as x tends

to 1, $f(x)$ being $(1-x)^{5/2}$. *(A.M.I.E., S-2009, W-1996)* **Ans. 0.36**

13. Obtain the Taylor's series expansion of $f(x) = \sin x$ about $x = 0$.

(A.M.I.E., S-2002)

Ans. $x - \dfrac{x^3}{3!} + \dfrac{x^5}{5!} - \dfrac{x^7}{7!} + \ldots\ldots + (-1)^{n-1} \dfrac{x^{2n-1}}{(2n-1)!} + \ldots\ldots$

14. Assuming the validity of expansion, show that

$$\tan^{-1} x = \tan^{-1} \frac{\pi}{4} + \frac{\left(x - \dfrac{\pi}{4}\right)}{\left(1 + \dfrac{\pi^2}{16}\right)} - \frac{\pi\left(x - \dfrac{\pi}{4}\right)^2}{4\left(1 + \dfrac{\pi^2}{16}\right)^2} + \ldots\ldots$$

15. Find the coefficient of x^3 in the expansion of $e^{\sin x}$. *(A.M.I.E., W-2010)*
Hint:

$$e^{\sin x} = e^{\left(x - \frac{x^3}{3!} + \ldots\right)} = 1 + \left(x - \frac{x^3}{3!} + \ldots\right) + \frac{1}{2!}\left(x - \frac{x^3}{3!} + \ldots\right)^2$$

$$+ \frac{1}{3!}\left(x - \frac{x^3}{3!} + \ldots\right)^3 + \frac{1}{4!}\left(x - \frac{x^3}{3!} + \ldots\right)^4 + \ldots$$

$$= 1 + x - \frac{x^3}{3!} + \frac{x^2}{2!} + \frac{x^3}{3!} - \frac{x^4}{3!} + \frac{x^4}{4!} + \ldots = 1 + x + \frac{x^2}{2} - \frac{x^4}{8} + \ldots$$

The coefficient of x^3 in the expansion of $e^{\sin x}$ is zero.]

Chapter 3

Definite Integrals

3.1 Definite Integral as the Limit of a Sum

We have so far looked upon the integration as an operation which is the inverse of differentiation, and we defined the integral of a function $f(x)$ as the function which when differentiated will give us $f(x)$. It will now be shown that a definite integral can also be represented as the limit of the sum of a finite series of numbers when the number of terms of the series tends to infinity while each term of the series tends to zero.

Let f be a continuous function defined on a closed interval $[a, b]$. Assume all the values taken by the function are non-negative, so that the graph of the function is a curve above the x-axis. Consider the region between this curve, the x-axis and the ordinates $x = a$ and $x = b$. The region R is shown by shaded portion in Fig. 3.1. Now, the problem is to find the area of this shaded region.

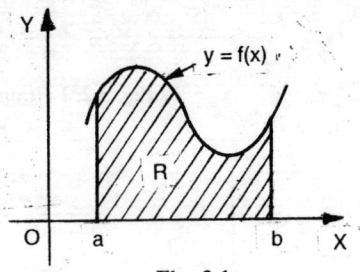

Fig. 3.1.

The "*rectangle method*" provides us with one aproach for determing the area between the graph of f and the interval $[a, b]$. The idea behind the rectangle method is as follows:

Divide the interval $[a, b]$ into n equal subintervals of length h so $h = (b-a)/n$. Take the values of f at the left end-points of the n sub-intervals. The values are $f(a), f(a+h), \cdots\cdots, f(a+\overline{n-1}\,h)$. Over each subinterval construct a rectangle. Refer Fig. 3.2, h and $f(a)$ are the adjacent sides of rectangle marked as 1. So $\left[h \cdot f(a)\right]$ is the area of rectangle marked as 1. Similarly $h \cdot f(a+h)$ is the area of the rectangle marked as 2 in the figure. Thus, $hf(a) + hf(a+h) + \cdots + hf(a+\overline{n-1}\,h)$ is the sum of the areas of these n rectangles marked in the figure. The union of these rectangles is approximately the region between the curve and the x-axis.

As *n* increases these approximations will get better and better and will approach the exact area as a limit (Fig. 3.3).

Thus if the function *f* is continuous on [*a, b*] and if $f(x) \geq 0$ for all *x* in [*a, b*], then the area of the region bounded by the curve *f*, *x*-axis, the ordinates $x = a$ and $x = b$ is given by

$$A = \lim_{n \to \infty} \Big[hf(a) + hf(a+h) + \cdots + hf\big(a + \overline{n-1}\, h\big) \Big], \text{ where } b - a = nh \qquad ...(3.1)$$

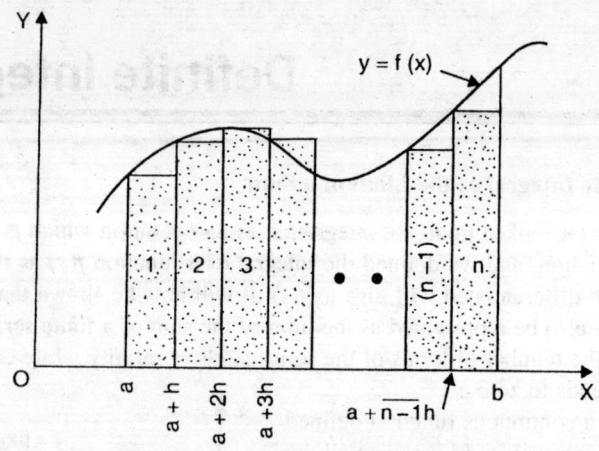

Fig. 3.2. Left end point approximation

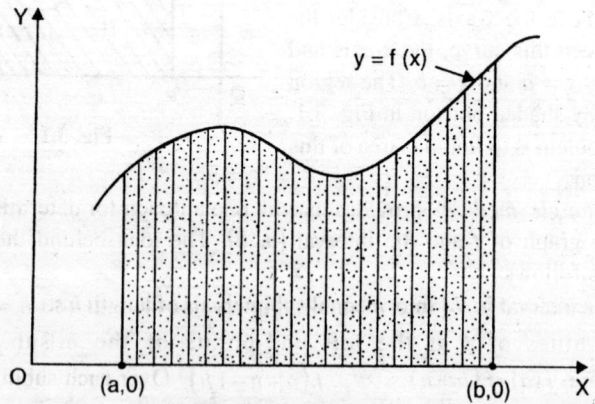

Fig, 3.3. As n increasing, the areas of retangles approaches the exact area under the curve

We take the expression on R.H.S of (3.1) as the definition of a definite integral.

This **definite integral** is denoted by $\int_a^b f(x)\,dx$ and is read as: 'integral of *f* from

a to b' or between the limits a and b. The numbers a and b are called the **lower limit or inferior limit of integration** and the **upper limit or superior limit of integration** respectively, and $f(x)$ is called the **integrand**.

The interval (a, b) is called the *range of integration*.

Thus, we have **Definition**

$$\int_a^b f(x)dx = \lim_{n \to \infty} \left[hf(a) + hf(a+h) + \cdots + hf\left(a + \overline{n-1}\,h\right) \right], \qquad ...(3.2)$$

where $h = (b-a)/n$.

Remark: 1. If we take the right end points instead of the left, then also, we obtain the same areas as the limit of areas of unions of some other rectangles (Fig. 3.4).

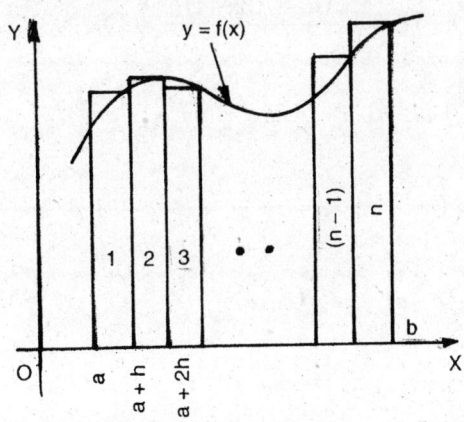

Fig. 3.4. Right end point approximation

Thus

$$A = \int_a^b f(x)dx = \lim_{n \to \infty} \left[h \cdot f(a+h) + h \cdot f(a+2h) + \cdots + h \cdot f(b) \right], \qquad ...(3.3)$$

where $h = (b-a)/n$.

It is true that (3.2) and (3.3) give the same limit.

2. Integration by summation is also called *integration from definition* or *integration from first principles* or *integration by ab-initio method*.

3.2 Illustrative Examples

Example 3.1 Evaluate $\int_a^b x^2 dx$ **as the limit of a sum.**

Solution. Here $f(x) = x^2$. Therefore $f(a) = a^2$, $f(a + h) = (a + h)^2$ etc.

By definition, we have

$$\int_a^b x^2 dx = \lim_{n \to \infty} h\left[f(a) + f(a+h) + \dots + f\left(a + \overline{n-1}h\right) \right],$$

where $b - a = nh$, when $n \to \infty, h \to 0$

$$= \lim_{n \to \infty} h\left[a^2 + (a+h)^2 + (a+2h)^2 + \dots + \left(a + \overline{n-1}h\right)^2 \right]$$

$$= \lim_{n \to \infty} h\left[\underbrace{a^2 + a^2 + \dots + a^2}_{n-times} + 2ah\left(1 + 2 + \dots + \overline{n-1}\right) + h^2\left\{1^2 + 2^2 + \dots + (n-1)^2\right\} \right]$$

$$= \lim_{n \to \infty} h\left[na^2 + n(n-1)ah + \frac{n(n-1)(2n-1)h^2}{6} \right]$$

$$= \lim_{n \to \infty} \left[nha^2 + nh \cdot (nh - h) \cdot a + \frac{1}{3}(nh - h)\left(nh - \frac{h}{2}\right)nh \right]$$

$$= \lim_{h \to 0} \left[(b-a)a^2 + (b-a)(b-a-h) \cdot a + \frac{1}{3}(b-a-h)\left(b-a-\frac{h}{2}\right)(b-a) \right]$$

$$= (b-a)a^2 + (b-a)^2 \cdot a + \frac{1}{3}(b-a)^3 \qquad (\because \text{ when } n \to \infty, h \to 0)$$

$$= \frac{1}{3}(b-a)\left\{3a^2 + 3(b-a)a + b^2 - 2ab + a^2\right\}$$

$$= (b-a)\left(a^2 + ab + b^2\right)/3 = \left(b^3 - a^3\right)/3.$$

Example 3.2 Evaluate from first principles $\int_a^b e^x dx$.

Solution. Here $f(x) = e^x$. Therefore $f(a) = e^a$, $f(a+h) = e^{a+h} \dots$, etc.

By definition, $\int_a^b e^x dx = \lim_{n \to \infty} h\left[e^a + e^{a+h} + e^{a+2h} + \dots + e^{a+\overline{n-1}h} \right]$,

where $b - a = nh$,

$$= \lim_{n \to \infty} he^a \left[1 + e^h + e^{2h} + \dots + e^{(n-1)h} \right]$$

$$= \lim_{n \to \infty} he^a \left[\frac{e^{nh} - 1}{e^h - 1} \right] \quad \text{(using the formula for sum to } n \text{ terms of a geometric progression)}$$

$$= \lim_{h \to 0} \left[h e^a \frac{e^{b-a} - 1}{e^h - 1} \right] = \left(e^b - e^a \right) \lim_{h \to 0} \frac{h}{e^h - 1} = e^b - e^a \cdot \left(\because \lim_{h \to 0} \frac{h}{e^h - 1} = 1 \right).$$

Example 3.3 Evaluate $\int_a^b \cos x \, dx$ **as the limit of a sum.**

Solution. Here $f(x) = \cos x$. Therefore

$$\int_a^b \cos x \, dx = \lim_{n \to \infty} h \left[\cos(a + h) + \cos(a + 2h) + \dots + \cos(a + nh) \right].$$

Let $S = \cos(a + h) + \cos(a + 2h) + \dots + \cos(a + nh)$.

Multiplying both sides by $2 \sin \left(\dfrac{h}{2} \right)$, we obtain

$$2 \sin \frac{h}{2} \cdot S = 2 \sin \frac{h}{2} \cos(a + h) + 2 \sin \frac{h}{2} \cos(a + 2h) + \dots + 2 \sin \frac{h}{2} \cos(a + nh)$$

$$= \sin \left(a + \frac{3h}{2} \right) - \sin \left(a + \frac{h}{2} \right) + \sin \left(a + \frac{5h}{2} \right) - \sin \left(a + \frac{3h}{2} \right)$$

$$+ \dots + \sin \left[a + \frac{(2n + 1)}{2} h \right] - \sin \left[a + \frac{(2n - 1)}{2} h \right]$$

$$= \sin \left[a + \frac{(2n + 1)h}{2} \right] - \sin \left[a + \frac{h}{2} \right] = \sin \left[b + \frac{h}{2} \right] - \sin \left[a + \frac{h}{2} \right], \text{ for } nh = b - a.$$

Thus
$$\int_a^b \cos x \, dx = \lim_{n \to \infty} \frac{h \left[\sin \left(b + \frac{h}{2} \right) - \sin \left(a + \frac{h}{2} \right) \right]}{2 \sin \frac{h}{2}}$$

$$= \lim_{n \to \infty} \frac{h/2}{\sin h/2} \left[\sin \left(b + \frac{h}{2} \right) - \sin \left(a + \frac{h}{2} \right) \right]$$

$$= \lim_{h \to 0} \frac{h/2}{\sin h/2} \lim_{h \to 0} \left[\sin \left(b + \frac{h}{2} \right) - \sin \left(a + \frac{h}{2} \right) \right] (\text{as } n \to \infty, h \to 0)$$

$$= 1 \cdot (\sin b - \sin a) = \sin b - \sin a.$$

3.3 The Mean-Value Theorem for Definite Integrals

The statement that *a continuous function on a closed interval assumes its average value at least once in the interval* is known as the Mean Value Theorem for Definite Integrals.

If $f(x)$ is continuous on a closed interval $[a, b]$, then at some point ξ in $[a, b]$,

$$\int_a^b f(x)\,dx = f(\xi)\cdot(b-a), \quad a < \xi < b.$$

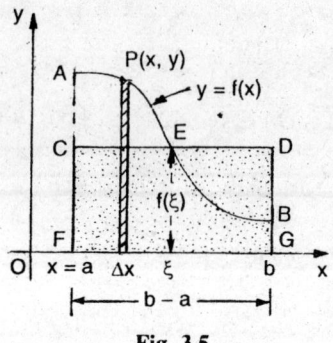

Fig. 3.5.

3.4 The Fundamental Theorem of Integral Calculus

The Fundamental Theorem of Calculus, Part I

If $f(x)$ be a continuous function defined on the closed interval $[a, b]$ and

$A(x) = \int_a^x f(t)\,dt,$ then $A(x)$ has a derivative at every point of $[a, b]$ and

$$A'(x) = \frac{d}{dx} A(x) = \frac{d}{dx} \int_a^x f(t)\,dt = f(x), \quad a \le x \le b. \qquad \text{...(3.4)}$$

In words, above Formula states:

If a definite integral has a variable upper limit of integration a constant lower limit of integration, and a continuous integrand, then the derivative of the integral with respect to its upper limit is equal to the integrand evaluated at the upper limit.

Proof: By the step rule for finding derivatives,

$$\frac{A(x+\Delta x) - A(x)}{\Delta x} = \frac{1}{\Delta x}\left[\int_a^{x+\Delta x} f(t)\,dt - \int_a^x f(t)\,dt \right]$$

$$\Rightarrow \quad \frac{\Delta A}{\Delta x} = \frac{1}{\Delta x}\left[\int_a^x f(t)\,dt + \int_x^{x+\Delta x} f(t)\,dt - \int_a^x f(t)\,dt \right] = \frac{1}{\Delta x} \int_x^{x+\Delta x} f(t)\,dt = f(\xi)$$

where ξ lies between x and $x + \Delta x$ (by the mean value theorem for integrals)

Then if ξ is any point interior to $[a, b]$, $A'(x) = \lim\limits_{\Delta x \to 0} \dfrac{\Delta A}{\Delta x} = \lim\limits_{\Delta x \to 0} f(\xi)$

But, as $\Delta x \to 0$, $x + \Delta x \to x$ and so $\lim\limits_{\Delta x \to 0} f(\xi) = \lim\limits_{\xi \to x} f(\xi)$,

and due to the continuity of the function $f(x)$, $\lim\limits_{\xi \to x} f(\xi) = f(x)$.

Thus $A'(x)$ exists and is equal to $f(x)$, and the theorem is proved.

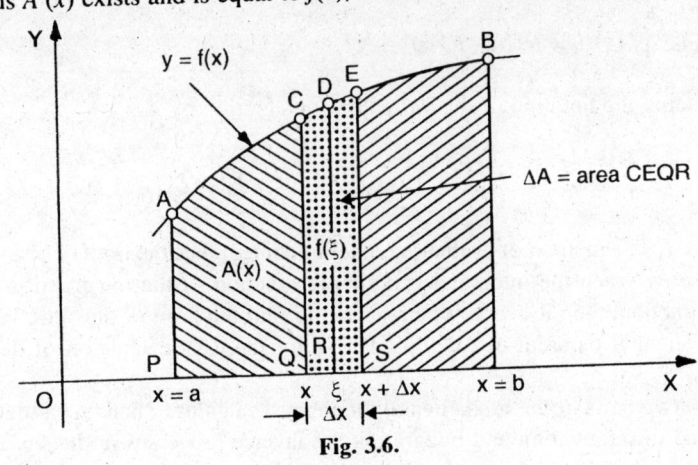

Fig. 3.6.

The Fundamental Theorem of Calculus, Part 2

If $f(x)$ be a continuous function defined on the closed interval $[a, b]$ and $F(x)$ be an anti-derivative of $f(x)$ on $[a, b]$. Then,

$$\int_a^b f(x)\,dx = F(b) - F(a).$$ ...(3.5)

In words, Formula (3.5) states:

The definite integral can be evaluated by finding any anti-derivative of the integrand and then subtracting the value of this anti-derivative at the lower limit of integration from its value at the upper limit of integration.

Proof: Let $F(x)$ be some anti-derivative of the function $f(x)$. By the first part of

the Fundamental Theorem, the function $\int_a^x f(t)\,dt$ is also an anti-derivative of $f(x)$.

We use the fact that two anti-derivatives of a given function differ by a constant. And so we can write

$$\int_a^x f(t)\,dt = F(x) + c, \text{ where } c \text{ is any constant.}$$

When the upper limit of integration is $x = a$, we have

$$\int_a^a f(t)\,dt = 0 = F(a) + c \text{ and } c = -F(a).$$

Hence, $\int_a^x f(t)\,dt = F(x) - F(a)$ and when the upper limit of integration is $x = b$,

this yields $\int_a^b f(t)\,dt = F(b) - F(a)$.

Replacing the notation of the variable of integration by x,

$$\int_a^b f(x)\,dx = F(b) - F(a).$$

Remarks.1. The method of evaluating a definite integral as the limit of a sum can be used only when the integrand is a simple function. If the integrand is not a simple function, then it may not be possible to find the limit of the sum. We use the fundamental theorem of calculus, Part 2 for the easy evaluation of definite integrals.

2. The two parts of the fundamental theorem of calculus, when taken together, tell us that differentiation and integration are inverse processes in the sense that each undoes the effect of the other.

3. If $f(x)$ and $g(x)$ are integrable on $[a, b]$ and if c is a constant, then $cf(x), f(x) + g(x)$ and $f(x) - g(x)$ are integrable on $[a, b]$ and

(i) $\int_a^b cf(x)\,dx = c\int_a^b f(x)\,dx,$

(ii) $\int_a^b \left[f(x) + g(x)\right]dx = \int_a^b f(x)\,dx + \int_a^b g(x)\,dx,$

(iii) $\int_a^b \left[f(x) - g(x)\right]dx = \int_a^b f(x)\,dx - \int_a^b g(x)\,dx.$

Part (ii) can be extended to more than two functions.

3.5 Leibnitz's Rule

Leibnitz's rule say that if f is continuous on $[a, b]$, and $u(x)$ and $v(x)$ are differentiable functions of x whose values lie in $[a, b]$, then

$$\frac{d}{dx}\int_{u(x)}^{v(x)} f(t)\,dt = f(v(x))\frac{dv}{dx} - f(u(x))\frac{du}{dx}. \qquad \text{...(3.6)}$$

3.6 Illustrative Examples

Example 3.4 Define F(x) by $F(x) = \int_1^x (t^3 + 1)\,dt.$ **Use the fundamental theorem of calculus to find F'(x). Check the result by first integrating and then differentiating.**

Solution. The integrand is a continuous function, so from the fundamental theorem of calculus

$$F'(x) = \frac{d}{dx}\left[\int_1^x (t^3 + 1)\,dt\right] = x^3 + 1$$

Alternative. Evaluating the integral and then differentiating yields

$$\int_1^x (t^3 + 1)\,dt = \left[\frac{t^4}{4} + t\right]_1^x = \left(\frac{x^4}{4} + x\right) - \left(\frac{1}{4} + 1\right) = \frac{x^4}{4} + x - \frac{5}{4}.$$

Therefore,
$$\frac{d}{dx}\left[\frac{x^4}{4} + x - \frac{5}{4}\right] = x^3 + 1.$$

Thus the two methods for differentiating the integral agree.

Example 3.5 Let $F(x) = \int_2^x \sqrt{3t^2 + 1}\,dt.$ **Find F'(2).**

Solution. The integrand is a continuous function, so from the fundamental theorem of calculus, we have

$$F'(x) = \frac{d}{dx}\left[\int_2^x \sqrt{3t^2 + 1}\,dt\right] = \sqrt{3x^2 + 1}.\ \text{Therefore,}\ F'(2) = \sqrt{3(2)^2 + 1} = \sqrt{13}.$$

Example 3.6 Find dy/dx if $y = \int_1^{x^2} \cos t\,dt.$

Solution. Here the upper limit of integration is not x but x^2. To find dy/dx we must therefore treat y as the composite of

$$y = \int_1^u \cos t\,dt \ \text{and}\ u = x^2. \quad \therefore \quad \frac{dy}{du} = \frac{d}{du}\int_1^u \cos t\,dt = \cos u \ \text{and} \ \frac{du}{dx} = 2x.$$

Hence $\dfrac{dy}{dx} = \dfrac{dy}{du}\cdot\dfrac{du}{dx} = (\cos x^2)\cdot 2x = 2x\cos x^2.$

Example 3.7 Without attempting to evaluate the integral, compute dy/dx if y is defined by the formula

(i) $y = \int_{x^3}^{x^2} \dfrac{t^6}{1+t^4}\,dt,$ (ii) $y = \int_{\sin x}^{x^2} (1+t)\,dt.$

Solution. (*i*) Using Leibnitz's Rule:

$$\frac{dy}{dx} = \frac{(x^2)^6}{1+(x^2)^4}\frac{d}{dx}(x^2) - \frac{(x^3)^6}{1+(x^3)^4}\frac{d}{dx}(x^3)$$

$$= 2x \cdot \frac{x^{12}}{1+x^8} - \frac{x^{18}}{1+x^{12}}(3x^2) = \frac{2x^{13}}{1+x^8} - \frac{3x^{20}}{1+x^{12}}.$$

(*ii*) $\dfrac{dy}{dx} = (1+x^2)\dfrac{d}{dx}(x^2) - (1+\sin x)\dfrac{d}{dx}(\sin x)$

$$= 2x(1+x^2) - \cos x(1+\sin x).$$

Example 3.8 A function f is defined for all real x by the formula

$$f(x) = 3 + \int_0^x \frac{1+\sin t}{2+t^2}\,dt.$$

Without attempting to evaluate this integral, find a quadratic polynomial $p(x) = a + bx + cx^2$ such that $p(0) = f(0)$, $p'(0) = f'(0)$, and $p''(0) = f''(0)$.

Solution. We have $f(x) = 3 + \int_0^x \dfrac{1+\sin t}{2+t^2}\,dt \Rightarrow f(0) = 3$.

$$\therefore \qquad f'(x) = \frac{1+\sin x}{2+x^2} \Rightarrow f'(0) = \frac{1}{2},$$

and $f''(x) = \dfrac{\cos x(2+x^2) - (1+\sin x)2x}{(2+x^2)^2} \Rightarrow f''(0) = \dfrac{1}{2}.$

Also $p(x) = a + bx + cx^2 \Rightarrow p(0) = a$

$\therefore \qquad p'(x) = b + 2cx \Rightarrow p'(0) = b \quad$ and $\quad p''(x) = 2c \Rightarrow p''(0) = 2c$

Now $a = p(0) = f(0) = 3$, $b = p'(0) = f'(0) = 1/2$,

and $c = \dfrac{1}{2}p''(0) = \dfrac{1}{2}f''(0) = \dfrac{1}{2}\left(\dfrac{1}{2}\right) = \dfrac{1}{4}.$

$\therefore$ Quadratic polynomial $p(x) = 3 + \dfrac{1}{2}x + \dfrac{1}{4}x^2.$

Example 3.9 Compute $f(2)$ if f is continuous and satisfies the given formula for $x \geq 0$.

$$\int_0^{x^2} f(t)\,dt = x^2(1+x).$$

Solution. Let $u = x^2$. The above formula may be written as

$$\frac{d}{du}\left(u + u^{3/2}\right) = f(u) \Rightarrow f(u) = 1 + \frac{3}{2}u^{1/2}. \text{ Therefore, } f(2) = 1 + \frac{3}{2}\sqrt{2}.$$

Example 3.10 Evaluate each of the following definite integrals:

(i) $\displaystyle\int_{-2}^{3} e^{-x/2}\,dx$, (ii) $\displaystyle\int_{1}^{2} \frac{dx}{\sqrt{x} - \sqrt{x-1}}$, (iii) $\displaystyle\int_{0}^{\pi/3} x^2 \sin 3x\,dx$,

(iv) $\displaystyle\int_{3}^{6} xy\,dx$, when $x = 6\cos\theta, y = 2\sin\theta$. (v) $\displaystyle\int_{0}^{a} \sqrt{a^2 - x^2}\,\cos^{-1}(x/a)\,dx$

Solution. (i) We have $\displaystyle\int_{-2}^{3} e^{-x/2}\,dx = \left[-2e^{-x/2}\right]_{-2}^{3} = -2\left(e^{-3/2} - e\right)$

$$= -2(0.22312 - 2.71828) = 4.99032.$$

(ii) We have $\displaystyle\int_{1}^{2} \frac{dx}{\sqrt{x} - \sqrt{x-1}} = \int_{1}^{2} \frac{\sqrt{x} + \sqrt{x-1}}{\left(\sqrt{x} - \sqrt{x-1}\right)\left(\sqrt{x} + \sqrt{x-1}\right)}\,dx$

$$= \int_{1}^{2} \frac{\sqrt{x} + \sqrt{x-1}}{x - (x-1)}\,dx = \int_{1}^{2}\left(\sqrt{x} + \sqrt{x-1}\right)dx = \frac{2}{3}\left[x^{3/2} + (x-1)^{3/2}\right]_{1}^{2}$$

$$= \frac{2}{3}\left[\left(2\sqrt{2} + 1\right) - (1 + 0)\right] = \frac{4\sqrt{2}}{3}.$$

(iii) We have $\displaystyle\int \underset{I}{x^2}\,\underset{II}{\sin 3x}\,dx = x^2 \frac{(-\cos 3x)}{3} - \int \frac{2x}{3}(-\cos 3x)\,dx$, (Integrating by

parts)

$$= -\frac{x^2}{3}\cos 3x + \frac{2}{3}\int x\cos 3x\,dx. \qquad \text{(Again integrating by parts).}$$

$$= -\frac{x^2}{3}\cos 3x + \frac{2}{3}\left[\frac{x\sin 3x}{3} - \frac{1}{3}\int \sin 3x\,dx\right]$$

$$= \frac{-x^2}{3}\cos 3x + \frac{2}{9}x\sin 3x + \frac{2}{27}\cos 3x.$$

$$\therefore \int_{0}^{\pi/3} x^2 \sin 3x\,dx = -\frac{1}{3}\left[x^2 \cos 3x\right]_{0}^{\pi/3} + \frac{2}{9}\left[x\sin 3x\right]_{0}^{\pi/3} + \frac{2}{27}\left[\cos 3x\right]_{0}^{\pi/3}$$

$$= -\frac{1}{3}\left(\frac{\pi^2}{9}\cos\pi - 0\right) + \frac{2}{9}\left(\frac{\pi}{3}\sin\pi - 0\right) + \frac{2}{27}(\cos\pi - \cos 0)$$

$$= -\frac{1}{3}\left(\frac{-\pi^2}{9}\right) + \frac{2}{27}(-1-1) = \frac{(\pi^2 - 4)}{27}.$$

(*iv*) Here we express x, y, and dx in terms of the parameter θ and $d\theta$. Change the limits of integration to corresponding values of the parameter and evaluate the resulting integral.

$dx = -6\sin\theta\, d\theta$. When $x = 6\cos\theta$, $\theta = 0$; and when $x = 6\cos\theta$, $\theta = \pi/3$. Hence

$$\int_3^6 xy\, dx = \int_{\pi/3}^0 (6\cos\theta)(2\sin\theta)(-6\sin\theta)d\theta = -72\int_{\pi/3}^0 \sin^2\theta\cos\theta\, d\theta$$

$$= -72\int_{\pi/3}^0 \sin^2\theta\frac{d}{d\theta}(\sin\theta)d\theta = -24\left[\sin^3\theta\right]_{\pi/3}^0 = -24\left[0 - \left(\sqrt{3}/2\right)^3\right] = 9\sqrt{3}.$$

(*v*) Put $x = a\cos\theta$ so that $dx = -a\sin\theta\, d\theta$ and $a^2 - x^2 = a^2\sin^2\theta$.

$$\therefore \quad \int_0^a \sqrt{(a^2 - x^2)}\cos^{-1}\left(\frac{x}{a}\right)dx = \int_0^{\pi/2}\theta\cdot(a\sin\theta)^2\, d\theta = \frac{a^2}{2}\int_0^{\pi/2}\theta(1-\cos 2\theta)d\theta$$

$$= \frac{a^2}{2}\left[\frac{\theta^2}{2} - \frac{\theta\sin 2\theta}{2} - \frac{\cos 2\theta}{4}\right]_0^{\pi/2} = \frac{a^2\pi^2}{16} + \frac{a^2}{4}.$$

Example 3.11 (*i*) Show that when $f(x)$ is of the form $a + bx + cx^2$,

$$\int_0^1 f(x)dx = \frac{1}{6}\left\{f(0) + 4f\left(\frac{1}{2}\right) + f(1)\right\}.$$

(*ii*) If $f(x)$ be a quadratic polynomial such that $f(0) = 2$, $f'(0) = -3$ and

$f''(0) = 4$, find the value of $\int_{-1}^1 f(x)dx$.

Solution. (*i*) Here $f(x) = a + bx + cx^2$.

Therefore, $f(0) = a$, $f(1/2) = a + \dfrac{b}{2} + \dfrac{c}{4}$, and $f(1) = a + b + c$.

R.H.S. $= \dfrac{1}{6}\left[a + 4\left\{a + \dfrac{b}{2} + \dfrac{c}{4}\right\} + a + b + c\right] = \dfrac{1}{6}(6a + 3b + 2c)$.

$$\text{L.H.S.} = \int_0^1 \left(a + bx + cx^2\right) dx = a\int_0^1 dx + b\int_0^1 xdx + c\int_0^1 x^2 dx$$

$$= a\left[x\right]_0^1 + b\left[\frac{x^2}{2}\right]_0^1 + c\left[\frac{x^3}{3}\right]_0^1 = a + \frac{b}{2} + \frac{c}{3}$$

$$= \frac{1}{6}(6a + 3b + 2c) = \text{R.H.S. Hence the result.}$$

(ii) Let $f(x) = ax^2 + bx + c$, so that $f'(x) = 2ax + b$, $f''(x) = 2a$.
Therefore, $a = 2$, $b = -3$, and $c = 2$.

Thus $f(x) = 2x^2 - 3x + 2$.

$$\therefore \int_{-1}^1 f(x)\,dx = \int_{-1}^1 \left(2x^2 - 3x + 2\right) dx = \left[2\frac{x^3}{3} - \frac{3x^2}{2} + 2x\right]_{-1}^1$$

$$= 2\left(\frac{1}{3} + \frac{1}{3}\right) - \frac{3}{2}(1 - 1) + 2(1 + 1) = \frac{16}{3}.$$

Exercise 3.1

1. Evaluate the following definite integrals as limits of sums:

   (i) $\int_a^b x\,dx$, (ii) $\int_2^3 x^3\,dx$, (iii) $\int_a^b e^{-x}\,dx$, (iv) $\int_\alpha^\beta \sin\theta\,d\theta$, (v) $\int_1^3 (2x^2 + 5x)\,dx$

   Ans. (i) $\left(b^2 - a^2\right)/2$, (ii) 65/4, (iii) $e^{-a} - e^{-b}$, (iv) $\cos\alpha - \cos\beta$, (v) 112/3.

2. Use the Fundamental Theorem of Calculus to find the derivatives of the following functions:

   (i) $\dfrac{d}{dx}\displaystyle\int_1^x \sin\left(\sqrt{t}\right) dt$, (ii) $\dfrac{d}{dx}\displaystyle\int_1^x \sqrt{1 + \cos^2 t}\, dt$,

   (iii) $\dfrac{d}{dx}\displaystyle\int_x^0 \dfrac{t}{\cos t}\,dt$, (iv) $\dfrac{d}{dt}\displaystyle\int_0^{t^4} \sqrt{u}\, du$, (v) $D_x\left(\displaystyle\int_x^0 t^2 dt\right)$.

   Ans. (i) $\sin\sqrt{x}$, (ii) $\sqrt{1 + \cos^2 x}$, (iii) $-\dfrac{x}{\cos x}$, (iv) $4t^5$, (v) $-x^2$.

3. Find dy/dx if

(i) $y = \int_0^x \dfrac{1}{1+t^2} dt$, (ii) $y = \int_0^{\sqrt{x}} \sin(t^2) dt$, (iii) $y = \int_0^{\sqrt{x}} \cos t \, dt$, (iv) $y = \int_0^{\tan x} \dfrac{dt}{1+t^2}$.

Ans. (i) $\dfrac{1}{1+x^2}$, (ii) $\dfrac{1}{2} x^{-1/2} \sin x$, (iii) $\left(\cos \sqrt{x}\right)\left(\dfrac{1}{2\sqrt{x}}\right)$, (iv) 1.

4. Use Leibnitz's rule to find the derivatives of the following functions:

(i) $f(x) = \int_{1/x}^x \dfrac{1}{t} dt$ (ii) $\phi(y) = \int_{\sqrt{y}}^{2\sqrt{y}} \sin t^2 \, dt$

(iii) $f(x) = \int_x^{x^2} \cos t^2 \, dt$, (iv) $\phi(x) = \int_{x^2}^{x^3} (t^2 + t + 1) dt$.

Ans. (i) $\dfrac{2}{x}$, (ii) $\dfrac{\sin 4y}{\sqrt{y}} - \dfrac{\sin y}{2\sqrt{y}}$ (iii) $2x \cos x^4 - \cos x^2$,

(iv) $3x^8 + x^5 - 2x^3 + 3x^2 - 2x$.

5. Without attempting to evaluate the following indefinite integrals, find the derivative $f'(x)$ in each case if $f(x)$ is equal to

(i) $\int_0^{x^2} (1+t^2)^{-3} dt$, (ii) $\int_{x^3}^{x^2} (1+t^2)^{-3} dt$, (iii) $\int_{x^2}^{4x} \cos t \, dt$.

Ans. (i) $2x(1+x^4)^{-3}$, (ii) $2x(1+x^4)^{-3} - 3x^2(1+x^6)^{-3}$,

(iii) $4\cos 4x - 2x \cos x^2$.

6. Let $\phi(x)$ be a differentiable function and $\phi(1) = 4$.

Prove that $\displaystyle\lim_{x \to 1} \int_4^{\phi(x)} \dfrac{2z}{x-1} dz = 8\phi'(1)$.

[Hint. $\displaystyle\lim_{x \to 1} \int_4^{\phi(x)} \dfrac{2z}{x-1} dz = \lim_{x \to 1}\left\{\dfrac{1}{x-1} \int_4^{\phi(x)} 2z \, dz\right\} = \lim_{x \to 1}\left\{\dfrac{1}{x-1}\left[z^2\right]_4^{\phi(x)}\right\}$

$= \displaystyle\lim_{x \to 1} \dfrac{\{\phi(x)\}^2 - 4^2}{x-1} = \lim_{x \to 1} \dfrac{\{\phi(x)+4\}\{\phi(x)-4\}}{x-1}$

$= \displaystyle\lim_{x \to 1}\{\phi(x)+4\} \cdot \lim_{x \to 1} \dfrac{\phi(x)-4}{x-1} = \{\phi(1)+4\} \cdot \lim_{x \to 1} \dfrac{\phi(x)-\phi(1)}{x-1}$

$= (4+4) \cdot \displaystyle\lim_{h \to 0} \dfrac{\phi(1+h)-\phi(1)}{h} = 8\phi'(1)].$

7. In each case, compute $f(2)$ if f is continuous and satisfies the given formula for all $x \geq 0$.

(i) $\int_0^x f(t)dt = x^2(1+x)$ (ii) $\int_0^{f(x)} t^2 dt = x^2(1+x)$,

(iii) $\int_0^{x^2(1+x)} f(t)dt = x$. **Ans.** (i) 16, (ii) $(36)^{1/3}$, (iii) $1/5$

8. Evaluate each of the following definite integrals

(i) $\int_3^4 \dfrac{dx}{25-x^2}$, (ii) $\int_{-1/2}^0 \dfrac{x^3 dx}{x^2+x+1}$, (iii) $\int_0^{2\pi/3} \dfrac{dx}{5+4\cos x}$, (iv) $\int_0^\infty \dfrac{dx}{x^2+a^2}$.

Ans. (i) $\dfrac{1}{5}\log\dfrac{3}{2}$, (ii) $\dfrac{\sqrt{3}\pi}{9}-\dfrac{5}{8}$, (iii) $\dfrac{\pi}{9}$, (iv) $\dfrac{\pi}{2a}$ (GGSIPU-2006)

9. Evaluate the following integrals:

(i) $\int_1^4 f(x)dx$, where $f(x) = \begin{cases} 7x+3, & \text{if } 1 \leq x \leq 3 \\ 8x, & \text{if } 3 \leq x \leq 4 \end{cases}$, (ii) $\int_{-5}^5 |x+2|dx$.

$\left[\textbf{Hint}\,(ii)\int_{-5}^5 |x+2|\,dx = \int_{-5}^{-2}(-x-2)dx + \int_{-2}^5 (x+2)dx \right]$ **Ans.** (i) 62, (ii) 29.

3.7 Properties of Definite Integrals

We consider below some important properties of the definite integrals. These will be useful in evaluating the definite integrals more easily.

Property I. (*Change of variables*)

$$\int_a^b f(x)\,dx = \int_a^b f(t)\,dt.$$

[**In words:** If the variables are changed without altering the form of the integrand, the value of the integral will be same provided the lower and upper limits are same in both cases.]

Proof. Let $\int f(x)dx = \phi(x)$. $\therefore \int_a^b f(x)dx = \left[\phi(x)\right]_a^b = \phi(b)-\phi(a)$.

Then, $\int f(t)dt = \phi(t)$, and $\int_a^b f(t)dt = \left[\phi(t)\right]_a^b = \phi(b)-\phi(a)$. Hence the result.

Thus the variable of integration in a definite integral plays no role in the end result, it is often referred to as a *dummy variable*.

Property II. (*Interchange of limits*)

$$\int_a^b f(x)\,dx = -\int_b^a f(x)\,dx$$

[In words: If the limits are interchanged then there is a change of sign in the definite integral.]

Proof. Let $\int f(x)\,dx = \phi(x)$.

$$\therefore \int_a^b f(x)\,dx = \left[\phi(x)\right]_a^b = \phi(b) - \phi(a) = -[\phi(a) - \phi(b)] = -\int_b^a f(x)\,dx.$$

Example 3.12 (*i*) **Prove that**

$$\int_a^b \frac{f(x-a)}{f(b-x)}\,dx = \int_a^b \frac{f(b-x)}{f(x-a)}\,dx,$$

provided f(x) remains finite when x vanishes.
(*ii*) **If f(a + b − x) = f(x), prove that**

$$\int_a^b x f(x)\,dx = \frac{a+b}{2}\int_a^b f(x)\,dx.$$

Solution. (*i*) Let $x - a = b - z \Rightarrow x = a + b - z$ so that $dx = -dz$. When $x = a$, $z = b$; and when $x = b$, $z = a$.

$$\therefore \int_a^b \frac{f(x-a)}{f(b-x)}\,dx = \int_b^a \frac{f(b-z)}{f(z-a)}(-dz) = -\int_b^a \frac{f(b-z)}{f(z-a)}\,dz$$

$$= \int_a^b \frac{f(b-z)}{f(z-a)}\,dz \text{ (by Property II)} = \int_a^b \frac{f(b-x)}{f(x-a)}\,dx. \qquad \text{(by Property I)}$$

(*ii*) Put $x = a + b - t$, so that $dx = -dt$. When $x = a$, $t = b$; and when $x = b$, $t = a$.

$$\therefore \int_a^b x f(x)\,dx = -\int_b^a (a+b-t)f(a+b-t)\,dt = \int_a^b (a+b-t)f(a+b-t)\,dt$$

$$= (a+b)\int_a^b f(a+b-t)\,dt - \int_a^b t f(a+b-t)\,dt$$

$$= (a+b)\int_a^b f(a+b-x)\,dx - \int_a^b x f(a+b-x)\,dx \qquad \text{(by Property I)}$$

$$= (a+b)\int_a^b f(x)dx - \int_a^b x f(x)dx \; (\because f(a+b-x) = f(x)).$$

$$\therefore \; 2\int_a^b x f(x)dx = (a+b)\int_a^b f(x)dx \Rightarrow \int_a^b x f(x)dx = \frac{a+b}{2}\int_a^b f(x)dx.$$

Property III. (*Decomposition of the range of integration*)

$$\int_a^b f(x)\,dx = \int_a^c f(x)\,dx + \int_c^b f(x)\,dx; \qquad \text{for } a < c < b$$

Proof. Let $\int f(x)\,dx = \phi(x)$. Then $\int_a^b f(x)dx = \left[\phi(x)\right]_a^b = \phi(b) - \phi(a)$. ...(3.7)

$$\int_a^c f(x)\,dx = \phi(c) - \phi(a) \qquad\qquad ...(3.8)$$

and

$$\int_c^b f(x)\,dx = \phi(b) - \phi(c) \qquad\qquad ...(3.9)$$

Adding (3.8) and (3.9) , we obtained

$$\int_a^c f(x)\,dx + \int_c^b f(x)\,dx = \phi(b) - \phi(a) = \int_a^b f(x)\,dx. \; \text{This proves the Property III.}$$

The generalised form of the above property

$$\int_a^b f(x)dx = \int_a^{c_1} f(x)dx + \int_{c_1}^{c_2} f(x)dx + \int_{c_2}^{c_3} f(x)dx + ... + \int_{c_{n-1}}^{c_n} f(x)dx + \int_{c_n}^b f(x)dx,$$

where $\qquad a < c_1 < c_2 < c_3 ... < c_{n-1} < c_n < b.$

The right-hand side is equal to $\phi(c_1) - \phi(a) + \phi(c_2) - \phi(c_1)$

$$+ \phi(c_3) - \phi(c_2) + ... + \phi(c_n) - \phi(c_{n-1}) + \phi(b) - \phi(c_n),$$

which is equal to $\phi(b) - \phi(a) = \int_a^b f(x)\,dx.$

Example 3.13 Prove that $\int_0^{2a} f(x)dx = \int_0^a f(x)dx + \int_0^a f(2a - x)dx.$

Solution. We have $\int\limits_{0}^{2a} f(x)\,dx = \int\limits_{0}^{a} f(x)\,dx + \int\limits_{a}^{2a} f(x)\,dx$, (by Property III) ...(i)

Let $z = 2a - x$ in the second integral on the right hand side. Then $dz = -dx$.
When $x = a, z = a$, and when $x = 2a, z = 0$. Also $x = 2a - z$.

$$\therefore \int\limits_{a}^{2a} f(x)\,dx = \int\limits_{a}^{0} f(2a-z)(-dz) = -\int\limits_{a}^{0} f(2a-z)\,dz = \int\limits_{0}^{a} f(2a-z)\,dz \quad \text{(by Prop.II)}$$

$$= \int\limits_{0}^{a} f(2a-x)\,dx, \qquad\qquad\qquad\qquad\qquad \text{(by Prop. I)}$$

Thus $\int\limits_{0}^{2a} f(x)\,dx = \int\limits_{0}^{a} f(x)\,dx + \int\limits_{0}^{a} f(2a-x)\,dx$.

Property IV. $\int\limits_{0}^{a} \mathbf{f(x)\,dx} = \int\limits_{0}^{a} \mathbf{f(a-x)\,dx}$.

[**In words.** If the lower limit of a definite integral is zero, then the definite integral remains unchanged if x is changed to (upper limit $-x$), the limits remaining the same.]

Proof. Let $z = a - x$, so $dz = -dx$. When $x = 0$, $z = a$; when $x = a$, $z = 0$.
Thus, as x varies from 0 to a, z varies from a to 0. Also $x = a - z$.

$$\therefore \int\limits_{0}^{a} f(x)\,dx = \int\limits_{a}^{0} f(a-z)(-dz) = -\int\limits_{a}^{0} f(a-z)\,dz = \int\limits_{0}^{a} f(a-z)\,dz \quad \text{(by Property II)}$$

$$= \int\limits_{0}^{a} f(a-x)\,dx \qquad\qquad \text{(by Property I)}.$$

Remark. The above property is very important and useful. It enables us, in certain cases, to evaluate a definite integral without first finding the indefinite integral which may be very difficult to find.

Example 3.14 Evaluate each of the following definite integrals

(i) $\int\limits_{0}^{\pi/2} \log\tan x\,dx$, (A.M.I.E., S-2005), (ii) $\int\limits_{0}^{\pi/2} \sin 2x \log\tan x\,dx$ (A.M.I.E.S-2000),

(iii) $\int\limits_{0}^{\pi/4} \log(1+\tan x)\,dx$ (GGSIPU2006; A.M.I.E. S-2003, Madras, 2000 PT),

(iv) $\displaystyle\int_0^{\pi/2}(2\log\sin x-\log\sin 2x)\,dx.$

Solution. (i) Let $I=\displaystyle\int_0^{\pi/2}\log\tan x\,dx.$

Then $I=\displaystyle\int_0^{\pi/2}\log\tan\left(\frac{\pi}{2}-x\right)dx,$ \hfill (by Property **IV**)

$$=\int_0^{\pi/2}\log\cot x\,dx=\int_0^{\pi/2}\log\left(\frac{1}{\tan x}\right)dx=\int_0^{\pi/2}\log(\tan x)^{-1}dx$$

$$=-\int_0^{\pi/2}\log\tan x\,dx=-I\Rightarrow 2I=0,\text{ or, }I=0.\text{ Hence }\int_0^{\pi/2}\log\tan x\,dx=0$$

(ii) Let $I=\displaystyle\int_0^{\pi/2}\sin 2x\log\tan x\,dx.$ Then $I=\displaystyle\int_0^{\pi/2}\sin 2\left(\frac{\pi}{2}-x\right)\log\tan\left(\frac{\pi}{2}-x\right)dx$

$$=\int_0^{\pi/2}\sin(\pi-2x)\log\cot x\,dx=\int_0^{\pi/2}\sin(\pi-2x)\log\left(\frac{1}{\tan x}\right)dx$$

$$=\int_0^{\pi/2}\sin 2x\log(\tan x)^{-1}dx=-\int_0^{\pi/2}\sin 2x\log\tan x\,dx=-I\Rightarrow 2I=0,\text{ or }I=0.$$

Hence $\displaystyle\int_0^{\pi/2}\sin 2x\log\tan x\,dx=0.$

(iii) Let $I=\displaystyle\int_0^{\pi/4}\log(1+\tan x)\,dx.$

Then $I=\displaystyle\int_0^{\pi/4}\log\left[1+\tan\left(\frac{\pi}{4}-x\right)\right]dx=\int_0^{\pi/4}\log\left[1+\frac{1-\tan x}{1+\tan x}\right]dx$

$$= \int_0^{\pi/4} \log\left(\frac{2}{1+\tan x}\right)dx = \log 2 \int_0^{\pi/4} dx - \int_0^{\pi/4} \log(1+\tan x)dx$$

$$= (\pi/4)\log 2 - I \Rightarrow 2I = (\pi/4)\log 2 \text{ or } I = (\pi/8)\log 2.$$

Hence $\int_0^{\pi/4} \log(1+\tan x)dx = \frac{\pi}{8}\log 2.$

(*iv*) We have $\int_0^{\pi/2} (2\log\sin x - \log\sin 2x)\,dx$

$$= \int_0^{\pi/2} [2\log\sin x - \log(2\sin x\cos x)]\,dx$$

$$= \int_0^{\pi/2} (2\log\sin x - \log 2 - \log\sin x - \log\cos x)\,dx$$

$$= \int_0^{\pi/2} (\log\sin x - \log 2 - \log\cos x)\,dx$$

$$= \int_0^{\pi/2} \log\sin x\,dx - \log 2 \int_0^{\pi/2} dx - \int_0^{\pi/2} \log\cos x\,dx$$

$$= \int_0^{\pi/2} \log\sin x\,dx - \log 2 [x]_0^{\pi/2} - \int_0^{\pi/2} \log\cos\left(\frac{\pi}{2}-x\right)dx$$

$$= \int_0^{\pi/2} \log\sin x\,dx - \frac{\pi}{2}\log 2 - \int_0^{\pi/2} \log\sin x\,dx = -\frac{\pi}{2}\log 2 = \frac{\pi}{2}\log 2^{-1} = \frac{\pi}{2}\log\frac{1}{2}.$$

Example 3.15 Evaluate the following definite integrals:

(*i*) $\displaystyle\int_0^{\pi/2} \frac{\sqrt{\sin x}}{\sqrt{\sin x}+\sqrt{\cos x}}\,dx$ (*IETE June 2009; A.M.I.E. S-2004; Kanpur, 1998*),

(*ii*) $\displaystyle\int_0^{\infty} \frac{x}{(1+x)(1+x^2)}\,dx$, (*iii*) $\displaystyle\int_0^{\pi/2} \frac{\sin^2 x}{\sin x+\cos x}\,dx$, (*A.M.I.E., W-2004*),

(iv) $\displaystyle\int_0^{\pi} \frac{x \tan x}{\sec x + \tan x}\,dx,$ (v) $\displaystyle\int_0^{\pi/2} \frac{x\,dx}{\sin x + \cos x}$

(A.M.I.E., W-97),   (A.M.I.E S-1996; Madras 1996).

Solution. (i) Let $I = \displaystyle\int_0^{\pi/2} \frac{\sqrt{\sin x}}{\sqrt{\sin x} + \sqrt{\cos x}}\,dx.$

Then $I = \displaystyle\int_0^{\pi/2} \frac{\sqrt{\sin\left(\dfrac{\pi}{2} - x\right)}}{\sqrt{\sin\left(\dfrac{\pi}{2} - x\right)} + \sqrt{\cos\left(\dfrac{\pi}{2} - x\right)}}\,dx = \int_0^{\pi/2} \frac{\sqrt{\cos x}}{\sqrt{\cos x} + \sqrt{\sin x}}\,dx.$ (by Prop. IV)

Adding the two values of I, we obtain

$$2I = \int_0^{\pi/2} \left[\frac{\sqrt{\sin x}}{\sqrt{\sin x} + \sqrt{\cos x}} + \frac{\sqrt{\cos x}}{\sqrt{\cos x} + \sqrt{\sin x}} \right] dx$$

$$= \int_0^{\pi/2} \frac{\sqrt{\sin x} + \sqrt{\cos x}}{\sqrt{\sin x} + \sqrt{\cos x}}\,dx = \int_0^{\pi/2} dx = \left[x\right]_0^{\pi/2} = \frac{\pi}{2}$$

Hence, $I = \displaystyle\int_0^{\pi/2} \frac{\sqrt{\sin x}}{\sqrt{\sin x} + \sqrt{\cos x}}\,dx = \frac{\pi}{4}.$

Remark. Each of the following definite integrals

$$\int_0^{\pi/2} \frac{\sin x}{\sin x + \cos x}\,dx, \quad \int_0^{\pi/2} \frac{\cos x}{\sin x + \cos x}\,dx, \quad \int_0^{\pi/2} \frac{\sqrt{\tan x}}{\sqrt{\tan x} + \sqrt{\cot x}}\,dx,$$

$$\int_0^{\pi/2} \frac{\sin^n x\,dx}{\sin^n x + \cos^n x}\,dx, \int_0^{\pi/2} \frac{dx}{1 + \tan x}, \int_0^{\pi/2} \frac{dx}{1 + \cot x}, \int_0^{\pi/2} \frac{\tan x}{1 + \tan x}\,dx \text{ is equal to } \pi/4.$$

(ii) Put $x = \tan\theta$ so that $dx = \sec^2\theta\,d\theta.$
When $x = 0, \theta = 0$; and when $x = \infty, \theta = \pi/2.$

Let $I = \displaystyle\int_0^{\infty} \frac{x\,dx}{(1+x)(1+x^2)} = \int_0^{\pi/2} \frac{\tan\theta\sec^2\theta\,d\theta}{(1+\tan\theta)(1+\tan^2\theta)}$

$$= \int_0^{\pi/2} \frac{\tan\theta}{1 + \tan\theta}\,d\theta = \int_0^{\pi/2} \frac{\sin\theta}{\sin\theta + \cos\theta}\,d\theta \qquad\qquad \ldots(i)$$

Then $I = \int\limits_0^{\pi/2} \dfrac{\sin(\pi/2-\theta)}{\sin\left(\dfrac{\pi}{2}-\theta\right)+\cos\left(\dfrac{\pi}{2}-\theta\right)}\,d\theta = \int\limits_0^{\pi/2}\dfrac{\cos\theta}{\cos\theta+\sin\theta}\,d\theta$ ...(ii)

Adding the two values of I, we obtain

$$2I = \int\limits_0^{\pi/2}\dfrac{\sin\theta+\cos\theta}{\sin\theta+\cos\theta}\,d\theta = \int\limits_0^{\pi/2}d\theta = \dfrac{\pi}{2} \quad\text{or}\quad I = \int\limits_0^{\infty}\dfrac{x\,dx}{(1+x)(1+x^2)} = \dfrac{\pi}{4}.$$

(*iii*) Let $\quad I = \int\limits_0^{\pi/2}\dfrac{\sin^2 x}{\sin x+\cos x}\cdot dx$ . ...(i)

Then $\quad I = \int\limits_0^{\pi/2}\dfrac{\sin^2\left(\dfrac{\pi}{2}-x\right)dx}{\sin\left(\dfrac{\pi}{2}-x\right)+\cos\left(\dfrac{\pi}{2}-x\right)} = \int\limits_0^{\pi/2}\dfrac{\cos^2 x\,dx}{\cos x+\sin x}$ . ...(ii)

Adding the two values of I, we obtain

$$2I = \int\limits_0^{\pi/2}\dfrac{(\sin^2 x+\cos^2 x)}{\sin x+\cos x}\,dx = \int\limits_0^{\pi/2}\dfrac{dx}{\sin x+\cos x}$$

or $\quad I = \dfrac{1}{2}\int\limits_0^{\pi/2}\dfrac{dx}{\sqrt2\left(\cos x\cdot\cos\dfrac{\pi}{4}+\sin x\cdot\sin\dfrac{\pi}{4}\right)} = \dfrac{1}{2\sqrt2}\int\limits_0^{\pi/2}\sec\left(x-\dfrac{\pi}{4}\right)dx$

$$= \dfrac{1}{2\sqrt2}\left[\log\left\{\sec\left(x-\dfrac{\pi}{4}\right)+\tan\left(x-\dfrac{\pi}{4}\right)\right\}\right]_0^{\pi/2} = \dfrac{1}{2\sqrt2}\log\dfrac{\sqrt2+1}{\sqrt2-1}$$

$$= \dfrac{1}{2\sqrt2}\log\left[\dfrac{(\sqrt2+1)(\sqrt2+1)}{(\sqrt2-1)(\sqrt2+1)}\right] = \dfrac{1}{2\sqrt2}\log(\sqrt2+1)^2 = \dfrac{1}{\sqrt2}\log(\sqrt2+1).$$

(*iv*) Let $I = \int\limits_0^{\pi}\dfrac{x\tan x}{\sec x+\tan x}\,dx = \int\limits_0^{\pi}\dfrac{x\sin x}{1+\sin x}\,dx$

Then $I = \int\limits_0^{\pi}\dfrac{(\pi-x)\sin(\pi-x)}{1+\sin(\pi-x)}\,dx = \int\limits_0^{\pi}\dfrac{(\pi-x)\sin x}{1+\sin x}\,dx$

$$= \pi \int_0^\pi \frac{\sin x}{1+\sin x} dx - \int_0^\pi \frac{x \sin x}{1+\sin x} dx = \pi \int_0^\pi \frac{\sin x}{1+\sin x} dx - I$$

$$\Rightarrow \quad 2I = \pi \int_0^\pi \frac{\sin x}{1+\sin x} dx \text{ or } I = \frac{\pi}{2} \int_0^\pi \frac{\sin x}{1+\sin x} dx = \frac{\pi}{2} \int_0^\pi \left(1 - \frac{1}{1+\sin x}\right) dx$$

$$= \frac{\pi}{2} \int_0^\pi \left(1 - \frac{1-\sin x}{\cos^2 x}\right) dx = \frac{\pi}{2} \int_0^\pi \left(1 - \sec^2 x + \sec x \tan x\right) dx$$

$$= \frac{\pi}{2} \left[x - \tan x + \sec x\right]_0^\pi = \frac{\pi}{2}(\pi - 1 - 1) = \frac{\pi}{2}(\pi - 2).$$

(v) Let $\quad I = \int_0^{\pi/2} \frac{x}{\sin x + \cos x} dx.$

Then $\quad I = \int_0^{\pi/2} \frac{\left(\frac{\pi}{2} - x\right)}{\sin\left(\frac{\pi}{2} - x\right) + \cos\left(\frac{\pi}{2} - x\right)} dx = \int_0^{\pi/2} \frac{\left(\frac{\pi}{2} - x\right)}{\cos x + \sin x} dx.$

Adding the two values of I, we obtain

$$2I = \frac{\pi}{2} \int_0^{\pi/2} \frac{dx}{\sin x + \cos x} = \frac{\pi}{2\sqrt{2}} \int_0^{\pi/2} \frac{dx}{\sin x \cos \frac{\pi}{4} + \cos x \sin \frac{\pi}{4}}$$

$$= \frac{\pi}{2\sqrt{2}} \int_0^{\pi/2} \frac{dx}{\sin\left(x + \frac{\pi}{4}\right)} = \frac{\pi}{2\sqrt{2}} \int_0^{\pi/2} \operatorname{cosec}\left(x + \frac{\pi}{4}\right) dx$$

$$= \frac{\pi}{2\sqrt{2}} \left[\log \tan\left(\frac{x}{2} + \frac{\pi}{8}\right)\right]_0^{\pi/2} = \frac{\pi}{2\sqrt{2}} \left[\log \tan\left(\frac{\pi}{4} + \frac{\pi}{8}\right) - \log \tan \frac{\pi}{8}\right]$$

or $I = \frac{\pi}{4\sqrt{2}} \left[\log \tan \frac{3\pi}{8} - \log \tan \frac{\pi}{8}\right]$

$$= \frac{\pi}{4\sqrt{2}} \left[\log \tan\left(\frac{\pi}{2} - \frac{\pi}{8}\right) - \log \tan \frac{\pi}{8}\right] = \frac{\pi}{4\sqrt{2}} \left[\log \cot \frac{\pi}{8} - \log \tan \frac{\pi}{8}\right]$$

$$= \frac{\pi}{4\sqrt{2}}\left[\log\frac{\cot \pi/8}{\tan \pi/8}\right] = \frac{\pi}{4\sqrt{2}}\log\left(\cot\frac{\pi}{8}\right)^2 = \frac{\pi}{2\sqrt{2}}\log\cot\frac{\pi}{8} = \frac{\pi}{2\sqrt{2}}\log\left(\sqrt{2}+1\right).$$

Example 3.16 Establish the following results

(i) $\displaystyle\int_0^1 \frac{\log(1+x)}{1+x^2}dx = \frac{\pi}{8}\log 2$, (*A.M.I.E., W-2001; D.C.E., 2001; GGSIPU 2006*)

(ii) $\displaystyle\int_0^\pi \frac{x\sin x}{1+\cos^2 x}dx = \frac{\pi^2}{4}$,   (*Kurukshetra, 1998; A.M.I.E., S-1998*)

(iii) $\displaystyle\int_0^\pi \frac{x\tan x}{\sec x+\cos x}dx = \frac{\pi^2}{4}$, (iv) $\displaystyle\int_0^1 \cot^{-1}(1-x+x^2)dx = \frac{\pi}{2}-\log 2.$

Solution. (*i*) Let $x = \tan\theta$, so that $dx = \sec^2\theta\, d\theta$. When $x = 0$, $\theta = \theta$ and when $x = 1$, $\theta = \pi/4$.

$$\therefore \int_0^1 \frac{\log(1+x)}{1+x^2}dx = \int_0^{\pi/4} \frac{\log(1+\tan\theta)}{1+\tan^2\theta}\sec^2\theta\, d\theta = \int_0^{\pi/4}\log(1+\tan\theta)\,d\theta = \frac{\pi}{8}\log 2.$$

(*ii*) Let $$I = \int_0^\pi \frac{x\sin x}{1+\cos^2 x}dx.$$

Then $I = \displaystyle\int_0^\pi \frac{(\pi-x)\sin(\pi-x)}{1+\cos^2(\pi-x)}dx = \int_0^\pi \frac{(\pi-x)\sin x}{1+\cos^2 x}dx = \pi\int_0^\pi \frac{\sin x}{1+\cos^2 x}dx - I$

$$\Rightarrow 2I = \pi\int_0^\pi \frac{\sin x}{1+\cos^2 x}dx.$$

Put $\cos x = z$, so that $\sin x\, dx = -dz$. When $x = 0$, $z = 1$; and when $x = \pi$, $z = -1$.

$$\therefore \int_0^\pi \frac{x\sin x}{1+\cos^2 x}dx = \frac{\pi}{2}\int_1^{-1}\frac{-dz}{1+z^2} = \frac{\pi}{2}\int_{-1}^1 \frac{dz}{1+z^2} = \pi\int_0^1 \frac{dz}{1+z^2}$$

$$= \pi\left[\tan^{-1}z\right]_0^1 = \pi\left(\tan^{-1}1 - \tan^{-1}0\right) = \pi\cdot\frac{\pi}{4} = \frac{\pi^2}{4}.$$

(*iii*) Let $I = \displaystyle\int_0^\pi \frac{x\tan x}{\sec x+\cos x}dx$. Then $I = \displaystyle\int_0^\pi \frac{(\pi-x)\tan(\pi-x)}{\sec(\pi-x)+\cos(\pi-x)}dx$

$$= \int_0^\pi \frac{(\pi - x)(-\tan x)}{-\sec x - \cos x} dx = \int_0^\pi \frac{(\pi - x)\tan x}{\sec x + \cos x} dx$$

$$= \pi \int_0^\pi \frac{\tan x}{\sec x + \cos x} dx - I \quad \Rightarrow \quad 2I = \pi \int_0^\pi \frac{\tan x}{\sec x + \cos x} dx$$

or, $\quad I = \dfrac{\pi}{2} \displaystyle\int_0^\pi \dfrac{\tan x}{\sec x + \cos x} dx = \dfrac{\pi}{2} \displaystyle\int_0^\pi \dfrac{\sin x}{1 + \cos^2 x} dx = \dfrac{\pi^2}{4}$, as proved above.

(iv) We have $\displaystyle\int_0^1 \cot^{-1}\left(1 - x + x^2\right) dx = \int_0^1 \tan^{-1} \frac{1}{1 - x + x^2} dx$

$$= \int_0^1 \tan^{-1}\left[\frac{x - (x - 1)}{1 + x(x - 1)}\right] dx = \int_0^1 \left[\tan^{-1}(x) - \tan^{-1}(x - 1)\right] dx$$

$$= \int_0^1 \tan^{-1} x \, dx - \int_0^1 \tan^{-1}(x - 1) dx = \int_0^1 \tan^{-1} x \, dx - \int_0^1 \tan^{-1}\left[(1 - x) - 1\right] dx$$
$$\text{(by Property IV)}$$

$$= \int_0^1 \tan^{-1} x \, dx - \int_0^1 \tan^{-1}(-x) dx = 2\int_0^1 \tan^{-1} x \, dx$$

$$= 2\left[\left(\tan^{-1} x\right)(x)\right]_0^1 - 2\int_0^1 \frac{1}{1 + x^2} \cdot x \, dx = 2\left(\frac{\pi}{4} - \frac{1}{2}\left[\log\left(1 + x^2\right)\right]_0^1\right) = \frac{\pi}{2} - \log 2.$$

Property V. $\quad \displaystyle\int_a^b f(x) dx = \int_a^b f(a + b - x) dx$

Proof: Let $z = a + b - x$. Then $dz = -dx$. When $x = a$, $z = b$ and when $x = b$, $z = a$. Thus, as x varies from a to b, z varies from b to a.

Also $x = a + b - z$. Therefore

$$\int_a^b f(x) dx = -\int_b^a f(a + b - z) dz = \int_a^b f(a + b - z) dz \qquad \text{(by Property II)}$$

$$= \int_a^b f(a + b - x) dx \qquad \text{(by Property I)}$$

Example 3.17 Evaluate each of the following definite integrals.

$$(i) \int_{2}^{3} \frac{\sqrt{x}\,dx}{\sqrt{5-x}+\sqrt{x}}, \qquad (ii) \int_{\pi/6}^{\pi/3} \frac{1}{1+\sqrt{\cot x}}\,dx.$$

Solution: (i) Let $I = \int_{2}^{3} \frac{\sqrt{x}\,dx}{\sqrt{5-x}+\sqrt{x}}$. ...(i)   Using the property V,

Then
$$I = \int_{2}^{3} \frac{\sqrt{5-x}\,dx}{\sqrt{5-(5-x)}+\sqrt{5-x}} = \int_{2}^{3} \frac{\sqrt{5-x}\,dx}{\sqrt{x}+\sqrt{5-x}}. \qquad ...(ii)$$

Adding the two values of I, we obtain

$$2I = \int_{2}^{3} \frac{\sqrt{x}+\sqrt{5-x}}{\sqrt{x}+\sqrt{5-x}}\,dx = \int_{2}^{3} dx = \left[x\right]_{2}^{3} = 1. \text{ Hence } I = \frac{1}{2}.$$

(ii) Let $I = \int_{\pi/6}^{\pi/3} \frac{1}{1+\sqrt{\cot x}}\,dx = \int_{\pi/6}^{\pi/3} \frac{\sqrt{\sin x}}{\sqrt{\sin x}+\sqrt{\cos x}}\,dx.$

Then
$$I = \int_{\pi/6}^{\pi/3} \frac{\sqrt{\sin\left(\frac{\pi}{3}+\frac{\pi}{6}-x\right)}}{\sqrt{\sin\left(\frac{\pi}{3}+\frac{\pi}{6}-x\right)}+\sqrt{\cos\left(\frac{\pi}{3}+\frac{\pi}{6}-x\right)}}\,dx \qquad \text{(by Property V)}$$

$$= \int_{\pi/6}^{\pi/3} \frac{\sqrt{\cos x}}{\sqrt{\cos x}+\sqrt{\sin x}}\,dx.$$

Adding the two values of I, we obtain

$$2I = \int_{\pi/6}^{\pi/3} \frac{\sqrt{\sin x}+\sqrt{\cos x}}{\sqrt{\cos x}+\sqrt{\sin x}}\,dx = \int_{\pi/6}^{\pi/3} dx = \left[x\right]_{\pi/6}^{\pi/3} = \frac{\pi}{3}-\frac{\pi}{6} = \frac{\pi}{6}. \text{ Hence } I = \frac{\pi}{12}.$$

Remarks. Some definite integrals can be easily evaluated when the integrand is either *even* or *odd* function of x. We therefore, give below the definition of these functions.

Even function. Def. *A function $f(x)$ is said to be an even if*

$f(-x) = f(x)$, *for all x.*

Since $\cos(-x) = \cos x$, therefore $\cos x$ is an even function. Similarly $\sec x$, x^2+3 etc. are even functions.

Odd function, Def. *A function f(x) is said to be an odd if*

$f(-x) = -f(x)$, *for all x.*

$\sin(-x) = -\sin x$, $\tan(-x) = -\tan x$. Therefore, $\sin x$ and $\tan x$ are odd functions. Similarly $x^3 + x$ is an odd function.

It may be easily verified that

(*i*) The product of *two even functions* is an even function.

(*ii*) The product of *two odd functions* is an even function.

(*iii*) The product of *an even and odd functions* is an odd function.

It may be pointed out that there are functions which are neither even nor odd. Thus $\sin x + \cos x$, $2x^2 + 6x + 8$, e^x etc. are neither even nor odd.

Property VI. $\displaystyle\int_{-a}^{a} f(x)\,dx = \begin{cases} 2\displaystyle\int_{0}^{a} f(x)\,dx, & \text{If f(x) is an even function} \\ 0, & \text{if f(x) is an odd, function} \end{cases}$

Proof. Using Property III, we have $\displaystyle\int_{-a}^{a} f(x)\,dx = \int_{-a}^{0} f(x)\,dx + \int_{0}^{a} f(x)\,dx$ ...(3.10)

Let $z = -x$ in the first integral on the right hand side. Then $dz = -dx$. When $x = -a$, $z = a$; and when $x = 0$, $z = 0$. Also $x = -z$.

$$\therefore \int_{-a}^{0} f(x)\,dx = \int_{a}^{0} f(-z)(-dz) = -\int_{a}^{0} f(-z)\,dz = \int_{0}^{a} f(-z)\,dz = \int_{0}^{a} f(-x)\,dx.$$

Now (3.10) may be rewritten as $\displaystyle\int_{-a}^{a} f(x)\,dx = \int_{0}^{a} f(-x)\,dx + \int_{0}^{a} f(x)\,dx$. ...(3.11)

(*i*) Now if $f(x)$ is an even function, $f(-x) = f(x)$,

then (3.11) becomes $\displaystyle\int_{-a}^{a} f(x)\,dx = \int_{0}^{a} f(x)\,dx + \int_{0}^{a} f(x)\,dx = 2\int_{0}^{a} f(x)\,dx.$

(*ii*) If $f(x)$ is an odd function, $f(-x) = -f(x)$, then (3.11) becomes

$$\int_{-a}^{a} f(x)\,dx = -\int_{0}^{a} f(x)\,dx + \int_{0}^{a} f(x)\,dx = 0.$$

Example 3.18 Show that (*i*) $\displaystyle\int_{-\pi/2}^{\pi/2} \sin^7 x\,dx = 0$, (*ii*) $\displaystyle\int_{-\pi/2}^{\pi/2} \cos^3\theta(1+\sin\theta)^2\,d\theta = \frac{8}{5}$.

Solution. (*i*) Let $f(x) = \sin^7 x$. Then

$$f(-x) = \{\sin(-x)\}^7 = (-\sin x)^7 = -\sin^7 x = -f(x). \text{ So } f(x) \text{ is an odd func-}$$

tion. Hence $\displaystyle\int_{-\pi/2}^{\pi/2} \sin^7 x\, dx = 0$, (by Property VI).

(*ii*) We have $\cos^3 \theta (1 + \sin \theta)^2 = \cos^3 \theta + \cos^3 \theta \sin^2 \theta + 2\cos^3 \theta \sin \theta$.

Therefore, $\displaystyle\int_{-\pi/2}^{\pi/2} \cos^3 \theta (1 + \sin \theta)^2 \, d\theta = \int_{-\pi/2}^{\pi/2} \cos^3 \theta \, d\theta$

$$+ \int_{-\pi/2}^{\pi/2} \cos^3 \theta \sin^2 \theta \, d\theta + 2 \int_{-\pi/2}^{\pi/2} \cos^3 \theta \sin \theta \, d\theta$$

$$= 2\int_{0}^{\pi/2} \cos^3 \theta \, d\theta + 2 \int_{0}^{\pi/2} \cos^3 \theta \sin^2 \theta \, d\theta + 0 \quad (\because \cos^3 \theta \text{ is an even function,}$$

$\cos^3 \theta \sin^2\theta$ is an even function and $\cos^3\theta \sin\theta$ is an odd function).

$$= 2 \cdot \frac{2}{3 \cdot 1} + 2 \cdot \frac{2}{5 \cdot 3 \cdot 1} = \frac{4}{3} + \frac{4}{15} = \frac{8}{5}.$$

Property VII: $\displaystyle\int_{0}^{2a} \mathbf{f(x)\,dx} = \left\{ \begin{array}{ll} 2\displaystyle\int_{0}^{a} \mathbf{f(x)\,dx}, & \text{if } \mathbf{f(2a-x)} = \mathbf{f(x)} \\[4mm] \mathbf{0}, & \text{if } \mathbf{f(2a-x)} = -\mathbf{f(x)} \end{array} \right\}$

Proof. Using Property III, we know $\displaystyle\int_{0}^{2a} f(x)\,dx = \int_{0}^{a} f(x)\,dx + \int_{a}^{2a} f(x)\,dx$ ...(i)

Let $z = 2a - x$ in the second integral on the right hand side.

Then $dz = -dx$. When $x = a, z = a$; and when $x = 2a, z = 0$. Also $x = 2a - z$.

$$\therefore \int_{0}^{2a} f(x)\,dx = \int_{a}^{0} f(2a - z)(-dz) = -\int_{a}^{0} f(2a - z)\,dz = \int_{0}^{a} f(2a - z)\,dx$$

$$= \int_{0}^{a} f(2a - x)\,dx. \qquad\qquad \text{(By Property I)}$$

Now, (*i*) becomes $\displaystyle\int_0^{2a} f(x)\,dx = \int_0^a f(x)\,dx + \int_0^a f(2a-x)\,dx$.

$$= \begin{cases} 2\displaystyle\int_0^a f(x)\,dx, \text{ when } f(2a-x)=f(x) \\ 0, \text{ when } f(2a-x)=-f(x). \end{cases}$$

Remark. The above property is generally used for evaluating a definite integral when the integrand is a trigonometric function of *x*, and the limits of integration are 0 and π (or 0 and a multiple of π).

Example 3.19 Show that (*i*) $\displaystyle\int_0^\pi x\, f(\sin x)\,dx = \pi \int_0^{\pi/2} f(\sin x)\,dx$ (*A.M.I.E., S-1999*),

(*ii*) $\displaystyle\int_0^\pi x\cos^4 x\,dx = \frac{3\pi^2}{16}$, (*iii*) $\displaystyle\int_0^{2\pi} \cos^5 x\,dx = 0$

Solution. (*i*) Let $\quad I = \displaystyle\int_0^\pi x\, f(\sin x)\,dx$.

Then $I = \displaystyle\int_0^\pi (\pi - x)\, f\big[\sin(\pi - x)\big]\,dx = \int_0^\pi (\pi - x)\, f(\sin x)\,dx$ (by Property IV)

$= \pi \displaystyle\int_0^\pi f(\sin x)\,dx - I \quad \Rightarrow \quad 2I = \pi \int_0^\pi f(\sin x)\,dx$ or $I = \frac{\pi}{2}\int_0^\pi f(\sin x)\,dx$.

Since $f(\sin x) = f\big[\sin(\pi - x)\big]$.

Therefore, $I = 2 \cdot \dfrac{\pi}{2}\displaystyle\int_0^{\pi/2} f(\sin x)\,dx = \pi \int_0^{\pi/2} f(\sin x)\,dx$.

(*ii*) Let $\quad I = \displaystyle\int_0^\pi x\cos^4 x\,dx$.

Then $I = \displaystyle\int_0^\pi (\pi - x)\cos^4(\pi - x)\,dx = \int_0^\pi (\pi - x)\cos^4 x\,dx = \pi \int_0^\pi \cos^4 x\,dx - I$

or, $2I = \pi \int\limits_0^\pi \cos^4 x\, dx = 2\pi \int\limits_0^{\pi/2} \cos^4 x\, dx = 2\pi \dfrac{3.1}{4.2} \cdot \dfrac{\pi}{2} = \dfrac{3\pi^2}{8}$. Therefore, $I = \dfrac{3\pi^2}{16}$.

(*iii*) Let $f(x) = \cos^5 x$. Then $f(2\pi - x) = \left[\cos(2\pi - x)\right]^5 = \cos^5 x$.

$\therefore \int\limits_0^{2\pi} \cos^5 x\, dx = 2 \int\limits_0^\pi \cos^5 x\, dx$. Now $f(\pi - x) = \left[\cos(\pi - x)\right]^5 = -\cos^5 x = -f(x)$.

$\therefore \int\limits_0^\pi \cos^5 x\, dx = 0$. (by Property VII). Hence, $\int\limits_0^{2\pi} \cos^5 x\, dx = 2 \int\limits_0^\pi \cos^5 x\, dx = 0$.

Example 3.20 Evaluate each of the following definite integrals:

(*i*) $\int\limits_0^\pi \theta \sin^6 \theta \cos^4 \theta\, d\theta$, (*ii*) $\int\limits_0^\pi \left(1 + \cos\theta + \cos^2\theta + \cos^3\theta\right) \sin^3\theta\, d\theta$,

(*iii*) $\int\limits_0^\pi \dfrac{x\, dx}{a^2 \cos^2 x + b^2 \sin^2 x}$. $\qquad\qquad$ (*A.M.I.E. W-2001*)

Solution. (*i*) Let $\quad I = \int\limits_0^\pi \theta \sin^6 \theta \cos^4 \theta\, d\theta$.

Then $\qquad\qquad I = \int\limits_0^\pi (\pi - \theta) \sin^6 (\pi - \theta) \cos^4 (\pi - \theta)\, d\theta \qquad$ (by Property IV)

$\qquad\qquad\qquad = \int\limits_0^\pi (\pi - \theta) \sin^6 \theta \cos^4 \theta\, d\theta = \pi \int\limits_0^\pi \sin^6 \theta \cos^4 \theta\, d\theta - I$

$\Rightarrow 2I = \pi \int\limits_0^\pi \sin^6 \theta \cos^4 \theta\, d\theta = 2\pi \int\limits_0^{\pi/2} \sin^6 \theta \cos^4 \theta\, d\theta, \qquad$ (by Property VII)

$\qquad = 2\pi \dfrac{5 \cdot 3 \cdot 1 \cdot 3 \cdot 1}{10 \cdot 8 \cdot 6 \cdot 4 \cdot 2} \cdot \dfrac{\pi}{2} = \dfrac{3\pi^2}{256}$ or $I = \dfrac{3\pi^2}{512}$, the required result.

(*ii*) We have $\cos(\pi - \theta) = -\cos\theta$, $\sin(\pi - \theta) = \sin\theta$.

$\therefore \int\limits_0^\pi \left(1 + \cos\theta + \cos^2\theta + \cos^3\theta\right) \sin^3\theta\, d\theta = \int\limits_0^\pi \sin^3\theta\, d\theta + \int\limits_0^\pi \cos\theta \sin^3\theta\, d\theta$

$$+\int_0^\pi \cos^2 \theta \cdot \sin^3 \theta \, d\theta + \int_0^\pi \cos^3 \theta \sin^3 \theta \, d\theta$$

$$= 2 \int_0^{\pi/2} \sin^3 \theta \, d\theta + 2 \int_0^{\pi/2} \cos^2 \theta \sin^3 \theta \, d\theta \qquad \text{(by Property VI)}$$

$$= 2 \int_0^{\pi/2} \sin^3 \theta \, d\theta + 2 \int_0^{\pi/2} \left(\sin^3 \theta - \sin^5 \theta\right) d\theta = 2 \cdot \frac{2}{3} + 2 \cdot \frac{2}{3} - 2 \cdot \frac{4.2}{5.3} = \frac{8}{5}.$$

(*iii*) Let $\qquad I = \displaystyle\int_0^\pi \frac{x \, dx}{a^2 \cos^2 x + b^2 \sin^2 x}.$

Then $\quad I = \displaystyle\int_0^\pi \frac{(\pi - x) \, dx}{a^2 \cos^2 (\pi - x) + b^2 \sin^2 (\pi - x)} = \int_0^\pi \frac{(\pi - x) \, dx}{a^2 \cos^2 x + b^2 \sin^2 x}$

$$= \pi \int_0^\pi \frac{dx}{a^2 \cos^2 x + b^2 \sin^2 x} - I \quad \Rightarrow \quad 2I = \pi \int_0^\pi \frac{dx}{a^2 \cos^2 x + b^2 \sin^2 x}$$

$$= 2\pi \int_0^{\pi/2} \frac{dx}{a^2 \cos^2 x + b^2 \sin^2 x} \quad \text{or} \quad I = \pi \int_0^{\pi/2} \frac{\sec^2 x \, dx}{a^2 + b^2 \tan^2 x}$$

$$= \pi \int_0^\infty \frac{dz}{a^2 + b^2 z^2} \; (\text{where } \tan x = z) = \frac{\pi}{b^2} \int_0^\infty \frac{dz}{z^2 + (a/b)^2} = \frac{\pi}{b^2} \cdot \frac{1}{a/b} \left[\tan^{-1} \frac{bz}{a} \right]_0^\infty$$

$$= \frac{\pi}{ab} \left[\tan^{-1} \infty - \tan^{-1} 0 \right] = \frac{\pi}{ab} \left(\frac{\pi}{2} \right) = \frac{\pi^2}{2ab}.$$

Example 3.21 Evaluate the following definite integrals

(*i*) $\displaystyle\int_0^\pi \sin^n x \, dx,$

(*ii*) $\displaystyle\int_0^\pi \cos^{2n+1} x \, dx,$

(*iii*) $\displaystyle\int_0^\pi \cos^{2n} x \, dx,$

(*iv*) $\displaystyle\int_0^{2\pi} \cos^{2n} x \, dx,$

(*v*) $\displaystyle\int_0^\pi \sin^m x \cos^n x \, dx,$

(*vi*) $\displaystyle\int_0^{2\pi} \sin^m x \cos^n x \, dx.$

Solution. (*i*) Since $\sin^n(\pi - x) = \sin^n x$.

$$\therefore \quad \int_0^\pi \sin^n x \, dx = 2 \int_0^{\pi/2} \sin^n x \, dx \qquad ...(i) \qquad \text{(by Property VII)}$$

Case I. If n is a + ve odd integer, then from (*i*),

$$\int_0^\pi \sin^n x \, dx = 2 \cdot \frac{(n-1)(n-3).....2}{n(n-2)......3}.$$

Case II. If n is a + ve even integer then from (*i*)

$$\int_0^\pi \sin^n x \, dx = 2 \cdot \frac{(n-1)(n-3)...1}{n(n-2)...2} \cdot \frac{\pi}{2} = \frac{(n-1)(n-3)...1}{n(n-2)...2} \pi.$$

(*ii*) We have $\left[\cos(\pi - x)\right]^{2n+1} = (-\cos x)^{2n+1} = -\cos^{2n+1} x.$

Here $\qquad\qquad\qquad\qquad f(2a - x) = -f(x)$

$$\therefore \quad \int_0^\pi \cos^{2n+1} x \, dx = 0. \qquad\qquad \text{(by Property VII)}$$

(*iii*) $\cos^{2n}(\pi - x) = (-\cos x)^{2n} = \cos^{2n} x. \quad \therefore \int_0^\pi \cos^{2n} x \, dx = 2 \int_0^{\pi/2} \cos^{2n} x \, dx$

$$= 2 \cdot \frac{(2n-1)(2n-3)......1}{2n(2n-2)......2} \cdot \frac{\pi}{2} = \frac{(2n-1)(2n-3)......1}{2n(2n-2)......2} \pi.$$

(*iv*) Since $\cos^{2n}(2\pi - x) = \cos^{2n} x$.

$$\therefore \quad \int_0^{2\pi} \cos^{2n} x \, dx = 2 \int_0^\pi \cos^{2n} x \, dx = 2 \cdot 2 \int_0^{\pi/2} \cos^{2n} x \, dx$$

$$\therefore \quad = 4 \cdot \frac{(2n-1)(2n-3)...1}{2n(2n-2)...2} \cdot \frac{\pi}{2} = 2\pi \frac{(2n-1)(2n-3)...1}{2n(2n-2)...2}.$$

(*v*) We have $f(x) = \sin^m x \cos^n x$. Therefore, $f(\pi - x) = \sin^m(\pi - x)\cos^n(\pi - x)$

$$= \sin^m x(-\cos x)^n = (-1)^n \sin^m x \cos^n x$$

$$= f(x) \text{ or } -f(x), \text{ according as } n \text{ is even or odd.}$$

Thus if n is even, $\displaystyle\int_0^\pi \sin^m x \cos^n x\, dx = 2\int_0^{\pi/2} \sin^m x \cos^n x\, dx$

and if n is odd, $\displaystyle\int_0^\pi \sin^m x \cos^n x\, dx = 0$.

(*vi*) We have $f(x) = \sin^m x \cos^n x$. $\therefore$ $f(2\pi - x) = \sin^m(2\pi - x)\cos^n(2\pi - x)$

$\qquad = (-1)^m \sin^m x \cos^n x = f(x)$ or $-f(x)$, according as m is even or odd.

Thus if m is odd, $\qquad \displaystyle\int_0^{2\pi} \sin^m x \cos^n x\, dx = 0$,

and if m is even, $\qquad \displaystyle\int_0^{2\pi} \sin^m x \cos^n x\, dx = 2\int_0^\pi \sin^m x \cos^n x\, dx$

$$= \begin{cases} 4\displaystyle\int_0^{\pi/2} \sin^m x \cos^n x\, dx, & \text{if } n \text{ is even.} \\[2mm] 0, & \text{if } n \text{ is odd.} \end{cases}$$

Example 3.22 Establish the following results

(*i*) $\displaystyle\int_0^{\pi/2} \log \sin x \ dx = \frac{\pi}{2}\log\frac{1}{2}$, (*A.M.I.E., S-2000; Kanpur, 1996*)

(*ii*) $\displaystyle\int_0^\pi \log(1 + \cos x)\,dx = \pi\log\frac{1}{2}$, (*A.M.I.E. W-1997*)

(*iii*) $\displaystyle\int_0^\infty \log\left(x + \frac{1}{x}\right)\frac{dx}{1 + x^2} = \pi\log 2$, (*A.M.I.E. W-2000*)

(*iv*) $\displaystyle\int_0^{\pi/2} \log(\tan\theta + \cot\theta)\,d\theta = \pi\log 2$, (*v*) $\displaystyle\int_0^1 \frac{\sin^{-1} x}{x}\,dx = \frac{\pi}{2}\log 2$

Solution. (*i*) Let $I = \displaystyle\int_0^{\pi/2} \log\sin x\, dx$.

Then $I = \int\limits_0^{\pi/2} \log\sin\left(\frac{\pi}{2}-x\right)dx = \int\limits_0^{\pi/2} \log\cos x \, dx.$

Adding the two values of I, we obtain

$$2I = \int\limits_0^{\pi/2}(\log\sin x + \log\cos x)dx = \int\limits_0^{\pi/2}\log(\sin x\cos x)dx = \int\limits_0^{\pi/2}\log\left(\frac{\sin 2x}{2}\right)dx$$

$$= \int\limits_0^{\pi/2}\log\sin 2x \, dx - \log 2\int\limits_0^{\pi/2}dx = \int\limits_0^{\pi/2}\log\sin 2x \, dx - \frac{\pi}{2}\log 2. \quad \ldots(i)$$

For the first integral on the right hand side, put $2x = z$ so that $dx = dz/2$.
When $x = 0$, $z = 0$; and when $x = \pi/2$, $z = \pi$.

$$\therefore \quad \int\limits_0^{\pi/2}\log\sin 2x \, dx = \frac{1}{2}\int\limits_0^{\pi}\log\sin z \, dz = \frac{1}{2}\int\limits_0^{\pi}\log\sin x \, dx \qquad \text{(by Property I)}$$

$$= \frac{1}{2}\cdot 2\int\limits_0^{\pi/2}\log\sin x \, dx, \text{(by Property VII)} = \int\limits_0^{\pi/2}\log\sin x \, dx = I.$$

From (i), we obtain $2I = I - (\pi/2)\log 2 \Rightarrow I = -(\pi/2)\log 2.$

Hence $\int\limits_0^{\pi/2}\log\sin x \, dx = -\frac{\pi}{2}\log 2 = \frac{\pi}{2}\log 2^{-1} = \frac{\pi}{2}\log\frac{1}{2}.$

Remark. $\int\limits_0^{\pi/2}\log\cos x \, dx = \int\limits_0^{\pi/2}\log\cos\left(\frac{\pi}{2}-x\right)dx = \int\limits_0^{\pi/2}\log\sin x \, dx = \frac{\pi}{2}\log\frac{1}{2}.$

(ii) Let $I = \int\limits_0^{\pi}\log(1+\cos x)dx = \int\limits_0^{\pi}\log\left(2\cos^2\frac{x}{2}\right)dx$

$$= \log 2\int\limits_0^{\pi}dx + 2\int\limits_0^{\pi}\log\cos\frac{x}{2}dx = \pi\log 2 + 2\int\limits_0^{\pi}\log\cos\frac{x}{2}dx. \quad \ldots(i)$$

For the integral on right hand side, put $x/2 = z$ so that $dx = 2dz$.
When $x = 0$, $z = 0$; and when $x = \pi$, $z = \pi/2$.

$$\therefore \quad \int\limits_0^{\pi}\log\cos\frac{x}{2}dx = 2\int\limits_0^{\pi/2}\log\cos z \, dz = 2\int\limits_0^{\pi/2}\log\cos x \, dx = \pi\log\frac{1}{2} = -\pi\log 2.$$

From (i), $I = \pi \log 2 - 2\pi \log 2 = -\pi \log 2 = \pi \log(1/2)$

(iii) Put $x = \tan \theta$, so that $dx = \sec^2 \theta \, d\theta$. When $x = 0$, $\theta = 0$, and when $x = \infty$, $\theta = \pi/2$.

$$\therefore \int_0^\infty \log\left(x + \frac{1}{x}\right) \frac{dx}{1+x^2} = \int_0^{\pi/2} \log(\tan \theta + \cot \theta) \, d\theta = \int_0^{\pi/2} \log \frac{1}{\sin \theta \cos \theta} \, d\theta$$

$$= -\int_0^{\pi/2} \log \sin \theta \, d\theta - \int_0^{\pi/2} \log \cos \theta \, d\theta = -\int_0^{\pi/2} \log \sin \theta \, d\theta - \int_0^{\pi/2} \log \cos\left(\frac{\pi}{2} - \theta\right) d\theta$$

$$= -\int_0^{\pi/2} \log \sin \theta \, d\theta - \int_0^{\pi/2} \log \sin \theta \, d\theta = -2 \int_0^{\pi/2} \log \sin \theta \, d\theta = -2\left(-\frac{\pi}{2}\log 2\right) = \pi \log 2.$$

(iv) Refer solution of part (iii).

(v) Put $x = \sin \theta$ so that $dx = \cos \theta \, d\theta$.

When $x = 0$, $\theta = 0$, and when $x = 1$, $\theta = \pi/2$.

Let $I = \int_0^1 \frac{\sin^{-1} x}{x} dx = \int_0^{\pi/2} \frac{\theta}{\sin \theta} \cos \theta \, d\theta = \int_0^{\pi/2} \theta \cot \theta \, d\theta$ (Integrating by parts)

$$= [\theta \log \sin \theta]_0^{\pi/2} - \int_0^{\pi/2} \log \sin \theta \, d\theta = -\int_0^{\pi/2} \log \sin \theta \, d\theta = \frac{\pi}{2} \log 2.$$

Exercise 3.2

1. Show that each of the following integrals vanish

(i) $\displaystyle\int_0^\pi \sin^6 x \cos^7 x \, dx$, (ii) $\displaystyle\int_0^{\pi/2} \frac{\sin x - \cos x}{1 + \sin x \cos x} dx$,   (A.M.I.E.. S-2004)

(iii) $\displaystyle\int_0^{\pi/2} (\sin x - \cos x) \log(\sin x + \cos x) dx$,   (A.M.I.E., W-1998)

(iv) $\displaystyle\int_0^1 \log\left(\frac{1}{x} - 1\right) dx$, (v) $\displaystyle\int_0^{\pi/2} \frac{\sin^2 x - \cos^2 x}{\sin^3 x + \cos^3 x} dx$.

2. Establish the following results.

(i) $\displaystyle\int_0^{\pi/2} x \cot x \, dx = \frac{\pi}{2}\log 2,$

(ii) $\displaystyle\int_0^{\pi} x \sin x \cos^4 x \, dx = \frac{\pi}{5},$

(iii) $\displaystyle\int_0^{\pi/2} \left(\frac{\theta}{\sin\theta}\right) d\theta = \pi \log 2,$

(iv) $\displaystyle\int_0^a \frac{dx}{x + \sqrt{a^2 - x^2}} = \frac{\pi}{4},$

(v) $\displaystyle\int_0^{\pi} \frac{x}{1 + \sin x} dx = \pi,$ *(Madras, 2000)*

(vi) $\displaystyle\int_0^{\pi} x \sin^7 x \, dx = \frac{16\pi}{35},$

(vii) $\displaystyle\int_{\pi/4}^{3\pi/4} \frac{\phi}{1 + \sin\phi} d\phi = \pi\left(\sqrt{2} - 1\right).$

3. Evaluate each of the following definite integrals

(i) $\displaystyle\int_0^{\pi} \frac{dx}{a^2 + b^2 \cos^2 x},$

(ii) $\displaystyle\int_0^{\pi/2} \frac{x + \sin x}{1 + \cos x} dx,$ *(AM.I.E., W-2004)*

(iii) $\displaystyle\int_0^{\pi} \sin\theta \cos^3 \frac{\theta}{2} d\theta$ *(A.M.I.E, S-2002),*

(iv) $\displaystyle\int_0^{\infty} \frac{\log(1 + x^2)}{1 + x^2} dx,$

(v) $\displaystyle\int_a^b \frac{f(x)}{f(a + b - x) + f(x)} dx,$

(vi) $\displaystyle\int_0^1 x(1 - x)^n \, dx,$

(vii) $\displaystyle\int_0^a \frac{dx}{1 + e^{f(x)}},$ given $f(a - x) + f(x) = 0.$

Ans: (i) $\dfrac{\pi}{a\sqrt{a^2 + b^2}}$, (ii) $\dfrac{\pi}{2}$, (iii) $\dfrac{4}{5}$, (iv) $\pi \log 2$, (v) $\dfrac{(b - a)}{2}$,

(vi) $\dfrac{1}{(n + 1)(n + 2)}$, (vii) $a/2$.

4. Evaluate the following

(i) $\displaystyle\int_0^{\pi/2} \frac{x \sin x \cos x}{\cos^4 x + \sin^4 x} dx,$

(ii) $\displaystyle\int_0^{\pi} \frac{x dx}{a^2 - \cos^2 x}.$ *(G.G.S.I.P.U-2001)*

[Hint (i) Let $I = \displaystyle\int_0^{\pi/2} \frac{x \sin x \cos x}{\cos^4 x + \sin^4 x} dx.$

Then $I = \int\limits_{0}^{\pi/2} \dfrac{(\pi/2 - x)\sin(\pi/2 - x)\cos(\pi/2 - x)}{\cos^4(\pi/2 - x) + \sin^4(\pi/2 - x)}\,dx = \int\limits_{0}^{\pi/2} \dfrac{(\pi/2 - x)\cos x \sin x}{\sin^4 x + \cos^4 x}\,dx.$

Adding the two values of I, we obtain $2I = \dfrac{\pi}{2} \int\limits_{0}^{\pi/2} \dfrac{\sin x \cos x}{\sin^4 x + \cos^4 x}\,dx$

$= \dfrac{\pi}{2} \int\limits_{0}^{\pi/2} \dfrac{\tan x \sec^2 x}{1 + \tan^4 x}\,dx$ (dividing the num. & denom. by $\cos^4 x$) Put $\tan^2 x = z$,

we get $I = \dfrac{\pi}{8} \int\limits_{0}^{\infty} \dfrac{dz}{1 + z^2} = \dfrac{\pi^2}{16}.$

(*ii*) Let $I = \int\limits_{0}^{\pi} \dfrac{x\,dx}{a^2 - \cos^2 x}.$ Then $I = \int\limits_{0}^{\pi} \dfrac{(\pi - x)\,dx}{a^2 - \cos^2(\pi - x)} = \int\limits_{0}^{\pi} \dfrac{(\pi - x)\,dx}{a^2 - \cos^2 x}.$

Adding the two values of I, we obtain $2I = \pi \int\limits_{0}^{\pi} \dfrac{dx}{a^2 - \cos^2 x}.$

or, $I = \pi \int\limits_{0}^{\pi/2} \dfrac{dx}{a^2 - \cos^2 x} = \pi \int\limits_{0}^{\pi/2} \dfrac{\sec^2 x\,dx}{a^2 \sec^2 x - 1} = \pi \int\limits_{0}^{\pi/2} \dfrac{\sec^2 x\,dx}{(a^2 - 1) + a^2 \tan^2 x}$

$= \dfrac{\pi}{a} \int\limits_{0}^{\infty} \dfrac{dz}{(a^2 - 1) + z^2},$ (put $a \tan x = z$)

$= \dfrac{\pi}{a\sqrt{a^2 - 1}} \left[\tan^{-1} \dfrac{z}{\sqrt{a^2 - 1}} \right]_{0}^{\infty} = \dfrac{\pi^2}{2a\sqrt{a^2 - 1}}, a > 1.]$

3.8 Summation of Series

The definition of integration as the limit of a sum enables us to find the limit of sums of series of a certain types by expressing them as definite integrals. The values of the required limit can be written down by the formula (3.2), viz.,

$$\int\limits_{a}^{b} f(x)\,dx = \lim_{n \to \infty} h\left[f(a) + f(a + h) + f(a + 2h) + \dots + f\left(a + \overline{n - 1}\,h\right) \right],$$

where $b - a = nh.$

When $n \to \infty, h \to 0$ and in that case

$$\int_a^b f(x)dx = \lim_{h\to 0} h\big[f(a)+f(a+h)+f(a+2h)+...+f\big(a+\overline{n-1}\,h\big)$$

$$= \lim_{h\to 0} h\sum_{r=0}^{n-1} f(a+rh), \text{ where } nh = b-a.$$

Putting $a = 0$ and $b = 1$, so that $h = 1/n$, we obtain

$$\int_0^1 f(x)dx = \lim_{n\to\infty}\frac{1}{n}\sum_{r=0}^{n-1} f\left(\frac{r}{n}\right) = \lim_{h\to 0} h\sum_{r=0}^{n-1} f(rh), \{nh = 1\}$$

(A) Rule to find the limit, as n → ∞, of the sum of the series of n terms, by integration.

Step 1. Write down the *r*th terms of the series and put it in the form $\frac{1}{n}\left\{f\left(\frac{r}{n}\right)\right\}$.

Then the series can be written as $\lim_{n\to\infty}\sum\frac{1}{n}\left\{f\left(\frac{r}{n}\right)\right\}$.

Step 2. Put $n = 1/h$ and write it so as to get h as a factor in the numerator.

Step 3. Replacing rh by x, h by dx and $\lim_{n\to\infty}\Sigma$ by $\int$.

Step 4. Integrate the resulting expression, taking the lower limit of the difinite integral $= \lim_{h\to 0}(rh)$, where r is as in the first term, and the upper limit $= \lim_{h\to 0}(rh)$, where r is as in the last term. This give the required limit. The procedure will be clear from the following solved examples.

(B) To find limit of a product by integration:

Let $P = \lim_{n\to\infty}$ (given product). Taking logarithms of both sides, we obtain

$\log P = \lim_{n\to\infty}$ (a series) $= k$ (say). Then $P = e^k$.

Example 3.23 Evaluate the following

(i) $\displaystyle\lim_{n\to\infty}\left[\frac{n}{n^2+1^2}+\frac{n}{n^2+2^2}+\frac{n}{n^2+3^2}+...+\frac{1}{2n}\right]$

(ii) $\displaystyle\lim_{n\to\infty}\left[\frac{1}{\sqrt{2n-1^2}}+\frac{1}{\sqrt{4n-2^2}}+\frac{1}{\sqrt{6n-3^2}}+...+\frac{1}{n}\right]$

(iii) $\displaystyle\lim_{n\to\infty}\left[\frac{1}{n}+\frac{n^2}{(n+1)^3}+\frac{n^2}{(n+2)^3}+...+\frac{1}{8n}\right],$

(iv) $\lim\limits_{n\to\infty}\left[\dfrac{1}{n}+\dfrac{\sqrt{n^2-1^2}}{n^2}+\dfrac{\sqrt{n^2-2^2}}{n^2}+...+\dfrac{\sqrt{n^2-(n-1)^2}}{n^2}\right]$,

(v) $\lim\limits_{n\to\infty}\left[\dfrac{\sqrt{n+1}+\sqrt{n+2}+\sqrt{n+3}+...+\sqrt{2n}}{n\sqrt{n}}\right]$,

(vi) $\lim\limits_{n\to\infty}\left[\dfrac{1}{n}+\dfrac{1}{n+1}+\dfrac{1}{n+2}+\dfrac{1}{n+3}+...+\dfrac{1}{3n}\right]$. *(IETE-June 2009)*

Solution. (i) $\lim\limits_{n\to\infty}\left[\dfrac{n}{n^2+1^2}+\dfrac{n}{n^2+2^2}+\dfrac{n}{n^2+3^2}+...+\dfrac{n}{n^2+n^2}\right]$

$=\lim\limits_{n\to\infty}\dfrac{1}{n}\left[\dfrac{1}{1+(1/n)^2}+\dfrac{1}{1+(2/n)^2}+\dfrac{1}{1+(3/n)^2}+...+\dfrac{1}{1+(n/n)^2}\right]$

$=\lim\limits_{n\to\infty}\dfrac{1}{n}\sum\limits_{r=1}^{n}\dfrac{1}{1+(r/n)^2}=\lim\limits_{h\to0}h\sum\limits_{r=1}^{n}\dfrac{1}{1+(rh)^2},\ nh=1$

$=\displaystyle\int_0^1\dfrac{dx}{1+x^2}\ \ [rh=x;\ \text{when } r=1,\ x=0\ as\ h\to0\ \text{and when } r=n,\ x=nh=1]$

$=\left[\tan^{-1}x\right]_0^1=\tan^{-1}1-\tan^{-1}0=\pi/4.$

(ii) $\lim\limits_{n\to\infty}\left[\dfrac{1}{\sqrt{2n-1^2}}+\dfrac{1}{\sqrt{4n-2^2}}+\dfrac{1}{\sqrt{6n-3^2}}+...+\dfrac{1}{\sqrt{2n\cdot n-n^2}}\right]$

$=\lim\limits_{n\to\infty}\dfrac{1}{n}\left[\dfrac{1}{\sqrt{\left(\dfrac{2}{n}\right)-\left(\dfrac{1}{n}\right)^2}}+\dfrac{1}{\sqrt{\left(\dfrac{4}{n}\right)-\left(\dfrac{2}{n}\right)^2}}+\dfrac{1}{\sqrt{\left(\dfrac{6}{n}\right)-\left(\dfrac{3}{n}\right)^2}}+...+\dfrac{1}{\sqrt{\dfrac{2n}{n}-\left(\dfrac{n}{n}\right)^2}}\right]$

$=\lim\limits_{n\to\infty}\dfrac{1}{n}\sum\limits_{r=1}^{n}\dfrac{1}{\sqrt{(2r/n)-(r/n)^2}}=\lim\limits_{h\to0}h\sum\limits_{r=1}^{n}\dfrac{1}{\sqrt{2rh-(rh)^2}},\ nh=1$

$=\displaystyle\int_0^1\dfrac{dx}{\sqrt{2x-x^2}}=\int_0^1\dfrac{dx}{\sqrt{1-(1-x)^2}}.$ Let $z=1-x$ so that $dz=-dx.$

When $x=0,\ z=1$; and when $x=1,\ z=0.$

$\therefore\ \ I=-\displaystyle\int_1^0\dfrac{dz}{\sqrt{1-z^2}}=-\left[\sin^{-1}z\right]_1^0=-\left[\sin^{-1}0-\sin^{-1}1\right]=-(0-\pi/2)=\pi/2.$

(*iii*) $\lim\limits_{n\to\infty}\left[\dfrac{1}{n}+\dfrac{n^2}{(n+1)^3}+\dfrac{n^2}{(n+2)^3}+...+\dfrac{1}{8n}\right]$

$=\lim\limits_{n\to\infty}\left[\dfrac{n^2}{(n+0)^3}+\dfrac{n^2}{(n+1)^3}+\dfrac{n^2}{(n+2)^3}+...+\dfrac{n^2}{(n+n)^3}\right]$

$=\lim\limits_{n\to\infty}\dfrac{n^2}{n^3}\left[\dfrac{1}{\{1+(0/n)\}^3}+\dfrac{1}{\{1+(1/n)\}^3}+\dfrac{1}{\{1+(2/n)\}^3}+....\dfrac{1}{\{1+(n/n)\}^3}\right]$

$=\lim\limits_{n\to\infty}\dfrac{1}{n}\sum\limits_{r=0}^{n}\dfrac{1}{\{1+(r/n)\}^3}=\lim\limits_{h\to0}h\sum\limits_{r=0}^{n}\dfrac{1}{(1+rh)^3},\ nh=1$

$=\int\limits_{0}^{1}\dfrac{dx}{(1+x)^3}=\left[-\dfrac{1}{2(1+x)^2}\right]_0^1=-\dfrac{1}{8}+\dfrac{1}{2}=\dfrac{3}{8}.$

(*iv*) $\lim\limits_{n\to\infty}\left[\dfrac{1}{n}+\dfrac{\sqrt{n^2-1^2}}{n^2}+\dfrac{\sqrt{n^2-2^2}}{n^2}+...+\dfrac{\sqrt{n^2-(n-1)^2}}{n^2}\right]$

$=\lim\limits_{n\to\infty}\left[\dfrac{\sqrt{n^2-0^2}}{n^2}+\dfrac{\sqrt{n^2-1^2}}{n^2}+\dfrac{\sqrt{n^2-2^2}}{n^2}+...+\dfrac{\sqrt{n^2-(n-1)^2}}{n^2}\right]$

$=\lim\limits_{n\to\infty}\dfrac{n}{n^2}\left[\sqrt{1-(0/n)^2}+\sqrt{1-(1/n)^2}+\sqrt{1-(2/n)^2}+......+\sqrt{1-\{(n-1)/n\}^2}\right]$

$=\lim\limits_{n\to\infty}\dfrac{1}{n}\sum\limits_{r=0}^{n-1}\sqrt{1-(r/n)^2}=\lim\limits_{h\to0}h\sum\limits_{r=0}^{n-1}\sqrt{1-(rh)^2},\ nh=1$

$=\int\limits_{0}^{1}\sqrt{1-x^2}\,dx=\left[\dfrac{x}{2}\sqrt{1-x^2}+\dfrac{1}{2}\sin^{-1}x\right]_0^1=\dfrac{1}{2}\sin^{-1}1=\dfrac{1}{2}\cdot\dfrac{\pi}{2}=\dfrac{\pi}{4}.$

(*v*) $\lim\limits_{n\to\infty}\left[\dfrac{\sqrt{n+1}+\sqrt{n+2}+\sqrt{n+3}+\cdots\cdots+\sqrt{2n}}{n\sqrt{n}}\right]$

$=\lim\limits_{n\to\infty}\dfrac{1}{n}\left[\dfrac{\sqrt{n+1}}{\sqrt{n}}+\dfrac{\sqrt{n+2}}{\sqrt{n}}+\dfrac{\sqrt{n+3}}{\sqrt{n}}+\cdots\cdots+\dfrac{\sqrt{n+n}}{\sqrt{n}}\right]$

$$= \lim_{n \to \infty} \frac{1}{n}\left[\sqrt{1+\frac{1}{n}}+\sqrt{1+\frac{2}{n}}+\sqrt{1+\frac{3}{n}}+\cdots\cdots+\sqrt{1+\frac{n}{n}}\right]$$

$$= \lim_{n \to \infty} \frac{1}{n}\sum_{r=1}^{n}\sqrt{1+(r/n)} = \lim_{h \to 0} h\sum_{r=1}^{n}\sqrt{1+rh}, \ nh = 1$$

$$= \int_{0}^{1}\sqrt{1+x}\ dx = \frac{2}{3}\left[(1+x)^{3/2}\right]_{0}^{1} = \frac{2}{3}\left(2\sqrt{2}-1\right).$$

(*vi*) The given expression $= \lim\limits_{n \to \infty}\left\{\sum\limits_{r=0}^{n-1}\dfrac{1}{n+r}+\sum\limits_{r=0}^{n}\dfrac{1}{2n+r}\right\}$

$$= \lim_{n \to \infty}\left\{\frac{1}{n}\sum_{r=0}^{n-1}\frac{1}{1+(r/n)}+\frac{1}{n}\sum_{r=0}^{n}\frac{1}{2+(r/n)}\right\} = \lim_{h \to 0}\left\{h\sum_{r=0}^{n-1}\frac{1}{1+rh}+h\sum_{r=0}^{n}\frac{1}{2+rh}\right\}, \ nh = 1$$

$$= \int_{0}^{1}\frac{dx}{1+x}+\int_{0}^{1}\frac{dx}{2+x} = \left[\log|1+x|\right]_{0}^{1}+\left[\log|2+x|\right]_{0}^{1} = \log_{e}2+\log_{e}3-\log_{e}2 = \log_{e}3.$$

Example 3.24 Evaluate the limit of the following, when *n* tends to infinity.

(*i*) $\displaystyle\sum_{r=1}^{n-1}\frac{1}{n}\sqrt{\frac{n+r}{n-r}},$ (*ii*) $\displaystyle\sum_{r=1}^{n}\frac{n+r}{n^{2}+r^{2}},$ (*iii*) $\displaystyle\sum_{r=0}^{n}\frac{n^{2}}{\left(n^{2}+r^{2}\right)^{3/2}},$

(*iv*) $\displaystyle\sum_{r=1}^{n}\frac{n}{(n+r)\sqrt{r(2n+r)}},$ (*v*) $\displaystyle\sum_{r=1}^{2n}\frac{1}{n}\cdot\frac{r}{\sqrt{n^{2}+r^{2}}}.$

Solution. (*i*) $\displaystyle\lim_{n \to \infty}\sum_{r=1}^{n-1}\frac{1}{n}\sqrt{\frac{n+r}{n-r}} = \lim_{n \to \infty}\sum_{r=1}^{n-1}\frac{1}{n}\sqrt{\frac{1+(r/n)}{1-(r/n)}} = \lim_{h \to 0} h\sum_{r=1}^{n-1}\sqrt{\frac{1+rh}{1-rh}}, \ nh = 1$

$$= \int_{0}^{1}\sqrt{\frac{1+x}{1-x}}\ dx = \int_{0}^{1}\frac{1+x}{\sqrt{1-x^{2}}}\ dx = \int_{0}^{\pi/2}(1+\sin\theta)\ d\theta, \ \text{where } x = \sin\theta$$

$$= \left[\theta-\cos\theta\right]_{0}^{\pi/2} = \frac{\pi}{2}-(-1) = \left(\frac{\pi}{2}\right)+1.$$

(*ii*) $\displaystyle\lim_{n \to \infty}\sum_{r=1}^{n}\frac{n+r}{n^{2}+r^{2}} = \lim_{n \to \infty}\frac{1}{n}\sum_{r=1}^{n}\frac{1+(r/n)}{1+(r/n)^{2}} = \lim_{h \to 0} h\sum_{r=1}^{n}\frac{1+rh}{1+(rh)^{2}}, \ nh = 1$

$$= \int_{0}^{1}\frac{1+x}{1+x^{2}}\ dx = \left[\tan^{-1}x+\frac{1}{2}\log\left(1+x^{2}\right)\right]_{0}^{1} = \tan^{-1}1+\frac{1}{2}\log 2 = \frac{\pi}{4}+\frac{1}{2}\log 2$$

(iii) $\displaystyle\lim_{n\to\infty}\sum_{r=0}^{n}\frac{n^2}{(n^2+r^2)^{3/2}}=\lim_{n\to\infty}\frac{1}{n}\sum_{r=0}^{n}\frac{1}{\left[1+(r/n)^2\right]^{3/2}}$

$\displaystyle=\lim_{h\to0}h\sum_{r=0}^{n}\frac{1}{\left[1+(rh)^2\right]^{3/2}},\ (nh=1)=\int_{0}^{1}\frac{dx}{(1+x^2)^{3/2}}=\int_{0}^{\pi/4}\frac{\sec^2\theta\,d\theta}{\sec^3\theta},$ where $x=\tan\theta$

$\displaystyle=\int_{0}^{\pi/4}\cos\theta\,d\theta=[\sin\theta]_0^{\pi/4}=\frac{1}{\sqrt{2}}.$

(iv) $\displaystyle\lim_{n\to\infty}\sum_{r=1}^{n}\frac{n}{(n+r)\sqrt{r(2n+r)}}=\lim_{n\to\infty}\frac{1}{n}\sum_{r=1}^{n}\frac{1}{\{1+(r/n)\}\{\sqrt{2(r/n)+(r/n)^2}\}}$

$\displaystyle=\lim_{h\to0}h\sum_{r=1}^{n}\frac{1}{(1+rh)\sqrt{2rh+(rh)^2}},\ nh=1$

$\displaystyle=\int_{0}^{1}\frac{dx}{(1+x)\sqrt{2x+x^2}}=\int_{0}^{1}\frac{dx}{(x+1)\sqrt{(x+1)^2-1}}=\left[\sec^{-1}(x+1)\right]_0^1$

$=\sec^{-1}2-\sec^{-1}1=\pi/3.$

(v) $\displaystyle\lim_{n\to\infty}\sum_{r=1}^{2n}\frac{1}{n}\cdot\frac{r}{\sqrt{n^2+r^2}}=\lim_{n\to\infty}\frac{1}{n}\sum_{r=1}^{2n}\frac{r/n}{\sqrt{1+(r/n)^2}}=\lim_{h\to0}h\sum_{r=1}^{2n}\frac{rh}{\sqrt{1+(rh)^2}},\ nh=1$

$\displaystyle=\int_{0}^{2}\frac{x}{\sqrt{1+x^2}}dx=\frac{1}{2}\int_{1}^{5}\frac{dz}{\sqrt{z}},\left(\text{where }1+x^2=z\right)=\left[\sqrt{z}\right]_1^5=\sqrt{5}-1.$

Example 3.25 Evaluate the following limit, where n tends to infinity, of the product:

(i) $\dfrac{[(n+1)(n+2)......(n+n)]^{1/n}}{n},$

(ii) $\left[(1+1/n)(1+2/n)^{1/2}(1+3/n)^{1/3}......(1+n/n)^{1/n}\right],$

(iii) $\left[\left(1+\dfrac{1}{n^2}\right)\left(1+\dfrac{2^2}{n^2}\right)\left(1+\dfrac{3^2}{n^2}\right).....\left(1+\dfrac{n^2}{n^2}\right)\right]^{1/n}.$

Solution. (*i*) Let $y = \lim\limits_{n \to \infty} \dfrac{[(n+1)(n+2)\cdots(2n)]^{1/n}}{n}$.

We have $\log y = \lim\limits_{n \to \infty} \left(\dfrac{1}{n}[\log(n+1) + \log(n+2) + \cdots + \log(2n)] - \log n\right)$

$= \lim\limits_{n \to \infty} \dfrac{1}{n}\left[\{\log(n+1) - \log n\} + \{\log(n+2) - \log n\} + \cdots + \{\log 2n - \log n\}\right]$

$= \lim\limits_{n \to \infty} \dfrac{1}{n}\left[\log\left(1 + \dfrac{1}{n}\right) + \log\left(1 + \dfrac{2}{n}\right) + \cdots + \log\left(1 + \dfrac{n}{n}\right)\right]$

$= \lim\limits_{n \to \infty} \dfrac{1}{n}\sum\limits_{r=1}^{n}\log\left(1 + \dfrac{r}{n}\right) = \lim\limits_{h \to 0} h\sum\limits_{r=1}^{n}\log(1 + rh),\ nh = 1$

$= \int\limits_{0}^{1}\log(1+x)\,dx = \left[x\log(1+x)\right]_0^1 - \int\limits_{0}^{1}\dfrac{x}{1+x}\,dx$

$= \log 2 - \left[x - \log(1+x)\right]_0^1 = \log 2 - (1 - \log 2)$

$= 2\log 2 - 1 = \log 4 - \log e = \log(4/e)$. Therefore, $y = 4/e$.

(*ii*) Let the required limit be A. We have

$\log A = \lim\limits_{n \to \infty}\left\{\log\left(1 + \dfrac{1}{n}\right) + \dfrac{1}{2}\log\left(1 + \dfrac{2}{n}\right) + \ldots + \left(\dfrac{1}{n}\right)\log\left(1 + \dfrac{n}{n}\right)\right\}$

$= \lim\limits_{n \to \infty}\sum\limits_{r=1}^{n}\dfrac{1}{r}\log\left(1 + \dfrac{r}{n}\right)$

$= \lim\limits_{n \to \infty}\dfrac{1}{n}\sum\limits_{r=1}^{n}\left\{\dfrac{1}{r/n}\log\left(1 + \dfrac{r}{n}\right)\right\} = \lim\limits_{h \to 0} h\sum\limits_{r=1}^{n}\dfrac{1}{rh}\log(1 + rh),\ nh = 1$

$= \int\limits_{0}^{1}\dfrac{1}{x}\log(1+x)\,dx = \int\limits_{0}^{1}\left(1 - \dfrac{x}{2} + \dfrac{x^2}{3} - \dfrac{x^3}{4} + \cdots\right)dx,\ \{\text{expanding } \log(1+x)\}$

$= \left[x - \dfrac{x^2}{2^2} + \dfrac{x^3}{3^2} - \dfrac{x^4}{4^2} + \cdots\right]_0^1 = 1 - \dfrac{1}{2^2} + \dfrac{1}{3^2} - \dfrac{1}{4^2} + \cdots = \dfrac{\pi^2}{12}$. Therefore, $A = e^{\pi^2/12}$.

(*iii*) Let P be the given expression, we have

$\log P = \lim\limits_{n \to \infty}\dfrac{1}{n}\sum\limits_{r=1}^{n}\log\left(1 + \dfrac{r^2}{n^2}\right) = \lim\limits_{h \to 0} h\sum\limits_{r=1}^{n}\log\{1 + (rh)^2\},\ nh = 1$

$$= \int_0^1 \log(1+x^2)\,dx = \left[x\log(1+x^2)\right]_0^1 - \int_0^1 \frac{2x^2}{1+x^2}\,dx$$

$$= \log 2 - 2\int_0^1\left(1-\frac{1}{1+x^2}\right)dx = \log 2 - 2\left[x-\tan^{-1}x\right]_0^1$$

$$= \log 2 - 2\left(1-\frac{\pi}{4}\right) = \log 2 + \frac{1}{2}(\pi-4), \text{ or, } \log P - \log 2 = \frac{1}{2}(\pi-4)$$

or $\log(P/2) = \dfrac{1}{2}(\pi-4)$ or $P/2 = e^{(\pi-4)/2}$ or $P = 2e^{(\pi-4)/2}$.

Exercise 3.3

1. Find the limit, when $n \to \infty$, of the series

(i) $\dfrac{1^m + 2^m + 3^m + \cdots + n^m}{n^{m+1}}$, $(m > -1)$ *(ii)* $\dfrac{1}{n+1} + \dfrac{1}{n+2} + \dfrac{1}{n+3} + \cdots + \dfrac{1}{2n}$,

(iii) $\dfrac{1}{n^2}\sec^2\dfrac{1}{n^2} + \dfrac{2}{n^2}\sec^2\dfrac{4}{n^2} + \dfrac{3}{n^2}\sec^2\dfrac{9}{n^2} + \cdots + \dfrac{1}{n}\sec^2 1$,

(iv) $\dfrac{1}{1+n^3} + \dfrac{4}{8+n^3} + \dfrac{9}{27+n^3} + \cdots + \dfrac{r^2}{r^3+n^3} + \cdots + \dfrac{1}{2n}$,

(v) $\dfrac{\sqrt{n}}{\sqrt{n^3}} + \dfrac{\sqrt{n}}{\sqrt{(n+3)^3}} + \dfrac{\sqrt{n}}{\sqrt{(n+6)^3}} + \cdots + \dfrac{\sqrt{n}}{\sqrt{\{n+3(n-1)\}^3}}$,

(vi) $\dfrac{n}{n^2} + \dfrac{n}{n^2+1^2} + \dfrac{n}{n^2+2^2} + \cdots + \dfrac{n}{n^2+(n+1)^2}$,

(vii) $\dfrac{1}{\sqrt{n^2}} + \dfrac{1}{\sqrt{n^2+n}} + \dfrac{1}{\sqrt{n^2+2n}} + \cdots + \dfrac{1}{\sqrt{n^2+(n-1)n}}$,

(viii) $\dfrac{n^2}{(n^2+1)^{3/2}} + \dfrac{n^2}{(n^2+2^2)^{3/2}} + \dfrac{n^2}{(n^2+3^2)^{3/2}} + \cdots + \dfrac{n^2}{\left[n^2+(n-1)^2\right]^{3/2}}$.

Ans: *(i)* $\dfrac{1}{m+1}$, *(ii)* $\log 2$, *(iii)* $\dfrac{1}{2}\tan 1$, *(iv)* $\dfrac{1}{3}\log 2$,

(v) $\dfrac{1}{3}$, *(vi)* $\pi/4$, *(vii)* $2(\sqrt{2}-1)$, *(viii)* $1/\sqrt{2}$.

2. Evaluate the limit of the following as $n \to \infty$.

(i) $\displaystyle\sum_{r=1}^{n} \frac{n}{n^2+r^2}$, (ii) $\displaystyle\sum_{r=1}^{n-1} \frac{1}{\sqrt{n^2+r^2}}$, (iii) $\displaystyle\sum_{r=1}^{n} \frac{r^3}{r^4+n^4}$,

(iv) $\displaystyle\sum_{r=1}^{n} \frac{n^3}{(n^2+r^2)(n^2+2r^2)}$, (v) $\displaystyle\sum_{r=1}^{n} \frac{1}{\sqrt{(4n^2-r^2)}}$,

(vi) $\displaystyle\sum_{r=1}^{n} \frac{\sqrt{n}}{\sqrt{r}(3\sqrt{r}+4\sqrt{n})^2}$, (vii) $\displaystyle\sum_{r=n+1}^{2n} \frac{n}{n^2+r^2}$.

Ans: (i) $\dfrac{\pi}{4}$, (ii) $\dfrac{\pi}{2}$, (iii) $\dfrac{1}{4}\log 2$, (iv) $\sqrt{2}\tan^{-1}\sqrt{2}-\dfrac{\pi}{4}$, (v) $\dfrac{\pi}{6}$,

(vi) $\dfrac{1}{14}$, (vii) $\tan^{-1}2-\dfrac{\pi}{4}$.

3. Evaluate $\displaystyle\lim_{n\to\infty}\left[\frac{n}{(n+1)\sqrt{2n+1}} + \frac{n}{(n+2)\sqrt{2(2n+2)}}\right.$

$\left.+\frac{n}{(n+3)\sqrt{3(2n+3)}}+\cdots+\frac{n}{2n\sqrt{n\cdot 3n}}\right]$ [**Hint:** The given limit

$= \displaystyle\lim_{n\to\infty}\frac{1}{n}\sum_{r=1}^{n} n\frac{1}{\left(1+\dfrac{r}{n}\right)\sqrt{\dfrac{r}{n}\left(2+\dfrac{r}{n}\right)}} = \lim_{h\to 0} h\sum_{r=1}^{n}\frac{1}{(1+rh)\sqrt{2rh+(rh)^2}}, nh=1$

$= \displaystyle\int_{0}^{1}\frac{dx}{(1+x)\sqrt{2x+x^2}} = \int_{0}^{1}\frac{dx}{(1+x)\sqrt{(1+x)^2-1}} = \left[\sec^{-1}(1+x)\right]_{0}^{1}$

$= \sec^{-1}2 - \sec^{-1}1 = \dfrac{\pi}{3} - 0 = \dfrac{\pi}{3}].$

4. Evaluate the following limit, as $n \to \infty$, of the product

(i) $\left\{\left(1+\dfrac{1}{n}\right)\left(1+\dfrac{2}{n}\right)\left(1+\dfrac{3}{n}\right)\cdots\left(1+\dfrac{n}{n}\right)\right\}^{\frac{1}{n}}$,

(ii) $\left\{\left(1+\dfrac{1}{n^2}\right)^{\frac{2}{n^2}}\left(1+\dfrac{2^2}{n^2}\right)^{\frac{4}{n^2}}\left(1+\dfrac{3^2}{n^2}\right)^{\frac{6}{n^2}}\cdots\left(1+\dfrac{n^2}{n^2}\right)^{\frac{2n}{n^2}}\right\}$, (iii) $\left(\dfrac{n!}{n^n}\right)^{\frac{1}{n}}$.

[**Hint.** (*ii*) Let P be the given expression. We have $\log P = \lim\limits_{n\to\infty} \sum\limits_{r=1}^{n} \dfrac{2r}{n^2} \log\left(1 + \dfrac{r^2}{n^2}\right)$

$$= \lim_{n\to\infty} \frac{1}{n} \sum_{r=1}^{n} \frac{2r}{n} \log\left(1 + \frac{r^2}{n^2}\right) = \lim_{h\to 0} h \sum_{r=1}^{n} 2rh \log\left\{1 + (rh)^2\right\}, nh = 1$$

$$= \int_{0}^{1} 2x \log\left(1 + x^2\right) dx = \int_{1}^{2} \log t \; dt, \left(\text{where } t = 1 + x^2\right)$$

$$= \left[t\log t - t\right]_{1}^{2} = 2\log 2 - 1 = \log 4 - \log e = \log\left(4/e\right). \text{ Therefore, } P = 4/e.]$$

Ans: (*i*) $4/e$, (*ii*) $4/e$, (*iii*) $1/e$.

Chapter 4

Reduction Formulae

4.1 Definition

A formula which connects an integral with another integral of the same form as the original but is of lower degree or order or is otherwise easier to integrate, is called a *reduction formula.*

Reduction formulae are generally obtained by the method of integration by parts.

In the following sections we shall see that many functions whose integrals cannot be written directly can be reduced, by repeated application of the same reduction formula, to a standard form or a form which can be easily integrated.

Example 4.1 Obtain the reduction formula for $\int \underset{\text{II}}{x^m} \underset{\text{I}}{(\log x)^n} \, dx$

Solution. Integrating by parts, taking x^m as second function, we obtain

$$\int x^m (\log x)^n \, dx = (\log x)^n \cdot \frac{x^{m+1}}{m+1} - \int n (\log x)^{n-1} \cdot \frac{1}{x} \cdot \frac{x^{m+1}}{m+1} \, dx$$

or $\quad \int x^m (\log x)^n dx = (\log x)^n \dfrac{x^{m+1}}{m+1} - \dfrac{n}{m+1} \int x^m (\log x)^{n-1} dx,$ \qquad ...(4.1)

which is the required reduction formula.

Remark. Here $\int x^m (\log x)^n \, dx$ is expressed in terms of $\int x^m (\log x)^{n-1} \, dx$. Thus the power of log x has been reduced from n to $(n-1)$.

By repeated application of this reduction formula, the given integral can ultimately be written in terms of $\int x^m dx$ which is a standard form.

4.2 Abridged Notation for Reduction Formulae

Sometimes reduction formulae are written using special notation. In the above example m and n are indices of x and $\log x$ respectively and the given integral may be written in the form

$$I_{m,n} = \int x^m (\log x)^n dx. \quad \text{Therefore, } I_{m,n-1} = \int \cdot x^m (\log x)^{n-1} dx.$$

Thus the reduction formula of the example 4.1 may be written as

$$\mathbf{I_{m,n}} = \frac{x^{m+1}}{m+1} \cdot (\log x)^n - \frac{n}{m+1} \mathbf{I_{m,n-1}}. \quad \quad ...(4.2)$$

Alternative notations are

$$U_{m,n} = \frac{x^{m+1}}{m+1} (\log x)^n - \frac{n}{m+1} U_{m,n-1},$$

or $\quad f(m,n) = \dfrac{x^{m+1}}{m+1} (\log x)^n - \dfrac{n}{m+1} f(m, n-1)$ etc.

4.3 Reduction formula for $\int \sin^n x\, dx$

and hence evaluate $\int \sin^6 x\, dx$. $\quad\quad$ (A.M.I.E., W-2005)

We write $\int \sin^n x\, dx = \int \sin^{n-1} x \cdot \sin x\, dx$. Integrating by parts, we get

$$\int \sin^n x\, dx = \sin^{n-1} x \cdot (-\cos x) - \int (n-1) \sin^{n-2} x \cdot \cos x (-\cos x)\, dx$$

$$= -\cos x \cdot \sin^{n-1} x + (n-1) \int \sin^{n-2} x \cdot \cos^2 x\, dx$$

$$= -\cos x \cdot \sin^{n-1} x + (n-1) \int \sin^{n-2} x (1 - \sin^2 x)\, dx$$

$$= -\cos x \cdot \sin^{n-1} x + (n-1) \int \sin^{n-2} x\, dx - (n-1) \int \sin^n x\, dx$$

or, $\quad (n-1+1) \displaystyle\int \sin^n x\, dx = -\cos x \cdot \sin^{n-1} x + (n-1) \int \sin^{n-2} x\, dx$

or, $\quad \displaystyle\int \mathbf{\sin^n x\, dx} = -\frac{\mathbf{\cos x \cdot \sin^{n-1} x}}{\mathbf{n}} + \frac{\mathbf{(n-1)}}{\mathbf{n}} \int \mathbf{\sin^{n-2} x\, dx} \quad\quad ...(4.3)$

which is the required reduction formula.

In abridged notation (4.3) may be written as

$$I_n = \int \sin^n x\, dx = -\frac{1}{n} \sin^{n-1} x \cos x + \left(\frac{n-1}{n}\right) I_{n-2} \quad\quad \text{(A.M.I.E., S-2003)}$$

Putting $n = 6, 4, 2$ successively in (4.3), we get

$$I_6 = \int \sin^6 x\,dx = -\frac{\cos x \sin^5 x}{6} + \frac{5}{6}\int \sin^4 x\,dx$$

$$I_4 = \int \sin^4 x\,dx = -\frac{\cos x \sin^3 x}{4} + \frac{3}{4}\int \sin^2 x\,dx$$

$$I_2 = \int \sin^2 x\,dx = -\frac{\cos x \sin x}{2} + \frac{1}{2}\int \sin^0 x\,dx = -\frac{\cos x \sin x}{2} + \frac{1}{2}x$$

Therefore, $I_4 = -\dfrac{\cos x \sin^3 x}{4} - \dfrac{3}{8}\cos x \sin x + \dfrac{3}{8}x$

and $\quad I_6 = -\dfrac{\cos x \sin^5 x}{6} - \dfrac{5}{24}\cos x \sin^3 x - \dfrac{5}{16}\cos x \sin x + \dfrac{5}{16}x.$

4.4 Reduction formula for $\int \cos^n x\,dx$

Let $\quad I_n = \int \cos^n x\,dx = \int \cos^{n-1} x \cdot \cos x\,dx$

$$= \cos^{n-1} x (\sin x) - \int (n-1)\cos^{n-2} x(-\sin x)\cdot \sin x\,dx \text{ (integrating by parts)}$$

$$= \sin x \cdot \cos^{n-1} x + \int (n-1)\cos^{n-2} x \cdot \left(1 - \cos^2 x\right)dx$$

$$= \sin x \cos^{n-1} x + (n-1)\int \cos^{n-2} x\,dx - (n-1)\int \cos^n x\,dx$$

or $\quad (n-1+1)\int \cos^n x\,dx = \sin x \cdot \cos^{n-1} x + (n-1)\int \cos^{n-2} x\,dx$

$$\therefore \quad \int \cos^n x\,dx = \frac{\sin x \cdot \cos^{n-1} x}{n} + \frac{n-1}{n}\int \cos^{n-2} x\,dx. \qquad \text{...(4.4)}$$

In abridged notation (4.4) may be written as, $nI_n = \sin x \cos^{n-1}x + (n-1)I_{n-2}$.

4.5 Evaluation of $\displaystyle\int_0^{\pi/2} \sin^n x\,dx$, n being a positive integer

By formula 4.3, we have

$$\int \sin^n x\,dx = -\frac{\cos x \sin^{n-1} x}{n} + \frac{n-1}{n}\int \sin^{n-2} x\,dx$$

Therefore, $\displaystyle\int_0^{\pi/2} \sin^n x\, dx = \left[\frac{-\cos x \sin^{n-1} x}{n}\right]_0^{\pi/2} + \frac{n-1}{n}\int_0^{\pi/2} \sin^{n-2} x\, dx$

$$= \frac{n-1}{n}\int_0^{\pi/2} \sin^{n-2} x\, dx$$

Denoting $\displaystyle\int_0^{\pi/2} \sin^n x\, dx$ by I_n we have $I_n = \dfrac{n-1}{n} I_{n-2}$.

Changing n to $n-2$ we get $I_{n-2} = \dfrac{n-3}{n-2} I_{n-4}$.

Similarly $I_{n-4} = \dfrac{n-5}{n-4} I_{n-6}$

... ...

... ...

$I_3 = (2/3) I_1$, if n is odd. $I_2 = (1/2) I_0$, if n is even.

Thus, we have $I_n = \begin{cases} \dfrac{n-1}{n}\cdot\dfrac{n-3}{n-2}\cdot\dfrac{n-5}{n-4}\cdots\dfrac{2}{3} I_1, & \text{when } n \text{ is odd,} \\[3mm] \dfrac{n-1}{n}\cdot\dfrac{n-3}{n-2}\cdot\dfrac{n-5}{n-4}\cdots\dfrac{1}{2} I_0, & \text{when } n \text{ is even.} \end{cases}$

Now $I_1 = \displaystyle\int_0^{\pi/2} \sin x\, dx = \left[-\cos x\right]_0^{\pi/2} = 1$, and $I_0 = \displaystyle\int_0^{\pi/2} \sin^0 x\, dx = \int_0^{\pi/2} 1\, dx = \pi/2$.

$\therefore\ I_n = \displaystyle\int_0^{\pi/2} \sin^n x\, dx = \begin{cases} \dfrac{n-1}{n}\cdot\dfrac{n-3}{n-2}\cdot\dfrac{n-5}{n-4}\cdots\dfrac{2}{3}, & \textbf{when n is odd.} \\[3mm] \dfrac{n-1}{n}\cdot\dfrac{n-3}{n-2}\cdot\dfrac{n-5}{n-4}\cdots\dfrac{3}{4}\cdot\dfrac{1}{2}\cdot\dfrac{\pi}{2}, & \textbf{when n is even.} \end{cases}$ ...(4.5)

Remark: The rule to write the value of $\displaystyle\int_0^{\pi/2} \sin^n x\, dx$ obviously, is as follows:

$$\int_0^{\pi/2} \sin^n x\, dx = \frac{(n-1)\times \textit{factors decreasing by 2}}{n \times \textit{factors decreasing by 2}} \times \frac{\pi}{2},$$

We should bear in mind that the factors in numerator and denominator are to be continued so long as they are positive that is, there should be no negative or zero factors. Thus if the factors (whether in the numerator or denominator) are even, the last factor will be 2 and if the factors are odd, the last factor is 1. We multiply by $\pi/2$ only if n is even.

4.6 Evaluation of $\int_0^{\pi/2} \cos^n x \, dx$, n being a positive integer

Putting, $x = (\pi/2) - y$ so that $dx = -dy$. When $x = 0$, $y = \pi/2$; and when

$x = \pi/2$, $y = 0$, we obtain $\displaystyle\int_0^{\pi/2} \cos^n x \, dx = -\int_{\pi/2}^0 \cos^n (\pi/2 - y) \, dy$

$$= \int_0^{\pi/2} \sin^n y \, dy \qquad \left[\because \int_a^b f(x) \, dx = -\int_b^a f(x) \, dx \right]$$

Therefore, $\displaystyle\int_0^{\pi/2} \cos^n x \, dx = \begin{cases} \dfrac{n-1}{n} \cdot \dfrac{n-3}{n-2} \cdot \dfrac{n-5}{n-4} \cdots \dfrac{2}{3}, & \text{when n is odd.} \\[3mm] \dfrac{n-1}{n} \cdot \dfrac{n-3}{n-2} \cdot \dfrac{n-5}{n-4} \cdots \dfrac{1}{2} \cdot \dfrac{\pi}{2}, & \text{when n is even.} \end{cases}$ ...(4.6)

[Proved in Art. 4.5]

Remark. From Formulas 4.5 and 4.6, we find that $\displaystyle\int_0^{\pi/2} \sin^n x \, dx = \int_0^{\pi/2} \cos^n x \, dx$.

Example 4.2 Evaluate the following definite integrals

(i) $\displaystyle\int_0^{\pi/2} \sin^7 x \, dx$. (ii) $\displaystyle\int_0^{\pi/2} \sin^6 x \, dx$. (A.M.I.E., S-2004)

(iii) $\displaystyle\int_0^{\pi/2} \cos^9 x \, dx$. (Madras 1998) (iv) $\displaystyle\int_0^{\pi/2} \cos^6 x \, dx$. (A.M.I.E.. W-2005)

(v) $\displaystyle\int_0^{\pi} (1 - \cos\theta)^3 \, d\theta$.

Solution. (i) $\displaystyle\int_0^{\pi/2} \sin^7 x \, dx = \frac{6}{7} \cdot \frac{4}{5} \cdot \frac{2}{3} = \frac{16}{35}$. (ii) $\displaystyle\int_0^{\pi/2} \sin^6 x \, dx = \frac{5}{6} \cdot \frac{3}{4} \cdot \frac{1}{2} \cdot \frac{\pi}{2} = \frac{5\pi}{32}$.

(iii) $\displaystyle\int_0^{\pi/2} \cos^9 x\,dx = \frac{8}{9}\cdot\frac{6}{7}\cdot\frac{4}{5}\cdot\frac{2}{3} = \frac{128}{315}$. *(iv)* $\displaystyle\int_0^{\pi/2} \cos^6 x\,dx = \frac{5}{6}\cdot\frac{3}{4}\cdot\frac{1}{2}\cdot\frac{\pi}{2} = \frac{5\pi}{32}$.

(v) We have $\displaystyle\int_0^{\pi}(1-\cos\theta)^3\,d\theta = \int_0^{\pi}\left(2\sin^2\frac{\theta}{2}\right)^3 d\theta = 8\int_0^{\pi}\sin^6\frac{\theta}{2}\,d\theta$

Put $\theta/2 = x$, then $d\theta = 2dx$. Also when $\theta = 0$, $x = 0$; when $\theta = \pi$, $x = \pi/2$.

$\therefore \displaystyle\int_0^{\pi}(1-\cos\theta)^3\,d\theta = 16\int_0^{\pi/2}\sin^6 x\,dx = 16\cdot\frac{5\cdot3\cdot1}{6\cdot4\cdot2}\cdot\frac{\pi}{2} = \frac{5\pi}{2}$.

4.7 Reduction formula for $\int \sin^p x\cdot\cos^q x\,dx$

(A) Let $\displaystyle I_{(p,q)} = \int \sin^p x\cdot\cos^q x\,dx = \int \cos^{q-1} x\cdot\left(\sin^p x\cdot\cos x\right)dx$

$\displaystyle = \cos^{q-1} x\cdot\frac{\sin^{p+1} x}{p+1} - \int (q-1)\cos^{q-2} x(-\sin x)\cdot\frac{\sin^{p+1} x}{p+1}\,dx$

$\displaystyle = \frac{\sin^{p+1} x\cdot\cos^{q-1} x}{p+1} + \frac{q-1}{p+1}\int \cos^{q-2} x\cdot\sin^{p+2} x\,dx$

$\displaystyle = \frac{\sin^{p+1} x\cdot\cos^{q-1} x}{p+1} + \frac{q-1}{p+1}\int \cos^{q-2} x\cdot\left(\sin^2 x\right)\sin^p x\,dx$

$\displaystyle = \frac{\sin^{p+1} x\cdot\cos^{q-1} x}{p+1} + \frac{q-1}{p+1}\int \cos^{q-2} x\cdot\left(1-\cos^2 x\right)\cdot\sin^p x\,dx$

or $\displaystyle\int \sin^p x\cos^q x\,dx = \frac{\sin^{p+1} x\cos^{q-1} x}{p+1} + \frac{q-1}{q+1}\int \cos^{q-2} x\sin^p x\,dx - \frac{q-1}{p+1}\int \cos^q x\sin^p x\,dx.$

Transposing the last integral to the left, we obtain

$\displaystyle\left(1+\frac{q-1}{p+1}\right)\int \sin^p x\cos^q x\,dx = \frac{\sin^{p+1} x\cos^{q-1} x}{p+1} + \frac{q-1}{p+1}\int \cos^{q-2} x\sin^p x\,dx$

$$...(4.7)$$

Now $1+\dfrac{q-1}{p+1} = \dfrac{p+q}{p+1}$. Multiplying both sides of (4.7) by $\dfrac{(p+1)}{(p+q)}$, we obtain

$$\int \sin^p x \cos^q x\, dx = \frac{\sin^{p+1} x \cos^{q-1} x}{p+q} + \frac{q-1}{p+q} \int \sin^p x \cos^{q-2} x\, dx$$

or $\quad I_{(p, q)} = \dfrac{\sin^{p+1} x \cos^{q-1} x}{p+q} + \dfrac{q-1}{p+q} I_{(p, q-2)},$ \hfill ...(4.8)

which is the required reduction formula.

(B) If we write $\displaystyle\int \sin^p x \cos^q x\, dx = \int \sin^{p-1} x \cdot \left(\cos^q x \sin x\right) dx$

and integrate by parts, proceeding as in (A), we get

$$\int \sin^p x \cos^q x\, dx = -\frac{\cos^{q+1} x \sin^{p-1} x}{p+q} + \frac{p-1}{p+q} \int \sin^{p-2} x \cos^q x\, dx$$

or $\qquad I_{(p, q)} = \dfrac{-\sin^{p-1} x \cos^{q+1} x}{p+q} + \dfrac{p-1}{p+q} I_{(p-2, q)},$ \hfill ...(4.9)

which is another form of the reduction formula.

4.8 Evaluation of $\displaystyle\int_0^{\pi/2} \sin^p x \cos^q x\, dx$, p and q being positive integers

(A.M.I.E., W-2009)

Using formula (4.9), we obtain the reduction formula

$$\int \sin^p x \cos^q x\, dx = -\frac{\cos^{q+1} x \sin^{p-1} x}{p+q} + \frac{p-1}{p+q} \int \sin^{p-2} x \cos^q x\, dx$$

$$\therefore \int_0^{\pi/2} \sin^p x \cos^q x\, dx = \left[-\frac{\cos^{q+1} x \sin^{p-1} x}{p+q}\right]_0^{\pi/2} + \frac{p-1}{p+q} \int_0^{\pi/2} \sin^{p-2} x \cos^q x\, dx$$

$$= \frac{p-1}{p+q} \int_0^{\pi/2} \sin^{p-2} x \cos^q x\, dx \hfill ...(4.10)$$

If we denote $\displaystyle\int_0^{\pi/2} \sin^p x \cos^q x\, dx$ by $I_{p, q}$ then (4.10) can be written in the form

$$I_{p, q} = \frac{p-1}{p+q} \cdot I_{p-2, q} \quad \text{Changing } p \text{ to } p-2, p-4 \text{ etc. we have}$$

$$I_{p-2, q} = \frac{p-3}{p+q-2} \cdot I_{p-4, q}$$

$$I_{p-4,\,q} = \frac{p-5}{p+q-4} I_{p-6,\,q}$$

… … … … … … … …
… … … … … … …, …

Finally
$$\begin{cases} I_{3,\,q} = \dfrac{2}{3+q} \cdot I_{1,\,q} \text{ when } p \text{ is odd} \\[3mm] I_{2,\,q} = \dfrac{1}{2+q} \cdot I_{0,\,q} \text{ when } p \text{ is even} \end{cases}$$

$$\therefore \quad I_{p,\,q} = \begin{cases} \dfrac{p-1}{p+q} \cdot \dfrac{p-3}{p+q-2} \,\cdots\cdots\, \dfrac{2}{q+3} \cdot I_{1,\,q} \text{ when } p \text{ is odd} \\[3mm] \dfrac{p-1}{p+q} \cdot \dfrac{p-3}{p+q-2} \,\cdots\cdots\, \dfrac{1}{q+2} \cdot I_{0,\,q} \text{ when } p \text{ is even} \end{cases}$$

Now,
$$I_{1,\,q} = \int_0^{\pi/2} \sin x \cos^q x \, dx = -\left[\frac{\cos^{q+1} x}{q+1} \right]_0^{\pi/2} = \frac{1}{q+1}$$

and
$$I_{0,\,q} = \int_0^{\pi/2} \sin^0 x \cos^q x \, dx = \int_0^{\pi/2} \cos^q x \, dx$$

$$\left. \begin{aligned} &= \frac{q-1}{q} \cdot \frac{q-3}{q-2} \cdots \frac{2}{3} \text{ when } q \text{ is odd.} \\[2mm] &= \frac{q-1}{q} \cdot \frac{q-3}{q-2} \cdots \frac{1}{2} \cdot \frac{\pi}{2} \text{ when } q \text{ is even} \end{aligned} \right] \text{(By Art 4.6)}$$

(*i*) Thus *when p is odd*, we have

$$\mathbf{I_{p,q}} = \frac{p-1}{p+q} \cdot \frac{p-3}{p+q-2} \cdot \frac{p-5}{p+q-4} \cdots \frac{2}{3+q} \cdot \frac{1}{q+1},$$

where *q* may be even or odd.

(*ii*) When *p is even and q is odd*, we have

$$\mathbf{I_{p,\,q}} = \frac{(p-1)(p-3)(p-5)\ldots\ldots1}{(p+q)(p+q-2)(p+q-4)\ldots\ldots2+q} \cdot \frac{(q-1)(q-3)\ldots\ldots2}{q(q-2)\ldots\ldots3}$$

(*iii*) When *p and q are both even*

$$\mathbf{I_{p,\,q}} = \frac{(p-1)(p-3)(p-5)\ldots\ldots1}{(p+q)(p+q-2)(p+q-4)\ldots\ldots2+q} \cdot \frac{(q-1)(q-3)\ldots\ldots1}{q(q-2)\ldots\ldots2} \cdot \frac{\pi}{2}.$$

Remark: *Just like the rule for* $\displaystyle\int_0^{\pi/2} \sin^n x \, dx$ *and* $\displaystyle\int_0^{\pi/2} \cos^n x \, dx$ *as mentioned*

in the Remark (Art. 4.5), we have the following rule for writing down the value

of the integral $\displaystyle\int_0^{\pi/2} \sin^p x \cos^q x \, dx = \dfrac{(p-1)(p-3)\ldots\ldots(q-1)(q-3)\ldots\ldots}{(p+q)(p+q-2)\ldots\ldots}$

$$= \dfrac{(p-1)\times \text{factors decreasing by } 2\ldots(q-1)\times \text{factors decreasing by } 2\ldots}{(p+q)\times \text{factors decreasing by } 2\ldots} \ldots(4.11)$$

ultimately followed by the factor $\pi/2$ *only when p and q are both even. We must bear in mind that no factor in numerator and denominator should be zero or negative.*

Example 4.3 **Evaluate the following definite integrals**

(i) $\displaystyle\int_0^{\pi/2} \sin^5 x \cos^6 x \, dx,$ (ii) $\displaystyle\int_0^{\pi/2} \sin^4 x \cos^3 x \, dx,$ (A.M.I.E., S-2004)

(iii) $\displaystyle\int_0^{\pi/2} \sin^7 x \cos^5 x \, dx,$ (iv) $\displaystyle\int_0^{\pi/2} \sin^6 x \cos^4 x \, dx,$ (v) $\displaystyle\int_0^2 x^3 \sqrt{2-x} \, dx.$

Solution. (i) $\displaystyle\int_0^{\pi/2} \sin^5 x \cos^6 x \, dx = \dfrac{4\cdot2\cdot5\cdot3\cdot1}{11\cdot9\cdot7\cdot5\cdot3\cdot1} = \dfrac{8}{693}.$

(Since both indices are not even, the result is not multiplied by $\pi/2$).

(ii) $\displaystyle\int_0^{\pi/2} \sin^4 x \cos^3 x \, dx = \dfrac{3\cdot1\cdot2}{7\cdot5\cdot3\cdot1} = \dfrac{2}{35}.$

(iii) $\displaystyle\int_0^{\pi/2} \sin^7 x \cos^5 x \, dx = \dfrac{6\cdot4\cdot2\cdot4\cdot2}{12\cdot10\cdot8\cdot6\cdot4\cdot2} = \dfrac{1}{120}.$

(iv) $\displaystyle\int_0^{\pi/2} \sin^6 x \cos^4 x \, dx = \dfrac{5\cdot3\cdot1\cdot3\cdot1}{10\cdot8\cdot6\cdot4\cdot2}\cdot\dfrac{\pi}{2} = \dfrac{3\pi}{512}.$

(v) Put $x = 2\sin^2\theta$ then $dx = 4\sin\theta\cos\theta \, d\theta.$ When $x = 0, \theta = 0$; when $x = 2, \theta = \pi/2.$ Therefore the given integral become

$$\int_0^{\pi/2} 2^3 \cdot \sin^6 \theta \cdot 2^{1/2} \cos\theta \cdot 4\sin\theta\cos\theta \, d\theta = 32\sqrt{2} \int_0^{\pi/2} \sin^7\theta \cos^2\theta \, d\theta$$

$$= 32\sqrt{2} \cdot \frac{6.4.2}{9.7.5.3} = \frac{512\sqrt{2}}{315}.$$

4.9 Reduction formula for $\int \tan^n x \, dx$ (A.M.I.E., S-2000, S-1996)

We have $\int \tan^n x \, dx = \int \tan^{n-2} x \, \tan^2 x \, dx = \int \tan^{n-2} x \left(\sec^2 x - 1 \right) dx$

$$= \int \tan^{n-2} x \sec^2 x \, dx - \int \tan^{n-2} x \, dx$$

or $\quad \int \tan^n x \, dx = \dfrac{\tan^{n-1} x}{n-1} - \int \tan^{n-2} x \, dx,$ ...(4.12)

which is the required reduction formula.
By using abridged notation, we have the reduction formula,

$$(n-1)I_n = \tan^{n-1} x - (n-1)I_{n-2} \qquad\qquad ...(4.13)$$

4.10 Reduction formula for $\int \cot^n x \, dx$ (A.M.I.E., W-2003, W-2002)

$$\int \cot^n x \, dx = \int \cot^{n-2} x \cdot \cot^2 x \, dx = \int \cot^{n-2} x \cdot \left(\operatorname{cosec}^2 x - 1 \right) dx$$

$$= \int \cot^{n-2} x \operatorname{cosec}^2 x \, dx - \int \cot^{n-2} x \, dx$$

$$= -\frac{\cot^{n-1} x}{n-1} - \int \cot^{n-2} x \, dx, \qquad\qquad ...(4.14)$$

which is the required reduction formula.

Example 4.4 (*a*) **Find a reduction formula for** $\int \tan^n x \, dx$ **and hence evaluate**

$\int \tan^5 x \, dx,$ *(A.M.I.E., S-1996)* $\qquad \int \tan^6 x \, dx.$

(*b*) **Evaluate** (*i*) $\displaystyle\int_0^{\pi/4} \tan^5 x \, dx$ *(A.M.I.E., W-2003)* (*ii*) $\displaystyle\int_0^{\pi/4} \tan^4 x \, dx$

(A.M.I.E., W-2005; W-2002)

Solution. (*a*) Refer Article 4.9.

$$\int \tan^n x \, dx = \frac{\tan^{n-1} x}{n-1} - \int \tan^{n-2} x \, dx \qquad ...(i)$$

Putting $n = 5, 3$ successively in (i), we obtain

$$\int \tan^5 x \, dx = \frac{\tan^4 x}{4} - \int \tan^3 x \, dx \qquad ...(ii)$$

$$\int \tan^3 x \, dx = \frac{\tan^2 x}{2} - \int \tan x \, dx = \frac{\tan^2 x}{2} - \log \sec x. \qquad ...(iii)$$

From (ii) and (iii), we obtain

$$\int \tan^5 x \, dx = \frac{\tan^4 x}{4} - \frac{\tan^2 x}{2} + \log \sec x = \frac{\tan^4 x}{4} - \frac{\tan^2 x}{2} - \log \cos x,$$

Now put $n = 6, 4, 2$ successively in (i), we obtain

$$I_6 = \frac{\tan^5 x}{5} - I_4, \qquad ...(iv) \qquad\qquad I_4 = \frac{\tan^3 x}{3} - I_2, \qquad ...(v)$$

$$I_2 = \tan x - I_0 \qquad ...(vi) \quad \text{But} \quad I_0 = \int (\tan x)^0 \, dx = \int dx = x.$$

Therefore, from (vi) $I_2 = \tan x - x$,

From (v) $\qquad I_4 = \dfrac{\tan^3 x}{3} - (\tan x - x) = \dfrac{\tan^3 x}{3} - \tan x + x,$

and from (iv), $\qquad I_6 = \dfrac{\tan^5 x}{5} - \dfrac{\tan^3 x}{3} + \tan x - x,$ the required result.

(b) $\quad (i)$ $\displaystyle\int_0^{\pi/4} \tan^5 x \, dx = \left[\frac{\tan^4 x}{4} - \frac{\tan^2 x}{2} - \log \cos x \right]_0^{\pi/4}$

$$= \left(\frac{1}{4} - \frac{1}{2} - \log \frac{1}{\sqrt{2}} \right) = -\frac{1}{4} + \frac{1}{2} \log 2.$$

(ii) $\displaystyle\int_0^{\pi/4} \tan^4 x \, dx = \left[\frac{\tan^3 x}{3} - \tan x + x \right]_0^{\pi/4} = \frac{3\pi - 8}{12}.$

4.11 Reduction formulae for $\int \sec^n x \, dx$ and $\int \operatorname{cosec}^n x \, dx$ (A.M.I.E., W-2001)

We have $\qquad \displaystyle\int \underset{\text{I}}{\sec^{n-2} x} \cdot \underset{\text{II}}{\sec^2 x} \, dx = \int \sec^{n-2} x \cdot \sec^2 x \, dx.$

Integrating by parts, we obtain

$$\int \sec^n x\, dx = \tan x \cdot \sec^{n-2} x - \int \tan x \cdot (n-2)\sec^{n-2} x \cdot \tan x\, dx$$

$$= \tan x \cdot \sec^{n-2} x - (n-2)\int \sec^{n-2} x\left(\sec^2 x - 1\right) dx$$

$$= \tan x \cdot \sec^{n-2} x - (n-2)\int \left[\sec^n x - \sec^{n-2} x\right] dx$$

Transposing $(n-2)\int \sec^n x\, dx$ to the left and dividing by

$1 + (n-2)$ i.e., $n - 1$, we obtain

$$\int \sec^n x\, dx = \frac{\tan x \cdot \sec^{n-2} x}{n-1} + \frac{n-2}{n-1}\int \sec^{n-2} x\, dx, \qquad \qquad ...(4.15)$$

which is the required reduction formula.

In abridged notation, this may be written as under:

$$\mathbf{(n-1)I_n = \tan x \sec^{n-2} x + (n-2)I_{n-2}, \quad n > 2} \; ...(4.16) \; (A.M.I.E., W\text{-}2003)$$

Similarly, it can be proved that

$$\int \mathbf{cosec}^n x\, dx = -\frac{\cot x\, \mathbf{cosec}^{n-2} x}{n-1} + \frac{n-2}{n-1}\int \mathbf{cosec}^{n-2} x\, dx. \qquad ...(4.17)$$

Example 4.5 **Evaluate** (a) $\displaystyle\int_0^{\pi/4} \sec^5 x\, dx$, (b) $\displaystyle\int_0^{\pi/4} \sec^6 x\, dx$. $\quad (A.M.I.E., S\text{-}2002)$

Solution. We first determine the indefinite integral $\int \sec^5 x\, dx$. Using formula
(4.15) and putting $n = 5, 3$, successively in it. We obtain,

$$\int \sec^5 x\, dx = \frac{\tan x \cdot \sec^3 x}{4} + \frac{3}{4}\int \sec^3 x\, dx \qquad ...(i)$$

$$\int \sec^3 x\, dx = \frac{\tan x \cdot \sec x}{2} + \frac{1}{2}\int \sec x\, dx = \frac{\tan x \sec x}{2} + \frac{1}{2}\log\tan\left(\frac{\pi}{4} + \frac{x}{2}\right) \qquad ...(ii)$$

From (i) and (ii), we obtain

$$\int \sec^5 x\, dx = \frac{\tan x \sec^3 x}{4} + \frac{3}{8}\tan x \sec x + \frac{3}{8}\log\tan\left(\frac{\pi}{4} + \frac{x}{2}\right).$$

$$\therefore \quad \int_0^{\pi/4} \sec^5 x\, dx = \frac{1}{4}\left[\tan x \sec^3 x + \frac{3}{2}\tan x \sec x + \frac{3}{2}\log\tan\left(\frac{\pi}{4} + \frac{x}{2}\right)\right]_0^{\pi/4}$$

$$= \frac{1}{4}\left[2\sqrt{2} + \frac{3}{2}\sqrt{2} + \frac{3}{2}\log\left(1 + \sqrt{2}\right)\right] = \frac{1}{4}\left[\frac{7}{2}\sqrt{2} + \frac{3}{2}\log\left(1 + \sqrt{2}\right)\right].$$

4.12 Reduction formula for $\int x^m \sin nx\, dx$. *(A.M.I.E., W-2004)*

Applying the rule of integration by parts, we obtain

$$\int x^m \sin nx\, dx = x^m\left(-\frac{\cos nx}{n}\right) - \int mx^{m-1}\left(-\frac{\cos nx}{n}\right)dx$$

$$= -\frac{x^m \cos nx}{n} + \frac{m}{n}\int x^{m-1}\cos nx\, dx \qquad \text{(Integrating again by parts)}$$

$$= -\frac{x^m \cos nx}{n} + \frac{m}{n}\left[x^{m-1}\frac{\sin nx}{n} - \int (m-1)x^{m-2}\frac{\sin nx}{n}dx\right]$$

$$= -\frac{x^m \cos nx}{n} + \frac{m}{n}\left[\frac{x^{m-1}\sin nx}{n} - \frac{m-1}{n}\int x^{m-2}\sin nx\, dx\right]$$

$$= -\frac{x^m \cos nx}{n} + \frac{m}{n^2}x^{m-1}\sin nx - \frac{m(m-1)}{n^2}\int x^{m-2}\sin nx\, dx,$$

By using abridged notation, we have the reduction formula,

$$I_{m,n} = \frac{-x^m \cos nx}{n} + \frac{m}{n^2}x^{m-1}\sin nx - \frac{m(m-1)}{n^2}I_{m-2},n. \qquad ...(4.18)$$

Similarly,

$$\int x^m \cos nx\, dx = \frac{nx^m \sin nx + m\, x^{m-1}\cos nx}{n^2} - \frac{m(m-1)}{n^2}\int x^{m-2}\cos nx\, dx.$$
$$...(4.19)$$

Example 4.6 If $I_n = \displaystyle\int_0^{\pi/2} x^n \sin x\, dx$, and $n > 1$, show that

$I_n + n(n-1)I_{n-2} = n(\pi/2)^{n-1}$. **Hence evaluate I_5.**

(GGSIPU 2006; AMIE, S-2001)

Solution. Integrating by parts

$$I_n = \left[x^n(-\cos x)\right]_0^{\pi/2} - \int_0^{\pi/2} nx^{n-1}(-\cos x)\, dx = n\int_0^{\pi/2} x^{n-1}\cos x\, dx.$$

Again, integrating by parts, we have

$$I_n = n\left[\left[x^{n-1}(\sin x)\right]_0^{\pi/2} - \int_0^{\pi/2}(n-1)x^{n-2}\cdot\sin x\,dx\right] = n\left[\left(\frac{\pi}{2}\right)^{n-1} - (n-1)I_{n-2}\right]$$

Therefore, $I_n + n(n-1)I_{n-2} = n(\pi/2)^{n-1}$ ...(i)

Putting $n = 5, 3, 1$ in (i) we have

$$I_5 + 5\cdot4\,I_3 = 5(\pi/2)^4 \;\;...(ii), \qquad I_3 + 3\cdot2I_1 = 3(\pi/2)^2 \;\;...(iii) \text{ and } I_1 = 1\cdot \;\;...(iv)$$

From (iii) and (iv) $\qquad\qquad I_3 = 3(\pi/2)^2 - 6$ ...(v)

From (ii) and (v) $I_5 + 20\left[3(\pi/2)^2 - 6\right] = 5\cdot(\pi/2)^4$

or $I_5 = 5\cdot\left(\pi^4/16\right) - 15\pi^2 + 120.$

4.13 Reduction formula for $\int e^{ax}\sin^n bx\,dx$.

$$\int e^{ax}\sin^n bx\,dx = \int \underset{I}{(\sin^n bx)}\cdot\underset{II}{e^{ax}}\,dx \qquad\qquad \text{(Integrating by parts)}$$

$$= (\sin^n bx)\cdot\frac{e^{ax}}{a} - \int(n\sin^{n-1}bx)(\cos bx)b\frac{e^{ax}}{a}\,dx$$

$$= \frac{e^{ax}\sin^n bx}{a} - \frac{nb}{a}\int \underset{I}{(\sin^{n-1}bx)}(\cos bx)\cdot\underset{II}{e^{ax}}\,dx \qquad \text{(Integrating again by parts)}$$

$$= \frac{e^{ax}\sin^n bx}{a} - \frac{nb}{a}\left[(\sin^{n-1}bx)(\cos bx)\cdot\frac{e^{ax}}{a}\right.$$

$$\left. -\int\left\{(n-1)\sin^{n-2}bx\cdot\cos bx\cdot b(\cos bx) + b\sin^{n-1}bx(-\sin bx)\right\}\frac{e^{ax}}{a}\,dx\right]$$

$$= \frac{e^{ax}\sin^n bx}{a} - \frac{nb}{a}\left[\frac{e^{ax}\sin^{n-1}bx\cdot\cos bx}{a}\right.$$

$$\left. -\int\left\{(n-1)b\sin^{n-2}bx\cdot\cos^2 bx - b\sin^n bx\right\}\frac{e^{ax}}{a}\,dx\right]$$

$$= \frac{e^{ax} \sin^n bx}{a} - \frac{nb}{a} \left[\frac{e^{ax} \sin^{n-1} bx \cdot \cos bx}{a} \right.$$

$$\left. - \int \left\{ (n-1) b \sin^{n-2} bx \left(1 - \sin^2 bx\right) - b \sin^n bx \right\} \frac{e^{ax}}{a} dx \right]$$

$$= \frac{e^{ax} \sin^n bx}{a} - \frac{nb}{a} \left[\frac{e^{ax} \sin^{n-1} bx \cdot \cos bx}{a} \right.$$

$$\left. - \int \left\{ (n-1) b \sin^{n-2} bx - nb \sin^n bx \right\} \frac{e^{ax}}{a} dx \right]$$

$$= \frac{e^{ax} \sin^n bx}{a} - \frac{nb}{a} \left[\frac{e^{ax} \sin^{n-1} bx \cdot \cos bx}{a} \right.$$

$$\left. - \frac{(n-1)b}{a} \int e^{ax} \sin^{n-2} bx \, dx + \frac{nb}{a} \int e^{ax} \sin^n bx \, dx \right]$$

$$= \frac{e^{ax} \sin^n bx}{a} - \frac{nb}{a^2} e^{ax} \sin^{n-1} bx \cos bx$$

$$+ \frac{n(n-1)b^2}{a^2} \int e^{ax} \sin^{n-2} bx \, dx - \frac{n^2 b^2}{a^2} \int e^{ax} \sin^n bx \, dx.$$

Transposing,

$$\left(1 + \frac{n^2 b^2}{a^2}\right) \int e^{ax} \sin^n bx \, dx$$

$$= \frac{a e^{ax} \sin^n bx - nb e^{ax} \sin^{n-1} bx \cdot \cos bx}{a^2} + \frac{n(n-1)b^2}{a^2} \int e^{ax} \sin^{n-2} bx \, dx$$

$$\therefore \int e^{ax} \sin^n bx \, dx$$

$$= \frac{e^{ax} \sin^{n-1} bx \left(a \sin bx - nb \cos bx\right)}{a^2 \left(1 + \dfrac{n^2 b^2}{a^2}\right)} + \frac{n(n-1)b^2}{a^2 \left(1 + \dfrac{n^2 b^2}{a^2}\right)} \int e^{ax} \sin^{n-2} bx \, dx$$

or
$$\int e^{ax} \sin^n bx \, dx = \frac{e^{ax} \sin^{n-1} bx \left(a \sin bx - nb \cos bx\right)}{a^2 + n^2 b^2}$$

$$+\frac{n(n-1)b^2}{a^2+n^2b^2}\int e^{ax}\sin^{n-2}bx\,dx, \qquad\qquad ...(4.20)$$

which is the required reduction formula.

4.14 Reduction formula for $\int \cos^m x \cos nx\,dx$

Applying the rule of integration by parts, we obtain

$$\int \underset{I}{\cos^m x}\,\underset{II}{\cos nx}\,dx = \cos^m x \cdot \frac{\sin nx}{n} + \frac{m}{n}\int \cos^{m-1} x \cdot \sin x \cdot \sin nx\,dx. \qquad ...(4.21)$$

Now, $\qquad\qquad\qquad \cos(n-1)x = \cos nx \cos x + \sin nx \sin x.$

Replacing $\sin nx \sin x$ by $\cos(n-1)x - \cos nx \cos x$ in the integral on the right of (4.21), we obtain

$$\int \cos^m x \cos nx\,dx = \frac{\cos^m x \sin nx}{n}$$

$$+\frac{m}{n}\int \cos^{m-1} x \cos(n-1)x\,dx - \frac{m}{n}\int \cos^m x \cos nx\,dx$$

or $\left(1+\dfrac{m}{n}\right)\displaystyle\int \cos^m x \cos nx\,dx = \dfrac{\cos^m x \sin nx}{n} + \dfrac{m}{n}\int \cos^{m-1} x \cos(n-1)x\,dx.$

$$\therefore \int \cos^m x \cos nx\,dx = \frac{\cos^m x \sin nx}{m+n} + \frac{m}{m+n}\int \cos^{m-1} x \cos(n-1)x\,dx, \quad ...(4.22)$$

which is the required reduction formula.
By using abridged notation, we can write (4.22) as

$$I_{m,n} = \frac{\cos^m x \sin nx}{m+n} + \frac{m}{m+n}I_{m-1,\,n-1} \qquad\qquad ...(4.23)$$

4.15 Reduction formula for $\int \dfrac{\sin nx}{\sin x}\,dx$

Let us connect $\displaystyle\int \frac{\sin nx}{\sin x}\,dx$ with $\displaystyle\int \frac{\sin(n-2)x}{\sin x}\,dx$

Now $\qquad \dfrac{\sin nx - \sin(n-2)x}{\sin x} = \dfrac{2\cos(n-1)x \sin x}{\sin x} = 2\cos(n-1)x$

That is.,
$$\frac{\sin nx}{\sin x} - \frac{\sin(n-2)x}{\sin x} = 2\cos(n-1)x.$$

Integrating both sides w.r.t. x, we obtain

$$\int \frac{\sin nx}{\sin x}\,dx - \int \frac{\sin(n-2)x}{\sin x}\,dx = 2\int \cos(n-1)x\,dx = 2\frac{\sin(n-1)x}{n-1}.$$

Transposing, $\displaystyle \int \frac{\sin nx}{\sin x}\,dx = 2\frac{\sin(n-1)x}{n-1} + \int \frac{\sin(n-2)x}{\sin x}\,dx,$ ...(4.24)

which is the required reduction formula.

Example 4.7 Prove that $\displaystyle \int_0^\pi \frac{\sin nx}{\sin x}\,dx = 0$ **or** π, **according as n is an even or odd integer.**

(A.M.I.E., W-2004)

Solution. Let $\displaystyle I_n = \int_0^\pi \frac{\sin nx}{\sin x}\,dx$. Using (4.24), we have

$$I_n = \left[2\frac{\sin(n-1)x}{n-1}\right]_0^\pi + I_{n-2} = I_{n-2} \quad \text{that is.,} \quad I_n = I_{n-2}.$$

Hence $\quad I_{n-2} = I_{n-4},\qquad\qquad I_{n-4} = I_{n-6}$ etc.

Finally $\begin{cases} I_4 = I_2 \text{ if } n \text{ is even,}\\ I_3 = I_1 \text{ if } n \text{ is odd.} \end{cases}$ Now $\displaystyle I_2 = \int_0^\pi \frac{\sin 2x}{\sin x}\,dx = \int_0^\pi 2\cos x\,dx = 2[\sin x]_0^\pi = 0.$

$$I_1 = \int_0^\pi \frac{\sin x}{\sin x}\,dx = \int_0^\pi dx = \pi. \quad \text{Hence the result.}$$

4.16 Illustrative Examples

Example 4.8 Evaluate the following definite integrals

(i) $\displaystyle \int_0^{\pi/2} \sin^4 x \cos^8 x\,dx,$ (ii) $\displaystyle \int_0^{\pi/2} \sin^5 x \cos^8 x\,dx,$ (iii) $\displaystyle \int_0^{\pi/4} \cos^3 2x \sin^4 4x\,dx.$

Solution. (i) Here both indices are even, therefore

$$\int_0^{\pi/2} \sin^4 x\cos^8 x\,dx = \frac{3\cdot1\cdot7\cdot5\cdot3\cdot1}{12\cdot10\cdot8\cdot6\cdot4\cdot2}\cdot\frac{\pi}{2} = \frac{7\pi}{2048}.$$

(ii) $\displaystyle\int_0^{\pi/2} \sin^5 x \cos^8 x\, dx = \frac{4\cdot2\cdot7\cdot5\cdot3\cdot1\cdot}{13\cdot11\cdot9\cdot7\cdot5\cdot3\cdot1} = \frac{1}{1287}.$

(iii) $\displaystyle\int_0^{\pi/4} \cos^3 2x \cdot \sin^4 4x\, dx = \int_0^{\pi/4} \cos^3 2x \cdot (2\sin 2x \cos 2x)^4\, dx = 16\int_0^{\pi/4} \cos^7 2x \cdot \sin^4 2x\, dx.$

Put $2x = t$ so that $2\,dx = dt$. When $x = 0$, $t = 0$ and when $x = \pi/4$, $t = \pi/2$.

$\therefore$ The given integral $= 8\displaystyle\int_0^{\pi/2} \cos^7 t \sin^4 t\, dt = 8\left[\frac{6\cdot4\cdot2\cdot3\cdot1}{11\cdot9\cdot7\cdot5\cdot3}\right] = \frac{128}{1155}.$

Example 4.9 Evaluate the following definite integrals

(i) $\displaystyle\int_0^1 x^2 (1-x)^{3/2}\, dx$, (ii) $\displaystyle\int_0^2 \frac{x^4}{\sqrt{4-x^2}}\, dx$, (iii) $\displaystyle\int_0^{2a} x^3 \sqrt{2ax - x^2}\, dx.$

(A.M.I.E., 2004)

Solution. (i) Put $x = \sin^2\theta$, then $dx = 2\sin\theta\cos\theta\, d\theta$; when $x = 0$, $\theta = 0$ and when $x = 1$, $\theta = \pi/2$.

$\therefore \displaystyle\int_0^1 x^2 (1-x)^{3/2}\, dx = 2\int_0^{\pi/2} \sin^5\theta \cos^4\theta\, d\theta = 2\cdot\frac{4\cdot2\cdot3\cdot1}{9\cdot7\cdot5\cdot3\cdot1} = \frac{16}{315}.$

(ii) Put $x = 2\sin\theta$, then $dx = 2\cos\theta\, d\theta$.

When $x = 0$, $\theta = 0$ and when $x = 2$, $\theta = \pi/2$.

$\therefore \displaystyle\int_0^2 \frac{x^4}{\sqrt{4-x^2}}\, dx = \int_0^{\pi/2} \frac{(2\sin\theta)^4}{\sqrt{4-4\sin^2\theta}} \cdot 2\cos\theta\, d\theta = 16\int_0^{\pi/2} \sin^4\theta\, d\theta = 3\pi.$

(iii) Let $x = 2a\sin^2\theta$; then $dx = 4a\sin\theta\cos\theta\, d\theta$.

Also $\theta = 0$ when $x = 0$ and $\theta = \pi/2$ when $x = 2a$.

$\therefore \displaystyle\int_0^{2a} x^3 \sqrt{2ax - x^2}\, dx = \int_0^{\pi/2} (2a)^3 \sin^6\theta\sqrt{4a^2\sin^2\theta - 4a^2\sin^4\theta}\cdot 4a\sin\theta\cos\theta\, d\theta$

$= 2^6 a^5 \displaystyle\int_0^{\pi/2} \sin^8\theta\cos^2\theta\, d\theta = 64a^5 \cdot \frac{7\cdot5\cdot3\cdot1}{10\cdot8\cdot6\cdot4\cdot2}\cdot\frac{\pi}{2} = \frac{7\pi a^5}{8}.$

Example 4.10 If $I_n = \int_0^{\pi/2} x \cos^n x\, dx$, **prove that** $n^2 I_n = n(n-1) I_{n-2} - 1$.

Hence evaluate I_4.

(D.C.E., 2004)

Solution. We have $I_n = \int_0^{\pi/2} x \cos^n x\, dx = \int_0^{\pi/2} \left(x \cos^{n-1} x \right) \cos x\, dx$.

Integrating by parts, the above integral

$$= \left[x \cos^{n-1} x \cdot \sin x \right]_0^{\pi/2} - \int_0^{\pi/2} \left[\cos^{n-1} x + x \cdot (n-1) \cos^{n-2} x (-\sin x) \right] \sin x\, dx$$

$$= -\int_0^{\pi/2} \cos^{n-1} x \sin x\, dx + (n-1) \int_0^{\pi/2} x \cos^{n-2} x \cdot \sin^2 x\, dx$$

$$= \left[\frac{\cos^{n-1+1} x}{n-1+1} \right]_0^{\pi/2} + (n-1) \int_0^{\pi/2} x \cos^{n-2} x \left(1 - \cos^2 x \right) dx$$

$$= -\frac{1}{n} + (n-1) \int_0^{\pi/2} x \cos^{n-2} x\, dx - (n-1) \int_0^{\pi/2} x \cos^n x\, dx = -\frac{1}{n} + (n-1) I_{n-2} - (n-1) I_n.$$

Transposing the last term to the left, we obtain

$$I_n + (n-1) I_n = -\frac{1}{n} + (n-1) I_{n-2}$$

or $\qquad I_n = -\frac{1}{n^2} + \frac{n-1}{n} I_{n-2}$ or $n^2 I_n = n(n-1) I_{n-2} - 1$.

Putting $n = 4, 2$ successively in the above result.

$$I_4 = -\frac{1}{4^2} + \frac{3}{4} \cdot I_2 \quad ...(i) \qquad\qquad \text{and} \qquad I_2 = -\frac{1}{2^2} + \frac{1}{2} I_0 \qquad ...(ii)$$

Also $I_0 = \int_0^{\pi/2} x\, dx = \left[\frac{x^2}{2} \right]_0^{\pi/2} = \frac{\pi^2}{8}$.

Putting the value of I_0 is (ii), we have $I_2 = -\frac{1}{4} + \frac{\pi^2}{16}$.

Substituting this value in (*i*)

Therefore, $\quad I_4 = -\dfrac{1}{16} + \dfrac{3}{4}\left(-\dfrac{1}{4} + \dfrac{\pi^2}{16}\right) = \dfrac{3\pi^2}{64} - \dfrac{1}{4}.$

Example 4.11 If $I_n = \displaystyle\int_0^{\pi/2} \theta \sin^n\theta \, d\theta$, and n > 1, prove that

$n^2 I_n = n(n-1)I_{n-2} + 1$. Hence deduce that $I_5 = 149/225$. (*A.M.I.E., S-2007*)

Solution. $I_n = \displaystyle\int_0^{\pi/2} \theta \sin^n\theta \, d\theta = \int_0^{\pi/2} \left(\theta \sin^{n-1}\theta\right) \sin\theta \, d\theta$

$= \left[\left(\theta \sin^{n-1}\theta\right)(-\cos\theta)\right]_0^{\pi/2} - \displaystyle\int_0^{\pi/2}\left[\sin^{n-1}\theta + \theta(n-1)\sin^{n-2}\theta\cos\theta\right](-\cos\theta)\,d\theta$

$= \displaystyle\int_0^{\pi/2} \sin^{n-1}\theta \cos\theta \, d\theta + (n-1)\int_0^{\pi/2} \theta\sin^{n-2}\theta\cos^2\theta \, d\theta$

$= \left[\dfrac{\sin^n\theta}{n}\right]_0^{\pi/2} + (n-1)\displaystyle\int_0^{\pi/2} \theta\sin^{n-2}\theta\left(1 - \sin^2\theta\right) d\theta$

$= \left(\dfrac{1}{n} - 0\right) + (n-1)\displaystyle\int_0^{\pi/2} \theta\sin^{n-2}\theta \, d\theta - (n-1)\int_0^{\pi/2} \theta\sin^n\theta \, d\theta$

$= \dfrac{1}{n} + (n-1)I_{n-2} - (n-1)I_n \quad \text{or} \quad (1 + n - 1)I_n = \dfrac{1}{n} + (n-1)I_{n-2}$

or $\quad nI_n = \dfrac{1}{n} + (n-1)I_{n-2} \quad$ or $\quad n^2 I_n = n(n-1)I_{n-2} + 1 \qquad \qquad …(i)$

Putting $n = 5, 3, 1$ successively in (*i*), we get

$$I_5 = \dfrac{4}{5}I_3 + \dfrac{1}{25}, \qquad\qquad I_3 = \dfrac{2}{3}I_1 + \dfrac{1}{9}.$$

and $I_1 = 1$. Therefore, $I_5 = \dfrac{4}{5}\left(\dfrac{2}{3}\cdot 1 + \dfrac{1}{9}\right) + \dfrac{1}{25} = \dfrac{149}{225}.$

Example 4.12 (*a*) If $I_n = \int\limits_0^\infty e^{-x} x^n dx$, where n is a positive integer, obtain a linear relation between I_n and I_{n-1}. Hence evaluate I_n.

(*b*) If $I_n = \int\limits_0^\infty e^{-x} x^{n-1} dx$ for n > 0 then show that $\dfrac{I_{n+1}}{I_n} = n$. (*AMIE S-2000*)

Solution: (*a*) Integrating by parts taking x^n as first function

$$I_n = \int\limits_0^\infty x^n e^{-x} dx = \left[x^n \left(-e^{-x}\right)\right]_0^\infty - \int\limits_0^\infty n x^{n-1}\left(-e^{-x}\right) dx$$

$$= -\left[x^n e^{-x}\right]_0^\infty + n \int\limits_0^\infty e^{-x} x^{n-1} dx.$$

Now, $\lim\limits_{x\to\infty}\left(x^n e^{-x}\right) = \lim\limits_{x\to\infty}\dfrac{x^n}{e^x} = \lim\limits_{x\to\infty}\dfrac{n x^{n-1}}{e^x}$. $\left(\text{By } L' \text{ Hospital's rule, } \dfrac{x}{x} \text{ form}\right)$.

Finally, we will have $\lim\limits_{x\to\infty}\dfrac{n(n-1)\dots 1}{e^x} = 0$, $\left[x^n e^{-x}\right]_0^\infty = 0$.

Therefore, $I_n = n\int\limits_0^\infty x^{n-1} e^{-x} dx = n I_{n-1}$, the required result.

Now $I_n = n I_{n-1}$, $I_{n-1} = (n-1) I_{n-2}$ etc.

$I_1 = I_0$ and $I_0 = \int\limits_0^\infty e^{-x} dx = \left[-e^{-x}\right]_0^\infty = 1$. $\therefore\ I_n = n(n-1)(n-2)\dots 1 = n!$

(*b*) We have $I_n = \int\limits_0^\infty e^{-x} x^{n-1} dx$ (n > 0).

Replacing n by (n + 1), we get

$$I_{n+1} = \int\limits_0^\infty e^{-x} x^n dx = \left[-x^n e^{-x}\right]_0^\infty + n\int\limits_0^\infty e^{-x} x^{n-1} dx = n I_n . \text{ Therefore, } \dfrac{I_{n+1}}{I_n} = n.$$

Example 4.13 If $I_{m,n} = \int \cos^m x \sin nx\, dx$, m > 1, n > 1 connect $I_{m,n}$ with $I_{m-1, n-1}$. Hence or otherwise evaluate $\int\limits_0^{\pi/2} \cos^5 x \sin 3x\, dx$.

(*AMIE; S-2009, W-2002*)

Solution. We have $I_{m,\,n} = \int \cos^m x \sin nx \; dx$

$$= \cos^m x \left(-\frac{\cos nx}{n} \right) - \int m \cos^{m-1} x (-\sin x) \left(\frac{-\cos nx}{n} \right) dx$$

$$= \frac{-\cos^m x \cos nx}{n} - \frac{m}{n} \int \cos^{m-1} x (\sin x \cos nx) \, dx. \qquad \ldots(i)$$

Now $\sin(n-1) x = \sin nx \cos x - \cos nx \sin x$

or $\sin x \cos nx = \sin nx \cos x - \sin(n-1) x$.

Substituting in (i),

$$I_{m,\,n} = \frac{-\cos^m x \cos nx}{n} - \frac{m}{n} \int \cos^{m-1} x [\sin nx \cos x - \sin(n-1)x] dx$$

$$= \frac{-\cos^m x \cos nx}{n} - \frac{m}{n} \int \cos^m x \sin nx \, dx + \frac{m}{n} \int \cos^{m-1} x \sin(n-1) x \, dx$$

$$= -\frac{\cos^m x \cos nx}{n} - \frac{m}{n} I_{m,\,n} + \frac{m}{n} I_{m-1,\,n-1}$$

or, $\left(1 + \dfrac{m}{n} \right) I_{m,\,n} = -\dfrac{\cos^m x \cos nx}{n} + \dfrac{m}{n} I_{m-1,\,n-1}$

or, $I_{m,\,n} = -\dfrac{\cos^m x \cos nx}{m+n} + \dfrac{m}{m+n} I_{m-1,\,n-1} \qquad \ldots(ii)$

or $(m+n) I_{m,n} = -\cos^m x \cos nx + m I_{m-1}, n-1,$

which is the required reduction formula.

Now $\left[I_{m,\,n} \right]_0^{\pi/2} = -\left[\dfrac{\cos^m x \cos nx}{m+n} \right]_0^{\pi/2} + \dfrac{m}{m+n} \left[I_{m-1,\,n-1} \right]_0^{\pi/2}$

$$= \frac{1}{m+n} + \frac{m}{m+n} \left[I_{m-1,\,n-1} \right]_0^{\pi/2} \qquad \ldots(iii)$$

Putting $m = 5, 4$ and $n = 3, 2$ successively in (iii) we have

$$\int_0^{\pi/2} \cos^5 x \sin 3x \, dx = \frac{1}{5+3} + \frac{5}{5+3} \int_0^{\pi/2} \cos^4 x \sin 2x \, dx$$

and $\displaystyle\int_0^{\pi/2} \cos^4 x \sin 2x \, dx = \frac{1}{4+2} + \frac{4}{4+2} \int_0^{\pi/2} \cos^3 x \sin x \, dx$

$$= \frac{1}{6} + \frac{4}{6} \left[\frac{-\cos^4 x}{4} \right]_0^{\pi/2} = \frac{1}{6} + \frac{1}{6}(1) = \frac{1}{3}.$$

$\therefore \quad \displaystyle\int_0^{\pi/2} \cos^5 x \sin 3x \, dx = \frac{1}{8} + \frac{5}{8}\left(\frac{1}{3}\right) = \frac{1}{3}.$

Example 4.14 Prove that, if n be a positive integer greater than unity,

$$\int_0^{\pi/2} \cos^{n-2} x \, \sin nx \, dx = \frac{1}{n-1}.$$

Solution. We have $\displaystyle\int_0^{\pi/2} \cos^{n-2} x \, \sin nx \, dx = \int_0^{\pi/2} \cos^{n-2} x \, \sin[(n-1)x + x] dx$

$$= \int_0^{\pi/2} \cos^{n-2} x \left[\sin(n-1)x \cos x + \cos(n-1)x \sin x \right] dx$$

$$= \int_0^{\pi/2} \underset{I}{\cos^{n-1} x} \sin(n-1)x \, dx + \int_0^{\pi/2} \underset{II}{\cos^{n-2} x} \sin x \cos(n-1)x \, dx$$

[Integrating only Ist integral on R.H.S. by parts]

$$= \left[\cos^{n-1} x \left\{ -\frac{\cos(n-1)x}{n-1} \right\} \right]_0^{\pi/2}$$

$$- \int_0^{\pi/2} (n-1) \cos^{n-2} x (-\sin x) \left[-\frac{\cos(n-1)x}{n-1} \right] dx + \int_0^{\pi/2} \cos^{n-2} x \sin x \cos(n-1)x \, dx$$

$$= -\frac{1}{n-1}(0-1) \, (\because \text{ last two integrals on R.H.S. cancel}) = \frac{1}{n-1}.$$

Example 4.15 Prove that $\displaystyle\int_0^{\pi/2} \cos^n x \, \cos nx \, dx = \frac{\pi}{2^{n+1}}$, n being +ve integer.

Solution. Let $I_{n,n} = \int\limits_0^{\pi/2} \cos^n x \, \cos nx \, dx$. Integrating by parts, taking $\cos^n x$ as the first function. We have

$$I_{n,n} = \left[\frac{1}{n}\cos^n x \sin nx\right]_0^{\pi/2} - \int\limits_0^{\pi/2} \sin nx \cos^{n-1} x(-\sin x)\,dx$$

$$= \int\limits_0^{\pi/2} \cos^{n-1} x\left\{\cos(n-1)x - \cos nx \cos x\right\}dx$$

$$= \int\limits_0^{\pi/2} \cos^{n-1} x \cos(n-1)x\,dx - \int\limits_0^{\pi/2} \cos^n x \cos nx\,dx = I_{n-1,n-1} - I_{n,n}.$$

Transposing $I_{n,n}$ from the right to the left and dividing by 2, we get

$$I_{n,n} = \frac{1}{2}I_{n-1,n-1} = \frac{1}{2^2}I_{n-2,n-2}. \qquad \left(\because I_{n-1,n-1} = \frac{1}{2}I_{n-2,n-2}\right)$$

Applying this reduction formula repeatedly, we have,

$$I_{n,n} = \frac{1}{2^n}I_{0,0} \text{ that is, } I_{n,n} = \left(\frac{1}{2^n}\right)\int\limits_0^{\pi/2} dx = \frac{\pi}{2^{n+1}}.$$

Example 4.16 Prove that

$$\int\limits_0^{\pi/2} \cos^m x \sin mx\,dx = \frac{1}{2^{m+1}}\left[2 + \frac{2^2}{2} + \frac{2^3}{3} + \dots + \frac{2^m}{m}\right].$$

Solution. In example 4.13 we have proved that

$$\int \cos^m x \sin nx\,dx = -\frac{\cos^m x \cos nx}{m+n} + \frac{m}{m+n}\int \cos^{m-1} x \sin(n-1)x\,dx.$$

$$\therefore \int\limits_0^{\pi/2} \cos^m x \sin nx\,dx = -\left[\frac{\cos^m x \cos nx}{m+n}\right]_0^{\pi/2} + \frac{m}{m+n}\int\limits_0^{\pi/2} \cos^{m-1} x \sin(n-1)x\,dx$$

$$= \frac{1}{m+n} + \frac{m}{m+n}\int\limits_0^{\pi/2} \cos^{m-1} x \sin(n-1)x\,dx.$$

Putting $n = m$ and writing $I_{m,m}$ for $\int\limits_0^{\pi/2} \cos^m x \sin mx\,dx$, we have

$$I_{m, m} = \frac{1}{2m} + \frac{1}{2} I_{m-1, m-1}.$$ ...(i)

Changing m into $m-1$, $I_{m-1, m-1} = \frac{1}{2(m-1)} + \frac{1}{2} I_{m-2, m-2}$ ...(ii)

From (i) and (ii)

$$I_{m, m} = \frac{1}{2m} + \frac{1}{2}\left[\frac{1}{2(m-1)} + \frac{1}{2} I_{m-2, m-2}\right] = \frac{1}{2m} + \frac{1}{2^2(m-1)} + \frac{1}{2^2} I_{m-2, m-2}.$$

In this way, by repeated application of (i), we have

$$I_{m, m} = \frac{1}{2m} + \frac{1}{2^2(m-1)} + \frac{1}{2^3(m-2)} + \cdots \text{ to } m \text{ terms} + \frac{1}{2^m} I_{0, 0}$$

$$= \frac{1}{2m} + \frac{1}{2^2(m-1)} + \frac{1}{2^3(m-2)} + \cdots + \frac{1}{2^m} + \frac{1}{2^m}\int_0^{\pi/2} \cos^0 x \sin 0 \, dx$$

$$= \frac{1}{2m} + \frac{1}{2^2(m-1)} + \frac{1}{2^3(m-2)} + \cdots + \frac{1}{2^m} + \frac{1}{2^m}(0)$$

$$= \frac{1}{2^{m+1}}\left[\frac{2^m}{m} + \frac{2^{m-1}}{m-1} + \cdots + \frac{2^3}{3} + \frac{2^2}{2} + 2\right].$$

Example 4.17 If $I_{m, n} = \displaystyle\int_0^{\pi/2} \cos^m x \cos nx \, dx$, prove that

$$(m^2 - n^2) I_{m, n} = m(m-1) I_{m-2, n} \qquad \text{(A.M.I.E., W-1997)}$$

Solution. Integrating by parts, $I_{m, n} = \displaystyle\int_0^{\pi/2} \underset{I}{\cos^m x} \cdot \underset{II}{\cos nx} \, dx$

$$= \left[\cos^m x \cdot \frac{\sin nx}{n}\right]_0^{\pi/2} + \frac{m}{n}\int_0^{\pi/2} \cos^{m-1} x \sin x \sin nx \, dx$$

$$= \frac{m}{n}\int_0^{\pi/2} \underset{I}{\left(\cos^{m-1} x \sin x\right)} \underset{II}{\sin nx} \, dx.$$

Integrating again by parts,

$$I_{m, n} = \frac{m}{n}\left[\left[\cos^{m-1} x \sin x \left(\frac{-\cos nx}{n}\right)\right]_0^{\pi/2}\right.$$

$$-\int_0^{\pi/2}\left\{\cos^{m-1}x\cos x-(m-1)\cos^{m-2}x\sin^2 x\right\}\left(\frac{-\cos nx}{n}\right)dx\Bigg]$$

$$=\frac{m}{n}\left[\frac{1}{n}\int_0^{\pi/2}\cos nx\left\{\cos^m x-(m-1)\cos^{m-2}x\left(1-\cos^2 x\right)\right\}dx\right]$$

$$=\frac{m}{n^2}\int_0^{\pi/2}\cos nx\left\{\cos^m x-(m-1)\cos^{m-2}x+(m-1)\cos^m x\right\}dx$$

$$=\frac{m}{n^2}\int_0^{\pi/2}\cos nx\left\{m\cos^m x-(m-1)\cos^{m-2}x\right\}dx=\frac{m}{n^2}\left[m I_{m,n}-(m-1)I_{m-2,n}\right]$$

or, $\quad I_{m,n}-\dfrac{m^2}{n^2}I_{m,n}=-\dfrac{m(m-1)}{n^2}I_{m-2,n}\quad$ or, $\quad\dfrac{m^2-n^2}{n^2}I_{m,n}=\dfrac{m(m-1)}{n^2}I_{m-2,n}$

or, $\quad I_{m,n}=\dfrac{m(m-1)}{m^2-n^2}I_{m-2,n}$, which is same as desired.

Example 4.18 If $\mathbf{I_m}=\displaystyle\int_0^{\pi/2}\dfrac{\sin(2m-1)\theta}{\sin\theta}d\theta$ and $\mathbf{I'_m}=\displaystyle\int_0^{\pi/2}\left(\dfrac{\sin m\theta}{\sin\theta}\right)^2 d\theta$, **m** being an integer, show that (*i*) $\mathbf{I_{m+1}-I_m=0}$, and (*ii*) $\mathbf{I'_{m+1}-I'_m=I_{m+1}}$.

Solution. (*i*) $I_{m+1}-I_m=\displaystyle\int_0^{\pi/2}\dfrac{\sin(2m+1)\theta-\sin(2m-1)\theta}{\sin\theta}d\theta$

$$=\int_0^{\pi/2}\frac{2\cos 2m\theta\cdot(\sin\theta)}{\sin\theta}\cdot d\theta=2\int_0^{\pi/2}\cos 2m\theta\, d\theta=\frac{1}{m}\left|\sin 2m\theta\right|_0^{\pi/2}=0.$$

(*ii*) $\quad I'_{m+1}-I'_m=\displaystyle\int_0^{\pi/2}\dfrac{\sin^2(m+1)\theta-\sin^2 m\theta}{\sin^2\theta}d\theta=\int_0^{\pi/2}\dfrac{\sin(2m+1)\theta\sin\theta}{\sin^2\theta}d\theta$

$$\left[\because\sin^2 A-\sin^2 B=\sin(A+B)\sin(A-B)\right]$$

$$=\int_0^{\pi/2}\frac{\sin(2m+1)\theta}{\sin\theta}d\theta=I_{m+1}.\text{ Hence proved.}$$

Example 4.19 If m and n are positive integers, n ≥ m and

$$I_{(m, n)} = \int_0^1 (\log x)^m \cdot x^{n-1} \, dx, \text{ prove that } I_{(m, n)} = -(m/n) I_{(m-1, n)}$$

Hence, show that $I_{(m, n)} = (-1)^m \, m!/n^{m+1}$. *(A.M.I.E., W-2001)*

Solution. We have

$$I_{(m, n)} = \int_0^1 \underset{I}{(\log x)^m} \cdot \underset{II}{x^{n-1}} \, dx = \left[(\log x)^m \cdot \frac{x^n}{n} \right]_0^1 - \int_0^1 m (\log x)^{m-1} \cdot \frac{1}{x} \cdot \frac{x^n}{n} \, dx$$

$$= -\frac{m}{n} \int_0^1 (\log x)^{m-1} \cdot x^{n-1} \, dx \quad \text{or,} \quad I_{(m, n)} = -\frac{m}{n} \cdot I_{(m-1, n)}. \qquad \qquad ...(i)$$

Changing m to $(m - 1)$, we get $I_{(m-1, n)} = -\dfrac{(m-1)}{n} I_{(m-2, n)}$ ...(ii)

From (i) and (ii) $I_{(m, n)} = (-1)^2 \dfrac{m(m-1)}{n^2} I_{(m-2, n)}$

$$= (-1)^3 \frac{m(m-1)(m-2)}{n^3} I_{(m-3, n)}$$

....
....

$$= \frac{(-1)^m m(m-1)(m-2)...3.2.1}{n^m} I_{(m-m, n)}$$

$$= (-1)^m \frac{m!}{n^m} \int_0^1 (\log x)^0 x^{n-1} \, dx = (-1)^m \frac{m!}{n^m} \left[\frac{x^n}{n} \right]_0^1 = \frac{(-1)^m m!}{n^m} \cdot \frac{1}{n} = \frac{(-1)^m m!}{n^{m+1}},$$

which is same as desired.

Example 4.20 By considering the value of $\int_0^1 (1-x^2)^n \, dx$, **show that**

$$1 - \frac{n}{1 \cdot 3} + \frac{n(n-1)}{1 \cdot 2 \cdot 5} - \frac{n(n-1)(n-2)}{1 \cdot 2 \cdot 3 \cdot 7} + ... = \frac{2}{3} \cdot \frac{4}{5} \cdot \frac{6}{7} ... \frac{2n}{2n+1}.$$

Solution. Let $I = \int_0^1 (1-x^2)^n \, dx$. Put $x = \sin\theta$, $dx = \cos\theta \, d\theta$.

$$\therefore I = \int_0^{\pi/2} \cos^{2n+1}\theta \, d\theta = \frac{2n}{2n+1} \cdot \frac{2n-2}{2n-1} \cdots \cdots \frac{4}{5} \cdot \frac{2}{3} = \frac{2}{3} \cdot \frac{4}{5} \cdot \frac{6}{7} \cdots \cdots \frac{2n}{2n+1} \quad ...(i)$$

Next expanding $\left(1 - x^2\right)^n$ by the binomial theorem, we have $I = \int_0^1 \left(1 - x^2\right)^n dx$

$$= \int_0^1 \left[1 - nx^2 + \frac{n(n-1)}{2!} x^4 - \frac{n(n-1)(n-2)}{3!} x^6 + \dots \right] dx$$

$$= 1 - n \cdot \frac{1}{3} + \frac{n(n-1)}{1 \cdot 2} \cdot \frac{1}{5} - \frac{n(n-1)(n-2)}{1 \cdot 2 \cdot 3} \cdot \frac{1}{7} + \dots \quad ...(ii)$$

From (i) and (ii), we have

$$1 - \frac{n}{1 \cdot 3} + \frac{n(n-1)}{1 \cdot 2 \cdot 5} - \frac{n(n-1)(n-2)}{1 \cdot 2 \cdot 3 \cdot 7} + \cdots \cdots = \frac{2}{3} \cdot \frac{4}{5} \cdot \frac{6}{7} \cdots \cdots \frac{2n}{2n+1}.$$

Example 4.21 If $I_n = \int \frac{t^n dt}{1+t^2}$, show that $I_{n+2} = \frac{t^{n+1}}{n+1} - I_n$ and hence

evaluate $\int_0^1 \frac{t^6}{1+t^2} dt$.

Solution. We have $I_n = \int \frac{t^n dt}{1+t^2}$. Now Replacing n by $(n + 2)$, we obtain

$$I_{n+2} = \int \frac{t^{n+2} dt}{1+t^2} = \int \frac{t^{n+2} + t^n - t^n}{1+t^2} dt = \int \frac{t^{n+2} + t^n}{1+t^2} dt - \int \frac{t^n}{1+t^2} dt$$

$$= \int \frac{t^n \left(t^2 + 1\right)}{1+t^2} dt - I_n \text{ or, } I_{n+2} = \frac{t^{n+1}}{n+1} - I_n, \text{ which is same as desired.}$$

Putting $n = 4, 2, 0$ successively in the above result, we obtain

$$I_6 = \left(t^5/5\right) - I_4, \quad ...(i), \qquad I_4 = \left(t^3/3\right) - I_2, \quad ...(ii), \qquad \text{and } I_2 = t - I_0 \quad ...(iii).$$

Also $I_0 = \int \frac{dt}{1+t^2} = \tan^{-1} t$. Putting the value of I_0 in (iii).

We have $I_2 = t - \tan^{-1} t$, $I_4 = \frac{t^3}{3} - t + \tan^{-1} t$, and $I_6 = \frac{t^5}{5} - \frac{t^3}{3} + t - \tan^{-1} t$.

$$\therefore \quad \int_0^1 t^6\, dt/(1+t^2) = \left[\frac{t^5}{5} - \frac{t^3}{3} + t - \tan^{-1} t\right]_0^1 = \frac{1}{5} - \frac{1}{3} + 1 - \frac{\pi}{4} = \frac{13}{15} - \frac{\pi}{4}.$$

Example 4.22 If $I_n = \int x^n (a^2 - x^2)^{1/2}\, dx$, **prove that**

$$I_n = -\frac{x^{n-1}(a^2 - x^2)^{3/2}}{n+2} + \frac{(n-1)}{n+2} a^2 I_{n-2}. \qquad \text{(D.C.E., 2001)}$$

Hence evaluate $\int x^5 (4 - x^2)^{1/2}\, dx.$ \qquad (A.M.I.E., W-1999)

Solution. Put $P = x^{n-1}(a^2 - x^2)^{3/2}$

$$\therefore \quad \frac{dP}{dx} = (n-1)x^{n-2}(a^2 - x^2)^{3/2} + x^{n-1} \cdot \frac{3}{2}(a^2 - x^2)^{1/2} \cdot (-2x)$$

$$= (n-1)x^{n-2}(a^2 - x^2)^{3/2} - 3x^n(a^2 - x^2)^{1/2}$$

$$= (n-1)x^{n-2}(a^2 - x^2)^{1/2}(a^2 - x^2) - 3x^n(a^2 - x^2)^{1/2}$$

$$= a^2(n-1)x^{n-2}(a^2 - x^2)^{1/2} - (n-1)x^n(a^2 - x^2)^{1/2} - 3x^n(a^2 - x^2)^{1/2}$$

$$= a^2(n-1)x^{n-2}(a^2 - x^2)^{1/2} - (n+2)x^n(a^2 - x^2)^{1/2}.$$

Integrating both sides, we get

$$\int dP = a^2(n-1)\int x^{n-2}(a^2 - x^2)^{1/2}\, dx - (n+2)\int x^n(a^2 - x^2)^{1/2}\, dx.$$

$$\therefore \quad P = a^2(n-1)I_{n-2} - (n+2)I_n$$

or $(n+2)I_n = -P + a^2(n-1)I_{n-2} = -x^{n-1}(a^2 - x^2)^{3/2} + a^2(n-1)I_{n-2}$

Therefore, $\quad I_n = -\dfrac{x^{n-1}(a^2 - x^2)^{3/2}}{n+2} + \dfrac{(n-1)}{n+2}a^2 I_{n-2}$ ...(i), the desired result.

Putting $n = 5$ and $a = 2$ in (i), we obtain $I_5 = -\dfrac{x^4(4-x^2)^{3/2}}{7} + \dfrac{4(4)}{7}I_3$ ...(ii).

Now putting $n = 3$, we have $I_3 = -\dfrac{x^2\left(4 - x^2\right)^{3/2}}{5} + \dfrac{4(2)}{5} I_1$

$$= -\dfrac{x^2\left(4 - x^2\right)^{3/2}}{5} + \dfrac{8}{5}\int x\left(4 - x^2\right)^{1/2} dx$$

$$= -\dfrac{x^2\left(4 - x^2\right)^{3/2}}{5} - \dfrac{8}{5}\cdot\dfrac{1}{2}\int\left(4 - x^2\right)^{1/2}(-2x)dx$$

$$= -\dfrac{x^2\left(4 - x^2\right)^{3/2}}{5} - \dfrac{4}{5}\dfrac{\left(4 - x^2\right)^{3/2}}{3/2} = -\dfrac{x^2\left(4 - x^2\right)^{3/2}}{5} - \dfrac{8}{15}\left(4 - x^2\right)^{3/2}.$$

Putting the value of I_3 in (ii), we get

$$I_5 = -\dfrac{x^4\left(4 - x^2\right)^{3/2}}{7} - \dfrac{16x^2\left(4 - x^2\right)^{3/2}}{35} - \dfrac{128\left(4 - x^2\right)^{3/2}}{105}$$

$$= -\left(4 - x^2\right)^{3/2}\left[\dfrac{x^4}{7} + \dfrac{16x^2}{35} + \dfrac{128}{105}\right] = \dfrac{-\left(4 - x^2\right)^{3/2}}{105}\left(15x^4 + 48x^2 + 128\right).$$

Example 4.23 **Obtain a reduction formula for** $I_m = \displaystyle\int_0^\infty e^{-x}\sin^m x\, dx$, **where**

$m \geq 2$, **in the form** $\left(1 + m^2\right)I_m = m(m - 1)I_{m-2}$. **Hence evaluate** I_4.

Solution. Integrating by parts, taking $\sin^m x$ as first function, we have

$$I_m = \int_0^\infty \underset{II}{e^{-x}}\ \underset{I}{\sin^m x}\, dx = \left[\sin^m x \cdot \dfrac{e^{-x}}{-1}\right]_0^\infty - m\int_0^\infty \sin^{m-1} x \cos x \dfrac{e^{-x}}{-1} dx$$

$$= m\int_0^\infty \left(\underset{I}{\sin^{m-1} x \cos x}\right)\cdot \underset{II}{e^{-x}}\, dx.$$

Again integrating by parts, taking $\sin^{m-1} x \cos x$ as first function, we have

$$I_m = m\left[\sin^{m-1} x \cos x \cdot \dfrac{e^{-x}}{-1}\right]_0^\infty - m\int_0^\infty \{(m-1)\sin^{m-2} x \cos^2 x\right.$$

$$\left. - \sin^{m-1} x \sin x\}\dfrac{e^{-x}}{-1} dx$$

$$= -m(m-1)\int_0^\infty \sin^{m-2} x \cdot \cos^2 x \cdot \frac{e^{-x}}{-1} dx + m\int_0^\infty \sin^m x \frac{e^{-x}}{-1} dx$$

$$= m(m-1)\int_0^\infty \sin^{m-2} x \cdot \left(1-\sin^2 x\right) e^{-x} dx - m\int_0^\infty \left(\sin^m x\right) e^{-x} dx$$

$$= m(m-1)\int_0^\infty \left(\sin^{m-2} x\right) e^{-x} dx - m(m-1)\int_0^\infty \left(\sin^m x\right) e^{-x} dx - m\int_0^\infty \left(\sin^m x\right) e^{-x} dx$$

or $\quad I_m = m(m-1)I_{m-2} - m(m-1)I_m - mI_m$

$$= m(m-1)I_{m-2} - \left(m^2 - m + m\right)I_m = m(m-1)I_{m-2} - m^2 I_m$$

or $\quad \left(1+m^2\right)I_m = m(m-1)I_{m-2}$, which is same as desired. $\qquad$ ...(i)

Putting $m = 4$ and 2 successively in (i), we get

$$(1+16)I_4 = 4(3)I_2 \text{ or } I_4 = (12\,I_2)/17 \qquad \text{...(ii)}$$

and $\quad 5I_2 = 2I_0$ or $I_2 = \dfrac{2}{5} I_0 = \dfrac{2}{5}\int_0^\infty e^{-x} \sin^0 x\, dx$

$$= \frac{2}{5}\int_0^\infty e^{-x} dx = \frac{2}{5}\left[\frac{e^{-x}}{-1}\right]_0^\infty = \frac{2}{5}.$$

Putting the value of I_2 in (ii), we get $I_4 = (12/17) \cdot (2/5) = 24/85$.

Example 4.24 If I_n denotes $\displaystyle\int_0^a \left(a^2 - x^2\right)^n dx$, and n > 0, prove that

$(2n+1)I_n = 2na^2 \cdot I_{n-1}$. **Hence evaluate** $\displaystyle\int_0^a \left(a^2 - x^2\right)^3 dx$ **and** I_4.

Solution. We have $I_n = \displaystyle\int_0^a \underset{I}{\left(a^2 - x^2\right)^n} \cdot \underset{II}{1\, dx}$

$$= \left[\left(a^2 - x^2\right)^n x\right]_0^a - \int_0^a n\left(a^2 - x^2\right)^{n-1} (-2x) x\, dx$$

$$= 2n \int_0^a \left(a^2 - x^2\right)^{n-1} x^2 \, dx = -2n \int_0^a \left(a^2 - x^2\right)^{n-1} \left(\overline{a^2 - x^2} - a^2\right) dx$$

$$= -2n \int_0^a \left(a^2 - x^2\right)^n dx + 2na^2 \int_0^a \left(a^2 - x^2\right)^{n-1} dx = -2n I_n + 2na^2 I_{n-1}$$

or $\quad (2n+1) I_n = 2na^2 I_{n-1}$ ...(i). Putting $n = 3$ in (i), we obtain

$$I_3 = \frac{6}{7} a^2 I_2 = \frac{6}{7} a^2 \left[\frac{4}{5} a^2 I_1\right], \text{ on putting } n = 2 \text{ in (i)}$$

$$= \frac{24}{35} a^4 I_1 = \frac{24}{35} a^4 \int_0^a \left(a^2 - x^2\right) dx$$

or $\quad \int_0^a \left(a^2 - x^2\right)^3 dx = \frac{24a^4}{35} \left[a^2 x - \frac{x^3}{3}\right]_0^a = \frac{24a^4}{35} \left[a^3 - \frac{a^3}{3}\right] = \frac{16a^7}{35}.$

Similarly, we can evaluate $\int_0^a \left(a^2 - x^2\right)^4 dx$ and its value is equal to $128a^9/315$.

Example 4.25 Obtain a reduction formula for $\int \left(a^2 + x^2\right)^{n/2} dx$ and hence

find the value of $\int \left(a^2 + x^2\right)^{5/2} dx$.

Solution. We have $\quad \underset{I}{\int \left(a^2 + x^2\right)^{n/2}} dx = \underset{II}{\int \left(a^2 + x^2\right)^{n/2} \cdot 1 \, dx}$

$$= x\left(a^2 + x^2\right)^{n/2} - \int \frac{n}{2} \left(a^2 + x^2\right)^{(n/2)-1} \cdot 2x \cdot x \, dx$$

$$= x\left(a^2 + x^2\right)^{n/2} - n \int \left(a^2 + x^2\right)^{(n/2)-1} x^2 \, dx$$

$$= x\left(a^2 + x^2\right)^{n/2} - n \int \left(a^2 + x^2\right)^{(n/2)-1} \left(\overline{a^2 + x^2} - a^2\right) dx$$

$$= x\left(a^2 + x^2\right)^{n/2} - n \int \left(a^2 + x^2\right)^{n/2} dx + na^2 \int \left(a^2 + x^2\right)^{(n/2)-1} dx$$

or, $(1+n) \int \left(a^2 + x^2\right)^{n/2} dx = x\left(a^2 + x^2\right)^{n/2} + na^2 \int \left(a^2 + x^2\right)^{(n/2)-1} dx$

or, $\int \left(a^2 + x^2\right)^{n/2} dx = \dfrac{x\left(a^2 + x^2\right)^{n/2}}{n+1} + \dfrac{na^2}{(n+1)} \int \left(a^2 + x^2\right)^{(n/2)-1} dx,$ ...(i),

the desired result.

Putting $n = 5$ in (i), we obtain

$$\int \left(a^2 + x^2\right)^{5/2} dx = \frac{x}{6}\left(a^2 + x^2\right)^{5/2} + \frac{5}{6}a^2 \int \left(a^2 + x^2\right)^{3/2} dx$$

$$= \frac{x}{6}\left(a^2 + x^2\right)^{5/2} + \frac{5}{6}a^2 \left\{ \frac{x}{4}\left(a^2 + x^2\right)^{3/2} + \frac{3}{4}a^2 \int \left(a^2 + x^2\right)^{1/2} dx \right\}$$

[On putting $n = 3$ in (i)]

$$= \frac{x}{6}\left(a^2 + x^2\right)^{5/2} + \frac{5}{24}a^2 x\left(a^2 + x^2\right)^{3/2} + \frac{5}{8}a^4 \int \sqrt{a^2 + x^2}\, dx$$

$$= \frac{x}{6}\left(a^2 + x^2\right)^{5/2} + \frac{5}{24}a^2 x\left(a^2 + x^2\right)^{3/2} + \frac{5}{8}a^4 \left[\frac{x}{2}\sqrt{a^2 + x^2} + \frac{a^2}{2}\sinh^{-1}\left(\frac{x}{a}\right) \right]$$

$$= \frac{x}{6}\left(a^2 + x^2\right)^{5/2} + \frac{5}{24}a^2 x\left(a^2 + x^2\right)^{3/2} + \frac{5}{16}a^4 x\sqrt{a^2 + x^2} + \frac{5}{16}a^6 \sinh^{-1}\left(\frac{x}{a}\right).$$

Example 4.26 Obtain a reduction formula for $\displaystyle\int \frac{dx}{\left(a^2 + x^2\right)^{n/2}}$ and hence find

the value of $\displaystyle\int \left(a^2 + x^2\right)^{-7/2} dx.$

Solution. Let $I_{n/2} = \displaystyle\int \frac{1}{\left(a^2 + x^2\right)^{n/2}} \cdot 1\ dx.$ Integrating by parts, we obtain

$$I_{n/2} = \frac{1}{\left(a^2 + x^2\right)^{n/2}} x + \int x \cdot \frac{n}{2}\left(a^2 + x^2\right)^{-(n/2)-1} \cdot 2x\, dx$$

$$= \frac{x}{\left(a^2 + x^2\right)^{n/2}} + n \int x^2 \left(a^2 + x^2\right)^{-(n/2)-1} dx$$

$$= \frac{x}{\left(a^2 + x^2\right)^{n/2}} + n \int \frac{x^2}{\left(a^2 + x^2\right)^{(n+2)/2}}\, dx = \frac{x}{\left(a^2 + x^2\right)^{n/2}} + n \int \frac{\left(a^2 + x^2\right) - a^2}{\left(a^2 + x^2\right)^{(n+2)/2}}\, dx$$

$$= \frac{x}{\left(a^2 + x^2\right)^{n/2}} + n \int \frac{1}{\left(a^2 + x^2\right)^{n/2}} dx - na^2 \int \frac{dx}{\left(a^2 + x^2\right)^{(n+2)/2}}$$

or, $(1-n) I_{n/2} = \dfrac{x}{\left(a^2 + x^2\right)^{n/2}} - na^2 \int \dfrac{dx}{\left(a^2 + x^2\right)^{(n+2)/2}}$

$$= \frac{x}{\left(a^2 + x^2\right)^{n/2}} - na^2 I_{(n/2)+1} \quad \text{or,} \quad na^2 I_{(n/2)+1} = \frac{x}{\left(a^2 + x^2\right)^{n/2}} + (n-1) I_{n/2}$$

or, $I_{(n/2)+1} = \dfrac{x}{na^2 \left(a^2 + x^2\right)^{n/2}} + \dfrac{(n-1)}{na^2} I_{n/2}.$

Changing $\left(\dfrac{n}{2}\right) + 1$ into $\dfrac{n}{2}$, we obtain

$$I_{n/2} = \frac{x}{(n-2)a^2 \left(a^2 + x^2\right)^{(n-2)/2}} + \frac{(n-3)}{(n-2)a^2} I_{(n/2)-1}, \qquad \text{...(i)}$$

the required reduction formula. Putting $n = 7$ in (i), we have

$$\int \frac{dx}{\left(a^2 + x^2\right)^{7/2}} = \frac{x}{(7-2)a^2 \left(a^2 + x^2\right)^{(7-2)/2}} + \frac{7-3}{(7-2)a^2} \int \frac{dx}{\left(a^2 + x^2\right)^{5/2}}$$

$$= \frac{x}{5a^2 \left(a^2 + x^2\right)^{5/2}} + \frac{4}{5a^2} \left(\frac{x}{(5-2)a^2 \left(a^2 + x^2\right)^{3/2}} + \frac{(5-3)}{(5-2)a^2} \int \frac{dx}{\left(a^2 + x^2\right)^{3/2}} \right)$$

$$= \frac{x}{5a^2 \left(a^2 + x^2\right)^{5/2}} + \frac{4}{5a^2} \left(\frac{x}{3a^2 \left(a^2 + x^2\right)^{3/2}} + \frac{2}{3a^2} \int \frac{dx}{\left(a^2 + x^2\right)^{3/2}} \right)$$

$$= \frac{x}{5a^2 \left(a^2 + x^2\right)^{5/2}} + \frac{4x}{15a^4 \left(a^2 + x^2\right)^{3/2}} + \frac{8}{15a^4} \int \frac{dx}{\left(a^2 + x^2\right)^{3/2}}$$

$$= \frac{x}{5a^2 \left(a^2 + x^2\right)^{5/2}} + \frac{4x}{15a^4 \left(a^2 + x^2\right)^{3/2}} + \frac{8}{15a^4} \left(\frac{x}{(3-2)a^2 \sqrt{a^2 + x^2}} \right).$$

$$= \frac{x}{5a^2 \left(a^2+x^2\right)^{5/2}} + \frac{4x}{15a^4 \left(a^2+x^2\right)^{3/2}} + \frac{8x}{15a^6 \left(a^2+x^2\right)^{1/2}}$$

$$= \frac{x}{5a^2 \left(a^2+x^2\right)^{1/2}} \left(\frac{1}{\left(a^2+x^2\right)^2} + \frac{4}{3a^2 \left(a^2+x^2\right)} + \frac{8}{3a^4} \right), \text{ the required result.}$$

Example 4.27 If $I_n = \int x^n (a-x)^{1/2} dx$, **prove that**

$$(2n+3)I_n = 2anI_{n-1} - 2x^n (a-x)^{3/2}.$$

Hence or otherwise evaluate $\int_0^a x^2 \sqrt{\left(ax-x^2\right)}dx$. (D.C.E., 2005, AMIE, S-99)

Solution: (i) We have $\qquad I_n = \int \underset{I}{x^n} \underset{II}{(a-x)^{1/2}} dx$

$$= x^n \cdot \left\{ -(2/3)(a-x)^{3/2} \right\} - \int nx^{n-1} \left\{ -(2/3)(a-x)^{3/2} \right\} dx$$

$$= -(2/3)x^n (a-x)^{3/2} + (2/3)n \int x^{n-1} (a-x)(a-x)^{1/2} dx$$

$$= -(2/3)x^n (a-x)^{3/2} + (2/3)na \int x^{n-1} (a-x)^{1/2} dx -(2/3)n \int x^n (a-x)^{1/2} dx$$

$$= -(2/3)x^n (a-x)^{3/2} + (2/3)na \, I_{n-1} - (2/3)nI_n$$

Transpoing the last integral on the right, to the left, we obtain

$$\left[1+(2/3)n\right]I_n = (2/3)na \, I_{n-1} - (2/3)x^n (a-x)^{3/2}$$

or, $\quad (2n+3)I_n = 2naI_{n-1} - 2x^n (a-x)^{3/2}$, the required result.

(ii) We have $I_n = \dfrac{2na}{(2n+3)} I_{n-1} - \dfrac{2x^n (a-x)^{3/2}}{(2n+3)}$

that is, $\displaystyle\int x^n (a-x)^{1/2} dx = \frac{2na}{(2n+3)} \int x^{n-1}(a-x)^{1/2} dx - \frac{2x^n (a-x)^{3/2}}{(2n+3)}$

$\therefore \displaystyle\int_0^a x^n (a-x)^{1/2} dx = \frac{2na}{(2n+3)} \int_0^a x^{n-1}(a-x)^{1/2} dx - \left[\frac{2x^n (a-x)^{3/2}}{2n+3} \right]_0^a$

or $\displaystyle\int_0^a x^n (a-x)^{1/2}\, dx = \frac{2na}{(2n+3)} \int_0^a x^{n-1}(a-x)^{1/2}\, dx$ ...(i)

Now the integral, in Ex. 4.27 part $(ii) = \displaystyle\int_0^a x^2 \left(ax - x^2\right)^{1/2} dx = \int_0^a x^{5/2}(a-x)^{1/2}\, dx$

$\displaystyle = \frac{2 \cdot (5/2)a}{2 \cdot (5/2)+3} \int_0^a x^{3/2}(a-x)^{1/2}\, dx, \left(\text{on putting } n = \frac{5}{2} \text{ in equation } (i)\right)$

$\displaystyle = \frac{5a}{8}\, \frac{2 \cdot (3/2)a}{2 \cdot (3/2)+3} \int_0^a x^{1/2}(a-x)^{1/2}\, dx, \left(\text{now on putting } n = \frac{3}{2} \text{ in } (i)\right)$

$\displaystyle = \frac{5a^2}{16} \int_0^{\pi/2} \left(a\sin^2\theta\right)^{1/2}\left(a\cos^2\theta\right)^{1/2} \cdot 2a\sin\theta\cos\theta\, d\theta, \quad (\text{putting } x = a\sin^2\theta)$

$\displaystyle = \frac{5a^4}{8} \int_0^{\pi/2} \sin^2\theta\cos^2\theta\, d\theta = \frac{5a^4}{8} \cdot \frac{1}{4\cdot 2} \cdot \frac{\pi}{2} = \frac{5\pi a^4}{128}.$

Exercise 4.1

1. Evaluate the following definite integrals:

(a) $\displaystyle\int_0^{\pi/2} \sin^4 x\, dx$, (b) $\displaystyle\int_0^{\pi/2} \cos^5 x\, dx$, (c) $\displaystyle\int_0^{\pi/4} \sin^7 2\theta\, d\theta$.

Ans. (a) $\dfrac{3\pi}{16}$, (b) $\dfrac{8}{15}$, (c) $\dfrac{8}{35}$.

2. Write down the values of

(a) $\displaystyle\int_0^{\pi/2} \sin^6 x\cos^5 x\, dx$, (b) $\displaystyle\int_0^{\pi/2} \sin^6 x\cos^8 x\, dx$, (c) $\displaystyle\int_0^{\pi/2} \sin^5 x\cos^5 x\, dx$.

Ans. (a) $\left(\dfrac{8}{693}\right)$, (b) $\dfrac{5\pi}{4096}$ (c) $\dfrac{1}{60}$.

3. Write down the values of (a) $\displaystyle\int_0^{\pi/6} \cos^6 3\theta \sin^2 6\theta\, d\theta$, (b) $\displaystyle\int_0^{\pi/8} \cos^3 4x\, dx$

Ans. (a) $\dfrac{7\pi}{384}$, (b) $\dfrac{1}{6}$

4. Evaluate: (*i*) $\int_0^1 x^2 \left(1-x^2\right)^{3/2} dx$, (*ii*) $\int_0^a x^4 \sqrt{a^2-x^2}\, dx$,   (*A.M.I.E., W-2005*)

(*iii*) $\int_0^\infty \frac{t^4}{\left(1+t^2\right)^4} dt$, (*iv*) $\int_0^4 x^3 \sqrt{4x-x^2}\, dx$, (*v*) $\int_0^1 x^{3/2}(1-x)^{3/2}\, dx$,

(*vi*) $\int_0^1 \frac{dx}{\left(1+x^2\right)^{3/2}}$, (*vii*) $\int_0^{2a} x^2 \sqrt{2ax-x^2}\, dx$, (*viii*) $\int_0^{2a} x^{3/2}(2a-x)^{1/2}\, dx$

Ans. (*i*) $\dfrac{\pi}{32}$, (*ii*) $\dfrac{\pi a^6}{32}$, (*iii*) $\dfrac{\pi}{32}$, (*iv*) 28π , (*v*) $\dfrac{3\pi}{128}$, (*vi*) $\dfrac{1}{\sqrt{2}}$

(*vii*) $\dfrac{5\pi a^4}{8}$, (*viii*) $\pi a^3/2$.

5. Prove that $\int_0^\pi \dfrac{\sin^4 \theta}{(1+\cos\theta)^2} d\theta = \dfrac{3\pi}{2}$.

6. Evaluate (*i*) $\int \cot^6 x\, dx$, (*ii*) $\int \cot^5 x\, dx$, (*iii*) $\int_{\pi/4}^{\pi/2} \cot^4 x\, dx$,

(*iv*) $\int \sec^6 x\, dx$, (*v*) $\int \text{cosec}^3 x\, dx$.   (*A.M.I.E., W-2001*)

Ans. (*i*) $-\dfrac{1}{5}\cot^5 x + \dfrac{1}{3}\cot^3 x - \cot x - x$, (*ii*) $\dfrac{-\cot^4 x}{4} + \dfrac{\cot^2 x}{2} + \log \sin x$,

(*iii*) $\dfrac{(3\pi-8)}{12}$, (*iv*) $\dfrac{1}{5}\tan x \sec^4 x + \dfrac{4}{15}\tan x \sec^2 x + \dfrac{8}{15}\tan x$,

(*v*) $\dfrac{-\text{cosec}\, x \cot x}{2} + \dfrac{1}{2}\log \tan\left(\dfrac{x}{2}\right)$.

7. If $I_n = \int_0^a \left(a^2-x^2\right)^n dx$, prove that $I_n = \dfrac{2n\, a^2}{2n+1} I_{n-1}$ for $n \geq 2$.

8. If $U_n = \int_0^{\pi/4} \tan^n x\, dx$, show that for $n > 4$, (i) $U_n + U_{n-2} = \dfrac{1}{n-1}$ and deduce the value of U_5 and U_6. *(GGSIPU 2006; A.M.I.E., W-2005)*

(ii) $n(U_{n+1} + U_{n-1}) = 1$. **Ans.** (i). $\dfrac{1}{2}\log 2 - \dfrac{1}{4}$; $\dfrac{13}{15} - \dfrac{\pi}{4}$.

[**Hint:** (ii) Replace $n - 1$ by n in the result of (i)]

9. Find a formula of reduction for $\int e^{ax} \cos^n bx\, dx$.

Ans. $\dfrac{a\cos bx + nb\sin bx}{a^2 + n^2 b^2} e^{ax} \cos^{n-1} bx + \dfrac{n(n-1)b^2}{a^2 + n^2 b^2} \int e^{ax} \cos^{n-2} bx\, dx$.

10. Prove that if $I_{m,n} = \int \cos^m x \sin nx\, dx$

then $(m+n) I_{m,n} = -\cos^m x \cos nx + m\, I_{m-1,\,n-1}$

and $\left[I_{m,n} \right]_0^{\pi/2} = \dfrac{1}{m+n} + \dfrac{m}{m+n} \left[I_{m-1,\,n-1} \right]_0^{\pi/2}$.

11. (a) Find a reduction formula for the integral $\int \dfrac{x^m\, dx}{(\log x)^n}$.

(b) If $I_{m,n} = \int x^m (\log x)^n\, dx$, show that $(m+1) I_{m,n} + n \cdot I_{m,n-1} = (\log x)^n x^{m+1}$.

Ans. (a) $(n-1) I_{m,n} = -\dfrac{x^{m+1}}{(\log x)^{n-1}} + (m+1) I_{(m,\,n-1)}$.

12. (a) If $I_{n,m} = \int x^n e^{mx}\, dx$, $(n > 0, m > 0)$ prove that

$I_{n,m} = \left(\dfrac{x^n e^{mx}}{m} \right) - \dfrac{n}{m} I_{n-1,m}$. Hence evaluate $\int x^4 e^{2x}\, dx$. *(A.M.I.E., W-2004)*

(b) If $I_n = \int x^n e^x\, dx$, show that $I_n + n I_{n-1} = x^n e^x$. Hence find I_4. *(Madras 2000)*

Ans. (a) $e^{2x}\left(x^4 - 2x^3 + 3x^2 - x + 1 \right)/2$; (b) $e^x\left(1 - x + x^2 - x^3 + x^4 \right)$.

13. If $U_n = \int_0^{\pi/4} \sin^{2n} x\, dx$, prove that $U_n = \left(1 - \dfrac{1}{2n} \right) U_{n-1} - \dfrac{1}{n \cdot 2^{n+1}}$.

14. If $I_n = \int x^n \sin ax\, dx$, show that

$$a^2 I_n = -ax^n \cos ax + nx^{n-1} \sin ax - n(n-1) I_{n-2}.$$

Hence, evaluate $\int_0^{\pi/4} x^4 \sin 2x\, dx$.　　　　**Ans.** $\left(\dfrac{\pi^4}{64} - \dfrac{3\pi}{8} + \dfrac{3}{4} \right)$

15. If $I_n = \int \dfrac{dx}{\left(x^2 + 1\right)^n}$, show that $2n\, I_{n+1} = (2n-1) I_n + \dfrac{x}{\left(x^2 + 1\right)^n}$

(A.M.I.E., S-2005)

16. If $I_n = \int_0^{\pi/2} x \cos^n x\, dx$, prove that $n^2 I_n = n(n-1) I_{n-1} - 1$. (A.M.I.E., S-2004)

17. If $U_n = \int \cos n\theta \, \operatorname{cosec} \theta \, d\theta$, prove that $U_n - U_{n-2} = \dfrac{2\cos(n-1)\theta}{n-1}.$

Hence, or otherwise, prove that: $\int_0^{\pi/2} \dfrac{\sin 3\theta \sin 5\theta}{\sin \theta} d\theta = \dfrac{71}{105}.$　　(D.U, 1998)

18. If $I_n = \int_0^1 x^m (1-x)^n\, dx$, prove that

$$(m+n+1) I_n = n\, I_{n-1}.$$

(A.M.I.E., W-2009)

19. Write a reduction formula for $\int x^n e^{2x}\, dx$.　　(A.M.I.E., W-2009)

Ans. $x^n \dfrac{e^{2x}}{2} - \dfrac{n}{2} \int x^{n-1} e^{2x}\, dx.$

OR

In abridged notation, $I_n = \dfrac{x^n e^{2x}}{2} - \dfrac{n}{2} I_{n-1}.$

Chapter 5

Elements of Curve Tracing and Standard Curves

5.1 Introduction

In the course of next few chapters, we shall be dealing with applications of integral calculus to areas of plane curves, lengths of curves, volumes and surfaces of solids of revolution etc. In these applications, which ultimately reduce to evaluation of definite integrals, we shall be required to write proper limits of integration. For this purpose it is essential that we have an idea of the general form of curves represented by given equations which in turn requires knowledge of curve tracing.

In this chapter we shall first discuss the general problem of curve tracing and then trace some important curves which frequently occur in engineering applications.

5.2 While tracing a curve, then along with other things, we should have a knowledge of

(A) *Symmetry.*
(B) *Tangents to the curve at the origin, if it passes through it.*
(C) *Asymptotes (particularly those that are parallel to the axes).*
(D) *Transformation of axes.*

We discuss these one by one.

5.2.1 Symmetry

(i) Symmetry about x-axis

If the equation of a curve remains unchanged when y is changed into −y, the curve is symmetrical about x-axis. In this case, if the point $P(x, y)$ lies on a curve, then $Q(x, -y)$ also lies on it.

For example, the curve $y^2 = 4\,ax$ is symmetrical about *x*-axis.

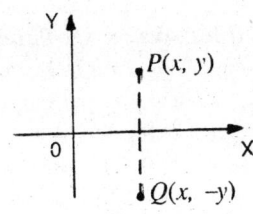

Fig. 5.1.

(ii) Symmetry about y-axis

If *the equation of a curve remains unchanged when x is changed into* $-x$, *the curve is symmetrical about y-axis.* In this case if (x, y) lies on the curve, then $(-x, y)$ also lies on its.

Obviously $x^2 = 4ay$ is symmetrical about y-axis.

Note. It follows from (*i*) and (*ii*) that in the case of algebraic curves, the curve is symmetrical about *x*-axis, if the equation of the curve involves only even powers of *y*. Similarly if only even powers of *x*, occur, the curve is symmetrical about *y-axis*.

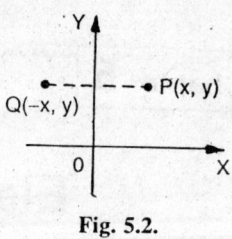

Fig. 5.2.

(iii) Symmetry in opposite quadrants

If *the equation of the curve remains unchanged when x is changed into* $-x$ *and y is changed into* $-y$, *then the curve is symmetrical in opposite quadrants.*

For example the curve $xy = c^2$ is symmetrical in opposite quadrants.

(iv) Symmetry about the line y = x

If *the equation remains unchanged when x and y are interchanged, then the curve is symmetrical about the line y = x.*

Clearly the curve $x^3 + y^3 = 3axy$ is symmetrical about the line $y = x$.

Imp. note: If we know that a curve is symmetrical about *x*-axis, then its form need be known above *x*-axis only. The part below *x*-axis, (on basis of symmetry) can be easily drawn.

Similarly in case of symmetry about *y*-axis, we have to known the shape on the one side of *y*-axis. The shape on the other side is known.

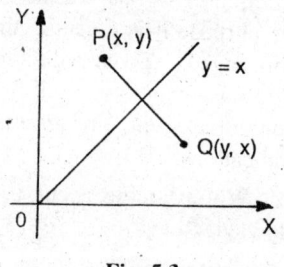

Fig. 5.3.

In case the curve is symmetrical about both axes, then it is sufficient to find the shape in the first quadrant. The remaining part can be, then, easily completed.

5.2.2 Tangents to the curve at the origin

Rule. *If a curve passes through the origin, then the tangents at the origin are obtained by equating to zero, the terms of the lowest degree in the equation.*

For example the parabola $y^2 = 4ax$ passes through the origin. The lowest degree term being $4ax$, the equation of the tangent to the curve at the origin is $4ax = 0$ or $x = 0$, (*i.e.*, the y-axis).

Similarly in the curve $x^2 = 4ay$, $y = 0$, (*i.e.*, the *x*-axis) is the tangent at the origin.

The curve $x^3 + y^3 - 3\,axy = 0$ also passes through the origin. Therefore the tangents there are given by $3\,axy = 0$ or $x = 0$ and $y = 0$. Thus both axes touch the curve at the origin.

When there are two tangents to a curve at a point, as in the last case above, it means two branches of the curve passes through that point. Such a point is called *double point*.

Types of double points

(i) *If the two tangents at a double point on a curve are real and different, the double point is called a NODE [Fig. 5.4 (a)].*

(ii) *If the two tangents coincide, the double point is called a CUSP [Fig. 5.4 (b) (i) & (ii).*

(iii) *In case the two tangents are imaginary, the double point is called a Conjugate Point or Isolated Point [Fig. 5.4 (c)].*

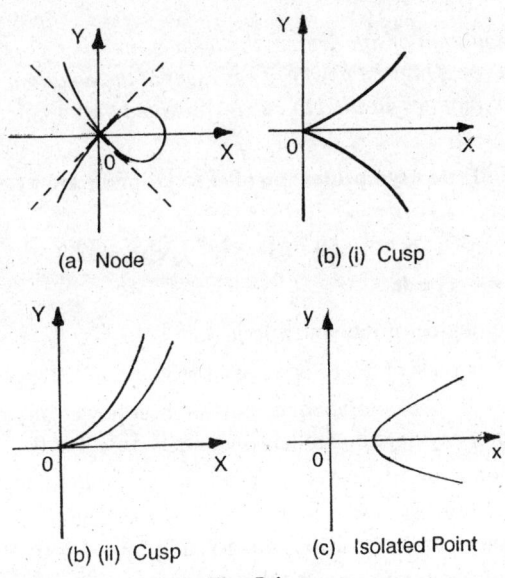

(a) Node   (b) (i) Cusp

(b) (ii) Cusp   (c) Isolated Point

Fig. 5.4.

If the curve passes through the origin, we can find the shape of the curve there by finding tangents at that point. If there are two real and different tangents, the origin is a node. In case there are two co-incident tangents, the origin is a cusp and if the two tangents at the origin are imaginary, the origin is on isolated point.

It should be noted that a point is called isolated point on the curve if there are no other points in its immediate neighbourhood which lie on the curve.

5.2.3 Asymptotes

Def. *An asymptote is a line which has the property that the distance from a point P on the curve to the line approaches zero as the distance from P to the origin increases without bound and P is on suitable piece of the curve.*

Asymptotes are usually classified as the following:

(*i*) Horizontal asymptotes, (*ii*) Vertical asymptotes, and (*iii*) Inclined or oblique asymptotes.

An asymptote which is not parallel to *y*-axis is called an **oblique asymptote**. We state below the rules for finding asymptotes to given curves.

Asymptotes parallel to *x*-axis

Rule. If, *in an equation of nth degree, the term containing x^n is absent, then the co-efficient of the next highest power of x present in the equation when equated to zero, gives the asymptotes parallel to x-axis (provided the coefficient is not merely a constant).*

Asymptotes parallel to *y*-axis.

Rule. *If, in an equation of nth degree, the term containing y^n is absent, then the co-efficient of the next highest power of y present in the equation when equated to zero, gives the asymptotes parallel to y-axis (provided the co-efficient is not merely a constant).*

Example 5.1 **Find the asymptotes parallel to co-ordinate axes, of the following curves:**

(i) $x^3 + 3xy^2 + y^2 + 2x + y = 0$ (ii) $x^2y^3 + x^3y^2 = x^3 + y^3$

(iii) $y^2x - a^2(x - a) = 0$.

Solution. (*i*) The degree of the equation is 3.

(*a*) Since x^3 is present, there is no asymptote parallel to *x*-axis.

(*b*) y^3 is absent. The coefficient of next highest power of *y i.e.*, y^2 is $3x + 1$. Therefore, asymptote parallel to *y*-axis is $3x + 1 = 0$.

(*ii*) The equation is of 5th degree. Here x^5 and y^5 are both absent.

(*a*) The next highest power of *x* is x^3. The coefficient of x^3 being $y^2 - 1$, asymptotes parallel to *x*-axis are given by $y^2 - 1 = 0$. These are $y = \pm 1$.

(*b*) Similarly asymptotes parallel to *y*-axis are given by $x^2 - 1 = 0$. These are $x = \pm 1$.

(*iii*) The degree of the equation is 3. Here both x^3 and y^3 are absent.

(*a*) The next highest power of *x* is *x*. The coefficient of *x* is $y^2 - a^2$ and, therefore, asymptotes parallel to *x*-axis are $y^2 - a^2 = 0$ *i.e.*, $y = \pm a$.

(*b*) Also the next highest power of *y* is y^2 and its coefficient is *x*. Therefore, $x = 0$ (*i.e.*, *y*-axis) is an asymptote.

Oblique asymptotes

Notation. Suppose the equation of the curve is of nth degree. Let $\phi_n(m)$ denote the *n*th degree terms when we put $x = 1$ and $y = m$ in them. Similarly $\phi_{n-1}(m)$ is obtained by putting $x = 1$ and $y = m$ in $(n - 1)$th degree terms and so on.

Rule to get oblique asymptotes

In the highest degree terms put $x = 1$, $y = m$ and get $\phi_n(m)$. Solve $\phi_n(m) = 0$. Let $m_1, m_2, ...$be its roots. These roots are the slopes of the asymptotes.

Next get $\phi_{n-1}(m)$ by putting $x = 1$, $y = m$ in $(n - 1)$th degree terms.

Put different values of m, i.e., m_1, m_2 ... in $c = -\dfrac{\phi_{n-1}(m)}{\phi'_n(m)}$

$$\left[\text{where } \phi'_n(m) = \frac{d}{dm}\phi_n(m) \right] \text{ and get } c_1, c_2, ...$$

Then $y = m_1 x + c_1$, $y = m_2 x + c_2$, ...are the oblique asymptotes.

Example 5.2 Find the asymptotes of $y^3 - x^2y + 2y^2 + 4y + 1 = 0$.

Solution. Putting $x = 1$ and $y = m$ in 3rd degree and second degree terms we get
$$\phi_3(m) = m^3 - m, \qquad \phi_2(m) = 2m^2.$$
The slopes of the asymptotes are roots of
$\phi_3(m) = m^3 - m = 0$. $\therefore$ $m = 0, 1, -1$ are the slopes of the asymptotes.

Now $$c = -\frac{\phi_2(m)}{\phi'_3(m)} = -\frac{2m^2}{3m^2 - 1} \qquad \qquad ...(i)$$

For $m = 0$, $c = -\dfrac{0}{-1} = 0$ $\left[\text{Putting } m = 0 \text{ in } (i)\right]$.

For $m = 1$, $c = -\dfrac{2}{3-1} = -1$. For $m = -1$, $c = -\dfrac{2}{3-1} = -1$.

Hence the three asymptotes are: $y = 0$, $y = x - 1$, $y = -x - 1$.

5.2.4 Shifting the origin to a point (h, k) without changing direction of axes

When the origin is shifted to a point (h, k) without changing the direction of axes, the equation $f(x, y) = 0$ to a curve becomes $f(x + h, y + k) = 0$.

Such a transformation is useful in finding the tangents to a curve at a given point.

Example 5.3 What will the equation $x^2 + y^2 - 6x + 4y - 12 = 0$ become when the origin of co-ordinates is shifted to $(3, -2)$?

Solution. Here $h = 3$, $k = -2$. Therefore, the transformed equation is
$$(x + 3)^2 + (y - 2)^2 - 6(x + 3) + 4(y - 2) - 12 = 0$$
or simplifying, $x^2 + y^2 - 25 = 0$.

5.3 Rules for curve tracing–Cartesian equations

We now give below some of the main rules which usually suffice to obtain the general form of curves from their equations.

1. **Symmetry**

   (*i*) If the equation remains unchanged when y is changed into $-y$, the curve is symmetrical about x-axis.

   (*ii*) If the equation does not change when x is changed into $-x$, the cuve is symmetrical about y-axis.

   (*iii*) If the equation does not change when signs of both x and y are changed, there is symmetry in opposite quadrants.

2. **Origin.** See whether the curve passes through the origin. If it does, write down the equations of the tangents there, by equating to zero the lowest degree terms.

3. **Intersection with axes.** Find the points of intersection with co-ordinate axes and, if necessary, the tangents there by shifting the origin to those points.

4. **Asymptotes.** Find the asymptotes to the curve, if any, particularly those which are parallel to the axes.

5. **Region.** If possible, express the equation of the curve in the form $y = f(x)$. See how y varies when x varies continuously. Find when x or y are imaginary and thus determine the regions where no part of the curve lies.

For example, consider the equation $y^2 = \dfrac{(x-1)^3}{(4-x)}$.

When $x < 1$, y^2 is negative so that y is imaginary.

Similarly when $x > 4$, y is imaginary. Thus the curve lie entirely in the region between the two parallel lines $x = 1$ and $x = 4$.

5.4 Illustrative Examples

Example 5.4. Trace the curve $y^2 = 4ax$. (Parabola)

Solution. We have the following information about the curve: (*i*) When y is changed into $-y$, the equation remains unchanged. Hence the curve is symmetrical about x-axis.

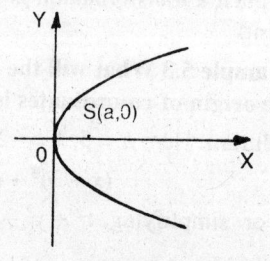

Fig. 5.5.

(*ii*) The curve passes through $(0, 0)$ and the tangent there is $4ax = 0$ or $x = 0$ *i.e.*, the y-axis.

(*iii*) When x is negative, y is imaginary. Thus no part of the curve lies to the left of y-axis.

(*iv*) Taking $y = \sqrt{4ax}$, we note that when x increases continuously y also increases continuously and when $x \to \infty$, y also tends to infinity.

When the shape above x-axis is known, then by symmetry, the shape below x-axis is also known. The graph of the curve is given in fig. 5.5.

Example 5.5 Trace the curve $a^2y^2 = x^2(a^2 - x^2)$

Solution. We have the following information about the curve: (*i*) Since there are only even powers of x and y in the equation, the curve is symmetrical about both the axes.

(*ii*) The curve passes through the origin and the tangents there are given by $a^2y^2 - a^2x^2 = 0$ or $y^2 = x^2$. So the tangents are $y = \pm x$. Hence the origin is a node.

(*iii*) Putting $x = 0$ we get $y^2 = 0$ or $y = 0$. Therefore, the curve meets the y-axis at $(0, 0)$ only.
Putting $y = 0$, we have $x^2(a^2 - x^2) = 0$
$\therefore x = 0, \pm a$. Hence the curve cuts the x-axis at $(0, 0)$, $(a, 0)$ $(-a, 0)$.
The curve being symmetrical about both the axes, an idea about the shape of the curve in the first quadrant is sufficient to trace the curve completely.

(*iv*) $ay = x\sqrt{a^2 - x^2}$. Now $y = 0$ when $x = 0$.

Also as x increases to a, y first increases and then decreases to zero.

Shifting the origin to $A(a, 0)$ the equation of the curve becomes
$$a^2y^2 = (x + a)^2 [a^2 - (x + a)^2]$$
or $\quad a^2y^2 = (x + a)^2 (-x^2 - 2xa)$.

The tangent to the curve at new origin *i.e.*, A is $x = 0$ which is new y-axis.

The shape of the curve in the first quadrant is given in Fig. 5.6.

Since the curve is symmetrical about both the axes, the graph of the curve is given in Fig. 5.7.

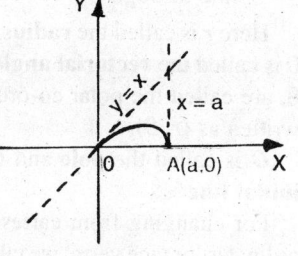

Fig. 5.6.

Example 5.6 Trace the curve $xy^2 = a^2 (a - x)$.

Solution. We have the following information about the curve: (*i*) The curve is symmetrical about x-axis.

(*ii*) It does not pass through the origin.

(*iii*) It meets the x-axis in $(a, 0)$ only. It does not intersect the y-axis.

(*iv*) y is imaginary when x is negative and also when $x > a$. Thus the curve lies entirely in the region between the parallel lines $x = 0$ and $x = a$.

(*v*) $x = 0$ *i.e.* y-axis is an asymptote to the curve.

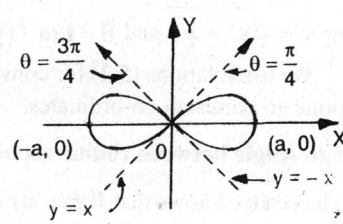

Fig. 5.7.

(*vi*) From $y^2 = \dfrac{a^2 (a - x)}{x}$, we note that $y = \infty$ when $x = 0$ and y decreases as x increases. It is zero when $x = a$.

(*vii*) Shifting the origin to the point $(a, 0)$, the equation reduces to

$$(x + a) y^2 = a^2 [a - (x + a)]$$

or $(x + a)y^2 = -a^2 x.$

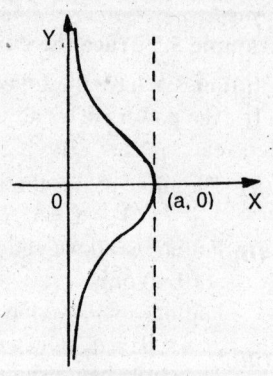

The tangent to the curve at the new origin (that is, $(a, 0)$] is $x = 0$ or the new y-axis.

Keeping in view all the above points, the graph of the curve is given in fig. 5.8.

5.5 Polar co-ordinates

We know how to represent points and curves in cartesian co-ordinates. Another system for

Fig. 5.8.

such representation is the **polar system**. In this system, the position of a point P on a plane is known, if we are given

(*i*) its distance r from a fixed point O.

(*ii*) the inclination θ of OP to OA, a fixed line through O.

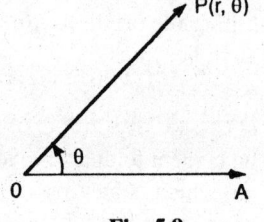

Here r is called the **radius vector** of P and θ is called the **vectorial angle**. The two r and θ, are called the polar co-ordinates of P and written as (r, θ).

O is called the **pole** and OA is called the **initial line**.

Fig. 5.9.

For changing from cartesian to polar co-ordinates, or vice versa, we take origin as pole, the x-axis as initial line.

Then **x = r cos θ, y = r sin θ** ...(5.1) are the formulae for conversion from cartesian to polar co-ordinates.

From (5.1) $x^2 + y^2 = r^2$ and $\tan \theta = y/x$

or $r = \sqrt{x^2 + y^2}$ and $\theta = \tan^{-1}(y/x)$ ...(5.2)

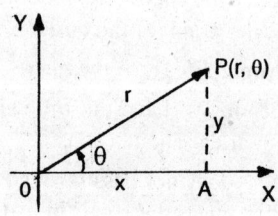

We use relations (5.2) for conversion from polar to cartesian co-ordinates.

Fig. 5.10.

5.6 Angle between radius vector and tangent

The reader knows that if $P(r, \theta)$ is any point on the curve represented by $f(r, \theta) = 0$ and PT is the tangent to the curve at $P(r, \theta)$, then ϕ, the angle between PT and OP produced is given by

$$\tan\phi = r \cdot (d\theta/dr) \qquad ...(5.3)$$

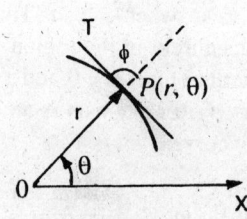

Fig. 5.11.

5.7 Rules for curve tracing–Polar co-ordinates

1. Symmetry

(*i*) If the equation of the curve remains unchanged when θ is changed into −θ, the curve is symmetrical about the initial line.

For example $r = a (1 - \cos\theta)$ remains unchanged when θ is changed to −θ. Hence the curve is symmetrical about the initial line.

(*ii*) If the powers of *r* in the equation of a curve are all even, then the curve is symmetrical about the pole.

The curve $r^2 = a^2 \cos 2\theta$, for example, is symmetrical about the pole.

(*iii*) If the equation remains unchanged when θ is changed into π − θ, the curve is symmetrical about the line through the pole and perpendicular to the initial line.

The curve $r = a \sin 3\theta$ is an example of this type.

(*iv*) If there is a value of θ for which $r = 0$, the curve passes through the pole.

2. **Region.** Find the regions in which a curve does not lie. This can be determined as follow:

(*i*) Find the value of θ which makes r^2 negative and therefore, *r* imaginary.

(*ii*) Find whether the values of r and θ are confined between certain limits.

For example, in the curve $r = a \sin 2\theta$, the values of *r* lie between the limits 0 and *a*. In this case the curve lies within the circle with centre 0 and radius *a*.

3. φ. Find tan φ which will give the angle that the tangent makes with the radius vector.

4. **Table.** If necessary, form a table of values of *r* for both positive and negative values of θ and note how *r* varies with θ.

Example 5.7 Trace the curve

(i) **r = a** (ii) **θ = α** (iii) **r = 2a cosθ.**

Solution. (*i*) If (*r*, θ) be the co-ordinates of a point, then *r* is the distance of the point from the pole.

Since the equation of the curve is $r = a$, the distance of any point on the curve from the pole is *a*.

Hence the curve is a circle with radius *a* and centre at the pole.

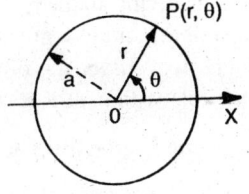

Remark. If $r = a$ then $r^2 = a^2$. Changing this equation in the cartesian form we get $x^2 + y^2 = a^2$ which, obviously, is a circle with centre at the pole and radius *a*.

Fig. 5.12.

(*ii*) θ, the vectorial angle of a point P(*r*, θ) is the angle which *OP* makes with the initial line.

In this case every point on the curve has the same vectorial angle α. Hence the curve is the line through the pole inclined at an angle α to the initial line.

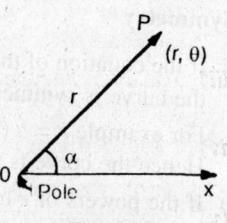

Note. Changing this equation in the cartesian form we get $\tan^{-1}(y/x) = \alpha$ or $y = x \tan \alpha$, which is a line through the origin with slope $\tan \alpha$.

(iii) (a) The equation $r = 2a \cos\theta$ remains unchanged when θ is changed into $-\theta$. Therefore, the curve is symmetrical about the initial line.

Fig. 5.13.

(b) $r = 0$ when $\theta = \pi/2$. Hence the curve passes through the pole.

Also when $\theta = 0$, $r = 2a$. This shows that the curve crosses the initial line at $A(2a, 0)$

(c) The greatest value of r is $2a$. Hence the curve lies entirely within a circle with centre at the pole and radius $2a$.

(d) When $\pi/2 < \theta < 3\pi/2$, r is negative. Hence no part of the curve lies to the left of the line perpendicular to the initial line through the pole.

(e) $\tan\phi = r\dfrac{d\theta}{dr} = (2a\cos\theta)\left(\dfrac{-1}{2a\sin\theta}\right) = -\cot\theta$, where ϕ is the angle between the tangent at (r, θ) and the radius vector.

When $\theta = 0$, $\phi = \pi/2$. Hence tangent at A is perpendicular to the initial line.

When $\theta = \dfrac{\pi}{2}$, $\phi = \pi$. Hence tangent at the pole coincides with the line perpendicular to the initial line.

Table 5.1

θ:	0	$\dfrac{\pi}{6}$	$\dfrac{\pi}{4}$	$\dfrac{\pi}{3}$	$\dfrac{\pi}{2}$
r:	$2a$	$\sqrt{3}a$	$\sqrt{2}a$	a	0

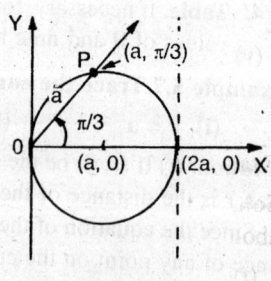

Considering all the points in Table 5.1, the graph of the curve is given in Fig. 5.14. From the graph of curve, it is obvious that the curve is a circle with centre at $(a, 0)$ and radius a.

Fig. 5.14.

5.8 Some Important Curves

5.8.1 Cartesian equations

Example 5.8 Trace the curve $y^2 = ax^3$ (Semi-cubical parabola).
Solution. We have the following information about the curve:

(i) The equation remains unchanged when y is changed to $-y$. Hence the curve is symmetrical about x-axis.

(*ii*) The curve passes through (0, 0), the origin and tangents there are $y^2 = 0$, *i.e.*, two co-incident tangents. Thus the origin is a cusp.

(*iii*) The curve meets the axes only in (0, 0), and no other point.

(*iv*) x cannot be negative. Therefore, no part of the curve lies to the left of y-axes.

(*v*) There are no asymptotes.

(*vi*) As x increases, y increases and when x is very large, y is also very large.

The graph of the curve is given in Fig. 5.15.

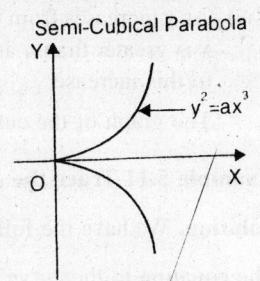

Semi-Cubical Parabola

$y^2 = ax^3$

Fig. 5.15.

Example 5.9 Trace the curve $y = ax^3$ (Cubical Parabola)

Solution. We have the following information about the curve:

(*i*) The equation remains unchanged when signs of both x and y are changed. Therefore, the curve is symmetrical in opposite quadrants.

(*ii*) The curve passes through the origin. The tangent there (by equating to zero the lowest degree term) is $y = 0$, *i.e.*, x-axis.

(*iii*) The curve meets the axes only at (0, 0).

(*iv*) When x is negative, y is negative and when x is positive, y is positive. Hence the curve lies in first and third quadrants only.

(*v*) When x increases y increases. There is no limit to the increase of either. The graph of the curve is given in Fig. 5.16.

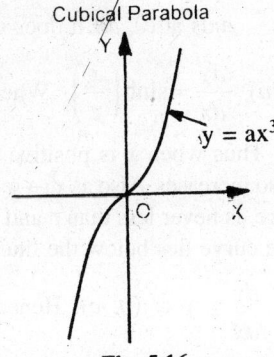

Cubical Parabola

$y = ax^3$

Fig. 5.16.

Example 5.10 Trace the curve $9ay^2 = x(x - a)^2$. (*MDU 2000*)

Solution. We have the following information about the curve:

(*i*) The curve is symmetrical about x-axis only.

(*ii*) It passes through the origin and tangent there is $x = 0$, *i.e.*, y-axis.

(*iii*) It meets the axes at (0, 0) and (a, 0).

(*iv*) The curve has no asymptotes.

(*v*) x cannot be negative. Hence no part of the curve lies to the left of y-axis.

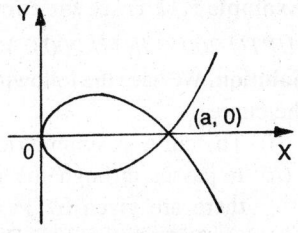

(a, 0)

Fig. 5.17.

(*vi*) As *x* increases from 0 to *a*, *y* first increase and then decreases to zero. When *x* is greater than *a* and increases then *y* also increases and there is no limit to this increase.

The graph of the curve is given in the figure 5.17.

Example 5.11 Trace the curve $y = c \cosh (x/c)$ (*Catenary*)

Solution. We have the following information about the curve:

The equation to the curve can be written in the form $y = c \cdot \left(e^{x/c} + e^{-x/c} \right) / 2$.

(*i*) The equation remains unchanged when *x* is changed into −*x*. Hence the curve is symmetrical about *y*-axis.

(*ii*) When $x = 0$, $y = c \cdot \dfrac{e^0 + e^0}{2} = c \cdot \dfrac{1+1}{2} = c$. Hence the curve intersects the *y*-axis at (0, *c*). It does not cut the *x*-axis.

(*iii*) $\dfrac{dy}{dx} = \sinh\left(\dfrac{x}{c}\right)$. When *x* is positive, $\dfrac{dy}{dx}$ is positive.

Thus when *x* is positive and increases, *y* also increases. Also as $x \to \infty$, $y \to \infty$. *y*, therefore, is never less than *c* and hence no part of the curve lies below the line $y = c$.

$\dfrac{dy}{dx} = 0$ at (0, *c*). Hence the tangent at

(0, *c*) is parallel to *x*-axis.

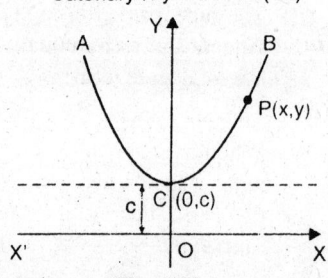

Catenary : y = c cosh (x/c)

Fig. 5.18.

The graph of the curve is given in Fig. 5.18.
Note. The curve represented by $y = c \cosh (x/c)$ is called **Catenary**. Point *C* is called the **vertex** of the catenary. In fact the curve in which a perfectly flexible and uniform cable hangs freely under gravity between two points (not in the same vertical line), is called a catenary.

Example 5.12 Trace the curve $y^2(2a - x) = x^3$ (**a > 0**)

(*UPTU 2005; JNTU 2003; VTU 2003S*)
Solution. We have the following information about the curve:

(*i*) The curve is symmetrical about *x*-axis.
(*ii*) It passes through the origin and tangents there are given by $y^2 = 0$. There are two coincident tangents. Hence the origin is a cusp.

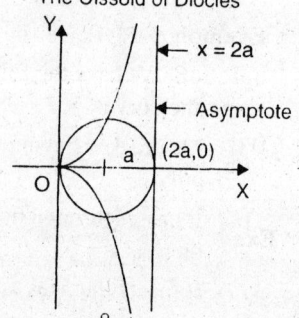

The Cissoid of Diocles

Fig. 5.19.

(*iii*) The curve cuts the axis only at the origin.

(*iv*) y is imaginary when x is negative or $x > 2a$.

Hence the entire curve lies between the line $x = 0$ and $x = 2a$.

(*v*) $x = 2a$ is an asymptote to the curve.

(*vi*) As x increases, y increases and as x approaches $2a$, y becomes very large. The graph of the curve is given in Fig. 5.19.

Example 5.13 Trace the curve $x(x^2 + y^2) = a(x^2 - y^2)$. (*V.T.U. 2000*)

Solution. The equation of the curve can be written in the form

$$y^2(a+x) = x^2(a-x) \text{ or } y^2 = \frac{a-x}{a+x}x^2.$$

We have the following information about the curve:

(*i*) The curve is symmetrical about x-axis.

(*ii*) It passes through the origin. Tangents there are given by $y^2 = x^2$. Hence the tangents are $y = \pm x$.

Since there are two real and different tangents, the origin is a node.

(*iii*) The curve meets the axes only at $(0, 0)$ and $(a, 0)$.

(*iv*) When $x > a$, y is imaginary. Also when $x < -a$, y is imaginary ($\because$ $x + a$ is negative). Hence the entire curve lies in the region between the parallel lines $x = -a$ and $x = a$.

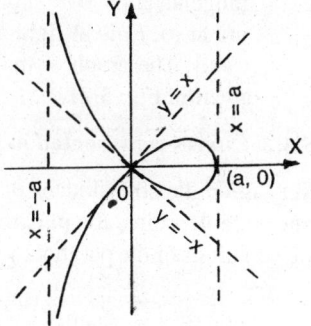

Fig. 5.20.

(*v*) By shifting the origin to the point $(a, 0)$, the equation becomes

$$y^2(x + 2a) = (x + a)^2[a - (x + a)]$$

or $$y^2(x + 2a) = -x(x + a)^2$$

or $$y^2(x + 2a) = -x(x^2 + 2ax + a^2).$$

The tangent at the new origin is $x = 0$, *i.e.*, the new y-axis.

(*vi*) $x + a = 0$ is an asymptote. (*vii*) When $x \to -a$, $y \to \infty$.

Considering the above particulars, the graph of the curve is given in Fig. 5.20.

Example 5.14 Trace the curve $(x/a)^{2/3} + (y/b)^{2/3} = 1$ (**Hypocycloid**)

(*GGSIPU 2006; VTU 2003; MDU 2003*)

Solution. The equation of the curve can be rewritten as

$$\left(x^2/a^2\right)^{1/3} + \left(y^2/b^2\right)^{1/3} = 1 \qquad \qquad \ldots(i)$$

We have the following information about the curve:

(*i*) The above form shows that there is symmetry about both the axes.

(*ii*) The curve does not passes through the origin.

(*iii*) Putting $y = 0$, we get $x = \pm a$ that is, the curve crosses the x-axis at $(a, 0)$ and $(-a, 0)$. Similarly the curve crosses the y-axis at $(0, b)$ and $(0, -b)$.

(*iv*) From the equation (*i*) of the curve we find that when $x > a$, y^2 is negative that is, y is imaginary so the curve does not exist for values of $x > a$. Similarly, the curve does not exist for values of $y > b$.

(*v*) There are no asymptotes.

(*vi*) For the given curve,

$$dy/dx = -b^{2/3} y^{1/3} / a^{2/3} x^{1/3}.$$

Tangents at $(a, 0)$ is along x-axis and at $(0, b)$ is at right angles to x-axis. The graph of the curve is given in Fig. 5.21.

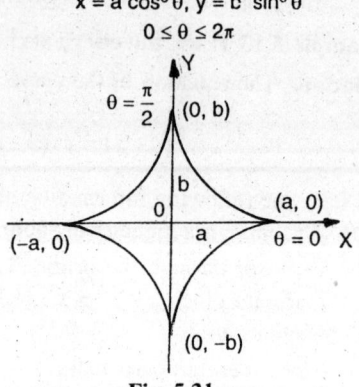

$x = a \cos^3 \theta, y = b \sin^3 \theta$
$0 \le \theta \le 2\pi$

Fig. 5.21.

5.8.2 Curves represented in parametric form

Very often the co-ordinates of any point on a curve can be expressed in terms of one variable, called the **parameter.** For example $x = at^2$, $y = 2at$ are the parametric equations of the parabola $y^2 = 4ax$. Here 't' is the parameter. We also know that

$x = a \cos \theta$, $y = b \sin \theta$ are parametric equations of the ellipse $\dfrac{x^2}{a^2} + \dfrac{y^2}{b^2} = 1$,

θ being the parameter in this case. We shall now trace some other curves when their parametric equations are given.

Example 5.15 Trace the *Cycloid*

$$x = a(\theta - \sin\theta), y = a(1 - \cos\theta).$$

Solution. We shall trace the curve when θ lies in the interval $(0, 2\pi)$.

We have the following information about the curve:

(*i*) **Symmetry.** When θ is changed into $-\theta$, x changes to $-x$ but y remains unchanged. Hence the curve is symmetrical about y-axis.

(*ii*) **Origin.** When $\theta = 0$, $x = 0$, $y = 0$. Therefore, the curve passes through the origin.

(*iii*) **Intersection with axes.** It intersects the x-axis where, putting $y = 0$ we get $1 - \cos\theta = 0$, or $\theta = 0$, 2π. For these values of θ, $x = 0$, $2\pi a$. Therefore, the point of intersection with x-axis are $(0, 0)$ $(2\pi a, 0)$.

To get points of intersection with y-axis, we put $x = 0$.

$$\therefore \qquad a(\theta - \sin\theta) = 0 \text{ or } \theta = 0.$$

For this value of θ, $y = 0$. Hence the point of intersection with y-axis is $(0, 0)$.

(iv) $\dfrac{dx}{d\theta} = a(1 - \cos\theta), \dfrac{dy}{d\theta} = a\sin\theta \cdot$

$$\therefore \quad \frac{dy}{dx} = \frac{dy}{d\theta} \cdot \frac{d\theta}{dx} = \frac{a\sin\theta}{a(1 - \cos\theta)} = \frac{2\sin(\theta/2)\cos(\theta/2)}{2\sin^2(\theta/2)} = \cot\frac{\theta}{2}.$$

Thus $\dfrac{dy}{dx} = 0$ when $\cot\dfrac{\theta}{2} = 0$ or $\theta = \pi$. For this value of θ

$x = \pi a, y = 2a$, Because tangent at $(\pi a, 2a)$ is parallel to x-axis.

$\dfrac{dy}{dx} = \infty$ when $\cot(\theta/2) = \infty$ or $\theta = 0, 2\pi$.

Hence tangent is perpendicular to x-axis at $(0, 0)$ and $(2\pi a, 0)$.

(v) **Region.** Now y cannot be negative ($\because \cos\theta \le 1$) and also y cannot be greater than $2a$. Therefore, the curve lies between the lines $y = 0$ and $y = 2a$. The graph of the curve is given in fig. 5.22.

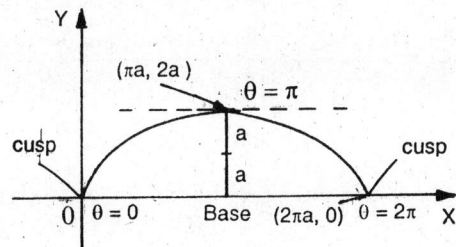

Fig. 5.22.

Remarks. 1. The complete curve consists of the portion from $\theta = 0$ to $\theta = 2\pi$ and endless repetitions of that part to the right and left.

2. The 'complete cycloid' means one arch of the cycloid between two consecutive cusps.

3. Depending upon the choice of axes, we have different forms of equations to the cycloid. The three types of equations with the forms of the curves they represent are given here:

(i) $x = a(\theta + \sin\theta), y = a(1 + \cos\theta)$ (ii) $x = a(\theta - \sin\theta), y = a(1 + \cos\theta)$

(iii) $x = a(\theta + \sin\theta), y = a(1 - \cos\theta)$ (cycloid, vertex at origin (Inverted cycloid)

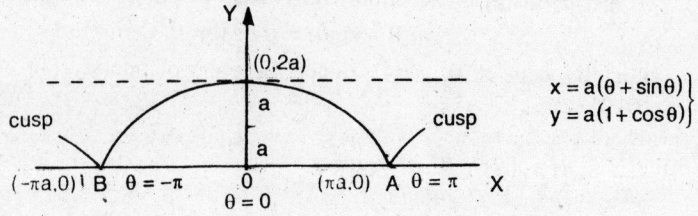

$$x = a(\theta + \sin\theta)$$
$$y = a(1 + \cos\theta)$$

Fig. 5.23.

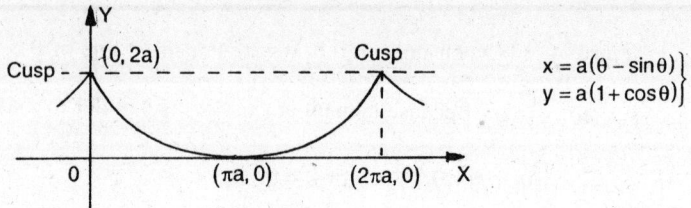

$$x = a(\theta - \sin\theta)$$
$$y = a(1 + \cos\theta)$$

Fig. 5.24.

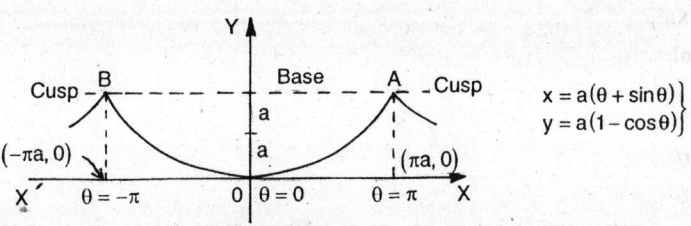

$$x = a(\theta + \sin\theta)$$
$$y = a(1 - \cos\theta)$$

Fig. 5.25.

Example 5.16 Trace the curve $x^{2/3} + y^{2/3} = a^{2/3}$ (Hypocycloid of four cusps – *Astroid*) *(MDU 2003; VTU 2003; Osmania 2002)*

Solution. The parametric equations of this curve are $x = a\cos^3\theta$, $y = a\sin^3\theta$. We have the following information about the curve:

(i) When θ is changed into $-\theta$, y changes to $-y$ but x remains unchanged. Hence the curve is symmetrical about x-axis.

(ii) When θ is changed to $\pi - \theta$ ($0 \le \theta \le \pi/2$), x changes into $-x$ but y remains unchanged. The curve, therefore, is symmetrical about y-axis also.

(iii) $y = 0$ when $\theta = 0$ and then $x = a$. Hence the curve intersects the x-axis at $A(a, 0)$.

$x = 0$ when $\theta = \pi/2$ and then $y = a$. Therefore, the curve cuts the y-axis at $(0, a)$.

(iv) $\dfrac{dy}{dx} = \dfrac{dy}{d\theta} \Big/ \dfrac{dx}{d\theta} = \left(3a\sin^2\theta\cos\theta\right) \Big/ \left(-3a\cos^2\theta\sin\theta\right) = -\tan\theta$.

Now $\dfrac{dy}{dx} = 0$, when $\theta = 0$. Thus the tangent at $A(a, 0)$ coincides with x-axis.

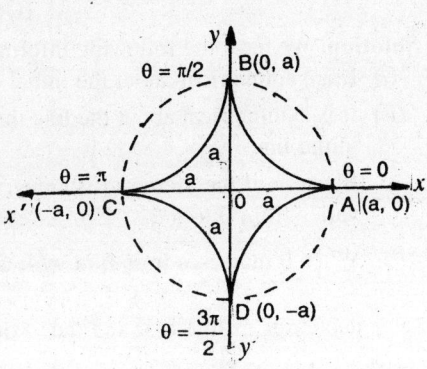

$\dfrac{dy}{dx} = -\infty$, when $\theta = \pi/2$.

Therefore the tangent is perpendicular to x-axis at $(0, a)$. In other words y-axis is the tangent.

(v) As θ increases from 0 to $\pi/2$, $\cos\theta$ and hence x decreases from a to 0 while y increases from 0 to a. The part of the curve in the first quadrant is now known.

Since there is symmetry about both axes, we can complete the curve in other quadrants.

Fig. 5.26.

The graph of the curve is given in Fig. 5.26.

5.8.3 Curve represented in polar form

Example 5.17 Trace the curve, $r = a (1 - \cos\theta)$ **(Cardioid)**

Solution. We have the following information about the curve:

(i) When θ is changed to $-\theta$, the equation remains unchanged. The curve, therefore, is symmetrical about the initial line.

(ii) $r = 0$ when $\theta = 0$. Hence the curve passes through the pole.

(iii) When θ increases from 0 to $\pi/2$, $\cos\theta$ decreases form 1 to 0 and, therefore, r increases from 0 to a. When θ increases from $\pi/2$ to π, $\cos\theta$ decreases from 0 to -1 and thus r increases from a to $2a$.

Now the part of the curve when θ increases from π to 2π is known because of symmetry about the initial line. The curve is given in Fig. 5.27.

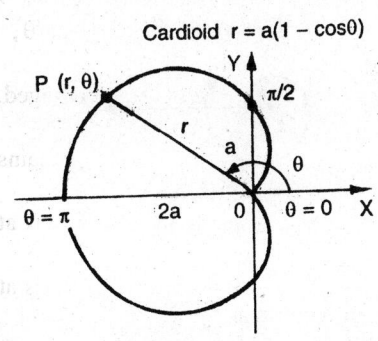

Cardioid $r = a(1 - \cos\theta)$

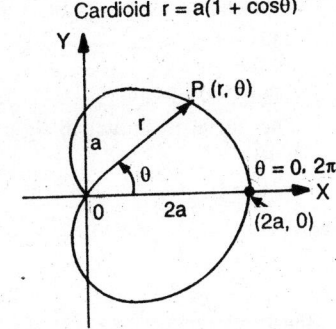

Cardioid $r = a(1 + \cos\theta)$

Fig. 5.27.

Fig. 5.28.

Note. The cardioid with equation in the form $r = a\,(1 + \cos\theta)$ can similarly be traced. The graph of the curve is given in the Fig. 5.28.

Example 5.18 Trace the curve $r^2 = a^2 \cos(2\theta)$ **(Lemniscate of Bernouilli)**

(GGSIPU 2006; JNTU 2003; UPTU 2001)

Solution. We have the following information about the curve:

(i) It is symmetrical about the initial line.

(ii) It is symmetrical about the line through the pole and perpendicular to the initial line.

Because of the aforesaid symmetry we need consider the change in r as θ varies from 0 to $\pi/2$.

(iii) When θ increases from 0 to $\pi/4$, 2θ changes from 0 to $\pi/2$ so that $\cos 2\theta$

decreases from 1 to 0 and thus r decreases from a to 0. For $\theta = \dfrac{\pi}{4}$, $r = 0$.

When θ increases from $\pi/4$ to $\pi/2$, $\cos 2\theta$ is negative and, therefore, r is imaginary. Thus there is no part of the curve between the lines $\theta = \pi/4$ and $\theta = \pi/2$.

The remaining part of the curve can be completed on the basis of symmetry.

The graph of the curve is given in figure 5.29.

Example 5.19 Trace the curve $r = ae^{b\theta}$ *(a, b > 0)* *(Equiangular spiral)*

(i) When $\theta = 0$, $r = a$.

(ii) When θ increases r also increases and as $\theta \to \infty$, $r \to \infty$.
Also when $\theta \to -\infty$, $r \to 0$.

(iii) r is always positive.
The curve is shown in figure 5.30.

Note. Since $\dfrac{dr}{d\theta} = abe^{b\theta}$.

The Lemniscate of Bernoulli

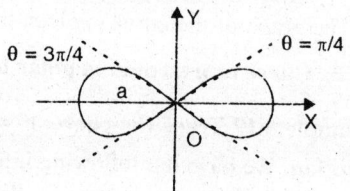

Rectangular coordinate form:
$$(x^2 + y^2)^2 = a^2(x^2 - y^2)$$
or
Polar form $r^2 = a^2 \cos(2\theta)$

Fig. 5.29.

Equiangular Spiral

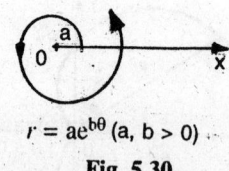

$r = ae^{b\theta}$ (a, b > 0)

Fig. 5.30.

$\therefore \quad \tan\phi = r\dfrac{d\theta}{dr} = \dfrac{ae^{b\theta}}{abe^{b\theta}} = \dfrac{1}{b}, \quad \Rightarrow \phi = \tan^{-1}\left(\dfrac{1}{b}\right)$, a constant.

Because the angle between the radius vector of any point and tangent there is always constant, the curve $r = ae^{b\theta}$ is called an equiangular spiral.

Example 5.20. Trace the curve r = a sin (3θ) (a > 0). *(Three leaved rose)*

Solution. We have the following information about the curve:

(i) $r = 0$ when $\theta = 0$. Hence the curve passes through the pole.

(ii) The value of r cannot exceed a. Therefore the curves lies inside the circle with center at the pole and radius a.

(iii) When θ increases from 0 to $\pi/6$, 3θ increases from 0 to $\pi/2$. Consequently r increases from 0 to a.

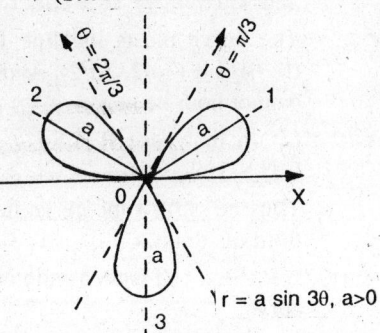

(Three leaved rose)

$r = a \sin 3\theta, a > 0$

Fig. 5.31.

When θ increases from $\pi/6$ to $\pi/3$, 3θ increases from $\pi/2$ to π and r decreases from a to 0.

When θ increases from $\pi/3$ to $\pi/2$, r is negative but its magnitude increases from 0 to a.

When θ increases from $\pi/2$ to $2\pi/3$, r is still negative and numerically decreases from a to 0.

Considering, similarly the variations of r. As θ increases from $2\pi/3$ to $5\pi/6$ and then from $5\pi/6$ to π, we get the second loop above the initial line.

As θ increases from π onward the same three loops of the curve are repeated.

The graph of the curve is given in fig. 5.31.

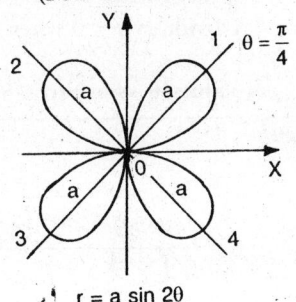

(Four leaved rose)

$\theta = \dfrac{\pi}{4}$

$r = a \sin 2\theta$

Fig. 5.32.

(Four leaved rose)

Example 5.21 Trace the curve r = a sin (2θ)

Proceeding as in the above example, we get the graph of curve given in Fig. 5.32.

Example 5.22 Trace the curve x³ + y³ = 3 axy *(Folium of Descartes).*

(Marthwada, 1998, UPTU 2003; MDU 2001)

Solution. To trace this curve, we require both Cartesian and Polar forms of its equation. We have the following information about the curve:

(i) The equation remains unchanged when x and y are interchanged. Hence the curve is symmetrical about the line $y = x$.

(ii) It passes through $(0, 0)$ that is, the origin. The tangents there are $xy = 0$ that is, $x = 0$ and $y = 0$. Thus the origin is a node on the curve.

(iii) It does not meet the axes at any other point.

(*iv*) Putting $y = x$ in $x^3 + y^3 = 3\,axy$, we get
$2x^3 - 3ax^2 = 0$ or $x^2(2x - 3a) = 0$.

(Folium of Descartes)

Therefore, $\qquad x = 0, 3a/2$.

Now when $\qquad x = 0, y = 0$;

and when $\qquad x = 3a/2, y = 3a/2$.

∴ The curve meets the line $y = x$ at $(0, 0)$ and $(3a/2, 3a/2)$. And x and y cannot both be negative.

(∵ in that case R.H.S. is +ve and L.H.S. is negative).

Thus no part of the curve lies in the third quadrant.

(*v*) $x + y + a = 0$ is an asymptote to the curve.

$x^3 + y^3 - 3axy = 0$

Fig. 5.33.

(*vi*) On transforming to polar co-ordinates, the polar form of the equation is
$r = (3a \sin\theta \cos\theta)/(\cos^3\theta + \sin^3\theta)$.
Obviously $r = 0$ when $\theta = 0$ and $\theta = \pi/2$.

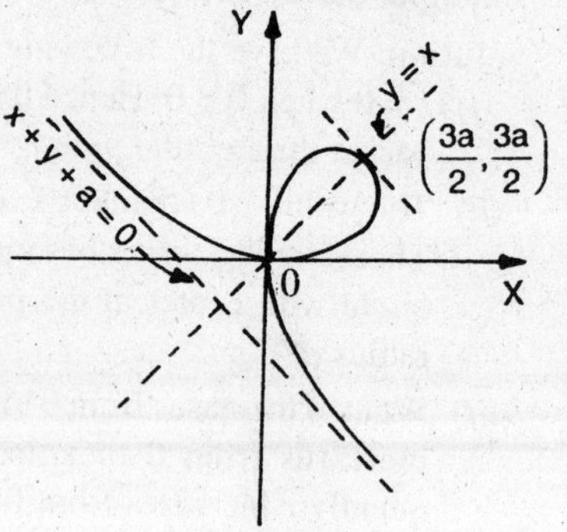

$$\frac{dr}{d\theta} = \frac{3a\left[\left(\cos^3\theta - \sin^3\theta\right)\left(\cos^2\theta - \sin^2\theta\right) - \sin\theta\cos\theta\left\{3\cos^2\theta(-\sin\theta) + 3\sin^2\theta\cos\theta\right\}\right]}{\left(\cos^3\theta + \sin^3\theta\right)^2}$$

$$= \frac{3a(\cos\theta - \sin\theta)\left(1 + \sin\theta\cos\theta + \sin^2\theta\cos^2\theta\right)}{\left(\cos^3\theta + \sin^3\theta\right)^2}$$

Now $\dfrac{dr}{d\theta} = 0$ when $\cos\theta - \sin\theta = 0$ that is, $\tan\theta = 1 \Rightarrow \theta = \dfrac{\pi}{4}$ or $\dfrac{5\pi}{4}$.

Since $\dfrac{dr}{d\theta}$ is + ve when θ increases from 0 to $\dfrac{\pi}{4}$, r increases from 0 to $3a/\sqrt{2}$.

As θ increases from $\pi/4$ to $\pi/2$, r decreases from $3a\sqrt{2}$ to 0. As θ increases from $\pi/2$ to $3\pi/4$, r is negative and numerical increases from 0 to ∞ so that the point (r, θ) describes the part of the curve shown in the fourth quadrant. Because of symmetry about $y = x$, the part in second quadrant can be drawn. The graph of the curve is given in fig. 5.33.

5.9 Curves for Reference

For the convenience of the readers a number of the more common curves are given here.

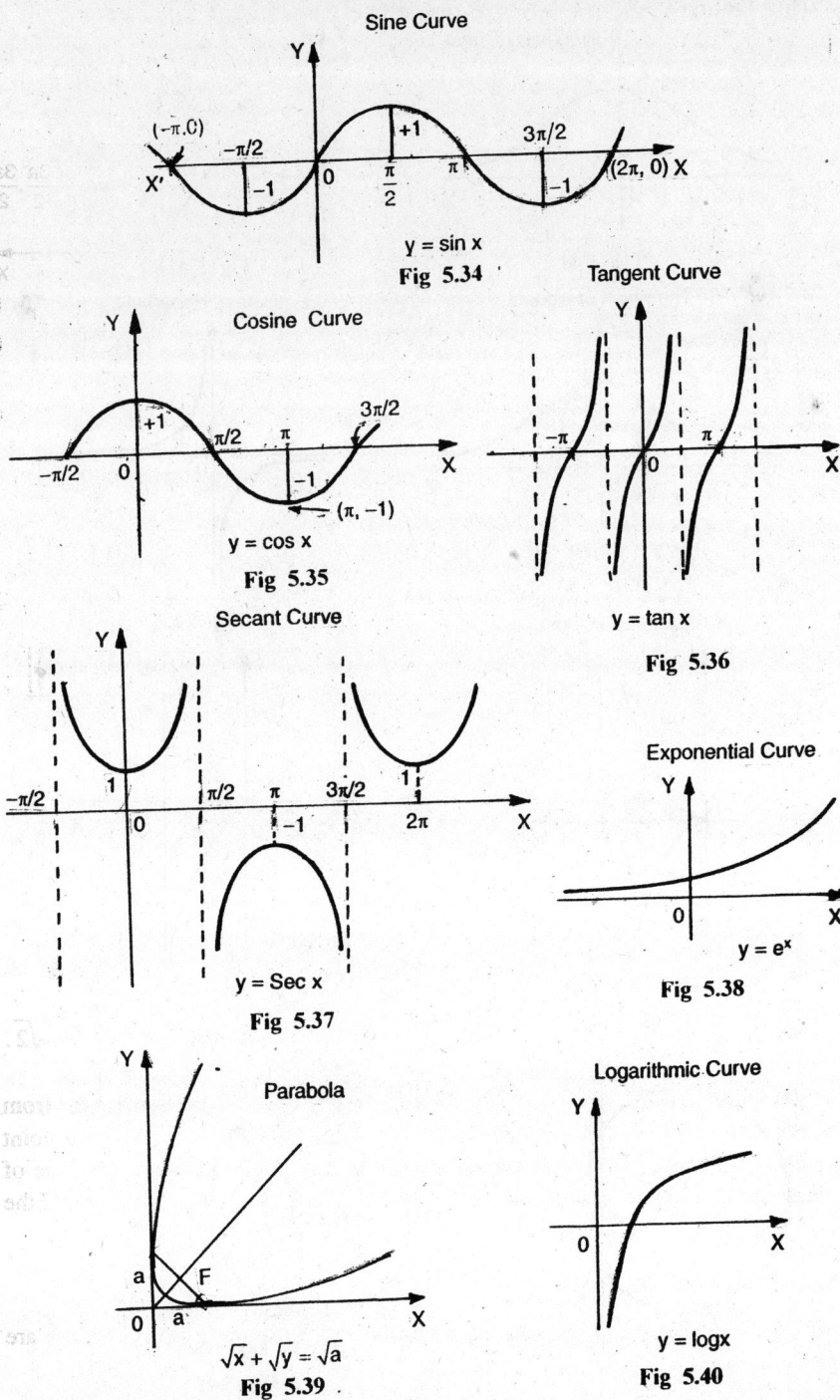

Sine Curve

$y = \sin x$

Fig 5.34

Cosine Curve

$y = \cos x$

Fig 5.35

Tangent Curve

$y = \tan x$

Fig 5.36

Secant Curve

$y = \sec x$

Fig 5.37

Exponential Curve

$y = e^x$

Fig 5.38

Parabola

$\sqrt{x} + \sqrt{y} = \sqrt{a}$

Fig 5.39

Logarithmic Curve

$y = \log x$

Fig 5.40

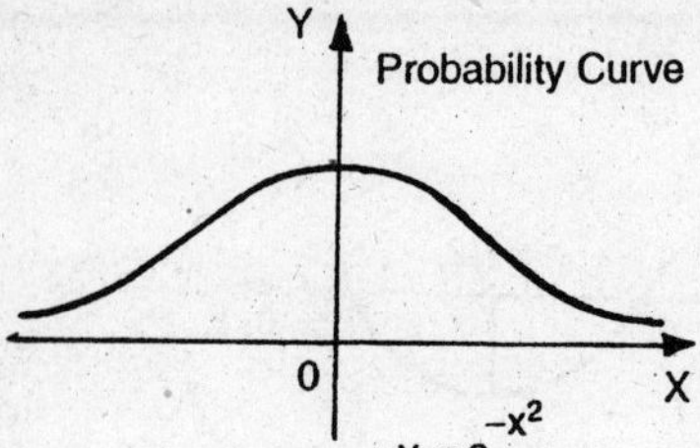

Y
Probability Curve
0
X

$y = e^{-x^2}$

Fig 5.41

Y
3ay² = x(x − a)²
(a, 0)
0
A
X

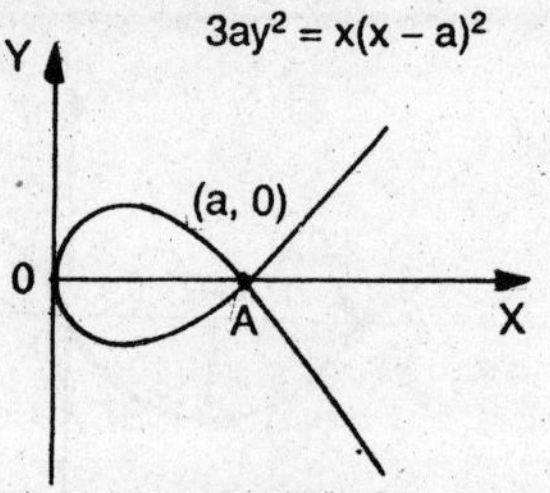

Fig 5.42

The Witch of Agnesi

Y
a
a
0
X

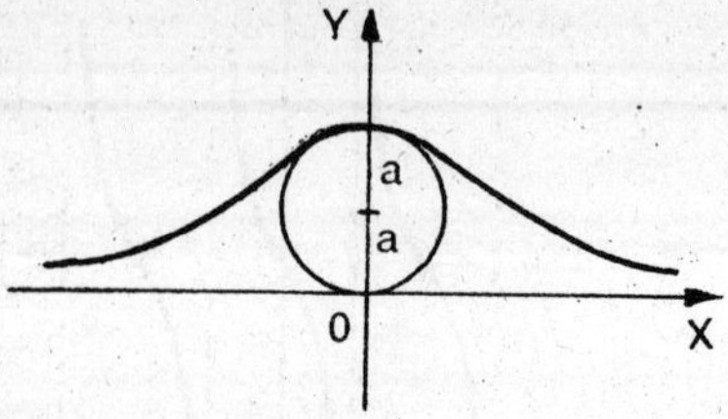

$x^2y = 4a^2(2a - y)$

Fig 5.43

Equilateral Hyperbola

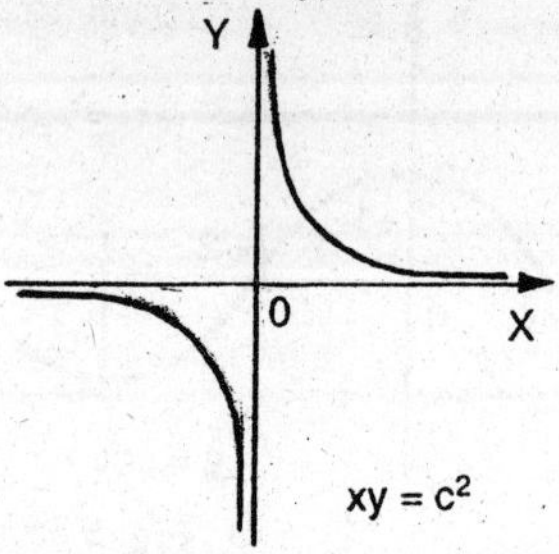

Y
0
X
xy = c²

Fig 5.45

The Cissoid of Diocles

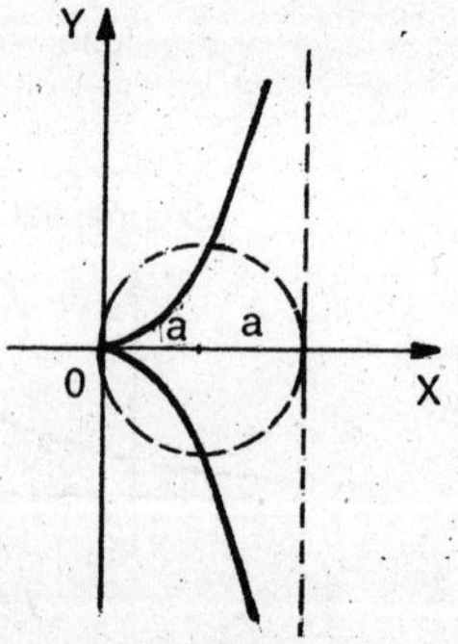

Y
a
a
0
X

$y^2 = x^3/(2a - x)$

Fig 5.44

Y
x = a
y = x
a
(−a, 0)
0
a
X
y = −x
Strophoid

$y^2 = x^2 \left(\dfrac{a + x}{a - x}\right)$

Fig 5.46

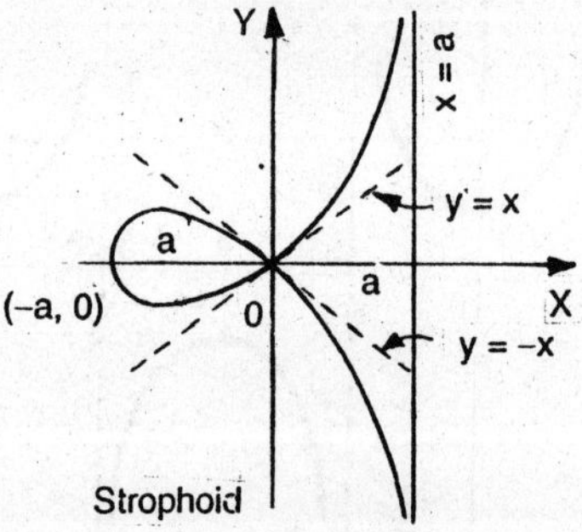

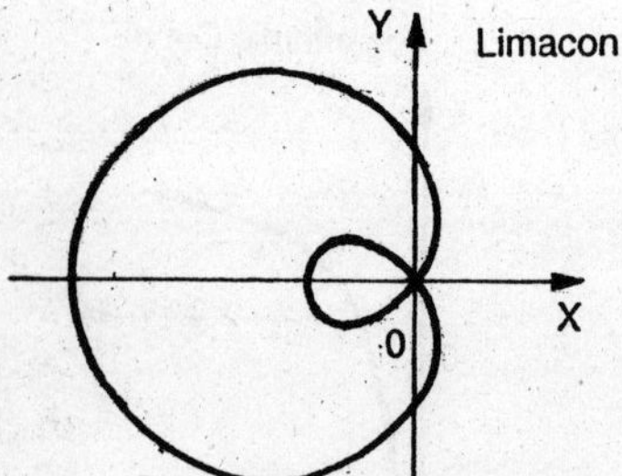

Y
Limacon
0
X

$r = b - a \cos\theta \ (b < a)$

Fig 5.47

Logarithmic or
Equiangular Spiral
0
X

$r = e^{a\theta}$, or $\log r = a\theta$

Fig 5.48

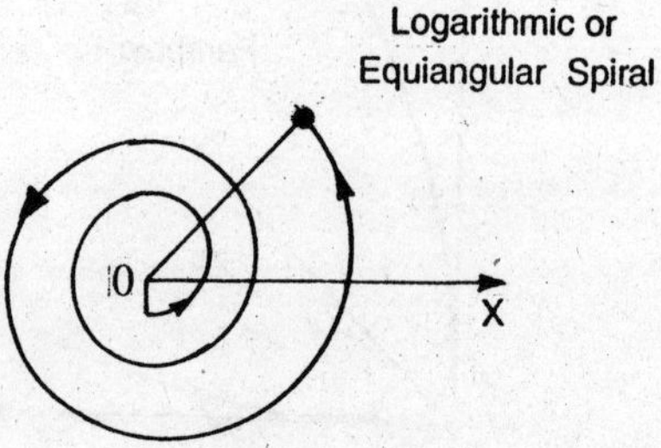

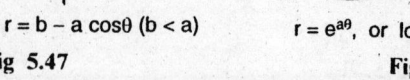

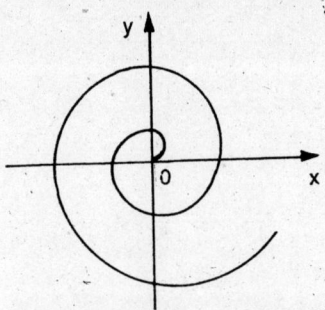

Spiral of Archimedes $r = a\theta$, $(a > 0)$

Fig 5.49

Cardioid: $r = 1 + \sin\theta$

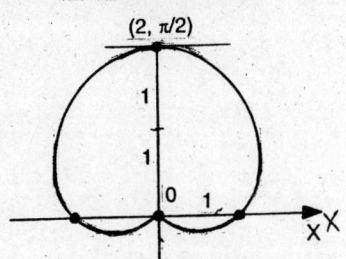

Fig 5.50

Four leaved rose, $r = a \cos(2\theta)$

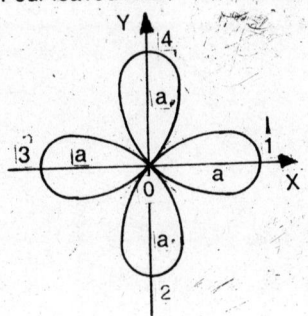

Fig 5.51

Three leaved rose, $r = a \cos(3\theta)$

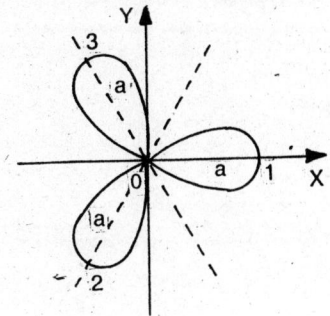

Fig 5.52

Chapter 6

Areas of Plane Curves

6.1 Areas of plane curves, rectangular co-ordinates

Let $y = f(x)$ be the equation to the curve.

The area of the region bounded by the curve, the x-axis, and the lines $x = a$, $x = b$ is given by the formula

$$\textbf{Area} = \int_a^b \mathbf{y}\,\mathbf{dx} = \int_a^b \mathbf{f(x)}\,\mathbf{dx}. \quad ...(6.1)$$

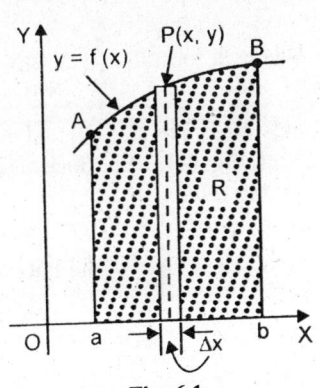

Fig. 6.1.

The value of y in terms of x being sub-stituted from the equation of the curve. The formula (6.1) is readily obtained by observing that the element of the area is a vertical rectangle of base Δx and altitude y. The required area is the limit of the sum of all such vertical rectangles between the lines $x = a$, $x = b$. (See Fig. 6.1.).

The area of the region bounded by the curve $x = \phi(y)$, the y-axis, and the lines $y = c$, $y = d$ is given by

$$\textbf{Area} = \int_c^d \mathbf{x}\,\mathbf{dy} = \int_c^d \boldsymbol{\phi}(\mathbf{y})\,\mathbf{dy}. \quad ...(6.2)$$

The value of x in terms of y is substituted from the equation of the curve. The formula (6.2) is obtained as the limit of the sum of all horizontal rectangles within the required area, x and Δy being, respectively the base and altitude of any elementary rectangle (Ref. Fig. 6.2).

Remarks. 1. The required area may be determined by dividing the region either into horizontal or vertical rectangles. Both type of rectangles should be considered before beginning to set up a definite integral and use only that which can easily determine the value of the definite integral.

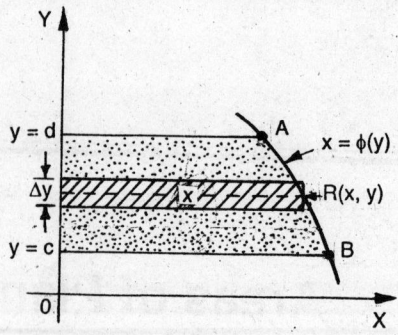

Fig. 6.2.

2. It is always necessary to trace the curve with reference to which the region whose area is to be determined, is given. The shape of the curve will enable the area to be expressed as a define integral or a suitable combination of such integrals.

3. The area formula $\int_a^b y\,dx$ gives a positive result if the area lie above the x-axis. [Ref. Fig. 6.3 (i)] and negative result if the area lies below the x-axis. [Ref. Fig. 6.3 (ii).]. Since area is a positive quantity, we take the magnitude of this value.

4. If the curve $y = f(x)$ is above the x-axis in the interval $a \leq x \leq c$ and below the x-axis in the interval $c \leq x \leq b$, then we write

$$\text{Area} = \int_a^c f(x)\,dx + \left| \int_c^b f(x)\,dx \right|$$

5. The process of finding the area of a plane region is known as *Quadrature*.

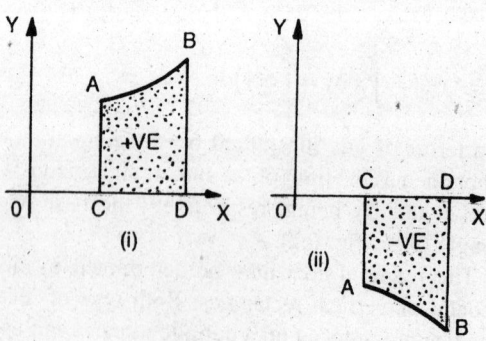

Fig. 6.3.

Example 6.1 Find the area bounded by the curve y = x², the x-axis and the ordinates x = 1 and x = 3.
(*AMIE, W-1999*)

Solution. Required area $A = \int_1^3 y\,dx = \int_1^3 x^2 dx = \left[\frac{x^3}{3}\right]_1^3 = 9 - \frac{1}{3} = \frac{26}{3}$ sq. units.

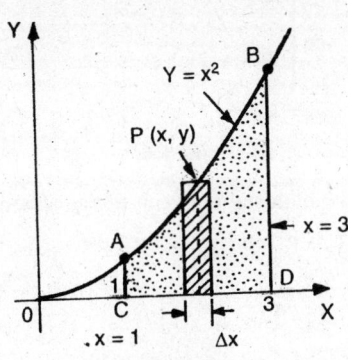

Fig. 6.4.

Example 6.2 Find the area bounded by the semi-cubical parabola ay² = x³, the y-axis, and the lines y = a and y = 2a.

Solution. Refer Fig. 6.5.

Required area $= \int_a^{2a} x\,dy = \int_a^{2a} a^{1/3} y^{2/3}\,dy$

$= \frac{3}{5} a^{1/3}\left[y^{5/3}\right]_a^{2a} = \frac{3}{5}a^2 \left(\sqrt[3]{32} - 1\right) = 1.304\,a^2.$

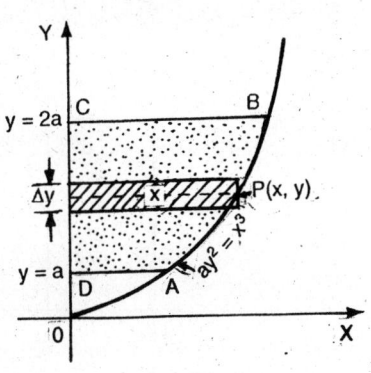

Fig. 6.5.

6.2 Areas between curves

Assume that f and g are continuous functions such that $g(x) \le f(x)$ for $a \le x \le b$. Then the curve $y = f(x)$ lies above the curve $y = g(x)$ between $x = a$ and $x = b$ as shown in the Fig. 6.6.

For the vertical rectangle the width is Δx, the height is {(value of y at the upper boundary) – (value of y at the lower boundary)} $= f(x) - g(x)$, and the area is $\{f(x) - g(x)\}\,\Delta x$.

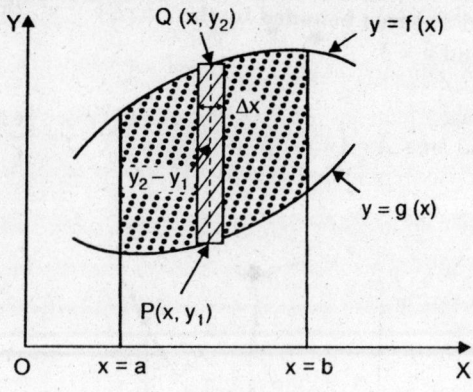

Fig. 6.6.

The area of the region enclosed between the curves $y = f(x), y = g(x)$ and the lines $x = a$ and $x = b$ is given by the formula

$$\text{Area} = \int_a^b \left[f(x) - g(x)\right] dx. \qquad ...(6.3)$$

Remark If $f(x) \geq g(x)$ in $[a,c]$ and $f(x) \leq g(x)$ in $[c,b] a < c < b$, then we write the area as (Fig. 6.7)

$$\text{Area} = \int_a^c \left[f(x) - g(x)\right] dx + \int_c^d \left[g(x) - f(x)\right] dx \qquad ...(6.4)$$

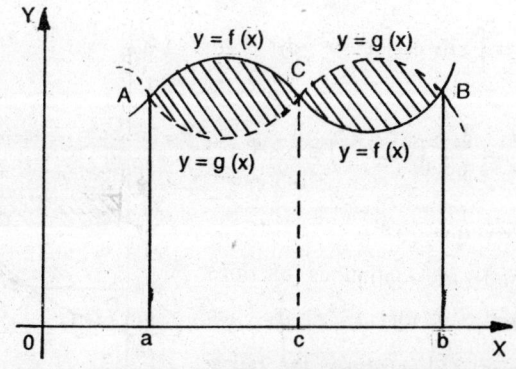

Fig. 6.7.

Example 6.3 Find the area bounded by the parabola $y^2 = 4x$ and the line $y = 2x - 4$.

Solution. The line intersects the parabola at the points $(1, -2)$ and $(4, 4)$. It will be seen from the two figures below that when vertical rectangle is used certain rectangle run from the line to the parabola and others from one branch of parabola to the other branch, while when horizontal rectangle is used, each rectangle runs from the parabola to the line. We give both solutions here to show the superiority of one over the other and to indicate that both types of rectangles should be considered before beginning to set up a definite integral.

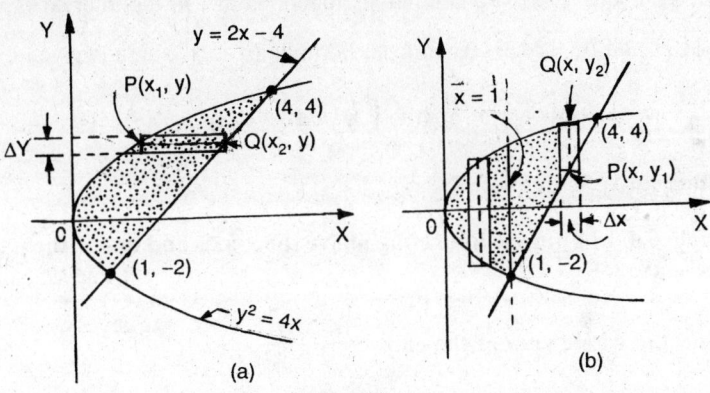

Fig. 6.8.

(a) *Using horizontal rectangle* [See Fig. 6.8(a)]. For the approximating rectangle, the width is Δy, the length is {(value of x on the line) – (value of x on the parabola)}

$$= \frac{(y+4)}{2} - \frac{y^2}{4} = 2 + \frac{y}{2} - \frac{y^2}{4} \text{ and the area of the elementary rectangle is}$$

$$\left(2 + \frac{y}{2} - \frac{y^2}{4}\right)\Delta y.$$

Required area $= \int_{-2}^{4}\left(2 + \frac{y}{2} - \frac{y^2}{4}\right)dy = \left[2y + \frac{y^2}{4} - \frac{y^3}{12}\right]_{-2}^{4} = 9$ sq. units.

(b) *Using Vertical rectangle.* See Fig 6.8 (b). Divide the region by the line $x = 1$. For the approximating rectangle to the right, the width is Δx, the height is $2\sqrt{x} - (2x - 4) = 2\sqrt{x} - 2x + 4$ and the area is $\left(2\sqrt{x} - 2x + 4\right)\Delta x$. For the approximating rectangle to the left of this line, the width is Δx, the height (making use of symmetry) is $2y = 4\sqrt{x}$, and the area is $4\sqrt{x}\,\Delta x$. Required area

$$= \int_{0}^{1} 4\sqrt{x}\,dx + \int_{1}^{4}\left(2\sqrt{x} - 2x + 4\right)dx = \left[\frac{8}{3}x^{3/2}\right]_{0}^{1} + \left[\frac{4}{3}x^{3/2} - x^2 + 4x\right]_{1}^{4}$$

$$= (8/3) + (19/3) = 9 \text{ square units.}$$

6.3 Area bounded by a curve represented in parametric form

Let the curve $y = f(x)$ be defined in parametric form as

$$x = \phi(t), y = \psi(t), \quad a \le t \le b$$

where $\phi(t)$ and $\psi(t)$ are continuous functions of t in the interval $[a, b]$. Let $x_0 = \phi(a)$ and $x_1 = \phi(b)$. Then from Formula (6.1), the area is given by

$$\text{Area} = \int_{x_0}^{x_1} y \, dx = \int_a^b \psi(t) \phi'(t) \, dt \qquad ...(6.5)$$

6.4 Illustrative Examples

Example 6.4 (a) **Find the area lying above the x-axis and under the parabola $y = 4x - x^2$**
(A.M.I.E., W-1998)

(b) **Find the whole area of the ellipse $\dfrac{x^2}{a^2} + \dfrac{y^2}{b^2} = 1$.**

(A.M.I.E. S-2005; MDU 2005)

Solution. (a) The given curve crosses the x-axis at $x = 0$ and $x = 4$. When vertical rectangle is used, these values become the limits of integration. For the elementary rectangle shown in Fig. 6.9. The width is Δx, the height is $y = 4x - x^2$, and the area is $(4x - x^2) \Delta x$. Then

$$\text{Required area} = \int_0^4 \left(4x - x^2\right) dx = \left[2x^2 - \frac{x^3}{3}\right]_0^4 = \frac{32}{3} \text{ sq. units.}$$

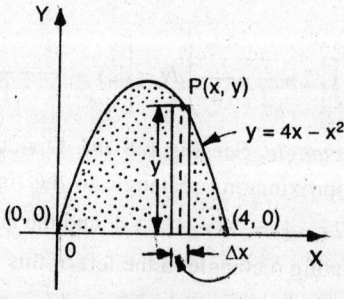

Fig. 6.9.

(b) The ellipse is symmetrical about both the co-ordinate axes, and two axes divide the area into four portions whose areas are equal.

Thus the area bounded by the ellipse = 4 times the area *OAB* (shown shaded. Refer Fig. 6.10)

$$\text{Area } OAB = \int_0^a y\, dx = \frac{b}{a} \int_0^a \sqrt{a^2 - x^2}\, dx$$

$$= \frac{b}{a} \left[\frac{x\sqrt{a^2 - x^2}}{2} + \frac{a^2}{2} \sin^{-1} \frac{x}{a} \right]_0^a = \frac{b}{a} \left[\frac{a^2}{2} \cdot \frac{\pi}{2} \right] = \frac{\pi}{4} ab.$$

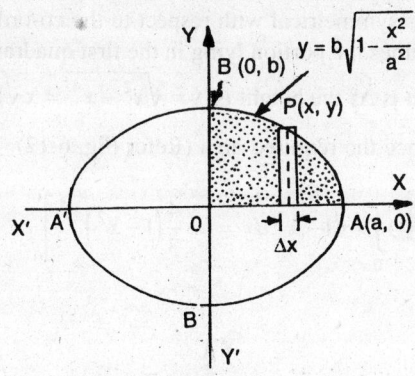

Fig. 6.10.

∴ Area of ellipse $= 4(\pi ab/4) = \pi\, ab.$

Example 6.5 (*a*) **Find the area between the curve $y = x^3 - 6x^2 + 8x$ and the x-axis.**

(*b*) **Find the area enclosed by the curve $y^2 = x^2(1 - x^2)$**   (*Nagpur, 1997*)

Solution. (*a*) We have $y = x^3 - 6x^2 + 8x = x(x^2 - 6x + 8) = x(x - 2)(x - 4)$.

The curve crosses the x-axis at $x = 0$, $x = 2$, and $x = 4$ as shown in Fig. 6.11. Using vertical rectangle, the area of the approximating rectangle with base on the interval $0 < x < 2$ is $(x^3 - 6x^2 + 8x)\Delta x$, and the area of the portion lying above the

x-axis is given by $\displaystyle\int_0^2 \left(x^3 - 6x^2 + 8x\right) dx$.

The area of the elementary rectangle with base on the interval $2 < x < 4$ is $|(x^3 - 6x^2 + 8x)|\, \Delta x$ and the area of the portion lying below the x-axis is given by

$$\left| \int_2^4 \left(x^3 - 6x^2 + 8x\right) dx \right| \text{ [Note this step]}$$

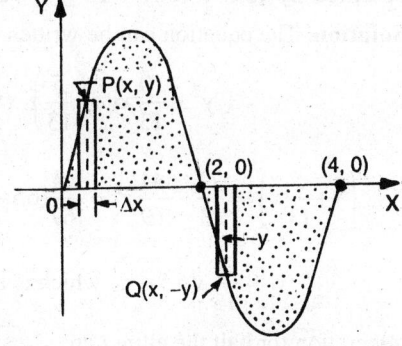

Fig. 6.11.

The required area is therefore,

$$A = \int_0^2 \left(x^3 - 6x^2 + 8x\right)dx + \left|\int_2^4 \left(x^3 - 6x^2 + 8x\right)dx\right|$$

$$= \left[\frac{x^4}{4} - 2x^3 + 4x^2\right]_0^2 + \left|\left[\frac{x^4}{4} - 2x^3 + 4x^2\right]_2^4\right| = 4 + 4 = 8 \text{ square units.}$$

(*b*) The curve is symmetrical with respect to the co-ordinate axes. Hence the required area is 4 times the portion lying in the first quadrant. For the elementary rectangle, the width is Δx the height is $y = \sqrt{x^2 - x^4} = x\sqrt{1 - x^2}$, and the area is $x\sqrt{1 - x^2} \cdot \Delta x$. Hence the required area (Refer Fig. 6.12)

$$= 4\int_0^1 x\sqrt{1 - x^2}\, dx = \left[-\frac{4}{3}\left(1 - x^2\right)^{3/2}\right]_0^1 = \frac{4}{3} \text{ sq. units.}$$

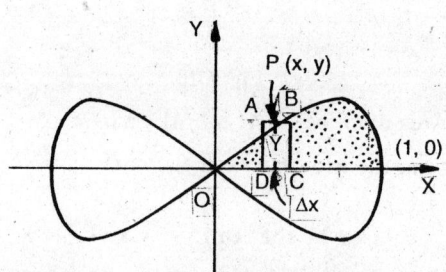

Fig. 6.12.

Example 6.6 Find the area of either of the halves into which the ellipse represented by $13x^2 + 10xy + 13 y^2 - 72 = 0$ is divided by the y-axis

Solution. The equation can be written in an equivalent form as:

$$y^2 + \frac{10}{13}xy + \left(\frac{5}{13}\right)^2 x^2 = \frac{72}{13} - x^2 + \frac{25}{169}x^2$$

i.e., $\left(y + \frac{5}{13}x\right)^2 = \frac{72}{13} - \frac{144}{169}x^2 = \frac{144}{169}(6.5 - x^2)$ or $y = -\frac{5}{13}x \pm \frac{12}{13}\sqrt{\left(6.5 - x^2\right)}$.

$\therefore\ y_2 - y_1 = \frac{24}{13}\sqrt{6.5 - x^2}$ which is zero when $x = \pm\sqrt{6.5}$. Thus the limits of integration for half the ellipse are $-\sqrt{6.5}$ and 0, or 0 and $\sqrt{6.5}$. Taking the latter limits, we have

$$A = \text{area of half ellipse} = \int_0^{\sqrt{6.5}} \frac{24}{13}\sqrt{6.5-x^2}\,dx.$$

Let $x = \sqrt{6.5}\sin\theta$, then $dx = \sqrt{6.5}\,\cos\theta\,d\theta$,

when $x = 0$, $\theta = 0$ and when $x = \sqrt{6.5}$, $\theta = \pi/2$.

Hence, substituting,

$$A = \int_0^{\pi/2} \frac{24}{13}\times 6.5\cos^2\theta\,d\theta = 6\left[\theta + \frac{\sin 2\theta}{2}\right]_0^{\pi/2} = 3\pi \text{ sq. units}$$

Example 6.7 (*a*) **Find the area of the region enclosed between the cycloid**

$$x = a(\theta - \sin\theta),\ y = a(1 - \cos\theta),\ \text{and its base i.e., x-axis.}$$

(A.M.I.E., S-1996, V.T.U., 2000)

(*b*) **Find the area bounded by the curve** $x = 3 + \cos\theta$, $y = 4\sin\theta$.

(*c*) **Find the area of the loop of the curve** $x = a(1 - t^2)$, $y = a\,t(1 - t^2)$.

(A.M.I.E; S-1998)

Solution. (a) An arch is described as θ varies from 0 to 2π, x varies from 0 to $2\pi a$ (Refer Fig. 6.13). $dx = a(1 - \cos\theta)\,d\theta$ Hence,

$$\text{Area} = \int_0^{2\pi a} y\,dx = \int_0^{2\pi} a(1-\cos\theta)\cdot a(1-\cos\theta)\,d\theta$$

$$= a^2 \int_0^{2\pi} \left(1 - 2\cos\theta + \cos^2\theta\right)d\theta$$

$$= a^2 \int_0^{2\pi} \left[1 - 2\cos\theta + \frac{1}{2}(1+\cos 2\theta)\right]d\theta$$

$$= a^2 \int_0^{2\pi} \left(\frac{3}{2} - 2\cos\theta + \frac{1}{2}\cos 2\theta\right)d\theta$$

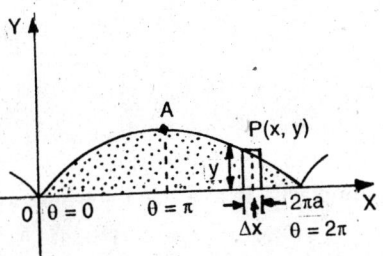

Fig. 6.13.

$$= a^2 \left[\frac{3}{2}\theta - 2\sin\theta + \frac{1}{4}\sin 2\theta\right]_0^{2\pi} = 3\pi a^2 \text{ square units.}$$

(*b*) The required area = 4 time the area of portion *QAB* (See Fig. 6.14).

$$\text{Area of } QAB = \int_3^4 y\,dx,$$

Here $dx = -\sin\theta\, d\theta$; when $x = 3$, $\theta = \pi/2$, when $x = 4$, $\theta = 0$.

∴ Area of QAB

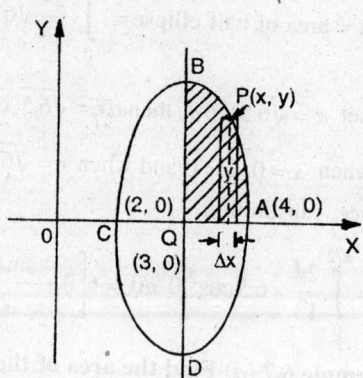

$$= \int_{\pi/2}^{0} 4\sin\theta\,(-\sin\theta)\,d\theta$$

$$= \int_{0}^{\pi/2} (4\sin\theta)(\sin\theta)\,d\theta$$

$$= 4\int_{0}^{\pi/2} \sin^2\theta\, d\theta$$

$$= 2\int_{0}^{\pi/2} (1-\cos 2\theta)\,d\theta = 2\left[\theta - \frac{1}{2}\sin 2\theta\right]_{0}^{\pi/2} = \pi$$

Fig. 6.14.

∴ Required area = 4π sq. units.

(c) Eliminating t, between the given equations of the curve, we obtain $ay^2 = x^2(a-x)$ as the Cartesian equation of the curve. The shape of the curve is shown in Fig. 6.15

Required area of the loop

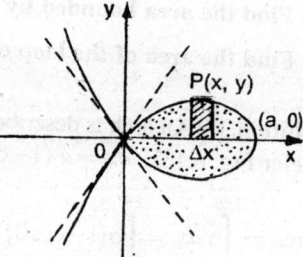

$$= 2\int_{0}^{a} y\,dx = 2\int_{0}^{a} \frac{x\sqrt{a-x}}{\sqrt{a}}\,dx, \text{ putting } x = a\sin^2\theta$$

Fig. 6.15.

and $dx = 2a\sin\theta\cos\theta\, d\theta$. When $x = 0$, $\theta = \pi$; when $x = a$, $\theta = \dfrac{\pi}{2}$.

$$\therefore \text{ Required area } = 4a^2 \int_{0}^{\pi/2} \sin^3\theta\cos^2\theta\, d\theta = 4a^2 \cdot \frac{2}{5\cdot 3} = \frac{8a^2}{15} \text{ sq. units}$$

Example 6.8 Prove that the whole area between the four infinite branches of

the tractrix $x = a\cos t + \dfrac{1}{2}a\log\tan^2\left(\dfrac{t}{2}\right)$, $y = a\sin t$ is πa^2.

Solution. Tracing the curve, we find that whole area between the four branches,

$$= 4\int_{0}^{\infty} y\,dx = 4\int_{\pi/2}^{\pi} y\frac{dx}{dt}\cdot dt. \text{ (Refer Fig. 6.16)}$$

$$(\because \text{ when } x = 0,\ t = \pi/2\,; \text{ when } x = \infty, t = \pi)$$

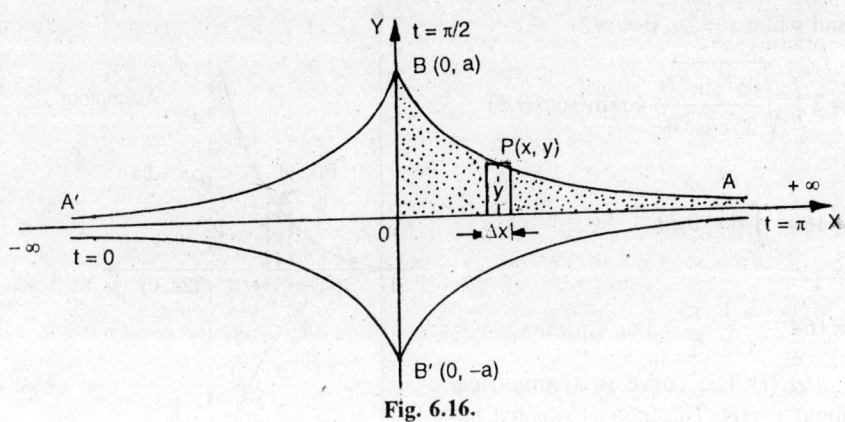

Fig. 6.16.

We have
$$\frac{dx}{dt} = -a\sin t + \frac{a}{2}\frac{1}{\tan^2 t/2} \cdot 2\tan\frac{t}{2} \cdot \sec^2\left(\frac{t}{2}\right) \cdot \frac{1}{2}$$

$$= -a\sin t + a\frac{1}{2\sin\dfrac{t}{2}\cdot\cos\dfrac{t}{2}} = -a\sin t + \frac{a}{\sin t} = \frac{a\cos^2 t}{\sin t}$$

$$\therefore \qquad A = 4\int_{\pi/2}^{\pi} a\sin t \cdot \frac{a\cos^2 t}{\sin t}\cdot dt = 4a^2\int_{\pi/2}^{\pi}\cos^2 t\, dt$$

$$= \frac{4a^2}{2}\int_{\pi/2}^{\pi}(1+\cos 2t)\,dt = 2a^2\left[t+\frac{\sin 2t}{2}\right]_{\pi/2}^{\pi} = 2a^2\left(\pi-\frac{\pi}{2}\right) = \pi a^2.$$

Example 6.9 (*a*) **Find the area between the curve**

$$y^2 = \frac{x^3}{(2a-x)}$$ **and its asymptote.** (*A.M.I.E., W-1995, Ranchi, 1998*)

(*b*) **Trace the curve** $x(x^2 + y^2) = a(x^2 - y^2)$ **and find the area enclosed by the curve and its asymptote. Also find the area of the loop.**

(*c*) **Find the area included between the curve** $xy^2 = a^2(a-x)$ **and it asymptote.**
(*A.M.I.E., S-1999, W-2001*)

Solution. (*a*) The shape of the curve is as shown in Fig. 6.17. The asymptote is the line $x = 2a$.

$$\text{Required area} = 2\int_0^{2a} y\,dx = 2\int_0^{2a}\sqrt{\frac{x^3}{2a-x}}\cdot dx$$

(Put $x = 2a \sin^2\theta$; when $x = 0$, $\theta = 0$
and when $x = 2a$, $\theta = \pi/2$)

$$= 2\int_0^{\pi/2} \sqrt{\frac{8a^3 \sin^6\theta}{2a\cos^2\theta}}\ 4a\sin\theta\cos\theta\, d\theta$$

$$= 16a^2 \int_0^{\pi/2} \sin^4\theta\, d\theta$$

$$= 16a^2 \cdot \frac{3 \cdot 1}{4 \cdot 2} \cdot \frac{\pi}{2} = 3\pi a^2 \text{ sq. units.}$$

(b) (i) The curve is symmetrical
about x-axis. The loop is situated between the lines $x = 0$, and $x = a$. The
line $x = -a$ is the asymptote of the curve.
The tangents at $(0, 0)$ are $y = \pm x$. The
graph of the curve is shown in Fig. 6.18.

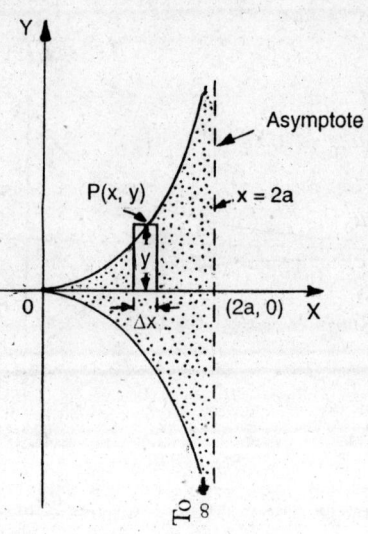

Fig. 6.17.

Required area = 2 area of portion $OCD = 2\int_{-a}^{0} y\, dx$

$$= 2\int_{-a}^{0} x\sqrt{\frac{a-x}{a+x}}\, dx \qquad \left(\because y^2 = \frac{x^2(a-x)}{(a+x)}\right)$$

$$= 2\int_{-\pi/2}^{0} a\sin\theta \cdot \sqrt{\frac{a-a\sin\theta}{a+a\sin\theta}} \cdot a\cos\theta\, d\theta = 2a^2 \int_{-\pi/2}^{0} \sqrt{\frac{1-\sin\theta}{1+\sin\theta}}\ \sin\theta\cos\theta\, d\theta$$
$$\text{(putting } x = a\sin\theta)$$

$$= 2a^2 \int_{-\pi/2}^{0} \frac{(1-\sin\theta)\sin\theta}{\sqrt{1-\sin^2\theta}}\cos\theta\, d\theta$$

$$= 2a^2 \int_{-\pi/2}^{0} (\sin\theta - \sin^2\theta)\, d\theta$$

$$= 2a^2 \left[-\cos\theta - \frac{\theta}{2} + \frac{\sin 2\theta}{4}\right]_{-\pi/2}^{0}$$

$$= 2a^2 \left[-1 - \frac{\pi}{4}\right] = (\pi + 4)\frac{a^2}{2}$$

(ignoring –ve sign)

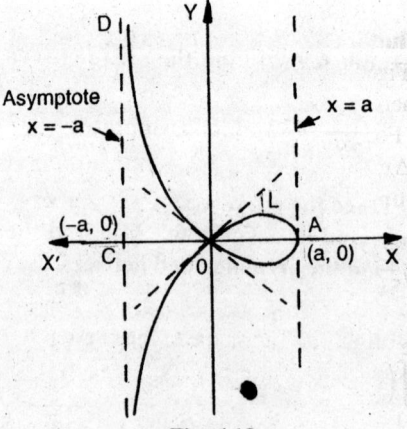

Fig. 6.18.

(*ii*) Area of loop $= 2\int_0^a y\, dx = 2\int_0^a x\sqrt{\dfrac{a-x}{a+x}}\, dx$

$= 2a^2\left[-\cos\theta - \dfrac{\theta}{2} + \dfrac{\sin 2\theta}{4}\right]_0^{\pi/2}$,

$= 2a^2\left[1-(\pi/4)\right] = a^2(4-\pi)/2$ square units.

(*c*) The curve is symmetrical about the *x*-axis and is shown in Fig. 6.19. It cuts the *x*-axis at $x = a$. The asymptote is the *y*-axis.
Required area

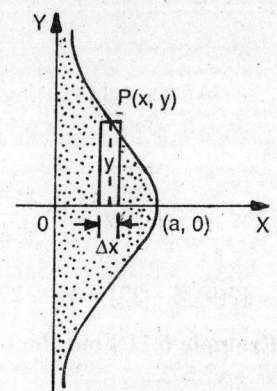

$= 2\int_0^a y\, dx = 2\int_0^a a\sqrt{\dfrac{a-x}{x}}\, dx$.

Let $x = a\sin^2\theta$ then $dx = 2a\sin\theta\cos\theta\, d\theta$.
When $x = 0$, $\theta = 0$; when $x = a$, $\theta = \pi/2$.

∴ Required Area $= 2a\displaystyle\int_0^{\pi/2}\sqrt{\dfrac{a\cos^2\theta}{a\sin^2\theta}}\; 2a\sin\theta\cos\theta\, d\theta$

$= 4a^2\displaystyle\int_0^{\pi/2}\cos^2\theta\, d\theta = 4a^2\cdot\dfrac{1}{2}\cdot\dfrac{\pi}{2} = \pi a^2$.

Fig. 6.19.

Example 6.10 (*a*) **Find the area bounded by the parabolas $y^2 = 5x + 6$ and $x^2 = y$.**

(*b*) **Find the area bounded by the curves $y^2 = 9x$ and $x^2 = 9y$.**

(*A.M.I.E. S-2003, S-2005*)

Solution. (*a*) The two parabolas are shown in Fig. 6.20. The curves intersect at the points where $x^4 = 4x + 6$. By trial $x = -1$ and $x = 2$.

For the elementary rectangle the width is Δx, the height is {(value of *y* at the upper boundary) − (value of *y* at the lower boundary)} $= \sqrt{5x+6} - x^2$, and its area is $\left(\sqrt{5x+6} - x^2\right)\cdot\Delta x$. Then required area

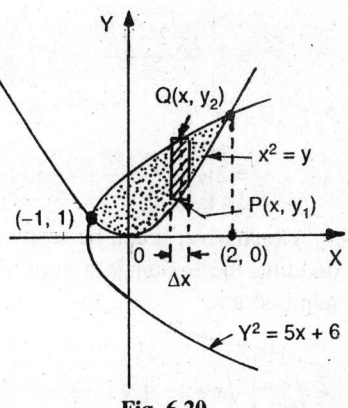

$= \displaystyle\int_{-1}^2\left(\sqrt{5x+6} - x^2\right)dx = \left[\dfrac{(5x+6)^{3/2}}{5\times 3/2} - \dfrac{x^3}{3}\right]_{-1}^2$

Fig. 6.20.

$$= \left(\frac{64 \times 2}{15} - \frac{8}{3} \right) - \left(\frac{2}{15} + \frac{1}{3} \right) = \frac{27}{5} \text{ sq. units.}$$

(b) The curves intersect at the points where $3\sqrt{x} = x^2/9$ or $x^4 - 729x = 0$ that is, at $x = 0$ and $x = 9$. Ref. Fig. 6.21.

Area of elementary rectangular strip

$$= \left(\sqrt{9x} - \frac{x^2}{9} \right) \Delta x \, \cdot$$

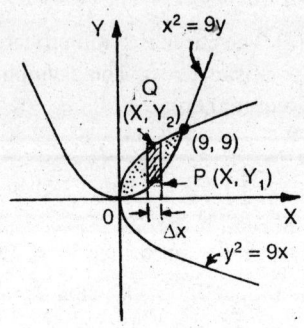

$$\therefore \quad \text{Area sought} = \int_0^9 \left(3\sqrt{x} - \frac{x^2}{9} \right) dx$$

$$= \left[3 \frac{x^{3/2}}{3/2} - \frac{x^3}{27} \right]_0^9$$

$$= [2(9)^{3/2} - 27] = 54 - 27 = 27 \text{ sq. units.}$$

Fig. 6.21.

Example 6.11 **Find the area common to the parabola $y^2 = x$ and the circle $x^2 + y^2 = 2$.**
(A.M.I.E., S-2001)

Solution. The two curves meet at $(1, 1)$ and $(1, -1)$ (Refer Fig. 6.22). The area required is twice the area in the first quadrant. We divide the region OAB into horizontal rectangles.

$$\text{Area sought} = 2\int_0^1 \left[\sqrt{2 - y^2} - y^2 \right] dy$$

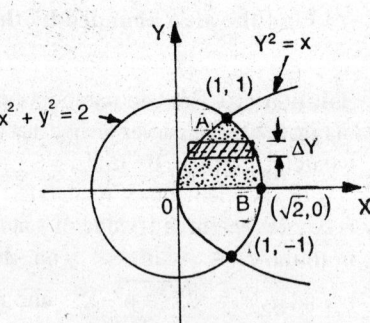

$$= 2 \left[\frac{y\sqrt{2 - y^2}}{2} + \sin^{-1} \frac{y}{\sqrt{2}} - \frac{y^3}{3} \right]_0^1$$

$$= 2 \left[\frac{1}{2} + \frac{\pi}{4} - \frac{1}{3} \right] = \frac{1}{6}(3\pi + 2) \text{ sq. units}$$

Alternative. It can be easily seen that by dividing the region into vertical rectangles, required area

Fig. 6.22.

$$A = 2 \left[\int_0^1 \sqrt{x} \, dx + \int_1^{\sqrt{2}} \sqrt{2 - x^2} \, dx \right] = 2 \left[\frac{2}{3} x^{3/2} \right]_0^1 + 2 \left[\frac{x\sqrt{2 - x^2}}{2} + \sin^{-1} \frac{x}{\sqrt{2}} \right]_1^{\sqrt{2}}$$

$$= 2\left[\left(\frac{2}{3}\right)+\left(0+\frac{\pi}{2}-\frac{1}{2}-\frac{\pi}{4}\right)\right] = 2\left(\frac{1}{6}+\frac{\pi}{4}\right) = \frac{1}{6}(3\pi+2).$$

Example 6.12 Find the area bounded by the curve $(x+y)^2 \doteq 4x$ and the line $y = -3$.

Solution. First trace the curve $(x+y)^2 = 4x$, we split it into two curves i.e.,

$x + y = \pm 2\sqrt{x}$ or $y = 2\sqrt{x} - x$ and $y = -2\sqrt{x} - x$.

For the curve $y = \sqrt{x}\left(2-\sqrt{x}\right)$; $y = 0$ when $x = 0$ and $y = 0$ when $x = 4$.

For $0 < x < 4$, y exists, first increases and then decreases.

For $x > 4$, y is negative. For $x < 0$, y does not exist.

For the curve $y = -\sqrt{x}\left(2+\sqrt{x}\right)$; $y = 0$ when $x = 0$ only.

For $x > 0$, y exists its values is negative and increases.

For $x < 0$, y does not exist.

Keeping in view the above points, the shape of the curve is shown in the figure, 6.23 which is a form of parabola. Tangent at the origin is $x = 0$ i.e. y-axis.

Consider rectangle parallel to x-axis. Area of elementary rectangle is $(x_2 - x_1)\Delta y$. The limit of integration will be from $y = -3$ to the tangent line to the curve at the highest point for which $dy/dx = 0$.

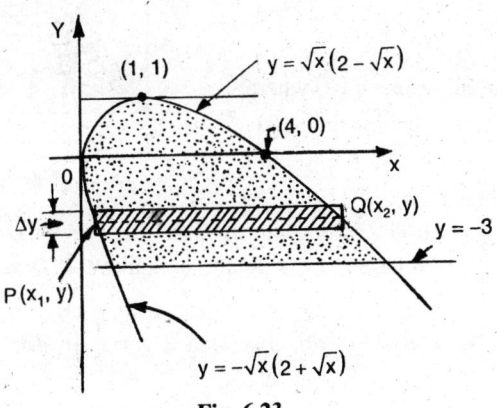

Fig. 6.23.

Differentiating the equation $(x+y)^2 = 4x$ w.r.t. x, we get $2(x+y)\left(1+\dfrac{dy}{dx}\right) = 4$.

Putting $dy/dx = 0$, we get the condition $x + y = 2$. The point where the tangent is parallel to x-axis is $(1, 1)$, which we obtain by putting $x + y = 2$ in the original equation. The area between the curve and the line $y = -3$ is

$$A = \int_{-3}^{1} (x_2 - x_1)\,dy \ . \text{ Since } x^2 + 2xy + y^2 = 4x \text{ or } x^2 + 2x(y-2) + y^2 = 0$$

$$\therefore \quad x = \frac{\left\{-2(y-2) \pm \sqrt{4(y-2)^2 - 4y^2}\right\}}{2} = -(y-2) \pm 2\sqrt{1-y}$$

From the above equation, it follows:

$$x_1 = -y + 2 + 2\sqrt{1-y} \text{ and } x_2 = -y + 2 - 2\sqrt{1-y}$$

Therefore, $(x_2 - x_1) = -4\sqrt{1-y}$.

Putting this value of $(x_2 - x_1)$ in the above integral, we obtain

$$A = -4\int_{-3}^{1} \sqrt{1-y}\,dy = \frac{8}{3}\Big[(1-y)^{3/2}\Big]_{-3}^{1} = \frac{8}{3}\left(-4^{3/2}\right)$$

$$= 64/3 \text{ sq. units (ignoring } -\text{ve sign)}$$

6.5 Area of plane curves; polar coordinates

To prove that the area bounded by the curve $r = f(\theta)$, $\alpha \le \theta \le \beta$ *and the radial lines* $\theta = \alpha$, *and* $\theta = \beta$ *is* $\dfrac{1}{2}\displaystyle\int_{\alpha}^{\beta} r^2\,d\theta.$

Let AB be the curve $r = f(\theta)$ where $f(\theta)$ is a continous function in (α, β). OA, OB the radial lines $\theta = \alpha$ and $\theta = \beta$. (Refer Fig. 6.24)

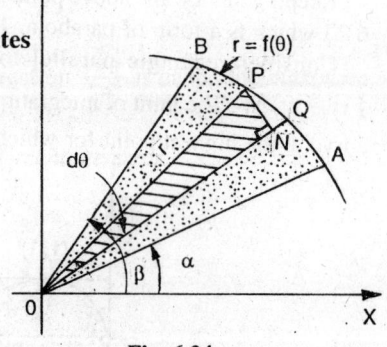

Fig. 6.24.

In an element area, OPQ we approximate area of the sector OPQ, by the area of the triangle OPN, with base $PN = r\,d\theta$ and height $ON \approx OP = r$. (PN is perpendicular to OQ). Then $dA = \dfrac{1}{2}r^2 d\theta$, and Area = $A = \dfrac{1}{2}\displaystyle\int_{\alpha}^{\beta} r^2 d\theta$. ...(6.6)

6.6 Illustrative Examples

Example 6.13 (*a*) **Find the area bounded by following curves**

(*i*) $r^2 = a^2 \cos 2\theta$ (*ii*) $r = a \cos 3\theta$; **the three leaved rose,** (*A.M.I.E., W-1999*)

(*iii*) $r = a\,(1 + \cos\theta)$. (*Pondicherry, 1998*)

(b) **Find the area of one of the loops of the curve r = a cos 4θ.**

Solution. (a) (i) From the Fig. 6.25 it is seen that the required area consists of four equal portions, one of which is swept over as θ varies from θ = 0 to θ = $\dfrac{\pi}{4}$. Thus, area

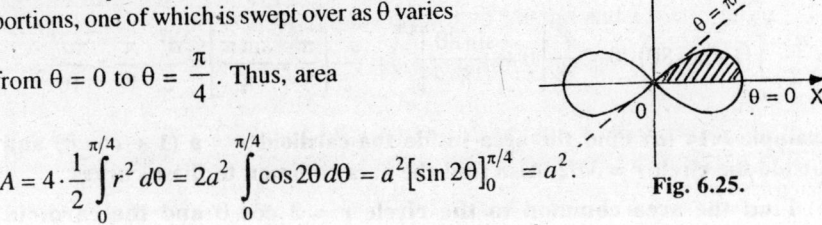

$r^2 = a^2 \cos 2\theta$.

Fig. 6.25.

$$A = 4 \cdot \frac{1}{2} \int_0^{\pi/4} r^2 \, d\theta = 2a^2 \int_0^{\pi/4} \cos 2\theta \, d\theta = a^2 \left[\sin 2\theta \right]_0^{\pi/4} = a^2.$$

(ii) The required area is 6 times the area shown shaded in Fig. 6.26 that is, the area swept over as θ varies from 0 to π/6. Hence

$$A = 6 \cdot \frac{1}{2} \int_\alpha^\beta r^2 \, d\theta = 6 \cdot \frac{1}{2} \int_0^{\pi/6} a^2 \cos^2 3\theta \, d\theta = 3a^2 \int_0^{\pi/6} \left(\frac{1}{2} + \frac{1}{2} \cos 6\theta \right) d\theta = \frac{1}{4} \pi a^2.$$

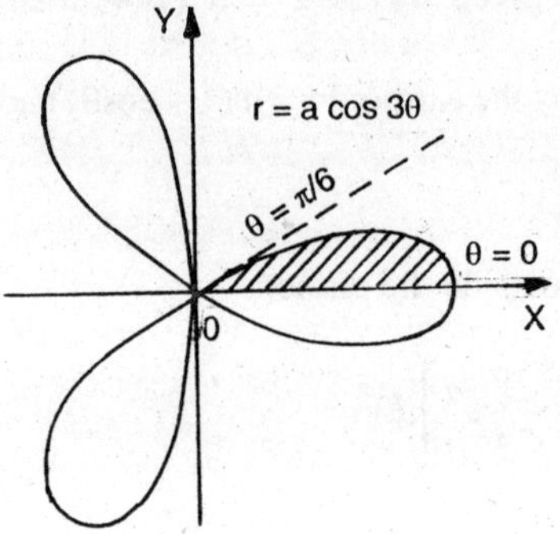

r = a cos 3θ

θ = π/6

θ = 0

Fig. 6.26.

r = a (1 + cos θ)

θ = π

θ = 0

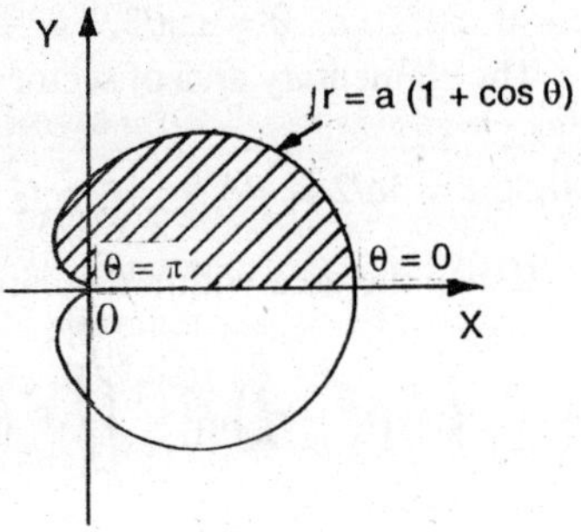

Fig. 6.27.

(iii) The required area = twice the area of shaded portion (Refer Fig. 6.27).

$$= 2 \cdot \frac{1}{2} \int_0^\pi a^2 (1 + \cos\theta)^2 \, d\theta = a^2 \int_0^\pi \left(1 + 2\cos\theta + \cos^2\theta \right) d\theta$$

$$= a^2 \int_0^\pi \left(\frac{3}{2} + 2\cos\theta + \frac{\cos 2\theta}{2} \right) d\theta = a^2 \left[\frac{3}{2}\theta + 2\sin\theta + \frac{\sin 2\theta}{4} \right]_0^\pi = \frac{3}{2} \pi a^2$$

(b) Put r = 0, then cos 4θ = 0. ∴ $4\theta = -\dfrac{\pi}{2}, \dfrac{\pi}{2}$ (two consecutive values),

or $\theta = -\dfrac{\pi}{8}, \dfrac{\pi}{8}$. Area of each loop

$$= \frac{1}{2}\int_{-\pi/8}^{\pi/8} r^2\, d\theta = \frac{1}{2}\int_{-\pi/8}^{\pi/8} a^2\cos^2 4\theta\, d\theta = 2\cdot\frac{1}{2}\int_{0}^{\pi/8} a^2\cos^2 4\theta\, d\theta = a^2\int_{0}^{\pi/8}\cos^2 4\theta\, d\theta$$

$$= \frac{a^2}{2}\int_{0}^{\pi/8}(1+\cos 8\theta)\, d\theta = \frac{a^2}{2}\left[\theta + \frac{\sin 8\theta}{8}\right]_{0}^{\pi/8} = \frac{a^2}{2}\left[\frac{\pi}{8}+\frac{\sin\pi}{8}\right] = \frac{a^2}{2}\cdot\frac{\pi}{8} = \frac{\pi a^2}{16}.$$

Example 6.14 (*a*) **Find the area inside the cardioid $r = a\,(1+\cos\theta)$ and outside the circle $r = 3a/2$. Also find the area common to these curves.**

(*b*) **Find the area common to the circle $r = 3\cos\theta$ and the cardioid $r = 1+\cos\theta$.**

(*c*) **Find the area common to the circles $r = a\sqrt{2}$ and $r = 2a\cos\theta$.**

Solution. (*a*) For the circle $r = 3a/2$ or $r^2 = 3ar/2$ or $x^2 + y^2 = 9a^2/4$, we have radius $3a/2$ and the centre at $(0, 0)$. The fig. 6.28 (a) shows the cardioid $r = a\,(1+\cos\theta)$ and the circle $r = 3a/2$.

These curves intersect at the points given by $3a/2 = a\,(1+\cos\theta)$ or $\cos\theta = 1/2$ i.e., $\theta = \pm\,\pi/3$.

The elementary area of sector bounded by the cardioid $r = a\,(1+\cos\theta)$ and the circle $r = 3a/2$ is $dA = \frac{1}{2}\left(r_1^2 - r_2^2\right)d\theta$.

Hence the area within the cardioid and outside the circle,

$$A = 2\int_{0}^{\pi/3}\frac{1}{2}\left(r_1^2 - r_2^2\right)d\theta = \int_{0}^{\pi/3}\left[a^2(1+\cos\theta)^2 - \frac{9}{4}a^2\right]d\theta$$

$$= a^2\int_{0}^{\pi/3}\left(1+2\cos\theta+\cos^2\theta - \frac{9}{4}\right)d\theta$$

$$= a^2\int_{0}^{\pi/3}\left[\left(1-\frac{9}{4}+\frac{1}{2}\right)+2\cos\theta+\frac{\cos 2\theta}{2}\right]d\theta$$

$$= a^2\int_{0}^{\pi/3}\left(2\cos\theta+\frac{\cos 2\theta}{2}-\frac{3}{4}\right)d\theta$$

$$= a^2\left[2\sin\theta+\frac{\sin 2\theta}{4}-\frac{3}{4}\theta\right]_{0}^{\pi/3}$$

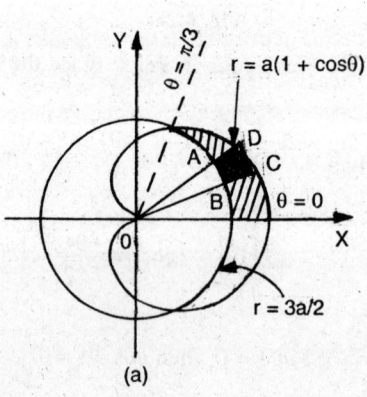

$r = a(1+\cos\theta)$

$r = 3a/2$

(a)

Fig. 6.28.

$$= a^2 \left(2 \cdot \frac{\sqrt{3}}{2} + \frac{\sqrt{3}}{8} - \frac{\pi}{4} \right)$$

$$= \frac{a^2}{8} \left(9\sqrt{3} - 2\pi \right) \text{ sq. units.}$$

(*ii*) Refer Fig.6.28(*b*) Required area

$$= 2 \int_0^{\pi/3} \frac{1}{2} \left(\frac{3a}{2} \right)^2 d\theta + 2 \int_{\pi/3}^{\pi} \frac{1}{2} \left\{ a(1+\cos\theta) \right\}^2 d\theta$$

$$= \frac{3\pi a^2}{4} + a^2 \int_{\pi/3}^{\pi} \left(1 + 2\cos\theta + \cos^2\theta \right) d\theta$$

$$= \frac{3\pi a^2}{4} + a^2 \int_{\pi/3}^{\pi} \left(1 + 2\cos\theta + \frac{1+\cos 2\theta}{2} \right) d\theta$$

$$= \frac{3\pi a^2}{4} + \frac{a^2}{2} \int_{\pi/3}^{\pi} (3 + \cos 2\theta + 4\cos\theta) \, d\theta$$

$$= \frac{3\pi a^2}{4} + \frac{a^2}{2} \left[3\theta + \frac{\sin 2\theta}{2} + 4\sin\theta \right]_{\pi/3}^{\pi}$$

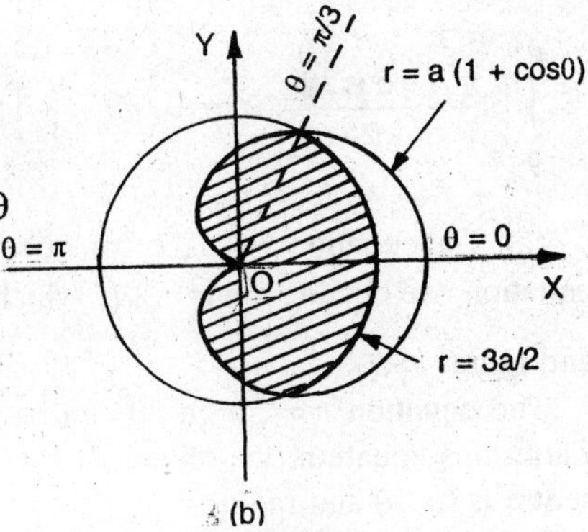

Fig. 6.28.

$$= \frac{3\pi a^2}{4} + \frac{a^2}{2} \left[3\left(\pi - \frac{\pi}{3} \right) + \frac{1}{2} \left(0 - \frac{\sqrt{3}}{2} \right) + 4\left(0 - \frac{\sqrt{3}}{2} \right) \right] = a^2 \left(14\pi - 9\sqrt{3} \right) / 8.$$

(*b*) The area common to the two curves and above the initial line consists of two portions one swept over by the radius vector $r = 1 + \cos\theta$ as θ varies from 0 to $\pi/3$ and the other by $r = 3\cos\theta$ as θ varies from $\pi/3$ to $\pi/2$ (Refer Fig. 6.29).

$$\text{Required area} = 2 \cdot \left[\frac{1}{2} \int_0^{\pi/3} (1+\cos\theta)^2 \, d\theta + \frac{1}{2} \int_{\pi/3}^{\pi/2} 9\cos^2\theta \, d\theta \right]$$

$$= \int_0^{\pi/3} \left(1 + \cos^2\theta + 2\cos\theta \right) d\theta + \int_{\pi/3}^{\pi/2} \frac{9}{2} (1+\cos 2\theta) \, d\theta$$

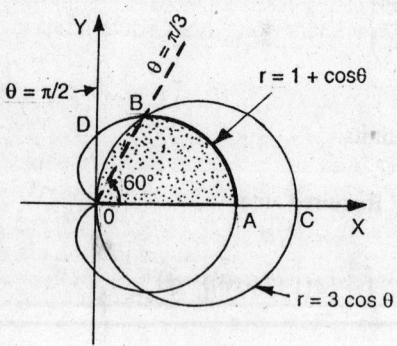

Fig. 6.29.

$$= \int_0^{\pi/3} \left(1 + \frac{1 + \cos 2\theta}{2} + 2\cos\theta\right) d\theta + \frac{9}{2} \int_{\pi/3}^{\pi/2} (1 + \cos 2\theta) d\theta = \frac{5\pi}{4} \text{ sq. units.}$$

(c) The equation $r = a\sqrt{2}$ can be written as $r^2 = 2a^2$. Changing to Cartesians equation, we obtain $x^2 + y^2 = 2a^2$, which represents a circle whose centre is $(0, 0)$ and radius $a\sqrt{2}$.

The equation $r = 2a \cos\theta$ can be written as $r^2 = 2ar \cos\theta$. Changing to Cartesians equation, we obtain $x^2 + y^2 = 2ax$, which represents a circle whose centre is $(a, 0)$ and radius is a.

The two circles which intersect at P are shown in the Fig. 6.30. At point P,

$$a\sqrt{2} = 2a\cos\theta, \text{ i.e., } \cos\theta = \frac{1}{\sqrt{2}}, \text{ or } \theta = \frac{\pi}{4}.$$

The required area = Twice the shaded area = $2 \times$ area $OAPQO$
       = 2 [area OAP + area $OPQO$].

Now area OAP is bounded by the circle $r = a\sqrt{2}$ and the radii vectors OA and OP corresponding to $\theta = 0$ and $\theta = \theta/4$.

$$\therefore \quad \text{Area } OPA = \frac{1}{2} \int_0^{\pi/4} r^2 \, d\theta = \frac{1}{2} \int_0^{\pi/4} 2a^2 \, d\theta = a^2 \cdot \frac{\pi}{4}.$$

The area $OPQO$ is bounded by the circle $r = 2a \cos\theta$, and the radii vectors corresponding to $\theta = \pi/4$ and $\theta = \pi/2$.

$$\therefore \quad \text{Area } OPQO = \frac{1}{2} \int_{\pi/4}^{\pi/2} (2a\cos\theta)^2 \, d\theta = 2a^2 \int_{\pi/4}^{\pi/2} \cos^2\theta \, d\theta$$

$$= a^2 \int_{\pi/4}^{\pi/2} (1+\cos 2\theta)\, d\theta = a^2 \left[\theta + \frac{\sin 2\theta}{2}\right]_{\pi/4}^{\pi/2}$$

$$= a^2\left[\frac{\pi}{2} - \left(\frac{\pi}{4}+\frac{1}{2}\right)\right] = a^2\left(\frac{\pi}{4}-\frac{1}{2}\right). \text{ Then}$$

$$\text{Required area } = 2\left[\frac{\pi}{4}a^2 + a^2\left(\frac{\pi}{4}-\frac{1}{2}\right)\right]$$

$= a^2(\pi - 1)$ sq. units.

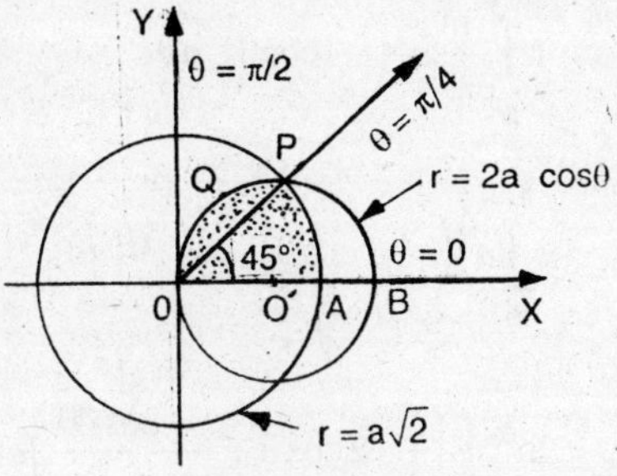

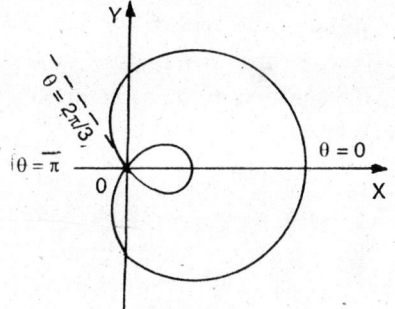

Fig. 6.30.

Example 6.15 Calculate the ratio of the area of the larger to the area of the smaller loop of the curve r = (1/2) + cos θ.

Solution. *Larger loop.* The required area is twice that swept over as θ varies from 0 to 2π/3. Refer Figure 6.31. Hence

$$A_1 = 2\cdot\frac{1}{2}\int_0^{2\pi/3}\left(\frac{1}{2}+\cos\theta\right)^2 d\theta = \int_0^{2\pi/3}\left(\frac{1}{4}+\cos\theta+\cos^2\theta\right)d\theta$$

$$= \frac{\pi}{2}+\frac{3\sqrt{3}}{8} \text{ sq. units.}$$

Smaller loop: The required area is twice that swept over as θ varies from 2π/3 to π. Hence,

$$A_2 = 2\cdot\frac{1}{2}\int_{2\pi/3}^{\pi}\left(\frac{1}{2}+\cos\theta\right)^2 d\theta$$

$$= \frac{\pi}{4}-\frac{3\sqrt{3}}{8} \text{ sq. units.}$$

Therefore, required ratio

$$A_1/A_2 = \left(4\pi+3\sqrt{3}\right)/\left(2\pi-3\sqrt{3}\right).$$

Fig. 6.31.

Example 6.16 Show that the area included between the cardioids r = a (1 + cos θ) and r = a (1 – cos θ) is $a^2(3\pi-8)/2$. *(Mangalore, 1999)*

Solution. The curves are shown in Fig. 6.32.

For the point of intersection, solve the given equations simultaneously. The polar coordinates of *P* are (*a*, π/2).

Required area = Four times the shaded part.

$$= 4 \int_0^{\pi/2} \frac{1}{2} \cdot a^2 (1 - \cos\theta)^2 \, d\theta$$

$$= 2 \int_0^{\pi/2} a^2 \left(1 - 2\cos\theta + \cos^2\theta\right) d\theta$$

$$= 2a^2 \int_0^{\pi/2} \left(1 - 2\cos\theta + \frac{1 + \cos 2\theta}{2}\right) d\theta$$

$$= 2a^2 \left[\theta - 2\sin\theta + \frac{\theta}{2} + \frac{1}{4}\sin 2\theta\right]_0^{\pi/2}$$

$$= 2a^2 \left[\frac{\pi}{2} - 2 + \frac{\pi}{4}\right] = 2a^2 \left(\frac{3\pi}{4} - 2\right) = \frac{a^2}{2}(3\pi - 8).$$

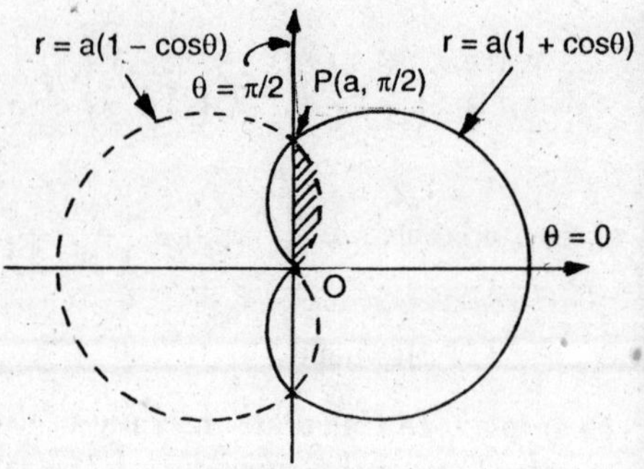

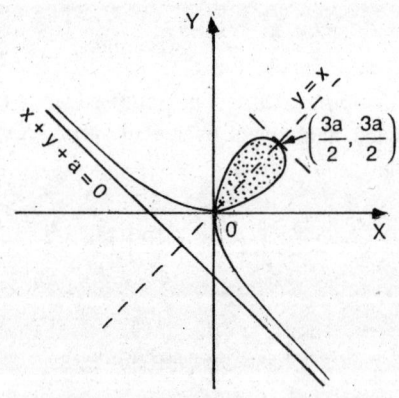

$r = a(1 - \cos\theta)$ $r = a(1 + \cos\theta)$

$\theta = \pi/2$ P(a, $\pi/2$)

$\theta = 0$

Fig. 6.32.

Example 6.17. Prove that the area of the loop of the curve $x^3 + y^3 = 3axy$ **is** $3a^2/2$. (*PTU 2006*)

Solution. To trace the loop of the curve $x^3 + y^3 = 3axy$, we change the equation to polar coordinates by putting $x = r\cos\theta$, and $y = r\sin\theta$.

The equation of the curve in polar co-ordinates is

$$r = \frac{3a\sin\theta\cos\theta}{\sin^3\theta + \cos^3\theta}.$$

Now r is zero, when $\theta = 0$ and $\pi/2$. Thus the loop of the curve lies between these two limits. (Refer Fig. 6.33)

Fig. 6.33.

$$\text{Required area} = \frac{1}{2}\int_0^{\pi/2} r^2 \, d\theta = \frac{1}{2}\int_0^{\pi/2} \left(\frac{3a\sin\theta\cos\theta}{\sin^3\theta + \cos^3\theta}\right)^2 d\theta$$

$$= \frac{9a^2}{2}\int_0^{\pi/2} \frac{\sin^2\theta\cos^2\theta}{\left(\sin^3\theta + \cos^3\theta\right)^2} \, d\theta$$

Dividing the numerator and denominator by $\cos^6\theta$, we get

$$A = \frac{9a^2}{2} \int_0^{\pi/2} \frac{\sec^2\theta \tan^2\theta}{\left(1+\tan^3\theta\right)^2} d\theta \quad \text{Put } z = 1+\tan^3\theta, \ dz = 3\sec^2\theta \tan^2\theta \, d\theta.$$

When $\theta = 0, z = 1$, and when $\theta = \pi/2, \ z = \infty$.

$$\therefore \quad A = \frac{3a^2}{2} \int_1^\infty \frac{dz}{z^2} = \frac{3a^2}{2}\left[-\frac{1}{z}\right]_1^\infty = \frac{3a^2}{2}.$$

Exercise 6.1

Find the area bounded by each of the following curves:

1.(a) The figure bounded by the lines and $y+6x=0$, $y=0$ and $x=4$.

(b) The region bounded by the lines $x-2y+4=0$; $x+y-5=0$ and $y=0$.

(c) The region in the first quadrant bounded by the parabola $y=4-x^2$.

(d) The region enclosed by the ellipse $x^2/4+y^2/9=1$.

(e) The region above the x-axis enclosed by the parabola $x^2=4y$ and the line $y=4$.

(f) The region included between the curve $y^2=x$ and the line $y=x$.

(g) The region bounded by the curve $y=x^2$ and the straight line $y=2x$.
   (A.M.I.E., S-2005) **Ans.** 4/3 sq. units.

(h) The region enclosed by the curves $x^2+y^2=a^2$ and $x+y=a$ in the first quadrant.

(i) The region bounded by the lines $x=0$, $y=0$, $x=\pi/2$, and curves $y=\cos x$ and $y=\sin x$.

(j) The area of the region bounded by the curves:
   $y=x, 0 \le x \le 1, y=2-x, 1 \le x \le 2$ and the x-axis.   **Ans.** 1 sq. unit.

(k) The region enclosed by the curve $y=\sin x$, and $y=0$, $x=0$ and $x=\pi$.
   [**Hint:** Refer Figures 6.34 (a to f and h, i).   **Ans.** 2 sq. units.

(a) Required area $= \left|\int_0^4 y \, dx\right| = \left|\int_0^4 -6x \, dx\right| = 48$ sq units.

(b) Required area $= \int_{-4}^2 \left(\frac{x+4}{2}\right) dx + \int_2^5 (5-x) \, dx = 13.5$ sq. units.

(c) Required area $= \int_0^2 (4 - x^2)\,dx = \left[4x - \frac{x^3}{3} \right]_0^2 = \frac{16}{3}$ sq. units.

(d) Required area

$$= 4 \cdot \frac{3}{2} \int_0^2 \sqrt{4 - x^2}\,dx = 6 \left[\frac{x\sqrt{4 - x^2}}{2} + \frac{4}{2} \sin^{-1} \frac{x}{2} \right]_0^2 = 6\pi \text{ sq. units} .$$

(e) Required area $= 2 \int_0^4 \left(4 - \frac{x^2}{4} \right) dx = 2 \left[4x - \frac{x^3}{12} \right]_0^4 = \frac{64}{3}$ sq. units.

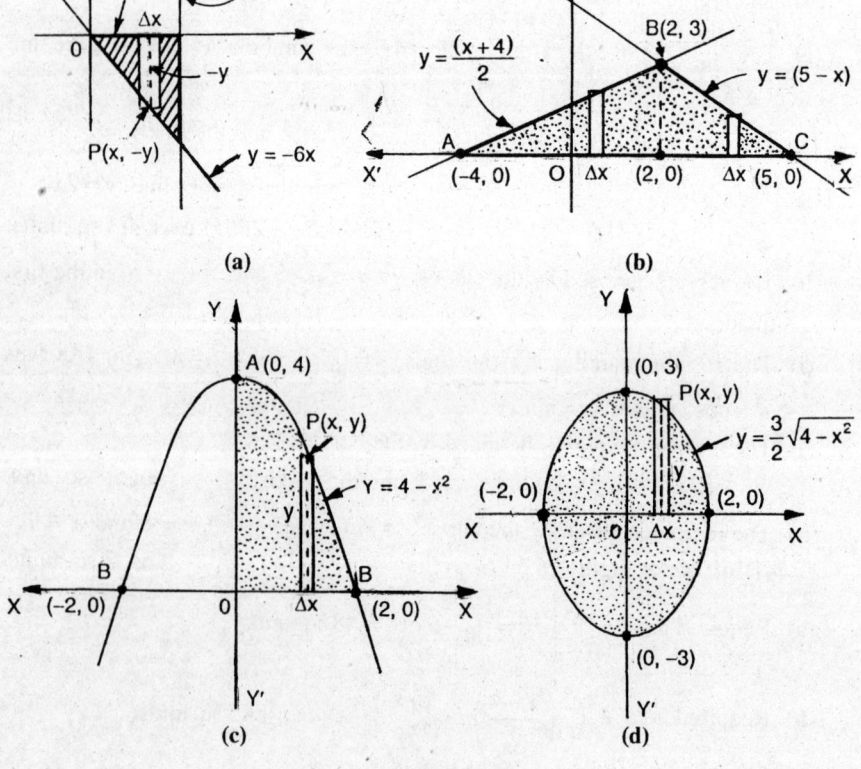

Fig. 6.34. *(Contd.)*

Y
P(x, y2)
y = 4
(−4, 4)
(4, 4)
4
y = x²/4
(x, y1)
X'
(0, 0)
O
Δx
X

(e)

Y
y = √x
P(x, y2)
A (1, 1)
y = x
Q(x, y1)
X
0
Δx

(f)

Y
B (0, a)
P(x, y2)
y = √a² − x²
y = a − x
Q(x, y1)
(a, 0)
X
0
Δx
A

(h)

Y
P(x, y2)
x = π/2
B
D
A
y = sin x
P (x, y1)
2π
C
0
π/4, 0
π/2
π
3π/2
X
y = cos x

(i)

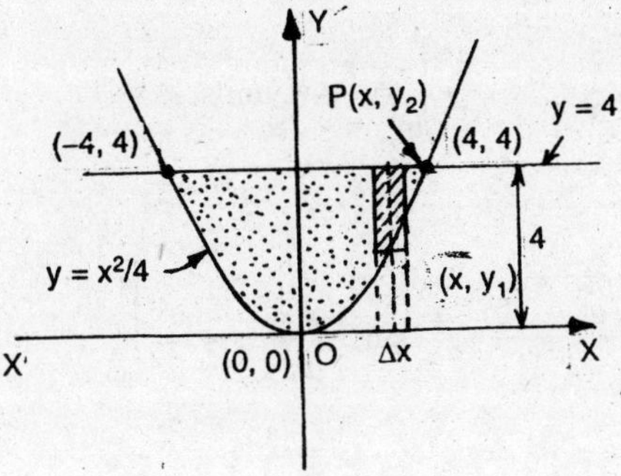

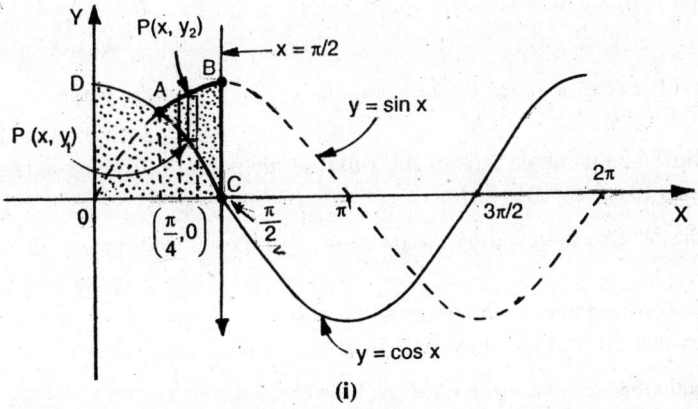

Fig. 6.34.

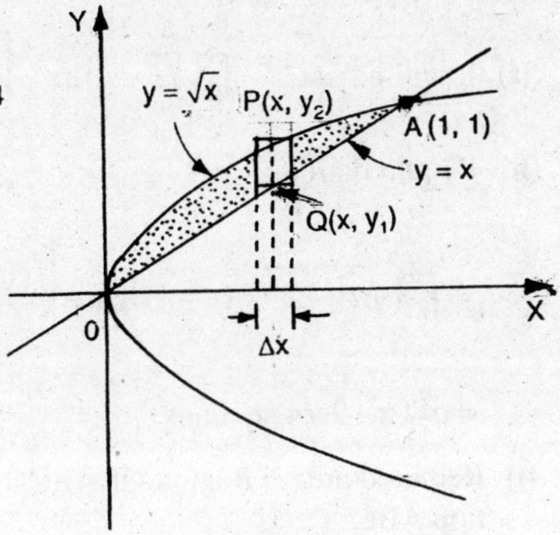

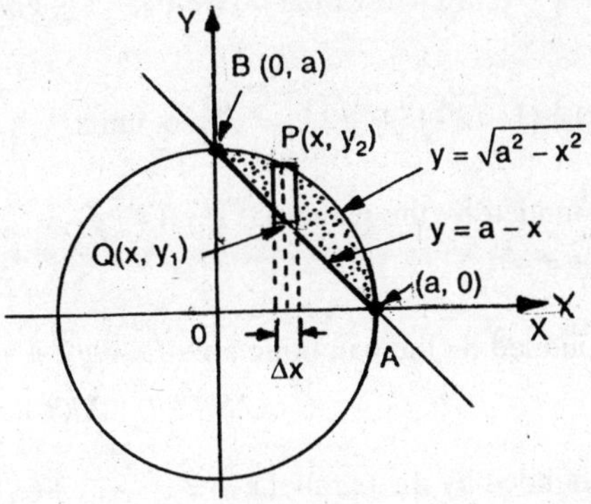

(f) Required area $= \int_0^1 \left(\sqrt{x} - x\right) dx = \left[\dfrac{2}{3}x^{3/2} - \dfrac{x^2}{2}\right]_0^1 = \dfrac{1}{6}$ sq. units.

(h) Required area

$$= \int_0^a \left\{\sqrt{a^2 - x^2} - (a - x)\right\} dx = \left[\dfrac{x\sqrt{a^2 - x^2}}{2} + \dfrac{a^2}{2}\sin^{-1}\dfrac{x}{a} - ax + \dfrac{x^2}{2}\right]_0^a$$

$= a^2(\pi - 2)/4$ sq. units.

(i) Required area = Region ODABCO = Area of portion ODC + Area of portion ABC

$$= \int_0^{\pi/2} \cos x\, dx + \int_{\pi/4}^{\pi/2}(\sin x - \cos x)\, dx = [\sin x]_0^{\pi/2} + [-\cos x - \sin x]_{\pi/4}^{\pi/2}$$

$= (1 - 0) + \left[0 - 1 + \left(1/\sqrt{2}\right) + \left(1/\sqrt{2}\right)\right] = \sqrt{2}$ sq. units.

2. Find the area bounded by the parabola $x = 8 + 2y - y^2$, the y-axis, and the lines $y = -1$ and $y = 3$. **Ans.** 92/3 sq. units.

3. Find the area bounded by the parabolic arc $\sqrt{x} + \sqrt{y} = \sqrt{a}$ and the coordinate axes. *(G.G.S.I.P.U., 2006).* **Ans.** $a^2/6$ sq. units.

4. Find the area bounded by the parabola $y = x^2 - 7x + 6$, the x-axis, and the lines $x = 2$ and $x = 6$. **Ans.** 56/3 sq. units.

5. Find the total area bounded by the curve $y = x^3 - 4x$ and the x-axix.

   Ans. 8 sq. units.

6. Find the area bounded by the parabola $x = 4 - y^2$ and the y-axis.

   Ans. 32/3 sq. units.
   [**Hint:** The parabola crosses the x-axis at the point (4, 0) the y-axis at the points (0, 2) and (0, −2)].

7. Find the area bounded by the parabolas $y = 6x - x^2$ and $y = x^2 - 2x$.

   Ans. 64/3 sq. units.
   [**Hint:** The parabolas intersect at the points (0, 0) and (4, 8). It is readily seen that vertical rectangle will yield the simpler solution].

8. Obtain the area above the x-axis included between the curves $y^2 = 2ax - x^2$ and $y^2 = ax$. *(A.M.I.E., S-2010)* **Ans.** $a^2(3\pi - 8)/12$ sq. units

9. Find the areas of the shaded regions. The curves are shown below in Fig. 6.35 (a) to (d). **Ans.** (a) $\pi/2$; (b) $1/12$; (c) $5/6$; (d) $5/6$.

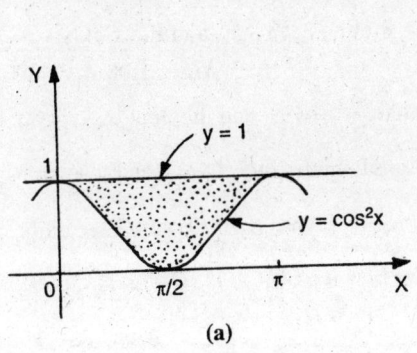

(a)

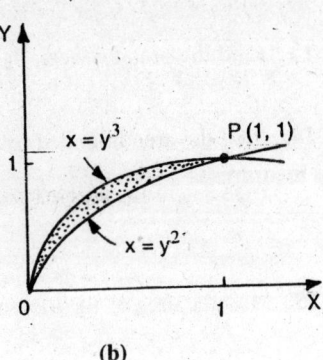

(b)

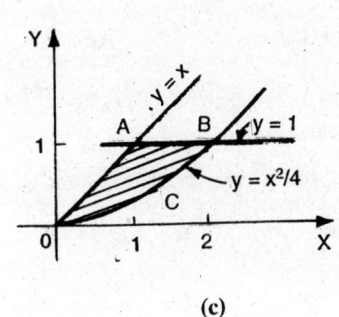

(c)

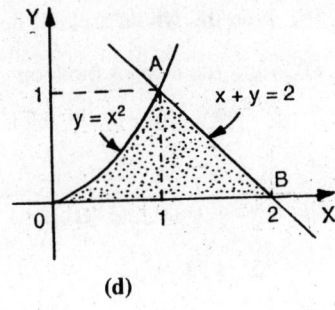

(d)

Fig. 6.35.

[**Hint:** (d) *Using Vertical rectangle:*

Required Area $= \int_0^1 x^2 dx + \int_1^2 (2-x) dx = \dfrac{5}{6}$ sq. units.

Alternative, *Using horizontal rectangle:*

Required Area $= \int_0^1 \left\{(2-y) - \sqrt{y}\right\} dy = \dfrac{5}{6}$ sq. units].

10. (a) Find the area common to the circles $x^2 + y^2 = 4$ and $x^2 + y^2 = 4x$.

Ans. $\left(11\pi - 6\sqrt{3}\right)/3$.

(b) Find the area of the circle $x^2 + y^2 = 16$ which is exterior to the parabola $y^2 = 6x$.

Ans. $\dfrac{4}{3}\left(8\pi - \sqrt{3}\right)$.

11. Prove that the area common to the parabolas $x^2 = 4ay$ and $y^2 = 4ax$ is $16a^2/3$.

(Kuvempu, 1998)

12. Find the smaller region bounded by the ellipse $x^2/9 + y^2/4 = 1$ and the straight line $x/3 + y/2 = 1$. (D.C.E., 2004) **Ans.** $3(\pi - 2)/2$.

13. Find the area between the curves $y = x(x-3)(x-5)$ and the x-axis.

    Ans. 21.08 sq. units.

14. Find the area bounded by the parabola $y^2 = 4ax$ and the line $x + y = 3a$. Also determine the area above the x-axis bounded by $y^2 = 4x$ and $x + y = 3$.

    Ans. $64a^2/3$; $10/3$ sq. units.

15. Find the area of the loop of the curve $ay^2 = x^2(a-x)$. (*Osmania, 2000S*)

    Ans. $8a^2/15$.

16. Find the whole area of the curve $a^2y^2 = x^3(2a - x)$. **Ans.** πa^2.

17. Find the area of the loop of the curve $y^2 = x^2\{(a+x)/(a-x)\}$. Also find the area between the curve and its asymptote.

    Ans. $a^2(4 - \pi)/2$; $a^2(4 + \pi)/2$.

18. Show that the area of a loop of the curve $x^4 = a^2(x^2 - y^2)$ is $2a^2/3$ sq. units.

19. Show that the curve $a^2y^2 = x^2(a^2 - x^2)$ consists of two loops and find the area of each loop. **Ans.** $2a^2/3$

20. Find the area cut off from the parabola $4y = 3x^2$ by the straight line $2y = 3x + 12$. **Ans.** 27 sq. units

21. Find the area between the following curves

    (a) $3y^2 = 25x$, $5x^2 = 9y$, **Ans.** 5 sq. units

    (b) $9xy = 4$, $2x + y = 2$, **Ans.** $(3 - 4\log 2)/9$ sq. units

    (c) $y = 3x^2 - x - 3$, $y = -2x^2 + 4x + 7$. **Ans.** 13.5 sq. units

    (d) $x = -2y^2$, $x = 1 - 3y^2$. (*A.M.I.E. W-2009*) **Ans.** 4/3 sq. units

22. Show that the area of the loop of the curve $ay^2 = (x - a)(x - 5a)^2$ is $128a^2/15$.

23. Find the area of the region bounded by the given curves $x^2/a^2 + y^2/b^2 = 1$ and $x^2/b^2 + y^2/a^2 = 1$, $0 < a < b$. **Ans.** $4ab \tan^{-1}(a/b)$.

24. Find the area bounded by each of the following curves

(*i*) The ellipse $x = a\cos t$, $y = b\sin t$,

(*ii*) $x = 2\cos\theta - \cos 2\theta - 1$, $y = 2\sin\theta - \sin 2\theta$,

(*iii*) $x = a\cos^3 t$, $y = a\sin^3 t$, or $x^{2/3} + y^{2/3} = a^{2/3}$,

(*iv*) $(x/a)^{2/3} + (y/b)^{2/3} = 1$ or $x = a\cos^3\theta$, $y = b\sin^3\theta$, $0 \le \theta \le 2\pi$.

Ans. (*i*) πab; (*ii*) 6π; (*iii*) $3\pi a^2/8$, (*iv*) $3\pi ab/8$.

25. Find the area included between the cycloid $x = a(\theta + \sin\theta)$,

$y = a(1 - \cos\theta)$ and its base. Also find the area between the curve and the

x-axis. [*A.M.I.E. S-2003, Gorakhpur, 1999*]. **Ans.** $3\pi a^2$; πa^2.

26. Find the area bounded by each of the following curves:

(*i*) $r^2 = 1 + \cos 2\theta$, (*ii*) $r = 4\cos\theta$, (*iii*) $r = 4(1 - \sin\theta)$,

(*iv*) $r = a(1 - \cos\theta)$, (*v*) $r = a(2 - \cos 2\theta)$, $0 \le \theta \le 2\pi$.

Ans. (*i*) π, (*ii*) 4π, (*iii*) 24π, (*iv*) $3\pi a^2/2$, (*v*) $9\pi a^2/2$.

27. Find the area bounded by the limacon $r = 2 + \cos\theta$. **Ans.** $9\pi/2$.

28. Find the area of a loop of the curve $r = a\sin 3\theta$. **Ans.** $\pi a^2/12$.

29. Find the area outside the circle $r = 2a\cos\theta$ and inside the cardioid $r = a(1 + \cos\theta)$. **Ans.** $\pi a^2/2$.

30. Find the area of the region that lies inside the circle $r = a\cos\theta$ and outside the cardioid $r = a(1 - \cos\theta)$.

[**Hint.** Refer Fig. 6.36. Area $= \dfrac{1}{2}\displaystyle\int_{-\pi/3}^{\pi/3}\left(r_1^2 - r_2^2\right)d\theta$

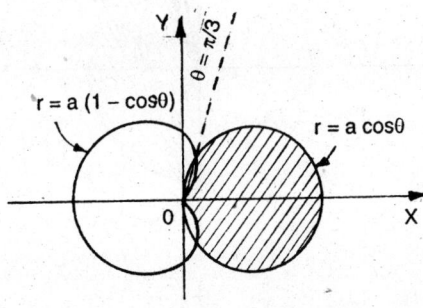

Fig. 6.36.

$$= \frac{a^2}{2} \int_{-\pi/3}^{\pi/3} \left[\cos^2\theta - (1-\cos\theta)^2\right] d\theta$$

$$= a^2 \int_0^{\pi/3} (2\cos\theta - 1) d\theta = a^2 \left[2\sin\theta - \theta\right]_0^{\pi/3} = \frac{a^2}{3}(3\sqrt{3} - \pi) \text{ sq. units}]$$

31. Find the area common to the two circles, $r = a\cos\theta$ and $r = a(\cos\theta + \sin\theta)$.

Ans. $a^2(\pi - 1)/4$ sq. units

32. Prove that the area between the equiangular spiral $r = ae^{\theta\cot\alpha}$ and two radii vectors of length r_1 and r_2 is equal to $\left\{(r_2^2 - r_1^2)\tan\alpha\right\}/4$.

33. (a) Find the area inside the smaller loop of the limacon, $r = 2\cos\theta + 1$.

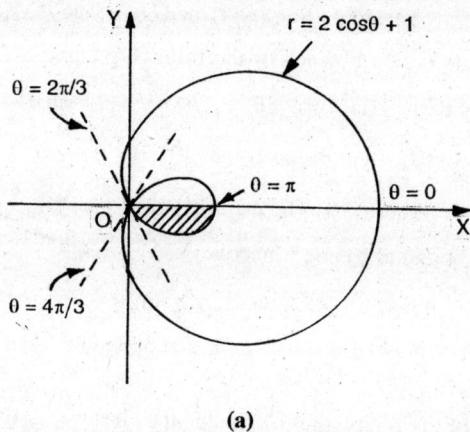

(a)

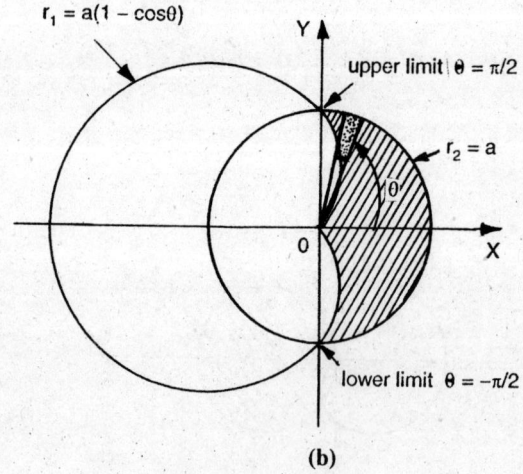

(b)

Fig. 6.37.

(b) Find the area of the region that lies inside the circle $r = a$ and outside the cardioid $r = a(1 - \cos\theta)$. [**Hint.** (a) Refer Fig 6.37(a)

$$A = 2\int_{2\pi/3}^{\pi} \frac{1}{2} r^2 d\theta = \int_{2\pi/3}^{\pi} (3 + 2\cos 2\theta + 4\cos\theta)\, d\theta = (2\pi - 3\sqrt{3})/2.$$

(b) Refer Fig. 6.37(b)

$$A = \int_{-\pi/2}^{\pi/2} \frac{1}{2}\left(r_2^2 - r_1^2\right) d\theta = a^2 \int_0^{\pi/2}\left[1 - \left(1 - 2\cos\theta + \cos^2\theta\right)\right] d\theta$$

$$= a^2 \int_0^{\pi/2} \left(2\cos\theta - \cos^2\theta\right) d\theta = a^2 \left[2\sin\theta - \frac{\theta}{2} - \frac{\sin 2\theta}{4}\right]_0^{\pi/2} = \frac{a^2}{4}(8 - \pi).$$

34. Prove that the area of the loop of the curve $x^3 + y^3 = 3axy$ is three times the area of one of loops of the curve $r^2 = a^2 \cos 2\theta$.

35. Find the area common to the curves $y^2 = ax$ and $x^2 + y^2 = 4ax$. (*A.M.I.E., S-1998; Raipur, 1998*) [**Hint.** Refer Fig. 6.38. Required area

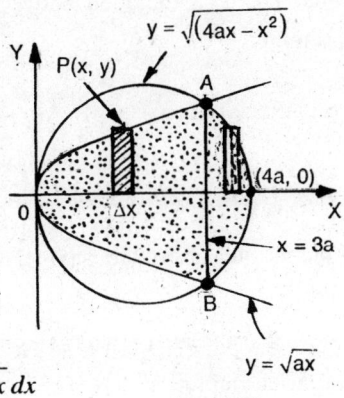

Fig. 6.38.

$$= 2\int_0^{3a} \sqrt{ax}\, dx + 2\int_{3a}^{4a} \sqrt{4ax - x^2}\, dx$$

$$= 2\sqrt{a}\left[\frac{2}{3} x^{3/2}\right]_0^{3a} + 2\int_{3a}^{4a} \sqrt{x}\sqrt{4a - x}\, dx$$

$$= \frac{4}{3}\sqrt{a}\,(3a)^{3/2} + 2\int_{\pi/3}^{\pi/2}\left(\sqrt{4a}\sin\theta\right)\left(\sqrt{4a}\cos\theta\right) 8a\sin\theta\cos\theta\, d\theta,$$

where $x = 4a\sin^2\theta$ where $x = 4a\sin^2\theta$

$$= 4\sqrt{3}a^2 + 64a^2 \int_{\pi/3}^{\pi/2} \sin^2\theta\cos^2\theta\, d\theta = 4\sqrt{3}a^2 + 8a^2\left[\theta - \frac{\sin 4\theta}{4}\right]_{\pi/3}^{\pi/2}$$

$$= 4\sqrt{3}a^2 + \left(4\pi a^2/3\right) - \sqrt{3}a^2 = 3\sqrt{3}a^2 + \left(4\pi a^2/3\right) = a^2\left(9\sqrt{3} + 4\pi\right)/3.]$$

36. Find by integration, the area of the ellipse $ax^2 + 2hxy + by^2 = 1$, $ab - h^2$ being positive. [**Hint.** The polar form of the ellipse is

$$r^2\left(a\cos^2\theta + 2h\cos\theta\sin\theta + b\sin^2\theta\right) = 1.$$

Since every diagonal of the ellipse bisects its area. Therefore,

$$\text{Area} = 2\int_{-\pi/2}^{\pi/2} \frac{1}{2}r^2 d\theta = \int_{-\pi/2}^{\pi/2} \frac{d\theta}{a\cos^2\theta + 2h\cos\theta\sin\theta + b\sin^2\theta}$$

$$= \int_{-\pi/2}^{\pi/2} \frac{\sec^2\theta\, d\theta}{a + 2h\tan\theta + b\tan^2\theta} = \int_{-\infty}^{\infty} \frac{dt}{a + 2ht + bt^2}, \text{ where } \tan\theta = t$$

$$= \frac{1}{b}\int_{-\infty}^{\infty} \frac{dt}{\{t + (h/b)\}^2 + \left(\sqrt{ab - h^2}/b\right)^2} = \frac{1}{b}\cdot\frac{b}{\sqrt{ab - h^2}}\left[\tan^{-1}\frac{(tb + h)}{\sqrt{ab - h^2}}\right]_{-\infty}^{\infty}$$

$$= \frac{1}{\sqrt{ab - h^2}}\left[\frac{\pi}{2} - \left(-\frac{\pi}{2}\right)\right] = \frac{\pi}{\sqrt{ab - h^2}}.$$

37. Find the area included between the curve $xy^2 = 4a^2(2a - x)$ and its asymptote. *(GGSIPU, 2006; AMIE, W-1997)*

[Hint. Required area $= 2\int_0^{2a} y\,dx = 4a\int_0^{2a} \frac{2a\cdot\sqrt{(2a - x)}}{\sqrt{x}}\,dx$, put $x = 2a\sin^2\theta$

$$= 16a^2\int_0^{\pi/2}\cos^2\theta\, d\theta = 4\pi a^2].$$

38. Find the area above the x-axis included between the curves $y^2 = x(2a - x)$ and $y^2 = ax$. **Ans.** $a^2(3\pi - 8)/12$.

39. Show that the area bounded by the Cissoid $x = a\sin^2 t$, $y = a\sin^3 t/\cos t$ and its asymptote is $3\pi a^2/4$.

40. Find the area included between the curve $x^2 y^2 = a^2(y^2 - x^2)$ and its asymptotes. *(GGSIPU, Dec. 2001)* **Ans.** $4a^2$.

[Hint. Required area $= 4\int_0^a \frac{ax}{\sqrt{a^2 - x^2}}\,dx = -4a\left[\sqrt{a^2 - x^2}\right]_0^a = 4a^2]$

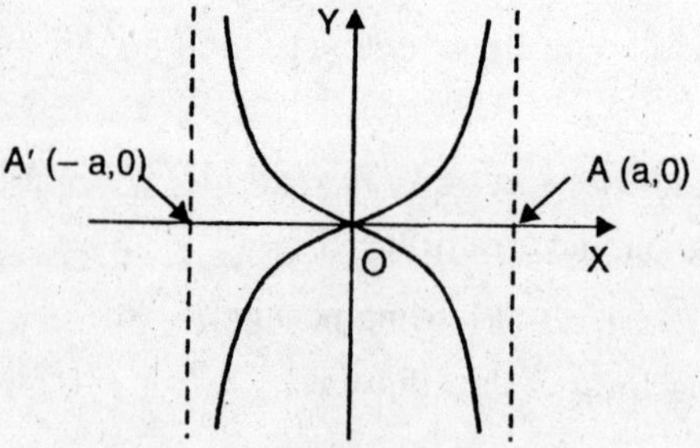

Fig. 6.39.

Chapter 7

Lengths of Plane Curves

7.1 Introduction

In this chapter, we shall determine the lengths of arcs of plane curves whose equations are given in Cartesian, Parametric or Polar forms. The process is known as *Rectification.*

The length of a curve is usually denoted by *s.* Any formula expressing the differential coefficient of *s* gives rise at once, by integration, to a formula in integral calculus for finding *s.* This will be clear from the following sections.

7.2 Arc Length of Cartesian Curves

To prove that the length of the portion of the curve $y = f(x)$ between the points A and B where $x = a$ and $x = b$ is given by

$$s = \int_a^b \sqrt{1 + \left(\frac{dy}{dx}\right)^2}\, dx = \int_a^b \sqrt{1 + \left(f'(x)\right)^2}\, dx.$$

Let *AB* be the curve $y = f(x)$, and *A*, *B* the points where $x = a$, $x = b$ (Refer Fig. 7.1). Let $P(x, y)$ be any point on the curve. If *s* denotes the length of the arc of the curve included between a fixed point *A* and a variable point *P* whose abscissa is *x* so that it is a function of *x*, we have by Differential Calculus

$$\frac{ds}{dx} = \sqrt{1 + \left(\frac{dy}{dx}\right)^2}$$

Therefore, $\displaystyle\int_a^b \sqrt{1 + \left(\frac{dy}{dx}\right)^2}\, dx$

$$= \int_a^b \frac{ds}{dx}\, dx = \left[s\right]_a^b$$

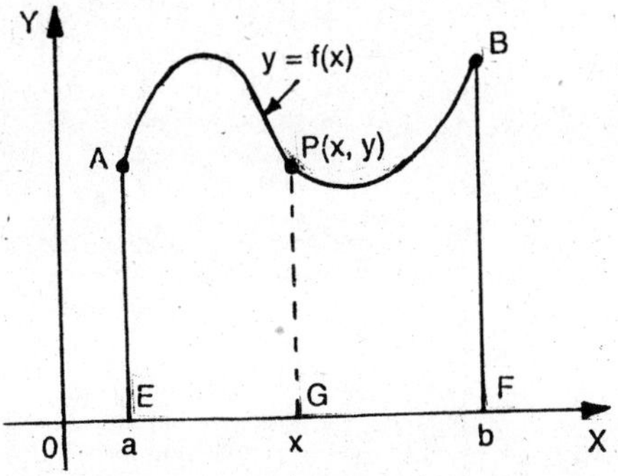

Fig. 7.1

= (value of s when x is equal to b) – (value of s when x is equal to a) = arc AB.

Hence $\quad s = \int_a^b \sqrt{1 + \left(\dfrac{dy}{dx}\right)^2}\ dx = \int_a^b \sqrt{1 + \left[f'(x)\right]^2}\ dx$. $\qquad$ ...(7.1)

Dealing with Discontinuities in *dy/dx*

At a point on a curve where dy/dx fails to exist, dx/dy may exist and we may be able to find the curve's length by expressing x as a function of y and applying the following formula.

Formula for the length of the portion of the curve defined by x = g (y), c ≤ y ≤ d is given by

$$s = \int_c^d \sqrt{1 + \left(\frac{dx}{dy}\right)^2}\ dy = \int_c^d \sqrt{1 + \left[g'(y)\right]^2}\ dy. \qquad ...(7.2)$$

7.3 Illustrative Examples

Example 7.1 Find the length of the arc of the curves

(a) $y = x^{3/2}$ from $x = 0$ to $x = 5$. (b) $x = 3y^{3/2} - 1$ from $y = 0$ to $y = 4$.

(c) $6xy = x^4 + 3$ from $x = 1$ to $x = 2$ (d) $y = (x/2)^{2/3}$ from $x = 0$ to $x = 2$.

Solution. (a) By differentiation, $\dfrac{dy}{dx} = \left(\dfrac{3}{2}\right)x^{1/2}$.

$$\therefore \quad s = \int_a^b \sqrt{1 + \left(\frac{dy}{dx}\right)^2}\ dx = \int_0^5 \sqrt{1 + \frac{9}{4}x}\ dx = \left[\frac{\{1 + (9/4)x\}^{3/2}}{(3/2)(9/4)}\right]_0^5$$

$$= \frac{8}{27}\left[\left(1 + \frac{9}{4}x\right)^{3/2}\right]_0^5 = \frac{8}{27}\left[\frac{343}{8} - 1\right] = \frac{335}{27} = 12.41 \text{ units.}$$

(b) Here $\dfrac{dx}{dy} = \dfrac{9}{2}y^{1/2}$. Therefore, $s = \int_c^d \sqrt{1 + \left(\dfrac{dx}{dy}\right)^2}\ dy$

$$= \int_0^4 \sqrt{1 + \frac{81}{4}y}\ dy = \frac{8}{243}\left[\left(1 + \frac{81}{4}y\right)^{3/2}\right]_0^4 = \frac{8}{243}\left(82\sqrt{82} - 1\right) \text{ units.}$$

(c)   By differentiation,

$$6\left[x\cdot\left(\frac{dy}{dx}\right)+y\right]=4x^3 \text{ or } \frac{dy}{dx}=\frac{\left(2x^3-3y\right)}{3x}=\frac{\left(2x^4-3xy\right)}{3x^2}$$

$$=\frac{2x^4-\left\{\left(x^4+3\right)/2\right\}}{3x^2}=\frac{x^4-1}{2x^2}, \text{ and } 1+\left(\frac{dy}{dx}\right)^2=\frac{1}{4}\left(\frac{x^4+1}{x^2}\right)^2.$$

$$\therefore \ s=\frac{1}{2}\int_1^2\left(x^2+\frac{1}{x^2}\right)dx=\frac{1}{2}\left[\frac{x^3}{3}-\frac{1}{x}\right]_1^2$$

$$=\frac{1}{2}\left[\left(\frac{8}{3}-\frac{1}{2}\right)-\left(\frac{1}{3}-1\right)\right]=\frac{1}{2}\left[\frac{17}{6}\right]=\frac{17}{12} \text{ units.}$$

(d)   Here the derivative, $\dfrac{dy}{dx}=\dfrac{2}{3}\left(\dfrac{x}{2}\right)^{-1/3}\cdot\dfrac{1}{2}=\dfrac{1}{3}\left(\dfrac{2}{x}\right)^{1/3}$ is not defined at $x=0$, so

we cannot find the curve's length with formula 7.1. We therefore rewrite the given equation to express x in term of y i.e., $x=2\,(y)^{3/2}$. The required length is obtained by using the equation of curve $x = 2\,(y)^{3/2}$ from $y = 0$ to $y = 1$.

The derivative, $\dfrac{dx}{dy}=3y^{1/2}$ is continuous on [0, 1]. We may therefore, use

formula (7.2) to find the curve's length.

Therefore, $$s=\int_0^1\sqrt{1+\left(\frac{dx}{dy}\right)^2}\,dy=\int_0^1\sqrt{1+9y}\,dy$$

$$=\frac{2}{27}\left[(1+9y)^{3/2}\right]_0^1=\frac{2}{27}\left(10\sqrt{10}-1\right)\approx 2.27 \text{ units.}$$

Example 7.2 (a) **Find the length of the arc of the catenary** $y = c\cosh\left(\dfrac{x}{c}\right)$

measured (i) **from the vertex (0, c) to the point (x, y)** (ii) **from x = 0, to x = c.
Also show that** $y^2 = c^2 + s^2$, **the arc s being measured from the vertex.**

(b)   **Find the length of the arc of parabola** $y^2 = 12x$ **cut-off by its latus rectum.**

Solution. (a) We have $y = c\cosh\left(\dfrac{x}{c}\right)$. $\therefore \dfrac{dy}{dx}=c\sinh\left(\dfrac{x}{c}\right)\cdot\dfrac{1}{c}=\sinh\left(\dfrac{x}{c}\right)$.

(*i*) The length of the arc of catenary $y = c \cosh\left(\dfrac{x}{c}\right)$ measured from the vertex to the point (x, y) (Refer Fig. 7.2) is

$$s = \int_0^x \sqrt{1 + \left(\frac{dy}{dx}\right)^2}\, dx = \int_0^x \sqrt{1 + \sinh^2\left(\frac{x}{c}\right)}\, dx$$

$$= \int_0^x \cosh\frac{x}{c}\, dx = c\left[\sinh\frac{x}{c}\right]_0^x = c \sinh\left(\frac{x}{c}\right).$$

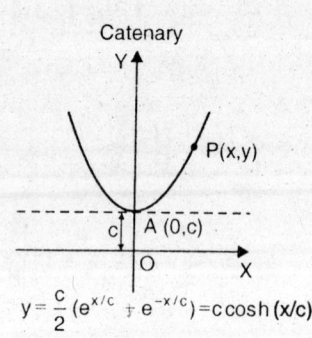

Catenary

$$y = \frac{c}{2}\left(e^{x/c} + e^{-x/c}\right) = c\cosh(x/c)$$

Fig. 7.2.

Further, $c^2 + s^2 = c^2 + c^2 \sinh^2\left(\dfrac{x}{c}\right)$

$$= c^2\left\{1 + \sinh^2\frac{x}{c}\right\} = c^2 \cosh^2\left(\frac{x}{c}\right) = y^2,$$

which is same as desired.

(*ii*) $s = \displaystyle\int_0^c \left(\dfrac{e^{x/c} + e^{-x/c}}{2}\right) dx = c\left[\dfrac{e^{x/c} - e^{-x/c}}{2}\right]_0^c = \dfrac{c}{2}\left(e - \dfrac{1}{e}\right).$

$$\left[\cosh\frac{x}{c} = \frac{e^{x/c} + e^{-x/c}}{2} \quad\text{and}\quad \sinh\frac{x}{c} = \frac{e^{x/c} - e^{-x/c}}{2}\right]$$

(*b*) The required length is twice that from the point $(0, 0)$ to the point $(3, 6)$

$$\frac{dx}{dy} = \frac{y}{6}, \quad\text{and}\quad 1 + \left(\frac{dx}{dy}\right)^2 = \frac{(36 + y^2)}{36}.$$

Then $s = 2\left(\dfrac{1}{6}\right)\displaystyle\int_0^6 \sqrt{36 + y^2}\, dy$

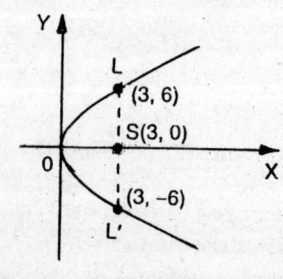

Fig. 7.3.

$$= \frac{1}{3}\left[\frac{1}{2}\cdot y\sqrt{36 + y^2} + \frac{36}{2}\log\left(y + \sqrt{36 + y^2}\right)\right]_0^6$$

$$= 6\left\{\sqrt{2} + \log\left(1 + \sqrt{2}\right)\right\}\text{ units.}$$

Example 7.3 (*a*) **Find the length of the curve $y^2 = (2x - 1)^3$ cut off by the line $x = 4$.** *(A.M.I.E., S-2002; V.T.U, 2000S)* (*b*) **Show that the length of the loop of the curve**

$3ay^2 = x(x-a)^2$ is $\dfrac{4a}{\sqrt{3}}$ **units.** (*D.C.E., 2007; Ranchi, 1998; A.M.I.E., S-1997*)

Solution: (*a*) The shape of the curve is shown in the Fig. 7.4.

$$\frac{dy}{dx} = \frac{3(2x-1)^2 \cdot 2}{2y} = \frac{3 \cdot (2x-1)^2}{(2x-1)^{3/2}} = 3\sqrt{2x-1}.$$

Employing, $s = \displaystyle\int_a^b \sqrt{1+\left(\frac{dy}{dx}\right)^2}\, dx$ to get

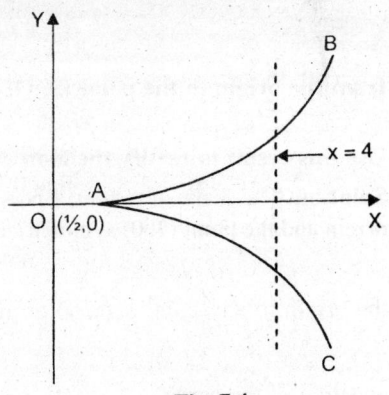

$$s = 2\int_{1/2}^4 \sqrt{1+\left(3\sqrt{2x-1}\right)^2}\, dx$$

$$= 2\int_{1/2}^4 (18x-8)^{1/2}\, dx$$

Fig. 7.4

$$= \frac{2}{18}\left[\frac{(18x-8)^{3/2}}{3/2}\right]_{1/2}^4 = \frac{2}{27}\left[(64)^{3/2}-1\right] = 37.85 \text{ units.}$$

(*b*) The shape of the curve is shown in Fig. 7.5. The required length of the loop

$$= 2\int_0^a \sqrt{1+\left(\frac{dy}{dx}\right)^2}\, dx.$$

We have $y = \dfrac{1}{\sqrt{3a}}\left[\sqrt{x}(x-a)\right]$

$$\therefore \quad \frac{dy}{dx} = \frac{1}{\sqrt{3a}}\left[\sqrt{x}+\frac{(x-a)}{2\sqrt{x}}\right] = \frac{3x-a}{2\sqrt{3ax}}$$

$$1+\left(\frac{dy}{dx}\right)^2 = 1+\frac{(3x-a)^2}{12ax} = \frac{(3x+a)^2}{12ax}.$$

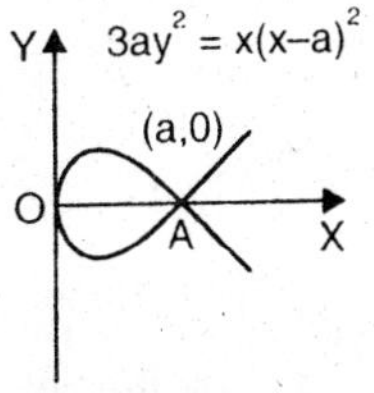

Therefore, $s = 2\displaystyle\int_0^a \sqrt{1+\left(\frac{dy}{dx}\right)^2}\, dx = 2\int_0^a \frac{3x+a}{\sqrt{12ax}}\, dx$

$3ay^2 = x(x-a)^2$

Fig. 7.5

$$= \frac{1}{\sqrt{3a}} \int_0^a \left(3\sqrt{x} + \frac{a}{\sqrt{x}} \right) dx = \frac{1}{\sqrt{3a}} \left[3 \cdot \frac{2}{3} x^{3/2} + 2a\sqrt{x} \right]_0^a$$

$$= \frac{1}{\sqrt{3a}} \left(2a^{3/2} + 2a^{3/2} \right) = \frac{4a}{\sqrt{3}} \text{ units.}$$

Example 7.4. Show that if 's' is the arc of the curve $y^2 = x \left[1 - \frac{x}{3} \right]^2$ measured

from the origin to the point (x, y), then $s^2 = y^2 + \frac{4x^2}{3}$.

Use this result to rectify the loop of the curve.

Solution. The shape of the curve is as shown in Fig. 7.6. It has a loop between the origin and the point (3, 0) symmetrical about the x-axis. The length of arc s, from

the origin to any point P (x, y) on the curve is $s = \int_0^x \sqrt{1 + \left(\frac{dy}{dx} \right)^2} \, dx$. From the

equation of the curve $y = x^{1/2} \left(1 - (x/3) \right)$,

we have $\dfrac{dy}{dx} = \dfrac{1}{2} x^{-1/2} - \dfrac{1}{3} \cdot \dfrac{3}{2} x^{1/2} = \dfrac{1-x}{2\sqrt{x}}$.

$$1 + \left(\frac{dy}{dx} \right)^2 = 1 + \frac{(1-x)^2}{4x} = \frac{(1+x)^2}{4x}.$$

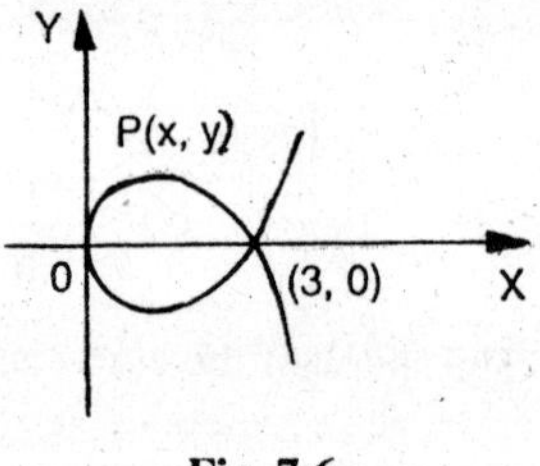

Fig. 7.6

$$\therefore \quad s = \int_0^x \sqrt{1 + \left(\frac{dy}{dx} \right)^2} \cdot dx = \int_0^x \frac{1+x}{2\sqrt{x}} dx = \frac{1}{2} \int_0^x \left(x^{-1/2} + x^{1/2} \right) dx$$

$$= \frac{1}{2} \left[\frac{x^{1/2}}{1/2} + \frac{x^{3/2}}{3/2} \right]_0^x = x^{1/2} + \frac{x^{3/2}}{3} = \frac{x^{1/2}}{3} (3+x).$$

Now $s^2 = \dfrac{x}{9}(3+x)^2 = x \left(1 + \dfrac{x}{3} \right)^2 = x \left(1 - \dfrac{x}{3} \right)^2 + \dfrac{4}{3} x^2$

$$= y^2 + \frac{4}{3} x^2, \text{ by the equation of the curve. Hence the result.}$$

The length of the loop lying above x-axis is obtained by putting $x = 3$, $y = 0$ in the above equation, which gives $s^2 = (4/3)\,9 = 12$. Therefore, $s = 2\sqrt{3}$. units. Thus, the length of the complete loop is $4\sqrt{3}$ units.

7.4 Arc length of parametric curves

To prove that the length of the arc of the curve $x = \phi(t)$, $y = \psi(t)$ between the points A and B corresponding to parametric values t_1 and t_2 of t, that is,

$t_1 \le t \le t_2$ is given by

$$s = \int_{t_1}^{t_2} \sqrt{\left(\frac{dx}{dt}\right)^2 + \left(\frac{dy}{dt}\right)^2}\, dt.$$

Let AB be the smooth curve $x = \phi(t)$, $y = \psi(t)$, and A, B the points, corresponds the values t_1 and t_2 of t. Let $P(x, y)$ be any point on the curve.

We know by Differential Calculus that

$$\frac{ds}{dt} = \sqrt{\left(\frac{dx}{dt}\right)^2 + \left(\frac{dy}{dt}\right)^2} \quad \therefore \quad \int_{t_1}^{t_2} \sqrt{\left(\frac{dx}{dt}\right)^2 + \left(\frac{dy}{dt}\right)^2}\, dt = \int_{t_1}^{t_2} \frac{ds}{dt}\, dt = [s]_{t_1}^{t_2}$$

$=$ (value of s for t equal to t_2) $-$ (value of s for t equal to t_1) $=$ Arc AB.

Thus $$s = \int_{t_1}^{t_2} \sqrt{\left(\frac{dx}{dt}\right)^2 + \left(\frac{dy}{dt}\right)^2}\, dt = \int_{t_1}^{t_2} \sqrt{\phi'^2(t) + \psi'^2(t)}\, dt \qquad \ldots(7.3)$$

Example 7.5 (*a*) **Find the length of the arc of the curve $x = t^2$, $y = t^3$ from $t = 0$ to $t = 4$.**

(*b*) **The position of a point at time t is given as $x = -t^2/2$, $y = (6t+9)^{3/2}/9$. Find the distance the point travels from $t = 0$ to $t = 4$.**

Solution. (*a*) We have $\dfrac{dx}{dt} = 2t$, and $\dfrac{dy}{dt} = 3t^2$.

$$\therefore \quad \sqrt{\left(\frac{dx}{dt}\right)^2 + \left(\frac{dy}{dt}\right)^2} = \sqrt{4t^2 + 9t^4} = 2t\sqrt{1 + (9t^2/4)}.$$

Hence, $$s = \int_0^4 \sqrt{1 + \frac{9}{4}t^2} \cdot 2t\, dt$$

$$= \frac{4}{9}\int_0^4 \sqrt{1 + \frac{9}{4}t^2} \cdot \frac{d}{dt}\left(\frac{9}{4}t^2\right) dt = \frac{4}{9}\left[\frac{\left\{1 + (9t^2/4)\right\}^{3/2}}{3/2}\right]_0^4 = \frac{8}{27}\left(37\sqrt{37} - 1\right) \text{ units.}$$

(b) $\dfrac{dx}{dt} = t, \dfrac{dy}{dt} = \sqrt{6t+9}$. $\therefore \left(\dfrac{dx}{dt}\right)^2 + \left(\dfrac{dy}{dt}\right)^2 = t^2 + 6t + 9 = (t+3)^2$.

Thus, $s = \displaystyle\int_0^4 \sqrt{\left(\dfrac{dx}{dt}\right)^2 + \left(\dfrac{dy}{dt}\right)^2}\, dt = \int_0^4 (t+3)\,dt = \left[\dfrac{(t+3)^2}{2}\right]_0^4 = \dfrac{1}{2}(49-9) = 20$ units.

Example 7.6 (a) **Find the length of one arch of the cycloid x = a (θ − sin θ), y = a (1 − cos θ).** (A.M.I.E., W-2005)

(b) **Find the length of one arch of the cycloid**

$$x = a\,(\theta + \sin\theta),\ y = a\,(1 - \cos\theta). \qquad \text{(A.M.I.E., S-1998)}$$

(c) **If $\theta \le \pi$ and s is the arc of the cycloid x = a (θ + sin θ), y = a (1 − cos θ) between the origin and the point (x, y) on the curve, show that $s^2 = 8ay$.**

Solution. (a) The shape of the curve is shown in Fig. 7.7. For the arch, θ varies from 0 to 2π.

We have $\dfrac{dx}{d\theta} = a(1 - \cos\theta)$, and $\dfrac{dy}{d\theta} = a\sin\theta$

$\left(\dfrac{dx}{d\theta}\right)^2 + \left(\dfrac{dy}{d\theta}\right)^2 = a^2(1 - \cos\theta)^2 + a^2\sin^2\theta$

$\qquad = 2a^2(1 - \cos\theta) = 4a^2 \sin^2\left(\dfrac{\theta}{2}\right).$

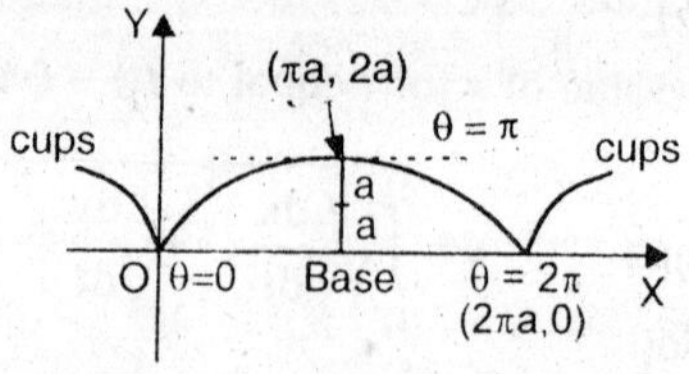

Fig. 7.7

Therefore, $s = 2a \displaystyle\int_0^{2\pi} \sin\dfrac{\theta}{2}\, d\theta = -4a\left[\cos\dfrac{\theta}{2}\right]_0^{2\pi} = 8a$ units.

(b) The shape of the curve is as shown in Fig. 7.8. For half of the cycloid θ varies from 0 to π. Hence the length of the complete cycloid,

$$s = 2\int_0^{\pi} \sqrt{\left(\dfrac{dx}{d\theta}\right)^2 + \left(\dfrac{dy}{d\theta}\right)^2}\, d\theta.$$

Differentiating the equations of curve, we have

$\dfrac{dx}{d\theta} = a(1 + \cos\theta)$ and $\dfrac{dy}{d\theta} = a\sin\theta$.

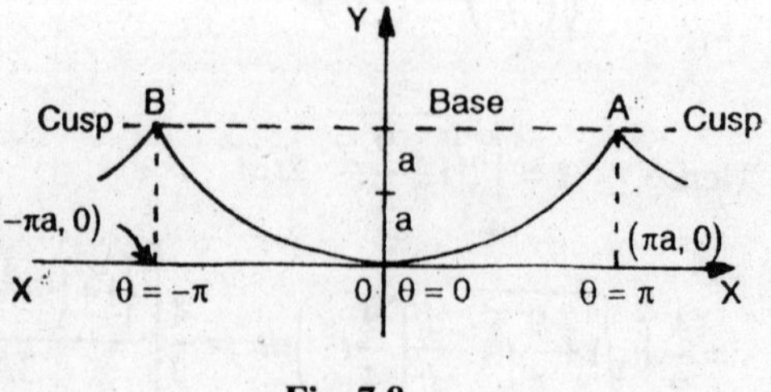

Then $\left(\dfrac{dx}{d\theta}\right)^2 + \left(\dfrac{dy}{d\theta}\right)^2$

Fig. 7.8

$$= a^2 (1+\cos\theta)^2 + a^2 \sin^2\theta = a^2 \left(1+2\cos\theta+\cos^2\theta+\sin^2\theta\right)$$

$$= 2a^2 (1+\cos\theta) = 4a^2 \cos^2\theta/2 \,.$$

$$\therefore s = 2\int_0^\pi \sqrt{4a^2 \cos^2\frac{\theta}{2}} \, d\theta = 4a \int_0^\pi \cos\frac{\theta}{2} d\theta. = 8a\left[\sin\frac{\theta}{2}\right]_0^\pi = 8a\left[\sin\frac{\pi}{2}\right] = 8a \text{ units.}$$

(c) We have $\left(\dfrac{dx}{d\theta}\right)^2 + \left(\dfrac{dy}{d\theta}\right)^2 = 4a^2 \cos^2\left(\dfrac{\theta}{2}\right)$

$$\therefore \quad s = \int_0^\theta \sqrt{\left(\frac{dx}{d\theta}\right)^2 + \left(\frac{dy}{d\theta}\right)^2} \, d\theta = \int_0^\theta 2a\cos\frac{\theta}{2} d\theta = 4a\left[\sin\frac{\theta}{2}\right]_0^\theta = 4a\sin\frac{\theta}{2}.$$

Squaring, $s^2 = 16a^2 \sin^2\left(\dfrac{\theta}{2}\right) = 8a^2 (1-\cos\theta)$

$$= 8a\{a(1-\cos\theta)\} = 8ay, \text{ the required result.}$$

Example 7.7 (a) Find the length of the involute of the circle given by $x = a\,(\cos\theta + \theta\sin\theta)$, $y = a\,(\sin\theta - \theta\cos\theta)$ between $\theta = 0$ to $\theta = 2\pi$.

(b) Show that the length of an arc of the curve

$$\left.\begin{array}{l} x\sin\theta + y\cos\theta = f'(\theta) \\ x\cos\theta - y\sin\theta = f''(\theta) \end{array}\right\} \text{ is given by } s = f(\theta) + f''(\theta) + c.$$

Solution. (a) Here $\dfrac{dx}{d\theta} = a(-\sin\theta + \sin\theta + \theta\cos\theta) = a\theta\cos\theta,$

and $\dfrac{dy}{d\theta} = a(\cos\theta - \cos\theta + \theta\sin\theta) = a\theta\sin\theta.$

Hence $s = \displaystyle\int_0^{2\pi} \sqrt{(a\theta\cos\theta)^2 + (a\theta\sin\theta)^2} \, d\theta$

$$= \int_0^{2\pi} a\theta \, d\theta = a\int_0^{2\pi} \theta \, d\theta = a\cdot\left[\frac{\theta^2}{2}\right]_0^{2\pi} = 2a\pi^2 \text{ units.}$$

(b) Solving the given equations for x and y, we have

$$x = f'(\theta)\sin\theta + f''(\theta)\cos\theta, \; y = f'(\theta)\cos\theta - f''(\theta)\sin\theta.$$

$$\therefore \quad \frac{dx}{d\theta} = \left[f'(\theta)\cos\theta + f''(\theta)\sin\theta\right] + \left[f''(\theta)(-\sin\theta) + f'''(\theta)\cos\theta\right]$$

$$= \left[f'(\theta) + f'''(\theta)\right]\cos\theta.$$

Similarly, $\dfrac{dy}{d\theta} = -\left[f'(\theta) + f'''(\theta)\right]\sin\theta$.

$\therefore\qquad s = \int \sqrt{\left(\dfrac{dx}{d\theta}\right)^2 + \left(\dfrac{dy}{d\theta}\right)^2}\, d\theta = \int \left[f'(\theta) + f'''(\theta)\right] d\theta$

$= f(\theta) + f''(\theta) + c$, where c is a constant.

Example 7.8 Show that the whole length of the curve

$$\left(\dfrac{x}{a}\right)^{2/3} + \left(\dfrac{y}{b}\right)^{2/3} = 1 \text{ is } 4\left(a^2 + ab + b^2\right)\big/(a+b).$$ (D.C.E, 2005)

Solution. Here we use the parametric equations for the curve,
that is, $x = a\cos^3\theta$, $y = b\sin^3\theta$.   ...(i)

Since the curve is symmetric about both coordinate axes, we shall first compute the length of a fourth part of it located in the first quadrant. We find

$$\left.\begin{aligned} \dfrac{dx}{d\theta} &= -3a\cos^2\theta\sin\theta, \\ \dfrac{dy}{d\theta} &= 3b\sin^2\theta\cos\theta. \end{aligned}\right]$$ ...(ii)

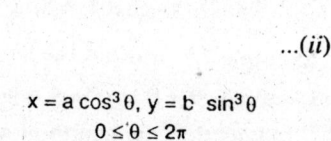

$x = a\cos^3\theta,\ y = b\ \sin^3\theta$
$0 \le \theta \le 2\pi$

The parameter will vary from 0 to $\pi/2$.
If s is the length of curve in first quadrant.

$$s = \int_0^{\pi/2} \sqrt{\left(\dfrac{dx}{d\theta}\right)^2 + \left(\dfrac{dy}{d\theta}\right)^2}\, d\theta$$

$$= \int_0^{\pi/2} \sqrt{9a^2\cos^4\theta\sin^2\theta + 9b^2\sin^4\theta\cos^2\theta}\cdot d\theta$$

$$= \int_0^{\pi/2} 3\sin\theta\cos\theta\sqrt{a^2\cos^2\theta + b^2\sin^2\theta}\, d\theta \quad ...(iii)$$

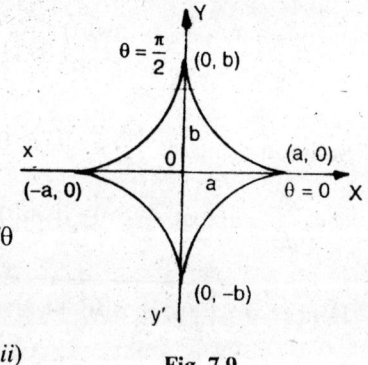

Fig. 7.9

To evaluate the integral, put $a^2\cos^2\theta + b^2\sin^2\theta = z^2$ so that $z = a$ when $\theta = 0$

and $z = b$ when $\theta = \dfrac{\pi}{2}$. Now $\sin\theta\cos\theta\, d\theta = z\, dz\big/\left(b^2 - a^2\right)$.

$$\therefore\quad s = \int_a^b 3z\cdot\dfrac{1}{b^2 - a^2} z\, dz = \dfrac{3}{b^2 - a^2}\int_a^b z^2\, dz = \dfrac{3}{b^2 - a^2}\left[\dfrac{z^3}{3}\right]_a^b$$

$$= \frac{b^3 - a^3}{b^2 - a^2} = \frac{b^2 + ab + a^2}{b + a}.$$

Therefore, whole length of the curve $= 4s = \dfrac{4\left(a^2 + ab + b^2\right)}{a + b}$ ...(iv)

Remarks. 1. Putting $b = a$ in (iv), the total length of the curve

$x^{2/3} + y^{2/3} = a^{2/3}$, that is $x = a\cos^3\theta$, $y = a\sin^3\theta$ is 6a units.

2. $x = a\cos^3\theta$, $y = a\sin^3\theta$ are the parametric equations of the curve

$x^{2/3} + y^{2/3} = a^{2/3}$. This curve is called a four-cusped hypocycloid (also called an astroid).

7.5 Arc length of polar curves

To prove that the length of the portion of the curve defined by $r = f(\theta)$ between two points as θ runs from α to β (that is, $\alpha \le \theta \le \beta$) is given by

$$s = \int_{\alpha}^{\beta} \sqrt{r^2 + \left(\frac{dr}{d\theta}\right)^2}\, d\theta.$$

Let AB be the curve $r = f(\theta)$, and A, B the points where $\theta = \alpha$, $\theta = \beta$ respectively (Refer Fig. 7.10).

Let $P(r, \theta)$ be any point on the curve.

We know by Differential Calculus that

$$\frac{ds}{d\theta} = \sqrt{r^2 + \left(\frac{dr}{d\theta}\right)^2}.$$

and $\dfrac{ds}{dr} = \dfrac{ds}{d\theta} \cdot \dfrac{d\theta}{dr} = \sqrt{r^2\left(\dfrac{d\theta}{dr}\right)^2 + 1}$

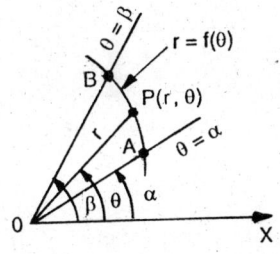

Fig. 7.10.

$$\therefore \quad \int_{\alpha}^{\beta} \sqrt{r^2 + \left(\frac{dr}{d\theta}\right)^2}\, d\theta = \int_{\alpha}^{\beta} \frac{ds}{d\theta}\, d\theta = \left[s\right]_{\alpha}^{\beta}$$

$= $ (value of s when θ equal to β) $-$ (value of s when θ equal to α) $=$ Arc AB

Thus $s = \displaystyle\int_{\alpha}^{\beta} \sqrt{r^2 + \left(\frac{dr}{d\theta}\right)^2}\, d\theta = \int_{\alpha}^{\beta} \sqrt{f^2(\theta) + f'^2(\theta)}\, d\theta$ ...(7.4) or $\qquad s = \int\limits_{r_1}^{r_2} \sqrt{\left[r^2 \left(\dfrac{d\theta}{dr}\right)^2 + 1 \right]}\, dr$ $\hspace{3cm}$ ...(7.5)

Example 7.9 (*a*) Find the length of the spiral $r = e^{2\theta}$ from $\theta = 0$ to $\theta = 2\pi$.
(*b*) Find the perimeter of the cardioid $r = a\,(1 - \cos\theta)$. $\quad$ (*A.M.I.E., S-2002*)
(*c*) Find the total length of the curve $r = a\,\sin^3(\theta/3)$. $\quad$ (*A.M.I.E., W-2010*)

Solution. (*a*) We have $\dfrac{dr}{d\theta} = 2e^{2\theta}$,

and $r^2 + \left(\dfrac{dr}{d\theta}\right)^2 = e^{4\theta} + 4e^{4\theta} = 5e^{4\theta}$.

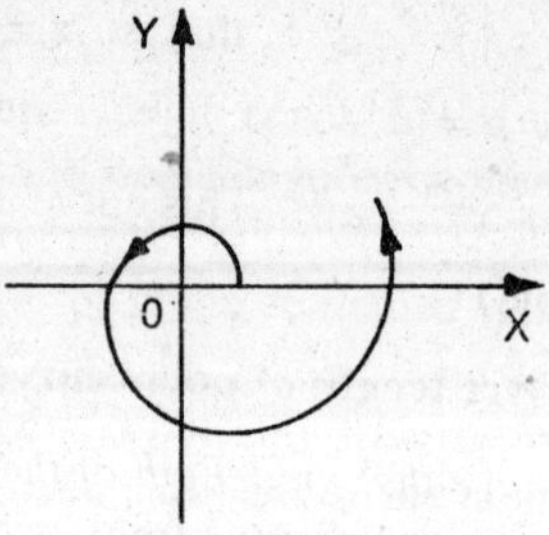

$\therefore \qquad s = \int\limits_{0}^{2\pi} \sqrt{r^2 + \left(\dfrac{dr}{d\theta}\right)^2}\, d\theta$

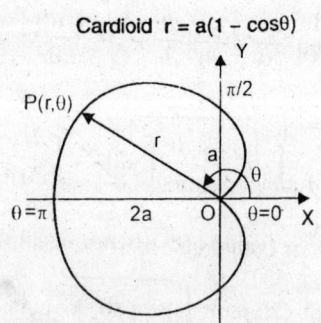

Fig. 7.11.

$\qquad = \sqrt{5}\int\limits_{0}^{2\pi} e^{2\theta}\, d\theta = \dfrac{\sqrt{5}}{2}\left[e^{4\pi} - 1\right]$ units.

(*b*) The cardioid is symmetrical about the initial line, and for the upper half, θ varies from 0 to π. Refer Fig. 7.12.

$\dfrac{dr}{d\theta} = a\sin\theta$, and $r^2 + \left(\dfrac{dr}{d\theta}\right)^2 = a^2(1 - \cos\theta)^2 + a^2\sin^2\theta$

$\qquad = 2a^2(1 - \cos\theta) = 4a^2\sin^2\left(\dfrac{\theta}{2}\right)$.

Therefore, perimeter $= 2\int\limits_{0}^{\pi} \sqrt{r^2 + \left(\dfrac{dr}{d\theta}\right)^2}\, d\theta$

Cardioid $r = a(1 - \cos\theta)$

$= 2\int\limits_{0}^{\pi} \sqrt{4a^2\sin^2(\theta/2)}\, d\theta = 4a\int\limits_{0}^{\pi} \sin\dfrac{\theta}{2}\, d\theta$

$= 4a\left[\dfrac{-\cos\theta/2}{1/2}\right]_{0}^{\pi} = -8a[0 - 1] = 8a$ units.

Fig. 7.12.

(c) Refer Fig. 7.13. The curve is defined when $0 \le \theta \le 3\pi$. We have

$dr/d\theta = a\sin^2(\theta/3)\cos(\theta/3)$ and $r^2 + (dr/d\theta)^2$

$= a^2\sin^6(\theta/3) + a^2\sin^4(\theta/3)\cos^2(\theta/3) = a^2\sin^4(\theta/3)$.

Therefore, $s = \int_0^{3\pi} \sqrt{r^2 + (dr/d\theta)^2}\, d\theta = a\int_0^{3\pi} \sin^2(\theta/3)\, d\theta$

$= 3a\int_0^{\pi} \sin^2 t\, dt$, where $\theta = 3t$.

$= \dfrac{3a}{2}\int_0^{\pi}(1-\cos 2t)\, dt = \dfrac{3a}{2}\left[t - \dfrac{\sin 2t}{2}\right]_0^{\pi} = \dfrac{3a\pi}{2}$ units.

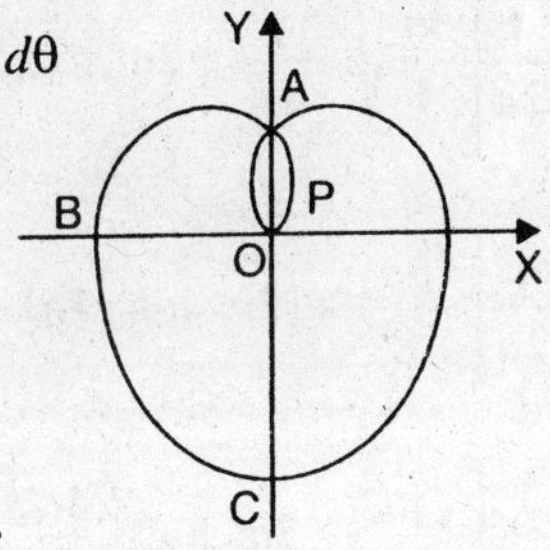

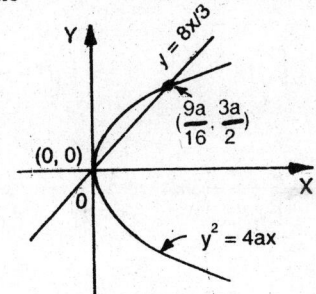

Fig. 7.13.

7.6 Miscellaneous Examples

Example 7.10. (a) **Find the length of the arc of the parabola $y^2 = 4ax$ cut-off by the line $3y = 8x$.** (*A.M.I.E., S-2003; Nagpur, 1998S*)

(b) **Find the perimeter of the loop of the curve**

$$9ay^2 = (x-2a)(x-5a)^2.$$ (*D.U. 1999*)

(c) **Show that in the curve $8a^2y^2 = x^2(a^2 - x^2)$, $s = a(2\theta + \sin\theta\cos\theta)/2\sqrt{2}$.**

where $x = a\sin\theta$ and that the perimeter of one of the loops is $\dfrac{\pi a}{\sqrt{2}}$. Show that

the whole length of the curve is equal to $\pi a\sqrt{2}$.

Solution. (a) Refer Fig. 7.14 The points of intersection of the curve with straight line

$y = \dfrac{8x}{3}$ are $(0,0)$ and $\left(\dfrac{9a}{16}, \dfrac{3a}{2}\right)$.

∴ Required length of the arc

$= \int_0^{3a/2} \sqrt{1 + \left(\dfrac{dx}{dy}\right)^2}\, dy = \int_0^{3a/2} \sqrt{1 + \dfrac{y^2}{4a^2}}\, dy$

$= \dfrac{1}{2a}\int_0^{3a/2} \sqrt{y^2 + 4a^2}\, dy$

Fig. 7.14.

$$= \frac{1}{2a}\left[\frac{y\sqrt{y^2+4a^2}}{2} + \frac{4a^2}{2}\cdot\log_e\left(y+\sqrt{y^2+4a^2}\right)\right]_0^{3a/2}$$

$$= \frac{1}{2a}\left[\frac{3a}{4}\sqrt{\frac{25a^2}{4}} + 2a^2\log_e\left(\frac{3a}{2}+\sqrt{\frac{25a^2}{4}}\right) - \frac{4a^2}{2}\log_e 2a\right]$$

$$= \frac{1}{2a}\left[\frac{15a^2}{8} + \left(2a^2\log_e 4a\right) - \left(2a^2\cdot\log_e 2a\right)\right]$$

$$= \frac{1}{2a}\left[\frac{15a^2}{8} + 2a^2\log_e\left(\frac{4a}{2a}\right)\right] = a\left[\frac{15}{16}+\log_e 2\right]\text{units.}$$

(b) The loop lies between the limits $x = 2a$ and $x = 5a$. The curve is symmetrical about x-axis and therefore, the perimeter of the loop is double of the length of its parts lying above x-axis. [Refer Fig. 7.15]

For any point on the arc lying above x-axis, we have

$$y = -\frac{(x-5a)\sqrt{(x-2a)}}{3\sqrt{a}}$$

$$\therefore \frac{dy}{dx} = -\frac{\sqrt{x-2a}+(x-5a)\cdot\frac{1}{2\sqrt{x-2a}}}{3\sqrt{a}}$$

$$= -\frac{x-3a}{2\sqrt{a}\sqrt{x-2a}}.$$

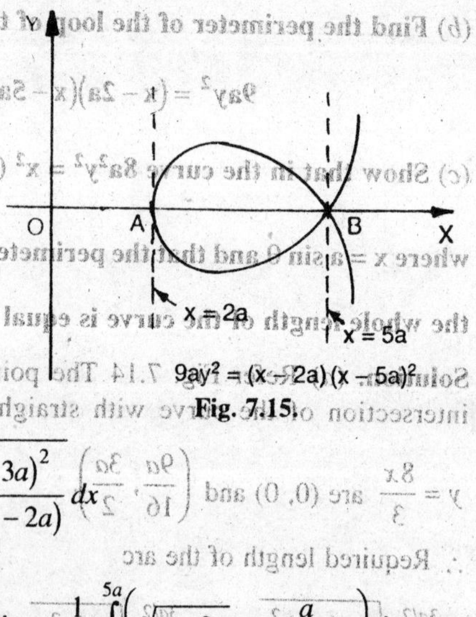

$\therefore$ The required perimeter

$$= 2\int_{2a}^{5a}\sqrt{1+\left(\frac{dy}{dx}\right)^2}\,dx = 2\int_{2a}^{5a}\sqrt{1+\frac{(x-3a)^2}{4a(x-2a)}}\,dx$$

$$= 2\int_{2a}^{5a}\frac{x-a}{2\sqrt{a}\sqrt{x-2a}}\,dx = \frac{1}{\sqrt{a}}\int_{2a}^{5a}\frac{x-2a+a}{\sqrt{x-2a}}\,dx = \frac{1}{\sqrt{a}}\int_{2a}^{5a}\left(\sqrt{x-2a}+\frac{a}{\sqrt{x-2a}}\right)dx$$

$$= \frac{1}{\sqrt{a}}\left[\frac{2}{3}(x-2a)^{3/2}+2a\sqrt{x-2a}\right]_{2a}^{5a} = \frac{1}{\sqrt{a}}\left[\frac{2}{3}(3a)^{3/2}+2a(3a)^{1/2}\right] = 4a\sqrt{3}\text{ units.}$$

Fig. 7.15.

(c) When $x = a \sin \theta$ is substituted in the equation of the curve, we have

$$8a^2 y^2 = \left(a^2 \sin^2 \theta\right) \cdot a^2 \cos^2 \theta$$

or, $\qquad y = \pm (a \sin \theta \cos \theta)/2\sqrt{2} = \pm (a \sin 2\theta)/4\sqrt{2}$.

Thus, $\qquad \dfrac{dx}{d\theta} = a \cos\theta$ and $\dfrac{dy}{d\theta} = \pm \dfrac{(a \cos 2\theta)}{2\sqrt{2}}$

$$\therefore \left(\frac{ds}{d\theta}\right)^2 = \left(\frac{dx}{d\theta}\right)^2 + \left(\frac{dy}{d\theta}\right)^2 = a^2 \cos^2\theta + \frac{1}{8}\left(a^2 \cos^2 2\theta\right)$$

$$= \frac{a^2}{8}\left[8\cos^2\theta + \left(2\cos^2\theta - 1\right)^2\right] = \frac{a^2}{8}\left(2\cos^2\theta + 1\right)^2, \text{ giving}$$

$$\frac{ds}{d\theta} = \frac{a(2\cos^2\theta + 1)}{2\sqrt{2}}$$

Hence $s = \dfrac{a}{2\sqrt{2}} \displaystyle\int_0^\theta \left(2\cos^2\theta + 1\right) d\theta$, measuring s from the point, where $\theta = 0$.

This gives $\quad s = \dfrac{a}{2\sqrt{2}} \displaystyle\int_0^\theta (2 + \cos 2\theta) d\theta = \dfrac{a}{2\sqrt{2}} \left[2\theta + \dfrac{\sin 2\theta}{2}\right]_0^\theta$

$$= \frac{a}{2\sqrt{2}}(2\theta + \sin\theta \cos\theta) \Big|_0^\theta \qquad \qquad \dots (i),$$

The shape of the curve is as shown in the figure 7.16. For the upper half of the loop on the right hand side, x varies from

0 to $\dfrac{\pi}{2}$, since $x = a \sin \theta$. Hence the

length of half the loop is obtained by

putting $\theta = \dfrac{\pi}{2}$ in (i). This gives

$s = \pi a/2\sqrt{2}$ units. The perimeter of one

loop is $\pi a/\sqrt{2}$. Since the curve consists

of two equal loops, thus the whole length

of the curve is $\pi a\sqrt{2}$.

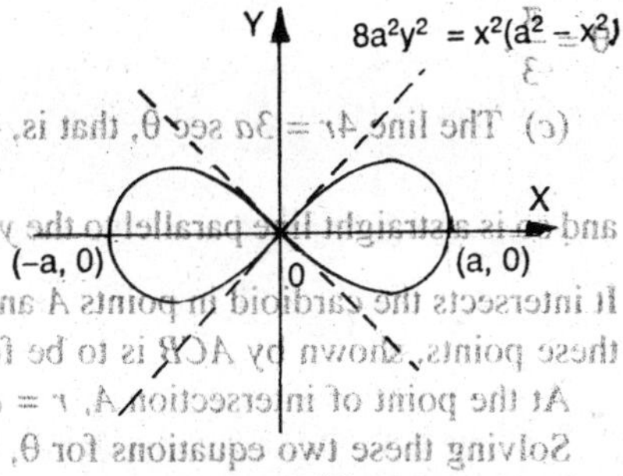

$$8a^2 y^2 = x^2(a^2 - x^2)$$

(−a, 0) $\qquad$ 0 $\qquad$ (a, 0)

Fig. 7.16.

Example 7.11 (*a*) **Find the entire length of the cardioid r = a (1 + cosθ).**

(*A.M.I.E., S-2004; Mysore, 1999*)

(*b*) **Show that the arc of the upper half of the above curve is bisected by**

$$\theta = \frac{\pi}{3}.$$
(*A.M.I.E., S-1999; Mangalore, 1999*)

(*c*) **Show that the length of the arc of that part of the cardioid r = a (1 + cosθ) which lies on the side of the line 4r = 3a sec θ remote from the pole is equal to 4a.**

Solution. (*a*) The curve is symmetrical about the initial line, therefore, the entire length is twice the length of the curve above the initial line. For the upper half, θ varies from 0 to π.

Hence required length $= 2\int_0^{\pi} \sqrt{r^2 + \left(\frac{dr}{d\theta}\right)^2}\, d\theta$

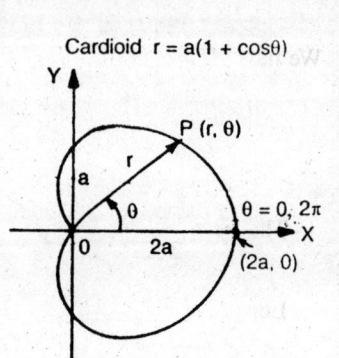

Cardioid r = a(1 + cosθ)

Fig. 7.17.

$$= 2\int_0^{\pi} \sqrt{a^2 (1 + \cos\theta)^2 + (-a\sin\theta)^2}\, d\theta$$

$$= 2\int_0^{\pi} 2a\cos\frac{\theta}{2}\, d\theta = 8a\left[\sin\frac{\theta}{2}\right]_0^{\pi} = 8a.$$

(*b*) The length of the upper half of the cardioid = 4*a*. The length of the arc between the points, where θ = 0 and θ = π/3

$$= \int_0^{\pi/3} \sqrt{r^2 + \left(\frac{dr}{d\theta}\right)^2}\, d\theta = \int_0^{\pi/3} 2a\cos\frac{\theta}{2}\, d\theta = \left[4a\sin\frac{\theta}{2}\right]_0^{\pi/3} = 2a$$

which is half of 4*a*. Hence the upper half of the cardioid is bisected by the line

$$\theta = \frac{\pi}{3}.$$

(*c*) The line 4*r* = 3*a* sec θ, that is, 4*r* cos θ = 3*a* in cartesian form is 4*x* = 3*a*

and so is a straight line parallel to the *y*-axis at a distance $\frac{3a}{4}$ unit from the origin.

It intersects the cardioid in points *A* and *B* and the length of the cardioid between these points, shown by *ACB* is to be found out (Refer Fig. 7.18).

At the point of intersection *A*, *r* = *a* (1+ cosθ) and 4*r* = 3*a* sec θ.

Solving these two equations for θ, we have

$4(1+\cos\theta) = 3\sec\theta$

or $4\cos^2\theta + 4\cos\theta - 3 = 0$

that is, $(2\cos\theta + 3)(2\cos\theta - 1) = 0$

or $\quad \cos\theta = -\dfrac{3}{2}$ or $\dfrac{1}{2}$.

Discarding the first one as untenable,

we have $\cos\theta = \dfrac{1}{2}$ or $\theta = \dfrac{\pi}{3}$.

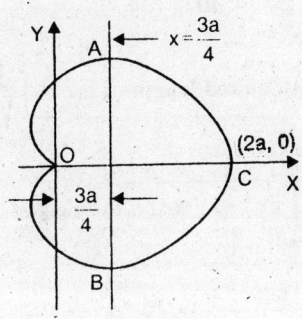

Fig. 7.18.

We have $\dfrac{dr}{d\theta} = -a\sin\theta$.

$$\therefore \quad \sqrt{r^2 + \left(\frac{dr}{d\theta}\right)^2} = \sqrt{a^2(1+\cos\theta)^2 + a^2\sin^2\theta} = 2a\cos\frac{\theta}{2}.$$

The length of the arc ACB = Twice the length of arc AC.

Length of the arc $AC = \displaystyle\int_0^{\pi/3} \sqrt{r^2 + \left(\frac{dr}{d\theta}\right)^2}\, d\theta = \int_0^{\pi/3} 2a\cos\frac{\theta}{2}\, d\theta$

$= 4a\left[\sin\dfrac{\theta}{2}\right]_0^{\pi/3} = 4a\sin\dfrac{\pi}{6} = 2a.$ Therefore, the length of the arc $ACB = 4a$ units.

Example 7.12 (*a*) **Find the length of the arc of the equiangular spiral $r = ae^{\theta\cot\alpha}$ between the points at which the radii vectors are r_1 and r_2.**

(*b*) **Find the length of the arc of the equiangular spiral $r = ae^{\theta\cot\alpha}$ from the pole to the point (r, θ).**

Solution. (*a*) The equation of the equiangular spiral is $r = ae^{\theta\cot\alpha}$.

Taking logarithms, $\log r = \log a + \theta\cot\alpha$

Differentiating w.r.t. θ, we have $\dfrac{1}{r}\cdot\dfrac{dr}{d\theta} = \cot\alpha$, or $r\left(\dfrac{d\theta}{dr}\right) = \tan\alpha$.

$$\therefore \text{ Length of arc } = \int_{r_1}^{r_2} \sqrt{1 + r^2\left(\frac{d\theta}{dr}\right)^2}\, dr = \int_{r_1}^{r_2} \sqrt{1 + \tan^2\alpha}\, dr$$

$$= \sec\alpha\cdot[r]_{r_1}^{r_2} = \sec\alpha\,(r_2 - r_1).$$

(b) Here $\dfrac{dr}{d\theta} = (\cot\alpha)ae^{\theta\cot\alpha} = r\cot\alpha$; also $\theta = -\infty$ when $r = 0$.

$\therefore$ Required length of arc $= \displaystyle\int_{-\infty}^{\theta}\sqrt{r^2 + \left(\dfrac{dr}{d\theta}\right)^2}\,d\theta$

$= \displaystyle\int_{-\infty}^{\theta}\sqrt{r^2 + r^2\cot^2\alpha}\,d\theta = \operatorname{cosec}\alpha\int_{-\infty}^{\theta} ae^{\theta\cot\alpha}\,d\theta$

$= \dfrac{a\,\operatorname{cosec}\,\alpha}{\cot\alpha}\left[e^{\theta\cot\alpha}\right]_{-\infty}^{\theta} = \dfrac{a}{\cos\alpha}\left(e^{\theta\cot\alpha} - 0\right) = r\sec\alpha.$

Exercise 7

1. Find the length of the arc of $24\,xy = x^4 + 48$ from $x = 2$ to $x = 4$.

   Ans. 17/6 units.

2. For the curve $ay^2 = x^3$, prove that if s be measured from the origin

   $$s = \frac{8a}{27}\left(1 + \frac{9x}{4a}\right)^{3/2} - \frac{8a}{27}.$$ Deduce that if the tangent at (x, y) makes an angle

   ψ with the x-axis, then $s = (8a/27)(\sec^3\psi - 1)$.

3. Find the arc of the parabola $y^2 = 4a\,(a - x)$ cut off by the y-axis.

   Ans. $a\left[2\sqrt{2} - \log\left(3 - 2\sqrt{2}\right)\right].$

4. Find the length of the arc of the parabola $y^2 = 4x$ between the points

   $(1, 2)$ and $(4, 4)$.

   Ans. $\dfrac{1}{2}\log\dfrac{2 + \sqrt{5}}{1 + \sqrt{2}}.$

5. Find the length of the curve $y = (1/3)\cosh 3x$ from $x = 0$ to $x = 1/3$.

   Ans. 0.3917

6. Find the length of the arc of the curve $y = \log\sec x$ from $x = 0$ to $x = \pi/3$.

   (A.M.I.E., W-2002) **Ans.** $\log\left(2 + \sqrt{3}\right)$

7. Find the length of the arc of the parabola $x^2 = 4ay$ measured from the vertex to one extremity of the latus rectum. *(A.M.I.E., S-2005)*

   Ans. $a\left[\sqrt{2} + \log\left(1 + \sqrt{2}\right)\right]$

8. Find the length of the arc of the parabola $y^2 = 4ax$, which is intercepted between the points of intersection of the line $y = 2x$.

   (A.M.I.E., S-1999) **Ans.** $a\left[\sqrt{2} + \log\left(1 + \sqrt{2}\right)\right].$

9. Prove that the length of the arc of the parabola $y^2 = 4ax$ cut off by the latus

rectum is $2a\left[\sqrt{2} + \log\left(1 + \sqrt{2}\right)\right]$. *(A.M.I.E. W-1996)*

10. Find the length of the parabola $y^2 = 4ax$ measured from the vertex to one extremity of the latus rectum. *(A.M.I.E., W-2002)*

Ans. $a\left[\sqrt{2} + \log\left(1 + \sqrt{2}\right)\right]$.

11. Show that length s of the curve, $x^{2/3} + y^{2/3} = a^{2/3}$, measured from $(0, a)$ to

the point (x, y) is given by $s = (3/2)\sqrt[3]{ax^2}$.

12. Show that $s^3 \propto x^2$, s being measured from the cusp which lies on the y-axis.

13. Find the perimeter of the loop of the following curves:

(*i*) $3ay^2 = x^2(a - x)$,

(*ii*) $9ay^2 = x(x - 3a)^2$ from $x = 0$ to $x = 3a$.
(A.M.I.E., W-2002)

(*iii*) $9y^2 = (x + 7)(x + 4)^2$.

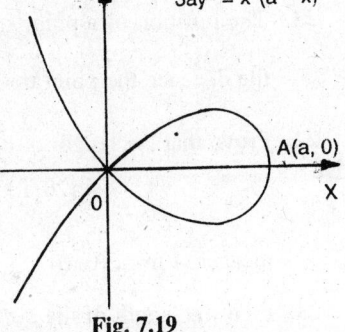

$3ay^2 = x^2(a - x)$

A(a, 0)

Fig. 7.19

Ans. (*i*) $4a/\sqrt{3}$, (*ii*) $4a\sqrt{3}$, (*iii*) $4\sqrt{3}$.

[**Hint:** (*i*) The shape of the curve is shown in Fig. 7.19. For the upper half of the loop, x varies from 0 to a.]

14. Find the length of the arc of catenary $y = c\cosh(x/c)$ measured from the vertex to the point (x, y). **Ans.** $c\sinh(x/c)$.

15. Prove that the whole length of the curve $x^2(a^2 - x^2) = 8a^2y^2$ is $\pi a\sqrt{2}$.

16. Find the length of the arc of the curve $y = \log\left\{\left(e^x - 1\right)/\left(e^x + 1\right)\right\}$

from $x = 1$ to $x = 2$. **Ans.** $\log\left(e + e^{-1}\right)$.

17. If 's' be the length of the arc of the curve $y^2 = x\left(1 - \dfrac{x}{3}\right)^2$ measured from

the origin to the ordinate, where $x = a$, show that $9s^2 = a(a + 3)^2$.

18. Show that the length of an arch of the cycloid

(*i*) $x = a(\theta - \sin\theta)$, $y = a(1 - \cos\theta)$ is $8a$.

(*ii*) $x = a(\theta + \sin\theta)$, $y = a(1 + \cos\theta)$ is $4a$.

(A.M.I.E., W-1999; Kuvempu, 1998).

19. In the curve $x = e^\theta \sin\theta$, $y = e^\theta \cos\theta$, show that $s = e^\theta \sqrt{2} + c$, where c is a constant. Hence rectify the curve from $\theta = 0$ to $\theta = \pi/2$.

Ans. $\sqrt{2}\left(e^{\pi/2} - 1\right)$.

20. Find the length of the loop of the curve

$$x = t^2, \; y = t\left(1 - \frac{t^2}{3}\right).$$

(Marathwada, 1998S)

[**Hint:** Eliminating t, the cartesian equation of the curve is seen to be

$$y^2 = x\left(1 - \frac{x}{3}\right)^2 .]$$

Ans. $4\sqrt{3}$ units.

21. The position of a point at time t is given as $x = \dfrac{t^2}{2}$, $y = \dfrac{(6t+9)^{3/2}}{9}$. Find the distance the point travels from $t = 0$ to $t = 9$.

Ans. 20 units.

22. Prove that the length of the arc of the curve

$$x = a\sin 2\theta(1 + \cos 2\theta), \; y = a\cos 2\theta(1 - \cos 2\theta),$$

measured from $(0, 0)$ to (x, y) is equal to $\dfrac{(4a \sin 3\theta)}{3}$.

23. Find the length of the curve having parametric equations:

(a) $x = e^{2t} - \dfrac{t}{8}$, $y = e^t$; $0 \le t \le \log 2$ **Ans.** $3 + \{(\log 2)/8\}$ units.

(b) $x = t^2/2$, $y = \dfrac{(2t+1)^{3/2}}{3}$, $0 \le t \le 4$.   **Ans.** 12 units.

(c) $x = e^t \cos t$, $y = e^t \sin t$ from $t = 0$ to $t = 4$.   **Ans.** $\sqrt{2}\left(e^4 - 1\right)$ units.

24. Find the length of arcs of the following curves

(i) $r = a\cos\theta$,   (ii) $r = ae^{m\theta}$,   (iii) $r = a\theta$,

(iv) $r = a\sin^2(\theta/2)$,   (v) $r^{1/3} = 8\cos(\theta/3)$,   (vi) $r = a\sec^2(\theta/2)$.

(A.M.I.E., W-1998)

Ans. (i) $a(\theta_2 - \theta_1)$, (ii) $(r_2 - r_1)\sqrt{1 + m^2}/m$, (iii) $\dfrac{a}{2}\left[\theta\sqrt{1 + \theta^2} + \sinh^{-1}\theta\right]_{\theta_1}^{\theta_2}$,

(iv) $2a\left(\cos\dfrac{\theta_1}{2} - \cos\dfrac{\theta_2}{2}\right)$, (v) $768\,\pi$, (vi) $2a\left[\sqrt{2} + \log\left(\sqrt{2} + 1\right)\right]$.

25. Find the lengths of the following curves.

(a) The spiral $r = \theta^2$, $0 \le \theta \le \sqrt{5}$, (b) $r = \cos^3(\theta/3)$, $0 \le \theta \le \pi/4$.

Ans. (a) 19/3 units (b) $\left[(\pi/8) + (3/8)\right]$ units.

26. Find the perimeter of the cardioid $r = a\,(1 - \cos\theta)$. Also show that the upper half is bisected by $\theta = 2\pi/3$.

(*A.M.I.E., W-2005; D.U. 1999*) **Ans.** 8a units.

27. Find the length of the cardiode $r = a(1 - \cos\theta)$ lying out-side the circle

$r = a\cos\theta.$

Ans. $4a\sqrt{3}$ units.

Chapter 8

Volumes and Surfaces of Solids of Revolution

Volumes of Solids of Revolution – Disks, Washers and Cylindrical Shells method

8.1 Solid of revolution

A *solid of revolution* is a solid that is generated by revolving a plane region about a line that does not intersect the region. The line about which the rotation takes place is called the *axis of revolution*. Many famliar solids of this type are shown in Fig. 8.1.

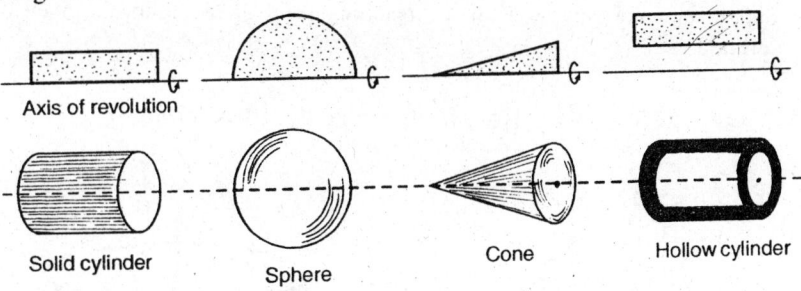

Fig. 8.1.

Some familiar solids of revolution

8.2 Volume of a solid of revolution — Disk Method

8.2.1 Rotation about the x-axis

Let f be a continuous function such that $f(x) > 0$ between $x = a$ and $x = b$ and let R be the region that is bounded by the arc AB of the curve $y = f(x)$, the x-axis, and the lines $x = a$ and $x = b$ (Fig. 8.2).

The volume of the solid of revolution that is generated by revolving the region *R* about the *x*-axis is given by

$$V = \pi \int_a^b \{f(x)\}^2 \, dx = \pi \int_a^b y^2 \, dx \quad \textbf{(disk formula)}. \qquad ...(8.1)$$

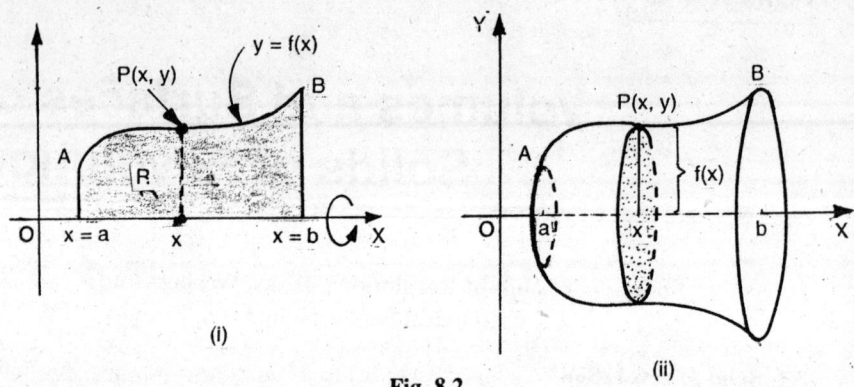

(i)

(ii)

Fig. 8.2.

8.2.2 Rotation about the y-axis

When the axis of rotation is the *y*-axis and the region *R* that is revolved lies between the *y*-axis and a curve $x = \phi(y)$ and between $y = c$ and $y = d$ [See fig. (8.3)], then the volume *V* of the resulting solid of revolution is given by the formula

$$V = \pi \int_c^d \{\phi(y)\}^2 \, dy = \pi \int_c^d x^2 \, dy \quad \textbf{(disk formula)}. \qquad ...(8.2)$$

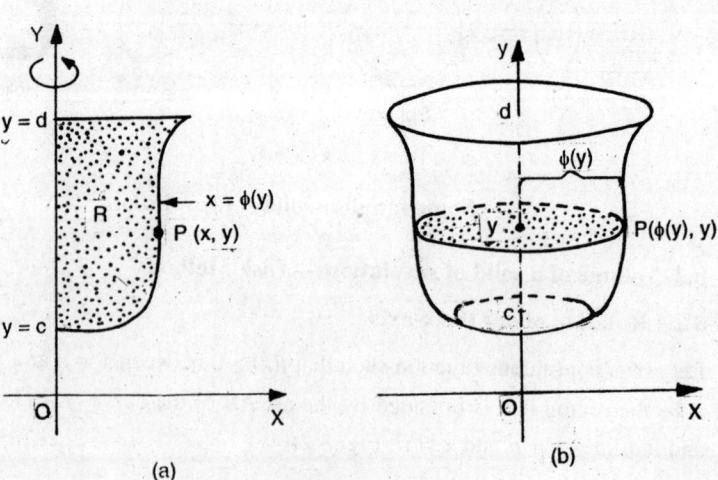

(a)

(b)

Fig. 8.3.

Remarks

(a) If the area bounded by the curve $y = f(x)$, the line $y = p$ and the lines $x = a$, $x = b$ is revolved about the line $y = p$ (a line parallel to the x-axis), then the volume of the solid of revolution is given by

$$V = \pi \int_a^b (y - p)^2 \, dx .$$ ...(8.3)

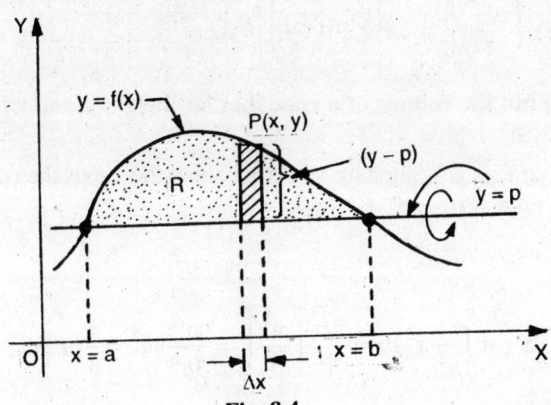

Fig. 8.4.

(b) If the area bounded by the curve $x = g(y)$, the line $x = q$ and the lines $y = c$, $y = d$ is revolved about the line $x = q$ (a line parallel to the y-axis), then the volume of solid of revolution is given by

$$V = \pi \int_c^d (x - q)^2 \, dy .$$ ...(8.4)

Example 8.1 Find the volume of a sphere of radius r.

Solution. The sphere is generated by revolving about the x-axis the region between the semicircle $y = \sqrt{r^2 - x^2}$ and the x-axis, between $x = -r$ and $x = r$.

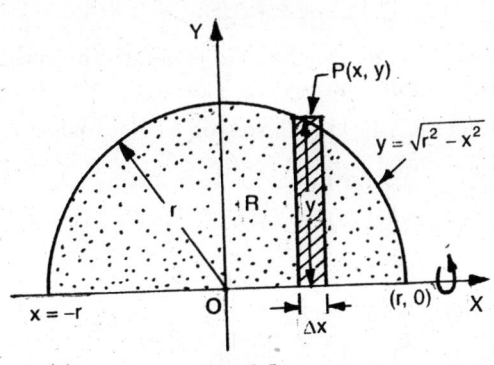

Fig. 8.5.

Hence by the disk formula, the volume of the sphere is

$$V = \pi \int_{-r}^{+r} y^2 \, dx = 2\pi \int_{0}^{r} y^2 \, dx = 2\pi \int_{0}^{r} \left(r^2 - x^2\right) dx$$

$$= 2\pi \left[r^2 x - \left(\frac{x^3}{3}\right)\right]_{0}^{r} = 2\pi \left[r^3 - \left(\frac{r^3}{3}\right)\right] = 4\pi r^3 / 3.$$

Example 8.2 Find the volume of a cone that has height h and whose base has radius r.

Solution. The cone is generated by revolving about the x-axis the region between the line $y = rx/h$ $\left[\because y = mx = (\tan\theta)x = rx/h\right]$ and the x-axis, between $x = 0$ and $x = h$. By the disk formula, the volume of the cone is

$$V = \pi \int_{0}^{h} y^2 \, dx = \pi \int_{0}^{h} \frac{r^2}{h^2} x^2 \, dx = \frac{\pi r^2}{h^2} \left[\frac{x^3}{3}\right]_{0}^{h} = \frac{\pi r^2}{3h^2} \cdot h^3 = \frac{1}{3} \pi r^2 h.$$

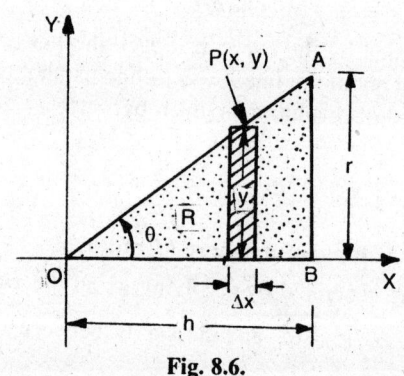

Fig. 8.6.

Example 8.3 Find the volume of the solid generated by revolving the region between the y-axis and the curve y = 3 – 2x, 0 ≤ y ≤ 2, about the y-axis.

Solution. Refer figure 8.7 showing the region R. The volume is

$$V = \int_{0}^{2} \pi \left(\phi(y)\right)^2 \, dy = \int_{0}^{2} \pi \left(\frac{3-y}{2}\right)^2 dy$$

$$= \frac{\pi}{4} \int_{0}^{2} \left(9 - 6y + y^2\right) dy = \frac{\pi}{4} \left[9y - 3y^2 + \frac{y^3}{3}\right]_{0}^{2} = \frac{\pi}{4} \left(18 - 12 + \frac{8}{3}\right) = \frac{13\pi}{6}. \text{ cubic units.}$$

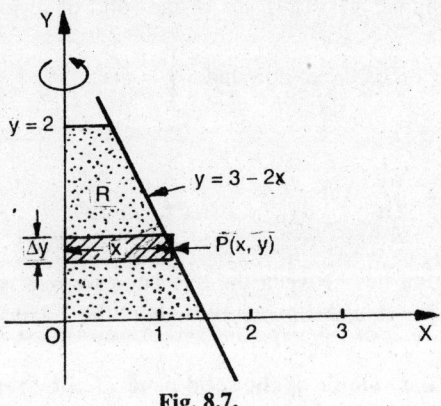

Fig. 8.7.

8.3 Volume of a solid of Revolution–Washer Method

Assume that $0 \le g(x) \le f(x)$ for $a \le x \le b$. Consider the region between $x = a$ and $x = b$ and lying between $y = g(x)$ and $y = f(x)$. (Ref. Fig. 8.8). Then the volume V of the solid of revolution obtained by revolving this region about the x-axis is given by the formula

$$V = \pi \int_{a}^{b} \left[(f(x))^{2} - (g(x))^{2} \right] dx \quad \text{(washer formula)} \qquad ...(8.5)$$

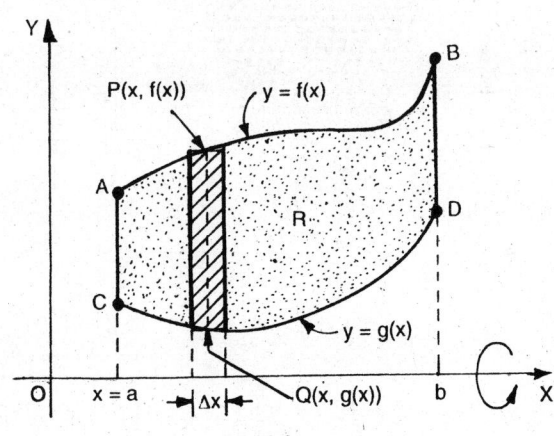

Fig. 8.8.

The desired volume is the difference of two volumes, the volume $\pi \int_{a}^{b} (f(x))^{2} dx$ of the solid of revolution generated by revolving about the x-axis the region under

$y = f(x)$ and the volume $\pi \int_a^b (g(x))^2 \, dx$ of the solid of revolution generated by revolving about the x-axis the region under $y = g(x)$.

A similar formula

$$V = \pi \int_c^d (\phi(y))^2 - (h(y))^2 \, dy \quad \text{(washer formula)} \qquad \qquad ...(8.6)$$

holds when the region lies between the two curves $x = \phi(y)$ and $x = h(y)$ and between $y = c$ and $y = d$, and it is revolved about the y-axis. (It is assumed that $0 \le h(y) \le \phi(y)$ for $c \le y \le d$.

Example 8.4 Find the volume of the solid generated by revolving the region bounded by the curves

(a) $y = x^2 + 1$, $y = 5$ about the line $x = 3$.
(b) $y = 3 - x^2$ and $y = -1$ about the line $y = -1$.

Solution. (a) The required region is shown in the Fig. 8.9. The volume is given by

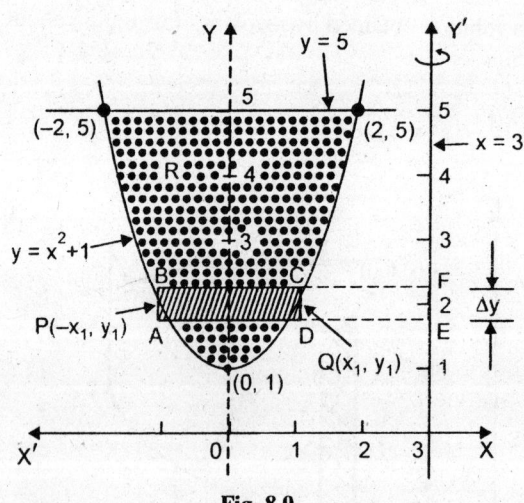

Fig. 8.9.

$$V = \pi \int_1^5 \left[(3+x_1)^2 - (3-x_1)^2 \right] dy = \pi \int_1^5 \left[(3+\sqrt{y-1})^2 - (3-\sqrt{y-1})^2 \right] dy$$

$$= 12\pi \int_1^5 \sqrt{y-1} \, dy = 12\pi (2/3) \left[(y-1)^{3/2} \right]_1^5 = 8\pi(8) = 64\pi \text{ cubic units.}$$

(*b*) The required region is given in Fig. 8.10. The region BAC is revolved about the line $y = -1$.

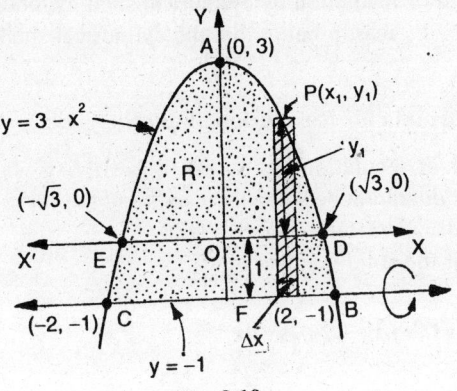

Fig. 8.10.

∴   Required volume $= \pi \int_{-2}^{2} (y_1 + 1)^2 \, dx = 2\pi \int_{0}^{2} (3 - x^2 + 1)^2 \, dx$

$$= 2\pi \int_{0}^{2} (16 - 8x^2 + x^4) \, dx = 2\pi \left[16x - \frac{8x^3}{3} + \frac{x^5}{5} \right]_0^2$$

$$= 2\pi \left(32 - \frac{64}{3} + \frac{32}{5} \right) = \frac{512\pi}{15} \text{ cubic units.}$$

8.4 Cylindrical Shells

A *cylindrical shell* is a solid enclosed by two concentric right circular cylinders. The volume V of a cylindrical shell with inner radius r_1, outer radius r_2, and height h can be written as

$V = $ (Area of cross section) · (height)

$$= \left(\pi r_2^2 - \pi r_1^2 \right) h$$

$$= \pi (r_2 + r_1)(r_2 - r_1) h$$

$$= 2\pi \left[\frac{1}{2} (r_2 + r_1) \right] \left[h \cdot (r_2 - r_1) \right]$$

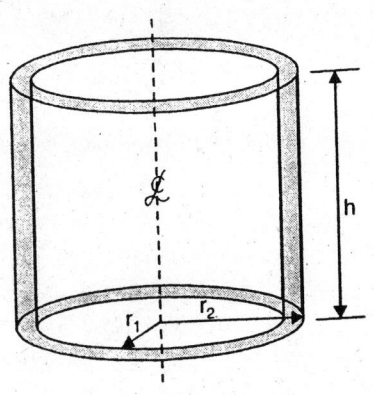

Fig. 8.11.

But $(r_1 + r_2)/2$ is the average radius of the shell and $(r_2 - r_1)$ is the thickness, so $V = 2\pi$ (average radius) · (height) · (thickness).

8.4.1 Volume of solid of revolution by Cylindrical Shells Method

The volume of solid of revolution by cylindrical shell method, some times work better than washers the reason being that the cylindrical shell formula does not require squaring.

8.4.2 The Shell Formula for Revolution about the y-axis

Consider the solid of revolution generated by revolving about the y-axis the region R in the first quardrant between the x-axis and the curve $y = f(x)$, and lying between $x = a$ and $x = b$ (Refer Fig. 8.12 (a)).
Then the volume of the solid is given by

$$V = \int_a^b 2\pi x\, f(x)\, dx = 2\pi \int_a^b xy\, dx \qquad \qquad ...(8.7)$$

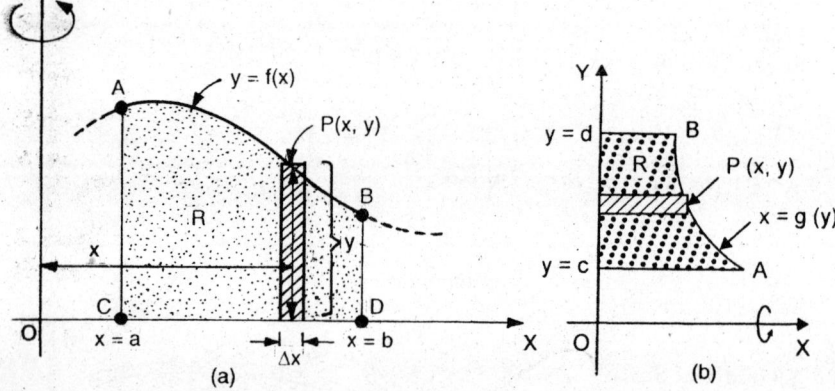

(a)   (b)

Fig. 8.12.

8.4.3 The Shell Formula for Revolution about the x-axis

The volume of the solid generated by revolving about the x-axis the region R between the y-axis and the curve $x = g(y)$, and lying between $y = c$ and $y = d$ (Refer Fig. 8.12b) is given by:

$$V = \int_c^d 2\pi y \cdot g(y) \cdot dy = 2\pi \int_c^d y\, x\, dy \qquad \qquad ...(8.8)$$

Example 8.5 Use the shell method to find the volumes of the solids generated by revolving the shaded region [Fig. 8.13 (i) to (iv)] about the indicated axis.

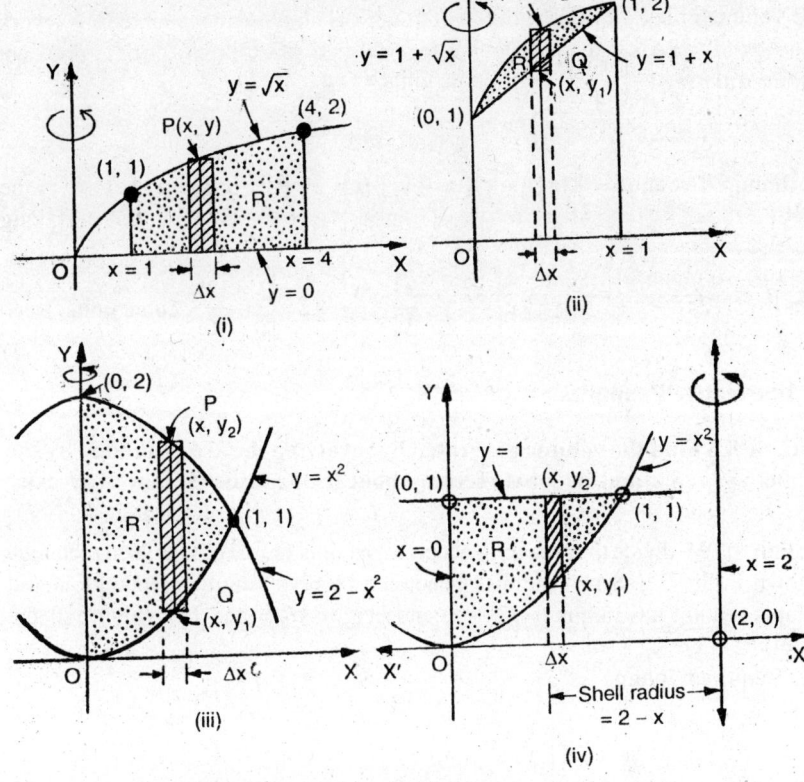

Fig. 8.13.

Solution. (*i*) Required volume, $V = \int\limits_{a}^{b} 2\pi(\text{shell radius})(\text{shell height})\, dx$

$$= \int\limits_{1}^{4} 2\pi\, xy\, dx = 2\pi \int\limits_{1}^{4} x\sqrt{x}\, dx = 2\pi\left[\frac{2}{5}x^{5/2}\right]_{1}^{4} = \frac{124\pi}{5} \text{ cubic units.}$$

(*ii*) By the difference of cylindrical shell formula, the volume is

$$V = \int\limits_{0}^{1} 2\pi x\left(y_2 - y_1\right) dx = 2\pi \int\limits_{0}^{1} x\left[\left(1+\sqrt{x}\right)-(1+x)\right] dx$$

$$= 2\pi \int\limits_{0}^{1} x\left(\sqrt{x} - x\right) dx = 2\pi\left[\frac{2}{5}x^{5/2} - \frac{x^3}{3}\right]_{0}^{1} = 2\pi\left[\frac{2}{5} - \frac{1}{3}\right] = \frac{2\pi}{15} \text{ cubic units.}$$

(*iii*) Required volume, $V = 2\pi \int_0^1 x\left[(2-x^2)-(x^2)\right]dx = 4\pi \int_0^1 (x-x^3)dx$

$$= 4\pi\left[\left(\frac{x^2}{2}\right)-\left(\frac{x^4}{4}\right)\right]_0^1 = \pi \text{ cubic units.}$$

(*iv*) Required volume $= \int_0^1 2\pi(2-x)(y_2 - y_1)dx = 2\pi \int_0^1 (2-x)(1-x^2)dx$

$$= 2\pi \int_0^1 (x^3 - 2x^2 - x + 2)dx = 2\pi\left[\frac{x^4}{4} - \frac{2x^3}{3} - \frac{x^2}{2} + 2x\right]_0^1 = \frac{13\pi}{6} \text{ cubic units.}$$

8.5 Illustrative Examples

Example 8.6 Find the volume generated by revolving the area bounded by the parabola $y^2 = 8x$ and its latus rectum about (i) the latus rectum (ii) y-axis.
(*A.M.I.E.;-W-1995*)

Solution. (*i*) We divide the area into rectangles parallel to x-axis. One such rectangle is shown in the Fig. 8.14. The disk generated by revolving this rectangle about the latus rectum has radius $= (2 - x)$ and height $AB = \Delta y$. The volume of the disk $= \pi(2 - x)^2 \Delta y$.

∴ Required volume

$$= \pi \int_{-4}^{4} (2-x)^2\, dy = \pi \int_{-4}^{4}\left(2 - \frac{y^2}{8}\right)^2 dy$$

$$= 2\pi \int_0^4 \left(2 - \frac{y^2}{8}\right)^2 dy$$

$$= 2\pi\left[4y + \frac{1}{64}\cdot\frac{y^5}{5} - \frac{1}{2}\cdot\frac{y^3}{3}\right]_0^4$$

Fig. 8.14.

$$= 2\pi\left[16 + \frac{1}{64}\cdot\frac{4^5}{5} - \frac{1}{2}\cdot\frac{4^3}{3}\right] = \frac{256\pi}{15} \text{ cubic units.}$$

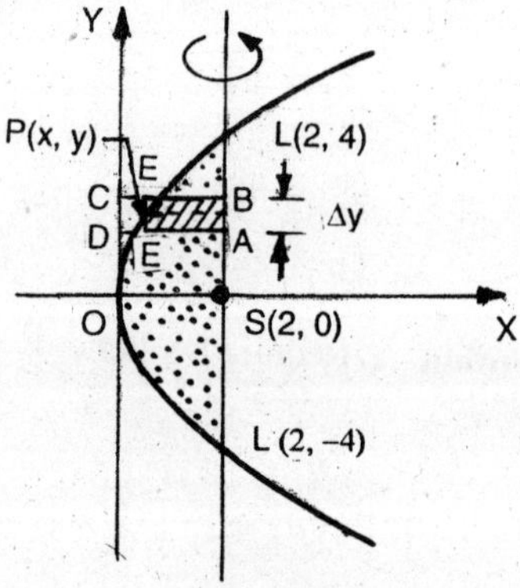

(*ii*) The revolution of the rectangle about y-axis generates a washer whose volume = (volume generated by revolving the rectangle *DABC*)–(volume generated by revolving the rectangle *DEFC*)

$$= \pi(2)^2 \Delta y - \pi(x)^2 \Delta y = \pi(4 - x^2)\Delta y.$$

$\therefore$ The required volume, $V = \pi \int_{-4}^{4} \left(4 - x^2\right) dy = \pi \int_{-4}^{4} \left(4 - \frac{y^4}{64}\right) dy$

$$= 2\pi \int_{0}^{4} \left(4 - \frac{y^4}{64}\right) dy = 2\pi \left[4y - \frac{y^5}{320}\right]_{0}^{4} = \frac{128}{5}\pi \text{ cubic units.}$$

Example 8.7 Find the volume generated by revolving about x-axis the area cut off from the parabola 9y = 4(9 – x²) by the line 4x + 3y = 12.

Solution. The parabola is symmetrical about y-axis and meets the x-axis at C(–3, 0) and B(3, 0) and the y-axis at A(0, 4).

The shape is shown is the Fig. 8.15. The line cuts the parabola at A(0, 4) and B(3, 0).

When the rectangle LMRK revolves about x-axis it generates a washer with volume

$$= \pi\left(y_1^2 - y_3^2\right)\Delta x.$$

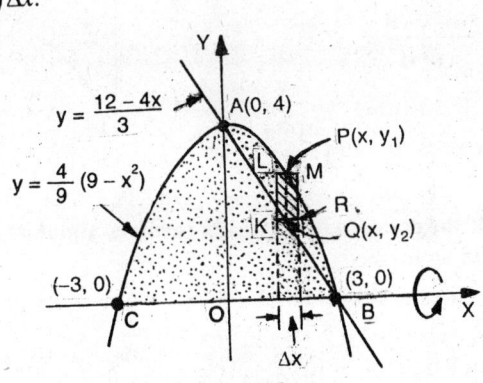

Fig. 8.15.

Hence the required volume

$$= \pi \int_{0}^{3} \left(y_1^2 - y_2^2\right) dx = \pi \int_{0}^{3} \left[\left\{ \frac{4}{9}\left(9 - x^2\right) \right\}^2 - \left(\frac{12 - 4x}{3}\right)^2 \right] dx$$

$$= \pi \int_{0}^{3} \left[\frac{16}{81}\left(81 + x^4 - 18x^2\right) - \frac{16}{9}\left(9 + x^2 - 6x\right)\right] dx = \frac{16\pi}{81} \int_{0}^{3} \left(x^4 - 27x^2 + 54x\right) dx$$

$$= \frac{16\pi}{81}\left[\frac{x^5}{5} - 9x^3 + 27x^2\right]_{0}^{3} = \frac{16\pi}{81}\left[\frac{243}{5} - 243 + 243\right] = \frac{48\pi}{5} \text{ cubic units.}$$

Example 8.8 Find the volume of the solid formed by the revolution of the curve $xy^2 = 4(2 - x)$ through four right angles about the y-axis.

Solution. The shape of the curve is shown in Fig. 8.16. The asymptote is the y-axis itself.

The required volume $= 2\int_0^\infty \pi x^2 dy = 2\pi \int_0^\infty \left(\frac{8}{y^2 + 4}\right)^2 dy$

$= 128\pi \int_0^\infty \frac{dy}{\left(y^2 + 4\right)^2}.$

Let $y = 2\tan\theta$ then $dy = 2\sec^2\theta\, d\theta.$

When $y = 0$, $\theta = 0$, when $y = \infty$, $\theta = \dfrac{\pi}{2}$.

$\therefore\ V = 128\pi \int_0^{\pi/2} \frac{2\sec^2\theta\, d\theta}{\left(4\tan^2\theta + 4\right)^2}$

$= 16\pi \int_0^{\pi/2} \frac{\sec^2\theta\, d\theta}{\sec^4\theta}$

$= 16\pi \int_0^{\pi/2} \cos^2\theta\, d\theta = 16\pi \cdot \frac{1}{2} \cdot \frac{\pi}{2} = 4\pi^2.$

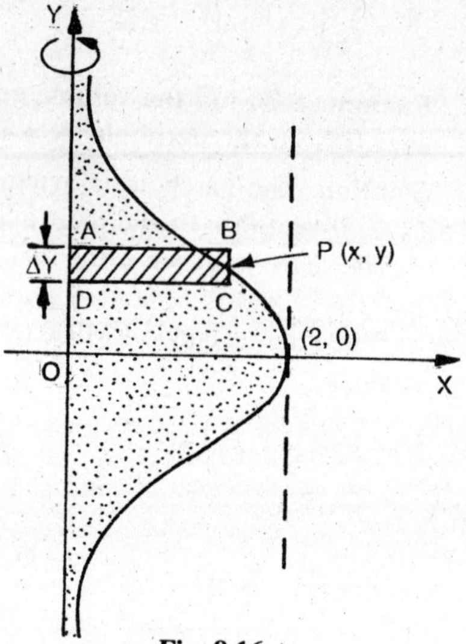

Fig. 8.16.

Example 8.9 If the curve $(a - x)\, y^2 = a^2 x$ revolves about its asymptote, find the volume of solid so formed.

Solution. The curve is shown in the Fig. 8.17. The asymptote is $x - a = 0$.

The required volume $= 2\int_0^\infty \pi(a - x)^2\, dy = 2\pi \int_0^\infty \left(a - \frac{ay^2}{a^2 + y^2}\right)^2 dy$

$= 2\pi \int_0^\infty \left(\frac{a^3}{a^2 + y^2}\right)^2 dy.$ Put $y = a\tan\theta$, then $dy = a\sec^2\theta\, d\theta.$

When $y = 0$, $\theta = 0$, when $y = \infty$, $\theta = \dfrac{\pi}{2}$.

$\therefore\quad$ Volume $= 2\pi \int_0^{\pi/2} \frac{a^6}{\left(a^2 + a^2\tan^2\theta\right)^2}\, a\sec^2\theta\, d\theta$

$$= 2\pi a^3 \int_0^{\pi/2} \cos^2\theta \, d\theta = 2\pi a^3 \cdot \frac{1}{2} \cdot \frac{\pi}{2} = \frac{\pi^2 a^3}{2} \quad \text{cubic units.}$$

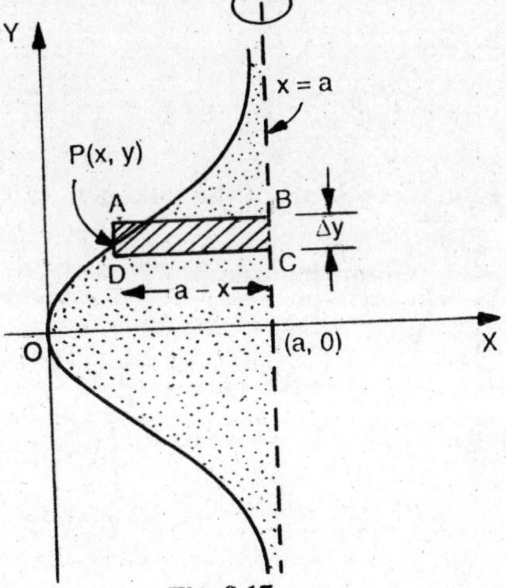

Fig. 8.17.

Example 8.10 Find the volume of the solid obtained by the revolution of the loop of the curve $y^2 = \{(a+x)/(a-x)\}x^2$ about the x-axis.

Solution. The curve is shown in the Fig. 8.18 with its loop symmetrical about x-axis.

$$\text{Required volume} = \pi \int_{-a}^{0} y^2 \, dx$$

$$= \pi \int_{-a}^{0} \frac{a+x}{a-x} x^2 \, dx.$$

Put $a - x = z$ so that $dx = -dz$; when $x = -a$, $z = 2a$, when $x = 0$, $z = a$.

$$\therefore \text{Volume} = -\pi \int_{2a}^{a} \frac{(2a-z)(a-z)^2 \, dz}{z}$$

$$= \pi \int_{a}^{2a} \frac{(a-z)^2 (2a-z)}{z} dz$$

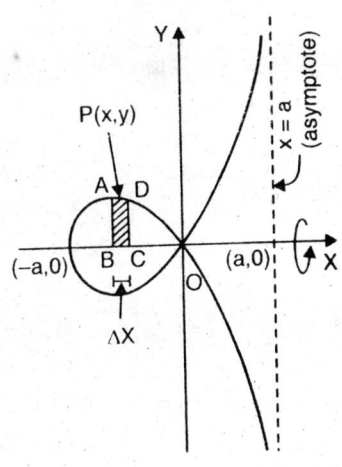

Fig. 8.18.

$$= \pi \int_{a}^{2a} \left(\frac{2a^3}{z} - 5a^2 + 4az - z^2 \right) dz = \pi \left[2a^3 \log z - 5a^2 z + 2az^2 - \frac{z^3}{3} \right]_{a}^{2a}$$

$$= \pi \left[\left(2a^3 \log 2a - 10a^3 + 8a^3 - \frac{8a^3}{3} \right) \right.$$

$$\left. - \left(2a^3 \log a - 5a^3 + 2a^3 - \frac{a^3}{3} \right) \right] = 2\pi a^3 \left(\log 2 - \frac{2}{3} \right).$$

Example 8.11 Find the volume of the solid generated by the revolution of the

Cissoid $y^2 = \dfrac{x^3}{(2a-x)}$ about its asymptote.   (*V.T.U., 2000*)

Solution. Refer Fig. 8.19.

$$\text{Reqd. Volume} = 2 \int_0^{\infty} \pi(2a-x)^2 \, dy = 2\pi \int_0^{2a} (2a-x)^2 \frac{dy}{dx} dx. \qquad ...(i)$$

Differentiating the given equation of the curve w.r.t. x, we obtain

$$2y \frac{dy}{dx} = \frac{3(2a-x) \cdot x^2 - x^3(-1)}{(2a-x)^2} = \frac{2x^2(3a-x)}{(2a-x)^2}$$

or $\quad \dfrac{dy}{dx} = \dfrac{x^2(3a-x)}{y(2a-x)^2}$

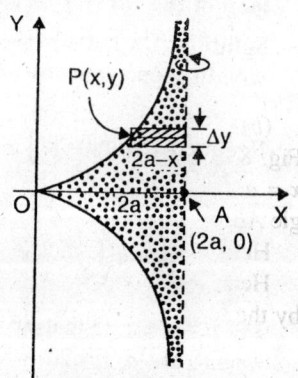

$$= \frac{x^2(3a-x)}{(2a-x)^2} \cdot \sqrt{\frac{2a-x}{x^3}} = \frac{(3a-x)}{(2a-x)^{3/2}} \cdot \sqrt{x}.$$

Therefore, from (*i*),

$$V = 2\pi \int_0^{2a} (2a-x)^2 \cdot \frac{3a-x}{(2a-x)^{3/2}} \cdot \sqrt{x} \, dx$$

$$= 2\pi \int_0^{2a} (3a-x) \sqrt{x(2a-x)} \, dx \qquad ...(ii)$$

Fig. 8.19.

On putting $x = 2a \sin^2 \theta$, $dx = 4a \sin \theta \cos \theta \, d\theta$ in (*ii*), we obtain

$$V = 2\pi \int_0^{\pi/2} (3a - 2a\sin^2 \theta) \cdot 2a \cos \theta \sin \theta \cdot 4a \sin \theta \cos \theta \, d\theta$$

$$= 16\pi a^3 \int_0^{\pi/2} (3\sin^2 \theta \cos^2 \theta - 2\sin^4 \theta \cos^2 \theta) \, d\theta$$

$$= 16\pi a^3 \left[\frac{3 \cdot 1}{4 \cdot 2} \cdot \frac{\pi}{2} - 2 \cdot \frac{3 \cdot 1}{6 \cdot 4 \cdot 2} \cdot \frac{\pi}{2} \right] = 2\pi^2 a^3 \text{ cubic units.}$$

Example 8.12 (a) A loop of the curve $y^2 = x^2(1 - x^2)$ is rotated about the y-axis. Find the volume generated.
(b) The loop of the curve $2ay^2 = x(x - a)^2$ revolves about the straight line $x = a$. Find the volume of the solid generated.

Solution. (a) The curve is symmetrical about the axes and forms two loops. Each loop is symmetrical about the x-axis.

Consider rectangle of width Δx and height y. When this rectangle revolves about y-axis, it generates a cylindrical shell of volume $V = 2\pi xy \, \Delta x$.

∴ Volume of solid generated by the loop is

$$V = 2 \int_0^1 2\pi xy \, dx = 4\pi \int_0^1 x \cdot x \sqrt{1 - x^2} \, dx$$

$$= 4\pi \int_0^{\pi/2} \sin^2 \theta \sqrt{1 - \sin^2 \theta} \cos\theta \, d\theta,$$

where $x = \sin\theta$, $dx = (\cos\theta) \cdot d\theta$

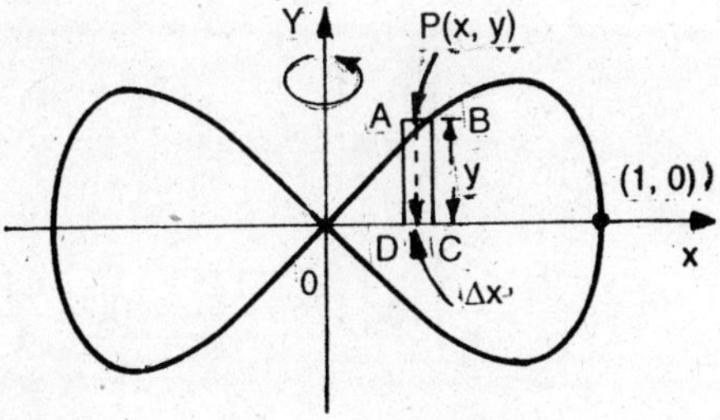

Fig. 8.20.

$$= 4\pi \int_0^{\pi/2} \sin^2 \theta \cos^2 \theta \, d\theta = 4\pi \cdot \frac{1}{4 \cdot 2} \cdot \frac{\pi}{2} = \frac{\pi^2}{4} \text{ cubic units.}$$

(b) The shape of the curve is shown in Fig. 8.21 having the loop between $x = 0$ to $x = a$. Consider the revolution of the rectangle *ABCD* about the line *EF*.

Here $\Delta V = 2\pi(a - x) y \cdot \Delta x$ [Note this step]
Hence the volume of the solid generated by the loop is

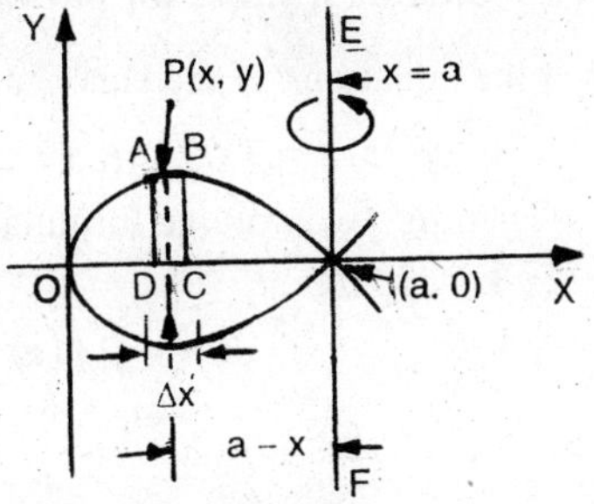

Fig. 8.21.

$$V = 2 \int_0^a 2\pi(a - x) y \, dx$$

$$= 4\pi \int_0^a (a - x) \frac{\sqrt{x}(x - a)}{\sqrt{2a}} \, dx = \frac{4\pi}{\sqrt{2a}} \int_0^a (2ax^{3/2} - a^2 x^{1/2} - x^{5/2}) \, dx$$

$$= \frac{4\pi}{\sqrt{2a}} \left[\frac{4a}{5} x^{5/2} - \frac{2}{3} a^2 x^{3/2} - \frac{2}{7} x^{7/2} \right]_0^a$$

$$= \frac{4\pi}{\sqrt{2a}} \left[\frac{4}{5} a^{7/2} - \frac{2}{3} a^{7/2} - \frac{2}{7} a^{7/2} \right] = \frac{32\sqrt{2}}{105} \pi a^3 \text{ cubic units.}$$

Example 8.13 Find the volume generated by revolving about the axis of x, the area bounded by the curves $x^2 + y^2 = 25$, $3x - 4y = 0$, $y = 0$ lying in the first quadrant. (*A.M.I.E., S-2009*)

Solution. The region is the area bounded by the circle $x^2 + y^2 = 25$, the line $3x - 4y = 0$ and $y = 0$, *i.e.*, x-axis. The co-ordinates of point C are (4, 3).

The required volume = volume generated by the right angled triangle OBC about OX + volume generated by the area bounded by the circle, the x-axis and line $x = 4$ about OX.

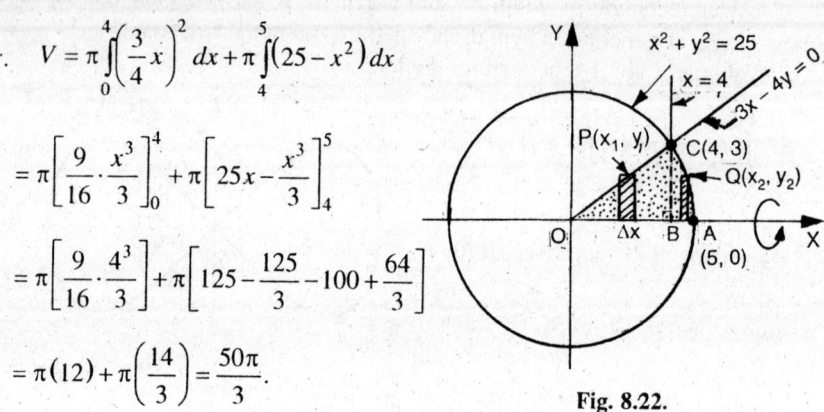

$$\therefore \quad V = \pi \int_0^4 \left(\frac{3}{4}x\right)^2 dx + \pi \int_4^5 (25 - x^2) dx$$

$$= \pi \left[\frac{9}{16} \cdot \frac{x^3}{3}\right]_0^4 + \pi \left[25x - \frac{x^3}{3}\right]_4^5$$

$$= \pi \left[\frac{9}{16} \cdot \frac{4^3}{3}\right] + \pi \left[125 - \frac{125}{3} - 100 + \frac{64}{3}\right]$$

$$= \pi(12) + \pi\left(\frac{14}{3}\right) = \frac{50\pi}{3}.$$

Fig. 8.22.

8.6 Volume formulae for parametric equations

Let the equations of the curve in the parametric form be

$$x = \phi(t), \, y = f(t) \quad t_1 \le t \le t_2.$$

Then the two volume formulae of Art. 8.2 can be written in the form

$$V = \int_{t_1}^{t_2} \pi y^2 \frac{dx}{dt} dt = \pi \int_{t_1}^{t_2} [f(t)]^2 \phi'(t) dt \qquad \qquad ...(8.9)$$

and $$V = \int_{t_1}^{t_2} \pi x^2 \frac{dy}{dt} dt = \pi \int_{t_1}^{t_2} \{\phi(t)\}^2 f'(t) dt \qquad \qquad ...(8.10)$$

Example 8.14 Find the volume of the spindle-shaped solid generated by revolving the hypocycloid $x^{2/3} + y^{2/3} = a^{2/3}$ about the x-axis. (*AMIE., W-2002*)

Solution. Using disk formula: $V = 2\int_0^a \pi y^2 dx$.

The parametric equations of the curve are:

$$x = a \cos^3\theta,$$
$$y = a \sin^3\theta.$$

$\therefore \quad dx = -3a \cos^2\theta \sin\theta \, d\theta.$

Also when $x = 0$, $\theta = \pi/2$ and when $x = a$, $\theta = 0$.

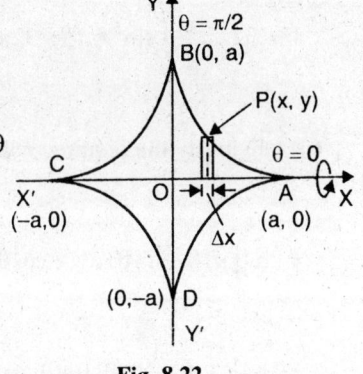

Fig. 8.22.

Hence $V = 2 \int_{\pi/2}^{0} \pi a^2 \sin^6\theta \left(-3a \cos^2\theta \sin\theta\right) d\theta$

$= 6\pi a^3 \int_{0}^{\pi/2} \sin^7\theta \cos^2\theta \, d\theta$

$= 6\pi a^3 \cdot \dfrac{6 \cdot 4 \cdot 2 \cdot 1}{9 \cdot 7 \cdot 5 \cdot 3} = \dfrac{32\pi a^3}{105}$ cubic units.

Example 8.15 Find the volume of the solid generated by revolving an arc of the cycloid $x = a\,(\theta - \sin\theta)$, $y = a\,(1 - \cos\theta)$ and x-axis about (i) the x-axis (ii) the y-axis, (iii) the tangent at its vertex. *(Osmania, 1995)*

Solution. (i) The volume generated by vertical strip when revolved about the x-axis is $\Delta V = \pi y^2 \cdot \Delta x$.

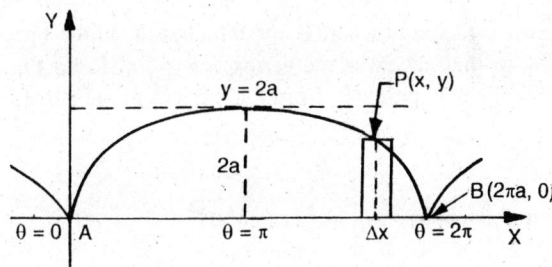

Fig. 8.24.

Hence the volume generated by the area of one arch of the cycloid when revolved about the x-axis is

$$V = \int_{0}^{2\pi a} \pi y^2 \cdot dx = \pi \int_{0}^{2\pi} a^2 (1 - \cos\theta)^2 \cdot \left[a(1 - \cos\theta) \right] d\theta$$

$$= \pi a^3 \int_{0}^{2\pi} (1 - \cos\theta)^3 \, d\theta = 2\pi a^3 \int_{0}^{\pi} (1 - \cos\theta)^3 \, d\theta$$

$$= 2\pi a^3 \int_0^\pi 8\sin^6\frac{\theta}{2}d\theta. \text{ Let } \frac{\theta}{2} = \phi, \text{ then } d\theta = 2d\phi.$$

$$\therefore \quad V = 32\pi a^3 \int_0^{\pi/2} \sin^6\phi\, d\phi = 32\pi a^3 \cdot \frac{5\cdot3\cdot1}{6\cdot4\cdot2}\cdot\frac{\pi}{2} = 5\pi^2 a^3 \text{ cubic units.}$$

(*ii*) Obviously the required volume is $V = 2\pi \int_0^{2\pi a} xy\, dx$ (using shell formula)

$$= 2\pi \int_0^{2\pi} a(\theta - \sin\theta)\, a(1 - \cos\theta)\, a(1 - \cos\theta)\, d\theta$$

$$= 2\pi a^3 \int_0^{2\pi} (\theta - 2\theta\cos\theta + \theta\cos^2\theta - \sin\theta + 2\sin\theta\cos\theta - \cos^2\theta\sin\theta)\, d\theta$$

$$= 2\pi a^3 \left[\frac{3}{4}\theta^2 - 2(\theta\sin\theta + \cos\theta) + \frac{1}{2}\left(\frac{1}{2}\theta\sin2\theta + \frac{1}{4}\cos2\theta\right)\right.$$

$$\left. + \cos\theta + \sin^2\theta + \frac{1}{3}\cos^3\theta\right]_0^{2\pi} = 6\pi^3 a^3 \text{ cu. units.}$$

(*iii*) The vertex is the point where $\theta = \pi$ (which is the highest point of the arch). The equation of the tangent at the vertex is $y = 2a$. (Refer Fig. 8.24). Now the distance of a point 'θ' from the tangent at the vertex

$$= 2a - y = 2a - a(1 - \cos\theta) = a(1 + \cos\theta) .$$

$$\therefore \quad \text{The required volume} = \int_0^{2\pi} \pi a^2 (1 + \cos\theta)^2 \cdot \left(\frac{dx}{d\theta}\right)\cdot d\theta$$

$$= \pi a^2 \int_0^{2\pi} (1 + \cos\theta)^2 \cdot a(1 - \cos\theta)\, d\theta = 2\pi a^3 \int_0^\pi (1 + \cos\theta)^2 (1 - \cos\theta)\, d\theta$$

$$= 2\pi a^3 \int_0^\pi 4\cos^4\frac{\theta}{2}\cdot 2\sin^2\frac{\theta}{2}d\theta = 16\pi a^3 \int_0^\pi \cos^4\frac{\theta}{2}\sin^2\frac{\theta}{2}d\theta, \left(\text{Put } \frac{\theta}{2} = \phi\right)$$

$$= 32\pi a^3 \int_0^{\pi/2} \cos^4\phi\sin^2\phi\cdot d\phi = 32\pi a^3 \cdot \frac{3\cdot1}{6\cdot4\cdot2}\cdot\frac{\pi}{2} = \pi^2 a^3 \text{ cu. units.}$$

8.7 Volume of solid of revolution–Rotation about any axis

Let the plane region $ABDC$ revolves about straight line AB which is not the x-or the y-axis, and CA and DB are the perpendiculars on AB from C and D respectively. Choose temporarily AB and AC as new axes of reference, say as the axes of ξ and η (Refer Fig. 8.25). Then the volume generated by the revolution of the area $ABDC$ about AB

$$= \pi \int \eta^2 d\xi = \pi \int \eta^2 \frac{d\xi}{dx} \cdot dx \qquad \qquad ...(8.11)$$

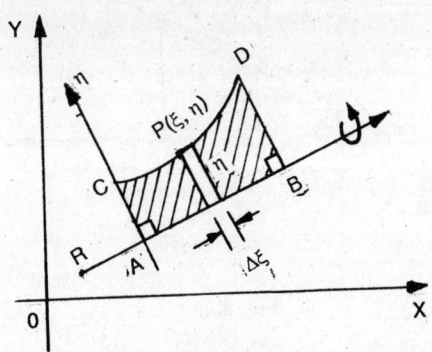

Fig. 8.25.

with proper limits of integration. The integral can be evaluated by the usual methods after expressing η^2 and $d\xi/dx$ in terms of x.

Example 8.16 The area cut off from the parabola $y^2 = 4ax$ by the chord joining the vertex to an end of the letus rectum is rotated through four right angles about the chord. Find the volume of the solid so formed.

Solution. Let A be the vertex, S the focus, SL the latus rectum. Let P be any point (x, y) on the arc AL, and PM the perpendicular on AL.

Let $AM = \xi$, $PM = \eta$.

The equation of chord AL is $y = 2x$.

$$\therefore \quad \eta = PM = (y - 2x)/\sqrt{5}.$$

Also $AM^2 = AP^2 - PM^2 = \left(x^2 + y^2\right) - \left\{(y - 2x)^2/5\right\}$

$$= \left(5x^2 + 5y^2 - y^2 - 4x^2 + 4xy\right)/5 = (x + 2y)^2/5.$$

$$\therefore \quad \xi = (x + 4\sqrt{ax})/\sqrt{5}.$$

Hence the required volume $= \pi \int_0^{AL} \eta^2 d\xi = \pi \int_0^a \eta^2 \frac{d\xi}{dx} \cdot dx$

$$= \frac{4\pi}{5\sqrt{5}} \int_0^a \left(\sqrt{ax} - x\right)^2 \left(1 + \frac{2\sqrt{a}}{\sqrt{x}}\right) dx = \frac{4\pi}{5\sqrt{5}} \int_0^a \left(x^2 - 3ax + 2a^{3/2}x^{1/2}\right) dx$$

$$= \frac{4\pi}{5\sqrt{5}} \left[\frac{x^3}{3} - \frac{3ax^2}{2} + \frac{4}{3}a^{3/2}x^{3/2}\right]_0^a = \frac{2\pi a^3}{15\sqrt{5}} = \left(2\sqrt{5}\right)\pi a^3 / 75 \text{ cubic units.}$$

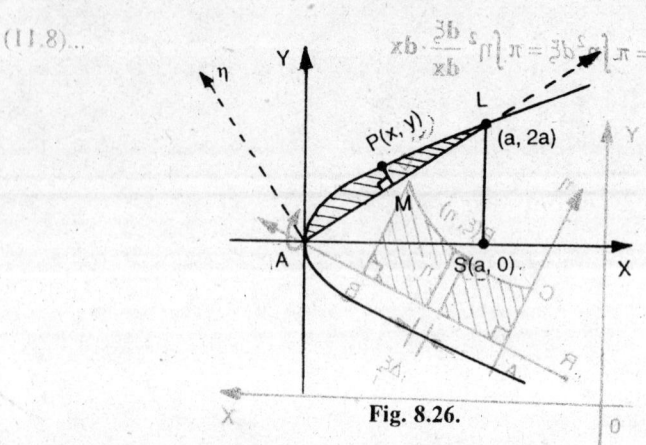

Fig. 8.26.

8.8 Volume formulae for polar equation

The volume of the solid generated by revolving the area bounded by the curve $r = f(\theta)$ and the radii vectors $\theta = \alpha$ and $\theta = \beta$

(i) about the initial line is $V = \frac{2}{3}\pi \int_\alpha^\beta r^3 \sin\theta \, d\theta$ ...(8.12)

(ii) about the line through the pole perpendicular to initial line is given by

$$V = \frac{2}{3}\pi \int_\alpha^\beta r^3 \cos\theta \, d\theta \qquad ...(8.13)$$

8.9 Illustrative Examples

Example 8.17 Find the volume of the solid formed by the revolution about the initial line of the loop of the curve $r = a(1 + \cos\theta)$ between $\theta = 0$ and $\theta = \pi$. *(A.M.I.E, S-2010, S-2001).*

Solution. The given curve is the cardioid shown in the Fig. 8.27. The required

$$\text{volume } = \frac{2}{3}\pi \int_0^\pi r^3 \sin\theta \, d\theta$$

$$= \frac{2}{3}\pi a^3 \int_0^\pi (1+\cos\theta)^3 \sin\theta \, d\theta$$

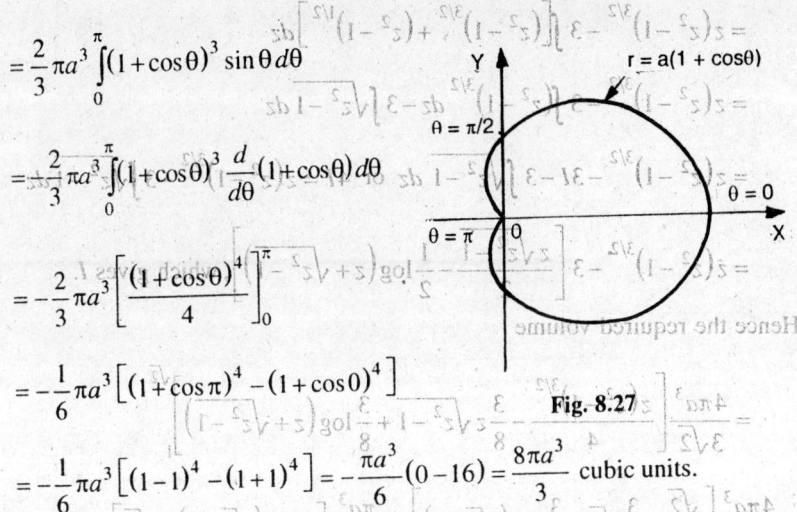

$$= \frac{2}{3}\pi a^3 \int_0^\pi (1+\cos\theta)^3 \frac{d}{d\theta}(1+\cos\theta)\, d\theta$$

$$= -\frac{2}{3}\pi a^3 \left[\frac{(1+\cos\theta)^4}{4} \right]_0^\pi$$

Fig. 8.27

$$= -\frac{1}{6}\pi a^3 \left[(1+\cos\pi)^4 - (1+\cos 0)^4 \right]$$

$$= -\frac{1}{6}\pi a^3 \left[(1-1)^4 - (1+1)^4 \right] = -\frac{\pi a^3}{6}(0-16) = \frac{8\pi a^3}{3} \text{ cubic units.}$$

Example 8.18 Find the volume of the solid generated by revolving the lemniscate $r^2 = a^2 \cos 2\theta$ about the initial line.

Solution. The required volume is twice that is generated by revolving the first quadrant area.

Therefore,
$$V = 2 \cdot \frac{2}{3}\pi \int_0^{\pi/4} r^3 \sin\theta \, d\theta$$

$$= \frac{4\pi}{3}\int_0^{\pi/4} a^3 (\cos 2\theta)^{3/2} \sin\theta \, d\theta$$

$$= \frac{4}{3}\pi a^3 \int_0^{\pi/4} (2\cos^2\theta - 1)^{3/2} \sin\theta \, d\theta$$

$$\left(\text{Putting } \sqrt{2}\cos\theta = z \right)$$

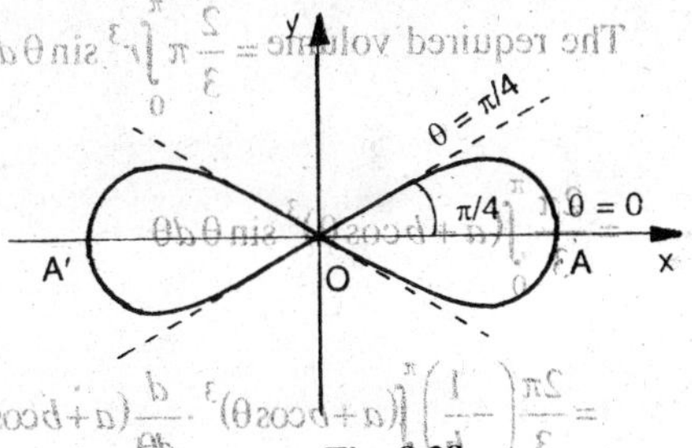

Fig. 8.28.

$$= \frac{4}{3}\pi a^3 \int_{\sqrt{2}}^{1} (z^2-1)^{3/2} \left(\frac{dz}{\sqrt{2}} \right) = \frac{4}{3\sqrt{2}}\pi a^3 \int_1^{\sqrt{2}} (z^2-1)^{3/2} \, dz$$

Let $I = \int (z^2-1)^{3/2} \, dz$. Intergrating by parts taking unity as second factor.

$$\therefore \quad I = (z^2-1)^{3/2} \cdot z - \int \frac{3}{2}(z^2-1)^{1/2} \cdot 2z \cdot z \, dz = z(z^2-1)^{3/2} - 3\int z^2 (z^2-1)^{1/2} \, dz$$

$$= z(z^2-1)^{3/2} - 3\int (z^2-1+1)(z^2-1)^{1/2} \, dz$$

$$= z(z^2-1)^{3/2} - 3\int\left[(z^2-1)^{3/2} + (z^2-1)^{1/2}\right]dz$$

$$= z(z^2-1)^{3/2} - 3\int(z^2-1)^{3/2}\,dz - 3\int\sqrt{z^2-1}\,dz$$

$$= z(z^2-1)^{3/2} - 3I - 3\int\sqrt{z^2-1}\,dz \quad \text{or} \quad 4I = z(z^2-1)^{3/2} - 3\int\sqrt{z^2-1}\,dz$$

$$= z(z^2-1)^{3/2} - 3\left[\frac{z\sqrt{z^2-1}}{2} - \frac{1}{2}\log\left(z+\sqrt{z^2-1}\right)\right], \text{ which gives } I.$$

Hence the required volume

$$= \frac{4\pi a^3}{3\sqrt{2}}\left[\frac{z(z^2-1)^{3/2}}{4} - \frac{3}{8}z\sqrt{z^2-1} + \frac{3}{8}\log\left(z+\sqrt{z^2-1}\right)\right]_1^{\sqrt{2}}$$

$$= \frac{4\pi a^3}{3\sqrt{2}}\left[\frac{\sqrt{2}}{4} - \frac{3}{8}\sqrt{2} + \frac{3}{8}\log\left(\sqrt{2}+1\right)\right] = \frac{\pi a^3}{6\sqrt{2}}\left[3\log\left(\sqrt{2}+1\right) - \sqrt{2}\right] \text{ cubic units.}$$

Example 8.19 Prove that the volume generated by the revolution about the initial line of the limacon $r = a + b\cos\theta$, $a > b$ is $4\pi a(a^2 + b^2)/3$.

Solution. The shape of the curve is shown in the Fig. 8.29.

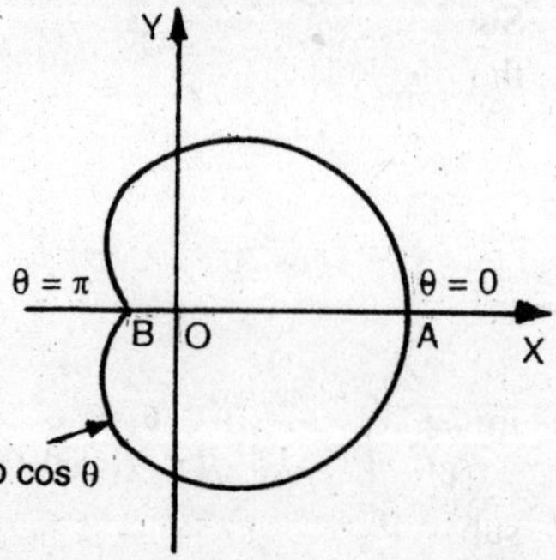

The required volume $= \dfrac{2}{3}\pi\displaystyle\int_0^\pi r^3 \sin\theta\,d\theta$

$$= \frac{2\pi}{3}\int_0^\pi (a+b\cos\theta)^3 \sin\theta\,d\theta$$

$$= \frac{2\pi}{3}\left(-\frac{1}{b}\right)\int_0^\pi (a+b\cos\theta)^3 \cdot \frac{d}{d\theta}(a+b\cos\theta)\,d\theta$$

$\theta = \pi$ B O   A $\theta = 0$

$r = a + b\cos\theta$

Fig. 8.29.

$$= \frac{-2\pi}{3b}\left[\frac{(a+b\cos\theta)^4}{4}\right]_0^\pi = \frac{-\pi}{6b}\left[(a-b)^4 - (a+b)^4\right]$$

$$= 4\pi a(a^2+b^2)/3 \text{ cubic units.}$$

Example 8.20 Show that the volume of the solid obtained by revolving one

loop of the curve $r^2 = a^2 \cos 2\theta$ about the line $\theta = \dfrac{\pi}{2}$ is $\dfrac{\pi^2 a^3}{4\sqrt{2}}$.

Solution. The required volume $= 2 \cdot \dfrac{2\pi}{3} \displaystyle\int_0^{\pi/4} r^3 \cos\theta\, d\theta$

$$= \frac{4\pi a^3}{3} \int_0^{\pi/4} (\cos 2\theta)^{3/2} \cos\theta\, d\theta$$

$$= \frac{4}{3}\pi a^3 \int_0^{\pi/4} \left(1 - 2\sin^2\theta\right)^{3/2} \cos\theta\, d\theta \qquad\qquad \left(\text{Let } \sqrt{2}\sin\theta = \sin\phi\right)$$

$$= \frac{4}{3}\pi a^3 \int_0^{\pi/2} \cos^3\phi \cdot \frac{1}{\sqrt{2}} \cos\phi\, d\phi \;=\; \frac{4}{3\sqrt{2}}\pi a^3 \int_0^{\pi/2} \cos^4\phi\, d\phi$$

$$= \frac{4}{3\sqrt{2}}\pi a^3 \cdot \frac{3\cdot 1}{4\cdot 2} \cdot \frac{\pi}{2} = \frac{\pi^2 a^3}{4\sqrt{2}}.$$

Example 8.21 Show that if the area lying within the cordioid $r = 2a\,(1 + \cos\theta)$ and without the parabola $r(1 + \cos\theta) = 2a$, revolves about the initial line, the volume generated is $18\pi a^3$.

Solution. The equations of the cardioid and the parabola are:

$$r = 2a(1 + \cos\theta), \qquad\qquad \dots(i)$$

$$r = \frac{2a}{1 + \cos\theta} \qquad\qquad \dots(ii)$$

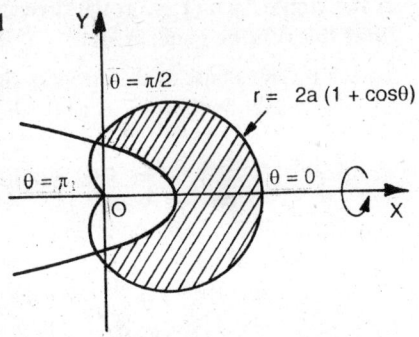

Fig. 8.30

They are both symmetrical about the inital line.

To find the points of inter-section, substituting the value of $r = 2a(1 + \cos\theta)$ from (i) in (ii), we have

$$2a(1 + \cos\theta) = \frac{2a}{1 + \cos\theta} \quad\text{ or }\quad 1 + 2\cos\theta + \cos^2\theta = 1.$$

$\therefore$ Either $\cos\theta = 0$ or $\cos\theta = -2$, which is impossible ($\because \cos\theta$ can not be numerically > 1.) Therefore, $\cos\theta = 0$, or $\theta = \dfrac{\pi}{2}$.

$$\therefore \qquad \text{Volume} = \frac{2}{3}\pi \int_0^{\pi/2} \left(r_1^3 - r_2^3\right) \sin\theta\, d\theta$$

$$= \frac{2}{3}\pi \int_0^{\pi/2} \left[\{2a(1+\cos\theta)\}^3 - \left(\frac{2a}{1+\cos\theta}\right)^3 \right] \sin\theta \, d\theta$$

$$= \frac{16\pi a^3}{3} \left[\int_0^{\pi/2} (1+\cos\theta)^3 \sin\theta \, d\theta - \int_0^{\pi/2} (1+\cos\theta)^{-3} \sin\theta \, d\theta \right]$$

$$= \frac{16\pi a^3}{3} \left[-\int_0^{\pi/2} (1+\cos\theta)^3 \frac{d}{d\theta}(\cos\theta) \, d\theta + \int_0^{\pi/2} (1+\cos\theta)^{-3} \frac{d}{d\theta}(\cos\theta) \, d\theta \right]$$

$$= \frac{16\pi a^3}{3} \left(-\left[\frac{(1+\cos\theta)^4}{4} \right]_0^{\pi/2} + \left[\frac{(1+\cos\theta)^{-2}}{-2} \right]_0^{\pi/2} \right)$$

$$= -\frac{4}{3}\pi a^3 \left[\left(1+\cos\frac{\pi}{2}\right)^4 - (1+\cos 0)^4 \right] - \frac{8}{3}\pi a^3 \left[\left(1+\cos\frac{\pi}{2}\right)^{-2} - (1+\cos 0)^{-2} \right]$$

$$= (20\pi a^3 - 2\pi a^3) = 18\pi a^3 \text{ cubic units.}$$

Example 8.22 A solid is formed by revolving the area between the two loops of the curve $r = a(1 + 2\cos\theta)$ through four right angles about the initial line. Find the volume generated.

Solution. The shape of the curve is shown in Fig. 8.31. The lined portion of the curve is traced for $0 \le \theta \le \pi$ while the dotted portion is traced for $\pi \le \theta \le 2\pi$.

At the pole, $\qquad \theta = \dfrac{2\pi}{3}$. The required volume

= (volume obtained by revolving the area OACO)–(volume obtained by revolving the area OBDO).

Now volume obtained by revolving the area *OACO*

$$= \frac{2}{3}\pi \int_0^{2\pi/3} r^3 \sin\theta \, d\theta = \frac{2\pi}{3} \int_0^{2\pi/3} a^3 (1+2\cos\theta)^3 \sin\theta \, d\theta$$

$$= -\frac{\pi a^3}{3} \int_0^{2\pi/3} (1+2\cos\theta)^3 \cdot (-2\sin\theta) \, d\theta$$

$$= -\frac{\pi a^3}{3 \cdot} \left[\frac{(1+2\cos\theta)^4}{4} \right]_0^{2\pi/3}$$

$$= -\frac{\pi a^3}{3} \left(0 - \frac{81}{4} \right) = \frac{27}{4} \pi a^3.$$

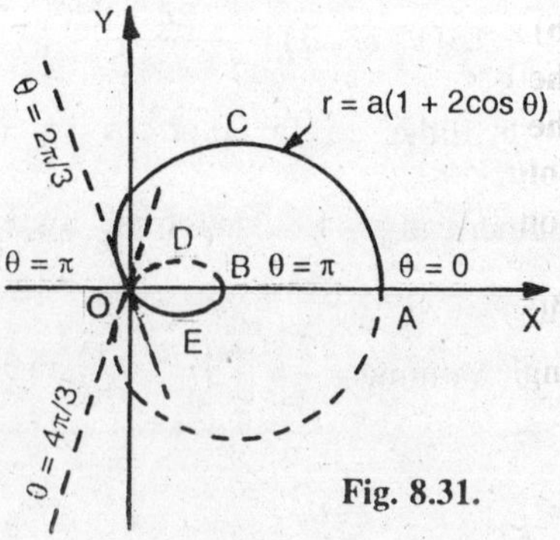

Fig. 8.31.

The volume obtained by revolving area *OBDO*

$$= \frac{2}{3}\pi \int_{\pi}^{4\pi/3} r^3 \sin\theta\, d\theta = -\frac{\pi a^3}{3}\left[\frac{(1+2\cos\theta)^4}{4}\right]_{\pi}^{4\pi/3}, \text{ as before}$$

$$= -\frac{\pi a^3}{3}\left(0 - \frac{1}{4}\right) = \frac{\pi a^3}{12}.$$

Hence, the required volume $= \frac{27}{4}\pi a^3 - \frac{1}{12}\pi a^3 = \frac{20}{3}\pi a^3$.

8.10 Finding Volumes of Solids by the Method of Slicing

Let a solid be bounded by two parallel planes perpendicular to the *x*-axis at $x = a$ and $x = b$. If , for each point *x* in the closed interval [*a*, *b*] the cross-sectional area of solid perpendicular to the *x*-axis is $A(x)$, then the volume of the solid of known integrable cross-section area $A(x)$ from $x = a$ to $x = b$ is given by:

$$V = \int_a^b A(x)\,dx, \qquad \text{(Cross-Section or Slicing formula)} \qquad ...(8.14)$$

There is a similar result for cross-sections perpendicular to the *y*-axis. If the solid is bounded by the planes $y = c$ and $y = d$, then volume of the solid can be written as

$$V = \int_c^d A(y)\,dy, \qquad \text{(Cross-Section or Slicing formula)} \qquad ...(8.15)$$

where $A(y)$ is the cross-sectional area.

Example 8.23 **The base of a certain solid is the region between the x-axis and the curve $y = 2\sqrt{\sin x}$ between x = 0 and x = π. Each cross-section of the solid cut off by a plane perpendicular to the x-axis is**
(a) an equilateral triangle with one side in the plane of the solid,
(b) a square with one side of the square in the base of the solid. Find the volume of the solid.

Solution. (*a*) Refer Fig. 8.32. The cross-section at *x* is an equilateral triangle of area

$A(x) = \sqrt{3}\left(2\sqrt{\sin x}\right)^2 / 4 = \sqrt{3}\sin x$. The triangles run from $x = 0$ to $x = \pi$.

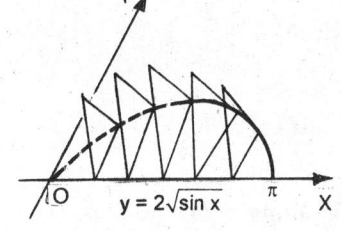

Fig. 8.32.

$$\therefore \quad V = \int_a^b A(x)\,dx = \int_0^\pi \sqrt{3}\sin x\, dx = \sqrt{3}[-\cos x]_0^\pi = \sqrt{3}(1+1) = 2\sqrt{3}.$$

(b) Required volume, $V = \int\limits_{0}^{\pi} \left(2\sqrt{\sin x}\right)^2 dx = 4 \int\limits_{0}^{\pi} \sin x \, dx = 4[-\cos x]_0^\pi = 8$ unit³.

Example 8.24 Find the volume of the solid whose base is enclosed by the circle $x^2 + y^2 = 16$ and whose cross-sections taken perpendicular to the base is an equilateral triangle.

Solution. The cross-section ABC of the solid is an equilateral triangle of side 2y and area

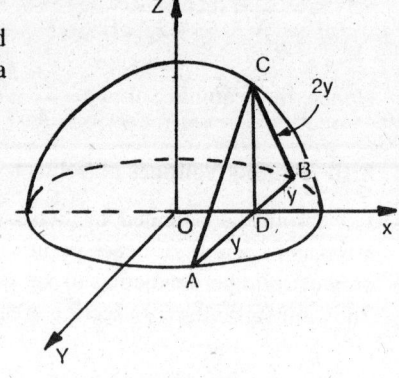

$A(x) = \dfrac{\sqrt{3}(2y)^2}{4} = \sqrt{3}y^2 = \sqrt{3}(16 - x^2)$.

Then, by the slicing formula,

$V = \sqrt{3} \int\limits_{-4}^{4} (16 - x^2)dx = 2\sqrt{3}\left[16x - \dfrac{x^3}{3}\right]_0^4$

$= 2\sqrt{3}\left[64 - \left(\dfrac{64}{3}\right)\right] = \dfrac{256\sqrt{3}}{3}$ cubic units.

Fig. 8.33.

Example 8.25 A solid has a base in the form of an ellipse with major axis 10 and minor axis 8. Find the volume if every section perpendicular to the major axis is an isosceles triangle with attitude 6.

Solution. Refer Fig. 8.34. The equation

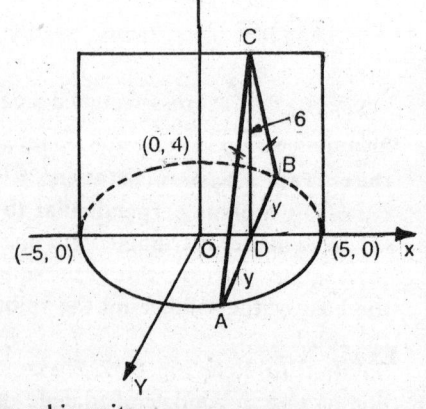

of the ellipse is $\dfrac{x^2}{25} + \dfrac{y^2}{16} = 1$. The section ABC an isosceles triangle of base 2y, altitude 6, and area

$A(x) = 2y \cdot 6/2 = 6y = 6\left(4\sqrt{25 - x^2}/5\right)$.

Hence $V = \dfrac{24}{5} \int\limits_{-5}^{5} \sqrt{25 - x^2} \, dx$

$= \dfrac{24}{5} \cdot 2\left[\dfrac{x}{2}\sqrt{25 - x^2} + \dfrac{25}{2}\sin^{-1}\dfrac{x}{5}\right]_0^5 = 60\pi$ cubic units.

Fig. 8.34.

Example 8.26 The cross sections of a certain solid made by planes perpendicular to the x axis are circles with diameters extending from the curve $y = 3x^2$ to the curve $y = 16 - x^2$. Find the volume of the solid which lies between the points of intersection of these curves.

Solution. Refer Fig. 8.35. At the points of intersection of the curves, we have $3x^2 = 16 - x^2 \Rightarrow x^2 = 4$, that is, $x = \pm 2$. Therefore, the points of intersection of the curves are $P(-2, 12)$ and $Q(2, -12)$.

Any point on the curve $y = 16 - x^2$ is $R\left(x, 16 - x^2\right)$.

Any point on the curve $y = 3x^2$ is $S\left(x, 3x^2\right)$.

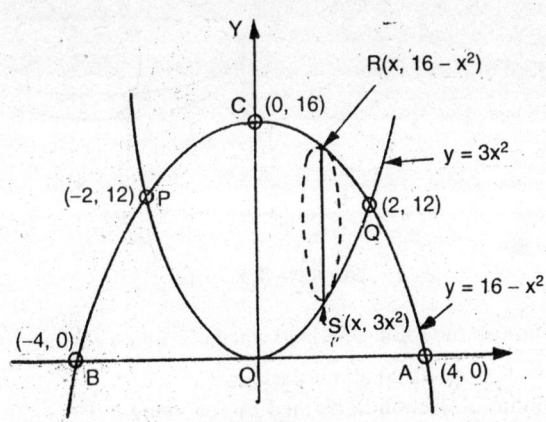

Fig. 8.35.

Diameter of the circle $= RS = (16 - x^2) - 3x^2 = 16 - 4x^2$. Area of the circle $=$

$$A(x) = \frac{\pi(RS)^2}{4} = \frac{\pi\left(16 - 4x^2\right)^2}{4} = 4\pi\left(4 - x^2\right)^2 .$$

Since the solid is symmetric about the y-axis, then, by the cross-section formula the required volume is,

$$V = 2\int_0^2 A(x)\,dx = 8\pi \int_0^2 \left(4 - x^2\right)^2 dx = 8\pi\left[16x - \frac{8x^3}{3} + \frac{x^5}{5}\right]_0^2$$

$$= 8\pi\left(32 - \frac{64}{3} + \frac{32}{5}\right) = \frac{2048\pi}{15} \text{ cubic units.}$$

Example 8.27 Find the volume of the ellipsoid $x^2/a^2 + y^2/b^2 + z^2/c^2 = 1$.

(I.E.T.E. June 2008)

Solution. Refer Fig. 8.36. The section of the ellipsoid made by a plane parallel to the yz-plane and at a distance x from it is an ellipse whose equation is

$$\frac{y^2}{b^2} + \frac{z^2}{c^2} = \left(1 - \frac{x^2}{a^2}\right) \quad \text{or} \quad \frac{y^2}{\left(b\sqrt{1 - \frac{x^2}{a^2}}\right)^2} + \frac{z^2}{\left(c\sqrt{1 - \frac{x^2}{a^2}}\right)^2} = 1$$

with semi axes $b_1 = b\sqrt{1 - \dfrac{x^2}{a^2}}$, $c_1 = c\sqrt{1 - \dfrac{x^2}{a^2}}$. The area of such an ellipse is

$\pi b_1 c_1$. Therefore $A(x) = \pi bc\left(1 - \dfrac{x^2}{a^2}\right)$.

The volume of the ellipsoid will be,

$$V = \pi bc \int_{-a}^{a} \left(1 - \frac{x^2}{a^2}\right) dx$$

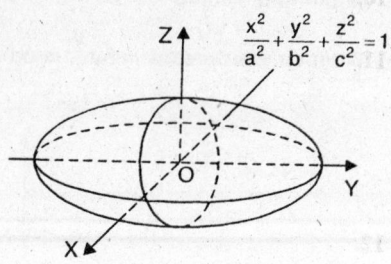

$$= \pi bc \left[x - \frac{x^3}{3a^2}\right]_{-a}^{a} = \frac{4}{3}\pi abc.$$

Fig. 8.36

Exercise 8.1

1. Find the volume of the solid generated when the region enclosed by $y = \sqrt{x}$, $y = 2$ and $x = 0$ is revolved about the y-axis. **Ans.** $32\,\pi/5$

2. Find the volume of the solid obtained by revolving
   (i) the ellipse $x^2/a^2 + y^2/b^2 = 1$ about the major axis.
   (MDU 2005; Bhopal 2002; A.M.I.E., W-1996)
   (ii) the area of the first quadrant bounded by the parabola $y^2 = 8x$ and its latus recturn about x-axis. *(MDU 2003)* **Ans.** (i) $4\pi ab^2/3$, (ii) 16π.

3. Find the volume of the solid generated when the region between the graphs of the equations $f(x) = x^2 + (1/2)$ and $g(x) = x$ over the interval $(0, 2)$ is revolved about the x-axis. **Ans.** $69\pi/10$

4. (a) Find the volume of a spherical segment of height h, the radius of whose base is c. **Ans.** $\pi h(3c^2 + h^2)/6$
   (b) Find the volume of a spherical segment of height h cut off from a sphere of radius r. *(A.M.I.E., W-1997)* **Ans.** $\pi h^2(3r - h)/3$

5. Show that the volume of the solid generated by revolving about x-axis the region bounded by $y = \log x$, $y = 0$ and $x = 2$ is $2\pi(1 - \log 2)^2$.

6. Show that the volume of solid generated by the revolution of the loop of $x(x^2 + y^2) = a(x^2 - y^2)$ about the x-axis is $2\pi a^3 \left\{\log 2 - \left(\dfrac{2}{3}\right)\right\}$.

7. The curve $y^2(a + x) = x^2(3a - x)$ revolves about the axis of x. Find the volume generated by the loop. **Ans.** $\pi a^3(8 \log 2 - 3)$

8. Show that the volume of the solid obtained by the revolution of the curve $a^2 y^2 = x^2(a^2 - x^2)$ about the axis of x is $4a^3/15$.

9. Find the volume of the solid generated by revolving the region bounded by the curves $y = 3 - x^2$ and $y = -1$ about the line $y = -1$.
(*A.M.I.E.-W-2003*) **Ans.** $512\pi/15$ cubic units.

10. Find the volume of the solid obtained by revolving the loop of the curve $a^2y^2 = x^2 (2a - x)(x - a)$ about the axis of x. **Ans.** $23\,\pi a^3/60$

11. Find the volume generated by revolving the area cut off from the parabola $y = 4x - x^2$ by the x-axis about the line $y = 6$.

[**Hint:** Required volume $= \pi \int\limits_{0}^{4} \left[6^2 - (6-y)^2 \right] dx = 1408\,\pi/15$ cubic units].

12. The area bounded by the parabola $y^2 = 4x$ and the straight line $4x - 3y + 2 = 0$ is rotated about the y-axis. Obtain the volume of the solid formed by this revolution. [**Hint.** $V = \pi \int\limits_{1}^{2} \left[\left(\dfrac{3y-2}{4} \right)^2 - \left(\dfrac{y^2}{4} \right)^2 \right] dy = \dfrac{\pi}{20}$] (*A.M.I.E.-S-1999*)

13. Prove that the volume of the solid generated by the revolution of the curve $y = \dfrac{a^3}{\left(a^2 + x^2 \right)}$ about its asymptote is $\dfrac{\pi^2 a^3}{2}$.

14. The part of the parabola $y^2 = 4ax$ cut off by the latus rectum revolves about the tangent at the vertex. Find the volume of the reel thus generated.
(*A.M.I.E., W-2004*) **Ans.** $4\pi a^3/5$

15. The segment of the parabola $y^2 = 4ax$ which is cut off by the latus rectum revolves about the directrix. Find the volume of the solid formed.

[**Hint:** $V = \pi \int\limits_{-2a}^{2a} \left[(2a)^2 - (a+x)^2 \right] dy = 128\,\pi a^3/15$]

16. Show that the volume of the solid generated by the revolution of the curve $(a - x)y^2 = a^2x$ about its asymptote is $\pi^2 a^3/2$.

17. Prove that the volume of the solid generated by the revolution of an ellipse round its minor axis, is a mean proportional between that generated by the revolution of the ellipse and of its auxiliary circle about the major axis.
[**Hint:** Volume of the sphere generated by the revolution of the auxiliary circle $(x^2 + y^2 = a^2)$ about major axis $= 4\pi a^3/3$].

18. Show that the volume of the solid generated by revolving the area included between the curves $y^2 = x^3$ and $x^2 = y^3$ about x-axis is $5\pi/28$.

19. Find the volume of the solid generated by the revolution of the plane area bounded by $y^2 = 9x$ and $y = 3x$ about the x-axis. **Ans.** $3\pi/2$ cubic units.

20. Find the area enclosed by the curve $xy^2 = 4(2 - x)$ and the y-axis and also the volume of the solid formed by the revolution of the curve through four right angles about the x-axis. **Ans.** 4π, $4\pi^2$

21. The area bounded by the hyperbola $xy = 4$ and the line $x + y = 5$ is revolved about the x-axis. Find the volume of the solid thus formed.

    [**Hint:** $V = \pi \int_1^4 \left(y_1^2 - y_2^2\right) dx = \pi \int_1^4 \left\{ (5-x)^2 - \dfrac{16}{x^2} \right\} dx$] **Ans.** 9π cubic units

22. Find the volume of the solid formed when the plane area bounded by $y = -x^2 - 3x + 6$ and $x + y - 3 = 0$ is revolved (*i*) about $y = 0$ (*ii*) about the line $x = 3$. **Ans.** (*i*) $1792\,\pi/15$ cubic units, (*ii*) $256\pi/3$ cubic units.

    [**Hint:** (*i*) $V = \pi \int_{-3}^1 \left\{ (y_c)^2 - (y_L)^2 \right\} dx = \pi \int_{-3}^1 \left\{ (-x^2 - 3x + 6)^2 - (3 - x^2) \right\} dx$

    (*ii*) $V = 2\pi \int_{-3}^1 (3 - x)(y_c - y_L) dx = 2\pi \int_{-3}^1 (x - 3)(x^2 + 2x - 3) dx$].

23. The ellipse $x^2/a^2 + y^2/b^2 = 1$ is divided into two parts by the line $x = a/2$ and the smaller part is rotated through four right angles about this line. Prove that the volume generated is $\pi a^2 b \left(9\sqrt{3} - 4\pi\right)/12$.

24. Trace the curves $y = \sin x$, $y = \cos x$ as x varies from 0 to $\pi/2$ and show that the volume of the solid obtained by revolving about x-axis the region enclosed by them and the x-axis is $\pi/4\,(\pi - 2)$.

25. The region bounded by the curve $y = x^2 + 1$ and the line $y = -x + 3$ is revolved about the x-axis to generate a solid. Find the volume of the solid generated.

    [**Hint:** $V = \pi \int_{-2}^1 \left[(-x + 3)^2 - (x^2 + 1)^2 \right] dx$ **Ans.** $117\pi/5$ cubic units

26. Use cylindrical shell method to find the volume of the solid generated when the region R under $y = x^2$ over the interval $[0, 2]$ is revolved about the x-axis.

    [**Hint:** $V = \int_0^4 2\pi y \left(2 - \sqrt{y}\right) dy$] **Ans.** $32\,\pi/5$ cubic units

27. Find the volume of the solid of revolution formed by revolving the curve $\sqrt{x} + \sqrt{y} = \sqrt{a}$ bounded by $x = 0$, $y = 0$ about the axis of x.

    (*A.M.I.E, W-2002*) **Ans.** $\pi a^3/12$

28. The area of the curve $x = a\cos^3\theta$, $y = a\sin^3\theta$ lying between $\theta = -\pi/2$ and $\theta = \pi/2$ rotates about the x-axis; find the volume of the solid so generated.

Ans. $16\ \pi a^3/105$

29. Show that the volume of the solid generated by revolution of the cycloid,
$x = a(\theta + \sin\theta)$, $y = a(1-\cos\theta)$ about the y-axis is $\pi a^3 (9\pi^2 - 16)/6$ and about the x-axis is $\pi^2 a^3$.

30. Show that the volume of the solid generated by the revolution of the curve $x = 2a\sin^2 t$, $y = 2a\sin^3 t/\cos t$ about its asymptote is $2\pi^2 a^3$.

31. Find the volume of the solid obtained by revolving one arc of the cycloid
$x = a(\theta + \sin\theta)$, $y = a(1 + \cos\theta)$ about x-axis.

(*D.C.E.*, 2004; VTU 2003) **Ans.** $5\pi^2 a^3$.

32. The smaller segment of a circle of radius 13 cm cut off by a chord 24 cm long is rotated about the chord. Find the volume of the solid generated.

Ans. $2\pi\left[1452 - 845\sin^{-1}(12/13)\right]$

33. Find the volume generated by revolving the area between the curve $(y+8)/x = x - 2$ and the x-axis about the line $x + 5 = 0$. (*D.C.E.*, 2005).

Ans. 432 π cubic units.

34. Find the volume of the solid generated by the revolution about y-axis of the area under the cuve $y = \sin x$ from $x = 0$ to $x = \pi$.
[**Hint:** Use shell formula] **Ans.** $2\pi^2$

35. A quadrant of a circle of radius a, revolves about its chord. Show that the volume of the spindle thus generated is $\pi(10 - 3\pi)a^3/6\sqrt{2}$.

[**Hint.** The chord AB has equation $x + y = a$. Therefore
$$\eta = PM = (x+y-a)/\sqrt{2}.$$
$$AM^2 = AP^2 - PM^2 = \left[(x-a)^2 + y^2\right]$$
$$-\left[(x+y-a)^2/2\right] = (x-a-y)^2/2.$$
$$\therefore \ \ \xi = (x-a-y)/\sqrt{2}.$$

Required volume $= \pi\int_0^{AB} \eta^2 \dfrac{d\xi}{dx}\cdot dx$

$$= \pi\int_0^a \frac{(x+y-a)^2}{2}\cdot\frac{1}{\sqrt{2}}\left(1 - \frac{dy}{dx}\right)dx.$$

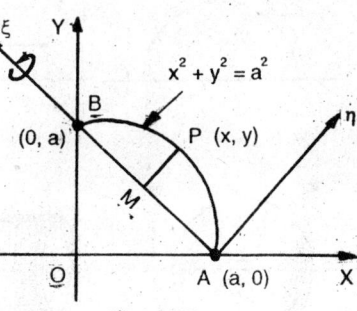

Fig. 8.37

For the circle $x = a\cos\theta$, $y = a\sin\theta$, $dy/dx = -\cos\theta/\sin\theta$. Substituting the values in the above definite integral, we obtain

$$W = \frac{\pi}{2\sqrt{2}} \int_0^{\pi/2} a^2 (\sin\theta + \cos\theta - 1)^2 \cdot \left(1 + \frac{\cos\theta}{\sin\theta}\right) \cdot a\sin\theta \, d\theta$$

$$= \frac{\pi a^3}{2\sqrt{2}} \int_0^{\pi/2} (\sin\theta + \cos\theta - 1)^2 \cdot (\sin\theta + \cos\theta) d\theta = \pi a^3 (10 - 3\pi)/6\sqrt{2} \;]$$

36. The curve $r = a(1 - \cos\theta)$ revolves about the initial line. Find the volume of the solid formed. **Ans.** $8\pi a^3/3$

37. The arc of the cardioid $r = a(1 + \cos\theta)$ included between $\theta = -\pi/2$ to $\pi/2$ is rotated about the line $\theta = \pi/2$. Show that the volume generated is $(16 + 5\pi)\pi a^3/4$.

38. The base of a certain solid is the region between the x-axis and the curve $y = \sin x$ between $x = 0$ and $x = \pi/2$. Each plane section of the solid perpendicular to the x-axis is an equilateral triagle with one side in the base of the solid. Find the volume.

[**Hint:** $V = \int_0^{\pi/2} A(x)\,dx = \int_0^{\pi/2} \frac{\sqrt{3}}{4}(\sin x)^2 \, dx = \frac{\sqrt{3}\,\pi}{16}$ cubic units.]

39. Find the volume of the solid whose base is enclosed by the circle $x^2 + y^2 = 1$ and whose cross-sections taken perpendicular to the base are
(*a*) semi circles, (*b*) squares, (*c*) an equilateral triangles.

Ans. (a) $2\pi/3$, (b) 16/3, (c) $4\sqrt{3}/3$.

8.11 Surface Area of a Solid of Revolution

A *surface area of a solid of revolution* is a surface that is generated by revolving a plane curve about an axis that does not intersect the curve. For example, the

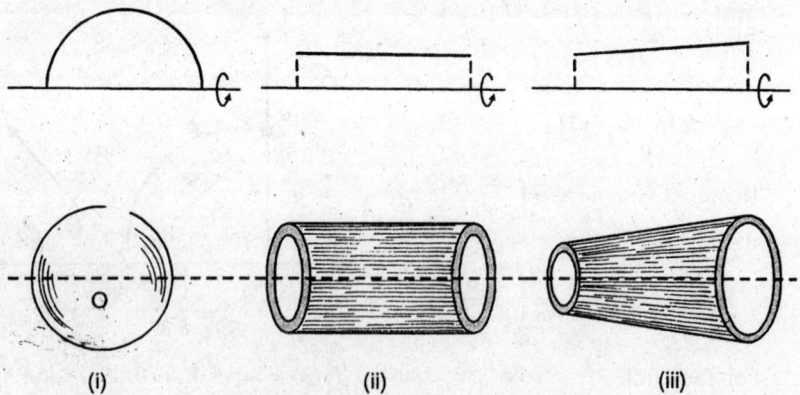

(i)   (ii)   (iii)

Fig. 8.38 Some surface of revolution

surface of a sphere can be generated by revolving a semi-circle arc about its diameter, and the lateral surface of a right circular cylinder can be generated by revolving a line segment about an axis that is parallel to it.

[Refer. Fig. 8.38 (*i*) to (*iii*)]

If an elementary arc PQ = *ds* of the curve $y = f(x)$ at P(x, y) revolves about the x-axis, then the curved surface of the disk formed is approximately given by $2\pi y ds$ as $2\pi y$ is the circumference of the disk and *ds* its breadth.

Summing up the surfaces of such disks, the surface of the solid formed by the revolution of the arc $y = f(x)$ between $x = a$ and $x = b$ about the x-axis is given by

$$S_x = \int_a^b 2\pi y \, ds \qquad \qquad ...(8.16)$$

8.11.1 Different forms of the surface area formula

The formula (8.16) has different forms depending upon the equation of the curve. The limits of integration are written corresponding to the values of the independent variable at the extremities A and B of the curve. (Refer Fig. 8.39).

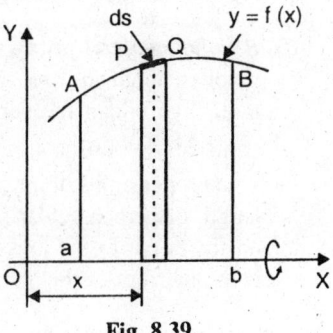

Fig. 8.39.

(*a*) If the curve $y = f(x)$, $f(x) > 0$ between $x = a$ and $x = b$ is revolved about the x-axis, the surface area generated is given by the formula

$$S_x = 2\pi \int_a^b y \sqrt{1+\left(\frac{dy}{dx}\right)^2} \, dx = 2\pi \int_a^b f(x)\sqrt{1+\left(f'(x)\right)^2} \, dx. \qquad ...(8.17)$$

If the region $x = g(y)$, $g(y) > 0$ between $y = c$ and $y = d$ is revolved about the y-axis, the surface area generated is given by the formula:

$$S_y = \int_c^d 2\pi x \, ds = 2\pi \int_c^d x \sqrt{1+\left(\frac{dx}{dy}\right)^2} \, dy = 2\pi \int_c^d g(y)\sqrt{1+\left(g'(y)\right)^2} \, dy \, ...(8.18)$$

(*b*) Similarly, if the curve is given by parametric equations $x = f(t)$, $y = \phi(t)$, $t_1 \le t \le t_2$ is revolved about the x-axis, then the surface area is given by the formula:

$$S_x = 2\pi \int_{t_1}^{t_2} y \sqrt{\left(\frac{dx}{dt}\right)^2 + \left(\frac{dy}{dt}\right)^2} \, dt, \qquad \text{where } y = \phi(t). \qquad ...(8.19)$$

The revolution around the y-axis will give

$$S_y = 2\pi \int_{t_1}^{t_2} x \sqrt{\left(\frac{dx}{dt}\right)^2 + \left(\frac{dy}{dt}\right)^2}\ dt, \quad \text{where } x = f(t). \qquad ...(8.20)$$

(c) If the curve is given in polar form $r = f(\theta)$, $\alpha \le \theta \le \beta$; the area of the surface generated by revolving the arc of the curve $r = f(\theta)$ about the polar axis is given by the formula:

$$S_x = \int_{\alpha}^{\beta} 2\pi y \frac{ds}{d\theta}\cdot d\theta = 2\pi \int_{\alpha}^{\beta} r\sin\theta \sqrt{r^2 + \left(\frac{dr}{d\theta}\right)^2}\ d\theta \qquad ...(8.21)$$

and about the 90° line is

$$S_y = \int_{\alpha}^{\beta} 2\pi x \frac{ds}{d\theta}\cdot d\theta = 2\pi \int_{\alpha}^{\beta} r\cos\theta \sqrt{r^2 + \left(\frac{dr}{d\theta}\right)^2}\ d\theta. \qquad ...(8.22)$$

(d) *Revolution about any axis:* If the curve *CD* revolves about any axis say *AB* (Refer Fig. 8.40), we obtain the surface area of the solid of revolution as $S = \int 2\pi (PN)ds$ with proper limits of integration. *PN* is the length of perpendicular from any point *P* on the curve to the axis of revolution, *N* being the foot of this perpendicular.

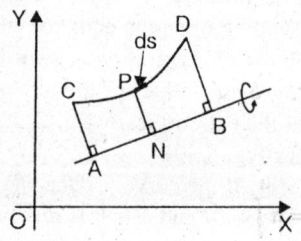

Fig. 8.40.

8.12 Illustrative Examples

Example 8.28 (a) Find the surface areas of solid generated by revolving about the x-axis, the arc of the parabola $y^2 = 12x$ from $x = 0$ to $x = 3$.
(b) A parabolic reflector of an automobile head light is 12 cm in diameter and 4 cm deep. Find the cost of plating of the front portion of the reflector if the cost of plating is Rs. 5.00 per cm².

Solution. (a) Using, $S_x = 2\pi \int_{0}^{P} y\,ds = 2\pi \int_{0}^{3} y \cdot \sqrt{1 + \left(\frac{dy}{dx}\right)^2}\ dx$

By differentiation,

$$\frac{dy}{dx} = \frac{6}{y} \quad \text{and} \quad 1 + \left(\frac{dy}{dx}\right)^2 = \frac{y^2 + 36}{y^2}.$$

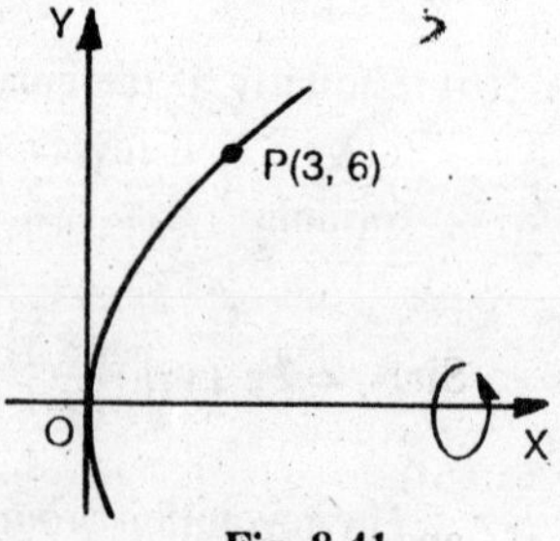

$$\therefore\ S_x = 2\pi \int_{0}^{3} y \frac{\sqrt{y^2 + 36}}{y}\,dx = 2\pi \int_{0}^{3} \sqrt{12x + 36}\ dx$$

$$= \frac{\pi}{9}\left[(12x + 36)^{3/2}\right]_{0}^{3} = 24\left(2\sqrt{2} - 1\right)\pi \text{ square units.}$$

Fig. 8.41.

Alternative Using, $S_x = 2\pi \int\limits_{O}^{P} y\,ds = 2\pi \int\limits_{0}^{6} y \sqrt{1+\left(\dfrac{dx}{dy}\right)^2}\,dy.$

Here $dx/dy = y/6$ and $1+(dx/dy)^2 = (y^2+36)/36$

$$\therefore\quad S_x = 2\pi \int\limits_{0}^{6} \frac{y\sqrt{36+y^2}}{6}\,dy = \frac{\pi}{9}\left[(36+y^2)^{3/2}\right]_0^6 = 24(2\sqrt{2}-1)\pi \text{ sq. units.}$$

(b) Let $y^2 = 4ax$ be the equation of the parabola. Since the point P(4, 6) lies on the parabola, therefore $36 = (4a)\,4$ or $4a = 9$. Thus the equation of the parabolic reflector is $y^2 = 9x$.

We have, $dy/dx = 9/2y$ and $1+\left(\dfrac{dy}{dx}\right)^2 = \dfrac{(4y^2+81)}{4y^2}.$

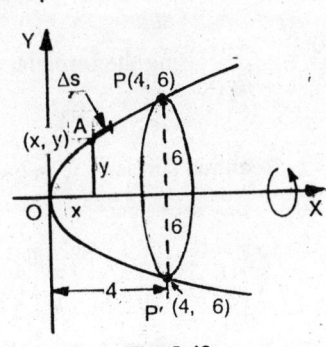

$$\therefore\quad S_x = 2\pi \int\limits_{0}^{4} y\cdot \frac{\sqrt{4y^2+81}}{2y}\,dx$$

$$= \pi \int\limits_{0}^{4}\sqrt{36x+81}\,dx = \pi\left[\frac{2}{3}\frac{(36x+81)^{3/2}}{36}\right]_0^4$$

$$= \frac{\pi}{2}\left[(4x+9)^{3/2}\right]_0^4 = \frac{\pi}{2}(125-27) = 154 \text{ sq. cm.}$$

Fig. 8.42.

Thus cost of plating = Rs. (154) 5 = Rs. 770/-

Example 8.29 Find the area of the surface of revolution that is generated by revolving the portion of curve $x = y^3$ from $y = 0$ to $y = 2$ about the y-axis.

Solution. The area of the surface generated by revolving the arc *OA* about the y-axis is given by

$$S_y = 2\pi \int\limits_{0}^{A} x\,ds = 2\pi \int\limits_{0}^{2} x\frac{ds}{dy}\,dy$$

$$= 2\pi \int\limits_{0}^{2} x \sqrt{1+\left(\frac{dx}{dy}\right)^2}\,dy$$

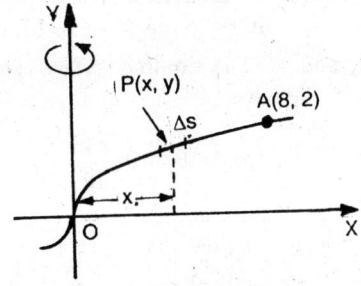

Since $x = y^3$, we have $\dfrac{dx}{dy} = 3y^2,$

and $1+\left(\dfrac{dx}{dy}\right)^2 = 1+9y^4.$

Fig. 8.43.

$$\therefore\; S_y = 2\pi \int_0^2 y^3 \sqrt{1+9y^4}\, dy = \frac{2\pi}{36} \int_0^2 (36 y^3)(1+9y^4)^{1/2}\, dy$$

$$= \frac{\pi}{18}\left[\frac{(1+9y^4)^{(1/2)+1}}{(1/2)+1} \right]_0^2 = \frac{\pi}{27}\left[(1+9y^4)^{3/2} \right]_0^2 = \frac{\pi(145\sqrt{145}-1)}{27} \text{ square units.}$$

Example 8.30 Find the area of the surface of revolution generated by revolving about the x-axis the ellipse whose semi-major and semi-minor axes are 4 units and 2 units respectively.   *(A.M.I.E., S-1999)*

Solution. The equation of the ellipse is $\dfrac{x^2}{16}+\dfrac{y^2}{4}=1$ or $x^2 + 4y^2 - 16 = 0$.

$\dfrac{dy}{dx} = -\dfrac{x}{4y}$. Using the formula, $S_x = 2\pi \displaystyle\int_{-4}^{4} y\,\dfrac{ds}{dx}\cdot dx.$

$\therefore$ Required surface, $S_x = 2\pi \displaystyle\int_{-4}^{4} \dfrac{y\sqrt{16y^2 + x^2}}{4y}\, dx$

$$= \pi \int_0^4 \sqrt{64-3x^2}\, dx = \frac{\pi}{\sqrt{3}}\left[\frac{x}{2}\sqrt{3}\sqrt{64-3x^2} + 32\sin^{-1}\left(\frac{x\sqrt{3}}{8} \right) \right]_0^4$$

$$= \frac{\pi}{\sqrt{3}}\left[8\sqrt{3} + 32\sin^{-1}\left(\frac{\sqrt{3}}{2} \right) \right] = \frac{\pi}{\sqrt{3}}\left(8\sqrt{3} + \frac{32\pi}{3} \right) = \frac{8\pi}{9}(9 + 4\sqrt{3}\pi).$$

Example 8.31 (a) Find the area of the surface of revolution generated by the revolution of the hypocycloid $x = a\cos^3\theta, y = a\sin^3\theta$ about the axis of x.

(b) Evaluate the surface area of the solid generated by revolving the cycloid $x = a(\theta - \sin\theta)$, $y = a(1 - \cos\theta)$, about the line $y = 0$.   *(V.T.U. 2000S)*

Solution. (a) The required surface is generated by revolving the arc from $\theta = 0$ to $\theta = \pi$. We have

$$\frac{dx}{d\theta} = -3a\cos^2\theta\sin\theta, \quad \frac{dy}{d\theta} = 3a\sin^2\theta\cos\theta \text{ and}$$

$$\sqrt{\left(\frac{dx}{d\theta}\right)^2 + \left(\frac{dy}{d\theta}\right)^2} = \sqrt{9a^2\cos^2\theta\sin^2\theta} = 3a\cos\theta\sin\theta.$$

$\therefore$ Required surface, $S_x = 2(2\pi) \int\limits_{0}^{\pi/2} y \cdot \sqrt{\left(\dfrac{dx}{d\theta}\right)^2 + \left(\dfrac{dy}{d\theta}\right)^2}\, d\theta$

$= 4\pi \int\limits_{0}^{\pi/2} \left(a\sin^3\theta\right) 3a\cos\theta\sin\theta\, d\theta$

$= 12\pi a^2 \int\limits_{0}^{\pi/2} \sin^4\theta\cos\theta\, d\theta = \dfrac{12\pi a^2}{5}$ square units.

(b) We have $\dfrac{dx}{d\theta} = a(1 - \cos\theta)$, $\dfrac{dy}{d\theta} = a\sin\theta$ and $\dfrac{ds}{d\theta} = \sqrt{\left(\dfrac{dx}{d\theta}\right)^2 + \left(\dfrac{dy}{d\theta}\right)^2}$

$= \sqrt{a^2(1 - \cos\theta)^2 + a^2\sin^2\theta}$

$= \sqrt{4a^2\sin^2\theta/2} = 2a\sin\theta/2$

Refer Fig. 8.44. For the portion
OPA, θ varies from 0 to 2π.

The required surface, $S_x = 2\pi \int\limits_{0}^{2\pi} y\dfrac{ds}{d\theta}\cdot d\theta$

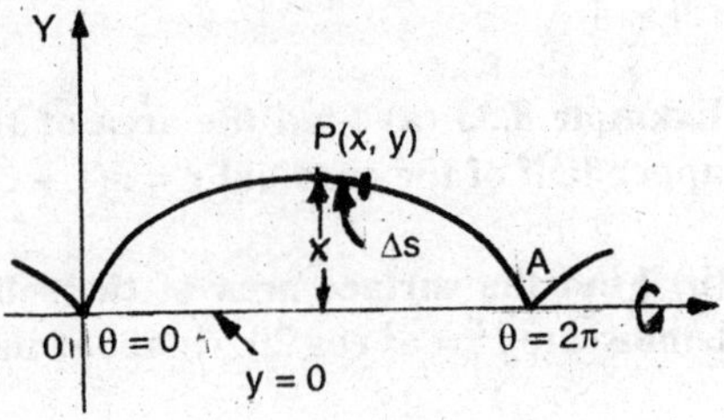

Fig. 8.44.

$= 2\pi \int\limits_{0}^{2\pi} a(1 - \cos\theta)\cdot 2a\sin\dfrac{\theta}{2}\, d\theta = 8\pi a^2 \int\limits_{0}^{2\pi} \sin^3\dfrac{\theta}{2}\, d\theta$

$= 16\pi a^2 \int\limits_{0}^{\pi} \sin^3 t\, dt, \left(\text{where } \dfrac{\theta}{2} = t\right) = 32\pi a^2 \int\limits_{0}^{\pi/2} \sin^3 t\, dt$

$= 32\pi a^2 \cdot \dfrac{2}{3} = \dfrac{64\pi a^2}{3}$ square units.

Example 8.32. Find the area of the surface of revolution generated by revolving about the x-axis, the cardioid

$$x = 2\cos\theta - \cos 2\theta, \quad y = 2\sin\theta - \sin 2\theta.$$

Solution. The required surface is generated by revolving the arc from $\theta = 0$
to $\theta = \pi$. We have

$$\dfrac{dx}{d\theta} = -2\sin\theta + 2\sin 2\theta, \quad \dfrac{dy}{d\theta} = 2\cos\theta - 2\cos 2\theta,$$

and $\dfrac{ds}{d\theta} = \sqrt{\left(\dfrac{dx}{d\theta}\right)^2 + \left(\dfrac{dy}{d\theta}\right)^2}$

$= \sqrt{8(1 - \sin\theta\sin 2\theta - \cos\theta\cos 2\theta)} = \sqrt{8(1 - \cos\theta)}$

The required surface,

$$S_x = 2\pi \int_0^\pi (2\sin\theta - \sin 2\theta)\cdot 2\sqrt{2}\sqrt{1 - \cos\theta}\, d\theta$$

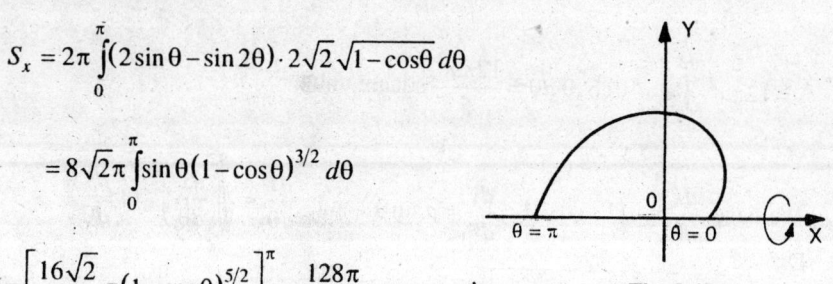

$$= 8\sqrt{2}\pi \int_0^\pi \sin\theta(1 - \cos\theta)^{3/2}\, d\theta$$

$$= \left[\dfrac{16\sqrt{2}}{5}\pi(1 - \cos\theta)^{5/2}\right]_0^\pi = \dfrac{128\pi}{5} \text{ square units.}$$

Fig. 8.45.

Example 8.33 (a) Find the area of the surface generated by revolving the upper half of the cardioid r = a(1 − cosθ) about the initial line.

(*VTU 2003; J.N.T.U., 2003*)

(b) Find the surface area of the solid generated by the revolution of the lemniscate r² = a² cos 2θ. (i) about the initial line (ii) about the line θ = π/2.

(*D.C.E., 2005; D.U. 1998*)

Solution. (a) The cardioid is described as θ varies from 0 to 2π. The required

surface $= \int_0^\pi 2\pi y \dfrac{ds}{d\theta}d\theta$. We have $\dfrac{ds}{d\theta} = \sqrt{r^2 + \left(\dfrac{dr}{d\theta}\right)^2} = \sqrt{a^2(1 - \cos\theta)^2 + a^2\sin^2\theta}$

$= \sqrt{4a^2\sin^2\dfrac{\theta}{2}} = 2a\sin\dfrac{\theta}{2}$. Therefore, $S = 2\pi\int_0^\pi r\sin\theta\dfrac{ds}{d\theta}\cdot d\theta$

$= 2\pi\int_0^\pi r\sin\theta\sqrt{r^2 + \left(\dfrac{dr}{d\theta}\right)^2}\, d\theta = 4\pi a^2\int_0^\pi(1 - \cos\theta)\sin\theta\cdot\sin\dfrac{\theta}{2}d\theta$

$= 16\pi a^2\int_0^\pi \sin^4\dfrac{\theta}{2}\cos\dfrac{\theta}{2}d\theta = \dfrac{32}{5}\pi a^2 \text{ square units.}$

(*b*) (*i*) The shape of the curve is shown in the Fig. 8.46. The required area is twice that generated by revolving the first quadrant arc about the initial line.

We have $\qquad r^2 = a^2\cos 2\theta.$

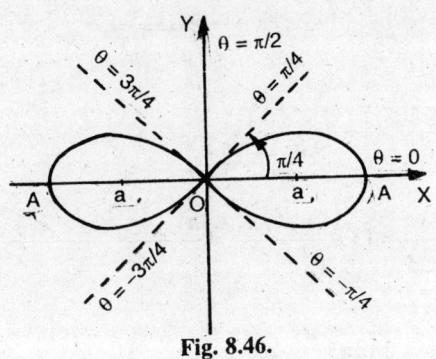

Fig. 8.46.

Differentiating, $\quad 2r \cdot \dfrac{dr}{d\theta} = -2a^2 \sin 2\theta \quad$ or $\quad \dfrac{dr}{d\theta} = -\dfrac{a^2 (\sin 2\theta)}{r}$

$\therefore \quad r^2 + \left(\dfrac{dr}{d\theta}\right)^2 = a^2 \cos 2\theta + \left(-\dfrac{a^2 \sin 2\theta}{r}\right)^2 = a^2 \cos 2\theta + \dfrac{a^4 \sin^2 2\theta}{a^2 \cos 2\theta} = \dfrac{a^4}{r^2}.$

The required area, $\quad S = 2 \cdot 2\pi \displaystyle\int_{\theta=0}^{\pi/4} (r \sin\theta) \cdot \dfrac{a^2}{r} d\theta = 4a^2\pi \int_0^{\pi/4} \sin\theta \, d\theta$

$\quad = 4\pi a^2 \left[-\cos\theta\right]_0^{\pi/4} = 2\pi a^2 \left(2 - \sqrt{2}\right)$ square units.

(*ii*) The required area is twice that generated by revolving the first quadrant arc about the line $\theta = \pi/2$

$\therefore \quad S = 2 \cdot 2\pi \displaystyle\int_0^{\pi/4} r \cos\theta \cdot \dfrac{a^2}{r} d\theta = 4\pi a^2 \int_0^{\pi/4} \cos\theta \, d\theta = 2\sqrt{2}\pi a^2$ square units.

Example 8.34 **Prove that the surface of the solid formed by the revolution of an ellipse of eccentricity e about its major axis is equal to**

$$2\pi ab \left[\sqrt{(1-e^2)} + (1/e)\sin^{-1} e\right]$$

Solution. Let the equation of the ellipse be $\quad \dfrac{x^2}{a^2} + \dfrac{y^2}{b^2} = 1 \qquad \qquad$ ...(1)

where $b^2 = a^2 (1 - e^2)$. From (1), we have, $\dfrac{y^2}{b^2} = 1 - \dfrac{x^2}{a^2} = \dfrac{a^2 - x^2}{a^2}$

or $\quad y = \dfrac{b}{a}\sqrt{a^2 - x^2}$ (Discarding $-$ve sign).

$$\therefore \quad \frac{dy}{dx} = \frac{b}{a} \cdot \frac{1}{2} \left(a^2 - x^2\right)^{-1/2} \cdot (-2x) = -\frac{bx}{a\sqrt{a^2 - x^2}},$$

and
$$\left(\frac{ds}{dx}\right)^2 = 1 + \left(\frac{dy}{dx}\right)^2 = 1 + \frac{b^2 x^2}{a^2 \left(a^2 - x^2\right)}$$

$$= 1 + \frac{a^2 \left(1 - e^2\right) x^2}{a^2 \left(a^2 - x^2\right)} = 1 + \frac{\left(1 - e^2\right) x^2}{a^2 - x^2} = \frac{a^2 - x^2 + x^2 - e^2 x^2}{a^2 - x^2} = \frac{a^2 - e^2 x^2}{a^2 - x^2}.$$

$$\therefore \quad \text{Required surface} = 2 \int_0^a 2\pi y \frac{ds}{dx} \cdot dx$$

$$= 4\pi \int_0^a \frac{b}{a} \sqrt{a^2 - x^2} \frac{\sqrt{a^2 - e^2 x^2}}{\sqrt{a^2 - x^2}} dx = 4\pi \frac{b}{a} \int_0^a \sqrt{a^2 - e^2 x^2} \, dx$$

$$= \frac{4\pi b e}{a} \int_0^a \sqrt{\left(\frac{a}{e}\right)^2 - x^2} \, dx = \frac{4\pi b e}{a} \left[\frac{x}{2} \sqrt{\left(\frac{a}{e}\right)^2 - x^2} + \frac{a^2}{2e^2} \sin^{-1}\left(\frac{ex}{a}\right)\right]_0^a$$

$$= \frac{4\pi b e}{a} \left[\frac{a}{2} \sqrt{\frac{a^2}{e^2} - a^2} + \frac{a^2}{2e^2} \sin^{-1} e\right] = \frac{4\pi b e}{a} \left[\frac{a^2}{2e} \sqrt{1 - e^2} + \frac{a^2}{2e^2} \left(\sin^{-1} e\right)\right]$$

$$= \frac{4\pi b e}{a} \times \frac{a^2}{2e} \left[\sqrt{1 - e^2} + \frac{\sin^{-1}(e)}{e}\right] = 2(\pi a b) \left[\sqrt{1 - e^2} + \frac{\sin^{-1} e}{e}\right] \text{ square units.}$$

Example 8.35 A quadrant of a circle of radius *a*, revolves round its chord. Show that the surface area of the spindle generated is $\sqrt{2}\,\pi a^2 (4 - \pi)/2$.

Solution. The equation of the circle is $x^2 + y^2 = a^2$, so the parametric equations for this circle are $x = a \cos\theta$, $y = a \sin\theta$.

The axis of revolution *i.e.*, the chord *AB* has its equation $x + y = a$. From any point $P(x, y)$ on the circle let *PN* be the perpendicular on *AB*, and its length equal to $PN = (x + y - a)/\sqrt{2}$ ...(1)

An elementary arc of the circle is $ds = a \cdot d\theta$ ...(2)

Employing $S = 2\pi \int_{AB} (PN) ds$ to get $S = 2\pi \int_0^{\pi/2} \frac{x + y - a}{\sqrt{2}} a \cdot d\theta$

As *P* is on the circle $x = a \cos\theta$, $y = a \sin\theta$, This gives

$$S = \sqrt{2}\pi a^2 \int_0^{\pi/2} (\cos\theta + \sin\theta - 1)\,d\theta$$

$$= \sqrt{2}\,\pi a^2 \left[\sin\theta - \cos\theta - \theta\right]_0^{\pi/2}$$

$$= \sqrt{2}\,\pi a^2 \left[1 + 1 - \left(\frac{\pi}{2}\right)\right] = 2\sqrt{2}\,\pi a^2 \left[1 - \left(\frac{\pi}{4}\right)\right]$$

$$= \sqrt{2}\pi a^2 (4-\pi)/2 \text{ square units.}$$

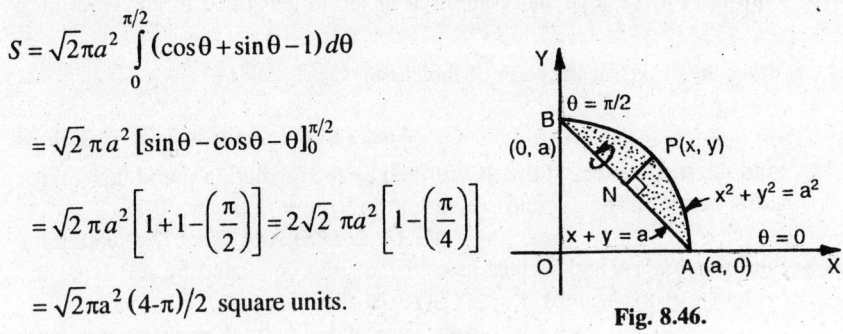

Fig. 8.46.

Exercise 8.2

1. Show that the area of the surface of sphere of radius r is $4\pi r^2$.
   (A.M.I.E., S-2002; Andhra, 1998)

2. Find the surface area of a cylinder of radius r and height h.
   [Hint: The surface is generated by revolving about the x-axis the curve
   $y = r$ from $x = 0$ to $x = h$. Here $dy/dx = 0$]. **Ans.** $2\pi rh$ sq. units

3. Find the surface area of the cup formed by the revolution about its axis of
   the smaller part of the parabola $y^2 = 4ax$ cut off by the line $x = 3a$.
   Ans. $56\pi a^2/3$

4. The arc of the curve $ay^2 = x^3$ between $x = 0$ and $x = a$ is revolved about
   x-axis. Find the surface area of the solid so generated.

   Ans. $\dfrac{\pi a^2 \left(10\sqrt{10} - 1\right)}{27}$ square units.

5. Find the surface area of the solid formed by the revolution of the loop of the
   curve $3ay^2 = x(x - a)^2$ about the line $y = 0$. **Ans.** $\pi a^2/3$.

6. Find the surface area of the solid generated by the revolution about the axis
   of y of the part of the curve $ay^2 = x^3$ from $x = 0$ to $x = 4a$ which is above the

   x-axis. **Ans.** $\dfrac{128\pi a^2 \left(125\sqrt{10} + 1\right)}{1215}$ square units.

7. Find the surface area of a cone generated by the revolution of a line segment

   $y = 2x$ from $x = 0$ to $x = 2$ about x-axis. **Ans.** $8\pi\sqrt{5}$ sq. units.

8. The quadrant of a circle of radius r revolves about the tangent at one extremity. Prove that the area of the curved surface generated is $\pi(\pi - 2)\,r^2$.

9. The part of parabola $y^2 = 4ax$ cut-off by the latus rectum revolves about the
   tangent at the vertex. Find the curved surface of the reel thus generated.

   (D.C.E., 2004) **Ans.** $\pi a^2 \left[3\sqrt{2} - \log\left(\sqrt{2} + 1\right)\right]$

10. Find the surface area and volume of the solid generated by the revolution about the x-axis of the loop of the curve $x = t^2$, $y = t - \left(\dfrac{t^3}{3}\right)$.

    Ans. 3π cubic units and $3\pi/4$ sq. units

11. Find the surface area of the reel formed by the revolution round the tangent at the vertex of the cycloid $x = a(\theta + \sin\theta)$, $y = a(1 - \cos\theta)$.

    (V.T.U., 2000S) **Ans.** $32\pi a^2/3$ square units

12. Find the volume and surface area of the solid generated by revolving the cycloid $x = a(\theta + \sin\theta)$, $y = a(1 + \cos\theta)$ about its base.

    Ans. $5\pi^2 a^3$ cubic units, $64\pi a^2/3$ square units

13. The arc of the curve $x^{2/3} + y^{2/3} = a^{2/3}$ in the first quadrant revolves about x-axis, find the surface area of the solid generated. *(Marathwada, 1998S)*

    Ans. $6\pi a^2/5$ square units.

14. A circular arc revolves about its chord. Prove that the area of the surface generated is $4\pi r^2 (\sin\alpha - \alpha\cos\alpha)$ where r is the radius of the circle and 2α the angle subtended by the arc at the centre.

    When the circular arc is a quadrant of a circle of radius r, show that the surface of the spindle generated is $\pi\sqrt{2}r^2 (4 - \pi)/2$ sq. units.

15. The curve $r = a(1 + \cos\theta)$ revolves about the initial line. Find the surface area of the solid so formed. *(A.M.I.E., S-1998)* **Ans.** $32\pi a^2/5$ square units.

16. The part of the lemniscate $r^2 = 2a^2 \cos 2\theta$, $0 \le \theta \le \pi/4$, is revolved about the x-axis. Find the surface area of the solid generated. *(A.M.I.E. W-2007)*

    Ans. $2\pi a^2 \left(2 - \sqrt{2}\right)$ square units.

    [Hint: Surface area, $S = \displaystyle\int_0^{\pi/4} 2\pi y \, ds = \int_0^{\pi/4} 2\pi (r\sin\theta) \sqrt{r^2 + \left(\dfrac{dr}{d\theta}\right)^2}\, d\theta$

    $= 2\pi \displaystyle\int_0^{\pi/4} r\sin\theta \sqrt{r^2 + \dfrac{4a^4 \sin^2 2\theta}{r^2}}\, d\theta = 4\pi a^2 \int_0^{\pi/4} \sin\theta\, d\theta = 4\pi a^2 \left[-\cos\theta\right]_0^{\pi/4}$ **]**.

17. Find the area of the surface of revolution formed by revolving the curve $r = 4\cos\theta$ about the initial line. **Ans.** 16π square units.

Chapter 9

Centroids, Centers of Gravity, Theorems of Pappus

9.1 Center of Gravity of a Body

A body comprises of several parts and its every part possesses weight. *Weight is the force of attraction between a body and the earth and is proportional to mass of the body.* The weights of all parts of a body can be considered as parallel forces directed towards the center of the earth. Therefore, they may be combined into a resultant force whose magnitude is equal to their algebraic sum. If a supporting force, equal and opposite to the resultant, is applied to the body along the line of action of the resultant, the body will be in equilibrium. This line of action will pass through the center of gravity of the body. Thus, center of gravity of the body may be defined as the *point through which the whole weight of a body may be assumed to act.* The center of gravity of a body or an object is usually denoted by c.g. or simply by *G*. The position of c.g. depends upon shape of the body and this may or may not necessarily be within the boundary of the body.

9.2 Determination of Center of Gravity of a Flat Plate by the Method of Moments

Consider the flat plate of irregular section shown in Fig. 9.1. The network shown in the figure divides the plate into small elements having weights w_1, w_2, w_3 etc., which act at the center of each element. These gravity forces form a parallel force system, the resultant of which is the total weight W of the plate. Let coordinates of the each elemental weight be (x_1, y_1), (x_2, y_2), (x_3, y_3) etc., and the coordinates of the resultant weight W be $(\bar{x}, \bar{y})$.

Taking moments of the weights about the Y-axis, we get

$$W\bar{x} = w_1 x_1 + w_2 x_2 + w_3 x_3 + ... = \Sigma w_i x_i \qquad ...(9.1)$$

Now taking moments of the weights about X-axis, we have

$$W\bar{y} = w_1 y_1 + w_2 y_2 + w_3 y_3 + ... = \Sigma w_i y_i \qquad ...(9.2)$$

The above equations state that the moment of a weight W about an axis is equal to the moment sum of its elemental weights.

From the above equations we can determine $\bar{x}$ and $\bar{y}$.

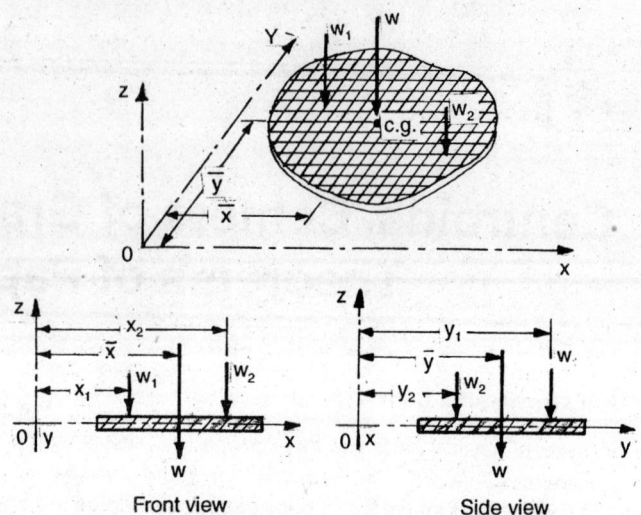

Fig. 9.1. Coordinates of the center of gravity.

9.2.1 Centroids of Areas and Lines

If the material of the plate in Fig. 9.1 is homogeneous, the weight W may be expressed as the product of density ρ multiplied by $t A$, where t is the thickness of the plate and A is its area. Similarly, the weight w of an element is given by $\rho t a$, where a is the cross-sectional area of the element. Substituting these values in above equations, we obtain

$$\rho t A \bar{x} = \rho t a_1 x_1 + \rho t a_2 x_2 + \rho t a_3 x_3 + \dots = \rho t \sum ax.$$

Cancelling the constant terms ρ and t, we get

and similarly
$$\left. \begin{array}{l} A\bar{x} = \sum a\,x \\ A\bar{y} = \sum a\,y \end{array} \right\} \qquad \dots(9.3)$$

The expression $A\bar{x}$ as well as $A\bar{y}$ is called the moment of area. It is equivalent to the sum of the moments of the elemental areas composing the total area. Moment of area is defined as the product of the area multiplied by the perpendicular distance from the center of area to the axis of moments.

The equation (9.3) is rewritten as $\bar{x} = \sum ax/A$, $\bar{y} = \sum ay/A$.

This gives coodinates of centroid of area.

The *centroid* of area is defined as the point corresponding to the center of gravity of a plate of infinitesimal thickness. The term "centroid" rather than "center of gravity" is used when referring to areas (as well as to lines and volumes) because such figures do not have weight. The term "center of gravity" should refer to the center of weight of actual physical bodies.

When referring to lines, we may determine the centroid by similar means. A line may be assumed to be the axis of a homogeneous slender wire.

Thus

$$\left. \begin{array}{l} L\,\overline{x} = \sum l\,x \\ \text{and} \quad L\,\overline{y} = \sum l\,y \end{array} \right\} \qquad \text{...(9.4)}$$

9.3 Centroids determined by Integration

Integration is the process of summing up infinitesimal quantities. If the area of an element had been expressed as the differential *dA* (*i.e.*, a small part of the total area *A*), the equations for determining the centroid of an area would have become

$$\left. \begin{array}{l} A\,\overline{x} = \int x\,d\,A \\ A\,\overline{y} = \int y\,d\,A \end{array} \right\} \qquad \text{...(9.5)}$$

The integral $\int xdA$ is known as the first moment of area with respect to the *Y*-axis. Similarly, the integral $\int ydA$ denotes the first moment of area with respect to the *X*-axis.

For determining the centroid of a line, we could have used

$$\left. \begin{array}{l} L\,\overline{x} = \int x\,d\,L \\ L\,\overline{y} = \int y\,d\,L \end{array} \right\} \qquad \text{...(9.6)}$$

When we determine the centroid by integration, the figure is divided into differential elements so that:

1. All points of the element are located the same distance from the axis of moments, or
2. The position of the centroid of the element is known so that the moment of the element about the axis of moments is the product of the element and the distance of its centroid from the axis.

If a plane figure has a line of symmetry, its centroid is located on that line. If the plane figure has two lines of symmetry, the centroid is located at the point of intersection of the lines.

9.4 Centroid of an Arc of a Curve

Let ρ be the mass per unit length of the curve. Take an elementary arc *ds* of the curve at the point $P(x, y)$ whose equation is $y = f(x)$. The mass of this element is $dm = \rho.ds$. If $(\overline{x}, \overline{y})$ denote the centroid of an element *PQ* then

$$\overline{x} = \frac{\int x \cdot \rho ds}{\int \rho ds}, \ \overline{y} = \frac{\int y \cdot \rho ds}{\rho ds}. \ \text{If } \rho \text{ is constant, we have}$$

$$\bar{x} = \frac{\int x\,ds}{\int ds}, \quad \bar{y} = \frac{\int y\,ds}{\int ds}. \quad \ldots(9.7)$$

To evaluate the integrals, ds must be

replaced by $\dfrac{ds}{dx}\,dx, \dfrac{ds}{dt}\,dt$ or $\dfrac{ds}{d\theta}\,d\theta$ de-

pending upon whether the equation of the curve is given in cartesian or in parametrical coordinates or in polar co-ordinates.

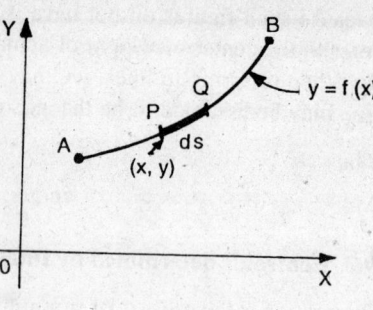

Fig. 9.2.

Derivative of Arc

(a) For the curve $y = f(x)$, we have $ds/dx = \sqrt{1 + (dy/dx)^2}$

(b) If the equation of the curve is $x = f(y)$, then $ds/dy = \sqrt{1 + (dx/dy)^2}$

(c) If the equation of the curve is in parametric form $x = f(t)$, $y = \phi(t)$, then

$$ds/dt = \sqrt{(dx/dt)^2 + (dy/dt)^2}$$

(d) For the curve $r = f(\theta)$, we have $ds/d\theta = \sqrt{r^2 + (dr/d\theta)^2}$.

In polar co-ordinates, replace x by $r\cos\theta$ and y by $r\sin\theta$.

Example 9.1 **Find the centroid of the first quadrant arc of the circle $x^2 + y^2 = r^2$.**

Solution. Differentiating, $x^2 + y^2 = r^2$, we have

$$dy/dx = -x/y, \quad \text{and} \quad 1 + \left(\frac{dy}{dx}\right)^2 = 1 + \left(\frac{x^2}{y^2}\right) = \frac{r^2}{y^2}.$$

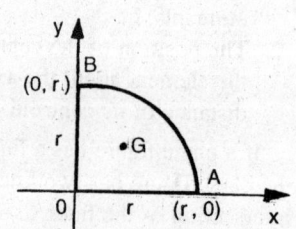

Since $s = \dfrac{2\pi r}{4} = \dfrac{\pi r}{2}$.

Fig. 9.3.

$$\therefore \quad \bar{y} = \frac{\int_0^r y \dfrac{ds}{dx} \cdot dx}{s} = \frac{\int_0^r y\sqrt{1 + \left(\dfrac{dy}{dx}\right)^2}\,dx}{s} = \frac{\int_0^r r\,dx}{\pi r/2} = \frac{2r^2}{\pi r} = \frac{2r}{\pi}.$$

By symmetry $\bar{x} = \bar{y}$ and the coordinates of the centroid are $\left(\dfrac{2r}{\pi}, \dfrac{2r}{\pi}\right)$.

Example 9.2 Find the centroid of the arc of a circle of radius r and central angle 2α..

Solution. Let the axis of symmetry be chosen as the X-axis. Then $\bar{y} = 0$.

The element of arc $dL = rd\theta$ and $x = r\cos\theta$.

Applying equation (9.6), we have

$$(2\alpha r)\bar{x} = \int_{-\alpha}^{+\alpha}(r\cos\theta)rd\theta = r^2\int_{-\alpha}^{+\alpha}\cos\theta\, d\theta = 2r^2\sin\alpha.$$

Therefore $\bar{x} = 2r^2\sin\alpha/2\alpha r = (r\sin\alpha)/\alpha.$

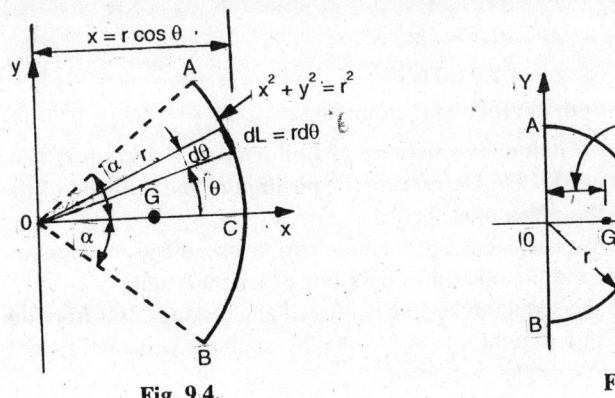

Fig. 9.4.   **Fig. 9.5.**

Thus, the centroid is on the bisecting radius at a distance $(r\sin\alpha)/\alpha$ from the center of the circle.

Remark. If the arc is a semi-circle, as in Fig. 9.5, $\alpha = 90° = \pi/2$ radians and $\sin\alpha = 1$. Substituting these values in the above result gives $\bar{x} = \dfrac{2r}{\pi}$.

9.5 Centroid of a Rod of Length $2l$

Choose the X-axis along the length of the rod with the origin at one end. Obviously the centroid lies on the X-axis and $\bar{y} = 0$.

Now consider an element of length dx at a distance x from the origin. Then $dm = \rho dx$.

$$\therefore \bar{x} = \int_{0}^{2l} x\rho\, dx \bigg/ \int_{0}^{2l} \rho\, dx = \left[x^2/2\right]_{0}^{2l}\bigg/\left[x\right]_{0}^{2l} = l.$$

Therefore, the centroid of a rod of uniform density lies at its mid-point.

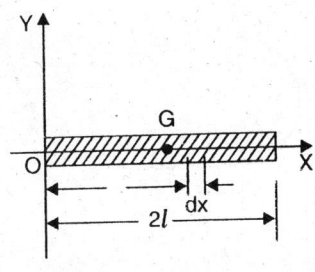

Fig. 9.6.

9.6 Centroids of Composite Figures

Many of the figures used in engineering are composed of combinations of the geometrical shapes. The centroid of the composite figure is determined by applying the following equations. In these equations the elemental areas become the areas of the geometrical shapes into which the entire area has been divided.

$$\left.\begin{array}{c} A\bar{x} = \Sigma ax \\ A\bar{y} = \Sigma ay \end{array}\right\}$$

A similar process may be applied to lines.

The given line may be divided into finite segments whose cenroids are known, and the following equations may be used:

$$\left.\begin{array}{c} L\bar{x} = \Sigma lx \\ L\bar{y} = \Sigma ly \end{array}\right\} \qquad \qquad ...(9.7)$$

This will be clear from the following example.

Example 9.3 A slender homogeneous wire of uniform cross section is bent into the form shown in Fig. 9.7. Determine the position of the centroid of the wire with respect to the given axes.

Solution. The wire is considered equivalent to two line segments, one a semicircle of radius 4 units and the other a straight line of length 8 units.

The centroid G_1 of the semicircular line is located at a distance $2r/\pi$ from the Y-axis and centroid G_2 of straight line at (4, −4). By applying equation(9.4)and paying careful attention to signs, we obtain

(i) $(4\pi+8)\bar{x} = 4\pi\left(-2\times\dfrac{4}{\pi}\right)+8(4)$

or $20.57\,\bar{x} = -32 + 32 = 0$. Therefore, $\bar{x} = 0$.

(ii) $(4\pi + 8)\,\bar{y} = 4\pi\,(0) + 8\,(-4)$ or $20.57\,\bar{y} = -32$ or $\bar{y} = -1.556$.

Thus the centroid of the wire lies on the Y-axis 1.556 unit below the X-axis. (Refer Fig. 9.8).

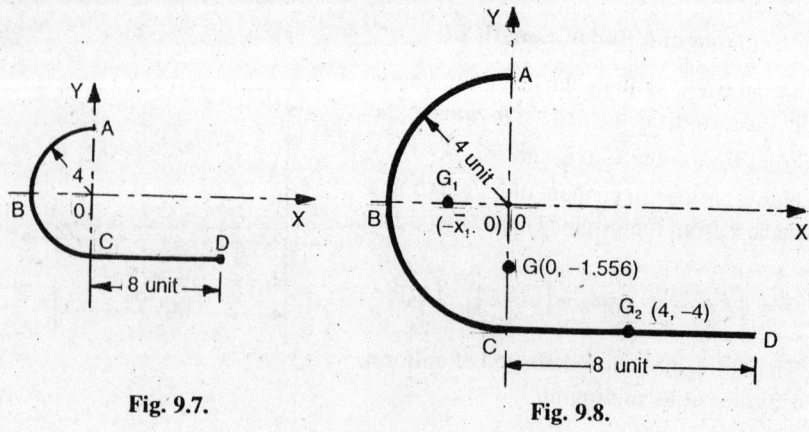

Fig. 9.7.

Fig. 9.8.

9.7 Centroid of a plane area—Cartesian Coordinates

Find the centroid of a plane area bounded by the curve $y = f(x)$, the x-axis and two ordinates at $x = a$ and $x = b$.

Use the approximating rectangle of the figure. Its area is $y.\Delta x$, its centroid is $(x, y/2)$, and its moment with respect to the X-axis is $(y/2) \cdot (y \cdot \Delta x)$.

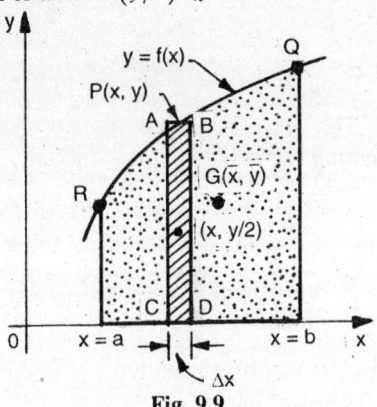

Then $M_x = \int\limits_{x=a}^{b} \frac{y^2}{2} dx = \frac{1}{2} \int\limits_{x=a}^{b} y^2 dx.$

Similarly, the moment of the approximating rectangle with respect to the Y-axis is $x (y.\Delta x)$ then

$$M_y = \int\limits_{x=a}^{b} xy \, dx.$$

Let $(\bar{x}, \bar{y})$ be the coordinates of the centroid, then

Fig. 9.9.

$$\bar{x} = M_y/A = \int\limits_{x=a}^{b} xy\,dx \Big/ \int\limits_{x=a}^{b} y\,dx, \text{ and } \bar{y} = M_x/A = \frac{1}{2} \int\limits_{x=a}^{b} y^2 dx \Big/ \int\limits_{x=a}^{b} y\,dx. \qquad ...(9.8)$$

Example 9.4 Locate the centroid of the triangular area with respect to the base. The triangle has a base b and an altitude h.
Solution. Refer Fig. 9.10. Take strip parallel to the base as the differential elements of area. Its area is $x \cdot \Delta y$, its moment with respect to the X-axis is $y(x \cdot \Delta y)$.

$$M_x = \int\limits_{0}^{h} xy\,dy. \qquad ...(i)$$

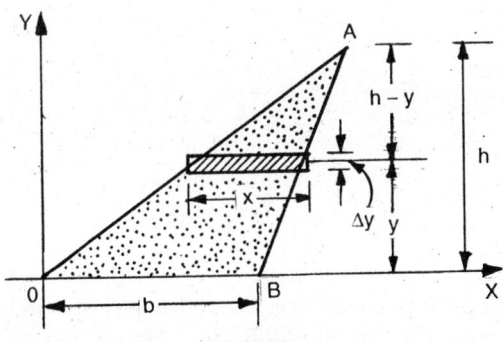

Fig. 9.10.

From, similar triangles, $\dfrac{x}{b} = \dfrac{h-y}{h}$ or $x = \dfrac{b}{h}(h-y)$ so that (*i*) becomes

$$M_x = \frac{b}{h}\int_0^h (h-y)\,y\,dy = \frac{b}{h}\left[\frac{hy^2}{2} - \frac{y^3}{3}\right]_0^h = \frac{bh^2}{6}.$$

Then $\bar{y} = M_x / A = \dfrac{bh^2/6}{bh/2} = h/3$. In any triangle the distance from the centroid to any side is equal to one-third of the altitude measured from that side. Furthermore, the centroid of a triangle is located on a median because the median to any side contains the centroids of all strips drawn parallel to that side. Therefore, the centroid is at the intersection of the medians.

9.7.1 Centroid of plane area—polar coordinates

The plane area bounded by the curve $r = f(\theta)$ and the radius vectors $\theta = \theta_1$ and

$\theta = \theta_2$ is given by $A = \dfrac{1}{2}\displaystyle\int_{\theta=\theta_1}^{\theta_2} r^2\,d\theta$. Proper

care must be taken to determine the proper limits of integration. The coordinates $(\bar{x}, \bar{y})$ of the centroid of the plane area bounded by the curve $r = f(\theta)$ and the radius vectors $\theta = \theta_1$ and $\theta = \theta_2$ are given by

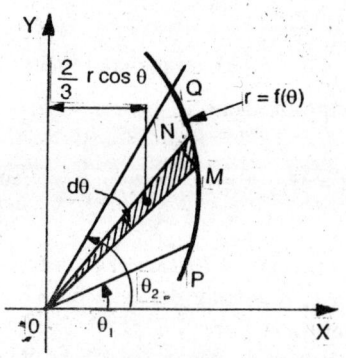

Fig. 9.11.

$$\bar{x} = \int_P^Q x\,dA \Big/ \int_P^Q dA = \int_{\theta_1}^{\theta_2} \frac{2}{3} r\cos\theta \left(\frac{1}{2} r^2 d\theta\right) \Big/ \int_{\theta_1}^{\theta_2} \frac{1}{2} r^2 d\theta = \int_{\theta_1}^{\theta_2} \frac{1}{3} r^3 \cos\theta\, d\theta \Big/ \int_{\theta_1}^{\theta_2} \frac{1}{2} r^2 d\theta$$

$$= \frac{2}{3}\int_{\theta=\theta_1}^{\theta_2} r^3 \cos\theta\, d\theta \Big/ \int_{\theta_1}^{\theta_2} r^2 d\theta \quad\text{and}\quad \bar{y} = \frac{2}{3}\int_{\theta=\theta_1}^{\theta_2} r^3 \sin\theta\, d\theta \Big/ \int_{\theta_1}^{\theta_2} r^2 d\theta \qquad ...(9.9)$$

[**Hint.** Take dm in the form of an elementary sector whose mass is

$\rho \cdot \dfrac{1}{2} r^2 d\theta$, with its centroid at $\left(\dfrac{2}{3} r\cos\theta, \dfrac{2}{3} r\sin\theta\right)$.]

Remarks. (1) For symmetrical bodies, the centre of gravity will always lie on the axis of symmetry. For examples, the centre of gravity of a circle is at its centre. The centre of gravity of a cylinder will be on its axis and at its mid-point. The centre of gravity of a cone is on the axis.

(2) The centre of gravity will always lie on the axis of revolution in case of volumes and surfaces of solids of revolution.

(3) If the equation of the curve does not change by interchanging x and y then $\bar{x} = \bar{y}$.

Example 9.5 Determine the location of the centroid of the area of the sector of the circle of radius r and subtended angle be 2α.

Solution. Choose the X-axis as the axis of symmetry, then $\bar{y} = 0$. Select as the element of area the shaded triangle the position of whose centriod is known to be $x = 2r\cos\theta/3$. The area of the ele-

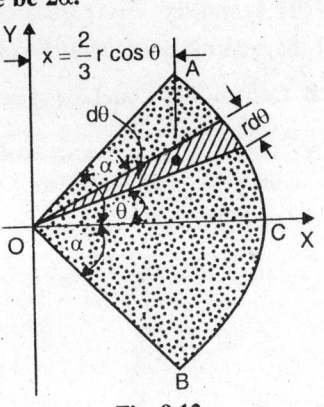

ment is $dA = \left(\dfrac{1}{2}r\right)r\,d\theta = \dfrac{1}{2}r^2 d\theta$, its moment with respect to the Y-axis is

$\dfrac{2}{3}r\cos\theta\left(\dfrac{1}{2}r^2 d\theta\right)$.

Fig. 9.12.

$$\therefore M_y = \int\limits_{\theta=-\alpha}^{+\alpha} \frac{2}{3}r\cos\theta\left(\frac{1}{2}r^2 d\theta\right) = \frac{1}{3}r^3 \int\limits_{\theta=-\alpha}^{+\alpha} \cos\theta\,d\theta = \frac{2}{3}r^3 \sin\alpha$$

$$A = \int\limits_{\theta=-\alpha}^{+\alpha} \frac{1}{2}r^2 d\theta = r^2\alpha.$$

Then $\bar{x} = M_y / A = \dfrac{(2r^3 \sin\alpha)/3}{r^2\alpha} = \dfrac{2}{3}r\,\dfrac{\sin\alpha}{\alpha}$.

Note. If the sector is a semi-circular area as in Fig. 9.13. $\alpha = 90° = \pi/2$ radians. We find the distance of the centroid from the diameter to be $\bar{x} = \dfrac{4r}{3\pi} = 0.424\,r$.

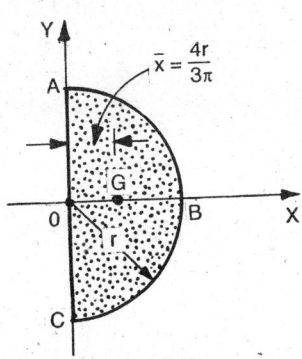

Fig. 9.13.

Example 9.6 Find the centroid of the area of the first quadrant loop of the rose $r = \sin 2\theta$.

Solution. Refer Fig. 9.14 $\bar{x} = \dfrac{2}{3}\int\limits_{0}^{\pi/2} (\sin^3 2\theta)\cos\theta\,d\theta \div \int\limits_{0}^{\pi/2} (\sin^2 2\theta)\,d\theta$

$$= \frac{2}{3} \cdot \frac{\displaystyle\int_0^{\pi/2} 8\sin^3\theta\cos^4\theta\, d\theta}{\dfrac{1}{2}\left[\theta - \dfrac{\sin 4\theta}{4}\right]_0^{\pi/2}} = \frac{32}{3} \cdot \frac{(2\cdot3\cdot1)/(7\cdot5\cdot3\cdot1)}{\pi/2} = \frac{128}{105\pi}$$

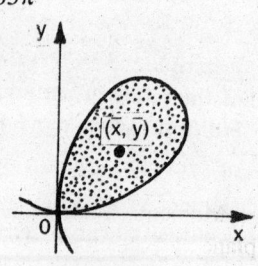

By symmetry $\bar{y} = 128/105\pi$. The co-ordinates of the centroid are $(128/105\,\pi,\ 128/105\,\pi)$.

9.8 Centroid of a surface of revolution

The coordinate $\bar{x}$ of the centroid of the surface, S generated by revolving an arc AB of a curve about the X-axis is given by

Fig. 9.14.

$$\therefore \qquad \bar{x} = \frac{2\pi\displaystyle\int_{AB} xy\, ds}{S} = \frac{2\pi\displaystyle\int_{AB} xy\sqrt{1+\left(\dfrac{dy}{dx}\right)^2}\, dx}{S},\ \bar{y} = 0 \qquad \dots(9.10)$$

Example 9.7 Find the centroid of the surface of revolution obtained by revolving a quadrant of a circle $x^2 + y^2 = r^2$ about x-axis.

Solution. We have

$$dy/dx = -x/y,\ \sqrt{1+(dy/dx)^2} = r/y \text{ and } S = \frac{1}{2}(4\pi r^2) = 2\pi r^2.$$

$$\therefore \quad \bar{x} = \left[2\pi\int_0^r xy\sqrt{1+(dy/dx)^2}\, dx\right] \div S = 2\pi\int_0^r xy\cdot\frac{r}{y}\, dx / 2\pi r^2 = \frac{1}{r}\left[\frac{x^2}{2}\right]_0^r = \frac{r}{2}.$$

The centroid of the surface generated is $(r/2, 0)$.

9.9 Centroid of a solid of revolution

The centroid of the solid, of volume V, generated by the revolution of the area bounded by the curve $y = f(x)$, the X-axis and the ordinates $x = a,\ x = b$, about the X-axis is situated at $(\bar{x}, 0)$, so that

$$\bar{x} = \pi\int_{x=a}^{b} xy^2 dx / V \qquad \dots(9.11)$$

Example 9.8 Find the centroid of the solid formed by the revolution of the area bounded by the parabola $y^2 = 4ax$, the axis of X and the latus-rectum about the latus-rectum.

Solution. Refer Fig. 9.15 the line $x = a$ is the geometrical axis of the solid. The centroid $(a, \bar{y})$ will then lie on this axis. Use the approximating rectangle and the shell method.

$$V = 2\pi \int_0^a (a-x) y \, dx = 2\pi \int_0^a (a-x) 2\sqrt{ax} \, dx = 4\pi\sqrt{a} \int_0^a \left(a\sqrt{x} - x^{3/2} \right) dx$$

$$= 4\pi\sqrt{a} \left[\frac{2}{3} ax^{3/2} - \frac{2}{5} x^{5/2} \right]_0^a = 4\pi\sqrt{a} \left(\frac{2}{3} a^{5/2} - \frac{2}{5} a^{5/2} \right) = \frac{16\pi}{15} a^3 .$$

Moment of the solid with respect to the plane through the point $(a, 0)$ and perpendicular to the Y-axis

$$M_x = 2\pi \int_0^a \frac{1}{2} y \cdot (a-x) y \, dx = 4\pi a \int_0^a (a-x) x \, dx$$

$$= 4\pi a \left[\frac{ax^2}{2} - \frac{x^3}{3} \right]_0^a = \frac{2\pi a^4}{3} .$$

$$\therefore \ \bar{y} = \frac{M_x}{V} = \frac{2\pi a^4}{3} \div \frac{16\pi a^3}{15} = \frac{5a}{8} .$$

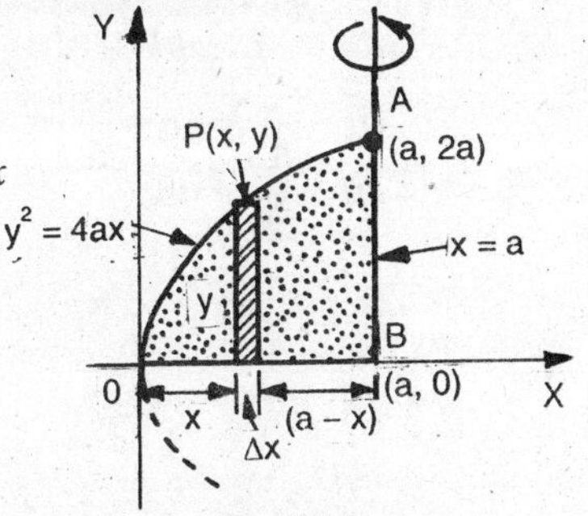

Fig. 9.15.

Hence, the coordinates of centroid of the solid of revolution are $(a, 5a/8)$.

9.10 Illustrative Examples

Example 9.9 Find the moments with respect to the coordinate axes of the plane area bounded by the parabola $y^2 = 4ax$ and the lines $y = 0$, $x = b$.

Solution. Use the approximating rectangle of the figure. Its area is $y \cdot \Delta x$, its centroid is $(x, y/2)$, and its moment with respect to the X-axis is $y(y \cdot \Delta x)/2$.

Then

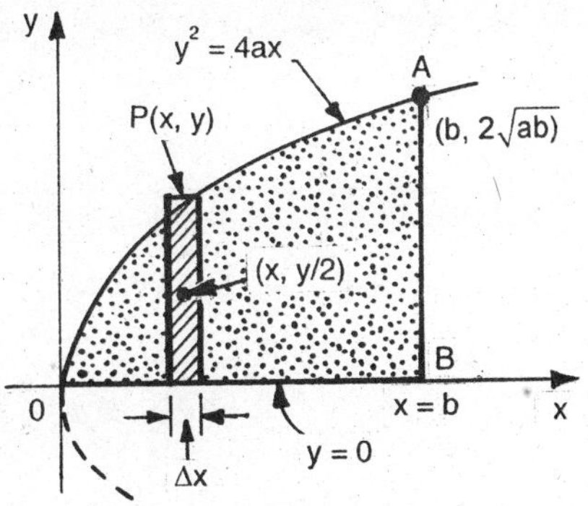

Fig. 9.16.

$$M_x = \frac{1}{2} \int_0^b y^2 \, dx = 2a \int_0^b x \, dx = 2a \cdot \left[\frac{x^2}{2} \right]_0^b = ab^2 .$$

Similarly, the moment of the approximating rectangle with respect to the Y-axis is $x(y \, \Delta x)$. Then

$$M_y = \int_0^b xy\,dx = 2\sqrt{a}\int_0^b x\sqrt{x}\,dx = 2\sqrt{a}\int_0^b x^{3/2}\,dx = \frac{4}{5}\sqrt{a}\,\left[x^{5/2}\right]_0^b$$

$$= \frac{4}{5}\sqrt{a}\,b^{5/2} = \frac{4}{5}b^2\sqrt{ab}.$$

Example 9.10 Find the centroid of an arc of the cycloid $x = a(\theta + \sin \theta)$, $y = a(1 - \cos \theta)$, **which lies in the positive quadrant.**

Solution. Ref. Fig. 9.17

$$\left.\begin{array}{l} \bar{x} = \displaystyle\int_0^\pi x\,\frac{ds}{d\theta}\cdot d\theta \bigg/ \int_0^\pi \frac{ds}{d\theta}\cdot d\theta, \\[4mm] \bar{y} = \displaystyle\int_0^\pi y\,\frac{ds}{d\theta}\cdot d\theta \bigg/ \int_0^\pi \frac{ds}{d\theta}\cdot d\theta. \end{array}\right\} \quad ...(i)$$

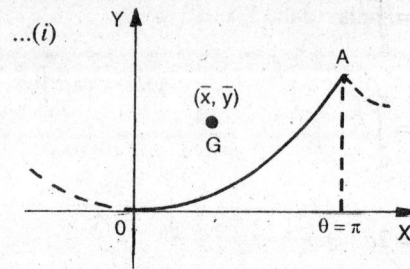

Now $\dfrac{ds}{d\theta} = \sqrt{\left(\dfrac{dx}{d\theta}\right)^2 + \left(\dfrac{dy}{d\theta}\right)^2}$

$$= \sqrt{a^2(1+\cos\theta)^2 + a^2\sin^2\theta} = 2a\cos\left(\frac{\theta}{2}\right)\,...(ii)$$

Fig. 9.17.

Substituting from the equation of the curve for x and y, and from (ii) for $\dfrac{ds}{d\theta}$

is (i), we have

$$\therefore\;\; \bar{x} = \frac{\displaystyle\int_0^\pi a(\theta+\sin\theta)\cdot 2a\cos\frac{\theta}{2}\,d\theta}{\displaystyle\int_0^\pi 2a\cos\frac{\theta}{2}\,d\theta} = \frac{2a^2\displaystyle\int_0^\pi\left(\theta\cos\frac{\theta}{2} + 2\sin\frac{\theta}{2}\cos^2\frac{\theta}{2}\right)d\theta}{2a\left[2\sin\frac{\theta}{2}\right]_0^\pi}$$

$$= 2a^2\left(2\pi - \frac{8}{3}\right)\bigg/4a = a\left(\pi - \frac{4}{3}\right).$$

Similarly, $\;\;\bar{y} = \dfrac{\displaystyle\int_0^\pi a(1-\cos\theta)\cdot 2a\cos\frac{\theta}{2}\cdot d\theta}{\displaystyle\int_0^\pi 2a\cos\frac{\theta}{2}\,d\theta} = \dfrac{4a^2\displaystyle\int_0^\pi \sin^2\frac{\theta}{2}\cos\frac{\theta}{2}\,d\theta}{4a}$

$$= \frac{8a^2}{3} \left[\sin^3 \frac{\theta}{2} \right]_0^\pi \Big/ 4a = \frac{8a^2}{3(4a)} = \frac{2a}{3}.$$

Thus the co-ordinates of the centroid are $\left(\dfrac{3a\pi - 4a}{3}, \dfrac{2a}{3} \right)$.

Example 9.11 Find the centroid of the arc of the cardioid $r = a(1 - \cos \theta)$.

Solution. Refer Fig. 9.18. By symmetry, $\bar{y} = 0$ and the abscissa of centroid of the entire arc is the same as that for the upper half.

Now, half the length of the cardioid

$$= \int_0^\pi \sqrt{r^2 + (dr/d\theta)^2} \, d\theta$$

$$= \int_0^\pi \sqrt{a^2 (1 - \cos \theta)^2 + a^2 \sin^2 \theta} \, d\theta$$

$$= 2a \int_0^\pi \sin(\theta/2) \, d\theta = 4a \left[-\cos \frac{\theta}{2} \right]_0^\pi = 4a.$$

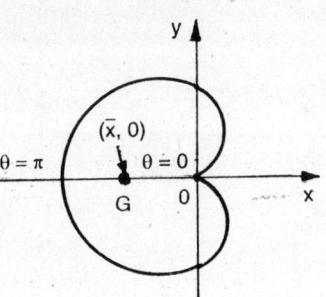

Fig. 9.18.

$$\bar{x} = \frac{\displaystyle\int_0^\pi (r \cos \theta) \sqrt{r^2 + (dr/d\theta)^2} \, d\theta}{4a} = \frac{2a^2 \displaystyle\int_0^\pi (1 - \cos \theta) \cos \theta \cdot \sin \frac{\theta}{2} \, d\theta}{4a}$$

$$= a \int_0^\pi \left(-2\cos^4 \frac{\theta}{2} + 3\cos^2 \frac{\theta}{2} - 1 \right) \sin \frac{\theta}{2} \, d\theta$$

$$= a \left[\frac{4}{5} \cos^5 \frac{\theta}{2} - 2\cos^3 \frac{\theta}{2} + 2\cos \frac{\theta}{2} \right]_0^\pi = -\frac{4a}{5}.$$

Thus, the co-ordinates of the centroid are $(-4a/5, 0)$.

Example 9.12 Find the centroid of the first quadrant area bounded by the parabola $y = x^2$ and line $y = x$.

Solution. Refer Fig. 9.19. The centroid of the approximating rectangle is

$$\left[x, \; y_1 + \frac{1}{2}(y_2 - y_1) \right] \quad i.e., \quad \left[x, \; x^2 + \frac{1}{2}(x - x^2) \right] \quad i.e., \quad \left[x, \; \frac{1}{2}(x + x^2) \right].$$

Area of shaded portion OABO $= \displaystyle\int_0^1 (y_2 - y_1) \, dx = \int_0^1 (x - x^2) \, dx = \left[\frac{x^2}{2} - \frac{x^3}{3} \right]_0^1 = \frac{1}{6}$

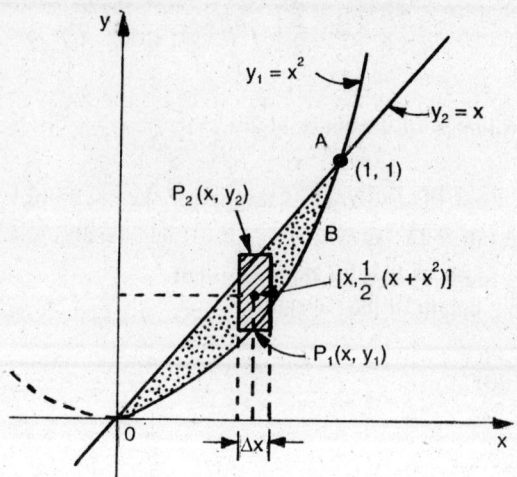

Fig. 9.19.

$$M_x = \int_0^1 \frac{1}{2}(x+x^2)(x-x^2)\,dx = \frac{1}{2}\int_0^1 (x^2-x^4)\,dx = \frac{1}{2}\left[\frac{x^3}{3}-\frac{x^5}{5}\right]_0^1 = \frac{1}{15}.$$

$$M_y = \int_0^1 x(x-x^2)\,dx = \int_0^1 (x^2-x^3)\,dx = \left[\frac{x^3}{3}-\frac{x^4}{4}\right]_0^1 = \frac{1}{12}.$$

Then $\bar{x} = \dfrac{M_y}{A} = \dfrac{1/12}{1/6} = \dfrac{1}{2}$, $\bar{y} = \dfrac{M_x}{A} = \dfrac{1/15}{1/6} = \dfrac{2}{5}$, and the co-ordinates of the

centroid are (1/2, 2/5).

Example 9.13 Find the centroid of the area bounded by the parabolas $y^2 = x$ and $x^2 = -8y$.

Solution. Refer Figure 9.20. The centroid of the approximating rectangle is

$$\left[x, y_2 + \frac{1}{2}(y_1 - y_2)\right] \text{ i.e., } \left[x, -\frac{x^2}{8} + \frac{1}{2}\left(-\sqrt{x} + \frac{x^2}{8}\right)\right]$$

$$i.e., \left[x, \frac{1}{2}\left(-\frac{x^2}{8} - \sqrt{x}\right)\right].$$

Area of shaded portion, *OABO*

$$= \int_0^4 \left(-\frac{x^2}{8} + \sqrt{x}\right) dx = \left[-\frac{x^3}{24} + \frac{2}{3}x^{3/2}\right]_0^4 = -\frac{8}{3} + \frac{16}{3} = \frac{8}{3}$$

$$M_x = \int_0^4 \frac{1}{2}\left(-\frac{x^2}{8} - \sqrt{x}\right)\left(-\frac{x^2}{8} + \sqrt{x}\right)dx$$

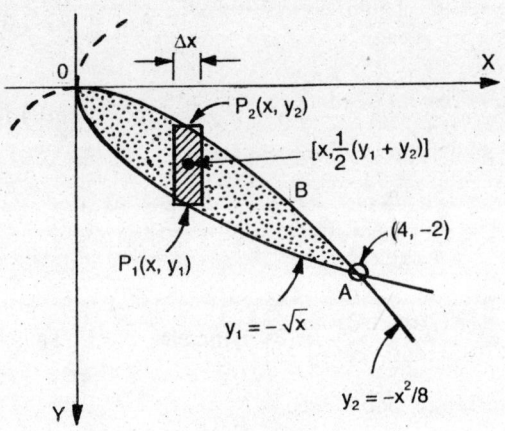

Fig. 9.20.

$$= \frac{1}{2}\int_0^4 \left(\frac{x^4}{64} - x\right)dx = \left[\frac{x^5}{640} - \frac{x^2}{4}\right]_0^4 = \frac{8}{5} - 4 = -\frac{12}{5}.$$

$$M_y = \int_0^4 x\left(-\frac{x^2}{8} + \sqrt{x}\right)dx = \int_0^4 \left(-\frac{x^3}{8} + x^{3/2}\right)dx = \left[-\frac{x^4}{32} + \frac{2}{5}x^{5/2}\right]_0^4 = \frac{24}{5}.$$

The co-ordinates of the centroid are $(M_y/A, M_x/A) = (9/5, -9/10)$.

Example 9.14 Find the centroid of the first quadrant area of the hypocycloid $x = a \cos^3 \theta$, $y = a \sin^3 \theta$.

Solution. Refer Fig. 9.21. The centroid of the approximating rectangle is $(x, y/2)$.

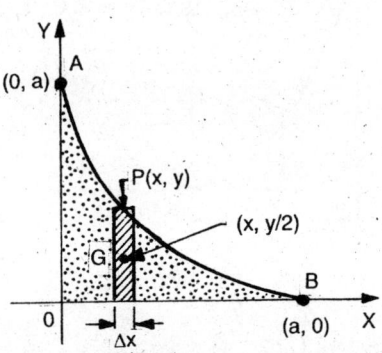

Fig. 9.21.

Area of the shaded portion, ABOA $= \int_0^a y\,dx = \int_{\pi/2}^0 a\sin^3\theta\left(-3a\cos^2\theta\sin\theta\right)d\theta$

$$= 3a^2 \int_0^{\pi/2} \sin^4\theta\cos^2\theta\,d\theta = 3a^2\cdot\frac{3\cdot1}{6\cdot4\cdot2}\cdot\frac{\pi}{2} = \frac{3a^2\pi}{32}.$$

$$M_x = \int_0^a \frac{y}{2}\cdot y\,dx = \frac{1}{2}\int_{\pi/2}^0 a^2\sin^6\theta\left(-3a\cos^2\theta\sin\theta\right)d\theta$$

$$= \frac{3a^3}{2}\int_0^{\pi/2}\sin^7\theta\cos^2\theta\,d\theta = \frac{3a^3}{2}\cdot\frac{6\cdot4\cdot2}{9\cdot7\cdot5\cdot3\cdot1} = \frac{8a^3}{105}.$$

Then $\bar{y} = M_x/A = \dfrac{8a^3/105}{3a^2\pi/32} = \dfrac{256a}{315\pi}$. By symmetry, $\bar{x} = \bar{y}$. Thus the centroid has

coordinates $(256a/315\pi,\ 256a/315\pi)$.

Example 9.15 Find the centroid of the surface generated by revolving the arc in the positive quadrant of the hypocycloid $x^{2/3} + y^{2/3} = a^{2/3}$ about the line x = a.

Solution. Refer Fig. 9.22. Consider the revolution of a portion of the arc $PQ = ds$, when revolved about the line $x =\ a$. The elementary surface generated is $2\pi(a - x)ds$. Its moment about the plane through the X-axis is $2\pi y(a - x)ds$.

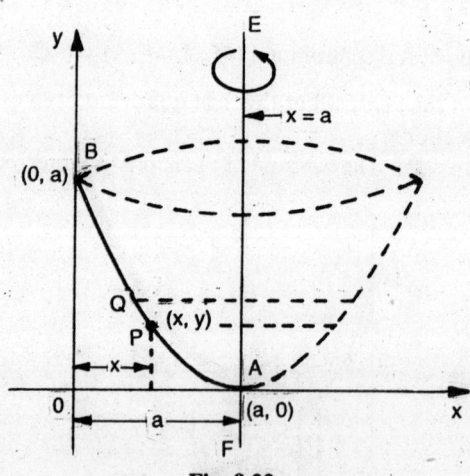

Fig. 9.22.

$$\bar{y} = 2\pi\int_A^B y(a-x)\,ds \div 2\pi\int_A^B (a-x)\,ds\ . \qquad \ldots(i)$$

Taking the equation of the hypocycloid as $x = a\cos^3\theta$, $y = a\sin^3\theta$.

$$\therefore \quad \frac{ds}{d\theta} = \sqrt{\left(\frac{dx}{d\theta}\right)^2 + \left(\frac{dy}{d\theta}\right)^2} = \sqrt{\left(-3a\cos^2\theta\sin\theta\right)^2 + \left(3a\sin^2\theta\cos\theta\right)^2}$$

$$= \sqrt{9a^2\cos^4\theta\sin^2\theta + 9a^2\sin^4\theta\cos^2\theta} = 3a\sin\theta\cos\theta.$$

Substituting the values in (i), we obtain

$$\bar{y} = \frac{\displaystyle\int_0^{\pi/2} a\sin^3\theta\left(a - a\cos^3\theta\right)3a\sin\theta\cos\theta\,d\theta}{\displaystyle\int_0^{\pi/2} \left(a - a\cos^3\theta\right)3a\sin\theta\cos\theta\,d\theta}$$

$$= a\frac{\displaystyle\int_0^{\pi/2}\left(\sin^4\theta\cos\theta - \sin^4\theta\cos^4\theta\right)d\theta}{\displaystyle\int_0^{\pi/2}\left(\sin\theta\cos\theta - \sin\theta\cos^4\theta\right)d\theta} = a\cdot\frac{(1/5)-(3\pi/256)}{1/2-1/5} \approx \frac{2a}{3}.$$

Thus the centroid has coordinates $(a, 2a/3)$.

Example 9.16 Find the centroid of the solid generated by revolving the area bounded by the curve $a^2y = x^3$, the x-axis and the line $x = 2a$ about the line $x = 2a$.

Solution. Refer Fig. 9.23. The line $AB(x = 2a)$ is the geometrical axis of the solid. The centroid $(2a, \bar{y})$ will then lie on this axis. Use the approximating rectangle and the disk method.

$$V = \pi\int_0^{8a}(2a - x)^2\,dy = \pi\int_0^{8a}\left(2a - a^{2/3}y^{1/3}\right)^2 dy$$

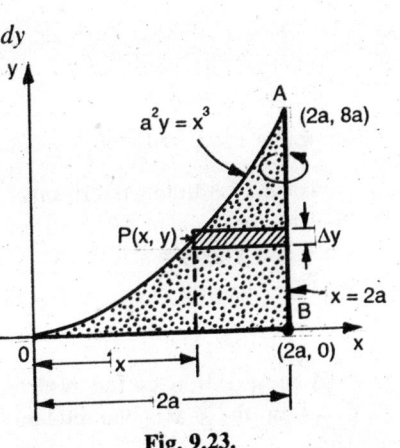

$$= \pi\int_0^{8a}\left(4a^2 - 4a^{5/3}y^{1/3} + a^{4/3}y^{2/3}\right)dy$$

$$= \pi\left[4a^2y - 3a^{5/3}y^{4/3} + \frac{3}{5}a^{4/3}y^{5/3}\right]_0^{8a}$$

$$= \pi\left(32a^3 - 48a^3 + \frac{96}{5}a^3\right) = \frac{16}{5}\pi a^3.$$

Moment with respect to the plane through the point $(2a, 0)$ and perpendicular to the Y-axis

Fig. 9.23.

$$= \pi \int_0^{8a} y \left(2a - a^{2/3} y^{1/3}\right)^2 dy = \pi \left[4a^2 \frac{y^2}{2} - \frac{12}{7} a^{5/3} y^{7/3} + \frac{3}{8} a^{4/3} y^{8/3}\right]_0^{8a}$$

$$= \pi \left(128a^4 - 128 \times \frac{12}{7} a^4 + 96a^4\right) = \frac{32\pi a^4}{7}.$$

$$\therefore \quad \bar{y} = \frac{32\pi a^4}{7} \div \frac{16\pi a^3}{5} = \frac{10a}{7}. \text{ Thus the centroid is at } (2a, 10a/7).$$

Exercise 9.1

1. Find the centroid of the arc of the cycloid $x = a\,(\theta - \sin\theta)$, $y = a\,(1 - \cos\theta)$ measured from cusp to cusp. **Ans.** $\left(\pi a,\ 4a/3\right)$.

2. Find the centroid of the arc in the positive quadrant of the curve $x^{2/3} + y^{2/3} = a^{2/3}$. **Ans.** $(2a/5, 2a/5)$.
   [**Hint.** Use parametric equations of the curve *i.e.* $x = a\cos^3\theta$, $y = a\sin^3\theta$. Take $s = 3a/2$].

3. The density at any point of the curve $x = a(\theta + \sin\theta)$, $y = a(1 - \cos\theta)$ varies as its distance from the x-axis, find the distance of the centroid from the X-axis.
   [**Hint.** We have to find the y-coordinate of the centroid of the arc between $\theta = -\pi$ to $\theta = \pi$. The density at the point (x, y) of the arc being λy, where λ is a constant.

   $$\therefore \quad \bar{y} = \int y\,dm / \int dm = \int y \cdot \lambda y\,ds / \int \lambda y\,ds = \int y^2\,ds / \int y\,ds$$

   $$= \int_{-\pi}^{\pi} a^2 (1 - \cos\theta)^2 \cdot 2a\cos\frac{\theta}{2} d\theta \div \int_{-\pi}^{\pi} a(1 - \cos\theta) \cdot 2a\cos\frac{\theta}{2} d\theta = 6a/5].$$

4. Show that the centroid of a wire bent into the form of a cardioide $r = a\,(1 + \cos\theta)$ and with line density $k\sec\left(\dfrac{\theta}{2}\right)$, k being constant, is on the axis of the cardioide at a distance $a/2$ from the cusp.

5. Find the position of the centroid of the area bounded by $y = \sin x$, $x = 0$, $x = \pi$ and X-axis. **Ans.** $(\pi/2, \pi/8)$.

6. Find the centroid of the area of the quadrant of a circle of radius r. **Ans.** $\left(4r/3\pi, 4r/3\pi\right)$.

7. Find the position of the centroid of the area bounded by the parabola $y^2 = 4\,ax$, the X-axis and the latus-rectum. **Ans.** $(3a/5, 3a/4)$

8. Find the centroid of a uniform lamina bounded by the co-ordinate axes and the arc of the ellipse $x^2/a^2 + y^2/b^2 = 1$ in the first quadrant.

    (*A.M.I.E-W-2009*) **Ans.** $\left(4a/3\pi, 4b/3\pi\right)$.

9. Find the centroid of the area of a figure bounded by the parabola $x^2 + 4y - 16 = 0$ and the X-axis. **Ans.** (0, 8/5).

10. Find the centroid of the area bounded by the co-ordinate axis and the parabola

    $\sqrt{x} + \sqrt{y} = \sqrt{a}.$ **Ans.** (a/5, a/5).

11. Find the coordinates of the centroid of the area between the parabola $y^2 = 4x$ and the straight line $y = 2x - 4$. **Ans.** (8/5, 1).

12. Find the centroid of the surface of revolution of the cardioide $r = a\,(1 + \cos\theta)$ about the initial line. **Ans.** (50a/63, 0).

13. The area bounded by the parabola $y^2 = 4\,ax$ and the latus-rectum revolves about the X-axis. Find the centroid of the volume of the solid generated.

    Ans. (2a/3, 0).

14. If the portion of the curve $ay^2 = x^3$ which is bounded by the curve, the X-axis and the ordinate $x = b$ rotates about the X-axis, find the centroid of the solid thus generated. **Ans.** (4b/5, 0).

15. Find the centroid of the solid generated by revolving the first quadrant area bounded by the parabola $x^2 = 4 - y$ about (*i*) the X-axis (*ii*) the Y-axis.

    Ans. (5/8, 0); (0, 4/3).

16. Find the centroid of the solid formed by revolving about the Y-axis the area in the first quadrant bounded by the lines $y = 0$, $x = a$, and the parabola $y^2 = 4ax$. **Ans.** (0, 5a/6).

17. Find centroid of uniform solid hemisphere of radius R. Take axis of symmetry as X-axis. **Ans.** (3R/8, 0).

18. Find centroid of a solid right circular cone of height h and semi-vertical angle α, and of its conical surface. Take axis of cone as Y-axis.

    Ans. (0, 3h/4); (0, 2h/3).

9.11 Two Theorems of Pappus

(A) First Theorem of Pappus for determining the surface area

Statement. If *a plane curve is revolved about a nonintersecting axis lying in its plane, the area of the surface generated is equal to the product of the length of the generating curve multiplied by the distance traveled by its centroid.*

Proof. Let the curve AB of length L (Ref. Fig. 9.24) be revolved about OX through an angle 2π radians. The differential length dL, which is at a distance y from the axis of revolution,

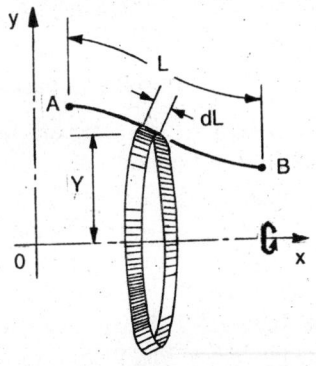

Fig. 9.24.

sweeps through the distance $2\pi y$ thereby generating a hoop whose area is $2\pi y dL$. The total area generated by AB is the area of all such hoops or

$$S = \int dS = \int 2\pi y\, dL = 2\pi \int y\, dL = 2\pi \bar{y}\, L = L(2\pi\bar{y}) \qquad ...(9.12)$$

If the generating line L is composed of several segments, the centroid $\bar{y}$ of that line need not be found since the product $L\bar{y}$ is equivalent to the sum of the moments of length (*i.e.*, Σly) of those segments.

(B) Second Theorem of Pappus for determining the volume generated by revolving a plane area

Statement. *If a bounded plane region, R is revolved once about a line that lies in the plane of R but is entirely on one side of R, then the volume of the solid of revolution thus generated is equal to the product of the plane area of R multiplied by the distance traveled by the centroid.*

Proof. Let the area of plane region R in Fig. 9.25 be A and this area is rotated about the axis OX through an angle of 2π radians. The differential area dA, which is at a distance y from the axis of revolution sweeps through the distance $2\pi y$ and generates a ring whose volume (dV) is $2\pi y dA$. The total volume generated is the sum of the volumes of all such rings or

$$V = \int dV = \int 2\pi y\, dA = 2\pi \int y\, dA = 2\pi\bar{y}A = A(2\pi\bar{y}) \qquad ...(9.13)$$

If the generating area is composed of several parts, the centroid $\bar{y}$ of that area need not be found since the product $A\bar{y}$ is equivalent to the sum of the moments of area (*i.e.* Σay) of the several parts.

Note. If the generating line or area is revolved through an angle θ less than 2π radians, the generated surface or volume may be found by substituting θ for 2π in equations (9.12) and (9.13).

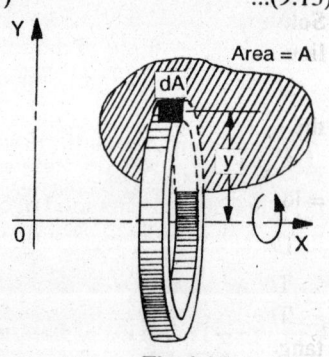

Fig. 9.25.

9.12 Solved Problems based on First Pappus' theorem.

Example 9.17 (*a*) **Obtain the area of the surface generated by revolving an equilateral triangle of side *a* about a line in its plane at a distance *c* from the centroid.**

(*b*) **Find the area of the surface generated by revolving the rectangle of dimensions *a* and *b* about an X-axis which is at a distance *c* (> *a*, *b*) units from the centroid.**

Solution. (*a*) The perimeter of the triangle is $3a$ and centroid describes a circle of radius c (see Fig. 9.26).

$$S = 3a \cdot 2\pi c = 6\pi\, ac \text{ sq. units.}$$

(*b*) The perimeter of the rectangle is $2(a + b)$ and the centroid describes a circle of radius c.

$$S = 2(a+b)\cdot 2\pi c = 4\pi(a+b)c \text{ square units.}$$

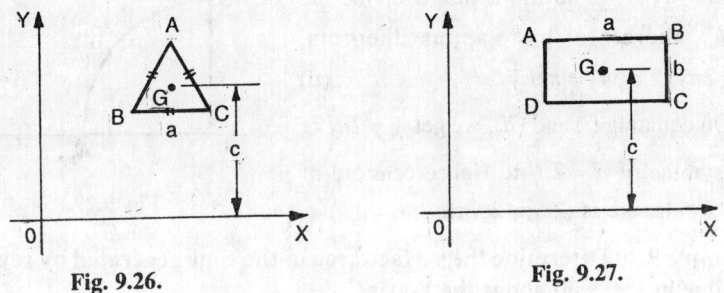

Fig. 9.26.   **Fig. 9.27.**

Example 9.18 The length of an arc of the cycloid

$$x = a(\theta - \sin\theta), \quad y = a(1 - \cos\theta)$$

is 8a, and the surface area generated by revolving it around the x-axis is $(64\pi a^2)/3$. Find the surface area generated by revolving the arc about the tangent at the highest point.

Solution. Refer Fig. 9.28 By symmetry the centroid of the arc-length lies on the line *BA*.

If $\overline{y}$ is the distance of centroid of the arch of the cycloid from the *X*-axis, then the length of the path of the centroid of arc $= 2\pi\overline{y}$.

Use Pappus' Theorem the surface area generated

= length of arc × length of path of centroid of arc = $8a \times 2\pi\overline{y} = 16a\pi\overline{y}$. ...(*i*)

Equating area (*i*) to $(64\pi a^2)/3$, we get $\overline{y} = 4a/3$.

The tangent is parallel to base of the cycloid at a distance $2a$ from it.

Thus the distance of centroid from tangent at the highest point of the cycloid $= 2a - (4a/3) = 2a/3$.

Hence it follows from Pappus' theorem that the surface area generated by revolving the arc about the tangent at the highest point

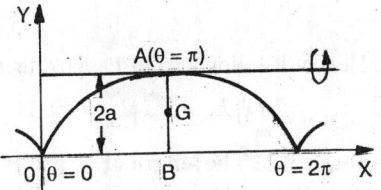

$$= (8a)\left(2\pi\cdot\frac{2a}{3}\right) = 32\pi a^2/3 \text{ square units.}$$

Fig. 9.28.

Example 9.19 Use the Theorem of Pappus to determine the position of centroid of the quadrant of a circular arc of radius r.

Solution. When the arc *ABC* revolves about *x*-axis, it generates hemi-sphere of surface area $2\pi r^2$ square units. ...(i)

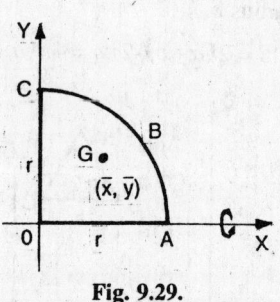

Let $G(\bar{x}, \bar{y})$ be the centroid of the arc *ABC*, therefore by Pappus theorem,

$$S = 2\pi\bar{y}(2\pi r/4) = \pi^2 r\bar{y} \qquad ...(ii)$$

On equating (i) and (ii), we get $\bar{y} = 2a/\pi$.

By symmetry $\bar{x} = 2a/\pi$. Hence centroid of the circular arc is $(2a/\pi, 2a/\pi)$.

Fig. 9.29.

Example 9.20 Determine the surface area of the cone generated by revolving the line in Fig. 9.30 about the Y-axis.

Solution. Two cones are generated by the line. One by the 4 unit length above O and the other by the 6 unit length below O. The *x*-coordinate of the centroid for each segment of the line is equal to $(L/2)\sin\theta = (L/2)\sin 30°$. Hence

$$\bar{x}_1 = \frac{4}{2}\times\frac{1}{2} = 1 \text{ unit and } \bar{x}_2 = \frac{6}{2}\times\frac{1}{2} = 1.5 \text{ units.}$$

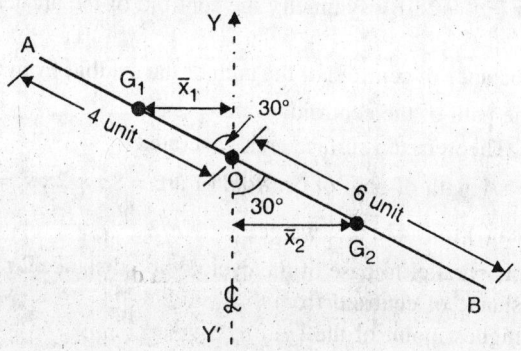

Fig. 9.30.

Hence, it follows from Pappus' theorem that total surface, area generated is

$$= 2\pi\bar{x}_1 L_1 + 2\pi\bar{x}_2 L_2 = 2\pi(1)(4) + 2\pi(1.5)(6) = 81.70 \text{ sq. units.}$$

Example 9.21 The length of an arc of a cycloid x = a (θ + sinθ), y = a (1 − cosθ) from θ = 0 to θ = π is 4a. Find the surface area generated by revolving this arc about x-axis.

Solution. The curve is shown in Fig. 9.31. Let *s* be the length of the arc OA

$$sy = \int y\,ds = \int_0^\pi y\frac{ds}{d\theta}\,d\theta = \int_0^\pi y\sqrt{\left(\frac{dx}{d\theta}\right)^2+\left(\frac{dy}{d\theta}\right)^2}\,d\theta$$

$$= \int_0^\pi a(1-\cos\theta)\sqrt{a^2(1+\cos\theta)^2+a^2\sin^2\theta}\,d\theta$$

$$= a^2\int_0^\pi (1-\cos\theta)\sqrt{2(1+\cos\theta)}\,d\theta = 4a^2\int_0^\pi \sin^2\frac{\theta}{2}\cos\frac{\theta}{2}\,d\theta$$

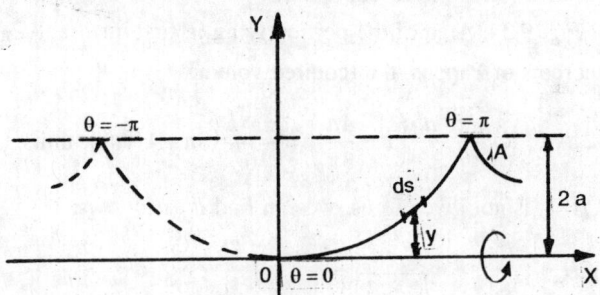

Fig. 9.31.

$$= 8a^2\int_0^\pi \sin^2\frac{\theta}{2}\,d\left(\sin\frac{\theta}{2}\right) = 8a^2\left[\frac{1}{3}\sin^3\frac{\theta}{2}\right]_0^\pi = \frac{8a^2}{3}.$$

∴ By Pappus' Theorem $S = 2\pi\bar{y}s = 2\pi(8a^2/3) = 16\pi a^2/3$ sq. units.

9.13 Problems based on Second Theorem of Pappus

Example 9.22 Find the volume of the torus generated by revolving the circle $x^2 + y^2 = 4$ about the line $x = 3$.
                    (A.M.I.E., Summer 1996)

Solution. Refer Fig. 9.32. The centroid of the area of the circle describes a circle of radius 3 units. The length of the path of centroid $= 2\pi \cdot 3 = 6\pi$ unit.

Area of circle $= \pi \cdot 2^2 = 4\pi$ square units. Hence volume of the torus generated

$= 4\pi(6\pi) = 24\pi^2$ cubic units.

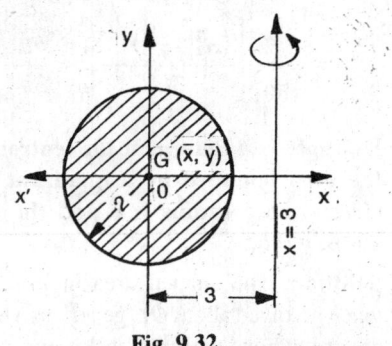

Fig. 9.32.

Example 9.23 (a) The ellipse $4x^2 + 36y^2 = 144$ is revolved about the line $y = 2$. Find the volume of the solid generated.   *(A.M.I.E., S-2002 S-2003)*

(*b*) **Find the volume of the solid generated by revolving the first quadrant of the ellipse** $x^2/a^2 + y^2/b^2 = 1$, **about the line x = a. Given** $\bar{x} = 4a/3\pi$.

Solution. (*a*) The equation of the ellipse is $x^2/36 + y^2/4 = 1$. Here $a = 6$, $b = 2$.

The area of the ellipse $= \pi(6)(2) = 12\pi$ square units.

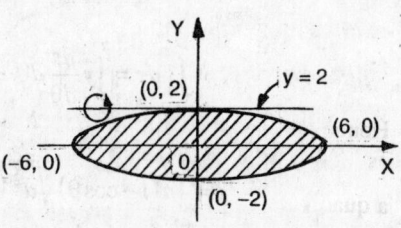

Fig. 9.33.

The centroid of the area of the ellipse, (*i.e.*, O) describes a circle of radius 2 units. Therefore, the length of the path of the centroid $= 2\pi (2) = 4\pi$ unit.

Hence $V = 12\ \pi\ (4\pi) = 48\ \pi^2$ cubic units.

(*b*) Refer Fig. 9.34. Area of the portion lying in first quadrant $= \pi ab/4$.

∴ By Theorems of Pappus, the required volume

$$= \frac{\pi ab}{4} \cdot 2\pi(a - \bar{x}) = \frac{\pi^2 ab}{2}\left(a - \frac{4a}{3\pi}\right) = \frac{\pi a^2 b}{6}(3\pi - 4)\,\text{cubic units.}$$

Remark. If $\bar{x}$ is not given to us, we can find it as follows:

$$\bar{x} = \frac{\displaystyle\int_0^a xy\,dx}{\pi ab/4} = \frac{4}{\pi ab}\int_0^a xb\sqrt{1 - \frac{x^2}{a^2}}\,dx$$

$$= \frac{4}{\pi a^2}\int_0^a x\sqrt{a^2 - x^2}\,dx$$

$$= -\frac{2}{\pi a^2}\int_0^a (-2x)(a^2 - x^2)^{1/2}\,dx$$

$$= -\frac{2}{\pi a^2}\left[\frac{(a^2 - x^2)^{3/2}}{3/2}\right]_0^a = \frac{4a}{3\pi}.$$

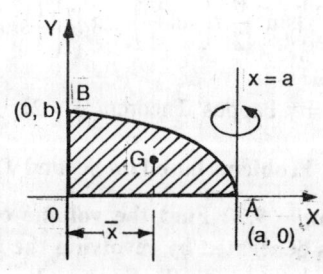

Fig. 9.34.

Example 9.24 Determine the centroid for the area bounded by a quadrant of a circle whose radius is r and the lines x = 0, y = 0.

Solution. The shaded area in Fig. 9.35 when revolved about *OY*, generates a hemisphere whose volume is known to be $2\pi r^3/3$. The length of the path of the

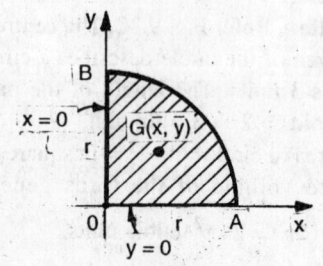

Fig. 9.35.

centroid is the volume divided by the area of the quadrant of a circle.

$$\therefore \quad L = \frac{2\pi r^3/3}{\pi r^2/4} = \frac{8\pi}{3}.$$ But the length of the path of the centroid is $2\pi\bar{x}$.

Hence $2\pi\bar{x} = 8r/3$ or $\bar{x} = 4r/3\pi$. By symmetry $\bar{y} = 4r/3\pi$. Hence centroid of

a quadrant of a circle is $\left(\dfrac{4r}{3\pi}, \dfrac{4r}{3\pi}\right)$.

Example 9.25 The figure bounded by a quadrant of a circle of radius r and the tangents at its extremities, revolves about one of these tangents. Prove that the volume of the solid thus generated is $(10 - 3\pi)\pi r^3/6$.

Solution. The revolving area is shown shaded in the Fig. 9.36. Let G be the centroid of the quadrant AOC. Then the distance of G from OA is $4r/3\pi$, and therefore from the tangent line CB is $r - (4r/3\pi)$.

Let BC be the axis of revolution. When the square $AOCB$ revolves about CB, we get a circular cylinder of radius r and height r, so that the volume generated is $\pi r^2 \cdot r = \pi r^3$. Also when the quadrant AOC rotates about CB the volume generated by Pappus' theorem is

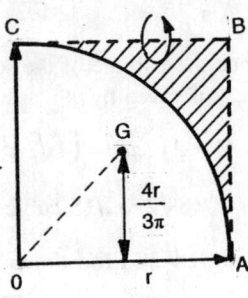

Fig. 9.36.

$$= \left(\frac{1}{4} \cdot \pi r^2\right) 2\pi \left(r - \frac{4r}{3\pi}\right) \text{cubic units.}$$

Hence the volume generated by the revolution of the arc $ACBA$ about CB = difference between the two volumes

$$= \pi r^3 - \frac{\pi}{4} r^2 \cdot 2\pi \left(r - \frac{4r}{3\pi}\right) = \left(\frac{5}{3} - \frac{\pi}{2}\right)\pi r^3 = \frac{(10 - 3\pi)\pi r^3}{6}.$$

Example 9.26 A semi-circular bend of lead pipe has a mean radius of 30 cm, the internal diameter of the pipe is 10 cm and the thickness of the lead is 12 mm. Find its weight given that 1 cm³ of lead weight 11.37 grams.

Solution. The area of a right section of the pipe = area included between two concentric circles of radii 6.2 cm and 5 cm
$= \pi [(6.2)^2 - 5^2] = \pi \cdot (11.2) (1.2)$ cm²
$= 13.44 \pi$ cm².

The centroid of the section describes the semi-circumference of a circle of radius 30 cm, so the length of its path = 30π cm.

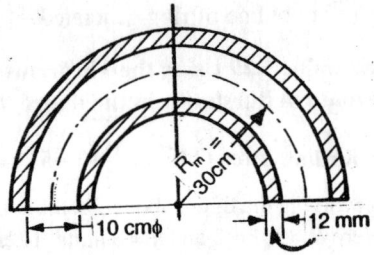

Fig. 9.37.

∴ Volume of lead in pipe = $13.44\,\pi \times 30\,\pi = 403.2\,\pi^2$ cm^3 and the weight of

the pipe $= \dfrac{403.2\pi^2 \times 11.37}{1000}$ kgm = 45.28 kgm.

Example 9.27 The loop of the curve $2ay^2 = x\,(a - x)^2$ revolves about the line $y = a$. Find the volume of the solid generated.

Solution. The given curve has a loop between $x = 0$, and $x = a$, symmetrical about the X-axis, not intersecting the line $y = a$. The centroid of the area of the loop is on X-axis and so is at a distance 'a' from the axis of revolution $y = a$.

The area of the loop,

$$A = 2\int_0^a y\,dx = 2\int_0^a \frac{x^{1/2}(a - x)}{\sqrt{2a}}\,dx \ [\text{as } a - x > 0]$$

$$= \sqrt{\frac{2}{a}} \int_0^a \left(ax^{1/2} - x^{3/2}\right)dx = \frac{4\sqrt{2}}{15}a^2 \ \text{square units.}$$

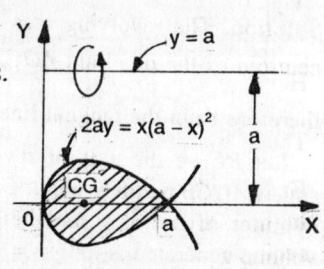

Hence the volume of the solid of revolution, V is given by

$$V = A(2\pi a) = \left(4\sqrt{2}\,a^2/15\right)2\pi a$$

$$= 8\sqrt{2}\,\pi a^3/15 \ \text{cubic units.}$$

Fig. 9.38.

Example 9.28 Find the volume of the ring generated by the revolution of an ellipse of eccentricity $1/\sqrt{2}$ and semi-minor axis length 1, about a straight line parallel to minor axis and situated at distance from the centre equal to six times the major axis. *(A.M.I.E. W-2004)*

Solution. We have $b = 1$, $e = 1/\sqrt{2}$. Employing $b^2 = a^2(1 - e^2)$ to get $a = \sqrt{2}$. The equation of ellipse is $x^2/2 + y^2/1 = 1$. Area of an ellipse $= = \pi(\sqrt{2})(1) = \sqrt{2}\,\pi$ sq. units. The centroid of the area of the ellipse describe a circle of radius $12\sqrt{2}$ units. The length of the path of the centroid $= 2\pi(12\sqrt{2}) = 24\pi\sqrt{2}$ units. Hence volume of ring generated $= (\sqrt{2}\pi)(24\pi\sqrt{2}) = 48\,\pi^2$ cubic units.

Example 9.29 Using the Theorems of Pappus to find the centroid of the trapezoid OABC shown in the figure 9.39.

Solution. Area $OABC = \dfrac{1}{2}(3 + 5)8 = 32$ sq. units.

Revolving the figure about OX, the solid formed is a frustum of a cone of revolution. Here $R = 5$, $r = 3$ and $h = 8$ units. Therefore, volume of frustum of a right circular cone,

$$V_x = \pi h \left(R^2 + Rr + r^2 \right)/3.$$

$$= \pi \cdot 8 \left(5^2 + 5 \times 3 + 3^2 \right)/3 = 392\,\pi/3 \text{ cubic units.}$$

Hence $\bar{y} = \dfrac{392\pi/3}{2\pi(32)} = 2.04$ units.

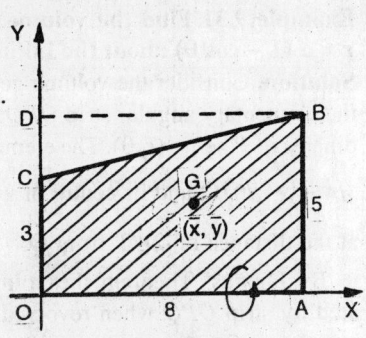

Now, revolving the figure about OY, the volume generated is the difference of the volumes of the cylinder generated by $ODBA$ and the cone generated by the triangle CDB. Hence

Fig. 9.38.

$$V_y = \left(\pi \cdot 8^2 \right) 5 - \left\{ \left(\pi \cdot 8^2 \right)(5-3)/3 \right\} = 320\pi - \left(128\pi/3 \right) = 832\pi/3 \text{ cubic units.}$$

Hence, $\bar{x} = \dfrac{V_y}{2\pi A} = \dfrac{832\,\pi/3}{2\pi(32)} = 4.33$ units.

Thus the centroid of the trapezoid $OABC$ is (4.33, 2.04).

Example 9.30 Use Theorems of Pappus to find the volume swept out by rotating the plane figure ABCDE in Fig. 9.40 (consisting of a rectangle and a sector of a circle), through one revolution about the side AE.

Solution. Refer Fig. 9.40, the centroid of the sector BCD is at G_1, where

$$DG_1 = \frac{2r}{3} \frac{\sin 30°}{(\pi/180)30} = \frac{2r}{\pi}, \text{ so that } FG_1 = \frac{2r}{\pi} \cdot \sin 30° = \frac{r}{\pi} = \frac{4}{\pi}, \text{ since } r = 4.$$

$\bar{x}_1 = $ the distance of G_1 from the line AE

is thus $2 + \left(\dfrac{4}{\pi} \right)$, and the area of the sec-

tor is $r^2 \cdot \left(\dfrac{\pi}{6} \right) = \dfrac{8\pi}{3}$ square units. The

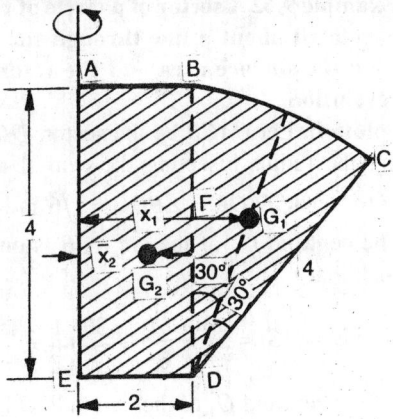

centroid of the rectangle ABDE is at G_2, where the distance of G_2 from AE is 1 unit, *i.e.*, $\bar{x}_2 = 1$, the area of rectangle is 8 square units.

∴ Required volume is, $V = 2\pi A \bar{x}$

$$= 2\pi \sum ax$$

Fig. 9.40.

$$= 2\pi \left[\frac{8\pi}{3} \left(2 + \frac{4}{\pi} \right) + 8(1) \right] = \frac{16\pi}{3}(2\pi + 7) \text{ cubic units.}$$

Example 9.31 Find the volume generated by revolving the cardioid
r = a (1 − cos θ) about the initial line.

Solution. Consider the volume generated by
the elementary angular strip *OPQ*. The coordinates of *P* being (r, θ). The elementary area

$dA = \left(r^2 d\theta\right)/2$ and its centre of gravity *G* is

at the distance of $2r/3$ from O.

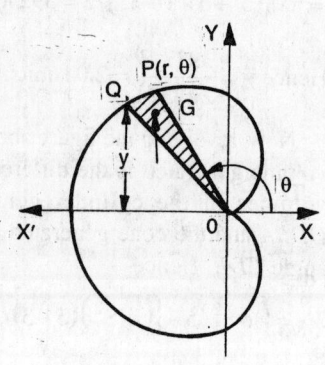

By Pappus' Theorem the volume generated by strip *OPQ* when revolved about the
initial line *OX* is

$$dV = 2\pi y \times dA = 2\pi \times \frac{2}{3} r \sin \theta \times \frac{1}{2} r^2 d\theta$$

$$= \frac{2}{3} \pi r^3 \sin \theta \, d\theta.$$

Fig. 9.41.

Therefore, the volume generated by the cardioid, when revolved about the
initial line is

$$V = \frac{2}{3} \pi \int_0^\pi r^3 \sin \theta \, d\theta = \frac{2}{3} \pi \int_0^\pi a^3 (1-\cos\theta)^3 \sin\theta \, d\theta$$

$$= \frac{2}{3} \pi a^3 \int_0^2 t^3 \cdot dt = \frac{2}{3} \pi a^3 \left[\frac{t^4}{4}\right]_0^2 = \frac{2}{3} \pi a^3 \cdot \frac{16}{4} = \frac{8}{3} \pi a^3 \text{ cubic units.}$$

Example 9.32 A sector of a circle of radius 5 units, the central angle being 60°
is rotated about a line through the centre parallel to the chord of the arc.
Find the surface area and the volume of the solid generated in a complete
revolution.

Solution. Let *OACB* be the sector, *OC* being the radius bisecting the central angle
AOB. Then $\angle AOC = \angle BOC = \pi/6$ radians.
The centroid G_1 of the arc *ACB* is on *OC*
such that

$$\bar{x}_1 = OG_1 = 5\frac{\sin \pi/6}{\pi/6} = \frac{30}{\pi} \cdot \frac{1}{2} = \frac{15}{\pi}.$$

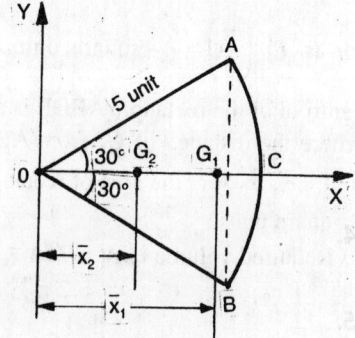

The centroid G_2 of the two radii *OA*, *OB*
is also on *OC* such that

$$\bar{x}_2 = OG_2 = (5/2)\cos(\pi/6) = 5\sqrt{3}/4.$$

Fig. 9.42.

The length of the arc $ACB = r\theta = 5\pi/3$.

Total surface area generated about Y-axis,

$$S = 2\pi \bar{x}L = 2\pi\Sigma xl = 2\pi\left[\frac{15}{\pi}\cdot\frac{5\pi}{3}+\frac{5\sqrt{3}}{4}\cdot 10\right]$$

$$= 25\left(2+\sqrt{3}\right)\pi = 293.23 \text{ square units.}$$

The area of the sector $= \dfrac{1}{2}\cdot 5^2 \cdot \dfrac{\pi}{3} = \dfrac{25\pi}{6}$ sq. unit, the centroid is on OC that is,

a distance $= \dfrac{2}{3}\cdot\dfrac{5\sin\pi/6}{\pi/6} = \dfrac{10}{3}\cdot\dfrac{6}{\pi}\cdot\dfrac{1}{2} = \dfrac{10}{\pi}$ from O.

$\therefore$ Volume of the solid generated about Y-axis

$$= \left(\frac{25\pi}{6}\right)\cdot\left(2\pi\cdot\frac{10}{\pi}\right) = 261.9 \text{ cubic units.}$$

Exercise 9.2

Use the Theorems of Pappus in the following problems

1. Find the surface area of an anchor ring formed by the revolution of a circle of radius a about a line in its plane at a distance b from the centre ($b > a$).
   Ans. $4\pi^2 ab$.

2. Find the surface area of a sphere of radius r. **Ans.** $4\pi r^2$.

3. In Fig. 9.43, determine the surface generated by revolving the arc of the semicircle of radius r about the Y-axis. Note that the centroid of the arc of the semi-circle is at a distance $\bar{x} = d + (2r/\pi)$. **Ans.** $(2\pi^2 rd + 4\pi r^2)$.

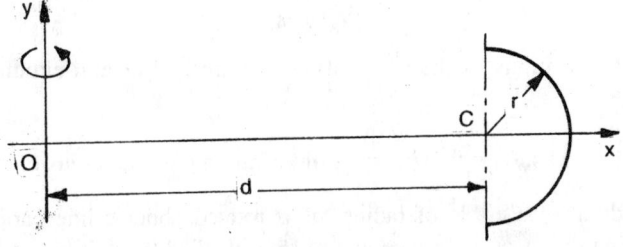

Fig. 9.43.

4. Show that the surface of the solid obtained by revolving the arc of the curve $y = \sin x$ from $x = 0$ to $x = \pi$ about the X-axis is $\pi^2\left[\sqrt{2}+\log\left(1+\sqrt{2}\right)\right]$ sq. units.

5. Find the surface of the reel formed by the revolution round the tangent at the vertex of the cycloid $x = a(\theta + \sin\theta)$, $y = a(1 - \cos\theta)$. *(V.T.U., 2005)*

   Ans. $32\pi a^2/3$ square units.

6. Find the area of the surface formed by revolving $r = 4 \cos \theta$ about the initial line.
   Ans. 16π Sq. units.

7. The curve $r = a(1 + \cos \theta)$ revolves about the initial line. Find the area of the surface formed.(*Nagpur 1998S; Marathwada, 1998; AMIE, S-1998*)
   Ans. $32\pi a^2/5$

8. Find the volume of the following:
   (*i*) A right circular cone of altitude h and base radius r.
   (*ii*) The ring obtained by revolving the ellipse $4(x - 6)^2 + 9(y - 5)^2 = 36$ about X-axis.   **Ans.** (*i*) $\pi r^2 h/3$ (*ii*) $60\pi^2$ cubic units.

9. Find the volume of the torus generated by revolving the circle $(x - 2)^2 + y^2 = 1$ about the Y-axis.   **Ans.** $4\pi^2$ cubic units

10. The co-ordinates of the vertices of a rectangular lamina ABCD are A(2, 2), B(6, 2), C(6, 4), and D(2, 4).
    Find the volume and surface generated by revolving this rectangle about the line $x = 9$.   **Ans.** 80π cubic units; 120π square units.

11. Determine the volume generated by revolving a semicircular area of radius r about the Y-axis. Refer to Fig. 9.44.

    Ans. $\left\{(\pi^2 r^2 d + (4\pi r^3/3))\right\}$ cubic units.

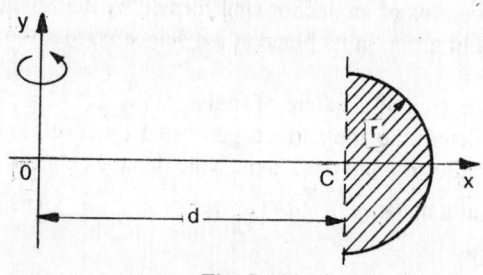

Fig. 9.44.

12. Find the positions of the centroids of the arc and a semi-circular area of radius r.

    Ans. (*i*) $\dfrac{2r}{\pi}$, (*ii*) $\dfrac{4r}{3\pi}$ from centre along the radius of symmetry.

13. A quadrant of a circle of radius 'a' is rotated about a line parallel to its bounding radius at a distance $d(>a)$ from it. Find the volume and the total surface of the solid generated by the revolution.

    Ans. $\pi a^2 (3\pi d - 4a)/6;\ \pi a(\pi d - 2a)$.

14. A circular quardrant of radius $a/2$ and centre B is removed from a square ABCD of side a. If the remaining portion is rotated through an angle of 360° about the side AD, prove that the volume generated is $\pi a^3 (104 - 12\pi)/96$ cubic units.

15. Find the volume of the solid generated by the revolution of the ellipse $x = a\cos\theta$, $y = b\sin\theta$ about the line $x = 2a$. **Ans.** $4\pi^2 a^2 b$ cubic unit

16. The ellipse $4x^2 + 9y^2 = 36$ is revolved about the straight line $2x + 3y = 12$. Find the volume so generated. **Ans.** $144\pi^2/\sqrt{13}$ cubic units.

17. The upper half of the area of the loop of the curve $ay^2 = x^2(a - x)$ is revolved about X-axis. Find the volume of the solid generated.

   Ans. $V = \pi a^3/12$ cubic units.

[**Hint:** $A = \int\limits_{0}^{a} y\, dx = \dfrac{4a^2}{15}$. $A\bar{y} = \dfrac{1}{2}\int\limits_{0}^{a} y^2\, dx = \dfrac{a^3}{24}$. Use Pappus' Theorem

$V = 2\pi A\bar{y} = \pi a^3/12.$]

18. Use the Theorems of Pappus to find the centroid of the arc of the cardioid $r = a(1 - \cos\theta)$ from $\theta = 0$ to $\theta = \pi$. **Ans.** $(-4a/5,\ 4a/5)$

   [**Hint.** Half length of the cardioid = $4\ a$ units, $S = 32\ \pi\ a^2/5$].

Example 9.33. Find the surface area of the solid generated by revolving the curve $x^2 + (y - b)^2 = a^2$ about x - axis. *(A.M.I.E., W-2010)*

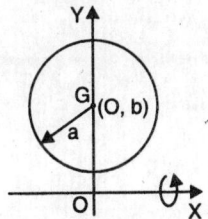

Fig. 1.

Solution: Ref. Fig. 1. The centroid of the circle describes a circle of radius b units. The length of the path of centroid = $2\pi b$ unit.

Circumference of the circle = $2\pi a$ unit.

Hence surface area of the solid generated, $S = 2\pi a(2\pi b) = 4\pi^2 ab$ square units.

Chapter 10

Infinite Series

10.1 Introduction

The subject of infinite series is of immense importance in solving many problems in applied mathematics. For example infinite series play an important role in the study of functions of complex variables, solution in series of some differential equations such as Bessel's equation and Legendre's equation, numerical solution of many important differential equations that occur in physical problems, evaluation of many non-elementary integrals, e.g. Elliptic integral

$$K(k) = \int_0^{\pi/2} \frac{dx}{\sqrt{1 - k^2 \sin^2 x}}, k^2 < 1.$$

Since an indiscriminate use of these series may lead to incorrect results, it is of vital importance that students of engineering should acquire an intelligent understanding of the behaviour of infinite series.

In the present chapter some fundamental notions and concepts of infinite series along with conditions in which their applications are valid, will be discussed.

10.2 Definitions

Sequence. *A sequence is a succession of numbers of terms formed according to some definite rule or law.*

For Example, 1, 4, 9, 16, 25 ...(*i*)

and $1, -x, \dfrac{x^2}{2}, \dfrac{-x^3}{3}, \dfrac{x^4}{4}, \dfrac{-x^5}{5}$ ...(*ii*)

are sequences.

The general term or *n*th term in a sequence denoted by u_n, is usually a function of *n* that indicates the rule of formation of the terms.

By giving different values to *n* in the expression for u_n, we get different terms of the sequence.

Thus if $u_n = n^2$, $u_1 = 1$, $u_2 = 4$, $u_3 = 9$, ... and the corresponding sequence is 1, 4, 9, ..., n^2, ...

Series. *A series is the indicated sum of the terms of a sequence.*

Thus from the sequences, (*i*) and (*ii*) we obtain the series $1 + 4 + 9 + 16 + 25$.

and
$$1 - x + \frac{x^2}{2} - \frac{x^3}{3} + \frac{x^4}{4} - \frac{x^5}{5}.$$

In general if the sequence is u_1, u_2, u_3, ...
then the sum $u_1 + u_2 + u_3 + ... + u_n + ...$ is called a series.

When the number of terms is limited, the sequence or series is said to be finite. When the number of terms is unlimited, the sequence or series is called an infinite sequence or infinite series.

An infinite series

$$u_1 + u_2 + u_3 + ... + u_n + ...$$

is denoted by $\sum\limits_{n=1}^{\infty} u_n$ or simply Σu_n.

10.3 Convergence, Divergence and Oscillation of Series

Consider an infinite series
$$\Sigma u_n = u_1 + u_2 + u_3 + ... + u_n + ... \to \infty$$
and let the sum of the first n terms be denoted by S_n so that
$$S_n = u_1 + u_2 + u_3 + ... + u_n.$$

If S_n tends to a finite limit as $n \to \infty$, the series is said to be *convergent*.

In case S_n tends to $+ \infty$ or $- \infty$ as $n \to \infty$, the series is called *divergent*.

It is said to be *oscillating* if as $n \to \infty$ the value of the sum changes from one value to a second or third etc., value according to whether an odd or even number of terms be taken. Thus $a - a + a - a + ... \to \infty$ is an oscillating series, the sum of an odd number of terms being a, and that of an even number of terms being zero.

10.4 Some General Properties of Infinite Series

With the help of above definitions, it can be easily verified that:

1. The behaviour of an infinite series as regards convergence, divergence or oscillation is not affected by
   (*i*) multiplication of all the terms by a non-zero constant; or by
   (*ii*) alteration, addition or deletion of a finite number of its terms.

2. If a series in which all the terms are positive is convergent the series will remain convergent even when some or all of its terms are made negative. This is because, the sum of the series will be clearly greater in the first case when all its terms positive.

3. An infinite series of positive terms can either converge or diverge to infinity but cannot oscillate.

4. If two series Σu_n and Σv_n both converges, then their sum/difference is $\Sigma(u_n \pm v_n)$.

that is,
$$\sum_{n=1}^{+\infty} (u_n \pm v_n) = \sum_{n=1}^{+\infty} u_n \pm \sum_{n=1}^{+\infty} v_n \text{ also converges.}$$

10.5 Illustrative Examples

Example 10.1 Discuss the nature of the following series:

(i) $1 + \dfrac{1}{2} + \dfrac{1}{4} + \dfrac{1}{8} + \dots \infty$, (ii) $1^2 + 2^2 + 3^2 + \dots \infty$,

(iii) $6 - 6 + 6 - 6 + 6 - \dots \infty$.

Solution. (i) The terms are in geometric progression with common ratio 1/2.

$$\text{Let } S_n = 1 + \frac{1}{2} + \frac{1}{2^2} + \frac{1}{2^3} + \dots + \frac{1}{2^{n-1}}.$$

The sum of the first n terms, $S_n = \dfrac{1 - \left(\dfrac{1}{2^n}\right)}{1 - \left(\dfrac{1}{2}\right)} = 2\left[1 - \left(\dfrac{1}{2^n}\right)\right]$

$$\therefore \lim_{n \to \infty} S_n = 2. \text{ Hence the series is convergent.}$$

(ii) $S_n = 1^2 + 2^2 + 3^2 + \dots + n^2 = n(n + 1)(2n + 1)/6$

Since $\lim_{n \to \infty} S_n = \infty$, the series diverges.

(iii) $S_n = 6 - 6 + 6 - 6 + \dots$ to n terms

$\qquad = 6$ or 0 according an n is odd or even.

Hence S_n does not tend to a unique limit and, therefore, the given series oscillates.

Example 10.2 Show that the geometric geometric series

$$1 + r + r^2 + r^3 + \dots \text{ to } \infty$$

(i) **Converges if $|r| < 1$** (ii) **Diverges if $r \geq 1$** and (iii) **Oscillates if $r \leq -1$.**
Solution. (i) Let $S_n = 1 + r + r^2 + \dots + r^{n-1}$.

$$\therefore \quad S_n = \frac{(1 - r^n)}{(1 - r)} = \frac{1}{1-r} - \frac{r^n}{1-r}.$$

If $|r| < 1$, $\lim_{n \to \infty} r^n = 0$. Therefore, $\lim_{n \to \infty} S_n = \lim_{n \to \infty} \dfrac{1 - r^n}{1 - r} = \dfrac{1}{1 - r}.$

Hence the series converges when $|r| < 1$.

(*ii*) When $|r| > 1$, $\lim_{n \to \infty} r^n \to \infty$. Therefore, $\lim_{n \to \infty} S_n = \lim_{n \to \infty} \dfrac{(r^n - 1)}{(r - 1)} \to \pm\infty$.

Thus the series diverges when $r > 1$.

If $r = 1$, the series becomes $1 + 1 + 1 + ...\infty$, whose sum is infinite

$\therefore$ $S_n = n$ **and** $\lim_{n \to \infty} S_n \to \infty$ and hence the series diverges.

(*iii*) When $r = -1$, the series is $1 - 1 + 1 - ...$

$\therefore$ $S_n = 1$ or 0 according as n is odd or even.

Hence the series is oscillatory. (In this case we say the series oscillates finitely).

When $r < -1$, let $r = -k$ where $k > 1$. Since $k^n \to \infty$ as $n \to \infty$.

$$\lim_{n \to \infty} S_n = \lim_{n \to \infty} \frac{1 - r^n}{1 - r} = \lim_{n \to \infty} \frac{1 - (-1)^n k^n}{1 + k} \to +\infty \text{ or } -\infty$$

according as n is odd or even. Hence the series oscillates.
(In this case we say the series oscillates infinitely).

Exercise 10.1

Determine the nature of the following series:

1. $1 + \dfrac{1}{3} + \dfrac{1}{3^2} + \dfrac{1}{3^3} + ... + \dfrac{1}{3^{n-1}} + ...$

2. $1 + \dfrac{2}{3} + \dfrac{4}{9} + \dfrac{8}{27} + ...\infty$

3. $\dfrac{3}{2} + \left(\dfrac{3}{2}\right)^2 + \left(\dfrac{3}{2}\right)^3 + ...\infty$

4. $1 + 2 + 3 + ...\infty$

5. $3 - 7 + 4 + 3 - 7 + 4 + ...\infty$

6. $\dfrac{1}{2!} + \dfrac{2}{3!} + \dfrac{3}{4!} + \dfrac{4}{5!} + ...\infty$

7. $\dfrac{1}{1 \cdot 3} + \dfrac{1}{3 \cdot 5} + \dfrac{1}{5 \cdot 7} + ... + \dfrac{1}{(2n - 1)(2n + 1)} + ...$

8. $\dfrac{1}{1 \cdot 2} + \dfrac{1}{2 \cdot 3} + \dfrac{1}{3 \cdot 4} + ...\infty$ *(Mangalore, 1999S)*

9. $\log 2 + \log \dfrac{3}{2} + \log \dfrac{4}{3} + \log \dfrac{5}{4} + ...\infty$ 10. $\displaystyle\sum_{n=1}^{\infty} \log \frac{n}{n + 1}$.

Ans. 1. Convergent **2.** Convergent **3.** Divergent **4.** Divergent **5.** $S_n = 0, 3, -4$
according as $n = 3m$, $3m + 1$, $3m + 2$. Hence the series oscillates.
6. Convergent **7.** Convergent **8.** Divergent **9.** Divergent **10.** Divergent

10.6 Theorem:

A necessary condition for a series $\sum u_n$ *to be convergent is* $\lim_{n \to \infty} u_n \ne 0$.

Proof. Let $S_n = u_1 + u_2 + u_3 + \ldots + u_n$.

Since Σu_n in convergent, therefore $\lim\limits_{n \to \infty} S_n = l$, a finite quantity.

Also $\qquad\qquad \lim\limits_{n \to \infty} S_{n-1} = l$.

Now $u_n = S_n - S_{n-1}$. Therefore, $\lim\limits_{n \to \infty} u_n = \lim\limits_{n \to \infty} S_n - \lim\limits_{n \to \infty} S_{n-1} = l - l = 0$.

Thus if Σu_n is convergent, $\lim\limits_{n \to \infty} u_n = 0$.

Important Note: It must clearly be understood that the converse of the above theorem is not true that is, $\lim\limits_{n \to \infty} u_n = 0$ does not prove that Σu_n is convergent.

There are series is which the nth term $\to 0$ as $n \to \infty$ but are divergent.

For example the harmonic series $1 + \dfrac{1}{2} + \dfrac{1}{3} + \ldots \infty$ is divergent even though

$$\lim\limits_{n \to \infty} u_n = \lim\limits_{n \to \infty} \frac{1}{n} = 0.$$

Thus $\lim\limits_{n \to \infty} u_n = 0$ *is necessary but not sufficient condition for convergence of* Σu_n.

10.7 Tests for Convergence and Divergence of Positive Term Series

To determine convergence or divergence of a series by definition we have to find S_n (the sum of first n terms). Since it is not possible to find S_n for every series, it becomes necessary to devise tests that enable us to determine the behaviour of a given series without finding S_n.

In the next few sections we shall consider tests for convergence of positive term series. Later on a test for series of positive and negative terms will be discussed.

10.8 Some Simple Tests

(A) *The nth Term Test for Divergence*

If $\lim\limits_{n \to \infty} u_n \neq 0$, *the series* Σu_n *diverges.*

As proved in Art 10.6 $\lim\limits_{n \to \infty} u_n = 0$ is necessary condition for convergence of the series Σu_n.

If $\lim\limits_{n \to \infty} u_n \neq 0$, the series can not converge. Since a positive term series either converges or diverges, it cannot oscillate.

Hence if Σu_n is not convergent, it must be divergent.

(B) *An infinite series of positive terms is divergent if every term after a particular terms, is greater than a fixed positive number however small.*

Ignoring the terms upto and including that fixed term. Let every term be greater than k, a fixed positive number. Then S_n is greater than nk.

By taking n sufficiently large, S_n can be made to exceed any finite number. Hence the series diverges.

(C) *An infinite series of positive terms is convergent if S_n, the sum of first n terms, is less than a fixed number for all values of n.*

A positive term series is either convergent or divergent, it cannot oscillate. Since S_n is less than a fixed number for all n, it cannot diverge. Hence it converges.

Example 10.3 Show that the following series does not converge.

(i) $\sqrt{\dfrac{1}{4}} + \sqrt{\dfrac{2}{6}} + \ldots + \sqrt{\dfrac{n}{2(n+1)}} + \ldots$ *(D.U., 1996)*   (ii) $\sum \cos(1/n)$ *(D.U., 1996)*

Solution. (*i*) We have $u_n = \sqrt{\dfrac{n}{2(n+1)}} = \dfrac{1}{\sqrt{2}}\sqrt{\dfrac{1}{1+(1/n)}}$.

$\therefore$ $\lim\limits_{n \to \infty} u_n = 1/\sqrt{2} \neq 0$. Hence the given series does not converge.

(*ii*) $\lim\limits_{n \to \infty} u_n = \lim\limits_{n \to \infty} \cos(1/n) = 1 \neq 0$. Hence the given series does not converge.

10.9 The p-Series:

$$\sum_{n=1}^{\infty} \frac{1}{n^p} = \frac{1}{1^p} + \frac{1}{2^p} + \frac{1}{3^p} + \ldots \infty, \, p > 0 \text{ is } p\text{-series}$$

convergent if $p > 1$ and divergent if $p \leq 1$.   *(Mysore, 1997S)*

Case I, p > 1

Since the convergence or divergence of a series of positive terms is not altered by a rearrangement of terms, we can write the infinite series in groups in the form

$$\sum_{n=1}^{\infty} \frac{1}{n^p} = \frac{1}{1^p} + \left(\frac{1}{2^p} + \frac{1}{3^p}\right) + \left(\frac{1}{4^p} + \frac{1}{5^p} + \frac{1}{6^p} + \frac{1}{7^p}\right) + \left(\frac{1}{8^p} + \frac{1}{9^p} + \ldots + \frac{1}{15^p}\right) + \ldots \text{ to } \infty$$

Now $\dfrac{1}{1^p} = 1$,

$$\frac{1}{2^p} + \frac{1}{3^p} < \frac{1}{2^p} + \frac{1}{2^p} = \frac{2}{2^p} = \frac{1}{2^{p-1}},$$

$$\frac{1}{4^p} + \frac{1}{5^p} + \frac{1}{6^p} + \frac{1}{7^p} < \frac{1}{4^p} + \frac{1}{4^p} + \frac{1}{4^p} + \frac{1}{4^p} = \frac{4}{4^p} = \frac{1}{4^{p-1}} = \left(\frac{1}{2^{p-1}}\right)^2,$$

$$\frac{1}{8^p}+\frac{1}{9^p}+...+\frac{1}{15^p}<\frac{1}{8^p}+\frac{1}{8^p}+...+\frac{1}{8^p}=\frac{8}{8^p}=\left(\frac{1}{2^{p-1}}\right)^3, \text{ and so on.}$$

Adding we have

$$\frac{1}{1^p}+\left(\frac{1}{2^p}+\frac{1}{3^p}\right)+\left(\frac{1}{4^p}+\frac{1}{5^p}+\frac{1}{6^p}+\frac{1}{7^p}\right)+\left(\frac{1}{8^p}+\frac{1}{9p}+...+\frac{1}{15^p}\right)+...$$

$$<\frac{1}{1^p}+\frac{2}{2^p}+\frac{4}{4^p}+\frac{8}{8^p}+...=\frac{1}{1^p}+\left(\frac{1}{2^{p-1}}\right)+\left(\frac{1}{2^{p-1}}\right)^2+\left(\frac{1}{2^{p-1}}\right)^3+......$$

which is a series in G.P. with common ratio, $r=1/2^{p-1}=2^{1-p}<1$.

Therefore, the series is *convergent* if $r=1/2^{p-1}<1$ or $p>1$.

Case 2, p = 1

The infinite series can be written as

$$1+\frac{1}{2}+\left(\frac{1}{3}+\frac{1}{4}\right)+\left(\frac{1}{5}+\frac{1}{6}+\frac{1}{7}+\frac{1}{8}\right)+...$$

Now $\frac{1}{3}+\frac{1}{4}>\frac{1}{4}+\frac{1}{4}=\frac{1}{2}$ and $\frac{1}{5}+\frac{1}{6}+\frac{1}{7}+\frac{1}{8}>\frac{1}{8}+\frac{1}{8}+\frac{1}{8}+\frac{1}{8}=\frac{1}{2}$ and so on.

$$\therefore \quad 1+\frac{1}{2}+\left(\frac{1}{3}+\frac{1}{4}\right)+\left(\frac{1}{5}+\frac{1}{6}+\frac{1}{7}+\frac{1}{8}\right)+...>1+\frac{1}{2}+\frac{1}{2}+\frac{1}{2}+...$$

The series on the right (omitting first term) is a G.P. with common ratio unity and hence divergent. Therefore the given series is *divergent* when $p=1$,

Case 3, p < 1

$$\frac{1}{1^p}+\frac{1}{2^p}+\frac{1}{3^p}+......\to\infty=\frac{1}{1^p}+\frac{1}{2^p}+\left(\frac{1}{3^p}+\frac{1}{4^p}\right)$$

$$+\left(\frac{1}{5^p}+\frac{1}{6^p}+\frac{1}{7^p}+\frac{1}{8^p}\right)+\text{(sum of next 8 terms)}+.......$$

$$>\frac{1}{1^p}+\left(\frac{1}{2^p}+\frac{2}{4^p}+\frac{4}{8^p}+.....\to\infty\right).$$

The last series inside the bracket is a G.P. whose common ratio $\frac{2}{2^p}=\frac{1}{2^{p-1}}>1$, and hence its sum to infinity is infinity. Therefore, the given series is *divergent* if $p<1$.

Note: The series $\sum\limits_{n=1}^{\infty} \dfrac{1}{n}$ is called *harmonic series and* $\sum \dfrac{1}{n^p}$ *is hyper harmonic series or p-series.*

Example 10.4 Test for convergence the following series

(a) $1 + \dfrac{3}{5} + \dfrac{8}{10} + \dfrac{15}{17} + ... + \dfrac{2^n - 1}{2^n + 1} + ... \infty$ (b) $\dfrac{1}{\sqrt{2}} + \dfrac{2}{\sqrt{5}} + \dfrac{4}{\sqrt{17}} + ... + \dfrac{2^n}{\sqrt{4^n + 1}} + ... \infty$

(c) (i) $\sum \dfrac{1}{n^2}$, (ii) $\sum \dfrac{1}{\sqrt{n}}$.

Solution. (a) $\lim\limits_{n \to \infty} u_n = \lim\limits_{n \to \infty} \dfrac{2^n - 1}{2^n + 1} = \lim\limits_{n \to \infty} \dfrac{1 - \left(1/2^n\right)}{1 + \left(1/2^n\right)} = 1 \neq 0.$

Hence by *n*th term test for divergence [Art 10.8(A)], the series is divergent.

Alternative

Obviously each term after the second is greater than 3/5. Hence by Art 10.8 (B), the series is divergent.

(b) $\lim\limits_{n \to \infty} u_n = \lim\limits_{n \to \infty} \dfrac{2^n}{\sqrt{4^n + 1}} = \lim\limits_{n \to \infty} \dfrac{2^n}{2^n \sqrt{1 + \left(1/4^n\right)}} = \lim\limits_{n \to \infty} \dfrac{1}{\sqrt{1 + \left(1/4^n\right)}} = 1 \neq 0.$

Hence the series is divergent

Alternative

Each term after the first is greater than $1/\sqrt{2}$. Hence by Art 10.8 (B), the series is divergent.

(c) (i) It is a *p*-series with $p = 2$. Since $p > 1$, the series is convergent.

(ii) It is a *p*-series with $p = 1/2$. As $p < 1$, the series is divergent.

10.10 Tests for Convergence

The following results are frequently used to test the convergence of an infinite series.

I. Comparison Tests

(a) Comparison Test for Convergence

If $\sum u_n$ and $\sum v_n$ are two positive terms series such that

(i) $\sum v_n$ is convergent (ii) $u_n \leq v_n$ for all values of *n*, then $\sum u_n$ is also convergent.

Proof. Since Σv_n is convergent, $\lim\limits_{n\to\infty} \Sigma v_n = \lambda$, a finite value. ...(10.1)

Now $\qquad\qquad\qquad\qquad u_1 \le v_1, u_2 \le v_2..., u_n \le v_n$.

Adding, $\qquad\qquad\qquad u_1 + u_2 +...+ u_n \le v_1 + v_2 +...+ v_n$

$$\therefore \lim_{n\to\infty} \left(u_1 + u_2 +...+ u_n\right) \le \lim_{n\to\infty} \left(v_1 + v_2 +...+ v_n\right)$$

that is, $\lim\limits_{n\to\infty} \Sigma u_n \le \lim\limits_{n\to\infty} \Sigma v_n = \lambda$ $\qquad\qquad\qquad\qquad$ [by(10.1)]

Hence the series Σu_n is also convergent.

(b) Comparison Test for Divergence

If Σu_n and Σv_n are two positive terms series such that

(i) Σv_n is divergent (ii) $u_n \ge v_n$ for all values of n, then Σu_n is also divegent.

Proof. Proceed as in part (a) above.

II. Limit Comparison Test

Statement: *Let Σu_n and Σv_n be two infinite series of positive terms and*

$$\lim_{n\to\infty} \frac{u_n}{v_n} = l, \ \ 0 < l < \infty \text{ that is, a finite number } (\ne 0)$$

then both the series Σu_n and Σv_n converge or diverge together.

Proof. By the definition of limit, for a given ε, however small, we can find a number m, such that

$$\left|\frac{u_n}{v_n} - l\right| < \varepsilon \text{ for } n > m \text{ that is, } -\varepsilon < \frac{u_n}{v_n} - l < +\varepsilon$$

or $\quad l - \varepsilon < \dfrac{u_n}{v_n} < l + \varepsilon$ for $n > m$.

Omitting the first m terms of both the series, we have

$$l - \varepsilon < \frac{u_n}{v_n} < l + \varepsilon \text{ for all } n \qquad\qquad ...(10.2)$$

Case I. When Σv_n is convergent, then

$$\lim_{n\to\infty} \left(v_1 + v_2 +...+v_n\right) = k\left(\text{say}\right), \text{ a finite number}$$

From (10.2) $u_n < (l + \varepsilon) v_n$ for all n.

$$\therefore \ \lim_{n\to\infty} \left(u_1 + u_2 +...+u_n\right) < \left(l+\varepsilon\right) \lim_{n\to\infty} \left(v_1 + v_2 +...+v_n\right) = \left(l+\varepsilon\right)k$$

Hence Σu_n is also convergent.

Case II. *When Σv_n is divergent, then in this case*

$$\lim_{n\to\infty}(v_1 + v_2 + ... + v_n) \to \infty \qquad\qquad ...(10.3)$$

Now from (10.2) $\quad l - \varepsilon < \dfrac{u_n}{v_n}$ or $u_n > (l - \varepsilon)\, v_n$ for all n.

Therefore, $\lim\limits_{n\to\infty}(u_1 + u_2 + ... + u_n) > (l - \varepsilon)\lim\limits_{n\to\infty}(v_1 + v_2 + ... + v_n) \to \infty$

Hence Σu_n is also divergent

Remark. For practical purposes, the limit form of comparison test is very useful. To determine convergence or divergence of a given series Σu_n by comparison test, we have to choose another suitable series Σv_n whose convergence or divergence is known and is such that

$$\lim_{n\to\infty}\frac{u_n}{v_n} \text{ is finite } (\neq 0).$$

Then Σu_n converges if Σv_n converges and Σu_n diverges if Σv_n diverges.

The series Σv_n is called *auxiliary series*. The geometric series (Art. 10.5 example 2) and *p*-series (Art. 10.9) will be found useful as auxiliary series.

While using *p*-series for this purpose, v_n is suggested by the form of u_n.

(*i*) If u_n is a fraction then $v_n = \dfrac{1}{n^{p-q}}$ where p and q are highest indices of n in

Denom. and Nume. respectively of u_n.

(*ii*) When u_n can be expanded in ascending powers of $1/n$,

say $\quad u_n = \dfrac{a_1}{n^k} + \dfrac{a_2}{n^{k+1}} + ...$ (a's being constants), then $v_n = \dfrac{1}{n^k}$.

The method will be clear from the following solved examples.

10.11 Illustrative Examples

Example 10.5 Test for convergence the series

(*i*) $\dfrac{1}{1.2.3} + \dfrac{3}{2.3.4} + \dfrac{5}{3.4.5} + ...\infty,$ *(Cochin 1999; Triputi, 1998S)*

(*ii*) $\displaystyle\sum_{n=1}^{\infty}\frac{\sqrt{n}}{n^2+1}$, *(Osmania, 2000S)*, (*iii*) $\dfrac{2}{1^p} + \dfrac{3}{2^p} + \dfrac{4}{3^p} + \dfrac{5}{4^p} + ...\infty.$

Solution. (*i*) We have $u_n = \dfrac{2n-1}{n(n+1)(n+2)}.$

The highest power of n in the denominator is 3 and in the numerator it is 1 that is, $p = 3$, $q = 1$. Hence $p - q = 2$.

Thus we take Σv_n the auxiliary series where $v_n = \dfrac{1}{n^2}$.

Now
$$\frac{u_n}{v_n} = \frac{2n-1}{n(n+1)(n+2)} \times n^2 = \frac{2n^2-n}{(n+1)(n+2)}$$

$$= \frac{2-(1/n)}{(1+1/n)(1+2/n)} \quad \text{(Dividing Nume and Denom. by } n^2\text{)}$$

$$\lim_{n\to\infty} \frac{u_n}{v_n} = \lim_{n\to\infty} \frac{2-(1/n)}{(1+1/n)(1+2/n)} = \frac{2-0}{(1+0)(1+0)} = 2$$

(which is finite and non-zero).

Therefore, both Σu_n and Σv_n converge or diverge together. Since Σv_n is convergent (p-series with $p = 2$). Hence, Σu_n is also convergent.

(ii) Here $u_n = \dfrac{\sqrt{n}}{n^2+1}$. $p = 2$, $q = 1/2$ or $p - q = 3/2$.

Taking Σv_n as the auxiliary series with $v_n = \dfrac{1}{n^{3/2}}$, we have

$$\frac{u_n}{v_n} = \frac{n^{1/2}}{n^2+1} \times n^{3/2} = \frac{n^2}{n^2+1} = \frac{1}{1+(1/n)^2}.$$

$$\therefore \quad \lim_{n\to\infty} \frac{u_n}{v_n} = \lim_{n\to\infty} \frac{1}{1+(1/n)^2} = 1 \text{ (finite and non-zero)}.$$

Since Σv_n is convergent, therefore Σu_n is also convergent.

(iii) We have $u_n = \dfrac{(n+1)}{n^p}$. We compare the series Σu_n with Σv_n where

$$v_n = \frac{1}{n^{p-1}}.$$

$$\frac{u_n}{v_n} = \frac{n+1}{n^p} \cdot n^{p-1} = \frac{n+1}{n} = 1 + \frac{1}{n}.$$

$$\therefore \quad \lim_{n\to\infty} \frac{u_n}{v_n} = \lim_{n\to\infty} \left(1 + \frac{1}{n}\right) = 1 \text{ (finite and non-zero)}.$$

Since Σv_n is convergent when $p - 1 > 1$ and divergent when $p - 1 \leq 1$.
Hence, Σu_n converges when $p - 1 > 1$ that is, $p > 2$ and diverges when
$p - 1 \leq 1$ that is, $p \leq 2$.

Example 10.6 Test for convergence the series:

(i) $\displaystyle\sum_{n=1}^{\infty} \left(\sqrt{n^2 + 1} - n \right)$, (ii) $1 + \dfrac{1}{2^2} + \dfrac{2^2}{3^3} + \dfrac{3^3}{4^4} + \dots \infty$, (iii) $\displaystyle\sum_{n=1}^{\infty} \dfrac{1}{3^n + 1}$.

(Osmania 2005)

Solution. (*i*) Here $u_n = \sqrt{n^2 + 1} - n$

$$= n\left[\left(1 + \frac{1}{n^2} \right)^{1/2} - 1 \right] = n\left[\left(1 + \frac{1}{2n^2} - \frac{1}{8n^4} + \dots \right) - 1 \right] = \frac{1}{2n} - \frac{1}{8n^3} + \dots$$

Let $v_n = \dfrac{1}{n}$. Then $\displaystyle\lim_{n \to \infty} \frac{u_n}{v_n} = \lim_{n \to \infty} \left[\left(\frac{1}{2n} - \frac{1}{8n^3} + \dots \right) \div \frac{1}{n} \right]$

$$= \lim_{n \to \infty} \left(\frac{1}{2} - \frac{1}{8n^2} + \text{terms involving higher powers of } \frac{1}{n} \right)$$

$$= \frac{1}{2} \text{ (finite and non-zero).} \quad \therefore \Sigma u_n \text{ and } \Sigma v_n \text{ converge or diverge together.}$$

But $v_n = 1/n$ diverges (*p*-series with $p = 1$). Hence, Σu_n also diverges.

(*ii*) We have $u_n = \dfrac{(n-1)^{n-1}}{n^n} = n^{n-1} \left(1 - \dfrac{1}{n} \right)^{n-1} \cdot \dfrac{1}{n^n} = \dfrac{1}{n} \left(1 - \dfrac{1}{n} \right)^{n-1}$.

Let $v_n = \dfrac{1}{n}$. Then $\displaystyle\lim_{n \to \infty} \frac{u_n}{v_n} = \lim_{n \to \infty} \left[\frac{1}{n} \left(1 - \frac{1}{n} \right)^{n-1} \div \frac{1}{n} \right]$

$$= \lim_{n \to \infty} \left(1 - \frac{1}{n} \right)^{n-1} = \lim_{n \to \infty} \left[\left(1 - \frac{1}{n} \right)^n \div \left(1 - \frac{1}{n} \right) \right] = e^{-1} \text{ (finite, and non-zero)}$$

$$\left[\because \lim_{n \to \infty} \left(1 - \frac{1}{n} \right)^n = e^{-1} \right]$$

But Σv_n is divergent, therefore Σu_n is also divergent.

(*iii*) Here $u_n = \dfrac{1}{3^n + 1}$. Let $v_n = \dfrac{1}{3^n}$.

$$\therefore \quad \lim_{n\to\infty}\frac{u_n}{v_n} = \lim_{n\to\infty}\left(\frac{1}{3^n+1}\times 3^n\right) = \lim_{n\to\infty}\frac{1}{1+3^{-n}} = 1 \text{ (finite and non-zero)}.$$

Since $\sum\dfrac{1}{3^n}$ is convergent (geometric series with common ratio < 1), therefore Σu_n is also convergent.

Exercise 10.2

Test the convergence or divergence of the following series:

1. $\dfrac{1}{1.2}+\dfrac{1}{3.4}+\dfrac{1}{5.6}+...+\dfrac{1}{(2n-1)2n}+......\infty.$

2. $\dfrac{1}{1.2}+\dfrac{2}{3.4}+\dfrac{3}{5.6}+......\infty.$ 3. $\dfrac{1}{1.3}+\dfrac{2}{3.5}+\dfrac{3}{5.7}+......\infty$ (*Mysore, 1997S*)

4. $\dfrac{1}{1+\sqrt{2}}+\dfrac{2}{1+2\sqrt{3}}+\dfrac{3}{1+3\sqrt{4}}+...\infty$ 5. $\sum_{n=1}^{\infty}\sin\dfrac{1}{n}$ 6. $\sum_{n=2}^{\infty}\left(\dfrac{1}{n\log n}\right).$

7. $\sum_{n=1}^{\infty}\dfrac{1}{n^{1+(1/n)}}.$ **[Hint.** Take $v_n=1/n$ and use $\lim_{n\to\infty}n^{1/n}=1$]

8. $\sum_{n=1}^{\infty}\dfrac{1}{\sqrt{n}+\sqrt{n+1}}$ (*A.M.I.E., W-2010*)   9. $\sum_{n=1}^{\infty}\dfrac{1}{x^n+x^{-n}}.$ (*VTU, 2005*)

10. $\sum\left(\sqrt[3]{n^3+1}-n\right).$ (*VTU 2001, Anthra 2000*)

11. $\sum_{n=1}^{\infty}\left(\sqrt{n^4+1}-\sqrt{n^4-1}\right).$ (*Raipur, 1998, DU 1999*)

12. $\sum_{n=0}^{\infty}\dfrac{2n^3+5}{4n^5+1}.$ 13. $1+\dfrac{2^2+1}{2^3+1}+\dfrac{3^2+1}{3^3+1}+\dfrac{4^2+1}{4^3+1}+......\infty.$

14. $\sum\dfrac{2n}{(n+1)(n+2)(n+3)}.$ 15. $\sum\dfrac{n^2+2^n}{2^n.n^2}$

16. $\dfrac{1.2}{3^2.4^2}+\dfrac{3.4}{5^2.6^2}+\dfrac{5.6}{7^2.8^2}+......+\dfrac{(2n-1)(2n)}{(2n+1)^2(2n+2)^2}+......\infty$

[Hint. (Q. 15) $\sum u_n = \sum\left(\dfrac{1}{2^n}+\dfrac{1}{n^2}\right)=\sum\dfrac{1}{2^n}+\sum\dfrac{1}{n^2}.$ Note that $\sum 1/2^n$ is geometric series and $\sum 1/n^2$ is p-series.]

Ans. 1. Convergent   **2.** Divergent   **3.** Divergent   **4.** Divergent
     5. Divergent   **6.** Divergent   **7.** Divergent   **8.** Divergent
     9. Convergent for $x \leqslant 1$ and Divergent for $x = 1$
      10. Convergent   **11.** Convergent   **12.** Convergent   **13.** Divergent
      14. Convergent   **15.** Convergent   **16.** Convergent

10.12 Comparison Test (Second Form)

Comparison of Ratios

If Σu_n and Σv_n are two positive terms series, then

(A) *Σu_n converges if (i) Σv_n converges, and (ii) from and after some particular*

     term $\dfrac{u_{n+1}}{u_n} < \dfrac{v_{n+1}}{v_n}$.

(B) *Σu_n diverges if (i) Σv_n diverges, and (ii) from and after some particular term*

$$\frac{u_{n+1}}{u_n} > \frac{v_{n+1}}{v_n}.$$

Let the inequality be satisfied from $n = m + 1$ onwards.

Omitting the first m terms of both series, let the series be

$$u_1 + u_2 + u_3 + \dots \text{ and } v_1 + v_2 + v_3 + \dots$$

(A) Since $\dfrac{u_2}{u_1} < \dfrac{v_2}{v_1}, \quad \dfrac{u_3}{u_2} < \dfrac{v_3}{v_2}, \dots\dots$

$$\therefore \quad u_1 + u_2 + u_3 + \dots = u_1\left(1 + \frac{u_2}{u_1} + \frac{u_3}{u_2}\cdot\frac{u_2}{u_1} + \dots\right)$$

$$< u_1\left(1 + \frac{v_2}{v_1} + \frac{v_3}{v_2}\cdot\frac{v_2}{v_1} + \dots\right) < \frac{u_1}{v_1}(v_1 + v_2 + v_3 + \dots)$$

Hence if Σv_n is convergent, Σu_n is also convergent.

(B) Proceed as in part (A) above.

Note. The above test may be stated in the following alternative form also.

If Σu_n and Σv_n and two positive terms series, then

(A) Σu_n converges if (i) Σv_n converges, and (ii) from and after some particular

     term $\dfrac{u_n}{u_{n+1}} > \dfrac{v_n}{v_{n+1}}$.

(B) $\sum u_n$ diverges if (i) $\sum v_n$ diverges, and (ii) from and after some particulor term $\dfrac{u_n}{u_{n+1}} < \dfrac{v_n}{v_{n+1}}$.

10.13 D'Alembert's Test (Ratio Test)

Let $\sum u_n$ be a positive terms series. Let $\lim\limits_{n\to\infty} \dfrac{u_{n+1}}{u_n} = l$.

Then, the series $\sum u_n$ is convergent if $l < 1$ and divergent if $l > 1$. (The test fails when $l = 1$. The theorem does not yield the convergence or divergence of the series).

Proof. By the definition of a limit, for a given ε, however small, we can find a number m such that

$$\left| \frac{u_{n+1}}{u_n} - l \right| < \varepsilon \text{ for } n \geq m \text{ that is, } -\varepsilon < \frac{u_{n+1}}{u_n} - l < +\varepsilon$$

or $\quad l - \varepsilon < \dfrac{u_{n+1}}{u_n} < l + \varepsilon \text{ for all } n \geq m.$

Omitting the first m terms, let the series be
$u_1 + u_2 + u_3 + \ldots \infty$ so that

$$l - \varepsilon < \frac{u_{n+1}}{u_n} < l + \varepsilon \text{ for all } n. \qquad \ldots(10.4)$$

Case I. Let $l < 1$

We can choose ε so small that $l + \varepsilon = k < 1$.

Now $\dfrac{u_{n+1}}{u_n} < k$ for all n. $\qquad\qquad$ [by (10.4)]

$\therefore \quad \dfrac{u_2}{u_1} < k, \ \dfrac{u_3}{u_2} < k, \ \dfrac{u_4}{u_3} < k, \ldots$ and so on.

Then $\qquad\qquad\qquad u_1 + u_2 + u_3 + \ldots \infty$

$$= u_1 \left(1 + \frac{u_2}{u_1} + \frac{u_3}{u_2} \cdot \frac{u_2}{u_1} + \frac{u_4}{u_3} \cdot \frac{u_3}{u_2} \cdot \frac{u_2}{u_1} + \ldots \infty \right)$$

$$< u_1 \left(1 + k + k^2 + k^3 + \ldots \infty \right)$$

$$= \frac{u_1}{1-k} \left(\text{a finite quantity}\right). \text{ Hence } \Sigma u_n \text{ is convergent.}$$

Case 2. Let $l > 1$

We choose ε sufficiently small so that $l - \varepsilon > 1$.

Thus $\dfrac{u_{n+1}}{u_n} > l - \varepsilon > 1$ for all n. [by (10.4)]. Therefore, $\dfrac{u_2}{u_1} > 1, \dfrac{u_3}{u_2} > 1, \ldots \ldots$ and

so on.

Then
$$S_n = u_1 + u_2 + u_3 + \ldots u_n = u_1 \left(1 + \frac{u_2}{u_1} + \frac{u_3}{u_2} \cdot \frac{u_2}{u_1} + \ldots\right)$$

$$> u_1 (1 + 1 + 1 + \ldots \text{to } n \text{ terms})$$

or $S_n > n u_1$. As $n \to \infty$, $S_n \to \infty$. Hence Σu_n is divergent.

Remark. (*i*) When $\lim\limits_{n \to \infty} \dfrac{u_{n+1}}{u_n} = 1$, the Ratio test fails that is, gives no information

about the convergence or divergence of the series.

For example consider the series $\Sigma \dfrac{1}{n}$ and $\Sigma \dfrac{1}{n^2}$.

In the first case $\lim\limits_{n \to \infty} \dfrac{u_{n+1}}{u_n} = \lim\limits_{n \to \infty} \dfrac{1/(n+1)}{1/n} = \lim\limits_{n \to \infty} \dfrac{n}{n+1} = 1.$

For the second series $\lim\limits_{n \to \infty} \dfrac{u_{n+1}}{u_n} = \lim\limits_{n \to \infty} \dfrac{n^2}{(n+1)^2} = 1.$

Thus in both cases $\lim\limits_{n \to \infty} \dfrac{u_{n+1}}{u_n} = 1$ but we know that the first series (*p*-series with

$p = 1$) is divergent and the second series (*p* series with $p = 2$) is convergent.

Hence when $\lim\limits_{n \to \infty} \dfrac{u_{n+1}}{u_n} = 1$, we cannot say whether Σu_n is convergent or diver-

gent. In this case some other test is required to find the nature of the series.

(ii) Alternative form of Ratio test

If Σu_n is a positive terms series such that $\lim\limits_{n \to \infty} \dfrac{u_n}{u_{n+1}} = l$, then the series Σu_n is

convergent if $l > 1$ and divergent if $l < 1$. The test fails when $l = 1$, the series may converge or diverge.

10.14 Illustrative Examples

Example 10.7 Test for convergence the series

$$\frac{1}{1+2}+\frac{2}{1+2^2}+\frac{3}{1+2^3}+...\infty. \qquad \text{(Andhra, 1999)}$$

Solution. We have $u_n = n/(1+2^n)$ and $u_{n+1} = (n+1)/(1+2^{n+1})$.

Now $\dfrac{u_{n+1}}{u_n} = \dfrac{n+1}{n}\cdot\dfrac{1+2^n}{1+2^{n+1}} = \left(1+\dfrac{1}{n}\right)\left(\dfrac{1}{2^n}+1\right)\Big/\left(\dfrac{1}{2^n}+2\right)..$

$\therefore \lim\limits_{n\to\infty}\dfrac{u_{n+1}}{u_n} = \lim\limits_{n\to\infty}\left[\left(1+\dfrac{1}{n}\right)\cdot\left(\dfrac{1}{2^n}+1\right)\Big/\left(\dfrac{1}{2^n}+2\right)\right] = (1+0)\left(\dfrac{0+1}{0+2}\right) = \dfrac{1}{2}$ which is

$<1.\left(\because \dfrac{1}{n} \text{ and } \dfrac{1}{2^n} \to 0 \text{ as } n\to\infty\right).$

Hence the given series is convergent.

Example 10.8 Test for convergence the series

$$1+\frac{2}{5}x+\frac{6}{9}x^2+\frac{14}{17}x^3+\frac{30}{33}x^4+...+\frac{2^n-2}{2^n+1}x^{n-1}+...(x>0)$$
$$\text{(AMIE., S-2007; Andhra 1999; Gulbarga, 1999; D.U. 1998)}$$

Solution. We have $u_n = \dfrac{2^n-2}{2^n+1}x^{n-1}$ and $u_{n+1} = \dfrac{2^{n+1}-2}{2^{n+1}+1}x^n.$

$\dfrac{u_{n+1}}{u_n} = \dfrac{2^{n+1}-2}{2^{n+1}+1}x^n\cdot\dfrac{2^n+1}{2^n-2}\cdot\dfrac{1}{x^{n-1}} = \dfrac{2-\dfrac{2}{2^n}}{2+\dfrac{1}{2^n}}\cdot\dfrac{1+\dfrac{1}{2^n}}{1-\dfrac{2}{2^n}}\cdot x$

$\therefore \lim\limits_{n\to\infty}\dfrac{u_{n+1}}{u_n} = \dfrac{2-0}{2+0}\cdot\dfrac{1+0}{1-0}x = x.$

Thus by Ratio test, the series Σu_n converges for $x<1$ and diverges for $x>1$.

For $x=1$, the test fails. In this case we note that

when $x=1,\ \lim\limits_{n\to\infty}u_n = \lim\limits_{n\to\infty}\dfrac{2^n-2}{2^n+1} = \lim\limits_{n\to\infty}\dfrac{1-(2/2^n)}{1+(1/2^n)} = 1\neq 0.$

$\therefore$ By nth term test, the series Σu_n diverges for $x=1$.

Hence the given series converges for $x<1$ and diverges for $x\geq 1$.

Example 10.9 Test for convergence the series:

(i) $\dfrac{1}{3} + \dfrac{2!}{3^2} + \dfrac{3!}{3^3} + \ldots \infty$, (ii) $1 + \dfrac{2^p}{2!} + \dfrac{3^p}{3!} + \dfrac{4^p}{4!} + \ldots \infty$

Solution. (i) Here $u_n = \dfrac{n!}{3^n}$ and $u_{n+1} = \dfrac{(n+1)!}{3^{n+1}}$.

$$\dfrac{u_{n+1}}{u_n} = \dfrac{(n+1)!}{3^{n+1}} \cdot \dfrac{3^n}{n!} = \dfrac{n+1}{3} . \quad \therefore \lim_{n \to \infty} \dfrac{u_{n+1}}{u_n} = \lim_{n \to \infty} \dfrac{n+1}{3} = \infty.$$

Hence the series diverges.

(ii) $u_n = \dfrac{n^p}{n!}$ and $u_{n+1} = \dfrac{(n+1)^p}{(n+1)!}$.

$$\therefore \quad \dfrac{u_{n+1}}{u_n} = \dfrac{(n+1)^p}{(n+1)!} \times \dfrac{n!}{n^p} = \left(1 + \dfrac{1}{n}\right)^p \cdot \dfrac{1}{n+1}$$

Since $\left(1 + \dfrac{1}{n}\right)^p \to 1$ and $\dfrac{1}{n+1} \to 0$ as $n \to \infty$. $\therefore \lim_{n \to \infty} \dfrac{u_{n+1}}{u_n} = 0.$

Hence the given series is convergent for all values of p.

Example 10.10 Test for convergence the series:

(i) $\displaystyle\sum_{n=1}^{\infty} \dfrac{a^n}{x^n + a^n}$, (ii) $\displaystyle\sum_{n=1}^{\infty} \dfrac{\sqrt{n}}{\sqrt{n^2 + 1}} x^n$ $(x > 0)$. *(Audhra, 1998)*

Solution. (i) Here $u_n = \dfrac{a^n}{\left(x^n + a^n\right)}$, $u_{n+1} = \dfrac{a^{n+1}}{\left(x^{n+1} + a^{n+1}\right)}$.

$$\therefore \quad \dfrac{u_{n+1}}{u_n} = \dfrac{a^{n+1}}{x^{n+1} + a^{n+1}} \cdot \dfrac{x^n + a^n}{a^n} = a \dfrac{x^n + a^n}{x^{n+1} + a^{n+1}} = \dfrac{a}{x}\left(\dfrac{1 + (a/x)^n}{1 + (a/x)^{n+1}}\right)$$

Now $\displaystyle \lim_{n \to \infty} \dfrac{u_{n+1}}{u_n} = \lim_{n \to \infty} \dfrac{a}{x} \cdot \left[\dfrac{1 + (a/x)^n}{1 + (a/x)^{n+1}}\right] = \dfrac{a}{x} < 1$, if $a < x$.

Thus the series converges if $x > a$.

Again $\displaystyle \lim_{n \to \infty} \dfrac{u_{n+1}}{u_n} = \lim_{n \to \infty} a \cdot \dfrac{x^n + a^n}{x^{n+1} + a^{n+1}} = \lim_{n \to \infty} \dfrac{(x/a)^n + 1}{(x/a)^{n+1} + 1} = 1$ if $x < a$.

Hence the ratio test fails.

Now $$\lim_{n\to\infty} \frac{a^n}{a^n + x^n} = \lim_{n\to\infty} \frac{1}{(x/a)^n + 1} = 1 \text{ when } x < a.$$

$\therefore$ By the nth term test, Σu_n diverges.

When $x = a$, the series becomes $\dfrac{1}{2} + \dfrac{1}{2} + \dfrac{1}{2} + \ldots$ which, obviously, is divergent.

Thus the series is divergent when $x \le a$ and convergent when $x > a$.

(ii) Here $$\frac{u_{n+1}}{u_n} = \frac{\sqrt{n+1}}{\sqrt{(n+1)^2 + 1}} x^{n+1} \cdot \frac{\sqrt{n^2 + 1}}{\sqrt{n}} \cdot \frac{1}{x^n}$$

$$= \sqrt{\left(\frac{n+1}{n} \cdot \frac{n^2 + 1}{n^2 + 2n + 2}\right)} \cdot x = \sqrt{\left[\left(1 + \frac{1}{n}\right)\frac{1 + (1/n^2)}{1 + (2/n) + (2/n^2)}\right]} x.$$

Hence $$\lim_{n\to\infty} \frac{u_{n+1}}{u_n} = \lim_{n\to\infty} \sqrt{\left(1 + \frac{1}{n}\right) \cdot \frac{1 + (1/n^2)}{1 + (2/n) + (2/n^2)}} x = x.$$

$\therefore \Sigma u_n$ is convergent for $x < 1$ and divergent for $x > 1$.

When $x = 1$, $u_n = \dfrac{\sqrt{n}}{\sqrt{n^2 + 1}} = \dfrac{n^{1/2}}{n\left(1 + 1/n^2\right)^{1/2}} = \dfrac{1}{n^{1/2}\left(1 + 1/n^2\right)^{1/2}}.$

Taking $v_n = \dfrac{1}{n^{1/2}}$, we have

$$\lim_{n\to\infty} \frac{u_n}{v_n} = \lim_{n\to\infty} \frac{1}{\left(1 + 1/n^2\right)^{1/2}} = 1 \text{ which is finite and non-zero.}$$

Since Σv_n is divergent (being p-series with $p = 1/2$), therefore Σu_n is also divergent.

Thus Σu_n converges for $x < 1$ and diverges for $x \ge 1$.

Exercise 10.3

Test for convergent following series:

1. $1 + \dfrac{1 \cdot 2}{1 \cdot 3} + \dfrac{1 \cdot 2 \cdot 3}{1 \cdot 3 \cdot 5} + \dfrac{1 \cdot 2 \cdot 3 \cdot 4}{1 \cdot 3 \cdot 5 \cdot 7} + \ldots$ 2. $2 + \dfrac{3}{2} \cdot \dfrac{1}{4} + \dfrac{4}{3} \cdot \dfrac{1}{4^2} + \dfrac{5}{4} \cdot \dfrac{1}{4^3} + \ldots$

3. $1 + \dfrac{3}{2!} + \dfrac{5}{3!} + \dfrac{7}{4!} + \ldots \infty$

4. $\dfrac{1}{2\sqrt{1}} + \dfrac{x^2}{3\sqrt{2}} + \dfrac{x^4}{4\sqrt{3}} + \dfrac{x^6}{5\sqrt{4}} + ...\infty$ *(AMIE-S-2010; ISM 2001; Andhra, 2000)*

5. $1 + 3x + 5x^2 + 7x^3 + ...\infty$ **6.** $x + \dfrac{3}{5}x^2 + \dfrac{8}{10}x^3 + \dfrac{15}{17}x^4 + ... + \dfrac{n^2 - 1}{n^2 + 1}x^n + ...\infty$

7. $1 + \dfrac{2!}{2^2} + \dfrac{3!}{3^3} + \dfrac{4!}{4^4} + ...\infty$ *(V.T.U; 2000S)*   **8.** $\sum_{n=1}^{\infty} \dfrac{x^n}{(2n)!}.$ *(Hamirpur, 1995)*

9. $\sum_{n=1}^{\infty} \dfrac{n! 2^n}{n^n}.$ *(Mangalore, 1997)*   **10.** $\sum \dfrac{n!}{3^n}.$

11. $\sum_{n=1}^{\infty} \dfrac{2^n + 1}{3^n + n}.$ **12.** $\sum \dfrac{2^n}{n^2}.$

13. $\sum_{n=1}^{\infty} \dfrac{n^2}{3^n}.$ *(Andhra, 1999)*   **14.** $\sum_{n=1}^{\infty} \dfrac{n^n x^n}{n!} (x > 0).$

15. $\sum_{n=1}^{\infty} \dfrac{x^n}{n(n+1)}.$ **16.** $\sum \dfrac{n^2}{e^n}.$

Ans. 1. Convergent **2.** Covergent **3.** Covergent **4.** Covergent for $x^2 \le 1$; Divergent for $x^2 > 1$ **5.** Convergent for $x < 1$; Divergent for $x \ge 1$ **6.** Convergent for $x < 1$; Divergent for $x \ge 1$ **7.** Convergent **8.** Convergent **9.** Convergent **10.** Divergent **11.** Convergent for $x < 1$; Divergent for $x \ge 1$ **12.** Divergent **13.** Convergent **14.** Convergent for $x < 1$; Divergent for $x \ge 1$ **15.** Convergent for $x \le 1$; Divergent for $x > 1$. **16.** Convergent.

10.15 Raabe's Test (Higher Ratio Test)

Let $\sum u_n$ be a positive terms series. Let

$$\lim_{n \to \infty} n\left(\dfrac{u_n}{u_{n+1}} - 1\right) = l.$$

Then the series $\sum u_n$ is convergent if $l > 1$, and divergent if $l < 1$. The test fails when $l = 1$.

Raabe's test may often give conclusive result, when the ratio test fails.

Proof. Case I *Let $l > 1$.* Choose a number p such that $l > p > 1$. Now compare the given series $\sum u_n$ with the p-series $\sum v_n = \sum \dfrac{1}{n^p}$ which we know converges if $p > 1$ and diverges if $p \le 1$.

Now Σu_n will converges, let from and after some term,

if $\qquad \dfrac{u_n}{u_{n+1}} > \dfrac{v_n}{v_{n+1}}$, for all n

or, if $\qquad \dfrac{u_n}{u_{n+1}} > \dfrac{(n+1)^p}{n^p}$ or $\left(1+\dfrac{1}{n}\right)^p$

or, if $\qquad \dfrac{u_n}{u_{n+1}} > \left(1+\dfrac{1}{n}\right)^p = 1+\dfrac{p}{n}+\dfrac{p(p-1)}{2!}\cdot\dfrac{1}{n^2}+\ldots\ldots$

or, if $\qquad n\left(\dfrac{u_n}{u_{n+1}}-1\right) > p+\dfrac{p(p-1)}{2n}+\ldots\ldots$.

Taking the limit we obtain

$$\lim_{n\to\infty} n\left(\dfrac{u_n}{u_{n+1}}-1\right) > \lim_{n\to\infty}\left[p+\dfrac{p(p-1)}{2n}+\ldots\right] > p$$

or if $l > p$ which is true. Hence Σu_n is convergent.

Case 2. Let $l < 1$. Choose a number p such that $l < p < 1$. Let Σv_n be divergent so that $p < 1$. Now Σu_n will diverges, let from and after some term,

if $\quad \dfrac{u_n}{u_{n+1}} < \dfrac{v_n}{v_{n+1}}$ or if $\dfrac{u_n}{u_{n+1}} < \dfrac{(n+1)^p}{n^p}$

or if $\dfrac{u_n}{u_{n+1}} < \left(1+\dfrac{1}{n}\right)^p = 1+\dfrac{p}{n}+\dfrac{p(p-1)}{2!}\left(\dfrac{1}{n}\right)^2+\ldots\ldots$

Therefore $n\left(\dfrac{u_n}{u_{n+1}}-1\right) < p+\dfrac{p(p-1)}{2n}+\ldots\ldots$

Taking the limit, we obtain $\lim_{n\to\infty} n\left(\dfrac{u_n}{u_{n+1}}-1\right) < p$ or if $l < p$.

But p itself < 1. Hence the given series Σu_n is divergent.

Remark. Raabe's test should be applied only when the D' Alembert's ratio test fails and when in the ratio test, u_n/u_{n+1} does not involve the number e.

Example 10.11. Test for convergence the series

(i) $\dfrac{\alpha}{\beta}+\dfrac{1+\alpha}{1+\beta}+\dfrac{(1+\alpha)(2+\alpha)}{(1+\beta)(2+\beta)}+\ldots\ldots$ $\qquad$ *(D.U. 1999) (ii)* $1 + \dfrac{1}{2}x + \dfrac{1 \cdot 3}{2 \cdot 4}x^2 + \dfrac{1 \cdot 3 \cdot 5}{2 \cdot 4 \cdot 6}x^3 + \dots \infty \ (x > 0)$ *(Andhra, 1998)*

Solution. *(i)* Here $u_n = \dfrac{(1+\alpha)(2+\alpha)(3+\alpha)\dots(n-1+\alpha)}{(1+\beta)(2+\beta)(3+\beta)\dots(n-1+\beta)}$,

and $u_{n+1} = \dfrac{(1+\alpha)(2+\alpha)(3+\alpha)\dots(n-1+\alpha)(n+\alpha)}{(1+\beta)(2+\beta)(3+\beta)\dots(n-1+\beta)(n+\beta)}$

$\therefore \quad \lim\limits_{n\to\infty} \dfrac{u_n}{u_{n+1}} = \lim\limits_{n\to\infty} \dfrac{n+\beta}{n+\alpha} = 1$. Hence the Ratio test fails.

Again $\lim\limits_{n\to\infty} n\left(\dfrac{u_n}{u_{n+1}} - 1\right) = \lim\limits_{n\to\infty} n\left(\dfrac{n+\beta}{n+\alpha} - 1\right) = \lim\limits_{n\to\infty} \dfrac{\beta-\alpha}{1+(\alpha/n)} = \beta - \alpha$.

Thus by Raabe's test, the series converges if $\beta - \alpha > 1$ that is, $\beta > \alpha + 1$, and diverges if $\beta < \alpha + 1$. This test fails for $\beta = \alpha + 1$.
Putting $\beta = 1 + \alpha$, the series becomes

$$\dfrac{\alpha}{1+\alpha} + \dfrac{1+\alpha}{2+\alpha} + \dfrac{1+\alpha}{3+\alpha} + \dfrac{1+\alpha}{4+\alpha} + \dots = \Sigma \dfrac{1+\alpha}{n+\alpha}$$

which diverges, by comparison with $\Sigma 1/n$.

Hence the series converges if $\beta > \alpha + 1$ and diverges if $\beta \le \alpha + 1$.

(ii) Neglecting the first term, we have

$$u_n = \dfrac{1 \cdot 3 \cdot 5 \dots (2n-1)}{2 \cdot 4 \cdot 6 \dots 2n} x^n \quad \text{and} \quad u_{n+1} = \dfrac{1 \cdot 3 \cdot 5 \dots (2n-1)(2n+1)}{2 \cdot 4 \cdot 6 \dots (2n)(2n+2)} x^{n+1}$$

$\therefore \quad \lim\limits_{n\to\infty} \dfrac{u_n}{u_{n+1}} = \lim\limits_{n\to\infty} \dfrac{2n+2}{2n+1} \cdot \dfrac{1}{x} = \dfrac{1}{x}$.

By ratio test, the series converges if $x < 1$ and diverges if $x > 1$.
Ratio test fails if $x = 1$ and we apply Raabe's test and get

$$\lim\limits_{n\to\infty} n\left(\dfrac{u_n}{u_{n+1}} - 1\right) = \lim\limits_{n\to\infty} n\left(\dfrac{2n+2}{2n+1} - 1\right)$$

$$= \lim\limits_{n\to\infty} n\left(\dfrac{1}{2n+1}\right) = \dfrac{1}{2} < 1$$

and so by Raabe's test Σu_n diverges if $x = 1$. Hence Σu_n converges if $x < 1$ and diverges if $x \ge 1$.

Example 10.12. Test for convergence the series

$$\sum \frac{4 \cdot 7 \dots (3n+1)}{1 \cdot 2 \dots n} x^n. \qquad\qquad (V.T.U.; 2000S)$$

Solution. Here $\dfrac{u_n}{u_{n+1}} = \dfrac{4.7\dots(3n+1)}{1.2\dots n} x^n \div \dfrac{4.7\dots(3n+1)(3n+4)}{1.2\dots n(n+1)} x^{n+1}$

$$= \frac{n+1}{3n+4} \cdot \frac{1}{x} = \left[\frac{1+1/n}{3+4/n}\right]\frac{1}{x}. \quad \text{Therefore, } \lim_{n\to\infty} \frac{u_n}{u_{n+1}} = \frac{1}{3x}.$$

Thus by **Ratio test**, the series converges for $\dfrac{1}{3x} > 1$ that is, for $x < \dfrac{1}{3}$ and

diverges for $x > 1/3$. But it fails for $x = 1/3$. Let us try the *Raabe's test*.

Now $\dfrac{u_n}{u_{n+1}} = \left(1 + \dfrac{1}{n}\right)\left(1 + \dfrac{4}{3n}\right)^{-1}.$ [Expand by Binomial Theorem]

$$= \left(1 + \frac{1}{n}\right)\left(1 - \frac{4}{3n} + \frac{16}{9n^2} - \dots\right) = \left(1 - \frac{1}{3n} + \frac{4}{9n^2} + \dots\right)$$

$$\therefore \quad n\left(\frac{u_n}{u_{n+1}} - 1\right) = -\frac{1}{3} + \frac{4}{9n} + \dots$$

$$\lim_{n\to\infty} n\left(\frac{u_n}{u_{n+1}} - 1\right) = \lim_{n\to\infty}\left(-\frac{1}{3} + \frac{4}{9n} + \dots\right) = -\frac{1}{3} \text{ which is } <1.$$

Thus by *Raabe's test*, the series diverges.
Hence the given series converges for $x < 1/3$ and diverges for $x \geq 1/3$.

Example 10.13. Examine the convergence of the series.

$$1 + \frac{2^2}{3^2} + \frac{2^2 \cdot 4^2}{3^2 \cdot 5^2} + \frac{2^2 \cdot 4^2 \cdot 6^2}{3^2 \cdot 5^2 \cdot 7^2} + \dots$$

Solution: Here $u_n = \dfrac{2^2 \cdot 4^2 \dots (2n-2)^2}{3^2 \cdot 5^2 \dots (2n-1)^2}$, and $u_{n+1} = \dfrac{2^2 \cdot 4^2 \dots (2n-2)^2 (2n)^2}{3^2 \cdot 5^2 \dots (2n-1)^2 (2n+1)^2}.$

$\dfrac{u_{n+1}}{u_n} = \dfrac{(2n)^2}{(2n+1)^2}.$ Therefore, $\lim\limits_{n\to\infty} \dfrac{u_{n+1}}{u_n} = \lim\limits_{n\to\infty} \dfrac{(2n)^2}{(2n+1)^2} = 1.$

Thus the ratio test fails.

Again $\lim\limits_{n\to\infty} n\left(\dfrac{u_n}{u_{n+1}}-1\right)=\lim\limits_{n\to\infty}\left[\dfrac{(2n+1)^2-(2n)^2}{(2n)^2}\right]=\lim\limits_{n\to\infty}\dfrac{4n+1}{4n}=1.$

∴ Raabe's test also fails.

Since $\dfrac{u_n}{u_{n+1}}$ does not involve n as an exponent or as a logarithm, the series diverges (See Remark Art 10.15).

Exercise 10.4

Examine the following series for convergence:

1. $1+a+\dfrac{a(a+1)}{1.2}+\dfrac{a(a+1)(a+2)}{1.2.3}+...$ to $\infty\,(a>0)$

2. $\dfrac{x}{1.2}+\dfrac{x^2}{2.3}+\dfrac{x^3}{3.4}+\dfrac{x^4}{4.5}+...\infty$ (*V.T.U., 2000, Gauhati, 1999*)

3. $1+\dfrac{3}{7}x+\dfrac{3.6}{7.10}x^2+\dfrac{3.6.9}{7.10.13}x^3+\dfrac{3.6.9.12}{7.10.13.16}x^4+...$ to ∞ (*D.U. 1998*)

4. $x+\dfrac{1}{2}\cdot\dfrac{x^3}{3}+\dfrac{1.3}{2.4}\cdot\dfrac{x^5}{5}+\dfrac{1.3.5}{2.4.6}\cdot\dfrac{x^7}{7}+...$ to $\infty\,(x>0)$

5. $1+\dfrac{1}{2}\cdot\dfrac{x^2}{4}+\dfrac{1.3.5}{2.4.6}\cdot\dfrac{x^4}{8}+\dfrac{1.3.5.7.9}{2.4.6.8.10}\cdot\dfrac{x^6}{12}+...$ to ∞ (*Roorkee, 2000*)

6. (a) $\sum\limits_{n=1}^{\infty}\dfrac{(2n)!}{(n!)^2}\cdot x^n,\ x>0$ (b) $\sum\dfrac{(n!)^2}{(2n)!}\cdot x^{2n}$

7. $\sum\limits_{n=1}^{\infty}\dfrac{1\cdot3\cdot5......(2n-1)}{2.4.6......2n}x^n,\ x>0$ (*D.U., 1996*)

8. $\sum\limits_{n=1}^{\infty}\dfrac{2\cdot4\cdot6......(2n+2)}{3\cdot5\cdot7......(2n+3)}x^{n-1}\,(x>0)$ (*D.U. 1996*)

9. $x^2(\log2)^p+x^3(\log3)^p+x^4(\log4)^p+...$ to ∞

10. $1+\dfrac{\alpha\cdot\beta}{1\cdot\gamma}x+\dfrac{\alpha(\alpha+1)\beta(\beta+1)}{1\cdot2\cdot\gamma(\gamma+1)}x^2+\dfrac{\alpha(\alpha+1)(\alpha+2)\beta(\beta+1)(\beta+2)}{1\cdot2\cdot3\gamma(\gamma+1)(\gamma+2)}x^3+....$to ∞ (*Ranchi, 1998*)

[**Hint.** Neglecting first term, we have $u_{n+1} = u_n \dfrac{(\alpha+n)(\beta+n)}{(n+1)(\gamma+n)} x$. Therefore

$$\lim_{n\to\infty} \frac{u_{n+1}}{u_n} = x.$$ Hence the series converges for $x < 1$ and diverges for $x > 1$.

When $x = 1$, ratio test fails. Applying Raabe's test

$$\lim_{n\to\infty} n\left(\frac{u_n}{u_{n+1}} - 1\right) = \lim_{n\to\infty} n\left[\frac{(n+1)(n+\gamma)}{(n+\alpha)(n+\beta)} - 1\right]$$

$$= \lim_{n\to\infty} n\left[\frac{n(1+\gamma-\alpha-\beta)+\gamma-\alpha\beta}{n^2 + n(\alpha+\beta)+\alpha\beta}\right] = 1+\gamma-\alpha-\beta.$$

Thus the series converges for $1+\gamma-\alpha-\beta > 1$ that is $\gamma > \alpha+\beta$ and diverges when $\gamma < \alpha+\beta$. But, when $\gamma = \alpha+\beta$, Raabe's test fails. Since u_n/u_{n+1} does not involve n as an exponent or as a logarithm, therefore the series Σu_n diverges for $\gamma = \alpha+\beta$. Hence, the series converges for $x < 1$ and diverges for $x > 1$. When $x = 1$, the series converges for $\gamma > \alpha+\beta$ and diverges for $\gamma \le \alpha+\beta$.]

Ans. 1. Convergent for $a \le 0$; Divergent for $a > 0$, **2.** Convergent for $x \le 1$; Divergent for $x > 1$, **3.** Convergent for $x \le 1$; Divergent for $x > 1$, **4.** Convergent for $x^2 \le 1$; Divergent for $x^2 > 1$, **5.** Convergent for $x^2 \le 1$; Divergent for $x^2 > 1$, **6.** (a) Convergent for $x < 1/4$; Divergent for $x \ge 1/4$ (b) Convergent for $x^2 < 4$; Divergent for $x^2 \ge 4$, **7.** Convergent for $x < 1$; Divergent for $x \ge 1$, **8.** Convergent for $x < 1$; Divergent for $x \ge 1$, **9.** Convergent for $x < 1$; Divergent for $x \ge 1$.

10.16 Cauchy's Root Test

Let Σu_n be a positive terms series. Let

$$\lim_{n\to\infty} (u_n)^{1/n} = l.$$

Then, the series Σu_n is convergent if $l < 1$ and divergent if $l > 1$. The test fails when $l = 1$.

Proof. By the definition of a limit, for a given ε, however small, we can find a number m such that

$$\left|(u_n)^{1/n} - l\right| < \varepsilon \text{ for } n > m$$

or $\quad l - \varepsilon < (u_n)^{1/n} < l + \varepsilon \text{ for } n > m.$

Omitting the first m terms of the series, we have

$$l - \varepsilon < \left(u_n\right)^{1/n} < l + \varepsilon \text{ for all } n.$$

Case I. Let $\lim_{n \to \infty} \left(u_n\right)^{1/n} = l < 1$ ...(10.5)

We can choose ε so small that $l + \varepsilon = k < 1$.   [from (10.5)]

Now $\left(u_n\right)^{1/n} < k$ for all n or $u_n < k^n$ for all n.

$\therefore \quad u_1 + u_2 + ... + \infty < k + k^2 + ... + k^n + ... \infty$ that is, $< \dfrac{1}{1-k}$ (a finite quantity).

Therefore, the series Σu_n is convergent.

Case 2. Let $\lim_{n \to \infty} \left(u_n\right)^{1/n} = l > 1.$

We can choose ε so small that $l - \varepsilon > 1.$

$\therefore \quad \left(u_n\right)^{1/n} > l - \varepsilon$ [from (10.5)]

That is, $\quad\quad\quad\quad\quad \left(u_n\right)^{1/n} > 1$ or $u_n > 1$.

$$S_n = u_1 + u_2 + u_3 + ... + u_n > n.$$

$\therefore \quad \lim_{n \to \infty} S_n \to \infty$. Hence, the given series Σu_n is divergent.

Remarks. 1. If $\lim_{n \to \infty} \left(u_n\right)^{1/n} = 1$, the root test fails that is, gives no information about the nature of the series as regards convergence or divergence.

For example consider the series $\Sigma \dfrac{1}{n^p}$.

Here $\lim_{n \to \infty} u_n^{1/n} = \lim_{n \to \infty} \left(\dfrac{1}{n^p}\right)^{1/n} = \lim_{n \to \infty} \left(\dfrac{1}{n^{1/n}}\right)^p = 1$ for all values of p.

But $\Sigma \dfrac{1}{n^p}$ converges when $p > 1$ and diverges when $p \le 1$.

Thus, when $\lim_{n \to \infty} \left(u_n\right)^{1/n} = 1$, we can not say whether Σu_n is convergent or divergent.

2. Try this test when u_n involves nth powers.

Example 10.14. Test the convergence of the following series

   (i) $\displaystyle\sum_{n=1}^{\infty} 1/n^n$, *(Gulbarga, 1999)*   (ii) $\Sigma \left(1 + \dfrac{1}{n}\right)^{n^2}$ *(Assam, 1998)* (*iii*) $\dfrac{1}{2}+\dfrac{2}{3}x+\left(\dfrac{3}{4}\right)^2 x^2+\left(\dfrac{4}{5}\right)^3 x^3+...\infty$ (*Assam, 1999; Mangalore, 1999*)

Solution. (*i*) We have $u_n = 1/n^n$. Therefore, $(u_n)^{1/n} = 1/n$.

$$\lim_{n\to\infty} u_n^{1/n} = \lim_{n\to\infty}\frac{1}{n} = 0 \text{ that is, } <1.$$

Hence the given series is convergent.

(*ii*) We have $u_n = \left(1+(1/n)\right)^{n^2}$.

Therefore $(u_n)^{1/n} = \left[\left(1+(1/n)\right)^{n^2}\right]^{1/n} = \left(1+(1/n)\right)^n$,

and $\lim_{n\to\infty} u_n^{1/n} = \lim_{n\to\infty}\left(1+1/n\right)^n = e$ which is >1.

Hence the series is divergent.

(*iii*) Omitting the first term, we obtain

$$u_n = \left(\frac{n+1}{n+2}\right)^n x^n. \quad \therefore \quad \lim_{n\to\infty} u_n^{1/n} = \lim_{n\to\infty}\frac{n+1}{n+2}x = \lim_{n\to\infty}\frac{(1+1/n)}{(1+2/n)}x = x.$$

Hence the series converges if $x < 1$ and diverges for $x > 1$.

When $x = 1$, $u_n = \left(\dfrac{n+1}{n+2}\right)^n = \dfrac{(1+1/n)^n}{(1+2/n)^n} = \left[1+(1/n)\right]^n \div \left[\{1+(2/n)\}^{n/2}\right]^2$

$\therefore \lim_{n\to\infty} u_n = \lim_{n\to\infty}\dfrac{(1+1/n)^n}{\left[(1+2/n)^{n/2}\right]^2} = \dfrac{e}{e^2} = \dfrac{1}{e}(\neq 0).$ The series is divergent, when $x = 1$

Hence, the given series converges if $x < 1$ and diverges if $x \geq 1$.

10.17 Cauchy's Integral Test

A *positive-terms series* $f(1)+f(2)+f(3)+...+f(n)+...$,
where *f(n)* does not increase with n, converges or diverges according as the integral

$$\int_1^\infty f(x)\,dx \quad ...(10.6)$$ is finite or infinite.

Proof. In the adjoining figure the area under the curve from $x = 1$ to $x = n + 1$ lies between the sum of the areas of the small rectangles and the sum of the large rectangles, all rectangles having unit bases.

Therefore, $f(1) + f(2) + ... + f(n)$

$$\geq \int_1^{n+1} f(x)\,dx$$

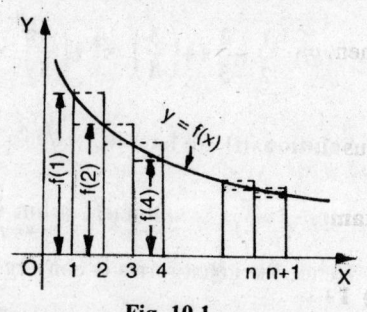

Fig. 10.1.

$$\geq f(2) + f(3) + ... + f(n+1)$$

or $\quad S_n \geq \int_1^{n+1} f(x)\,dx \geq S_{n+1} - f(1)$.

As $n \to \infty$, we find from the second inequality that if the integral has a finite value then $\lim\limits_{n \to \infty} S_{n+1}$ is also finite and, therefore, $\Sigma f(n)$ converges.

Similarly if the integral is infinite then we see from the first inequality that $\lim\limits_{n \to \infty} S_n \to \infty$ and, therefore, the series diverges.

Remark. This test only applies to series that have positive terms. Try this test when $f(x)$ is easy to integrate.

Example 10.15. Apply the integral test to determine the convergence of the p-series.

$$\sum_{n=1}^{\infty} \frac{1}{n^p} = \frac{1}{1^p} + \frac{1}{2^p} + ... + \frac{1}{n^p} + ... \infty. \qquad \textit{(Andhra, 1999; Mysore, 1999S)}$$

Solution. Here $f(x) = \dfrac{1}{x^p}$ which decreases as x increases $(x \geq 1)$.

$\therefore$ By the integral test the given series is convergent or divergent according as the integral

$$\int_1^{+\infty} \frac{1}{x^p}\,dx \quad \text{is finite or infinite.}$$

When $p \neq 1$, $\displaystyle\int_1^{+\infty} \frac{1}{x^p}\,dx = \lim_{u \to +\infty} \int_1^u \frac{1}{x^p}\,dx = \lim_{u \to +\infty} \left| \frac{x^{1-p}}{1-p} \right|_1^u = \frac{1}{p-1}\left[\lim_{u \to +\infty} u^{1-p} - 1 \right].$

Now if $p > 1$, $\dfrac{1}{1-p}\left[\lim\limits_{u \to +\infty} u^{1-p} - 1 \right] = \dfrac{1}{1-p}\left[\lim\limits_{u \to +\infty} \dfrac{1}{u^{p-1}} - 1 \right] = \dfrac{1}{p-1}$,

which is finite. Hence the series converges when $p > 1$.

When $p < 1$, $\dfrac{1}{1-p}\left[\lim\limits_{u \to +\infty} u^{1-p} - 1 \right] = +\infty$. Therefore, the series diverges.

When $p = 1$, $\int_1^{+\infty} \dfrac{1}{x} dx = \lim_{u \to +\infty} |\log u| = +\infty$. $\therefore$ The series diverges.

Thus $\Sigma \dfrac{1}{n^p}$ converges when $p > 1$ and diverges when $p \le 1$.

Example 10.16. **Examine the given series for convergence**

(a) $1 + \dfrac{1}{\sqrt{3}} + \dfrac{1}{\sqrt{5}} + \dfrac{1}{\sqrt{7}} + \dots,$ (b) $\dfrac{1}{2\log 2} + \dfrac{1}{3\log 3} + \dfrac{1}{4\log 4} + \dots.$

(c) $\sum \dfrac{(\log n)}{n}.$ *Alternative* $\sum_{n=2}^{\infty} \dfrac{1}{n \log n}$

Solution. (a) Here $S_n = \dfrac{1}{\sqrt{2n-1}}$. Let $f(x) = \dfrac{1}{\sqrt{2x-1}}$.

On $[1, +\infty]$, $f(x) > 0$ and f is decreasing.

$$\int_1^{+\infty} \frac{1}{\sqrt{2x-1}} dx = \lim_{u \to +\infty} \int_1^u \frac{dx}{\sqrt{2x-1}} = \lim_{u \to +\infty} \frac{1}{2} \int_1^u (2x-1)^{-1/2}(2) dx$$

$$= \lim_{u \to +\infty} \frac{1}{2}(2) \Big|(2x-1)^{1/2}\Big|_1^u = \lim_{u \to +\infty} \left((2u-1)^{1/2} - 1\right) = +\infty$$

Hence, the series diverges.

(b) Here $S_n = \dfrac{1}{n \log n}$ is defined for $n \ge 2$.

$$\int_2^{+\infty} \frac{dx}{x \log x} = \lim_{u \to +\infty} \int_2^u \frac{dx}{x \log x} = \lim_{u \to +\infty} \Big|\log(\log u)\Big|_2^u$$

$$= \lim_{u \to +\infty} \left(\log(\log u) - \log(\log 2)\right) = +\infty$$

Hence, the series diverges by the integral test.

(c) Let $f(x) = \dfrac{\log x}{x}$. Now $\int_1^{+\infty} \dfrac{\log x}{x} dx = \lim_{u \to +\infty} \int_1^u \dfrac{\log x}{x} dx$

$$= \lim_{u \to +\infty} \frac{1}{2} \Big[(\log x)^2\Big]_1^u = \lim_{u \to +\infty} \frac{1}{2}\left((\log u)^2 - 0\right) = +\infty$$

Hence, by the integral test, $\sum \dfrac{\log n}{n}$ diverges.

Example 10.17. Examine the convergence, using the integral test.

(i) $\sum \dfrac{1}{n^2}$,   (ii) $\sum\limits_{n=1}^{\infty} ne^{-n^2}$.

Solution. (i) Let $f(x) = \dfrac{1}{x^2}$. Now $\int\limits_{1}^{+\infty} \dfrac{1}{x^2}\,dx = \lim\limits_{u \to +\infty} \int\limits_{1}^{u} \dfrac{1}{x^2}\,dx = \lim\limits_{u \to +\infty} \left| -\dfrac{1}{x} \right|_{1}^{u}$

$= \lim\limits_{u \to +\infty} -\left(\dfrac{1}{u} - 1 \right) = 1$.

Hence, by the integral test, $\sum \dfrac{1}{n^2}$ converges.

(ii) Let $f(x) = xe^{-x^2}$ which decreases as x increases. Now

$\int\limits_{1}^{+\infty} xe^{-x^2}\,dx = \lim\limits_{u \to +\infty} \int\limits_{1}^{u} xe^{-x^2}\,dx = \lim\limits_{u \to +\infty} \left[-\dfrac{1}{2} e^{-x^2} \right]_{1}^{u}$

$= -\dfrac{1}{2} \cdot \lim\limits_{u \to +\infty} \left(e^{-u^2} - e^{-1} \right) = \dfrac{e^{-1}}{2} = \dfrac{1}{2e}$, which is finite.

Hence the given series converges.

10.18 Logarithmic Test

Let $\sum u_n$ be a positive terms series. Let

$$\lim\limits_{n \to \infty} \left(n \log \dfrac{u_n}{u_{n+1}} \right) = l.$$

Then the series $\sum u_n$ convergent if $l > 1$ and divergent if $l < 1$. The test fails when $l = 1$.

Proof: Case 1. Let $l > 1$. Choose a number p such that $l > p > 1$. Compare the given series $\sum u_n$ with the auxiliary series $\sum v_n = \sum \left(1/n^p \right)$ which is convergent if $p > 1$.

Now $\sum u_n$ will be convergent if, from and after a particular term,

$$\dfrac{u_n}{u_{n+1}} > \dfrac{v_n}{v_{n+1}} \text{ for all values of } n$$

or if $\dfrac{u_n}{u_{n+1}} > \dfrac{(n+1)^p}{n^p}$ or $\left(1 + \dfrac{1}{n} \right)^p$.

Taking logarithms of both sides

$$\log \frac{u_n}{u_{n+1}} > p \log\left(1+\frac{1}{n}\right) \quad \text{or if} \quad \log \frac{u_n}{u_{n+1}} > p\left(\frac{1}{n} - \frac{1}{2n^2} + \frac{1}{3n^3} - \cdots\right)$$

[Expanding log [1+1/n] in powers of 1/n]

or if $n \log \dfrac{u_n}{u_{n+1}} > p\left(1 - \dfrac{1}{2n} + \dfrac{1}{3n^2} - \cdots\right)$

or if $\lim\limits_{n\to\infty} n \log \dfrac{u_n}{u_{n+1}} > p$ that is, $l > p$ which is true since $l > p > 1$.

Hence Σu_n is convergent.

Case 2. *Let $l < 1$.*

Let Σv_n be divergent, so that $p \le 1$.

Then Σu_n will also diverges if $\dfrac{u_n}{u_{n+1}} < \dfrac{v_n}{v_{n+1}}$ or if $\log \dfrac{u_n}{u_{n+1}} < \log \dfrac{v_n}{v_{n+1}}$

or if $\log \dfrac{u_n}{u_{n+1}} < \log\left(1+\dfrac{1}{n}\right)^p = p \log\left(1+\dfrac{1}{n}\right)$

or if $\log \dfrac{u_n}{u_{n+1}} < p\left(\dfrac{1}{n} - \dfrac{1}{2n^2} + \dfrac{1}{3n^3} - \cdots\right)$

or if $n \log \dfrac{u_n}{u_{n+1}} < p\left[1 - \dfrac{1}{2n} + \dfrac{1}{3n^2} - \cdots\right]$

or if $\lim\limits_{n\to\infty} n \log \dfrac{u_n}{u_{n+1}} < p$. That is, $l < p$ which is true since $l < p < 1$.

Hence Σu_n is divergent.

Remarks. 1. This test is an alternative to Raabe's test and should be applied when

D' Alembert's ratio test fails and when either (*i*) n occurs as an exponent in $\dfrac{u_n}{u_{n+1}}$

or (*ii*) taking logarithm of $\dfrac{u_n}{u_{n+1}}$ makes the evaluation of limits easier.

2. If $\dfrac{u_n}{u_{n+1}}$ does not involve n as an exponent or a logarithm, the series

Σu_n diverges.

Example 10.18. **Test for convergence and divergence the following series**

(i) $\quad 1 + \dfrac{2x}{2!} + \dfrac{3^2 x^2}{3!} + \dfrac{4^3 x^3}{4!} + \dfrac{5^4 x^4}{5!} + ...$

(ii) $\quad x + \dfrac{2^2 x^2}{2!} + \dfrac{3^3 x^3}{3!} + \dfrac{4^4 x^4}{4!} + \dfrac{5^5 x^5}{5!} + ...$ $\qquad$ *(J.N.T.U., 1998, Ranchi, 1998)*

Solution. (i) Let $\sum u_n$ be the given series, then

$$u_n = \frac{n^{n-1} x^{n-1}}{n!} \text{ and } u_{n+1} = \frac{(n+1)^n x^n}{(n+1)!} .$$

$$\frac{u_n}{u_{n+1}} = \frac{n^{n-1}}{(n+1)^n} \cdot \frac{n+1}{x} = \frac{(1+1/n) \cdot 1}{\left[1 + (1/n)\right]^n} \cdot \frac{1}{x}$$

Therefore, $\qquad \lim_{n \to \infty} \dfrac{u_n}{u_{n+1}} = \dfrac{1}{ex}.$ $\qquad \left[\because \lim_{n \to \infty} \left(1 + \dfrac{1}{n}\right)^n \to e \right]$

From Ratio test, the series $\sum u_n$ is convergent or divergent according as $x <$ or $>$

$(1/e)$. If $x = 1/e$, then $\lim_{n \to \infty} \dfrac{u_n}{u_{n+1}} = 1$, and so Ratio test fails and we apply logarithmic

test.

Now $\lim_{n \to \infty} n \log \dfrac{u_n}{u_{n+1}} = \lim_{n \to \infty} n \log \left[\dfrac{e}{(1 + 1/n)^{n-1}} \right]$

$$= \lim_{n \to \infty} n \left\{ \log e - (n-1) \log \left(1 + \frac{1}{n}\right) \right\}$$

$$= \lim_{n \to \infty} n \left\{ 1 - \frac{n-1}{n} + \frac{n-1}{2n^2} - \frac{n-1}{3n^3} + ... \right\}, \qquad \left[\text{on expanding } \log\{1 + (1/n)\} \right]$$

$$= \lim_{n \to \infty} \left\{ n - (n-1) + \left(\frac{1}{2} - \frac{1}{2n}\right) - \frac{(n-1)}{3n^2} + ... \right\}$$

$= 1 + \dfrac{1}{2} = \dfrac{3}{2}$, which is greater than 1, Thus from logarithmic test the series $\sum u_n$

is convergent when $x = 1/e$.

We conclude that the given series is convergent if $x \le (1/e)$ and divergent if

$x > (1/e)$.

(*ii*) Let Σu_n be the given series, then we have

$$u_n = \frac{n^n x^n}{n!} \text{ and } u_{n+1} = \frac{(n+1)^{n+1} x^{n+1}}{(n+1)!}.$$

Therefore $\lim\limits_{n \to \infty} \dfrac{u_n}{u_{n+1}} = \lim\limits_{n \to \infty} \dfrac{(n+1)!}{(n+1)^{n+1} x^{n+1}} \times \dfrac{n^n x^n}{n!}$

$$= \lim\limits_{n \to \infty} \frac{n^n}{(n+1)^n x} = \lim\limits_{n \to \infty} \frac{1}{(1+1/n)^n} \cdot \frac{1}{x} = \frac{1}{ex}.$$

Thus by Ratio test, the series converges for $x < 1/e$, and diverges for $x > 1/e$. If $x = 1/e$, the ratio test fails and we apply logarithmic test. We thus have

$$\lim\limits_{n \to \infty} n \log \frac{u_n}{u_{n+1}} = \lim\limits_{n \to \infty} n \log \left\{ \left(1+\frac{1}{n}\right)^{-n} e \right\} = \lim\limits_{n \to \infty} n \left\{ \log e - n \log \left(1+\frac{1}{n}\right) \right\}$$

$$= \lim\limits_{n \to \infty} n \left\{ 1 - n \left(\frac{1}{n} - \frac{1}{2n^2} + \frac{1}{3n^3} - \cdots \right) \right\} = \lim\limits_{n \to \infty} \left\{ \frac{1}{2} - \frac{1}{3n} + \cdots \right\}$$

$$= \frac{1}{2} \text{ which is } < 1. \text{ Thus by the } logarithmic \text{ } test, \text{ the series diverges if } x = 1/e.$$

Hence the given series converges for $x < 1/e$ and diverges for $x \geq 1/e$.

Exercise 10.5

Examine the convergence of the following series using root test:

1. $\Sigma \left(1 + \left(1/\sqrt{n} \right) \right)^{-n^{3/2}}$ (*V.T.U., 2000S; Hamirpur, 1996S*)

2. $\sum\limits_{n=1}^{\infty} \dfrac{n!}{n^n}$ (*AMIE, W-2006*)   **3.** $\sum n^{n^2} \div (n+1)^{n^2}$

4. $\sum 2^{-n-(-1)^n}$ **5.** $\sum\limits_{n=1}^{\infty} n/5^n$ **6.** $1 + \dfrac{x}{2} + \dfrac{x^2}{3^2} + \dfrac{x^3}{4^3} + \cdots \infty \ (x > 0)$

7. $\left(\dfrac{2^2}{1^2} - \dfrac{2}{1} \right)^{-1} + \left(\dfrac{3^3}{2^3} - \dfrac{3}{2} \right)^{-2} + \left(\dfrac{4^4}{3^4} - \dfrac{4}{3} \right)^{-3} + \cdots \infty$

8. $\Sigma \left(\dfrac{n}{n+1} \right)^{n^2}$ or $\Sigma \left(1 + \dfrac{1}{n} \right)^{-n^2}$ **9.** $\Sigma \dfrac{(n+1)^n x^n}{n^{n+1}}$

10. (a) $\dfrac{1}{\sqrt{3}}+\dfrac{1}{\sqrt{5}}+\dfrac{1}{\sqrt{7}}+\dfrac{1}{\sqrt{9}}+...$ (b) $\dfrac{1}{4}+\dfrac{1}{16}+\dfrac{1}{36}+\dfrac{1}{64}+...$

(c) $\sin\pi+\dfrac{1}{4}\sin\dfrac{\pi}{2}+\dfrac{1}{9}\sin\dfrac{\pi}{3}+\dfrac{1}{16}\sin\dfrac{\pi}{4}+......$

[Hint: (a) $S_n=\dfrac{1}{\sqrt{2n+1}}$; take $f(x)=\dfrac{1}{\sqrt{2x+1}}$.

(b) $S_n=\dfrac{1}{4n^2}$; take $f(x)=\dfrac{1}{4x^2}$.

(c) $S_n=\dfrac{1}{n^2}\sin\dfrac{\pi}{n}$; take $f(x)=\dfrac{1}{x^2}\sin\dfrac{\pi}{x}$.

On the interval $0>x>2$, $f(x)>0$ and decreases as x increases.

Now $\displaystyle\int_2^{+\infty}f(x)dx=\lim_{a\to+\infty}\int_2^a\dfrac{1}{x^2}\sin\dfrac{\pi}{x}dx=\dfrac{1}{\pi}\lim_{a\to+\infty}\left|\cos\dfrac{\pi}{x}\right|_2^a=\dfrac{1}{\pi}$.

11. $\displaystyle\sum\dfrac{50}{n(n+1)}$.   **12.** $\displaystyle\sum\dfrac{1}{\sqrt{n}}$.   **13.** $\displaystyle\sum_{n=1}^{\infty}\dfrac{1}{n^2+1}$.

14. $\displaystyle\sum_{n=1}^{\infty}\dfrac{n^2}{2n^3-1}$.   **15.** $\displaystyle\sum_{n=2}^{\infty}\dfrac{1}{n\log n}$.   **16.** $\displaystyle\sum_{n=1}^{\infty}\dfrac{\log n}{n}$.

17. $\displaystyle\sum n^2/e^n$.

18. Test for convergence, using logarithmic test

$$1+\dfrac{x}{2}+\dfrac{2!}{3^2}x^2+\dfrac{3!}{4^3}x^3+......$$

Ans. 1. Convergent **2.** Convergent **3.** Convergent **4.** Convergent **5.** Convergent **6.** Convergent **7.** Convergent **8.** Convergent **9.** Convergent for $x<1$, Divergent for $x\geq1$ **10.** (a) Divergent (b) Convergent (c) Convergent **11.** Convergent **12.** Divergent **13.** Convergent **14.** Divergent **15.** Divergent **16.** Divergent **17.** Convergent **18.** Convergent for $x<e$; Divergent for $x\geq e$.

10.19 Alternating Series

Def. *A series whose terms are alternately positive and negative is said to be an alternating series. It can be written in the form*

$$\sum_{n=1}^{\infty} (-1)^{n-1} u_n = u_1 - u_2 + u_3 - u_4 + u_5 - \dots (u_n > 0)$$

where $u_1, u_2, \dots u_n$ are all positive.

Thus $1 - \dfrac{1}{2} + \dfrac{1}{3} - \dfrac{1}{4} + \dots$ to ∞, $\qquad\qquad 1 - \dfrac{1}{\sqrt{2}} + \dfrac{1}{\sqrt{3}} - \dfrac{1}{\sqrt{4}} + \dots$ to ∞,

$\dfrac{1}{\log 2} - \dfrac{1}{\log 3} + \dfrac{1}{\log 4} - \dfrac{1}{\log 5} + \dots$ to ∞, etc. are some examples of an alternating series.

10.20 Leibnitz's Test for Alternating Series

If an alternating series

$$u_1 - u_2 + u_3 - u_4 + \dots\dots (u_n > 0) \text{ is such that}$$

(i) each term is numerically less than its preceeding term that is, $|u_{n+1}| < |u_n|$

for all values of n and (ii) $\lim\limits_{n \to \infty} u_n = 0$, then the series is convergent.

Proof. Consider S_{2n}, the sum of an even number of terms of the series

$$S_{2n} = (u_1 - u_2) + (u_3 - u_4) + \dots + (u_{2n-1} - u_{2n}) \qquad \dots(10.7)$$

This can be written in the form

$$S_{2n} = u_1 - (u_2 - u_3) - (u_4 - u_5) - \dots - (u_{2n-2} - u_{2n-1}) - u_{2n}$$
$$= u_1 - \left[(u_2 - u_3) + (u_4 - u_5) + \dots + (u_{2n-2} - u_{2n-1}) + u_{2n} \right] \qquad \dots(10.8)$$

From (10.7) S_{2n} is positive and it increases with n.

($\because u_1 > u_2 > u_3 > \dots$ and hence $u_1 - u_2$, $u_3 - u_4$ etc are all + ve).

From (10.8) S_{2n} is always less than u_1.

Therefore, S_{2n} must tends to a finite limit as $n \to \infty$.

Now consider S_{2n+1}, the sum of an odd number of terms.

Since $S_{2n+1} = S_{2n} + u_{2n+1}$.

Therefore, $\qquad \lim\limits_{n \to \infty} S_{2n+1} = \lim\limits_{n \to \infty} S_{2n} + \lim\limits_{n \to \infty} u_{2n+1} \qquad \dots(10.9)$

$\qquad = \lim\limits_{n \to \infty} S_{2n} \qquad \left(\because \lim\limits_{n \to \infty} u_{2n+1} = 0 \right)$

Thus S_n tends to the same limit whether n is even or odd.

Hence the given series is convergent.

Note. It should be noted that $\lim\limits_{n \to \infty} u_n = 0$ is necessary for convergence. If $\lim u_n \neq 0$ then from (10.9), S_{2n} and S_{2n+1} tend to two different limits according as n is even or odd and therefore, the alternating series is oscillatory.

Example 10.19. Show that the series

$$1 - \frac{1}{2^2} + \frac{1}{3^2} - \frac{1}{4^2} + \dots \text{converges.}$$

Solution. If the given series is $u_1 - u_2 + u_3 - u_4 + \dots$

then $|u_n| = \frac{1}{n^2}$ and $|u_{n+1}| = \frac{1}{(n+1)^2}$.

Clearly $\quad \frac{1}{n^2} > \frac{1}{(n+1)^2}$. Also $\quad \lim_{n \to \infty} |u_n| = \lim_{n \to \infty} \frac{1}{n^2} = 0$.

By Leibnitz test, the series converges.

Example 10.20 Discuss the convergence of the series

$$\sum_{n=1}^{\infty} \frac{(-1)^{n-1} \cdot n}{2n-1}.$$

Solution. The given series is $\sum_{n=1}^{\infty} \frac{(-1)^{n-1} \cdot n}{2n-1} = 1 - \frac{2}{3} + \frac{3}{5} - \frac{4}{7} + \dots$

It is an alternating series.

Now $\quad |u_n| - |u_{n+1}| = \frac{n}{2n-1} - \frac{n+1}{2n+1} = \frac{1}{(2n-1)(2n+1)} > 0$.

$\therefore$ $|u_n| > |u_{n+1}|$ for all n. But $\lim_{n \to \infty} |u_n| = \lim_{n \to \infty} \frac{n}{2n-1} = \frac{1}{2} \neq 0$.

The second condition of Leibnitz's test is not satisfied.
Hence the given alternating series is not convergent. It is oscillatory.

10.21 Absolute Convergence, Conditional Convergence

Consider the series $\Sigma u_n = u_1 + u_2 + u_3 + \dots$ $\qquad \qquad$ ...(10.10)
in which any term may be either positive or negative.

Let $|u_n|$ denotes the absolute value of u_n that is,

$|u_n| = u_n$ if u_n is positive and $|u_n| = -u_n$ if u_n is negative. Then

$$\Sigma |u_n| = |u_1| + |u_2| + |u_3| + \dots\dots$$

is a positive term series each term of which is numerically equal to the corresponding term of (10.10). If the series Σu_n is convergent, it is not necessary that
$\Sigma |u_n|$ be also convergent.

For example if $\quad \Sigma u_n = 1 - \dfrac{1}{2} + \dfrac{1}{3} - \dfrac{1}{4} + \dfrac{1}{5} - ...$ $\qquad\qquad$ ...(10.11)

then $\qquad\quad \Sigma|u_n| = 1 + \dfrac{1}{2} + \dfrac{1}{3} + \dfrac{1}{4} + \dfrac{1}{5} + ...$ $\qquad\qquad$ ...(10.12)

Here Σu_n is convergent but $\Sigma|u_n|$ is divergent.

Again if $\qquad\quad \Sigma u_n = 1 - \dfrac{1}{2^2} + \dfrac{1}{3^2} - \dfrac{1}{4^2} + ...$ $\qquad\qquad$ ...(10.13)

then $\qquad\quad \Sigma|u_n| = 1 + \dfrac{1}{2^2} + \dfrac{1}{3^2} + \dfrac{1}{4^2} + ...$ $\qquad\qquad$ ...(10.14)

In this case both Σu_n and $\Sigma|u_n|$ are convergent.

Thus if Σu_n is convergent $\Sigma|u_n|$ may be convergent or divergent.

To distinguish between the two cases that arise in this way, we have the following definitions.

Def. (*a*) Consider an arbitrary series Σu_n which contains positive as well as negative terms is said to be *absolutely* or *unconditionally* convergent if the series $\Sigma|u_n|$ is convergent and is said to diverge absolutely if the series of absolute values diverges.

(*b*) A series Σu_n is said to be *conditionally convergent if Σu_n is convergent but $\Sigma|u_n|$ is divergent.*

For example, the series (10.11) is conditionally convergent since it converges but $1 + 1/2 + 1/3 + 1/4 + ...$ diverges while (10.13) is absolutely convergent.

It should be noted that by property (2) Art 10.4, if $\Sigma|u_n|$ is convergent then Σu_n is also convergent.

The Ratio Test for Absolute Convergence

A series Σu_n with mixed terms is absolutely convergent if $\displaystyle\lim_{n\to\infty}\left|\dfrac{u_{n+1}}{u_n}\right| < 1$ and

is divergent if $\displaystyle\lim_{n\to\infty}\left|\dfrac{u_{n+1}}{u_n}\right| > 1$ or if equal to $+\infty$. If the limit is 1, the test gives no conclusion about convergence or absolute convergence.

We compare the series with some series which we know to be convergent, as

$a + ar + ar^2 + ar^3 + ..., (r < 1)$ $\qquad\qquad\qquad$ (geometric series)

$1 + \dfrac{1}{2^p} + \dfrac{1}{3^p} + \dfrac{1}{4^p} + ..., (p > 1)$ $\qquad\qquad\qquad$ (*p*-sereis)

or compare the given series with some series which is known to be divergent, as

$$1 + \frac{1}{2} + \frac{1}{3} + \frac{1}{4} + \dots\dots,$$ (harmonic series)

$$1 + \frac{1}{2^p} + \frac{1}{3^p} + \frac{1}{4^p} + \dots\dots, (p < 1)$$ (*p*-sereis)

Example 10.21 Test the following series for convergence and absolute convergence

(*i*) $1 - \frac{1}{2\sqrt{2}} + \frac{1}{3\sqrt{3}} - \frac{1}{4\sqrt{4}} + \dots$ to ∞ *(V.T.U., 2000)*

(*ii*) $1 - 2x + 3x^2 - 4x^3 + \dots \infty$, (*iii*) $\sum_{n=1}^{\infty} (-1)^n \frac{(2n-1)!}{3^n}$.

Solution. (*i*) Here the terms are alternately positive and negative, each term is numerically less then the preceding term, and

$$\lim_{n \to \infty} |u_n| = \lim_{n \to \infty} \left(\frac{1}{n\sqrt{n}} \right) = 0.$$ Hence the series is convergent.

Also $\Sigma |u_n| = \Sigma \frac{1}{n^{3/2}}$, which is convergent. Hence the given series is abso-

lutely convergent.

(*ii*) Applying ratio test to $\Sigma |u_n|$, we get

$$\lim_{n \to \infty} \frac{|u_{n+1}|}{|u_n|} = \lim_{n \to \infty} \frac{n+1}{n} |x| = |x|.$$

Hence the given series is absolutely convergent (and therefore convergent) if $|x| < 1$.

If $|x| \geq 1$, u_n increases numerically as n increases. Hence the series will be oscillatory if $x \geq 1$. If $x \leq -1$, the terms will be all positive and the series will be divergent.

(*iii*) **Use the ratio test for absolute convergence.**

Taking the absolute value of the general term u_n, we obtain

$$|u_n| = \left| (-1)^n \frac{(2n-1)!}{3^n} \right| = \frac{(2n-1)!}{3^n}$$

Thus $\lim_{n \to +\infty} \frac{|u_{n+1}|}{|u_n|} = \lim_{n \to +\infty} \frac{[2(n+1)-1]!}{3^{n+1}} \cdot \frac{3^n}{(2n-1)!}$

$$= \lim_{n \to \infty} \frac{1}{3} \cdot \frac{(2n+1)!}{(2n-1)!} = \frac{1}{3} \lim_{n \to +\infty} (2n)(2n+1) = +\infty,$$

which imples that the series diverges.

Example 10.22 Find the interval of convergence of the series

$$x - \frac{x^2}{\sqrt{2}} + \frac{x^3}{\sqrt{3}} - \frac{x^4}{\sqrt{4}} + \frac{x^5}{\sqrt{5}} - \ldots\ldots$$

Solution. The given series is $\sum_{n=1}^{\infty} u_n = \sum_{n=1}^{\infty} \frac{(-1)^{n-1} x^n}{\sqrt{n}}$

Here $|u_n| = \frac{|x^n|}{\sqrt{n}} = \frac{|x|^n}{\sqrt{n}}$ and $|u_{n+1}| = \frac{|x|^{n+1}}{\sqrt{n+1}}$

$\therefore \quad \frac{|u_n|}{|u_{n+1}|} = \sqrt{\frac{n+1}{n}} \cdot \frac{1}{|x|} = \sqrt{1+\frac{1}{n}} \cdot \frac{1}{|x|}$

$\lim\limits_{n \to \infty} \frac{|u_n|}{|u_{n+1}|} = \frac{1}{|x|}.$

$\therefore$ By ratio test, the series $\Sigma |u_n|$ is convergent if $\frac{1}{|x|} > 1$

i.e., if $|x| < 1$ *i.e., if* $-1 < x < 1$ and divergent if $\frac{1}{|x|} < 1$ *i.e., if* $|x| > 1$ *i.e.*

if $x > 1$ or $x < -1$. .

Ratio test fails when $|x| = 1$, that is, when $x = 1$ or -1. When $x = 1$, the series

becomes $1 - \frac{1}{\sqrt{2}} + \frac{1}{\sqrt{3}} - \frac{1}{\sqrt{4}} + \frac{1}{\sqrt{5}} - \ldots$ which is an alternating series and is

convergent.

When $x = -1$, the series becomes $-\left(1 + \frac{1}{\sqrt{2}} + \frac{1}{\sqrt{3}} + \frac{1}{\sqrt{4}} + \frac{1}{\sqrt{5}} + \ldots\right)$ which is di-

vergent by *p*-series test.
Hence the given series converges for $-1 < x \le 1$.

Exercise 10.6

Test the convergence of the following series:

1. $1 - \frac{1}{2} + \frac{1}{3} - \frac{1}{4} + \frac{1}{5} - \ldots \infty$

2. $1 - \dfrac{1}{\sqrt{2}} + \dfrac{1}{\sqrt{3}} - \dfrac{1}{\sqrt{4}} + ...\infty$
(*Triputi, 1998*)

3. $1 - \dfrac{1}{2^p} + \dfrac{1}{3^p} - \dfrac{1}{4^p} + ...\infty$ for $p > 0$

4. (a) $\sum \dfrac{(-1)^n n}{2n+3}$, (b) $\displaystyle\sum_{n=1}^{\infty} (-1)^{n-1} \dfrac{n}{n^2+1}$.
(*Mysore, 1997*)

5. (a) $\dfrac{1}{n} - \dfrac{1}{n+a} + \dfrac{1}{n+2a} - \dfrac{1}{n+3a} + ...$ to $\infty\, (a>0)$, (b) $\displaystyle\sum_{n=1}^{\infty} (-1)^{n-1} \dfrac{n+1}{n}$.

Examine the following convergent series for absolute or conditional convergence.

6. $1 - \dfrac{1}{2} + \dfrac{1}{4} - \dfrac{1}{8} + \dfrac{1}{16} - ...\infty$
7. $1 - \dfrac{1}{\sqrt{2}} + \dfrac{1}{\sqrt{3}} - \dfrac{1}{\sqrt{4}} + ...\infty$

8. $1 - \dfrac{2}{3} + \dfrac{3}{3^2} - \dfrac{4}{3^3} + ...\infty$
[**Hint.** $u_n = (-1)^{n+1} \dfrac{n}{3^{n-1}}$. Apply the ratio test.]

9. (a) $\dfrac{(-1)^{n+1}}{\sqrt{n(n+1)}}$; (b) $\dfrac{(-1)^{n-1}}{(n!)^3}$.

10. Prove that the series $\dfrac{\sin x}{1^3} - \dfrac{\sin 2x}{2^3} + \dfrac{\sin 3x}{3^3} - ...$ converges absolutely.

11. Discuss the absolute convergence of the series $\displaystyle\sum_{n=0}^{\infty} \dfrac{(-1)^n x^n}{n+1}$.
(*Mangalore, 1997*)

12. Test for convergence the series

$$\dfrac{1}{1^p} + \dfrac{x}{3^p} + \dfrac{x^2}{5^p} + ... + \dfrac{x^{n-1}}{(2n-1)^p} + ...\infty$$
(*AMIE, W-2009; JNTU, 2006*)

Ans. 1. Convergent **2.** Convergent **3.** Convergent **4.** (a) Divergent (b) Convergent **5.** (a) Convergent (b) Divergent **6.** Absolute Convergent **7.** Conditional Convergent **8.** Absolute Convergent **9.** (a) Conditional Convergent (b) Absolute Convergent **11.** Absolute Convergent for $0 < x < 1$. **12.** Convergent for $x < 1$, Divergent for $x > 1$; Convergent for $p > 1$ and Divergent for $p \leq 1$.

Chapter 11

Partial Differentiation

11.1 Introduction

In many applied problems we come across functions which depend on two or more independent variables. For example the area of a rectangle depend on its sides of length x and y; volume of a rectangular parallelopiped depends on its edges of length x, y, z; and the ideal gas law in physics states that under appropriate conditions the pressure exerted by a gas is a function of the volume of the gas and its temperature.

If the value of a quantity z depends upon the values of two variables x and y then z is called a function of x and y and, as in the case of function of single variable defined by $y = f(x)$, we write function of two variables as

$$z = f(x, y). \qquad ...(11.1)$$

The variables x and y are called independent variables while z is called the dependent variable. One can inquire how the value of z changes if y is held fixed and x is allowed to vary, or if x is held fixed and y is allowed to vary.

A similar definition can be given for functioning of more than two variables. In general, we define a real valued function of n variables as

$$z = f(x_1, x_2, ..., x_n). \qquad ...(11.2)$$

The function as defined by equation (11.2) is called an *explicit* function, whereas a function defined by $\phi(z, x_1, x_2, ..., x_n) = 0$ is called an *implicit* function.

We shall discuss the calulus of the functions of two variables in detail and then generalize to the case of several variables.

11.2 Functions of Two Variables

Consider the function of two variables $z = f(x, y)$. The set of points (x, y) in the xy-plane for which function $f(x, y)$ is defined is called the *domain* of definition or simply *domain* of the function and is denoted by D. The domain may be the entire xy plane or a part of the xy plane. The collection of the corresponding values of z is called the *range* of the function.

For example if $z = \sqrt{1 - (x^2 + y^2)}$, the domain for which z is real consists of the set of points (x, y) such that $x^2 + y^2 \leq 1$, *i.e.*, the set of points inside and on the circle in the xy-plane having centre at $(0, 0)$ and radius 1 unit.

Neighborhood of a point

For a function of two independent variables, two kinds of neighbourhoods are in common use – the *circular* and the *square* neighbourhoods. A circular neighbourhood of the point $P_0(x_0, y_0)$ is the set of all points (x, y) which lie inside a circle of radius $\delta > 0$ with centre at the point (x_0, y_0). The number δ is arbitrary and can be chosen as small as we like but not zero. We usually denote this neighborhood by $N_\delta(P)$. Thus the δ-neighbourhood of $P_0(x_0, y_0)$ is the circular region defined by

$$N_\delta(P) = \left\{(x, y): \sqrt{(x - x_0)^2 + (y - y_0)^2} < \delta\right\}. \qquad ...(11.3)$$

A square neighbourhood of the point $P(x_0, y_0)$ can also be defined as

$$N_\delta(P) = \left\{(x, y): |x - x_0| < \delta \text{ and } |y - y_0| < \delta\right\} \qquad ...(11.4)$$

that is, the set of all points which lie inside a square of side 2δ with centre at (x_0, y_0) and sides parallel to the coordinate axes (Fig. 11.1).

If the point $P(x_0, y_0)$ is not included in the set, then it is called the *deleted* δ-*neighborhood* of the point, that is, the set of points which satisfy

$$0 < \sqrt{(x - x_0)^2 + (y - y_0)^2} < \delta \qquad ...(11.5)$$

is called the deleted neighborhood of $P(x_0, y_0)$.

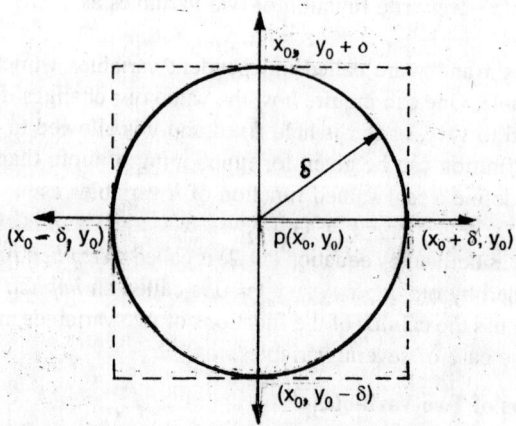

Fig. 11.1. Neighborhood of a point $P(x_0, y_0)$

11.3 Limits

Let $z = f(x, y)$ be a function of two variables defined in a deleted δ neighborhood of (x_0, y_0) [that is, $f(x, y)$ may be undefined at (x_0, y_0)]. We say that L is the limit of $f(x, y)$ as (x, y) approaches (x_0, y_0) and write $\lim\limits_{(x, y) \to (x_0, y_0)} f(x, y) = L$ if for

every number $\varepsilon > 0$ there exists a corresponding number $\delta > 0$ such that for all (x, y) in the domain of f satisfies

$|f(x, y) - L| < \varepsilon$, whenever the distance between (x, y) and (x_0, y_0) satisfies

$$0 < \sqrt{(x - x_0)^2 + (y - y_0)^2} < \delta . \qquad \text{...(11.6)}$$

Remark

1. As in the case of function of one variable, the existence of the limit of $f(x, y)$ as $(x, y) \to (x_0, y_0)$ is in no way dependent on the existence of a value of $f(x, y)$ at (x_0, y_0).

2. *It must be noted that in order for* $\lim\limits_{(x, y) \to (x_0, y_0)} f(x, y)$ or $\lim\limits_{\substack{x \to x_0 \\ y \to y_0}} f(x, y)$ *to*

   exist, it must have the same value regardless of the approach of (x, y) *to* (x_0, y_0)*. The limit is same along all the paths, that is, limit is independent of the path. Thus if two different approaches give different values, the limit can not exist. In other words if the limit is dependent on a path, then the limit does not exist.*

3. Sometimes it is possible to determine the limit by changing the given function to polars form.

11.4 Properties of Limits of Functions of Two Variables

Let $u = f(x, y)$ and $v = g(x, y)$ be two real valued functions defined in a domain D. Let

$$\lim\limits_{(x, y) \to (x_0, y_0)} f(x, y) = L_1 \text{ and } \lim\limits_{(x, y) \to (x_0, y_0)} g(x, y) = L_2.$$

Then, the following rules hold

1. **Sum Rule:** $\lim\limits_{(x, y) \to (x_0, y_0)} [f(x, y) + g(x, y)] = L_1 + L_2.$

2. **Difference Rule:** $\lim\limits_{(x, y) \to (x_0, y_0)} [f(x, y) - g(x, y)] = L_1 - L_2.$

3. **Constant multiple Rule:** $\lim\limits_{(x, y) \to (x_0, y_0)} [k f(x, y)] = kL_1$ for any real constant k.

4. **Product Rule:** $\lim\limits_{(x, y) \to (x_0, y_0)} [f(x, y) \cdot g(x, y)] = L_1 L_2.$

5. **Quotient Rule:** $\lim\limits_{(x, y) \to (x_0, y_0)} [f(x, y)/g(x, y)] = L_1/L_2, \ L_2 \neq 0.$

11.5 Illustrative Examples

Example 11.1 Evaluate the following limits:

(i) $\lim\limits_{(x,y)\to(3,1)}\left(\dfrac{2xy^2}{6+y^2}+\dfrac{1}{2}xy\right)$, (ii) $\lim\limits_{(x,y)\to(0,0)}\dfrac{x\sin\left(x^2+y^2\right)}{x^2+y^2}$,

(iii) $\lim\limits_{(x,y)\to(2,1)}\dfrac{\sin^{-1}\left(xy-2\right)}{\tan^{-1}\left(3xy-6\right)}$, (iv) $\lim\limits_{(x,y)\to(0,0)}\dfrac{y-x}{y+x}$.

Solution: (i) Using Standard laws for limits, we see that

(i) $\lim\limits_{(x,\,y)\to(3,\,1)}\left(\dfrac{2xy^2}{6+y^2}+\dfrac{1}{2}xy\right)=\lim\limits_{(x,\,y)\to(3,\,1)}\dfrac{2xy^2}{6+y^2}+\lim\limits_{(x,\,y)\to(3,\,1)}\dfrac{1}{2}xy$

$$=\dfrac{2(3)(1)}{6+1}+\dfrac{1}{2}(3)(1)=\dfrac{6}{7}+\dfrac{3}{2}=\dfrac{33}{14}.$$

(ii) $\lim\limits_{(x,\,y)\to(0,0)}\dfrac{x\sin\left(x^2+y^2\right)}{x^2+y^2}=\lim\limits_{(x,\,y)\to(0,0)}x\cdot\lim\limits_{(x,\,y)\to(0,0)}\dfrac{\sin\left(x^2+y^2\right)}{x^2+y^2}=0\cdot1=0.$

(iii) Let $xy-2=t$. Therefore $t\to0$ as $(x,\,y)\to(2,\,1)$.   We can now write

$\lim\limits_{(x,\,y)\to(2,\,1)}f\left(x,\,y\right)=\lim\limits_{t\to0}\dfrac{\sin^{-1}t}{\tan^{-1}3t}=\lim\limits_{t\to0}\left[\dfrac{\left(\sin^{-1}t\right)/t}{\tan^{-1}\left(3t\right)/\left(3t\right)}\right]\left[\dfrac{t}{3t}\right]=1\cdot\dfrac{1}{3}=\dfrac{1}{3}.$

(iv) In some cases the Standard laws donot suffice. The limit does not exist, if it is not finite or if it depends on a particular path.

Consider the path $y=mx$. As $(x,\,y)\to(0,\,0)$, we get $x\to0$. Therefore

$\lim\limits_{(x,y)\to(0,0)}\dfrac{y-x}{y+x}=\lim\limits_{x\to0}\dfrac{mx-x}{mx+x}=\dfrac{m-1}{m+1}$, which depends on m. For different

values of m, we obtain different limits. Hence, the limit does not exist.

Example 11.2 Show that the following limits

(i) $\lim\limits_{(x,y)\to(0,0)}\dfrac{x}{\sqrt{x^2+y^2}}$, (ii) $\lim\limits_{(x,y)\to(0,0)}\dfrac{x+\sqrt{y}}{x^2+y}$,

(iii) $\lim\limits_{(x,y)\to(0,1)}\tan^{-1}\left(y/x\right)$, *(IETE Dec. 2006)* (iv) $\lim\limits_{(x,y)\to(0,0)}\dfrac{xy^3}{x^2+y^6}$

do not exist.

Solution. (i) Consider the path $y=mx$. As $(x,\,y)\to(0,\,0)$ we get $x\to0$.

Therefore, $\lim\limits_{(x,\,y)\to(0,\,0)} \dfrac{x}{\sqrt{x^2+y^2}} = \lim\limits_{x\to 0}\dfrac{x}{x\sqrt{1+m^2}} = \dfrac{1}{\sqrt{1+m^2}}$ which depends on m.

For different values of m, we obtain different limits. Hence, the limit does not exist.

Alternative. Setting $x = r\cos\theta$, $y = r\sin\theta$, we obtain

$$\lim\limits_{(x,\,y)\to(0,\,0)} \dfrac{x}{\sqrt{x^2+y^2}} = \lim\limits_{r\to 0}\dfrac{r\cos\theta}{r} = \cos\theta,\ \text{which depends on } \theta.\ \text{Hence, the limit}$$

is dependent on different radial paths θ = constant, the limit does not exist.

(*ii*) Choose the path $y = mx^2$. As $(x, y) \to (0, 0)$, we get $x \to 0$. Therefore,

$$\lim\limits_{(x,\,y)\to(0,\,0)} \dfrac{x+\sqrt{y}}{x^2+y} = \lim\limits_{x\to 0}\dfrac{x(1+\sqrt{m})}{x^2(1+m)} = \lim\limits_{x\to 0}\dfrac{1+\sqrt{m}}{x(1+m)} = \infty.$$

Since the limit is not finite, so limit does not exist.

(*iii*) We have $\lim\limits_{(x,\,y)\to(0,\,1)} \tan^{-1}(y/x) = \tan^{-1}(\pm\infty) = \pm\pi/2$ depending on

whether the point $(0, 1)$ is approached from left or from right along the line $y = 1$. If we approach from left, we obtain the limit as $-\pi/2$ and if we approach from right, we obtain the limit as $\pi/2$. Since the limit is not unique, the limit does not exist as $(x, y) \to (0.1)$.

(*iv*) Choose the path $x = my^3$. As $(x, y) \to (0, 0)$, we get $y \to 0$.

Therefore $\lim\limits_{(x,\,y)\to(0,\,0)} \dfrac{xy^3}{x^2+y^6} = \lim\limits_{y\to 0}\dfrac{my^6}{(1+m^2)y^6} = \dfrac{m}{1+m^2}$

which depends on m. For different values of m, we obtain different limits. Hence, the limit does not exist.

Example 11.3 Examine the limits of $f(x, y) = x^2y/(x^4 + y^2)$, $(x, y) \neq (0, 0)$ as $(x, y) \to (0, 0)$ along the line $y = mx$ and along the parabola $y = x^2$. Does $f(x, y)$ have a limit as $(x, y) \to (0, 0)$? *(I.E.T.E; W-2004, AMIE., S-1998)*

Solution. *First consider the path $y = mx$. As $(x, y) \to (0, 0)$,* we get $x \to 0$. Therefore,

$$\lim\limits_{(x,\,y)\to(0,\,0)} \dfrac{x^2y}{x^4+y^2} = \lim\limits_{x\to 0}\dfrac{x^2(mx)}{x^4+m^2x^2} = \lim\limits_{x\to 0}\dfrac{mx}{x^2+m^2} = 0.$$

Now consider the path $y = x^2$, when $(x, y) \to (0, 0)$, we get $x \to 0$.

Therefore, $\lim\limits_{(x,\,y)\to(0,\,0)} \dfrac{x^2y}{x^4+y^2} = \lim\limits_{x\to 0}\dfrac{x^2(x^2)}{x^4+x^4} = \lim\limits_{x\to 0}\dfrac{1}{2} = \dfrac{1}{2}$.

Since the limits along the two paths are different, so limit does not exist.

11.6 Continuity

Def. A function $z = f(x, y)$ of the two real independent variable x and y is said to be *continuous* at a point (x_0, y_0), if

(i) $f(x, y)$ is defined at the point (x_0, y_0), (ii) $\lim\limits_{(x, y) \to (x_0, y_0)} f(x, y)$ exists, and

(iii) $\lim\limits_{(x, y) \to (x_0, y_0)} f(x, y) = f(x_0, y_0)$.

A function is *continuous* if it is continuous at every point of its domain.

Therefore, a function $f(x, y)$ is continuous at (x_0, y_0) if

$$\left| f(x, y) - f(x_0, y_0) \right| < \varepsilon, \text{ whenever } \sqrt{(x - x_0)^2 + (y - y_0)^2} < \delta. \quad \dots(11.7)$$

If any one of the above conditions is not satisfied, then the function is said to be *discontinuous* at the point (x_0, y_0) which is then called a *point of discontinuity*.

Example 11.4(a). Investigate for continuity at the point (1, 2).

$$f(x, y) = \begin{cases} x^2 + 2y, & (x, y) \neq (1, 2) \\ 0 & (x, y) = (1, 2) \end{cases}.$$

(b) Show that the given function is continuous at the point (0, 0).

$$f(x, y) = \begin{cases} \dfrac{\sin^{-1}(x + 2y)}{\tan^{-1}(2x + 4y)}, & (x, y) \neq (0, 0) \\ 1/2, & (x, y) = (0, 0). \end{cases}$$

Solution. (a) $\lim\limits_{(x, y) \to (1, 2)} f(x, y) = \lim\limits_{(x, y) \to (1, 2)} \left(x^2 + 2y \right)$

$$= \lim\limits_{(x, y) \to (1, 2)} x^2 + \lim\limits_{(x, y) \to (1, 2)} 2y = 1 + 4 = 5.$$

Since $\lim\limits_{(x, y) \to (1, 2)} f(x, y) \neq f(1, 2)$,

The function is not continuous at $(1, 2)$. The point $(1, 2)$ is a *point of removable discontinuity*.

(b) Let $x + 2y = t$. Therefore, $t \to 0$ as $(x, y) \to (0, 0)$. We can now write

$$\lim\limits_{(x,y) \to (0,0)} f(x, y) = \lim\limits_{t \to 0} \frac{\sin^{-1} t}{\tan^{-1} 2t} = \lim\limits_{t \to 0} \left[\frac{(\sin^{-1} t)/t}{(\tan^{-1} 2t)/2t} \right] \left[\frac{t}{2t} \right] = \frac{1}{2}.$$

Since $\lim\limits_{(x,y) \to (0,0)} f(x, y) = f(0, 0) = 1/2$, the given function is continuous at $(x, y) = (0, 0)$.

Example 11.5 Show that the following functions are discontinuous at the given points

(i) $f(x, y) = \begin{cases} \dfrac{x^2 - y^2}{x^2 + y^2}, & (x, y) \neq (0, 0) \\ 0, & (x, y) = (0, 0) \end{cases}$ at the point (0, 0).

(ii) $f(x, y) = \begin{cases} \dfrac{x^2 - x\sqrt{y}}{x^2 + y}, & (x, y) \neq (0, 0) \\ 0, & (x, y) = (0, 0) \end{cases}$ at the point (0, 0).

Solution. (i) Choose the path $y = mx$. As $(x, y) \to (0, 0)$, we get $x \to 0$.

Therefore, $\displaystyle \lim_{(x, y) \to (0,0)} \frac{x^2 - y^2}{x^2 + y^2} = \lim_{x \to 0} \frac{x^2(1 - m^2)}{x^2(1 + m^2)} = \frac{1 - m^2}{1 + m^2}$

which depends on m. Since, the limit does not exist, the function is not continuous at (0. 0).

(ii) Choose the path $y = m^2 x^2$. As $(x, y) \to (0, 0)$, we get $x \to 0$.

Therefore, $\displaystyle \lim_{(x, y) \to (0,0)} \frac{x^2 - x\sqrt{y}}{x^2 + y} = \lim_{x \to 0} \frac{x^2(1 - m)}{x^2(1 + m^2)} = \frac{1 - m}{1 + m^2}$

which depends on m. Since the limit does not exist, the function is not continuous at (0, 0).

Exercise 11.1

1. Evaluate each of the following limits where they exists.

(i) $\displaystyle \lim_{(x, y) \to (\pi, 0)} x \cos\left(\frac{x - y}{4}\right)$,

(ii) $\displaystyle \lim_{(x, y) \to (4, \pi)} x^2 \sin\frac{y}{x}$,

(iii) $\displaystyle \lim_{(x, y) \to (1,1)} \frac{x^3 - y^3}{x^2 - y^2}$, (I.E.T.E., S-2005) [**Hint:** Factorize and cancel $(x - y)$]

(iv) $\displaystyle \lim_{\substack{x \to 0 \\ y \to 0}} \frac{y}{x^2 + y^2}$,

(v) $\displaystyle \lim_{(x,y) \to (0,0)} \frac{(1 + x^2)\sin y}{y}$,

(vi) $\displaystyle \lim_{(x,y) \to (0,0)} \frac{x^2}{x^2 + y^2}$,

(vii) $\displaystyle \lim_{(x, y) \to (0,0)} \frac{xy}{\sqrt{(x^2 + y^2)}}$, (IETE June 2007)

(viii) $\displaystyle \lim_{(x,y) \to (0,0)} \frac{2x^2 y}{x^4 + y^2}$ (I.E.T.E., June 2006 Dec., 2004),

(ix) $\lim_{(x,y)\to(0,0)} \dfrac{x^3 y}{x^6 + y^2}$

[**Hint:** Choose the path $y = mx^3$]

(x) $\lim_{(x,y)\to(0,0)} \dfrac{x^2 - y^2}{x^2 + y^2}$,

(xi) $\lim_{(x,y)\to(0,0)} \dfrac{x^2 - xy}{\sqrt{x} - \sqrt{y}}$

(xii) $\lim_{(x,y)\to(0,0)} (x^2 + y^2) \log(x^2 + y^2)$ [**Hint:** Use polar coordinates (r, θ) of the point (x, y) with $r \geq 0$.

$$\lim_{(x,y)\to(0,0)} (x^2 + y^2) \log(x^2 + y^2) = \lim_{r\to 0^+} r^2 \log r^2$$

$$= \lim_{r\to 0^+} \frac{2\log r}{1/r^2} \left(\frac{\infty}{\infty} \text{ form}\right) = \lim_{r\to 0^+} \frac{2/r}{-2/r^3} = \lim_{r\to 0^+} (-r^2) = 0.]$$

(xiii) $\lim_{(x,y)\to(2,-2)} \dfrac{x^2 + xy + x + y}{x + y}$

(IETE S-2008)

(xiv) $\lim_{(x,y)\to(0,0)} \dfrac{x + \sqrt{y}}{\sqrt{x^2 + y}}$

(IETE Dec. 2007)

Ans. (i) $\pi\sqrt{2}/2$ (ii) $8\sqrt{2}$, (iii) $3/2$. (iv) limit does not exist, (v) 1, (vi) no limit, (vii) 0, (viii) no limit, (ix) limit does not exist, (x) limit does not exist, (xi) 0, (xii) 0, (xiii) 3, (xiv) limit does not exist.

2. Discuss the continuity of the following functions at the point (0, 0).

(i) $f(x, y) = \begin{cases} \dfrac{xy}{x^2 + y^2}, & (x, y) \neq (0, 0) \\ 0, & (x, y) = (0, 0), \end{cases}$

(ii) $f(x, y) = \begin{cases} \dfrac{2x - y}{x + y}, & (x, y) \neq (0, 0) \\ 0, & (x, y) = (0, 0), \end{cases}$

(iii) $f(x, y) = \begin{cases} \dfrac{2xy}{\sqrt{x^2 + y^2}}, & (x, y) \neq (0, 0) \\ 0, & (x, y) = (0, 0), \end{cases}$

(iv) $f(x, y) = \begin{cases} \dfrac{x^2 + y^2}{\tan xy}, & (x, y) \neq (0, 0) \\ 0, & (x, y) = (0, 0). \end{cases}$ [**Hint:** (iv) Choose the path $y = mx$]

Ans. (i) Discontinuous, (ii) Discontinuous, (iii) Continuous, (iv) Discontinuous.

3. Show that the given function is discontinuous at the point $(2, -2)$.

$$f(x, y) = \begin{cases} \dfrac{x^2 + xy + x + y}{x + y}, & (x, y) \neq (2, -2) \\ 4, & (x, y) = (2, -2) \end{cases}$$ (*IETE, June 2009*)

$\left[\textbf{Hint :} \text{ Since } \lim\limits_{(x, y) \to (2, -2)} f(x, y) \neq f(2, -2) \text{ the function is not continuous at } (2, -2)\right].$

11.7 Partial Derivatives

Let $z = f(x, y)$ be a function of two independent variables x and y. If x varies while y is held fixed then z becomes a function of x. The derivative of f with respect to x, treating y as constant is called first partial derivative of f with respect to x and is denoted by symbols $\dfrac{\partial z}{\partial x}\left(\text{or } \dfrac{\partial f}{\partial x} \text{ or } f_x(x, y)\right)$.

Thus $\dfrac{\partial z}{\partial x} = \lim\limits_{\Delta x \to 0} \dfrac{f(x + \Delta x, y) - f(x, y)}{\Delta x}$,   ...(11.8)

provided the limit exists.

Similarly, if y varies while x is held fixed, the (*first*) *partial derivative* of f with respect to y is $f_y(x, y) = \dfrac{\partial z}{\partial y} = \dfrac{\partial f}{\partial y} = \lim\limits_{\Delta y \to 0} \dfrac{f(x, y + \Delta y) - f(x, y)}{\Delta y}$ ...(11.9)

provided the limit exists.

In general, if z is a function of a number of independent variables, then the partial derivative of z with respect to any one of those variables is obtained by differentiating z with respect to that variable while all other variables are held constant.

11.8 Illustrative Examples

Example 11.6 Find the first order partial derivatives of the following functions at the specified point from the first principles.

   (i) $f(x, y) = 2x^2 - xy + y^2$ at (x_0, y_0),

   (ii) $f(x, y) = ye^{-x}$ at $(4, 2)$,

   (iii) $f(x, y) = \sin(3x + 2y)$ at $(0, \pi/3)$.

Solution. We have

(*i*) $\dfrac{\partial f}{\partial x}\bigg|_{(x_0,\,y_0)} = f_x(x_0,\,y_0) = \lim\limits_{\Delta x \to 0} \dfrac{f(x_0 + \Delta x,\,y_0) - f(x_0,\,y_0)}{\Delta x}$

$\qquad\qquad = \lim\limits_{\Delta x \to 0} \dfrac{\left[2(x_0 + \Delta x)^2 - (x_0 + \Delta x)y_0 + y_0^2\right] - \left[2x_0^2 - x_0 y_0 + y_0^2\right]}{\Delta x}$

$\qquad = \lim\limits_{\Delta x \to 0} \dfrac{(4x_0 - y_0)\Delta x + 2(\Delta x)^2}{\Delta x} = \lim\limits_{\Delta x \to 0} 4x_0 - y_0 + 2\Delta x = 4x_0 - y_0.$

$\dfrac{\partial f}{\partial y}\bigg|_{(x_0,\,y_0)} = f_y(x_0,\,y_0) = \lim\limits_{\Delta y \to 0} \dfrac{f(x_0,\,y_0 + \Delta y) - f(x_0,\,y_0)}{\Delta y}$

$\qquad = \lim\limits_{\Delta y \to 0} \dfrac{\left[2x_0^2 - x_0(y_0 + \Delta y) + (y_0 + \Delta y)^2\right] - \left[2x_0^2 - x_0 y_0 + y_0^2\right]}{\Delta y}$

$\qquad = \lim\limits_{\Delta y \to 0} \dfrac{(-x_0 + 2y_0)\Delta y + (\Delta y)^2}{\Delta y} = \lim\limits_{\Delta y \to 0} -x_0 + 2y_0 + \Delta y = -x_0 + 2y_0.$

(*ii*) $\dfrac{\partial f}{\partial x} = \lim\limits_{\Delta x \to 0} \dfrac{ye^{-(x+\Delta x)} - ye^{-x}}{\Delta x} = \lim\limits_{\Delta x \to 0} -\dfrac{ye^{-x}\left(1 - e^{-\Delta x}\right)}{\Delta x}$

$\qquad = -ye^{-x} \lim\limits_{\Delta x \to 0} \dfrac{1 - e^{-\Delta x}}{\Delta x} = -ye^{-x}.$

$\dfrac{\partial f}{\partial y} = \lim\limits_{\Delta y \to 0} \dfrac{(y + \Delta y)e^{-x} - ye^{-x}}{\Delta y} = e^{-x}.$

$\therefore \quad \dfrac{\partial f}{\partial x}\bigg|_{(4,2)} = -2e^{-4}; \dfrac{\partial f}{\partial x}\bigg|_{(4,2)} = e^{-4}.$

(*iii*) $\dfrac{\partial f}{\partial x} = \lim\limits_{\Delta x \to 0} \dfrac{\sin\{3(x + \Delta x) + 2y\} - \sin(3x + 2y)}{\Delta x}$

$\qquad = \lim\limits_{\Delta x \to 0} \dfrac{2\cos\left[3x + 2y + (3\Delta x/2)\right]\sin(3\Delta x/2)}{\Delta x}$

$\qquad = \lim\limits_{\Delta x \to 0} \dfrac{3\cos\left[3x + 2y + (3\Delta x/2)\right]\sin(3\Delta x/2)}{3\Delta x/2}$

$$= 3 \lim_{\Delta x \to 0} \cos\left[3x + 2y + (3\Delta x/2)\right] \cdot \lim_{\Delta x \to 0} \frac{\sin(3\Delta x/2)}{(3\Delta x/2)} = 3\cos(3x + 2y).$$

$$\frac{\partial f}{\partial y} = \lim_{\Delta y \to 0} \frac{\sin\{3x + 2(y + \Delta y)\} - \sin(3x + 2y)}{\Delta y}$$

$$= \lim_{\Delta y \to 0} \frac{2\cos(3x + 2y + \Delta y)\sin \Delta y}{\Delta y} = 2 \lim_{\Delta y \to 0} \cos(3x + 2y + \Delta y) \cdot \lim_{\Delta y \to 0} \frac{\sin \Delta y}{\Delta y}$$

$$= 2\cos(3x + 2y).$$

$$\therefore \quad \left. \frac{\partial f}{\partial x} \right|_{(0,\, \pi/3)} = 3\cos(2\pi/3) = -\frac{3}{2}, \text{ and } \left. \frac{\partial f}{\partial y} \right|_{(0,\, \pi/3)} = 2\cos(2\pi/3) = -1.$$

Example 11.7 Show that the function

$$f(x, y) = \begin{cases} \dfrac{x^2 - y^2}{x - y}, & (x, y) \neq (1, -1) \\ 0 & ,(x, y) = (1, -1) \end{cases}$$

(*i*) **is continuous at (0, 0),**
(*ii*) **possesses partial derivatives f$_x$ (1, −1) and f$_y$ (1, −1).**

Solution. (*i*) We have

$$\lim_{(x, y) \to (1, -1)} \frac{x^2 - y^2}{x - y} = \lim_{(x, y) \to (1, -1)} (x + y) = 0 = f(1, -1).$$

Therefore, the function is continuous at (1, −1).

(*ii*) The partial derivatives are given by

$$f_x(1, -1) = \lim_{\Delta x \to 0} \frac{f(1 + \Delta x, -1) - f(1, -1)}{\Delta x} = \lim_{\Delta x \to 0} \frac{1}{\Delta x}\left[\frac{(1 + \Delta x)^2 - 1}{(1 + \Delta x) + 1} - 0\right]$$

$$= \lim_{\Delta x \to 0} \frac{2 + \Delta x}{2 + \Delta x} = 1.$$

$$f_y(1, -1) = \lim_{\Delta y \to 0} \frac{f(1, -1 + \Delta y) - f(1, -1)}{\Delta y} = \lim_{\Delta y \to 0} \frac{1}{\Delta y}\left[\frac{1 - (-1 + \Delta y)^2}{1 - (-1 + \Delta y)} - 0\right]$$

$$= \lim_{\Delta y \to 0} \frac{2 - \Delta y}{2 - \Delta y} = 1.$$

Therefore, the first order partial derivatives exist at (1, −1).

Example 11.8 Show that the function

$$f(x, y) = \begin{cases} \dfrac{xy}{x^2 + 2y^2}, & (x, y) \neq (0, 0) \\ 0 & , (x, y) = (0, 0) \end{cases}$$

is discontinuous at (0, 0) but its partial derivatives f_x and f_y exist at (0, 0).

Solution. Choose the path $y = mx$. We see that

$$\lim_{(x,y) \to (0,0)} f(x, y) = \lim_{x \to 0} \frac{mx^2}{(1 + 2m^2)x^2} = \frac{m}{1 + 2m^2}.$$

Since the limit depends on the value of m, that is, on the path of approach and is different for the different paths followed and therefore the limit does not exist. Hence, the function $f(x, y)$ is not continuous at (0, 0). We now have

$$f_x(0, 0) = \lim_{\Delta x \to 0} \frac{f(\Delta x, 0) - f(0, 0)}{\Delta x} = \lim_{\Delta x \to 0} \frac{0 - 0}{\Delta x} = 0.$$

$$f_y(0, 0) = \lim_{\Delta y \to 0} \frac{f(0, \Delta y) - f(0, 0)}{\Delta y} = \lim_{\Delta y \to 0} \frac{0 - 0}{\Delta y} = 0.$$

Therefore, the partial derivatives f_x and f_y exist at (0, 0).

Example 11.9 Show that for the function $f(x, y) = \sqrt{|xy|}$, partial derivatives f_x and f_y both exist at the origin and have the value 0. Also show that these two partial derivatives are continuous except at the origin. *(I.E.T.E., Dec. 2005)*

Solution. For $(x, y) = (0, 0)$

$$f_x(0, 0) = \lim_{\Delta x \to 0} \frac{f(\Delta x, 0) - f(0, 0)}{\Delta x} = \lim_{\Delta x \to 0} \frac{0 - 0}{\Delta x} = 0$$

$$f_y(0, 0) = \lim_{\Delta y \to 0} \frac{f(0, \Delta y) - f(0, 0)}{\Delta y} = \lim_{\Delta y \to 0} \frac{0 - 0}{\Delta y} = 0.$$

For $(x, y) \neq (0, 0)$

$$f_x(x, y) = \lim_{\Delta x \to 0} \frac{f(x + \Delta x, y) - f(x, y)}{\Delta x} = \lim_{\Delta x \to 0} \frac{\sqrt{|x + \Delta x| \, |y|} - \sqrt{|x| \, |y|}}{\Delta x}$$

$$= \lim_{\Delta x \to 0} \sqrt{|y|} \, \frac{|x + \Delta x| - |x|}{\Delta x \left[\sqrt{|x + \Delta x|} + \sqrt{|x|} \right]}.$$

Now as $\Delta x \to 0$, we can take $x + \Delta x > 0$, i.e., $|x + \Delta x| = x + \Delta x$, when $x > 0$.
And $\quad x + \Delta x < 0$ or $|x + \Delta x| = -(x + \Delta x)$, when $x < 0$.

$$\therefore \quad f_x(x, y) = \begin{cases} \sqrt{|y|}/2\sqrt{|x|}, & \text{when } x > 0, \\ -\sqrt{|y|}/2\sqrt{|x|}, & \text{when } x < 0. \end{cases}$$

Similarly, $\qquad f_y(x, y) = \begin{cases} \sqrt{|x|}/2\sqrt{|y|}, & \text{when } y > 0, \\ -\sqrt{|x|}/2\sqrt{|y|}, & \text{when } y < 0 \end{cases}$

which are not continuous at the origin.

Example 11.10 Find the first order partial derivatives of (i) $z = \log(x^2 + y^2)$, (ii) $u = 3\,x^2y - x \sin xy$.

Solution. (i) we have $z = \log(x^2 + y^2)$. Treating y as a constant and differentiating w.r.t. x, then

$$\frac{\partial z}{\partial x} = \frac{1}{x^2 + y^2} \cdot 2x = \frac{2x}{x^2 + y^2}.$$

Now treating x as constant and differentiating w.r.t y, we have

$$\frac{\partial z}{\partial y} = \frac{1}{x^2 + y^2} \cdot 2y = \frac{2y}{x^2 + y^2}.$$

(ii) We have $u = 3\,x^2y - x \sin xy$.

$$\therefore \quad \frac{\partial u}{\partial x} = 6\,xy - [xy \cos(xy) + \sin xy] = 6\,xy - xy \cos(xy) + \sin(xy),$$

and $\partial u/\partial y = 3x^2 - x^2 \cos(xy)$.

Example 11.11 (a) If $z = e^{2x^2 + xy + 3y^2}$, find $\partial z/\partial x$, $\partial z/\partial y$.

(b) If $f(x, y) = x^3y - xy^3$, find $\left[\dfrac{1}{\partial f/\partial x} + \dfrac{1}{\partial f/\partial y} \right]_{\substack{x=1 \\ y=2}}$.

(c) The relation $\left(P + \dfrac{a}{v^2} \right)(v - b) = RT$, in which a, b and R are constants is

given. Find an expression for $\left(\dfrac{\partial P}{\partial T} \right)_v$ and show that $T \left(\dfrac{\partial P}{\partial T} \right)_v - P = \dfrac{a}{v^2}$.

(A.M.I.E; W-2003)

Solution. (a) We have $\dfrac{\partial z}{\partial x} = (4x + y)e^{2x^2 + xy + 3y^2}$ (Here y is treated as a constant)

And $\dfrac{\partial z}{\partial y} = (x + 6y)e^{2x^2 + xy + 3y^2}$ (Here x has been treated as a constant).

(b) We have $f(x, y) = x^3y - xy^3$.

$\therefore \quad \partial f/\partial x = 3x^2y - y^3; \; \partial f/\partial y = x^3 - 3xy^2,$

and $\left[\dfrac{1}{\partial f/\partial x} + \dfrac{1}{\partial f/\partial y}\right]_{\substack{x=1 \\ y=2}} = \left[\dfrac{1}{3x^2y - y^3} + \dfrac{1}{x^3 - 3xy^2}\right]_{\substack{x=1 \\ y=2}} = \left[\dfrac{1}{6-8} + \dfrac{1}{1-12}\right] = -\dfrac{13}{22}.$

(c) We have $\left(P + \dfrac{a}{v^2}\right)(v - b) = RT.$ ...(i)

Differentiating partially w.r.t T, treating v as constant.

$\left\{\left(\dfrac{\partial P}{\partial T}\right)_v + 0\right\}(v - b) + \left(P + \dfrac{a}{v^2}\right)0 = R$ or $\left(\dfrac{\partial P}{\partial T}\right)_v (v - b) = R \Rightarrow \left(\dfrac{\partial P}{\partial T}\right)_v = \dfrac{R}{v - b}.$

    ...(ii)

Now from (ii), $T\left(\dfrac{\partial P}{\partial T}\right)_v = \dfrac{R \cdot T}{v - b} = \left(P + \dfrac{a}{v^2}\right)$, from (i) or $T\left(\dfrac{\partial P}{\partial T}\right)_v - P = \dfrac{a}{v^2}.$

Example 11.12 If $u = (1 - 2xy + y^2)^{-1/2}$, prove that $x\dfrac{\partial u}{\partial x} - y\dfrac{\partial u}{\partial y} = y^2u^3$.

(MDU-2001)

Solution. We have

$\dfrac{\partial u}{\partial x} = -\dfrac{1}{2}(1 - 2xy + y^2)^{-3/2} \cdot (-2y) = y\big/(1 - 2xy + y^2)^{3/2},$

$\dfrac{\partial u}{\partial y} = -\dfrac{1}{2}(1 - 2xy + y^2)^{-3/2} \cdot (-2x + 2y) = (x - y)\big/(1 - 2xy + y^2)^{3/2}.$

$\therefore \quad x\dfrac{\partial u}{\partial x} - y\dfrac{\partial u}{\partial y} = (xy - xy + y^2)\big/(1 - 2xy + y^2)^{3/2}$

$= y^2(1 - 2xy + y^2)^{-3/2} = y^2u^3$, which is same as required.

Example 11.13 If $x = r\cos\theta$, $y = r\sin\theta$, find

(a) $(\partial x/\partial r)_\theta$, $(\partial x/\partial r)_y$, $(\partial r/\partial x)_y$, $(\partial\theta/\partial y)_x$

(b) $\partial^2\theta/\partial x^2 + \partial^2\theta/\partial y^2$.   (UPTU 2001; MDU-2001)

Note: Before we solve the problem let us clearly understand the meaning of $(\partial x/\partial r)_\theta$, $(\partial x/\partial r)_y$ *etc.*

In the present problem we have four variables x, y, r, θ connected by two relations x = r cos θ, y = r sin θ. Any one of the four variables can be expressed in terms of any two of the remaining three. For example x can be expressed in terms of r, θ or r, y or θ, y.

Now $(\partial x/\partial r)_\theta$ *means the partial derivative of x w.r.t. r treating θ as constant in a relation expressing x as function of r and θ.*

Solution. To get $(\partial x/\partial r)_\theta$, we have to express x in terms of r and θ which in fact is already given to us because $x = r\cos\theta$. Therefore, $(\partial x/\partial r)_\theta = \cos\theta$.   ...(i)

To find $(\partial x/\partial r)_y$, we have to express x as a function of r and y.

Since $x = r\cos\theta$, $y = r\sin\theta$. Therefore, $x^2 + y^2 = r^2$. Thus

$$x = \sqrt{r^2 - y^2}\ .$$

$$\therefore \qquad \left(\frac{\partial x}{\partial r}\right)_y = \frac{r}{\sqrt{r^2 - y^2}} = \frac{r}{x} = \sec\theta\ . \qquad\qquad ...(ii)$$

[From (i) and (ii) we note that $\left(\dfrac{\partial x}{\partial r}\right)_\theta \neq \left(\dfrac{\partial x}{\partial r}\right)_y$]

To determine $(\partial r/\partial x)_y$ we know that $r = \sqrt{x^2 + y^2}\ .$

$$\therefore \qquad \left(\frac{\partial r}{\partial x}\right)_y = \frac{x}{\sqrt{x^2 + y^2}} = \frac{x}{r} = \cos\theta.$$

As $\qquad \theta = \tan^{-1}(y/x),\ \ ...(iii).\quad \therefore\quad \left(\dfrac{\partial\theta}{\partial y}\right)_x = \dfrac{x}{x^2 + y^2}\ . \qquad ...(iv)$

(b) From (iii), $\qquad \left(\dfrac{\partial\theta}{\partial x}\right)_y = -\dfrac{y}{x^2 + y^2}\ .$

$$\therefore\quad \frac{\partial^2\theta}{\partial x^2} = \frac{\partial}{\partial x}\left(\frac{\partial\theta}{\partial x}\right) = \frac{\partial}{\partial x}\left(-\frac{y}{x^2 + y^2}\right) = \frac{2xy}{\left(x^2 + y^2\right)^2}\ .$$

From (iv), $\dfrac{\partial^2\theta}{\partial y^2} = \dfrac{\partial}{\partial y}\left(\dfrac{x}{x^2 + y^2}\right) = -\dfrac{2xy}{\left(x^2 + y^2\right)^2}$. Hence, $\dfrac{\partial^2\theta}{\partial x^2} + \dfrac{\partial^2\theta}{\partial y^2} = 0.$

Exercise 11.2

1. Find f_x and f_y when
   (i) $f(x, y) = x^4 + x^2 y^2 + y^4$,   (ii) $f(x, y) = e^{ax} \sin by$,

(iii) $f(x, y) = \tan^{-1}\left(\dfrac{x^2 + y^2}{x + y}\right)$,   (iv) $f(x, y) = x^y$.   *(IETE Dec 2007)*

Ans. (i) $4x^3 + 2xy^2$, $2x^2 y + 4y^3$. (ii) $ae^{ax} \sin by$, $be^{ax} \cos by$.
(iii) $(x^2 + 2xy - y^2/\{(x + y)^2 + (x^2 + y^2)^2\}$,
$(y^2 + 2xy - x^2)/\{(x + y)^2 + (x^2 + y^2)^2\}$.   (iv) yx^{y-1}, $x^y \log x$.

2. If $f(x, y) = \begin{cases} \dfrac{xy}{x^2 + y^2}, & (x, y) \neq (0, 0) \\ 0, & (x, y) = (0, 0) \end{cases}$

   Show that both the partial derivatives exist at $(0, 0)$ but the function is not continuous thereat.

3. Show that the function $f(x, y) = \begin{cases} \dfrac{x^2 + y^2}{x - y}, & (x, y) \neq (0, 0) \\ 0, & (x, y) = (0, 0) \end{cases}$

   possesses partial derivatives at $(0, 0)$, though it is not continuous at $(0, 0)$.

4. If $f(x, y) = 2x^2 - 3xy + 4y^2$, show that $f_x(2, 3) = -1$, $f_y(2, 3) = 18$.

5. (a) If $u = \sin^{-1}(x/y)$, show that $x\dfrac{\partial u}{\partial x} + y\dfrac{\partial u}{\partial y} = 0$.

   (b) If $z = \log\sqrt{x^2 + y^2}$, Show that $x\dfrac{\partial z}{\partial x} + y\dfrac{\partial z}{\partial y} = 1$ *(GGSIPU 2006)*

6. Given that $x = r\cos\theta$, $y = r\sin\theta$, find $r_x, r_y, \theta_x, \theta_y$.
   [**Hint:** $r^2 = x^2 + y^2$. Note

   $r_x \equiv \left(\dfrac{\partial r}{\partial x}\right), r_y \equiv \left(\dfrac{\partial r}{\partial y}\right), \theta_x \equiv \left(\dfrac{\partial \theta}{\partial x}\right), \theta_y \equiv \left(\dfrac{\partial \theta}{\partial y}\right)$]

Ans. $r_x = \cos\theta$, $r_y = \sin\theta$, $\theta_x = (\sin\theta)/r$, $\theta_y = (\cos\theta)/r$.

7. If $z = f(ax + by)$, show that $b\dfrac{\partial z}{\partial x} - a\dfrac{\partial z}{\partial y} = 0$.

8. If $z(x + y) = x^2 + y^2$, show that $\left(\dfrac{\partial z}{\partial x} - \dfrac{\partial z}{\partial y}\right)^2 = 4\left(1 - \dfrac{\partial z}{\partial x} - \dfrac{\partial z}{\partial y}\right)$.   *(VTU 2003)*

9. Find the first order partial derivatives with respect to each variable, if

$$f(r, \theta, z) = \frac{r(r - \cos 2\theta)}{r^2 + z^2}. \qquad \text{Ans. } \frac{\partial f}{\partial r} = \frac{2rz^2 + (r^2 - z^2)\cos 2\theta}{(r^2 + z^2)^2},$$

$$\frac{\partial f}{\partial \theta} = \frac{2r\sin 2\theta}{(r^2 + z^2)}; \quad \frac{\partial f}{\partial z} = -\frac{2z(r^2 - r\cos 2\theta)}{(r^2 + z^2)^2}.$$

10. If $z = e^{ax + by}$. $f(ax - by)$, prove that $b\dfrac{\partial z}{\partial x} + a\dfrac{\partial z}{\partial y} = 2abz$.

(GGSIPU 2006; PTU 2006) (A.M.I.E., S-2004; I.E.T.E., S-2004; V.T.U.-2000)

11. If $f(x, y) = (1 - 2xy + y^2)^{-1/2}$, show that

$$\frac{\partial}{\partial x}\left\{(1 - x^2)\frac{\partial f}{\partial x}\right\} + \frac{\partial}{\partial y}\left(y^2\frac{\partial f}{\partial y}\right) = 0. \qquad \text{(AMIE, S-2009; Hamirpur, 1995)}$$

[**Hint** $\dfrac{\partial f}{\partial x} = y(1 - 2xy + y^2)^{-3/2}, \dfrac{\partial f}{\partial y} = (x - y)(1 - 2xy + y^2)^{-3/2}$.

$\therefore \quad \dfrac{\partial}{\partial x}\left[(1 - x^2)y(1 - 2xy + y^2)^{-3/2}\right] = 3y^2(1 - x^2)(1 - 2xy + y^2)^{-5/2}$

$$-2xy(1 - 2xy + x^2)^{-3/2},$$

and $\dfrac{\partial}{\partial y}\left[(y^2 x - y^3)(1 - 2xy + y^2)^{-3/2}\right] = 3y^2(x - y)^2(1 - 2xy + y^2)^{-5/2}$

$$+(2xy - 3y^2)(1 - 2xy + y^2)^{-3/2}.$$

Adding, we get $(1 - 2xy + y^2)^{-5/2}\left[3y^2(1 - x^2) - 2xy(1 - 2xy + y^2)\right.$

$$\left. +3y^2(x - y)^2 + (2xy - 3y^2)(1 - 2xy + y^2) = 0\right].$$

12. If $\phi = x\{1 + (a^2/r^2)\}$, $\psi = y\{1 - (a^2/r^2)\}$, where $r^2 = x^2 + y^2$. Show that $\partial\phi/\partial x = -\partial\psi/\partial y$ and $\partial\phi/\partial y = -\partial\psi/\partial x$. *(A.M.I.W., S-2003)*

13. If $\dfrac{x^2}{a^2 + u} + \dfrac{y^2}{b^2 + u} + \dfrac{z^2}{c^2 + u} = 1$, show that

$$\left(\frac{\partial u}{\partial x}\right)^2 + \left(\frac{\partial u}{\partial y}\right)^2 + \left(\frac{\partial u}{\partial z}\right)^2 = 2\left(x\frac{\partial u}{\partial x} + y\frac{\partial u}{\partial y} + z\frac{\partial u}{\partial z}\right).$$

(UPTU 2003; A.M.I.E., S-1997)

[**Hint.** We have $x^2 \cdot (a^2 + u)^{-1} + y^2(b^2 + u)^{-1} + z^2(c^2 + u)^{-1} = 1$, ...(i) where u is a function of x, y, z. Differentiating (i) partially w.r.t. x, we obtain

$$2x \cdot (a^2 + u)^{-1} - \left\{ x^2 (a^2 + u)^{-2} + y^2 (b^2 + u)^{-2} + z^2 (c^2 + u)^{-2} \right\} \frac{\partial u}{\partial x} = 0$$

$$\Rightarrow \frac{\partial u}{\partial x} = 2x (a^2 + u)^{-1} \div \left[\frac{x^2}{(a^2 + u)^2} + \frac{y^2}{(b^2 + u)^2} + \frac{z^2}{(c^2 + u)^2} \right] = \frac{2x}{a^2 + u} \cdot \frac{1}{Q}$$

where $Q = \sum \dfrac{x^2}{(a^2 + u)^2}$. Similarly $\dfrac{\partial u}{\partial y} = \dfrac{2y}{b^2 + u} \cdot \dfrac{1}{Q}$ and $\dfrac{\partial u}{\partial z} = \dfrac{2z}{c^2 + u} \cdot \dfrac{1}{Q}$.

Therefore, $\left(\dfrac{\partial u}{\partial x} \right)^2 + \left(\dfrac{\partial u}{\partial y} \right)^2 + \left(\dfrac{\partial u}{\partial z} \right)^2 = \dfrac{4}{Q^2} \left[\dfrac{x^2}{(a^2 + u)^2} + \dfrac{y^2}{(b^2 + u)^2} + \dfrac{z^2}{(c^2 + u)^2} \right]$

$$= \frac{4}{Q^2} \cdot Q = \frac{4}{Q}.$$

Now $2 \left(x \dfrac{\partial u}{\partial x} + y \dfrac{\partial u}{\partial y} + z \dfrac{\partial u}{\partial z} \right) = 2 \left[\dfrac{2x^2}{Q(a^2 + u)} + \dfrac{2y^2}{Q(b^2 + u)} + \dfrac{2z^2}{Q(c^2 + u)} \right]$

$$= \frac{4}{Q} \left[\frac{x^2}{a^2 + u} + \frac{y^2}{b^2 + u} + \frac{z^2}{c^2 + u} \right] = \frac{4}{Q} ..(1) \quad = \frac{4}{Q} = \left(\frac{\partial u}{\partial x} \right)^2 + \left(\frac{\partial u}{\partial y} \right)^2 + \left(\frac{\partial u}{\partial z} \right)^2.$$

14. If $u = \log(x^3 + y^3 - x^2 y - xy^2)$. Show that $\dfrac{\partial u}{\partial x} + \dfrac{\partial u}{\partial y} = \dfrac{2}{x + y}$.

(AMIE., S-2007)

[**Hint.** $u = \log(x^3 + y^3 - x^2 y - xy^2) = \log \left[x^2 (x - y) - y^2 (x - y) \right]$

$= \log \left[(x^2 - y^2)(x - y) \right] = \log \left[(x - y)^2 (x + y) \right]$

$= 2 \log(x - y) + \log(x + y)].$

15. If $u = f(y/x)$. Show that $x \dfrac{\partial u}{\partial x} + y \dfrac{\partial u}{\partial y} = 0$.　　*(IETE., S-2008)*

16. If $f(x, y) = \tan^{-1}(y/x)$, find $\partial f / \partial x$ and $\partial f / \partial y$　　*(AMIE., W-2009)*

Ans. $-y/(x^2 + y^2), \ x/(x^2 + y^2).$

11.9 Higher Order Partial Derivatives

Let the first-order partial derivatives of a function of two variables $z = f(x, y)$ that is, $\partial z/\partial x$ and $\partial z/\partial y$ exist at all the points in the domain D. Then, the first order partial derivatives are also functions of x and y and thus can be further differentiated partially with respect to these independent variables that is. w.r.t. x or w.r.t. y. This gives rise to four possible **second order** partial derivatives of z or f, which are defined as

$$\frac{\partial^2 f}{\partial x^2} = \frac{\partial}{\partial x}\left(\frac{\partial f}{\partial x}\right) = f_{xx}(x, y) = \lim_{\Delta x \to 0}\left[\frac{f_x(x+\Delta x, y) - f_x(x, y)}{\Delta x}\right]$$

$$\frac{\partial^2 f}{\partial y \partial x} = \frac{\partial}{\partial y}\left(\frac{\partial f}{\partial x}\right) = f_{yx}(x, y) = \lim_{\Delta y \to 0}\left[\frac{f_x(x, y+\Delta y) - f_x(x, y)}{\Delta y}\right]$$

(differentiate partially first with respect to x and then with respect to y)

$$\frac{\partial^2 f}{\partial x \partial y} = \frac{\partial}{\partial x}\left(\frac{\partial f}{\partial y}\right) = f_{xy}(x, y) = \lim_{\Delta x \to 0}\left[\frac{f_y(x+\Delta x, y) - f_y(x, y)}{\Delta x}\right]$$

(differentiate partially first with respect to y and then with respect to x)

$$\frac{\partial^2 f}{\partial y^2} = \frac{\partial}{\partial y}\left(\frac{\partial f}{\partial y}\right) = f_{yy}(x, y) = \lim_{\Delta y \to 0}\left[\frac{f_y(x, y+\Delta y) - f_y(x, y)}{\Delta y}\right],$$

provided the limits exists. The derivatives f_{xy} and f_{yx} are called *mixed* derivatives. If f_{xy} and f_{yx} are continuous at a point $P(x, y)$, then at this point $f_{xy} = f_{yx}$. If all the second order partial derivatives exists at all points in D, then these derivatives are also functions of x and y can be further differentiated.

11.10 Illustrative Examples

Example 11.14 Find all the second order partial derivatives of the function $f(x, y) = x^2 y^3 + x^4 y$.

Solution. We have

$$f_x(x, y) = \frac{\partial f}{\partial x} = 2xy^3 + 4x^3 y, \quad f_y(x, y) = \frac{\partial f}{\partial y} = 3x^2 y^2 + x^4.$$

$$f_{yx}(x, y) = \frac{\partial}{\partial y}\left(\frac{\partial f}{\partial x}\right) = \frac{\partial}{\partial y}(f_x) = \frac{\partial}{\partial y}(2xy^3 + 4x^3 y) = 6xy^2 + 4x^3,$$

$$f_{xy}(x, y) = \frac{\partial}{\partial x}\left(\frac{\partial f}{\partial y}\right) = \frac{\partial}{\partial x}(f_y) = \frac{\partial}{\partial x}(3x^2 y^2 + x^4) = 6xy^2 + 4x^3,$$

$$f_{xx}(x, y) = \frac{\partial}{\partial x}\left(\frac{\partial f}{\partial x}\right) = \frac{\partial}{\partial x}(f_x) = \frac{\partial}{\partial x}(2xy^3 + 4x^3 y) = 2y^3 + 12x^2 y,$$

$$f_{yy}(x, y) = \frac{\partial}{\partial y}\left(\frac{\partial f}{\partial y}\right) = \frac{\partial}{\partial y}(f_y) = \frac{\partial}{\partial y}(3x^2 y^2 + x^4) = 6x^2 y.$$

We note that $f_{xy} = f_{yx}$.

Example 11.15 For the function

$$f(x, y) = \begin{cases} \dfrac{xy(x^2 - y^2)}{x^2 + y^2}, & (x,y) \neq (0, 0) \\ 0 & (x,y) = (0, 0) \end{cases}$$

show that $f_{xy}(0, 0) \neq f_{yx}(0, 0)$.

Solution. We obtain the required derivatives as

$$f_x(0, 0) = \lim_{\Delta x \to 0} \frac{f(\Delta x, 0) - f(0, 0)}{\Delta x} = 0, \; f_y(0, 0) = \lim_{\Delta y \to 0} \frac{f(0, \Delta y) - f(0, 0)}{\Delta y} = 0$$

$$f_x(0, y) = \lim_{\Delta x \to 0} \frac{f(\Delta x, y) - f(0, y)}{\Delta x} = \lim_{\Delta x \to 0} \frac{y\left[(\Delta x)^2 - y^2\right]\Delta x}{\left[(\Delta x)^2 + y^2\right]\Delta x} = -y.$$

$$f_y(x, 0) = \lim_{\Delta y \to 0} \frac{f(x, \Delta y) - f(x, 0)}{\Delta y} = \lim_{\Delta y \to 0} \frac{x\left[x^2 - (\Delta y)^2\right]\Delta y}{\left[x^2 + (\Delta y)^2\right]\Delta y} = x,$$

Now, $\quad f_{xy}(0, 0) = \dfrac{\partial}{\partial x}\left(\dfrac{\partial f}{\partial y}\right)_{(0,0)} = \lim_{\Delta x \to 0} \dfrac{f_y(\Delta x, 0) - f_y(0, 0)}{\Delta x} = \lim_{\Delta x \to 0} \dfrac{\Delta x - 0}{\Delta x} = 1.$

$$f_{yx}(0, 0) = \frac{\partial}{\partial y}\left(\frac{\partial f}{\partial x}\right)_{(0,0)} = \lim_{\Delta y \to 0} \frac{f_x(0, \Delta y) - f_x(0, 0)}{\Delta y} = \lim_{\Delta y \to 0} \frac{-\Delta y - 0}{\Delta y} = -1.$$

Hence, $\quad f_{xy}(0, 0) \neq f_{yx}(0, 0)$.

Example 11.16 Compute $f_{xy}(0, 0)$ **and** $f_{yx}(0, 0)$ **for the function**

$$f(x,y) = \begin{cases} \dfrac{xy^3}{x + y^2}, & (x, y) \neq (0, 0) \\ 0 & , (x,y) = (0, 0) \end{cases}$$

(I.E.T.E., June 2008; Dec. 2004)

Also discuss the continuity of f_{xy} **and** f_{yx} **at** (0, 0).

Solution. We have

$$f_x(0, 0) = \lim_{\Delta x \to 0} \frac{f(\Delta x, 0) - f(0, 0)}{\Delta x} = 0, \; f_y(0, 0) = \lim_{\Delta y \to 0} \frac{f(0, \Delta y) - f(0, 0)}{\Delta y} = 0,$$

$$f_x(0, y) = \lim_{\Delta x \to 0} \frac{f(\Delta x, y) - f(0, y)}{\Delta x} = \lim_{\Delta x \to 0} \frac{y^3 \Delta x}{\left[\Delta x + y^2\right]\Delta x} = y$$

$$f_y(x,0) = \lim_{\Delta y \to 0} \frac{f(x,\Delta y) - f(x,0)}{\Delta y} = \lim_{\Delta y \to 0} \frac{x(\Delta y)^3}{\left[x + (\Delta y)^2\right]\Delta y} = 0.$$

$$f_{xy}(0,0) = \lim_{\Delta x \to 0} \frac{f_y(\Delta x, 0) - f_y(0,0)}{\Delta x} = 0.$$

$$f_{yx}(0,0) = \lim_{\Delta y \to 0} \frac{f_x(0,\Delta y) - f_x(0,0)}{\Delta y} = \lim_{\Delta y \to 0} \frac{\Delta y}{\Delta y} = 1.$$

Since $f_{xy}(0,0) \neq f_{yx}(0,0)$, f_{xy} and f_{yx} are not continuous at $(0,0)$.

Alternative. We have $f_x(x,y) = \dfrac{y^3(x+y^2) - xy^3}{(x+y^2)^2} = \dfrac{y^5}{(x+y^2)^2}$.

We find that for $(x,y) \neq (0,0)$

$$f_{yx}(x,y) = \frac{y^6 + 5xy^4}{(x+y^2)^3} = f_{xy}(x,y).$$

Along the path $x = my^2$, we obtain

$$\lim_{(x,y) \to (0,0)} f_{yx}(x,y) = \lim_{y \to 0} \frac{y^6(1+5m)}{y^6(1+m)^3} = \frac{1+5m}{(1+m)^3}.$$ Since the limit does not

exist, f_{yx} is not continuous at $(0,0)$.

Example 11.17 Verify that $\dfrac{\partial^2 u}{\partial y\, \partial x} = \dfrac{\partial^2 u}{\partial x\, \partial y}$ for the following cases:

(*i*) $u = ax^3 + 3\,bx^2y + 3\,cxy^2 + dy^3$, (*ii*) $u = \tan^{-1}(x/y)$.

Solution. (*i*) We have $u = ax^3 + 3\,bx^2y + 3\,cxy^2 + dy^3$.

$$\frac{\partial u}{\partial x} = 3ax^2 + 6bxy + 3cy^2, \quad \frac{\partial u}{\partial y} = 3bx^2 + 6cxy + 3dy^2.$$

$$\therefore \quad \frac{\partial^2 u}{\partial y\, \partial x} = \frac{\partial}{\partial y}\left(\frac{\partial u}{\partial x}\right) = \frac{\partial}{\partial y}\left(3ax^2 + 6bxy + 3cy^2\right) = 6bx + 6cy, \qquad \text{...}(a)$$

and $\quad \dfrac{\partial^2 u}{\partial x\, \partial y} = \dfrac{\partial}{\partial x}\left(\dfrac{\partial u}{\partial y}\right) = \dfrac{\partial}{\partial x}\left(3bx^2 + 6cxy + 3dy^2\right) = 6bx + 6cy. \qquad \text{...}(b)$

From (*a*) and (*b*), follows the required result.

(*ii*) We have $u = \tan^{-1}\left(\dfrac{x}{y}\right)$. $\therefore \dfrac{\partial u}{\partial x} = \dfrac{1}{1+(x/y)^2} \cdot (1/y) = \dfrac{y}{x^2+y^2}$,

$\dfrac{\partial u}{\partial y} = \dfrac{1}{1+(x/y)^2} \dfrac{-x}{y^2} = \dfrac{-x}{x^2+y^2}$.

$$\dfrac{\partial^2 u}{\partial y \partial x} = \dfrac{\partial}{\partial y}\left(\dfrac{\partial u}{\partial x}\right) = \dfrac{\partial}{\partial y}\left(\dfrac{y}{x^2+y^2}\right) = \dfrac{(x^2+y^2)-y(2y)}{(x^2+y^2)^2} = \dfrac{x^2-y^2}{(x^2+y^2)^2}, \qquad ...(c)$$

and $\dfrac{\partial^2 u}{\partial x \partial y} = \dfrac{\partial}{\partial x}\left(\dfrac{\partial u}{\partial y}\right) = \dfrac{\partial}{\partial x}\left(\dfrac{-x}{x^2+y^2}\right) = \dfrac{(x^2+y^2)(-1)-(-x)(2x)}{(x^2+y^2)^2}$

$$= \dfrac{x^2-y^2}{(x^2+y^2)^2}. \qquad ...(d)$$

From (*c*) and (*d*) follows the required result.

Example 11.18 If u = x^y,

show that $\dfrac{\partial^3 u}{\partial x^2 \partial y} = \dfrac{\partial^3 u}{\partial x \partial y \partial x}$. *(UPTU-2001; PTU 1999; A.M.I.E., S-1995)*

Solution. We have $u = x^y$. Therefore, $\dfrac{\partial u}{\partial y} = x^y \log_e x$.

$$\dfrac{\partial^2 u}{\partial x \partial y} = \dfrac{\partial}{\partial x}\left(\dfrac{\partial u}{\partial y}\right) = yx^{y-1}\log_e x + x^y \cdot \dfrac{1}{x} = x^{y-1}\left(y\log_e x + 1\right).$$

$\therefore \qquad \dfrac{\partial^3 u}{\partial x^2 \partial y} = \dfrac{\partial}{\partial x}\left(\dfrac{\partial^2 u}{\partial x \partial y}\right) = \dfrac{\partial}{\partial x}\left[x^{y-1}\left(y\log_e x + 1\right)\right]. \qquad ...(i)$

Also $\qquad \dfrac{\partial u}{\partial x} = yx^{y-1}$.

Therefore, $\dfrac{\partial^2 u}{\partial y \partial x} = \dfrac{\partial}{\partial y}\left(\dfrac{\partial u}{\partial x}\right) = x^{y-1} + y \cdot x^{y-1}\log_e x = x^{y-1}\left(1 + y\log_e x\right)$,

and $\qquad \dfrac{\partial^3 u}{\partial x \partial y \partial x} = \dfrac{\partial}{\partial x}\left(\dfrac{\partial^2 u}{\partial y \partial x}\right) = \dfrac{\partial}{\partial x}\left\{x^{y-1}\left(y\log_e x + 1\right)\right\}. \qquad ...(ii)$

From (*i*) and (*ii*) follows the required result.

Example 11.19 (*a*) **Show that at the point on the surface $x^x \, y^y \, z^z = c$, where x = y = z, we have**

$$\frac{\partial^2 z}{\partial x \partial y} = -\left[x \log(ex)\right]^{-1}.$$

(*A.M.I.E., W-2005; DCE 2004, I.E.T.E., S-2004; UPTU 2001, PTU 1999*)

(*b*) **If $z = x^2 \tan^{-1}(y/x) - y^2 \tan^{-1}(x/y)$, prove that**

$$\partial^2 z / \partial y \partial x = (x^2 - y^2)/(x^2 + y^2).$$

(*A.M.I.E., S-2003; Mysore, 1997*)

Solution. (*a*) We have $x^x y^y z^z = c$.

Taking logarithms, we get $\log x^x + \log y^y + \log z^z = \log c$

or $\quad x \log x + y \log y + z \log z = \log c$.

Differentiating partially w.r.t. *y*, we obtain

$$y \cdot \frac{1}{y} + \log y + \left(z \cdot \frac{1}{z} + \log z\right)\frac{\partial z}{\partial y} = 0 \text{ or } \frac{\partial z}{\partial y} = -\frac{1 + \log y}{1 + \log z}.$$

$$\therefore \quad \frac{\partial z}{\partial y} = -\frac{\log e + \log y}{\log e + \log z} = -\frac{\log ey}{\log ez}. \text{ Similarly } \frac{\partial z}{\partial x} = -\frac{\log ex}{\log ez} \qquad \ldots(i)$$

Now $\dfrac{\partial^2 z}{\partial x \partial y} = \dfrac{\partial}{\partial x}\left(\dfrac{\partial z}{\partial y}\right) = \dfrac{\partial}{\partial x}\left(-\dfrac{\log ey}{\log ez}\right) = -\log ey \cdot \dfrac{\partial}{\partial x}(\log ez)^{-1}$

$= \log ey (\log ez)^{-2} \cdot \dfrac{e}{ez} \cdot \dfrac{\partial z}{\partial x} = \dfrac{-\log ey}{(\log ez)^2} \cdot \dfrac{1}{z} \cdot \dfrac{\log ex}{\log ez}.$ [Using (*i*)]

Thus at $y = z = x$ we obtain

$$\frac{\partial^2 z}{\partial x \partial y} = -\frac{\log ex}{(\log ex)^2} \cdot \frac{1}{x} \cdot \frac{\log(ex)}{\log(ex)} = -\frac{1}{x\log(ex)}.$$

Hence $\qquad \dfrac{\partial^2 z}{\partial x \partial y} = -\left[x\log(ex)\right]^{-1}.$

(*b*) We have $z = x^2 \tan^{-1}\left(\dfrac{y}{x}\right) - y^2 \tan^{-1}(x/y).$

$$\frac{\partial z}{\partial x} = 2x \tan^{-1}\left(\frac{y}{x}\right) + x^2 \cdot \frac{1}{1 + (y/x)^2} \cdot \left(\frac{-y}{x^2}\right) - y^2 \cdot \frac{1}{1 + (x/y)^2} \cdot \frac{1}{y}$$

$$= 2x\tan^{-1}\left(\frac{y}{x}\right) - \frac{x^2 y}{x^2 + y^2} - \frac{y^3}{x^2 + y^2} = 2x\tan^{-1}\left(\frac{y}{x}\right) - y.$$

Therefore, $\dfrac{\partial^2 z}{\partial y \partial x} = \dfrac{\partial}{\partial y}\left(\dfrac{\partial z}{\partial x}\right) = \dfrac{\partial}{\partial y}\left(2x\tan^{-1}\left(\dfrac{y}{x}\right) - y\right) = 2x \cdot \dfrac{1}{1 + (y/x)^2} \cdot \dfrac{1}{x} - 1$

$$= \frac{2x^2}{x^2 + y^2} - 1 = \frac{x^2 - y^2}{x^2 + y^2}.$$

Example 11.20 If $v = (x^2 + y^2 + z^2)^{-1/2}$, show that

$$\frac{\partial^2 v}{\partial x^2} + \frac{\partial^2 v}{\partial y^2} + \frac{\partial^2 v}{\partial z^2} = 0.$$

(J.N.T.U., 2000, A.M.I.E., W-1999)

Solution. $\quad \dfrac{\partial v}{\partial x} = -\dfrac{1}{2}\left(x^2 + y^2 + z^2\right)^{-3/2} \cdot 2x = -x\left(x^2 + y^2 + z^2\right)^{-3/2}.$

Therefore, $\dfrac{\partial^2 v}{\partial x^2} = \dfrac{\partial}{\partial x}\left(\dfrac{\partial v}{\partial x}\right) = \dfrac{\partial}{\partial x}\left[-x\left(x^2 + y^2 + z^2\right)^{-3/2}\right]$

$$= -\left[x \cdot (-3/2)\left(x^2 + y^2 + z^2\right)^{-5/2} \cdot 2x + \left(x^2 + y^2 + z^2\right)^{-3/2}\right]$$

$$= -\left[-3x^2\left(x^2 + y^2 + z^2\right)^{-5/2} + \left(x^2 + y^2 + z^2\right)^{-3/2}\right]$$

$$= -\left(x^2 + y^2 + z^2\right)^{-5/2}\left[-3x^2 + \left(x^2 + y^2 + z^2\right)\right]$$

$$= -\left(x^2 + y^2 + z^2\right)^{-5/2}\left(-2x^2 + y^2 + z^2\right).$$

Similarly $\quad \dfrac{\partial^2 v}{\partial y^2} = -\left(x^2 + y^2 + z^2\right)^{-5/2}\left(x^2 - 2y^2 + z^2\right),$

and $\quad \dfrac{\partial^2 v}{\partial z^2} = -\left(x^2 + y^2 + z^2\right)^{-5/2}\left(x^2 + y^2 - 2z^2\right).$

$\therefore \quad \dfrac{\partial^2 v}{\partial x^2} + \dfrac{\partial^2 v}{\partial y^2} + \dfrac{\partial^2 v}{\partial z^2} = -\left(x^2 + y^2 + z^2\right)^{-5/2}(0) = 0.$

Example 11.21 If $z = f(x + ct) + \phi(x - ct)$ where f and ϕ are twice differentiable functions and c is a real constant. Show that $\dfrac{\partial^2 z}{\partial t^2} = c^2 \cdot \dfrac{\partial^2 z}{\partial x^2}.$

(A.M.I.E., W-2000; I.E.T.E., June 1995)

Solution. We have $\dfrac{\partial z}{\partial t} = cf'(x+ct) - c\phi'(x-ct),$

and $\qquad \dfrac{\partial^2 z}{\partial t^2} = \dfrac{\partial}{\partial t}\left(\dfrac{\partial z}{\partial t}\right) = c^2 f''(x+ct) + c^2 \phi''(x-ct)$

$$= c^2\left[f''(x+ct) + \phi''(x-ct)\right]. \qquad \text{...(i)}$$

Similarly, $\qquad \dfrac{\partial^2 z}{\partial x^2} = f''(x+ct) + \phi''(x-ct). \qquad \text{...(ii)}$

From (i) and (ii) it follows that $\dfrac{\partial^2 z}{\partial t^2} = c^2 \cdot \dfrac{\partial^2 z}{\partial x^2}.$

Example 11.22 If $\theta = t^n\, e^{-r^2/4t}$, **what value of n will make**

$$\frac{1}{r^2}\frac{\partial}{\partial r}\left(r^2\frac{\partial\theta}{\partial r}\right) = \frac{\partial\theta}{\partial t}\ ? \qquad (I.E.T.E.,\ S\text{-}2003)$$

Solution. We have $\dfrac{\partial\theta}{\partial r} = t^n \cdot e^{-r^2/4t}\left(\dfrac{-r}{2t}\right).$

$$\therefore \qquad r^2\frac{\partial\theta}{\partial r} = \frac{-r^3}{2t}t^n \cdot e^{-r^2/4t} = \frac{-r^3}{2}t^{n-1}\cdot e^{-r^2/4t},$$

and $\dfrac{\partial}{\partial r}\left(r^2\dfrac{\partial\theta}{\partial r}\right) = \dfrac{\partial}{\partial r}\left(\dfrac{-r^3}{2}t^{n-1}\cdot e^{-r^2/4t}\right) = \dfrac{-t^{n-1}}{2}\left[3r^2\,e^{-r^2/4t} + r^3\cdot e^{-r^2/4t}\cdot\left(\dfrac{-r}{2t}\right)\right]$

$$= \frac{-t^{n-1}}{2}r^2 e^{-r^2/4t}\left(3 - \frac{r^2}{2t}\right).$$

$$\therefore \qquad \frac{1}{r^2}\frac{\partial}{\partial r}\left(r^2\frac{\partial\theta}{\partial r}\right) = \frac{-t^{n-1}}{2}e^{-r^2/4t}\left(3 - \frac{r^2}{2t}\right) = t^{n-2}\cdot e^{-r^2/4t}\left(\frac{-3t}{2} + \frac{r^2}{4}\right). \qquad \text{...(i)}$$

Also $\quad \dfrac{\partial\theta}{\partial t} = nt^{n-1}e^{-r^2/4t} + t^n\cdot e^{-r^2/4t}\left(\dfrac{r^2}{4t^2}\right) = t^{n-2}e^{-r^2/4t}\left(nt + \dfrac{r^2}{4}\right). \qquad \text{...(ii)}$

Comparing (i) and (ii), we see that n = –3/2 will make them equal.

Example 11.23 If **u = f(r) and x = r cos θ, y = r sin θ, prove that**

$$\frac{\partial^2 u}{\partial x^2} + \frac{\partial^2 u}{\partial y^2} = f''(r) + \frac{1}{r}f'(r).$$

(UPTU 2001; MDU 2001; Mangalore 1997; A.M.I.E., W-1995)

Solution. We have $u = f(r)$ and $r^2 = x^2 + y^2$. Therefore,

$$\frac{\partial u}{\partial x} = f'(r) \cdot \frac{\partial r}{\partial x} = f'(r) \cdot \frac{x}{r}, \quad \frac{\partial^2 u}{\partial x^2} = \frac{\partial}{\partial x}\left(\frac{\partial u}{\partial x}\right) = \frac{\partial}{\partial x}\left[f'(r) \cdot \frac{x}{r}\right]$$

$$= f''(r) \cdot \frac{\partial r}{\partial x} \cdot \frac{x}{r} + f'(r)\left[\frac{r - x(\partial r/\partial x)}{r^2}\right] = f''(r) \cdot \frac{x^2}{r^2} + f'(r) \cdot \frac{r - x(x/r)}{r^2}$$

$$= f''(r) \cdot \frac{x^2}{r^2} + f'(r) \cdot \frac{r^2 - x^2}{r^3}. \qquad \qquad \ldots(i)$$

Similarly, $\qquad \dfrac{\partial^2 u}{\partial y^2} = f''(r) \cdot \dfrac{y^2}{r^2} + f'(r) \cdot \dfrac{r^2 - y^2}{r^3}. \qquad \ldots(ii)$

Adding (i) and (ii), we get

$$\frac{\partial^2 u}{\partial x^2} + \frac{\partial^2 u}{\partial y^2} = f''(r)\left(\frac{x^2 + y^2}{r^2}\right) + f'(r) \cdot \frac{2r^2 - (x^2 + y^2)}{r^3}$$

$$= f''(r) + \frac{1}{r} f'(r), \text{ the required result.}$$

Example 11.24 If $x = e^{r \cos \theta} \cos(r \sin \theta)$ and $y = e^{r \cos \theta} \sin(r \sin \theta)$ **prove that**

$$\frac{\partial x}{\partial r} = \frac{1}{r} \cdot \frac{\partial y}{\partial \theta}, \frac{\partial y}{\partial r} = -\frac{1}{r} \frac{\partial x}{\partial \theta}. \text{ Hence deduce that } \frac{\partial^2 x}{\partial r^2} + \frac{1}{r} \frac{\partial x}{\partial r} + \frac{1}{r^2} \frac{\partial^2 x}{\partial \theta^2} = 0.$$

(DCE 2004)

Solution. We have $\dfrac{\partial x}{\partial r} = \cos\theta \, e^{r\cos\theta} \cos(r\sin\theta) - \sin\theta \, e^{r\cos\theta} \sin(r\sin\theta)$

$$= e^{r\cos\theta} \cos(\theta + r\sin\theta). \qquad \qquad \ldots(i)$$

$$\frac{\partial x}{\partial \theta} = -r\sin\theta \, e^{r\cos\theta} \cos(r\sin\theta) - r\cos\theta \, e^{r\cos\theta} \sin(r\sin\theta)$$

$$= -re^{r\cos\theta} \sin(\theta + r\sin\theta). \qquad \qquad \ldots(ii)$$

$$\frac{\partial y}{\partial r} = \cos\theta \, e^{r\cos\theta} \sin(r\sin\theta) + \sin\theta \, e^{r\cos\theta} \cos(r\sin\theta)$$

$$= e^{r\cos\theta} \sin(\theta + r\sin\theta). \qquad \qquad \ldots(iii)$$

and $\dfrac{\partial y}{\partial \theta} = -r\sin\theta \, e^{r\cos\theta} \sin(r\sin\theta) + r\cos\theta \, e^{r\cos\theta} \cos(r\sin\theta)$

$$= re^{r\cos\theta} \cos(\theta + r\sin\theta). \qquad \qquad ,..(iv)$$

From (*i*) and (*iv*) $\dfrac{\partial x}{\partial r} = \dfrac{1}{r}\dfrac{\partial y}{\partial \theta}$, and from (*ii*) and (*iii*) $\dfrac{\partial y}{\partial r} = -\dfrac{1}{r}\dfrac{\partial x}{\partial \theta}$.

Now
$$\frac{\partial^2 x}{\partial r^2} = \frac{\partial}{\partial r}\left(\frac{\partial x}{\partial r}\right) = \frac{\partial}{\partial r}\left(\frac{1}{r}\frac{\partial y}{\partial \theta}\right) = -\frac{1}{r^2}\frac{\partial y}{\partial \theta} + \frac{1}{r}\frac{\partial^2 y}{\partial r \partial \theta},$$

$$\frac{\partial^2 x}{\partial \theta^2} = \frac{\partial}{\partial \theta}\left(\frac{\partial x}{\partial \theta}\right) = \frac{\partial}{\partial \theta}\left(-r\frac{\partial y}{\partial r}\right) = -r\frac{\partial^2 y}{\partial \theta \partial r}.$$

Therefore, $\dfrac{\partial^2 x}{\partial r^2} + \dfrac{1}{r}\dfrac{\partial x}{\partial r} + \dfrac{1}{r^2}\dfrac{\partial^2 x}{\partial \theta^2} = -\dfrac{1}{r^2}\dfrac{\partial y}{\partial \theta} + \dfrac{1}{r}\dfrac{\partial^2 y}{\partial r \partial \theta} + \dfrac{1}{r^2}\dfrac{\partial y}{\partial \theta} - \dfrac{1}{r}\dfrac{\partial^2 y}{\partial \theta \partial r} = 0.$

11.11 Homogeneous Functions

A function $f(x, y)$ is said to be *homogeneous* of degree n in x and y, if it can be written in any one of the following forms

(i) $f(\lambda x, \lambda y) = \lambda^n f(x, y), \lambda > 0.$ ...(11.10)

(ii) $f(x, y) = x^n g(y/x).$ ...(11.11)

(iii) $f(x, y) = y^n h(x/y).$ ...(11.12)

Similarly, a function $f(x, y, z)$ of three variables is said to be homogeneous, of degree n, if it can be written as

$f(\lambda x, \lambda y, \lambda z) = \lambda^n f(x, y, z)$ or $f(x, y, z) = x^n g(y/x, z/x)$ etc.

Some examples of homogeneous functions are the following:

Function, *f*	Degree of homogeneity
1. $x^3 + 3x^2 y$	3
2. $x^2 \cos(y/x)$	2
3. $\sin^{-1}(y/x)$	0
4. $\dfrac{x+y}{\sqrt{x}+\sqrt{y}}$	1/2
5. $\sqrt{ax^2 + 2hxy + by^2}$	1
6. $\dfrac{1}{xy+x^2}$	−2
7. $1/(x^3 + y^3 + z^3)$	−3

The function $f(x, y) = (x + y^2)/(y + x^2)$ is not homogeneous.

Example 11.25 Determine the degree of homogeneity of each of the following functions

(i) $f(x,y) = \dfrac{x^3 + y^3 + 3x^2 y}{2x + 3y}$, (ii) $f(l, k) = (al^4 + bk^4)^{1/2}$, (*iii*) $\mathbf{f(x, y) = (x^2 \, y)^{1/3}}$,   (*iv*) $\mathbf{f(x, y) = Ax^\alpha \, y^\beta}$.

Solution. (*i*) We have $f(x, y) = \dfrac{x^3 + y^3 + 3x^2 y}{2x + 3y}$. Replacing x by λx and y by λy, $\lambda > 0$, we obtain

$$f(\lambda x, \lambda y) = \frac{(\lambda x)^3 + (\lambda y)^3 + 3(\lambda x)^2 (\lambda y)}{2\lambda x + 3\lambda y} = \lambda^2 \left(\frac{x^3 + y^3 + 3x^2 y}{2x + 3y} \right) = \lambda^2 f(x, y).$$

Hence the degree of homogeneity is 2.

(*ii*) Setting $l = \lambda l$, $k = \lambda k$, $\lambda > 0$, we obtain

$$f(\lambda l, \lambda k) = \left[a(\lambda l)^4 + b(\lambda k)^4 \right]^{1/2} = \lambda^2 \left(a l^4 + b k^4 \right)^{1/2} = \lambda^2 f(l, k).$$

Hence the degree of homogeneity is 2.

(*iii*) We have $f(x, y) = (x^2 y)^{1/3}$.

$\therefore \ f(\lambda x, \lambda y) = [(\lambda x)^2 (\lambda y)]^{1/3} = \lambda (x^2 y)^{1/3} = \lambda f(x, y).$

Hence the degree of homogeneity is 1.

(*iv*) $f(x, y) = A \, x^\alpha y^\beta$.

$\therefore$ $f(\lambda x, \lambda y) = A \, (\lambda x)^\alpha \, (\lambda y)^\beta = \lambda^{\alpha+\beta} \left(A x^\alpha y^\beta \right) = \lambda^{\alpha+\beta} f(x, y).$

So that the degree of homogeneity is $\alpha + \beta$.

11.12 Euler's Theorem on Homogeneous Functions

If $f(x, y)$ is a homogeneous function of degree n in x and y and has continuous first and second order partial derivatives, then

(*a*) $x\dfrac{\partial f}{\partial x} + y\dfrac{\partial f}{\partial y} = nf.$ ...(11.13) (*I.E.T.E., June 2009; A.M.I.E., W-2003, W-2002*)

(*b*) $x^2 \dfrac{\partial^2 f}{\partial x^2} + 2xy \dfrac{\partial^2 f}{\partial x \partial y} + y^2 \dfrac{\partial^2 f}{\partial y^2} = n(n-1) f.$ ...(11.14)

(*I.E.T.E., June 2009; UPTU 2005; Mangalore 1997*)

Proof. (*a*) Since $f(x, y)$ is homogeneous function of degree n in x and y, we can write $f(x, y) = x^n g\left(\dfrac{y}{x}\right)$. Differentiating partially with respect to x and y, we obtain

$$\frac{\partial f}{\partial x} = nx^{n-1} g\left(\frac{y}{x}\right) + x^n \cdot g'\left(\frac{y}{x}\right) \cdot \left(-\frac{y}{x^2}\right) = nx^{n-1} g\left(\frac{y}{x}\right) - yx^{n-2} g'\left(\frac{y}{x}\right).$$

$$\frac{\partial f}{\partial y} = x^n g'\left(\frac{y}{x}\right) \cdot \frac{1}{x} = x^{n-1} g'\left(\frac{y}{x}\right).$$

Hence, we obtain

$$x\frac{\partial f}{\partial x}+y\frac{\partial f}{\partial y}=nx^{n}g\left(\frac{y}{x}\right)-yx^{n-1}g'\left(\frac{y}{x}\right)+yx^{n-1}g'\left(\frac{y}{x}\right)=nx^{n}g\left(\frac{y}{x}\right)=nf.$$

Note. [In general if $f(x_1, x_2, ..., x_k)$ be a homogeneous function of order n, then

$$x_1\frac{\partial f}{\partial x_1}+x_2\frac{\partial f}{\partial x_2}+......+x_k\frac{\partial f}{\partial x_k}=nf.]$$

(b) Differentiating (11.13) partially w.r.t x and y, we get

$$x\frac{\partial^2 f}{\partial x^2}+\frac{\partial f}{\partial x}+y\frac{\partial^2 f}{\partial x\partial y}=n\frac{\partial f}{\partial x}, \qquad ...(11.15)$$

and

$$x\frac{\partial^2 f}{\partial y\partial x}+\frac{\partial f}{\partial y}+y\frac{\partial^2 f}{\partial y^2}=n\frac{\partial f}{\partial y}. \qquad ...(11.16)$$

Multiplying (11.15) by x and (11.16) by y and adding, we obtain.

$$x^2\frac{\partial^2 f}{\partial x^2}+\left(x\frac{\partial f}{\partial x}+y\frac{\partial f}{\partial y}\right)+xy\left(\frac{\partial^2 f}{\partial x\partial y}+\frac{\partial^2 f}{\partial y\partial x}\right)+y^2\frac{\partial^2 f}{\partial y^2}=n\left(x\frac{\partial f}{\partial x}+y\frac{\partial f}{\partial y}\right),$$

or

$$x^2\frac{\partial^2 f}{\partial x^2}+2xy\frac{\partial^2 f}{\partial x\partial y}+y^2\frac{\partial^2 f}{\partial y^2}=n(n-1)f.$$

11.13 Illustrative Examples

Example 11.26 Verify Euler's theorem for the following functions:

(i) $f(x, y) = ax^2 + 2hxy + by^2$,   *(Osmania, 1995)*

(ii) $f(x, y) = \dfrac{\left(x^{1/4} + y^{1/4}\right)}{\left(x^{1/5} + y^{1/5}\right)}$,   *(UPTU 2005)*

(iii) $f(x, y, z) = 3x^2yz + 5xy^2z + 4z^4$.   *(PTU 2006; J.N.T.U., 1999)*

Solution. (i) We have $\dfrac{\partial f}{\partial x}=2ax+2hy, \dfrac{\partial f}{\partial y}=2hx+2by.$

Therefore, $x\dfrac{\partial f}{\partial x}+y\dfrac{\partial f}{\partial y}=x(2ax+2hy)+y(2hx+2by)$

$= 2(ax^2 + 2hxy + by^2) = 2f(x, y).$

Since $f(x, y)$ is homogeneous of degree 2, Euler's theorem is verified.

(ii) $f(x, y) = \left(x^{1/4} + y^{1/4}\right)/\left(x^{1/5} + y^{1/5}\right).$

Both the numerator and denominator are homogeneous. Their degrees are 1/4 and 1/5 respectively. So the degree of $f(x, y)$ is 1/4 − 1/5 = 1/20.

Now $\dfrac{\partial f}{\partial x} = \dfrac{\dfrac{1}{4} x^{-3/4}\left(x^{1/5}+y^{1/5}\right)-\dfrac{1}{5}x^{-4/5}\left(x^{1/4}+y^{1/4}\right)}{\left(x^{1/5}+y^{1/5}\right)^2}$,

and $\dfrac{\partial f}{\partial y} = \dfrac{\dfrac{1}{4} y^{-3/4}\left(x^{1/5}+y^{1/5}\right)-\dfrac{1}{5}y^{-4/5}\left(x^{1/4}+y^{1/4}\right)}{\left(x^{1/5}+y^{1/5}\right)^2}$.

Therefore, $x\dfrac{\partial f}{\partial x}+y\dfrac{\partial f}{\partial y} = \dfrac{\left(\dfrac{1}{4}-\dfrac{1}{5}\right)\left(x^{1/5}+y^{1/5}\right)\left(x^{1/4}+y^{1/4}\right)}{\left(x^{1/5}+y^{1/5}\right)^2}$

$= \dfrac{1}{20}\cdot\dfrac{x^{1/4}+y^{1/4}}{x^{1/5}+y^{1/5}} = \dfrac{1}{20}f(x,y)$, which verifies Euler's theorem.

(*iii*) $\dfrac{\partial f}{\partial x} = 6xyz+5y^2z,\ \dfrac{\partial f}{\partial y} = 3x^2z+10xyz$ and $\dfrac{\partial f}{\partial z} = 3x^2y+5xy^2+16z^3$.

Therefore, $x\dfrac{\partial f}{\partial x}+y\dfrac{\partial f}{\partial y}+z\dfrac{\partial f}{\partial z} = 6x^2yz+5y^2zx+3x^2zy$

$+10xy^2z+3x^2yz+5xy^2z+16z^4$

$= 12\ x^2yz + 20\ xy^2z + 16\ z^4 = 4\left[3x^2yz+5xy^2z+4z^4\right] = 4f(x,\ y,\ z)$. Since the degree of homogeneity is 4 so Euler's theorem is verified.

Example 11.27 Find the value of $x\dfrac{\partial u}{\partial x}+y\dfrac{\partial u}{\partial y}$ when

(*i*) $u = \sin^{-1}\left(\dfrac{x}{y}\right)+\tan^{-1}\left(\dfrac{y}{x}\right)$,   (*I.E.T.E., June 2010, A.M.I.E., S-2003*)

(*ii*) $u = x^3\log\dfrac{\sqrt[3]{y}-\sqrt[3]{x}}{\sqrt[3]{y}+\sqrt[3]{x}}$,   (*iii*) $u = \sin\left\{(x-y)/(x+y)\right\}^{1/2}$,

(*iv*) $u = 3\log x - \log y^3 + \tan^{-1}(y/x)$.

Solution. (*i*) Since u is a homogeneous function of degree 0, thus by Euler's theorem $x\left(\dfrac{\partial u}{\partial x}\right)+y\left(\dfrac{\partial u}{\partial y}\right) = 0.u = 0$.

(ii) $u = x^3 \log \left[\dfrac{(y/x)^{1/3} - 1}{(y/x)^{1/3} + 1} \right]$ is homogeneous function of degree 3. Therefore,

by Euler's theorem $x \dfrac{\partial u}{\partial x} + y \dfrac{\partial u}{\partial y} = nu = 3u = 3x^3 \log \dfrac{\sqrt[3]{y} - \sqrt[3]{x}}{\sqrt[3]{y} + \sqrt[3]{x}}$.

(iii) $u = \sin \left(\dfrac{x-y}{x+y} \right)^{1/2} = \sin \left(\dfrac{1-(y/x)}{1+(y/x)} \right)^{1/2}$ is a homogeneous function of degree

zero. Therefore, $x \left(\dfrac{\partial u}{\partial x} \right) + y \left(\dfrac{\partial u}{\partial y} \right) = 0$.

(iv) $u = 3 \log x - \log y^3 + \tan^{-1}(y/x) = \log(x^3/y^3) + \tan^{-1}(y/x)$ is a homogeneous function of degree 0. Therefore $x \left(\dfrac{\partial u}{\partial x} \right) + y \left(\dfrac{\partial u}{\partial y} \right) = 0$.

Example 11.28 If $F(x, y) = x^4 y^2 \sin^{-1}(y/x)$, then find the value of

$x \left(\dfrac{\partial F}{\partial x} \right) + y \left(\dfrac{\partial F}{\partial y} \right)$ by actual differentiation. (*I.E.T.E. Dec. 2001*)

Solution. We have $F(x, y) = x^4 y^2 \sin^{-1}(y/x)$.

Differentiating partially w.r.t. x, we obtain

$$\frac{\partial F}{\partial x} = y^2 \left[4x^3 \sin^{-1}(y/x) + x^4 \cdot \frac{1}{\sqrt{1-(y/x)^2}} \cdot \left(-\frac{y}{x^2} \right) \right].$$

Therefore, $\quad x \dfrac{\partial F}{\partial x} = 4x^4 y^2 \sin^{-1}(y/x) - \dfrac{x^4 y^3}{\sqrt{x^2 - y^2}}.$ ...(i)

Now, differentiating partially w.r.t. y, we obtain

$$\frac{\partial F}{\partial y} = x^4 \left[2y \sin^{-1}(y/x) + y^2 \cdot \frac{1}{\sqrt{1-(y/x)^2}} \cdot \frac{1}{x} \right].$$

Therefore, $\quad y \dfrac{\partial F}{\partial y} = 2x^4 y^2 \sin^{-1}(y/x) + \dfrac{x^4 y^3}{\sqrt{x^2 - y^2}}.$ ...(ii)

Adding (i) and (ii), we obtain

$$x \frac{\partial F}{\partial x} + y \frac{\partial F}{\partial y} = 6 \left[x^4 y^2 \sin^{-1}(y/x) \right] = 6F(x, y).$$

Example 11.29 (*a*) If $u = \sin^{-1}\left(\dfrac{x^2 + y^2}{x + y}\right)$, show that $x\dfrac{\partial u}{\partial x} + y\dfrac{\partial u}{\partial y} = \tan u$.

(*IETE, Dec. 2009; GGSIPU 2006; VTU 2004; A.M.I.E.; W-2004*)

(*b*) If $f(x, y) = \dfrac{1}{x^2} + \dfrac{1}{xy} + \dfrac{\log x - \log y}{x^2 + y^2}$, show that

$$x\frac{\partial f}{\partial x} + y\frac{\partial f}{\partial y} + 2f(x, y) = 0. \quad (A.M.I.E., S\text{-}2007; PTU\ 2004)$$

(*c*) If $u = \cos\left(\dfrac{xy + yz + zx}{x^2 + y^2 + z^2}\right)$, prove that $x\dfrac{\partial u}{\partial x} + y\dfrac{\partial u}{\partial y} + z\dfrac{\partial u}{\partial z} = 0$.

(*d*) If $u = \log\left(\dfrac{x^4 + y^4}{x + y}\right)$, show that $x\dfrac{\partial u}{\partial x} + y\dfrac{\partial u}{\partial y} = 3$. (*MDU 2004; UPTU 2001*)

Solution. (*a*) The given function can be written as

$$\sin u = \frac{x^2 + y^2}{x + y} = x\left\{\frac{1 + (y/x)^2}{1 + (y/x)}\right\}.$$

Therefore, $\sin u$ is a homogeneous function of degree 1. Using the Euler's theorem for $f = \sin u$ and $n = 1$, we obtain

$$x\frac{\partial}{\partial x}(\sin u) + y\frac{\partial}{\partial y}(\sin u) = \sin u \qquad \ldots(i)$$

or $\quad x\cos u\dfrac{\partial u}{\partial x} + y\cos u\dfrac{\partial u}{\partial y} = \sin u$ or $x\dfrac{\partial u}{\partial x} + y\dfrac{\partial u}{\partial y} = \tan u$.

(*b*) $f(x, y) = \dfrac{1}{x^2}\left\{1 + \dfrac{x}{y} + \dfrac{\log(x/y)}{1 + (y^2/x^2)}\right\} = x^{-2}\left\{1 + (y/x)^{-1} - \dfrac{\log(y/x)}{1 + (y^2/x^2)}\right\}.$

The degree of homogeneity is -2. Thus by Euler's theorem, we have

$$x\frac{\partial f}{\partial x} + y\frac{\partial f}{\partial y} = -2f(x, y) \text{ or } x\frac{\partial f}{\partial x} + y\frac{\partial f}{\partial y} + 2f(x, y) = 0.$$

(*c*) $u = f(x, y, z) = \cos\left(\dfrac{xy + yz + zx}{x^2 + y^2 + z^2}\right).$

The degree of homogeneity is 0. Thus by Euler's theorem, we have

$$x\frac{\partial u}{\partial x} + y\frac{\partial u}{\partial y} + z\frac{\partial u}{\partial z} = 0(u) = 0.$$

(d) Let $z = e^u = \dfrac{x^4 + y^4}{x+y} = x^3\dfrac{\left[1+(y/x)^4\right]}{\left[1+(y/x)\right]}$. The degree of homogeneity is 3.

∴ By Euler's theorem, $x\left(\dfrac{\partial z}{\partial x}\right) + y\left(\dfrac{\partial z}{\partial y}\right) = 3z$

or $\quad x\cdot\dfrac{\partial z}{\partial u}\cdot\dfrac{\partial u}{\partial x} + y\dfrac{\partial z}{\partial u}\cdot\dfrac{\partial u}{\partial y} = 3z$ or $x\cdot e^u \cdot\dfrac{\partial u}{\partial x} + y\cdot e^u\cdot\dfrac{\partial u}{\partial y} = 3e^u$

or $\quad x\left(\dfrac{\partial u}{\partial x}\right) + y\left(\dfrac{\partial u}{\partial y}\right) = 3.$

Example 11.30 Use Euler's theorem to prove

$$x\frac{\partial v}{\partial x} + y\frac{dv}{dy} + \frac{1}{2}\cot v = 0, \text{ where } v = \cos^{-1}\frac{x+y}{\sqrt{x}+\sqrt{y}}.$$

(GGSIPU 2006; VTU 2004; A.M.I.E.; W-2004; Gorakhpur S-1994)

Solution. The given function can be written as

$$\cos v = \frac{x+y}{\sqrt{x}+\sqrt{y}} = \frac{x\{1+(y/x)\}}{\sqrt{x}\{1+(\sqrt{y/x})\}} = x^{1/2}\left\{\frac{1+(y/x)}{1+(\sqrt{y/x})}\right\}.$$

Therefore, $\cos v$ is a homogeneous function of degree 1/2. Using the Euler Theorem for $f = \cos v$ and $n = 1/2$, we obtain

$$x\frac{\partial}{\partial x}(\cos v) + y\frac{\partial}{\partial y}(\cos v) = \frac{1}{2}\cos v \text{ or } -x\sin v\frac{\partial v}{\partial x} - y\sin v\frac{\partial v}{\partial y} = \frac{1}{2}\cos v.$$

or $\quad x\dfrac{\partial v}{\partial x} + y\dfrac{\partial v}{\partial y} + \dfrac{1}{2}\cot v = 0,$ which is same as required.

Example 11.31 If z is a homogeneous function of x, y of degree n and z = f (u) then

$$(a) \quad x\frac{\partial u}{\partial x} + y\frac{\partial u}{\partial y} = n\cdot\frac{f(u)}{f'(u)},$$

$$(b) \quad x^2\frac{\partial^2 u}{\partial x^2} + 2xy\frac{\partial^2 u}{\partial x\partial y} + y^2\frac{\partial^2 u}{\partial y^2} = g(u)\{g'(u)-1\} \text{ where } g(u) = \frac{nf(u)}{f'(u)}.$$

Solution. (*a*) Since z is a homogeneous function of x, y of degree n, we have, by Euler's theorem

$$x\frac{\partial z}{\partial x} + y\frac{\partial z}{\partial y} = nz \qquad \qquad ...(i)$$

As $z = f(u)$, we have $\dfrac{\partial z}{\partial x} = \dfrac{\partial z}{\partial u}\dfrac{\partial u}{\partial x} = f'(u)\dfrac{\partial u}{\partial x}$.

Similarly $\dfrac{\partial z}{\partial y} = f'(u)\dfrac{\partial u}{\partial y}$. Substituting in (*i*), we obtain

$$x\frac{\partial u}{\partial x}f'(u) + y\frac{\partial u}{\partial y}f'(u) = n \cdot f(u) \quad \text{or} \quad x\frac{\partial u}{\partial x} + y\frac{\partial u}{\partial y} = \frac{nf(u)}{f'(u)}.$$

(*ii*) Set $\qquad x\dfrac{\partial u}{\partial x} + y\dfrac{\partial u}{\partial y} = \dfrac{nf(u)}{f'(u)} = g(u).$ \qquad\qquad ...(*ii*)

Differentiating (*ii*) partially w.r.t., x we obtain

$$x\frac{\partial^2 u}{\partial x^2} + \frac{\partial u}{\partial x} + y\frac{\partial^2 u}{\partial x \partial y} = g'(u)\frac{\partial u}{\partial x} \quad \text{or} \quad x\frac{\partial^2 u}{\partial x^2} + y\frac{\partial^2 u}{\partial x \partial y} = \{g'(u) - 1\}\frac{\partial u}{\partial x}. \qquad ...(iii)$$

Similarly $\qquad\qquad y\dfrac{\partial^2 u}{\partial y^2} + x\dfrac{\partial^2 u}{\partial y \partial x} = \{g'(u) - 1\}\dfrac{\partial u}{\partial y}.$ \qquad\qquad ...(*iv*)

Multiplying (*iii*) by x and (*iv*) by y and then adding, we obtain

$$x^2\frac{\partial^2 u}{\partial x^2} + 2xy\frac{\partial^2 u}{\partial x \partial y} + y^2\frac{\partial^2 u}{\partial y^2} = \{g'(u) - 1\}\left(x\frac{\partial u}{\partial x} + y\frac{\partial u}{\partial y}\right)$$

$$= \{g'(u) - 1\}g(u). \qquad \{\text{by } (ii)\}$$

Example 11.32 If $\mathbf{U} = \sec^{-1}\left(\dfrac{\mathbf{x^3 - y^3}}{\mathbf{x - y}}\right)$, then evaluate

$$\mathbf{x^2 U_{xx} + 2xy\, U_{xy} + y^2 U_{yy}}. \qquad\qquad (I.E.T.E.\ Dec.\ 2001)$$

Solution. The given function can be written as

$$\sec U = \frac{x^3 - y^3}{x - y} = \frac{x^3\left[1 - (y/x)^3\right]}{x\left[1 - (y/x)\right]} = x^2\left\{\frac{1 - (y/x)^3}{1 - (y/x)}\right\}.$$

Therefore sec U is a homogeneous function of degree 2. Using Euler's theorem for $f = \sec U$ and $n = 2$, we obtain

$$x\frac{\partial}{\partial x}(\sec U) + y\frac{\partial}{\partial y}(\sec U) = 2\sec U$$

or $\quad x\sec U\tan U\frac{\partial U}{\partial x} + y\sec U\tan U\frac{\partial U}{\partial y} = 2\sec U$

or $\quad x\frac{\partial U}{\partial x} + y\frac{\partial U}{\partial y} = \frac{2\sec U}{\sec U\tan U} = 2\cot U.$

In the above result, if we set $2\cot U = G(u)$, then

$$x^2\frac{\partial^2 U}{\partial x^2} + 2xy\frac{\partial^2 U}{\partial x\partial y} + y^2\frac{\partial^2 U}{\partial y^2} = G(U)[G'(U) - 1] = 2\cot U\left[-2\csc^2 U - 1\right]$$

$$= 2\cot U\left[-2(1 + \cot^2 U) - 1\right] = -2\cot U[2\cot^2 U + 3].$$

Thus the value of $x^2 U_{xx} + 2xy\,U_{xy} + y^2 U_{yy}$ is equal to $-2\cot U\,(2\cot^2 U + 3)$.

Example 11.33 If $u = \sin^{-1}\left\{\dfrac{x^{1/3} + y^{1/3}}{x^{1/2} + y^{1/2}}\right\}^{1/2}$ or $\csc^{-1}\left\{\dfrac{x^{1/2} + y^{1/2}}{x^{1/3} + y^{1/3}}\right\}^{1/2}$, **prove**

that $x^2 u_{xx} + 2xy\,u_{xy} + y^2 u_{yy} = \dfrac{\tan u}{144}(13 + \tan^2 u).$

(DCE, 2004; MDU 2003; A.M.I.E., S-2000)

Solution. The given function can be written as

$$\sin u = \left\{\frac{x^{1/3} + y^{1/3}}{x^{1/2} + y^{1/2}}\right\}^{1/2} = \frac{x^{1/6}}{x^{1/4}}\left\{\frac{1 + (y/x)^{1/3}}{1 + (y/x)^{1/2}}\right\}^{1/2} = x^{-1/12}\left\{\frac{1 + (y/x)^{1/3}}{1 + (y/x)^{1/2}}\right\}^{1/2}.$$

Therefore, $\sin u$ is a homogeneous function of degree $-1/12$. Using Euler's theorem for $f = \sin u$ and $n = -1/12$, we obtain

$$x\frac{\partial}{\partial x}(\sin u) + y\frac{\partial}{\partial y}(\sin u) = -\frac{1}{12}\sin u \quad\text{or}\quad x\cos u\frac{\partial u}{\partial x} + y\cos u\frac{\partial u}{\partial y} = -\frac{1}{12}\sin u$$

or $\quad x\dfrac{\partial u}{\partial x} + y\dfrac{\partial u}{\partial y} = -\dfrac{1}{12}\tan u.$ $\qquad\qquad$ ...(i)

We know that if $x\dfrac{\partial u}{\partial x} + y\dfrac{\partial u}{\partial y} = G(u)$ then

$$x^2 \frac{\partial^2 u}{\partial x^2} + 2xy \frac{\partial^2 u}{\partial x \partial y} + y^2 \frac{\partial^2 u}{\partial y^2} = G(u)\{G'(u) - 1\}. \qquad \ldots(ii)$$

Set $G(u) = -\dfrac{1}{12}\tan u$ and $G'(u) = -\dfrac{1}{12}\sec^2 u = -\dfrac{1}{12}(1 + \tan^2 u)$, we obtain

$$x^2 \frac{\partial^2 u}{\partial x^2} + 2xy \frac{\partial^2 u}{\partial x \partial y} + y^2 \frac{\partial^2 u}{\partial y^2} = -\frac{1}{12}\tan u \left\{ -\frac{1}{12}(1 + \tan^2 u) - 1 \right\}$$

or $\quad x^2 u_{xx} + 2xy\, u_{xy} + y^2 u_{yy} = \dfrac{\tan u}{144}(13 + \tan^2 u)$, which is same as required.

Example 11.34 (*a*) **If** $z = x\phi\left(\dfrac{y}{x}\right) + \psi\left(\dfrac{y}{x}\right)$, **where ϕ and ψ are differentiable**

functions, then prove that (*a*) $\quad x^2 \dfrac{\partial^2 z}{\partial x^2} + 2xy \dfrac{\partial^2 z}{\partial x \partial y} + y^2 \dfrac{\partial^2 z}{\partial y^2} = 0.$

(*A.M.I.E. W-2009, S-2008; IETE S-2002*)

(*b*) **If** $z = x^m f(y/x) + x^n g(x/y)$, **show that**

$$x^2 \frac{\partial^2 z}{\partial x^2} + 2xy \frac{\partial^2 z}{\partial x \partial y} + y^2 \frac{\partial^2 z}{\partial y^2} + mnz = (m+n-1)\left(x \frac{\partial z}{\partial x} + y \frac{\partial z}{\partial y} \right).$$

(*A.M.I.E. S-2000*)

Solution: (*a*) We have $z = x\phi(y/x) + \psi(y/x)$, then

$$\frac{\partial z}{\partial x} = \phi(y/x) + \{x\phi'(y/x) + \psi'(y/x)\}(-y/x^2),$$

and $\quad \dfrac{\partial z}{\partial y} = \{x\phi'(y/x) + \psi'(y/x)\}(1/x).$

Therefore, $\qquad x\dfrac{\partial z}{\partial x} + y\dfrac{\partial z}{\partial y} = x\phi\left(\dfrac{y}{x}\right).$

Differentiating partially with respect to x and y, we obtain

$$x\frac{\partial^2 z}{\partial x^2} + \frac{\partial z}{\partial x} + y\frac{\partial^2 z}{\partial x \partial y} = x\phi'\left(\frac{y}{x}\right)\cdot\left(\frac{-y}{x^2}\right) + \phi\left(\frac{y}{x}\right), \qquad \ldots(i)$$

and $\quad x\dfrac{\partial^2 z}{\partial y \partial x} + \dfrac{\partial z}{\partial y} + y\dfrac{\partial^2 z}{\partial y^2} = x\phi'\left(\dfrac{y}{x}\right)\cdot\dfrac{1}{x} \qquad \ldots(ii)$

Multiplying (*i*) by *x* and (*ii*) by *y* and then adding, we get

$$x^2 \frac{\partial^2 z}{\partial x^2} + 2xy \frac{\partial^2 z}{\partial x \partial y} + y^2 \frac{\partial^2 z}{\partial y^2} + \left\{ x \frac{\partial z}{\partial x} + y \frac{\partial z}{\partial y} \right\} = x\phi\left(\frac{y}{x}\right)$$

or $\quad x^2 \frac{\partial^2 z}{\partial x^2} + 2xy \frac{\partial^2 z}{\partial x \partial y} + y^2 \frac{\partial^2 z}{\partial y^2} = 0, \quad \left[\because x \frac{\partial z}{\partial x} + y \frac{\partial z}{\partial y} = x\phi\left(\frac{y}{x}\right) \right].$

(*b*) Let $x^m f(y/x) = u$ and $x^n g(x/y) = v$

Then $z = u + v$.   ...(i)

Now $u = x^m f(y/x)$ is a homogeneous function of degree *m*. We know that

$$x^2 \frac{\partial^2 u}{\partial x^2} + 2xy \frac{\partial^2 u}{\partial x \partial y} + y^2 \frac{\partial^2 u}{\partial y^2} = m(m-1)u. \qquad \text{...(ii)}$$

Also $v = x^n g(x/y)$ is a homogeneous function of degree *n*, so we have

$$x^2 \frac{\partial^2 v}{\partial x^2} + 2xy \frac{\partial^2 v}{\partial x \partial y} + y^2 \frac{\partial^2 v}{\partial y^2} = n(n-1)v. \qquad \text{...(iii)}$$

Adding (*ii*) and (*iii*), we have

$$x^2 \frac{\partial^2}{\partial x^2}(u+v) + 2xy \frac{\partial^2}{\partial x \partial y}(u+v) + y^2 \frac{\partial^2}{\partial y^2}(u+v) = m(m-1)u + n(n-1)\,v,$$

or $\qquad x^2 \frac{\partial^2 z}{\partial x^2} + 2xy \frac{\partial^2 z}{\partial x \partial y} + y^2 \frac{\partial^2 z}{\partial y^2} = m(m-1)u + n(n-1)v$ ...(iv)

Since *u* and *v*, are homogeneous functions is *x* and *y* of degree *m* and *n* respectively, therefore by Euler's theorem

$$x \frac{\partial u}{\partial x} + y \frac{\partial u}{\partial y} = mu \text{ and } x \frac{\partial v}{\partial x} + y \frac{\partial v}{\partial y} = nv.$$

Adding these and with the help of (*i*), we obtain

$$x \frac{\partial z}{\partial x} + y \frac{\partial z}{\partial y} = mu + nv. \qquad \text{...(v)}$$

Now $m(m-1)u + n(n-1)\,v = (m^2 u + n^2 v) - (mu + nv)$
$= m\,(m+n)\,u + n(m+n)\,v - mn\,(u+v) - (mu+nv)$
$= (m+n)\,(mu+nv) - mn\,(u+v) - (mu+nv)$
$= (m+n-1)\,(mu+nv) - mn\,(z), \text{ from (i)}.$

Substituting the value in (*iv*), we obtain

$$x^2 \frac{\partial^2 z}{\partial x^2} + 2xy \frac{\partial^2 z}{\partial x \partial y} + y^2 \frac{\partial^2 z}{\partial y^2} = (m+n-1)(mu+nv) - mnz$$

or $\quad x^2 \dfrac{\partial^2 z}{\partial x^2} + 2xy \dfrac{\partial^2 z}{\partial x \partial y} + y^2 \dfrac{\partial^2 z}{\partial y^2} + mnz = (m+n-1)(mu+nv)$

$$= (m+n-1)\left(x \dfrac{\partial z}{\partial x} + y \dfrac{\partial z}{\partial y} \right), \text{ from } (v). \text{ Hence proved.}$$

Example 11.35 If $u(x,y) = \dfrac{(x^2+y^2)^n}{2n(2n-1)} + xf\left(\dfrac{y}{x}\right) + g\left(\dfrac{y}{x}\right)$, where f, g and arbitrary functions, prove by using Euler's theorem that

$$\left(x\dfrac{\partial}{\partial x} + y\dfrac{\partial}{\partial y} \right)^2 u(x,y) = (x^2+y^2)^n. \qquad \text{(A.M.I.E. S-2003)}$$

Solution. Set $\dfrac{(x^2+y^2)^n}{2n(2n-1)} = v$, $xf\left(\dfrac{y}{x}\right) = w$ and $g\left(\dfrac{y}{x}\right) = s$.

Then $\qquad\qquad\qquad\qquad u = v + w + s \qquad\qquad\qquad\qquad \text{...(i)}$

$v = \dfrac{(x^2+y^2)^n}{2n(2n-1)} = \dfrac{x^{2n}\left\{1+(y/x)^2\right\}^n}{2n(2n-1)}$ is a homogeneous function of degree 2 n.

We know that If v be a homogeneous function of degree $2n$, then

$$x^2 \dfrac{\partial^2 v}{\partial x^2} + 2xy \dfrac{\partial^2 v}{\partial x \partial y} + y^2 \dfrac{\partial^2 v}{\partial y^2} = 2n(2n-1)v. \qquad\qquad \text{...(ii)}$$

Also $w = xf(y/x)$ is a homogeneous function of degree one

$\therefore \quad x^2 \dfrac{\partial^2 w}{\partial x^2} + 2xy \dfrac{\partial^2 w}{\partial x \partial y} + y^2 \dfrac{\partial^2 w}{\partial y^2} = 1(1-1)w = 0, \qquad\qquad \text{...(iii)}$

and $s = g(y/x)$ is a homogeneous function of degree zero, so we have

$$x^2 \dfrac{\partial^2 s}{\partial x^2} + 2xy \dfrac{\partial^2 s}{\partial x \partial y} + y^2 \dfrac{\partial^2 s}{\partial y^2} = 0. \qquad\qquad \text{...(iv)}$$

Adding (ii), (iii) and (iv), we obtain

$$x^2 \dfrac{\partial^2}{\partial x^2}(v+w+s) + 2xy\dfrac{\partial^2}{\partial x \partial y}(v+w+s) + y^2 \dfrac{\partial^2}{\partial y^2}(v+w+s) = 2n(2n-1)v$$

or $\quad \left(x^2 \dfrac{\partial^2}{\partial x^2} + 2xy \dfrac{\partial^2}{\partial x \partial y} + y^2 \dfrac{\partial^2}{\partial y^2} \right) u(x,y) = \dfrac{2n(2n-1)(x^2+y^2)^n}{2n(2n-1)}$

or
$$\left(x\frac{\partial}{\partial x} + y\frac{\partial}{\partial y} \right)^2 u(x, y) = \left(x^2 + y^2 \right)^n.$$

Exercise 11.3

1. If $f(x, y) = x^4 - 4 x^3 y + 8 xy^3 - y^4$, show that f_{xx} $(2, -1) = 96$, f_{xy} $(2, -1) = -24$, f_{yy} $(2, -1) = -108$.

2. If $z = \tan^{-1}(y/x)$, show that $\dfrac{\partial^2 z}{\partial x^2} + \dfrac{\partial^2 z}{\partial y^2} = 0$.

3. If $u = \dfrac{1}{2}\log\left(x^2 + y^2\right)$ then show that $\dfrac{\partial^2 u}{\partial x^2} + \dfrac{\partial^2 u}{\partial y^2} = 0$.

4. If $f(x, y) = \log (x^2 + y^2) + \tan^{-1} (y/x)$, then show that $\dfrac{\partial^2 f}{\partial x^2} + \dfrac{\partial^2 f}{\partial y^2} = 0$.

5. If $u = x^3 - 3xy^2$, show that $\dfrac{\partial^2 u}{\partial x^2} + \dfrac{\partial^2 u}{\partial y^2} = 0$.

6. Show that the function $u = \log 1/r$, where $r = \sqrt{(x-a)^2 + (y-b)^2}$ satisfies

the partial differential equation $\dfrac{\partial^2 u}{\partial x^2} + \dfrac{\partial^2 u}{\partial y^2} = 0$.   *(DCE, 2004)*

7. If $v = (ax + by)^2 - (x^2 + y^2)$ where $a^2 + b^2 = 2$, show that $\dfrac{\partial^2 v}{\partial x^2} + \dfrac{\partial^2 v}{\partial y^2} = 0$.

8. If $V = \sqrt{x^2 + y^2 + z^2}$, show that $V_{xx} + V_{yy} + V_{zz} = 2/v$ *(A.M.I.E.; W-2000)*

9. Find the value of a, if $V = x^3 + axy^2$ satisfies the equation $\dfrac{\partial^2 V}{\partial x^2} + \dfrac{\partial^2 V}{\partial y^2} = 0$.

Taking this value of V, show that If $u = r^n V$, and $r^2 = x^2 + y^2$, then

$\dfrac{\partial^2 u}{\partial x^2} + \dfrac{\partial^2 u}{\partial y^2} = n(n+6)r^{n-2}V$.   **Ans.** $a = -3$.

10. If $u = x^2\tan^{-1} (y/x) - y^2 \tan^{-1}(x/y)$, show that $\dfrac{\partial^2 u}{\partial x \partial y} = \dfrac{x^2 - y^2}{x^2 + y^2} = \dfrac{\partial^2 u}{\partial y \partial x}$.

(A.M.I.E., 2003; 2000, Madras 2000)

11. Verify that $\dfrac{\partial^2 u}{\partial x\,\partial y} = \dfrac{\partial^2 u}{\partial y\,\partial x}$, when u is equal to

(*i*) $ax^2 + 2\,hxy + by^2$,   (*ii*) $\log\dfrac{x^2 + y^2}{xy}$,

(*iii*) $\sin^{-1}(x/y)$,   (*iv*) $\log(y \sin x + x \sin y)$.

12. If $u = \log_e r^2$, where $r^2 = x^2 + y^2 + z^2$ show that $r^2(u_{xx} + u_{yy} + u_{zz}) = 1$.

13. For the function $f(x, y) = \begin{cases} x^2 y\left(\dfrac{x-y}{x^2+y^2}\right), & (x, y) \neq (0, 0) \\ 0, & (x, y) = (0, 0) \end{cases}$

Show that $f_{xy}(0, 0) \neq f_{yx}(0, 0)$.

14. For the function

$$f(x, y) = \begin{cases} x^2 \tan^{-1}\left(\dfrac{y}{x}\right) - y^2 \tan^{-1}\left(\dfrac{x}{y}\right), & x \neq 0,\, y \neq 0 \\ 0, & \text{elsewhere.} \end{cases}$$

Show that $f_{xy} \neq f_{yx}$ at $(0, 0)$.

15. If $V = x \log_e(x + r) - r$, where $r^2 = x^2 + y^2$, show that

$$\frac{\partial^2 V}{\partial x^2} + \frac{\partial^2 V}{\partial y^2} = \frac{1}{x+r}.$$

16. Let $r = \sqrt{x^2 + y^2}$ be the radius vector from the origin to the origin to the point (x, y) in the xy-plane and θ be the angle which r makes with x-axis. Show that

(*i*) $\dfrac{\partial^2 r}{\partial y^2} = \dfrac{\cos^2 \theta}{r}$, (*ii*) $\dfrac{\partial^2 r}{\partial x^2} + \dfrac{\partial^2 r}{\partial y^2} = \dfrac{1}{r}\left\{\left(\dfrac{\partial r}{\partial x}\right)^2 + \left(\dfrac{\partial r}{\partial y}\right)^2\right\}$. (A.M.I.E., S-2008)

17. If $u = x/r^3$ and $r^2 = x^2 + y^2 + z^2$, show that

$$\frac{\partial^2 u}{\partial x^2} + \frac{\partial^2 u}{\partial y^2} + \frac{\partial^2 u}{\partial z^2} = 0.$$

18. If $f(x, y) = \dfrac{1}{\sqrt{y}} e^{-(x-a)^2/4y}$, show that $f_{xy}(x, y) = f_{yx}(x, y)$.

19. If $V = r^m$ where $r^2 = x^2 + y^2 + z^2$, show that $V_{xx} + V_{yy} + V_{zz} = m(m + 1)\, r^{m-2}$. (AMIE., S-2007)

20. If $V = f(r)$ and $r^2 = x^2 + y^2 + z^2$, show that $V_{xx} + V_{yy} + V_{zz} = f''(r) + \dfrac{2}{r}f'(r)$, where the notations have their usual meaning.

21. If $u = Ae^{-gx} \sin(nt - gx)$, where A, g, n are positive constants, satisfies the heat conduction equation $\dfrac{\partial u}{\partial t} = \mu \dfrac{\partial^2 u}{\partial x^2}$, then show that $g = \sqrt{n/2\mu}$.

22. If $u = \tan^{-1} \dfrac{xy}{\sqrt{(1 + x^2 + y^2)}}$, show that $\dfrac{\partial^2 u}{\partial x \, \partial y} = \dfrac{1}{\left(1 + x^2 + y^2\right)^{3/2}}$.

23. If $z(z^2 + 3x) + 3y = 0$, show that $\dfrac{\partial^2 z}{\partial x^2} + \dfrac{\partial^2 z}{\partial y^2} = \dfrac{2z(x-1)}{\left(z^2 + x\right)^3}$.

24. If z is a function of x and y determined by the equation $z^3 - 3yz - 3x = 0$, show that $z\dfrac{dz}{dx} = \dfrac{dz}{dy}$ and $z\left\{\dfrac{\partial^2 z}{\partial x \, \partial y} + \left(\dfrac{\partial z}{\partial x}\right)^2\right\} = \dfrac{\partial^2 z}{\partial y^2}$.

25. If $u = e^{xyz}$, show that $\dfrac{\partial^3 u}{\partial x \, \partial y \, \partial z} = \left(1 + 3xyz + x^2 y^2 z^2\right)e^{xyz}$.

(A.M.I.E., W-2009, W-2000; Gauhati 1999)

[**Hint.** $\dfrac{\partial u}{\partial z} = xye^{xyz}$, $\dfrac{\partial^2 u}{\partial y \, \partial z} = x^2 yze^{xyz} + xe^{xyz} = \left(x^2 yz + x\right)e^{xyz}$.

$\therefore \quad \dfrac{\partial^3 u}{\partial x \, \partial y \, \partial z} = yz\left(x^2 yz + x\right)e^{xyz} + \left(2xyz + 1\right)e^{xyz} = \left(x^2 y^2 z^2 + 3xyz + 1\right)e^{xyz}$].

26. If $u = \log(e^x + e^y + e^z)$, show that $\dfrac{\partial^3 u}{\partial x \, \partial y \, \partial z} = 2e^{x+y+z-3u}$.

27. If $u = e^{xy} \sin z$, show that
$$\dfrac{\partial^3 u}{\partial x \, \partial y \, \partial z} = \dfrac{\partial^3 u}{\partial y \, \partial z \, \partial x} = e^{xy}(1 + xy)\cos z.$$

28. If $u = \log(x^3 + y^3 + z^3 - 3xyz)$, show that

(a) $\dfrac{\partial u}{\partial x} + \dfrac{\partial u}{\partial y} + \dfrac{\partial u}{\partial z} = \dfrac{3}{x+y+z}$, (b) $\left(\dfrac{\partial}{\partial x} + \dfrac{\partial}{\partial y} + \dfrac{\partial}{\partial z}\right)^2 u = \dfrac{-9}{(x+y+z)^2}$,

(IETE, Dec. 2009; PTU 2005; UPTU 2004; MDU 2004, JNTU 2003, Delhi 1997; AMIE, W-1996)

(c) $\dfrac{\partial^2 u}{\partial x^2} + \dfrac{\partial^2 u}{\partial y^2} + \dfrac{\partial^2 u}{\partial z^2} + 2\dfrac{\partial^2 u}{\partial x\, \partial y} + 2\dfrac{\partial^2 u}{\partial y\, \partial z} + 2\dfrac{\partial^2 u}{\partial z\, \partial x} = \dfrac{-9}{(x+y+z)^2}$.

[**Hint.** (c) $\dfrac{\partial u}{\partial x} = \dfrac{3x^2 - 3yz}{x^3 + y^3 + z^3 - 3xyz}$; $\dfrac{\partial u}{\partial y} = \dfrac{3y^2 - 3xz}{x^3 + y^3 + z^3 - 3xyz}$;

$\dfrac{\partial u}{\partial z} = \dfrac{3z^2 - 3xy}{x^3 + y^3 + z^3 - 3xy}$. Adding, $\sum \dfrac{\partial u}{\partial x} = \dfrac{3}{x+y+z}$.

$\left(\because x^3 + y^3 + z^3 - 3xyz\right) = (x+y+z)\left(x^2 + y^2 + z^2 - xy - yz - zx\right)$

Now $\left(\dfrac{\partial}{\partial x} + \dfrac{\partial}{\partial y} + \dfrac{\partial}{\partial z}\right)^2 u = \left(\dfrac{\partial}{\partial x} + \dfrac{\partial}{\partial y} + \dfrac{\partial}{\partial z}\right)\left(\dfrac{\partial}{\partial x} + \dfrac{\partial}{\partial y} + \dfrac{\partial}{\partial z}\right) u$

$= \left(\dfrac{\partial}{\partial x} + \dfrac{\partial}{\partial y} + \dfrac{\partial}{\partial z}\right)\left(\dfrac{\partial u}{\partial x} + \dfrac{\partial u}{\partial y} + \dfrac{\partial u}{\partial z}\right) = \left(\dfrac{\partial}{\partial x} + \dfrac{\partial}{\partial y} + \dfrac{\partial}{\partial z}\right)\left(\dfrac{3}{x+y+z}\right)$

$= \dfrac{-3}{(x+y+z)^2} - \dfrac{3}{(x+y+z)^2} - \dfrac{3}{(x+y+z)^2} = \dfrac{-9}{(x+y+z)^2}$].

29. If $z = \log(e^x + e^y)$, show that $rt - s^2 = 0$,

where $r \equiv \dfrac{\partial^2 z}{\partial x^2}$, $t \equiv \dfrac{\partial^2 z}{\partial y^2}$, $s \equiv \dfrac{\partial^2 z}{\partial y\, \partial x}$.

30. If $u = e^{xyz} f(yx/z)$, show that $x\dfrac{\partial u}{\partial x} + z\dfrac{\partial u}{\partial z} = 2xyzu$; $y\dfrac{\partial u}{\partial y} + z\dfrac{\partial u}{\partial z} = 2xyzu$.

Also deduce that $x\dfrac{\partial^2 u}{\partial z\, \partial x} = y\dfrac{\partial^2 u}{\partial z\, \partial y}$. *(A.M.I.E., S-2001)*

31. If $u(x, y) = \dfrac{x^3 + y^3}{x+y}$, $(x, y) \neq 0$, then evaluate $x\dfrac{\partial^2 u}{\partial x^2} + y\dfrac{\partial^2 u}{\partial x\, \partial y} - \dfrac{\partial u}{\partial x}$. **Ans.** 0

32. Show that $\dfrac{\partial^2 z}{\partial x^2} - \dfrac{2\partial^2 z}{\partial x\, \partial y} + \dfrac{\partial^2 z}{\partial y^2} = 0$, where $z = xf(x+y) + yg(x+y)$.

33. Verify Euler's theorem in the following cases:

(i) $f(x, y, z) = axy + byz + czx$, (ii) $u = \dfrac{(x^3 + y^3)}{(x+y)}$, (*iii*) $z = e^{x/y}$, (*iv*) $f(x,y) = \dfrac{(x^2 + y^2)}{\sqrt{x+y}}$. (*A.M.I.E., S-2002*)

34. If $u = f\left(\dfrac{x}{y}, \dfrac{y}{z}, \dfrac{z}{x}\right)$, show that $x\dfrac{\partial u}{\partial x} + y\dfrac{\partial u}{\partial y} + z\dfrac{\partial u}{\partial z} = 0$. (*A.M.I.E., S-2001*)

35. (*a*) If $z = \log (x^2 + xy + y^2)$, show that $x\dfrac{\partial z}{\partial x} + y\dfrac{\partial z}{\partial y} = 2$. (*A.M.I.E., S-2005*)

(*b*) Show that $x\dfrac{\partial u}{\partial x} + y\dfrac{\partial u}{\partial y} + z\dfrac{\partial u}{\partial z} = 2\tan u$, where $u = \sin^{-1}\left(\dfrac{x^3 + y^3 + z^3}{ax + by + cz}\right)$.

(*UPTU 2003*)

36. If $u = \tan^{-1}\left(\dfrac{x^3 + y^3}{x - y}\right)$, show that (*i*) $x\dfrac{\partial u}{\partial x} + y\dfrac{\partial u}{\partial y} = \sin 2u$.

(*GGSIPU 2006; MDU 2002, Bhopal 2002, UPTU 2001, A.M.I.E., W-2001*)

(*ii*) $x^2 u_{xx} + 2xy u_{xy} + y^2 u_{yy} = \sin 4u - \sin 2u = 2 \sin u \cos 3u$.

(*DCE, 2004; PTU-2002; Andhra, 2000; Assam, 1998; Ranchi 1998*)

37. If $z = x^4 y^2 \sin^{-1} (x/y) + \log x - \log y$, then show that

$$x\dfrac{\partial z}{\partial x} + y\dfrac{\partial z}{\partial y} = 6x^4 y^2 \sin^{-1}(x/y).$$ (*I.E.T.E., June 2003; UPTU 2003*)

38. Show that $x\dfrac{\partial u}{\partial x} + y\dfrac{\partial u}{\partial y} = 2u \log u$ where $\log u = \dfrac{x^3 + y^3}{3x + 4y}$. (*Mysore; 1998*)

39. If $u(x, y) = x^2 \tan^{-1} (y/x) - y^2 \tan^{-1} (x/y)$, $x > 0$, $y > 0$, then show that

$$x^2 \dfrac{\partial^2 u}{\partial x^2} + 2xy \dfrac{\partial^2 u}{\partial x \partial y} + y^2 \dfrac{\partial^2 u}{\partial y^2} = 2u.$$

(*A.M.I.E., S-2009, W-2002; I.E.T.E., W-2005*)

40. If $u = \tan^{-1}\left(\dfrac{x^2 - y^2}{x - y}\right)$, show that $x\dfrac{\partial u}{\partial x} + y\dfrac{\partial u}{\partial y} = \dfrac{1}{2}\sin 2u$ and

$$x^2 \dfrac{\partial^2 u}{\partial x^2} + 2xy \dfrac{\partial^2 u}{\partial x \partial y} + y^2 \dfrac{\partial^2 u}{\partial y^2} = \dfrac{1}{4}(\sin 4u - 2\sin 2u).$$ (*I.E.T.E., June-1995*)

41. If $u = \sin^{-1}\left(\dfrac{x+y}{\sqrt{x} + \sqrt{y}}\right)$, show that

$$x^2 \frac{\partial^2 u}{\partial x^2} + 2xy \frac{\partial^2 u}{\partial x \partial y} + y^2 \frac{\partial^2 u}{\partial y^2} = -\frac{\sin u \cos 2u}{4 \cos^3 u}.$$

(DCE 2005; MDU 2004; UPTU 2001)

42. If $u = \dfrac{x^2 y^2}{x^2 + y^2} \log \dfrac{y}{x}$ and $v = \cos^{-1}\left(\dfrac{xy}{x^2 + y^2}\right)$ and if $z = u + v$, prove that

$$x \frac{\partial z}{\partial x} + y \frac{\partial z}{\partial y} = 2u.$$

(I.E.T.E., Dec.-2003)

43. If $x = e^u \tan v$, $y = e^u \sec v$, find the value of

$$\left(x \frac{\partial u}{\partial x} + y \frac{\partial u}{\partial y}\right)\left(x \frac{\partial v}{\partial x} + y \frac{\partial v}{\partial y}\right).$$ *(I.E.T.E., June 2001)* **Ans. 0.**

44. Let $u(x, y) = \dfrac{x^3 + y^3}{x + y} + x \tan^{-1}\left(\dfrac{y}{x}\right)$, $(x, y) \neq (0, 0)$, then

$$x \frac{\partial u}{\partial x} + y \frac{\partial u}{\partial y} = \frac{2(x^3 + y^3)}{x + y} + x \tan^{-1} \frac{y}{x}.$$ *(I.E.T.E., June-2005)*

45. If $u = \tan^{-1}(y^2/x)$, show that

$$x^2 \frac{\partial^2 u}{\partial x^2} + 2xy \frac{\partial^2 u}{\partial x \partial y} + y^2 \frac{\partial^2 u}{\partial y^2} = -\sin^2 u \sin 2u.$$ *(PTU 2006)*

46. Given that $u(x, y) = 5 + x \sin(x/y) + y^2 \sin(y/x)$.

Compute $x^2 \dfrac{\partial^2 u}{\partial x^2} + 2xy \dfrac{\partial^2 u}{\partial x \partial y} + y^2 \dfrac{\partial^2 u}{\partial y^2}$. **Ans.** $2y^2 \sin(y/x)$.

47. Find $x^2 u_{xx} + 2xy\, u_{xy} + y^2 u_{yy}$ when
    (a) $u = \sin^{-1}(x^3 + y^3)^{2/5}$, (b) $u = \log(x^3 + y^3 - x^2 y - xy^2)$,

    (c) $u = \log(x^4 + x^2 y^2 + y^4)$, *(AMIE, S-2000)* (d) $u = \tan^{-1}\left\{\dfrac{(x^3 + y^3)^{1/2}}{\sqrt{x} + \sqrt{y}}\right\}$,

    (e) $u = \sin^{-1}\left((x^2 + y^2)/(x + y)\right)$, (f) $u = (x^2 + y^2)^{1/3}$

Ans. (a) $\dfrac{6}{5} \tan u\left(\dfrac{6}{5} \sec^2 u - 1\right)$, (b) -3, (c) -4, (d) $-2 \sin^3 u \cos u$,

(e) $\tan^3 u$, (f) $-2u/9$.

48. If $u = \dfrac{x^2 y^2}{x+y}$, show that $x\dfrac{\partial^2 u}{\partial x^2} + y\dfrac{\partial^2 u}{\partial x \partial y} = 2\dfrac{\partial u}{\partial x}$. (*GGSIPU 2007*)

49. If $e^u = x^3 + y^3 + z^3 - 3xyz$, then prove that

$$\left(\frac{\partial}{\partial x} + \frac{\partial}{\partial y} + \frac{\partial}{\partial z}\right)^3 u = \frac{54}{(x+y+z)^3}.$$

11.14 Total Differential

Let a function of two variables $z = f(x, y)$ be defined in some domain D in the xy-plane. Let $P(x, y)$ be any point in D and $(x + \Delta x, y + \Delta y)$ be a point in the neighborhood of (x, y) in D. Then $\Delta z = f(x + \Delta x, y + \Delta y) - f(x, y)$.
is called the *total increment* in z corresponding to the increments Δx in x and Δy in y. Adding and subtracting $f(x, y + \Delta y)$ in the second member, we have

$$\Delta z = \{f(x + \Delta x, y + \Delta y) - f(x, y + \Delta y)\} + \{f(x, y + \Delta y) - f(x, y)\}$$

Applying the mean value theorem for functions of one variable, we obtain

$$\Delta z = \Delta x f_x(x + \theta_1 \Delta x, y + \Delta y) + \Delta y f_y(x, y + \theta_2 \Delta y)\ 0 < \theta_1 < 1, 0 < \theta_2 < 1.$$

Since f_x and f_y are continuous at (x, y), it follows that

$$f_x(x + \theta_1 \Delta x, y + \Delta y) = f_x(x, y) + \varepsilon_1, f_y(x, y + \theta_2 \Delta y) = f_y(x, y) + \varepsilon_2$$

where $\varepsilon_1 \to 0, \varepsilon_2 \to 0$ as $\Delta x \to 0$ and $\Delta y \to 0$.
Thus,

$$\Delta z = f_x \Delta x + f_y \Delta_y + \varepsilon_1 \Delta_x + \varepsilon_2 \Delta y.$$

Defining $\Delta x = dx, \Delta y = dy$, we have $\Delta z = f_x dx + f_y dy + \varepsilon_1 dx + \varepsilon_2 dy.$

We call $\qquad \mathbf{dz = f_x dx + f_y dy}$ ...(11.17)

The differential of z (or f) or the principal part of Δz (or Δf).

Remark. For a function of n variables $z = f(x_1, x_2, ..., x_n)$, we write the total differential as

$$dz = f_{x_1} dx_1 + f_{x_2} dx_2 + ... + f_{x_n} dx_n.$$ ...(11.18)

11.15 Second Order Total Differentials

For the function $z = f(x, y)$, the first order total differential of z is $dz = f_x dx + f_y dy$.
The second order total differential of z, denoted by $d^2 z$, is given by

$$d^2 z = d(dz) = d(f_x dx + f_y dy)$$

$$= \frac{\partial}{\partial x}\left(f_x dx + f_y dy\right)dx + \frac{\partial}{\partial y}\left(f_x dx + f_y dy\right)dy$$

$$= \left[f_{xx}dx + f_x \frac{\partial}{\partial x}(dx) + f_{yx}dy + f_y \frac{\partial}{\partial x}(dy) \right] dx$$

$$+ \left[f_{yx}dx + f_x \frac{\partial}{\partial y}(dx) + f_{yy}dy + f_y \frac{\partial}{\partial y}(dy) \right] dy .$$

Since dx and dy are considered as constants so

$$\frac{\partial}{\partial x}(dx) = 0, \frac{\partial}{\partial x}(dy) = 0, \frac{\partial}{\partial y}(dx) = 0, \frac{\partial}{\partial y}(dy) = 0 .$$

Therefore, $d^2z = \left(f_{xx}dx + f_{yx}dy \right)dx + \left(f_{yx}dx + f_{yy}dy \right)dy$

or $\qquad \mathbf{d^2z = f_{xx}(dx)^2 + 2f_{xy}dx\,dy + f_{yy}(dy)^2}$. $\quad (\because f_{xy} = f_{yx})$. $\quad$ ...(11.19)

In abbreviated notation, it may be written as $d^2z = \left(\dfrac{\partial}{\partial x}dx + \dfrac{\partial}{\partial y}dy \right)^2 z$.

Similarly, $\qquad d^3z = \left(\dfrac{\partial}{\partial x}dx + \dfrac{\partial}{\partial y}dy \right)^3 z.$

Example 11.36 Find the total differential of the following functions

(i) $z = xy^2 \log(y/x)$, $\quad$ (ii) $u = x^2 e^{y/x}$, $\quad$ (iii) $u = \left(xz + \dfrac{x}{z} \right)^y$, $z \neq 0$.

Solution. (i) We have, $f(x, y) = xy^2 \log(y/x)$

$$f_x = y^2 \log\left(\frac{y}{x} \right) + xy^2 \cdot \frac{x}{y}\left(-yx^{-2} \right) = y^2 \log\left(\frac{y}{x} \right) - y^2 ,$$

and $f_y = 2xy \log\left(\dfrac{y}{x} \right) + xy^2 \cdot \dfrac{x}{y} \cdot \dfrac{1}{x} = 2xy \log\left(\dfrac{y}{x} \right) + xy.$

Therefore, we obtain the total differential as

$$dz = f_x dx + f_y dy = \left\{ y^2 \log(y/x) - y^2 \right\} dx + \left\{ 2xy \log(y/x) + xy \right\} dy.$$

(ii) $f(x, y) = x^2 e^{y/x} \cdot f_x = x^2 e^{y/x} \cdot \left(-yx^{-2} \right) + 2xe^{y/x}, \; f_y = x^2 e^{y/x} \cdot \left(\dfrac{1}{x} \right) = xe^{y/x}.$

Therefore, we obtain the total differential as $\quad du = f_x \, dx + f_y \, dy$

$$= \left(2xe^{y/x} - ye^{y/x} \right) dx + xe^{y/x} dy .$$

(iii) $\qquad f(x, y, z) = \left(xz + \dfrac{x}{z} \right)^y, \; f_x = y\left(xz + \dfrac{x}{z} \right)^{y-1} \left(z + \dfrac{1}{z} \right),$

$$f_y = \left(xz + \frac{x}{z}\right)^y \log\left(xz + \frac{x}{z}\right), \ f_z = y\left(xz + \frac{x}{z}\right)^{y-1} \cdot \left(x - \frac{x}{z^2}\right).$$

Therefore, we obtain the total differential as

$$du = \left(xz + \frac{x}{z}\right)^{y-1}\left[y\left(z + \frac{1}{z}\right)dx + xy\left(1 - \frac{1}{z^2}\right)dz\right]$$

$$+ \left[\left(xz + \frac{x}{z}\right)^y \log\left(xz + \frac{x}{z}\right)\right]dy.$$

Example 11.37 If z = f(x, y) = x²y − 3y, find (a) Δz, (b) dz, (c) f (5.12, 6.85) without direct computation.

Solution. (a) $\Delta z = f(x + \Delta x, y + \Delta y) - f(x, y)$

$$= \left\{(x + \Delta x)^2 (y + \Delta y) - 3(y + \Delta y)\right\} - \left(x^2 y - 3y\right)$$

$$= \underbrace{2xy\,\Delta x + (x^2 - 3)\Delta y}_{(A)} + \underbrace{(\Delta x)^2\, y + 2x\Delta x\Delta y + (\Delta x)^2\, \Delta y}_{(B)}.$$

(b) The sum (A) is the principal part of Δz and is the differential of z that is, dz. Thus $dz = 2xy\,\Delta x + (x^2 - 3)\,\Delta y = 2\,xy\,dx + (x^2 - 3)\,dy.$

Alternative

$$dz = \frac{\partial f}{\partial x}dx + \frac{\partial f}{\partial y}dy = f_x dx + f_y dy = 2\,xy\,dx + (x^2 - 3)\,dy.$$

(c) We have to find $f(x + \Delta x, y + \Delta y)$, when $x + \Delta x = 5.12$ and $y + \Delta y = 6.85$. Choosing $x = 5$, $\Delta x = 0.12$, $y = 7$, $\Delta y = -0.15$. Since Δx and Δy are small, we use the fact that

$f(x + \Delta x, y + \Delta y) = f(x, y) + \Delta z \approx f(x, y) + dz$ that is, $z + dz$.

Now $z = f(x, y) = f(5, 7) = 5^2(7) - 3(7) = 154.$

$$dz = f_x dx + f_y\, dy = 2xy\,dx + (x^2 - 3)\,dy$$

$$= 2(5)\,(7)\,(0.12) + (5^2 - 3)\,(-0.15) = 5.1.$$

Then the required value is $154 + 5.1 = 159.1$ (approx.)

Example 11.38 Find the second order total differentials of the following function

(i) z = 7y log (1 + x),   (ii) z = xeˣʸ.

Solution. (i) We have $z = f(x, y) = 7y \log(1 + x)$.

$f_x = 7y/(1 + x), f_y = 7 \log(1 + x)$.

Also $f_{xx} = -7y/(1 + x)^2$, $f_{yy} = 0$, $f_{xy} = 7/(1 + x)$.

Therefore, the second order total differential of z is

$$d^2z = f_{xx}(dx)^2 + 2f_{xy}\,dx\,dy + f_{yy}(dy)^2 = -\frac{7y}{(1+x)^2}(dx)^2 + \frac{14}{1+x}dxdy.$$

(ii) We have $f(x,y) = xe^{xy}$. $f_x = \dfrac{\partial f}{\partial x} = e^{xy} + xye^{xy}$,

$$f_y = \frac{\partial f}{\partial y} = x^2 e^{xy};\quad f_{xx} = \frac{\partial^2 f}{\partial x^2} = ye^{xy} + y\{e^{xy} + xye^{xy}\},$$

$$f_{yy} = \frac{\partial^2 f}{\partial y^2} = \frac{\partial}{\partial y}\left(\frac{\partial f}{\partial y}\right) = \frac{\partial}{\partial y}(x^2 e^{xy}) = x^3 e^{xy},$$

$$f_{xy} = \frac{\partial}{\partial y} = \left(\frac{\partial f}{\partial x}\right) = \frac{\partial}{\partial y}(e^{xy} + xye^{xy}) = xe^{xy} + x(e^{xy} + yxe^{xy}).$$

Therefore, the second order total differential of z is
$$d^2z = f_{xx}(dx)^2 + 2f_{xy}\,dx\,dy + f_{yy}(dy)^2$$
$$= \{ye^{xy} + ye^{xy} + xy^2\,e^{xy}\}(dx)^2 + \{2xe^{xy} + 2x(e^{xy} + xye^{xy}\}dxdy + x^3\,e^{xy}(dy)^2$$
or $d^2z = (xy + 2)\,ye^{xy}(dx)^2 + 2(xy + 2)\,xe^{xy}dxdy + x^3e^{xy}(dy)^2.$

Example 11.39 If $x^2 + y^2 + z^2 - 2xyz = 1$, show that

$$\frac{dx}{\sqrt{1-x^2}} + \frac{dy}{\sqrt{1-y^2}} + \frac{dz}{\sqrt{1-z^2}} = 0.$$ (A.M.I.E., W-1996; Hamirpur, 1996)

Solution. We have
$$x^2 + y^2 + z^2 - 2xyz = 1.$$
$$\therefore\qquad 2xdx + 2\,ydy + 2\,zdz - 2(xydz + yzdx + zxdy) = 0$$
or $\qquad (x - yz)\,dx + (y - zx)\,dy + (z - xy)\,dz = 0 \qquad\qquad …(i)$
Now $\qquad (x - yz)^2 = x^2 + y^2z^2 - 2\,xyz = (x^2 - 2\,xyz) + y^2z^2$
$$= (1 - y^2 - z^2) + y^2z^2 = 1 - y^2 - z^2 + y^2z^2 = (1 - y^2)(1 - z^2)$$
$$\therefore\quad x - yz = \sqrt{(1-y^2)(1-z^2)}.$$

Similarly $\qquad y - zx = \sqrt{(1-x^2)(1-z^2)}$, and $z - xy = \sqrt{(1-x^2)(1-y^2)}$.

Putting these values of $x - yz$, $y - zx$ and $z - xy$ in (i), we get

$$\sqrt{(1-y^2)(1-z^2)}\,dx + \sqrt{(1-x^2)(1-z^2)}\,dy + \sqrt{(1-x^2)(1-y^2)}\,dz = 0.$$

Dividing throughout by $\sqrt{(1-x^2)(1-y^2)(1-z^2)}$, we obtain

$$\frac{dx}{\sqrt{1-x^2}} + \frac{dy}{\sqrt{1-y^2}} + \frac{dz}{\sqrt{1-z^2}} = 0,\text{ the required result.}$$

11.16 Exact Differential

We know that if a function $f(x, y)$ of two variables has continuous first order partial derivatives, then the *total or exact differential, df,* of $f(x, y)$ is expressed as

$$df = \frac{\partial f}{\partial x} dx + \frac{\partial f}{\partial y} dy. \qquad \qquad ...(11.20)$$

A differential expression $M(x, y)dx + N(x, y)dy$ is an *exact differential* in a region R of the xy-plane if it corresponds to the differential of some function $f(x, y)$.

For simplicity let us assume that $M(x, y)$ and $N(x, y)$ have continuous first partial derivatives for all (x, y). Now of the expression $M(x, y) dx + N(x, y) dy$ is exact, there exists some function f such that for all x in a region R,

$$M(x, y)dx + N(x, y)dy = \frac{\partial f}{\partial x} dx + \frac{\partial f}{\partial y} dy.$$

Therefore, $M(x, y) = \dfrac{\partial f}{\partial x}, N(x, y) = \dfrac{\partial f}{\partial y},$

and $\quad \dfrac{\partial M}{\partial y} = \dfrac{\partial}{\partial y}\left(\dfrac{\partial f}{\partial x}\right) = \dfrac{\partial^2 f}{\partial y \partial x} = \dfrac{\partial}{\partial x}\left(\dfrac{\partial f}{\partial y}\right) = \dfrac{\partial N}{\partial x}.$

The equality of the mixed partials is a comequence of the continuity of the first partial derivatives of $M(x, y)$ and $N(x, y)$.

Thus a necessary and sufficient condition that the expression

$$M(x, y)dx + N(x, y)dy \text{ be an exact differential is } \frac{\partial M}{\partial y} = \frac{\partial N}{\partial x}. \qquad ...(11.21)$$

Example 11.40 State whether following expressions are exact differential or not

(i) $2x \sin y \, dx + x^2 \cos y \, dy$, (ii) $(x^2y + 2y^3)dx - (2x^3 + 3xy^2)dy$,

(iii) $\left(\dfrac{x^2}{y} - \dfrac{y^2}{x}\right)dx - \left(\dfrac{x^3}{3y^2} + 2y \log x\right)dy.$

Solution. (i) Here $M = 2x \sin y$ and $N = x^2 \cos y$,

$\dfrac{\partial M}{\partial y} = 2x \cos y, \dfrac{\partial N}{\partial x} = 2x \cos y$ Since $\dfrac{\partial M}{\partial y} = \dfrac{\partial N}{\partial x}.$

Therefore, the given expression is an exact differential.

(ii) $M = x^2y + 2y^3, N = -(2x^3 + 3 xy^2),$

$$\frac{\partial M}{\partial y} = x^2 + 6y^2, \frac{\partial N}{\partial x} = -\left(6x^2 + 3y^2\right). \text{ Since} \frac{\partial M}{\partial y} \neq \frac{\partial N}{\partial x}.$$

Thus, the given expression is not an exact differential.

(*iii*) $M = \dfrac{x^2}{y} - \dfrac{y^2}{x}, \quad N = -\left(\dfrac{x^3}{3y^2} + 2y \log x\right).$

$$\frac{\partial M}{\partial y} = -\frac{x^2}{y^2} - \frac{2y}{x}; \ \frac{\partial N}{\partial x} = -\frac{x^2}{y^2} - \frac{2y}{x}. \text{ Since } \frac{\partial M}{\partial y} = \frac{\partial N}{\partial x}.$$

Thus, the given expression is exact.

Example 11.41 Show that expression $(3x^2y - 2y^2)\, dx + (x^3 - 4\, xy + 6\, y^2)\, dy$ can be written as an exact differential of a function $\phi(x, y)$ and find this function.

Solution. Suppose that $(3x^2y - 2y^2)\, dx + (x^3 - 4\, xy + 6y^2)dy$

$$= d\phi = \frac{\partial \phi}{\partial x} dx + \frac{\partial \phi}{\partial y} dy .$$

Comparing, we get $\dfrac{\partial \phi}{\partial x} = 3x^2y - 2y^2$ ...(*i*) and $\dfrac{\partial \phi}{\partial y} = x^3 - 4xy + 6y^2$ ...(*ii*).

Integrating (*i*) w.r.t. x keeping y constant, we have $\phi = x^3y - 2xy^2 + F(y)$, where $F(y)$ is the constant of integration. Substituting this into (*ii*), we obtain $x^3 - 4xy + F'(y) = x^3 - 4xy + 6y^2$ from which $F'(y) = 6y^2$ that is, $F(y) = 2y^3 + c.$

Hence, the required function is $\phi(x, y) = x^3y - 2xy^2 + 2y^3 + c$, where c is an arbitrary constant.

11.17 Two Variable Chain Rule for Derivatives

Let $z = f(x, y)$ be a function of two independent variables x and y. Suppose that x and y are in turn functions of a single variable t, say $x = \phi(t)$, $y = \psi(t)$. Then $z = f[\phi(t), \psi(t)]$ is a composite function of the independent variable t.

Now, assume that the partial derivatives f_x, f_y are continuous functions of x, y and $\phi(t), \psi(t)$ are differentiable functions of t.

Let $\Delta x, \Delta y$ and Δz be the increments respectively in x, y and z corresponding to the increment Δt in t. Then we have

$$\Delta z = \frac{dz}{dx} \Delta x + \frac{\partial z}{\partial y} \Delta y + \varepsilon_1 \Delta x + \varepsilon_2 \Delta y.$$

Dividing both sides by Δt, we obtain

$$\frac{\Delta z}{\Delta t} = \frac{\partial z}{\partial x}\frac{\Delta x}{\Delta t} + \frac{\partial z}{\partial y}\cdot\frac{\Delta y}{\Delta t} + \varepsilon_1\frac{\Delta x}{\Delta t} + \varepsilon_2\frac{\Delta y}{\Delta t}. \qquad \text{...(11.22)}$$

Chain Rule for Functions of Three Independent Variables

If $w = f(x, y, z)$ ...(11.23) is differentiable and x, y, and z are differentiable functions of t, then w is differentiable function of t and

$$\frac{dw}{dt} = \frac{\partial w}{\partial x}\frac{dx}{dt} + \frac{\partial w}{\partial y}\frac{dy}{dt} + \frac{\partial w}{\partial z}\frac{dz}{dt}, \qquad \text{...(11.24)}$$

and so on for any number of variables.

11.18 Illustrative Examples

Example 11.42 If $z = e^{xy^2}$, $x = t\cos t$, $y = t\sin t$, compute dz/dt at $t = \pi/2$.
(AMIE., S-2006)

Solution. When $t = \pi/2$, we get $x = 0$, $y = \pi/2$. Using the chain rule, we obtain

$$\frac{dz}{dt} = \frac{\partial z}{\partial x}\frac{dx}{dt} + \frac{\partial z}{\partial y}\frac{dy}{dt}$$

$$= y^2 e^{xy^2}(\cos t - t\sin t) + 2xy\,e^{xy^2}(\sin t + t\cos t).$$

Substituting $t = \pi/2$, $x = 0$ and $y = \pi/2$, we obtain $(dz/dt)_{t=\pi/2} = -\pi^3/8$.

Example 11.43 (a) If $u = \sin^{-1}(x - y)$, $x = 3t$, $y = 4t^3$, show that

$$\frac{du}{dt} = 3(1 - t^2)^{-1/2} \qquad \text{(PTU 2005; Gauhati 1999; Kuvempu, 1998S)}$$

(b) If $u = x^3 + y^3$, where $x = a\cos t$, $y = b\sin t$. Find $\dfrac{du}{dt}$ and verify the result by direct substitution.

(c) If $u = xe^y z$, where $y = \sqrt{a^2 - x^2}$, $z = \sin^2 x$. Find du/dx.

Solution. (a) We have $\dfrac{\partial u}{\partial x} = \dfrac{1}{\sqrt{1 - (x - y)^2}}$, $\qquad\qquad \dfrac{dx}{dt} = 3$;

$$\frac{\partial u}{\partial y} = \frac{-1}{\sqrt{1 - (x - y)^2}}, \qquad\qquad \frac{dy}{dt} = 12t^2.$$

Using the chain rule, we obtain $\dfrac{du}{dt} = \dfrac{\partial u}{\partial x}\dfrac{dx}{dt} + \dfrac{\partial u}{\partial y}\dfrac{dy}{dt}$

$$= \frac{1}{\sqrt{1-(x-y)^2}} \cdot 3 - \frac{1}{\sqrt{1-(x-y)^2}} \cdot 12t^2$$

$$= \frac{3}{\sqrt{1-(3t-4t^3)^2}} - \frac{12t^2}{\sqrt{1-(3t-4t^3)^2}} \text{, (Substitute for the intermediate variables)}$$

$$= \frac{3(1-4t^2)}{\sqrt{1-(3t-4t^3)^2}} = \frac{3(1-4t^2)}{\sqrt{1-9t^2-16t^6+24t^4}} = \frac{3(1-4t^2)}{\sqrt{(1-t^2)(1-8t^2+16t^4)}} = \frac{3}{\sqrt{1-t^2}}.$$

Alternative

$u = \sin^{-1}(x-y) = \sin^{-1}(3t-4t^3)$. Setting $t = \sin\theta$, we obtain
$u = \sin^{-1}(3\sin\theta - 4\sin^3\theta) = \sin^{-1}(\sin 3\theta) = 3\theta = 3\sin^{-1}t$.

Therefore, $\qquad \dfrac{du}{dt} = \dfrac{3}{\sqrt{1-t^2}}$.

(b) We have $\qquad \dfrac{\partial u}{\partial x} = 3x^2, \dfrac{\partial u}{\partial y} = 3y^2$ and $\dfrac{dx}{dt} = -a\sin t, \dfrac{dy}{dt} = b\cos t$.

Using the chain rule, we obtain $\dfrac{du}{dt} = \dfrac{\partial u}{\partial x}\dfrac{dx}{dt} + \dfrac{\partial u}{\partial y}\dfrac{dy}{dt}$

$= 3x^2(-a\sin t) + 3y^2(b\cos t) = 3a^2\cos^2 t(-a\sin t) + 3b^2\sin^2 t(b\cos t)$

(Substitute for the intermediate variables.)

$\qquad\qquad = -3a^3\cos^2 t\sin t + 3b^3\sin^2 t\cos t$.

Verification: $\qquad u = x^3 + y^3 = a^3\cos^3 t + b^3\sin^3 t$.

Therefore, $\qquad \dfrac{du}{dt} = 3a^3\cos^2 t(-\sin t) + 3b^3\sin^2 t(\cos t)$

$\qquad = -3a^3\cos^2 t\sin t + 3b^3\sin^2 t\cos t$.

(c) $\dfrac{\partial u}{\partial x} = e^y z, \dfrac{\partial u}{\partial y} = xe^y z, \dfrac{\partial u}{\partial z} = xe^y$ and $\dfrac{dy}{dx} = \dfrac{1}{2}(a^2-x^2)^{-1/2}\cdot(-2x)$

$$= \frac{-x}{\sqrt{a^2-x^2}}, \qquad\qquad \frac{dz}{dx} = 2\sin x\cos x.$$

Using the chain rule, we obtain $\dfrac{du}{dx} = \dfrac{\partial u}{\partial x} + \dfrac{\partial u}{\partial y}\cdot\dfrac{dy}{dx} + \dfrac{\partial u}{\partial z}\cdot\dfrac{dz}{dx}$

$$= e^y z + x e^y z \left(\frac{-x}{\sqrt{a^2 - x^2}} \right) + x e^y (\sin 2x) = e^y \left[z - \frac{x^2 z}{\sqrt{a^2 - x^2}} + x \sin 2x \right].$$

Example 11.44 If $u = \sin(x/y)$, $x = e^t$ and $y = t^2$ then find $\dfrac{du}{dt}$ as function

of t.

(PTU, 2006)

Solution. $\qquad \dfrac{\partial u}{\partial x} = \dfrac{1}{y} \cos \dfrac{x}{y}, \qquad \dfrac{\partial u}{\partial y} = -\dfrac{x}{y^2} \cos \dfrac{x}{y}; \qquad \dfrac{dx}{dt} = e^t, \qquad \dfrac{dy}{dt} = 2t.$

Using the chain rule, we obtain $\dfrac{du}{dt} = \dfrac{\partial u}{\partial x} \dfrac{dx}{dt} + \dfrac{\partial u}{\partial y} \dfrac{dy}{dt}$

$$= \cos\left(\frac{x}{y}\right) \cdot \frac{1}{y} \cdot (e^t) + \cos\left(\frac{x}{y}\right) \cdot \left(-\frac{x}{y^2}\right) \cdot 2t,$$

$$= \frac{e^t}{t^2} \cos\left(\frac{e^t}{t^2}\right) - \frac{2e^t}{t^3} \cos\left(\frac{e^t}{t^2}\right) \qquad \text{(Substitute for the intermediate variables).}$$

$$= (t - 2) \frac{e^t}{t^3} \cos\left(\frac{e^t}{t^2}\right).$$

Example 11.45 Given $u = e^{ax} (y - z)$, $y = a \sin x$, $z = \cos x$; find du/dx.

Solution. $\qquad \dfrac{\partial u}{\partial x} = a e^{ax} (y - z), \dfrac{\partial u}{\partial y} = e^{ax}, \dfrac{\partial u}{\partial z} = -e^{ax};$

$$\frac{dy}{dx} = a \cos x, \frac{dz}{dx} = -\sin x.$$

Using the chain rule, we obtain $\dfrac{du}{dx} = \dfrac{\partial u}{\partial x} + \dfrac{\partial u}{\partial y} \cdot \dfrac{dy}{dx} + \dfrac{\partial u}{\partial z} \cdot \dfrac{dz}{dx}$

$$= a e^{ax}(y - z) + a e^{ax} \cos x + e^{ax} \sin x$$
$$= a e^{ax}(a \sin x - \cos x) + a e^{ax} \cos x + e^{ax} \sin x$$
$$\qquad\qquad \text{(Substitute for the intermediate variables)}$$

$$= e^{ax} (a^2 + 1) \sin x.$$

11.18.1 Related Rates Problems

Example 11.46. At a certain instant the three dimensions of a rectangular parallelopiped are 15 cm, 20 cm, 25 cm, and these are increasing at the respective rates of 5 mm per second, 8 mm per second, 3 mm per second. How fast is the volume increasing?

Solution: Let x, y, and z be the edges of a rectangular parallelopiped. Then $V = xyz$. Considering x, y and z as functions of time t, we have

$$\frac{dV}{dt} = \frac{\partial V}{\partial x}\frac{dx}{dt} + \frac{\partial V}{\partial y}\cdot\frac{dy}{dt} + \frac{\partial V}{\partial z}\cdot\frac{dz}{dt} = yz\frac{dx}{dt} + zx\frac{dy}{dt} + xy\frac{dz}{dt}.$$

It is given that $x = 15$, $\dfrac{dx}{dt} = 0.5$, $y = 20$, $\dfrac{dy}{dt} = 0.8$, $z = 25$ and $\dfrac{dz}{dt} = 0.3$,

we have $\dfrac{dV}{dt} = (20)(25)(0.5) + (25)(15)(0.8) + (15)(20)(0.3)$

$$= 250 + 300 + 90 = 640 \text{ cm}^3 \text{ per sec.}$$

Thus the volume is increasing at the rate of 640 cu cm per sec.

Example 11.47 The altitude of a right circular cone is 15 units and is increasing at the rate of 0.2 unit per min. The radius of the base is 10 units and is decreasing at the rate of 0.3 unit per min. At what rate is the volume changing?

Solution. Let r be the radius and h the altitude of the cone. Then $V = \pi r^2 h/3$, considering r and h as functions of time t. We have

$$\frac{\partial V}{\partial r} = \frac{2}{3}\pi rh, \quad \frac{\partial V}{\partial h} = \frac{\pi r^2}{3}, \quad h = 15, \frac{dh}{dt} = +0.2, \quad r = 10, \frac{dr}{dt} = -0.3.$$

Therefore, $\dfrac{dV}{dt} = \dfrac{\partial V}{\partial r}\cdot\dfrac{dr}{dt} + \dfrac{\partial V}{\partial h}\cdot\dfrac{dh}{dt} = \dfrac{\pi}{3}\left(2rh\dfrac{dr}{dt} + r^2\dfrac{dh}{dt}\right)$

$$= \frac{\pi}{3}\left[2(10)(15)(-0.3) + 10^2(0.2)\right] = -\frac{70\pi}{3}\text{ units}^3/\text{min}.$$ The negative sign indicate that volume is decreasing.

Example 11.48 A point is moving along the curve of intersection of $x^2 + 3xy + 3y^2 = z^2$ and the plane $x - 2y + 4 = 0$. When $x = 2$ and is increasing at 3 units/sec, find (*a*) how *y* is changing, (*b*) how *z* is changing, and (*c*) the speed of the point.

Solution. From $z = \pm(x^2 + 3xy + 3y^2)^{1/2}$, we obtain $\dfrac{dz}{dt} = \dfrac{\partial z}{\partial x}\dfrac{dx}{dt} + \dfrac{\partial z}{\partial y}\dfrac{dy}{dt}$

$$= \frac{1}{2}\frac{(2x+3y)}{\sqrt{x^2+3xy+3y^2}}\frac{dx}{dt} + \frac{1}{2}\frac{(3x+6y)}{\sqrt{x^2+3xy+3y^2}}\frac{dy}{dt}.$$

Since $x - 2y + 4 = 0$, $y = 3$, when $x = 2$; also differentiation yields $\dfrac{dx}{dt} = 2\cdot\dfrac{dy}{dt}$.

(*a*) when $x = 2$, $\dfrac{dx}{dt} = +3$ then $\dfrac{dy}{dt} = \dfrac{3}{2}$ units/sec, increasing.

(b) $z = \pm(4 + 18 + 27)^{1/2} = \pm 7$. At $(2, 3, 7)$, $\dfrac{dz}{dt} = \dfrac{13}{14} \cdot 3 + \dfrac{24}{14} \cdot \dfrac{3}{2} = \dfrac{75}{14}$

units/sec, increasing.

At $(2, 3, -7)$, $\dfrac{dz}{dt} = -\dfrac{39}{14} - \dfrac{36}{14} = -\dfrac{75}{14}$ units/sec that is, decreasing.

(c) Speed of the point $= \sqrt{\left(\dfrac{dx}{dt}\right)^2 + \left(\dfrac{dy}{dt}\right)^2 + \left(\dfrac{dz}{dt}\right)^2}$

$$= \sqrt{9 + \dfrac{9}{4} + \left(\dfrac{75}{14}\right)^2} = \sqrt{39.9489} = 6.32 \text{ units/sec.}$$

Example 11.49 The radius of the base of a certain cone is increasing at the rate of 8 cm/minute and the altitude is decreasing at the rate of 10 cm/minute. Find the rate of change of the total surface of the cone when the radius is 18 cm and the altitude 60 cm.

Solution. Let r be the radius, and h the altitude of the cone. Then the total surface area of the cone is

$$S = \pi r^2 + \pi r l = \pi r^2 + \pi r \sqrt{h^2 + r^2} = f(r, h).$$

Considering r and h as functions of time t,

$$\therefore \quad \frac{dS}{dt} = \frac{\partial S}{\partial r} \cdot \frac{dr}{dt} + \frac{\partial S}{\partial h} \cdot \frac{dh}{dt}.$$

Now $\dfrac{\partial S}{\partial r} = 2\pi r + \pi \left[\sqrt{h^2 + r^2} + \dfrac{r^2}{\sqrt{h^2 + r^2}} \right]$

$$= 2\pi r + \frac{\pi(h^2 + 2r^2)}{\sqrt{h^2 + r^2}} \quad \text{and} \quad \frac{\partial S}{\partial h} = \frac{\pi r h}{\sqrt{h^2 + r^2}}.$$

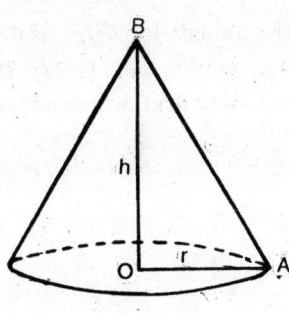

Fig. 11.2

Therefore, $\dfrac{dS}{dt} = 2\pi r \dfrac{dr}{dt} + \dfrac{\pi(h^2 + 2r^2)}{\sqrt{h^2 + r^2}} \dfrac{dr}{dt} + \dfrac{\pi r h}{\sqrt{h^2 + r^2}} \dfrac{dh}{dt}.$ ...(i)

When $r = 18$, $h = 60$, $dr/dt = +8$ and $dh/dt = -10$, then

$$\frac{dS}{dt} = 2\pi(18)(8) + \frac{\pi(4248)}{62.64}(8) + \frac{\pi(18)(60)}{62.64}(-10)$$

$$= 288\,\pi + 370\,\pi = 658\,\pi \text{ cm}^2/\text{min.}$$

Thus the total surface of the cone is increasing at the rate of $658\,\pi$ cm^2/minute.

11.19 Two-Variables Chain Rule for Partial Derivatives

In Article 11.17, the variables x and y are each functions of a single variable t. We now consider the case where x and y are each functions of two variables. Let

$$z = f(x, y) \qquad \qquad ...(11.25)$$

and suppose that x and y are functions of r and s, say $x = \phi(r,s)$, $y = \psi(r,s)$. We could then express z as a function of the two variables r and s with the composite function $z = f\big(\phi(r,s), \psi(r,s)\big)$. If all the three functions are differentiable, then z has partial derivatives with respect to r and s, given by the formulas

$$\frac{\partial z}{\partial r} = \frac{\partial z}{\partial x} \cdot \frac{\partial x}{\partial r} + \frac{\partial z}{\partial y} \cdot \frac{\partial y}{\partial r} \qquad \qquad ...(11.26)$$

and

$$\frac{\partial z}{\partial s} = \frac{\partial z}{\partial x} \cdot \frac{\partial x}{\partial s} + \frac{\partial z}{\partial y} \cdot \frac{\partial y}{\partial s} \qquad \qquad ...(11.27)$$

If s is held fixed, then $x = \phi(r,s)$ and $y = \psi(r,s)$ become functions of r alone. The chain rule (11.23) will apply with r in place of t, and if we used ∂ rather than d to indicate that the variable s is fixed, we obtain

$$\frac{\partial z}{\partial r} = \frac{\partial z}{\partial x} \cdot \frac{\partial x}{\partial r} + \frac{\partial z}{\partial y} \cdot \frac{\partial y}{\partial r}.$$

The formula for $\partial z/\partial s$ is derived similarly. Figure 11.3 shows tree diagrams for Eqs. (11.26) and (11.27). The above formulae can be easily extended to the case of more than two variables.

Chain Rule

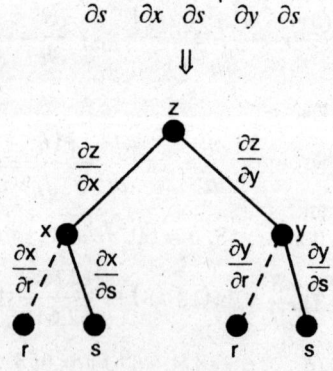

Tree diagram for Eq. (11.26)   Tree diagram for Eq. (11.27)

Fig. 11.3.

11.20 Illustrative Examples

Example 11.50 (*a*) If $F(x, y) = (2x + y)/(y - 2x)$, $x = 2u - 3v$, $y = u + 2v$, find (*i*) $\partial F/\partial u$, (*ii*) $\partial F/\partial v$, where $u = 2$, $v = 1$.

(*b*) If $z = f(x,y)$, $x = e^u + e^{-v}$ and $y = e^{-u} - e^v$, then show that

$$\frac{\partial f}{\partial u} - \frac{\partial f}{\partial v} = x\frac{\partial f}{\partial x} - y\frac{\partial f}{\partial y}. \qquad (A.M.I.E., \ W\text{-}2001; \ MDU \ 2000)$$

(*c*) If $u = f(r, s)$, $r = x + at$, $s = y + bt$ and x, y, t are independent variables, show that

$$\frac{\partial u}{\partial t} = a\frac{\partial u}{\partial x} + b\frac{\partial u}{\partial y}. \qquad (Madras, \ 1996S)$$

Solution. (*a*) Using the chain rule, we obtain $\dfrac{\partial F}{\partial u} = \dfrac{\partial F}{\partial x}\cdot\dfrac{\partial x}{\partial u} + \dfrac{\partial F}{\partial y}\cdot\dfrac{\partial y}{\partial u}$

$$= \left\{\frac{2(y-2x)-(2x+y)(-2)}{(y-2x)^2}\right\}\cdot 2 + \left\{\frac{(y-2x)-(2x+y)}{(y-2x)^2}\right\}\cdot 1 = \frac{4(2y-x)}{(y-2x)^2},$$

$$\frac{\partial F}{\partial v} = \frac{\partial F}{\partial x}\frac{\partial x}{\partial v} + \frac{\partial F}{\partial y}\frac{\partial y}{\partial v} = \frac{4y}{(y-2x)^2}(-3) + \frac{(-4x)}{(y-2x)^2}\cdot 2 = -\frac{4(2x+3y)}{(y-2x)^2}.$$

When $u = 2$, $v = 1$; then $x = 1$, $y = 4$.

Therefore, $\left(\dfrac{\partial F}{\partial u}\right)_{\substack{x=1\\y=4}} = 4\left[\dfrac{(2y-x)}{(y-2x)^2}\right]_{\substack{x=1\\y=4}} = 7$ and $\left(\dfrac{\partial F}{\partial V}\right)_{\substack{x=1\\y=4}} = -14.$

(*b*) We have $\dfrac{\partial x}{\partial u} = e^u$, $\dfrac{\partial x}{\partial v} = -e^{-v}$, $\dfrac{\partial y}{\partial u} = -e^{-u}$ and $\dfrac{\partial y}{\partial v} = -e^v$.

Using the chain rule, we obtain $\dfrac{\partial f}{\partial u} = \dfrac{\partial f}{\partial x}\cdot\dfrac{\partial x}{\partial u} + \dfrac{\partial f}{\partial y}\cdot\dfrac{\partial y}{\partial u} = e^u\dfrac{\partial f}{\partial x} - e^{-u}\dfrac{\partial f}{\partial y}$, ...(*i*)

and $\dfrac{\partial f}{\partial v} = \dfrac{\partial f}{\partial x}\cdot\dfrac{\partial x}{\partial v} + \dfrac{\partial f}{\partial y}\cdot\dfrac{\partial y}{\partial v} = -e^{-v}\dfrac{\partial f}{\partial x} - e^v\dfrac{\partial f}{\partial y}.$ ...(*ii*)

Therefore, $\dfrac{\partial f}{\partial u} - \dfrac{\partial f}{\partial v} = (e^u + e^{-v})\dfrac{\partial f}{\partial x} - (e^{-u} - e^v)\dfrac{\partial f}{\partial y} = x\dfrac{\partial f}{\partial x} - y\dfrac{\partial f}{\partial y}.$

(*c*) We have by the chain rule

$$\frac{\partial u}{\partial x} = \frac{\partial u}{\partial r}\cdot\frac{\partial r}{\partial x} + \frac{\partial u}{\partial s}\cdot\frac{\partial s}{\partial x} = 1\cdot\frac{\partial u}{\partial r} + 0\cdot\frac{\partial u}{\partial s} = \frac{\partial u}{\partial r} \qquad ...(i)$$

$$\frac{\partial u}{\partial y} = \frac{\partial u}{\partial r} \cdot \frac{\partial r}{\partial y} + \frac{\partial u}{\partial s} \cdot \frac{\partial s}{\partial y} = 0 \cdot \frac{\partial u}{\partial r} + 1 \cdot \frac{\partial u}{\partial s} = \frac{\partial u}{\partial s} \qquad \text{...(ii)}$$

$$\frac{\partial u}{\partial t} = \frac{\partial u}{\partial r} \cdot \frac{\partial r}{\partial t} + \frac{\partial u}{\partial s} \cdot \frac{\partial s}{\partial t} = a \frac{\partial u}{\partial r} + b \frac{\partial u}{\partial s} . \qquad \text{...(iii)}$$

From (*i*), (*ii*) and (*iii*), we find

$$\frac{\partial u}{\partial t} = a \frac{\partial u}{\partial x} + b \frac{\partial u}{\partial y}.$$

Example 11.51 (*a*) **If w = f (u, v), where u = $\sqrt{x^2 + y^2}$ and v = tan^{-1}(y/x), find $\partial f/\partial x$ and $\partial f/\partial y$.**

(*b*) **If f(x, y, z) = z sin(y/x), where x = 3r^2 + 2s, y = 4r – 2s^3 and z = 2r^2 – 3 s^2, find $\partial f/\partial r$ and $\partial f/\partial s$.**

Solution. (*a*) We have $u = \sqrt{x^2 + y^2}$ and $v = \tan^{-1}(y/x)$. We obtain

$$\frac{\partial u}{\partial x} = \frac{x}{\sqrt{x^2 + y^2}}, \frac{\partial u}{\partial y} = \frac{y}{\sqrt{x^2 + y^2}}, \frac{\partial v}{\partial x} = \frac{-y}{x^2 + y^2} \text{ and } \frac{\partial v}{\partial y} = \frac{x}{x^2 + y^2}.$$

Using the chain rule, we obtain

$$\frac{\partial f}{\partial x} = \frac{\partial f}{\partial u} \cdot \frac{\partial u}{\partial x} + \frac{\partial f}{\partial v} \cdot \frac{\partial v}{\partial x}$$

$$= \frac{x}{\sqrt{x^2 + y^2}} \frac{\partial f}{\partial u} - \frac{y}{x^2 + y^2} \frac{\partial f}{\partial v} = \frac{1}{x^2 + y^2} \left(x\sqrt{x^2 + y^2} \frac{\partial f}{\partial u} - y \frac{\partial f}{\partial v} \right),$$

and

$$\frac{\partial f}{\partial y} = \frac{\partial f}{\partial u} \cdot \frac{\partial u}{\partial y} + \frac{\partial f}{\partial v} \cdot \frac{\partial v}{\partial y} = \frac{y}{\sqrt{x^2 + y^2}} \frac{\partial f}{\partial u} + \frac{x}{x^2 + y^2} \frac{\partial f}{\partial v}$$

$$= \frac{1}{x^2 + y^2} \left(y\sqrt{x^2 + y^2} \frac{\partial f}{\partial u} + x \frac{\partial f}{\partial v} \right).$$

(*b*) Using the chain rule, we obtain $\dfrac{\partial f}{\partial r} = \dfrac{\partial f}{\partial x} \cdot \dfrac{\partial x}{\partial r} + \dfrac{\partial f}{\partial y} \cdot \dfrac{\partial y}{\partial r} + \dfrac{\partial f}{\partial z} \cdot \dfrac{\partial z}{\partial r}$

$$= \left\{ -\frac{y}{x^2} z \cos(y/x) \right\} 6r + \left\{ \frac{z}{x} \cos(y/x) \right\} 4 + (\sin y/x)(4r)$$

$$= \frac{-6ryz}{x^2} \cos(y/x) + \frac{4z}{x} \cos(y/x) + 4r \sin(y/x),$$

and

$$\frac{\partial f}{\partial s} = \frac{\partial f}{\partial x} \cdot \frac{\partial x}{\partial s} + \frac{\partial f}{\partial y} \cdot \frac{\partial y}{\partial s} + \frac{\partial f}{\partial z} \cdot \frac{\partial z}{\partial s}$$

$$= \left\{ -\frac{yz}{x^2} \cos(y/x) \right\} 2 + \left\{ \frac{z}{x} \cos(y/x) \right\} (-6s^2) + \sin(y/x) \cdot (-6s)$$

$$= \frac{-2yz}{x^2} \cos(y/x) - 6 \frac{s^2 z}{x} \cos(y/x) - 6s \sin(y/x).$$

Example 11.52 If $u = f(y - z, z - x, x - y)$, show that $\dfrac{\partial u}{\partial x} + \dfrac{\partial u}{\partial y} + \dfrac{\partial u}{\partial z} = 0.$

(A.M.I.E. W-2009, W-2006; UPTU 2005; IETE June 2002, Andhra-1998)

Solution. Let $r = y - z$, $s = z - x$, $t = x - y$, so that $u = f(r, s, t)$. Then u is a function of the three variables r, s, t, which in turn are functions of the three variables x, y, z.

Using the chain rule, we obtain $\dfrac{\partial u}{\partial x} = \dfrac{\partial u}{\partial r} \cdot \dfrac{\partial r}{\partial x} + \dfrac{\partial u}{\partial s} \cdot \dfrac{\partial s}{\partial x} + \dfrac{\partial u}{\partial t} \cdot \dfrac{\partial t}{\partial x}$

$$= 0 \cdot \frac{\partial u}{\partial r} + (-1)\frac{\partial u}{\partial s} + (1)\frac{\partial u}{\partial t} = -\frac{\partial u}{\partial s} + \frac{\partial u}{\partial t}. \qquad \qquad \text{...(i)}$$

$$\frac{\partial u}{\partial y} = \frac{\partial u}{\partial r} \cdot \frac{\partial r}{\partial y} + \frac{\partial u}{\partial s} \cdot \frac{\partial s}{\partial y} + \frac{\partial u}{\partial t} \cdot \frac{\partial t}{\partial y} = (1)\frac{\partial u}{\partial r} + 0\left(\frac{\partial u}{\partial s}\right) + (-1)\frac{\partial u}{\partial t} = \frac{\partial u}{\partial r} - \frac{\partial u}{\partial t}, \qquad \text{...(ii)}$$

and $\dfrac{\partial u}{\partial z} = \dfrac{\partial u}{\partial r} \cdot \dfrac{\partial r}{\partial z} + \dfrac{\partial u}{\partial s} \cdot \dfrac{\partial s}{\partial z} + \dfrac{\partial u}{\partial t} \cdot \dfrac{\partial t}{\partial z} = (-1)\dfrac{\partial u}{\partial r} + (1)\dfrac{\partial u}{\partial s} + (0)\dfrac{\partial u}{\partial t} = -\dfrac{\partial u}{\partial r} + \dfrac{\partial u}{\partial s}$ ...(iii)

Adding (i), (ii) and (iii), we obtain

$$\frac{\partial u}{\partial x} + \frac{\partial u}{\partial y} + \frac{\partial u}{\partial z} = 0.$$

Example 11.53 If $\phi(cx - az, cy - bz) = 0$, show that $ap + bq = c$, where

$p \equiv \dfrac{\partial z}{\partial x}$ and $q \equiv \dfrac{\partial z}{\partial y}$.

$$\text{...(i)}$$

Solution. Let $u = cx - az$ and $v = cy - bz$.
Then the given function is $\phi(u, v) = 0$

$$\therefore \quad \frac{\partial \phi}{\partial u} \frac{\partial u}{\partial x} + \frac{\partial \phi}{\partial v} \frac{\partial v}{\partial x} = 0, \qquad \frac{\partial \phi}{\partial u} \frac{\partial u}{\partial y} + \frac{\partial \phi}{\partial v} \frac{\partial v}{\partial y} = 0.$$

Eliminating $\dfrac{\partial \phi}{\partial u}$ and $\dfrac{\partial \phi}{\partial v}$, we obtain $\begin{vmatrix} \dfrac{\partial u}{\partial x} & \dfrac{\partial v}{\partial x} \\ \dfrac{\partial u}{\partial y} & \dfrac{\partial v}{\partial y} \end{vmatrix} = 0.$ ...(ii)

From (i) $\dfrac{\partial u}{\partial x} = c - ap, \dfrac{\partial u}{\partial y} = -aq; \dfrac{\partial v}{\partial x} = -bp, \dfrac{\partial v}{\partial y} = c - bq$.

Substituting these values in (ii), we obtain

$$\begin{vmatrix} c - ap & -bp \\ -aq & c - bq \end{vmatrix} = 0 \text{ or } (c - ap)(c - bq) - abpq = 0 \text{ or } ap + bq = c.$$

Example 11.54 **Let u be a function of (x, y) and x, y are functions of (θ, ϕ) defined by $x + y = 2e^{\theta} \cos \phi$, and $x - y = 2i\, e^{\theta} \sin \phi$, where $i = \sqrt{-1}$. Show that**

(i) $x\dfrac{\partial u}{\partial x} + y\dfrac{\partial u}{\partial y} = \dfrac{\partial u}{\partial \theta}$, *(I.E.T.E. Dec. 2004)*

(ii) $\dfrac{\partial^2 u}{\partial \theta^2} + \dfrac{\partial^2 u}{\partial \phi^2} = 4xy\dfrac{\partial^2 u}{\partial x \partial y}.$

(I.E.T.E. June 2010, Dec. 2009, 2007, MDU 2002)

Solution. We have $x = e^{\theta}(\cos \phi + i \sin \phi) = e^{\theta} \cdot e^{i\phi}$,
and $y = e^{\theta}(\cos \phi - i \sin \phi) = e^{\theta} \cdot e^{-i\phi}$.

Here u is a composite function of θ and ϕ.

$\therefore \quad \dfrac{\partial u}{\partial \theta} = \dfrac{\partial u}{\partial x} \cdot \dfrac{\partial x}{\partial \theta} + \dfrac{\partial u}{\partial y} \cdot \dfrac{\partial y}{\partial \theta} = \dfrac{\partial u}{\partial x}\left(e^{\theta} \cdot e^{i\phi}\right) + \dfrac{\partial u}{\partial y}\left(e^{\theta} \cdot e^{-i\phi}\right)$ ...(i)

$= x\dfrac{\partial u}{\partial x} + y\dfrac{\partial u}{\partial y},$ the required result.

(i) may be written in operator form

$$\dfrac{\partial}{\partial \theta} = e^{\theta}\left(\dfrac{\partial}{\partial x}e^{i\phi} + \dfrac{\partial}{\partial y}e^{-i\phi}\right).$$ ...(ii)

Also $\dfrac{\partial u}{\partial \phi} = \dfrac{\partial u}{\partial x} \cdot \dfrac{\partial x}{\partial \phi} + \dfrac{\partial u}{\partial y} \cdot \dfrac{\partial y}{\partial \phi} = \dfrac{\partial u}{\partial x}\left(e^{\theta} \cdot ie^{i\phi}\right) + \dfrac{\partial u}{\partial y}\left(e^{\theta} \cdot -ie^{-i\phi}\right)$

or $$\dfrac{\partial}{\partial \phi} = ie^{\theta}\left(\dfrac{\partial}{\partial x} \cdot e^{i\phi} - \dfrac{\partial}{\partial y}e^{-i\phi}\right).$$ ...(iii)

Using the operator (*ii*), we have

$$\frac{\partial^2 u}{\partial \theta^2} = \frac{\partial}{\partial \theta}\left(\frac{\partial u}{\partial \theta}\right) = e^{2\theta}\left(\frac{\partial}{\partial x}e^{i\phi} + \frac{\partial}{\partial y}e^{-i\phi}\right)\left(\frac{\partial u}{\partial x}e^{i\phi} + \frac{\partial u}{\partial y}e^{-i\phi}\right)$$

$$= e^{2\theta}\left(e^{2i\phi}\frac{\partial^2 u}{\partial x^2} + 2\frac{\partial^2 u}{\partial x \partial y} + e^{-2i\phi}\frac{\partial^2 u}{\partial y^2}\right), \qquad \qquad ...(iv)$$

and

$$\frac{\partial^2 u}{\partial \phi^2} = \frac{\partial}{\partial \phi}\left(\frac{\partial u}{\partial \phi}\right) = (ie^{\theta})^2\left(\frac{\partial}{\partial x}e^{i\phi} - \frac{\partial}{\partial y}e^{-i\phi}\right)\left(\frac{\partial u}{\partial x}e^{i\phi} - \frac{\partial u}{\partial y}e^{-i\phi}\right)$$

$$= -e^{2\theta}\left(e^{2i\phi}\frac{\partial^2 u}{\partial x^2} - 2\frac{\partial^2 u}{\partial x \partial y} + e^{-2i\phi}\frac{\partial^2 u}{\partial y^2}\right). \qquad \qquad ...(v)$$

Adding (*iv*) and (*v*), we get

$$\frac{\partial^2 u}{\partial \theta^2} + \frac{\partial^2 u}{\partial \phi^2} = e^{2\theta}\left[4\frac{\partial^2 u}{\partial x \partial y}\right] = 4xy\frac{\partial^2 u}{\partial x \partial y}. \qquad \qquad [\because xy = e^{2\theta}]$$

Example 11.55 If $x = e^v \sec u$, $y = e^v \tan u$ and $\phi = \phi(x, y)$, **show that**

$$\cos u\left(\frac{\partial^2 \phi}{\partial u \partial v} - \frac{\partial \phi}{\partial u}\right) = xy\left(\frac{\partial^2 \phi}{\partial x^2} + \frac{\partial^2 \phi}{\partial y^2}\right) + (x^2 + y^2)\frac{\partial^2 \phi}{\partial x \partial y}. \qquad (A.M.I.E.\ W\text{-}2000)$$

Solution. We have $x = e^v \sec u$, $y = e^v \tan u$ and $\phi = \phi(x, y)$.
Using the chain rule, we obtain

$$\frac{\partial \phi}{\partial u} = \frac{\partial \phi}{\partial x}\cdot\frac{\partial x}{\partial u} + \frac{\partial \phi}{\partial y}\cdot\frac{\partial y}{\partial u} = \frac{\partial \phi}{\partial x}(e^v \sec u \tan u) + \frac{\partial \phi}{\partial y}(e^v \sec^2 u)$$

$$= (\sec u)\, y\frac{\partial \phi}{\partial x} + (\sec u)\, x\frac{\partial \phi}{\partial y} \quad \text{or} \quad \cos u \cdot \frac{\partial \phi}{\partial u} = y\frac{\partial \phi}{\partial x} + x\frac{\partial \phi}{\partial y}$$

$$\Rightarrow \qquad \qquad \cos u\frac{\partial}{\partial u} = y\frac{\partial}{\partial x} + x\frac{\partial}{\partial y} \qquad \qquad ...(i)$$

$$\frac{\partial \phi}{\partial v} = \frac{\partial \phi}{\partial x}\cdot\frac{\partial x}{\partial v} + \frac{\partial \phi}{\partial y}\cdot\frac{\partial y}{\partial v} = \frac{\partial \phi}{\partial x}(e^v \sec u) + \frac{\partial \phi}{\partial y}(e^v \tan u) = x\frac{\partial \phi}{\partial x} + y\frac{\partial \phi}{\partial y}. \qquad (ii)$$

Now $\cos u\left(\dfrac{\partial^2 \phi}{\partial u \partial v}\right) = \cos u\dfrac{\partial}{\partial u}\left(\dfrac{\partial \phi}{\partial v}\right) = \left(y\dfrac{\partial}{\partial x} + x\dfrac{\partial}{\partial y}\right)\left(x\dfrac{\partial \phi}{\partial x} + y\dfrac{\partial \phi}{\partial y}\right)$ from (*i*) & (*ii*)

$$= y\frac{\partial}{\partial x}\left(x\frac{\partial \phi}{\partial x} + y\frac{\partial \phi}{\partial y}\right) + x\frac{\partial}{\partial y}\left(x\frac{\partial \phi}{\partial x} + y\frac{\partial \phi}{\partial y}\right)$$

$$= y\left[1\cdot\frac{\partial\phi}{\partial x}+x\frac{\partial^2\phi}{\partial x^2}+y\frac{\partial^2\phi}{\partial x\partial y}\right]+x\left(x\frac{\partial^2\phi}{\partial y\,\partial x}+1\cdot\frac{\partial\phi}{\partial y}+y\frac{\partial^2\phi}{\partial y^2}\right)$$

$$= xy\left(\frac{\partial^2\phi}{\partial x^2}+\frac{\partial^2\phi}{\partial y^2}\right)+\left(x^2+y^2\right)\frac{\partial^2\phi}{\partial x\,\partial y}+\left(x\frac{\partial\phi}{\partial y}+y\frac{\partial\phi}{\partial x}\right),$$

$$= xy\left(\frac{\partial^2\phi}{\partial x^2}+\frac{\partial^2\phi}{\partial y^2}\right)+\left(x^2+y^2\right)\frac{\partial^2\phi}{\partial x\,\partial y}+\cos u\frac{\partial\phi}{\partial u}, \qquad \text{from } (i)$$

or $\cos u\left(\dfrac{\partial^2\phi}{\partial u\,\partial v}-\dfrac{\partial\phi}{\partial u}\right)=xy\left(\dfrac{\partial^2\phi}{\partial x^2}+\dfrac{\partial^2\phi}{\partial y^2}\right)+\left(x^2+y^2\right)\dfrac{\partial^2\phi}{\partial x\,\partial y}$, which is same as required.

Example 11.56 If $x=\dfrac{\cos\theta}{u},\ y=\dfrac{\sin\theta}{u}$ and $z=f(x,y)$, then show that

$$\frac{\partial^2 z}{\partial x^2}+\frac{\partial^2 z}{\partial y^2}=u^4\frac{\partial^2 z}{\partial u^2}+u^3\frac{\partial z}{\partial u}+u^2\frac{\partial^2 z}{\partial\theta^2}. \qquad (I.E.T.E.\ Dec.\ 2002)$$

Solution. We have $\dfrac{\partial x}{\partial u}=\dfrac{-\cos\theta}{u^2},\ \dfrac{\partial x}{\partial\theta}=\dfrac{-\sin\theta}{u}$, and $\dfrac{\partial y}{\partial u}=\dfrac{-\sin\theta}{u^2}$ and $\dfrac{\partial y}{\partial\theta}=\dfrac{\cos\theta}{u}$.

Using the chain rule, we obtain $\dfrac{\partial z}{\partial u}=\dfrac{\partial z}{\partial x}\cdot\dfrac{\partial x}{\partial u}+\dfrac{\partial z}{\partial y}\cdot\dfrac{\partial y}{\partial u}=-\dfrac{\cos\theta}{u^2}\dfrac{\partial z}{\partial x}-\dfrac{\sin\theta}{u^2}\dfrac{\partial z}{\partial y}$,

and $\dfrac{\partial z}{\partial\theta}=\dfrac{\partial z}{\partial x}\cdot\dfrac{\partial x}{\partial\theta}+\dfrac{\partial z}{\partial y}\cdot\dfrac{\partial y}{\partial\theta}=\dfrac{-\sin\theta}{u}\dfrac{\partial z}{\partial x}+\dfrac{\cos\theta}{u}\dfrac{\partial z}{\partial y}$.

Solving these equations, we get

$$\frac{\partial z}{\partial x}=-\left(u^2\cos\theta\frac{\partial z}{\partial u}+u\sin\theta\frac{\partial z}{\partial\theta}\right)\Rightarrow\frac{\partial}{\partial x}=-\left(u^2\cos\theta\frac{\partial}{\partial u}+u\sin\theta\frac{\partial}{\partial\theta}\right),$$

(in operator form)

and $\dfrac{\partial z}{\partial y}=\left(u\cos\theta\dfrac{\partial z}{\partial\theta}-u^2\sin\theta\dfrac{\partial z}{\partial u}\right)\Rightarrow\dfrac{\partial}{\partial y}=\left(u\cos\theta\dfrac{\partial}{\partial\theta}-u^2\sin\theta\dfrac{\partial}{\partial u}\right).$

(in operator form)

Hence $\dfrac{\partial^2 z}{\partial x^2}=\dfrac{\partial}{\partial x}\left(\dfrac{\partial z}{\partial x}\right)=\left(u^2\cos\theta\dfrac{\partial}{\partial u}+u\sin\theta\dfrac{\partial}{\partial\theta}\right)\left(u^2\cos\theta\dfrac{\partial z}{\partial u}+u\sin\theta\dfrac{\partial z}{\partial\theta}\right)$

$$= u^4\cos^2\theta\frac{\partial^2 z}{\partial u^2}+2u^3\sin\theta\cos\theta\frac{\partial^2 z}{\partial u\,\partial\theta}+u^2\sin^2\theta\frac{\partial^2 z}{\partial\theta^2}$$

$$+\left(2u^3\cos^2\theta - u^3\sin^2\theta\right)\frac{\partial z}{\partial u} + 2u^2\cos\theta\sin\theta\frac{\partial z}{\partial\theta}. \qquad \text{...}(i)$$

Similarly
$$\frac{\partial^2 z}{\partial y^2} = -2u^3\sin\theta\cos\theta\frac{\partial^2 z}{\partial u\,\partial\theta} + u^2\cos^2\theta\frac{\partial^2 z}{\partial\theta^2} + u^4\sin^2\theta\frac{\partial^2 z}{\partial u^2}$$

$$+\left(2u^3\sin^2\theta - u^3\cos^2\theta\right)\frac{\partial z}{\partial u} - 2u^2\sin\theta\cos\theta\frac{\partial z}{\partial\theta}. \qquad \text{...}(ii)$$

Adding (*i*) and (*ii*), we obtain

$$\frac{\partial^2 z}{\partial x^2} + \frac{\partial^2 z}{\partial y^2} = u^4\frac{\partial^2 z}{\partial u^2} + u^3\frac{\partial z}{\partial u} + u^2\frac{\partial^2 z}{\partial\theta^2}, \text{ the required result.}$$

Example 11.57 If $u = f(x, y)$, $x = r\cos\theta$, $y = r\sin\theta$, show that $\dfrac{\partial^2 u}{\partial x^2} + \dfrac{\partial^2 u}{\partial y^2} = 0$

is transformed to the form $\dfrac{\partial^2 u}{\partial r^2} + \dfrac{1}{r}\dfrac{\partial u}{\partial r} + \dfrac{1}{r^2}\dfrac{\partial^2 u}{\partial\theta^2} = 0$.

(GGSIPU 2007; DCE 2004; UPTU 2001; Delhi 1997)

Solution. We have $x = r\cos\theta$, $y = r\sin\theta$. Therefore,

$$r = \sqrt{\left(x^2 + y^2\right)}, \text{ and } \theta = \tan^{-1}\left(y/x\right).$$

$$\frac{\partial r}{\partial x} = \frac{x}{\sqrt{\left(x^2 + y^2\right)}} = \cos\theta \text{ and } \frac{\partial\theta}{\partial x} = -\frac{y}{x^2 + y^2} = -\frac{\sin\theta}{r}.$$

Using the chain rule, we obtain $\dfrac{\partial u}{\partial x} = \dfrac{\partial u}{\partial r}\cdot\dfrac{\partial r}{\partial x} + \dfrac{\partial u}{\partial\theta}\cdot\dfrac{\partial\theta}{\partial x}$

$$= \cos\theta\frac{\partial u}{\partial r} - \frac{\sin\theta}{r}\frac{\partial u}{\partial\theta}. \text{ Therefore, } \frac{\partial}{\partial x} = \cos\theta\frac{\partial}{\partial r} - \frac{\sin\theta}{r}\frac{\partial}{\partial\theta}.$$

Similarly
$$\frac{\partial}{\partial y} = \sin\theta\frac{\partial}{\partial r} + \frac{\cos\theta}{r}\frac{\partial}{\partial\theta}.$$

$$\therefore \quad \frac{\partial^2 u}{\partial x^2} = \frac{\partial}{\partial x}\left(\frac{\partial u}{\partial x}\right) = \left(\cos\theta\frac{\partial}{\partial r} - \frac{\sin\theta}{r}\frac{\partial}{\partial\theta}\right)\left(\cos\theta\frac{\partial u}{\partial r} - \left(\frac{\sin\theta}{r}\right)\frac{\partial u}{\partial\theta}\right)$$

$$= \cos^2\theta\frac{\partial^2 u}{\partial r^2} - \frac{2\sin\theta\cos\theta}{r}\frac{\partial^2 u}{\partial r\,\partial\theta} + \frac{\sin^2\theta}{r^2}\frac{\partial^2 u}{\partial\theta^2} + 2\frac{\sin\theta\cos\theta}{r^2}\frac{\partial u}{\partial\theta} + \frac{\sin^2\theta}{r}\frac{\partial u}{\partial r}, \qquad \text{...}(i)$$

Now
$$\frac{\partial^2 u}{\partial y^2} = \frac{\partial}{\partial y}\left(\frac{\partial u}{\partial y}\right) = \left(\sin\theta\frac{\partial}{\partial r} + \frac{\cos\theta}{r}\frac{\partial}{\partial\theta}\right)\left(\sin\theta\frac{\partial u}{\partial r} + \frac{\cos\theta}{r}\frac{\partial u}{\partial\theta}\right)$$

$$= \sin^2\theta \frac{\partial^2 u}{\partial r^2} + \frac{2\sin\theta\cos\theta}{r}\frac{\partial^2 u}{\partial r\,\partial\theta} \div \frac{\cos^2\theta}{r^2}\frac{\partial^2 u}{\partial\theta^2} + \frac{\cos^2\theta}{r}\frac{\partial u}{\partial r} - \frac{2\sin\theta\cos\theta}{r^2}\frac{\partial u}{\partial\theta}. \quad ...(ii)$$

Adding (*i*) and (*ii*), we obtain

$$\frac{\partial^2 u}{\partial x^2} + \frac{\partial^2 u}{\partial y^2} = \frac{\partial^2 u}{\partial r^2} + \frac{1}{r^2}\frac{\partial^2 u}{\partial\theta^2} + \frac{1}{r}\frac{\partial u}{\partial r}.$$

$$\therefore \quad \frac{\partial^2 u}{\partial x^2} + \frac{\partial^2 u}{\partial y^2} = 0 \quad \text{transforms into} \quad \frac{\partial^2 u}{\partial r^2} + \frac{1}{r}\frac{\partial u}{\partial r} + \frac{1}{r^2}\frac{\partial^2 u}{\partial\theta^2} = 0.$$

11.21 Differentiation of Implicit Functions

An equation of the form $f(x, y) = 0$ ...(11.28)
in which y is not expressible directly in terms of x, is known as an implicit function of x and y. The above equation defines either x or y as an implicit function of the other.

To find dy/dx for the implicit function $f(x, y) = 0$, we proceed as follows:
Let $z = f(x, y)$, where $z = 0$ for all (x, y).

Then $dz = \dfrac{\partial f}{\partial x}dx + \dfrac{\partial f}{\partial y}dy$ or $dz = f_x\,dx + f_y\,dy$.

On the other hand, since $z = 0$, for all (x, y) implies $dz = 0$ and hence

$$f_x\,dx + f_y\,dy = 0 \qquad \text{or} \qquad \frac{\mathbf{dy}}{\mathbf{dx}} = -\frac{\partial f/\partial x}{\partial f/\partial y} = -\frac{\mathbf{f_x}}{\mathbf{f_y}}\,\left(f_y \neq 0\right) \quad ...(11.29)$$

We can find d^2y/dx^2 by differentiating the result (11.29) with respect to x

$$\frac{d^2 y}{dx^2} = \frac{d}{dx}\left(\frac{dy}{dx}\right) = -\frac{d}{dx}\left(\frac{f_x}{f_y}\right) = -\frac{f_y\dfrac{d}{dx}(f_x) - f_x\dfrac{d}{dx}(f_y)}{(f_y)^2}$$

$$= \left[-f_y\left(f_{xx} + (f_{yx})\frac{dy}{dx}\right) + f_x\left(f_{xy} + (f_{yy})\frac{dy}{dx}\right)\right]\Big/(f_y)^2$$

$$= -\frac{(f_y f_{xx} - f_x f_{xy}) + (f_y f_{yx} - f_x f_{yy})(-f_x/f_y)}{(f_y)^2}$$

$$= -\frac{(f_x)^2 f_{yy} - 2f_x f_y f_{xy} + (f_y)^2 f_{xx}}{(f_y)^3}. \qquad ...(11.30)$$

...(11.31)

The equation $F(x, y, z) = 0$ defines z as an implicit function of the two independent variables x and y. To find the partial derivatives of z with respect to x and to y, proceed as follows:

Let $u = F(x, y, z)$.

Then $du = \dfrac{\partial F}{\partial x}dx + \dfrac{\partial F}{\partial y}dy + \dfrac{\partial F}{\partial z}dz$, and this holds no matter what the independent variables are. Now let z be chosen as that function of the independent variables x and y which satisfied (11.31). Then $u = 0$, $du = 0$, and we have

$$\frac{\partial F}{\partial x}dx + \frac{\partial F}{\partial y}dy + \frac{\partial F}{\partial z}dz = 0.$$ ...(11.32)

But now $dz = \dfrac{\partial z}{\partial x}dx + \dfrac{\partial z}{\partial y}dy.$

Substituting this value in (11.32) and simplifying gives

$$\left(\frac{\partial F}{\partial x} + \frac{\partial F}{\partial z}\frac{\partial z}{\partial x}\right)dx + \left(\frac{\partial F}{\partial y} + \frac{\partial F}{\partial z}\frac{\partial z}{\partial y}\right)dy = 0.$$

Since x and y are independent, we have

$$\frac{\partial F}{\partial x} + \frac{\partial F}{\partial z}\frac{\partial z}{\partial x} = 0 \ and \ \frac{\partial F}{\partial y} + \frac{\partial F}{\partial z}\frac{\partial z}{\partial y} = 0,$$

from which $\quad \dfrac{\partial z}{\partial x} = -\dfrac{\partial F/\partial x}{\partial F/\partial z} \ and \ \dfrac{\partial z}{\partial y} = -\dfrac{\partial F/\partial y}{\partial F/\partial z}.$

This also can be written as $\dfrac{\partial z}{\partial x} = -\dfrac{F_x}{F_z}$ ...(11.33) and $\dfrac{\partial z}{\partial y} = -\dfrac{F_y}{F_z},$ ...(11.34)

where $F_z \neq 0$.

11.22 Illustrative Examples

Example 11.58 Find dy/dx, when

(i) $f(x, y) = e^{xy} + y^2 - \cos\sqrt{x} = 0.$

(ii) $f(x, y) = \log(x^2 + y^2) + \tan^{-1}(y/x) = 0.$ (A.M.I.E., S-2002)

Solution. (i) $\dfrac{\partial f}{\partial x} = ye^{xy} + \dfrac{1}{2}\dfrac{\sin\sqrt{x}}{\sqrt{x}} = \dfrac{\left(2\sqrt{x}ye^{xy} + \sin\sqrt{x}\right)}{2\sqrt{x}}$

and $\dfrac{\partial f}{\partial y} = xe^{xy} + 2y.$ Therefore, $\dfrac{dy}{dx} = -\dfrac{\partial f/\partial x}{\partial f/\partial y} = -\left[\dfrac{2\sqrt{x}\,ye^{xy} + \sin\sqrt{x}}{2\sqrt{x}\left(xe^{xy} + 2y\right)}\right].$

(ii) $\dfrac{\partial f}{\partial x} = \dfrac{2x}{x^2+y^2} + \dfrac{1}{1+(y/x)^2}\left(\dfrac{-y}{x^2}\right) = \dfrac{2x}{x^2+y^2} - \dfrac{y}{x^2+y^2} = \dfrac{2x-y}{x^2+y^2}$,

and $\dfrac{\partial f}{\partial y} = \dfrac{2y}{x^2+y^2} + \dfrac{1}{1+(y/x)^2}\left(\dfrac{1}{x}\right) = \dfrac{2y}{x^2+y^2} + \dfrac{x}{x^2+y^2} = \dfrac{x+2y}{x^2+y^2}$.

Therefore, $\dfrac{dy}{dx} = -\dfrac{\partial f/\partial x}{\partial f/\partial y} = -\left(\dfrac{2x-y}{x+2y}\right) = \dfrac{y-2x}{2y+x}$.

Example 11.59 If $f(x, y) = x^3 + y^3 - 3\,axy = 0$, find dy/dx and d^2y/dx^2.

(A.M.I.E., W-2005; J.N.T.U., 1998)

Solution. We have $f(x, y) = x^3 + y^3 - 3axy = 0$.

$$f_x = \frac{\partial f}{\partial x} = 3x^2 - 3ay \text{ and } f_y = \frac{\partial f}{\partial y} = 3y^2 - 3ax .$$

∴ $$\frac{dy}{dx} = -\frac{\partial f/\partial x}{\partial f/\partial y} = -\left(\frac{f_x}{f_y}\right) = -\left(\frac{3x^2-3ay}{3y^2-3ax}\right) = \frac{ay-x^2}{y^2-ax}.$$

Now $$f_{xx} = 6x,\ f_{yy} = 6y,\ f_{xy} = f_{yx} = -3a.$$

∴ $$\frac{d^2y}{dx^2} = -\left\{\frac{\left(f_y\right)^2 f_{xx} - 2f_x f_y f_{xy} + \left(f_x\right)^2 f_{yy}}{\left(f_y\right)^3}\right\}$$

$$= -\left\{\frac{\left(3y^2-3ax\right)^2 6x - 2\left(3x^2-3ay\right)\left(3y^2-3ax\right)(-3a) + \left(3x^2-3ay\right)^2 6y}{\left(3y^2-3ax\right)^3}\right\}$$

$$= -\left\{\frac{2x\left(y^2-ax\right)^2 + 2a\left(x^2-ay\right)\left(y^2-ax\right) + 2y\left(x^2-ay\right)^2}{\left(y^2-ax\right)^3}\right\} = -\frac{2a^3xy}{\left(y^2-ax\right)^3}.$$

Example 11.60 If x increases at the rate of 2 cm per second at the instant when x = 3 cm and y = 1 cm, at what rate must y be changing in order that the function $2xy^2 - 3x^2y$ shall be neither increasing nor decreasing?

Solution. Let $u = 2xy^2 - 3x^2y$. Using the chain rule, we obtain

$$\frac{du}{dt} = \frac{\partial u}{\partial x}\frac{dx}{dt} + \frac{\partial u}{\partial y}\cdot\frac{dy}{dt} = \left(2y^2 - 6xy\right)\frac{dx}{dt} + \left(4xy - 3x^2\right)\frac{dy}{dt}. \qquad \text{...(i)}$$

As u is neither increasing nor decreasing so $du/dt = 0$.

Thus,
$$\frac{dy}{dt} = -\left(\frac{2y^2 - 6xy}{4xy - 3x^2}\right)\frac{dx}{dt}.$$

When $x = 3$, $y = 1$, $\dfrac{dx}{dt} = +2$, then $\dfrac{dy}{dt} = -\left(\dfrac{2-18}{12-27}\right)2 = -2\dfrac{2}{15}$ cm/sec.

Thus y is decreasing at the rate of $2\dfrac{2}{15}$ cm/sec.

Example 11.61 If $u = x^3y$, find du/dt if $x^5 + y = t$ ...(i)

and $\qquad\qquad\qquad\qquad x^2 + y^3 = t^2.$ ...(ii)

Solution. Equations (*i*) and (*ii*) define x and y as implicit functions of t. Now differentiating these equations w.r.t. t, we obtain

$$5x^4\left(\frac{dx}{dt}\right) + \left(\frac{dy}{dt}\right) = 1 \;...(iii) \quad \text{and} \quad 2x\left(\frac{dx}{dt}\right) + 3y^2\left(\frac{dy}{dt}\right) = 2t. \qquad ...(iv)$$

Solving (*iii*) & (*iv*) simultaneously for dx/dt and dy/dt, we obtain

$$\frac{dx}{dt} = \frac{\begin{vmatrix} 1 & 1 \\ 2t & 3y^2 \end{vmatrix}}{\begin{vmatrix} 5x^4 & 1 \\ 2x & 3y^2 \end{vmatrix}} = \frac{3y^2 - 2t}{15x^4y^2 - 2x}, \; \frac{dy}{dt} = \frac{\begin{vmatrix} 5x^4 & 1 \\ 2x & 2t \end{vmatrix}}{\begin{vmatrix} 5x^4 & 1 \\ 2x & 3y^2 \end{vmatrix}} = \frac{10x^4t - 2x}{15x^4y^2 - 2x}.$$

Using the chain rule, we obtain $\dfrac{du}{dt} = \dfrac{\partial u}{\partial x}\dfrac{dx}{dt} + \dfrac{\partial u}{\partial y}\dfrac{dy}{dt}$

$$= 3x^2y\left(\frac{3y^2 - 2t}{15x^4y^2 - 2x}\right) + x^3\left(\frac{10x^4t - 2x}{15x^4y^2 - 2x}\right).$$

Example 11.62 If $u = x \log(xy)$, where $x^3 + y^3 + 3xy = 1$, find du/dx.

$\qquad\qquad\qquad\qquad\qquad\qquad\qquad\qquad\qquad$ *(UPTU 2001)*

Solution. We have $f(x, y) = x^3 + y^3 + 3xy - 1 = 0$.

$$\frac{\partial f}{\partial x} = 3x^2 + 3y \text{ and } \frac{\partial f}{\partial y} = 3y^2 + 3x.$$

Therefore, $\qquad \dfrac{dy}{dx} = \dfrac{-\partial f/\partial x}{\partial f/\partial y} = -\dfrac{x^2 + y}{y^2 + x}$. Given $u = x\log(xy)$.

Therefore, $\dfrac{\partial u}{\partial x} = \log(xy) + \dfrac{xy}{xy} = 1 + \log(xy)$, and $\dfrac{\partial u}{\partial y} = \dfrac{x^2}{xy} = \dfrac{x}{y}$.

Using the chain rule, we obtain $\dfrac{du}{dx} = \dfrac{\partial u}{\partial x} + \dfrac{\partial u}{\partial y} \cdot \dfrac{dy}{dx} = 1 + \log(xy) - \dfrac{x(x^2 + y)}{y(x + y^2)}$.

Example 11.63 Use implicit differentiation to find $\dfrac{\partial z}{\partial x}$ and $\dfrac{\partial z}{\partial y}$, given

(a) $F(x, y, z) = x^2 + 3xy - 2y^2 + 3xz + z^2 = 0$ (b) $z = e^x \cos(y + z)$.

Solution. (a) $F_x = \dfrac{\partial F}{\partial x} = 2x + 3y + 3z$, $F_y = \dfrac{\partial F}{\partial y} = 3x - 4y$,

and $\qquad F_z = \dfrac{\partial F}{\partial z} = 3x + 2z$.

$$\frac{\partial z}{\partial x} = -\frac{F_x}{F_z} = -\frac{2x + 3y + 3z}{3x + 2z} \text{ and } \frac{\partial z}{\partial y} = -\frac{F_y}{F_z} = -\frac{3x - 4y}{3x + 2z}.$$

(b) Set $F(x, y, z) = z - e^x \cos(y + z)$; then

$$\frac{\partial F}{\partial x} = -e^x \cos(y + z), \quad \frac{\partial F}{\partial y} = e^x \sin(y + z),$$

and $\qquad \dfrac{\partial F}{\partial z} = 1 + e^x \sin(y + z)$.

Therefore, $\dfrac{\partial z}{\partial x} = -\dfrac{\partial F/\partial x}{\partial F/\partial z} = \dfrac{e^x \cos(y + z)}{1 + e^x \sin(y + z)} = \dfrac{z}{1 + e^x \sin(y + z)}$,

and $\qquad \dfrac{\partial z}{\partial y} = -\dfrac{F_y}{F_z} = -\dfrac{e^x \sin(y + z)}{1 + e^x \sin(y + z)}$.

Example 11.64 By the equation $x + 2y + z - 2\sqrt{xyz} = 10$, z is defined as an implicit function of x and y. Find the partial derivatives of this function.

Solution. Set $F(x, y, z) = x + 2y + z - 2\sqrt{xyz} - 10 = 0$.

$$\frac{\partial F}{\partial x} = 1 - \frac{yz}{\sqrt{xyz}} = \frac{\sqrt{xyz} - yz}{\sqrt{xyz}}; \quad \frac{\partial F}{\partial y} = 2 - \frac{zx}{\sqrt{xyz}} = \frac{2\sqrt{xyz} - zx}{\sqrt{xyz}},$$

$$\frac{\partial F}{\partial z} = 1 - \frac{xy}{\sqrt{xyz}} = \frac{\sqrt{xyz} - xy}{\sqrt{xyz}}.$$

Therefore, $\dfrac{\partial z}{\partial x} = -\dfrac{\partial F/\partial x}{\partial F/\partial z} = \dfrac{yz - \sqrt{xyz}}{\sqrt{xyz} - xy}$, $\dfrac{\partial z}{\partial y} = -\dfrac{\partial F/\partial y}{\partial F/\partial z} = \dfrac{zx - 2\sqrt{xyz}}{\sqrt{xyz} - xy}$

Example 11.65 If u and v are defined as functions of x and y by the equations $2u - v = - x^2 - xy$ and $u + 2v = y^2 - xy$, find

(a) $\partial u/\partial x$, (b) $\partial v/\partial x$, (c) $\partial u/\partial y$, (d) $\partial v/\partial y$.

Solution. Differentiate the given equations with respect to x, considering u and v as functions of x and y. Then

$$2\frac{\partial u}{\partial x} - \frac{\partial v}{\partial x} = -2x - y, \ ...(i) \qquad \frac{\partial u}{\partial x} + 2\frac{\partial v}{\partial x} = -y. \quad ...(ii)$$

Solving (i) and (ii) simultaneously for du/dx and dv/dx.

$$\frac{\partial u}{\partial x} = \frac{\begin{vmatrix} -2x-y & -1 \\ -y & 2 \end{vmatrix}}{\begin{vmatrix} 2 & -1 \\ 1 & 2 \end{vmatrix}} = \frac{1}{5}(4x+3y), \frac{\partial v}{\partial x} = \frac{\begin{vmatrix} 2 & -2x-y \\ 1 & -y \end{vmatrix}}{\begin{vmatrix} 2 & -1 \\ 1 & 2 \end{vmatrix}} = \frac{1}{5}(2x-y).$$

Differentiating with respect to y, we have

$$2\frac{\partial u}{\partial y} - \frac{\partial v}{\partial y} = -x, \quad ...(iii) \qquad \frac{\partial u}{\partial y} + 2\frac{\partial v}{\partial y} = 2y - x.$$

Solving, $\dfrac{\partial u}{\partial y} = \dfrac{\begin{vmatrix} -x & -1 \\ 2y-x & 2 \end{vmatrix}}{\begin{vmatrix} 2 & -1 \\ 1 & 2 \end{vmatrix}} = \dfrac{1}{5}(2y-3x), \dfrac{\partial v}{\partial y} = \dfrac{\begin{vmatrix} 2 & -x \\ 1 & 2y-x \end{vmatrix}}{\begin{vmatrix} 2 & -1 \\ 1 & 2 \end{vmatrix}} = \dfrac{1}{5}(4y-x).$

Exercises 11.4

1. Find the total differential of the following functions

   (i) $f(x, y, z) = x^2z - yz^3 + x^4$, (ii) $u = (x^2 + y^2 + z^2)^{-1/2}$,
   (iii) $w = xyz + (xyz)^{-1}$, (iv) $z = \log_e(\cos y/x)$, (v) $z = x^{\log y}$.

   Ans. (i) $df = (2xz + 4x^3)dx - z^3dy + (x^2 - 3yz^2)\,dz$.
   (ii) $du = -(x^2 + y^2 + z^2)^{-3/2}(x\,dx + y\,dy + z\,dz)$,
   (iii) $dw = [1 - (xyz)^{-2}]\,[yz\,dx + zx\,dy + xy\,dz]$,

   (iv) $dz = \dfrac{1}{x^2}\tan\dfrac{y}{x}(y\,dx - x\,dy)$, (v) $dz = \dfrac{x^{(\log y - 1)}}{y}(y \log y\,dx + x \log x\,dy)$.

2. (a) If $f(x, y) = e^{xy^2}$, find the total differential of the function at the point (1, 2). (I.E.T.E. W-2005) **Ans.** $e^4(dx + dy)$.
   (b) If $f(x, y) = \tan^{-1}(x/y)$, find the total differential of the function at the point (1, 1). (I.E.T.E. S-2005) **Ans.** $(dx - dy)/2$.

3. Find the second order total differentials of the following functions

   (i) $z = xy(x + y)$, (ii) $u = xyz$, (AMIE, W 1996) (iii) $z = x + y + \log(xy)$.

   Ans. (i) $d^2z = 2y(dx)^2 + 2x(dy)^2 + 4(x + y)\, dxdy$,

   (ii) $d^2u = 2(z\, dx\, dy + xdydz + ydzdx)$, (iii) $d^2z = -\dfrac{1}{x^2}(dx)^2 - \dfrac{1}{y^2}(dy)^2$.

4. Compute Δu and du for the function $u = 2x^2 + 3y^2$, when $x = 10$, $y = 8$, $\Delta x = 0.2$, $\Delta y = 0.3$, and compare the results.

   Ans. $\Delta u = 22.75$, $du = 22.4$, $\Delta u - du = 1.6\%$ of Δu.

5. Determine whether each of the following are exact differentials of a function and if so, find the function

   (a) $(2xy + y^2)\, dx + (x^2 + 2\, xy)\, dy$,

   (b) $(2xy^2 + 3y\cos 3x)\, dx + (2x^2y + \sin 3x)\, dy$,

   (c) $(5xy - y^2)\, dx + (3\, xe^y - x^2)\, dy$,

   Ans. (a) $x^2y + xy^2 + c$, (b) $x^2y^2 + y\sin 3x + c$, (c) not exact.

6. (i) If $z = u^2 + v^2$ and $u = at^2$ and $v = 2at$, find dz/dt. **Ans.** $4a^2t(t^2 + 2)$

   (ii) Find dz/dt, when $z = xy^2 + yx^2$ and $x = at^2$, $y = 2\,at$. Verify by direct substitution. **Ans.** $a^3(16\, t^3 + 10\, t^4)$

   (iii) If $u = x^4y^5$, where $x = t^2$ and $y = t^3$, find du/dt. **Ans.** $23\, t^{22}$.

   (iv) If $u = x^2y^3$, $x = \log t$, $y = e^t$, find du/dt. **Ans.** $2\log t \cdot \dfrac{e^{3t}}{t} + 3(\log t)^2 e^{3t}$

7. Find du/dx of the following functions

   (i) $u = \sin(x^2 + y^2)$, where $a^2x^2 + b^2y^2 = c^2$.

   Ans. $\dfrac{2x}{b^2}(b^2 - a^2)\cos(x^2 + y^2)$.

   (ii) $u = y^2 + yz + z^2$, where $z = \sin x$, $y = e^x$.

   Ans. $e^x(2e^x + \cos x) + \sin x(e^x + 2\cos x)$.

8. If $u = x^2 - y^2 + \sin yz$, where $y = e^x$ and $z = \log x$, find du/dx. Verify by direct substitution.(*UPTU 2005*) **Ans.** $2(x - e^{2x}) + e^x\cos(e^x\log x)[\log x + x^{-1}]$.

9. Find dw/dt when

   (i) $w = x^2 + y^2 + z^2$, $x = e^t\cos t$, $y = e^t\sin t$, $z = e^t$.

   (ii) $w = e^x\sin yz$, $x = t^2$, $y = t - 1$, $z = 1/t$.

   (iii) $w = e^{x+y}\cos 2z$, $x = \log t$, $y = \log(t^2 + 1)$, $z = t$.

   (iv) $w = \dfrac{x}{z} + \dfrac{y}{z}$, $x = \cos^2 t$, $y = \sin^2 t$, $z = 1/t; t = 3$.

   Ans. (i) $4\, e^{2t}$, (ii) $e^{t^2}\left(2t\sin\dfrac{t-1}{t} + \dfrac{1}{t^2}\cos\dfrac{t-1}{t}\right)$,

   (iii) $(3t^2 + 1)\cos 2t - 2(t^3 + t)\sin 2t$. (iv) $dw/dt = 1$; dw/dt at $t = 3$ is 1.

10. The length, width and height of a rectangular box are increasing at rates of 1 unit/sec, 2 unit/sec and 3 units/sec respectively.
    (a) At what rate is the volume increasing when the length is 2 units, the width is 3 units and the height is 6 units?
    (b) At what rate is the length of the diagonal increasing at that instant?
    (c) At what rate is the surface area of the box increasing at the given instant?
    Ans. (a) 60 cubic units/sec (b) 26/7 unit per sec.

11. A point is moving on the curve of intersection of the surface $x^2 + xy + y^2 - z^2 = 0$ and the plane $x - y + 2 = 0$. When x is 3 and is increasing 2 units per second, find (a) the rate at which y is changing, (b) the rate at which z is changing, (c) the speed with which the point is moving.
    Ans. (a) 2 units per second; (b) 24/7 units per sec; (c) 4.44 units per second.

12. At a certain instant the radius of a right circular cylinder is 6 units and is increasing at the rate 0.2 unit/sec, while the altitude is 8 units and is decreasing at the rate of 0.4 unit/sec. Find the time rate of change (a) of the volume and (b) of the surface at that instant.
    Ans. (a) $4.8\,\pi$ unit^3/sec; (b) $3.2\,\pi$ units^2/sec

13. Find the total differential coefficient of $u = x^2 y$ with respect to x, when x and y are connected by the relation $x^2 + xy + y^2 = 1$. *(D.C.E. 2004)*

$$\textbf{Ans.}\quad \frac{du}{dx} = 2xy - \frac{x^2(2x+y)}{x+2y}.$$

14. (a) If $u = f(r, s)$, $r = x + y$, $s = x - y$, show that $\dfrac{\partial u}{\partial x} + \dfrac{\partial u}{\partial y} = 2\dfrac{\partial u}{\partial r}$.

    (b) If $f(u) = \sin u$ and $u = \sqrt{x^2 + y^2}$, show that $\left(\dfrac{\partial f}{\partial x}\right)^2 + \left(\dfrac{\partial f}{\partial y}\right)^2 = \cos^2 u$.
    (A.M.I.E., S-2001)

15. (i) Given $V = f(x, y, z)$ where $x = r \cos \theta$, $y = r \sin \theta$, $z = t$; obtain expressions for $\dfrac{\partial V}{\partial r}, \dfrac{\partial V}{\partial \theta}, \dfrac{\partial V}{\partial t}$ in terms of $\dfrac{\partial V}{\partial x}, \dfrac{\partial V}{\partial y}, \dfrac{\partial V}{\partial z}$.

    (ii) If $u = f(r, s, t)$, and $r = x/y$, $s = y/z$, $t = z/x$, then show that

$$x\frac{\partial u}{\partial x} + y\frac{\partial u}{\partial y} + z\frac{\partial u}{\partial z} = 0. \qquad \textit{(A.M.I.E., S-1999)}$$

    (iii) If $z = x^3 - xy + y^3$, and $x = r \cos \theta$, $y = r \sin \theta$, find $\dfrac{\partial z}{\partial r}$ and $\dfrac{\partial z}{\partial \theta}$.

    Ans. $(3x^2 - y)\cos \theta + (3y^2 - x)\sin \theta$; $(3x^2 - y)(-r \sin\theta) + (3y^2 - x)r\cos \theta$.

[**Hint.** (*i*) $\dfrac{\partial x}{\partial r} = \cos\theta,\ \dfrac{\partial x}{\partial \theta} = -r\sin\theta,\ \dfrac{\partial x}{\partial t} = 0,\ \dfrac{\partial y}{\partial r} = \sin\theta,\ \dfrac{\partial y}{\partial \theta} = r\cos\theta,\ \dfrac{\partial y}{\partial t} = 0$

$\dfrac{\partial z}{\partial r} = \dfrac{\partial z}{\partial \theta} = 0,\ \dfrac{\partial z}{\partial t} = 1.$

$\dfrac{\partial V}{\partial r} = \dfrac{\partial V}{\partial x}\dfrac{\partial x}{\partial r} + \dfrac{\partial V}{\partial y}\dfrac{\partial y}{\partial r} + \dfrac{\partial V}{\partial z}\dfrac{\partial z}{\partial r} = \cos\theta\dfrac{\partial V}{\partial x} + \sin\theta\dfrac{\partial V}{\partial y}$

$\dfrac{\partial V}{\partial \theta} = \dfrac{\partial V}{\partial x}\dfrac{\partial x}{\partial \theta} + \dfrac{\partial V}{\partial y}\dfrac{\partial y}{\partial \theta} + \dfrac{\partial V}{\partial z}\dfrac{\partial z}{\partial \theta} = -r\sin\theta\dfrac{\partial V}{\partial x} + r\cos\theta\dfrac{\partial V}{\partial y}$

$\dfrac{\partial V}{\partial t} = \dfrac{\partial V}{\partial x}\dfrac{\partial x}{\partial t} + \dfrac{\partial V}{\partial y}\dfrac{\partial y}{\partial t} + \dfrac{\partial V}{\partial z}\cdot\dfrac{\partial z}{\partial t} = \dfrac{\partial V}{\partial z}$].

16. If $u = f\left[\dfrac{y-x}{xy},\ \dfrac{z-x}{xz}\right]$, show that $x^2\dfrac{\partial u}{\partial x} + y^2\dfrac{\partial u}{\partial y} + z^2\dfrac{\partial u}{\partial z} = 0.$

(DCE 2004; UPTU 2004; I.E.T.E., June 1996)

17. If $x^2 = au + bv,\ y^2 = au - bv$ and V is a function of $x,\ y$, then show that

$$x\dfrac{\partial V}{\partial x} + y\dfrac{\partial V}{\partial y} = 2\left(u\dfrac{\partial V}{\partial u} + v\dfrac{\partial V}{\partial v}\right).$$

18. If $\phi = f(x,\ y,\ z)$ and $x = \sqrt{vw},\ y = \sqrt{wu},\ z = \sqrt{uv}$, then show that

$$u\dfrac{\partial \phi}{\partial u} + v\dfrac{\partial \phi}{\partial v} + w\dfrac{\partial \phi}{\partial w} = x\dfrac{\partial \phi}{\partial x} + y\dfrac{\partial \phi}{\partial y} + z\dfrac{\partial \phi}{\partial z}.$$

19. If $u = \dfrac{x^2 - y^2}{x^2 + y^2}$ and $z = \sin^{-1}\sqrt{u}$, find $dz,\ \dfrac{\partial z}{\partial x}$ and $\dfrac{\partial z}{\partial y}$.

Ans. $\sqrt{2}\cdot x(y\,dx - x\,dy)\big/\left(x^2 + y^2\right)\left(\sqrt{(x^2 - y^2)}\right);$

$\sqrt{2}\,xy\big/\left(x^2 + y^2\right)\sqrt{(x^2 - y^2)};\ -\sqrt{2}\,x^2\big/\left(x^2 + y^2\right)\sqrt{(x^2 - y^2)}.$

20. If $x = u + v + w,\ y = vw + wu + uv,\ z = uvw$ and F is a function for $x,\ y,\ z$, then show that

$$u\dfrac{\partial F}{\partial u} + v\dfrac{\partial F}{\partial v} + w\dfrac{\partial F}{\partial w} = x\dfrac{\partial F}{\partial x} + 2y\dfrac{\partial F}{\partial y} + 3z\dfrac{\partial F}{\partial z}.$$

21. Given $z = xy^2 + x^3y$, where $x = r^2 - 2s,\ y = 2r - s^2$, find $\partial z/\partial r$ and $\partial z/\partial s$.

Ans. $2\left\{\left(y^2 + 3x^2y\right)r + \left(2xy + x^3\right)\right\};\ -2\left\{\left(y^2 + 3x^2y\right) + s\left(2xy + x^3\right)\right\}$

22. If $u = x^2 + y^2$, $x = s + 3t$ and $y = 2s-t$, obtain the values of $\partial^2 u/\partial s^2$ and $\partial^2 u/\partial t^2$
    (A.M.I.E., W-2003) **Ans.** 10, 20.

23. Find $\partial w/\partial r$ when $w = (x+y+z)^2$, $x = r-s$, $y = \cos(r+s)$,

    $z = \sin(r+s)$ at $r = 1, s = -1$.   **Ans.** 12.

24. If $f(x, y) = \phi(u, v)$ and $u = x^2 - y^2$ and $v = 2xy$ then show that

$$\frac{\partial^2 f}{\partial x^2} + \frac{\partial^2 f}{\partial y^2} = 4(x^2 + y^2)\left(\frac{\partial^2 \phi}{\partial u^2} + \frac{\partial^2 \phi}{\partial v^2}\right).$$

25. If $z = f(x, y)$, $x = e^u \cos v$, $y = e^u \sin v$, show that

$$\frac{\partial^2 z}{\partial x^2} + \frac{\partial^2 z}{\partial y^2} = e^{-2u}\left(\frac{\partial^2 z}{\partial u^2} + \frac{\partial^2 z}{\partial v^2}\right).$$

26. If $u = f(x^2 + 2yz, y^2 + 2zx)$, prove that

$$(y^2 - zx)\frac{\partial u}{\partial x} + (x^2 - yz)\frac{\partial u}{\partial y} + (z^2 - xy)\frac{\partial u}{\partial z} = 0.$$

27. (a) If $\phi(x + y + z, x^2 + y^2 + z^2) = 0$, show that
    $(y - z) p + (z - x) q = x - y$, where $p = \partial z/\partial x$ and $q = \partial z/\partial y$.

28. If $\phi\left(\dfrac{z}{x^3}, \dfrac{y}{x}\right) = 0$, show that $px + qy = 3z$.

29. Find p and q if

$$x = \sqrt{a}\,(\sin u + \cos v), \quad y = \sqrt{a}\,(\cos u - \sin v), \quad z = -1 + \sin(u - v)$$

    where p and q mean $\dfrac{\partial z}{\partial x}$ and $\dfrac{\partial z}{\partial y}$ respectively.   (A.W.I.E., W-1998)

[Hint. $\dfrac{x^2}{a} = \sin^2 u + \cos^2 v + 2\sin u \cos v$,   ...(i)

                  $\dfrac{y^2}{a} = \cos^2 u + \sin^2 v - 2\cos u \sin v$.   ...(ii)

Adding (i) and (ii), we obtain

$$\left(\frac{x^2}{a}\right) + \left(\frac{y^2}{a}\right) = 2 + 2\sin(u-v) = 2 + 2(1+z) \qquad [\because\ z = -1 + \sin(u-v)]$$

or $z = \dfrac{1}{2a}(x^2 + y^2) - 2$. Therefore, $\dfrac{\partial z}{\partial x} = \dfrac{x}{a}$ and $\dfrac{\partial z}{\partial y} = \dfrac{y}{a}$].

30. Using implicit differentiation, obtain the following
   (*i*) dy/dx, when $x^y + y^x = a^b$, a and b are constant,
   (*ii*) dy/dx, when $x^3 + 3\,x^2y + 6\,xy^2 + y^3 = 1$,
   (*iii*) dy/dx, when $(\cos x)^y = (\sin y)^x$,
   (*iv*) dy/dx, when $\tan^{-1}(x/y) + y^3 + 1 = 0$; $x > 0$, $y > 0$.   (*A.M.I.E., W-2002*)

Ans. (*i*) $-\dfrac{\left(yx^{y-1} + y^x \log y\right)}{\left(xy^{x-1} + x^y \log x\right)}$,

(*ii*) $-\dfrac{\left(x^2 + 2xy + 2y^2\right)}{x^2 + 4xy + y^2}$,

(*iii*) $\dfrac{y \tan x + \log \sin y}{\log \cos x - x \cot y}$,

(*iv*) $\dfrac{y}{\left(x - 3x^2y^2 - 3y^4\right)}$.

31. If $f(x, y) = 0$ and $\phi(y, z) = 0$ then show that

$$\frac{\partial f}{\partial y} \cdot \frac{\partial \phi}{\partial z} \cdot \frac{dz}{dx} = \frac{\partial f}{\partial x} \cdot \frac{\partial \phi}{\partial y}$$

(*A.M.I.E., W-2005; UPTU 2005*)

32. Find $\partial w/\partial v$, when $u = 0$, $v = 0$ if $w = (x^2 + y - 2)^4 + (x - y + 2)^3$,
   $x = u - 2v + 1$ and $y = 2u + v - 2$.   **Ans. 99.**

[**Hint:** Using the chain rule: $\dfrac{\partial w}{\partial v} = \dfrac{\partial w}{\partial x} \cdot \dfrac{\partial x}{\partial v} + \dfrac{\partial w}{\partial y} \cdot \dfrac{\partial y}{\partial v}$

When $u = v = 0$, then $x = 1$, $y = -2$. Find $\left(\dfrac{\partial w}{\partial v}\right)_{\substack{x=1 \\ y=-2}}$].

33. If $u = f(x, y)$, $x = r \cos \theta$, $y = r \sin \theta$, then show that

$$\left(\frac{\partial u}{\partial x}\right)^2 + \left(\frac{\partial u}{\partial y}\right)^2 = \left(\frac{\partial u}{\partial r}\right)^2 + \frac{1}{r^2}\left(\frac{\partial u}{\partial \theta}\right)^2.$$

(*IETE, June 2008; V.T.U., 2000; A.M.I.E., W-1997*)

34. If z be a function of x and y, and u and v be two other variables such that
$u = lx + my$, $v = ly - mx$,

show that $\dfrac{\partial^2 z}{\partial x^2} + \dfrac{\partial^2 z}{\partial y^2} = \left(l^2 + m^2\right)\left(\dfrac{\partial^2 z}{\partial u^2} + \dfrac{\partial^2 z}{\partial v^2}\right).$ (*PTU, 2004*)

[**Hint.** Using the chain rule, we obtain $\dfrac{\partial z}{\partial x} = \dfrac{\partial z}{\partial u} \cdot \dfrac{\partial u}{\partial x} + \dfrac{\partial z}{\partial v} \cdot \dfrac{\partial v}{\partial x} = l\dfrac{\partial z}{\partial u} - m\dfrac{\partial z}{\partial v}$

$\Rightarrow \dfrac{\partial}{\partial x} = l\dfrac{\partial}{\partial u} - m\dfrac{\partial}{\partial v}$. Similarly, $\dfrac{\partial}{\partial y} = m\dfrac{\partial}{\partial u} + l\dfrac{\partial}{\partial v}$.

$\therefore \quad \dfrac{\partial^2 z}{\partial x^2} = \dfrac{\partial}{\partial x}\left(\dfrac{\partial z}{\partial x}\right) = \left(l\dfrac{\partial}{\partial u} - m\dfrac{\partial}{\partial v}\right)\left(l\dfrac{\partial z}{\partial u} - m\dfrac{\partial z}{\partial v}\right)$

$$= l^2 \frac{\partial^2 z}{\partial u^2} - 2lm \frac{\partial^2 z}{\partial u \partial v} + m^2 \frac{\partial^2 z}{\partial v^2}. \qquad ...(i)$$

Similarly, $\qquad \dfrac{\partial^2 z}{\partial y^2} = m^2 \dfrac{\partial^2 z}{\partial u^2} + 2lm \dfrac{\partial^2 z}{\partial u \partial v} + l^2 \dfrac{\partial^2 z}{\partial v^2}. \qquad ...(ii)$

Adding (i) and (ii), we get the required result.]

35. (i) If $x = u + v$, $y = uv$ and z be any function of x and y, show that

$$u \frac{\partial z}{\partial u} + v \frac{\partial z}{\partial v} = x \frac{\partial z}{\partial x} + 2y \frac{\partial z}{\partial y}.$$

(ii) If $z = uv$ and $u^2 + v^2 - x - y = 0$, $u^2 - v^2 + 3x + y = 0$. Find $\dfrac{\partial z}{\partial x}$.

[Hint. (ii) Solving the last two equations for u^2 and v^2, we get
$u^2 = -x$, $v^2 = 2x + y$.

Now $\qquad \dfrac{\partial z}{\partial x} = u \dfrac{\partial v}{\partial x} + v \dfrac{\partial u}{\partial x} = u \left(\dfrac{1}{v} \right) + v \left(-\dfrac{1}{2u} \right) = \dfrac{2u^2 - v^2}{2uv}$

$$\left(\because \frac{\partial u}{\partial x} = -\frac{1}{2u} \text{ and } \frac{\partial v}{\partial x} = \frac{1}{v} \right).\Bigg]$$

36. If u and v are defined as functions of x and y by the equations

$u^2 - v = 3x + y$ and $u - 2v^2 = x - 2y$, find $\dfrac{\partial u}{\partial x}, \dfrac{\partial v}{\partial x}, \dfrac{\partial u}{\partial y}$ and $\dfrac{\partial v}{\partial y}$.

Ans. $\dfrac{1-12v}{1-8uv}, \dfrac{2u-3}{1-8uv}, -\dfrac{(2+4v)}{1-8uv}, -\dfrac{1+4u}{1-8uv}$ assumed that $1-8uv \neq 0$.

37. Suppose that u and v are two functions of x and y and satisfy the relations

$u^2 - v^2 = x$, $2uv = y$. Then compute $\dfrac{\partial u}{\partial x}, \dfrac{\partial u}{\partial y}, \dfrac{\partial v}{\partial x}$ and $\dfrac{\partial v}{\partial y}$, assuming that

they exist. (*AMIE.; S-2008*)

[Hint. $\dfrac{2u\partial u}{\partial x} - 2v \dfrac{\partial v}{\partial x} = 1$, $2v \dfrac{\partial u}{\partial x} + 2u \dfrac{\partial v}{\partial x} = 0$. Solving simultaneously for

$\partial u/\partial x$ and $\partial v/\partial x$. We have

$$\frac{\partial u}{\partial x} = \frac{\begin{vmatrix} 1 & -2v \\ 0 & 2u \end{vmatrix}}{\begin{vmatrix} 2u & -2v \\ 2v & 2u \end{vmatrix}} = \frac{u}{2(u^2 + v^2)},$$

$$\frac{\partial v}{\partial x} = \frac{\begin{vmatrix} 2u & 1 \\ 2u & 0 \\ 2u & -2v \\ 2v & 2u \end{vmatrix}}{} = -\frac{v}{2(u^2 + v^2)}.$$ Similarly, the other partial derivatives

are obtained. **Ans.** $\dfrac{u}{2(u^2 + v^2)}, \dfrac{v}{2(u^2 + v^2)}, -\dfrac{v}{2(u^2 + v^2)}, \dfrac{u}{2(u^2 + v^2)}.$

38. If $z^3 - xz - y = 0$, prove that $\dfrac{\partial^2 z}{\partial x \partial y} = -\dfrac{x + 3z^2}{(3z^2 - x)^3}.$

39. If $u = f(x, y)$ and $x = \xi \cos \alpha - \eta \sin \alpha$, $y = \xi \sin \alpha + \eta \cos \alpha$, where α is a constant, show that

$$\frac{\partial^2 u}{\partial x^2} + \frac{\partial^2 u}{\partial y^2} = \frac{\partial^2 u}{\partial \xi^2} + \frac{\partial^2 u}{\partial \eta^2}.$$

(A.M.I.E. S-2000)

[**Hint:** Using the chain rule, we obtain

$$\frac{\partial u}{\partial \xi} = \frac{\partial u}{\partial x} \cdot \frac{\partial x}{\partial \xi} + \frac{\partial u}{\partial y} \cdot \frac{\partial y}{\partial \xi} = \frac{\partial u}{\partial x} \cos \alpha + \frac{\partial u}{\partial y} \sin \alpha$$

$$= \left(\cos \alpha \frac{\partial}{\partial x} + \sin \alpha \frac{\partial}{\partial y} \right) u \Rightarrow \frac{\partial}{\partial \xi} = \left(\cos \alpha \frac{\partial}{\partial x} + \sin \alpha \frac{\partial}{\partial y} \right).$$

$$\therefore \quad \frac{\partial^2 u}{\partial \xi^2} = \left(\cos \alpha \frac{\partial}{\partial x} + \sin \alpha \frac{\partial}{\partial y} \right)^2 u$$

$$= \cos^2 \alpha \frac{\partial^2 u}{\partial x^2} + 2 \sin \alpha \cos \alpha \frac{\partial^2 u}{\partial x \partial y} + \sin^2 \alpha \frac{\partial^2 u}{\partial y^2}. \qquad ...(i)$$

Similarly $\dfrac{\partial^2 u}{\partial \eta^2} = \sin^2 \alpha \dfrac{\partial^2 u}{\partial x^2} - 2 \sin \alpha \cos \alpha \dfrac{\partial^2 u}{\partial x \partial y} + \cos^2 \alpha \dfrac{\partial^2 u}{\partial y^2}. \qquad ...(ii)$

Adding (*i*) and (*ii*), we obtain

$$\frac{\partial^2 u}{\partial \xi^2} + \frac{\partial^2 u}{\partial \eta^2} = \frac{\partial^2 u}{\partial x^2} + \frac{\partial^2 u}{\partial y^2}.$$

40. If $x + y = (u + v)^2$, $x - y = (u - v)^2$, show that

$$4(x^2 - y^2) \left(\frac{\partial^2 f}{\partial x^2} - \frac{\partial^2 f}{\partial y^2} \right) = (u^2 - v^2) \left(\frac{\partial^2 f}{\partial u^2} - \frac{\partial^2 f}{\partial v^2} \right).$$

41. If $x = r \cos \theta$ and $y = r \sin \theta$, then for any function $u = f(x, y)$, prove that

$$\frac{\partial u}{\partial x} = \cos\theta \frac{\partial u}{\partial r} - \sin\frac{\theta}{r}\frac{\partial u}{\partial \theta}.$$ *(A.M.I.E., S-2002; I.E.T.E. June 1998)*

[**Hint.** We have $r = \sqrt{x^2 + y^2}$ and $\theta = \tan^{-1}(y/x)$. Using the chain rule,

$$\frac{\partial u}{\partial x} = \frac{\partial u}{\partial r}\cdot\frac{\partial r}{\partial x} + \frac{\partial u}{\partial \theta}\cdot\frac{\partial \theta}{\partial x} = \frac{x}{\sqrt{x^2+y^2}}\frac{\partial u}{\partial r} + \left(\frac{-y}{x^2+y^2}\right)\frac{\partial u}{\partial \theta}$$

$$= \cos\theta \frac{\partial u}{\partial r} - \frac{\sin\theta}{r}\frac{\partial u}{\partial \theta}].$$

42. If $u = x + ay$ and $v = x + by$ transform the equation

$$2\frac{\partial^2 z}{\partial x^2} - 5\frac{\partial^2 z}{\partial x \partial y} + 3\frac{\partial^2 z}{\partial y^2} = 0 \text{ into } \frac{\partial^2 z}{\partial u \partial v} = 0,$$

and find the possible values of 'a' and b. *(AMIE, W-2004)*

[**Hint:** We have $\frac{\partial u}{\partial x} = 1, \frac{\partial u}{\partial y} = a, \frac{\partial v}{\partial x} = 1$ and $\frac{\partial v}{\partial y} = b$. Using the chain rule, we

obtain $\quad \frac{\partial z}{\partial x} = \frac{\partial z}{\partial u}\cdot\frac{\partial u}{\partial x} + \frac{\partial z}{\partial v}\cdot\frac{\partial v}{\partial x} = \frac{\partial z}{\partial u} + \frac{\partial z}{\partial v} \Rightarrow \frac{\partial}{\partial x} = \frac{\partial}{\partial u} + \frac{\partial}{\partial v}.$

Similarly $\qquad\qquad \frac{\partial}{\partial y} = a\frac{\partial}{\partial u} + b\frac{\partial}{\partial v}.$

$$\therefore \quad \frac{\partial^2 z}{\partial x^2} = \frac{\partial}{\partial x}\left(\frac{\partial z}{\partial x}\right) = \left(\frac{\partial}{\partial u} + \frac{\partial}{\partial v}\right)\left(\frac{\partial z}{\partial u} + \frac{\partial z}{\partial v}\right) = \frac{\partial^2 z}{\partial u^2} + 2\frac{\partial^2 z}{\partial u \partial v} + \frac{\partial^2 z}{\partial v^2},$$

$$\frac{\partial^2 z}{\partial y^2} = a^2\frac{\partial^2 z}{\partial u^2} + 2ab\frac{\partial^2 z}{\partial u \partial v} + b^2\frac{\partial^2 z}{\partial v^2},$$

and $\qquad \frac{\partial^2 z}{\partial x \partial y} = \frac{\partial}{\partial x}\left(\frac{\partial z}{\partial y}\right) = a\frac{\partial^2 z}{\partial u^2} + (a+b)\frac{\partial^2 z}{\partial u \partial v} + b\frac{\partial^2 z}{\partial v^2}.$

Thus $\qquad 2\frac{\partial^2 z}{\partial x^2} - 5\frac{\partial^2 z}{\partial x \partial y} + 3\frac{\partial^2 z}{\partial y^2} = \frac{\partial^2 z}{\partial u^2}(2 - 5a + 3a^2)$

$$+ \frac{\partial^2 z}{\partial u \partial v}(4 - 5a - 5b + 6ab) + \frac{\partial^2 z}{\partial v^2}(2 - 5b + 3b^2).$$

It is given that left hand expression be zero, and only term $\dfrac{\partial^2 z}{\partial u \, \partial v}$ on right

hand expression should exist so coefficients of $\dfrac{\partial^2 z}{\partial u^2}$ and $\dfrac{\partial^2 z}{\partial v^2}$ must be equal

to 0. Therefore $2 - 5a + 3a^2 = 0$ and $2 - 5b + 3b^2 = 0$.

On solving, we get the solution sets (a, b) either $(1, 2/3)$ or $(2/3, 1)$.]

43. (a) If $z = f(x, y)$, where $x = e^u \cos v$ and $y = e^u \sin v$, show that

$$y\frac{\partial z}{\partial u} + x\frac{\partial z}{\partial v} = e^{2u}\frac{\partial z}{\partial y}.$$

(b) If $u = u(x, y)$ and $x = e^r \cos \theta$, $y = e^r \sin \theta$, show that

$$\left(\frac{\partial u}{\partial x}\right)^2 + \left(\frac{\partial u}{\partial y}\right)^2 = e^{-2r}\left[\left(\frac{\partial u}{\partial r}\right)^2 + \left(\frac{\partial u}{\partial \theta}\right)^2\right].$$

44. If $z = f(u, v)$ where $u = x^2 - y^2$ and $v = 2xy$, with $(x, y) \neq (0, 0)$. Show

that the differential equation $x\dfrac{\partial z}{\partial x} - y\dfrac{\partial z}{\partial y} = 0$ is equivalent to $\partial z / \partial u = 0$.

(A.M.I.E., S-2007)

11.23 Jacobians

Def. If $u = f(x, y)$ and $v = \phi(x, y)$ be two continuous functions of the independent variables x and y such that $\partial u/\partial x$, $\partial u/\partial y$, $\partial v/\partial x$, $\partial v/\partial y$ are also continuous in x and y, then the *Jacobian* of u and v with respect to x and y is defined by the

determinant $\begin{vmatrix} \partial u/\partial x & \partial u/\partial y \\ \partial v/\partial x & \partial v/\partial y \end{vmatrix}$.

The above determinant is often denoted by the symbol $\dfrac{\partial(u, v)}{\partial(x, y)}$ or $J\left(\dfrac{u, v}{x, y}\right)$.

Similarly if u, v, w are functions of three independent variables x, y, z then the determinant

$$\begin{vmatrix} \partial u/\partial x & \partial u/\partial y & \partial u/\partial z \\ \partial v/\partial x & \partial v/\partial y & \partial v/\partial z \\ \partial w/\partial x & \partial w/\partial y & \partial w/\partial z \end{vmatrix}$$

is called the Jacobian of u, v, w with respect to x, y, z.. This determinant is denoted by the symbol

$$\frac{\partial(u, v, w)}{\partial(x, y, z)} \text{ or } J\left(\frac{u, v, w}{x, y, z}\right).$$

In general if $x_1, x_2, ..., x_n$ be n functions of the n variables $y_1, y_2 ... y_n$ then

$$\frac{\partial(x_1, x_2, ..., x_n)}{\partial(y_1, y_2, ..., y_n)} = \begin{vmatrix} \partial x_1/\partial y_1 & \partial x_1/\partial y_2 & \partial x_1/\partial y_n \\ \partial x_2/\partial y_1 & \partial x_2/\partial y_2 & \partial x_2/\partial y_n \\ ... & ... & ... \\ \partial x_n/\partial y_1 & \partial x_n/\partial y_2 & \partial x_n/\partial y_n \end{vmatrix}$$

is the Jacobian of $x_1, x_2, ..., x_n$ with respect to $y_1, y_2, ..., y_n$.

An important application of Jacobians is in connection with the change of variables in multiple integrals.

11.24 Properties of Jacobians

1. *If J_1 is the Jacobian of u, v with respect to x, y and J_2 the Jacobian of x, y with respect to u, v then $J_1 J_2 = 1$.* (IETE, June-2005; A.M.I.E., W-2003)

Proof.

$$J_1 J_2 = \begin{vmatrix} \partial u/\partial x & \partial u/\partial y \\ \partial v/\partial x & \partial v/\partial y \end{vmatrix} \times \begin{vmatrix} \partial x/\partial u & \partial x/\partial v \\ \partial y/\partial u & \partial y/\partial v \end{vmatrix} = \begin{vmatrix} \partial u/\partial x & \partial u/\partial y \\ \partial v/\partial x & \partial v/\partial y \end{vmatrix} \times \begin{vmatrix} \partial x/\partial u & \partial y/\partial u \\ \partial x/\partial v & \partial y/\partial v \end{vmatrix}$$

(interchanging rows and columns of the second determinant)

$$= \begin{vmatrix} \dfrac{\partial u}{\partial x} \cdot \dfrac{\partial x}{\partial u} + \dfrac{\partial u}{\partial y} \cdot \dfrac{\partial y}{\partial u} & \dfrac{\partial u}{\partial x} \cdot \dfrac{\partial x}{\partial v} + \dfrac{\partial u}{\partial y} \cdot \dfrac{\partial y}{\partial v} \\ \dfrac{\partial v}{\partial x} \cdot \dfrac{\partial x}{\partial u} + \dfrac{\partial v}{\partial y} \cdot \dfrac{\partial y}{\partial u} & \dfrac{\partial v}{\partial x} \cdot \dfrac{\partial x}{\partial v} + \dfrac{\partial v}{\partial y} \cdot \dfrac{\partial y}{\partial v} \end{vmatrix} \qquad ...(11.35)$$

Let $u = f(x, y)$ and $v = g(x, y)$. Suppose, on solving for x and y, we get $x = \phi(u, v)$ and $y = \psi(u, v)$. Then

$$\left. \begin{array}{ll} \dfrac{\partial u}{\partial u} = 1 = \dfrac{\partial u}{\partial x} \cdot \dfrac{\partial x}{\partial u} + \dfrac{\partial u}{\partial y} \cdot \dfrac{\partial y}{\partial u} & \dfrac{\partial u}{\partial v} = 0 = \dfrac{\partial u}{\partial x} \cdot \dfrac{\partial x}{\partial v} + \dfrac{\partial u}{\partial y} \cdot \dfrac{\partial y}{\partial v} \\[2mm] \dfrac{\partial v}{\partial u} = 0 = \dfrac{\partial v}{\partial x} \cdot \dfrac{\partial x}{\partial u} + \dfrac{\partial v}{\partial y} \cdot \dfrac{\partial y}{\partial u} & \dfrac{\partial v}{\partial v} = 1 = \dfrac{\partial v}{\partial x} \cdot \dfrac{\partial x}{\partial v} + \dfrac{\partial v}{\partial y} \cdot \dfrac{\partial y}{\partial v} \end{array} \right\} \qquad ...(11.36)$$

With the help of Eq. (11.36) we can write Eq. (11.35) in the form

$$\begin{vmatrix} 1 & 0 \\ 0 & 1 \end{vmatrix} = 1. \quad \therefore \quad J_1 J_2 = 1 \text{ that is, } \frac{\partial(u, v)}{\partial(x, y)} \cdot \frac{\partial(x, y)}{\partial(u, v)} = 1.$$

Note. The result can be easily extended.

Thus
$$\frac{\partial(u, v, w, ...)}{\partial(x, y, z, ...)} \cdot \frac{\partial(x, y, z, ...)}{\partial(u, v, w, ...)} = 1.$$

2. *If u, v are functions of r, s where r, s are functions of x, y, then*

$$\frac{\partial(u, v)}{\partial(x, y)} = \frac{\partial(u, v)}{\partial(r, s)} \times \frac{\partial(r, s)}{\partial(x, y)}.$$

Proof. R.H.S. $= \dfrac{\partial(u, v)}{\partial(r, s)} \times \dfrac{\partial(r, s)}{\partial(x, y)} = \begin{vmatrix} \partial u/\partial r & \partial u/\partial s \\ \partial v/\partial r & \partial v/\partial s \end{vmatrix} \times \begin{vmatrix} \partial r/\partial x & \partial r/\partial y \\ \partial s/\partial x & \partial s/\partial y \end{vmatrix}$

$$= \begin{vmatrix} \partial u/\partial r & \partial u/\partial s \\ \partial v/\partial r & \partial v/\partial s \end{vmatrix} \times \begin{vmatrix} \partial r/\partial x & \partial s/\partial x \\ \partial r/\partial y & \partial s/\partial y \end{vmatrix}$$

(interchanging rows and columns in second determinant)

$$= \begin{vmatrix} \dfrac{\partial u}{\partial r} \cdot \dfrac{\partial r}{\partial x} + \dfrac{\partial u}{\partial s} \cdot \dfrac{\partial s}{\partial x} & \dfrac{\partial u}{\partial r} \cdot \dfrac{\partial r}{\partial y} + \dfrac{\partial u}{\partial s} \cdot \dfrac{\partial s}{\partial y} \\ \dfrac{\partial v}{\partial r} \cdot \dfrac{\partial r}{\partial x} + \dfrac{\partial v}{\partial s} \cdot \dfrac{\partial s}{\partial x} & \dfrac{\partial v}{\partial r} \cdot \dfrac{\partial r}{\partial y} + \dfrac{\partial v}{\partial s} \cdot \dfrac{\partial s}{\partial y} \end{vmatrix} = \begin{vmatrix} \partial u/\partial x & \partial u/\partial y \\ \partial v/\partial x & \partial v/\partial y \end{vmatrix} = \frac{\partial(u, v)}{\partial(x, y)} = \text{L.H.S.}$$

3. *If functions u, v, w of three independent variables x, y, z are not independent, then the jacobian of u, v, w with respect to x, y, z vanishes, i.e.*

$$\frac{\partial(u, v, w)}{\partial(x, y, z)} = 0.$$

Proof. In case u, v, w are not independent, then there must be a relation $f(u, v, w) = 0$ connecting them.

Differentiating $f(u, v, w) = 0$ with respect to x, y, z respectively, we get

$$\left. \begin{aligned} \frac{\partial f}{\partial u} \frac{\partial u}{\partial x} + \frac{\partial f}{\partial v} \frac{\partial v}{\partial x} + \frac{\partial f}{\partial w} \frac{\partial w}{\partial x} &= 0, \\ \frac{\partial f}{\partial u} \frac{\partial u}{\partial y} + \frac{\partial f}{\partial v} \frac{\partial v}{\partial y} + \frac{\partial f}{\partial w} \frac{\partial w}{\partial y} &= 0, \\ \frac{\partial f}{\partial u} \frac{\partial u}{\partial z} + \frac{\partial f}{\partial v} \frac{\partial v}{\partial z} + \frac{\partial f}{\partial w} \frac{\partial w}{\partial z} &= 0. \end{aligned} \right\}$$

Eliminating $\dfrac{\partial f}{\partial u}, \dfrac{\partial f}{\partial v}, \dfrac{\partial f}{\partial w}$ from the above equations, we have

$$\begin{vmatrix} \partial u/\partial x & \partial v/\partial x & \partial w/\partial x \\ \partial u/\partial y & \partial v/\partial y & \partial w/\partial y \\ \partial u/\partial z & \partial v/\partial z & \partial w/\partial z \end{vmatrix} = 0 \quad \text{or} \quad \frac{\partial(u, v, w)}{\partial(x, y, z)} = 0.$$

11.25 Illustrative Examples

Example 11.66 If x = r cos θ, y = r sin θ, find the Jacobians $J = \dfrac{\partial(x,y)}{\partial(r,\theta)}$ and

$J' = \dfrac{\partial(r,\theta)}{\partial(x,y)}$. Show that JJ' = 1.

(IETE, S-2008; A.M.I.E., W-2002; Andhra 2000)

Solution. We have $x = r\cos\theta, y = r\sin\theta$.

$$\frac{\partial x}{\partial r} = \cos\theta, \ \frac{\partial x}{\partial \theta} = -r\sin\theta, \ \frac{\partial y}{\partial r} = \sin\theta, \ \frac{\partial y}{\partial \theta} = r\cos\theta.$$

$$\therefore \ J = \frac{\partial(x,y)}{\partial(r,\theta)} = \begin{vmatrix} \partial x/\partial r & \partial x/\partial \theta \\ \partial y/\partial r & \partial y/\partial \theta \end{vmatrix} = \begin{vmatrix} \cos\theta & -r\sin\theta \\ \sin\theta & r\cos\theta \end{vmatrix} = r\left(\cos^2\theta + \sin^2\theta\right) = r.$$

Now
$$r^2 = x^2 + y^2, \ \theta = \tan^{-1}(y/x).$$

$$\frac{\partial r}{\partial x} = \frac{x}{r}, \ \frac{\partial r}{\partial y} = \frac{y}{r}, \ \frac{\partial \theta}{\partial x} = \frac{-y}{x^2+y^2} = \frac{-y}{r^2}, \ \frac{\partial \theta}{\partial y} = \frac{x}{x^2+y^2} = \frac{x}{r^2}.$$

$$\therefore J' = \frac{\partial(r,\theta)}{\partial(x,y)} = \begin{vmatrix} \partial r/\partial x & \partial r/\partial y \\ \partial \theta/\partial x & \partial \theta/\partial y \end{vmatrix} = \begin{vmatrix} x/r & y/r \\ -y/r^2 & x/r^2 \end{vmatrix} = \frac{x^2}{r^3} + \frac{y^2}{r^3} = \frac{r^2}{r^2} = \frac{1}{r}.$$

Hence $JJ' = r(1/r) = 1.$

Example 11.67 If u = x² – 2y, v = x + y, prove that

$$\frac{\partial(u,v)}{\partial(x,y)} = 2x + 2.$$

Solution. We have $\dfrac{\partial u}{\partial x} = 2x, \ \dfrac{\partial u}{\partial y} = -2, \ \dfrac{\partial v}{\partial x} = 1, \ \dfrac{\partial v}{\partial y} = 1.$

$$\therefore \quad \frac{\partial(u,v)}{\partial(x,y)} = \begin{vmatrix} \dfrac{\partial u}{\partial x} & \dfrac{\partial u}{\partial y} \\ \dfrac{\partial v}{\partial x} & \dfrac{\partial v}{\partial y} \end{vmatrix} = \begin{vmatrix} 2x & -2 \\ 1 & 1 \end{vmatrix} = 2x + 2.$$

Example 11.68 If in cylindrical co-ordinates x = r cos θ, y = r sin θ, z = z, find

$$\frac{\partial(x,y,z)}{\partial(r,\theta,z)}.$$

Solution. We have $\dfrac{\partial x}{\partial r} = \cos\theta$, $\dfrac{\partial x}{\partial \theta} = -r\sin\theta$, $\dfrac{\partial x}{\partial z} = 0$;

$$\dfrac{\partial y}{\partial r} = \sin\theta,\ \dfrac{\partial y}{\partial \theta} = r\cos\theta,\ \dfrac{\partial y}{\partial z} = 0\ ;\ \dfrac{\partial z}{\partial r} = 0,\ \dfrac{\partial z}{\partial \theta} = 0,\ \dfrac{\partial z}{\partial z} = 1.$$

$$\therefore \dfrac{\partial(x,y,z)}{\partial(r,\theta,z)} = \begin{vmatrix} \cos\theta & -r\sin\theta & 0 \\ \sin\theta & r\cos\theta & 0 \\ 0 & 0 & 1 \end{vmatrix} = 1\left(r\cos^2\theta + r\sin^2\theta\right) = r.$$

Example 11.69 If $F = xu + v - y$, $G = u^2 + vy + w$, $H = zu - v + vw$.

Compute $\dfrac{\partial(F,G,H)}{\partial(u,w,v)}$. *(Kuvempu, 1996)*

Solution. $\dfrac{\partial(F,G,H)}{\partial(u,w,v)} = \begin{vmatrix} \partial F/\partial u & \partial F/\partial w & \partial F/\partial v \\ \partial G/\partial u & \partial G/\partial w & \partial G/\partial v \\ \partial H/\partial u & \partial H/\partial w & \partial H/\partial v \end{vmatrix} = \begin{vmatrix} x & 0 & 1 \\ 2u & 1 & y \\ z & v & -1+w \end{vmatrix}$

$$= x(w-1-vy) + 1(2uv-z) = xw - x - xyv + 2uv - z.$$

Example 11.70 In spherical polar coordinates,

$$x = \rho \sin\phi \cos\theta,\ y = \rho\sin\phi\sin\theta,\ \text{and}\ z = \rho\cos\phi,$$

show that $J\left(\dfrac{x,y,z}{\rho,\phi,\theta}\right) = \rho^2\sin\phi.$ *(A.M.I.E. S-2004, Gauhati, 1999)*

Solution. We have $\dfrac{\partial x}{\partial \rho} = \sin\phi\cos\theta$, $\dfrac{\partial x}{\partial \phi} = \rho\cos\phi\cos\theta$, $\dfrac{\partial x}{\partial \theta} = -\rho\sin\phi\sin\theta$,

$$\dfrac{\partial y}{\partial \rho} = \sin\phi\sin\theta,\ \dfrac{\partial y}{\partial \phi} = \rho\cos\phi\sin\theta,\ \dfrac{\partial y}{\partial \theta} = \rho\sin\phi\cos\theta,$$

$$\dfrac{\partial z}{\partial \rho} = \cos\phi,\ \dfrac{\partial z}{\partial \phi} = -\rho\sin\phi,\ \dfrac{\partial z}{\partial \theta} = 0.$$

$$\therefore\ J\left(\dfrac{x,y,z}{\rho,\phi,\theta}\right) = \dfrac{\partial(x,y,z)}{\partial(\rho,\phi,\theta)} = \begin{vmatrix} \partial x/\partial \rho & \partial x/\partial \phi & \partial x/\partial \theta \\ \partial y/\partial \rho & \partial y/\partial \phi & \partial y/\partial \theta \\ \partial z/\partial \rho & \partial z/\partial \phi & \partial z/\partial \theta \end{vmatrix}$$

$$= \begin{vmatrix} \sin\phi\cos\theta & \rho\cos\phi\cos\theta & -\rho\sin\phi\sin\theta \\ \sin\phi\sin\theta & \rho\cos\phi\sin\theta & \rho\sin\phi\cos\theta \\ \cos\phi & -\rho\sin\phi & 0 \end{vmatrix}$$

$$= \frac{\rho^2}{\cos\theta} \begin{vmatrix} \sin\phi\cos^2\theta & \cos\phi\cos^2\theta & -\sin\phi\sin\theta\cos\theta \\ \sin\phi\sin\theta & \cos\phi\sin\theta & \sin\phi\cos\theta \\ \cos\phi & -\sin\phi & 0 \end{vmatrix} \begin{matrix} (R_1 \rightarrow R_1 + R_2 \sin\theta) \\ \\ \\ \end{matrix}$$

$$= \frac{\rho^2}{\cos\theta} \begin{vmatrix} \sin\phi & \cos\phi & 0 \\ \sin\phi\sin\theta & \cos\phi\sin\theta & \sin\phi\cos\theta \\ \cos\phi & -\sin\phi & 0 \end{vmatrix}$$

$$= \frac{\rho^2}{\cos\theta}(-\sin\phi\cos\theta)(-\sin^2\phi - \cos^2\phi) = \rho^2\sin\phi.$$

Example 11.71 If x = a cosh ξ cos η, y = a sinh ξ sin η, show that

$$\frac{\partial(x,y)}{\partial(\xi,\eta)} = \frac{a^2}{2}(\cosh 2\xi - \cos 2\eta).$$ *(Ranchi, 1998)*

Solution. $\dfrac{\partial(x,y)}{\partial(\xi,\eta)} = \begin{vmatrix} \partial x/\partial\xi & \partial x/\partial\eta \\ \partial y/\partial\xi & \partial y/\partial\eta \end{vmatrix} = \begin{vmatrix} a\sinh\xi\cos\eta & -a\cosh\xi\sin\eta \\ a\cosh\xi\sin\eta & a\sinh\xi\cos\eta \end{vmatrix}$

$$= a^2 [\sinh^2\xi\cos^2\eta + \cosh^2\xi\sin^2\eta]$$
$$= a^2[(\cosh^2\xi - 1)\cos^2\eta + \cosh^2\xi(1 - \cos^2\eta)] = a^2[\cosh^2\xi - \cos^2\eta]$$

$$= a^2\left[\frac{1 + \cosh 2\xi}{2} - \frac{1 + \cos 2\eta}{2}\right] = \frac{a^2}{2}(\cosh 2\xi - \cos 2\eta).$$

Example 11.72 If $y_1 = x_2 x_3/x_1$, $y_2 = x_3 x_1/x_2$, $y_3 = x_1 x_2/x_3$. Show that the Jacobian of y_1, y_2, y_3 with respect to x_1, x_2, x_3 is 4.

(A.M.I.E., W-2001, J.N.J.U., 1998)

Solution. $\dfrac{\partial(y_1, y_2, y_3)}{\partial(x_1, x_2, x_3)} = \begin{vmatrix} \partial y_1/\partial x_1 & \partial y_1/\partial x_2 & \partial y_1/\partial x_3 \\ \partial y_2/\partial x_1 & \partial y_2/\partial x_2 & \partial y_2/\partial x_3 \\ \partial y_3/\partial x_1 & \partial y_3/\partial x_2 & \partial y_3/\partial x_3 \end{vmatrix}$

$$= \begin{vmatrix} \dfrac{-x_2 x_3}{x_1^2} & \dfrac{x_3}{x_1} & \dfrac{x_2}{x_1} \\[2mm] \dfrac{x_3}{x_2} & \dfrac{-x_3 x_1}{x_2^2} & \dfrac{x_1}{x_2} \\[2mm] \dfrac{x_2}{x_3} & \dfrac{x_1}{x_3} & \dfrac{-x_1 x_2}{x_3^2} \end{vmatrix} = \frac{1}{x_1^2 x_2^2 x_3^2} \begin{vmatrix} -x_2 x_3 & x_3 x_1 & x_2 x_1 \\ x_3 x_2 & -x_3 x_1 & x_1 x_2 \\ x_2 x_3 & x_3 x_1 & -x_1 x_2 \end{vmatrix}$$

$$= \frac{x_1^2 x_2^2 x_3^2}{x_1^2 x_2^2 x_3^2} \begin{vmatrix} -1 & 1 & 1 \\ 1 & -1 & 1 \\ 1 & 1 & -1 \end{vmatrix} \begin{matrix} \\ (R_2 \leftarrow R_2 + R_1) \\ (R_3 \leftarrow R_3 + R_1) \end{matrix} = \begin{vmatrix} -1 & 1 & 1 \\ 0 & 0 & 2 \\ 0 & 2 & 0 \end{vmatrix} = -1(0-4) = 4.$$

Example 11.73 If $u = 2xy$, $v = x^2 - y^2$, $x = r \cos\theta$ and $y = r \sin\theta$,

evaluate $\dfrac{\partial(u,v)}{\partial(r,\theta)}$.

Solution. We know that

$$\frac{\partial(u,v)}{\partial(r,\theta)} = \frac{\partial(u,v)}{\partial(x,y)} \cdot \frac{\partial(x,y)}{\partial(r,\theta)} \quad \text{[Art. 11.24, Property 2]}$$

$$= \begin{vmatrix} \dfrac{\partial u}{\partial x} & \dfrac{\partial u}{\partial y} \\ \dfrac{\partial v}{\partial x} & \dfrac{\partial v}{\partial y} \end{vmatrix} \begin{vmatrix} \dfrac{\partial x}{\partial r} & \dfrac{\partial x}{\partial \theta} \\ \dfrac{\partial y}{\partial r} & \dfrac{\partial y}{\partial \theta} \end{vmatrix} = \begin{vmatrix} 2y & 2x \\ 2x & -2y \end{vmatrix} \begin{vmatrix} \cos\theta & -r\sin\theta \\ \sin\theta & r\cos\theta \end{vmatrix}$$

$$= (-4y^2 - 4x^2)(r\cos^2\theta + r\sin^2\theta) = -4(x^2 + y^2)\, r = -4r^3.$$

Example 11.74 If $u = xyz$, $v = xy + yz + zx$ and, $w = x + y + z$.

Compute $\dfrac{\partial(u,v,w)}{\partial(x,y,z)}$.

Solution. $J\left(\dfrac{u,v,w}{x,y,z}\right) = \begin{vmatrix} \partial u/\partial x & \partial u/\partial y & \partial u/\partial z \\ \partial v/\partial x & \partial v/\partial y & \partial v/\partial z \\ \partial w/\partial x & \partial w/\partial y & \partial w/\partial z \end{vmatrix}$

$$= \begin{vmatrix} yz & xz & xy \\ y+z & x+z & y+x \\ 1 & 1 & 1 \end{vmatrix} \begin{matrix} C_1 \to C_1 - C_3 \\ C_2 \to C_2 - C_3 \end{matrix} = \begin{vmatrix} yz - xy & xz - xy & xy \\ z - x & z - y & x+y \\ 0 & 0 & 1 \end{vmatrix}$$

$$= \begin{vmatrix} y(z-x) & x(z-y) & xy \\ z-x & z-y & x+y \\ 0 & 0 & 1 \end{vmatrix} = (z-x)(z-y)\begin{vmatrix} y & x & xy \\ 1 & 1 & x+y \\ 0 & 0 & 1 \end{vmatrix}$$

$$= (z-x)(z-y)(y-x) = (x-y)(y-z)(z-x).$$

Example 11.75 (a) **For the transformation** $x = e^u \cos v$, $y = e^u \sin v$, **prove that**

$$\frac{\partial(x,y)}{\partial(u,v)} \cdot \frac{\partial(u,v)}{\partial(x,y)} = 1.$$

(b) **For the transformations x = a(u + v), y = b(u − v)**

and u = r² cos 2θ, v = r² sin 2θ, find $\dfrac{\partial(x, y)}{\partial(r, \theta)}$.

Solution. (a) The inverse transformation for $x = e^u \cos v$, $y = e^u \sin v$,

are $u = \dfrac{1}{2} \log(x^2 + y^2)$, $v = \tan^{-1} \dfrac{y}{x}$.

$$\therefore \quad \frac{\partial(x, y)}{\partial(u, v)} = \begin{vmatrix} \partial x/\partial u & \partial x/\partial v \\ \partial y/\partial u & \partial y/\partial v \end{vmatrix} = \begin{vmatrix} e^u \cos v & -e^u \sin v \\ e^u \sin v & e^u \cos v \end{vmatrix} = e^{2u}. \qquad ...(i)$$

$$\frac{\partial(u, v)}{\partial(x, y)} = \begin{vmatrix} \partial u/\partial x & \partial u/\partial y \\ \partial v/\partial x & \partial v/\partial y \end{vmatrix} = \begin{vmatrix} \dfrac{x}{x^2 + y^2} & \dfrac{y}{x^2 + y^2} \\ \dfrac{-y}{x^2 + y^2} & \dfrac{x}{x^2 + y^2} \end{vmatrix} = \frac{1}{x^2 + y^2} = e^{-2u}. \qquad ...(ii)$$

Thus, from (i) and (ii), we have

$$\frac{\partial(x, y)}{\partial(u, v)} \cdot \frac{\partial(u, v)}{\partial(x, y)} = e^{2u} \cdot e^{-2u} = e^{2u-2u} = e^0 = 1.$$

(b) $\dfrac{\partial(x, y)}{\partial(u, v)} = \begin{vmatrix} a & a \\ b & -b \end{vmatrix} = -2ab$, $\qquad ...(iii)$

and $\dfrac{\partial(u, v)}{\partial(r, \theta)} = \begin{vmatrix} 2r \cos 2\theta & -2r^2 \sin 2\theta \\ 2r \sin 2\theta & 2r^2 \cos 2\theta \end{vmatrix} = 4r^3$. $\qquad ...(iv)$

∴ Using the Property 2, we get

$$\frac{\partial(x, y)}{\partial(r, \theta)} = \frac{\partial(x, y)}{\partial(u, v)} \cdot \frac{\partial(u, v)}{\partial(r, \theta)} = (-2ab) \cdot (4r^3) = -8abr^3.$$

Exercise 11.5

1. Given $x = u^2 - v^2$, $y = 2uv$, calculate $J\left(\dfrac{x, y}{u, v}\right)$. **Ans.** $4(u^2 + v^2)$.

2. If $x = r \sin^2 \theta$, $y = r \cos^2 \theta$, find $J\left(\dfrac{x, y}{r, \theta}\right)$. **Ans.** $2r \sin \theta \cos \theta$.

3. Find $\dfrac{\partial(u, y)}{\partial(x, y)}$, if $u = x - y^{2/3}$ and $v = x + y^2$ **Ans.** $2y + \left(2/(3y^{1/3})\right)$.

4. If $x = \cos u$, $y = \cos u \sin u$, $z = \cos w \sin v \sin u$, show that

$$\frac{\partial(x,y,z)}{\partial(u,v,w)} = -\sin^3 u \sin^2 v \sin w.$$

5. If $x = u(1 - v)$, $y = uv$, evaluate $J = \dfrac{\partial(x,y)}{\partial(u,v)}$ and $J' = \dfrac{\partial(u,v)}{\partial(x,y)}$

and hence verify the result $JJ' = 1$.

(A.W.I.E., W-2004; V.T.U. 2000S; Andhra 1998)

[**Hint.** $x = u - uv = u - y \Rightarrow u = x + y$ and $v = y/u = y/(x + y)$, $J = u$, $J' = 1/(x + y) = 1/u$. $\therefore$ $JJ' = 1$].

6. (a) If $u = x + y + z$, $uv = y + z$, $uvw = z$ then show that

$$\frac{\partial(x,y,z)}{\partial(u,v,w)} = u^2v.$$

(A.M.I.E., W-2003, V.T.U., 2000)

(b) If $u = yz/x$, $v = zx/y$ and $w = xy/z$. Show that $\dfrac{\partial(u,v,w)}{\partial(x,y,z)} = 4$.

(A.M.I.E. W-2001; Mysore, 1997; Kottayam, 1996S)

[**Hint.** (a) $x = u - uv$, $y = uv - uvw$, $z = uvw$.

$$\therefore \quad \frac{\partial(x,y,z)}{\partial(u,v,w)} = \begin{vmatrix} 1-v & -u & 0 \\ v-vw & u-uw & -uv \\ vw & uw & uv \end{vmatrix} \quad (\text{operate } R_2 \leftarrow R_2 + R_3)$$

$$= \begin{vmatrix} 1-v & -u & 0 \\ v & u & 0 \\ vw & uw & uv \end{vmatrix} = u^2v].$$

7. If $u = x(1 - r^2)^{-1/2}$, $v = y(1 - r^2)^{-1/2}$, $w = z(1 - r^2)^{-1/2}$, where $r^2 = x^2 + y^2 + z^2$, then show that

$$\frac{\partial(u,v,w)}{\partial(x,y,z)} = (1 - r^2)^{-5/2}$$

8. If $x = u^2 - v^2$, $y = 2uv$, find $\dfrac{\partial(u,v)}{\partial(x,y)}$. **Ans.** $1/4(u^2 + v^2)$.

Remark. While solving the following problems, we should remember that

if $\dfrac{\partial(u,v)}{\partial(x,y)} = 0$ then there exists a functional relationship between u and v.

9. Show that the functions

$f(x, y) = x^2 + xy + y^2 = 0$; $\phi(x, y) = x^4 + 2x^3y + 3x^2y^2 + 2xy^3 + y^4 = 0$ are functionally dependent.

[**Hint.** Show that $\begin{vmatrix} f_x & f_y \\ \phi_x & \phi_y \end{vmatrix} = 0$].

10. If $u = \dfrac{x+y}{x-y}$ and $v = \dfrac{xy}{(x-y)^2}$, find $\dfrac{\partial(u,v)}{\partial(x,y)}$. Are u and v functionally

    related? If so, find the relationship. **Ans.** $u^2 = 1 + 4v$.

11. If $u = y + z$, $v = x + 2z^2$, $w = x - 4yz - 2y^2$, find $\dfrac{\partial(u,v,w)}{\partial(x,y,z)}$. Are u, v, w

    functionally related? If so find relationship. **Ans.** $w = v - u^2$.

12. If $u = xy + yz + zx$, $v = x^2 + y^2 + z^2$ and $w = x + y + z$, determine whether
    there is a functional relationship between u, u, w and if so, find it.
    Ans. $w^2 = v + 2u$.

13. If $u = \dfrac{x+y}{1-xy}$ and $v = \tan^{-1} x + \tan^{-1} y$. Evaluate $\dfrac{\partial(u,v)}{\partial(x,y)}$ when $xy \neq 1$. State

    whether u and v are functionally related. If so find the relationship between
    them. *(AMIE, S-2005, S-2003)*

    [**Hint.** $\dfrac{\partial(u,v)}{\partial(x,y)} = \begin{vmatrix} \dfrac{1+y^2}{(1-xy)^2} & \dfrac{1+x^2}{(1-xy)^2} \\ \dfrac{1}{1+x^2} & \dfrac{1}{1+y^2} \end{vmatrix} = 0.$ $(xy \neq 1)$.

    Therefore, u, v are functionally related.

    $\tan v = \tan\left(\tan^{-1} x + \tan^{-1} y\right) = \dfrac{x+y}{1-xy} = u.$]

14. If $u = \sin^{-1} x + \sin^{-1} y$ and $v = x\sqrt{1-y^2} + y\sqrt{1-x^2}$, determine whether
    there is a functional relationship between u and v, if so find it.
    Ans. $\sin u = v$. *(A.M.I.E. W-2001)*

    [**Hint.** We find $\dfrac{\partial(u,v)}{\partial(x,y)} = 0.$ Since Jacobian u, v with respect to x, y

    vanishes, therefore u and v are functionally related.
    We have $u = \sin^{-1} x + \sin^{-1} y$. $\therefore$ $\sin u = \sin(\sin^{-1}x + \sin^{-1}y)$
    $= \sin(\sin^{-1}x) \cos(\sin^{-1}y) + \cos(\sin^{-1}x) \sin(\sin^{-1}y)$

$$= x\sqrt{1-\sin^2\left(\sin^{-1}y\right)} + \sqrt{1-\sin^2\left(\sin^{-1}x\right)} \cdot y$$

$$= x\sqrt{1-\left\{\sin\left(\sin^{-1}y\right)\right\}^2} + \sqrt{1-\left\{\sin\left(\sin^{-1}x\right)\right\}^2} \cdot y = x\sqrt{1-y^2} + y\sqrt{1-x^2} = v.$$

$\therefore v = \sin u$ is the required relationship between u and v].

15. Let $u = x+y-z$, $v = x-y+z$, $w = x^2+y^2+z^2+ayz$. Find the value of 'a' if u, v, w are functionally dependent. Hence find the relation.

Ans. $a = 2$; $u^2 + v^2 = 2w$.

Chapter 12

Applications of Partial Derivatives

12.1 Geometrical Interpretation of partial derivatives of a function of two variables

The partial derivatives have simple geometric interpretations. Consider the surface S represented by $z = f(x, y)$ in Fig. 12.1. Through the point $P(x, y, z)$, there is a curve APB that is, the intersection with the surface of the plane through P parallel to the xz plane. Similarly CPD is the curve through P i.e., the intersection with the surface of the plane through P parallel to the yz plane. As x varies while y is held fixed, P moves along the curve APB and the value of $\partial z/\partial x$ at (x, y) is the slope of the tangent line to the curve APB at P. Similarly, as y varies while x is held fixed, P moves along the curve CPD, and the value of $\partial z/\partial y$ at (x, y) is the slope of the tangent line to the curve CPD at P.

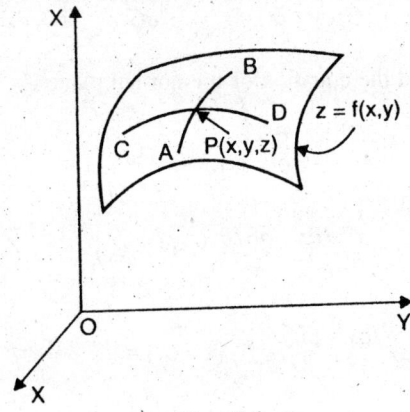

Fig. 12.1.

12.2 Applications to Geometry

12.2.1 Tangent Plane and Normal Line to a Surface

Let the equation of the given surface be $F(x, y, z) = 0$, and let $P_0(x_0, y_0, z_0)$ be the point on the given surface then direction ratios of a normal to the tangent plane at any point $P_0(x_0, y_0, z_0)$ on the surface are given by $(\partial F/\partial x)_{P_0}$, $(\partial F/\partial y)_{P_0}$, $(\partial F/\partial z)_{P_0}$.

The equation of the tangent plane to the surface $F(x, y, z) = 0$ at one of its points $P_0(x_0, y_0, z_0)$ is

$$\left.\frac{\partial F}{\partial x}\right|_{P_0} (x - x_0) + \left.\frac{\partial F}{\partial y}\right|_{P_0} (y - y_0) + \left.\frac{\partial F}{\partial z}\right|_{P_0} (z - z_0) = 0, \qquad ...(12.1)$$

and the equations of the normal line that is, perpendicular to the tangent plane at the point $P_0(x_0, y_0\ z_0)$ are

$$\frac{x - x_0}{\left.\partial F/\partial x\right|_{P_0}} = \frac{y - y_0}{\left.\partial F/\partial y\right|_{P_0}} = \frac{z - z_0}{\left.\partial F/\partial z\right|_{P_0}} \qquad ...(12.2)$$

with the understanding that the partial derivatives have been evaluated at the point P_0.

Setting each of these ratios equal to a parameter (such as t or u) and solving for x, y, and z yields the *parametric equations* of the normal line.

12.2.2 Tangent Line and Normal Plane to a Space Curve

A space curve may also be defined by the pair of equations.

$$F(x, y, z) = 0, \quad G(x, y, z) = 0.$$

At the point $P_0(x_0, y_0, z_0)$ of the curve, the equations of the tangent line are

$$\frac{x - x_0}{\left.\begin{vmatrix} \partial F/\partial y & \partial F/\partial z \\ \partial G/\partial y & \partial G/\partial z \end{vmatrix}\right|_{P_0}} = \frac{y - y_0}{\left.\begin{vmatrix} \partial F/\partial z & \partial F/\partial x \\ \partial G/\partial z & \partial G/\partial x \end{vmatrix}\right|_{P_0}} = \frac{z - z_0}{\left.\begin{vmatrix} \partial F/\partial x & \partial F/\partial y \\ \partial G/\partial x & \partial G/\partial y \end{vmatrix}\right|_{P_0}} \qquad ...(12.3)$$

and the equation of the normal plane is

$$\left.\begin{vmatrix} \partial F/\partial y & \partial F/\partial z \\ \partial G/\partial y & \partial G/\partial z \end{vmatrix}\right|_{P_0} (x - x_0)$$

$$+ \left.\begin{vmatrix} \partial F/\partial z & \partial F/\partial x \\ \partial G/\partial z & \partial G/\partial x \end{vmatrix}\right|_{P_0} (y - y_0)$$

$$+ \left.\begin{vmatrix} \partial F/\partial x & \partial F/\partial y \\ \partial G/\partial x & \partial G/\partial y \end{vmatrix}\right|_{P_0} (z - z_0) = 0 \quad ...(12.4)$$

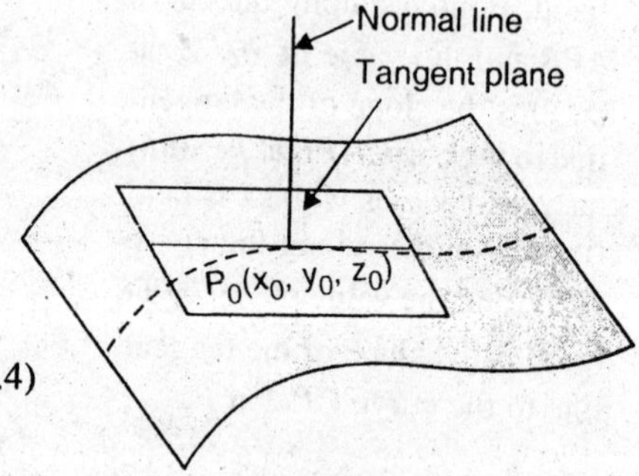

Normal line
Tangent plane
$P_0(x_0, y_0, z_0)$

Fig. 12.2.

Note: All partial derivatives are to be evaluated at the point P_0.

Example 12.1 Find the equations of the tangent plane and normal line to the surface $2x^2 + y^2 = 3 - 2z$ at the point $(2, 1, -3)$.

Solution. Here $F(x, y, z) = 2x^2 + y^2 + 2z - 3 = 0$

and $\dfrac{\partial F}{\partial x} = 4x, \dfrac{\partial F}{\partial y} = 2y, \dfrac{\partial F}{\partial z} = 2$

that is, [4x, 2y, 2] are direction ratios of the normal at (x, y, z).

At the point (2, 1, –3), $\dfrac{\partial F}{\partial x} = 8, \dfrac{\partial F}{\partial y} = 2, \dfrac{\partial F}{\partial z} = 2$.

Thus at (2, 1, –3), direction ratios of the normal are (8, 2, 2) or (4, 1, 1). Hence the equation of the tangent plane at (2, 1, –3) is

$4(x - 2) + (y - 1) + (z + 3) = 0$ or $4x + y + z = 6$.

The normal at (2, 1, –3) is a line through (2, 1, –3) perpendicular to the above plane. The equations of the normal line at (2, 1, –3) are thus

$$\frac{x-2}{4} = \frac{y-1}{1} = \frac{z+3}{1}.$$

Remark. By setting each of these ratios equal to the parameter t, we have $x = 2 + 4t, y = 1 + t, z = t - 3$ called the *parametric equations* for the line.

Example 12.2 Show that the surfaces

$F(x, y, z) = x^2 + 4y^2 - 4z^2 - 4 = 0$ and $G(x, y, z) = x^2 + y^2 + z^2 - 6x - 6y + 2z + 10 = 0$ are tangent at the point (2, 1, 1).

Solution. We have to prove that the two surfaces have the same tangent plane at the given point.

At (2, 1, 1), $\dfrac{\partial F}{\partial x} = 2x = 4, \dfrac{\partial F}{\partial y} = 8y = 8, \dfrac{\partial F}{\partial z} = -8z = -8$

and $\dfrac{\partial G}{\partial x} = 2x - 6 = -2, \dfrac{\partial G}{\partial y} = 2y - 6 = -4, \dfrac{\partial G}{\partial z} = 2z + 2 = 4.$

Since the sets of direction numbers [4, 8, –8] and [–2, –4, 4] of the normal lines of the two surfaces are proportional, the surfaces have the common tangent plane

$1(x - 2) + 2(y - 1) - 2(z - 1) = 0$ or $x + 2y - 2z = 2$.

Example 12.3 Show that the surface $f(x, y, z) = x^2 - 2yz + y^3 - 4 = 0$ is perpendicular to any member of the family of surfaces

$F(x, y, z) = x^2 + 1 - (2 - 4a) y^2 - az^2 = 0$ at the point of intersection (1, –1, 2).

Solution. We have $f(x, y, z) = x^2 - 2yz + y^3 - 4 = 0$.

Therefore, $\dfrac{\partial f}{\partial x} = 2x, \dfrac{\partial f}{\partial y} = -2z + 3y^2, \dfrac{\partial f}{\partial z} = -2y.$

At the point (1, –1, 2) $\dfrac{\partial f}{\partial x} = 2, \dfrac{\partial f}{\partial y} = -1, \dfrac{\partial f}{\partial z} = 2.$

Also, we have $F(x, y, z) = x^2 + 1 - (2 - 4a)y^2 - az^2 = 0$.

$$\therefore \quad \frac{\partial F}{\partial x} = 2x, \frac{\partial F}{\partial y} = -2(2-4a)y, \frac{\partial F}{\partial z} = -2az.$$

At the point $(1, -1, 2)$, the values of $\partial F/\partial x$, $\partial F/\partial y$ and $\partial F/\partial z$ are respectively $2, 4 - 8a, -4a$.

Now applying $\left(\dfrac{\partial f}{\partial x}\right)\left(\dfrac{\partial F}{\partial x}\right) + \left(\dfrac{\partial f}{\partial y}\right)\left(\dfrac{\partial F}{\partial y}\right) + \left(\dfrac{\partial f}{\partial z}\right)\left(\dfrac{\partial F}{\partial z}\right) = 0,$

we obtain $\qquad (2)(2) + (-1)(4 - 8a) + (2)(-4a) = 0,$

it follows that given surfaces are perpendicular for all 'a', and so the required result follows.

Example 12.4 Find the equations of the tangent line and the normal plane to the curve $x^2 + 2y^2 + 2z^2 = 5$, $3x - 2y - z = 0$ at the point $(1, 1, 1)$.

Solution. The equations of the surfaces intersecting in the curve are

$$F(x, y, z) = x^2 + 2y^2 + 2z^2 - 5 = 0, \quad G(x, y, z) = 3x - 2y - z = 0.$$

At $(1, 1, 1)$,

$$\begin{vmatrix} \partial F/\partial y & \partial F/\partial z \\ \partial G/\partial y & \partial G/\partial z \end{vmatrix} = \begin{vmatrix} 4y & 4z \\ -2 & -1 \end{vmatrix} = \begin{vmatrix} 4 & 4 \\ -2 & -1 \end{vmatrix} = 4,$$

$$\begin{vmatrix} \partial F/\partial z & \partial F/\partial x \\ \partial G/\partial z & \partial G/\partial x \end{vmatrix} = \begin{vmatrix} 4z & 2x \\ -1 & 3 \end{vmatrix} = \begin{vmatrix} 4 & 2 \\ -1 & 3 \end{vmatrix} = 14,$$

$$\begin{vmatrix} \partial F/\partial x & \partial F/\partial y \\ \partial G/\partial x & \partial G/\partial y \end{vmatrix} = \begin{vmatrix} 2x & 4y \\ 3 & -2 \end{vmatrix} = \begin{vmatrix} 2 & 4 \\ 3 & -2 \end{vmatrix} = -16.$$

With $[2, 7, -8]$ as a set of direction numbers of the tangent, its equations are

$\dfrac{x-1}{2} = \dfrac{y-1}{7} = \dfrac{z-1}{-8}$. The equation of the normal plane is $2(x-1) + 7(y-1)$

$-8(z-1) = 2x + 7y - 8z - 1 = 0$.

Exercise 12.1

1. Find the equations of the tangent plane and normal line to the given surface at the given point:

   (a) $x^2 + 2y^2 + 3z^2 = 12$, $(1, 2, -1)$. (b) $z = xy$, $(3, -4, -12)$.
   (c) $2x^2 + 2xy + y^2 + z + 1 = 0$, $(1, -2, -3)$.
   (d) $x^2 + 3y^2 - 4z^2 + 3xy - 10yz + 4x - 5z - 22 = 0$, $(1. -2, 1)$.

   Ans. (a) $x + 4y - 3z = 12$; $x - 1 = \dfrac{(y-2)}{4} = \dfrac{(z+1)}{-3}$.

(b) $4x - 3y + z = 12$; $\dfrac{(x-3)}{4} = \dfrac{(y+4)}{-3} = z + 12$.

(c) $z - 2y = 1$; $x - 1 = 0$, $\dfrac{(y+2)}{2} = \dfrac{(z+3)}{-1}$.

(d) $19y - 7z + 45 = 0$; $x = 1$, $7y + 19z - 5 = 0$.

2. Find the equations of the tangent planes and the normal line to the following surfaces at the specified points:

 (i) $x^2 + y^2 = 4z$ at $(2, -4, 5)$.

 (ii) $z = 3x^2 + 2y^2 - 11$ at $(2, 1, 3)$.

 (iii) $x^2yz + 3y^2 = 2xz^2 - 8z$ at $(1, 2, -1)$.

 (iv) $x^2y + xz^2 = z - 1$ at $(1, -3, 2)$. (AMIE; W-2008)

Ans. (i) $x - 2y - z = 5$; $\dfrac{(x-2)}{-1} = \dfrac{(y+4)}{2} = z - 5$.

 (ii) $12x + 4y - z = 25$; $\dfrac{(x-2)}{12} = \dfrac{(y-1)}{4} = \dfrac{(z-3)}{-1}$.

 (iii) $6x - 11y - 14z + 2 = 0$; $\dfrac{(x-1)}{-6} = \dfrac{(y-2)}{11} = \dfrac{(z+1)}{14}$.

 (iv) $2x - y - 3y + 1 = 0$; $\dfrac{x-1}{-2} = y + 3 = \dfrac{z-2}{3}$.

3. Show that the surfaces $F(x, y, z) = xy + yz - 4zx = 0$ and $G(x, y, z) = 3z^2 - 5x + y = 0$ intersect at right angles at the point $(1, 2, 1)$.

4. Show that the surfaces $F(x, y, z) = 3x^2 + 4y^2 + 8z^2 - 36 = 0$ and $G(x, y, z) = x^2 + 2y^2 - 4z^2 - 6 = 0$ intersect at right angles.

5. The surfaces $x^2y^2 + 2x + z^3 = 16$ and $3x^2 + y^2 - 2z = 9$ intersect in a curve which passes through the point $(2, 1, 2)$. What are the equations of the respective tangent planes to the two surfaces at that point?

 Ans. $3x + 4y + 6z = 22$; $6x + y - z = 11$.

6. Find the condition that the plane $ax + by + cz + d = 0$ should touch the surface $px^2 + qy^2 + 2z = 0$. **Ans.** $a^2/p + b^2/q + 2cd = 0$.

7. Find equations for the (a) tangent line and (b) normal plane to the curve

$$3x^2y + y^2z = -2, \quad 2xz - x^2y = 3 \text{ at the point } (1, -1, 1).$$

 Ans. (a) $\dfrac{(x-1)}{3} = \dfrac{(y+1)}{16} = \dfrac{(z-1)}{2}$ or $x = 1 + 3t$, $y = 16t - 1$, $z = 1 + 2t$;

 (b) $3x + 16y + 2z + 11 = 0$.

8. Find the equations of the tangent line and normal plane to the curve,

$$x^2 + 2y^2 + 3z^2 = 3, \ 2x + 3y + 4z = 5 \ \text{at} \ (1, 1, 0). \qquad (AMIE., W-2007)$$

$$\textbf{Ans.} \ \frac{(x-1)}{-8} = \frac{(y-1)}{4} = z, \ 8x - 4y - z - 4 = 0.$$

12.3 Gradient of a Scalar Field and Directional Derivative

Before defining Gradient of a scalar field and directional derivative, we shall introduce the concepts of a scalar functions and vector functions.

Point Functions. A variable quantity the value of which at any point in a region of space depends on the position of the point, is called a *point function* or *function of position*.

There are two types of point functions.

(a) *Scalar point function* (b) *Vector point function*.

12.3.1 Scalar point function

If to each point P in a region of space R, there corresponds a scalar $f(P)$, then f is called a *scalar function of position* or *scalar point function* and we say that scalar field f has been defined in R. Examples are:

(i) The distance $f(P)$ of any point P from a fixed point P_0 in space is a scalar field. This is certainly independent of the choice of the coordinate system.

(ii) The temperature T at any point on the earth's surface at a certain time defines a scalar field.

(iii) The density at any point of a certain body occupying given region is a scalar field.

(iv) $\phi(x, y, z) = 3xy^2z + x^2y$ defines a scalar field.

12.3.2 Vector point function

If to each point P in a region R of space there corresponds a vector $\mathbf{v}(P)$, then $\mathbf{v}$ is called a *vector function of position* or *vector point function* and we say that a vector field $\mathbf{v}$ has been defined in R. Examples are:

(i) the velocity at each point of a fluid in motion is a vector field.

(ii) $\mathbf{v} = \mathbf{v}(P) = v_1\mathbf{i} + v_2\mathbf{j} + v_3\mathbf{k}$ defined at each point P is called a vector function.

(iii) gravitational and electromagnetic force fields etc.

If we introduce Cartesian co-ordinates x, y, z then the value of the scalar point function f at a point $P(x, y, z)$ may be written as $f(x, y, z)$ but we should bear in mind that this value is independent of particular choice of axes of co-ordinates and depends only on the position of P. In order to indicate this fact we sometimes write $f(P)$ instead $f(x, y, z)$.

Similarly at a point $P(x, y, z)$ the vector point function $\mathbf{v}$ may be written in the form:

$\mathbf{v}(x, y, z) = v_1(x, y, z)\mathbf{i} + v_2(x, y, z)\mathbf{j} + v_3(x, y, z)\mathbf{k}$ bearing in mind that $\mathbf{v}$ depends only on the position of P and defines the same vector for every choice of co-ordinate system. This fact is indicated symbolically by writing $\mathbf{v}(P)$ in place of $\mathbf{v}(x, y, z)$.

If the scalar and vector fields depend on time also, then we denote them as $f(P, t)$ and $\mathbf{v}(P, t)$ respectively. Both the fields are independent of the choice of the coodinate systems.

Differentiation of these scalar and vector point functions follow the same rules as those of ordinary calculus. Thus if $f(x, y, z)$ and $\mathbf{v}(x, y, z)$ are respectively scalar and vector point functions, then

$$df = \frac{\partial f}{\partial x}dx + \frac{\partial f}{\partial y}dy + \frac{\partial f}{\partial z}dz , \qquad \frac{df}{dt} = \frac{\partial f}{\partial x}\frac{dx}{dt} + \frac{\partial f}{\partial y}\frac{dy}{dt} + \frac{\partial f}{\partial z}\frac{dz}{dt} ,$$

and $$d\mathbf{v} = \frac{\partial \mathbf{v}}{\partial x}dx + \frac{\partial \mathbf{v}}{\partial y}dy + \frac{\partial \mathbf{v}}{\partial z}dz, \qquad \frac{d\mathbf{v}}{dt} = \frac{\partial \mathbf{v}}{\partial x}\frac{dx}{dt} + \frac{\partial \mathbf{v}}{\partial y}\frac{dy}{dt} + \frac{\partial \mathbf{v}}{\partial z}\frac{dz}{dt} .$$

Some of the vector fields can be obtained from scalar fields. The relation between the two types of fields is accomplished by the "*gradient*". To define the gradient of a scalar field, we first introduce a vector differential operator called *del* operator denoted by ∇.

12.3.3 The vector differential operator Del

We define the vector differential operator ∇ (based on partial differentiations) in two and three dimensions as

$$\nabla = \mathbf{i}\frac{\partial}{\partial x} + \mathbf{j}\frac{\partial}{\partial y} \quad \text{and} \quad \nabla = \mathbf{i}\frac{\partial}{\partial x} + \mathbf{j}\frac{\partial}{\partial y} + \mathbf{k}\frac{\partial}{\partial z} .$$

The del operator is analogous to the derivative operator d/dx, which when applied to $f(x)$ produces the derivative $f'(x)$. The symbol ∇, an inverted capital Greek delta is also known as *nabla*.

Gradient of a scalar field. *When the vector differential operator ∇ is applied to a differentiable function $z = f(x, y)$ or $w = f(x, y, z)$, defining a scalar field we say that the vectors*

$$\nabla f(x, y) = \left(\frac{\partial}{\partial x}\mathbf{i} + \frac{\partial}{\partial y}\mathbf{j} \right) f = \frac{\partial f}{\partial x}\mathbf{i} + \frac{\partial f}{\partial y}\mathbf{j} \qquad \qquad ...(12.5)$$

and $$\nabla f(x, y, z) = \left(\frac{\partial}{\partial x}\mathbf{i} + \frac{\partial}{\partial y}\mathbf{j} + \frac{\partial}{\partial z}\mathbf{k} \right) f = \frac{\partial f}{\partial x}\mathbf{i} + \frac{\partial f}{\partial y}\mathbf{j} + \frac{\partial f}{\partial z}\mathbf{k} \qquad ...(12.6)$$

are gradient of the respective functions.

Note that the del operator ∇ operates on a "scalar field and produce a vector field. The vector ∇f is usually read "grad f" as well as "gradient of f" and "del f".

12.3.4 Geometrical Interpretation of Gradient

Let $f(P) = f(x, y, z)$ be a differential scalar function. Let $f(x, y, z) = c = $ constant be a level surface* through P. Vector normal to tangent plane, to the surface at the point P, is called the normal vector to the surface at this point.

Let $x = x(t)$, $y = y(t)$, $z = z(t)$ be the parametric representation of the smooth curve C on the surface passing through a point P on the surface. The curve C in the space can be represented by a vector function.

$\mathbf{r}(t) = x(t)\mathbf{i} + y(t)\mathbf{j} + z(t)\mathbf{k}$. Since the curve lie on the surface, we have

$$f(x(t), y(t), z(t)) = c. \text{ Then } \frac{d}{dt} f(x(t), y(t), z(t)) = 0.$$

By chain rule, we have $\dfrac{\partial f}{\partial x}\dfrac{dx}{dt} + \dfrac{\partial f}{\partial y}\dfrac{dy}{dt} + \dfrac{\partial f}{\partial z}\dfrac{dz}{dt} = 0$ ...(12.7)

or $\left(\mathbf{i}\dfrac{\partial}{\partial x} + \mathbf{j}\dfrac{\partial}{\partial y} + \mathbf{k}\dfrac{\partial}{\partial z}\right) f \cdot \left(\mathbf{i}\dfrac{\partial x}{\partial t} + \mathbf{j}\dfrac{dy}{dt} + \mathbf{k}\dfrac{dz}{dt}\right) = 0$ or $\nabla f \cdot \mathbf{r}'(t) = 0$ ...(12.8)

$\mathbf{r}'(t)$ is a tangent vector to curve C at the point P and lies in the tangent plane. Hence grad f is orthogonal to all the tangent vector at P. Thus gradient as surface normal vector.

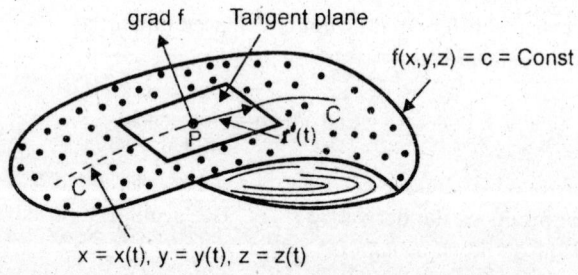

x = x(t), y = y(t), z = z(t)

Fig. 12.3. Gradient as surface normal vector

Illustrative Examples

Example 12.5 Find the gradient of the following scalar fields
(i) $f(x, y) = x^3 - 3x^2y^2 + y^3$ at $(1, 2)$.

---

* If a surface $f(x, y, z) = c = $ constant, is drawn through any point P of the region such that at each point on the surface, the function has the same value as at P then such a surface is known as *level surface* through P. For different values of c, we obtain different surfaces. For example, if $f(x, y, z)$ represents temperature in a medium, then $f(x, y, z) = c$ represents a surface on which the temperature is a constant c. Such surfaces are called *isothermal* surfaces.

(ii) $f(x, y, z) = xy^2 + 3x^2 - z^3$ at $(2, -1, 4)$.

Solution. (i) We have $\nabla f(x, y) = \left(\mathbf{i} \dfrac{\partial}{\partial x} + \mathbf{j} \dfrac{\partial}{\partial y} \right) \left(x^3 - 3x^2 y^2 + y^3 \right)$

$$= \mathbf{i} \dfrac{\partial}{\partial x} \left(x^3 - 3x^2 y^2 + y^3 \right) + \mathbf{j} \dfrac{\partial}{\partial y} \left(x^3 - 3x^2 y^2 + y^3 \right)$$

$$= \left(3x^2 - 6xy^2 \right) \mathbf{i} + \left(-6x^2 y + 3y^2 \right) \mathbf{j}.$$

At $(1, 2)$, we obtain $\nabla f(x, y) = -21\,\mathbf{i}$.

(ii) We have $\nabla f(x, y, z) = \left(\mathbf{i} \dfrac{\partial}{\partial x} + \mathbf{j} \dfrac{\partial}{\partial y} + \mathbf{k} \dfrac{\partial}{\partial z} \right) \left(xy^2 + 3x^2 - z^3 \right)$

$$= (y^2 + 6x)\,\mathbf{i} + 2\,xy\,\mathbf{j} - 3z^2\mathbf{k}.$$

At $(2, -1, 4)$ we obtain $\nabla f(2, -1, 4) = 13\,\mathbf{i} - 4\mathbf{j} - 48\mathbf{k}$.

Example 12.6 (a). If $\mathbf{r} = x\mathbf{i} + y\mathbf{j} + z\mathbf{k}$, $|\mathbf{r}| = r$ and $\hat{\mathbf{r}} = \mathbf{r}/r$, then show that

(i) grad $r = \hat{\mathbf{r}}$, (ii) grad $(1/r) = -\hat{\mathbf{r}}/r^2 = -\dfrac{\mathbf{r}}{r^3}$.   (*PTU 2003; UPTU 2002*)

(b) Find $\nabla \phi$ if $\phi = \log |\mathbf{r}|$.   (*A,M.I.E., W-2008*)

Solution. (a) (i) We have $r^2 = x^2 + y^2 + z^2$. Therefore grad r, that is,

$$\nabla r = \mathbf{i} \dfrac{\partial r}{\partial x} + \mathbf{j} \dfrac{\partial r}{\partial y} + \mathbf{k} \dfrac{\partial r}{\partial z} = \mathbf{i} \dfrac{x}{r} + \mathbf{j} \dfrac{y}{r} + \mathbf{k} \dfrac{z}{r} = \dfrac{x\mathbf{i} + y\mathbf{j} + z\mathbf{k}}{r} = \dfrac{\mathbf{r}}{r} = \hat{\mathbf{r}}.$$

(ii) grad $\dfrac{1}{r} = \left(\mathbf{i} \dfrac{\partial}{\partial x} + \mathbf{j} \dfrac{\partial}{\partial y} + \mathbf{k} \dfrac{\partial}{\partial z} \right) \left(\dfrac{1}{r} \right) = \mathbf{i} \left(-\dfrac{1}{r^2} \dfrac{\partial r}{\partial x} \right) + \mathbf{j} \left(-\dfrac{1}{r^2} \dfrac{\partial r}{\partial y} \right) + \mathbf{k} \left(-\dfrac{1}{r^2} \dfrac{\partial r}{\partial z} \right)$

$$= -\dfrac{1}{r^2} \left(\dfrac{x}{r} \mathbf{i} + \dfrac{y}{r} \mathbf{j} + \dfrac{z}{r} \mathbf{k} \right) = -\dfrac{1}{r^2} \left(\dfrac{\mathbf{r}}{r} \right) = -\dfrac{\mathbf{r}}{r^3} = -\dfrac{\hat{\mathbf{r}}}{r^2}.$$

(b) We have $|\mathbf{r}| = \sqrt{x^2 + y^2 + z^2}$ and $\phi = \log|\mathbf{r}| = \dfrac{1}{2} \log \left(x^2 + y^2 + z^2 \right)$..

$$\therefore \ \nabla \phi = \dfrac{1}{2} \nabla \log \left(x^2 + y^2 + z^2 \right) = \dfrac{1}{2} \left\{ \mathbf{i} \dfrac{\partial}{\partial x} \log \left(x^2 + y^2 + z^2 \right) + \right.$$

$$\left. \mathbf{j} \dfrac{\partial}{\partial y} \log \left(x^2 + y^2 + z^2 \right) + \mathbf{k} \dfrac{\partial}{\partial z} \log \left(x^2 + y^2 + z^2 \right) \right\}$$

$$= \dfrac{1}{2} \left\{ \mathbf{i} \dfrac{2x}{x^2 + y^2 + z^2} + \mathbf{j} \dfrac{2y}{x^2 + y^2 + z^2} + \mathbf{k} \dfrac{2z}{x^2 + y^2 + z^2} \right\} = \dfrac{x\mathbf{i} + y\mathbf{i} + z\mathbf{k}}{x^2 + y^2 + z^2} = \dfrac{\mathbf{r}}{r^2}.$$

Example 12.7 Find $\nabla\phi$ if

(i) $\phi = f(r)$ where $r = \sqrt{x^2 + y^2 + z^2}$.

(ii) $\phi = r^n$.   (AMIE; W 2007; UPTU 2004; Delhi 2002; MDU 2001)

Solution. (i) $\nabla\phi = \nabla f(r) = \left(\mathbf{i}\dfrac{\partial}{\partial x} + \mathbf{j}\dfrac{\partial}{\partial y} + \mathbf{k}\dfrac{\partial}{\partial z}\right) f(r)$

$$= \mathbf{i}\frac{\partial}{\partial x} f(r) + \mathbf{j}\frac{\partial}{\partial y} f(r) + \mathbf{k}\frac{\partial}{\partial z} f(r)$$

$$= \mathbf{i} f'(r)\frac{\partial r}{\partial x} + \mathbf{j} f'(r)\frac{\partial r}{\partial y} + \mathbf{k} f'(r)\frac{\partial r}{\partial z}, \text{ where } f'(r) = \frac{d}{dr} f(r).$$

Since $r = \sqrt{x^2 + y^2 + z^2}$. $\therefore \dfrac{\partial r}{\partial x} = \dfrac{\partial}{\partial x}\left(x^2 + y^2 + z^2\right)^{1/2} = \dfrac{x}{\sqrt{x^2 + y^2 + z^2}} = \dfrac{x}{r}$.

Similarly $\dfrac{\partial r}{\partial y} = \dfrac{y}{r}$ and $\dfrac{\partial r}{\partial z} = \dfrac{z}{r}$.

$\therefore \quad \nabla\phi = \mathbf{i} f'(r)\dfrac{x}{r} + \mathbf{j} f'(r)\dfrac{y}{r} + \mathbf{k} f'(r)\dfrac{z}{r} = \dfrac{f'(r)}{r}\left(x\mathbf{i} + y\mathbf{j} + z\mathbf{k}\right) = \dfrac{f'(r)}{r}\mathbf{r}$.

(ii) Using the result of part (i), we have $f(r) = r^n$. Therefore $f'(r) = nr^{n-1}$.

Hence $\nabla\phi = \nabla r^n = \dfrac{nr^{n-1}}{r}\mathbf{r} = nr^{n-2}\mathbf{r}$.

Alternative $\nabla r^n = \mathbf{i}\dfrac{\partial}{\partial x} r^n + \mathbf{j}\dfrac{\partial}{\partial y} r^n + \mathbf{k}\dfrac{\partial}{\partial z} r^n$.

Now $\dfrac{\partial}{\partial x} r^n = \dfrac{\partial}{\partial x}\left(x^2 + y^2 + z^2\right)^{n/2} = \dfrac{n}{2}\left(x^2 + y^2 + z^2\right)^{(n-2)/2} \cdot 2x$

$$= nx\left(x^2 + y^2 + z^2\right)^{(n-2)/2} = nxr^{n-2}.$$

Similarly $\dfrac{\partial}{\partial y} r^n = nyr^{n-2}$ and $\dfrac{\partial}{\partial z} r^n = nzr^{n-2}$.

$\therefore \quad \nabla r^n = nxr^{n-2}\mathbf{i} + nyr^{n-2}\mathbf{j} + nzr^{n-2}\mathbf{k} = nr^{n-2}\left(x\mathbf{i} + y\mathbf{j} + z\mathbf{k}\right) = nr^{n-2}\mathbf{r}$.

Example 12.8 (i) If $\mathbf{r} = x\mathbf{i} + y\mathbf{j} + z\mathbf{k}$ and $\phi = |\mathbf{r}|^3$, then find $\nabla|\mathbf{r}|^3$,

(ii) Find $\nabla\phi$ if $\phi = \left(x^2 + y^2 + z^2\right)e^{-\sqrt{(x^2+y^2+z^2)}}$.

Solution. (i) $|\mathbf{r}| = \sqrt{(x^2 + y^2 + z^2)}$, therefore $|\mathbf{r}|^3 = (x^2 + y^2 + z^2)^{3/2}$ and

$$\nabla|\mathbf{r}|^3 = \left(\mathbf{i}\frac{\partial}{\partial x} + \mathbf{j}\frac{\partial}{\partial y} + \mathbf{k}\frac{\partial}{\partial z}\right)(x^2 + y^2 + z^2)^{3/2}$$

$$= \mathbf{i}\frac{\partial}{\partial x}(x^2 + y^2 + z^2)^{3/2} + \mathbf{j}\frac{\partial}{\partial y}(x^2 + y^2 + z^2)^{3/2} + \mathbf{k}\frac{\partial}{\partial z}(x^2 + y^2 + z^2)^{3/2}$$

$$= \mathbf{i}\left[\frac{3}{2}(x^2 + y^2 + z^2)^{1/2} \cdot 2x\right] + \mathbf{j}\left[\frac{3}{2}(x^2 + y^2 + z^2)^{1/2} \cdot 2y\right]$$

$$+ \mathbf{k}\left[\frac{3}{2}(x^2 + y^2 + z^2)^{1/2} \cdot 2z\right] = 3(x^2 + y^2 + z^2)^{1/2}[x\mathbf{i} + y\mathbf{j} + z\mathbf{k}] = 3r\,\mathbf{r}.$$

(ii) $\mathbf{r} = x\mathbf{i} + y\mathbf{j} + z\mathbf{k}$, $r = |\mathbf{r}| = \sqrt{(x^2 + y^2 + z^2)}$.

$$\therefore \quad \nabla\phi = \left(\mathbf{i}\frac{\partial}{\partial x} + \mathbf{j}\frac{\partial}{\partial y} + \mathbf{k}\frac{\partial}{\partial z}\right)(x^2 + y^2 + z^2)e^{-\sqrt{(x^2 + y^2 + z^2)}}$$

$$= \sum \mathbf{i}\frac{\partial}{\partial x}\left[(x^2 + y^2 + z^2)e^{-\sqrt{(x^2 + y^2 + z^2)}}\right]$$

$$= \sum \mathbf{i}\left[2xe^{-\sqrt{(x^2 + y^2 + z^2)}} - (x^2 + y^2 + z^2)e^{-\sqrt{(x^2 + y^2 + z^2)}} \cdot \frac{1}{2}(x^2 + y^2 + z^2)^{-1/2} \cdot 2x\right]$$

$$= \sum \mathbf{i}\left[e^{-\sqrt{(x^2 + y^2 + z^2)}} \cdot x\left\{2 - \sqrt{(x^2 + y^2 + z^2)}\right\}\right]$$

$$= e^{-\sqrt{(x^2 + y^2 + z^2)}}\left\{2 - \sqrt{(x^2 + y^2 + z^2)}\right\}(x\mathbf{i} + y\mathbf{j} + z\mathbf{k}) = e^{-r}(2 - r)\mathbf{r}.$$

Example 12.9 Find a unit normal vector to the surface $x^2y + 2xz = 4$ at the point $(2, -2, 3)$.
(A.M.I.E., W-2005)

Solution. Let us regard the given surface as a particular level surface of the function $\phi = x^2y + 2xz$. Then the gradient of this function at the point $(2, -2, 3)$ will be perpendicular to the level surface through $(2, -2, 3)$, which is the given surface.

Now $\phi(x, y, z) = x^2y + 2xz - 4$, $\dfrac{\partial\phi}{\partial x} = 2xy + 2z$, $\dfrac{\partial\phi}{\partial y} = x^2$ and $\dfrac{\partial\phi}{\partial z} = 2x$.

Then, the normal vector is given by

$$\nabla\phi = \mathbf{i}\frac{\partial\phi}{\partial x} + \mathbf{j}\frac{\partial\phi}{\partial y} + \mathbf{k}\frac{\partial\phi}{\partial z} = (2xy + 2z)\mathbf{i} + x^2\mathbf{j} + 2x\mathbf{k}.$$

At $(2, -2, 3)$, we obtain the normal vector as $\nabla\phi(2, -2, 3) = -2\mathbf{i} + 4\mathbf{j} + 4\mathbf{k}$. Hence a unit vector normal to the given surface is given by

$$\mathbf{n} = \frac{\nabla\phi}{|\nabla\phi|} = \frac{(-2\mathbf{i}+4\mathbf{j}+4\mathbf{k})}{\sqrt{4+16+16}} = -\frac{1}{3}\mathbf{i}+\frac{2}{3}\mathbf{j}+\frac{2}{3}\mathbf{k}.$$

Example 12.10 Find the normal vector and the equation of the tangent plane to the surface $xz^2 + x^2y - z + 1 = 0$ at the point $(1, -3, 2)$.

Solution. Let $f(x, y, z) = xz^2 + x^2y - z + 1 = 0$ be the surface. Then, the normal vector is given by $\mathbf{N} = $

$$\nabla f = \left(\mathbf{i}\frac{\partial}{\partial x} + \mathbf{j}\frac{\partial}{\partial y} + \mathbf{k}\frac{\partial}{\partial z}\right)\left(xz^2 + x^2y - z + 1\right) = \left(z^2 + 2xy\right)\mathbf{i} + x^2\mathbf{j} + (2xz - 1)\mathbf{k}.$$

At $(1, -3, 2)$, the normal vector is given by $\mathbf{N} = \nabla f(1, -3, 2) = -2\mathbf{i} + \mathbf{j} + 3\mathbf{k}$. The tangent plane at the point $(1, -3, 2)$ is given by
$-2(x - 1) + 1(y + 3) + 3(z - 2) = 0$ or $2x - y - 3z + 1 = 0$.

Example 12.11 Find the angle between the surfaces $x^2 + y^2 + z^2 = 9$ and $z + 3 = x^2 + y^2$ at the point $(2, -1, 2)$. *(UPTU 2003; Madras 2003; MDU, 2001)*

Solution. The angle between two surfaces at a common point is the angle between their normals at that point. We have
$$f_1(x, y, z) = x^2 + y^2 + z^2 - 9 = 0, \quad \nabla f_1(x, y, z) = 2x\mathbf{i} + 2y\mathbf{j} + 2z\mathbf{k}.$$
$\nabla f_1(2, -1, 2) = 4\mathbf{i} - 2\mathbf{j} + 4\mathbf{k}$, which is normal $\mathbf{N}_1$ to $x^2 + y^2 + z^2 = 9$ at $(2, -1, 2)$.
$$f_2(x, y, z) = x^2 + y^2 - z - 3 = 0, \quad \nabla f_2(x, y, z) = 2x\mathbf{i} + 2y\mathbf{j} - \mathbf{k}.$$
$\mathbf{N}_2 = \nabla f_2(2, -1, 2) = 4\mathbf{i} - 2\mathbf{j} - \mathbf{k}$. If the required angle is θ then

$$\cos\theta = \frac{\mathbf{N}_1 \cdot \mathbf{N}_2}{|\mathbf{N}_1||\mathbf{N}_2|} = \frac{(4\mathbf{i}-2\mathbf{j}+4\mathbf{k})\cdot(4\mathbf{i}-2\mathbf{j}-\mathbf{k})}{\sqrt{(4)^2+(-2)^2+(4)^2}\sqrt{(4)^2+(-2)^2+(-1)^2}}$$

$$= \frac{16+4-4}{6\sqrt{21}} = \frac{8\sqrt{21}}{63}, \text{ or } \theta = \cos^{-1}\left(8\sqrt{21}/63\right) = 54°25'.$$

Example 12.12 If $\nabla f = (x + 2y + 4z)\mathbf{i} + (2x - 3y - z)\mathbf{j} + (4x - y + 2z)\mathbf{k}$, find the scalar function f.

Solution. Since $\nabla f = \frac{\partial f}{\partial x}\mathbf{i} + \frac{\partial f}{\partial y}\mathbf{j} + \frac{\partial f}{\partial z}\mathbf{k}$. Comparing, we obtain

$$\frac{\partial f}{\partial x} = x + 2y + 4z \cdots(i), \quad \frac{\partial f}{\partial y} = 2x - 3y - z \cdots(ii), \quad \frac{\partial f}{\partial z} = 4x - y + 2z \cdots(iii).$$

Integrating the first equation, we obtain

$$f(x, y, z) = \frac{x^2}{2} + 2xy + 4zx + g(y, z) \ldots(A). \text{ Substituting in the second equation,}$$

we get $\dfrac{\partial f}{\partial y} = 2x - 3y - z = 2x + \dfrac{\partial g}{\partial y}$, or $\dfrac{\partial g}{\partial y} = -3y - z.$

Therefore $g(y, z) = -3\dfrac{y^2}{2} - yz + g(z).$

Now (A) become $f(x, y, z) = (x^2/2) + 2xy + 4zx + (-3/2)y^2 - yz + g(z)$ ...(B)

Substituting in the third equation, we get

$\dfrac{\partial f}{\partial z} = 4x - y + 2z = 4x - y + \dfrac{dg}{dz}$ or $\dfrac{dg}{dz} = 2z$. Integrating, we get

$g = z^2 + k$, constant.

Thus from (B), we obtain

$$f(x, y, z) = (x^2/2) + 2xy + 4xz - yz + (-3/2)y^2 + z^2 + k.$$

Remark: A vector field **F** derived from a scalar field f such that $\mathbf{F} = \nabla f$ is called *a conservative vector field* and f is called the *scalar potential*.

Example 12.13 Show that the vector field defined by the vector function
$\mathbf{v} = xyz(yz\mathbf{i} + zx\mathbf{j} + xy\mathbf{k})$ is conservative. *(IETE, June 2008)*

Solution. If the given vector field is conservative, then it can be expressed as the gradient of a scalar function $f(x, y, z)$. Therefore

$$\nabla f = \left(\mathbf{i}\frac{\partial f}{\partial x} + \mathbf{j}\frac{\partial f}{\partial y} + \mathbf{k}\frac{\partial f}{\partial z} \right) = \mathbf{v} = xyz(yz\mathbf{i} + zx\mathbf{j} + xy\mathbf{k}).$$

Comparing, we obtain

$$\frac{\partial f}{\partial x} = xy^2z^2 \ \ ...(i), \qquad \frac{\partial f}{\partial y} = x^2yz^2 \ \ ...(ii), \qquad \frac{\partial f}{\partial z} = x^2y^2z \ \ ...(iii).$$

Integrating the first equation, we obtain

$$f(x, y, z) = \frac{1}{2}x^2y^2z^2 + g(y, z). \ \ ...(A) \ \text{Substituting in the second equation,}$$

we get

$$\frac{\partial f}{\partial y} = x^2yz^2 = x^2yz^2 + \frac{\partial g}{\partial y} \ \text{or} \ \frac{\partial g}{\partial y} = 0 \Rightarrow g = g(z).$$

Now (A) become $f(x, y, z) = \dfrac{1}{2}x^2y^2z^2 + g(z)$ ...(B)

Substituting in the third equation, we get

$$\frac{\partial f}{\partial z} = x^2 y^2 z = x^2 y^2 z + \frac{dg}{dz} \quad \text{or} \quad \frac{dg}{dz} = 0 \implies g = k_1, \text{ constant}$$

Hence, $f(x, y, z) = x^2 y^2 z^2 + k$. Therefore, there exists a scalar function $f(x, y, z)$ such that $\nabla f = \mathbf{v}$ and the vector field $\mathbf{v}$ is conservative.

12.3.5 Directional derivative

Let $f(P) = f(x, y, z)$ be a differential scalar field. Then $\partial f / \partial x, \partial f / \partial y, \partial f / \partial z$ denote the rates of change of f in the directions of x, y and z axis, respectively. Let $\hat{\mathbf{b}} = b_1 \mathbf{i} + b_2 \mathbf{j} + b_3 \mathbf{k}$ be any unit vector. Let P_o *be any point, its position vector is* $\mathbf{a} = a_1 \mathbf{i} + a_2 \mathbf{j} + a_3 \mathbf{k}$. Refer Fig. 12.4. Let $P_o C$ is parallel to $\mathbf{b}$ and Q is a variable point on the directed line segment $P_o C$. Then the position vector of point Q is given by

$$\mathbf{r}(s) = \mathbf{a} + s\hat{\mathbf{b}}, \quad s \geq 0$$

$$= \left(a_1 + b_1 s\right)\mathbf{i} + \left(a_2 + b_2 s\right)\mathbf{j} + \left(a_3 + b_3 s\right)\mathbf{k} = x(s)\mathbf{i} + y(s)\mathbf{j} + z(s)\mathbf{k}.$$

The vector $\overline{P_o Q}$ is given by $s\,\hat{\mathbf{b}}$. Since $\left|\hat{\mathbf{b}}\right| = 1$, so the distance from P_0 to Q is s.

Then $\dfrac{\partial f}{\partial s} = \lim_{s \to 0} \dfrac{f(Q) - f(P_0)}{s}$

if it exists, is called the *directional derivative* of f at the point P_0 in the direction of $\hat{\mathbf{b}}$. It is the rate of change of f in the direction of $\mathbf{b}$.

Therefore, $\dfrac{\partial}{\partial s} f\left(x(s), y(s), z(s)\right)$ is the rate of change of f with respect to the distance s.

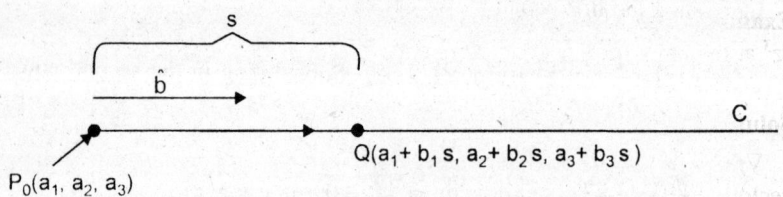

Fig. 12.4. Directional derivative

We have by the chain rule $\dfrac{\partial f}{\partial s} = \dfrac{\partial f}{\partial x}\dfrac{dx}{ds} + \dfrac{\partial f}{\partial y}\dfrac{dy}{ds} + \dfrac{\partial f}{\partial z}\dfrac{dz}{ds}$ ...(i)

where dx/ds, dy/ds, dz/ds are calculated at the point P_0 (where $s = 0$). We write (i) as

$$\frac{\partial f}{\partial s} = \left(\mathbf{i}\frac{\partial}{\partial x} + \mathbf{j}\frac{\partial}{\partial y} + \mathbf{k}\frac{\partial}{\partial z}\right) f \cdot \left(\mathbf{i}\frac{dx}{ds} + \mathbf{j}\frac{dy}{ds} + \mathbf{k}\frac{dz}{ds}\right) = \nabla f \cdot \frac{d\mathbf{r}}{ds}$$

But $d\mathbf{r}/ds = \hat{\mathbf{b}}$ (a unit vector). Therefore, the directional derivative of f in the direction of $\hat{\mathbf{b}}$, denoted by $D_b\ (f)$, is given by

$$D_b\left(f\right) = \nabla f \cdot \hat{\mathbf{b}} = grad\left(f\right)\cdot \hat{\mathbf{b}} \qquad \qquad ...(12.9)$$

Remark If the direction is specified by a vector $\mathbf{u}$, then we have $\hat{\mathbf{b}} = \mathbf{u}/|\mathbf{u}|$ and

$$D_b\left(f\right) = \nabla f \cdot \mathbf{u}/|\mathbf{u}|.$$

Maximum/Minimum value of the Directional derivative

Let f represent a function of either two or three variables. Since the directional derivative is a dot product, we see from the definition of dot or scalar product that

$$D_b\left(f\right) = \nabla f \cdot \hat{\mathbf{b}} = |\nabla f|\ |\hat{\mathbf{b}}|\cos\alpha = |\nabla f|\cos\alpha, \qquad \text{as}\ \left(|\hat{\mathbf{b}}| = 1\right),$$

where α is the angle between the vector ∇f and $\mathbf{b}$. Because $0 \le \alpha \le \pi$, we have $-1 \le \cos \alpha \le 1$ and consequently $-|\nabla f| \le D_b\ (f) \le |\nabla f|$.

Thus the maximum value of the directional derivative is $|\nabla f|$ and it occurs when $\alpha = 0$ *i.e.*; $\hat{\mathbf{b}}$ has the same direction as $\nabla\phi$. This direction is the direction of the normal vector. The minimum value of the directional derivative is $-|\nabla\phi|$ and it occurs when $\theta = \pi$ that is $\hat{\mathbf{b}}$ and ∇f have opposite directions. We may also say that the gradient vector ∇f points in the direction in which f increases most rapidly, whereas $-\nabla f$ points in the direction of the most rapid decrease of f.

12.4 Illustrative Examples

Example 12.14 What is the directional derivative of the function $f(x, y, z) = xy^2 + yz^3$ at the point $(2, -1, 1)$ in the direction of the vector $\mathbf{i} + 2\mathbf{j} + 3\mathbf{k}$?

(I.E.T.E, Dec. 2009; A.M.I.E., S-2005; Karnataka 1995)

Solution. We have $\nabla f = \nabla(xy^2 + yz^3) = y^2\mathbf{i} + (2xy + z^3)\mathbf{j} + 3yz^2\mathbf{k}$.

$\nabla f (2, -1, 1) = \mathbf{i} - 3\mathbf{j} - 3\mathbf{k}$. The projection of this in the direction of the given vector will be the required directional derivative.

The unit vector in the given direction is $\hat{\mathbf{b}} = \dfrac{(\mathbf{i}+2\mathbf{j}+3\mathbf{k})}{\sqrt{1+4+9}} = \dfrac{\mathbf{i}+2\mathbf{j}+3\mathbf{k}}{\sqrt{14}}$.

Therefore, $\qquad D_b\left(2, -1, 1\right) = \nabla f \cdot \hat{\mathbf{b}} = \left(\mathbf{i} - 3\mathbf{j} - 3\mathbf{k}\right)\cdot\dfrac{\left(\mathbf{i} + 2\mathbf{j} + 3\mathbf{k}\right)}{\sqrt{14}}$

$$= \frac{(1)(1)+(-3)(2)+(-3)(3)}{\sqrt{14}} = \frac{-14}{\sqrt{14}} = -\sqrt{14}.$$

The negative sign, of course, indicates that f decreases in the given direction.

Example 12.15. Find the directional derivative of $f(x, y) = 2x^2y^3 + 6xy$ at $(1, 1)$ in the direction of a unit vector whose angle with the positive x-axis is $\pi/6$.

Solution. We have $\nabla f(x, y) = (4xy^3 + 6y)\ \mathbf{i} + (6x^2y^2 + 6x)\mathbf{j}$. At the point $(1, 1)$, we

have $\nabla f(1, 1) = 10\mathbf{i} + 12\mathbf{j}$. Now, at $\theta = \pi/6$, $\hat{\mathbf{b}} = (\cos\theta)\ \mathbf{i} + (\sin\theta)\mathbf{j} = \dfrac{\sqrt{3}}{2}\mathbf{i} + \dfrac{1}{2}\mathbf{j}$.

Therefore $D_{\mathbf{b}}(1, 1) = \nabla f \cdot \hat{\mathbf{b}} = (10\mathbf{i}+12\mathbf{j}) \cdot \left(\dfrac{\sqrt{3}}{2}\mathbf{i}+\dfrac{1}{2}\mathbf{j}\right) = 5\sqrt{3}+6.$

Example 12.16 (a) In what direction from the point $(3, 1, -2)$ is the directional derivative of $f(x, y, z) = x^2y^2z^4$ maximum and what is its magnitude?

(A.M.I.E., W-2005; MDU 2004; Bhopal 1998; Kerala 1997)

Solution. We have $\nabla f = \nabla(x^2y^2z^4) = 2xy^2z^4\mathbf{i} + 2x^2yz^4\mathbf{j} + 4x^2y^2z^3\mathbf{k}$.

At the point $(3, 1, -2)$, we have $\nabla f = 96\mathbf{i} + 288\mathbf{j} - 288\mathbf{k}$.

(a) the directional derivative is a maximum in the direction $\nabla f = 96\mathbf{i} + 288\mathbf{j} - 288\mathbf{k}$. (b) the magnitude of this maximum directional derivative is

$$|\nabla f| = \sqrt{(96)^2 + (288)^2 + (-288)^2} = \sqrt{175104} = 96\sqrt{19}.$$

Example 12.17 (a) Find the directional derivative of the function $f(x, y, z) = 2x^3y - 3y^2z$ at $P(1, 2, -1)$ in a direction toward $Q(3, -1, 5)$.

(b) In what direction from P is the directional derivative a maximum?

(c) What is the magnitude of the maximum directional derivative?

Solution. (a) We have $\nabla f = 6x^2y\mathbf{i} + (2x^3 - 6yz)\mathbf{j} - 3y^2\mathbf{k}$.

At the point $(1, 2, -1)$, we have $\nabla f = 12\mathbf{i} + 14\mathbf{j} - 12\mathbf{k}$.

The vector from P to $Q = (3 - 1)\mathbf{i} + (-1-2)\mathbf{j} + [5 -(-1)]\mathbf{k} = 2\mathbf{i} - 3\mathbf{j} + 6\mathbf{k}$.

The unit vector from P to $Q = \hat{\mathbf{b}} = \dfrac{2\mathbf{i}-3\mathbf{j}+6\mathbf{k}}{\sqrt{(2)^2 +(-3)^2 +(6)^2}} = \dfrac{2\mathbf{i}-3\mathbf{j}+6\mathbf{k}}{7}$

Therefore, directional derivative at P, $D_{\mathbf{b}}f = \nabla f \cdot \hat{\mathbf{b}}$

$$= (12\mathbf{i}+14\mathbf{j}-12\mathbf{k}) \cdot \left(\dfrac{2\mathbf{i}-3\mathbf{j}+6\mathbf{k}}{7}\right) = \dfrac{24-42-72}{7} = -\dfrac{90}{7},$$

that is, f is decreasing in the given direction.

(b) The directional derivative is a maximum in the direction $\nabla f = 12\mathbf{i} + 14\mathbf{j} - 12\mathbf{k}$.

(c) The value of the maximum directional derivative is

$$|\nabla f| = |12\mathbf{i}+14\mathbf{j}-12\mathbf{k}| = \sqrt{144+196+144} = 22.$$

Example 12.18 If the directional derivative of $f(x, y, z) = axy^2 + byz + cz^2x^3$ at $(1, 2, -1)$ has a maximum magnitude of 64 in the direction parallel to z-axis, find the values of constants a, b, c.

(I.E.T.E., June 2001)

Solution. The maximum directional derivative of f at any point is the value of ∇f at the point.

Now $\nabla f = \dfrac{\partial f}{\partial x}\mathbf{i} + \dfrac{\partial f}{\partial y}\mathbf{j} + \dfrac{\partial f}{\partial z}\mathbf{k} = \left(ay^2 + 3cz^2x^2\right)\mathbf{i} + \left(2axy + bz\right)\mathbf{j} + \left(by + 2czx^3\right)\mathbf{k}.$

$\therefore \quad \nabla f_{(1, 2, -1)} = (4a + 3c)\mathbf{i} + (4a - b)\mathbf{j} + (2b - 2c)\mathbf{k}.$

Since its magnitude is 64 and direction is parallel to z-axis, we have

$(4a + 3c)\mathbf{i} + (4a - b)\mathbf{j} + (2b - 2c)\mathbf{k} = 64\mathbf{k}.$

$\Rightarrow 4a + 3c = 0,\ 4a - b = 0,\ 2b - 2c = 64.$ Solving, we get $a = 6, b = 24, c = -8.$

Example 12.19 Find the directional derivative of $f(x, y, z) = x^2yz^3$ along the curve $x = e^{-u}, y = 2\sin u + 1, z = u - \cos u$ at the point P where $u = 0$.

Solution. The point P corresponding to $u = 0$ is $(1, 1, -1)$.

$\nabla f = 2xyz^3\mathbf{i} + x^2z^3\mathbf{j} + 3x^2yz^2\mathbf{k}. \quad \therefore \ \nabla f(1, 1, -1) = -2\mathbf{i} - \mathbf{j} + 3\mathbf{k}.$

The position vector $\mathbf{r}$ of any point on the given curve is given by

$\mathbf{r} = e^{-u}\mathbf{i} + (2\sin u + 1)\mathbf{j} + (u - \cos u)\mathbf{k}.$

A tangent vector to the curve is

$\mathbf{r}'(u) = \dfrac{d\mathbf{r}}{du} = \dfrac{d}{du}\left\{e^{-u}\mathbf{i} + (2\sin u + 1)\mathbf{j} + (u - \cos u)\mathbf{k}\right\}$

$= -e^{-u}\mathbf{i} + 2\cos u\,\mathbf{j} + (1 + \sin u)\mathbf{k}. \quad \therefore \ (d\mathbf{r}/du)_{\text{at } P}\,\mathbf{r}'(0) = -\mathbf{i} + 2\mathbf{j} + \mathbf{k}.$

The unit tangent vector in this direction $= \dfrac{-\mathbf{i} + 2\mathbf{j} + \mathbf{k}}{\sqrt{1 + 4 + 1}} = \dfrac{-\mathbf{i} + 2\mathbf{j} + \mathbf{k}}{\sqrt{6}}.$

Then, directional derivative $= \nabla f \cdot$ (unit tangent vector in this direction)

$= (-2\mathbf{i} - \mathbf{j} + 3\mathbf{k}) \cdot \left(\dfrac{-\mathbf{i} + 2\mathbf{j} + \mathbf{k}}{\sqrt{6}}\right) = \dfrac{1}{\sqrt{6}}(2 - 2 + 3) = \dfrac{3}{\sqrt{6}} = \dfrac{\sqrt{6}}{2}.$

Since this is positive, f is increasing in this direction.

12.5 Properties of Gradient

If f and g are differentiable scalar functions of position (x, y, z), then

(i) $\nabla(k\,f) = k\,\nabla f$ (any number k),

(ii) $\nabla(f \pm g) = \nabla f \pm \nabla g,$

(iii) $\nabla(c_1 f + c_2 g) = c_1\,\nabla f + c_2\,\nabla g,\ c_1, c_2$ arbitrary constants, (*iv*) $\nabla(fg) = f \nabla g + g \nabla f,$

(*v*) $\nabla\left(\dfrac{f}{g}\right) = \dfrac{g\nabla f - f\nabla g}{g^2}, \; g \neq 0.$

The proof of properties (*ii*), (*iv*) and (*v*) is given below.

(*ii*)
$$\nabla(f \pm g) = \left(\mathbf{i}\frac{\partial}{\partial x} + \mathbf{j}\frac{\partial}{\partial y} + \mathbf{k}\frac{\partial}{\partial z}\right)(f \pm g)$$

$$= \mathbf{i}\frac{\partial}{\partial x}(f \pm g) + \mathbf{j}\frac{\partial}{\partial y}(f \pm g) + \mathbf{k}\frac{\partial}{\partial z}(f \pm g)$$

$$= \left(\mathbf{i}\frac{\partial f}{\partial x} + \mathbf{j}\frac{\partial f}{\partial y} + \mathbf{k}\frac{\partial f}{\partial z}\right) \pm \left(\mathbf{i}\frac{\partial g}{\partial x} + \mathbf{j}\frac{\partial g}{\partial y} + \mathbf{k}\frac{\partial g}{\partial z}\right) = \nabla f \pm \nabla g.$$

(*iv*)
$$\nabla(fg) = \left(\mathbf{i}\frac{\partial}{\partial x} + \mathbf{j}\frac{\partial}{\partial y} + \mathbf{k}\frac{\partial}{\partial z}\right)fg$$

$$= \mathbf{i}\left(f\frac{\partial g}{\partial x} + g\frac{\partial f}{\partial x}\right) + \mathbf{j}\left(f\frac{\partial g}{\partial y} + g\frac{\partial f}{\partial y}\right) + \mathbf{k}\left(f\frac{\partial g}{\partial z} + g\frac{\partial f}{\partial z}\right)$$

$$= f\left(\mathbf{i}\frac{\partial g}{\partial x} + \mathbf{j}\frac{\partial g}{\partial y} + \mathbf{k}\frac{\partial g}{\partial z}\right) + g\left(\mathbf{i}\frac{\partial f}{\partial x} + \mathbf{j}\frac{\partial f}{\partial y} + \mathbf{k}\frac{\partial f}{\partial z}\right) = f\nabla g + g\nabla f.$$

(*v*)
$$\nabla\left(\frac{f}{g}\right) = \left(\mathbf{i}\frac{\partial}{\partial x} + \mathbf{j}\frac{\partial}{\partial y} + \mathbf{k}\frac{\partial}{\partial z}\right)\left(\frac{f}{g}\right)$$

$$= \mathbf{i}\frac{\left(g\frac{\partial f}{\partial x} - f\frac{\partial g}{\partial x}\right)}{g^2} + \mathbf{j}\frac{\left(g\frac{\partial f}{\partial y} - f\frac{\partial g}{\partial y}\right)}{g^2} + \mathbf{k}\frac{\left(g\frac{\partial f}{\partial z} - f\frac{\partial g}{\partial z}\right)}{g^2}, \text{ if } g \neq 0$$

$$= \frac{1}{g^2}\left[g\left(\mathbf{i}\frac{\partial f}{\partial x} + \mathbf{j}\frac{\partial f}{\partial y} + \mathbf{k}\frac{\partial f}{\partial z}\right) - f\left(\mathbf{i}\frac{\partial g}{\partial x} + \mathbf{j}\frac{\partial g}{\partial y} + \mathbf{k}\frac{\partial g}{\partial z}\right)\right] = \frac{g\nabla f - f\nabla g}{g^2}.$$

Exercise 12.2

1. Compute the gradient of the following scalar fields

(*i*) $f(x, y) = y^2 - 4xy,$ (*ii*) $f(x, y) = 5y - x^3y^2,$

(*iii*) $f(x, y, z) = xy^2 + 3x^2 - z^3,$ (*iv*) $f(x, y, z) = xy^2/z^3,$

(*v*) $f(x, y, z) = x^2 + yz,$ (*vi*) $\phi(x, y, z) = x^3 + y^3 + z^3 - 3xyz,$

(*vii*) $\phi(x, y, z) = \log(x^2 + y^2 + z^2)$.

Ans. (*i*) $-4y\mathbf{i} + (2y - 4x)\mathbf{j}$ (*ii*) $-3x^2y^2\mathbf{i} + (5 - 2x^3y)\mathbf{j}$

(*iii*) $(y^2 + 6x)\mathbf{i} + 2xy\mathbf{j} - 3z^2\mathbf{k}$ (*iv*) $(y^2/z^3)\mathbf{i} + (2xy/z^3)\mathbf{j} - (3xy^2/z^4)\mathbf{k}$

(*v*) $2x\mathbf{i} + z\mathbf{j} + y\mathbf{k}$ (*vi*) $3(x^2 - yz)\mathbf{i} + 3(y^2 - zx)\mathbf{j} + 3(z^2 - xy)\mathbf{k}$

[**Hint:** (*vii*) $\nabla \log\left(x^2 + y^2 + z^2\right) = \nabla \log r^2 = \dfrac{2}{r} \cdot \dfrac{\mathbf{r}}{r} = \dfrac{2\mathbf{r}}{r^2} = \dfrac{2(x\mathbf{i} + y\mathbf{j} + 3\mathbf{k})}{x^2 + y^2 + z^2}.$]

2. Find the gradient of the given function at the indicated point
   (*i*) $f(x, y) = x^2 - 4y^2$, (2, 4). (*ii*) $f(x, y, z) = 3x^2y - y^3z^2$, (1, -2, -1).
   (*iii*) $f(x, y, z) = 2xz^4 - x^2y$, (2, -2, -1).
   (*iv*) $f(x, y, z) = x^2y + y^2x + z^2$, (1, 1, 1).
   (*v*) $f(x, y, z) = x^3 - y^3 + xz^2$, (1, 1, 2).

   Ans.(*i*) $4\mathbf{i} - 32\mathbf{j}$, (*ii*) $-12\mathbf{i} - 9\mathbf{j} - 16\mathbf{k}$, (*iii*) $10\mathbf{i} - 4\mathbf{j} - 16\mathbf{k}$,
   (*iv*) $3\mathbf{i} + 3\mathbf{j} + 2\mathbf{k}$, (*v*) $7\mathbf{i} - 3\mathbf{j} + 4\mathbf{k}$.

3. (*a*) Evaluate $\nabla\left(3r^2 - 4\sqrt{r} + \dfrac{6}{\sqrt[3]{r}}\right)$. (*b*) Show that $\text{grad}\left(\dfrac{1}{r^2}\right) = \dfrac{-2\mathbf{r}}{r^4}$.

   (*MDU 2002*) **Ans.** $(a)\left(6 - 2r^{-3/2} - 2r^{-7/3}\right)\mathbf{r}$

4. (*a*) Find the normal vector and the unit normal vector to the given curve/
   surface at the indicated point
   (*i*) $x^2 - y^2 = 12$, (4, 2)　　　　　　(*ii*) $z = x^2 + y^2$, (1, 2, 5).
   (*iii*) $xy^3z^2 = 4$, (-1, -1, 2).　　　　　　(*Madras 1999S*)
   (*iv*) $x^4 - 3xyz + z^2 + 1 = 0$, (1, 1, 1).
   (*v*) $x^3 + y^3 + 3xyz = 3$, (1, 2, -1).　　　　　　(*I.E.T.E., June 2002*)
   (*vi*) $x^3 - xyz = 1$, (1, 1, 1).　　　　　　(*I.E.T.E., Dec. 2004*)
   (*vii*) $f(x, y, z) = xy^2 + 2yz - 8 = 0$ at (3, -2, 1)　　　　　　(*A.M.I.E., W-2007*)

   Ans.(*i*) $4(2\mathbf{i} - \mathbf{j})$, $(2\mathbf{i} - \mathbf{j})/\sqrt{5}$ (*ii*) $(2\mathbf{i} + 4\mathbf{j} - \mathbf{k})$; $(2\mathbf{i} + 4\mathbf{j} - \mathbf{k})/\sqrt{21}$
   (*iii*) $4(\mathbf{i} + 3\mathbf{j} - \mathbf{k})$; $-(\mathbf{i} + 3\mathbf{j} - \mathbf{k})/\sqrt{11}$ (*iv*) $\mathbf{i} - 3\mathbf{j} - \mathbf{k}$; $(\mathbf{i} - 3\mathbf{j} - \mathbf{k})/\sqrt{11}$,
   (*v*) $(-\mathbf{i} + 3\mathbf{j} + 2\mathbf{k})$; $(-\mathbf{i} + 3\mathbf{j} + 2\mathbf{k})/\sqrt{14}$ (*vi*) $2\mathbf{i} - \mathbf{j} - \mathbf{k}$; $(2\mathbf{i} - \mathbf{j} - \mathbf{k})/\sqrt{6}$,
   (*vii*) $4i - 10j - 4k$; $(2\mathbf{i} - 5\mathbf{j} - 2\mathbf{k})/\sqrt{33}$.

   (*b*) Find the magnitude of the vector drawn perpendicular to the surface
   $x^2 + 2y^2 + z^2 = 7$ at the point (1, -1, 2). (*A.M.I.E., S 2009*) **Ans.** 6.

5. If $u = x + y + z$, $v = x^2 + y^2 + z^2$, $w = xy + yz + zx$, show that
   $\nabla u \cdot [\nabla v \times \nabla w] = 0$. (*UPTU2002; Patna 1999*)

6. What is the angle between the normals to the surface $xy = z^2$ at the points
   (1, 4, 2) and (-3, -3, 3)? **Ans.** $\cos^{-1}(1/\sqrt{22})$.

7. Find the angle between the two surfaces at the indicated point of intersection.

   (i) $xy^2z = 3x + z^2$, $3x^2 - y^2 + 2z = 1$; $(1, -2, 1)$.   (*PU 1999*)

   (ii) $x \log z = y^2 - 1$, $x^2y = 2 - z$, $(1, 1, 1)$.   (*IETE, Dec. 2006, JNTU 2003*)

   Ans. (i) $\cos^{-1}(\sqrt{6}/14)$, (ii) $\cos^{-1}(1/\sqrt{30})$.

8. (a) Find the normal vector and the equation to tangent plane to surface

   $z = \sqrt{x^2 + y^2}$ at point $(3, 4, 5)$.   (*I.E.T.E., Dec. 2005*)

   Ans. $-\dfrac{3}{5}\mathbf{i} - \dfrac{4}{5}\mathbf{j} + \mathbf{k}$; $3x + 4y - 5z = 0$.

   (b) Find the equation of the tangent plane to the ellipsoid $x^2 + 4y^2 + z^2 = 18$ at the point $(1, 2, 1)$, and determine the acute angle that this plane makes with the xy-plane.   **Ans.** $x + 8y + z = 18$; $\approx 83°$.

9. Find the value of λ and μ so that the surfaces $\lambda x^2 - \mu yz = (\lambda + 2)x$ and $4x^2y + z^3 = 4$ intersect orthogonally at $(1, -1, 2)$.

   (*I.E.T.E., S-2009, W-2008, W-2005; MDU 2005*) **Ans.** $\lambda = 5/2$, $\mu = 1$.

10. Find the constants a and b such that the surface $ax^2 - byz = (a + 2)x$ will be orthogonal to the surface $4x^2y + z^3 = 4$ at the point $(1, -1, 2)$.

    (*AMIE, S-2007*) **Ans.** $a = 5/2$, $b = 1$

    [Hint: Since $(1, -1, 2)$ lies on $ax^2 - 2byz = (a + 4)x$, we get $b = 1$. Use the condition of orthogonality to get a.]

11. If $\nabla f = 2xy\,z^3\mathbf{i} + x^2z^3\,\mathbf{j} + 3x^2yz^2\,\mathbf{k}$, find $f(x, y, z)$ if $f(1, -2, 2) = 4$.

    Ans. $f(x, y, z) = x^2yz^3 + 20$.

12. Find the directional derivative of

    (i) $f(x, y) = xe^y + \cos(xy)$ at $(2, 0)$ in the direction of $3\mathbf{i} - 4\mathbf{j}$.

    (ii) $f(x, y, z) = 4xz^3 - 3x^2y^2z^2$ at $(2, -1, 2)$ along z-axis.

    (iii) $f(x, y, z) = x^2yz + 4xz^2$ at $(1, 2, -1)$ in the direction $2\mathbf{i} - \mathbf{j} - 2\mathbf{k}$.

    (*AMIE., W-2008; MDU 2006; VTU 2001*)

    (iv) $\phi(x, y, z) = xy^2 - 4x^2y + z^2$ at $(1, -1, 2)$ in the direction of $6\mathbf{i} + 2\mathbf{j} + 3\mathbf{k}$.

    (v) $f(x, y, z) = 4xz^3 - 3x^2y^2z$ at $(2, -1, 2)$ in the direction $2\mathbf{i} - 3\mathbf{j} + 6\mathbf{k}$.

    (vi) $\phi(x, y, z) = 4\,e^{2x - y + z}$ at $(1, 1, -1)$ in the direction of vector $\mathbf{j}$.

    (vii) $\phi = 3e^{2x - y + z}$ at $A(1, 1, -1)$ in the direction $\overrightarrow{AB}$ where B is the point $(-3, 5, 6)$.   (*IETE, Dec. 2001*)

    (viii) $\phi(x, y) = x/(x^2 + y^2)$ in the direction of a line making an angle of $30°$ with the positive x-axis at the point $(0, 1)$.   (*IETE, Dec. 2002*)

    Ans. (i) -1, (ii) 48, (iii) 37/3, (iv) 54/7, (v) 376/7, (vi) -4 (vii) $-5/3$, (viii) $\sqrt{3}/2$.

13. What is the directional derivative of the scalar function $f = xy^2 + yz^2$ at the point $(2, -1, 1)$ in the direction of normal to the surface $x \log z - y^2 + 4 = 0$ at $(-1, 2, 1)$? (*MDU 2007; IETE, W-2004; Andhra, 2000*) **Ans.** $15/\sqrt{17}$.

14. Find the directional derivative of $\vec{V}^2$ where $\vec{V} = xy^2\mathbf{i} + zy^2\mathbf{j} + xz^2\mathbf{k}$ at the point $(2, 0, 3)$ in the direction of the outward normal to the sphere $x^2 + y^2 + z^2 = 14$ at the point $(3, 2, 1)$. (*IETE Dec. 2001*) **Ans.** $702\sqrt{14}/7$

15. (*a*) Find the directional derivative of the function $\phi = 4e^{2x-y+z}$ at the point $A(1, 1, -1)$ in the direction towards the point $B(-3, 5, 6)$. **Ans.** $-20/9$.

(*b*) In what direction from the point $(1, 1, -1)$ is the directional derivative of $\phi = x^2 - 2y^2 + 4z^2$ a maximum? Also find the value of this maximum directional derivative. **Ans.** $2\mathbf{i} - 4\mathbf{j} - 8\mathbf{k}$; $2\sqrt{21}$.

16. What is the greatest rate of increase of

(*a*) $\phi(x, y, z) = x^2 + yz^2$ at $(1, -1, 3)$? (*A.M.I.E., S-2004; Madurai 1998*)

(*b*) $f(x, y, z) = xyz^2$ at $(1, 0, 3)$? **Ans.** (*a*) 11; (*b*) 9

17. Find the directions in which $\phi(x, y) = (x^2 + y^2)/2$ increases most rapidly and decreases most rapidly at the point $(1, 1)$.

Ans. $\dfrac{1}{\sqrt{2}}\mathbf{i} + \dfrac{1}{\sqrt{2}}\mathbf{j}$; $-\dfrac{1}{\sqrt{2}}\mathbf{i} - \dfrac{1}{\sqrt{2}}\mathbf{j}$.

18. (*a*) Find the directional derivative of $f(x, y, z) = 2xy - z^2$ at $(2, -1, 1)$ in a direction toward $(3, 1, -1)$. (*b*) In what direction is the directional derivative a maximum? (*c*) What is the value of this maximum?

Ans. (*a*) 10/3, (*b*) $-2\mathbf{i} + 4\mathbf{j} - 2\mathbf{k}$ (*c*) $2\sqrt{6}$.

19. (*a*) Given that the directional derivative of $f(x, y, z)$ at the point $(3, -2, 1)$ in the direction of $\mathbf{a} = 2\mathbf{i} - \mathbf{j} - 2\mathbf{k}$ is -5 and that $|\nabla f(3, -2, 1)| = 5$, find $\nabla f(3, -2, 1)$. **Ans.** $-5(2\mathbf{i} - \mathbf{j} - 2\mathbf{k})/3$.

(*b*) Find the directional derivative of $\phi = 5x^2y - 5y^2z + 2.5z^2x$ at the point $P(1,1,1)$ in the direction of the line $\dfrac{x-1}{2} = \dfrac{y-3}{-2} = z$.

(*UPTU 2004*) **Ans.** 35/3.

20. The temperature at a point (x, y, z) in a space is given by $T(x, y, z) = x^2 + y^2 - z$. A mosquito located at $(1, 1, 2)$ desires to fly in such a direction that it will get warm as soon as possible. In what direction should it fly?

(*A.M.I.E., S-2005*) **Ans.** $(2\mathbf{i} + 2\mathbf{j} - \mathbf{k})/3$

21. The temperature at a point (x, y, z) in a space is given by $T(x, y, z) = x^2 + y^2 - z$. A fly located at $(4, 4, 2)$ desires to fly in a direction that gets cooler faster. Find the direction in which it should fly. Also find the rate decrease of temperature in the direction of flight. (*I.E.T.E., June 2005*)

Ans. In the direction of maximum rate of decrease, $-(8\mathbf{i} + 8\mathbf{j} - \mathbf{k})$

22. Find the values of a, b, c so that the directional derivative of $\phi = ax^2 + by^2 + cz^2$ at the point $(1, 1, 2)$ has a maximum magnitude 4 in the direction of y-axis. [**Hint:** $\nabla\phi$ at $(1, 1, 2) = 4\mathbf{j}$]   **Ans.** $a = 0, b = 2, c = 0$.

23. If the directional derivative of $\phi = ax^2y + by^2z + cz^2x$ at $(1, 1, 1)$ has maximum magnitude 15 in the direction parallel to line $\dfrac{x-1}{2} = \dfrac{y-3}{-2} = z$, find the value of a, b, c.   *(IETE June 2003; UPTU 2002)*

[**Hint:** The maximum directional derivative of ϕ at any point is the value of $\nabla\phi$ at the point. $\nabla\phi = (2axy + cz^2)\mathbf{i} + (2byz + ax^2)\mathbf{j} + (by^2 + 2czx)\mathbf{k}$.
$\nabla\phi(1, 1, 1) = (2a + c)\mathbf{i} + (2b + a)\mathbf{j} + (b + 2c)\mathbf{k}$

$$= 15\frac{(2\mathbf{i} - 2\mathbf{j} + \mathbf{k})}{\sqrt{4+4+1}} = 10\mathbf{i} - 10\mathbf{j} + 5\mathbf{k}.$$

$\Rightarrow 2a + c = 10,\ 2b + a = -10,\ b + 2c = 5$. Solving, we get $a = 20/9$; $b = -55/9, c = 50/9$].

24. Find $\phi(r)$ such that $\nabla\phi = \mathbf{r}/r^5$ and $\phi(1) = 0$ **Ans.** $\dfrac{1}{3}\left(1 - \dfrac{1}{r^3}\right)$

25. If $f(x, y, z) = x^2z + e^{y/x}$ and $g(x, y, z) = 2 z^2y - xy^2$, find (a) $\nabla(f + g)$ and (b) $\nabla(fg)$ at the point $(1, 0, -2)$.   **Ans.** (a) $-4\mathbf{i} + 9\mathbf{j} + \mathbf{k}$, (b) $-8\mathbf{j}$

26. If $f(x, y) = \log_e \sqrt{x + y^2}$, show that

$$\text{grad } f = \frac{\mathbf{r} - (\mathbf{k} \cdot \mathbf{r})\mathbf{k}}{\{\mathbf{r} - (\mathbf{k} \cdot \mathbf{r})\mathbf{k}\} \cdot \{\mathbf{r} - (\mathbf{k} \cdot \mathbf{r})\mathbf{k}\}}, \text{ where } \mathbf{r} = x\mathbf{i} + y\mathbf{j} + z\mathbf{k}.$$

[**Hint.** We have $\mathbf{k} \cdot \mathbf{r} = z$.

Therefore, $\mathbf{r} - (\mathbf{k} \cdot \mathbf{r})\mathbf{k} = (x\mathbf{i} + y\mathbf{j} + z\mathbf{k}) - z\mathbf{k} = x\mathbf{i} + y\mathbf{j} = \mathbf{r}_1$ say)

We have to show that $\nabla f = \dfrac{\mathbf{r}_1}{|\mathbf{r}_1^2|} = \dfrac{x\mathbf{i} + y\mathbf{j}}{x^2 + y^2}$. Now $f(x, y) = \dfrac{1}{2}\log(x^2 + y^2)$.

Therefore, $\nabla f = \left(\mathbf{i}\dfrac{\partial}{\partial x} + \mathbf{j}\dfrac{\partial}{\partial y}\right)\dfrac{1}{2}\log(x^2 + y^2)$

$$= \frac{1}{2}\left[\frac{2x\mathbf{i}}{x^2 + y^2} + \frac{2y\mathbf{j}}{x^2 + y^2}\right] = \frac{x\mathbf{i} + y\mathbf{j}}{x^2 + y^2}.$$]

12.6 Taylor's Theorem for a function of two variables

We know by Taylor's theorem that for a function $f(x)$ of single variable x,

$$f(x+h) = f(x) + hf'(x) + \frac{h^2}{2!}f''(x) + \frac{h^3}{3!}f'''(x) + \ldots\ldots$$

Now let $f(x, y)$ be a function of two independent variables x and y. *If y is kept constant*, then by Taylor's theorem for a function of single variable x, we have

$$f(x+h, y+k) = f(x, y+k) + h\frac{\partial}{\partial x} f(x, y+k) + \frac{h^2}{2!}\frac{\partial^2}{\partial x^2} f(x, y+k) +$$

$$\frac{h^3}{3!}\frac{\partial^3}{\partial x^3} f(x, y+k) + ... \qquad ...(12.10)$$

Now *keeping x constant* and applying Taylor's theorem for a function of single variable y, we have

$$f(x, y+k) = f(x, y) + k\frac{\partial}{\partial y} f(x, y) + \frac{k^2}{2!}\frac{\partial^2}{\partial y^2} f(x, y) + \frac{k^3}{3!}\frac{\partial^3}{\partial y^3} f(x,y) + ...$$

$$...(12.11)$$

Using (12.11), we can write (12.10) in the form

$$f(x+h, y+k) = \left\{ f(x, y) + k\frac{\partial}{\partial y} f(x, y) + \frac{k^2}{2!}\frac{\partial^2}{\partial y^2} f(x, y) \right.$$

$$\left. + \frac{k^3}{3!}\frac{\partial^3}{\partial y^3} f(x, y) + ... \right\}$$

$$+ h\frac{\partial}{\partial x}\left\{ f(x, y) + k\frac{\partial}{\partial y} f(x, y) + \frac{k^2}{2!}\cdot\frac{\partial^2}{\partial y^2} f(x, y) + ... \right\}$$

$$+ \frac{h^2}{2!}\frac{\partial^2}{\partial x^2}\left\{ f(x, y) + k\frac{\partial}{\partial y} f(x, y) + ... \right\} + \frac{h^3}{3!}\frac{\partial^3}{\partial x^3}\left[f(x, y) + ... \right] + ...$$

$$= \left\{ f(x, y) + k\frac{\partial}{\partial y} f(x, y) + \frac{k^2}{2!}\frac{\partial^2}{\partial y^2} f(x, y) + \frac{k^3}{3!}\frac{\partial^3}{\partial y^3} f(x, y) + ... \right\}$$

$$+ \left\{ h\frac{\partial}{\partial x} f(x, y) + hk\frac{\partial^2}{\partial x \partial y} f(x, y) + \frac{hk^2}{2!}\frac{\partial^3}{\partial x \partial y^2} f(x, y) + ... \right\}$$

$$+ \left\{ \frac{h^2}{2!}\frac{\partial^2}{\partial x^2} f(x, y) + \frac{h^2 k}{2!}\frac{\partial^3}{\partial x^2 \partial y} f(x, y) + ... \right\} + \frac{h^3}{3!}\frac{\partial^3}{\partial x^3} f(x, y) + ...$$

$$= f(x,y) + \left\{ h\frac{\partial}{\partial x} f(x,y) + k\frac{\partial}{\partial y} f(x,y) \right\}$$

$$+\left\{\frac{h^2}{2!}\frac{\partial^2}{\partial x^2}f(x,y)+hk\frac{\partial^2}{\partial x \partial y}f(x,y)+\frac{k^2}{2!}\frac{\partial^2}{\partial y^2}f(x,y)\right\}$$

$$+\left\{\frac{h^3}{3!}\frac{\partial^3}{\partial x^3}f(x,y)+\frac{h^2 k}{2!}\frac{\partial^3}{\partial x^2 \partial y}f(x,y)+\frac{hk^2}{2!}\frac{\partial^3}{\partial x \partial y^2}f(x,y)\right.$$

$$\left.+\frac{k^3}{3!}\frac{\partial^3}{\partial y^3}f(x,y)\right\}+...$$

$$=f(x,y)+\left\{h\frac{\partial}{\partial x}f(x,y)+k\frac{\partial f}{\partial y}(x,y)\right\}$$

$$+\frac{1}{2!}\left\{h^2\frac{\partial^2}{\partial x^2}f(x,y)+2hk\frac{\partial^2}{\partial x \partial y}f(x,y)+k^2\frac{\partial^2}{\partial y^2}f(x,y)\right\}$$

$$+\frac{1}{3!}\left\{h^3\frac{\partial^3}{\partial x^3}f(x,y)+3h^2 k\frac{\partial^3}{\partial x^2 \partial y}f(x,y)+3hk^2\frac{\partial^3}{\partial x \partial y^2}\right.$$

$$\left. f(x,y)+k^3\frac{\partial^3}{\partial y^3}f(x,y)\right\}+... \qquad \qquad ...(12.12)$$

The above result can be written symbolically as

$$f(x+h,\,y+k)=f(x,y)+\left(h\frac{\partial}{\partial x}+k\frac{\partial}{\partial y}\right)f(x,y)$$

$$+\frac{1}{2!}\left(h\frac{\partial}{\partial x}+k\frac{\partial}{\partial y}\right)^2 f(x,y)+\frac{1}{3!}\left(h\frac{\partial}{\partial x}+k\frac{\partial}{\partial y}\right)^3 f(x,.y)+...$$

where $\left(h\dfrac{\partial}{\partial x}+k\dfrac{\partial}{\partial y}\right)^2$ stands for the operator $h^2\dfrac{\partial^2}{\partial x^2}+2hk\dfrac{\partial^2}{\partial x \partial y}+k^2\dfrac{\partial^2}{\partial y^2}.$

Similarly for $\left(h\dfrac{\partial}{\partial x}+k\dfrac{\partial}{\partial y}\right)^3$ and etc.

Remarks. 1. Putting $x = a$ and $y = b$ and denoting the value of $\dfrac{\partial f}{\partial x}$ for $x = a$,

$y = b$ by $f_x(a, b)$, and etc. we get from (12.12)

$f(a + h, b + k) = f(a, b) + [hf_x(a, b) + kf_y (a, b)]$

$$+ \frac{1}{2!}\ [h^2 f_{xx}\ (a,\ b) + 2hk f_{xy}(a,\ b) + k^2 f_{yy}(a,\ b)]$$

$$+ \frac{1}{3!}\Big[h^3 f_{xxx}\left(a,\ b\right)+3h^2 kf_{xxy}\left(a,\ b\right)+3hk^2 f_{xyy}\left(a,\ b\right)+k^3 f_{yyy}\left(a,\ b\right)\Big]+... \quad (12.13)$$

2. If we put $a + h = x$ and $b + k = y$ so that $h = (x - a)$ and $k = (y - b)$, the result is *Taylor's Theorem for a function of two variables,*

$$f(x,\ y) = f(a,\ b) + [(x - a)f_x\ (a,\ b) + (y - b)f_y\ (a,\ b)] +$$

$$+ \frac{1}{2!}\Big[(x-a)^2 f_{xx}\left(a,\ b\right)+2(x-a)(y-b)f_{xy}\left(a,\ b\right)+(y-b)^2 f_{yy}\left(a,\ b\right)\Big]+...$$

$$...(12.14)$$

3. Setting $a = b = 0$ in eqn. (12.14), we obtain

$$f\left(x,\ y\right) = f\left(0,\ 0\right)+\left\{x f_x\left(0,\ 0\right)+y f_y\left(0,\ 0\right)\right\}+\frac{1}{2!}\left\{x^2 f_{xx}\left(0,\ 0\right)\right.$$

$$+2xy f_{xy}\left(0,\ 0\right)+y^2 f_{yy}\left(0,\ 0\right)\right\}$$

$$+\frac{1}{3!}\left\{x^3 f_{xxx}\left(0,\ 0\right)+3x^2 yf_{xxy}\left(0,\ 0\right)+3xy^2 f_{xyy}\left(0,\ 0\right)+y^3 f_{yyy}\left(0,\ 0\right)\right\}+......(12.15)$$

This expansion is known as Maclaurin's expansion for functions of two variables $f(x,\ y)$.

4. The expansion (12.14) is called Taylor's expansion of $f(x,\ y)$ at $(a,\ b)$ or in the neighbourhood of $(a,\ b)$ or in powers of $(x - a)$ and $(y - b)$. Obviously (12.15) is expansion of $f(x,\ y)$ at $(0,\ 0)$ (or in the neighbourhood of $(0,\ 0)$ or in powers of x and y).

12.7 Illustrative Examples

Example 12.20 Expand $e^x \sin y$ by Taylor's series in powers of x and y as the terms of third degree. *(A.M.I.E., S-2002, I.E.T.E., Dec. 2001, June 2002)*

Solution. We have $f(x,\ y) = e^x \sin y$; $\qquad\qquad f(0,\ 0) = 0$

$f_x(x,\ y) = e^x \sin y$; $\qquad\qquad f_x(0,\ 0) = 0$

$f_y(x,\ y) = e^x \cos y$; $\qquad\qquad f_y(0,\ 0) = 1$

$f_{xx}(x,\ y) = e^x \sin y$; $\qquad\qquad f_{xx}(0,\ 0) = 0$

$f_{xy}(x,\ y) = e^x \cos y$; $\qquad\qquad f_{xy}(0,\ 0) = 1$

$f_{yy}(x,\ y) = -e^x \sin y$; $\qquad\qquad f_{yy}(0,\ 0) = 0$

$f_{xxx}\left(x,y\right)\left\{i.e.,f_{x^3}\left(x,y\right)\right\} = e^x \sin y$; $\qquad fx^3\left(0,0\right) = 0$

$$f_{xxy}(x,y)\{i.e., f_{x^2y}(x,y)\} = e^x \cos y; \qquad f_{x^2y}(0,0) = 1$$

$$fxy^2(x,y) = -e^x \sin y; \qquad\qquad f_{xy^2}(0,0) = 0$$

$$fy^3(x,y) = -e^x \cos y, \qquad\qquad f_{y^3}(0,0) = -1$$

....

....

Substituting these values in Taylor's series, we obtain

$$e^x \sin y = 0 + x(0) + y(1) + \frac{1}{2!}\left[x^2(0) + 2xy(1) + y^2(0)\right]$$

$$+ \frac{1}{3!}\left[x^3(0) + 3x^2y(1) + 3xy^2(0) + y^3(-1)\right] + ...$$

$$= y + xy + \frac{1}{2}x^2y - \frac{1}{6}y^3 + ...$$

Example 12.21 Expand $\dfrac{(x+h)(y+k)}{x+h+y+k}$ in powers of h and k upto and inclusive

of the second degree terms. (A.M.I.E., S-2001, Nagpur 1995)

Solution. We have $f(x+h, y+k) = \dfrac{(x+h)(y+k)}{x+h+y+k}$. Therefore $f(x, y) = \dfrac{xy}{x+y}$.

$$f_x = \frac{(x+y)y - xy}{(x+y)^2} = \frac{y^2}{(x+y)^2}; \ f_y = \frac{x^2}{(x+y)^2}; \ f_{xx} = -\frac{2y^2}{(x+y)^3};$$

$$f_{xy} = \frac{(x+y)^2 \cdot 2x - x^2 \cdot 2(x+y)}{(x+y)^4} = \frac{2xy}{(x+y)^3}; \ f_{yy} = -\frac{2x^2}{(x+y)^3}.$$

$$\therefore \quad \frac{(x+h)(y+k)}{x+h+y+k} = f(x+h, y+k) = f(x, y) + (hf_x + kf_y)$$

$$+ \frac{1}{2!}\left(h^2 f_{xx} + 2hk f_{xy} + k^2 f_{yy}\right) + ...$$

$$= \frac{xy}{x+y} + \left(\frac{hy^2}{(x+y)^2} + \frac{kx^2}{(x+y)^2}\right) + \frac{1}{2!}\left(\frac{-2h^2y^2}{(x+y)^3} + \frac{4hkxy}{(x+y)^3} - \frac{2k^2x^2}{(x+y)^3}\right) + ...$$

$$= \frac{xy}{x+y} + \frac{1}{(x+y)^2}\left(hy^2 + kx^2\right) - \frac{1}{(x+y)^3}\left(h^2y^2 - 2hkxy + k^2x^2\right) + ...$$

Example 12.22 Expand $f(x, y) = x^2y + 3y - 2$ in powers of $(x - 1)$ and $(y + 2)$ upto 3rd degree terms.   (*A.M.I.E., S-2004; I.E.T.E., W-2003; Assam 1998*)

Solution. The expansion of $f(x, y)$ at the point (a, b) or in powers of $(x - a)$ and $(y - b)$ is given by

$$f(x, y) = f(a, b) + \{(x-a)f_x(a, b) + (y-b)f_y(a, b)\}$$

$$+ \frac{1}{2!}\left[(x-a)^2 f_{xx}(a, b) + 2(x-a)(y-b)f_{xy}(a, b) + (y-b)^2 f_{yy}(a,b)\right]$$

$$+ \frac{1}{3!}\left[(x-a)^3 f_{xxx}(a, b) + 3(x-a)^2(y-b)f_{xxy}\right.$$

$$\left. + 3(x-a)(y-b)^2 f_{xyy}(a, b) + (y-b)^3 f_{yyy}\right] + ... \qquad ...(1)$$

We have

$f(x, y) = x^2y + 3y - 2; a = 1, b = -2,$	$f(1, -2) = -10$
$f_x(x, y) = 2xy;$	$f_x(1, -2) = -4$
$f_y(x, y) = x^2 + 3;$	$f_y(1, -2) = 4$
$f_{xx}(x, y) = 2y;$	$f_{xx}(1, -2) = -4$
$f_{xy}(x, y) = 2x;$	$f_{xy}(1, -2) = 2$
$f_{yy}(x, y) = 0;$	$f_{yy}(1, -2) = 0$
$f_{xxx}(x, y) = 0;$	$f_{xxx}(1, -2) = 0$
$f_{xxy}(x, y) = 2;$	$f_{xxy}(1, -2) = 2$
$f_{xyy}(x, y) = 0;$	$f_{xyy}(1, -2) = 0$
$f_{yyy}(x, y) = 0;$	$f_{yyy}(1, -2) = 0.$

Substituting these values in (1), we obtain

$$f(x, y) = x^2y + 3y - 2 = -10 + \left[(x-1)(-4) + (y+2)(4)\right]$$

$$+ \frac{1}{2!}\left[(x-1)^2(-4) + 2(x-1)(y+2)(2) + (y+2)^2(0)\right]$$

$$+ \frac{1}{3!}\left[0 + 3(x-1)^2(y+2)(2) + 0 + 0\right]$$

$$= -10 - 4(x-1) + 4(y+2) - 2(x-1)^2 + 2(x-1)(y+2) + (x-1)^2(y+2).$$

Example 12.23 Expand $f(x, y) = \tan^{-1}(y/x)$ about $(1, 1)$ up to the third degree terms. Hence compute $f(1.1, 0.9)$ approximately.   (*A.M.I.E., W-1996; S-2002*)

Solution. We have $f(x, y) = \tan^{-1}(y/x)$;

	$f(1, 1) = \pi/4$
$f_x(x, y) = -y/(x^2 + y^2);$	$f_x(1, 1) = -1/2$
$f_y(x, y) = x/(x^2 + y^2);$	$f_y(1, 1) = 1/2$
$f_{xx}(x, y) = 2xy/(x^2 + y^2)^2;$	$f_{xx}(1, 1) = 1/2$
$f_{xy}(x, y) = (y^2 - x^2)/(x^2 + y^2)^2;$	$f_{xy}(1, 1) = 0$

$$f_{yy}(x, y) = -2xy/(x^2 + y^2)^2; \qquad\qquad f_{yy}(1, 1) = -1/2$$
$$f_{xxx}(x, y) = (2y^3 - 6x^2y/(x^2 + y^2)^3; \qquad f_{xxx}(1, 1) = -1/2$$
$$f_{xxy}(x, y) = (2x^3 - 6xy^2/(x^2 + y^2)^3; \qquad f_{xxy}(1, 1) = -1/2$$
$$f_{xyy}(x, y) = (6x^2y - 2y^3)/(x^2 + y^2)^3; \qquad f_{xyy}(1, 1) = 1/2$$
$$f_{yyy}(x, y) = (6xy^2 - 2x^3)/(x^2 + y^2)^3; \qquad f_{yyy}(1, 1) = 1/2$$

Taylor's expansion of $f(x, y)$ in powers of $(x - 1)$ and $(y - 1)$ is given by

$$f(x, y) = f(1, 1) + [(x - 1)f_x(1, 1) + (y - 1)f_y(1, 1)]$$

$$+ \frac{1}{2!}\left[(x-1)^2 f_{xx}(1, 1) + 2(x-1)(y-1)f_{xy}(1, 1) + (y-1)^2 f_{yy}(1, 1)\right]$$

$$+ \frac{1}{3!}\left[(x-1)^3 f_{xxx}(1, 1) + 3(x-1)^2(y-1)f_{xxy}(1, 1)\right.$$

$$\left. + 3(x-1)(y-1)^2 f_{xyy}(1, 1) + (y-1)^3 f_{yyy}(1, 1)\right] + \dots$$

$$\therefore \quad \tan^{-1}\left(\frac{y}{x}\right) = \frac{\pi}{4} + \left[(x-1)\left(-\frac{1}{2}\right) + (y-1)\left(\frac{1}{2}\right)\right]$$

$$+ \frac{1}{2!}\left[(x-1)^2\left(\frac{1}{2}\right) + 2(x-1)(y-1)(0) + (y-1)^2\left(-\frac{1}{2}\right)\right]$$

$$+ \frac{1}{3!}\left[(x-1)^3\left(-\frac{1}{2}\right) + 3(x-1)^2(y-1)\left(-\frac{1}{2}\right) + 3(x-1)\right.$$

$$\left. (y-1)^2\left(\frac{1}{2}\right) + (y-1)^3\left(\frac{1}{2}\right)\right] + \dots$$

$$= \frac{\pi}{4} - \frac{1}{2}\left[(x-1) - (y-1)\right] + \frac{1}{4}\left[(x-1)^2 - (y-1)^2\right]$$

$$- \frac{1}{12}\left[(x-1)^3 + 3(x-1)^2(y-1) - 3(x-1)(y-1)^2 - (y-1)^3\right] + \dots$$

Putting $x = 1.1$ and $y = 0.9$, we obtain

$$f(1.1, 0.9) \approx \frac{\pi}{4} - \frac{1}{2}(0.2) + \frac{1}{4}(0) - \frac{1}{12}\left[(0.1)^3 - 3(0.1)^3 - 3(0.1)^3 - (-0.1)^3\right]$$

$$\approx 0.7854 - 0.1 + 0.0003 \approx 0.6857.$$

Example 12.24 (*a*) Expand $f(x, y) = \sin xy$ in powers of $(x - 1)$ and $(y - \pi/2)$ upto second degree terms. *(A.M.I.E., W-2000, 2004; I.E.T.E., June 1996)*

(b) Find the linear and the quadratic Taylor's series polynomial approximations to the function $f(x, y) = 2x^3 - 4x^2y + 3y^3$, about the point $(1, 2)$

(I.E.T.E, Dec. 2006; A.M.I.E., W-2002)

Solution. (a) We have $f(x, y) = \sin xy$;

$f_x(x, y) = y \cos (xy)$;

$f_y(x, y) = x \cos (xy)$;

$f_{xx} (x, y) = -y^2 \sin(xy)$;

$f_{xy} = \cos(xy) + xy (-\sin xy)$;

$f_{yy} = -x^2 \sin (xy)$;

....

....

$f(1, \pi/2) = 1$

$f_x(1, \pi/2) = 0$

$f_y(1, \pi/2) = 0$

$f_{xx}(1, \pi/2) = -\pi^2/4$

$f_{xy}(1, \pi/2) = -\pi/2$

$f_{yy}(1, \pi/2) = -1$.

....

Substituting the values of partial derivatives obtained above in Taylor's series expansion, we obtain

$$\sin(xy) = 1 + (x-1)(0) + \left(y - \frac{\pi}{2}\right)(0)$$

$$+ \frac{1}{2!}\left[(x-1)^2\left(-\frac{\pi^2}{4}\right) + 2(x-1)\left(y-\frac{\pi}{2}\right)\left(-\frac{\pi}{2}\right) + \left(y-\frac{\pi}{2}\right)^2(-1)\right] + ...$$

or $$\sin(xy) = 1 - \frac{\pi^2}{8}(x-1)^2 - \frac{\pi}{2}(x-1)\left(y-\frac{\pi}{2}\right) - \frac{1}{2}\left(y-\frac{\pi}{2}\right)^2 + ...$$

(b) We have

$f(x, y) = 2x^3 - 4x^2y + 3y^3$;

$f_x(x, y) = 6x^2 - 8xy$;

$f_y(x, y) = -4x^2 + 9y^2$;

$f_{xx}(x, y) = 12x - 8y$;

$f_{xy}(x, y) = -8x$;

$f_{yy}(x, y) = 18y$;

$f(1, 2) = 2 - 8 + 24 = 18$

$f_x(1, 2) = 6 - 16 = -10$

$f_y(1, 2) = -4 + 36 = 32$

$f_{xx}(1, 2) = 12 - 16 = -4$

$f_{xy}(1, 2) = -8$

$f_{yy}(1, 2) = 36$

$f_{xxx}(x,y) = 12, \quad f_{xxy}(x,y) = -8, \quad f_{xyy}(x,y) = 0, \quad f_{yyy}(x,y) = 18$.

The linear approximation in given by

$$f(x,y) \approx f(1,2) + \left[(x-1)f_x(1,2) + (y-2)f_y(1,2)\right]$$

$$= 18 + (x-1)(-10) + (y-2)(32) = 18 - 10(x-1) + 32(y-2).$$

The quadratic approximation is given by

$$f(x,y) \approx f(1,2) + \left[(x-1)f_x(1,2) + (y-2)f_y(1,2)\right]$$

$$+ \frac{1}{2}\left[(x-1)^2 f_{xx}(1,2) + 2(x-1)(y-2)f_{xy}(1,2) + (y-2)^2 f_{yy}(1,2)\right]$$

$$= 18 - 10(x-1) + 32(y-2) + \frac{1}{2}\left[-4(x-1)^2 - 16(x-1)(y-2) + 36(y-2)^2\right]$$

$$= 18 - 10(x-1) + 32(y-2) - 2\left[(x-1)^2 + 4(x-1)(y-2) - 9(y-2)^2\right].$$

Example 12.25 If $f(x, y) = \tan^{-1}(xy)$, find an approximate value of $f(1.1, 0.8)$ using the Taylor's series quadratic approximation. *(IETE, June 2007)*

Solution. Let $(a, b) = (1.0, 1.0)$, $h = 0.1$, $k = -0.2$. Then

$$f(1.1, 0.8) = f(1 + 0.1, 1 - 0.2).$$

Using the Taylor's series quadratic approximation, we have

$$f(1.1, 0.8) \approx f(1, 1) + [hf_x(1, 1) + kf_y(1, 1)] + \frac{1}{2!}[h^2 f_{xx}(1, 1)$$

$$+ 2hk f_{xy}(1, 1) + k^2 f_{yy}(1, 1)] \qquad \qquad ...(1)$$

We have

$$f(x, y) = \tan^{-1}(xy); \qquad\qquad f(1, 1) = \tan^{-1}(1) = \pi/4 \approx 0.7854$$

$$f_x(x, y) = \frac{y}{1 + x^2 y^2}; \qquad\qquad f_x(1, 1) = 0.5$$

$$f_y(x, y) = \frac{x}{1 + x^2 y^2}; \qquad\qquad f_y(1, 1) = 0.5$$

$$f_{xx}(x, y) = -\frac{2xy^3}{\left(1 + x^2 y^2\right)^2}; \qquad\qquad f_{xx}(1, 1) = -0.5$$

$$f_{xy}(x, y) = \frac{\left(1 + x^2 y^2\right) - 2x^2 y^2}{\left(1 + x^2 y^2\right)^2} = \frac{1 - x^2 y^2}{\left(1 + x^2 y^2\right)^2}; \qquad\qquad f_{xy}(1, 1) = 0$$

$$f_{yy}(x, y) = -\frac{2x^3 y}{\left(1 + x^2 y^2\right)^2}, \qquad\qquad f_{yy}(1, 1) = -0.5$$

Therefore by (1), we obtain,

$$f(1.1, 0.8) \approx 0.7854 + \{(0.5)(0.1) + (0.5)(-0.2)\} + \frac{1}{2}\{(0.01)(-0.5)$$

$$+ 2(0.1)(-0.2)(0) + (0.04)(-0.5)\} + \ldots\ldots = 0.7854 - 0.05 - 0.0125 = 0.7229.$$

Exercise 12.3

1. (a) Expand $e^x \cos y$ by Taylor's theorem is powers of x and y as far as terms of the third degree.

Alternative Expand $e^x \cos y$ at the point $(0, 0)$ up to terms of the third degree.

$$\text{Ans. } 1+x+\frac{1}{2!}\left(x^2-y^2\right)+\frac{1}{3!}\left(x^3-3xy^2\right)+...$$

(b) Find the Taylor's series for the function $f(x, y) = 1/xy$ at the point $(2, -1)$ up to terms of the second degree.

$$\text{Ans. } \left(-\frac{1}{2}\right)+\frac{1}{4}(x-2)-\frac{1}{2}(y+1)-\frac{1}{8}(x-2)^2+\frac{1}{4}(x-2)(y+1)-\frac{1}{2}(y+1)^2+...$$

(c) Find a quadratic $f(x, y) = \sin x \sin y$ near the origin. **Ans.** $\sin x \sin y = xy.$

2. Show that $e^y \log(1+x) = x+xy-\frac{x^2}{2}+\frac{1}{2}\left(xy^2-x^2y\right)+\frac{1}{3}x^3+...$

3. (a) Expand $e^x \cos y$ in the neighbourhood of $(1, \pi/4)$. *(A.M.I.E., W-1997)*

$$\text{Ans. } \frac{e}{\sqrt{2}}\left[1+(x-1)-\left(y-\frac{\pi}{4}\right)+\frac{1}{2}(x-1)^2-(x-1)\left(y-\frac{\pi}{4}\right)-\frac{1}{2}\left(y-\frac{\pi}{4}\right)^2+...\right]$$

(b) Expand $e^x \cos y$ in powers of x and $\left(y-\frac{\pi}{2}\right)$ as far as 3rd degree terms, using Taylor's series expansion. *(I.E.T.E., S-2005)*

4. (a) Expand y^2/x^3 at the point $(1, -1)$.

   Ans. $1 - 3(x - 1) - 2(y + 1) + 6(x - 1)^2 + 6(x - 1)(y + 1) + (y + 1)^2 +....$

   (b) Expand $f(x, y) = e^{xy}$ at the point $(1, 1)$ up to terms of the third degree.

$$\text{Ans. } e+(x-1)e+(y-1)e+\frac{1}{2}(x-1)^2 e+2(x-1)(y-1)e+\frac{1}{2}(y-1)^2 e$$

$$+\frac{1}{6}(x-1)^3 e+\frac{3}{2}(x-1)^2(y-1)e+\frac{3}{2}(x-1)(y-1)^2 e+\frac{1}{6}(y-1)^3 e+...$$

5. Find the first six terms of the expansion of the function $e^x \log(1 + y)$ in a Taylor series in the neighbourhood of point $(0, 0)$.

   (A.M.I.E., 1999; Andhra 1999) **Ans.** $y+xy-\frac{y^2}{2}+\frac{1}{2}x^2y-\frac{1}{2}xy^2+\frac{1}{3}y^3+...$

6. (a) Expand $f(x, y) = y^x$ in neighbourhood of $(1, 1)$ upto the terms of second degree. **Ans.** $1 + (y - 1) + (x - 1)(y - 1) + ...$

   (b) Expand the function $f(x, y) = x^y$ in powers of $(x - 1)$ and $(y - 1)$, finding the terms upto the third order included. Use the result to compute (without tables) $(1.02, 1.1)$ and $1.1^{1.02}$. *(I.E.T.E., W-2004)*

   Ans. $1 + (x - 1) + (x - 1)(y - 1) + 1/2(x - 1)^2$
   $(y - 1) +...; f(1.02, 1.1) \approx 1.022; 1.1^{1.02} = 1.1021.$

7. Verify the following expansion.

(a) $\sin(x+y) = x + y - \dfrac{(x+y)^3}{3!} + ...$

(b) $a^x \log(1+y) = y + \dfrac{1}{2}\left(2xy \log a - y^2 + x^2 y \log^2 a - xy^2 \log a\right) + \dfrac{1}{3}y^3 + ...$

8. If $f(x, y) = \tan^{-1}xy$, compute $f(0.9, -1.2)$ approximately. **Ans. 2.306**

9. Expand $xy^2 + \sin xy$ at the point $(1, \pi/2)$ upto terms of the third degree.

Ans. $1 + \dfrac{\pi^2}{4} + \dfrac{\pi^2}{4}(x-1) + \pi\left(y - \dfrac{\pi}{2}\right) + \dfrac{1}{2!}$

$$\left[-\dfrac{\pi^2}{4}(x-1)^2 + \pi(x-1)\left(y - \dfrac{\pi}{2}\right) + \left(y - \dfrac{\pi}{2}\right)^2\right] + ...$$

10. Expand $x^2 y + \sin y + e^x$ in powers of $(x-1)$ and $(y-\pi)$ upto quadratic terms by using Taylor's series. *(A.M.I.E., S-1996)*

Ans. $(\pi + e) + (2\pi + e)(x-1) + \dfrac{1}{2!}(x-1)^2(2\pi + e) + 2(x-1)(y-\pi)$.

11. (a) Expand $\sin\left(\dfrac{x+\pi}{y+2}\right)$ in powers of x and y up to the terms of second

degree and hence find the approximate value of $f(x, y)$ at $x = 0.1$ and $y = 0.1$ by taking $\pi = 3.1416$. *(A.M.I.E., W-1997)*

Ans. $1 - \dfrac{1}{8}x^2 + \pi\dfrac{xy}{8} - \dfrac{\pi^2}{32}y^2 + ...$; 0.9996

(b) Expand $f(x,y) = e^{2x} \cos 3y$, in Taylor's power series of x and y upto quadratic terms. *(A.M.I.E., S-2004)* **Ans.** $1 + 2x + 2x^2 - 9y^2/2 + ...$

12. Expand $f(x,y) = 21 + x - 20y + 4x^2 + xy + 6y^2$ in Taylor series of maximum order about the point $(-1, 2)$.

Ans. $f(x,y) = 6 - 5(x+1) + 3(y-2) + 4(x+1)^2 + (x+1)(y-2) + 6(y-2)^2$.

13. Obtain the second order Taylor's series approximation to the function $f(x,y) = xy^2 + y\cos(x-y)$ about the point $(1, 1)$.

Ans. $f(x,y) = 2 + (x-1) + 3(y-1) + \dfrac{1}{2}\left[-(x-1)^2 + 6(x-1)(y-1) + (\dot{y}-1)^2\right]$

12.8 Maximum and Minimum Values of a Function

Let $z = f(x, y)$ be a function of two independent variables x and y which is defined and continuous in some closed and bounded region R. Let (a, b) be an interior point of R and $(a + h, b + k)$ be a point in its neighbourhood and lies inside R. We define the following:

Maximum Value. *A function* $f(x, y)$ *is said to have a relative (or local) maximum value for* $x = a, y = b$, *if there exists a small neighbourhood of* (a, b) *such that for every point* $(a + h, b + k)$ *of this neighbourhood,*

$$f(a + h, b + k) \leq f(a, b) \quad \text{for all } h \text{ and } k.$$ The point (a, b) is called a point of *relative (or local) maximum.*

Minimum value. *A function* $f(x, y)$ *is said to have a relative (or local) minimum value for* $x = a, y = b$ *if there exists a small neighbourhood of* (a, b) *such that for every point* $(a + h, b + k)$ *of this neighbourhood*

$$f(a+h,b+k) \geq f(a,b) \quad \text{for all } h \text{ and } k.$$ The point (a, b) is called a point of *relative (or local) minimum.*

The points at which maximum/minimum values of the function occur are also called *point of extrema* or the *stationary points* and the maximum and the minimum values taken together are called the *extreme* or *extremum values* of the function.

We can represent $z = f(x, y)$ graphically by surface. From the figure 12.5 (*a*) it is clear that $z = f(x, y)$ has a maximum value for $x = a, y = b$ because $f(a, b)$ *i.e.,* PQ is greater than $f(x, y)$ for all points (x, y) close to (a, b). Thus the maximum value of $z = f(x, y)$ occurs at the top of an elevation.

Similarly, we can easily see from figure 12.5 (*b*) that minimum value occurs at the bottom of a depression.

Thus at a local maximum point the surface *descends* in every direction while at a local minimum point it *ascends* in every direction.

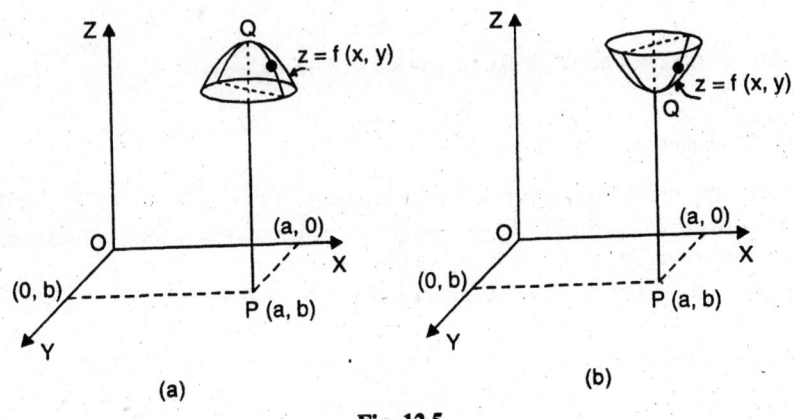

Fig. 12.5.

12.8.1 Absolute maximum and absolute minimum

A function $f(x, y)$ may also attain its maximum or minimum values on the boundary of the region. The greatest and the least values attained by a function over the entire region including the boundary are called the *absolute maximum and absolute minimum* values respectively.

12.9 Conditions for a function to have an Extremum

By definition, a function $f(x, y)$ has an extreme value at (x, y) if $f(x + h, y + k) - f(x, y)$ does not change sign for small values of h, k. If this sign is negative $f(x, y)$ is maximum and in case it is positive $f(x, y)$ is minimum.

By Taylor's Theorem

$$f\left(x+h,\ y+k\right) = f\left(x,\ y\right) + \left(h\frac{\partial f}{\partial x} + k\frac{\partial f}{\partial y} \right)$$

$$+ \frac{1}{2!}\left(h^2\frac{\partial^2 f}{\partial x^2} + 2hk\frac{\partial^2 f}{\partial x \partial y} + k^2\frac{\partial^2 f}{\partial y^2} \right) + \ldots$$

$$\therefore \quad f\left(x+h,\ y+k\right) - f\left(x,\ y\right) = \left(h\frac{\partial f}{\partial x} + k\frac{\partial f}{\partial y} \right)$$

$$+ \frac{1}{2!}\left(h^2\frac{\partial^2 f}{\partial x^2} + 2hk\frac{\partial^2 f}{\partial x \partial y} + k^2\frac{\partial^2 f}{\partial y^2} \right) + \ldots \qquad \ldots(12.16)$$

When h and k are small, the sign on the R.H.S. and, therefore of the L.H.S. is governed by $h\dfrac{\partial f}{\partial x} + k\dfrac{\partial f}{\partial y}$ which changes sign with h and k.

Thus, (x, y) is a point of extreme value if $h\dfrac{\partial f}{\partial x} + k\dfrac{\partial f}{\partial y} = 0$.

Since h and k are independent $\dfrac{\partial f}{\partial x} = 0,\ \dfrac{\partial f}{\partial y} = 0,$ $\qquad \ldots(12.17)$

which is the *necessary condition* for extreme value.

By solving equations (12.17) we get points which may be points of maximum or minimum value. Let (a, b) be one such point.

Now with $x = a, y = b$ and using (12.17) we have from (12.16)

$$f\left(a+h,\ b+k\right) - f\left(a,b\right) = \frac{1}{2!}\left[h^2 f_x^2\left(a,\ b\right) + 2hk\, f_{xy}\left(a,\ b\right) + k^2 f_y^2\left(a,b\right) \right] + \ldots$$

$$= \frac{1}{2!}\left(rh^2 + 2shk + tk^2 \right) + \ldots \qquad \ldots(12.18)$$

where $$r = f_x^2(a, b), s = f_{xy}(a, b), t = f_y^2(a, b).$$

For sufficiently small values of h and k, the sign on the R.H.S. of (12.18) is the same as that of $(rh^2 + 2shk + tk^2)$.

Case I. $rt - s^2 > 0$

In this case $r \neq 0$, $t \neq 0$ and, we have

$$rh^2 + 2shk + tk^2 = \frac{1}{r}\left[(rh + sk)^2 + (rt - s^2)k^2\right].$$

Since $rt - s^2 > 0$, the sign of $rh^2 + 2shk + tk^2$ is always the same as that of r. In this case $f(a, b)$ has an extreme value which is maximum if r is negative and minimum if r is positive.

Case II. $rt - s^2 < 0$.

In case $r \neq 0$, then $rh^2 + 2shk + tk^2 = \frac{1}{r}\left[(rh + sk)^2 + (rt - s^2)k^2\right]$.

Since $rt - s^2 < 0$, the expression has different signs when $k = 0$, and when $rh + sk = 0$. Hence $f(a, b)$ is not an extreme value.

The proof when $t \neq 0$ is similar.

When $r = 0$, $t = 0$, then $rh^2 + 2shk + tk^2 = 2shk$ which assume values with different signs. Hence $f(a, b)$ is not an extreme value in this case.

Case III. $rt - s^2 = 0$.

In this case $rh^2 + 2shk + tk^2 = \frac{1}{r}\left[(rh + sk)^2 + (rt - s^2)k^2\right] = (rh + sk)^2 / r$

which is zero for values of h, k such that $h/k = -s/r$.

This is, therefore, a doubtful case. To determine the nature of sign, further investigation is required.

Working rule to find extremum values of a function z = f(x, y).

To determine the extremum values for $z = f(x, y)$, we proceed as follows:

(i) *Find* $\dfrac{\partial z}{\partial x}$ *and* $\dfrac{\partial z}{\partial y}$ *and equate each to zero. Solve the equations thus ob-*

tained, i.e., $\dfrac{\partial z}{\partial x} = 0$ *and* $\dfrac{\partial z}{\partial y} = 0$ *simultaneously. Let (x_1, y_1), (x_2, y_2) ..., be*

the pairs of roots of these equations.

(ii) *Find* $r = \dfrac{\partial^2 f}{\partial x^2}$, $s = \dfrac{\partial^2 f}{\partial x \partial y}$ *and* $t = \dfrac{\partial^2 f}{\partial y^2}$ *for each of pair of roots.*

(iii) (a) *If $rt - s^2 > 0$, and $r < 0$, then $z = f(x, y)$ has a relative maximum value for that pair.*

(b) *If $rt - s^2 > 0$ and $r > 0$ then the function f has a relative minimum value at that point.*

(c) *If $rt - s^2 < 0$, the function f has no local minimum or local maximum value at that point. In this case, the point $P(a, b)$ is called a saddle point. At a saddle point, the tangent plane to the surface $z = f(x, y)$ is horizontal, but near the point the surface is partly above and partly below this tangent plane.*

(d) *If $rt - s^2 = 0$ no conclusion about an extremum can be drawn and further investigation is needed.*

Remark: The points (x_1, y_1), (x_2, y_2)... which are the roots of $\dfrac{\partial f}{\partial x} = 0$ and $\dfrac{\partial f}{\partial y} = 0$ are called **stationary points**. The function $f(x, y)$ is said to be stationary at (x_1, y_1), (x_2, y_2) etc. and its values there are called **stationary values**. Thus every extreme value is a stationary value but the converse may not be true.

12.10 Illustrative Examples

Example 12.26 Examine $f(x, y) = x^3 + y^3 - 3xy$ for maximum and minimum values.   (*A.M.I.E., W-1996; J.N.T.U. 1999*)

Solution. We have $\dfrac{\partial f}{\partial x} = 3x^2 - 3y$, $\dfrac{\partial f}{\partial y} = 3y^2 - 3x$,

$r = \dfrac{\partial^2 f}{\partial x^2} = 6x$, $s = \dfrac{\partial^2 f}{\partial x \partial y} = -3$, $t = \dfrac{\partial^2 f}{\partial y^2} = 6y$. Now $\dfrac{\partial f}{\partial x} = 0$ and $\dfrac{\partial f}{\partial y} = 0$ give

$$3x^2 - 3y = 0 \ \text{ or } \ x^2 - y = 0, \qquad ...(1)$$
$$\text{and } 3y^2 - 3x = 0 \ \text{ or } \ y^2 - x = 0. \qquad ...(2)$$

We solve (1) and (2) to get the stationary points.

Putting $y = x^2$ from (1) in (2), we get $x^4 - x = 0$ or $x(x - 1)(x^2 + x + 1) = 0$. The real roots of this equation are $x = 0$, $x = 1$.

Putting these values of x in (1) we get $y = 0$ and $y = 1$.

$\therefore$ (0, 0) and (1, 1) are the stationary points.

At (0, 0) $r = 0$, $s = -3$ and $t = 0$.

Since $rt - s^2 = -9 < 0$, there is no extremum value at (0, 0).

At (1, 1) $r = 6$, $s = -3$, $t = 6$. Now $rt - s^2 = 36 - 9 = 27 > 0$. Also $r > 0$.

Therefore (1, 1) is a point of minimum value.

The minimum value is $f(1, 1) = 1 + 1 - 3 = -1$.

Example 12.27 Find the extreme values of $xy(a - x - y)$ (*A.M.I.E., S-2003*)

Alternative: Decompose a positive number 'a' into three parts so that their sum is 'a' and product is maximum.   (*A.M.I.E., W-2009; I.E.T.E., June 1998*)

Solution. Let x = first part, y = second part then $(a - x - y)$ = third part, and function to be examined is $f(x, y) = xy(a - x - y) = axy - x^2y - xy^2$.

$\therefore \; \partial f/\partial x = ay - 2xy - y^2, \; \partial f/\partial y = ax - x^2 - 2xy, \; r = \partial^2 f/\partial x^2 = -2y,$

$s = \partial^2 f/\partial x \partial y = a - 2x - 2y, \; t = \partial^2 f/\partial y^2 = -2x.$

We now solve the simultaneous equations $\partial f/\partial x = 0$ and $\partial f/\partial y = 0$.
We get as one pair of values $x = a/3, \; y = a/3$.
Now $rt - s^2 = 4xy - (a - 2x - 2y)^2$. When $x = a/3$ and $y = a/3$,
$rt - s^2 = a^2/3 > 0$ and $\partial^2 f/\partial x^2 = -2a/3 < 0$.

It is seen that product is a maximum when $x = a/3$ and $y = a/3$. Therefore, the third part is also $a/3$, and the maximum value of the product $= a^3/27$.

Example 12.28 Find the extreme values of $u = x^2y^2 - 5x^2 - 8xy - 5y^2$.

Solution. We have $\dfrac{\partial u}{\partial x} = 2xy^2 - 10x - 8y, \; \dfrac{\partial u}{\partial y} = 2x^2y - 8x - 10y.$

Solving these two simultaneous equations $\dfrac{\partial u}{\partial x} = 0$ and $\dfrac{\partial u}{\partial y} = 0$ i.e.,

$$xy^2 - 5x - 4y = 0 \qquad \qquad ...(1),$$
and $$x^2y - 4x - 5y = 0 \qquad \qquad ...(2).$$

Subtracting (1) from (2) we get $xy(x - y) + x - y = 0$ or $(x - y)(xy + 1) = 0$
Substituting $x = y$ in (1) we get $y^3 - 9y = 0 \; y = 0, \pm 3$.

Again putting $x = -1/y$ in (1) we have $-y^2 + 1 = 0$ or $y = \pm 1$.

Hence $(0, 0), (3, 3), (-3, -3), (-1, 1), (1, -1)$ are five stationary points.

Now, $$r = \frac{\partial^2 u}{\partial x^2} = 2y^2 - 10, \; s = \frac{\partial^2 u}{\partial y \partial x} = 4xy - 8, \; t = \frac{\partial^2 u}{\partial y^2} = 2x^2 - 10.$$

At $(0, 0)$, $r = -10, \; s = -8, \; t = -10$. $\therefore rt - s^2 = 100 - 64 > 0$ and $r < 0$.
Hence $(0, 0)$ is a point of maximum value. The maximum value, obviously, is zero.
At $(3, 3)$, $r = 8, \; s = 28, \; t = 8$. $\therefore \; rt - s^2 < 0$.
Thus there is no extreme value at $(3, 3)$.
Similarly there is no extreme value at $(-3, -3), (-1, 1), (1, -1)$,

Example 12.29 Determine the points where the function $x^4 + y^4 - 2x^2 + 4xy - 2y^2$ has a maximum or minimum. *(A.M.I.E., W-96, Delhi 1997)*

Solution. Let $u = x^4 + y^4 - 2x^2 + 4xy - 2y^2$.

$\therefore \; \dfrac{\partial u}{\partial x} = 4x^3 - 4x + 4y, \quad \dfrac{\partial u}{\partial y} = 4y^3 + 4x - 4y,$

$r = \dfrac{\partial^2 u}{\partial x^2} = 12x^2 - 4, \quad s = \dfrac{\partial^2 u}{\partial x \partial y} = 4, \quad t = \dfrac{\partial^2 u}{\partial y^2} = 12y^2 - 4.$

Now the points of maximum or minimum values are given by

$$\frac{\partial u}{\partial x} = 4x^3 - 4x + 4y = 0, \quad ...(1) \qquad \frac{\partial u}{\partial y} = 4y^3 + 4x - 4y = 0. \quad ...(2)$$

Adding (1) and (2) we get $x^3 + y^3 = 0$ or $x = -y$.
Substituting $x = -y$ in (2) we have $y^3 - 2y = 0$.

$\therefore \quad y = 0, \pm \sqrt{2}$ and the coresponding values of x are $0, \mp \sqrt{2}$.

Hence the stationary points are $(0, 0), \left(-\sqrt{2}, \sqrt{2}\right)$, and $\left(\sqrt{2}, -\sqrt{2}\right)$.

$At\left(-\sqrt{2}, \sqrt{2}\right)$, $r = 20$, $s = 4$, $t = 20$. As $rt - s^2 = 384 > 0$ and $r > 0$.

$\therefore At\left(-\sqrt{2}, \sqrt{2}\right)$, u has a minimum value.

Similarly at $\left(\sqrt{2}, -\sqrt{2}\right)$, the function has a minimum value.

Now at $(0, 0)$, $r = -4$, $s = 4$, $t = -4$, therefore $rt - s^2 = 0$.

Hence at $(0, 0)$ further investigation is needed.

Now $u = x^4 + y^4 - 2x^2 + 4xy - 2y^2 = x^4 + y^4 - 2(x - y)^2$.

At $(0, 0)$, $u = 0$. When x, y are small, $u = -2(x - y)^2$ approximately which is negative when $x \neq y$. When $x = y$, $u = x^4 + y^4$ which is + ve.

Thus in a small neighbourhood of $(0, 0)$ there are points where the function is -ve *i.e.*, $<u (0, 0)$ and also there are points where the function is +ve *i.e.*, $> u (0, 0)$.

Hence at $(0, 0)$ there is no exterme value. $(0, 0)$ is a saddle point.

Example 12.30 Find the absolute maxima and minima of the function $f(x, y) = x^2 + xy + y^2 - 6x$ on the rectangular plate $0 \leq x \leq 5$, $-3 \leq y \leq 3$.

Solution. The function f can attain maximum/minimum values at the critical points or on the boundary of the rectangle ABCD (Refer Fig. 12.6).

We have $f_x = 2x + y - 6 = 0$, $f_y = x + 2y = 0$. The critical point is $(4, -2)$. Now $r = f_{xx} = 2, s = f_{xy} = 1, t = f_{yy} = 2, rt - s^2 = 3$. Since $rt - s^2 > 0$ and $r > 0$, the point $(4, -2)$ is a point of relative minimum. The minimum value is $f(4, -2) = -12$.

On the boundary line AB, we have $y = -3$ and $f(x, y) = f(x, -3) = g(x) = x^2 - 9x + 9$, which is a function of one variable. Setting $dg/dx = 0$, we get $2x - 9 = 0$ or $x = 9/2$. Now $d^2 g/dx^2 = 2 > 0$. Therefore, at $x = 9/2$, the function has a minimum. The minimum value is $g(9/2) = (81/4) - (81/2) + 9 = -45/4$.

Also at the corners $(0, -3), (5, -3)$, we have $f(0, -3) = 9$, $f(5, -3) = -11$.

Similarly, along the other boundary lines, we have the following results.

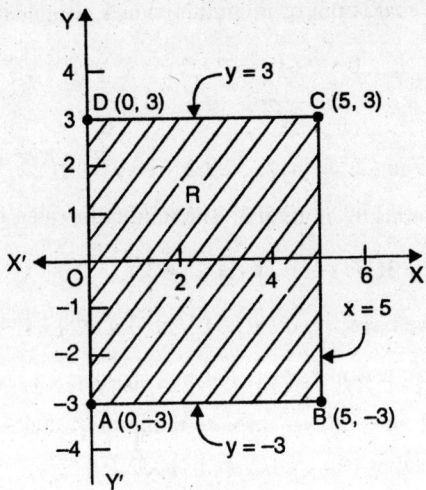

Fig. 12.6.

At $x = 5$: $h(y) = y^2 + 5y - 5$; $dh/dy = 2y + 5 = 0$, gives $y = -5/2$; $d^2h/dy^2 = 2 > 0$. Therefore $y = -5/2$ is a point of minimum. The minimum value is $f(5, -5/2) = -45/4$. At the corner $(5, 3)$ we have $f(5, 3) = 19$.

At $y = 3$: $g(x) = x^2 - 3x + 9$; $dg/dx = 2x - 3 = 0$ gives $x = 3/2$; $d^2g/dx^2 = 2 > 0$. Therefore $x = 3/2$ is a point of minimum. The minimum value is $f(3/2, 3) = 27/4$. At the corner point $(0, 3)$, we have $f(0, 3) = 9$.

At $x = 0$: $h(y) = y^2$, $dh/dy = 2y = 0$, gives $y = 0$; $d^2h/dy^2 = 2 > 0$. Therefore $y = 0$ is a point of minimum. The minimum value is $f(0, 0) = 0$. Therefore, the absolute minimum value is -12 which occur at $(4, 2)$ and the absolute maximum value is 19 which occurs at the point $(5, 3)$.

Example 12.31 Find the absolute maximum and minimum values of the function $f(x, y) = 3x^2 + y^2 - x$ over the region $2x^2 + y^2 \leq 1$.

(A.M.I.E.-W-2007; IETE, W-2008)

Solution. We have $f_x = 6x - 1 = 0$ and $f_y = 2y = 0$. Therefore, the critical point is $(1/6, 0)$. We find that $r = f_{xx} = 6$, $s = f_{xy} = 0$, $t = f_{yy} = 2$ and $rt - s^2 = 12 > 0$. Therefore, $(1/6, 0)$ is a point of minimum. The minimum value at this point is

$$f(1/6, 0) = 3(1/6)^2 - (1/6) = -1/12.$$

On the **boundary**, we have $y^2 = 1 - 2x^2$, $-1/\sqrt{2} \leq x \leq 1/\sqrt{2}$. Substituting in $f(x, y)$, we obtain

$f(x,y) = 3x^2 + (1-2x^2) - x = 1 - x + x^2 = h(x)$, which is a function of one variable. Setting $dh/dx = 0$. $dh/dx = -1 + 2x = 0$ or $x = 1/2$.

Also $d^2h/dx^2 = 2 > 0$.

For $x = 1/2$, we get $y^2 = 1 - 2x^2 = 1/2$ or $y = \pm 1/\sqrt{2}$. Hence, the points $(1/2, \pm 1/\sqrt{2})$ are points of minimum. The minimum value is

$$f(1/2, \pm 1/\sqrt{2}) = 3(1/4) + (1/2) - (1/2) = 3/4.$$

At the vertices, we have $f(1/\sqrt{2}, 0) = (3/2) - (\sqrt{2}/2) = (3 - \sqrt{2})/2,$

$f(-1/\sqrt{2}, 0) = (3/2) + (\sqrt{2}/2) = (3 + \sqrt{2})/2,$ $f(0, \pm 1) = 1.$

Therefore, the given function has absolute minimum value −1/12 at (1/6, 0) and absolute maximum value $(3 + \sqrt{2})/2$ at $(-1/\sqrt{2}, 0)$.

Example 12.32 Given $f(x, y, z) = \dfrac{5xyz}{x + 2y + 4z}$, find the values of x, y, z for which $f(x, y, z)$ is maximum subject to $xyz = 8$.   (*A.M.I.E., W-2002*)

Solution. Since $xyz = 8$, we obtain

$$f(x, y, z) = \frac{5xyz}{x + 2y + 4z} = \frac{5 \times 8}{x + 2y + (32/xy)} \quad \text{or} \quad \frac{40}{f} = x + 2y + (32/xy).$$

Now f is a maximum when $40/f$ is a minimum.

$$\frac{\partial(40/f)}{\partial x} = 1 - \frac{32}{x^2 y} = 0 \text{ and } \frac{\partial(40/f)}{\partial y} = 2 - \frac{32}{xy^2} = 0.$$

Solving these two simultaneous equations, we obtain $x = 4$, $y = 2$.

At (4, 2) We obtain $r = \dfrac{\partial^2}{\partial x^2}\left(\dfrac{40}{f}\right) = \dfrac{64}{x^3 y} = \dfrac{64}{4^3 \times 2} = \dfrac{1}{2},$

$t = \dfrac{\partial^2}{\partial y^2}\left(\dfrac{40}{f}\right) = \dfrac{64}{xy^3} = \dfrac{64}{4 \times 2^3} = 2,$ and $s = \dfrac{\partial^2}{\partial x \partial y}\left(\dfrac{40}{f}\right) = \dfrac{32}{x^2 y^2} = \dfrac{32}{4^2 \times 2^2} = \dfrac{1}{2}.$

Therefore $rt - s^2 = \dfrac{1}{2} \cdot 2 - \left(\dfrac{1}{2}\right)^2 = \dfrac{3}{4} > 0, r > 0.$ Hence

$x = 4$, $y = 2$ and $z = \dfrac{8}{xy} = \dfrac{8}{4 \times 2} = 1$ that is, (4, 2, 1) gives a minimum $40/f$ and a maximum $f(x, y, z)$.

Example 12.33 A long piece of tin 12 unit wide is to be made into a trough by bending up two sides to form equal angles with the base. Find the amount to be bent up and the angle of inclination of each side that will make the carrying capacity a maximum.

Solution. Let x unit be the amount bent up and θ the inclination of the sides.

The carrying capacity will be maximum when the area K of the cross section shown in the figure must be a maximum. The cross-section is a trapezoid of upper base $EF = 12 - 2x + 2x\cos\theta$, lower base $CD = 12 - 2x$ and altitude $= x\sin\theta$. The area, K is given by

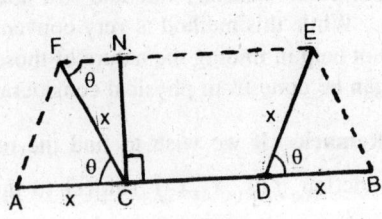

Fig. 12.7

$$K = 12\,x\sin\theta - 2x^2\sin\theta + x^2\sin\theta\cos\theta.$$

By differentiation, we have

$$\partial K/\partial\theta = 12x\cos\theta - 2x^2\cos\theta + x^2(\cos^2\theta - \sin^2\theta)$$
$$= x[12\cos\theta - 2x\cos\theta + x\cos^2\theta - x\sin^2\theta]$$
$$\partial K/\partial x = 12\sin\theta - 4x\sin\theta + 2x\sin\theta\cos\theta = 2\sin\theta\,(6 - 2x + x\cos\theta).$$

Setting the partial derivatives equal to zero, we have the two equations

$$x[12\cos\theta - 2x\cos\theta + x(\cos^2\theta - \sin^2\theta)] = 0 \qquad \text{...(1)}$$

and

$$2\sin\theta\,(6 - 2x + x\cos\theta) = 0 \qquad \text{...(2)}$$

One solution of this system is $\theta = 0$, $x = 0$, which has no meaning in the physical problem. Assuming $\theta \neq 0$, $x \neq 0$, and solving the equations, we get $\cos\theta = 1/2 = \cos 60°$ that is, $\theta = 60°$ and $x = 4$.

A consideration of the physical problem shows that there must exist a maximum value of the area. Hence the maximum value occurs when $\theta = 60°$ and $x = 4$ units.

12.11 Lagrange's method of Multipliers

This method is useful when we have to find the extremum of the function of a number of variables which are not independent but are connected by some given relation or relations.

Supposing we have to find the extremum of the function $f(x, y, z)$ under the condition

$$\phi\,(x, y, z) = 0 \qquad \text{...(12.19)}$$

In such cases we construct an auxiliary function of the form

$$F(x, y, z, \lambda) = f(x, y, z) + \lambda\phi(x, y, z) \qquad \text{...(12.20)}$$

where λ is an undetermined parameter and is known as *Lagrange multipliers.*

Then, to determine the stationary points of F, we have the necessary conditions

$$\frac{\partial F}{\partial x} = 0,\ \frac{\partial F}{\partial y} = 0,\ \frac{\partial F}{\partial z} = 0,\ \text{which give the equations}$$

$$\frac{\partial f}{\partial x} + \lambda \frac{\partial \phi}{\partial x} = 0, \quad \frac{\partial f}{\partial y} + \lambda \frac{\partial \phi}{\partial y} = 0, \quad \frac{\partial f}{\partial z} + \lambda \frac{\partial \phi}{\partial z} = 0.$$
...(12.21)

From these equations and $\phi(x, y, z) = 0$ determine x, y, z and λ, the latter only played an auxiliary role and will not be needed any more.

While this method is very convenient in finding the stationary points, it does not help in finding the nature of those stationary points. In applied problems this can be done from physical considerations.

Remark. If we wish to find the relative maximum or minimum values of a function $f(x_1, x_2, x_3)$ subject to the constraint conditions $\phi_1(x_1, x_2, x_3) = 0$, $\phi_2(x_1, x_2, x_3) = 0$, we form the auxiliary function

$$F(x_1, x_2, x_3, \lambda_1, \lambda_2) = f(x_1, x_2, x_3) + \lambda_1 \phi_1(x_1, x_2\ x_3) + \lambda_2 \phi_2(x_1, x_2\ x_3)$$

i.e., $F(x_1, x_2, x_3, \lambda_1, \lambda_2) = f + \lambda_1 \phi_1 + \lambda_2 \phi_2$ subject to the necessary conditions $\partial F / \partial x_1 = 0, \partial F / \partial x_2 = 0, \partial F / \partial x_3 = 0$, where λ_1, λ_2 are independent of x_1, x_2, x_3 are the *Lagrange multipliers*.

12.12 Illustrative Examples

Example 12.34 A rectangular box open at the top is to have a given capacity. Find the dimensions of the box requiring least material for its construction.

(IETE, June 2009; A.M.I.E., S-2001, W-2004; J.N.T.U 1998)

Solution. Let x, y, z be the dimensions of the box, S its surface area and V its capacity.

Now $V = xyz$ and $S = xy + 2yz + 2xz$.

Consider the auxiliary function $F(x, y, z, \lambda)$
$= xy + 2yz + 2zx + \lambda(xyz - V)$.

The necessary conditions for extremum as

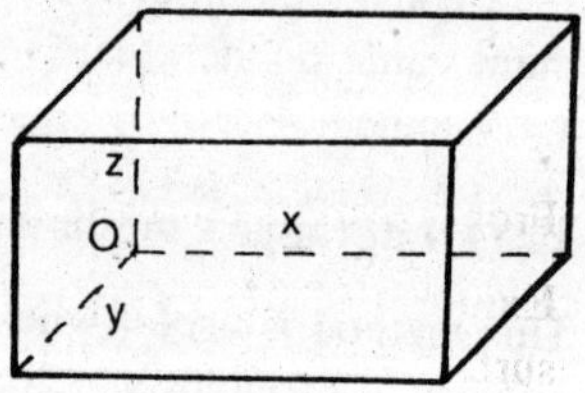

$$\frac{\partial F}{\partial x} = y + 2z + \lambda(yz) = 0, \qquad ...(1)$$

Fig. 12.8.

$$\frac{\partial F}{\partial y} = x + 2z + \lambda(xz) = 0, \qquad ...(2)$$

$$\frac{\partial F}{\partial z} = 2y + 2x + \lambda(xy) = 0. \qquad ...(3)$$

and $\qquad xyz = V.$ ...(4)

Multiplying (1) by x and (2) by y and substracting we get

$2zx - 2zy = 0$ or $2z(x - y) = 0$. Hence $x = y$. $[z \neq 0]$.

Similarly from (2) and (3) we get $y = 2z$.

Substituting $x = y = 2z$ in (4) we get $2z = (2V)^{1/3}$.

$\therefore$ $x = y = 2z = (2V)^{1/3}$ are the required dimensions which, from physical considerations, give the minimum surface.

Note. When $V = 32$ cubic feet, $x = y = 4$ ft, $z = 2$ ft.

Example 12.35 A tent of a given volume has a square base of side 2 a, has its four sides vertical of height b and is surmounted by a regular pyramid of height h. Find the values of a and b in terms of h such that the canvas required for its construction is minimum.

Solution. Let V be the volume and S the surface area of the tent.

Now $V = 4a^2 b + \left\{ (4a^2) h / 3 \right\}$, and $S = 8ab + 4a \sqrt{a^2 + h^2}$.

Consider the auxiliary function

$$F(a, b, h, \lambda) = 8ab + 4a\sqrt{a^2 + h^2} + \lambda\left[4a^2 b + (4a^2 h/3) - V \right].$$

For the extremum, we have the necessary conditions

$$\frac{\partial F}{\partial a} = 8b + 4\sqrt{a^2 + h^2} + \frac{4a^2}{\sqrt{a^2 + h^2}} + \lambda\left[8ab + \frac{8ah}{3} \right] = 0, \qquad ...(1)$$

$$\frac{\partial F}{\partial b} = 8a + 4\lambda a^2 = 0, \quad ...(2) \qquad \frac{\partial F}{\partial h} = \frac{4ah}{\sqrt{a^2 + h^2}} + \frac{4\lambda}{3} a^2 = 0 . \qquad ...(3)$$

From (2) $\lambda a + 2 = 0$. ...(4). From (3), $3h + \lambda a \sqrt{a^2 + h^2} = 0$ $(\because a \neq 0)$. ...(5)

Substituting the value of λa from (4) in (5) and simplifying, we get $a = \sqrt{5} h / 2$.

Substituting $\lambda a = -2$ and $a = \sqrt{5} h / 2$ in (1) and simplifying, we get $b = h/2$.

Thus when $a = \sqrt{5} h / 2$ and $b = h/2$, we get the stationary value of S.

From physical considerations this stationary value is a minimun.

Example 12.36 A torpedo has the shape of a cylinder with conical ends. For given surface area S, show that the dimensions which give maximum volume are $l = h = 2r/\sqrt{5}$, where l is the length of the cylinder, r its radius and h the altitude of the cone.

Solution. Here $S = 2\pi r l + 2\pi r \sqrt{r^2 + h^2}$,   ...(a)

and $$V = \pi r^2 l + 2\pi r^2 h / 3. \qquad ...(b)$$

We have to find the maximum value of V subject to the condition (a).

Considering the auxiliary function

$$F(l, h, r, \lambda) = \pi r^2 l + (2\pi r^2 h/3) + \lambda\left[2\pi r l + 2\pi r \sqrt{r^2 + h^2} - S \right].$$

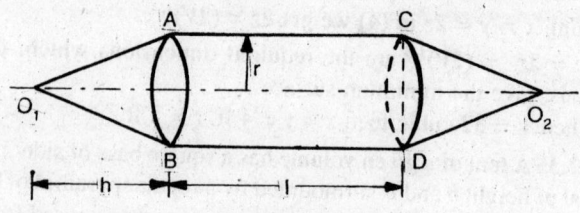

Fig. 12.9.

For the extremum, we have the necessary conditions

$$\frac{\partial F}{\partial l} = \pi r^2 + \lambda (2\pi r) = 0, \qquad \qquad ...(1)$$

$$\frac{\partial F}{\partial h} = \frac{2}{3}\pi r^2 + \frac{2\pi r h \lambda}{\sqrt{r^2 + h^2}} = 0, \qquad \qquad ...(2)$$

$$\frac{\partial F}{\partial r} = 2\pi r l + \frac{4}{3}\pi r h + 2\pi\lambda \left[l + \sqrt{r^2 + h^2} + \frac{r^2}{\sqrt{r^2 + h^2}} \right] = 0. \qquad \qquad ...(3)$$

From (1) $r + 2\lambda = 0$ ($\because r \neq 0$), or $2\lambda = -r$.

Putting $2\lambda = -r$ in (2) and simplifying, we get $h = 2r/\sqrt{5}$.

Substituting $h = 2r/\sqrt{5}$ and $2\lambda = -r$ in (3) and simplifying, we get $l = 2r/\sqrt{5}$.

The maximum value of V is when $l = h = 2r/\sqrt{5}$.

Example 12.37. Find the volume of the greatest rectangular parallelopiped with edges parallel to the coordinate axes that can be inscribed inside the ellipsoid

$$\frac{x^2}{a^2} + \frac{y^2}{b^2} + \frac{z^2}{c^2} = 1. \qquad \qquad \textit{(I.E.T.E., S-2004, A.M.I.E., W-2008)}$$

Solution. Let the edges of the parallelopiped which are parallel to the coordinate axes be $2x$, $2y$ and $2z$. Then its volume is $V = 8xyz$.

Now we have to find the maximum value of $V = 8xyz$ subject to the condition

$$\frac{x^2}{a^2} + \frac{y^2}{b^2} + \frac{z^2}{c^2} = 1. \qquad \qquad ..(i)$$

Consider the auxiliary function

$$F(x, y, z, \lambda) = 8xyz + \lambda \left(\frac{x^2}{a^2} + \frac{y^2}{b^2} + \frac{z^2}{c^2} - 1 \right).$$

For the extremum, we have the necessary conditions

$$\frac{\partial F}{\partial x} = 8yz + \lambda\left(\frac{2x}{a^2}\right) = 0, \quad ...(ii)$$

$$\frac{\partial F}{\partial y} = 8zx + \lambda\left(\frac{2y}{b^2}\right) = 0, \quad ...(iii)$$

$$\frac{\partial F}{\partial z} = 8xy + \lambda\left(\frac{2z}{c^2}\right) = 0, \quad ...(iv)$$

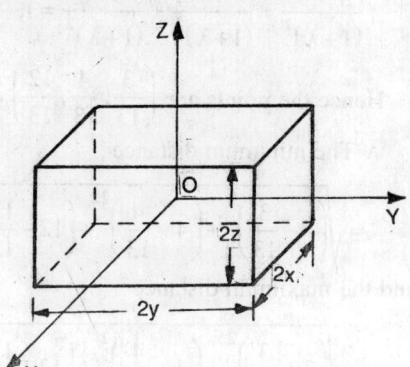

Fig. 12.10

Equating the values of λ from (ii) and (iii), we get $x^2/a^2 = y^2/b^2$.
Similarly from (iii) and (iv) we obtain

$$y^2/b^2 = z^2/c^2. \quad \therefore \quad \frac{x^2}{a^2} = \frac{y^2}{b^2} = \frac{z^2}{c^2}.$$

Substituting $\dfrac{x^2}{a^2} = \dfrac{y^2}{b^2} = \dfrac{z^2}{c^2}$ in (1) we get $\dfrac{z^2}{c^2} = \dfrac{1}{3}$.

$$\therefore \quad \frac{x^2}{a^2} = \frac{y^2}{b^2} = \frac{z^2}{c^2} = \frac{1}{3} \text{ which gives } x = \frac{a}{\sqrt{3}}, y = \frac{b}{\sqrt{3}}, z = \frac{c}{\sqrt{3}}.$$

Hence the greatest volume $= 2a/\sqrt{3} \cdot 2b/\sqrt{3} \cdot 2c/\sqrt{3} = 8abc/3\sqrt{3}$.

Example 12.38 Find the maximum and minimum distances of the point $(3, 4, 12)$ from the sphere $x^2 + y^2 + z^2 = 1$. *(A.M.I.E., S-1997)*

Solution. Let (x, y, z) be any point on the sphere, then its distance from $(3, 4, 12)$

is given by $\sqrt{(x-3)^2 + (y-4)^2 + (z-12)^2}$.

Obviously the 'distance' and 'the square of the distance have extremum values for the same values of x, y, z.

Consider the auxiliary function

$F(x, y, z, \lambda) = \{(x - 3)^2 + (y - 4)^2 + (z - 12)^2\} + \lambda(x^2 + y^2 + z^2 - 1)$.

The necessary conditions for extremum as $\dfrac{\partial F}{\partial x} = 2(x-3) + 2\lambda x = 0$,

$$\frac{\partial F}{\partial y} = 2(y-4) + 2\lambda y = 0, \frac{\partial F}{\partial z} = 2(z-12) + 2\lambda z = 0 \text{ give}$$

$$x = \frac{3}{1+\lambda}, y = \frac{4}{1+\lambda}, z = \frac{12}{1+\lambda}.$$

Substituting in the equation $x^2 + y^2 + z^2 = 1$, we obtain

$$\frac{9}{(1+\lambda)^2} + \frac{16}{(1+\lambda)^2} + \frac{144}{(1+\lambda)^2} = 1, \text{ or } 1 + \lambda = \pm 13 \Rightarrow \lambda = 12, -14.$$

Hence the points are $\left(\frac{3}{13}, \frac{4}{13}, \frac{12}{13}\right)$ and $\left(-\frac{3}{13}, -\frac{4}{13}, -\frac{12}{13}\right)$.

∴ The minimum distance

$$= \sqrt{\left(3 - \frac{3}{13}\right)^2 + \left(4 - \frac{4}{13}\right)^2 + \left(12 - \frac{12}{13}\right)^2} = 12 \text{ units},$$

and the maximum distance

$$= \sqrt{\left(3 + \frac{3}{13}\right)^2 + \left(4 + \frac{4}{13}\right)^2 + \left(12 + \frac{12}{13}\right)^2} = 14 \text{ units}.$$

Example 12.39 A pentagon is made by mounting an issoceles triangle on top of a rectangle. What dimensions minimise the perimeter P for a given area K?

Solution. Let x and y be the length and breadth of rectangle and z, the altitude of an issoceles triangle, as shown in the Figure 12.11.

Then area of pentagon $K = xy + (xz)/2$.

The perimeter, P of the pentagon is given by

$$P = x + 2y + 2\sqrt{(x^2/4) + z^2} = x + 2y + \sqrt{x^2 + 4z^2}.$$

Consider the auxiliary function

$$F(x, y, z, \lambda) = x + 2y + \sqrt{x^2 + 4z^2} + \lambda\left(xy + \frac{1}{2}xz - K\right).$$

For the extremum, we have the necessary conditions

$$\frac{\partial F}{\partial x} = 1 + \frac{x}{\sqrt{x^2 + 4z^2}} + \lambda\left(y + \frac{z}{2}\right) = 0, \dots(1)$$

$$\frac{\partial F}{\partial y} = 2 + \lambda x = 0, \qquad \dots(2)$$

$$\frac{\partial F}{\partial z} = \frac{4z}{\sqrt{x^2 + 4z^2}} + \frac{\lambda x}{2} = 0. \qquad \dots(3)$$

From (2) $\lambda = -2/x$. Putting this value is (3), we obtain

$$\frac{4z}{\sqrt{x^2 + 4z^2}} - 1 = 0 \Rightarrow 2\sqrt{3}z = x \qquad \dots(4)$$

From (1), we obtain $(\sqrt{3} + 1)z = y$ $\qquad \dots(5)$

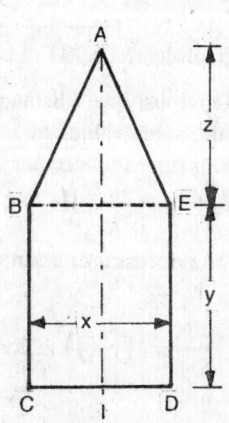

Fig. 12.11.

From (4) and (5), we have $\dfrac{x}{2\sqrt{3}} = \dfrac{y}{\sqrt{3}+1} = \dfrac{z}{1}$

or $\quad x : y : z = 2\sqrt{3} : 1+\sqrt{3} : 1$, for the perimeter to be minimum.

Example 12.40. Find the shortest distance from $(0, 0)$ to hyperbola $x^2 + 8xy + 7y^2 = 225$ in XY plane. $\qquad$ *(A.M.I.E., S-2006; I.E.T.E., W-2005)*

Solution. We have to find the minimum value of $x^2 + y^2$ (the square of the distance from the origin to any point in the XY plane) subject to the constraint $x^2 + 8xy + 7y^2 = 225$.

Consider the auxiliary function, $F(x, y, \lambda) = (x^2 + y^2) + \lambda (x^2 + 8xy + 7y^2 - 225)$.

For the extremum, we have the necessary conditions:

$$\frac{\partial F}{\partial x} = 2x + 2\lambda x + 8\lambda y = 0 \text{ or } (1 + \lambda)x + 4\lambda y = 0 \qquad ...(1)$$

$$\frac{\partial F}{\partial y} = 2y + 8\lambda x + 14\lambda y = 0 \text{ or } 4\lambda x + (1 + 7\lambda) y = 0 \qquad ...(2)$$

From (1) and (2), Since $(x, y) \neq (0, 0)$, we must have

$$\begin{vmatrix} 1+\lambda & 4\lambda \\ 4\lambda & 1+7\lambda \end{vmatrix} = 0 \text{ i.e., } 9\lambda^2 - 8\lambda - 1 = 0 \text{ or } \lambda = 1, -1/9.$$

Case I: $\lambda = 1$. From (1) or (2), $x = -2y$ and substitution in $x^2 + 8xy + 7y^2 = 225$ gives $y^2 = -45$, for which no real solution exists.

Case 2: $\lambda = -1/9$. From (1) or (2), $y = 2x$ and substitution in $x^2 + 8xy + 7y^2 = 225$ gives $x^2 = 5$. Then $y^2 = 4x^2 = 4 \bullet 5 = 20$ and so $x^2 + y^2 = 25$. Thus the required shortest distance is $\sqrt{25} = 5$ units.

Example 12.41 If r is the distance of a point on the conic $ax^2 + by^2 + cz^2 = 1$, $lx + my + nz = 0$ from the origin, show that the stationary values of r are given by

$$\frac{l^2}{1-ar^2} + \frac{m^2}{1-br^2} + \frac{n^2}{1-cr^2} = 0. \qquad \text{(IETE, Dec. 2002)}$$

Solution. We have to find the stationary values of
$$r^2 = x^2 + y^2 + z^2 \qquad ...(i)$$
subject to the conditions that $\quad ax^2 + by^2 + cz^2 - 1 = 0 \qquad ...(ii)$
and $\qquad\qquad\qquad\qquad\qquad lx + my + nz = 0 \qquad ...(iii)$

In this case, we use two Lagrange multipliers λ_1, λ_2. Consider the auxiliary function

$$F(x, y, z, \lambda_1, \lambda_2) = (x^2 + y^2 + z^2) + \lambda_1(ax^2 + by^2 + cz^2 - 1) + \lambda_2(lx + my + nz).$$

For the extremum, we have the necessary conditions, set the partial derivatives of F w.r.t. x, y, z equal to zero. Thus, we obtain

$$\frac{\partial F}{\partial x} = 2x + 2\lambda_1 ax + l\lambda_2 = 0, \qquad ...(iv)$$

$$\frac{\partial F}{\partial y} = 2y + 2\lambda_1 by + m\lambda_2 = 0, \qquad \text{...(v)}$$

$$\frac{\partial F}{\partial z} = 2z + 2\lambda_1 \; cz + n\lambda_2 = 0. \qquad \text{...(vi)}$$

On multiplying (iv), (v), (vi) by x, y, and z respectively and then on adding, we get

$$2(x^2 + y^2 + z^2) + 2\lambda_1 (ax^2 + by^2 + cz^2) + \lambda_2 (lx + my + nz) = 0$$

or $\qquad 2(r^2) + 2\lambda_1(1) + \lambda_2(0) = 0$ or $\lambda_1 = -r^2$.

Substituting for λ_1 in (iv), we get

$$2x - r^2(2\;ax) + \lambda_2(l) = 0 \text{ or } 2x(1 - ar^2) = -\lambda_2 l$$

or $\quad \dfrac{l}{1 - ar^2} = -\dfrac{2x}{\lambda^2}$ or $\dfrac{l^2}{1 - ar^2} = -\dfrac{2lx}{\lambda_2}$.

Similarly, $\quad \dfrac{m^2}{1 - br^2} = -\dfrac{2my}{\lambda_2}$ and $\dfrac{n^2}{1 - cr^2} \mp -\dfrac{2nz}{\lambda_2}$. Adding these, we get

$$\frac{l^2}{1 - ar^2} + \frac{m^2}{1 - br^2} + \frac{n^2}{1 - cr^2} = -\frac{2}{\lambda_2} (lx + my + nz) = \frac{-2}{\lambda_2}(0) = 0, \text{ the required result.}$$

Example 12.42 Find the maximum and minimum values of $x^2 + y^2 + z^2$ subject to

the constraint conditions $\dfrac{x^2}{4} + \dfrac{y^2}{5} + \dfrac{z^2}{25} = 1$, and $z = x + y$.

Solution. We must find the extrema of $f(x, y, z) = x^2 + y^2 + z^2$ subject to the

constraint conditions $\phi_1 = \dfrac{x^2}{4} + \dfrac{y^2}{5} + \dfrac{z^2}{25} - 1 = 0$ and $\phi_2 = x + y - z = 0$.

Consider the auxiliary function

$$F\left(x, y, z, \lambda_1, \lambda_2\right) = x^2 + y^2 + z^2 + \lambda_1 \left(\frac{x^2}{4} + \frac{y^2}{5} + \frac{z^2}{25} - 1\right) + \lambda_2 \left(x + y - z\right).$$

For the extremum, we have the necessary conditions

$$\left.\begin{aligned}
\frac{\partial F}{\partial x} &= 2x + \frac{\lambda_1}{2} x + \lambda_2 = 0; \\
\frac{\partial F}{\partial y} &= 2y + \frac{2\lambda_1}{5} y + \lambda_2 = 0, \\
\frac{\partial F}{\partial z} &= 2z + \frac{2\lambda_1}{25} z - \lambda_2 = 0.
\end{aligned}\right\} \qquad \text{...(1)}$$

Solving these equations for x, y, z, we find

$$x = \frac{-2\lambda_2}{\lambda_1 + 4}, \quad y = \frac{-5\lambda_2}{2\lambda_1 + 10}, \quad z = \frac{25\lambda_2}{2\lambda_1 + 50} \qquad \ldots (2)$$

Substituting in second constraint condition, $x + y - z = 0$, we get

$$\frac{2}{\lambda_1 + 4} + \frac{5}{2\lambda_1 + 10} + \frac{25}{2\lambda_1 + 50} = 0, \lambda_2 \neq 0 \qquad \ldots (3)$$

For if, $\lambda_2 = 0$, $x = y = z = 0$, but $(0, 0, 0)$ does not satisfy the other condition of constraint. Hence, from (3), $17\lambda_1^2 + 245\lambda_1 + 750 = 0$

or $(\lambda_1 + 10)(17\lambda_1 + 75) = 0$ from which $\lambda_1 = -10$ or $\dfrac{-75}{17}$.

Case 1: $\lambda_1 = -10$. From (2), $x = \dfrac{\lambda_2}{3}, y = \dfrac{\lambda_2}{2}, z = \dfrac{5\lambda_2}{6}$.

Substituting in the first constraint condition, $\dfrac{x^2}{4} + \dfrac{y^2}{5} + \dfrac{z^2}{25} = 1$, yields

$\lambda_2^2 = \dfrac{180}{19}$ or $\lambda_2 = \pm 6\sqrt{5/19}$. This gives the two critical points.

$$\left(2\sqrt{5/19}, 3\sqrt{5/19}, 5\sqrt{5/19}\right), \left(-2\sqrt{5/19}, -3\sqrt{5/19}, -5\sqrt{5/19}\right).$$

The value of $x^2 + y^2 + z^2$ corresponding to these critical points is
$$(20 + 45 + 125)/19 = 10.$$

Case 2: $\lambda_1 = -75/17$.

From (2), $x = \dfrac{34}{7}\lambda_2, y = -\dfrac{17}{4}\lambda_2, z = \dfrac{17}{28}\lambda_2$. Substituting in the first constraint

condition, $\dfrac{x^2}{4} + \dfrac{y^2}{5} + \dfrac{z^2}{25} = 1$ yield $\lambda_2 = \pm\dfrac{140}{17\sqrt{646}}$ which gives the critical points

$$\left(\frac{40}{\sqrt{646}}, -\frac{35}{\sqrt{646}}, \frac{5}{\sqrt{646}}\right), \left(-\frac{40}{\sqrt{646}}, \frac{35}{\sqrt{646}}, \frac{-5}{\sqrt{646}}\right).$$ The value of $x^2 + y^2 + z^2$

corresponding to these is $(1600 + 1225 + 25)/646 = 75/17$. Thus, the required maximum value is 10 and the minimum value is 75/17.

Exercise 12.4

1. Discuss for maxima and minima the following functions:
(i) $x^3 + y^3 + 3xy$.   (ii) $x^2 + y^2 + 6x + 12$.
(iii) $x^3 + 3xy^2 - 15x^2 + 15y^2 + 72x$. (A.M.I.E., S-2001; Andhra, 1999;)

   (*iv*) $3x^2 + xy + y^2 + x + y.$ (*A.M.I.E., S-2002*)

   (*v*) $x^3 y^2 (1 - x - y), x > 0, y > 0$ (*I.E.T.E., Dec. 1995; A.M.I.E., S-2004*)

  (*vi*) $\sin x \sin y \sin(x + y).$ $(0 \le x \le \pi, 0 \le y \le \pi).$ (*A.M.I.E., W-1998*)

(*vii*) $x^2 y + xy^2 - axy.$ (*A.M.I.E., S-2003*)   (*viii*) $x^3 + y^3 - 3x - 12y + 20.$

Ans. (*i*) Maximum value 1 at $(-1, -1)$.

  (*ii*) Minimum value 3 at $(-3, 0)$.

 (*iii*) Maximum value 112 at $(4, 0)$, Minimum value 108 at $(6, 0)$.

 (*iv*) Minimum value $-33/121$ at $(-1/11, -5/11)$.

  (*v*) Maximum value $1/432$ at $(1/2, 1/3)$; no extreme value at $(0, 0)$.

 (*vi*) Maximum value $3\sqrt{3}/8$ at $(\pi/3, \pi/3)$; Minimum value $-3\sqrt{3}/8$ at $(2\pi/3, 2\pi/3)$.

(*vii*) Maximum or Minimum exist at $(a/3, a/3)$; Minimum value $= -a^3/27$; Maximum value $= a^3/27$.

(*viii*) Minimum value 2 at $(1, 2)$; Maximum value 38 at $(-1, -2)$; $(-1, 2)$ and $(1, -2)$ are saddle points.

  2. (*a*) Test for local maxima or minima, given

$$f(x, y) = x^2 + 2xy + 2y^2 + 2x + y.$$

   (*b*) Discuss the maxima or minima of the function

$$z = xy + \frac{a^3}{x} + \frac{a^3}{y}.$$ (*A.M.I.E., W-2004, Marathwada, 1995*)

              Ans. (*a*) Minimum at $(-3/2, 1/2)$; (*b*) Minimum at (a, a).

  3. (*a*) Find the stationary points of $f(x, y) = x^3 + y^2 + 3x^2 + 4xy$. Examine them for the extreme value of the function.

   (*b*) Test for maximum and minimum the function: $z = x^2 + xy + y^2 + x^{-1} + y^{-1}$.

Ans. (*a*) $(0, 0)$, $(2/3, -4/3)$; Minimum value $= -4/27$.

  (*b*) Minimum z at $x = y = 1/\sqrt[3]{3}$.

  4. Show that the maximum value of

$$\sin x + \sin y + \sin(x + y) \left(0 \le x \le \frac{\pi}{2}, \ 0 \le y \le \frac{\pi}{2}\right) \text{ is } 3\sqrt{3}/\sqrt{2}.$$

  5. Find the absolute maximum and minimum values of the function $f(x, y) = 4x^2 + 9y^2 - 8x - 12y + 4$ over the rectangle in first quadrant bounded by the lines $x = 2, y = 3$ and the coordinate axes.

                                          (*IETE, June-2007*)

Ans. Absolute minimum value is -4 which occurs at $(1, 2/3)$ and the absolute maximum value is 49 which occurs at the point $(2, 3)$ and $(0, 3)$.

  6. (*a*) Find the point in the plane $2x + y + 2z = 16$ nearest to the origin.

   (*b*) Find the point $P(x, y, z)$ on the plane $2x - y - z - 5 = 0$ that lies closest to the origin.

(c) Find the point on the plane $x + 2y + 3z = 13$ closest to the point $(1, 1, 1)$.

Ans. (a) $\left(\dfrac{32}{9}, -\dfrac{16}{9}, \dfrac{32}{9}\right)$; (b) $\left(\dfrac{5}{3}, \dfrac{5}{6}, -\dfrac{5}{6}\right)$; (c) $\left(\dfrac{3}{2}, 2, \dfrac{5}{2}\right)$.

7. (a) Find points on the surface $z^2 = xy + 1$ that is nearest to the origin.
   (*A.M.I.E., S-1999; Andhra 2000;*)

   (b) Find the points on the surface $z^2 = xy + 4$ closest to the origin.
   Ans. (a) $(0, 0, \pm 1)$; (b) $(2, -2, 0)$ and $(-2, 2, 0)$

8. (a) Find the minimum distance from the surface $x^2 + y^2 - z^2 = 1$ to the origin.

   (b) Find the minimum distance from the origin to the surface $x^2 - z^2 - 1 = 0$.

   (c) Find the shortest and largest distance from the point $(1, 2, -1)$ to the sphere $x^2 + y^2 + z^2 = 24$.

   Ans. (a) Minimum distance $= 1$. (b) points on the surface closest to the origin $(\pm 1, 0, 0)$, Minimum distance $= 1$. (c) Minimum distance $= \sqrt{6}$; Maximum distance $= \sqrt{54}$.

9. Find the stationary values of $x^2 + y^2 + z^2$ given that $ax + by + cz = p$.
   (*A.M.I.E., S-1999; Mangalore, 1999*) **Ans.** $p^2/(a^2 + b^2 + c^2)$.

10. Prove that the rectangular solid of maximum value which can be inscribed in a sphere is cube. (*A.M.I.E., W-1999; Delhi 1997*)

11. Divide 120 into 3 parts such that the sum of the products taken two at a time shall be minimum. (*Nagpur 1995*) **Ans.** 40, 40, 40.

12. Given $x + y + z = a$, find the minimum value of $x^m y^n z^p$. (*Nagpur 1995*)

13. Find the maximum value of $u = x^p y^q z^r$ when the variable x, y, z are subject to the condition $ax + by + cz = p + q + r$. **Ans.** $(p^p q^q r^r)/a^p b^q c^r$.

14. In a plane triangle ABC, find the maximum value of $\cos A \cos B \cos C$.
    (*Raipur 1998; Nagpur 1995*) **Ans.** 1/8.

    [Hint: Let $F = \cos A \cos B \cos C + \lambda (A + B + C - \pi)$.

    $\partial F/\partial A = -\sin A \cos B \cos C + \lambda$. Also find $\partial F/\partial B$ and $\partial F/\partial C$ and equate them to zero, we get $\sin A \cos B \cos C = \sin B \cos A \cos C = \sin C \cos A \cos B$

    $\Rightarrow \qquad \tan A = \tan B = \tan C \Rightarrow A \pm B = C = \pi/3$.

    Maximum value of $\cos A \cos B \cos C = (1/2)(1/2)(1/2) = 1/8$].

15. Find the dimensions of the rectangular parallelopiped of maximum volume which has three faces in the co-ordinates planes and one vertex in the plane $x/a + y/b + z/c = 1$. **Ans.** $\left(a/3, b/3, c/3\right)$; $V = abc/27$.

16. The temperature T at any point (x, y, z) in space is $T = 400\, xyz^2$. Find the highest temperature on the surface of the unit sphere $x^2 + y^2 + z^2 = 1$.
    (*A.M.I.E., W-2005*) **Ans.** 50.

17. Find the dimensions of the rectangular box open at the top of maximum capacity whose surface is 432 sq. cm. (*Madras, 2005*) **Ans.** (12, 12, 6 cm)

18. A tent having the form of a cylinder surmounted by a cone is to contain a given volume. Find its dimensions if the convas required is a minimum.

    Ans. radius: cylinder height: cone height $= \sqrt{5} : 1 : 2$.

19. Find the largest product the positive numbers x, y and z can have if $x + y + z^2 = 16$.

20. If $3x^{-1} + 4y^{-1} + 5z^{-1} = 6$. find the values of x, y, z which make $x + y + z$ a minimum. **Ans.** $x = \lambda\sqrt{3}$, $y = 2\lambda$, $z = \lambda\sqrt{5}$, where $\lambda = \left(\sqrt{3} + 2 + \sqrt{5}\right)/6$.

21. A solid consists of a cylinder surmounted by a cone. Let r = radius, h = altitude of the cylinder, H = altitude of the cone; S = surface area, and V = volume of the solid. Show that the minimum value of S for a given value of V is given by

$$ r^3 = \frac{3V}{\pi\left(3 + \sqrt{5}\right)}, \quad H = \frac{2r}{\sqrt{5}}, \quad S = \frac{3V}{r}. $$

22. Find the stationary value of $u = xyz$ subject to $x + y + z = 5$, $xy + yz + zx = 8$.   (*A.M.I.E., S-1995*) **Ans.** $x = 4/3$, $y = 4/3$, $z = 7/3$.
    [**Hint:** Find the stationary value of $u = xy(5 - x - y)$ subject to condition $yx + (x + y)(5 - x - y) = 8$].

23. Using Lagrange's method of multipliers, find the critical points (stationary values) of the function $f(x, y, z) = x^2 + y^2 + z^2$ given that $z^2 = xy + 1$.

    Ans. $(0, 0, \pm 1)$.

24. Find the extreme values of $f(x, y, z) = 2x + 3y + z$ such that $x^2 + y^2 = 5$ and $x + z = 1$.   (*IETE, Dec. 2007*)

    Ans. Maximum value is $1 + 5\sqrt{2}$ and minimum value $1 - 5\sqrt{2}$.

25. If $u = ax^2 + by^2 + cz^2$ where $x^2 + y^2 + z^2 = 1$ and $lx + my + nz = 0$, prove that the stationary values of u satisfy the equation:

$$ \frac{l^2}{a - u} + \frac{m^2}{b - u} + \frac{n^2}{c - u} = 0. $$

26. Find the relative maximum and minimum values of the function.

    $f(x, y) = x^4 - y^4 - 2x^2 + 2y^2$.   (*I.E.T.E., June 2005*)
    Ans. $(0, 1)$ and $(0, -1)$ are points of relative minimum and minimum value at each point is -1; $(-1, 0)$ and $(1, 0)$ are points of relative maximum and maximum value at each point is 1; Points $(0, 0)$, $(\pm 1, \pm 1)$ are neither the points of maximum nor minimum.

27. Divide 24 into three parts such that the continued product of the first, the square of the second and the cube of the third may be maximum.

    (*AMIE., S-2007*) **Ans.** $x = 4$, $y = 8$, $z = 12$.
    [**Hint:** $x + y + z = 24$ and xy^2z^3 is a maximum.]

28. Find the stationary value of $a^3x^2 + b^3y^2 + c^3z^2$ subject to the fulfilment of the condition $x^{-1} + y^{-1} + z^{-1} = 1$, given a, b, c are not zero.

$$\text{(I.E.T.E., June 2003)} \quad \textbf{Ans.} \ x = \Sigma\, a/a, \ y = \Sigma\, a/b, \ z = \Sigma\, a/c.$$

29. A point P moves on a plane containing a triangle ABC. Prove that $(AP)^2 + (BP)^2 + (CP)^2$ will be minimum when P is at the centroid of the triangle ABC. *(A.M.I.E., W-2009, S-2005)*

30. Find the extreme value of $x^2 + y^2 + z^2 + xy + yz + zx$ subject to the conditions $x + y + z = 1$ and $x + 2y + 3z = 3$ *(I.E.T.E., Dec. 2004)*

Ans. Extreme value is 11/12 at (−1/6, 1/3, 5/6).

31. Show that the stationary values of $u = \dfrac{x^2}{a^4} + \dfrac{y^2}{b^4} + \dfrac{z^2}{c^4}$, where

$lx + my + nz = 0$ and $\dfrac{x^2}{a^2} + \dfrac{y^2}{b^2} + \dfrac{z^2}{c^2} = 1$ are the roots of the equation

$$\frac{l^2a^4}{1-a^2u} + \frac{m^2b^4}{1-b^2u} + \frac{n^2c^4}{1-c^2u} = 0. \qquad \text{(I.E.T.E., Dec 2003)}$$

32. Show that the maximum and minimum distances from the origin to section of the surface $(x^2 + y^2 + z^2)^2 = x^2 + 2y^2 + 3z^2$ made by the plane $x + y + z = 0$ are the roots of the equation $3r^4 - 12r^2 + 11 = 0$, where r is the distance from the origin to any point on the section.

(I.E.T.E., Dec. 1999)

33. Using Lagrange's multipliers, find a point (x, y, z) on the unit sphere $x^2 + y^2 + z^2 = 1$ which minimizes the function $x + y^2 + yz + 2z^2$. **Ans.** (−1, 0, 0)

34. Find the minimum value of $x^2 + y^2 + z^2$ subject to the condition $xyz = a^3$.

(IETE, June 2009, June 2008; AMIE., S-2008) **Ans.** $3a^2$.

35. Find the shortest distance between the line $y = 10 - 2x$ and the ellipse $x^2/4 + y^2/9 = 1$. *(IETE, Dec. 2006)* **Ans.** $\sqrt{5}$ units.

36. Find the extreme values of $f(x,y,z) = x^2 + 2xy + z^2$ subject to the constraints $g(x,y,z) = 2x + y = 0$ and $h(x,y,z) = x + y + z = 1$.

(IETE., June 2006). **Ans.** Extreme value is 3/2 at (1/2, −1, 3/2).

37. Show that $f(x,y,z) = (x+y+z)^3 - 3(x+y+z) - 24xyz + a^3$ has a minima at (1, 1, 1) and a maxima at (−1, −1, −1). *(AMIE., W-2007)*

38. Find the greatest and the least value of the function $f(x,y) = x^2 + 2xy - 4x + 8y$ in the rectangle bounded by the straight lines $x = 0$, $y = 0$, $x = 1$, $y = 2$.

Ans. The greatest value $f(x, y) = 17$ is at the point (1, 2); the least value $f(x,y) = -3$ at the point (1, 0); the stationary point (−4, 6) lies outside the given domain.

39. Find the absolute maxima and minima of the $f(x,y) = 2x^2 - 4x + y^2$
$- 4y + 1$ on the closed triangular plate bounded by the lines $x = 0$, $y = 0$,
$y = 2x$ in the first quadrant.

Ans. The absolute maxima value $f(x,y) = 1$ is at the point (0, 0); the
absolute minima value $f(x,y) = -5$ at the point (1, –2).

40. Find the absolute maximum and minimum values of $f(x,y) = 3xy - 6x$
$- 3y + 7$ on the closed triangular region R with vertices (0, 0), (3, 0), and
(0, 5).

Ans. The absolute maximum value of $f(x,y) = 7$ is at the point (0, 0); the
absolute minimum value $f(x,y) = -11$ at the point (3, 0).

12.13 Approximation by Total Differentials

Let the function $z = f(x, y)$ be differentiable at the point (x, y). The total increment
Δz of the function z corresponding to the increments Δx in x and Δy in y is
defined by

$$\Delta z = f(x + \Delta x, y + \Delta y) - f(x, y),$$

or $\qquad f(x + \Delta x, y + \Delta y) = f(x, y) + \Delta z,$ ...(12.22)

We know the approximate formula $\Delta z \approx dz,$ ...(12.23)

where $\qquad dz = \dfrac{\partial f}{\partial x}\Delta x + \dfrac{\partial f}{\partial y}\Delta y.$

Substituting into (12.22) the expanded expression for dz in place of Δz, we get
the approximate formula

$$f(x + \Delta x,\ y + \Delta y) \approx f(x,\ y) + \frac{\partial}{\partial x}f(x,\ y)\cdot \Delta x + \frac{\partial}{\partial y}f(x,\ y)\cdot \Delta y \qquad ...(12.24)$$

This result has applications in estimating errors in calculations.

Example 12.43 A certain function $z = f(x, y)$ has values $f(2, 3) = 5, f_x(2, 3) = 3$ and
$f_y(2, 3) = 7$. Find an approximate value of $f(1.98, 3.01)$.

Solution. We have $x = 2$, $y = 3$, $\Delta x = -0.02$, $\Delta y = 0.01$.

$\therefore\ f(1.98, 3.01) \approx f(2, 3) + f_x(2, 3) \cdot \Delta x + f_y(2, 3) \cdot \Delta y$
$\qquad\qquad = 5 + (3)(-0.02) + (7)(0.01) = 5.01.$

Example 12.44 Using differentials, find an approximate value of

(i) $f(4.1, 4.9)$, where $f(x,\ y) = \sqrt{x^3 + x^2 y}$. (A.M.I.E., W-2005)

(ii) $f(1.94, 0.98, 2.01)$, where $f(x,\ y,\ z) = \sqrt{x^2 + y^2 + z^2}$.

(iii) $f(4.05, 7.97)$, where $f(x,\ y) = (x)^{1/2}\,(y)^{1/3}$.

Solution. (i) Let $(x, y) = (4, 5)$, $\Delta x = 0.1$, $\Delta y = -0.1$. We have

$$f(x, y) = \sqrt{x^3 + x^2 y}, \qquad\qquad f(4, 5) = \sqrt{64 + 16 \times 5} = 12,$$

$$f_x(x, y) = \frac{3x^2 + 2xy}{2\sqrt{x^3 + x^2 y}}, \qquad f_x(4, 5) = \frac{48 + 40}{2\sqrt{64 + 80}} = \frac{11}{3},$$

$$f_y(x, y) = \frac{x^2}{2\sqrt{x^3 + x^2 y}}, \qquad f_y(4, 5) = \frac{16}{2 \times 12} = \frac{2}{3}.$$

Therefore, $f(4.1, 4.9) \approx f(4, 5) + f_x(4, 5)\, \Delta x + f_y(4, 5)\, \Delta y$

$$= 12 + \left(\frac{11}{3}\right)(0.1) + \left(\frac{2}{3}\right)(-0.1) = 12.3.$$

(*ii*) Let $(x, y, z) = (2, 1, 2)$, $\Delta x = -0.06$, $\Delta y = -0.02$, $\Delta z = 0.01$.

Take $\quad f(x, y, z) = \sqrt{x^2 + y^2 + z^2}, \qquad f(2, 1, 2) = 3,$

$$f_x(x, y, z) = \frac{x}{\sqrt{x^2 + y^2 + z^2}}, \qquad f_x(2, 1, 2) = \frac{2}{3},$$

$$f_y(x, y, z) = \frac{y}{\sqrt{x^2 + y^2 + z^2}}, \qquad f_y(2, 1, 2) = \frac{1}{3},$$

$$f_z(x, y, z) = \frac{z}{\sqrt{x^2 + y^2 + z^2}}, \qquad f_z = \frac{2}{3}.$$

Therefore, $f(1.94, 0.98, 2.01) \approx f(2, 1, 2)$
$$+ f_x(2, 1, 2)\, \Delta x + f_y(2, 1, 2)\, \Delta y + f_z(2, 1, 2)\Delta z$$

$$= 3 + \left(\frac{2}{3}\right)(-0.06) + \left(\frac{1}{3}\right)(-0.02) + \left(\frac{2}{3}\right)(0.01) \approx 2.96.$$

(*iii*) Let $(x, y) = (4, 8)$, $\Delta x = 0.05$, $\Delta y = -0.03$, we have

$$f(x, y) = x^{1/2}. y^{1/3}, \qquad\qquad f(4, 8) = 4^{1/2}. 8^{1/3} = 4,$$

$$f_x(x, y) = \frac{y^{1/3}}{2x^{1/2}}, \qquad\qquad f_x(4, 8) = \frac{1}{2},$$

$$f_y(x, y) = \frac{x^{1/2}}{3y^{2/3}}, \qquad\qquad f_y(4, 8) = \frac{1}{6}.$$

Therefore $f(4.05, 7.97) \approx f(4, 8) + f_x(4, 8) \cdot \Delta x + f_y(4, 8) \cdot \Delta y$

$$= 4 + \left(\frac{1}{2}\right)(0.05) + \left(\frac{1}{6}\right)(-0.03) = 4.02.$$

12.14 Use of a differentials to estimate errors in calculations

Consider a function of n variables $x_1, x_2, ..., x_n$. Let the function $z = f(x_1, x_2, ..., x_n)$ be differentiable at the point $P(x_1, x_2, ..., x_n)$.

Let there be errors $\Delta x_1, \Delta x_2, ..., \Delta x_n$ in measuring the values of $x_1, x_2, ..., x_n$ respectively. Then, the computed value of z using the inexact values of the arguments will be obtained with an error

$$\Delta z = f(x_1 + \Delta x_1, x_2 + \Delta x_2, ..., x_n + \Delta x_n) - f(x_1, x_2..., x_n).$$

When the errors $\Delta x_1, \Delta x_2, ..., \Delta x_n$ are small in magnitude, we can replace, approximately, the total increment by the total differential, we obtain

$$f(x_1 + \Delta x_1, x_2 + \Delta x_2, ...x_n + \Delta x_n) \approx f(x_1, x_2, ..., x_n) + f_{x_1} \Delta x_1 + f_{x_2} \Delta x_2 + ... + f_{x_n} \Delta x_n,$$

where the partial derivatives are evaluated at the point $(x_1, x_2, ..., x_n)$. This is the generalization of the result for functions of two variables.

Since the values of the partial derivatives and the errors in arguments may be either positive or negative, we define the *absolute error* as

$$|\Delta z| \approx |dz| = |df| = \left| f_{x_1} \Delta x_1 + f_{x_2} \Delta x_2 + ... f_{x_n} \Delta x_n \right|.$$

Then $|df| \leq |f_{x_1}||\Delta x_1| + |f_{x_2}||\Delta x_2| + ... + |f_{x_n}||\Delta x_n|$ gives the *maximum absolute error* in z.

Relative and Percentage error. The expression $\dfrac{|df|}{|f|}$ is called the *maximum relative error* and $\dfrac{|df|}{|f|} \times 100$ is called the *percentage error*.

The maximum relative error can also be written as

$$\frac{|df|}{|f|} \leq \left| \frac{\partial f / \partial x_1}{f} \right| |\Delta x_1| + \left| \frac{\partial f / \partial x_2}{f} \right| |\Delta x_2| + ... + \left| \frac{\partial f / \partial x_n}{f} \right| |\Delta x_n|.$$

But $\dfrac{\partial f / \partial x_1}{f} = \dfrac{\partial}{\partial x_1} \log|f|, \dfrac{\partial f / \partial x_2}{f} = \dfrac{\partial}{\partial x_2} \log|f|, ...$ etc.

$$\therefore \quad \frac{|df|}{|f|} \leq \left| \frac{\partial}{\partial x_1} \left[\log|f| \right] \right| |\Delta x_1| + \left| \frac{\partial}{\partial x_2} \left[\log|f| \right] \right| |\Delta x_2| + ... + \left| \frac{\partial}{\partial x_n} \left[\log|f| \right] \right| |\Delta x_n|.$$

Remarks.

(1) We may write $\Delta x_1, \Delta x_2, ...$as dx_1, dx_2 respectively.

(2) While solving problems, relative error may be found directly by logarithmic differentiation.

(3) In practical problems, error in calculation are due to small errors in the data upon which calculation is based. The latter may arise from lack of precision in the measurements or from other causes.

12.15 Illustrative Examples

Example 12.45 The period T of a simple pendulum is $T = 2\pi\sqrt{l/g}$. Find the maximum error in T due to possible error upto 1% in l and 2.5% in g.

Solution. We have $T = 2\pi\sqrt{l/g}$. Taking logarithms, we get

$$\log T = \log 2\pi + \frac{1}{2}\log l - \frac{1}{2}\log g. \quad \therefore \quad \frac{1}{T}dT = 0 + \frac{1}{2}\cdot\frac{dl}{l} - \frac{1}{2}\cdot\frac{dg}{g}$$

or $\quad 100\dfrac{dT}{T} = \dfrac{1}{2}\left[100\left(\dfrac{dl}{l}\right) - 100\left(\dfrac{dg}{g}\right)\right].$

The percentage error will be maximum when

$$100\left(\frac{dl}{l}\right) = 1 \text{ and } 100\left(\frac{dg}{g}\right) = -2.5$$

that is, the error in l is positive and that in g is negative.

$\therefore \quad 100\dfrac{dT}{T} = \dfrac{1}{2}[1-(-2.5)] = 1.75$. Hence maximum error in $T = 1.75\%$.

Example 12.46 The work that must be done to propel a ship of displacement D for a distance s in time t is proportional to $s^2 D^{2/3} t^2$. Find approximately the percentage increase of work necessary when the displacement is increased by 1%, the time is diminished by 1% and the distance is increased by 3%.

Solution. Let the work done be W. Therefore, $W = ks^2 D^{2/3} t^2$, where k is a constant of proprotionality.

Taking logarithms, we obtain $\log W = \log k + 2\log s + \left(\dfrac{2}{3}\right)\log D + 2\log t.$

$\therefore \quad \dfrac{dW}{W} = 2\dfrac{ds}{s} + \dfrac{2}{3}\dfrac{dD}{D} + 2\dfrac{dt}{t}.$ Multiplying both sides by 100, we obtain

$$100\frac{dW}{W} = 2\left(100\frac{ds}{s}\right) + \frac{2}{3}\left(100\frac{dD}{D}\right) + 2\left(100\frac{dt}{t}\right).$$

Since $\qquad 100\left(\dfrac{ds}{s}\right) = 3,\ 100\left(\dfrac{dD}{D}\right) = 1 \text{ and } 100\left(\dfrac{dt}{t}\right) = -1.$

$\therefore \quad 100\dfrac{dW}{W} = $ Percentage increase of work $= 2(3) + \dfrac{2}{3}(1) + 2(-1) = 4\dfrac{2}{3}\%.$

Example 12.47 The indicated horse power I of an engine is calculated from the formula $I = PLAN/33000$, where $A = \pi D^2/4$. Assuming that error of λ percent may have been made in measuring *P, L, N* and *D*, find the greatest possible error in *I*.

Solution. We have $I = PLAN/33000$. Putting $A = \pi D^2/4$, we get

$$I = \frac{\pi PLD^2 N}{4 \times 33000} = \pi PL \frac{D^2 N}{132000}.$$

$\therefore \log I = \log \pi + \log P + \log L + 2 \log D + \log N - \log 132000.$

Differentiating and multiplying by 100, we obtain

$$100 \frac{dI}{I} = 100 \frac{dP}{P} + 100 \frac{dL}{L} + 2 \times 100 \frac{dD}{D} + 100 \frac{dN}{N}.$$

Putting $100\left(\frac{dP}{P}\right) = 100\left(\frac{dL}{L}\right) = 100\left(\frac{dD}{D}\right) = 100\left(\frac{dN}{N}\right) = \lambda$ $\therefore$ $100\left(\frac{dI}{I}\right) = 5\lambda.$

Thus the greatest possible error in $I = 5\lambda\%$.

Example 12.48 With the usual meaning for *a, b, c* and *s*, if Δ be the area of a triangle prove that error in Δ resulting from a small error in the measurement of *c*, is given by

$$d\Delta = \frac{\Delta}{4}\left[\frac{1}{s} + \frac{1}{s-a} + \frac{1}{s-b} - \frac{1}{s-c}\right]dc.$$

Solution. We know that

$$\Delta = \sqrt{s(s-a)(s-b)(s-c)}, \text{ where } s = (a+b+c)/2.$$

$\therefore$ $\log \Delta = \frac{1}{2}\left[\log s + \log(s-a) + \log(s-b) + \log(s-c)\right].$

Hence $\quad \dfrac{d\Delta}{\Delta} = \dfrac{1}{2}\left[\dfrac{ds}{s} + \dfrac{ds}{s-a} + \dfrac{ds}{s-b} + \dfrac{ds-dc}{s-c}\right].$ ...(*i*)

Now $\quad s = (a+b+c)/2$, therefore $ds = dc/2$.

Hence puting $ds = dc/2$ in (*i*) we get

$$\frac{d\Delta}{\Delta} = \frac{1}{2}\left[\frac{dc}{2s} + \frac{dc}{2(s-a)} + \frac{dc}{2(s-b)} - \frac{dc}{2(s-c)}\right],$$

or $\quad d\Delta = \dfrac{\Delta}{4}\left[\dfrac{1}{s} + \dfrac{1}{s-a} + \dfrac{1}{s-b} - \dfrac{1}{s-c}\right]dc.$

Example 12.49 In estimating the cost of a pile of bricks measured as 2 m × 15 m × 1.2 m, the tape is stretched 1 per cent beyond the standard length. If the count is 450 bricks to 1 cu. m, and bricks cost Rs. 530.00 per thousand, find the aproximate error in the cost. *(A.M.I.E., S-2004, V.T.U.-2001)*

Solution. Let x, y and z be the length, breadth and height of the pile so that its volume, $V = xyz$.

Therefore, by logarithmic differentiation, we have $\dfrac{dV}{V} = \dfrac{dx}{x} + \dfrac{dy}{y} + \dfrac{dz}{z}$.

Given $\dfrac{dx}{x} = \dfrac{dy}{y} = \dfrac{dz}{z} = \dfrac{1}{100}$, and $V = 2 \times 15 \times 1.2 = 36 \text{ m}^3$.

Therefore $dV = 36\left(\dfrac{3}{100}\right) = 1.08 \text{ m}^3$. Number of bricks in $dV = 1.08 \times 450 = 486$.

$\therefore$ Error in the cost $= \text{Rs. } 486 \times \left(\dfrac{530}{1000}\right) = \text{Rs. } 257.58$,

which is a loss to the seller of bricks.

Example 12.50 In a plane triangle ABC, if the sides a, b, be kept constant, show that the variations of its angles are given by the relation

$$\frac{dA}{\sqrt{a^2 - b^2 \sin^2 A}} = \frac{dB}{\sqrt{b^2 - a^2 \sin^2 B}} = -\frac{dC}{c},$$

the letters having their usual significance. *(I.E.T.E., W-2005)*

Solution. In any triangle ABC, $\dfrac{\sin A}{a} = \dfrac{\sin B}{b}$ or $b \sin A = a \sin B$.

Differentiating, $b \cos A \dfrac{dA}{dB} = a \cos B$, ($a$, b being constants.)

or $b \cos A\, dA - a \cos B\, dB + 0.\, dC = 0$. ...(i)

But $A + B + C = \pi$ Therefore, $dA + dB + dC = 0$. ...(ii)

From (i) and (ii) by cross-multiplication, we obtain

$$\frac{dA}{-a \cos B} = \frac{dB}{-b \cos A} = \frac{dC}{b \cos A + a \cos B}$$

or
$$\frac{dA}{a \cos B} = \frac{dB}{b \cos A} = -\frac{dC}{b \cos A + a \cos B}.$$

Using $\cos\theta = \sqrt{1 - \sin^2\theta}$ and projection formula $c = a \cos B + b \cos A$, we

obtain $\dfrac{dA}{\sqrt{a^2 - a^2 \sin^2 B}} = \dfrac{dB}{\sqrt{b^2 - b^2 \sin^2 A}} = -\dfrac{dC}{c}$. Since $a \sin B = b \sin A$,

$$\therefore \quad \frac{dA}{\sqrt{a^2 - b^2 \sin^2 A}} = \frac{dB}{\sqrt{b^2 - a^2 \sin^2 B}} = -\frac{dC}{c}.$$

Example 12.51 The area of a triangle is calculated from angles A and C and the side b. If dA is the error in measuring $\angle A$, show that the relative error in measuring area is $\dfrac{\sin C}{\sin A \sin(C+A)} dA$.

(A.M.I.E., W-1999)

Solution. Let K be the area of the triangle $= \dfrac{1}{2} ab \sin C$ (Refer Fig. 12.12)

Using the Sine Laws: $\dfrac{a}{\sin A} = \dfrac{b}{\sin B} = \dfrac{c}{\sin C}$. We have

$$a = \frac{b \sin A}{\sin B} \text{ and } \angle B = \pi - (\angle C + \angle A).$$

$\therefore \quad K = \dfrac{1}{2} \cdot \dfrac{b^2 \sin A \sin C}{\sin B} = \dfrac{t \sin A}{\sin(C+A)}$, where $t = \dfrac{1}{2} b^2 \sin C$.

Taking logarithms on both sides, we have

$\log K = \log t + \log \sin A - \log \sin (C + A)$. Differentiating, we get

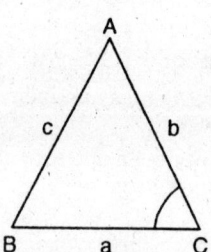

$$\frac{dK}{K} = \left[\frac{\cos A}{\sin A} - \frac{\cos(C+A)}{\sin(C+A)} \right] dA$$

$$= \left[\frac{\cos A \sin(C+A) - \cos(C+A) \sin A}{\sin A \sin(C+A)} \right] dA$$

$$= \left\{ \frac{\sin(\overline{C+A} - A)}{\sin A \sin(C+A)} \right\} dA = \frac{\sin C}{\sin A \sin(C+A)} dA.$$

$\therefore \quad$ Relative error $= \dfrac{\sin C}{\sin A \sin(C+A)} dA$.

Fig. 12.12.

Example 12.52 Find the percentage error in the computed area of an ellipse when an error of +1 percent is made in measuring the major and minor axes.

(IETE, June 2009; V.T.U., 2005; A.M.I.E., W-1998)

Solution. Let the major and minor axes of the ellipse be $2a$ and $2b$ respectively. The errors Δa and Δb in computing the lengths of the semi-major and semi-minor axes are, $\Delta a = a(0.01) = 0.01a$ and $\Delta b = b(0.01) = 0.01 b$.

The area of the ellipse is given by $A = \pi ab$. The error in computing the area of the ellipse is

$$dA = \left| \frac{\partial A}{\partial a} \right| \cdot \Delta a + \left| \frac{\partial A}{\partial b} \right| \cdot \Delta b = \pi b (0.01a) + \pi a (0.01b) = 0.02 \pi ab.$$

Relative error is $\left| \dfrac{dA}{A} \right| = \dfrac{0.02 \pi ab}{\pi ab} = 0.02$.

Percentage error in $A = \left| \dfrac{dA}{A} \right| \times 100 = 0.02 \times 100 = 2\%$.

Example 12.53 In measuring a rectangular block of wood, the dimensions were found to be 10, 12 and 20 units with a possible error of 0.05 unit in each of the measurements. Find (approximately) the greatest error in the surface area of the block and the percentage error in the area caused by the errors in the individual measurements.

Solution. The surface area $S = 2(xy + yz + zx)$, where x, y, z are the lengths of the edges. Using the relation

$$dS = \frac{\partial S}{\partial x}\Delta x + \frac{\partial S}{\partial y}\Delta y + \frac{\partial S}{\partial z}\Delta z \text{ to get } dS = 2(y+z)\Delta x + 2(x+z)\Delta y + 2(y+x)\Delta z.$$

The greatest error in S occurs when the errors in the lengths are of the same sign, say positive. Then

$$dS = 2(12+20)(0.05) + 2(10+20)(0.05) + 2(12+10)(0.05) = 8.4 \text{ unit}^2. \text{ The}$$
surface area, $S = 2(120 + 240 + 200) = 1120 \text{ unit}^2$.

$\therefore$ The percentage error in the surface area $= \dfrac{dS}{S} \times 100 = \dfrac{8.4}{1120} \times 100 = 0.75\%$

Example 12.54 The power consumed in an electrical resistor is given by $P = E^2/R$ (in watts). If $E = 200$ volts and $R = 8$ ohms, by how much does the power change if E is decreased by 5 volts and R is decreased by 0.2 ohm?

Solution. We have $\dfrac{\partial P}{\partial E} = \dfrac{2E}{R}, \dfrac{\partial P}{\partial R} = -\dfrac{E^2}{R^2}$. Therefore $dP = \dfrac{2E}{R} \cdot \Delta E - \dfrac{E^2}{R^2}\Delta R.$

When $E = 200$, $R = 8$, $\Delta E = d\dot{E} = -5$ and $\Delta R = dR = -0.2$, then

$$dP = \frac{2(200)}{8}(-5) - \left(\frac{200}{8}\right)^2(-0.2) = -250 + 125 = -125.$$

Thus the power is reduced by approximately 125 watts.

Example 12.55 In a triangle ABC, right angled at C, leg $a = 32$ units and leg $c = 75$ units, determined with maximum absolute errors of 0.2 unit and 0.1 unit in measuring c and a respectively. Determine the angle A from the formula $\sin A = a/c$; and compute the maximum absolute error in the calculation of angle A.

Solution. We have $\sin A = a/c$ or $\angle A = \sin^{-1}(a/c)$.

Differentiating partially, we obtain $\dfrac{\partial A}{\partial a} = \dfrac{1}{\sqrt{c^2 - a^2}}$ and $\dfrac{\partial A}{\partial c} = -\dfrac{a}{c\sqrt{c^2 - a^2}}.$

Employing the relation $|dA| = \left|\dfrac{\partial A}{\partial a}\right||\Delta a| + \left|\dfrac{\partial A}{\partial c}\right||\Delta c|$ and substituting the proper values, we get

$$|dA| = \left|\frac{1}{\sqrt{75^2 - 32^2}}\right| |0.1| + \left|\frac{-32}{75\sqrt{75^2 - 32^2}}\right| |0.2| = 0.00273 \text{ radian} = 9'24''$$

Thus
$$\angle A = \sin^{-1}(32/75) \pm 9'24''$$

Example 12.56 Suppose $u = xe^y + y \sin z$ and that x, y, z can be measured with maximum possible errors of ± 0.2, ± 0.6 and $\pm \dfrac{\pi}{180}$ respectively. Estimate the maximum possible error in calculating u from the measured values $x = 2$, $y = \log 3$, $z = \pi/2$.

Solution. We have $u = xe^y + y \sin z$. Therefore $\dfrac{\partial u}{\partial x} = e^y$,

$$\frac{\partial u}{\partial y} = xe^y + \sin z \text{ and } \frac{\partial u}{\partial z} = y \cos z.$$

Employing $du = \dfrac{\partial u}{\partial x} \cdot \Delta x + \dfrac{\partial u}{\partial y} \cdot \Delta y + \dfrac{\partial u}{\partial z} \cdot \Delta z$, we obtain

$$du = e^y \cdot \Delta x + (xe^y + \sin z)\Delta y + (y \cos z)\Delta z. \qquad \ldots(i)$$

Substituting the values of x, y, z, Δx, Δy and Δz in (i), we get

$$du = e^{\log 3}(\pm 0.2) + \left[2 \cdot (e^{\log 3}) + \sin \frac{\pi}{2}\right](\pm 0.6) + (\log 3)\left(\cos \frac{\pi}{2}\right)\left(\pm \frac{\pi}{180}\right)$$

$= 3(\pm 0.2) + (6 + 1)(\pm 0.6) + 0 = \pm 4.8$, which are the maximum possible positive and negative errors in u.

Example 12.57 The sides of a triangle are measured as 12 cm and 15 cm and the angle included between them as 60°. If the lengths can be measured within 1% accuracy while the angle can be measured within 2% accuracy, find the percentage error in determining (i) area of the triangle (ii) length of the opposite side of the triangle.   *(I.E.T.E.; Dec. 2002)*

Solution. (i) Refer Fig. 12.13 Area of a triangle, $K = \dfrac{1}{2}bc \sin A$. Differentiating partially, we get

$$\frac{\partial K}{\partial b} = \frac{1}{2}c \sin A, \frac{\partial K}{\partial c} = \frac{1}{2}b \sin A \text{ and } \frac{\partial K}{\partial A} = \frac{1}{2}bc \cos A.$$

Using the formula $dK = \dfrac{\partial k}{\partial b}\Delta b + \dfrac{\partial k}{\partial c}\Delta c + \dfrac{\partial k}{\partial A}\Delta A$, we get

$$dK = \frac{1}{2}c(\sin A) \cdot \Delta b + \frac{1}{2}b(\sin A) \cdot \Delta c + \frac{1}{2}bc(\cos A) \cdot \Delta A. \qquad \ldots(i)$$

Substituting $b = 12$, $c = 15$, $\angle A = 60°$,

$\Delta b = 0.12$, $\Delta c = 0.15$ and $\Delta A = \dfrac{\pi^c}{150}$ in (i), we obtain

$$dK = \frac{1}{2}(15)(\sin 60°)(0.12) + \frac{1}{2}(12)(\sin 60°)(0.15)$$

$$+ \frac{1}{2}(12)(15)\cos 60° \left(\frac{\pi}{150}\right)$$

$$= \frac{15\sqrt{3}\,(0.12)}{4} + 3\sqrt{3}\,(0.15) + \frac{3\pi}{10}$$

$$= 0.7794 + 0.7794 + 0.9429 = 2.5017 \text{ cm}^2.$$

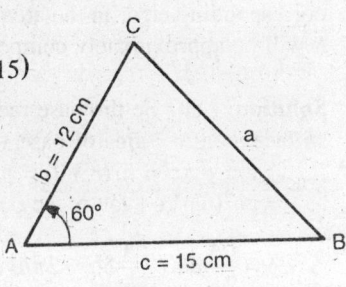

Fig. 12.13.

Area of triangle, $K = \dfrac{1}{2}(15)(12)\sin 60° = 77.94 \text{ cm}^2.$

$\therefore$ Percentage error in computing area $= \dfrac{dK}{K} \times 100 = \dfrac{2.5017}{77.94} \times 100 = 3.21\%.$

(ii) Using the law of cosines,

$$\cos A = \frac{b^2 + c^2 - a^2}{2bc} \implies a^2 = b^2 + c^2 - 2bc \cos A \qquad \text{...}(ii)$$

Differentiating (ii) partially, we get

$$\frac{\partial a}{\partial b} = \frac{b - c\cos A}{a}, \quad \frac{\partial a}{\partial c} = \frac{c - b\cos A}{a}, \quad \frac{\partial a}{\partial A} = \frac{bc \sin A}{a}.$$

$\therefore \quad da = \dfrac{\partial a}{\partial b} \cdot \Delta b + \dfrac{\partial a}{\partial c} \cdot \Delta c + \dfrac{\partial a}{\partial A} \cdot \Delta A$

$$= \frac{(b - c\cos A)\Delta b + (c - b\cos A)\Delta c + (bc \sin A)\Delta A}{a}. \qquad \text{...}(iii)$$

Substituting $b = 12$, $c = 15$, $\angle A = 60°$, $\Delta b = 0.12$, $\Delta c = 0.15$ and $\Delta A = \pi/150$ in (iii), we obtain.

$$da = \left\{ \left(12 - 15 \cdot \frac{1}{2}\right)0.12 + (15 - 6)0.15 + (12)(15)\left(\frac{\sqrt{3}}{2}\right) \cdot \frac{\pi}{150} \right\} 13.748$$

$= (0.54 + 1.35 + 3.266)/13.748 = 5.156/13.748 = 0.375$

Side 'a' is determined by employing the formula $a^2 = b^2 + c^2 - 2\,bc \cos A$ to get $a^2 = 189$ or $a = 13.748$.

Therefore % error in third side 'a' $= \dfrac{da}{a} \times 100 = \dfrac{0.375 \times 100}{13.748} = 2.73\%$

Example 12.58 The height h and semi-vertical angle α of a cone are measured and the total area A of surface of cone including that of base is calculated in terms of h, α. If h and α are in error by small quantities δh and $\delta \alpha$ respectively, find the corresponding error in the area. Show further that if $\alpha = \pi/6$, an error of $+1\%$ in h will be approximately compensated by an error of -0.33 degree in α.

(I.E.T.E., June 2003, Delhi 1997)

Solution. Let r be the base radius and l the slant height of the cone, then total surface area A = area of base + area of curved surface

$$= \pi r^2 + \pi r l = \pi r(r + l) = \pi h \tan \alpha(h \tan \alpha + h \sec \alpha)$$
$$= \pi h^2(\tan^2\alpha + \tan \alpha \sec \alpha).$$

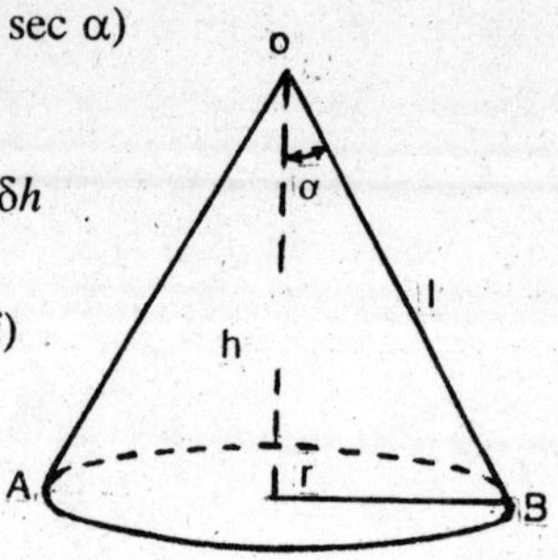

$$\therefore \; dA = \frac{\partial A}{\partial h}\delta h + \frac{\partial A}{\partial \alpha}\delta \alpha = 2\pi h\left(\tan^2\alpha + \tan \alpha \sec \alpha\right)\delta h$$

$$+\pi h^2\left(2 \tan \alpha \sec^2\alpha + \sec^3\alpha + \tan^2\alpha \sec \alpha\right)\delta \alpha \quad ...(i)$$

which is the error dA in A corresponding to errors δh and $\delta \alpha$ in h and α respectively.

Putting $\alpha = \dfrac{\pi}{6}$ and $\delta h = \dfrac{h}{100}$ in (i), we have

Fig. 12.14.

$$dA = \frac{2\pi h^2}{100}\left(\frac{1}{3} + \frac{1}{\sqrt{3}}\cdot\frac{2}{\sqrt{3}}\right) + \pi h^2\left(\frac{2}{\sqrt{3}}\cdot\frac{4}{3} + \frac{8}{3\sqrt{3}} + \frac{1}{3}\cdot\frac{2}{\sqrt{3}}\right)\delta \alpha$$

$$= \frac{2\pi h^2}{100}(1) + \pi h^2\left(\frac{8}{3\sqrt{3}} + \frac{8}{3\sqrt{3}} + \frac{2}{3\sqrt{3}}\right)\delta \alpha = 2\pi h^2(0.01) + \pi h^2(3.464)\,\delta \alpha.$$

The error in h will be compensated by the error in α, when dA = 0 that is,
$2\pi h^2(0.01) + \pi h^2(3.464)\,\delta \alpha = 0$

or $\quad \delta \alpha = \dfrac{-2(0.01)}{3.464}$ radian $= -\dfrac{0.02}{3.464}\times 57.3^\circ = -0.33^\circ.$

Exercise 12.5

1. Evaluate $\log\left(\sqrt[3]{1.03} + \sqrt[4]{0.98} - 1\right)$.

   [**Hint:** Let $z = \log\left(x^{1/3} + y^{1/4} - 1\right)$.

   Take $x = 1$, $y = 1$, $dx = 0.03$, $dy = -0.02$. Then $z = \log 1 = 0$.

   $\therefore \; dz = \left(\dfrac{1}{3}x^{-2/3}dx + \dfrac{1}{4}y^{-3/4}dy\right)\Big/\left(x^{1/3} + y^{1/4} - 1\right)$. Substituting the proper

   values, we get $dz = 3^{-1}(0.03) + 4^{-1}(-0.02) = 0.005$, which is the approximate value required].

2. (a) If $f = x^2 y^3 z^{1/10}$, find the approximate value of f, when $x = 1.99$, $y = 3.01$ and $z = 0.98$.

   (b) Find an approximate value of $f(2.1, 3.2)$, where $f(x,y) = x^y$.
   Given $\log 2 = 0.3010$.   **Ans.** (a) 107.784, (b) 9.6816.

3. Using differential, compute $\sqrt[5]{(3.8)^2 + 2(2.1)^3}$ approximately.
   (I.E.T.E.; Dec. 1998). **Ans.** 0.01

4. If the power, P required to propel a steamer varies as the cube of the velocity and the square of length, prove that a 3% increase in velocity and 4% increase in length will require an increase of about 17% H.P.
   (A.M.I.E., W-2000)

   [**Hint:** $P = cv^3 l^2$. Therefore, $\log P = \log c + 3 \log v + 2 \log l$.

   Thus $100\dfrac{dP}{P} = 3\left(100\dfrac{dv}{v}\right) + 2\left(100\dfrac{dl}{l}\right) = 3 \times 3 + 2 \times 4 = 17\%.$]

5. The period T of a simple pendulum of length l with small oscillations is given by $T = 2\pi\sqrt{l/g}$. If T is computed using $l = 2$ m and $g = 9.8$ m/sec^2, find the approximate error in T if the true length are $l = 2.015m$ and $g = 9.82$ m/sec^2. Find also the percentage error.

   [**Hint:** $dT = T\left(\dfrac{dl}{2l} - \dfrac{dg}{2g}\right)$. Substituting $l = 2$, $g = 9.8$, $dl = 0.015$,

   $dg = 0.02$ and $T = 2\pi\sqrt{\dfrac{2}{9.8}} = 2.84$, we get $dT = 0.00775$ sec. to be added to

   computed value and $\left(\dfrac{dT}{T}\right)(100) = 0.273\%$]

6. If $\theta = \lambda/v^2$ and $\phi = \mu/v$, where λ, μ are constants, show that the percentage error of θ is nearly double of that of ϕ.   *(A.M.I.E., W-2003)*

7. Show that the relative error of a quotient does not exceed the sum of the relative errors of the dividend and the divisor.   *(A.M.I.E., W-2001)*

8. If the kinetic energy T is given by the formula $T = mv^2/2$. Use total differential to find approximately the change in T as m changes from 49 to 49.5 and v changes from 1600 to 1590.   *(Mysore, 1995; I.E.T.E., June 2002)*

   [**Hint:** $dT = \left(v^2 dm + 2mv\, dv\right)/2 = mv\, dv + \left(v^2 dm\right)/2$. Here $m = 49$,

   $v = 1600$, $dm = 0.5$, $dv = -10$. $\therefore$ $dT = -144000$].

9. The diameter and altitude of a can in the shape of a right circular cylinder are measured as 4 cm and 6 cm respectively. The possible error in each

measurements is 0.1 cm. Find approximately the maximum possible error in the values computed for the volume and the lateral surface.

$$\text{(A.M.I.E., W-2004; Mangalore 1999; IETE, Dec. 2008)}$$

[**Hint:** $V = \dfrac{\pi x^2 y}{4}$, $S = \pi xy$, where x is diameter and y is altitude of the

cylinder $dV = \pi\left(2xydx + x^2dy\right)/4$, $dS = \pi\left(xdy + ydx\right)$.

Since $x = 4$, $y = 6$, $dx = dy = 0.1$, we get $dV = 1.6\,\pi$ cm^3 and $dS = \pi$ cm^2].

10. At a distance of 50 metres from foot of a tower, the elevation of its top is 30°. If the possible errors in measuring the distance and the elevation are 2 cm and 0.05 degree respectively, find the approximate error in the calculated height.

[**Hint:** Let $h = x \tan \alpha$. $\therefore$ $dh = (\tan \alpha)\,dx + (x \sec^2\alpha)\,d\alpha$.
Given $\alpha = \pi/6$, $x = 50$, $dx = 1/50$, $d\alpha = 0.05 \times \pi/180$ radian.

$$\therefore dh = \frac{1}{\sqrt{3}} \cdot \frac{1}{50} + 50\left(\frac{2}{\sqrt{3}}\right)^2 \times 0.05 \times \frac{\pi}{180} = \frac{\sqrt{3}}{150} + \frac{11}{189} = 0.07 \text{ m} = 7 \text{ cm.}]$$

11. A triangle ABC is inscribed in a circle of radius R. Prove that if the triangle be slightly varied so as to remain inscribed in the same circle, then

$$\frac{da}{\cos A} + \frac{db}{\cos B} + \frac{dc}{\cos C} = 0. \qquad \text{(Hamirpur, 1995; Mysore, 1995S)}$$

[**Hint:** $R = \dfrac{a}{2\sin A} = \dfrac{b}{2\sin B} = \dfrac{c}{2\sin C}$.

$$a = 2R \sin A, \ b = 2R \sin B, \ c = 2R \sin C.$$

Hence $da = 2R \cos A\, dA$, $db = 2R \cos B\, dB$, $dc = 2R \cos C\, dC$

i.e., $dA = da/2R \cos A$, $dB = db/2R \cos B$, $dC = \dfrac{dc}{2R\cos C}$. In any triangle,

ABC, $\angle A + \angle B + \angle C = \pi$ or $dA + dB + dC = 0$. Hence etc.]

12. A metal box is 4 unit long, 3 unit wide and 2 unit deep. The thickness of the metal is 0.05 unit. Find approximately the volume of the metal in the box.

Ans. 1.3 cubic unit.

13. The voltage V across a resistor is measured with an error h, and resistance R is measured with an error k. Show that the error in calculating the $W(V, R) = V^2/R$ generated in the resistor is $V(2Rh - Vk)/R^2$.

14. The time of swing t of a pendulum of length l under certain condition is given by $t = 2\pi\sqrt{l/g'}$, where $g' = g\left(\dfrac{r}{r+h}\right)^2$. Find the percentage error in t due to the errors of p percent in h and q per cent in l **Ans.** $\dfrac{1}{2}\left(q + \dfrac{2ph}{r+h}\right)$.

15. Approximate the area of rectangle of dimensions 35.02 by 24.97 units.

    [Hint: Let area of rectangle be denoted by K.

    $$K = xy, \ dK = \frac{\partial K}{\partial x} dx + \frac{\partial K}{\partial y} dy = y dx + x dy.$$

    Take $x = 35$, $y = 25$, $dx = 0.02$, $dy = -0.03$. Therefore,

    $K = (35)(25) = 875$. $dK = 25(0.02) + 35(-0.03) = -0.55$

    The area is approx. $(K + dK) = 875 - 0.55 = 874.45 \ \text{unit}^2$]

16. For the formula $R = E/C$, find the maximum error and the percentage error in R if $C = 20$ Amp. with a possible error of 0.1 Amp and $E = 120$ volts with a possible error of 0.05 volt. **Ans. 0.0325 ohm, 0.54%.**

    [Hint: The maximum error will occur when $dE = 0.05$ and $dC = -0.1$. Find dR, using $dR = \frac{E}{C}\left(\frac{dE}{E} - \frac{dC}{C}\right)$ and hence find $\frac{dR}{R}(100)$.]

    Ans. 0.0325 ohm, 0.54%

17. Approximate the change in the hypotenuse of a right triangle of legs 6 and 8 units when the shorter leg is lengthened by 1/4 unit and the longer leg is shortened by 1/8 unit.

    [Hint: Let x, y, and z be the shorter leg, the longer leg and the hypotenuse of the triangle. Then

    $$z = \sqrt{x^2 + y^2}, \ \partial z/\partial x = x\Big/\sqrt{x^2+y^2}, \ \partial z/\partial y = y\Big/\sqrt{x^2+y^2}.$$

    $$\therefore \ dz = \frac{\partial z}{\partial x} dx + \frac{\partial z}{\partial y} dy = \frac{x\,dx + y\,dy}{\sqrt{x^2+y^2}}.$$

    When $x = 6$, $y = 8$, $dx = 1/4$, $dy = -1/8$, then

    $$dz = \frac{6(1/4) + 8(-1/8)}{\sqrt{6^2 + 8^2}} = \frac{1}{20} \ \text{unit. Thus the hypotenuse is lengthened by}$$

    approximetely 1/20 unit].

18. A quantity u can be computed by the formula $u = 4\alpha^{1/2}\beta^{3/4}\gamma^{-1/5}$. If an error of $\pm 0.5\%$, $\pm 1\%$ and $\pm 0.2\%$ is made in measuring α, β and r respectively, compute maximum percentage error in u, when the above formula is used.
    (I.E.T.E., Dec. 1996) **Ans. 1.04%**

19. If specific gravity is determined by the formula $s = \dfrac{A}{A - W}$, where A is the weight in the air and W the weight in water. What is (a) approximately the largest error in s if A can be read within 0.01 gm and W within 0.02 gm, the actual readings being $A = 1.1$ gm; $W = 0.6$ gm? (b) the largest relative error?
    (A.M.I.E., S-1999)

[**Hint:** $ds = \dfrac{\partial s}{\partial A} dA + \dfrac{\partial s}{\partial W} dW = -\dfrac{WdA}{(A-W)^2} + \dfrac{AdW}{(A-W)^2}$. The greatest error in

s will occur when the error in A is of negative sign. Put $dA = -0.01$, $dW = 0.02$, $A = 1.1$, $W = 0.6$, we obtain

$ds = +\dfrac{0.6(0.01)}{0.25} + \dfrac{1.1(0.02)}{0.25} = 0.112$, hence $\dfrac{ds}{s} = \dfrac{0.112 \times 5}{11} = \dfrac{56}{1100}$].

20. Two sides of a triangle are found by measurement to be 18.9 m and 23.4 m and the included angle to be 60°. These measurements were subject to errors whose maximum values are 3 cm in each length and 1° in the angle.

(a) Find the approximate maximum error and the percentage error made in calculating the third side from these measurements.

(b) What is approximately the greatest possible error in the computed value of the area?   **Ans.** (a) 0.34, 1.58%, (b) 8.22 m^2.

[**Hint:** (a) Differentiating $x^2 = a^2 + b^2 - 2ab \cos \alpha$ partially with respect to a, b and α respectively, where a, b are the given sides and α the included angle, we get

$$\frac{\partial x}{\partial \alpha} = \frac{a - b\cos\alpha}{x}, \quad \frac{\partial x}{\partial b} = \frac{b - a\cos\alpha}{x}, \quad \frac{\partial x}{\partial \alpha} = \frac{ab\sin\alpha}{x}$$

$$\therefore \quad dx = \frac{\partial x}{\partial a} \cdot da + \frac{\partial x}{\partial b} \cdot db + \frac{\partial x}{\partial \alpha} \cdot d\alpha$$

$$= \frac{(a - b\cos\alpha)da + (b - a\cos\alpha)db + ab\sin\alpha d\alpha}{x}.$$

Put $a = 18.9$, $b = 23.4$, $\alpha = 60°$, $da = db = 0.03$. Therefore,

$d\alpha = 1° = \pi/180$. We get $dx = 0.34$ and hence $\dfrac{dx}{x}(100) = 1.58\%$.

(b) $\Delta = \dfrac{1}{2} ab \sin \alpha$, $\dfrac{\partial \Delta}{\partial a} = \dfrac{1}{2} b \sin \alpha$, $\dfrac{\partial \Delta}{\partial b} = \dfrac{1}{2} a \sin \alpha$, $\dfrac{\partial \Delta}{\partial \alpha} = \dfrac{1}{2} ab \cos \alpha$.

$\therefore \quad d\Delta = \dfrac{\partial \Delta}{\partial a} da + \dfrac{\partial \Delta}{\partial b} \cdot db + \dfrac{\partial \Delta}{\partial \alpha} \cdot d\alpha$]

21. Two sides of a triangle were measured as 150 and 200 units and the included angle is 60°. If the possible errors are 0.2 unit in measuring the sides and 1° in the angle, what is the greatest possible error in the computed area?   **Ans.** 161.21 units².

22. Show that the approximate change in the angle A of a triangle ABC due to small changes δa, δb, δc in the sides a, b, c respectively is given by

$\delta A = \dfrac{a}{2\Delta}(\delta a - \delta b \cos C - \delta c \cos B)$, where Δ is the area of the triangle ABC.

Verify that $\delta A + \delta B + \delta C = 0$. *(I.E.T.E. Dec 2007, June 2001).*

[**Hint:** Using cosine and projection formula. We know that

$\cos A = (b^2 + c^2 - a^2)/2bc$ or $2\, bc \cos A = b^2 + c^2 - a^2.$

$\therefore$ $(\delta b)c \cos A + (\delta c) \cdot b \cos A - bc \sin A(\delta A) = b(\delta b) + c(\delta c) - a(\delta a)$

$= (c \cos A + a \cos C)\delta b + [a \cos B + b \cos A]\delta c - a\,\delta\,a,$

or $\quad bc\,(\sin A)\,\delta A = a[(\delta a) - (\delta b)\cos C - (\delta\,c)\cos B)]$

or $\quad 2\Delta\,\delta A = a[\delta a - \delta b \cos C - \delta\,c \cos B]$ $\qquad \left(\because \Delta = \dfrac{1}{2}bc \sin A\right)$

or $\quad \delta A = \dfrac{a}{2\Delta}(\delta a - \delta b \cos C - \delta c \cos B).$

Similarly, $\delta B = \dfrac{b}{2\Delta}(\delta b - \delta c \cos A - \delta a \cos C)$ and δC etc.

Thus $\delta A + \delta B + \delta C = 0].$

23. The deflection at the centre of a rod of length l and diameter d supported at its ends and loaded at the centre with a weight w varies as wl^3d^{-4}. What is the increase in the deflection corresponding to $p\%$ increase in w, $q\%$ decrease in l and $r\%$ increase in d? **Ans. $(p - 3q - 4r)\%$.**

Example 12.59. *Find the shortest distance between the line $2x + y = 10$ and the ellipse $9x^2 + 4y^2 = 36$.* *(A.M.I.E., W-2010)*

Solution: Let (x, y) be a point on the ellipse and (a, b) be a point on the line. Then the shortest distance between the line and the ellipse is the square root of the minimum value of

$f(x, y, a, b) = (x - a)^2 + (y - b)^2$ subject to the constraints

$\phi_1(x, y) = 9x^2 + 4y^2 - 36 = 0$ and $\phi_2(a, b) = 2a + b - 10 = 0.$

Consider the auxiliary function

$F(x, y, a, b, \lambda_1, \lambda_2) = (x - a)^2 + (y - b)^2 + \lambda_1(9x^2 + 4y^2 - 36) + \lambda_2(2a + b - 10).$

For extremum, we have the necessary conditions

$\dfrac{\partial F}{\partial x} = 2(x - a) + 18x\,\lambda_1 = 0,$ or $\lambda_1 x = (a - x)/9$

$\dfrac{\partial F}{\partial y} = 2(y - b) + 8y\,\lambda_1 = 0,$ or $\lambda_1 y = (b - y)/4$

$\dfrac{\partial F}{\partial a} = -2(x - a) + 2\lambda_2 = 0,$ or $\lambda_2 = x - a$

$\dfrac{\partial F}{\partial b} = -2(y - b) + \lambda_2 = 0,$ or $\lambda_2 = 2(y - b)$

Eliminating λ_1 and λ_2 from the above equations, we obtain
$4(a-x)y = 9(b-y)x$ and $x - a = 2(y-b)$.

Dividing the two equations, we get $8y = 9x$. Substituting in the equation of the ellipse, we obtain

$$9x^2 + 4\left(\frac{9x}{8}\right)^2 = 36, \text{ or } x^2 = 64/25 \Rightarrow x = \pm 8/5 \text{ and } y = \pm 9/5.$$

Corresponding to $x = 8/5$, $y = 9/5$, we obtain

$$\frac{8}{5} - a = 2\left(\frac{9}{5} - b\right) \text{ or } a = 2b - 2.$$

Substituting in the equation of the line $2a + b - 10 = 0$, we get $a = 18/5$ and $b = 14/5$.

Hence, an extremum is obtained when $(x, y) = (8/5, 9/5)$ and $(a, b) = (18/5, 14/5)$. The distance between the two points is

$$\sqrt{\left(\frac{18}{5} - \frac{8}{5}\right)^2 + \left(\frac{14}{5} - \frac{9}{5}\right)^2} = \sqrt{5} \text{ units.}$$

Corresponding to $x = -8/5$, $y = -9/5$, we get $a - 2b = 2$. Substituting in the equation $2a + b - 10 = 0$, we get $a = 22/5$, $b = 6/5$. Hence another extremum is obtained when $(x, y) = (-8/5, -9/5)$ and $(a, b) = (22/5, 6/5)$. The distance

between these two points $= \sqrt{\left(\frac{22}{5} + \frac{8}{5}\right)^2 + \left(\frac{6}{5} + \frac{9}{5}\right)^2} = 3\sqrt{5}$ units. Hence

the shortest distance between the line and the ellipse is $\sqrt{5}$ units.

Chapter 13

Multiple Integrals

13.1 Double Integrals

Notation: The symbol $\int\limits_{a}^{b}\int\limits_{c}^{d} f(x,y)\,dy\,dx$ is called a double integral. It is evaluated as follows:

We first integrate $f(x, y)$ with respect to y (considering x as a constant) between the limits c and d. The result is then integrated with respect to x between the limits a and b.

Thus the double integral

$$\int\limits_{a}^{b}\int\limits_{c}^{d} f(x,y)\,dy\,dx = \int\limits_{a}^{b}\left(\int\limits_{c}^{d} f(x,y)\,dy\right) dx = \int\limits_{a}^{b}dx \int\limits_{c}^{d} f(x,y)\,dy.$$

The limits c and d may be constants or functions of x.

Example 13.1 Evaluate (i) $\int\limits_{0}^{3}\int\limits_{1}^{2}\left(x^2 + 3y^2\right)dy\,dx$, **(ii)** $\int\limits_{x=0}^{1} \int\limits_{y=0}^{2}\left(x^2 + 3xy^2\right)dx\,dy$.

(IETE June 2009)

Solution. (*i*) The given integral $= \int\limits_{0}^{3}\left(\int\limits_{1}^{2}(x^2 + 3y^2)\,dy\right)dx$

$$= \int\limits_{0}^{3}\left[x^2 y + 3\cdot\frac{y^3}{3}\right]_{1}^{2} dx \quad \text{(integrating w.r. to } y, \text{ treating } x \text{ as a constant)}$$

$$= \int\limits_{0}^{3}\left[(2x^2 + 2^3) - (x^2 + 1)\right]dx = \int\limits_{0}^{3}(x^2 + 7)\,dx = \left[\frac{x^3}{3} + 7x\right]_{0}^{3} = \left(\frac{3^3}{3} + 7(3)\right) = 30.$$

(ii) $\displaystyle\int_{x=0}^{1}\int_{y=0}^{2}\left(x^2+3xy^2\right)dxdy = \int_{x=0}^{1}dx\int_{y=0}^{2}\left(x^2+3xy^2\right)dy = \int_0^1\left[x^2y+xy^3\right]_0^2 dx$

$\displaystyle = 2\int_0^1\left(x^2+4x\right)dx = 2\left[\frac{x^3}{3}+2x^2\right]_0^1 = \frac{14}{3}.$

Example 13.2 Evaluate $\displaystyle\int_0^1\int_0^{\sqrt{1+x^2}}\frac{dydx}{1+x^2+y^2}.$

(PTU 2006; Rajasthan, 2005; UPTU 2001)

Solution. Since the limits of y are functions of x, the integration will first be performed with respect to y (treating x as a constant). Thus

$\displaystyle I = \int_0^1\left[\int_0^{\sqrt{1+x^2}}\frac{1}{\left(1+x^2\right)+y^2}dy\right]dx = \int_0^1\frac{1}{\sqrt{1+x^2}}\left[\tan^{-1}\frac{y}{\sqrt{1+x^2}}\right]_0^{\sqrt{1+x^2}}dx$

$\displaystyle = \int_0^1\frac{1}{\sqrt{1+x^2}}\left[\tan^{-1}1-\tan^{-1}0\right]dx = \frac{\pi}{4}\int_0^1\frac{dx}{\sqrt{1+x^2}}$

$\displaystyle = \frac{\pi}{4}\left[\log_e\left(x+\sqrt{1+x^2}\right)\right]_0^1 = \frac{\pi}{4}\left[\log_e\left(1+\sqrt{2}\right)-\log 1\right] = \frac{\pi}{4}\log_e\left(\sqrt{2}+1\right).$

Example 13.3 Evaluate $\displaystyle\int_0^2\int_0^{x^2}x\left(x^2+y^2\right)dy\,dx.$

Solution. The given integral $\displaystyle= \int_0^2\left[\int_0^{x^2}\left(x^3+xy^2\right)dy\right]dx$

$\displaystyle = \int_0^2\left[x^3y+x\frac{y^3}{3}\right]_0^{x^2}dx = \int_0^2\left(x^3\cdot x^2+x\cdot\frac{\left(x^2\right)^3}{3}\right)dx$

$\displaystyle = \int_0^2\left(x^5+\frac{x^7}{3}\right)dx = \left[\frac{x^6}{6}+\frac{x^8}{24}\right]_0^2 = \frac{2^6}{6}+\frac{2^8}{24} = \frac{64}{3}.$

Example 13.4 Evaluate $\int\limits_{0}^{\pi}\int\limits_{0}^{x} x\sin(x+y)\,dy\,dx$.

Solution. The given integral $= \int\limits_{0}^{\pi} x\left[-\cos(x+y)\right]_{0}^{x} dx$

$$= -\int\limits_{0}^{\pi} x(\cos 2x - \cos x)\,dx = -\left[\left[x\left(\frac{\sin 2x}{2} - \sin x\right)\right]_{0}^{\pi} - \int\limits_{0}^{\pi}\left(\frac{\sin 2x}{2} - \sin x\right)dx\right]$$

$$= \int\limits_{0}^{\pi}\left(\frac{\sin 2x}{2} - \sin x\right)dx = \left[-\frac{\cos 2x}{4} + \cos x\right]_{0}^{\pi} = \left[\left(-\frac{1}{4}-1\right) - \left(-\frac{1}{4}+1\right)\right] = -2.$$

Definition

We know the definition of the definite integral $\int\limits_{a}^{b} f(x)\,dx$ as the limit of a sum.

We shall now define the double integral of a function $f(x, y)$ over a closed region R, assuming that $f(x, y)$ is given for all points of the region.

Sub-divide the region R into rectangles by drawing lines parallel to co-ordinate axes. Number the rectangles which lie entirely inside R, from 1 to n.

Let $P(x_i, y_i)$ be any point inside ith rectangle whose area is ΔA_i and form the sum

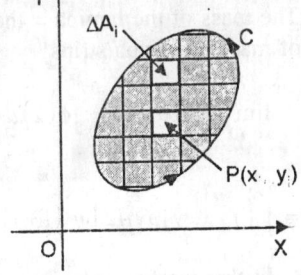

Fig. 13.1 Subdivision of R

$$S_n = \sum_{i=1}^{n} f(x_i, y_i)\Delta A_i.$$

Let $n \to \infty$, such that the length of the largest diagonal of the rectangles approaches zero. If $\lim\limits_{n\to\infty} S_n$ exists, independent of the choice of the subdivision and the point (x_i, y_i), then we say that $f(x, y)$ is integrable over R.

This limit is called the *double integral* of $f(x, y)$ over the region R and is denoted by $\iint\limits_{R} f(x, y)\,dA$. Other symbols used for this limit are

$$\iint\limits_{R} f(x, y)\,dy\,dx \text{ or } \iint\limits_{R} f(x, y)\,dx\,dy.$$

A Useful Example

We shall now find the mass of plane lamina with variable density in the form of a double integral. This will indicate how double integrals are evaluated in practice.

Let the plane lamina be bounded by the curves $y = f_1(x)$, $y = f_2(x)$ and the ordinates $x = a$ and $x = b$. Let the density σ at any point $P(x, y)$ be given by $\sigma = f(x, y)$.

We divide the area of the lamina into strips parallel to y-axis. One such strip is *ABCD*. We divide the strip into elementary rectangular meshes like *PQRS* with dimensions Δx and Δy (as shown).

The mass of this mesh $= f(x, y)\, \Delta y\, \Delta x$.

The mass of the strip = Sum of masses of such meshes $= \lim\limits_{\Delta y \to 0} \sum f(x, y)\Delta y\, \Delta x$

(x and Δx remain constant for this summation)

$$= \left[\int_{f_1(x)}^{f_2(x)} f(x, y)\, dy \right]\Delta x = \phi(x)\,\Delta x,$$

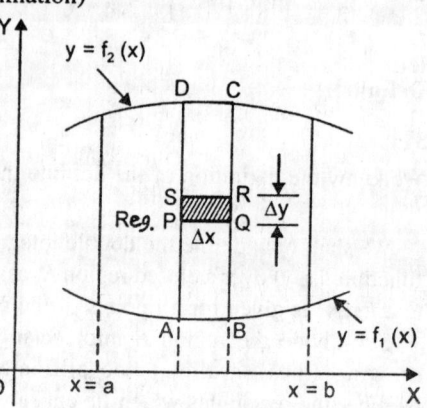

where $\phi(x) = \int_{f_1}^{f_2} f(x, y)\, dy.$ ...(13.1)

The mass of the lamina = the sum of masses of such strips

$$= \lim\limits_{\Delta x \to 0} \sum \phi(x)\Delta x = \int_{a}^{b}\phi(x)\, dx$$

$$= \int_{a}^{b}\int_{f_1}^{f_2} f(x, y)\, dy\, dx \text{ by } (13.1).\ ...(13.2)$$

Fig. 13.2.

In this we first integrate $f(x, y)$ w.r. to y, treating x as a constant and the resulting expression is integrated with respect to x.

Similarly, if the lamina is bounded by the curves $x = \Psi_1(y)$, $x = \psi_2(y)$ and the lines $y = c$, $y = d$, we can divide the area, thus bounded, into strips parallel to x-axis. Proceeding as above, we can show that the mass of the lamina

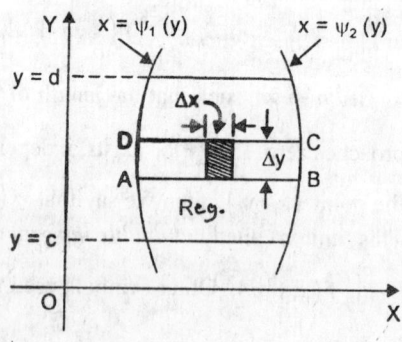

$$= \int_{c}^{d}\int_{\psi_1}^{\psi_2} f(x, y)\, dx\, dy.\ ...(13.3)$$

Fig. 13.3.

In this we first integrate $f(x, y)$ w.r. to x, treating y as a constant and then the resulting expression is integrated w.r.t. y.

Remarks. 1. The double integral

$$\int_a^b \int_{f_1(x)}^{f_2(x)} f(x,y)\,dy\,dx$$ is also called the

value of $\iint f(x, y)\,dy\,dx$ over the region bounded by the curves $y = f_1(x)$, $y = f_2(x)$, and the lines $x = a$ and $x = b$.

(2) The area of the adjoining figure may be considered as bounded by the curves, $y = f_1(x)$, $y = f_2(x)$ and the lines $x = a$, $x = b$.

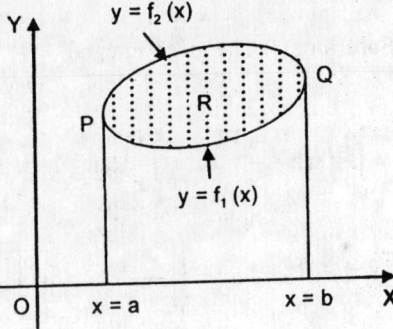

Fig. 13.4.

Hence $\iint f(x, y)\,dy\,dx$ over this region $= \int_a^b \int_{f_1}^{f_2} f(x,y)\,dy\,dx.$

(3) We should carefully note that when the area is divided into strips parallel to y-axis, then the first integral is w.r. to y, treating x as a constant.

The limits of integration for y are the values of y (as functions of x) at the lower end of the strip and the upper end of the strip. These values are given by the equations of the curves on which the lower extremity and the upper extremity lie.

Then the integration w.r. to x is between the limits which are values of x at the extreme left and extreme right of the region.

Similar remarks apply when the region is divided into strips parallel to x-axis.

(4) Some writers interpret the double integral $\int_a^b \int_{f_1}^{f_2} f(x,y)\,dx\,dy$ to mean integral of $f(x, y)$ w.r. to y, treating x as a constant and then integrating the resulting expression w.r.t. x. We should bear in mind that if f_1 and f_2 are functions of x then it is understood that we integrate w.r. to y first and then w.r. to x and if f_1 and f_2 are functions of y then we integrate w.r. to x first and then w.r. to y.

13.2 Illustrative Examples

Example 13.5 Evaluate the following double integrals

(i) $\displaystyle \int_1^2 \int_0^x \frac{dy\,dx}{y^2 + x^2}$,

(ii) $\displaystyle \int_0^4 dx \int_0^{x^2} e^{y/x}\,dy$, (*Osmania, 1999S*)

(iii) $\displaystyle \int_0^\pi \int_0^{\cos y} x \sin y\,dx\,dy$,

(iv) $\displaystyle \int_0^{\log 3} \int_0^{\log 2} e^{x+y}\,dy\,dx$,

(v) $\displaystyle \int_0^1 dy \int_0^1 \frac{x-y}{(x+y)^3}\,dx$.

Solution. (*i*) The given integral $= \int\limits_{1}^{2}\left(\int\limits_{0}^{x}\frac{dy}{y^2+x^2}\right)dx$

$= \int\limits_{1}^{2}\frac{1}{x}\left[\tan^{-1}\frac{y}{x}\right]_{0}^{x}dx = \int\limits_{1}^{2}\frac{1}{x}\left(\tan^{-1}1-\tan 0\right)dx = \frac{\pi}{4}\int\limits_{1}^{2}\frac{1}{x}dx = \frac{\pi}{4}\left[\log x\right]_{1}^{2} = \frac{\pi}{4}\log 2.$

(*ii*) $\int\limits_{0}^{4}dx\int\limits_{0}^{x^2}e^{y/x}\,dy = \int\limits_{0}^{4}\left[\frac{e^{y/x}}{1/x}\right]_{0}^{x^2}dx = \int\limits_{0}^{4}\left[xe^{y/x}\right]_{0}^{x^2}dx$

$= \int\limits_{0}^{4}\left(\underset{I}{x\,e^x}-\underset{II}{x}\right)dx = \left[xe^x\right]_{0}^{4}-\int\limits_{0}^{4}e^x dx-\left[\frac{x^2}{2}\right]_{0}^{4} = \left[xe^x-e^x\right]_{0}^{4}-\frac{1}{2}\left[x^2\right]_{0}^{4} = \underline{3e^4-7}$

(*iii*) $\int\limits_{0}^{\pi}\int\limits_{0}^{\cos y}x\sin y\,dx\,dy = \int\limits_{0}^{\pi}\left(\int\limits_{0}^{\cos y}x\sin y\,dx\right)dy = \int\limits_{0}^{\pi}\left[\frac{x^2}{2}\sin y\right]_{x=0}^{\cos y}dy$

$= \int\limits_{0}^{\pi}\frac{1}{2}\cos^2 y\sin y\,dy = \left[-\frac{1}{6}\cos^3 y\right]_{0}^{\pi} = \frac{1}{3}.$

(*iv*) $\int\limits_{0}^{\log 3}\int\limits_{0}^{\log 2}e^{x+y}\,dy\,dx = \int\limits_{0}^{\log 3}\left[\int\limits_{0}^{\log 2}e^{x+y}\,dy\right]dx = \int\limits_{0}^{\log 3}\left[e^{x+y}\right]_{0}^{\log 2}dx$

$= \int\limits_{0}^{\log 3}\left(e^{x+\log 2}-e^x\right)dx = \int\limits_{0}^{\log 3}\left(2e^x-e^x\right)dx = \int\limits_{0}^{\log 3}e^x\,dx = \left[e^x\right]_{0}^{\log 3} = 3-1 = 2.$

(*v*) $\int\limits_{0}^{1}dy\int\limits_{0}^{1}\frac{x-y}{(x+y)^3}dx = \int\limits_{0}^{1}dy\int\limits_{0}^{1}\left\{\frac{1}{(x+y)^2}-\frac{2y}{(x+y)^3}\right\}dx$

$= \int\limits_{0}^{1}\left[-\frac{1}{x+y}+\frac{y}{(x+y)^2}\right]_{0}^{1}dy$

$= \int\limits_{0}^{1}\left[-\frac{1}{1+y}+\frac{y}{(1+y)^2}\right]dy = -\int\limits_{0}^{1}\frac{dy}{(1+y)^2} = \left[\frac{1}{1+y}\right]_{0}^{1} = -\frac{1}{2}.$

Example 13.6 (*i*) Evaluate $\iint\limits_{R}(x^2+y^2)\,dy\,dx$, where **R** is the region in the positive quadrant for which $x+y \le 1$.

(ii) **Evaluate** $\iint\limits_{R} xy\, dx\, dy$**, where R is the quadrant of the circle** $x^2 + y^2 = a^2$

where $x \ge 0, y \ge 0$**.** *(Rajasthan, 2006; V.T.U., 2001; Madras 2000; IETE S-1999)*

(iii) **Evaluate** $\iint\limits_{S} \sqrt{xy - y^2}\, dx\, dy$ **over the region S, triangle with vertices (0, 0),**

(10, 1) and (1, 1).

Solution. *(i)* Refer Fig. 13.5 Here the region of integration is the area bounded by the two axes and straight line $x + y = 1$. We can consider it as the area bounded by the lines, $y = 0$, $y = 1 - x$, $x = 0$, and $x = 1$.

Hence, $\iint\limits_{R} (x^2 + y^2)\, dy\, dx = \int\limits_{x=0}^{1} \int\limits_{y=0}^{1-x} (x^2 + y^2)\, dy\, dx = \int\limits_{0}^{1} \left[x^2 y + \frac{y^3}{3} \right]_{0}^{1-x} dx$

$= \int\limits_{0}^{1} \left[x^2 (1-x) + \frac{1}{3}(1-x)^3 \right] dx$

$= \left[\frac{x^3}{3} - \frac{x^4}{4} + \frac{1}{3} \frac{(1-x)^4}{4(-1)} \right]_{0}^{1}$

$= \left[\frac{1}{3}(1-0) - \frac{1}{4}(1-0) - \frac{1}{12}(0-1) \right]$

$= \frac{1}{3} - \frac{1}{4} + \frac{1}{12} = \frac{1}{6}.$

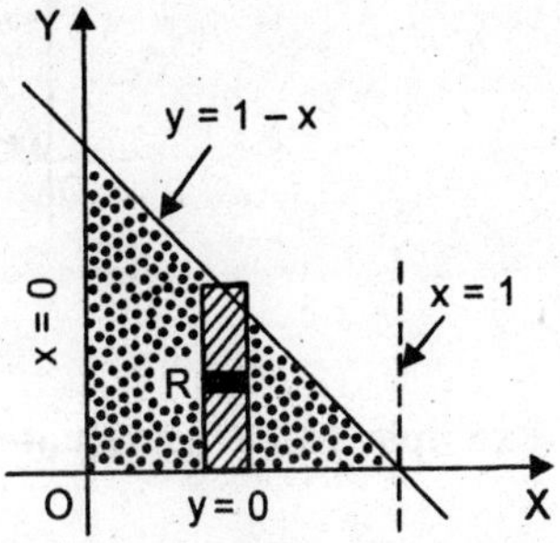

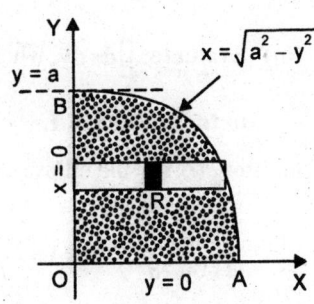

Fig. 13.5.

(ii) Refer Fig. 13.6 The region R is bounded by the curves $x = 0$, $x = \sqrt{a^2 - y^2}$, $y = 0$, and $y = a$

Hence $\iint\limits_{R} xy\, dx\, dy = \int\limits_{y=0}^{a} \int\limits_{x=0}^{\sqrt{(a^2-y^2)}} xy\, dx\, dy$

$= \int\limits_{0}^{a} y \left[\frac{x^2}{2} \right]_{0}^{\sqrt{(a^2-y^2)}} dy$

$= \int\limits_{0}^{a} y \cdot \frac{1}{2}(a^2 - y^2)\, dy = \frac{1}{2} \int\limits_{0}^{a} (a^2 y - y^3)\, dy$

Fig. 13.6.

$$= \frac{1}{2}\left[\frac{a^2 y^2}{2} - \frac{y^4}{4}\right]_0^a = \frac{1}{2}\left[\frac{a^4}{2} - \frac{a^4}{4}\right] = \frac{a^4}{8}.$$

(*iii*) The region S is bounded by the lines, $y = 0$, $y = 1$, $x = y$ and $x = 10y$.

$$\iint_S \sqrt{xy - y^2}\, dx\, dy = \int_{y=0}^{1} dy \int_{x=y}^{10y} \left(xy - y^2\right)^{1/2} dx = \int_0^1 \left[\frac{2}{3y}\left(xy - y^2\right)^{3/2}\right]_y^{10y} dy$$

$$= \int_0^1 \frac{2}{3y}\left[\left(10y^2 - y^2\right)^{3/2}\right] dy = 18\int_0^1 y^2 dy = 18\left[\frac{y^3}{3}\right]_0^1 = 6.$$

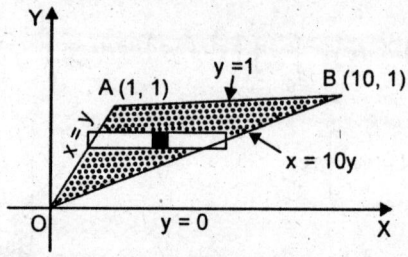

Fig. 13.7.

Example 13.7 (*i*) Evaluate $\iint(x+y)^2\, dy\, dx$ over the area bounded by the

ellipse $\dfrac{x^2}{a^2} + \dfrac{y^2}{b^2} = 1$. *(U.P.T.U., 2004S; Andhra, 1995)*

(*ii*) Evaluate $\iint_R x^2 dx\, dy$, where R is the region in the first quadrant bounded

by the hyperbola $xy = 16$ and the lines $y = x$, $y = 0$ and $x = 8$.
(UPTU-2002; AMIE., S-2001)

(*iii*) Evaluate $\iint_R dx\, dy$, where R is the region between $y = 2x$ and $y = x^2$ lying

to the left of $x = 1$.

Solution. (*i*) For the ellipse

$y/b = \pm\sqrt{\left(1 - \dfrac{x^2}{a^2}\right)}$, the region of integration can be considered as bounded by

the curves

$$y = -b\sqrt{\left[1-\left(x^2/a^2\right)\right]}, \ y = b\sqrt{\left[1-\left(x^2/a^2\right)\right]}, \ x = -a \text{ and } x = a.$$

Therefore, the given double integral

$$= \int_{-a}^{a} \int_{-b\sqrt{1-x^2/a^2}}^{b\sqrt{(1-x^2/a^2)}} \left(x^2 + 2xy + y^2\right) dy \, dx = 2 \int_{-a}^{a} \int_{0}^{b\sqrt{(1-x^2/a^2)}} \left(x^2 + y^2\right) dy \, dx$$

($\because xy$ is an odd function of y and $x^2 + y^2$ is an even function of y)

$$= 2 \int_{-a}^{a} \left[x^2 y + \left(y^3/3\right)\right]_{0}^{b\sqrt{(1-x^2/a^2)}} dx = 4 \int_{0}^{a} \left\{ x^2 b \sqrt{1 - \frac{x^2}{a^2}} + \frac{1}{3} b^3 \left(1 - \frac{x^2}{a^2}\right)^{3/2} \right\} dx$$

$$= 4b \int_{0}^{\pi/2} \left\{ a^2 \sin^2 \theta \cos\theta + \frac{1}{3} b^2 \cos^3 \theta \right\} a \cos\theta \, d\theta$$

[On putting $x = a \sin\theta$ and $dx = a \cos\theta \, d\theta$,]

$$= 4ab \int_{0}^{\pi/2} \left(a^2 \sin^2 \theta \cos^2 \theta + \frac{1}{3} b^2 \cos^4 \theta \right) d\theta$$

$$= 4ab \left[a^2 \cdot \frac{1}{4} \cdot \frac{1}{2} \cdot \frac{\pi}{2} + \frac{1}{3} b^2 \cdot \frac{3}{4} \cdot \frac{1}{2} \cdot \frac{\pi}{2} \right] = 4ab \left[\frac{\pi a^2}{16} + \frac{\pi b^2}{16} \right] = \frac{\pi ab}{4} \left(a^2 + b^2 \right).$$

(*ii*) Refer Fig. 13.8 Here the region R is separated into two regions. Let R_1 denote the part of R lying below the line $y = 2$ and R_2 the part above that line. Then

$$\iint_{R} x^2 \, dx \, dy = \iint_{R_1} x^2 \, dx \, dy + \iint_{R_2} x^2 \, dx \, dy = \int_{y=0}^{2} \int_{x=y}^{8} x^2 \, dx \, dy + \int_{2}^{4} \int_{x=y}^{16/y} x^2 \, dx \, dy$$

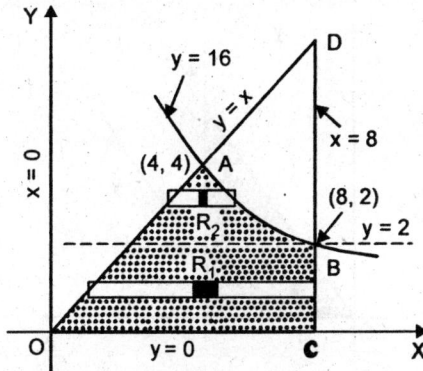

Fig. 13.8.

$$= \int_0^2 \left[\frac{x^3}{3}\right]_y^8 dy + \int_2^4 \left[\frac{x^3}{3}\right]_y^{16/y} dy = \frac{1}{3}\int_0^2 (8^3 - y^3)\,dy + \frac{1}{3}\int_2^4 \left(\frac{16^3}{y^3} - y^3\right) dy$$

$$= \frac{1}{3}\left[8^3 y - \frac{y^4}{4}\right]_0^2 + \frac{1}{3}\left[-\frac{(16)^3}{2y^2} - \frac{y^4}{4}\right]_2^4$$

$$= \frac{1}{3}\left(8^3 \cdot 2 - \frac{2^4}{4}\right) + \frac{1}{3}\left[-\frac{(16)^3}{2}\cdot\left(\frac{1}{16} - \frac{1}{4}\right) - \frac{1}{4}(4^4 - 2^4)\right]$$

$$= \frac{1020}{3} + \frac{1}{3}\left[\frac{3}{2}(16)^2 - 60\right] = 340 + 108 = 448.$$

[**Alternative.** Using the vertical strips, separate R with the line $x = 4$ and we

obtain $\iint\limits_R x^2\,dy\,dx = \int_0^4 \int_{y=0}^x x^2\,dy\,dx + \int_4^8 \int_{y=0}^{16/x} x^2\,dy\,dx.$]

(*iii*) Refer Fig. 13.9, it is obvious that the region R is separated into two regions and an iterated integral is evaluated for each. Let R_1 denote the part R lying below CD and R_2 the part above CD. Then

$$\iint\limits_R dx\,dy = \iint\limits_{R_1} dx\,dy + \iint\limits_{R_2} dx\,dy = \int_{y=0}^1 \int_{x=y/2}^{\sqrt{y}} dx\,dy + \int_{y=1}^2 \int_{x=y/2}^1 dx\,dy$$

$$= \int_0^1 \left(\sqrt{y} - \frac{y}{2}\right)dy + \int_1^2 \left(1 - \frac{y}{2}\right)dy = \left[\frac{2}{3}y^{3/2} - \frac{y^2}{4}\right]_0^1 + \left[y - \frac{y^2}{4}\right]_1^2$$

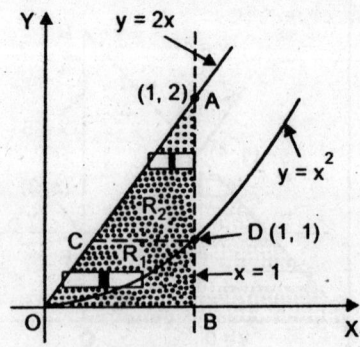

Fig. 13.9.

$$= \left(\frac{2}{3}-\frac{1}{4}\right)+\left\{(2-1)-\frac{1}{4}(4-1)\right\}=\frac{5}{12}+\frac{1}{4}=\frac{2}{3}.$$

13.3 Change of order of Integration

In Art 13.1 we have seen that a double integral may be evaluated with respect to y first and then with respect to x or with respect to x first and then with respect to y. Sometimes one order of evaluation may be more convenient as compared to the other. For this we may have to change the order of integration to evaluate a given integral.

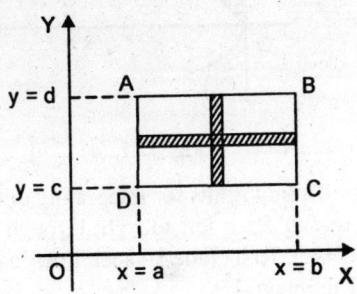

1. In case the region of integration is a rectangle with sides parallel to axes, then all the limits of integration are constants. In this case the change in the order of integration does not, ordinarily, require change in the limits of integration. From Fig. 13.10 it is clear that

Fig. 13.10.

$$\int\limits_{x=a}^{b}\int\limits_{y=c}^{d} f(x,y)\,dy\,dx = \int\limits_{y=c}^{d}\int\limits_{x=a}^{b} f(x,y)\,dx\,dy.$$

2. But if the limits of integration are variable, a change in the order of integration requires change in the limits of integration. In such cases a rough sketch of the region of integration is essential. This helps in fixing the new limits of integration. Sometimes in changing the order of integration we may have to split up the region of integration in two or more parts and the given integral will be expressed as the sum of two or more integrals. The following solved examples will make the ideas clear.

13.4 Illustrative Examples

Example 13.8 Change the order of integration and then evaluate it

$$\int\limits_{0}^{a}\int\limits_{x/a}^{\sqrt{x/a}} \left(x^2 + y^2\right)\,dy\,dx. \qquad\qquad (I.E.T.E.\ June\ 2002)$$

Solution. Refer Fig. 13.11(a) the region is bounded by the curves

$$y = \frac{x}{a} \text{ and } y = \sqrt{\frac{x}{a}} \text{ between } x = 0 \text{ and } x = a.$$

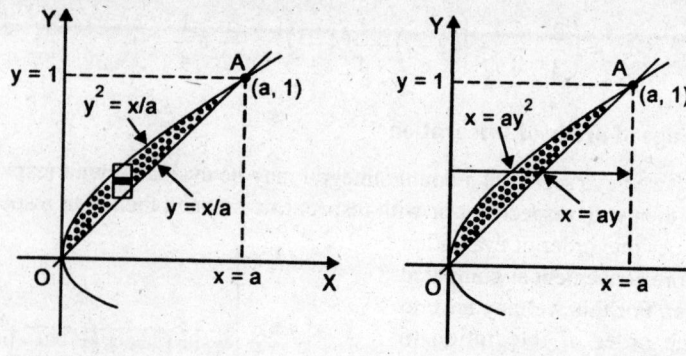

Fig. 13.11(a)   **Fig. 13.11(b)**

To find limits for integrating in the reverse order, we imagine a horizontal line passing from left to right through the region. It enters at $x = ay^2$ and leaves at $x = ay$. To include all such lines, we let y run from $y = 0$ to $y = 1$ [Fig. 13.11 (b)]. The integral is

$$\int_{y=0}^{1} \int_{x=ay^2}^{ay} (x^2 + y^2) \, dx \, dy = \int_0^1 \left[\frac{x^3}{3} + xy^2 \right]_{ay^2}^{ay} dy = \int_0^1 \left(\frac{a^3 y^3}{3} + ay^3 - \frac{a^3 y^6}{3} - ay^4 \right) dy$$

$$= \left[\frac{a^3 y^4}{12} + \frac{ay^4}{4} - \frac{a^3 y^7}{21} - \frac{ay^5}{5} \right]_0^1 = \frac{a^3}{12} + \frac{a}{4} - \frac{a^3}{21} - \frac{a}{5} = \frac{a}{4} \left(\frac{a^2}{7} + \frac{1}{5} \right).$$

Example 13.9 Change the order of integration in

$$\int_{y=0}^{a} \int_{x=y}^{a} \frac{x \, dx \, dy}{x^2 + y^2}, \text{ and hence evaluate the same.}$$

Solution. From the limits of integration, it is clear that the region of integration is bounded by $x = y$, $x = a$, $y = 0$ and $y = a$ that is, the triangle $0AB$.

The given order of integration indicates that the region is supposed to be divided into horizontal strips.

For changing the order of integration, the area is supposed to be divided into vertical strips which are bounded from below by $y = 0$ and from above $y = x$. Then the values of x at the extreme left and right are 0 and a respectively.

Hence on changing the order of integration the given integral becomes

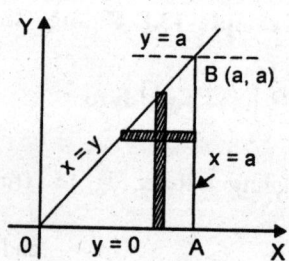

Fig. 13.12.

$$\int_{x=0}^{a} \int_{y=0}^{x} \frac{x\,dy\,dx}{x^2+y^2}. \quad \text{Therefore,} \quad = \int_{y=0}^{a} \int_{x=y}^{a} \frac{x}{x^2+y^2}\,dx\,dy = \int_{x=0}^{a} \int_{y=0}^{x} \frac{x}{x^2+y^2}\,dy\,dx$$

$$= \int_{0}^{a} x \cdot \left| \frac{1}{x} \tan^{-1} \frac{y}{x} \right|_{0}^{x} dx = \int_{0}^{a} \frac{\pi}{4} dx = \frac{\pi a}{4}.$$

Example 13.10 Evaluate the following integrals by changing the order of integration.

(i) $\displaystyle\int_{0}^{4a} \int_{x^2/4a}^{2\sqrt{(ax)}} dy\,dx$, (*PTU 2001; MDU-2001; Bangalore, 1998S; Mangalore, 1997*)

(ii) $\displaystyle\int_{0}^{a} \int_{x^2/a}^{2a-x} xy\,dy\,dx$, (*PTU 2006; DCE, 2005; IETE June 2005; MDU 2003*)

(iii) $\displaystyle\int_{0}^{1} \int_{x}^{\sqrt{(2-x^2)}} \frac{x\,dy\,dx}{\sqrt{(x^2+y^2)}}$. (*UPTU, 2006; JNTU, 2005; MDU 2004*)

Solution. (*i*) The given limits show that the region of integration is the area bounded by the curves $y = x^2/4a$, $y = 2\sqrt{(ax)}$, $x = 0$, and $x = 0$, and $x = 4a$. (See Fig. 13.13). The first two curves are parabolas $x^2 = 4ay$ and $y^2 = 4ax$, which intersect at $(0, 0)$ and $(4a, 4a)$. We can also consider the same area enclosed by $x = y^2/4a$, $x = 2\sqrt{(ay)}$, $y = 0$, and $y = 4a$.

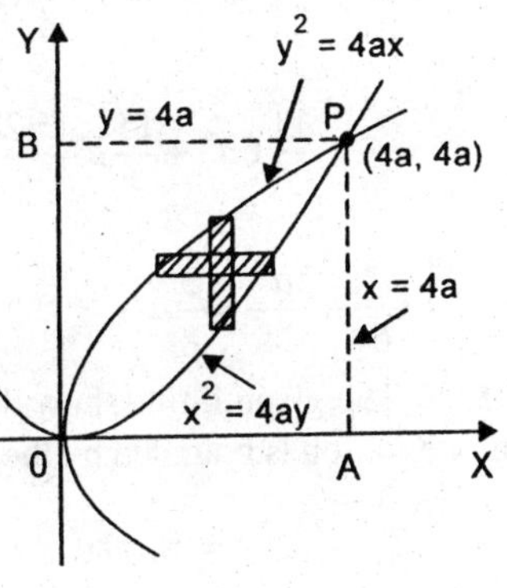

Hence, $\displaystyle\int_{0}^{4a} \int_{x^2/4a}^{2\sqrt{(ax)}} dy\,dx = \int_{0}^{4a} \int_{y^2/4a}^{2\sqrt{(ay)}} dx\,dy$

Fig. 13.13.

$$= \int_{0}^{4a} [x]_{y^2/4a}^{2\sqrt{(ay)}} dy = \int_{0}^{4a} \left(2\sqrt{(ay)} - \frac{y^2}{4a} \right) dy$$

$$= \left[2\sqrt{a}\, \frac{y^{3/2}}{3/2} - \frac{y^3}{3 \cdot 4a} \right]_{0}^{4a} = \frac{4}{3}\sqrt{a}\,(4a)^{3/2} - \frac{(4a)^3}{12a}$$

$$= \frac{32}{3}a^2 - \frac{16}{3}a^2 = \frac{16a^2}{3}.$$

(*ii*) From the limits of integration it is clear that the region of integration is bounded by the curves $x^2 = ay$, $x + y = 2a$, $x = 0$ and $x = a$ (shown shaded).

To cover this area by horizontal strips we have to split it into two parts, one lying below the line *BA* and the other above it.

(This is necessitated by the fact that the right extremities of strips below *BA* lie on the parabola $x^2 = ay$ while in the case of strips above *BA*, the right extremities lie on the line $x + y = 2a$)

Therefore,

$$\int_{x=0}^{a}\int_{y=x^2/a}^{2a-x} xy\,dy\,dx = \int_{y=0}^{a}\int_{x=0}^{\sqrt{ay}}(xy\,dx\,dy) + \int_{y=a}^{2a}\int_{x=0}^{2a-y} xy\,dx\,dy$$

$$= \int_{0}^{a} y\left[\frac{x^2}{2}\right]_{0}^{\sqrt{ay}} dy + \int_{a}^{2a} y\left[\frac{x^2}{2}\right]_{0}^{2a-y} dy$$

$$= \frac{a}{2}\int_{0}^{a} y^2\,dy + \frac{1}{2}\int_{a}^{2a} y(2a-y)^2\,dy$$

$$= \frac{a}{2}\left|\frac{y^3}{3}\right|_{0}^{a} + \frac{1}{2}\int_{a}^{2a}(4a^2y + y^3 - 4ay^2)\,dy$$

$$= \frac{a^4}{6} + \frac{1}{2}\left[2a^2y^2 + \frac{y^4}{4} - \frac{4ay^3}{3}\right]_{a}^{2a}$$

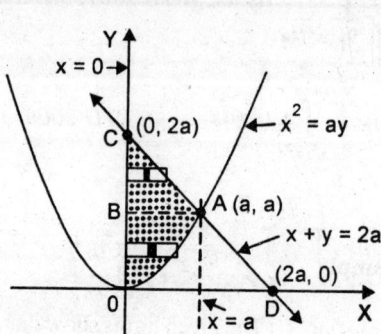

Fig. 13.14.

$$= \frac{a^4}{6} + \frac{1}{2}\left[2a^2(3a^2) + \frac{1}{4}(16a^4 - a^4) - \frac{4a}{3}(8a^3 - a^3)\right]$$

$$= \frac{a^4}{6} + \frac{1}{2}\left(6a^4 + \frac{15}{4}a^4 - \frac{28a^4}{3}\right)$$

$$= \frac{a^4}{6} + \frac{5a^4}{24} = \frac{3}{8}a^4.$$

(*iii*) The given limits show that the region of integration is bounded by the curves $y = x$, $y = \sqrt{(2-x^2)}$, $x = 0$, and $x = 1$ (See Fig. 13.15).

Thus the area *OABCO* lies between the lines $y = x$, $x = 0$, and the circle $x^2 + y^2 = 2$.

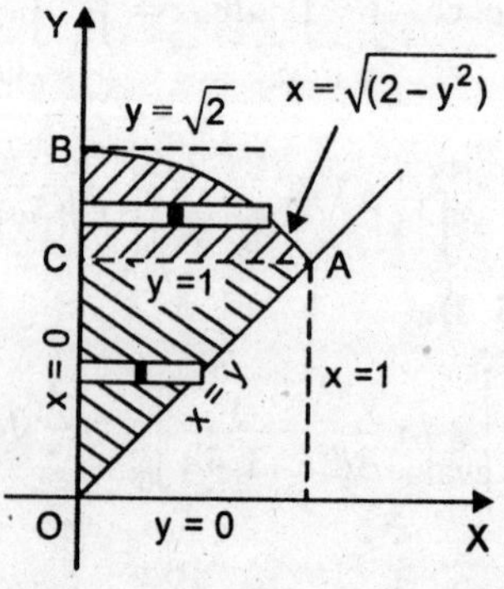

Fig. 13.15.

To cover this area by horizontal strips, we shall split it into two parts; one lying below the line *CA* and the other above *CA*; then

$$\int_{x=0}^{1}\int_{y=x}^{\sqrt{(2-x^2)}}\frac{x\,dy\,dx}{\sqrt{(x^2+y^2)}} = \int_{y=0}^{1}\int_{x=0}^{y}\frac{x\,dx\,dy}{\sqrt{(x^2+y^2)}} + \int_{y=1}^{\sqrt{2}}\int_{x=0}^{\sqrt{(2-y^2)}}\frac{x\,dx\,dy}{\sqrt{(x^2+y^2)}}$$

$$= \int_0^1 \left[\sqrt{(x^2+y^2)}\right]_0^y dy + \int_1^{\sqrt{2}}\left[\sqrt{(x^2+y^2)}\right]_0^{\sqrt{(2-y^2)}} dy$$

$$= \int_0^1\left[\sqrt{(y^2+y^2)}-\sqrt{(y^2)}\right]dy + \int_1^{\sqrt{2}}\left[\sqrt{2}-y\right]dy = (\sqrt{2}-1)\left[\frac{y^2}{2}\right]_0^1 + \left[\sqrt{2}y-\frac{y^2}{2}\right]_1^{\sqrt{2}}$$

$$= \frac{1}{2}(\sqrt{2}-1) + \left[\sqrt{2}(\sqrt{2}-1)-\frac{1}{2}(2-1)\right] = \left(1-\frac{1}{2}\sqrt{2}\right).$$

Example 13.11 Change the order of integration and evaluate $\displaystyle\int_0^{\infty}\int_x^{\infty}\frac{e^{-y}}{y}dy\,dx.$

(V.T.U., 2007; DCE, 2004; Madras 2003; Triputi, 1998S)

Solution. The first integration is by vertical strips extending from $y = x$ to $y = \infty$; the strips starting from $x = 0$ and going to $x = \infty$. So the region of integration is the upper half of the first quadrant. If the same region is covered by horizontal strips, their ends will be on $x = 0$ and $x = y$, while the strips will start from $y = 0$ and go to $y = \infty$. Hence

$$\int_0^{\infty}\int_x^{\infty}\left(e^{-y}/y\right)dy\,dx = \int_0^{\infty}\int_0^{y}\left(e^{-y}/y\right)dx\,dy = \int_0^{\infty}\left(e^{-y}/y\right)[x]_0^y\,dy$$

$$= \int_0^{\infty}e^{-y}dy = \left[-e^{-y}\right]_0^{\infty} = 1.$$

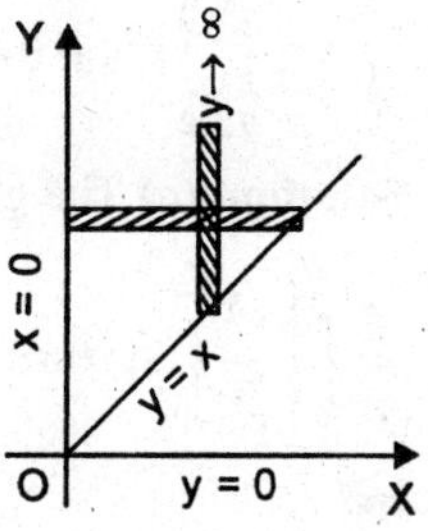

Fig. 13.16.

13.5 Double Integrals in Polar Form

The evaluation of the integral $\displaystyle\int_{\theta_1}^{\theta_2}\int_{r_1}^{r_2} f(r,\theta)\,dr\,d\theta$ is done exactly as in the case of double integrals in Cartesian co-ordinates i.e., we first integrate $f(r,\theta)$ w.r.t. r (treating θ as a constant), between the limits r_1 and r_2. The resulting expression is integrated w.r.t. θ between the limits θ_1 and θ_2. Here r_1 and r_2 may be constants or functions of θ.

Similarly to evaluate $\int\limits_{r_1}^{r_2}\int\limits_{\theta_1}^{\theta_2} f(r, \theta)\,d\theta\,dr$, we first integrate w.r.t. θ between the

limits θ_1 and θ_2. The resulting expression is integrated w.r.t. r between the limits r_1 and r_2. The limits θ_1, θ_2 may be constants or functions of r.

13.5.1 Geometrical Interpretation of $\int\limits_{\theta_1}^{\theta_2}\int\limits_{r_1}^{r_2} f(r, \theta)\,dr\,d\theta$.

Refer Fig. 13.17, the region *ACDB* is bounded by the curves $r = f_1(\theta)$ and $r = f_2(\theta)$ and the radii vectors $\theta = \theta_1$ and $\theta = \theta_2$.

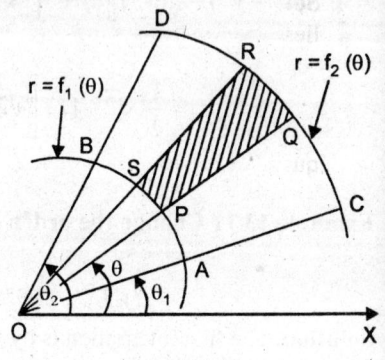

Now $\int\limits_{r_1}^{r_2} f(r, \theta)\,dr$ shows that the in-

tegration is along the shaded portion from *PS* to *QR* and then integration w.r.t. θ shows that the shaded portion slides from *AC* to *BD*.

Fig. 13.17.

Example 13.12 Evaluate

(a) $\int\limits_{0}^{\pi/2} \int\limits_{0}^{a} r^2 \sin\theta\,dr\,d\theta$, (b) $\int\limits_{0=\theta}^{\pi} \int\limits_{r=0}^{a\sin\theta} r\,dr\,d\theta$.

Solution. (a) The given integral

$$= \int\limits_{0}^{\pi/2} \left[\frac{r^3}{3}\right]_0^a \sin\theta\,d\theta = \frac{a^3}{3}\int\limits_{0}^{\pi/2}\sin\theta\,d\theta = \frac{a^3}{3}[-\cos\theta]_0^{\pi/2} = \frac{a^3}{3}.$$

(b) $\int\limits_{0}^{\pi} \int\limits_{0}^{a\sin\theta} r\,dr\,d\theta = \int\limits_{0}^{\pi}\left[\frac{r^2}{2}\right]_0^{a\sin\theta} d\theta = \frac{1}{2}\int\limits_{0}^{\pi} a^2\sin^2\theta\,d\theta = \frac{a^2}{4}\int\limits_{0}^{\pi}(1-\cos 2\theta)\,d\theta$

$$= \frac{a^2}{4}\left[\theta - \frac{\sin 2\theta}{2}\right]_0^{\pi} = \frac{a^2}{4}(\pi) = \frac{\pi a^2}{4}.$$

13.6 Changing Cartesian Integrals into Polar Integrals

From the limits of the given integral sketch the region of integration. The equations of the bounding curves can now be expressed in polar co-ordinates. Then change $dy\, dx$ into $r\, dr\, d\theta$ and put $x = r \cos \theta$ and $y = r \sin \theta$ in the integrand. The limits of integration can be determined from the equations of the bounding curves.

The following solved examples will clarify the procedure.

Example 13.13 Evaluate the following integrals by changing to polar co-ordinates

(a) $\displaystyle\int_{x=0}^{a} \int_{y=0}^{\sqrt{a^2-x^2}} y^2 \sqrt{x^2+y^2}\; dy\, dx,$ (A.M.I.E., S-2001) (b) $\displaystyle\int_{0}^{2} \int_{0}^{\sqrt{2x-x^2}} \frac{x\, dy\, dx}{\sqrt{x^2+y^2}}$

Solution. (a) The given limits of integration show that the region of integration lies between the curves

$$y = 0,\ y = \sqrt{(a^2 - x^2)},\ x = 0 \text{ and } x = a. \qquad \text{[See Fig. 13.18}(a)].$$

Thus the region of integration is the part of the circle $x^2 + y^2 = a^2$ in the first quadrant. In polar co-ordinates, the region of integration is bounded by the curves:

$$r = 0,\ r = a,\ \theta = 0,\ \theta = \frac{\pi}{2}. \qquad \text{[See Fig. 13.18}(b)]$$

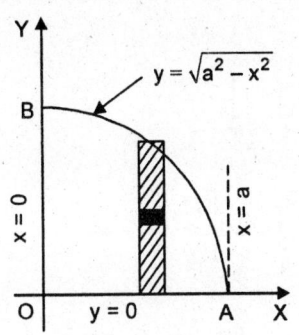

Fig. 13.18(a)

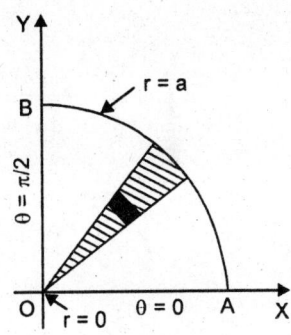

Fig. 13.18(b)

Therefore, $\displaystyle\int_{0}^{a} \int_{0}^{\sqrt{a^2-x^2}} y^2 \sqrt{(x^2+y^2)}\; dy\, dx = \int_{0}^{\pi/2} \int_{0}^{a} r^2 \sin^2 \theta \cdot r \cdot r\, dr\, d\theta$

$$= \int_{0}^{\pi/2} \sin^2 \theta \left[\frac{r^5}{5}\right]_{0}^{a} d\theta = \frac{1}{5} a^5 \int_{0}^{\pi/2} \sin^2 \theta\, d\theta = \frac{1}{5} a^5 \cdot \frac{1}{2} \cdot \frac{\pi}{2} = \frac{\pi a^5}{20}.$$

(b) The limits of integration show that the region of integration lies between the curves $y = 0,\ y = \sqrt{(2x - x^2)}, x = 0$ and $x = 2$.

Thus the region of integration is the semi-circular area shown in the Fig. 13.19(a), The circle $y = \sqrt{2x - x^2}$, that is, $x^2 + y^2 = 2x$, transforms into $r^2 = 2r \cos \theta$, or $r = 2 \cos\theta$.

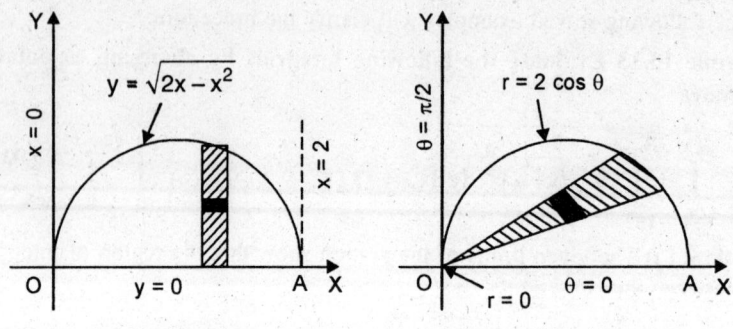

Fig. 13.19(a)   **Fig. 13.19(b)**

Hence, in polar co-ordinates, the region of integration is bounded by the curves

$$r = 0, \ r = 2\cos\theta, \ \theta = 0, \ \theta = \frac{\pi}{2}.$$

Therefore,

$$\int\limits_{0}^{2} \int\limits_{0}^{\sqrt{(2x-x^2)}} \frac{x \, dy \, dx}{\sqrt{(x^2 + y^2)}} = \int\limits_{0}^{\pi/2} \int\limits_{0}^{2\cos\theta} \frac{r\cos\theta}{r} \cdot r \, dr \, d\theta$$

$$= \int\limits_{0}^{\pi/2} \int\limits_{0}^{2\cos\theta} r\cos\theta \, dr \, d\theta = \int\limits_{0}^{\pi/2} \cos\theta \left[\frac{r^2}{2}\right]_{0}^{2\cos\theta} d\theta$$

$$= \frac{1}{2} \int\limits_{0}^{\pi/2} \left(4\cos^2\theta\right)\cos\theta \, d\theta = 2 \int\limits_{0}^{\pi/2} \cos^3\theta \, d\theta = 2 \cdot \frac{2}{3} = \frac{4}{3}.$$

Example 13.14 Change into polar co-ordinates and evaluate.

(a) $\displaystyle\int\limits_{0}^{a}\int\limits_{y}^{a}\frac{x \, dx \, dy}{x^2 + y^2}$, (b) $\displaystyle\int\limits_{0}^{\infty}\int\limits_{0}^{\infty}e^{-(x^2+y^2)} \, dy \, dx$. **Hence show that** $\displaystyle\int\limits_{0}^{\infty}e^{-x^2} \, dx = \frac{\sqrt{\pi}}{2}$.

(MDU 2005; UPTU 2003; Madras 2003; JNTU 2000)

Solution. (a) The given limits of integration show that the region of integration lies between the curves $x = y$, $x = a$, $y = 0$, and $y = a$.

Thus the region of integration is the triangular area shown in the diagram 13.20(a).

The given area can be covered by radial strips as in the Fig. 13.20(*b*). One end of these strips is at $r = 0$ and the other end at $x = a$ i.e., $r\cos\theta = a$ or $r = a\sec\theta$.

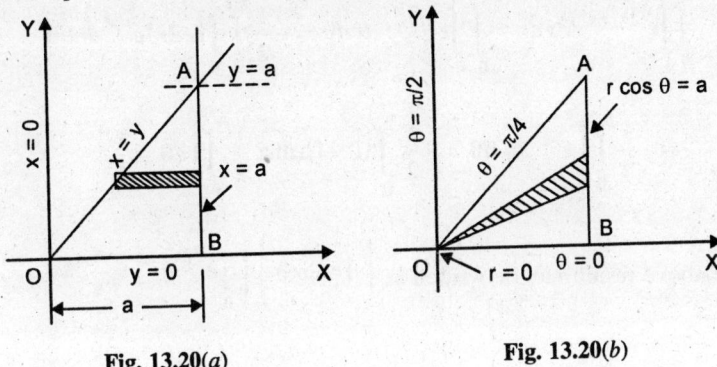

Fig. 13.20(*a*)   Fig. 13.20(*b*)

The strips start from $\theta = 0$ and end on $\theta = \dfrac{\pi}{4}$. Hence

$$\int_0^a \int_y^a \frac{x\,dx\,dy}{x^2+y^2} = \int_0^{\pi/4} \int_0^{a/\cos\theta} \frac{r\cos\theta}{r^2}\, r\,dr\,d\theta = \int_0^{\pi/4} \int_0^{a\sec\theta} \cos\theta\, dr\,d\theta$$

$$= \int_0^{\pi/4} \cos\theta\,[r]_0^{a\sec\theta}\,d\theta = \int_0^{\pi/4} \cos\theta\, a\sec\theta\,d\theta = \int_0^{\pi/4} a\,d\theta = a[\theta]_0^{\pi/4} = \frac{\pi a}{4}.$$

(*b*) From the limits of integration we find that the first integration is along a vertical strip extending from $y = 0$ to $y = \infty$. The strip slides from $x = 0$ and goes to ∞. Thus the region of integration is the whole of first quadrant.

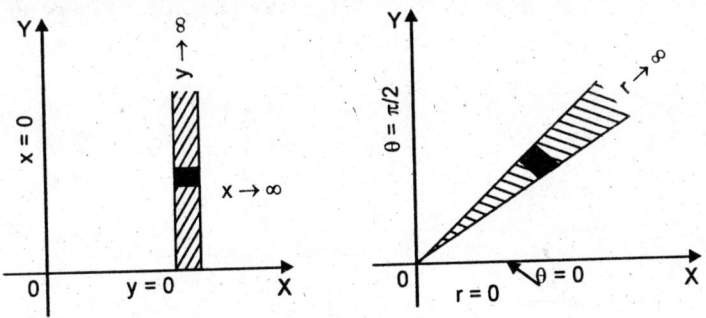

Fig. 13.21(*a*)   Fig. 13.21(*b*)

This region can be covered by radial strips extending from $r = 0$ to $r = \infty$. The strips start from $\theta = 0$ and go upto $\theta = \pi/2$.

Hence $\displaystyle\int_0^\infty\int_0^\infty e^{-(x^2+y^2)}dy\,dx = \int_0^{\pi/2}\int_0^\infty e^{-r^2}\cdot r\,dr\,d\theta = -\frac{1}{2}\int_0^{\pi/2}\int_0^\infty(-2r)e^{-r^2}dr\,d\theta$

$$= -\frac{1}{2}\int_0^{\pi/2}\left[e^{-r^2}\right]_0^\infty d\theta = -\frac{1}{2}\int_0^{\pi/2}(0-1)d\theta = \frac{1}{2}\int_0^{\pi/2}1\,d\theta = \frac{\pi}{4}.$$

The above result may be written as $\left(\displaystyle\int_0^\infty e^{-x^2}dx\right)\cdot\left(\int_0^\infty e^{-y^2}dy\right) = \frac{\pi}{4}$

or $\left(\displaystyle\int_0^\infty e^{-x^2}dx\right)\left(\int_0^\infty e^{-x^2}dx\right) = \frac{\pi}{4}$ or $\left(\displaystyle\int_0^\infty e^{-x^2}dx\right)^2 = \frac{\pi}{4}$ $\Rightarrow \displaystyle\int_0^\infty e^{-x^2}dx = \frac{\sqrt\pi}{2}.$

Exercise 13.1

1. Evaluate the following double integrals.

(i) $\displaystyle\int_0^2\int_1^2(x^2+y^2)dx\,dy$, (ii) $\displaystyle\int_{-1}^2\int_{-3}^3(y^2-3xy)dx\,dy$, (iii) $\displaystyle\int_1^2\int_0^2 e^{x+y}dx\,dy$

(iv) $\displaystyle\iint_R f(x,y)dx\,dy$ for $f(x,y)=1-6x^2y$ and $R:0\le x\le 2,\,-1\le y\le 1.$

(v) $\displaystyle\int_0^1\int_x^{2x}(x^2+y^2)dy\,dx$, (vi) $\displaystyle\int_0^{\pi/2}\int_0^\pi \cos(x+y)dx\,dy$, *(AMIE, W-2009)*

(vii) $\displaystyle\int_0^1\int_x^{\sqrt x}(x^2+y^2)dy\,dx$ *(V.T.U., 2000)*, (viii) $\displaystyle\int_0^1\int_{x^2}^{2-x}xy\,dx\,dy$, *(IETE, W-2009)*,

(ix) $\displaystyle\int_0^1\int_0^1\frac{dx\,dy}{\sqrt{(1-x^2)(1-y^2)}}$, (x) $\displaystyle\int_0^a\int_0^{\sqrt{a^2-x^2}}\sqrt{a^2-x^2-y^2}\,dy\,dx$

[**Hint.** (x) The given integral may be re-written as $\displaystyle\int_0^a dx\int_0^{\sqrt{a^2-x^2}}\sqrt{(a^2-x^2)-y^2}\,dy\,$]

(xi) $\displaystyle\int_0^2 dx \int_{\sqrt{2x}}^2 \frac{y}{\sqrt{x^2+y^2+1}}\,dy$, (xii) $\displaystyle\int_0^{2a}\int_0^{\sqrt{2ax-x^2}} x^2\,dy\,dx$.

Ans. (i) 22/3, (ii) 18, (iii) $e(e+1)(e-1)^2$, (iv) 4, (v) 5/6, (vi) −2, (vii) 3/35,

$(viii)$ 3/8, (ix) $\pi^2/4$, (x) $\pi a^3/6$, (xi) $(5/4)\log 5 - 1$, (xii) $5\pi a^4/8$.

2. Show that $\displaystyle\int_0^1 dx \int_0^1 \frac{x-y}{(x+y)^3}\,dy \ne \int_0^1 dy \int_0^1 \frac{x-y}{(x+y)^3}\,dx.$ (A.M.I.E., W-2009)

3. Sketch the region R in the xy plane bounded by $y = x^2$, $x = 2$, $y = 1$, and

evaluate the double integral $\displaystyle\iint_R (x^2 + y^2)\,dx\,dy.$ **Ans.** 1006/105.

[**Hint.** The given integral $= \displaystyle\int_{y=1}^4 \int_{x=\sqrt{y}}^2 (x^2+y^2)\,dx\,dy.$]

4. Evaluate $\displaystyle\iint_R xy\,dy\,dx$ over the region R enclosed between $y = x/2$,

$y = \sqrt{x}$, $x = 2$ and $x = 4$. **Ans.** 11/6.

5. Evaluate the double integral $\displaystyle\iint_D (4 - x^2 - y^2)\,dx\,dy$, if the domain D is

bounded by the straight lines $x = 0$, $x = 1$, $y = 0$ and $y = 3/2$. **Ans.** 35/8.

6. Calculate $\displaystyle\iint_R \frac{\sin x}{x}\,dy\,dx$, where R is the triangle in the xy-plane bounded by

the x-axis, the line $y = x$ and the line $x = 1$. **Ans.** 0.46.

7. Evaluate the double integral of the function $f(x, y) = 1 + x + y$ over the
region bounded by the lines $y = -x$, $x = \sqrt{y}$, $y = 2$ and $y = 0$.

Ans. $(44\sqrt{2} + 65)/15.$

8. Evaluate $\displaystyle\iint e^{2x+3y}\,dx\,dy$ over the triangle bounded by $x = 0$, $y = 0$ and

$x + y = 1$. **Ans.** $(e-1)^2(2e+1)/6.$

9. Find the value of $\displaystyle\iint_R dy\,dx$ and $\displaystyle\iint_R dx\,dy$ where R is the region in the first

quadrant bounded by the semi-cubical parabola $y^2 = x^3$ and the line $y = x$.

Ans. 1/10, 1/10.

10. Evaluate $\iint_A xy\,dx\,dy$, where A is the region bounded by x-axis, ordinate

$x = 2a$ and the curve $x^2 = 4\,ay$. (*A.M.I.E., S-2006, MDU 2001*)**Ans.** $a^4/3$.

11. Evaluate $\iint xy\,dy\,dx$, over the region in the positive quadrant for which

$x + y \le 1$.   (*A.M.I.E., W-1999*) **Ans.** 1/24

12. Find the value of $\iint (x^2 + y^2)\,dx\,dy$ over the region bounded by the curves

$y = 4x, x + y = 3, y = 0$ and $y = 2$.   **Ans.** 9.646.

13. (*a*) Evaluate $\iint \dfrac{xy}{\sqrt{1-y^2}}\,dy\,dx$, over the positive quadrant of the circle

$x^2 + y^2 = 1$.

(*b*) Evaluate $\iint xy(x+y)\,dx\,dy$ over the area between $y = x^2$ and $y = x$.

(*I.E.T.E., June 2003*) **Ans.** (*a*) 1/6, (*b*) 3/56.

14. Evaluate $\iint_R (2x - y^2)\,dx\,dy$ over the triangular region R enclosed between

the lines $y = -x + 1, y = x + 1$ and $y = 3$.   **Ans.** 68/3.

15. Evaluate the double integral $\iint_R xy\,dy\,dx$ where R is the region bounded by

the x-axis, the line $y = 2x$ and the parabola $y = x^2/4a$.

(*IETE, S-2009, W-2006*).

[**Hint.** $\iint_R xy\,dy\,dx = \int_{x=0}^{8a} \int_{y=x^2/4a}^{2x} xy\,dy\,dx = \dfrac{2048\,a^4}{3}$].

Alternative. $\iint_R xy\,dx\,dy = \int_0^{16a} \int_{x=y/2}^{\sqrt{4ay}} xy\,dx\,dy = \dfrac{2048}{3}a^4$].

16. Evaluate $\int\int_R (x^2 + y^2)\,dy\,dx$, where R is the region bounded by the x-axis,

the line $y = 2x$ and the parabola $x^2 = 4y$. (*A.M.I.E.,S-2004*)

Ans. 165888/105.

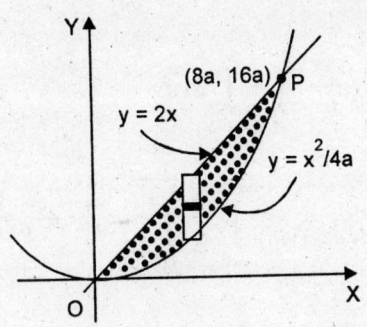

Fig. 13.22.

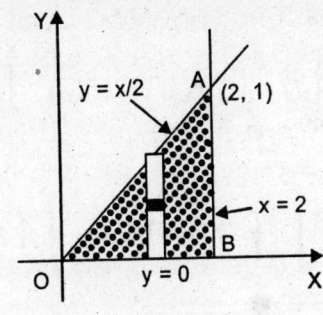

Fig. 13.23

17. Evaluate the double integral $\iint\limits_{R} e^{x^2}\, dy\, dx$, where the region R is given by

$R: 2y \le x \le 2$ and $0 \le y \le 1$.   *(A.M.I.E., Dec., 2010; IETE, Dec., 2009)*

[Hint. $\iint\limits_{R} e^{x^2}\, dy\, dx = \int\limits_{x=0}^{2} \int\limits_{y=0}^{x/2} e^{x^2}\, dy\, dx = \frac{1}{4}(e^4 - 1).$]

18. Sketch the region of integration for the following integrals and write an equivalent double integral with the order of integration reversed.

(i) $\displaystyle\int\limits_{0}^{4} \int\limits_{x}^{2\sqrt{x}} f(x, y)\, dy\, dx$, (ii) $\displaystyle\int\limits_{0}^{2a} \int\limits_{y^2/4a}^{3a-y} f(x, y)\, dx\, dy$, (iii) $\displaystyle\int\limits_{0}^{2} \int\limits_{x^2/4}^{3-x} \phi(x, y)\, dy\, dx$,

(iv) $\displaystyle\int\limits_{0}^{2} \int\limits_{x^2}^{2x} (4x+2)\, dy\, dx$, (v) $\displaystyle\int\limits_{-1}^{1} dx \int\limits_{0}^{\sqrt{1-x^2}} f(x, y)\, dy$, (vi) $\displaystyle\int\limits_{0}^{5} \int\limits_{2-x}^{2+x} f(x, y)\, dy\, dx$

(vii) $\displaystyle\int\limits_{0}^{a} \int\limits_{\sqrt{a^2-y^2}}^{y+a} f(x, y)\, dx\, dy$, *(IETE, June 1999)* (viii) $\displaystyle\int\limits_{x=-a}^{a} \int\limits_{y=0}^{\sqrt{a^2-x^2}} (a-x)^2\, dy\, dx$.

Ans. (i) $\displaystyle\int\limits_{0}^{4} \int\limits_{y^2/4}^{y} f(x, y)\, dx\, dy$, (ii) $\displaystyle\int\limits_{0}^{a} \int\limits_{0}^{2\sqrt{ax}} f(x, y)\, dy\, dx + \int\limits_{0}^{3a} \int\limits_{0}^{3a-x} f(x, y)\, dy\, dx$,

(iii) $\displaystyle\int\limits_{0}^{1} \int\limits_{0}^{2\sqrt{y}} \phi(x, y)\, dx\, dy + \int\limits_{1}^{3} \int\limits_{0}^{3-y} \phi(x, y)\, dx\, dy$, (iv) $\displaystyle\int\limits_{0}^{4} \int\limits_{y/2}^{\sqrt{y}} (4x+2)\, dx\, dy$,

(v) $\displaystyle\int_0^1 dy \int_{-\sqrt{1-y^2}}^{\sqrt{1-y^2}} f(x, y)\,dx$, (vi) $\displaystyle\int_{y=-3}^{2}\int_{x=2-y}^{5} dx\,dy + \int_{y=2}^{7}\int_{x=y-2}^{5} dx\,dy$

(vii) $\displaystyle\int_0^a \int_{\sqrt{a^2-x^2}}^{a} f(x, y)\,dy\,dx + \int_a^{2a}\int_{x-a}^{a} f(x, y)\,dy\,dx$ (viii) $\displaystyle\int_{y=0}^{a}\int_{x=-\sqrt{a^2-y^2}}^{\sqrt{a^2-y^2}} (a-x)^2\,dx\,dy$

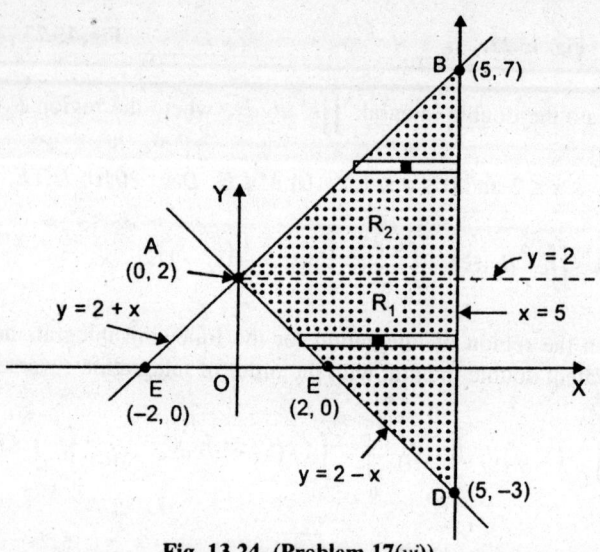

Fig. 13.24. (Problem 17(vi))

19. Evaluate the following integrals by changing the order of integration.

(i) $\displaystyle\int_0^1 \int_{x^2}^{2-x} xy\,dy\,dx$, (S.V.T.U., 2007; UPTU, 2005), (ii) $\displaystyle\int_0^a \int_{\sqrt{ax}}^{a} \frac{y^2\,dy\,dx}{\sqrt{y^4 - a^2 x^2}}$,

(S.V.T.U., 2006; Delhi, 1997)

(iii) $\displaystyle\int_0^1 \int_{3y}^{3} e^{x^2}\,dx\,dy$, (iv) $\displaystyle\int_0^1 \int_{e^x}^{e} \frac{dy\,dx}{\log y}$ (v) $\displaystyle\int_0^1 \int_{x}^{2-x} \frac{x}{y}\,dy\,dx$, (I.E.T.E., June 2004)

[**Hint.** (v) The given integral $= \displaystyle\int_0^1 \int_0^y \frac{x}{y}\,dx\,dy + \int_1^2 \int_0^{2-y} \frac{x}{y}\,dx\,dy$

$= \dfrac{1}{4} + 2\log_e 2 - \dfrac{5}{4} = 2\log_e 2 - 1$]

(vi) $\displaystyle\int_0^\infty\int_0^x xe^{-x^2/y}\,dy\,dx$ (MDU 2005; VTU-2004; Delhi 2002; A.M.I.E., W-2000)

(vii) $\displaystyle\int_0^1\int_y^1 x^2e^{xy}\,dx\,dy$, (I.E.T.E., June 2006) (viii) $\displaystyle\int_0^1\int_{y^2}^y \frac{y\,dx\,dy}{(1-x)(x-y^2)^{1/2}}$

(UPTU 2001)

[**Hint.** (viii) The given integral $= \displaystyle\int_0^1\int_x^{\sqrt{x}} \frac{y\,dy\,dx}{(1-x)(x-y^2)^{1/2}}$

$= \displaystyle\int_0^1\left[\frac{-(x-y^2)^{1/2}}{1-x}\right]_x^{\sqrt{x}}dx = \int_0^1\frac{(x-x^2)^{1/2}}{1-x}dx = \int_0^1\left(\frac{x}{1-x}\right)^{1/2}dx$

$= \displaystyle\int_0^{\pi/2}\left(\frac{\sin^2\theta}{1-\sin^2\theta}\right)^{1/2}\cdot 2\sin\theta\cos\theta\,d\theta \text{ (on putting } x=\sin^2\theta) = \int_0^{\pi/2} 2\sin^2\theta\,d\theta = \frac{\pi}{2}].$

(ix) $\displaystyle\int_0^{2a}\int_{x^2/4a}^{3a-x}(x^2+y^2)\,dy\,dx$, (x) $\displaystyle\int_0^a\int_0^{a-\sqrt{a^2-y^2}}\frac{xy\log(x+a)}{(x-a)^2}dx\,dy$

(DCE-2004)   (DCE-2004)

(xi) $\displaystyle\int_0^1\int_0^{\sqrt{1-x^2}} y^2\,dy\,dx$ (PTU 2004)

Ans. (i) 3/8, (ii) $\pi a^2/6$, (iii) $\dfrac{1}{6}(e^9-1)$, (iv) $(e-1)$, (v) $2\log_e 2-1$, (vi) 1/2,

(vii) $\displaystyle\int_{x=0}^1\int_{y=0}^x x^2e^{xy}\,dy\,dx; (e-2)/2$, (viii) $\dfrac{\pi}{2}$, (ix) $\dfrac{314a^4}{35}$,

(x) $\dfrac{a^2}{8}(1+2\log a), (xi)\dfrac{\pi}{16}.$

20. Change the order of integration for the integral $\displaystyle\int_0^{2a}\int_{\sqrt{2ax-x^2}}^{\sqrt{2ax}} dy\,dx$, and hence

evaluate it.

(I.E.T.E., Dec. 2003) **Ans.** $(16-3\pi)a^2/6$.

[**Hint.** The region of integration is bounded by the circle $x^2 + y^2 = 2ax$, the parabola $y^2 = 2ax$, $x = 0$ and $x = 2a$. The region of integration is divided into three parts OAB, BCD and AEC. The equation of the circle is $x^2 - 2ax + y^2 = 0$. Therefore $x = a \pm \sqrt{(a^2 - y^2)}$. Out of the two values of x the lessor one is $a - \sqrt{a^2 - y^2}$ and the greater one is $a + \sqrt{a^2 - y^2}$.

Hence, the given integral $= \iint\limits_{R_1} dx\, dy + \iint\limits_{R_2} dx\, dy + \iint\limits_{R_3} dx\, dy$

$$= \int\limits_{y=0}^{a} \int\limits_{x=y^2/2a}^{a-\sqrt{a^2-y^2}} dx\, dy + \int\limits_{y=0}^{a} \int\limits_{x=a+\sqrt{a^2-y^2}}^{2a} dx\, dy + \int\limits_{y=a}^{2a} \int\limits_{x=y^2/2a}^{2a} dx\, dy \,].$$

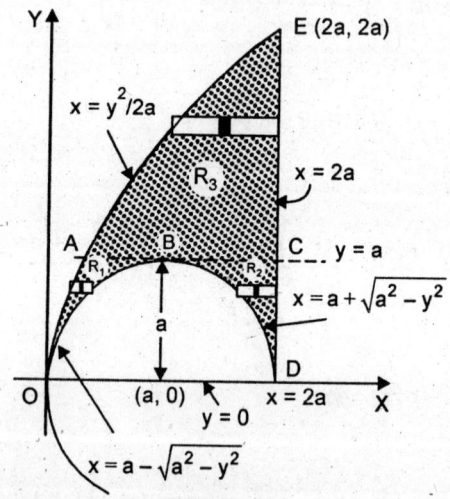

Fig. 13.25.

21. Change the order of integration in the double integral

$$\int\limits_0^a \int\limits_0^x \frac{\phi'(y)\, dy\, dx}{\sqrt{(a-x)(x-y)}} \quad \text{and hence find its value.} \qquad \text{(A.M.I.E., S-2007)}$$

[**Hint.** The region of integration is bounded by $y = 0$, $y = x$ and $x = a$. Thus OAB is the region of integration. Consider an elementary strip parallel to x-axis. The double integral after changing the order of integration is

$$J = \int\limits_{y=0}^{a} \int\limits_{x=y}^{a} \frac{\phi'(y)\, dx\, dy}{\sqrt{(a-x)(x-y)}} . \quad \text{To find its value, proceed as under:}$$

Put $x = a\sin^2\theta + y\cos^2\theta$,

$dx = 2(a-y)\sin\theta\cos\theta\,d\theta$.

$a - x = (a-y)\cos^2\theta$,

$x - y = (a-y)\sin^2\theta$.

When $x = y, \sin^2\theta = 0$, that is $\theta = 0$

and when $x = a$, $\theta = \pi/2$. Thus the

limits of x are from 0 to $\pi/2$.

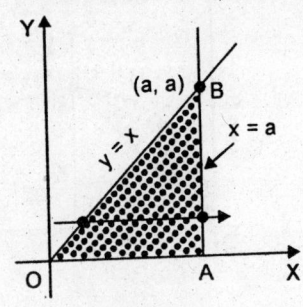

Fig. 13.26.

$$\therefore \int_y^a \frac{dx}{\sqrt{(a-x)(x-y)}} = \int_0^{\pi/2} \frac{2(a-y)\sin\theta\cos\theta\,d\theta}{(a-y)\sin\theta\cos\theta} = 2\int_0^{\pi/2} d\theta = \pi.$$

Hence $I = \int_0^a \pi\phi'(y)\,dy = \pi\big[\phi(y)\big]_0^a = \pi\big[\phi(a) - \phi(0)\big].$

22. Evaluate the following integrals

(i) $\displaystyle\int_0^{2\pi}\int_0^{1-\cos\theta} r^3\cos^2\theta\,dr\,d\theta,$ (ii) $\displaystyle\int_0^{\pi/4}\int_0^{\tan\theta\sec\theta} r^3\cos^2\theta\,dr\,d\theta,$

(iii) $\displaystyle\int_0^{\pi}\int_0^{a(1+\cos\theta)} r^2\sin\theta\,dr\,d\theta,$ (iv) $\displaystyle\int_0^{\pi/2}\int_0^{2a\cos\theta} \frac{r^3}{a}\,dr\,d\theta,$ (v) $\displaystyle\int_0^{\pi/2}\int_{a\cos\theta}^{a} r^4\,dr\,d\theta.$

Ans. (i) $49\pi/32$, (ii) $1/20$, (iii) $4a^3/3$, (iv) $3\pi a^3/4$, (v) $\left(\pi - \dfrac{16}{15}\right)\dfrac{a^5}{10}.$

23. Compute the following integrals by changing to polar co-ordinates:

(i) $\displaystyle\int_0^1\int_0^x \sqrt{x^2+y^2}\,dy\,dx,$ (ii) $\displaystyle\int_0^a\int_y^a \frac{x^2\,dx\,dy}{\sqrt{x^2+y^2}},$ (iii) $\displaystyle\int_0^a\int_0^{\sqrt{a^2-x^2}} (x^2+y^2)\,dx\,dy,$

(*Andhra, 1998*)

(iv) $\displaystyle\int_0^{2a}\int_0^{\sqrt{2ax-x^2}} dy\,dx,$ (v) $\displaystyle\int_0^1\int_x^{\sqrt{2x-x^2}} (x^2+y^2)\,dy\,dx,$

[**Hint.** (v) In polar co-ordinates, the region of integration is bounded by $r = 0$, $r = 2\cos\theta$, $\theta = \pi/4$, $\theta = \pi/2$. Refer Figure 13.27 (a) and (b).]

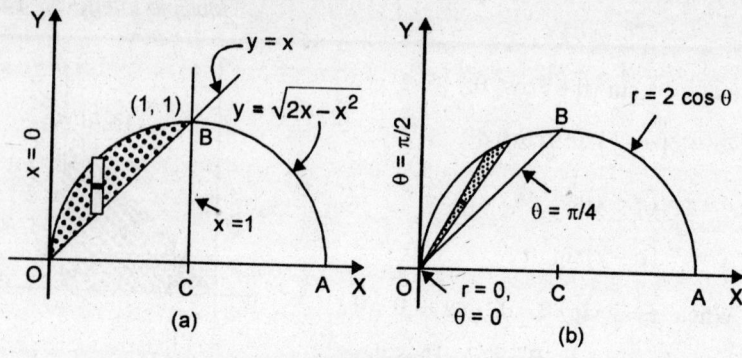

Fig. 13.27.

(*vi*) $\displaystyle\int_{0}^{2a}\int_{0}^{\sqrt{2ax-x^2}}\left(x^2+y^2\right)dy\,dx$, [**Hint.** the given integral $=\displaystyle\int_{0}^{\pi/2}\int_{0}^{2a\cos\theta}r^2\cdot r\,dr\,d\theta$]

(*A.M.I.E., S-2003*)

Ans. (*i*) $\dfrac{1}{6}\left[\sqrt{2}+\log\left(1+\sqrt{2}\right)\right]$, (*ii*) $\dfrac{1}{3}a^3\log\left(\sqrt{2}+1\right)$, (*iii*) $\pi a^4/8$, (*iv*) $\pi a^2/2$,

(*v*) $\left(\dfrac{3}{8}\pi-1\right)$, (*vi*) $3\pi a^4/4$.

24. Let $\displaystyle\int_{0}^{1}\int_{1}^{2}\dfrac{1}{x^2+y^2}\,dx\,dy+\int_{1}^{2}\int_{y}^{2}\dfrac{1}{x^2+y^2}\,dx\,dy=\iint_{R}\dfrac{1}{x^2+y^2}\,dy\,dx$. Recognise the

region R of integration on the r.h.s. and then evaluate the integral on the right in the order indicated. (*I.E.T.E., Dec. 2004*)

Ans. $\displaystyle\iint_{R}\dfrac{1}{x^2+y^2}\,dy\,dx=\int_{x=1}^{2}\int_{y=0}^{x}\dfrac{1}{x^2+y^2}\,dy\,dx=\dfrac{\pi}{4}\log_{e}2$.

25. Express $\displaystyle\int_{0}^{a/\sqrt{2}}\int_{0}^{x}x\,dy\,dx+\int_{a/\sqrt{2}}^{a}\int_{0}^{\sqrt{a^2-x^2}}x\,dy\,dx$, as a single integral and then

evaluate it. (*I.E.T.E., Dec. 2005*) **Ans.** $\displaystyle\int_{0}^{\pi/4}\int_{0}^{a}r^2\cos\theta\,dr\,d\theta$.

26. Express the sum of the following integrals as a single integral and hence evaluate by changing the order of integration:

$$\int\limits_{0}^{a/\sqrt{2}}\int\limits_{0}^{x}\cos\left[k\left(x^2+y^2\right)\right]dy\,dx + \int\limits_{a/\sqrt{2}}^{a}\int\limits_{0}^{\sqrt{a^2-x^2}}\cos\left[k\left(x^2+y^2\right)\right]dy\,dx .$$

[**Hint.** The given integral $= \int\limits_{\theta=0}^{\pi/4}d\theta\int\limits_{r=0}^{a}r\cos kr^2\,dr = \dfrac{\pi}{8k}\sin\left(ka^2\right)$].

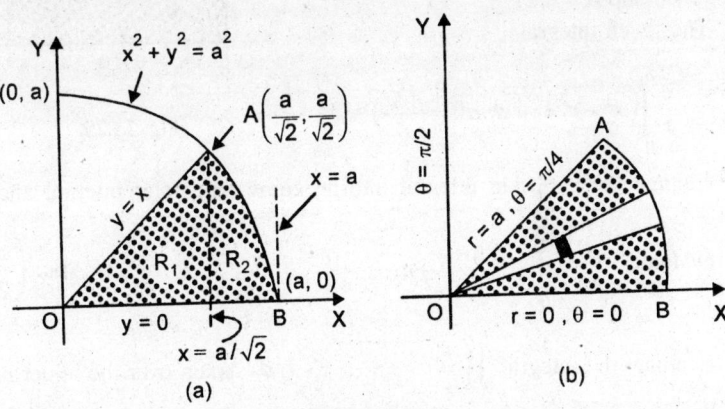

Fig. 13.28.

27. Transform each of the given integrals to one or more iterated integrals in polar coordinates

(a) $\displaystyle\int\limits_{0}^{1}\int\limits_{0}^{x}f(x,y)\,dy\,dx$, (b) $\displaystyle\int\limits_{0}^{1}dx\int\limits_{0}^{1}f(x,y)\,dy$, (c) $\displaystyle\int\limits_{0}^{1}\left[\int\limits_{0}^{x^2}f(x,y)\,dy\right]dx$.

Ans. (a) $\displaystyle\int\limits_{0}^{\pi/4}\left[\int\limits_{0}^{\sec\theta}f(r\cos\theta,r\sin\theta)r\,dr\right]d\theta$, (b) $\displaystyle\int\limits_{0}^{\pi/4}\left[\int\limits_{0}^{\sec\theta}f(r\cos\theta,r\sin\theta)r\,dr\right]d\theta$

$+\displaystyle\int\limits_{\pi/4}^{\pi/2}\left[\int\limits_{0}^{\csc\theta}f(r\cos\theta,r\sin\theta)r\,dr\right]d\theta$, (c) $\displaystyle\int\limits_{0}^{\pi/4}\left[\int\limits_{\tan\theta\sec\theta}^{\sec\theta}f(r\cos\theta,r\sin\theta)r\,dr\right]d\theta$

28. Compute the integral $\displaystyle\int\limits_{0}^{a}\int\limits_{0}^{\sqrt{a^2-x^2}}\sqrt{a^2-x^2-y^2}\,dy\,dx$ by changing to polar coordinates.

[**Hint.** The region of integration lies between the curves

$$y = 0, \ y = \sqrt{a^2 - x^2}, \ x = 0 \text{ and } x = a$$

(Fig. 13.29).
In polar coordinates, the region is bounded by the curves $r = 0$, $r = a$, $\theta = 0$ and $\theta = \pi/2$.

∴ The given integral

$$= \int_0^{\pi/2} \int_0^a (a^2 - r^2)^{1/2} r \, dr \, d\theta = \frac{\pi a^3}{6}].$$

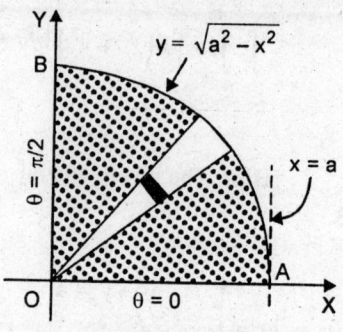

Fig. 13.29.

29. Change the cartesian integral into an equivalent polar integral and then

evaluate $\displaystyle\int_0^2 \int_0^{\sqrt{1-(x-1)^2}} \frac{x+y}{x^2+y^2} \, dy \, dx.$ **Ans.** $\left(\dfrac{\pi}{2}+1\right)$

30. Evaluate the integral $\displaystyle\iint \sqrt{4a^2 - x^2 - y^2} \, dy \, dx$ taken over the upper half of the circle $x^2 + y^2 - 2ax = 0$, by changing to polar co-ordinates.

[**Hint.** We have to evaluate $\displaystyle\int_0^{2a} \int_0^{\sqrt{2ax-x^2}} \sqrt{4a^2 - x^2 - y^2} \, dy \, dx$

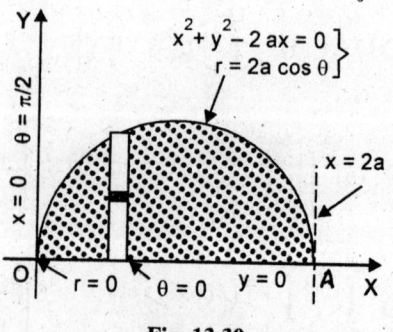

$$x^2 + y^2 - 2ax = 0 \left.\begin{array}{c}\\ \\\end{array}\right\}$$
$$r = 2a \cos\theta$$

Fig. 13.30

Changing to polar co-ordinates, we have the given integral

$$= \int_{\theta=0}^{\pi/2} \int_{r=0}^{2a\cos\theta} \sqrt{4a^2 - r^2} \cdot r \, dr \, d\theta$$

$$= \frac{4a^3}{9}(3\pi - 4).$$

31. Evaluate the integral $\displaystyle\iint_R (a^2 - x^2 - y^2) \, dy \, dx$, where R is the region

$x^2 + y^2 \le a^2$.   **Ans.** $\pi a^4/2$

32. Evaluate the integral $\displaystyle\iint_R \sqrt{x^2 + y^2} \, dx \, dy$ by changing to polar coordinates,

where R is the region in the xy-plane bounded by the circles
$x^2 + y^2 = a^2$ and $x^2 + y^2 = b^2$, $b > a$.
(*I.E.T.E., Dec. 1997*)

[**Hint.** The given integral $= \int\limits_0^{2\pi} \int\limits_a^b r(r \, dr \, d\theta) = \dfrac{2\pi}{3}(b^3 - a^3)$].

33. Evaluate the integral $\iint\limits_R e^{-x^2 - y^2} \, dx \, dy$, where R is the annulus bounded by

$x^2 + y^2 = 1$ and $x^2 + y^2 = 4$ by changing to polar coordinates.
(*I.E.T.E., June 1997*)] **Ans.** $\pi(e^3 - 1)/e^4$.

13.7 Applications of Double Integrals

Double integrals have large number of applications. We state some of them.

A. Area by double integration

If we put $f(x, y) = 1$ in equation (13.2) of Art 13.1, we find that the area A of the
region bounded by $y = f_1(x)$, $y = f_2(x)$, $x = a$ and $x = b$ is given by

$$A = \int\limits_a^b \int\limits_{f_1}^{f_2} dy \, dx.$$

In polar co-ordinates the area bounded
by the curves $r = f_1(\theta)$, $r = f_2(\theta)$, the radii
vectors $\theta = \theta_1$, $\theta = \theta_2$ is given by

$$\int\limits_{\theta_1}^{\theta_2} \int\limits_{f_1}^{f_2} r \, dr \, d\theta.$$

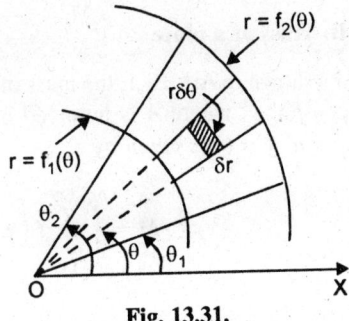

Fig. 13.31.

Example 13.15. Find the area bounded
by the parabola $y = x^2$ and the line
$y = 2x + 3$.

Solution. Ref. Fig. 13.32, using vertical
strips.

$$\text{Required area} = \int\limits_{-1}^{3} \int\limits_{x^2}^{2x+3} dy \, dx$$

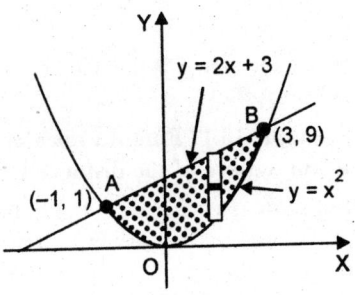

Fig. 13.32.

$$= \int\limits_{-1}^{3} (2x + 3 - x^2) \, dx = \left[x^2 + 3x - \frac{x^3}{3} \right]_{-1}^{3} = 32/3 \text{ sq. units.}$$

Example 13.16. Find the area outside the circle r = 2 and inside the cardioid r = 2(1 + cosθ).

Solution. Due to symmetry, the required area is twice that swept over as θ varies from θ = 0 to θ = π/2. Thus, area,

$$= 2 \int_{0}^{\pi/2} \int_{2}^{2(1+\cos\theta)} r \, dr \, d\theta$$

$$= 2 \int_{0}^{\pi/2} \left[\frac{1}{2} r^2 \right]_{2}^{2(1+\cos\theta)} d\theta$$

$$= 4 \int_{0}^{\pi/2} \left(2\cos\theta + \cos^2\theta \right) d\theta$$

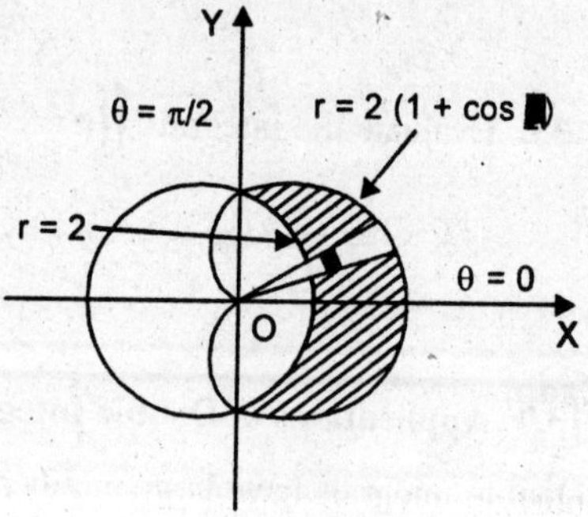

Fig. 13.33.

$$= 4 \left[2\sin\theta + \frac{\theta}{2} + \frac{\sin 2\theta}{4} \right]_{0}^{\pi/2} = (\pi + 8) \text{ square units.}$$

B. Mass of a plate

As shown in Art. 13.1, the mass of plate occupying an area A and having a density $\rho = f(x, y)$ where A is bounded by the curves $y = f_1(x)$, $y = f_2(x)$ and the lines $x = a$, $x = b$, is given by

$$M = \int_{a}^{b} \int_{f_1(x)}^{f_2(x)} f(x, y) \, dy \, dx.$$

In polar coordinates, the mass is given by $M = \int_{\theta_1}^{\theta_2} \int_{f_1(\theta)}^{f_2(\theta)} F(r, \theta) r \, dr \, d\theta$,

where the area A is bounded by $r = f_1(\theta)$, $r = f_2(\theta)$, $\theta = \theta_1$ and $\theta = \theta_2$ and density $\rho = F(r, \theta)$.

Example 13.17 Find the mass of a square plate of side 'a' if the density varies as the square of the distance from a vertex.

Solution. (Ref. Fig. 13.34). Let the vertex from which distances are measured be at the origin. Then $\rho(x, y) = k(x^2 + y^2)$

and $\qquad M = \iint_{A} \rho(x, y) \, dy \, dx = \int_{0}^{a} \int_{0}^{a} k\left(x^2 + y^2\right) dy \, dx$

$$= k \int_0^a \left[x^2 y + \frac{y^3}{3} \right]_0^a dx$$

$$= k \int_0^a \left(x^2 a + \frac{a^3}{3} \right) dx$$

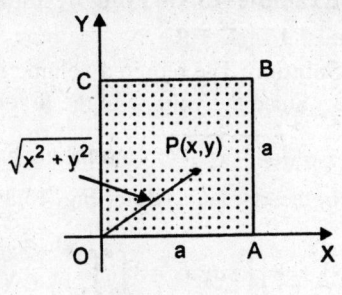

Fig. 13.34.

$$= k \left[\frac{ax^3}{3} + \frac{a^3}{3} x \right]_0^a = \frac{2}{3} k a^4 \text{ units.}$$

Example 13.18 Find the mass of a plate in the form of a cardioid $r = a(1 + \cos\theta)$, if the density of the mass varies as the distance from the pole.

Solution. At the point $P(r, \theta)$ the density $\rho = kr$. Since total mass M is twice the mass above the initial line,

$$M = 2 \int_0^\pi \int_0^{a(1+\cos\theta)} (kr) \cdot r \, dr \, d\theta$$

$$= 2k \frac{a^3}{3} \int_0^\pi (1 + \cos\theta)^3 \, d\theta$$

$$= \frac{2}{3} ka^3 \int_0^\pi 2^3 \cdot \cos^6 \frac{\theta}{2} \, d\theta$$

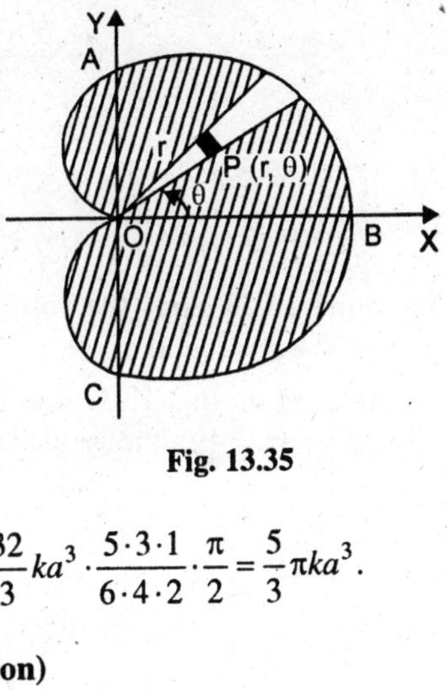

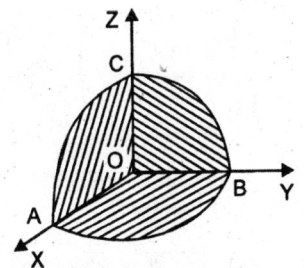

Fig. 13.35

$$= \frac{16}{3} ka^3 \int_0^{\pi/2} 2 \cos^6 t \, dt \left(\text{where } \frac{\theta}{2} = t \right) = \frac{32}{3} ka^3 \cdot \frac{5 \cdot 3 \cdot 1}{6 \cdot 4 \cdot 2} \cdot \frac{\pi}{2} = \frac{5}{3} \pi ka^3.$$

C. Volume under a surface (Double integration)

The volume of a vertical column whose upper end is in the surface $z = f(x, y)$ and whose lower base is in the XOY plane is defined by

the double integral $V = \iint_R z \, dy \, dx$ or

$\iint_R f(x, y) \, dy \, dx$ the region R being the lower

base of the column.

Fig. 13.36.

Example 13.19 Find by double integration the volume of the sphere
$x^2 + y^2 + z^2 = a^2$.

Solution. The required volume is eight times the volume in the first octant.

Since the volume in the first octant is bounded from above by the surface

$z = \sqrt{a^2 - x^2 - y^2}$ and has its base R as the quadrant of the circle $x^2 + y^2 = a^2$
in the xy-plane. (Refer Fig. 13.36)

$$V = 8 \iint_R z \, dy \, dx = 8 \int_0^a \int_0^{\sqrt{a^2-x^2}} \sqrt{(a^2 - x^2) - y^2} \, dy \, dx$$

$$= 8 \int_0^a \left[\frac{y}{2} \sqrt{(a^2 - x^2) - y^2} + \frac{(a^2 - x^2)}{2} \sin^{-1} \frac{y}{\sqrt{a^2 - x^2}} \right]_0^{\sqrt{a^2-x^2}} dx$$

$$= 8 \int_0^a \left[0 + \frac{a^2 - x^2}{2} \cdot \frac{\pi}{2} \right] dx = 2\pi \int_0^a (a^2 - x^2) \, dx$$

$$= 2\pi \left[a^2 x - \frac{x^3}{3} \right]_0^a = 2\pi \left[a^3 - \frac{a^3}{3} \right] = \frac{4\pi a^3}{3} \text{ cubic units.}$$

Example 13.20 Find the volume bounded by the cylinder $x^2 + y^2 = 4$, the planes $y + z = 3$ and $z = 0$. *(Madras 2000S; A.M.I.E., W-2000)*

Solution. From Fig. 13.37 it is evident that $z = 3 - y$ is to be integrated over the circle $x^2 + y^2 = 4$ in the xy-plane. Hence

$$V = \int_{-2}^{2} \int_{-\sqrt{4-y^2}}^{\sqrt{4-y^2}} (3 - y) \, dx \, dy$$

$$= 2 \int_{-2}^{2} \int_0^{\sqrt{4-y^2}} (3 - y) \, dx \, dy = 2 \int_{-2}^{2} (3 - y) \sqrt{4 - y^2} \, dy$$

$$= 6 \int_{-2}^{2} \sqrt{4 - y^2} \, dy - 2 \int_{-2}^{2} y \sqrt{4 - y^2} \, dy = 12 \int_0^{2} \sqrt{4 - y^2} \, dy$$

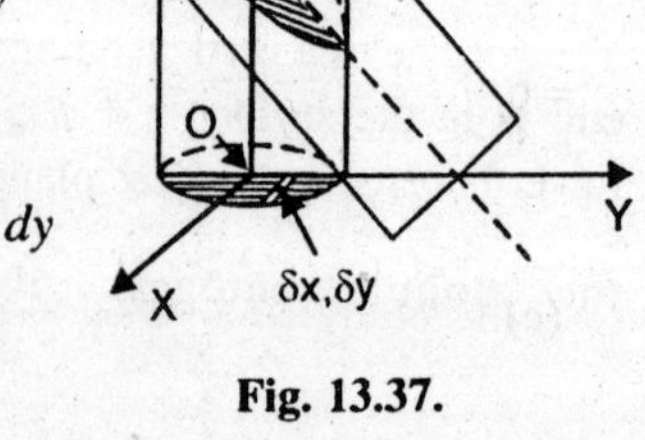

(The second term vanishes as the integrand
is an odd function)

Fig. 13.37.

$$= 12 \left[\frac{y\sqrt{4 - y^2}}{2} + \frac{4}{2} \sin^{-1} \frac{y}{2} \right]_0^{2} = 12 \cdot 2 \cdot \frac{\pi}{2} = 12\pi \text{ cubic units.}$$

13.8 Illustrative Examples

Example 13.21 Find by double integration

(a) the area lying between the parabola $y = 4x - x^2$ and the line $y = x$,
(DCE, 2004; MDU 2003)

(b) the area bounded by the parabolas $y^2 = 4 - x$ and $y^2 = 4 - 4x$,

(c) the area included between the curves $y = x^2 - 6x + 3$ and $y = 2x - 9$,
(I.E.T.E.; June 2001)

(d) the area included between the curves $y^2 = 4a(x + a)$ and $y^2 = 4b(b - x)$.
(I.E.T.E., June 2001)

Solution. (a) The two curves intersect at points whose abscissae are given by $4x - x^2 = x$, that is, $x = 0$ or 3.

The area under consideration lies between the curves $y = x$, $y = 4x - x^2$, $x = 0$ and $x = 3$. Using vertical strips,

required area $A = \displaystyle\int_0^3 \int_{y=x}^{4x-x^2} dy\,dx = \int_0^3 \left[y\right]_x^{4x-x^2} dx$

$= \displaystyle\int_0^3 \left(4x - x^2 - x\right)dx = \int_0^3 \left(3x - x^2\right)dx$

$= \left[\dfrac{3}{2}x^2 - \dfrac{x^3}{3}\right]_0^3 = \dfrac{27}{2} - 9 = 4.5$ square units.

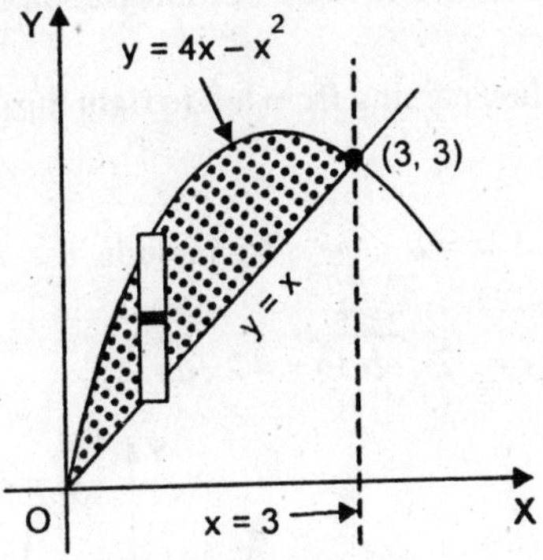

Fig. 13.38.

(b) Using horizontal strips and taking advantage of symmetry, we have

$A = 2\displaystyle\int_0^2 \int_{1-(y^2/4)}^{4-y^2} dx\,dy = 2\int_0^2 \left[(4 - y^2) - \left(1 - \dfrac{y^2}{4}\right)\right] dy$

$= 6\displaystyle\int_0^2 \left(1 - \dfrac{y^2}{4}\right) dy = 6\left[y - \dfrac{y^3}{12}\right]_0^2 = 8$ sq. units.

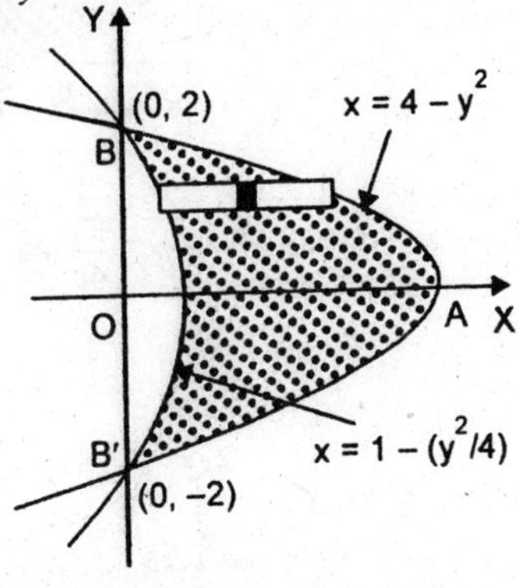

(c) Required area, $A = \displaystyle\int_{x=2}^6 \int_{y=x^2-6x+3}^{2x-9} dy\,dx$

(Refer Fig. 13.40)

$= \displaystyle\int_2^6 \left(2x - 9 - x^2 + 6x - 3\right)dx$

Fig. 13.39.

$$= -\int_2^6 \left(x^2 - 8x + 12\right) dx = \int_6^2 \left(x^2 - 8x + 12\right) dx$$

$$= \left[\frac{x^3}{3} - 4x^2 + 12x\right]_6^2$$

$$= \frac{1}{3}(8 - 216) - 4(4 - 36) + 12(2 - 6)$$

$$= -\frac{1}{3}(208) + 128 - 48 = \frac{32}{3} \text{ square units.}$$

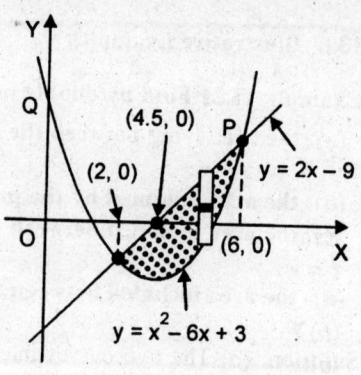

Fig. 13.40.

(d) The region of integration is shown in Fig. 13.41. We imagine a horizontal line passing from left to right through the region. It enters at $x = \dfrac{y^2}{4a} - a$ and leaves at $x = b - \dfrac{y^2}{4b}$. To include the area between the curves, we let y run from $y = -2\sqrt{ab}$ to $y = 2\sqrt{ab}$.

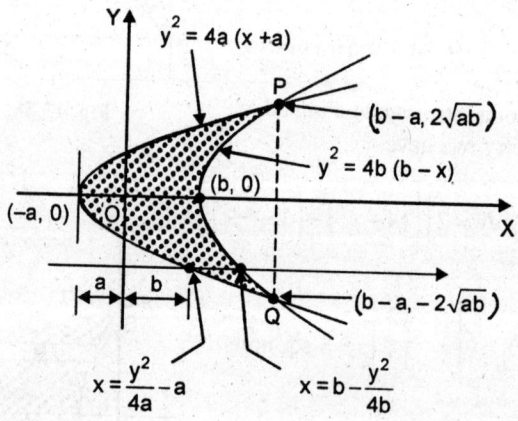

Fig. 13.41.

$$\therefore \text{ Reqd. Area} = \int_{y=-2\sqrt{ab}}^{2\sqrt{ab}} \int_{x=(y^2/4a)-a}^{b-(y^2/4b)} dx \, dy = \int_{-2\sqrt{ab}}^{2\sqrt{ab}} \left(b - \frac{y^2}{4b} - \frac{y^2}{4a} + a\right) dy$$

$$= \left[(b+a)y - \frac{1}{12ab}(a+b) \cdot y^3 \right]_{-2\sqrt{ab}}^{2\sqrt{ab}} = (b+a)4\sqrt{ab} - \frac{(a+b)}{12ab} \cdot 16ab\sqrt{ab}$$

$$= \frac{8}{3}(a+b)\sqrt{ab} \text{ square units.}$$

Example 13.22 (a) Find by double integration the area lying inside the circle $r = a \sin \theta$ and outside the cardioid $r = a(1 - \cos\theta)$.

(DCE 2005, MDU 2004, Assam, 1998)

(b) Integrate $r \sin \theta$ over the area of the cardioid $r = a(1 + \cos\theta)$ above the initial line. *(MDU, 2004)*

(c) Compute the area included between the curve $r = a(\sec\theta + \cos\theta)$ and its asymptote $r = a \sec\theta$. *(Bhopal, 1998; Hamirpur 1995S)*

Solution. *(a)* It is obvious from the diagram that the circle $r = a \sin \theta$ and the cardioid $r = a(1 - \cos\theta)$ cut at $\theta = 0$ and $\theta = \pi/2$.

The area A between the two curves can be covered by radial strips whose ends are at $r = a(1 - \cos\theta)$ and $r = a \sin\theta$. The strips start at $\theta = 0$ and end at $\theta = \pi/2$. Hence the desired area

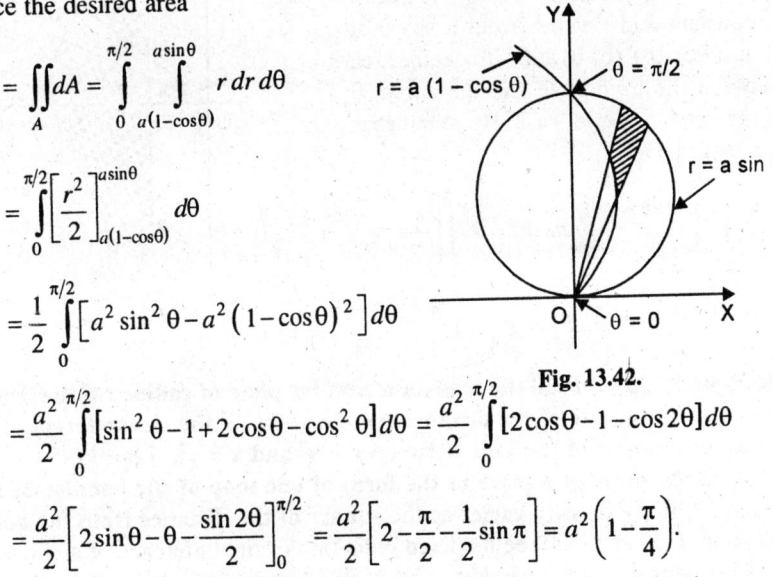

Fig. 13.42.

$$= \iint_A dA = \int_0^{\pi/2} \int_{a(1-\cos\theta)}^{a\sin\theta} r \, dr \, d\theta$$

$$= \int_0^{\pi/2} \left[\frac{r^2}{2} \right]_{a(1-\cos\theta)}^{a\sin\theta} d\theta$$

$$= \frac{1}{2} \int_0^{\pi/2} \left[a^2 \sin^2 \theta - a^2 (1-\cos\theta)^2 \right] d\theta$$

$$= \frac{a^2}{2} \int_0^{\pi/2} [\sin^2 \theta - 1 + 2\cos\theta - \cos^2 \theta] d\theta = \frac{a^2}{2} \int_0^{\pi/2} [2\cos\theta - 1 - \cos 2\theta] d\theta$$

$$= \frac{a^2}{2} \left[2\sin\theta - \theta - \frac{\sin 2\theta}{2} \right]_0^{\pi/2} = \frac{a^2}{2} \left[2 - \frac{\pi}{2} - \frac{1}{2}\sin \pi \right] = a^2 \left(1 - \frac{\pi}{4} \right).$$

(b) Here the region of integration A can be covered by radial strips, whose ends are at $r = 0$ and $r = a(1 + \cos\theta)$. The strips start from $\theta = 0$ and end at $\theta = \pi$. Therefore the required integral is

$$= \iint_A r \sin \theta \, dA = \int_0^{\pi} \int_0^{a(1+\cos\theta)} r \sin\theta \, r \, dr \, d\theta = \int_0^{\pi} \sin\theta \left[\frac{r^3}{3} \right]_0^{a(1+\cos\theta)} d\theta$$

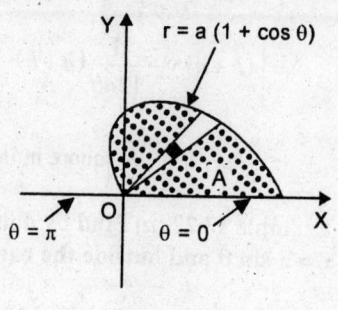

Fig. 13.43.

$$= \frac{1}{3}a^3 \int_0^\pi \sin\theta (1+\cos\theta)^3 \, d\theta$$

$$= \frac{1}{3}a^3 \int_0^\pi 2\sin\frac{\theta}{2}\cos\frac{\theta}{2} \cdot 8\cos^6\frac{\theta}{2} \, d\theta$$

$$= \frac{16}{3}a^3 \int_0^{\pi/2} (\sin t \cos^7 t) 2 \, dt$$

(on putting $\theta = 2t$ and $d\theta = 2dt$) $= \frac{32}{3}a^3 \left[-\frac{\cos^8 t}{8} \right]_0^{\pi/2} = \frac{4}{3}a^3$.

(c) The curve is symmetrical about the initial line and has an asymptote $r = a\sec\theta$. Taking any line OB cutting the curve at B and its asymptote at B'. Along this line, θ is constant and r varies from $a \sec\theta$ at B' to $a(\sec\theta + \cos\theta)$ at B. To get the area between the asymptote and the curve, θ varies from $-\pi/2$ to $+\pi/2$. By symmetry, the required area

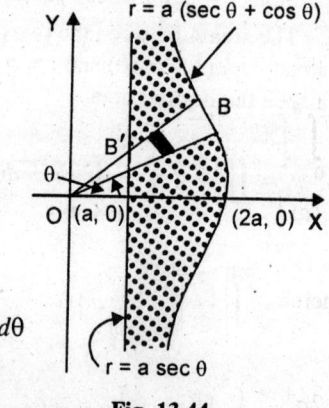

Fig. 13.44.

$$= 2\int_0^{\pi/2} \int_{a\sec\theta}^{a(\sec\theta+\cos\theta)} r \, dr \, d\theta = 2\int_0^{\pi/2} \left[\frac{r^2}{2} \right]_{a\sec\theta}^{a(\sec\theta+\cos\theta)} d\theta$$

$$= a^2 \int_0^{\pi/2} (2+\cos^2\theta) \, d\theta = a^2 \left[\pi + \frac{\pi}{4} \right] = \frac{5\pi a^2}{4} \text{ square units.}$$

Example 13.23 (*a*) **Find the mass of a circular plate of radius r if the density varies as the square of the distance from a point on the circumference.**
(*b*) **Find the mass of the area between $y = x^3$ and $x = y^2$, if $\rho = k(x^2 + y^2)$.**
(*c*) **Find the mass of a plate in the form of one loop of the lemniscate $r^2 = a^2\cos 2\theta$, if the density varies as the square of the distance from the pole.**
Solution. (*a*) Let $A(r, 0)$ be the fixed point on the circumference of a circle. Then density at $P(x, y) = k[(x-r)^2 + y^2]$ and therefore,

$$M = \iint_A \rho(x, y) \, dA = 2\int_{-r}^{r} \int_0^{\sqrt{r^2-x^2}} k\left[(x-r)^2 + y^2 \right] dy \, dx = \frac{3}{2}k\pi r^4 \text{ units.}$$

(*b*) The curves $y = x^3$ and $x = y^2$ cut at $x = 0$ and $x = 1$. Hence the region of integration can be considered to lie between the curves

$$y = x^3, \ y = \sqrt{x}, \ x = 0 \text{ and } x = 1.$$

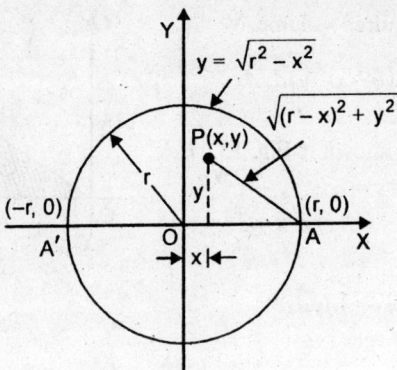

Fig. 13.45.

Hence the mass of the area between the curves

$$= \iint_A \rho(x, y)\,dx\,dy = \iint_A k(x^2 + y^2)\,dx\,dy$$

$$= \int_0^1 \int_{x^3}^{\sqrt{x}} k(x^2 + y^2)\,dy\,dx, \text{ which on simplification reduces to } \frac{23k}{105}.$$

(c) At the point $P(r, \theta)$ the density $\rho = kr^2$.

Therefore, $M = 2\displaystyle\int_0^{\pi/4} \int_0^{a\sqrt{\cos 2\theta}} kr^2 \cdot r\,dr\,d\theta = 2k\int_0^{\pi/4} d\theta = 2k\int_0^{\pi/4} \int_0^{a\sqrt{\cos 2\theta}} r^3\,dr\,d\theta$

$$= \frac{2a^4}{4} k \int_0^{\pi/4} \cos^2 2\theta\,d\theta = \frac{a^4 k}{2} \int_0^{\pi/2} \frac{1}{2}\cos^2 t\,dt$$

(Putting $2\theta = t$)

$$= \frac{1}{4} a^4 k \frac{1}{2} \cdot \frac{\pi}{2} = \frac{a^4 k\pi}{16}.$$

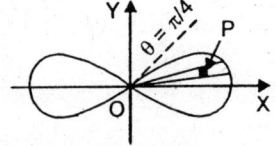

Fig. 13.46.

Example 13.24 (a) Find the volume under the plane $x + y + z = 6$ and above the triangle in the xy-plane bounded by $2x = 3y$, $y = 0$, $x = 3$.

   (b) Find the volume bounded by the plane $z = 0$, surface $z = x^2 + y^2 + 2$ and the cylinder $x^2 + y^2 = 4$.

   (c) Find the volume in the first octant bounded by the circular cylinder $x^2 + y^2 = 2$ and the planes $z = x + y$, $y = x$, $z = 0$ and $x = 0$.

   (d) Find the volume cut off from the paraboloid $x^2 + (y^2/4) + z = 1$ by the plane $z = 0$.

Solution. (*a*) The required volume, V

$$= \iint_A z\,dy\,dx = \iint_A (6-x-y)\,dy\,dx,$$

where A is the region, shown in Fig. 13.47 bounded by the curves

$$y = 0,\ y = 2x/3,\ x = 0 \text{ and } x = 3.$$

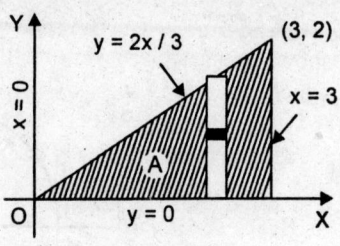

Fig. 13.47.

$$\therefore\ V = \int_0^3 \int_0^{2x/3} (6-x-y)\,dy\,dx$$

$$= \int_0^3 \left[6y - xy - \frac{y^2}{2} \right]_0^{2x/3} dx = \int_0^3 \left(4x - \frac{2}{3}x^2 - \frac{2}{9}x^2 \right) dx$$

$$= \int_0^3 \left(4x - \frac{8x^2}{9} \right) dx = \left[2x^2 - \frac{8}{27}x^3 \right]_0^3 = 10 \text{ cubic units.}$$

(*b*) The region A in the positive quadrant of *XOY* plane is bounded by the curves $y = 0$, $y = \sqrt{(4 - x^2)}$, $x = 0$ and $x = 2$.

Required volume = 4 times the volume lying in the first octant

$$= 4 \int_0^2 \int_0^{\sqrt{(4-x^2)}} (x^2 + y^2 + 2)\,dy\,dx = 4 \int_0^2 \left[x^2 y + (y^3/3) + 2y \right]_0^{\sqrt{(4-x^2)}} dx$$

$$= 4 \int_0^2 \left[x^2 \sqrt{(4-x^2)} + (1/3)(4-x^2)^{3/2} + 2\sqrt{(4-x^2)} \right] dx$$

$$= 4 \int_0^{\pi/2} \left[4\sin^2\theta \cdot 2\cos\theta + (1/3)8\cos^3\theta + 2\cdot 2\cos\theta \right] 2\cos\theta\,d\theta$$

(on putting $x = 2\sin\theta$, $\sqrt{(4-x^2)} = \sqrt{\{4(1-\sin^2\theta)\}} = 2\cos\theta$ and $dx = 2\cos\theta\,d\theta$).

$$= 32 \int_0^{\pi/2} \left[2\sin^2\theta\cos^2\theta + \frac{2}{3}\cos^4\theta + \cos^2\theta \right] d\theta$$

$$= 32 \left[2 \cdot \frac{1}{4} \cdot \frac{1}{2} \cdot \frac{\pi}{2} + \frac{2}{3} \cdot \frac{3\cdot 1}{4\cdot 2} \cdot \frac{\pi}{2} + \frac{1}{2} \cdot \frac{\pi}{2} \right] = 32 \left(\frac{\pi}{8} + \frac{\pi}{8} + \frac{\pi}{4} \right) = 16\pi.$$

(c) The region of integration A is shown by shaded lines in the diagram, which is bounded by the curves

$$y = x, \quad y = \sqrt{(2-x^2)},$$

$$x = 0 \text{ and } x = 1.$$

Hence the required volume,

$$V = \int_0^1 \int_x^{\sqrt{(2-x^2)}} (x+y) \, dy \, dx, \text{ which}$$

on simplification reduces to $\dfrac{2\sqrt{2}}{3}$.

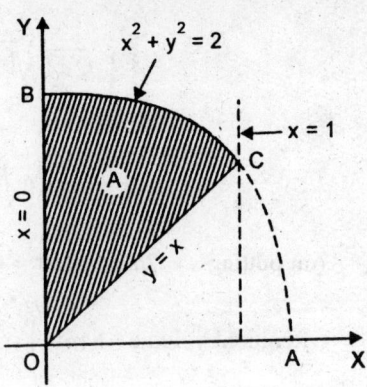

Fig. 13.48.

Example 13.25 (a) **Find the volume common to the cylinders $x^2 + y^2 = a^2$, $x^2 + z^2 = a^2$.** (S.V.T.U., 2006)

(b) **Prove that the volume enclosed between the cylinders $x^2 + y^2 = 2ax$ and $z^2 = 2ax$ is $\dfrac{128a^3}{15}$.** (Delhi, 1997)

Solution. (a) Figure 13.49 shows one-eighth of the volume common to the cylinders.

Required volume = 8 times volume of region shown in Fig. 13.49.

$$= 8 \int_{x=0}^{a} \int_{y=0}^{\sqrt{(a^2-x^2)}} z \, dy \, dx = 8 \int_{x=0}^{a} \int_{y=0}^{\sqrt{(a^2-x^2)}} \sqrt{(a^2-x^2)} \, dy \, dx$$

$$= 8 \int_0^a \sqrt{a^2 - x^2} \, [y]_0^{\sqrt{(a^2-x^2)}} \, dx$$

$$= 8 \int_0^a (a^2 - x^2) \, dx = 8 \left[a^2 x - \frac{x^3}{3} \right]_0^a$$

$$= 8 \left(a^3 - \frac{a^3}{3} \right) = \frac{16a^3}{3} \text{ cubic units.}$$

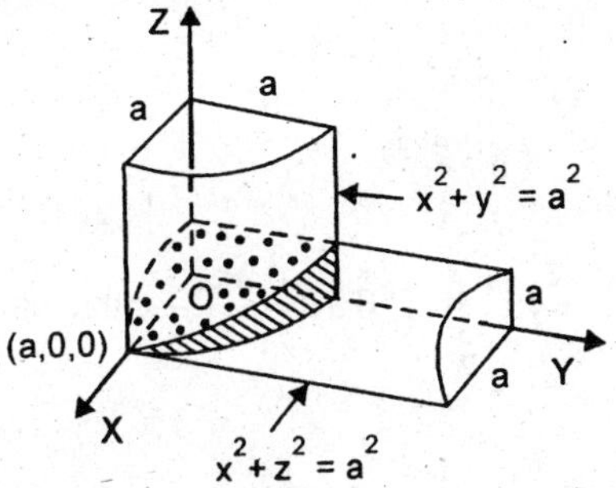

Fig. 13.49.

(b) Here $z = \pm\sqrt{2ax}$. Thus half of the volume lies above the xy-plane and half below it.

Therefore, $V = 2 \int_0^{2a} \int_{-\sqrt{(2ax-x^2)}}^{\sqrt{(2ax-x^2)}} \sqrt{2ax} \, dy \, dx = 4 \int_0^{2a} \int_0^{\sqrt{(2ax-x^2)}} \sqrt{2ax} \, dy \, dx$

$$= 4 \int_0^{2a} \sqrt{2ax} \cdot \sqrt{(2ax - x^2)}\, dx$$

$$= 4 \int_0^{\pi/2} 2a \sin\theta \sqrt{\left\{2a \sin^2\theta \left(2a - 2a \sin^2\theta\right)\right\}} \cdot 4a \sin\theta \cos\theta\, d\theta$$

(on putting $x = 2a \sin^2\theta$, $dx = 4a \sin\theta \cos\theta\, d\theta$).

Required volume $= 64a^3 \int_0^{\pi/2} \sin^3\theta \cos^2\theta\, d\theta = 64a^3 \dfrac{2}{5 \cdot 3 \cdot 1} = \dfrac{128a^3}{15}$.

Example 13.26. Find the volume bounded by the paraboloid $x^2 + y^2 = az$, the cylinder $x^2 + y^2 = 2\,ay$ and the plane $z = 0$. *(Mysore 1995)*

Solution. The required volume is obtained by integrating $z = (x^2 + y^2)/a$ over the circle $x^2 + y^2 - 2ay = 0$. Changing to polar coordinates in the xy-plane, we have $x = r \cos\theta$ $y = r \sin\theta$ so that $z = r^2/a$ and the polar equation of the circle is $r = 2a \sin\theta$. To cover the circle, r varies from 0 to $2a \sin\theta$ and θ varies from 0 to π. (Refer Fig. 13.50).

Hence the required volume, $V = \displaystyle\iint_R z\, dA$

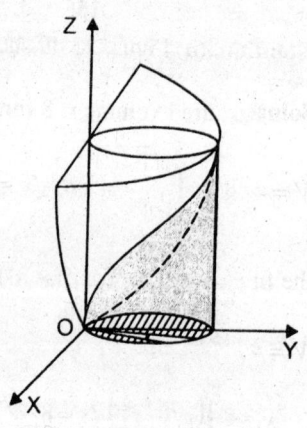

$$= \int_{\theta=0}^{\pi} \int_{r=0}^{2a\sin\theta} z \cdot r\, dr\, d\theta = \frac{1}{a}\int_0^{\pi} d\theta \int_0^{2a\sin\theta} r^3\, dr$$

$$= \frac{1}{a}\int_0^{\pi} \left[\frac{r^4}{4}\right]_0^{2a\sin\theta} d\theta$$

$$= 4a^3 \int_0^{\pi} \sin^4\theta\, d\theta = \frac{3\pi a^3}{2} \text{ cubic units.}$$

Fig. 13.50.

Example 13.27 Find the volume of the ellipsoid $\dfrac{x^2}{a^2} + \dfrac{y^2}{b^2} + \dfrac{z^2}{c^2} = 1$.

(PTU 2006; DCE 2004; MDU 2001; J.N.T.U., 1998; Mysore, 1997)

Solution. On account of symmetry, the required volume is 8 times the volume of the ellipsoid in the first octant. The projection of the surface $z = c\sqrt{1 - \dfrac{x^2}{a^2} - \dfrac{y^2}{b^2}}$

in the xy-plane is the region in the first quadrant of the ellipse $\dfrac{x^2}{a^2} + \dfrac{y^2}{b^2} = 1$.

Therefore, $V = 8 \displaystyle\int_0^a dx \int_0^{b\sqrt{1-x^2/a^2}} c\sqrt{1 - \dfrac{x^2}{a^2} - \dfrac{y^2}{b^2}}\, dy = 8c \int_0^a dx \int_0^{bk} \sqrt{k^2 - \dfrac{y^2}{b^2}}\, dy$,

where $k^2 = 1 - (x^2/a^2)$. Putting $y = bk \sin\theta$ and $dy = bk \cos\theta\, d\theta$, we obtain

$$V = 8c \int_0^a dx \int_0^{\pi/2} \sqrt{k^2 - k^2 \sin^2\theta}\,(bk\cos\theta)\,d\theta = 8bc \int_0^a dx \int_0^{\pi/2} k^2 \cos^2\theta\, d\theta$$

$$= 8bc \cdot \frac{1}{2} \cdot \frac{\pi}{2} \int_0^a \left(1 - \frac{x^2}{a^2}\right) dx = \frac{2\pi bc}{a^2} \int_0^a (a^2 - x^2)\, dx = \frac{2\pi bc}{a^2}\left[a^2 x - \frac{x^3}{3}\right]_0^a$$

$$= \frac{4\pi abc}{3}\ \text{units}^3.$$

Example 13.28 Find the volume of the solid bounded by the conical surface

$$(z-2)^2 = \frac{x^2}{3} + \frac{y^2}{2}\ \text{and the plane } z = 0.$$

Solution. By symmetry, volume required is given by

$$V = 4 \iint_R \left(2 + \sqrt{\frac{x^2}{3} + \frac{y^2}{2}}\right) dy\, dx,\ \text{where } R \text{ is the part of the ellipse } \frac{x^2}{12} + \frac{y^2}{8} = 1 \text{ in}$$

the first quadrant of the xy plane. Putting $x = u\sqrt{3}, y = v\sqrt{2}$, we obtain

$$V = 4 \iint_{R_1} \left(2 + \sqrt{u^2 + v^2}\right) \sqrt{6}\, du\, dv,\ \text{where } R_1 \text{ is the quadrant of the circle } u^2 + v^2 = 4.$$

Again putting $u = r\cos\theta,\ v = r\sin\theta$, we obtain

$$V = 4\sqrt{6} \int_0^{\pi/2} \int_0^2 (2+r)r\, dr\, d\theta = 4\sqrt{6} \int_0^{\pi/2}\left[r^2 + \frac{r^3}{3}\right]_0^2 d\theta = 4\sqrt{6} \cdot \frac{20}{3} \int_0^{\pi/2} 1\, d\theta$$

$$= 4\sqrt{6} \times \frac{20}{3} \times \frac{\pi}{2} = \frac{40}{3}\pi\sqrt{6}\ \text{cubic units.}$$

Exercise 13.2

1. Find the area bounded by the curves $y = x,\ y = 2 - x^2$. **Ans.** 4.5 units^2.

2. (*a*) Find the smaller of the areas bounded by $y = 2 - x$ and $x^2 + y^2 = 4$.

Ans. $(\pi - 4)$ unit².

(*b*) Find the area enclosed by the ellipse $x^2/a^2 + y^2/b^2 = 1$.

(VTU 2001; Osmania 2000S; MDU 2000) **Ans.** πab sq. units.

(*c*) Evaluate $\iint xy \, dx \, dy$ over the area between $y = x^2$ and $y = x$.

(A.M.I.E., S-2009) **Ans.** 1/24.

3. Find the area in the first quadrant bounded by the *x*-axis and the curves $x^2 + y^2 = 10$ and $y^2 = 9x$. **Ans.** 6.75 square units.

4. Find the area of the region enclosed by the curves $\sqrt{x} + \sqrt{y} = \sqrt{a}$ and $x + y = a$. **Ans.** $a^2/3$ square units.

5. Find the area enclosed by the curves $y^2 = 4ax$ and $x^2 = 4 \, ay$

(Kerala 2005, Rohtak 2003; MDU 2003) **Ans.** $16a^2/3$ square units.

6. Find the area of a figure bounded by the curves $y = \sin x$, $y = \cos x$, $x = 0$.

Ans. $(\sqrt{2} - 1)$ square units.

7. Find the area enclosed between the parabola $y = x(4 - x)$ and the axis of *x*.

Ans. 10.667 square units.

8. Evaluate the integral $\iint\limits_R \sqrt{x^2 + y^2} \, dx \, dy$ by changing to polar coordinates,

where R is the region in the *xy*-plane bounded by the circles $x^2 + y^2 = 4$ and $x^2 + y^2 = 9$. *(IETE Dec. 2009, June 2007)*

[**Hint.** Using $x = r\cos\theta$, $y = r\sin\theta$, $dx \, dy = r \, dr \, d\theta$, we obtain

$$I = \int\limits_0^{2\pi} \int\limits_2^3 r(r \, dr \, d\theta) = \int\limits_0^{2\pi} \left(\frac{r^3}{3}\right)_2^3 d\theta = \frac{19}{3}\int\limits_0^{2\pi} d\theta = \frac{38\pi}{3}.\,]$$

9. Evaluate $\iint\limits_A r^3 dr \, d\theta$, where A is the area included between the circles

$r = 2 \sin\theta$ and $r = 4 \sin\theta$.

[**Hint.** The given integral $= \int\limits_{\theta=0}^{\pi} d\theta \int\limits_{r=2\sin\theta}^{4\sin\theta} r^3 dr$] *(MDU-2003, JNTU, 1999)* **Ans.** 22.5 π

10. Find the area in the *xy*-plane bounded by the lemniscate $r^2 = a^2 \cos 2\theta$.

(Madras, 2000S; Hamirpur, 1995S) **Ans.** a^2 square units

11. (*a*) Find the area inside the circle $r = 2a \cos\theta$ and outside the circle $r = a$.

Ans. $2a^2 \left(\pi/6 + \sqrt{3}/4\right)$ square units

(b) Find the area lying inside a cardioide $r = 1 + \cos\theta$ and outside the parabola $r(1 + \cos\theta) = 1$. *(DCE 2005)* **Ans.** $(9\pi + 16)/12$.

12. Find the area common to the cardioids $r = a(1 - \cos\theta)$ and $r = a(1 + \cos\theta)$.
Ans. $a^2(3\pi - 8)/2$ square units.

13. Find the area inside the circle $r = 4\sin\theta$ and outside the lemniscate $r^2 = 8\cos 2\theta$. **Ans.** $(8\pi + 12\sqrt{3} - 12)/3$ square units.

[Hint. Area $= 2\displaystyle\int_{\pi/6}^{\pi/4}\int_{2\sqrt{2\cos 2\theta}}^{4\sin\theta} r\, dr\, d\theta + 2\int_{\pi/4}^{\pi/2}\int_{0}^{4\sin\theta} r\, dr\, d\theta.$]

14. Find the mass of a plate in the form of a right triangle with legs 'a' and 'b', if the density varies as the sum of distances from the legs.
Ans. $kab(a + b)/6$ units.

15. Find the mass of a plate in the form of an ellipse $b^2x^2 + a^2y^2 = a^2b^2$, if the density varies as the sum of the distances from the axes.
Ans. $4kab(a + b)/3$ units.

16. Find the mass of a plate in the form of a quadrant of an ellipse $x^2/a^2 + y^2/b^2 = 1$, whose density per unit area is given by $\rho = kxy$.
Ans. $ka^2b^2/8$ units.

[Hint. The quadrant of the ellipse $x^2/a^2 + y^2/b^2 = 1$ is enclosed by the curves $y = 0$, $y = b\sqrt{(1 - x^2/a^2)}$, $x = 0$ and $x = a$. This gives the region of integration R].

17. A plate has its edge the curve $y = e^x$, the line $x = 1$ and the coordinate axes. If the density varies as the square of the distance from the origin, find the mass of the plate. **Ans.** $2.84\,k$ units.

[Hint. $M = \displaystyle\int_{0}^{1}\int_{0}^{e^x} k(x^2 + y^2)\,dy\,dx$]

18. A plate is in the form of a parabolic segment bounded by parabola $y^2 = 8x$ and its latus rectum $x = 2$. If the density varies as the distance from the latus rectum, find the mass of the plate. **Ans.** $64k/15$ units.

19. Determine the mass of a circular plate of radius r if the surface density $f(x, y)$ of the material at each point $P(x, y)$ is proportional to the distance of the point (x, y) from the centre of the circle. **Ans.** $2k\pi r^3/3$ units.

Use double integration to find the volume of each solid (20-29)

20. The solid bounded by the elliptic paraboloid $4z = 16 - 4x^2 - y^2$ and the XOY-plane. **Ans.** 16π cubic units

21. The solid bounded by the cylinder $x^2 + y^2 = 9$ and the planes $y + z = 9$ and $z = 0$. *(I.E.T.E., Dec. 1995)* **Ans.** 81π cubic units.

22. The solid enclosed by the surfaces $y^2 + z^2 = 4x$ and $x = 5$. **Ans.** 50π units^3.

[**Hint.** $V = 4 \int_0^5 \int_0^{2\sqrt{x}} \sqrt{4x - y^2}\, dy\, dx$.]

23. The solid bounded under $z = 3x$ and above the first quadrant area enclosed by $x = 0$, $y = 0$, $x = 4$ and $x^2 + y^2 = 25$. **Ans.** 98 cubic units

24. The solid bounded under the plane $z = x + y$ and above the area cut from the first quadrant by the ellipse $4x^2 + 9y^2 = 36$. **Ans.** 10 cubic units.

[**Hint.** Reqd. Volume $= \int_0^3 \int_0^{2\sqrt{(9-x^2)}/3} (x+y)\, dy\, dx$].

25. The solid bounded by the cylinders $y = x^2$, $y^2 = x$ and the planes $z = 0$ and $x + y + z = 2$. **Ans.** 11/30 cubic units.

[**Hint.** Volume $= \int_0^1 \int_{x^2}^{\sqrt{x}} (2 - x - y)\, dy\, dx$]

26. The solid bounded under the plane $x + z = 2$, above the plane $z = 0$ and within the cylinder $x^2 + y^2 = 4$. **Ans.** 8π cubic units.

27. The solid bounded by the cylinders $y^2 = z$ and $x^2 + y^2 = a^2$ and the plane $z = 0$ **Ans.** $\pi a^4/4$ cubic units

28. The solid bounded by the co-ordinate planes and that portion of the plane which lie in the first quadrant.
    (a) $lx + my + nz = 1$ *(I.E.T.E., Dec. 2001)* **Ans.** 1/6 *lmn* cubic units
    (b) $x + 2y + 3z = 4$ **Ans.** 16/9 cubic units

29. The solid (tetrahedron) bounded by the coordinate planes $x = 0$, $y = 0$, $z = 0$ and the plane $\dfrac{x}{a} + \dfrac{y}{b} + \dfrac{z}{c} = 1$. *(A.M.I.E., S-2004; Mysore, 1998)*

[**Hint.** Volume $= \int_0^a \int_0^{b(1-x/a)} c\left(1 - \dfrac{x}{a} - \dfrac{y}{b}\right) dy\, dx$

$= c \int_0^a \int_0^{b(a-x)/y} \left[\dfrac{1}{a}(a-x) - \dfrac{y}{b}\right] dy\, dx = c \int_0^a \left[\dfrac{1}{a}(a-x)y - \dfrac{1}{b} \cdot \dfrac{y^2}{2}\right]_0^{b(a-x)/a} dx$

$= c \int_0^a \left[\dfrac{b}{a^2}(a-x)^2 - \dfrac{b(a-x)^2}{2a^2}\right] dx = \dfrac{bc}{2a^2} \int_0^a (a-x)^2\, dx = \dfrac{abc}{6}$ cubic units].

30. A triangular prism is formed by the planes, whose equations are $ay = bx$, $y = 0$ and $x = a$. Show that the volume of this prism between the plane $z = 0$ and the surface $z = c + xy$ is $ab(4c + ab)/8$.

[**Hint.** $V = \int_0^a \int_0^{bx/a} (c + xy)\, dy\, dx$].

31. Show that the volume common to the sphere $x^2 + y^2 + z^2 = a^2$ and the cylinder $x^2 + y^2 = ay$ is $2a^3(3\pi - 4)/9$ cubic units. *(DCE, 2004)*
[**Hint.** The required volume is twice the volume obtained by integrating

$z = \sqrt{a^2 - x^2 - y^2}$ over the circle $x^2 + y^2 = ay$ in the xy-plane.

$$\text{Required volume} = 2\int_0^\pi \int_0^{a\sin\theta} \sqrt{a^2 - r^2}\cdot r\, dr\, d\theta]$$

32. Show that the volume cut off the sphere $x^2 + y^2 + z^2 = a^2$ by the cone $x^2 + y^2 = z^2$ is $\pi(2 - \sqrt{2})\, a^3/3$ cubic units.

33. Find the volume bounded by the surface $z = c\left(1 - \dfrac{x}{a}\right)\left(1 - \dfrac{y}{b}\right)$ and the positive quadrant of the elliptic cylinder $x^2/a^2 + y^2/b^2 = 1$, $z = 0$.

(I.E.T.E., Dec. 2005) **Ans.** $\dfrac{abc}{4}\left(\pi - \dfrac{13}{16}\right)$ cubic units .

34. Find the volume of the paraboloid of revolution $x^2 + y^2 = 4z$ cut off by the plane $z = 4$. *(D.C.E. 2005)*

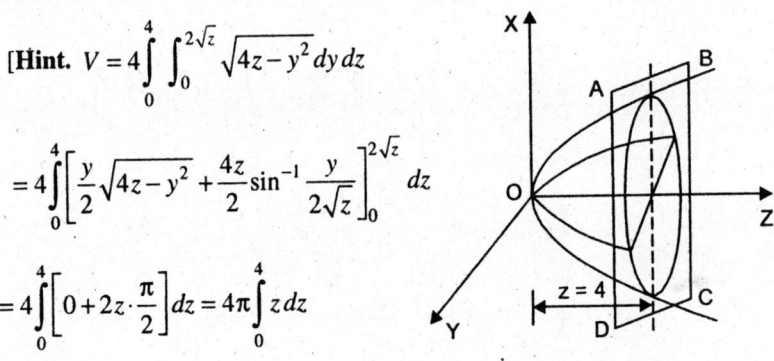

[**Hint.** $V = 4\int_0^4 \int_0^{2\sqrt{z}} \sqrt{4z - y^2}\, dy\, dz$

$= 4\int_0^4 \left[\dfrac{y}{2}\sqrt{4z - y^2} + \dfrac{4z}{2}\sin^{-1}\dfrac{y}{2\sqrt{z}}\right]_0^{2\sqrt{z}} dz$

$= 4\int_0^4 \left[0 + 2z\cdot\dfrac{\pi}{2}\right] dz = 4\pi\int_0^4 z\, dz$

$= 32\pi$ cubic units.]

Fig. 13.51.

35. Find the volume of the cylindrical column standing on the area common to the parabolas $y^2 = x$, $x^2 = y$ and cut off by the surface $z = 12 + y - x^2$.
(DCE-2003; UPTU 2001) **Ans.** 49/140.

13.9 Triple integrals

Notation: The symbol $\int\limits_{a}^{b}\int\limits_{c}^{d}\int\limits_{h}^{k}f(x,\,y,\,z)\,dz\,dy\,dx$ is called a *triple integral*.

We integrate $f(x, y, z)$ with respect to z (treating x and y as constants), between the limits h and k. The resulting expression is then integrated with respect to y (treating x as a constant), between the limits c and d. The result is integrated with respect to x between the limits a and b.

Thus

$$\int\limits_{a}^{b}\int\limits_{c}^{d}\int\limits_{h}^{k}f(x,\,y,\,z)\,dz\,dy\,dx = \int\limits_{a}^{b}\left[\int\limits_{c}^{d}\left(\int\limits_{h}^{k}f(x,\,y,\,z)\,dz\right)dy\right]dx$$

or

$$\int\limits_{a}^{b}dx\int\limits_{c}^{d}dy\int\limits_{h}^{k}f(x,\,y,\,z)\,dz$$

Remark. h, k may be constants or functions of x and y.

Similarly c, d may be constants or functions of x.

Example 13.29 Evaluate $\int\limits_{0}^{1}\int\limits_{0}^{1}\int\limits_{0}^{x}(x-2y+z)\,dz\,dy\,dx$.

Solution. The integral $= \int\limits_{0}^{1}\int\limits_{0}^{1}\left[(x-2y)z+\dfrac{z^2}{2}\right]_{0}^{x}dy\,dx$

$$= \int\limits_{0}^{1}\int\limits_{0}^{1}\left[(x-2y)x+\frac{x^2}{2}\right]dy\,dx = \int\limits_{0}^{1}\int\limits_{0}^{1}\left(\frac{3x^2}{2}-2xy\right)dy\,dx$$

$$= \int\limits_{0}^{1}\left[\frac{3x^2}{2}\cdot y-2x\frac{y^2}{2}\right]_{0}^{1}dx = \int\limits_{0}^{1}\left[\frac{3x^2}{2}\cdot 1-2x\cdot\frac{1}{2}\right]dx$$

$$= \int\limits_{0}^{1}\left(\frac{3x^2}{2}-x\right)dx = \left[\frac{3}{2}\cdot\frac{x^3}{3}-\frac{x^2}{2}\right]_{0}^{1} = 0.$$

Example 13.30 Evaluate $\int_{0}^{1}\int_{0}^{1-x}\int_{0}^{1-x^2-y^2}dz\,dy\,dx$.   (A.M.I.E. W-1998)

Solution. The given integral $= \int_0^1 \int_0^{1-x} \left(1-x^2-y^2\right) dy\, dx.$

$$= \int_0^1 \left[y - x^2 y - \frac{y^3}{3} \right]_0^{1-x} dx = \int_0^1 \left\{ (1-x) - x^2(1-x) - \frac{1}{3}(1-x)^3 \right\} dx$$

$$= \left[x - \frac{x^2}{2} - \frac{x^3}{3} + \frac{x^4}{4} + \frac{1}{3}\frac{(1-x)^4}{4} \right]_0^1 = 1 - \frac{1}{2} - \frac{1}{3} + \frac{1}{4} - \frac{1}{12} = \frac{1}{3}.$$

Example 13.31 Evaluate $\displaystyle\int_0^{\pi/2} \int_0^{a\sin\theta} \int_0^{(a^2-r^2)/a} r\, dz\, dr\, d\theta.$

(V.T.U. 2007; Gulbarga, 1999)

Solution. The given integral $= \displaystyle\int_0^{\pi/2} \left[\int_0^{a\sin\theta} \left(\int_0^{(a^2-r^2)/a} r\, dz \right) dr \right] d\theta$

$$= \int_0^{\pi/2} \left[\int_0^{a\sin\theta} r[z]_0^{(a^2-r^2)/a}\, dr \right] d\theta = \int_0^{\pi/2} \left[\int_0^{a\sin\theta} r\left(\frac{a^2-r^2}{a}\right) dr \right] d\theta$$

$$= \frac{1}{a} \int_0^{\pi/2} \left[\frac{a^2 r^2}{2} - \frac{r^4}{4} \right]_0^{a\sin\theta} d\theta = \frac{1}{a} \int_0^{\pi/2} \frac{a^4}{4}\left(2\sin^2\theta - \sin^4\theta\right) d\theta$$

$$= \frac{a^3}{4}\left[2\cdot\frac{1}{2}\cdot\frac{\pi}{2} - \frac{3}{4}\cdot\frac{1}{2}\frac{\pi}{2} \right] = \frac{5\pi a^3}{64}.$$

13.10 Volume under a surface (Triple integration)

The volume of a vertical column (that is, a column whose lateral surface is generated by lines parallel to z-axis) whose upper end is in the surface $z = f(x, y)$ and whose lower end is in the *XOY* plane is also given by the triple integral

$$\iint_R \left[\int_0^{f(x,y)} dz \right] dy\, dx,$$ where *R* is the lower base of the column. Obviously *R* is

the orthogonal projection of the upper end *S* of the column in *XOY* plane.

If the upper end of the vertical column lies in the surface $z = f_2(x, y)$ and lower end in $z = f_1(x, y)$ then the volume is given by $V = \displaystyle\iint_R \left[\int_{f_1}^{f_2} dz \right] dy\, dx,$ where *R* is the projection of the upper end (or lower end) of the column in the *XOY* plane.

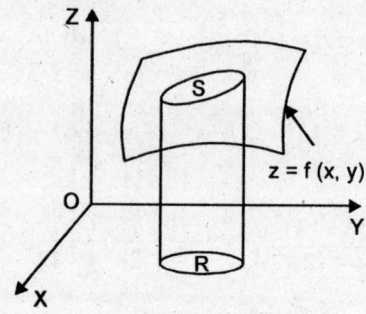

Fig. 13.52.

Note. In case the lateral surface of the column is generated by lines parallel to x-axis and one end lies in the surface $x = \phi(y, z)$ and base R in the YOZ plane then

$$V = \iint_R \left[\int_0^{\phi(y, z)} dx \right] dy\, dz.$$

When the base of the column lies in XOZ plane and the other end in the surface

$$y = F(x, z) \text{ then } V = \iint_R \left[\int_0^{F(x, z)} dy \right] dx\, dz.$$

13.11 Cylindrical and Spherical Coordinates

In this section we will discuss two new types of coordinate systems in 3-space that are often more useful than rectangular coordinate systems. These are: the cylindrical coordinate system and the spherical coordinate system. Cylindrical coordinates simplify the equations of cylinders. Spherical coordinates simplify the equations of spheres and cones.

Cylindrical Coordinates (r, θ, z). See Fig. 13-53.

Cylindrical coordinates represent a point P in space by ordered triples (r, θ, z) in which r and θ are polar coordinates for the vertical projection of P on the xy-plane, and z is the rectangular vertical coordinate.

The values of x, y, r and θ in rectangular and cylindrical coordinates are related by the following equations.

$$x = r\cos\theta,\ y = r\sin\theta,\ z = z,\ r^2 = x^2 + y^2,\ \tan\theta = y/x.$$

In cylindrical co-ordinates, the equation $r = r_0$ is a right circular cylinder of radius r_0 centered on the z-axis, where $0 \le \theta < 2\pi$.

Transformation equations: $x = r\cos\theta,\ y = r\sin\theta,\ z = z;$

Element of Volume: $dV = r\, dr\, d\theta\, dz.$

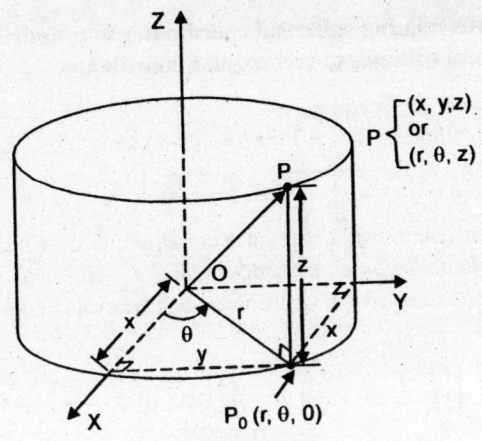

Fig. 13.53.

Sometimes a triple integral that is difficult to integrate in rectangular coordinates, can be evaluated more easily by making the substitution $x = r \cos\theta$, $y = r \sin\theta$, $z = z$ to convert it to an integral in cylindrical coordinates.

Under such a substitution, a rectangular triple integral can be expressed as an iterated integral in cylindrical coordinates as

$$\iiint_R f(x, y, z)\,dV = \iiint_{\substack{\text{appropriate}\\ \text{limits}}} f(r\cos\theta, r\sin\theta, z)\,dz \cdot r\,dr\,d\theta$$

Spherical Coordinates (ρ, ϕ, θ). See Fig. 13.54

Spherical coordinates represent a point P in space by an ordered triples (ρ, ϕ, θ) in which the first coordinate, ρ is the distance from P to the origin, the second coordinate, ϕ, is the angle $\overrightarrow{OP}$ makes with the positive z-axis $(0 \le \phi \le \pi)$, and the third coordinate is the angle θ as measured in cylindrical coordinates.

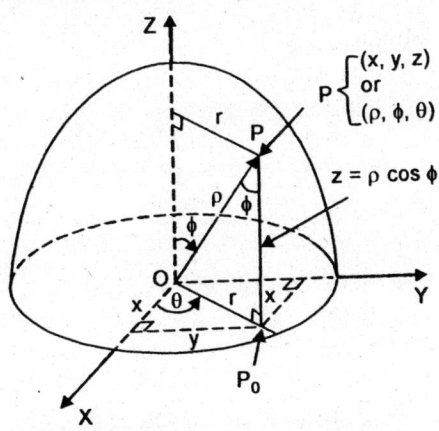

$z = \rho \cos\phi$

The equation $\rho = \rho_0$ describes the sphere of radius ρ_0 centered at the origin. The equation $\phi = \phi_0$ describes a single cone whose vertex lies at the origin and whose axis lies along the z-axis. If ϕ_0 is greater than $\pi/2$, the cone $\phi = \phi_0$ opens downward.

Fig. 13.54. Spherical coordinates are measured with a distance and two angle $(0 \le \theta \le 2\pi, 0 \le \phi \le \pi)$

13.11.1 Equations relating spherical coordinates to cylindrical coordinates and cylindrical coordinates to rectangular coordinates

$r = \rho \sin\phi,\ \theta = \theta,\ z = \rho \cos\phi$

$x = r \cos\theta = \rho \sin\phi \cos\theta,\ y = r \sin\theta = \rho \sin\phi \sin\theta$

$$\rho = \sqrt{x^2 + y^2 + z^2} = \sqrt{r^2 + z^2}.$$

Transformation equations: $x = \rho \sin\phi \cos\theta$, $y = \rho \sin\phi \sin\theta$, $z = \rho \cos\phi$.
Element of volume: $dV = \rho^2 \sin\phi\, d\rho\, d\phi\, d\theta$.

A triple integrals in spherical coordinates can be evaluated as an iterated integral of the form

$$\iiint_D f(\rho, \phi, \theta)\, dV = \iiint_{\substack{appropriate \\ limits}} f(\rho, \phi, \theta)\rho^2 \sin\phi\, d\rho\, d\phi\, d\theta$$

To evaluate $\displaystyle\iiint_D f(\rho, \phi, \theta)\, dV$ over a region D in space in spherical coordinates, integrating first with respect to ρ, then with respect to ϕ, and finally with respect to θ.

13.12 Illustrative Example

Example 13.32. Evaluate the triple integral $\displaystyle\iiint_R (x^2 + y^2 + z^2)\, dx\, dy\, dz$, the region R bounded by $x + y + z = a\ (a > 0)$, $x = 0$, $y = 0$, $z = 0$. *(AMIE. W-2002)*

Solution. $\displaystyle\iiint_R (x^2 + y^2 + z^2)\, dz\, dy\, dx = \int_{x=0}^{a} \int_{y=0}^{a-x} \int_{z=0}^{a-x-y} (x^2 + y^2 + z^2)\, dz\, dy\, dx$

$$= \int_{x=0}^{a} \int_{y=0}^{a-x} \left[x^2 z + y^2 z + \frac{z^3}{3} \right]_{z=0}^{a-x-y} dy\, dx$$

$$= \int_{x=0}^{a} \int_{y=0}^{a-x} \left\{ x^2(a-x) - x^2 y + (a-x)y^2 - y^3 + \frac{(a-x-y)^3}{3} \right\} dy\, dx$$

$$= \int_{x=0}^{a} \left[x^2(a-x)y - \frac{x^2 y^2}{2} + \frac{(a-x)y^3}{3} - \frac{y^4}{4} - \frac{(a-x-y)^4}{12} \right]_{y=0}^{a-x} dx$$

$$= \int_0^a \left\{ x^2 (a-x)^2 - \frac{x^2 (a-x)^2}{2} + \frac{(a-x)^4}{3} - \frac{(a-x)^4}{4} + \frac{(a-x)^4}{12} \right\} dx$$

$$= \int_0^a \left\{ \frac{x^2 (a-x)^2}{2} + \frac{(a-x)^4}{6} \right\} dx = \frac{1}{2} \int_0^a (a^2 x^2 - 2ax^3 + x^4) dx + \frac{1}{6} \int_0^a (a-x)^4 dx$$

$$= \frac{1}{2} \left[\frac{a^2 x^3}{3} - \frac{ax^4}{2} + \frac{x^5}{5} \right]_0^a - \frac{1}{6} \left[\frac{(a-x)^5}{5} \right]_0^a = \frac{a^5}{6} - \frac{a^5}{4} + \frac{a^5}{10} + \frac{a^5}{30} = \frac{a^5}{20}.$$

Example 13.33 Evaluate the integral $\iiint\limits_T z\, dx\, dy\, dz$, where **T** is the region

bounded by the cone $x^2 \tan^2 \alpha + y^2 \tan^2 \beta = z^2$ and the planes $z = 0$ to $z = h$
in the first octant. *(IETE, Dec. 2009, Dec. 2007)*

Solution. The required region can be written as

$$0 \le z \le \sqrt{x^2 \tan^2 \alpha + y^2 \tan^2 \beta},\ 0 \le y \le \left(\sqrt{h^2 - x^2 \tan^2 \alpha}\right) \cot \beta,\ 0 \le x \le h \cot \alpha.$$

Therefore, $I = \int_0^{h\cot\alpha} dx \left[\int_0^{\left(\sqrt{h^2 - x^2 \tan^2 \alpha}\right)\cot\beta} \frac{1}{2} (x^2 \tan^2 \alpha + y^2 \tan^2 \beta) dy \right.$

$$= \frac{1}{2} \int_0^{h\cot\alpha} \left[x^2 \sqrt{h^2 - x^2 \tan^2 \alpha}\ \tan^2 \alpha + \frac{1}{3} (h^2 - x^2 \tan^2 \alpha)^{3/2} \right] \cot \beta\, dx \;.$$

Substituting $x \tan \alpha = h \sin t,\ dx = h \cot \alpha \cos t\, dt$. When $x = 0, t = 0$ and when
$x = h \cot \alpha, t = \pi/2$. We obtain

$$I = \frac{\cot \beta}{2} \int_{t=0}^{\pi/2} \left[h^2 \sin^2 t (h\cos t) + \frac{1}{3} (h^3 \cos^3 t) \right] h \cot \alpha \cos t\, dt$$

$$= \frac{h^4}{2} \cot \alpha \cot \beta \left[\int_0^{\pi/2} \left(\sin^2 t \cos^2 t + \frac{1}{3} \cos^4 t \right) dt \right]$$

$$= \frac{h^4}{2} \cot \alpha \cot \beta \left[\int_0^{\pi/2} \left(\sin^2 t - \sin^4 t + \frac{1}{3} \cos^4 t \right) dt \right]$$

$$= \frac{h^4}{2} \cot \alpha \cot \beta \left[\frac{1}{2} \cdot \frac{\pi}{2} - \frac{3 \cdot 1}{4 \cdot 2} \cdot \frac{\pi}{2} + \frac{1}{3} \cdot \frac{3 \cdot 1}{4 \cdot 2} \cdot \frac{\pi}{2} \right] = \frac{h^4}{2} \cot \alpha \cot \beta \left(\frac{\pi}{4} - \frac{3\pi}{16} + \frac{\pi}{16} \right)$$

$$= \frac{\pi h^4}{16} \cot \alpha \cot \beta.$$

Example 13.34 Using triple integral, find the

(a) **volume of the region R bounded by the parabolic cylinder $z = 4 - x^2$ and the plane $x = 0$, $y = 0$, $y = 6$, $z = 0$.**

(b) **volume in the first octant bounded by the co-ordinate planes and the plane $x + 2y + 3z = 4$.**

(c) **volume of the sphere $x^2 + y^2 + z^2 = a^2$.** *(IETE, June 2009)*

Solution. (a) The region R is shown in Fig. 13.55.

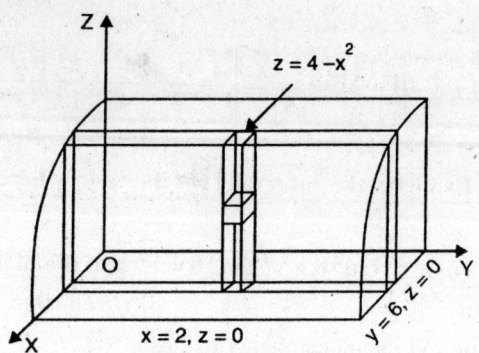

Fig. 13.55.

$$\text{Required volume} = \iiint_R dz\,dy\,dx = \int_{x=0}^{2}\int_{y=0}^{6}\int_{z=0}^{4-x^2} dz\,dy\,dx = \int_{0}^{2}\int_{0}^{6}(4-x^2)\,dy\,dx$$

$$= \int_{0}^{2}\left[4y - x^2 y\right]_{0}^{6} dx = \int_{0}^{2}(24 - 6x^2)\,dx = \left[24x - 2x^3\right]_{0}^{2} = 48 - 16 = 32 \text{ cubic units.}$$

(b) Here $z = (4 - x - 2y)/3$ and R the triangle OAB.

$\therefore$ Volume required $= \iint_R\left[\int_{0}^{z} dz\right] dy\,dx$

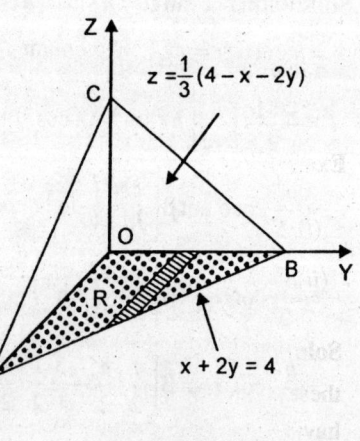

$$= \iint_R z\,dy\,dx = \frac{1}{3}\int_{0}^{4}\int_{0}^{(4-x)/2}(4 - x - 2y)\,dy\,dx$$

$$= \frac{1}{3}\int_{0}^{4}\left[(4-x)y - y^2\right]_{0}^{(4-x)/2} dx$$

Fig. 13.56.

$$= \frac{1}{3} \int_0^4 \left[\frac{1}{2}(4-x)^2 - \frac{1}{4}(4-x)^2 \right] dx$$

$$= \frac{1}{12} \int_0^4 (4-x)^2 \, dx = -\frac{1}{12} \left[\frac{(4-x)^3}{3} \right]_0^4 = \frac{16}{9} \text{ cubic units.}$$

(c) Let the equation of the sphere be $x^2 + y^2 + z^2 = a^2$.

By symmetry the volume of the sphere is 8 times the volume in the first octant.

$$\therefore \quad \text{Volume} = 8 \int_0^a \int_0^{\sqrt{a^2-x^2}} \int_0^{\sqrt{a^2-x^2-y^2}} dz \, dy \, dx$$

(Here R is the quadrant of the circle $x^2 + y^2 = a^2$ in the XY plane).

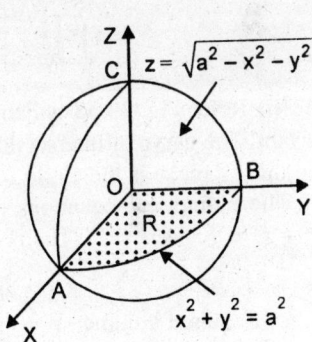

Fig. 13.57.

$$= 8 \int_0^a \int_0^{\sqrt{a^2-x^2}} \sqrt{a^2 - x^2 - y^2} \, dy \, dx$$

$$= 8 \int_0^a \int_0^t \sqrt{t^2 - y^2} \, dy \, dx, \text{ where } t = \sqrt{a^2 - x^2} \qquad \dots(i)$$

$$= 8 \int_0^a \left[\frac{t}{2}\sqrt{t^2 - y^2} + \frac{t^2}{2}\sin^{-1}\frac{y}{t} \right]_0^t dx = 8 \int_0^a \left(\frac{t^2}{2}\sin^{-1} 1 \right) dx = 4 \int_0^a \left(t^2 \cdot \frac{\pi}{2} \right) dx$$

$$= 2\pi \int_0^a (a^2 - x^2) \, dx = 2\pi \left[a^2 x - \frac{x^3}{3} \right]_0^a = \frac{4\pi a^3}{3}.$$

Example 13.35 Find the volume of the solid

(i) **bounded by the surfaces** $z = 4 - x^2 - \frac{1}{4}y^2$, $z = 3x^2 + \frac{1}{4}y^2$,

(ii) **whose upper surface is on the sphere** $x^2 + y^2 + z^2 = 8$ **and whose lower surface is on the paraboloid of revolution** $x^2 + y^2 = 2z$.

Solution. (i) The surfaces are the elliptic paraboloids. Eliminating z between these equations, we obtain $4x^2 + (y^2/2) = 4$ which is the equation of the cylinder having generator parallel to z-axis.

$$\text{Required volume} = \int_{x=-1}^{1} \int_{y=-2\sqrt{2(1-x^2)}}^{2\sqrt{2(1-x^2)}} \int_{z=3x^2+\frac{1}{4}y^2}^{4-x^2-\frac{1}{4}y^2} dz\,dy\,dx$$

$$= 4 \int_{x=0}^{1} \int_{y=0}^{2\sqrt{2(1-x^2)}} \int_{z=3x^2+\frac{1}{4}y^2}^{4-x^2-\frac{1}{4}y^2} dz\,dy\,dx$$

$$= 4 \int_{0}^{1} \int_{0}^{2\sqrt{2(1-x^2)}} \left(4 - 4x^2 - \frac{1}{2}y^2\right) dy\,dx = 4\pi\sqrt{2} \text{ cubic units.}$$

(*ii*) Refer Fig. 13.58 which shows the sphere and the paraboloid in the first octant. The curve of intersection *AB* lies in the plane $z = 2$ and its projection *DE* on the *XY* plane is the circle $x^2 + y^2 = 4$.

The cylindrical equations of given sphere and paraboloid are respectively $r^2 + z^2 = 8$ and $r^2 = 2z$. The polar equation of the circle, $x^2 + y^2 = 4$ is $r = 2$.

$$\therefore \text{ Required volume, } V = \int_{\theta=0}^{2\pi} \int_{r=0}^{2} \int_{z=r^2/2}^{\sqrt{8-r^2}} dz \cdot r\,dr\,d\theta$$

$$= \int_{0}^{2\pi} \int_{0}^{2} [z]_{r^2/a}^{\sqrt{(8-r^2)}} r\,dr\,d\theta$$

$$= \int_{0}^{2\pi} \int_{0}^{2} \left[\sqrt{(8-r^2)} - \frac{r^2}{2} \right] r\,dr\,d\theta$$

$$= \int_{0}^{2\pi} \left[\frac{(8-r^2)^{3/2}}{-2\cdot 3/2} - \frac{r^4}{8} \right]_{0}^{2} d\theta$$

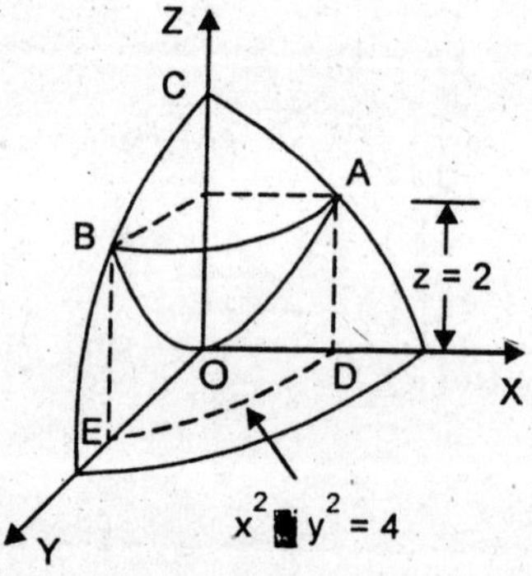

Fig. 13.58.

$$= \int_{0}^{2\pi} \left[-\frac{1}{3}(8) + \frac{16\sqrt{2}}{3} - 2 \right] d\theta = \int_{0}^{2\pi} \frac{2}{3}(8\sqrt{2}-7)d\theta = \frac{4\pi}{3}(8\sqrt{2}-7) \text{ cubic units.}$$

Example 13.35 Use triple integration in spherical coordinates to find the volume of the solid G bounded above by the sphere $x^2 + y^2 + z^2 = 16$ and below by the cone $z = \sqrt{x^2 + y^2}$.

Solution. The solid *G* is shown in Fig. 13.59. In spherical coordinates, the equation of the sphere $x^2 + y^2 + z^2 = 16$ is $\rho = 4$ and the equation of the cone

$z = \sqrt{x^2 + y^2}$ is $\rho \cos\phi = \sqrt{\rho^2 \sin^2\phi \cos^2\theta + \rho^2 \sin^2\phi \sin^2\theta}$, which simplifies to $\rho \cos\phi = \rho \sin\phi$. On dividing both sides by $\rho \cos\phi$, we get $\tan\phi = 1 = \tan \pi/4$. Thus $\phi = \pi/4$. The volume of G is

$$V = \iiint_G dV = \int_{\theta=0}^{2\pi} \int_{\phi=0}^{\pi/4} \int_{\rho=0}^{4} \rho^2 \sin\phi\, d\rho\, d\phi\, d\theta$$

$$= \int_0^{2\pi} \int_0^{\pi/4} \left[\frac{\rho^3}{3}\sin\phi\right]_0^4 d\phi\, d\theta = \int_0^{2\pi}\int_0^{\pi/4} \frac{64}{3}\sin\phi\, d\phi\, d\theta$$

$$= \frac{64}{3}\int_0^{2\pi} [-\cos\phi]_{\phi=0}^{\pi/4}\, d\theta = \frac{64}{3}\int_0^{2\pi}\left(1 - \frac{\sqrt2}{2}\right) d\theta$$

$$= \frac{64\pi}{3}(2 - \sqrt2)\text{ cubic units.}$$

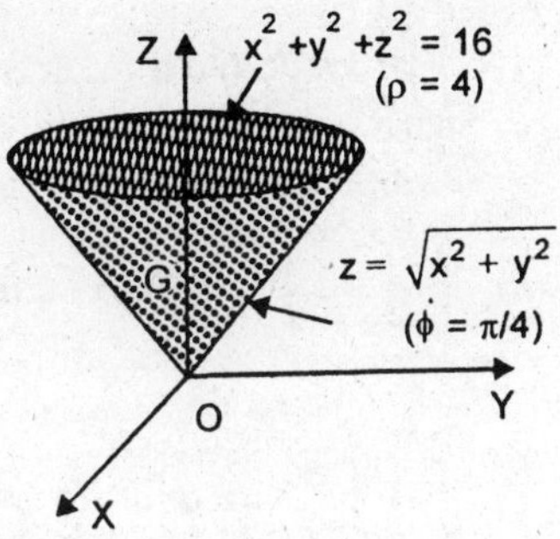

Fig. 13.59.

Exercise 13.3

1. Evaluate the following triple integrals:

(i) $\displaystyle\int_2^3 \int_1^2 \int_2^5 xy^2\, dz\, dy\, dx$,

(ii) $\displaystyle\int_1^2 \int_0^1 \int_{-1}^1 (x^2 + y^2 + z^2)\, dx\, dy\, dz$,

(iii) $\displaystyle\int_{x=0}^1 \int_{y=0}^1 \int_{z=\sqrt{x^2+y^2}}^2 xyz\, dz\, dy\, dx$, (iv) $\displaystyle\int_{-1}^1 dz \int_0^z dx \int_{x-z}^{x+z}(x+y+z)\, dy$, (*IETE, Dec.'08; JNTU*, 2006)

(v) $\displaystyle\int_0^4 \int_0^{2\sqrt z} \int_0^{\sqrt{4z-x^2}} dy\, dx\, dz$, (vi) $\displaystyle\int_0^a \int_0^x \int_0^{x+y} e^{x+y+z}\, dz\, dy\, dx$,

(*Gauhati, 1999*), (*JNTU, 2005*)

(vii) $\displaystyle\iiint_R (x - 2y + z)\, dz\, dy\, dx$, where R is the region determined by

$0 \le x \le 1, 0 \le y \le x^2, 0 \le z \le x + y$

[**Hint.** $\displaystyle\iiint_R f(x, y, z)\, dz\, dy\, dx = \int_0^1 \int_0^{x^2} \int_0^{x+y}(x - 2y + z)\, dz\, dy\, dx$].

(viii) $\displaystyle\int_0^1\int_{y^2}^1\int_0^{1-x} x\,dz\,dx\,dy$,   (ix) $\displaystyle\int_0^{\pi/2}\int_0^{\cos\theta}\int_{r^2}^{r\cos\theta} r\,dz\,dr\,d\theta$,

(x) $\displaystyle\int_0^a\int_0^x\int_0^y xyz\,dz\,dy\,dx$,   (xi) $\displaystyle\int_0^1\int_0^{\sqrt{1-x^2}}\int_0^{\sqrt{1-x^2-y^2}} xyz\,dz\,dy\,dx$,   *(V.T.U., 2003S)*

(xii) $\displaystyle\int_0^1\int_0^{1-y}\int_0^{1-x^2-y^2} dz\,dy\,dx$, (xiii) $\displaystyle\int_1^e\int_1^{\log y}\int_1^{e^x} \log z\,dz\,dx\,dy$,   *(Rohtak, 2005)*

(xiv) $\displaystyle\int_0^{2\pi}\int_0^{\pi/4}\int_0^a r^2\sin\theta\,dr\,d\theta\,d\phi$, where $a=$ constant.

[Hint: *(xiv)* The given integral $=\displaystyle\int_0^{2\pi} d\phi\int_0^{\pi/4} d\theta\int_0^a r^2\sin\theta\,dr$

$=\dfrac{a^3}{3}\displaystyle\int_0^{2\pi} d\phi\int_0^{\pi/4}\sin\theta\,d\theta=\dfrac{a^3}{3}\int_0^{2\pi}\left(-\dfrac{1}{\sqrt 2}+1\right)d\phi=\dfrac{a^3}{3}\cdot\dfrac{\sqrt 2-1}{\sqrt 2}\cdot 2\pi=\dfrac{\pi a^3}{3}\left(2-\sqrt 2\right)].$

Ans. *(i)* 35/2, *(ii)* 6, *(iii)* 3/8, *(iv)* 0, *(v)* 8π, *(vi)* $\dfrac{e^{4a}}{8}-\dfrac{3}{4}e^{2a}+e^a-\dfrac{3}{8}$, *(vii)* 8/35,

(viii) 4/35, *(ix)* $\pi/64$, *(x)* $a^6/48$, *(xi)*1/48, *(xii)*1/3, *(xiii)* $\dfrac{1}{4}\left(e^2-8e+13\right)$.

2. Find the volume of the solid cut off by the surface $z=(x+y)^2$ from the right prism whose base, in the plane $z=0$, is the triangle bounded by the lines $x=0$, $y=0$, $x+y=1$.   **Ans.** 1/4 cubic units.

[Hint. Required volume $=\displaystyle\int_{x=0}^1\int_{y=0}^{1-x}\int_{z=0}^{(x+y)^2} dz\,dy\,dx$]

3. Find the volume bounded by the paraboloid $x^2+y^2=1+z$ and $z=0$

[Hint. Required volume $=\displaystyle\int_{x=-1}^1\int_{y=-\sqrt{(1-x^2)}}^{\sqrt{(1-x^2)}}\int_{z=0}^{x^2+y^2-1} dz\,dy\,dx$]   **Ans.** $\pi/2$ unit^3.

4. Evaluate the triple integral of the function $f(x, y, z) = x^2$ over the region V enclosed by the planes $x = 0$, $y = 0$, $z = 0$, and $x + y + z = a$. **Ans.** $a^5/60$

5. Evaluate $\iiint\limits_R xyz\,dz\,dy\,dx$ (i) R: Region bounded by $x + y + z = 1$ and the

coordinate planes.

(ii) Over the positive octant of the sphere $x^2 + y^2 + z^2 = a^2$.

Ans. (i) $1/720$, (ii) $a^6/48$.

6. Compute $\iiint\limits_D \dfrac{dz\,dy\,dx}{(x+y+z+1)^3}$ if the domain D, of integration is bounded by

the coordinate planes and the plane $x + y + z = 1$. **Ans.** $\dfrac{\log 2}{2} - \dfrac{5}{16}$.

(A.M.I.E.; S-2003; I.E.T.E., Dec. 2003)

7. Find the volume of the solid within the cylinder $x^2 + y^2 = 9$ and between the planes $z = 1$ and $x + z = 5$. **Ans.** $36\,\pi$ cubic units.

8. (a) Prove that $\iiint\limits_V (1+x+y+z)^2\,dz\,dy\,dx = \dfrac{31}{60}$, where V is the region of the

tetrahedron bounded by $x = 0$, $y = 0$, $z = 0$ and $x + y + z = 1$.

(A.M.I.E., W-2003)

(b) Evaluate the integral $\iiint\limits_T y\,dx\,dy\,dz$ where T is the region bounded by the

surfaces $x = y^2$, $x = y+2$, $4z = x^2 + y^2$ and $z = y+3$.

(IETE Dec. 2008)

[**Hint.** Limits of z are from $(x^2+y^2)/4$ to $y + 3$, limits of x from y^2 to $y + 2$ and limits of y from -1 to 2.

$\therefore \iiint\limits_T y\,dx\,dy\,dz = \int\limits_{-1}^{2}\left[\int\limits_{y^2}^{(y+2)}\left[\int\limits_{(x^2+y^2)/4}^{(y+3)} y\,dz\right]dx\right]dy = \dfrac{837}{160}.$]

9. Evaluate $\iiint\limits_D x\,dx\,dy\,dz$, where D is the domain bounded by the tetrahedron

with faces, $x/a + y/b + z/c = 1$, $x = 0$, $y = 0$, $z = 0$. *(A.M.I.E., S-2004)* **Ans.** $a^2bc/24$.

10. Find the volume of that portion of the ellipsoid $x^2/a^2 + y^2/b^2 + z^2/c^2 = 1$ which lies in the first octant. **Ans.** $\pi abc/6$ cubic units.

[**Hint.** $V = \int\limits_0^a \int_0^{b\sqrt{1-x^2/a^2}} \int_0^{c\sqrt{1-x^2/a^2-y^2/b^2}} dz\,dy\,dx.$]

11. Find the volume of the tetrahedron bounded by the coordinate planes $x = 0$,
    $y = 0$, $z = 0$ and the plane $\dfrac{x}{a} + \dfrac{y}{b} + \dfrac{z}{c} = 1$.   (*A.M.I.E. S-2004*)

[**Hint.** Volume, $V = \int\limits_0^a \int_0^{b(1-x/a)} \int_0^{c(1-x/a-y/b)} dz\,dy\,dx$

$= \int\limits_0^a \int_0^{b(1-x/a)} c\left(1 - x/a - y/b\right) dy\,dx = c\int\limits_0^a \int_0^{b(a-x)/a} \left\{\dfrac{1}{a}(a-x) - \dfrac{y}{b}\right\} dy\,dx$

$= c\int\limits_0^a \left[\dfrac{1}{a}(a-x)y - \dfrac{1}{b}\dfrac{y^2}{2}\right]_0^{b(a-x)/a} dx = \dfrac{bc}{2a^2}\int\limits_0^a (a-x)^2\,dx = abc/6$ cubic units.]

12. Evaluate $\iiint \dfrac{dx\,dy\,dz}{\sqrt{1 - x^2 - y^2 - z^2}}$ throughout the volume of the sphere
    $x^2 + y^2 + z^2 = 1$.
                                                                    Ans. π^2.

13. Find the volume of the region above the xy-plane bounded by the parabo-
    loid $z = x^2 + y^2$ and the cylinder $x^2 + y^2 = a^2$.   **Ans.** $\pi a^4/2$ cubic units
    [**Hint.** The volume is most easily found by using cylindrical co-ordinates.
    The equations of the paraboloid and cylinder in cylindrical co-ordinates are
    respectively $z = r^2$ and $r = a$. Required volume = 4 times volume in the first
    octant

$$= 4\int\limits_{\theta=0}^{\pi/2} \int\limits_{r=0}^{a} \int\limits_{z=0}^{r^2} r\,dz\,dr\,d\phi = \pi a^4/2.\big]$$

14. Find the volume of the region bounded by $z = x^2 + y^2$ and $z = 2x$.
                                                                    Ans. $\pi/2$.

15. Find the volume of the paraboloid of revolution $x^2 + y^2 = 4z$ cut off by the
    plane $z = 4$.   **Ans.** 32π cubic units.
    [**Hint.** Required volume

$$= \int\limits_{x=-4}^{4} \int\limits_{y=-\sqrt{16-x^2}}^{\sqrt{16-x^2}} \int\limits_{z=(x^2+y^2)/4}^{z=4} dz\,dy\,dx = 4\int\limits_{0}^{4} \int\limits_{0}^{\sqrt{16-x^2}} \int\limits_{(x^2+y^2)/4}^{4} dz\,dy\,dx.]$$

16. Find the volume of the region bounded by the paraboloids $z = x^2 + y^2$ and

$z = 6 - \dfrac{x^2 + y^2}{2}$. **Ans.** 12π cubic units.

[**Hint.** $V = \displaystyle\int_{x=-2}^{2} \int_{y=-\sqrt{4-x^2}}^{\sqrt{4-x^2}} \int_{x^2+y^2}^{6-\frac{1}{2}(x^2+y^2)} dz\,dy\,dx$

$= \displaystyle\int_{-2}^{2} \int_{-\sqrt{4-x^2}}^{\sqrt{4-x^2}} \left[6 - \frac{1}{2}(x^2 + y^2) - (x^2 + y^2) \right] dy\,dx$

$= 6 \displaystyle\int_{0}^{2} \int_{0}^{\sqrt{4-x^2}} \left[4 - (x^2 + y^2) \right] dy\,dx = 6 \int_{0}^{\pi/2} \int_{0}^{2} (4 - r^2) r\,dr\,d\theta = 12\pi.\,]$

17. Evaluate $\displaystyle\iiint (x + y + z)\,dz\,dy\,dx$ over the tetrahedron bounded by the planes $x = 0,\ y = 0,\ z = 0$ and $x + y + z = 1$. (*Bangalore, 1998S*) **Ans.** 1/4.

18. Evaluate $\displaystyle\iiint z\,dx\,dy\,dz$, over the volume enclosed between the cone $x^2 + y^2 = z^2$ and the sphere $x^2 + y^2 + z^2 = 1$ on the positive side of the xy-plane. [**Hint.** Integrate w.r.t. z and then change to polars]. **Ans.** $\pi/8$.

19. Find the volume of the solid bounded by the sphere $x^2 + y^2 + z^2 = 4$ and the surface of the paraboloid $x^2 + y^2 = 3z$. **Ans.** 19 $\pi/6$ cubic units

[**Hint.** The two surfaces intersect at $z = 1$.

$V = \displaystyle\iiint_{D} dz\,dy\,dx = \iint_{R} dy\,dx \int_{z=\frac{1}{3}(x^2+y^2)}^{\sqrt{(4-x^2-y^2)}} dz = \iint_{R} \left[\sqrt{4 - x^2 - y^2} - \frac{1}{3}(x^2 + y^2) \right] dy\,dx$

Changing to polars, R being the circle $x^2 + y^2 \le 3$.

Therefore, $V = \displaystyle\int_{0}^{2\pi} d\theta \int_{0}^{\sqrt{3}} \left(\sqrt{4 - r^2} - \frac{r^2}{3} \right) r\,dr = \frac{19\pi}{6}.\,]$

20. Find the volume of the solid bounded by the surfaces $z = 0,\ 3z = x^2 + y^2$ and $x^2 + y^2 = 9$. (*IETE June 2005*) **Ans.** 27 $\pi/2$.

13.13 Change of Variables in Multiple Integrals; Jacobian

Sometimes it may not be possible to evaluate a double integral with the help of methods discussed so far. Those methods fail because the integrand and/or, the region of integration happen to be complicated. In such cases certain substitutions of the type $x = f_1(u, v)$, $y = f_2(u, v)$ may

(i) reduce the integrand to a simple form.

(ii) map a complicated region R in xy-plane into a simpler region R' (say a rectangle) in uv-plane, over which the integral can be easily evaluated.

The procedure for evaluation of double integral and triple integral without proof, is given below:

A. Double Integrals

Consider the double integral $\displaystyle\iint_R f(x, y)\,dx\,dy$ ...(13.4)

and let the substitutions be $x = f_1(u, v)$, $y = f_2(u, v)$.   ...(13.5)

Then $dx\,dy$ is given by $dx\,dy = \left|\dfrac{\partial(x, y)}{\partial(u, v)}\right| du\,dv.$

Thus, under the transformation (13.5), (13.4) is transformed as

$$\iint_R f(x, y)\,dx\,dy = \iint_{R'} F(u, v)\left|\dfrac{\partial(x, y)}{\partial(u, v)}\right| du\,dv,$$

that is, the integrand is expressed in terms of u and v, and $dx\,dy$ is replaced by $du\,dv$ times the absolute value of the **Jacobian**.

Remark: Sometimes we use the transformation $u = \phi_1(x, y)$, $v = \phi_2(x, y)$. In some cases it may not be possible to obtain the inverse transformation that is, $x = x(u, v)$, $y = y(u, v)$. In such cases, we get $\dfrac{\partial(u, v)}{\partial(x, y)}$. Then with the help of

the property $\dfrac{\partial(x, y)}{\partial(u, v)} \cdot \dfrac{\partial(u, v)}{\partial(x, y)} = 1$, we obtain $\dfrac{\partial(x, y)}{\partial(u, v)} = 1 \Big/ \dfrac{\partial(u, v)}{\partial(x, y)}.$

B. Triple Integrals

Consider the triple integral $\displaystyle\iiint_V f(x, y, z)\,dx\,dy\,dz.$ Let the transformation be

$$x = f_1(u, v, w),\ y = f_2(u, v, w),\ z = f_3(u, v, w).$$

Then $dx\,dy\,dz = \left|\dfrac{\partial(x, y, z)}{\partial(u, v, w)}\right| du\,dv\,dz,$

and $\displaystyle\iiint_V f(x,\,y,\,z)\,dx\,dy\,dz = \iiint_{V'} F(u,\,v,\,w)\left|\frac{\partial(x,\,y,\,z)}{\partial(u,\,v,\,w)}\right| du\,dv\,dw,$

where V' is the new region of integration.

Note: The remark after case (A) above is applicable in this case also that is, if we find convenient to use the transformation

$$u = \phi_1(x,\,y,\,z),\; v = \phi_2(x,\,y,\,z),\; w = \phi_3(x,\,y,\,z),\; \text{then we get } \frac{\partial(u,\,v,\,w)}{\partial(x,\,y,\,z)}.$$

The value of $\dfrac{\partial(x,\,y,\,z)}{\partial(u,\,v,\,w)}$ is given by $\dfrac{\partial(x,\,y,\,z)}{\partial(u,\,v,\,w)} = 1 \Big/ \dfrac{\partial(u,\,v,\,w)}{\partial(x,\,y,\,z)}.$

13.14 Illustrative Examples

Example 13.37 Evaluate $\displaystyle\iint xy\,dy\,dx$ over the area bounded by $y^2 = 4x$, $y^2 = 8x$, $x^2 = 4y$, $x^2 = 8y$.

Solution. Obviously the region of integration R in the xy-plane is very compli-

cated. By the substitutions $\dfrac{y^2}{x} = u$, $\dfrac{x^2}{y} = v$, the region R maps into region R' of

uv-plane which is bounded by the lines $u = 4$, $u = 8$ and $v = 4$ and $v = 8$. Thus R' is a square.

Also $xy = \left(\dfrac{y^2}{x}\right)\left(\dfrac{x^2}{y}\right) = uv$ and $\dfrac{\partial(u,\,v)}{\partial(x,\,y)} = \begin{vmatrix} -y^2/x^2 & 2y/x \\ 2x/y & -x^2/y^2 \end{vmatrix} = 1 - 4 = -3.$

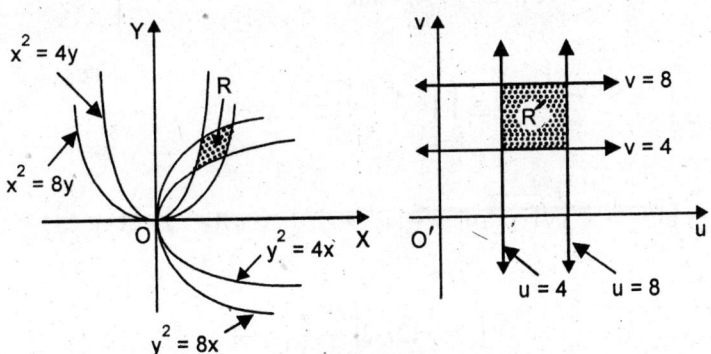

Fig. 13.60.

Therefore, $\left|\dfrac{\partial(u, v)}{\partial(x, y)}\right| = 3 \Rightarrow \left|\dfrac{\partial(x, y)}{\partial(u, v)}\right| = \dfrac{1}{3}$. Hence, $dx\,dy = \dfrac{1}{3}\,du\,dv$.

Thus $\displaystyle\iint_{R} xy\,dy\,dx = \int_{4}^{8}\int_{4}^{8} uv\,\dfrac{1}{3}\,du\,dv = \dfrac{1}{3}\int_{4}^{8}\left[\dfrac{u^2}{2}\right]_{4}^{8} v\,dv$

$$= \dfrac{1}{3}\int_{4}^{8}\dfrac{1}{2}\cdot 48\,v\,dv = 4\left[v^2\right]_{4}^{8} = 192.$$

Example 13.38 Use the transformation

$$x = \dfrac{u}{2}(1+v),\ y = \dfrac{u}{2}(1-v)\ \text{to evaluate}\ \int_{0}^{\infty}\int_{0}^{\infty} e^{-(x+y)}\sqrt{xy}\,dy\,dx.$$

Solution. We have $x + y = u,\ \sqrt{xy} = u\sqrt{1-v^2}\big/2$.

Also $\dfrac{\partial(x, y)}{\partial(u, v)} = \begin{vmatrix} \dfrac{1+v}{2} & \dfrac{u}{2} \\[2mm] \dfrac{1-v}{2} & \dfrac{-u}{2} \end{vmatrix} = -\dfrac{u}{2}$. Therefore, $dy\,dx = \left|\dfrac{\partial(x, y)}{\partial(u, v)}\right| dv\,du = \dfrac{u}{2}dv\,du$.

Obviously, the region of integration R in xy-plane is in the first quadrant.

The line $x = 0$ transforms into $u = 0$, $v = -1$. Also the transform of $y = 0$ is $u = 0$, $v = 1$.

As $x \to \infty$, $y \to \infty$ we find $u \to \infty$, while $-1 < v < 1$.

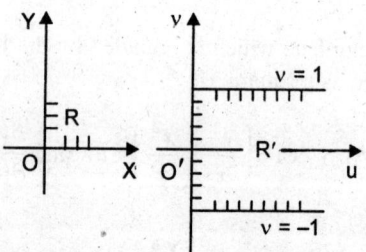

Fig. 13.61.

Hence $\displaystyle\int_{0}^{\infty}\int_{0}^{\infty} e^{-(x+y)}\sqrt{xy}\,dy\,dx = \int_{0}^{\infty}\int_{-1}^{1} e^{-u}\cdot\dfrac{u}{2}\sqrt{1-v^2}\,\dfrac{u}{2}\,dv\,du$

$$= \dfrac{1}{4}\int_{0}^{\infty} u^2 e^{-u}\,du \int_{-1}^{1}\sqrt{1-v^2}\,dv = \dfrac{1}{4}\left[u^2(-e^{-u}) - (2u)(e^{-u}) + 2(-e^{-u})\right]_{0}^{\infty}$$

$$2\left[\dfrac{v}{2}\sqrt{1-v^2} + \dfrac{1}{2}\sin^{-1} v\right]_{0}^{1} = \dfrac{1}{4}(2)\cdot 2\left[\dfrac{1}{2}\cdot\dfrac{\pi}{2}\right] = \dfrac{\pi}{4}.$$

Example 13.39 Evaluate $\displaystyle\int_{y=0}^{2}\int_{x=0}^{y}\left[(x-y)^2+2(x+y)+1\right]^{-1/2}dx\,dy$ **using the**

substitutions $x = u(1 + v)$, $y = v(1 + u)$, $u \geq 0$, $v \geq 0$.

Solution. Substituting $x = u(1 + v)$, $y = v(1 + u)$ in the integrand, we get

$$\left[(x-y)^2+2(x+y)+1\right]^{-1/2}=\left[(u-v)^2+2(u+v+2uv)+1\right]^{-1/2}$$

$$=\left[(u+v+1)^2\right]^{-1/2}=1/(u+v+1).$$

Also $dx\,dy=\left|\dfrac{\partial(x,y)}{\partial(u,v)}\right|dv\,du.$ Now $\dfrac{\partial(x,y)}{\partial(u,v)}=\begin{vmatrix}1+v & u\\ v & 1+u\end{vmatrix}=1+u+v.$

The region of integration R of the xy-plane is bounded by the curves $x = 0$, $x = y$, $y = 0$ and $y = 2$. When $x = 0$, $u = 0$ or $v = -1$. But $v \geq 0 \therefore x = 0 \Rightarrow u = 0$.
When $x = y$, we have $u = v$.
Also $y = 0 \Rightarrow v = 0$ and $y = 2 \Rightarrow v(1 + u) = 2$, which is a rectangular hyperbola having $v = 0$ and $u = -1$ as asymptotes.
The region R' into which R is mapped under the given transformation is as shown.

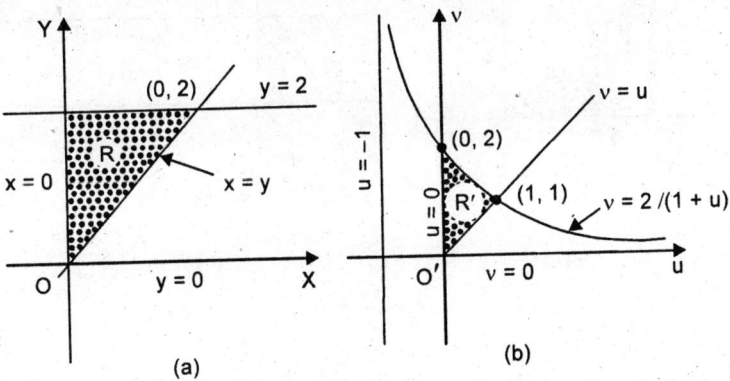

Fig. 13.62.

$$\int_{0}^{2}\int_{0}^{y}\left[(x-y)^2+2(x+y)+1\right]^{-1/2}dx\,dy=\int_{u=0}^{1}\int_{v=u}^{2/(1+u)}\frac{1}{u+v+1}(u+v+1)\,dv\,du$$

$$=\int_{0}^{1}\int_{u}^{2/(1+u)}1\,dv\,du=\int_{0}^{1}\left(\frac{2}{1+u}-u\right)du=\left[2\log(1+u)-\frac{u^2}{2}\right]_{0}^{1}=2\log 2-\frac{1}{2}.$$

Example 13.40 Using the transformation $x + y = u$, $y = uv$,

show that $\displaystyle\int\limits_{x=0}^{1}\int\limits_{y=0}^{1-x} e^{y/(x+y)}dy\,dx = \frac{1}{2}(e-1)$. (*P.T.U 2003*)

Solution. The region R of integration of xy-plane is as shown in Fig. 13.63(a). It

is bounded by the curves $y = 0$, $y = 1 - x$, $x = 0$, $x = 1$. $e^{y/(x+y)} = e^{uv/u} = e^{v}$.

Since $y = uv$, $x = u - y = u - uv$ that is, $x = u(1 - v)$.

Now, $\dfrac{\partial(x, y)}{\partial(u, v)} = \begin{vmatrix} 1-v & -u \\ v & u \end{vmatrix} = u$, Therefore, $dy\,dx = u\,du\,dv$.

When $y = 0$, $u = 0$, $v = 0$. $y = 1 - x$ that is, $x + y = 1$, we have $u = 1$

$x = 0 \Rightarrow u = 0$, $1 - v = 0$ or $v = 1$. $x = 1 \Rightarrow u(1 - v) = 1$, a rectangular hyperbola.

The R' in the uv-plane is bounded by $u = 0$, $v = 0$, $u = 1$, $v = 1$

and $u(1 - v) = 1$. It is a square as shown in the figure 13.63(b).

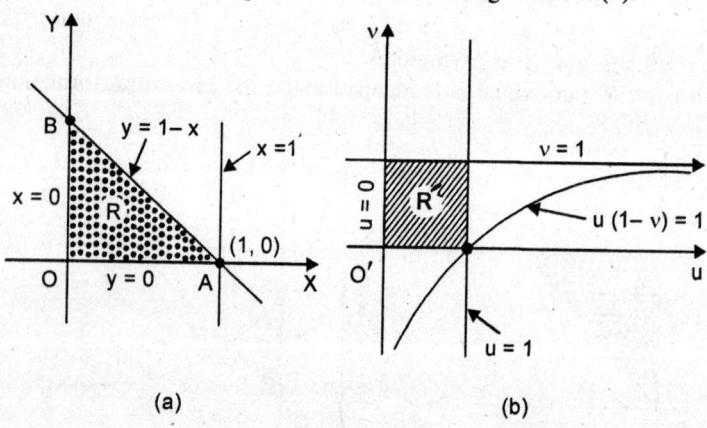

(a)   (b)

Fig. 13.63.

$\therefore \displaystyle\int\limits_{x=0}^{1}\int\limits_{y=0}^{1-x} e^{y/(x+y)}dy\,dx = \int\limits_{v=0}^{1}\int\limits_{u=0}^{1} e^{v}u\,du\,dv = \int\limits_{0}^{1} e^{v}\left[\frac{u^2}{2}\right]_0^1 dv = \frac{1}{2}\int\limits_{0}^{1} e^{v}dv = \frac{1}{2}(e-1)$.

Example 13.41 Evaluate $\displaystyle\iiint\limits_{R}\frac{dx\,dy\,dz}{\left(x^2 + y^2 + z^2\right)^{3/2}}$, where R is the region bounded

by the spheres $x^2 + y^2 + z^2 = a^2$ and $x^2 + y^2 + z^2 = b^2$ ($a > b > 0$).

Solution. Transforming to spherical polar co-ordinates (ρ, ϕ, θ) that is using the
substitutions $x = \rho \sin\phi \cos\theta$, $y = \rho \sin \phi \sin \theta$, and $z = \rho \cos \phi$.

We have by Ex 11.70 page 11.82, $\dfrac{\partial(x, y, z)}{\partial(\rho, \phi, \theta)} = \rho^2 \sin\phi$. Also $x^2 + y^2 + z^2 = \rho^2$.

Therefore, the given integral reduces to

$$\int_{\theta=0}^{2\pi} \int_{\phi=0}^{\pi} \int_{\rho=b}^{a} \frac{1}{\rho^3}\rho^2 \sin\phi\, d\rho\, d\phi\, d\theta = \int_0^{2\pi}\int_0^{\pi}[\log\rho]_b^a \sin\phi\, d\phi\, d\theta = \log\frac{a}{b}\int_0^{2\pi}[-\cos\phi]_0^{\pi}\, d\theta$$

$$= \log\left(\frac{a}{b}\right)2\cdot 2\pi = 4\pi\log\left(\frac{a}{b}\right).$$

Exercise 13.4

Evaluate the following problems 1 to 3.

1. $\displaystyle\iint(x^4 - y^4)\,dx\,dy$ in the first quadrant in which $1 \le x^2 - y^2 \le 2;\ 1 \le xy \le 2$.

   [**Hint.** Let $x^2 - y^2 = u,\ xy = v$]. **Ans. 3/4.**

2. $\displaystyle\iint_R \cos\left(\frac{x-y}{x+y}\right)dx\,dy$, where R be the region bounded by $x + y = 1,\ x = 0$,

   $y = 0$. [**Hint.** Let $x - y = u,\ x + y = v$]   **Ans.** $(\sin 1)/2$.

3. $\displaystyle\iint y^3\, dx\,dy$ over the area enclosed by $y^2 = x,\ y^2 = 2x,\ x^2 = y,\ x^2 = 3y$

   Ans. 28/9.

4. Using the transformation $x + y = u,\ y = uv$, compute the integral

   $\displaystyle\iint \sqrt{xy(1-x-y)}\,dx\,dy$, taken over the area of the triangle with sides $x = 0$,

   $y = 0$, and $x + y = 1$.   (*IETE Dec. 2009*) **Ans.** $2\pi/105$.

5. Using the transformation $u = (x + y)/2,\ v = (x - y)/2$, evaluate the integral

   $\displaystyle\iint_R \sin\frac{1}{2}(x+y)\cos\frac{1}{2}(x-y)\,dx\,dy$, over the triangular region R with

   vertices $(0, 0),\ (2, 0)$ and $(1, 1)$.   **Ans.** $(2-\sin 2)/2$.

6. Evaluate the integral $\displaystyle\int_0^{\infty}\int_0^{\infty} e^{-(x+y)}\sin\frac{\pi y}{x+y}\,dx\,dy$, by means of transformation

   $u = x + y$ and $v = y$.   (*I.E.T.E., Dec. 1996*) **Ans.** $2/\pi$.

7. Evaluate $\displaystyle\iint(x^2 + y^2)\,dx\,dy$, over the area bounded by $x^2 - y^2 = 1$,

   $x^2 - y^2 = 9,\ xy = 2$ and $xy = 4$.   **Ans. 8.**

8. Using the transformation $u = y - x$, $v = y + x$, evaluate the integral $\iint_S e^{(y-x)/(y+x)} dx\, dy$, where S is the triangle bounded by the line $x + y = 2$ and the two coordinate axes.   **Ans.** $(e - e^{-1})$.

[**Hint.** Solving for x and y, we find $x = (v-u)/2$ and $y = (v+u)/2$.

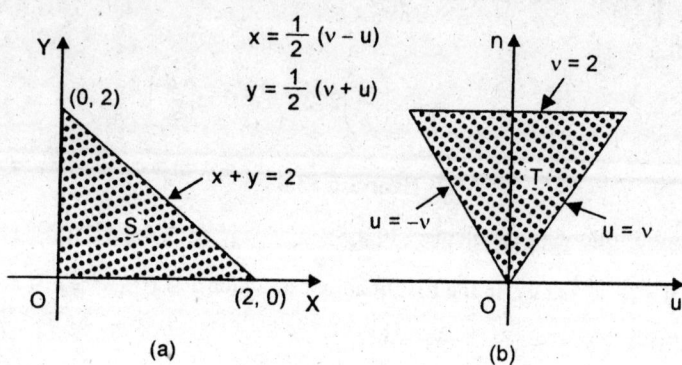

Fig. 13.64. Mapping by a linear transformation

The Jacobian determinant is $J(u, v) = -1/2$, $|J| = 1/2$.

$$\iint_S e^{(y-x)/(y+x)} dx\, dy = \frac{1}{2} \iint_T e^{u/v} du\, dv = \frac{1}{2} \int_{v=0}^{2} \int_{u=-v}^{v} e^{u/v}\, du\, dv = e - e^{-1}.]$$

9. Use a suitable linear transformation to evaluate the double integral $\iint_S (x-y)^2 \sin^2(x+y) dx\, dy$, where S is the rhombus with successive vertices at $(\pi, 0)$, $(2\pi, \pi)$, $(\pi, 2\pi)$, $(0, \pi)$.   **Ans.** $\pi^4/3$.

[**Hint.** The region S is given in fig. 13.65. The equations of the sides AB, BC, CD and DA are respectively $x - y = \pi$, $x + y = 3\pi$, $x - y = -\pi$ and $x + y = \pi$. Substitute $y - x = u$ and $y + x = v$, we obtain

$$I = \iint_S (x-y)^2 \sin^2(x+y) dx\, dy$$

$$= \frac{1}{2} \int_{\pi}^{3\pi} \int_{-\pi}^{\pi} u^2 \sin^2 v\, du\, dv.]$$

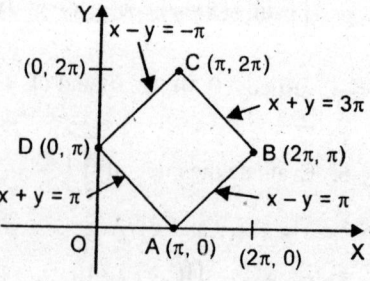

Fig. 13.65.

10. Compute the double integral $\iint\limits_{D}(y-x)dx\,dy$ over the region D in the xy-plane bounded by the straight lines

$y = x+1, \; y = x-3, \; y = (-x/3)+(7/3), \; y = (-x/3)+5.$ **Ans. −8.**

[**Hint.** Put $u = y - x, \; v = y+(x/3), |J| = 3/4.$

Therefore, $\iint\limits_{D}(y-x)dx\,dy = \int\limits_{7/3}^{5}\int\limits_{-3}^{1}\dfrac{3}{4}u\,du\,dv = -8$].

11. Evaluate $\int\limits_{y=0}^{4}\int\limits_{x=y/2}^{(y/2)+1}\dfrac{2x-y}{2}dx\,dy$ by applying the transformation

$u = (2x-y)/2, \; v = y/2.$ **Ans. 2.**

12. Evaluate $\int_{0}^{1}\int_{0}^{1-x}\sqrt{x+y}\,(y-2x)^2\,dy\,dx,$ by applying the transformation

$u = x + y$ and $v = y - 2x.$ **Ans. 2/9.**

13. Evaluate $\iint\limits_{R}e^{xy}dy\,dx,$ where R is the region enclosed by the lines

$y = \dfrac{x}{2}, \; y = x$ and the hyperbolas $y = \dfrac{1}{x}, \; y = \dfrac{2}{x}.$

[**Hint.** Use $u = \dfrac{y}{x}, v = xy.$] **Ans.** $\left[(e^2 - e)\log 2\right]/2.$

14. Using the transformation $u = x+y$ and $v = x-2y,$ evaluate

$\iint\limits_{R}(x+y)^2\,dx\,dy,$ where R is the parallelogram in the xy-plane with vertices

$(1, 0), (3, 1), (2, 2)$ and $(0, 1)$. *(U.P.T.U-2004; Andhra, 1999)* **Ans. 21.**

15. Evaluate the integral $\iiint\left(x^2 + y^2 + z^2\right)dx\,dy\,dz$ taken over the volume

enclosed by the sphere $x^2 + y^2 + z^2 = 1.$ **Ans. 4π/5.**
[**Hint.** Changing to spherical polar coordinates].

16. Evaluate $\iiint\limits_{V}dz\,dy\,dx,$ where V is the volume bounded by the ellipsoid

$(2x + y + z)^2 + (3x - y + z)^2 + 4(x + 2y)^2 = 16$ **Ans. 32π/3.**

Chapter 14

Vector Calculus

14.1 Introduction

In this chapter, we shall study the vector differential and integral calculus. We call this study as vector analysis. We first introduce few concepts..

Scalar function: If to each point P in a region of space R, there corresponds a scalar f(P), then f is called a *scalar function of position* or *scalar point function* and we say that scalar field f has been defined in R. Examples, are

(i) The distance $f(P)$ of any point P from a fixed point P_0 in space is a scalar field. This is certainly independent of the choice of the coordinate system.

(ii) The temperature T at any point on the earth's surface at a certain time defines a scalar field.

(iii) The density at any point of a certain body occupying given region is a scalar field.

(iv) $\phi\,(x,\,y,\,z) = 3xy^2z + x^2y$ defines a scalar field.

Vector function: If to each point P in a region R of space there corresponds a vector $\mathbf{v}(P)$, then $\mathbf{v}$ is called a *vector function of position* or *vector point function* and we say that a vector field $\mathbf{v}$ has been defined in R. Examples are

(i) the velocity at each point of a fluid in motion is a vector field.

(ii) $\mathbf{v} = \mathbf{v}(P) = v_1\mathbf{i} + v_2\mathbf{j} + v_3\mathbf{k}$ defined at each point P is called a vector function.

(iii) gravitational and electromagnetic force fields etc.

If we introduce Cartesian co-ordinates x, y, z then the value of the scalar point function f at a point $P(x, y, z)$ may be written as $f(x, y, z)$ but we should bear in mind that this value is independent of particular choice of axes of co-ordinates and depends only on the position of P. In order to indicate this fact we sometimes write $f(P)$ instead $f(x, y, z)$.

Similarly at a point $P(x, y, z)$ the vector point function $\mathbf{v}$ may be written in the form:

$\mathbf{v}(x, y, z) = v_1(x, y, z)\mathbf{i} + v_2(x, y, z)\mathbf{j} + v_3(x, y, z)\mathbf{k}$ bearing in mind that $\mathbf{v}$ depends only on the position of P and defines the same vector for every choice of coordinate system. This fact in indicated symbolically by writing $\mathbf{v}(P)$ in place of $\mathbf{v}(x, y, z)$.

If the scalar and vector fields depend on time also, then we denote them as $f(P, t)$ and $\mathbf{v}(P, t)$ respectively. Both the fields are independent of the choice of the co-ordinate systems.

Level Surfaces: If a surface $f(x, y, z) = c = $ constant, is drawn through any point P of the region such that at each point on the surface, the function has the same value as at P then such a surface is known as *level surface* through P. For different values of c, we obtain different surfaces. For example, if $f(x, y, z)$ represents temperature in a medium, then $f(x, y, z) = c$ represents a surface on which the temperature is a constant c. Such surfaces are called *iso thermal* surfaces.

14.2 Parametric Representation of Curves

A curve C in the two dimensional xy-plane can be parametrised by $x = x(t)$, $y = y(t)$, $a \le t \le b$. Then, the position vector of a point P on the curve C can be written as

$$\mathbf{r}(t) = x(t)\mathbf{i} + y(t)\mathbf{j}, \qquad ...(14.1)$$

where t is called the parameter of the representation. The parameter t may be time or something else. Therefore, the position vector of a point on a curve defines a vector function, (Fig. 14.1). Similarly, a three-dimensional curve or a space curve C can be parametrised as

$$\mathbf{r}(t) = x(t)\mathbf{i} + y(t)\mathbf{j} + z(t)\mathbf{k}, \ a \le t \le b \qquad ...(14.2)$$

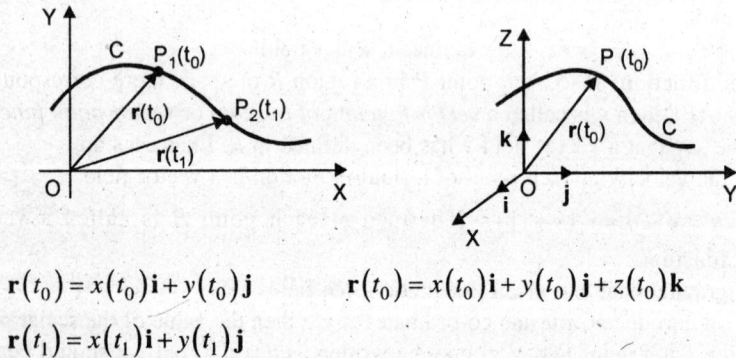

$$\mathbf{r}(t_0) = x(t_0)\mathbf{i} + y(t_0)\mathbf{j} \qquad \mathbf{r}(t_0) = x(t_0)\mathbf{i} + y(t_0)\mathbf{j} + z(t_0)\mathbf{k}$$
$$\mathbf{r}(t_1) = x(t_1)\mathbf{i} + y(t_1)\mathbf{j}$$

Fig. 14.1. Position vector of a point.

The following are the parametric forms of some of the curves:

1. Straight line: A straight line L which passes through the point A with position vector $\mathbf{a}$ in the direction of a constant vector $\mathbf{b}$ can be represented in the form:

$$\mathbf{r}(t) = \mathbf{a} + t\mathbf{b} = (a_1 + tb_1)\mathbf{i} + (a_2 + tb_2)\mathbf{j} + (a_3 + tb_3)\mathbf{k} \qquad ...(14.3)$$

2. Ellipse, Circle: The vector function

$$r(t) = a\cos t\,\mathbf{i} + b\sin t\,\mathbf{j}, \; 0 \le t \le 2\pi$$

$$...(14.4)$$

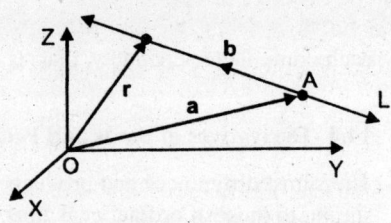

represents an ellipse, $\dfrac{x^2}{a^2} + \dfrac{y^2}{b^2} = 1$, $z = 0$,

in the xy- plane with centre at the origin and principal axes in the direction of the x and y axes.

Fig. 14.2. Parametric representation of a straight line.

If $b = a$, then (14.4) represent a circle of radius 'a'.

3. Parabola: The vector function $\mathbf{r}(t) = at^2\mathbf{i} + 2at\mathbf{j}$ ...(14.5)

represent a parabola in the xy-plane in standard form.

4. Circle in a plane in 3-dimensions: Consider the circle having centre at origin and radius 'a', which lies in the plane $z = d$. The parametric representation of the vector is

$$\mathbf{r}(t) = a\cos t\,\mathbf{i} + b\sin t\,\mathbf{j} + d\,\mathbf{k}. \qquad ....(14.6)$$

14.3 Derivative of a Vector Function

If to each value of a scalar variable t there corresponds a value of a vector $\mathbf{r}$, then $\mathbf{r}$ is called a *vector function* of t and is denoted by $\mathbf{r}(t)$.

Let the variable t be given a small increment Δt and let $\Delta \mathbf{r}$ be the corresponding change in $\mathbf{r}$ so that

$$\Delta \mathbf{r} = \mathbf{r}(t + \Delta t) - \mathbf{r}(t).$$

Then the derivative of a vector function $\mathbf{r}(t)$ with respect to t is defiend by,

$$\mathbf{r}'(t) = \frac{d\mathbf{r}}{dt} = \lim_{\Delta t \to 0} \frac{\Delta \mathbf{r}}{\Delta t} = \lim_{\Delta t \to 0} \frac{\mathbf{r}(t + \Delta t) - \mathbf{r}(t)}{\Delta t},$$

$$...(14.7)$$

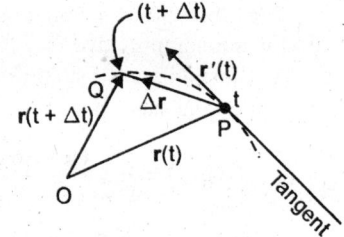

provided that this limit exists. In this case, we also say that $\mathbf{r}(t)$ is *differentiable* or derivable. See Fig. 14.3, the vector $\mathbf{r}'$ (t) is called the derivative of $\mathbf{r}(t)$.

Fig. 14.3. Derivative of a vector function

The derivative of dr/dt with respect to t is defined exactly in same way and is denoted by $\dfrac{d\mathbf{r}'(t)}{dt}$ or $\dfrac{d^2\mathbf{r}}{dt^2}$ and is called the **second derivative** of $\mathbf{r}$. Similarly, higher order derivatives of any kind can be defined.

In three dimensions, we can write $\mathbf{r}(t) = f(t)\mathbf{i} + g(t)\mathbf{j} + h(t)\mathbf{k}$, f, g and h are differentiable functions of t. The derivative $\mathbf{r}'(t)$ is obtained by differentiating each component separately, that is $r'(t) = \left(\dfrac{df}{dt}\right)i + \left(\dfrac{dg}{dt}\right)j + \left(\dfrac{dh}{dt}\right)k.$ ...(14.8)

14.4 Derivatives of Sums and Products of Vectors

The sums, differences and products of vectors can be differentiated by formulas similar to those of ordinary calculus. Thus if f is a scalar function of t and $\mathbf{u}$, $\mathbf{v}$ and $\mathbf{w}$ are vector functions of t then

1. $\dfrac{d}{dt}(\mathbf{u} \pm \mathbf{v}) = \dfrac{d\mathbf{u}}{dt} \pm \dfrac{d\mathbf{v}}{dt}$

2. $\dfrac{d}{dt}(f\mathbf{u}) = f\dfrac{d\mathbf{u}}{dt} + \dfrac{df}{dt}\mathbf{u}$

3. $\dfrac{d}{dt}(\mathbf{u} \cdot \mathbf{v}) = \mathbf{u} \cdot \dfrac{d\mathbf{v}}{dt} + \dfrac{d\mathbf{u}}{dt} \cdot \mathbf{v}$

4. $\dfrac{d}{dt}(\mathbf{u} \times \mathbf{v}) = \mathbf{u} \times \dfrac{d\mathbf{v}}{dt} + \dfrac{d\mathbf{u}}{dt} \times \mathbf{v}$

5. $\dfrac{d}{dt}(\mathbf{u} \cdot \mathbf{v} \times \mathbf{w}) = \dfrac{d\mathbf{u}}{dt} \cdot \mathbf{v} \times \mathbf{w} + \mathbf{u} \cdot \dfrac{d\mathbf{v}}{dt} \times \mathbf{w} + \mathbf{u} \cdot \mathbf{v} \times \dfrac{d\mathbf{w}}{dt}$

6. $\dfrac{d}{dt}[\mathbf{u} \times (\mathbf{v} \times \mathbf{w})] = \dfrac{d\mathbf{u}}{dt} \times (\mathbf{v} \times \mathbf{w}) + \mathbf{u} \times \left(\dfrac{d\mathbf{v}}{dt} \times \mathbf{w}\right) + \mathbf{u} \times \left(\mathbf{v} \times \dfrac{d\mathbf{w}}{dt}\right).$

The order of the vectors must be carefully observed because cross multiplication is not commutative.

As illustrations we shall prove formulae (3), (4) and (5) only. Others can be proved on the same lines.

3. $\dfrac{d}{dt}(\mathbf{u} \cdot \mathbf{v}) = \lim\limits_{\Delta t \to 0} \dfrac{(\mathbf{u} + \Delta\mathbf{u}) \cdot (\mathbf{v} + \Delta\mathbf{v}) - \mathbf{u} \cdot \mathbf{v}}{\Delta t} = \lim\limits_{\Delta t \to 0} \dfrac{\mathbf{u} \cdot \Delta\mathbf{v} + \Delta\mathbf{u} \cdot \mathbf{v} + \Delta\mathbf{u} \cdot \Delta\mathbf{v}}{\Delta t}$

$= \lim\limits_{\Delta t \to 0}\left(\mathbf{u} \cdot \dfrac{\Delta\mathbf{v}}{\Delta t} + \dfrac{\Delta\mathbf{u}}{\Delta t} \cdot \mathbf{v} + \dfrac{\Delta\mathbf{u}}{\Delta t} \cdot \Delta\mathbf{v}\right) = \mathbf{u} \cdot \dfrac{d\mathbf{v}}{dt} + \dfrac{d\mathbf{u}}{dt} \cdot \mathbf{v}$

4. $\dfrac{d}{dt}(\mathbf{u} \times \mathbf{v}) = \lim\limits_{\Delta t \to 0} \dfrac{(\mathbf{u} + \Delta\mathbf{u}) \times (\mathbf{v} + \Delta\mathbf{v}) - \mathbf{u} \times \mathbf{v}}{\Delta t}$

$= \lim\limits_{\Delta t \to 0} \dfrac{\mathbf{u} \times \Delta\mathbf{v} + \Delta\mathbf{u} \times \mathbf{v} + \Delta\mathbf{u} \times \Delta\mathbf{v}}{\Delta t} = \lim\limits_{\Delta t \to 0}\left(\mathbf{u} \times \dfrac{\Delta\mathbf{v}}{\Delta t} + \dfrac{\Delta\mathbf{u}}{\Delta t} \times \mathbf{v} + \dfrac{\Delta\mathbf{u}}{\Delta t} \times \Delta\mathbf{v}\right)$

$= \mathbf{u} \times \dfrac{d\mathbf{v}}{dt} + \dfrac{d\mathbf{u}}{dt} \times \mathbf{v}$

5. $\dfrac{d}{dt}(\mathbf{u}\cdot\mathbf{v}\times\mathbf{w}) = \mathbf{u}\cdot\dfrac{d}{dt}(\mathbf{v}\times\mathbf{w}) + \dfrac{d\mathbf{u}}{dt}\cdot(\mathbf{v}\times\mathbf{w})$ {by formula (3)}

$\qquad\qquad = \mathbf{u}\cdot\left(\dfrac{d\mathbf{v}}{dt}\times\mathbf{w}\right) + \mathbf{u}\cdot\left(\mathbf{v}\times\dfrac{d\mathbf{w}}{dt}\right) + \dfrac{d\mathbf{u}}{dt}\cdot(\mathbf{v}\times\mathbf{w})$ {by formula (4)}

14.5 Derivative of a Scalar Product

Consider $f(t) = \mathbf{a}(t) \cdot \mathbf{b}(t)$, where $\mathbf{a}$ and $\mathbf{b}$ are vector functions of the scalar t. Then we have $f(t + \Delta t) = \mathbf{a}(t + \Delta t) \cdot \mathbf{b}(t + \Delta t)$. Divide both sides of the identity

$$f(t+\Delta t) - f(t) = \mathbf{a}(t+\Delta t)\cdot\{\mathbf{b}(t+\Delta t) - \mathbf{b}(t)\} + \{\mathbf{a}(t+\Delta t) - \mathbf{a}(t)\}\cdot\mathbf{b}(t)$$

by Δt and proceed to the limit as Δt tends to zero. It follows that

$$\frac{d}{dt}(\mathbf{a}\cdot\mathbf{b}) = \mathbf{a}\cdot\frac{d\mathbf{b}}{dt} + \frac{d\mathbf{a}}{dt}\cdot\mathbf{b}.$$

In particular, we have

$$\frac{d}{dt}\mathbf{a}^2 = \frac{d}{dt}(\mathbf{a}\cdot\mathbf{a}) = 2\mathbf{a}\cdot\frac{d\mathbf{a}}{dt}.$$

But $\mathbf{a}^2 = a^2$ and so $\dfrac{d}{dt}a^2 = 2a\dfrac{da}{dt}$. Hence we have $\mathbf{a}\cdot\dfrac{d\mathbf{a}}{dt} = a\dfrac{da}{dt}$.

Further, if $\mathbf{a}$ is a vector of *constant magnitude*, $da/dt = 0$ and so $\mathbf{a} \cdot da/dt = 0$. That is, the derivative of a vector of constant magnitude is either orthogonal to it or zero.

Remarks. A vector changes if either its magnitude or direction changes or both magnitude and direction changes. The necessary and sufficient condition for a vector function to remain constant in magnitude or in direction or in both is given below:

(i) For the vector $\mathbf{v}(t)$ to be constant, $d\mathbf{v}/dt = \mathbf{O}$. That is, the derivative of a constant vector is the zero vector.

(ii) For the vector $\mathbf{v}(t)$ to have a constant magnitude, $\mathbf{v}\cdot\dfrac{d\mathbf{v}}{dt} = O$.

(iii) For the vector $\mathbf{v}(t)$ to have a constant direction, $\mathbf{v}\times\dfrac{d\mathbf{v}}{dt} = \mathbf{O}.$

Example 14.1 (i) Differentiate $\dfrac{t\mathbf{i}+t^2\mathbf{j}}{\left|t\mathbf{i}+t^2\mathbf{j}\right|}$ with respect to t.

(ii) If $\dfrac{d\mathbf{u}}{dt} = \mathbf{w}\times\mathbf{u}$ and $\dfrac{d\mathbf{v}}{dt} = \mathbf{w}\times\mathbf{v}$, show that $\dfrac{d}{dt}(\mathbf{u}\times\mathbf{v}) = \mathbf{w}\times(\mathbf{u}\times\mathbf{v})$.

Solution. (*i*) Since $\left| t\mathbf{i}+t^2\mathbf{j} \right| = \sqrt{t^2+t^4}$.

$$\therefore \quad \frac{d}{dt}\left\{ \frac{t\mathbf{i}+t^2\mathbf{j}}{\left| t\mathbf{i}+t^2\mathbf{j} \right|} \right\} = \frac{d}{dt}\left[\frac{t\mathbf{i}+t^2\mathbf{j}}{\sqrt{t^2+t^4}} \right] = \frac{d}{dt}\left[\frac{\mathbf{i}+t\mathbf{j}}{\sqrt{1+t^2}} \right]$$

$$= \frac{\sqrt{1+t^2}\,(\mathbf{j}) - (\mathbf{i}+t\mathbf{j})\dfrac{t}{\sqrt{1+t^2}}}{1+t^2} = \frac{\left(1+t^2\right)\mathbf{j} - t\mathbf{i} - t^2\mathbf{j}}{\left(1+t^2\right)^{3/2}} = \frac{\mathbf{j} - t\mathbf{i}}{\left(1+t^2\right)^{3/2}}.$$

(*ii*) Now $\dfrac{d}{dt}(\mathbf{u}\times\mathbf{v}) = \dfrac{d\mathbf{u}}{dt}\times\mathbf{v} + \mathbf{u}\times\dfrac{d\mathbf{v}}{dt} = (\mathbf{w}\times\mathbf{u})\times\mathbf{v} + \mathbf{u}\times(\mathbf{w}\times\mathbf{v})$

$$= (\mathbf{w}\cdot\mathbf{v})\mathbf{u} - (\mathbf{u}\cdot\mathbf{v})\mathbf{w} + (\mathbf{u}\cdot\mathbf{v})\mathbf{w} - (\mathbf{u}\cdot\mathbf{w})\mathbf{v}$$

$$[\because \mathbf{A}\times(\mathbf{B}\times\mathbf{C}) = (\mathbf{A}\cdot\mathbf{C})\mathbf{B} - (\mathbf{A}\cdot\mathbf{B})\mathbf{C}]$$

$$= (\mathbf{w}\cdot\mathbf{v})\mathbf{u} - (\mathbf{u}\cdot\mathbf{w})\mathbf{v} = \mathbf{w}\times(\mathbf{u}\times\mathbf{v}).$$

Example 14.2 Find $\mathbf{v}'(t)$ in each of the following cases

(*i*) $\mathbf{v}(t) = (\sin t)\left(5t^2\mathbf{i} + t\mathbf{j} + t^3\mathbf{k}\right)$

(*ii*) $\mathbf{v}(t) = (\sin 2t\,\mathbf{i} - \cos 2t\,\mathbf{j} + t\,\mathbf{k})\cdot(\cos 2t\,\mathbf{i} - \sin 2t\,\mathbf{j} + t^2\mathbf{k})$

(*iii*) $\mathbf{v}(t) = \left(6t^2\mathbf{i} - t\mathbf{j} + 3t^2\mathbf{k}\right)\times\left(t\mathbf{i} + t^2\mathbf{j} + 2t\,\mathbf{k}\right)$.

Solution. (*i*) Using formula 2, Article 14.4, we obtain

$$\mathbf{v}'(t) = (\sin t)\left(5t^2\mathbf{i} + t\mathbf{j} + t^3\mathbf{k}\right)' + (\sin t)'\left(5t^2\mathbf{i} + t\mathbf{j} + t^3\mathbf{k}\right)$$

$$= \sin t\left(10t\mathbf{i} + \mathbf{j} + 3t^2\mathbf{k}\right) + (\cos t)\left(5t^2\mathbf{i} + t\mathbf{j} + t^3\mathbf{k}\right)$$

$$= \left(10t\sin t + 5t^2\cos t\right)\mathbf{i} + (\sin t + t\cos t)\mathbf{j} + \left(3t^2\sin t + t^3\cos t\right)\mathbf{k}$$

(*ii*) Using formula 3, Article 14.4, we obtain

$$\mathbf{v}'(t) = (\sin 2t\,\mathbf{i} - \cos 2t\,\mathbf{j} + t\,\mathbf{k})\cdot\left(\cos 2t\,\mathbf{i} - \sin 2t\,\mathbf{j} + t^2\mathbf{k}\right)'$$

$$+ (\sin 2t\,\mathbf{i} - \cos 2t\,\mathbf{j} + t\,\mathbf{k})'\cdot\left(\cos 2t\,\mathbf{i} - \sin 2t\,\mathbf{j} + t^2\mathbf{k}\right)$$

$$= (\sin 2t\,\mathbf{i} - \cos 2t\,\mathbf{j} + t\,\mathbf{k})\cdot(-2\sin 2t\,\mathbf{i} - 2\cos 2t\,\mathbf{j} + 2t\,\mathbf{k})$$

$$+ (2\cos 2t\,\mathbf{i} + 2\sin 2t\,\mathbf{j} + \mathbf{k})\cdot\left(\cos 2t\,\mathbf{i} - \sin 2t\,\mathbf{j} + t^2\mathbf{k}\right)$$

$$= \left(-2\sin^2 2t + 2\cos^2 2t + 2t^2\right) + \left(2\cos^2 2t - 2\sin^2 2t + t^2\right)$$

$$= 4\left(\cos^2 2t - \sin^2 2t\right) + 3t^2 = 4\cos 4t + 3t^2.$$

(iii) Using formula 4, Article 14.4, we obtain

$$\mathbf{v}'(t) = \left(6t^2\mathbf{i} - t\,\mathbf{j} + 3t^2\mathbf{k}\right) \times \left(t\mathbf{i} + t^2\mathbf{j} + 2t\,\mathbf{k}\right)' + \left(6t^2\mathbf{i} - t\,\mathbf{j} + 3t^2\mathbf{k}\right)' \times \left(t\mathbf{i} + t^2\mathbf{j} + 2t\,\mathbf{k}\right)$$

$$= \left(6t^2\mathbf{i} - t\,\mathbf{j} + 3t^2\mathbf{k}\right) \times \left(\mathbf{i} + 2t\,\mathbf{j} + 2\mathbf{k}\right) + \left(12t\,\mathbf{i} - \mathbf{j} + 6t\,\mathbf{k}\right) \times \left(t\mathbf{i} + t^2\mathbf{j} + 2t\,\mathbf{k}\right)$$

$$= \begin{vmatrix} \mathbf{i} & \mathbf{j} & \mathbf{k} \\ 6t^2 & -t & 3t^2 \\ 1 & 2t & 2 \end{vmatrix} + \begin{vmatrix} \mathbf{i} & \mathbf{j} & \mathbf{k} \\ 12t & -1 & 6t \\ t & t^2 & 2t \end{vmatrix} = \begin{aligned} &\mathbf{i}\left(-2t - 6t^3\right) - \mathbf{j}\left(12t^2 - 3t^2\right) + \mathbf{k}\left(12t^3 + t\right) \\ &+\mathbf{i}\left(-2t - 6t^3\right) - \mathbf{j}\left(24t^2 - 6t^2\right) + \mathbf{k}\left(12t^3 + t\right) \end{aligned}$$

$$= -4t\left(1 + 3t^2\right)\mathbf{i} - 27t^2\mathbf{j} + 2t\left(1 + 12t^2\right)\mathbf{k}.$$

14.6 Geometric Interpretation of r'(t)

Let the vector $\mathbf{r}'(t) \neq \mathbf{0}$ at the point P, then we may draw tangent to the curve C,

at P. The vectors $\Delta\mathbf{r} = \mathbf{r}(t + \Delta t) - \mathbf{r}(t)$ and $\dfrac{\Delta\mathbf{r}}{\Delta t} = \dfrac{\mathbf{r}(t + \Delta t) - \mathbf{r}(t)}{\Delta t}$ are parallel.

(See Fig. 14.4) Now, let $\Delta t \to 0$, the vectors $\mathbf{r}(t)$ and $\mathbf{r}(t + \Delta t)$ become close. If

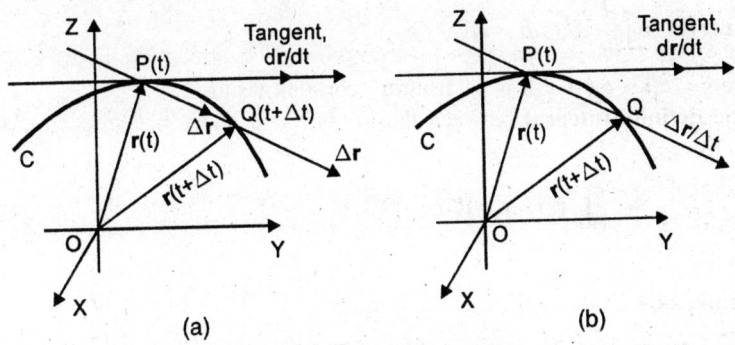

Fig. 14.4. Tangent to a curve

$\lim\limits_{\Delta t \to 0} \dfrac{\Delta\mathbf{r}}{\Delta t}$ exists, then the limiting position of the vector $\dfrac{\Delta\mathbf{r}}{\Delta t}$ is the tangent line to

the curve C at the point P. Thus $\dfrac{d\mathbf{r}}{dt} = \mathbf{r}'(t)$ ...(14.9)

represents the tangent vector to the curve C.

Example 14.3 Find the tangent vector to the curve whose parametric representation is $x = t^2 - 1$, $y = t + 1$, $z = t/(t + 1)$, at $t = 2$. Hence, find the unit tangent vector.

Solution. The vector function that gives the position of a point P on the curve is

$$\mathbf{r}(t) = \left(t^2 - 1\right)\mathbf{i} + (t + 1)\mathbf{j} + \left(\frac{t}{t+1}\right)\mathbf{k}. \text{ Therefore, the tangent vector is}$$

$$\mathbf{r}'(t) = 2t\,\mathbf{i} + \mathbf{j} + \left[1\big/(1+t)^2\right]\mathbf{k} \text{ and } \mathbf{r}'(2) = 4\mathbf{i} + \mathbf{j} + \frac{1}{9}\mathbf{k}.$$

$$|\mathbf{r}'(2)| = \sqrt{16 + 1 + \frac{1}{81}} = \frac{\sqrt{1378}}{9}.$$

Hence, the unit tangent vector at $t = 2$ is $\hat{\mathbf{r}}'(2) = \left[36\mathbf{i} + 9\mathbf{j} + \mathbf{k}\right]\big/\sqrt{1378}$.

14.7 Integrals of Vector Functions

Let $\mathbf{r}(t) = x(t)\mathbf{i} + y(t)\mathbf{j} + z(t)\mathbf{k}$, where its component functions $x(t)$, $y(t)$, and $z(t)$ are each continuous in a specified interval I, then

$$\int \mathbf{r}(t)\,dt = \left(\int x(t)\,dt\right)\mathbf{i} + \left(\int y(t)\,dt\right)\mathbf{j} + \left(\int z(t)\,dt\right)\mathbf{k} \qquad \text{...(14.10)}$$

is called an indefinite integral of $\mathbf{r}(t)$.

If there exists a vector $\mathbf{R}(t)$ such that $\mathbf{r}(t) = \dfrac{d}{dt}\mathbf{R}(t)$ then

$$\int \mathbf{r}(t)\,dt = \int \frac{d}{dt}\mathbf{R}(t)\,dt = \mathbf{R}(t) + \mathbf{c},$$

where $\mathbf{c} = c_1\mathbf{i} + c_2\mathbf{j} + c_3\mathbf{k}$ is an arbitrary constant vector.

The **definite integral** between the limits $t = a$ and $t = b$ in such cases is given by

$$\int_a^b \mathbf{r}(t)\,dt = \mathbf{R}(b) - \mathbf{R}(a) \qquad \text{...(14.11)}$$

Example 14.4 Evaluate the indefinite integral $\displaystyle\int \left(t^2\mathbf{i} - 2t\,\mathbf{j} + \frac{1}{t}\mathbf{k}\right)dt.$

Solution. We have $\displaystyle\int \left(t^2\mathbf{i} - 2t\mathbf{j} + \frac{1}{t}\mathbf{k}\right)dt = \left(\int t^2\,dt\right)\mathbf{i} - \left(\int 2t\,dt\right)\mathbf{j} + \left(\int \frac{1}{t}\,dt\right)\mathbf{k}$

$= \dfrac{t^3}{3}\mathbf{i} - t^2\mathbf{j} + \log_e|t|\,\mathbf{k} + \mathbf{c}$, where $\mathbf{c} = c_1\mathbf{i} + c_2\mathbf{j} + c_3\mathbf{k}$ is an arbitrary vector constant

of integration.

Example 14.5 Evaluate the definite integral $\displaystyle\int_0^1 \left\{t^2\mathbf{i} + e^t\mathbf{j} - (2\cos\pi t)\mathbf{k}\right\}dt.$

Solution. We have

$$\int_0^1 \left\{t^2\mathbf{i} + e^t\mathbf{j} - (2\cos\pi t)\mathbf{k}\right\}dt = \left(\int_0^1 t^2\,dt\right)\mathbf{i} + \left(\int_0^1 e^t\,dt\right)\mathbf{j} - \left(\int_0^1 2\cos\pi t\,dt\right)\mathbf{k}$$

$$= \left[\frac{t^3}{3}\right]_0^1 \mathbf{i} + \left[e^t\right]_0^1 \mathbf{j} - \left[\frac{2}{\pi}\sin \pi t\right]_0^1 \mathbf{k} = \frac{1}{3}\mathbf{i} + (e-1)\mathbf{j}.$$

Example 14.6 Solve the vector initial-value problem for $\mathbf{y}(t)$ by integrating and using the initial conditions to find the constants of integration.

$$\mathbf{y}''(t) = \mathbf{i} + e^t\mathbf{j}, \quad \mathbf{y}(0) = 2\mathbf{i}, \quad \mathbf{y}'(0) = \mathbf{j}.$$

Solution. We have $\mathbf{y}''(t) = \mathbf{i} + e^t\mathbf{j}$. Integrating, both sides of the differential equation with respect to t gives

$$\mathbf{y}'(t) = t\mathbf{i} + e^t\mathbf{j} + \mathbf{A}. \quad ...(i) \qquad \text{As } \mathbf{y}'(t) = \mathbf{j} \text{ for } t = 0, \text{ we get } \mathbf{A} = 0.$$

Therefore (i) takes the form $\mathbf{y}'(t) = t\mathbf{i} + e^t\mathbf{j}$. Integrating again, we have

$$\mathbf{y}(t) = \frac{t^2}{2}\mathbf{i} + e^t\mathbf{j} + \mathbf{B} \quad....(ii)$$

Use initial condition. As $\mathbf{y}(t) = 2\mathbf{i}$, when $t = 0$, we get $\mathbf{B} = 2\mathbf{i} - \mathbf{j}$.

Therefore (ii) takes the form: $\mathbf{y}(t) = \left(\frac{t^2}{2} + 2\right)\mathbf{i} + (e^t - 1)\mathbf{j}$, the required solution of the given differential equation.

Example 14.7 Given that $\mathbf{r}(t) = \begin{cases} 2\mathbf{i} - \mathbf{j} + 2\mathbf{k} & \text{when } t = 2, \\ 4\mathbf{i} - 2\mathbf{j} + 3\mathbf{k} & \text{when } t = 3. \end{cases}$

Show that $\displaystyle\int_2^3 \mathbf{r} \cdot \frac{d\mathbf{r}}{dt}\,dt = 10.$

Solution. Since $\mathbf{r} \cdot \mathbf{r} = r^2$. Therefore, $\dfrac{d}{dt}(\mathbf{r} \cdot \mathbf{r}) = \dfrac{d}{dt}(r^2)$

$$\Rightarrow 2\mathbf{r} \cdot \frac{d\mathbf{r}}{dt} = \frac{d}{dt}(r^2) \text{ or } \mathbf{r} \cdot \frac{d\mathbf{r}}{dt} = \frac{1}{2}\frac{d}{dt}(r^2). \text{ Hence } \int_2^3 \mathbf{r} \cdot \frac{d\mathbf{r}}{dt}\,dt = \frac{1}{2}[r^2]_2^3.$$

Now $r^2 = |\mathbf{r}|^2 = (4)^2 + (-2)^2 + (3)^2 = 29$, when $t = 3$

and $r^2 = (2)^2 + (-1)^2 + (2)^2 = 9$, when $t = 2$.

$$\therefore \quad \int_2^3 \mathbf{r} \cdot \frac{d\mathbf{r}}{dt}\,dt = \frac{1}{2}[29-9] = 10.$$

14.8 Motion of a Body or Particle on a Curve

If $\mathbf{r}(t)$ is the position vector of a body or particle moving along a smooth curve in space, then

$v(t) = \dfrac{d\mathbf{r}}{dt} = \mathbf{r}'(t)$ is the particle's **velocity vector**, tangent to the curve. At any time t, the direction of $\mathbf{v}$ is the **direction of motion**, the magnitude of $\mathbf{v}(t)$ is the particle's **speed**, and the derivative, $d\mathbf{v}/dt$, when it exists, is the particle's **acceteration vector** $\mathbf{a}(t)$. In short

1. Velocity is the derivative of position: $\mathbf{v} = \dfrac{d\mathbf{r}}{dt}$.

2. Speed is the magnitude of velocity: speed $= |\mathbf{v}(t)|$.

3. Acceleration is the derivative of velocity: $\mathbf{a}(t) = \dfrac{d\mathbf{v}}{dt} = \dfrac{d^2\mathbf{r}}{dt^2}$.

4. The vector $\mathbf{v}/|\mathbf{v}|$ is the direction of motion at time t. The velocity of a moving particle can be expressed as the product of its speed and direction.

   Velocity $= |\mathbf{v}|\left(\dfrac{\mathbf{v}}{|\mathbf{v}|}\right) =$ (speed) (direction).

Example 14.8 The vector $\mathbf{r}(t) = \{2\log(t+1)\}\mathbf{i} + t^2\mathbf{j} + (t^2/2)\mathbf{k}$ gives the position of a particle in space at time t. Find the particle's velocity and acceleration vectors. Then find the particle's speed and direction of motion when $t = 1$. Write the particle's velocity at $t = 1$ as the product of its speed and direction.

Solution. We have $\mathbf{r}(t) = \{2\log(t+1)\}\mathbf{i} + t^2\mathbf{j} + (t^2/2)\mathbf{k}$.

Therefore, $\mathbf{v}(t) = \dfrac{d\mathbf{r}}{dt} = \left(\dfrac{2}{t+1}\right)\mathbf{i} + 2t\,\mathbf{j} + t\,\mathbf{k}$,

and $\mathbf{a}(t) = \dfrac{d\mathbf{v}}{dt} = \dfrac{d^2\mathbf{r}}{dt^2} = \left(\dfrac{-2}{(t+1)^2}\right)\mathbf{i} + 2\mathbf{j} + \mathbf{k}$.

At $t = 1$, the particle's speed and direction are:

Speed: $|\mathbf{v}(1)| = |\mathbf{i} + 2\mathbf{j} + \mathbf{k}| = \sqrt{1+4+1} = \sqrt{6}$;

Direction: $\dfrac{\mathbf{v}(1)}{|\mathbf{v}(1)|} = \dfrac{1}{\sqrt{6}}\mathbf{i} + \dfrac{2}{\sqrt{6}}\mathbf{j} + \dfrac{1}{\sqrt{6}}\mathbf{k}$.

Therefore, $\mathbf{v}(1) = |\mathbf{v}(1)|\left(\dfrac{\mathbf{v}(1)}{|\mathbf{v}(1)|}\right) = \sqrt{6}\left(\dfrac{1}{\sqrt{6}}\mathbf{i} + \dfrac{2}{\sqrt{6}}\mathbf{j} + \dfrac{1}{\sqrt{6}}\mathbf{k}\right)$.

14.9 Illustrative Examples

Example 14.9 Find the angle between the tangents to the curve

$\mathbf{r}(t) = t^2\mathbf{i} - 2t\mathbf{j} + t^3\mathbf{k}$ at points $t = 1$ and $t = 2$.

Solution. We have $\dfrac{d\mathbf{r}}{dt} = 2t\mathbf{i} - 2\mathbf{j} + 3t^2\mathbf{k}$ a vector along the tangent at any point 't'.

If $\mathbf{T}_1$ and $\mathbf{T}_2$ are vectors along the tangents at $t = 1$ and $t = 2$ repectively then
$\mathbf{T}_1 = 2\mathbf{i} - 2\mathbf{j} + 3\mathbf{k}$, and $\mathbf{T}_2 = 4\mathbf{i} - 2\mathbf{j} + 12\mathbf{k}$.
If θ is the angle between the tangents we have

$$\mathbf{T}_1 \cdot \mathbf{T}_2 = |\mathbf{T}_1||\mathbf{T}_2|\cos\theta \Rightarrow (2)(4) + (-2)(-2) + (3)(12)$$

$$= \sqrt{4+4+9}\sqrt{16+4+144}\,\cos\theta$$

or $\cos\theta = \dfrac{48}{\sqrt{17}\sqrt{164}}$ or $\theta = \cos^{-1}\left(\dfrac{24}{\sqrt{697}}\right)$.

Example 14.10 A particle moves along the curve $x = 2t^2$, $y = t^2 - 4t$, $z = 3t - 5$, where t is the time. Find the components of its velocity and acceteration at time $t = 1$ in the direction $\mathbf{i} - 3\mathbf{j} + 2\mathbf{k}$.

Solution. The position vector of the particle at time 't' is

$$\mathbf{r}(t) = 2t^2\mathbf{i} + (t^2 - 4t)\mathbf{j} + (3t - 5)\mathbf{k}.$$

The velocity, $\mathbf{v}(t) = \dfrac{d\mathbf{r}}{dt} = 4t\mathbf{i} + (2t - 4)\mathbf{j} + 3\mathbf{k}$.

The acceleration, $\mathbf{a}(t) = \dfrac{d^2\mathbf{r}}{dt^2} = 4\mathbf{i} + 2\mathbf{j}$.

When $t = 1$, $\mathbf{v}(1) = 4\mathbf{i} - 2\mathbf{j} + 3\mathbf{k}$, and $\mathbf{a}(1) = 4\mathbf{i} + 2\mathbf{j}$.

$\therefore$ At $t = 1$, component of velocity $\mathbf{v}(1)$ in the direction $\mathbf{i} - 3\mathbf{j} + 2\mathbf{k}$

$$= (4\mathbf{i} - 2\mathbf{j} + 3\mathbf{k}) \cdot \dfrac{\mathbf{i} - 3\mathbf{j} + 2\mathbf{k}}{\sqrt{1+9+4}} = \dfrac{(4)(1) + (-2)(-3) + (3)(2)}{\sqrt{14}} = \dfrac{16}{\sqrt{14}} = \dfrac{8\sqrt{14}}{7}.$$

Similarly, the component of acceleration $\mathbf{a}(1)$ in the given direction

$$= (4\mathbf{i} + 2\mathbf{j}) \cdot \dfrac{\mathbf{i} - 3\mathbf{j} + 2\mathbf{k}}{\sqrt{14}} = \dfrac{(4)(1) + (2)(-3)}{\sqrt{14}} = \dfrac{-2}{\sqrt{14}} = \dfrac{-\sqrt{14}}{7}.$$

Example 14.11 Find the tangent vector to the curve whose parametric representation is $x = t$, $y = 2t^2$, $z = 3t^3$ at $t = 2$. Hence, find the parametric representation of the tangent vector.

Solution. The vector function that gives the position of a point P on the given curve is given by $\mathbf{r}(t) = t\mathbf{i} + 2t^2\mathbf{j} + 3t^3\mathbf{k}$.

Therefore, the tangent vector is $\mathbf{r}'(t) = \mathbf{i} + 4t\mathbf{j} + 9t^2\mathbf{k}$, and so $\mathbf{r}'(2) = \mathbf{i} + 8\mathbf{j} + 36\mathbf{k}$ which is tangent to curve C at the point whose position vector is $\mathbf{r}(2) = 2\mathbf{i} + 8\mathbf{j} + 24\mathbf{k}$, that is P (2, 8, 24).

Thus the parametric equations of the tangent line are

$$\frac{x-2}{1} = \frac{y-8}{8} = \frac{z-24}{36} = t \,(\text{say}) \, i.e., \, x = 2 + t, \, y = 8 + 8t, \, z = 24 + 36t.$$

Example 14.12 A particle moves along the curve

$$\mathbf{r}(t) = (t^3 - 4t)\mathbf{i} + (t^2 + 4t)\mathbf{j} + (8t^2 - 3t^3)\mathbf{k},$$

where t is the time. Find the magnitudes of the tangential and normal components of its acceleration where $t = 2$. *(V.T.U, 2003S)*

Solution. Velocity, $\mathbf{v}(t) = \dfrac{d\mathbf{r}}{dt} = (3t^2 - 4)\mathbf{i} + (2t + 4)\mathbf{j} + (16t - 9t^2)\mathbf{k}$.

Acceleration, $\mathbf{a}(t) = \dfrac{d^2\mathbf{r}}{dt^2} = 6t\mathbf{i} + 2\mathbf{j} + (16 - 18t)\mathbf{k}$.

$\mathbf{v}(t)$, the velocity at $t = 2$ is $\mathbf{v}(2) = 8\mathbf{i} + 8\mathbf{j} - 4\mathbf{k}$.

$\mathbf{a}(t)$, the acceleration at $t = 2$ is $\mathbf{a}(2) = 12\mathbf{i} + 2\mathbf{j} - 20\mathbf{k}$.

Since the direction of velocity is along the tangent to the curve, therefore the tangential component of acc, that is, $\overline{|PD|}$ (Refer Fig. 14.5)

$$= \mathbf{a} \cdot \frac{\mathbf{v}}{|\mathbf{v}|} = (12\mathbf{i} + 2\mathbf{j} - 20\mathbf{k}) \cdot \frac{8\mathbf{i} + 8\mathbf{j} - 4\mathbf{k}}{\sqrt{64 + 64 + 16}}$$

$$= \frac{(12)(8) + (2)(8) + (-20)(-4)}{12} = \frac{192}{12} = 16.$$

Fig. 14.5.

$$\overline{|PF|} = |\mathbf{a}| = \sqrt{144 + 4 + 400} = \sqrt{548}.$$

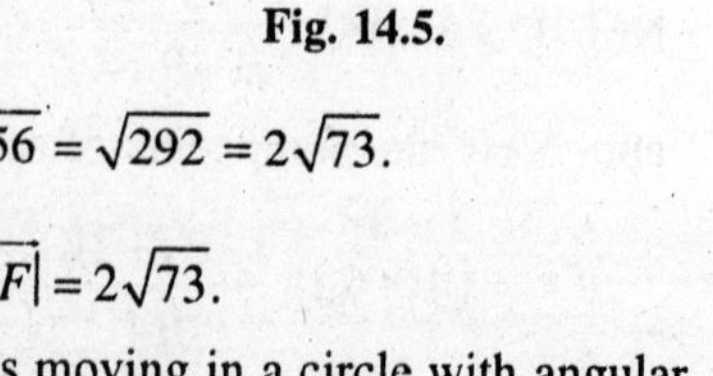

Therefore, $\overline{DF} = \sqrt{\overline{PF}^2 - \overline{PD}^2} = \sqrt{548 - 256} = \sqrt{292} = 2\sqrt{73}.$

$\therefore$ Normal component of acc, $\mathbf{a}(2) = \overline{|PE|} = \overline{|DF|} = 2\sqrt{73}.$

Example 14.13 A particle (position vector $\mathbf{r}$) is moving in a circle with angular velocity ω. Show by vector methods that the acceleration is equal to $-\omega^2\mathbf{r}$.

Solution. Let P be the position of the particle moving along a circle of radius c with constant angular speed $\omega = d\theta/dt$.

The position vector of P is $\mathbf{r}(\theta) = (c\cos\theta)\mathbf{i} + (c\sin\theta)\mathbf{j}$.

Velocity, $\mathbf{v}(t) = \dfrac{d\mathbf{r}}{dt} = (-c\sin\theta\,\mathbf{i} + c\cos\theta\,\mathbf{j})\dfrac{d\theta}{dt} = (-c\sin\theta\,\mathbf{i} + c\cos\theta\,\mathbf{j})\omega$.

Acceleration, $\mathbf{a}(t) = \dfrac{d^2\mathbf{r}}{dt^2} = (-c\cos\theta\,\mathbf{i} - c\sin\theta\,\mathbf{j})\omega\left(\dfrac{d\theta}{dt}\right)$

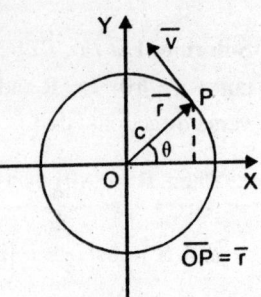

$= -(c\cos\theta\,\mathbf{i} + c\sin\theta\,\mathbf{j})\omega^2 = -\omega^2\mathbf{r}$

Alternatively: $\mathbf{r} = c\cos\omega t\,\mathbf{i} + c\sin\omega t\,\mathbf{j}$.

Therefore, $\mathbf{v}(t) = \dfrac{d\mathbf{r}}{dt} = -c\omega\sin\omega t\,\mathbf{i} + c\omega\cos\omega t\,\mathbf{j}$,

and $acc.,\mathbf{a}(t) = \dfrac{d^2\mathbf{r}}{dt^2} = -c\omega^2\cos\omega t\,\mathbf{i} - c\,\omega^2\sin\omega t\,\mathbf{j}$.

$\overline{OP} = \overline{r}$

$= -\omega^2(c\cos\omega t\,\mathbf{i} + c\sin\omega t\,\mathbf{j}) = -\omega^2\mathbf{r}$.

Fig. 14.6.

Example 14.14 Find by vector methods, the tangential and normal components of acceleration of a particle moving in a plane curve.

Solution. Let $\hat{\mathbf{T}}$ and $\hat{\mathbf{N}}$ respectively denote unit vectors along tangent and normal at any point on the curve at any time t.

If $\mathbf{v}$ denotes velocity at P along the tangent then $\mathbf{v} = v\hat{\mathbf{T}}$ where $v = |\mathbf{v}|$.

Acceleration along tangent

$= \dfrac{d\mathbf{v}}{dt} = \dfrac{dv}{dt}\hat{\mathbf{T}} + v\dfrac{d\hat{\mathbf{T}}}{dt} = \dfrac{dv}{dt}\hat{\mathbf{T}} + v\cdot\dfrac{d\hat{\mathbf{T}}}{d\psi}\cdot\dfrac{d\psi}{ds}\cdot\dfrac{ds}{dt}$

$= \dfrac{dv}{dt}\hat{\mathbf{T}} + v^2\dfrac{1}{\rho}\dfrac{d\hat{\mathbf{T}}}{d\psi}\cdots\left(\because \dfrac{d\psi}{ds} = \dfrac{1}{\rho},\text{where } \rho \text{ is radius of circle and } \dfrac{ds}{dt} = v\right)$ ...(1)

Now $\hat{\mathbf{T}} = \cos\psi\,\mathbf{i} + \sin\psi\,\mathbf{j}$,

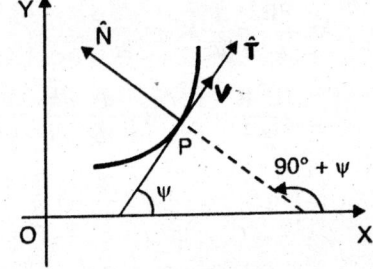

and $\hat{\mathbf{N}} = -\sin\psi\,\mathbf{i} + \cos\psi\,\mathbf{j}$.

$\therefore \dfrac{d\hat{\mathbf{T}}}{d\psi} = -\sin\psi\,\mathbf{i} + \cos\psi\,\mathbf{j} = \hat{\mathbf{N}}$.

Hence from (1) we have

$\dfrac{d\mathbf{v}}{dt} = \dfrac{dv}{dt}\hat{\mathbf{T}} + \dfrac{v^2}{\rho}\hat{\mathbf{N}}$.

Fig. 14.7.

Thus tangential and normal components of acceleration respectively are

$$\frac{dv}{dt} \text{ and } \frac{v^2}{\rho}.$$

Example 14.15 Find the radial and transverse components of velocity and acceleration of a particle describing a plane curve.

(Kurukshetra, 2006; Rajasthan, 2006, Punjab, 1997)

Solution. Let $P(r, \theta)$ be the position of a moving particle at any time t along a curve $r = f(\theta)$. Let $\hat{R}$ and $\hat{S}$ respectively denote unit vectors in the radial and transverse directions, that is, along and perpendicular to OP.

Then $\hat{R} = \cos\theta \mathbf{i} + \sin\theta \mathbf{j}$,

and $\hat{S} = \cos\left(\dfrac{\pi}{2} + \theta\right)\mathbf{i} + \sin\left(\dfrac{\pi}{2} + \theta\right)\mathbf{j}$

$= -\sin\theta \mathbf{i} + \cos\theta \mathbf{j}$.

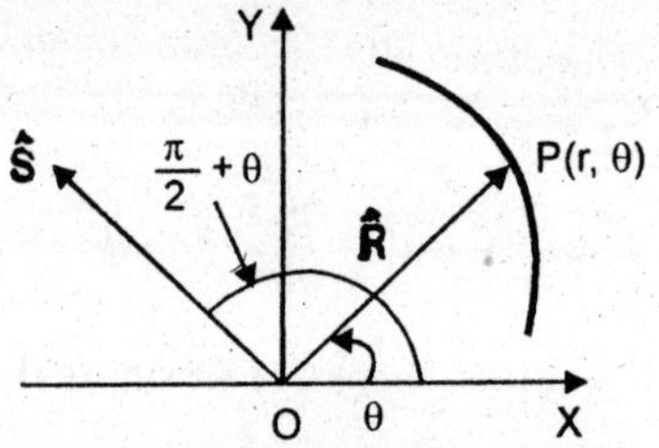

Therefore, $\dfrac{d\hat{R}}{dt} = (-\sin\theta \mathbf{i} + \cos\theta \mathbf{j})\dfrac{d\theta}{dt} = \hat{S}\dfrac{d\theta}{dt}$ ...(i)

and $\dfrac{d\hat{S}}{dt} = (-\cos\theta \mathbf{i} - \sin\theta \mathbf{j})\dfrac{d\theta}{dt} = -\hat{R}\dfrac{d\theta}{dt}$ ...(ii)

Now **R**, the position vector of P is given by $\mathbf{R} = r\hat{R}$ [where $r = |\mathbf{R}|$].

Therefore $\dfrac{d\mathbf{R}}{dt} = \dfrac{dr}{dt}\hat{R} + r\dfrac{d\hat{R}}{dt} = \dfrac{dr}{dt}\hat{R} + r\dfrac{d\theta}{dt}\hat{S}$ ...{By (i)}

Hence *radial component* of velocity is $\dfrac{dr}{dt}$ and *transverse component* is $r\left(\dfrac{d\theta}{dt}\right)$.

As $\dfrac{d\mathbf{R}}{dt} = \dfrac{dr}{dt}\hat{R} + r\dfrac{d\theta}{dt}\hat{S}$.

$\therefore \dfrac{d^2\mathbf{R}}{dt^2} = \dfrac{d^2r}{dt^2}\hat{R} + \dfrac{dr}{dt}\cdot\dfrac{d\hat{R}}{dt} + \dfrac{dr}{dt}\cdot\dfrac{d\theta}{dt}\hat{S} + r\dfrac{d^2\theta}{dt^2}\hat{S} + r\cdot\dfrac{d\theta}{dt}\cdot\dfrac{d\hat{S}}{dt}$

$= \dfrac{d^2r}{dt^2}\hat{R} + \dfrac{dr}{dt}\left(\hat{S}\dfrac{d\theta}{dt}\right) + \dfrac{dr}{dt}\cdot\dfrac{d\theta}{dt}\cdot\hat{S} + r\dfrac{d^2\theta}{dt^2}\hat{S} + r\dfrac{d\theta}{dt}\left(-\hat{R}\dfrac{d\theta}{dt}\right)$

$= \left[\dfrac{d^2r}{dt^2} - r\left(\dfrac{d\theta}{dt}\right)^2\right]\hat{R} + \left[r\dfrac{d^2\theta}{dt^2} + 2\dfrac{dr}{dt}\dfrac{d\theta}{dt}\right]\hat{S}.$

Fig. 14.8.

Hence the *radial component of acceleration* is $\dfrac{d^2r}{dt^2} - r\left(\dfrac{d\theta}{dt}\right)^2$,

and *transverse component of acceleration* is $r\dfrac{d^2\theta}{dt^2} + 2\dfrac{dr}{dt}\dfrac{d\theta}{dt}$.

Remark: We can also write radial and transverse components of acceleration in the form:

$\ddot{r} - r\dot{\theta}^2$ and $r\ddot{\theta} + 2\dot{r}\dot{\theta}$ respectively,

where dots denote derivative w.r.t. '*t*', that is $\dot{r} = \dfrac{dr}{dt}$ etc.

Exercise 14.1

1. If $r(t) = \sin t\, i + \cos t\, j + t\, k$, evaluate

   (i) dr/dt,   (ii) d^2r/dt^2,   (iii) $|dr/dt|$,   (iv) $|d^2r/dt^2|$.

     Ans. (i) $\cos t\, i - \sin t\, j + k$, (ii) $-\sin t\, i - \cos t\, j$, (iii) $\sqrt{2}$, (iv) 1.

2. If $r = e^{-t} i + \log(t^2 + 1)j - \tan t\, k$, find (i) dr/dt, (ii) d^2r/dt^2, (iii) $|dr/dt|$,

   (iv) $\left|d^2r/dt^2\right|$ at $t = 0$.   **Ans.** (i) $-i-k$, (ii) $i + 2j$, (iii) $\sqrt{2}$, (iv) $\sqrt{5}$.

3. Differentiate the following with respect to t. (**a** and **b** are constant vectors)

   (i) $\dfrac{r}{r^2} + \dfrac{rb}{a \cdot r}$,   (ii) $\dfrac{r \times a}{r \cdot a}$.

    Ans. (i) $\dfrac{r'}{r^2} - 2\dfrac{r'r}{r^3} + \dfrac{br'}{a \cdot r} - r\dfrac{(a \cdot r')b}{(a \cdot r)^2}$, (ii) $\dfrac{r' \times a}{r \cdot a} - \dfrac{(r' \cdot a)(r \times a)}{(r \cdot a)^2}$.

4. If $A = t^3 i + 2t\, j + e^t k$ and $B = \sin t\, i - \cos t\, j + t\, k$, find (i) $\dfrac{d}{dt}(A \cdot B)$,

   (ii) $\dfrac{d}{dt}(A \times B)$.   **Ans.** (i) $(t^3 - 2)\cos t + (3t + 2)t \sin t + (t + 1)e^t$,

         (ii) $(4t + e^t \cos t - e^t \sin t)i + (e^t \sin t + e^t \cos t - 4t^3)j$

             $+ (t^3 \sin t - 3t^2 \cos t - 2t \cos t - 2 \sin t)k$.

5. Evaluate the following:

   (i) $\dfrac{d}{dt}\left(r \times \dfrac{dr}{dt}\right)$, (ii) $\dfrac{d}{dt}\left(r \cdot \dfrac{dr}{dt} \times \dfrac{d^2r}{dt^2}\right)$, (iii) $\dfrac{d}{dt}\left(\dfrac{F(t)}{|F(t)|}\right)$.

    Ans. (i) $r \times \dfrac{d^2r}{dt^2}$, (ii) $r \cdot \dfrac{dr}{dt} \times \dfrac{d^3r}{dt^3}$, (iii) $\left[(F \cdot F)F' - F(F \cdot F')\right]/(F \cdot F)^{3/2}$.

Hint. (iii) $\dfrac{d}{dt}\left[\dfrac{F}{|F|}\right] = \dfrac{d}{dt}\left[\dfrac{F}{(F \cdot F)^{1/2}}\right]$.

6. If $\mathbf{u} = 5t^2\mathbf{i} + t\mathbf{j} - t^3\mathbf{k}, \mathbf{v} = \sin t\,\mathbf{i} - \cos t\,\mathbf{j}$. Verify the formulae

   (i) $\dfrac{d}{dt}(\mathbf{u} \cdot \mathbf{v}) = \dfrac{d\mathbf{u}}{dt} \cdot \mathbf{v} + \mathbf{v} \cdot \dfrac{d\mathbf{v}}{dt}$,

   (ii) $\dfrac{d}{dt}(\mathbf{u} \times \mathbf{v}) = \mathbf{u} \times \dfrac{d\mathbf{v}}{dt} + \dfrac{d\mathbf{u}}{dt} \times \mathbf{v}$.

7. If $\mathbf{r}(t) = (a\cos t)\mathbf{i} + (a\sin t)\mathbf{j} + (at\tan\alpha)\mathbf{k}$, show that

   $$\dfrac{d\mathbf{r}}{dt} \cdot \dfrac{d^2\mathbf{r}}{dt^2} \times \dfrac{d^3\mathbf{r}}{dt^3} = a^3 \tan\alpha.$$

8. If $\mathbf{r}(t) = 5t^2\,\mathbf{i} + t\mathbf{j} - t^3\,\mathbf{k}$, find $\displaystyle\int_1^2 \mathbf{r} \times \dfrac{d^2\mathbf{r}}{dt^2}\,dt$.   **Ans.** $-14\mathbf{i} + 75\mathbf{j} - 15\mathbf{k}.$

9. Given $\mathbf{v}(t) = t\mathbf{i} + (t^2 - 2t)\mathbf{j} + (3t^2 + t^3)\mathbf{k}$, find $\displaystyle\int_0^1 \mathbf{v}(t)\,dt$.

   $$\textbf{Ans. } \dfrac{1}{2}\mathbf{i} - \dfrac{2}{3}\mathbf{j} + \dfrac{5}{4}\mathbf{k}.$$

10. Find $\displaystyle\int_a^b \mathbf{v}(t)\,dt$, given $\mathbf{v}(t) = u^3\mathbf{i} + (3u^2 - 2u)\mathbf{j} + 3\mathbf{k}; a = 0, b = 2$

    Ans. $4\mathbf{i} + 4\mathbf{j} + 6\mathbf{k}.$

11. If $\mathbf{r} = a\cos t\,\mathbf{i} + a\sin t\,\mathbf{j} + bt\,\mathbf{k}$, prove that

    (i) $\left[\dfrac{d\mathbf{r}}{dt}\,\dfrac{d^2\mathbf{r}}{dt^2}\,\dfrac{d^3\mathbf{r}}{dt^3}\right] = a^2 b$, (ii) $\left(\dfrac{d\mathbf{r}}{dt}\right)^2 = a^2 + b^2$, (iii) $\left(\dfrac{d\mathbf{r}}{dt} \times \dfrac{d^2\mathbf{r}}{dt^2}\right)^2 = a^2(a^2 + b^2)$.

12. Find the unit tangent vector on the curve $x = 2t + 1, y = t^2 - t, z = t^3$ at the point $t = 2$.   **Ans.** $(2\mathbf{i} + 3\mathbf{j} + 12\mathbf{k})/157$.

13. Find the angle between the tangents to the curve $y = x^2, z = x^3$ at $(1, 1, 1)$ and $(-1, 1, -1)$.   **Ans.** $\cos^{-1}(3/7)$

    [**Hint.** The parametric equations of the curve are $x = t, y = t^2, z = t^3$. The points $(1, 1, 1)$ and $(-1, 1, -1)$ correspond to $t = 1$ and $t = -1$ etc.]

14. A particle moves along the curve $x = e^{-t}, y = 2\cos 3t, z = 2\sin 3t$, where is the time variable. Find the (i) velocity (ii) acceleration, at $t = 0$.

    (*P.T.U, 2003; VTU 2003S*) **Ans.** (i) $\mathbf{i} + 6\mathbf{k}$ (ii) $\mathbf{i} - 18\mathbf{j}$.

15. If $\mathbf{r} = t^2\mathbf{i} - t^3\mathbf{j} + t^4\mathbf{k}$ is the position vector of a moving particle at time t. What is the component of its velocity in the direction of the vector $8\mathbf{i} - \mathbf{j} + 4\mathbf{k}$ at $t = 1$?

    What is the acceleration of the particle at $t = 1$? What are the tangential and normal components of its acceleration?

    $$\textbf{Ans. } 35/9, 2\mathbf{i} - 6\mathbf{j} + 12\mathbf{k}, 70/\sqrt{29}, \sqrt{(436/29)}.$$

16. The acceleration of a particle at any time $t \geq 0$ is given by $12\cos 2t\,\mathbf{i} - 8\sin 2t\,\mathbf{j} + 16t\,\mathbf{k}$. The velocity and displacement are zero at $t = 0$. Find the velocity and displacement at any time t.

    Ans. $\mathbf{v}(t) = 6\sin 2t\,\mathbf{i} + 4(\cos 2t - 1)\mathbf{j} + 8t^2\,\mathbf{k};$
    $\mathbf{r}(t) = 3(1 - \cos 2t)\mathbf{i} + 2(\sin 2t - 2t)\mathbf{j} + 8t^3\,\mathbf{k}/3.$

17. A particle moves so that its position vector is given by $\mathbf{r}(t) = (\cos \omega t)\mathbf{i} + (\sin \omega t)\mathbf{j}$. Show that the velocity, $\mathbf{r}'(t)$ of the particle is perpendicular to $\mathbf{r}(t)$ and $\mathbf{r}(t) \times \mathbf{r}'(t)$ is a constant vector.

18. A particle moves along the curve $x = 3t^2$, $y = t^2 - 2t$, $z = t^3$, where t is time. Find (*i*) the magnitudes of its velocity and acceleration at time $t = 1$; (*ii*) the components of velocity and acceleration at time $t = 1$ in the direction $\mathbf{b} = 4\mathbf{i} - 2\mathbf{j} + 4\mathbf{k}$.   **Ans.** (*i*) $|\mathbf{v}(1)| = 3\sqrt{5}, |\mathbf{a}(1)| = 2\sqrt{19}$; (*ii*) 6, 22/3.

19. The position vector of a particle at time t is
    $$\mathbf{r}(t) = \cos(t - 1)\,\mathbf{i} + \sinh(t - 1)\,\mathbf{j} + \alpha t^3\mathbf{k}.$$
    Find the condition imposed on α by requiring that at $t = 1$, the acceleration is normal to the position vector.   **Ans.** $\alpha = \pm 1/\sqrt{6}.$

20. The acceleration of a particle at any time $t \geq 0$ is given by $\mathbf{a}(t) = d\mathbf{v}/dt = e^t\mathbf{i} + e^{2t}\mathbf{j} + \mathbf{k}$. If at $t = 0$, the displacement is $\mathbf{r}(0) = 0$ and the velocity is $\mathbf{v}(0) = \mathbf{i} + \mathbf{j}$, find $\mathbf{r}(t)$ and $\mathbf{v}(t)$ at any time t.

    Ans. $\mathbf{r}(t) = (e^t - 1)\mathbf{i} + \left(\dfrac{1}{4}e^{2t} + \dfrac{1}{2}t - \dfrac{1}{4}\right)\mathbf{j} + \dfrac{1}{2}t^2\mathbf{k};$ $\mathbf{v}(t) = e^t\mathbf{i} + \dfrac{1}{2}(e^{2t} + 1)\mathbf{j} + t\,\mathbf{k}.$

21. The acceleration of a particle at any time t is given by $\mathbf{a} = d\mathbf{v}/dt = (t + 1)\mathbf{i} + t^2\mathbf{j} + (t^2 - 2)\mathbf{k}$. If at $t = 0$, the displacement is $\mathbf{r}(0) = 0$ and the velocity is $\mathbf{v}(0) = \mathbf{i} - \mathbf{k}$, find $\mathbf{v}$ and $\mathbf{r}$ at any time t.

    Ans. $\mathbf{v}(t) = \left(\dfrac{1}{2}t^2 + t + 1\right)\mathbf{i} + \dfrac{1}{3}t^3\mathbf{j} + \left(\dfrac{1}{3}t^3 - 2t - 1\right)\mathbf{k};$

    $\mathbf{r}(t) = \left(\dfrac{1}{6}t^3 + \dfrac{1}{2}t^2 + t\right)\mathbf{i} + \dfrac{1}{12}t^4\mathbf{j} + \left(\dfrac{1}{12}t^4 - t^2 - t\right)\mathbf{k}.$

14.10 Gradient of a Scalar Field and Directional Derivative

Let $f(x, y, z)$ be a real valued function defining a scalar field. To define the gradient of a scalar field, we first introduce a vector differential operator called *del* operator denoted by ∇.

We define the vector differential operator ∇ (based on partial differentiations) in two and three dimensions as

$$\nabla = \mathbf{i}\frac{\partial}{\partial x} + \mathbf{j}\frac{\partial}{\partial y} \quad \text{and} \quad \nabla = \mathbf{i}\frac{\partial}{\partial x} + \mathbf{j}\frac{\partial}{\partial y} + \mathbf{k}\frac{\partial}{\partial z}.$$

The del operator is analogous to the derivative operator d/dx, which when applied to $f(x)$ produces the derivative $f'(x)$. The symbol ∇, an inverted capital Greek delta is also known as "*nabla*".

Gradient of a scalar field. *When the vector differential operator ∇ is applied to a differentiable function $z = f(x, y)$ or $w = f(x, y, z)$, defining a scalar field, we say that the vectors*

$$\nabla f(x, y) = \left(\frac{\partial}{\partial x}\mathbf{i} + \frac{\partial}{\partial y}\mathbf{j}\right)f = \frac{\partial f}{\partial x}\mathbf{i} + \frac{\partial f}{\partial y}\mathbf{j} \qquad \qquad ...(14.12)$$

and $\quad \nabla f(x, y, z) = \left(\frac{\partial}{\partial x}\mathbf{i} + \frac{\partial}{\partial y}\mathbf{j} + \frac{\partial}{\partial z}\mathbf{k}\right)f = \frac{\partial f}{\partial x}\mathbf{i} + \frac{\partial f}{\partial y}\mathbf{j} + \frac{\partial f}{\partial z}\mathbf{k} \qquad ...(14.13)$

are gradient of the respective functions.

Note that the del operator ∇ operates on a "scalar field and produce a vector field. The vector ∇f is usually read "grad f" as well as "gradient of f" and "del f".

Example 14.16 Find the gradient of the following scalar fields

(i) $f(x, y) = x^3 - 3x^2y^2 + y^3$ at $(1, 2)$.

(ii) $f(x, y, z) = 3x^2y - y^3z^2$ at $(1, -2, -1)$. *(IETE Dec 2008)*

(iii) $\phi(x, y, z) = xy^2 + 3x^2 - z^3$ at $(2, -1, 4)$.

Solution. (i) We have $\nabla f(x, y) = \left(\mathbf{i}\dfrac{\partial}{\partial x} + \mathbf{j}\dfrac{\partial}{\partial y}\right)\left(x^3 - 3x^2y^2 + y^3\right)$

$$= \mathbf{i}\frac{\partial}{\partial x}\left(x^3 - 3x^2y^2 + y^3\right) + \mathbf{j}\frac{\partial}{\partial y}\left(x^3 - 3x^2y^2 + y^3\right)$$

$$= \left(3x^2 - 6xy^2\right)\mathbf{i} + \left(-6x^2y + 3y^2\right)\mathbf{j}. \text{ At } (1, 2), \text{ we obtain } \nabla f(1, 2) = -21\,\mathbf{i}.$$

(ii) We have $\nabla f(x, y, z) = \left(\mathbf{i}\dfrac{\partial}{\partial x} + \mathbf{j}\dfrac{\partial}{\partial y} + \mathbf{k}\dfrac{\partial}{\partial z}\right)\left(3x^2y - y^3z^2\right)$

$$= 6xy\mathbf{i} + \left(3x^2 - 3y^2z^2\right)\mathbf{j} - 2y^3z\mathbf{k}.$$

At $(1, -2, -1)$, we obtain $\nabla f(1, -2, -1) = -\left(12\mathbf{i} + 9\mathbf{j} + 16\mathbf{k}\right)$.

(iii) We have $\nabla\phi(x, y, z) = \left(\mathbf{i}\dfrac{\partial}{\partial x} + \mathbf{j}\dfrac{\partial}{\partial y} + \mathbf{k}\dfrac{\partial}{\partial z}\right)\left(xy^2 + 3x^2 - z^3\right)$

$$= \left(y^2 + 6x\right)\mathbf{i} + 2xy\mathbf{j} - 3z^2\mathbf{k}.$$

At $(2, -1, 4)$, we obtain $\nabla\phi(2, -1, 4) = 13\mathbf{i} - 4\mathbf{j} - 48\mathbf{k}$.

Example 14.17 (a). *If* $\mathbf{r} = x\mathbf{i} + y\mathbf{j} + z\mathbf{k}$, $|\mathbf{r}| = r$ *and* $\hat{\mathbf{r}} = \mathbf{r}/r$, *then show that*

(i) grad $r = \hat{\mathbf{r}}$, (ii) grad $(1/r) = -\hat{\mathbf{r}}/r^2 = -\mathbf{r}/r^3$. *(PTU 2003; UPTU 2002)*

(b) Find $\nabla\phi$, if $\phi = \log|\mathbf{r}|$. *(A.M.I.E., W-2008)*

Solution. (*a*) (*i*) We have $r^2 = x^2 + y^2 + z^2$. Therefore grad r, that is

$$\nabla r = \mathbf{i}\frac{\partial r}{\partial x} + \mathbf{j}\frac{\partial r}{\partial y} + \mathbf{k}\frac{\partial r}{\partial z} = \mathbf{i}\frac{x}{r} + \mathbf{j}\frac{y}{r} + \mathbf{k}\frac{z}{r} = \frac{x\mathbf{i} + y\mathbf{j} + z\mathbf{k}}{r} = \frac{\mathbf{r}}{r} = \hat{\mathbf{r}}.$$

(*ii*) grad $\dfrac{1}{r} = \left(\mathbf{i}\dfrac{\partial}{\partial x} + \mathbf{j}\dfrac{\partial}{\partial y} + \mathbf{k}\dfrac{\partial}{\partial z}\right)\left(\dfrac{1}{r}\right) = \mathbf{i}\left(-\dfrac{1}{r^2}\dfrac{\partial r}{\partial x}\right) + \mathbf{j}\left(-\dfrac{1}{r^2}\dfrac{\partial r}{\partial y}\right) + \mathbf{k}\left(-\dfrac{1}{r^2}\dfrac{\partial r}{\partial z}\right)$

$$= -\frac{1}{r^2}\left(\frac{x}{r}\mathbf{i} + \frac{y}{r}\mathbf{j} + \frac{z}{r}\mathbf{k}\right) = -\frac{1}{r^2}\left(\frac{\mathbf{r}}{r}\right) = -\frac{\mathbf{r}}{r^3} = -\frac{\hat{\mathbf{r}}}{r^2}.$$

(*b*) We have $r = |\mathbf{r}| = \sqrt{x^2 + y^2 + z^2}$ and $\phi = \log|\mathbf{r}| = \dfrac{1}{2}\log\left(x^2 + y^2 + z^2\right)$.

$$\therefore \ \nabla\phi = \frac{1}{2}\nabla\log\left(x^2 + y^2 + z^2\right) = \frac{1}{2}\left\{\mathbf{i}\frac{\partial}{\partial x}\log\left(x^2 + y^2 + z^2\right) + \right.$$

$$\left. \mathbf{j}\frac{\partial}{\partial y}\log\left(x^2 + y^2 + z^2\right) + \mathbf{k}\frac{\partial}{\partial z}\log\left(x^2 + y^2 + z^2\right)\right\}$$

$$= \frac{1}{2}\left\{\mathbf{i}\frac{2x}{x^2 + y^2 + z^2} + \mathbf{j}\frac{2y}{x^2 + y^2 + z^2} + \mathbf{k}\frac{2z}{x^2 + y^2 + z^2}\right\} = \frac{x\mathbf{i} + y\mathbf{i} + z\mathbf{k}}{x^2 + y^2 + z^2} = \frac{\mathbf{r}}{r^2}.$$

Example 14.18 Find $\nabla\phi$ if

(*i*) $\phi = f(r)$ where $r = \sqrt{x^2 + y^2 + z^2}$.

(*ii*) $\phi = r^n$.(*AMIE; W 2007; UPTU 2004; Delhi 2002; MDU 2001*)

Solution. (*i*) $\nabla\phi = \nabla f(r) = \left(\mathbf{i}\dfrac{\partial}{\partial x} + \mathbf{j}\dfrac{\partial}{\partial y} + \mathbf{k}\dfrac{\partial}{\partial z}\right)f(r)$

$$= \mathbf{i}\frac{\partial}{\partial x}f(r) + \mathbf{j}\frac{\partial}{\partial y}f(r) + \mathbf{k}\frac{\partial}{\partial z}f(r)$$

$$= \mathbf{i}f'(r)\frac{\partial r}{\partial x} + \mathbf{j}f'(r)\frac{\partial r}{\partial y} + \mathbf{k}f'(r)\frac{\partial r}{\partial z}, \text{ where } f'(r) = \frac{d}{dr}f(r).$$

Since $r = \sqrt{x^2 + y^2 + z^2}$. $\quad \therefore \ \dfrac{\partial r}{\partial x} = \dfrac{\partial}{\partial x}\left(x^2 + y^2 + z^2\right)^{1/2} = \dfrac{x}{\sqrt{x^2 + y^2 + z^2}} = \dfrac{x}{r}.$

Similarly $\dfrac{\partial r}{\partial y} = \dfrac{y}{r}$ and $\dfrac{\partial r}{\partial z} = \dfrac{z}{r}.$

$$\therefore \quad \nabla\phi = \mathbf{i} f'(r)\frac{x}{r} + \mathbf{j} f'(r)\frac{y}{r} + \mathbf{k} f'(r)\frac{z}{r} = \frac{f'(r)}{r}(x\mathbf{i} + y\mathbf{j} + z\mathbf{k}) = \frac{f'(r)}{r}\mathbf{r}.$$

(ii) Using the result of part (i), we have $f(r) = r^n$. Therefore $f'(r) = nr^{n-1}$.

Hence
$$\nabla\phi = \nabla r^n = \frac{nr^{n-1}}{r}\mathbf{r} = nr^{n-2}\mathbf{r}.$$

Alternative
$$\nabla r^n = \mathbf{i}\frac{\partial}{\partial x}r^n + \mathbf{j}\frac{\partial}{\partial y}r^n + \mathbf{k}\frac{\partial}{\partial z}r^n.$$

Now
$$\frac{\partial}{\partial x}r^n = \frac{\partial}{\partial x}\left(x^2 + y^2 + z^2\right)^{n/2} = \frac{n}{2}\left(x^2 + y^2 + z^2\right)^{(n-2)/2} \cdot 2x$$

$$= nx\left(x^2 + y^2 + z^2\right)^{(n-2)/2} = nxr^{n-2}.$$

Similarly $\dfrac{\partial}{\partial y}r^n = nyr^{n-2}$ and $\dfrac{\partial}{\partial z}r^n = nzr^{n-2}$.

$$\therefore \quad \nabla r^n = nxr^{n-2}\mathbf{i} + nyr^{n-2}\mathbf{j} + nzr^{n-2}\mathbf{k} = nr^{n-2}(x\mathbf{i} + y\mathbf{j} + z\mathbf{k}) = nr^{n-2}\mathbf{r}.$$

Example 14.19 (i) If $\mathbf{r} = x\mathbf{i} + y\mathbf{j} + z\mathbf{k}$ and $\phi = |\mathbf{r}|^3$, then find $\nabla|\mathbf{r}|^3$,

(ii) Find $\nabla\phi$ if $\phi = \left(x^2 + y^2 + z^2\right)e^{-\sqrt{\left(x^2+y^2+z^2\right)}}$.

Solution. (i) $|\mathbf{r}| = \sqrt{\left(x^2 + y^2 + z^2\right)}$, therefore $|\mathbf{r}|^3 = \left(x^2 + y^2 + z^2\right)^{3/2}$ and

$$\nabla|\mathbf{r}|^3 = \left(\mathbf{i}\frac{\partial}{\partial x} + \mathbf{j}\frac{\partial}{\partial y} + \mathbf{k}\frac{\partial}{\partial z}\right)\left(x^2 + y^2 + z^2\right)^{3/2}$$

$$= \mathbf{i}\frac{\partial}{\partial x}\left(x^2 + y^2 + z^2\right)^{3/2} + \mathbf{j}\frac{\partial}{\partial y}\left(x^2 + y^2 + z^2\right)^{3/2} + \mathbf{k}\frac{\partial}{\partial z}\left(x^2 + y^2 + z^2\right)^{3/2}$$

$$= \mathbf{i}\left[\frac{3}{2}\left(x^2 + y^2 + z^2\right)^{1/2} \cdot 2x\right] + \mathbf{j}\left[\frac{3}{2}\left(x^2 + y^2 + z^2\right)^{1/2} \cdot 2y\right]$$

$$+ \mathbf{k}\left[\frac{3}{2}\left(x^2 + y^2 + z^2\right)^{1/2} \cdot 2z\right] = 3\left(x^2 + y^2 + z^2\right)^{1/2}\left[x\mathbf{i} + y\mathbf{j} + z\mathbf{k}\right] = 3r\,\mathbf{r}.$$

(ii) $\mathbf{r} = x\mathbf{i} + y\mathbf{j} + z\mathbf{k}$, $r = |\mathbf{r}| = \sqrt{\left(x^2 + y^2 + z^2\right)}$.

$$\therefore \quad \nabla\phi = \left(\mathbf{i}\frac{\partial}{\partial x} + \mathbf{j}\frac{\partial}{\partial y} + \mathbf{k}\frac{\partial}{\partial z}\right)\left(x^2 + y^2 + z^2\right)e^{-\sqrt{\left(x^2+y^2+z^2\right)}}$$

$$= \sum \mathbf{i}\frac{\partial}{\partial x}\left[\left(x^2 + y^2 + z^2\right)e^{-\sqrt{\left(x^2+y^2+z^2\right)}}\right]$$

$$= \sum \mathbf{i} \left[2xe^{-\sqrt{(x^2+y^2+z^2)}} - \left(x^2+y^2+z^2\right)e^{-\sqrt{(x^2+y^2+z^2)}} \cdot \frac{1}{2}\left(x^2+y^2+z^2\right)^{-1/2} \cdot 2x \right]$$

$$= \sum \mathbf{i} \left[e^{-\sqrt{(x^2+y^2+z^2)}} \cdot x \left\{ 2 - \sqrt{\left(x^2+y^2+z^2\right)} \right\} \right]$$

$$= e^{-\sqrt{(x^2+y^2+z^2)}} \left\{ 2 - \sqrt{\left(x^2+y^2+z^2\right)} \right\} \left(x\mathbf{i} + y\mathbf{j} + z\mathbf{k} \right) = e^{-r}(2-r)\mathbf{r}.$$

Geometrical Interpretation of Gradient

Let $f(P) = f(x, y, z)$ be a differential scalar function. Let $f(x, y, z) = c$ = constant be a level surface through P. Vector normal to tangent plane, to the surface at the point P, is called the normal vector to the surface at this point.

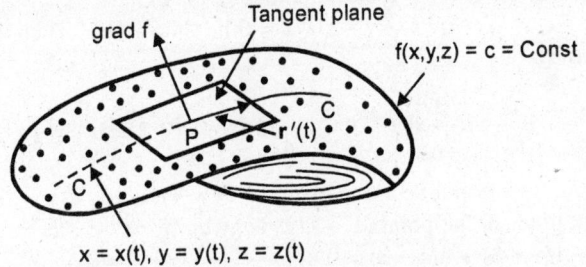

Fig. 14.9. Gradient as surface normal vector

Let $x = x(t)$, $y = y(t)$, $z = z(t)$ be the parametric representation of the smooth curve C on the surface passing through a point P on the surface. The curve C in space can be represented by a vector function $\mathbf{r}(t) = x(t)\mathbf{i} + y(t)\mathbf{j} + z(t)\mathbf{k}$.

Since the curve lie on the surface, we have

$$f(x(t), y(t), z(t)) = c. \text{ Then } \frac{d}{dt} f(x(t), y(t), z(t)) = 0.$$

By chain rule, we have $\dfrac{\partial f}{\partial x}\dfrac{dx}{dt} + \dfrac{\partial f}{\partial y}\dfrac{dy}{dt} + \dfrac{\partial f}{\partial z}\dfrac{dz}{dt} = 0$

or $\left(\mathbf{i}\dfrac{\partial}{\partial x} + \mathbf{j}\dfrac{\partial}{\partial y} + \mathbf{k}\dfrac{\partial}{\partial z} \right) f \cdot \left(\mathbf{i}\dfrac{d}{dt} + \mathbf{j}\dfrac{dy}{dt} + \mathbf{k}\dfrac{dz}{dt} \right) = 0$ or $\nabla f \bullet \mathbf{r}'(t) = 0.$ ...(14.14)

$\mathbf{r}'(t)$ is tangent vector to curve C at the point P and lies in the tangent plane. Hence grad f is orthogonal to all the tangent vector at P. Thus grad as surface normal vector.

Example 14.20 Find a unit normal vector to the surface $x^2y + 2xz = 4$ at the point $(2, -2, 3)$. *(A.M.I.E., W-2005)*

Solution. Let us regard the given surface as a particular level surface of the function $\phi = x^2y + 2xz$. Then the gradient of this function at the point $(2, -2, 3)$ will be perpendicular to the level surface through $(2, -2, 3)$, which is the given surface.

Now $\phi(x, y, z) = x^2y + 2xz - 4$. $\dfrac{\partial\phi}{\partial x} = 2xy + 2z$, $\dfrac{\partial\phi}{\partial y} = x^2$ and $\dfrac{\partial\phi}{\partial z} = 2x$.

Then, the normal vector is given by

$$\nabla\phi = i\frac{\partial\phi}{\partial x} + j\frac{\partial\phi}{\partial y} + k\frac{\partial\phi}{\partial z} = (2xy + 2z)i + x^2j + 2xk.$$

At $(2, -2, 3)$, we obtain the normal vector as $\nabla\phi(2, -2, 3) = -2i + 4j + 4k$. Hence a unit vector normal to the given surface is given by

$$n = \frac{\nabla\phi}{|\nabla\phi|} = \frac{(-2i + 4j + 4k)}{\sqrt{4 + 16 + 16}} = -\frac{1}{3}i + \frac{2}{3}j + \frac{2}{3}k.$$

Example 14.21. Find the normal vector and the equation of the tangent plane to the surface $xz^2 + x^2y - z + 1 = 0$ at the point $(1, -3, 2)$.

Solution. Let $f(x, y, z) = xz^2 + x^2y - z + 1 = 0$ be the surface. Then, the normal vector is given by

$$N = \nabla f = \left(i\frac{\partial}{\partial x} + j\frac{\partial}{\partial y} + k\frac{\partial}{\partial z}\right)(xz^2 + x^2y - z + 1) = (z^2 + 2xy)i + x^2j + (2xz - 1)k.$$

At $(1, -3, 2)$, the normal vector is given by $N = \nabla f(1, -3, 2) = -2i + j + 3k$. The tangent plane at the point $(1, -3, 2)$ is given by
$-2(x - 1) + 1(y + 3) + 3(z - 2) = 0$ or $2x - y - 3z + 1 = 0$.

Example 14.22 Find the angle between the surfaces $x^2 + y^2 + z^2 = 9$ and $z + 3 = x^2 + y^2$ at the point $(2, -1, 2)$. *(UPTU 2003; Madras 2003; MDU, 2001)*

Solution. The angle between two surfaces at a common point is the angle between their normals at that point. We have
$$f_1(x, y, z) = x^2 + y^2 + z^2 - 9 = 0, \quad \nabla f_1(x, y, z) = 2xi + 2yj + 2zk.$$
$\nabla f_1(2, -1, 2) = 4i - 2j + 4k$, which is normal N_1 to $x^2 + y^2 + z^2 = 9$ at $(2, -1, 2)$.
$$f_2(x, y, z) = x^2 + y^2 - z - 3 = 0, \quad \nabla f_2(x, y, z) = 2xi + 2yj - k.$$
$$N_2 = \nabla f_2(2, -1, 2) = 4i - 2j - k.$$
If the required angle is θ then

$$\cos\theta = \frac{N_1 \cdot N_2}{|N_1||N_2|} = \frac{(4i - 2j + 4k)\cdot(4i - 2j - k)}{\sqrt{(4)^2 + (-2)^2 + (4)^2}\ \sqrt{(4)^2 + (-2)^2 + (-1)^2}}$$

$$= \frac{16 + 4 - 4}{6\sqrt{21}} = \frac{8\sqrt{21}}{63}, \text{ or } \theta = \cos^{-1}\left(8\sqrt{21}/63\right) = 54°25'.$$

Directional derivative

Let $f(P) = f(x, y, z)$ be a differential scalar field. Then, $\partial f/\partial x, \partial f/\partial y, \partial f/\partial z$ denote the rates of change of f in the directions of x, y and z axis, respectively. Let

$\hat{\mathbf{b}} = b_1\mathbf{i} + b_2\mathbf{j} + b_3\mathbf{k}$ be any unit vector. Let P_0 be any point, its position vector is $\mathbf{a} = a_1\mathbf{i} + a_2\mathbf{j} + a_3\mathbf{k}$. Refer Fig. 14.10. Let P_0C is parallel to $\mathbf{b}$ and Q is a variable point on the directed line segment P_0C. Then the position vector of point Q is given by $\mathbf{r}(s) = \mathbf{a} + s\hat{\mathbf{b}}$,

$$= \left(a_1 + b_1 s\right)\mathbf{i} + \left(a_2 + b_2 s\right)\mathbf{j} + \left(a_3 + b_3 s\right)\mathbf{k} = x(s)\mathbf{i} + y(s)\mathbf{j} + z(s)\mathbf{k}.$$

The vector $\overrightarrow{P_0Q}$ is given by $s\hat{\mathbf{b}}$. Since $|\hat{\mathbf{b}}| = 1$, so the distance from P_0 to Q is s. Then

$$\frac{\partial f}{\partial s} = \lim_{s \to 0} \frac{f(Q) - f(P_0)}{s} \quad \text{if it exists,}$$

is called the *directional derivative* of f at the point P_0 in the direction of $\hat{\mathbf{b}}$. It is the rate of change of f in the direction of $\mathbf{b}$.

Therefore, $\dfrac{\partial}{\partial s} f\big(x(s), y(s), z(s)\big)$ is the rate of change of f with respect to the distance s.

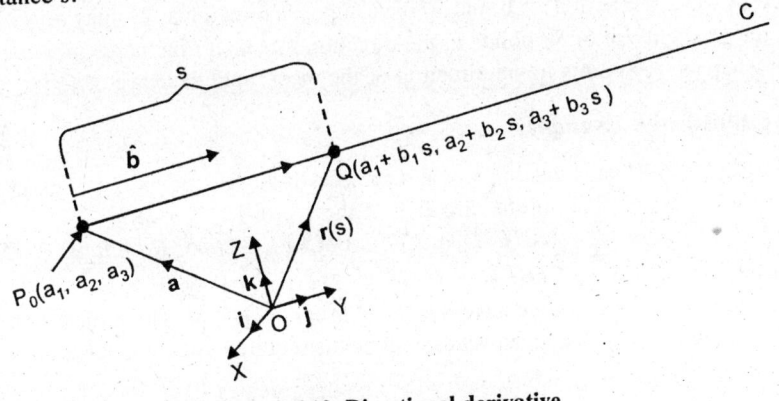

Fig. 14.10. Directional derivative

We have (by chain rule) $\dfrac{\partial f}{\partial s} = \dfrac{\partial f}{\partial x}\dfrac{dx}{ds} + \dfrac{\partial f}{\partial y}\dfrac{dy}{ds} + \dfrac{\partial f}{\partial z}\dfrac{dz}{ds}$ ...(14.15)

where dx/ds, dy/ds, dz/ds are calculated at the point P_0 (where $s = 0$). We write equation (14.15) as

$$\frac{\partial f}{\partial s} = \left(\mathbf{i}\frac{\partial}{\partial x} + \mathbf{j}\frac{\partial}{\partial y} + \mathbf{k}\frac{\partial}{\partial z}\right)f\cdot\left(\mathbf{i}\frac{dx}{ds} + \mathbf{j}\frac{dy}{ds} + \mathbf{k}\frac{dz}{ds}\right) = \nabla f\cdot\frac{d\mathbf{r}}{ds}.$$

But $d\mathbf{r}/ds = \hat{\mathbf{b}}$ (a unit vector). Therefore, the directional derivative of f in the direction of $\hat{\mathbf{b}}$, which is denoted by $D_{\mathbf{b}}(f)$, is given by

$$D_{\mathbf{b}}(f) = \nabla f \cdot \hat{\mathbf{b}} = \mathbf{grad}(f) \cdot \hat{\mathbf{b}} \qquad \ldots(14.16)$$

Remark: If the direction is specified by a vector **u**, then we have

$$\hat{\mathbf{b}} = \mathbf{u}/|\mathbf{u}| \text{ and } D_{\mathbf{b}}(f) = \nabla f \cdot \frac{\mathbf{u}}{|\mathbf{u}|}.$$

Maximum/Minimum value of the Directional derivative

Let f represent a function of either two or three variables. Since the directional derivative is a dot product, we see from the definition of dot or scalar product that

$$D_{\mathbf{b}}(f) = \nabla f \cdot \hat{\mathbf{b}} = |\nabla f||\hat{\mathbf{b}}|\cos\alpha = |\nabla f|\cos\alpha, \qquad \text{as } \left(|\hat{\mathbf{b}}| = 1\right),$$

where α is the angle between ∇f and $\hat{\mathbf{b}}$. Because $0 \le \alpha \le \pi$, we have $-1 \le \cos\alpha \le 1$ and, consequently $-|\nabla f| \le D_{\mathbf{b}}(f) \le |\nabla f|$.

Thus the maximum value of the directional derivative is $|\nabla f|$ and it occurs when $\alpha = 0$, that is, when $\hat{\mathbf{b}}$ has the same direction as ∇f. This direction is the direction of the normal vector. The minimum value of the directional derivative is $-|\nabla f|$ and it occurs when $\theta = \pi$, that is, $\hat{\mathbf{b}}$ and ∇f have opposite directions. We may also say that the gradient vector ∇f points in the direction in which f increases most rapidly, whereas $-\nabla f$ points in the direction of the most rapid decrease of f.

14.11 Illustrative Examples

Example 14.23 What is the directional derivative of the function $f(x, y, z) = xy^2 + yz^3$ at the point $(2, -1, 1)$ in the direction of the vector $\mathbf{i} + 2\mathbf{j} + 3\mathbf{k}$?

(IETE, Dec. 2009; A.M.I.E., S-2005; Karnataka 1995)

Solution. We have $\nabla f = \nabla(xy^2 + yz^3) = y^2\mathbf{i} + (2xy + z^3)\mathbf{j} + 3yz^2\mathbf{k}$.

At the point $(2, -1, 1)$, we have $\nabla f(2, -1, 1) = \mathbf{i} - 3\mathbf{j} - 3\mathbf{k}$. The projection of this in the direction of the given vector will be the required directional derivative.

The unit vector in the given direction is $\hat{\mathbf{b}} = \dfrac{(\mathbf{i}+2\mathbf{j}+3\mathbf{k})}{\sqrt{1+4+9}} = \dfrac{\mathbf{i}+2\mathbf{j}+3\mathbf{k}}{\sqrt{14}}$.

Therefore,
$$D_{\mathbf{b}}(2,-1,1) = \nabla f \cdot \hat{\mathbf{b}} = (\mathbf{i}-3\mathbf{j}-3\mathbf{k}) \cdot \frac{(\mathbf{i}+2\mathbf{j}+3\mathbf{k})}{\sqrt{14}}$$

$$= \frac{(1)(1)+(-3)(2)+(-3)(3)}{\sqrt{14}} = \frac{-14}{\sqrt{14}} = -\sqrt{14}.$$

The negative sign, of course, indicates that f decreases in the given direction.

Example 14.24 Find the directional derivative of $f(x, y) = 2x^2y^3 + 6xy$ at $(1, 1)$ in the direction of a unit vector whose angle with the positive x-axis is $\pi/6$.

Solution. We have $\nabla f(x, y) = (4xy^3 + 6y)\,\mathbf{i} + (6x^2y^2 + 6x)\mathbf{j}$. At the point $(1, 1)$, we have $\nabla f(1, 1) = 10\mathbf{i} + 12\mathbf{j}$. Now, at $\theta = \pi/6$, $\hat{\mathbf{b}} = (\cos\theta)\,\mathbf{i} + (\sin\theta)\mathbf{j} = \dfrac{\sqrt{3}}{2}\mathbf{i} + \dfrac{1}{2}\mathbf{j}$.

Therefore $D_b(1, 1) = \nabla f \cdot \hat{\mathbf{b}} = (10\mathbf{i} + 12\mathbf{j}) \cdot \left(\dfrac{\sqrt{3}}{2}\mathbf{i} + \dfrac{1}{2}\mathbf{j} \right) = 5\sqrt{3} + 6$.

Example 14.25 (*a*) In what direction from the point $(3, 1, -2)$ is the directional derivative of $f(x, y, z) = x^2y^2z^4$ maximum and what is its magnitude?

(A.M.I.E., W-2005; MDU 2004; Bhopal 1998; Kerala 1997)

Solution. We have $\nabla f = \nabla(x^2y^2z^4) = 2xy^2z^4\mathbf{i} + 2x^2yz^4\mathbf{j} + 4x^2y^2z^3\mathbf{k}$.

At the point $(3, 1, -2)$, we have $\nabla f = 96\mathbf{i} + 288\mathbf{j} - 288\mathbf{k}$.

(*a*) the directional derivative is a maximum in the direction
$\nabla f = 96\mathbf{i} + 288\mathbf{j} - 288\mathbf{k}$.

(*b*) the magnitude of this maximum directional derivative is

$$|\nabla f| = \sqrt{(96)^2 + (288)^2 + (-288)^2} = \sqrt{175104} = 96\sqrt{19}\,.$$

Example 14.26 (*a*) Find the directional derivative of the function $f(x, y, z) = 2x^3y - 3y^2z$ at $P(1, 2, -1)$ in a direction toward $Q(3, -1, 5)$.

(*b*) In what direction from P is the directional derivative a maximum?

(*c*) what is the magnitude of the maximum directional derivative?

Solution. (*a*) We have $\nabla f = 6x^2y\mathbf{i} + (2x^3 - 6yz)\mathbf{j} - 3y^2\mathbf{k}$.

At the point $(1, 2, -1)$, we have $\nabla f = 12\mathbf{i} + 14\mathbf{j} - 12\mathbf{k}$.

The vector from P to $Q = (3 - 1)\mathbf{i} + (-1-2)\mathbf{j} + [5 -(-1)]\mathbf{k} = 2\mathbf{i} - 3\mathbf{j} + 6\mathbf{k}$.

The unit vector from P to $Q = \hat{\mathbf{b}} = \dfrac{2\mathbf{i} - 3\mathbf{j} + 6\mathbf{k}}{\sqrt{(2)^2 + (-3)^2 + (6)^2}} = \dfrac{2\mathbf{i} - 3\mathbf{j} + 6\mathbf{k}}{7}$.

Therefore, directional derivative at P, $D_b f = \nabla f \cdot \hat{\mathbf{b}}$

$$= (12\mathbf{i} + 14\mathbf{j} - 12\mathbf{k}) \cdot \left(\dfrac{2\mathbf{i} - 3\mathbf{j} + 6\mathbf{k}}{7} \right) = \dfrac{24 - 42 - 72}{7} = -\dfrac{90}{7},$$

that is, f is decreasing in the given direction.

(*b*) The directional derivative is a maximum in the direction $\nabla f = 12\mathbf{i} + 14\mathbf{j} - 12\mathbf{k}$.

(*c*) The value of the maximum directional derivative is

$$|\nabla f| = |12\mathbf{i} + 14\mathbf{j} - 12\mathbf{k}| = \sqrt{144 + 196 + 144} = 22.$$

Example 14.27 If the directional derivative of $f(x, y, z) = axy^2 + byz + cz^2x^3$ at $(1, 2, -1)$ has a maximum magnitude of 64 in the direction parallel to z-axis, find the values of constants a, b, c.

(I.E.T.E., June 2001)

Solution. The maximum directional derivative of f at any point is the value of ∇f at the point.

Now $\nabla f = \dfrac{\partial f}{\partial x}\mathbf{i} + \dfrac{\partial f}{\partial y}\mathbf{j} + \dfrac{\partial f}{\partial z}\mathbf{k} = \left(ay^2 + 3cz^2x^2\right)\mathbf{i} + \left(2axy + bz\right)\mathbf{j} + \left(by + 2czx^3\right)\mathbf{k}.$

$\therefore \quad \nabla f(1,2,-1) = (4a + 3c)\mathbf{i} + (4a - b)\mathbf{j} + (2b - 2c)\mathbf{k}.$

Since its magnitude is 64 and direction is parallel to z-axis, we have

$(4a + 3c)\mathbf{i} + (4a - b)\mathbf{j} + (2b - 2c)\mathbf{k} = 64\mathbf{k}$

$\Rightarrow 4a + 3c = 0,\ 4a - b = 0,\ 2b - 2c = 64.$ Solving, we get $a = 6,\ b = 24,\ c = -8.$

Example 14.28 Find the directional derivative of $f(x, y, z) = x^2yz^3$ along the curve $x = e^{-u},\ y = 2\sin u + 1,\ z = u - \cos u$ at the point P where $u = 0$.

Solution. The point P corresponding to $u = 0$ is $(1, 1, -1)$.

$\nabla f = 2xyz^3\mathbf{i} + x^2z^3\mathbf{j} + 3x^2yz^2\mathbf{k}.$ Therefore, $\nabla f(1, 1, -1) = -2\mathbf{i} - \mathbf{j} + 3\mathbf{k}.$

The position vector $\mathbf{r}$ of any point on the given curve is given by

$\mathbf{r} = e^{-u}\mathbf{i} + (2\sin u + 1)\mathbf{j} + (u - \cos u)\mathbf{k}.$

A tangent vector to the curve is

$\mathbf{r}'(u) = \dfrac{d\mathbf{r}}{du} = \dfrac{d}{du}\left\{e^{-u}\mathbf{i} + (2\sin u + 1)\mathbf{j} + (u - \cos u)\mathbf{k}\right\}$

$= -e^{-u}\mathbf{i} + 2\cos u\,\mathbf{j} + (1 + \sin u)\,\mathbf{k}.\quad \therefore \left(\dfrac{d\mathbf{r}}{du}\right)_{\text{at } P} = \mathbf{r}'(0) = -\mathbf{i} + 2\mathbf{j} + \mathbf{k}.$

The unit tangent vector in this direction $= \dfrac{-\mathbf{i} + 2\mathbf{j} + \mathbf{k}}{\sqrt{1 + 4 + 1}} = \dfrac{-\mathbf{i} + 2\mathbf{j} + \mathbf{k}}{\sqrt{6}}.$

Then, directional derivative $= \nabla f \bullet$ (unit tangent vector in this direction)

$= (-2\mathbf{i} - \mathbf{j} + 3\mathbf{k}) \bullet \left(\dfrac{-\mathbf{i} + 2\mathbf{j} + \mathbf{k}}{\sqrt{6}}\right) = \dfrac{1}{\sqrt{6}}(2 - 2 + 3) = \dfrac{3}{\sqrt{6}} = \dfrac{\sqrt{6}}{2}.$

Since this is positive, f is increasing in this direction.

14.12 Properties of Gradient

If f and g are differentiable scalar functions of position (x, y, z), then

(i) $\nabla(kf) = k\,\nabla f$ (any number k),

(ii) $\nabla(f \pm g) = \nabla f \pm \nabla g,$

(iii) $\nabla(c_1 f + c_2 g) = c_1\,\nabla f + c_2\,\nabla g,\ c_1, c_2$ arbitrary constants,

(iv) $\nabla(fg) = f\nabla g + g\,\nabla f,$

(v) $\nabla\left(\dfrac{f}{g}\right) = \dfrac{g\nabla f - f\nabla g}{g^2},\ g \neq 0.$

The proof of properties (*ii*), (*iv*) and (*v*) is given below.

(*ii*) $\nabla(f \pm g) = \left(\mathbf{i}\dfrac{\partial}{\partial x} + \mathbf{j}\dfrac{\partial}{\partial y} + \mathbf{k}\dfrac{\partial}{\partial z} \right)(f \pm g)$

$$= \mathbf{i}\dfrac{\partial}{\partial x}(f \pm g) + \mathbf{j}\dfrac{\partial}{\partial y}(f \pm g) + \mathbf{k}\dfrac{\partial}{\partial z}(f \pm g)$$

$$= \left(\mathbf{i}\dfrac{\partial f}{\partial x} + \mathbf{j}\dfrac{\partial f}{\partial y} + \mathbf{k}\dfrac{\partial f}{\partial z} \right) \pm \left(\mathbf{i}\dfrac{\partial g}{\partial x} + \mathbf{j}\dfrac{\partial g}{\partial y} + \mathbf{k}\dfrac{\partial g}{\partial z} \right) = \nabla f \pm \nabla g.$$

(*iv*) $\nabla(fg) = \left(\mathbf{i}\dfrac{\partial}{\partial x} + \mathbf{j}\dfrac{\partial}{\partial y} + \mathbf{k}\dfrac{\partial}{\partial z} \right)fg$

$$= \mathbf{i}\left(f\dfrac{\partial g}{\partial x} + g\dfrac{\partial f}{\partial x} \right) + \mathbf{j}\left(f\dfrac{\partial g}{\partial y} + g\dfrac{\partial f}{\partial y} \right) + \mathbf{k}\left(f\dfrac{\partial g}{\partial z} + g\dfrac{\partial f}{\partial z} \right)$$

$$= f\left(\mathbf{i}\dfrac{\partial g}{\partial x} + \mathbf{j}\dfrac{\partial g}{\partial y} + \mathbf{k}\dfrac{\partial g}{\partial z} \right) + g\left(\mathbf{i}\dfrac{\partial f}{\partial x} + \mathbf{j}\dfrac{\partial f}{\partial y} + \mathbf{k}\dfrac{\partial f}{\partial z} \right) = f\nabla g + g\nabla f.$$

(*v*) $\nabla\left(\dfrac{f}{g} \right) = \left(\mathbf{i}\dfrac{\partial}{\partial x} + \mathbf{j}\dfrac{\partial}{\partial y} + \mathbf{k}\dfrac{\partial}{\partial z} \right)\left(\dfrac{f}{g} \right)$

$$= \mathbf{i}\dfrac{\left(g\dfrac{\partial f}{\partial x} - f\dfrac{\partial g}{\partial x} \right)}{g^2} + \mathbf{j}\dfrac{\left(g\dfrac{\partial f}{\partial y} - f\dfrac{\partial g}{\partial y} \right)}{g^2} + \mathbf{k}\dfrac{\left(g\dfrac{\partial f}{\partial z} - f\dfrac{\partial g}{\partial z} \right)}{g^2}, \text{ if } g \neq 0$$

$$= \dfrac{1}{g^2}\left[g\left(\mathbf{i}\dfrac{\partial f}{\partial x} + \mathbf{j}\dfrac{\partial f}{\partial y} + \mathbf{k}\dfrac{\partial f}{\partial z} \right) - f\left(\mathbf{i}\dfrac{\partial g}{\partial x} + \mathbf{j}\dfrac{\partial g}{\partial y} + \mathbf{k}\dfrac{\partial g}{\partial z} \right) \right] = \dfrac{g\nabla f - f\nabla g}{g^2}.$$

Conservative Vector Field

A vector field **v** is said to be conservative if the vector function can be written as the gradient of a scalar function f, that is $\mathbf{v} = \nabla f$. In such a vector field, the work done in moving a particle from a point A to a point B depends only on the end point A and B and is independent of path joining the two points.

Remark. It may be noted that not every vector field is conservative.

Example 14.29 If $\nabla f = (x + 2y + 4z)\,\mathbf{i} + (2x - 3y - z)\mathbf{j} + (4x - y + 2z)\,\mathbf{k}$, find the scalar function f.

Solution. Since $\nabla f = \dfrac{\partial f}{\partial x}\mathbf{i} + \dfrac{\partial f}{\partial y}\mathbf{j} + \dfrac{\partial f}{\partial z}\mathbf{k}$. Comparing, we obtain

$$\frac{\partial f}{\partial x} = x + 2y + 4z \ ...(i) \quad \frac{\partial f}{\partial y} = 2x - 3y - z \ ...(ii) \quad \frac{\partial f}{\partial z} = 4x - y + 2z \ ...(iii).$$

Integrating the first equation, we obtain

$$f(x, y, z) = \left(x^2/2\right) + 2xy + 4zx + g(y, z). \qquad ...(A)$$

Substituting in the second equation, we get

$$\frac{\partial f}{\partial y} = 2x - 3y - z = 2x + \frac{\partial g}{\partial y}, \text{ or } \frac{\partial g}{\partial y} = -3y - z. \text{ Therefore}$$

$$g(y, z) = (-3/2)y^2 - yz + g(z).$$

Now (A) become $f(x, y, z) = \left(x^2/2\right) + 2xy + 4zx - (3/2)y^2 - yz + g(z)$ ...(B)

Substituting in the third equation, we get

$$\frac{\partial f}{\partial z} = 4x - y + 2z = 4x - y + \frac{dg}{dz} \text{ or } \frac{dg}{dz} = 2z. \text{ Integrating, we get}$$

$g = z^2 + k$, constant.

Thus from (B), we obtain $f(x, y, z) = \left(x^2/2\right) + 2xy + 4xz - yz - \left(3y^2/2\right) + z^2 + k.$

Example 14.30 Show that the vector field defined by the vector function
$\mathbf{v} = xyz(yz\mathbf{i} + zx\mathbf{j} + xy\mathbf{k})$ is conservative. *(IETE, June 2008)*
Solution. If the given vector field is conservative, then it can be expressed as the gradient of a scalar function $f(x, y, z)$. Therefore

$$\nabla f = \left(\mathbf{i}\frac{\partial f}{\partial x} + \mathbf{j}\frac{\partial f}{\partial y} + \mathbf{k}\frac{\partial f}{\partial z} \right) = \mathbf{v} = xyz\left(yz\mathbf{i} + zx\mathbf{j} + xy\mathbf{k}\right).$$

Comparing, we obtain

$$\frac{\partial f}{\partial x} = xy^2z^2 \ ...(i), \quad \frac{\partial f}{\partial y} = x^2yz^2 \ ...(ii), \quad \frac{\partial f}{\partial z} = x^2y^2z \ ...(iii).$$

Integrating the first equation, we obtain

$$f(x, y, z) = \frac{1}{2}x^2y^2z^2 + g(y, z). \qquad ...(A)$$

Substituting in the second equation, we get

$$\frac{\partial f}{\partial y} = x^2yz^2 = x^2yz^2 + \frac{\partial g}{\partial y} \text{ or } \frac{\partial g}{\partial y} = 0 \Rightarrow g = g(z).$$

Now (A) become $f(x, y, z) = (1/2)x^2y^2z^2 + g(z)$ ...(B)

Substituting in the third equation, we get

$$\frac{\partial f}{\partial z} = x^2y^2z = x^2y^2z + \frac{dg}{dz} \text{ or } \frac{dg}{dz} = 0 \Rightarrow g = k_1, \text{ constant}$$

Hence, $f(x, y, z) = x^2y^2z^2 + k$. Therefore, there exists a scalar function $f(x, y, z)$ such that $\nabla f = \mathbf{v}$ and the vector field $\mathbf{v}$ is conservative.

Exercise 14.2

1. Compute the gradient of the following scalar fields.

   (i) $f(x, y) = y^2 - 4xy$,　　　　(ii) $f(x, y) = 5y - x^3y^2$,

   (iii) $f(x, y, z) = xy^2 + 3x^2 - z^3$,　　(iv) $f(x, y, z) = xy^2/z^3$,

   (v) $f(x, y, z) = x^2 + yz$,　　　(vi) $\phi(x, y, z) = x^3 + y^3 + z^3 - 3xyz$,

   (vii) $\phi(x, y, z) = \log(x^2 + y^2 + z^2)$.

   Ans. (i) $-4y\mathbf{i} + (2y - 4x)\mathbf{j}$, (ii) $-3x^2y^2\mathbf{i} + (5 - 2x^3y)\mathbf{j}$,
   (iii) $(y^2 + 6x)\mathbf{i} + 2xy\mathbf{j} - 3z^2\mathbf{k}$, (iv) $(y^2/z^3)\mathbf{i} + (2xy/z^3)\mathbf{j} - (3xy^2/z^4)\mathbf{k}$,

   (v) $2x\mathbf{i} + z\mathbf{j} + y\mathbf{k}$, (vi) $3(x^2 - yz)\mathbf{i} + 3(y^2 - zx)\mathbf{j} + 3(z^2 - xy)\mathbf{k}$.

[Hint: (vii) $\nabla \log\left(x^2 + y^2 + z^2\right) = \nabla \log r^2 = \dfrac{2}{r} \cdot \dfrac{\mathbf{r}}{r} = \dfrac{2\mathbf{r}}{r^2} = \dfrac{2\left(x\mathbf{i} + y\mathbf{j} + 3\mathbf{k}\right)}{x^2 + y^2 + z^2}.$]

2. Find the gradient of the given function at the indicated point

   (i) $f(x, y) = x^2 - 4y^2$, $(2, 4)$. (ii) $f(x, y, z) = 3x^2y - y^3z^2$, $(1, -2, -1)$,

   (iii) $f(x, y, z) = 2xz^4 - x^2y$, $(2, -2, -1)$.

   (iv) $f(x, y, z) = x^2y + y^2x + z^2$, $(1, 1, 1)$,

   (v) $f(x, y, z) = x^3 - y^3 + xz^2$, $(1, 1, 2)$.

   Ans. (i) $4\mathbf{i} - 32\mathbf{j}$, (ii) $-12\mathbf{i} - 9\mathbf{j} - 16\mathbf{k}$, (iii) $10\mathbf{i} - 4\mathbf{j} - 16\mathbf{k}$,
   (iv) $3\mathbf{i} + 3\mathbf{j} + 2\mathbf{k}$, (v) $7\mathbf{i} - 3\mathbf{j} + 4\mathbf{k}$.

3. If $f = 3x^2y$, $\phi = xz^2 - 2y$; evaluate grad $[(\text{grad } f) \cdot (\text{grad } \phi)]$.

   Ans. $\left(6yz^2 - 12x\right)\mathbf{i} + 6xz^2\mathbf{j} + 12xyz\mathbf{k}$.

4. (a) Evaluate $\nabla\left(3r^2 - 4\sqrt{r} + \dfrac{6}{\sqrt[3]{r}}\right)$, (b) Show that grad $\left(\dfrac{1}{r^2}\right) = \dfrac{-2\mathbf{r}}{r^4}$.

   (*MDU 2002*) **Ans.** (a) $\left(6 - 2r^{-3/2} - 2r^{-7/3}\right)\mathbf{r}$.

5. Find the normal vector and the unit normal vector to the given curve/surface at the indicated point.

   (i) $x^2 - y^2 = 12$, $(4, 2)$,　　(ii) $z = x^2 + y^2$, $(1, 2, 5)$,

   (iii) $xy^3z^2 = 4$, $(-1, -1, 2)$,

   (*Madras 1999S*)

(iv) $x^4 - 3xyz + z^2 + 1 = 0$, $(1, 1, 1)$,

(v) $x^3 + y^3 + 3xyz = 3$, $(1, 2, -1)$, *(I.E.T.E., June 2002)*

(vi) $x^3 - xyz = 1$, $(1, 1, 1)$, *(I.E.T.E., Dec. 2004)*

(vii) $f(x, y, z) = xy^2 + 2yz - 8 = 0$, $(3, -2, 1)$, *(A.M.I.E., W-2007)*

(viii) $(x-1)^2 + y^2 + (z+2)^2 = 9$, $(3, 1, -4)$.

Ans. (i) $4(2\mathbf{i} - \mathbf{j})$, $(2\mathbf{i} - \mathbf{j})/\sqrt{5}$ (ii) $(2\mathbf{i} + 4\mathbf{j} - \mathbf{k})$; $(2\mathbf{i} + 4\mathbf{j} - \mathbf{k})/\sqrt{21}$

(iii) $4(\mathbf{i} + 3\mathbf{j} - \mathbf{k})$; $(\mathbf{i} + 3\mathbf{j} - \mathbf{k})/\sqrt{11}$. (iv) $\mathbf{i} - 3\mathbf{j} - \mathbf{k}$; $(\mathbf{i} - 3\mathbf{j} - \mathbf{k})/\sqrt{11}$,

(v) $(-\mathbf{i} + 3\mathbf{j} + 2\mathbf{k})$, $(-\mathbf{i} + 3\mathbf{j} + 2\mathbf{k})/\sqrt{14}$.

(vi) $2\mathbf{i} - \mathbf{j} - \mathbf{k}$; $(2\mathbf{i} - \mathbf{j} - \mathbf{k})/\sqrt{6}$. (vii) $4\mathbf{i} - 10\mathbf{j} - 4\mathbf{k}$; $(2\mathbf{i} - 5\mathbf{j} - 2\mathbf{k})/\sqrt{33}$.

(viii) $4\mathbf{i} + 2\mathbf{j} - 4\mathbf{k}$; $(2\mathbf{i} + \mathbf{j} - 2\mathbf{k})/3$.

6. If $u = x + y + z$, $v = x^2 + y^2 + z^2$, $w = xy + yz + zx$, show that

$$\nabla u \cdot [\nabla v \times \nabla w] = 0.$$ *(UPTU 2002; Patna 1999)*

7. (a) What is the angle between the normals to the surface $xy = z^2$ at the points $(1, 4, 2)$ and $(-3, -3, 3)$? **Ans.** $\cos^{-1}(1/\sqrt{22})$.

(b) Find the magnitude of the vector drawn perpendicular to the surface $x^2 + 2y^2 + z^2 = 7$ at the point $(1, -1, 2)$. *(AMIE, S-2009)* **Ans. 6.**

8. Find the angle between the two surfaces at the indicated point of intersection.

(i) $xy^2z = 3x + z^2$, $3x^2 - y^2 + 2z = 1$; $(1, -2, 1)$. *(PU 1999)*

(ii) $x \log z = y^2 - 1$, $x^2y = 2 - z$; $(1, 1, 1)$. *(IETE, Dec. 2006; JNTU 2003)*

Ans. (i) $\cos^{-1}(\sqrt{6}/14)$, (ii) $\cos^{-1}(1/\sqrt{30})$.

9. (a) Find the normal vector and the equation of tangent plane to surface $z = \sqrt{x^2 + y^2}$ at point $(3, 4, 5)$. *(I.E.T.E., Dec. 2005)*

Ans. $(-3/5)\mathbf{i} - (4/5)\mathbf{j} + \mathbf{k}$; $3x + 4y - 5z = 0$.

(b) Find the equation of the tangent plane to the ellipsoid $x^2 + 4y^2 + z^2 = 18$ at the point $(1, 2, 1)$, and determine the acute angle that this plane makes with the xy-plane. **Ans.** $x + 8y + z = 18$; $\approx 83°$.

10. Find the values of constants λ and μ so that the surfaces $\lambda x^2 - \mu yz = (\lambda + 2) x$ and $4x^2y + z^3 = 4$ intersect orthogonally at $(1, -1, 2)$.

(I.E.T.E., Dec 2008, 2005; MDU 2005; JNTU 2001) **Ans.** $\lambda = 5/2$, $\mu = 1$.

11. Find the constants a and b such that the surface $ax^2 - byz = (a + 2)x$ will be orthogonal to the surface $4x^2y + z^3 = 4$ at the point $(1, -1, 2)$.

    (AMIE, S-2007; Madras, 2004; IETE Dec 2003) **Ans.** $a = 5/2, b = 1$.

    [**Hint:** Since $(1, -1, 2)$ lies on $ax^2 - 2byz = (a + 4)x$, we get $b = 1$. Use the condition of orthogonality to get a.]

12. If $\nabla f = 2xy\, z^3 \mathbf{i} + x^2z^3\, \mathbf{j} + 3x^2yz^2\, \mathbf{k}$, find $f(x, y, z)$ if $f(1, -2, 2) = 4$.

    Ans. $f(x, y, z) = x^2yz^3 + 20$.

13. Find the directional derivative of

    (i) $f(x, y) = xe^y + \cos(xy)$ at $(2, 0)$ in the direction of $3\mathbf{i} - 4\mathbf{j}$.

    (ii) $f(x, y, z) = 4xz^3 - 3x^2y^2z^2$ at $(2, -1, 2)$ along z-axis.

    (iii) $f(x, y, z) = x^2yz + 4xz^2$ at $(1, 2, -1)$ in the direction $2\mathbf{i} - \mathbf{j} - 2\mathbf{k}$.

    (AMIE., W-2008; MDU 2006; VTU 2001)

    (iv) $\phi(x, y, z) = xy^2 - 4x^2y + z^2$ at $(1, -1, 2)$ in the direction of $6\mathbf{i} + 2\mathbf{j} + 3\mathbf{k}$.

    (v) $f(x, y, z) = 4xz^3 - 3x^2y^2z$ at $(2, -1, 2)$ in the direction $2\mathbf{i} - 3\mathbf{j} + 6\mathbf{k}$.

    (vi) $\phi(x, y, z) = 4\, e^{2x - y + z}$ at $(1, 1, -1)$ in the direction of vector $\mathbf{j}$.

    (vii) $\phi = 3e^{2x-y+z}$ at $A(1,1,-1)$ in the direction $\overrightarrow{AB}$ where B is the point $(-3, 5, 6)$. *(IETE, Dec. 2001)*

    (viii) $\phi(x,y) = x/(x^2 + y^2)$ in the direction of a line making an angle of 30° with the positive x-axis at the point $(0, 1)$. *(IETE, Dec. 2002)*

    Ans. (i) -1, (ii) 48, (iii) 37/3, (iv) 54/7, (v) 376/7, (vi) -4, (vii) $-5/3$, (viii) $\sqrt{3}/2$.

14. What is the directional derivative of the scalar function $f = xy^2 + yz^3$ at the point $(2, -1, 1)$ in the direction of the normal to the surface $x \log z - y^2 + 4 = 0$ at $(-1, 2, 1)$?

    (MDU, 2007; J.N.T.U, 2005; V.T.U, 2004; Andhra, 2000) **Ans.** $15/\sqrt{17}$.

15. Find the directional derivative of $\vec{V}^2$ where $\vec{V} = xy^2\mathbf{i} + zy^2\mathbf{j} + xz^2\mathbf{k}$ at the point $(2, 0, 3)$ in the direction of the outward normal to the sphere $x^2 + y^2 + z^2 = 14$ at the point $(3, 2, 1)$. *(IETE Dec 2007)* **Ans.** $702\sqrt{14}/7$.

[**Hint.** $\phi(x,y,z) = x^2y^4 + z^2y^4 + x^2z^4$. At the point $(2, 0, 3)$, $\nabla\phi = 324\mathbf{i} + 432\mathbf{k}$. The unit vector in the given direction is $\hat{b} = (3\mathbf{i} + 2\mathbf{j} + \mathbf{k})/\sqrt{14}$.

Therefore, $D_b(2,0,3) = \nabla\phi \cdot \hat{b} = (324\mathbf{i} + 432\mathbf{k}) \cdot \dfrac{3\mathbf{i} + 2\mathbf{j} + \mathbf{k}}{\sqrt{14}} = 702\sqrt{14}/7$.]

16. (a) Find the directional derivative of the function $\phi = 4e^{2x - y + z}$ at the point $A(1, 1, -1)$ in the direction towards the point $B(-3, 5, 6)$. **Ans.** $-20/9$.

    (b) In what direction from the point $(1, 1, -1)$ is the directional derivative of $\phi = x^2 - 2y^2 + 4z^2$ a maximum? Also find the value of this maximum directional derivative.

    Ans. $2\mathbf{i} - 4\mathbf{j} - 8\mathbf{k}; 2\sqrt{21}$.

17. Find the directional derivative of the function $\phi(x,y,z) = x^2 - y^2 + 2z^2$ at the point $P(1, 2, 3)$ in the direction of the line PQ, where Q has coordinates $(5, 0, 4)$.   *(MDU 2004; UPTU 2001; DU 2001)* **Ans.** $4\sqrt{21}/3$.

18. Find the directional derivative of $\phi(x,y) = x^2y^3 + xy$ at $(2, 1)$ in the direction of a unit vector which makes an angle of $\pi/3$ with x-axis.

    [Hint. $\nabla f = (2xy^3 + y)\mathbf{i} + (3x^2y^2 + x)\mathbf{j}$. At $(2,1)$, $\nabla f = 5\mathbf{i} + 14\mathbf{j}$.

    $\hat{\mathbf{b}} = \cos\theta\,\mathbf{i} + \sin\theta\,\mathbf{j} = \mathbf{i}/2 + (\sqrt{3}/2)\mathbf{j}$. Therefore direction derivative

    $= (5\mathbf{i} + 14\mathbf{j})\cdot(\mathbf{i}/2 + (\sqrt{3}/2)\mathbf{j}) = (5 + 14\sqrt{3})/2$.]

19. Find the directional derivative of $1/r$ in the direction $\mathbf{r}$ where $\mathbf{r} = x\mathbf{i} + y\mathbf{j} + z\mathbf{k}$.
    (UPTU 2003, 2002) **Ans.** $D_b f = -1/(x^2 + y^2 + z^2)$.

20. (a) Given that the directional derivative of $f(x, y, z)$ at the point $(3, -2, 1)$ in the direction of $\mathbf{a} = 2\mathbf{i} - \mathbf{j} - 2\mathbf{k}$ is -5 and that $|\nabla f(3, -2, 1)| = 5$, find $\nabla f(3, -2, 1)$.   **Ans.** $-5(2\mathbf{i} - \mathbf{j} - 2\mathbf{k})/3$.

    (b) Find the directional derivative of $\phi = 5x^2y - 5y^2z + 2.5z^2x$ at the point $P(1,1,1)$ in the direction of the line $\dfrac{x-1}{2} = \dfrac{y-3}{-2} = z$.   *(UPTU 2004)* **Ans.** 35/3.

21. What is the greatest rate of increase of

    (a) $\phi(x, y, z) = x^2 + yz^2$ at $(1, -1, 3)$?   *(A.M.I.E., S-2004; Madurai 1998)*
    (b) $f(x, y, z) = xyz^2$ at $(1, 0, 3)$?   **Ans.** (a) 11; (b) 9.

22. Find the directions in which $\phi(x,y) = (x^2 + y^2)/2$ increases most rapidly and decreases most rapidly at the point $(1, 1)$.

    Ans. $(1/\sqrt{2})\mathbf{i} + (1/\sqrt{2})\mathbf{j}$; $(-1/\sqrt{2})\mathbf{i} - (1/\sqrt{2})\mathbf{j}$.

23. (a) Find the directional derivative of $f(x, y, z) = 2xy - z^2$ at $(2, -1, 1)$ in a direction toward $(3, 1, -1)$. (b) In what direction is the directional derivative a maximum? (c) What is the value of this maximum?

    Ans. (a) 10/3, (b) $-2\mathbf{i} + 4\mathbf{j} - 2\mathbf{k}$ (c) $2\sqrt{6}$.

24. The temperature at a point (x, y, z) in a space is given by $T(x, y, z) = x^2 + y^2 - z$. A mosquito located at $(1, 1, 2)$ desires to fly in such a direction that it will get warm as soon as possible. In what direction should it fly?
    (A.M.I.E., S-2005) **Ans.** $(2\mathbf{i} + 2\mathbf{j} - \mathbf{k})/3$.

25. The temperature at a point (x, y, z) in a space is given by $T(x, y, z) = x^2 + y^2 - z$. A fly located at $(4, 4, 2)$ desires to fly in a direction that gets cooler

faster. Find the direction in which it should fly. Also find the rate decrease of temperature in the direction of flight. (*I.E.T.E., June 2005*)

Ans. In the direction of maximum rate of decrease, $-(8\mathbf{i} + 8\mathbf{j} - \mathbf{k})$.

26. Find the values of a, b, c so that the directional derivative of $\phi = ax^2 + by^2 + cz^2$ at the point $(1, 1, 2)$ has a maximum magnitude 4 in the direction of y-axis. [**Hint:** $\nabla\phi$ at $(1, 1, 2) = 4\mathbf{j}$] **Ans.** $a = 0, b = 2, c = 0$.

27. If the directional derivative of $\phi = ax^2y + by^2z + cz^2x$ at the point $(1, 1, 1)$

has maximum magnitude 15 in the direction parallel to line $\dfrac{x-1}{2} = \dfrac{y-3}{-2} = z$,

find the value of a, b, c. (*IETE June 2003; UPTU 2002*)

[**Hint:** The maximum directional derivative of ϕ at any point is the value of $\nabla\phi$ at the point. $\nabla\phi = (2axy + cz^2)\mathbf{i} + (2byz + ax^2)\mathbf{j} + (by^2 + 2czx)\mathbf{k}$.

$$\nabla\phi(1, 1, 1) = (2a + c)\mathbf{i} + (2b + a)\mathbf{j} + (b + 2c)\mathbf{k} = 15\dfrac{(2\mathbf{i}-2\mathbf{j}+\mathbf{k})}{\sqrt{4+4+1}}.$$

$= 10\mathbf{i} - 10\mathbf{j} + 5\mathbf{k}. \Rightarrow 2a + c = 10, 2b + a = -10, b + 2c = 5.$ Solving, we get $a = 20/9; b = -55/9, c = 50/9$].

28. Find the directional derivative of ϕ at the point $(1, -2, 1)$ in the direction of the normal to the surface $xy^2z = 3x + z^2$, where $\phi(x,y,z) = 2x^3y^2z^4$.

Ans. $1724/\sqrt{21}$.

29. Find the directional derivative of the scalar point function $\phi = x^2y + y^2z + z^2x$ at the point $(2, 2, 2)$ in the direction of the normal to the surface $4x^2y + 2z^2 = 2$ at the point $(2, -1, 3)$. (*IETE Dec 2006*)

30. Find $\phi(r)$ such that $\nabla\phi = \mathbf{r}/r^5$ and $\phi(1) = 0$. **Ans.** $\dfrac{1}{3}\left(1 - \dfrac{1}{r^3}\right)$.

31. If $f(x, y, z) = x^2z + e^{y/x}$ and $g(x, y, z) = 2\,z^2y - xy^2$, find (a) $\nabla(f + g)$ and (b) $\nabla(fg)$ at the point $(1, 0, -2)$. **Ans.** (a) $-4\mathbf{i} + 9\mathbf{j} + \mathbf{k}$, (b) $-8\mathbf{j}$.

32. Show that the field of force given by $\mathbf{F}(x,y,z) = \left(y^2\cos x + z^3\right)\mathbf{i} + \left(2y\sin x - 4\right)\mathbf{j} + (3xz^2 + 2)\mathbf{k}$ is conservative and find its scalar potential.

Ans. $\phi(x,y,z) = y^2\sin x + xz^3 - 4y + 2z + k_1$

33. If $f(x,y) = \log_e \sqrt{x^2 + y^2}$, show that

$$\text{grad } f = \dfrac{\mathbf{r} - (\mathbf{k}\cdot\mathbf{r})\mathbf{k}}{\{\mathbf{r} - (\mathbf{k}\cdot\mathbf{r})\mathbf{k}\}\cdot\{\mathbf{r} - (\mathbf{k}\cdot\mathbf{r})\mathbf{k}\}}, \text{ where } \mathbf{r} = x\mathbf{i} + y\mathbf{j} + z\mathbf{k}.$$

[**Hint.** We have $\mathbf{k} \cdot \mathbf{r} = z$.

Therefore, $\mathbf{r} - (\mathbf{k} \cdot \mathbf{r}) \mathbf{k} = (x\mathbf{i} + y\mathbf{j} + z\mathbf{k}) - z\mathbf{k} = x\mathbf{i} + y\mathbf{j} = \mathbf{r}_1 \text{ (say)}$.

We have to show that $\nabla f = \dfrac{\mathbf{r}_1}{\left|\mathbf{r}_1^2\right|} = \dfrac{x\mathbf{i} + y\mathbf{j}}{x^2 + y^2}$.

Now $f(x, y) = \dfrac{1}{2} \log(x^2 + y^2)$.

$\therefore \quad \nabla f = \left(\mathbf{i}\dfrac{\partial}{\partial x} + \mathbf{j}\dfrac{\partial}{\partial y}\right)\dfrac{1}{2} \log(x^2 + y^2) = \dfrac{1}{2}\left[\dfrac{2x\mathbf{i}}{x^2 + y^2} + \dfrac{2y\mathbf{j}}{x^2 + y^2}\right] = \dfrac{x\mathbf{i} + y\mathbf{i}}{x^2 + y^2}.]$

14.13 Divergence and Curl of a Vector Field

Let $\mathbf{v} = v_1(x, y, z)\mathbf{i} + v_2(x, y, z)\mathbf{j} + v_3(x, y, z)\mathbf{k}$ define a vector field.

Divergence of vector field v

The Divergence of $\mathbf{v}$, denoted by div $\mathbf{v}$, is defined as the scalar field

$$\text{div } \mathbf{v} = \frac{\partial v_1}{\partial x} + \frac{\partial v_2}{\partial y} + \frac{\partial v_3}{\partial z}. \qquad \text{...(14.17)}$$

Another common notation for the divergence of $\mathbf{v}$ is $\nabla \cdot \mathbf{v}$ that is del dot $\mathbf{v}$.

$$\text{div } \mathbf{v} = \nabla \cdot \mathbf{v} = \left(\mathbf{i}\frac{\partial}{\partial x} + \mathbf{j}\frac{\partial}{\partial y} + \mathbf{k}\frac{\partial}{\partial z}\right) \cdot (v_1\mathbf{i} + v_2\mathbf{j} + v_3\mathbf{k}) = \frac{\partial v_1}{\partial x} + \frac{\partial v_2}{\partial y} + \frac{\partial v_3}{\partial z}.$$

Note that $\nabla \cdot \mathbf{v}$ is a notation for div $\mathbf{v}$ and it is not a scalar product in the usual sense. Also note that $\nabla \cdot \mathbf{v} \neq \mathbf{v} \cdot \nabla$

Example 14.31 Find the divergence of the vector field

$\mathbf{v} = 3xyz^2\,\mathbf{i} + 2xy^3\,\mathbf{j} - x^2yz\,\mathbf{k}$ at the point $(1, -1, 1)$.

Solution. We have $\text{div } \mathbf{v} = \left(\mathbf{i}\dfrac{\partial}{\partial x} + \mathbf{j}\dfrac{\partial}{\partial y} + \mathbf{k}\dfrac{\partial}{\partial z}\right) \cdot \left(3xyz^2\,\mathbf{i} + 2xy^3\mathbf{j} - x^2yz\,\mathbf{k}\right)$

$= 3yz^2 + 6xy^2 - x^2y = 3(-1)(1)^2 + 6(1)(-1)^2 - (1)^2(-1) = 4, \text{ at}(1, -1, 1).$

Curl of vector field v

The *curl* of a vector field $\mathbf{v}$ denoted by curl $\mathbf{v}$, is defined as the vector field.

$$\text{curl } \mathbf{v} = \left(\frac{\partial v_3}{\partial y} - \frac{\partial v_2}{\partial z}\right)\mathbf{i} + \left(\frac{\partial v_1}{\partial z} - \frac{\partial v_3}{\partial x}\right)\mathbf{j} + \left(\frac{\partial v_2}{\partial x} - \frac{\partial v_1}{\partial y}\right)\mathbf{k}. \qquad \text{...(14.18)}$$

Another common notation for the curl of **v** is $\nabla \times \mathbf{v}$ that is del cross **v**.

$$\text{curl } \mathbf{v} = \nabla \times \mathbf{v} = \left(\mathbf{i}\frac{\partial}{\partial x} + \mathbf{j}\frac{\partial}{\partial y} + \mathbf{k}\frac{\partial}{\partial z} \right) \times \left(v_1 \mathbf{i} + v_2 \mathbf{j} + v_3 \mathbf{k} \right)$$

$$= \begin{vmatrix} \mathbf{i} & \mathbf{j} & \mathbf{k} \\ \partial/\partial x & \partial/\partial y & \partial/\partial z \\ v_1 & v_2 & v_3 \end{vmatrix} \qquad \qquad \text{...(14.19)}$$

$$= \begin{vmatrix} \partial/\partial y & \partial/\partial z \\ v_2 & v_3 \end{vmatrix} \mathbf{i} - \begin{vmatrix} \partial/\partial x & \partial/\partial z \\ v_1 & v_3 \end{vmatrix} \mathbf{j} + \begin{vmatrix} \partial/\partial x & \partial/\partial y \\ v_1 & v_2 \end{vmatrix} \mathbf{k}$$

$$= \left(\frac{\partial v_3}{\partial y} - \frac{\partial v_2}{\partial z} \right) \mathbf{i} + \left(\frac{\partial v_1}{\partial z} - \frac{\partial v_3}{\partial x} \right) \mathbf{j} + \left(\frac{\partial v_2}{\partial x} - \frac{\partial v_1}{\partial y} \right) \mathbf{k}.$$

Note that $\nabla \times \mathbf{v}$ is just a notation for curl **v** and it is not a cross or vector product in the usual sense. $\nabla \times \mathbf{v}$ is not equal to $\mathbf{v} \times \nabla$.

Sometimes curl **v** is also written as

$$\text{curl } \mathbf{v} = \Sigma \left(\frac{\partial v_3}{\partial y} - \frac{\partial v_2}{\partial z} \right) \mathbf{i}, \text{ where } \Sigma \text{ denotes summation obtained by the cyclic}$$

rotation of the unit vectors **i, j, k,** the components v_1, v_2, v_3 and the independent variables x, y, z respectively.

Example 14.32 Find the curl of the vector field $\mathbf{v} = xyz\,\mathbf{i} + 2x^2 y\,\mathbf{j} + \left(xz^2 - y^2 z \right) \mathbf{k}$ and verify that div (curl **v**) = 0.

Solution. We have

$$\text{curl } \mathbf{v} = \nabla \times \mathbf{v} = \begin{vmatrix} \mathbf{i} & \mathbf{j} & \mathbf{k} \\ \partial/\partial x & \partial/\partial y & \partial/\partial z \\ xyz & 2x^2 y & xz^2 - y^2 z \end{vmatrix}$$

$$= \left[\frac{\partial}{\partial y}\left(xz^2 - y^2 z \right) - \frac{\partial}{\partial z}\left(2x^2 y \right) \right] \mathbf{i} + \left[\frac{\partial}{\partial z}(xyz) - \frac{\partial}{\partial x}\left(xz^2 - y^2 z \right) \right] \mathbf{j} + \left[\frac{\partial}{\partial x}\left(2x^2 y \right) - \frac{\partial}{\partial y}(xyz) \right] \mathbf{k}$$

$$= -2yz\mathbf{i} + \left(xy - z^2 \right) \mathbf{j} + \left(4xy - xz \right) \mathbf{k} = - \left[2yz\mathbf{i} + \left(z^2 - xy \right) \mathbf{j} + \left(xz - 4xy \right) \mathbf{k} \right].$$

$$\text{div}\left(\text{curl } \mathbf{v}\right) = \nabla \cdot \left(\nabla \times \mathbf{v} \right) = \nabla \cdot \left[-\left\{ 2yz\,\mathbf{i} + \left(xy - z^2 \right) \mathbf{j} + \left(4xy - xz \right) \mathbf{k} \right\} \right]$$

$$= \frac{\partial}{\partial x}\left(-2yz \right) + \frac{\partial}{\partial y}\left(-xy + z^2 \right) + \frac{\partial}{dz}\left(-4xy + xz \right) = 0 - x + x = 0.$$

Example 14.33 If **a** is constant vector and $\mathbf{r} = x\mathbf{i} + y\mathbf{j} + z\mathbf{k}$, show that

(i) grad(**a** · **r**) = **a**. *(IETE, June 2002; P.U. 1999)* (*ii*) $\nabla \cdot (\mathbf{a} \times \mathbf{r}) = 0$

(*iii*) curl $(\mathbf{a} \times \mathbf{r}) = 2\mathbf{a}$.   (*IETE Dec. 2009, 2005; UPTU 2002*)

Solution. (*i*) Let $\mathbf{a} = a_1\mathbf{i} + a_2\mathbf{j} + a_3\mathbf{k}$, where a_1, a_2, a_3 are constants. We have
$\mathbf{a} \cdot \mathbf{r} = (a_1\mathbf{i} + a_2\mathbf{j} + a_3\mathbf{k}) \cdot (x\mathbf{i} + y\mathbf{j} + z\mathbf{k}) = a_1 x + a_2 y + a_3 z$.

Now $\nabla(\mathbf{a} \cdot r) = \left(\mathbf{i}\dfrac{\partial}{\partial x} + \mathbf{j}\dfrac{\partial}{\partial y} + \mathbf{k}\dfrac{\partial}{\partial z} \right)(a_1 x + a_2 y + a_3 z) = a_1\mathbf{i} + a_2\mathbf{j} + a_3\mathbf{k} = \mathbf{a}$.

(*ii*) $\mathbf{a} \times \mathbf{r} = \begin{vmatrix} \mathbf{i} & \mathbf{j} & \mathbf{k} \\ a_1 & a_2 & a_3 \\ x & y & z \end{vmatrix} = \mathbf{i}(a_2 z - a_3 y) + \mathbf{j}(a_3 x - a_1 z) + \mathbf{k}(a_1 y - a_2 x)$

Therefore,

$$\nabla \cdot (\mathbf{a} \times \mathbf{r}) = \left(\mathbf{i}\frac{\partial}{\partial x} + \mathbf{j}\frac{\partial}{\partial y} + \mathbf{k}\frac{\partial}{\partial z} \right) \cdot \left[\mathbf{i}(a_2 z - a_3 y) + \mathbf{j}(a_3 x - a_1 z) + \mathbf{k}(a_1 y - a_2 x) \right]$$

$$= \frac{\partial}{\partial x}(a_2 z - a_3 y) + \frac{\partial}{\partial y}(a_3 x - a_1 z) + \frac{\partial}{\partial z}(a_1 y - a_2 x) = 0.$$

(*iii*) We have $\nabla \times (\mathbf{a} \times \mathbf{r}) = \left(\mathbf{i}\dfrac{\partial}{\partial x} + \mathbf{j}\dfrac{\partial}{\partial y} + \mathbf{k}\dfrac{\partial}{\partial z} \right) \times (\mathbf{a} \times \mathbf{r})$

$$= \mathbf{i} \times \frac{\partial}{\partial x}(\mathbf{a} \times \mathbf{r}) + \mathbf{j} \times \frac{\partial}{\partial y}(\mathbf{a} \times \mathbf{r}) + \mathbf{k} \times \frac{\partial}{\partial z}(\mathbf{a} \times \mathbf{r})$$

$$= \mathbf{i} \times \left(\mathbf{a} \times \frac{\partial \mathbf{r}}{\partial x} \right) + \mathbf{j} \times \left(\mathbf{a} \times \frac{\partial \mathbf{r}}{\partial y} \right) + \mathbf{k} \times \left(\mathbf{a} \times \frac{\partial \mathbf{r}}{\partial z} \right) \quad \text{(Since } \mathbf{a} \text{ is a constant vector)}$$

$$= \mathbf{i} \times (\mathbf{a} \times \mathbf{i}) + \mathbf{j} \times (\mathbf{a} \times \mathbf{j}) + \mathbf{k} \times (\mathbf{a} \times \mathbf{k})$$

$$= (\mathbf{i} \cdot \mathbf{i})\mathbf{a} - (\mathbf{i} \cdot \mathbf{a})\mathbf{i} + (\mathbf{j} \cdot \mathbf{j})\mathbf{a} - (\mathbf{j} \cdot \mathbf{a})\mathbf{j} + (\mathbf{k} \cdot \mathbf{k})\mathbf{a} - (\mathbf{k} \cdot \mathbf{a})\mathbf{k}$$

$$= 3\mathbf{a} - (a_1\mathbf{i} + a_2\mathbf{j} + a_3\mathbf{k}) = 2\mathbf{a}.$$

Example 14.34 If $\mathbf{A} = 2xz^2\mathbf{i} - yz\mathbf{j} + 3xz^3\mathbf{k}$ and $\phi = x^2 yz$, find curl curl $\mathbf{A}$, and curl grad ϕ at the point $(1, 1, 1)$.

Solution. (*i*) Curl curl $\mathbf{A} = \nabla \times (\nabla \times \mathbf{A})$

$$= \nabla \times \begin{vmatrix} \mathbf{i} & \mathbf{j} & \mathbf{k} \\ \partial/\partial x & \partial/\partial y & \partial/\partial z \\ 2xz^2 & -yz & 3xz^3 \end{vmatrix} = \nabla \times \left[y\mathbf{i} - (3z^3 - 4xz)\mathbf{j} \right]$$

$$= \begin{vmatrix} \mathbf{i} & \mathbf{j} & \mathbf{k} \\ \partial/\partial x & \partial/\partial y & \partial/\partial z \\ y & -3z^3 + 4xz & 0 \end{vmatrix} = (9z^2 - 4x)\mathbf{i} + (4z - 1)\mathbf{k} = 5\mathbf{i} + 3\mathbf{k} \quad \text{at} \quad (1,1,1).$$

(*ii*) We have $\nabla\phi = \left(i\dfrac{\partial}{\partial x} + j\dfrac{\partial}{\partial y} + k\dfrac{\partial}{\partial z} \right) x^2 yz = 2xyz\,i + x^2 z\,j + x^2 y\,k.$

$\therefore$ Curl grad $\phi =$

$$\nabla\times\nabla\phi = \begin{vmatrix} i & j & k \\ \partial/\partial x & \partial/\partial y & \partial/\partial z \\ 2xyz & x^2 z & x^2 y \end{vmatrix} = (x^2 - x^2)i - (2xy - 2xy)j + (2xz - 2xz)k = 0.$$

Example 14.35 If $r = x i + y j + z k$ and $r = |r|$, show that div $(r/r^3) = 0$.

Solution. We have $\nabla\cdot\left(\dfrac{r}{r^3}\right) = \left(i\dfrac{\partial}{\partial x} + j\dfrac{\partial}{\partial y} + k\dfrac{\partial}{\partial z} \right)\cdot\left(\dfrac{x}{r^3}i + \dfrac{y}{r^3}j + \dfrac{z}{r^3}k \right)$

$$= \dfrac{\partial}{\partial x}\left(\dfrac{x}{r^3}\right) + \dfrac{\partial}{\partial y}\left(\dfrac{y}{r^3}\right) + \dfrac{\partial}{\partial z}\left(\dfrac{z}{r^3}\right) = \dfrac{3}{r^3} - \dfrac{3}{r^4}\left(x\dfrac{\partial r}{\partial x} + y\dfrac{\partial r}{\partial y} + z\dfrac{\partial r}{\partial z} \right).$$

Since $r^2 = x^2 + y^2 + z^2$, we obtain $\dfrac{\partial r}{\partial x} = \dfrac{x}{r}, \dfrac{\partial r}{\partial y} = \dfrac{y}{r}$ and $\dfrac{\partial r}{\partial z} = \dfrac{z}{r}$.

Therefore, $\qquad \nabla\cdot\left(\dfrac{r}{r^3}\right) = \dfrac{3}{r^3} - \dfrac{3}{r^5}\left(x^2 + y^2 + z^2 \right) = \dfrac{3}{r^3} - \dfrac{3}{r^3} = 0.$

Example 14.36 If $A = x^2 y i + y^2 x^3 j - 3\,x^2 y\,k$, $B = 2xz^2 i - yz j + x^2 y^3 k$
and $\phi(x, y, z) = x^2 yz$, find the values of the following at $(1, 2, 1)$.

(*i*) $(A \cdot \nabla)\phi, \qquad$ (*ii*) $(A \cdot \nabla)\,B.$

Solution. (*i*) At $(1, 2, 1)$ $A = 2i + 4j - 6\,k.$

$$A\cdot\nabla = (2i + 4j - 6k)\cdot\left(i\dfrac{\partial}{\partial x} + j\dfrac{\partial}{\partial y} + k\dfrac{\partial}{\partial z} \right) = 2\dfrac{\partial}{\partial x} + 4\dfrac{\partial}{\partial y} - 6\dfrac{\partial}{\partial z}.$$

Hence $(A\cdot\nabla)\phi = \left(2\dfrac{\partial}{\partial x} + 4\dfrac{\partial}{\partial y} - 6\dfrac{\partial}{\partial z} \right) x^2 yz$

$$= 4xyz + 4x^2 z - 6x^2 y = 8 + 4 - 12 = 0 \text{ at } (1,2,1).$$

(*ii*) $(A\cdot\nabla)B = \left(2\dfrac{\partial}{\partial x} + 4\dfrac{\partial}{\partial y} - 6\dfrac{\partial}{\partial z} \right)\left(2xz^2\,i - yz\,j + x^2 y^3\,k \right)$

$$= 2\left(2z^2\,i + 2xy^3\,k \right) + 4\left(-z\,j + 3x^2 y^2\,k \right) - 6\left(4xz\,i - y\,j \right)$$

$$= \left(4z^2 - 24xz \right)i + \left(6y - 4z \right)j + \left(4xy^3 + 12x^2 y^2 \right)k = -20i + 8j + 80k \text{ at } (1, 2, 1).$$

Physical interpretation of divergence

We shall present an interpretation in fluid mechanics. Consider the steady-state motion (*i.e.*, independent of time) of a compressible fluid, an infinitesimal volume element (parrallelopiped with edges Δx, Δy, Δz) is placed in the fluid as shown in Fig. 14.11.

Let $\mathbf{v} = v_1\mathbf{i} + v_2\mathbf{j} + v_3\mathbf{k}$ be the velocity vector of the motion.

Since the fluid is assumed to be compressible, therefore the density ρ, as well as the velocity may vary from point to point. The fluid enters the elemental volume through the faces: rear, left and bottom and goes out from the front, right and top faces respectively.

We now compute the loss of fluid as it flows through the element in time Δt. The volume of fluid which passes through one face of the element Δv in time Δt is approximately equal to the component of the fluid velocity normal to the face × area of the face × Δt.

The corresponding mass flow = volume of this fluid × density of fluid ρ

Now, we compute the loss of fluid in time Δt through each face in turn. We have

$$\begin{cases} \text{Front face:} & \left[\rho v_1 + \dfrac{\partial(\rho v_1)}{\partial x}\Delta x\right]\Delta y\, \Delta z\, \Delta t & (i) \\[3mm] \text{Rear face:} & -\rho v_1\, \Delta y\, \Delta z\, \Delta t & (ii) \end{cases}$$

$$\begin{cases} \text{Right face:} & \left[\rho v_2 + \dfrac{\partial(\rho v_2)}{\partial y}\Delta y\right]\Delta x\, \Delta z\, \Delta t & (iii) \\[3mm] \text{Left face:} & -\rho v_2\, \Delta x\, \Delta z\, \Delta t & (iv) \end{cases}$$

$$\begin{cases} \text{Top face:} & \left[\rho v_3 + \dfrac{\partial(\rho v_3)}{\partial z}\Delta z\right]\Delta x\, \Delta y\, \Delta t & (v) \\[3mm] \text{Bottom face:} & -\rho v_3\, \Delta x\, \Delta y\, \Delta t & (vi) \end{cases}$$

Adding the above equations.

Therefore, the total loss of mass of fluid from Δv ($= \Delta x\, \Delta y\, \Delta z$) in the interval Δt

$$= \left[\frac{\partial(\rho v_1)}{\partial x} + \frac{\partial(\rho v_2)}{\partial y} + \frac{\partial(\rho v_3)}{\partial z}\right]\Delta x\, \Delta y\, \Delta z\, \Delta t \qquad ...(14.20)$$

Dividing eqn. (14.20) by $\Delta v\, \Delta t$, we obtain in the limit the loss per unit volume per unit time that is,

$$\text{Rate of loss of fluid/unit volume} = \frac{\partial(\rho v_1)}{\partial x} + \frac{\partial(\rho v_2)}{\partial y} + \frac{\partial(\rho v_3)}{\partial z} \qquad ...(14.21)$$

which is precisely the divergence of the vector ρv.

If the fluid is incompressible, there can be neither gain nor loss of fluid in a general element.

Hence for an incompressible fluid, (the density ρ is constant), we have

$$\nabla \cdot \rho v = \rho \nabla \cdot v = 0 \quad \text{or} \quad \nabla \cdot v = 0$$
...(14.22)

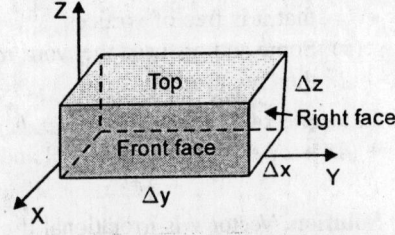

Fig. 14.11. Flow through parallelopiped.

which is known as the equation of continuity for incompressible fluids.

Thus, when the fluid is incompressible, the divergence of the velocity vector vanishes.

Remark. Let **v** denote the velocity of a fluid, if div(**v**) = 0, then the fluid is said to be *incompressible*. In electromagnetic theory, if div(**v**) = 0, then the vector field **v** is said to be *solenoidal*.

Physical Interpretation of curl

Consider a rigid body rotate with the uniform angular velocity $\Omega = a\mathbf{i} + b\mathbf{j} + c\mathbf{k}$ about a fixed axis *l* through O, the origin of co-ordinates. Let the position vector of any point $P(x, y, z)$ on the rotating body be $\mathbf{r} = x\mathbf{i} + y\mathbf{j} + z\mathbf{k}$. The tangential (linear) velocity **v** of the point $P(x, y, z)$ is given by

$$\mathbf{v} = \Omega \times \mathbf{r} = (a\mathbf{i} + b\mathbf{j} + c\mathbf{k}) \times (x\mathbf{i} + y\mathbf{j} + z\mathbf{k})$$

$$= \begin{vmatrix} \mathbf{i} & \mathbf{j} & \mathbf{k} \\ a & b & c \\ x & y & z \end{vmatrix} = (bz - cy)\mathbf{i} + (cx - az)\mathbf{j} + (ay - bx)\mathbf{k}.$$

Now $\text{curl } \mathbf{v} = \nabla \times \mathbf{v} = \begin{vmatrix} \mathbf{i} & \mathbf{j} & \mathbf{k} \\ \partial/\partial x & \partial/\partial y & \partial/\partial z \\ bz - cy & cx - az & ay - bx \end{vmatrix}$

$$= 2(a\mathbf{i} + b\mathbf{j} + c\mathbf{k}) = 2\Omega \quad \text{or} \quad \Omega = (1/2) \text{ curl } \mathbf{v}.$$

Thus the angular velocity of a uniformly rotating rigid body is equal to one-half the curl of the linear velocity of any point of the body.

Remark. A force field **F** is said to be *conservative* if it is obtained from a potential function *f*, that is **F** = grad *f*. Then curl (**F**) = curl (grad *f*) = **0**. Thus, if **F** is conservative then curl (**F**) = **0** and there exists a scalar potential function *f* such as **F** = grad *f*.

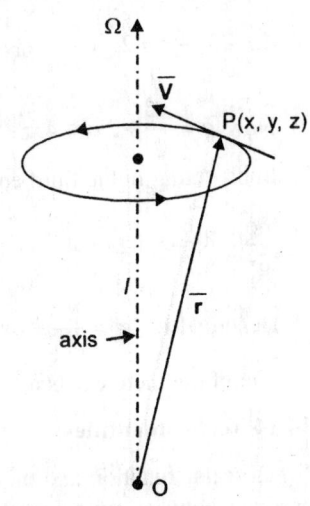

Fig. 14.12.

(a) If curl $\mathbf{v} = \mathbf{0}$, then the flow of the fluid is said to be irrotational, which means that it is free of vortices.

(b) Some authors used the word *rotation* instead of curl. The symbol curl $\mathbf{F}$ is then replaced by rot $\mathbf{F}$.

Example 14.37 Find constants a, b, c so that $\mathbf{v} = (x + 2y + az)\,\mathbf{i} + (bx - 3y - z)\mathbf{j} + (4x + cy + 2z)\mathbf{k}$ is irrotational and hence find its scalar potential $\phi(x, y, z)$.

(AMIE, W-2008)

Solution. Vector $\mathbf{v}$ is irrotational if curl $\mathbf{v} = \mathbf{O}$.

$$\text{curl } \mathbf{v} = \nabla \times \mathbf{v} = \begin{vmatrix} \mathbf{i} & \mathbf{j} & \mathbf{k} \\ \partial/\partial x & \partial/\partial y & \partial/\partial z \\ x+2y+az & bx-3y-z & 4x+cy+2z \end{vmatrix}$$

$$= (c+1)\mathbf{i} + (a-4)\mathbf{j} + (b-2)\mathbf{k}.$$

This equals zero when $a = 4$, $b = 2$, $c = -1$ and

$$\mathbf{v} = (x + 2y + 4z)\mathbf{i} + (2x - 3y - z)\mathbf{j} + (4x - y + 2z)\mathbf{k}.$$

Assume $\mathbf{v} = \nabla\phi = \dfrac{\partial\phi}{\partial x}\mathbf{i} + \dfrac{\partial\phi}{\partial y}\mathbf{j} + \dfrac{\partial\phi}{\partial z}\mathbf{k}.$

Then $\dfrac{\partial\phi}{\partial x} = x + 2y + 4z$ ...(i), $\dfrac{\partial\phi}{\partial y} = 2x - 3y - z$ ...(ii), $\dfrac{\partial\phi}{\partial z} = 4x - y + 2z$ ...(iii).

Integrating (i) partially with respect to x, keeping y and z constant,

$\phi = \dfrac{x^2}{2} + 2xy + 4xz + f(y, z)$. Substituting in the second equation, we get

$2x - 3y - z = 2x + \dfrac{\partial f}{\partial y}$, or $\dfrac{\partial f}{\partial y} = -3y - z$ or $f = \dfrac{-3y^2}{2} - yz + f(z)$.

Now $\phi = \dfrac{x^2}{2} + 2xy + 4xz - \dfrac{3y^2}{2} - yz + f(z)$.

Substituting in the third equation, we get

$\dfrac{\partial\phi}{\partial z} = 4x - y + 2z = 4x - y + \dfrac{\partial f}{\partial z}$, or $\dfrac{\partial f}{\partial z} = 2z$ or $f = z^2 + c$, (c being constant)

Hence $\phi(x, y, z) = \dfrac{x^2}{2} + 2xy + 4xz - \dfrac{3y^2}{2} - yz + z^2 + c.$

Therefore, there exists a scalar function $\phi(x, y, z)$ such that $\nabla\phi = \mathbf{v}$.

14.14 Vector Identities

If f is scalar function and $\mathbf{u}$, $\mathbf{v}$ are vector functions, then the possible forms of products between these functions are $f\mathbf{u}$, $\mathbf{u} \cdot \mathbf{v}$ and $\mathbf{u} \times \mathbf{v}$.

Since we have defined gradient of scalar point functions and divergence, curl of vector point functions. Now the cases to be considered are:

$$\nabla(\mathbf{u \cdot v}), \nabla\cdot(f\mathbf{v}), \nabla\cdot(\mathbf{u\times v}), \nabla\times(f\,\mathbf{v}), \nabla\times(\mathbf{u\times v}).$$

We shall now prove the following expansion formulae:

1. $\nabla(\mathbf{u \cdot v}) = \mathbf{u}\times(\nabla\times\mathbf{v})+\mathbf{v}\times(\nabla\times\mathbf{u})+(\mathbf{u}\cdot\nabla)\mathbf{v}+(\mathbf{v}\cdot\nabla)\mathbf{u}$

    that is grad $(\mathbf{u}\cdot\mathbf{v}) = (\mathbf{u}\times\text{curl } \mathbf{v})+(\mathbf{v}\times\text{curl } \mathbf{u})+(\mathbf{u}\cdot\nabla)\mathbf{v}+(\mathbf{v}\cdot\nabla)\mathbf{u}.$

2. $\nabla\cdot(f\,\mathbf{v}) = f(\nabla\cdot\mathbf{v})+(\nabla f)\cdot\mathbf{v}$ *(UPTU 2004)*

    that is $\text{div}(f\,\mathbf{v}) = f \text{ div } \mathbf{v}+(\text{ grad} f)\cdot\mathbf{v}$

3. $\nabla\cdot(\mathbf{u\times v}) = \mathbf{v}\cdot(\nabla\times\mathbf{u})-\mathbf{u}\cdot(\nabla\times\mathbf{v})$

    that is, $\text{div}(\mathbf{u\times v}) = \mathbf{v}\cdot\text{curl } \mathbf{u}-\mathbf{u}\cdot\text{curl } \mathbf{v}$ *(IETE, Dec. 2009; UPTU 2003)*

4. $\nabla\times(f\,\mathbf{v}) = f(\nabla\times\mathbf{v})+(\nabla f)\times\mathbf{v},$

    that is, $\text{curl}(f\,\mathbf{v}) = f \text{ curl } \mathbf{v}+(\text{grad} f)\times\mathbf{v}$ *(UPTU 2003)*

5. $\nabla\times(\mathbf{u\times v}) = \mathbf{u}(\nabla\cdot\mathbf{v})-\mathbf{v}(\nabla\cdot\mathbf{u})+(\mathbf{v}\cdot\nabla)\mathbf{u}-(\mathbf{u}\cdot\nabla)\mathbf{v},$

    that is, $\text{curl }(\mathbf{u\times v}) = \mathbf{u}(\text{div } \mathbf{v})-\mathbf{v}(\text{div } \mathbf{u})+(\mathbf{v}\cdot\nabla)\mathbf{u}-(\mathbf{u}\cdot\nabla)\mathbf{v}.$

Proof:

1. $\nabla(\mathbf{u}\cdot\mathbf{v}) = \sum \mathbf{i}\dfrac{\partial}{\partial x}(\mathbf{u}\cdot\mathbf{v}) = \sum\mathbf{i}\left(\dfrac{\partial\mathbf{u}}{\partial x}\cdot\mathbf{v}+\mathbf{u}\cdot\dfrac{\partial\mathbf{v}}{\partial x}\right)$

    $$= \sum\mathbf{i}\left(\dfrac{\partial\mathbf{u}}{\partial x}\cdot\mathbf{v}\right)+\sum\mathbf{i}\left(\mathbf{u}\cdot\dfrac{\partial\mathbf{v}}{\partial x}\right). \qquad \text{...}(i)$$

    We know that $\mathbf{a}\times(\mathbf{b\times c}) = (\mathbf{a}\cdot\mathbf{c})\mathbf{b}-(\mathbf{a}\cdot\mathbf{b})\mathbf{c}$

    $\Rightarrow (\mathbf{a}\cdot\mathbf{c})\mathbf{b} = \mathbf{a}\times(\mathbf{b\times c})+(\mathbf{a}\cdot\mathbf{b})\mathbf{c}.$

    Now $\mathbf{v}\times\left(\mathbf{i}\times\dfrac{\partial\mathbf{u}}{\partial x}\right) = \mathbf{i}\left(\dfrac{\partial\mathbf{u}}{\partial x}\cdot\mathbf{v}\right)-\dfrac{\partial\mathbf{u}}{\partial x}(\mathbf{v}\cdot\mathbf{i})$

    or $\mathbf{i}\left(\dfrac{\partial\mathbf{u}}{\partial x}\cdot\mathbf{v}\right) = \mathbf{v}\times\left(\mathbf{i}\times\dfrac{\partial\mathbf{u}}{\partial x}\right)+(\mathbf{v}\cdot\mathbf{i})\dfrac{\partial\mathbf{u}}{\partial x}$

    $\therefore$ $\sum\mathbf{i}\left(\dfrac{\partial\mathbf{u}}{\partial x}\cdot\mathbf{v}\right) = \mathbf{v}\times\left[\sum\mathbf{i}\times\dfrac{\partial\mathbf{u}}{\partial x}\right]+\sum(\mathbf{v}\cdot\mathbf{i})\dfrac{\partial\mathbf{u}}{\partial x}$

    $$= \mathbf{v}\times(\nabla\times\mathbf{u})+(\mathbf{v}\cdot\nabla)\mathbf{u} \qquad \text{...}(ii)$$

    Interchanging $\mathbf{v}$ and $\mathbf{u}$, we have

    $$\sum\mathbf{i}\left[\dfrac{\partial\mathbf{v}}{\partial x}\cdot\mathbf{u}\right] = \mathbf{u}\times(\nabla\times\mathbf{v})+(\mathbf{u}\cdot\nabla)\mathbf{v} \qquad \text{...}(iii)$$

Using (*ii*) and (*iii*), we can write (*i*) in the form

$$\nabla(\mathbf{u}\cdot\mathbf{v})=\mathbf{u}\times(\nabla\times\mathbf{v})+\mathbf{v}\times(\nabla\times\mathbf{u})+(\mathbf{u}\cdot\nabla)\mathbf{v}+(\mathbf{v}\cdot\nabla)\mathbf{u}.$$

2. We have $\nabla\cdot(f\mathbf{v})=\left(\mathbf{i}\dfrac{\partial}{\partial x}+\mathbf{j}\dfrac{\partial}{\partial y}+\mathbf{k}\dfrac{\partial}{\partial z}\right)\cdot\left(f v_1\mathbf{i}+f v_2\mathbf{j}+f v_3\mathbf{k}\right)$

$$=\frac{\partial}{\partial x}(f v_1)+\frac{\partial}{\partial y}(f v_2)+\frac{\partial}{\partial z}(f v_3)$$

$$=\left(v_1\frac{\partial f}{\partial x}+f\frac{\partial v_1}{\partial x}\right)+\left(v_2\frac{\partial f}{\partial y}+f\frac{\partial v_2}{\partial y}\right)+\left(v_3\frac{\partial f}{\partial z}+f\frac{\partial v_3}{\partial z}\right)$$

$$=f\left(\frac{\partial v_1}{\partial x}+\frac{\partial v_2}{\partial y}+\frac{\partial v_3}{\partial z}\right)+\left(v_1\mathbf{i}+v_2\mathbf{j}+v_3\mathbf{k}\right)\cdot\left(\mathbf{i}\frac{\partial f}{\partial x}+\mathbf{j}\frac{\partial f}{\partial y}+\mathbf{k}\frac{\partial f}{\partial z}\right)$$

$$=f(\nabla\cdot\mathbf{v})+\mathbf{v}\cdot\nabla f.$$

3. $\nabla\cdot(\mathbf{u}\times\mathbf{v})=\sum\mathbf{i}\cdot\dfrac{\partial}{\partial x}(\mathbf{u}\times\mathbf{v})$

$$=\sum\mathbf{i}\cdot\left[\frac{\partial\mathbf{u}}{\partial x}\times\mathbf{v}+\mathbf{u}\times\frac{\partial\mathbf{v}}{\partial x}\right]=\sum\mathbf{i}\cdot\frac{\partial\mathbf{u}}{\partial x}\times\mathbf{v}+\sum\mathbf{i}\cdot\mathbf{u}\times\frac{\partial\mathbf{v}}{\partial x}$$

$$=\sum\mathbf{i}\times\frac{\partial\mathbf{u}}{\partial x}\cdot\mathbf{v}+\sum\mathbf{i}\times\mathbf{u}\cdot\frac{\partial\mathbf{v}}{\partial x}\qquad\text{(interchanging dot and cross)}$$

$$=\left[\sum\mathbf{i}\times\frac{\partial\mathbf{u}}{\partial x}\right]\cdot\mathbf{v}-\left[\sum\mathbf{i}\times\frac{\partial\mathbf{v}}{\partial x}\right]\cdot\mathbf{u}=(\nabla\times\mathbf{u})\cdot\mathbf{v}-(\nabla\times\mathbf{v})\cdot\mathbf{u}$$

4. $\operatorname{curl}(f\mathbf{v})=\nabla\times(f\mathbf{v})=\nabla(f v_1\mathbf{i}+f v_2\mathbf{j}+f v_3\mathbf{k})$

$$=\sum\left[\frac{\partial}{\partial y}(f v_3)-\frac{\partial}{\partial z}(f v_2)\right]\mathbf{i}=\sum\left[f\left(\frac{\partial v_3}{\partial y}-\frac{\partial v_2}{\partial z}\right)+v_3\frac{\partial f}{\partial y}-v_2\frac{\partial f}{\partial z}\right]\mathbf{i}$$

$$=f\left[\left(\frac{\partial v_3}{\partial y}-\frac{\partial v_2}{\partial z}\right)\mathbf{i}+\left(\frac{\partial v_1}{\partial z}-\frac{\partial v_3}{\partial x}\right)\mathbf{j}+\left(\frac{\partial v_2}{\partial x}-\frac{\partial v_1}{\partial y}\right)\mathbf{k}\right]$$

$$+\left[\left(v_3\frac{\partial f}{\partial y}-v_2\frac{\partial f}{\partial z}\right)\mathbf{i}+\left(v_1\frac{\partial f}{\partial z}-v_3\frac{\partial f}{\partial x}\right)\mathbf{j}+\left(v_2\frac{\partial f}{\partial x}-v_1\frac{\partial f}{\partial y}\right)\mathbf{k}\right]$$

$$=f(\operatorname{curl}\mathbf{v})+\left(\mathbf{i}\frac{\partial f}{\partial x}+\mathbf{j}\frac{\partial f}{\partial y}+\mathbf{k}\frac{\partial f}{\partial z}\right)\times(\mathbf{i}v_1+\mathbf{j}v_2+\mathbf{k}v_3)$$

$$=f\operatorname{curl}\mathbf{v}+(\operatorname{grad}f)\times\mathbf{v}.$$

5. $\nabla \times (\mathbf{u} \times \mathbf{v}) = \sum \mathbf{i} \times \dfrac{\partial}{\partial x}(\mathbf{u} \times \mathbf{v}) = \sum \mathbf{i} \times \left[\dfrac{\partial \mathbf{u}}{\partial x} \times \mathbf{v} + \mathbf{u} \times \dfrac{\partial \mathbf{v}}{\partial x} \right]$

$$= \sum \mathbf{i} \times \left[\dfrac{\partial \mathbf{u}}{\partial x} \times \mathbf{v} \right] + \sum \mathbf{i} \times \left[\mathbf{u} \times \dfrac{\partial \mathbf{v}}{\partial x} \right]$$

$$= \sum \left[(\mathbf{i}.\mathbf{v}) \dfrac{\partial \mathbf{u}}{\partial x} - \left(\mathbf{i} \cdot \dfrac{\partial \mathbf{u}}{\partial x} \right) \mathbf{v} \right] + \sum \left[\left(\mathbf{i} \cdot \dfrac{\partial \mathbf{v}}{\partial x} \right) \mathbf{u} - (\mathbf{i} \cdot \mathbf{u}) \dfrac{\partial \mathbf{v}}{\partial x} \right]$$

$$\left[\because \ \mathbf{a} \times (\mathbf{b} \times \mathbf{c}) = (\mathbf{a} \cdot \mathbf{c})\mathbf{b} - (\mathbf{a} \cdot \mathbf{b})\mathbf{c} \right]$$

$$= \sum (\mathbf{i} \cdot \mathbf{v}) \dfrac{\partial \mathbf{u}}{\partial x} - \sum \left(\mathbf{i} \cdot \dfrac{\partial \mathbf{u}}{\partial x} \right) \mathbf{v} + \sum \left(\mathbf{i} \cdot \dfrac{\partial \mathbf{v}}{\partial x} \right) \mathbf{u} - \sum (\mathbf{i} \cdot \mathbf{u}) \dfrac{\partial \mathbf{v}}{\partial x}$$

$$= (\mathbf{v} \cdot \nabla)\mathbf{u} - (\nabla \cdot \mathbf{u})\mathbf{v} + (\nabla \cdot \mathbf{v})\mathbf{u} - (\mathbf{u} \cdot \nabla)\mathbf{v}$$

14.15 Illustrative Examples

Example 14.38 Prove the following identities:

(i) Curl(grad f) = $\mathbf{O}$ or $\nabla \times (\nabla f) = \mathbf{O}$.   *(PTU 2000)*

(ii) div(curl $\mathbf{v}$) = 0 or $\nabla \cdot (\nabla \times \mathbf{v}) = 0$ *(MDU 2003; PTU 1999)*

(iii) div(grad f) = $\nabla^2 f$, where $\nabla^2 = \dfrac{\partial^2}{\partial x^2} + \dfrac{\partial^2}{\partial y^2} + \dfrac{\partial^2}{\partial z^2}$ is the Laplacian operator.

(iv) curl(curl $\mathbf{v}$) i.e., $\nabla \times (\nabla \times \mathbf{v}) = \nabla(\nabla \cdot \mathbf{v}) - \nabla^2 \mathbf{v}$,

or grad (div $\mathbf{v}$) = curl (curl $\mathbf{v}$) + $\nabla^2 \mathbf{v}$, where $\mathbf{v}$, a differentiable vector field.

(PTU 2006; IETE Dec 2004; UPTU 2003; MDU 2002)

Solution. (i) From the definition, we have

$$\nabla \times (\nabla f) = \begin{vmatrix} \mathbf{i} & \mathbf{j} & \mathbf{k} \\ \partial/\partial x & \partial/\partial y & \partial/\partial z \\ \partial f/\partial x & \partial f/\partial y & \partial f/\partial z \end{vmatrix}$$

$$= \mathbf{i} \left(\dfrac{\partial^2 f}{\partial y \partial z} - \dfrac{\partial^2 f}{\partial y \partial z} \right) + \mathbf{j} \left(\dfrac{\partial^2 f}{\partial x \partial z} - \dfrac{\partial^2 f}{\partial x \partial z} \right) + \mathbf{k} \left(\dfrac{\partial^2 f}{\partial x \partial y} - \dfrac{\partial^2 f}{\partial x \partial y} \right) = \mathbf{O}.$$

(ii) We have $\mathbf{v} = v_1 \mathbf{i} + v_2 \mathbf{j} + v_3 \mathbf{k}$. From the definition

$$\text{div(curl } \mathbf{v}) = \nabla \cdot (\nabla \times \mathbf{v}) = \nabla \cdot \begin{vmatrix} \mathbf{i} & \mathbf{j} & \mathbf{k} \\ \partial/\partial x & \partial/\partial y & \partial/\partial z \\ v_1 & v_2 & v_3 \end{vmatrix}$$

$$= \nabla \cdot \left[\left(\frac{\partial v_3}{\partial y} - \frac{\partial v_2}{\partial z} \right) \mathbf{i} + \left(\frac{\partial v_1}{\partial z} - \frac{\partial v_3}{\partial x} \right) \mathbf{j} + \left(\frac{\partial v_2}{\partial x} - \frac{\partial v_1}{\partial y} \right) \mathbf{k} \right].$$

$$= \frac{\partial}{\partial x} \left(\frac{\partial v_3}{\partial y} - \frac{\partial v_2}{\partial z} \right) + \frac{\partial}{\partial y} \left(\frac{\partial v_1}{\partial z} - \frac{\partial v_3}{\partial x} \right) + \frac{\partial}{\partial z} \left(\frac{\partial v_2}{\partial x} - \frac{\partial v_1}{\partial y} \right)$$

$$= \frac{\partial^2 v_3}{\partial x \partial y} - \frac{\partial^2 v_2}{\partial x \partial z} + \frac{\partial^2 v_1}{\partial y \partial z} - \frac{\partial^2 v_3}{\partial y \partial x} + \frac{\partial^2 v_2}{\partial z \partial x} - \frac{\partial^2 v_1}{\partial z \partial y} = 0.$$

(*iii*) $\text{div}(\text{grad} f) = \nabla \cdot (\nabla f) = \left(\mathbf{i} \frac{\partial}{\partial x} + \mathbf{j} \frac{\partial}{\partial y} + \mathbf{k} \frac{\partial}{\partial z} \right) \cdot \left(\mathbf{i} \frac{\partial f}{\partial x} + \mathbf{j} \frac{\partial f}{\partial y} + \mathbf{k} \frac{\partial f}{\partial z} \right)$

$$= \frac{\partial^2 f}{\partial x^2} + \frac{\partial^2 f}{\partial y^2} + \frac{\partial^2 f}{\partial z^2} = \nabla^2 f.$$ The operator ∇^2 is called Laplacian operator.

Remarks. (*i*) $\nabla^2 f = 0$ is called *Laplace's equation*.

(*ii*) A function f satisfying $\nabla^2 f = 0$ is said to be a *harmonic function*.

(*iv*) $\text{curl}(\text{curl } \mathbf{v}) = \nabla \times (\nabla \times \mathbf{v}) = \left(\sum \mathbf{i} \frac{\partial}{\partial x} \right) \times \left[\sum \mathbf{i} \left(\frac{\partial v_3}{\partial y} - \frac{\partial v_2}{\partial z} \right) \right]$

$$= \sum \mathbf{i} \left[\frac{\partial}{\partial y} \left(\frac{\partial v_2}{\partial x} - \frac{\partial v_1}{\partial y} \right) - \frac{\partial}{\partial z} \left(\frac{\partial v_1}{\partial z} - \frac{\partial v_3}{\partial x} \right) \right]$$

$$= \sum \mathbf{i} \left[\frac{\partial^2 v_2}{\partial y \partial x} + \frac{\partial^2 v_3}{\partial z \partial x} - \left(\frac{\partial^2 v_1}{\partial y^2} + \frac{\partial^2 v_1}{\partial z^2} \right) \right]$$

$$= \sum \mathbf{i} \left[\frac{\partial}{\partial x} \left(\frac{\partial v_2}{\partial y} + \frac{\partial v_3}{\partial z} \right) - \left(\frac{\partial^2 v_1}{\partial y^2} + \frac{\partial^2 v_1}{\partial z^2} \right) \right]$$

$$= \sum \mathbf{i} \left[\frac{\partial}{\partial x} \left(\frac{\partial v_1}{\partial x} + \frac{\partial v_2}{\partial y} + \frac{\partial v_3}{\partial z} \right) - \left(\frac{\partial^2 v_1}{\partial x^2} + \frac{\partial^2 v_1}{\partial y^2} + \frac{\partial^2 v_1}{\partial z^2} \right) \right]$$

$$= \left(\sum \mathbf{i} \frac{\partial}{\partial x} \right) (\nabla \cdot \mathbf{v}) - \left(\frac{\partial^2}{\partial x^2} + \frac{\partial^2}{\partial y^2} + \frac{\partial^2}{\partial z^2} \right) \left(\sum \mathbf{i} v_1 \right) = \nabla (\nabla \cdot \mathbf{v}) - \nabla^2 \mathbf{v}.$$

Hence $\text{grad}(\text{div } \mathbf{v}) = \text{curl}(\text{curl } \mathbf{v}) + \nabla^2 \mathbf{v}.$

Example 14.39 Show that $\nabla^2(r^n) = n(n+1)r^{n-2}$. Hence show that $\nabla^2\left(\dfrac{1}{r}\right) = 0$.

(S.V.T.U, 2006; J.N.T.U 2006; UPTU 2005)

Solution. $\nabla^2(r^n) = \nabla\bullet(\nabla r^n) = \nabla\cdot\left(nr^{n-1}\dfrac{\mathbf{r}}{r}\right) = n\nabla\cdot\left(r^{n-2}\mathbf{r}\right)$

$$= n\left[(\nabla r^{n-2})\cdot\mathbf{r} + r^{n-2}(\nabla\cdot\mathbf{r})\right] \qquad \text{[by Art. 14.14, formula (2)]}$$

$$= n\left[(n-2)r^{n-3}\cdot\dfrac{\mathbf{r}}{r}\cdot\mathbf{r} + r^{n-2}(3)\right] = n\left[(n-2)r^{n-2} + 3r^{n-2}\right] = n(n+1)r^{n+2}.$$

Setting $n = -1$ in the above result, we obtain $\nabla^2(r^{-1}) = 0$ or $\nabla^2\left(\dfrac{1}{r}\right) = 0$.

Example 14.40 If $\nabla\cdot\mathbf{E} = 0$, $\nabla\cdot\mathbf{H} = 0$, $\nabla\times\mathbf{E} = -\dfrac{1}{c}\dfrac{\partial\mathbf{H}}{\partial t}$, $\nabla\times\mathbf{H} = \dfrac{1}{c}\dfrac{\partial\mathbf{E}}{\partial t}$,

Show that the vectors $\mathbf{E}$ and $\mathbf{H}$ satisfy the equation $\nabla^2\psi = \dfrac{1}{c^2}\dfrac{\partial^2\psi}{\partial t^2}$.

Solution. $\nabla\times(\nabla\times\mathbf{E}) = \nabla\times\left(-\dfrac{1}{c}\dfrac{\partial\mathbf{H}}{\partial t}\right) = -\dfrac{1}{c}\dfrac{\partial}{\partial t}(\nabla\times\mathbf{H})$

$$= -\dfrac{1}{c}\dfrac{\partial}{\partial t}\left(\dfrac{1}{c}\dfrac{\partial\mathbf{E}}{\partial t}\right) = -\dfrac{1}{c^2}\dfrac{\partial^2\mathbf{E}}{\partial t^2}.$$

We also know that $\nabla\times(\nabla\times\mathbf{E}) = -\nabla^2\mathbf{E} + \nabla(\nabla\cdot\mathbf{E}) = -\nabla^2\mathbf{E}$, since $\nabla\cdot\mathbf{E} = 0$.

Then $\qquad\qquad\qquad\qquad \nabla^2\mathbf{E} = \dfrac{1}{c^2}\dfrac{\partial^2\mathbf{E}}{\partial t^2}.$ ...(i)

Similarly, $\nabla\times(\nabla\times\mathbf{H}) = \nabla\times\left(\dfrac{1}{c}\dfrac{\partial\mathbf{E}}{\partial t}\right) = \dfrac{1}{c}\dfrac{\partial}{\partial t}(\nabla\times\mathbf{E})$

$$= \dfrac{1}{c}\dfrac{\partial}{\partial t}\left(-\dfrac{1}{c}\dfrac{\partial\mathbf{H}}{\partial t}\right) = -\dfrac{1}{c^2}\dfrac{\partial^2\mathbf{H}}{\partial t^2}.$$

But $\nabla\times(\nabla\times\mathbf{H}) = -\nabla^2\mathbf{H} + \nabla(\nabla\cdot\mathbf{H}) = -\nabla^2\mathbf{H}.$ $\qquad (\because \nabla\cdot\mathbf{H} = 0)$

Then $\nabla^2 H = \dfrac{1}{c^2}\dfrac{\partial^2\mathbf{H}}{\partial t^2}.$ ...(ii)

From (*i*) and (*ii*) it is obvious that **E** and **H** satisfy the equation $\nabla^2 \psi = \dfrac{1}{c^2}\dfrac{\partial^2 \psi}{\partial t^2}$.

Example 14.41 If r is the distance of a point (x, y, z) from the origin, prove that $\text{curl}\left(\mathbf{k}\times \text{gard }\dfrac{1}{r}\right) + \text{grad}\left(\mathbf{k}\cdot\text{grad }\dfrac{1}{r}\right) = 0$, where **k** is a unit vector in the direction of *OZ*.

(*IETE, Dec. 2009; UPTU 2001; PTU 1999*)

Solution. We have $\text{curl}\left(\mathbf{k}\times \text{grad }\dfrac{1}{r}\right) = \nabla \times \left(\mathbf{k}\times \nabla\dfrac{1}{r}\right)$

$$= \mathbf{k}\left(\nabla\cdot\nabla\dfrac{1}{r}\right) - (\mathbf{k}\cdot\nabla)\nabla\dfrac{1}{r} + \left(\nabla\cdot\nabla\dfrac{1}{r}\right)\mathbf{k} - (\nabla\cdot\mathbf{k})\nabla\dfrac{1}{r}$$

$$= \mathbf{k}\left(\nabla\cdot\nabla\dfrac{1}{r}\right) - (\mathbf{k}\cdot\nabla)\nabla\dfrac{1}{r} + 0 - 0 = \mathbf{k}\left(\nabla\cdot\nabla\dfrac{1}{r}\right) - (\mathbf{k}\cdot\nabla)\nabla\dfrac{1}{r} \qquad ...(i)$$

Now $\text{grad}\left(\mathbf{k}\cdot\text{grad }\dfrac{1}{r}\right) = \nabla\left(\mathbf{k}\cdot\nabla\dfrac{1}{r}\right)$

$$= \mathbf{k}\times\left(\nabla\times\nabla\dfrac{1}{r}\right) + \nabla\dfrac{1}{r}\times(\nabla\times\mathbf{k}) + (\mathbf{k}\cdot\nabla)\nabla\dfrac{1}{r} + \left(\nabla\dfrac{1}{r}\cdot\nabla\right)\mathbf{k}$$

$$= 0 + 0 + (\mathbf{k}\cdot\nabla)\nabla\dfrac{1}{r} + 0 = (\mathbf{k}\cdot\nabla)\nabla\dfrac{1}{r}. \qquad ...(ii)$$

Adding (*i*) and (*ii*) we obtain $\text{curl}\left(\mathbf{k}\times\text{grad }\dfrac{1}{r}\right) + \text{grad}\left(\mathbf{k}\cdot\text{grad }\dfrac{1}{r}\right)$

$$= \mathbf{k}\left(\nabla\cdot\nabla\dfrac{1}{r}\right) = \mathbf{k}\left[\nabla\cdot\left(-\dfrac{\mathbf{r}}{r^3}\right)\right] = -\mathbf{k}\left[\nabla\cdot\left(\dfrac{\mathbf{r}}{r^3}\right)\right] = -\mathbf{k}\left[\dfrac{1}{r^3}\nabla\cdot\mathbf{r} + \nabla\dfrac{1}{r^3}\cdot\mathbf{r}\right]$$

$$= -\mathbf{k}\left[\dfrac{3}{r^3} - \dfrac{3}{r^4}\dfrac{\mathbf{r}}{r}\cdot\mathbf{r}\right] = -\mathbf{k}\left[\dfrac{3}{r^3} - \dfrac{3}{r^3}\right] = 0. \text{ Hence proved.}$$

Example 14.42 If r and **r** have their usual meanings and **a** is a constant vector, prove that

$$\text{curl}\left(\dfrac{\mathbf{a}\times\mathbf{r}}{r^n}\right) = \dfrac{2-n}{r^n}\mathbf{a} + nr^{-(n+2)}(\mathbf{a}\cdot\mathbf{r})\mathbf{r}.$$

(*IETE June 2008; JNTU, 2005; Andhra 2000*)

Solution. $\text{curl}\left(\dfrac{\mathbf{a}\times\mathbf{r}}{r^n}\right) = \nabla\times\left[r^{-n}(\mathbf{a}\times\mathbf{r})\right]$

$$= r^{-n}\left[\nabla\times(\mathbf{a}\times\mathbf{r})\right]+\nabla r^{-n}\times(\mathbf{a}\times\mathbf{r}) \qquad \text{[Art. 14.14 formula 4]}$$

$$= r^{-n}\left[(\nabla\cdot\mathbf{r})\mathbf{a}-(\mathbf{a}\cdot\nabla)\mathbf{r}\right]+\left[-nr^{-n-1}\frac{\mathbf{r}}{r}\right]\times(\mathbf{a}\times\mathbf{r})$$

$$= r^{-n}(3\mathbf{a}-\mathbf{a})-nr^{-(n+2)}\mathbf{r}\times(\mathbf{a}\times\mathbf{r}) \qquad \left[\because \nabla\cdot\mathbf{r}=3 \text{ and } (\mathbf{a}\cdot\nabla)\mathbf{r}=\mathbf{a}\right]$$

$$= 2\mathbf{a}r^{-n}-nr^{-(n+2)}\left\{(\mathbf{r}\cdot\mathbf{r})\mathbf{a}-(\mathbf{a}\cdot\mathbf{r})\mathbf{r}\right\} \quad \text{(expanding the triple vector product)}$$

$$= \frac{2\mathbf{a}}{r^n}-nr^{-(n+2)}\left[\mathbf{a}r^2-(\mathbf{a}\cdot\mathbf{r})\mathbf{r}\right]=\frac{2\mathbf{a}}{r^n}-\frac{n\mathbf{a}}{r^n}+nr^{-(n+2)}(\mathbf{a}\cdot\mathbf{r})\mathbf{r}$$

$$= \frac{2-n}{r^n}\mathbf{a}+nr^{-(n+2)}(\mathbf{a}\cdot\mathbf{r})\mathbf{r}.$$

Example 14.43 A central field **A** in space is defined by $\mathbf{A}=\mathbf{r}f(r)$, where $\mathbf{r}=x\mathbf{i}+y\mathbf{j}+z\mathbf{k}$ and $r=(x^2+y^2+z^2)^{1/2}$. Determine $f(r)$ so that the field may be irrotational and solenoidal. *(IETE June 2005)*

Solution. $\nabla\times\left[\mathbf{r}f(r)\right]=f(r)(\nabla\times\mathbf{r})+\nabla f(r)\times\mathbf{r}$ (by Art. 14.14, formula 4)

$$= f(r)(\nabla\times\mathbf{r})+f'(r)\cdot\frac{\mathbf{r}}{r}\times\mathbf{r}=0 \qquad (\because \nabla\times\mathbf{r}=0 \text{ and } \mathbf{r}\times\mathbf{r}=\mathbf{0})$$

Hence $\mathbf{r}f(r)$ is irrotational for all $f(r)$. Again since $\mathbf{r}f(r)$ is solenoidal, therefore, $\nabla\cdot[\mathbf{r}f(r)]=0$

or $f(r)(\nabla\cdot\mathbf{r})+\mathbf{r}\cdot\nabla f(r)=0$ [by Art. 14.14, formula 2]

or $3f(r)+\mathbf{r}\cdot f'(r)\dfrac{\mathbf{r}}{r}=0$ $(\because \nabla\cdot\mathbf{r}=3)$

or $3f(r)+f'(r)\,r=0$ $(\because \mathbf{r}\cdot\mathbf{r}=r^2)$

or $\dfrac{f'(r)}{f(r)}+\dfrac{3}{r}=0$ or $\log f(r)+3\log r=\log c$

or $f(r)(r^3)=c$. Therefore $f(r)=c/r^3$, where c is an arbitrary constant.

Hence if $f(r)=\dfrac{c}{r^3}$, then $\mathbf{r}\,f(r)$ is both irrotational and solenoidal.

Example 14.44 (*a*) If $u=f(r)$ and $r^2=x^2+y^2+z^2$, Show that

$$\nabla^2 u=f''(r)+\frac{2}{r}f'(r).$$

(*b*) Also show that $u=A+\dfrac{B}{r}$ is a possible solution of $\nabla^2 u=0$, where A and B are arbitrary constants. *(IETE Dec 2003)*

Solution. (a) We have $\nabla^2 u = \nabla^2 f(r) = \nabla \cdot \nabla [f(r)]$

$$= \nabla \cdot \left(\frac{f'(r)}{r} \mathbf{r} \right) = \frac{f'(r)}{r} (\nabla \cdot \mathbf{r}) + \nabla \left(\frac{f'(r)}{r} \right) \cdot \mathbf{r}$$

$$= \frac{3f'(r)}{r} + \mathbf{r} \cdot \left[\frac{rf''(r) - f'(r)}{r^2} \right] \frac{\mathbf{r}}{r} = \frac{3f'(r)}{r} + \frac{rf''(r) - f'(r)}{r} = \frac{2f'(r)}{r} + f''(r).$$

(b) From the result of the part (a), we have

$$\nabla^2 \left(A + \frac{B}{r} \right) = \frac{d^2}{dr^2} \left(A + \frac{B}{r} \right) + \frac{2}{r} \frac{d}{dr} \left(A + \frac{B}{r} \right)$$

$$= \frac{2B}{r^3} + \frac{2}{r} \left(-\frac{B}{r^2} \right) = 0, \text{ showing that } A + \frac{B}{r} \text{ is a solution of Laplace's equation.}$$

Example 14.45 Prove that $div(r^n \mathbf{r}) = (n+3)r^n$. Hence show that $\mathbf{r}/r^3$ is solenoidal. (V.T.U, 2006; U.P.T.U, 2006; P.T.U, 2005)

Solution. We have $\mathbf{r} = x\mathbf{i} + y\mathbf{j} + z\mathbf{k}$ and $r = \sqrt{x^2 + y^2 + z^2}$.

$$\therefore \quad div(r^n \mathbf{r}) = \nabla \cdot (x^2 + y^2 + z^2)^{n/2} (x\mathbf{i} + y\mathbf{j} + z\mathbf{k})$$

$$= \left(\frac{\partial}{\partial x}\mathbf{i} + \frac{\partial}{\partial y}\mathbf{j} + \frac{\partial}{\partial z}\mathbf{k} \right) \cdot (x\mathbf{i} + y\mathbf{j} + z\mathbf{k})(x^2 + y^2 + z^2)^{n/2}$$

$$= \frac{\partial}{\partial x}\left[x(x^2 + y^2 + z^2)^{n/2} \right] + \frac{\partial}{\partial y}\left[y(x^2 + y^2 + z^2)^{n/2} \right] + \frac{\partial}{\partial z}\left[z(x^2 + y^2 + z^2)^{n/2} \right]$$

$$= \Sigma \left[1 \cdot (x^2 + y^2 + z^2)^{n/2} + x\frac{n}{2}(x^2 + y^2 + z^2)^{\frac{n}{2}-1} 2x \right]$$

$$= \Sigma r^n + n\Sigma x^2 (x^2 + y^2 + z^2)^{\frac{n}{2}-1} = 3r^n + n(r^2) \cdot r^{2\left(\frac{n}{2}-1\right)} = r^n(n+3).$$

Thus $div(r^n \mathbf{r}) = (n+3)r^n$. Substitute $n = -3$ in above equation, we get $div(\mathbf{r}/r^3) = 0$ that is, $\mathbf{r}/r^3$ is solenoidal.

Exercise 14.3

1. If $\mathbf{r} = x\mathbf{i} + y\mathbf{j} + z\mathbf{k}$ and $r = |\mathbf{r}|$, show that
(i) $\nabla \cdot \mathbf{r} = 3$, (ii) $\nabla \times \mathbf{r} = 0$. (UPTU, 2006; P.T.U, 2006; Madras 1998S)
2. Compute divF where $\mathbf{F} = 2x^2 z\mathbf{i} - xy^2 z\mathbf{j} + 3yz^2\mathbf{k}$ **Ans.** $4xz - 2xyz + 6yz$.

3. Find div **F** and curl **F** where
   $\mathbf{F} = \nabla(x^3 + y^3 + z^3 - 3xyz)$.   (*Kurukshetra, 2006*) **Ans.** $6(x + y + z)$; **0**

4. If $\mathbf{v} = xyz\,\mathbf{i} + 3x^2y\,\mathbf{j} + (xz^2 - y^2z)\mathbf{k}$, find (*i*) div **V**, (*ii*) curl **V**.

     Ans. (*i*) $yz + 3x^2 + 2xz - y^2$, (*ii*) $-2yz\,\mathbf{i} - (z^2 - xy)\mathbf{j} + (6xy - xz)\mathbf{k}$.

5. If $\mathbf{A} = yz^2\mathbf{i} + zx^2\mathbf{j} + xy^2\mathbf{k}$, evaluate curl **A** and show that

   $\mathbf{A} \cdot \text{curl } \mathbf{A} = xyz(xy + yz + zx)$.

6. Find the divergence and curl of the vector field given by

   (*i*) $xy^2\,\mathbf{i} + 2x^2yz\,\mathbf{j} - 3yz^2\mathbf{k}$,

   (*ii*) $(x^2 - y^2)\mathbf{i} + 2xy\,\mathbf{j} + (y^2 - xy)\mathbf{k}$.

     Ans. (*i*) $y^2 + 2x^2z - 6yz$, $-(3z^2 + 2x^2y)\mathbf{i} + (4xyz - 2xy)\mathbf{k}$;

          (*ii*) $4x$, $(2y - x)\mathbf{i} + y\mathbf{j} + 4y\mathbf{k}$.

7. If $\mathbf{A} = x^2z\,\mathbf{i} + yz^3\,\mathbf{j} - 3xy\,\mathbf{k}$, $\mathbf{B} = y^2\mathbf{i} - yz\,\mathbf{j} + 2x\,\mathbf{k}$ and $\phi = 2x^2 + yz$, find

   (*i*) $(\mathbf{A} \cdot \nabla)\phi$ (*ii*) $\mathbf{A} \cdot \nabla\phi$ (*iii*) $(\mathbf{A} \cdot \nabla)\mathbf{B}$.

Ans. (*i*) $4x^3z + 4z^4 - 3xy^2$, (*ii*) same as (*i*), (*iii*) $2y^2z^3\mathbf{i} + (3xy^2 - yz^4\mathbf{j}) + 2x^2z\mathbf{k}$.

8. If $\mathbf{v} = (x + z)\mathbf{i} - 3x^2y\,\mathbf{j} + 2yz\,\mathbf{k}$,   find div **v**, curl **v**, grad(div **v**) and

   div(curl **v**).   **Ans.** $1 - 3x^2 + 2y$; $2z\mathbf{i} + \mathbf{j} - 6xy\,\mathbf{k}$; $-6x\mathbf{i} + 2\mathbf{j}$; 0.

9. Evaluate curl grad $(x^2 + y^2 + z^2)$.   **Ans. O.**

10. If $\mathbf{A} = 3xyz^2\mathbf{i} + 2xy^3\mathbf{j} - x^2yz\,\mathbf{k}$ and $\phi = 3x^2 - yz$, find

   (*i*) $\nabla \times \mathbf{A}$, (*ii*) $\nabla \cdot (\nabla \times \mathbf{A})$, (*iii*) $\nabla \times (\nabla\phi)$.

     Ans. (*i*) $-x^2z\,\mathbf{i} + 8xyz\,\mathbf{j} + (2y^3 - 3xz^2)\mathbf{k}$, (*ii*) 0, (*iii*) **O.**

11. If $\mathbf{F}(x, y, z) = 3xz^2\mathbf{i} - yz\,\mathbf{j} + (x + 2z)\mathbf{k}$, find curl(curl **F**).

                  Ans. $-6x\mathbf{i} + (6z - 1)\mathbf{k}$.

12. Given $f(x, y, z) = 2x^3y^2z^4$. Find (*a*) div(grad f) (*b*) $\nabla^2 f$.

        Ans. (*a*) and (*b*) $12xy^2z^4 + 4x^3z^4 + 24x^3y^2z^2$.

13. Prove that $\nabla^2(1/r) = 0$.

   [Hint. $\nabla^2(1/r) = \left(\dfrac{\partial^2}{\partial x^2} + \dfrac{\partial^2}{\partial y^2} + \dfrac{\partial^2}{\partial z^2}\right)\left(\dfrac{1}{\sqrt{x^2 + y^2 + z^2}}\right)$

$$\frac{\partial}{\partial x}\left(\frac{1}{\sqrt{x^2+y^2+z^2}}\right)=-x\left(x^2+y^2+z^2\right)^{-3/2}.$$

$$\therefore \frac{\partial^2}{\partial x^2}=\frac{\partial}{\partial x}\left(\frac{\partial}{\partial x}\right)=\frac{\partial}{\partial x}\left[-x\left(x^2+y^2+z^2\right)^{-3/2}\right]=\frac{2x^2-y^2-z^2}{\left(x^2+y^2+z^2\right)^{-5/2}}.$$

Thus $\displaystyle\sum\frac{\partial^2}{\partial x^2}\left(\frac{1}{\sqrt{x^2+y^2+z^2}}\right)=0.]$

14. Show that the vector $\mathbf{v}=(x+3y)\mathbf{i}+(y-3y)\mathbf{j}+(x-2z)\mathbf{k}$ is solenoidal.

15. Show that the vector

$\mathbf{F}=\left(x^2-yz\right)\mathbf{i}+\left(4y+z^2x\right)\mathbf{j}-(2xz+4z)\mathbf{k}$ is solenoidal.

16. Find the value of the constant a so that the vector

$\mathbf{A}=(x+3y)\mathbf{i}+(y-2z)\mathbf{j}+(x+az)\mathbf{k}$ is solenoidal.　　　**Ans.** -2.

17. Show that the vector field $\mathbf{A}$ where $\mathbf{A}=\left(x^2-y^2+x\right)\mathbf{i}-(2xy+y)\mathbf{j}$ is irrotational; and find a scalar function ϕ such that $\mathbf{A}=\operatorname{grad}\phi$.

Ans. $\left(2x^3-6xy^2+3x^2-3y^2\right)/6$.

18. (a) Prove that $\mathbf{v}=2xyz\mathbf{i}+x^2z\mathbf{j}+x^2y\mathbf{k}$ is irrotational and find $u(x,y,z)$ so that $\mathbf{v}=\operatorname{grad}u$.　　　**Ans.** $u=x^2yz$

(b) Show that the vector field defined by $\mathbf{F}=2xyz^3\mathbf{i}+x^2z^3\mathbf{j}+3x^2yz^2\mathbf{k}$ is irrotational. Find the scalar potential ϕ such that $\mathbf{F}=\operatorname{grad}\phi$.

Ans. $\phi(x,y,z)=x^2yz^3$.

19. Show that $\mathbf{v}(x,y,z)=\mathbf{i}(z-y)+\mathbf{j}(x-z)+\mathbf{k}(y-x)$ is solenoidal and find $\mathbf{w}(x,y,z)$ such that $\mathbf{v}=\operatorname{curl}\mathbf{w}$.

Ans. $\mathbf{w}(x,y,z)=x(y+z)\mathbf{i}+y(z+x)\mathbf{j}+z(x+y)\mathbf{k}.$

20. A vector field $\mathbf{F}$ is given by $\mathbf{F}=\left(x^2+xy^2\right)\mathbf{i}+\left(y^2+x^2y\right)\mathbf{j}$.

Show that the field is irrotational (or conservative) and find the scalar potential ϕ.　　　**Ans.** $\phi=\dfrac{x^3}{3}+\dfrac{x^2y^2}{2}+\dfrac{y^3}{3}+c.$

21. A fluid motion is given by

$\overline{v}=(y\sin z-\sin x)\hat{\mathbf{i}}+(x\sin z+2yz)\hat{\mathbf{j}}+\left(xy\cos z+y^2\right)\hat{\mathbf{k}}.$

Is the motion irrotational? If so, find the velocity potential.

(*IETE, Dec 2007; PTU 2006*)   **Ans.** Yes. $\phi = xy\sin z + \cos x + zy^2 + c$.

22. Show that $\mathbf{F} = \left(y^2 + 2xz^2\right)\mathbf{i} + \left(2xy - z\right)\mathbf{j} + \left(2x^2z - y + 2z\right)\mathbf{k}$ is irrotational and hence find its scalar potential.   (*IETE Dec 2004*)

23. Show that vector field represented by

    $\mathbf{F} = \left(z^2 + 2x + 3y\right)\mathbf{i} + \left(3x + 2y + z\right)\mathbf{j} + \left(y + 2zx\right)\mathbf{k}$ is irrotational but not solenoidal. Also obtain a scalar ϕ function such that grad $\phi = \mathbf{F}$.

    (*IETE Dec 2003*)

24. A vector field is given by $F = \left(x^2 - y^2 + x\right)\mathbf{i} - \left(2xy + y\right)\mathbf{j}$. Show that the field is irrotational and find its scalar potential.   (*IETE June 2003*)

    Ans. $\phi = \dfrac{1}{3}x^3 - xy^2 + \dfrac{1}{2}\left(x^2 - y^2\right) + c$.

25. Show that the vector field given by

    $\mathbf{A} = 3x^2y\mathbf{i} + \left(x^3 - 2yz^2\right)\mathbf{j} + \left(3z^2 - 2y^2z\right)\mathbf{k}$ is irrotational but not solenoidal. Also find a scalar function ϕ such that grad $\phi = \mathbf{A}$.   (*IETE Dec. 2002*)

26. If $\nabla\phi = \left(y^2 - 2xyz^3\right)\mathbf{i} + \left(3 + 2xy - x^2z^3\right)\mathbf{j} + \left(6z^3 - 3x^2yz^2\right)\mathbf{k}$, find ϕ.

    Ans. $\phi = xy^2 - x^2yz^3 + 3y + \dfrac{3}{2}z^4 + c$.

27. If ϕ and ψ are scalar functions, prove that $\nabla\phi \times \nabla\psi$ is solenoidal.

28. If $\phi\mathbf{F} = \nabla\psi$, where ϕ, ψ are scalar functions and $\mathbf{F}$ is vector function, show that $\mathbf{F}\cdot(\text{curl }\mathbf{F}) = 0$, that is $\mathbf{F}\cdot(\nabla \times \mathbf{F}) = 0$.

29. If $\mathbf{r} = x\mathbf{i} + y\mathbf{i} + z\mathbf{k}$ and $r = |\mathbf{r}|$ show that

    (*i*) $\text{div}\left(\text{grad } r^m\right) = m(m+1)r^{m-2}$, (*ii*) $\text{div}\left(\mathbf{r}\phi\right) = 3\phi + \mathbf{r}\cdot\text{grad }\phi$.

    (*UPTU 2005; Osmania 2003; Madras 2003; Delhi 2002*)

30. If $r = |\mathbf{r}| = |x\mathbf{i} + y\mathbf{j} + z\mathbf{k}|$, prove that $\nabla\cdot\left(r^3\mathbf{r}\right) = 6r^3$.

31. If $\mathbf{a}$ and $\mathbf{b}$ are constant vectors and $\mathbf{r}$ is the position vector with modulus r, prove that

    (*i*) $\nabla\times(\mathbf{a}\cdot\mathbf{r})\mathbf{a} = 0$, (*ii*) $\nabla\times\left[\mathbf{a}\times(\mathbf{b}\times\mathbf{r})\right] = \mathbf{a}\times\mathbf{b}$

    (*iii*) $\nabla\times\left(\dfrac{\mathbf{a}\times\mathbf{r}}{r}\right) = \dfrac{\mathbf{a}}{r} + \dfrac{(\mathbf{a}\cdot\mathbf{r})\mathbf{r}}{r^3}$, (*iv*) $\text{curl}\left(\dfrac{\mathbf{r}\times\mathbf{a}}{r^3}\right) = \dfrac{\mathbf{a}}{r^3} - \dfrac{3r}{r^5}(\mathbf{r}.\mathbf{a})$ (*MDU 2007*)

    (*v*) $\nabla\cdot\left(\dfrac{\mathbf{a}\times\mathbf{r}}{r^n}\right) = 0$, (*vi*) $\nabla\left(\dfrac{\mathbf{a}\cdot\mathbf{r}}{r^n}\right) = \dfrac{\mathbf{a}}{r^n} - \dfrac{n(\mathbf{a}\cdot\mathbf{r})\mathbf{r}}{r^{n+2}}$.

(vii) $\mathbf{b} \cdot \nabla\left[\mathbf{a} \cdot \nabla\left(\dfrac{1}{r}\right)\right] = \dfrac{3(\mathbf{a} \cdot \mathbf{r})(\mathbf{b} \cdot \mathbf{r})}{r^5} - \dfrac{\mathbf{a} \cdot \mathbf{b}}{r^3}.$ *(IETE Dec 2006)*

[Hint. *(vii)* Let $\mathbf{a} = a_1\mathbf{i} + a_2\mathbf{j} + a_3\mathbf{k}$, $\mathbf{r} = x\mathbf{i} + y\mathbf{j} + z\mathbf{k}$.

$$\mathbf{a} \cdot \nabla\left(\frac{1}{r}\right) = \mathbf{a} \cdot \left(-\frac{\mathbf{r}}{r^3}\right) = -\frac{\mathbf{a} \cdot \mathbf{r}}{r^3} = -\left(\frac{a_1 x + a_2 y + a_3 z}{r^3}\right).$$

Now $\nabla\left\{\mathbf{a} \cdot \nabla\left(\dfrac{1}{r}\right)\right\} = \nabla\left\{-\left(\dfrac{a_1 x + a_2 y + a_3 z}{r^3}\right)\right\} = \sum \mathbf{i}\dfrac{\partial}{\partial x}\left(-\dfrac{a_1 x + a_2 y + a_3 z}{r^3}\right)$

$$= \sum \mathbf{i}\left[-\frac{a_1}{r^3} + \frac{3(a_1 x + a_2 y + a_3 z)}{r^4}\left(\frac{x}{r}\right)\right]$$

$$= -\frac{1}{r^3}\sum \mathbf{i}\,a_1 + \frac{3(a_1 x + a_2 y + a_3 z)}{r^5}\sum \mathbf{i}x = -\frac{\mathbf{a}}{r^3} + \frac{3(\mathbf{a} \cdot \mathbf{r})}{r^5}\mathbf{r}.$$

Therefore, $\mathbf{b} \cdot \nabla\left\{\mathbf{a} \cdot \nabla\left(\dfrac{1}{r}\right)\right\} = \dfrac{3(\mathbf{a} \cdot \mathbf{r})(\mathbf{b} \cdot \mathbf{r})}{r^5} - \dfrac{(\mathbf{a} \cdot \mathbf{b})}{r^3}$, on taking dot product

of $\mathbf{b}$ and $\nabla\left\{\mathbf{a} \cdot \nabla\left(\dfrac{1}{r}\right)\right\}$].

32. If $u = x^2 + y^2 + z^2$, $\overline{v} = x\mathbf{i} + y\mathbf{j} + z\mathbf{k}$ then find the value of $\nabla \cdot (u\overline{v})$.

    (IETE June 2008, 2007, 2004; PTU 2005) **Ans.** $5u$.

    [Hint. div $u\overline{v} = u$ div $\overline{v} + (\text{grad } u)\bullet\overline{v}$].

33. If **E** and **H** are irrotational fields, prove that $\mathbf{E} \times \mathbf{H}$ is solenoidal.

    [Hint. As **E** and **H** are irrotational. Therefore $\nabla \times \mathbf{E} = \mathbf{0}$ and $\nabla \times \mathbf{H} = \mathbf{0}$.
    Now $\nabla \cdot (\mathbf{E} \times \mathbf{H}) = (\nabla \times \mathbf{E}) \cdot \mathbf{H} - (\nabla \times \mathbf{H}) \cdot \mathbf{E} = 0$.]

34. If $\nabla \cdot \left\{\dfrac{f(r)\mathbf{r}}{r}\right\} = 0$, show that $f(r) = A/r^2$, where A is an arbitrary constant.

35. Show that $r^\alpha \mathbf{r}$ is any irrotational vector for any value of α but is solonoidal
    if $\alpha + 3 = 0$ where $\mathbf{r} = x\mathbf{i} + y\mathbf{i} + z\mathbf{k}$ and r is the magnitude of $\mathbf{r}$.

    (V.T.U., 2006; Kottayam, 2005)

    [Hint: Let $\mathbf{A} = r^\alpha \mathbf{r} = \left(x^2 + y^2 + z^2\right)^{\alpha/2}\left(x\mathbf{i} + y\mathbf{j} + z\mathbf{k}\right) = \sum x\left(x^2 + y^2 + z\right)^{\alpha/2}\mathbf{i}$

$$\therefore \quad \text{curl } \mathbf{A} = \begin{vmatrix} \mathbf{i} & \mathbf{j} & \mathbf{k} \\ \partial/\partial x & \partial/\partial y & \partial/\partial z \\ x\left(x^2+y^2+z\right)^{\alpha/2} & y\left(x^2+y^2+z\right)^{\alpha/2} & y\left(x^2+y^2+z\right)^{\alpha/2} \end{vmatrix}.$$

$$= \sum \mathbf{i}\left\{\frac{\alpha z}{2}\left(x^2+y^2+z\right)^{\frac{\alpha}{2}-1}(2y) - \frac{\alpha y}{2}\left(x^2+y^2+z\right)^{\frac{\alpha}{2}-1}(2z)\right\} = 0. \text{ Hence } \mathbf{A}$$

is irrotational for any value of α.

Now $\text{div } \mathbf{A} = \nabla\cdot\left(r^\alpha \mathbf{r}\right) = (\alpha+3)r^\alpha$ which is zero when $\alpha+3=0$, that is $\mathbf{A}$ is solenoidal if $\alpha+3=0$.]

36. Prove that:

(i) $\nabla^2\left(\phi f\right) = \phi\nabla^2 f + 2\nabla\phi\cdot\nabla f + f\nabla^2\phi,$

(ii) $\nabla\cdot\left(\phi\nabla f - f\nabla\phi\right) = \phi\nabla^2 f - f\nabla^2\phi.$

14.16 Line Integrals

Let the parametric representation of simple curve C be

$$x = x(t), y = y(t), z = z(t), \qquad (a \le t \le b). \qquad \qquad ...(14.23)$$

Therefore, the position vector of a point on the curve C can be written as

$$\mathbf{r}(t) = x(t)\mathbf{i} + y(t)\mathbf{j} + z(t)\mathbf{k}, \qquad (a \le t \le b). \qquad \qquad ...(14.24)$$

The integral of a vector field $\mathbf{F}$ defined on a curve segment C is called a line integral.

Definition. A **line integral** of a vector function $\mathbf{F}$ over a curve C is defined by

$$\int_C \mathbf{F}\cdot d\mathbf{r} = \int_a^b \mathbf{F}\cdot\frac{d\mathbf{r}}{dt}\,dt \qquad \qquad ...(14.25)$$

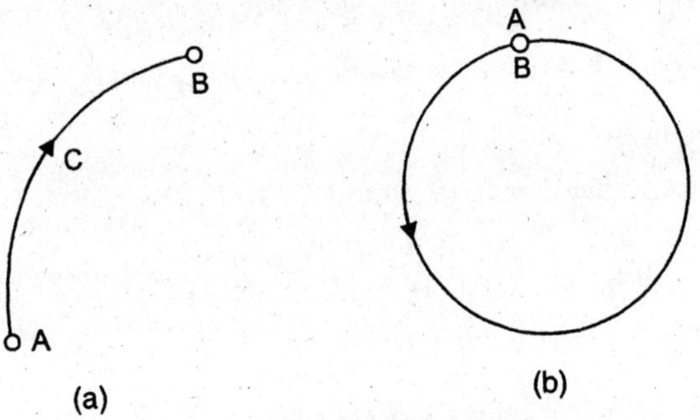

(a)

(b)

Fig. 14.13.

If $\mathbf{F}(x,y,z) = F_1(x,y,z)\mathbf{i} + F_2(x,y,z)\mathbf{j} + F_3(x,y,z)\mathbf{k}$ and $\mathbf{r} = x\mathbf{i} + y\mathbf{j} + z\mathbf{k}$, then equation (14.25) becomes

$$\int_C \mathbf{F} \cdot d\mathbf{r} = \int_C (F_1 dx + F_2 dy + F_3 dz) = \int_a^b \left(F_1 \frac{dx}{dt} + F_2 \frac{dy}{dt} + F_3 \frac{dz}{dt} \right) dt.$$

We call C the path of integration.

If the path of integration C in (14.25) is a *closed curve*, (Fig. 14.13(b)) then the integral around C is denoted by $\oint_C \mathbf{F} \cdot d\mathbf{r}$ instead of $\int_C \mathbf{F} \cdot d\mathbf{r}$.

If the curve C is piecewise smooth Fig. (14.14) containing the curve $C_1, C_2,...$ C_n joined end to end, we define a line integral along a piecewise smooth curve C to be the sum of the integrals along the sections.

$$\int_C \mathbf{F} \cdot d\mathbf{r} = \int_{C_1} \mathbf{F} \cdot d\mathbf{r} + \int_{C_2} \mathbf{F} \cdot d\mathbf{r} + ... + \int_{C_n} \mathbf{F} \cdot d\mathbf{r} \qquad ...(14.26)$$

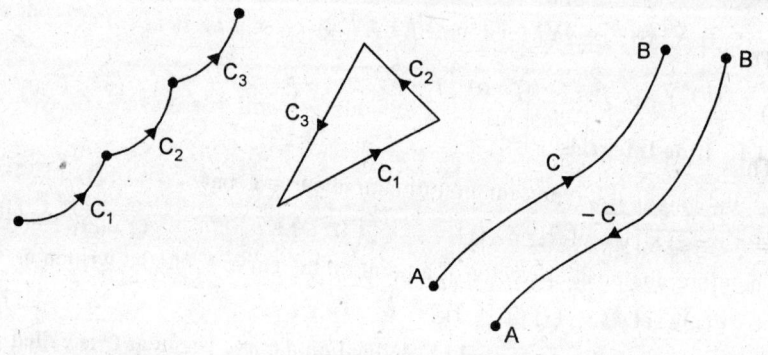

Fig. 14.14.   **Fig. 14.15.**

Remark: Reversal of the path of integration changes the sign of the line integral. (Fig. 14.15)

14.17 Illustrative Examples

Example 14.46 If $\mathbf{F} = x^2\mathbf{i} + xy\mathbf{j}$, evaluate $\int_C \mathbf{F} \cdot d\mathbf{r}$ along the curve C in the xy plane consisting of

(i) the straight line joining (0, 0) to (1, 1).

(ii) the parabola $y^2 = x$ joining the same points.

(iii) the straight lines from (0, 0) to (1, 0), then to (1, 1).

Solution. Since $\mathbf{r} = \mathbf{i}x + \mathbf{j}y, d\mathbf{r} = \mathbf{i}dx + \mathbf{j}dy$.

Therefore, $\int_C \mathbf{F} \cdot d\mathbf{r} = \int_C (x^2\mathbf{i} + xy\,\mathbf{j}) \cdot (\mathbf{i}\,dx + \mathbf{j}\,dy) = \int_C x^2 dx + xy\,dy.$

(i) Along the straight line joining $(0, 0)$ to $(1, 1)$ $y = x \Rightarrow dy = dx$, and x varies from 0 to 1. [Fig. 14.16 (a)]

$$\therefore \int_C \mathbf{F} \cdot d\mathbf{r} = \int_0^1 x^2 dx + x(x) dx = \int_0^1 (x^2 + x^2) dx = 2 \int_0^1 x^2 dx = \frac{2}{3}.$$

(ii) Along the parabola $y^2 = x$, Fig. 14.16 (b) we have $dx = 2y \, dy$ and y goes from 0 to 1. Therefore, $\int_C \mathbf{F} \cdot d\mathbf{r} = \int_C x^2 dx + xy \, dy = \int_0^1 y^4 (2y \, dy) + y^2 \cdot y \, dy$

$$= \int_0^1 (2y^5 + y^3) dy = \left[\frac{y^6}{3} + \frac{y^4}{4} \right]_0^1 = \frac{7}{12}.$$

Alternative. Let $y = t$ in $y^2 = x$. Then the parametric equations of C are $y = t, x = t^2$.

Points $(0, 0)$, $(1, 1)$ correspond to $t = 0$ and $t = 1$ respectively.

Then $\int_C \mathbf{F} \cdot d\mathbf{r} = \int_0^1 (t^2)^2 \, 2t \, dt + t^2 (t) \, dt = \int_0^1 (2t^5 + t^3) dt = \frac{7}{12}.$

(iii) Refer Fig. 14.16(c). Along C_1, $y = 0$, $dy = 0$. Along C_2, $x = 1$, $dx = 0$.

Thus along the path OMP, $\int_C \mathbf{F} \cdot d\mathbf{r} = \int_{C_1} \mathbf{F} \cdot d\mathbf{r} + \int_{C_2} \mathbf{F} \cdot d\mathbf{r}$

$$= \int_0^1 x^2 dx + \int_0^1 y \, dy = \frac{1}{3} + \frac{1}{2} = \frac{5}{6}.$$

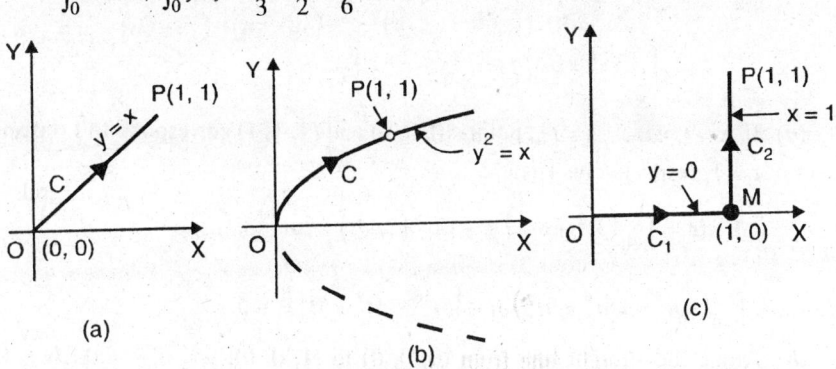

(a)   (b)   (c)

Fig. 14.16.

Example 14.47 Evaluate the line integral of $\mathbf{v} = x^2 \mathbf{i} - 2y \mathbf{j} + z^2 \mathbf{k}$ over the straight line path from $(-1, 2, 3)$ to $(2, 3, 5)$.

Solution. We have

$$\int_C \mathbf{v} \cdot d\mathbf{r} = \int_C (x^2 \mathbf{i} - 2y \mathbf{j} + z^2 \mathbf{k}) \cdot (dx \, \mathbf{i} + dy \, \mathbf{j} + dz \, \mathbf{k}) = \int_C x^2 dx - 2y \, dy + z^2 dz.$$

The equations of the straight line from $(-1, 2, 3)$ to $(2, 3, 5)$ are

$$\frac{x-(-1)}{2-(-1)} = \frac{y-2}{3-2} = \frac{z-3}{5-3} = t \text{ (say), that is, } \frac{x+1}{3} = y-2 = \frac{z-3}{2} = t.$$

Therefore $x = 3t-1$, $y = t+2$, $z = 2t+3$ are its parametric equations, where t varies from 0 to 1. Then the line integral equals

$$\int_{t=0}^{1} (3t-1)^2 \, 3dt - 2(2+t) \, dt + (2t+3)^2 \, 2dt$$

$$= \int_{t=0}^{1} (35t^2 + 4t + 17) \, dt = \left[\frac{35}{3}t^3 + 2t^2 + 17t\right]_0^1 = \frac{92}{3}.$$

Example 14.48 If $F = (3x^2 + 6y)i - 14yz \, j + 20xz^2 \, k$, evaluate $\int_C F \cdot dr$ from $(0, 0, 0)$ to $(1, 1, 1)$ along the following paths C:

(a) $x = t, y = t^2, z = t^3$.   (*V.T.U. 2001; Bhopal 1999*)

(b) the straight lines from $(0, 0, 0)$ to $(1, 0, 0)$, then to $(1, 1, 0)$, and then to $(1, 1, 1)$.

(c) the straight line joining $(0, 0, 0)$ to $(1, 1, 1)$.   (*AMIE, S-2006*)

Solution. We have

$$\int_C F \cdot dr = \int_C \left[(3x^2 + 6y)i - 14yz \, j + 20xz^2 k \right] \cdot (dx \, i + dy \, j + dz \, k)$$

$$= \int_C (3x^2 + 6y) \, dx - 14yz \, dy + 20xz^2 \, dz.$$

(a) If $x = t, y = t^2, z = t^3$, points $(0, 0, 0)$ and $(1, 1, 1)$ correspond to $t = 0$ and $t = 1$, respectively. Then

$$\int_C F \cdot dr = \int_{t=0}^{1} (3t^2 + 6t^2) \, dt - 14t^5 (2t \, dt) + 20t^7 (3t^2 dt)$$

$$= \int_{t=0}^{1} (60t^9 - 28t^6 + 9t^2) \, dt = [6t^{10} - 4t^7 + 3t^3]_0^1 = 5.$$

(b) Along the straight line from $(0, 0, 0)$ to $(1, 0, 0)$, $y = 0$, $z = 0$, $dy = 0$, $dz = 0$ while x varies from 0 to 1. Then the integral over this part of the path is

$$\int_{x=0}^{1} \{3x^2 + 6(0)\} \, dx = \int_0^1 3x^2 dx = [x^3]_0^1 = 1.$$

Along the straight line from $(1, 0, 0)$ to $(1, 1, 0)$, $x = 1$, $z = 0$, $dx = 0$, $dz = 0$ while y varies from 0 to 1. Then the integral over this part of the path is

$$\int_{y=0}^{1} -14y(0) \, dy = 0.$$

Along the straight line from $(1, 1, 0)$ to $(1, 1, 1)$, $x = 1$, $y = 1$, $dx = 0$, $dy = 0$ while z varies from 0 to 1. Then the integral over this part of the path is

$$\int_{z=0}^{1} 20(1)z^2 dz = \frac{20}{3}\left[z^3\right]_0^1 = \frac{20}{3}. \text{ Therefore, } \int_C \mathbf{F}\cdot d\mathbf{r} = 1+0+(20/3) = 23/3.$$

(c) The straight line joining $(0, 0, 0)$ and $(1, 1, 1)$ is given in parametric form by $x = t$, $y = t$, $z = t$ then $\int_C \mathbf{F}\cdot d\mathbf{r} = \int_{t=0}^{1}\left(3t^2 + 6t - 14t^2 + 20t^3\right)dt$

$$= \int_{t=0}^{1}\left(20t^3 - 11t^2 + 6t\right)dt = \left[5t^4 - \frac{11}{3}t^3 + 3t^2\right]_0^1 = \frac{13}{3}.$$

Example 14.49 Evaluate $\int_C (x+y)dx - x^2 dy + (y+z)dz$, where $C: x^2 = 4y$,

$z = x, 0 \le x \le 2$. (A.M.I.E., W-2010; IETE, June 2008)

Solution. We parametrise C as $x = t$, $y = t^2/4$, $z = t$, $0 \le t \le 2$.

$$\therefore \int_C (x+y)dx - x^2 dy + (y+z)dz = \int_0^2 \left(t + \frac{t^2}{4}\right)dt - t^2\cdot d\left(\frac{t^2}{4}\right) + \left(\frac{t^2}{4} + t\right)dt$$

$$= \int_0^2 \left(t + \frac{t^2}{4} - \frac{t^3}{2} + \frac{t^2}{4} + t\right)dt = \int_0^2 \left(2t + \frac{t^2}{2} - \frac{t^3}{2}\right)dt$$

$$= \left[t^2 + \frac{t^3}{6} - \frac{t^4}{8}\right]_0^2 = 4 + \frac{4}{3} - 2 = \frac{10}{3}.$$

Example 14.50 Evaluate $\oint_C \left(3x^2 - 8y^2\right)dx + (4y - 6xy)dy$ on the closed curve C defined by $x^2 = y$ and $y^2 = x$.

Solution. Since C is piecewise smooth, we express the integral as a sum of integrals. We write

$$\oint_C Pdx + Qdy = \int_{C_1} Pdx + Qdy + \int_{C_2} Pdx + Qdy,$$

where C_1, C_2 are the curves shown in Fig. 14.17.

On C_1, we use x as a parameter. Since

$y = x^2$, $dy = 2xdx$, we have

$$\int_{C_1}\left(3x^2 - 8y^2\right)dx + (4y - 6xy)dy$$

$$= \int_0^1 (3x^2 - 8x^4)dx + (4x^2 - 6x^3)2xdx$$

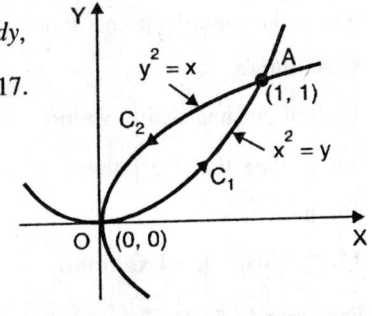

Fig. 14.17.

$$= \int_0^1 \left(-20x^4 + 8x^3 + 3x^2\right) dx = \left[-4x^5 + 2x^4 + x^3\right]_0^1 = -1.$$

On C_2, we use y as a parameter. From $x = y^2$, we get $dx = 2y\, dy$ and so

$$\int_{C_2} \left(3x^2 - 8y^2\right) dx + \left(4y - 6xy\right) dy = \int_1^0 \left(3y^4 - 8y^2\right) 2y\, dy + \left(4y - 6y^3\right) dy$$

$$= \int_1^0 \left(6y^5 - 22y^3 + 4y\right) dy = \left[y^6 - \frac{11}{2} y^4 + 2y^2\right]_1^0 = \frac{5}{2}.$$

Hence $\oint_C \left(3x^2 - 8y^2\right) dx + \left(4y - 6xy\right) dy = -1 + \frac{5}{2} = \frac{3}{2}.$

14.18 Applications of line integrals

Work done by a force (*i*) If **F** is the variable force acting on a particle which is moving along the curve C. Then, the work done by the force **F** in displacing the particle from the point P to the point Q along the curve C is given by

$$W = \int_P^Q \mathbf{F} \cdot d\mathbf{r} = \int_{C^*} \mathbf{F} \cdot d\mathbf{r}, \qquad \ldots(14.27)$$

where C^* is the part of C, whose initial and terminal points are P and Q.
(*ii*) Suppose that **F** is a *conservative* vector field. Then **F** can be written as $\mathbf{F} = \nabla\phi$, where ϕ is a scalar potential (field). Then, the work done

$$W = \int_{C^*} \mathbf{F} \cdot d\mathbf{r} = \int_{C^*} (\nabla\phi) \cdot d\mathbf{r} = \int_{C^*} \left(\frac{\partial\phi}{\partial x} dx + \frac{\partial\phi}{\partial y} dy + \frac{\partial\phi}{\partial z} dz\right)$$

$$= \int_P^Q d\phi = \left[\phi(x, y, z)\right]_P^Q \qquad \ldots(14.28)$$

Therefore, work done depends only on the initial and terminal points of the curve C^*, that is the work done is independent of the path of integration.
Remark. If a particle moving in a conservative vector field traverses a closed path C that begins and ends at (x_0, y_0), then the work performed by the field is zero. This is because the points P and Q are same.

Circulation

In fluid mechanics, Let **v** represent the velocity field of a fluid and C be a closed curve, then the line integral of vector field $\oint_C \mathbf{v} \cdot d\mathbf{r}$ is called the *circulation* around C,

14.19 Illustrative Examples

Example 14.51 Find the work done in moving a particle once round the circle $x^2 + y^2 = 9$, $z = 0$ if the vector field is given by

$$\mathbf{F} = (2x - y - z)\mathbf{i} + (x + y - z^2)\mathbf{j} + (3x - 2y + 4z)\mathbf{k}.$$

Solution. Since the particle moves in the plane $z = 0$

(*i.e., xy*-plane), $\mathbf{F} = (2x - y)\mathbf{i} + (x + y)\mathbf{j} + (3x - 2y)\mathbf{k}$ and $d\mathbf{r} = dx\,\mathbf{i} + dy\,\mathbf{j}$.

Work done $= \oint_C \mathbf{F} \cdot d\mathbf{r} = \oint_C \left[(2x - y)\mathbf{i} + (x + y)\mathbf{j} + (3x - 2y)\mathbf{k} \right] \cdot (dx\,\mathbf{i} + dy\,\mathbf{j})$

$$= \oint_C (2x - y)\,dx + (x + y)\,dy.$$

Now C is the circle $x^2 + y^2 = 9$.

The parametric equations of the circle are
$x = 3\cos\theta$, $y = 3\sin\theta$, where θ varies from
0 to 2π (See Figure 14.18).

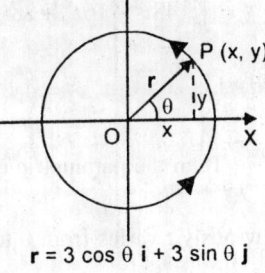

$r = 3\cos\theta\,\mathbf{i} + 3\sin\theta\,\mathbf{j}$

Fig. 14.18.

$$\therefore \oint_C \mathbf{F} \cdot d\mathbf{r} = \int_0^{2\pi} \left[2(3\cos\theta) - 3\sin\theta \right]$$

$$(-3\sin\theta)\,d\theta + [3\cos\theta + 3\sin\theta](3\cos\theta)\,d\theta$$

$$= \int_0^{2\pi} (9 - 9\sin\theta\cos\theta)\,d\theta = \left[9\theta - 9\frac{\sin^2\theta}{2} \right]_0^{2\pi} = 18\pi.$$

Example 14.52 Find the work done is moving the particle in the force field
$\mathbf{F}(x, y, z) = 3x^2\mathbf{i} + (2xz - y)\mathbf{j} + z\mathbf{k}$ along.

(*i*) the straight line from $A(0, 0, 0)$ to $B(2, 1, 3)$. (*SVTU, 2007; JNTU, 2002*)

(*ii*) the space curve $x = 2t^2$, $y = t$, $z = 4t^2 - t$ from $t = 0$ to $t = 1$.

(*iii*) the curve defined by $x^2 = 4y, 3x^3 = 8z$ from $x = 0$ to $x = 2$.

(*IETE, June 2006; MDU 2002*)

Solution. $\int_C \mathbf{F} \cdot d\mathbf{r} = \int_C \left[3x^2\mathbf{i} + (2xz - y)\mathbf{j} + z\mathbf{k} \right] \cdot (dx\,\mathbf{i} + dy\,\mathbf{j} + dz\,\mathbf{k})$

$$= \int_C 3x^2\,dx + (2xz - y)\,dy + z\,dz.$$

(*i*) The equations of line AB are $\dfrac{x}{2} = y = \dfrac{z}{3} = t$ (say).

$\therefore$ $x = 2t$, $y = t$ and $z = 3t$ are the parametric equations of the line AB.

The points $(0, 0, 0)$ to $(2, 1, 3)$ correspond to $t = 0$ and $t = 1$ respectively.

Work done $= \int_C \mathbf{F} \cdot d\mathbf{r}$

$$= \int_0^1 3(2t)^2\,d(2t) + [2(2t)(3t) - t]\,dt + 3t\,d(3t) = \int_0^1 24t^2\,dt + (12t^2 - t)\,dt + 9t\,dt$$

$$= \int_0^1 (36t^2 + 8t)\,dt = 16.$$

(*ii*) Work done $= \int_C \mathbf{F} \cdot d\mathbf{r} = \int_0^1 3(2t^2)^2 \, d(2t^2) + \left[2(2t^2)(4t^{2^\bullet} - t) - t \right] dt$

$$+ (4t^2 - t) \, d(4t^2 - t)$$

$$= \int_0^1 48t^5 \, dt + (16t^4 - 4t^3 - t) \, dt + (32t^3 - 12t^2 + t) \, dt$$

$$= \int_0^1 (48t^5 + 16t^4 + 28t^3 - 12t^2) \, dt = \left[8t^6 + \frac{16}{5}t^5 + 7t^4 - 4t^3 \right]_0^1 = 14.2.$$

(*iii*) Let $x = t$ in $x^2 = 4y$, $3x^3 = 8z$.

Then the parametric equations of C are $x = t$, $y = \dfrac{t^2}{4}$, $z = \dfrac{3t^3}{8}$.

Obviously t varies from 0 to 2. Hence the work done $= \int_C \mathbf{F} \cdot d\mathbf{r}$

$$= \int_0^2 3t^2 \, dt + \left[2(t) \left(\frac{3t^3}{8} \right) - \frac{t^2}{4} \right] d\left(\frac{t^2}{4} \right) + \frac{3t^3}{8} \, d\left(\frac{3t^3}{8} \right)$$

$$= \int_0^2 \left(3t^2 - \frac{t^3}{8} + \frac{51}{64} t^5 \right) dt = 16.$$

Example 14.53 Find the work done by the force $\mathbf{F} = x \, \mathbf{i} - z \, \mathbf{j} + 2y \, \mathbf{k}$ in the displacement along the closed path C consisting of the segments C_1, C_2 and C_3 where on C_1, $0 \le x \le 1$, $y = x$, $z = 0$, on C_2, $0 \le z \le 1$, $x = 1$, $y = 1$ on C_3, $1 \ge x \ge 0$, $y = z = x$.

Solution. Total work done $= \oint_C \mathbf{F} \cdot d\mathbf{r}$

$$= \oint_C (x \, \mathbf{i} - z \, \mathbf{j} + 2y \, \mathbf{k}) \cdot (dx \, \mathbf{i} + dy \, \mathbf{j} + dz \, \mathbf{k})$$

$$= \oint_C x \, dx - z \, dy + 2y \, dz.$$

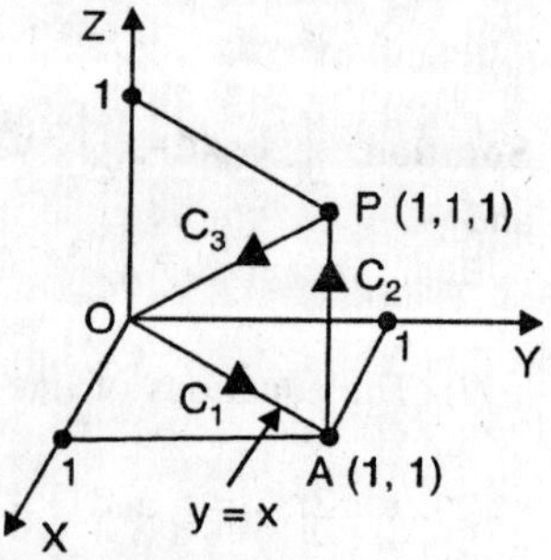

The closed path C consisting of segments C_1, C_2 and C_3 is shown in the Fig. 14.19.

Let W_1, W_2, W_3 be the work done in displacement along C_1, C_2 and C_3 respectively.

Since on C_1, $0 \le x \le 1$, $y = x$, $z = 0$, $dy = dx$, $dz = 0$.

Fig. 14.19.

$$\therefore \ W_1 = \int_{C_1} x \, dx = \int_0^1 x \, dx = 1/2.$$

On C_2, $0 \le z \le 1$, $x = 1$, $y = 1$, $dx = 0$, $dy = 0$. $\therefore \ W_2 = \int_{C_2} 2 \, dz = 2 \int_0^1 dz = 2.$

On $C_3, 1 \ge x \ge 0, y = z = x, dy = dz = dx$.

$$\therefore \qquad W_3 = \int_{C_3} x\,dx - x\,dx + 2x\,dx = 2\int_1^0 x\,dx = -1.$$

Thus. total work done $= W_1 + W_2 + W_3 = (1/2) + 2 - 1 = 3/2$.

Example 14.54 Find the constant 'a' so that the vector field $\mathbf{v} = (axy - z^3)\mathbf{i} + (a-2)x^2\mathbf{j} + (1-a)xz^2\mathbf{k}$ is conservative. Calculate its scalar potential and the work done in moving a particle from $P: (1, 2, -3)$ to $Q(1, -4, 2)$ in the field.

Solution. $\mathbf{v}$ is a conservative field if curl $(\mathbf{v}) = \mathbf{O}$.

$$\text{curl}(\mathbf{v}) = \begin{vmatrix} \mathbf{i} & \mathbf{j} & \mathbf{k} \\ \partial/\partial x & \partial/\partial y & \partial/\partial z \\ axy - z^3 & (a-2)x^2 & (1-a)xz^2 \end{vmatrix}$$

$$= \mathbf{i}(0) - \mathbf{j}\left[z^2(1-a) + 3z^2\right] + \mathbf{k}\left[2x(a-2) - ax\right] = \mathbf{O}$$

or $z^2(1-a+3) = 0$ and $x(a-4) = 0$, whence we have $a = 4$.

Now $\qquad \mathbf{v} = \left(4xy - z^3\right)\mathbf{i} + 2x^2\mathbf{j} - 3xz^2\mathbf{k}$.

We have $\qquad \mathbf{v} = \text{grad } \phi = \dfrac{\partial \phi}{\partial x}\mathbf{i} + \dfrac{\partial \phi}{\partial y}\mathbf{j} + \dfrac{\partial \phi}{\partial z}\mathbf{k}$.

Hence $\qquad \dfrac{\partial \phi}{\partial x} = 4xy - z^3 \quad ...(i) \qquad \dfrac{\partial \phi}{\partial y} = 2x^2 \quad ...(ii) \qquad \dfrac{\partial \phi}{\partial z} = -3xz^2 \quad ...(iii)$

Integrating we get from (i), (ii) and (iii) respectively,

$$\phi(x, y, z) = 2x^2y - z^3x + f_1(y, z), \qquad\qquad ...(iv)$$

$$\phi(x, y, z) = 2x^2y + f_2(x, z), \qquad\qquad ...(v)$$

and $\qquad\qquad \phi(x, y, z) = -xz^3 + f_3(x, z). \qquad\qquad ...(vi)$

The results (iv), (v) and (vi) agree if we choose

$$f_1(y, z) = 0, f_2(x, z) = -xz^3, f_3(x, y) = 2x^2y.$$

Therefore, the scalar potential is given by

$$\phi(x, y, z) = 2x^2y - xz^3 + c.$$

Hence, the work done by $\mathbf{v}$ in moving a particle from $P(1, 2, -3)$ to $Q(1, -4, 2)$ is

$$W = \int_P^Q \mathbf{v} \cdot d\mathbf{r} = \left[\phi(x, y, z)\right]_P^Q = \left[2x^2y - xz^3 + c\right]_{(1,2,-3)}^{(1,-4,2)}$$

$$= (-8 - 8 + c) - (4 + 27 + c) = -16 - 31 = -47.$$

Example 14.55 Show that the vector field $\mathbf{F} = 2x(y^2 + z^3)\,\mathbf{i} + 2x^2y\,\mathbf{j} + 3\,x^2z^2\mathbf{k}$ is conservative. Find its scalar potential and the work done by it in moving a particle from $P:(-1, 2, 1)$ to $Q: (2, 3, 4)$.   (*IETE., June 2009, Dec 2008, 2005*)

Solution. We have

$$\operatorname{curl}(\mathbf{F}) = \begin{vmatrix} \mathbf{i} & \mathbf{j} & \mathbf{k} \\ \partial/\partial x & \partial/\partial y & \partial/\partial z \\ 2x(y^2 + z^3) & 2x^2y & 3x^2z^2 \end{vmatrix}$$

$$= \mathbf{i}(0) - \mathbf{j}(6xz^2 - 6xz^2) + \mathbf{k}(4xy - 4xy) = \mathbf{O}.$$

Therefore, the vector field $\mathbf{F}$ is conservative. We have $\mathbf{F} = \operatorname{grad} f$. Hence

$$\frac{\partial f}{\partial x} = 2x(y^2 + z^3) \;...(i), \quad \frac{\partial f}{\partial y} = 2x^2y \;...(ii), \quad \frac{\partial f}{\partial z} = 3x^2z^2 \;...(iii)$$

Integrating, we get from (*i*), (*ii*) and (*iii*) respectively,

$$f(x,y,z) = x^2(y^2 + z^3) + f_1(y,z) \qquad\qquad ...(iv)$$

$$f(x,y,z) = x^2y^2 + f_2(x,z) \qquad\qquad ...(v)$$

$$f(x,y,z) = x^2z^3 + f_3(x,y) \qquad\qquad ...(vi)$$

The results (*iv*), (*v*) and (*vi*) agree if we choose

$$f_1(y,z) = 0,\ f_2(x,z) = x^2z^3,\ f_3(x,y) = x^2y^2.$$

Therefore, the scalar potential is given by $f(x,y,z) = x^2(y^2 + z^3) + c.$

Hence, the work done by $\mathbf{F}$ in moving a particle from $P(-1, 2, 1)$ to $Q(2, 3, 4)$ is

$$W = \int_P^Q \mathbf{F} \cdot d\mathbf{r} = \left[f(x,y,z) \right]_P^Q = \left[x^2y^2 + x^2z^3 \right]_{(-1,2,1)}^{(2,3,4)} = (36 + 256 - 4 - 1) = 287.$$

14.20 Line Integrals Independent of the (i) Path in plane (ii) path in space

(*i*) We know that the value of $\int \mathbf{F} \cdot d\mathbf{r}$ or $\int P\,dx + Q\,dy$ depends not only on the end points A and B of the curve but also on the path of C. The condition under which the line integral $\int_C P\,dx + Q\,dy$ is independent of the path of integration, that is, it depends only on the end points A and B of the curve C if and only if $P\,dx + Q\,dy$ is an exact differential.

If there exists a scalar function $\phi(x, y)$ such that $d\phi = P(x, y)dx + Q(x, y)dy$, then components, $P = \dfrac{\partial \phi}{\partial x}, Q = \dfrac{\partial \phi}{\partial y}$. A line integral $\int_C P\,dx + Q\,dy$, which is independ-

ent of the path between the endpoints A and B, is often written $\int_A^B P\,dx + Q\,dy$.

(ii) If C is a space curve, then a line integral $\int_C \mathbf{F} \cdot d\mathbf{r}$ is independent of the path whenever the differential expression $P(x, y, z)dx + Q(x, y, z)dy + R(x, y, z)dz$ is an exact differential.

$\int_C P\,dx + Q\,dy + R\,dz$ is independent of path C if and only if

$$\frac{\partial P}{\partial y} = \frac{\partial Q}{\partial x}, \quad \frac{\partial P}{\partial z} = \frac{\partial R}{\partial x}, \quad \text{and} \quad \frac{\partial Q}{\partial z} = \frac{\partial R}{\partial y}.$$

Remark. If the line integral is path independent, then $\mathbf{F} = \text{grad}(\phi)$.
Hence curl $(\mathbf{F}) = \text{curl}(\text{grad } \phi) = \mathbf{O}$.   ...(14.29)
    The given vector field $\mathbf{F}$ is called a *gradient field* and the function ϕ is called the potential function for $\mathbf{F}$. Therefore, in a gradient force field, the work done by the force $\mathbf{F}$ in moving a particle from a position P to a position Q is independent of the path of integration. Such a force field is also called a *conservative field*.

Example 14.56 Show that the integral $\int_{(1,2)}^{(3,4)} \left(xy^2 + y^3\right)dx + \left(x^2y + 3xy^2\right)dy$ is in-

dependent of the path joining the points (1, 2) and (3, 4). Hence evaluate the integral.   *(IETE, June 2007)*

Solution. We have $P(x, y) = xy^2 + y^3$ and $Q(x, y) = x^2y + 3xy^2$. Now,

$$\frac{\partial P}{\partial y} = 2xy + 3y^2 \quad \text{and} \quad \frac{\partial Q}{\partial x} = 2xy + 3y^2.$$

    Since $\partial P/\partial y = \partial Q/\partial x$, the integral is independent of the path. Also the integrand is an exact differential. Therefore there exists a function $\phi(x, y)$ such that

$$\partial\phi/\partial x = P(x, y) = xy^2 + y^3 \quad \text{and} \quad \partial\phi/\partial y = Q(x, y) = x^2y + 3xy^2.$$

    Integrating the first equation with respect to x, we get

$\phi(x, y) = \left(x^2 y^2 /2\right) + xy^3 + g(y)$, where $g(y)$ is the constant of integration. Taking the partial derivative of this last equation with respect to y and equating the result equal to $Q(x, y)$, we have,

$\dfrac{\partial\phi}{\partial y} = x^2y + 3xy^2 + g'(y) = x^2y + 3xy^2$, which implies $g'(y) = 0$ and so $g(y) = k$,

constant.

Hence, $\phi(x, y) = \dfrac{x^2 y^2}{2} + xy^3 + k.$

Therefore, $\int_C \left(xy^2 + y^3\right)dx + \left(x^2y + 3xy^2\right)dy = \int_{(1,2)}^{(3,4)} d\left(\dfrac{x^2 y^2}{2} + xy^3\right)$

$$= \left[\frac{x^2 y^2}{2} + xy^3 \right]_{(1,2)}^{(3,4)} = (72 + 192) - (2 + 8) = 264 - 10 = 254.$$

Alternative. Since the integral is independent of the path, we can integrate along any convenient curve connecting the given points. In particular

$y - 2 = \dfrac{4-2}{3-1}(x-1)$ *i.e.*, $y = x + 1$ is such a curve. Using x as a parameter we get

$$\int_C (xy^2 + y^3) dx + (x^2 y + 3xy^2) dy$$

$$= \int_1^3 \left[x(x+1)^2 + (x+1)^3 \right] dx + \left[x^2 (x+1) + 3x(x+1)^2 \right] dx$$

$$= \int_1^3 (6x^3 + 12x^2 + 7x + 1) dx = \left[\frac{3}{2} x^4 + 4x^3 + \frac{7}{2} x^2 + x \right]_1^3$$

$$= \frac{3}{2}(80) + 4(26) + \frac{7}{2}(8) + 2 = 120 + 104 + 28 + 2 = 254.$$

Example 14.57 Show that the following line integral is independent of path C from points $P(-1, 2, 3)$ to $Q(2, 2, 4)$ and hence evaluate the integral

$$\int_C (2xz + y) dx + (x+z) dy + (x^2 + y) dz. \qquad \text{(IETE, June 2005)}$$

Solution. We have $P(x, y, z) = 2xz + y$, and $Q(x, y, z) = x + z$, $R(x, y, z) = x^2 + y$.

We find $\dfrac{\partial P}{\partial y} = 1 = \dfrac{\partial Q}{\partial x}, \dfrac{\partial P}{\partial z} = 2x = \dfrac{\partial R}{\partial x}$, and $\dfrac{\partial Q}{\partial z} = 1 = \dfrac{\partial R}{\partial y}$.

The integral is independent of path of integration. Also, the integrand is an exact differential. Therefore, there exists a function ϕ (x, y, z) such that

$$\frac{\partial \phi}{\partial x} = 2xz + y, \quad \frac{\partial \phi}{\partial y} = x + z, \quad \text{and} \quad \frac{\partial \phi}{\partial z} = x^2 + y.$$

Integrating the first equation with respect to x, we obtain

$$\phi(x, y, z) = x^2 z + xy + g(y, z).$$

Substituting in the second equation, we get

$$\frac{\partial \phi}{\partial y} = x + z = x + \frac{\partial g}{\partial y}, \quad \text{or} \quad \frac{\partial g}{\partial y} = z.$$

Integrating, we get $g(y, z) = yz + h(z)$ and $\phi(x, y, z) = x^2 z + xy + yz + h(z)$.

Substituting in the third equation, we get

$\frac{\partial \phi}{\partial z} = x^2 + y = x^2 + y + h'(z)$, or $h'(z) = 0$ or $h(z) = k$, constant.

Therefore, $\phi(x,y,z) = x^2 z + xy + yz + k$.

The value of the integral is $\int_C (2xz + y)dx + (x + z)dy + (x^2 + y)dz$

$$= \int_{(-1,2,3)}^{(2,2,4)} d(x^2 z + xy + yz) = \left[x^2 z + xy + yz\right]_{(-1,2,3)}^{(2,2,4)} = 28 - 7 = 21$$

Example 14.58 Show that $\int_C (yz - 1)dx + (z + xz + z^2)dy + (y + xy + 2yz)dz$ is independent of the path of integration from (1, 2, 2) to (2, 3, 4) and hence evaluate the integral. *(IETE, June 2009, PTU 2006)*

Solution. We have

$P(x,y,z) = yz - 1, Q(x,y,z) = z + xz + z^2$, and $R(x,y,z) = y + xy + 2yz$.

We find $\frac{\partial P}{\partial y} = z = \frac{\partial Q}{\partial x}, \frac{\partial P}{\partial z} = y = \frac{\partial R}{\partial x}$, and $\frac{\partial Q}{\partial z} = 1 + x + 2z = \frac{\partial R}{\partial y}$.

The integral is independent of the path of integration. Also, the integrand is an exact differential and so there exists a function $\phi(x, y, z)$ such that

$\frac{\partial \phi}{\partial x} = yz - 1, \frac{\partial \phi}{\partial y} = z + xz + z^2$, and $\frac{\partial \phi}{\partial z} = y + xy + 2yz$.

Integrating the first equation with respect to x, we obtain

$$\phi(x,y,z) = xyz - x + g(y, z).$$

Substituting in the second equation, we get

$$\frac{\partial \phi}{\partial y} = z + xz + z^2 = xz + \frac{\partial g}{\partial y}, \text{ or } \frac{\partial g}{\partial y} = z + z^2.$$

Integrating, we get $g(y,z) = yz + yz^2 + h(z)$ and

$\phi(x,y,z) = xyz - x + yz + yz^2 + h(z)$. Substituting in the third equation, we obtain.

$$\frac{\partial \phi}{\partial z} = y + xy + 2yz = xy + y + 2yz + h'(z).$$

From this we get $h'(z) = 0$, or $h(z) = k$, constant.

Therefore, $\phi(x,y,z) = xyz - x + yz + yz^2 + k$.

The value of the integral $= \int_C (yz - 1)dx + (z + xz + z^2)dy + (y + xy + 2yz)dz$

$$= \int_{(1,2,2)}^{(2,3,4)} d\left(xyz - x + yz + yz^2\right) = \left[xyz - x + yz + yz^2\right]_{(1,2,2)}^{(2,3,4)}$$

$$= (24 - 2 + 12 + 48) - (4 - 1 + 4 + 8) = 82 - 15 = 67.$$

Exercise 14.4

1. A vector field is given by
$$\mathbf{F} = (\sin y)\mathbf{i} + x\,(1 + \cos y)\mathbf{j}.$$
   Evaluate the line integral $\int_C \mathbf{F} \cdot d\mathbf{r}$, where C is a circular path given by $x^2 + y^2 = a^2, z = 0$. *(IETE Dec. 2007; MDU 2006; PTU 2003)* **Ans.** πa^2.

2. Show that $\int_C \mathbf{F} \cdot d\mathbf{r} = 3\pi$, given that $\mathbf{F} = z\mathbf{i} + x\mathbf{j} + y\mathbf{k}$ and C being the arc of the curve $\mathbf{r} = \cos t\,\mathbf{i} + \sin t\,\mathbf{j} + t\,\mathbf{k}$ from $t = 0$ to $t = 2\pi$.

3. Evaluate the line integral $\oint_C \left(x^2 + xy\right)dx + \left(x^2 + y^2\right)dy$, where C is the square formed by the lines $y = \pm 1$ and $x = \pm 1$. *(Kerala 2001)*

   [**Hint.** $\oint_C \left(x^2 + xy\right)dx + \left(x^2 + y^2\right)dy,$

   $= \int_{AB} \left(x^2 + xy\right)dx + \left(x^2 + y^2\right)dy + \int_{BC} \left(x^2 + xy\right)dx + \left(x^2 + y^2\right)dy$

   $+ \int_{CD} \left(x^2 + xy\right)dx + \left(x^2 + y^2\right)dy + \int_{DA} \left(x^2 + xy\right)dx + \left(x^2 + y^2\right)dy$

   $= \int_{-1}^{1}\left(x^2 - x\right)dx + \int_{-1}^{1}\left(1 + y^2\right)dy + \int_{1}^{-1}\left(x^2 + x\right)dx + \int_{1}^{-1}\left(1 + y^2\right)dy = 0.$]

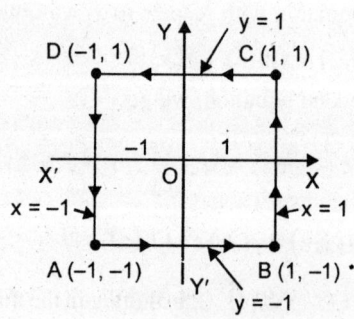

Fig. 14.20.

4. Evaluate $\oint_c \left(y^2 dx - x^2 dy\right)$, on the given closed curve C:

   triangle with vertices $A(1, 0)$, $B(0, 1)$ and $C(-1, 0)$ traversed counter clockwise. *(PTU 1999)* **Ans.** $-2/3$.

   [**Hint.** Refer Fig. 14.21. The equation of curve C_1: Line AB is $x + y = 1$
   $\Rightarrow dx + dy = 0.$

$$\therefore \int_{C_1}\left(y^2 dx - x^2 dy\right) = \int_{C_1}(1-x)^2\, dx - x^2(-dx)$$

$$= \int_{C_1}\left(2x^2 - 2x + 1\right)dx = \int_1^0\left(2x^2 - 2x + 1\right)dx = \frac{-2}{3}.$$

The equation of curve C_2 : line BC is $-x + y = 1 \Rightarrow dx = dy.$

$$\int_{C_2}\left(y^2 dx - x^2 dy\right) = \int_0^{-1}(1+x)^2\, dx - x^2 dx = \int_0^{-1}(2x+1)\,dx = 0.$$

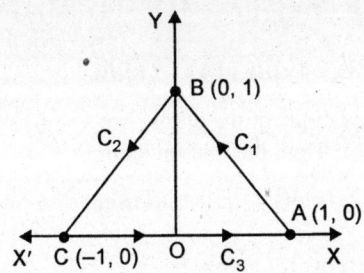

Fig. 14.21.

The equation of curve C_3: line CA is $y = 0 \Rightarrow dy = 0.$ $\int_{C_3}\left(y^2 dx - x^2 dy\right) = 0.$

Hence $\oint_c\left(y^2 dx - x^2 dy\right) = \int_{C_1}\left(y^2 dx - x^2 dy\right) + \int_{C_2}\left(y^2 dx - x^2 dy\right)$

$$+\int_{C_3}\left(y^2 dx - x^2 dy\right) = (-2/3) + 0 + 0 = -2/3. \Big]$$

5. Evaluate $\oint_C \mathbf{F} \cdot d\mathbf{r}$ where $\mathbf{F} = xy\mathbf{i} + \left(x^2 + y^2\right)\mathbf{j}$ and C is the rectangle in the xy-plane bounded by the lines $y = 2,\ x = 4,\ y = 10$ and $x = 1.$ **Ans. 60.**

6. Evaluate the line integral of $\mathbf{F} = \left(2x^2 y + y + z^2\right)\mathbf{i} + (2 + 2yz)\mathbf{j}$

$$+\left(2z + 3y^2 z^2\right)\mathbf{k} \text{ along the curve } C : a^2 - y^2 - z^2 = 0,\ x = 0. \quad \textbf{Ans. zero.}$$

7. Find the total work done in moving a particle in a force field given by $\mathbf{F} = 3xy\mathbf{i} - 5z\mathbf{j} + 10x\mathbf{k}$ along the curve $x = t^2 + 1,\ y = 2t^2, z = t^3$ from $t = 1$ to $t = 2.$

[**Hint.** Total work done $= \int_C \mathbf{F} \cdot d\mathbf{r} = \int_C 3xy\, dx - 5z\, dy + 10x\, dz$

$$= \int_{t=1}^2 3(t^2+1)(2t^2)d(t^2+1) - 5(t^3)d(2t^2) + 10(t^2+1)d(t^3)$$

$$= \int_1^2\left(12t^5 + 10t^4 + 12t^3 + 30t^2\right)dt = 303. \,]$$

8. Find the work done in moving a particle once round an ellipse $x^2/16 + y^2/9 = 1$ in the xy plane (traced in the anticlockwise direction) if the force field is given by $\mathbf{F} = (3x - 4y + 2z)\mathbf{i} + (4x + 2y - 3z^2)\mathbf{j} + (2xz - 4y^2 + z^3)\mathbf{k}$. **Ans. 96π.**

[**Hint.** In the plane $z = 0$, $\mathbf{F} = (3x - 4y)\mathbf{i} + (4x + 2y)\mathbf{j} - 4y^2\mathbf{k}$ and $d\mathbf{r} = dx\mathbf{i} + dy\mathbf{j}$.

Work done $= \oint_C \mathbf{F} \cdot d\mathbf{r} = \int_C \{(3x - 4y)\mathbf{i} + (4x + 2y)\mathbf{j} - 4y^2\mathbf{k}\}$

$\cdot (dx\mathbf{i} + dy\mathbf{j}) = \int_C (3x - 4y)dx + (4x + 2y)dy.$

The parametric equations of the ellipse are $x = 4\cos t$, $y = 3\sin t$, where t varies from 0 to 2π. Then the line integral

$= \int_{t=0}^{2\pi} (12\cos t - 12\sin t)(-4\sin t)dt + (16\cos t + 6\sin t)(3\cos t)dt$

$= \int_{t=0}^{2\pi} (48 - 30\sin t \cos t)dt = 96\pi.]$

9. Calculate the work done when a force $\mathbf{F} = 3xy\mathbf{i} - y^2\mathbf{j}$ moves a particle in the xy-plane from (0, 0) to (1, 2) along the parabola $y = 2x^2$. **Ans. –7/6.**

10. Find the work done when a force $\mathbf{F} = (x^2 - y^2 + x)\mathbf{i} - (2xy + y)\mathbf{j}$ moves a particle in the xy-plane from (0, 0) to (1, 1) along the parabola $y^2 = x$. Is the work done different when the path is the straight line $y = x$?
    Ans. –2/3, work done is same.

11. Find the work done in moving a particle once round the circle $x^2 + y^2 = 9$ in the xy plane if the field of force is $\mathbf{F} = (2x - y - z)\mathbf{i} + (x + y - z^2)\mathbf{j} + (3x - 2y + 4z)\mathbf{k}$. Also find the scalar potential of $\mathbf{F}$. **Ans. 18π.**

12. Compute the work done by the force $\mathbf{F} = (2y + 3)\mathbf{i} + xz\,\mathbf{j} + (yz - x)\mathbf{k}$ when it moves a particle from the point (0, 0, 0) to the point (2, 1, 1) along the curve $x = 2t^2$, $y = t$, $z = t^3$. (*Madras 2000; Bhopal 1999*) **Ans. 288/35.**

13. Evaluate $\int_C (y + 3z)dx + (2z + x)dy + (3x + 2y)dz$, where C is the arc of the helix $x = a\cos\theta$, $y = a\sin\theta$, $z = (2a\theta)/\pi$ between the points $(a, 0, 0)$ and $(0, a, a)$. **Ans. $2a^2$.**

14. Find the circulation of $\mathbf{F}$ around the curve C, where
    $\mathbf{F} = e^x[(\sin y)\mathbf{i} + (\cos y)\mathbf{j}]$ and C is the rectangle whose vertices are (0, 0), (1, 0), (1, π/2) and (0, π/2). **Ans. zero.**

15. Find the circulation of $\mathbf{F}$ round the curve C, where
    $\mathbf{F} = (2x + y^2)\mathbf{i} + (3y - 4x)\mathbf{j}$ and C is the curve $y = x^2$ from (0, 0) to (1, 1) and the curve $y^2 = x$ from (1, 1) to (0, 0). **Ans. –49/30.**

16. Show that vector field defined by the vector function
    $\mathbf{v} = xyz(yz\,\mathbf{i} + xz\,\mathbf{j} + xy\,\mathbf{k})$ is conservative. *(IETE June 2008)*

17. Show that $\mathbf{F} = \left(2xy + z^3\right)\mathbf{i} + x^2\mathbf{j} + 3xz^2\mathbf{k}$ is a conservative force field. Find
    its scalar potential. Find also the work done in moving an object in this field
    from (1, –2, 1) to (3, 1, 4).   **Ans.** $\phi = x^2 y + xz^3 + c$; 202 .

18. Prove that $\mathbf{F} = \left(2xy + z^3\right)\mathbf{i} + x^2\mathbf{j} + 3xz^2\mathbf{k}$ is a conservative force field. Also,

    find the scalar potential.   *(A.M.I.E., W-2007)* **Ans.** $x^2 y + xz^3 + c$.

19. Show that $\mathbf{F} = \left(e^x \cos y + yz\right)\mathbf{i} + \left(xz - e^x \sin y\right)\mathbf{j} + \left(xy + z\right)\mathbf{k}$ is conservative

    and find a potential function for it.   **Ans.** $e^x \cos y + xyz + \left(z^2/2\right) + c$.

20. Show that the integral $\int_{(1,2)}^{(3,4)} \left(6xy^2 - y^3\right)dx + \left(6x^2 y - 3xy^2\right)dy$ is inde-

    pendent of the path joining (1, 2) and (3, 4). Hence evaluate the integral.
    Ans. 236.

21. Show that the line integral $\int_C \left(3x^2 dx + 2yz\,dy + y^2 dz\right)$ is independent of the

    path in any domain in space and find its value, if C has the initial point
    $A(0, 1, 2)$ and terminal point $B(1, –1, 7)$.   *(AMIE S-2009)* **Ans.** 6.

    [Hint. $\int_C d\left(x^3 + y^2 z\right) = x^3 + y^2 z$. The value of the integral is

    $\left[x^3 + y^2 z\right]_{(0,1,2)}^{(1,-1,7)} = (1 + 7) - (2) = 6.]$

22. Show that the integral $\int_C \left(x^2 - 2y^3\right)dx + \left(x + 5y\right)dy$ is not independent of
    the path C.

    [Hint. From, $P(x, y) = x^2 - 3y^3$ and $Q(x, y) = 3x + 6y$, we find

    $\dfrac{\partial P}{\partial y} = -9y^2$ and $\dfrac{\partial Q}{\partial x} = 3$. Since $\partial P/\partial y \neq \partial Q/\partial x$, therefore the integral is

    not independent of the path.]

23. Determine whether the line integral

    $\int 2xyz^2 dx + \left(x^2 z^2 + z \cos yz\right)dy + \left(2x^2 yz + y\cos yz\right)dz$ is independent of

    the path of integration? If so, then evaluate it from (1, 0, 1) to (0, π/2, 1).
    Ans. Independent of path; 1.

    $\left[\text{Hint. } \int_C \left(2xyz^2 dx + x^2 z^2 dy + 2x^2 yz\,dz\right) + \left(z\cos yz\,dy + y\cos yz\,dz\right)\right.$

$$= \int_C d\left(x^2 yz^2\right) + \left\{(\cos yz) \cdot d\left(yz\right)\right\} = \int_C d\left(x^2 yz^2 + \sin yz\right) = x^2 yz^2 + \sin yz.$$

The value of the integral is $\left[x^2 yz^2 + \sin yz\right]_{(1,0,1)}^{(0,\pi/2,1)} = 1$.

24. Show that $ydx + xdy + 4dz$ is exact, and evaluate the integral

$\int_{(1,1,1)}^{(2,3,-1)} ydx + xdy + 4dz$ over the line segment from (1, 1, 1) to (2, 3, –1).

Ans. –3.

25. Calculate the integral

$$\oint_C \left[\left(y^2 + 2xz^2 - 1\right)dx + 2xydy + 2x^2 zdz\right], \text{ if } C \text{ is the ellipse}$$

$x^2/a^2 + y^2/b^2 = 1; z = 0.$

Ans. zero.

26. If $\mathbf{F} = 2y\mathbf{i} - z\mathbf{j} + x\mathbf{k}$, evaluate $\int_C \mathbf{F} \times d\mathbf{r}$ along the curve $x = \cos t, \; y = \sin t,$

$z = 2\cos t$ from $t = 0$ to $t = \pi/2.$

$$\left[\textbf{Hint. } \mathbf{F} \times d\mathbf{r} = \begin{vmatrix} \mathbf{i} & \mathbf{j} & \mathbf{k} \\ 2y & -z & x \\ dx & dy & dz \end{vmatrix} \right.$$

$= -\left(z\,dz + x\,dy\right)\mathbf{i} - \left(2y\,dz - x\,dx\right)\mathbf{j} + \left(2y\,dy + z\,dx\right)\mathbf{k}$

$= -\left(-4\cos t \sin t + \cos^2 t\right)dt\,\mathbf{i} - \left(\cos t \sin t - 4\sin^2 t\right)dt\,\mathbf{j}$

$+ \left(2\sin t \cos t - 2\cos t \sin t\right)dt\,\mathbf{k}.$

$\therefore \int_C \mathbf{F} \times d\mathbf{r} = \int_0^{\pi/2} \left\{\left(4\cos t \sin t - \cos^2 t\right)\mathbf{i} + \left(4\sin^2 t - \cos t \sin t\right)\mathbf{j}\right\}dt$

$\left. = \left(2 - \frac{\pi}{4}\right)\mathbf{i} + \left(\pi - \frac{1}{2}\right)\mathbf{j}\right].$

14.21 Green's Theorem in the Plane
(Transformation between double integrals and line integrals)

Let R be a closed bounded region in the xy-plane whose boundary C consists of finitely many smooth curves. Let $P(x, y)$ and $Q(x, y)$ be functions that are continuous and possess continuous partial derivatives in a region R bounded by simple closed curve C, then

$$\oint_C Pdx + Qdy = \iint_R \left(\frac{\partial Q}{\partial x} - \frac{\partial P}{\partial y}\right)dy\,dx, \qquad \ldots(14.30)$$

where C is traversed in the counterclockwise direction.

Remark. A curve "traversed in counterclockwise direction" is also called traversed in positive direction. Unless otherwise stated, integral along a closed curve C will imply integral along C described in positive direction.

14.22 Illustrative Examples

Example 14.59 Verify the Green's theorem in the plane for $\oint_C (xy + y^2) dx + x^2 dy$, where C is the closed curve of the region bounded by $y = x$ and $y = x^2$.

(S.V.T.U, 2006S; PTU 2005, VTU 2005, MDU 2003; IETE, Dec 2001)

Solution. The curves $y = x$ and $y = x^2$ intersect at $(0, 0)$ and $(1, 1)$. The positive direction in traversing C is as shown in the adjacent diagram.

Now $\oint_C = \int_{C_1} + \int_{C_2}$ as shown in the figure 14.22.

Along $y = x^2$ from $(0, 0)$ to $(1, 1)$, the line integral

$$\int_{C_1} (xy + y^2) dx + x^2 dy$$

$$= \int_0^1 \left[x(x^2) + x^4 \right] dx + x^2 (2x) dx$$

$$= \int_0^1 (3x^3 + x^4) dx = \frac{19}{20}.$$

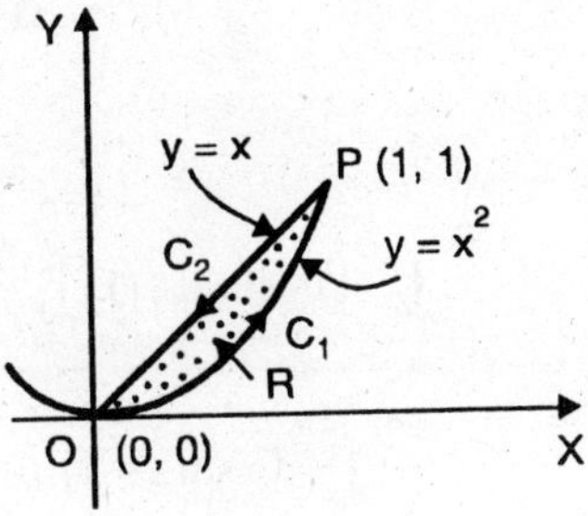

Line integral along C_2 is along $y = x$ from $(1, 1)$ to $(0, 0)$ Fig. 14.22.

Fig. 14.22.

$$\therefore \int_{C_2} (xy + y^2) dx + x^2 dy = \int_1^0 (x(x) + x^2) dx + x^2 dx = \int_1^0 3x^2 dx = -1.$$

Hence, the required line integral $= (19/20) - 1 = -(1/20)$.

Also $\iint_R \left(\frac{\partial Q}{\partial x} - \frac{\partial P}{\partial y} \right) dx dy = \iint_R \left[\frac{\partial}{\partial x} (x^2) - \frac{\partial}{\partial y} (xy + y^2) \right] dx dy = \iint_R (x - 2y) dx dy$

$$= \int_{x=0}^1 \int_{y=x^2}^x (x - 2y) dy dx = \int_0^1 \left[xy - y^2 \right]_{x^2}^x dx = \int_0^1 (x^4 - x^3) dx = -\frac{1}{20}.$$

Hence, the Green's theorem is verified.

Example 14.60 Use Green's theorem to evaluate $\oint_C (y - \sin x) dx + \cos x \, dy$, where C is a plane triangle enclosed by the lines $y = 0$, $x = \pi/2$ and $y = 2x/\pi$.

(IETE, Dec. 2009; MDU 2007; PTU 2006; AMIE-2005; JNTU 2005)

Solution. Here $P(x, y) = y - \sin x$, $Q(x, y) = \cos x$.

$$\therefore \oint_C (y - \sin x)\,dx + \cos x\,dy = \oint_C P(x,y)\,dx + Q(x,y)\,dy$$

$$= \iint_R \left(\frac{\partial Q}{\partial x} - \frac{\partial P}{\partial y} \right) dx\,dy = \iint_R \left[\frac{\partial}{\partial x}(\cos x) - \frac{\partial}{\partial y}(y - \sin x) \right] dy\,dx$$

(By the Green's theorem)

$$= \int_{x=0}^{\pi/2} \left[\int_{y=0}^{2x/\pi} (-\sin x - 1)\,dy \right] dx \qquad \left(\because \text{equation of } OB \text{ is } y = \frac{2x}{\pi} \right)$$

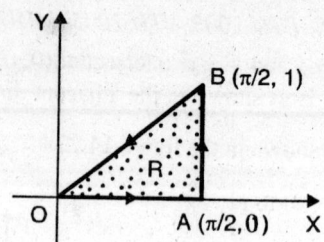

Fig. 14.23.

$$= \int_0^{\pi/2} \left[(-\sin x - 1) y \right]_0^{2x/\pi} dx = \int_0^{\pi/2} (-\sin x - 1)\frac{2x}{\pi}\,dx$$

$$= \frac{-2}{\pi} \int_0^{\pi/2} (x \sin x + x)\,dx = \frac{-2}{\pi} \left[x(-\cos x) + \sin x + \frac{x^2}{2} \right]_0^{\pi/2}$$

$$= -\left(\frac{2}{\pi} + \frac{\pi}{4} \right), \text{ the required value.}$$

Example 14.61 Verify the Green's theorem for $f(x, y) = e^{-x} \sin y$, $g(x,y) = e^{-x} \cos y$ and C is the square with vertices at $(0, 0)$, $(\pi/2, 0)$, $(\pi/2, \pi/2)$, $(0, \pi/2)$.   (AMIE S-2008; IETE-June 2004)

Solution. We can write the line integral as

$$\oint_C f(x,y)\,dx + g(x,y)\,dy = \left[\int_{C_1} + \int_{C_2} + \int_{C_3} + \int_{C_4} \right] (f(x,y)\,dx + g(x,y)\,dy)$$

where C_1, C_2, C_3 and C_4 are the boundary lines shown in Fig. 14.24.
Along C_1: $y = 0$, $0 \le x \le \pi/2$ and

$$\int_{C_1} e^{-x} (\sin y\,dx + \cos y\,dy) = 0,$$

along C_2: $x = \pi/2$, $0 \le y \le \pi/2$ and

$$\int_{C_2} e^{-x} (\sin y\,dx + \cos y\,dy) = \int_0^{\pi/2} e^{-\pi/2} \cos y\,dy = e^{-\pi/2},$$

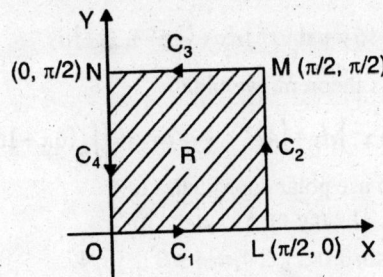

Fig. 14.24.

along $C_3 : y = \pi/2, \pi/2 \le x \le 0$ and $\int_{C_3} e^{-x} \left(\sin y dx + \cos y dy \right)$

$$= \int_{\pi/2}^0 e^{-x} dx = \left[-e^{-x} \right]_{\pi/2}^0 = \left[e^{-x} \right]_0^{\pi/2} = e^{-\pi/2} - 1,$$

along $C_4 : x = 0, \pi/2 \le y \le 0$ and

$$\int_{C_4} e^{-x} \left(\sin y \, dx + \cos y \, dy \right) = \int_{\pi/2}^0 \cos y dy = -1.$$

Therefore, $\oint_C f(x,y) dx + g(x,y) dy = e^{-\pi/2} + e^{-\pi/2} - 1 - 1 = 2 \left(e^{-\pi/2} - 1 \right).$

Using the Green's theorem, we obtain

$$\oint_C f(x,y) dx + g(x,y) dy = \iint_R \left(\frac{\partial g}{\partial x} - \frac{\partial f}{\partial y} \right) dx \, dy$$

$$= \iint_R \left(-e^{-x} \cos y - e^{-x} \cos y \right) dx dy = \int_{x=0}^{\pi/2} \int_{y=0}^{\pi/2} -2e^{-x} \cos y \, dy \, dx$$

$$= -2e^{-x} \int_{x=0}^{\pi/2} \left[\sin y \right]_0^{\pi/2} dx = -2 \int_0^{\pi/2} e^{-x} dx = 2 \left[e^{-x} \right]_0^{\pi/2} = 2 \left(e^{-\pi/2} - 1 \right).$$

Hence, the Green's theorem is verified.

Example 14.62 Find the work done by the force

$$\mathbf{F} = \left(-16y + \sin x^2 \right) \mathbf{i} + \left(4e^y + 3x^2 \right) \mathbf{j} \text{ acting}$$

along the simple closed curve C shown in Fig. 14.25.

Solution. The work done by the force is given by

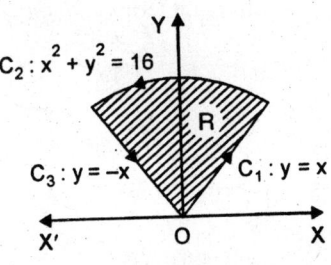

Fig. 14.25.

$$W = \oint_c F \cdot d\mathbf{r} = \oint \left(-16y + \sin x^2\right) dx + \left(4e^y + 3x^2\right) dy.$$

Using the Green's theorem, we obtain

$$\oint_c \left(-16y + \sin x^2\right) dx + \left(4e^y + 3x^2\right) dy = \iint_R (6x + 16) dx dy.$$

It is convenient to use polar coordinates to evaluate the integral. The region R is given by

$$R : x = r\cos\theta, y = r\sin\theta, 0 \le r \le 4, \pi/4 \le \theta \le 3\pi/4.$$

Therefore, $W = \iint_R (6x + 16) dx dy = \int_{\pi/4}^{3\pi/4} \int_0^4 (6r\cos\theta + 16) r\, dr\, d\theta$

$$= \int_{\pi/4}^{3\pi/4} \left[\int_0^4 \left(6r^2 \cos\theta + 16r\right) dr \right] d\theta = \int_{\pi/4}^{3\pi/4} \left[2r^3 \cos\theta + 8r^2\right]_0^4 d\theta$$

$$= \int_{\pi/4}^{3\pi/4} (128\cos\theta + 128) d\theta = [128\sin\theta + 128\theta]_{\pi/4}^{3\pi/4} = 64\pi.$$

Remarks. 1. From the statement of the Green's theorem, it is clear that

$\oint_C P dx + Q dy$ is equal to zero around every simple closed curve in a simply

connected region, if and only if $\dfrac{\partial P}{\partial y} = \dfrac{\partial Q}{\partial x}$.

2. Simple closed curve, Simply and Multiply connected Regions

A curve is called a *simple closed curve* if it does not cross itself. For example the curve in Fig. 14.26 (a) is simple closed while the curve in Fig. 14.26(b) is not simple closed and is known as *multiple curve*.

A region is called *simply connected* if every closed curve in the region encloses points of the region only. A region which is not simply connected is called *multiply connected*. For example the region bounded by a circle or a square is simply connected region while the region between two concentric circles that is the region $r_1 \le |z - z_0| \le r_2$ [Fig. 14.26(c)] is an example of a multiply connected region.

In plain terms a *simply connected region is one which has no holes.*

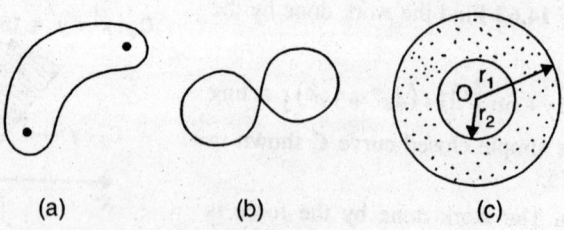

(a)   (b)   (c)

Fig. 14.26.

A region with one hole is called *doubly connected* and a region with two holes is called *triply connected* and so on.

Example 14.63 Find the work done by moving a particle once around the circle $x^2 + y^2 = a^2, z = b$ in the force field

$$\mathbf{F} = (xy + \sin xy)\mathbf{i} + \left[\frac{x^2}{2} + \frac{x}{y}\sin xy + \frac{1}{y^2}\cos xy\right]\mathbf{j} + e^x \cos xy\mathbf{k}$$

Solution. Work done $= \oint_C \mathbf{F} \cdot d\mathbf{r}$

$$= \oint_C (xy + \sin xy)\mathbf{i} + \left(\frac{x^2}{2} + \frac{x}{y}\sin xy + \frac{1}{y^2}\cos xy\right)\mathbf{j} + (e^x \cos xy)\mathbf{k} \cdot (dx\,\mathbf{i} + dy\,\mathbf{j})$$

$$= \oint_C (xy + \sin xy)\,dx + \left(\frac{x^2}{2} + \frac{x}{y}\sin xy + \frac{1}{y^2}\cos xy\right)dy$$

Here $P(x, y) = xy + \sin xy$ and $Q(x, y) = \dfrac{x^2}{2} + \dfrac{x}{y}\sin xy + \dfrac{\cos xy}{y^2}$

Now $\dfrac{\partial P}{\partial y} = x + x\cos xy$, and $\dfrac{\partial Q}{\partial x} = x + \dfrac{1}{y}\sin xy + x\cos xy - \dfrac{\sin xy}{y} = x + x\cos xy.$

Since $\dfrac{\partial P}{\partial y} = \dfrac{\partial Q}{\partial x}$, then by the Green's theorem

$$\oint_C P(x, y)\,dx + Q(x, y)\,dy = \iint_R \left(\frac{\partial Q}{\partial x} - \frac{\partial P}{\partial y}\right)dx\,dy = 0,$$

where R is the region bounded by C. (See Remarks 1 and 2.)

Example 14.64 Apply the Green's theorem to show that the area bounded by a simple closed curve C is given by $\dfrac{1}{2}\oint_C x\,dy - y\,dx$. Hence, find the area of an ellipse whose semi-major and minor axes are of lengths a and b.

(*PTU 2006; Kerala 2005; Madras 2003; MDU 2002; VTU 2000S*).

Solution. In the Green's theorem, put $P = -y$, $Q = x$. The

$$\oint_C x\,dy - y\,dx = \iint_R \left(\frac{\partial}{\partial x}(x) - \frac{\partial}{\partial y}(-y)\right)dx\,dy = 2\iint_R dx\,dy = 2A,$$

where A is the required area. Thus $A = \dfrac{1}{2}\oint_C x\,dy - y\,dx$. The parametric equations of an ellipse are $x = a\cos\theta$, $y = b\sin\theta$ and θ varies from 0 to 2π.

$$\text{Area} = \frac{1}{2}\oint_C x\,dy - y\,dx = \frac{1}{2}\int_0^{2\pi}(a\cos\theta)(b\cos\theta)\,d\theta - (b\sin\theta)(-a\sin\theta)\,d\theta$$

$$= \frac{1}{2}\int_0^{2\pi} ab\left(\cos^2\theta + \sin^2\theta\right)d\theta = \frac{1}{2}\int_0^{2\pi} ab\,d\theta = \pi ab.$$

Exercise 14.5

1. Verify the Green's Theorem in the plane for

(a) $\oint_C\left(3x^2 - 8y^2\right)dx + \left(4y - 6xy\right)dy$, where C is the boundary of the region defined by:

   (i) $y = \sqrt{x},\ y = x^2$;   (*AMIE W-2005*)

   (ii) $x = 0,\ y = 0,\ x + y = 1$.   (*Kerala, 2005; Anna 2003S; Madras 2000*)

   Ans. (i) Common value = 3/2, (ii) Common value = 5/3.

(b) $\oint_C\left(2x - y^3\right)dx - xy\,dy$, where C is the boundary of the region enclosed by

   the circles $x^2 + y^2 = 1$ and $x^2 + y^2 = 9$.   **Ans.** Common value = $60\,\pi$.

   [**Hint.** The region is multiply connected. Therefore, $\oint_C = \oint_{C_2} - \oint_{C_1}$, where

   C_2 and C_1 are circumferences of outer circle and inner circle respectively, traversed in the positive sense.

   $\oint_{C_2} = \dfrac{243\pi}{4}$ and $\oint_{C_1} = \dfrac{3\pi}{4}$. Therefore, $\oint_C \dfrac{243\pi}{4} - \dfrac{3\pi}{4} = 60\pi$].

2. Evaluate $\oint\left(x^2 - 2xy\right)dx + \left(x^2 y + 3\right)dy$ around the boundary of the region

   defined by $y^2 = 8x$ and $x = 2$, using the Green's theorem.   **Ans.** 128/5.

3. Evaluate by the Green's theorem

   $$\oint_C e^{-x}\left(\sin y\,dx + \cos y\,dy\right)$$

   where C is the boundary of the rectangle with vertices (0, 0), (π, 0), (π, $\pi/2$), (0, $\pi/2$).   (*AMIE S-2006*) **Ans.** $2\left(e^{-\pi} - 1\right)$.

4. Using the Green's theorem in a plane to evaluate the integral

   $\oint_C\left(2x^2 - y^2\right)dx + \left(x^2 + y^2\right)dy$, where C is the boundary in the xy-plane of

   the area enclosed by the x-axis and the semi-circle $x^2 + y^2 = 1$ in the upper half xy-plane.   (*AMIE S-2010*) **Ans.** 4/3.

5. Verify the Green's theorem in the xy-plane for

$$\oint_C\left[\left(xy^2-2xy\right)dx+\left(x^2y+3\right)dy\right]$$

around the boundary C of the region enclosed by $y^2=8x$ and $x=2$.

6. Evaluate:

$\int_{(0,0)}^{(2,1)}\left(10x^4-2xy^3\right)dx-3x^2y^2dy$ along the path $x^4-6xy^3=4y^2$. **Ans. 60.**

[**Hint.** It is difficult to evaluate directly. We note that $P(x,y)=10x^4-2xy^3$

and $Q(x,y)=-3x^2y^2$ and $\dfrac{\partial P}{\partial y}=-6xy^2=\dfrac{\partial Q}{\partial x}$. Hence the integral is inde-

pendent of path *i.e.* we can use any path. One simple path consists of the straight line segments from (0, 0) to (2, 0) and then from (2, 0) to (2, 1). Along the straight line path from (0, 0) to (2, 0), the integral is equal to

$\int_0^2 10x^4dx=64.$

Along the straght line path from (2, 0) to (2, 1) the integral

$=\int_0^1-12y^2dy=-4.$

∴ Required value of line integral = 64 – 4 = 60.]

7. Apply the Green's theorem to evaluate $\oint_C\left(2x^2-y^2\right)dx+\left(x^2+y^2\right)dy$,

where C is the boundary of the area enclosed by the x-axis and the upper half of circle $x^2+y^2=a^2$. *(UTTU 2005)* **Ans. $4a^3/3$.**

8. Using the Green's theorem to evaluate $\int_C\left(x^2ydx+x^2dy\right)$, where C is the boundary described counter clockwise of the triangle with vertices (0, 0), (1, 0), (1, 1). *(UPTU 2004)* **Ans. 5/12.**

9. Evaluate $\oint_C\left(2xy^3-y^2\cos x\right)dx+\left(1-2y\sin x+3x^2y^2\right)dy$ around a paral-

lelogram with vertices at (0, 0), (3, 0), (5, 2), (2, 2). **Ans. zero.**
[**Hint.** Here $P(x,y)=2xy^3-y^2\cos x$, and $Q(x,y)=1-2y\sin x+3x^2y^2$.
Since $\partial P/\partial y=6xy^2-2y\cos x=\partial Q/\partial x$, so the line integral around any closed path, in particular the parallelogram is zero.]

10. Prove that the line integral $\int_C\dfrac{xdy-ydx}{x^2+y^2}$, taken in the positive direction

over any closed contour with the origin inside it is equal to 2π.
(AMIE, W-2008)

11. Use Green's theorem to evaluate the line integral

$$\oint_C \left(e^{-x^2} + y^2\right)dx + \left(\log y - x^2\right)dy,$$ where C is the square with vertices
$(0, 1)$, $(1, 1)$ $(1, 2)$, and $(0, 2)$. *(AMIE, W-2006)* **Ans.** -4.

12. Find the work done by the force field $\mathbf{F}(x, y) = \left(e^x - y^3\right)\mathbf{i} + \left(\cos y + x^3\right)\mathbf{j}$
on a particle that travels once around the unit circle $x^2 + y^2 = 1$ in the
counterclockwise direction. **Ans.** $3\pi/2$.

13. Apply the Green's theorem to evaluate the following integral:

$$\oint_c \left(y^2 dx + x^2 dy\right)$$ C: the triangle bounded by $x = 0$, $x + y = 1$, $y = 0$.

Ans. Zero.

14.23 Surface Integrals

Let F be a vector function and S be a two sided* surface having projection R on
the xy plane. We consider any one side of S as positive (This is the outer side if S
is a closed surface). Denote by **n** the unit vector normal to any point of the positive
side.

Divide the surface S into k parts δS_1, δS_2,... δS_k. Let P_i be any point is δS_i and
$\hat{\mathbf{n}}_i$ a unit vector normal to the surface at P_i.

Form the sum $\sum\limits_{i=1}^{k} \mathbf{F}(P_i) \cdot \hat{\mathbf{n}}_i \delta S_i.$

The limit of this sum as $k \to \infty$ and area of each $\delta S_i \to 0$ is called the normal

surface integral of vector $\mathbf{F}$ over S and is denoted by $\iint\limits_{S} \mathbf{F} \cdot \mathbf{n}\, dA$.

---

* An example of one sided surface is the Mobius strip. A model of this surface can be
made by taking a long piece of rectangular paper and sticking the shorter sides so that
the two points of A and the two points of B coincide.

If we tried to colour only one side of the surface, we would find the whole thing
coloured.

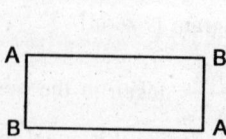

(a)

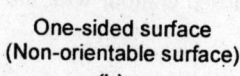

One-sided surface
(Non-orientable surface)
(b)

Now
$$\iint_S \mathbf{F}\cdot\mathbf{n}\,dA = \iint_R \mathbf{F}\cdot\mathbf{n}\,\frac{dx\,dy}{\mathbf{n}\cdot\mathbf{k}}, \qquad \text{...(14.31)}$$

where R is the projection of S on the xy-plane.

$(\because dx\,dy = dA\cos\gamma = (\mathbf{n}\cdot\mathbf{k})\,dA \Rightarrow dA = dx\,dy/\mathbf{n}\cdot\mathbf{k})$

If R is the projection of S on the yz-plane, then

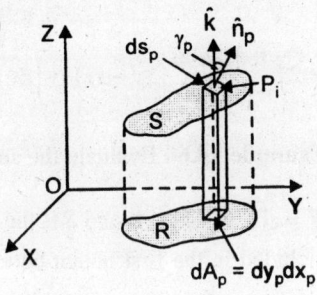

$$\iint_S \mathbf{F}\cdot\mathbf{n}\,dA = \iint_R \mathbf{F}\cdot\mathbf{n}\,\frac{dy\,dz}{\mathbf{n}\cdot\mathbf{i}} \qquad \text{...(14.32)}$$

If R is the projection of a surface S on the zx-plane, then

$$\iint_S \mathbf{F}\cdot\mathbf{n}\,dA = \iint_R \mathbf{F}\cdot\mathbf{n}\,\frac{dz\,dx}{\mathbf{n}\cdot\mathbf{j}}. \qquad \text{...(14.33)}$$

$dA_p = dy_p dx_p$

Fig. 14.27. Surface Integral

While evaluating surface integrals, we should choose the most convenient of the results (14.31), (14.32), (14.33).

14.24 Illustrative Examples

Example 14.65 Evaluate the surface integral $\iint_S \mathbf{F}\cdot\mathbf{n}\,dA$ where

$\mathbf{F} = (x+y^2)\mathbf{i} - 2x\mathbf{j} + 2yz\mathbf{k}$ and S is the portion of the plane $2x+y+2z=6$ which is in the first octant.

Solution. Let $f(x, y, z) = 2x + y + 2z - 6 = 0$ be the surface. Then grad $f = 2\mathbf{i}+\mathbf{j}+2\mathbf{k}$. A unit vector normal to the same surface is

$$\mathbf{n} = \frac{\text{grad }f}{|\text{grad }f|} = \frac{2\mathbf{i}+\mathbf{j}+2\mathbf{k}}{\sqrt{4+1+4}} = \frac{1}{3}(2\mathbf{i}+\mathbf{j}+2\mathbf{k}).$$

Refer Fig. 14.28. The projection of the portion of the plane ABC in the first octant is the triangle bounded by $x = 0$, $y = 0$ and $2x + y = 6$. We have

$$dA = \frac{dx\,dy}{\mathbf{n}\cdot\mathbf{k}} = \frac{dx\,dy}{(1/3)(2\mathbf{i}+\mathbf{j}+2\mathbf{k})\cdot\mathbf{k}} = \frac{dx\,dy}{2/3}.$$

Therefore,

$$\iint_S \mathbf{F}\cdot\mathbf{n}\,dA = \iint_S \frac{2(x+y^2)-2x+4yz}{3}\,dA$$

$$= \frac{2}{3}\iint_S (y^2+2yz)\,dA.$$

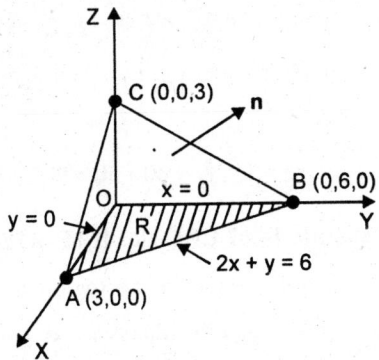

Fig. 14.28.

From the equation of the surface, we get $2z = 6 - 2x - y$. Hence,

$$\iint_S \mathbf{F} \cdot \mathbf{n}\, dA = \frac{2}{3} \iint_S \left(y^2 + 6y - 2xy - y^2 \right) dA = 2 \iint_R (3y - xy)\, dx\, dy$$

$$= 2 \int_{x=0}^{3} \left[\int_{y=0}^{6-2x} (3y - xy)\, dy \right] dx = 2 \int_0^3 \left[(3 - x)\frac{y^2}{2} \right]_0^{6-2x} dx = 4 \int_0^3 (3 - x)^3\, dx = 81.$$

Example 14.66 Evaluate the surface integral $\iint_S \mathbf{F} \cdot \mathbf{n}\, dA$, where

$\mathbf{F} = z\mathbf{i} + x\mathbf{j} - 3y^2 z\mathbf{k}$ and S is the portion of the surface of the cylinder $x^2 + y^2 = 16$ included in the first octant between $z = 0$ and $z = 5$. *(IETE Dec 2003)*

Solution. Let $f(x, y, z) = x^2 + y^2 - 16 = 0$ be the surface. Then

$$\operatorname{grad} f = 2x\mathbf{i} + 2y\mathbf{j},\ \mathbf{n} = \frac{\operatorname{grad} f}{|\operatorname{grad} f|} = \frac{2(x\mathbf{i} + y\mathbf{j})}{\sqrt{4(x^2 + y^2)}} = \frac{1}{4}(x\mathbf{i} + y\mathbf{j}).$$

The projection of S on xy plane cannot be consider here. Project S on the yz-plane. The projection is a rectangle with sides of lengths 4 and 5. We have

$$dA = \frac{dy\, dz}{\mathbf{n} \cdot \mathbf{i}} = \frac{dy\, dz}{x/4}.$$

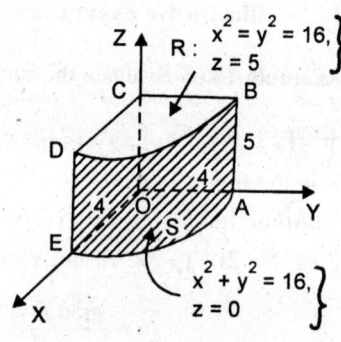

Fig. 14.29.

Therefore, $\iint_S \mathbf{F} \cdot \mathbf{n}\, dA = \iint_S \frac{1}{4}(zx + xy)\, dA$

$$= \int_{z=0}^{5} \int_{y=0}^{4} \frac{1}{4} x(y + z) \frac{dy\, dz}{(x/4)}$$

$$= \int_0^5 \left[\int_0^4 (y + z)\, dy \right] dz = \int_0^5 \left[\frac{y^2}{2} + yz \right]_0^4 dz = \int_0^5 (8 + 4z)\, dz$$

$$= \left[8z + 2z^2 \right]_0^5 = 40 + 50 = 90.$$

Example 14.67 Evaluate $\iint_S (\mathbf{F} \cdot \mathbf{n})\, dA$, where $\mathbf{F} = 2x^2 y\mathbf{i} - y^2\mathbf{j} + 4xz^2\mathbf{k}$ and S is the closed surface of the region in the first octant bounded by the cylinder $y^2 + z^2 = 9$ and the planes $x = 0$, $x = 2$, $y = 0$ and $z = 0$. *(IETE, Dec 2006)*

Solution. Here $\iint_S (\mathbf{F} \cdot \mathbf{n})dA = \iint_{S_1} (\mathbf{F} \cdot \mathbf{n})dA + \iint_{S_2} (\mathbf{F} \cdot \mathbf{n})\, dA$

$$+ \iint_{S_3} (\mathbf{F} \cdot \mathbf{n}) dA + \iint_{S_4} (\mathbf{F} \cdot \mathbf{n}) dA + \iint_{S_5} (\mathbf{F} \cdot \mathbf{n}) dA,$$

where S_1 is the rectangular face in xy-plane; S_2 the rectangular face in xz-plane; faces S_3 and S_4 are quadrants of circles in yz-plane and the plane $x = 2$ respectively; S_5 is the curved surface to the cylinder in the first octant.

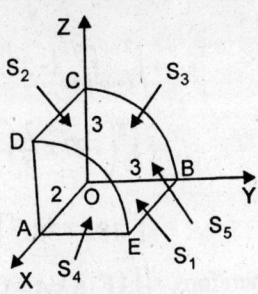

Fig. 14.30.

Now $\iint_{S_1} (\mathbf{F} \cdot \mathbf{n} \, dA) = \iint_{S_1} \{ \mathbf{F} \cdot (-\mathbf{k}) \} dA$

$= \iint_{S_1} -4xz^2 \, dA = 0,$ because $z = 0$ in the xy-plane.

$\iint_{S_2} (\mathbf{F} \cdot \mathbf{n}) dA = \iint_{S_2} \mathbf{F} \cdot (-\mathbf{j}) dA = \iint_{S_2} y^2 dA = 0,$ because $y = 0$ in the xz-plane.

$\iint_{S_3} (\mathbf{F} \cdot \mathbf{n}) dS = \iint_{S_3} \mathbf{F} \cdot (-\mathbf{i}) dA = \iint_{S_3} -2x^2 y dA = 0$ because $x = 0$ in yz-plane.

$\iint_{S_4} \mathbf{F} \cdot \mathbf{n} \, dS = \iint_{S_4} \mathbf{F} \cdot \mathbf{i} \, dA = \iint_{S_4} 2x^2 y \, dA$

$$= \int_{z=0}^{3} \int_{y=0}^{\sqrt{9-z^2}} 8y \, dy \, dz = 4 \int_0^3 (9 - z^2) dz = 72.$$

To get $\mathbf{n}$ in S_5 we note that $\nabla f(y^2 + z^2) = 2y\mathbf{i} + 2z\mathbf{k}.$

Thus $\mathbf{n} = \dfrac{2y\mathbf{j} + 2z\mathbf{k}}{\sqrt{4y^2 + 4z^2}} = \dfrac{y\mathbf{j} + z\mathbf{k}}{3}$ $\qquad (\because y^2 + z^2 = 9)$

$\therefore \mathbf{F} \cdot \mathbf{n} = (2x^2 y\mathbf{i} - y^2\mathbf{j} + 4xz^2\mathbf{k}) \cdot \dfrac{y\mathbf{j} + z\mathbf{k}}{3} = \dfrac{-y^3 + 4xz^3}{3},$ and $\mathbf{n} \cdot \mathbf{k} = z/3.$

Hence $\iint_{S_5} (\mathbf{F} \cdot \mathbf{n}) dA = \dfrac{1}{3} \int_0^2 \int_0^3 (-y^3 + 4xz^3) \cdot \dfrac{dy \, dx}{z/3}$

$$= \int_{x=0}^2 \int_{y=0}^3 \left(\dfrac{-y^3}{z} + 4xz^2 \right) dy \, dx,$$

$$= \int_{x=0}^2 \int_{y=0}^3 \left[\dfrac{-y^3}{\sqrt{9 - y^2}} + 4x(9 - y^2) \right] dy \, dx \quad (\because z^2 = 9 - y^2)$$

$$= \int_{x=0}^2 \int_{\theta=0}^{\pi/2} \left[\dfrac{-27 \sin^3 \theta}{3 \cos \theta} + 4x(9 \cos^2 \theta) \right] 3 \cos \theta \, d\theta \, dx \quad \text{(Putting } y = 3 \sin \theta \text{)}$$

$$= \int_{x=0}^{2}\int_{\theta=0}^{\pi/2}\left(-27\sin^3\theta+108x\cos^3\theta\right)d\theta\,dx$$

$$= \int_{0}^{2}\left(-27\times\frac{2}{3}+108x\times\frac{2}{3}\right)dx = \int_{0}^{2}(-18+72x)dx$$

$$= \left[-18x+36x^2\right]_{0}^{2} = -36+144 = 108.$$

Therefore, $\iint_{S}(\mathbf{F}\cdot\mathbf{n})dA = 0+0+0+72+108 = 180$.

Exercise 14.6

1. Evaluate $\iint_{S}\mathbf{F}\cdot\mathbf{n}\,dA$, where $\mathbf{F}=18z\mathbf{i}-12\mathbf{j}+3y\mathbf{k}$ and S is that part of the plane $2x+3y+6z=12$ which is located in the first octant.

   (M.D.U. 2001) **Ans. 24.**

   [**Hint.** $\mathbf{n}=\dfrac{1}{7}(2\mathbf{i}+3\mathbf{j}+6\mathbf{k})$, $\mathbf{F}\cdot\mathbf{n}=\dfrac{36z-36+18y}{7}=\dfrac{36-12x}{7}$,

   $\mathbf{n}\cdot\mathbf{k}=\left(\dfrac{2}{7}\mathbf{i}+\dfrac{3}{7}\mathbf{j}+\dfrac{6}{7}\mathbf{k}\right)\cdot\mathbf{k}=\dfrac{6}{7}$. Using $z=(12-2x-3y)/6$,

   from the equation of the surface.

   $\therefore \iint_{S}\mathbf{F}\cdot\mathbf{n}\,dA = \iint_{S}\dfrac{36-12x}{7}dS = \iint_{R}\left(\dfrac{36-12x}{7}\right)\dfrac{dx\,dy}{\mathbf{n}\cdot\mathbf{k}} = \iint_{R}(6-2x)dx\,dy$

   $= \int_{x=0}^{6}\left[\int_{y=0}^{(12-2x)/3}(6-2x)dy\right]dx = 24.$]

2. If $\mathbf{F}=2y\mathbf{i}-z\mathbf{j}+x^2\mathbf{k}$ and S is surface of the parabolic cylinder $y^2=8x$ in the first octant bounded by the planes $y=4$, $z=6$, evaluate the surface integral $\iint_{S}\mathbf{F}\cdot\mathbf{n}\,dA$.

   [**Hint.** $\mathbf{n}=\dfrac{8\mathbf{i}-2y\mathbf{j}}{\sqrt{64+4y^2}}$, $\mathbf{F}\cdot\mathbf{n}=\dfrac{16y+2yz}{\sqrt{64+4y^2}}$, $\mathbf{n}\cdot\mathbf{i}=\dfrac{8}{\sqrt{64+4y^2}}$. Projecting the

   surface S on yz plane, we obtain

   $\therefore \iint_{S}\mathbf{F}\cdot\mathbf{n}\,dA = \iint_{R}\mathbf{F}\cdot\mathbf{n}\dfrac{dy\,dz}{\mathbf{n}\cdot\mathbf{i}}$, where R is the rectangular region in the yz plane bounded by the lines $y=0$, $z=0$, $y=4$ and $z=6$.

   $= \dfrac{1}{4}\int_{z=0}^{6}\int_{y=0}^{4}(8y+yz)dy\,dz = 132.$]

3. Evaluate $\iint_S \mathbf{F} \cdot \mathbf{n} \, dA$, where $\mathbf{F} = xy\mathbf{i} - x^2\mathbf{j} + (x+z)\mathbf{k}$, S is that portion of the plane $2x + 2y + z = 6$ included in the first octant and $\mathbf{n}$ is a unit normal to S. *(IETE June 2004, Dec 2001)* **Ans. 27/4.**

4. Show that $\iint_S \mathbf{F} \cdot \mathbf{n} \, dA = \dfrac{3}{2}$, where $\mathbf{F} = 4xz\mathbf{i} - y^2\mathbf{j} + yz\mathbf{k}$ and S is the surface of the cube bounded by the planes, $x = 0$, $x = 1$, $y = 0$, $y = 1$, $z = 0$, $z = 1$.

5. Evaluate $\iint_S \mathbf{F} \cdot \mathbf{n} \, dA$, where $\mathbf{F}(x,y,z) = xy\mathbf{i} + z^2\mathbf{j} + 2yz\mathbf{k}$, S is the surface of the plane $x + y + z = 1$ in the first octant and $\mathbf{n}$ is unit outward normal to the surface S. **Ans. 1/8**

6. Show that $\iint_S (yz\mathbf{i} + zx\mathbf{j} + xy\mathbf{k}) \cdot \mathbf{n} \, dA = 3/8$, where S is the surface of the sphere $x^2 + y^2 + z^2 = 1$ in the first octant.

14.25 Divergence theorem of Gauss
(Transformation between Volume Integrals and Surface Integrals)

Let D be a closed and bounded region in 3-space whose boundary is a piecewise smooth orientable surface S. Let $\mathbf{F}(x, y, z)$ be a vector function that is continuous and has continuous first partial derivatives in a region of 3-space containing D. Then

$$\iiint_D div(\mathbf{F}) dV = \iiint_D \nabla \cdot \mathbf{F} dV = \iint_S (\mathbf{F} \cdot \mathbf{n}) \, dA \qquad \ldots(14.34)$$

where $\mathbf{n}$ is the outer unit normal vector to S.

Expressed in words the divergence theorem states that "the surface integral of the normal component of a vector point function $\mathbf{F}$, taken over a closed surface is equal to the integral of the divergence of $\mathbf{F}$, taken over the volume enclosed by the surface".

14.25.1 Cartesian Form of Divergence Theorem

Let vector function $\mathbf{F}(x, y, z)$ has components (F_1, F_2, F_3).

Therefore, $\mathbf{F} = F_1\mathbf{i} + F_2\mathbf{j} + F_3\mathbf{k}$. Now $div \, \mathbf{F} = \nabla \cdot \mathbf{F} = \dfrac{\partial F_1}{\partial x} + \dfrac{\partial F_2}{\partial y} + \dfrac{\partial F_3}{\partial z}$ and

$(\mathbf{F} \cdot \mathbf{n}) dA = \left[F_i (\mathbf{n} \cdot \mathbf{i}) + F_2 (\mathbf{n} \cdot \mathbf{j}) + F_3 (\mathbf{n} \cdot \mathbf{k}) \right] dA$.

If α, β, γ are angles which $\mathbf{n}$ the unit normal makes with the positive directions of x, y, z axis respectively or $\mathbf{i}$, $\mathbf{j}$, $\mathbf{k}$ directions then
$\mathbf{n} \cdot \mathbf{i} = \cos\alpha, \mathbf{n} \cdot \mathbf{j} = \cos\beta$, and $\mathbf{n} \cdot \mathbf{k} = \cos\gamma$. Thus $\mathbf{n} = (\cos\alpha)\mathbf{i} + (\cos\beta)\mathbf{j} + (\cos\gamma)\mathbf{k}$.

Therefore $(\mathbf{F} \cdot \mathbf{n}) dA = F_1 \cos\alpha \, dA + F_2 \cos\beta \, dA + F_3 \cos\gamma \, dA \qquad \ldots(14.35)$

$$= F_1 dy \, dz + F_2 dz \, dx + F_3 dx \, dy.$$

Therefore, $\iiint_D \nabla \cdot \mathbf{F}\, dV = \iint_S (\mathbf{F} \cdot \mathbf{n})\, dA$ may be written as

$$\iiint_D \left(\frac{\partial F_1}{\partial x} + \frac{\partial F_2}{\partial y} + \frac{\partial F_3}{\partial z} \right) dx\, dy\, dz = \iint_S F_1 dy\, dz + F_2 dz\, dx + F_3 dx\, dy,$$

which is the Cartesian form of the divergence theorem.

14.25.2 If ℓ, m, n are direction cosines of $\mathbf{n}$ then (14.35) can be written as

$$\left(F_1 \ell + F_2 m + F_3 n \right) dA.$$

Then the divergence theorem may be written in the form

$$\iiint_D \left(\frac{\partial F_1}{\partial x} + \frac{\partial F_2}{\partial y} + \frac{\partial F_3}{\partial z} \right) dx\, dy\, dz = \iint_S \left(\ell F_1 + m F_2 + n F_3 \right) dA \qquad \dots(14.36)$$

14.26 Illustrative Examples

Example 14.68 Use divergence theorem to evaluate:

(i) $\iint_S (\mathbf{F} \cdot \mathbf{n})\, dA$ where $\mathbf{F} = 4x\mathbf{i} - 2y^2 \mathbf{j} + z^2 \mathbf{k}$ and S is the surface bounding the region $x^2 + y^2 = 4$, $z = 0$, $z = 3$. *(SVTU 2007S; JNTU '06; Kerala '05)*

(ii) $\iint_S (\mathbf{r} \cdot \mathbf{n})\, dA$ where $\mathbf{r} = x\mathbf{i} + y\mathbf{j} + z\mathbf{k}$ and S is the surface of the sphere $x^2 + y^2 + z^2 = 9$. *(UPTU-2003)*

(iii) $\iint_S \left(\nabla r^2 \right) \cdot \mathbf{n}\, dA$ where S is a closed surface enclosing the volume V and $r^2 = x^2 + y^2 + z^2$. *(AMIE S-2006)*

Solution. (i) By the divergence theorem

$$\iint_S (\mathbf{F} \cdot \mathbf{n})\, dA = \iiint_D (\nabla \cdot \mathbf{F})\, dV.$$

$$= \iiint_D (\nabla \cdot \mathbf{F})\, dV = \iiint_D \left[\frac{\partial}{\partial x}(4x) + \frac{\partial}{\partial y}\left(-2y^2\right) + \frac{\partial}{\partial z}\left(z^2\right) \right] dV$$

$$= \iiint_D (4 - 4y + 2z)\, dV$$

$$= \int_{x=-2}^{2} \int_{y=-\sqrt{4-x^2}}^{\sqrt{4-x^2}} \int_{z=0}^{3} (4 - 4y + 2z)\, dz\, dy\, dx$$

$$= \int_{x=-2}^{2} \int_{y=-\sqrt{4-x^2}}^{\sqrt{4-x^2}} \left[4z - 4yz + z^2 \right]_0^3 dy\, dx = \int_{x=-2}^{2} \int_{y=-\sqrt{4-x^2}}^{\sqrt{4-x^2}} (21 - 12y)\, dy\, dx$$

$$= 42 \int_{x=-2}^{x} \sqrt{4-x^2}\, dx = 84 \int_0^2 \sqrt{4-x^2}\, dx = 84 \left[\frac{x}{2}\sqrt{4-x^2} + \frac{4}{2}\sin^{-1}\frac{x}{2} \right]_0^2 = 84\pi.$$

(ii) $\iint_S (\mathbf{r} \cdot \mathbf{n}) \, dA = \iiint_D (\nabla \cdot \mathbf{r}) \, dV$ (divergence theorem)

$$= \iiint_D \left(\frac{\partial}{\partial x}\mathbf{i} + \frac{\partial}{\partial y}\mathbf{j} + \frac{\partial}{\partial z}\mathbf{k} \right) \cdot (x\mathbf{i} + y\mathbf{j} + z\mathbf{k}) \, dV = \iiint_V \left(\frac{\partial x}{\partial x} + \frac{\partial y}{\partial y} + \frac{\partial z}{\partial z} \right) dV$$

$$= 3 \iiint_D dV = 3V, \text{ where } V \text{ is the volume of the sphere } x^2 + y^2 + z^2 = 9,$$

$$= 3(4/3)\pi 3^3 = 108\pi.$$

(iii) Using the divergence theorem

$$\iint_S (\nabla r^2) \cdot \mathbf{n} \, dA = \iiint_D (\nabla \cdot \nabla r^2) \, dV = \iiint_D \nabla \cdot \left(2r\frac{\mathbf{r}}{r} \right) dV \quad \left[\because \nabla f(r) = f'(r)\frac{\mathbf{r}}{r} \right]$$

$$= 2 \iiint_D \nabla \cdot \mathbf{r} \, dV = 2 \iiint_D \left[\frac{\partial}{\partial x}(x) + \frac{\partial}{\partial y}(y) + \frac{\partial}{\partial z}(z) \right] dV,$$

since $\mathbf{r} = x\mathbf{i} + y\mathbf{j} + z\mathbf{k} = 6 \iiint_D dV = 6V.$

Example 14.69 Evaluate $\iint_S (\mathbf{F} \cdot \mathbf{n}) \, dA$, where

$$\mathbf{F} = (2x + 3z)\mathbf{i} - (xz + y)\mathbf{j} + (y^2 + 2z)\mathbf{k}$$

and S is the surface of the sphere having center at $(3, -1, 2)$ and radius 3 units.
(IETE, Dec 2007; UPTU 2001).

Solution. Let V be the volume enclosed by the surface S. Thus by divergence theorem, we have

$$\iint_S (\mathbf{F} \cdot \mathbf{n}) \, dA = \iiint_D (\nabla \cdot \mathbf{F}) \, dV = \iiint_D \left[\frac{\partial}{\partial x}(2x + 3z) + \frac{\partial}{\partial y}(-xz - y) \right.$$

$$\left. + \frac{\partial}{\partial z}(y^2 + 2z) \right] dV = \iiint_D (2 - 1 + 2) \, dV = 3 \iiint_D dV = 3V.$$

The volume of a sphere of radius 3 units $= (4/3)\pi(3)^3 = 36\pi.$

Hence $\iint_S (\mathbf{F} \cdot \mathbf{n}) \, dA = 3(36\pi) = 108\pi.$

Example 14.70 Verify divergence theorem for $\mathbf{F} = 4xz\mathbf{i} - y^2\mathbf{j} + yz\mathbf{k}$ taken over the cube bounded by the planes $x = 0$, $x = 2$, $y = 0$, $y = 2$, $z = 0$, $z = 2$.
(MDU 2002; Madras 1998)

Solution. We have $\nabla \cdot \mathbf{F} = \frac{\partial}{\partial x}(4xz) + \frac{\partial}{\partial y}(-y^2) + \frac{\partial}{\partial z}(yz) = 4z - 2y + y = 4z - y.$

By the divergence theorem, the required integral is equal to

$$\iiint_D (\nabla \cdot \mathbf{F})\,dV = \iiint_D (4z - y)\,dV = \int_{x=0}^2 \int_{y=0}^2 \int_{z=0}^2 (4z - y)\,dz\,dy\,dx$$

$$= \int_{x=0}^2 \int_{y=0}^2 \left[2z^2 - yz\right]_0^2 dy\,dx = 2\int_{x=0}^2 \int_{y=0}^2 (4 - y)\,dy\,dx$$

$$= 2\int_{x=0}^2 \left[4y - \frac{y^2}{2}\right]_0^2 dx = 12\int_0^2 dx = 24.$$

To evaluate surface integral $\iint_S (\mathbf{F} \cdot \mathbf{n})\,dA$, we note that S consists of six faces.

Over the surface $ABCO$: $\mathbf{n} = \mathbf{i}$, $x = 0$. Then

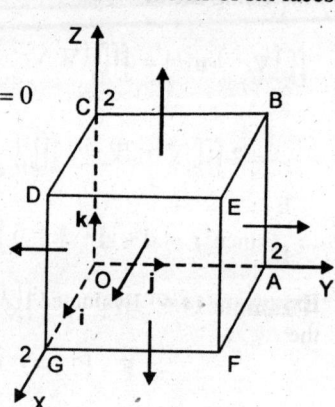

$$\iint_{ABCD} (\mathbf{F} \cdot \mathbf{n})\,dA = \int_0^2 \int_0^2 (-y^2\mathbf{j} + yz\mathbf{k}) \cdot (-\mathbf{i})\,dy\,dz = 0$$

Over the face $OGDC$: $\mathbf{n} = -\mathbf{j}$, $y = 0$. Then

$$\iint_{OGDC} (\mathbf{F} \cdot \mathbf{n})\,dA = \int_0^2 \int_0^2 (4xz\mathbf{i}) \cdot (-\mathbf{j})\,dx\,dz = 0$$

Over the face $AFGO$: $\mathbf{n} = -\mathbf{k}$, $z = 0$. Then

$$\iint_{AFGO} (\mathbf{F}.\mathbf{n})\,dA = \int_0^2 (-y^2\mathbf{j}) \cdot (-\mathbf{k})\,dx\,dy = 0.$$

Over the face $DEFG$: $\mathbf{n} = \mathbf{i}$, $x = 2$. Then

$$\iint_{DEFG} (\mathbf{F}.\mathbf{n})\,dA$$

Fig. 14.31.

$$= \int_0^2 \int_0^2 (8z\mathbf{i} - y^2\mathbf{j} + yz\mathbf{k}) \cdot \mathbf{i}\,dy\,dz = 8\int_0^2 \int_0^2 z\,dy\,dz = 32.$$

Over the face $ABEF$: $\mathbf{n} = \mathbf{j}$, $y = 2$. Then

$$\iint_{ABEF} (\mathbf{F}.\mathbf{n})\,dA = \int_0^2 \int_0^2 (4xz\mathbf{i} - 4\mathbf{j} + 2z\mathbf{k}) \cdot \mathbf{j}\,dx\,dz = -4\int_0^2 \int_0^2 dx\,dz = -16.$$

Over the face $BCDE$: $\mathbf{n} = \mathbf{k}$, $z = 2$. Then

$$\iint_{BCDE} (\mathbf{F} \cdot \mathbf{n})\,dA = \int_0^2 \int_0^2 (8x\mathbf{i} - y^2\mathbf{j} + 2y\mathbf{k}) \cdot \mathbf{k}\,dx\,dy = 2\int_0^2 \int_0^2 y\,dx\,dy = 8$$

Adding, we see that over the whole surface

$$\iint_S (\mathbf{F} \cdot \mathbf{n})\,dA = 0 + 0 + 0 + 32 - 16 + 8 = 24.$$

Hence $\iint_S (\mathbf{F} \cdot \mathbf{n})\,dA = \iiint_D (\nabla \cdot \mathbf{F})\,dV.$

Example 14.71 Verify the divergence theorem for the function

$$\mathbf{F} = 2x^2 y\mathbf{i} - y^2\mathbf{j} + 4xz^2\mathbf{k}$$

taken over the region in the first octant bounded by $y^2 + z^2 = 9$ and $x = 2$.

Solution. Volume integral

$$= \iiint_D (\nabla \cdot \mathbf{F}) dV = \iiint_D \left[\frac{\partial}{\partial x}(2x^2 y) + \frac{\partial}{\partial y}(-y^2) + \frac{\partial}{\partial z}(4xz^2) \right] dV$$

$$= \iiint_D (4xy - 2y + 8xz) dV = \int_0^3 \int_0^{\sqrt{9-y^2}} \int_0^2 (4xy - 2y + 8xz) dx\, dz\, dy$$

$$= \int_0^3 \int_0^{\sqrt{9-y^2}} \left[2x^2 y - 2yx + 4x^2 z\right]_0^2 dz\, dy = 4\int_0^3 \int_0^{\sqrt{9-y^2}} (y + 4z) dz\, dy$$

$$= 4\int_0^3 \left[yz + 2z^2\right]_0^{\sqrt{9-y^2}} dy = 4\int_0^3 \left\{ y\sqrt{9-y^2} + 2(9 - y^2) \right\} dy$$

$$= 4\left[-\frac{1}{3}(9 - y^2)^{3/2} + 18y - \frac{2}{3}y^3 \right]_0^3 = 180.$$

By Example 14.65 Surface integral, $\iint_S (\mathbf{F} \cdot \mathbf{n}) dA = 180.$

Hence, the divergence theorem is verified in this case.

Example 14.72 Verify the divergence theorem for $\mathbf{F} = 4x\mathbf{i} - 2y^2\mathbf{j} + z^2\mathbf{k}$ taken over the region bounded by the cylinder $x^2 + y^2 = 4$, $z = 0$ and $z = 3$.

(*MDU 2004; IETE-June 2003, June 2001*)

Solution. By Example 14.68(*i*), volume integral $= \iiint_D (\nabla \cdot \mathbf{F}) dV = 84\pi.$

The surface S of the cylinder consists of a base $S_1 (z = 0)$, the top $S_2 (z = 3)$ and the convex portion $S_3 (x^2 + y^2 = 4)$. Refer Fig. 14.32. Then,

surface integral $\iint_S (\mathbf{F} \cdot \mathbf{n}) dA = \iint_{S_1} (\mathbf{F} \cdot \mathbf{n}) dA + \iint_{S_2} (\mathbf{F} \cdot \mathbf{n}) dA + \iint_{S_3} (\mathbf{F} \cdot \mathbf{n}) dA.$

Over the surface S_1: $z = 0$, $\mathbf{n} = -\mathbf{k}$, $\mathbf{F} = 4x\mathbf{i} - 2y^2\mathbf{j}$ and $\mathbf{F} \cdot \mathbf{n} = 0.$

Therefore $\iint_{S_1} (\mathbf{F} \cdot \mathbf{n}) dA = 0.$

Over the surface S_2: $z = 3$, $\mathbf{n} = \mathbf{k}$,

$\mathbf{F} = 4x\mathbf{i} - 2y^2\mathbf{j} + 9\mathbf{k}$ so that $\mathbf{F} \cdot \mathbf{n} = 9.$

$\therefore \iint_{S_2} (\mathbf{F} \cdot \mathbf{n}) dA = \iint_{S_2} 9 dA = 9$ (area of

circular region with radius 2) $= 36\pi.$

Over the surface S_3: $x^2 + y^2 = 4$. A vector normal to $x^2 + y^2 - 4 = 0$ is given by

$\nabla(x^2 + y^2 - 4) = 2x\mathbf{i} + 2y\mathbf{j}.$

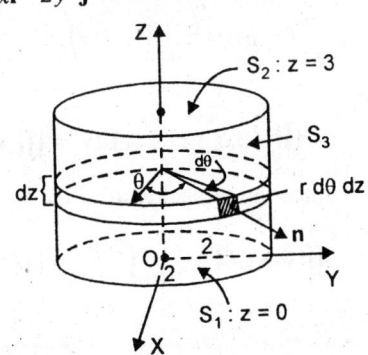

Fig. 14.32.

A unit vector normal to S_3 is

$$\mathbf{n} = \frac{2x\mathbf{i} + 2y\mathbf{j}}{\sqrt{4x^2 + 4y^2}} = \frac{x\mathbf{i} + y\mathbf{j}}{2}, \quad \text{since } x^2 + y^2 = 4.$$

$$\mathbf{F} \cdot \mathbf{n} = \left(4x\mathbf{i} - 2y^2\mathbf{j} + z^2\mathbf{k}\right) \cdot \left(\frac{x\mathbf{i} + y\mathbf{j}}{2}\right) = 2x^2 - y^3.$$

Using the cylindrical coordinates, we write $x = r \cos\theta = 2 \cos\theta$, $y = r \sin\theta = 2 \sin\theta$, $dA = r\, d\theta\, dz = 2\, d\theta\, dz$ and so

$$\iint_{S_3} (\mathbf{F} \cdot \mathbf{n}) dA = \int_{\theta=0}^{2\pi} \int_{z=0}^{3} \left[2(2\cos\theta)^2 - (2\sin\theta)^3\right] 2 dz\, d\theta$$

$$= 16 \int_{\theta=0}^{2\pi} \int_{z=0}^{3} \left(\cos^2\theta - \sin^3\theta\right) dz\, d\theta = 48 \int_{0}^{2\pi} \left(\cos^2\theta - \sin^3\theta\right) d\theta$$

$$= 48 \int_{0}^{2\pi} \left[\frac{1}{2}(1 + \cos 2\theta) - \frac{1}{4}(3\sin\theta - \sin 3\theta)\right] d\theta = 48\pi.$$

$$\left[\because \int_{0}^{2\pi} \sin nx\, dx = \int_{0}^{2\pi} \cos nx\, dx = 0, \ n \text{ being an integer}\right].$$

Then the surface integral $= 0 + 36\pi + 48\pi = 84\pi$.

Therefore $\iiint_D (\nabla \cdot \mathbf{F}) dV = \iint_S (\mathbf{F} \cdot \mathbf{n}) dA$ and the divergence theorem is verified.

Example 14.73 Use the divergence theorem to evaluate $\iint_S (\mathbf{v} \cdot \mathbf{n}) dA$; where $\mathbf{v} = x^2 z\mathbf{i} + y\mathbf{j} - xz^2\mathbf{k}$ and S is the boundary of the region bounded by the paraboloid $z = x^2 + y^2$ and the plane $z = 4y$. *(IETE June 2009, June 2008)*
Solution. We have

$$\iint_S (\mathbf{v} \cdot \mathbf{n}) dA = \iiint_D (\nabla \cdot \mathbf{v}) dV = \iiint_D \left[\frac{\partial}{\partial x}(x^2 z) + \frac{\partial}{\partial y}(y) + \frac{\partial}{\partial z}(-xz^2)\right] dV$$

$$= \iiint_D (2xz + 1 - 2xz) dV = \iiint_D dV = \int_{y=0}^{4} \int_{x=-\sqrt{4y-y^2}}^{\sqrt{4y-y^2}} \int_{z=x^2+y^2}^{4y} dz\, dx\, dy,$$

since the projection of S on the xy-plane is $x^2 + y^2 = 4y$. Hence,

$$\iint_S (\mathbf{v} \cdot \mathbf{n}) dA = \int_{y=0}^{4} \int_{x=-\sqrt{4y-y^2}}^{\sqrt{4y-y^2}} (4y - x^2 - y^2) dx\, dy$$

$$= \int_{y=0}^{4} \left[2(4y - y^2)(4y - y^2)^{\frac{1}{2}} - \frac{2}{3}(4y - y^2)^{3/2}\right] dy$$

$$= \int_{y=0}^{4} \frac{4}{3}\left(4y - y^2\right)^{3/2} dy = \frac{4}{3} \int_{y=0}^{4} \left[4 - (y-2)^2\right]^{3/2} dy.$$ Put $y - 2 = 2\sin\theta$, we obtain

$$\iint_S (\mathbf{v} \cdot \mathbf{n}) \, dA = \frac{4}{3} \int_{-\pi/2}^{\pi/2} 16\cos^4\theta \, d\theta = \frac{4}{3}(32)\frac{3}{4}\cdot\frac{1}{2}\cdot\frac{\pi}{2} = 8\pi.$$

Example 14.74 By transforming to triple integral, evaluate

$$I = \iint_S x^3 dy\,dz + x^2 y\,dz\,dx + x^2 z\,dx\,dy$$ where S is the closed surface consisting of

the cylinder $x^2 + y^2 = a^2$ $(0 \leq z \leq b)$ and the circular discs $z = 0$ and $z = b$.

(Burdwan, 2003)

Solution. Since $dy\,dz = \cos\alpha \, dA$, $dz\,dx = \cos\beta \, dA$, $dx\,dy = \cos\gamma \, dA$ the given integral can be written as

$$I = \iint_S \left(x^3 \cos\alpha + x^2 y\cos\beta + x^2 z\cos\gamma\right) dA = \iint_S (\mathbf{F} \cdot \mathbf{n}) \, dA;$$

where $\mathbf{F} = x^3\mathbf{i} + x^2 y\mathbf{j} + x^2 z\mathbf{k}$ and $\mathbf{n} = (\cos\alpha)\mathbf{i} + (\cos\beta)\mathbf{j} + (\cos\gamma)\mathbf{k}$, the outward drawn unit normal.

Then by divergence theorem, the integral equals

$$I = \iiint_D (\nabla \cdot \mathbf{F}) \, dV = \iiint_D \left\{ \frac{\partial}{\partial x}(x^3) + \frac{\partial}{\partial y}(x^2 y) + \frac{\partial}{\partial z}(x^2 z) \right\} dV$$

$$= \iiint_D (3x^2 + x^2 + x^2) \, dV = \iiint_D 5x^2 \, dx\,dy\,dz,$$

where D is the region bounded by the cylinder and the circular discs.

Making use of the symmetry of the region D

$$I = 4\int_{z=0}^{b} \int_{y=0}^{a} \int_{x=0}^{\sqrt{a^2 - y^2}} 5x^2 \, dx\,dy\,dz = 20 \int_0^b \int_0^a \left[\frac{x^3}{3}\right]_0^{\sqrt{a^2-y^2}} dy\,dz$$

$$= \frac{20}{3} \int_0^b \int_0^a \left(a^2 - y^2\right)^{3/2} dy\,dz.$$ Set $y = a\sin\theta$, we obtain

$$\int_0^a \left(a^2 - y^2\right)^{3/2} dy = \int_0^{\pi/2} \left(a^3 \cos^3\theta\right) a\cos\theta \, d\theta = a^4 \int_0^{\pi/2} \cos^4\theta \, d\theta = \frac{3\pi a^4}{16}.$$

Therefore, $I = \dfrac{20}{3} \displaystyle\int_0^b \dfrac{3\pi a^4}{16} dz = \dfrac{5}{4}\pi a^4 b.$

Example 14.75 Use the divergence theorem to evaluate $\iint_S \left(\ell x^2 + my^2 + nz^2\right) dA$

taken over the sphere $(x - a)^2 + (y - b)^2 + (z - c)^2 = \rho^2$ and ℓ, m, n are the direction cosines of the external normal to the sphere. (IETE June 2009, W 2008)

14.90 *Mathematics for Engineers*

Solution. *Employing Coordinates Conversion Formulas*: Spherical to rectangular of the given sphere having centre at (a, b, c) and radius ρ.

We have, $x = a + \rho \sin\phi \cos\theta$, $y = b + \rho \sin\phi \sin\theta$, $z = c + \rho \cos\phi$.

Corresponding volume elements $dV = \rho^2 \sin\phi \, d\rho \, d\phi \, d\theta$,

radius varies from 0 to ρ, θ varies from 0 to 2π and ϕ varies from 0 to π.

Now $\iint_S (\ell x^2 + my^2 + nz^2) dA$ may be written in the form

$$\iiint_V \left\{ \frac{\partial}{\partial x}(x^2) + \frac{\partial}{\partial y}(y^2) + \frac{\partial}{\partial z}(z^2) \right\} dx\,dy\,dz = 2\iiint_V (x+y+z)\,dV$$

$$= 2\int_{\theta=0}^{2\pi}\int_{\phi=0}^{\pi}\int_{\rho=0}^{\rho}\left[(a+b+c)+\rho(\sin\phi\cos\theta+\sin\phi\sin\theta+\cos\phi)\right]\rho^2\sin\phi\,d\rho\,d\phi\,d\theta$$

$$= 2\int_{\theta=0}^{2\pi}\int_{\phi=0}^{\pi}\int_{\rho=0}^{\rho}(a+b+c)\rho^2\sin\phi\,d\rho\,d\phi\,d\theta + 2\int_{\theta=0}^{2\pi}\int_{\phi=0}^{\pi}\int_{\rho=0}^{\rho}\{\sin^2\phi$$

$$(\cos\theta+\sin\theta)+\sin\phi\cos\phi\}\rho^3\,d\rho\,d\phi\,d\theta$$

$$= \frac{2}{3}\int_{\theta=0}^{2\pi}\int_{\phi=0}^{\pi}\rho^3(a+b+c)\sin\phi\,d\phi\,d\theta + \frac{\rho^4}{2}\int_{\theta=0}^{2\pi}\int_{\phi=0}^{\pi}(\cos\theta+\sin\theta)$$

$$\sin^2\phi\,d\phi\,d\theta + \frac{\rho^4}{4}\int_{\theta=0}^{2\pi}\int_{\phi=0}^{\pi}\sin 2\phi\,d\phi\,d\theta$$

$$= \frac{2}{3}\rho^3(a+b+c)\int_{\theta=0}^{2\pi}[-\cos\phi]_0^\pi\,d\theta + \frac{\rho^4}{4}(\cos\theta+\sin\theta)\int_{\theta=0}^{2\pi}\left[\phi-\frac{\sin 2\phi}{2}\right]_0^\pi\,d\theta$$

$$-\frac{\rho^4}{8}\int_{\theta=0}^{2\pi}[\cos 2\phi]_0^\pi\,d\theta = \frac{4}{3}\rho^3(a+b+c)[\theta]_0^{2\pi} +$$

$$\frac{\pi\rho^4}{4}\int_{\theta=0}^{2\pi}(\cos\theta+\sin\theta)d\theta, \quad \left(\because [\cos 2\theta]_0^\pi = 0.\right)$$

$$= \frac{8}{3}\pi\rho^3(a+b+c)+0 = \frac{8}{3}\pi\rho^3(a+b+c).$$

Exercise 14.7

1. Show that for the vector $\mathbf{F} = y\cos(2x+y^2)\mathbf{i} - \cos(2x+y^2)\mathbf{j}$,

$\iint_S (\mathbf{F}\cdot\mathbf{n})dA = 0$, where S is a closed surface bounding a volume V.

[**Hint.** $\iint_S (\mathbf{F}\cdot\mathbf{n})dA = \iiint_D (\nabla\cdot\mathbf{F})dV = 0$

$\because \nabla\cdot\mathbf{F} = -2y\sin(2x+y^2)+2y\sin(2x+y^2) = 0].$

2. Verify the divergence theorem for
$$\mathbf{F} = \left(x^2 - yz\right)\mathbf{i} + \left(y^2 - zx\right)\mathbf{j} + \left(z^2 - xy\right)\mathbf{k}$$
taken over the rectangular parallelopiped $0 \le x \le a,\, 0 \le y \le b,\, 0 \le z \le c$.

Ans. Common value $= abc(a + b + c)$

(MDU 2006; PTU 2003; Madras 2003; UPTU 2002; VTU 2000s)

3. Use divergence theorem to evaluate $\iint_S (\mathbf{F} \cdot \mathbf{n})\, dA$, where

$\mathbf{F} = x^3\mathbf{i} + y^3\mathbf{j} + z^3\mathbf{k}$ and S is the surface of the sphere $x^2 + y^2 + z^2 = a^2$.

(IETE, Dec. 2009; PTU 2004; Gauhati 1999; Mysore 1999)

[**Hint.** $\iint_S (\mathbf{F} \cdot \mathbf{n})\, dA = \iiint_V (\nabla \cdot \mathbf{F})\, dV$, (where V is the volume of the sphere)

$= \iiint_V \left(3x^2 + 3y^2 + 3z^2\right) dx\, dy\, dz = 3\iiint_V \left(x^2 + y^2 + z^2\right) dx\, dy\, dz$

$= 3\iiint_V r^2 \cdot r^2 \sin\theta \; dr\, d\theta\, d\phi$ (by transforming to spherical polar coordinates)

$= 3\int_0^{2\pi} \int_0^\pi \int_0^a r^4 \sin\theta \, dr\, d\theta\, d\phi = \dfrac{12}{5}\pi a^5.\,]$

4. Show that if $\mathbf{r}$ is the position vector of any point of a closed surface S bounding a volume V, then $\iint_S (\mathbf{r} \cdot \mathbf{n})\, dA = 3V$.

[**Hint.** $\iint_S (\mathbf{r} \cdot \mathbf{n})\, dA = \iiint_V (\nabla \cdot \mathbf{r})\, dV$ (divergence theorem) $= 3\iiint_V dV = 3V.\,]$

5. Find the value of $\iint_S (\mathbf{F} \cdot \mathbf{n})\, dA$ where

$\mathbf{F} = r^2\left(\mathbf{i}\, x + \mathbf{j}\, y + \mathbf{k}\, z\right)$ and $r^2 = x^2 + y^2 + z^2$ and S is the surface of the sphere $x^2 + y^2 + z^2 = a^2$. Compute the integral with the aid of divergence theorem.

[**Hint.** $\nabla \cdot \mathbf{F} = \nabla \cdot \left(r\mathbf{r}^2\right) = r^2 \nabla \cdot \mathbf{r} + \mathbf{r} \cdot \nabla r^2 = 3r^2 + \mathbf{r} \cdot 2r\dfrac{\mathbf{r}}{r} = 3r^2 + 2r^2 = 5r^2$.

$\therefore \iint_S (\mathbf{F} \cdot \mathbf{n})\, dA = \iiint_V (\nabla \cdot \mathbf{F})\, dV = \iiint_V 5r^2\, dV = 5\left(\dfrac{4}{5}\pi a^5\right)$ by problem (3)

above $= 4\pi a^5.\,]$

6. The vector field $\mathbf{F} = x^2\mathbf{i} + z\mathbf{j} + yz\mathbf{k}$ is defined over the volume of the cuboid

given by $0 \le x \le a$, $0 \le y \le b$, $0 \le z \le c$. Evaluate the surface integral $\iint_S (\mathbf{F} \cdot \mathbf{n}) dA$, where S is the surface of the cuboid.

(IETE June 2007, Dec 2006; UPTU 2002) **Ans.** $abc(2a+b)/2$.

7. Verify Gauss divergence theorem for $\mathbf{F} = x^2 \mathbf{i} + y^2 \mathbf{j} + z^2 \mathbf{k}$ on the surface S of the cuboid formed by the planes $x = 0$, $x = a$, $y = 0$, $y = b$, $z = 0$ and $z = c$. *(IETE Dec 2004)*

8. Verify the divergence theorem for $\mathbf{F} = (2x-z)\mathbf{i} + x^2 y \mathbf{j} - xz^2 \mathbf{k}$ taken over the region bounded by $x = 0$, $x = 1$, $y = 0$, $y = 1$, $z = 0$, $z = 1$.

Ans. Common value = 11/6.

9. Find the value of the surface integral

$$\iint_S \left(2x^2 y \, dy \, dz - y^2 \, dz \, dx + 4xz^2 \, dx \, dy \right),$$ where S is the curved surface of

the cylinder $y^2 + z^2 = 9$ bounded by the planes $x = 0$, $x = 2$.

(IETE Dec 2006) **Ans.** 180.

10. Verify the divergence theorem for $\mathbf{F}(x, y, z) = x\mathbf{i} + y\mathbf{j} + z\mathbf{k}$ on the surface S of the sphere $x^2 + y^2 + z^2 = a^2$. *(IETE June 2006)*

11. Verify the divergence theorem for the function $F = y\mathbf{i} + x\mathbf{j} + z^2 \mathbf{k}$ taken over the cylindrical region bounded by $x^2 + y^2 = a^2$, $z = 0$, $z = h$.

(MDU 2005; Mangalore 1999) **Ans.** Common value = $\pi a^2 h^2$.

12. Verify the divergence theorem for $\mathbf{F} = \dfrac{x}{r}\mathbf{i} + \dfrac{y}{r}\mathbf{j} + \dfrac{z}{r}\mathbf{k}$, where

$r = \sqrt{x^2 + y^2 + z^2}$, over the region $x^2 + y^2 + z^2 \le a^2$.

[**Hint.** $\mathbf{F} = \mathbf{r}/r$. Therefore, $\nabla \cdot \mathbf{F} = \dfrac{1}{r} \nabla \cdot \mathbf{r} + \nabla \dfrac{1}{r} \cdot \mathbf{r} = \dfrac{2}{r}$. Therefore,

$$\iiint_V (\nabla \cdot \mathbf{F}) dV = \iiint_V \dfrac{2}{r} dV = \int_0^{2\pi} \int_0^{\pi} \int_0^a \dfrac{2}{r} \cdot r^2 \sin\theta \, dr \, d\theta \, d\phi = 4\pi a^2.$$

$(\because$ in spherical coordinates $dV = r^2 \sin\theta \, dr \, d\theta \, d\phi)$

$\mathbf{N} = \nabla f = \nabla \left(x^2 + y^2 + z^2 - a^2 \right) = 2x\mathbf{i} + 2y\mathbf{j} + 2z\mathbf{k}$.

Also $\mathbf{n} = \dfrac{\nabla f}{|\nabla f|} = \dfrac{x\mathbf{i} + y\mathbf{j} + z\mathbf{k}}{a}$ and $\mathbf{F} \cdot \mathbf{n} = 1$ $(\because$ on the surface $r = a)$

$\therefore \quad \iint_S (\mathbf{F} \cdot \mathbf{n}) dA = \iint_S 1 \, dA = 4\pi a^2$. Hence divergence theorem is verified.]

13. Evaluate:

(i) $\iint_S x^2 \, dy \, dz + y^2 \, dz \, dx + z^2 \, dx \, dy$, where S is the surface of the cube $0 \le x \le 1, 0 \le y \le 1, 0 \le z \le 1$.

(ii) $\iint_S yz \, dy \, dz + zx \, dz \, dx + xy \, dx \, dy$, S being the surface of the sphere $x^2 + y^2 + z^2 = 4$. **Ans.** (*i*) 3 (*ii*) zero.

14. Evaluate $\iint_S xz^2 \, dy \, dz + (x^2 y - z^3) \, dz \, dx + (2xy + y^2 z) \, dx \, dy$, where S is the

entire surface of the hemispherical region bounded by $z = \sqrt{a^2 - x^2 - y^2}$

and $z = 0$ by using the divergence theorem. **Ans.** $2\pi a^5/5$.

[**Hint.** $\iint_S xz^2 \, dy \, dz + (x^2 y - z^3) \, dz \, dx + (2xy + y^2 z) \, dx \, dy$

$= \iint_S \left[\{ xz^2 \mathbf{i} + (x^2 y - z^3) \mathbf{j} + (2xy + y^2 z) \mathbf{k} \} \cdot \{ (\cos\alpha) \mathbf{i} + (\cos\beta) \mathbf{j} + (\cos\gamma) \mathbf{k} \} \right] dA$

$= \iint_S (\mathbf{F} \cdot \mathbf{n}) \, dA = \iiint_D (\nabla \cdot \mathbf{F}) \, dV$ (by divergence theorem), where $\mathbf{n}$ is outward drawn unit normal,

$= \iiint_D \left\{ \frac{\partial}{\partial x}(xz^2) + \frac{\partial}{\partial y}(x^2 y - z^3) + \frac{\partial}{\partial z}(2xy + y^2 z) \right\} dV = \iiint_D (x^2 + y^2 + z^2) \, dV,$

where D is the region bounded by the hemisphere and the *xy*-plane.

By use of spherical coordinates, above integral is equal to

$$4 \int\limits_{\theta=0}^{\pi/2} \int\limits_{\phi=0}^{\pi/2} \int\limits_{\rho=0}^{a} \rho^2 \cdot \rho^2 \sin\phi \, d\rho \, d\phi \, d\theta = 2\pi a^5/5.]$$

15. If $\mathbf{F} = \nabla\phi$ and $\nabla^2 \phi = -4\pi\rho$, prove that

$$\iint_S (\mathbf{F} \cdot \mathbf{n}) \, dA = -4\pi\rho \iiint_V dV, \text{ where the symbols have their usual meanings.}$$

16. Prove that $\iiint_V \frac{dV}{r^2} = \iint_S \left(\frac{\mathbf{r} \cdot \mathbf{n}}{r^2} \right) dA$, where $r = |\mathbf{r}| = |x\mathbf{i} + y\mathbf{j} + z\mathbf{k}|$, other

symbols having their usual meanings.

[**Hint.** R.H.S. $= \iint_S \left(\frac{\mathbf{r} \cdot \mathbf{n}}{r^2} \right) dA = \iiint_V \left(\nabla \cdot \frac{\mathbf{r}}{r^2} \right) dV.$

Now $\nabla \cdot \frac{\mathbf{r}}{r^2} = \frac{1}{r^2} \nabla \cdot \mathbf{r} + \mathbf{r} \cdot \nabla \frac{1}{r^2} = \frac{3}{r^2} + \mathbf{r} \cdot \left(-2r^{-3} \frac{\mathbf{r}}{r} \right) = \frac{1}{r^2}$ etc.]

17. Evaluate the surface integral

$\iint_S (yz\mathbf{i} + zx\mathbf{j} + xy\mathbf{k}) \cdot dA$, where S is the

surface of the sphere $x^2 + y^2 + z^2 = 1$
in the first octant.

[**Hint.** Here $\mathbf{F} = yz\mathbf{i} + zx\mathbf{j} + xy\mathbf{k}$.

$\iiint_D (\nabla \cdot \mathbf{F})dV = \iint_{S_1} (\mathbf{F} \cdot \mathbf{n})dA$, where

S_1 is the closed surface bounding the
volume of the sphere in first octant.
(Refer Fig. 14.33)

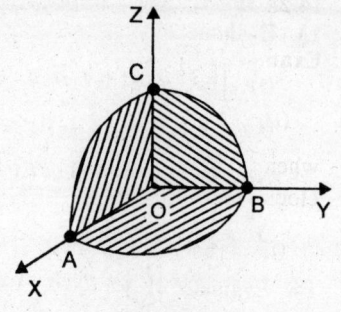

Fig. 14.33.

Since $\nabla \cdot \mathbf{F} = 0$. Therefore, $\iiint_D (\nabla \cdot \mathbf{F})dV = 0$

By divergence theorem, $\iiint_D \operatorname{div} \mathbf{F}\, dV = \iint_{S_1} (\mathbf{F} \cdot \mathbf{n})dA = \iint_{OBC} (\mathbf{F} \cdot \mathbf{n})dA$

$+ \iint_{OAC} (\mathbf{F} \cdot \mathbf{n})dA + \iint_{OAB} (\mathbf{F} \cdot \mathbf{n})dA + \iint_S (\mathbf{F} \cdot \mathbf{n})dA,$

where S is the curved surface of the sphere in first octant.

$\Rightarrow 0 = -\dfrac{1}{8} - \dfrac{1}{8} - \dfrac{1}{8} + \iint_S (\mathbf{F} \cdot \mathbf{n})dA.$ Therefore $\iint_S (\mathbf{F} \cdot \mathbf{n})dA = \dfrac{3}{8}.$]

14.27 Stokes's Theorem
(Transformation between surface integrals and line integrals)

Let S be a piecewise smooth orientable surface in space and let the boundary of
S be a piecewise smooth simple closed curve C. Let $\mathbf{F}(x, y, z)$ be a continuous
vector function that has continuous first partial derivatives in a region of 3-space
containing S. Then

$$\iint_S (\operatorname{curl} \mathbf{F}) \cdot \mathbf{n}\, dA = \iint_S (\nabla \times \mathbf{F}) \cdot \mathbf{n}\, dA = \oint_C \mathbf{F} \cdot d\mathbf{r} \qquad \text{...(14.37)}$$

where $\mathbf{n}$ is the unit normal vector at any point of S drawn in the sense in which a
right handed screw would move when rotated in the sense of description of C.

Remarks:

1. Expressed in words, the Stokes's theorem states that "The line integral of
   the tangential component of a vector $\mathbf{F}$ taken around a simple closed curve
   C is equal to the surface integral of the normal component of curl $\mathbf{F}$ taken
   over any surface S having C as its boundary".
2. The Cartesian form of Stokes's theorem is

$$\iint_S \left[\left(\frac{\partial F_3}{\partial y} - \frac{\partial F_2}{\partial z} \right) \cos \alpha + \left(\frac{\partial F_1}{\partial z} - \frac{\partial F_3}{\partial x} \right) \cos \beta + \left(\frac{\partial F_2}{\partial x} - \frac{\partial F_1}{\partial y} \right) \cos \gamma \right] dA$$

$$= \oint_C \left(F_1 dx + F_2 dy + F_3 dz \right) \qquad \text{...(14.38)}$$

14.28 Illustrative Examples

Example 14.76 Apply Stokes's theorem to evaluate:

(*i*) $\oint_C \sin z\, dx - \cos x\, dy + \sin y\, dz$,

where C is the boundary of the rectangle $0 \le x \le \pi$, $0 \le y \le 1$, $z = 3$ oriented
clockwise as viewed from the origin. *(AMIE S-2009)*

(*ii*) $\oint_C (x+y)\,dx + (2x-z)\,dy + (y+z)\,dz$,

where C is the boundary of the triangle with vertices $(2, 0, 0)$, $(0, 3, 0)$ and
$(0, 0, 6)$. *(Nagpur 1997)*

Solution. (*i*) $\oint_C \sin z\, dx - \cos x\, dy + \sin y\, dz$

$= \oint_C (\sin z\, \mathbf{i} - \cos x\mathbf{j} + \sin y\mathbf{k}) \cdot d\mathbf{r}$

$(\because d\mathbf{r} = dx\,\mathbf{i} + dy\,\mathbf{j} + dz\,\mathbf{k})$

$= \oint_C \mathbf{F} \cdot d\mathbf{r}$, where $\mathbf{F} = \sin z\,\mathbf{i} - \cos x\,\mathbf{j} + \sin y\,\mathbf{k}$

$= \iint_S (\nabla \times \mathbf{F}) \cdot \mathbf{n}\, dA$, (by Stokes's theorem)

where S is the surface of the rectangle
$0 \le x \le \pi$, $0 \le y \le 1$, $z = 3$.

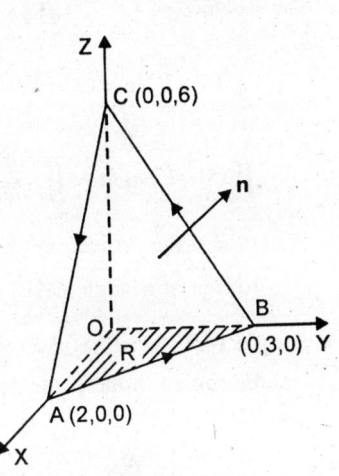

Fig. 14.34.

Now $\nabla \times \mathbf{F} = \begin{vmatrix} \mathbf{i} & \mathbf{j} & \mathbf{k} \\ \partial/\partial x & \partial/\partial y & \partial/\partial z \\ \sin z & -\cos x & \sin y \end{vmatrix}$

$= \mathbf{i} \cos y + \mathbf{j} \cos z + \mathbf{k} \sin x.$

Also for the surface of the rectangle, $\mathbf{n} = \mathbf{k}$.

$\therefore (\nabla \times \mathbf{F}) \cdot \mathbf{n} = (\mathbf{i} \cos y + \mathbf{j} \cos z + \mathbf{k} \sin x) \cdot \mathbf{k} = \sin x.$

Thus $\iint_S (\nabla \times \mathbf{F}) \cdot \mathbf{n}\, dA = \iint_S \sin x\, dA = \int_0^\pi \int_0^1 \sin x\, dy\, dx = \int_0^\pi \sin x\, dx = 2.$

(*ii*) $\oint_C (x+y)\,dx + (2x-z)\,dy + (y+z)\,dz$

$= \oint_C [(x+y)\mathbf{i} + (2x-z)\mathbf{j} + (y+z)\mathbf{k}] \cdot d\mathbf{r}$

$= \oint_C \mathbf{F} \cdot d\mathbf{r}$ where $\mathbf{F} = (x+y)\mathbf{i} + (2x-z)\mathbf{j} + (y+z)\mathbf{k}$

$= \iint_S (\nabla \times \mathbf{F}) \cdot \mathbf{n}\, dA.$ (By Stokes's Theorem),

S being the surface of the triangle ABC (see Fig. 14.34.).

Now $\nabla \times \mathbf{F} = \begin{vmatrix} \mathbf{i} & \mathbf{j} & \mathbf{k} \\ \partial/\partial x & \partial/\partial y & \partial/\partial z \\ x+y & 2x-z & y+z \end{vmatrix} = 2\mathbf{i} + \mathbf{k}.$

To obtain $\mathbf{n}$ we note that a vector normal to the plane of the triangle ABC, that

is, $\dfrac{x}{2} + \dfrac{y}{3} + \dfrac{z}{6} = 1$, *i.e.*, $3x + 2y + z - 6 = 0$ is $\nabla(3x + 2y + z - 6) = 3\mathbf{i} + 2\mathbf{j} + \mathbf{k}.$

A unit normal at any point of the triangle ABC is

$$\mathbf{n} = \frac{3\mathbf{i} + 2\mathbf{j} + \mathbf{k}}{\sqrt{9+4+1}} = \frac{3\mathbf{i} + 2\mathbf{j} + \mathbf{k}}{\sqrt{14}}.$$

Now $(\nabla \times \mathbf{F}) \cdot \mathbf{n} = (2\mathbf{i} + \mathbf{k}) \cdot \dfrac{3\mathbf{i} + 2\mathbf{j} + \mathbf{k}}{\sqrt{14}} = \dfrac{7}{\sqrt{14}}.$

$$\therefore \iint_S (\nabla \times \mathbf{F}) \cdot \mathbf{n}\, dA = \iint_R \frac{7}{\sqrt{14}} \frac{dx\, dy}{\mathbf{n} \cdot \mathbf{k}} = \iint_R 7\, dx\, dy \left(\because \mathbf{n} \cdot \mathbf{k} = \frac{1}{\sqrt{14}} \right),$$

where R is the projection of S on the xy-plane, *i.e.*, triangle OAB

$= 7$ (area of triangle OAB) $= 7 \left(1/2 \right)(2)(3) = 21.$

Example 14.77 Verify Stoke's theorem for the function $\mathbf{F} = x^2\mathbf{i} + xy\mathbf{j}$ integrated round the square in the plane $z = 0$ and bounded by the lines $x = 0$, $y = 0$, $x = a$ and $y = a$. (PTU 2002)

Solution. curl $\mathbf{F} = \nabla \times \mathbf{F} = \begin{vmatrix} \mathbf{i} & \mathbf{j} & \mathbf{k} \\ \partial/\partial x & \partial/\partial y & \partial/\partial z \\ x^2 & xy & 0 \end{vmatrix} = \mathbf{k}\, y.$

$$\therefore \iint_S (\nabla \times \mathbf{F}) \cdot \mathbf{n}\, dA = \int_0^a \int_0^a \mathbf{k}\, y \cdot \mathbf{k}\, dy\, dx = \int_{x=0}^a \int_{y=0}^a y\, dy\, dx = \frac{a^3}{2}.$$

Now $\oint_C \mathbf{F} \cdot d\mathbf{r} = \int_{OA} \mathbf{F} \cdot d\mathbf{r} + \int_{AB} \mathbf{F} \cdot d\mathbf{r}$

$+ \int_{BC} \mathbf{F} \cdot d\mathbf{r} + \int_{CO} \mathbf{F} \cdot d\mathbf{r}.$

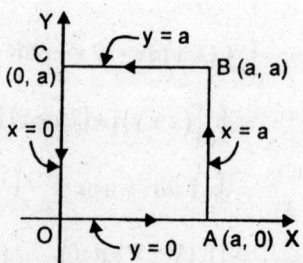

Along OA, $y = 0$, $dy = 0$. $\mathbf{F} = x^2\mathbf{i}$, $d\mathbf{r} = dx\, \mathbf{i}$ and x varies from 0 to a.

$$\therefore \int_{OA} \mathbf{F} \cdot d\mathbf{r} = \int_0^a x^2\mathbf{i} \cdot dx\, \mathbf{i} = \int_0^a x^2\, dx = \frac{a^3}{3}.$$

Along AB, $x = a$, $dx = 0$,

Fig. 14.35.

$\mathbf{F} = a^2\mathbf{i} + ay\,\mathbf{j}, d\mathbf{r} = dy\,\mathbf{j}$ and y changes from o to a.

$$\therefore \int_{AB} \mathbf{F} \cdot d\mathbf{r} = \int_0^a \left(a^2\mathbf{i} + ay\mathbf{j}\right) \cdot dy\,\mathbf{j} = a\int_0^a y\,dy = \frac{a^3}{2}.$$

Similarly $\int_{BC} \mathbf{F} \cdot d\mathbf{r} = \int_a^0 \left(x^2\mathbf{i} + xa\mathbf{j}\right) \cdot dx\,\mathbf{i} = \int_a^0 x^2\,dx = -\frac{a^3}{3}.$

$\int_{CO} \mathbf{F} \cdot d\mathbf{r} = 0$ ($\because$ along CO, $x = 0$. Therefore, $\mathbf{F} = 0$)

$$\therefore \oint_C \mathbf{F} \cdot d\mathbf{r} = \frac{a^3}{3} + \frac{a^3}{2} - \frac{a^3}{3} + 0 = \frac{a^3}{2}.$$

Hence Stoke's theorem is verified.

Example 14.78 Verify Stokes's theorem for the vector field
$\mathbf{F} = (2x - y)\mathbf{i} - yz^2\mathbf{j} - y^2z\,\mathbf{k}$ over the upper half of the surface of $x^2 + y^2 + z^2 = 1$ bounded by its projection on the xy-plane. (*Madras 2006; SVTU 2006; PTU 2001*)

Solution. Here the curve C bounding the surface of the sphere is the circle in the xy-plane with radius one and centre at $(0, 0)$. The parametric equations of the circle are: $x = \cos\theta, y = \sin\theta, 0 \le \theta \le 2\pi$.

$$\therefore \oint_C \mathbf{F} \cdot d\mathbf{r} = \oint_C (2x - y)\,dx - yz^2\,dy - y^2z\,dz$$

$$= \int_0^{2\pi} (2\cos\theta - \sin\theta)(-\sin\theta\,d\theta) = \pi.$$

Since $\nabla \times \mathbf{F} = \begin{vmatrix} \mathbf{i} & \mathbf{j} & \mathbf{k} \\ \partial/\partial x & \partial/\partial y & \partial/\partial z \\ 2x - y & -yz^2 & -y^2z \end{vmatrix} = \mathbf{k}$

$$\therefore \iint_S (\nabla \times \mathbf{F}) \cdot \mathbf{n}\,dA = \iint_S \mathbf{k} \cdot \mathbf{n}\,dA = \iint_D dy\,dx = \pi(1)^2 = \pi.$$

(where D is the projection of S on the xy-plane, *i.e.*, a circle with radius unity)

$$\therefore \iint_S (\mathbf{F} \cdot \mathbf{n})\,dA = \oint_C \mathbf{F} \cdot d\mathbf{r}$$

and the Stokes's theorem is verified.

Example 14.79 Evaluate $\iint_S (\nabla \times \mathbf{F}) \cdot \mathbf{n}\,dA$, where $\mathbf{F} = y\mathbf{i} + z\mathbf{j} + x\mathbf{k}$ and S the portion of the surface $x^2 + y^2 - 2ax + az = 0$ above the xy-plane.

Also Verify Stokes's theorem

Solution. We have $\iint_S (\nabla \times \mathbf{F}) \cdot \mathbf{n}\,dA = \oint_C \mathbf{F} \cdot d\mathbf{r}$ \hfill (By Stokes's theorem) where C is the circle given by $x^2 + y^2 - 2ax + az = 0, z = 0,$

that is, $x^2 + y^2 - 2ax = 0$ in the xy-plane.

The equation of the circle can be written in the form $(x-a)^2 + y^2 = a^2$

or $x = a + a\cos\theta, y = a\sin\theta,$ the parametric equations of the circle.

In the xy-plane *i.e.*, $z = 0,$ $\mathbf{F} \cdot d\mathbf{r} = (y\mathbf{i} + x\mathbf{k}) \cdot (\mathbf{i}\, dx + \mathbf{j}\, dy) = y dx.$

$$\therefore \oint_C \mathbf{F} \cdot d\mathbf{r} = \int_0^{2\pi} (a\sin\theta)(-a\sin\theta)\, d\theta = -a^2 \int_0^{2\pi} \sin^2\theta\, d\theta = -\pi a^2.$$

Now we obtain the surface integral, we note that

$$\iint_S (\nabla \times \mathbf{F}) \cdot \mathbf{n}\, dA = \iint_{S_1} (\nabla \times \mathbf{F}) \cdot \mathbf{n}\, dA,$$ where S_1 is the plane region bounded by

the circle C.

Here $\mathbf{n} = \mathbf{k}.$ $(\nabla \times \mathbf{F}) \cdot \mathbf{n} = -(\mathbf{i} + \mathbf{j} + \mathbf{k}) \cdot \mathbf{k} = -1.$

$$\therefore \iint_{S_1} (\nabla \times \mathbf{F}) \cdot \mathbf{n}\, dA = -\iint_{S_1} 1 dA = -\pi a^2.$$

Hence the Stokes's theorem is verified.

Remark.

$x^2 + y^2 - 2ax + az = 0$ can be written in the form $(x-a)^2 + y^2 = -a(z-a).$
Thus the surface is a paraboloid with vertex at (a, o, a), axis parallel to z-axis and turned towards negative direction of z-axis.

Example 14.80 Apply Stokes's theorem to prove that

$$\oint_C y\, dx + z\, dy + x\, dz = -2\sqrt{2}\, \pi a^2,$$

where C is the curve given by $x^2 + y^2 + z^2 - 2ax - 2ay = 0, x + y = 2a$
and begins at the point $(2a, 0, 0)$ and goes at first below the z-plane.
Solution. Obviously the curve C is a great circle of the sphere $x^2 + y^2 + z^2$
$- 2ax - 2ay = 0$ (the circle is shown dotted in Fig. 14.36) in the plane $x + y = 2a$
which passes through A and B and is perpendicular to the z-plane.

Now $\oint_C y\, dx + z\, dy + x\, dz = \int_C (y\mathbf{i} + z\mathbf{j} + x\mathbf{k}) \cdot d\mathbf{r} = \iint_S \{\nabla \times (y\mathbf{i} + z\mathbf{j} + x\mathbf{k})\} \cdot \mathbf{n} dA$

where S is the surface of the circle referred to above.

Now $\nabla \times (y\mathbf{i} + z\mathbf{j} + x\mathbf{k}) = -(\mathbf{i} + \mathbf{j} + \mathbf{k}),$ $\phi = x + y - 2a = 0$

$\therefore \mathbf{N} = \nabla\phi = \mathbf{i} + \mathbf{j}.$ Hence $\mathbf{n} = \dfrac{\mathbf{N}}{|\mathbf{N}|} = \dfrac{\mathbf{i} + \mathbf{j}}{\sqrt{2}}.$ Thus the given integral

$$= \iint_S \left[-(i+j+k) \cdot \frac{i+j}{\sqrt{2}} \right] dA$$

$$= \frac{-2}{\sqrt{2}} \iint_S dA = \frac{-2}{\sqrt{2}} \text{ (area of the circle)}$$

$$= \frac{-2}{\sqrt{2}} (2\pi a^2) = -2\sqrt{2}\pi a^2$$

($\because$ radius of circle $= \sqrt{2} \cdot a$)

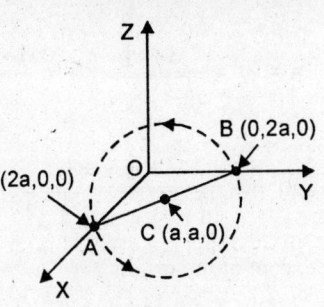

Fig. 14.36.

Example 14.81 Verify Stokes's theorem for the function $F = 2y^3 i + x^3 j + z k$,

where C is the curve of intersection of cone $z = \sqrt{x^2 + y^2}$ by the plane $z = 4$ and S is surface of cone below $z = 4$. (*PTU 2006; IETE Dec 2005*)

Solution. Refer Figure 14.37. The bounding curve C of S is a circle with equations $x^2 + y^2 = 16$, $z = 4$ setting the parametric equations of circle $x = 4 \cos\theta$, $y = 4 \sin\theta$, we obtain

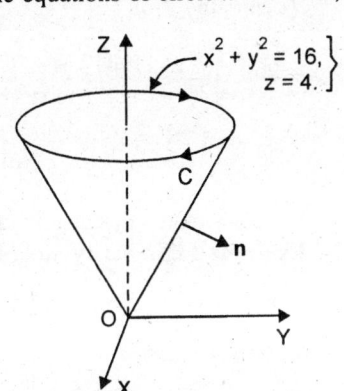

$$\oint_C F \cdot dr = \oint_C 2y^3 dx + x^3 dy + z dz$$

$$= \oint_C 2y^3 dx + x^3 dy$$

$$= \int_{2\pi}^0 64 \left[2\sin^3 \theta (-4\sin\theta) + \cos^3 \theta (4\cos\theta) \right] d\theta$$

$$= -256 \int_0^{2\pi} (\cos^4 \theta - 2\sin^4 \theta) d\theta$$

$$= -1024 \int_0^{\pi/2} (\cos^4 \theta - 2\sin^4 \theta) d\theta$$

$$= -1024 \left[\frac{3}{4} \cdot \frac{1}{2} \cdot \frac{\pi}{2} - 2\left(\frac{3}{4} \cdot \frac{1}{2} \cdot \frac{\pi}{2} \right) \right] = 192\pi.$$

Fig. 14.37.

Also curl $F = \nabla \times F = \begin{vmatrix} i & j & k \\ \partial/\partial x & \partial/\partial y & \partial/\partial z \\ 2y^3 & x^3 & z \end{vmatrix} = i(0) - j(0) + k(3x^2 - 6y^2)$
$= k(3x^2 - 6y^2).$

If the outward normal to S is taken, then it points downwards. The orientation of C is shown in Fig. 14.34. Let $f(x, y, z) = \sqrt{x^2 + y^2} - z = 0$ be taken as the

equation of the surface. Then unit normal is given by $n = \dfrac{N}{|N|}$.

$$\mathbf{N} = \nabla f = \frac{x\mathbf{i} + y\mathbf{j}}{\sqrt{x^2 + y^2}} - \mathbf{k} = \frac{x\mathbf{i} + y\mathbf{j} - z\mathbf{k}}{z},$$

and $\mathbf{n} = \dfrac{(x\mathbf{i} + y\mathbf{j} - z\mathbf{k})/z}{\sqrt{(x^2 + y^2 + z^2)/z^2}} = \dfrac{x\mathbf{i} + y\mathbf{j} - z\mathbf{k}}{\sqrt{2}\, z},$

except at the origin.

We have $\displaystyle\iint_S (\nabla \times \mathbf{F}) \cdot \mathbf{n}\, dA = \iint_S -\frac{(3x^2 - 6y^2)}{\sqrt{2}}\, dA = -\iint_R \frac{(3x^2 - 6y^2)}{\sqrt{2}}\, \frac{dx\, dy}{\mathbf{n} \cdot \mathbf{k}}$

$$= -\iint_R \frac{(3x^2 - 6y^2)}{\sqrt{2}}\, \frac{dx\, dy}{(-1/\sqrt{2})} = \iint_R (3x^2 - 6y^2)\, dx\, dy.$$

In polar coordinates, this become

$$\int_{\theta=2\pi}^{0} \int_{r=0}^{4} (3\cos^2\theta - 6\sin^2\theta)\, r^3\, dr\, d\theta = 64\int_{2\pi}^{0} (3\cos^2\theta - 6\sin^2\theta)\, d\theta$$

$$= 96\int_{2\pi}^{0} \left[(1 + \cos 2\theta) - 2(1 - \cos 2\theta)\right] d\theta$$

$$= 96\int_{2\pi}^{0} (3\cos 2\theta - 1)\, d\theta = 96\left[\frac{3}{2}\sin 2\theta - \theta\right]_{2\pi}^{0} = 96(2\pi) = 192\pi.$$

Hence, Stokes's theorem is verified.

Example 14.82 Verify Stokes's theorem for the vector field

$\mathbf{F}(x, y, z) = xy\mathbf{i} + yz\mathbf{j} + zx\mathbf{k}$, taking S to be part of the cylinder $z = 1 - x^2$ for which $0 \le x \le 1, -2 \le y \le 2$.

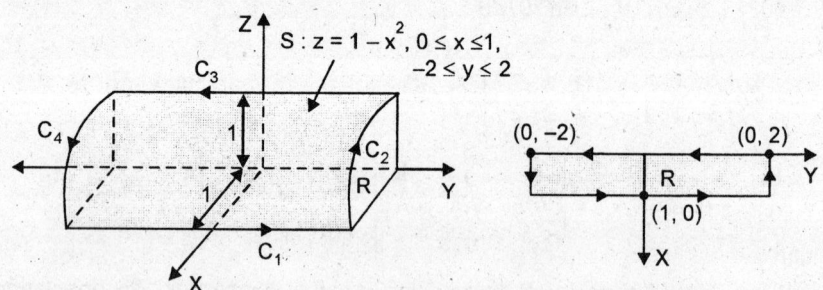

Fig. 14.38.

Solution. The surface S, the curve C (which is composed of the union of C_1, C_2, C_3 and C_4); and the region R are shown in Fig. 14.38.

We write $\oint_C = \int_{C_1} + \int_{C_2} + \int_{C_3} + \int_{C_4}$.

We have $\oint_C F \cdot dr = \oint_C xy\,dx + yz\,dy + zx\,dz$. On C_1: $x = 1, z = 0, dx = 0, dz = 0,$

so $\int_{C_1} y(0) + y(0)dy + 0 = 0.$

On C_2: $y = 2, z = 1 - x^2, dy = 0, dz = -2x\,dx,$ so

$$\int_{C_2} 2x\,dx + 2(1 - x^2)0 + x(1 - x^2)(-2x)dx = \int_1^0 (2x - 2x^2 + 2x^4)dx$$

$$= \int_0^1 (2x^2 - 2x - 2x^4)dx = \left[\frac{2}{3}x^3 - x^2 - 2\frac{x^5}{5}\right]_0^1 = \frac{2}{3} - 1 - \frac{2}{5} = -\frac{11}{15}.$$

On C_3: $x = 0, z = 1, dx = 0, dz = 0,$ so

$$\int_{C_3} 0 + y\,dy + 0 = \int_2^{-2} y\,dy = 0.$$

On C_4: $y = -2, z = 1 - x^2, dy = 0, dz = -2x\,dx,$ so

$$\int_{C_4} -2x\,dx - 2(1 - x^2)0 + x(1 - x^2)(-2x\,dx) = \int_0^1 (-2x - 2x^2 + 2x^4)dx$$

$$= \left[-x^2 - \frac{2}{3}x^3 + \frac{2}{5}x^5\right]_0^1 = -\frac{19}{15}.$$

Hence, $\oint_C xy\,dx + yz\,dy + zx\,dz = 0 - \frac{11}{15} + 0 - \frac{19}{15} = -2.$ ...(i)

$$\text{curl } \mathbf{F} = \nabla \times \mathbf{F} = \begin{vmatrix} \mathbf{i} & \mathbf{j} & \mathbf{k} \\ \partial/\partial x & \partial/\partial y & \partial/\partial z \\ xy & yz & zx \end{vmatrix}$$

$$= \mathbf{i}(0 - y) - \mathbf{j}(z - 0) + \mathbf{k}(0 - x) = -y\mathbf{i} - z\mathbf{j} - x\mathbf{k}.$$

Let $f(x, y, z) = x^2 + z - 1 = 0$ be taken as the equation of the surface. Then

unit normal is given by $\mathbf{n} = \dfrac{\nabla f}{|\nabla f|} = \dfrac{2x\mathbf{i} + \mathbf{k}}{\sqrt{4x^2 + 1}}.$

Therefore, $\displaystyle\iint_S (\nabla \times \mathbf{F}) \cdot \mathbf{n}\,dA = \iint_S \frac{-2xy - x}{\sqrt{4x^2 + 1}}dA = \iint_R \frac{-2xy - x}{\sqrt{4x^2 + 1}} \cdot \frac{dx\,dy}{\mathbf{n} \cdot \mathbf{k}}$

$$= \int_{x=0}^{1} \int_{y=-2}^{2} (-2xy - x)\, dy\, dx = \int_{0}^{1} \left[-xy^2 - xy\right]_{-2}^{2} dx = \int_{0}^{1} (-4x)\, dx = -2,$$

which agrees with (i). Hence Stokes's theorem is verified.

Exercise 14.8

1. Express Stokes's theorem in words and write it in Cartesian form.

    Evaluate $\oint_C \mathbf{F} \cdot d\mathbf{r}$ by Stokes's theorem, where $\mathbf{F} = y\mathbf{i} + xz^3\mathbf{j} - zy^3\mathbf{k}$,

    C is the circle $x^2 + y^2 = 4, z = 1.5$.　　　　　(*Delhi, 1999*) **Ans.** $19\pi/2$.
    What is the physical meaning of this integral if $\mathbf{F}$ defines a field of force.

    [**Hint.** $\oint_C \mathbf{F} \cdot d\mathbf{r} = \iint_S (\nabla \times \mathbf{F}) \cdot \mathbf{n}\, dA$

    $\nabla \times \mathbf{F} = -\left(3y^2z + 3xz^2\right)\mathbf{i} + \left(z^3 - 1\right)\mathbf{k}$ and $\mathbf{n} = \mathbf{k}$

    Therefore, $(\nabla \times \mathbf{F}) \cdot \mathbf{n} = \left(z^3 - 1\right)\mathbf{k} \cdot \mathbf{k} = z^3 - 1$.

    Thus $\iint_S \nabla \times \mathbf{F} \cdot \mathbf{n}\, dA = \iint_S \left(z^3 - 1\right) dA = \dfrac{19}{8} \iint_S dA = \dfrac{19}{8} \times 4\pi = \dfrac{19\pi}{2}$.

    ($\because z = 1.5$ on the surface of the given circle)]

2. Verify Stokes's theorem for a vector field defined by $\mathbf{F} = \left(x^2 - y^2\right)\mathbf{i} + 2xy\mathbf{j}$
    in the rectangular region in the xy-plane bounded by the lines
    $x = 0, x = a, y = 0$ and $y = b$　.　　　　　**Ans.** Common value = $2\ ab^2$.

3. Verify Stokes's theorem for the vector field $\mathbf{F} = 4xz\mathbf{i} - y^2\mathbf{j} + yz\mathbf{k}$ over the
    area in the plane $z = 0$ bounded by $x = 0, y = 0$ and $x^2 + y^2 = 1$.

4. Apply Stokes's theorem to evaluate $\iint_S (\nabla \times \mathbf{F}) \cdot \mathbf{n}\, dA$, where $\mathbf{F} = z\mathbf{i} + x\mathbf{j} + y\mathbf{k}$
    and S is the sphere $x^2 + y^2 + z^2 = 1$ above the xy-plane. Also verify Stokes's
    theorem.　　　　　　　　　　　　　　　　　　　　**Ans.** π.

5. Verify Stokes's theorem for the vector function $\mathbf{F}(x, y, z) = z\mathbf{i} + x\mathbf{j} + y\mathbf{k}$.
    Assume that the surface S is oriented upward. S is the portion of the plane
    $2x + y + 2z = 6$ in the first octant.

6. Verify Stokes's theorem for $\mathbf{F} = \left(x^2 + y^2\right)\mathbf{i} - 2xy\mathbf{j}$ taken round the rectan-
    gle bounded by the lines $x = \pm a, y = 0, y = b$.
    　　　　　(*VTU 2007; Raipur 2005; UPTU 2003; PTU 2003; JNTU 2003*)
    　　　　　　　　　　　　　　　　　　　Ans. Common value = $-4ab^2$.

7. Use Stokes's theorem or otherwise, evaluate

$\oint_C (2x-y)dx - yz^2 dy - y^2 z\, dz$, where C is the circle $x^2 + y^2 = 1$, corresponding to the surface of sphere of unit radius.

(MDU 2007; PTU 2003; UPTU 2002) **Ans.** π.

8. Evaluate $\oint_C \mathbf{F} \cdot d\mathbf{r}$ by Stokes's theorem, where $\mathbf{F} = y^2 \mathbf{i} + x^2 \mathbf{j} - (x+z)\mathbf{k}$ and C is the boundary of the triangle with vertices at $(0, 0, 0)$, $(1, 0, 0)$ and $(1, 1, 0)$. *(AMIE, S-2005, UPTU 2001)* **Ans.** $1/3$.

[**Hint.** $\oint \mathbf{F} \cdot d\mathbf{r} = \iint_S \operatorname{curl} \mathbf{F} \cdot \mathbf{n}\, dA = 2\int_0^1 \int_{y=0}^x (x-y)\, dy\, dx = \dfrac{1}{3}.$]

9. Compute the circulation of vector field $\mathbf{v} = x^2 y^3 \mathbf{i} + \mathbf{j} + z\mathbf{k}$ along the circle $x^2 + y^2 = a^2, z = 0$ with the help of Stokes's theorem.

[**Hint.** Circulation of the field $= \oint \mathbf{v} \cdot d\mathbf{r} = \iint_S (\nabla \times \mathbf{v}) \cdot \mathbf{n}\, dA$ (by Stokes's theorem)

Now $\nabla \times \mathbf{v} = \begin{vmatrix} \mathbf{i} & \mathbf{j} & \mathbf{k} \\ \partial/\partial x & \partial/\partial y & \partial/\partial z \\ x^2 y^3 & 1 & z \end{vmatrix} = -3x^2 y^2 \mathbf{k},$

and $(\nabla \times \mathbf{v}) \cdot \mathbf{n} = -3x^2 y^2 \mathbf{k} \cdot \mathbf{k} = -3x^2 y^2.$

$\therefore \iint_S (\nabla \times \mathbf{v}) \cdot \mathbf{n}\, dA = -4\int_0^a \int_0^{\sqrt{a^2 - x^2}} 3x^2 y^2\, dy\, dx = -\dfrac{\pi a^6}{8}.$]

10. State Stokes's theorem. Hence or otherwise evaluate $\iint_S (\operatorname{curl} \mathbf{v}) \cdot \mathbf{n}\, dA.$. If $\mathbf{v} = y\mathbf{i} + z\mathbf{j} + x\mathbf{k}$ and S being part of surface of paraboloid $z = 1 - x^2 - y^2$, $z \geq 0$ and $(\operatorname{curl} \mathbf{v}) \cdot \mathbf{n}$ denotes the normal component of curl $\mathbf{v}$ on surface element. **Ans.** $-\pi$.

[**Hint.** $\iint_S (\operatorname{curl} \mathbf{v}) \cdot \mathbf{n}\, dA = \iint_{S_1} (\operatorname{curl} \mathbf{v}) \cdot \mathbf{n}\, dA$, where S_1 is the area of circle $x^2 + y^2 = 1$. It can be easily shown that $\iint (\operatorname{curl} \mathbf{v}) \cdot \mathbf{n}\, dA = -\pi.$. Because $\nabla \times \mathbf{v} = -\mathbf{j} - \mathbf{k}, \mathbf{n} = \mathbf{k}$ therefore $(\nabla \times \mathbf{v}) \cdot \mathbf{n} = -1.$]

11. Verify the Stokes's theorem for the function $\mathbf{F} = (x - y - z)\mathbf{i} + (y - z - x)\mathbf{j}$ $+ (z - x - y)\mathbf{k}$ over the enclosed surface of the cylinder $\dfrac{x^2}{a^2} + \dfrac{y^2}{b^2} = 1,$

bounded by the plane $z = h$ and open at the end $z = 0$.

Ans. Common value $= 0$

12. Find the value of $\iint_S (\nabla \times \mathbf{F}) \cdot \mathbf{n} \, dA$ taken over the upper portion of the surface $x^2 + y^2 - 2ax + az = 0$ and the bounding curve lies in the plane $z = 0$, where $\mathbf{F} = \left(y^2 + z^2 - x^2\right)\mathbf{i} + \left(z^2 + x^2 - y^2\right)\mathbf{j} + \left(x^2 + y^2 - z^2\right)\mathbf{k}$.

(IETE Dec 2002) **Ans.** $2\pi a^3$.

[**Hint.** By Stokes's Theorem $\iint_S (\nabla \times \mathbf{F}) \cdot \mathbf{n} \, dA = \oint_C \mathbf{F} \cdot d\mathbf{r}$ where C is the circle $x^2 + y^2 - 2ax = 0$, that is, $(x - a)^2 + y^2 = a^2$ with parametric equations $x = a + a\cos\theta$, $y = a\sin\theta$. Also

$$\mathbf{F} \cdot d\mathbf{r} = \left\{\left(y^2 - x^2\right)\mathbf{i} + \left(x^2 - y^2\right)\mathbf{j}\right\} \cdot \left(\mathbf{i} \, dx + \mathbf{j} \, dy\right)]$$

13. Apply Stokes's theorem to evaluate $\oint_C y \, dx + z \, dy + x \, dz$, where C is the curve of intersection of $x^2 + y^2 + z^2 = a^2$ and $x + z = a$.

(JNTU 1999) **Ans.** $-\pi a^2 / \sqrt{2}$.

14. If $\mathbf{F} = 2yz\mathbf{i} - (x + 3y - 2)\mathbf{j} + \left(x^2 + z\right)\mathbf{k}$, evaluate $\iint_S (\nabla \times \mathbf{F}) \cdot \mathbf{n} \, dA$ over the surface of intersection of the cylinders $x^2 + y^2 = a^2, x^2 + z^2 = a^2$ which is included in the first octant.

[**Hint.** Refer Fig. 14.39. $\iint_S (\nabla \times \mathbf{F}) \cdot \mathbf{n} \, dA = \oint_C \mathbf{F} \cdot d\mathbf{r}$

$= \int_{C_1} \mathbf{F} \cdot d\mathbf{r} + \int_{C_2} \mathbf{F} \cdot d\mathbf{r} + \int_{C_3} \mathbf{F} \cdot d\mathbf{r} + \int_{C_4} \mathbf{F} \cdot d\mathbf{r}$

$= \dfrac{a^2}{2} + \left(\dfrac{3a^2}{2} - 2a\right) + \left(-\dfrac{2a^3}{3} - \dfrac{a^2}{2}\right)$

$+ \left(-\dfrac{\pi a^2}{4} - \dfrac{3}{2}a^2 + 2a\right)$

$= -\dfrac{a^2}{12}(3\pi + 8a)].$

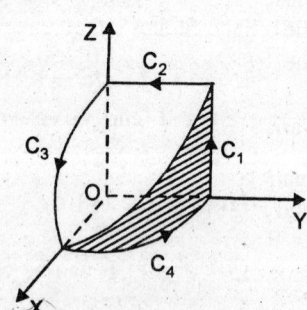

Fig. 14.39.

Chapter 15

Linear Algebra: Matrices*, Vectors, Determinants, Linear Systems of Equations

15.1 Introduction

Linear algebra includes the theory and applications of linear systems of equations, linear transformations and eigenvalue problems. Linear algebra makes systematic use of vectors and matrices and, to a lesser extent, determinants. These days, matrix theory occupies an important place and has applications in almost all branches of Engineering and physical sciences. Certain mathematical operations and results can be expressed in elegant and compact form by using matrix algebra.

15.2 Definition

A **matrix** is a rectangular array of numbers (real or complex) or functions enclosed within square brackets. These numbers (real or complex) or functions in the array are called **entries** or **elements** of the matrix. Each element or entry of the matrix can be a real or a complex number or a function of one or more variables or any other object. For example $\mathbf{A} = \begin{bmatrix} 4 & -7 & 5 \\ 8 & 1 & 6 \end{bmatrix}$ is a 2 × 3 matrix or matrix of order 2 × 3 (read as 2 by 3 matrix). It has two *"rows"* (horizontal lines) and three *columns* (vertical lines).

If a matrix has m rows and n columns, we say that its **size** is m by n (written $m \times n$).

General notations and concepts

The matrices are usually denoted by single capital boldface letters $\mathbf{A}, \mathbf{B}, \mathbf{C},...$ etc. Thus, in general

---

* The theory of matrices was the invention of the eminent English mathematician Arthur Cayley (1821-1895).

$$A = \begin{bmatrix} a_{11} & a_{12} & a_{13} & \cdots & a_{1n} \\ a_{21} & a_{22} & a_{23} & \cdots & a_{2n} \\ \vdots & & & & \vdots \\ a_{m1} & a_{m2} & a_{m3} & \cdots & a_{mn} \end{bmatrix} \qquad ...(15.1)$$

Minor diagonal —————— Principal diagonal

is an $m \times n$ matrix

Here each element has two subscripts. The first subscript indicates the row and the second indicates the column in which the element is located. For example a_{12} is the element in the first row and the second column and a_{ij} is the element in the ith row and jth column. This element is also known as the (i, j)th element. All elements in the second row have 2 as first subscript and each element in the 3rd column has 3 as second subscript. An $m \times n$ matrix **A** above is abbreviated as $A = (a_{ij})_{m \times n}$. Sometimes double bars ‖ ‖ or pair of parentheses ( ) are used in place of square brackets in indicating a matrix.

If all the elements of a matrix are real, it is called a **real matrix**, whereas if one or more elements of a matrix are complex, it is called a **complex matrix**.

Remark. A matrix, by definition, is simply an arrangement of numbers and has no numerical value.

Some physical examples of matrices

(a) The coordinates of a point can be written in the form of a 1×3 matrix $[x_1 \ y_1 \ z_1]$.

(b) The direction-ratios of a line can be written in the form of a matrix $[a \ b \ c]$.

(c) The co-efficients of x, y, z in the equations:

$$a_1x + b_1y + c_1z = d_1, \ a_2x + b_2y + c_2z = d_2, \ a_3x + b_3y + c_3z = d_3$$

can be written in the form of a 3×3 matrix $\begin{bmatrix} a_1 & b_1 & c_1 \\ a_2 & b_2 & c_2 \\ a_3 & b_3 & c_3 \end{bmatrix}$,

which is called coefficient matrix.

15.3 Some types of matrices

Row vector A matrix of order $1 \times n$, that is, it has exactly one row and n columns is called a *row vector* or a *row matrix* of order n and is written as

$[a_{11} \ a_{12} \ a_{13}... \ a_{1n}]$, or $[a_1 \ a_2 \ a_3... \ a_n]$ in which a_{1j} (or a_j) is the jth element.

Column vector A matrix of order $m \times 1$, that is, it has m rows and exactly one column is called a *column vector* or a *column matrix* of order m and is written as $\begin{bmatrix} b_{11} \\ b_{21} \\ \vdots \\ b_{m1} \end{bmatrix}$, or $\begin{bmatrix} b_1 \\ b_2 \\ \vdots \\ b_m \end{bmatrix}$ in which b_{j1} (or b_j) is the jth element.

The number of elements in a row/column vector is called its *order*. The vectors are usually denoted by boldface lower case letters such as **a, b, c,**... etc.

Rectangular matrix A matrix **A** of order $m \times n$, $m \neq n$ is called a *rectangular matrix*

Square matrix A matrix **A** of order $m \times n$ in which $m = n$ that is, number of rows is equal to the number of columns is called a *square matrix* of order n.

For example $\mathbf{B} = \begin{bmatrix} 4 & 3 \\ 1 & 2 \end{bmatrix}$ and $\mathbf{C} = \begin{bmatrix} a & b & c \\ d & e & f \\ g & h & i \end{bmatrix}$ are square matrices of order 2 and 3 respectively.

The elements a_{ii}, that is the elements a_{11}, a_{22}, ..., a_{nn} are called the *diagonal elements* and the line on which these elements lie is called the **Principal diagonal** or the **main** or **leading** diagonal of the matrix. The sum of the diagonal elements of a square matrix is called the **trace** of the matrix. The *trace* of **A**, written $tr\,(\mathbf{A})$, is the sum of the diagonal elements. Namely $tr(\mathbf{A}) = a_{11} + a_{22} + a_{33} + ... + a_{nn}$. In the above examples $tr(\mathbf{B}) = 4 + 2 = 6$, $tr(\mathbf{C}) = a + e + i$.

Null matrix A matrix **A** of order $m \times n$ in which all the elements are zero is called a *null matrix* or a *zero matrix* and is denoted by **O**. For example,

$\mathbf{O} = \begin{bmatrix} 0 & 0 & 0 \\ 0 & 0 & 0 \end{bmatrix}$ is a 2×3 null matrix, $\mathbf{O} = \begin{bmatrix} 0 & 0 \\ 0 & 0 \\ 0 & 0 \end{bmatrix}$ is a 3×2 null matrix, etc.

Diagonal matrix A square matrix **A** having non-zero elements in the principal diagonal positions and zero in all other positions, is called a *diagonal matrix*. A diagonal matrix of order n with diagonal elements a_{11}, a_{22}, ..., a_{nn} is denoted by $\mathbf{D} = \text{diag } [a_{11}, a_{22}, a_{33} ... a_{nn}]$.

If all the elements of a diagonal matrix of order n are equal, that is $a_{ii} = k$ for all i, (k being some number) then the matrix

$\mathbf{A} = \begin{bmatrix} k & 0 & \cdots & 0 \\ 0 & k & & 0 \\ & & \ddots & \\ 0 & 0 & & k \end{bmatrix}$ is called a **scalar matrix** of order n.

If all the elements of a diagonal matrix of order n are 1, then the matrix

$$A = \begin{bmatrix} 1 & & & 0 \\ & 1 & & \\ & & \ddots & \\ 0 & & & 1 \end{bmatrix}$$ is called an **unit matrix** or an **identity matrix** of order n.

An identity matrix of order n is denoted by $\mathbf{I_n}$ or more simply by $\mathbf{I}$.

For example, $\begin{bmatrix} 1 & 0 \\ 0 & 1 \end{bmatrix}$ is a unit matrix of order 2 and is denoted by $\mathbf{I_2}$, whereas

$\begin{bmatrix} 1 & 0 & 0 \\ 0 & 1 & 0 \\ 0 & 0 & 1 \end{bmatrix}$ is a unit matrix of order 3 and is denoted by $\mathbf{I_3}$.

Triangular matrices   A square matrix $\mathbf{A} = (a_{ij})$ is called a *lower triangular matrix* if $a_{ij} = 0$ for $i < j$, that is all elements above the principal diagonal are zero and an *upper triangular matrix* if $a_{ij} = 0$ for $i > j$, that is all the elements below the principal diagonal are zero.

Examples of lower and upper triangular matrices are given below:

$$\begin{pmatrix} 4 & 0 \\ -2 & 3 \end{pmatrix}, \quad \begin{pmatrix} 3 & 0 & 0 \\ 8 & -2 & 0 \\ 7 & 5 & 4 \end{pmatrix}, \quad \begin{pmatrix} 3 & 0 & 0 & 0 \\ 8 & -4 & 0 & 0 \\ 3 & 0 & 2 & 0 \\ 2 & 4 & 3 & 6 \end{pmatrix}$$

Lower triangular matrices

$$\begin{pmatrix} 1 & 4 \\ 0 & 3 \end{pmatrix}, \quad \begin{pmatrix} 1 & 2 & 8 \\ 0 & 4 & 2 \\ 0 & 0 & 5 \end{pmatrix}, \quad \begin{pmatrix} 4 & 2 & 2 & 0 \\ 0 & -5 & 4 & 1 \\ 0 & 0 & 3 & -6 \\ 0 & 0 & 0 & -7 \end{pmatrix}$$

Upper triangular matrices

It is to be noted that there is no restriction on the elements on or below the principal diagonal for an lower triangular matrix; they may be any number, even zero. The elements above the principal diagonal must be zero. Similarly, for upper triangular matrix, there is no restriction on the elements on or above the principal diagonal. The null matrix, the diagonal matrix, and the identity matrix, defined above, are in the upper triangular form. The identity matrix is both upper as well as lower triangular.

Submatrix   If a finite number of rows or columns are deleted in a matrix the resulting matrix is known as a *submatrix*.

Consider the matrix $A = \begin{bmatrix} 6 & 7 & 8 & 5 \\ 4 & 3 & 2 & -1 \\ 1 & 0 & 7 & 5 \end{bmatrix}$. A is a 3×4 matrix. If we delete the

first column, we get $\begin{bmatrix} 7 & 8 & 5 \\ 3 & 2 & -1 \\ 0 & 7 & 5 \end{bmatrix}$, which is a 3×3 matrix. If we delete the

second row, we get $\begin{bmatrix} 6 & 7 & 8 & 5 \\ 1 & 0 & 7 & 5 \end{bmatrix}$, which is a 2×4 matrix. If we delete the

second column and third row, we get $\begin{bmatrix} 6 & 8 & 5 \\ 4 & 2 & -1 \end{bmatrix}$ which is a 2×3 matrix. Thus

by deleting some rows and columns of a given matrix, we get different matrices. Each of these is called a *sub-matrix* of A.

Remark. The matrix A itself is also a submatrix of A obtained by omitting no rows or columns.

Equal matrices Two matrices $A = (a_{ij})_{m \times n}$ and $B = (b_{ij})_{p \times q}$ are said to be equal, when

(*i*) they are of the same order, that is $m = p$, $n = q$ and
(*ii*) their corresponding elements are equal, that is $a_{ij} = b_{ij}$ for each i, and j.
Matrices of different sizes cannot be equal.

Two matrices are said to be *conformable* if they are of the same order, that is if they have the same number of rows and the same number of columns.

Example. Find the values of x, y, z, t which satisfy the matrix equation.

(A.M.I.E W 2005)

$$\begin{bmatrix} x+3 & 2y+x \\ z-1 & 4t-6 \end{bmatrix} = \begin{bmatrix} 0 & -7 \\ 3 & 2t \end{bmatrix}.$$

Solution. By the definition of equality of matrices, the four corresponding entries must be equal. Thus:

$x + 3 = 0$, $2y + x = -7$, $z - 1 = 3$, and $4t - 6 = 2t$.

Solving the above system of equations yields $x = -3$, $y = -2$, $z = 4$, and $t = 3$.

15.4 Matrix Algebra—Operations on matrices

(a) Scalar multiplication (multiplication of a matrix by a number)

If $A = (a_{ij})$ be a given matrix of order $m \times n$ and k be a scalar (real or complex number), then B is the matrix kA of order $m \times n = (ka_{ij})$, obtained by multiplying each entry in **A** by k. The order of the new matrix **B** is same as that of the matrix **A**.

We also define $-\mathbf{A} = (-1)\mathbf{A}$. Similarly $(-k)\mathbf{A}$ is written $-k\mathbf{A}$. The matrix $-\mathbf{A}$ is called the *negative* of the matrix **A**.

(b) Addition/subtraction of two matrices

If **A** and **B** are two $m \times n$ matrices then the sum of **A** and **B**, written $\mathbf{A} + \mathbf{B}$, is defined as an $m \times n$ matrix **C** such that each element of **C** is the sum of the corresponding elements of **A** and **B**. The sum of matrices with different sizes is not defined.

Thus, if $\mathbf{A} = \begin{bmatrix} 6 & 5 & 4 \\ -2 & 3 & 7 \end{bmatrix}$, $\mathbf{B} = \begin{bmatrix} 4 & 1 & -5 \\ 3 & 9 & 2 \end{bmatrix}$

then $\mathbf{C} = \mathbf{A} + \mathbf{B} = \begin{bmatrix} 6+4 & 5+1 & 4-5 \\ -2+3 & 3+9 & 7+2 \end{bmatrix} = \begin{bmatrix} 10 & 6 & -1 \\ 1 & 12 & 9 \end{bmatrix}$.

Simlarly the difference $\mathbf{A}-\mathbf{B}$ of the matrices **A** and **B** is a matrix **C** each element of which is obtained by subtracting the elements of **B** from the corresponding elements of **A**.

Thus if $\mathbf{A} = \begin{bmatrix} 6 & 2 \\ 7 & -5 \end{bmatrix}$, $\mathbf{B} = \begin{bmatrix} 8 & 1 \\ 3 & 4 \end{bmatrix}$

then $\mathbf{C} = \mathbf{A} - \mathbf{B} = \begin{bmatrix} 6 & 2 \\ 7 & -5 \end{bmatrix} - \begin{bmatrix} 8 & 1 \\ 3 & 4 \end{bmatrix} = \begin{bmatrix} 6-8 & 2-1 \\ 7-3 & -5-4 \end{bmatrix} = \begin{bmatrix} -2 & 1 \\ 4 & -9 \end{bmatrix}$

Matrices of the same order are said to be *conformable* for addition/subtraction.

Properties of the matrix addition and scalar multiplication

Let **A**, **B**, **C** be the matrices which are conformable for addition and k, λ be scalars we state:

1. $\mathbf{A} + \mathbf{B} = \mathbf{B} + \mathbf{A}$ (*i.e.*, Commutative law holds good)
2. $(\mathbf{A} + \mathbf{B}) + \mathbf{C} = \mathbf{A} + (\mathbf{B} + \mathbf{C})$ (Associative law holds good)
3. $\mathbf{A} + \mathbf{O} = \mathbf{A}$ (**O** is the null matrix of the same order as **A**)
4. $\mathbf{A} + (-\mathbf{A}) = \mathbf{O}$
5. $k(\mathbf{A} + \mathbf{B}) = k\mathbf{A} + k\mathbf{B} = (\mathbf{A} + \mathbf{B})k$ (*i.e.*, Distributive law holds good)
6. $(k + \lambda)\mathbf{A} = k\mathbf{A} + \lambda\mathbf{A}$.
7. $k(\lambda\mathbf{A}) = (k\lambda)\mathbf{A} = k\lambda\mathbf{A}$.   (Associative law for multiplication by scalars)
8. $1 \times \mathbf{A} = \mathbf{A}$ and $0 \times \mathbf{A} = \mathbf{O}$

(c) Multiplication of two matrices

The product **AB** of two matrices **A** and **B** is defined only when the number of columns in **A** is equal to the number of rows in **B**. Such matrices are said to be *conformable* for multiplication.

Let $\mathbf{A} = (a_{ij})$ be an $m \times p$ matrix and $\mathbf{B} = (b_{ij})$ be an $p \times n$ matrix.

Then the product matrix $\mathbf{C} = (c_{ij}) = \mathbf{AB}$ is the $m \times n$ matrix whose ij-entry is obtained by multiplying the ith rows of $\mathbf{A}$ by the jth column of $\mathbf{B}$. Figure 15.1 illustrates this.

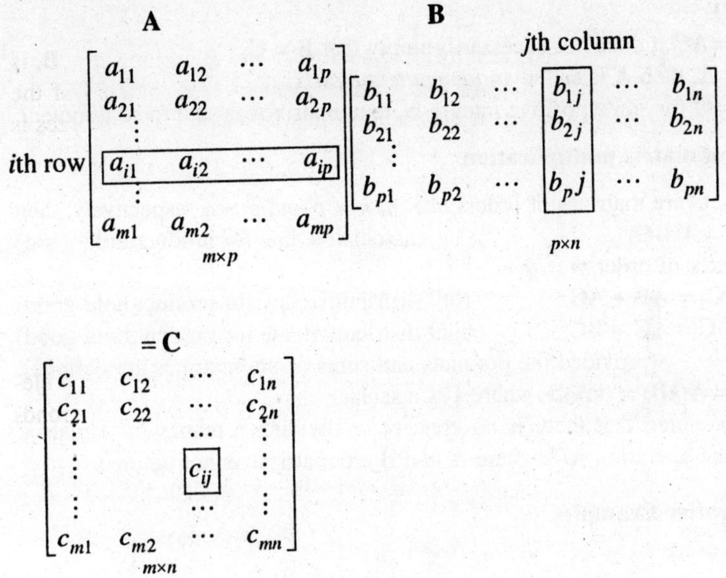

Fig. 15.1 Matrix multiplication $\mathbf{AB} = \mathbf{C}$

where $c_{ij} = \left(\displaystyle\sum_{k=1}^{p} a_{ik} b_{kj} \right)_{m \times n} = a_{i1} b_{1j} + a_{i2} b_{2j} + \cdots + a_{ip} b_{pj}.$ ...(15.2)

The product $\mathbf{AB}$ is not defined if $\mathbf{A}$ is an $m \times p$ matrix and $\mathbf{B}$ is a $q \times n$ matrix, where $p \neq q$.

In the product $\mathbf{AB}$, the matrix $\mathbf{B}$ is said to be pre-multiplied by $\mathbf{A}$ or $\mathbf{A}$ is said to be post-multiplied by $\mathbf{B}$.

The product $\mathbf{AB}$ of a row matrix $\mathbf{A}$ of order $1 \times n$ and column matrix $\mathbf{B}$ of order $n \times 1$ is a matrix of order 1×1, that is a single element, and $\mathbf{BA}$ is a matrix of order $n \times n$.

Imp. Notes:

1. It is possible that for two given matrices $\mathbf{A}$ and $\mathbf{B}$, the product matrices $\mathbf{AB}$ and $\mathbf{BA}$ may or may not exist and may be equal or different.

   For example, if $\mathbf{A}$ is a $m \times n$ matrix and $\mathbf{B}$ is $n \times k$ matrix then the product matrix $\mathbf{AB}$ is defined and is a matrix of order $m \times k$, whereas the product matrix $\mathbf{BA}$ is not defined.

2. If both the product matrices **AB** and **BA** are defined, then both the matrices **AB** and **BA** are square matrices. In general **AB** ≠ **BA**. Thus, the matrix product is not commutative.

   If **AB** = **BA**, then the matrices **A** and **B** are said to *commute* with each other.

3. If **AB** = **O**, then it does not necessarily imply that either **A** = **O** or **B** = **O** or **BA** = **O**.

4. If **AB** = **AC** it does not necessarily imply that **B** = **C**.

5. If $A^2 = A$, then **A** is called an *idempotent matrix*.

6. If $A^m = O$ for some positive integer m, then matrix **A** is said to be nilpotent.

Properties of matrix multiplication

1. If **A, B, C** are matrices of orders $m \times n$, $n \times p$ and $p \times q$ respectively, then
   $$(AB)C = A(BC) \qquad \text{(associative law for product hold good)}$$
   is a matrix of order $m \times q$.

2. $A(B + C) = AB + AC$ (left distributive law for product hold good)

3. $(A + B)C = AC + BC$ (right distributive law for product hold good)
   (provided the products and sums of the matrices are defined).

4. $k(AB) = A(kB) = (kA)B$, where k is a scalar.

   It may be noted that there is no concept of dividing a matrix by a matrix. Therefore, the operation **A/B** where **A** and **B** are matrices is not defined.

15.5 Illustrative Examples

Example. 15.1 (*a*) If $A = \begin{bmatrix} 10 \\ 3 \\ 12 \end{bmatrix}, B = \begin{bmatrix} 5 \\ -7 \\ 9 \end{bmatrix}$ and $C = \begin{bmatrix} -9 \\ 4 \\ -8 \end{bmatrix}$ **find A − B − C.**

(*b*) If $A = \begin{bmatrix} 1 & 2 & -1 \\ 0 & -1 & 3 \end{bmatrix}$ and $B = \begin{bmatrix} 2 & 3 & 1 \\ 0 & -1 & 5 \end{bmatrix}$, find $2A - 3B$.

Solution. (**a**) We have $A - B - C = \begin{bmatrix} 10 \\ 3 \\ 12 \end{bmatrix} - \begin{bmatrix} 5 \\ -7 \\ 9 \end{bmatrix} - \begin{bmatrix} -9 \\ 4 \\ -8 \end{bmatrix}$.

First perform the scalar multiplication and then a matrix addition.

$$A - B - C = A + (-B) + (-C) = \begin{bmatrix} 10 \\ 3 \\ 12 \end{bmatrix} + \begin{bmatrix} -5 \\ 7 \\ -9 \end{bmatrix} + \begin{bmatrix} 9 \\ -4 \\ 8 \end{bmatrix} = \begin{bmatrix} 10-5+9 \\ 3+7-4 \\ 12-9+8 \end{bmatrix} = \begin{bmatrix} 14 \\ 6 \\ 11 \end{bmatrix}.$$

(**b**) First perform the scalar multiplication and then a matrix addition.

$$2A - 3B = 2A + (-3B) = \begin{bmatrix} 2 & 4 & -2 \\ 0 & -2 & 6 \end{bmatrix} + \begin{bmatrix} -6 & -9 & -3 \\ 0 & 3 & -15 \end{bmatrix} = \begin{bmatrix} -4 & -5 & -5 \\ 0 & 1 & -9 \end{bmatrix}$$

Note: The matrix $2A - 3B$ i.e., $2A + (-3B)$ is called a *linear combination* of matrices **A** and **B**.

Example 15.2 If $P = \begin{bmatrix} 3 & 5 & 7 \\ 9 & 3 & 2 \\ 1 & 2 & 5 \end{bmatrix}$, and $Q = \begin{bmatrix} 3 & 5 & 4 \\ 4 & 7 & 8 \\ 3 & 2 & 1 \end{bmatrix}$, find the matrix R such

that $3P + 2Q - 2R$ is a null matrix.

Solution.

We have $3P + 2Q - 2R = \begin{bmatrix} 9 & 15 & 21 \\ 27 & 9 & 6 \\ 3 & 6 & 15 \end{bmatrix} + \begin{bmatrix} 6 & 10 & 8 \\ 8 & 14 & 16 \\ 6 & 4 & 2 \end{bmatrix} - 2R = \begin{bmatrix} 0 & 0 & 0 \\ 0 & 0 & 0 \\ 0 & 0 & 0 \end{bmatrix}$

$\Rightarrow 2R = \begin{bmatrix} 15 & 25 & 29 \\ 35 & 23 & 22 \\ 9 & 10 & 17 \end{bmatrix}$ or $R = 1/2 \begin{bmatrix} 15 & 25 & 29 \\ 35 & 23 & 22 \\ 9 & 10 & 17 \end{bmatrix}$.

Example 15.3 Find x, y, z, w where

$$3 \begin{bmatrix} x & y \\ z & w \end{bmatrix} = \begin{bmatrix} x & 6 \\ -1 & 2w \end{bmatrix} + \begin{bmatrix} 4 & x+y \\ z+w & 3 \end{bmatrix}.$$

Solution. Write each side as a single equation we have

$$\begin{bmatrix} 3x & 3y \\ 3z & 3w \end{bmatrix} = \begin{bmatrix} x+4 & x+y+6 \\ z+w-1 & 2w+3 \end{bmatrix}.$$

By the definition of equality of matrices, the four corresponding entries must be equal. Thus, we obtain the following four equations:

$3x = x + 4$, $3y = x + y + 6$, $3z = z + w - 1$, $3w = 2w + 3$.

Solving the above system of equations yields $x = 2$, $y = 4$, $z = 1$, $w = 3$.

Example 15.4 Let $A = \begin{bmatrix} 1 & 2 & 3 \end{bmatrix}$, $B = \begin{bmatrix} 1 & 2 \\ -1 & 0 \\ 1 & 3 \end{bmatrix}$, and $C = \begin{bmatrix} 3 \\ 1 \\ 5 \end{bmatrix}$.

Compute the ones which are defined: CA, $AC + B$, $AB + C$.

Solution. Since C is a matrix of order 3×1 and A is a matrix of order 1×3, CA is defined and is a matrix of order 3×3.

$$CA = \begin{bmatrix} 3 \\ 1 \\ 5 \end{bmatrix} \begin{bmatrix} 1 & 2 & 3 \end{bmatrix} = \begin{bmatrix} 3 & 6 & 9 \\ 1 & 2 & 3 \\ 5 & 10 & 15 \end{bmatrix}.$$

A is a 1×3 matrix and C is a 3×1 matrix, AC is defined and is a 1×1 matrix. But B is 3×2 matrix, therefore $AC + B$ is not defined.

Further, AB is 1×2 matrix but C is 3×1 matrix, therefore $AB + C$ is not defined.

Example 15.5 If $A = \begin{bmatrix} 1 & 2 \\ 3 & 4 \end{bmatrix}$, show that $A^2 - 5A = 2I$, where I is the unit matrix

of order 2. Hence, or otherwise determine A^4.

Solution. We have $A^2 = AA = \begin{bmatrix} 1 & 2 \\ 3 & 4 \end{bmatrix}\begin{bmatrix} 1 & 2 \\ 3 & 4 \end{bmatrix} = \begin{bmatrix} 7 & 10 \\ 15 & 22 \end{bmatrix}$.

$\therefore A^2 - 5A = A^2 + (-5A) = \begin{bmatrix} 7 & 10 \\ 15 & 22 \end{bmatrix} + \begin{bmatrix} -5 & -10 \\ -15 & -20 \end{bmatrix} = \begin{bmatrix} 2 & 0 \\ 0 & 2 \end{bmatrix} = 2\begin{bmatrix} 1 & 0 \\ 0 & 1 \end{bmatrix} = 2I$

We know $A^2 = 5A + 2I$. Hence $A^4 = A^2 A^2 = (5A + 2I)(5A + 2I)$

$\qquad = 25A^2 + 10AI + 10IA + 4I^2$

$\qquad = 25(5A + 2I) + 20A + 4I$, since $AI = IA = A$ and $I^2 = I$.

$\qquad = 145A + 54I$

$\qquad = 145\begin{bmatrix} 1 & 2 \\ 3 & 4 \end{bmatrix} + 54\begin{bmatrix} 1 & 0 \\ 0 & 1 \end{bmatrix} = \begin{bmatrix} 199 & 290 \\ 435 & 634 \end{bmatrix}$

Example 15.6 If $A = \begin{bmatrix} 2 & -1 \\ 1 & 0 \\ -3 & 4 \end{bmatrix}$ and $B = \begin{bmatrix} 1 & -2 & -5 \\ 3 & 4 & 0 \end{bmatrix}$, evaluate **AB** and **BA**

(AMIE S-2001)

Solution. Since **A** is a 3×2 matrix and **B** a 2×3 matrix, the product **AB** is defined and is a 3×3 matrix. Multiply the rows of **A** by the columns of **B** to obtain.

$AB = \begin{bmatrix} 2 & -1 \\ 1 & 0 \\ -3 & 4 \end{bmatrix}\begin{bmatrix} 1 & -2 & -5 \\ 3 & 4 & 0 \end{bmatrix} = \begin{bmatrix} 2(1)+(-1)3 & 2(-2)+(-1)4 & 2(-5)+0 \\ 1(1)+0 & 1(-2)+0 & 1(-5)+0 \\ (-3)(1)+4(3) & (-3)(-2)+4(4) & (-3)(-5)+0 \end{bmatrix}$

$= \begin{bmatrix} 2-3 & -4-4 & -10 \\ 1 & -2 & -5 \\ -3+12 & 6+16 & 15 \end{bmatrix} = \begin{bmatrix} -1 & -8 & -10 \\ 1 & -2 & -5 \\ 9 & 22 & 15 \end{bmatrix}$.

Further, product **BA** is defined and is a 2×2 matrix. Multiply the rows of **B** by the columns of **A** to obtain

$BA = \begin{bmatrix} 1 & -2 & -5 \\ 3 & 4 & 0 \end{bmatrix}\begin{bmatrix} 2 & -1 \\ 1 & 0 \\ -3 & 4 \end{bmatrix} = \begin{bmatrix} 2-2+15 & -1+0-20 \\ 6+4+0 & -3+0+0 \end{bmatrix} = \begin{bmatrix} 15 & -21 \\ 10 & -3 \end{bmatrix}$.

Note: $AB \neq BA$.

Example 15.7 Evaluate $A^2 - 3A + 9I$ if **I** is the unit matrix of order 3 and

$$A = \begin{bmatrix} 1 & -2 & 3 \\ 2 & 3 & -1 \\ -3 & 1 & 2 \end{bmatrix}. \qquad \text{(A.M.I.E., S-1999)}$$

Solution. $A^2 = AA = \begin{bmatrix} 1 & -2 & 3 \\ 2 & 3 & -1 \\ -3 & 1 & 2 \end{bmatrix}\begin{bmatrix} 1 & -2 & 3 \\ 2 & 3 & -1 \\ -3 & 1 & 2 \end{bmatrix}$

$$= \begin{bmatrix} 1(1)+(-2)2+3(-3) & 1(-2)+(-2)3+3(1) & 1(3)+(-2)(-1)+3(2) \\ 2(1)+3(2)+(-1)(-3) & 2(-2)+3(3)+(-1)1 & 2(3)+3(-1)+(-1)2 \\ (-3)1+1(2)+2(-3) & (-3)(-2)+1(3)+2(1) & (-3)3+1(-1)+2(2) \end{bmatrix}$$

$$= \begin{bmatrix} 1-4-9 & -2-6+3 & 3+2+6 \\ 2+6+3 & -4+9-1 & 6-3-2 \\ -3+2-6 & 6+3+2 & -9-1+4 \end{bmatrix} = \begin{bmatrix} -12 & -5 & 11 \\ 11 & 4 & 1 \\ -7 & 11 & -6 \end{bmatrix}$$

$$\therefore A^2 - 3A + 9I = \begin{bmatrix} -12 & -5 & 11 \\ 11 & 4 & 1 \\ -7 & 11 & -6 \end{bmatrix} + \begin{bmatrix} -3 & 6 & -9 \\ -6 & -9 & 3 \\ 9 & -3 & -6 \end{bmatrix} + \begin{bmatrix} 9 & 0 & 0 \\ 0 & 9 & 0 \\ 0 & 0 & 9 \end{bmatrix} = \begin{bmatrix} -6 & 1 & 2 \\ 5 & 4 & 4 \\ 2 & 8 & -3 \end{bmatrix}.$$

Example 15.8 Let $A = \begin{bmatrix} 2 & -5 \\ 3 & 1 \end{bmatrix}$ and let $f(x) = x^3 - 2x^2 - 5$. Find $f(A)$.

Solution. $A^2 = AA = \begin{bmatrix} 2 & -5 \\ 3 & 1 \end{bmatrix}\begin{bmatrix} 2 & -5 \\ 3 & 1 \end{bmatrix} = \begin{bmatrix} 4-15 & -10-5 \\ 6+3 & -15+1 \end{bmatrix} = \begin{bmatrix} -11 & -15 \\ 9 & -14 \end{bmatrix}$

$A^3 = A^2A = \begin{bmatrix} -11 & -15 \\ 9 & -14 \end{bmatrix}\begin{bmatrix} 2 & -5 \\ 3 & 1 \end{bmatrix} = \begin{bmatrix} -22-45 & 55-15 \\ 18-42 & -45-14 \end{bmatrix} = \begin{bmatrix} -67 & 40 \\ -24 & -59 \end{bmatrix}.$

Substitute A for x and $5I$ for constant term in $f(x)$, we obtain

$$f(A) = A^3 - 2A^2 - 5I = \begin{bmatrix} -67 & 40 \\ -24 & -59 \end{bmatrix} + \begin{bmatrix} 22 & 30 \\ -18 & 28 \end{bmatrix} + \begin{bmatrix} -5 & 0 \\ 0 & -5 \end{bmatrix} = \begin{bmatrix} -50 & 70 \\ -42 & -36 \end{bmatrix}.$$

Example 15.9 Find the diagonal and trace of each matrix:

(a) $P = \begin{bmatrix} 2 & -5 & 8 \\ 3 & -6 & -7 \\ 4 & 0 & -1 \end{bmatrix}$, (b) $Q = \begin{bmatrix} 4 & 3 & -6 \\ 2 & -5 & 0 \end{bmatrix}$.

Solution. (a) The diagonal elements of matrix P are 2, –6 and –1. The trace of P is the sum of these diagonal elements Thus $tr(P) = 2 - 6 - 1 = -5$.

(b) The diagonal and trace are only defined for square matrices, whereas the given matrix is rectangle of order 2 × 3.

Example 15.10 If $\begin{bmatrix} 4 \\ 1 \\ 3 \end{bmatrix} X = \begin{bmatrix} -4 & 8 & 4 \\ -1 & 2 & 1 \\ -3 & 6 & 3 \end{bmatrix}$, find X.

Solution. Let $X = \begin{bmatrix} a & b & c \end{bmatrix}$.

We have $\begin{bmatrix} 4 \\ 1 \\ 3 \end{bmatrix}\begin{bmatrix} a & b & c \end{bmatrix} = \begin{bmatrix} 4a & 4b & 4c \\ a & b & c \\ 3a & 3b & 3c \end{bmatrix} = \begin{bmatrix} -4 & 8 & 4 \\ -1 & 2 & 1 \\ -3 & 6 & 3 \end{bmatrix}$ (given)

By the definition of equality of matrices, we obtain $a = -1$, $b = 2$ and $c = 1$. Therefore, $X = [-1 \ 2 \ 1]$.

Example 15.11 If e^A is defined as $I + A + \dfrac{A^2}{2!} + \dfrac{A^3}{3!} + \cdots$ show that

$$e^A = e^x \begin{bmatrix} \cosh x & \sinh x \\ \sinh x & \cosh x \end{bmatrix} \text{ when } A = \begin{bmatrix} x & x \\ x & x \end{bmatrix}. \qquad (A.M.I.E., W\text{-}2002)$$

Solution. Here $A = \begin{bmatrix} x & x \\ x & x \end{bmatrix}$, Therefore,

$$A^2 = AA = \begin{bmatrix} x & x \\ x & x \end{bmatrix}\begin{bmatrix} x & x \\ x & x \end{bmatrix} = \begin{bmatrix} 2x^2 & 2x^2 \\ 2x^2 & 2x^2 \end{bmatrix} \text{ and }$$

$$A^3 = A^2 A = \begin{bmatrix} 2x^2 & 2x^2 \\ 2x^2 & 2x^2 \end{bmatrix}\begin{bmatrix} x & x \\ x & x \end{bmatrix} = \begin{bmatrix} 4x^3 & 4x^3 \\ 4x^3 & 4x^3 \end{bmatrix} \text{ etc.}$$

In general $A^n = \begin{bmatrix} 2^{n-1}\,x^n & 2^{n-1}\,x^n \\ 2^{n-1}\,x^n & 2^{n-1}\,x^n \end{bmatrix} = \dfrac{1}{2}\begin{bmatrix} (2x)^n & (2x)^n \\ (2x)^n & (2x)^n \end{bmatrix}$

Hence $e^A = I + A + \dfrac{A^2}{2!} + \dfrac{A^3}{3!} + \cdots$

$$= \begin{bmatrix} 1 & 0 \\ 0 & 1 \end{bmatrix} + \begin{bmatrix} x & x \\ x & x \end{bmatrix} + \frac{1}{2!}\begin{bmatrix} (2x)^2 & (2x)^2 \\ (2x)^2 & (2x)^2 \end{bmatrix} + \frac{1}{3!}\begin{bmatrix} 4x^3 & 4x^3 \\ 4x^3 & 4x^3 \end{bmatrix} + \cdots$$

$$= \frac{1}{2}\begin{bmatrix} 2 + 2x + \dfrac{(2x)^2}{2!} + \dfrac{(2x)^3}{3!} + \cdots & 2x + \dfrac{(2x)^2}{2!} + \dfrac{(2x)^3}{3!} + \cdots \\[2mm] 2x + \dfrac{(2x)^2}{2!} + \dfrac{(2x)^3}{3!} + \cdots & 2 + 2x + \dfrac{(2x)^2}{2!} + \dfrac{(2x)^3}{3!} + \cdots \end{bmatrix}$$

$$= \frac{1}{2}\begin{bmatrix} 1 + e^{2x} & e^{2x} - 1 \\ e^{2x} - 1 & 1 + e^{2x} \end{bmatrix} = e^x \begin{bmatrix} \dfrac{e^x + e^{-x}}{2} & \dfrac{e^x - e^{-x}}{2} \\[2mm] \dfrac{e^x - e^{-x}}{2} & \dfrac{e^x + e^{-x}}{2} \end{bmatrix} = e^x \begin{bmatrix} \cosh x & \sinh x \\ \sinh x & \cosh x \end{bmatrix}.$$

Exercise 15.1

1. (*a*) Evaluate $3\begin{bmatrix} 2 & 9 & -6 \\ 4 & -5 & 3 \end{bmatrix} - 2\begin{bmatrix} 1 & 6 & 2 \\ 4 & 3 & -5 \end{bmatrix}$,

(*b*) Given $A = \begin{bmatrix} 1 & 2 & -3 \\ 5 & 0 & 2 \\ 1 & -1 & 1 \end{bmatrix}$, $B = \begin{bmatrix} 3 & -1 & 2 \\ 4 & 2 & 5 \\ 2 & 0 & 3 \end{bmatrix}$, and $C = \begin{bmatrix} 4 & 1 & 2 \\ 0 & 3 & 2 \\ 1 & -2 & 3 \end{bmatrix}$,

Compute (*i*) $A + B$ and $A - C$ (*ii*) verify $A + (B - C) = (A + B) - C$.

Ans. (*a*) $\begin{bmatrix} 4 & 15 & -22 \\ 4 & -21 & 19 \end{bmatrix}$ (*b*) (*i*) $\begin{bmatrix} 4 & 1 & -1 \\ 9 & 2 & 7 \\ 3 & -1 & 4 \end{bmatrix}; \begin{bmatrix} -3 & 1 & -5 \\ 5 & -3 & 0 \\ 0 & 1 & -2 \end{bmatrix}$.

2. (*a*) If $A = \begin{bmatrix} 2 & -1 \\ 4 & 3 \end{bmatrix}$, $B = \begin{bmatrix} -1 & 1 \\ 2 & -4 \end{bmatrix}$, and $C = \begin{bmatrix} 1 & 4 \\ -2 & -1 \end{bmatrix}$, find $3A + 2B - 4C$.

(*b*) Find *x, y, z, t* such that $\begin{bmatrix} x+2 & 2y+x \\ z-1 & 4t+1 \end{bmatrix} = \begin{bmatrix} 0 & -6 \\ 3 & 2t \end{bmatrix}$

Ans. (*a*) $\begin{bmatrix} 0 & -17 \\ 24 & 5 \end{bmatrix}$ (*b*) $x = -2, y = -2, z = 4, t = -1/2$.

3. If $A = \begin{bmatrix} 1 & 2 \\ 3 & 4 \\ 5 & 6 \end{bmatrix}$, $B = \begin{bmatrix} -3 & -2 \\ 1 & -5 \\ 4 & 3 \end{bmatrix}$, find $D = \begin{bmatrix} p & q \\ r & s \\ t & u \end{bmatrix}$ such that $A + B - D = O$.

Ans. $\begin{bmatrix} -2 & 0 \\ 4 & -1 \\ 9 & 9 \end{bmatrix}$.

4. If $A = \begin{bmatrix} 0 & 1 & 2 \\ 1 & 2 & 3 \\ 2 & 3 & 4 \end{bmatrix}$, and $B = \begin{bmatrix} 1 & -2 \\ -1 & 0 \\ 2 & -1 \end{bmatrix}$, obtain the product

AB. Explain why BA is not defined? **Ans.** $\begin{bmatrix} 3 & -2 \\ 5 & -5 \\ 7 & -8 \end{bmatrix}$.

5. If $A = \begin{bmatrix} 1 & 0 & 0 \\ 0 & -2 & 3 \\ 2 & 1 & -1 \end{bmatrix}$, $B = \begin{bmatrix} 0 & 2 & -1 \\ 3 & 1 & 0 \\ 0 & -1 & 2 \end{bmatrix}$ and $C = \begin{bmatrix} 1 & 2 \\ -4 & -2 \\ 0 & 1 \end{bmatrix}$,

find (*a*) AB (*b*) BA (*c*) BC (*d*) CB (*e*) AC (*f*) CA.

Ans. (*a*) $\begin{bmatrix} 0 & 2 & -1 \\ -6 & -5 & 6 \\ 3 & 6 & -4 \end{bmatrix}$ (*b*) $\begin{bmatrix} -2 & -5 & 7 \\ 3 & -2 & 3 \\ 4 & 4 & -5 \end{bmatrix}$, (*c*) $\begin{bmatrix} -8 & -5 \\ -1 & 4 \\ 4 & 4 \end{bmatrix}$,

(*d*) Not defined (*e*) $\begin{bmatrix} 1 & 2 \\ 8 & 7 \\ -2 & 1 \end{bmatrix}$ (*f*) Not defined.

6. (*a*) If $A = \begin{bmatrix} 1 & -2 & 3 \\ 2 & 3 & -1 \\ -3 & 1 & 2 \end{bmatrix}$ and $B = \begin{bmatrix} 1 & 0 & 2 \\ 0 & 1 & 2 \\ 1 & 2 & 0 \end{bmatrix}$, obtain the products AB and

BA and show that $AB \neq BA$.

(b) Evaluate $\begin{bmatrix} 1 \\ -2 \\ 3 \end{bmatrix} \times \begin{bmatrix} 4 & 5 & 2 \end{bmatrix} \times \begin{bmatrix} 2 \\ -3 \\ 5 \end{bmatrix} \times \begin{bmatrix} 3 & 2 \end{bmatrix}$

Ans. (a) $AB = \begin{bmatrix} 4 & 4 & -2 \\ 1 & 1 & 10 \\ -1 & 5 & -4 \end{bmatrix}, BA = \begin{bmatrix} -5 & 0 & 7 \\ -4 & 5 & 3 \\ 5 & 4 & 1 \end{bmatrix}$; (b) $\begin{bmatrix} 9 & 6 \\ -18 & 2 \\ 27 & 18 \end{bmatrix}$.

7. (a) If $A = \begin{bmatrix} 2 & 1 & 1 \\ -1 & 1 & 4 \end{bmatrix}, B = \begin{bmatrix} 1 & 5 \\ 7 & 8 \\ -1 & 0 \end{bmatrix}$ and $C = \begin{bmatrix} -1 & 1 \\ 2 & -1 \end{bmatrix}$, verity that

$A(B + C) = AB + AC$ and $(AB) C = A (BC)$.

(b) If $E = \begin{bmatrix} 0 & 1 & 0 \\ 0 & 0 & 1 \\ 0 & 0 & 0 \end{bmatrix}$ and $F = \begin{bmatrix} 0 & 0 & 0 \\ 1 & 0 & 0 \\ 0 & 1 & 0 \end{bmatrix}$, calculate the matrices products

EF and FE and show that $E^2F + FE^2 = E$.

Ans. (a) Each is $\begin{bmatrix} 28 & -10 \\ 4 & -1 \end{bmatrix}$ (b) $EF = \begin{bmatrix} 1 & 0 & 0 \\ 0 & 1 & 0 \\ 0 & 0 & 0 \end{bmatrix}$ and $FE = \begin{bmatrix} 0 & 0 & 0 \\ 0 & 1 & 1 \\ 0 & 0 & 0 \end{bmatrix}$.

8. If $A = \begin{bmatrix} 2 & 1 & -1 \\ 1 & -2 & 3 \\ -2 & 1 & 2 \end{bmatrix}$ and $B = \begin{bmatrix} 1 & -1 & 2 \\ -2 & 1 & 3 \\ 2 & -1 & 1 \end{bmatrix}$, show that

$(A + B)^2 = A^2 + AB + BA + B^2$.

9. If $A = \begin{bmatrix} 3 & -2 \\ 1 & -1 \end{bmatrix}$ and $B = \begin{bmatrix} 1 & 2 \\ -1 & 1 \end{bmatrix}$, verify that

$(A + B)(A - B) = A^2 + BA - AB - B^2$.

10. Find x and y such that

$\begin{bmatrix} 2 & -1 \\ -3 & 4 \end{bmatrix} \begin{bmatrix} x \\ y \end{bmatrix} + \begin{bmatrix} 8 \\ 1 \end{bmatrix} = \begin{bmatrix} 0 \\ 0 \end{bmatrix}$

Ans. $x = -33/5, y = -26/5$.

11. If $A = \begin{bmatrix} 1 & -2 & 3 \\ 2 & 3 & -1 \\ -3 & 1 & 2 \end{bmatrix}, B = \begin{bmatrix} 6 & -1 & -2 \\ -5 & -4 & -4 \\ -2 & -8 & 3 \end{bmatrix}$ and I is a unit matrix of order 3,

evaluate $A^2 - 3A + I + B$.

Ans. $\begin{bmatrix} -8 & 0 & 0 \\ 0 & -8 & 0 \\ 0 & 0 & -8 \end{bmatrix}$.

12. Find matrix A, such that

$A \begin{bmatrix} 0 & 1 \\ 2 & -1 \end{bmatrix} = \begin{bmatrix} 2 & 1 \\ -1 & 0 \end{bmatrix}$ (*IETE, June 2008*)

Ans. $\begin{bmatrix} 2 & 1 \\ -1/2 & -1/2 \end{bmatrix}$.

13. Prove that the product of the two matrices

$$\begin{bmatrix} \cos^2\theta & \cos\theta\sin\theta \\ \cos\theta\sin\theta & \sin^2\theta \end{bmatrix} \text{ and } \begin{bmatrix} \cos^2\phi & \cos\phi\sin\phi \\ \cos\phi\sin\phi & \sin^2\phi \end{bmatrix}$$

is a null matrix when θ and ϕ differ by an odd multiple of $\pi/2$.

14. (*a*) Given $A = \begin{bmatrix} 3 & -4 \\ 1 & -1 \end{bmatrix}$ show that $A^n = \begin{bmatrix} 1+2n & -4n \\ n & 1-2n \end{bmatrix}$.

(*b*) If $A = \begin{bmatrix} k & 1 \\ 0 & k \end{bmatrix}$, prove that $A^n = \begin{bmatrix} k^n & nk^{n-1} \\ 0 & k^n \end{bmatrix}$ for every positive integer n.

(**Hint.** Use the method of mathematical Induction)   (*A.M.I.E., S-2001*)

15. Given $A = \begin{bmatrix} \cos\theta & \sin\theta \\ -\sin\theta & \cos\theta \end{bmatrix}$, prove that $A^2 = \begin{bmatrix} \cos 2\theta & \sin 2\theta \\ -\sin 2\theta & \cos 2\theta \end{bmatrix}$. Also find

the value of A^n.

(*A.M.I.E., S-2002*)

Ans. $\begin{bmatrix} \cos n\theta & \sin n\theta \\ -\sin n\theta & \cos n\theta \end{bmatrix}$.

16. If $A = \begin{bmatrix} 0 & 1 & 0 \\ 0 & 0 & 1 \\ p & q & r \end{bmatrix}$ and I is a unit matrix of order 3, show that

$A^3 = pI + qA + rA^2$.

17. If $A = \begin{bmatrix} 0 & 1 \\ -1 & 0 \end{bmatrix}$, choose α and β so that $(\alpha I + \beta A)^2 = A$.

Ans. $\alpha = \beta = 1/\sqrt{2}$.

18. Show that the matrix $A = \begin{bmatrix} 1 & 3 & 2 \\ 2 & 0 & -1 \\ 1 & 2 & 3 \end{bmatrix}$ satisfies the matrix equation

$A^3 - 4A^2 - 3A + 11I = 0$, where I is an identity matrix of order 3.

(*A.M.I.E., W-1997*)

19. Let A, B be the indicated matrices. Show that

$tr(AB) = tr(BA)$.

(*a*) $A = \begin{pmatrix} 1 & -1 & 1 \\ 2 & 4 & 1 \\ 3 & 0 & 1 \end{pmatrix}$, $B = \begin{pmatrix} 3 & 1 & 2 \\ 1 & 1 & 0 \\ -1 & 2 & 1 \end{pmatrix}$.

(*b*) $A = \begin{pmatrix} 1 & -1 & 1 \\ 2 & 4 & 1 \\ 3 & 0 & 1 \end{pmatrix}$, $B = \begin{pmatrix} 3 & 1 & 2 \\ 1 & 1 & 0 \\ -1 & 2 & 1 \end{pmatrix}$.

20. If $A = \begin{bmatrix} 1 & -4 & 2 \\ -1 & 4 & -2 \end{bmatrix}$, $B = \begin{bmatrix} 1 & 2 \\ -1 & 3 \\ 5 & -2 \end{bmatrix}$, $C = \begin{bmatrix} 2 & 2 \\ 1 & -1 \\ 1 & -3 \end{bmatrix}$,

compute $A(2B - 3C)$. **Ans.** $\begin{bmatrix} 30 & -28 \\ -30 & 28 \end{bmatrix}$.

21. Calculate $AB - BA$ in each case

(a) $A = \begin{bmatrix} 1 & 2 & 2 \\ 2 & 1 & 2 \\ 1 & 2 & 3 \end{bmatrix}$, $B = \begin{bmatrix} 4 & 1 & 1 \\ -4 & 2 & 0 \\ 1 & 2 & 1 \end{bmatrix}$.

(b) $A = \begin{pmatrix} 2 & 0 & 0 \\ 1 & 1 & 2 \\ -1 & 2 & 1 \end{pmatrix}$, $B = \begin{pmatrix} 3 & 1 & -2 \\ 3 & -2 & 4 \\ -3 & 5 & 11 \end{pmatrix}$.

Ans. $(a) \begin{pmatrix} -9 & -2 & -10 \\ 6 & 14 & 8 \\ -7 & 5 & -5 \end{pmatrix}$, $(b) \begin{pmatrix} -3 & 5 & -4 \\ 0 & 3 & 24 \\ 12 & -27 & 0 \end{pmatrix}$.

15.6 Transpose of a matrix

A matrix obtained by interchanging the corresponding rows and columns of a given matrix A is called the *transpose of matrix* A and is denoted by A^T or A'. That is, if

$$A = \begin{bmatrix} a_{11} & a_{12} & \cdots & a_{1n} \\ a_{21} & a_{22} & \cdots & a_{2n} \\ \vdots & & & \\ a_{m1} & a_{m2} & \cdots & a_{mn} \end{bmatrix}, \text{ then } A^T \text{ or } A' = \begin{bmatrix} a_{11} & a_{21} & \cdots & a_{m1} \\ a_{12} & a_{22} & & a_{m2} \\ \vdots & \vdots & & \vdots \\ a_{1n} & a_{2n} & & a_{mn} \end{bmatrix}.$$

If A is an $m \times n$ matrix, then A^T is an $n \times m$ matrix.

For example if $A = \begin{bmatrix} 6 & 5 & 4 \\ 3 & 2 & -7 \end{bmatrix}$ then $A^T = \begin{bmatrix} 6 & 3 \\ 5 & 2 \\ 4 & -7 \end{bmatrix}$.

Also, both the product matrices $A^T A$ and AA^T are defined and $A^T A = (n \times m)$ $(m \times n)$ is an $n \times n$ square matrix and $AA^T = (m \times n)(n \times m)$ is an $m \times m$ square

matrix. A row vector $\mathbf{a}$ can also be written as $\begin{bmatrix} a_1 \\ a_2 \\ \vdots \\ a_n \end{bmatrix}^T$. The transpose of a row

matrix is a column matrix and the transpose of a column matrix is a row matrix.

Properties of Transpose

If A^T and B^T be the transposes respectively of matrices A and B and if k a scalar then whenever the sum and product defined.

1. $(A^T)^T = A$ that is, the transpose of the transpose of a matrix is the matrix itself.
2. $(kA)^T = kA^T$
3. $(A + B)^T = A^T + B^T$ that is, the transpose of the sum of two matrices is equal to the sum of their transposes.
4. $(AB)^T = B^T A^T$ that is, the transpose of a product of two matrices is equal to the product of the transposes taken in the reverse order.

Proof

1. Let $A = (a_{ij})$ is an $m \times n$ matrix, then A^T, is an $n \times m$ matrix and $(A^T)^T$, the transpose of A^T, is an $m \times n$ matrix. Thus the matrices $(A^T)^T$ and A are of the same order.
   Further, the (i, j)-th element of $(A^T)^T = (j, i)$-th element of $A^T = (i, j)$-th element of A.
   Since $(A^T)^T$ and A are comparable and their corresponding elements are equal, $(A^T)^T = A$

2. Let $A = (a_{ij})$ is an $m \times n$ matrix. Thus kA is an $m \times n$ matrix, so that $(kA)^T$ is an $n \times m$ matrix.
   Further A^T is an $n \times m$ matrix and, therefore, kA^T is also an $n \times m$ matrix. Thus both the matrices $(kA)^T$ and kA^T are of the same order.
   Also (i, j)-th element of $(kA)^T = (j, i)$-th element of $kA = k$ (j, i)-th element of $A = k$ (i, j)-th element of $A^T = (i, j)$-th element of (kA^T).
   Since the matrices $(kA)^T$ and kA^T are of the same order the their corresponding elements are equal, they are equal.

3. Let $A = (a_{ij})$, $B = (b_{ij})$ be $m \times n$ matrices.
   Since each of the matrices A and B is $m \times n$ matrix, $(A + B)$ is an $m \times n$ matrix and, therefore, $(A + B)^T$ is an $n \times m$ matrix.
   Also A^T and B^T are both $n \times m$ matrices. So that $(A^T + B^T)$ is an $n \times m$ matrix.
   Thus $(A + B)^T$ and $(A^T + B^T)$ are of the same order.
   Further (i, j)-th element of $(A + B)^T$
   $= (j, i)$-th element of $(A + B) = a_{ji} + b_{ji} = (j, i)$-th element of $A + (j, i)$-th element of $B = (i, j)$-th element of $A^T + (i, j)$-th element of $B^T = (i, j)$-th element of $(A^T + B^T)$.
   Since $(A + B)^T$ and $A^T + B^T$ are of the same order and their corresponding elements are equal,
   Thus, $(A + B)^T + A^T + B^T$.

4. Let $A = (a_{ij})$ is an $m \times n$ matrix and $B = (b_{ij})$ is an $n \times p$ matrix.
   Since AB is an $m \times p$ matrix, $(AB)^T$ is a $p \times m$ matrix.
   Further B^T is $p \times n$ matrix and A^T is an $n \times m$ matrix and therefore $B^T A^T$ is a $p \times m$ matrix.

Thus $(AB)^T$ and $B^T A^T$ are matrices of the same order.

Now the (j, i)-th element of $(AB)^T = (i, j)$-th element of AB

$$= \sum_{k=1}^{n} a_{ik} b_{kj} \qquad \ldots(15.3)$$

Also the jth row of B^T is $b_{1j}, b_{2j}, \ldots, b_{nj}$ and ith column of A^T is $a_{i1}, a_{i2}, \ldots, a_{in}$.

$\therefore$ (j, i)-th element of $B^T A^T = \sum_{k=1}^{n} a_{kj} b_{ik}$ $\qquad \ldots(15.4)$

From (15.3) and (15.4), we have

$\therefore$ (j, i)-th element of $(AB)^T = (j, i)$-th element of $B^T A^T$.

Since the matrices $(AB)^T$ and $B^T A^T$ are of the same order and their corresponding elements are equal, we have $(AB)^T = B^T A^T$.

15.7 Symmetric and skew-symmetric matrices

Def. A real square matrix $A = (a_{ij})$ is said to be *symmetric*, if $a_{ij} = a_{ji}$ for all $i\, j$, that is $A^T = A$. This means that the entries in a symmetric matrix are symmetric with respect to the main diagonal of the matrix.

Skew-symmetric, if $a_{ij} = -a_{ji}$ for all i and j, that is $A^T = -A$.

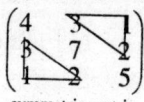 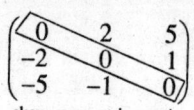

symmetric matrix   skew-symmetric matrix

Properties of symmetric and skew-symmetric matrices

1. In a skew-symmetric matrix $A = (a_{ij})$, all the elements in the principal diagonal are zero.

2. If A is a real square matrix, then matrix $A + A^T$ is symmetric and matrix $A - A^T$ is skew-symmetric or anti-symmetric matrix. Therefore any real square matrix can always be expressed as the sum of a real symmetric matrix and a real skew-symmetric matrix. That is

$$A = \frac{1}{2}\left(A + A^T\right) + \frac{1}{2}\left(A - A^T\right)$$

Proof: $(A + A^T)^T = A^T + A = A + A^T$.

$\therefore \qquad A + A^T$ and hence $\frac{1}{2}\left(A + A^T\right)$ is symmetric.

Also, since $(A - A^T)^T = A^T - A = -(A - A^T)$, it follows that $\frac{1}{2}\left(A - A^T\right)$ is skew-symmetric. The result is thus proved.

3. If **A** and **B** be two symmetric matrices of the same order, then the product matrix **AB** is symmetric if and only if **AB** = **BA**, that is the matrices **A** and **B** commute.

Proof. Since the matrices **A** and **B** are symmetric we have

$A^T = A$ and $B^T = B$.

Let **AB** be symmetric. Then $(AB)^T = AB$, or $B^T A^T = AB$, $\Rightarrow BA = AB$. Now, let **AB** = **BA**. Taking transpose on both sides, we obtain

$(AB)^T = (BA)^T = A^T B^T = AB$. Hence, the result.

Example 15.12 Given $A = \begin{bmatrix} 1 & 2 & 3 \\ 4 & 5 & 6 \\ 7 & 8 & 9 \end{bmatrix}$ and $B = \begin{bmatrix} 1 & 1 & 1 \\ 1 & 1 & 1 \\ 1 & 1 & 1 \end{bmatrix}$.

Show that $(AB)^T = B^T A^T$. *(A.M.I.E., S-2002)*

Solution. We have $AB = \begin{bmatrix} 1 & 2 & 3 \\ 4 & 5 & 6 \\ 7 & 8 & 9 \end{bmatrix}\begin{bmatrix} 1 & 1 & 1 \\ 1 & 1 & 1 \\ 1 & 1 & 1 \end{bmatrix}$

$= \begin{bmatrix} 1+2+3 & 1+2+3 & 1+2+3 \\ 4+5+6 & 4+5+6 & 4+5+6 \\ 7+8+9 & 7+8+9 & 7+8+9 \end{bmatrix} = \begin{bmatrix} 6 & 6 & 6 \\ 15 & 15 & 15 \\ 24 & 24 & 24 \end{bmatrix}$

$\therefore$ $(AB)^T = \begin{bmatrix} 6 & 6 & 6 \\ 15 & 15 & 15 \\ 24 & 24 & 24 \end{bmatrix}^T = \begin{bmatrix} 6 & 15 & 24 \\ 6 & 15 & 24 \\ 6 & 15 & 24 \end{bmatrix}$

Now $B^T A^T = \begin{bmatrix} 1 & 1 & 1 \\ 1 & 1 & 1 \\ 1 & 1 & 1 \end{bmatrix}\begin{bmatrix} 1 & 4 & 7 \\ 2 & 5 & 8 \\ 3 & 6 & 9 \end{bmatrix} = \begin{bmatrix} 1+2+3 & 4+5+6 & 7+8+9 \\ 1+2+3 & 4+5+6 & 7+8+9 \\ 1+2+3 & 4+5+6 & 7+8+9 \end{bmatrix}$

$= \begin{bmatrix} 6 & 15 & 24 \\ 6 & 15 & 24 \\ 6 & 15 & 24 \end{bmatrix}$. Hence $(AB)^T = B^T A^T$.

Example 15.13 Express the following matrix **A** as a sum of a symmetric and skew-symmetric matrix

$$A = \begin{bmatrix} -1 & 7 & 1 \\ 2 & 3 & 4 \\ 5 & 0 & 5 \end{bmatrix}.$$ *(A.M.I.E., W-1998)*

Solution. We have $A^T = \begin{bmatrix} -1 & 2 & 5 \\ 7 & 3 & 0 \\ 1 & 4 & 5 \end{bmatrix}$.

Then $A + A^T = \begin{bmatrix} -1 & 7 & 1 \\ 2 & 3 & 4 \\ 5 & 0 & 5 \end{bmatrix} + \begin{bmatrix} -1 & 2 & 5 \\ 7 & 3 & 0 \\ 1 & 4 & 5 \end{bmatrix} = \begin{bmatrix} -2 & 9 & 6 \\ 9 & 6 & 4 \\ 6 & 4 & 10 \end{bmatrix}$.

$$\therefore \quad \frac{1}{2}(A+A^T) = \begin{bmatrix} -1 & 9/2 & 3 \\ 9/2 & 3 & 2 \\ 3 & 2 & 5 \end{bmatrix}.$$

Similarly, $\frac{1}{2}(A-A^T) = \begin{bmatrix} 0 & 5/2 & -2 \\ -5/2 & 0 & 2 \\ 2 & -2 & 0 \end{bmatrix}$

$$\therefore A = \frac{1}{2}(A+A^T) + \frac{1}{2}(A-A^T) = \underbrace{\begin{bmatrix} -1 & 9/2 & 3 \\ 9/2 & 3 & 2 \\ 3 & 2 & 5 \end{bmatrix}}_{\substack{\text{symmetrical} \\ \text{matrix}}} + \underbrace{\begin{bmatrix} 0 & 5/2 & -2 \\ -5/2 & 0 & 2 \\ 2 & -2 & 0 \end{bmatrix}}_{\substack{\text{skew-symmetric} \\ \text{matrix}}}.$$

Example 15.14 If $2A + B^T = \begin{bmatrix} 1 & 2 \\ 0 & 3 \end{bmatrix}$ ...(i)

and $-A + 2B = \begin{bmatrix} 4 & 3 \\ 4 & 6 \end{bmatrix}$, ...(ii)

find the matrices **A** and **B**. (A.M.I.E., S-2003)

Solution. Multiplying (ii) by 2 and adding, we get

$$4B + B^T = \begin{bmatrix} 9 & 8 \\ 8 & 15 \end{bmatrix} \qquad \text{...(iii)}$$

Let $B = \begin{bmatrix} p & q \\ r & s \end{bmatrix}$, $\qquad \therefore B^T = \begin{bmatrix} p & r \\ q & s \end{bmatrix}$.

Matrix equation (iii) may be re-written as $\begin{bmatrix} 5p & 4q+r \\ 4r+q & 5s \end{bmatrix} = \begin{bmatrix} 9 & 8 \\ 8 & 15 \end{bmatrix}$.

By the definition of equality of matrices, we have

$5p = 9, \; 4q + r = 8, \; 4r + q = 8, \; 5s = 15$

On solving these equations, we obtain $p = 9/5$, $q = \dfrac{8}{5}$, $r = \dfrac{8}{5}$, $s = 3$.

$$\therefore B = \begin{bmatrix} 9/5 & 8/5 \\ 8/5 & 3 \end{bmatrix} \text{ and } A = 2B - \begin{bmatrix} 4 & 3 \\ 4 & 6 \end{bmatrix} = \begin{bmatrix} \dfrac{18}{5} & \dfrac{16}{5} \\ \dfrac{16}{5} & 6 \end{bmatrix} + \begin{bmatrix} -4 & -3 \\ -4 & -6 \end{bmatrix} = \begin{bmatrix} \dfrac{-2}{5} & \dfrac{1}{5} \\ \dfrac{-4}{5} & 0 \end{bmatrix}.$$

Exercise 15.2

1. If $A = \begin{bmatrix} 1 & 2 & 0 \\ 3 & -1 & 4 \end{bmatrix}$, find AA^T and A^TA.   **Ans.** $\begin{bmatrix} 5 & 1 \\ 1 & 26 \end{bmatrix}$, $\begin{bmatrix} 10 & -1 & 12 \\ -1 & 5 & -4 \\ 12 & -4 & 16 \end{bmatrix}$.

2. If $A = \begin{bmatrix} 1 & 2 & -1 \\ 3 & 0 & 2 \\ 4 & 5 & 0 \end{bmatrix}$, $B = \begin{bmatrix} 1 & 0 & 0 \\ 2 & 1 & 0 \\ 0 & 1 & 3 \end{bmatrix}$ verify that $(AB)^T = B^TA^T$.

3. If $A = \begin{bmatrix} 4 \\ 8 \\ -10 \end{bmatrix}$ and $B = [2\ 4\ 5]$, find (a) $A^T A$, (b) $B^T B$, (c) $A + B^T$.

Ans. (a) 180; (b) $\begin{bmatrix} 4 & 8 & 10 \\ 8 & 16 & 20 \\ 10 & 20 & 25 \end{bmatrix}$; (c) $\begin{bmatrix} 6 \\ 12 \\ -5 \end{bmatrix}$.

4. Find (a) A^TB, (b) A^TC, where $A = \begin{bmatrix} 1 & -1 & 2 \\ 0 & 3 & 4 \end{bmatrix}$, $B = \begin{bmatrix} 4 & 0 & -3 \\ -1 & -2 & 3 \end{bmatrix}$,

$C = \begin{bmatrix} 2 & -3 & 0 & 1 \\ 5 & -1 & -4 & 2 \\ -1 & 0 & 0 & 3 \end{bmatrix}$.   Ans. (a) $\begin{bmatrix} 4 & 0 & -3 \\ -7 & -6 & 12 \\ 4 & -8 & 6 \end{bmatrix}$; (b) not defined.

5. (a) Express the matrix $\begin{bmatrix} 4 & 5 & 6 \\ -1 & 0 & 1 \\ 2 & 1 & 2 \end{bmatrix}$ as the sum of a symmetric and a skew-symmetric matrix.

Ans. $\begin{bmatrix} 4 & 2 & 4 \\ 2 & 0 & 1 \\ 4 & 1 & 2 \end{bmatrix} + \begin{bmatrix} 0 & 3 & 2 \\ -3 & 0 & 0 \\ -2 & 0 & 0 \end{bmatrix}$.

(b) If $A = \begin{pmatrix} 3 & 1 \\ -4 & -1 \end{pmatrix}$, represent it as $A = B + C$, where B is a symmetric and C is a skew-symmetric matrix.

Ans. $B = \begin{bmatrix} 3 & -3/2 \\ -3/2 & -1 \end{bmatrix}$; $C = \begin{bmatrix} 0 & 5/2 \\ -5/2 & 0 \end{bmatrix}$.

6. If A and B are symmetric matrices, prove that $AB - BA$ is a skew-symmetric matrix.

15.8 Determinants

Corresponding to every square matrix A of order n, that is for every $n \times n$ matrix there is associated an expression, called the *determinant* of A denoted by either det (A) or |A|. The determinant has a value and this value is real if the matrix is real and may be real or complex, if the matrix is complex. A determinant of order n is defined as

$$\det(\mathbf{A}) \text{ or } |\mathbf{A}| = \begin{vmatrix} a_{11} & a_{12} & \cdots & a_{1n} \\ a_{21} & a_{22} & \cdots & a_{2n} \\ \vdots & & & \vdots \\ a_{n1} & a_{n2} & \cdots & a_{nn} \end{vmatrix}. \qquad \ldots(15.5)$$

The n^2 quantities $a_{11}, a_{12}, \ldots, a_{nn}$ are called **elements** of the determinant. The horizontal lines of elements are called **rows** and the vertical lines of elements and called **columns**. A determinant of nth order has n rows and n columns.

We now discuss methods to find the value of a determinant. For a, 1×1 matrix $\mathbf{A} = [a]$, we have $\det (\mathbf{A}) = |a| = a$.

A determinant of 2×2 matrix is written as $|\mathbf{A}| = \begin{vmatrix} a_{11} & a_{12} \\ a_{21} & a_{22} \end{vmatrix}$ and its value is

given by $|\mathbf{A}| = a_{11}a_{22} - a_{12}a_{21}$.

We evaluate higher order determinants using minors and co-factors.

Minors, Co-factors and Laplace Expansion

Minor Let a_{ij} be the general element of a determinant. If we delete the i-th row and the j-th column from the determinant, we obtain a new determinant of order $(n - 1)$ which is called the *minor* of the element a_{ij}. We denote the minor by M_{ij}.

For example, the minor corresponding to the element 3 in the 3rd row and 2nd column of the fourth order det $(\mathbf{B})$

$$= \begin{vmatrix} 2 & 1 & -1 & 4 \\ -3 & 2 & 5 & 6 \\ 1 & 3 & -2 & 2 \\ 4 & 5 & -3 & 1 \end{vmatrix} \text{ is } M_{32} = \begin{vmatrix} 2 & -1 & 4 \\ -3 & 5 & 6 \\ 4 & -3 & 1 \end{vmatrix} \text{ which is obtained by deleting}$$

the elements shown in the blocks.

Co-factor If we multiply the minor of a_{ij} by $(-1)^{i+j}$, the result is called the co-factor of a_{ij} and is denoted by A_{ij}, that is $A_{ij} = (-1)^{i+j} \cdot M_{ij}$. $\qquad \ldots(15.6)$

For example the co-factor A_{32} corresponding to the element 3 in the det $(\mathbf{B})$ is $(-1)^{3+2}$ times its minor, that is

$$A_{32} = -\begin{vmatrix} 2 & -1 & 4 \\ -3 & 5 & 6 \\ 4 & -3 & 1 \end{vmatrix}.$$ We can expand a determinant of order n through the

elements of any row or any column.

The Laplace expansion The value of a determinant is defined as the sum of the products of the elements of the ith row (or jth column) by their corresponding cofactors and is called the *Laplace expansion*. Thus, we have

$$\det(A) \text{ or } |A| = \sum_{j=1}^{n}(-1)^{i+j}a_{ij}M_{ij} = \sum_{j=1}^{n}a_{ij}A_{ij}, \qquad \qquad ...(15.7a)$$

when we expand through the elements of the ith row or

$$\det(A) \text{ or } |A| = \sum_{i=1}^{n}(-1)^{i+j}a_{ij}M_{ij} = \sum_{i=1}^{n}a_{ij}A_{ij} \qquad \qquad ...(15.7b)$$

when we expand through the elements of the jth column.

Remark Note that, the sum of the products formed by multiplying the elements of any row (or column) of an n-squared matrix **A** by their corresponding co-factors gives the value of the determinant. However, the sum of the products of the elements of any row (or column) of an n-square matrix **A** by the corresponding co-factors of **any other** row (or column) is zero.

A **determinant of third order** can by defined by

$$\det(A) = \begin{vmatrix} a_{11} & a_{12} & a_{13} \\ a_{21} & a_{22} & a_{23} \\ a_{31} & a_{32} & a_{33} \end{vmatrix} = a_{11}\begin{vmatrix} a_{22} & a_{23} \\ a_{32} & a_{33} \end{vmatrix} - a_{12}\begin{vmatrix} a_{21} & a_{23} \\ a_{31} & a_{33} \end{vmatrix} + a_{13}\begin{vmatrix} a_{21} & a_{22} \\ a_{31} & a_{32} \end{vmatrix}$$

$$= a_{11}\begin{vmatrix} a_{22} & a_{23} \\ a_{32} & a_{33} \end{vmatrix} + a_{12}\left(-\begin{vmatrix} a_{21} & a_{23} \\ a_{31} & a_{33} \end{vmatrix}\right) + a_{13}\begin{vmatrix} a_{21} & a_{22} \\ a_{31} & a_{32} \end{vmatrix}$$

$$= a_{11}A_{11} + a_{12}A_{12} + a_{13}A_{13} \text{ [Co-factor expansion of |A| along the first row].}$$

A 3×3 matrix has nine co-factors:

$$A_{11} = M_{11} \qquad A_{12} = -M_{12} \qquad A_{13} = M_{13}$$
$$A_{21} = -M_{21} \qquad A_{22} = M_{22} \qquad A_{23} = -M_{23}$$
$$A_{31} = M_{31} \qquad A_{32} = -M_{32} \qquad A_{33} = M_{33}$$

Similarly, it can be shown that

det $(A) = a_{12}A_{12} + a_{22}A_{22} + a_{32}A_{32}$, which is the co-factor expansion of |A| along the second column.

Generally, we expand a determinant through that row or column which has a number of zeroes.

Note that determinants are defined only for square matrices.

Example 15.15 Evaluate the determinant of $A = \begin{bmatrix} 3 & -2 & 2 \\ 1 & 2 & -3 \\ 4 & 1 & 2 \end{bmatrix}$, taking expansion along the first row.

Solution. $\det(A) = \begin{vmatrix} 3 & -2 & 2 \\ 1 & 2 & -3 \\ 4 & 1 & 2 \end{vmatrix} = 3A_{11} + (-2)A_{12} + 2A_{13}.$

Now, the co-factors of the entries/elements in the first row of **A** are:

$$A_{11} = (-1)^{1+1} \begin{vmatrix} 3 & -2 & 2 \\ 1 & 2 & -3 \\ 4 & 1 & 2 \end{vmatrix} = (-1)^2 \begin{vmatrix} 2 & -3 \\ 1 & 2 \end{vmatrix} = 4+3 = 7,$$

$$A_{12} = (-1)^{1+2} \begin{vmatrix} 3 & -2 & 2 \\ 1 & 2 & -3 \\ 4 & 1 & 2 \end{vmatrix} = (-1)^3 \begin{vmatrix} 1 & -3 \\ 4 & 2 \end{vmatrix} = -(2+12) = -14,$$

$$A_{13} = (-1)^{1+3} \begin{vmatrix} 3 & -2 & 2 \\ 1 & 2 & -3 \\ 4 & 1 & 2 \end{vmatrix} = (-1)^4 \begin{vmatrix} 1 & 2 \\ 4 & 1 \end{vmatrix} = (1-8) = -7.$$

Thus det (**A**) = 3(7) + (-2) (-14) + 2(-7) = 21 + 28 - 14 = 35.

Note In place of first row we could expand along any other row or column, the value of the determinant remain same. Let us take co-factor expansion along the third column. The co-factors of elements of third column *i.e.* 2, -3 and 2 are

respectively $(-1)^{1+3} \begin{vmatrix} 1 & 2 \\ 4 & 1 \end{vmatrix}$, $(-1)^{2+3} \begin{vmatrix} 3 & -2 \\ 4 & 1 \end{vmatrix}$, $(-1)^{3+3} \begin{vmatrix} 3 & -2 \\ 1 & 2 \end{vmatrix}$

That is, -7, -11 and 8. Therefore expansion along third column gives det (**A**) = 2(-7)-3(-11)+2(8) = -14 + 33 + 16 = 35, which is same as above.

Example 15.16 Solve the equation $\begin{vmatrix} x-1 & -6 & 2 \\ -6 & x-2 & -4 \\ 2 & -4 & x-6 \end{vmatrix} = 0$.

Solution. Expansion along the first row gives

$$(x-1) \begin{vmatrix} x-2 & -4 \\ -4 & x-6 \end{vmatrix} - (-6) \begin{vmatrix} -6 & -4 \\ 2 & x-6 \end{vmatrix} + 2 \begin{vmatrix} -6 & x-2 \\ 2 & -4 \end{vmatrix} = 0$$

or $(x-1)(x^2 - 8x + 12 - 16) + 6(-6x + 36 + 8) + 2(24 - 2x + 4) = 0$

Simplifying, we obtain $x^3 - 9x^2 - 36x + 324 = 0$ ...(i)

By Remainder Theorem, $x = 6$ is a root of this equation. Hence we can write (i) as

$(x - 6)(x^2 - 3x - 54) = 0$ or $(x - 6)(x + 6)(x - 9) = 0$.

∴ $x = 6, -6, 9$ are the required roots.

Exercise 15.3

1. Find the co-factors of the elements of the second row and third column of

$$|A| = \begin{vmatrix} 1 & 2 & 3 \\ 2 & 3 & 2 \\ 1 & 2 & 2 \end{vmatrix}.$$

Ans. 2, -1, 0; 1, 0, -1.

2. Show that the co-factor of each element of

$$\begin{vmatrix} -1/3 & -2/3 & -2/3 \\ 2/3 & 1/3 & -2/3 \\ 2/3 & -2/3 & 1/3 \end{vmatrix} \text{ is that element itself.}$$

3. Show that the co-factor of an element of any row of $\begin{vmatrix} -4 & -3 & -3 \\ 1 & 0 & 1 \\ 4 & 4 & 3 \end{vmatrix}$ is the

corresponding element of the same numbered column *i.e.*, $A_{ij} = a_{ji}$.

4. Compute the determinant of each of the following matrices:

(a) $\begin{bmatrix} 2 & -1 & 1 \\ 3 & 2 & 4 \\ -1 & 0 & 3 \end{bmatrix}$,   (b) $\begin{bmatrix} 3 & 4 & 5 \\ 5 & 3 & 4 \\ 4 & 5 & 3 \end{bmatrix}$,

(c) $\begin{bmatrix} 3 & -4 & -3 \\ 2 & 7 & -31 \\ 5 & -9 & 2 \end{bmatrix}$,   (d) $\begin{bmatrix} -2 & -1 & 4 \\ 6 & -3 & -2 \\ 4 & 1 & 2 \end{bmatrix}$.

Ans. (a) 27, (b) 36, (c) zero, (d) 100

5. Solve the following equations:

(a) $\begin{vmatrix} 2x & 5 \\ 9 & x+3 \end{vmatrix} = \begin{vmatrix} 5 & 4 \\ 13 & 3x \end{vmatrix}$,   (b) $\begin{vmatrix} 2x^2 & -3 & -16 \\ x & 5 & 5 \\ 11 & 20 & -15 \end{vmatrix} = 0$,

(c) $\begin{vmatrix} x & 2 & 3 \\ 6 & x+4 & 4 \\ 7 & 8 & x+8 \end{vmatrix} = 0$,   (d) $\begin{vmatrix} x-4 & 3 \\ 2 & x-9 \end{vmatrix} = 0$.

Ans. (a) 3.5, 1 (b) 1, $-2\dfrac{3}{70}$ (c) 1, 1.39, -14.39 (d) 3, 10.

6. Find values of λ for which $\begin{vmatrix} 3-\lambda & 0 & 0 \\ 0 & 4-\lambda & \sqrt{3} \\ 0 & \sqrt{3} & 6-\lambda \end{vmatrix} = 0$.   **Ans.** 3, 7.

15.9 In applied mathematics very often we are required to evaluate determinants whose elements are numbers. Such determinants can be evaluated without much difficulty by the method of Laplace expansion when the determinants are of second or third order. In case the determinants are of higher order, this process becomes very cumbersome. The expansion in such cases can be effected very conveniently with the help of the following properties of determinants.

Properties of determinants

1. If all the elements in a row (or column) of a determinant are zero, then the value of the determinant is zero.

2. The value of a determinant is not changed if its corresponding rows and columns are interchanged, that is $|A| = |A^T|$.

3. If any two rows (or columns) of a determinant are interchanged, then the value of the determinant is multiplied by (-1).

4. If the corresponding elements of two rows (or columns) are same, that is two rows (or columns) are identical, then the value of the determinant is zero.

5. If the corresponding elements of two rows (or columns) are proportional to each other, then the value of the determinant is zero.

6. If each element of a row (or column) of a determinant is multiplied by a scalar k, then the determinant is multiplied by the scalar k.

7. If each element of any row (or column) of a determinant can be expressed as the sum of two (or more) terms, then the determinant can be expressed as sum of two (or more) determinants.

8. If we multiply the elements of any row (or column) by a non-zero number and add to corresponding elements of any other row (or column), then the value of the determinant remains the same.

9. The value of the determinant of a diagonal or a lower triangular or an upper triangular matrix is the product of its diagonal elements.

Remark: When the elements of the ith row are multiplied by a non-zero constant k and added to the corresponding elements of the jth row, we denote this operation as $R_j \leftarrow R_j + kR_i$, where R_j is the jth row of $|A|$. The elements of the ith row remain unchanged whereas the elements of the jth row get changed. This operation is called an *elementary row operation*. Similarly, the operation $C_j \leftarrow C_j + kC_i$, where C_j is the jth column of $|A|$, is called the *elementary column operation*. Therefore, under elementary row (or column) operations, the value of the determinant is unchanged.

15.10 Illustrative Examples

Example 15.17 Evaluate:

$$(a)\ |A| = \begin{vmatrix} 28 & 25 & 38 \\ 42 & 38 & 65 \\ 56 & 47 & 83 \end{vmatrix},\ (b)\ |A| = \begin{vmatrix} 1 & a & b+c \\ 1 & b & c+a \\ 1 & c & a+b \end{vmatrix}.$$

Solution. (a) Taking 14 as common factor, we obtain

$$|A| = 14 \begin{vmatrix} 2 & 25 & 38 \\ 3 & 38 & 65 \\ 4 & 47 & 83 \end{vmatrix}\ (\text{operate } C_2 \leftarrow C_2 - 12C_1 \text{ and } C_3 \leftarrow C_3 - 19C_1)$$

$$= 14 \begin{vmatrix} 2 & 1 & 0 \\ 3 & 2 & 8 \\ 4 & -1 & 7 \end{vmatrix}\ (\text{operate } C_1 \leftarrow C_1 - 2C_2) = 14 \begin{vmatrix} 0 & 1 & 0 \\ -1 & 2 & 8 \\ 6 & -1 & 7 \end{vmatrix}.$$

Expanding the determinant by using the first row, we have

$$|A| = 14(-1) \begin{vmatrix} -1 & 8 \\ 6 & 7 \end{vmatrix} = 14 \times 55 = 770.$$

(b) operating $C_3 \leftarrow C_3 + C_2$, the given determinant becomes

$$|A| = \begin{vmatrix} 1 & a & a+b+c \\ 1 & b & a+b+c \\ 1 & c & a+b+c \end{vmatrix} = (a+b+c) \begin{vmatrix} 1 & a & 1 \\ 1 & b & 1 \\ 1 & c & 1 \end{vmatrix} = 0,$$

because two columns are identical.

Example 15.18 Prove, without expanding, that each of the following determinants vanishes.

$$(a)\ |A| = \begin{vmatrix} 1 & 15 & 14 & 4 \\ 12 & 6 & 7 & 9 \\ 8 & 10 & 11 & 5 \\ 13 & 3 & 2 & 16 \end{vmatrix}, \qquad (b)\ |A| = \begin{vmatrix} 1 & a & a^2 & a^3 + bcd \\ 1 & b & b^2 & b^3 + cda \\ 1 & c & c^2 & c^3 + dab \\ 1 & d & d^2 & d^3 + abc \end{vmatrix}.$$

Solution. (a) Operating $C_1 \leftarrow C_1 + C_2 + C_3 + C_4$ and taking out 34, the common factor from C_1, the given determinant

$$|A| = 34 \begin{vmatrix} 1 & 15 & 14 & 4 \\ 1 & 6 & 7 & 9 \\ 1 & 10 & 11 & 5 \\ 1 & 3 & 2 & 16 \end{vmatrix} \quad \text{(operate } R_4 \to R_4 - R_1,\ R_3 \to R_3 - R_1, \text{and} \\ R_2 \to R_2 - R_1)$$

$$= 34 \begin{vmatrix} 1 & 15 & 14 & 4 \\ 0 & -9 & -7 & 5 \\ 0 & -5 & -3 & 1 \\ 0 & -12 & -12 & 12 \end{vmatrix} \begin{array}{l} \text{(taking out 12 as common} \\ \text{factor from } R_4) \end{array} = 408 \begin{vmatrix} 1 & 15 & 14 & 4 \\ 0 & -9 & -7 & 5 \\ 0 & -5 & -3 & 1 \\ 0 & -1 & -1 & 1 \end{vmatrix}.$$

Expanding the determinant by using the first column, we have

$$|A| = 408 \begin{vmatrix} -9 & -7 & 5 \\ -5 & -3 & 1 \\ -1 & -1 & 1 \end{vmatrix} \quad \text{(operate } C_3 \leftarrow C_3 + C_2 \text{ and } C_2 \leftarrow C_2 - C_1)$$

$$= 408 \begin{vmatrix} -9 & 2 & -2 \\ -5 & 2 & -2 \\ -1 & 0 & 0 \end{vmatrix} = 0. \quad \text{(because } C_2 \text{ and } C_3 \text{ are proportional)}$$

(b) By property 7, the given determinant

$$|A| = \begin{vmatrix} 1 & a & a^2 & a^3 \\ 1 & b & b^2 & b^3 \\ 1 & c & c^2 & c^3 \\ 1 & d & d^2 & d^3 \end{vmatrix} + \begin{vmatrix} 1 & a & a^2 & bcd \\ 1 & b & b^2 & cda \\ 1 & c & c^2 & dab \\ 1 & d & d^2 & abc \end{vmatrix} = |B| + |C| \text{ (say)}$$

Taking $a^{-1}, b^{-1}, c^{-1}, d^{-1}$ common from R_1, R_2, R_3 and R_4 respectively in $|C|$, we have

$$|\mathbf{C}| = \frac{1}{abcd} \begin{vmatrix} a & a^2 & a^3 & abcd \\ b & b^2 & b^3 & bcda \\ c & c^2 & c^3 & cdab \\ d & d^2 & d^3 & dabc \end{vmatrix} \quad \text{(taking } abcd \text{ common from } C_4\text{)}$$

$$= \begin{vmatrix} a & a^2 & a^3 & 1 \\ b & b^2 & b^3 & 1 \\ c & c^2 & c^3 & 1 \\ d & d^2 & d^3 & 1 \end{vmatrix} \quad \text{(passing } C_4 \text{ over first three columns)}$$

$$= (-1)^3 \begin{vmatrix} 1 & a & a^2 & a^3 \\ 1 & b & b^2 & b^3 \\ 1 & c & c^2 & c^3 \\ 1 & d & d^2 & d^3 \end{vmatrix} = -|\mathbf{B}|.$$

$\therefore |\mathbf{A}| = |\mathbf{B}| - |\mathbf{B}| = 0$. Hence the result.

Example 15.19 Prove that

$$|\mathbf{A}| = \begin{vmatrix} 1+a & 1 & 1 & 1 \\ 1 & 1+b & 1 & 1 \\ 1 & 1 & 1+c & 1 \\ 1 & 1 & 1 & 1+d \end{vmatrix} = abcd\left(1 + a^{-1} + b^{-1} + c^{-1} + d^{-1}\right).$$

Solution. Taking a, b, c, d common from R_1, R_2, R_3, R_4 respectively, we obtain

$$|\mathbf{A}| = abcd \begin{vmatrix} a^{-1}+1 & a^{-1} & a^{-1} & a^{-1} \\ b^{-1} & b^{-1}+1 & b^{-1} & b^{-1} \\ c^{-1} & c^{-1} & c^{-1}+1 & c^{-1} \\ d^{-1} & d^{-1} & d^{-1} & d^{-1}+1 \end{vmatrix}$$

[operating $R_1 \leftarrow R_1 + (R_2 + R_3 + R_4)$ and taking out the common factor from R_1]

$$= abcd\left(1 + a^{-1} + b^{-1} + c^{-1} + d^{-1}\right) \begin{vmatrix} 1 & 1 & 1 & 1 \\ b^{-1} & b^{-1}+1 & b^{-1} & b^{-1} \\ c^{-1} & c^{-1} & c^{-1}+1 & c^{-1} \\ d^{-1} & d^{-1} & d^{-1} & d^{-1}+1 \end{vmatrix}$$

(operating $C_2 \leftarrow C_2 - C_1$, $C_3 \leftarrow C_3 - C_1$ and $C_4 \leftarrow C_4 - C_1$)

$$= abcd\left(1 + a^{-1} + b^{-1} + c^{-1} + d^{-1}\right) \begin{vmatrix} 1 & 0 & 0 & 0 \\ b^{-1} & 1 & 0 & 0 \\ c^{-1} & 0 & 1 & 0 \\ d^{-1} & 0 & 0 & 1 \end{vmatrix}.$$

Expanding the determinant by using the first row we obtain
$|A| = abcd\,(1 + a^{-1} + b^{-1} + c^{-1} + d^{-1}).$

Example 15.20 Show that

$$
\begin{vmatrix}
a-b-c & 2b & 2c \\
2a & b-c-a & 2c \\
2a & 2b & c-a-b
\end{vmatrix} = (a+b+c)^3.
$$

Solution. Let $|A|$ be the given determinant. Operating $C_1 \leftarrow C_1 + (C_2 + C_3)$, we obtain

$$
|A| = \begin{vmatrix}
a+b+c & 2b & 2c \\
a+b+c & b-c-a & 2c \\
a+b+c & 2b & c-a-b
\end{vmatrix}
$$

$$
= (a+b+c)\begin{vmatrix}
1 & 2b & 2c \\
1 & b-c-a & 2c \\
1 & 2b & c-a-b
\end{vmatrix} \quad (\text{operate } R_2 \leftarrow R_2 - R_1, \text{ and } R_3 \leftarrow R_3 - R_1)
$$

$$
= (a+b+c)\begin{vmatrix}
1 & 2b & 2c \\
0 & -b-c-a & 0 \\
0 & 0 & -c-a-b
\end{vmatrix}.
$$

Expanding the determinant by using the first column, we have
$|A| = (a + b + c)\,(a + b + c)^2 = (a + b + c)^3.$

Example 15.21 Prove that

$$
|A| = \begin{vmatrix}
1+a^2+b^2 & 2ab & -2b \\
2ab & 1+a^2+b^2 & 2a \\
2b & -2a & 1-a^2-b^2
\end{vmatrix} = (1+a^2+b^2)^3.
$$

Solution. Operating $C_1 \leftarrow C_1 - bC_3$ and $C_2 \leftarrow C_2 + aC_3$, the det (A)

$$
= \begin{vmatrix}
1+a^2+b^2 & 0 & -2b \\
0 & 1+a^2+b^2 & 2a \\
b+a^2b+b^3 & -a-a^3-ab^2 & 1-a^2-b^2
\end{vmatrix} \quad
\begin{array}{l}
[\text{taking out }(1+a^2+b^2) \\
\text{common from } C_1 \text{ and } C_2]
\end{array}
$$

$$
= (1+a^2+b^2)^2\begin{vmatrix}
1 & 0 & -2b \\
0 & 1 & 2a \\
b & -a & 1-a^2-b^2
\end{vmatrix} \quad (\text{operate } C_3 \leftarrow C_3 + 2b\, C_1)
$$

$$
= (1+a^2+b^2)^2\begin{vmatrix}
1 & 0 & 0 \\
0 & 1 & 2a \\
b & -a & 1-a^2+b^2
\end{vmatrix}.
$$

Expanding the determinant by using the first row, we have

$$|A| = (1+a^2+b^2)^2 \begin{vmatrix} 1 & 2a \\ -a & 1-a^2+b^2 \end{vmatrix}$$

$$= (1+a^2+b^2)^2[1-a^2+b^2+2a^2] = (1+a^2+b^2)^3.$$

Example 15.22 Show that

$$|A| = \begin{vmatrix} x^2+1 & xy & xz & xu \\ xy & y^2+1 & yz & yu \\ zx & zy & z^2+1 & zu \\ ux & uy & uz & u^2+1 \end{vmatrix} = x^2+y^2+z^2+u^2+1.$$

Solution. Multiplying C_1, C_2, C_3, C_4 by x, y, z, u respectively and dividing the determinant by $xyzu$, we obtain

$$|A| = \frac{1}{xyzu} \begin{vmatrix} x(x^2+1) & xy^2 & xz^2 & xu^2 \\ x^2y & y(y^2+1) & yz^2 & yu^2 \\ x^2z & zy^2 & z(z^2+1) & zu^2 \\ x^2u & uy^2 & uz^2 & u(u^2+1) \end{vmatrix}$$

(taking x, y, z, u common from R_1, R_2, R_3, R_4 respectively)

$$= \frac{xyzu}{xyzu} \begin{vmatrix} x^2+1 & y^2 & z^2 & u^2 \\ x^2 & y^2+1 & z^2 & u^2 \\ x^2 & y^2 & z^2+1 & u^2 \\ x^2 & y^2 & z^2 & u^2+1 \end{vmatrix}$$

[operating $C_1 \leftarrow C_1 + (C_2 + C_3 + C_4)$ and taking $x^2 + y^2 + z^2 + u^2 + 1$ common from C_1)]

$$= (x^2+y^2+z^2+u^2+1) \begin{vmatrix} 1 & y^2 & z^2 & u^2 \\ 1 & y^2+1 & z^2 & u^2 \\ 1 & y^2 & z^2+1 & u^2 \\ 1 & y^2 & z^2 & u^2+1 \end{vmatrix}$$

(operate $R_2 \leftarrow R_2 - R_1$, $R_3 \leftarrow R_3 - R_1$ and $R_4 \leftarrow R_4 - R_1$)

$$= (x^2+y^2+z^2+u^2+1) \begin{vmatrix} 1 & y^2 & z^2 & u^2 \\ 0 & 1 & 0 & 0 \\ 0 & 0 & 1 & 0 \\ 0 & 0 & 0 & 1 \end{vmatrix} = x^2+y^2+z^2+u^2+1.$$

Example 15.23 If $|A| = \begin{vmatrix} a & a^2 & a^3-1 \\ b & b^2 & b^3-1 \\ c & c^2 & c^3-1 \end{vmatrix} = 0$

in which a, b, c are all different, show that $abc = 1$.

Solution. Since each term of C_3 in the given det (**A**) consists of two terms, the given determinant can be expressed as a sum of two determinants. (Property 7).

Therefore, $\begin{vmatrix} a & a^2 & a^3-1 \\ b & b^2 & b^3-1 \\ c & c^2 & c^3-1 \end{vmatrix} = \begin{vmatrix} a & a^2 & a^3 \\ b & b^2 & b^3 \\ c & c^2 & c^3 \end{vmatrix} + \begin{vmatrix} a & a^2 & -1 \\ b & b^2 & -1 \\ c & c^2 & -1 \end{vmatrix}$

(taking a, b, c common from R_1, R_2, R_3 respectively of the first determinant and -1 from C_3 of the second determinant). The given det (**A**)

$$= abc \begin{vmatrix} 1 & a & a^2 \\ 1 & b & b^2 \\ 1 & c & c^2 \end{vmatrix} - \begin{vmatrix} a & a^2 & 1 \\ b & b^2 & 1 \\ c & c^2 & 1 \end{vmatrix} \text{ (passing } C_3 \text{ over } C_2 \text{ and } C_1 \text{ in the second determinant)}$$

$$= abc \begin{vmatrix} 1 & a & a^2 \\ 1 & b & b^2 \\ 1 & c & c^2 \end{vmatrix} - (-1)^2 \begin{vmatrix} 1 & a & a^2 \\ 1 & b & b^2 \\ 1 & c & c^2 \end{vmatrix} = (abc-1) \begin{vmatrix} 1 & a & a^2 \\ 1 & b & b^2 \\ 1 & c & c^2 \end{vmatrix} = 0.$$

(as the given determinant is equal to zero).

Since a, b, c are all different, so

$$\begin{vmatrix} 1 & a & a^2 \\ 1 & b & b^2 \\ 1 & c & c^2 \end{vmatrix} \neq 0. \text{ Hence } abc - 1 = 0 \text{ or } abc = 1.$$

Example 15.24 Solve the equation $\begin{vmatrix} a+x & b+x & c+x \\ b+x & c+x & a+x \\ c+x & a+x & b+x \end{vmatrix} = 0.$

Solution. Operating $C_1 \leftarrow C_1 + (C_2 + C_3)$ and taking $(a + b + c + 3x)$ as common from first column, we obtain

$$(a+b+c+3x) \begin{vmatrix} 1 & b+x & c+x \\ 1 & c+x & a+x \\ 1 & a+x & b+x \end{vmatrix} = 0. \text{ (operating } R_2 \leftarrow R_2 - R_1 \text{ and } R_3 \leftarrow R_3 - R_1)$$

$$(a+b+c+3x) \begin{vmatrix} 1 & b+x & c+x \\ 0 & c-b & a-c \\ 0 & a-b & b-c \end{vmatrix} = 0.$$

i.e., $(a + b + c + 3x) [-(b - c)^2 + (c - a)(a - b)] = 0.$

∴ $x = -(a + b + c)/3$, provided that $(b - c)^2 \neq (a - b)(c - a)$.

Example 15.25 Solve the equation

$$\begin{vmatrix} x+2 & 2x+3 & 3x+4 \\ 2x+3 & 3x+4 & 4x+5 \\ 3x+5 & 5x+8 & 10x+17 \end{vmatrix} = 0.$$

Solution. Operating $R_3 \leftarrow R_3 - (R_1 + R_2)$, we obtain

$$\begin{vmatrix} x+2 & 2x+3 & 3x+4 \\ 2x+3 & 3x+4 & 4x+5 \\ 0 & 1 & 3x+8 \end{vmatrix} = 0 \quad \text{(operating } R_2 \leftarrow R_2 - R_1 \text{ and } R_1 \leftarrow R_1 + R_3)$$

$$\begin{vmatrix} x+2 & 2x+4 & 6x+12 \\ x+1 & x+1 & x+1 \\ 0 & 1 & 3x+8 \end{vmatrix} = 0$$

or $\quad (x+1)(x+2)\begin{vmatrix} 1 & 2 & 6 \\ 1 & 1 & 1 \\ 0 & 1 & 3x+8 \end{vmatrix} = 0 \quad \text{(operate } R_1 \leftarrow R_1 - R_2)$

$$(x+1)(x+2)\begin{vmatrix} 0 & 1 & 5 \\ 1 & 1 & 1 \\ 0 & 1 & 3x+8 \end{vmatrix} = 0.$$

Expanding the determinant by using the first column, we have
$-(x+1)(x+2)(3x+8-5) = 0$ or $(x+1)(x+2)(x+1) = 0$
$\therefore \quad x = -1, -1, -2.$

Exercise 15.4

1. Compute the determinant of each of the following matrices:

(i) $\begin{bmatrix} 2 & 3 & -2 & 4 \\ 3 & -2 & 1 & 2 \\ 3 & 2 & 3 & 4 \\ -2 & 4 & 0 & 5 \end{bmatrix}$,

(ii) $\begin{bmatrix} 3 & 5 & 8 & 3 \\ 2 & -3 & -5 & -10 \\ 4 & 1 & 6 & 7 \\ 5 & -4 & -2 & -4 \end{bmatrix}$,

(iii) $\begin{bmatrix} 2 & -5 & -7 & -5 \\ 8 & 3 & -11 & -27 \\ 3 & -9 & 4 & 2 \\ -5 & 2 & 3 & 13 \end{bmatrix}$,

(iv) $\begin{bmatrix} 5 & -3 & 7 & 2 \\ 3 & 8 & -6 & -9 \\ 8 & -1 & 1 & -10 \\ 1 & 3 & 5 & 12 \end{bmatrix}$.

Ans. (i) −286 (ii) 0, (iii) −868 (iv) −1368.

2. For the matrix

$A = \begin{bmatrix} 3 & 9 & 5 \\ 4 & 6 & 0 \\ -1 & -3 & 2 \end{bmatrix}$, show that det (A^T) = det (A).

3. Evaluate the following determinants:

(i) $\begin{vmatrix} a+b & a & b \\ a & c+a & c \\ b & c & b+c \end{vmatrix}$, (ii) $\begin{vmatrix} 1+i & 1-i & i \\ 1-i & i & 1+i \\ i & 1+i & 1-i \end{vmatrix}$ where $i = \sqrt{-1}$.

Ans. (i) 4 abc (ii) 4 + 7i.

4. Prove, without expanding, that each of the following determinants vanishes:

(i) $\begin{vmatrix} -3 & 1 & 1 & 1 \\ 1 & -3 & 1 & 1 \\ 1 & 1 & -3 & 3 \\ 1 & 1 & 1 & -3 \end{vmatrix}$, (ii) $\begin{vmatrix} 1 & a & a^2-bc \\ 1 & b & b^2-ca \\ 1 & c & c^2-ab \end{vmatrix}$, (iii) $\begin{vmatrix} a-b & b-c & c-a \\ b-c & c-a & a-b \\ c-a & a-b & b-c \end{vmatrix}$,

(iv) $\begin{vmatrix} 1^2 & 2^2 & 3^2 & 4^2 \\ 2^2 & 3^2 & 4^2 & 5^2 \\ 3^2 & 4^2 & 5^2 & 6^2 \\ 4^2 & 5^2 & 6^2 & 7^2 \end{vmatrix}$, (v) $\begin{vmatrix} 3 & 5 & 8 & 3 \\ 2 & -3 & -5 & -10 \\ 4 & 1 & 6 & 7 \\ 5 & -4 & -2 & -4 \end{vmatrix}$, (vi) $\begin{vmatrix} 1 & bc & a(b+c) \\ 1 & ca & b(c+a) \\ 1 & ab & c(a+b) \end{vmatrix}$.

Prove the following:

5. $\begin{vmatrix} a & b & c \\ l & m & n \\ p & q & r \end{vmatrix} = \begin{vmatrix} m & b & q \\ l & a & p \\ n & c & r \end{vmatrix} = \begin{vmatrix} l & m & n \\ p & q & r \\ a & b & c \end{vmatrix}$.

6. $\begin{vmatrix} a+b & b+c & c+a \\ b+c & c+a & a+b \\ c+a & a+b & b+c \end{vmatrix} = 2\begin{vmatrix} a & b & c \\ b & c & a \\ c & a & b \end{vmatrix}$.

7. $\begin{vmatrix} a & a+b & a+b+c \\ 2a & 3a+2b & 4a+3b+2c \\ 3a & 6a+3b & 10a+6b+3c \end{vmatrix} = a^3$. **8.** $\begin{vmatrix} b+c & a-c & a-b \\ b-c & c+a & b-a \\ c-b & c-a & a+b \end{vmatrix} = 8abc$.

9. $\begin{vmatrix} a & b & c \\ a-b & b-c & c-a \\ b+c & c+a & a+b \end{vmatrix} = a^3+b^3+c^3-3abc$.

10. $\begin{vmatrix} 1 & 1 & 1 \\ \alpha & \beta & r \\ \alpha^2 & \beta^2 & r^2 \end{vmatrix} = (\alpha-\beta)(\beta-\gamma)(\gamma-\alpha)$.

11. $\begin{vmatrix} (b+c)^2 & a^2 & a^2 \\ b^2 & (c+a)^2 & b^2 \\ c^2 & c^2 & (a+b)^2 \end{vmatrix} = 2abc(a+b+c)^3$. (*JNTU-1998*)

12. $\begin{vmatrix} 1 & \alpha & \beta\gamma \\ 1 & \beta & \gamma\alpha \\ 1 & \gamma & \alpha\beta \end{vmatrix} = (\alpha-\beta)(\beta-\gamma)(\gamma-\alpha)$.

13. $\begin{vmatrix} 1 & 1 & 1 \\ a & b & c \\ a^3 & b^3 & c^3 \end{vmatrix} = (a-b)(b-c)(c-a)(a+b+c)$.

14. $\begin{vmatrix} a & a^2 & b+c \\ b & b^2 & c+a \\ c & c^2 & a+b \end{vmatrix} = (a-b)(b-c)(c-a)(a+b+c).$

15. $\begin{vmatrix} a+b+c & -c & -b \\ -c & a+b+c & -a \\ -b & -a & a+b+c \end{vmatrix} = 2(b+c)(c+a)(a+b).$

16. $\begin{vmatrix} a & a^2 & bc \\ b & b^2 & ca \\ c & c^2 & ab \end{vmatrix} = \begin{vmatrix} 1 & a^2 & a^3 \\ 1 & b^2 & b^3 \\ 1 & c^2 & c^3 \end{vmatrix} = (a-b)(b-c)(c-a)(ab+bc+ca).$

17. $\begin{vmatrix} 1+a & 2 & 3 \\ 1 & 2+a & 3 \\ 1 & 2 & 3+a \end{vmatrix} = a^2(a+6).$

18. $\begin{vmatrix} 1+a & 1 & 1 & 1 \\ 1 & 1+a & 1 & 1 \\ 1 & 1 & 1+a & 1 \\ 1 & 1 & 1 & 1+a \end{vmatrix} = a^4 + 4a^3.$

19. $\begin{vmatrix} a^2+\lambda & ab & ac & ad \\ ab & b^2+\lambda & bc & bd \\ ac & bc & c^2+\lambda & cd \\ ad & bd & cd & d^2+\lambda \end{vmatrix} = \lambda^3(a^2+b^2+c^2+d^2+\lambda).$

[**Hint.** Operate $C_1 \leftarrow aC_1$, $C_2 \leftarrow bC_2$, $C_3 \leftarrow cC_3$, and $C_4 \leftarrow dC_4$, divide the determinant, by *abcd*.

Now take *a, b, c, d* common from R_1, R_2, R_3, R_4 respectively. Next operate $C_1 \leftarrow C_1 + (C_2 + C_3 + C_4)$ and take out $(\Sigma a^2 + \lambda)$ as a common factor from C_1. Finally, operating $R_2 \leftarrow R_2 - R_1, R_3 \leftarrow R_3 - R_1$ and $R_4 \leftarrow R_4 - R_1$, we obtain the desired result].

20. $\begin{vmatrix} a^3 & 3a^2 & 3a & 1 \\ a^2 & a^2+2a & 2a+1 & 1 \\ a & 2a+1 & a+2 & 1 \\ 1 & 3 & 3 & 1 \end{vmatrix} = (a-1)^6.$

Solve the following equations:

21. $\begin{vmatrix} x-3 & x+2 & x-1 \\ x+2 & x-4 & x+4 \\ x-1 & x+4 & x-5 \end{vmatrix} = 0.$ **22.** $\begin{vmatrix} 4x+5 & 4x+7 & 4x+9 \\ 4x+9 & 4x+5 & 4x+7 \\ 4x+7 & 4x+9 & 4x+5 \end{vmatrix} = 0.$

Ans 2/33   **Ans.** $x = -1.75$

23. $\begin{vmatrix} x+1 & 2x+1 & 3x+1 \\ 2x & 4x+3 & 6x+3 \\ 4x+1 & 6x+4 & 8x+4 \end{vmatrix} = 0.$ **Ans.** $x = 0, -1/2$

15.11 Product of two determinants

The determinant of a product of two square matrices of the same order **A** and **B** is the product of their determinants, that is det **(AB)** = det **(A)** det **(B)** or |**AB**| = |**A**||**B**|.

Since |**A**| = |**A**^{T}|, we can multiply two determinants in any one of the following ways:

(*i*) row by row, (*ii*) column by column, (*iii*) row by column, (*iv*) column by row. The value of the determinant is same in each case.

15.12 Illustrative Examples

Example 15.26 Given the matrices

$$A = \begin{bmatrix} 2 & 3 & 1 \\ 3 & 2 & -2 \\ 4 & -4 & 3 \end{bmatrix}, B = \begin{bmatrix} 1 & 3 & -2 \\ 2 & 1 & 3 \\ -1 & 2 & 2 \end{bmatrix}. \text{ Verify that } |AB| = |A||B|.$$

Solution. We have

$$AB = \begin{bmatrix} 2 & 3 & 1 \\ 3 & 2 & -2 \\ 4 & -4 & 3 \end{bmatrix} \begin{bmatrix} 1 & 3 & -2 \\ 2 & 1 & 3 \\ -1 & 2 & 2 \end{bmatrix} = \begin{bmatrix} 2+6-1 & 6+3+2 & -4+9+2 \\ 3+4+2 & 9+2-4 & -6+6-4 \\ 4-8-3 & 12-4+6 & -8-12+6 \end{bmatrix}$$

$$= \begin{bmatrix} 7 & 11 & 7 \\ 9 & 7 & -4 \\ -7 & 14 & -14 \end{bmatrix}$$

Therefore, $|AB| = \begin{vmatrix} 7 & 11 & 7 \\ 9 & 7 & -4 \\ -7 & 14 & -14 \end{vmatrix} = 7(-98+56) - 11(-126-28) + 7(126+49)$

$= 7(-42) - 11(-154) + 7(175) = -294 + 1694 + 1225 = 2625.$

The product $|A||B| = \begin{vmatrix} 2 & 3 & 1 \\ 3 & 2 & -2 \\ 4 & -4 & 3 \end{vmatrix} \begin{vmatrix} 1 & 3 & -2 \\ 2 & 1 & 3 \\ -1 & 2 & 2 \end{vmatrix}$

$= \begin{vmatrix} 2+9-2 & 4+3+3 & -2+6+2 \\ 3+6+4 & 6+2-6 & -3+4-4 \\ 4-12-6 & 8-4+9 & -4-8+6 \end{vmatrix}$ (multiplying determinants row by row)

$= \begin{vmatrix} 9 & 10 & 6 \\ 13 & 2 & -3 \\ -14 & 13 & -6 \end{vmatrix} \begin{matrix} \text{(Operate } R_1 \leftarrow R_1 + 2R_2) \\ \\ \text{(Operate } R_3 \leftarrow R_3 - 2R_2) \end{matrix} = \begin{vmatrix} 35 & 14 & 0 \\ 13 & 2 & -3 \\ -40 & 9 & 0 \end{vmatrix}.$

Expanding the determinant by using the **third** column, we have

$$|\mathbf{A}||\mathbf{B}| = 3\begin{vmatrix} 35 & 14 \\ -40 & 9 \end{vmatrix} = 3(315 + 560) = 2625, \text{ which is same as } |\mathbf{AB}|.$$

Note: We can also find $|\mathbf{A}|$ and $|\mathbf{B}|$ and then multiply.

Example 15.27 Express $\begin{vmatrix} 0 & z & y \\ z & 0 & x \\ y & x & 0 \end{vmatrix}^2$ as a determinant. Hence evaluate the determinant.

Solution.
$$\begin{vmatrix} 0 & z & y \\ z & 0 & x \\ y & x & 0 \end{vmatrix}^2 = \begin{vmatrix} 0 & z & y \\ z & 0 & x \\ y & x & 0 \end{vmatrix}\begin{vmatrix} 0 & z & y \\ z & 0 & x \\ y & x & 0 \end{vmatrix}$$

$$= \begin{vmatrix} 0+z^2+y^2 & 0+0+xy & 0+zx+0 \\ 0+0+xy & z^2+0+x^2 & zy+0+0 \\ 0+xz+0 & yz+0+0 & y^2+x^2+0 \end{vmatrix} \quad \text{(multiplying determinants row by row)}$$

$$= \begin{vmatrix} y^2+z^2 & xy & zx \\ xy & z^2+x^2 & yz \\ zx & yz & x^2+y^2 \end{vmatrix}.$$

Since $\begin{vmatrix} 0 & z & y \\ z & 0 & x \\ y & x & 0 \end{vmatrix} = -z\begin{vmatrix} z & x \\ y & 0 \end{vmatrix} + y\begin{vmatrix} z & 0 \\ y & x \end{vmatrix} = 2xyz.$ $\therefore \begin{vmatrix} 0 & z & y \\ z & 0 & x \\ y & x & 0 \end{vmatrix}^2 = 4x^2y^2z^2.$

Example 15.28 Show that the

$$\det{(\mathbf{A})} = \begin{vmatrix} 2b_1+c_1 & c_1+3a_1 & 2a_1+3b_1 \\ 2b_2+c_2 & c_2+3a_2 & 2a_2+3b_2 \\ 2b_3+c_3 & c_3+3a_3 & 2a_3+3b_3 \end{vmatrix} \text{ is a multiple of the}$$

determinant $\begin{vmatrix} a_1 & b_1 & c_1 \\ a_2 & b_2 & c_2 \\ a_3 & b_3 & c_3 \end{vmatrix}.$

Solution. The given determinant can be written in the form

$$|\mathbf{A}| = \begin{vmatrix} a_1.0+b_1.2+c_1.1 & a_1.3+b_1.0+c_1.1 & a_1.2+b_1.3+c_1.0 \\ a_2.0+b_2.2+c_2.1 & a_2.3+b_2.0+c_2.1 & a_2.2+b_2.3+c_2.0 \\ a_3.0+b_3.2+c_3.1 & a_3.3+b_3.0+c_3.1 & a_3.2+b_3.3+c_3.0 \end{vmatrix}.$$

From the rule for writing the elements of the product determinant, we can write

$$|A| = \begin{vmatrix} a_1 & b_1 & c_1 \\ a_2 & b_2 & c_2 \\ a_3 & b_3 & c_3 \end{vmatrix} \begin{vmatrix} 0 & 2 & 1 \\ 3 & 0 & 1 \\ 2 & 3 & 0 \end{vmatrix} \text{ which proves the result.}$$

Example 15.29 Prove that

$$\begin{vmatrix} 2ll' & lm' + ml' & ln' + nl' \\ lm' + ml' & 2mm' & mn' + nm' \\ ln' + nl' & mn' + nm' & 2nn' \end{vmatrix} = 0.$$

Solution. The given determinant can be written as

$$\begin{vmatrix} l.l' + l'.l + 0.0 & lm' + l'.m + 0.0 & l.n' + l'.n + 0.0 \\ m.l' + m'.l + 0.0 & m.m' + m'.m + 0.0 & m.n' + m'.n + 0.0 \\ n.l' + n'.l + 0.0 & n.m' + n'.m + 0.0 & n.n' + n'.n + 0.0 \end{vmatrix}$$

$$= \begin{vmatrix} l & l' & 0 \\ m & m' & 0 \\ n & n' & 0 \end{vmatrix} \begin{vmatrix} l' & l & 0 \\ m' & m & 0 \\ n' & n & 0 \end{vmatrix} = 0,$$

because each of the determinants on the left is a zero-determinant.

Example 15.30 Show that

$$\begin{vmatrix} A_1 & B_1 & C_1 \\ A_2 & B_2 & C_2 \\ A_3 & B_3 & C_3 \end{vmatrix} = \begin{vmatrix} a_1 & b_1 & c_1 \\ a_2 & b_2 & c_2 \\ a_3 & b_3 & c_3 \end{vmatrix}^2$$

where capital letters denote the co-factors of the corresponding small letters of the determinant on the right.

Solution. Let $|P| = \begin{vmatrix} a_1 & b_1 & c_1 \\ a_2 & b_2 & c_2 \\ a_3 & b_3 & c_3 \end{vmatrix}$ and $|Q| = \begin{vmatrix} A_1 & B_1 & C_1 \\ A_2 & B_2 & C_2 \\ A_3 & B_3 & C_3 \end{vmatrix}.$

Then multiplication of determinants row by row gives

$$|P||Q| = \begin{vmatrix} a_1A_1 + b_1B_1 + c_1C_1 & a_1A_2 + b_1B_2 + c_1C_2 & a_1A_3 + b_1B_3 + c_1C_3 \\ a_2A_1 + b_2B_1 + c_2C_1 & a_2A_2 + b_2B_2 + c_2C_2 & a_2A_3 + b_2B_3 + c_2C_3 \\ a_3A_1 + b_3B_1 + c_3C_1 & a_3A_2 + b_3B_2 + c_3C_2 & a_3A_3 + b_3B_3 + c_3C_3 \end{vmatrix}$$

$$= \begin{vmatrix} |P| & 0 & 0 \\ 0 & |P| & 0 \\ 0 & 0 & |P| \end{vmatrix} = \{|P|\}^3. \quad \text{Therefore, } |Q| = \{|P|\}^2, \text{ if } |P| \neq 0.$$

Note. $|Q|$ is called the *adjugate* or *reciprocal* determinant of **P**.

Example 15.31 Show that the

$$\det(\mathbf{D}) = \begin{vmatrix} 1 & \cos(\beta - \alpha) & \cos(\gamma - \alpha) \\ \cos(\alpha - \beta) & 1 & \cos(\gamma - \beta) \\ \cos(\alpha - \gamma) & \cos(\beta - \gamma) & 1 \end{vmatrix} \text{ is a perfect square (of a deter-}$$

minant) and find its value.

Solution. The give det **(D)** =

$$\begin{vmatrix} \cos\alpha\cos\alpha + \sin\alpha\sin\alpha & \cos\alpha\cos\beta + \sin\alpha\sin\beta & \cos\alpha\cos\gamma + \sin\alpha\sin\gamma \\ \cos\beta\cos\alpha + \sin\beta\sin\alpha & \cos\beta\cos\beta + \sin\beta\sin\beta & \cos\beta\cos\gamma + \sin\beta\sin\gamma \\ \cos\gamma\cos\alpha + \sin\gamma\sin\alpha & \cos\gamma\cos\beta + \sin\gamma\sin\beta & \cos\gamma\cos\gamma + \sin\gamma\sin\gamma \end{vmatrix}$$

$$= \begin{vmatrix} \cos\alpha & \sin\alpha & 0 \\ \cos\beta & \sin\beta & 0 \\ \cos\gamma & \sin\gamma & 0 \end{vmatrix}\begin{vmatrix} \cos\alpha & \sin\alpha & 0 \\ \cos\beta & \sin\beta & 0 \\ \cos\gamma & \sin\gamma & 0 \end{vmatrix} = \begin{vmatrix} \cos\alpha & \sin\alpha & 0 \\ \cos\beta & \sin\beta & 0 \\ \cos\gamma & \sin\gamma & 0 \end{vmatrix}^2 = 0 \times 0 = 0.$$

Exercise 15.5

1. Express $\begin{vmatrix} 1 & 4 & -3 \\ -3 & -2 & 1 \\ 1 & 3 & 5 \end{vmatrix}\begin{vmatrix} 3 & 4 & 2 \\ 2 & 5 & 4 \\ -1 & 6 & 1 \end{vmatrix}$

   as a single determinant and hence find its value.   **Ans.** −3384

2. Find the value of the det (M) if
   $$|\mathbf{M}| = 3\{|\mathbf{A}|\}^2 + |\mathbf{AB}| + \{|\mathbf{B}|\}^2$$

   where $|\mathbf{A}| = \begin{vmatrix} 2 & 1 & 1 \\ 1 & 2 & 1 \\ 0 & -1 & 0 \end{vmatrix}, |\mathbf{B}| = \begin{vmatrix} 1/2 & 0 & -1 \\ -1 & 0 & 1 \\ 1 & 2 & 3 \end{vmatrix},$

   without evaluating |A| and |B| independently.
3. Show that   **Ans.** 5

   $$\begin{vmatrix} (x-a)^2 & (y-a)^2 & (z-a)^2 \\ (x-b)^2 & (y-b)^2 & (z-b)^2 \\ (x-c)^2 & (y-c)^2 & (z-c)^2 \end{vmatrix} = 2\begin{vmatrix} x^2 & x & 1 \\ y^2 & y & 1 \\ z^2 & z & 1 \end{vmatrix} \times \begin{vmatrix} a^2 & a & 1 \\ b^2 & b & 1 \\ c^2 & c & 1 \end{vmatrix}$$
   $$= 2(x-y)(y-z)(z-x)(a-b)(b-c)(c-a).$$

   $$\left[\textbf{Hint.} \text{R.H.S.} = -2\begin{vmatrix} x^2 & x & 1 \\ y^2 & y & 1 \\ z^2 & z & 1 \end{vmatrix}\begin{vmatrix} 1 & a & a^2 \\ 1 & b & b^2 \\ 1 & c & c^2 \end{vmatrix} = \begin{vmatrix} x^2 & x & 1 \\ y^2 & y & 1 \\ z^2 & z & 1 \end{vmatrix}\begin{vmatrix} 1 & -2a & a^2 \\ 1 & -2b & b^2 \\ 1 & -2c & c^2 \end{vmatrix}\right]$$

4. Evaluate $\begin{vmatrix} \lambda & c & -b \\ -c & \lambda & a \\ b & -a & \lambda \end{vmatrix}\begin{vmatrix} a^2 + \lambda^2 & ab + \lambda c & ca - \lambda b \\ ab - \lambda c & b^2 + \lambda^2 & bc + \lambda a \\ ca + \lambda b & bc - \lambda a & c^2 + \lambda^2 \end{vmatrix}.$

   [**Hint.** The second determinant is adjugate of first.]

   Ans. $\lambda^3(a^2 + b^2 + c^2 + \lambda^2)^3.$

5. Express
$$\begin{vmatrix} 2bc-a^2 & c^2 & b^2 \\ c^2 & 2ca-b^2 & a^2 \\ b^2 & a^2 & 2ab-c^2 \end{vmatrix}$$

as the square of a determinant and hence find its value.

[**Hint.** Proceeding as in examples 15.28 and 15.29 the given determinant

$$= \begin{vmatrix} a & b & c \\ b & c & a \\ c & a & b \end{vmatrix} \begin{vmatrix} -a & c & b \\ -b & a & c \\ -c & b & a \end{vmatrix} = \begin{vmatrix} a & b & c \\ b & c & a \\ c & a & b \end{vmatrix}^2 \Bigg].$$

Ans. $(a^3 + b^3 + c^3 - 3\,abc)^2$.

6. Show that

$$\begin{vmatrix} b^2+c^2 & ab & ca \\ ab & c^2+a^2 & bc \\ ca & bc & a^2+b^2 \end{vmatrix} = \begin{vmatrix} 0 & c & b \\ c & 0 & a \\ b & a & 0 \end{vmatrix}^2 = 4a^2b^2c^2.$$

7. If $|\mathbf{P}| = \begin{vmatrix} a & b & c \\ b & c & a \\ c & a & b \end{vmatrix}$, show that $|\mathbf{Q}| = \begin{vmatrix} bc-a^2 & ca-b^2 & ab-c^2 \\ ca-b^2 & ab-c^2 & bc-a^2 \\ ab-c^2 & bc-a^2 & ca-b^2 \end{vmatrix} = \{|\mathbf{P}|\}^2$.

[**Hint.** $|\mathbf{Q}| = adj|\mathbf{P}|$].

8. Show that

$$\begin{vmatrix} \sin(A+P) & \sin(B+P) & \sin(C+P) \\ \sin(A+Q) & \sin(B+Q) & \sin(C+Q) \\ \sin(A+R) & \sin(B+R) & \sin(C+R) \end{vmatrix} = 0.$$

15.13 Rank of a matrix

Def. *A matrix* **A** *is said to be of rank r when it has at least one* non-zero *minor of* $|A|$ *of order r, all minors of order* $(r + 1)$ *are zero.*

The rank of a matrix **A** is denoted symbolically by r or $r(A)$ or $\rho(A)$.

We know that any determinant can be expressed in terms of determinants of next lower order. As such if all minors of order $(r + 1)$ are zero then all minors of order $(r + 2)$ and higher orders (if any) are also zero. *Thus the rank of matrix* **A** *is equal to the order of its largest non-zero minor of* $|A|$.

Since as a result of elementary operations the zero or non-zero character of a minor is not changed, the rank of a matrix remains unaltered by these operations.

From the definition of the rank of a matrix it follows that:

(*i*) The rank of a non-singular matrix of order n is n. If the matrix is singular its rank is less than n.

(*ii*) The rank of an $m \times n$ matrix can at most be equal to the smaller of numbers m and n but it may be less.

(*iii*) If there is a non-zero minor of order r, then rank is $\geq r$.

(*iv*) The rank of a null matrix is zero and if the rank of a matrix is zero, then it must be a null matrix. The rank of a non-zero matrix is ≥ 1.

(*v*) The rank of $I_n = n$.

In practice, while determining the rank of a matrix, we may start with the largest order minor or minors. If they are all zero we go to the minors of next lower order and so on till we get a non-zero minor. The order of that minor is the rank of the given matrix.

Example 15.32 Determine the rank of the following matrices:

(*i*) $\begin{bmatrix} 1 & 2 & 3 \\ -4 & 0 & 5 \end{bmatrix}$, (*ii*) $\begin{bmatrix} 0 & 2 & 3 \\ 0 & 4 & 6 \\ 0 & 6 & 9 \end{bmatrix}$, (*iii*) $\begin{bmatrix} 1 & 2 & 3 \\ 2 & 4 & 7 \\ 3 & 6 & 10 \end{bmatrix}$, (*iv*) $\begin{bmatrix} 1 & 2 & 3 \\ 1 & 4 & 2 \\ 2 & 6 & 5 \end{bmatrix}$.

(*IETE, June 2009*)

Solution. (*i*) It is a 2×3 matrix, therefore the rank cannot exceed 2.

Since $\begin{vmatrix} 1 & 2 \\ -4 & 0 \end{vmatrix} \neq 0$, the rank is 2.

(*ii*) Obviously the given matrix is singular. Therefore its rank is less than 3. Also each of the nine minors of order 2 is zero. Hence the rank is less than 2.

Since it is non-zero matrix, the rank is 1.

(*iii*) Let the given matrix be denoted by **A**.

$$\therefore |A| = \begin{vmatrix} 1 & 2 & 3 \\ 2 & 4 & 7 \\ 3 & 6 & 10 \end{vmatrix} \text{ (Operate } R_2 \leftarrow R_2 - 2R_1 \text{ and } R_3 \leftarrow R_3 - 3R_1)$$

$$= \begin{vmatrix} 1 & 2 & 3 \\ 0 & 0 & 1 \\ 0 & 0 & 1 \end{vmatrix} = 0. \text{ Hence the matrix } \mathbf{A} \text{ is singular and therefore, the rank is less}$$

than 3.

Since $\begin{vmatrix} 2 & 3 \\ 4 & 7 \end{vmatrix} = 2 \neq 0$, the rank of the given matrix is 2.

(*iv*) Let the given matrix be denoted by **A**.

$$\therefore |A| = \begin{vmatrix} 1 & 2 & 3 \\ 1 & 4 & 2 \\ 2 & 6 & 5 \end{vmatrix} \text{ (operate } R_2 \leftarrow R_2 - R_1 \text{ and } R_3 \leftarrow R_3 - 2R_1)$$

$$= \begin{vmatrix} 1 & 2 & 3 \\ 0 & 2 & -1 \\ 0 & 2 & -1 \end{vmatrix} = 0. \text{ Matrix } \mathbf{A} \text{ is singular and therefore the rank is less than 3.}$$

Since $\begin{vmatrix} 1 & 2 \\ 1 & 4 \end{vmatrix} \neq 0$, the rank is 2.

Example 15.33 Find all values of μ for which rank of the matrix

$$A = \begin{bmatrix} \mu & -1 & 0 & 0 \\ 0 & \mu & -1 & 0 \\ 0 & 0 & \mu & -1 \\ -6 & 11 & -6 & 1 \end{bmatrix} \text{ is equal to 3.}$$

(I.E.T.E. Dec. 2005)

Solution. The matrix **A** is of order 4. Rank (**A**) ≤ 4. Now Rank (**A**) = 3, if $|A|$ = 0 and there is atleast one submatrix of order 3 whose determinant is not zero. Expanding the determinant through the elements of first row, we get

$$|A| = \mu \begin{vmatrix} \mu & -1 & 0 \\ 0 & \mu & -1 \\ 11 & -6 & 1 \end{vmatrix} + 1 \begin{vmatrix} 0 & -1 & 0 \\ 0 & \mu & -1 \\ -6 & -6 & 1 \end{vmatrix} = \mu \left[\mu(\mu - 6) + 1(11) \right] - 6$$

$$= \mu^3 - 6\mu^2 + 11\mu - 6 = (\mu - 1)(\mu - 2)(\mu - 3).$$

Setting $|A| = 0$, we obtain $\mu = 1, 2, 3$. The given matrix **A** possesses a minor of order 3

$$viz \begin{vmatrix} \mu & -1 & 0 \\ 0 & \mu & -1 \\ 0 & 0 & \mu \end{vmatrix} = \mu^3 \neq 0. \text{ (For } \mu = 1, 2, 3.)$$

Hence $r(\mathbf{A}) = 3$, when $\mu = 1$ or 2 or 3. For other values of μ, $r(\mathbf{A}) = 4$.

15.14 Adjoint of a square matrix

Definition. Let $A = (a_{ij})$ be a square matrix of order n and let A_{ij} be the cofactor of elements a_{ij} in the det(**A**). Then the transpose of the cofactor matrix (A_{ij}) is defined as the adjoint of **A**, and is denoted by *adj*A.

Thus if $A = \begin{bmatrix} a_{11} & a_{12} & a_{13} \\ a_{21} & a_{22} & a_{23} \\ a_{31} & a_{32} & a_{33} \end{bmatrix}$ is a square matrix of order 3,

then $adj\ A = \begin{bmatrix} A_{11} & A_{12} & A_{13} \\ A_{21} & A_{22} & A_{23} \\ A_{31} & A_{32} & A_{33} \end{bmatrix}^T = \begin{bmatrix} A_{11} & A_{21} & A_{31} \\ A_{12} & A_{22} & A_{32} \\ A_{13} & A_{23} & A_{33} \end{bmatrix}.$

If **A** is a square matrix of order 2, then adjoint **A** can be obtained directly by simply interchanging the diagonal elements and changing the signs of non-diagonal elements as is illustrated below:

If $A = \begin{bmatrix} 1 & 4 \\ 2 & 10 \end{bmatrix}$, then $adj\ A = \begin{bmatrix} 10 & -4 \\ -2 & 1 \end{bmatrix}.$

Properties of the Adjoint of a matrix

1. If **A** is a square matrix of order n, then $A(adjA) = (adjA)A = |A|I_n$, where I_n is a unit or identity matrix of order n.

2. If **A** is a square matrix of order n, then $adj(\mathbf{A}^T) = (adj\mathbf{A})^T$.

3. If **A** and **B** are two square matrices of the same order, then
$adj(\mathbf{AB}) = adj(\mathbf{B})\, adj(\mathbf{A})$.

We prove the property **1** when $n = 3$.

$$\mathbf{A}(adj\,\mathbf{A}) = \begin{bmatrix} a_{11} & a_{12} & a_{13} \\ a_{21} & a_{22} & a_{23} \\ a_{31} & a_{32} & a_{33} \end{bmatrix}\begin{bmatrix} A_{11} & A_{21} & A_{31} \\ A_{12} & A_{22} & A_{32} \\ A_{13} & A_{23} & A_{33} \end{bmatrix}.$$

$$= \begin{bmatrix} a_{11}A_{11}+a_{12}A_{12}+a_{13}A_{13} & a_{11}A_{21}+a_{12}A_{22}+a_{13}A_{23} \\ a_{21}A_{11}+a_{22}A_{12}+a_{23}A_{13} & a_{21}A_{21}+a_{22}A_{22}+a_{23}A_{23} \\ a_{31}A_{11}+a_{32}A_{12}+a_{33}A_{13} & a_{31}A_{21}+a_{32}A_{22}+a_{33}A_{23} \end{bmatrix}$$

$$\begin{bmatrix} a_{11}A_{31}+a_{12}A_{32}+a_{13}A_{33} \\ a_{21}A_{31}+a_{22}A_{32}+a_{23}A_{33} \\ a_{31}A_{31}+a_{32}A_{32}+a_{33}A_{33} \end{bmatrix}$$

$$= \begin{bmatrix} \det(\mathbf{A}) & 0 & 0 \\ 0 & \det(\mathbf{A}) & 0 \\ 0 & 0 & \det(\mathbf{A}) \end{bmatrix} = \det(\mathbf{A})\begin{bmatrix} 1 & 0 & 0 \\ 0 & 1 & 0 \\ 0 & 0 & 1 \end{bmatrix} = |\mathbf{A}|I_3 = |\mathbf{A}|I \quad ...(15.8)$$

Similarly, it can be shown in exactly the same manner that $(adj\mathbf{A})\mathbf{A} = |\mathbf{A}|\mathbf{I}$.

$$...(15.9)$$

Example 15.34 If $\mathbf{A} = \begin{bmatrix} 1 & 1 & 1 \\ 1 & 2 & -3 \\ 2 & -1 & 3 \end{bmatrix}$, verify that $\mathbf{A}(adj\mathbf{A}) = |\mathbf{A}|I_3 = (adj\mathbf{A})\mathbf{A}$.

Solution. We have $|\mathbf{A}| = \begin{vmatrix} 1 & 1 & 1 \\ 1 & 2 & -3 \\ 2 & -1 & 3 \end{vmatrix}$. The cofactors corresponding to the elements in $|\mathbf{A}|$ are

$$A_{11} = \begin{vmatrix} 2 & -3 \\ -1 & 3 \end{vmatrix} = 3, \; A_{12} = -\begin{vmatrix} 1 & -3 \\ 2 & 3 \end{vmatrix} = -9, \; A_{13} = \begin{vmatrix} 1 & 2 \\ 2 & -1 \end{vmatrix} = -5,$$

$$A_{21} = -\begin{vmatrix} 1 & 1 \\ -1 & 3 \end{vmatrix} = -4, \; A_{22} = \begin{vmatrix} 1 & 1 \\ 2 & 3 \end{vmatrix} = 1, \; A_{23} = -\begin{vmatrix} 1 & 1 \\ 2 & -1 \end{vmatrix} = 3,$$

$$A_{31} = \begin{vmatrix} 1 & 1 \\ 2 & -3 \end{vmatrix} = -5, \; A_{32} = -\begin{vmatrix} 1 & 1 \\ 1 & -3 \end{vmatrix} = 4, \; A_{33} = \begin{vmatrix} 1 & 1 \\ 1 & 2 \end{vmatrix} = 1.$$

$\therefore |\mathbf{A}| = (1)(3) + (1)(-9) + (1)(-5) = -11.$

The matrix of cofactors $= \begin{bmatrix} 3 & -9 & -5 \\ -4 & 1 & 3 \\ -5 & 4 & 1 \end{bmatrix}$.

$$\therefore adj\ \mathbf{A} = \begin{bmatrix} 3 & -9 & -5 \\ -4 & 1 & 3 \\ -5 & 4 & 1 \end{bmatrix}^T = \begin{bmatrix} 3 & -4 & -5 \\ -9 & 1 & 4 \\ -5 & 3 & 1 \end{bmatrix}.$$

Hence

$$\mathbf{A}(adj\ \mathbf{A}) = \begin{bmatrix} 1 & 1 & 1 \\ 1 & 2 & -3 \\ 2 & -1 & 3 \end{bmatrix}\begin{bmatrix} 3 & -4 & -5 \\ -9 & 1 & 4 \\ -5 & 3 & 1 \end{bmatrix} = \begin{bmatrix} 3-9-5 & -4+1+3 & -5+4+1 \\ 3-18+15 & -4+2-9 & -5+8-3 \\ 6+9-15 & -8-1+9 & -10-4+3 \end{bmatrix}$$

$$= \begin{bmatrix} -11 & 0 & 0 \\ 0 & -11 & 0 \\ 0 & 0 & -11 \end{bmatrix} = -11\begin{bmatrix} 1 & 0 & 0 \\ 0 & 1 & 0 \\ 0 & 0 & 1 \end{bmatrix} = |\mathbf{A}|I_3 .$$

Similarly, it can be shown that $(adj\mathbf{A})\mathbf{A} = |\mathbf{A}|I_3$. Hence the result.

15.15. Singular and the Non-singular Matrices

A square matrix $\mathbf{A}$ of order n is said to be **singular** if $|\mathbf{A}| = 0$, and it is called **non-singular** if $|\mathbf{A}| \neq 0$.

In other words, a square matrix of order n is singular if its rank $\mathbf{A} < n$ and non-singular if its rank $\mathbf{A} = n$. (Refer Article 15.13).

15.16 Inverse (or reciprocal) of a Square Matrix

Def. If $\mathbf{A}$ and $\mathbf{B}$ are square matrices of order n such that $\mathbf{AB} = \mathbf{BA} = \mathbf{I}$, ...(15.10) than $\mathbf{B}$ is called the inverse of $\mathbf{A}$ and is denoted by $\mathbf{A}^{-1}$, that is $\mathbf{B} = \mathbf{A}^{-1}$.

From the symmetry of (15.10) it follows that if $\mathbf{B}$ is inverse of $\mathbf{A}$ then $\mathbf{A}$ is inverse of $\mathbf{B}$.

Thus, it follows that $\mathbf{A} = \mathbf{B}^{-1} = (\mathbf{A}^{-1})^{-1}$

If $|\mathbf{A}| \neq 0$, we have from results (15.8) and (15.9) of Art. 15.14.

$$\mathbf{A}.\frac{adj\ \mathbf{A}}{|\mathbf{A}|} = \frac{adj\ \mathbf{A}}{|\mathbf{A}|}.\mathbf{A} = \mathbf{I}.$$

From the definition of inverse of $\mathbf{A}$ it follows that

$$\mathbf{A}^{-1} = \frac{adj\ \mathbf{A}}{|\mathbf{A}|}. \qquad ...(15.11)$$

Note. If a matrix has an inverse, both the matrix and its inverse must be non-singular. When a matrix $\mathbf{A}$ is singular, that is $|\mathbf{A}| = 0$, its inverse is not defined that is, a singular matrix has no inverse.

Properties of the inverse of a matrix

1. The inverse of a non-sigular matrix, is unique.
2. The inverse of the inverse is the original matrix itself, that is $(\mathbf{A}^{-1})^{-1} = \mathbf{A}$.
3. The inverse of the transpose of a matrix is the transpose of its inverse, that is

$$(A^T)^{-1} = (A^{-1})^T.$$

4. Let $D = diag(d_{11}, d_{22}, ..., d_{nn})$, $d_{ii} \neq 0$. Then $D^{-1} = diag(1/d_{11}, 1/d_{22}, ..., 1/d_{nn})$.

5. If A and B are two invertible matrices of the same order, then
   $(AB)^{-1} = B^{-1}A^{-1}$.
   Proof. We have $(AB)(AB)^{-1} = I$.
   Premultiplying both sides first by A^{-1} and then by B^{-1}, we obtain
   $B^{-1}A^{-1}(AB)(AB)^{-1} = B^{-1}A^{-1}$
   or $B^{-1}(A^{-1}A)B(AB)^{-1} = B^{-1}A^{-1}$ or $(AB)^{-1} = B^{-1}A^{-1}$.
   Hence for more than two factors,
   $(AB ...PQ)^{-1} = Q^{-1}P^{-1}...B^{-1}A^{-1}$.

6. If a matrix A is symmetric and invertible, then its inverse matrix, that is A^{-1} is symmetric matrix.

15.17 Illustrative Examples

Example 15.35 Find the inverse of the matrix $A = \begin{bmatrix} -1 & 1 & 2 \\ 3 & -1 & 1 \\ -1 & 3 & 4 \end{bmatrix}$ and check the answer by direct multiplication.

Solution. $\det(A) = -1 \begin{vmatrix} -1 & 1 \\ 3 & 4 \end{vmatrix} - 1 \begin{vmatrix} 3 & 1 \\ -1 & 4 \end{vmatrix} + 2 \begin{vmatrix} 3 & -1 \\ -1 & 3 \end{vmatrix}$

$= -1(-4-3) - 1(12+1) + 2(9-1) = 10 \neq 0.$

Thus A^{-1} exists and is given by $A^{-1} = \dfrac{1}{|A|} adj\, A$.

The cofactors corresponding to the elements in det (A) are

$A_{11} = \begin{vmatrix} -1 & 1 \\ 3 & 4 \end{vmatrix} = -7,\ A_{12} = -\begin{vmatrix} 3 & 1 \\ -1 & 4 \end{vmatrix} = -13,\ A_{13} = \begin{vmatrix} 3 & -1 \\ -1 & 3 \end{vmatrix} = 8,$

$A_{21} = -\begin{vmatrix} 1 & 2 \\ 3 & 4 \end{vmatrix} = 2,\ A_{22} = \begin{vmatrix} -1 & 2 \\ -1 & 4 \end{vmatrix} = -2,\ A_{23} = -\begin{vmatrix} -1 & 1 \\ -1 & 3 \end{vmatrix} = 2,$

$A_{31} = \begin{vmatrix} 1 & 2 \\ -1 & 1 \end{vmatrix} = 3,\ A_{32} = -\begin{vmatrix} -1 & 2 \\ 3 & 1 \end{vmatrix} = 7,\ A_{33} = \begin{vmatrix} -1 & 1 \\ 3 & -1 \end{vmatrix} = -2.$

$\therefore adj\, A = \begin{bmatrix} -7 & -13 & 8 \\ 2 & -2 & 2 \\ 3 & 7 & -2 \end{bmatrix}^T = \begin{bmatrix} -7 & 2 & 3 \\ -13 & -2 & 7 \\ 8 & 2 & -2 \end{bmatrix}.$

Hence $A^{-1} = \dfrac{1}{|A|} adj\, A = \dfrac{1}{10} \begin{bmatrix} -7 & 2 & 3 \\ -13 & -2 & 7 \\ 8 & 2 & -2 \end{bmatrix} = \begin{bmatrix} -0.7 & 0.2 & 0.3 \\ -1.3 & -0.2 & 0.7 \\ 0.8 & 0.2 & -0.2 \end{bmatrix}.$

$$AA^{-1} = \begin{bmatrix} -1 & 1 & 2 \\ 3 & -1 & 1 \\ -1 & 3 & 4 \end{bmatrix} = \begin{bmatrix} -0.7 & 0.2 & 0.3 \\ -1.3 & -0.2 & 0.7 \\ 0.8 & 0.2 & -0.2 \end{bmatrix}$$

$$= \begin{bmatrix} 0.7-1.3+1.6 & -0.2-0.2+0.4 & -0.3+0.7-0.4 \\ -2.1+1.3+0.8 & 0.6+0.2+0.2 & 0.9-0.7-0.2 \\ 0.7-3.9+3.2 & -0.2-0.6+0.8 & -0.3+2.1-0.8 \end{bmatrix} = \begin{bmatrix} 1 & 0 & 0 \\ 0 & 1 & 0 \\ 0 & 0 & 1 \end{bmatrix} = I.$$

We can also show that $A^{-1}A = I$. This gives the required check.

Example 15.36 Show that the matrix $A = \begin{bmatrix} 2 & -1 & 1 \\ -1 & 2 & -1 \\ 1 & -1 & 2 \end{bmatrix}$ satisfies the matrix

equation $A^3 - 6A^2 + 9A - 4I = O$ where I is an identity matrix of order 3. Hence, find the matrix (*i*) A^{-1} and (*ii*) A^{-2}.

Solution. We have

$$A^2 = AA = \begin{bmatrix} 2 & -1 & 1 \\ -1 & 2 & -1 \\ 1 & -1 & 2 \end{bmatrix} \begin{bmatrix} 2 & -1 & 1 \\ -1 & 2 & -1 \\ 1 & -1 & 2 \end{bmatrix}$$

$$= \begin{bmatrix} 4+1+1 & -2-2-1 & 2+1+2 \\ -2-2-1 & 1+4+1 & -1-2-2 \\ 2+1+2 & -1-2-2 & 1+1+4 \end{bmatrix} = \begin{bmatrix} 6 & -5 & 5 \\ -5 & 6 & -5 \\ 5 & -5 & 6 \end{bmatrix}.$$

$$A^3 = A^2A = \begin{bmatrix} 6 & -5 & 5 \\ -5 & 6 & -5 \\ 5 & -5 & 6 \end{bmatrix} \begin{bmatrix} 2 & -1 & 1 \\ -1 & 2 & -1 \\ 1 & -1 & 2 \end{bmatrix}$$

$$= \begin{bmatrix} 12+5+5 & -6-10-5 & 6+5+10 \\ -10-6-5 & 5+12+5 & -5-6-10 \\ 10+5+6 & -5-10-6 & 5+5+12 \end{bmatrix} = \begin{bmatrix} 22 & -21 & 21 \\ -21 & 22 & -21 \\ 21 & -21 & 22 \end{bmatrix}.$$

Substituting in $C = A^3 - 6A^2 + 9A - 4I$, we get

$$C = \begin{bmatrix} 22 & -21 & 21 \\ -21 & 22 & -21 \\ 21 & -21 & 22 \end{bmatrix} - \begin{bmatrix} 36 & -30 & 30 \\ -30 & 36 & -30 \\ 30 & -30 & 36 \end{bmatrix} + \begin{bmatrix} 18 & -9 & 9 \\ -9 & 18 & -9 \\ 9 & -9 & 18 \end{bmatrix}$$

$$- \begin{bmatrix} 4 & 0 & 0 \\ 0 & 4 & 0 \\ 0 & 0 & 4 \end{bmatrix} = \begin{bmatrix} 0 & 0 & 0 \\ 0 & 0 & 0 \\ 0 & 0 & 0 \end{bmatrix} = O.$$

(*i*) Premultiplying $A^3 - 6A^2 + 9A - 4I = O$ by A^{-1}, we get
$A^{-1}A^3 - 6A^{-1}A^2 + 9A^{-1}A - 4A^{-1} = O$ or $4A^{-1} = A^2 - 6A + 9I$

$$= \begin{bmatrix} 6 & -5 & 5 \\ -5 & 6 & -5 \\ 5 & -5 & 6 \end{bmatrix} - \begin{bmatrix} 12 & -6 & 6 \\ -6 & 12 & -6 \\ 6 & -6 & 12 \end{bmatrix} + \begin{bmatrix} 9 & 0 & 0 \\ 0 & 9 & 0 \\ 0 & 0 & 9 \end{bmatrix} = \begin{bmatrix} 3 & 1 & -1 \\ 1 & 3 & 1 \\ -1 & 1 & 3 \end{bmatrix}.$$

Hence $\mathbf{A}^{-1} = 1/4 \begin{bmatrix} 3 & 1 & -1 \\ 1 & 3 & 1 \\ -1 & 1 & 3 \end{bmatrix}$.

(ii) $\mathbf{A}^{-2} = \left(\mathbf{A}^{-1}\right)^2 = \dfrac{1}{16} \begin{bmatrix} 3 & 1 & -1 \\ 1 & 3 & 1 \\ -1 & 1 & 3 \end{bmatrix} \begin{bmatrix} 3 & 1 & -1 \\ 1 & 3 & 1 \\ -1 & 1 & 3 \end{bmatrix}$

$= \dfrac{1}{16} \begin{bmatrix} 9+1+1 & 3+3-1 & -3+1-3 \\ 3+3-1 & 1+9+1 & -1+3+3 \\ -3+1-3 & -1+3+3 & 1+1+9 \end{bmatrix} = \dfrac{1}{16} \begin{bmatrix} 11 & 5 & -5 \\ 5 & 11 & 5 \\ -5 & 5 & 11 \end{bmatrix}$.

Example 15.37 If $\mathbf{A} = \begin{bmatrix} 1 & 1 & -2 \\ 1 & 9 & -11 \\ 1 & -1 & 2 \end{bmatrix}$ and $\mathbf{B} = \begin{bmatrix} 1 & 2 & 0 \\ 2 & 3 & -1 \\ 1 & 1 & -3 \end{bmatrix}$, find $(\mathbf{AB})^{-1}$.

Solution. We have $\mathbf{AB} = \begin{bmatrix} 1 & 1 & -2 \\ 1 & 9 & -11 \\ 1 & -1 & 2 \end{bmatrix}\begin{bmatrix} 1 & 2 & 0 \\ 2 & 3 & -1 \\ 1 & 1 & -3 \end{bmatrix} = \begin{bmatrix} 1 & 3 & 5 \\ 8 & 18 & 24 \\ 1 & 1 & -5 \end{bmatrix}$.

Therefore, $|\mathbf{AB}| = \begin{vmatrix} 1 & 3 & 5 \\ 8 & 18 & 24 \\ 1 & 1 & -5 \end{vmatrix} = 1(-90-24) - 3(-40-24) + 5(8-18)$

$$= -114 + 192 - 50 = 28 \neq 0.$$

Hence $(\mathbf{AB})^{-1}$ exists and is given by $(\mathbf{AB})^{-1} = \dfrac{1}{|\mathbf{AB}|} adj(\mathbf{AB})$.

It can be easily seen that $adj(\mathbf{AB}) = \begin{bmatrix} -114 & 20 & -18 \\ 64 & -10 & 16 \\ -10 & 2 & -6 \end{bmatrix}$.

$\therefore (\mathbf{AB})^{-1} = \dfrac{1}{28}\begin{bmatrix} -114 & 20 & -18 \\ 64 & -10 & 16 \\ -10 & 2 & -6 \end{bmatrix} = \begin{bmatrix} -57/14 & 5/7 & -9/14 \\ 16/7 & -5/14 & 4/7 \\ -5/14 & 1/14 & -3/14 \end{bmatrix}$.

Example 15.38 Find matrix $\mathbf{A}$ such that $\begin{bmatrix} 2 & 1 \\ 3 & 2 \end{bmatrix}\mathbf{A}\begin{bmatrix} -3 & 2 \\ 5 & -3 \end{bmatrix} = \begin{bmatrix} -2 & 4 \\ 3 & -1 \end{bmatrix}$.

Solution. Let $\mathbf{B} = \begin{bmatrix} 2 & 1 \\ 3 & 2 \end{bmatrix}$, $\mathbf{C} = \begin{bmatrix} -3 & 2 \\ 5 & -3 \end{bmatrix}$ and $\mathbf{D} = \begin{bmatrix} -2 & 4 \\ 3 & -1 \end{bmatrix}$. Then the given matrix equation is

$$\mathbf{BAC} = \mathbf{D}. \qquad \qquad ...(i)$$

Now $|\mathbf{B}| = 4 - 3 = 1 \neq 0$. Thus $\mathbf{B}$ is non-singular and hence $\mathbf{B}^{-1}$ exists. Premultiplying both sides of equation (i) by $\mathbf{B}^{-1}$ gives $\mathbf{AC} = \mathbf{B}^{-1}\mathbf{D}$.

But $\mathbf{B}^{-1} = \dfrac{1}{|\mathbf{B}|} adj \; \mathbf{B} = \begin{bmatrix} 2 & -1 \\ -3 & 2 \end{bmatrix}$.

Therefore $\mathbf{AC} = \mathbf{B}^{-1}\mathbf{D} = \begin{bmatrix} 2 & -1 \\ -3 & 2 \end{bmatrix}\begin{bmatrix} -2 & 4 \\ 3 & -1 \end{bmatrix} = \begin{bmatrix} -7 & 9 \\ 12 & -14 \end{bmatrix}$.   ...(*ii*)

Now $|\mathbf{C}| = 9 - 10 = -1 \neq 0$. Thus $\mathbf{C}$ is non-singular and hence $\mathbf{C}^{-1}$ exists.

But $\mathbf{C}^{-1} = \dfrac{1}{|\mathbf{C}|} adj \mathbf{C} = -\begin{bmatrix} -3 & -2 \\ -5 & -3 \end{bmatrix} = \begin{bmatrix} 3 & 2 \\ 5 & 3 \end{bmatrix}$.

Now post-multiplying both sides of equation (*ii*) by $\mathbf{C}^{-1}$ gives

$\mathbf{A} = \begin{bmatrix} -7 & 9 \\ 12 & -14 \end{bmatrix}\begin{bmatrix} 3 & 2 \\ 5 & 3 \end{bmatrix} = \begin{bmatrix} 24 & 13 \\ -34 & -18 \end{bmatrix}$, the required result.

Example 15.39 Given the inverse of the matrix $\mathbf{A}$ as:

$$\mathbf{A}^{-1} = \begin{bmatrix} 2/21 & 1/7 & -13/21 \\ -1/7 & 2/7 & 3/7 \\ 5/21 & -1/7 & -1/21 \end{bmatrix}, \; obtain \; \mathbf{A}^2 + 2\mathbf{A}.$$

Solution. We have

$$|\mathbf{A}^{-1}| = \begin{vmatrix} 2/21 & 1/7 & -13/21 \\ -1/7 & 2/7 & 3/7 \\ 5/21 & -1/7 & -1/21 \end{vmatrix} \quad \text{(Operating } R_1 \leftarrow 2R_1 - R_2 \text{ and } R_3 \leftarrow 2R_3 + R_2\text{)}$$

$$= 1/4 \begin{vmatrix} 1/3 & 0 & -35/21 \\ -1/7 & 2/7 & 3/7 \\ 1/3 & 0 & 1/3 \end{vmatrix} = -\frac{1}{4\times 63}\begin{vmatrix} 0 & 1 & -35/7 \\ 2 & -1 & 3 \\ 0 & 1 & 1 \end{vmatrix} = \frac{1\times 2}{4\times 63}\left(1 + \frac{35}{7}\right) = \frac{1}{21} \neq 0.$$

It can be easily seen that $adj\left(\mathbf{A}^{-1}\right) = \begin{bmatrix} 1/21 & 2/21 & 5/21 \\ 2/21 & 3/21 & 1/21 \\ -1/21 & 1/21 & 1/21 \end{bmatrix}$.

$$\therefore \; \mathbf{A} = \left(\mathbf{A}^{-1}\right)^{-1} = \frac{1}{|\mathbf{A}^{-1}|} adj\left(\mathbf{A}^{-1}\right) = 21\begin{bmatrix} 1/21 & 2/21 & 5/21 \\ 2/21 & 3/21 & 1/21 \\ -1/21 & 1/21 & 1/21 \end{bmatrix} = \begin{bmatrix} 1 & 2 & 5 \\ 2 & 3 & 1 \\ -1 & 1 & 1 \end{bmatrix}.$$

Now $\mathbf{A}^2 = \mathbf{AA} = \begin{bmatrix} 1 & 2 & 5 \\ 2 & 3 & 1 \\ -1 & 1 & 1 \end{bmatrix}\begin{bmatrix} 1 & 2 & 5 \\ 2 & 3 & 1 \\ -1 & 1 & 1 \end{bmatrix} = \begin{bmatrix} 0 & 13 & 12 \\ 7 & 14 & 14 \\ 0 & 2 & -3 \end{bmatrix}$,

and hence $\mathbf{A}^2 + 2\mathbf{A} = \begin{bmatrix} 0 & 13 & 12 \\ 7 & 14 & 14 \\ 0 & 2 & -3 \end{bmatrix} + \begin{bmatrix} 2 & 4 & 10 \\ 4 & 6 & 2 \\ -2 & 2 & 2 \end{bmatrix} = \begin{bmatrix} 2 & 17 & 22 \\ 11 & 20 & 16 \\ -2 & 4 & -1 \end{bmatrix}$.

15.18 Systems of $n \times n$ linear equations and their solution

Consider a system of n equations in the n unknowns $x_1, x_2, ...x_n$

$$\left.\begin{array}{l} a_{11}x_1 + a_{12}x_2 + \quad \cdots\cdots \quad +a_{1n}x_n = b_1 \\ a_{21}x_1 + a_{22}x_2 + \quad \cdots\cdots \quad +a_{2n}x_n = b_2 \\ \quad\vdots \qquad\qquad\qquad\qquad\qquad \vdots \\ a_{n1}x_1 + a_{n2}x_2 + \quad \cdots\cdots \quad +a_{nn}x_n = b_n \end{array}\right\} \qquad ...(15.12)$$

The coefficients of the unknowns in the linear system (15.12) can be abbreviated as a_{ij}, where i denotes the row and j denotes the column in which the coefficient appears. Thus, $i = j = 1, 2, 3, ..., n$. The numbers $b_1, b_2, ..., b_n$ are called the **constants** of the system.

Matrix representation of a system of equations

In matrix form, we can write the system of equations (15.12) as

$$\mathbf{Ax = b} \qquad\qquad ...(15.13)$$

where $\mathbf{A} = \begin{bmatrix} a_{11} & a_{12} & \cdots\cdots & a_{1n} \\ a_{21} & a_{22} & \cdots\cdots & a_{2n} \\ \vdots & \vdots & & \vdots \\ a_{n1} & a_{n2} & \cdots\cdots & a_{nn} \end{bmatrix}$, $\mathbf{b} = \begin{bmatrix} b_1 \\ b_2 \\ \vdots \\ b_n \end{bmatrix}$, $\mathbf{x} = \begin{bmatrix} x_1 \\ x_2 \\ \vdots \\ x_n \end{bmatrix}$

and $\mathbf{A}$, $\mathbf{b}$, $\mathbf{x}$ are respectively called the *coefficient matrix*, the right hand side column vector and the solution vector.

If $\mathbf{b} = 0$, that is all elements $b_1, b_2, ..., b_n$ is 0, the system (15.12) is said to be **homogeneous**. If $b \neq 0$, that is, atleast one of the elements $b_1, b_2, ..., b_n$ is not zero, the system is said to be **non homogeneous**.

A **solution** of a linear system (15.12) is a set of n numbers $x_1, x_2, ..., x_n$ that satisfies each equation in the system.

The system of equations is said to be **consistent** if it has atleast one solution and **inconsistent**, if it has no solution. If a linear system is consistent, it has either a unique solution (that is, precisely one solution), or infinitely many solutions.

Non-homogeneous system of equations

The non-homogeneous system of equation $\mathbf{Ax = b}$ can be solved by the following methods.

Matrix method

Let coefficient matrix $\mathbf{A}$ be non-singular, that is $|\mathbf{A}| \neq 0$, or the rank of the matrix $\mathbf{A}$ is n so that $\mathbf{A}^{-1}$ exists. Premultiplying $\mathbf{Ax = b}$ by $\mathbf{A}^{-1}$, we obtain

$$\mathbf{x = A^{-1}b}. \qquad\qquad ...(15.14)$$

The system of equations is consistent and has a unique solution. If $\mathbf{b = O}$, then $\mathbf{x = O}$ (trivial solution) is the only solution.

Cramer's rule

Let **A** be non-singular matrix. The Cramer's rule for the solution of $\mathbf{Ax} = \mathbf{b}$ is given by

$$x_k = \frac{|\mathbf{A}_k|}{|\mathbf{A}|}, k = 1, 2, ..., n \qquad ...(15.15)$$

where $|\mathbf{A}_k|$ is the determinant of the matrix $\mathbf{A}_k$ obtained from $|\mathbf{A}|$ by removing the kth column and replacing it by the right hand side column vector **b**. We discuss the following cases.

Case 1. When $|\mathbf{A}| \neq 0$, the system of equations is consistent and the unique solution is obtained by using equation (15.15).

Case 2. When $|\mathbf{A}| = 0$ and all $|\mathbf{A}_k| = 0$, $k = 1, 2, ..., n$, then the system of equations is consistent and has infinite number of solutions.

Case 3. When $|\mathbf{A}| = 0$, and one or more of $|\mathbf{A}_k|$, $k = 1, 2, ...n$, are not zero, then the system of equations has no solution, that is the system is inconsistent.

Homogeneous system of equations

Consider the homogeneous system of equations $\mathbf{Ax} = \mathbf{O}$ ...(15.16)

If the coefficient matrix **A** is non-singular, then $\mathbf{x} = \mathbf{A}^{-1}\mathbf{O} = \mathbf{O}$. That is, a homogeneous system has only the trivial solution $(x_1 = 0, x_2 = 0, ..., x_n = 0)$ if and only if **A** is non-singular. However, if the coefficient matrix **A** is singular, the homogeneous system of equations has a nontrivial solution. A homogeneous system of equations is always consistent.

In view of above, we can conclude that a homogeneous system of n linear equations in n unknowns $\mathbf{Ax} = \mathbf{O}$ possesses

only the trivial solution if and only if $|\mathbf{A}| \neq 0$, and

a non trivial solution if and only if $|\mathbf{A}| = 0$.

In this case, the homogeneous system of equations has infinite number of solutions.

Example 15.40 Solve the following system of equations with the help of matrix inversion.

$x + y + z = 3$, $x + 2y + 3z = 4$, $x + 4y + 9z = 6$.

Solution. The given system can be written in the matrix form as $\mathbf{Ax} = \mathbf{b}$,

where $\mathbf{A} = \begin{bmatrix} 1 & 1 & 1 \\ 1 & 2 & 3 \\ 1 & 4 & 9 \end{bmatrix}$, $\mathbf{x} = \begin{bmatrix} x \\ y \\ z \end{bmatrix}$, and $\mathbf{b} = \begin{bmatrix} 3 \\ 4 \\ 6 \end{bmatrix}$.

We find that $|\mathbf{A}| = \begin{vmatrix} 1 & 1 & 1 \\ 1 & 2 & 3 \\ 1 & 4 & 9 \end{vmatrix} = 1(18-12) - 1(9-3) + 1(4-2) = 2 \neq 0.$

Therefore, the coefficient matrix **A** is non-singular and the given system of equations has a unique solution.

We find $adj\ \mathbf{A} = \begin{bmatrix} 6 & -5 & 1 \\ -6 & 8 & -2 \\ 2 & -3 & 1 \end{bmatrix}$ and obtain $\mathbf{A}^{-1} = \frac{1}{|\mathbf{A}|} adj\ \mathbf{A} = \frac{1}{2}\begin{bmatrix} 6 & -5 & 1 \\ -6 & 8 & -2 \\ 2 & -3 & 1 \end{bmatrix}$.

Therefore, $\mathbf{x} = \mathbf{A}^{-1}\mathbf{b} = \frac{1}{2}\begin{bmatrix} 6 & -5 & 1 \\ -6 & 8 & -2 \\ 2 & -3 & 1 \end{bmatrix}\begin{bmatrix} 3 \\ 4 \\ 6 \end{bmatrix} = \frac{1}{2}\begin{bmatrix} 4 \\ 2 \\ 0 \end{bmatrix} = \begin{bmatrix} 2 \\ 1 \\ 0 \end{bmatrix}$

Hence, $x = 2$, $y = 1$ and $z = 0$.

Example 15.41 If $\mathbf{A} = \begin{bmatrix} 3 & -3 & 4 \\ 2 & -3 & 4 \\ 0 & -1 & 1 \end{bmatrix}$, find $\mathbf{A}^{-1}$ and hence solve the following

system of linear equations:

$x - y = 0$, $-2x + 3y - 4z = 9$, $-2x + 3y - 3z = -5$. \hfill *(A.M.I.E. S. 2003)*

Solution. We find $|\mathbf{A}| = \begin{vmatrix} 3 & -3 & 4 \\ 2 & -3 & 4 \\ 0 & -1 & 1 \end{vmatrix} = 3(-3+4) + 3(2) + 4(-2) = 1 \neq 0$.

Therefore $\mathbf{A}^{-1}$ exists and in given by $\mathbf{A}^{-1} = \frac{1}{|\mathbf{A}|} adj\ \mathbf{A}$.

The Cofactors A_{ij} of $\mathbf{A}$ are:

$A_{11} = \begin{vmatrix} -3 & 4 \\ -1 & 1 \end{vmatrix} = 1,\ A_{12} = -\begin{vmatrix} 2 & 4 \\ 0 & 1 \end{vmatrix} = -2,\ A_{13} = \begin{vmatrix} 2 & -3 \\ 0 & -1 \end{vmatrix} = -2,$

$A_{21} = -\begin{vmatrix} -3 & 4 \\ -1 & 1 \end{vmatrix} = -1,\ A_{22} = \begin{vmatrix} 3 & 4 \\ 0 & 1 \end{vmatrix} = 3,\ A_{23} = -\begin{vmatrix} 3 & -3 \\ 0 & -1 \end{vmatrix} = 3,$

$A_{31} = \begin{vmatrix} -3 & 4 \\ -3 & 4 \end{vmatrix} = 0,\ A_{32} = -\begin{vmatrix} 3 & 4 \\ 2 & 4 \end{vmatrix} = -4,\ A_{33} = \begin{vmatrix} 3 & -3 \\ 2 & -3 \end{vmatrix} = -3.$

$adj\ \mathbf{A} = \begin{bmatrix} 1 & -2 & -2 \\ -1 & 3 & 3 \\ 0 & -4 & -3 \end{bmatrix}^{T} = \begin{bmatrix} 1 & -1 & 0 \\ -2 & 3 & -4 \\ -2 & 3 & -3 \end{bmatrix}$

We obtain $\mathbf{A}^{-1} = \frac{adj\ \mathbf{A}}{|\mathbf{A}|} = \begin{bmatrix} 1 & -1 & 0 \\ -2 & 3 & -4 \\ -2 & 3 & -3 \end{bmatrix}$.

Now the given system of equations can be written in the matrix form as

$\begin{bmatrix} 1 & -1 & 0 \\ -2 & 3 & -4 \\ -2 & 3 & -3 \end{bmatrix}\begin{bmatrix} x \\ y \\ z \end{bmatrix} = \begin{bmatrix} 0 \\ 9 \\ -5 \end{bmatrix}$

i.e., $\mathbf{A}^{-1}\mathbf{x} = \mathbf{b} \Rightarrow \mathbf{x} = \mathbf{Ab}$, where $\mathbf{x} = [x\ y\ z]^{T}$ and $\mathbf{b} = (0\ 9\ -5)^{T}$.

Therefore $\mathbf{x} = \begin{bmatrix} 3 & -3 & 4 \\ 2 & -3 & 4 \\ 0 & -1 & 1 \end{bmatrix} \begin{bmatrix} 0 \\ 9 \\ -5 \end{bmatrix} = \begin{bmatrix} -47 \\ -47 \\ -14 \end{bmatrix}$. Hence $x = -47$, $y = -47$, and $z = -14$.

Example 15.42 Solve the homogeneous system of equations
$$x + y - z = 0, \ x - 2y + z = 0, \ 3x + 6y - 5z = 0.$$
Solution. In matrix form, we can write the given system of equations as $\mathbf{Ax = O}$.

$$\begin{bmatrix} 1 & 1 & -1 \\ 1 & -2 & 1 \\ 3 & 6 & -5 \end{bmatrix} \begin{bmatrix} x \\ y \\ z \end{bmatrix} = \begin{bmatrix} 0 \\ 0 \\ 0 \end{bmatrix}.$$

We find that $|\mathbf{A}| = \begin{vmatrix} 1 & 1 & -1 \\ 1 & -2 & 1 \\ 3 & 6 & -5 \end{vmatrix} = 1(10-6) - 1(-5-3) - 1(6+6) = 0$. The system

possesses non-trivial solutions. Hence, the given system has infinite number of solutions. Solving the first two equations

$$\begin{bmatrix} 1 & 1 \\ 1 & -2 \end{bmatrix} \begin{bmatrix} x \\ y \end{bmatrix} = \begin{bmatrix} z \\ -z \end{bmatrix}$$

$$\Rightarrow \begin{bmatrix} x \\ y \end{bmatrix} = \begin{bmatrix} 1 & 1 \\ 1 & -2 \end{bmatrix}^{-1} \begin{bmatrix} z \\ -z \end{bmatrix} = -1/3 \begin{bmatrix} -2 & -1 \\ -1 & 1 \end{bmatrix} \begin{bmatrix} z \\ -z \end{bmatrix} = -1/3 \begin{bmatrix} -z \\ -2z \end{bmatrix} = \begin{bmatrix} z/3 \\ 2z/3 \end{bmatrix}.$$

We obtain $x = z/3$, $y = 2z/3$, where z is arbitrary. This solution satisfies the third equation.

Example 15.43 Show that the system of equations $2x + y + 4z = 2$, $x + 3y = 2z + 7$, $5z - 8 = 5x + 3y$ has a unique solution. Solve this system using Cramer's rule.

Solution. We have $|\mathbf{A}| = \begin{vmatrix} 2 & 1 & 4 \\ 1 & 3 & -2 \\ 5 & 3 & -5 \end{vmatrix} = 2(-15+6) - 1(-5+10) + 4(3-15)$

$= -71 \neq 0$. Therefore, the coefficient matrix $\mathbf{A}$ is non-singular and the given system of equations has a unique solution.
We find that

$$|\mathbf{A}_1| = \begin{vmatrix} 2 & 1 & 4 \\ 7 & 3 & -2 \\ -8 & 3 & -5 \end{vmatrix} = 2(-15+6) - 1(-35-16) + 4(21+24) = 213.$$

$$|\mathbf{A}_2| = \begin{vmatrix} 2 & 2 & 4 \\ 1 & 7 & -2 \\ 5 & -8 & -5 \end{vmatrix} = 2(-35-16) - 2(-5+10) + 4(-8-35) = -284.$$

$$|\mathbf{A}_3| = \begin{vmatrix} 2 & 1 & 2 \\ 1 & 3 & 7 \\ 5 & 3 & -8 \end{vmatrix} = 2(-24-21) - 1(-8-35) + 2(3-15) = -71.$$

Therefore, $x = \dfrac{|A_1|}{|A|} = -3$, $y = \dfrac{|A_2|}{|A|} = 4$, $z = \dfrac{|A_3|}{|A|} = 1$.

Hence $x = -3$, $y = 4$ and $z = 1$ is the required solution.

Example 15.44 Show that the system of equations $2x - y + 3z = 4$, $x + y - 3z = -1$, $5x - y + 3z = 7$ has infinite number of solution. Hence, find the solutions.

Solution. We find that

$$|A| = \begin{vmatrix} 2 & -1 & 3 \\ 1 & 1 & -3 \\ 5 & -1 & 3 \end{vmatrix} = 2(3-3) + 1(3+15) + 3(-1-5) = 0,$$

$$|A_1| = \begin{vmatrix} 4 & -1 & 3 \\ -1 & 1 & -3 \\ 7 & -1 & 3 \end{vmatrix} = 4(3-3) + 1(-3+21) + 3(1-7) = 0,$$

$$|A_2| = \begin{vmatrix} 2 & 4 & 3 \\ 1 & -1 & -3 \\ 5 & 7 & 3 \end{vmatrix} = 2(-3+21) - 4(3+15) + 3(7+5) = 0,$$

$$|A_3| = \begin{vmatrix} 2 & -1 & 4 \\ 1 & 1 & -1 \\ 5 & -1 & 7 \end{vmatrix} = 2(7-1) + 1(7+5) + 4(-1-5) = 0.$$

The system is consistent and has infinite number of solution. Taking the first two equations.

$2x - y = 4 - 3z$, $x + y = -1 + 3z$ and solving, we obtain $x = 1$ and $y = 3z - 2$ where z is arbitrary. This solution satisfies the third equation.

Example 15.45 Show that the system of equations
$x + 2y + z = 6$, $x + 4y = 10 - 3z$, $x + 4y + 3z = 8$ is inconsistent.

Solution. We find that $|A| = \begin{vmatrix} 1 & 2 & 1 \\ 1 & 4 & 3 \\ 1 & 4 & 3 \end{vmatrix} = 0, |A_1| = \begin{vmatrix} 6 & 2 & 1 \\ 10 & 4 & 3 \\ 8 & 4 & 3 \end{vmatrix}$

$= 6(12-12) - 2(30-24) + 1(40-32) = -12 + 8 = -4 \neq 0,$

Since $|A| = 0$ and $|A_1| = -4 \neq 0$, the system of equations is inconsistent, that is the system has no solution.

Exercise 15.6

1. Find the adjoint of the matrix $A = \begin{bmatrix} 1 & 2 & 0 \\ -1 & 3 & 4 \\ -2 & 5 & 6 \end{bmatrix}$. **Ans.** $\begin{bmatrix} -2 & -12 & 8 \\ -2 & 6 & -4 \\ 1 & -9 & 5 \end{bmatrix}$.

2. Find the adjoint of the matrix $A = \begin{bmatrix} 1 & 1 & 1 \\ 2 & 2 & 3 \\ 1 & 4 & 9 \end{bmatrix}$, verify that

$A(adj\ A) = |A|I_3 = (adj\ A)A.$ **Ans.** $\begin{bmatrix} 6 & -5 & 1 \\ -15 & 8 & -1 \\ 6 & -3 & 0 \end{bmatrix}$.

3. If matrix $C = \begin{bmatrix} -1 & -2 & -2 \\ 2 & 1 & -2 \\ 2 & -2 & 1 \end{bmatrix}$, show that $adjC = 3C^T$.

4. Given two matrices $A = \begin{bmatrix} 2 & 3 & 4 \\ 1 & 2 & 3 \\ -4 & -5 & 7 \end{bmatrix}$ and $B = \begin{bmatrix} 2 & 1 & 3 \\ 0 & -1 & 3 \\ 4 & 2 & 1 \end{bmatrix}$.

Prove that $adj(AB) = adjB \cdot adjA$

5. For what values of x is the matrix $\begin{bmatrix} 3-x & 2 & 2 \\ 2 & 4-x & 2 \\ -2 & -4 & -1-x \end{bmatrix}$ singular?

(I.E.T.E. June 2004) **Ans.** 0 and 3.

6. Find the inverse of each of the following matrices:

(i) $\begin{bmatrix} 5 & 1 \\ -3 & 4 \end{bmatrix}$, *(ii)* $\begin{bmatrix} 3 & 5 \\ 7 & -11 \end{bmatrix}$, *(iii)* $\begin{bmatrix} 3 & 2 & 2 \\ 2 & 5 & 3 \\ 1 & 2 & 1 \end{bmatrix}$,

(iv) $\begin{bmatrix} 1 & 3 & 7 \\ 4 & 2 & 3 \\ 1 & 2 & 1 \end{bmatrix}$, *(v)* $\begin{bmatrix} 1 & 2 & 5 \\ 3 & 1 & 4 \\ 1 & 1 & 2 \end{bmatrix}$, *(vi)* $\begin{bmatrix} 1 & 0 & -1 \\ 3 & 4 & 5 \\ 0 & -6 & -7 \end{bmatrix}$,

(vii) $\begin{bmatrix} 2 & 5 & 3 \\ 3 & 1 & 2 \\ 1 & 2 & 1 \end{bmatrix}$, *(viii)* $\begin{bmatrix} 3 & -3 & 4 \\ 2 & -3 & 4 \\ 0 & -1 & 1 \end{bmatrix}$, *(ix)* $\begin{bmatrix} 1 & -2 & 3 \\ 0 & 2 & -1 \\ -4 & 5 & 2 \end{bmatrix}$,

(x) $\begin{bmatrix} 1 & 0 & 2 \\ 2 & -1 & 3 \\ 4 & 1 & 8 \end{bmatrix}$, *(xi)* $\begin{bmatrix} 2 & 4 & 3 \\ 0 & 1 & 1 \\ 2 & 2 & -1 \end{bmatrix}$. *(IETE, Dec. 2001)*

Ans. *(i)* $\begin{bmatrix} 4/23 & -1/23 \\ 3/23 & 5/23 \end{bmatrix}$, *(ii)* $\begin{bmatrix} 11/68 & 5/68 \\ 7/68 & -3/68 \end{bmatrix}$, *(iii)* $1/3\begin{bmatrix} 1 & -2 & 4 \\ -1 & -1 & 5 \\ 1 & 4 & -11 \end{bmatrix}$,

(iv) $1/35\begin{bmatrix} -4 & 11 & -5 \\ -1 & -6 & 25 \\ 6 & 1 & -10 \end{bmatrix}$, *(v)* $1/4\begin{bmatrix} -2 & 1 & 3 \\ -2 & -3 & 11 \\ 2 & 1 & -5 \end{bmatrix}$, *(vi)* $1/20\begin{bmatrix} 2 & 6 & 4 \\ 21 & -7 & -8 \\ -18 & 6 & 4 \end{bmatrix}$,

(vii) $1/4\begin{bmatrix} -3 & 1 & 7 \\ -1 & -1 & 5 \\ 5 & 1 & -13 \end{bmatrix}$, (viii) $\begin{bmatrix} 1 & -1 & 0 \\ -2 & 3 & -4 \\ -2 & 3 & -3 \end{bmatrix}$, (ix) $1/25\begin{bmatrix} 9 & 19 & -4 \\ 4 & 14 & 1 \\ 8 & 3 & 2 \end{bmatrix}$,

(x) $\begin{bmatrix} -11 & 2 & 2 \\ -4 & 0 & 1 \\ 6 & -1 & -1 \end{bmatrix}$, (xi) $1/4\begin{bmatrix} 3 & -10 & -1 \\ -2 & 8 & 2 \\ 2 & -4 & -2 \end{bmatrix}$.

7. For any three non-singular matrices **A**, **B**, **C** each of order n, show that $(\mathbf{ABC})^{-1} = \mathbf{C}^{-1}\mathbf{B}^{-1}\mathbf{A}^{-1}$. (A.M.I.E., W-2005)

8. Does the inverse of an identity matrix exist? If so, what is it?

   If $\mathbf{A} = 1/9\begin{bmatrix} -8 & 1 & 4 \\ 4 & 4 & 7 \\ 1 & -8 & 4 \end{bmatrix}$, prove that $\mathbf{A}^{-1} = \mathbf{A}^T$. **Ans. $\mathbf{I}_3$**

9. The inverse of $\begin{bmatrix} 1 & 2 & -2 \\ -1 & 3 & 0 \\ 0 & -2 & 1 \end{bmatrix}$ is $\begin{bmatrix} 3 & 2 & 6 \\ 1 & 1 & k \\ 2 & 2 & 5 \end{bmatrix}$, then the value of k is...

   Ans. 2.

10. For the matrix $\mathbf{A} = \begin{bmatrix} 3 & -3 & 4 \\ 2 & -3 & 4 \\ 0 & -1 & 1 \end{bmatrix}$, verify that $\mathbf{A}^3 = \mathbf{A}^{-1}$.

11. Show that $\begin{bmatrix} \cos\theta & -\sin\theta \\ \sin\theta & \cos\theta \end{bmatrix} = \begin{bmatrix} 1 & -\tan\theta/2 \\ \tan\theta/2 & 1 \end{bmatrix}\begin{bmatrix} 1 & \tan\theta/2 \\ -\tan\theta/2 & 1 \end{bmatrix}^{-1}$.

12. If $\mathbf{A}^T$ denotes the transpose of matrix **A** and $\mathbf{A} = \begin{bmatrix} 1 & -2 & 3 \\ 0 & -1 & 4 \\ -2 & 2 & 1 \end{bmatrix}$,

    obtain $(\mathbf{A}^T)^{-1}$. (A.M.I.E., S-2003) **Ans.** $\begin{bmatrix} -9 & -8 & -2 \\ 8 & 7 & 2 \\ -5 & -4 & -1 \end{bmatrix}$.

13. Find the matrix **A** satisfying the matrix equation:

    $\begin{bmatrix} 5 & 4 \\ 1 & 1 \end{bmatrix}\mathbf{A} = \begin{bmatrix} 1 & -2 \\ 1 & 3 \end{bmatrix}$. **Ans.** $\begin{bmatrix} -3 & -14 \\ 4 & 17 \end{bmatrix}$.

14. Find matrix **X** such that

    $\mathbf{X}\begin{bmatrix} 1 & -2 & 3 \\ 0 & 2 & -1 \\ -4 & 5 & 2 \end{bmatrix} = \begin{bmatrix} 1 & -2 & 3 \\ -4 & 5 & 2 \\ 0 & 5 & -1 \end{bmatrix}$. **Ans.** $\begin{bmatrix} 1 & 0 & 0 \\ 0 & 0 & 1 \\ 0 & 1 & 0 \end{bmatrix}$.

15. If a matrix **A** satisfies a relation $\mathbf{A}^2 + \mathbf{A} - \mathbf{I} = \mathbf{O}$, prove that $\mathbf{A}^{-1}$ exists and that $\mathbf{A}^{-1} = \mathbf{I} + \mathbf{A}$, **I** being an identity matrix. (A.M.I.E., W-2003)

[**Hint.** If **A** and **B** are square matrices of the same order, then $|AB| = |A||B|$.

We have, $\qquad A^2 + A = I$ or $A(A + I) = I$. $\qquad\qquad$ ...(*i*)

$\therefore$ $|A||A + I| = |I| = 1$. Thus $|A| \neq 0$ and therefore A^{-1} exists. Pre- multiplying
(*i*) by A^{-1}, we obtain $A + I = A^{-1}$].

16. Show that the matrix $A = \begin{bmatrix} 1 & 0 & 0 \\ 2 & 1 & 0 \\ 3 & 2 & 1 \end{bmatrix}$ satisfies the equation $A^3 - 3A^2$

    $+ 3A - I = O$. Hence obtain A^{-1}. (*A.M.I.E., S-2001*) **Ans.** $\begin{bmatrix} 1 & 0 & 0 \\ -2 & 1 & 0 \\ 1 & -2 & 1 \end{bmatrix}$.

17. If $A = \begin{bmatrix} 1 & 1 & -2 \\ -1 & 2 & 1 \\ 0 & 1 & -1 \end{bmatrix}$ and $B = \begin{bmatrix} 1 & 1 & 3 \\ 0 & 3 & 2 \\ 1 & 1 & 1 \end{bmatrix}$ be two square matrices of order

    3, verify that $B^{-1}AB = \begin{bmatrix} -1 & 0 & 0 \\ 0 & 2 & 0 \\ 0 & 0 & 1 \end{bmatrix}$.

18. Given two matrixes $A = \begin{bmatrix} 1 & 1 & 1 \\ 2 & 4 & 1 \\ 2 & 3 & 1 \end{bmatrix}$ and $B = \begin{bmatrix} 2 & 3 \\ 3 & 4 \end{bmatrix}$. Find **X** such that

    $BXA = \begin{bmatrix} 1 & 0 & 1 \\ 0 & 1 & 0 \end{bmatrix}$. $\qquad\qquad$ **Ans.** $\begin{bmatrix} -4 & 7 & -7 \\ 3 & -5 & 5 \end{bmatrix}$.

19. If $A = \begin{bmatrix} 1 & 1 & 1 \\ 1 & 2 & 3 \\ 1 & 4 & 9 \end{bmatrix}$ and $B = \begin{bmatrix} 2 & 5 & 3 \\ 3 & 1 & 2 \\ 1 & 2 & 1 \end{bmatrix}$ be two square matrices of order 3,

    verify that $(AB)^{-1} = B^{-1} A^{-1}$.

20. Given $A^{-1} = \begin{bmatrix} 5/7 & 1/7 \\ 3/7 & 2/7 \end{bmatrix}$, evaluate $A^2 + 2A$. $\qquad$ **Ans.** $\begin{bmatrix} 11 & -9 \\ -27 & 38 \end{bmatrix}$.

21. If $A = \begin{bmatrix} 1 & 2 & 3 \\ 2 & 3 & 1 \\ 3 & 1 & 2 \end{bmatrix}$, verify that $(A^2)^{-1} = (A^{-1})^2$.

22. If $A = \begin{bmatrix} 2 & -1 \\ -2 & 3 \end{bmatrix}$ and $B = \begin{bmatrix} 7 & 6 \\ 9 & 8 \end{bmatrix}$, find 2×2 matrices **C** and **D** such that

    $AC = B$ and $DA = B$. $\qquad$ **Ans.** $C = \begin{bmatrix} 15/2 & 13/2 \\ 8 & 7 \end{bmatrix}, D = \begin{bmatrix} 33/4 & 19/4 \\ 43/4 & 25/4 \end{bmatrix}$.

    [**Hint.** $C = A^{-1} B$, $D = BA^{-1}$].

23. If $A = \begin{bmatrix} 1 & 2 & -1 \\ 3 & 8 & 2 \\ 4 & 9 & -1 \end{bmatrix}$, find A^{-1}. Using A^{-1}, solve the following system of linear equations: $x + 2y - z = 2$, $3x + 8y + 2z = 10$, $4x + 9y - z = 12$.

Ans. $\begin{bmatrix} -26 & -7 & 12 \\ 11 & 3 & -5 \\ -5 & -1 & 2 \end{bmatrix}$; 22, –8, 4.

24. Solve the following system of equations by (*i*) matrix method (*ii*) Cramer's rule
    (*a*) $x + 2y + 3z = 1$, $2x + 3y + 2z = 2$, $3x + 3y + 4z = 1$.
    (*b*) $3x - 2y + 2z = 10$, $x + 2y - 3z = -1$, $4x + y + 2z = 3$.
    (*c*) $x - y + z = 4$, $2x + y = 3z$, $x + y + z = 2$.
    (*d*) $x + y + z = 3$, $x + 2y + 3z = 4$, $x + 4y + 9z = 6$.
    (*e*) $x + 2y + 3y = 2$, $2x + 3y + 5z = 3$, $3x + 5y + z = 12$.   (*AMIE., W-2002*)

    Ans. (*a*) –3/7, 8/7, –2/7 (*b*) 2, –3, –1,
    (*c*) 2, –1, 1, (*d*) 2, 1, 0, (*e*) 1, 2, –1.

25. The currents I_1, I_2, I_3 and I_4 in an electric network satisfy the system of equations.
    $3I_1 + 2I_3 - I_4 = 60$, $2I_1 - I_2 + 4I_3 = 160$, $4I_2 + I_3 - 2I_4 = 20$, $5I_1 - I_2 - 2I_3 + I_4 = 0$. Find I_3.   **Ans.** 40.

26. Find the value of λ for which the equations
    $(2 - \lambda) x + 2y + 3 = 0$, $2x + (4 - \lambda) y + 7 = 0$, $2x + 5y + (6 - \lambda) = 0$ are consistent and find the values of x and y corresponding to each of these value of λ. (*Bhopal, 1998*)
    Ans. $\lambda = -1$, (–1/11, –15/11); 1, (–5, 1); 12, (1/2, 1).

27. Investigate the following system of equations. Find the solution if it exists.

    $\begin{bmatrix} 2 & 1 & -1 \\ 1 & -2 & 1 \\ 4 & -3 & 1 \end{bmatrix} \begin{bmatrix} x \\ y \\ z \end{bmatrix} = \begin{bmatrix} 1 \\ 3 \\ 5 \end{bmatrix}$.   **Ans.** Inconsistent.

15.19 Vector Spaces

Let *V* denote a non-empty set of certain objects, which may be vectors, matrices, functions or some other objects. Each object is an element of *V* and is called a vector. The elements of V are denoted by lowercase letters **a, b, c, u, v.** etc.

Definition of a vector space

A vector space V is a non-empty set of objects which can be added and multiplied by numbers, in such a way that the sum of two elements of V is again an element of **V,** the product of an element of V by a number is an element of V and the following ten properties, (axioms) are satisfied.

Properties (axioms) for vector addition

1. If **a** and **b** are in V, then **a** + **b** is in V.
2. For any two vectors **a** and **b** of V, **a** + **b** = **b** + **a**. (Commutative law)
3. For any three vectors **a**, **b**, **c** of V, **a** + (**b** + **c**) = (**a** + **b**) + **c** (associative law)
4. There is a unique vector **0** in V such that
   a + **0** = **0** + **a** = **a** for all **a** in V (existence of zero vector)
5. For every **a** in V, there exists a vector −**a** such that
   a + (−**a**) = (−**a**) + **a** = **0** (existence of additive inverse or negative vector in V)

Properties (axioms) for scalar multiplication

6. If k_1 is any scalar and **a** is in v, then k_1**a** is in V.
7. k_1(**a** + **b**) = k_1**a** + k_1**b** (right distributive law)
8. $(k_1 + k_2)$ **a** = k_1**a** + k_2**a** (left distributive law)
9. $k_1(k_2$**a**$) = (k_1 k_2)$ **a**.
10. 1**a** = **a** (existence of multiplicative inverse)

The properties defined in **1** and **6** are called the *closure* properties. When these two properties are satisfied, we say that the vector space V is closed under the vector addition and scalar multiplication.

If the elements of V are real, then it is called a *real vector space* when the scalars k_1, k_2 are real numbers. If *real number* is replaced by *complex number* in properties **6, 7, 8, 9**, the resulting structure is called a *complex vector space*. When we use the term vector space without further designation, it is to be understood that the space can be real or complex.

The following are some examples of vector spaces under the usual operations of vector addition and scalar multiplication.

1. The set V of real or complex numbers.
2. The set V of n-tuples of numbers in R^{n}* or $\mathbb{C}^n$ **.
3. The set of polynomials P_n of degree less than or equal to n.
4. The set of real valued continuous functions f on any closed interval $[a, b]$.
5. The set V of all $m \times n$ matrices.

Example 15.46 Let V be the set of all ordered pairs (x, y), where x, y are real numbers. Let $\mathbf{u} = (x_1, y_1)$ and $\mathbf{v} = (x_2, y_2)$ be two elements in V. Define the addition as
$$\mathbf{u} + \mathbf{v} = (x_1, y_1) + (x_2, y_2) = (2x_1 - 3x_2, y_1 - y_2)$$
and the scalar multiplication as
$$k(x_1, y_1) = (kx_1/3, ky_1/3).$$
Show that V is not a vector space.

Solution: We shall go through properties that are not satisfied.

---

* The set of all n-tuples of real numbers, denoted by R^n *is called* n-*space*.

** The set of all n-tuples of complex numbers, denoted by $\mathbb{C}^n$, *is called* complex n-space.

(i) $(x_2, y_2) + (x_1, y_1) = (2x_2 - 3x_1, y_2 - y_1) \neq (x_1, y_1) + (x_2, y_2)$

Therefore, **property 2** (commutative law) does not hold.

(ii) $((x_1, y_1) + (x_2, y_2)) + (x_3, y_3) = (2x_1 - 3x_2, y_1 - y_2) + (x_3, y_3) = (4x_1 - 6x_2 - 3x_3, y_1 - y_2 - y_3)$.

$(x_1, y_1) + ((x_2, y_2) + (x_3, y_3)) = (x_1, y_1) + (2x_2 - 3x_3, y_2 - y_3) = 2x_1 - 6x_2 + 9x_3, y_1 - y_2 + y_3)$.

Therefore, **property 3** (associative law) is not satisfied.

(iii) $1(x_1, y_1) = (x_1/3, y_1/3) \neq (x_1, y_1)$.

Therefore, **property 10.** (existence of multiplicative inverse) is not satisfied.

Hence, V is not a vector space.

Example 15.47 Let V be the set of all positive real numbers where addition is defined by $x + y = xy$ and scalar multiplication is defined by $kx = x^k$. Determine whether V is a vector space.

Solution. We shall go through all ten properties.

1. For $x = x > 0$ and $y = y > 0$, $x + y = xy > 0$. Thus, the sum $x + y$ is in V; V is closed under addition.

2. For all $x = x$ and $y = y$ in V, $x + y = xy = yx = y + x$. Thus addition is commutative.

3. For all $x = x$, $y = y$, $z = z$ in V, $x + (y + z) = x(yz) = (xy)z = (x + y) + z$. Thus addition is associative.

4. Since $1 + x = 1x = x = x$ and $x + 1 = x1 = x = x$, the zero vector O is $1 = 1$.

5. If we define $-x = 1/x$, then $x + (-x) = x\dfrac{1}{x} = 1 = 1 = 0$ and

$(-x) + x = \dfrac{1}{x}x = 1 = 1 = 0$. The negative of a vector is its reciprocal.

6. If k is any scalar and $x = x > 0$ is any vector, then $kx = x^k > 0$. Hence V is closed under scalar multiplication.

7. For k is any scalar, then $k(x + y) = (xy)^k = x^k y^k = kx + ky$

8. For scalars k_1 and k_2, $(k_1 + k_2)x = x^{(k_1 + k_2)} = x^{k_1} x^{k_2} = k_1 x + k_2 x$

9. For scalars k_1 and k_2, $k_1(k_2 x) = (x^{k_2})^{k_1} = x^{k_1 k_2} = (k_1 k_2)x$

10. $1x = x^1 = x = x$.

Since all the properties of vector space are satisfied, we conclude that V is a vector space.

15.19.1 Sub space of a vector space

If a subset W of a vector space V is itself a vector space under the operations of vector addition and scalar multiplication defined on V. then W is called a **subspace** of V. Every vector space V has atleast two subspaces: V itself and the zero subspace

{0}; {0} is a subspace since the zero vector must be an element in every vector space. The vector space V is also taken as a subspace of V.

To show that a subset W of a vector space V is a subspace, it is not necessary to show that all the ten properties are satisifed.

Criteria for a subspace

A non empty subset W of a vector space V is a subspace of V if and only if W is closed under vector addition and scalar multiplication defined on V:

(*i*) If **a** and **b** are in W, then **a** + **b** is in W.

(*ii*) If **a** is in W and k is any scalar, then k**a** is W.

Example 1. Let $V = R^n$ and let W be the set of vectors in V whose last coordinate is equal to 0. Then W is a subspace of V, which we could identify with R^{n-1}.

Example 2. Let U and W be subspaces of a vector space V. We denote by $U \cap W$ the intersection of U and W, *i.e.*, the set of elements which lie both in U and W. Then $U \cap W$ is a subspace of V.

Linear Combinations, Spanning Set

Let V be a vector space and let $\mathbf{v}_1, \mathbf{v}_2, ..., \mathbf{v}_m$ be any m elements in V under usual vector addition and scalar multiplication. Then, the set of all linear combinations of these elements, that is the set of all elements of the form

$$k_1\mathbf{v}_1 + k_2\mathbf{v}_2 + ... + k_m\mathbf{v}_m \qquad ...(15.17)$$

is a subspace of V, where $k_1, k_2 ..., k_m$ are scalars.

Example 15.48 Express $\mathbf{x} = (2, -5, 3)$ in R^3 as a linear combination of the vectors $\mathbf{u} = (1, -3, 2)$, $\mathbf{v} = (2, -4, -1)$ and $\mathbf{w} = (1, -5, 7)$.

Solution. Let k_1, k_2, k_3 be scalars, such that

$\mathbf{x} = k_1\mathbf{u} + k_2\mathbf{v} + k_3\mathbf{w}$. Thus, we have

$$\begin{pmatrix} 2 \\ -5 \\ 3 \end{pmatrix} = k_1 \begin{pmatrix} 1 \\ -3 \\ 2 \end{pmatrix} + k_2 \begin{pmatrix} 2 \\ -4 \\ -1 \end{pmatrix} + k_3 \begin{pmatrix} 1 \\ -5 \\ 7 \end{pmatrix} \text{ or } \begin{array}{l} k_1 + 2k_2 + k_3 = 2, \\ 3k_1 - 4k_2 - 5k_3 = -5, \\ 2k_1 - k_2 + 7k_3 = 3. \end{array}$$

We write the augmented matrix and reduce it to *row-echelon form by applying elementary row-operations.

$$(\mathbf{A}\,|\,\mathbf{b}) = \begin{bmatrix} 1 & 2 & 1 & 2 \\ -3 & -4 & -5 & -5 \\ 2 & -1 & 7 & 3 \end{bmatrix} \sim \begin{bmatrix} 1 & 2 & 1 & 2 \\ 0 & 2 & -2 & 1 \\ 0 & -5 & 5 & -1 \end{bmatrix} \sim \begin{bmatrix} 1 & 2 & 1 & 2 \\ 0 & 2 & -2 & 1 \\ 0 & 0 & 0 & 3/2 \end{bmatrix}.$$

Using the back substitution method, we obtain.

$k_1 + 2k_2 + k_3 = 2$, $2k_2 - 2k_3 = 1$, $0 = 3/2$.

The system is inconsistent and has no solution. Thus $\mathbf{x}$ cannot be written as a linear combination of **u, v, w**.

---

* Refer Article 15.21, Example 15.66.

Example 15.49 Express $x = (4, 3, 10)$ in R^3 as a linear combination of the vectors

$$u = (1, 2, -1), \; v = (2, 3, 4) \text{ and } w = (1, 5, -3).$$

Solution. Let k_1, k_2, k_3 be scalars, such that
$x = k_1 u + k_2 v + k_3 w$; that is

$$\begin{bmatrix} 4 \\ 3 \\ 10 \end{bmatrix} = k_1 \begin{bmatrix} 1 \\ 2 \\ -1 \end{bmatrix} + k_2 \begin{bmatrix} 2 \\ 3 \\ 4 \end{bmatrix} + k_3 \begin{bmatrix} 1 \\ 5 \\ -3 \end{bmatrix} \text{ or } \begin{aligned} k_1 + 2k_2 + k_3 &= 4, \\ 2k_1 + 3k_2 + 5k_3 &= 3, \\ -k_1 + 4k_2 - 3k_3 &= 10. \end{aligned}$$

We write the augmented matrix and reduce it to row-echelon form by applying elementary row-operations.

$$(A \mid b) = \begin{bmatrix} 1 & 2 & 1 & 4 \\ 2 & 3 & 5 & 3 \\ -1 & 4 & -3 & 10 \end{bmatrix} \text{ (operate } R_2 \leftarrow R_2 - 2R_1 \text{ and } R_3 \leftarrow R_3 + R_1)$$

$$\sim \begin{bmatrix} 1 & 2 & 1 & 4 \\ 0 & -1 & 3 & -5 \\ 0 & 6 & -2 & 14 \end{bmatrix} \text{ (operate } R_3 \leftarrow R_3 + 6R_2) \sim \begin{bmatrix} 1 & 2 & 1 & 4 \\ 0 & -1 & 3 & -5 \\ 0 & 0 & 16 & -16 \end{bmatrix}.$$

Using the back substitution method, we obtain the solution

$$16k_3 = -16 \text{ or } k_3 = -1$$
$$-k_2 + 3k_3 = -5 \text{ or } k_2 = 2$$
$$k_1 + 2k_2 + k_3 = 4 \text{ or } k_1 = 1 \quad \text{Thus } x = u + 2v - w.$$

Spanning Set. Let S be a subset of a vector space V and suppose that every element in V can be obtained as a linear combination of the elements taken from S, then S is said to be the *spanning set* for V. We may also say that S spans V.

Example 15.50 Let V be the set of all 2×2 real matrices. Show that the set

$$S = \left\{ \begin{pmatrix} 2 & 1 \\ 1 & -2 \end{pmatrix}, \begin{pmatrix} 1 & 1 \\ 1 & 0 \end{pmatrix}, \begin{pmatrix} 0 & 0 \\ 1 & 1 \end{pmatrix}, \begin{pmatrix} 0 & 2 \\ 0 & -1 \end{pmatrix} \right\} \text{ Spans } V.$$

Solution. Let $Y = \begin{bmatrix} a & b \\ c & d \end{bmatrix}$ be an arbitrary element of V. We are required to determine the scalars $\alpha_1, \alpha_2, \alpha_3, \alpha_4$ so that

$$\begin{bmatrix} a & b \\ c & d \end{bmatrix} = \alpha_1 \begin{bmatrix} 2 & 1 \\ 1 & -2 \end{bmatrix} + \alpha_2 \begin{bmatrix} 1 & 1 \\ 1 & 0 \end{bmatrix} + \alpha_3 \begin{bmatrix} 0 & 0 \\ 1 & 1 \end{bmatrix} + \alpha_4 \begin{bmatrix} 0 & 2 \\ 0 & -1 \end{bmatrix}.$$

Equating the corresponding elements, we obtain the system of equations

$$2\alpha_1 + \alpha_2 = a, \qquad\qquad \alpha_1 + \alpha_2 + 2\alpha_4 = b,$$
$$\alpha_1 + \alpha_2 + \alpha_3 = c, \qquad\qquad -2\alpha_1 + \alpha_3 - \alpha_4 = d.$$

We write the augmented matrix and reduce it to row-echelon form by applying elementary row-operations.

$$(A|b) = \begin{bmatrix} 2 & 1 & 0 & 0 & | & a \\ 1 & 1 & 0 & 2 & | & b \\ 1 & 1 & 1 & 0 & | & c \\ -2 & 0 & 1 & -1 & | & d \end{bmatrix} \sim \begin{bmatrix} 2 & 1 & 0 & 0 & | & a \\ 0 & 1 & 0 & 4 & | & 2b-a \\ 0 & 0 & 1 & -2 & | & c-b \\ 0 & 1 & 1 & -1 & | & d+a \end{bmatrix}$$

$$\sim \begin{bmatrix} 2 & 1 & 0 & 0 & | & a \\ 0 & 1 & 0 & 4 & | & 2b-a \\ 0 & 0 & 1 & -2 & | & c-b \\ 0 & 0 & 1 & -5 & | & 2a-2b+d \end{bmatrix} \sim \begin{bmatrix} 2 & 1 & 0 & 0 & | & a \\ 0 & 1 & 0 & 4 & | & 2b-a \\ 0 & 0 & 1 & -2 & | & c-b \\ 0 & 0 & 0 & -3 & | & 2a-b-c+d \end{bmatrix}.$$

Using the back substitution method, we obtain the solution

$-3\alpha_4 = 2a - b - c + d$ or $\alpha_4 = (-2a + b + c - d)/3$

$\alpha_3 - 2\alpha_4 = c - b$ or $\alpha_3 = c - b + \dfrac{2b + 2c - 4a - 2d}{3} = (-4a - b + 5c - 2d)/3$

$\alpha_2 + 4\alpha_4 = 2b - a$ or

$\alpha_2 = 2b - a + (8a - 4b - 4c + 4d)/3 = (5a + 2b - 4c + 4d)/3$

$2\alpha_1 + \alpha_2 = a$ or $\alpha_1 = \dfrac{a}{2} + \dfrac{4c - 4d - 5a - 2b}{6} = \dfrac{(-a - b + 2c - 2d)}{3}.$

Therefore, we can write

$$\begin{bmatrix} a & b \\ c & d \end{bmatrix} = \frac{(-a - b + 2c - 2d)}{3}\begin{bmatrix} 2 & 1 \\ 1 & -2 \end{bmatrix} + \frac{(5a + 2b - 4c + 4d}{3}\begin{bmatrix} 1 & 1 \\ 1 & 0 \end{bmatrix}$$

$$+ \frac{(-4a - b + 5c - 2d)}{3}\begin{bmatrix} 0 & 0 \\ 1 & 1 \end{bmatrix} + \frac{(-2a + b + c - d)}{3}\begin{bmatrix} 0 & 2 \\ 0 & -1 \end{bmatrix}.$$

Since every element of V can be written as a linear combination of the element of S, the set S spans the vector space V.

Example 15.51 Let V be the set of all 3×1 real matrices. Show that the set

$$S = \left\{ \begin{pmatrix} 1 \\ 1 \\ 0 \end{pmatrix}, \begin{pmatrix} 1 \\ -1 \\ 0 \end{pmatrix}, \begin{pmatrix} 0 \\ 0 \\ 1 \end{pmatrix} \right\} \text{ spans } V.$$

Solution. Let $\mathbf{X} = [a\ b\ c]^T$ be an arbitrary element of V. We are required to determine the scalars α_1, α_2, α_3 so that

$$\begin{bmatrix} a \\ b \\ c \end{bmatrix} = \alpha_1\begin{bmatrix} 1 \\ 1 \\ 0 \end{bmatrix} + \alpha_2\begin{bmatrix} 1 \\ -1 \\ 0 \end{bmatrix} + \alpha_3\begin{bmatrix} 0 \\ 0 \\ 1 \end{bmatrix}.$$

Equating the corresponding elements, we obtain the system of equations

$\alpha_1 + \alpha_2 = a, \qquad \alpha_1 - \alpha_2 = b, \qquad \alpha_3 = c$

The solution of this system of equations is

$\alpha_1 = (a + b)/2$, $\alpha_2 = (a - b)/2$, and $\alpha_3 = c$.

Therefore, we can write $\begin{bmatrix} a \\ b \\ c \end{bmatrix} = \dfrac{(a+b)}{2}\begin{bmatrix} 1 \\ 1 \\ 0 \end{bmatrix} + \dfrac{(a-b)}{2}\begin{bmatrix} 1 \\ -1 \\ 0 \end{bmatrix} + c\begin{bmatrix} 0 \\ 0 \\ 1 \end{bmatrix}.$

Since every element of V can be written as a linear combination of the elements of S, the set S spans the vector space V.

15.19.2 Linear Independence of vectors

Let V be a vector space. A finite set $\{v_1, v_2, ..., v_n\}$ of the elements of V is said to be *linearly dependent* if there exists scalars $k_1, k_2, ..., k_n$, not all of them zero, such that

$$k_1 v_1 + k_2 v_2 + ... + k_n v_n = O \qquad ...(15.18)$$

where O on the right hand side denotes a null vector. Otherwise, we say that the vectors are *linearly independent*. In this case the equation (15.18) will be satisfied only when

$$k_1 = k_2 = ... = k_n = 0$$

The above definition of linear dependence of $v_1, v_2, ..., v_n$ can be written alternately as follows.

Theorem: The set of vectors $\{v_1, v_2, ..., v_n\}$ is linearly dependent if and only if at least one element of the set is a linear combination of the remaining elements.

Proof: Let the elements $v_1, v_2, ..., v_n$ be linearly dependent. Then, there exists scalars $k_1, k_2, ..., k_n$, not all zero such that

$$k_1 v_1 + k_2 v_2 + ... k_{i-1}v_{i-1} + k_i v_i + k_{i+1}v_{i+1} ... + k_n v_n = 0.$$

Let $k_i \neq 0$. Then, we can write

or $\quad v_i = -\left(\dfrac{k_1}{k_i}\right)v_1 - \left(\dfrac{k_2}{k_i}\right)v_2 - \cdots - \left(\dfrac{k_{i-1}}{k_i}\right)v_{i-1} - \left(\dfrac{k_{i+1}}{k_i}\right)v_{i+1} - \cdots - \left(\dfrac{k_n}{k_i}\right)v_n$

$$= c_1 v_1 + c_2 v_2 + c_{i-1}v_{i-1} + c_{i+1}v_{i+1} + ... + c_n v_n$$

where $c_1, c_2, ...c_n$ are some scalars. Hence, the vector v_i is a linear combination of the vectors $v_1, v_2, ..., v_{i-1}, v_{i+1}, ..., v_n$.

Now let v_i be a linear combination of $v_1, v_2, ..., v_{i-1}, v_{i+1}, ..., v_n$. Therefore, we have

$$v_i = a_1 v_1 + a_2 v_2 + ... + a_{i-1} v_{i-1} + a_{i+1} v_{i+1} + ... + a_n v_n.$$

where a_i's are scalars. Then

$$a_1 v_1 + a_2 v_2 + ... + a_{i-1}v_{i-1} + (-1)v_i + a_{i+1}v_{i+1} + ... + a_n v_n = 0.$$

Since the coefficient of v_i is not zero, the elements are linearly dependent.

Remark: Equation (15.18) gives a homogeneous system of algebraic equations. Non-trivial solution exist if det (coefficient matrix) $= 0$, that is the vectors are linearly dependent in this case. If the det (coefficient matrix) $\neq 0$, then by Cramer's rule, $k_1 = k_2 = ... = k_n = 0$ and the vectors are linearly independent.

Example. Let $V = R^n$ and consider the vectors

$v_1 = (1, 0, ..., 0)$, $v_2 = (0, 1, ..., 0)$, $...v_n = (0, 0, ..., 1)$, then $v_1, v_2, ...v_n$ are linearly independent.

Solution. Let $k_1, k_2, ..., k_n$ be scalars. Consider the vector equation

$k_1 v_1 + k_2 v_2 + ... + k_n v_n = 0$.

Substituting for $v_1, v_2, ... v_n$ we obtain

$k_1(1, 0, ..., 0) + k_2(0, 1, ..., 0) + ... k_n(0, 0, ..., 1) = 0$

or $(k_1, k_2, ..., k_n) = 0$

Comparing, we obtain $k_i = 0$. Therefore, the given set of vectors is linearly independent.

Alternative

$$\det (v_1, v_2, \cdots v_n) = \begin{vmatrix} 1 & 0 & 0 & \cdots & 0 \\ 0 & 1 & 0 & \cdots & 0 \\ \cdots & \cdots & \cdots & \cdots & \cdots \\ \cdots & \cdots & \cdots & \cdots & \cdots \\ 0 & 0 & 0 & \cdots & 1 \end{vmatrix} = 1 \neq 0.$$

Therefore, the given vectors are linearly independent.

Rule for testing linear dependence. Let the number of vectors be m.

(i) Write the given vectors as row vectors.

(ii) Add such multiples of one vector to the other vectors so that the resulting vectors have zero as first component. In this way, we are left with $(m-1)$ vectors each with first component zero.

(iii) Again add suitable multiples of one from the $(m-1)$ vectors to the remaining vectors so that the resulting vectors have their second component zero. We now have $(m-2)$ vectors with first two components zero.

(iv) Repeat this process till we are left with only one vector.

If all the components of this vector are zero, the given m vectors are linearly dependent, otherwise they are linearly independent.

Example 15.52 Investigate the linear dependence or independence of the vectors.

(a) $v_1 = (2, -1, 3, 2)$, $v_2 = (1, 3, 4, 2)$ and $v_3 = (3, -5, 2, 2)$.   *(MDU 2001)*

(b) $v_1 = (1, 1, 1)$, $v_2 = (1, 2, 3)$ and $v_3 = (2, 3, 8)$

(c) $v_1 = (1, -1, 0)$, $v_2 = (0, 1, -1)$, $v_3 = (0, 2, 1)$ and $v_4 = (1, 0, 3)$
    (AMIE, S-2008)

Solution. (a) Using v_2 to reduce the first component of resulting vectors to zero, we have

$$v_1 - 2v_2 = (0, -7, -5, -2) \qquad ...(i)$$

and

$$v_3 - 3v_2 = (0, -14, -10, -4) \qquad ...(ii)$$

Multiplying (i) by 2 and subtracting from (ii), we get

$(v_3 - 3v_2) - 2(v_1 - 2v_2) = (0, 0, 0, 0)$ or $2v_1 - v_2 - v_3 = 0$.

Therefore, the given vectors are linearly dependent.

(b)   We have $v_2 - v_1 = (0, 1, 2)$, $v_3 - 2v_1 = (0, 1, 6)$

$(v_3 - 2v_1) - (v_2 - v_1) = (0, 0, 4)$ or $v_3 - v_1 - v_2 = (0, 0, 4) \neq 0$.

Since the resulting vector is not a null vector, so the given vectors are linearly independent.

(c) We have $v_1 - v_4 = (0, -1, -3)$, $v_2 = (0, 1, -1)$ and $v_3 = (0, 2, 1)$

$(v_1 - v_4) + v_2 = (0, 0, -4)$, ... (i)   and $2v_2 - v_3 = (0, 0, -3)$ ...(ii)

Multiplying (*i*) by 3, (*ii*) by –4 and adding, we obtain

$$3(v_1 - v_4) + 3v_2 - 8v_2 + 4v_3 = (0, 0, 0) \text{ or } v_1 - \frac{5}{3}v_2 + \frac{4}{3}v_3 - v_4 = 0$$

Therefore, the set of vectors $\{v_1, v_2, v_3, v_4\}$ is linearly dependent.

Example 15.53 Are the following vectors linearly dependent
$v_1 = (1, 2, 4)$, $v_2 = (2, -1, 3)$, $v_3 = (0, 1, 2)$ and $v_4 = (-3, 7, 2)$?
If so, find the relation between them.   (*AMIE., S-2007*)

Solution. Using v_1 to reduce the first component to zero we obtain
$v_2 - 2v_1 = (0, -5, -5)$, $v_3 = (0, 1, 2)$, $v_4 + 3v_1 = (0, 13, 14)$.

Using v_3 (the simplest of the above three vectors) to reduce the second component to zero, we obtain

$(v_2 - 2v_1) + 5v_3 = (0, 0, 5)$, $(v_4 + 3v_1) - 13v_3 = (0, 0, -12)$.

Multiplying the first equation by 12 and the second by 5, and adding, we obtain

$$12(v_2 - 2v_1 + 5v_3) + 5(v_4 + 3v_1 - 13v_3) = 0$$

or $-9v_1 + 12v_2 - 5v_3 + 5v_4 = 0$ ...(*i*)

Therefore, the given vectors are linearly dependent and are connected by relation (*i*).

Example 15.54 Let $v_1 = (2, 2, 1)$, $v_2 = (1, -1, 1)$ and $v_3 = (1, 0, 1)$ be elements in R^3. Show that the set of vectors $\{v_1, v_2, v_3\}$ is linearly independent.

Solution. Consider the vector equation
$k_1 v_1 + k_2 v_2 + k_3 v_3 = 0$.

Substituting for v_1, v_2, v_3, we obtain
$k_1(2, 2, 1) + k_2(1, -1, 1) + k_3(1, 0, 1) = 0$,
or $(2k_1 + k_2 + k_3; 2k_1 - k_2, k_1 + k_2 + k_3) = 0$

Comparing, we obtain $2k_1 + k_2 + k_3 = 0$, $2k_1 - k_2 = 0$ and $k_1 + k_2 + k_3 = 0$. The solution of these equations is $k_1 = k_2 = k_3 = 0$. Therefore, the given set of vectors is linearly independent.

Alternative.

$$\det (v_1, v_2, v_3) = \begin{vmatrix} 2 & 2 & 1 \\ 1 & -1 & 1 \\ 1 & 0 & 1 \end{vmatrix} = 2(-1) - 2(1-1) + 1(1) = -2 + 1 = -1 \neq 0.$$

Therefore, the given vectors are linearly independent.

15.19.3 Dimension and Basis

If in a vector space V there exists a set S of n linearly independent elements of V and if every set of $n + 1$ or more elements in V is linearly dependent, then V is said to have *dimension n*. Thus, the maximum number of linearly independent elements in V is the *dimension* of V and is denoted by dim (V). We write dim $(V) = n$. The vector space $\{0\}$ is defined to have dimension zero.

A linearly independent set in V consisting of a maximum possible number of elements in V is called a *basis* for V. Thus the number of elements of a basis for V equals dim (V).

Some of the standard basis are listed below:

1. If V consists of n-tuples in $\mathbf{R}^n$, then

$$\mathbf{e}_1 = (1, 0, 0, ..., 0), \mathbf{e}_2 = (0, 1, 0, ..., 0), ...\mathbf{e}_n = (0, 0, ..., 0, 1) \text{ is called a}$$

standard basis in $\mathbf{R}^n$.

2. If V consists of all polynomials $P(t)$ of degree $\leq n$, then $\{ 1, t, t^2, ..., t^n \}$ is taken as its standard basis.

3. The real 2×2 matrices form a four-dimensional real vector space. The following four matrices form a basis:

$$\mathbf{E}_{11} = \begin{bmatrix} 1 & 0 \\ 0 & 0 \end{bmatrix}, \mathbf{E}_{12} = \begin{bmatrix} 0 & 1 \\ 0 & 0 \end{bmatrix}, \mathbf{E}_{21} = \begin{bmatrix} 0 & 0 \\ 1 & 0 \end{bmatrix}, \mathbf{E}_{22} = \begin{bmatrix} 0 & 0 \\ 0 & 1 \end{bmatrix}.$$

Any matrix $\begin{bmatrix} l & m \\ p & q \end{bmatrix}$ in V can be written as

$$\begin{bmatrix} l & m \\ p & q \end{bmatrix} = l\mathbf{E}_{11} + m\mathbf{E}_{12} + p\mathbf{E}_{21} + q\mathbf{E}_{22}.$$

Example 15.55 Determine whether or not each of the following set of vectors $\{\mathbf{u}, \mathbf{v}, \mathbf{w}\}$ forms a basis in $\mathbf{R}^3$, where

(i) $\mathbf{u} = (1, 1, 1)$, $\mathbf{v} = (1, 2, 3)$, $\mathbf{w} = (2, -1, 1)$

(ii) $\mathbf{u} = (1, 2, 5)$, $\mathbf{v} = (2, 5, 1)$, $\mathbf{w} = (1, 5, 2)$.

Solution. If the set $\{\mathbf{u}, \mathbf{v}, \mathbf{w}\}$ forms a basis in $\mathbf{R}^3$, then $\mathbf{u}, \mathbf{v}, \mathbf{w}$ must be linearly independent. Let k_1, k_2, k_3 be scalars. Then, the only solution of the equation

$$k_1\mathbf{u} + k_2\mathbf{v} + k_3\mathbf{w} = 0 \qquad ...(1)$$

must be $k_1 = k_2 = k_3 = 0$.

(i) Using equation (1), we obtain the system of equations

$k_1 + k_2 + 2k_3 = 0$, $k_1 + 2k_2 - k_3 = 0$ and $k_1 + 3k_2 + k_3 = 0$.

The solution of this system of equations is $k_1 = k_2 = k_3 = 0$. Therefore, $\mathbf{u}, \mathbf{v}, \mathbf{w}$ are linearly independent and they form a basis in $\mathbf{R}^3$.

(ii) Using equation (1), we obtain the system of equations

$k_1 + 2k_2 + k_3 = 0$, $2k_1 + 5k_2 + 5k_3 = 0$ and $5k_1 + k_2 + 2k_3 = 0$.

The solution of this system of equations is $k_1 = k_2 = k_3 = 0$. Therefore, $\mathbf{u}, \mathbf{v}, \mathbf{w}$ are linearly independent and they form a basis in $\mathbf{R}^3$.

15.19.4 Linear Transformations

First we introduce some notation and terminology concerning arbitrary functions. A function is a rule which assigns to each number some number.

Let A and B be two arbitrary non-empty sets. Suppose to each element in A there is assigned a unique element of B; the collection f of such assignments is called a *mapping* or a transformation from A into B. Thus, a transformation maps the elements of A into the elements of B. The set A is called the *domain* of the transformation, and B is called the *target set*. We denote transformation by capital letters T, S etc. If T is the transformation from A into B, we write

$$T : A \rightarrow B. \qquad \text{...(15.19)}$$

For each element $\mathbf{a}$ in A, the element $T(\mathbf{a})$ in $\mathbf{b}$ is called the *image of* $\mathbf{a}$ under the mapping and we say that T maps $\mathbf{a}$ onto $T(\mathbf{a})$. The collection of all such images in B is called the *range* or the image set of the transformation T.

Now we assume that V and W be two vector spaces, both real or complex, over the same field F of scalars. Let T be a mapping from V into W. The mapping T is said to be a *Linear transformation* or a *linear mapping*, if it satisfies the following two properties:

(*i*) For any two elements $\mathbf{v}_1$, $\mathbf{v}_2$ in V

$$T(\mathbf{v}_1 + \mathbf{v}_2) = T(\mathbf{v}_1) + T(\mathbf{v}_2) \qquad \text{...(15.20)}$$

(*ii*) For every scalar α and every element $\mathbf{v}$ in V

$$T(k\mathbf{v}) = kT(\mathbf{v}) \qquad \text{...(15.21)}$$

The two properties can be combined into one formula which states that

$$T(k_1\mathbf{v}_1 + k_2\mathbf{v}_2) = T(k_1\mathbf{v}_1) + T(k_2\mathbf{v}_2) = k_1T(\mathbf{v}_1) + k_2T(\mathbf{v}_2)$$

for $\mathbf{v}_1$ and $\mathbf{v}_2$ in V and any scalars k_1 and k_2.

Matrix Mappings

Let $\mathbf{A}$ be an $m \times n$ real (or complex) matrix. Let the rows of $\mathbf{A}$ represent the elements in R^n (or $\mathbb{C}^n$) and the columns of $\mathbf{A}$ represent the elements in R^m (or $\mathbb{C}^m$). If $\mathbf{x}$ is in R^n, then $\mathbf{Ax}$ is in R^m.

Thus, an $m \times n$ matrix maps the element in R^n into the element in R^m. We write

$T = \mathbf{A} : R^n \rightarrow R^m$, and $\mathbf{Tx} = \mathbf{Ax}$. The mapping $\mathbf{A}$ is a linear transformation.

Example 15.56 Let $L: R^2 \rightarrow R^2$ be a linear map, having the following effect on the indicated vectors: $L(1, 1) = (1, 4)$ and $L(2, -1) = (-2, 3)$. Compute $L(3, -1)$.

Solution. We write $(3, -1)$ as a linear combination of $(1, 1)$ and $(2, -1)$. Thus, we have

$$(3, -1) = k_1 (1, 1) + k_2 (2, -1) .$$

i.e., $$k_1 + 2k_2 = 3, \quad k_1 - k_2 = -1.$$

The solution is $k_1 = 1/3$, $k_2 = 4/3$. Hence

$$L(3,-1) = k_1 L(1,1) + k_2 L(2,-1)$$

$$= \frac{1}{3}(1,4) + \frac{4}{3}(-2,3) = \left(-\frac{7}{3}, \frac{16}{3}\right).$$

Example 15.57 Let T be a linear transformation from R^3 into R^2 defined by the relations

$TX = AX, A = \begin{bmatrix} 3 & 4 & 5 \\ 6 & 7 & 8 \end{bmatrix}.$ Find TX when **X** is given by $[5\ 6\ 7]^T$.

Solution. We have

$$TX = AX = \begin{bmatrix} 3 & 4 & 5 \\ 6 & 7 & 8 \end{bmatrix}\begin{bmatrix} 5 \\ 6 \\ 7 \end{bmatrix} = \begin{bmatrix} 15+24+35 \\ 30+42+56 \end{bmatrix} = \begin{bmatrix} 74 \\ 128 \end{bmatrix}.$$

Example 15.58 Let T be a linear transformation defined by

$$T\left[\begin{pmatrix} 1 & 1 \\ 1 & 1 \end{pmatrix}\right] = \begin{bmatrix} 1 \\ 2 \\ 3 \end{bmatrix}, T\left[\begin{pmatrix} 0 & 1 \\ 1 & 1 \end{pmatrix}\right] = \begin{bmatrix} 1 \\ -2 \\ 3 \end{bmatrix}, T\left[\begin{pmatrix} 0 & 0 \\ 1 & 1 \end{pmatrix}\right] = \begin{bmatrix} 1 \\ -2 \\ -3 \end{bmatrix}, T\left[\begin{pmatrix} 0 & 0 \\ 0 & 1 \end{pmatrix}\right] = \begin{bmatrix} -1 \\ 2 \\ 3 \end{bmatrix}.$$

Find $T\left[\begin{pmatrix} 4 & 5 \\ 3 & 8 \end{pmatrix}\right].$ *(I.E.T.E., W, 2005)*

Solution. The matrices $\begin{bmatrix} 1 & 1 \\ 1 & 1 \end{bmatrix}, \begin{bmatrix} 0 & 1 \\ 1 & 1 \end{bmatrix}, \begin{bmatrix} 0 & 0 \\ 1 & 1 \end{bmatrix}, \begin{bmatrix} 0 & 0 \\ 0 & 1 \end{bmatrix}$ are linearly independent and hence form a basis in the space of 2×2 matrices. We write for any scalars k_1, k_2, k_3, k_4, not all zero

$$\begin{bmatrix} 4 & 5 \\ 3 & 8 \end{bmatrix} = k_1\begin{bmatrix} 1 & 1 \\ 1 & 1 \end{bmatrix} + k_2\begin{bmatrix} 0 & 1 \\ 1 & 1 \end{bmatrix} + k_3\begin{bmatrix} 0 & 0 \\ 1 & 1 \end{bmatrix} + k_4\begin{bmatrix} 0 & 0 \\ 0 & 1 \end{bmatrix}$$

$$= \begin{bmatrix} k_1 & k_1+k_2 \\ k_1+k_2+k_3 & k_1+k_2+k_3+k_4 \end{bmatrix}.$$

Comparing the elements and solving the resulting system of equations, we obtain $k_1 = 4, k_2 = 1, k_3 = -2, k_4 = 5$. Since T is a linear transformation, we get

$$T\left[\begin{pmatrix} 4 & 5 \\ 3 & 8 \end{pmatrix}\right] = k_1 T\left[\begin{pmatrix} 1 & 1 \\ 1 & 1 \end{pmatrix}\right] + k_2 T\left[\begin{pmatrix} 0 & 1 \\ 1 & 1 \end{pmatrix}\right] + k_3 T\left[\begin{pmatrix} 0 & 0 \\ 1 & 1 \end{pmatrix}\right] + k_4 T\left[\begin{pmatrix} 0 & 0 \\ 1 & 1 \end{pmatrix}\right]$$

$$= 4\begin{bmatrix} 1 \\ 2 \\ 3 \end{bmatrix} + 1\begin{bmatrix} 1 \\ -2 \\ 3 \end{bmatrix} + (-2)\begin{bmatrix} 1 \\ -2 \\ -3 \end{bmatrix} + 5\begin{bmatrix} -1 \\ 2 \\ 3 \end{bmatrix} = \begin{bmatrix} 4+1-2-5 \\ 8-2+4+10 \\ 12+3+6+15 \end{bmatrix} = \begin{bmatrix} -2 \\ 20 \\ 36 \end{bmatrix}.$$

Example 15.59 The set of vectors $\{x_1, x_2\}$, where $x_1 = (1,3)^T, x_2 = (4,6)^T$ is a basis in R^2. Find a linear transformation $T : R^2 \rightarrow R^3$, such that $Tx_1 = (-2, 2, -7)^T$ and $Tx_2 = (-2, -4, -10)^T$. *(I.E.T.E., W, 2007)*

Solution. The transformation T maps column vectors in R^2 into column vectors in R^3. Therefore T must be a matrix **A** of order 3×2. Let $A = \begin{bmatrix} a_1 & b_1 \\ a_2 & b_2 \\ a_3 & b_3 \end{bmatrix}.$

Therefore, we have $\begin{bmatrix} a_1 & b_1 \\ a_2 & b_2 \\ a_3 & b_3 \end{bmatrix}_{3\times2} \begin{bmatrix} 1 \\ 3 \end{bmatrix}_{2\times1} = \begin{bmatrix} -2 \\ 2 \\ -7 \end{bmatrix}_{3\times1}$ and $\begin{bmatrix} a_1 & b_1 \\ a_2 & b_2 \\ a_3 & b_3 \end{bmatrix} \begin{bmatrix} 4 \\ 6 \end{bmatrix} = \begin{bmatrix} -2 \\ -4 \\ -10 \end{bmatrix}$.

Multiplying and comparing the corresponding elements, we get

$$a_1 + 3b_1 = -2 \qquad\qquad 4a_1 + 6b_1 = -2$$
$$a_2 + 3b_2 = 2 \qquad\qquad 4a_2 + 6b_2 = -4$$
$$a_3 + 3b_3 = -7 \qquad\qquad 4a_3 + 6b_3 = -10$$

Solving these equations, we obtain $\mathbf{A} = \begin{bmatrix} 1 & -1 \\ -4 & 2 \\ 2 & -3 \end{bmatrix}$.

Example 15.60 Find a linear transformation T from R^3 into R^3 such that

$$T\begin{bmatrix} 1 \\ 1 \\ 1 \end{bmatrix} = \begin{bmatrix} 6 \\ 2 \\ 4 \end{bmatrix}, T\begin{bmatrix} 1 \\ -1 \\ 1 \end{bmatrix} = \begin{bmatrix} 2 \\ -4 \\ 2 \end{bmatrix}, T\begin{bmatrix} 1 \\ -2 \\ 3 \end{bmatrix} = \begin{bmatrix} 6 \\ 6 \\ 5 \end{bmatrix}. \qquad (I.E.T.E., June 2007)$$

Solution. The transformation T maps elements in R^3 into R^3. Therefore, the transformation is a matrix of order 3×3. Let this matrix be written as

$$T = \mathbf{A} = \begin{bmatrix} a_1 & b_1 & c_1 \\ a_2 & b_2 & c_2 \\ a_3 & b_3 & c_3 \end{bmatrix}.$$ We determine the elements of the matrix $\mathbf{A}$ such as

$$\begin{bmatrix} a_1 & b_1 & c_1 \\ a_2 & b_2 & c_2 \\ a_3 & b_3 & c_3 \end{bmatrix}\begin{bmatrix} 1 \\ 1 \\ 1 \end{bmatrix} = \begin{bmatrix} 6 \\ 2 \\ 4 \end{bmatrix}, \begin{bmatrix} a_1 & b_1 & c_1 \\ a_2 & b_2 & c_2 \\ a_3 & b_3 & c_3 \end{bmatrix}\begin{bmatrix} 1 \\ -1 \\ 1 \end{bmatrix} = \begin{bmatrix} 2 \\ -4 \\ 2 \end{bmatrix}, \begin{bmatrix} a_1 & b_1 & c_1 \\ a_2 & b_2 & c_2 \\ a_3 & b_3 & c_3 \end{bmatrix}\begin{bmatrix} 1 \\ -2 \\ 3 \end{bmatrix} = \begin{bmatrix} 6 \\ 6 \\ 5 \end{bmatrix}.$$

Multiplying and comparing the corresponding elements, we obtain

$$a_1 + b_1 + c_1 = 6 \qquad a_1 - b_1 + c_1 = 2 \qquad a_1 - 2b_1 + 3c_1 = 6$$
$$a_2 + b_2 + c_2 = 2 \qquad a_2 - b_2 + c_2 = -4 \qquad a_2 - 2b_2 + 3c_2 = 6$$
$$a_3 + b_3 + c_3 = 4 \qquad a_3 - b_3 + c_3 = 2 \qquad a_3 - 2b_3 + 3c_3 = 5$$

Solving these equations, we obtain $\mathbf{A} = \begin{bmatrix} 1 & 2 & 3 \\ -15/2 & 3 & 13/2 \\ 1 & 1 & 2 \end{bmatrix}$.

Matrix representation of a linear transformation

Let T be a linear transformation from a n dimensional vector space V into another m-dimensional vector space W over the same field F, such that $T : V \to W$. Let
$$\mathbf{X} = \{v_1, v_2, ...v_n\}, \quad \mathbf{Y} = \{w_1, w_2, ..., w_m\}$$
be the ordered basis of V and W respectively. Since T has values in W, each element $T(v_i)$ can be expressed uniquely as a linear combination of the basis elements $w_1, w_2, ..., w_m$ in W. That is there exist scalars a_{ij} $i = 1, 2 ..., n, j = 1, 2, ..., m$ not all zero, such that

$$T(v_i) = a_{1i}w_1 + a_{2i} w_2 + ...+ a_{mi} w_m$$
$$= [w_1, w_2, ..., w_m] [a_{1i}, a_{2i}, ...a_{mi}] \qquad i = 1, 2, ..., n$$

where $a_{1i}, a_{2i}, ..., a_{mi}$ are the components of $T(v_i)$ relative of the ordered basis $(w_1, w_2, ..., w_m)$.

Hence, we can write

$$T\left[v_1, v_2, \cdots, v_n\right] = \left[w_1, w_2, \cdots, w_m\right] \begin{bmatrix} a_{11} & a_{12} & \cdots & a_{1n} \\ a_{21} & a_{22} & \cdots & a_{2n} \\ \vdots & & & \vdots \\ a_{m1} & a_{m2} & \cdots & a_{mn} \end{bmatrix} \qquad ...(15.22)$$

or $TX = YA$ $\qquad\qquad$ where $\mathbf{A}$ is the m by n matrix

Thus, every linear transformation T of an n-dimensional space V into an m-dimensional space W give rise to an $m \times n$ matrix, $\mathbf{A} = (a_{ij})$ whose columns consist of the components of $T(v_1), T(v_2), ..., T(v_n)$ relative to the basis $(w_1, w_2, ..., w_m)$. We call this the matrix representation of T relative to the given choice of ordered basis $(v_1, v_2, ..., v_n)$ for V and $(w_1, w_2, ..., w_m)$ for W.

Example 15.61 Let $T : R^3 \to R^2$ be a linear transformation defined by

$$T\begin{bmatrix} x \\ y \\ z \end{bmatrix} = \begin{pmatrix} x+y \\ x-z \end{pmatrix}.$$

Find the matrix representation of T with respect to the ordered basis.

(i) $\mathbf{X} = \left\{ \begin{pmatrix} 1 \\ 0 \\ 1 \end{pmatrix}, \begin{pmatrix} 1 \\ 1 \\ 0 \end{pmatrix}, \begin{pmatrix} 0 \\ 1 \\ 1 \end{pmatrix} \right\}$ in R^3 and $\mathbf{Y} = \left\{ \begin{pmatrix} 1 \\ 0 \end{pmatrix}, \begin{pmatrix} 0 \\ 1 \end{pmatrix} \right\}$ in R^2.

(ii) $\mathbf{X} = \left\{ \begin{pmatrix} 1 \\ 0 \\ 1 \end{pmatrix}, \begin{pmatrix} 1 \\ 1 \\ 0 \end{pmatrix}, \begin{pmatrix} 0 \\ 1 \\ 1 \end{pmatrix} \right\}$ in R^3 and $\mathbf{Y} = \left\{ \begin{pmatrix} 1 \\ 3 \end{pmatrix}, \begin{pmatrix} 2 \\ 5 \end{pmatrix} \right\}$ in R^2.

Solution. Let $V = R^3$, $W = R^2$. Let $\mathbf{X} = \{v_1, v_2, v_3\}$, $\mathbf{Y} = \{w_1, w_2\}$.

(*i*)   We have $\mathbf{v}_1 = \begin{bmatrix} 1 \\ 0 \\ 1 \end{bmatrix}$, $\mathbf{v}_2 = \begin{bmatrix} 1 \\ 1 \\ 0 \end{bmatrix}$, $\mathbf{v}_3 = \begin{bmatrix} 0 \\ 1 \\ 1 \end{bmatrix}$, $\mathbf{w}_1 = \begin{bmatrix} 1 \\ 0 \end{bmatrix}$, $\mathbf{w}_2 = \begin{bmatrix} 0 \\ 1 \end{bmatrix}$.

We obtain $T\begin{bmatrix} 1 \\ 0 \\ 1 \end{bmatrix} = \begin{bmatrix} 1 \\ 0 \end{bmatrix} = \begin{bmatrix} 1 \\ 0 \end{bmatrix}(1) + \begin{bmatrix} 0 \\ 1 \end{bmatrix}(0)$, $T\begin{bmatrix} 1 \\ 1 \\ 0 \end{bmatrix} = \begin{bmatrix} 2 \\ 1 \end{bmatrix} = \begin{bmatrix} 1 \\ 0 \end{bmatrix}(2) + \begin{bmatrix} 0 \\ 1 \end{bmatrix}(1)$,

$$T\begin{bmatrix} 0 \\ 1 \\ 1 \end{bmatrix} = \begin{bmatrix} 1 \\ -1 \end{bmatrix} = \begin{bmatrix} 1 \\ 0 \end{bmatrix}(1) + \begin{bmatrix} 0 \\ 1 \end{bmatrix}(-1).$$

Using the notation $T\mathbf{X} = \mathbf{YA}$, we write

$$T[\mathbf{v}_1, \mathbf{v}_2, \mathbf{v}_3] = [\mathbf{w}_1, \mathbf{w}_2]\begin{bmatrix} 1 & 2 & 1 \\ 0 & 1 & -1 \end{bmatrix}$$

or $$T\left[\begin{pmatrix} 1 \\ 0 \\ 1 \end{pmatrix}, \begin{pmatrix} 1 \\ 1 \\ 0 \end{pmatrix}, \begin{pmatrix} 0 \\ 1 \\ 1 \end{pmatrix}\right] = \left[\begin{pmatrix} 1 \\ 0 \end{pmatrix}, \begin{pmatrix} 0 \\ 1 \end{pmatrix}\right]\begin{bmatrix} 1 & 2 & 1 \\ 0 & 1 & -1 \end{bmatrix}.$$

Therefore, the matrix of the linear transformation T with respect to the given basis vectors is given by

$$A = \begin{bmatrix} 1 & 2 & 1 \\ 0 & 1 & -1 \end{bmatrix}.$$

(*ii*)   We have $\mathbf{v}_1 = \begin{bmatrix} 1 \\ 0 \\ 1 \end{bmatrix}$, $\mathbf{v}_2 = \begin{bmatrix} 1 \\ 1 \\ 0 \end{bmatrix}$, $\mathbf{v}_3 = \begin{bmatrix} 0 \\ 1 \\ 1 \end{bmatrix}$, $\mathbf{w}_1 = \begin{bmatrix} 1 \\ 3 \end{bmatrix}$, $\mathbf{w}_2 = \begin{bmatrix} 2 \\ 5 \end{bmatrix}$.

We obtain $T\begin{bmatrix} 1 \\ 0 \\ 1 \end{bmatrix} = \begin{bmatrix} 1 \\ 0 \end{bmatrix} = \begin{bmatrix} 1 \\ 3 \end{bmatrix}(-5) + \begin{bmatrix} 2 \\ 5 \end{bmatrix}(3)$,

$$T\begin{bmatrix} 1 \\ 1 \\ 0 \end{bmatrix} = \begin{bmatrix} 2 \\ 1 \end{bmatrix} = \begin{bmatrix} 1 \\ 3 \end{bmatrix}(-8) + \begin{bmatrix} 2 \\ 5 \end{bmatrix}(5), \quad T\begin{bmatrix} 0 \\ 1 \\ 1 \end{bmatrix} = \begin{bmatrix} 1 \\ -1 \end{bmatrix} = \begin{bmatrix} 1 \\ 3 \end{bmatrix}(-7) + \begin{bmatrix} 2 \\ 5 \end{bmatrix}(4).$$

Using $TX = YA$, we write

$$T\left[\begin{pmatrix}1\\0\\1\end{pmatrix},\begin{pmatrix}1\\1\\0\end{pmatrix},\begin{pmatrix}0\\1\\1\end{pmatrix}\right]=\left[\begin{pmatrix}1\\3\end{pmatrix},\begin{pmatrix}2\\5\end{pmatrix}\right]\begin{bmatrix}-5 & -8 & -7\\3 & 5 & 4\end{bmatrix}.$$

Therefore, the matrix of the linear transformation T with respect to the given basis vectors is given by

$$A=\begin{bmatrix}-5 & -8 & -7\\3 & 5 & 4\end{bmatrix}.$$

Example 15.62 Let V and W be two vector spaces in R^3. Let $T : V \to W$ be a linear transformation defined by

$$T\begin{bmatrix}x\\y\\z\end{bmatrix}=\begin{bmatrix}x+z\\x+y\\x+y+z\end{bmatrix}.$$

Find the matrix representation of T with respect to the ordered basis

$$X=\left\{\begin{bmatrix}-1\\1\\1\end{bmatrix},\begin{bmatrix}1\\-1\\1\end{bmatrix},\begin{bmatrix}1\\1\\-1\end{bmatrix}\right\}\text{ in }V\text{ and }Y=\left\{\begin{bmatrix}1\\-1\\-1\end{bmatrix},\begin{bmatrix}-1\\1\\-1\end{bmatrix},\begin{bmatrix}-1\\-1\\1\end{bmatrix}\right\}\text{ in }W.$$

Solution. Let $X = \{v_1, v_2, v_3\}$, $Y = \{w_1, w_2, w_3\}$.

We have

$$v_1=\begin{bmatrix}-1\\1\\1\end{bmatrix}, v_2=\begin{bmatrix}1\\-1\\1\end{bmatrix}, v_3=\begin{bmatrix}1\\1\\-1\end{bmatrix}, w_1=\begin{bmatrix}1\\-1\\-1\end{bmatrix}, w_2=\begin{bmatrix}-1\\1\\-1\end{bmatrix}, w_3=\begin{bmatrix}-1\\-1\\1\end{bmatrix}.$$

We obtain $T\begin{bmatrix}-1\\1\\1\end{bmatrix}=\begin{bmatrix}0\\0\\1\end{bmatrix}=\begin{bmatrix}1\\-1\\-1\end{bmatrix}\left(-\dfrac{1}{2}\right)+\begin{bmatrix}1\\1\\-1\end{bmatrix}\left(-\dfrac{1}{2}\right)+\begin{bmatrix}-1\\-1\\1\end{bmatrix}(0)$

$$T\begin{bmatrix}1\\-1\\1\end{bmatrix}=\begin{bmatrix}2\\0\\1\end{bmatrix}=\begin{bmatrix}1\\-1\\-1\end{bmatrix}\left(-\dfrac{1}{2}\right)+\begin{bmatrix}-1\\1\\-1\end{bmatrix}\left(-\dfrac{3}{2}\right)+\begin{bmatrix}-1\\-1\\1\end{bmatrix}(-1)$$

$$T\begin{bmatrix} 1 \\ 1 \\ -1 \end{bmatrix} = \begin{bmatrix} 0 \\ 2 \\ 1 \end{bmatrix} = \begin{bmatrix} 1 \\ -1 \\ -1 \end{bmatrix}\left(-\frac{3}{2}\right) + \begin{bmatrix} -1 \\ 1 \\ -1 \end{bmatrix}\left(-\frac{1}{2}\right) + \begin{bmatrix} -1 \\ -1 \\ 1 \end{bmatrix}(-1)$$

Using $TX = YA$, we write

$$T\begin{bmatrix} \begin{bmatrix} -1 \\ 1 \\ 1 \end{bmatrix} & \begin{bmatrix} 1 \\ -1 \\ 1 \end{bmatrix} & \begin{bmatrix} 1 \\ 1 \\ -1 \end{bmatrix} \end{bmatrix} = \begin{bmatrix} \begin{bmatrix} 1 \\ -1 \\ -1 \end{bmatrix} & \begin{bmatrix} -1 \\ 1 \\ -1 \end{bmatrix} & \begin{bmatrix} -1 \\ -1 \\ 1 \end{bmatrix} \end{bmatrix}\begin{bmatrix} -1/2 & -1/2 & -3/2 \\ -1/2 & -3/2 & -1/2 \\ 0 & -1 & -1 \end{bmatrix}.$$

Therefore, the matrix of the linear transformation T with respect to the given basis vectors is given by

$$A = \begin{bmatrix} -1/2 & -1/2 & -3/2 \\ -1/2 & -3/2 & -1/2 \\ 0 & -1 & -1 \end{bmatrix}.$$

Example 15.63 Let $T : R^3 \to R^2$ be a linear transformation. Let $A = \begin{bmatrix} 1 & 2 & 1 \\ 2 & -3 & -4 \end{bmatrix}$ be the matrix representation of the linear transformation with respect to the ordered basis vectors $v_1 = [1, -1, 1]^T$, $v_2 = [2, 3, -1]^T$, $v_3 = [1, 1, 1]^T$ in R^3 and $w_1 = [1, 1]^T$, $w_2 = [2, 3]^T$ in R^2. Then determine the linear transformation T.

Solution. We have $T\begin{bmatrix} v_1, & v_2, & v_3 \end{bmatrix} = \begin{bmatrix} w_1, & w_2 \end{bmatrix}A = \begin{bmatrix} 1 & 2 \\ 1 & 3 \end{bmatrix}\begin{bmatrix} 1 & 2 & 1 \\ 2 & -3 & -4 \end{bmatrix}$

$$= \begin{bmatrix} 5 & -4 & -7 \\ 7 & -7 & -11 \end{bmatrix}.$$

Now, any vector $x = (x_1, x_2, x_3)^T$ in R^3 with respect to the given basis can be written as

$$\begin{bmatrix} x_1 \\ x_2 \\ x_3 \end{bmatrix} = k_1\begin{bmatrix} 1 \\ -1 \\ 1 \end{bmatrix} + k_2\begin{bmatrix} 2 \\ 3 \\ -1 \end{bmatrix} + k_3\begin{bmatrix} 1 \\ 1 \\ -1 \end{bmatrix}.$$

We obtain $k_1 = \dfrac{2x_1 - x_2 + x_3}{4}$, $k_2 = \dfrac{x_2 + x_3}{2}$, $k_3 = \dfrac{2x_1 - 3x_2 - 5x_3}{4}$. Hence, we have

$$Tx = k_1 Tv_1 + k_2 Tv_2 + k_3 Tv_3$$

$$\text{or } Tx = k_1 \begin{bmatrix} 5 \\ 7 \end{bmatrix} + k_2 \begin{bmatrix} -4 \\ -7 \end{bmatrix} + k_3 \begin{bmatrix} -7 \\ -11 \end{bmatrix} = \begin{bmatrix} 5k_1 - 4k_2 - 7k_3 \\ 7k_1 - 7k_2 - 11k_3 \end{bmatrix} = \begin{bmatrix} -x_1 + 2x_2 + 8x_3 \\ -2x_1 + 3x_2 + 12x_3 \end{bmatrix}.$$

Exercise 15.7

1. Let V be the set of all positive real numbers where addition is defined by $x + y = xy$ and scalar multiplication is defined by $kx = x$. Determine whether V is a vector space. **Ans. yes**.

2. The vectors $\mathbf{u}_1 = (1, 0, 0)$, $\mathbf{u}_2 = (1, 1, 0)$ and $\mathbf{u}_3 = (1, 1, 1)$ form a basis for the vector space R^3. Express the vector $x = (3, -4, 8)$ as a linear combination of $\mathbf{u}_1$, $\mathbf{u}_2$ and $\mathbf{u}_3$. **Ans. $x = 7\mathbf{u}_1 - 12\mathbf{u}_2 + 8\mathbf{u}_3$.**

3. Express $\mathbf{M}$ as a linear combination of the matrices $\mathbf{A}$, $\mathbf{B}$, $\mathbf{C}$ where

$$\mathbf{M} = \begin{bmatrix} 4 & 7 \\ 7 & 9 \end{bmatrix}, \text{ and } \mathbf{A} = \begin{bmatrix} 1 & 1 \\ 1 & 1 \end{bmatrix}, \mathbf{B} = \begin{bmatrix} 1 & 2 \\ 3 & 4 \end{bmatrix}, \mathbf{C} = \begin{bmatrix} 1 & 1 \\ 4 & 5 \end{bmatrix}.$$

Ans. $\mathbf{M} = 2\mathbf{A} + 3\mathbf{B} - \mathbf{C}$.

4. Let $u = (1, 2, -1)$, $v = (2, 3, 4)$ and $w = (1, 5, -3)$. Determine whether or not x is a linear combinations of $\mathbf{u}$, $\mathbf{v}$, $\mathbf{w}$ where $x = (3, 2, 5)$ **Ans. $2\mathbf{u} + \mathbf{v} - \mathbf{w}$**

5. Investigate the linear dependence or independence of the following set of vectors:

(a) $\mathbf{v}_1 = (1, 2, 1)$, $\mathbf{v}_2 = (4, 1, 2)$, $\mathbf{v}_3 = (6, 5, 4)$ and $\mathbf{v}_4 = (-3, 8, 1)$.

(b) $\mathbf{v}_1 = (2, -1, 4)$, $\mathbf{v}_2 = (0, 1, 2)$, $\mathbf{v}_3 = (6, -1, 16)$, $v_4 = (4, 0, 12)$.

(c) $\mathbf{v}_1 = (1, 2, -1, 0)$, $\mathbf{v}_2 = (1, 3, 1, 2)$, $\mathbf{v}_3 = (6, 1, 0, 1)$ and $\mathbf{v}_4 = (4, 1, 2, 0)$.

(d) $\mathbf{v}_1 = (1, 2, 2)$, $\mathbf{v}_2 = (2, 1, -2)$ and $\mathbf{v}_3 = (2, -2, 1)$.

Ans. (a) Dependent, (b) Dependent, (c) Independent, (d) Independent.

6. Are the following vectors linearly dependent? If so, find the relation between them.

(i) $\mathbf{v}_1 = (1, 2, 1)$, $\mathbf{v}_2 = (2, 1, 4)$, $\mathbf{v}_3 = (4, 5, 6)$, and $\mathbf{v}_4 = (1, 8, -3)$

(ii) $\mathbf{v}_1 = (3, 1, -4)$, $\mathbf{v}_2 (2, 2, -3)$, $\mathbf{v}_3 = (0, -4, 1)$ and $\mathbf{v}_4 = (-4, -4, 6)$

Ans. (i) yes, $\mathbf{v}_3 = 2\mathbf{v}_1 + \mathbf{v}_2$ and $\mathbf{v}_4 = 5\mathbf{v}_1 - 2\mathbf{v}_2$ (ii) yes, $2\mathbf{v}_1 - \mathbf{v}_2 - \mathbf{v}_3 + \mathbf{v}_4 = 0$.

7. Let $\mathbf{v}_1 = (1, -1, 0)$, $\mathbf{v}_2 = (0, 1, -1)$ and $\mathbf{v}_3 = (0, 0, 1)$ be elements of R^3. Show that the set of vectors $\{\mathbf{v}_1, \mathbf{v}_2, \mathbf{v}_3\}$ is linearly independent.

(I.E.T.E., Dec. 2005)

8. Examine whether the transformation T given is linear or not. If not linear, state why?

$$T : R^2 \to R^2; \; T\begin{bmatrix} x \\ y \end{bmatrix} = \begin{bmatrix} x+y \\ x \end{bmatrix}. \qquad \textbf{Ans. Linear}.$$

9. Let T be a transformation from R^3 into R^2 defined by

$$T(x_1, x_2, x_3) = x_1^2 + x_2^2 + x_3^2.$$

Show that T is not a linear transformation.

(P.T.U., 2005)

[**Hint:** Let $x = (x_1, x_2, x_3)$ and $y = (y_1, y_2, y_3)$ be any two elements in R^3.

Then $x + y = (x_1 + y_1, x_2 + y_2, x_3 + y_3)$.

$T(x) = x_1^2 + x_2^2 + x_3^2, T(y) = y_1^2 + y_2^2 + y_3^2$

$T(x + y) = (x_1 + y_1)^2 + (x_2 + y_2)^2 + (x_3 + y_3)^2 \neq T(x) + T(y)$.

Therefore, T is not a linear transformation.]

10. Let T be a linear transformation from R^3 into R^2 defined by the relations

$Tx = Ax$, $A = \begin{bmatrix} 1 & -4 & 5 \\ 2 & 3 & -6 \end{bmatrix}$. Find Tx, when x is given by $[1 \ 3 \ -5]^T$.

Ans. $\begin{bmatrix} -36 \\ 41 \end{bmatrix}$.

11. Let $T : R^3 \to R^2$ be a linear transformation defined by $T\begin{bmatrix} x \\ y \\ z \end{bmatrix} = \begin{pmatrix} y + z \\ y - z \end{pmatrix}$.

Taking $\left\{ \begin{bmatrix} 1 \\ 1 \\ 0 \end{bmatrix}, \begin{bmatrix} 0 \\ 1 \\ 1 \end{bmatrix}, \begin{bmatrix} 1 \\ 0 \\ 1 \end{bmatrix} \right\}$ as a basis in R^3 and $\left\{ \begin{pmatrix} 1 \\ 0 \end{pmatrix}, \begin{pmatrix} 0 \\ 1 \end{pmatrix} \right\}$ as a basis in R^2,

determine the matrix of linear transformation. (*I.E.T.E., Dec. 2006*)

Ans. $\begin{bmatrix} 1 & 2 & 1 \\ 1 & 0 & -1 \end{bmatrix}$.

12. Let $T : R^3 \to R^2$ be a linear transformation defined by

$$T\begin{bmatrix} x \\ y \\ z \end{bmatrix} = \begin{bmatrix} 2x + 3y - z \\ 4x - y + 2z \end{bmatrix}.$$

Find the matrix representation of T with respect to the ordered basis

$$X = \left\{ \begin{bmatrix} 1 \\ 1 \\ 0 \end{bmatrix}, \begin{bmatrix} 1 \\ 2 \\ 3 \end{bmatrix}, \begin{bmatrix} 1 \\ 3 \\ 5 \end{bmatrix} \right\} \text{ in } R^3 \text{ and } Y = \left\{ \begin{bmatrix} 1 \\ 2 \end{bmatrix}, \begin{bmatrix} 2 \\ 3 \end{bmatrix} \right\} \text{ in } R^2.$$

Ans. $\begin{bmatrix} -9 & 1 & 4 \\ 7 & 2 & 1 \end{bmatrix}$.

15.20 Elementary Transformation (or operation) of a matrix

Any one of the following three operations applied on the rows (or columns) of a matrix is called an *elementary transformation* (or *operation*).

(*i*) The interchange of any two rows (or columns).

(*ii*) The multiplication of each element of any row (or column) by a non-zero constant.

(*iii*) The addition to the elements of any row (or column), the constant times the corresponding elements of any other row (or column).

Notation. *The following symbols are used for elementary transformations:*

(*a*) $R_i \leftrightarrow R_j$ *for the interchange of ith row with j^{th} row.*

$C_i \leftrightarrow C_j$ *for the interchange of ith column with j^{th} column.*

(*b*) $R_i \rightarrow kR_i$ *for the multiplication of ith row by non-zero constant k.*

$C_i \rightarrow kC_i$ *for the multiplication of ith column by non-zero constant k.*

(*c*) $R_i \rightarrow R_i + kR_j$ *for addition to i^{th} row, k times the elements of the jth row.*

$C_i \rightarrow C_i + kC_j$ *for addition to i^{th} column, k times the elements of the jth column.*

Two matrices are said to be equivalent if one is obtained from the other by elementary transformations. The symbol ~ is used the **equivalence**.

Any elementary row transformation of a matrix **A** can be accomplished by **premultiplying A** by a unit matrix on whose rows the same elementary transformation has been performed. Similarly, any elementary column transformation of a matrix **A** can be accomplished by **post multiplying A** by a unit matrix on whose columns the same elementary transformation has been performed.

15.20.1 Inverse of a matrix by elementary operations

If **A** is a matrix such that A^{-1} exists, then to find A^{-1} using elementary row operations, write **A = IA** till we get **I = BA**. The matrix **B** will be the inverse of **A**. Similarly, if we wish to find A^{-1} using column operations, then write **A = AI** and apply a sequence of column operations on **A = AI** till we get, **I = AB**.

Remark

In case, after applying one or more elementary row (column) operations on **A = IA** (**A = AI**), if we obtain all zeros in one or more rows of the matrix **A** on L.H.S., then A^{-1} does not exist.

Note. A matrix obtained from a unit matrix by subjecting it to any of the elementary transformations, is called an **elementary matrix.**

15.21 Echelon form of a matrix

An $m \times n$ matrix A is called a row *echelon matrix*, or in row *echelon form* if the number of zeros preceeding the first non-zero element of a row increases row by row until a row having all zero elements is obtained. Therefore, a matrix is in row echelon form if the following are satisfied:

(i) if the ith row contains all zero, it is true for all subsequent rows.

(ii) If a column contains a non-zero element of any row, then every subsequent entry in this column is zero.

(iii) Rows containing all zeros occur only after all non-zero rows.

For example, the following matrices are in row echelon form whose pivots have been circled.

$$\begin{bmatrix} ②&3&7&9\\ 0&⑤&4&1\\ 0&0&0&6 \end{bmatrix} \quad \begin{bmatrix} ①&-4&5&6\\ 0&0&③&5\\ 0&0&0&0 \end{bmatrix} \quad \begin{bmatrix} ①&-3&4&6\\ 0&②&-3&-3\\ 0&0&⑥&0\\ 0&0&0&0 \end{bmatrix}$$

Remarks: (1) If A is a square matrix, then the row-echelon form is an upper triangular matrix and column echelon form is a lower-triangular matrix.

(2) From echelon form of a matrix, we examine whether a given set of vectors are linearly independent or not. We form the matrix with each vector as its row (or column) and reduce it to the row (column) echelon form. The given vectors are *linearly independent*, if the row echelon form has *no row with all its elements as zeros.* The number of non-zero rows is the *dimension* of the given set of vectors and the set of vectors consisting of the non-zero rows is the *basis.*

Rank of a matrix A To find the rank of a matrix from definition, lot of computational work is involved because we have to evaluate several determinants. The given matrix reduces to row-echelon form give us the rank, as per following norms.

Let $A = (a_{ij})$ be a given $m \times n$ matrix. Assume that $a_{11} \neq 0$. If $a_{11} = 0$, we interchange the first row with some other row to make the element in the $(1, 1)$ position as non-zero. Using elementary row operations, we reduce the matrix A to its row-echelon form, elements of first column below a_{11} are made zero, then elements in the second column below a_{22} are made zero and so on. The number of non-zero rows in the row echelon form of matrix A gives the *rank of the matrix* A.

The following examples will make the procedure clear.

Example 15.64 Find the rank of the following matrices:

(i) $\mathbf{A} = \begin{bmatrix} 1 & 3 & 4 & 5 \\ 1 & 2 & 6 & 7 \\ 1 & 5 & 0 & 10 \end{bmatrix}$, (ii) $\mathbf{A} = \begin{bmatrix} 1 & 2 & 3 & 0 \\ 2 & 4 & 3 & 2 \\ 3 & 2 & 1 & 3 \\ 6 & 8 & 7 & 5 \end{bmatrix}$, (iii) $\mathbf{A} = \begin{bmatrix} 6 & 1 & 3 & 8 \\ 4 & 2 & 6 & -1 \\ 10 & 3 & 9 & 7 \\ 16 & 4 & 12 & 15 \end{bmatrix}$.

(*Osmania 2003, MDU 2003, Delhi 1997*)

Solution. (*i*) Since it is a 3×4 matrix, the rank is ≤ 3.

(*i*) Applying the operations $R_2 \leftarrow R_2 - R_1$, and $R_3 \leftarrow R_3 - R_1$, we obtain

$$\mathbf{A} \sim \begin{bmatrix} 1 & 3 & 4 & 5 \\ 0 & -1 & 2 & 2 \\ 0 & 2 & -4 & 5 \end{bmatrix} \text{(operate } R_3 \leftarrow R_3 + 2R_2) \sim \begin{bmatrix} 1 & 3 & 4 & 5 \\ 0 & -1 & 2 & 2 \\ 0 & 0 & 0 & 9 \end{bmatrix}.$$

This is the row echelon form of **A**. Since the number of non-zero rows in the row echelon form is 3, we get rank (**A**) = 3.

(*ii*) Applying the operations $R_2 \leftarrow R_2 - 2R_1, R_3 \leftarrow R_3 - 3R_1$ and $R_4 \leftarrow R_4 - 6R_1$, we obtain

$$\mathbf{A} \sim \begin{bmatrix} 1 & 2 & 3 & 0 \\ 0 & 0 & -3 & 2 \\ 0 & -4 & -8 & 3 \\ 0 & -4 & -11 & 5 \end{bmatrix} \text{(operate } R_3 \leftarrow -R_3/4) \sim \begin{bmatrix} 1 & 2 & 3 & 0 \\ 0 & 0 & -3 & 2 \\ 0 & 1 & 2 & -3/4 \\ 0 & -4 & -11 & 5 \end{bmatrix}$$

$$\text{(operate } R_4 \leftarrow R_4 + 4R_3)$$

$$\sim \begin{bmatrix} 1 & 2 & 3 & 0 \\ 0 & 0 & -3 & 2 \\ 0 & 1 & 2 & -3/4 \\ 0 & 0 & -3 & 2 \end{bmatrix} \text{(performing } R_2 \leftrightarrow R_3) \sim \begin{bmatrix} 1 & 2 & 3 & 0 \\ 0 & 1 & 2 & -3/4 \\ 0 & 0 & -3 & 2 \\ 0 & 0 & -3 & 2 \end{bmatrix}$$

$$\text{(operate } R_4 \leftarrow R_4 - R_3)$$

$$\sim \begin{bmatrix} 1 & 2 & 3 & 0 \\ 0 & 1 & 2 & -3/4 \\ 0 & 0 & -3 & 2 \\ 0 & 0 & 0 & 0 \end{bmatrix}. \text{ Since the number of non-zero rows in the row echelon form of } \mathbf{A} \text{ is 3, we get rank } (\mathbf{A}) = 3.$$

(*iii*) $\mathbf{A} = \begin{bmatrix} 6 & 1 & 3 & 8 \\ 4 & 2 & 6 & -1 \\ 10 & 3 & 9 & 7 \\ 16 & 4 & 12 & 15 \end{bmatrix} R_1 \leftrightarrow R_2 \sim \begin{bmatrix} 4 & 2 & 6 & -1 \\ 6 & 1 & 3 & 8 \\ 10 & 3 & 9 & 7 \\ 16 & 4 & 12 & 15 \end{bmatrix} \begin{array}{l} R_2 \leftarrow R_2 - \dfrac{3}{2}R_1 \\[2mm] R_3 \leftarrow R_3 - \dfrac{5}{2}R_1 \\[2mm] R_4 \leftarrow R_4 - 4R_1 \end{array}$

$$\sim \begin{bmatrix} 4 & 2 & 6 & -1 \\ 0 & -2 & -6 & 19/2 \\ 0 & -2 & -6 & 19/2 \\ 0 & -4 & -12 & 19 \end{bmatrix} \begin{matrix} \\ \\ R_3 \leftarrow R_3 - R_2 \\ R_4 \leftarrow R_4 - 2R_2 \end{matrix} \sim \begin{bmatrix} 4 & 2 & 6 & -1 \\ 0 & -2 & -6 & 19/2 \\ 0 & 0 & 0 & 0 \\ 0 & 0 & 0 & 0 \end{bmatrix}$$

since the number of non-zero rows in the row echelon form of **A** is 2, we get rank (**A**) = 2.

Example 15.65 Find the rank of following matrices:

$$(i)\ \mathbf{A} = \begin{bmatrix} 2 & 3 & -1 & -1 \\ 1 & -1 & -2 & -4 \\ 3 & 1 & 3 & -2 \\ 6 & 3 & 0 & -7 \end{bmatrix}, \begin{matrix} (UPTU2005, MDU2001, \\ A.M.I.E., W - 1998) \end{matrix} (ii)\ \mathbf{B} = \begin{bmatrix} 1 & 2 & -2 & 3 & 1 \\ 1 & 3 & -2 & 3 & 0 \\ 2 & 4 & -3 & 6 & 4 \\ 1 & 1 & -1 & 4 & 6 \end{bmatrix}$$

Solution. (*i*) Performing the operation $R_1 \leftrightarrow R_2$, we get

$$\mathbf{A} \sim \begin{bmatrix} 1 & -1 & -2 & -4 \\ 2 & 3 & -1 & -1 \\ 3 & 1 & 3 & -2 \\ 6 & 3 & 0 & -7 \end{bmatrix} \begin{matrix} \text{(applying the operations } R_2 \leftarrow R_2 - 2R_1, \\ R_3 \leftarrow R_3 - 3R_1 \text{ and } R_4 \leftarrow R_4 - 6R_1) \end{matrix}$$

$$\sim \begin{bmatrix} 1 & -1 & -2 & -4 \\ 0 & 5 & 3 & 7 \\ 0 & 4 & 9 & 10 \\ 0 & 9 & 12 & 17 \end{bmatrix} \begin{matrix} \text{(applying the operations } R_3 \leftarrow R_3 - \dfrac{4}{5}R_2 \\ \text{and } R_4 \leftarrow R_4 - \dfrac{9}{5}R_2) \end{matrix}$$

$$\sim \begin{bmatrix} 1 & -1 & -2 & -4 \\ 0 & 5 & 3 & 7 \\ 0 & 0 & 33/5 & 22/5 \\ 0 & 0 & 33/5 & 22/5 \end{bmatrix} \text{(applying } R_4 \leftarrow R_4 - R_3) \sim \begin{bmatrix} 1 & -1 & -2 & -4 \\ 0 & 5 & 3 & 7 \\ 0 & 0 & 33/5 & 22/5 \\ 0 & 0 & 0 & 0 \end{bmatrix}.$$

This is the row echelon form of **A**. Since the number of non-zero rows in the row echelon form of **A** is 3. We get rank (**A**) = 3.

(*ii*) By applying the operations $R_2 \leftarrow R_2 - R_1$, $R_3 \leftarrow R_3 - 2R_1$, and $R_4 \leftarrow R_4 - R_1$, we obtain

$$B \sim \begin{bmatrix} 1 & 2 & -2 & 3 & 1 \\ 0 & 1 & 0 & 0 & -1 \\ 0 & 0 & 1 & 0 & 2 \\ 0 & -1 & 1 & 1 & 5 \end{bmatrix} \text{ (operate } R_4 \leftarrow R_4 + R_2)$$

$$\sim \begin{bmatrix} 1 & 2 & -2 & 3 & 1 \\ 0 & 1 & 0 & 0 & -1 \\ 0 & 0 & 1 & 0 & 2 \\ 0 & 0 & 1 & 1 & 4 \end{bmatrix} \text{ (operate } R_4 \leftarrow R_4 - R_3) \sim \begin{bmatrix} 1 & 2 & -2 & 3 & 1 \\ 0 & 1 & 0 & 0 & -1 \\ 0 & 0 & 1 & 0 & 2 \\ 0 & 0 & 0 & 1 & 2 \end{bmatrix}.$$

This is the row echelon form of **B**. Since the number of non-zero rows in the row echelon form of **B** is 4, we get rank (**B**) = 4.

Normal form of a matrix

Every non-zero matrix **A** of rank r, can be reduced by a sequence of elementary transformations to one of the following four forms, called the **normal form of matrix A.**

(i) $[\mathbf{I}_r]$, (ii) $[\mathbf{I}_r \ \mathbf{O}]$, (iii) $\begin{bmatrix} \mathbf{I}_r \\ \mathbf{O} \end{bmatrix}$, (iv) $\begin{bmatrix} \mathbf{I}_r & \mathbf{O} \\ \mathbf{O} & \mathbf{O} \end{bmatrix}$, where $\mathbf{I}_r$ is $r \times r$ identity matrix

and **O** is null matrix of any order.

By a combination of rows transformations followed by column transformations a non-zero matrix **A** can be easily reduced to normal form which gives the rank of the matrix **A**.

Example 15.66 Find the ranks of the following matrices by reducing them to the normal form:

(i) $\mathbf{A} = \begin{bmatrix} 3 & 2 & 5 & 7 & 12 \\ 1 & 1 & 2 & 3 & 5 \\ 3 & 3 & 6 & 9 & 15 \end{bmatrix}$, (ii) $\mathbf{B} = \begin{bmatrix} 0 & 1 & -3 & -1 \\ 1 & 0 & 1 & 1 \\ 3 & 1 & 0 & 2 \\ 1 & 1 & -2 & 0 \end{bmatrix}$. *(A.M.I.E., W - 2003; IETE Dec 1995)*

Solution. (i) By performing the operation $R_1 \leftrightarrow R_2$, the given matrix

$$A \sim \begin{bmatrix} 1 & 1 & 2 & 3 & 5 \\ 3 & 2 & 5 & 7 & 12 \\ 3 & 3 & 6 & 9 & 15 \end{bmatrix} \text{ (perform the operations } (R_2 \leftarrow R_2 - 3R_1 \text{ and } R_3 \leftarrow R_3 - 3R_1)$$

$$\sim \begin{bmatrix} 1 & 1 & 2 & 3 & 5 \\ 0 & -1 & -1 & -2 & -3 \\ 0 & 0 & 0 & 0 & 0 \end{bmatrix} \begin{array}{l} \text{(perform the operations} \\ C_2 \leftarrow C_2 - C_1, C_3 \leftarrow C_3 - 2C_1 \text{ and} \\ C_4 \leftarrow C_4 - 3C_1 \text{ and } C_5 \leftarrow C_5 - 5C_1) \end{array}$$

$$\sim \begin{bmatrix} 1 & 0 & 0 & 0 & 0 \\ 0 & -1 & -1 & -2 & -3 \\ 0 & 0 & 0 & 0 & 0 \end{bmatrix} \text{(perform the operation } R_2 \leftarrow -R_2)$$

$$\sim \begin{bmatrix} 1 & 0 & 0 & 0 & 0 \\ 0 & 1 & 1 & 2 & 3 \\ 0 & 0 & 0 & 0 & 0 \end{bmatrix} \begin{array}{l} \text{(perform the operations } (C_3 \leftarrow C_3 - C_2, \\ C_4 \leftarrow C_4 - 2C_2, \text{ and } C_5 \leftarrow C_5 - 3C_2) \end{array}$$

$$\sim \begin{bmatrix} 1 & 0 & 0 & 0 & 0 \\ 0 & 1 & 0 & 0 & 0 \\ 0 & 0 & 0 & 0 & 0 \end{bmatrix} \sim \left[\begin{array}{cc|ccc} 1 & 0 & 0 & 0 & 0 \\ 0 & 1 & 0 & 0 & 0 \\ \hline 0 & 0 & 0 & 0 & 0 \end{array}\right] \sim \begin{bmatrix} \mathbf{I_2} & \mathbf{O} \\ \mathbf{O} & \mathbf{O} \end{bmatrix},$$

which is the required normal form. Hence the rank $(\mathbf{A}) = 2$.

(ii) Applying the operations $C_3 \leftarrow C_3 - C_1$ and $C_4 \leftarrow C_4 - C_1$, we get

$$B \sim \begin{bmatrix} 0 & 1 & -3 & -1 \\ 1 & 0 & 0 & 0 \\ 3 & 1 & -3 & -1 \\ 1 & 1 & -3 & -1 \end{bmatrix} \begin{array}{l} \\ \\ R_3 \leftarrow R_3 - R_1 \\ R_4 \leftarrow R_4 - R_1 \end{array}$$

$$\sim \begin{bmatrix} 0 & 1 & -3 & -1 \\ 1 & 0 & 0 & 0 \\ 3 & 0 & 0 & 0 \\ 1 & 0 & 0 & 0 \end{bmatrix} \begin{array}{l} \text{(Operating } C_3 \leftarrow C_3 + 3C_2 \\ \text{and } C_4 \leftarrow C_4 + C_2) \end{array}$$

$$\sim \begin{bmatrix} 0 & 1 & 0 & 0 \\ 1 & 0 & 0 & 0 \\ 3 & 0 & 0 & 0 \\ 1 & 0 & 0 & 0 \end{bmatrix} \begin{array}{l} (R_3 \leftarrow R_3 - 3R_2 \\ R_4 \leftarrow R_4 - R_2) \end{array}$$

$$\sim \begin{bmatrix} 0 & 1 & 0 & 0 \\ 1 & 0 & 0 & 0 \\ 0 & 0 & 0 & 0 \\ 0 & 0 & 0 & 0 \end{bmatrix} \text{(operating } R_1 \leftrightarrow R_2)$$

$$\sim \begin{bmatrix} 1 & 0 & 0 & 0 \\ 0 & 1 & 0 & 0 \\ \hline 0 & 0 & 0 & 0 \\ 0 & 0 & 0 & 0 \end{bmatrix} \sim \begin{bmatrix} I_2 & O \\ O & O \end{bmatrix}. \text{ Hence Rank of matrix } \mathbf{B} = 2.$$

Example 15.67 For the matrix $\mathbf{A} = \begin{bmatrix} 1 & 2 & 3 \\ 3 & 1 & 2 \end{bmatrix}$, find non-singular matrices $\mathbf{P}$ and

$\mathbf{Q}$ such that $\mathbf{PAQ}$ is in the normal form.

Solution. The given matrix $\mathbf{A}$ is of order 2×3. we start with $I_2 A I_3$, while I_2 is meant for row transformations only, I_3 is meant for column transformations only, which are performed on $\mathbf{A}$ to reduce it to normal form. The procedure is as follows.

$$I_2 A I_3 = \begin{bmatrix} 1 & 0 \\ 0 & 1 \end{bmatrix} \begin{bmatrix} 1 & 2 & 3 \\ 3 & 1 & 2 \end{bmatrix} \begin{bmatrix} 1 & 0 & 0 \\ 0 & 1 & 0 \\ 0 & 0 & 1 \end{bmatrix} \text{(Applying the operation } R_2 \leftarrow R_2 - 3R_1)$$

$$\sim \begin{bmatrix} 1 & 0 \\ -3 & 1 \end{bmatrix} \begin{bmatrix} 1 & 2 & 3 \\ 0 & -5 & -7 \end{bmatrix} \begin{bmatrix} 1 & 0 & 0 \\ 0 & 1 & 0 \\ 0 & 0 & 1 \end{bmatrix} \begin{array}{l} \text{(Applying the operations } C_2 \leftarrow C_2 - 2C_1 \\ \text{and } C_3 \leftarrow C_3 - 3C_1) \end{array}$$

$$\sim \begin{bmatrix} 1 & 0 \\ -3 & 1 \end{bmatrix} \begin{bmatrix} 1 & 0 & 0 \\ 0 & -5 & -7 \end{bmatrix} \begin{bmatrix} 1 & -2 & -3 \\ 0 & 1 & 0 \\ 0 & 0 & 1 \end{bmatrix} \text{(Applying the operation } C_2 \leftarrow - C_2 /5)$$

$$\sim \begin{bmatrix} 1 & 0 \\ -3 & 1 \end{bmatrix} \begin{bmatrix} 1 & 0 & 0 \\ 0 & 1 & -7 \end{bmatrix} \begin{bmatrix} 1 & 2/5 & -3 \\ 0 & -1/5 & 0 \\ 0 & 0 & 1 \end{bmatrix} \text{(Applying the operation } C_3 \leftarrow C_3 + 7C_2)$$

$$\sim \begin{bmatrix} 1 & 0 \\ -3 & 1 \end{bmatrix} \begin{bmatrix} 1 & 0 & 0 \\ 0 & 1 & 0 \end{bmatrix} \begin{bmatrix} 1 & 2/5 & -1/5 \\ 0 & -1/5 & -7/5 \\ 0 & 0 & 1 \end{bmatrix} \text{ which may be written as}$$

$$\begin{bmatrix} 1 & 0 \\ -3 & 1 \end{bmatrix} [I_2 \ 0] \begin{bmatrix} 1 & 2/5 & -1/5 \\ 0 & -1/5 & -7/5 \\ 0 & 0 & 1 \end{bmatrix}. \text{ Hence } \mathbf{P} = \begin{bmatrix} 1 & 0 \\ -3 & 1 \end{bmatrix} \text{ and } \mathbf{Q} = \frac{1}{5} \begin{bmatrix} 5 & 2 & -1 \\ 0 & -1 & -7 \\ 0 & 0 & 5 \end{bmatrix}$$

Notes:

1. It is clear that **P** and **Q** are not unique.
2. As is clear from the procedure, we start with unit matrices of suitable order. For example if **A** is $m \times n$ then we start with $I_m AI_n$, while I_m is meant for row transformations only, I_n is meant for column transformations only, which are performed on **A** to reduce it to normal form.

Example 15.68 Examine whether the following set of vectors is linearly independent. Find the dimension and the basis of the given set of vectors.

(i) $\{(1, 1, 0, 1), (1, 1, 1, 1), (4, 4, 1, 1), (1, 0, 0, 1)\}$. *(I.E.T.E., June 2006)*

(ii) $\{(2, 2, 0, 2), (4, 1, 4, 1), (3, 0, 4, 0)\}$.

(iii) $\{(2, 3, 6, -3, 4), (4, 2, 12, -3, 6), (4, 10, 12, -9, 10)\}$.

Solution. Let each given vector represent a row of a matrix **A**. We reduce **A** to row echelon form. If all the rows of the row echelon form have some non-zero elements, then the given set of vectors are linearly independent.

(i) $A = \begin{bmatrix} 1 & 1 & 0 & 1 \\ 1 & 1 & 1 & 1 \\ 4 & 4 & 1 & 1 \\ 1 & 0 & 0 & 1 \end{bmatrix} \begin{matrix} R_2 \leftarrow R_2 - R_1 \\ R_3 \leftarrow R_3 - 4R_1 \\ R_4 \leftarrow R_4 - R_1 \end{matrix} \sim \begin{bmatrix} 1 & 1 & 0 & 1 \\ 0 & 0 & 1 & 0 \\ 0 & 0 & 1 & -3 \\ 0 & -1 & 0 & 0 \end{bmatrix}$ (operate $R_2 \leftrightarrow R_4$)

$\sim \begin{bmatrix} 1 & 1 & 0 & 1 \\ 0 & -1 & 0 & 0 \\ 0 & 0 & 1 & -3 \\ 0 & 0 & 1 & 0 \end{bmatrix} (R_4 \leftarrow R_4 - R_3) \sim \begin{bmatrix} 1 & 1 & 0 & 1 \\ 0 & -1 & 0 & 0 \\ 0 & 0 & 1 & -3 \\ 0 & 0 & 0 & 3 \end{bmatrix}.$

Since all the rows in the row echelon form of **A** are non-zero, the given set of vectors are linearly independent and the dimension of the given set of vectors is 4. The set of vectors $[(1, 1, 0, 1), (0, -1, 0, 0), (0, 0, 1, -3), (0, 0, 0, 3)$ or the given set itself forms the basis.

(ii) $A = \begin{bmatrix} 2 & 2 & 0 & 2 \\ 4 & 1 & 4 & 1 \\ 3 & 0 & 4 & 0 \end{bmatrix} \begin{matrix} \text{operate } R_2 \leftarrow R_2 - 2R_1 \\ \text{and } R_3 \leftarrow R_3 - \dfrac{3}{2}R_1 \end{matrix}$

$\sim \begin{bmatrix} 2 & 2 & 0 & 2 \\ 0 & -3 & 4 & -3 \\ 0 & -3 & 4 & -3 \end{bmatrix} \text{(operate } R_3 \leftarrow R_3 - R_2) \sim \begin{bmatrix} 2 & 2 & 0 & 2 \\ 0 & -3 & 4 & -3 \\ 0 & 0 & 0 & 0 \end{bmatrix}.$

Since all the rows in the row echelon form of **A** are not non-zero, the given set of vectors are linearly dependent. Since the number of non-zero rows is

2, the dimension of the given set of vectors is 2 and its basis can be taken as the set $\{(2, 2, 0, 2), (0, -3, 4, -3)\}$

(*iii*) $\quad \mathbf{A} = \begin{bmatrix} 2 & 3 & 6 & -3 & 4 \\ 4 & 2 & 12 & -3 & 6 \\ 4 & 10 & 12 & -9 & 10 \end{bmatrix}$ (operate $R_2 \leftarrow R_2 - 2R_1$ and $R_3 \leftarrow R_3 - 2R_1$)

$$\sim \begin{bmatrix} 2 & 3 & 6 & -3 & 4 \\ 0 & -4 & 0 & 3 & -2 \\ 0 & 4 & 0 & -3 & 2 \end{bmatrix} R_3 \leftarrow R_3 + R_2 \sim \begin{bmatrix} 2 & 3 & 6 & -3 & 4 \\ 0 & -4 & 0 & 3 & -2 \\ 0 & 0 & 0 & 0 & 0 \end{bmatrix}.$$

Since all the rows in the row echelon form of **A** are not non-zero, the given set of vectors are linearly dependent. Since the number of non-zero rows is 2, the dimension of the given set of vectors is 2 and its basis can be taken as the set $\{(2, 3, 6, -3, 4), (0, -4, 0, 3, -2)\}$.

15.22 Solution of General linear system of equations

In Article 15.18, we have discussed the matrix method and the Cramer's rule for solving a system of n linear equations in n unknowns, $\mathbf{Ax} = \mathbf{b}$. We assumed that the coefficient matrix **A** is non-singular, that is $|\mathbf{A}| \neq 0$, or the rank of the matrix **A** is n. The matrix method requires evaluation of n^2 determinants each of order $(n - 1)$ to generate the cofactor matrix, and one determinant of order n, whereas the Cramer's rule requires evaluation of $(n + 1)$ determinants each of order n. Since the evaluation of high order determinants is very time consuming, these methods are not used for large value of n, say greater than four.

A system of m linear equations in n unknowns $x_1, x_2, ..., x_n$ has the general form

$$\left. \begin{array}{l} a_{11}x_1 + a_{12}x_2 + \cdots\cdots + a_{1n}x_n = b_1 \\ a_{21}x_1 + a_{22}x_2 + \cdots\cdots + a_{2n}x_n = b_2 \\ \quad\quad\vdots \\ a_{m1}x_1 + a_{m2}x_2 + \cdots\cdots + a_{mn}x_n = b_m \end{array} \right\} \quad\quad ...(15.23)$$

can be written compactly as a matrix equation $\mathbf{Ax} = \mathbf{b}$ $\quad\quad ...(15.24)$

where $\quad \mathbf{A} = \begin{bmatrix} a_{11} & a_{12} & \cdots & a_{1n} \\ a_{21} & a_{22} & \cdots & a_{2n} \\ \vdots & & & \vdots \\ a_{m1} & a_{m2} & \cdots & a_{mn} \end{bmatrix}, \mathbf{b} = \begin{bmatrix} b_1 \\ b_2 \\ \vdots \\ b_m \end{bmatrix}, \mathbf{x} = \begin{bmatrix} x_1 \\ x_2 \\ \vdots \\ x_n \end{bmatrix}$

are respectively called the *coefficient matrix*, *right hand side column vector* and the *solution vector*. The orders of the matrices **A, b, x** are respectively $m \times n$, $m \times 1$ and $n \times 1$. The matrix

$$(A|b) = \begin{bmatrix} a_{11} & a_{12} & \cdots & \cdots & a_{1n} & b_1 \\ a_{21} & a_{22} & \cdots & \cdots & a_{2n} & b_2 \\ \vdots & & & & \vdots & \vdots \\ \vdots & & & & \vdots & \vdots \\ a_{m1} & a_{m2} & \cdots & \cdots & a_{mn} & b_m \end{bmatrix} \qquad ...(15.25)$$

is called the *augmented matrix* of the given system of equations and has m rows and $(n + 1)$ columns that is, of order $m \times (n + 1)$.

To solve a system of m linear equations in n unknowns using an augmented matrix $(A|b)$, we may use either **Gauss elimination method** or the **Gauss-Jordan method**.

Gauss Elimination method for non-homogeneous system

Consider a non-homogeneous system of m equations in n unknown

$$Ax = b. \qquad \text{(Refer equation 15.24)}$$

We write the augmented matrix $(A|b)$ of order $m \times (n + 1)$ as

$$A|b = \begin{bmatrix} a_{11} & a_{12} & \cdots & a_{1n} & b_1 \\ a_{21} & a_{22} & \cdots & a_{2n} & b_2 \\ \vdots & \vdots & & & \vdots \\ a_{m1} & a_{m2} & \cdots & a_{mn} & b_m \end{bmatrix}$$

and reduce it to the row echelon form by using elementary row operations. We need a maximum of $(m - 1)$ stages of elimination to reduce the given augmented matrix to the equivalent row echelon form. This process may terminate at an earlier stage. We then obtained an equivalent system of the form

$$(A|b) = \begin{bmatrix} a_{11} & a_{12} & \cdots & \cdots & a_{1r} & \cdots & a_{1n} & b_1 \\ 0 & \bar{a}_{22} & \cdots & \cdots & \bar{a}_{2r} & \cdots & \bar{a}_{2n} & \bar{b}_2 \\ \vdots & & & & & & & \\ 0 & 0 & \cdots & \cdots & a_{rr}^* & \cdots & a_{rn}^* & b_r^* \\ 0 & 0 & \cdots & \cdots & 0 & \cdots & 0 & b_{r+1}^* \\ \vdots & & & & & & & \vdots \\ 0 & 0 & \cdots & \cdots & 0 & \cdots & 0 & b_m^* \end{bmatrix}$$

where $r \le m$ and $a_{11} \ne 0$, $\bar{a}_{22} \ne 0$, ..., $a_{rr}^* \ne 0$ are called pivots. We have the following cases:

(i) **No solution.** If $r < m$ and one or more of the elements $b_{r+1}^*, b_{r+2}^*, \ldots, b_m^*$ are not zero. Then rank $(A) \neq$ rank $(A|b)$.

(ii) **Unique solution i.e., precisely one solution.** Let $m \geq n$ and $r = n$ and $b_{r+1}^*, b_{r+2}^*, \ldots, b_m^*$ are all zeros. In this case rank $(A) =$ rank $(A|b) = n$. We solve the nth equations for x_n, the $(n-1)^{th}$ equation for x_{n-1} and so on. This procedure is called the *back substitution method*.

(iii) **Infinitely many solutions.** If $r < n$ and $b_{r+1}^*, b_{r+2}^*, \ldots, b_m^*$ are all zeros. In this case, r unknown, $x_1, x_2, \ldots, x_r$ can be obtained in terms of the remaining $(n-r)$ unknown $x_{r+1}, x_{r+2}, \ldots, x_r$ by solving the rth equation for x_r, $(r-1)$ th equation for x_{r-1} and so on. In this case, the solution of the system contains $(n-r)$ parameters.

Remarks

1. We donot, normally use column elementary operations in solving the linear system of equations.
2. Gauss elimination method may be written as

$$(A|b) \xrightarrow[\text{row operations}]{\text{Elementary}} (B|c).$$

The matrix **B** is the row echelon form of the matrix **A** and **c** is the new right hand side column vector. We obtain the solution vector (if it exists) using the back substitution method.

Gauss-Jordon method to solve system of equations

In the Gauss-Jordan method, we reduce the augmented matrix $(A|b)$ of the given system of equations to the form $(I|d)$ by using elementary row transformations only, where **I** is the unit matrix and matrix **d** is the new right hand side column vector. The solution of the system of equations is given by column vector **d**. This method does not require back-substitution. The solution of the system will be apparent by inspection of the final matrix.

Gauss-Jordan method involves the following sequence of elementary row operations:

(a) In matrix **A**, make the element in the first row first column, pivot a_{11}, to unity by some suitable elementary row operation.
(b) Reduce all other elements in the first column to zero with the help of unity obtained in first step.
(c) Reduce the element in the second row second column, pivot $\bar{a}_{22}$, to unity by suitable elementary row operations.
(d) Reduce all other elements in the second column to zero with the help of unity obtained in third step.
(e) At the kth step, all the elements above and below the pivot a_{kk} are made

zero. The pivot in the (k, k) position be made 1. This process is continued until an identity matrix is obtained.

Remarks

1. Whenever a row of zeros exists to the left of the vertical line and a non-zero number appears to the right, the system is inconsistent that is it has no solution.

2. Whenever a complete row of zeros occurs (that is zeros on both sides of the vertical line), the system is dependent that is it has infinite number of solutions.

3. Otherwise, the given system has a unique solution given in the column to the right of the vertical line.

Example 15.69 Solve the following systems of equations (if possible) using Gauss elimination methods.

(i) $x - y + z = 0$, $-x + y - z = 0$, $10y + 25z = 90$, $20x + 10y = 80$,

(ii) $2x_1 + x_2 + 2x_3 + x_4 = 6$, $6x_1 - 6x_2 + 6x_3 + 12x_4 = 36$, $4x_1 + 3x_2 + 3x_3 - 3x_4 = -1$, $2x_1 + 2x_2 - x_3 + x_4 = 10$.

(iii) $4y + 3z = 8$, $2x - z = 2$, $3x + 2y = 5$

(iv) $5x + 5y - 10z = 0$, $2w - 3x - 3y + 6z = 2$, $4w + x + y - 2z = 4$.

Solution. We write the augmented matrix and reduce it to row echelon form by applying elementary row operations.

$$(i)\ (A|b)\begin{bmatrix} 1 & -1 & 1 & | & 0 \\ -1 & 1 & -1 & | & 0 \\ 0 & 10 & 25 & | & 90 \\ 20 & 10 & 0 & | & 80 \end{bmatrix} \begin{matrix} \\ R_2 \leftarrow R_2 + R_1 \\ R_4 \leftarrow R_4 - 20R_1 \\ \\ \end{matrix} \sim \begin{bmatrix} 1 & -1 & 1 & | & 0 \\ 0 & 0 & 0 & | & 0 \\ 0 & 10 & 25 & | & 90 \\ 0 & 30 & -20 & | & 80 \end{bmatrix} (R_2 \leftrightarrow R_4)$$

$$\sim \begin{bmatrix} 1 & -1 & 1 & | & 0 \\ 0 & 30 & -20 & | & 80 \\ 0 & 10 & 25 & | & 90 \\ 0 & 0 & 0 & | & 0 \end{bmatrix} (R_3 \leftarrow R_3 - R_2/3) \sim \begin{bmatrix} 1 & -1 & 1 & | & 0 \\ 0 & 30 & -20 & | & 80 \\ 0 & 0 & 95/3 & | & 190/3 \\ 0 & 0 & 0 & | & 0 \end{bmatrix}.$$

Using the back substitution method, we obtain the solution as

$$95z/3 = 190/3, \text{ or } z = 2,$$
$$30y - 20z = 80, \text{ or } y = (2z + 8)/3 = 4,$$
$$x - y + z = 0, \text{ or } x = y - z = 2.$$

Therefore, the system of equations has the unique solution $x = 2$, $y = 4$, $z = 2$.

(ii) We have

$$(A|b) = \begin{bmatrix} 2 & 1 & 2 & 1 & | & 6 \\ 6 & -6 & 6 & 12 & | & 36 \\ 4 & 3 & 3 & -3 & | & -1 \\ 2 & 2 & -1 & 1 & | & 10 \end{bmatrix} \begin{matrix} \text{(operating } R_2 \leftarrow R_2 - 3R_1, R_3 \leftarrow R_3 - 2R_1 \\ \text{and } R_4 \leftarrow R_4 - R_1) \end{matrix}$$

$$\sim \begin{bmatrix} 2 & 1 & 2 & 1 & 6 \\ 0 & -9 & 0 & 9 & 18 \\ 0 & 1 & -1 & -5 & -13 \\ 0 & 1 & -3 & 0 & 4 \end{bmatrix} R_2 \leftarrow -R_2/9 \sim \begin{bmatrix} 2 & 1 & 2 & 1 & 6 \\ 0 & 1 & 0 & -1 & -2 \\ 0 & 1 & -1 & -5 & -13 \\ 0 & 1 & -3 & 0 & 4 \end{bmatrix} \begin{matrix} \\ R_3 \leftarrow R_3 - R_2 \\ R_4 \leftarrow R_4 - R_2 \\ \\ \end{matrix}$$

$$\sim \begin{bmatrix} 2 & 1 & 2 & 1 & 6 \\ 0 & 1 & 0 & -1 & -2 \\ 0 & 0 & -1 & -4 & -11 \\ 0 & 0 & -3 & 1 & 6 \end{bmatrix} R_4 \leftarrow R_4 - 3R_3 \sim \begin{bmatrix} 2 & 1 & 2 & 1 & 6 \\ 0 & 1 & 0 & -1 & -2 \\ 0 & 0 & -1 & -4 & -11 \\ 0 & 0 & 0 & 13 & 39 \end{bmatrix}.$$

Using the back substitution method, we obtain the solution as

$$13x_4 = 39, \text{ or } x_4 = 3,$$
$$-x_3 - 4x_4 = -11, \text{ or } x_3 = -1,$$
$$x_2 - x_4 = -2, \text{ or } x_2 = 1,$$
$$2x_1 + x_2 + 2x_3 + x_4 = 6, \text{ or } x_1 = 2.$$

Therefore, the system of equations has the unique solution

$$x_1 = 2, x_2 = 1, x_3 = -1, x_4 = 3.$$

(*iii*) $(\mathbf{A}|\mathbf{b}) = \begin{bmatrix} 0 & 4 & 3 & 8 \\ 2 & 0 & -1 & 2 \\ 3 & 2 & 0 & 5 \end{bmatrix}$ (perform $R_1 \leftrightarrow R_3$)

$$\sim \begin{bmatrix} 3 & 2 & 0 & 5 \\ 2 & 0 & -1 & 2 \\ 0 & 4 & 3 & 8 \end{bmatrix} \left(\text{operate } R_2 \leftarrow R_2 - \frac{2}{3}R_1 \right)$$

$$\sim \begin{bmatrix} 3 & 2 & 0 & 5 \\ 0 & -4/3 & -1 & -4/3 \\ 0 & 4 & 3 & 8 \end{bmatrix} (\text{operate } R_3 \leftarrow R_3 + 3R_2) \sim \begin{bmatrix} 3 & 2 & 0 & 5 \\ 0 & -4/3 & -1 & -4/3 \\ 0 & 0 & 0 & 4 \end{bmatrix}.$$

The third row of the last augmented matrix means $0x + 0y + 0z = 4$ (or $0 = 4$). Since no numbers x, y and z can satisfy this equation, we conclude that the system has no solution—inconsistent system.

Alternately We find that rank $(\mathbf{A}) = 2$ and rank $(\mathbf{A}|\mathbf{b}) = 3$. Therefore, the system of equations has no solution.

(*iv*) $(\mathbf{A}|\mathbf{b}) = \begin{bmatrix} 5 & 5 & -10 & 0 & 0 \\ -3 & -3 & 6 & 2 & 2 \\ 1 & 1 & -2 & 4 & 4 \end{bmatrix}$ (perform $(R_1 \leftrightarrow R_3)$

$$\sim \begin{bmatrix} 1 & 1 & -2 & 4 & | & 4 \\ -3 & -3 & 6 & 2 & | & 2 \\ 5 & 5 & -10 & 0 & | & 0 \end{bmatrix} \text{(operate } R_2 \leftarrow R_2 + 3R_1 \text{ and } R_3 \leftarrow R_3 - 5R_1)$$

$$\sim \begin{bmatrix} 1 & 1 & -2 & 4 & | & 4 \\ 0 & 0 & 0 & 14 & | & 14 \\ 0 & 0 & 0 & -20 & | & -20 \end{bmatrix} \left(\text{operate } R_3 \leftarrow R_3 + \frac{10}{7}R_2 \right) \sim \begin{bmatrix} 1 & 1 & -2 & 4 & | & 4 \\ 0 & 0 & 0 & 14 & | & 14 \\ 0 & 0 & 0 & 0 & | & 0 \end{bmatrix}.$$

The system is consistent and has infinite number of solutions. From the second equation, we obtain $w = 1$. From the first equation, we obtain $x = 4 - y + 2z - 4w = 2z - y$

Thus, we obtain a two parameter family of solutions:

$x = 2z - y$ and $w = 1$, where y and z are arbitrary.

Example 15.70 Use Gauss-Jordon elimination to solve the given system or show that no solution exists.

(*i*) $x - y - z = -3$, $2x + 3y + 5z = 7$, $x - 2y + 3z = -11$

(*ii*) $x - y - z = 8$, $x - y + z = 3$, $-x + y + z = 4$

(*iii*) $x + 3y - 2z = -7$, $4x + y + 3z = 5$, $2x - 5y + 7z = 19$

Solution. (*i*) We have

$$(A|b) = \begin{bmatrix} 1 & -1 & -1 & | & -3 \\ 2 & 3 & 5 & | & 7 \\ 1 & -2 & 3 & | & -11 \end{bmatrix} \text{(operate } R_2 \leftarrow R_2 - 2R_1 \text{ and } R_3 \leftarrow R_3 - R_1)$$

$$\sim \begin{bmatrix} 1 & -1 & -1 & | & -3 \\ 0 & 5 & 7 & | & 13 \\ 0 & -1 & 4 & | & -8 \end{bmatrix} \text{(operate } R_2 \leftarrow R_2/5)$$

$$\sim \begin{bmatrix} 1 & -1 & -1 & | & -3 \\ 0 & 1 & 7/5 & | & 13/5 \\ 0 & -1 & 4 & | & -8 \end{bmatrix} \text{(operate } R_1 \leftarrow R_1 + R_2 \text{ and } R_3 \leftarrow R_3 + R_2)$$

$$\sim \begin{bmatrix} 1 & 0 & 2/5 & | & -2/5 \\ 0 & 1 & 7/5 & | & 13/5 \\ 0 & 0 & 27/5 & | & -27/5 \end{bmatrix} \left(\text{operate } R_3 \leftarrow \frac{5R_3}{27} \right)$$

$$\sim \begin{bmatrix} 1 & 0 & 2/5 & | & -2/5 \\ 0 & 1 & 7/5 & | & 13/5 \\ 0 & 0 & 1 & | & -1 \end{bmatrix} \left(\text{operate } R_1 \leftarrow R_1 - \frac{2}{5}R_3 \text{ and } R_2 \leftarrow R_2 - \frac{7}{5}R_3 \right)$$

$$\sim \begin{bmatrix} 1 & 0 & 0 & | & 0 \\ 0 & 1 & 0 & | & 4 \\ 0 & 0 & 1 & | & -1 \end{bmatrix}.$$ The last matrix is in reduced row-echelon form. We see that

the solution of the system is $x = 0$, $y = 4$, $z = -1$.

(ii) $(A|b) = \begin{bmatrix} 1 & -1 & -1 & | & 8 \\ 1 & -1 & 1 & | & 3 \\ -1 & 1 & 1 & | & 4 \end{bmatrix}$ (operate $R_2 \leftarrow R_2 - R_1$ and $R_3 \leftarrow R_3 + R_1$)

$$\sim \begin{bmatrix} 1 & -1 & -1 & | & 8 \\ 0 & 0 & 2 & | & -5 \\ 0 & 0 & 0 & | & 12 \end{bmatrix}.$$ This shows that the system has no solution-inconsistent

(iii) $(A|b) = \begin{bmatrix} 1 & 3 & -2 & | & -7 \\ 4 & 1 & 3 & | & 5 \\ 2 & -5 & 7 & | & 19 \end{bmatrix}$ (operate $R_2 \leftarrow R_2 - R_1$ and $R_3 \leftarrow R_3 - 2R_1$)

$$\sim \begin{bmatrix} 1 & 3 & -2 & | & -7 \\ 0 & -11 & 11 & | & 33 \\ 0 & -11 & 11 & | & 33 \end{bmatrix}$$ (operate $R_2 \leftarrow -R_2/11$ and $R_3 \leftarrow -R_3/11$)

$$\sim \begin{bmatrix} 1 & 3 & -2 & | & -7 \\ 0 & 1 & -1 & | & -3 \\ 0 & 1 & -1 & | & -3 \end{bmatrix} \begin{matrix} R_1 \leftarrow R_1 - 3R_2 \\ R_3 \leftarrow R_3 - R_2 \end{matrix} \sim \begin{bmatrix} 1 & 0 & 1 & | & 2 \\ 0 & 1 & -1 & | & -3 \\ 0 & 0 & 0 & | & 0 \end{bmatrix}.$$

The system of equations is consistent and has infinite number of solutions. Choose z as arbitrary. From the second equation, we obtain
$$y - z = -3 \text{ or } y = z - 3.$$
From the first equation, we obtain $x + z = 2$ or $x = 2 - z$.

Example 15.71 Test for consistency and solve the system of equation (if possible) using Gauss elimination method—
$$x - 2y + 3z = 2, \ 2x - 3z = 3, \ x + y + z = 0.$$
Solution. We write the augmented matrix of the system and reduce it to row echelon form by applying elementary row operations.

$$(A|b) = \begin{bmatrix} 1 & -2 & 3 & | & 2 \\ 2 & 0 & -3 & | & 3 \\ 1 & 1 & 1 & | & 0 \end{bmatrix}$$ (operate $R_2 \leftarrow R_2 - 2R_1$ and $R_3 \leftarrow R_3 - R_1$)

$$\sim \begin{bmatrix} 1 & -2 & 3 & | & 2 \\ 0 & 4 & -9 & | & -1 \\ 0 & 3 & -2 & | & -2 \end{bmatrix} R_2 \leftarrow R_2/4 \sim \begin{bmatrix} 1 & -2 & 3 & | & 2 \\ 0 & 1 & -9/4 & | & -1/4 \\ 0 & 3 & -2 & | & -2 \end{bmatrix} R_3 \leftarrow R_3 - 3R_2$$

$$\sim \begin{bmatrix} 1 & -2 & 3 & | & 2 \\ 0 & 1 & -9/4 & | & -1/4 \\ 0 & 0 & 19/4 & | & -5/4 \end{bmatrix}$$ We find that rank $(\mathbf{A})$ = rank $(\mathbf{A}|\mathbf{b})$ = 3 = n.

Therefore, the system of equations is consistent.

Using the back substitution method, we obtain the solution as

$$\frac{19}{4}z = \frac{-5}{4}, \text{ or } z = \frac{-5}{19}, \quad y - \frac{9}{4}z = -\frac{1}{4}, \text{ or } y = \frac{-16}{19},$$

$$x - 2y + 3z = 2, \text{ or } x = \frac{21}{19}.$$

Therefore, the system of equation has the unique solution
$$x = 21/19, \ y = -16/19, \ z = -5/19.$$

Example 15.72 Show that the system of equations
$2x - y + z = 4, \ 3x - y + z = 6, \ 4x - y + 2z = 7, \ -x + y - z = 9$ is inconsistent.

Solution. We write the augmented matrix and reduce it to row echelon form by applying elementary row operations.

$$(\mathbf{A}|\mathbf{b}) = \begin{bmatrix} 2 & -1 & 1 & | & 4 \\ 3 & -1 & 1 & | & 6 \\ 4 & -1 & 2 & | & 7 \\ -1 & 1 & -1 & | & 9 \end{bmatrix} \text{(operate } R_1 \leftrightarrow R_4)$$

$$\sim \begin{bmatrix} -1 & 1 & -1 & | & 9 \\ 3 & -1 & 1 & | & 6 \\ 4 & -1 & 2 & | & 7 \\ 2 & -1 & 1 & | & 4 \end{bmatrix} \begin{matrix} \\ R_2 \leftarrow R_2 + 3R_1 \\ R_3 \leftarrow R_3 + 4R_1 \\ R_4 \leftarrow R_4 + 2R_1 \end{matrix} \sim \begin{bmatrix} -1 & 1 & -1 & | & 9 \\ 0 & 2 & -2 & | & 33 \\ 0 & 3 & -2 & | & 43 \\ 0 & 1 & -1 & | & 22 \end{bmatrix} \text{(operate } R_2 \leftrightarrow R_4)$$

$$\sim \begin{bmatrix} -1 & 1 & -1 & | & 9 \\ 0 & 1 & -1 & | & 22 \\ 0 & 3 & -2 & | & 43 \\ 0 & 2 & -2 & | & 33 \end{bmatrix} \begin{matrix} \\ \\ R_3 \leftarrow R_3 - 3R_2 \\ R_4 \leftarrow R_4 - 2R_2 \end{matrix} \sim \begin{bmatrix} -1 & 1 & -1 & | & 9 \\ 0 & 1 & -1 & | & 22 \\ 0 & 0 & 1 & | & -23 \\ 0 & 0 & 0 & | & -11 \end{bmatrix}$$

We find that rank $(A) = 3$ and rank $(A|b) = 4$. Therefore, the given system of equations is inconsistent and has no solution.

Example 15.73 Test the following system of equations for consistent and find the solution for the consistent system.

$$\begin{bmatrix} 1 & 4 & 7 \\ 2 & 5 & 8 \\ 1 & 2 & 3 \end{bmatrix} \begin{bmatrix} x \\ y \\ z \end{bmatrix} = \begin{bmatrix} 1 \\ 2 \\ 1 \end{bmatrix}.$$

Solution. We write the augmented matrix and reduce it to row echelon form by applying elementary row operations.

$$(A|b) = \begin{bmatrix} 1 & 4 & 7 & | & 1 \\ 2 & 5 & 8 & | & 2 \\ 1 & 2 & 3 & | & 1 \end{bmatrix} \text{(operate } R_2 \leftarrow R_2 - 2R_1 \text{ and } R_3 \leftarrow R_3 - R_1)$$

$$\sim \begin{bmatrix} 1 & 4 & 7 & | & 1 \\ 0 & -3 & -6 & | & 0 \\ 0 & -2 & -4 & | & 0 \end{bmatrix} R_3 \leftarrow R_3 - 2R_2/3 \sim \begin{bmatrix} 1 & 4 & 7 & | & 1 \\ 0 & -3 & -6 & | & 0 \\ 0 & 0 & 0 & | & 0 \end{bmatrix}.$$

We find that rank $(A) = $ rank $(A|b) = 2 < $ number of unknowns, the system is consistent and has infinite number of solutions. We find that the last equation is satisfied for all values of x, y, z.

From the second equation, we get $y = -2z$. From the first equation, we get $x + 4y + 7z = 1$. or $x = 1 + z$. Therefore, we obtain the solution $x = 1 + z, y = -2z$ and z is arbitrary.

Example 15.74 Determine for what values of λ and μ the following equations have (*i*) a unique solution, (*ii*) infinite number of solutions and (*iii*) no solution.

$$x + y + z = 6, \; x + 2y + 3z = 10, \; x + 2y + \lambda z = \mu$$

(AMIE, W-2007; I.E.T.E., June 2002; UPTU 2002; PTU 2002)

Solution. We write the augmented matrix and reduce it to row echelon form by applying elementary row operation.

$$(A|b) = \begin{bmatrix} 1 & 1 & 1 & | & 6 \\ 1 & 2 & 3 & | & 10 \\ 1 & 2 & \lambda & | & \mu \end{bmatrix} \text{(operate } R_2 \leftarrow R_2 - R_1 \text{ and } R_3 \leftarrow R_3 - R_1)$$

$$\sim \begin{bmatrix} 1 & 1 & 1 & | & 6 \\ 0 & 1 & 2 & | & 4 \\ 0 & 1 & \lambda-1 & | & \mu-6 \end{bmatrix} R_3 \leftarrow R_3 - R_2 \sim \begin{bmatrix} 1 & 1 & 1 & | & 6 \\ 0 & 1 & 2 & | & 4 \\ 0 & 0 & \lambda-3 & | & \mu-10 \end{bmatrix}.$$

(*i*) There is a unique solution if rank (**A**) = rank (**A**|**b**) = 3 that is, if $\lambda \neq 3$, μ may have any value.

(*ii*) There are infinite number of solutions if

rank (**A**) = rank (**A**|**b**) < 3 that is $\lambda = 3$, $\mu = 10$.

(*iii*) There is no solution if rank (**A**) $\neq$ rank (**A**|**b**) that is $\lambda = 3$, $\mu \neq 10$.

15.22.1 Homogeneous System of $m \times n$ linear equations

Consider the homogeneous system of equations **Ax = O**   ...(15.26)
where **A** is an $m \times n$ matrix. The homogeneous system is always *consistent*, since $x_1 = 0$, $x_2 = 0$, ...$x_n = 0$ will satisfy each equation in the system. The solution consisting of all zeros is called the **trivial solution**. In this case, rank(**A**) = rank (**A**|**O**) = number of unknowns.

Non-trivial solution exists if and only if rank (**A**) < n. If rank **A** = $r < n$, we obtain an $(n - r)$ parameter family of solutions which form a vector space of dimension $(n - r)$ called the **solution space** of homogeneous system. The solution space of homogeneous system is called the **null space** of **A** because **Ax = O** for every **x** in the solution space and its dimension is called the **nullity** of **A**. In view of above, we obtain the result

rank (**A**) + nullity (**A**) = n.

Remarks: (1) A homogeneous linear system with fewer equations than unknowns always possesses non-trivial solution.

(2) The following diagram 15.2 outlines the connection between the concept of rank of a matrix and the solution of m linear equations in n unknown **Ax = b**. Let rank (**A**) = r.

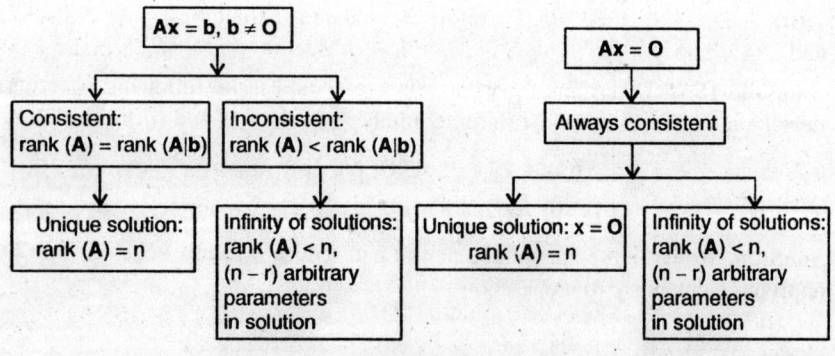

Fig. 15.2

Example 15.75 Solve the following homogeneous system of equations **Ax = O**, where **A** is given by

(i) $\begin{bmatrix} 2 & 1 \\ 1 & -1 \\ 3 & 2 \end{bmatrix}$, (ii) $\begin{bmatrix} 1 & 2 & 3 \\ 3 & 4 & 4 \\ 7 & 10 & 12 \end{bmatrix}$ (*Osmania, 1999*)

(iii) $\begin{bmatrix} 2 & -4 & 3 \\ 1 & 1 & -2 \end{bmatrix}$ (iv) $\begin{bmatrix} 4 & 2 & 1 & 3 \\ 6 & 3 & 4 & 7 \\ 2 & 1 & 0 & 1 \end{bmatrix}$. Find the rank (**A**) and nullity (**A**)

Solution. We write the augmented matrix (**A|O**) and reduce it to row echelon form.

$(i)\,(\mathbf{A|O}) = \begin{bmatrix} 2 & 1 & | & 0 \\ 1 & -1 & | & 0 \\ 3 & 2 & | & 0 \end{bmatrix}$ (perform $R_2 \leftarrow R_2 - R_1/2$ and $R_3 \leftarrow R_3 - 3R_1/2$)

$\sim \begin{bmatrix} 2 & 1 & | & 0 \\ 0 & -3/2 & | & 0 \\ 0 & 1/2 & | & 0 \end{bmatrix} R_3 \leftarrow R_3 + R_2/3 \sim \begin{bmatrix} 2 & 1 & | & 0 \\ 0 & -3/2 & | & 0 \\ 0 & 0 & | & 0 \end{bmatrix}$. Since, rank (**A**) = 2 = number

of unknown, the system has only a trivial solution.
    Hence nullity (**A**) = **0**.

$(ii)\,(\mathbf{A|O}) = \begin{bmatrix} 1 & 2 & 3 & | & 0 \\ 3 & 4 & 4 & | & 0 \\ 7 & 10 & 12 & | & 0 \end{bmatrix}$ (operate $R_2 \leftarrow R_2 - 3R_1$ and $R_3 \leftarrow R_3 - 7R_1$)

$\sim \begin{bmatrix} 1 & 2 & 3 & | & 0 \\ 0 & -2 & -5 & | & 0 \\ 0 & -4 & -9 & | & 0 \end{bmatrix} R_3 \leftarrow R_3 - 2R_2 \sim \begin{bmatrix} 1 & 2 & 3 & | & 0 \\ 0 & -2 & -5 & | & 0 \\ 0 & 0 & 1 & | & 0 \end{bmatrix}$.

    Since rank (**A**) = 3 = number of unknown, the homogeneous system has only a trivial solution. Therefore, nullity (**A**) = **0**.
    (iii) Since the number of equations is less than the number of unknowns, we know that the given system has non trivial solution (See Remark of Art. 15.22.1). Using Gauss-Jordan elimination, we find

$(\mathbf{A|O}) = \begin{bmatrix} 2 & -4 & 3 & | & 0 \\ 1 & 1 & -2 & | & 0 \end{bmatrix} R_1 \leftrightarrow R_2 \sim \begin{bmatrix} 1 & 1 & -2 & | & 0 \\ 2 & -4 & 3 & | & 0 \end{bmatrix} R_2 \leftarrow R_2 - 2R_1$

$\sim \begin{bmatrix} 1 & 1 & -2 & | & 0 \\ 0 & -6 & 7 & | & 0 \end{bmatrix}$ (perform $R_2 \leftarrow - R_2/6) \sim \begin{bmatrix} 1 & 1 & -2 & | & 0 \\ 0 & 1 & -7/6 & | & 0 \end{bmatrix} R_1 \leftarrow R_1 - R_2$

$$\begin{bmatrix} 1 & 0 & -5/6 & | & 0 \\ 0 & 1 & -7/6 & | & 0 \end{bmatrix}.$$ Here, rank (A) = 2 and the number of unknowns is 3, we

obtain a one parameter family of solutions as $x = 5z/6$, $y = 7z/6$, where z is arbitrary. Therefore, nullity (A) = 1.

(iv) $(A|O) = \begin{bmatrix} 4 & 2 & 1 & 3 & | & 0 \\ 6 & 3 & 4 & 7 & | & 0 \\ 2 & 1 & 0 & 1 & | & 0 \end{bmatrix} \begin{matrix} \\ R_2 \leftarrow R_2 - 3R_1/2 \\ R_3 \leftarrow R_3 - R_1/2 \end{matrix}$

$\sim \begin{bmatrix} 4 & 2 & 1 & 3 & | & 0 \\ 0 & 0 & 5/2 & 5/2 & | & 0 \\ 0 & 0 & -1/2 & -1/2 & | & 0 \end{bmatrix}$ (operate $R_3 \leftarrow R_3 + R_2/5$) $\sim \begin{bmatrix} 4 & 2 & 1 & 3 & | & 0 \\ 0 & 0 & 5/2 & 5/2 & | & 0 \\ 0 & 0 & 0 & 0 & | & 0 \end{bmatrix}.$

Rank (A) = 2 and the number of unknowns is 4. Hence, we obtain a two parameter family of solutions as $x_3 = -x_4$, $x_2 = (-4x_1 - x_3 - 3x_4)/2 = -(2x_1 + x_4)$, where x_1 and x_4 are arbitrary. Therefore, nullity (A) = 2.

15.23 Gauss-Jordon method to find the inverse of a matrix

Any non-singular matrix can be reduced to a unit matrix by successive-elementary row (or column) transformations. Assuming this result, we have the following theorem.

Theorem: *The elementary row transformations which reduce a given matrix* **A** *to a unit matrix, when applied to the unit matrix give the inverse matrix* **A**$^{-1}$.

Let **A** be a non-singular matrix of order n. We may write **A** = **IA** ...(15.27).

Now we apply elementary row operations to **A** on the left hand side (15.27) so that **A** is reduced to **I**.

Simultaneously, apply the same elementary row operations to the prefactor **I** on the right side. Let **I** reduce to **B**, so that (15.27) becomes **I** = **BA**.

Post multiplying by **A**$^{-1}$, we obtain **IA**$^{-1}$ = **BAA**$^{-1}$ = **BI** = **B**

$$\Rightarrow A^{-1} = B.$$ Thus **B** is the inverse of **A**.

Similarly, the elementary column transformations that reduce a given matrix **A** to a unit matrix, when applied to the unit matrix give inverse matrix **A**$^{-1}$.

In practice the two matrices **A** and **I** are written side by side and the same row operations are performed on them. The aim is to reduce **A** to **I**. As soon as this is achieved, the other matrix gives **A**$^{-1}$.

$$(A|I) \xrightarrow[\text{row operations}]{\text{Elementary}} (I|B). \text{ Hence } B = A^{-1}.$$

The procedure will be clear from the following solved example.

Example 15.76 Using the Gauss-Jordan method, find the inverses of the following matrices:

(i) $\mathbf{A} = \begin{bmatrix} 2 & 3 & 4 \\ 4 & 3 & 1 \\ 1 & 2 & 4 \end{bmatrix}$, (ii) $\mathbf{A} = \begin{bmatrix} 4 & 2 & -1 \\ 3 & -2 & 1 \\ 2 & 4 & -3 \end{bmatrix}$, (iii) $\mathbf{A} = \begin{bmatrix} 1 & -1 & 1 \\ -3 & 2 & -1 \\ -2 & 1 & -1 \end{bmatrix}$.

Solution. (i) Writing the given matrix side by side with unit matrix of order 3, that is A|I we have

$$\mathbf{A}|\mathbf{I} = \left[\begin{array}{ccc|ccc} 2 & 3 & 4 & 1 & 0 & 0 \\ 4 & 3 & 1 & 0 & 1 & 0 \\ 1 & 2 & 4 & 0 & 0 & 1 \end{array}\right] \text{(operate } R_1 \leftrightarrow R_3\text{)}$$

$$\sim \left[\begin{array}{ccc|ccc} 1 & 2 & 4 & 0 & 0 & 1 \\ 4 & 3 & 1 & 0 & 1 & 0 \\ 2 & 3 & 4 & 1 & 0 & 0 \end{array}\right] \text{(operate } R_2 \leftarrow R_2 - 4R_1 \text{ and } R_3 \leftarrow R_3 - 2R_1\text{)}$$

$$\sim \left[\begin{array}{ccc|ccc} 1 & 2 & 4 & 0 & 0 & 1 \\ 0 & -5 & -15 & 0 & 1 & -4 \\ 0 & -1 & -4 & 1 & 0 & -2 \end{array}\right] \left(\text{operate } R_2 \leftarrow \frac{-R_2}{5} \text{ and } R_3 \leftarrow -R_3\right)$$

$$\sim \left[\begin{array}{ccc|ccc} 1 & 2 & 4 & 0 & 0 & 1 \\ 0 & 1 & 3 & 0 & -1/5 & 4/5 \\ 0 & 1 & 4 & -1 & 0 & 2 \end{array}\right] \text{(operate } R_1 \leftarrow R_1 - 2R_2 \text{ and } R_3 \leftarrow R_3 - R_2\text{)}$$

$$\sim \left[\begin{array}{ccc|ccc} 1 & 0 & -2 & 0 & 2/5 & -3/5 \\ 0 & 1 & 3 & 0 & -1/5 & 4/5 \\ 0 & 0 & 1 & -1 & 1/5 & 6/5 \end{array}\right] \text{(operate } R_1 \leftarrow R_1 + 2R_3 \text{ and } R_2 \leftarrow R_2 - 3R_3\text{)}$$

$$\sim \left[\begin{array}{ccc|ccc} 1 & 0 & 0 & -2 & 4/5 & 9/5 \\ 0 & 1 & 0 & 3 & -4/5 & -14/5 \\ 0 & 0 & 1 & -1 & 1/5 & 6/5 \end{array}\right]. \text{ Hence, } \mathbf{A}^{-1} = \frac{1}{5}\begin{bmatrix} -10 & 4 & 9 \\ 15 & -4 & -14 \\ -5 & 1 & 6 \end{bmatrix}.$$

(ii) We have $\mathbf{A}|\mathbf{I} = \left[\begin{array}{ccc|ccc} 4 & 2 & -1 & 1 & 0 & 0 \\ 3 & -2 & 1 & 0 & 1 & 0 \\ 2 & 4 & -3 & 0 & 0 & 1 \end{array}\right] \left(\text{operate } R_1 \leftarrow \frac{R_1}{4}\right)$

$$\sim \left[\begin{array}{ccc|ccc} 1 & 1/2 & -1/4 & 1/4 & 0 & 0 \\ 3 & -2 & 1 & 0 & 1 & 0 \\ 2 & 4 & -3 & 0 & 0 & 1 \end{array}\right] \text{(operate } R_2 \leftarrow R_2 - 3R_1 \text{ and } R_3 \leftarrow R_3 - 2R_1\text{)}$$

$$\sim \begin{bmatrix} 1 & 1/2 & -1/4 & | & 1/4 & 0 & 0 \\ 0 & -7/2 & 7/4 & | & -3/4 & 1 & 0 \\ 0 & 3 & -5/2 & | & -1/2 & 0 & 1 \end{bmatrix} \text{(operate } R_2 \leftrightarrow R_3)$$

$$\sim \begin{bmatrix} 1 & 1/2 & -1/4 & | & 1/4 & 0 & 0 \\ 0 & 3 & -5/2 & | & -1/2 & 0 & 1 \\ 0 & -7/2 & 7/4 & | & -3/4 & 1 & 0 \end{bmatrix} \left(\text{operate } R_2 \leftarrow \frac{1}{3} R_2 \right)$$

$$\sim \begin{bmatrix} 1 & 1/2 & -1/4 & | & 1/4 & 0 & 0 \\ 0 & 1 & -5/6 & | & -1/6 & 0 & 1/3 \\ 0 & -7/2 & 7/4 & | & -3/4 & 1 & 0 \end{bmatrix} \left(\begin{array}{l} \text{operate } R_1 \leftarrow R_1 - \dfrac{1}{2} R_2 \\ \text{and } R_3 \leftarrow R_3 + \dfrac{7}{2} R_2 \end{array} \right)$$

$$\sim \begin{bmatrix} 1 & 0 & 1/6 & | & 1/3 & 0 & -1/6 \\ 0 & 1 & -5/6 & | & -1/6 & 0 & 1/3 \\ 0 & 0 & -7/6 & | & -4/3 & 1 & 7/6 \end{bmatrix} \left(\text{operate } R_3 \leftarrow -\frac{6}{7} R_3 \right)$$

$$\sim \begin{bmatrix} 1 & 0 & 1/6 & | & 1/3 & 0 & -1/6 \\ 0 & 1 & -5/6 & | & -1/6 & 0 & 1/3 \\ 0 & 0 & 1 & | & 8/7 & -6/7 & -1 \end{bmatrix} \left(\begin{array}{l} \text{operate } R_1 \leftarrow R_1 - \dfrac{1}{6} R_3 \\ \text{and } R_2 \leftarrow R_2 + \dfrac{5}{6} R_3 \end{array} \right)$$

$$\sim \begin{bmatrix} 1 & 0 & 0 & | & 1/7 & 1/7 & 0 \\ 0 & 1 & 0 & | & 11/14 & -5/7 & -1/2 \\ 0 & 0 & 1 & | & 8/7 & -6/7 & -1 \end{bmatrix} . \text{ Hence } \mathbf{A}^{-1} = \frac{1}{14} \begin{bmatrix} 2 & 2 & 0 \\ 11 & -10 & -7 \\ 16 & -12 & -14 \end{bmatrix}.$$

(*iii*) We have $\mathbf{A} | \mathbf{I} = \begin{bmatrix} 1 & -1 & 1 & | & 1 & 0 & 0 \\ -3 & 2 & -1 & | & 0 & 1 & 0 \\ -2 & 1 & -1 & | & 0 & 0 & 1 \end{bmatrix}$ (operating $R_2 \leftarrow R_2 + 3R_1$, and $R_3 \leftarrow R_3 + 2R_1$)

$$\sim \begin{bmatrix} 1 & -1 & 1 & | & 1 & 0 & 0 \\ 0 & -1 & 2 & | & 3 & 1 & 0 \\ 0 & -1 & 1 & | & 2 & 0 & 1 \end{bmatrix} \text{(operate } R_2 \leftarrow -R_2 \text{ and } R_3 \leftarrow -R_3)$$

$$\sim \begin{bmatrix} 1 & -1 & 1 & | & 1 & 0 & 0 \\ 0 & 1 & -2 & | & -3 & -1 & 0 \\ 0 & 1 & -1 & | & -2 & 0 & -1 \end{bmatrix} \begin{array}{l} R_1 \leftarrow R_1 + R_2 \\ R_3 \leftarrow R_3 - R_2 \end{array}$$

$$\sim \begin{bmatrix} 1 & 0 & -1 & -2 & -1 & 0 \\ 0 & 1 & -2 & -3 & -1 & 0 \\ 0 & 0 & 1 & 1 & 1 & -1 \end{bmatrix} \begin{matrix} R_1 \leftarrow R_1 + R_3 \\ R_2 \leftarrow R_2 + 2R_3 \\ \end{matrix}$$

$$\sim \begin{bmatrix} 1 & 0 & 0 & -1 & 0 & -1 \\ 0 & 1 & 0 & -1 & 1 & -2 \\ 0 & 0 & 1 & 1 & 1 & -1 \end{bmatrix}. \text{ Hence, } A^{-1} = \begin{bmatrix} -1 & 0 & -1 \\ -1 & 1 & -2 \\ 1 & 1 & -1 \end{bmatrix}.$$

15.24 Inverse Transformation

The relation

$$\left. \begin{aligned} y_1 &= a_{11}x_1 + a_{12}x_2 + a_{13}x_3 \\ y_2 &= a_{21}x_1 + a_{22}x_2 + a_{23}x_3 \\ y_3 &= a_{31}x_1 + a_{32}x_2 + a_{33}x_3 \end{aligned} \right\} \qquad \text{...(15.28)}$$

is a linear transformation in three dimensions which sends the point with coordinates (x_1, x_2, x_3) into a point with coordinates (y_1, y_2, y_3).

The relation (15.28) may be written in the matrix form $\mathbf{y} = \mathbf{Ax}$

where $\qquad \mathbf{y} = \begin{pmatrix} y_1 \\ y_2 \\ y_3 \end{pmatrix}, \mathbf{x} = \begin{pmatrix} x_1 \\ x_2 \\ x_3 \end{pmatrix}$ and $\mathbf{A} = \begin{pmatrix} a_{11} & a_{12} & a_{13} \\ a_{21} & a_{22} & a_{23} \\ a_{31} & a_{32} & a_{33} \end{pmatrix}.$

Here $\mathbf{A}$ is the matrix of transformation.

In general $\mathbf{y} = \mathbf{Ax}$ ....(15.29)

where $\qquad \mathbf{y} = \begin{pmatrix} y_1 \\ y_2 \\ \vdots \\ y_n \end{pmatrix}, \mathbf{A} = \begin{pmatrix} a_{11} & a_{12} & \cdots & a_{1n} \\ a_{21} & a_{22} & \cdots & a_{2n} \\ \cdots & \cdots & \cdots & \cdots \\ a_{n1} & a_{n2} & \cdots & a_{nn} \end{pmatrix}$ and $\mathbf{x} = \begin{pmatrix} x_1 \\ x_2 \\ \vdots \\ x_n \end{pmatrix}$

is a transformation of n-component vector $\mathbf{x}$ into n-component vector $\mathbf{y}$ by the matrix of transformation $\mathbf{A}$.

The determinant $|\mathbf{A}|$ is called the *modulus of transformation.*

If $|\mathbf{A}| = 0$, the transformation matrix $\mathbf{A}$ is called *singular* otherwise *non-singular or regular.*

If $\mathbf{A}$ is a non-singular matrix, $\mathbf{A}^{-1}$ exists. Then from (15.29), we have

$$\mathbf{A}^{-1}\mathbf{y} = \mathbf{A}^{-1}\mathbf{Ax} = \mathbf{x} \text{ or } \mathbf{x} = \mathbf{A}^{-1}\mathbf{y} \text{ which gives the inverse transformation.}$$

Remark. If a transformation from $(x_1, x_2, ..., x_n)$ to $(y_1, y_2, ..., y_n)$ is given by $\mathbf{y} = \mathbf{Ax}$ and another transformation from $(y_1, y_2, ..., y_n)$ to $(z_1, z_2, ..., z_n)$ is given by

$z = By$, then the transformation from $(x_1, x_2, ..., x_n)$ to $(z_1, z_2, ..., z_n)$ is given by
$z = By = B(Ax) = (BA)x$.

Example 15.77 Show that the transformation

$$y_1 = x_1 - x_2 + x_3,\ y_2 = 3x_1 - x_2 + 2x_3,\ y_3 = 2x_1 - 2x_2 + 3x_3$$

is non-singular. Find the inverse transformation. *(IETE, June 2008)*

Solution. In matrix notation, the given transformation is $y = Ax$,

where
$$y = \begin{pmatrix} y_1 \\ y_2 \\ y_3 \end{pmatrix},\ A = \begin{pmatrix} 1 & -1 & 1 \\ 3 & -1 & 2 \\ 2 & -2 & 3 \end{pmatrix},\ x = \begin{pmatrix} x_1 \\ x_2 \\ x_3 \end{pmatrix}.$$

Since $|A| = \begin{vmatrix} 1 & -1 & 1 \\ 3 & -1 & 2 \\ 2 & -2 & 3 \end{vmatrix} \begin{matrix} R_2 \leftarrow R_2 - 3R_1 \\ R_3 \leftarrow R_3 - 2R_1 \end{matrix} \approx \begin{vmatrix} 1 & -1 & 1 \\ 0 & 2 & -1 \\ 0 & 0 & 1 \end{vmatrix} = 1 \begin{vmatrix} 2 & -1 \\ 0 & 1 \end{vmatrix} = 2 \neq 0,$

the matrix A is non-singular and hence, the given transformation is non-singular or regular.

We compute A^{-1} using the Gauss-Jordan method. We have

$$A|I = \begin{bmatrix} 1 & -1 & 1 & | & 1 & 0 & 0 \\ 3 & -1 & 2 & | & 0 & 1 & 0 \\ 2 & -2 & 3 & | & 0 & 0 & 1 \end{bmatrix} \begin{matrix} R_2 \leftarrow R_2 - 3R_1 \\ R_3 \leftarrow R_3 - 2R_1 \end{matrix} \sim \begin{bmatrix} 1 & -1 & 1 & | & 1 & 0 & 0 \\ 0 & 2 & -1 & | & -3 & 1 & 0 \\ 0 & 0 & 1 & | & -2 & 0 & 1 \end{bmatrix} R_2 \leftarrow R_2/2$$

$$\sim \begin{bmatrix} 1 & -1 & 1 & | & 1 & 0 & 0 \\ 0 & 1 & -1/2 & | & -3/2 & 1/2 & 0 \\ 0 & 0 & 1 & | & -2 & 0 & 1 \end{bmatrix} R_1 \leftarrow R_1 + R_2$$

$$\sim \begin{bmatrix} 1 & 0 & 1/2 & | & -1/2 & 1/2 & 0 \\ 0 & 1 & -1/2 & | & -3/2 & 1/2 & 0 \\ 0 & 0 & 1 & | & -2 & 0 & 1 \end{bmatrix} \begin{matrix} R_1 \leftarrow R_1 - \dfrac{1}{2}R_3 \\ R_2 \leftarrow R_2 + \dfrac{1}{2}R_3 \end{matrix}$$

$$\sim \begin{bmatrix} 1 & 0 & 0 & | & 1/2 & 1/2 & -1/2 \\ 0 & 1 & 0 & | & -5/2 & 1/2 & 1/2 \\ 0 & 0 & 1 & | & -2 & 0 & 1 \end{bmatrix}. \text{ Hence } A^{-1} = \frac{1}{2}\begin{pmatrix} 1 & 1 & -1 \\ -5 & 1 & 1 \\ -4 & 0 & 2 \end{pmatrix}.$$

The inverse transformation is given by $x = A^{-1}y$

i.e.
$$\begin{pmatrix} x_1 \\ x_2 \\ x_3 \end{pmatrix} = \frac{1}{2}\begin{pmatrix} 1 & 1 & -1 \\ -5 & 1 & 1 \\ -4 & 0 & 2 \end{pmatrix}\begin{pmatrix} y_1 \\ y_2 \\ y_3 \end{pmatrix}$$

$$\Rightarrow x_1 = \frac{1}{2}(y_1 + y_2 - y_3),\ x_2 = \frac{1}{2}(-5y_1 + y_2 + y_3),\ x_3 = -2y_1 + y_3.$$

Example 15.78 Let
$$y_1 = 5x_1 + 3x_2 + 3x_3$$
$$y_2 = 3x_1 + 2x_2 - 2x_3$$
$$y_3 = 2x_1 - x_2 + 2x_3$$

be a linear transformation from (x_1, x_2, x_3) to (y_1, y_2, y_3) and $z_1 = 4x_1 + 2x_3$,

$z_2 = x_2 + 4x_3$, $z_3 = 5x_3$ be a linear transformation from (x_1, x_2, x_3) to (z_1, z_2, z_3). Find the linear transformation from (z_1, z_2, z_3) to (y_1, y_2, y_3) by inverting appropriate matrix and matrix multiplication. *(IETE, Dec. 2004)*

Solution. The transformation from the variables x_1, x_2, x_3 to y_1, y_2, y_3 is given by

$$y = Ax, \text{ where } A = \begin{bmatrix} 5 & 3 & 3 \\ 3 & 2 & -2 \\ 2 & -1 & 2 \end{bmatrix}. \text{ Another transformation from } x_1, x_2, x_3 \text{ to}$$

z_1, z_2, z_3 is given by z = Bx, where $B = \begin{bmatrix} 4 & 0 & 2 \\ 0 & 1 & 4 \\ 0 & 0 & 5 \end{bmatrix}$.

The required transformation is $y = Ax = (AB^{-1})z$. We now determine B^{-1} by using Gauss-Jordan elimination method.

We have

$$B|I = \begin{bmatrix} 4 & 0 & 2 & | & 1 & 0 & 0 \\ 0 & 1 & 4 & | & 0 & 1 & 0 \\ 0 & 0 & 5 & | & 0 & 0 & 1 \end{bmatrix} R_1 \leftarrow R_1/4 \sim \begin{bmatrix} 1 & 0 & 1/2 & | & 1/4 & 0 & 0 \\ 0 & 1 & 4 & | & 0 & 1 & 0 \\ 0 & 0 & 5 & | & 0 & 0 & 1 \end{bmatrix} R_3 \leftarrow R_3/5$$

$$\sim \begin{bmatrix} 1 & 0 & 1/2 & | & 1/4 & 0 & 0 \\ 0 & 1 & 4 & | & 0 & 1 & 0 \\ 0 & 0 & 1 & | & 0 & 0 & 1/5 \end{bmatrix} \begin{matrix} R_1 \leftarrow R_1 - \dfrac{1}{2}R_3 \\ R_2 \leftarrow R_2 - 4R_3 \end{matrix} \sim \begin{bmatrix} 1 & 0 & 0 & | & 1/4 & 0 & -1/10 \\ 0 & 1 & 0 & | & 0 & 1 & -4/5 \\ 0 & 0 & 1 & | & 0 & 0 & 1/5 \end{bmatrix}$$

$$\therefore \quad B^{-1} = \begin{bmatrix} 1/4 & 0 & -1/10 \\ 0 & 1 & -4/5 \\ 0 & 0 & 1/5 \end{bmatrix}.$$

Hence $AB^{-1} = \begin{bmatrix} 5 & 3 & 3 \\ 3 & 2 & -2 \\ 2 & -1 & 2 \end{bmatrix} \begin{bmatrix} 1/4 & 0 & -1/10 \\ 0 & 1 & -4/5 \\ 0 & 0 & 1/5 \end{bmatrix} = \begin{bmatrix} 5/4 & 3 & -1/2 - 12/5 + 3/5 \\ 3/4 & 2 & -3/10 - 8/5 - 2/5 \\ 1/2 & -1 & -2/10 + 4/5 + 2/5 \end{bmatrix}$

$$= \begin{bmatrix} 5/4 & 3 & -23/10 \\ 3/4 & 2 & -23/10 \\ 1/2 & -1 & 1 \end{bmatrix} = \frac{1}{20} \begin{bmatrix} 25 & 60 & -46 \\ 15 & 40 & -46 \\ 10 & -20 & 20 \end{bmatrix}.$$

Thus $y_1 = \frac{1}{20}(25z_1 + 60z_2 - 46z_3)$; $y_2 = \frac{1}{20}(15z_1 + 40z_2 - 46z_3)$;

$y_3 = \frac{1}{2}(z_1 - 2z_2 + 2z_3)$ is the required transformation.

Exercise 15.8

1. Using the elementary row operations, determine the ranks of the following matrices.

(i) $\begin{bmatrix} 1 & -2 & 3 \\ -2 & 4 & -1 \\ -1 & 2 & 7 \end{bmatrix}$,　　(ii) $\begin{bmatrix} 3 & -1 & 2 \\ -6 & 2 & 4 \\ -3 & 1 & 2 \end{bmatrix}$,　　(*MDU 2005, Andhra, 2000*)

(iii) $\begin{bmatrix} 0 & 1 & 2 & -2 \\ 4 & 0 & 2 & 6 \\ 2 & 1 & 3 & 1 \end{bmatrix}$,　(iv) $\begin{bmatrix} 1 & -1 & 2 & -3 \\ 4 & 1 & 0 & 2 \\ 0 & 3 & 0 & 4 \\ 0 & 1 & 0 & 2 \end{bmatrix}$, (v) $\begin{bmatrix} 1 & -1 & 1 & -1 \\ 4 & 2 & -1 & 2 \\ 2 & 2 & -2 & 2 \end{bmatrix}$,

(vi) $\begin{bmatrix} 1 & 3 & 4 & 3 \\ 3 & 9 & 12 & 9 \\ -1 & -3 & -4 & -3 \end{bmatrix}$, (vii) $\begin{bmatrix} 1 & 3 & 4 & 7 \\ 2 & 4 & 5 & 8 \\ 3 & 1 & 2 & 4 \end{bmatrix}$, (viii) $\begin{bmatrix} 1 & 0 & 2 & 1 \\ 0 & 1 & -2 & 1 \\ 1 & -1 & 4 & 0 \\ -2 & 2 & 8 & 0 \end{bmatrix}$

(*PTU, 2004*)

(ix) $\begin{bmatrix} 9 & 3 & 1 & 0 \\ 3 & 0 & 1 & -6 \\ 1 & 1 & 1 & 1 \\ 0 & -6 & 1 & 9 \end{bmatrix}$ (*IETE Dec. 2004*), (x) $\begin{bmatrix} 1 & 2 & 3 & 4 \\ 2 & 3 & 4 & 5 \\ 3 & 4 & 5 & 6 \\ 4 & 5 & 6 & 7 \end{bmatrix}$ (*IETE June 2002*)

Ans. (i) 2, (ii) 2, (iii) 2, (iv) 4, (v) 3, (vi) 1, (vii) 3, (viii) 3, (ix) 4, (x) 2

2. Reduce each of the following matrices to normal form and hence, find their ranks.

(i) $\begin{bmatrix} 8 & 1 & 3 & 6 \\ 0 & 3 & 2 & 2 \\ -8 & -1 & -3 & 4 \end{bmatrix}$ (*JNTU-2000*), (ii) $\begin{bmatrix} 2 & 1 & -3 & -6 \\ 3 & -3 & 1 & 2 \\ 1 & 1 & 1 & 2 \end{bmatrix}$ (*UPTU-2003*)

(iii) $\begin{bmatrix} 2 & 3 & -1 & -1 \\ 1 & -1 & -2 & -4 \\ 3 & 1 & 3 & -2 \\ 6 & 3 & 0 & -7 \end{bmatrix}$, (*AMIE - W 2007; UPTU - 2005; MDU 2000*)

(iv) $\begin{bmatrix} 1 & 2 & -1 & 4 \\ 2 & 4 & 3 & 4 \\ 1 & 2 & 3 & 4 \\ -1 & -2 & 6 & -7 \end{bmatrix}$ (*UPTU-2005, 2002*) **Ans.** (*i*) 3, (*ii*) 3, (*iii*) 3, (*iv*) 3

3. Find the non-singular matrices **P** and **Q** such that **PAQ** is in the normal form, where

$$A = \begin{bmatrix} 1 & 1 & 1 \\ 1 & -1 & -1 \\ 3 & 1 & 1 \end{bmatrix} \quad \textbf{Ans. } P = \begin{bmatrix} 1/2 & 1/2 & 0 \\ 1/2 & -1/2 & 0 \\ -2 & -1 & 1 \end{bmatrix}, Q = \begin{bmatrix} 1 & 0 & 0 \\ 0 & 1 & -1 \\ 0 & 0 & 1 \end{bmatrix}.$$

4. If $A = \begin{bmatrix} 3 & -3 & 4 \\ 2 & -3 & 4 \\ 0 & -1 & 1 \end{bmatrix}$, find A^{-1}.

Also find non-singular matrices **P** and **Q** such that **PAQ = I**, where **I** is a unit matrix and verify that $A^{-1} = QP$.

$$\textbf{Ans. } P = \begin{bmatrix} 1 & -1 & 0 \\ 0 & 0 & 1 \\ -2 & 3 & -3 \end{bmatrix}, Q = \begin{bmatrix} 1 & 0 & 0 \\ 0 & -1 & 1 \\ 0 & 0 & 1 \end{bmatrix}.$$

5. Find the non-singular matrices **R** and **S**, such that **RAS** is in the normal

form, where $A = \begin{bmatrix} 2 & 2 & -6 \\ -1 & 2 & 2 \end{bmatrix}$. **Ans.** $R = \begin{bmatrix} 1/2 & 0 \\ 1/2 & 1 \end{bmatrix}, S = \begin{bmatrix} 1 & -1/3 & 8/3 \\ 0 & 1/3 & 1/3 \\ 0 & 0 & 1 \end{bmatrix}.$

6. Find the non-singular matrices **P** and **Q** such that **PAQ** is in the normal form where

$$A = \begin{bmatrix} 1 & -1 & -1 \\ 1 & 1 & 1 \\ 3 & 1 & 1 \end{bmatrix}, (ii)\ A = \begin{bmatrix} 1 & 1 & 2 \\ 1 & 2 & 3 \\ 0 & -1 & -1 \end{bmatrix}$$

(UPTU 2001, MDU 2001)

Ans. (i) $P = \begin{bmatrix} 1 & 0 & 0 \\ -1/2 & 1/2 & 0 \\ 1/2 & 1 & -1/2 \end{bmatrix}$, $Q = \begin{bmatrix} 1 & 1 & 0 \\ 0 & 1 & -1 \\ 0 & 0 & 1 \end{bmatrix}$.

(ii) $P = \begin{bmatrix} 1 & 0 & 0 \\ -1 & 1 & 0 \\ -1 & 1 & 0 \end{bmatrix}$, $Q = \begin{bmatrix} 1 & -1 & -1 \\ 0 & 1 & -1 \\ 0 & 0 & 1 \end{bmatrix}$.

7. Test for consistency and hence solve:
   (i) $x + y + z = 6, x - y + 2z = 5, 3x + y + z = 8, 2x - 2y + 3z = 7$.
   (ii) $5x + 3y + 7z = 4, 3x + 26y + 2z = 9, 7x + 2y + 10z = 5$.

   (IETE June 2009, 2008; PTU 2005; VTU 2004)

   (iii) $2x - 3y + 7z = 5, 3x + y - 3z = 13, 2x + 19y - 47z = 32$.

   (GGSIPU 2005; MDU 2003, D.U. 2002, A.M.I.E., S-2002, Andhra, 2000)

   Ans. (i) 1, 2, 3 (ii) $x = (7 - 16z)/11, y = (z + 3)/11$, (iii) inconsistent and possesses no solution.

8. Investigate for consistency of the following systems and find the solution if it exists.
   (i) $4x + 7y - 5z + 3 = 0, 9x - 11y + 5z + 1 = 0, 11x - 9y + 7z - 5 = 0$, $3y + 10z - 23 = 0$.
   (ii) $3x + 3y + 2z = 1, x + 2y = 4, 10y + 3z = -2, 2x - 3y - z = 5$.

   (UPTU 2001) **Ans.** (i) 0, 1, 2; (ii) 2, 1, –4.

   (iii) $4x - 2y + 6z = 8, x + y - 3z = -1, 15x - 3y + 9z = 21$.

   (IETE, June 2009) **Ans.** Consistent; $x = 1, y = 3k - 2, z = k$ for all k.

9. Show that the equations
   $$3x + 4y + 5z = a, 4x + 5y + 6z = b, 5x + 6y + 7z = c$$
   do not have a solution unless $a + c = 2b$. Solve the equations when $a = b = c = -1$. *(Kerala, 1999)*

   Ans. $x = 1 + z, y = -(1 + 2z)$ and z is arbitrary.

10. Investigate the values of μ and λ, so that the equations
    $$2x + 3y + 5z = 9, 7x + 3y - 2z = 8, 2x + 3y + \lambda z = \mu \text{ has}$$
    (i) a unique solution; (ii) an infinite number of solutions (iii) no solution.

    (I.E.T.E. June 2007; Ranchi 2000)

    Ans. (i) $\lambda \neq 5, \mu$ may have any value, (ii) $\lambda = 5, \mu = 9$; (iii) $\lambda = 5$ and $\mu \neq 9$.

11. For what value of λ the equations $x + y + z = 1, x + 2y + 4z = \lambda$, $x + 4y + 10z = \lambda^2, 3x + 7y + 15z = 3$ are consistent. Solve the set of equations for such a value of λ.
    (*A.M.I.E., S-2003*)
    Ans. $\lambda = 1, x = 1 + 2z, y = -3y$, and z is arbitrary.

12. Find the values of the parameter λ for which the system of equation
    $x + y + 4z = 1, x + 2y - 2z = 1, \lambda x + y + z = 1$
    will have (*i*) a unique solution (*ii*) no solution. Also find the solution for $\lambda = 1/2$ by applying elementary row transformations to the augmented matrix. (*A.M.I.E., W-2000*)   **Ans.** (*i*) $\lambda \neq 7/10$ (*ii*) $\lambda = 7/10$; $\left[-\dfrac{3}{2}, \dfrac{3}{2}, \dfrac{1}{4} \right]$.

13. For what values of k the equations $x + y + z = 1, 2x + y + 4z = k$, $4x + y + 10z = k^2$ have a solution and solve them completely in each case.
    (*PTU 2005; VTU 2004*) **Ans.** $\begin{pmatrix} k = 1, x = -3\lambda, y = 1 + 2\lambda, z = \lambda; \\ k = 2, x = 1 - 3\lambda, y = 2\lambda, z = \lambda. \end{pmatrix}$.

14. Determine the values of λ for which the system of equations
    $x - \lambda y + z = 0, \lambda x + 3y - \lambda z = 0, 3x + y - z = 0$
    has (*i*) only trivial solution, (*ii*) non-trivial solution.
    Ans. (*i*) $\lambda \neq 2$ and $\lambda \neq -3$, (*ii*) $\lambda = 2$, or $\lambda = -3$

15. For what values of k will be system
    $2x + ky + z = 0, (k - 1)x - y + 2z = 0, 4x + y + 4z = 0$
    have non-trivial solutions?
    Ans. $k = 1, 9/4.$

16. Find the values of λ for which the equations
    $(\lambda - 1)x + (3\lambda + 1) y + 2\lambda z = 0, (\lambda - 1) x + (4\lambda - 2)y + (\lambda + 3) z = 0$,
    $2x + (3\lambda + 1)y + 3(\lambda - 1)z = 0$ are consistent, and hence find the ratios of $x: y: z$ when λ has the smallest of these values.
    (*A.M.I.E., S-2007; IETE., Dec. 2007; D.U. 2002*)   **Ans.** $\lambda = 0, 3$; $1:1:1$.

17. Consider the system of equations:
    $x + 2y + z = 3, \lambda y + 5z = 10, 2x + 7y + \lambda z = \mu$.
    Find (*a*) those values of λ for which the system has a unique solution (*b*) those pairs of values (λ, μ) for which the system has more than one solution.   **Ans.** (*a*) $\lambda \neq 5$ and $\lambda \neq -3$ (*b*) $(5, 12)$ and $(-3, -4)$

18. Discuss the consistency of the following system of equations for various values of λ:
    $2x_1 - 3x_2 + 6x_3 - 5x_4 = 3, x_2 - 4x_3 + x_4 = 1, 4x_1 - 5x_2 + 8x_3 - 9x_4 = \lambda$ and if consistent, solve it. (*I.E.T.E., Dec. 2006*)
    Ans. $\lambda = 7$; two parameter family of solutions $x_1 = 3 + 3x_3 + x_4$ and $x_2 = 1 + 4x_3 - x_4$, where x_3 and x_4 are arbitrary.

19. (*a*) Solve system of equations by matrix method
    $5x + 3y + 14z = 4, y + 2z = 1, 2x + y + 6z = 2, x + y + 2z = 0$.
    (*I.E.T.E., June 2006*) **Ans.** $[-1, -1/2, 3/4]$.

(b) Solve the following system of linear equations by Gauss-elimination method.

$$2x + y + z = 10, 3x + 2y + 3z = 18, x + 4y + 9z = 16$$

(*AMIE, S-2008, W-2009, Punjab 1997*) **Ans.** $x = 7, y = -9, z = 5.$

20. Examine for non-trivial solution the system:

$$7x_1 + x_2 - 2x_3 = 0, x_1 + 5x_2 - 4x_3 = 0, 3x_1 - 2x_2 + x_3 = 0$$

Ans. $x_1 = 3t, x_2 = 13t, x_3 = 17t$, where t is arbitrary.

21. Investigate the values of λ and μ so that the equations

$$x + 2y + 3z = 6, 3x + 4y - 10z = 1, 3x + 4y + \lambda z = \mu$$

have (*i*) unique solution, (*ii*) infinite number of solutions (*iii*) no solution.

Ans. (*i*) $\lambda \neq -10$, μ may have any value, (*ii*) $\lambda = -10$, $\mu = 1$, (*iii*) $\lambda = -10$, $\lambda \neq 1$.

22. Determine the condition so that the equations

$$x + y + z = 1, x + 2y + 3z = 2, 2x + 3y + \lambda z = \mu$$

have (*i*) unique solution, (*ii*) infinite many solutions, (*iii*) no solution

Ans. (*i*) $\lambda \neq 4$, μ may have any value, (*ii*) $\lambda = 4$, $\mu = 3$, (*iii*) $\lambda = 4$, $\mu \neq 3$.

23. Using the Gauss-Jordan elimination method, to find the inverses of the following matrices.

(*i*) $\begin{bmatrix} 0 & 1 & 2 \\ 1 & 2 & 3 \\ 3 & 1 & 1 \end{bmatrix}$ (*UPTU – 2001, A.M.I.E., S - 1999*) (*ii*) $\begin{bmatrix} 1 & 1 & 3 \\ 1 & 3 & -3 \\ -2 & -4 & -4 \end{bmatrix}$ (*Andhra, 1998*)

(*iii*) $\begin{bmatrix} 2 & 0 & 1 \\ -2 & 3 & 4 \\ -5 & 5 & 6 \end{bmatrix}$ (*iv*) $\begin{bmatrix} -1 & 1 & 2 \\ 3 & -1 & 1 \\ -1 & 3 & 4 \end{bmatrix}$ (*v*) $\begin{bmatrix} 1 & 1 & 1 & 1 \\ -2 & 1 & -1 & -2 \\ -4 & -2 & -3 & 1 \\ 1 & 1 & 1 & 0 \end{bmatrix}$

(*vi*) $\begin{bmatrix} 8 & 4 & -3 \\ 2 & 1 & 1 \\ 1 & 2 & 1 \end{bmatrix}$ (*GGSIPU-2005*) (*vii*) $\begin{bmatrix} 1 & 2 & 3 \\ 2 & 4 & 5 \\ 3 & 5 & 6 \end{bmatrix}$ (*IETE, June 2009*)

Ans. (*i*) $\frac{1}{2}\begin{bmatrix} 1 & -1 & 1 \\ -8 & 6 & -2 \\ 5 & -3 & 1 \end{bmatrix}$, (*ii*) $\frac{1}{4}\begin{bmatrix} 12 & 4 & 6 \\ -5 & -1 & -3 \\ -1 & -1 & -1 \end{bmatrix}$, (*iii*) $\begin{bmatrix} -2 & 5 & -3 \\ -8 & 17 & -10 \\ -5 & -10 & 6 \end{bmatrix}$,

(*iv*) $\frac{1}{10}\begin{bmatrix} -7 & 2 & 3 \\ -13 & -2 & 7 \\ 8 & 2 & -2 \end{bmatrix}$, (*v*) $\begin{bmatrix} 4 & 1 & -2 & -9 \\ 3 & 1 & -1 & -5 \\ -7 & -2 & 3 & 15 \\ -1 & 0 & 0 & -1 \end{bmatrix}$, (*vi*) $\frac{1}{21}\begin{bmatrix} 1 & 10 & -7 \\ 1 & -11 & 14 \\ -3 & 12 & 0 \end{bmatrix}$,

$$(vii) \begin{bmatrix} 1 & -3 & 2 \\ -3 & 3 & -1 \\ 2 & -1 & 0 \end{bmatrix}.$$

24. Using matrix method, show that the equations

$$3x_1 + 2x_2 + 2x_3 - 5x_4 = 8$$
$$2x_1 + 5x_2 + 5x_3 - 18x_4 = 9$$
$$4x_1 - x_2 - x_3 + 8x_4 = 7$$

are consistent and hence solve them. *(A.M.I.E., S-2009)*

[Hint: The corresponding augmented matrix is

$$\begin{bmatrix} 3 & 2 & 2 & -5 & | & 8 \\ 2 & 5 & 5 & -18 & | & 9 \\ 4 & -1 & -1 & 8 & | & 7 \end{bmatrix}.$$ Applying the row operations. Finally we arrive at the augmented matrix (i).

$$\begin{bmatrix} 3 & 2 & 2 & -5 & | & 8 \\ 2 & 5 & 5 & -18 & | & 9 \\ 4 & -1 & -1 & 8 & | & 7 \end{bmatrix} \sim \begin{bmatrix} 1 & 2/3 & 2/3 & -5/3 & | & 8/3 \\ 2 & 5 & 5 & -18 & | & 9 \\ 4 & -1 & -1 & 8 & | & 7 \end{bmatrix} \sim \begin{bmatrix} 1 & 2/3 & 2/3 & -5/3 & | & 8/3 \\ 0 & 11/3 & 11/3 & -44/3 & | & 11/3 \\ 0 & -11/3 & -11/3 & 44/3 & | & -11/3 \end{bmatrix}$$

$$\begin{bmatrix} 1 & 2/3 & 2/3 & -5/3 & | & 8/3 \\ 0 & 1 & 1 & -4 & | & 1 \\ 0 & -1 & -1 & 4 & | & -1 \end{bmatrix} \sim \begin{bmatrix} 1 & 2/3 & 2/3 & -5/3 & | & 8/3 \\ 0 & 1 & 1 & -4 & | & 1 \\ 0 & 0 & 0 & 0 & | & 0 \end{bmatrix}$$

$$\begin{bmatrix} 1 & 0 & 0 & 1 & | & 2 \\ 0 & 1 & 1 & -4 & | & 1 \\ 0 & 0 & 0 & 0 & | & 0 \end{bmatrix} \cdots (i)$$ The corresponding system of equatons can be solved for x_1, x_2 in terms of x_3 and x_4, giving us

$$x_1 = 2 - x_4, x_2 = 1 - x_3 + 4x_4$$

If we let $x_3 = t_1$ and $x_4 = t_2$, where t_1 and t_2 are arbitrary real numbers, the vector (x_1, x_2, x_3, x_4) in V_4 is given by

$(x_1, x_2, x_3, x_4) = (2 - t_2, 1 - t_1 + 4t_2, t_1, t_2)$ is a solution. By separating the parts involving t_1 and t_2, we can rewrite this as follows:

$(x_1, x_2, x_3, x_4) = (2, 1, 0, 0) + t_1(0, -1, 1, 0) + t_2(-1, 4, 0, 1).$

This equation gives the general solution of the system. The vector (2, 1, 0, 0) is a particular solution of the non-homogeneous system. The two vectors (0, −1, 1, 0) and (−1, 4, 0, 1) are solution of the corresponding homogeneous system.**]**

25. Show that the transformation

$$y_1 = 2x_1 + x_2 + x_3, \ y_2 = x_1 + x_2 + 2x_3, \ y_3 = x_1 - 2x_3$$

is regular. Write down the inverse transformation. *(Kerala, 1995)*

Ans. $x_1 = 2y_1 - 2y_2 - y_3; x_2 = -4y_1 + 5y_2 + 3y_3; x_3 = y_1 - y_2 - y_3.$

26. A transformation from the variables x_1, x_2, x_3 to y_1, y_2, y_3 is given by $\mathbf{Y} = \mathbf{AX}$, and another transformation from y_1, y_2, y_3 to z_1, z_2, z_3 is given by $\mathbf{Z} = \mathbf{BY}$, where

$$\mathbf{A} = \begin{bmatrix} 2 & 1 & 0 \\ 0 & 1 & -2 \\ -1 & 2 & 1 \end{bmatrix}, \quad \mathbf{B} = \begin{bmatrix} 1 & 1 & 1 \\ 1 & 2 & 3 \\ 1 & 3 & 5 \end{bmatrix}$$

obtain the transformation from x_1, x_2, x_3 to z_1, z_2, z_3.

Hint. The required transformation is $\mathbf{Z} = \mathbf{BY} = \mathbf{B(AX)} = \mathbf{(BA)X}$,

where $\mathbf{BA} = \begin{bmatrix} 1 & 1 & 1 \\ 1 & 2 & 3 \\ 1 & 3 & 5 \end{bmatrix}\begin{bmatrix} 2 & 1 & 0 \\ 0 & 1 & -2 \\ -1 & 2 & 1 \end{bmatrix} = \begin{bmatrix} 1 & 4 & -1 \\ -1 & 9 & -1 \\ -3 & 14 & -1 \end{bmatrix}$.

Ans. $z_1 = x_1 + 4x_2 - x_3$; $z_2 = -x_1 + 9x_2 - x_3$; $z_3 = -3x_1 + 14x_2 - x_3$.

27. Represent each of the transformations

$$x_1 = 3y_1 + 2y_2; y_1 = z_1 + 2z_2 \text{ and } x_2 = -y_1 + 4y_2, y_2 = 3z_1$$

by the use of matrices and find the composite transformation which expresses x_1, x_2 in terms of z_1, z_2.   *(Kerala, 1995)*

Ans. $\begin{pmatrix} x_1 \\ x_2 \end{pmatrix} = \begin{pmatrix} 3 & 2 \\ -1 & 4 \end{pmatrix}\begin{pmatrix} y_1 \\ y_2 \end{pmatrix}$; $\begin{pmatrix} y_1 \\ y_2 \end{pmatrix} = \begin{pmatrix} 1 & 2 \\ 3 & 0 \end{pmatrix}\begin{pmatrix} z_1 \\ z_2 \end{pmatrix}$; $\begin{pmatrix} x_1 \\ x_2 \end{pmatrix} = \begin{pmatrix} 9 & 6 \\ 11 & -2 \end{pmatrix}\begin{pmatrix} z_1 \\ z_2 \end{pmatrix}$.

Example 15.79. Find the value(s) of k for which the vectors (1, k, 5), (1, –3, 2), (2, –1, 1) will form a basis in R^3.   *(A.M.I.E., W-2010)*

Solution: The three vectors form a basis if and only if they are linearly independent. Thus form the matrix whose rows are the given vectors, and row reduce the matrix to echelon form:

$$\begin{bmatrix} 1 & -3 & 2 \\ 2 & -1 & 1 \\ 1 & k & 5 \end{bmatrix} \sim \begin{bmatrix} 1 & -3 & 2 \\ 0 & 5 & -3 \\ 0 & k+3 & 3 \end{bmatrix} \sim \begin{bmatrix} 1 & -3 & 2 \\ 0 & 5 & -3 \\ 0 & k+8 & 0 \end{bmatrix}.$$

If the value of $k \neq -8$, then the echelon matrix has no zero rows. Hence the three vectors are linearly independent, and so they do form a basis of R^3.

Chapter 16

Linear Algebra: Matrix Eigenvalue Problems

16.1 Eigenvalues and Eigenvectors

Let $A = \begin{bmatrix} a_{ij} \end{bmatrix}$ be a given square matrix of order n. Consider the vector equation

$$Ax = \lambda x \text{ or in matrix notation, } (A - \lambda I) x = 0 \qquad ...(16.1)$$

where λ is a scalar, I is an identity matrix of order n, x is an solution vector. If we let $x = [x_1, x_2, ...x_n]^T$ then (16.1) is the same as

$$\left. \begin{array}{l} (a_{11} - \lambda) x_1 + a_{12} x_2 + ... + a_{1n} x_n = 0 \\ a_{21} x_1 + (a_{22} - \lambda) x_2 + ... + a_{2n} x_n = 0 \\ .... \\ a_{n1} x_1 + a_{n2} x_2 + ... + (a_{nn} - \lambda) x_n = 0 \end{array} \right\} \qquad ...(16.2)$$

Although an obvious solution of homogeneous system (16.2) is $x_1 = 0, x_2 = 0,$ $...x_n = 0$, for any value of λ. This is of no practical interest. We are seeking only non-trivial solutions. We know that a homogeneous system of n linear equations in n unknowns has a nontrivial solution if and only if the determinant of the coefficient matrix is equal to zero. Thus to find a non zero solution of x for (16.1), we must have

$$|A - \lambda I| = 0. \qquad ...(16.3)$$

The expansion of $|A - \lambda I|$ by cofactors results in an nth degree polynomial in λ, which is of the form $\lambda^n - k_1 \lambda^{n-1} + k_2 \lambda^{n-2} - ... + (-1)^n k_n = 0$, where $k_1, k_2, ..., k_n$ are functions of the a's. The equation (16.3) is called **characteristic equation** of the matrix A. The polynomial equation of degree n in λ will have n roots (λ_1, $\lambda_2, ...\lambda_n$) which can be real or complex, simple or repeated. Thus, the **eigenvalues** or the **characteristic values** (or *latent roots*) of the matrix A are the roots of the **characteristic equation.** The corresponding nontrivial solution vectors x are called the eigenvectors or the characteristic vectors of A corresponding to that eigenvalue λ.

If **x** is a nontrivial solution of the homogeneous system (16.2), then $k\mathbf{x}$, where k is any constant is also a solution of the homogeneous system. Hence, an eigenvector is unique only upto a constant multiple. Thus the eigenvector corresponding to an eigenvalue is not unique.

The set of the eigenvalues is called the **spectrum** of **A**. The largest of the absolute values of the eigenvalues of **A** is called the **spectral radius** of **A** and is denoted by $\rho(\mathbf{A})$. After determining the eigenvalues λ_i's, we solve the homogeneous system $(\mathbf{A} - \lambda_i\mathbf{I})\,\mathbf{x} = \mathbf{0}$ for each λ_i, $i = 1, 2, ..., n$ to obtain the corresponding eigenvectors.

The set of all eigenvectors corresponding to an eigenvalue of **A**, together with **O**, forms a vector space called the **eigenspace** of **A** corresponding to this eigenvalue. The problem of determining the eigenvalues and eigenvectors of a matrix is called an *eigenvalue problem*.

Remark

A square matrix **A** of order n has always n linearly independent eigenvectors when its eigenvalues are distinct. The matrix **A** may also have n linearly independent eigenvectors even when some eigenvalues are repeated.

Properties of eigenvalues

We give below, without proof, some important properties of eigenvalues.

1. The sum of the eigenvalues of a matrix **A** is equal to the *trace* of the matrix **A**.

2. The product of the eigenvalues of a matrix **A** is equal to the determinant of **A**. Note that matrix **A** is singular if and only if at least one of its eigenvalues is zero.

3. If **A** is a square matrix, **A** and $\mathbf{A}^T$ have the same eigenvalues, since a determinant can be expanded by rows or columns.

4. The eigenvalues of an upper triangular, lower triangular, or diagonal matrix are the main diagonal entries.

5. If $\lambda_1, \lambda_2, ..., \lambda_n$ are the eigenvalues of **A**, then the eigenvalues of
   (*i*) $k\mathbf{A}$ are $k\lambda_1, k\lambda_2, ..., k\lambda_n$.

   (*ii*) $\mathbf{A}^m$ are $\lambda_1^m, \lambda_2^m, ..., \lambda_n^m$

   (*iii*) $\mathbf{A}^{-1}$ (if it exists) are $\dfrac{1}{\lambda_1}, \dfrac{1}{\lambda_2}, ..., \dfrac{1}{\lambda_n}$.

6. $\mathbf{A} - k\mathbf{I}$ has the eigenvalue $\lambda_1 - k, ..., \lambda_n - k$, for any scalar k.

7. $(\mathbf{A} - k\mathbf{I})^{-1}$ has the eigenvalue $1/(\lambda_1 - k), ..., 1/(\lambda_n - k)$.

8. A real matrix **A** may have complex eigenvalues. That is if $\alpha + i\beta$ is an eigenvalue, then its conjugate $\alpha - i\beta$ is also an eigenvalue.

9. The eigenvalues of a symmetric matrix are real.

10. The eigenvalues of a skew-symmetric matrix are pure imaginary or zero.

16.2 Cayley-Hamilton Theorem

Statement. Every square matrix **A** satisfies its own characteristic equation.

Proof: Let $\mathbf{A} = \begin{bmatrix} a_{11} & a_{12} & \cdots & a_{1n} \\ a_{21} & a_{22} & \cdots & a_{2n} \\ \cdots & \cdots & \cdots & \cdots \\ a_{n1} & a_{n2} & \cdots & a_{nn} \end{bmatrix}$.

Then characteristic equation of **A** is
$$|\mathbf{A} - \lambda\mathbf{I}| = 0.$$
Let $|\mathbf{A} - \lambda\mathbf{I}| = p_0 + p_1\lambda + p_2\lambda^2 + \ldots\ldots + p_n\lambda^n$ where $p_n = (-1)^n$.

$\therefore$ Characteristic equation is
$$p_0 + p_1\lambda + p_2\lambda^2 + \ldots + p_n\lambda^n = 0.$$
Now we have to prove that
$$p_0\mathbf{I} + p_1\mathbf{A} + p_2\mathbf{A}^2 + \ldots + p_n\mathbf{A}^n = \mathbf{0}.$$
Let $\mathbf{B} = Adj(\mathbf{A} - \lambda\mathbf{I})$.

Since co-factors of elements in $|\mathbf{A} - \lambda\mathbf{I}|$ are polynomials in λ of degree $(n-1)$ or less, the elements of **B** are also such polynomials.

We can, therefore break up **B** into a number of matrices each containing the same power of λ and write
$$\mathbf{B} = \mathbf{C}_0 + \mathbf{C}_1\lambda + \mathbf{C}_2\lambda^2 + \ldots + \mathbf{C}_{n-1}\lambda^{n-1}$$
where $\mathbf{C}_0, \mathbf{C}_1, \mathbf{C}_2, \ldots, \mathbf{C}_{n-1}$ are square matrices of order n whose elements are functions of the elements of **A**.

Then $(\mathbf{A} - \lambda\mathbf{I})\mathbf{B} = |\mathbf{A} - \lambda\mathbf{I}|\mathbf{I}$
$$\Rightarrow (\mathbf{A} - \lambda\mathbf{I})(\mathbf{C}_0 + \mathbf{C}_1\lambda + \mathbf{C}_2\lambda^2 + \ldots\ldots + \mathbf{C}_{n-1}\lambda^{n-1}) = (p_0 + p_1\lambda + p_2\lambda^2 + \ldots + p_n\lambda^n)\mathbf{I}.$$
Equating Coeff. of different powers of λ, we get
$$\mathbf{AC}_0 = p_0\mathbf{I}$$
$$\mathbf{AC}_1 - \mathbf{IC}_0 = p_1\mathbf{I}$$
$$\mathbf{AC}_2 - \mathbf{IC}_1 = p_2\mathbf{I}$$
$$\cdots \quad \cdots \quad \cdots \quad \cdots \quad \cdots$$
$$\mathbf{AC}_{n-1} - \mathbf{IC}_{n-2} = p_{n-1}\mathbf{I}$$
$$-\mathbf{IC}_{n-1} = p_n\mathbf{I}.$$
Pre-multiplying the above equations by $1, \mathbf{A}, \mathbf{A}^2, \ldots \mathbf{A}^n$ respectively and adding, we get
$$p_0\mathbf{I} + p_1\mathbf{A} + p_2\mathbf{A}^2 + \ldots + p_n\mathbf{A}^n$$
$$= \mathbf{AC}_0 + (\mathbf{A}^2\mathbf{C}_1 - \mathbf{AC}_0) + (\mathbf{A}^3\mathbf{C}_2 - \mathbf{A}^2\mathbf{C}_1) + \ldots$$
$$+ (\mathbf{A}^n\mathbf{C}_{n-1} - \mathbf{A}^{n-1}\mathbf{C}_{n-2}) - \mathbf{A}^n\mathbf{C}_{n-1} = \mathbf{0}$$
i.e.,
$$p_0\mathbf{I} + p_1\mathbf{A} + p_2\mathbf{A}^2 + \ldots + p_n\mathbf{A}^n = \mathbf{0} \qquad \ldots(16.4)$$
which proves the theorem.

Note: We can use (16.4) to find $\mathbf{A}^{-1}$ (if it exists) in terms of the powers of the matrix $\mathbf{A}$. Pre-multiplying equation (16.4) by $\mathbf{A}^{-1}$, we have

$$p_0 \mathbf{A}^{-1} + p_1 \mathbf{I} + p_2 \mathbf{A} + \ldots + p_n \mathbf{A}^{n-1} = 0$$

$$\Rightarrow \quad \mathbf{A}^{-1} = \frac{1}{p_0}\left(-p_1 \mathbf{I} - p_2 \mathbf{A} \ldots \ldots - p_n \mathbf{A}^{n-1}\right).$$

16.3 Illustrative Examples

Example 16.1 Verify that $\mathbf{x} = [1, -1, 1]^T$ is an eigenvector of the matrix.

$$\mathbf{A} = \begin{bmatrix} 0 & -1 & -3 \\ 2 & 3 & 3 \\ -2 & 1 & 1 \end{bmatrix}.$$

Solution. By carrying out the multiplication $\mathbf{Ax}$, we obtain

$$\mathbf{Ax} = \begin{bmatrix} 0 & -1 & -3 \\ 2 & 3 & 3 \\ -2 & 1 & 1 \end{bmatrix}\begin{bmatrix} 1 \\ -1 \\ 1 \end{bmatrix} = \begin{bmatrix} 1-3 \\ 2-3+3 \\ -2-1+1 \end{bmatrix} = \begin{bmatrix} -2 \\ 2 \\ -2 \end{bmatrix} = (-2)\begin{bmatrix} 1 \\ -1 \\ 1 \end{bmatrix} \begin{array}{c} \text{eigenvalue} \\ \downarrow \\ = (-2)\mathbf{x}. \end{array}$$

We see from the definition that $\lambda = -2$ is an eigenvalue of $\mathbf{A}$.

Example 16.2 Verify Cayley-Hamilton theorem for the matrix

$$\mathbf{A} = \begin{bmatrix} 7 & -1 & 3 \\ 6 & 1 & 4 \\ 2 & 4 & 8 \end{bmatrix} \qquad\qquad \textit{(MDU 2003)}$$

Also obtain $\mathbf{A}^{-1}$ and $\mathbf{A}^3$.

Solution. The characteristic equation of $\mathbf{A}$ is given by

$$|\mathbf{A} - \lambda \mathbf{I}| = \begin{vmatrix} 7-\lambda & -1 & 3 \\ 6 & 1-\lambda & 4 \\ 2 & 4 & 8-\lambda \end{vmatrix}$$

$$= (7-\lambda)\left[(1-\lambda)(8-\lambda)-16\right]+1\left[6(8-\lambda)-8\right]+3\left[24-2(1-\lambda)\right]$$
$$= (7-\lambda)\left[\lambda^2 - 9\lambda - 8\right] + (40 - 6\lambda) + 3(22 + 2\lambda)$$
$$= -\lambda^3 + 16\lambda^2 - 55\lambda + 50 = 0$$

Now

$$\mathbf{A}^2 = \begin{bmatrix} 7 & -1 & 3 \\ 6 & 1 & 4 \\ 2 & 4 & 8 \end{bmatrix}\begin{bmatrix} 7 & -1 & 3 \\ 6 & 1 & 4 \\ 2 & 4 & 8 \end{bmatrix} = \begin{bmatrix} 49 & 4 & 41 \\ 56 & 11 & 54 \\ 54 & 34 & 86 \end{bmatrix}.$$

$$A^3 = A^2 A = \begin{bmatrix} 49 & 4 & 41 \\ 56 & 11 & 54 \\ 54 & 34 & 86 \end{bmatrix} \begin{bmatrix} 7 & -1 & 3 \\ 6 & 1 & 4 \\ 2 & 4 & 8 \end{bmatrix} = \begin{bmatrix} 449 & 119 & 491 \\ 566 & 171 & 644 \\ 754 & 324 & 986 \end{bmatrix}.$$

We have $-A^3 + 16A^2 - 55A + 50I = -\begin{bmatrix} 449 & 119 & 491 \\ 566 & 171 & 644 \\ 754 & 324 & 986 \end{bmatrix} + 16 \begin{bmatrix} 49 & 4 & 41 \\ 56 & 11 & 54 \\ 54 & 34 & 86 \end{bmatrix}$

$$-55 \begin{bmatrix} 7 & -1 & 3 \\ 6 & 1 & 4 \\ 2 & 4 & 8 \end{bmatrix} + 50 \begin{bmatrix} 1 & 0 & 0 \\ 0 & 1 & 0 \\ 0 & 0 & 1 \end{bmatrix} = \begin{bmatrix} 0 & 0 & 0 \\ 0 & 0 & 0 \\ 0 & 0 & 0 \end{bmatrix} = 0 \qquad \ldots(i)$$

Hence, A satisfies the characteristic equation $-\lambda^3 + 16\lambda^2 - 55\lambda + 50 = 0$.
From equation (i), we obtain

$$A^{-1} = \frac{1}{50}[A^2 - 16A + 55I]$$

$$= \frac{1}{50}\left[\begin{pmatrix} 49 & 4 & 41 \\ 56 & 11 & 54 \\ 54 & 34 & 86 \end{pmatrix} - \begin{pmatrix} 112 & -16 & 48 \\ 96 & 16 & 64 \\ 32 & 64 & 128 \end{pmatrix} + \begin{pmatrix} 55 & 0 & 0 \\ 0 & 55 & 0 \\ 0 & 0 & 55 \end{pmatrix} \right] = \frac{1}{50}\begin{bmatrix} -8 & 20 & -7 \\ -40 & 50 & -10 \\ 22 & -30 & 13 \end{bmatrix}$$

From equation (i), we get

$$A^3 = 16A^2 - 55A + 50I = \begin{bmatrix} 784 & 64 & 656 \\ 896 & 176 & 864 \\ 864 & 544 & 1376 \end{bmatrix} - \begin{bmatrix} 385 & -55 & 165 \\ 330 & 55 & 220 \\ 110 & 220 & 440 \end{bmatrix} + \begin{bmatrix} 50 & 0 & 0 \\ 0 & 50 & 0 \\ 0 & 0 & 50 \end{bmatrix}$$

$$= \begin{bmatrix} 449 & 119 & 491 \\ 566 & 171 & 644 \\ 754 & 324 & 986 \end{bmatrix}.$$

Example 16.3 If $A = \begin{bmatrix} 2 & 2 & 0 \\ 2 & 1 & 1 \\ -7 & 2 & -3 \end{bmatrix}$, find the eigenvalues of the matrix A, A^2 and

verify that eigenvalues of A^2 are squares of those of A. Also determine the spectral radius of A.

Solution. The characteristic equation of A is given by

$$|A - \lambda I| = \begin{vmatrix} 2-\lambda & 2 & 0 \\ 2 & 1-\lambda & 1 \\ -7 & 2 & -3-\lambda \end{vmatrix}$$

$$= (2-\lambda)\left[(1-\lambda)(-3-\lambda)-2\right]-2\left[2(-3-\lambda+7)\right]=0$$

or $(2 - \lambda)(\lambda^2 + 2\lambda - 5) -2(1 - 2\lambda) = 0$ or $-\lambda^3 + 13\lambda - 12 = 0$

We have
$$\mathbf{A}^2 = \begin{bmatrix} 2 & 2 & 0 \\ 2 & 1 & 1 \\ -7 & 2 & -3 \end{bmatrix}\begin{bmatrix} 2 & 2 & 0 \\ 2 & 1 & 1 \\ -7 & 2 & -3 \end{bmatrix} = \begin{bmatrix} 8 & 6 & 2 \\ -1 & 7 & -2 \\ 11 & -18 & 11 \end{bmatrix}.$$

Eigenvalues of $\mathbf{A}$ are the roots of $\lambda^3 - 13\lambda + 12 = 0$ ...(i)
By trial $\lambda = 1$ is a root of this equation. Now (i) can be written as
$(\lambda - 1)(\lambda^2 + \lambda - 12) = 0$ or $(\lambda - 1)(\lambda - 3)(\lambda + 4) = 0$ or $\lambda = 1, 3, -4$.
The characteristic equation of $\mathbf{A}^2$ is given by

$$\begin{vmatrix} 8-\lambda & 6 & 2 \\ -1 & 7-\lambda & -2 \\ 11 & -18 & 11-\lambda \end{vmatrix}$$

$$= (8-\lambda)\left[(7-\lambda)(11-\lambda)-36\right]-6\left[\lambda-11+22\right]+2\left[18-11(7-\lambda)\right]=0$$

or $(8 - \lambda)(\lambda^2 - 18\lambda + 41) -6\lambda - 66 + 22\lambda - 118 = 0$
or $\lambda^3 - 26\lambda^2 + 169\lambda - 144 = 0$
or $(\lambda - 1)(\lambda^2 - 25\lambda + 144) = 0$ or $(\lambda - 1)(\lambda - 9)(\lambda - 16) = 0$.
The eigenvalues of $\mathbf{A}^2$ are 1, 9, 16 which are the squares of the eigenvalues of $\mathbf{A}$.
The spectral radius of $\mathbf{A}$ is given by
$\rho(\mathbf{A}) = $ largest eigenvalue in magnitude $= \max|\lambda_i| = |-4| = 4$.

Example 16.4 If $\mathbf{A} = \begin{bmatrix} 1 & 0 & 0 \\ 1 & 0 & 1 \\ 0 & 1 & 0 \end{bmatrix}$, then show that $\mathbf{A}^n = \mathbf{A}^{n-2} + \mathbf{A}^2 - \mathbf{I}$ for $n \geq 3$.

Hence, find $\mathbf{A}^{50}$.
                                                                    (I.E.T.E., W-2006)
Solution. The characteristic equation of $\mathbf{A}$ is given by

$$|\mathbf{A} - \lambda\mathbf{I}| = \begin{vmatrix} 1-\lambda & 0 & 0 \\ 1 & -\lambda & 1 \\ 0 & 1 & -\lambda \end{vmatrix} = (1-\lambda)(\lambda^2 - 1) = 0, \text{ or } \lambda^3 - \lambda^2 - \lambda + 1 = 0$$

and thus by the Cayley-Hamilton theorem, we get
$\mathbf{A}^3 - \mathbf{A}^2 - \mathbf{A} + \mathbf{I} = \mathbf{0}$ or $\mathbf{A}^3 - \mathbf{A}^2 = \mathbf{A} - \mathbf{I}$
Premultiplying both sides successively by $\mathbf{A}$, we obtain
$\mathbf{A}^4 - \mathbf{A}^3 = \mathbf{A}^2 - \mathbf{A}$
....
....
$\mathbf{A}^{n-1} - \mathbf{A}^{n-2} = \mathbf{A}^{n-3} - \mathbf{A}^{n-4}$
$\mathbf{A}^n - \mathbf{A}^{n-1} = \mathbf{A}^{n-2} - \mathbf{A}^{n-3}$.

Adding these equations, we get

$$A^n - A^2 = A^{n-2} - I, \text{ or } A^n = A^{n-2} + A^2 - I, \, n \geq 3. \qquad \ldots(i)$$

Replacing n by $n-2$, $n-4$, etc., we obtain

$$A^{n-2} = A^{n-4} + A^2 - I, \, A^{n-4} = A^{n-6} + A^2 - I$$

Now (i) may be re-written as

$$A^n = (A^{n-4} + A^2 - I) + A^2 - I = A^{n-4} + 2(A^2 - I)$$
$$= (A^{n-6} + A^2 - I) + 2(A^2 - I) = A^{n-6} + 3(A^2 - I)$$

....

....

$$= A^{n-(n-2)} + \frac{(n-2)}{2}(A^2 - I) = \frac{n}{2}A^2 - \left(\frac{n-2}{2}\right)I$$

Substituting $n = 50$, we get

$$A^{50} = 25A^2 - 24I = 25\begin{bmatrix} 1 & 0 & 0 \\ 1 & 1 & 0 \\ 1 & 0 & 1 \end{bmatrix} - 24\begin{bmatrix} 1 & 0 & 0 \\ 0 & 1 & 0 \\ 0 & 0 & 1 \end{bmatrix} = \begin{bmatrix} 1 & 0 & 0 \\ 25 & 1 & 0 \\ 25 & 0 & 1 \end{bmatrix}.$$

Example 16.5 If $A = \begin{bmatrix} 1 & -1 \\ 2 & 3 \end{bmatrix}$. Find A^8.

Solution. The characteristic equation of A is given by

$$|A - \lambda I| = \begin{vmatrix} 1-\lambda & -1 \\ 2 & 3-\lambda \end{vmatrix} = (1-\lambda)(3-\lambda) + 2 = \lambda^2 - 4\lambda + 5 = 0$$

and thus by the Cayley-Hamilton theorem, we get

$$A^2 - 4A + 5I = 0 \Rightarrow A^2 = 4A - 5I.$$

$$A^4 = 16A^2 - 40A + 25I, \text{ using the Commutative rule,}$$
$$= 16(4A - 5I) - 40A + 25I = 24A - 55I;$$
$$A^8 = 24^2 A^2 - (48)(55) A + 55^2 I = 24^2(4A - 5I) - (48)(55) A + 55^2 I$$

$$= -336A + 145I = -336\begin{bmatrix} 1 & -1 \\ 2 & 3 \end{bmatrix} + 145\begin{bmatrix} 1 & 0 \\ 0 & 1 \end{bmatrix} = \begin{bmatrix} -191 & 336 \\ -672 & -863 \end{bmatrix}.$$

Example 16.6 Verify Cayley-Hamilton theorem for the matrix $A = \begin{bmatrix} 1 & 4 \\ 2 & 3 \end{bmatrix}$ and find its inverse. Also express $A^5 - 4A^4 - 7A^3 + 11A^2 - A - 10I$ as a linear polynomial in A.

(MDU - 2003; IETE, June 2003)

Solution. The characteristic equation of A is given by

$$|A - \lambda I| = \begin{vmatrix} 1-\lambda & 4 \\ 2 & 3-\lambda \end{vmatrix} = (1-\lambda)(3-\lambda) - 8 = \lambda^2 - 4\lambda - 5 = 0 \qquad ...(i)$$

Now
$$A^2 = \begin{bmatrix} 1 & 4 \\ 2 & 3 \end{bmatrix}\begin{bmatrix} 1 & 4 \\ 2 & 3 \end{bmatrix} = \begin{bmatrix} 9 & 16 \\ 8 & 17 \end{bmatrix}.$$

We have
$$A^2 - 4A - 5I = \begin{bmatrix} 9 & 16 \\ 8 & 17 \end{bmatrix} - \begin{bmatrix} 4 & 16 \\ 8 & 12 \end{bmatrix} - \begin{bmatrix} 5 & 0 \\ 0 & 5 \end{bmatrix} = \begin{bmatrix} 0 & 0 \\ 0 & 0 \end{bmatrix} = 0 \qquad ...(ii)$$

Hence, A satisfies the characteristic equation $\lambda^2 - 4\lambda + 5 = 0$.
From equation (ii), we get

$$A^{-1} = \frac{1}{5}[A - 4I] = \frac{1}{5}\left[\begin{pmatrix} 1 & 4 \\ 2 & 3 \end{pmatrix} - \begin{pmatrix} 4 & 0 \\ 0 & 4 \end{pmatrix}\right] = \frac{1}{5}\begin{bmatrix} -3 & 4 \\ 2 & -1 \end{bmatrix}.$$

Now dividing the polynomial
$\lambda^5 - 4\lambda^4 - 7\lambda^3 + 11\lambda^2 - \lambda - 10$ by the polynomial $\lambda^2 - 4\lambda - 5$, we obtain
$$\lambda^5 - 4\lambda^4 - 7\lambda^3 + 11\lambda^2 - \lambda - 10 = (\lambda^2 - 4\lambda - 5)(\lambda^3 - 2\lambda + 3) + \lambda + 5$$
$$= \lambda + 5 \qquad \text{[by } (i)\text{]}$$
Hence $A^5 - 4A^4 - 7A^3 + 11A^2 - A - 10I = A + 5I$, which is a linear polynomial in A.

Example 16.7 Find the eigenvalues and the corresponding eigenvectors of the following matrices

$$(i) \ A = \begin{bmatrix} 4 & 2 \\ 6 & 5 \end{bmatrix}, \qquad\qquad (ii) \ A = \begin{bmatrix} 6 & -1 \\ 5 & 4 \end{bmatrix}.$$

Solution. (i) The characteristic equation of A is given by

$$|A - \lambda I| = \begin{vmatrix} 4-\lambda & 2 \\ 6 & 5-\lambda \end{vmatrix} = 0 \ \text{ or } (4 - \lambda)(5 - \lambda) - 12 = 0$$

or $\quad \lambda^2 - 9\lambda + 8 = 0$ or $\lambda = 1, 8$.

Corresponding to the eigenvalue $\lambda = 1$, we have

$$(A - I)x = \begin{bmatrix} 3 & 2 \\ 6 & 4 \end{bmatrix}\begin{bmatrix} x_1 \\ x_2 \end{bmatrix} = \begin{bmatrix} 0 \\ 0 \end{bmatrix} \text{ or } \begin{matrix} 3x_1 + 2x_2 = 0 \\ 6x_1 + 4x_2 = 0 \end{matrix} \bigg\} \text{ or } \begin{matrix} 3x_1 + 2x_2 = 0, \\ \text{ or } x_1 = -2x_2/3. \end{matrix}$$

Hence, the eigenvector x is given by

$$x = \begin{bmatrix} x_1 \\ x_2 \end{bmatrix} = \begin{bmatrix} -2x_2/3 \\ x_2 \end{bmatrix} = x_2\begin{bmatrix} -2/3 \\ 1 \end{bmatrix}.$$

Since an eigenvector is unique upto a constant multiple, we can take the eigenvector as $[-2, 3]^T$.

Corresponding to the eigenvalue $\lambda = 8$, we have

$$(A - 8I)x = \begin{bmatrix} -4 & 2 \\ 6 & -3 \end{bmatrix} \begin{bmatrix} x_1 \\ x_2 \end{bmatrix} = 0 \text{ or } \left. \begin{array}{c} -4x_1 + 2x_2 = 0 \\ 6x_1 - 3x_2 = 0 \end{array} \right\} \text{ or } x_1 = \frac{x_2}{2}.$$

Choose $x_2 = 2$, $x_1 = 1$. Therefore, the eigenvector is given by
$x = (x_1, x_2)^T = x_1(1, 2)^T$ or simply $(1, 2)^T$.

(*ii*) The characteristic equation of **A** is given by

$$|A - \lambda I| = \begin{vmatrix} 6 - \lambda & -1 \\ 5 & 4 - \lambda \end{vmatrix} = 0 \text{ or } \lambda^2 - 10\lambda + 29 = 0, \text{ or } \lambda = 5 \pm 2i.$$

Corresponding to the eigenvalue $\lambda = 5 + 2i$, we have

$$\left[A - (5 + 2i)I \right]x = \begin{bmatrix} 1 - 2i & -1 \\ 5 & -1 - 2i \end{bmatrix} \begin{bmatrix} x_1 \\ x_2 \end{bmatrix} = \begin{bmatrix} 0 \\ 0 \end{bmatrix}$$

or $(1 - 2i)x_1 - x_2 = 0$ and $5x_1 - (1 + 2i)x_2 = 0$. On solving, we get
$x_2 = (1 - 2i)x_1$. Choosing $x_1 = 1$, we get $x_2 = 1 - 2i$.
Therefore, the eigenvector is $x = [1, 1 - 2i]^T$.

Corresponding to the eigenvalue $\lambda = 5 - 2i$, we have

$$\left[A - (5 - 2i)I \right]x = \begin{bmatrix} 1 + 2i & -1 \\ 5 & -1 + 2i \end{bmatrix} \begin{bmatrix} x_1 \\ x_2 \end{bmatrix} = \begin{bmatrix} 0 \\ 0 \end{bmatrix}$$

or $(1 + 2i)x_1 - x_2 = 0$ and $5x_1 + x_2(-1 + 2i) = 0$. On solving, we get
$x_2 = (1 + 2i)x_1$. Choosing $x_1 = 1$, we get $x_2 = (1 + 2i)$.
Therefore, the eigenvector is $x = [1, 1 + 2i]^T$.

Remark. For a real matrix **A**, the eigenvalues and the corresponding eigenvectors can be complex.

Example 16.8 Find the eigenvalues and the corresponding eigenvectors of the matrix

$$A = \begin{bmatrix} 3 & -4 & 4 \\ 1 & -2 & 4 \\ 1 & -1 & 3 \end{bmatrix} \qquad\qquad (AMIE, S\text{-}2005)$$

Solution. The characteristic equation of **A** is given by

$$|A - \lambda I| = \begin{vmatrix} 3 - \lambda & -4 & 4 \\ 1 & -2 - \lambda & 4 \\ 1 & -1 & 3 - \lambda \end{vmatrix} = 0.$$

Operating $C_2 \leftarrow C_2 + C_3$, we get

$$|\mathbf{A} - \lambda \mathbf{I}| = \begin{vmatrix} 3-\lambda & 0 & 4 \\ 1 & 2-\lambda & 4 \\ 1 & 2-\lambda & 3-\lambda \end{vmatrix} = 0 \text{ or } (2-\lambda) \begin{vmatrix} 3-\lambda & 0 & 4 \\ 1 & 1 & 4 \\ 1 & 1 & 3-\lambda \end{vmatrix} = 0.$$

Expanding the determinant along R_1, we obtain

$$(2 - \lambda)[(3 - \lambda)(3 - \lambda - 4)] + 4(1 - 1) = 0$$

or $(2 - \lambda)(3 - \lambda)(-\lambda - 1) = 0$ or $\lambda = 2, 3, -1$.

The eigenvalues are 2, 3, –1.

We have the $(\mathbf{A} - \lambda \mathbf{I})\,\mathbf{x} = \mathbf{0}$,

that is,
$$\begin{bmatrix} 3-\lambda & -4 & 4 \\ 1 & -2-\lambda & 4 \\ 1 & -1 & 3-\lambda \end{bmatrix} \begin{bmatrix} x_1 \\ x_2 \\ x_3 \end{bmatrix} = \begin{bmatrix} 0 \\ 0 \\ 0 \end{bmatrix}$$

that is,
$$\left. \begin{array}{l} (3-\lambda)x_1 - 4x_2 + 4x_3 = 0 \\ x_1 - (2+\lambda)x_2 + 4x_3 = 0 \\ x_1 - x_2 + (3-\lambda)x_3 = 0 \end{array} \right\}. \qquad \dots(i)$$

Putting the different values of λ (*i.e.*, eigenvalues) in (*i*) we get the corresponding eigenvectors which are non-zero solutions of (*i*).

(*a*) When $\lambda = 2$, the system of equations (*i*) is

$$\left. \begin{array}{l} x_1 - 4x_2 + 4x_3 = 0 \\ x_1 - 4x_2 + 4x_3 = 0 \\ x_1 - x_2 + x_3 = 0 \end{array} \right\} \text{ or } \left. \begin{array}{l} x_1 - 4x_2 + 4x_3 = 0, \\ x_1 - x_2 + x_3 = 0. \end{array} \right\}$$

Obviously only two equations are independent. Solving the last two equations, we get

$$\frac{x_1}{-4+4} = \frac{x_2}{4-1} = \frac{x_3}{-1+4} \text{ or } \frac{x_1}{0} = \frac{x_2}{3} = \frac{x_3}{3} \text{ or } \frac{x_1}{0} = \frac{x_2}{1} = \frac{x_3}{1}.$$

Hence the corresponding eigenvector is $(0, 1, 1)^T$.

(*b*) When $\lambda = 3$, equations (*i*) are

$$0 \cdot x_1 - 4x_2 + 4x_3 = 0,$$
$$x_1 - 5x_2 + 4x_3 = 0,$$
$$x_1 - x_2 + 0 \cdot x_3 = 0.$$

Since only two equations are independent we can omit one of them. From first two equations, we obtain

$$\frac{x_1}{-16+20} = \frac{x_2}{4-0} = \frac{x_3}{0+4} \text{ or } \frac{x_1}{1} = \frac{x_2}{1} = \frac{x_3}{1}.$$

Thus the corresponding eigenvector is $(1, 1, 1)^T$.

(c) When $\lambda = -1$, equations (i) are

$$\left.\begin{array}{c} 4x_1 - 4x_2 + 4x_3 = 0 \\ x_1 - x_2 + 4x_3 = 0 \\ x_1 - x_2 + 4x_3 = 0 \end{array}\right\} \text{ or } \begin{array}{c} x_1 - x_2 + x_3 = 0 \\ x_1 - x_2 + 4x_3 = 0 \end{array}$$

Since only two equations are independent, from first two equations we have

$$\frac{x_1}{-4+1} = \frac{x_2}{1-4} = \frac{x_3}{-1+1} \text{ or } \frac{x_1}{1} = \frac{x_2}{1} = \frac{x_3}{0}.$$

The corresponding eigenvector is $(1, 1, 0)^T$

Hence the eigenvectors of a 3×3 matrix **A** corresponding to the eigenvalues 2, 3, -1 are $[0, 1, 1]^T$, $[1, 1, 1]^T$ and $[1, 1, 0]^T$ respectively.

Example 16.9 Find the eigenvalues and eigenvectors of the matrix

$$\mathbf{A} = \begin{bmatrix} 6 & -2 & 2 \\ -2 & 3 & -1 \\ 2 & -1 & 3 \end{bmatrix}. \quad (DU\ 2002;\ MDU\ 2002;\ Madras\ 2002)$$

Solution. The characteristic equation of **A** is given by

$$|\mathbf{A} - \lambda\mathbf{I}| = \begin{vmatrix} 6-\lambda & -2 & 2 \\ -2 & 3-\lambda & -1 \\ 2 & -1 & 3-\lambda \end{vmatrix} = 0$$

Operating $C_2 \leftarrow C_2 + C_3$, we get

$$\begin{vmatrix} 6-\lambda & 0 & 2 \\ -2 & 2-\lambda & -1 \\ 2 & 2-\lambda & 3-\lambda \end{vmatrix} = 0 \text{ or } (2-\lambda) \begin{vmatrix} 6-\lambda & 0 & 2 \\ -2 & 1 & -1 \\ 2 & 1 & 3-\lambda \end{vmatrix} = 0$$

or $(2-\lambda)[(6-\lambda)(3-\lambda+1) + 2(-2-2)] = 0$

or $(2-\lambda)(\lambda^2 - 10\lambda + 16) = 0$ *i.e.*, $(\lambda - 2)(\lambda - 2)(\lambda - 8) = 0$ or $\lambda = 2, 2, 8$

We have $(\mathbf{A} - \lambda\mathbf{I})\mathbf{x} = \mathbf{0}$,

that is, $\left.\begin{array}{c} (6-\lambda)x_1 - 2x_2 + 2x_3 = 0 \\ -2x_1 + (3-\lambda)x_2 - x_3 = 0 \\ 2x_1 - x_2 + (3-\lambda)x_3 = 0 \end{array}\right\}$ ...(i)

(a) When $\lambda = 8$, the equations (i) are

$$\left.\begin{array}{c} -2x_1 - 2x_2 + 2x_3 = 0 \\ -2x_1 - 5x_2 - x_3 = 0 \\ 2x_1 - x_2 - 5x_3 = 0 \end{array}\right\} \text{ or } \begin{array}{c} x_1 + x_2 - x_3 = 0, \\ x_2 + x_3 = 0. \end{array}$$

Choosing $x_3 = 1$, we get $x_2 = -1$, $x_1 = 2$.

Hence $(2, -1, 1)^T$ or any multiple is a characteristic vector.

(b) When $\lambda = 2$, (the equal root) the equations (i) reduce to

$$\begin{aligned} 4x_1 - 2x_2 + 2x_3 &= 0 \\ -2x_1 + x_2 - x_3 &= 0 \\ 2x_1 - x_2 + x_3 &= 0 \end{aligned} \right\} \quad \text{or} \quad 2x_1 - x_2 + x_3 = 0.$$

Clearly only one equation is independent, the other two being multiples of that. Here we can get any number of eigenvectors by giving arbitrary values to any two of the three variables. To get two linearly independent vectors we put $x_2 = 0$ and $x_3 = 0$ successively is any one equation.

For $x_2 = 0$, $x_1 = 1$, we get $x_3 = -2$. The eigenvector is $(1, 0, -2)^T$.

For $x_3 = 0$, $x_1 = 1$, we get $x_2 = 2$. The eigenvector is $(1, 2, 0)^T$.

It can be verified that any other eigenvector corresponding to $\lambda = 2$ is a linear combination of these two.

Hence the three eigenvectors are

$(2, -1, 1)^T$, $(1, 0, -2)^T$ and $(1, 2, 0)^T$.

Example 16.10 Find the eigenvalues and eigenvectors of the matrix

$$A = \begin{bmatrix} 2 & -2 & 2 \\ 1 & 1 & 1 \\ 1 & 3 & -1 \end{bmatrix}. \qquad (AMIE, \ W\text{-}2009)$$

Solution. The characteristic equation of **A** is given by

$$|A - \lambda I| = \begin{vmatrix} 2-\lambda & -2 & 2 \\ 1 & 1-\lambda & 1 \\ 1 & 3 & -1-\lambda \end{vmatrix} = 0.$$

On expanding the determinant, we obtain

$(2 - \lambda)\,[(1 - \lambda)(-1 - \lambda) - 3] + 2\,[-1 - \lambda - 1] + 2\,[3 - (1 - \lambda)] = 0$

or $(2 - \lambda)\,[\lambda^2 - 4] - 2\lambda - 4 + 2\lambda + 4 = 0$

or $(2 - \lambda)(\lambda - 2)(\lambda + 2) = 0$ or $\lambda = -2, 2, 2$

The eigenvalues are $-2, 2, 2$.

we have $(A - \lambda I)\,x = 0$,

that is, $$\begin{aligned} (2-\lambda)x_1 - 2x_2 + 2x_3 &= 0 \\ x_1 + (1-\lambda)x_2 + x_3 &= 0 \\ x_1 + 3x_2 - (1+\lambda)x_3 &= 0 \end{aligned} \right\} \qquad \text{...}(i)$$

(a) When $\lambda = -2$, equations (i) are

$$\left.\begin{array}{r} 4x_1 - 2x_2 + 2x_3 = 0, \\ x_1 + 3x_2 + x_3 = 0, \\ x_1 + 3x_2 + x_3 = 0. \end{array}\right\} \text{ or } \left.\begin{array}{r} 2x_1 - x_2 + x_3 = 0, \\ x_1 + 3x_2 + x_3 = 0. \end{array}\right\}$$

Obviously only first two equations are independent. Solving these equations

$$\frac{x_1}{-1-3} = \frac{x_2}{1-2} = \frac{x_3}{6+1} \text{ i.e., } \frac{x_1}{-4} = \frac{x_2}{-1} = \frac{x_3}{7}.$$

$\therefore$ Corresponding eigenvector is $(-4, -1, 7)^T$.

(b) When $\lambda = 2$ (the equal root) equations (i) become

$$\left.\begin{array}{r} 0 \cdot x_1 - 2x_2 + 2x_3 = 0 \\ x_1 - x_2 + x_3 = 0 \\ x_1 + 3x_2 - 3x_3 = 0 \end{array}\right\} \text{ or } \left.\begin{array}{r} x_2 - x_3 = 0, \\ x_1 - x_2 + x_3 = 0, \\ x_1 + 3x_2 - 3x_3 = 0. \end{array}\right\}$$

Here only two equations are independent. Solving first two equations, we get

$$\frac{x_1}{-1+1} = \frac{x_2}{1} = \frac{x_3}{1} \text{ i.e., } \frac{x_1}{0} = \frac{x_2}{1} = \frac{x_3}{1}.$$

The eigenvector corresponding to the eigenvalues 2 is $(0, 1, 1)^T$.
In this case we have only two independent eigenvectors.

Remarks

(1) Eigenvectors corresponding to distinct eigenvalues are linearly independent.

(2) If λ is an eigenvalue of multiplicity m of a square matrix $\mathbf{A}$ of order n, then the number of linearly independent eigenvectors associated with λ is given by

$$p = n - r, \text{ where } r = \text{rank } (\mathbf{A} - \lambda\mathbf{I}), 1 \leq p \leq m.$$

For example let a 3×3 matrix has the eigenvalue $\lambda = 1$ of multiplicity 3.

(a) If the rank of the matrix $\mathbf{A} - \mathbf{I}$ is 2, we obtain one linearly independent eigenvector.

(b) If the rank of matrix $\mathbf{A} - \mathbf{I}$ is 1, we obtain two linearly independent eigenvectors.

(c) If the rank of matrix $\mathbf{A} - \mathbf{I}$ is 0, we obtain three linearly independent eigenvectors.

Example 16.11 Let a 4×4 matrix $\mathbf{A}$ have eigenvalues $1, -1, 2, -2$. Find the value of the determinant of the matrix $\mathbf{B} = 2\mathbf{A} + \mathbf{A}^{-1} - \mathbf{I}$ and trace of matrix $\mathbf{B}$.

(I.E.T.E., June 2005)

Solution. Eigenvalues of $\mathbf{B}$ are $2\lambda_i + (1/\lambda_i) - 1, i = 1, 2, 3, 4$.

That is, $2(1) + 1 - 1, 2(-1) - 1 - 1, 2(2) + (1/2) - 1, 2(-2) - (1/2) - 1$
or $2, -4, 7/2, -11/2$.

$|\mathbf{B}|$ = Product of eigenvalues of $\mathbf{B}$ = $2(-4)(7/2)(-11/2)$ = 154.

Trace of $\mathbf{B}$ = Sum of eigenvalues of $\mathbf{B}$ = $2 - 4 + (7/2) - (11/2) = -4$.

Example 16.12 Let a 3×3 matrix $\mathbf{A}$ have eigenvalues 1, 2, -1. Find the trace of the matrix $\mathbf{B} = \mathbf{A} - \mathbf{A}^{-1} + \mathbf{A}^2$.

Solution. Eigenvalues of $\mathbf{B}$ are $\lambda_i - (1/\lambda_i) + \lambda_i^2$, $i = 1, 2, 3$.

That is, $1 - 1 + 1$, $2 - (1/2) + 2^2$, $-1 + 1 + 1$ or $1, 11/2, 1$.

Trace of $\mathbf{B}$ = sum of eigenvalues of $\mathbf{B}$ = $1 + (11/2) + 1 = 15/2$.

16.4 Similar Matrices

Suppose $\mathbf{A}$ and $\mathbf{B}$ are square matrices of the same order. The matrix $\mathbf{B}$ is said to be similar to the matrix $\mathbf{A}$ if there exists an invertible matrix $\mathbf{P}$ such that

$$\mathbf{B} = \mathbf{P}^{-1}\mathbf{AP} \text{ or } \mathbf{PB} = \mathbf{AP}. \qquad \ldots(16.5)$$

That is, $\mathbf{B}$ is said to be obtained from $\mathbf{A}$ by similarity transformation. Post multiplying both sides in (16.5) by $\mathbf{P}^{-1}$, we get

$$\mathbf{PBP}^{-1} = \mathbf{A}$$

Thus, $\mathbf{B}$ is similar to $\mathbf{A}$, if and only if $\mathbf{A}$ is similar to $\mathbf{B}$. The matrix $\mathbf{P}$ is called the *similarity matrix*.

Important Results

1. Two similar matrices have the same characteristic equation and hence the same eigenvalues. However, the converse is not true. Two matrices which have the same characteristic equation need not always be similar.

2. If $\mathbf{x}$ is an eigenvector of $\mathbf{A}$ corresponding to the eigenvalue λ, then $\mathbf{y} = \mathbf{P}^{-1}\mathbf{x}$ is an eigenvector of $\mathbf{B}$ corresponding to the eigenvalue λ, where $\mathbf{P}$ is the similarity matrix. Thus similarity transformation retains same eigenvalues but changed eigenvectors.

3. If $\mathbf{A}$ is similar to $\mathbf{B}$ and $\mathbf{B}$ is similar to $\mathbf{C}$, then $\mathbf{A}$ is similar to $\mathbf{C}$.
   In practice, it is usually difficult to obtain a non-singular matrix $\mathbf{P}$ which satisfies the equation $\mathbf{B} = \mathbf{P}^{-1}\mathbf{AP}$ for any two matrices $\mathbf{A}$ and $\mathbf{B}$. However, it is possible to obtain the matrix $\mathbf{P}$ when $\mathbf{A}$ or $\mathbf{B}$ is a diagonal matrix.

Example 16.13 Examine whether $\mathbf{A}$ is similar to $\mathbf{B}$, where

(i) $\mathbf{A} = \begin{bmatrix} 1 & 1 \\ 2 & -3 \end{bmatrix}$ and $\mathbf{B} = \begin{bmatrix} -34 & 57 \\ -19 & 32 \end{bmatrix}$,

(ii) $\mathbf{A} = \begin{bmatrix} 1 & 0 \\ 0 & 1 \end{bmatrix}$ and $\mathbf{B} = \begin{bmatrix} 1 & 1 \\ 0 & 1 \end{bmatrix}$.

Solution. The given matrices are similar if there exists a non-singular matrix $\mathbf{P}$ such that $\mathbf{B} = \mathbf{P}^{-1}\mathbf{AP}$ or $\mathbf{PB} = \mathbf{AP}$.

Let $P = \begin{bmatrix} a & b \\ c & d \end{bmatrix}$. We shall find a, b, c, d such that $PB = AP$ and then check whether P is non-singular.

(i) $\begin{bmatrix} a & b \\ c & d \end{bmatrix} \begin{bmatrix} -34 & 57 \\ -19 & 32 \end{bmatrix} = \begin{bmatrix} 1 & 1 \\ 2 & -3 \end{bmatrix} \begin{bmatrix} a & b \\ c & d \end{bmatrix}$

or $\begin{bmatrix} -34a - 19b & 57a + 32b \\ -34c - 19d & 57c + 32d \end{bmatrix} = \begin{bmatrix} a+c & b+d \\ 2a-3c & 2b-3d \end{bmatrix}$.

Equating the corresponding elements, we obtain the system of equations

$-34a - 19b = a + c$,   or $35a + 19b + c = 0$

$57a + 32b = b + d$,   or $57a + 31b - d = 0$

$-34c - 19d = 2a - 3c$,   or $2a + 31c + 19d = 0$

$57c + 32d = 2b - 3d$,   or $2b - 57c - 35d = 0$

On solving these equations, we obtain $a = 1$, $b = -2$, $c = 3$, $d = -5$.

Therefore, we get $P = \begin{bmatrix} 1 & -2 \\ 3 & -5 \end{bmatrix}$, which is a non-singular matrix. Hence the matrices A and B are similar.

(ii) $\begin{bmatrix} a & b \\ c & d \end{bmatrix} \begin{bmatrix} 1 & 1 \\ 0 & 1 \end{bmatrix} = \begin{bmatrix} 1 & 0 \\ 0 & 1 \end{bmatrix} \begin{bmatrix} a & b \\ c & d \end{bmatrix}$, or $\begin{bmatrix} a & a+b \\ c & c+d \end{bmatrix} = \begin{bmatrix} a & b \\ c & d \end{bmatrix}$.

Equating the corresponding elements, we obtain

$a + b = b$, $c + d = d$ or $a = c = 0$. Therefore, $P = \begin{bmatrix} 0 & b \\ 0 & d \end{bmatrix}$, which is singular matrix. Since an invertible matrix P does not exist the matrices A and B are not similar.

16.5 Diagonalizable matrices

A matrix A is diagonalizable if it is similar to a diagonal matrix, that is there exists an invertible matrix P such that $P^{-1}AP = D$, where D is a diagonal matrix. Since, similar matrices have the same eigenvalues, the diagonal elements of matrix D are the eigenvalues of A.

Theorem. A square matrix A of order n is diagonalizable if and only if it has n linearly independent eigenvectors.

Proof: Let x_1, x_2, ..., x_n be n linearly independent eigenvectors, corresponding to the eigenvalues λ_1, λ_2, ..., λ_n of the matrix A in the same order, that is the eigenvector x_i corresponds to the eigenvalues λ_i, $i = 1, 2,..., n$. Then $Ax_i = \lambda_i x_i$.

Denote by P the square matrix whose columns are x_1, x_2, ..., x_n. We write

$P = [x_1, x_2, ..., x_n]$. Then

$$AP = A[\mathbf{x}_1, \mathbf{x}_2, ..., \mathbf{x}_n] = [A\mathbf{x}_1, A\mathbf{x}_2, ..., A\mathbf{x}_n]$$
$$= [\lambda_1 \mathbf{x}_1, \lambda_2 \mathbf{x}_2, ..., \lambda_n \mathbf{x}_n] = [\mathbf{x}_1, \mathbf{x}_2, ..., \mathbf{x}_n]\, D = PD, \qquad \qquad ...(16.6)$$

where
$$D = \begin{bmatrix} \lambda_1 & 0 & 0 & \cdots & 0 \\ 0 & \lambda_2 & 0 & \cdots & 0 \\ \vdots & \vdots & \vdots & \vdots & \vdots \\ 0 & 0 & 0 & \cdots & \lambda_n \end{bmatrix}.$$

Premultiplying both sides of (16.6) by $\mathbf{P}^{-1}$, we obtain

$\mathbf{P}^{-1}A\mathbf{P} = \boxed{\mathbf{P}^{-1}\mathbf{P}}\, D = D,$ which implies that A is similar to D.

The matrix of eigenvectors $\mathbf{P}$ reduces a matrix A to its diagonal form. The matrix $\mathbf{P}$ is called a *modal matrix* of A and D is called the *spectral matrix* of A.

Post multiplying both sides of (16.6) by $\mathbf{P}^{-1}$, we obtain
$$A = PDP^{-1}. \qquad \qquad ...(16.7)$$

Remarks

1. The reduction to diagonal form enables us to calculate powers of A. We have
   $$\mathbf{P}^{-1}\, A\mathbf{P} = D \qquad \qquad ...(i)$$
   Premultiplying both sides of (i) by $\mathbf{P}$, we obtain $A\mathbf{P} = \mathbf{P}D \qquad ...(ii)$
   Post multiplying both sides of (ii) by $\mathbf{P}^{-1}$, we obtain
   $$A = PDP^{-1}.$$
   $\therefore \quad A^2 = AA = (PDP^{-1})(PDP^{-1}) = PD(P^{-1}P)DP^{-1} = PD^2P^{-1}.$

   Repeating the pre-multiplication (post-multiplication) m times, we obtain
   $A^m = PD^mP^{-1}$, for any positive integer m.

2. If D is a diagonal matrix of order n, and
   $$D = \begin{bmatrix} \lambda_1 & 0 & 0 \\ 0 & \lambda_2 & 0 \\ \vdots & \vdots & \vdots \\ 0 & 0 & \lambda_n \end{bmatrix}, \text{ then } D^m = \begin{bmatrix} \lambda_1^m & 0 & 0 \\ 0 & \lambda_2^m & 0 \\ \vdots & \vdots & \vdots \\ 0 & 0 & \lambda_n^m \end{bmatrix}$$

   for any positive integer m. If $f(D)$ is a polynomial in D, then we obtain

   $$f(D) = \begin{bmatrix} f(\lambda_1) & 0 & \cdots & 0 \\ 0 & f(\lambda_2) & \cdots & 0 \\ \cdots & \cdots & \ddots & \cdots \\ 0 & 0 & \cdots & f(\lambda_n) \end{bmatrix}.$$

Let a matrix **A** be diagonalizable. Then we have $\mathbf{A} = \mathbf{PDP}^{-1}$ and $\mathbf{A}^m = \mathbf{PD}^m\mathbf{P}^{-1}$, for any positive integer m. Hence, we obtain

$$f(\mathbf{A}) = f(\mathbf{PDP}^{-1}) = \mathbf{P}f(\mathbf{D})\mathbf{P}^{-1}, \text{ for any matrix polynomial } f(\mathbf{A}).$$

16.6 Illustrative Examples

Example 16.14 Show that the matrix

$$\mathbf{A} = \begin{bmatrix} 2 & 2 & 1 \\ 1 & 3 & 1 \\ 1 & 2 & 2 \end{bmatrix} \qquad (AMIE\ S\text{-}2005,\ IETE.,\ Dec.\ 2003)$$

is diagonalizable. Hence, find **P** such that $\mathbf{P}^{-1}\mathbf{AP}$ is a diagonal matrix.

Solution. The characteristic equation of **A** is given by

$$|\mathbf{A} - \lambda\mathbf{I}| = \begin{vmatrix} 2-\lambda & 2 & 1 \\ 1 & 3-\lambda & 1 \\ 1 & 2 & 2-\lambda \end{vmatrix}$$

$$= (2-\lambda)[(3-\lambda)(2-\lambda)-2]-2(2-\lambda-1)+1(2-3+\lambda)$$

$$= (2-\lambda)(\lambda^2 - 5\lambda + 4) - 2 + 2\lambda + \lambda - 1$$

$$= \lambda^3 - 7\lambda^2 + 11\lambda - 5 = 0 \text{ or } (\lambda - 5)(\lambda - 1)^2 = 0 \qquad \text{or } \lambda = 5, 1, 1.$$

We first find the eigenvectors corresponding to the repeated eigenvalue $\lambda = 1$. We have the system

$$(\mathbf{A} - \mathbf{I})\mathbf{x} = \begin{bmatrix} 1 & 2 & 1 \\ 1 & 2 & 1 \\ 1 & 2 & 1 \end{bmatrix} \begin{bmatrix} x_1 \\ x_2 \\ x_3 \end{bmatrix} = \begin{bmatrix} 0 \\ 0 \\ 0 \end{bmatrix} \text{ or } x_1 + 2x_2 + x_3 = 0$$

The rank of the coefficient matrix is 1. Therefore, the system has two linearly independent solutions. We use the equation $x_1 + 2x_2 + x_3 = 0$ to find two linearly independent eigenvectors. Taking $x_3 = 0$, $x_2 = -1$, we obtain the eigenvector $[2, -1, 0]^T$ and taking $x_2 = 0$, $x_3 = -1$, we obtain the eigenvector $[1, 0, -1]^T$.

Eigenvector corresponding to the eigenvalue $\lambda = 5$ is the solution of the system.

$$(\mathbf{A} - 5\mathbf{I})\mathbf{x} = \begin{bmatrix} -3 & 2 & 1 \\ 1 & -2 & 1 \\ 1 & 2 & -3 \end{bmatrix} \begin{bmatrix} x_1 \\ x_2 \\ x_3 \end{bmatrix} = \begin{bmatrix} 0 \\ 0 \\ 0 \end{bmatrix}$$

A solution of this system is $[1, 1, 1]^T$.

The given 3×3 matrix has three linearly independent eigenvectors. Therefore, the matrix **A** is diagonalizable. The modal matrix **P** is given by

$$P = \begin{bmatrix} 1 & 2 & 1 \\ 1 & -1 & 0 \\ 1 & 0 & -1 \end{bmatrix} \text{ and } P^{-1} = \frac{1}{4} \begin{bmatrix} 1 & 2 & 1 \\ 1 & -2 & 1 \\ 1 & 2 & -3 \end{bmatrix}.$$

We shall now verify that $P^{-1}AP = D = \text{diag } (5, 1, 1)$.

$$P^{-1}AP = \frac{1}{4} \begin{bmatrix} 1 & 2 & 1 \\ 1 & -2 & 1 \\ 1 & 2 & -3 \end{bmatrix} \begin{bmatrix} 2 & 2 & 1 \\ 1 & 3 & 1 \\ 1 & 2 & 2 \end{bmatrix} \begin{bmatrix} 1 & 2 & 1 \\ 1 & -1 & 0 \\ 1 & 0 & -1 \end{bmatrix}$$

$$= \frac{1}{4} \begin{bmatrix} 5 & 10 & 5 \\ 1 & -2 & 1 \\ 1 & 2 & -3 \end{bmatrix} \begin{bmatrix} 1 & 2 & 1 \\ 1 & -1 & 0 \\ 1 & 0 & -1 \end{bmatrix} = \frac{1}{4} \begin{bmatrix} 20 & 0 & 0 \\ 0 & 4 & 0 \\ 0 & 0 & 4 \end{bmatrix} = \begin{bmatrix} 5 & 0 & 0 \\ 0 & 1 & 0 \\ 0 & 0 & 1 \end{bmatrix} = \text{diag } (5, 1, 1).$$

Example 16.15 Examine whether the matrix **A**, where **A** is given by

(i) $A = \begin{bmatrix} 1 & 3 & 3 \\ 1 & 4 & 3 \\ -1 & 3 & 4 \end{bmatrix}$, (ii) $A = \begin{bmatrix} -2 & 2 & -3 \\ 2 & 1 & -6 \\ -1 & -2 & 0 \end{bmatrix}$, *(I.E.T.E., Dec. 2008, June 2007)*

(iii) $A = \begin{bmatrix} 3 & 2 & 1 \\ 0 & 2 & 0 \\ 1 & 2 & 3 \end{bmatrix}$. *(I.E.T.E., June 2006)*

is diagonalizable. If so, obtain the matrix **P** such that $P^{-1}AP$ is a diagronal matrix.

Solution. (i)The characteristic equation of the matrix **A** is given by

$$|A - \lambda I| = \begin{vmatrix} 1-\lambda & 3 & 3 \\ 1 & 4-\lambda & 3 \\ -1 & 3 & 4-\lambda \end{vmatrix}$$

$$= (1 - \lambda)[(4 - \lambda)^2 - 9] - 3(4 - \lambda + 3) + 3(3 + 4 - \lambda)$$
$$= (\lambda - 1)^2 (\lambda - 7) = 0, \text{ or } \lambda = 1, 1, 7.$$

We first find the eigenvectors corresponding to the repeated eigenvalue $\lambda = 1$. We have the system

$$(A - I)x = \begin{bmatrix} 0 & 3 & 3 \\ 1 & 3 & 3 \\ -1 & 3 & 3 \end{bmatrix} \begin{bmatrix} x_1 \\ x_2 \\ x_3 \end{bmatrix} = \begin{bmatrix} 0 \\ 0 \\ 0 \end{bmatrix}.$$

Since the rank of the coefficient matrix is 2, it has one linearly independent eigenvector. We obtain another linearly independent eigenvector corresponding to the eigenvalue $\lambda = 7$. Since the matrix **A** has only two linearly independent eigenvectors, the matrix is not diagonalizable.

(*ii*) The characteristic equation of the matrix **A** is given by

$$|A - \lambda I| = \begin{vmatrix} -2 - \lambda & 2 & -3 \\ 2 & 1 - \lambda & -6 \\ -1 & -2 & 0 - \lambda \end{vmatrix} = 0 \text{ or } \lambda^3 + \lambda^2 - 21\lambda - 45 = 0$$

or $(\lambda - 5)(\lambda + 3)^2 = 0$ or $\lambda = 5, -3, -3$.

Eigenvector corresponding to the eigenvalue $\lambda = 5$ is the solution of the system

$$(A - 5I)x = \begin{bmatrix} -7 & 2 & -3 \\ 2 & -4 & -6 \\ -1 & -2 & -5 \end{bmatrix} \begin{bmatrix} x_1 \\ x_2 \\ x_3 \end{bmatrix} = \begin{bmatrix} 0 \\ 0 \\ 0 \end{bmatrix}.$$

A solution of this system is $[1, 2, -1]^T$.

Eigenvectors corresponding to $\lambda = -3$ are the solutions of the system

$$(A + 3I)x = \begin{bmatrix} 1 & 2 & -3 \\ 2 & 4 & -6 \\ -1 & -2 & 3 \end{bmatrix} \begin{bmatrix} x_1 \\ x_2 \\ x_3 \end{bmatrix} = \begin{bmatrix} 0 \\ 0 \\ 0 \end{bmatrix} \text{ or } x_1 + 2x_2 - 3x_3 = 0.$$

The rank of the coefficient matrix is 1. Therefore, the system has two linearly independent solutions. We use the equation $x_1 + 2x_2 - 3x_3 = 0$ to find two linearly independent eigenvectors. Choosing $x_3 = 0$, $x_2 = -1$, we obtain the eigenvector $[2, -1, 0]^T$ and taking $x_2 = 0$, $x_3 = 1$, we obtain the eigenvector $[3, 0, 1]^T$. The given 3×3 matrix has three linearly independent eigenvectors. Therefore, the matrix **A** is diagonalizable. The modal matrix **P** is given by

$$P = \begin{bmatrix} 1 & 2 & 3 \\ 2 & -1 & 0 \\ -1 & 0 & 1 \end{bmatrix} \text{ and } P^{-1} = \frac{1}{8} \begin{bmatrix} 1 & 2 & -3 \\ 2 & -4 & -6 \\ 1 & 2 & 5 \end{bmatrix}$$

Now we have to verified that $P^{-1}AP = \text{diag}[5, -3, -3]$

$$P^{-1}AP = \frac{1}{8} \begin{bmatrix} 1 & 2 & -3 \\ 2 & -4 & -6 \\ 1 & 2 & 5 \end{bmatrix} \begin{bmatrix} -2 & 2 & -3 \\ 2 & 1 & -6 \\ -1 & -2 & 0 \end{bmatrix} \begin{bmatrix} 1 & 2 & 3 \\ 2 & -1 & 0 \\ -1 & 0 & 1 \end{bmatrix}$$

$$= \frac{1}{8} \begin{bmatrix} 5 & 10 & -15 \\ -6 & 12 & 18 \\ -3 & -6 & -15 \end{bmatrix} \begin{bmatrix} 1 & 2 & 3 \\ 2 & -1 & 0 \\ -1 & 0 & 1 \end{bmatrix} = \frac{1}{8} \begin{bmatrix} 40 & 0 & 0 \\ 0 & -24 & 0 \\ 0 & 0 & -24 \end{bmatrix} = \text{diag}[5, -3, -3].$$

(*iii*) The characteristic equation of the matrix A is given by

$$|\mathbf{A} - \lambda \mathbf{I}| = \begin{vmatrix} 3-\lambda & 2 & 1 \\ 0 & 2-\lambda & 0 \\ 1 & 2 & 3-\lambda \end{vmatrix} = (3-\lambda)^2(2-\lambda) + 1(\lambda - 2)$$

$$= (\lambda - 2)^2(\lambda - 4) = 0 \text{ or } \lambda = 2, 2, 4.$$

Eigenvector corresponding to the eigenvalue $\lambda = 2$ is the solution of the system

$$(\mathbf{A} - 2\mathbf{I})\mathbf{x} = \begin{bmatrix} 1 & 2 & 1 \\ 0 & 0 & 0 \\ 1 & 2 & 1 \end{bmatrix} \begin{bmatrix} x_1 \\ x_2 \\ x_3 \end{bmatrix} = \begin{bmatrix} 0 \\ 0 \\ 0 \end{bmatrix} \text{ or } x_1 + 2x_2 + x_3 = 0.$$

The rank of the coefficient matrix is 1. Therefore, the system has two linearly independent solutions. We use the equation $x_1 + 2x_2 + x_3 = 0$ to find two linearly independent eigenvectors. Choosing $x_2 = 0$, $x_3 = -1$, we obtain the eigenvector $[1, 0, -1]^T$ and taking $x_3 = 0$, $x_2 = -1$, we obtain the eigen vector $[2, -1, 0]^T$.

Eigenvector corresponding to the eigenvalue $\lambda = 4$ is the solution of the system

$$(\mathbf{A} - 4\mathbf{I})\mathbf{x} = \begin{bmatrix} -1 & 2 & 1 \\ 0 & -2 & 0 \\ 1 & 2 & -1 \end{bmatrix} \begin{bmatrix} x_1 \\ x_2 \\ x_3 \end{bmatrix} = \begin{bmatrix} 0 \\ 0 \\ 0 \end{bmatrix}.$$

The solution of this system is $[1, 0, 1]^T$.

The given 3×3 matrix has three linearly independent eigenvectors. Therefore, the matrix A is diagonalizable.

The modal matrix **P** is given by

$$\mathbf{P} = \begin{bmatrix} 1 & 2 & 1 \\ 0 & -1 & 0 \\ -1 & 0 & 1 \end{bmatrix} \text{ and } \mathbf{P}^{-1} = \frac{1}{2} \begin{bmatrix} 1 & 2 & -1 \\ 0 & -2 & 0 \\ 1 & 2 & 1 \end{bmatrix}.$$

It can be verified that $\mathbf{P}^{-1}\mathbf{A}\mathbf{P} = \dfrac{1}{2} \begin{bmatrix} 1 & 2 & -1 \\ 0 & -2 & 0 \\ 1 & 2 & 1 \end{bmatrix} \begin{bmatrix} 3 & 2 & 1 \\ 0 & 2 & 0 \\ 1 & 2 & 3 \end{bmatrix} \begin{bmatrix} 1 & 2 & 1 \\ 0 & -1 & 0 \\ -1 & 0 & 1 \end{bmatrix}$

$$= \frac{1}{2} \begin{bmatrix} 2 & 4 & -2 \\ 0 & -4 & 0 \\ 4 & 8 & 4 \end{bmatrix} \begin{bmatrix} 1 & 2 & 1 \\ 0 & -1 & 0 \\ -1 & 0 & 1 \end{bmatrix} = \frac{1}{2} \begin{bmatrix} 4 & 0 & 0 \\ 0 & 4 & 0 \\ 0 & 0 & 8 \end{bmatrix} = \begin{bmatrix} 2 & 0 & 0 \\ 0 & 2 & 0 \\ 0 & 0 & 4 \end{bmatrix} = \text{diag } (2, 2, 4).$$

Example 16.16 The eigenvectors of a 3×3 matrix **A** corresponding to the eigenvalues 1, 2, 3 are $[1, 2, 1]^T$, $[2, 3, 4]^T$, $[1, 4, 9]^T$ respectively. Find the matrix **A**.

Solution. We have

$$\text{modal matrix } \mathbf{P} = \begin{bmatrix} 1 & 2 & 1 \\ 2 & 3 & 4 \\ 1 & 4 & 9 \end{bmatrix} \text{ and the spectral matrix } \mathbf{D} = \begin{bmatrix} 1 & 0 & 0 \\ 0 & 2 & 0 \\ 0 & 0 & 3 \end{bmatrix}.$$

We find that $\mathbf{P}^{-1} = \dfrac{1}{12} \begin{bmatrix} -11 & 14 & -5 \\ 14 & -8 & 2 \\ -5 & 2 & 1 \end{bmatrix}.$

Therefore, $\mathbf{A} = \mathbf{PDP}^{-1} = \dfrac{1}{12} \begin{bmatrix} 1 & 2 & 1 \\ 2 & 3 & 4 \\ 1 & 4 & 9 \end{bmatrix} \begin{bmatrix} 1 & 0 & 0 \\ 0 & 2 & 0 \\ 0 & 0 & 3 \end{bmatrix} \begin{bmatrix} -11 & 14 & -5 \\ 14 & -8 & 2 \\ -5 & 2 & 1 \end{bmatrix}$

$$= \dfrac{1}{12} \begin{bmatrix} 1 & 4 & 3 \\ 2 & 6 & 12 \\ 1 & 8 & 27 \end{bmatrix} \begin{bmatrix} -11 & 14 & -5 \\ 14 & -8 & 2 \\ -5 & 2 & 1 \end{bmatrix} = \dfrac{1}{12} \begin{bmatrix} 30 & -12 & 6 \\ 2 & 4 & 14 \\ -34 & 4 & 38 \end{bmatrix}.$$

Example 16.17 Show that the matrix

$$\mathbf{A} = \begin{bmatrix} 1 & 2 & 0 \\ 2 & 1 & -6 \\ 2 & -2 & 3 \end{bmatrix}$$

is diagonalizable. Hence, find $\mathbf{P}$ such that $\mathbf{P}^{-1}\mathbf{AP}$ is a diagonal matrix. Then, obtain the matrix $\mathbf{B} = \mathbf{A}^2 + 6\mathbf{A} + 4\mathbf{I}$.

Solution. The characteristic equation of $\mathbf{A}$ is given by

$$|\mathbf{A} - \lambda\mathbf{I}| = \begin{vmatrix} 1-\lambda & 2 & 0 \\ 2 & 1-\lambda & -6 \\ 2 & -2 & 3-\lambda \end{vmatrix} = \lambda^3 - 5\lambda^2 - 9\lambda + 45 = (\lambda-5)(\lambda-3)(\lambda+3) = 0, \text{ or } \lambda = 5, 3, -3.$$

Since the matrix $\mathbf{A}$ has three distinct eigenvalues, it has three linearly independent eigenvectors and hence it is diagonalizable.

The eigenvector corresponding to the eigenvalue $\lambda = 5$ is the solution of the system

$$(\mathbf{A} - 5\mathbf{I})\mathbf{x} = \begin{bmatrix} -4 & 2 & 0 \\ 2 & -4 & -6 \\ 2 & -2 & -2 \end{bmatrix} \begin{bmatrix} x_1 \\ x_2 \\ x_3 \end{bmatrix} = \begin{bmatrix} 0 \\ 0 \\ 0 \end{bmatrix}. \text{ The solution of } \mathbf{x}_1 = \begin{bmatrix} 1 \\ 2 \\ -1 \end{bmatrix}.$$

The eigenvector corresponding to the eigenvalue $\lambda = 3$ is the solution of the system

$$(A - 3I)x = \begin{bmatrix} -2 & 2 & 0 \\ 2 & -2 & -6 \\ 2 & -2 & 0 \end{bmatrix} \begin{bmatrix} x_1 \\ x_2 \\ x_3 \end{bmatrix} = \begin{bmatrix} 0 \\ 0 \\ 0 \end{bmatrix}. \text{ The solution of } x_2 = \begin{bmatrix} 1 \\ 1 \\ 0 \end{bmatrix}.$$

The eigenvector corresponding to the eigenvalue $\lambda = -3$ is the solution of the system

$$(A + 3I)x = \begin{bmatrix} 4 & 2 & 0 \\ 2 & 4 & -6 \\ 2 & -2 & 6 \end{bmatrix} \begin{bmatrix} x_1 \\ x_2 \\ x_3 \end{bmatrix} = \begin{bmatrix} 0 \\ 0 \\ 0 \end{bmatrix}. \text{ The solution of } x_3 = \begin{bmatrix} -1 \\ 2 \\ 1 \end{bmatrix}.$$

Hence, the modal matrix is given by

$$P = \begin{bmatrix} x_1, x_2, x_3 \end{bmatrix} = \begin{bmatrix} 1 & 1 & -1 \\ 2 & 1 & 2 \\ -1 & 0 & 1 \end{bmatrix}.$$

We shall verify that $P^{-1}AP = D = \text{diag }(5, 3, -3)$.

The characteristic equation of P is $\lambda^3 - 3\lambda^2 + 4 = 0$.

Hence, by the Cayley-Hamilton theorem

$P^3 - 3P^2 + 4I = 0$, where I is the unit matrix of order 3. Premultiplying by P^{-1}, we obtain

$$P^2 - 3P + 4P^{-1} = 0 \text{ or } P^{-1} = (-P^2 + 3P)/4$$

$$= -\frac{1}{4} \begin{bmatrix} 1 & 1 & -1 \\ 2 & 1 & 2 \\ -1 & 0 & 1 \end{bmatrix} \begin{bmatrix} 1 & 1 & -1 \\ 2 & 1 & 2 \\ -1 & 0 & 1 \end{bmatrix} + \frac{3}{4} \begin{bmatrix} 1 & 1 & -1 \\ 2 & 1 & 2 \\ -1 & 0 & 1 \end{bmatrix}$$

$$= -\frac{1}{4} \begin{bmatrix} 4 & 2 & 0 \\ 2 & 3 & 2 \\ -2 & -1 & 2 \end{bmatrix} + \frac{3}{4} \begin{bmatrix} 1 & 1 & -1 \\ 2 & 1 & 2 \\ -1 & 0 & 1 \end{bmatrix} = \begin{bmatrix} -1/4 & 1/4 & -3/4 \\ 1 & 0 & 1 \\ -1/4 & 1/4 & 1/4 \end{bmatrix}.$$

$$AP = \begin{bmatrix} 1 & 2 & 0 \\ 2 & 1 & -6 \\ 2 & -2 & 3 \end{bmatrix} \begin{bmatrix} 1 & 1 & -1 \\ 2 & 1 & 2 \\ -1 & 0 & 1 \end{bmatrix} = \begin{bmatrix} 5 & 3 & 3 \\ 10 & 3 & -6 \\ -5 & 0 & -3 \end{bmatrix}$$

$$\therefore P^{-1}AP = \begin{bmatrix} -1/4 & 1/4 & -3/4 \\ 1 & 0 & 1 \\ -1/4 & 1/4 & 1/4 \end{bmatrix} \begin{bmatrix} 5 & 3 & 3 \\ 10 & 3 & -6 \\ -5 & 0 & -3 \end{bmatrix} = \begin{bmatrix} 5 & 0 & 0 \\ 0 & 3 & 0 \\ 0 & 0 & -3 \end{bmatrix} = \text{diag }(5, 3, -3).$$

We have $D = \text{diag}(5, 3, -3)$, $D^2 = \text{diag}(25, 9, 9)$.

$\therefore \quad A^2 + 6A + 4I = P(D^2 + 6D + 4I)P^{-1}$

Now $\quad D^2 + 6D + 4I = \begin{bmatrix} 25 & 0 & 0 \\ 0 & 9 & 0 \\ 0 & 0 & 9 \end{bmatrix} + \begin{bmatrix} 30 & 0 & 0 \\ 0 & 18 & 0 \\ 0 & 0 & -18 \end{bmatrix} + \begin{bmatrix} 4 & 0 & 0 \\ 0 & 4 & 0 \\ 0 & 0 & 4 \end{bmatrix}$

$= \begin{bmatrix} 59 & 0 & 0 \\ 0 & 31 & 0 \\ 0 & 0 & -5 \end{bmatrix}$. Hence, we obtain

$A^2 + 6A + 4I = \begin{bmatrix} 1 & 1 & -1 \\ 2 & 1 & 2 \\ -1 & 0 & 1 \end{bmatrix} \begin{bmatrix} 59 & 0 & 0 \\ 0 & 31 & 0 \\ 0 & 0 & -5 \end{bmatrix} \begin{bmatrix} -1/4 & 1/4 & -3/4 \\ 1 & 0 & 1 \\ -1/4 & 1/4 & 1/4 \end{bmatrix}$

$= \begin{bmatrix} 15 & 16 & -12 \\ 4 & 27 & -60 \\ 16 & -16 & 43 \end{bmatrix}$.

Example 16.18 Find A^n, given $A = \begin{bmatrix} -1 & 2 \\ -10 & 8 \end{bmatrix}$.

Solution. The characteristic equation of A is given by

$|A - \lambda I| = \begin{vmatrix} -1-\lambda & 2 \\ -10 & 8-\lambda \end{vmatrix} = 0$ or $(-1-\lambda)(8-\lambda) + 20 = 0$

or $\lambda^2 - 7\lambda + 12 = 0$, or $\lambda = 3, 4$.

Corresponding to the eigenvalue $\lambda = 3$, we have

$(A - 3I)x = \begin{bmatrix} -4 & 2 \\ -10 & 5 \end{bmatrix} \begin{bmatrix} x_1 \\ x_2 \end{bmatrix} = \begin{bmatrix} 0 \\ 0 \end{bmatrix}$ or $2x_1 - x_2 = 0$ or $x_1 = \dfrac{x_2}{2}$.

Hence, the eigenvector x is given by

$x = \begin{bmatrix} x_1 \\ x_2 \end{bmatrix} = \begin{bmatrix} x_2/2 \\ x_2 \end{bmatrix} = x_2 \begin{bmatrix} 1/2 \\ 1 \end{bmatrix}$.

Since an eigenvector is unique upto a constant multiple, we can take the eigenvector as $[1, 2]^T$.

Corresponding to the eigenvalue $\lambda = 4$, we have

$(A - 4I)x = \begin{bmatrix} -5 & 2 \\ -10 & 4 \end{bmatrix} \begin{bmatrix} x_1 \\ x_2 \end{bmatrix} = \begin{bmatrix} 0 \\ 0 \end{bmatrix}$ or $5x_1 - 2x_2 = 0$ or $x_1 = \dfrac{2}{5}x_2$.

Therefore, the eigenvector is given by $\mathbf{x} = (x_1, x_2)^T = x_2(2, 5)^T$ or simply $(2, 5)^T$
Hence, the modal matrix is given by

$$\mathbf{P} = \begin{bmatrix} 1 & 2 \\ 2 & 5 \end{bmatrix} \text{ and } \mathbf{P}^{-1} = \begin{bmatrix} 5 & -2 \\ -2 & 1 \end{bmatrix}.$$

Now $\mathbf{AP} = \begin{bmatrix} -1 & 2 \\ -10 & 8 \end{bmatrix}\begin{bmatrix} 1 & 2 \\ 2 & 5 \end{bmatrix} = \begin{bmatrix} 3 & 8 \\ 6 & 20 \end{bmatrix}$ and $\mathbf{P}^{-1}\mathbf{AP} = \begin{bmatrix} 5 & -2 \\ -2 & 1 \end{bmatrix}\begin{bmatrix} 3 & 8 \\ 6 & 20 \end{bmatrix} = \begin{bmatrix} 3 & 0 \\ 0 & 4 \end{bmatrix}.$

The reduction to diagonal form enables us to calculate powers of $\mathbf{A}$. For example,

$$\begin{bmatrix} 3^3 & 0 \\ 0 & 4^3 \end{bmatrix} = (\mathbf{P}^{-1}\mathbf{AP})^3 = (\mathbf{P}^{-1}\mathbf{AP})(\mathbf{P}^{-1}\mathbf{AP})(\mathbf{P}^{-1}\mathbf{AP})$$

$$= \mathbf{P}^{-1}\mathbf{A}(\mathbf{PP}^{-1})\mathbf{A}(\mathbf{PP}^{-1})\mathbf{AP} = \mathbf{P}^{-1}\mathbf{A}^3\mathbf{P}. \text{ Therefore, } \mathbf{A}^3 = \mathbf{P}\begin{bmatrix} 3^3 & 0 \\ 0 & 4^3 \end{bmatrix}\mathbf{P}^{-1}.$$

More generally, we have $\mathbf{A}^n = \begin{bmatrix} 1 & 2 \\ 2 & 5 \end{bmatrix}\begin{bmatrix} 3^n & 0 \\ 0 & 4^n \end{bmatrix}\begin{bmatrix} 5 & -2 \\ -2 & 1 \end{bmatrix}$

$$= \begin{bmatrix} 3^n & 2 \cdot 4^n \\ 2 \cdot 3^n & 5 \cdot 4^n \end{bmatrix}\begin{bmatrix} 5 & -2 \\ -2 & 1 \end{bmatrix} = \begin{bmatrix} 3^n \cdot 5 - 2^2 \cdot 4^n & -2 \cdot 3^n + 2 \cdot 4^n \\ 10 \cdot 3^n - 5 \cdot 4^n \cdot 2 & -4 \cdot 3^n + 5 \cdot 4^n \end{bmatrix}$$

$$= \begin{bmatrix} 5 \cdot 3^n - 2^{2n+2} & -2 \cdot 3^n + 2^{2n+1} \\ 10 \cdot 3^n - 5 \cdot 2^{2n+1} & -4 \cdot 3^n + 5 \cdot 2^{2n} \end{bmatrix}.$$

16.7 Inner Product (dot product) of vectors

Let $\mathbf{x} = (x_1, x_2, ..., x_n)^T$ and $\mathbf{y} = (y_1, y_2, ..., y_n)^T$ are two vectors of dimensions n in R^n or $\mathbb{C}^n$. We define the following:

Inner Product or dot product of two vectors $\mathbf{x}$ and $\mathbf{y}$ in R^n is denoted and defined by

$$\mathbf{x} \cdot \mathbf{y} = \mathbf{x}^T\mathbf{y} = x_1 y_1 + x_2 y_2 + \text{........} + x_n y_n = \sum_{i=1}^{n} x_i y_i \qquad ...(16.8)$$

and is a scalar. The inner product is also denoted by $\langle \mathbf{x}, \mathbf{y} \rangle$. In this case $\mathbf{x} \cdot \mathbf{y} = \mathbf{y} \cdot \mathbf{x}$. Note that

$$\left. \begin{array}{l} \mathbf{x} \cdot \mathbf{x} \geq 0 \\ \mathbf{x} \cdot \mathbf{x} = 0 \text{ if and only if } \mathbf{x} = \mathbf{0} \end{array} \right\}.$$

Example 16.19 Consider vectors $x = (2, 3, -4)^T$ and $y = (3, -1, -2)^T$. find $\langle x, y \rangle$.

Solution. $\langle x, y \rangle = x^T y = (2)\,(3) + (3)\,(-1) + (-4)\,(-2) = 11$.

If x and y are in $\mathbb{C}^n$, then the inner product of these vectors is defined as

$$x \cdot y = x^T \bar{y} = \sum_{i=1}^{n} x_i \, \bar{y}_i \text{ and } y \cdot x = y^T \bar{x} = \sum_{i=1}^{n} y_i \, \bar{x}_i , \qquad \ldots(16.9)$$

where $\bar{x}$ and $\bar{y}$ are complex conjugate vectors of x and y respectively.

Note that $x \cdot y = \overline{y \cdot x}$. $\qquad \ldots(16.10)$

Example 16.20 Consider vectors $x = (5 + i, 2 - 3i, 7 + 2i)^T$ and $y = (3 - 2i, 4i, 1 + 6i)^T$, find $\langle x, y \rangle$.

Solution. $x \cdot y = x^T \bar{y} = (5+i)\overline{(3-2i)} + (2-3i)\overline{(4i)} + (7+2i)\overline{(1+6i)}$
$= (5 + i)\,(3 + 2i) + (2 - 3i)\,(-4i) + (7 + 2i)(1 - 6i)$
$= (13 + 13i) + (-12 - 8i) + (19 - 40i) = 20 - 35i$.

It can be easily verified that for any vectors x, y, z and scalars k_1 and k_2

$$\langle k_1 x + k_2 y, z \rangle = k_1 \langle x, z \rangle + k_2 \langle y, z \rangle = k_1 (x \cdot z) + k_2 (y \cdot z)$$

Problems based on Inner products

Example 16.21 Expand:

  (i) $\langle 3u_1 - 4u_2, 2v_1 - 5v_2 + 6v_3 \rangle$

  (ii) $\langle 3u + 5v, 4u - 6v \rangle$,

Solution. (i) $\langle 3u_1 - 4u_2, 2v_1 - 5v_2 + 6v_3 \rangle = 6\langle u_1, v_1 \rangle - 15 \langle u_1, v_2 \rangle$
$+ 18 \langle u_1, v_3 \rangle - 8 \langle u_2, v_1 \rangle + 20 \langle u_2, v_2 \rangle - 24 \langle u_2, v_3 \rangle$

  (ii) $\langle 3u + 5v, 4u - 6v \rangle = 12\langle u, u \rangle - 18 \langle u, v \rangle + 20 \langle v, u \rangle - 30 \langle v, v \rangle$
$= 12 \langle u, u \rangle + 2 \langle u, v \rangle - 30 \langle v, v \rangle$, [use symmetry, $\langle u, v \rangle = \langle v, u \rangle$]

Example 16.22 Let $\langle u, v \rangle = 3 + 2i$ in a complex inner product space v. Find:

  (i) $\langle (4 - 6i) u, v \rangle$;   (ii) $\langle u, (8 + 3i) v \rangle$,   (iii) $\langle (3 - 6i)u, (5 - 2i) v \rangle$.

Solution. (i) $\langle (4 - 6i) u, v \rangle = (4 - 6i) \langle u, v \rangle$
$= (4 - 6i)\,(3 + 2i) = 24 - 10i$

  (ii) $\langle u, (8+3i)v \rangle = \overline{(8+3i)}\langle u, v \rangle = (8 - 3i)(3 + 2i) = 30 + 7i$

  (iii) $\langle (3 - 6i)u, (5 - 2i)v \rangle = (3 - 6i)\overline{(5 - 2i)} \langle u, v \rangle$
$= (3 - 6i)(5 + 2i)(3 + 2i)$
$= (27 - 24i)(3 + 2i) = 81 + 48 - 18i = 129 - 18i$.

Example 16.23 Let $\mathbf{u} = (1, 3, -4, 2)$, $\mathbf{v} = (4, -2, 2, 1)$, $\mathbf{w} = (5, -1, -2, 6)$ in R^4.

Show $\langle 3\mathbf{u} - 2\mathbf{v}, \mathbf{w} \rangle = 3\langle \mathbf{u}, \mathbf{w} \rangle - 2\langle \mathbf{v}, \mathbf{w} \rangle$.

Solution. We have

$3\mathbf{u} - 2\mathbf{v} = (3, 9, -12, 6) - (8, -4, 4, 2) = (-5, 13, -16, 4)$.

$\therefore \langle 3\mathbf{u} - 2\mathbf{v}, \mathbf{w} \rangle = (-5)(5) + (13)(-1) + (-16)(-2) + 4(6)$

$= -25 - 13 + 32 + 24 = 18$.

$\langle \mathbf{u}, \mathbf{w} \rangle = 5 - 3 + 8 + 12 = 22$ and $\langle \mathbf{v}, \mathbf{w} \rangle = 20 + 2 - 4 + 6 = 24$

$\therefore 3\langle \mathbf{u}, \mathbf{w} \rangle - 2\langle \mathbf{v}, \mathbf{w} \rangle = 3(22) - 2(24) = 18 = \langle 3\mathbf{u} - 2\mathbf{v}, \mathbf{w} \rangle$.

Example 16.24 Consider $\mathbf{u} = (1 + i, 3, 4 - i)$ and $\mathbf{v} = (3 - 4i, 1 + i, 2i)$ in $\mathbb{C}^3$. Find

(i) $\langle \mathbf{u}, \mathbf{v} \rangle$, (ii) $\langle \mathbf{v}, \mathbf{u} \rangle$.

Solution. (i) $\langle \mathbf{u}, \mathbf{v} \rangle = (1+i)(\overline{3-4i}) + 3(\overline{1+i}) + (4-i)(\overline{2i})$

$= (1 + i)(3 + 4i) + 3(1 - i) + (4 - i)(-2i) = -1 + 7i + 3 - 3i - 8i - 2 = -4i$.

(ii) $\langle \mathbf{v}, \mathbf{u} \rangle = (3-4i)(\overline{1+i}) + (1+i)(3) + (2i)(\overline{4-i})$

$= (3 - 4i)(1 - i) + 3 + 3i + 2i(4 + i) = -1 - 7i + 3 + 3i + 8i - 2 = 4i$.

Alternatively. $\langle \mathbf{v}, \mathbf{u} \rangle = \overline{\langle \mathbf{u}, \mathbf{v} \rangle} = \overline{-4i} = 4i$.

Norm (Length) of a vector

The *norm* or *length* of a vector $\mathbf{x}$ in R^n or $\mathbb{C}^n$, denoted by $\|\mathbf{x}\|$, is defined to be the non-negative square root of $\mathbf{x} \cdot \mathbf{x}$. Thus

$$\|\mathbf{x}\| = \sqrt{\mathbf{x} \cdot \mathbf{x}} = \sqrt{\mathbf{x}^T \mathbf{x}} = \sqrt{x_1^2 + x_2^2 + \dots\dots + x_n^2} \qquad \dots(16.11)$$

is called the *length* or the *norm* of the vector $\mathbf{x}$

Example 16.25 Find $\|(3, -4, 12)\|$.

Solution. $\|3, -4, 12\| = \sqrt{(3)^2 + (-4)^2 + 12^2} = \sqrt{9 + 16 + 144} = \sqrt{169} = 13$

Example 16.26 Prove $\|k\mathbf{x}\| = |k| \|\mathbf{x}\|$, for real number k.

Solution: $\|k\mathbf{x}\| = \|k(x_1, x_2, \dots, x_n)\| = \|(kx_1, kx_2, \dots\dots, kx_n)\|$

$= \sqrt{(kx_1)^2 + (kx_2)^2 + \dots\dots + (kx_n)^2} = \sqrt{k^2(x_1^2 + x_2^2 + \dots + x_n^2)}$

$= \sqrt{k^2} \sqrt{x_1^2 + x_2^2 + \dots + x_n^2} = |k| \|\mathbf{x}\|$.

Unit Vector

The vector $\mathbf{x}$ is called a unit vector if $\|\mathbf{x}\| = 1$. For any non zero vector $\mathbf{x}$ in R^n, the vector $x/\|x\|$ is always a unit vector. Vectors of unit length are called **unit or normalized vectors.**

Orthogonal Vectors

The vectors x and y are said to be *orthogonal vectors* if and only if $x \cdot y = 0$. That is, inner product of these vectors is zero.

A non empty set of mutually orthogonal nonzero vector is called an **orthogonal set of vectors.**

Example 16.27. Show that vectors $x = [1/3, -2/3, -2/3]^T$ and $y = [2/3, -1/3, 2/3]^T$ are orthogonal.

Solution. $x \cdot y = x^T y = [1/3, -2/3, -2/3] \begin{bmatrix} 2/3 \\ -1/3 \\ 2/3 \end{bmatrix} = \begin{matrix} (1/3)(2/3) + (-2/3)(-1/3) \\ + (-2/3)(2/3) \end{matrix}$

$= 2/9 + 2/9 - 4/9 = 0$ and the vectors are orthogonal.

Example 16.28 Consider the vectors $x = (1, 1, 1)$, $y = (1, 2, -3)$, $z = (1, -4, 3)$ in R^3. Show that x is orthogonal to both y and z. But y and z are not orthogonal.

Solution. $\langle x, y \rangle = (1)(1) + (1)(2) + (1)(-3) = 1 + 2 - 3 = 0$,

$\langle x, z \rangle = (1)(1) + (1)(-4) + (1)(3) = 1 - 4 + 3 = 0$,

$\langle y, z \rangle = (1)(1) + (2)(-4) + (-3)(3) = 1 - 8 - 9 = -16.$

Thus x is orthogonal to vectors y and z, but vectors y and z are not orthogonal.

Orthonormal Vectors. The vectors x and y for which
$$x \cdot y = 0 \text{ and } \|x\| = 1, \|y\| = 1$$
are called *orthonormal* vectors. If x, y are any vector and $x \cdot y = 0$, then $x/\|x\|$, $y/\|y\|$ are orthonormal.

Orthonormal System of Vectors. Let $x_1, x_2, ..., x_m$ be m vectors in R^n. Then this set of vectors forms an orthonormal system of vectors, if

$$x_i \cdot x_j = x_i^T x_j = \begin{cases} 0, & i \neq j \\ 1, & i = J. \end{cases}$$

16.8 Orthogonal matrix

A real matrix A is called an orthogonal matrix if A is non-singular and its transpose is the same as its inverse.
That is, $A^T = A^{-1}$ or $A^T A = AA^T = I$.
A simple example is
$$A = \begin{bmatrix} \sin\theta & \cos\theta \\ -\cos\theta & \sin\theta \end{bmatrix}.$$

∴ Consider an orthogonal matrix $\mathbf{B} = \begin{bmatrix} 1/\sqrt{3} & 1/\sqrt{6} & -1/\sqrt{2} \\ 1/\sqrt{3} & -2/\sqrt{6} & 0 \\ 1/\sqrt{3} & 1/\sqrt{6} & 1/\sqrt{2} \end{bmatrix}$.

The column vectors and also the (row vectors) of an orthogonal matrix form an orthonormal system of vectors.

Remarks

1. The product of two or more orthogonal matrices is orthogonal.
2. The inverse and the transpose of an orthogonal matrix are orthogonal.
3. The determinant of an orthogonal matrix has the value ± 1.
4. The eigenvalues of an orthogonal matrix are real or complex conjugates in pairs and have absolute value 1.

Example 16.29 Verify that the following matrices are orthogonal

(i) $\mathbf{A} = \begin{bmatrix} 1/\sqrt{3} & 0 & -2/\sqrt{6} \\ 1/\sqrt{3} & 1/\sqrt{2} & 1/\sqrt{6} \\ 1/\sqrt{3} & -1/\sqrt{2} & 1/\sqrt{6} \end{bmatrix}$, (ii) $\mathbf{A} = \dfrac{1}{3}\begin{bmatrix} -1 & 2 & -2 \\ -2 & 1 & 2 \\ 2 & 2 & 1 \end{bmatrix}$ (A.M.I.E., W-2003)

Solution. (*i*) We need only verify that $\mathbf{A}^T\mathbf{A} = \mathbf{I}$

$$\mathbf{A}^T\mathbf{A} = \begin{bmatrix} 1/\sqrt{3} & 1/\sqrt{3} & 1/\sqrt{3} \\ 0 & 1/\sqrt{2} & -1/\sqrt{2} \\ -2/\sqrt{6} & 1/\sqrt{6} & 1/\sqrt{6} \end{bmatrix}\begin{bmatrix} 1/\sqrt{3} & 0 & -2/\sqrt{6} \\ 1/\sqrt{3} & 1/\sqrt{2} & 1/\sqrt{6} \\ 1/\sqrt{3} & -1/\sqrt{2} & 1/\sqrt{6} \end{bmatrix}$$

$$= \begin{bmatrix} \dfrac{1}{3}+\dfrac{1}{3}+\dfrac{1}{3} & \dfrac{1}{\sqrt{6}}-\dfrac{1}{\sqrt{6}} & -\dfrac{2}{3\sqrt{2}}+\dfrac{1}{3\sqrt{2}}+\dfrac{1}{3\sqrt{2}} \\ 0+\dfrac{1}{\sqrt{6}}-\dfrac{1}{\sqrt{6}} & \dfrac{1}{2}+\dfrac{1}{2} & 0+\dfrac{1}{2\sqrt{3}}-\dfrac{1}{2\sqrt{3}} \\ -\dfrac{2}{3\sqrt{2}}+\dfrac{1}{3\sqrt{2}}+\dfrac{1}{3\sqrt{2}} & \dfrac{1}{2\sqrt{3}}-\dfrac{1}{2\sqrt{3}} & \dfrac{2}{3}+\dfrac{1}{6}+\dfrac{1}{6} \end{bmatrix} = \begin{bmatrix} 1 & 0 & 0 \\ 0 & 1 & 0 \\ 0 & 0 & 1 \end{bmatrix} = \mathbf{I}.$$

(*ii*) $\mathbf{A}^T\mathbf{A} = \dfrac{1}{3}\cdot\dfrac{1}{3}\begin{bmatrix} -1 & -2 & 2 \\ 2 & 1 & 2 \\ -2 & 2 & 1 \end{bmatrix}\begin{bmatrix} -1 & 2 & -2 \\ -2 & 1 & 2 \\ 2 & 2 & 1 \end{bmatrix} = \dfrac{1}{9}\begin{bmatrix} 9 & 0 & 0 \\ 0 & 9 & 0 \\ 0 & 0 & 9 \end{bmatrix} = \begin{bmatrix} 1 & 0 & 0 \\ 0 & 1 & 0 \\ 0 & 0 & 1 \end{bmatrix} = \mathbf{I}.$

Example 16.30 Show that the vectors $\mathbf{x}_1 = [\cos\theta, \sin\theta, 0]^T$, $\mathbf{x}_2 = [-\sin\theta, \cos\theta, 0]^T$, $\mathbf{x}_3 = [0, 0, 1]^T$ form an orthogonal set or system of vectors.

Solution. Since the vectors are real, we must show that

$$\mathbf{x}_i \cdot \mathbf{x}_j = \mathbf{x}_i^T \mathbf{x}_j = \begin{cases} 0, & i \neq j \\ 1, & i = j, \end{cases}$$

If $i = j = 1$, we have

$$\mathbf{x}_1^T \mathbf{x}_1 = [\cos\theta, \sin\theta, 0] \begin{bmatrix} \cos\theta \\ \sin\theta \\ 0 \end{bmatrix} = \cos^2\theta + \sin^2\theta = 1.$$

Similarly, we find if $i = j = 2$ and $i = j = 3$, $\mathbf{x}_2^T \mathbf{x}_2 = 1$, $\mathbf{x}_3^T \mathbf{x}_3 = 1$.

Thus $\mathbf{x}_1$, $\mathbf{x}_2$, $\mathbf{x}_3$ are unit vectors.
Take $i = 1, j = 2$, the orthogonality of any two of the vectors.

$$\mathbf{x}_1^T \mathbf{x}_2 = [\cos\theta \quad \sin\theta \quad 0] \begin{bmatrix} -\sin\theta \\ \cos\theta \\ 0 \end{bmatrix} = 0.$$

Similarly $\mathbf{x}_2^T \mathbf{x}_3 = 0$, $\mathbf{x}_1^T \mathbf{x}_3 = 0$. The vectors are mutually orthogonal. Thus the vectors form an orthonormal system.

Example 16.31 Let $A = \begin{bmatrix} 1 & 1 & -1 \\ 1 & 3 & 4 \\ 7 & -5 & 2 \end{bmatrix}$. Determine whether or not

(*i*) the rows of A are orthogonal;
(*ii*) A is an orthogonal matrix;
(*iii*) the columns of A are orthogonal;
(*iv*) normalizing each row of matrix A and find matrix B;
(*v*) Is B an orthogonal matrix?
(*vi*) Are the columns of matrix B orthogonal?

Solution. (*i*) yes, since $(1, 1, -1) \cdot (1, 3, 4) = (1)(1) + (1)(3) + (-1)(4) = 0$,
$(1, 3, 4) \cdot (7, -5, 2) = 7 - 15 + 8 = 0$, and $(1, 1, -1) \cdot (7, -5, 2) = 7 - 5 - 2 = 0$.

(*ii*) No, since the rows of A are not unit vector.

$$\|(1, 1, -1)\| = \sqrt{1^2 + 1^2 + (-1)^2} = \sqrt{3}.$$

(*iii*) No, for example $(1, 1, 7) \cdot (1, 3, -5) = 1 + 3 - 35 = -31 \neq 0$.

(*iv*) $\|(1, 1, -1)\| = \sqrt{1+1+1} = \sqrt{3}$, $\|(1, 3, 4)\| = \sqrt{1+9+16} = \sqrt{26}$,

$$\|(7, -5, 2)\| = \sqrt{49+25+4} = \sqrt{78}.$$

$$\text{Thus } \mathbf{B} = \begin{bmatrix} 1/\sqrt{3} & 1/\sqrt{3} & -1/\sqrt{3} \\ 1/\sqrt{26} & 3/\sqrt{26} & 4/\sqrt{26} \\ 7/\sqrt{78} & -5/\sqrt{78} & 2/\sqrt{78} \end{bmatrix}.$$

(*v*) Yes, since the rows of matrix **B** are still orthogonal and are now unit vectors.

.(*vi*) Yes, since the rows of matrix **B** form an orthonormal set of vectors, the columns of **B** must automatically form an orthonormal set.

Example 16.32 Show that the matrices **A** and **A**^{T} have the same eigenvalues. Further if λ, μ are two distinct eigenvalues, then show that the eigenvector corresponding to λ for **A** is orthogonal to eigenvector corresponding to μ for **A**^{T}.

[*I.E.T.E., Dec. 2005*]

Solution. We have $|\mathbf{A} - \lambda\mathbf{I}| = |(\mathbf{A}^T)^T - \lambda\mathbf{I}^T| = |[\mathbf{A}^T - \lambda\mathbf{I}]^T| = |\mathbf{A}^T - \lambda\mathbf{I}|$.

Since **A** and **A**^{T} have the same characteristic equation, they have the same eigenvalues.

Let **x** be the eigenvector corresponding to the eigenvalue λ for **A** and **y** be the eigenvector corresponding to the eigenvalue μ for **A**^{T}. We have $\mathbf{Ax} = \lambda\mathbf{x}$.

Premultiplying by **y**^{T}, we obtain

$$\mathbf{y}^T\mathbf{Ax} = \lambda\mathbf{y}^T\mathbf{x} \qquad \qquad ...(i)$$

we also have $\mathbf{A}^T\mathbf{y} = \mu\mathbf{y}$, or $(\mathbf{A}^T\mathbf{y})^T = (\mu\mathbf{y})^T$ or $\mathbf{y}^T\mathbf{A} = \mu\mathbf{y}^T$.

Postmultiplying by **x**, we get

$$\mathbf{y}^T\mathbf{Ax} = \mu\mathbf{y}^T\mathbf{x}. \qquad \qquad ...(ii)$$

From (*i*) and (*ii*), we obtain $(\lambda - \mu)\mathbf{y}^T\mathbf{x} = 0$.

Since $\lambda \neq \mu$, we obtain $\mathbf{y}^T\mathbf{x} = \mathbf{0}$. Therefore, the vectors **x** and **y** are orthogonal.

Exercise 16

Verify the Cayley-Hamilton theorem for the matrix **A**. Find **A**$^{-1}$, if it exists, where **A** is as given in Problems 1 (*i*) to (*viii*).

1. (i) $\begin{bmatrix} 2 & 3 \\ 3 & 5 \end{bmatrix}$,   (ii) $\begin{bmatrix} 2 & -1 & 1 \\ -1 & 2 & -1 \\ 1 & -1 & 2 \end{bmatrix}$,

(*D.U. 1999; PU 1999*)   (*AMIE, W-2005; JNTU '02; MDU '02*)

(iii) $\begin{bmatrix} 1 & -2 & 2 \\ 1 & 2 & 3 \\ 0 & -1 & 2 \end{bmatrix}$,   (iv) $\begin{bmatrix} 1 & 2 & 1 \\ -1 & 0 & 3 \\ 2 & -1 & 1 \end{bmatrix}$,   (v) $\begin{bmatrix} 1 & 3 & 7 \\ 4 & 2 & 3 \\ 1 & 2 & 1 \end{bmatrix}$,

(*Madras, 2003; MDU 2002*)

(vi) $\begin{bmatrix} 1 & 1 & 3 \\ 1 & 3 & -3 \\ -2 & -4 & -4 \end{bmatrix}$, (I.E.T.E., June 2002) (vii) $\begin{bmatrix} 1 & 0 & 3 \\ 2 & 1 & -1 \\ 1 & -1 & 1 \end{bmatrix}$ (Osmania, 2000S)

(viii) $\begin{bmatrix} 7 & 2 & -2 \\ -6 & -1 & 2 \\ 6 & 2 & -1 \end{bmatrix}$. (Coimbatore 2001; Madras 2000S)

Ans. (i) $\begin{bmatrix} 5 & -3 \\ -3 & 2 \end{bmatrix}$, (ii) $\dfrac{1}{4}\begin{bmatrix} 3 & 1 & -1 \\ 1 & 3 & 1 \\ -1 & 1 & 3 \end{bmatrix}$, (iii) $\dfrac{1}{9}\begin{bmatrix} 7 & 2 & -10 \\ -2 & 2 & -1 \\ -1 & 1 & 4 \end{bmatrix}$,

(iv) $\dfrac{1}{18}\begin{bmatrix} 3 & -3 & 6 \\ 7 & -1 & -4 \\ 1 & 5 & 2 \end{bmatrix}$, (v) $\dfrac{1}{35}\begin{bmatrix} -4 & 11 & -5 \\ -1 & -6 & 25 \\ 6 & 1 & -10 \end{bmatrix}$, (vi) $\dfrac{1}{4}\begin{bmatrix} 12 & 4 & 6 \\ -5 & -1 & -3 \\ -1 & -1 & -1 \end{bmatrix}$,

(vii) $\dfrac{1}{9}\begin{bmatrix} 0 & 3 & 3 \\ 3 & 2 & -7 \\ 3 & -1 & -1 \end{bmatrix}$, (viii) $\dfrac{1}{3}\begin{bmatrix} -3 & -2 & 2 \\ 6 & 5 & -2 \\ -6 & -2 & 5 \end{bmatrix}$

2. If $A = \begin{bmatrix} 3 & 1 \\ -1 & 2 \end{bmatrix}$, express $2A^5 - 3A^4 + A^2 - 4I$ as a linear polynomial in A.

Ans. $138A - 403I$

3. Evaluate the matrix $A^5 - 27A^3 + 65A^2$,

where $A = \begin{bmatrix} 0 & 0 & 1 \\ 3 & 1 & 0 \\ -2 & 1 & 4 \end{bmatrix}$.   **Ans.** $\begin{bmatrix} -40 & 0 & 73 \\ 219 & 33 & 0 \\ -146 & 73 & 252 \end{bmatrix}$.

4. Given $A = \begin{bmatrix} 1 & 1 & 2 \\ 3 & 1 & 1 \\ 2 & 3 & 1 \end{bmatrix}$. Use the fact that A is satisfies its characteristic

equation to compute A^3 and A^4. Also compute A^{-1} and A^{-2} since A is non-singular matrix.

[**Hint.** The characteristic equation of A is $\lambda^3 - 3\lambda^2 - 7\lambda - 11 = 0$.

$$A^3 = 3A^2 + 7A + 11I = \begin{bmatrix} 42 & 31 & 29 \\ 45 & 39 & 31 \\ 53 & 45 & 42 \end{bmatrix}$$

$$A^4 = 3A^3 + 7A^2 + 11A = \begin{bmatrix} 193 & 160 & 144 \\ 224 & 177 & 160 \\ 272 & 224 & 193 \end{bmatrix}$$

$$A^{-1} = \frac{1}{11}(A^2 - 3A - 7I) = \frac{1}{11}\begin{bmatrix} -2 & 5 & -1 \\ -1 & -3 & 5 \\ 7 & -1 & -2 \end{bmatrix};$$

$$A^{-2} = \frac{1}{11}[A - 3I - 7A^{-1}] = \frac{1}{121}\begin{bmatrix} -8 & -24 & 29 \\ 40 & -1 & -24 \\ -27 & 40 & -8 \end{bmatrix}].$$

5. (a) Find the characteristic equation of the matrix $A = \begin{bmatrix} 4 & 3 & 1 \\ 2 & 1 & -2 \\ 1 & 2 & 1 \end{bmatrix}$. Hence find A^{-1}.

(UPTU-2001)

Ans. $\lambda^3 - 6\lambda^2 + 6\lambda - 11 = 0$, $A^{-1} = 1/11\begin{bmatrix} 5 & -1 & -7 \\ -4 & 3 & 10 \\ 3 & -5 & -2 \end{bmatrix}$.

(b) Show that the characteristic equation of the matrix

$A = \begin{bmatrix} 5 & 7 & 3 \\ 1 & 5 & 2 \\ 3 & 2 & 1 \end{bmatrix}$ is $\lambda^3 - 11\lambda^2 + 15\lambda - 1 = 0$. Deduce that A is non-singular

and that $A^{-1} = A^2 - 11A + 15I$.

6. If $A = \begin{bmatrix} 3 & -7 \\ -4 & 1 \end{bmatrix}$, find $A^3 + 3A^2 + 12A$. **Ans.** $\begin{bmatrix} 370 & -455 \\ -260 & 240 \end{bmatrix}$

7. Find $A^4 + A^2 + I$ if $A = \begin{bmatrix} 3 & 5 \\ -1 & -2 \end{bmatrix}$. **Ans.** $\begin{bmatrix} 16 & 20 \\ -4 & -4 \end{bmatrix}$.

8. If $A = \begin{bmatrix} 1 & 1 & 2 \\ 0 & 2 & 1 \\ 1 & 0 & 2 \end{bmatrix}$, show that $A^3 = (5A - I)(A - I)$ and

$A^{-1} = (A - 3I)(A - 2I)$. Deduce explicit forms for A^3 and A^{-1}.

Ans. $A^3 = \begin{bmatrix} 10 & 9 & 23 \\ 5 & 9 & 14 \\ 9 & 5 & 19 \end{bmatrix}$, $A^{-1} = \begin{bmatrix} 4 & -2 & -3 \\ 1 & 0 & -1 \\ -2 & 1 & 2 \end{bmatrix}$.

9. The matrix A is defined by $A = \begin{bmatrix} 1 & 2 & 3 \\ 3 & 1 & 2 \\ 2 & 3 & 1 \end{bmatrix}$. Show that $A^3 - 3A^2 - 16A -$

$16I = 2I - A$ and express $(2I - A)^{-1}$ as a quadratic polynomial in A.
Ans. $(17I + A - A^2)/52$.

10. If $A = \begin{bmatrix} 1 & 2 \\ -1 & 3 \end{bmatrix}$. Verify Cayley-Hamilton theorem. Hence express

$A^6 - 4A^5 + 8A^4 - 12A^3 + 14A^2$, as a linear polynomial in A.
Ans. $-4A + 5I$.

11. Using Cayley-Hamilton theorem, find A^8, if $A = \begin{bmatrix} 1 & 2 \\ 2 & -1 \end{bmatrix}$.

(PTU 2003; Anna 2003; MDU 2001) **Ans. 625I.**

12. Find the characteristic equation of the matrix $A = \begin{bmatrix} 2 & 1 & 1 \\ 0 & 1 & 0 \\ 1 & 1 & 2 \end{bmatrix}$ and hence

find the matrix represented by $A^8 - 5A^7 + 7A^6 - 3A^5 + A^4 - 5A^3 + 8A^2 - 2A + I$.   *(I.E.T.E., June 2004; UPTU 2003, MDU 2002)*

Ans. $\lambda^3 - 5\lambda^2 + 7\lambda - 3 = 0$, $A^2 + A + I = \begin{bmatrix} 8 & 5 & 5 \\ 0 & 3 & 0 \\ 5 & 5 & 8 \end{bmatrix}$.

13. If $A = \begin{bmatrix} 1 & 0 & 2 \\ 0 & -1 & 1 \\ 0 & 1 & 0 \end{bmatrix}$. Verify Cayley-Hamilton theorem.

Compute $2A^8 - 3A^5 + A^4 + A^2 - 4I$.

[**Hint:** The characteristic equation of A is $\lambda^3 - 2\lambda + 1 = 0$. By Cayley-Hamilton theorem $A^3 - 2A + I = 0$.

$2A^8 - 3A^5 + A^4 + A^2 - 4I = 2A^5(A^3 - 2A + I) + 4A^6 - 5A^5 + A^4 + A^2 - 4I$

$= 4A^3(A^3 - 2A + I) - 5A^5 + 9A^4 - 4A^3 + A^2 - 4I$

$= -5A^2(A^3 - 2A + I) + 9A^4 - 14A^3 + 6A^2 - 4I$

$= 9A(A^3 - 2A + I) - 14A^3 + 24A^2 - 9A - 4I$

$= -14(A^3 - 2A + I) + 24A^2 - 37A + 10I = 24A^2 - 37A + 10I$

$$= 24\begin{bmatrix} 1 & 2 & 2 \\ 0 & 2 & -1 \\ 0 & -1 & 1 \end{bmatrix} - 37\begin{bmatrix} 1 & 0 & 2 \\ 0 & -1 & 1 \\ 0 & 1 & 0 \end{bmatrix} + \begin{bmatrix} 10 & 0 & 0 \\ 0 & 10 & 0 \\ 0 & 0 & 10 \end{bmatrix} = \begin{bmatrix} -3 & 48 & -26 \\ 0 & 95 & -61 \\ 0 & -61 & 34 \end{bmatrix}.]$$

14. Show that $A = \begin{bmatrix} 1 & 2 & 0 \\ 2 & -1 & 0 \\ 0 & 0 & -1 \end{bmatrix}$ satisfies its own characteristic equation. Hence

    or otherwise evaluate A^{-2}. **Ans.** $\dfrac{1}{5}\begin{bmatrix} 1 & 0 & 0 \\ 0 & 1 & 0 \\ 0 & 0 & 5 \end{bmatrix}$.

15. If $A = \begin{bmatrix} 1 & 0 & 0 \\ 1 & 0 & 1 \\ 0 & 1 & 0 \end{bmatrix}$, then show that $A^n = A^{n-2} + A^3 - A$ for $n \geq 4$. Hence,

    find A^{20}. **Ans.** $\begin{bmatrix} 1 & 0 & 0 \\ 10 & 1 & 0 \\ 10 & 0 & 1 \end{bmatrix}$,

16. Find eigenvalues and corresponding eigenvectors for each of the following matrices.

    (a) $\begin{bmatrix} 2 & 2 \\ -1 & 5 \end{bmatrix}$,   (b) $\begin{bmatrix} \cos\theta & -\sin\theta \\ \sin\theta & \cos\theta \end{bmatrix}$,

    (c) $\begin{bmatrix} 8 & -6 & 2 \\ -6 & 7 & -4 \\ 2 & -4 & 3 \end{bmatrix}$ (*AMIE, 2005; I.E.T.E., 2005; VTU 2004; MDU 2003; PTU 2003;*)

(d) $\begin{bmatrix} 1 & 1 & 3 \\ 1 & 5 & 1 \\ 3 & 1 & 1 \end{bmatrix}$ *(IETE; June 2009)* (e) $\begin{bmatrix} 2 & 2 & -7 \\ 2 & 1 & 2 \\ 0 & 1 & -3 \end{bmatrix}$,

(f) $\begin{bmatrix} -2 & 2 & -3 \\ 2 & 1 & -6 \\ -1 & -2 & 0 \end{bmatrix}$, *(MDU 2005; AMIE S-2010, 2005; I.E.T.E., 2004)*

(g) $\begin{bmatrix} 3 & 1 & 4 \\ 0 & 2 & 6 \\ 0 & 0 & 5 \end{bmatrix}$, *(UPTU 2005; IETE, June 2009; Madras, 1998)*

(h) $\begin{bmatrix} 6 & -2 & 2 \\ -2 & 3 & -1 \\ 2 & -1 & 3 \end{bmatrix}$;

(i) $\begin{bmatrix} 2 & 2 & 1 \\ 1 & 3 & 1 \\ 1 & 2 & 2 \end{bmatrix}$, (j) $\begin{bmatrix} 3 & 10 & 5 \\ -2 & -3 & -4 \\ 3 & 5 & 7 \end{bmatrix}$, (k) $\begin{bmatrix} 1 & 2 & 1 \\ 6 & -1 & 0 \\ -1 & -2 & -1 \end{bmatrix}$,

(l) $\begin{bmatrix} 2 & 2 & 1 \\ -4 & 8 & 1 \\ -1 & -2 & 0 \end{bmatrix}$, (m) $\begin{bmatrix} 3 & 2 & 4 \\ 2 & 0 & 2 \\ 4 & 2 & 3 \end{bmatrix}$, (n) $\begin{bmatrix} 3 & -4 & 0 \\ 4 & 3 & 0 \\ 0 & 0 & 5 \end{bmatrix}$,

(o) $\begin{bmatrix} 1 & 0 & 0 \\ 0 & 2 & 1 \\ 2 & 0 & 3 \end{bmatrix}$ *(AMIE, S-2008)* (p) $\begin{bmatrix} 2 & 0 & 1 \\ 0 & 2 & 0 \\ 1 & 0 & 2 \end{bmatrix}$ *(GGSIPU 2004; MDU 2002; JNTU 2002)*

Ans. (a) 3, 4; $[2, 1]^T$, $[1, 1]^T$ (b) $e^{\pm i\theta}$; $[1, -i]^T$, $[1, i]^T$

(c) 0, 3, 15; $[1, 2, 2]^T$, $[2, 1, -2]^T$, $[2, -2, 1]^T$

(d) -2, 3, 6; $[-1, 0, 1]^T$, $[1, -1, 1]^T$, $[1, 2, 1]^T$

(e) 1, 3, -4; $[-1, 4, 1]^T$, $[5, 6, 1]^T$, $[3, -2, 2]^T$

(f) 5, -3, -3; $[1, 2, -1]^T$, $[2, -1, 0]^T$, $[3, 0, 1]^T$

(g) 2, 3, 5; $[1, -1, 0]^T$, $[1, 0, 0]^T$, $[3, 2, 1]^T$

(h) 2, 2, 8; $[1, 0, -2]^T$, $[1, 2, 0]^T$, $[2, -1, 1]^T$

(i) 5, 1, 1; $[1, 1, 1]^T$, $[1, 0, -1]^T$, $[2, -1, 0]^T$; (j) 3, 2, 2; $[1, 1, -2]^T$, $[5, 2, -5]^T$

(k) 0, 3, -4; $[1, 6, -13]^T$, $[2, 3, -2]^T$, $[-1, 2, 1]^T$

(l) 1, 3, 6; $[1, 1, -3]^T$, $[1, 1, -1]^T$, $[2, 5, -2]^T$

(m) 8, -1, -1; $[2, 1, 2]^T$, $[1, -2, 0]^T$, $[0, -2, 1]^T$

(n) 5, 3 ± 4i; $[0, 0, 1]^T$, $[1, -i, 0]^T$, $[1, i, 0]^T$.

(o) 1, 2, 3; $[-1, -1, 1]^T$, $[0, 1, 0]^T$, $[0, -1, 1]^T$

(p) 1, 2, 3; $[1, 0, -1]^T$, $[0, 1, 0]^T$, $[1, 0, 1]^T$

17. Examine whether **A** is similar to **B**, where

$$A = \begin{bmatrix} 5 & 5 \\ -2 & 0 \end{bmatrix} \text{ and } B = \begin{bmatrix} 1 & 2 \\ -3 & 4 \end{bmatrix}$$

(I.E.T.E., Dec-2006)

Ans. Matrices **A** and **B** are similar. $P = \begin{bmatrix} 1 & 1 \\ 1 & 2 \end{bmatrix}$.

18. Show that the following matrices **A** are diagonalizable. Find the matrix **P** such that $P^{-1}AP$ is a diagonal matrix.

(i) $\begin{bmatrix} 5 & 4 \\ 1 & 2 \end{bmatrix}$, *(IETE Dec. 2004)*, (ii) $\begin{bmatrix} -5 & 9 \\ -6 & 10 \end{bmatrix}$, (iii) $\begin{bmatrix} -1 & 1 & 2 \\ 0 & -2 & 1 \\ 0 & 0 & -3 \end{bmatrix}$,

(iv) $\begin{bmatrix} 5 & 7 & -5 \\ 10 & 4 & -1 \\ 2 & 8 & -3 \end{bmatrix}$, (v) $\begin{bmatrix} 0 & 2 & 1 \\ -4 & 6 & 1 \\ -1 & -2 & -2 \end{bmatrix}$, (vi) $\begin{bmatrix} 5 & -6 & -6 \\ -1 & 4 & 2 \\ 3 & -6 & -4 \end{bmatrix}$,

(vii) $\begin{bmatrix} 3 & 1 & -1 \\ -2 & 1 & 2 \\ 0 & 1 & 2 \end{bmatrix}$ *(IETE, Dec. 2007)*, (viii) $\begin{bmatrix} 3 & 2 & 1 \\ 0 & 2 & 0 \\ 1 & 2 & 3 \end{bmatrix}$ *(IETE June 2006)*,

(ix) $\begin{bmatrix} -1 & 2 & -2 \\ 1 & 2 & 1 \\ -1 & -1 & 0 \end{bmatrix}$ *(IETE, June 2001)*

Ans. (i) $\lambda = 6$: $[4, 1]^T$; $\lambda = 1$: $[1, -1]^T$. $P = \begin{bmatrix} 4 & 1 \\ 1 & -1 \end{bmatrix}$; $P^{-1} = \dfrac{1}{-5}\begin{bmatrix} -1 & -1 \\ -1 & 4 \end{bmatrix}$

(ii) $\lambda = 1$: $[3, 2]^T$; $\lambda = 4$: $[1, 1]^T$ $P = \begin{bmatrix} 3 & 1 \\ 2 & 1 \end{bmatrix}$; $P^{-1} = \begin{bmatrix} 1 & -1 \\ -2 & 3 \end{bmatrix}$

(iii) $\lambda = -1$: $[1, 0, 0,]^T$; $\lambda = -2$: $[1, -1, 0]^T$; $\lambda = -3$: $[1, 2, -2]^T$.

$$P = \begin{bmatrix} 1 & 1 & 1 \\ 0 & -1 & 2 \\ 0 & 0 & -2 \end{bmatrix}; \quad P^{-1} = \frac{1}{2}\begin{bmatrix} 2 & 2 & 3 \\ 0 & -2 & -2 \\ 0 & 0 & -1 \end{bmatrix},$$

(iv) $\lambda = 1$: $[2, 1, 3]^T$; $\lambda = 2$: $[1, 1, 2]^T$; $\lambda = 3$: $[-1, 1, 1]^T$.

$$P = \begin{bmatrix} 2 & 1 & -1 \\ 1 & 1 & 1 \\ 3 & 2 & 1 \end{bmatrix}; \quad P^{-1} = \begin{bmatrix} -1 & -3 & 2 \\ 2 & 5 & -3 \\ -1 & -1 & 1 \end{bmatrix}.$$

(v) $\lambda = 4$: $[2, 5, -2]^T$; $\lambda = -1$: $[1, 1, -3]^T$; $\lambda = 1$: $[1, 1, -1]^T$.

$$P = \begin{bmatrix} 2 & 1 & 1 \\ -5 & 1 & 1 \\ -2 & -3 & -1 \end{bmatrix}; \quad P^{-1} = 1/6\begin{bmatrix} -2 & 2 & 0 \\ -3 & 0 & -3 \\ 13 & -4 & 3 \end{bmatrix}.$$

(vi) $\lambda = 1$: $[3, -1, 3]^T$; $\lambda = 2, 2$: $[2, 1, 0]^T, [2, 0, 1]^T$.

$$P = \begin{bmatrix} 3 & 2 & 2 \\ -1 & 1 & 0 \\ 3 & 0 & 1 \end{bmatrix}; \quad P^{-1} = \begin{bmatrix} -1 & 2 & 2 \\ -1 & 3 & 2 \\ 3 & -6 & -5 \end{bmatrix}.$$

(vii) $\lambda = 1$: $[1, -1, 1]^T$; $\lambda = 2$: $[1, 0, 1]^T$; $\lambda = 3$: $[0, 1, 1]^T$

$$P = \begin{bmatrix} 1 & 1 & 0 \\ -1 & 0 & 1 \\ 1 & 1 & 1 \end{bmatrix}; \quad P^{-1} = \begin{bmatrix} -1 & -1 & 1 \\ 2 & 1 & -1 \\ -1 & 0 & 1 \end{bmatrix}.$$

(viii) $\lambda = 2, 2; [1, 0, -1]^T, [-2, 1, 0]^T; \lambda = 4: [1, 0, 1]^T$.

$$P = \begin{bmatrix} 1 & -2 & 1 \\ 0 & 1 & 0 \\ -1 & 0 & 1 \end{bmatrix}; \quad P^{-1} = 1/2\begin{bmatrix} 1 & 2 & -1 \\ 0 & 2 & 0 \\ 1 & 2 & 1 \end{bmatrix}.$$

(ix) $\lambda = 1: [1, 0, -1]^T; \lambda = \sqrt{5}: [\sqrt{5}-1, 1, -1]^T; \lambda = -\sqrt{5}: [\sqrt{5}+1, -1, 1]^T$

$$P = \begin{bmatrix} 1 & \sqrt{5}-1 & \sqrt{5}+1 \\ 0 & 1 & -1 \\ -1 & -1 & 1 \end{bmatrix}; P^{-1} = \begin{bmatrix} 0 & -1 & -1 \\ 1/2\sqrt{5} & (2+\sqrt{5})/2\sqrt{5} & 1/2\sqrt{5} \\ 1/2\sqrt{5} & (2-\sqrt{5})/2\sqrt{5} & 1/2\sqrt{5} \end{bmatrix}$$

19. Find the matrix **A** whose eigenvalues and the corresponding eigenvectors are as given below:

   (*i*) Eigenvalues: 1, 1, 3; Eigenvectors: $[1, 0, -1]^T$, $[0, 1, -1]^T$, $[1, 1, 0]^T$.

(*IETE, June 2008*)

   (*ii*) Eigenvalues: 0, -4, 3; Eigenvectors: $[1, 6, -13]^T$, $[-1, 2, 1]^T$, $[2, 3, -2]$.

   (*iii*) Eigenvalues: 1, 2, 3; Eigenvectors: $[-1, -1, 1]^T$, $[0, 1, 0]^T$, $[0, -1, 1]^T$.

(*A.M.I.E., S-2009*)

Ans. (*i*) $P = \begin{bmatrix} 1 & 0 & 1 \\ 0 & 1 & 1 \\ -1 & -1 & 0 \end{bmatrix}$; $P^{-1} = \dfrac{1}{2}\begin{bmatrix} 1 & -1 & -1 \\ -1 & 1 & -1 \\ 1 & 1 & 1 \end{bmatrix}$; $A = PDP^{-1} = \begin{bmatrix} 2 & 1 & 1 \\ 1 & 2 & 1 \\ 0 & 0 & 1 \end{bmatrix}$

(*ii*) $P = \begin{bmatrix} 1 & -1 & 2 \\ 6 & 2 & 3 \\ -13 & 1 & -2 \end{bmatrix}$; $P^{-1} = \dfrac{1}{84}\begin{bmatrix} -7 & 0 & -7 \\ -27 & 24 & 9 \\ 32 & 12 & 8 \end{bmatrix}$; $A = PDP^{-1} = \begin{bmatrix} 1 & 2 & 1 \\ 6 & -1 & 0 \\ -1 & -2 & -1 \end{bmatrix}$.

(*iii*) $A = \begin{bmatrix} 1 & 0 & 0 \\ 0 & 2 & 1 \\ 2 & 0 & 3 \end{bmatrix}$.

20. If λ be an eigen value of a non-singular matrix **A**, show that $|A|/\lambda$ is an eigen value of the matrix adj $\cdot$ **A**. (*PTU 2003; UPTU 2001*).

21. Verify that $x_1 = [4, 1, -1]^T$, $x_2 = [1, 0, 4]^T$ and $x_3 = [1, -4, 0]^T$ are eigenvectors for the symmetric matrix

$$A = \begin{bmatrix} 7 & 4 & -4 \\ 4 & -8 & -1 \\ -4 & -1 & -8 \end{bmatrix}$$ corresponding to the eigenvalues $\lambda_1 = 9$ and

$\lambda_2 = \lambda_3 = -9$, respectively.

22. Find A^n in the following cases:

   (*i*) $A = \begin{pmatrix} -3 & 10 \\ -3 & 8 \end{pmatrix}$; (*ii*) $A = \begin{pmatrix} -13 & 6 \\ -35 & 16 \end{pmatrix}$.

Ans. (*i*) $\begin{bmatrix} 3.2^{n+1} - 5.3^n & -5.2^{n+1} + 10.3^n \\ 3.2^n - 3^{n+1} & -5.2^n + 2.3^{n+1} \end{bmatrix}$; (*ii*) $\begin{bmatrix} -7.2^{n+1} + 15 & 3.2^{n+1} - 6 \\ -35.2^n + 35 & 15.2^n - 14 \end{bmatrix}$.

23. Show that the transformation

$$y_1 = x_1 \cos\theta + x_2 \sin\theta$$
$$y_2 = -x_1 \sin\theta + x_2 \cos\theta$$

is orthogonal.

[**Hint:** The given transformation can be written as

$$\begin{pmatrix} y_1 \\ y_2 \end{pmatrix} = \begin{pmatrix} \cos\theta & \sin\theta \\ -\sin\theta & \cos\theta \end{pmatrix} \begin{pmatrix} x_1 \\ x_2 \end{pmatrix}.$$

Here the matrix of transformation $A = \begin{pmatrix} \cos\theta & \sin\theta \\ -\sin\theta & \cos\theta \end{pmatrix}$.

Since $A^{-1} = A^T$, the transformation in orthogonal.]

24. Determine the values of α, β, γ when

$$A = \begin{bmatrix} 0 & 2\beta & r \\ \alpha & \beta & -r \\ \alpha & -\beta & r \end{bmatrix}$$ is orthogonal. **Ans.** $\alpha = \pm 1/\sqrt{2}, \beta = \pm 1/\sqrt{6}, \gamma = \pm 1/\sqrt{3}$.

25. Prove that the inverse of an orthogonal matrix is orthogonal and its transpose is also orthogonal.

26. Show that the following matrices are orthogonal

(i) $A = \begin{bmatrix} \cos\theta & -\sin\theta & 0 \\ \sin\theta & \cos\theta & 0 \\ 0 & 0 & 1 \end{bmatrix}$, (ii) $B = \dfrac{1}{9}\begin{bmatrix} -8 & 4 & 1 \\ 1 & 4 & -8 \\ 4 & 7 & 4 \end{bmatrix}$ (*IETE, W-2008*)

(iii) $A = \dfrac{1}{3}\begin{bmatrix} 2 & 2 & 1 \\ -2 & 1 & 2 \\ 1 & -2 & 2 \end{bmatrix}$, (iv) $B = \dfrac{1}{7}\begin{bmatrix} 6 & -3 & 2 \\ -3 & -2 & 6 \\ 2 & 6 & 3 \end{bmatrix}$

27. Let $A = \begin{bmatrix} 1 & 0 \\ -1 & 1 \end{bmatrix}$. Prove that $A^2 = 2A - I$ and compute A^{100}.

Ans. $\begin{bmatrix} 1 & 0 \\ -100 & 1 \end{bmatrix}$.

28. If $A = \begin{bmatrix} 1 & 2 \\ 5 & 4 \end{bmatrix}$, find a non singular matrix

P such that $P^{-1}AP = \begin{bmatrix} 6 & 0 \\ 0 & -1 \end{bmatrix}$. **Ans.** $P = \begin{bmatrix} 2 & 1 \\ 5 & -1 \end{bmatrix}$.

29. Two eigenvalues of the matrix $A = \begin{bmatrix} 2 & 2 & 1 \\ 1 & 3 & 1 \\ 1 & 2 & 2 \end{bmatrix}$ are equal to 1 each. Find the

eigenvalues of A^{-1}. (*A.M.I.E., S-2009*) **Ans.** $1, 1, 1/5$.

30. Find the eigen values and corresponding eigen vectors of the matrix

$$A = \begin{bmatrix} 1 & 0 & 0 \\ 0 & 2 & 1 \\ 2 & 0 & 3 \end{bmatrix}.$$

(A.M.I.E., W-2010)

Ans. 1, 2, 3, $[-1, -1, 1]^T$, $[0, 1, 0]^T$, $[0, -1, 1]^T$.

31. Verify whether the matrix $A = \begin{bmatrix} 2 & 1 \\ 1 & 0 \end{bmatrix}$ is diagonalizable.

(A.M.I.E., W-2010)

Hint. The characteristic equation of **A** is given by

$$|A - \lambda I| = \begin{vmatrix} 2 - \lambda & 1 \\ 1 & -\lambda \end{vmatrix} = 0 \text{ or } \lambda^2 - 2\lambda - 1 = 0 \text{ or } \lambda = 1 \pm \sqrt{2}.$$

Since the matrix **A** has two distinct eigen values, it has two linearly independent eigen vectors and hence it is diagonalizable.

Corresponding to the eigen value $\lambda = 1 + \sqrt{2}$, we have eigen vector $[1 + \sqrt{2}, 1]^T$.

Corresponding to the eigen value $\lambda = 1 - \sqrt{2}$, we have eigen vector $[1 - \sqrt{2}, 1]^T$.

The model matrix **P** is given by

$$P = \begin{bmatrix} 1 + \sqrt{2} & 1 - \sqrt{2} \\ 1 & 1 \end{bmatrix} \text{ and } P^{-1} = \frac{1}{2\sqrt{2}} \begin{bmatrix} 1 & \sqrt{2} - 1 \\ -1 & 1 + \sqrt{2} \end{bmatrix}.$$

If can be verified that

$$P^{-1}AP = \begin{bmatrix} 1 + \sqrt{2} & 0 \\ 0 & 1 - \sqrt{2} \end{bmatrix} = \text{diag} (1 + \sqrt{2}, 1 - \sqrt{2})].$$

Chapter 17

Ordinary Differential Equations of First Order

17.1 Introduction

The study of differential equations is of paramount importance in engineering mathematics. This is primarily due to the reason that many physical laws and relations are expressible mathematically in the form of differential equations. It is from these equations that we obtain relationship among the variables of the problems. The interpretation and analysis of the results provides us with the information concerning the manner in which the physical quantities involved depend upon one another. Before we classify and examine different kinds of differential equations, let us consider the following differential equations:

(i) $\dfrac{dy}{dx} + 2y + x = 0$ (ii) $(2x + y)dx + (\sin y + x)dy = 0$

(iii) $\dfrac{d^2 y}{dx^2} + c^2 y = 0$ (iv) $\rho = \left\{ 1 + \left(\dfrac{dy}{dx}\right)^2 \right\}^{3/2} \Big/ \dfrac{d^2 y}{dx^2}$

(v) $\dfrac{d^3 y}{dx^3} + 5\dfrac{d^2 y}{dx^2} + 5\dfrac{dy}{dx} + y = 0$ (vi) $\dfrac{\partial^2 u}{\partial t^2} = c^2 \dfrac{\partial^2 u}{\partial x^2}$

(vii) $x\dfrac{\partial u}{\partial x} + y\dfrac{\partial u}{\partial y} = nu$ (viii) $\dfrac{dx}{dt} + wy = a\cos pt, \dfrac{dy}{dt} + wx = b\sin pt$

From the above equations, it is clear that all of them contain either the differential coefficients or differentials. A *differential equation* can be defined as an equation containing derivatives of various orders and the variables. Symbolically a differential equation may be written as follows:

$$f\left(x, y, y', y'', \ldots, y^{(n)}\right) = 0 \text{ or } f\left(x, y, \dfrac{dy}{dx}, \dfrac{d^2 y}{dx^2}, \ldots, \dfrac{d^n y}{dx^n}\right) = 0.$$

Differential equations which involve one independent variable are called *ordinary differential equations*.

Equations (*i*) to (*v*) are all examples of ordinary differential equations.

If the differential equation involves more than one independent variable and partial derivatives of the dependent variable with respect to them, then it is called a *partial differential equation*.

Equations (*vi*) and (*vii*) are partial differential equations.

The differential equations which have two or more dependent variables and a single independent variable are called *simultaneous differential equations*. Equations (*viii*) is an example of simultaneous equations.

Other Definitions

(1) **Order:** The order of a differential equation is the order of the highest order derivative occurring in the equation. Equations (*i*) and (*ii*) are of the first order while equation (*iii*) is of the second order and equation (*v*) is of the three order.

(2) **Degree:** The degree of a differential equation is the degree (or power) of the highest order derivative occurring in the equation after the equations has been made free from fractions and the radicals as far as derivatives are concerned. Equations (*i*) and (*ii*) are ordinary, first order and first degree equations; equation (*iii*) is an ordinary, second order and first degree; equation (*v*) is an ordinary, third order and first degree; equation (*iv*) is an ordinary, second order and second degree as it can be written as:

$$\rho^2 \left(\frac{d^2 y}{dx^2} \right)^2 = \left\{ 1 + \left(\frac{dy}{dx} \right)^2 \right\}^3 .$$

Similarly equation (*vi*) is a second order, first degree partial differential equation and so on.

(3) **Linear and Non-linear differential equation**

Linear: A differential equation is linear, when the dependent variable and all its derivatives are of the first degree (that is, the power of each term involving y is 1) and no products of the dependent variable and its derivatives or of various order derivatives occur. The form of a linear ordinary differential equation, in general, is

$$c_0 y^{(n)} + c_1 y^{(n-1)} + c_2 y^{(n-2)} + \ldots + c_{n-1} y' + c_n y = g(x),$$

where $c_i = c_i(x)$ are some functions of x or are constants.
The following equations

$$y' = 4x^2, \quad y'' + 3y' + 4y = 2x, \quad x^2 \frac{d^2 y}{dx^2} + x \frac{dy}{dx} + 3y = \log x$$

and $y''' + 6y'' + 3y' + y = 0$ are some examples of linear differential equations.

Non-linear: A differential equation which is not linear, is called a non-linear differential equation. Some examples of non-linear ordinary differential equations are given below.

$$(y')^2 + 3xy' + y = 0, \left[1 + (y')^2\right]^{1/2} = 8y, \ y'' + 4yy' + 2y = \cos x, \ y'' + \sin y = 0.$$

Example 17.1 Find the order and the degree of the following differential equations. State also whether they are linear or non-linear.

(a) $y'' + 4xy' + 2y = \cos x.$ (b) $y''' + xy'' + 2y \ (y')^2 + xy = 0.$

(c) $y'' + y' + \cos y = 0.$ (d) $\dfrac{d(xy')}{dx} + x^2 y = 0.$

(e) $y' = \sqrt{x} + \sqrt{y}.$ (f) $y'y'' + y' + 5y = \sin x.$ (g) $y + \dfrac{dy}{dx} = \dfrac{1}{2} \int y \, dx.$

Ans. (a) Order two; degree one; linear.

(b) Order three; degree one; non-linear (because of the presence of first derivative having degree two).

(c) Order two; degree one; non-linear (because of the presence of cos y, which is a non-linear function of y).

(d) $\dfrac{d(xy')}{dx} + x^2 y = 0 \Rightarrow xy'' + y' + x^2 y = 0.$

Order two; degree one; linear.

(e) Order one; degree one; non-linear.

(f) Order two; degree one; non-linear (because of the occurrence of the product of first and second order derivatives).

(g) Differentiating w.r.t. x, we get $dy/dx + d^2y/dx^2 = y/2.$

Order two; degree one; linear.

There is no discipline in engineering and other branches of science where differential equations are not finding place. Problems dealing with the vibration of strings, drying of wood and baking of bread by the simultaneous heat and mass transfer, conduction of heat in solids, vibrations of electrical and mechanical systems, bending of beams and velocity of chemical reactions are a few of the many areas where differential equations are extensively used. It is therefore, imperative for an engineering student to know the differential equations in depth and in detail. However, an engineers' approach to the study of differential equations is different from that of a student of mathematics. An engineer is not much concerned with the mathematical rigor of the solution but views the differential equation as a tool by which he derives the solution to authenticate the practical results. The study of differential equations from an engineer's view point can thus be classified into the following stages.

(a) Setting up of the differential equation from the given physical problem.

(b) To obtain the solution and evaluate the arbitrary constants with the help of the initial and boundary conditions.

(c) To analyse and interpret the results from the mathematical solution.

17.2 Formation of Differential Equations

An ordinary differential equation is formed by the elimination of arbitrary constants which occur in the relationship between the variables.

Let y and x be the dependent and the independent variables respectively. The equation
$$f(x, y, c) = 0 \qquad \qquad ...(17.1)$$
containing one arbitrary constant c, represents a family of curves. The equation
$$\phi(x, y, c, d) = 0 \qquad \qquad ...(17.2)$$
containing two arbitrary constants c and d also represents a family of curves. We often say that it represents a two parameter family of curves.

To eliminate the arbitrary constant c in equation (17.1), we need two equations. One equation is given by (17.1) itself and the second equation is obtained by differentiating equation (17.1) with respect to x. On eliminating c from the two equations, we obtain an equation containing x, y and y' which is a first order differential equation.

Similarly, to eliminate the arbitrary constants c and d in equation (17.2), we need three equations. One equation is given by equation (17.2), and the remaining two equations are obtained by differentiating equation (17.2) with respect to x two times.

On eliminating c and d from these three equations, we obtain a second order differential equation.

Example 17.2 **Form the differential equation not containing a, from the relation $y^2 = 4\,ax$.**

Solution. We have $y^2 = 4ax$. $\qquad \qquad (i)$

Differentiating (i) w.r.t. x, we get $2y\left(\dfrac{dy}{dx}\right) = 4a = \dfrac{y^2}{x}$. $\qquad$ [using (i)].

Therefore, since $y \neq 0$, required equation is $\dfrac{dy}{dx} = \dfrac{y}{2x}$. $\qquad \qquad ...(ii)$

Equation (i) defines a family of parabolas for various values of a, while equation (ii) gives the property of these parabolas, namely, that the slope of the tangent at any point to (i) is the ratio of the ordinate to twice the abscissa of that point.

Example 17.3 **Find the differential equation of the family of curves given by the equation $x^2 - y^2 + 2axy = 1$, where a is a parameter.**

Solution. Differentiating the equation of the curve w.r.t. x, we obtain

$$2x - 2y\frac{dy}{dx} + 2a\left(y + x\frac{dy}{dx}\right) = 0. \qquad \qquad ...(i)$$

Substituting for 2*a* from the original equation in (*i*), we get

$$2\left(x-y\frac{dy}{dx}\right)+\left(\frac{1-x^2+y^2}{xy}\right)\left(y+x\frac{dy}{dx}\right)=0$$

or
$$(x^2+y^2)\left(y-x\frac{dy}{dx}\right)+\left(x\frac{dy}{dx}+y\right)=0. \qquad\qquad ...(ii)$$

Relation (*ii*) is the required differential equation.

Example 17.4 Form the differential equation not involving A, B, C from the equation 2x² + 3y² + Ax + By + C = 0.

Solution. This equation has three arbitrary constants. Differentiating successively three times w.r.t. *x*, we obtain

$$4x+6y\frac{dy}{dx}+A+B\frac{dy}{dx}=0 \;\;...(i), \;\; 4+6y\frac{d^2y}{dx^2}+6\left(\frac{dy}{dx}\right)^2+B\frac{d^2y}{dx^2}=0. \qquad ...(ii)$$

$$6y\frac{d^3y}{dx^3}+6\left(\frac{dy}{dx}\right)\frac{d^2y}{dx^2}+12\left(\frac{dy}{dx}\right)\left(\frac{d^2y}{dx^2}\right)+B\frac{d^3y}{dx^3}=0. \qquad\qquad ...(iii)$$

Eliminating (B) from (*ii*) and (*iii*), the eliminant is given by

$$\frac{d^3y}{dx^3}\left\{2+3\left(\frac{dy}{dx}\right)^2\right\}=9\frac{dy}{dx}\left(\frac{d^2y}{dx^2}\right)^2.$$

17.3 Solution of a Differential Equation

By the solution of the differential equation is meant the relation between the dependent and the independent variables not involving derivatives. We have seen in the examples of Art. 17.2 that when we eliminate arbitrary constants from a given relation, the order of the resulting differential equation is the same as the number of arbitrary constants. It is therefore, logical that the *solution(integral)* of the differential equation must contain, in general, the same number of arbitrary constants as the order of the equation. Such a solution is known as the *general solution* or the *complete solution* of differential equation. The general solution is also called the *complete integral* or *complete primitive* of the differential equation. For example *y* = A cos α*t* + B sin α*t*

is the complete solution of the differential equation $\dfrac{d^2y}{dt^2}+\alpha^2 y=0$ because the

number of arbitrary constants A and B corresponds exactly with the order of the equation. It is important to note that the general solution of a differential equation can be expressed in different (but equivalent) forms.

A *particular solution* is one which is obtained from the general solution by assigning particular values (definite values) to the arbitrary constants. Thus

$y^2 = 4x$ is a particular solution of $dy/dx = 2/y$ as it can be derived from the general solution $y^2 = 4Ax$ by assuming that the solution curve passes through the point (1, 2).

A differential equation may sometimes have an additional solution that cannot be obtained from the general solution (by assigning specific values to its arbitrary constant) and is then called a **singular solution**. This is not of great engineering interest.

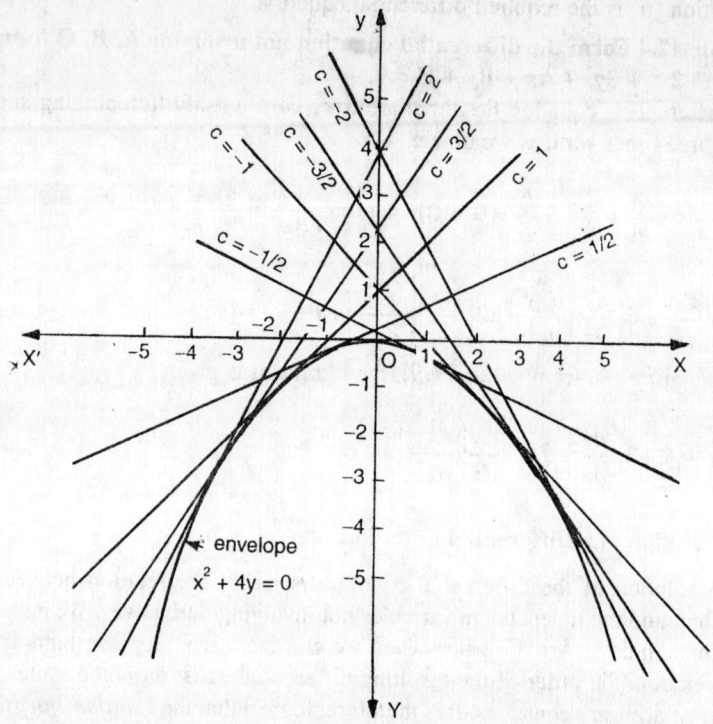

Fig. 17.1 Singular solution of $y'^2 + xy' - y = 0$.

For example the differential equation $y'^2 + xy' - y = 0$ has the general solution $y = cx + c^2$, which may be verified by differentiation and substitution. This represents a family of straight lines, one straight line for each value of c. These are the particular solutions shown in Fig. 17.1. However, we find that $x^2 = -4y$, which is the equation of the parabola, is also a solution of the differential equation. This is a singular solution of above differential equation because we cannot find it from $y = cx + c^2$ by choosing a suitable value of c. It may be noted that this singular solution is the envelope of the family of the straight lines represented by the general solution.

A differential equation together with the condition is called an *initial value problem* (I.V.P.) and condition is called the *initial condition*. It may be noted that initial condition for example $y = \pi/4$ when $x = \sqrt{2}$ may be written as $y(\sqrt{2}) = \pi/4$. The given initial condition is used to determine a value of arbitrary constant in the general solution and hence to find a particular solution.

Example 17.5 Verify that $y = Ae^{3x} + Bxe^{3x}$ is a general solution of the differential equation $y'' - 6y' + 9y = 0$.

Solution. We have $y = Ae^{3x} + Bxe^{3x}$.

Therefore, $\qquad y' = 3Ae^{3x} + Be^{3x} + 3Bxe^{3x}, \quad y'' = 9Ae^{3x} + 6Be^{3x} + 9Bxe^{3x}$.

Then $y'' - 6y' + 9y = \left(9Ae^{3x} + 6Be^{3x} + 9Bxe^{3x}\right) - 6\left(3Ae^{3x} + Be^{3x} + 3Bxe^{3x}\right)$

$$+ 9\left(Ae^{3x} + Bxe^{3x}\right) = 0,$$

and so that the given relation is a general solution since the order of the differential equation and the number of arbitrary constants in the solution are both equal to 2.

Example 17.6 Show that $y = c_1 e^x + c_2 e^{-3x} + \sin x$ is a general solution of the differential equation $y'' + 2y' - 3y = 2\cos x - 4\sin x$. Also find the particular solution of the differential equation satisfying the conditions

$$y(0) = 2, \; y'(0) = -5.$$

Solution. We have $y = c_1 e^x + c_2 e^{-3x} + \sin x$.

$\therefore \quad y' = c_1 e^x - 3c_2 e^{-3x} + \cos x, \; y'' = c_1 e^x + 9c_2 e^{-3x} - \sin x$.

Substitute y, y', y'' in the differential equation, we obtain

$$y'' + 2y' - 3y = \left(c_1 e^x + 9c_2 e^{-3x} - \sin x\right) + 2\left(c_1 e^x - 3c_2 e^{-3x} + \cos x\right)$$

$$-3\left(c_1 e^x + c_2 e^{-3x} + \sin x\right) = 2\cos x - 4\sin x$$

and so the given relation is a general solution since the order of the differential equation and the number of arbitrary constants in the solution are both equal to 2.

The general solution is $y = y(x) = c_1 e^x + c_2 e^{-3x} + \sin x$. $\qquad$ ...(*i*)

At $x = 0$, $\qquad\qquad y(0) = c_1 + c_2 = 2$ that is, $c_1 + c_2 = 2$.

Differentiating (*i*) w.r.t. x, we obtain $y'(x) = c_1 e^x - 3c_2 e^{-3x} + \cos x$

At $x = 0$, $\qquad\qquad y'(0) = c_1 - 3c_2 + 1 = -5$ that is, $c_1 - 3c_2 = -6$. $\qquad$ ...(*ii*)

Solving (*i*) and (*ii*) simultaneously, we find $c_1 = 0, c_2 = 2$. The required particular solution is $y_p(x) = 2e^{-3x} + \sin x$.

Exercise 17.1

Describe each of the following differential equations giving its order and degree. State also whether they are linear or non-linear.

1. $y''' + 6y'' + 4y' + y = e^x$ **2.** $y'' + 4yy' + 2y = \cos x$

3. $(y')^2 + 3xy' + 6y = 0$ **4.** $L\dfrac{d^2Q}{dt^2} + R\dfrac{dQ}{dt} + \dfrac{Q}{C} = 0$

5. $(y'')^2 + (y')^3 + 3y = x^2$ **6.** $y''' + 2(y'')^2 + y' = \cos x$

7. $(dy/dx)^2 + 5y^{1/3} = x$ **8.** $(dy/dx)^{1/2} + y = x^{1/3}$

Ans. 1. Three, one, linear.   2. Two, one, non-linear.
3. One, two; non-linear.   4. Two, one, linear
5. Two, two, non-linear.   6. Three, one, non-linear.
7. One, two, non-linear.   8. One, one, non-linear.

Form the differential equations from the following equations:

9. $x^2 + y^2 + 2gx + 2fy + c = 0$, not containing f, g, c.

$$\textbf{Ans.}\ \left\{1 + \left(\frac{dy}{dx}\right)^2\right\}\frac{d^3y}{dx^3} - 3\frac{dy}{dx}\left(\frac{d^2y}{dx^2}\right)^2 = 0.$$

10. $y^2 = m(a^2 - x^2)$, not containing m and a

$$\textbf{Ans.}\ x\left\{(dy/dx)^2 + y(d^2y/dx^2)\right\} - y(dy/dx) = 0.$$

11. $x^2 + y^2 - 2ky = k^2$, not containing k

$$\textbf{Ans.}\ (x^2 - 2y^2)(dy/dx)^2 - 4xy(dy/dx) - x^2 = 0.$$

12. $x = A\sin(\omega t + \alpha)$, not containing A and α $\textbf{Ans.}\ \dfrac{d^2x}{dt^2} + \omega^2 x = 0.$

13. $(x - h)^2 + (y - k)^2 = a^2$, not containing h and k
Alternative form: Obtain the differential equation of all circles of radius 'a' and centre at (h, k).   *(Andhra, 1999; Mysore 1997)*

$$\textbf{Ans.}\ (1 + y'^2)^3 = a^2 y''^2.$$

14. $y = ae^{2x} + be^{-3x} + ce^x$, not containing a, b, and c.   *(Karnatake, 1995)*

$$\textbf{Ans.}\ (d^3y/dx^3) - 7(dy/dx) + 6y = 0.$$

15. $xy = Ae^x + Be^{-x} + x^2$, not containing A and B.   *(A.M.I.E., S-2005)*
$$\textbf{Ans.}\ xy'' + 2y' + x^2 - xy - 2 = 0$$

16. $y = e^x(A\cos x + B\sin x)$, not containing A and B.
(PTU 2003; Andhra, 1998) **Ans.** $y^2 - 2y' + 2y = 0.$

17. Find the differential equation of the family of circles of fixed radius r with centres on the x-axis. $\textbf{Ans.}\ (y')^2 y^2 + y^2 = r^2.$

18. Form the differential equation of all parabolas with principal axis along the x-axis.
    Ans. $yy'' + y'^2 = 0$.

19. Form the differential equation of all parabolas whose axes are parallel to y-axis.
    Ans. $d^3y/dx^3 = 0$.

20. Find the differential equation of the family of parabolas with foci at the origin and axes along the x-axis.
    Ans. $yy'^2 + 2xy' - y = 0$.

21. Find the differential equation of all straight lines at a unit distance from the origin.
    Ans. $(xy' - y)^2 = 1 + (y')^2$

22. Find the differential equation of the family of cardiods $r = a(1 - \cos\theta)$.
    Ans. $(1 - \cos\theta)dr = r\sin\theta\, d\theta$.

23. Verify that $y = ce^{-x} + x^2 - 2x$ is solution of the differential equation $dy/dx + y = x^2 - 2$.

24. Show that $y = e^{-x}(a\cos x + b\sin x)$ is the solution of the differential equation $y'' + 2y' + 2y = 0$.

25. Show that $y = 2(x^2 - 1) + ce^{-x^2}$ is a solution of $y' + 2xy = 4x^3$.

26. Show that $Ax^2 + By^2 = 1$ is a solution of the differential equation
    $$x(yy'' + y'^2) = yy'.$$

27. Show that $y = a\cos(\log x) + b\sin(\log x)$ is a solution of the differential equation $x^2 y'' + xy' + y = 0$.
    Ans. $\left(1 + y'^2\right)^3 = a^2 y''^2$.

17.4 Analytical Solution of a Differential Equation of the first Order and first Degree

A differential equation of the first order and fist degree can be written as

$$f\left(x, y, \frac{dy}{dx}\right) = 0 \quad \text{or} \quad \frac{dy}{dx} = \phi(x, y) \quad \text{or} \quad M(x, y)dx + N(x, y)dy = 0 \quad ...(17.3)$$

It may be remarked that it is not always possible to solve every first order differential equation in a closed form. We will discuss the solutions of a few types of equations which can be classified into any one of the following forms.
(a) Equations in the variables separable form.
(b) Equations which are homogeneous.
(c) Equations which are linear.
(d) Equations which are exact.

17.5 Variables Separable Form

In this case differential equation can be written in the form $M(x)\, dx + N(y)\, dy = 0$ that is x appears only in the co-efficient of dx and y appears only in the co-efficient of dy.

Integrating, we have the general solution.

$$\int M(x)\,dx + \int N(y)\,dy = c, \text{ a constant.}$$

Example 17.6 Solve $\dfrac{dy}{dx} + \sqrt{\dfrac{1-y^2}{1-x^2}} = 0.$

Solution. The equation can be written in the form

$$\dfrac{dx}{\sqrt{1-x^2}} + \dfrac{dy}{\sqrt{1-y^2}} = 0. \text{ Integrating, we get } \sin^{-1} x + \sin^{-1} y = c_1 \qquad ...(1)$$

or $\sin^{-1}\left\{ x\sqrt{1-y^2} + y\sqrt{1-x^2} \right\} = c_1$

$$\left[\because \sin^{-1} A + \sin^{-1} B = \sin^{-1}\left(A\sqrt{1-B^2} + B\sqrt{1-A^2} \right) \right]$$

or $x\sqrt{1-y^2} + y\sqrt{1-x^2} = c$, where $c = \sin c_1$ is an arbitrary constant.

$\therefore \quad x\sqrt{1-y^2} + y\sqrt{1-x^2} = c$, is the required solution.

Example 17.7 Solve the initial value problem $x\dfrac{dy}{dx} + \cot y = 0, \ y\left(\sqrt{2}\right) = \dfrac{\pi}{4}.$

Solution. The equation can be written as $\dfrac{\sin y}{\cos y}\,dy + \dfrac{dx}{x} = 0.$ Integrating, we ob-

tain $-\log |(\cos y)| + \log |x| = c_1$ or $\log\left(\dfrac{x}{\cos y}\right) = c_1$ or $\dfrac{x}{\cos y} = c$, where $c = e^{c_1}.$

$\therefore \ x = c \cos y$ is the general solution of the given differential equation.

Using the given conditions $y = \pi/4$, when $x = \sqrt{2}$, we get $c = 2$.

The particular solution of initial value problem is $x = 2 \cos y$.

Example 17.8 Solve the differential equation

$3e^{2x} \sec^2 3y\,dy + 2\,(e^{2x} - 1)\tan 3y\,dx = 0.$

Solution. The given equation may be written as $\dfrac{3\sec^2 3y}{\tan 3y}\,dy + \dfrac{2\left(e^{2x}-1\right)}{e^{2x}}\,dx = 0.$

Integrating, we obtain $\log|\tan 3y| + 2\left(x + \dfrac{e^{-2x}}{2} \right) = c$

or $\log |\tan 3y| + 2x + e^{-2x} = c$, the required solution.

17.5.1 Equations Reducible to Separable Form

The following form of the differential equation can be reduced to separable form by substitution.

Equation of the form $\dfrac{dy}{dx} = f(ax + bx + c)$

Substituting $ax + by + c = u$, we get $a + b\dfrac{dy}{dx} = \dfrac{du}{dx}$ or $\dfrac{dy}{dx} = \dfrac{1}{b}\left(\dfrac{du}{dx} - a\right)$.

The above differential equation simplifies to

$$\frac{1}{b}\left(\frac{du}{dx} - a\right) = f(u) \text{ or } \frac{du}{dx} = a + bf(u) \text{ or } \frac{du}{a + bf(u)} = dx.$$

Integrating, we obtain $\displaystyle\int \frac{du}{a + bf(u)} = x + c.$

Example 17.9 Solve the initial-value problem $\dfrac{dy}{dx} = (y - 2x)^2 - 7,\ y(0) = 0.$

Solution. Let $u = y - 2x$, then $\dfrac{du}{dx} = \left(\dfrac{dy}{dx}\right) - 2$ so the given differential equation is

transformed into $\dfrac{du}{dx} + 2 = u^2 - 7$ or $\dfrac{du}{dx} = u^2 - 9$ or $\dfrac{du}{(u+3)(u-3)} = dx.$ Using

partial fractions,

$\dfrac{1}{6}\left[\dfrac{1}{u-3} - \dfrac{1}{u+3}\right] du = dx.$ Integrating, we obtain

$\dfrac{1}{6}\log\left|\dfrac{u-3}{u+3}\right| = x + c_1$ or $\dfrac{u-3}{u+3} = e^{6x+6c_1} = ce^{6x}$ or $u = \dfrac{3(1+ce^{6x})}{1-ce^{6x}}.$

Resubstituting $u = y - 2x$, we get

$y = 2x + \dfrac{3(1+ce^{6x})}{1-ce^{6x}}$. Now applying the initial condition $y(0) = 0$, we get

$c = -1$. Therefore, the particular solution is $y = 2x + 3\left\{(1 - e^{6x})/(1 + e^{6x})\right\}.$

Example 17.10 Solve $x^4 \dfrac{dy}{dx} + x^3 y + \operatorname{cosec}(xy) = 0.$

Solution. The given differential equation can be written as

$$x^3\left(x\frac{dy}{dx}+y\right)=-\operatorname{cosec} xy \qquad\qquad ...(i)$$

Substituting $xy=u$ and $x\left(\dfrac{dy}{dx}\right)+y=\dfrac{du}{dx}$ in (i), we obtain.

$$x^3\frac{du}{dx}=-\operatorname{cosec} u \quad\text{or}\quad x^{-3}dx=-\sin u\,du.$$

Integrating, we obtain $-\dfrac{1}{2x^2}=\cos u+c_1$ or $\cos xy+\dfrac{1}{2x^2}=c$, the required solution of the given equation.

Exercise 17.2

Find the general solution of the following differential equations.

1. (a) $(1+2x)y'=1-x$. **Ans.** $2x+4y=3\log(2x+1)+c$.

2. $dy/dx=(1+y^2)/(1+x^2)$. **Ans.** $y=(x+c)/(1-cx)$.

3. $(x+1)y'+1=2e^{-y}$. (*Madras, 2005*) **Ans.** $(x+1)(2-e^y)=c$.

4. $(1-y)\,y'=1+y^2$. **Ans.** $x=\tan^{-1}y-\dfrac{1}{2}\log\left(1+y^2\right)+c$.

5. $\sec^2 x\tan y\,dx+\sec^2 y\cdot\tan x\,dy=0$. (*PTU 2002*) **Ans.** $\tan x\tan y=c$.

6. $(x^2-yx^2)\,dy+(y^2+xy^2)\,dx=0$. **Ans.** $\log(x/y)-(y+x)/xy=c$.

7. $xdx+ye^{-x^2}dy=0$. **Ans.** $e^{x^2}+y^2=c$.

8. $\dfrac{y}{x}\cdot y'=\sqrt{1+x^2+y^2+x^2y^2}$. **Ans.** $\sqrt{1+y^2}=\dfrac{2(1+x^2)^{3/2}}{3}+c$.

9. $dy/dx=e^{x-y}+x^2e^{-y}$. (*GGSIPU 2006; A.M.I.E., S-2002*)

Ans. $e^y=e^x+\left(\dfrac{x^3}{3}\right)+c$.

10. $y\log y\,dx+x\log x\,dy=0$. (*UPTU-2004*) **Ans.** $y=e^{c/\log x}$.

11. $(1+e^x)yy'-e^x(1+y)=0$. **Ans.** $(1+e^x)(1+y)=ce^y$.

12. $(x+y)^2y'=a^2$. (*A.M.I.E., S-2005*) **Ans.** $x+y=a\tan\left(\dfrac{y+c}{a}\right)$.

13. $y' = \sin(x+y) + \cos(x+y)$.   **Ans.** $\log\left|1 + \tan\left(\dfrac{x+y}{2}\right)\right| = x + c.$

14. $xy^3 y' = 1 - x^2 + y^2 - x^2 y^2$.   **Ans.** $\log x - \dfrac{x^2}{2} - \dfrac{y^2}{2} + \dfrac{1}{2}\log(1+y^2) = c.$

15. $(x+y)^2 (xy' + y) = xy(1 + y')$.   **Ans.** $\dfrac{1}{x+y} + \log xy = c.$

16. $\dfrac{dy}{dx} + \dfrac{1+y^3}{xy^2(1+x^2)} = 0$.   **Ans.** $\dfrac{x^6(1+y^3)^2}{(1+x^2)^3} = c.$

17. $\log\left(\dfrac{dy}{dx}\right) = ax + by$. *(I.E.T.E.; W-2002)*   **Ans.** $\dfrac{e^{ax}}{a} + \dfrac{e^{-by}}{b} = c.$

18. $\cos(x+y)\,dy = dx$. *(Patna 1997S)*   **Ans.** $y = c + \tan\left(\dfrac{x+y}{2}\right).$

19. $x \cos x \cos y + (\sin y)\,dy/dx = 0$. *(PTU 2002)*

                     Ans. $x \sin x + \cos x = \log \cos y + c.$

20. $xy\,dy + \dfrac{1+y^2}{1+x^2}\,dx = 0$.   **Ans.** $x^2(1+y^2) = c(1+x^2).$

21. $xyy' = 1 + x + y + xy$. *(PTU 2003)* $x - y + \log x(1+y) = c.$

22. $(x + y + 1)y' = 1$.   **Ans.** $x = ce^y - y - 2.$

23. $y - xy' = a(y^2 + y')$. *(AMIE., S-2007)*   **Ans.** $cy = (x + a)(1 - ay).$

24. $3e^x \tan y\,dx + (1 - e^x)\sec^2 y\,dy = 0$.   **Ans.** $\tan y = c(1 - e^x)^3$

25. $x^4 y' + x^3 y = -\sec(xy)$ *(I.E.T.E., Dec. 2003)*   **Ans.** $\sin(xy) = (1/2\,x^2) + c.$

26. $xy\dfrac{dy}{dx} = \dfrac{1+y^2}{1+x^2}(1 + x + x^2)$.   *(AMIE, W-2007)*

                     Ans. $1 + y^2 = (cx^2)e^{2\tan^{-1}x}.$

Solve the following initial value problems.

27. $(x \log x)\,y' = 2y$, given $y(2) = (\log 2)^2$.   **Ans.** $y = (\log x)^2.$

28. $y^2 y' = \cos^2 x$, given that $y = 0$, when $x = 0$.   **Ans.** $\dfrac{y^3}{3} = \dfrac{1}{2}\left(x + \dfrac{\sin 2x}{2}\right).$

29. $\cos x\left(e^{2y} - y\right)y' = e^y \sin 2x, y(0) = 0.$ **Ans.** $e^y + ye^{-y} + e^{-y} = 4 - 2\cos x.$

30. $\dfrac{dy}{dx} = \dfrac{x(2\log x + 1)}{\sin y + y\cos y}$, given that $y = 0$, when $x = 1$. (*V.T.U. 2000*)

   Ans. $y \sin y = x^2 \log x.$

31. $y - x\dfrac{dy}{dx} = a\left(y^2 + \dfrac{dy}{dx}\right),$ $y(0) = a.$ **Ans.** $(1 - a^2)y = (x + a)(1 - ay).$
   (*AMIE, W-2010*)

32. $\dfrac{dy}{dx} = (4x + y + 1)^2$, $y(0) = 1.$ (*A.M.I.E., S-2009; S-2002; Mangalore 1999*)

   Ans. $(4x + y + 1) = 2\tan\left[2x + \left(\dfrac{\pi}{4}\right)\right].$

33. $\dfrac{dy}{dx} = \dfrac{x}{y} - \dfrac{x}{1+y}, y(0) = 2.$ **Ans.** $2y^3 + 3y^2 = 3x^2 + 28.$

34. $3e^x \tan y\, dx + (1 + e^x) \sec^2 y\, dy = 0$, given $y(0) = \pi/4.$
   Ans. $\tan y(1 + e^x)^3 = 8.$

35. $e^x\left(\cos y\, dx - \sin y\, dy\right) = 0, y(0) = 0.$ (*G.G.SIPU 2006*)

   Ans. $y = \cos^{-1}\left(e^{-x}\right).$

36. Find the equation of the curve which satisfies the differential equation
   $(1 + y^2)\, dx - xy\, dy = 0$ and passes through the point $\left(1, \sqrt{3}\right)$. Also name the
   curve. (*AMIE., S-2005*) **Ans.** $4x^2 - y^2 = 1$; Hyperbola.

17.6 Homogeneous Equations

If a function $f(x, y)$ possesses the property $f(\lambda x, \lambda y) = \lambda^n f(x,y)$ for some real
number n, then $f(x, y)$ is said to be a **homogeneous function** of degree n.
For example, $f(x, y) = x^4 + y^4$ is a homogeneous function of degree 4 since
$f(\lambda x, \lambda y) = (\lambda x)^4 + (\lambda y)^4 = \lambda^4(x^4 + y^4) = \lambda^4 f(x, y),$
whereas $f(x, y) = x^2 + y^2 + 4$ is seen not to be homogeneous.

A first order differential equation in differential form

$$M(x, y)dx + N(x, y)\, dy = 0 \qquad \qquad ...(17.4)$$

is said to be **homogeneous** if both coefficients M and N are homogeneous functions of the *same* degree. In other words, (17.4) is homogeneous if

$$M(\lambda x, \lambda y) = \lambda^n M(x, y) \text{ and } N(\lambda x, \lambda y) = \lambda^n N(x, y).$$

If M and N are homogeneous functions of degree n, we can also write

$$M(x, y) = x^n M(1, v) \text{ and } N(x, y) = x^n N(1,v) \text{ where } v = y/x, \qquad ...(17.5)$$

and $M(x, y) = y^n M(u, 1)$ and $N(x, y) = y^n N(u, 1)$ where $u = x/y.$ $...(17.6)$

Either of the substitution $y = vx$ or $x = uy$, where v and u are new dependent variables will reduce a homogeneous equation to a *separable* first-order differential equation.

As a consequence of (17.5) a homogeneous equation

$M(x, y) \, dx + N(x, y) \, dy = 0$ can be rewritten as

$$x^n M(1, v) \, dx + x^n N(1, v) \, dy = 0 \text{ or } M(1, v) \, dx + N(1, v) \, dy = 0 \qquad \text{...(17.7)}$$

where $v = y/x$ or $y = vx$. By substituting the differential $dy = v \, dx + x \, dv$ in (17.7) and gathering terms, we obtain a separable differential equation in the variables v and x:

$$M(1, v) \, dx + N(1, v) \, [v \, dx + x \, dv] = 0$$

or

$$[M(1, v) + vN(1, v)] \, dx + xN(1, v) \, dv = 0$$

or

$$\frac{dx}{x} + \frac{N(1, v) \, dv}{M(1, v) + vN(1, v)} = 0.$$

Integrating, we obtain

$$\int \frac{N(1, v) \, dv}{M(1, v) + vN(1, v)} + \log|x| = \log|c|.$$

After integrating, we replace v by y/x to obtain the general solution of the given differential equation. To solve the homogeneous equation

$M(u, 1) \, dx + N(u, 1) \, dy = 0$, we use the substitution

$$x = uy \text{ and } dx = u \, dy + y \, du.$$

Example 17.11 Find the solution of the differential equation

$$x\frac{dy}{dx} = y(\log y - \log x + 1). \qquad (I.E.T.E., \text{ June } 2001)$$

Solution. The given differential equation

$$\frac{dy}{dx} = \frac{y}{x}\left\{ \log\left(\frac{y}{x}\right) + 1 \right\} \text{ is homogeneous.}$$

Substituting $y = vx$ and $dy/dx = v + x\left(\dfrac{dv}{dx}\right)$, we obtain

$$v + x\left(\frac{dv}{dx}\right) = v \log v + v \text{ or } x\left(\frac{dv}{dx}\right) = v \log v.$$

Separating the variables, we obtain

$$\frac{dv}{v \log v} = \frac{dx}{x}. \text{ Integrating, we obtain}$$

$\log|\log v| = \log|x| + \log|c|$ *or* $\log v = cx$.

Replacing v by y/x, we obtain $\log(y/x) = cx$ or $y = xe^{cx}$.

Example 17.12 Find the solution of differential equation

$$x(y - x)\, dy = y(x + y)\, dx.$$

Solution. The given differential equation $\dfrac{dy}{dx} = \dfrac{\left(xy + y^2\right)}{\left(xy - x^2\right)}$, is homogeneous.

Substituting $y = vx$ and $\dfrac{dy}{dx} = v + x\left(\dfrac{dv}{dx}\right)$, we obtain

$$v + x\frac{dv}{dx} = \frac{v + v^2}{v - 1} \quad \text{or} \quad x\frac{dv}{dx} = \frac{2v}{v - 1}.$$

Separating the variables, we get $\dfrac{v-1}{2v}\, dv = \dfrac{dx}{x}$ or $\left(\dfrac{1}{2} - \dfrac{1}{2v}\right) dv = \dfrac{dx}{x}$.

Integrating, we obtain $\dfrac{1}{2}v - \dfrac{1}{2}\log|v| = \log|x| + \log|c_1|$,

or $v - \log|v| = 2\log|x| + 2\log(c_1)$ or $v = \log vx^2 c$, where $c = c_1^2$.
Replacing v by y/x, we get $y = x(\log xyc)$ or $y/x = \log xyc$.

Example 17.13 Solve the initial value problem

$$(3xy + y^2)\, dx + (x^2 + xy)\, dy = 0,\ y(1) = 1.$$

Solution. The given differential equation

$$\frac{dy}{dx} = -\frac{3xy + y^2}{x^2 + xy} \quad \text{is homogeneous.}$$

Substituting $y = vx$ and $\dfrac{dy}{dx} = v + x\dfrac{dv}{dx}$, we obtain

$$v + x\frac{dv}{dx} = -\frac{3v + v^2}{1 + v} \quad \text{or} \quad x\frac{dv}{dx} = -\frac{3v + v^2}{1 + v} - v = -\frac{2v(v + 2)}{v + 1}.$$

Separating the variables, we get

$$\frac{(v + 1)}{v(v + 2)}\, dv = -2\frac{dx}{x} \quad \text{or} \quad \frac{1}{2}\left[\frac{1}{v} + \frac{1}{v + 2}\right] dv = -\frac{2}{x}\, dx.$$

Integrating, we obtain

$$\frac{1}{2}\left[\log|v| + \log|v + 2|\right] = -2\log|x| + \log|c_1|$$

or $\log|v(v + 2)| + 4\log|x| = 2\log|c_1|$,

or $v(v+2)x^4 = c$, where $c = c_1^2$. Replacing, v by y/x, we get $x^2y(2x + y) = c$.

Using the initial condition $y(1) = 1$, we obtain $c = 3$.
The particular solution is $x^2y(2x + y) = 3$.

Example 17.14 Solve the differential equation

$$(1 + 3e^{x/y})\ dx + 3e^{x/y}\ \{1 - (x/y)\}dy = 0.$$

Solution. The given equation is homogeneous. The appearance of x/y throughout

the equation suggests the use of the transformation $x = uy$, $\dfrac{dx}{dy} = u + y\dfrac{du}{dy}$ to

obtain $(1+3e^u)\left[u+y\left(\dfrac{du}{dy}\right)\right]+3e^u(1-u)=0$

or $\quad u+y\dfrac{du}{dy} = \dfrac{3e^u(u-1)}{1+3e^u}$ or $y\dfrac{du}{dy} = \dfrac{-3e^u - u}{1+3e^u}$.

Separating the variables, we get

$\dfrac{1+3e^u}{u+3e^u}du = -\dfrac{dy}{y}$. Integrating, we obtain $\log|u + 3e^u| = -\log|y| + \log|c|$

or $\quad u+3e^u = \left(\dfrac{c}{y}\right)$. Replacing u by $\dfrac{x}{y}$, we obtain

$$\dfrac{x}{y}+3e^{x/y} = \dfrac{c}{y} \qquad \text{or } x+3ye^{x/y} = c.$$

Example 17.15 find the solution of differential equation $xy \log (x/y)\ dx +$ $\{y^2 - x^2 \log (x/y)\}\ dy = 0$, given that $y(1) = e$. *(I.E.T.E., June 2001)*

Solution. The given equation can be rewritten as

$$-xy \log (y/x)\ dx + \{y^2 + x^2 \log (y/x)\}\ dy = 0,$$

or $\quad \dfrac{dy}{dx} = \dfrac{xy\log(y/x)}{y^2 + x^2\log(y/x)}$, which is homogeneous.

Using the transformation $y = vx$ and $\dfrac{dy}{dx} = v + x\dfrac{dv}{dx}$ to obtain

$$v+x\dfrac{dv}{dx} = \dfrac{v\log v}{v^2+\log v} \quad \text{or } x\dfrac{dv}{dx} = \dfrac{-v^3}{v^2+\log v}. \text{ Separating the variables,}$$

$\dfrac{dx}{x}+\dfrac{(v^2+\log v)}{v^3}dv = 0$. Integrating and replacing v by y/x, we obtain

$$\log|x|+\log|v|-\left(\frac{\log v}{2}+\frac{1}{4}\right)\frac{1}{v^2}=c \quad \text{or} \quad \log y-\left[\frac{1}{2}\log\left(\frac{y}{x}\right)+\frac{1}{4}\right]\left(\frac{x^2}{y^2}\right)=c.$$

Using the initial condition $y(1)=e$, we obtain $c=(4e^2-3)/4e^2$.

The particular solution is $\log y-\dfrac{x^2}{4y^2}\left\{2\log\left(\dfrac{y}{x}\right)+1\right\}=\dfrac{4e^2-3}{4e^2}.$

Example 17.16 Solve the differential equation $\{x\tan(y/x)-y\sec^2(y/x)\}\,dx +$ $\{x\sec^2(y/x)\}\,dy=0$ (A.M.I.E., W-2005; PTU-2003; Mysore, 1999)

Solution. The given equation is homogeneous and can be rewritten as

$$\{\tan(y/x)-(y/x)\sec^2(y/x)\}+\{\sec^2(y/x)\}\frac{dy}{dx}=0.$$

Using the transformation $y=vx$ and $\dfrac{dy}{dx}=v+x\left(\dfrac{dv}{dx}\right)$ to obtain

$$\tan v-v\sec^2 v+\sec^2 v\left[v+x\left(\frac{dv}{dx}\right)\right]=0 \quad \text{or} \quad \tan v+x\sec^2 v\left(\frac{dv}{dx}\right)=0.$$

Separating the variables, we obtain $\dfrac{dx}{x}+\dfrac{\sec^2 v}{\tan v}\,dv=0.$

Integrating, we obtain $\log|x|+\log|\tan v|=\log|c|$ or $x\tan v=c.$

Replacing v by y/x, we get $x\tan(y/x)=c$, the required solution.

Exercise 17.3

Find the general solution of the following differential equations.

1. $(x+y)dx+(y-x)dy=0.$ *(PTU 2004)*

                       Ans. $\log(x^2+y^2)=2\tan^{-1}(y/x)+c.$

2. $(2x+y)\,dx+(2y+x)\,dy=0.$ *(I.E.T.E., 2002)*   **Ans.** $x^2+xy+y^2=c.$

3. $(x-y)\,dy-(2x-y)\,dx=0.$ **Ans.** $2x^2-2xy+y^2=c.$

4. $y^2-(xy+x^2)\,y'=0.$ **Ans.** $cx-y=x\log y.$

5. $xy'+\left(\dfrac{y^2}{x}\right)=y.$ **Ans.** $cx=e^{x/y}.$

6. $xy^2y'-y^3=x^3.$ **Ans.** $\log x=\left(\dfrac{y^3}{3x^3}\right)+c.$

7. $(y^2+2xy)dx+(2x^2+3xy)\,dy=0.$ **Ans.** $xy^2(x+y)=c.$

8. $2x^2y'=y(x+y).$ **Ans.** $(x-y)^2=cxy^2.$

9. $x(x-y)y'=y(x+y).$ *(GGSIPU 2006)*   **Ans.** $(x/y)+\log(xy)=c.$

10. $y^2 + x^2y' = xyy'$. *(Triputi, 1998-S)*   **Ans.** $y = x \log (cy)$.

11. $(1 + e^{x/y}) dx + e^{x/y} \{1 - (x/y)\} dy = 0$.   *(IETE, June 2009, VTU 2001)*
       Ans. $x + ye^{x/y} = c$.

12. $\left\{ x\cos\left(\dfrac{y}{x}\right) + y\sin\left(\dfrac{y}{x}\right) \right\} y\,dx + \left\{ x\cos\left(\dfrac{y}{x}\right) - y\sin\left(\dfrac{y}{x}\right) \right\} x\,dy = 0.$

Alternative form: $x\cos\left(\dfrac{y}{x}\right)(y\,dx + x\,dy) = y\sin\left(\dfrac{y}{x}\right)(x\,dy - y\,dx) \cdot$

[Hint: The given equation may be rewritten as

$$\frac{dy}{dx} = \frac{y\left(y\sin\left(\dfrac{y}{x}\right) + x\cos\left(\dfrac{y}{x}\right)\right)}{x\left\{ y\sin\left(\dfrac{y}{x}\right) - x\cos\left(\dfrac{y}{x}\right) \right\}} \,].$$
       Ans. $xy \cos (y/x) = c$.

13. $x \sin (y/x)\{y\,dx + x\,dy\} + \cos(y/x) \{x\,dy - y\,dx\} = 0$.   **Ans.** $xy \sin (y/x) = c$.

14. $(x^2 - y^2)dx - xy\,dy = 0$. *(Assam, 1998)*   **Ans.** $x^2(x^2 - 2y^2) = c$.

15. $x^3y' = y^3 + y^2\sqrt{y^2 - x^2}$.   **Ans.** $xy = c\left(y + \sqrt{y^2 - x^2}\right)$.

16. $xdy - ydx = \sqrt{x^2 + y^2}\, dx$.   *(PTU 2003; Mysore 1997)*

       Ans. $y + \sqrt{x^2 + y^2} = cx^2$.

17. $\dfrac{y}{x}\cos\dfrac{y}{x}\,dx - \left(\dfrac{x}{y}\sin\dfrac{y}{x} + \cos\dfrac{y}{x}\right)dy = 0.$ **Ans.** $y\sin\left(\dfrac{y}{x}\right) = c.$

18. $\{x\tan(y/x) - y\sec^2(y/x)\} dx + x\sec^2(y/x)dy = 0.$ *(PT.U. 2003)*

       Ans. $x\tan(y/x) = c.$

19. $2xy\,dy = (x^2 + y^2)\,dx.$ *(AMIE, W-2010)* **Ans.** $x^2 - y^2 = cx.$

20. $xy' - y = x \sin (y/x)$ *(V.T.U. 2000S)* **Ans.** $y = 2x\tan^{-1}(cx).$

21. $(y^3 - 3xy^2) dx + (2x^2y - xy^2) dy = 0.$ *(I.E.T.E.-2001)* **Ans.** $x^3e^{y/x} = cy^2.$

22. $(2x - y)e^{y/x}dx + (y + xe^{y/x})dy = 0.$ **Ans.** $y^2 + 2x^2e^{y/x} = c.$

23. $(2xy + x^2)y' = 3y^2 + 2xy.$ *(IETE., Dec. 2006)* **Ans.** $y(x + y) = cx^3.$

Solve the following initial value problems.

24. $(x^2 + 2y^2) dx - xy\,dy = 0, \ y(1) = 0.$ **Ans.** $y^2 = x^4 - x^2.$

25. $(x^3 - 3y^3)\, dx + 3xy^2\, dy = 0$, $y(1) = 2$.   **Ans.** $y^3 = x^3\,(8 - \log|x|)$

26. $(x + ye^{y/x})\, dx - xe^{y/x}\, dy = 0$, $y(1) = 0$.   **Ans.** $\left|\log|x|\right| = e^{y/x} - 1$.

27. $(3xy + y^2)\, dx + (x^2 + xy)\, dy = 0$, $y(1) = 1$.   **Ans.** $x^2 y(2x + y) = 3$.

28. $x^2 dy + y(x + y)\, dx = 0$ given that $y = 1$ when $x = 1$.   *(UPTU 2006)*

Ans. $x^2 y = (2x + y)/3$.

29. $\left(x^2 - y^2\right)dx + 2xy\, dy = 0$, $y(1) = 1$.   **Ans.** $x^2 + y^2 = 2x$.

17.7 Equations of the form: $dy/dx = (a_1 x + b_1 y + c_1)/(a_2 x + b_2 y + c_2)$

The differential equation

$$\frac{dy}{dx} = \frac{a_1 x + b_1 y + c_1}{a_2 x + b_2 y + c_2}, \qquad \text{...(17.8) is not homogeneous.}$$

Case I. If $\dfrac{a_1}{a_2} = \dfrac{b_1}{b_2} = t$, then equation (17.8) can be written as

$$\frac{dy}{dx} = \frac{t(a_2 x + b_2 y) + c_1}{a_2 x + b_2 y + c_2}. \qquad \text{...(17.9)}$$

Substituting $a_2 x + b_2 y = v$, we obtain $a_2 + b_2 y' = v'$. Then Eq. (17.9) becomes

$$\frac{dy}{dx} = \frac{1}{b_2}(v' - a_2) = \frac{tv + c_1}{v + c_2} \quad \text{or} \quad v' = a_2 + \frac{b_2(tv + c_1)}{v + c_2} = \frac{a_2(v + c_2) + b_2(tv + c_1)}{v + c_2}$$

or $\dfrac{(v + c_2)\, dv}{a_2(v + c_2) + b_2(tv + c_1)} = dx$, which is in separable form.

Integrating and replacing v by $a_2 x + b_2 y$, we obtain the general solution.

Case II. If $\dfrac{a_1}{a_2} \neq \dfrac{b_1}{b_2}$, then we substitute $x = X + h$, $y = Y + k$ in the differential equation to get

$$\frac{dy}{dx} = \frac{dY}{dX} = \frac{a_1(X + h) + b_1(Y + k) + c_1}{a_2(X + h) + b_2(Y + k) + c_2} = \frac{a_1 X + b_1 Y + (a_1 h + b_1 k + c_1)}{a_2 X + b_2 Y + (a_2 h + b_2 k + c_2)} \quad \text{...(17.10)}$$

Choose h and k such that $a_1 h + b_1 k + c_1 = 0$ and $a_2 h + b_2 k + c_2 = 0$. Then, equation (17.10) simplifies to

$$\frac{dY}{dX} = \frac{a_1 X + b_1 Y}{a_2 X + b_2 Y},$$

which is a homogeneous equation in the variables, X, Y. We solve this equation and substitute $X = x - h$ and $Y = y - k$ to obtain the general solution.

Example 17.17 Find the solution of the differential equation

$$(4x + 2y + 5)\, dy - (2x + y - 1)\, dx = 0.$$

Solution. We have $\dfrac{dy}{dx} = \dfrac{2x+y-1}{2(2x+y)+5}$.

Substituting $2x + y = v$, we get $2 + y' = v'$. Hence

$$y' = v' - 2 = \frac{v-1}{2v+5} \text{ or } v' = 2 + \frac{v-1}{2v+5} = \frac{5v+9}{2v+5}.$$

Separating the variables, we obtain

$$\frac{(2v+5)}{5v+9}\, dv = dx \text{ or } \frac{2}{5}\left[1 + \frac{7/10}{v+(9/5)}\right] dv = dx.$$

Integrating, we obtain $\dfrac{2}{5}(v) + \dfrac{7}{25}\log\left|v + \dfrac{9}{5}\right| = x + c.$

Replacing v by $(2x + y)$, we obtain $\dfrac{2}{5}(2x+y) + \dfrac{7}{25}\log\left|2x + y + \dfrac{9}{5}\right| - x = c.$

Example 17.18 Find the solution of the differential equation

$$(x - 2y - 3)\, dy + (x - y - 2)\, dx = 0.$$

Solution. The given differential equation is $\dfrac{dy}{dx} = \dfrac{y-x+2}{x-2y-3}$.

To convert it into a homogeneous equation, make the substitution $x = X + h$, $y = Y + k$. Then

$$\frac{dY}{dX} = \frac{Y - X + (k - h + 2)}{X - 2Y + (h - 2k - 3)}.$$

Choose h, k such that $k - h + 2 = 0$, $h - 2k - 3 = 0$. Solving, we obtain $h = 1$ and $k = -1$. For this choice of h, k, we get

$$\frac{dY}{dX} = \frac{Y - X}{X - 2Y}. \text{ Substituting } Y = vX, \text{ we obtain}$$

$$\frac{dY}{dX} = v + X\frac{dv}{dX} = \frac{vX - X}{X - 2vX} = \frac{v-1}{1-2v} \text{ or } X\frac{dv}{dX} = \frac{v-1}{1-2v} - v = \frac{2v^2 - 1}{1-2v}.$$

Separating the variables, we get $\dfrac{dX}{X} = \dfrac{2v-1}{1-2v^2}\, dv = \dfrac{(-1/2)(-4v)-1}{1-2v^2}\, dv.$

Integrating, we obtain $\log |X| = -\dfrac{1}{2}\log|1-2v^2| - \displaystyle\int \dfrac{dv}{1-2v^2} + \log|c_1|$

$$= -\frac{1}{2}\log|1-2v^2| - \frac{1}{2\sqrt{2}}\log\left|\frac{1+v\sqrt{2}}{1-v\sqrt{2}}\right| + \log|c_1|,$$

or $\qquad X^2\left(1-2v^2\right)\left(\dfrac{1+v\sqrt{2}}{1-v\sqrt{2}}\right)^{1/\sqrt{2}} = c$, where $c = c_1^2$.

Substituting $v = \dfrac{Y}{X} = \dfrac{y-k}{x-h} = \dfrac{y+1}{x-1}$ and simplifying, we get the solution as

$$(x-1)^2\left[1 - 2\left(\frac{y+1}{x-1}\right)^2\right]\left[\frac{1+\{(y+1)/(x-1)\}\sqrt{2}}{1-\{(y+1)/(x-1)\}\sqrt{2}}\right]^{1/2} = c$$

or $\quad (x-1)^2 - 2(y+1)^2 = c\left(\dfrac{x-y\sqrt{2}-\sqrt{2}-1}{x+y\sqrt{2}+\sqrt{2}-1}\right)^{1/\sqrt{2}}.$

Exercise 17.4

Find the general solution of the following differential equations.

1. $(2x + y + 1)dx + (4x + 2y - 1)\,dy = 0.$ **Ans.** $x + 2y + \log(2x + y - 1) = c.$
2. $(2x - y + 2)\,dx + (4x - 2y - 1)\,dy = 0.$ **Ans.** $x + 2y + \log(2x - y) = c.$
3. $(6x + 9y + 6)y' = (2x + 3y - 1).$ **Ans.** $2x + 3y + 1 = c \exp(x - 3y).$
4. $(2x + 4y + 3)y' = x + 2y + 1.$ **Ans.** $\log(4x + 8y + 5) = 4x - 8y + c.$
5. $(6x - 4y + 1)\,y' = 3x - 2y + 1.$ *(IETE., W-2006)*

$$\textbf{Ans. } x - 2y + \frac{1}{4}\log\left|3x - 2y + \frac{1}{4}\right| = c.$$

6. $(2x + 3y + 4)dx - (4x + 6y + 5)\,dy = 0.$ *(Madras 2000S)*

$$\textbf{Ans. } 7x - 14y + 3\log(14x + 21y + 22) = c.$$

7. $(3x - 6y + 2)dx - (x - 2y + 1)\,dy = 0.$

$$\textbf{Ans. } \frac{1}{5}(x - 2y) + \frac{2}{25}\log\left|x - 2y + \frac{3}{5}\right| + x = c.$$

8. $y' + \dfrac{2x+3y}{y+2} = 0.$ **Ans.** $(2x + y - 4)^2 = c(x + y - 1).$

9. $(3x - y - 2)\,y' = x + 3y - 4.$

$$\textbf{Ans. } \log\left[(x-1)^2 + (y-1)^2\right] = 6\tan^{-1}\left(\frac{y-1}{x-1}\right) + c.$$

10. $y' = \dfrac{7x-3y-7}{7y-3x+3}$. *(I.E.T.E. June 2004)* **Ans.** $(x + y - 1)^5 (x - y - 1)^2 = c$.

11. $(x + 2y - 3)dx = (2x + y - 3) \, dy$. **Ans.** $(x + y - 2) = c(x - y)^3$.
(Madras, 2000 PT; V.T.U, 2000; Andhra 1998)

12. $y'(x+y-2) = x - y + 1$. **Ans.** $x^2 - 2xy - y^2 + 2x + 4y = c$.

13. $y' = (y + x - 2)/(y - x - 4)$. **Ans.** $x^2 + 2xy - y^2 - 4x + 8y = c$.
(J.N.T.U., 1999; Mangalore, 1999; Patna 1997S)

14. $(x + 2y)(dx - dy) = dx + dy$. *(V.T.U. 2001; Rewa 1995)*
Ans. $(3x + 6y - 1)^{2/3} = c_1 e^{x-y}$

15. $(y - x + 1)dy - (y + x + 2)dx = 0$. *(IETE, June 2007)*

Ans. $2(y - x + 1)^2 - (2x + 1)^2 = c_1$.

17.8 Linear First Order Differential Equations

Definition. *When the dependent variable y and its derivative dy/dx occur in the differential equation in the first degree and are not multiplied together, the equation is known as a linear differential equation of the first order.*

A typical linear equation of the first order is

$$\frac{dy}{dx} + yP(x) = Q(x) \qquad\qquad ...(17.11)$$

where $P(x)$ and $Q(x)$ are given continuous functions of x (or are constants). Eq. (17.11) is also sometimes called **Leibnitz linear equation.**

If $Q(x) = 0$, then the equation is easily solved, as it is in the separable form. In this case, the equation $y' + yP(x) \, y = 0$ is also called a *homogeneous first order equation.*

To find the solution of (17.11), we multiply both sides by $e^{\int P(x)dx}$ and obtain

$$\frac{dy}{dx} e^{\int P(x)dx} + yP(x) e^{\int P(x)dx} = Q(x) \cdot e^{\int P(x)dx}.$$

or $$\frac{d}{dx}\left(y e^{\int P(x)dx} \right) = Q(x) e^{\int P(x)dx},$$

$e^{\int P(x)dx}$ is an *integrating factor* (I.F.) of (17.11) and its solution is

$$y(x) e^{\int P(x)dx} = \int Q(x) \cdot e^{\int P(x)dx} dx + c$$

Hence we write the solution of (17.11) as

$$y(x) \times I.F. = \int [Q(x) \times (I.F.)] dx + c \text{ where c is a constant.}$$

Example 17.19 Solve the initial value problem

$$y' + y \tan x = \sin 2x, \ y(0) = 1.$$

Solution. An integrating factor is $I.F. = e^{\int \tan x \, dx} = e^{\log|\sec x|} = \sec x.$

The general solution of equation is $y \cdot \sec x = \int (\sin 2x) \sec x \, dx + c$

or $\qquad y(x) = \cos x \left[2 \int \sin x \, dx + c \right] = c \cos x - 2 \cos^2 x.$

Applying the initial condition, $y(0) = 1$, we get $1 = c - 2$ or $c = 3$.

Hence, the solution of IVP is $y = 3 \cos x - 2 \cos^2 x$.

Example 17.20 Solve: $(1 + y^2) \, dx = (\tan^{-1} y - x) \, dy$.

(A.M.I.E., W-2010, S-2004; IETE., W-2007)

Solution. This equation is linear if we take y as independent and x as dependent variable. Thus it is linear differential equation when written in the form

$$\frac{dx}{dy} + \frac{x}{1 + y^2} = \frac{\tan^{-1} y}{1 + y^2}.$$

An integrating factor is I.F. $= e^{\int dy/(1+y^2)} = e^{\tan^{-1} y}$.

The general solution of equation is $x \, e^{\tan^{-1} y} = \int \frac{\tan^{-1} y \left(e^{\tan^{-1} y} \right)}{1 + y^2} \, dy + c.$

Putting $\tan^{-1} y = t$ and $\dfrac{dy}{1 + y^2} = dt$, we get

$$\int \left(\frac{\tan^{-1} y}{1 + y^2} \right) \cdot e^{\tan^{-1} y} \, dy = \int t \, e^t \, dt = t \, e^t - e^t = e^t (t - 1) = \left(\tan^{-1} y - 1 \right) e^{\tan^{-1} y}.$$

Hence the solution is

$$x \, e^{\tan^{-1} y} = \left(\tan^{-1} y - 1 \right) e^{\tan^{-1} y} + c \quad \text{or} \quad x = \tan^{-1} y - 1 + c \, e^{-\tan^{-1} y}.$$

Example 17.21 Solve $y \log y \, dx + (x - \log y) \, dy = 0$.

Solution. The equation, with x taken as dependent variable, may be put in the form $\dfrac{dx}{dy} + \dfrac{1}{y \log y} x = \dfrac{1}{y}.$

An integrating factor is $I.F. = e^{\int dy/(y \log y)} = e^{\log(\log y)} = \log y$.

The general solution of equation is

$$x \cdot \log y = \int \frac{1}{y} \cdot \log y \, dy + c_1 = \frac{1}{2} \log^2 y + c_1 \quad \text{or} \quad 2x \log y = (\log y)^2 + c.$$

Example 17.22 Solve $2(1+x)y' - (1+2x)y = x^2 \sqrt{1+x}$

Solution. The given equation is

$$\frac{dy}{dx} - \frac{(1+2x)}{2(1+x)} y = \frac{x^2}{2\sqrt{1+x}}. \quad \text{An integrating factor is}$$

$$I.F. = e^{\int P \, dx} = e^{\int -\frac{(1+2x)}{2(1+x)} dx} = e^{\int \left\{-1 + \frac{1}{2(1+x)}\right\} dx} = e^{\left\{-x + \frac{1}{2} \log(1+x)\right\}} = \sqrt{1+x} \cdot e^{-x}$$

The general solution of the equation is

$$y \cdot \sqrt{1+x} \cdot e^{-x} = \int \frac{x^2}{2\sqrt{1+x}} \cdot \sqrt{1+x} \cdot e^{-x} dx + c = \frac{1}{2} \int x^2 e^{-x} dx + c$$

$$= \frac{1}{2} \left[x^2 (-e^{-x}) - (2x) e^{-x} + 2(-e^{-x}) \right] + c \quad \text{or} \quad y\sqrt{1+x} = ce^x - \frac{1}{2}(x^2 + 2x + 2).$$

Exercise 17.5

Find the general solution of the following first order linear differential equations.

1. $xy' - 3y = x - 1.$
  Ans. $y = cx^3 - x/2 + 1/3.$

2. $(x^2 + 1) y' + 2xy = 4x^2.$
  Ans. $(x^2 + 1) y = (4x^3/3) + c.$

3. $(x^2 + 3) y' + 4xy = x.$ *(IETE., 1995)*
  Ans. $y(x^2 + 3)^2 = \frac{x^4}{4} + \frac{3x^2}{2} + c.$

4. $(x \log x) \dfrac{dy}{dx} + y = \log x^2.$
  (Andhra 1998; A.M.I.E., W-1999)
  Ans. $y \log x = (\log x)^2 + c.$

5. $(x + 2y^3) y' = y \cdot$
  Ans. $x = y^3 + cy.$

  [**Hint:** Take x as the dependent variable. The equation can be written as

  $$\left(\frac{dx}{dy} \right) - \left(\frac{x}{y} \right) = 2y^2 \,].$$

6. $y' + \dfrac{2y}{x} = \sin x.$
  Ans. $x^2 y = (2 - x^2) \cos x + 2x \sin x + c.$

7. $e^{-y} \sec^2 y \, dy = dx + x \, dy.$
  (A.M.I.E., S-2002; I.E.T.E. Dec. 2002)
  Ans. $xe^y = \tan y + c.$

8. $\left(1-x^2\right)\dfrac{dy}{dx} - xy = 1$.

Ans. $y\sqrt{\left(1-x^2\right)} = \sin^{-1} x + c$.

9. $dr + (2r \cot \theta + \sin 2\theta)\, d\theta = 0$.

Ans. $2r \sin^2\theta + \sin^4\theta = c$.

10. $x^2 y' = 3x^2 - 2xy + 1$.

Ans. $y = x + x^{-1} + cx^{-2}$.

11. $\dfrac{dy}{dx} = \dfrac{y}{2y \log y + y - x}$.

Ans. $x = y \log y + cy^{-1}$.

12. $\sqrt{1-y^2}\, dx = \left(\sin^{-1} y - x\right) dy$.

Ans. $x = \sin^{-1} y - 1 + ce^{-\sin^{-1} y}$.

13. $\dfrac{dy}{dx} = \dfrac{\log x - 3x^3 y}{x^4}$.

Ans. $x^3 y = c + \dfrac{1}{2}\left(\log x\right)^2$.

14. $ye^y\, dx = (y^3 + 2xe^y)dy$. *(Andhra, 1997)*

Ans. $xy^{-2} = c - e^{-y}$.

15. $(x + 1)y' - y = e^{3x}(x + 1)^2$.

Ans. $y = \left(\dfrac{1}{3}e^{3x} + c\right)(x + 1)$.

(AMIE, S-2005; I.E.T.E.; June 2002; Andhra, 1998)

16. $\left(1+y^2\right)dx + \left(x - e^{\tan^{-1} y}\right)dy = 0$.

(Delhi 1997, Patna 1997)

Ans. $xe^{\tan^{-1} y} = \dfrac{1}{2}e^{2\tan^{-1} y} + c$.

17. $\cot 3x\, \dfrac{dy}{dx} - 3y = \cos 3x + \sin 3x, 0 < x < \pi/2$.

(AMIE., S-2008; I.E.T.E. June 2007)

Ans. $y \cos 3x = \dfrac{1}{12}\left[6x - \cos 6x - \sin 6x\right] + c$.

18. $y' \sec x = y + \sin x$. *(I.E.T.E.; W-2005)*

Ans. $y = ce^{\sin x} - \sin x - 1$.

19. $y' + y \cot x = 2\cos x$. *(I.E.T.E., June 2003)*

Ans. $y \sin x = c - \dfrac{1}{2}\cos 2x$.

20. $\left(1+y^2\right)dx + \left(x - e^{-\tan^{-1} y}\right)dy = 0$.

Ans. $xe^{\tan^{-1} y} = \tan^{-1} y + c$.

Solve the following initial value problems.

21. $y' + \dfrac{2xy}{1+x^2} = \dfrac{1}{\left(1+x^2\right)^2}$, $\qquad y(1) = 0$.

(Andhra, 1998)

Ans. $y(1 + x^2) = \tan^{-1} x - (\pi/4)$.

22. $\left(1-x^2\right)y' + 2xy = x\sqrt{1-x^2}$, $y(0) = 0$.

Ans. $y = \sqrt{1-x^2} + x^2 - 1$.

23. $x^2 y' + 2xy - x + 1 = 0$, $y(1) = 0$.   **Ans.** $yx^2 = \dfrac{x^2}{2} - x + \dfrac{1}{2}$.

24. $y' + y \cot x = 5\, e^{\cos x}$, $y(\pi/2) = -4$.   **Ans.** $y \sin x + 5\, e^{\cos x} = 1$.

25. $2 \cos x \cdot \dfrac{dy}{dx} + 4y \sin x = \sin 2x$, $y\left(\pi/3\right) = 0$. Show that the maximum

   value of $y = 1/8$.   **Ans.** $y = -2 \cos^2 x + \cos x$.

26. $y' + y \cot x = 4\, x \operatorname{cosec} x$, $y(\pi/2) = 0$.   **Ans.** $y \sin x = 2\, x^2 - (\pi^2/2)$.

27. $xy' - 3y = x - 1$, $y(1) = 0$.   **Ans.** $y = \left(x^3 - 3x + 2\right)/6$.

17.9 Equations Reducible to Linear Form

The equation $\dfrac{dy}{dx} + yP(x) = y^n Q(x)$, $\quad (n \neq 0,1)$ ...(17.12)

is known as **Bernoulle's equation** after the swiss mathematician JAKOB BERNOULLI (1654-1705). This equation is non-linear, but it can be changed to a linear form by suitable substitution and the solution of the transformed equation is obtained by the above method.

To find the solution of (17.12), consider the following transformation
$$z(x) = [y(x)]^{1-n}. \qquad ...(17.13)$$

We obtain $\dfrac{dz}{dx} = (1-n)y^{-n}\dfrac{dy}{dx} = (1-n)y^{-n}\left[y^n Q(x) - yP(x) \right]$

$$= (1-n)\left[Q(x) - y^{1-n}P(x) \right] = (1-n)\left[Q(x) - zP(x) \right]$$

using equation (17.13). Hence, equation (17.12) reduces to the linear equation

$$\dfrac{dz}{dx} + (1-n)zP(x) = (1-n)Q(x) \qquad ...(17.14)$$

When the solution of this linear equation is obtained and $z(x)$ is replaced by y^{1-n}, we get the general solution of the Bernoulli equation.

Remark: An equation of the form $f'(y)\dfrac{dy}{dx} + f(y)P(x) = Q(x)$ is a linear equa-

tion of the first order. To solve this equation, we put $f(y) = z$ etc.

Example 17.23 Solve the differential equation $y' - 4y - xy^{3/2} = 0$.

Solution. Write the given equation as

$\dfrac{dy}{dx} - 4y = xy^{3/2}$, which is a Bernoulli equation with $n = 3/2$. Consider the trans-

formation $z = y^{1-(3/2)} = y^{-1/2}$, we obtain

$$\frac{dz}{dx} = -\frac{1}{2}y^{-3/2}\frac{dy}{dx} = -\frac{1}{2}y^{-3/2}\left[4y + xy^{3/2}\right] = -2y^{-1/2} - \frac{1}{2}x \quad \text{or} \quad \frac{dz}{dx} + 2z = -\frac{x}{2}.$$

An integrating factor of this equation is

$$I.F. = e^{\int 2dx} = e^{2x}.$$

The general solution of the transformed equation is obtained as

$$z \cdot e^{2x} = -\frac{1}{2}\int xe^{2x}\,dx + c = -\frac{1}{2}\left(x\frac{e^{2x}}{2} - \frac{1}{4}e^{2x}\right) + c = \frac{1}{8}e^{2x} - \frac{1}{4}xe^{2x} + c$$

or $\quad y^{-1/2}e^{2x} = \dfrac{1}{8}e^{2x} - \dfrac{1}{4}xe^{2x} + c_1 \quad$ or $\quad y = \dfrac{64e^{4x}}{\left(e^{2x} - 2xe^{2x} + c\right)^2}.$

Example 17.24 Solve the differential equation

$$xy\left(x^2y^2 + 1\right)\frac{dy}{dx} = 1. \qquad \text{(A.M.I.E. W-2005)}$$

Solution. Write the given equation as

$$\frac{dx}{dy} - yx = x^3y^3. \qquad \qquad ...(1)$$

Here the dependent variable is x and the independent variable is y. The equation (1) is Bernoulli equation with $n = 3$. Set $z = x^{1-3} = x^{-2}$, we obtain

$$\frac{dz}{dy} = -2x^{-3}\frac{dx}{dy} = -2x^{-3}\left(xy + x^3y^3\right) = -2x^{-2}y - 2y^3 \quad \text{or} \quad \frac{dz}{dy} + 2y \cdot z = -2y^3.$$

An integrating factor is $I.F. = e^{\int 2y \cdot dy} = e^{y^2}.$
The general solution of the transformed equation is obtained as

$$z \cdot e^{y^2} = \int\left(-2y^3\right)e^{y^2}\,dy + c = -\int te^t\,dt + c \qquad \text{(put } y^2 = t \text{ and } 2ydy = dt)$$

$$= c - e^t(t-1) = c - e^{y^2}\left(y^2 - 1\right), \quad \text{or} \quad x^{-2}e^{y^2} = c - e^{y^2}\left(y^2 - 1\right).$$

or $x^2 = \dfrac{e^{y^2}}{c - e^{y^2}\left(y^2 - 1\right)}.$

Example 17.25 Solve $\dfrac{dz}{dx} + \left(\dfrac{z}{x}\right)\log z = \dfrac{z}{x}(\log z)^2.$ \qquad (I.E.T.E, June 2003)

Solution. The given differential equation is $\dfrac{1}{z}\dfrac{dz}{dx} + \dfrac{1}{x}\log z = \dfrac{1}{x}(\log z)^2 \qquad ...(i)$

Put log $z = t$ so that $\dfrac{1}{z}\dfrac{dz}{dx} = \dfrac{dt}{dx}$. Therefore (*i*) become,

$$\dfrac{dt}{dx} + \dfrac{t}{x} = \dfrac{t^2}{x} \quad \text{or} \quad \dfrac{1}{t^2}\dfrac{dt}{dx} + \dfrac{1}{xt} = \dfrac{1}{x} \qquad \text{...(ii)}$$

Now set $\dfrac{1}{t} = v$ so that $-\dfrac{1}{t^2}\dfrac{dt}{dx} = \dfrac{dv}{dx}$. The equation (*ii*) reduces to

$$-\dfrac{dv}{dx} + \dfrac{v}{x} = \dfrac{1}{x} \quad \text{or} \quad \dfrac{dv}{dx} - \dfrac{v}{x} = -\dfrac{1}{x}. \text{ This is Leibnitz's linear equation in } v.$$

An integrating factor of this equation is I.F $= e^{\int -\frac{1}{x}dx} = e^{-\log x} = e^{\log x^{-1}} = \dfrac{1}{x}$

The solution of the transformed equation is $v \cdot \dfrac{1}{x} = -\int \dfrac{1}{x} \cdot \dfrac{1}{x} dx + c = \dfrac{1}{x} + c.$

Replacing v by $1/\log z$, we obtain

$(x \log z)^{-1} = x^{-1} + c$ or $(\log z)^{-1} = 1 + cx$, the required solution.

Example 17.26 Solve the differential equation

$$x\, dy - [y + xy^3 (1 + \sin x)]\, dx = 0.$$

Solution. Write the given equation as

$$\dfrac{dy}{dx} - \dfrac{y}{x} = y^3(1 + \sin x), \text{ which is a Bernoulli equation with } n = 3.$$

Set $\qquad z = y^{1-3} = y^{-2}.$

Therefore, $\dfrac{dz}{dx} = -2y^{-3}\dfrac{dy}{dx} = -2y^{-3}\left[\dfrac{y}{x} + y^3(1 + \sin x)\right] = \dfrac{-2y^{-2}}{x} - 2(1 + \sin x)$

or $\qquad \dfrac{dz}{dx} + \dfrac{2}{x}z = -2(1 + \sin x)$. An integrating factor of this equation is

$$\text{I.F} = e^{\int \frac{2}{x}dx} = e^{2\log x} = e^{\log x^2} = x^2. \text{ The solution is}$$

$$z\,x^2 = -2\int (1 + \sin x)x^2 dx = -2\int \left(\underset{I}{x^2} + \underset{II}{x^2 \sin x}\right)dx$$

$$= -\dfrac{2x^3}{3} - 2\left[x^2(-\cos x) - \int 2x(-\cos x)dx\right]$$

$$= -\dfrac{2x^3}{3} + 2x^2 \cos x - 4\int x \cos x\, dx$$

$$= -\frac{2x^3}{3} + 2x^2 \cos x - 4\left[x \sin x - \int \sin x\, dx\right]$$

$$= -\frac{2x^3}{3} + 2x^2 \cos x - 4x \sin x - 4\cos x + c$$

or $\quad\dfrac{x^2}{y^2} = -\dfrac{2}{3}x^3 + 2x^2 \cos x - 4x \sin x - 4\cos x + c.$

Exercise 17.6

Find the solution of the following equations.

1. $y' + \dfrac{y}{x} = \dfrac{y^2}{x^2}.$

Ans. $\dfrac{1}{y} = cx + \dfrac{1}{2x}.$

2. $y' + \dfrac{xy}{1-x^2} = xy^{1/2}.$

(A.M.I.E.-S-2005; Delhi 1999)

Ans. $y^{1/2} = c(1-x^2)^{1/4} - \{(1-x^2)/3\}.$

3. $x\left(\dfrac{dy}{dx}\right) + y = x^3 y^6.$ (Andhra, 1998; Mangalore, 1997; Delhi 1997)

Ans. $1 = (2.5 + cx^2)x^3 y^5.$

4. $xy(1 + xy^2)y' = 1.$ (AMIE., S-2006; I.E.T.E., W-1999; Gulbarga 1999)

Ans. $1/x = (2 - y^2) + ce^{-y^2/2}$

[**Hint:** Rewrite the given equation as $\dfrac{dx}{dy} - yx = y^3 x^2$]

5. $3\dfrac{dy}{dx} + 3\dfrac{y}{x} = 2x^4 y^4.$

Ans. $y^{-3} = -x^5 + cx^3.$

6. $y' + y \tan x = y^3 \cos x.$

Ans. $\cos^2 x = y^2\left(c - 2\sin x + \dfrac{2}{3}\sin^3 x\right).$

7. $y' - \dfrac{\tan y}{1+x} = (1+x)e^x \sec y.$ (ISM 2001) **Ans.** $\sin y = (1+x)(e^x + c).$

8. $r \sin\theta - \dfrac{dr}{d\theta}\cos\theta = r^2.$

Ans. $\dfrac{1}{r} = \sin\theta + c\cos\theta.$

9. $y' + x \sin 2y = x^3 \cos^2 y.$ (PTU 2002; Mysore, 1999; A.M.I.E. W-1995)

Alternative form: $\sec^2 y \dfrac{dy}{dx} + 2x \tan y = x^3$.   **Ans.** $\tan y = \dfrac{1}{2}(x^2 - 1) + ce^{-x^2}$.

10. $y^2 + x^2 \dfrac{dy}{dx} = xy \dfrac{dy}{dx}$.   **Ans.** $y/x = \log y + c$.

11. $(1 + x)y' + 1 = 2e^{-y}$.   (*PTU 2001*) **Ans.** $(x + 1)e^y = 2x + c$.

12. $2xy' = y^2(y^3 + x^2)$.   **Ans.** $x^2 = ce^{y^2/3} - y^3 - 3$.

13. $\tan y \dfrac{dy}{dx} + \tan x = \cos y \cos^3 x$.   (*A.M.I.E., W-1997; Mysore 1997S*)

   Ans. $\sec y \sec x = (\sin 2x)/4 + x/2 + c$.

14. $\tan x \cos y \, dy + \sin y \, dx + e^{\sin x} dx = 0$.

Alternative form: $\dfrac{dy}{\cot x} + \dfrac{dx}{\cot y} + \dfrac{e^{\sin x}}{\cos y} dx = 0$.   **Ans.** $\sin x \sin y = c - e^{\sin x}$.

15. $3y^2 \dfrac{dy}{dx} + 2xy^3 = 4xe^{-x^2}$.   **Ans.** $y^3 e^{x^2} = 2x^2 + c$.

16. $\dfrac{dy}{dx} + \dfrac{y}{x} \log y = \dfrac{y}{x^2}(\log y)^2$.   (*I.E.T.E. Dec 2003, 2001*)

   [**Hint:** The given equation may be written as

   $\dfrac{1}{y(\log y)^2} y' + \dfrac{1}{x(\log y)} = \dfrac{1}{x^2}$. Put $\dfrac{1}{\log y} = z$.]   **Ans.** $\dfrac{1}{x \log y} = \dfrac{1}{2x^2} + c$.

17. $xy \dfrac{dy}{dx} + (x^2 + x + y^2) = 0$.   (*I.E.T.E. June 1996*)

   [**Hint:** The given equation may be written as $yy' + (y^2/x) = -(x + 1)$.

   Put $y^2 = z$]   **Ans.** $6x^2 y^2 + 3x^4 + 4x^3 = c$.

18. $dx - xy (1 + xy^2) \, dy = 0$. (*AMIE W-2005*) **Ans.** $1 + x\left(y^2 - 2 + ce^{-y^2/2}\right) = 0$.

   [**Hint:** The equation may written as $\dfrac{1}{x^2} \dfrac{dx}{dy} - \dfrac{y}{x} = y^3$]

19. $y' + \sin y + x \cos y + x = 0$. [**Hint:** Write the equation in the form :

   $y' + 2\sin\dfrac{y}{2}\cos\dfrac{y}{2} + x\left(2\cos^2\dfrac{y}{2}\right) = 0 \Rightarrow y' \dfrac{\sec^2(y/2)}{2} + \tan\dfrac{y}{2} = -x$. Put $\tan y/2 = t$]

   Ans. $\tan\dfrac{y}{2} = 1 - x + ce^{-x}$.

20. $e^y(y' + 1) = e^x \cdot$ *(PTU 2002; V.T.U. 2000S; I.E.T.E., S-1997)*

Ans. $2e^y = e^x + ce^{-x}$.

21. $xy' + y = y^2 \log x$. **Ans.** $1 = (1 + \log x)y - cxy$.
22. $(x - y^2)\, dx + 2\, xy\, dy = 0$. **Ans.** $y^2/x = c - \log x$.

23. $\left(xy^2 - e^{1/x^3}\right) dx - x^2 y\, dy = 0$. *(PTU-2001)* **Ans.** $3y^2 = 2x^2\, e^{1/x^3} + cx^2$.

24. $y' \tan y + \tan x = \cos y \cos^2 x$. **Ans.** $\sec y = (c + \sin x)\cos x$.

25. $y\,dx - x\,dy + 3x^2 y^2 e^{x^3}\,dx = 0$. **Ans.** $(x/y) + e^{x^3} = c$.

26. Find the general solution of linear differential equation of first order

$$\frac{dx}{dy} + Px = Q \text{ (where } P \text{ and } Q \text{ are constants or functions of } y\text{).}$$

(IETE, June 2009) **Ans.** $x \cdot e^{\int Pdy} = \int Q \cdot e^{\int Pdy}\,dy + c$.

27. Solve $3\dfrac{dy}{dx} + xy = xy^{-2}$.   *(IETE, June 2009)* **Ans.** $y = \left[1 + ce^{-x^2/2}\right]^{1/3}$

17.10 Exact Differential Equations

The differential of a function of one or more variables is called an exact differential. For example $3x^2 y\, dx + x^3\, dy$ is an exact differential, being the differential of $x^3 y$.

Definition. *An equation obtained by equating an exact differential to zero, without any operation of reduction or elimination, is called an exact differential equation.*

Thus $\qquad 3x^2 y\, dx + x^3\, dy = 0 \qquad\qquad$ ...(i)

is an exact differential equation because it has been obtained by equating an exact differential to zero while $3y\, dy + x\, dy = 0$ is not because it has been obtained from (i) by dividing both sides by x^2.

17.11 Condition for M (x, y) dx + N (x, y) dy = 0 to be Exact

Necessary Condition. In case $M(x, y)dx + N(x, y)dy = 0$ is an exact differential equation then $M(x, y)\, dx + N(x, y)dy$ is an exact differential of some function u of x and y i.e.$u(x, y)$.

Hence $\qquad\qquad du = M(x, y)dx + N(x, y)dy.$ ...(17.15a)

The total or exact differential of $u(x, y)$ is expressed as

$$du = \frac{\partial u}{\partial x}dx + \frac{\partial u}{\partial y}dy \qquad\qquad \text{...(17.15b)}$$

Comparing Eqs. (17.15a) and (17.15b), we get

$$\frac{\partial u}{\partial x} = M(x, y) \text{ and } \frac{\partial u}{\partial y} = N(x, y) \qquad ...(17.16)$$

Now from Eq. (17.16), we obtain

$$\frac{\partial^2 u}{\partial y \, \partial x} = \frac{\partial M}{\partial y} \text{ and } \frac{\partial^2 u}{\partial x \partial y} = \frac{\partial N}{\partial x}.$$

Since $\quad \dfrac{\partial^2 u}{\partial y \partial x} = \dfrac{\partial^2 u}{\partial x \partial y}$, we have $\dfrac{\partial M}{\partial y} = \dfrac{\partial N}{\partial x} \qquad ...(17.17)$

which is, therefore, the *necessary* condition of exactness.

Sufficient Condition. If $\dfrac{\partial M}{\partial y} = \dfrac{\partial N}{\partial x}$, we shall prove that

$M(x, y) \, dx + N(x, y) \, dy = 0$ is exact.

Set $\int M(x, y) \, dx = u$, then $\dfrac{\partial u}{\partial x} = M(x, y)$ and hence

$$\frac{\partial^2 u}{\partial y \partial x} = \frac{\partial M}{\partial y} = \frac{\partial N}{\partial x} \text{ by equation (17.17), } i.e., \frac{\partial N}{\partial x} = \frac{\partial}{\partial x}\left(\frac{\partial u}{\partial y}\right).$$

Integration, we get $N = \dfrac{\partial u}{\partial y} + \phi(y)$, where $\phi(y)$ is a function of y alone.

Hence $\qquad M(x, y) \, dx + N(x, y) \, dy = \dfrac{\partial u}{\partial x} dx + \left[\dfrac{\partial u}{\partial y} + \phi(y)\right] dy$

$= \left(\dfrac{\partial u}{\partial x} dx + \dfrac{\partial u}{\partial y} dy\right) + \phi(y) dy = du + \phi(y) dy = d\left[u + \int \phi(y) dy\right]$ which shows that

$M(x, y)dx + N(x, y)dy$ is an exact differential and hence $M(x, y)dx + N(x, y)dy = 0$ is an exact differential equation. By proving the sufficiency part of the condition, we have found out the method of solution of an exact equation.

For when $\dfrac{\partial M}{\partial y} = \dfrac{\partial N}{\partial x}$, $M(x, y)dx + N(x, y)dy = d\left[u + \int \phi(y)dy\right]$.

$\therefore \quad M(x, y)dx + N(x, y)dy = 0$ gives $d\left[u + \int \phi(y)dy\right] = 0.$

Integrating, $\quad u + \int \phi(y)dy = c \quad$ or $\quad \int M(x, y)dx + \int \phi(y)dy = c, \qquad ...(17.18)$

which is the required solution.

Here $\int M(x, y)dx$ stands for integral of M w.r.t. x regarding y as a constant and $\phi(y)$ stands for those terms of N that donot involve x.

Method to solve an exact equation $Mdx + Ndy = 0$

(i) *Integrate M w.r.t. x, regarding y as a constant.*

(ii) *Integrate w.r.t. y those terms in N that don't contain x.*

(iii) *Equate the sum of integrals thus obtained to a constant which is the required solution.*

Example 17.27 Check the equation

$$(1 + 6y^2 - 3x^2y)\, dy = (3xy^2 - x^3)dx$$

for exactness. If it is exact, find the solution.

Solution. The given equation may be written as

$$\left(3xy^2 - x^3\right)dx - \left(1 + 6y^2 - 3x^2 y\right)dy = 0.$$

We have $M(x, y) = 3xy^2 - x^3$, $N(x, y) = -(1 + 6y^2 - 3x^2y)$.

$$\frac{\partial M}{\partial y} = 6xy, \frac{\partial N}{\partial x} = 6xy. \text{ Since } \frac{\partial M}{\partial y} = \frac{\partial N}{\partial x}.$$

Therefore the given equation is exact and its solution is given by

$$\int_{(y \text{ const.})} M dx + \int (\text{terms in } N \text{ not containing } x)\, dy = c.$$

i.e., $\int_{y \text{ const.}} \left(3xy^2 - x^3\right)dx - \int \left(1 + 6y^2\right)dy = c$ or $3x^2y^2/2 - x^4/4 - y - 2y^3 = c.$

Example 17.28 Solve the equation

$$(2y \sin x - \cos y)dx + (x \sin y - 2 \cos x)\, dy = 0.$$

Solution. We have $M(x, y) = 2y \sin x - \cos y$, $N(x, y) = x \sin y - 2 \cos x$.

$$\frac{\partial M}{\partial y} = 2 \sin x + \sin y, \frac{\partial N}{\partial x} = \sin y + 2 \sin x.$$

Since $\dfrac{\partial M}{\partial y} = \dfrac{\partial N}{\partial x}$, the given equation is exact. Therefore its solution is given by

$$\int_{(y \text{ const.})} (2y \sin x - \cos y)\, dx + \int (\text{terms in } N \text{ not containing } x)\, dy = c$$

i.e., $-2y \cos x - x \cos y = \text{const. or, } 2y \cos x + x \cos y = c.$

Example 17.29 Solve the equation $\left(y^2 e^{xy^2} + 4x^3\right)dx + \left(2xye^{xy^2} - 3y^2\right)dy = 0.$

Solution. We have $M(x, y) = y^2 e^{xy^2} + 4x^3$, $N = 2xye^{xy^2} - 3y^2$.

$$\frac{\partial M}{\partial y} = 2ye^{xy^2} + y^2 e^{xy^2} \cdot 2xy, \quad \frac{\partial N}{\partial x} = 2ye^{xy^2} + 2xye^{xy^2} \cdot y^2.$$

Since $\partial M/\partial y = \partial N/\partial x$, the given equation is exact. Therefore its solution is

$$\int\limits_{(y\,\text{const.})} M\,dx + \int(\text{terms in } N \text{ not containing } x)\,dy = c$$

i.e., $\int\limits_{(y\,\text{const.})}\left(y^2 e^{xy^2} + 4x^3\right)dx + \int\left(-3y^2\right)dy = c$ or $e^{xy^2} + x^4 - y^3 = c.$

Example 17.30 Determine for what values of *a* and *b*, the following differential equation is exact and obtain the general solution of the exact equation $(y + x^3)\,dx + (ax + by^3)dy = 0.$

Solution. We have $M(x, y) = y + x^3$, $N(x, y) = ax + by^3$.

$\dfrac{\partial M}{\partial y} = 1$, $\dfrac{\partial N}{\partial x} = a$. Hence if $a = 1$, the equation is exact *i.e.*, the equation is exact for $a = 1$, irrespective of the value of *b*. The solution is given by

$$\int\limits_{(y\,\text{const.})} M\,dx + \int(\text{terms in } N \text{ not containing } x)\,dy = c \;\; i.e., \;\; \int\left(y+x^3\right)dx + b\int y^3\,dy = c$$

or $xy + \dfrac{x^4}{4} + \dfrac{by^4}{4} = c$ for all *b* and *c* is the arbitrary constant.

Exercise 17.7

Solve the following differential equations:

1. $(x^2 + y^2 - a^2)x\,dx + (x^2 - y^2 - b^2)\,y\,dy = 0.$ *(MDU 2003, Kanpur, 1996)*
        Ans. $x^4 + 2\,x^2y^2 - 2a^2x^2 - y^4 - 2b^2y^2 = c$

2. $\dfrac{dy}{dx} = \dfrac{y+1}{(y+2)e^y - x}.$ **Ans.** $(y + 1)(x - e^y) = c$

3. $(2x^2 + 6xy - y^2)dx + (3x^2 - 2xy + y^2)dy = 0.$
            Ans. $2x^3 + 9x^2y - 3xy^2 + y^3 = c.$

4. $(x^2 - ay)dx = (ax - y^2)dy.$ *(Kurukshetra, 1998)* **Ans.** $x^3 + y^3 - 3\,axy = c.$
5. $(1 + e^{x/y})\,dx + e^{x/y}\,[1 - (x/y)]dy = 0.$ **Ans.** $x + ye^{x/y} = c.$
                  (V.T.U. 2001, Bhopal 1998, Triputi 1998s, AMIE S-2005)
6. $\{y\,(1 + x^{-1}) + \cos y\}\,dx + (x + \log x - x \sin y)dy = 0.$
                              (VTU 2000S; MDU 2002)
               Ans. $y(x + \log x) + x \cos y = c.$
7. $(\sec x \tan x \tan y - e^x)\,dx + \sec x \sec^2y\,dy = 0.$
          (AMIE, S-1997; MDU 2003) **Ans.** $\sec x \tan y - e^x = c.$
8. $y \sin 2x\,dx - (y^2 + \cos^2x)dy = 0.$ **Ans.** $3y \cos 2x + 2y^3 + 3y = c.$
9. $(5x^4 + 3x^2y^2 - 2xy^3)dx + (2x^3y - 3x^2y^2 - 5y^4)dy = 0.$ *(Gulbarga, 1999S)*
               Ans. $x^5 + x^3y^2 - x^2y^3 - y^5 = c.$

10. $(x^2 - 4xy - 2y^2)dx + (y^2 - 4xy - 2x^2)\, dx = 0.$

   (*Madras, 1998S, AMIE; S-1998*) **Ans.** $x^3 - 6x^2y - 6xy^2 + y^3 = c.$

11. $(2xy + y - \tan y)\, dx + (x^2 - x\tan^2 y + \sec^2 y)\, dy = 0.$ (*Osmania 2000S*)

   Ans. $x^2y + xy - x\tan y + \tan y = c.$

12. $(y \cos x + \sin y + y)\, dx + (\sin x + x \cos y + x)\, dy = 0,\ y = 0$ when $x = 0$

   (*A.M.I.E., W-2010*)

   Ans. $y \sin x + x \sin y + xy = 0.$

13. $(e^{2y} - y \cos xy)\, dx + (2xe^{2y} - x \cos xy + 2y)\, ay = 0.$

   Ans. $xe^{2y} - \sin xy + y^2 + c = 0.$

14. $\dfrac{dy}{dx} = \dfrac{xy^2 - \cos x \sin x}{y(1 - x^2)},\ y(0) = 2.$ **Ans.** $y^2(1 - x^2) - \cos^2 x = 3.$

15. $(xe^{xy} + 2y)y' + ye^{xy} = 0.$ **Ans.** $e^{xy} + y^2 = c.$

16. $\left(\dfrac{y}{x}\sec y - \tan y\right)dx + (\sec y \log x - x)dy = 0.$ (*I.E.T.E., W-2005*)

   [**Hint:** We may rewrite the given equation as

   $\left(\dfrac{y}{x} - \sin y\right)dx + (\log x - x \cos y)dy = 0$.]   **Ans.** $y \log x - x \sin y = c.$

17. Find the value of λ for which the differential equation $(xy^2 + \lambda x^2 y)dx$

   $+ (x + y)x^2 dy = 0$ is exact.   (*GGSIPU 2006*) **Ans.** $\lambda = 3.$

17.17 Integrating factor found by inspection

Many times when the equation $M(x, y)\, dx + N(x, y)\, dy = 0$ is not exact, it can be changed into an exact form by multiplying it by a suitable factor. Such a multiplier is called an *integrating factor* (I.F.) of the differential equation. For example equation $y\, dx - x\, dy + y^3\, dy = 0$ is not exact. However, if we divide it by y^2, we get

$$\frac{y\, dx - x\, dy}{y^2} + y\, dy = 0 \ \text{ or } \ \frac{d}{dx}\left(\frac{x}{y}\right) + \frac{d}{dy}\left(\frac{y^2}{2}\right) = 0.$$

This is an exact equation and the solution is given by $\dfrac{x}{y} + \dfrac{y^2}{2} = \text{constant}$. In this case $1/y^2$ is called an integrating factor (I.F.).

Thus it may be possible to guess an integrating factor in the given equation by grouping properly the terms of the equation. A few combinations of most common occurence which will be of immense help in reducing a non-exact equation to an exact form are:

(i) $x\, dy + y\, dx = d(xy);$ (ii) $\dfrac{x\, dy - y\, dx}{x^2} = d\left(\dfrac{y}{x}\right)$

(iii) $\dfrac{xdy - ydx}{y^2} = -d\left(\dfrac{x}{y}\right)$ (iv) $\dfrac{xdy - ydx}{x^2 + y^2} = \dfrac{(xdy - ydx)/x^2}{1 + (y/x)^2} = d\left[\tan^{-1}(y/x)\right]$

(v) $\dfrac{xdy - ydx}{x^2 + y^2} = \dfrac{(xdy - ydx)/y^2}{1 + (x/y)^2} = -d\left[\tan^{-1}(x/y)\right]$

(vi) $\dfrac{xdy - ydx}{xy} = d\left(\log\left|\dfrac{y}{x}\right|\right)$ (vii) $\dfrac{ydx - xdy}{xy} = d\left(\log|x/y|\right)$

(viii) $\dfrac{xdx + ydy}{x^2 + y^2} = \dfrac{1}{2}d\left[\log(x^2 + y^2)\right]$ (ix) $\dfrac{2xydx - x^2 dy}{y^2} = d\left(\dfrac{x^2}{y}\right)$

(x) $\dfrac{2xy\,dy - y^2 dx}{x^2} = d\left(\dfrac{y^2}{x}\right)$ (xi) $\dfrac{x\,dy + y\,dx}{xy} = d\left[\log(xy)\right]$

(xii) $\dfrac{xdy - ydx}{x^2 - y^2} = \dfrac{1}{2}d\left(\log\dfrac{x+y}{x-y}\right)$ (xiii) $d\left(\dfrac{e^x}{y}\right) = \dfrac{ye^x dx - e^x dy}{y^2}$

Example 17.31 Solve the following differential equations by finding an integrating factor by inspection.

(a) y dx − x dy + e^{1/x}dx = 0.

(b) $xdx + ydy = \dfrac{a^2(xdy - ydx)}{x^2 + y^2}$. *(AMIE S-2001; PTU 2005)*

(c) (1 + xy)y dx + (1 − xy) x dy = 0.
(d) y(2xy + e^x)dx = e^x dy. *(AMIE., W-2008)* (e)x dy − y dx − (1 − x^2) dx = 0.

Solution. (a) Multiplying the differential equation throughout by an integrating factor $1/x^2$, we obtain

$$\dfrac{ydx - xdy}{x^2} + \dfrac{e^{1/x}}{x^2}dx = 0 \quad \text{or} \quad -d(y/x) - d(e^{1/x}) = 0.$$

Integrating, we obtain $\left(\dfrac{y}{x}\right) + e^{1/x} = c$ or $y + xe^{1/x} = cx$.

(b) The given differential equation may be rewritten as:

$$xdx + ydy - \left\{a^2 \dfrac{(xdy - ydx)}{x^2} \middle/ 1 + \left(\dfrac{y}{x}\right)^2\right\} = 0$$

or $x\,dx + y\,dy - a^2 \cdot d(\tan^{-1} y/x) = 0.$

Integrating, we get $\dfrac{x^2}{2}+\dfrac{y^2}{2}-a^2\tan^{-1}(y/x)=c_1$

or $x^2+y^2-2a^2\tan^{-1}(y/x)=c$, the required result.

(c) The given equation may be rewritten as:

$$(y\,dx+x\,dy)+xy(y\,dx-x\,dy)=0. \qquad \qquad \text{...}(i)$$

If we multiply equation (i) by $1/x^2y^2$, the equation becomes exact.

$$\frac{(y\,dx+x\,dy)}{x^2y^2}+\frac{1}{xy}(y\,dx-x\,dy)=0$$

or $\dfrac{d(xy)}{x^2y^2}+\dfrac{y\,dx-x\,dy}{xy}=0$ or $d\!\left(-\dfrac{1}{xy}\right)+\left(\dfrac{dx}{x}-\dfrac{dy}{y}\right)=0$.

Integrating, we get the solution as $-\dfrac{1}{xy}+\log x-\log y=c$

or $-\dfrac{1}{xy}+\log\!\left(\dfrac{x}{y}\right)=c$ or $\log\!\left(\dfrac{x}{y}\right)=c+\left(\dfrac{1}{xy}\right)$.

(d) Dividing each term by y^2, we get $2x\,dx+\left(e^x/y\right)dx-\left(e^x/y^2\right)dy=0$

or $2x\,dx+\left(\dfrac{ye^x\,dx-e^x\,dy}{y^2}\right)=0$ or $2x\,dx+d(e^x/y)=0$.

Integrating, we get the solution as $x^2+(e^x/y)=c$,

or $x^2y+e^x=cy$ is the required solution.

(e) Here $1/x^2$ an integrating factor. Upon introducing it, the equation becomes

$$\frac{x\,dy-y\,dx}{x^2}-\left(\frac{1}{x^2}-1\right)dx=0.$$

Integrating, we get the solution as $\left(\dfrac{y}{x}\right)+\left(\dfrac{1}{x}\right)+x=c$ or $y+x^2+1=cx$.

17.13 Rules for Finding Integrating Factors

Rule 1. If the functions $M(x,\,y)$ and $N(x,\,y)$ in the equation $M(x,\,y)\,dx+N(x,\,y)\,dy=0$ are homogeneous functions of degree n and $Mx+Ny\neq 0$, then

$\dfrac{1}{Mx+Ny}$ is an integrating factor. If $Mx+Ny=0$, then $1/(xy)$, or $1/x^2$ or $1/y^2$ is an integrating factor.

Example 17.32 Solve $y^2dx + (x^2 - xy - y^2)dy = 0$.

Solution. The equation is homogeneous and $Mx + Ny \neq 0$,

then $\dfrac{1}{Mx + Ny} = \dfrac{1}{y(x^2 - y^2)}$ is an integrating factor.

Using this I, F, the given differential equation becomes

$$\frac{y}{x^2 - y^2}dx + \frac{x^2 - xy - y^2}{y(x^2 - y^2)}dy = 0, \text{ which is exact.}$$

Integrating it, the solution is

$$\int \frac{y}{x^2 - y^2}dx + \int \frac{1}{y}dx = c_2 \qquad \left(\because \frac{x^2 - xy - y^2}{y(x^2 - y^2)} = \frac{1}{y} - \frac{x}{x^2 - y^2} \right)$$
(y const.)

or $\dfrac{1}{2}\displaystyle\int \left(\dfrac{1}{x - y} - \dfrac{1}{x + y} \right)dx + \int \dfrac{1}{y}dx = c_2$ or $\dfrac{1}{2}\log \dfrac{x - y}{x + y} + \log y = \log c_1$.

or $(x - y)y^2 = c(x + y)$, the required result.

Example 17.33 Solve: $(x^2y - 2xy^2) dx - (x^3 - 3x^2y)dy = 0$. (PTU 2003)

Solution. The equation is homogeneous with $M(x, y) = x^2y - 2xy^2$ and $N(x, y) = -(x^3 - 3x^2y)$.

An *I.F.* is $\dfrac{1}{Mx + Ny} = \dfrac{1}{x(x^2y - 2xy^2) - y(x^3 - 3x^2y)} = \dfrac{1}{x^2y^2}$.

Multiplying the differential equation throughout by an $I.F = 1/x^2y^2$ the given differential equation becomes

$$\left(\frac{1}{y} - \frac{2}{x} \right)dx - \left(\frac{x}{y^2} - \frac{3}{y} \right)dy = 0 \text{ which can be seen as exact. Integrating it,}$$

the solution is $\dfrac{x}{y} - 2\log x + 3\log y = \log c$ or $\log \dfrac{y^3}{cx^2} = -\dfrac{x}{y}$ or $y^3 = cx^2e^{-x/y}$.

Rule 2. For the non-exact equation $M(x, y) dx + N(x, y) dy = 0$,

if $\left(\dfrac{\partial M}{\partial y} - \dfrac{\partial N}{\partial x} \right) \Big/ N$ is a function of x alone, say $\phi(x)$, then an integrating factor is

given by $I.F. = e^{\int \phi(x)dx}$.

Example 17.34 Solve $\left(xy^2 - e^{1/x^3}\right)dx - x^2y\,dy = 0$. (A.M.I.E., W-2005)

Solution. We have $M(x, y) = xy^2 - e^{1/x^3}$, $N(x, y) = -x^2 y$.

$\dfrac{\partial M}{\partial y} = 2xy$, $\dfrac{\partial N}{\partial x} = -2xy$. The equation is not exact.

Since, $\dfrac{\left(\dfrac{\partial M}{\partial y}\right) - \left(\dfrac{\partial N}{\partial x}\right)}{N} = \dfrac{2xy - (-2xy)}{-x^2 y} = \dfrac{-4}{x} = f(x)$ a function of x alone,

$e^{\int f(x) dx}$ is an integration factor.

We have $I.F. = e^{\int f(x) dx} = e^{\int \frac{-4}{x} dx} = e^{-4 \log x} = x^{-4}$.

Multiplying the differential equation throughout by an $I.F.$ $1/x^4$, we obain

$\left(\dfrac{y^2}{x^3} - \dfrac{1}{x^4} e^{1/x^3}\right) dx - \dfrac{y}{x^2} dy = 0$, which is an exact equation.

The solution is $\displaystyle\int \left(\dfrac{y^2}{x^3} - \dfrac{1}{x^4} e^{1/x^3}\right) dx + 0 = c$

or $-\dfrac{y^2 x^{-2}}{2} + \dfrac{1}{3} \displaystyle\int e^{x^{-3}} (-3x^{-4}) dx = c$ or $\dfrac{1}{3} e^{x^{-3}} - \dfrac{1}{2} \dfrac{y^2}{x^2} = c$.

Example 17.35 Solve: $(x^2 + y^2 + x) dx + xy\, dy = 0$.   *(A.M.I.E. W-2005)*

Solution. We have $M(x, y) = x^2 + y^2 + x$, $N(x, y) = xy$.

$\dfrac{\partial M}{\partial y} = 2y$, $\dfrac{\partial N}{\partial x} = y$, the equation is not exact.

Since, $\dfrac{\left(\dfrac{\partial M}{\partial y}\right) - \left(\dfrac{\partial N}{\partial x}\right)}{N} = \dfrac{2y - y}{xy} = \dfrac{1}{x} = f(x)$, a function of x alone, $e^{\int f(x) dx}$ is an

integrating factor.

We have $I.F. = e^{\int f(x) dx} = e^{\int \frac{1}{x} dx} = e^{\log x} = x$.

Multiplying the differential equation throughout by the $I.F.$, we get

$x(x^2 + y^2 + x) dx + x^2 y\, dy = 0$, which being exact. Integrating, we obtain the

solution as $\dfrac{x^4}{4} + \dfrac{x^2 y^2}{2} + \dfrac{x^3}{3} = c_1$ or $3x^4 + 4x^3 + 6x^2 y^2 = c$.

Rule 3. For the non-exact equation $M(x, y)\ dx + N(x, y)\ dy = 0$, if

$\dfrac{\left(\dfrac{\partial N}{\partial x}\right)-\left(\dfrac{\partial M}{\partial y}\right)}{M}$ is a function of y alone, say, $g(y)$, then the integrating factor is

given by $I.F. = e^{\int g(y)\,dy}.$

Example 17.36 Solve the equation:

$$(3x^2y^4 + 2\,xy)\,dx + (2x^3y^3 - x^2)\,dy = 0.$$

Solution. We have $M(x, y) = 3x^2 y^4 + 2xy,\ \dfrac{\partial M}{\partial y} = 12x^2 y^3 + 2x.$

$N(x,y) = 2x^3 y^3 - x^2,\ \dfrac{\partial N}{\partial x} = 6x^2 y^3 - 2x.$

The equation is not exact. We have

$$\dfrac{\left(\dfrac{\partial N}{\partial x}\right)-\left(\dfrac{\partial M}{\partial y}\right)}{M} = \dfrac{6x^2 y^3 - 2x - 12x^2 y^3 - 2x}{xy\left(3xy^3 + 2\right)} = -\dfrac{2}{y} = g(y)$$

which is a function of y alone. The integrating factor is

$$I.F. = e^{\int g(y)dy} = e^{\int (-2/y)dy} = \dfrac{1}{y^2}.$$

Multiplying the given differential equation throughout by the integrating

factor $\dfrac{1}{y^2}$, we get $\left(3x^2 y^2 + \dfrac{2x}{y}\right)dx + \left(2x^3 y - \dfrac{x^2}{y^2}\right)dy = 0.$

This equation is exact and its solution is

$$x^3 y^2 + \left(\dfrac{x^2}{y}\right) = c, \text{ a constant or } x^3 y^3 + x^2 = cy.$$

Example 17.37 Solve $(xy^3 + y)dx + 2(x^2 y^2 + x + y^4)\,dy = 0.$ (A.M.I.E. W-2005)
Solution. We have $M(x, y) = xy^3 + y$, $N(x, y) = 2(x^2 y^2 + x + y^4)$.

$\partial M/\partial y = 3xy^2 + 1$, $\partial N/\partial x = 4xy^2 + 2$. The equation is not exact.

Since $\qquad \dfrac{1}{M}\left(\dfrac{\partial N}{\partial x} - \dfrac{\partial M}{\partial y}\right) = \dfrac{1}{y(xy^2 + 1)}\left(4xy^2 + 2 - 3xy^2 - 1\right) = \dfrac{1}{y}$

which is a function of y alone. The integrating factor is

$$I.F. = e^{\int g(y)dy} = e^{\int \frac{1}{y}dy} = e^{\log y} = y.$$

Multiplying the given differential equation throughout by the integrating factor y, we get

$(xy^4 + y^2) dx + (2x^2y^3 + 2 xy + 2y^5) dy = 0$, which is an exact equation.

Integrating, we obtain the solution as $3x^2y^4 + 6xy^2 + 2y^6 = c$.

Rule 4. If the non-exact equation $M(x, y) dx + N(x, y) dy = 0$ is of the form

$yf_1(xy) dx + xf_2(xy) dy = 0$, then $\dfrac{1}{Mx - Ny}$ is an integrating factor, provided

$Mx - Ny \neq 0$.

Example 17.38 Solve the equation:

$$(x^3y^3 + x^2y^2 + xy + 1)y\, dx + (x^3y^3 - x^2y^2 - xy + 1)x dy = 0.$$

Solution. The equation is of the form $yf_1(xy)dx + xf_2(xy)dy = 0$ and $1/(Mx - Ny)$ is an integrating factor.

We have $M(x, y) = y(x^3y^3 + x^2y^2 + xy + 1)$, $N(x, y) = x(x^3y^3 - x^2y^2 - xy + 1)$.

An integrating factor is

$$\frac{1}{Mx - Ny} = \frac{1}{xy(x^3y^3 + x^2y^2 + xy + 1) - (x^3y^3 - x^2y^2 - xy + 1)xy}$$

$$= \frac{1}{2x^2y^2(xy + 1)}.$$

Multiplying the given differential equation throughout by this factor, we obtain

$$\left(\frac{x^2y^2 + 1}{x^2y^2}\right)ydx + \left[\frac{(x^2y^2 - xy + 1) - xy}{x^2y^2}\right]xdy = 0$$

or $\quad (ydx + xdy) + \dfrac{ydx + xdy}{x^2y^2} - \dfrac{2x^2y}{x^2y^2}dy = 0$

or $\quad d(xy) + \dfrac{d(xy)}{x^2y^2} - \dfrac{2}{y}dy = 0$ or $dz + \dfrac{dz}{z^2} - \dfrac{2}{y}dy = 0$, where $z = xy$.

Integrating, we obtain $z - (1/z) - 2\log y = c$ or $xy - (1/xy) - 2\log y = c$,

Example 17.39 Solve $(1 + xy)y\, dx + (1 - xy) x\, dy = 0$.

Solution. The given equation is of the form

$yf_1(xy) dx + xf_2(xy) dy = 0$. and $1/(Mx - Ny)$ is an integrating factor.

We have $\quad M(x, y) = (1 + xy)y$, $N(x, y) = (1 - xy)x$.

The integrating factor , I.F. $= \dfrac{1}{Mx - Ny} = \dfrac{1}{(1 + xy)xy - (1 - xy)xy} = \dfrac{1}{2x^2y^2}$.

Multiplying the differential equation throughout by the *I.F.* $1/2\, x^2 y^2$, we obtain

$$\left(\frac{1}{2x^2 y}+\frac{1}{2x}\right)dx+\left(\frac{1}{2xy^2}-\frac{1}{2y}\right)dy=0, \text{ which is an exact equation.}$$

The solution is $\dfrac{1}{2y}\left(-\dfrac{1}{x}\right)+\dfrac{1}{2}\log x-\dfrac{1}{2}\log y=c_1$ or $\log\dfrac{x}{y}-\dfrac{1}{xy}=c.$

Example 17.40 Solve: $(xy \sin xy + \cos xy)\, y\, dx + (xy \sin xy - \cos xy)\, x\, dy = 0$

Solution. The given equation is of the form

$f_1(xy)\, y\, dx + f_2(xy)\, x\, dy = 0$ and $1/(Mx - Ny)$ is an integrating factor.

We have $M(x, y) = (xy \sin xy + \cos xy)y,$ $\qquad N(x, y) = (xy \sin xy - \cos xy)x.$
The integrating factor is

$$I.F. = \frac{1}{Mx-Ny}=\frac{1}{(xy\sin xy+\cos xy)xy-(xy\sin xy-\cos xy)xy}=\frac{1}{2xy\cos xy}.$$

Multiplying the given differential equation by the integrating factor $1/2xy\cos xy$, we get

$$\frac{1}{2}\left(\tan xy+\frac{1}{xy}\right)ydx+\frac{1}{2}\left(\tan xy-\frac{1}{xy}\right)xdy=0$$

or $\quad \tan xy\,(ydx+xdy)+\dfrac{1}{x}dx-\dfrac{1}{y}dy=0 \quad$ or $\quad (\tan xy)d(xy)+\dfrac{1}{x}dx-\dfrac{1}{y}dy=0$

or $\quad \tan z\,dz+\dfrac{1}{x}dx-\dfrac{1}{y}dy=0$, where $z = xy.$

Integrating it, the solution is
$\log |\sec z| + \log x - \log y = \log c$

or $\quad \log\left\{\dfrac{x\sec z}{y}\right\}=\log c$ or $\dfrac{x}{y}\sec z=c$ or $x\sec(xy)=cy.$

Rule 5. For the equation of the type

$x^a y^b(my\, dx + nx\, dy) + x^{a'} y^{b'}(m'y\, dx + n'x\, dy) = 0$, an integrating factor is $x^h y^k$

where $\qquad \dfrac{a+h+1}{m}=\dfrac{b+k+1}{n},\ \dfrac{a'+h+1}{m'}=\dfrac{b'+k+1}{n'}.$

Example 17.41 Solve $x^3 y^3(2y\, dx + x\, dy) - (5\,y\, dx + 7\,x\, dy) = 0$

Solution. Comparing the given equation with

$$x^a y^b\,(mydx+nxdy)+x^{a'}y^{b'}\,(m'ydx+n'xdy)=0.$$

we have $a = b = 3,\ m = 2,\ n = 1;\ a' = b' = 0,\ m' = -5,\ n' = -7.$

The integrating factor is $I.F = x^h y^k$

where
$$\frac{a+h+1}{m} = \frac{b+k+1}{n}, \frac{a'+h+1}{m'} = \frac{b'+k+1}{n'}$$

i.e.,
$$\frac{3+h+1}{2} = \frac{3+k+1}{1}, \frac{0+h+1}{-5} = \frac{0+k+1}{-7}$$

or
$$h - 2k = 4, 7h - 5k = -2.$$

Solving these, we get $h = -8/3$ and $k = -10/3$. $\therefore$ $I.F. = \dfrac{1}{x^{8/3} y^{10/3}}$.

Multiplying the given differential equation by the integrating factor $1/x^{8/3} y^{10/3}$, we get

$$x^{1/3} y^{-1/3} (2ydx + xdy) - \frac{1}{x^{8/3} y^{10/3}} (5ydx + 7x\,dy) = 0$$

or $(2x^{1/3} y^{2/3} - 5x^{-8/3}y^{-7/3})dx + (x^{4/3}y^{-1/3} - 7x^{-5/3}y^{-10/3})dy = 0$
which is an exact equation. The solution is

$$\frac{3}{2} x^{4/3} y^{2/3} + 3x^{-5/3} y^{-7/3} = c, \text{ or } x^3 y^3 + 2 = cx^{5/3}y^{7/3}.$$

Example 17.42 Solve $(y^2 + 2x^2y)\,dx + (2x^3 - xy)\,dy = 0$ *(VTU 2001, DU 1998)*

Solution. Rewriting the equation as

$y(y\,dx - x\,dy) + 2x^2 (y\,dx + x\,dy) = 0$ and comparing with

$x^a y^b (mydx + nxdy) + x^{a'} y^{b'} (m'ydx + n'xdy) = 0,$

we have $a = 0, b = 1, m = 1, n = -1; a' = 2, b' = 0, m' = n' = 1$.
The integrating factor is $I.F. = x^h y^k$

where
$$\frac{a+h+1}{m} = \frac{b+k+1}{n}, \frac{a'+h+1}{m'} = \frac{b'+k+1}{n'}$$

i.e.,
$$\frac{0+h+1}{1} = \frac{1+k+1}{-1}, \frac{2+h+1}{1} = \frac{0+k+1}{1} \text{ or } h + k = -3 \text{ and } h - k = -2.$$

Solving these, we get $h = -5/2; k = -1/2$. $\therefore$ $I.F. = 1/x^{5/2}y^{1/2}$.

Multiplying the given differential equation by the integrating factor $1/x^{5/2}y^{1/2}$, we get

$$(x^{-5/2}y^{3/2} + 2x^{-1/2}y^{1/2})dx + (2x^{1/2}y^{-1/2} - x^{-3/2}y^{1/2})dy = 0,$$
which is an exact equation.

The solution is $-\dfrac{2}{3} x^{-3/2} y^{3/2} + 4x^{1/2} y^{1/2} = c$ or $6(xy)^{1/2} - (y/x)^{3/2} = c.$

Exercise 17.8

1. Find the value of α so that $e^{\alpha x^2}$ is an integrating factor of following differential equations:

   (a) $x(1-y)dx - dy = 0$. (I.E.T.E. June 2005) Ans. $1/2$.

   (b) $\left(e^{-y^2/2} - xy\right)dy - dx = 0$. (I.E.T.E. Dec. 2005) Ans. $1/2$.

2. Given that the integrating factor of the equation

   $$y\sec^2 x\,dx + \left[3\tan x - \left(\frac{\sec y}{y}\right)^2\right]dy = 0 \text{ is of the form } y^n. \text{ Find } n \text{ and hence}$$

   solve the equation. Ans. $n=2; y^3 \tan x - \tan y = c$.

   Solve the following differential equations:

3. $x^2 y\,dx - (x^3 + y^3)\,dy = 0$ Ans. $x^3 = 3y^3 (\log y - c)$.

4. $(3xy^2 - y^3)dx - (2x^2y - xy^2)\,dy = 0$. Ans. $3\log x - 2\log y + \left(\dfrac{y}{x}\right) = c$.

5. $(3xy - 2ay^2)\,dx + (x^2 - 2\,axy)dy = 0$. (Andhra 2000) Ans. $x^2(ay^2 - xy) = c$.

6. $(x^4 + y^4)\,dx - xy^3 dy = 0$. Ans. $y^4 = 4x^4 \log x + cx^4$.

7. $(x + y^2)\,dx - 2xy\,dy = 0$. Ans. $x = ce^{y^2/x}$.

8. $(x^2 + y^2 + 2x)dx + 2y\,dy = 0$. Ans. $(x^2 + y^2)e^x = c$.

9. $(2xy^2 - 3y^3)\,dx + (7 - 3xy^2)\,dy = 0$. Ans. $x^2 - (7/y) - 3xy = c$.

10. $2xy\log y\,dx + \left(x^2 + y^2\sqrt{y^2+1}\right)dy = 0$. Ans. $x^2\log y + \left\{(y^2 + 1)^{3/2}/3\right\} = c$.

11. $(xy^2 - x^2)\,dx + (3x^2y^2 + x^2y - 2x^3 + y^2)\,dy = 0$.
    Ans. $e^{6y}(54x^2y^2 - 36x^3 + 18y^2 - 6y + 1) = c$.

12. $(x^2y^2 + xy + 1)y\,dx + (x^2y^2 - xy + 1)x\,dy = 0$.
    Ans. $xy + \log(x/y) - (1/xy) = c$.

13. $y(xy + 2x^2y^2)dx + x(xy - x^2y^2)dy = 0$. Ans. $x^2 = cye^{1/xy}$.

14. $(y\,dx + 3x\,dy) + 2y(3y\,dx + 4x\,dy) = 0$. Ans. $6y + 1 = c\,x^{3/5}\,y^{9/5}$.

15. $(3x + 2y^2)\cdot y\,dx + 2x(2x + 3y^2)\,dy = 0$. Ans. $x^3y^4 + x^2y^6 = c$.

16. $(x^4e^x - 2m\,xy^2)dx + 2\,mx^2y\,dy = 0$. (JNTU 1998) Ans. $e^x + m(y/x)^2 = c$.

17. $(x^4y^4 + x^2y^2 + xy)y\,dx + (x^4y^4 - x^2y^2 + xy)\,x\,dy = 0$. (AMIE W-2004)
    Ans. $\{x^2y^2/2\} + \log(x/y) - (1/xy) = c$.

18. $(3x^2y^3e^y + y^3 + y^2)\,dx + (x^3y^3e^y - xy)\,dy = 0$. (IETE, June 2008)
    Ans. $y(x^3e^y + x) + x = cy$.

19. $\left(1 + xy + x^2 y^2\right) y \, dx + \left(1 - xy + x^2 y^2\right) x \, dy = 0.$ *(MDU-2005, 2003)*

Ans. $(-1/xy) + \log x + xy - \log y = c.$

20. $(2x^2 y^2 + y) dx - (x^3 y - 3x) \, dy = 0.$ *(MDU 2002)* **Ans.** $4x^2 y = 5 + cx^{4/7} y^{12/7}.$

21. $(8y \, dx + 8x \, dy) + x^2 y^3 \, (4y \, dx + 5 \, x \, dy) = 0.$ **Ans.** $4x^2 y^2 + x^4 y^5 = c.$

22. $\sin y \left(dy/dx\right) = \cos x \left(2 \cos y - \sin^2 x\right).$ *(I.E.T.E. Dec. 1996)*

Ans. $\cos y = \dfrac{\sin^2 x}{2} - \dfrac{\sin x}{2} + \dfrac{1}{4} + \lambda e^{-2 \sin x}.$

23. $x dx + y dy + \left\{(x dy - y dx)/\left(x^2 + y^2\right)\right\} = 0.$ *(AMIE S-2010)*

Ans. $x^2 - 2 \tan^{-1}\left(x/y\right) + y^2 = c.$

24. $\left(y^3 - 2x^2 y\right) dx + \left(2xy^2 - x^3\right) dy = 0.$

Ans. $x^2 y^2 \left(y^2 - x^2\right) = c.$

25. $\left(5x^3 + 12x^2 + 6y^2\right) dx + 6xy \, dy = 0.$ *(AMIE., S-2008)*

Ans. $x^5 + 3x^4 + 3x^2 y^2 = c.$

17.14 Differential Equations of the First Order and Higher Degree

Introduction. In the preceeding sections we have discussed the solution of the differential equations of the first order and first degree. Here we study the solution of the differential equations of the first order but with a degree higher than the first. For convenience, we will represent $\dfrac{dy}{dx}$ by p.

A most general form of the differential equation of the first order and nth degree is

$$p^n + P_1 p^{n-1} + P_2 \, p^{n-2} + \dots + P_n = 0, \qquad \dots(17.19)$$

where $P_1, P_2, \dots$ are functions of x and y.

We shall discuss the solution of Eq. (17.19) under the following cases:

(*a*) Equations solvable for p.

(*b*) Equations solvable for y.

(*c*) Equations solvable for x.

17.15 Equations Solvable for p

Since Eq. (17.19) is a polynomial of degree n in p, we resolve it into n linear factors of the form

$$(p - F_1)(p - F_2) \dots (p - F_n) = 0, \qquad \dots(17.20)$$

where $F_1, F_2, \dots F_n$ are functions of x and y.

Equation (17.20) is equivalent to
$$p - F_1 = 0, p - F_2 = 0, \dots p - F_n = 0$$

or
$$\frac{dy}{dx} = F_1; \frac{dy}{dx} = F_2; \dots \frac{dy}{dx} = F_n. \qquad \dots(17.21)$$

Each of the equations in (17.21) is a differential equation of the first order and first degree and therefore, can be easily integrated.

Let the solutions of these equations be
$$f_1(x, y, c_1) = 0; f_2(x, y, c_2) = 0 \dots \text{ and } f_n(x, y, c_n) = 0 \qquad \dots(17.22)$$

where $c_1, c_2, \dots c_n$ are arbitrary constants.

Combining all the solutions found out in Eq. (17.22), the solution of Eq. (17.19) is given by
$$f_1(x, y, c) \cdot f_2(x, y, c) \dots f_n(x, y, c) = 0. \qquad \dots(17.23)$$

There is no loss of generality by replacing n arbitrary constants $c_1, c_2, \dots c_n$ by a single constant c as every particular solution can be obtained by giving suitable value to c.

Example 17.43 Solve the equation $p^2 - 5p + 6 = 0$ $\left(p = \dfrac{dy}{dx}\right)$.

Solution. The given equation can be written as $(p - 2)(p - 3) = 0$.

Hence $p = 2$ or 3. When $p = 2$ i.e., $\dfrac{dy}{dx} = 2$. Integrating this, we get $y = 2x + c$.

Similarly $p = 3$ gives $y = 3x + c$. Hence the complete solution of the equation is given by
$$(y - 2x - c)(y - 3x - c) = 0.$$

Example 17.44 Solve equation $x^2\left(\dfrac{dy}{dx}\right)^2 - 2xy\left(\dfrac{dy}{dx}\right) + 2y^2 - x^2 = 0$.

Solution. The given equation can be written in terms of p as
$x^2 p^2 - 2xyp + 2y^2 - x^2 = 0$. Solving for p, we get
$$p = \frac{y \pm \sqrt{x^2 - y^2}}{x} = \frac{dy}{dx}.$$

This is a homogeneous equation and therefore putting $y = vx$, we get
$$v + x\left(\frac{dv}{dx}\right) = v \pm \sqrt{1 - v^2} \quad \text{or} \quad x\left(\frac{dv}{dx}\right) = \pm\sqrt{1 - v^2} \quad \text{or} \quad \frac{dv}{\sqrt{1 - v^2}} = \pm\frac{dx}{x}.$$

Integration gives $\sin^{-1}v = \pm \log x + c$ or $\sin^{-1}\left(\dfrac{y}{x}\right) = \pm \log x + c$.

Hence the solution of the given equation is

$$\left(\sin^{-1}\frac{y}{x} - \log x - c\right)\left(\sin^{-1}\frac{y}{x} + \log x - c\right) = 0.$$

Exercise 17.9

Solve the following differential equations:

1. $p^2 - 9p + 18 = 0.$ **Ans.** $(y - 3x - c)(y - 6x - c) = 0.$

2. $p^3 + 2xp^2 - y^2p^2 - 2xy^2p = 0.$ **Ans.** $(y - c)(y + x^2 - c)(xy - cy + 1) = 0.$

3. $p^2 + 2py \cot x = y^2.$ *(Kanpur, 1998)* **Ans.** $\left(y - \dfrac{c}{1+\cos x}\right)\left(y - \dfrac{c}{1-\cos x}\right) = 0.$

4. $y = x\left(p + \sqrt{1+p^2}\right).$ **Ans.** $x^2 + y^2 = \lambda x.$

5. $p^2 + x^3y - x^3p - yp = 0.$ **Ans.** $(y - ce^x)(4y - x^4 + c) = 0.$

6. $p - \dfrac{1}{p} = \dfrac{x}{y} - \dfrac{y}{x}.$ *(Raipur, 1998)*   **Ans.** $(xy - c)(x^2 - y^2 - c) = 0.$

7. $p(p + y) = x\,(x + y).$ **Ans.** $(2y - x^2 + c)(x + y - 1 + ce^{-x}) = 0.$

8. $xyp^2 + p(3x^2 - 2y^2) - 6xy = 0.$ **Ans.** $(y - cx^2)(y^2 + 3x^2 - c) = 0.$

9. $xyp^2 - (x^2 + y^2)\,p + xy = 0.$ **Ans.** $(y - cx)(y^2 - x^2 - c) = 0.$

10. $xy^2(p^2 + 2) = 2py^3 + x^3.$ **Ans.** $(x^2 - y^2 + c)(x^2 - y^2 + cx^4) = 0.$

17.16 Equations Solvable for y

Sometimes it is possible to express y explicitly as a function of x and p in the given equation $F(x, y, p) = 0$, so that it can be put in the form

$$y = f(x, p). \qquad\qquad ...(17.24)$$

Differentiation with respect to x leads to an equation of the form

$$\frac{dy}{dx} = p = \frac{\partial f}{\partial x} + \frac{\partial f}{\partial p}\frac{\partial p}{\partial x}. \qquad\qquad ...(17.25)$$

Equation (17.25) is a first order differential equation in x and p. Let the solution of this equation be $\phi(x, p, c) = 0$ ...(17.26)

Elimination of p between equation (17.24) and (17.26) yields the solution of the given equation.

In certain problems it may be tedious to eliminate p between (17.24) and (17.26). In such cases, (17.24) and (17.26) together constitute the solution giving x and y in terms of the parameter p.

Example 17.45 Solve the differential equation $y = 2\,px + p^4x^2$.

Solution. Differentiating with respect to x and denoting $\dfrac{dy}{dx}$ by p, we get

$$p = 2p + 2x\left(\frac{dp}{dx}\right) + 2xp^4 + 4x^2p^3\left(\frac{dp}{dx}\right) \quad \text{or} \quad \left[p + 2x\left(\frac{dp}{dx}\right)\right](1 + 2xp^3) = 0.$$

This gives $p + 2x\dfrac{dp}{dx} = 0$ and $1 + 2xp^3 = 0$.

Case I. $p + 2x\left(\dfrac{dp}{dx}\right) = 0$ or $2\left(\dfrac{dp}{p}\right) + \left(\dfrac{dx}{x}\right) = 0.$

Integration gives $2 \log p + \log x = \log \lambda$ or $p^2 = \dfrac{\lambda}{x}.$

Eliminating p from this relation and the given equation, we get

$$y = 2x\sqrt{\frac{\lambda}{x}} + \lambda^2 \quad \text{or} \quad \left(y - \lambda^2\right)^2 = 4\lambda x \, .$$

Case II. $1 + 2xp^3 = 0$. This gives $p = \left(-\dfrac{1}{2x}\right)^{1/3}.$

Eliminating p from this relation and the given equation, we get

$$y = \left(-\frac{1}{2x}\right)^{1/3}\left\{2x - \frac{x}{2}\right\}.$$

This also gives a solution but as it does not contain any arbitrary constant, it is called a singular solution.

Example 17.46 Solve the differential equation y = (1 + p)x + p².

Solution. Differentiating with respect to x, we get

$$\frac{dy}{dx} = p = 1 + p + x\left(\frac{dp}{dx}\right) + 2p\left(\frac{dp}{dx}\right).$$

This gives $\quad (x + 2p)\dfrac{dp}{dx} = -1 \quad$ or $\quad \dfrac{dx}{dp} + x = -2p$

This is a linear equation. An integrating factor is $I.F. = e^{\int dp} = e^p$.
The solution of this equation is

$$xe^p = -2\int pe^p \, dp + \lambda = -2\left(pe^p - e^p\right) + \lambda$$

or, $\qquad x = \lambda e^{-p} + 2(1 - p) \qquad \qquad \text{...(i)}$

Substituting this value of x in the given equation, we get

$$y = (1 + p)\{2(1 - p) + \lambda e^{-p}\} + p^2. \qquad \text{...(ii)}$$

Equations (*i*) and (*ii*) together with p as parameter constitute the solution of the given equation.

Note. The above equation is a particular case of more general class of equation $y = xf(p) + g(p)$ which is called Lagrange's equation.

Example 17.47 Solve y $-$ 2px = tan^{-1}(xp^2). (Bhopal, 1998)

Solution. We have $y = 2px + \tan^{-1}(xp^2)$. ...(1)

Differentiating both sides with respect to x,

$$\frac{dy}{dx} = p = 2\left[p + x\frac{dp}{dx}\right] + \frac{1}{1 + x^2 p^4}\cdot\left[p^2 + 2xp\frac{dp}{dx}\right]$$

or

$$\left(p + 2x\frac{dp}{dx}\right) + \left(p + 2x\frac{dp}{dx}\right)\frac{p}{1 + x^2 p^4} = 0$$

or

$$\left(p + 2x\frac{dp}{dx}\right)\left(1 + \frac{p}{1 + x^2 p^4}\right) = 0.$$ This gives $p + 2x\dfrac{dp}{dx} = 0.$

Separating the variables and integrating, we have $\displaystyle\int\frac{dx}{x} + 2\int\frac{dp}{p} = a$ constant.

or $\log x + 2\log p = \log\lambda$ or $\log xp^2 = \log\lambda$ or $xp^2 = \lambda$ or $p = \sqrt{\dfrac{\lambda}{x}}.$ ...(ii)

Eliminating p from (i) and (ii), we get $y = 2\sqrt{\lambda x} + \tan^{-1}\lambda,$

which is the general solution of (i).

Note: The factor $\left(1 + \dfrac{p}{1 + x^2 p^4}\right) = 0,$ has not been considered as it concerns 'singular solution' of (i), whereas we are interested only in finding *general solution*.

Exercise 17.10

Solve the following differential equations:

1. $y = 3x + \log p.$ **Ans.** $y = 3x + \log\dfrac{3}{1 - \lambda e^{3x}}.$

2. $p^3 + mp^2 = a(y + mx).$ **Ans.** The required solution is the p-eliminant of

$$ax = \lambda + \frac{3p^2}{2} - mp + m^2\log(p + m)\ \text{ and }\ p^3 + mp^2 = a(y + mx).$$

3. $xp + y = x^4 p^2.$ (D.U. 1998) **Ans.** $xy = \lambda^2 x + \lambda.$

4. $y = 2px + f(xp^2).$ **Ans.** $y = 2\lambda\sqrt{x} + f(\lambda^2).$

5. $y = p\tan p + \log\cos p.$ **Ans.** $x = \tan p + \lambda;\ y = p\tan p + \log\cos p.$

6. $y = \dfrac{1}{\sqrt{1 + p^2}} + b.$ **Ans.** $(x + \lambda)^2 + (y - b)^2 = 1.$

7. $y = x + a \tan^{-1} p$.

> **Ans.** $x = \lambda + \dfrac{a}{2}\left\{ \log \dfrac{p-1}{\sqrt{1+p^2}} - \tan^{-1} p \right\}$ with the given relation.

8. $y = (1 + p)x + e^p$. **Ans.** $x = \lambda e^{-p} - \dfrac{1}{2}e^p$; $y = \lambda(1+p)e^{-p} + \dfrac{1}{2}(1-p)e^p$.

9. $y = p^2x + p$.   **Ans.** $x = (\log p - p + \lambda)(p-1)^{-2}$, with the given relation.

10. $y = p \sin p + \cos p$.   **Ans.** $x = \sin p + \lambda$, with the given relation.

17.17 Equations Solvable for x

Let $F(x, y, p) = 0$ be the given equation. Suppose it is possible to solve it for x in terms of y and p so that $\qquad x = f(y, p)$. ...(17.27)

Differentiate (17.27) with respect to y and denote $\dfrac{dx}{dy} = \dfrac{1}{p}$, we get

$$\frac{dx}{dy} = \frac{1}{p} = \frac{\partial f}{\partial y} + \frac{\partial f}{\partial p}\frac{dp}{dy}. \qquad \qquad ...(17.28)$$

Equation (17.28) is a differential equation of first order and first degree in y and p. Let its solution be

$$\phi \,(y, p, c) = 0 \qquad \qquad ...(17.29)$$

Elimination of p from (17.27) and (17.29) gives the solution of the given equation. In case it is not possible to eliminate p then (17.27) and (17.29) together constitute the solution with p as a parameter.

Example 17.48 Solve the differential equation x = y + p².

Solution. Differentiating with respect to y and denoting $\dfrac{dx}{dy}$ by $\dfrac{1}{p}$, we get

$$\frac{dx}{dy} = \frac{1}{p} = 1 + 2p\left(\frac{dp}{dy}\right) \quad \text{or} \quad \frac{1-p}{p} = 2p\frac{dp}{dy}. \text{ Separating the variables, we get}$$

$-\dfrac{2p^2}{p-1}dp = dy$. Integration gives $-2 \log (p-1) -2p - p^2 = y + c$. ...(i)

Since it is not possible to eliminate p from (i) and the given equation, we denote the solution by

$$y = \lambda - 2\left[\log(p-1) + p + \frac{p^2}{2} \right] \text{ and } x = y + p^2.$$

Exercise 17.11

Solve the following differential equations:

1. $x + \left(\dfrac{p}{\sqrt{1+p^2}} \right) = a.$ **Ans.** $(x - a)^2 + (y + \lambda)^2 = 1.$

2. $p^3 - 2xyp + 4y^2 = 0.$ *(D.U. 1999)*   **Ans.** $16y = \lambda(2x - \lambda)^2.$

3. $p = \tan\left(x - \dfrac{p}{1+p^2} \right).$ **Ans.** $y + \left(1 + p^2\right)^{-1} = c,$ with the given relation.

4. $p^2 y + 2px - y = 0.$ **Ans.** $y^2 = 2cx + c^2.$
5. $p^3 - 4xyp + 8y^2 = 0.$ *(Kanpur, 1996)*   **Ans.** $y = c(x - c)^2.$
6. $y = 2px + y^2 p^3.$ **Ans.** $y^2 = 2cx + c^3.$

17.18 Clairaut's Equation

A non-linear differential equation of the form $y = px + f(p)$ is called the *Clairaut's equation.* Differentiating with respect to x, we get

$$\frac{dy}{dx} = p = x\left(\frac{dp}{dx}\right) + p + f'(p)\left(\frac{dp}{dx}\right) \qquad \qquad ...(17.30)$$

or $$\left[x + f'(p)\right]\frac{dp}{dx} = 0 \qquad \qquad ...(17.31)$$

Now either $\dfrac{dp}{dx} = 0$ or $x + f'(p) = 0.$

Let $\dfrac{dp}{dx} = 0.$ Integration gives $p = \lambda.$ Elimination of p between this relation and the given equation yields $y = \lambda x + f(\lambda)$ $...(17.32)$
The other factor $x + f'(p) = 0$ leads to a singular solution and is therefore not discussed.

Rule. In order to solve Clairaut's equation, we replace p by λ and this gives the solution.

Example 17.49 Solve $y = (x - a)\, p - p^2.$

Solution. The given equation is written as $y = px - ap - p^2.$ $...(i)$
Since (i) is a Clairaut's equation, its solution is
$$y = (x - a)\, \lambda - \lambda^2.$$

Example 17.50 Solve $p = \sin(y - px).$

Solution. The given equation may be written as
$\sin^{-1}p = y - px$ or $y = px + \sin^{-1}p$ which is the Clairaut's equation.
$\therefore$ Its solution is $y = \lambda x + \sin^{-1}\lambda.$

17.19 Equations Reducible to Clairaut's Form

There are certain equations of the first order and degree greater than one which can be solved by transforming them into Clairaut's form by appropriate substitutions.

Example 17.51 Solve the differential equation: $x^2(y - px) = p^2y$.

Solution. Put $x^2 = u$ and $y^2 = v$. This gives $2x\,dx = du$; $2y\,dy = dv$

or $\quad \dfrac{y}{x} \cdot \dfrac{dy}{dx} = \dfrac{dv}{du}$ i.e., $\dfrac{y}{x} p = \dfrac{dv}{du}$ or $p = \dfrac{x}{y} \cdot \dfrac{dv}{du}$.

Substituting for p in the given equation, we get

$$x^2 \left(y - \frac{x^2}{y} \frac{dv}{du} \right) = \frac{x^2}{y^2} \left(\frac{dv}{du} \right)^2 \cdot y$$

or $\quad \left(y^2 - x^2 \dfrac{dv}{du} \right) = \left(\dfrac{dv}{du} \right)^2$ or $v = u \dfrac{dv}{du} + \left(\dfrac{dv}{du} \right)^2$.

This is a Clairaut's equation and so its solution is

$$v = u\lambda + \lambda^2 \text{ or } y^2 = \lambda x^2 + \lambda^2$$

Exercise 17.12

Find the general solution of the following differential equations:

1. $(px - y)(py + x) = a^2p$; (use $x^2 = u$ and $y^2 = v$).

   Ans. $y^2 = \lambda x^2 - a^2\lambda/(\lambda+1)$.

2. $a\,x\,y\,p^2 + (x^2 - ay^2 - b)\,p - xy = 0$; use $x^2 = u$ and $y^2 = v$.

   Ans. $y^2 = \lambda x^2 - b\lambda/(a\lambda+1)$.

3. $xyp^2 - (x^2 + y^2 - 1)\,p + xy = 0$; use $x^2 = u$ and $y^2 = v$.

   Ans. $y^2 = \lambda x^2 + \left(\dfrac{\lambda}{\lambda-1} \right)$.

4. $y = 2px + y^2p^3$; use $y^2 = v$.

   Ans. $y^2 = \lambda x + \left(\dfrac{\lambda^3}{8} \right)$.

5. $4yp^2 + 2xp - y = 0$; Put $y^2 = v$.

   Ans. $y^2 = \lambda x + \lambda^2$.

6. $(px + y)^2 = py^2$; Put $xy = v$.

   Ans. $xy = \lambda y - \lambda^2$.

7. $e^{3x}(p - 1) + p^3e^{2y} = 0$; Put $e^x = u$ and $e^y = v$.

   Ans. $e^y = \lambda e^x + \lambda^3$.

8. $y = px + \sqrt{1 + p^2}$. *(Bangalore, 1998S)*   **Ans.** $y = \lambda x + \sqrt{1 + \lambda^2}$.

9. $y = px + \left(\dfrac{ap}{\sqrt{1 + p^2}}\right)$.   **Ans.** $y = \lambda x + \left(\dfrac{a\lambda}{\sqrt{1 + \lambda^2}}\right)$.

10. $p^3 x - p^2 y - 1 = 0$.   **Ans.** $y = \lambda x - \left(\dfrac{1}{\lambda^2}\right)$.

11. $(x - a)(dy/dx)^2 + (x - y)(dy/dx) - y = 0$. **Ans.** $(y - \lambda x)(1 + \lambda) + a\lambda^2 = 0$.

12. $(y - px)(p - 1) = p$.   *(Kanpur, 1996)* **Ans.** $(y - \lambda x)(\lambda - 1) = \lambda$.

13. $\sin(px - y) = p$.   *(Marathwada, 1998)*

Ans. General solution: $y = \lambda x - \sin^{-1}\lambda$.

14. $p = \log(px - y)$ **Ans.** General solution: $y = \lambda x - e^\lambda$.

Example 17.52. Solve

$$\frac{dy}{dx} + \frac{y \cos + \sin y + y}{\sin x + x \cos y + x} = 0, \; y = 0 \text{ when } x = 0.$$ *(A.M.I.E., W-2010)*

Solution. The given equation may be written as

$(y \cos x + \sin y + y)dx + (\sin x + x \cos y + x)dy = 0$.

We have $M(x, y) = y \cos x + \sin y + y$, $N(x, y) = \sin x + x \cos y + x$.

$\partial M / \partial y = \cos x + \cos y + 1$, $\partial N / \partial x = \cos x + \cos y + 1$.

Since $\partial M / \partial y = \partial N / \partial x$, the given equation is exact.

Therefore, its solution is given by

$$\int (y \cos x + \sin y + y)dx + \int (\text{terms in } N \text{ not containing } x)dy = c.$$
$$(y \text{ const.})$$

That is, $y \sin x + x \sin y + xy = c$

Using the initial condition $y(0) = 0$, we get $c = 0$.

Hence $y \sin x + x \sin y + xy = 0$ is the solution of initial value problem.

Chapter 18

Linear Differential Equations

18.1 Definitions

A differential equation in which the dependent variable and its derivatives occur in the first degree and are also not multiplied together is known as a linear differential equation. The most general form of a linear differential equation of nth order is given by

$$P_0 \frac{d^n y}{dx^n} + P_1 \frac{d^{n-1} y}{dx^{n-1}} + P_2 \frac{d^{n-2} y}{dx^{n-2}} + \dots + P_{n-1} \frac{dy}{dx} + P_n y = X, \; P_0 \neq 0 \qquad \dots(18.1)$$

where the function X and the coefficients $P_0, P_1, P_2, \dots, P_{n-1}, P_n$ are all functions of x. If X = 0, the resulting equation is called a *homogeneous linear equation,* otherwise *non-homogeneous linear differential equation.*

If the coefficients $P_0, P_1, P_2, \dots, P_{n-1}, P_n$ are all constants, eqn. (18.1) can be written in the form

$$a_0 \frac{d^n y}{dx^n} + a_1 \frac{d^{n-1} y}{dx^{n-1}} + a_2 \frac{d^{n-2} y}{dx^{n-2}} + \dots + a_{n-1} \frac{dy}{dx} + a_n y = X, \; a_0 \neq 0. \qquad \dots(18.2)$$

Eq. (18.2) is called a non-homogeneous linear differential equation of nth order with constant coefficients. Such equations are of common occurrence in many engineering disciplines like electrical circuits, mechanical vibrations etc.

18.2 Differential Operator D

The symbol D is called a *differential operator* because it transforms a differentiable function into another function. We have

$$\frac{d}{dx} \equiv D, \; \frac{d^2}{dx^2} \equiv D^2, \; \dots \frac{d^n}{dx^n} \equiv D^n,$$

so that $\dfrac{dy}{dx} = Dy, \; \dfrac{d^2 y}{dx^2} = D^2 y, \dots$ and in general $\dfrac{d^n y}{dx^n} = D^n y.$

Thus in terms of the operator notation D, the differential equation (18.2) can be written as

$$a_0 D^n y + a_1 D^{n-1} y + a_2 D^{n-2} y + ... + a_{n-1} Dy + a_n y = X$$

or

$$(a_0 D^n + a_1 D^{n-1} + a_2 D^{n-2} + ... + a_{n-1} D + a_n)y = X$$

or

$$f(D)y = X \qquad ...(18.3)$$

where

$$f(D) = a_0 D^n + a_1 D^{n-1} + a_2 D^{n-2} + ... + a_{n-1} D + a_n.$$

It is obvious that $f(D)$ is a polynomial of degree n in D. In many ways the operator D can be treated as an algebraical symbol and it obeys almost all laws of algebra. *i.e.*,

(*i*) $D(u + v) = Du + Dv$; (Distributive law)

(*ii*) $D^p D^q u = D^q D^p u$; (Commutative law)

(*iii*) $D(\lambda u) = \lambda Du$; (Commutative law for constant λ)

(*iv*) $D^p D^q u = D^{p+q} u$; (the Index law).

It may be mentioned that the commutative law does not hold good in respect of variables *i.e.*, $Du \neq uD$ and $D(uv) \neq uDv$.

Keeping in view the above properties of symbol D, the function $f(D)$ in Eq. (18.3) can be factorised into n linear factors which can be written in any order. For example

$$\frac{d^2 y}{dx^2} - 5\frac{dy}{dx} + 6y = (D^2 - 5D + 6)\, y = (D - 3)\,(D - 2)y = (D - 2)\,(D - 3)y.$$

Eq. (18.3) with $X = 0$ can be written as

$$(a_0 D^n + a_1 D^{n-1} + a_2 D^{n-2} + ... + a_{n-1}D + a_n)y = 0 \text{ or } f(D)y = 0 \qquad ...(18.4)$$

18.3 Theorem

If $y_1(x)$ and $y_2(x)$ are two independent solutions of Eq. (18.4), then $c_1 y_1 + c_2 y_2$ is also a solution of Eq. (18.4), where c_1 and c_2 are constants.

Proof. Since $y_1(x)$ and $y_2(x)$ are two independent solutions of Eq. (18.4), they must satisfy this equation. This yields

$$f(D)y_1 = 0 \qquad ...(18.5)$$

and

$$f(D)y_2 = 0 \qquad ...(18.6)$$

Multiplying Eq. (18.5) by c_1 and Eq. (18.6) by c_2 and adding, we obtain.

$$c_1 f(D)\, y_1 + c_2 f(D)\, y_2 = 0$$

or

$$f(D)[c_1 y_1 + c_2 y_2] = 0 \qquad ....(18.7)$$

Eq. (18.7) shows that $y = c_1 y_1 + c_2 y_2$ satisfies Eq. (18.4) and hence this is also a solution of Eq. (18.4).

The result of the above theorem can now be extended to any finite number of independent solutions of Eq. (18.4). Thus if $y = y_1$, $y = y_2$, ...and $y = y_n$ are n independent solutions of Eq.(18.4), then its complete solution $y = y_c$ is given by

$$y_c = c_1 y_1 + c_2 y_2 + ... + c_n y_n \qquad ...(18.8)$$

and
$$f(D)y_c = 0 \qquad ...(18.9)$$

where $c_1, c_2, ...c_n$ are arbitrary constants.

Again let $y = y_p$ be any function of x, not containing any arbitrary constant, which satisfies Eq. (18,3). Then

$$f(D)y_p = X \qquad ...(18.10)$$

Adding Eqs. (18.9) and (18.10), we have $f(D)[y_c + y_p] = X$. $\qquad ...(18.11)$

We observe from Eq. (18.11) that $y(x) = y_c(x) + y_p(x)$ satisfies Eq. (1833). Since $y_c(x)$ contains n arbitrary constants, the solution $y(x) = y_c(x) + y_p(x)$ also contains n arbitrary constants. Hence $y(x) = y_c(x) + y_p(x)$ is the general or complete solution of Eq. (18.3).

The part $y_c(x)$ is called the *complementary function* (C.F.) of Eq. (18.3) and is obtained from the solution of this equation after putting $X = 0$. The part $y_p(x)$ is called the *particular integral* (P.I.) of Eq. (18.3). It is some function of x, not involving any constant, which satisfies Eq. (18.3). Hence the general or complete solution (C.S.) of Eq. (18.3) is given by
$$y(x) = \text{C.F.} + \text{P.I.}$$

18.4 The Complementary Function (C.F.)

To find the complementary function, we solve the homogeneous equation

$$a_0 \frac{d^n y}{dx^n} + a_1 \frac{d^{n-1} y}{dx^{n-1}} + ... + a_{n-1}\frac{dy}{dx} + a_n y = 0, \ a_0 \neq 0. \qquad ...(18.12)$$

In operator notation, Eq. (18.12) can be written as
$$(a_0 D^n + a_1 D^{n-1} + a_2 D^{n-2} + ... + a_{n-1}D + a_n)y = 0$$

or
$$f(D)y = 0. \qquad ...(18.13)$$

Since $f(D)$ is a polynomial of degree n in D and D obeys the algebraical laws, we can in general factorise $f(D)$ into n linear factors and Eq. (18.13) can be written as
$$(D - m_1)(D - m_2)(D - m_3)...(D - m_n)y = 0 \qquad ...(18.14)$$

where $m_1, m_2, ...m_n$ are real or complex numbers.

The linear factors in Eq. (18.14) can be written in any order and the solution corresponding to each one of them can be obtained from the linear equations
$$(D - m_1)\, y = 0; \ (D - m_2)\, y = 0; \ ...(D - m_n)y = 0.$$

The differential equation $(D - m)y = 0$ for any m, has $y = e^{mx}$ as its solution. Substituting e^{mx} for y and its derivatives in Eq. (18.13), we get
$$(a_0 m^n + a_1 m^{n-1} + a_2 m^{n-2} + ... + a_n)e^{mx} = 0. \qquad ...(18.15)$$

Since $e^{mx} \neq 0$. Eq. (18.15) gives
$$a_0 m^n + a_1 m^{n-1} + a_2 m^{n-2} + ... + a_n = 0 \qquad ...(18.16)$$

Eq. (18.16) is called the *characteristic equation or the auxiliary equation* of Eq. (18.13) and is obtained by putting $D = m$ in $f(D) = 0$. Being an algebraic equation of degree n in m, Eq. (18.16) determines n roots $m_1, m_2 \ldots m_n$ as needed in Eq. (18.14).

These roots may be (*i*) real and different (*ii*) equal or (*iii*) imaginary.

Case I. *Characteristic equation having real and different roots.*

Let $m_1, m_2, \ldots, m_n$ be n real and different roots of Eq. (18.16). The solution corresponding to each one of them is given by $e^{m_1 x}, e^{m_2 x}, \ldots, e^{m_n x}$. Associating n arbitrary constants, the general solution of Eq. (18.13) is given by

$$y = c_1 e^{m_1 x} + c_2 e^{m_2 x} + \ldots + c_n e^{m_n x} \qquad \ldots(18.17)$$

Case II. *Characteristic equation having equal roots*

Let two of the n roots m_1 and m_2 of Eq. (18.13) be equal. Its solution as given in (18.17) above can be written as

$$y = \left(c_1 + c_2\right) e^{m_1 x} + c_3 e^{m_3 x} + \ldots + c_n e^{m_n x}$$

or $\quad y = c e^{m_1 x} + c_3 e^{m_3 x} + \ldots c_n e^{m_n x} \ldots(18.18)$ $\quad$ where $\quad c = c_1 + c_2$.

This solution contains only $(n - 1)$ arbitrary constants and is therefore, not a complete solution of the given equation. To obtain complete solution in this case, the given equation can be written as

$$\left(D - m_1\right)\left(D - m_1\right)\left(D - m_3\right)\ldots\left(D - m_n\right) y = 0. \qquad \ldots(18.19)$$

For the repeated factor, the corresponding part of the equation is given by

$$(D - m_1)^2 y = 0 \quad \text{or} \quad (D - m_1)(D - m_1) y = 0 \qquad \ldots(18.20)$$

Let $\qquad (D - m_1) y = v \qquad \ldots(18.21)$

Eq. (18.20) can be written as $(D - m_1)v = 0 \qquad$ or $\quad \dfrac{dv}{dx} - m_1 v = 0 \ldots(18.22)$

Eq. (18.22) is a linear first order differential equation and its solution is given by $v = A e^{m_1 x}$. Substituting for v in Eq. (18.21), we get

$$\left(D - m_1\right) y = A e^{m_1 x} \quad \text{or} \quad \dfrac{dy}{dx} - m_1 y = A e^{m_1 x} \qquad \ldots(18.23)$$

Eq. (18.23) is a linear differential equation of the first order. The integrating factor is $e^{-m_1 x}$. Hence the solution of (18.23) is given by

or $\quad y e^{-m_1 x} = \int A\,dx + B = Ax + B \qquad$ or $\quad y = (Ax + B) e^{m_1 x}. \qquad \ldots(18.24)$

Thus the complete solution of Eq. (18.13) is given by

$$y(x) = (Ax + B) e^{m_1 x} + c_3 e^{m_3 x} + \ldots + c_n e^{m_n x} \qquad \ldots(18.25)$$

It can similarly be shown that if the given equation $f(D)y = 0$ has three equal roots, each equal to m_1, then the solution corresponding to the repeated factor is given by the equation

$$(D - m_1)^3 y = 0. \qquad \text{...(18.26)}$$

Solving Eq. (18.26) on similar lines as for double roots above, its solution is given by

$$y(x) = (Ax^2 + Bx + C)e^{m_1 x}. \qquad \text{...(18.27)}$$

The complete solution of the given equation is given by

$$y(x) = \left(Ax^2 + Bx + C\right)e^{m_1 x} + c_4 e^{m_4 x} + ... + c_n e^{m_n x} \qquad \text{...(18.28)}$$

Case III. *Characteristic equation having imaginary roots*

We know that imaginary roots of an algegraic equation with real coefficients always occur in conjugate pairs. Let the two roots m_1 and m_2 of the characteristic equation (18.16) be denoted by $m_1 = \alpha + i\beta$ and $m_2 = \alpha - i\beta$. Since the roots are different, using case I, we can write the solution

$$y(x) = c_1 e^{(\alpha + i\beta)x} + c_2 e^{(\alpha - i\beta)x} + c_3 e^{m_3 x} + ... + c_n e^{m_n x}$$

$$= e^{\alpha x}\left[c_1 e^{i\beta x} + c_2 e^{-i\beta x}\right] + c_3 e^{m_3 x} + ... + c_n e^{m_n x}$$

$$= e^{\alpha x}\left[(c_1 + c_2)\cos\beta x + i(c_1 - c_2)\sin\beta x\right] + c_3 e^{m_3 x} + ... + ... c_n e^{m_n x}$$

$$\text{[Using Euler's theorem } e^{i\theta} = \cos\theta + i\sin\theta]$$

$$= e^{\alpha x}\left(A_1 \cos\beta x + A_2 \sin\beta x\right) + c_3 e^{m_3 x} + ... + c_n e^{m_n x} \qquad \text{...(18.29)}$$

where $\qquad A_1 = c_1 + c_2; \qquad A_2 = i(c_1 - c_2).$

Note: The form $e^{\alpha x}(A_1 \cos\beta x + A_2 \sin\beta x)$ can also be represented by $A_1 e^{\alpha x} \cos(\beta x + A_2)$ or by $A_1 e^{\alpha x} \sin(\beta x + A_2)$.

Case IV. *Characteristic equation having repeated imaginary roots*

Let $m_1 = \alpha + i\beta$ be repeated twice and $m_2 = \alpha - i\beta$ be also repeated twice. Since this is case of two repeated roots. Using case II we can write the solution as

$$y(x) = \left(A_1 x + A_2\right)e^{(\alpha + i\beta)x} + \left(A_3 x + A_4\right)e^{(\alpha - i\beta)x} + c_5 e^{m_5 x} + ... + c_n e^{m_n x}$$

$$= e^{\alpha x}\left\{\left(A_1' x + A_2'\right)\cos\beta x + \left(A_3' x + A_4'\right)\sin\beta x\right\} + c_5 e^{m_5 x} + ... + c_n e^{m_n x}$$

where $A_1' = A_1 + A_3,\ A_2' = A_2 + A_4,\ A_3' = i(A_1 - A_3),\ A_4' = i(A_2 - A_4).$

Working Procedure to find the C.F. of

$$\frac{d^n y}{dx^n} + a_1 \frac{d^{n-1} y}{dx^{n-1}} + ... + a_n y = X$$

Step 1. Write the characteristic equation of the corresponding homogeneous equation.

$$m^n + a_1 m^{n-1} + a_2 m^{n-2} + ... + a_n = 0 \text{ and solve it for } m.$$

Step 2. Then write the complementary function according to the following scheme.

Table 18.1

Nature of the roots of the characteristic equation	Complementary Function (C.F)
1. All roots real and distinct $m_1, m_2, m_3, \ldots \ldots m_n$	$c_1 e^{m_1 x} + c_2 e^{m_2 x} + c_3 e^{m_3 x} + \ldots \ldots$
2. Two roots real and equal and remaining distinct $m_1, m_1, m_3, m_4, \ldots \ldots m_n$	$\left(c_1 + c_2 x\right) e^{m_1 x} + c_3 e^{m_3 x} + \ldots \ldots \ldots$
3. Three roots real and equal and remaining distinct $m_1, m_1, m_1, m_4, \ldots \ldots m_n$	$\left(c_1 + c_2 x + c_3 x^2\right) e^{m_1 x} + c_1 e^{m_4 x} + \ldots$
4. A pair of roots complex and remaining distinct $\alpha \pm i\beta, m_3, m_4, \cdots \cdots m_n$	$e^{\alpha x}\left(c_1 \cos \beta x + c_2 \sin \beta x\right) + c_3 e^{m_3 x} + \ldots \ldots$
5. Two pairs of complex roots equal and remaining distinct. $\alpha \pm i\beta, \alpha \pm i\beta, m_5, m_6 \ldots \ldots m_n$	$e^{\alpha x}\{(c_1 + c_2 x)\cos \beta x$ $+ (c_3 + c_4 x)\sin \beta x\} + c_5 e^{m_5 x} + \ldots \ldots$

18.5 Illustrative Examples

Example 18.1 Solve the differential equation

$$\frac{d^2 y}{dx^2} + 3a\frac{dy}{dx} - 4a^2 y = 0.$$

Solution. The given homogeneous equation in operator notation is
$$(D^2 + 3aD - 4a^2)y = 0.$$

The characteristic equation of the differential equation is $m^2 + 3am - 4a^2 = 0$ or $(m - a)(m + 4a) = 0$. Its roots are $m = a, -4a$.

Hence a complete solution is $y(x) = c_1 e^{ax} + c_2 e^{-4ax}$.

Example 18.2 Solve the differential equation
$$y''' - 3y'' + 4y = 0.$$

Solution. The given homogeneous equation in operator notation is
$$(D^3 - 3D^2 + 4) \, y = 0.$$

The characteristic equation of the differential equation is $m^3 - 3m^2 + 4 = 0$ or $(m + 1)(m - 2)^2 = 0$. Its roots are $m = -1, 2, 2$. Hence a complete solution of the equation is $y(x) = c_1 e^{-x} + (c_2 + c_3 x)e^{2x}$.

Example 18.3 Solve the differential equation $y^{iv} + 32y'' + 256y = 0.$

(I.E.T.E., Dec. 2004)

Solution. The given homogeneous equation in operator notation is
$$(D^4 + 32D^2 + 256) \, y = 0.$$

The characteristic equation of the differential equation is $m^4 + 32\, m^2 + 256 = 0$ or $(m^2 + 16)^2 = 0$. The roots of this equation are $m = \pm 4i, \pm 4$.

Hence a complete solution is $y(x) = (c_1 + c_2 \, x) \cos 4x + (c_3 + c_4 x) \sin 4x.$

Example 18.4 Solve the differential equation

$$y^{iv} + a^4 y = 0.$$ *(Karnataka, 1995)*

Solution. The given homogeneous equation in operator notation is $(D^4 + a^4)y = 0.$
The characteristic equation of the differential equation is $m^4 + a^4 = 0$

or $\quad m^4 + 2m^2 a^2 + a^4 - 2m^2 a^2 = 0 \quad$ or $\quad \left(m^2 + a^2\right)^2 - \left(\sqrt{2}\,am\right)^2 = 0$

or $\left(m^2 + a^2 - \sqrt{2}\,am\right)\left(m^2 + a^2 + \sqrt{2}\,am\right) = 0.$

Its roots are $m = \dfrac{a(1 \pm i)}{\sqrt{2}}, \ \dfrac{a(-1 \pm i)}{\sqrt{2}}.$

Hence a complete solution of the equation is

$$y(x) = e^{ax/\sqrt{2}}\left[c_1 \cos\left(ax/\sqrt{2}\right) + c_2 \sin\left(ax/\sqrt{2}\right)\right]$$

$$+ e^{-ax/\sqrt{2}}\left[c_3 \cos\left(ax/\sqrt{2}\right) + c_4 \sin\left(ax/\sqrt{2}\right)\right].$$

Example 18.5 Solve the differential equation $\dfrac{d^3 s}{dt^3} + 9\dfrac{ds}{dt} = 0.$

Solution. The given homogeneous equation in operator notation is
$$(D^3 + 9D) \, s = 0.$$

The characteristic equation of the differential equation is $m^3 + 9m = 0$ or $m(m^2 + 9) = 0$. Its roots are $m = 0, \pm 3i$. Hence a complete solution of the equation is $s(t) = c_1 + c_2 \cos 3t + c_3 \sin 3t.$

Example 18.6 Solve the following initial value problems

(i) $4y'' - 8y' + 3y = 0$; $y(0) = 1$, $y'(0) = 3$.

(ii) $y'' + 4y' + 13y = 0$; $y(0) = 0$, $y'(0) = 1$.

(iii) $y''' + 3y'' - 4y = 0$; $y(0) = 1$, $y'(0) = 0$, $y''(0) = 1/2$.

Solution. (*i*) The characteristic equation of the given differential equation is $4m^2 - 8m + 3 = (2m - 3)(2m - 1) = 0$. Its roots are $m = 3/2$, $1/2$. Hence a complete solution is $y(x) = c_1 e^{(3x)/2} + c_2 e^{x/2}$.

By differentiating this, we find $y'(x) = \dfrac{3}{2} c_1 e^{(3x)/2} + \dfrac{1}{2} c_2 e^{x/2}$.

Substituting the given data into the equations for $y(x)$ and $y'(x)$ respectively, we have

$$y(0) = 1 = c_1 + c_2 \text{ and } y'(0) = 3 = \frac{3}{2} c_1 + \frac{c_2}{2}.$$

The solution of this system is $c_1 = 5/2$, $c_2 = -3/2$. Hence, the particular solution is $y(x) = [5e^{(3x)/2} - 3e^{x/2}]/2$.

(*ii*) The characteristic equation of the given differential equation is given by

$$m^2 + 4m + 13 = 0, \text{ or } m = \frac{-4 \pm \sqrt{16 - 52}}{2} = -2 \pm 3i = \alpha \pm i\beta.$$

The complete solution is given by

$$y(x) = (c_1 \cos\beta x + c_2 \sin\beta x)e^{\alpha x} = (c_1 \cos 3x + c_2 \sin 3x)e^{-2x}.$$

The first initial condition gives $y(0) = 0 = c_1$, and so

$$y'(x) = c_2 e^{-2x}(3 \cos 3x - 2 \sin 3x).$$

The second initial condition gives $y'(0) = 1 = 3 c_2$ or $c_2 = 1/3$.
Hence, the particular solution is

$$y(x) = (e^{-2x} \sin 3x)/3.$$

(*iii*) The characteristic equation of the given differential equation is $m^3 + 3m^2 - 4 = 0$ or $(m - 1)(m^2 + 4m + 4) = 0$ or $(m - 1)(m + 2)^2 = 0$. The roots of this equation are $m = 1, -2, -2$. The complete solution is

$$y(x) = c_1 e^x + (c_2 + c_3 x)e^{-2x} \qquad \qquad ...(i)$$

By successive differentiating (*i*), we find $y'(x) = c_1 e^x + c_3 e^{-2x} - 2(c_2 + c_3 x)e^{-2x}$, and $y''(x) = c_1 e^x - 4c_3 e^{-2x} + 4(c_3 x + c_2)e^{-2x}$.

Substituting the given data into the equations for $y(x)$, $y'(x)$ and $y''(x)$ respectively, we have

$y(0) = 1 = c_1 + c_2$, $y'(0) = 0 = c_1 - 2c_2 + c_3$, and $y''(0) = 1/2 = c_1 + 4c_2 - 4c_3$.
The solution of this system is $c_1 = c_2 = c_3 = 1/2$.
Hence, the particular solution is $y(x) = [e^x + (x + 1) e^{-2x}]/2$.

Example 18.7 Find a complete solution of each of the following equations.

(i) $\dfrac{d^6y}{dx^6} - 64y = 0$, (ii) $(D^8 + 6D^6 - 32D^2)y = 0$. *(I.E.T.E., June 2005)*

Solution. (*i*) The characteristic equaion of the given differential equation is
$m^6 - 64 = 0$ or $(m^2 - 4)(m^4 + 4m^2 + 16) = 0$.
Solving $m^4 + 4m^2 + 16 = 0$ for m^2, we get

$$m^2 = \frac{-4 \pm \sqrt{16-64}}{2} = -2 \pm 2\sqrt{3}\, i.$$

Let $m = \alpha + i\beta \Rightarrow m^2 = (\alpha + i\beta)^2 = \alpha^2 - \beta^2 + 2i\alpha\beta = -2 + 2\sqrt{3}\, i$.

By comparing the real and imaginiary parts, we have $\alpha^2 - \beta^2 = -2$ and $\alpha\beta = \sqrt{3}$.
Now $(\alpha^2 + \beta^2)^2 = (\alpha^2 - \beta^2)^2 + 4\alpha^2\beta^2 \Rightarrow \alpha^2 + \beta^2 = 4$.
Soliving the equations $\alpha^2 + \beta^2 = 4$ and $\alpha^2 - \beta^2 = -2$, we get

$\alpha = \pm 1$ and $\beta = \pm\sqrt{3}$. Therefore, the roots of the auxiliary equation are

$m = \pm 2, 1 \pm i\sqrt{3}, -1 \pm i\sqrt{3}$.

A complete solution of the given equation is.

$$y(x) = c_1 e^{2x} + c_2 e^{-2x} + \left(c_3 \cos\sqrt{3}x + c_4 \sin\sqrt{3}x\right)e^x$$

$$+ \left(c_5 \cos\sqrt{3}x + c_6 \sin\sqrt{3}x\right)e^{-x}.$$

(*ii*) The characteristic equation is
$$m^8 + 6m^6 - 32m^2 = m^2(m^6 + 6m^4 - 32) = 0.$$

when $m^2 = 2$, the polynomial inside the bracket vanishes, it has $m^2 - 2$ as a factor. Dividing $m^6 + 6m^4 - 32$ by $m^2 - 2$ to obtain $m^4 + 8m^2 + 16$. The characteristic equation can be re-written as

$$m^2\left(m^2 - 2\right)\left(m^4 + 8m^2 + 16\right) = m^2\left(m + \sqrt{2}\right)\left(m - \sqrt{2}\right)\left(m^2 + 4\right)^2 = 0,$$

which has unrepreated roots $m_1 = -\sqrt{2}$ and $m_2 = \sqrt{2}$ and the repeated roots $m_3 = 0$, $m_4 = 0 \pm 2i$.
A complete solution of the given equation is

$$y(x) = c_1 e^{-\sqrt{2}x} + c_2 e^{\sqrt{2}x} + c_3 + c_4 x + \left(c_5 + c_6 x\right)\cos 2x + \left(c_7 + c_8 x\right)\sin 2x.$$

Exercise 18.1

Find a complete solution of each of the following differential equations.

1. $2\dfrac{d^2y}{dx^2} + 5\dfrac{dy}{dx} + 2y = 0$.   **Ans.** $y(x) = c_1 e^{-x/2} + c_2 e^{-2x}$.

2. $4y'' + 12y' + 9y = 0$.   **Ans.** $y(x) = (c_1 + c_2 x)e^{-3x/2}$.

3. $y''' - 2y'' - 5y' + 6y = 0$.   **Ans.** $y(x) = c_1 e^x + c_2 e^{-2x} + c_3 e^{3x}$.

4. $y''' - 3y'' + 3y' - y = 0$. (*Kanpur, 1998*)   **Ans.** $y(x) = (c_1 + c_2 x + c_3 x^2)e^x$.

5. $y''' + y'' + 4y' + 4 = 0$.   **Ans.** $y(x) = c_1 e^{-x} + c_2 \cos 2x + c_3 \sin 2x$.

6. $4y''' + 4y'' + y' = 0$. (*Mysore; 1997S*)   **Ans.** $y(x) = c_1 + (c_2 + c_3 x)e^{-x/2}$.

7. (a) $\dfrac{d^3 y}{dx^3} + y = 0$.   **Ans.** $y(x) = c_1 e^{-x} + e^{x/2}\left(c_2 \cos \dfrac{\sqrt{3}}{2}x + c_3 \sin \dfrac{\sqrt{3}}{2}x\right)$.

   (b) $\left(D^3 - 8\right)y = 0$. (*IETE., Dec 2008*)

   Ans. $y = e^{-x}\left(c_1 \cos \sqrt{3}x + c_2 \sin \sqrt{3}x\right) + c_3 e^{2x}$.

8. $4y^{iv} - 12y''' - y'' + 27y' - 18y = 0$.

   Ans. $y(x) = c_1 e^x + c_2 e^{2x} + c_3 e^{-3x/2} + c_4 e^{3x/2}$.

9. $y^{iv} + 8y'' + 16y = 0$.   **Ans.** $y(x) = (Ax + B)\cos 2x + (Cx + D)\sin 2x$.

10. $(D^4 + 2D^3 + 3D^2 + 2D + 1)y = 0$.

   Ans. $y(x) = e^{-x/2}\left[(c_1 + c_2 x)\cos \dfrac{\sqrt{3}}{2}x + (c_3 + c_4 x)\sin \dfrac{\sqrt{3}}{2}x\right]$.

11. $\left(D^4 - 6D^3 + 12D^2 - 8D\right)y = 0$. (*IETE., Dec 2008*)

   Ans. $y(x) = a + (b + cx + dx^2)e^{2x}$.

Solve the following initial value problems.

12. $y'' - 4y' + 4y = 0;\ y(0) = 3,\ y'(0) = 4$ **Ans.** $y(x) = (3 - 2x)e^{2x}$.

13. $y'' - 10y' + (25 + \pi^2)y = 0;\ y(0) = 0,\ y'(0) = \pi e$. **Ans.** $y(x) = e^{5x+1}\sin \pi x$.

14. $(4D^2 + 16D + 17)y = 0;\ y = 1$ when $t = 0$ and $y = 0$ when $t = \pi$.

   Ans. $y(t) = e^{-2t}\left[c_1 \cos \dfrac{t}{2} + c_2 \sin \dfrac{t}{2}\right]$.

15. $y''' - 6y'' + 11y' - 6y = 0;\ y(0) = 0,\ y'(0) = -4,\ y''(0) = -18$.

   [**Hint:** The auxiliary equation is $m^3 - 6m^2 + 11m - 6 = 0$, or $(m-1)(m-2)$ $(m-3) = 0$. The general solution is $y(x) = c_1 e^x + c_2 e^{2x} + c_3 e^{3x}$. Find $y'(x)$, $y''(x)$ and substituting the initial conditions and then solving the equations obtained to get c_1, c_2 etc.]   **Ans.** $y(x) = e^x + 2e^{2x} - 3\,e^{3x}$.

18.6 A. Particular Integral (P.I.)

We have seen that particular integral y_p is that value of y which does not involve any arbitrary constant and satisfies the entire equation $f(D)y = X$. To make this concept further clear we consider the differential equation

$\dfrac{d^2 y}{dx^2} + 2\dfrac{dy}{dx} + 5y = 2 + 5x$. It is obvious that $y = x$ is a solution of this equation.

This is called the particular integral. Similarly the equation

$$\frac{d^3y}{dx^3} - 7\frac{d^2y}{dx^2} + 5\frac{dy}{dx} + 6y = 5e^x$$

is satisfied by $y = e^x$ which is the particular integral of the equation.

18.7 The inverse operator $\frac{1}{f(D)}$

Definition: $\frac{1}{f(D)}X$ defines that function of x which when operated on by $f(D)$

gives X i.e. $f(D)\left[\frac{1}{f(D)}X\right] = X$ ...(18.30)

In other words $\frac{1}{f(D)}X$ is the solution of $f(D)y = X$ and is therefore, called the

particular integral of this equation.

· Thus the particular integral y_p of the equation $f(D)y = X$ is symbolically given

by $y_p = \frac{1}{f(D)}X$.

It must be noted that $\frac{1}{f(D)}$ is an operator and must always precede the func-

tion operated on. It is incorrect to write y_p as $X \cdot \frac{1}{f(D)}$ or $\frac{X}{f(D)}$.

$f(D)$ and $\frac{1}{f(D)}$ are called inverse operators.

As particular cases, we will show that

(i) $\frac{1}{D}X = \int X \, dx$; **(ii)** $\frac{1}{D-\alpha}X = e^{\alpha x}\int e^{-\alpha x}X \, dx$ ...(18.31)

Case. *(i)* $\frac{1}{D}X$.   Let $\frac{1}{D}X = y$.

Operating on both sides by D and remembering D and $\frac{1}{D}$ are inverse opera-

tors, we get

$D\left[\frac{1}{D}X\right] = Dy$ or $X = \frac{dy}{dx}$. Therefore, $y = \int X \, dx$. Thus, $\frac{1}{D}X = \int X \, dx$.

Case (*ii*) $\dfrac{1}{D-\alpha}X$. Let $\dfrac{1}{D-\alpha}X = y$. Operating on both sides by $D - \alpha$, we get

$$(D-\alpha)\left[\frac{1}{D-\alpha}X\right] = (D-\alpha)y \quad \text{or} \quad X = (D-\alpha)y = \frac{dy}{dx} - \alpha y.$$

This is a linear first order differential equation. Its I.F. is $e^{-\alpha x}$. Hence its solution is

$$ye^{-\alpha x} = \int e^{-\alpha x} X\, dx \text{ or } y = e^{\alpha x}\int e^{-\alpha x} X\, dx \qquad \text{(no constant to be added).}$$

18.8 Methods of Finding the Particular Integrals (P.I.)

Let us consider the differential equation

$$\left(a_0 D^n + a_1 D^{n-1} + a_2 D^{n-2} + \ldots\ldots + a_{n-1}D + a_n\right)y = X \text{ or } f(D)y = X.$$

Its particular integral is given by $y_p(x) = \dfrac{1}{f(D)}X.$

We now discuss the methods for obtaining $y_p(x)$.

Case I. General Methods.

   (*i*) *Method of Factors.* Since $f(D)$ is a polynomial of degree n in D, it can be factorised into n linear factors *i.e.*, $f(D) = (D - m_1)(D - m_2) \ldots (D - m_n)$.

$$\therefore \quad y_p(x) = \frac{1}{(D-m_1)(D-m_2)\ldots(D-m_n)}X.$$

Now $\dfrac{1}{D-m_n}X = e^{m_n x}\int e^{-m_n x}X\, dx.$ \qquad (by Eq. (18.31))

The above result is next operated by $\dfrac{1}{D-m_{n-1}}$. Performing integration successively by taking the symbolic factors in the denominator from right to left, we get $y_p(x)$.

   (*ii*) *Method of Partial Fractions.* It is sometimes possible to express $\dfrac{1}{f(D)}$ into partial fractions so that

$$y_p(x) = \frac{1}{f(D)}X = \left(\frac{A_1}{x-m_1} + \frac{A_2}{x-m_2} + \ldots + \frac{A_n}{x-m_n}\right)X.$$

Using result (*ii*) of Eq. (18.31), we get

$$y_p(x) = A_1 e^{m_1 x}\int e^{-m_1 x}X\, dx + \ldots + A_n e^{m_n x}\int e^{-m_n x}X\, dx.$$

It may be mentioned that the use of second method is advisable when the factors are all distinct.

Example 18.8 Solve the differential equation $y'' - 3y' + 2y = e^{3x}$.

Solution. The given equation in operator notation is $(D^2 - 3D + 2)y = e^{3x}$.

The characteristic equation of the corresponding homogeneous equation is $m^2 - 3m + 2 = 0$ or $(m - 1)(m - 2) = 0$. Its roots are $m = 1, 2$. The complementary function is $y_c(x) = c_1 e^x + c_2 e^{2x}$.

The particular integral is $y_p(x) = \dfrac{1}{(D-1)(D-2)} \cdot e^{3x} = \dfrac{1}{(D-1)} \cdot \dfrac{1}{(D-2)} e^{3x}$.

Using the method of factors

Here we first evaluate $\dfrac{1}{D-2} e^{3x} = e^{2x} \int e^{3x} e^{-2x}\, dx = e^{2x} \int e^x dx = e^{3x}$.

$\therefore \quad \dfrac{1}{D-1}\left\{\dfrac{1}{D-2} e^{3x}\right\} = \dfrac{1}{D-1}\{e^{3x}\} = e^x \int e^{3x} e^{-x} dx = \dfrac{e^{3x}}{2}$.

Thus $\dfrac{1}{(D-1)} \cdot \dfrac{1}{(D-2)} e^{3x} = \dfrac{e^{3x}}{2}$.

Using the method of partial fractions

$y_p(x) = \dfrac{1}{(D-1)(D-2)} e^{3x} = \left(\dfrac{1}{D-2} - \dfrac{1}{D-1}\right) e^{3x} = \dfrac{1}{D-2} e^{3x} - \dfrac{1}{D-1} e^{3x}$.

$= e^{2x} \int e^{-2x} e^{3x}\, dx - e^x \int e^{-x} e^{3x} dx = e^{3x} - \dfrac{e^{3x}}{2} = \dfrac{e^{3x}}{2}$, which is same as above.

Hence the general solution of equation is $y(x) = y_c(x) + y_p(x) = c_1 e^x + c_2 e^{2x} + \dfrac{e^{3x}}{2}$.

Case II. Special Methods.
Depending upon the particular form of the function X, we shall now develop the methods of obtaining the particular integral, $y_p(x)$, of the differential equation $f(D)y = X$.

(i) $X = e^{\alpha x}$.
Since $D\, e^{\alpha x} = \alpha e^{\alpha x}$; $D^2 e^{\alpha x} = \alpha^2 e^{\alpha x}$,...., $D^{n-1} e^{\alpha x} = \alpha^{n-1} e^{\alpha x}$, $D^n e^{\alpha x} = \alpha^n e^{\alpha x}$.
Therefore $f(D)e^{\alpha x} = (a_0 \alpha^n + a_1 \alpha^{n-1} + a_2 \alpha^{n-2} + ... + a_{n-1}\alpha + a_n)e^{\alpha x}$.

Operating on both sides by $\dfrac{1}{f(D)}$, we obtain

$$\frac{1}{f(D)}\left[f(D)e^{\alpha x}\right] = \frac{1}{f(D)}\left[f(\alpha)e^{\alpha x}\right] \quad \text{or} \quad e^{\alpha x} = f(\alpha)\frac{1}{f(D)}e^{\alpha x}.$$

Thus
$$\frac{1}{f(D)}e^{\alpha x} = \frac{1}{f(\alpha)}e^{\alpha x}, \text{ provided } f(\alpha) \neq 0. \qquad \text{...(18.32)}$$

Case of Failure. If $f(a) = 0$, then $f(D)$ has $(D - a)$ as a factor, and the above method fails. To calculate $y_p(x)$ in this case, we write
$$f(D) = (D - \alpha)\,\phi(D) \text{ where } \phi(D) \neq 0 \text{ at } D = \alpha.$$

$$\therefore \quad y_p(x) = \frac{1}{f(D)}e^{\alpha x} = \frac{1}{\phi(D)(D-\alpha)}e^{\alpha x} = \frac{1}{\phi(\alpha)(D-\alpha)}e^{\alpha x}$$

$$= \frac{1}{\phi(\alpha)}e^{\alpha x}\int e^{-\alpha x}\cdot e^{\alpha x}dx = \frac{1}{\phi(\alpha)}e^{\alpha x}.$$

[Since $f(D) = (D - \alpha)\,\phi(D)$, differentiating both sides w.r.t. D, we get
$$f'(D) = (D - \alpha)\phi'(D) + \phi(D)$$
$$\therefore \qquad f'(D)|_{D=\alpha} = \phi(\alpha), \therefore f'(\alpha) = \phi(\alpha)].$$

$$\therefore \quad \frac{1}{f(D)}e^{\alpha x} = x\frac{1}{\phi(\alpha)}e^{\alpha x} = x\frac{1}{f'(\alpha)}e^{\alpha x}, \text{ provided } f'(\alpha) \neq 0. \qquad \text{...(18.33)}$$

If $f'(\alpha) = 0$, then $(D - \alpha)^2$ is a factor of $f(D)$ and the above method fails. Then we write

$$f(D) = (D - \alpha)^2 \,\phi(D).$$

Hence the particular integral $y_p(x)$ in this case is given by
$$y_p(x) = \frac{1}{\phi(D)(D-\alpha)^2}e^{\alpha x} = \frac{1}{\phi(\alpha)(D-\alpha)^2}e^{\alpha x}$$

$$= x^2\frac{1}{2!\phi(\alpha)}e^{\alpha x} = x^2\frac{1}{f''(\alpha)}e^{\alpha x}, \text{provided } f''(\alpha) \neq 0.$$

In general if $(D - \alpha)^r$ occur as a factor in $f(D)$, then
$f(D) = (D - \alpha)^r\,\phi(D)$ and $y_p(x)$ is given by

$$y_p(x) = \frac{1}{f(D)}e^{\alpha x} = \frac{1}{(D-\alpha)^r\,\phi(D)}e^{\alpha x} = \frac{x^r e^{\alpha x}}{r!\,\phi(\alpha)} = x^r\frac{1}{f^r(\alpha)}e^{\alpha x} \qquad \text{...(18.34)}$$

Example 18.9 Find the particular integral of the following differential equations

(i) $(D^3 - 2D^2 - 5D + 6)y = e^{4x}$,   (ii) $(D^3 - 3D^2 + 4)y = e^{2x}$.

Solution. (*i*) The particular integral is $y_p(x) = \dfrac{1}{D^3 - 2D^2 - 5D + 6} \cdot e^{4x}$

$$= \frac{1}{(D-1)(D-3)(D+2)} e^{4x} = \frac{1}{(4-1)(4-3)(4+2)} e^{4x} = \frac{1}{18} e^{4x}.$$

(*ii*)   The particular integral is $y_p(x) = \dfrac{1}{D^3 - 3D^2 + 4} e^{2x}$

$f(D) = D^3 - 3D^2 + 4$ vanishes when D is replaced by 2. It is a case of failure. We multiply the numerator by x and differentiate the denominator w.r.t. D.

$$\therefore y_p(x) = x \cdot \frac{1}{3D^2 - 6D} e^{2x}.$$

It is again a case of failure. We multiply the numerator by x and differentiate the denominator w.r.t. D.

$$y_p(x) = x^2 \frac{1}{6D - 6} e^{2x} = \frac{x^2}{6(2)-6} e^{2x} = \frac{x^2}{6} e^{2x}.$$

(*ii*) **When X = sin(ax + b) or cos (ax + b)**

We have $D \sin (ax + b) = a \cos (ax + b)$,

$D^2 \sin (ax + b) = -a^2 \sin (ax + b)$,

$D^4 \sin (ax + b) = (-a^2)^2 \sin (ax + b)$.

In general $(D^2)^k \sin (ax + b) = (-a^2)^k \sin (ax + b)$.

Since $f(D^2)$ is a polynomial expression in D^2, we have

$f(D^2) \sin (ax + b) = f(-a^2) \sin (ax + b)$.

Operating on both sides of the above relation by $\dfrac{1}{f(D^2)}$, we obtain

$$\frac{1}{f(D^2)}\left[f(D^2)\sin(ax+b)\right] = \frac{1}{f(D^2)}\left[f(-a^2)\sin(ax+b)\right]$$

or $\sin(ax+b) = f(-a^2)\dfrac{1}{f(D^2)}\sin(ax+b)$.

On rewriting the terms, we get

$$\frac{1}{f(D^2)}\sin(ax+b) = \frac{1}{f(-a^2)}\sin(ax+b), \text{ provided } f(-a^2) \neq 0 \qquad ...(18.35)$$

Case of Failure. If $f(-a^2) = 0$, then $D^2 + a^2$ is a factor of $f(D)$ and the above method fails and we proceed as follows.

Since $e^{i(ax + b)} = \cos(ax + b) + i \sin(ax + b)$.

$$\therefore \quad \frac{1}{f(D^2)} \sin(ax+b) = \frac{1}{f(D^2)} \text{ imaginary part of } e^{i(ax+b)}$$

$$= x \cdot \frac{1}{f'(D^2)}, \text{ imaginary part of } e^{i(ax+b)}$$

where $D^2 = -a^2$ and $f'(D^2) = \dfrac{d}{dD} f(D^2)$ $\qquad\qquad [\because f(-a^2) = 0]$

$$\frac{1}{f(D^2)} \sin(ax+b) = x \frac{1}{f'(-a^2)} \sin(ax+b), \quad \textbf{provided } \mathbf{f'(-a^2) \neq 0}$$

If $f'(-a^2) = 0$, then the particular integral is given by

$$\frac{1}{f(D)^2} \sin(ax+b) = x^2 \frac{1}{f''(-a^2)} \sin(ax+b), \textbf{ provided } \mathbf{f''(-a^2) \neq 0}.$$

Similarly $\dfrac{1}{f(D^2)} \cos(ax+b) = \dfrac{1}{f(-a^2)} \cos(ax+b),$ $\qquad\qquad$ ...(18.36)

$$\text{provided } f(-a^2) \neq 0$$

If $f(-a^2) = 0$, then $\dfrac{1}{f(D)^2} \cos(ax+b) = x \dfrac{1}{f'(-a^2)} \cos(ax+b),$ $\quad$ ...(18.37)

provided $f'(-a^2) \neq 0$ and so on.

Illustration: The particular integral of $(D^2 + a^2)y = \sin ax$ is $\dfrac{-x}{2a} \cos ax$.

(IETE., June 2009)

Example 18.10 Find the particular integral of the following differential equations

$\quad$ (i) $(D^3 + 1)y = \cos(2x + 3)$. $\qquad$ (ii) $(D^3 + 4D)y = \sin 2x$.

$\quad$ (iii) $\dfrac{d^2y}{dx^2} + 4y = \cos 2x$. $\qquad\qquad\qquad$ *(A.M.I.E., W-2005)*

Solution. (*i*) The particular integral is

$$y_p(x) = \frac{1}{D^3 + 1}\cos(2x+3) = \frac{1}{D \cdot D^2 + 1}\cos(2x+3)$$

$$= \frac{1}{D(-2^2)+1}\cos(2x+3) = \frac{1}{1-4D}\cos(2x+3) = \frac{1+4D}{1-16D^2}\cos(2x+3)$$

$$= \frac{1+4D}{1-16(-2^2)}\cos(2x+3) = \frac{1}{65}[\cos(2x+3) - 8\sin(2x+3)].$$

(*ii*) The particular integral is $y_p(x) = \dfrac{1}{D(D^2+4)}\sin 2x$

$$= x\frac{1}{3D^2+4}\sin 2x = x\frac{1}{3(-2^2)+4}\sin 2x = -\frac{x}{8}\sin 2x.$$

(*iii*) The particular integral is $y_p(x) = \dfrac{1}{D^2+4}\cdot\cos 2x$

$$= x\cdot\frac{1}{2D}\cos 2x = \frac{x}{2}\int\cos 2x\,dx = \frac{x\sin 2x}{4}.$$

(*iii*) **when X = x^m, m being a positive integer.**

The particular integral $y_p(x) = \dfrac{1}{f(D)}x^m = [f(D)]^{-1}x^m.$

We expand $[f(D)]^{-1}$ by Binomial theorem in ascending powers of D upto the *m*th power and evaluate the result by direct differentiation of x^m. The powers of D beyond m need not be considered as their derivatives will all be zero. This is illustrated in the following problem.

Example 18.11 Find the particular integral of $(D^2 + 5D + 4)y = x^2 + 7x + 9$.

Solution. The particular integral is $y_p(x) = \dfrac{1}{D^2+5D+4}(x^2+7x+9)$

$$= \frac{1}{4\left(1+\dfrac{5D}{4}+\dfrac{D^2}{4}\right)}(x^2+7x+9) = \frac{1}{4}\left[1+\left(\frac{5D}{4}+\frac{D^2}{4}\right)\right]^{-1}(x^2+7x+9)$$

$$= \frac{1}{4}\left[1-\left(\frac{5D}{4}+\frac{D^2}{4}\right)+\left(\frac{5D}{4}+\frac{D^2}{4}\right)^2 -...\right](x^2+7x+9)$$

$$= \frac{1}{4}\left[1 - \frac{5D}{4} - \frac{D^2}{4} + \frac{25D^2}{16} + \ldots\ldots\right](x^2 + 7x + 9)$$

$$= \frac{1}{4}\left[1 - \frac{5D}{4} + \frac{21D^2}{16} + \ldots\ldots\right](x^2 + 7x + 9)$$

$$= \frac{1}{4}\left[(x^2 + 7x + 9) - \frac{5}{4}D(x^2 + 7x + 9) + \frac{21}{16}D^2(x^2 + 7x + 9)\right].$$

$$= \frac{1}{4}\left[(x^2 + 7x + 9) - \frac{5}{4}(2x + 7) + \frac{21}{16}(2)\right] = \frac{1}{4}\left(x^2 + \frac{9}{2}x + \frac{23}{8}\right).$$

(iv) $\mathbf{X = e^{\alpha x} V}$, **where V is a function of** x.
Since when $D(e^{\alpha x}V) = \alpha e^{\alpha x}V + e^{\alpha x}DV = e^{\alpha x}(D + \alpha)V$.

$$D^2(e^{\alpha x}V) = \alpha^2 e^{\alpha x}V + \alpha e^{\alpha x}DV + \alpha e^{\alpha x}DV + e^{\alpha x}D^2V$$

$$= e^{\alpha x}(D^2 + 2D\alpha + \alpha^2)V = e^{\alpha x}(D + \alpha)^2 V.$$

Similarly $D^3(e^{\alpha x}V) = e^{\alpha x}(D + \alpha)^3 V$

In general, $D^n(e^{\alpha x}V) = e^{\alpha x}(D + \alpha)^n V$.

Hence $f(D)(e^{\alpha x}V) = e^{\alpha x}f(D + \alpha)V.$ \hfill ...(18.38)

Now let $f(D + \alpha)V = V_1$.

Operating on both sides by $\dfrac{1}{f(D + \alpha)}$, we get

$$\frac{1}{f(D + \alpha)}\left[f(D + \alpha)V\right] = \frac{1}{f(D + \alpha)}V_1 \text{ or } V = \frac{1}{f(D + \alpha)}V_1. \qquad ...(18.39)$$

Substituting for V from Eq. (18.39) in Eq. (18.38), we get

$$f(D)\left(e^{\alpha x}\frac{1}{f(D + \alpha)}V_1\right) = e^{\alpha x}V_1.$$

Operating on both sides by $\dfrac{1}{f(D)}$, we have

$$\frac{1}{f(D)}\left[f(D)\left(e^{\alpha x}\frac{1}{f(D + \alpha)}V_1\right)\right] = \frac{1}{f(D)}e^{\alpha x}V_1$$

Rewriting the terms, we obtain

$$\frac{1}{f(D)}e^{\alpha x}V_1 = e^{\alpha x}\frac{1}{f(D + \alpha)}V_1$$

Now since V_1 is any function of x, we have the rule

$$\frac{1}{f(D)}e^{\alpha x}V = e^{\alpha x}\frac{1}{f(D+\alpha)}V. \qquad ...(18.40)$$

Example 18.12 Find the particular integral of $(D^2 - 2D + 5)y = e^{2x}\sin x$.

Solution. The particular integral is

$$y_p(x) = \frac{1}{D^2 - 2D + 5}e^{2x}\sin x = e^{2x}\frac{1}{(D+2)^2 - 2(D+2) + 5}\sin x$$

$$= e^{2x}\frac{1}{D^2 + 2D + 5}\sin x = e^{2x}\frac{1}{-1 + 2D + 5}\sin x$$

$$= \frac{e^{2x}}{2}\frac{1}{D+2}\sin x = \frac{e^{2x}}{2}\frac{(D-2)}{D^2 - 4}\sin x = \frac{e^{2x}}{2}\frac{(D-2)}{-5}\sin x = -\frac{e^{2x}}{10}(\cos x - 2\sin x).$$

(v) When $X = xV$, where V is a function of x.

We have by successive differentiation, $D(xV) = xDV + V$,

$$D^2(xV) = xD^2V + 2DV,$$
$$D^3(xV) = xD^3V + 3D^2V,$$

$$............................$$
$$............................$$

In general, $\qquad D^n(xV) = xD^nV + nD^{n-1}V = xD^nV + \dfrac{d}{dD}(D^n)V.$

Hence $f(D)[xV] = xf(D)V + f'(D)V$ ...(18.41), where $f'(D) = \dfrac{d}{dD}f(D)$.

Now let $V_1 = f(D)V$ so that $V = \dfrac{1}{f(D)}V_1$. Substituting in (18.41), we get

$$f(D)\left[x\cdot\frac{1}{f(D)}V_1\right] = xf(D)\frac{1}{f(D)}V_1 + f'(D)\frac{1}{f(D)}V_1,$$

or $\qquad xV_1 = f(D)\left[x\cdot\dfrac{1}{f(D)}V_1\right] - f'(D)\dfrac{1}{f(D)}V_1$

and $\qquad \dfrac{1}{f(D)}xV_1 = x\dfrac{1}{f(D)}V_1 - \dfrac{1}{f(D)}f'(D)\dfrac{1}{f(D)}V_1 = x\dfrac{1}{f(D)}V_1 - \dfrac{f'(D)}{\left[f(D)\right]^2}V_1$

and since V_1 is any function of x, we have

$$\frac{1}{f(D)}[xV] = x\frac{1}{f(D)}V - \frac{f'(D)}{\left[f(D)\right]^2}V. \qquad ...(18.42$$

Example 18.13 Find the particular integral of $(D^2 - 2D + 1)y = x \sin x$.

Solution. The particular integral is.

$$y_p(x) = \frac{1}{D^2 - 2D + 1} \cdot x \sin x = x \frac{1}{D^2 - 2D + 1} \sin x - \frac{(2D-2)}{\left(D^2 - 2D + 1\right)^2} \sin x$$

$$= x \frac{1}{-1 - 2D + 1} \sin x - \frac{(2D-2)}{\left(-1 - 2D + 1\right)^2} \sin x$$

$$= -\frac{1}{2} x \int \sin x \, dx - \frac{1}{2D^2}(D-1)\sin x = \frac{1}{2} x \cos x - \frac{1}{2} \cdot \frac{1}{D} \int (\cos x - \sin x) \, dx$$

$$= \frac{1}{2} x \cos x - \frac{1}{2} \cdot \frac{1}{D}(\sin x + \cos x) = \frac{1}{2} x \cos x - \frac{1}{2} \int (\sin x + \cos x) \, dx .$$

$$\frac{1}{2} x \cos x - \frac{1}{2}(-\cos x + \sin x) = \frac{1}{2}(x \cos x + \cos x - \sin x).$$

Working Procedure to find the P.I.

The particular integral, $y_p(x)$, of the differential equation $f(D) y = X$ is symbolically written as $y_p(x) = \dfrac{1}{f(D)} X$.

Table 18.2

Form of X	Particular integral
(1) e^{ax}	

$$y_{p(x)} = \frac{1}{f(D)} e^{ax} = \frac{1}{f(a)} e^{ax}, \quad \text{provided } f(a) \ne 0.$$

$$y_p(x) = x \cdot \frac{1}{f'(a)} e^{ax}, \quad \text{when } f(a) = 0, \text{ provided } f'(a) \ne 0,$$

$$\text{where } f'(a) = \frac{d}{dD} f(D)\Big|_{D=a}.$$

$$y_p(x) = x^2 \cdot \frac{1}{f''(a)} e^{ax}, \quad \text{when } f(a) = f'(a) = 0, \ f''(a) \ne 0.$$

(Contd.)

(2) $\sin(ax + b)$ or $\cos(ax + b)$

$$y_p(x) = \frac{1}{f(D^2)}{\sin \atop \cos}(ax+b) = \frac{1}{f(-a^2)}{\sin \atop \cos}(ax+b), \text{ provided } f(-a^2) \neq 0$$

$$= x\frac{1}{f'(-a^2)}{\sin \atop \cos}(ax+b), \quad f(-a^2)=0, f'(-a^2) \neq 0$$

$$= x^2\frac{1}{f''(-a^2)}{\sin \atop \cos}(ax+b), \quad f(-a^2)=f'(-a^2)=0, f''(-a^2) \neq 0.$$

(3) x^m (m = positive integer)

$y_p(x) = [f(D)]^{-1}x^m$. Expand $[f(D)]^{-1}$ in ascending powers of D upto D^m and differentiate term by term.

(4) $e^{ax}V$

$$y_p(x) = \frac{1}{f(D)}e^{ax}V = e^{ax} \cdot \frac{1}{f(D+a)}V$$

(5) xV

$$y_p(x) = \frac{1}{f(D)}[xV] = x\frac{1}{f(D)}V - \frac{f'(D)}{[f(D)]^2}V$$

18.9 Illustrative Examples

Example 18.14 Find the general solution of the following differential equations.

(i) $y''' - 2y'' - 5y' + 6y = e^{4x} + 2$,   (ii) $y''' - y' = e^x + e^{-x}$,

(iii) $\dfrac{d^3y}{dx^3} - 5\dfrac{d^2y}{dx^2} + 8\dfrac{dy}{dx} - 4y = e^{2x} + 3e^{-x}$, (iv) $(2D + 1)^2y = 4e^{-x/2}$,

(v) $\dfrac{d^3y}{dx^3} + \dfrac{d^2y}{dx^2} - \dfrac{dy}{dx} - y = \cosh x$.

Solution. (*i*) The given equation in operator notation is

$(D^3 - 2D^2 - 5D + 6)y = e^{4x} + 2$.

The characteristic equation of the correponding homogeneous equation is

$m^3 - 2m^2 - 5m + 6 = (m - 1)(m - 3)(m + 2) = 0$. Its roots are $m = 1, 3, -2$.

The complementary function is given by $y_c(x) = c_1e^x + c_2e^{3x} + c_3e^{-2x}$.

The particular integral is

$$y_p(x) = \frac{1}{(D-1)(D-3)(D+2)} e^{4x} + 2 = \frac{1}{(D-1)(D-3)(D+2)} e^{4x}$$

$$+ 2 \cdot \frac{1}{(D-1)(D-3)(D+2)} e^{ox}$$

$$= \frac{1}{(4-1)(4-3)(4+2)} e^{4x} + 2 \cdot \frac{1}{(-1)(-3)(2)} e^{0x} = \frac{e^{4x}}{18} + \frac{1}{3}.$$

The general solution is

$$y(x) = y_c(x) + y_p(x) = c_1 e^x + c_2 e^{3x} + c_3 e^{-2x} + \frac{e^{4x}}{18} + \frac{1}{3} \cdots$$

(*ii*) The characteristic equation of the corresponding homogeneous equation $y''' - y' = 0$ is $m^3 - m = 0$ or $m(m+1)(m-1) = 0$.

The roots of this equation are $m = 0, -1, 1$.

The complementary function is $y_c(x) = c_1 + c_2 e^{-x} + c_3 e^x$.

The particular integral is $y_p(x) = \dfrac{1}{D(D+1)(D-1)} (e^x + e^{-x})$

$$= \frac{1}{D(D+1)(D-1)} e^x + \frac{1}{D(D+1)(D-1)} e^{-x}$$

$$= \frac{1}{2(D-1)} e^x + \frac{1}{2(D+1)} e^{-x} = \frac{xe^x}{2} + \frac{xe^{-x}}{2}.$$

The general solution of the equation is

$$y(x) = y_c(x) + y_p(x) = c_1 + c_2 e^{-x} + c_3 e^x + \frac{xe^x}{2} + \frac{xe^{-x}}{2}.$$

(*iii*) The given equaiton in operator notation is

$$(D^3 - 5D^2 + 8D - 4) y = e^{2x} + 3e^{-x}.$$

The characteristic equation of the corresponding homogeneous equation $y''' - 5y'' + 8y' - 4y = 0$ is $m^3 - 5m^2 + 8m - 4 = 0$ or $(m-1)(m-2)^2 = 0$.

Its roots are $m = 1, 2, 2$.

The complementary function is $y_c(x) = c_1 e^x + (c_2 + c_3 x)e^{2x}$.

The particular integral is $y_p(x) = \dfrac{1}{(D-1)(D-2)^2} (e^{2x} + 3e^{-x})$

$$= \frac{1}{(D-1)(D-2)^2} e^{2x} + 3 \frac{1}{(D-1)(D-2)^2} e^{-x}$$

$$= \frac{1}{(2-1)(D-2)^2}e^{2x}+3\frac{1}{(-1-1)(-1-2)^2}e^{-x} = \frac{1}{(D-2)^2}e^{2x} - \frac{e^{-x}}{6}$$

$$= \frac{x^2e^{2x}}{2} - \frac{e^{-x}}{6}. \text{ The general solution of the equation is}$$

$$y'(x) = y_c(x) + y_p(x) = c_1e^x + (c_2+c_3x)e^{2x} + \frac{(x^2e^{2x})}{2} - \frac{(e^{-x})}{6}.$$

(*iv*) The characteristic equation of the corresponding homogeneous equation **is** $(2m+1)^2 = 0$. The roots of this equation are $m = -1/2, -1/2$.
The complementary function is $y_c(x) = (c_1 + c_2x)e^{-x/2}$.
The particular integral is

$$y_p(x) = \frac{1}{(2D+1)^2}4e^{-x/2} = 4\frac{1}{(2D+1)^2}e^{-x/2}$$

$$= 4\cdot\frac{1}{4}\frac{1}{(D+(1/2))^2}e^{-x/2} = \frac{(x^2e^{-x/2})}{2}.$$

The general solution of the equation is $y(x) = y_c(x) + y_p(x)$

$$= (c_1+c_2x)e^{-x/2} + \frac{(x^2e^{-x/2})}{2}.$$

(*v*) The given equation in operator notation is
$$(D^3+D^2-D-1)y = \cosh x = (e^x+e^{-x})/2.$$
The characteristic equation of the corresponding homogeneous equation is $m^3 + m^2 - m - 1 = 0$ or $(m-1)(m+1)^2 = 0$. Its roots are $m = 1, -1, -1$.
The complementary function is $y_c(x) = c_1e^x + (c_2+c_3x)e^{-x}$.

The particular integral is $y_p(x) = \frac{1}{(D-1)(D+1)^2}\cosh x$

$$= \frac{1}{(D-1)(D+1)^2}\cdot\frac{e^x+e^{-x}}{2} = \frac{1}{2}\left[\frac{1}{(D-1)(D+1)^2}e^x + \frac{1}{(D+1)^2(D-1)}e^{-x}\right]$$

$$= \frac{1}{2}\left[\frac{1}{(D-1)(1+1)^2}e^x + \frac{1}{(D+1)^2(-1-1)}e^{-x}\right]$$

$$= \frac{1}{2}\left[\frac{1}{4}\cdot\frac{1}{D-1}e^x - \frac{1}{2}\cdot\frac{1}{(D+1)^2}e^{-x}\right] = \frac{1}{8}xe^x - \frac{1}{4}\frac{x^2}{2}e^{-x} = \frac{1}{8}xe^x - \frac{1}{8}x^2e^{-x}.$$

The general solution of the equation is $y(x) = y_c(x) + y_p(x)$

$$= c_1 e^x + (c_2 + c_3 x) e^{-x} + \frac{x e^x}{8} - \frac{x^2 e^{-x}}{8}.$$

Example 18.15 Find the general solution of the following differential equations.

(i) $y'' - 5y' + 4y = 65 \sin 2x.$ *(I.E.T.E., June 2005)*

(ii) $y'' + 3y' + 2y = \cos 2x,$ (iii) $y''' + 6y'' + 11y' + 6y = 2 \sin x.$

(iv) $(D - 1)^2 (D^2 + 1)y = e^x + \sin^2 \dfrac{x}{2}.$

(v) $(D^4 - 3D^2 - 4)y = 24 \sin 2x - 40 e^{-2x}.$

Solution. (i) The given equation in operator notation is
$(D^2 - 5D + 4)y = 65 \sin 2x.$ The characteristic equation of the corresponding homogeneous equation is $m^2 - 5m + 4 = 0$ or $(m - 1)(m - 4) = 0.$ Its roots are $m = 1, 4.$

The complementary function is $y_c(x) = c_1 e^x + c_2 e^{4x}.$

The particular integral is $y_p(x) = \dfrac{1}{D^2 - 5D + 4} \cdot 65 \sin 2x$

$$= 65 \cdot \frac{1}{-2^2 - 5D + 4} \sin 2x = -13 \frac{\sin 2x}{D} = \frac{13}{2} \cos 2x.$$

The general solution of the equation is $y(x) = y_c(x) + y_p(x)$

$$y(x) = c_1 e^x + c_2 e^{4x} + \frac{(13 \cos 2x)}{2}.$$

(ii) The given equation in operator notation is $(D^2 + 3D + 2)y = \cos 2x.$ The characteristic equation of the corresponding homogeneous equation is $m^2 + 3m + 2 = (m + 1)(m + 2) = 0.$ Its roots are $m = -1, -2.$
The complementary function is $y_c(x) = c_1 e^{-x} + c_2 e^{-2x}.$

The particular integral is $y_p(x) = \dfrac{1}{D^2 + 3D + 2} \cos 2x.$

$$= \frac{1}{(-2^2) + 3D + 2} \cos 2x = \frac{\cos 2x}{3D - 2} = \frac{(3D + 2)}{9D^2 - 4} \cos 2x = \frac{(3D + 2) \cos 2x}{9(-4) - 4}$$

$$= -\frac{1}{40}(-6 \sin 2x + 2 \cos 2x) = \frac{1}{20}(3 \sin 2x - \cos 2x).$$

The general solution of the equation is

$$y(x) = y_c(x) + y_p(x) = c_1 e^{-x} + c_2 e^{-2x} + \frac{(3 \sin 2x - \cos 2x)}{20}.$$

(*iii*) The given equation in operator notation is $(D^3 + 6D^2 + 11D + 6)y = 2\sin x$. The characteristic equation of the corresponding homogeneous equation is

$$m^3 + 6m^2 + 11m + 6 = 0 \quad \text{or} \quad (m+1)(m+2)(m+3) = 0.$$

Its root are $m = -1, -2, -3$.
The complementary function is $y_c(x) = c_1 e^{-x} + c_2 e^{-2x} + c_3 e^{-3x}$.

The particular integral is $y_p(x) = \dfrac{1}{D^3 + 6D^2 + 11D + 6} 2\sin x$

$$= \frac{1}{D(D^2+11)+6(D^2+1)} 2\sin x = \frac{1}{D(-1^2+11)+6(-1^2+1)} 2\sin x$$

$$= \frac{2\sin x}{10D} = -\frac{1}{5}\cos x.$$

The general solution of the equation is

$$y(x) = y_c(x) + y_p(x) = c_1 e^{-x} + c_2 e^{-2x} + c_3 e^{-3x} - \frac{(\cos x)}{5}.$$

(*iv*) The characteristic equation of the correstponding homogeneous equation is $(m-1)^2(m^2+1) = 0$. Its roots are $m = 1, 1, \pm i$.
The complementary function is $y_c(x) = (c_1 + c_2 x)e^x + c_3 \cos x + c_4 \sin x$.

The particular integral is $y_p(x) = \dfrac{1}{(D-1)^2(D^2+1)}\left(e^x + \sin^2\dfrac{x}{2}\right)$.

Now $\dfrac{1}{(D-1)^2(D^2+1)} e^x = \dfrac{1}{2}\cdot\dfrac{1}{(D-1)^2} e^x = \dfrac{1}{2}\cdot\dfrac{x^2}{2} e^x = \dfrac{x^2}{4} e^x$,

and $\dfrac{1}{(D-1)^2(D^2+1)} \sin^2\dfrac{x}{2} = \dfrac{1}{2}\dfrac{1}{(D^2+1)(D-1)^2}(1-\cos x)$

$$= \frac{1}{2}\left[\frac{1}{(D^2+1)(D-1)^2} e^{0x} - \frac{1}{(D^2+1)(D-1)^2}\cos x\right]$$

$$= \frac{1}{2}\left[1 - \frac{1}{(D^2+1)}\cdot\frac{1}{-2D}\cos x\right] = \frac{1}{2}\left[1 + \frac{1}{2}\frac{1}{(D^2+1)}\sin x\right]$$

$$= \frac{1}{2} + \frac{1}{4}\cdot\frac{1}{D^2+1}\sin x = \frac{1}{2} + \frac{1}{4}\cdot x\cdot\frac{\sin x}{2D} = \frac{1}{2} - \frac{1}{8} x\cos x.$$

The general solution of the equation is $y(x) = y_c(x) + y_p(x)$

$$= (c_1 + c_2 x)e^x + (c_3 \cos x + c_4 \sin x) + \left(x^2 e^x/4\right) + 1/2 - (x\cos x)/8.$$

(v) The characteristic equation of the corresponding homogenous equation is $m^4 - 3m^2 - 4 = 0$ or $(m^2 - 4)(m^2 + 1) = 0$. Its roots are $m = \pm 2, \pm i$.

The complementary function $y(x) = c_1 e^{2x} + c_2 e^{-2x} + c_3 \cos x + c_4 \sin x$.

The particular integral is $y_p(x) = \dfrac{1}{D^4 - 3D^2 - 4} 24 \sin 2x - 40 e^{-2x}$.

Now $\dfrac{1}{D^4 - 3D^2 - 4} \sin 2x = \dfrac{1}{\left(-2^2\right)^2 - (3)\left(-2^2\right) - 4} \sin 2x = \dfrac{\sin 2x}{24}$,

and $\dfrac{1}{D^4 - 3D^2 - 4} e^{-2x} = x \cdot \dfrac{1}{4D^3 - 6D} e^{-2x} = x \dfrac{1}{4(-2)^3 - 6(-2)} e^{-2x} = \dfrac{x e^{-2x}}{-20}$.

$\therefore \quad y_p(x) = \sin 2x + 2x e^{-2x}$.

The general solution of the equation is $y(x) = y_c(x) + y_p(x)$

$= c_1 e^{2x} + c_2 e^{-2x} + c_3 \cos x + c_4 \sin x + \sin 2x + 2x e^{-2x}$.

Example 18.16 Find the general solution of the following differential equations.

   (i) $(D^3 - 2D + 4) y = 3x^2 - 5x + 2$.

  (ii) $(D^3 + 4D)y = 24x^2 + 12 + 8 \sin 2x$.

 (iii) $(D^2 - 4D + 3) y = x^3$.

 (iv) $(D - 1)^2 (D + 1)^2 y = \sin^2 \dfrac{x}{2} + e^x + x$.   (MDU 2005)

Solution. (i) The characteristic equation of the corresponding homogeneous equation is $m^3 - 2m + 4 = 0$ or $(m + 2)(m^2 - 2m + 2) = 0$.

It roots are $m = -2, 1 \pm i$.

The complementary function is $y_c(x) = c_1 e^{-2x} + e^x (c_2 \cos x + c_3 \sin x)$.

The particular integral is $y_p(x) = \dfrac{1}{D^3 - 2D + 4}(3x^2 - 5x + 2)$

$= \dfrac{1}{4} \left\{ 1 - \left(\dfrac{D}{2} - \dfrac{D^3}{4} \right) \right\}^{-1} (3x^2 - 5x + 2) = \dfrac{1}{4} \left(1 + \dfrac{D}{2} + \dfrac{D^2}{4} + \ldots \right)(3x^2 - 5x + 2)$

$= \dfrac{1}{4} \left[(3x^2 - 5x + 2) + \left(3x - \dfrac{5}{2} \right) + \dfrac{3}{2} \right] = \dfrac{1}{4}(3x^2 - 2x + 1)$.

The general solution of the equation is $y(x) = y_c(x) + y_p(x)$

$= c_1 e^{-2x} + e^x (c_2 \cos x + c_3 \sin x) + \dfrac{(3x^2 - 2x + 1)}{4}$.

(*ii*) The characteristic equation of the corresponding homogeneous equation is $m^3 + 4m = 0$ or $m(m^2 + 4) = 0$. Its roots are $m = 0, \pm 2i$.
The complementary function is $y_c(x) = c_1 + c_2 \cos 2x + c_3 \sin 2x$.

The particular integral is $y_p(x) = \dfrac{1}{D^3 + 4D}(24x^2 + 12 + 8\sin 2x)$

$$= \frac{1}{D^3 + 4D}(24x^2 + 12) + \frac{1}{D^3 + 4D} 8\sin 2x$$

$$= \frac{12}{4D}\left(1 + \frac{D^2}{4}\right)^{-1}(2x^2 + 1) + 8x\frac{1}{3D^2 + 4}\sin 2x$$

$$= \frac{3}{D}\left(1 - \frac{D^2}{4} + \ldots\ldots\right)(2x^2 + 1) + 8x \cdot \frac{1}{3(-2^2) + 4}\sin 2x$$

$$= \frac{3}{D}(2x^2 + 1 - 1) - x\sin 2x = 2x^3 - x\sin 2x \,.$$

The general solution of the equation is $y(x) = y_c(x) + y_p(x)$
$= c_1 + c_2 \cos 2x + c_3 \sin 2x + 2x^3 - x \sin 2x$.

(*iii*) The characteristic equation of the corresponding homogeneous equation
is $m^2 - 4m + 3 = 0$ or $(m - 1)(m - 3) = 0$. Its roots are $m = 1, 3$.
The complementary function is $y_c(x) = c_1 e^x + c_2 e^{3x}$.

The particular integral is $y_p(x) = \dfrac{1}{D^2 - 4D + 3}x^3 = \dfrac{1}{3\left(1 - \dfrac{4D - D^2}{3}\right)}x^3$

$$= \frac{1}{3}\left(1 - \frac{4D - D^2}{3}\right)^{-1}x^3 = \frac{1}{3}\left[1 + \frac{4D - D^2}{3} + \frac{16D^2 - 8D^3}{9} + \frac{64D^3}{27} + \ldots\right]x^3$$

$$= \frac{1}{3}\left[x^3 + \frac{12x^2 - 6x}{3} + \frac{96x - 48}{9} + \frac{128}{9}\right] = \frac{1}{3}\left(x^3 + 4x^2 + \frac{26x}{3} + \frac{80}{9}\right)$$

$$= (9x^3 + 36x^2 + 78x + 80)/27 \,.$$

The general solution of the equation is $y(x) = y_c(x) + y_p(x)$

$$= c_1 e^x + c_2 e^{3x} + \frac{(9x^3 + 36x^2 + 78x + 80)}{27} \,.$$

(*iv*) The characteristic equation of the corresponding homogeneous equation
is $(m - 1)^2 (m + 1)^2 = 0$. Its roots are $m = 1, 1, -1, -1$.

The complementary function is $y_c(x) = (c_1 + c_2 x)e^x + (c_3 + c_4 x)e^{-x}$,

The particular integral is $y_p(x) = \dfrac{1}{(D-1)^2(D+1)^2}\sin^2\dfrac{x}{2} + e^x + x$

$$= \frac{1}{(D-1)^2(D+1)^2}\sin^2\frac{x}{2} + \frac{1}{(D-1)^2(D+1)^2}e^x + \frac{1}{(D-1)^2(D+1)^2}x$$

Now $\dfrac{1}{(D-1)^2(D+1)^2}\sin^2\dfrac{x}{2} = \dfrac{1}{(D-1)^2(D+1)^2}\dfrac{(1-\cos x)}{2}$

$$= \frac{1}{2}\frac{1}{(D-1)^2(D+1)^2}e^{0x} - \frac{1}{2}\frac{1}{(D^2-2D+1)(D^2+2D+1)}\cos x$$

$$= \frac{1}{2} - \frac{1}{2}\cdot\frac{1}{4}\cos x = \frac{1}{2} - \frac{1}{8}\cos x,$$

$$\frac{1}{(D-1)^2(D+1)^2}e^x = \frac{1}{4}\cdot\frac{1}{(D-1)^2}e^x = \frac{1}{4}x\cdot\frac{1}{2(D-1)}e^x$$

$$= \frac{1}{8}x\frac{1}{(D-1)}e^x = \frac{1}{8}x^2 e^x, \text{ and } \frac{1}{(D-1)^2(D+1)^2}x = \frac{1}{D^4-2D^2+1}x$$

$$= \left[1-(2D^2-D^4)\right]^{-1}x = (1+2D^2-D^4+....)x = x.$$

$$\therefore \quad y_p(x) = \frac{1}{2} - \frac{1}{8}\cos x + \frac{x^2 e^x}{8} + x.$$

The general solution of the equation is $y(x) = y_c(x) + y_p(x)$

$$= (c_1 + c_2 x)e^x + (c_3 + c_4 x)e^{-x} + \frac{1}{2} - \frac{1}{8}\cos x + \frac{x^2 e^x}{8} + x.$$

Example 18.17 Find the general solution of the following differential equations.

(i) $\dfrac{d^2 y}{dx^2} - 4\dfrac{dy}{dx} + 4y = 8x^2 e^{2x}\sin 2x$,   *(UPTU 2004; DU 2002; IETE, Dec. 99)*

(ii) $(D^2 - 4D + 3)y = 2xe^{3x} + 3e^x \cos 2x$,

(iii) $(D^3 - 3D^2 - 6D + 8)y = xe^{-3x}$,

(iv) $(D^2 - 2D + 1)y = xe^x \sin x$.   *(A.M.I.E., W-2010, S-2005; UPTU 2005)*

Solution. (*i*) The given equation in operator notation is

$$(D^2 - 4D + 4)y = 8x^2 e^{2x} \sin 2x.$$

The characteristic equation of the corresponding homogeneous equation is $m^2 - 4m + 4 = 0$ or $(m - 2)^2 = 0$. Its roots are $m = 2, 2$.

The complementary function is $y_c(x) = (c_1 + c_2 x)e^{2x}$.

The particular integral is $y_p(x) = \dfrac{1}{(D-2)^2} 8x^2 e^{2x} \sin 2x$

$$= 8e^{2x} \frac{1}{(D+2-2)^2} x^2 \sin 2x = 8e^{2x} \frac{1}{D^2} x^2 \sin 2x .$$

Now $\dfrac{1}{D^2} x^2 \sin 2x = \dfrac{1}{D} \underset{\underset{\text{II}}{\text{I}}}{\int x^2 \sin 2x \, dx} = \dfrac{1}{D}\left[\left(\dfrac{1}{4} - \dfrac{x^2}{2}\right)\cos 2x + \dfrac{x}{2}\sin 2x\right]$

$$= \int\left[\left(\frac{1}{4} - \frac{x^2}{2}\right)\cos 2x + \frac{x}{2}\sin 2x\right]dx = \frac{3-2x^2}{8}\sin 2x - \frac{x\cos 2x}{2} .$$

$$\therefore \ y_p(x) = e^{2x}\left[(3-2x^2)\sin 2x - 4x\cos 2x\right].$$

The general solution of the equation is $y(x) = y_c(x) + y_p(x)$

$$= (c_1 + c_2 x)e^{2x} + \left[(3-2x^2)\sin 2x - 4x\cos 2x\right]e^{2x} .$$

(*ii*) The characteristic equation of the corresponding homogeneous equation is $m^2 - 4m + 3 = 0$. or $(m-1)(m-3) = 0$. Its roots are $m = 1, 3$.

The complementary function is $y_c(x) = c_1 e^x + c_2 e^{3x}$.

The particular integral is $y_p(x) = \dfrac{1}{D^2 - 4D + 3}\left(2xe^{3x} + 3e^x \cos 2x\right)$

$$= 2\frac{1}{D^2 - 4D + 3} xe^{3x} + 3\frac{1}{D^2 - 4D + 3} e^x \cos 2x$$

$$= 2e^{3x} \frac{1}{(D+3)^2 - 4(D+3)+3} x + 3e^x \frac{1}{(D+1)^2 - 4(D+1)+3}\cos 2x$$

$$= 2e^{3x} \frac{1}{D^2 + 2D} x + 3e^x \frac{1}{D^2 - 2D}\cos 2x$$

$$= \left(2e^{3x} \cdot \frac{1}{D}\cdot\frac{1}{D+2} x\right) + \left(3e^x \frac{1}{-4-2D}\cos 2x\right)$$

$$= 2e^{3x} \frac{1}{D}\left(\frac{1}{2} - \frac{1}{4}D\right)x - \frac{3}{2}e^x \frac{D-2}{D^2 - 4}\cos 2x$$

$$= \frac{1}{2}e^{3x}\frac{1}{D}(2x-1) + \frac{3}{16}e^{x}(D-2)\cos 2x$$

$$= \frac{1}{2}e^{3x}\left(x^2-x\right) - \frac{3}{8}e^{x}\left(\cos 2x + \sin 2x\right).$$

The general solution of the equation is $y(x) = y_c(x) + y_p(x)$

$$= c_1 e^x + c_2 e^{3x} + e^{3x}\left(x^2-x\right)/2 - 3e^{x}\left(\cos 2x + \sin 2x\right)/8.$$

(*iii*) The characteristic equation of the corresponding homogeneous equation is $m^3 - 3m^2 - 6m + 8 = 0$, or $(m-1)(m+2)(m-4) = 0$. Its roots are $m = 1, -2, 4$. The complementary function is $y_c(x) = c_1 e^x + c_2 e^{-2x} + c_3 e^{4x}$.

The particular integral is $y_p(x) = \dfrac{1}{D^3 - 3D^2 - 6D + 8} \cdot xe^{-3x}$

$$= e^{-3x}\frac{1}{(D-3)^3 - 3(D-3)^2 - 6(D-3)+8}x$$

$$= e^{-3x}\frac{1}{D^3 - 12D^2 + 39D - 28}x = e^{-3x}\left(-\frac{1}{28} - \frac{39}{784}D\right)x = e^{-3x}\left(-\frac{1}{28}x - \frac{39}{784}\right).$$

The general solution of the equation is $y(x) = y_c(x) + y_p(x)$

$$= c_1 e^x + c_2 e^{-2x} + c_3 e^{4x} - e^{-3x}(28x+39)/784.$$

(*iv*) The characteristic equation of the corresponding homogeneous equation is $m^2 - 2m + 1 = 0$ or $(m-1)^2 = 0$. Its roots are 1, 1.

The complementary function is $y_c(x) = (c_1 + c_2 x)e^x$.

The particular integral is $y_p(x) = \dfrac{1}{(D-1)^2}e^x \cdot x\sin x = e^x \cdot \dfrac{1}{(D+1-1)^2}x\sin x$

$$= e^x \frac{1}{D^2}x\sin x = e^x \cdot \frac{1}{D}\int\limits_{I}x\sin\underset{II}{x}dx$$

$$= e^x \cdot \frac{1}{D}\left[x(-\cos x) - \int 1(-\cos x)dx\right] = e^x\frac{1}{D}(-x\cos x + \sin x)$$

$$= e^x\int(-x\cos x + \sin x)dx = e^x\left[-\left(x\sin x - \int\sin x dx\right) - \cos x\right]$$

$$= e^x(-x\sin x - \cos x - \cos x) = -e^x(x\sin x + 2\cos x).$$

The general solution of the equation is $y(x) = y_c(x) + y_p(x)$

$$= (c_1 + c_2 x)e^x - e^x(x\sin x + 2\cos x).$$

Example 18.18 Find the general solution of the following differential equations.

(i) $y'' + 3y' + 2y = x \sin 2x,$ (ii) $y'' - y = x^2 \sin 3x,$

(iii) $y'' + 4y = x \sin x + 2^x,$ *(A.M.I.E.; W-2005)*

(iv) $\left(D^4 + 2D^2 + 1\right)y = x^2 \cos x.$ *(A.M.I.E.; W-2008)*

Solution. (*i*) The given equation in operator notation is $(D^2 + 3D + 2)\,y = x \sin 2x.$ The characteristic equation of the corresponding homogeneous equation is $m^2 + 3m + 2 = 0$ or $(m + 1)(m + 2) = 0.$ Its roots are $m = -1, -2.$ The complementary function is $y_c(x) = c_1 e^{-x} + c_2 e^{-2x}.$

The particular integral is $y_p(x) = \dfrac{1}{D^2 + 3D + 2} x \sin 2x$

$$= x\frac{1}{D^2 + 3D + 2}\sin 2x - \frac{2D + 3}{\left(D^2 + 3D + 2\right)^2}\sin 2x$$

$$= x\frac{1}{3D - 2}\sin 2x - \frac{2D + 3}{D^4 + 6D^3 + 13D^2 + 12D + 4}\sin 2x$$

$$= x\frac{1}{3D - 2}\sin 2x - \frac{2D + 3}{(-4)^2 + 6(-4)D + 13(-4) + 12D + 4}\sin 2x$$

$$= x\frac{3D + 2}{9D^2 - 4}\sin 2x + \frac{1}{4}\frac{(2D + 3)(3D - 8)}{9D^2 - 64}\sin 2x$$

$$= \frac{-x(3\cos 2x + \sin 2x)}{20} + \frac{24\sin 2x + 7\cos 2x}{200}.$$

The general solution of the equation is $y(x) = y_c(x) + y_p(x)$

$$= c_1 e^{-x} + c_2 e^{-2x} - \left(\frac{30x - 7}{200}\right)\cos 2x - \left(\frac{5x - 12}{100}\right)\sin 2x.$$

(*ii*) The given equation in operator notation is $(D^2 - 1)y = x^2 \sin 3x.$ The characteristic equation of the corresponding homogeneous equation is $m^2 - 1 = 0.$ Its roots are $m = 1, -1.$ The complementary function is $y_c(x) = c_1 e^x + c_2 e^{-x}.$

The particular integral is $y_p(x) = \dfrac{1}{D^2 - 1} x^2 \sin 3x$

$$= x\frac{1}{D^2 - 1}x\sin 3x - \frac{2D}{\left(D^2 - 1\right)^2}x\sin 3x = \left\{x^2\frac{1}{D^2 - 1}\sin 3x - x\frac{2D}{\left(D^2 - 1\right)^2}\sin 3x\right\}$$

$$-2D\left\{ x\frac{1}{D^4-2D^2+1}\sin 3x - \frac{4D^3-4D}{\left(D^4-2D^2+1\right)^2}\sin 3x \right\}$$

$$= x^2\frac{1}{D^2-1}\sin 3x - x\frac{2D}{\left(D^2-1\right)^2}\sin 3x - 2D\left\{ x\frac{1}{\left(D^2-1\right)^2}\sin 3x \right\} + \frac{8D^2}{\left(D^2-1\right)^3}\sin 3x$$

$$= -\frac{1}{10}x^2\sin 3x - \frac{3}{50}x\cos 3x - \frac{1}{50}D(x\sin 3x) + \frac{9}{125}\sin 3x$$

$$= -\frac{1}{10}x^2\sin 3x - \frac{3}{25}x\cos 3x + \frac{13}{250}\sin 3x .$$

The general solution of the equation is $y(x) = y_c\,(x) + y_p\,(x)$

$$= c_1e^x + c_2e^{-x} - \frac{25x^2-13}{250}\sin 3x - \frac{3}{25}x\cos 3x.$$

(*iii*) The given equation in operator notation is $(D^2+4)y = x\sin x + 2^x$.

The characteristic equation of the corresponding homogeneous equation is $m^2 + 4 = 0$. Its roots are $m = 2i, -2i$.

The complementary function is $y_c(x) = c_1\cos 2x + c_2\sin 2x$.

The particular integral is $y_p\,(x) = \dfrac{1}{D^2+4}\left(x\sin x + 2^x\right)$

$$= \frac{1}{D^2+4}x\sin x + \frac{1}{D^2+4}2^x = x\frac{1}{D^2+4}\sin x - \frac{2D}{\left(D^2+4\right)^2}\sin x + \frac{1}{D^2+4}e^{\log 2^x}$$

$$= \frac{x}{3}\sin x - \frac{2}{9}\cos x + \frac{1}{D^2+4}e^{x\log 2} = \frac{x}{3}\sin x - \frac{2}{9}\cos x + \frac{1}{\left(\log 2\right)^2+4}2^x .$$

The general solution of the equation is $y(x) = y_c(x) + y_p(x)$

$$= c_1\cos 2x + c_2\sin 2x + \frac{x}{3}\sin x - \frac{2}{9}\cos x + \frac{1}{\left(\log 2\right)^2+4}\cdot 2^x .$$

(*iv*) The characteristic equation of the corresponding homogeneous equation is

$m^4 + 2m^2 + 1 = 0$ or $\left(m^2+1\right)^2 = 0$, which has roots $\pm i, \pm i$.

The complementary function is $y_c\,(x) = \left(c_1+c_2x\right)\cos x + \left(c_3+c_4x\right)\sin x$.

The particular integral is $y_p\,(x) = \dfrac{1}{D^4+2D^2+1}x^2\cos x$

$$= \frac{1}{\left(D^2+1\right)^2} x^2 \left(\frac{e^{ix}+e^{-ix}}{2}\right) = \frac{1}{2}\left[\frac{1}{\left(D^2+1\right)^2} x^2 e^{ix} + \frac{1}{\left(D^2+1\right)^2} x^2 e^{-ix}\right].$$

Now $\dfrac{1}{\left(D^2+1\right)^2} x^2 e^{ix} = e^{ix} \dfrac{1}{\left\{(D+i)^2+1\right\}^2} x^2$

$$= e^{ix} \frac{1}{\left(D^2+2Di\right)^2} x^2 = e^{ix} \frac{1}{4i^2 D^2} \frac{1}{\left(1+\dfrac{D}{2i}\right)^2} x^2$$

$$= -\frac{1}{4} e^{ix} \cdot \frac{1}{D^2} \left(1+\frac{D}{2i}\right)^{-2} x^2 = -\frac{1}{4} e^{ix} \cdot \frac{1}{D^2}\left(1-\frac{D}{i}-\frac{3}{4}D^2+...\right) x^2$$

$$= -\frac{1}{4} e^{ix} \cdot \frac{1}{D^2}\left(x^2 - \frac{2x}{i} - \frac{3}{2}\right) = -\frac{1}{4} e^{ix} \cdot \frac{1}{D}\left(\frac{x^3}{3} - \frac{x^2}{i} - \frac{3x}{2}\right)$$

$$= -\frac{1}{4} e^{ix}\left(\frac{x^4}{12} - \frac{x^3}{3i} - \frac{3x^2}{4}\right).$$

Similarly, or on changing i to $-i$ in the above, we obtain

$$\frac{1}{\left(D^2+1\right)^2} x^2 e^{-ix} = -\frac{1}{4} e^{-ix}\left(\frac{x^4}{12} + \frac{x^3}{3i} - \frac{3x^2}{4}\right).$$

Therefore, the particular integral is

$$y_p(x) = -\frac{1}{8}\left[\frac{x^4}{6} \cdot \frac{e^{ix}+e^{-ix}}{2} - \frac{2x^3}{3} \frac{e^{ix}-e^{-ix}}{2i} - \frac{3x^2}{2} \cdot \frac{e^{ix}+e^{-ix}}{2}\right]$$

$$= -\frac{1}{8}\left[\frac{x^4}{6}\cos x - \frac{2x^3}{3}\sin x - \frac{3x^2}{2}\cos x\right].$$

The general solution of the equation is $y(x) = y_c(x) + y_p(x)$

$$= \left(c_1+c_2 x\right)\cos x + \left(c_3+c_4 x\right)\sin x + \frac{1}{12} x^3 \sin x + \frac{1}{48}\left(9x^2 - x^4\right)\cos x.$$

Example 18.19 Find the general solution of the following differential equations
(i) $y'' + y = \text{cosec } x.$ (AMIE, W-2007; MDU-2002)

(ii) **y″ + a²y = sec ax.**

(iii) **(D² + 4)y = tan 2x.**

(AMIE-2005; VTU, 2001; IETE, 1999)

(MDU 2002; DU 1997)

Solution. The given equation in operator notation is $(D^2 + 1)y = \operatorname{cosec} x$.

The characteristic equation of the corresponding homogeneous equation is $(m^2 + 1) = 0$. Its roots are $m = i, -i$.

The complementary function is $y_c(x) = c_1 \cos x + c_2 \sin x$.

The particular integral is $y_p(x) = \dfrac{1}{D^2 + 1} \operatorname{cosec} x = \dfrac{1}{(D+i)(D-i)} \operatorname{co sec} x$

$$= \frac{1}{2i}\left(\frac{1}{D-i} - \frac{1}{D+i}\right)\cos ec\, x = \frac{1}{2i}\left(\frac{1}{D-i}\cos ec\, x - \frac{1}{D+i}\cos ec\, x\right)$$

Now $\dfrac{1}{D-i}\cos ec\, x = \dfrac{1}{D-i}e^{ix}\dfrac{e^{-ix}}{\sin x} = e^{ix}\dfrac{1}{D+i-i}\dfrac{e^{-ix}}{\sin x} = e^{ix}\dfrac{1}{D}\dfrac{\cos x - i\sin x}{\sin x}$

$= e^{ix}\int(\cot x - i)\,dx = e^{ix}\left[\log|\sin x| - ix\right].$

Changing i to $-i$, we have $\dfrac{1}{D+i}\cos ec\, x = e^{-ix}\left[\log|\sin x| + ix\right].$

$\therefore \quad y_p(x) = \dfrac{1}{2i}\left[e^{ix}\left(\log|\sin x| - ix\right) - e^{-ix}\left(\log|\sin x| + ix\right)\right]$

$$= \log|\sin x|\left(\frac{e^{ix} - e^{-ix}}{2i}\right) - \frac{ix}{i}\left(\frac{e^{ix} + e^{-ix}}{2}\right) = \sin x \log|\sin x| - x\cos x.$$

The general solution of the equation is $y(x) = y_c(x) + y_p(x)$

$= c_1 \cos x + c_2 \sin x - x \cos x + (\sin x) \log|\sin x|$.

(ii) The given equation in operator notation is $(D^2 + a^2)y = \sec ax$.

The characteristic equation of the corresponding homogeneous equation is $m^2 + a^2 = 0$. Its roots are $m = \pm ia$.

The complementary function is $y_c(x) = c_1 \cos ax + c_2 \sin ax$.

The particular integral is $y_p(x) = \dfrac{1}{D^2 + a^2}\sec ax$

$$= \frac{1}{2ia}\left[\frac{1}{D-ia} - \frac{1}{D+ia}\right]\sec ax = \frac{1}{2ia}\left[\frac{1}{D-ia}\sec ax - \frac{1}{D+ia}\sec ax\right].$$

Now $\dfrac{1}{D-ia}\sec ax = e^{iax}\int \sec ax \cdot e^{-iax}\,dx$

$$= e^{iax}\int \frac{\cos ax - i\sin ax}{\cos ax}\,dx = e^{iax}\int(1 - i\tan ax)\,dx = e^{iax}\left(x + \frac{i}{a}\log\cos ax\right).$$

Changing i to $-i$, we have $\dfrac{1}{D+ia}\sec ax = e^{-iax}\left(x-\dfrac{i}{a}\log\cos ax\right)$.

$\therefore$ The particular integral is

$$y_p(x)=\frac{1}{2ia}\left[e^{iax}\left(x+\frac{i}{a}\log\cos ax\right)-e^{-iax}\left(x-\frac{i}{a}\log\cos ax\right)\right]$$

$$=\frac{x}{a}\frac{e^{iax}-e^{-iax}}{2i}+\frac{1}{a^2}\log\cos ax\cdot\frac{e^{iax}+e^{-iax}}{2}=\frac{x}{a}\sin ax+\frac{1}{a^2}(\cos ax\cdot\log\cos ax).$$

The general solution of the equation is
$y(x)=c_1\cos ax+c_2\sin ax+(1/a)\,x\sin ax+(1/a^2)\cos ax\,\log(\cos ax)$.
(*iii*)   The characteristic equation is $m^2+4=0$. Its roots are $m=2i,-2i$.
The complementary function is $y_c(x)=c_1\cos 2x+c_2\sin 2x$.
The particular integral is

$$y_p(x)=\frac{1}{D^2+4}\cdot\tan 2x=\frac{1}{(D+2i)(D-2i)}\tan 2x$$

$$=\frac{1}{4i}\left[\frac{1}{D-2i}-\frac{1}{D+2i}\right]\tan 2x=\frac{1}{4i}\left[\frac{1}{D-2i}\tan 2x-\frac{1}{D+2i}\tan 2x\right]$$

Now $\dfrac{1}{D-2i}\tan 2x=e^{2ix}\displaystyle\int\tan 2x\cdot e^{-2ix}\,dx$

$=e^{2ix}\displaystyle\int\tan 2x(\cos 2x-i\sin 2x)\,dx=e^{2ix}\displaystyle\int\left[\sin 2x-i(\sin^2 2x/\cos 2x)\right]dx$

$=e^{2ix}\displaystyle\int\left[\sin 2x-i\big((1-\cos^2 2x)/\cos 2x\big)\right]dx=e^{2ix}\displaystyle\int[\sin 2x-i(\sec 2x-\cos 2x)]dx$

$=\dfrac{e^{2ix}}{2}\left[-\cos 2x-i\log|\sec 2x+\tan 2x|+i\sin 2x\right]$

$=-\dfrac{e^{2ix}}{2}\left[(\cos 2x-i\sin 2x)+i\log|\sec 2x+\tan 2x|\right]$

$=-\dfrac{e^{2ix}}{2}\left[e^{-2ix}+i\log|\sec 2x+\tan 2x|\right]=-\dfrac{1}{2}\left[1+ie^{2ix}\log|\sec 2x+\tan 2x|\right].$

Changing i to $-i$, we have

$\dfrac{1}{D+2i}\tan 2x=-\dfrac{1}{2}\left[1-ie^{-2ix}\log|\sec 2x+\tan 2x|\right].$

Thus $y_p(x) = \dfrac{1}{4i}\left[\dfrac{-1}{2}\{1 + ie^{2ix} \log|\sec 2x + \tan 2x|\} \right.$

$\left. + \dfrac{1}{2}\{1 - ie^{-2ix} \log|\sec 2x + \tan 2x|\} \right]$

$= -\dfrac{1}{4}\left(\dfrac{e^{2ix} + e^{-2ix}}{2} \right) \log|\sec 2x + \tan 2x| = -\dfrac{1}{4}\cos 2x \log|\sec 2x + \tan 2x|.$

The general solution of the equation is $y(x) = y_c(x) + y_p(x)$

$= c_1 \cos 2x + c_2 \sin 2x - \left[\cos 2x \log|\sec 2x + \tan 2x| \right]/4.$

Exercise 18.2

Find the general solution of the following differential equations.

1. $y'' - 3y' + 2y = e^{3x}$. *(IETE, W-2007)* **Ans.** $y(x) = c_1 e^x + c_2 e^{2x} + \left(e^{3x}/2 \right)$.

2. $y''' - 6y'' + 11y' - 6y = e^{-x}$. **Ans.** $y(x) = c_1 e^x + c_2 e^{2x} + c_3 e^{3x} - \left(e^{-x}/24 \right)$.

3. $y'' + 4y' + 4y = 12e^{-2x}$. **Ans.** $y(x) = (c_1 + c_2 x)e^{-2x} + 6x^2 e^{-2x}$.

4. $y'' - 2y' - 3y = 3e^{2x}$. **Ans.** $y(x) = c_1 e^{-x} + c_2 e^{3x} - e^{2x}$.

5. $y''' - 2y'' - 5y' + 6y = 4e^{-x} - e^{2x}$.

Ans. $y(x) = c_1 e^x + c_2 e^{-2x} + c_3 e^{3x} + \dfrac{e^{-x}}{2} + \dfrac{e^{2x}}{4}$.

6. $y'' - 9y = 54e^{3x}$. **Ans.** $y(x) = c_1 e^{3x} + c_2 e^{-3x} + 9xe^{3x}$.

7. $y'' + 5y' + 6y = 3e^{-2x} + e^{3x}$. **Ans.** $y(x) = c_1 e^{-2x} + c_2 e^{-3x} + 3xe^{-2x} + \dfrac{e^{3x}}{30}$.

8. $\dfrac{d^2 y}{dx^2} - \dfrac{dy}{dx} - 6y = e^x \cosh 2x$. **Ans.** $y(x) = c_1 e^{3x} + c_2 e^{-2x} + \dfrac{xe^{3x}}{10} - \dfrac{e^{-x}}{8}$.

[**Hint:** $e^x \cosh 2x = e^x \left(\dfrac{e^{2x} + e^{-2x}}{2} \right) = \dfrac{e^{3x} + e^{-x}}{2}$]

9. $4y'' - 4y' + y = e^{x/2}$. **Ans.** $y(x) = (c_1 + c_2 x)e^{x/2} + (x^2 e^{x/2})/8$.

10. $(D^3 - 5D^2 + 8D - 4)y = e^{2x} + 2e^x + 3e^{-x} + 2$.

Ans. $y(x) = c_1 e^x + (c_2 + c_3 x)e^{2x} + \dfrac{x^2 e^{2x}}{2} + 2xe^x - \dfrac{e^{-x}}{6} - \dfrac{1}{2}$.

[Hint: P.I. of $\dfrac{1}{D^3-5D^2+8D-4}2=\dfrac{1}{(D-1)(D-2)^2}2$

$$=2\dfrac{1}{(D-1)(D-2)^2}\cdot e^{0x}=2\dfrac{1}{(-1)(-2)^2}=-\dfrac{1}{2}.]$$

11. $(D+2)(D-1)^2y=e^{-2x}+2\sinh x.$

Ans. $y(x)=c_1e^{-2x}+(c_2+c_3x)e^x+\dfrac{(xe^{-2x})}{9}+\left(\dfrac{x^2e^x}{6}\right)+\dfrac{e^{-x}}{4}.$

12. $\dfrac{d^2y}{dx^2}+6\dfrac{dy}{dx}+9y=50e^{2x}.$ Ans. $y(x)=(c_1+c_2x)e^{-3x}+2e^{2x}.$

13. $\dfrac{d^3y}{dx^3}-5\dfrac{d^2y}{dx^2}+8\dfrac{dy}{dx}-4y=e^{2x}+3e^{-x}.$

Ans. $y(x)=c_1e^x+(c_2+c_3x)e^{2x}+\left(\dfrac{x^2e^{2x}}{2}\right)-\dfrac{e^{-x}}{6}$

14. $y''+4y'+8y=(1+e^x)^2.$

Ans. $y(x)=(c_1\cos 2x+c_2\sin 2x)e^{-2x}+\left(\dfrac{2e^x}{13}\right)+\dfrac{e^{2x}}{20}+\dfrac{1}{8}.$

(GGSIPU 2006)

15. $(D^2+4)y=\cos 2x.$

Ans. $y(x)=c_1\cos 2x+c_2\sin 2x+(x\sin 2x)/4.$

(I.E.T.E., June 2002)

16. $y''+3y'+2y=\sin 2x.$

Ans. $y(x)=c_1e^{-x}+c_2e^{-2x}-(6\cos 2x+2\sin 2x)/40.$

17. $(D^3+D^2+D+1)y=\sin 2x.$

Ans. $y(x)=c_1e^{-x}+(c_2\cos x+c_3\sin x)+(2\cos 2x-\sin 2x)/15.$

18. $\dfrac{d^2y}{dx^2}+5\dfrac{dy}{dx}+6y=e^{-2x}+\sin x.$

Ans. $y(x)=c_1e^{-2x}+c_2e^{-3x}+xe^{-2x}+(\sin x-\cos x)/10.$

(UPTU 2006)

19. $(D^3-3D^2+4D-2)y=e^x+\cos x.$

Ans. $y(x)=c_1e^x+e^x(c_2\cos x+c_3\sin x)+xe^x+(3\sin x+\cos x)/10.$

20. $\dfrac{d^3y}{dx^3} + 2\dfrac{d^2y}{dx^2} + \dfrac{dy}{dx} = e^{-x} + \sin 2x.$ *(VTU 2004)*

Ans. $y(x) = c_1 + (c_2 + c_3x)e^{-x} - (x^2 e^{-x}/2) + (3\cos 2x - 4\sin 2x)/50.$

21. $(D^2 + 4)y = e^x + \sin 3x.$

(MDU 2003)

Ans. $y(x) = c_1 \cos 2x + c_2 \sin 2x + (e^x - \sin 3x)/5.$

22. $y''' + y = \sin 3x - \cos^2(x/2).$

(MDU 2003; Madras 2000)

Ans. $y(x) = c_1 e^{-x} + e^{x/2}\left(c_2 \cos \dfrac{\sqrt{3}}{2}x + c_3 \sin \dfrac{\sqrt{3}}{2}x \right)$

$+\dfrac{1}{730}(\sin 3x + 27\cos 3x) - \dfrac{1}{2} - \dfrac{1}{4}(\cos x - \sin x).$

23. $\dfrac{d^2y}{dx^2} + 2\dfrac{dy}{dx} + y = e^{2x} - \cos^2 x.$

(Delhi, 2002)

$\left[\text{Hint : } \cos^2 x = (1 + \cos 2x)/2 \right]$

Ans. $y(x) = (c_1 + c_2x)e^{-x} + \dfrac{1}{9}e^{2x} - \dfrac{1}{2} + \dfrac{1}{50}(3\cos 2x - 4\sin 2x).$

24. $y''' + 2y'' + y' = e^{2x} + \sin 2x.$

(Andhra, 1999)

Ans. $y(x) = c_1 + (c_2 + c_3x)e^{-x} + \dfrac{e^{2x}}{18} + (6\cos 2x - 8\sin 2x)/100.$

25. $(D^2 - 4D + 3)\,y = \sin 3x \cos 2x.$ *(MDU 2003; Madras, 2000)*
    [Hint: We may write $\sin 3x \cos 2x = (\sin 5x + \sin x)/2$].

Ans. $y(x) = c_1 e^x + c_2 e^{3x} + (10\cos 5x - 11\sin 5x)/884 + (\sin x + 2\cos x)/20.$

26. $y'' + a^2 y = \cos ax.$ Ans. $y(x) = c_1 \cos ax + c_2 \sin ax + (x \sin ax)/2a.$

27. $y'' + 2y' + 401y = \sin 20x + 40 \cos 20\,x.$

Ans. $y(x) = e^{-x}(c_1 \cos 20x + c_2 \sin 20x) + \sin 20x.$

28. $(2D^2 + 2D + 3)\,y = x^2 + 2x - 1.$

Ans. $y(x) = e^{-x/2}\left(c_1 \cos \dfrac{1}{2}\sqrt{5}x + c_2 \sin \dfrac{1}{2}\sqrt{5}x \right) + \dfrac{1}{3}x^2 + \dfrac{2}{9}x - \dfrac{25}{27}.$

29. $(D^4 - 2D^3 + D^2)y = x^3.$

(AMIE, W-2008)

Ans. $y(x) = c_1 + c_2x + (c_3 + c_4x)e^x + (x^5 + 10x^4 + 60x^3 + 240x^2)/20.$

30. $(D^2 - 4D + 4)y = 8(e^{2x} + \sin 2x + x^2)$. *(GGSIPU-2006; IETE-June 2004)*

Ans. $y(x) = (c_1 + c_2 x)e^{2x} + 4x^2 e^{2x} + \cos 2x + 2x^2 + 4x + 3$.

31. $y'' + 4y = e^x + \sin 2x + x^2$.

Ans. $y(x) = c_1 \cos 2x + c_2 \sin 2x + \dfrac{1}{5}e^x - \dfrac{x}{4}\cos 2x + \dfrac{1}{4}\left(x^2 - \dfrac{1}{2}\right)$.

32. $(D^2 + 3D + 2)y = e^{2x} + x^2 + \sin x$. *(D.C.E., 2004)*

Ans. $y(x) = c_1 e^{-x} + c_2 e^{-2x} + e^{2x}/12 + (2x^2 - 6x + 7)/4 - (3\cos x - \sin x)/10$.

33. $y''' + 2y'' + y' = e^{2x} + x^2 + x$.

Ans. $y(x) = c_1 + (c_2 + c_3 x)e^{-x} + (e^{2x} + 6x^3 - 27x^2 + 72x)/18$.

34. $y''' + y = 5e^{2x} + x^3 + 3$.

Ans. $y(x) = c_1 e^{-x} + e^{x/2}\left(c_2 \cos \dfrac{\sqrt{3}}{2}x + c_3 \sin \dfrac{\sqrt{3}}{2}x\right) + \dfrac{5}{9}e^{2x} + x^3 - 3$.

35. $y''' + 2y'' + y' = x^2 + x$.

Ans. $y(x) = c_1 + (c_2 + c_3 x)e^{-x} + x^3/3 - 3x^2/2 + 4x$.

36. $(D^4 + 2D^3 - 3D^2)y = x^2$.

Ans. $y(x) = c_1 + c_2 x + c_3 e^x + c_4 e^{-3x} - x^2(3x^2 + 8x + 28)/108$.

37. $y^{iv} + 4y'' = 96x^2$. **Ans.** $y(x) = c_1 + c_2 x + c_3 \cos 2x + c_4 \sin 2x + 2x^4 - 6x^2$.

38. $y^{iv} + y''' + y'' = 5x^2 + \cos x$.

Ans. $y(x) = c_1 + c_2 x + e^{-x/2}\left(c_3 \cos \dfrac{\sqrt{3}}{2}x + c_4 \sin \dfrac{\sqrt{3}}{2}x\right)$

$$+ \dfrac{5}{12}x^4 - \dfrac{5}{3}x^3 + 10x - 10 - \sin x.$$

39. $y'' - 2y' + 5y = e^{2x}\sin x$.

Ans. $y(x) = e^x(c_1 \cos 2x + c_2 \sin 2x) - e^{2x}(\cos x - 2\sin x)/10$.

40. $(D^2 - 4D + 3)y = e^x \cos 2x$. *(Osmania 1999)*

Ans. $y(x) = c_1 e^x + c_2 e^{3x} - [e^x(\cos 2x + \sin 2x)]/8$.

41. $y'' - 2y' = e^x \sin x$. *(VTU 2000)* **Ans.** $y(x) = c_1 + c_2 e^{2x} - (e^x \sin x)/2$.

42. $y''' + 2y'' + y' = x^2 e^{2x} + \sin^2 x$.

Ans. $y(x) = c_1 + (c_2 + c_3 x)e^{-x} + \dfrac{e^{2x}}{18}\left(x^2 - \dfrac{7x}{3} + \dfrac{11}{6}\right) + \dfrac{x}{2} + \dfrac{(3\sin 2x + 4\cos 2x)}{100}$.

43. $y'' - 2y' + 2y = x + e^x \cos x.$ $\hspace{2cm}$ *(UPTU 2002)*

$\hspace{1.5cm}$ **Ans.** $y(x) = e^x (c_1 \cos x + c_2 \sin x) + (x+1)/2 + (xe^x \sin x)/2$.

44. $y'' + y = e^{-x} + x^3 + e^x \sin x + 7.$

Ans. $y(x) = c_1 \cos x + c_2 \sin x + e^{-x}/2 + (x^3 - 6x) - e^x (2\cos x - \sin x)/5 + (7).$

45. $16y'' + 8y' + y = 48xe^{-x/4}.$ $\hspace{2cm}$ *(I.E.T.E., June 2006)*

$\hspace{2.5cm}$ **Ans.** $y(x) = (c_1 + c_2 x)e^{-x/4} + (x^3 e^{-x/4})/2.$

46. $y'' - 4y = x^2 e^{3x}.$ $\hspace{3cm}$ *(I.E.T.E., Dec. 1998)*

$\hspace{1.5cm}$ **Ans.** $y(x) = c_1 e^{2x} + c_2 e^{-2x} + e^{3x} (25x^2 - 60x + 62)/125.$

47. $\dfrac{d^3 y}{dx^3} - 3\dfrac{d^2 y}{dx^2} + 3\dfrac{dy}{dx} - y = (x+1)e^x.$

$\hspace{2.5cm}$ **Ans.** $y(x) = \left(c_1 + c_2 x + c_3 x^2\right)e^x + \dfrac{x^3 e^x}{6}\left(1 + \dfrac{x}{4}\right).$

48. $(D^2 - 4D + 13)y = 18e^{2x} \sin 3x.$ $\hspace{1cm}$ *(IETE Dec. 2008, Dec 2007)*

$\hspace{2cm}$ **Ans.** $y(x) = e^{2x}(c_1 \cos 3x + c_2 \sin 3x) - 3x\, e^{2x}\cos 3x.$

49. $y''' + 2y'' + y' = x^2 e^{2x} + \sin^2 x.$ $\hspace{2cm}$ *(PTU 2003)*

$\hspace{1cm}$ **Ans.** $y(x) = c_1 + (c_2 + c_3 x)e^{-x} + e^{2x}(6x^2 - 14x + 11)/108 + (x/2)$

$\hspace{5cm} + (3\sin 2x + 4\cos 2x)/100.$

50. $(D^2 + 4D + 3)y = e^{-x} \sin x + xe^{3x}.$ $\hspace{2cm}$ *(Anna 2002S)*

$\hspace{1cm}$ **Ans.** $y(x) = c_1 e^{-x} + c_2 e^{-3x} - \left[e^{-x}(2\cos x + \sin x)\right]/5 - e^{-3x}(x^2 + x)/4$

51. $y'' + 3y' + 2y = \cosh 2x \sin x.$

$\hspace{2cm}$ **Ans.** $y(x) = c_1 e^{-x} + c_2 e^{-2x} - e^{2x}\ (7\cos x - 11\sin x)/340$

$\hspace{4cm} + e^{-2x}(\cos x - \sin x)/4.$

52. $(D^2 - 4D + 1)y = e^{2x} \sin 2x.$ $\hspace{2cm}$ *(I.E.T.E., Dec. 2005)*

$\hspace{2cm}$ **Ans.** $y(x) = c_1 e^{(2+\sqrt{3})x} + c_2 e^{(2-\sqrt{3})x} - (e^{2x} \sin 2x)/7.$

53. $(D^2 + 4)y = x \sin x.$ $\hspace{3cm}$ *(A.M.I.E., S-2010)*

$\hspace{1.5cm}$ **Ans.** $y(x) = c_1 \cos 2x + c_2 \sin 2x + (3x\sin x - 2\cos x)/9.$

54. $y'' + 4y' + 3y = x \sin 2x.$ $\hspace{2cm}$ *(IETE., June 2009, 2008)*

$\hspace{1.5cm}$ **Ans.** $y(x) = Ae^{-x} + Be^{-3x} - \dfrac{1}{4225}[65x(8\cos 2x + \sin 2x)$

$\hspace{4cm} - 188\cos 2x - 316\sin 2x]$

55. $y'' + 4y = x\sin^2 x$.

$$\left[\textbf{Hint}: x\sin^2 x = x(1 - \cos 2x)/2 = (x/2) - (x\cos 2x)/2\right].$$

Ans. $y(x) = c_1 \cos 2x + c_2 \sin 2x + (x/8) - (x\cos 2x)/128 - (x^2 \sin 2x)/16$.

56. $y^{iv} - y = x \sin x$.

Ans. $y(x) = c_1 e^x + c_2 e^{-x} + c_3 \cos x + c_4 \sin x + (x^2 \cos x - 3x \sin x)/8$.

57. $y^{iv} + 2y'' + y = x^2 \cos x$. *(AMIE., W-2008; Osmania, 2003S)*

Ans. $y(x) = (c_1 + c_2 x) \cos x + (c_3 + c_4 x) \sin x + [4x^3 \sin x - x^2(x^2 - 9) \cos x]/48$.

58. $y'' + 4y = x \cos x$. *(DCE, 2005)*

Ans. $y(x) = c_1 \cos 2x + c_2 \sin 2x + (3x \cos x + 2 \sin x)/9$.

59. $\left(D^2 + 5D + 6\right) y = e^{-2x}\left(\sec^2 x\right)(1 + 2\tan x)$. *(IETE., June 2003)*

Ans. $y(x) = c_1 e^{-3x} + c_2 e^{-2x} + e^{-2x} \tan x$.

Find the solution of the following differential equations satisfying the given conditions.

60. $y''' - 6y'' + 11y' - 6y = 0$, $y(0) = 0$, $y'(0) = -4$, $y''(0) = -18$.

Ans. $y(x) = e^x + 2e^{2x} - 3e^{3x}$.

61. $y'' - 7y' + 6y = e^{2x}$, $y(0) = 0$. *(GGSIPU 2006)*

Ans. $y(x) = c\left(e^x - e^{6x}\right) + \left\{e^{2x}\left(e^{4x} - 1\right)\right\}/4$.

62. $y'' + 2y' + 10y + 37 \sin 3x = 0$, $y(0) = 3$; $y'(0) = 0$. Also find $y(\pi/2)$.

Ans. $y(x) = -3e^{-x} \cos 3x + 6\cos 3x - \sin 3x$; 1.

63. $y'' + 4y' + 5y = -2 \cosh x$, $y(0) = 0$, $y'(0) = 1$.

Ans. $y(x) = 3e^{-2x}(\cos x + 3\sin x)/5 - e^x/10 - e^{-x}/2$.

64. $y^{iv} + 13y'' + 36y = 0$, $y(0) = 0$, $y''(0) = 0$, $y(\pi/2) = -1$, $y'(\pi/2) = -4$.

Ans. $y(x) = 2 \sin 2x + \sin 3x$.

65. $(D^2 + 6D + 10)y = 50x$; $y(0) = 0$, $y'(0) = 1$.

Ans. $y(x) = 5x - 3 + e^{-3x}(3 \cos x + 5 \sin x)$.

66. $(D^2 + 1)^2 y = 24x \cos x$, $y(0) = 0$, $y'(0) = y''(0) = 0$ and $y'''(0) = 12$.

Ans. $y(x) = 3x^2 \sin x - x^3 \cos x$.

67. Find the general solution of the differential equation

$$\frac{d^3 y}{dx^3} + \frac{d^2 y}{dx^2} - \frac{dy}{dx} - y = \sin(2x - 3).$$ *(A.M.I.E., S-2009)*

Ans. $y(x) = c_1 e^x + (c_2 + c_3 x)e^{-x} + \frac{1}{25}[2\cos(2x - 3) - \sin(2x - 3)]$.

18.10 Two important methods for determining a particular integral of $f(D)y = X$ are available.

I. Method of Variation of Parameters

This is a very elegant method for finding the particular integral (P.I.) of a linear differential equation whose complementary function (C.F.) is known. Though the method is general, we will illustrate it by applying it to a second order linear differential equation with constant coefficients. Consider the equation.

$$\frac{d^2y}{dx^2} + p\frac{dy}{dx} + qy = X \qquad \qquad ...(18.43)$$

where p, q, are all constants and X is functions of x. It gives

$$\text{P.I.} = -y_1 \int \frac{y_2 X}{W} dx + y_2 \int \frac{y_1 X}{W} dx \qquad \qquad ...(18.44)$$

where y_1 and y_2 are the solutions of differential equation

$$\frac{d^2y}{dx^2} + p\frac{dy}{dx} + qy = 0 \qquad \qquad ...(18.45)$$

and $W = \begin{vmatrix} y_1 & y_2 \\ y_1' & y_2' \end{vmatrix}$ is called the Wronskian of (y_1, y_2).

Proof. Let the complementary function of Eq. (18.43) be

$$y_c = c_1 y_1 + c_2 y_2 \qquad \qquad ...(18.46)$$

Replacing c_1, c_2 (regarded as parameters) by unknown functions $u(x)$ and $v(x)$, let the P.I. be

$$y = u y_1 + v y_2 \qquad \qquad ...(18.47)$$

Differentiating Eq. (18.47) w.r.t. x we get

$$y' = u y_1' + v y_2' + u' y_1 + v' y_2$$
$$= u y_1' + v y_2', \qquad \qquad ...(18.48)$$

on assuming that u, v satisfy the equation

$$u' y_1 + v' y_2 = 0. \qquad \qquad ...(18.49)$$

Differentiating Eq. (18.46) w.r.t., x, we obtain

$$y'' = u y_1'' + u' y_1' + v y_2'' + v' y_2'.$$

Substituting the values of y, y' and y'' in Eq. (18.42), we get

$$(u y_1'' + u' y_1' + v y_2'' + v' y_2') + p(u y_1' + v y_2') + q(u y_1 + v y_2) = X$$
or $u(y_1'' + p y_1' + q y_1) + v(y_2'' + p y_2' + q y_2)$
$$+ u' y_1' + v' y_2' = X \text{ or } u' y_1' + v' y_2' = X \qquad \qquad ...(18.50)$$

Solving Eq. (18.48) and Eq. (18.49) we get

$$u' = \begin{vmatrix} 0 & y_2 \\ X & y_2' \end{vmatrix} \div \begin{vmatrix} y_1 & y_2 \\ y_1' & y_2' \end{vmatrix} = -\frac{y_2 X}{W} \quad \text{and} \quad v' = \begin{vmatrix} y_1 & 0 \\ y_1' & X \end{vmatrix} \div \begin{vmatrix} y_1 & y_2 \\ y_1' & y_2' \end{vmatrix} = \frac{y_1 X}{W}$$

where $W = \begin{vmatrix} y_1 & y_2 \\ y_1' & y_2' \end{vmatrix}$ is called the Wronskian of (y_1, y_2).

Integrating, $\qquad u = -\int \frac{y_2 X}{W} dx, \qquad v = \int \frac{y_1 X}{W} dx.$

Substituting these in Eq. (18.47), the P.I. is known.

Remark. As the solution is obtained by varying the arbitrary constants c_1, c_2 of the complementary function, the method is known as *variation of parameters*.

Example 18.20 Solve the following differential equations by the method of variation of parameters

(i) $(D^2 - 1)y = 2/(1 + e^x).$ *(I.E.T.E., June 2001; UPTU-2001)*

(ii) $(D^2 - 1)y = e^{-x} \sin(e^{-x}) + \cos(e^{-x}).$ *(I.E.T.E., Dec. 2002)*

(iii) $y'' - 2y' + y = e^x \log x.$ *(V.T.U., 2000S)*

(iv) $y'' + 4y = \tan 2x.$ *(J.N.T.U., 1999; Mangalore, 1997).*

Solution. (*i*) The characteristic equation of the corresponding homogeneus equation is $m^2 - 1 = (m - 1)(m + 1) = 0$. Its roots are $m_1 = 1$ and $m_2 = -1$. The complementary function is

$$y_c(x) = c_1' e^x + c_2' e^{-x}.$$

To determine $y_p(x)$, proceed as under:

Here $\qquad y_1 = e^x, \ y_2 = e^{-x} \text{ and } X = 2/(1 + e^x).$

The Wronskian $W(y_1, y_2)$ is given by $W\left(e^x, e^{-x}\right) = \begin{vmatrix} e^x & e^{-x} \\ e^x & -e^{-x} \end{vmatrix} = -2.$

Thus $\quad y_p(x) = -y_1 \int \frac{y_2 X}{W} dx + y_2 \int \frac{y_1 X}{W} dx = e^x \int \frac{e^{-x}}{1 + e^x} dx - e^{-x} \int \frac{e^x}{1 + e^x} dx.$

To evaluate $\qquad \int \frac{dx}{e^x (1 + e^x)}$, put $e^x = t, \ dx = \frac{dt}{t}.$

$\therefore \qquad \int \frac{dx}{e^x (1 + e^x)} = \int \frac{dt}{t^2 (1 + t)} = \int \left(\frac{1}{t^2} - \frac{1}{t} + \frac{1}{1 + t} \right) dt$

$$= -\frac{1}{t} - \log t + \log(1 + t) = -\frac{1}{e^x} + \log \frac{(1 + e^x)}{e^x} + c_3,$$

and $\int \dfrac{e^x}{1+e^x} dx = \log(1+e^x) + c_4.$

$\therefore \quad y_p(x) = e^x\left[-\dfrac{1}{e^x} + \log(1+e^x) - x + c_3\right] - e^{-x}\left[\log(1+e^x) + c_4\right].$

The general solution of the equation is

$$y(x) = c_1' e^x + c_2' e^{-x} + e^x\left[-\dfrac{1}{e^x} + \log(1+e^x) - x + c_3\right] - e^{-x}\left[\log(1+e^x) + c_4\right]$$

$$= e^x\left[-\dfrac{1}{e^x} + \log(1+e^x) - x + c_1\right] + e^{-x}\left[c_2 - \log(1+e^x)\right].$$

(*ii*) The characteristic equation of the corresponding homogeneous equation is $(m^2 - 1) = (m+1)(m-1) = 0$. Its roots are $m_1 = 1$ and $m_2 = -1$. The complementary function is $y_c(x) = c_1 e^x + c_2 e^{-x}$.

To determine $y_p(x)$, proceed as under:
Here $y_1 = e^x$, $y_2 = e^{-x}$ and $X = e^{-x}\sin(e^{-x}) + \cos(e^{-x})$

The Wronskian $W(y_1, y_2)$ is given by $W(e^x, e^{-x}) = \begin{vmatrix} e^x & e^{-x} \\ e^x & -e^{-x} \end{vmatrix} = -2.$

Thus $y_p(x) = -y_1 \int \dfrac{y_2 X}{W} dx + y_2 \int \dfrac{y_1 X}{W} dx$

$= \dfrac{e^x}{2} \int e^{-x}\left[e^{-x}\sin(e^{-x}) + \cos(e^{-x})\right] dx - \dfrac{e^{-x}}{2} \int e^x\left[e^{-x}\sin(e^{-x}) + \cos(e^{-x})\right] dx$

$= \dfrac{e^x}{2} \int \left[e^{-2x}\sin(e^{-x}) + e^{-x}\cos(e^{-x})\right] dx - \dfrac{e^{-x}}{2} \int \left[\sin(e^{-x}) + e^x\cos(e^{-x})\right] dx.$

Now $\int \left[e^{-2x}\sin(e^{-x}) + e^{-x}\cos(e^{-x})\right] dx = e^{-x}\cos(e^{-x}) - \sin(e^{-x}) - \sin(e^{-x})$

$= e^{-x}\cos(e^{-x}) - 2\sin(e^{-x})$ and $\int \left[\sin e^{-x} + e^x\cos(e^{-x})\right] dx = \dfrac{\cos e^{-x}}{e^{-x}}.$

$\therefore \quad y_p(x) = \dfrac{e^x}{2}\left[e^{-x}\cos(e^{-x}) - 2\sin(e^{-x})\right] - \dfrac{e^{-x}}{2}\left[\dfrac{\cos(e^{-x})}{e^{-x}}\right]$

$= \dfrac{\cos(e^{-x})}{2} - e^x\sin(e^{-x}) - \dfrac{\cos(e^{-x})}{2} = -e^x\sin(e^{-x}).$

The general solution of the equation is

$$y(x) = c_1 e^x + c_2 e^{-x} - e^x \sin(e^{-x}).$$

(*iii*) The characteristic equation of the corresponding homogeneous equation is $m^2 - 2m + 1 = (m-1)^2 = 0$. and its roots are $m = 1, 1$, which is a repeated root.

The complementary function is $y_c = (c_1 + c_2 x)e^x$.

To find $y_p(x)$, proceed as under:

Here $\qquad y_1 = e^x, \ y_2 = xe^x, \ X = e^x \log x.$

The wronskian $W(y_1, y_2)$ is given by $W(e^x, xe^x) = \begin{vmatrix} e^x & xe^x \\ e^x & (1+x)e^x \end{vmatrix} = e^{2x}.$

Thus $y_p(x) = -y_1 \int \dfrac{y_2 X}{W} dx + y_2 \int \dfrac{y_1 X}{W} dx$

$= -e^x \int \dfrac{xe^x \cdot e^x \log x}{e^{2x}} dx + xe^x \int \dfrac{e^x \cdot e^x \log x}{e^{2x}} dx = -e^x \underset{II}{\int} x \log x \, dx + xe^x \underset{I}{\int} \log x \, dx$

$= -e^x \left(\dfrac{x^2}{2} \log x - \int \dfrac{1}{x} \cdot \dfrac{x^2}{2} dx \right) + x \cdot e^x \left(x \log x - \int \dfrac{1}{x} \cdot x \, dx \right)$

$= -e^x \left(\dfrac{x^2}{2} \log x - \dfrac{x^2}{4} \right) + xe^x (x \log x - x) = \dfrac{1}{4} x^2 e^x (2 \log x - 3).$

Hence the general solution of the equation is

$$y(x) = (c_1 + c_2 x)e^x + \dfrac{1}{4} x^2 e^x (2 \log x - 3).$$

(*iv*) The characteristic equation of the corresponding homogeneous equation is $m^2 + 4 = 0$. Its roots are $m = \pm 2i$.

The complementary function is $y_c(x) = c_1 \cos 2x + c_2 \sin 2x$.

To find $y_p(x)$, proceed as under:

Here $y_1 = \cos 2x, \ y_2 = \sin 2x$ and $X = \tan 2x$.

The Wronskian $W(y_1, y_2)$ is given by

$$W(\cos 2x, \sin 2x) = \begin{vmatrix} \cos 2x & \sin 2x \\ -2\sin 2x & 2\cos 2x \end{vmatrix} = 2.$$

Thus $y_p(x) = -y_1 \int \dfrac{y_2 X}{W} dx + y_2 \int \dfrac{y_1 X}{W} dx$

$= -\cos 2x \int \dfrac{\sin 2x \tan 2x}{2} dx + \sin 2x \int \dfrac{\cos 2x \tan 2x}{2} dx$

$$= -\frac{1}{2}\cos 2x \int (\sec 2x - \cos 2x)\, dx + \frac{1}{2}\sin 2x \int \sin 2x\, dx$$

$$= -\frac{1}{4}\cos 2x \left[\log|\sec 2x + \tan 2x| - \sin 2x\right] - \frac{1}{4}\sin 2x \cos 2x$$

$$= -\frac{1}{4}\cos 2x \log|\sec 2x + \tan 2x|.$$

The general solution of the equation is

$$y(x) = c_1 \cos 2x + c_2 \sin 2x - \frac{1}{4}\cos 2x \log|\sec 2x + \tan 2x|.$$

II. Method of Undetermined Coefficients

In this method we assume a *trial solution* (assumed particular solution) containing unknown constants (indicated by A, B, C, E...) which are to be determined by substitution in the given equation. The trial solution to be assumed in each case depends on the special form of X containing constants, polynomials, exponentials, sines and cosines functions, sums or products of these functions. These set of functions has the remarkable property that derivatives of their sums and products are again sums and products of constants polynomials, exponentials, sines and cosines. The method of underdetermined coefficients is not applicable to the sets of functions that contains $\log x$, x^{-1}, $\tan x$, $\sec x$, $\sin^{-1}x$ etc. since the number of terms obtained by differenting each of these function is infinite. The special form of X and assumed trial solution is shown in the following table.

Table 18.3. Trial Particular Solutions

X	Form of $y_p(x)$
1. 4 (any constant)	A
2. $6x + 9$	$Ax + B$
3. $2x^2 - 3$	$Ax^2 + Bx + C$
4. $x^3 - x + 1$	$Ax^3 + Bx^2 + Cx + E$
5. $\sin 3x$	$A\cos 3x + B\sin 3x$
6. $\cos 3x$	$A\cos 3x + B\sin 3x$
7. e^{4x}	$A e^{4x}$
8. $(9x - 3)e^{4x}$	$(Ax + B)e^{4x}$
9. $x^2 e^{4x}$	$(Ax^2 + Bx + C)e^{4x}$
10. $e^{3x}\sin 4x$	$e^{3x}(A\cos 4x + B\sin 4x)$
11. $6x^2 \sin 3x$	$(Ax^2 + Bx + C)\cos 3x + (Ex^2 + Fx + G)\sin 3x$
12. $xe^{2x}\cos 3x$	$(Ax + B)e^{2x}\cos 3x + (Cx + E)e^{2x}\sin 3x$

The above method holds in case no terms in the assumed trial solution appears in the complementary solution. If any term of the assumed trial solution does appear in the complementary solution, we must multiply this trial solution by the smallest positive integral power of x which is large enough so that none of the terms which are then present appear in the complementary solution.

Example 18.21 Find the general solution of the following differential equations by the method of undetermined coefficients.

(i) $y'' + 2y' + 4y = 2x^2 + 3e^{-x}$.   (ii) $y'' - 2y' + 3y = x^3 + \sin x$.

(iii) $y'' - 2y' - 3y = 4x - 5 + 6xe^{2x}$.   (*VTU-2000*)

(iv) $y'' - 2y' = e^x \sin x$.

Solution. (*i*) The characteristic equation of the homogeneous equation is $m^2 + 2m + 4 = 0$. The roots of this equation are

$$m = -2 \pm \sqrt{4 - 16}/2 = -1 \pm i\sqrt{3}.$$

The complementary function is $y_c(x) = e^{-x}\left(c_1 \cos \sqrt{3}x + c_2 \sin \sqrt{3}x\right)$.

Since $X = 2x^2 + 3e^{-x}$, we choose the particular integral as

$$y_p(x) = (Ax^2 + Bx + C) + Ee^{-x}.$$

Substituting y_p and the derivatives $y_p'(x) = 2Ax + B - Ee^{-x}$ and $y_p''(x) = 2A + Ee^{-x}$ into the given differential equation, we obtain

$$y_p'' + 2y_p' + 4y_p = 4Ax^2 + (4A + 4B)x + (2A + 2B + 4C) + 3Ee^{-x} = 2x^2 + 3e^{-x}.$$

Equating coefficients of like terms, we obtain $A = 1/2$, $B = -1/2$, $C = 0$ and

$E = 1$. Hence, the particular integral is $y_p(x) = \dfrac{1}{2}x^2 - \dfrac{1}{2}x + e^{-x}$. The general

solution of the given equation is

$$y(x) = e^{-x}\left(c_1 \cos \sqrt{3}x + c_2 \sin \sqrt{3}x\right) + \frac{1}{2}x^2 - \frac{1}{2}x + e^{-x}.$$

(*ii*) The characteristic equation of the homogeneous equation is $m^2 - 2m + 3 = 0$. The roots of this equation are $m = \left\{2 \pm \sqrt{4 - 12}\right\}/2 = 1 \pm i\sqrt{2}$.

The complementary function is $y_c(x) = e^x\left(c_1 \cos \sqrt{2}x + c_2 \sin \sqrt{2}x\right)$.

We choose the particular integral as $y_p(x) = (Ax^3 + Bx^2 + Cx + E) +$ ($F \sin x + G \cos x$).

Then $y_p'(x) = 3Ax^2 + 2Bx + C + F \cos x - G \sin x$,

$y_p''(x) = 6Ax + 2B - F \sin x - G \cos x$.

Substituting in the given equation, we obtain

$$y_p'' - 2y_p' + 3y_p = 3Ax^3 + 3(B - 2A)x^2 + (3C - 4B + 6A)x + (3E - 2C + 2B)$$
$$+ 2(F + G)\sin x + 2(G - F)\cos x = x^3 + \sin x.$$

Equating coefficients of like terms, we obtain
$3A = 1$ and $A = 1/3$; $B - 2A = 0$ and $B = 2/3$; $3C - 4B + 6A = 0$ and
$C = 2/9$; $3E - 2C + 2B = 0$ and $E = -8/27$; $2(F + G) = 1$, $G - F = 0$ and
$F = G = 1/4$.

Thus, the particular integral is

$$y_p = \frac{1}{3}x^3 + \frac{2}{3}x^2 + \frac{2}{9}x - \frac{8}{27} + \frac{1}{4}(\sin x + \cos x).$$

The general solution of the given equation is

$$y(x) = e^x \left(c_1 \cos \sqrt{2}x + c_2 \sin \sqrt{2}x \right)$$

$$+ \frac{1}{27}(9x^3 + 18x^2 + 6x - 8) + \frac{1}{4}(\sin x + \cos x).$$

(*iii*) The characteristic equation of the homogeneous equation is
$m^2 - 2m - 3 = 0$ or $(m + 1)(m - 3) = 0$. Its roots are $m_2 = 3$ and $m_1 = -1$
The complementary function is
$y_c(x) = c_2 e^{3x} + c_1 e^{-x}$. We choose the particular integral as
$y_p(x) = (Ax + B) + (Cx + E)e^{2x}$.
Then $y'_p(x) = A + Ce^{2x} + 2(Cx + E)e^{2x}$,
  $y''_p(x) = 2Ce^{2x} + 2Ce^{2x} + 4(Cx + E)e^{2x}$,

and $y''_p - 2y'_p - 3y_p = -3Ax - 2A - 3B - 3Cxe^{2x} + (2C - 3E)e^{2x}$
  $= 4x - 5 + 6\,xe^{2x}$.

Equating coefficients of like terms, we get $-3A = 4$ and $A = -4/3$;
$2A + 3B = 5$ and $B = 23/9$; $-3C = 6$ and $C = -2$; $2C - 3E = 0$ and $E = -4/3$
Hence a particular integral is

$$y_p(x) = -\frac{4}{3}x + \frac{23}{9} - 2xe^{2x} - \frac{4}{3}e^{2x}.$$

The general solution of the given equation is

$$y(x) = c_2 e^{3x} + c_1 e^{-x} - \frac{4}{3}x + \frac{23}{9} - \left(2x + \frac{4}{3}\right)e^{2x}.$$

(*iv*) The characteristic equation of the homogeneous equation is
$m^2 - 2m = 0$. Its roots are $m_1 = 0$ and $m_2 = 2$.
The complementary function is $y_c(x) = c_1 + c_2 e^{2x}$. We choose the particular
integral as
$y_p(x) = e^x(A \sin x + B \cos x)$.
Then $y'_p = (A - B)e^x \sin x + (A + B)e^x \cos x$,
  $y''_p = -2B\,e^x \sin x + 2A\,e^x \cos x$,

and $y''_p - 2y'_p = -2Ae^x \sin x - 2Be^x \cos x = e^x \sin x$.

Equating coefficients of like terms, we obtain $-2A = 1$ and $A = -1/2$; $-2B = 0$ and $B = 0$. Hence, a particular integral of the differential equation is

$$y_p(x) = -\left(e^x \sin x\right)/2.$$

The general solution of the given equation is $y(x) = c_1 + c_2 e^{2x} - \left(e^x \sin x\right)/2$.

Example 18.22. Using the method of undetermined coefficients, solve the given initial value problems

(i) $5y'' + y' = -6x$, $\quad y(0) = 0$, $\quad y'(0) = -10$,

(ii) $y'' + 2y' + y = e^{-x}$, $\quad y(0) = -1$, $\quad y'(0) = 1$,

(iii) $y'' - y' - 2y = 10 \sin x$, $\quad y(0) = 1$, $\quad y'(0) = 3$.

Solution. (*i*) The characteristic equation of the homogeneous equation is $5m^2 + m = 0$ or $m(5m+1) = 0$ and its roots are $0, -1/5$.

The complementary function is $y_c(x) = c_1 + c_2 e^{-x/5}$.

We choose the particular integral as

$y_p(x) = x(Ax + B) = Ax^2 + Bx$. We have $y_p'(x) = 2Ax + B$, $y_p'' = 2A$.

Substituting in the given equation, we obtain

$5y_p'' + y_p' = 5(2A) + (2Ax + B) = 2Ax + (10A + B) = -6x$.

Equating coefficients of like terms, we get

$2A = -6$ and $A = -3$; $\quad 10A + B = 0$ and $B = 30$.

Thus, the particular integral is $y_p(x) = -3x^2 + 30x$.

The general solution is $y(x) = c_1 + c_2 e^{-x/5} - 3x^2 + 30x$.

Thus $y(0) = c_1 + c_2 = 0$, from the first initial condition.

By differentiating and from the second initial condition

$y'(x) = (-1/5)c_2 e^{-x/5} - 6x + 30$, $\quad y'(0) = (-c_2/5) + 30 = -10 \Rightarrow c_2 = 200$.

$\therefore \quad c_1 = -200$. The solution of the given initial value problem is

$y(x) = -200 + 200 e^{-x/5} - 3x^2 + 30x$.

(*ii*) The characteristic equation of the homogeneous equation is $m^2 + 2m + 1 = 0$

or $(m+1)^2 = 0$ and its roots are $-1, -1$:

The complementary function is $y_c(x) = (c_1 + c_2 x)e^{-x}$.

We choose the particular integral as $y(x) = Ax^2 e^{-x}$.

We have $y_p'(x) = A(2x - x^2)e^{-x}$, $y_p''(x) = A(2 - 4x + x^2)e^{-x}$. Substituting

in the given equation, we get $y_p'' + 2y_p' + y_p = A(2 - 4x + x^2)e^{-x}$

$+2A(2x - x^2)e^{-x} + Ax^2e^{-x} = 2Ae^{-x} = e^{-x}$. Hence A = 1/2.
This gives the general solution of the given equation,
$y(x) = (c_1 + c_2 x)e^{-x} + (1/2)x^2 e^{-x}$.

Thus $y(0) = c_1 = -1$, from the first initial condition. By differentiation and from

the second initial condition $y'(x) = (c_2 - c_1 - c_2 x)e^{-x} + \left(x - \dfrac{x^2}{2}\right)e^{-x}$,

$y'(0) = c_2 - c_1 = 1 \Rightarrow c_2 = 0$.

This gives the answer $y(x) = \left(\dfrac{x^2}{2} - 1\right)e^{-x}$.

(*iii*) The characteristic equation of the homogeneous equation is $m^2 - m - 2 = 0$
or $(m+1)(m-2) = 0$ and its roots are $-1, 2$. The complementary function is
$y_c(x) = c_1 e^{-x} + c_2 e^{2x}$. We choose the partcular integral as
$y_p(x) = A\cos x + B\sin x$.

We have $y_p'(x) = -A\sin x + B\cos x$, $y_p''(x) = -A\cos x - B\sin x$.
Substituting in the given equation, we get
$y_p'' - y_p' - 2y_p = -(A\cos x + B\sin x) + (A\sin x - B\cos x) - (2A\cos x + 2B\sin x)$

$= (A - 3B)\sin x - (3A + B)\cos x = 10\sin x$.
Equating coefficients of like terms, we get $A - 3B = 10$; $3A + B = 0$. On solving, we obtain $A = 1$ and $B = -3$. Thus, the particular integral is
$y_p(x) = \cos x - 3\sin x$.

The general solution is $y(x) = c_1 e^{-x} + c_2 e^{2x} + \cos x - 3\sin x$ (*i*)
Substituting the initial conditions in (*i*) we get
$y(0) = 1 = c_1 + c_2 + 1$, $y'(x) = -c_1 e^{-x} + 2c_2 e^{2x} - \sin x - 3\cos x$,

$y'(0) = 3 = -c_1 + 2c_2 - 3$. The solution is $c_1 = -2, c_2 = 2$. The solution of the
given initial value problem is $y(x) = -2e^{-x} + 2e^{2x} + \cos x - 3\sin x$.

Example 18.23 Find the general solution of the following differential equations by the method of undetermined coefficients

(i) $y'' - 6y' + 9y = 6x^2 + 2 - 12e^{3x}$,

(ii) $y''' - 6y'' + 12y' - 8y = 12e^{2x} + 27e^{-x}$,

(iii) $y'' + 9y = \cos 3x$,   *(IETE., Dec 2008)*

(iv) $y'' - 4y' + 13y = 12e^{2x} \sin 3x$.   *(IETE June 2007)*

Solution. (*i*) The characteristic equation of the homogeneous equation is $m^2 - 6m + 9 = 0$, or $(m - 3)^2 = 0$ and its roots are $m_1 = m_2 = 3$.

The complementary function is $y_c(x) = (c_1 + c_2 x)e^{3x}$.

We note that e^{3x} and xe^{3x} are present in the complementary function (due to the double root $m = 3$) and e^{3x} is also a term on the right hand side X. Therefore, we choose the particular integral as $y_p(x) = (Ax^2 + Bx + C) + Ex^2 e^{3x}$.

We have $\qquad y_p'(x) = 2Ax + B + 2Exe^{3x} + 3Ex^2 e^{3x}$,

$\qquad\qquad y_p''(x) = 2A + 2Ee^{3x} + 12Exe^{3x} + 9Ex^2 e^{3x}$.

Substituting in the given equation, we get

$y_p'' - 6y_p' + 9y_p = 9Ax^2 + (-12A + 9B)x + 2A - 6B + 9C + 2Ee^{3x}$

$\qquad\qquad\qquad = 6x^2 + 2 - 12e^{3x}$.

Equating coefficients of like terms, we get

$9A = 6$ and $A = 2/3$; $-12A + 9B = 0$ and $B = 8/9$;

$2A - 6B + 9C = 2$ and $C = 2/3$; $2E = -12$ and $E = -6$.

Thus, the particular integral is

$$y_p(x) = \frac{2}{3}x^2 + \frac{8}{9}x + \frac{2}{3} - 6x^2 e^{3x}.$$

Hence the general solution is $y(x) = y_c(x) + y_p(x)$

$$= \left(c_1 + c_2 x\right)e^{3x} + \frac{2}{3}x^2 + \frac{8}{9}x + \frac{2}{3} - 6x^2 e^{3x}.$$

(*ii*) The characteristic equation of the homogeneous equation is $m^3 - 6m^2 + 12m - 8 = (m - 2)^3 = 0$. Its roots are $m_1 = m_2 = m_3 = 2$.

The complementary function is $y_c(x) = (c_1 + c_2 x + c_3 x^2)e^{2x}$.

We note that e^{2x}, xe^{2x} and $x^2 e^{2x}$ are present in the complementary function (due to the triple roots $m = 2$) and e^{2x} is also a term on the right hand side X. Therefore, we choose the particular integral as $y_p(x) = Ax^3 e^{2x} + Be^{-x}$.

We have $y_p'(x) = A(3x^2 + 2x^3)e^{2x} - Be^{-x}$,

$\qquad\qquad y_p''(x) = A(6x + 12x^2 + 4x^3)e^{2x} + Be^{-x}$,

$\qquad\qquad y_p'''(x) = A(6 + 36x + 36x^2 + 8x^3)e^{2x} - Be^{-x}$.

Substituting in the given equation, we get

$y_p''' - 6y_p'' + 12y_p' - 8y_p = Ae^{2x}[(6 + 36x + 36x^2 + 8x^3)$

$-6(6x + 12x^2 + 4x^3) + 12(3x^2 + 2x^3) - 8x^3] + Be^{-x}[-1 - 6 - 12 - 8]$

$= 6Ae^{2x} - 27Be^{-x} = 12e^{2x} + 27e^{-x}$.

Equating coefficients of like terms, we get $6A = 12$ and $A = 2$;

$-27B = 27$ and $B = -1$.

Therefore, the particular integral is $y_p(x) = 2x^3 e^{2x} - e^{-x}$.

The general solution is

$$y(x) = (c_1 + c_2 x + c_3 x^2) e^{2x} + 2x^3 e^{2x} - e^{-x}.$$

(*iii*) The characteristic equation of the homogeneous equation is $m^2 + 9 = 0$ and its roots are $m = \pm 3i$. The complementary function is

$$y_c(x) = c_1 \cos 3x + c_2 \sin 3x.$$

We note that $\cos 3x$ appears as a term both in complementary function and the right hand side X. Therefore, we choose the particular integral as

$$y_p(x) = x(A \cos 3x + B \sin 3x).$$

We have $y'_p(x) = A \cos 3x + B \sin 3x + 3x(-A \sin 3x + B \cos 3x)$,

$y''_p(x) = 6(-A \sin 3x + B \cos 3x) + 9x(-A \cos 3x - B \sin 3x)$.

Substituting in the given equation, we get

$$y''_p + 9y_p = \sin 3x[-6A - 9xB + 9xB] + \cos 3x[6B - 9xA + 9xA] = \cos 3x$$

or $-6A \sin 3x + 6B \cos 3x = \cos 3x$.

Equating coefficients of like terms, we get $A = 0$; $6B = 1$ and $B = 1/6$.

The particular integral is $y_p(x) = (x \sin 3x)/6$.

The general solution is

$$y(x) = c_1 \cos 3x + c_2 \sin 3x + (x \sin 3x)/6.$$

(*iv*) The characteristic equation of the homogeneous equation is $m^2 - 4m + 13 = 0$. The roots of this equation are

$$m = 4 \pm \sqrt{16 - 52}/2 = 2 \pm 3i.$$

The complementary function is $y_c(x) = e^{2x}(c_1 \cos 3x + c_2 \sin 3x)$.

We note that $e^{2x} \sin 3x$ appears both in the complementary function and the right hand side X. Therefore, we choose

$$y_p(x) = x e^{2x}(A \cos 3x + B \sin 3x).$$

We have $y'_p(x) = (1 + 2x)e^{2x} (A \cos 3x + B \sin 3x)$

$+ 3xe^{2x}(-A \sin 3x + B \cos 3x)$,

$y''_p(x) = (4 + 4x)e^{2x} (A \cos 3x + B \sin 3x)$

$+ 6(1 + 2x)e^{2x} (-A \sin 3x + B \cos 3x) + 9xe^{2x}(-A \cos 3x - B \sin 3x)$.

Substituting in the given equation, we get

$y''_p - 4y'_p + 13y_p = e^{2x} \cos 3x[A(4 + 4x) + 6B(1 + 2x)$

$-9Ax - 4A(1 + 2x) - 12xB + 13Ax] + e^{2x} \sin 3x[B(4 + 4x) - 6A(1 + 2x)$

$-9Bx - 4B(1 + 2x) + 12Ax + 13xB] = 12e^{2x} \sin 3x$

or $6Be^{2x} \cos 3x - 6Ae^{2x} \sin 3x = 12e^{2x} \sin 3x$.

Equating coefficients of like terms, we get $-6A = 12$ and $A = -2$; $6B = 0$ and $B = 0$. Therefore, the particular integral is $y_p(x) = -2xe^{2x} \cos 3x$.

The general solution is

$$y(x) = e^{2x}(c_1 \cos 3x + c_2 \sin 3x - 2x \cos 3x).$$

Exercise 18.3

Find the general solution of the following differential equations, using the method of variation of parameters.

1. $y'' + a^2y = \sec ax.$ (*V.T.U., 2001; Osmania 1999*)

Ans. $y(x) = c_1 \cos ax + c_2 \sin ax + (1/a^2) \cos ax \log |\cos ax|) + (1/a) x\sin ax.$

2. $(D^2 + 4)y = \cos 2x.$ (*Osmania, 2000S; Madras, 1998*)

Ans. $y(x) = c_1 \cos 2x + c_2 \sin 2x + (x/4) \sin 2x.$

3. $y'' + y = \tan x.$ (*Pondicherry, 1998*)

Ans. $y(x) = c_1 \cos x + c_2 \sin x - \cos x \log |\sec x + \tan x|.$

4. $y'' + y = 2x \sin x.$ (*Andhra, 1998*)

Ans. $y(x) = c_1 \cos x + c_2 \sin x + \dfrac{x\sin x}{2} - \dfrac{x^2 \cos x}{2}.$

5. $y'' + 4y = 4\sec^2 2x.$

Ans. $y(x) = c_1 \cos 2x + c_2 \sin 2x - 1 + \sin 2x \log |\sec 2x + \tan 2x|.$

6. $y'' - 6y' + 9y = \left(e^{3x}/x^2\right).$ (*I.E.T.E., S-2009; A.M.I.E., W-2005*)

Ans. $y(x) = (c_1 + c_2 x)e^{3x} - e^{3x} \log x.$

7. $(D^2 - 1)y = (1 + e^{-x})^{-2}.$ **Ans.** $y(x) = c_1 e^x + c_2 e^{-x} - 1 + e^{-x} \log |1 + e^x|.$

8. $y'' - 2y' + 2y = e^x \tan x + 3x.$

Ans. $y(x) = \left\{c_1 \sin x + \left[c_2 - \log|\sec x + \tan x|\right]\cos x\right\}e^x + (3x/2) + (3/2).$

9. $y'' - 4y' + 4y = (x + 1) e^{2x}.$

Ans. $y(x) = \left(c_1 + c_2 x\right)e^{2x} + (1/6)x^3 e^{2x} + (1/2)x^2 e^{2x}$

10. $y'' + 2y' + y = e^{-x} \cos x.$ (*A.M.I.E., S-2006*)

Ans. $y(x) = (c_1 + c_2 x - \cos x) e^{-x}$

11. $y'' + 16y = 32\sec 2x.$ (*I.E.T.E., Dec. 2007*)

Ans. $y(x) = c_1 \cos 4x + c_2 \sin 4x + 8\cos 2x - 4\sin 4x \log|\sec 2x + \tan 2x|.$

12. $y'' + 2y' + y = e^{-x} \log x.$

Ans. $y(x) = c_1 e^{-x} + c_2 x e^{-x} + \left(\dfrac{x^2}{2}\log x - \dfrac{3}{4}x^2\right)e^{-x}.$

13. $y'' + 3y' + 2y = 2e^x$ (*A.M.I.E., Dec. 2010; IETE., June 2009, 2008*)

Ans. $y(x) = c_1 e^{-x} + c_2 e^{-2x} + \left(e^x/3\right).$

14. $y'' - 2y' + y = e^x/x^3.$ **Ans.** $y(x) = \left[c_1 + c_2 x + (1/2x)\right]e^x.$

15. $y'' + y = 3x - 8\cot x.$

Ans. $y(x) = c_1 \sin x + c_2 \cos x + 8\sin x \log|\text{cosec } x + \cot x| + 3x.$

16. $y'' - 4y' + 4y = e^{2x} \sec^2 x.$ **Ans.** $y(x) = \left(c_1 + c_2 x - \log \cos x\right) e^{2x}.$

17. $y'' - 8y' + 16y = 12 e^{4x}/x^4.$ **Ans.** $y(x) = \left(c_1 + c_2 x + 2x^{-2}\right) e^{4x}.$

18. $y'' + y = \cos ax.$ *(Osmania, 2000S; Madras, 1998)*

Ans. $y(x) = c_1 \cos x + c_2 \sin x - x \cos x + \sin x \log |\sin x|.$

Find the general solution of the following differential equations by the method of undetermined coefficients.

19. $y'' + 4y = 8x^2;$ **Ans.** $y(x) = c_1 \cos 2x + c_2 \sin 2x + 2x^2 - 1.$

20. $y'' + y = x^2 + x.$ **Ans.** $y(x) = c_1 \cos x + c_2 \sin x + x^2 + x - 2.$

21. $y'' + 4y' + 3y = 5e^{2x}.$ **Ans.** $y(x) = c_1 e^{-3x} + c_2 e^{-x} + (e^{2x}/3).$

22. $y'' - 3y' + 2y = e^x.$ **Ans.** $y(x) = c_1 e^x + c_2 e^{2x} - xe^x.$

23. $y^2 + 5y' + 6y = 3e^{-2x} + e^{3x}.$ **Ans.** $y(x) = c_1 e^{-2x} + c_2 e^{-3x} + 3xe^{-2x} + (e^{3x}/30).$

24. $y'' + 2y' + y = e^{-x}.$ [**Hint:** Choose $y_p = Ax^2 e^{-x}$]

Ans. $y(x) = (c_1 + c_2 x)e^{-x} + (x^2 e^{-x}/2).$

25. $y'' - 2y' - 3y = 6e^{-x} - 8e^x.$ **Ans.** $y(x) = c_1 e^{-x} + c_2 e^{3x} - (3xe^{-x}/2) + 2e^x.$

26. $y'' + 4y' + 4y = 12e^{-2x}.$ **Ans.** $y(x) = (c_1 + c_2 x)e^{-2x} + 6x^2 e^{-2x}.$

27. $y'' - y' - 2y = 4 \sin 3x.$

Ans. $y(x) = c_1 e^{2x} + c_2 e^{-x} + (6 \cos 3x - 22 \sin 3x)/65.$

28. $y'' + y = 3 \sin x.$ **Ans.** $y(x) = c_1 \cos x + c_2 \sin x - (3x \cos x)/2.$

29. $y'' + y = 2 \cos x.$ *(V.T.U., 2000S)* **Ans.** $y(x) = c_1 \cos x + c_2 \sin x - x \sin x.$

30. $y'' + 2y' + 4y = 2x^2 + 3e^{-x}.$

Ans. $y(x) = e^{-x}\left(c_1 \cos \sqrt{3}x + c_2 \sin \sqrt{3}x\right) + \frac{1}{2}x^2 - \frac{1}{2}x + e^{-x}.$

31. $y'' + 3y' + 2y = 10 e^{3x} + 4x^2.$

Ans. $y(x) = c_1 e^{-2x} + c_2 e^{-x} + (e^{3x})/2 + \left(2x^2 - 6x + 7\right).$

32. $y'' + 2y' + 5y = \frac{5}{4} e^{x/2} + 40 \cos 4x - 55 \sin 4x.$

[**Hint:** Choose $y_p = Ae^{x/2} + B \cos 4x + C \sin 4x$].

Ans. $y(x) = e^{-x}(c_1 \cos 2x + c_2 \sin 2x) + (e^{x/2}/5) + 5 \sin 4x.$

33. $y'' - 2y' + 2y = 2e^x \cos x.$ **Ans.** $y(x) = (c_1 \cos x + c_2 \sin x) e^x + xe^x \sin x.$

34. $y'' - 2y' = e^x \sin x.$ *(V.T.U., 2000)*   **Ans.** $y(x) = c_1 + c_2 e^{2x} - (e^x \sin x)/2.$

18.11 Linear Equations with Variable Coefficients—The Euler-Cauchy and Legendre Linear Equations

We will now study two types of differential equations with variable coefficients which can be transformed into differential equations with constant coefficients by appropriate substitutions.

(A) The Euler-Cauchy linear equation

An equation of the form

$$a_0 x^n \frac{d^n y}{dx^n} + a_1 x^{n-1} \frac{d^{n-1} y}{dx^{n-1}} + a_2 x^{n-2} \frac{d^{n-2} y}{dx^{n-2}} + \ldots\ldots + a_{n-1} x \frac{dy}{dx} + a_n y = X \quad \ldots(18.51)$$

where $a_0, a_1, a_2, ..., a_n$ are constants and X is a function of x is called *Euler-Cauchy's equation*. Eq. (18.51) can be solved by changing it into a linear equation with constant coefficients by the following transformation.

Under the transformation $x = e^z$ or $z = \log x$, we have

$$y' = \frac{dy}{dx} = \frac{dy}{dz} \cdot \frac{dz}{dx} = \frac{1}{x} \frac{dy}{dz}. \qquad \qquad ...(18.52)$$

$$y'' = \frac{d(y')}{dx} = \frac{d}{dx}\left(\frac{dy}{dx}\right) = \frac{d}{dx}\left(\frac{1}{x}\frac{dy}{dz}\right) = -\frac{1}{x^2}\frac{dy}{dz} + \frac{1}{x}\frac{d}{dx}\left(\frac{dy}{dz}\right)$$

$$= -\frac{1}{x^2}\frac{dy}{dz} + \frac{1}{x}\frac{d}{dz}\left(\frac{dy}{dz}\right)\frac{dz}{dx} = -\frac{1}{x^2}\frac{dy}{dz} + \frac{1}{x^2}\frac{d^2y}{dz^2} = \frac{1}{x^2}\left(\frac{d^2y}{dz^2} - \frac{dy}{dz}\right) \qquad ...(18.53)$$

Similarly

$$y''' = -\frac{2}{x^3}\left(\frac{d^2y}{dz^2} - \frac{dy}{dz}\right) + \frac{1}{x^3}\left(\frac{d^3y}{dz^3} - \frac{d^2y}{dz^2}\right)$$

$$= \frac{1}{x^3}\left(\frac{d^3y}{dz^3} - 3\frac{d^2y}{dz^2} + 2\frac{dy}{dz}\right) \qquad \qquad ...(18.54)$$

and so on.

In operator notation, set $D = d/dx$, $D^2 = d^2/dx^2$, $D^3 = d^3/dx^3$,

$\theta = d/dz$, $\theta^2 = d^2/dz^2$, $\theta^3 = d^3/dz^3$ etc. Then, Eqs. (18.52), (18.53) and (18.54) can be written as

$$y' = \frac{dy}{dx} = Dy = \frac{1}{x}\theta y, \quad \text{or} \quad xDy = \theta y$$

$$y'' = \frac{d^2y}{dx^2} = D^2 y = \frac{1}{x^2}\left(\theta^2 y - \theta y\right), \quad \text{or} \quad x^2 D^2 y = \theta(\theta-1)y$$

and $\quad y''' = \frac{d^3y}{dx^3} = D^3 y = \frac{1}{x^3}\left(\theta^3 y - 3\theta^2 y + 2\theta y\right), \quad \text{or} \quad x^3 D^3 y = \theta(\theta-1)(\theta-2)y.$

These suggest the generalization

$$x^n D^n y = \theta(\theta-1)......[\theta-(n-1)]y.$$

Substituting for $x\dfrac{dy}{dx}$, $x^2\dfrac{d^2y}{dx^2}$, ..., Eq. (18.51) is reduced to linear equation with constant coefficients and thus can be solved by the methods already discussed.

(B) The Legendre's Linear Equation

This equation is of the form

$$a_0 (ax+b)^n \frac{d^n y}{dx^n} + a_1 (ax+b)^{n-1} \frac{d^{n-1} y}{dx^{n-1}} + \dots + a_{n-1}(ax+b)\frac{dy}{dx} + a_n y = X \ \dots(18.55)$$

where $a_0, a_1, a_2, \dots, a_n$ are constants and X is a function of x.

Such equation can straight away be reduced to linear equation with constant coefficients by putting $ax + b = e^z$ or $z = \log (ax + b)$.

$$y' = \frac{dy}{dx} = Dy = \frac{dy}{dz} \cdot \frac{dz}{dx} = \frac{a}{ax+b}\frac{dy}{dz}, \text{ or } (ax+b)Dy = a\frac{dy}{dz} = a\,\theta\, y,$$

$$y'' = \frac{d^2 y}{dx^2} = D^2 y = \frac{a^2}{(ax+b)^2}\left(\frac{d^2 y}{dz^2} - \frac{dy}{dz}\right) \text{ or } (ax+b)^2 D^2 y = a^2\,\theta\,(\theta-1)\,y,$$

and so on.

Substituting for $x\dfrac{dy}{dx}, x^2\dfrac{d^2 y}{dx^2}\dots\dots$, Eq. (18.55) is reduced to linear equation with constant coefficients.

18.12 Illustrative Examples

Example 18.24 Find the general solution of the following homogeneous differential equations

(i) $2x^2 y'' + xy' - 6y = 0.$ (ii) $(x^3 D^3 + 3x^2 D^2 - 2xD + 2)y = 0.$
(iii) $x^2 y'' + xy' + 9y = 0.$

Solution: (i) Using the transformation $x = e^z$, so that $z = \log x$. This reduces the given equation (in operator notation) to $\{2\theta(\theta-1)+ \theta-6\} y = 0$ or

$(2\theta^2 - \theta - 6)y = 0$ or $(\theta-2)(2\theta+3)y = 0$, where $\theta = d/dz$.

The solution is $y(z) = c_1 e^{2z} + c_2 e^{-3z/2}$.

Substituting $e^z = x$; we get the general solution as $y(x) = c_1 x^2 + \left(c_2/x\sqrt{x}\right)$.

(ii) The transformation $x = e^z$ reduces the given equation to

$\{\theta(\theta-1)(\theta-2) + 3\theta(\theta-1) - 2\theta + 2\} y = (\theta^3 - 3\theta + 2)y$

$= (\theta-1)(\theta^2 + \theta - 2)y = (\theta-1)^2(\theta+2)y = 0$, (where $\theta = d/dz$), whose solution is $y(z) = (c_1 + c_2 z)e^z + c_3 e^{-2z}$.

Substituting $e^z = x$, we obtain the general solution as

$y(x) = c_1 x + c_2 x \log x + \left(c_3/x^2\right)$.

(*iii*) The transformation $x = e^z$ reduces the given equation to
$$\{\theta\,(\theta - 1) + \theta + 9\}\, y = (\theta^2 + 9)\, y = 0,$$

whose solution is $y(z) = c_1 \cos 3z + c_2\sin 3z$.

Replacing z by $\log x$, the general solution of the given equation is
$$y\,(x) = c_1 \cos (3 \log x) + c_2 \sin (3 \log x)\cdot$$

Example 18.25 Find the general solution of the following differential equations

   (i) $x^2y'' - 2xy' - 4y = x^2 + 2 \log x$,   (*Bhopal 2003S*)

   (ii) $x^2y'' - 2xy' + 2y = x^3 + \sin(5 \log x)$,

   (iii) $x^2y'' + 3xy' + y = (1 - x)^{-2}$.   (*P.T.U.-2003*)

Solution. (*i*) Using the transformation $x = e^z$ so that $z = \log x$. We obtain the equation (in operator notation) as $\{\theta(\theta - 1) - 2\theta - 4\}y = e^{2z} + 2z$
or $(\theta^2 - 3\theta - 4)\, y = e^{2z} + 2z$, where $\theta = d/dz$.

The characteristic equation of the corresponding homogeneous equation is

$$m^2 - 3m - 4 = 0, \quad \text{or} \quad (m+1)(m-4) = 0, \quad \text{or} \quad m = -1, 4.$$

The complementary function is given by $y_c(z) = c_1 e^{-z} + c_2 e^{4z}$.

The particular integral is

$$y_p\,(z) = \frac{1}{\theta^2 - 3\theta - 4}\left(e^{2z} + 2z\right) = \frac{1}{\theta^2 - 3\theta - 4}e^{2z} + 2\frac{1}{\theta^2 - 3\theta - 4}z$$

$$= \frac{1}{2^2 - 3(2) - 4}e^{2z} - \frac{1}{2}\left(1 + \frac{3\theta}{4} - \frac{\theta^2}{4}\right)^{-1} z = -\frac{1}{6}e^{2z} - \frac{1}{2}\left(1 - \frac{3\theta}{4} + \ldots\ldots\right)z$$

$$= -\frac{1}{6}e^{2z} - \frac{1}{2}\left(z - \frac{3}{4}\right).$$

The general solution is $y(z) = y_c + y_p = c_1 e^{-z} + c_2 e^{4z} - \dfrac{e^{2z}}{6} - \dfrac{z}{2} + \dfrac{3}{8}$.

Substituting $e^z = x$, we obtain $y(x) = c_1 x^{-1} + c_2 x^4 - \dfrac{x^2}{6} - \dfrac{1}{2}\log x + \dfrac{3}{8}$.

(*ii*) The transformation $x = e^z$ reduces the equation (in operator notation) to
$$\{\theta\,(\theta - 1) - 2\theta + 2\}\, y = e^{3z} + \sin 5z$$

or $(\theta^2 - 3\theta + 2)\, y = (\theta - 1)(\theta - 2)\, y = e^{3z} + \sin 5z$, where $\theta = d/dz$.

The characteristic equation of the corresponding homogeneous equation is
$$(m - 1)(m - 2) = 0 \quad \text{or} \quad m = 1, 2.$$

The complementary function is $y_c\,(z) = c_1 e^z + c_2\, e^{2z}$. The particular integral is

$$y_p\,(z) = \frac{1}{\theta^2 - 3\theta + 2}\left(e^{3z} + \sin 5z\right) = \frac{1}{\theta^2 - 3\theta + 2}e^{3z} + \frac{1}{\theta^2 - 3\theta + 2}\sin 5z$$

$$= \frac{1}{3^2-3.3+2}e^{3z} + \frac{1}{-25-30+2}\sin 5z = \frac{1}{2}e^{3z} - \frac{1}{23+30}\cdot\frac{(23-30)}{(23-30)}\sin 5z$$

$$= \frac{1}{2}e^{3z} - \frac{(23-30)\sin 5z}{529-90^2} = \frac{1}{2}e^{3z} - \frac{(23\sin 5z-15\cos 5z)}{529-9(-25)}$$

$$= \frac{1}{2}e^{3z} - \frac{(23\sin 5z-15\cos 5z)}{754} .$$

The general solution is $y(z) = y_c(z) + y_p(z)$

$$= c_1 e^z + c_2 e^{2z} + \frac{1}{2}e^{3z} - \frac{1}{754}(23\sin 5z-15\cos 5z)$$

Substituting $e^z = x$, we obtain

$$y(x) = c_1 x + c_2 x^2 + \frac{1}{2}x^3 - \frac{1}{754}\left[23\sin(5\log x) - 15\cos(5\log x)\right].$$

(*iii*) The transformation $x = e^z$ reduces the equation (in operator notation) to

$$\{\theta(\theta-1)+3\theta+1\} y = (1-e^z)^{-2}$$

or $\{\theta^2 + 2\theta + 1\} y = (\theta+1)^2 y = \dfrac{1}{(1-e^z)^2}$, where $\theta = d/dz$.

The characteristic equation of the corresponding homogeneous equation is $(m+1)^2 = 0$ or $m = -1, -1$.

The complementary function is $y_c(z) = (c_1 + c_2 z)e^{-z}$.

The particular integral is $y_p(z) = \dfrac{1}{(\theta+1)^2}\cdot\dfrac{1}{(1-e^z)^2} = \dfrac{1}{\theta+1}\cdot\dfrac{1}{\theta+1}\cdot\dfrac{1}{(1-e^z)^2}$

$$= \frac{1}{\theta+1}\cdot e^{-z}\int\frac{e^z}{(1-e^z)^2}dz = \frac{1}{\theta+1}e^{-z}\left(\frac{1}{1-e^z}\right)$$

$$= e^{-z}\cdot\int e^{-z}\frac{e^z}{1-e^z}dz = e^{-z}\int\frac{dt}{t(1-t)}, \text{where } t = e^z,$$

$$= e^{-z}\int\left(\frac{1}{t}+\frac{1}{1-t}\right)dt = e^{-z}\left[\log t - \log(1-t)\right] = e^{-z}\left[z - \log(1-e^z)\right].$$

The term ze^{-z} occurring in the particular integral is absorbable in the complementary function.

The general solution is $y(z) = y_c(z) + y_p(z)$

$$= (c_1 + c_2 z)e^{-z} - e^{-z} \log(1 - e^z) = \left[c_1 + c_2 z - \log(1 - e^z)\right]e^{-z}$$

Substituting $e^z = x$, we obtain

$$y(x) = [c_1 + c_2 \log x - \log(1 - x)]x^{-1}.$$

Example 18.26 Find the general solution of the following differential equations

(i) $x^2 y'' - xy' + 4y = \cos(\log x) + x \sin(\log x)$,

(ii) $x^3 y''' + 2xy' - 2y = x^2 \log x + 3x$,

(iii) $x^2 y'' - 3xy' + y = \log x \dfrac{\sin(\log x) + 1}{x}$ (*I.E.T.E., June 2007*)

(iv) $x^3 y''' + 5x^2 y'' + 5xy' + y = x^2 + \log x, \ x > 0.$ (*I.E.T.E., Dec. 2007*)

Solution. (*i*) Using the transformation $x = e^z$, we get (in operator notation)

$$\{\theta(\theta - 1) - \theta + 4\}y = \cos z + e^z \sin z$$

or $(\theta^2 - 2\theta + 4)y = \cos z + e^z \sin z$, where $\theta = d/dz$.

The characteristic equation of the corresponding homogeneous equation is

$$m^2 - 2m + 4 = 0 \text{ or } m = 1 \pm i\sqrt{3}.$$

The complementary function is $y_c(z) = e^z\left(c_1 \cos \sqrt{3}z + c_2 \sin \sqrt{3}z\right)$.

The particular integral is

$$y_p(z) = \frac{1}{\theta^2 - 2\theta + 4}\cos z + \frac{1}{\theta^2 - 2\theta + 4}\cdot e^z \sin z$$

$$= \frac{1}{3 - 2\theta}\cos z + e^z \frac{1}{(\theta + 1)^2 - 2(\theta + 1) + 4}\sin z$$

$$= \frac{1}{3 - 2\theta}\cos z + e^z \frac{1}{\theta^2 + 3}\sin z = \frac{3 + 2\theta}{9 - 4\theta^2}\cos z + e^z \cdot \frac{1}{-1^2 + 3}\sin z$$

$$= \frac{(3 + 2\theta)}{9 - 4(-1)}\cos z + \frac{1}{2}e^z \sin z = \frac{1}{13}(3\cos z - 2\sin z) + \frac{1}{2}e^z \sin z.$$

The general solution is $y(z) = e^z\left(c_1 \cos \sqrt{3}z + c_2 \sin \sqrt{3}z\right)$

$$+ \frac{1}{13}(3\cos z - 2\sin z) + \frac{1}{2}e^z \sin z. \text{ Substituting } e^z = x, \text{ we obtain}$$

$$y(x) = x\left[c_1 \cos \sqrt{3}(\log x) + c_2 \sin \sqrt{3}(\log x)\right]$$

$$+ \frac{1}{13}[3\cos(\log x) - 2\sin(\log x)] + \frac{1}{2}x\sin(\log x).$$

(*ii*) Using the transformation $x = e^z$, we get (in operator notation)
$$\{\theta(\theta - 1)(\theta - 2) + 2\theta - 2\}y = ze^{2z} + 3e^z.$$

or $\quad (\theta - 1)(\theta^2 - 2\theta + 2)y = ze^{2z} + 3e^z$, where $\theta = d/dz$.

The characteristic equation of the corresponding homogeneous equation is

$$(m-1)(m^2 - 2m + 2) = 0 \quad \text{or} \quad m = 1, 1 \pm i.$$

The complementary function is
$$y_c(z) = c_1 e^z + e^z(c_2 \cos z + c_3 \sin z).$$

The particular integral is $y_p(z) = \dfrac{1}{\theta^3 - 3\theta^2 + 4\theta - 2} ze^{2z} + 3e^z$

$$= e^{2z} \frac{1}{(\theta+2)^3 - 3(\theta+2)^2 + 4(\theta+2) - 2} z + 3 \cdot \frac{1}{\theta^3 - 3\theta^2 + 4\theta - 2} e^z$$

$$= e^{2z} \frac{1}{\theta^3 + 3\theta^2 + 4\theta + 2} z + 3z \frac{1}{3\theta^2 - 6\theta + 4} e^z$$

$$= \frac{e^{2z}}{2}(1 + 2\theta + \cdots)^{-1} z + 3ze^z = \frac{e^{2z}}{2}(1 - 2\theta - \cdots)z + 3ze^z = \frac{e^{2z}}{2}(z - 2) + 3ze^z$$

The general solution is $y(z) = y_c(z) + y_p(z)$

$$= c_1 e^z + e^z(c_2 \cos z + c_3 \sin z) + \frac{1}{2}e^{2z}(z - 2) + 3ze^z$$

Substituting $e^z = x$, we obtain

$$y(x) = c_1 x + x[c_2 \cos(\log x) + c_3 \sin(\log x)] + \frac{1}{2}x^2(\log x - 2) + 3x\log x.$$

(*iii*) Using the transformation $x = e^z$, we get (in operator notation)
$$\{\theta(\theta-1) - 3\theta + 1\}y = z\frac{(\sin z + 1)}{e^z}$$

or $\quad (\theta^2 - 4\theta + 1)y = e^{-z}z(\sin z + 1)$, where $\theta = d/dz$.

The characteristic equation of the corresponding homogeneous equation is
$$m^2 - 4m + 1 = 0 \text{ or } m = 2 \pm \sqrt{3}.$$

The complementary function is $y_c(z) = c_1 e^{(2+\sqrt{3})z} + c_2 e^{(2-\sqrt{3})z}$

$$= e^{2z}\left(c_1 e^{\sqrt{3}z} + c_2 e^{-\sqrt{3}z}\right).$$

The particular integral is $y_p(z) = \dfrac{1}{\theta^2 - 4\theta + 1} e^{-z} z(\sin z + 1)$

$$= e^{-z} \frac{1}{(\theta-1)^2 - 4(\theta-1)+1} z(\sin z+1) = e^{-z}\left\{\frac{1}{\theta^2 - 6\theta + 6} z + \frac{1}{\theta^2 - 6\theta + 6} z \sin z\right\}.$$

Now $\dfrac{1}{\theta^2 - 6\theta + 6} z = \dfrac{1}{6}\left(1 - \dfrac{6\theta - \theta^2}{6}\right)^{-1} z = \dfrac{1}{6}(1+\theta)z = \dfrac{1}{6}(z+1)$, ...(i)

and $\dfrac{1}{\theta^2 - 6\theta + 6} z \sin z = z \dfrac{1}{\theta^2 - 6\theta + 6} \sin z - \dfrac{2\theta - 6}{\left(\theta^2 - 6\theta + 6\right)^2} \sin z$

$$= z \cdot \frac{1}{5 - 6\theta} \cdot \frac{5 + 6\theta}{5 + 6\theta} \sin z - \frac{2\theta - 6}{\theta^4 - 12\theta^3 + 48\theta^2 - 72\theta + 36} \sin z$$

$$= z \cdot \frac{(5 + 6\theta)\sin z}{25 - 36\theta^2} - \frac{(2\theta - 6)}{\left(\theta^2\right)^2 - 12\theta^2(\theta) + 48\theta^2 - 72\theta + 36} \sin z$$

$$= \frac{z(5\sin z + 6\cos z)}{61} + \frac{(2\cos z - 6\sin z)}{600 + 11}$$

$$= \frac{1}{61} z(5\sin z + 6\cos z) + \frac{2}{600 + 11} \cdot \frac{600 - 11}{600 - 11}\cos z - \frac{6}{600 + 11} \cdot \frac{600 - 11}{600 - 11}\sin z$$

$$= \frac{1}{61} z(5\sin z + 6\cos z) + \frac{120\sin z + 22\cos z}{3721} + \frac{360\cos z - 66\sin z}{3721}$$

$$= \frac{1}{61} z(5\sin z + 6\cos z) + \frac{2}{3721}(27\sin z + 191\cos z) \qquad ...(ii).$$

Adding (i) and (ii), we get

$$y_p(z) = e^{-z}\left[\frac{1}{6}(z+1) + \frac{z}{61}(5\sin z + 6\cos z) + \frac{2}{3721}(27\sin z + 191\cos z)\right].$$

The general solution of the given equation $y(z) = y_c(z) + y_p(z)$

$$= e^{2z}\left(c_1 e^{\sqrt{3}z} + c_2 e^{-\sqrt{3}z}\right) + e^{-z}\left[\frac{1}{6}(z+1) + \frac{z}{61}(5\sin z + 6\cos z)\right.$$

$$\left. + \frac{2}{3721}(27\sin z + 191\cos z)\right]$$

Substituting $e^z = x$, we get the general solution as

$$y(x) = x^2 \left(c_1 x^{\sqrt{3}} + c_2 x^{-\sqrt{3}} \right) + \frac{1}{x} \left[\frac{1}{6} (\log x + 1) \right.$$

$$+ \frac{\log x}{61} \left\{ 5 \sin (\log x) + 6 \cos (\log x) \right\} + \frac{2}{3721} \left\{ 27 \sin (\log x) + 191 \cos (\log x) \right\} \right].$$

(*iv*) Using the transformation $x = e^z$, get (in operator notation)

$$\{ \theta(\theta - 1)(\theta - 2) + 5\theta(\theta - 1) + 5\theta + 1 \} y = e^{2z} + z$$

or $\quad \{ \theta^3 + 2\theta^2 + 2\theta + 1 \} y = e^{2z} + z$, where $\theta = d/dz$.

The characteristic equation of the corresponding homogeneous equation is

$$m^3 + 2m^2 + 2m + 1 = 0, \text{ or } (m+1)(m^2 + m + 1) = 0.$$

Its roots are $m = -1$, $\left(-1 \pm i\sqrt{3} \right)/2$. The complementary function is

$$y_c (z) = c_1 e^{-z} + e^{-z/2} \left\{ c_2 \cos \left(\sqrt{3} \, z/2 \right) + c_3 \sin \left(\sqrt{3} \, z/2 \right) \right\}$$

The particular integral is $y_p (z) = \dfrac{1}{\theta^3 + 2\theta^2 + 2\theta + 1} e^{2z} + \dfrac{1}{\theta^3 + 2\theta^2 + 2\theta + 1} z$

$$= \frac{1}{8 + 8 + 4 + 1} e^{2z} + (1 + 2\theta + \cdots)^{-1} z = \frac{1}{21} e^{2z} + (1 - 2\theta + \cdots) z = \frac{1}{21} e^{2z} + z - 2.$$

The general solution is

$$y(z) = c_1 e^{-z} + e^{-z/2} \left\{ c_2 \cos \left(\sqrt{3} \, z/2 \right) + c_3 \sin \left(\sqrt{3} \, z/2 \right) \right\} + \frac{1}{21} e^{2z} + z - 2.$$

Substituting $e^z = x$, we get

$$y(x) = \frac{c_1}{x} + \frac{1}{\sqrt{x}} \left[c_2 \cos \left(\sqrt{3} \log x / 2 \right) + c_3 \sin \left(\sqrt{3} \log x / 2 \right) \right] + \frac{x^2}{21} + \log x - 2.$$

Example 18.27 Solve the differential equation $x^2 y'' - 3xy' + 3y = 2x^4 e^x$, by variation of parameters.

Solution. Using the transformation $x = e^z$, we get (in operator notation)

$$\{ \theta(\theta - 1) - 3\theta + 3 \} y = 2e^{4z} \cdot e^{e^z} = 2e^{4z + e^z}$$

or $\quad (\theta^2 - 4\theta + 3) y = 2e^{4z + e^z}$, where $\theta = d/dz$.

The characteristic equation of the corresponding homogeneous equation is

$m^2 - 4m + 3 = 0$ or $(m-1)(m-3) = 0$. Its roots are $m = 1, 3$.

The complementary function is $y_c(z) = c_1 e^z + c_2 e^{3z}$.

To determine $y_p(z)$, *proceed as under:*

Here $y_1 = e^z$, $y_2 = e^{3z}$, $X = 2e^{4z+e^z}$

The Wronskian $W(y_1, y_2)$ is given by

$$W(e^z, e^{3z}) = \begin{vmatrix} e^z & e^{3z} \\ e^z & 3e^{3z} \end{vmatrix} = 3e^{4z} - e^{4z} = 2e^{4z}.$$

Employing the formula:

$$y_p(x) = -y_1 \int \frac{y_2 X}{W} dx + y_2 \int \frac{y_1 X}{W} dx, \text{ we obtain}$$

$$y_p(z) = -e^z \int \frac{e^{3z} \cdot 2e^{4z+e^z}}{2e^{4z}} dz + e^{3z} \int \frac{e^z \cdot 2e^{4z+e^z}}{2e^{4z}} dz$$

$$= -e^z \int e^{3z} \cdot e^{e^z} dz + e^{3z} \int e^z \cdot e^{e^z} dz$$

$$= -e^z \left(e^{2z} \cdot e^{e^z} - 2e^z e^{e^z} + 2e^{e^z} \right) + e^{3z} \cdot e^{e^z} = 2e^{e^z} \left(e^{2z} - e^z \right).$$

The general solution is

$$y(z) = y_c(z) + y_p(z) = c_1 e^z + c_2 e^{3z} + 2e^{e^z} \left(e^{2z} - e^z \right).$$

Substituting $e^z = x$, we obtain $y(x) = c_1 x + c_2 x^3 + 2x^2 e^x - 2x e^x$.

Example 18.28 Solve the following differential equations:

(i) $(3x+2)^2 \dfrac{d^2 y}{dx^2} + 3(3x+2) \dfrac{dy}{dx} - 36y = 3x^2 + 4x + 1$, *(Sambalpur 2002)*

(ii) $(1+x)^2 \dfrac{d^2 y}{dx^2} + (1+x) \dfrac{dy}{dx} + y = 4\cos\left[\log(1+x)\right]$. *(MDU 2005)*

Solution. (*i*) Put $3x + 2 = e^z$ so that $z = \log(3x+2)$ and $dz/dx = 3/(3x+2)$.

Then $\dfrac{dy}{dx} = \dfrac{dy}{dz} \cdot \dfrac{dz}{dx} = \dfrac{3}{3x+2} \cdot \dfrac{dy}{dz} \Rightarrow (3x+2)\dfrac{dy}{dx} = 3\theta y$, where $\theta = \dfrac{d}{dz}$.

Now $\dfrac{d^2 y}{dx^2} = \dfrac{d}{dx}\left(\dfrac{3}{3x+2} \cdot \dfrac{dy}{dz} \right) = -\dfrac{3^2}{(3x+2)^2} \dfrac{dy}{dz} + \dfrac{3}{3x+2} \cdot \dfrac{d}{dx}\left(\dfrac{dy}{dz} \right)$

$= \dfrac{-3^2}{(3x+2)^2} \dfrac{dy}{dz} + \dfrac{3}{3x+2} \cdot \dfrac{d}{dz}\left(\dfrac{dy}{dz} \right) \dfrac{dz}{dx} = -\dfrac{3^2}{(3x+2)^2} \dfrac{dy}{dz} + \dfrac{3^2}{(3x+2)^2} \dfrac{d^2 y}{dz^2}$

or $(3x+2)^2 \dfrac{d^2y}{dx^2} = 3^2 \left[\theta(\theta-1)y\right].$

The differential equation in operator notation is

$$\left[3^2\theta(\theta-1)+3^2\theta-36\right]y = \frac{\left(e^z-2\right)^2}{3} + \frac{4}{3}\left(e^z-2\right)+1 = \frac{e^{2z}-1}{3}$$

or $9\left(\theta^2-4\right)y = \dfrac{\left(e^{2z}-1\right)}{3}.$

The characteristic equation of the corresponding homogeneous equation is $9\left(m^2-4\right)=0.$ Its roots are $m=\pm 2.$ The complementary function is $y_c(z) = c_1 e^{2z} + c_2 e^{-2z}.$ The particular integral is

$$y_p(z) = \frac{1}{27}\left(\frac{1}{\theta^2-4}e^{2z} - \frac{1}{\theta^2-4}e^{0z}\right) = \frac{1}{27}\cdot z\frac{e^{2z}}{4} + \frac{1}{27}\cdot\frac{1}{4} = \frac{1}{108}\left(ze^{2z}+1\right).$$

The general solution is $y(z) = y_c(z) + y_p(z)$

$$= c_1 e^{2z} + c_2 e^{-2z} + \frac{1}{108}\left(ze^{2z}+1\right).$$

Substituting $e^z = 3x+2$, we obtain

$$y(x) = c_1(3x+2)^2 + c_2(3x+2)^{-2} + \frac{1}{108}\left[(3x+2)^2\log(3x+2)+1\right].$$

(ii)   Put $1+x = e^z$ or $z = \log(1+x)$, so that

$$(1+x)\frac{dy}{dx} = \theta y \text{ and } (1+x)^2\frac{d^2y}{dx^2} = \theta(\theta-1)y, \text{ where } \theta = \frac{d}{dz}.$$

The given equation in operator notation is $\theta(\theta-1)y + \theta y + y = 4\cos z$
or $(\theta^2+1)y = 4\cos z.$
The characteristic equation of the corresponding homogeneous equation is

$$\left(m^2+1\right)=0.$$ Its roots are $m = \pm i$

The complementary function is $y_c(z) = c_1\cos z + c_2\sin z.$
The particular integral is

$$y_p(z) = 4\cdot\frac{1}{\theta^2+1}\cos z = 4z\cdot\frac{1}{2\theta}\cos z = 2z\sin z.$$

The general solution is $y(z) = y_c(z) + y_p(z) = c_1\cos z + c_2\sin z + 2z\sin z.$
Substituting $e^z = 1+x$, we obtain

$y(x) = c_1\cos[\log(1+x)] + c_2\sin[\log(1+x)] + 2\log(1+x)\sin[\log(1+x)],$

Exercise 18.4

Find the general solution of the following homogeneous differential equations.

1. $x^2 y'' + xy' - y = 0$.　　　　　　　　　**Ans.** $y = c_1 x + c_2 x^{-1}$

2. $x^2 y'' - 2xy' - 4y = 0$.　　　　　　　　**Ans.** $y = c_1 x^4 + c_2 x^{-1}$

3. $4x^2 y'' + 8xy' + y = 0$.　　　　　**Ans.** $y = c_1 x^{-1/2} + c_2 x^{-1/2} \log x$

4. $x^2 y'' + 7xy' + 13y = 0$.　　　**Ans.** $y = x^{-3}[c_1 \cos(2 \log x) + c_2 \sin (2 \log x)]$

5. $x^3 y''' + 5x^2 y'' + 7xy' + 8y = 0$

　　　　　　　Ans. $y = c_1 x^{-2} + c_2 \cos (2 \log x) + c_3 \sin(2 \log x)$.

6. $y''' = 6y/x^3$.　　　　**Ans.** $y = c_1 x^3 + c_2 \cos\left(\sqrt{2} \log x + c_3\right)$.

   [**Hint:** The given equation is $x^3 y''' - 6y = 0$].

Find the general solution of the following differential equations

7. $2x^2 y'' + 7xy' - 3y = x^2 + x^{-2}$.　　**Ans.** $y = c_1 \sqrt{x} + \dfrac{c_2}{x^3} + \dfrac{x^2}{15} - \dfrac{1}{5x^2}$.

8. $x^3 y''' - x^2 y'' + 2xy' - 2y = x^3 + 3x$.　　　　　*(I.E.T.E., June 2006)*

　　　　　Ans. $y = (c_1 + c_2 \log x) x + c_3 x^2 + \dfrac{x^3}{4} - \dfrac{3}{2} x (\log x)^2$.

9. $x^3 y''' + 2x^2 y'' + 2y = 10(x + x^{-1})$.　*(GGSIPU-2006; DCE, 2005; PTU 03)*

　　Ans. $y = c_1 x^{-1} + x[c_2 \cos (\log x) + c_3 \sin (\log x)] + 5x + 2x^{-1} \log x$.

10. $2x^2 y'' + 3xy' - 3y = x^3$. *(I.E.T.E., Dec. 2005)*　　**Ans.** $y = c_1 x + \dfrac{c_2}{x\sqrt{x}} + \dfrac{x^3}{18}$.

11. $x^2 D^2 y - xDy + y = \log x$, where $D = d/dx$.

   (Pondicherry, 1998S; Mangalore, 1997) **Ans.** $y = (c_1 + c_2 \log x) x + \log x + 2$.

12. $x^3 y''' + 3x^2 y'' + xy' + y = x + \log x$.　　　　　　　*(UPTU 2001)*

　　Ans. $y = c_1 x^{-1} + \sqrt{x}\left[c_2 \cos\left(\dfrac{\sqrt{3}}{2} \log x\right) + c_3 \sin\left(\dfrac{\sqrt{3}}{2} \log x\right)\right] + \dfrac{1}{2} x + \log x$.

13. $x^2 y'' - 2xy' - 4y = x^2 + 2 \log x$. **Ans.** $y = c_1 x^{-1} + c_2 x^4 - \dfrac{x^2}{6} - \dfrac{1}{2} \log x + \dfrac{3}{8}$.

14. $x^2 y'' - xy' - 3y = x^2 \log x$　　　　　　　　　　*(A.M.I.E., S-2010)*

　　　　　Ans. $y = c_1 x^3 + c_2 x^{-1} - \dfrac{x^2}{9}(3 \log x + 2)$.

15. $x^2 y'' + 2xy' - 12y = x^3 \log x$.

　　　　　Ans. $y = c_1 x^3 + c_2 x^{-4} + \left\{x^3 \log x (7 \log x - 2)\right\}/98$.

16. $x^2 y'' + xy' + y = \log x \cdot \sin(\log x).$ *(PTU 2003; UPTU 2002)*

Ans. $y = c_1 \cos(\log x) + c_2 \sin(\log x) + \dfrac{\log x}{4}\sin(\log x) - \dfrac{(\log x)^2}{4}\cos(\log x).$

17. $y'' + \dfrac{1}{x}y' = \dfrac{12\log x}{x^2}.$ *(AMIE, S-1997)* **Ans.** $y = c_1 \log x + c_2 + 2(\log x)^3.$

18. $x^2 y'' - 2y = x^2 + x^{-1}.$ **Ans.** $y = c_1 x^2 + c_2 x^{-1} + \left\{\log x \left(x^2 - x^{-1}\right)\right\}/3.$

19. $x^2 y'' + 2xy' - 20y = (x+1)^2$ **Ans.** $y = c_1 x^{-5} + c_2 x^4 - \left(\dfrac{x^2}{14} + \dfrac{x}{9} + \dfrac{1}{20}\right).$

20. $x^2 y'' - 3xy' + y = \dfrac{\sin(\log x)}{x}.$

Ans. $y = x^2\left(c_1 x^{\sqrt{3}} + c_2 x^{-\sqrt{3}}\right) + \dfrac{1}{61x}\left(5\sin\log x + 6\cos\log x\right)$

21. $x^2 y'' - 3xy' + 3y = x \log x.$ *(I.E.T.E., June 1995)*

Ans. $y = c_1 x + c_2 x^3 - \dfrac{x}{4}\left[(\log x)^2 + \log x\right].$

22. $x^2 y'' - xy' + 2y = x \log x.$

Ans. $y = x\left(c_1 \cos\log x + c_2 \sin\log x\right) + x\log x.$

23. $x^2 y'' - 2xy' + 2y = x + x^2 \log x + x^3.$

Ans. $y = c_1 x + c_2 x^2 - x\log x + \dfrac{x^3}{2} + x^2\left[\dfrac{1}{2}(\log x)^2 - \log x\right].$

24. $x^2 y'' - 3xy' + 5y = x^2 \sin(\log x).$ *(MDU2002)*

Ans. $y = x^2\left[c_1 \cos(\log x) + c_2 \sin(\log x)\right] - \left\{x^2 \log x \cdot \cos(\log x)\right\}/2.$

Using the method of variation of parameters, find the general solution of the following differential equations:

25. $x^2 D^2 y + xDy - y = x^2 e^x.$ *(I.E.T.E., Dec. 2005)*

Ans. $y = c_1 x + c_2 x^{-1} + e^x - x^{-1}e^x.$

26. $x^2 y'' - 2xy' + 2y = x^4.$ **Ans.** $y = c_1 x + c_2 x^2 + x^4/6.$

27. $x^2 y'' + 3xy' + y = (1 - x^2).$ **Ans.** $y = \left(c_1 + c_2 \log x\right)x^{-1} + 1 - \dfrac{1}{2}x + \dfrac{1}{9}x^2.$

28. $x^2 y'' + xy' - y = x^3/(1 + x^2).$ *(IETE., Dec. 2006)*

Ans. $y = c_1 x + c_2 x^{-1} + \dfrac{x}{4}\log\left(1 + x^2\right) + \dfrac{x}{4} - \dfrac{1}{4x}\log\left(1 + x^2\right).$

Find the solutions of the following differential equations, which satisfy the given conditions.

29. $x^2 y'' - 3xy' + 4y = 0$, $y(1) = 0$ and $y'(1) = 3$.   **Ans.** $y(x) = 3x^2 \log x$.

30. $x^3 y''' - 3x^2 y'' + 6xy' - 6y = 0$, $y(1) = 2$, $y'(1) = 1$, $y''(1) = -4$.

                Ans. $y = 2x + x^2 - x^3$.

31. $x^3 y'' - x^2 y' + xy = 1$; $y(1) = 1/4$, $y(e) = e + (1/4e)$.   *(I.E.T.E., Dec. 2004)*

                Ans. $y = (1/4x) + x \log x$, $x > 0$.

32. $(x^3 D^2 + x^2 D - x)y = 3x^3$; $y(1) = 1$, $y'(1) = 2$.   **Ans.** $y = x^2$, $x > 0$.

33. $(x^2 D^2 - 3xD + 3)y = \log x$; $y(1) = 1$, $y'(1) = 2$.

$$\text{Ans. } y = \frac{1}{9}\left(5x^3 + 4 + \log x^3\right), \ x > 0.$$

Find the general solution of the following differential equations.

34. $(x + 1)^2 y'' - 3(x + 1)y' + 4y = x^2$.

$$\text{Ans. } y = \left[c_1 + c_2 \log(x+1)\right](x+1)^2 + \frac{1}{2}(x+1)^2 \left\{\log(x+1)\right\}^2 - 2(x+1) + \frac{1}{4}.$$

35. $(x + 2)^2 y'' - (x + 2) y' + y = 3x + 4$.

$$\text{Ans. } y = (x+2)\left[c_1 + c_2 \log|x+2| + \frac{3}{2}\log^2 |x+2|\right] - 2.$$

36. $(2x - 1)^2 y'' - 4(2x - 1) y' + 8y = 8x$.

        Ans. $y = c_1(2x - 1) + c_2(2x - 1)^2 - (2x - 1) \log (2x - 1) + 1/2$.

37. $(x + 1)^2 y'' + (x + 1) y' = (2x + 3)(2x + 4)$.

        Ans. $y = c_1 + c_2 \log (x + 1) + [\log (x + 1)]^2 + x^2 + 8x$.

38. $(2x + 3)^2 y'' - 2(2x + 3)y' - 12y = 6x$.   *(VTU 2003S; Raipur 1998)*

        Ans. $y = c_1(2x + 3)^{-1} + c_2(2x + 3)^3 - (3x/2) + (3/4)$.

39. $(1+x)^2 \dfrac{d^2 y}{dx^2} + (1+x)\dfrac{dy}{dx} + y = \sin\left[2\log(1+x)\right]$.   *(V.T.U., 2004)*

$$\text{Ans. } y = c_1 \cos[\log (1 + x)] + c_2 \sin[\log (1 + x)] - \frac{1}{3}\sin\left\{2\log(1+x)\right\}.$$

40. $(1+x)^2 y'' + (1+x)y' + y = 2\sin\left[\log(1+x)\right]$.   *(VTU 2001)*

$$\text{Ans. } y = c_1 \cos\left[\log(1+x)\right] + c_2 \sin\left[\log(1+x)\right] - \log(1+x)\cos\left[\log(1+x)\right].$$

41. $(1+2x)^2 \dfrac{d^2 y}{dx^2} - 6(1+2x)\dfrac{dy}{dx} + 16y = 8(1+2x)^2$.

        Ans. $y = (1 + 2x)^2[\{\log (1 + 2x)\}^2 + c_1 \log (1 + 2x) + c_2]$.

18.13 Simultaneous Linear Equations

Many problems in engineering give rise to differential equations in which there are two or more dependent variables and a single independent variable usually the time t. Such equations are termed as *simultaneous linear equations*. We consider here a system of linear differential equations with constant coefficients only. We will discuss the solution of a system of two linear equations in two linear dependent variables x and y and one independent variable t. The solution of such system can be obtained by eliminating one of the variables and solving the resulting linear equation for the second variable. We illustrate the method of obtaining the solution through the following solved examples.

Example 18.29 Solve the system of equations

$$dx/dt + 4x + 3y = t, \ dy/dt + 2x + 5y = e^t.$$

Solution. The given equations in the operator notation are given by

$$(D + 4) x + 3y = t, \qquad \text{...(i)}$$

$$2x + (D + 5) y = e^t. \qquad \text{...(ii)}$$

Operate on (i) by $(D + 5)$ and multiply (ii) by 3 and subtract, we obtain

$$(D^2 + 9D + 14)x = 1 + 5t - 3e^t. \qquad \text{...(iii)}$$

Eq. (iii) is a linear differential equation of 2nd order. Its characteristic equation is given by $m^2 + 9m + 14 = 0$, gives $m_1 = -2$ and $m_2 = -7$.

The complementary function is given by $x_c(t) = c_1 e^{-2t} + c_2 e^{-7t}$

The particular integral is

$$x_p(t) = \frac{1}{D^2 + 9D + 14}\left(e^{ot} + 5t - 3e^t\right) = \frac{1}{14} + \frac{5}{14}\left(t - \frac{9}{14}\right) - \frac{3}{24}e^t$$

$$= -\frac{31}{196} + \frac{5t}{14} - \frac{1}{8}e^t.$$

Hence the general solution of the equation is

$$x(t) = c_1 e^{-2t} + c_2 e^{-7t} + \frac{5t}{14} - \frac{1}{8}e^t - \frac{31}{196}. \qquad \text{...(iii)}$$

Differentiating (iii), $dx/dt = -2c_1 e^{-2t} - 7c_2 e^{-7t} + (5/14) - (e^t/8)$.

Substituting for dx/dt and x in the first equation, we get

$$-2c_1 e^{-2t} - 7c_2 e^{-7t} + 5/14 - (e^t/8) + 4c_1 e^{-2t} + 4c_2 e^{-7t}$$

$$-(31/49) + (10t/7) - (e^t/2) - t = -3y.$$

$$\therefore \quad y(t) = -\frac{2}{3}c_1 e^{-2t} + c_2 e^{-7t} - \frac{1}{7}t + \frac{5}{24}e^t + \frac{9}{98}. \qquad \text{...(iv)}$$

The required solution is given by equation (iii) and (iv).

Example 18.30 Solve the simultaneous equations

$$\frac{dx}{dt}+5x-2y=t, \quad \frac{dy}{dt}+2x+y=0 \text{ subject to } x(0)=0, y(0)=0.$$

(IETE., June 2009; Mangalore, 1999; Andhra, 1997)

Solution. The given equations in the operator notation are given by

$$(D+5)x-2y=t, \qquad \qquad ...(i)$$
$$2x+(D+1)y=0. \qquad \qquad ...(ii)$$

Multiplying (*i*) by 2 and operate on (*ii*) by $(D+5)$ and then subtracting, we obtain
$(D^2+6D+9)y=-2t.$ Its characteristic equation is $m^2+6m+9=0$
$\Rightarrow (m+3)^2=0$, giving $m_1=m_2=-3$.
The complementary function is given by $y_c(t)=(c_1+c_2 t)e^{-3t}$.
The particular integral is

$$y_p(t)=\frac{1}{(D+3)^2}(-2t)$$

$$=-\frac{2}{9}\left(1+\frac{D}{3}\right)^{-2}t=-\frac{2}{9}\left(1-\frac{2D}{3}+...\right)t \ =-\frac{2}{9}t+\frac{4}{27}.$$

Hence
$$y(t)=(c_1+c_2 t)e^{-3t}-\frac{2}{9}t+\frac{4}{27}. \qquad \qquad ...(iii)$$

Now to find x, substitute the value of y in (*ii*). From (*iii*)

$$D_y(t)=c_2 e^{-3t}+(c_1+c_2 t)(-3)e^{-3t}-\frac{2}{9}.$$

Substituting for y and Dy in (*ii*), we obtain

$$x(t)=-\frac{1}{2}[Dy+y]=\left\{\left(c_1-\frac{1}{2}c_2\right)+c_2 t\right\}e^{-3t}+\frac{t}{9}+\frac{1}{27}. \qquad \qquad ...(iv)$$

The stipulated initial conditions then apply, the equation (*iii*) and (*iv*) gives

$$0=c_1+\frac{4}{27} \text{ and } c_1-\frac{1}{2}c_2+\frac{1}{27}=0 \text{ whence } c_1=-4/27 \text{ and } c_2=-2/9.$$

Thus the solution of the initial-value problem is

$$x(t)=-\frac{1}{27}(1+6t)e^{-3t}+\frac{1}{27}(1+3t), \quad y(t)=-\frac{2}{27}(2+3t)e^{-3t}+\frac{2}{27}(2-3t).$$

Example 18.31 Find the solution of the system of equations

$$\frac{d^2x}{dt^2}-3x-4y=0, \quad \frac{d^2y}{dt^2}+x+y=0. \qquad (A.M.I.E., W-2008)$$

Solution. The given equations in the operator notation are written as:

$$(D^2 - 3)x - 4y = 0, \ x + (D^2 + 1)y = 0.$$

Operate on the first by $D^2 + 1$ and multiply second by 4 and adding, we get

$$(D^4 - 2D^2 + 1)x = 0 \ \text{or} \ (D^2 - 1)^2 = 0.$$

The characteristic equation is $(m^2 - 1)^2 = 0$, giving $m_1, m_2 = 1$, and $m_3, m_4 = -1$.

Hence the solution is $x = (c_1 + c_2 t)e^t + (c_3 + c_4 t)e^{-t}$.

As done in the above example, we calculate d^2x/dt^2 and substitute for x, d^2x/dt^2 and get y. But this is laborious and we calculate y directly.

Operate on the second equation by $(D^2 - 3)$ and subtract first from the result of the second, we get $(D^4 - 2D^2 + 1)y = 0$.

The characteristic equation is $m^4 - 2m^2 + 1 = 0$ giving $m_1, m_2 = 1$, and $m_3, m_4 = -1$.

$$\therefore \quad y = (k_1 + k_2 t)e^t + (k_3 + k_4 t)e^{-t}.$$

Substituting for y and x in the first equation, we get

$$k_1 = \frac{c_2 - c_1}{2}, \ k_2 = -\frac{c_2}{2}, \ k_3 = -\frac{c_4 + c_3}{2}, \ k_4 = -\frac{c_4}{2},$$

Hence $y = \left(\dfrac{c_2 - c_1}{2} - \dfrac{c_2 t}{2}\right)e^t - \left(\dfrac{c_3 + c_4}{2} + \dfrac{c_4 t}{2}\right)e^{-t}.$

It may be noted that by solving two equations we got eight constants. These constants are not all independent as is obvious from above. The number of independent arbitrary constants is equal to the exponent of the highest powers of D obtained in the expansion of the determinant formed by the operator coefficients in the dependent variable. In this case, the determinant is

$$\begin{vmatrix} D^2 - 3 & -4 \\ 1 & D^2 + 1 \end{vmatrix} = D^4 - 2D^2 + 1$$

and the exponent of the highest power of D is 4. So there are 4 independent constants.

Exercise 18.5

Solve the following simultaneous equations

1. $\dfrac{dx}{dt} - 7x + y = 0, \quad \dfrac{dy}{dt} - 2x - 5y = 0.$ (P.T.U., 1999)

Ans. $x = e^{6t}(c_1 \cos t + c_2 \sin t), \ y = e^{6t}[(c_1 - c_2) \cos t + (c_1 + c_2) \sin t].$

2. $(D+5)x-2y=t, \quad 2x+(D+1)y=0;$ given that $x=0=y$ when $t=0.$

Ans. $x=-\dfrac{1}{27}(1+6t)e^{-3t}+\dfrac{1}{27}(1+3t), \quad y=-\dfrac{2}{27}(2+3t)e^{-3t}+\dfrac{2}{27}(2-3t).$

3. $\dfrac{dx}{dt}+\dfrac{dy}{dt}+2x+y=0, \quad \dfrac{dy}{dt}+5x+3y=0.$ $\hspace{2cm}$ *(GGSIPU-2006)*

Ans. $x=c_1 \cos t + c_2 \sin t, \quad y=\dfrac{1}{2}\Big[\big(c_2-3c_1\big)\cos t-\big(c_1+3c_2\big)\sin t\Big].$

4. $(D+5)x+y=e^t, \quad (D+3)y-x=e^{2t}.$

Ans. $x=e^{-4t}\big(c_1+c_2t\big)+\dfrac{4e^t}{25}-\dfrac{1}{36}e^{2t}, y=-c_2e^{-4t}-e^{-4t}\big(c_1+c_2t\big)+\dfrac{1}{25}e^t+\dfrac{7}{36}e^{2t}.$

5. $4\dfrac{dx}{dt}+9\dfrac{dy}{dt}+2x+31y=e^t, \quad 3\dfrac{dx}{dt}+7\dfrac{dy}{dt}+x+24y=3.$

Ans. $x=-\dfrac{e^{-4t}}{2}\big(c_1\cos t+c_2\sin t\big)-\dfrac{e^{-4t}}{2}\big(c_1\sin t-c_2\cos t\big)+\dfrac{31}{26}e^t-\dfrac{93}{17},$

$$y=e^{-4t}\big(c_1\cos t+c_2\sin t\big)+\dfrac{6}{17}-\dfrac{2e^t}{13}.$$

6. $4\dfrac{dx}{dt}+9\dfrac{dy}{dt}+44x+49y=t, \quad 3\dfrac{dx}{dt}+7\dfrac{dy}{dt}+34x+38y=e^t.$

Ans. $x=c_1e^{-t}+c_2e^{-6t}+\dfrac{19}{3}t-\dfrac{56}{9}-\dfrac{29}{7}e^t, y=-c_1e^{-t}+4c_2e^{-6t}-\dfrac{17}{3}t+\dfrac{55}{9}+\dfrac{24}{7}e^t$

7. $\dfrac{dx}{dt}+\dfrac{dy}{dt}-2y=2\cos t-7\sin t, \quad \dfrac{dx}{dt}-\dfrac{dy}{dt}+2x=4\cos t-3\sin t.$

Ans. $x=c_1 e^{\sqrt{2}t}+c_2e^{-\sqrt{2}t}+3\cos t, \quad y=\big(\sqrt{2}+1\big)c_1e^{\sqrt{2}t}+\big(1-\sqrt{2}\big)c_2e^{-\sqrt{2}t}+2\sin t.$

8. $\dfrac{dx}{dt}=y, \quad \dfrac{dy}{dt}=-x, x(0)=0, y(0)=0. \text{ (UPTU 2002) }$ **Ans.** $x(t)=0=y(t).$

9. $\dfrac{dx}{dt}-\dfrac{dy}{dt}+2y=\cos 2t, \quad \dfrac{dx}{dt}+\dfrac{dy}{dt}-2x=\sin 2t.$ $\hspace{2cm}$ *(D.C.E. 2004)*

Ans. $x=e^t\big(c_1\cos t+c_2\sin t\big)-\dfrac{1}{2}\cos 2t, \quad y=e^t\big(c_1\sin t-c_2\cos t\big)-\dfrac{1}{2}\sin 2t.$

10. $Dx+2y=e^t, \quad Dy-2x=e^{-t}.$ $\hspace{2cm}$ *(Pondicherry, 1998S)*

Ans. $x=\dfrac{1}{5}e^t+\dfrac{2}{5}e^{-t}-c_1\sin 2t+c_2\cos 2t, \quad y=\dfrac{2}{5}e^t+\dfrac{1}{5}e^{-t}+c_1\cos 2t+c_2\sin 2t.$

11. $(3D+1)x+3Dy = 3t+1, \quad (D-3)x+Dy = 2t.$

$$\text{Ans. } x = \frac{(1-3t)}{10}, \, y = \frac{11}{20}t^2 + \frac{6}{10}t + c.$$

12. $\dfrac{d^2x}{dt^2}+y=\sin t, \quad \dfrac{d^2y}{dt^2}+x=\cos t.$

$$\text{(UPTU 2004; D.U. 1995)}$$

$$\text{Ans. } x = c_1 e^t + c_2 e^{-t} + c_3 \cos t + c_4 \sin t + \frac{t}{4}(\sin t - \cos t),$$

$$y = -c_1 e^t - c_2 e^{-t} + c_3 \cos t + c_4 \sin t + \frac{1}{4}(2+t)(\sin t - \cos t).$$

13. $\dfrac{dx}{dt}+2x+3y=0, \quad 3x+\dfrac{dy}{dt}+2y=2e^{2t}.$ $\quad$ (DU 2002; Nagarjuna, 1998S)

$$\text{Ans. } x = c_1 e^t + c_2 e^{-5t} - \frac{6}{7}e^{2t}, \, y = c_2 e^{-5t} - c_1 e^t + \frac{8}{7}e^{2t}.$$

14. A mechanical system with two degrees of freedom satisfies the equations.

$$2\frac{d^2x}{dt^2}+3\frac{dy}{dt}=4, \quad 2\frac{d^2y}{dt^2}-3\frac{dx}{dt}=0.$$

Obtain expressions for x and y in terms of t, given x, y, dx/dt, dy/dt all vanish

at $t = 0$. $\qquad$ $\text{Ans. } x = \dfrac{8}{9}\left(1-\cos\dfrac{3t}{2}\right), \, y = \dfrac{4}{3}t - \dfrac{8}{9}\sin\dfrac{3}{2}t.$

15. $\dfrac{d^2x}{dt^2}+4x+5y=t^2, \quad \dfrac{d^2y}{dt^2}+5x+4y=t+1.$

$$\text{(DCE 2004)}$$

$$\text{Ans. } x(t) = c_1 e^t + c_2 e^{-t} + c_3 \cos 3t + c_4 \sin 3t - \frac{1}{9}\left(4t^2 - 5t + \frac{37}{9}\right),$$

$$y(t) = -c_1 e^t - c_2 e^{-t} + c_3 \cos 3t + c_4 \sin 3t + \frac{1}{9}\left(5t^2 - 4t + \frac{44}{9}\right).$$

16. $\dfrac{dx}{dt}+\dfrac{dy}{dt}+3x=\sin t, \quad \dfrac{dx}{dt}+y-x=\cos t.$ $\qquad$ (U.P.T.U. 2003)

$$\text{Ans. } x = c_1 e^{3t} + c_2 e^{-t} + \frac{1}{5}(2\sin t - \cos t), \, y = \frac{1}{5}(2\cos t + \sin t) - 2c_1 e^{3t} + 2c_2 e^{-t}.$$

17. $\dfrac{dx}{dt}+2x-3y=t, \quad \dfrac{dy}{dt}-3x+2y=e^{2t}.$

$$\text{Ans. } x = c_1 e^{-5t} + c_2 e^t + \frac{3}{7}e^{2t} - \frac{2t}{5} - \frac{13}{25}, \, y = -c_1 e^{-5t} + c_2 e^t + \frac{4}{7}e^{2t} - \frac{3t}{5} - \frac{12}{25}.$$

Chapter 19

Partial Differential Equations

19.1 Introduction

A differential equation which involves partial derivatives is called a partial differential equation.

Such an equation involves one dependent variable and two or more than two independent variables.

For instance

$$\frac{\partial z}{\partial x} + 2\frac{\partial z}{\partial y} = z, \qquad ...(1), \qquad \frac{\partial^2 z}{\partial x^2} + \frac{\partial^2 z}{\partial y^2} = 0, \qquad ...(2)$$

$$\frac{\partial u}{\partial t} = c^2 \left(\frac{\partial^2 u}{\partial x^2} + \frac{\partial^2 u}{\partial y^2} \right), \quad ...(3) \qquad \frac{\partial^2 u}{\partial x \, \partial y} = \left(\frac{\partial u}{\partial z} \right)^3, \qquad ...(4)$$

are examples of partial differential equations.

The *order* of a partial differential equation is the order of the highest partial derivative in the equation. The *degree* of an equation is the degree of the highest order derivative occurring in the equation.

Thus equation (1) is a partial differential equation of the first order while equations (2), (3) and (4) are partial differential equations of second order. The degree of all the above equations is one.

The *solutions* of partial differential equations are of the form $f(x, y, z) = c$, which represents a surface in three dimensions, where c is a constant.

The *general solution* of a partial differential equation contains, arbitrary constants or arbitrary functions and sometimes both.

In the ensuing discussion, we shall restrict ourselves to partial differential equations involving one dependent variable z and only two independent variables x and y so that $z = f(x, y)$ and we shall use the following notations:

The first order partial derivatives $p = \partial z / \partial x = z_x$ and $q = \partial z / \partial y = z_y$, and the second order partial derivatives

$$r = \partial^2 z / \partial x^2 = z_{xx} \, , \ s = \partial^2 z / \partial x \partial y = z_{xy}, \ t = \partial^2 z / \partial y^2 = z_{yy} \ \text{etc.}$$

An equaiton containing x, y, z, p, q defines a first order partial differential equation, that is $f(x, y, z, p, q) = 0$. ...(19.1)

An equation containing x, y, z, p, q, r, s, t defines a second order partial differential equation, that is $g(x, y, z, p, q, r, s, t) = 0$. ...(19.2)

19.2 Formation of Partial Differential Equations

Just like ordinary differential equations, partial differential equations also arise as a result of elimination of arbitrary constants from a given relation between the variables. Besides, partial differential equations are also formed by elimination of arbitrary functions from given functional relations between variables. We shall note from the solved examples of next section that if the number of constants to be eliminated is equal to the number of independent variables, the partial differential equations obtained are of first order. In case number of constants is more than the number of independent variables, the differential equations that arise are of second or higher order. When partial differential equations are obtained by eliminating arbitrary functions the order of differential equations is, in general, equal to the number of arbitrary functions eliminated.

19.3 Illustrative Examples

Example 19.1 Derive partial differential equations by eliminating the arbitrary constants a and b, from the following equations

(*i*) **z = ax + by + ab,**   (*ii*) **z = ax + a²y² + b.**

Solution: (*i*) Differentiating z partially with respect to x and y, we have

$$p = \frac{\partial z}{\partial x} = a, \quad q = \frac{\partial z}{\partial y} = b.$$

Substituting for a and b in $z = ax + by + ab$, we obtain $z = px + qy + pq$, a partial differential equation of first order.

(*ii*) Differentiating z partially w.r.t x and y, we get $p = \dfrac{\partial z}{\partial x} = a, \ q = \dfrac{\partial z}{\partial y} = 2a^2 y$.

Eliminating 'a' between these results, we obtain

$q = 2p^2 y$, the required partial differential equation of first order.

Example 19.2 Form a partial differential equation by eliminating the arbitrary constant from $z = \left(x^2 + a\right)\left(y^2 + b\right)$.

Solution: Differentiating z partially w.r.t. x and y respectively, we get

$$p = \frac{\partial z}{\partial x} = 2x\left(y^2 + b\right) \qquad \qquad \text{...(1)}$$

and
$$q = \frac{\partial z}{\partial y} = 2y\left(x^2 + a\right). \qquad \ldots(2)$$

Multiplying (1) and (2), we get

$pq = 4xy\left(x^2 + a\right)\left(y^2 + b\right)$ or $pq = 4xyz$, which is a partial differential equation of the first order.

Example 19.3 Form the partial differential equation by eliminating the arbitrary functions f and g from z = f(2x + 3y) +g(2x + y).

Solution: Differentiating z partially with respect to x and y, we obtain

$$p = \frac{\partial z}{\partial x} = 2f'(2x+3y) + 2g'(2x+y), \qquad \ldots(1)$$

and
$$q = \frac{\partial z}{\partial y} = 3f'(2x+3y) + g'(2x+y). \qquad \ldots(2)$$

Now differentiating (1) partially w.r.t. x

$$r = \frac{\partial^2 z}{\partial x^2} = 4f''(2x+3y) + 4g''(2x+y), \qquad \ldots(3)$$

Differentiating (2) partially w.r.t. x

$$s = \frac{\partial^2 z}{\partial x\, \partial y} = 6f''(2x+3y) + 2g''(2x+y), \qquad \ldots(4)$$

Differentiating (2) partially w.r.t. y

$$t = \frac{\partial^2 z}{\partial y^2} = 9f''(2x+3y) + g''(2x+y). \qquad \ldots(5)$$

Solving (3) and (4), we get

$$f''(2x+3y) = (2s-r)/8, \; g''(2x+y) = (3r-2s)/8.$$

Substituting these values in (5), we have

$$t = \frac{9(2s-r)}{8} + \frac{(3r-2s)}{8} = 2s - \frac{3r}{4}$$

or $3r - 8s + 4t = 0$, i.e., $3\dfrac{\partial^2 z}{\partial x^2} - 8\dfrac{\partial^2 z}{\partial x\, \partial y} + 4\dfrac{\partial^2 z}{\partial y^2} = 0$, the required equation.

Example 19.4 Find the partial differential equation by eliminating ϕ from ϕ (u, v) = 0 where u, v are known functions of x, y and z.

Solution: Differentiating $\phi(u,v) = 0$ partially w.r.t. x and y respectively, we get

$$\frac{\partial \phi}{\partial u}\left(\frac{\partial u}{\partial x}+\frac{\partial u}{\partial z}p\right)+\frac{\partial \phi}{\partial v}\left(\frac{\partial v}{\partial x}+\frac{\partial v}{\partial z}p\right)=0, \qquad \text{...(1)}$$

and $$\frac{\partial \phi}{\partial u}\left(\frac{\partial u}{\partial y}+\frac{\partial u}{\partial z}q\right)+\frac{\partial \phi}{\partial v}\left(\frac{\partial v}{\partial y}+\frac{\partial v}{\partial z}q\right)=0. \qquad \text{...(2)}$$

Eliminating $\dfrac{\partial \phi}{\partial u}, \dfrac{\partial \phi}{\partial v}$ from (1) and (2), we have

$$\begin{vmatrix} \dfrac{\partial u}{\partial x}+p\dfrac{\partial u}{\partial z} & \dfrac{\partial v}{\partial x}+p\dfrac{\partial v}{\partial z} \\[2mm] \dfrac{\partial u}{\partial y}+q\dfrac{\partial u}{\partial z} & \dfrac{\partial v}{\partial y}+q\dfrac{\partial v}{\partial z} \end{vmatrix}=0$$

or $$\left(\frac{\partial u}{\partial x}+p\frac{\partial u}{\partial z}\right)\left(\frac{\partial v}{\partial y}+q\frac{\partial v}{\partial z}\right)-\left(\frac{\partial v}{\partial x}+p\frac{\partial v}{\partial z}\right)\left(\frac{\partial u}{\partial y}+q\frac{\partial u}{\partial z}\right)=0$$

or $$p\left(\frac{\partial u}{\partial z}\frac{\partial v}{\partial y}-\frac{\partial u}{\partial y}\frac{\partial v}{\partial z}\right)+q\left(\frac{\partial u}{\partial x}\frac{\partial v}{\partial z}-\frac{\partial u}{\partial z}\frac{\partial v}{\partial x}\right)=\left(\frac{\partial u}{\partial y}\frac{\partial v}{\partial x}-\frac{\partial u}{\partial x}\frac{\partial v}{\partial y}\right) \qquad \text{...(3)}$$

The above equation can be written as a first order partial differential equation given by

$$p\frac{\partial(u,\,v)}{\partial(y,\,z)}+q\frac{\partial(u,\,v)}{\partial(z,\,x)}=\frac{\partial(u,\,v)}{\partial(x,\,y)} \text{ or } P(x,y.z)p+Q(x,y.z)q=R(x,y,z)$$

where $P(x,y,z)=\dfrac{\partial(u,\,v)}{\partial(y,\,z)},\, Q(x,y,z)=\dfrac{\partial(u,\,v)}{\partial(z,\,x)},\, R(x,y,z)=\dfrac{\partial(u,\,v)}{\partial(x,\,y)}$

are the Jacobians.

Remarks 1. In example 19.1(i), the given relation $z = ax + by + ab$ is called the primitive or general solution of $z = px + qy + pq$, a differential equation of first order.

Similarly in example 19.1 (ii) $z = ax + a^2y^2 + b$ is the general solution of $q = 2p^2y$ a differential equation of first order. *Thus the general solution of a partial differential equation of first order contains, in general, two arbitrary constants.*

2. In example 19.3, $z = f(2x + 3y) + g(2x + y)$ is the general solution of

$$3\frac{\partial^2 z}{\partial x^2}-8\frac{\partial^2 z}{\partial x\,\partial y}+4\frac{\partial^2 z}{\partial y^2}=0, \text{ (a partial differential equation of second order). This}$$

solution contains two arbitrary functions f and g. *In general, the general solution of a differential equation of nth order contains n arbitrary functions.*

Example 19.5 Obtain the partial differential equation governing the equations $\phi(u, v) = 0$, $u = xyz$, $v = x + y + z$.

Solution. We can use $Pp + Qq = R$ directly to obtain the partial differential equation.

We have $P(x, y, z) = \dfrac{\partial(u, v)}{\partial(y, z)} = \begin{vmatrix} \partial u/\partial y & \partial u/\partial z \\ \partial v/\partial y & \partial v/\partial z \end{vmatrix} = \begin{vmatrix} xz & xy \\ 1 & 1 \end{vmatrix} = xz - xy = x(z - y),$

$Q(x, y, z) = \dfrac{\partial(u, v)}{\partial(z, x)} = \begin{vmatrix} \partial u/\partial z & \partial u/\partial x \\ \partial v/\partial z & \partial v/\partial x \end{vmatrix} = \begin{vmatrix} xy & yz \\ 1 & 1 \end{vmatrix} = xy - yz = y(x - z),$

$R(x, y, z) = \dfrac{\partial(u, v)}{\partial(x, y)} = \begin{vmatrix} \partial u/\partial x & \partial u/\partial y \\ \partial v/\partial x & \partial v/\partial y \end{vmatrix} = \begin{vmatrix} yz & xz \\ 1 & 1 \end{vmatrix} = yz - xz = z(y - x).$

Hence, we obtain the linear partial differential equation

$$x(y - z)p + y(z - x)q = z(x - y).$$

Exercise 19.1

Form the partial differential equations (by eliminating the arbitrary constants) from the following equations, 1 to 5.

1. $z = ax + by + a^2 + b^2$ **Ans.** $z = px + qy + p^2 + q^2$.

2. $(x - a)^2 + (y - b)^2 + z^2 = c^2$. (IETE, June, 2008; Osmania, 1999S)

   Ans. $z^2(p^2 + q^2 + 1) = c^2$.

3. $z = xy + y\sqrt{x^2 - a^2} + b$. **Ans.** $px + qy = pq$.

4. $z = \dfrac{x^2}{a^2} + \dfrac{y^2}{b^2} + \dfrac{z^2}{c^2} - 1$. **Ans.** $xz(\partial^2 z/\partial x^2) + x(\partial z/\partial x)^2 - z(\partial z/\partial x) = 0$.

5. $z = a(x + y) + b(x - y) + abt + c$ **Ans.** $\left(\dfrac{\partial z}{\partial x}\right)^2 - \left(\dfrac{\partial z}{\partial y}\right)^2 = 4\dfrac{\partial z}{\partial t}$.

6. Find the differential equation of all spheres of radius 3 units having their centres in the xy-plane. (*Madras, 2000S*) **Ans.** $z^2(p^2 + q^2 + 1) = 9$.

7. Find the differential equation of all planes which are at a constant distance a from the origin. **Ans.** $z = px + qy + a\sqrt{\left(1 + p^2 + q^2\right)}$.

In the following problems, eliminate the arbitrary functions to obtain a partial differential equation.

8. (i) $\phi\left(x+y+z, x^2+y^2-z^2\right)=0.$ **Ans.** $p(y+z)-q(z+x)=x-y.$

(ii) $f\left(xy+z^2, x+y+z\right)=0$ **Ans.** $p(x-2z)+q(2z-y)=y-x$

(iii) $z=f(y/x).$ **Ans.** $px+qy=0$

(iv) $z=yf(x)+xg(y).$ **Ans.** $xy\dfrac{\partial^2 z}{\partial x\,\partial y}=x\dfrac{\partial z}{\partial x}+y\dfrac{\partial z}{\partial y}-z.$

(v) $u=f(x+4t)+g(x-4t)$ **Ans.** $\dfrac{\partial^2 u}{dt^2}=16\dfrac{\partial^2 u}{\partial x^2}.$

(vi) $xyz=\phi(x+y+z).$ **Ans.** $px(y-z)+qy(z-x)=z(x-y).$

(vii) $z=y^2+2f\left(x^{-1}+\log y\right).$ *(Madras 2000; VTU 2000S)*

[**Hint:** $p=\partial z/\partial x=2f'\left(x^{-1}+\log y\right)\left(-x^{-2}\right)\Rightarrow -px^2=2f'\left(x^{-1}+\log y\right)$ ...(i)

$q=\partial z/\partial y=2y+2f'\left(x^{-1}+\log y\right)y^{-1}\Rightarrow qy-2y^2=2f'\left(x^{-1}+\log y\right)$ ...(ii)

From (i) and (ii), $px^2+qy=2y^2$].

(viii) $z=f\left(x^2-y^2\right)$ *(IETE; June 2008)* **Ans.** $yp+xq=0.$

Partial differential equations in which p and q occur in the first degree and they are also not multiplied together are defined as the *linear partial differential equations*.

Differential equations which are not linear are called *non-linear differential equations*.

In what follows, we shall discuss methods for solving some important types of linear and non-linear differential equations of the first and second order.

19.4 Linear Partial Differential Equations of the First Order.

A linear partial differential equation of the first order, commonly known *as Lagrange's linear equation* is the form

$$Pp + Qq = R,$$

where P, Q and R are functions of x, y, z or constants, Basically speaking this equation is quasi-linear equation by virtue of the presence of z in P, Q and R. When P, Q and R do not involve z, the equation is known as *linear equation*.

The procedure for solving this differential equation will be clear from the following theorem.

Theorem. *If $u(x, y, z) = a$ and $v(x, y, z) = b$ are two independent solutions of*

$\dfrac{dx}{P} = \dfrac{dy}{Q} = \dfrac{dz}{R}$ *, where P, Q, R are functions of x, y, z then $\phi(u, v) = 0$ with ϕ arbitrary, is the general solution of $Pp + Qq = R$.*

Such an equation is obtained by the elimination of an arbitrary function ϕ from the equation $\phi(u, v) = 0$ where u and v are functions of x, y, z.

Proof. Taking the differential of $u(x, y, z) = a$ and $v(x, y, z) = b$, we get

$$\frac{\partial u}{\partial x} dx + \frac{\partial u}{\partial y} dy + \frac{\partial u}{\partial z} dz = 0, \qquad \frac{\partial v}{\partial x} dx + \frac{\partial v}{\partial y} dy + \frac{\partial v}{\partial z} dz = 0.$$

As u and v are independent functions, we solve for the ratios and get

$$\frac{dx}{\dfrac{\partial u}{\partial y}\dfrac{\partial v}{\partial z} - \dfrac{\partial u}{\partial z}\dfrac{\partial v}{\partial y}} = \frac{dy}{\dfrac{\partial u}{\partial z}\dfrac{\partial v}{\partial x} - \dfrac{\partial u}{\partial x}\dfrac{\partial v}{\partial z}} = \frac{dz}{\dfrac{\partial u}{\partial x}\dfrac{\partial v}{\partial y} - \dfrac{\partial u}{\partial y}\dfrac{\partial v}{\partial x}}$$

or

$$\frac{dx}{P} = \frac{dy}{Q} = \frac{dz}{R}. \qquad \qquad ...(1)$$

Since the denominators are the relations defining P, Q, R in the equation $Pp + Qq = R$ whose general solution is $\phi(u, v) = 0$ (See solved Example 19.4, Art. 19.3), the result follows.

Rule to solve Pp + Qq = R Lagrange's Equation

(i) Form the *auxiliary or subsidiary equations*

$$\frac{dx}{P} = \frac{dy}{Q} = \frac{dz}{R}. \qquad \qquad ...(19.3)$$

(ii) Find any two independent solutions of the subsidiary equations say $u = a$, $v = b$, (a, b being arbitrary constants).

(iii) Then $\phi(u, v) = 0$ or $u = \phi(v)$ is a solution of the equation $Pp + Qq = R$. Here $\phi(v)$ is an arbitrary function of v.

The method of solving the Lagrange's equation will be clear from the solved examples of next section. In this process we have to get two independent solutions ($u = a$, $v = b$) of the simultaneous equations (19.3) and this may pose some difficulty. The readers are advised to note carefully the methods used to get these solutions of the system (19.3), given below.

(a) Method of grouping. We take two terms from the simultaneous equations

(19.3), say $\dfrac{dx}{P} = \dfrac{dy}{Q}$ when z is absent or can be cancelled. Thus we have a differ-

ential equation in x and y only which can be solved and we get one solution.

---

* These equations are called auxiliary or subsidiary equations

Similarly it may be convenient to solve $\dfrac{dx}{P} = \dfrac{dz}{R}$ or $\dfrac{dy}{Q} = \dfrac{dz}{R}$ and get the second solution.

(b) Method of multipliers. We use multipliers l, m, n (which need not always be constants) and get $\dfrac{dx}{P} = \dfrac{dy}{Q} = \dfrac{dz}{R} = \dfrac{ldx + mdy + ndz}{lP + mQ + nR}$. ...(19.4)

These multipliers are so chosen that $lP + mQ + nR = 0$. Then
$$ldx + mdy + ndz = 0.$$

$\left[\because \text{ If } \dfrac{ldx + mdy + ndz}{lP + mQ + nR} = k, \text{ then } ldx + mdy + ndz = k(lP + mQ + nR). \right.$

Now if $lP + mQ + nR = 0$, then $ldx + mdy + ndz = 0 \big]$.

Integrating this we get one solution.

It may be possible to get another set of multipliers l, m, n and get another solution.

(c) Combination of methods (a) and (b).

19.5 Illustrative Examples

Example 19.6 Find the general solution of following partial differential equations

(a) $2p + 3q = 1$,   (b) $yq - xp = z$.

Solution: (a) Comparing the given equation with $Pp + Qq = R$, we obtain $P = 2$, $Q = 3$, $R = 1$. The subsidiary simultaneous equations are $\dfrac{dx}{2} = \dfrac{dy}{3} = \dfrac{dz}{1}$.

From $\dfrac{dx}{2} = \dfrac{dy}{3}$, we get $3dx = 2dy$, whose solution is $3x - 2y = c_1$, a constant.

Similarly, solving $\dfrac{dy}{3} = \dfrac{dz}{1}$, we get $y - 3z = c_2$, a constant.

The general solution is given by $\phi(c_1, c_2) = 0$, or $\phi(3x - 2y, y - 3z) = 0$.

Remark. The general solution can also be written in the alternative forms $3x - 2y = \psi(y - 3z)$ or $y - 3z = f(3x - 2y)$.

(b) Here $P = -x$, $Q = y$, $R = z$. The subsidiary equations are $\dfrac{dx}{-x} = \dfrac{dy}{y} = \dfrac{dz}{z}$.

Taking the first pair of terms, we get $\dfrac{dx}{-x} = \dfrac{dy}{y}$, whose solution is

or
$$\log x + \log y = \log c_1 \text{ that is, } xy = c_1 \qquad \qquad ...(i)$$

Taking the second and third terms, we get $\dfrac{dy}{y} = \dfrac{dz}{z}$, whose solution is or $\qquad \log \dfrac{y}{z} = \log c_2$ that is, $\dfrac{y}{z} = c_2$. $\qquad\qquad$ ...(ii)

From (i) and (ii), the general solution is $\phi\left(xy, \dfrac{y}{z}\right) = 0$.

Example 19.7 Solve the following partial differential equations

$\quad$ (a) $y^2 p - xyq = x(z - 2y)$ $\qquad$ (A.M.I.E., S-2001; Osmania 1999)

$\quad$ (b) $x(y - z)p + y(z - x)q = z(x - y)$ $\qquad$ (A.M.I.E., W-2001; Calicut, 1995)

$\quad$ **Solution:** (a) The subsidiary equations are

$$\frac{dx}{y^2} = \frac{dy}{-xy} = \frac{dz}{x(z - 2y)}.$$

Taking the first pair of terms, we get

$$\frac{dx}{y^2} = \frac{dy}{-xy} \qquad \text{or} \qquad \frac{dx}{y} = -\frac{dy}{x} \qquad \text{or} \qquad x\,dx + y\,dy = 0$$

whose solution is $\dfrac{x^2}{2} + \dfrac{y^2}{2} = c$ or $x^2 + y^2 = c_1$, where $c_1 = 2c$.

Taking the second and third term, we get

$$\frac{dy}{-xy} = \frac{dz}{x(z - 2y)} \qquad \text{or} \qquad \frac{-dy}{y} = \frac{dz}{z - 2y}$$

or $\quad (z - 2y)dy + ydz = 0 \qquad$ that is, $(zdy + ydz) - 2ydy = 0$

This can be written in the form $d(yz) - 2y\,dy = 0$.

Solving this differential equation, we have $yz - y^2 = c_2$

The general solution is given by $\phi(x^2 + y^2, yz - y^2) = 0$

(b) The subsidiary equations are given by

$$\frac{dx}{x(y - z)} = \frac{dy}{y(z - x)} = \frac{dz}{z(x - y)} \qquad\qquad ...(1)$$

Taking the multipliers 1, 1, 1, we note that

$$\text{each term} = \frac{dx + dy + dz}{x(y - z) + y(z - x) + z(x - y)}.$$

Since $x(y - z) + y(z - x) + z(x - y) = 0$, therefore, $dx + dy + dz = 0$.

Integration gives, $x + y + z = c_1$.

The equations (1) can be written in the form

$$\frac{dx/x}{y-z} = \frac{dy/y}{z-x} = \frac{dz/z}{x-y}.$$ Again using the multipliers 1, 1, 1 we have

$$\text{each term} = \frac{dx/x + dy/y + dz/z}{y-z+z-x+x-y}.$$

Since $y-z+z-x+x-y = 0$, Therefore, $\dfrac{dx}{x} + \dfrac{dy}{y} + \dfrac{dz}{z} = 0$. Integration gives,

$$\log x + \log y + \log z = \log b \ \text{ or } \ xyz = c_2.$$

The general solution is given by $\phi(c_1, c_2) = 0$, or $\phi(x+y+z, xyz) = 0$.

Example 19.8 **Find the solution of the partial differential equation**

$$(\mathbf{mz - ny})\frac{\partial \mathbf{z}}{\partial \mathbf{x}} + (\mathbf{nx - \ell z})\frac{\partial \mathbf{z}}{\partial \mathbf{y}} = \boldsymbol{\ell}\mathbf{y - mx} \qquad \textit{(A.M.I.E. S-2002; W-2001)}$$

Solution: The subsidiary equations are

$$\frac{dx}{mz - ny} = \frac{dy}{nx - lz} = \frac{dz}{ly - mx}.$$

Using the multipliers l, m, n, each term $= \dfrac{ldx + mdy + ndz}{l(mz-ny) + m(nx-lz) + n(ly-mx)}$.

Since $\qquad l(mz-ny) + m(nx-lz) + n(ly-mx) = 0$ therefore,

$\therefore \quad ldx + mdy + ndz = 0$. Integration gives $lx + my + nz = c_1$. ...(i)

Similarly, using multipliers x, y, z, we have each term $= \dfrac{xdx + ydy + zdz}{0}$.

$\therefore \quad xdx + ydy + zdz = 0$, which on integration gives $\dfrac{x^2}{2} + \dfrac{y^2}{2} + \dfrac{z^2}{2} = k$

or $\quad x^2 + y^2 + z^2 = c_2$, where $c_2 = 2k$.

The general solution is given by

$$\phi(c_1, c_2) = 0, \text{ or } \phi\left(lx + my + nz, \ x^2 + y^2 + z^2\right) = 0.$$

Example 19.9 Solve the following partial differential equations

(a) $\left(x^2 - y^2 - z^2\right)p + 2xyq = 2xz$,   *(A.M.I.E., S-2004; Madras, 2000S)*

(b) $x^2p + y^2q = (x+y)z$.   *(A.M.I.E. W-2006; S-2002)*

Solution: (*a*) The subsidiary equations are

$$\frac{dx}{x^2 - y^2 - z^2} = \frac{dy}{2xy} = \frac{dz}{2xz}. \qquad ...(1)$$

Taking the last pair of terms, we get

$\frac{dy}{2xy} = \frac{dz}{2xz}$ that is, $\frac{dy}{y} = \frac{dz}{z}$. The solution is $y/z = c_2$.

Using multipliers x, y, z, we obtain

$$\text{each term} = \frac{xdx + ydy + zdz}{x(x^2 - y^2 - z^2) + y(2xy) + z(2xz)} = \frac{xdx + ydy + zdz}{x(x^2 + y^2 + z^2)}. \qquad ...(2)$$

Taking the second term of term eqn. (1) and last term of eqn. (2), we get

$$\frac{dy}{2xy} = \frac{xdx + ydy + zdz}{x(x^2 + y^2 + z^2)} \quad \text{or} \quad \frac{dy}{y} = \frac{2xdx + 2ydy + 2zdz}{x^2 + y^2 + z^2}.$$

Integration gives, $\log y = \log(x^2 + y^2 + z^2) + \log c$

or $\quad \log \frac{x^2 + y^2 + z^2}{y} = -\log c = \log \frac{1}{c} \quad$ or $\quad \frac{x^2 + y^2 + z^2}{y} = c_1$, where $c_1 = c^{-1}$.

The general solution is given by $\phi(c_1, c_2) = 0$, or $\phi\left(\frac{x^2 + y^2 + z^2}{y}, \frac{y}{z}\right) = 0$.

(*b*) The subsidiary equations are $\quad \frac{dx}{x^2} = \frac{dy}{y^2} = \frac{dz}{(x+y)z}. \qquad ...(1)$

Taking the first pairs of terms, we get $\frac{dx}{x^2} = \frac{dy}{y^2}$ which on integration gives

$$-\frac{1}{x} = -\frac{1}{y} + a \quad \text{or} \quad \frac{1}{y} - \frac{1}{x} = a.$$

Equations (1) can be written in the form

$$\frac{dx/x}{x} = \frac{dy/y}{y} = \frac{dz/z}{x+y} = \frac{dx/x + dy/y - dz/z}{x + y - x - y} \quad \text{(using 1, 1, −1 as multipliers)}$$

$$\therefore \quad \frac{dx}{x} + \frac{dy}{y} - \frac{dz}{z} = 0 \quad \text{or} \quad \log x + \log y - \log z = \log b \quad \text{or} \quad \frac{xy}{z} = b.$$

The general solution is given by $\phi(a,b) = 0$, or $\phi\left[\dfrac{1}{y} - \dfrac{1}{x}, \dfrac{xy}{z}\right] = 0$.

Example 19.10 Find the general solution of the following partial differential equations

(a) $(y+z)p+(z+x)q = x+y$,

(b) $x\dfrac{\partial z}{\partial x} + y\dfrac{\partial z}{\partial y} + t\dfrac{\partial z}{\partial t} = xyt$, z being the dependent variable.

Solution: (a) The subsidiary equations are

$$\frac{dx}{y+z} = \frac{dy}{z+x} = \frac{dz}{x+y}.$$

Now, each term $= \dfrac{dx-dy}{y-x} = \dfrac{dy-dz}{z-y} = \dfrac{dx+dy+dz}{2(x+y+z)}.$ ...(1)

From the first pair of terms, we get $\dfrac{dx-dy}{(x-y)} = \dfrac{dy-dz}{(y-z)}.$

Integration gives, $\log(x-y) = \log(y-z) + \log c$ or $\dfrac{(x-y)}{(y-z)} = c.$ ...(2)

From the first and last two terms in (1), we have

$$-\frac{dx-dy}{x-y} = \frac{dx+dy+dz}{2(x+y+z)}.$$

Integration gives, $-\log(x-y) = \dfrac{1}{2}\log(x+y+z) - \log b_1$ ...(3)

or $(x-y)^2(x+y+z) = b.$ The general solution is given by

$$\phi(b,c) = 0, \text{ or } \phi\left((x-y)^2(x+y+z), \frac{x-y}{y-z}\right) = 0.$$

(b) In the differential equation $Pp + Qq = R$, there are two independent variables x and y. To solve this we have to find two independent solutions satisfying

the equations $\dfrac{dx}{P} = \dfrac{dy}{Q} = \dfrac{dz}{R}.$

This procedure can be extended to first order first degree differential equations involving more than two independent variables. In the present case the number of independent variables is three *i.e.*, *x, y, t*.

$\therefore$ The subsidiary equations are

$$\frac{dx}{x} = \frac{dy}{y} = \frac{dt}{t} = \frac{dz}{xyt}. \qquad \ldots(1)$$

Taking the first pair of eqn. (1), $\dfrac{dx}{x} = \dfrac{dy}{y}$.

This gives $\log x = \log y + \log a$ or $\dfrac{x}{y} = a$.

Similarly $\quad \dfrac{dy}{y} = \dfrac{dt}{t}$ gives $\dfrac{y}{t} = b$.

Using the multipliers yt, xt, xy, -3, we get

$$\text{each term} = \frac{ytdx + xtdy + xydt - 3dz}{ytx + xty + xyt - 3xyt}.$$

Since $ytx + xty + xyt - 3xyt = 0$ therefore, $ytdx + xtdy + xydt - 3dz = 0$

Integration gives $xyt - 3z = c$.

Hence the general solution is $\phi\left(\dfrac{x}{y}, \dfrac{y}{t}, xyt - 3z\right) = 0$.

Example 19.11 Solve the partial differential equation $x\dfrac{\partial z}{\partial y} = y\dfrac{\partial z}{\partial x} + xe^{(x^2+y^2)}$.

Solution: The given equation is $-yp + xq = xe^{x^2+y^2}$.

The subsidiary equations are

$$\frac{dx}{-y} = \frac{dy}{x} = \frac{dz}{xe^{x^2+y^2}}. \text{ From } \frac{dx}{-y} = \frac{dy}{x}, \text{ we get } xdx + ydy = 0.$$

Integration gives $\dfrac{x^2}{2} + \dfrac{y^2}{2} = \dfrac{a}{2}$ (say), $\therefore$ $x^2 + y^2 = a$ $\qquad\qquad \ldots(1)$

From $\dfrac{dy}{x} = \dfrac{dz}{xe^{x^2+y^2}}$, we have $e^a dy = dz$ $\qquad\qquad \left(\text{putting } x^2 + y^2 = a\right)$

Solving $\qquad e^a y = z + b \quad$ or $\quad e^a y - z = b \quad$ or $\quad ye^{x^2+y^2} - z = b$.

Therefore, solution required is $\phi\left(x^2 + y^2, ye^{x^2+y^2} - z\right) = 0$.

Example 19.12 Find the general solution of the partial differential equation
$$\left(x^2 - yz\right)\frac{\partial z}{\partial x} + \left(y^2 - zx\right)\frac{\partial z}{\partial y} = z^2 - xy.$$

(A.M.I.E. S-2009, 2003, 2001; Madras, 2000)

Solution. The subsidiary equations are

$$\frac{dx}{x^2 - yz} = \frac{dy}{y^2 - zx} = \frac{dz}{z^2 - xy}. \qquad \qquad ...(1)$$

Now, each term $= \dfrac{dx - dy}{x^2 - y^2 - (y - x)z} = \dfrac{dy - dz}{y^2 - z^2 - x(z - y)}$

i.e., $\dfrac{d(x - y)}{(x - y)(x + y + z)} = \dfrac{d(y - z)}{(y - z)(x + y + z)}$ or $\dfrac{d(x - y)}{x - y} = \dfrac{d(y - z)}{y - z}.$

Integration gives, $\log(x - y) = \log(y - z) + \log a$ or $\dfrac{x - y}{y - z} = a$ ...(2)

Now use multipliers x, y, z. Each term of subsidiary equations (1)

$$= \frac{xdx + ydy + zdz}{x^3 + y^3 + z^3 - 3xyz} = \frac{xdx + ydy + zdz}{(x + y + z)\left(x^2 + y^3 + z^3 - xy - yz - zx\right)} \qquad ...(3)$$

Also each term of the subsidiary equations $(1) = \dfrac{dx + dy + dz}{x^2 + y^2 + z^2 - xy - yz - zx}$

...(4)

Equating (3) and (4) and cancelling the common factor, we get

$$\frac{xdx + ydy + zdz}{x + y + z} = dx + dy + dz$$

or $\displaystyle\int (xdx + ydy + zdz) = \int (x + y + z)d(x + y + z) + c_1$

or $x^2 + y^2 + z^2 = (x + y + z)^2 + 2c_1$ or $xy + yz + zx = b$ ...(5)

Combining (2) and (5), the general solution is $\dfrac{x - y}{y - z} = f(xy + yz + zx)$.

Exercise 19.2

Find the general solution of the following partial differential equations.

1. $3p + 4q = 2$.   **Ans.** $\phi(3y - 4x, 3z - 2x) = 0$.

2. $px^2 + qy^2 = z^2$.   **Ans.** $x - y = xy\phi(1/x - 1/z)$.

3. $p\cos(x+y) + q\sin(x+y) = z$.

$$\textbf{Ans. } \left[\cos(x+y) + \sin(x+y)\right]e^{y-x} = \phi\left[z^{-\sqrt{2}}\tan\left(\frac{x+y}{2} + \frac{\pi}{8}\right)\right].$$

4. $p - q = \log(x+y)$ **Ans.** $x\log(x+y) - z = f(x+y)$.

5. $px - qy = y^2 - x^2$.

[**Hint.** The auxiliary equations are $\dfrac{dx}{x} = \dfrac{dy}{-y} = \dfrac{dz}{\left(y^2 - x^2\right)}$. Use multipliers

$x, y, 1$]   **Ans.** $x^2 + y^2 + 2z = f\left[\log(xy)\right]$.

6. $p\tan x + q\tan y = \tan z$ **Ans.** $\phi\left(\dfrac{\sin x}{\sin y}, \dfrac{\sin y}{\sin z}\right) = 0$.

7. $\dfrac{y-z}{yz}p + \dfrac{z-x}{zx}q = \dfrac{x-y}{xy}$ (*A.M.I.E. S-2003, W-2002, W-2001*)

[**Hint.** Multiplying throughout by xyz we get

$x(y-z)p + y(z-x)q = z(x-y)$.

Now proceed as in solved Example 19.7(*b*)]   **Ans.** $\phi(x+y+z,\ xyz) = 0$.

8. $2yzp + zxq = 3xy$ **Ans.** $\phi\left(x^2 - 2y^2,\ 3y^2 - z^2\right) = 0$.

9. $x\left(z^2 - y^2\right)\dfrac{\partial z}{\partial x} + y\left(x^2 - z^2\right)\dfrac{\partial z}{\partial y} = z\left(y^2 - x^2\right)$

(*AMIE S-2008, 2007, W-2003; Madras 2000; Osmania 2000S; V.T.U. 2000S*)

[**Hint.** The subsidiary equations are $\dfrac{dx}{x\left(z^2 - y^2\right)} = \dfrac{dy}{y\left(x^2 - z^2\right)} = \dfrac{dz}{z\left(y^2 - x^2\right)}$.

Using multipliers x, y, z, each term

$$= \frac{xdx + ydy + zdz}{x^2\left(z^2 - y^2\right) + y^2\left(x^2 - z^2\right) + z^2\left(y^2 - x^2\right)}.$$

Since $x^2\left(z^2 - y^2\right) + y^2\left(x^2 - z^2\right) + z^2\left(y^2 - x^2\right) = 0$, therefore

$xdx + ydy + zdz = 0$. Integration gives $x^2 + y^2 + z^2 = c_1$ where $c_1 = 2a$.

Now, Using multipliers $\dfrac{1}{x}, \dfrac{1}{y}, \dfrac{1}{z}$ we get $\dfrac{dx}{x} + \dfrac{dy}{y} + \dfrac{dz}{z} = 0$ which on inte-

gration gives $xyz = c_2$. The general solution is $\phi\left(x^2 + y^2 + z^2, \, xyz\right) = 0$].

10. $x^2(y-z)p + y^2(z-x)q = z^2(x-y)$ **Ans.** $\phi\left(\dfrac{1}{x} + \dfrac{1}{y} + \dfrac{1}{z}, \, xyz\right) = 0$.

11. $x\left(y^2 + z\right)p - y\left(x^2 + z\right)q = z\left(x^2 - y^2\right)$.   **Ans.** $\phi\left(x^2 + y^2 - 2z, \, xyz\right) = 0$.

[**Hint.** Use multipliers x, y, -1 and then $\dfrac{1}{x}, \dfrac{1}{y}, \dfrac{1}{z}$]

12. $xy^2 p + y^3 q = \left(zxy^2 - 4x^3\right)$.   **Ans.** $\phi\left[\dfrac{y}{x}, \, x - \log\left(z - \dfrac{4x^2}{y^2}\right)\right] = 0$.

13. $(y-z)p + (z-x)q = x - y$.   **Ans.** $f\left(x + y + z, \, x^2 + y^2 + z^2\right) = 0$.

[**Hint.** Use multipliers 1, 1, 1 and then x, y, z]

14. $px\left(z - 2y^2\right) = (z - qy)\left(z - y^2 - 2x^3\right)$.   (*A.M.I.E., S-2000, Nagpur 1995*)

[**Hint.** The given equation is

$$px\left(z - 2y^2\right) + qy\left(z - y^2 - 2x^3\right) = \left(z - y^2 - 2x^3\right)z.$$

The subsidiary equations are

$$\frac{dx}{x\left(z - 2y^2\right)} = \frac{dy}{y\left(z - y^2 - 2x^3\right)} = \frac{dz}{z\left(z - y^2 - 2x^3\right)}.$$

From second and third terms, we get $\dfrac{dy}{y} = \dfrac{dz}{z}$.

Integration gives $\dfrac{y}{z} = a$.   ...(1)

Now taking first and third terms $\dfrac{dx}{x\left(z - 2y^2\right)} = \dfrac{dz}{z\left(z - y^2 - 2x^3\right)}$

or $\dfrac{dx}{x\left(z - 2a^2 z^2\right)} = \dfrac{dz}{z\left(z - a^2 z^2 - 2x^3\right)}$ [Putting $y = az$ from (1)]

or $\dfrac{dx}{x\left(1 - 2a^2 z\right)} = \dfrac{dz}{z - a^2 z^2 - 2x^3}$

or
$$\left(z - a^2 z^2 - 2x^3\right)dx - x\left(1 - 2a^2 z\right)dz = 0$$

or
$$\left(zdx - xdz\right) - a^2\left(z^2 dx - 2xzdz\right) - 2x^3 dx = 0$$

or
$$\frac{xdz - zdx}{x^2} - a^2\left(\frac{2xzdz - z^2 dx}{x^2}\right) + 2xdx = 0 .$$ Integration gives

$$\frac{z}{x} - a^2\left(\frac{z^2}{x}\right) + x^2 = b \text{ or } \frac{z}{x} - \frac{y^2}{x} + x^2 = b. \quad \therefore \quad f\left(\frac{y}{z}, \frac{z}{x} - \frac{y^2}{x} + x^2\right) = 0].$$

15. $zx\dfrac{\partial z}{\partial x} - zy\dfrac{\partial z}{\partial y} = y^2 - x^2 .$ (*A.M.I.E. S-2005*) **Ans.** $\phi\left(xy, x^2 + y^2 + z^2\right) = 0.$

16. $(y - z)p + (x - y)q = z - x.$ **Ans.** $\phi\left(x + y + z, x^2 + 2yz\right) = 0.$

17. $(y + z)p - (x + z)q = x - y.$ **Ans.** $\phi\left[x + y + z, (x + z)(y + z)\right] = 0.$

18. $\left(z^2 - 2yz - y^2\right)p + (xy + zx)q = xy - zx.$ (*A.M.I.E., W-2000; Madras 1997S*)

[**Hint.** The subsidiary equations are $\dfrac{dx}{z^2 - 2yz - y^2} = \dfrac{dy}{xy + zx} = \dfrac{dz}{xy - zx}.$

Using x, y, z as multipliers, we have $xdx + ydy + zdz = 0$, which on integra-

tion gives $x^2 + y^2 + z^2 = a$, where $a = 2c_1.$ ...(1)

Now taking the last two terms, we have $\dfrac{dy}{y + z} = \dfrac{dz}{y - z}$

$\Rightarrow (y - z)dy = (y + z)dz \Rightarrow ydy - (zdy + ydz) - zdz = 0$

or, $ydy - d(yz) - zdz = 0$, which on integration gives

$$y^2 - 2yz - z^2 = b. \quad ...(2)$$

From (1) and (2), the general solution is

$$\phi\left(x^2 + y^2 + z^2, y^2 - 2yz - z^2\right) = 0].$$

19. $2xzp + 2yzq = z^2 - x^2 - y^2 .$

[**Hint.** The subsidiary equations are $\dfrac{dx}{2xz} = \dfrac{dy}{2yz} = \dfrac{dz}{z^2 - x^2 - y^2}.$

From the first pair of terms, we get $\log y = \log ax$ or $y = ax.$ Using
multipliers x, y, z. Now each term is equal to

$$\frac{xdx + ydy + zdz}{2x^2 z + 2y^2 z + z^3 - x^2 z - y^2 z} = \frac{xdx + ydy + zdz}{z\left(x^2 + y^2 + z^2\right)}.$$

From the first term and above term, we get $\dfrac{dx}{x} = \dfrac{2\left(xdx + ydy + zdz\right)}{x^2 + y^2 + z^2}$.

Integration gives $\log\left(x^2 + y^2 + z^2\right) = \log x + \log b$ or $x^2 + y^2 + z^2 = bx$.

The general solution is given by $\phi\left(\dfrac{y}{x}, \dfrac{x^2 + y^2 + z^2}{x}\right) = 0.$]

20. $z = px + qy + a\sqrt{x^2 + y^2 + z^2}$. **Ans.** $\dfrac{z + \sqrt{x^2 + y^2 + z^2}}{x^{1-a}} = \phi\left(\dfrac{y}{x}\right)$.

21. $z(x+y)\dfrac{\partial z}{\partial x} + z(x-y)\dfrac{\partial z}{\partial y} = x^2 + y^2$. **Ans.** $f\left(x^2 - y^2 - z^2, 4xy - z^2\right)$.

19.6 Non-linear Equations of the First Order

In this section we shall discuss special methods for solving four important types (or forms) of non-linear differential equations of first order. These will be followed by a general method for solving this type of equations. The complete solution of such equations contains only two arbitrary constants, *i.e.*, equal to the number of independent variables and the particular solution is obtained by assigning particular values to the constants.

Type 1. $f(p, q) = 0$ *i.e., equations involving p and q only*

This type of equation does not contain x, y, z, explicitly. For example $p^2 + q^2 = 9$ is an equation of this type.

Method of Solution. We assume $z = ax + by + c$ as a solution. Then

$$p = \frac{\partial z}{\partial x} = a, \, q = \frac{\partial z}{\partial y} = b.$$

$\therefore z = ax + by + c$ is a solution where $f(a, b) = 0$ or $b = \phi(a)$

or $z = ax + \phi(a)y + c$ is the complete solution, where a and c are arbitrary constants.

Example 19.13 Solve $p^2 + q^2 = 9$.

Solution. $z = ax + by + c$ is a solution where $a^2 + b^2 = 9$.

(by putting $p = a$ and $q = b$ in $p^2 + q^2 = 9$)

Now $b = \sqrt{9 - a^2}$. $\therefore z = ax + \sqrt{9 - a^2}\, y + c$ is the complete solution required.

Type 2. $f(z, p, q) = 0$ *that is, equations containing p, q, z only.*
Method of Solution. We assume $z = \phi(y + ax) = \phi(u)$ as a trial solution.

Then $p = \dfrac{\partial z}{\partial x} = \phi'(y + ax) \cdot a = a\phi'(u) = a\dfrac{dz}{du}, \quad q = \dfrac{\partial z}{\partial y} = \phi'(y + ax) = \phi'(u) = \dfrac{dz}{du}.$

This type of equation does not contain x, y explicitly.
Substituting these values of p and q in the given equation, we get

$$f\left(z, a\dfrac{dz}{du}, \dfrac{dz}{du}\right) = 0,$$ which is an ordinary differential equation in u and z.

Let $z = F(u, b)$ be its solution. Then $z = F(y + ax, b)$ is the complete solution required.

Example 19.14 Solve $p(1 + q^2) = q(z - a)$.

Solution: Let $\quad z = \phi(y + ax) = \phi(u)$.

Then $\qquad p = \dfrac{\partial z}{\partial x} = a\phi'(y + ax) = a\phi'(u) = a\dfrac{dz}{du},$

$$q = \dfrac{\partial z}{\partial y} = \phi'(y + ax) = \phi'(u) = \dfrac{dz}{du}.$$

Substituting the values of p and q in the given equation, we get

$$a\dfrac{dz}{du}\left[1 + \left(\dfrac{dz}{du}\right)^2\right] = \dfrac{dz}{du}(z - a).$$

Now either $\dfrac{dz}{du} = 0$ which gives $z = c$ (*a* constant)

or $\quad a\left[1 + \left(\dfrac{dz}{du}\right)^2\right] = z - a,$ that is, $\left(\dfrac{dz}{du}\right)^2 = \dfrac{(z - 2a)}{a}.$

$\therefore \quad \dfrac{dz}{du} = \dfrac{1}{\sqrt{a}}\sqrt{z - 2a}.$ Integration gives $\sqrt{a}\displaystyle\int\dfrac{dz}{\sqrt{z - 2a}} = u + b.$

or $\quad 2\sqrt{a}\sqrt{z - 2a} = y + ax + b,$ or $\quad 4a(z - 2a) = (y + ax + b)^2,$ is the required solution.

Type III. $f_1(x, p) = f_2(y, q)$ *that is, variables separable form.*
This type of equation does not contains z explicitly and the terms involving p and x can be separated from those involving q and y and the equation can be written in the form $f_1(x, p) = f_2(y, q)$.

Method of Solution

Let $f_1(x, p) = f_2(y, q) = a$(an arbitrary constant).
  Let $p = F_1(x, a)$ be the solution of $f_1(x, p) = a$
and $q = F_2(y, a)$ be that of $f_2(y, q) = a$.
  Since z is a function of x and y, we have

$$dz = \frac{\partial z}{\partial x} dx + \frac{\partial z}{\partial y} dy \text{ or } dz = pdx + qdy.$$

Putting the values of p and q, we get $dz = F_1(x, a)dx + F_2(y, a)dy$.

Integration, gives $z = \int F_1(x, a)dx + \int F_2(y, a)dy + b$, which is the required solution.

Example 19.15 Solve $p - x^2 = q + y^2$.

Solution: Let $\quad p - x^2 = a$ and $q + y^2 = a$.

∴ $\qquad\qquad p = a + x^2$ and $q = a - y^2$.

Putting values of p and q in $dz = pdx + qdy$, we obtain

$$dz = \left(a + x^2\right)dx + \left(a - y^2\right)dy.$$

Solving, we get $\quad z = ax + \dfrac{x^3}{3} + ay - \dfrac{y^3}{3} + b.$

Type IV. $z = px + qy + f(p, q)$ *(Clairaut's form)*

Method of Solution. It can be easily verified that $z = ax + by + f(a, b)$ is a solu-

tion, because $\dfrac{\partial z}{\partial x} = p = a$ and $\dfrac{\partial z}{\partial y} = q = b$.

Example 19.16 Solve $z = px + qy + 4p^2q^2$.

Solution: The equation is of Clairaut's form. Hence the complete solution is
$z = ax + by + 4a^2b^2$.

Example 19.17 Solve $p^2q^2(px + qy - z) = 2$.

Solution: We write the given equation as $z = px + qy - \left(2/p^2q^2\right)$, which is a

Clairaut's equation. Hence the solution is $z = ax + by - \left(2/a^2b^2\right)$.

19.7 Illustrative Examples

Example 19.18 Solve

(a) $\sqrt{p} + \sqrt{q} = 1$,   (b) $pq + p + q = 0$.

Solution: (a) It is of the type $f(p, q) = 0$.

Hence the complete solution is $z = ax + by + c$,

where $\sqrt{a} + \sqrt{b} = 1$ or $b = \left(1 - \sqrt{a}\right)^2$.

Thus $z = ax + \left(1 - \sqrt{a}\right)^2 y + c$, is the required solution.

(b) The complete solution is $z = ax + by + c$,

where $ab + a + b = 0$ or $b = \left(\dfrac{-a}{a+1}\right)$.

$\therefore$ $z = ax - \dfrac{a}{a+1} y + c$ is the required solution.

Example 19.19 Solve $x^2p^2 + y^2q^2 = z^2$.   *(Osmania, 2000S; Madras, 1998)*

Solution: This equation can be reduced to the type I by putting

$X = \log x$, $Y = \log y$ and $Z = \log z$,

$\therefore$ $\dfrac{\partial Z}{\partial X} = \dfrac{\partial Z}{\partial z} \cdot \dfrac{\partial z}{\partial X} = \dfrac{1}{z} \dfrac{\partial z}{\partial X}$ or $z \dfrac{\partial Z}{\partial X} = \dfrac{\partial z}{\partial X} = \dfrac{\partial z}{\partial x} \cdot \dfrac{\partial x}{\partial X} = px$.

Similarly $z \dfrac{\partial Z}{\partial Y} = qy$.

Substituting in the given equation, we get

$$\left(\dfrac{\partial Z}{\partial X}\right)^2 + \left(\dfrac{\partial Z}{\partial Y}\right)^2 = 1 \text{ (cancelling } z^2 \text{ from both sides)}.$$

Now $Z = aX + bY + c$ is its solution, where $a^2 + b^2 = 1$ or $b = \sqrt{1 - a^2}$.

$\therefore$ Complete solution is

$Z = aX + \sqrt{1 - a^2} \, Y + c$ or $\log z = a \log x + \sqrt{1 - a^2} \, \log y + c$.

Example 19.20 Solve

(a) $z = p^2 + q^2$, *(Madras, 2000)*   (b) $p(1 + q) = qz$.   *(Madras, 2000S)*

Solution: (a) Let $z = \phi(y + ax) = \phi(u)$ be a solution.

Then $p = \dfrac{\partial z}{\partial x} = a\phi'(u) = a\dfrac{dz}{du}$ and $q = \dfrac{\partial z}{\partial y} = \phi'(u) = \dfrac{dz}{du}$.

Substituting in the given equation, we have

$$z = a^2\left(\frac{dz}{du}\right)^2 + \left(\frac{dz}{du}\right)^2 = \left(a^2+1\right)\left(\frac{dz}{du}\right)^2 \text{ or } \frac{dz}{du} = \frac{1}{\sqrt{a^2+1}} z^{1/2}.$$

Solving, we get $2\sqrt{a^2+1}\ z^{1/2} = u+b$

or $4\left(a^2+1\right)z = (u+b)^2 = (y+ax+b)^2$, as the required solution.

(b) Proceeding as in part (a), we get

$$a \cdot \frac{dz}{du}\left(1 + \frac{dz}{du}\right) = \frac{dz}{du}\cdot z.$$

∴ Either $\dfrac{dz}{du} = 0$ which gives $z = c$ (a constant)

or $a\left(1 + \dfrac{dz}{du}\right) = z$ or $\dfrac{dz}{du} = \dfrac{z-a}{a}$.

Solving, $a\log(z-a) = u+b$ or $a\log(z-a) = y + ax + b$ is the solution.

Example 19.21 Solve $z^2\left(p^2x^2 + q^2\right) = 1$. *(Madras 2000S)*

Solution: This question can be reduced to the type II by putting $X = \log x$, then

$$p = \frac{\partial z}{\partial x} = \frac{\partial z}{\partial X}\cdot\frac{\partial X}{\partial x} = \frac{1}{x}\cdot\frac{\partial z}{\partial X} \text{ so that } px = \frac{\partial z}{\partial X}.$$

Now the given equation reduces to $z^2\left\{\left(\dfrac{\partial z}{\partial X}\right)^2 + \left(\dfrac{\partial z}{\partial y}\right)^2\right\} = 1.$ ...(1)

Put $\qquad z = \phi(y + aX) = \phi(u).$

Then $\qquad \dfrac{\partial z}{\partial X} = a\phi'(u) = a\dfrac{dz}{du}$ and $\dfrac{\partial z}{\partial y} = \phi'(u) = \dfrac{dz}{du}.$

Now (1) becomes

$$z^2\left\{a^2\left(\frac{dz}{du}\right)^2 + \left(\frac{dz}{du}\right)^2\right\} = 1 \text{ or } z^2\left(a^2+1\right)\left(\frac{dz}{du}\right)^2 = 1.$$

$$\therefore \qquad \left(a^2+1\right)^{1/2} z \frac{dz}{du} = 1 \qquad \text{that is, } \sqrt{a^2+1} \cdot z\,dz = du.$$

Solving, $\qquad \sqrt{a^2+1} \dfrac{z^2}{2} = u+b \quad$ or, $\sqrt{a^2+1}\ z^2 = 2(y+aX+b)$

or $\sqrt{a^2+1}\ z^2 = 2(y+a\log x+b)$ is the required solution.

Example 19.22 Solve yp + xq + pq = 0.

Solution. The given equation can be written in the form

$$\frac{y}{q}+\frac{x}{p}+1 = 0 \ \left(\text{Variable separable form}\right).$$

Now let $\qquad \dfrac{y}{q} = -\left(\dfrac{x}{p}+1\right) = \dfrac{1}{a}$, so that $q = ay$ and $p = -ax/(a+1)$.

Substituting in $dz = pdx + qdy$ we get $dz = \dfrac{-ax}{a+1}dx + ay\,dy$.

Solving $z = -\dfrac{ax^2}{2(a+1)}+\dfrac{a}{2}y^2 +b$ or $2z = -\dfrac{ax^2}{(a+1)}+ay^2 +2b$.

Example 19.23 Solve : $z = px+qy + \sqrt{\left(1+p^2 +q^2\right)}$.

Solution: This is Clairauts' form.

Hence the solution is $z = ax+by+\sqrt{1+a^2 +b^2}$.

Exercise 19.3

Obtain the complete solution of the following equations

1. $p^3 - q^3 = 0.$ 　　　　　　　　　　　　　**Ans.** $z = a(x+y)+c.$

2. $p^2 +q^2 = 2.$ 　　　　　　　　　　**Ans.** $z = ax+\sqrt{2-a^2}\ y+c.$

3. $z^2 = 1+p^2 +q^2.$ 　　　　　**Ans.** $z = \cosh\left[(y+ax+b)/\sqrt{a^2 +1}\right].$

4. $z^2\left(p^2 +q^2 +1\right)= 1.$ 　　　**Ans.** $\left(a^2 +1\right)\left(1-z^2\right) = (y+ax+b)^2.$

5. $p + q = \sin x + \sin y.$ 　　　**Ans.** $z = a(x-y)-(\cos x +\cos y)+b.$

6. $\sqrt{p}+\sqrt{q} = x-y.$ 　　　　**Ans.** $3z = (x+a)^3 +(y+a)^3 +b.$

7. $z = px + qy + \sqrt{\dfrac{pq}{(p+q)}}.$ **Ans.** $z = ax + by + \sqrt{\dfrac{ab}{(a+b)}}.$

8. $p^2 - q^2 = x - y.$ (*Madras, 1998*) **Ans.** $z = \dfrac{2}{3}\left[(x+a)^{3/2} + (y+a)^{3/2}\right] + b.$

9. $p^2 y(1 + x^2) = qx^2.$ **Ans.** $z = a\sqrt{1 + x^2} + \dfrac{a}{2}y^2 + b.$

10. $p(1 - q^2) = q(1 - z).$ (*A.M.I.E. W-2004*) **Ans.** $4(1 - a + az) = (x + ay + b)^2.$

11. $z^2 = pqxy.$ **Ans.** $z = ax^b y^{1/b}.$

[**Hint.** The given equation is $1 = \left(\dfrac{x}{z}\dfrac{\partial z}{\partial x}\right)\left(\dfrac{y}{z}\dfrac{\partial z}{\partial y}\right).$ Put $Z = \log z$, $X = \log x$,

$Y = \log y$. Then equation reduces to $1 = PQ$. This is of the standard type I,

Putting therefore $P = b$ and $Q = 1/b$, we have $z = bX + \dfrac{1}{b}Y + a$

$\Rightarrow \log z = b \log x + \dfrac{1}{b}\log y + \log a \Rightarrow z = ax^b y^{1/b}.$]

19.8 Charpit's Method

This is a general method for finding the complete solution of non-linear partial differential equations of first order.

Let $\qquad f(x, y, z, p, q) = 0$ $\qquad\qquad$ ...(a)

be the given differential equation.

If we can find another relation between p, q, x, y and z say,

$\qquad F(x, y, z, p, q) = 0$ $\qquad\qquad$ ...(b)

then we can solve (a) and (b) for p and q and substitute them in

$\qquad dz = p \, dx + q \, dy.$ $\qquad\qquad$ ...(19.5)

Solution of (19.5), if it exists, is the complete solution of (a).

We shall now discuss a method to determine the relation (b).

Let us assume the relation (b) between x, y, z, p and q.

Differentiating (a) and (b) partially w.r.t. x and y we get

$$\frac{\partial f}{\partial x} + \frac{\partial f}{\partial z}p + \frac{\partial f}{\partial p}\frac{\partial p}{\partial x} + \frac{\partial f}{\partial q}\frac{\partial q}{\partial x} = 0, \qquad\qquad \text{...(19.6)}$$

$$\frac{\partial F}{\partial x} + \frac{\partial F}{\partial z}p + \frac{\partial F}{\partial p}\frac{\partial p}{\partial x} + \frac{\partial F}{\partial q}\frac{\partial q}{\partial x} = 0, \qquad\qquad \text{...(19.7)}$$

$$\frac{\partial f}{\partial y} + \frac{\partial f}{\partial z}q + \frac{\partial f}{\partial p}\frac{\partial p}{\partial y} + \frac{\partial f}{\partial q}\frac{\partial q}{\partial y} = 0, \qquad\qquad \text{...(19.8)}$$

$$\frac{\partial F}{\partial y} + \frac{\partial F}{\partial z}q + \frac{\partial F}{\partial p}\frac{\partial p}{\partial y} + \frac{\partial F}{\partial q}\frac{\partial q}{\partial y} = 0, \qquad\qquad \text{...(19.9)}$$

Eliminating $\dfrac{\partial p}{\partial x}$ between the equations (19.6) and (19.7), $\dfrac{\partial q}{\partial y}$ between the

equations (19.8) and (19.9), we get

$$\left(\frac{\partial f}{\partial x}\frac{\partial F}{\partial p}-\frac{\partial F}{\partial x}\frac{\partial f}{\partial p}\right)+\left(\frac{\partial f}{\partial z}\frac{\partial F}{\partial p}-\frac{\partial F}{\partial z}\frac{\partial f}{\partial p}\right)p+\left(\frac{\partial f}{\partial q}\frac{\partial F}{\partial p}-\frac{\partial F}{\partial q}\frac{\partial f}{\partial p}\right)\frac{\partial q}{\partial x}=0 \quad ...(19.10)$$

and $\left(\dfrac{\partial f}{\partial y}\dfrac{\partial F}{\partial q}-\dfrac{\partial F}{\partial y}\dfrac{\partial f}{\partial q}\right)+\left(\dfrac{\partial f}{\partial z}\dfrac{\partial F}{\partial q}-\dfrac{\partial F}{\partial z}\dfrac{\partial f}{\partial q}\right)q$

$$+\left(\frac{\partial f}{\partial p}\frac{\partial F}{\partial q}-\frac{\partial F}{\partial p}\frac{\partial f}{\partial q}\right)\frac{\partial p}{\partial y}=0 \qquad\qquad ...(19.11)$$

Adding the above equations and noting that

$$\frac{\partial q}{\partial x}=\frac{\partial^2 z}{\partial x\,\partial y}=\frac{\partial p}{\partial y}, \quad \text{we get}$$

$$\left(\frac{\partial f}{\partial x}+p\frac{\partial f}{\partial z}\right)\frac{\partial F}{\partial p}+\left(\frac{\partial f}{\partial y}+q\frac{\partial f}{\partial z}\right)\frac{\partial F}{\partial q}-\frac{\partial f}{\partial p}\frac{\partial F}{\partial x}-\frac{\partial f}{\partial q}\frac{\partial F}{\partial y}-\left(p\frac{\partial f}{\partial p}+q\frac{\partial f}{\partial q}\right)\frac{\partial F}{\partial z}=0.$$
$$...(19.12)$$

This is a Lagrange's form of linear differential equation of first order in which F is the dependent variable and p, q, z, x, y are independent variables.

The auxiliary equations are

$$\frac{dp}{\dfrac{\partial f}{\partial x}+p\dfrac{\partial f}{\partial z}}=\frac{dq}{\dfrac{\partial f}{\partial y}+q\dfrac{\partial f}{\partial z}}=\frac{dx}{-\dfrac{\partial f}{\partial p}}=\frac{dy}{-\dfrac{\partial f}{\partial q}}=\frac{dz}{-p\dfrac{\partial f}{\partial p}-q\dfrac{\partial f}{\partial q}}=\frac{dF}{0}. \quad ...(19.13)$$

Any integral of (19.13) will satisfy (19.12) and if that integral involves p or q or both it can be taken as the relation (b). In practice we choose the simplest of the integrals of (19.13). The relation (19.13) should be committed to memory to solve problems by Charpit's method.

19.9 Illustrative Examples

Example 19.24 Solve $(p^2 + q^2)\,y = qz$.   *(Osmania, 2000S)*

Solution: Let $f(x, y, z, p, q) = (p^2 + q^2)\,y - qz = 0$.   ...(1)

The charpit's equations become

$$\frac{dx}{-2py}=\frac{dy}{z-2qy}=\frac{dz}{-qz}=\frac{dp}{-pq}=\frac{dq}{p^2}.$$

The last two ratios give $pdp + qdq = 0$. Integration gives, $p^2 + q^2 = c^2$. ...(2)

Now to solve (1) and (2), put $p^2 + q^2 = c^2$ in (1), so that $q = c^2 y/z$.

Substituting this value of q in (2), we get $p = c\sqrt{\left(z^2 - c^2 y^2\right)}/z$.

Hence
$$dz = pdx + qdy = \frac{c}{z}\sqrt{\left(z^2 - c^2 y^2\right)}\,dx + \frac{c^2 y}{z}\,dy$$

or $\quad z\,dz - c^2 y\,dy = c\sqrt{\left(z^2 - c^2 y^2\right)}\,dx \quad$ or $\quad \dfrac{\frac{1}{2}d\left(z^2 - c^2 y^2\right)}{\sqrt{\left(z^2 - c^2 y^2\right)}} = cdx$.

Integration gives $\sqrt{\left(z^2 - c^2 y^2\right)} = cx + a$

or $z^2 = \left(a + cx\right)^2 + c^2 y^2$, where a and c are arbitrary constant.

Example 19.25 Solve $2zx - px^2 - 2qxy + pq = 0$. ...(1)

Solution. We have $f = 2zx - px^2 - 2qxy + pq$.

Therefore, $\quad \dfrac{\partial f}{\partial x} = 2z - 2px - 2qy, \quad \dfrac{\partial f}{\partial y} = -2qx, \quad \dfrac{\partial f}{\partial z} = 2x,$

$$\frac{\partial f}{\partial p} = -x^2 + q, \quad \frac{\partial f}{\partial q} = -2xy + p.$$

The charpit's equations are

$$\frac{dp}{\dfrac{\partial f}{\partial x} + p\dfrac{\partial f}{\partial z}} = \frac{dq}{\dfrac{\partial f}{\partial y} + q\dfrac{\partial f}{\partial z}} = -\frac{dx}{\dfrac{\partial f}{\partial p}} = \frac{dy}{-\dfrac{\partial f}{\partial q}} = \frac{dz}{-p\dfrac{\partial f}{\partial p} - q\dfrac{\partial f}{\partial q}},$$

which become $\quad \dfrac{dp}{2z - 2qy} = \dfrac{dq}{0} = \dfrac{dx}{x^2 - q} = \dfrac{dy}{2xy - p} = \dfrac{dz}{px^2 + 2xyq}$.

From first two ratios, $dq = 0$ or $q = a$.

Putting $q = a$ in (1), we get $p = \dfrac{2x(z - ay)}{x^2 - a}$. Substituting for p and q in the

relation $dz = pdx + qdy$, we get

$$dz = \frac{2x(z - ay)\,dx}{x^2 - a} + a\,dy \quad \text{or} \quad \frac{dz - a\,dy}{z - ay} = \frac{2x\,dx}{x^2 - a}.$$

Integration gives, $\log(z - ay) = \log(x^2 - a) + \log b = \log b(x^2 - b)$

or $\quad z = ay + b(x^2 - a)$ is the complete solution.

Example 19.26 Solve $\quad 2z + p^2 + qy + 2y^2 = 0$. $\hspace{2cm}$ ...(1)

Solution. We have $\hspace{2cm} f = 2z + p^2 + qy + 2y^2$.

The charpit's equations become

$$\frac{dp}{2p} = \frac{dq}{4y + 3q} = \frac{dx}{-2p} = \frac{dy}{-y} = \frac{dz}{-(2p^2 + qy)} .$$

From first and third ratios, we obtain $dp = -dx$ or $-p = x + a$.

Putting $p = -(x + a)$ in (1), we get

$$q = \frac{1}{y}\left[-2z - 2y^2 - (x + a)^2 \right].$$

$$\therefore \quad dz = p\,dx + q\,dy = -(x + a)\,dx + \frac{1}{y} \cdot \left[-2z - 2y^2 - (x + a)^2 \right] dy$$

or $\hspace{2cm} y\,dz = -y(x + a)\,dx - 2z\,dy - 2y^2\,dy - (x + a)^2\,dy.$

Multiplying both sides by $2y$, we obtain

$$2y^2\,dz + 4yz\,dy = -2y^2(x + a)\,dx - 2y(x + a)^2\,dy - 4y^3\,dy.$$

Integration gives, $2zy^2 = -\left[y^2(x + a)^2 + y^4 \right] + b$

or $y^2\left[(x + a)^2 + 2z + y^2 \right] = b$, the required solution.

Exercise 19.4

Solve the following equations

1. $z = pq$. $\hspace{2cm}$ **Ans.** $4az = (ax + y + b)^2$.

2. $p - 3x^2 = q^2 - y$. $\hspace{1cm}$ **Ans.** $z = -(a - x)^3/3 + x^3 + ay - xy + b$.

3. $p^2 + qy - z = 0$. $\hspace{1.5cm}$ **Ans.** $z = ay + \dfrac{1}{4}(x + b)^2$.

4. $p(q^2 + 1) + (b - z)q = 0$. $\hspace{0.5cm}$ **Ans.** $4[a(z - b) - 1] = (x + ay + b)^2$.

5. $1 + p^2 = qz$.

Ans. $\dfrac{z^2}{2} \pm \left[\dfrac{z}{2}\sqrt{z^2 - 4a^2} - 2a^2 \log\left(z + \sqrt{z^2 - 4a^2} \right) \right] = 2ax + 2y + b$.

6. $pq + 2x(y+1)p + y(y+2)q - 2(y+1)z = 0$. **Ans.** $z = ax + b\left(y^2 + 2y + a \right)$.

7. $p^2 x + q^2 y = z$. **Ans.** $\sqrt{(1+a)z} = \sqrt{ax} + \sqrt{y} + b$.

8. $pxy + pq + qy = yz$. **Ans.** $z = ax + be^y / (y+a)^a$.

9. $\left(p^2 + q^2 \right) x = pz$. **Ans.** $z^2 - a^2 x^2 = (ay + b)^2$.

10. $z^2 = pq\, xy$. **Ans.** $z = ax^b y^{1/b}$.

19.10 Linear Homogeneous Equations with Constant Coefficients

An equation of the form

$$\frac{\partial^n z}{\partial x^n} + a_1 \frac{\partial^n z}{\partial x^{n-1}\partial y} + a_2 \frac{\partial^n z}{\partial x^{n-2}\partial y^2} + \ldots + a_n \frac{\partial^n z}{\partial y^n} = f(x,\, y) \qquad \ldots(19.14)$$

where a's are constants and $f(x,y)$ is a function of x and y is called a *linear homogeneous partial differential equation of nth order with constant coefficients*. The equation is called homogeneous because all derivatives involved are of the same order.

If we write $D^r = \dfrac{\partial^r}{\partial x^r}$ and $D'^r = \dfrac{\partial^r}{\partial y^r}$ then (19.14) can be written in the form

$$D^n z + a_1 D^{n-1} D' z + a_2 D^{n-2} D'^2 z \ldots + a_n D'^n z = f(x,\, y)$$

or

$$\left(D^n + a_1 D^{n-1} D' + a_2 D^{n-2} D'^2 \ldots + a_n D'^n \right) z = f(x,\, y)$$

or briefly $\quad F\left(D,\, D' \right) z = f(x,\, y)$.

The readers already know how to solve ordinary linear differential equations with constant coefficients. As in the case of ordinary differential equations, the complete solution of (19.14) consists of two parts, *complementary function* (C.F.) and the *particular integral* (P.I.).

(*a*) The complementary function is the solution of $F(D,\, D')z = 0$. It must contain n arbitrary functions where n is the order of the differential equation.

(*b*) The particular integral is the particular solution of

$$F(D, D')z = f(x, y).$$

The complete solution of (19.14) is z = C.F. + P.I.

19.11 Rules for finding the complementary function (C.F.)

The procedure for finding the complementary function will be explained by considering a differential equations of second order. The rule can be easily extended to equations of higher order.

Let us consider the equation

$$\frac{\partial^2 z}{\partial x^2} + a_1 \frac{\partial^2 z}{\partial x \partial y} + a_2 \frac{\partial^2 z}{\partial y^2} = 0 \qquad ...(1)$$

This equation can be written in the symbolic notation (or operator form) as

$$\left(D^2 + a_1 DD' + a_2 D'^2\right)z = 0 \qquad ...(2)$$

The equation obtained by equating the operator on z equal to zero that is,

$$D^2 + a_1 DD' + a_2 D'^2 = 0 \qquad ...(3)$$

is called the *auxiliary equation* or (A.E.).

Case I. *Roots of A.E. real and different*

Let m_1, m_2 be roots of (3) considered as a quadratic in D/D'.

Then (2) can be put in the form $(D - m_2 D')(D - m_1 D')z = 0.$ ...(4)

Now the solution of $(D - m_1 D')z = 0$ is also a solution of (4),

But $\left(D - m_1 D'\right)z = 0$ is equivalent to $\dfrac{\partial z}{\partial x} - m_1 \dfrac{\partial z}{\partial y} = 0$

that is, $p - m_1 q = 0$ (Lagrange's form).

The subsidiary equations for this are

$$\frac{dx}{1} = \frac{dy}{-m_1} = \frac{dz}{0} \qquad \text{(Art. 19.4)}$$

From these equations, we get $y + m_1 x = a$ and $z = b$.

$\therefore \ z = \phi_1(y + m_1 x)$, a solution of $(D - m_1 D') z = 0$ is also a solution of (4).

Similarly $z = \phi_2(y + m_2 x)$ is a solution of (4).

Hence $z = \phi_1(y + m_1 x) + \phi_2(y + m_2 x)$ is the complete solution of (4).

Case II. *Roots of A.E. equal*

Let the two roots of A.E. be each equal to m.

Then (2) becomes

$$\left(D - mD'\right)\left(D - mD'\right)z = 0 \quad \text{or} \quad \left(D - mD'\right)^2 z = 0 \qquad ...(5)$$

Let $(D - mD')z = u$, then (5) becomes

$(D - mD')u = 0$ which has $u = \phi(y + mx)$ as its solution.

Hence $(D - mD')z = \phi(y + mx)$ or $p - mq = \phi(y + mx)$.   [Lagrange's form]
The auxiliary system for this is

$$\frac{dx}{1} = \frac{dy}{-m} = \frac{dz}{\phi(y + mx)} . \text{ Now } \frac{dx}{1} = \frac{dy}{-m} \text{ gives } y + mx = a. \qquad ...(6)$$

Then $\dfrac{dx}{1} = \dfrac{dz}{\phi(a)}$ or $z = x\phi(a) + b$ or $z = x\phi(y + mx) + b$.   ...(7)

From (6) and (7), $z - x\phi(y + mx) = \psi(y + mx)$

or $z = x\phi(y + mx) + \psi(y + mx)$ **is the required solution.**

Remark. If $\dfrac{\partial^2 z}{\partial x^2} + a_1 \dfrac{\partial^2 z}{\partial x \partial y} + a_2 \dfrac{\partial^2 z}{\partial y^2} = 0$ that is, $\left(D^2 + a_1 DD' + a_2 D'^2\right)z = 0$

be the differential equation, then $D^2 + a_1 DD' + a_2 D'^2 = 0$ is the auxiliary equation.
If we put $D = m$, $D' = 1$ in this, we get $m^2 + a_1 m + a_2 = 0$.
Now roots of $D^2 + a_1 DD' + a_2 D'^2 = 0$, considered as quadratic in D/D', are the
same as those of $m^2 + a_1 m + a_2 = 0$. Hence this quadratic in m may also be called
the A.E. Henceforth we shall write the A.E. in this form.

19.12 Rule to write the complementary function

Let the given differential equation be

$$\frac{\partial^2 z}{\partial x^2} + a_1 \frac{\partial^2 z}{\partial x \partial y} + a_2 \frac{\partial^2 z}{\partial y^2} = 0.$$

In the symbolic form, this can be written as $\left(D^2 + a_1 DD' + a_2 D'^2\right)z = 0.$ ...(1)

1. *Write the A.E., that is,* $m^2 + a_1 m + a_2 = 0$, *that is, replace D by m and D' by 1.*

2. *If the roots of the A.E. are real and distinct, say m_1 and m_2
   then $z = \phi_1(y + m_1 x) + \phi_2(y + m_2 x)$ is the C.F.*

3. *If the roots of the A.E. are equal, say m each, then*

$$z = \phi_1(y + mx) + x\phi_2(y + mx) \text{ is the C.F.}$$

Remarks. 1. The rule for finding the C.F. can be easily extended to homogeneous differential equations of higher order.

For example if the differential equation is of 3rd order and m_1, m_2, m_3 are roots of A.E., then

$$z = \phi_1(y + m_1 x) + \phi_2(y + m_2 x) + \phi_3(y + m_3 x) \text{ is the C.F.}$$

In case all the three roots are equal and each equal to m (say), then

$$z = \phi_1(y + mx) + x\phi_2(y + mx) + x^2\phi_3(y + mx) \text{ is the C.F. etc.}$$

2. If the given differential equation is $f(D, D')z = 0$, then C.F. is the complete solution.

Example 19.27 Find the complementary function of following partial differential equations

(a) $\dfrac{\partial^2 z}{\partial x^2} - a^2 \dfrac{\partial^2 z}{\partial y^2} = 0$, $\quad (b)$ $\dfrac{\partial^2 z}{\partial x^2} - 2\dfrac{\partial^2 z}{\partial x \partial y} + \dfrac{\partial^2 z}{\partial y^2} = 0$.

Solution. (a) The given equation in the operator form is $(D^2 - a^2 D'^2)z = 0$.

The A.E. is $\quad m^2 - a^2 = 0$. Its roots are $\quad m = \pm a$.

Therefore, C.F. is $\quad z = \phi_1(y + ax) + \phi_2(y - ax)$.

(b) The A.E. is $m^2 - 2m + 1 = 0$ or $(m - 1)^2 = 0$ $\therefore$ $m = 1, 1$.

Since each root is 1, the C.F. is $z = \phi_1(y + x) + x\phi_2(y + x)$.

19.13 Particular integral (P.I.) (*Special Methods*)

If the given differential equation is F(D, D') z = f (x, y) ...(1)

then its particular integral is denoted by $\dfrac{1}{F(D, D')} f(x, y)$.

$\left[\because \text{If } z \text{ is replaced by } \dfrac{1}{F(D, D')} f(x, y) \text{ in (1), then} \right.$

$\text{L.H.S.} = F(D, D') \dfrac{1}{F(D, D')} f(x, y) = f(x, y) = \text{R.H.S.,}$

$\text{bearing in mind that } \dfrac{1}{F(D, D')} \text{ and } F(D, D') \text{ are inverse operators}]$

We shall give below, without proof, methods for obtaining P.I. of (1) for different forms of $f(x, y)$. These are analogous to the short methods for finding P.I. in ordinary differential equations.

(i) If $F(D, D')z = e^{ax+by}$

then

$$P.I. = \frac{1}{F(D, D')} e^{ax+by} = \frac{1}{F(a,b)} e^{ax+by},$$

provided $F(a, b) \neq 0$. $F(a, b) = 0$ is called a case of failure, See Art 19.16 Example 19.37.

(ii) If $F(D^2, DD', D'^2)z = \sin(ax + by)$

then $P.I. = \dfrac{1}{F(-a^2, -ab, -b^2)} \sin(ax + by)$ provided $F(-a^2, -ab, -b^2) \neq 0.$

The same is the rule when f(x, y) = cos (ax + by). For the case of failure, *i.e.*, $F(-a^2, -ab, -b^2) = 0$, see Art 19.16 Example 19.37.

(iii) If $F(D, D')z = x^p y^q$, **(p, q being positive integers), then**

$$P.I. = \frac{1}{F(D, D')} x^p y^q = \left[F(D, D')\right]^{-1} x^p y^q$$

We find the particular integral by expanding $[F(D, D')]^{-1}$ in ascending powers of D or D' and then operate on $x^p y^q$. The procedure is illustrated below.

Example 19.28 Find the P.I. of following partial differential equations

(i) $\dfrac{\partial^2 z}{\partial x^2} + 3\dfrac{\partial^2 z}{\partial x \partial y} + \dfrac{\partial^2 z}{\partial y^2} = e^{2x+3y}$, **(ii)** $\left(D^2 - 3DD' + 2D'^2\right)z = \sin(x - 2y)$,

(iii) $\left(D^2 - DD' - 6D'^2\right)z = x + y.$

Solution: (*i*) The operator form of the differential equation is
$(D^2 + 3DD' + D'^2)z = e^{2x + 3y}.$

$\therefore$

$$P.I. = \frac{1}{D^2 + 3DD' + D'^2} e^{2x+3y}.$$

Here $a = 2, b = 3$. Putting $D = 2, D' = 3$, we get

$$P.I. = \frac{e^{2x+3y}}{2^2 + 3\cdot 2\cdot 3 + 3^2} = \frac{e^{2x+3y}}{31}.$$

(ii)

$$P.I. = \frac{1}{D^2 - 3DD' + 2D'^2} \sin(x - 2y)$$

Now $a = 1, b = -2$. Putting $D^2 = -1^2$, $DD' = -(1)(-2)$, $D'^2 = -(-2)^2$, we get

$$P.I. = \frac{1}{-1 - 3(2) + 2(-4)} \sin(x - 2y) = -\frac{1}{15} \sin(x - 2y).$$

(*iii*) P.I.$= \dfrac{1}{D^2 - DD' - 6D'^2}(x+y) = \dfrac{1}{D^2}\left(1 - \dfrac{D'}{D} - 6\dfrac{D'^2}{D^2}\right)^{-1}(x+y)$

$= \dfrac{1}{D^2}\left[1 - \left(\dfrac{D'}{D} + 6\dfrac{D'^2}{D^2}\right)\right]^{-1}(x+y) = \dfrac{1}{D^2}\left(1 + \dfrac{D'}{D} + ...\right)(x+y)$

(we expand by Binomial Theorem upto first degree term in $\dfrac{D'}{D}$, because $x + y$

contains first degree term in y)

$= \dfrac{1}{D^2}(x+y+x)\left[\because \dfrac{D'}{D}(x+y) = \dfrac{1}{D}(0+1) = \dfrac{1}{D}1 = \int 1\, dx = x\right]$

$= \dfrac{1}{D^2}(2x+y) = 2\cdot\dfrac{x^3}{6} + y\cdot\dfrac{x^2}{2} = \dfrac{x^3}{3} + \dfrac{x^2 y}{2}.$

$\left[\because \dfrac{1}{D^2}x = \dfrac{1}{D}\int x\, dx = \dfrac{1}{D}\cdot\dfrac{x^2}{2} = \int \dfrac{x^2}{2}dx = \dfrac{x^3}{6}\right]$

Notes. (*i*) In part (*iii*) above we could take D'^2 common in place of D^2. In that case the result will be different but the difference can be included in C.F.

(2) To evaluate $\dfrac{1}{f(D, D')}x^p y^q$ we should expand $[f(D, D')]^{-1}$ in powers of

$\dfrac{D'}{D}$, if power of y that is, q is less than power of x (that is, p.). We expand in

powers of $\dfrac{D}{D'}$ if $p < q$.

19.14 Illustrative Examples

Example 19.29 Find the solutions of the following partial differential equations

(*a*) $\dfrac{\partial^2 z}{\partial x^2} + \dfrac{\partial^2 z}{\partial x \partial y} - 2\dfrac{\partial^2 z}{\partial y^2} = 0,$ (*b*) $\left(2D^2 + 5DD' + 2D'^2\right)z = 0.$

Solution: (*a*) The given differential equation is $\left(D^2 + DD' - 2D'^2\right)z = 0$.

The auxiliary equation is $m^2 + m - 2 = 0$ whose roots are $m = 1, -2$.

The solution required is $z = \phi_1(y + x) + \phi_2(y - 2x)$.

(b)   The A.E. is $2m^2 + 5m + 2 = 0$. Its roots are $m = -2, -\dfrac{1}{2}$.

Hence the solution is $z = \phi_1(y - 2x) + \phi_2\left(y - \dfrac{x}{2}\right)$

or $z = \phi_1(y - 2x) + \phi_2(2y - x)$.

Example 19.30 Solve the following partial differential equations

(a) $(D - 2D')(D + 3D')^2 z = 0$,   (b) $\dfrac{\partial^3 z}{\partial x^3} - 3\dfrac{\partial^3 y}{\partial x^2 \partial y} + 3\dfrac{\partial^3 z}{\partial x \partial y^2} - \dfrac{\partial^3 z}{\partial y^3} = 0$.

Solution: (a) The A.E. is $(m - 2)(m + 3)^2 = 0$. $\therefore$ $m = 2, -3, -3$.
Here two roots of A.E. are equal. Therefore, the solution is

$z = \phi_1(y + 2x) + \phi_2(y - 3x) + x\phi_3(y - 3x)$.

(b) The given equation can be written as $\left(D^3 - 3D^2 D' + 3DD'^2 - D'^3\right)z = 0$.

The A.E. is $m^3 - 3m^2 + 3m - 1 = 0$ or $(m - 1)^3 = 0$. $\therefore$ $m = 1, 1, 1$.
Since all the three roots are equal, the solution is

$z = \phi_1(y + x) + x\phi_2(y + x) + x^2\phi_3(y + x)$.

Example 19.31. Solve the following partial differential equations

(a) $\dfrac{\partial^2 z}{\partial x^2} - 5\dfrac{\partial^2 z}{\partial x \partial y} + 6\dfrac{\partial^2 z}{\partial y^2} = e^{x+y}$,   *(A.M.I.E W-2002)*

(b) $\dfrac{\partial^3 z}{\partial x^3} - 3\dfrac{\partial^3 z}{\partial x^2 \partial y} + 4\dfrac{\partial^3 z}{\partial y^3} = e^{x+2y}$.   *(A.M.I.E W-1998)*

Solution: (a) The given differential equation is $\left(D^2 - 5DD' + 6D'^2\right)z = e^{x+y}$.
The auxiliary equation in $m^2 - 5m + 6 = 0$. $\therefore$ $m = 2, 3$.
Hence the C.F. $= \phi_1(y + 2x) + \phi_2(y + 3x)$.

$$\text{P.I.} = \dfrac{1}{D^2 - 5DD' + 6D'^2}e^{x+y} = \dfrac{1}{1^2 - 5(1)(1) + 6(1)^2}e^{x+y} = e^{x+y}/2.$$

(Putting $D = 1$, $D' = 1$)

$\therefore$ Complete solution is $z = \text{C.F.} + \text{P.I.} = \phi_1(y + 2x) + \phi_2(y + 3x) + e^{x+y}/2$.

(b) The given equation can be written in the symbolic form

$$\left(D^3 - 3D^2D' + 4D'^3\right)z = e^{x+2y}.$$

A.E. is $m^3 - 3m^2 + 4 = 0$ or $(m-2)^2(m+1) = 0$. ∴ $m = 2, 2, -1.$

Hence the C.F. $= \phi_1(y+2x) + x\phi_2(y+2x) + \phi_3(y-x)$.

$$\text{P.I.} = \frac{1}{D^3 - 3D^2D' + 4D'^3}e^{x+2y}$$

$$= \frac{1}{1^3 - 3 \cdot 1^2 \cdot 2 + 4 \cdot 2^3}e^{x+2y} \qquad \left(\text{Putting } D = 1, D' = 2\right)$$

$$= \frac{e^{x+2y}}{27}. \text{ The complete solution is}$$

$$z = \phi_1(y+2x) + x\phi_2(y+2x) + \phi_3(y-x) + e^{(x+2y)}/27.$$

Example 19.32 Solve the following partial differential equations

(a) $\left(D^3 - 4D^2D' + 4DD'^2\right)z = 6\sin(3x + 2y),$ \hfill (AMIE., W-2008)

(b) $\dfrac{\partial^2 z}{\partial x^2} + \dfrac{\partial^2 z}{\partial x\, \partial y} - 6\dfrac{\partial^2 z}{\partial y^2} = \cos(3x + y),$ (c) $\dfrac{\partial^2 z}{\partial x^2} - \dfrac{\partial^2 z}{\partial x\, \partial y} = \sin x \cos 2y.$

Solution: (a) The A.E. is $m^3 - 4m^2 + 4m = 0$ or $m\left(m^2 - 4m + 4\right) = 0$

that is, $\qquad m(m-2)^2 = 0$. Therefore, $m = 0, 2, 2.$

Hence $\qquad$ C.F. $= \phi_1(y) + \phi_2(y+2x) + x\phi_3(y+2x).$

$$\text{P.I.} = \frac{1}{D^3 - 4D^2D' + 4DD'^2}6\sin(3x+2y)$$

$$= \frac{1}{D\left(D^2 - 4DD' + 4D'^2\right)}6\sin(3x+2y).$$

Putting $D^2 = -3^2 = -9$, $DD' = -(3)(2) = -6$, $D'^2 = -2^2 = -4$, we get

$$\text{P.I.} = \frac{1}{D\left[-9 - 4(-6) + 4(-4)\right]}6\sin(3x+2y) = -\frac{1}{D}6\sin(3x+2y)$$

$$= -6\left[\frac{-\cos(3x+2y)}{3}\right] = 2\cos(3x+2y).$$

The complete solution is $z = \phi_1(y) + \phi_2(y+2x) + x\phi_3(y+2x) + 2\cos(3x+2y)$.

(b) The symbolic form of the equation is

$$\left(D^2 + DD' - 6D'^2\right)z = \cos(3x+y).$$

The A.E. is $m^2 + m - 6 = 0$. Therefore, $m = -3, 2$.

$$\text{C.F.} = \phi_1(y-3x) + \phi_2(y+2x)$$

$$\text{P.I.} = \frac{1}{D^2 + DD' - 6D'^2}\cos(3x+y)$$

$$= \frac{1}{-3^2 - (3)(1) - 6\left(-1^2\right)}\cos(3x+y) = -\cos(3x+y)/6.$$

Therefore, required solution is $z = \phi_1(y-3x) + \phi_2(y+2x) - \cos(3x+y)/6$.

(c) The given equation is $(D^2 - DD')\,z = \sin x \cos 2y$.

A.E. is $m^2 - m = 0$ or $m = 0, 1$. C.F. $= \phi_1(y) + \phi_2(y+x)$.

$$\text{P.I.} = \frac{1}{D^2 - DD'}\sin x \cos 2y = \frac{1}{2}\frac{1}{D^2 - DD'}\left[\sin(x+2y) + \sin(x-2y)\right]$$

$$= \frac{1}{2}\left[\frac{1}{D^2 - DD'}\sin(x+2y) + \frac{1}{D^2 - DD'}\sin(x-2y)\right]$$

$$= \frac{1}{2}\left[\frac{1}{-1^2 + 1.2}\sin(x+2y) + \frac{1}{-1^2 - (-1)(-2)}\sin(x-2y)\right]$$

$$= \frac{1}{2}\left[\sin(x+2y) - \frac{1}{3}\sin(x-2y)\right] = \frac{1}{2}\sin(x+2y) - \frac{1}{6}\sin(x-2y).$$

∴ Complete solution is

$$z = \phi_1(y) + \phi_2(y+x) + \frac{1}{2}\sin(x+2y) - \frac{1}{6}\sin(x-2y).$$

Example 19.33 Solve the following equations

(a) $\dfrac{\partial^2 z}{\partial x^2} - 3\dfrac{\partial^2 z}{\partial x\,\partial y} + 2\dfrac{\partial^2 z}{\partial y^2} = e^{2x+3y} + \sin(x-2y),$

(b) $\left(D^3 - 7DD'^2 - 6D'^3\right)z = \sin(x+2y) + e^{2x+y}$. (A.M.I.E; W-2008, S-2001)

Solution: (a) The given equation in operator form is

$$\left(D^2 - 3DD' + 2D'^2\right)z = e^{2x+3y} + \sin(x-2y).$$

A.E. is $m^2 - 3m + 2 = 0$, therefore $m = 1, 2$. C.F. $= \phi_1(y+x) + \phi_2(y+2x)$.

$$\text{P.I.} = \frac{1}{D^2 - 3DD' + 2D'^2}\left[e^{2x+3y} + \sin(x-2y)\right]$$

$$= \frac{1}{D^2 - 3DD' + 2D'^2}e^{2x+3y} + \frac{1}{D^2 - 3DD' + 2D'^2}\sin(x-2y)$$

$$= \frac{1}{2^2 - 3(2)(3) + 2(3)^2}e^{2x+3y} + \frac{\sin(x-2y)}{-1^2 - 3(-1)(-2) + 2\left[-(-2)^2\right]}$$

$$= \frac{1}{4}e^{2x+3y} - \frac{1}{15}\sin(x-2y).$$

The complete solution is

$$z = \phi_1(y+x) + \phi_2(y+2x) + \frac{1}{4}e^{2x+3y} - \frac{1}{15}\sin(x-2y).$$

(b) A.E. is $m^3 - 7m - 6 = 0$. Solving we get $m = -1, -2, 3$.

$\therefore \qquad$ C.F. $= \phi_1(y-x) + \phi_2(y-2x) + \phi_3(y+3x)$.

$$\text{P.I.} = \frac{1}{D^3 - 7DD'^2 - 6D'^3}\left[\sin(x+2y) + e^{2x+y}\right]$$

$$= \frac{1}{D^3 - 7DD'^2 - 6D'^3}\sin(x+2y) + \frac{1}{D^3 - 7DD'^2 - 6D'^3}e^{2x+y}.$$

P.I. of first term $= \dfrac{1}{D^3 - 7DD'^2 - 6D'^3}\sin(x+2y)$

$$= \frac{1}{D \cdot D^2 - 7\left(D \cdot D'^2\right) - 6D' \cdot D'^2}\sin(x+2y),$$

(writing $\qquad D^2 = -(1)^2 = -1, \; D'^2 = -2^2 = -4$)

$$= \frac{1}{D(-1) - 7D(-4) - 6D'(-4)}\sin(x+2y) = \frac{1}{3(9D + 8D')}\sin(x+2y) \quad \text{(A)}$$

$$= \frac{9D - 8D'}{3\left(81D^2 - 64D'^2\right)}\sin(x+2y) = \frac{9D - 8D'}{3[81(-1) - 64(-4)]}\sin(x+2y)$$

$$= \frac{9D - 8D'}{3(175)}\sin(x+2y) = \frac{1}{525}\left[9\cos(x+2y) - 16\cos(x+2y)\right]$$

$$= \frac{-7}{525}\cos(x+2y) = -\frac{1}{75}\cos(x+2y).$$

P.I. of second term

$$= \frac{1}{D^3 - 7DD'^2 - 6D'^3}e^{2x+y} = \frac{1}{2^3 - 7(2)(1^2) - 6(1)^3}e^{2x+y} = -\frac{1}{12}e^{2x+y}$$

∴ Complete solution is

$$z = \phi_1(y-x) + \phi_2(y-2x) + \phi_3(y+3x) - \frac{1}{75}\cos(x+2y) - \frac{1}{12}e^{2x+y}.$$

Important Note. An alternate and simpler method for evaluation of P.I. after step (A) in the above solution is as follows:

$$\text{P.I.} = \frac{1}{3(9D+8D')}\sin(x+2y) = \frac{D}{3(9D^2 + 8DD')}\sin(x+2y) \qquad \text{(note this step)}$$

$$= \frac{D}{3[9(-1)+8(-2)]}\sin(x+2y) = -\frac{1}{75}\cos(x+2y).$$

Example 19.34 Solve the following equations

(a) $\dfrac{\partial^2 z}{\partial x^2} + \dfrac{\partial^2 z}{\partial x \partial y} - 6\dfrac{\partial^2 z}{\partial y^2} = x + y,$ *(A.M.I.E S-2005, S-2003)*

(b) $\dfrac{\partial^2 z}{\partial x^2} + 3\dfrac{\partial^2 z}{\partial x \partial y} + 2\dfrac{\partial^2 z}{\partial y^2} = x + y.$ *(A.M.I.E W-2004)*

Solution. (a) The given equation is $\left(D^2 + DD' - 6D'^2\right)z = x+y$.

The A.E. is $m^2 + m - 6 = 0$ so that $m = 2, -3$.

$$\text{C.F.} = \phi_1(y+2x) + \phi_2(y-3x).$$

$$\text{P.I.} = \frac{1}{D^2 + DD' - 6D'^2}(x+y) = \frac{1}{D^2}\left[1 + \left(\frac{D'}{D} - 6\frac{D'^2}{D^2}\right)\right]^{-1}(x+y)$$

$$= \frac{1}{D^2}\left(1 - \frac{D'}{D}\right)(x+y) = \frac{1}{D^2}\left[x+y - \frac{1}{D}(0+1)\right]$$

$$= \frac{1}{D^2}(x+y-x) = \frac{1}{D^2}y = \frac{x^2 y}{2}.$$

$$\left[\because \frac{D'}{D}(x+y) = \frac{1}{D}\frac{\partial}{\partial y}(x+y) = \frac{1}{D}(1) = \int 1\,dx = x\right]$$

∴ Complete solution is $z =$ C.F. + P.I. $= \phi_1(y+2x) + \phi_2(y-3x) + \dfrac{x^2 y}{2}$.

(b) The operator form of the differential equation is

$$\left(D^2 + 3DD' + 2D'^2\right)z = x+y.$$

The A.E. is $m^2 + 3m + 2 = 0$ so that $m = -1, -2$.

C.F. $= \phi_1(y-x) + \phi_2(y-2x)$.

$$\text{P.I.} = \frac{1}{D^2 + 3DD' + 2D'^2}(x+y) = \frac{1}{D^2}\left[1 + \frac{3D'}{D} + 2\frac{D'^2}{D^2}\right]^{-1}(x+y)$$

$$= \frac{1}{D^2}\left(1 - \frac{3D'}{D}\right)(x+y) = \frac{1}{D^2}(x+y-3x) = \frac{1}{D^2}(y-2x) = \frac{x^2}{2}y - \frac{x^3}{3}$$

∴ Complete solution is $z =$ C.F. + P.I $= \phi_1(y-x) + \phi_2(y-2x) + \dfrac{x^2 y}{2} - \dfrac{x^3}{3}$.

Example 19.35 Solve the following partial differential equations

(a) $\dfrac{\partial^2 z}{\partial x^2} - a^2 \dfrac{\partial^2 z}{\partial y^2} = x^2$, (b) $\dfrac{\partial^3 z}{\partial x^3} - 2\dfrac{\partial^3 z}{\partial x^2 \partial y} = 2e^{2x} + 3x^2 y$.

(A.M.I.E. S-2004, S-2001; Madras, 1999)

Solution: (a) The given equation is $\left(D^2 - a^2 D'^2\right)z = x^2$.

A.E. is $\qquad m^2 - a^2 = 0, \; \therefore m = \pm a$. C.F. $= \phi_1(y+ax) + \phi_2(y-ax)$.

$$\text{P.I.} = \frac{1}{D^2 - a^2 D'^2}x^2 = \frac{1}{D^2}\left(1 - a^2 \frac{D'^2}{D^2}\right)^{-1}x^2$$

$$= \frac{1}{D^2}\left(1 + a^2 \frac{D'^2}{D^2}\right)x^2 = \frac{1}{D^2}(x^2) = \frac{x^4}{12}.$$

∴ Complete solution is $z = \phi_1(y+ax) + \phi_2(y-ax) + \left(\dfrac{x^4}{12}\right)$.

(b) A.E. is $m^3 - 2m^2 = 0$ or $m = 0, 0, 2$.

$$C.F = \phi_1(y) + x\phi_2(y) + \phi_3(y+2x).$$

$$P.I = \cdot \frac{1}{D^3 - 2D^2 D'}(2e^{2x} + 3x^2 y) = \frac{1}{D^3 - 2D^2 D'} \cdot 2e^{2x} + \frac{1}{D^3 - 2D^2 D'} 3x^2 y$$

$$= \frac{1}{2^3 - 2 \cdot 2^2 \cdot 0} 2e^{2x} + \frac{3}{D^3}\left(1 - 2\frac{D'}{D}\right)^{-1} x^2 y$$

$$= \frac{e^{2x}}{4} + \frac{3}{D^3}\left(1 + 2\frac{D'}{D}\right) x^2 y = \frac{e^{2x}}{4} + \frac{3}{D^3}\left(x^2 y + 2\frac{1}{D}x^2\right)$$

$$= \frac{e^{2x}}{4} + \frac{3}{D^3}\left(x^2 y + \frac{2x^3}{3}\right) = \frac{e^{2x}}{4} + 3y\frac{1}{D^3}x^2 + 2 \cdot \frac{1}{D^3}x^3$$

$$= \frac{e^{2x}}{4} + 3y \cdot \frac{x^5}{3 \cdot 4 \cdot 5} + 2 \cdot \frac{x^{6 \bullet}}{4 \cdot 5 \cdot 6} = \frac{e^{2x}}{4} + \frac{x^5 y}{20} + \frac{x^6}{60}.$$

$\therefore$ Complete solution is $z = $ C.F. + P.I.

$$= \phi_1(y) + x\phi_2(y) + \phi_3(y+2x) + \frac{e^{2x}}{4} + \frac{x^5 y}{20} + \frac{x^6}{60}.$$

19.15 General method for finding P.I.

So far we know how to find the P.I. of $F(D, D')z = f(x, y)$ when $f(x, y)$ is of the form e^{ax+by}, sin or $\cos(ax+b)$, $x^p y^q$.

When $f(x, y)$ is of a different form, the P.I. can be found by using the general method, *viz.*

$$\frac{1}{D - mD'}f(x, y) = \int f(x, c - mx)\,dx$$

in which c is to be replaced by $y + mx$ *after integration.*

In fact this method can be used in all cases discussed in Art. 19.14 including cases of failure, *i.e.,* in the case

$$\frac{1}{F(D, D')}e^{ax+by} \text{ when } F(a, b) = 0$$

or $\quad \dfrac{1}{F(D^2, DD', D'^2)}$ sin or $\cos(ax+by)$ when $F(-a^2, -ab, -b^2) = 0$.

The following examples will make the procedure clear.

19.16 Illustrative Examples

Example 19.36 Solve the following equations

(a) $\left(D^2 - DD' - 2D'^2\right)z = (y-1)e^x$,

(b) $\dfrac{\partial^2 z}{\partial x^2} + \dfrac{\partial^2 z}{\partial x \partial y} - 6\dfrac{\partial^2 z}{\partial y^2} = y \cos x$ or $y(\sec x)^{-1}$.

(*AMIE, S-2003; S-2000, Assam 1999*)

Alternative $r + s - 6t = y \cos x$ or $y(\sec x)^{-1}$.

Solution. (a) The A.E. is $m^2 - m - 2 = 0$ with roots $m = -1, 2$.

$\therefore$ C.F. $= \phi_1(y-x) + \phi_2(y+2x)$.

$$\text{P.I.} = \frac{1}{D^2 - DD' - 2D'^2}(y-1)e^x = \frac{1}{(D-2D')(D+D')}(y-1)e^x$$

$$= \frac{1}{D-2D'} \cdot \frac{1}{D+D'}(y-1)e^x = \frac{1}{D-2D'}\int (c+x-1)e^x\,dx$$

$$\left[\because \frac{1}{D-mD'}f(x,y) = \int f(x, c-mx)\,dx, \right.$$

where c is replaced by $y + mx$ after integration].

$$= \frac{1}{D-2D'}\left[(c+x-1)e^x - e^x\right] = \frac{1}{D-2D'}\left[(y-x+x-1)e^x - e^x\right]$$

(replacing c by $y-x$)

$$= \frac{1}{D-2D'}\left[(y-2)e^x\right] = \int (c-2x-2)e^x\,dx \qquad \text{(replacing } y \text{ by } c-2x)$$

$$= (c-2x-2)e^x - (-2)e^x = (c-2x)e^x \qquad \text{(replacing } c \text{ by } y+2x)$$

$$= (y+2x-2x)e^x = ye^x.$$

Hence complete solution is $z = \phi_1(y-x) + \phi_2(y+2x) + ye^x$.

(b) Given equation in *symbolic form* is $(D^2 + DD' - 6D'^2)z = y \cos x$ and the A.E. is $m^2 + m - 6 = 0$. Solving the roots are $m = 2, -3$.

$\therefore$ C.F. $= \phi_1(y+2x) + \phi_2(y-3x)$.

$$\text{P.I.} = \frac{1}{D^2 + DD' - 6D'^2}y\cos x = \frac{1}{(D+3D')(D-2D')}y\cos x$$

$$= -\frac{1}{D+3D'} \cdot \frac{1}{D-2D'} y \cos x \qquad \text{(replacing } y \text{ by } (c-2x))$$

$$= \frac{1}{D+3D'} \int (c-2x) \cos x \, dx$$

$$= \frac{1}{D+3D'} \big[(c-2x)(\sin x) - (-2)(-\cos x)\big]$$

$$= \frac{1}{D+3D'} \big[(c-2x)\sin x - 2\cos x\big] \qquad \text{(replacing } c \text{ by } y+2x)$$

$$= \frac{1}{D+3D'} \big[y\sin x - 2\cos x\big] \qquad \text{(replacing } y \text{ by } c+3x)$$

$$= \int \big[(c+3x)\sin x - 2\cos x\big] dx$$

$$= (c+3x)(-\cos x) - (3)(-\sin x) - 2\sin x = (c+3x)(-\cos x) + \sin x$$

$$= -y\cos x + \sin x . \qquad (\because \text{ on replacing } c \text{ by } y-3x)$$

Hence complete solution is $z = \phi_1(y+2x) + \phi_2(y-3x) - y\cos x + \sin x$.

Example 19.37 Obtain the solutions of following equations

(a) $\left(D-2D'\right)\left(D-D'\right)^2 z = e^{x+y}$,

(b) $r+s-6t = \cos(2x+y)$.

Solution: (a) The auxiliary equation $(m-2)(m-1)^2 = 0$ has $m = 2, 1, 1$ as its roots.

$\therefore$ C.F. $= \phi_1(y+2x) + \phi_2(y+x) + x\phi_3(y+x)$.

Here $\qquad\qquad F(D, D') = \left(D-2D'\right)\left(D-D'\right)^2$.

Since $a = 1, b = 1, F(a, b) = 0$, and, therefore, it is a case of failure and we cannot find P.I. by the method of Art. 19.14.

Applying the general method

$$\text{P.I.} = \frac{1}{\left(D-2D'\right)\left(D-D'\right)^2}\left(e^{x+y}\right) = \frac{1}{(1-2)\left(D-D'\right)^2} e^{x+y}$$

$$= -\frac{1}{D-D'} \cdot \frac{1}{D-D'} e^{x+y} = -\frac{1}{D-D'} \int e^{x+c-x} dx$$

$$= -\frac{1}{D-D'} \int e^c \, dx = -\frac{1}{D-D'} \cdot xe^c = -\frac{1}{D-D'} xe^{x+y}$$

$$= -\int xe^{x+c-x}dx = -e^c\int x\,dx = -\frac{x^2}{2}e^c = -\frac{x^2}{2}e^{x+y}.$$

$\therefore$ Complete solution is $z = \phi_1(y+2x)+\phi_2(y+x)+x\phi_3(y+x)-\dfrac{x^2}{2}e^{x+y}.$

(b) We know that $p = \dfrac{\partial z}{\partial x}, q = \dfrac{\partial z}{\partial y}; r = \dfrac{\partial^2 z}{\partial x^2}, s = \dfrac{\partial^2 z}{\partial x\,\partial y}, t = \dfrac{\partial^2 z}{\partial y^2}.$

$\therefore$ The given equation is

$$\frac{\partial^2 z}{\partial x^2}+\frac{\partial^2 z}{\partial y\,\partial x}-6\frac{\partial^2 z}{\partial y^2}=\cos(2x+y).$$

A.E. is $m^2 + m - 6 = 0.$ $\therefore$ $m = -3, 2.$

   C.F.$= \phi_1(y-3x)+\phi_2(y+2x).$

Here $a = 2, b = 1$

and $f(D^2, DD', D'^2) = D^2 + DD' - 6D'^2.$

Now $f(-a^2, -ab, -b^2) = -2^2 -(2)(1)-6(-1^2) = 0\cdot$

Therefore, it is a case of failure and we apply the general method.

Now $\dfrac{1}{D^2 + DD' - 6D'^2}\cos(2x+y) = \dfrac{1}{(D+3D')(D-2D')}\cos(2x+y)$

$$= \frac{1}{D+3D'}\int \cos(2x+c-2x)dx = \frac{1}{D+3D'}\int \cos c\,dx = \frac{1}{D+3D'}x\cos c$$

$$= \frac{1}{D+3D'}x\cos(y+2x)= \int x\cos(c+3x+2x)dx$$

$$= \int x\cos(5x+c)dx = \frac{x\sin(5x+c)}{5}+\frac{\cos(5x+c)}{25}$$

$$= \frac{x}{5}\sin(5x+y-3x)+\frac{\cos(5x+y-3x)}{25} = \frac{x}{5}\sin(2x+y)+\frac{\cos(2x+y)}{25}.$$

Hence complete solution is

$$z = \phi_1(y-3x)+\phi_2(y+2x)+\frac{x}{5}\sin(2x+y)+\frac{1}{25}\cos(2x+y).$$

Exercise 19.5

Solve the following partial differential equations

1. $\dfrac{\partial^2 z}{\partial x^2} + 7\dfrac{\partial^2 z}{\partial x\, \partial y} + 12\dfrac{\partial^2 z}{\partial y^2} = 0$.

   Ans. $z = \phi_1(y-3x) + \phi_2(y-4x)$.

2. $\left(2D^2 + 5DD' + 2D'^2\right)z = 0$.

   Ans. $z = \phi_1(y-2x) + \phi_2(2y-x)$.

3. $\dfrac{\partial^3 z}{\partial x^3} - 3\dfrac{\partial^3 z}{\partial x^2 \partial y} + 4\dfrac{\partial^3 z}{\partial y^3} = 0$. **Ans.** $y = \phi_1(y-x) + \phi_2(y+2x) + x\phi_3(y+2x)$.

4. $\left(D^2 - 2DD' + D'^2\right)z = e^{x+2y}$.

   Ans. $z = \phi_1(y+x) + x\phi_2(y+x) + e^{x+2y}$.

5. $\left(D^2 - 7DD' + 12D'^2\right)z = e^{x-y}$. **Ans.** $z = \phi_1(y+3x) + \phi_2(y+4x) + \dfrac{e^{x-y}}{20}$.

6. $\dfrac{\partial^2 z}{\partial x^2} - \dfrac{\partial^2 z}{\partial x\, \partial y} = \cos(x+2y)$.

   Ans. $z = \phi_1(y) + \phi_2(y+x) + \cos(x+2y)$.

7. $\left(2D^2 - 5DD' + 2D'^2\right)z = 5\sin(2x-y)$.

   Ans. $z = \phi_1(y+2x) + \phi_2(2y+x) - \dfrac{1}{4}\sin(2x-y)$.

8. $\dfrac{\partial^2 z}{\partial x^2} - 2\dfrac{\partial^2 z}{\partial x \partial y} + \dfrac{\partial^2 z}{\partial y^2} = \sin x$.

   Ans. $z = \phi_1(y+x) + x\phi_2(y+x) - \sin x$.

9. $\dfrac{\partial^2 z}{\partial x^2} + 4\dfrac{\partial^2 z}{\partial x\, \partial y} - 5\dfrac{\partial^2 z}{\partial y^2} = y^2 + x$.

   Ans. $z = \phi_1(y+x) + \phi_2(y-5x) + \dfrac{y^2 x^2}{2} + \dfrac{x^3}{6} - \dfrac{4x^3 y}{3} + \dfrac{7x^4}{4}$.

10. (a) $\dfrac{\partial^2 z}{\partial x^2} - \dfrac{\partial^2 z}{\partial y^2} = x^2 + y^2$. **Ans.** $z = \phi_1(y-x) + \phi_2(y+x) + \left(\dfrac{x^4}{6} + \dfrac{x^2 y^2}{2}\right)$.

    (b) $\dfrac{\partial^2 z}{\partial x^2} + 3\dfrac{\partial^2 z}{\partial x \partial y} + 2\dfrac{\partial^2 z}{\partial x^2} = 12xy$.

    (A.M.I.E., S-1998)

    Ans. $z = \phi_1(y-x) + \phi_2(y-2x) + 2x^3\left(y - \dfrac{3}{4}x\right)$.

11. $\left(D^2 - 6DD' + 9D'^2\right)z = 6x + 2y.$ *(AMIE W 2003)*

Ans. $z = \phi_1(y+3x) + x\phi_2(y+3x) + x^2(y+3x).$

12. $\left(2D^2 - 5DD' + 2D'^2\right)z = 5\sin(2x+y).$

Ans. $z = \phi_1(y+2x) + \phi_2(2y+x) - (5/3)x\cos(2x+y).$

13. $\left(D - 2D'\right)^2\left(D + 3D'\right)z = e^{2x+y}.$

Ans. $z = \phi_1(y+2x) + x\phi_2(y+2x) + \phi_3(y-3x) + \dfrac{1}{10}x^2 e^{2x+y}.$

14. $\dfrac{\partial^2 z}{\partial x^2} + \dfrac{\partial^2 z}{\partial y^2} = 12(x+y).$ **Ans.** $z = \phi_1(y+ix) + \phi_2(y-ix) + 2\left(x^3 + 3x^2 y\right).$

15. $\left(D^2 + 2DD' + D'^2\right)z = 2\cos y - x\sin y.$

Ans. $z = \phi_1(y-x) + x\phi_2(y-x) + x\sin y.$

16. $r + 2s + t = 2(y-x) + \sin(x-y).$

Ans. $z = \phi_1(y-x) + x\phi_2(y-x) + x^2 y - x^3 + \dfrac{x^2}{2}\sin(x-y).$

17. $\left(D^2 - 6DD' + 9D'^2\right)z = 12x^2 + 36xy.$ *(A.M.I.E. W-2001)*

Ans. $z = \phi_1(y+3x) + x\phi_2(y+3x) + 10x^4 + 6x^3 y$

18. $r - 4s + 4t = e^{2x+y}.$ **Ans.** $z = f_1(y+2x) + xf_2(y+2x) + \dfrac{1}{2}x^2 e^{y+2x}.$

19. (a) $\dfrac{\partial^2 z}{\partial x^2} - \dfrac{\partial^2 z}{\partial x\,\partial y} = \cos x\cos 2y,$ (b) $\dfrac{\partial^2 z}{\partial x^2} - \dfrac{\partial^2 z}{\partial y^2} = x - y.$

Ans. (a) $z = \phi_1(y) + \phi_2(y+x) + \dfrac{1}{2}\cos(x+2y) - \dfrac{1}{6}\cos(x-2y),$

(b) $z = \phi_1(y+x) + \phi_2(y-x) + \dfrac{x^3}{3} - \dfrac{x^2 y}{2}.$

20. (a) $\left(D^3 + D^2 D' - DD'^2 - D'^3\right)z = e^x\cos 2y.$

Ans. $z = \phi_1(y+x) + \phi_2(y-x) + x\phi_3(y-x) + \dfrac{e^x}{25}(\cos 2y + 2\sin 2y).$

$$\left[\text{Hint: } \frac{1}{f(D, D')} e^{ax} X = e^{ax} \frac{1}{f(D+a, D')} X, \text{ etc.}\right]$$

(b) $\dfrac{\partial^2 z}{\partial x^2} - \dfrac{\partial^2 z}{\partial y^2} = e^{x+2y}$. **Ans.** $z = \phi_1(y+x) + \phi_2(y-x) - \dfrac{1}{3} e^{x+2y}$.

19.17 Non-homogeneous Linear Equations with constant Coefficients

In the equation $\qquad F(D, D')z = f(x, y) \qquad \qquad ...(19.15)$

if the polynomial expression $F(D, D')$ is not homogeneous then (19.15) is defined as non-homogeneous linear partial differential equation. As in the case of homogeneous linear partial differential equations, the complete solution (C.S.) of (19.15) is given by

$$\text{C.S.} = \text{C.F.} + \text{P.I.}$$

The procedure of getting the P.I. is the same as in the case of homogeneous linear equations. However, to evaluate the C.F., we factorise the polynomial expression $f(D, D')$ into factors of the type $(D - mD' - c)$. This gives us an equation $(D - mD' - c)z = 0$. This equation written as

$$p - mq = cz \qquad \qquad ...(19.16)$$

is the Lagrange's first order partial differential equations. The subsidiary equations are:

$$\frac{dx}{1} = \frac{dy}{-m} = \frac{dz}{cz}.$$

The integrals are given by $y + mx = a$ and $z = be^{cx}$.

Taking $b = \phi(a)$, the solution of (19.16) is given by $z = \phi(a)e^{cx} = \phi(y + mx)e^{cx}$.

Such solutions obtained from various factors when added, give the C.F. of (19.15).

Example 19.38 Solve $\left(D^2 + DD' + D' - 1\right)z = e^{-x}$.

Solution: Here $F(D, D')z = (D^2 + DD' + D' - 1)z = (D+1)(D + D' - 1)z$.

For C.F., we write $(D+1)(D + D' - 1)z = 0$. $\qquad \qquad ...(1)$

As given above, the solution corresponding to the factor $(D - mD' - c)$ is known to be $z = e^{cx}\phi(y + mx)$.

$\therefore \qquad \qquad \text{C.F.} = e^x \phi(y - x) + e^{-x}\psi(y)$.

$$\text{P.I.} = \frac{1}{(D+1)(D+D'-1)} e^{-x} . = -\frac{1}{2}\frac{1}{D+1} e^{-x} = -\frac{x}{2} e^{-x}.$$

Hence the complete solution is $z = e^x \phi(y-x) + e^{-x} \psi(y) - \dfrac{x}{2} e^{-x}$.

Example 19.39 Solve $(D^2 - DD' + D' - 1) z = \cos(x + 2y)$.

Solution: Here $F(D, D')z = (D^2 - DD' + D' - 1)z = (D-1)(D-D'+1)z$.

Since the solution corresponding to the factor $(D - mD' - c)$ is known to be

$z = \phi(y + mx)e^{cx}$, the C.F. of the given equation is $z = e^x \phi(y) + e^{-x} \psi(y+x)$.

$$\text{P.I.} = \frac{1}{D^2 - DD' + D' - 1} \cos(x+2y) = \frac{1}{-1+2+D'-1} \cos(x+2y)$$

$$= \frac{1}{D'} \cos(x+2y) = \frac{1}{2} \sin(x+2y).$$

$\therefore$ The complete solution is given by $z = e^x \cdot \phi(y) + e^{-x} \psi(y+x) + \dfrac{1}{2} \sin(x + 2y)$.

Example 19.40 Solve: $(2DD' + D'^2 - 3D')z = 3 \cos (3x - 2y)$.

Solution: Here $f(D, D')z = D'(D' + 2D - 3)$.

Since the solution corresponding to the factor $D' - mD - c$ is known to be $z = e^{cy}\phi(x + my)$.

$\therefore$ C.F. $= \phi(x) + e^{3y} \psi(x-2y)$. P.I. $= \dfrac{1}{2DD' + D'^2 - 3D'} 3\cos(3x - 2y)$

$$= \frac{1}{-2(3)(-2)-(-2)^2 - 3D'} 3\cos(3x-2y)$$

$$= 3 \cdot \frac{1}{8 - 3D'} \cos(3x-2y) = 3 \cdot \frac{8 + 3D'}{64 - 9D'^2} \cos(3x-2y)$$

$$= 3 \cdot \frac{8 + 3D'}{64 - 9\left\{-(-2)^2\right\}} \cos(3x-2y)$$

$$= \frac{3}{100}\left[8\cos(3x-2y) + 6\sin(3x-2y)\right] = \frac{3}{50}\left[4\cos(3x-2y) + 3\sin(3x-2y)\right].$$

Hence the complete solution is

$$z = \phi(x) + e^{3y} \psi(x-2y) + \frac{3}{50}\left[4\cos(3x-2y) + 3\sin(3x-2y)\right].$$

Exercise 19.6

Solve the following partial differential equations

1. $(D - D' - 1)(D - D' - 2)z = 0.$ **Ans.** $z = e^x \phi(y + x) + e^{2x} \psi(y + x).$

2. $(2D^2 - DD' - D'^2 + 6D + 3D')z = 0.$ **Ans.** $z = \phi(x - 2y) + e^{-3x} \psi(y + x).$

3. $(D^2 + 2DD' + D'^2 - 2D - 2D')z = \sin(x + 2y).$

    Ans. $z = \phi(y - x) + e^{2x} \psi(y - x) + \dfrac{1}{39}\{2\cos(x + 2y) - 3\sin(x + 2y)\}.$

4. $(D^2 - DD' - 2D'^2 + 2D + 2D')z = e^{2x+3y} + \sin(2x + y) + xy.$

    Ans. $z = \phi(y - x) + e^{-2x} \psi(y + 2x) - \dfrac{1}{10}e^{2x+3y} - \dfrac{1}{6}\cos(2x + y)$

                                                       $+ \dfrac{x}{24}(6xy - 6y - 2x^2 + 9x - 12).$

5. $\dfrac{\partial^2 z}{\partial x^2} - \dfrac{\partial^2 z}{\partial x \partial y} + \dfrac{\partial z}{\partial x} = x^2 + y^2.$ *(Madras 2000S)*

    Ans. $z = \phi(y) + e^{-x} \psi(y + x) + \dfrac{1}{3}x^3 - x^2 + xy^2 + 2xy + 4x.$

6. $(D - D' - 1)(D - D' - 2)z = e^{2x-y} + x.$

    Ans. $z = \phi(y + x)e^x + \psi(y + x)e^{2x} + \dfrac{1}{2}e^{2x-y} + \dfrac{x}{2} + \dfrac{3}{4}.$

7. $(D^2 - DD' + D' - 1)z = \cos(x + 2y) + e^y.$

    Ans. $z = e^x \phi(y) + e^{-x} \psi(y + x) + \dfrac{1}{2}\sin(x + 2y) - xe^y.$

8. $\dfrac{\partial^2 z}{\partial x^2} - \dfrac{\partial^2 z}{\partial y^2} + \dfrac{\partial z}{\partial x} + 3\dfrac{\partial z}{\partial y} - 2z = e^{x-y} - x^2 y.$

    Ans. $z = e^{-2x} \phi(y + x) + e^x \psi(y - x) - \dfrac{1}{4}e^{x-y} + \dfrac{1}{2}\left(x^2 y + xy + \dfrac{3}{2}x^2 + \dfrac{3}{2}y + 3x + \dfrac{21}{4}\right).$

9. $(D + D' - 1)(D + 2D' - 3)z = 4 + 3x + 6y.$

    Ans. $z = e^x \phi(y - x) + e^{3x} \psi(y - 2x) + x + 2y + 6.$

10. $(D^2 - D'^2 - 3D + 3D')z = xy + e^{x+2y}.$

Ans. $z = e^{3x} \phi(y - x) + \psi(y + x) - ye^{x+2y} - \dfrac{1}{54}(9x^2 y + 6xy + 9x^2 + 4x).$

Chapter 20

Applications of Partial Differential Equations

20.1 Introduction

The mathematical formulation of many problems in Science and Engineering gives rise to partial differential equations. Some of the important partial differential equations thus obtained are:

1.
$$\frac{\partial^2 y}{\partial t^2} = c^2 \cdot \frac{\partial^2 y}{\partial x^2}.$$

This is known as *wave equation* or *vibrating string equation*. This is the simplest example of hyperbolic equation.

2.
$$\frac{\partial u}{\partial t} = c^2 \cdot \frac{\partial^2 u}{\partial x^2}.$$

This equation occurs in problems of *one-dimensional heat flow*. This is a parabolic equation and is also known as *diffusion equation*.

3.
$$\frac{\partial^2 u}{\partial x^2} + \frac{\partial^2 u}{\partial y^2} = 0.$$

This is known as *Laplace equation* in two dimensions. It occurs in problems of two-dimensional heat flow in steady state, that is, when temperature does not change with time. This is an example of two dimensional elliptic equation.

4. *Transmission Line Equations*

(a) *Telephone Equations*

(i)
$$\frac{\partial V}{\partial x} = -L \frac{\partial i}{\partial t} - Ri \; ; \quad \frac{\partial i}{\partial x} = -C \frac{\partial V}{\partial t} - GV .$$

(ii) $\quad \dfrac{\partial^2 V}{\partial x^2} = LC \dfrac{\partial^2 V}{\partial t^2} + (RC + LG) \dfrac{\partial V}{\partial t} + RGV.$

(iii) $\quad \dfrac{\partial^2 i}{\partial x^2} = LC \dfrac{\partial^2 i}{\partial t^2} + (RC + LG) \dfrac{\partial i}{\partial t} + RGi$

where i and V denote the current and voltage respectively and L, C, R and G are the inductance, capacitance, Resistance and conductance respectively.

    (b) *Telegraph Equations:* If L and G are small ($L = G = 0$) which is precisely the case in submarine cables, the telephone equations reduces to telegraph equations and the resulting equations takes the form

(i) $\quad \dfrac{\partial V}{\partial x} = -Ri \;\; ; \;\; \dfrac{\partial i}{\partial x} = -C \dfrac{\partial V}{\partial t}.$

(ii) $\quad \dfrac{\partial^2 V}{\partial x^2} = RC \dfrac{\partial V}{\partial t}.$

(iii) $\quad \dfrac{\partial^2 i}{\partial x^2} = RC \dfrac{\partial i}{\partial t}.$

    Equations (ii) and (iii) are similar to one-dimensional heat flow equations.

    (c) *Radio Equations:* In the case of high frequency lines where the effects of conductance and resistance are negligible ($G = R = 0$), the telephone equations assume the form:

(i) $\quad \dfrac{\partial V}{\partial x} = -L \dfrac{\partial i}{\partial t} \; ; \; \dfrac{\partial i}{\partial x} = -C \dfrac{\partial V}{\partial t}.$

(ii) $\quad \dfrac{\partial^2 V}{\partial x^2} = LC \dfrac{\partial^2 V}{\partial t^2}.$

(iii) $\quad \dfrac{\partial^2 i}{\partial x^2} = LC \dfrac{\partial^2 i}{\partial t^2}.$

    The last two of the above equations are similar to the *wave equation*.

20.2 In the case of physical problems giving rise to partial differential equations, we have to find solutions which satisfy certain initial[*] and boundary[**] conditions. Such problems are called *boundary value problems*. In the previous chapter we found solutions of differential equations which involved arbitrary functions. These

---

[*]   The conditions which are given for time $t = 0$, are called *initial conditions*.
[**] The conditions given at the boundary of the region or interval are called *boundary conditions*.

arbitrary functions are difficult to adjust so as to satisfy given boundary conditions. A method suitable for such problems is the *method of separation of variables* also known as *product method*.

20.3 Method of Separation of Variables (Product Method)

This method describes the solution which expresses the dependent variable as a product of functions each of which depends only on one variable. For example if z is the dependent variable and depends on two independent variables x and y that is, $z(x, y)$, then we assume the solution in the form

$$z = X(x)\, Y(y). \qquad \ldots(20.1)$$

where X is a function of x alone and Y is a function of y alone. Substitution for z from (20.1) in the given differential equation reduces it to ordinary differential equations which can be conveniently solved for the solution. The procedure will be clear from the following solved examples.

Example 20.1 Solve $\dfrac{\partial^2 z}{\partial x^2} - 2\dfrac{\partial z}{\partial x} + \dfrac{\partial z}{\partial y} = 0$ **by the method of separation of variables.**

(*I.E.T.E., June 2009; Mysore; 1997; Osmania 1995*)

Solution. Let $\qquad z = X(x) \cdot Y(y)$ $\qquad\qquad\qquad$ ...(*i*)

be the solution of the given differential equation, where X is a function of x and Y a function of y only.

Let $\qquad \dfrac{dX}{dx} = X',\ \dfrac{d^2 X}{dx^2} = X'',\ \dfrac{dY}{dy} = Y'$ and $\dfrac{d^2 Y}{dy^2} = Y''.$

Substituting (*i*) in the differential equation we get

$$X''Y - 2X'Y + XY' = 0, \quad \text{or, } (X'' - 2X')Y + XY' = 0. \qquad \ldots(ii)$$

Dividing (*ii*) by XY and separating the variables, this equation becomes

$$\frac{X'' - 2X'}{X} = -\frac{Y'}{Y}. \qquad \ldots(iii)$$

Now x and y are independent variables.

If we vary y, the right side of (*iii*) will change while the left hand side will remain unaffected and vice-versa. In order that (*iii*) holds, each side must be equal to a constant because X and Y are respectively functions of x and y only.

Let $\dfrac{X'' - 2X'}{X} = -\dfrac{Y'}{Y} = k$, a constant.

This gives $\qquad\qquad X'' - 2X' - kX = 0 \qquad\qquad\qquad$ ...(*iv*)

$$Y' + kY = 0. \qquad\qquad\qquad \ldots(v)$$

These are two ordinary differential equations.

To solve (*iv*), we know that $m^2 - 2m - k = 0$ is the auxiliary equation.

Its roots are $$m = \frac{2 \pm \sqrt{4 + 4k}}{2} = 1 \pm \sqrt{1+k}.$$

Let $$1 + \sqrt{1+k} = \alpha \text{ and } 1 - \sqrt{1+k} = \beta.$$

$\therefore$ $$X = c_1 e^{\alpha x} + c_2 e^{\beta x}.$$

From (*v*), we have

$$\frac{Y'}{Y} = -k \text{ or } \log Y = -ky + c \qquad \qquad \text{...(vi)}$$

$\therefore$ $Y = e^{-ky+c} = c_3 e^{-ky} \left(\text{where } c_3 = e^c \right)$ $\qquad \qquad$ ...(*vii*)

From (*vi*) and (*vii*) the required solution is

$$z(x, y) = XY = \left(c_1 e^{\alpha x} + c_2 e^{\beta x} \right) c_3 e^{-ky} = \left(A e^{\alpha x} + B e^{\beta x} \right) e^{-ky},$$

where $c_1 c_3$ and $c_2 c_3$ are replaced by A and B respectively.

Example 20.2 **Using the method of separation of variables, solve the partial**

differential equation $\dfrac{\partial u}{\partial x} = 2 \dfrac{\partial u}{\partial t} + u$ **, under the condition** $u(x, 0) = 6 e^{-3x}$.

(IETE, June 2009; V.T.U. 2001; A.M.I.E., S-2001)

[Before we solve the problem we should note that u is a function of x and t and therefore may be written as $u(x, t)$. The value of u when $t = 0$ is written as $u(x, 0)$ and its value when $x = 0$ is denoted by $u(0, t)$].

Solution. Let $u(x, t) = X(x) \cdot T(t)$ be the solution.

Substituting in the given equation, we have

$$X'T = 2XT' + XT \qquad \qquad \text{...(20.2)}$$

or $$(X' - X)T = 2XT'$$

or $$\frac{X' - X}{2X} = \frac{T'}{T} = k \text{ (say) (after dividing by XT)}$$

$\therefore$ $X' - X - 2kX = 0$ or $X' - (1 + 2k)X = 0$ $\qquad \qquad$ ...(*i*)

and $$T' - kT = 0. \qquad \qquad \text{...(ii)}$$

From (*i*) $\dfrac{X'}{X} = 1 + 2k$. Solving $\log X = (1 + 2k)x + c'$

or $$X = c_1 e^{(1+2k)x}. \qquad \left(\text{where } c_1 = e^{c'} \right). \qquad \text{...(iii)}$$

From (*ii*)
$$\frac{T'}{T} = k \text{ or } \log T = kt + c''$$

$\therefore$ $\qquad T = c_2 e^{kt} \text{ where } (c_2 = e^{c''}).$ ...(*iv*)

From (*iii*) and (*iv*) we get
$$u = XT = c_1 c_2 e^{(1 + 2k)x} \cdot e^{kt}$$
$$= A e^{(1 + 2k)x} \cdot e^{kt}.$$ ...(*v*)

Now $\qquad u(x, 0) = A e^{(1 + 2k)x}.$ $\qquad (\because e^{ok} = e^0 = 1)$

But $u(x, 0) = 6e^{-3x}$ (given). Therefore, $A e^{(1 + 2k)x} = 6e^{-3x}$.

Comparing the two sides, we get $A = 6$ and $1 + 2k = -3$ or $k = -2$.

Putting these values in (*v*). Thus $u(x, t) = 6 e^{-3x} e^{-2t} = 6 e^{-(3x + 2t)}$ is the required solution of the given equation.

Remark: Equation (20.2) can also be written in the form $X'T = (2T' + T)X$ that is,

$$\frac{X'}{X} = \frac{2T' + T}{T} = k.$$

Solving $X' - kX = 0$ and $2T' + (1 - k)T = 0$, we get the same result.

20.4 Equation of Vibrating String

$$\frac{\partial^2 y}{\partial t^2} = c^2 \cdot \frac{\partial^2 y}{\partial x^2}.$$

Consider a perfectly flexible homogeneous string of length *l* tightly stretched between two points 0 and *A*. We assume the tension in the string to be so large that gravity may be neglected in comparison with it.

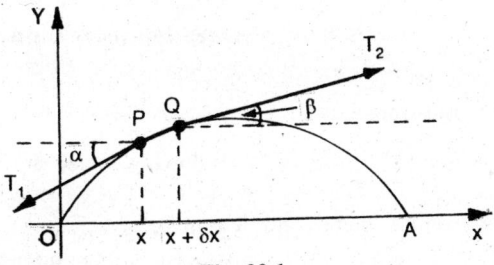

Fig. 20.1.

Take 0 as origin and *OA* as *X*-axis and let *XY*-plane be the vertical plane. We shall determine the differential equation governing the motion when the string is set vibrating in the vertical plane, each point moving along a straight line at right angles to the string *i.e.*, we will investigate the transverse vibrations of the string.

Consider the forces acting on a small portion between *P*(*x*, *y*) and $Q(x + \delta x, y + \delta y)$ moving vertically, with acceleration $\partial^2 y/\partial t^2$ at any time *t*.

The only forces acting on PQ are the tensions T_1 and T_2 along the tangents at P and Q inclined at angles α and β respectively with the horizontal. Since there is no motion in the horizontal direction, the horizontal components of tension must be constant.

$$\therefore \qquad T_1 \cos \alpha = T_2 \cos \beta = T \text{ (a constant).} \qquad \ldots(20.3)$$

For vertical motion of PQ, we have by Newton's second law of motion

$$m \, \delta x \frac{\partial^2 y}{\partial t^2} = T_2 \sin \beta - T_1 \sin \alpha$$

(m is the mass per unit length of the string)

or
$$\frac{m \delta x}{T} \frac{\partial^2 y}{\partial t^2} = \frac{T_2 \sin \beta}{T_2 \cos \beta} - \frac{T_1 \sin \alpha}{T_1 \cos \alpha} \qquad \ldots\text{[By (20.3)]}$$

$$= \tan \beta - \tan \alpha. \qquad \ldots(20.4)$$

Now $\tan \alpha$ and $\tan \beta$ are slopes of the tangents at $P(x, y)$ and $Q(x + \delta x, y + \delta y)$.

Therefore,
$$\tan \alpha = \left(\frac{\partial y}{\partial x}\right)_x \text{ and } \tan \beta = \left(\frac{\partial y}{\partial x}\right)_{x+\delta x}.$$

Thus (20.4) can be written in the form

$$\frac{m}{T} \delta x \frac{\partial^2 y}{\partial t^2} = \left(\frac{\partial y}{\partial x}\right)_{x+\delta x} - \left(\frac{\partial y}{\partial x}\right)_x \quad \text{or} \quad \frac{m}{T} \frac{\partial^2 y}{\partial t^2} = \frac{1}{\delta x}\left[\left(\frac{\partial y}{\partial x}\right)_{x+\delta x} - \left(\frac{\partial y}{\partial x}\right)_x\right].$$

As $\delta x \to 0$, we get

$$\frac{m}{T} \frac{\partial^2 y}{\partial t^2} = \frac{\partial^2 y}{\partial x^2} \quad \text{or} \quad \frac{\partial^2 y}{\partial t^2} = c^2 \frac{\partial^2 y}{\partial x^2}, \ldots(20.5) \qquad\qquad \text{where } c^2 = T/m.$$

Note: $\dfrac{\partial^2 y}{\partial t^2} = c^2 \dfrac{\partial^2 y}{\partial x^2}$ is also known as **one-dimensional wave equation**.

20.5 Boundary Conditions

Equation (20.5) is obliged to satisfy the following initial and boundary conditions.

(1) At the ends O and A, there is no motion i.e.,

(i) $y = 0$ at $x = 0$; and (ii) $y = 0$ at $x = l$. These conditions must hold for all values of time t i.e. $t \geq 0$.

Alternately $y(o, t) = y(l, t) = 0$ for $t \geq 0$.

(2) If the string is initially held in the form of the curve $y = f(x)$ and then released from this position, the initial condition of vibration will be

(i) $y = f(x)$ at $t = 0$ and (ii) $\dfrac{\partial y}{\partial t} = 0$ at $t = 0$.

(3) If every element of the string is moving with a definite velocity and the motion is reckoned when the string is at the equilibrium position, then the initial conditions are represented by

(i) $\dfrac{\partial y}{\partial t} = \phi(x)$ at $t = 0$ and (ii) $y = 0$ at $t = 0$.

(4) If the string is moving with a specified velocity and is also given a displacement, the conditions can be specified accordingly.
In the following examples we shall solve this equation by the method of separation of variables.

20.6 Illustrative Examples

Example 20.3 (a) Obtain the solution of the wave equation $\dfrac{\partial^2 y}{\partial t^2} = c^2 \cdot \dfrac{\partial^2 y}{\partial x^2}$, **using the method of separation of variables.** *(A.M.I.E., W-1998)*

(b) A string is stretched and fastened to two points at a distance l apart. Motion is started by displacing the string in the form $y = a \sin\left(\dfrac{\pi x}{l}\right)$ from which it is released at a time $t = 0$. Show that the displacement of any point at distance x from one end at time t is given by

$$y(x,t) = a \sin\left(\frac{\pi x}{l}\right) \cos\left(\frac{\pi ct}{l}\right).$$ *(A.M.I.E., S-2002)*

Solution. (a) Let $Y(x, t) = X(x) \cdot T(t)$ ...(i)
be the solution of the given equation. Substituting (i) in the equation, we get

$$XT'' = c^2 X''T$$

or $\dfrac{X''}{X} = \dfrac{1}{c^2}\dfrac{T''}{T} = h\,(\text{say})$ (after dividing both sides of above by XT).

$\therefore$ $$X'' - hX = 0$$ ...(ii)

and $$T'' - c^2 hT = 0.$$ ...(iii)

Now three cases arise.

Case I ($h > 0$): Let $h = k^2$.

Then (ii) and (iii) become respectively

$$X'' - k^2 X = 0 \text{ and } T'' - c^2 k^2 T = 0.$$

Solving these ordinary equations, we get

$$X = c_1 e^{kx} + c_2 e^{-kx} \text{ and } T = c_3 e^{ckt} + c_4 e^{-ckt}.$$ ...(iv)

Case II. $h = 0$.

Then (ii) and (iii) become $X'' = 0$ and $T'' = 0$.

Solving, we get $X = c_5 x + c_6$ and $T = c_7 t + c_8$. ...(v)

Case III. ($h < 0$), Let $h = -k^2$.

From (ii) and (iii) we get

$$X'' + k^2 X = 0 \text{ and } T'' + c^2 k^2 T = 0.$$

Solving, we get $\qquad X = c_9 \cos kx + c_{10} \sin kx,$

and $\qquad\qquad\qquad T = c_{11} \cos ckt + c_{12} \sin ckt.$ ...(vi)

Thus depending upon the value of h we have from (iv), (v) and (vi) the solution as

$$y = (c_1 e^{kx} + c_2 e^{-kx})(c_3 e^{ckt} + c_4 e^{-ckt}) \qquad\qquad ...(vii)$$

$$y = (c_5 x + c_6)(c_7 t + c_8) \qquad\qquad\qquad\qquad ...(viii)$$

$$y = (c_9 \cos kx + c_{10} \sin kx)(c_{11} \cos ckt + c_{12} \sin ckt). \qquad ...(ix)$$

Since it is a problem on vibrating string every position of which is repeated as time passes, the solution should be a periodic function of t. Hence only (ix) can be taken as the solution of the given equation.

(b) The equation of the vibrating string is

$$\frac{\partial^2 y}{\partial t^2} = c^2 \frac{\partial^2 y}{\partial x^2} \qquad\qquad ...(i)$$

Let O and A be the points to which the ends are fastened so that $OA = l$ (Refer Fig. 20.1).

Since the ends of the string are fixed, $y = 0$ when $x = 0$ and $x = l$, for all values of t.

$\therefore$ *Boundary conditions* are

$$y(0, t) = 0 \text{ and } y(l, t) = 0. \qquad\qquad ...(ii)$$

The string is in the shape $y = a \sin\left(\dfrac{\pi x}{l}\right)$ and at rest when $t = 0$.

$\therefore$ The *initial conditions* are

$$y(x, 0) = a \sin\frac{\pi x}{l}, \quad \left(\frac{\partial y}{\partial t}\right)_{t=0} = 0. \qquad\qquad ...(iii)$$

Now we have to solve (i) subject to the conditions (ii) and (iii).

As shown in part (a), the solution is of the form

$$y(x, t) = (a_1 \cos kx + a_2 \sin kx)(a_3 \cos ckt + a_4 \sin ckt). \qquad ...(iv)$$

Since $y(0, t) = 0$ for all t (first boundary condition), therefore,

$0 = a_1(a_3 \cos ckt + a_4 \sin ckt)$. Hence $a_1 = 0$.

Therefore, (iv) reduces to

$$y(x, t) = a_2 \sin kx(a_3 \cos ckt + a_4 \sin ckt)$$

$$= \sin kx(b_1 \cos ckt + b_2 \sin ckt). \quad \text{(where } b_1 = a_2 a_3, \ b_2 = a_2 a_4).$$

From the second boundary condition $y(l, t) = 0$ for all t, we get

$$0 = \sin kl(b_1 \cos ckt + b_2 \sin ckt).$$

$\therefore \quad \sin kl = 0 \quad$ or $\quad kl = n\pi$ (n being an integer) or $k = n\pi/l$.

$$\therefore \quad y(x,t) = \sin\frac{n\pi x}{l}\left(b_1 \cos\frac{cn\pi}{l}t + b_2 \sin\frac{cn\pi}{l}t\right).$$

Now
$$\frac{\partial y}{\partial t} = \sin\frac{n\pi x}{l}\left(-b_1\frac{cn\pi}{l}\sin\frac{cn\pi t}{l} + b_2\frac{cn\pi}{l}\cos\frac{cn\pi t}{l}\right).$$

By second initial condition $\left(\dfrac{\partial y}{\partial t}\right)_{t=0} = 0.$

$$\therefore \quad 0 = \sin\frac{n\pi x}{l}\left(b_2\frac{cn\pi}{l}\right) \text{ which gives } b_2 = 0.$$

Hence
$$y(x, t) = b_1 \sin\frac{n\pi x}{l}\cos\frac{cn\pi}{l}t. \qquad\qquad ...(v)$$

From this equation $y(x, 0) = b_1 \sin\dfrac{n\pi x}{l}$ at $t = 0$.

But $y(x, 0) = a\sin\dfrac{\pi x}{l}$ (first initial condition).

$\therefore \quad a\sin\dfrac{\pi x}{l} = b_1 \sin\dfrac{n\pi x}{l}$. This holds when $b_1 = a$ and $n = 1$.

Hence $y(x,t) = a\sin\dfrac{\pi x}{l}\cos\dfrac{\pi ct}{l}$ is the required solution.

Important note. We know that if $y_1, y_2,..., y_n$ are solutions of a differential equation, then
$$A_1 y_1 + A_2 y_2 + ...A_n y_n$$ where A's are arbitrary constants, is also a solution. This is known as **superposition principle**.

From (v) above $y = b_1 \sin\dfrac{n\pi x}{l}\cos\dfrac{cn\pi t}{l}$

or $y = \sin\dfrac{n\pi x}{l}\cos\dfrac{cn\pi t}{l}$ is a solution of the wave equation, n being an integer.

Putting $n = 1, 2, 3,...$

$$y_1 = \sin\frac{\pi x}{l}\cos\frac{c\pi t}{l}, y_2 = \sin\frac{2\pi x}{l}\cos\frac{2c\pi t}{l},... \text{ are also solutions. Hence by}$$

superposition principle

$$y = A_1 \sin\frac{\pi x}{l}\cos\frac{c\pi t}{l} + A_2 \sin\frac{2\pi x}{l}\cos\frac{2c\pi t}{l}$$

$$+ \ldots + A_n \sin\frac{n\pi x}{l}\cos\frac{nc\pi t}{l} + \ldots$$

or $\qquad y = \sum_{n=1}^{\infty} A_n \sin\frac{n\pi x}{l}\cos\frac{cn\pi t}{l}$ is also a solution.

We could have taken this to be the solution in the above example. Then using the first initial condition,

that is, $\qquad\qquad y(x,0) = a\sin\frac{\pi x}{l}$, we get

$$a\sin\frac{\pi x}{l} = \sum A_n \sin\frac{n\pi x}{l} = A_1 \sin\frac{\pi x}{l} + A_2 \sin\frac{2\pi x}{l} + \ldots$$

which gives $A_1 = a, A_2 = A_3 = \ldots = 0.$

$\therefore \quad y(x,t) = a\sin\frac{\pi x}{l}\cos\frac{c\pi t}{l}$ is a solution.

Example 20.4 A tightly stretched string with fixed ends at x = 0 and x = *l* is initially in equilibrium position. It is set vibrating by giving to each of its points an initial velocity $\left(\dfrac{\partial y}{\partial t}\right)_{t=0} = b\sin^3\left(\dfrac{\pi x}{l}\right)$. Find the displacement y(x, t).

(I.E.T.E. June 2009; A.M.I.E., S-2001; Madras 1995)

Solution. The equation of the vibrating string is

$$\frac{\partial^2 y}{\partial t^2} = c^2 \frac{\partial^2 y}{\partial x^2}.$$

We have to solve this equation subject to the *boundary conditions:*
$$y(0, t) = 0 \text{ and } y(l, t) = 0$$
and subject to the *initial conditions.*

$$y(x, 0) = 0 \quad \text{and} \quad \left(\frac{\partial y}{\partial t}\right)_{t=0} = b\sin^3\frac{\pi x}{l}.$$

Assuming the solution to be of the form y(x, t) = X(x)T(t), we get
$$XT'' = c^2 X'' T$$

or $\qquad\qquad \dfrac{X''}{X} = \dfrac{1}{c_2}\cdot\dfrac{T''}{T} = -k^2 \text{ (a constant)}.$

(As is clear from Example 20.3 we have taken the constant as $-k^2$ so that solution is a periodic function of *t*)

$$\therefore \qquad X'' + k^2 X = 0 \text{ and } T'' + c^2 k^2 T = 0.$$

Solving the differential equations we get

$$X = c_1 \cos kx + c_2 \sin kx \text{ and } T = c_3 \cos ckt + c_4 \sin ckt.$$

Hence the solution is of the form

$$y = (c_1 \cos kx + c_2 \sin kx)(c_3 \cos ckt + c_4 \sin ckt).$$

Now $y(0, t) = 0$ (first boundary condition).

$$\therefore \qquad 0 = c_1(c_3 \cos ckt + c_4 \sin ckt).$$

This is true for all t and this gives $c_1 = 0$.

$$\therefore \qquad y(x, t) = c_2 \sin kx(c_3 \cos ckt + c_4 \sin ckt)$$
$$= \sin kx(a_1 \cos ckt + a_2 \sin ckt)$$

$$\text{(where } a_1 = c_2 c_3 \text{ and } a_2 = c_2 c_4).$$

Also $\qquad\qquad y(l, t) = 0$ (second boundary condition).

$$\therefore \qquad 0 = \sin kl(a_1 \cos ckt + a_2 \sin ckt) \text{ for all } t.$$

This gives $kl = n\pi$ or $k = n\pi/l$. (n being an integer).

Hence $\qquad y(x,t) = \sin \dfrac{n\pi x}{l}\left(a_1 \cos \dfrac{cn\pi}{l} t + a_2 \sin \dfrac{cn\pi}{l} t \right).$

By first initial condition $y(x, 0) = 0$.

$$\therefore \qquad 0 = a_1 \sin \dfrac{n\pi x}{l} \quad \text{for all } x \quad \text{or } a_1 = 0.$$

Thus $\qquad\qquad y(x,t) = a_2 \sin \dfrac{n\pi x}{l} \sin \dfrac{cn\pi}{l} t.$

Since this is a solution for $n = 1, 2, 3, \ldots$, therefore the sum of these solutions is also a solution.

Hence $y(x,t) = \sum A_n \sin \dfrac{n\pi x}{l} \sin \dfrac{n\pi ct}{l}$ $\qquad\qquad\qquad$ ...(i)

is also a solution (superposition principle).

Now $\qquad\qquad \dfrac{\partial y}{\partial t} = \sum A_n \sin \dfrac{n\pi x}{l} \cdot \dfrac{cn\pi}{l} \cos \dfrac{cn\pi t}{l}.$

$$\therefore \quad \left(\dfrac{\partial y}{\partial t} \right)_{t=0} = \sum A_n \dfrac{cn\pi}{l} \sin \dfrac{n\pi x}{l}. \quad \text{But } \left(\dfrac{\partial y}{\partial t} \right)_{t=0} = b \sin^3 \dfrac{\pi x}{l} \text{(given)}.$$

$$\therefore \quad b \sin^3 \dfrac{\pi x}{l} = \sum A_n \dfrac{cn\pi}{l} \sin \dfrac{n\pi x}{l}. \quad \text{Now } \sin^3 \dfrac{\pi x}{l} = \dfrac{1}{4}\left(3\sin \dfrac{\pi x}{l} - \sin \dfrac{3\pi x}{l} \right)$$

$$(\because \sin 3\theta = 3\sin\theta - 4\sin^3\theta).$$

$\therefore$ $$\frac{3b}{4}\sin\frac{\pi x}{l} - \frac{b}{4}\sin\frac{3\pi x}{l} = \sum A_n \frac{cn\pi}{l}\sin\frac{n\pi x}{l}$$

$$= A_1\frac{c\pi}{l}\sin\frac{\pi x}{l} + A_2\frac{2c\pi}{l}\sin\frac{2\pi x}{l} + A_3\frac{3c\pi}{l}\sin\frac{3\pi x}{l} + ...$$

Equating cofficients, $A_1\dfrac{c\pi}{l} = \dfrac{3b}{4}$, $A_2 = 0$, $A_3\dfrac{3c\pi}{l} = -\dfrac{b}{4}$, $\qquad A_4 = A_5 = ...0$,

$\therefore$ $$A_1 = \frac{3bl}{4c\pi}, \quad A_3 = -\frac{bl}{12c\pi}.$$

Substituting in (*i*), the required solution is

$$y(x,t) = \frac{3bl}{4c\pi}\sin\frac{\pi x}{l}\sin\frac{c\pi}{l}t - \frac{bl}{12c\pi}\sin\frac{3\pi x}{l}\sin\frac{3c\pi}{l}t$$

$$= \frac{bl}{12c\pi}\left[9\sin\frac{\pi x}{l}\sin\frac{\pi ct}{l} - \sin\frac{3\pi x}{l}\sin\frac{3\pi ct}{l}\right].$$

Important Note: In the following solved examples and others that we shall encounter in this chapter, we shall be required to determine constants $b_n (n = 1, 2, 3, 4,...)$ subject to the condition

$$f(x) = \sum_{n=1}^{\infty} b_n \sin\frac{n\pi x}{l} \quad (0 < x < l).$$

We know that the series on the right is a sine series in which $f(x)$ can be expressed in the interval $(0, l)$ and is known *half-range Fourier sine series.*

Therefore b_n can be determined by the formula

$$b_n = \frac{2}{l}\int_0^l f(x)\sin\frac{n\pi x}{l}\,dx.$$

The value of the integral on the right is a function of *n*.

By putting $n = 1, 2, 3, ...$, we can get b_1, b_2, b_3 etc. Similarly, if we have

$$\phi(x) = \sum_{n=0}^{\infty} b_n \cos\frac{n\pi x}{l} \quad (0 < x < l)$$

then $\qquad b_0 = \dfrac{2}{l}\int_0^l \phi(x)\,dx$ and $\quad b_n = \dfrac{2}{l}\int_0^l \phi(x)\cos\dfrac{n\pi x}{l}\,dx.$

Example 20.5 A string is stretched and fastened to two points *l* apart. Motion is started by displacing the string into the form $y = k(lx - x^2)$ from which it is released at time $t = 0$. Find the displacement of any point on the string at a distance x from one end at time t.

Solution. The equation of the vibrating string is

$$\frac{\partial^2 y}{\partial t^2} = c^2 \frac{\partial^2 y}{\partial x^2}. \qquad \qquad ...(i)$$

Since the end points are fixed, $y = 0$ when $x = 0$ and $x = l$ for all t.

$\therefore$ $y(0, t) = 0$ and $y(l \cdot t) = 0$ are the *boundary conditions.* ...(ii)

Also at $t = 0$, the string is at rest in the shape $y = k(lx - x^2)$.

$\therefore$ *Initial conditions* are

$$\left(\frac{\partial y}{\partial t}\right)_{t=0} = 0 \text{ and } y(x,0) = k\left(lx - x^2\right). \qquad \qquad ...(iii)$$

Now we have to solve (i) subject to the conditions (ii) and (iii). Proceeding as

in example 20.4 and using the conditions (ii) and also $\left(\dfrac{\partial y}{\partial t}\right)_{t=0} = 0$, we get the

solution in the form

$$y(x, t) = \sum_{n=1}^{\infty} A_n \sin\frac{n\pi x}{l}\cos\frac{c n\pi t}{l}. \qquad \qquad ...(iv)$$

Since $y(x, 0) = k(lx - x^2)$, we have

$$k\left(lx - x^2\right) = \sum_{n=1}^{\infty} A_n \sin\frac{n\pi x}{l}. \qquad \qquad ...(v)$$

Now (v) is the half-range sine series representing $k(lx - x^2)$ in the interval (o, l).

$$\therefore \quad A_n = \frac{2}{l}\int_0^l k\left(lx - x^2\right)\sin\frac{n\pi x}{l}\,dx$$

$$= \frac{2k}{l}\left[\left(lx - x^2\right)\left(-\frac{l}{n\pi}\cos\frac{n\pi x}{l}\right)\right.$$

$$\left. -(l - 2x)\left(-\frac{l^2}{n^2\pi^2}\sin\frac{n\pi x}{l}\right) + (-2)\left(\frac{l^3}{n^3\pi^3}\cos\frac{n\pi x}{l}\right)\right]_0^l$$

$$= \begin{cases} \dfrac{4kl^2}{n^3\pi^3}(1 - \cos n\pi) = 0, \text{ when } n \text{ is even} \\[4mm] \dfrac{8kl^2}{n^3\pi^3}, \text{ when } n \text{ is odd}. \end{cases}$$

Putting $n = 2m - 1$, the solution (iv) can be written in the form

$$y = \frac{8kl^2}{\pi^3}\sum_{m=1}^{\infty}\left[\frac{1}{(2m-1)^3}\sin(2m-1)\frac{\pi x}{l}\cos(2m-1)\frac{c\pi t}{l}\right].$$

Example 20.6 A tightly stretched string with fixed end points $x = 0$ and $x = l$ is initially at rest in its equilibrium position. If it is set vibrating by giving to each of its points a velocity $\lambda x(l - x)$, find the displacement of the string at any distance x from one end at any time t. (*A.M.I.E., S-1998*)

Solution. We have to solve the equation

$$\frac{\partial^2 y}{\partial t^2} = c^2 \frac{\partial^2 y}{\partial x^2}$$

subject to the *boundary conditions* $y(0, t) = 0$, $y(l, t) = 0$ and *initial conditions*

$$y = 0 \text{ and } \frac{\partial y}{\partial t} = \lambda x(l - x) \text{ when } t = 0.$$

Proceeding as in example 20.3, the solution satisfying the boundary conditions is

$$y = \sin \frac{n\pi x}{l} \left(b_1 \cos \frac{cn\pi}{l} t + b_2 \sin \frac{cn\pi}{l} t \right).$$

Since $y = 0$ when $t = 0$, we have $b_1 = 0$.

Hence $\qquad y = b_n \sin \frac{n\pi x}{l} \sin cn\pi \frac{t}{l} \left(\text{replacing } b_2 \text{ by } b_n \right).$

Adding up the solutions for different values of n.

$$y = \sum_{n=1}^{\infty} b_n \sin \frac{n\pi x}{l} \sin \frac{cn\pi t}{l}. \qquad \ldots(i)$$

$$\therefore \qquad \frac{\partial y}{\partial t} = \sum_{n=1}^{\infty} \frac{c\pi}{l} n b_n \sin \frac{n\pi x}{l} \cos \frac{cn\pi t}{l}.$$

As $\frac{\partial y}{\partial t} = \lambda x(l - x)$ when $t = 0$, $\therefore \lambda x(l - x) = \frac{c\pi}{l} \sum n b_n \sin \frac{n\pi x}{l}. \qquad \ldots(ii)$

Since (*ii*) is half-range sine-series, we have

$$\frac{c\pi}{l} n b_n = \frac{2}{l} \int_0^l \lambda x(l - x) \sin \frac{n\pi x}{l} dx = \frac{4\lambda l^2}{n^3 \pi^3}(1 - \cos n\pi) = \begin{cases} 0 \text{ when n is even} \\ \frac{8\lambda l^2}{n^3 \pi^3} \text{ when } n \text{ is odd} \end{cases}$$

or $\qquad b_n = \frac{8\lambda l^3}{c\pi^4 n^4}$ when n is odd.

Putting the value of b_n in (*i*) we get the solution

$$y = \frac{8\lambda l^3}{c\pi^4} \sum_{n=1,3,5,\ldots}^{\infty} \left[\frac{1}{n^4} \sin \frac{n\pi x}{l} \sin \frac{n\pi ct}{l} \right].$$

Exercise 20.1

Using the method of separation of variables find the solution of the equations 1 to 3

1. $u_x + u_y = 2(x + y)u.$ **Ans.** $u = ce^{x^2 + y^2 + kx - ky}.$

2. $2x\dfrac{\partial z}{\partial x} - 3y\dfrac{\partial z}{\partial y} = 0$ **Ans.** $z = Ax^{h/2}\,y^{h/3}.$

3. $x^2 u_{xy} + 3y^2 u = 0$ **Ans.** $u = Ae^{-(k/x + y^3/k)}.$

Solve each of the following boundary-value problems 4 to 9 by the method of separation of variables.

4. $\dfrac{\partial u}{\partial x} = 4\dfrac{\partial u}{\partial y}, \quad u(0, y) = 8e^{-3y}.$ **Ans.** $u(x, y) = 8e^{-12x - 3y}.$

5. $\dfrac{\partial u}{\partial x} = 4\dfrac{\partial u}{\partial y}, \quad u(0, y) = 8e^{-3y} + 4e^{-5y}.$ **Ans.** $u(x, y) = 8e^{-12x-3y} + 4e^{-20x-5y}.$

6. $3\dfrac{\partial u}{\partial x} + 2\dfrac{\partial u}{\partial y} = 0,\ u(x, 0) = 4e^{-x}.$ **Ans.** $u(x, y) = 4e^{(3y-2x)/2}.$

7. $\dfrac{\partial^2 u}{\partial x^2} = \dfrac{\partial u}{\partial y} + 2u,$ given that $u = 0$ and $\dfrac{\partial u}{\partial x} = 1 + e^{-3y},$ when $x = 0$ for all

values of y. (*Andhra 2000*) **Ans.** $u = \dfrac{1}{\sqrt{2}}\sinh\sqrt{2}x + e^{-3y}\sin x.$

8. $\dfrac{\partial u}{\partial x} - 2\dfrac{\partial u}{\partial y} = u,$ given that $u(x, 0) = 3e^{-5x} + 2e^{-3x}.$ (*I.E.T.E., June 2001*)

Ans. $u(x, y) = 3e^{-5x - 3y} + 2e^{-3x - 2y}.$

9. $4\dfrac{\partial u}{\partial x} + \dfrac{\partial u}{\partial y} = 3u;$ given that $u = 3e^{-y} - e^{-5y}$ at $x = 0.$ (*I.E.T.E., Dec. 2002*)

Ans. $u(x, y) = 3\,e^{x-y} - e^{2x - 5y}.$

10. A tightly stretched string with fixed end points $x = 0$ and $x = l$ is initially in a position given by

$$y = y_0 \sin^3\left(\frac{\pi x}{l}\right).$$

If it is released from rest from this position, find the displacement $y(x, t).$

(*Andhra 2000*) **Ans.** $y(x, t) = \dfrac{y_0}{4}\left[3\sin\dfrac{\pi x}{l}\cos\dfrac{\pi ct}{l} - \sin\dfrac{3\pi x}{l}\cos\dfrac{3\pi ct}{l}\right].$

11. Solve the boundary value problem

$$\frac{\partial^2 y}{\partial t^2} = 4\frac{\partial^2 y}{\partial x^2}, y(0, t) = y(5, t) = 0, \ y(x, 0) = 0, \left(\frac{\partial y}{\partial t}\right)_{t=0} = f(x).$$

If (a) $f(x) = 5 \sin \pi x$ (b) $f(x) = 3 \sin 2\pi x - 2 \sin 5\pi x$ (*A.M.I.E., S-2004*)

Ans. (a) $y(x, t) = (5 \sin \pi x \sin 2\pi t)/2\pi$.

(b) $y (x, t) = (15 \sin 2\pi x \sin 4 \ \pi t - 4 \sin 5\pi x \sin 10\pi t)/20\pi$.

12. Show how the wave equation

$$\frac{\partial^2 y}{\partial t^2} = c^2 \frac{\partial^2 y}{\partial x^2}$$

can be solved by the method of separation of variables. If the initial displacement and velocity of a string stretched between the points $x = 0$ and

$x = l$ are given by $y = f(x)$ and $\dfrac{\partial y}{\partial t} = g(x)$,

determine the displacement of any point x of the string at any time $t > 0$.

Ans. $y = \displaystyle\sum_{n=1}^{\infty} \sin\frac{n\pi x}{l}\left(A_n \cos\frac{cn\pi t}{l} + B_n \sin\frac{cn\pi t}{l}\right),$

where $A_n = \dfrac{2}{l}\displaystyle\int_0^l f(x)\sin\frac{n\pi x}{l}dx$ and $B_n = \dfrac{2}{n\pi c}\displaystyle\int_0^l g(x)\sin\frac{n\pi x}{l}dx.$

13. A tightly stretched string with fixed end points $x = 0$ and $x = L$ is initially at rest in its equilibrium position. If it is set vibrating by giving to each of its points a velocity.

$$\left(\frac{\partial y}{\partial t}\right)_{t=0} = 3\left(Lx - x^2\right), \text{ find } y(x, t).$$

Ans. $y(x, t) = \dfrac{24L^3}{c\pi^4}\displaystyle\sum_{m=1}^{\infty}\left[\frac{1}{(2m-1)^4}\sin(2m-1)\frac{\pi x}{L} \times \sin(2m-1)\frac{c\pi t}{L}\right].$

14. A tightly stretched string of length l is fastened at both ends. The mid- point of the string is taken to a height b and then released from rest in that position. Show that the displacement at any time t is given by

$$y = \frac{8b}{\pi^2}\left[\frac{1}{1^2}\sin\frac{\pi x}{l}\cos\frac{\pi ct}{l} - \frac{1}{3^2}\sin\frac{3\pi x}{l}\cos\frac{3\pi ct}{l} + ...\right].$$

15. A string of length l is fastened at both ends. One end is taken at the origin and at a distance b from this end, the string is displaced a distance d

transversely and is released from rest from this position. Find the equation of subsequent motion.

$$\text{Ans. } y(x,t) = \frac{2dl^2}{b(l-b)\pi^2} \sum_{n=1}^{\infty} \frac{1}{n^2} \sin\frac{n\pi b}{l} \sin\frac{n\pi x}{l} \cos\frac{n\pi ct}{l}.$$

20.7 One-Dimensional Heat flow: $\dfrac{\partial u}{\partial t} = c^2 \dfrac{\partial^2 u}{\partial x^2}.$

Consider a homogeneous bar of uniform cross section with its sides insulated so that heat flows in the direction perpendicular to the cross-section of area A(shown shaded). We take the origin at one end O of the bar and X-axis along the direction of heat flow.

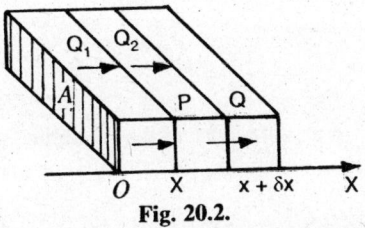

Fig. 20.2.

The temperature u at any point of the bar depends on the distance x from one end and time t. Moreover the temperature of all points of any cross-section is the same.

Experimentally we know that the rate of flow of heat across any area A is proportional to the area and the temperature gradient *i.e.*, rate of change of temperature w.r.t. distance normal to the area.

Hence Q_1, the quantity of heat flowing across the section at a distance x is given by

$$Q_1 = -KA\left(\frac{\partial u}{\partial x}\right)_x \quad \text{cal per sec,}$$

where K is coefficient of conductivity, (also known as thermal conductivity), A the area of cross section; the negative sign on the right is attached because as x increases, u decreases.

Similarly Q_2, the quantity of heat flowing across the section at a distance $x + \delta x$ is

$$Q_2 = -KA\left(\frac{\partial u}{\partial x}\right)_{x+\delta x} \quad \text{cal per sec.}$$

Hence the amount of heat gained per second by the slab with thickness δx is

$$Q_1 - Q_2 = KA\left[\left(\frac{\partial u}{\partial x}\right)_{x+\delta x} - \left(\frac{\partial u}{\partial x}\right)_x\right]. \qquad \qquad ...(20.6)$$

But the rate of increase of heat in the slab $= sp A \delta x \dfrac{\partial u}{\partial t}$ ...(20.7)

where s is the specific heat and ρ, the density of the material.

$\therefore$ From (20.6) and (20.7)

$$sp A \delta x \frac{\partial u}{\partial t} = KA \left[\left(\frac{\partial u}{\partial x} \right)_{x+\delta x} - \left(\frac{\partial u}{\partial x} \right)_x \right]$$

or

$$sp \frac{\partial u}{\partial t} = K \cdot \frac{1}{\delta x} \left[\left(\frac{\partial u}{\partial x} \right)_{x+\delta x} - \left(\frac{\partial u}{\partial x} \right)_x \right].$$

As

$$\delta x \to 0 \qquad sp \frac{\partial u}{\partial t} = K \cdot \frac{\partial^2 u}{\partial x^2}$$

or

$$\frac{\partial u}{\partial t} = c^2 \frac{\partial^2 u}{\partial x^2}, \qquad \qquad ...(20.8)$$

where $c^2 = k/sp$, is known as *diffusivity* of the material.

We shall solve this equation by the method of separation of variables.

20.8 Illustrative Examples

Example 20.7 (a) Obtain the solution of heat flow equation $\dfrac{\partial u}{\partial t} = c^2 \dfrac{\partial^2 u}{\partial x^2}$ by the method of separation of variables.

(b) A rod of length l with insulated sides is initially at a uniform temperature u_0. Its ends are suddenly cooled to $0°C$ and are kept at that temperature. Find the temperature function $u(x, t)$.

Solution. (a) Let $u(x, t) = X(x)\, T(t)$ be the solution of the equation

$$\frac{\partial u}{\partial t} = c^2 \frac{\partial^2 u}{\partial x^2}. \qquad \qquad ...(1)$$

Substituting $u = XT$ in this we get $XT' = c^2 X''T$

or

$$\frac{X''}{X} = \frac{1}{c^2} \frac{T'}{T} = h\,(\text{say}), \quad \text{where } h \text{ is a constant. This gives}$$

$$X'' - hX = 0 \qquad \qquad ...(2)$$

and

$$T' - c^2 hT = 0. \qquad \qquad ...(3)$$

Three cases arise according as h is positive, zero or negative.

Case I $(h > 0)$: Let $h = k^2$.

Then (2) and (3) become

$$X'' - k^2 X = 0 \quad \text{and} \quad T' - c^2 k^2 T = 0.$$

Solving these differential equations, we get

$$X = c_1 e^{kx} + c_2 e^{-kx} \quad \text{and} \quad T = c_3 e^{c^2 k^2 t}. \qquad \text{...(4)}$$

Case II. $h = 0$.

Then (2) and (3) become $X'' = 0$ and $T' = 0$

Solving these differential equations, we get

$$X = c_4 x + c_5 \quad \text{and} \quad T = c_6. \qquad \text{...(5)}$$

Case III. $(h < 0)$: Let $h = -k^2$.

Then (2) and (3) reduce to

$$X'' + k^2 X = 0 \quad \text{and} \quad T' + c^2 k^2 T = 0.$$

Solving these differential equations, we get

$$X = (c_7 \cos kx + c_8 \sin kx) \quad \text{and} \quad T = c_9 e^{-c^2 k^2 t}. \qquad \text{...(6)}$$

Thus depending upon the values of h, we have from (4), (5) and (6), the solution of (1) in the form

$$u(x, t) = \left(b_1 e^{kx} + b_2 e^{-kx}\right) e^{c^2 k^2 t} \qquad \text{...(7)}$$

$$u(x, t) = b_3 x + b_4 \qquad \text{...(8)}$$

and

$$u(x, t) = \left(b_5 \cos kx + b_6 \sin kx\right) e^{-c^2 k^2 t}. \qquad \text{...(9)}$$

Solution (7) will have to be rejected as $u(x, t)$ will become infinitely large for large values of t and this will violate the physical requirement of the problem.

Solution (8) is a trivial solution as it gives a non-zero temperature for all times. Hence solution (9) can justificably be taken as a correct viable solution of the heat conduction equation (1). However, in certain situations in heat flow problems, the combination of (8) and (9) may give a desired solution, as will be seen later in solved example.

(*b*) Let OA be the rod with length *l*. We take 0 as the origin and OA as X-axis. The temperature $u(x, t)$ at any point satisfies the differential equation

$$\frac{\partial u}{\partial t} = c^2 \frac{\partial^2 u}{\partial x^2}.$$

By part (a) the solution of this differential equation is of the form

$$u(x, t) = (A \cos kx + B \sin kx) e^{-c^2 k^2 t}. \qquad \text{...(1)}$$

Since the ends are cooled and at 0° C and kept at that temperature throughout, the *boundary conditions* are

$$u(0, t) = 0, \quad u(l, t) = 0. \qquad \text{...(2)}$$

The rod is initially at temperature u_o, therefore the *initial condition* is

$$u(x, 0) = u_0. \qquad \text{...(3)}$$

Now we have to find the constants in (1) subject to the boundary conditions (2) and initial condition (3).

Since $u(0, t) = 0$ (first boundary condition), we have from (1)

$$0 = Ae^{-c^2k^2t}, \text{ or } A = 0 \cdot$$

Thus (1) becomes $\qquad u(x, t) = B \sin kx \, e^{-c^2k^2t}$. $\qquad$ ...(4)

Also $u(l, t) = 0$ (second boundary condition), therefore, from (4)

$0 = B \sin kl \, e^{-c^2k^2t}$. Since $B \neq 0$,

$$\therefore \qquad\qquad kl = n\pi \text{ or } k = \frac{n\pi}{l} \text{ (n being an integer)}.$$

The solution (4) now reduces to

$$u(x, t) = B \sin \frac{n\pi x}{l} e^{-c^2 n^2 \pi^2 t/l^2}.$$

Since this is a solution holding for all values of n, *i.e.*, $n = 1, 2, 3, \ldots$, the sum of these is also a solution (superposition principle). Hence

$$u(x, t) = \sum_{n=1}^{\infty} b_n \sin \frac{n\pi x}{l} e^{-c^2 n^2 \pi^2 t/l^2}, \text{ is the general solution.}$$

Now by (3) $\quad u(x, 0) = u_o$. $\qquad$ ...(5)

$$\therefore \qquad\qquad u_0 = \sum_{n=1}^{\infty} b_n \sin \frac{n\pi x}{l}. \qquad\qquad ...(6)$$

Since this is half-range sine series for u_o,

$$b_n = \frac{2}{l} \int_0^l u_0 \sin \frac{n\pi x}{l} dx = \begin{cases} 0, & \text{when n is even} \\ 4u_0/n\pi, & \text{when n is odd} \end{cases}$$

Hence $\qquad u(x,t) = \frac{4u_0}{\pi} \sum_{n=1,3,5}^{\infty} \frac{1}{n} \sin \frac{n\pi x}{l} \exp \left\{ \frac{-c^2 n^2 \pi^2 t}{l^2} \right\}.$

Example 20.8 Solve $\dfrac{\partial^2 u}{\partial x^2} = h^2 \dfrac{\partial u}{\partial t}$ when $u(0, t) = u(l, t) = 0$ and

$u(x, 0) = \sin\left(\dfrac{\pi x}{l}\right)$, by the method of separation of variables.

Solution. Proceeding as in Example 20.7, the solution satisfying the condition $u(0, t) = 0$ and $u(l, t) = 0$ is

$$u(x,t) = \sum_{n=1}^{\infty} b_n \sin\frac{n\pi x}{l} e^{-n^2\pi^2 t/h^2 l^2} \quad \left(\because \text{ here } c^2 = \frac{1}{h^2} \right).$$

Since $u(x,0) = \sin\dfrac{\pi x}{l}$, we get $\sin\dfrac{\pi x}{l} = \sum_{n=1}^{\infty} b_n \sin\dfrac{n\pi x}{l}$.

Comparing coefficients of different terms on both sides,
$b_1 = 1, b_2 = b_3 = \ldots = 0.$

Hence $u(x,t) = \sin\dfrac{\pi x}{l} e^{-\pi^2 t/h^2 l^2}$ is the required solution.

Example 20.9 A rod of length *l* has its ends A and B kept at 0°C and 100°C respectively, until, steady state conditions prevail. If the temperature at B is reduced suddenly to 0°C and kept so, while that of A is maintained, find the temperature u(x, t) at a distance x from A and at time t. *(IETE-June 2002)*

Solution. Before the temperature at B is reduced to 0°C, the temperature at any point of the rod did not change with time (steady state conditions prevailed). When it depends on x only, the equation of heat conduction, that is,

$$\frac{\partial u}{\partial t} = c^2 \frac{\partial^2 u}{\partial x^2} \qquad \ldots(1)$$

reduces to $\dfrac{d^2 u}{dx^2} = 0$ whose general solution is $u = c_1 x + c_2.$ $\qquad \ldots(2)$

At that time we had
$$u = 0 \text{ when } x = 0 \text{ and } u = 100 \text{ when } x = l.$$

Putting these values in (2) we get $c_2 = 0$ and $c_1 = \dfrac{100}{l}.$

$$\therefore \qquad u = \frac{100x}{l} \text{ when } t = 0, \text{ that is, } u(x,0) = \frac{100x}{l}.$$

Thus we have to solve (1) subject to the *boundary conditions*
$$u(0, t) = 0, \quad u(l, t) = 0$$

and the *initial condition* $u(x,0) = \dfrac{100x}{l}.$

Proceeding as in example 20.7 the solution satisfying the boundary conditions is

$$u(x,t) = \sum b_n \sin\frac{n\pi x}{l} e^{-\frac{n^2\pi^2 c^2 t}{l^2}}.$$

Since $u(x,0) = \dfrac{100x}{l}$ we get $\dfrac{100x}{l} = \sum_{n=1}^{\infty} b_n \sin\dfrac{n\pi x}{l},$

where $b_n = \dfrac{2}{l} \int_0^l \dfrac{100x}{l} \sin \dfrac{n\pi x}{l} dx = -\dfrac{200}{n\pi} \cos n\pi = \dfrac{200}{n\pi}(-1)^{n+1}$.

Hence $\qquad u(x, t) = \dfrac{200}{\pi} \sum_{n=1}^{\infty} \dfrac{(-1)^{n+1}}{n} \sin \dfrac{n\pi x}{l} e^{-n^2\pi^2 c^2 t/l^2}$.

Rod with non-zero end temperatures

In the following example we shall consider the case when the ends of the bar are maintained at non-zero temperatures, that is, $u(0, t)$ and $u(l, t) \neq 0$.

Example 20.10 **The ends A and B of a rod 20 cm long have their temperatures at 30°C and 80°C until steady state conditions prevails. The temperatures of the ends are changed to 40°C and 60°C respectively. Find the temperature distribution in the rod at any time t.** *(I.E.T.E., Dec. 2003; Kerala, 1995)*

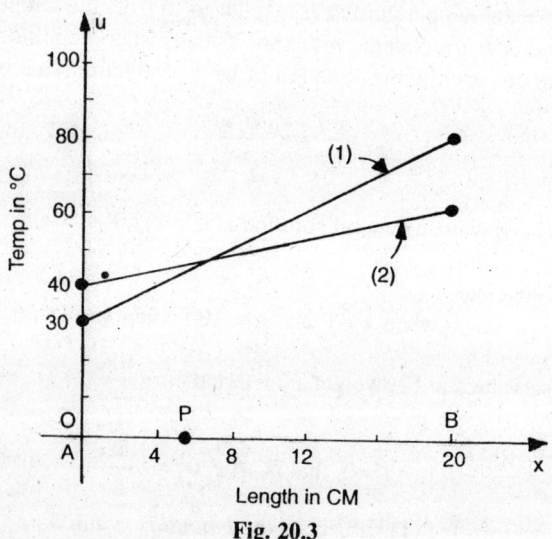

Fig. 20.3

Solution. When the steady-state conditions prevail, the temperature at any point P of the rod depends only on its distance from the point A and is independent of the time. The heat equation becomes

$$\frac{d^2u}{dx^2} = 0. \qquad \qquad ...(1)$$

The solution of (1) is given by $u = ax + b$. Using the conditions at the end points that is, $u = 30°C$ at A and $u = 80°C$ at B, the values of a and b are given by $b = 30$ and $a = 5/2$. Hence when the steady-state conditions prevail, the temperature distribution is given by

$$u = 30 + (5x/2). \qquad \text{...(2)}$$

The temperature distribution given in (2) acts as the initial temperature for the next stage of the problem. When the temperatures at A and B are suddenly changed; let the temperature at any point of the rod at a distance x from the end A be $u(x, t)$. Then $u(x, t)$ satisfies the differential equation

$$\frac{\partial u}{\partial t} = c^2 \frac{\partial^2 u}{\partial x^2} \qquad \text{...(3)}$$

subject to the *boundary* and *initial conditions* as

$$u(0, t) = 40 \qquad \text{...(4)}$$
$$u(20, t) = 60 \qquad \text{...(5)}$$
$$u(x, 0) = 30 + (5x/2). \qquad \text{...(6)}$$

Now since the temperatures at the ends of the rod are not zero as is borne out from (4) and (5), we express $u(x, t)$ as

$$u(x, t) = v(x, t) + g(x). \qquad \text{...(7)}$$

where $v(x, t)$ denotes the transient part of the solution that is, a function of both x and t and $g(x)$ denotes the steady state part of the solution which is a function of x alone and is independent of t.

Substituting from above in (3), we get

$$\frac{\partial v}{\partial t} = c^2 \frac{\partial^2 v}{\partial x^2} + c^2 g''(x)$$

Now choose $g(x)$ such that

(i) $g''(x) = 0$ (ii) $v(0, t) = 0$ (iii) $v(20, t) = 0$

From (i) on integration, we get

$$g(x) = ax + b \qquad \text{(From 7)}$$

and
$$u(0, t) = v(0, t) + g(0)$$

or
$$40 = 0 + b \therefore b = 40.$$

$$u(20, t) = v(20, t) + g(20)$$

or
$$60 = 0 + g(20) = 20a + 40. \quad \therefore a = 1$$

This gives us
$$g(x) = 40 + x.$$

Hence
$$u(x, t) = v(x, t) + 40 + x \qquad \text{...(8)}$$

and $v(x, t)$ satisfies the differential equation

$$\frac{\partial v}{\partial t} = c^2 \frac{\partial^2 v}{\partial x^2}. \qquad \text{...(9)}$$

With the *boundary conditions*

$$v(0, t) = 0 \qquad \text{...(10)}$$
$$v(20, t) = 0 \qquad \text{...(11)}$$

and initial condition
$$v(x, 0) = u(x, 0) - 40 - x$$
$$= 30 + (5x/2) - 40 - x = (3x/2) - 10. \qquad \text{...(12)}$$

The solution of (9) subject to boundary conditions (10) and (11) is given by

$$v(x,t) = \sum_{n=1}^{\infty} B_n \sin \frac{n\pi x}{20} e^{-\left(c^2 \pi^2 n^2 t/400\right)}.$$ [Refer example 20.7 (b)]

Using the initial condition (12), we get

$$\frac{3}{2}x - 10 = \sum_{n=1}^{\infty} B_n \sin \frac{n\pi x}{20}.$$ ...(13)

(13) is a Fourier half range sine series where

$$B_n = \frac{2}{20} \int_0^{20} \left(\frac{3}{2}x - 10\right) \sin \frac{n\pi x}{20} dx.$$

$$= \frac{1}{10}\left[\frac{-20}{n\pi}(10 + 20\cos n\pi)\right] = \frac{-20}{\pi}\left(\frac{1 + 2\cos n\pi}{n}\right).$$

∴ The complete solution $u(x, t) = v(x, t) + 40 + x$ is given by

$$u(x, t) = 40 + x - \frac{20}{\pi}\sum_{n=1}^{\infty}\left(\frac{1 + 2\cos n\pi}{n}\right)\sin \frac{n\pi x}{20} e^{-\left(c^2 \pi^2 n^2 t/400\right)}.$$

Example 20.11 An insulated metal rod of length *l* has one end A kept at 0°C and the other end B at 100°C until steady state is attained. At t = 0, the temperature at A is then suddenly raised to 50°C and thereafter maintained while at the same time *t* = 0, the end B is insulated. Find the temperature at any point of the rod at any time t.

Solution. When the steady state temperature is reached, the temperature at any point of the rod depends upon its distance from the point A. The heat equation

reduces to $\dfrac{d^2 u}{dx^2} = 0$. Its solution is $u = ax + b$. With the *boundary conditions*

$u = 0$ at A and $u = 100$ at B, the values of a and b are 100/*l* and 0 respectively. The initial temperature distribution defining the *initial condition* is given by

$$u(x, 0) = \frac{100x}{l}.$$ ...(1)

Since one of the boundary conditions subsequent to temperature raise is not zero, we consider the solution $u(x, t)$ as the sum of the steady state solution and the transient solution that is,

$$u(x, t) = v(x, t) + u_s(x)$$ ...(2)

The equation of heat conduction together with the boundary and initial conditions is given by

$$\frac{\partial u}{\partial t} = c^2 \frac{\partial^2 u}{\partial x^2}$$ ...(3)

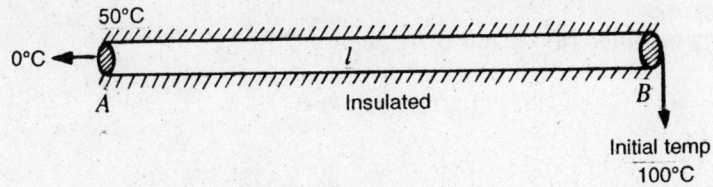

Fig.20.4

$$u(0, t) = 50 \qquad \ldots(4)$$

$$\frac{\partial u(l, t)}{\partial x} = 0 \qquad \ldots(5)$$

$$u(x, 0) = \frac{100x}{l} \qquad \ldots(6)$$

Equations (2) and (3) together give

$$\frac{\partial v}{\partial t} = c^2 \left(\frac{\partial^2 v}{\partial x^2} + \frac{d^2 u_0(x)}{dx^2} \right). \qquad \ldots(7)$$

Now choosing $u_s(x)$ such that *(i)* $\dfrac{d^2 u_s(x)}{dx^2} = 0$

(ii) $v(0, t) = 0$ *(iii)* $\dfrac{\partial v}{\partial x}(l, t) = 0$.

Equation *(i)* on integration gives $u_s(x) = ax + b$. The constants a and b are

evaluated from the boundary conditions *i.e.* $u_s(0) = 50$; $\dfrac{du_s(x)}{dx} = 0$ at $x = l$. This

gives $a = 0$, $b = 50$. Thus the steady-state part $u_s(x) = 50$.

The transient part $v(x, t)$ satisfies the differential equation

$$\frac{\partial v}{\partial t} = c^2 \frac{\partial^2 v}{\partial x^2} \qquad \ldots(8)$$

and the *initial* and *boundary* conditions given by

(i) $v(x, 0) = u(x, 0) - u_s(x) = \dfrac{100x}{l} - 50$.

(ii) $v(0, t) = 0$ and *(iii)* $\dfrac{\partial v(l, t)}{\partial x} = 0$

The solution of (8) is given by

$$v(x, t) = (A\cos kx + B\sin kx)e^{-c^2k^2t}.$$

Using condition (ii) i.e., $x = 0$, we get $A = 0$. Hence

$$v(x, t) = B\sin kx\, e^{-c^2k^2t}.$$

Using condition (iii) i.e., $\left.\dfrac{\partial v}{\partial x}\right|_{x=l} = 0 = Bk\cos kl\, e^{-c^2k^2t}$, we get

$$\cos kl = 0 \;(B \text{ connot be zero as } A \text{ is already zero})$$

This gives $\quad kl = \dfrac{(2n-1)\pi}{2}$ or $k = \dfrac{(2n-1)\pi}{2l},\; n = 1, 2, 3, \dots$

$$\therefore\quad v(x, t) = \sum_{n=1}^{\infty} B_n \sin\frac{(2n-1)\pi x}{2l}\, e^{-c^2\frac{(2n-1)^2\pi^2 t}{4l^2}}. \qquad \dots(9)$$

To evaluate B_n, we put $t = 0$, This gives

$$\frac{100x}{l} - 50 = \sum_{n=1}^{\infty} B_n \sin\frac{(2n-1)\pi x}{2l}. \qquad \dots(10)$$

This equation is a Fourier half range sine series in $(0, l)$. B_n is given by

$$B_n = \frac{2}{l}\int_0^l \left(\frac{100x}{l} - 50\right)\sin\frac{(2n-1)\pi x}{2l}\, dx$$

$$= \frac{2}{l}\left[\left(\frac{100x}{l} - 50\right)\left(-\cos\frac{(2n-1)\pi x}{2l} \times \frac{2l}{(2n-1)\pi}\right)\right.$$

$$\left.-\frac{100}{l}\left(-\sin\frac{(2n-1)\pi x}{2l} \times \frac{4l^2}{(2n-1)^2\pi^2}\right)\right]_0^l$$

$$= \frac{2}{l}\left[\frac{400l}{(2n-1)^2\pi^2}\sin\frac{(2n-1)\pi}{2} - \frac{100l}{(2n-1)\pi}\right].$$

Substituting for B_n is (9), we get

$$v(x, t) = \sum_{n=1}^{\infty}\left(\frac{800(-1)^n}{(2n-1)^2\pi^2} - \frac{200}{(2n-1)\pi}\right)\sin\frac{(2n-1)\pi x}{2l}\, e^{\frac{-c^2(2n-1)^2\pi^2 t}{4l^2}} \qquad \dots(11)$$

Combining (2) and (11), the complete-solution of the problem is given by

$$u(x, t) = 50 + \sum_{n=1}^{\infty} \left(\frac{(-1)^n 800}{(2n-1)^2 \pi^2} - \frac{200}{(2n-1)\pi} \right) \sin \frac{(2n-1)\pi x}{2l} e^{\frac{-c^2(2n-1)^2 \pi^2 t}{4l^2}}.$$

Example 20.12 A bar with insulated sides is initially at temperature 0°C throughout. The end $x = 0$ is kept at 0°C for all time and the heat is suddenly applied so that $\dfrac{\partial u}{\partial x} = 10$ at $x = l$ for all time. Find the temperature function $u(x, t)$.

Solution. The differential equation governing the flow of heat in the bar is given by

$$\frac{\partial u}{\partial t} = c^2 \frac{\partial^2 u}{\partial x^2} \qquad \qquad \text{...(1)}$$

subject to the *boundary* and *initial conditions* presscribed as

$$u(0, t) = 0 \qquad \qquad \text{...(2)}$$

$$\frac{\partial u}{\partial x}(l, t) = 10 \qquad \qquad \text{...(3)}$$

$$u(x, 0) = 0 \qquad \qquad \text{...(4)}$$

Since one of the conditions prescribed on boundary is other than zero, we express the solution as the combination of the steady-state solution and the transient solution, the later ultimately vanishing with large time. Therefore the solution of (1) is given by

$$u(x, t) = v(x, t) + u_s(x). \qquad \qquad \text{...(5)}$$

As in the previous problem, the steady-state solution $u_s(x)$ is given as
$$u_s(x) = ax + b.$$
Using boundary conditions (2) and (3), $b = 0$, $a = 10$. $\therefore$ $u_s(x) = 10x$ ...(6)
The transient solution $v(x, t)$ is obtained by solving the differential equation.

$$\frac{\partial v}{\partial t} = c^2 \cdot \frac{\partial^2 v}{\partial x^2} \qquad \qquad \text{...(7)}$$

with the *boundary conditions*

$$v(0, t) = 0 \qquad \qquad \text{...(8)}$$

$$\frac{\partial v}{\partial t}(0, t) = 0 \qquad \qquad \text{...(9)}$$

$$v(x, 0) = 0. \qquad \qquad \text{...(10)}$$

The solution of (7) is given by

$$v = (A \cos kx + B \sin kx) e^{-c^2 k^2 t} \quad \text{(Refer example 20.7(b)).}$$

Using conditions (8) and (9), $A = 0$, $k = \dfrac{(2n-1)\pi}{2l}$ (refer previous example)

$$\therefore \quad v = \sum_{n=1}^{\infty} B_n \sin \frac{(2n-1)\pi x}{2l} e^{\frac{-c^2(2n-1)^2\pi^2 t}{4l^2}} \qquad\qquad \text{...(11)}$$

The complete solution is

$$u = 10x + \sum_{n=1}^{\infty} B_n \sin \frac{(2n-1)\pi x}{2l} e^{\frac{-c^2(2n-1)^2\pi^2 t}{4l^2}}.$$

Now $u(x, 0) = v(x, 0) = 0$. Therefore, B_n, using half range sine series, is given by

$$B_n = \frac{2}{l} \int_0^l -10x \sin \frac{(2n-1)\pi x}{2l} dx = \frac{-80l}{(2n-1)^2 \pi^2} \sin \frac{(2n-1)\pi}{2}.$$

$\therefore$ The complete solution $u(x, t)$ is given by

$$u(x, t) = 10x - \frac{80l}{\pi^2} \sum_{n=1}^{\infty} \frac{(-1)^n}{(2n-1)^2} \sin \frac{(2n-1)\pi x}{2l} \cdot e^{-\frac{c^2(2n-1)^2\pi^2 t}{4l^2}}.$$

Exercise 20.2

1. Using the method of separation of variables show that one set of solution of
$$\frac{\partial u}{\partial t} = c^2 \frac{\partial^2 u}{\partial x^2} \text{ is } u = \left(A\cos px + B\sin px\right)e^{-c^2 p^2 t}, \text{ where } A, B, p \text{ are constants.}$$

2. Solve the equation
$$\frac{\partial u}{\partial t} = 4 \frac{\partial^2 u}{\partial x^2},$$
$u(0, t) = 0$, $u(\pi, t) = 0$, $u(x, 0) = 2 \sin 3x - 4 \sin 5x$ by the method of separation of variables. **Ans.** $u = 2e^{-36t} \sin 3x - 4e^{-100t} \sin 5x$.

3. Find by the method of separation of variables the solution $u(x, t)$ of the boundary value problem
$$\frac{\partial u}{\partial t} = 3 \frac{\partial^2 u}{\partial x^2}, t > 0, 0 < x < 2.$$
$u(0, t) = 0$, $u(2, t) = 0$, $t > 0 \qquad u(x, 0) = x$, $0 < x < 2$.

$$\textbf{Ans. } u(x, t) = -\frac{4}{\pi} \sum_{n=1}^{\infty} \frac{(-1)^n}{n} \sin \frac{n\pi x}{2} e^{-3k^2 t}.$$

4. Solve the boundary value problem

$$\frac{\partial u}{\partial t} = \frac{\partial^2 u}{\partial x^2}, \left(\frac{\partial u}{\partial x}\right)_{x=0} = 0, \ u(2, t) = 0, \ u(x, 0) = 8\cos\frac{3\pi x}{4} - 6\cos\frac{9\pi x}{4}$$

by the method of separation of variables.

Ans. $u = 8 \exp(-9\pi^2 t/16) \cos\dfrac{3\pi x}{4} - 6\exp\left(-\dfrac{81\pi^2 t}{16}\right)\cos\dfrac{9\pi x}{4}.$

5. Solve the differential equation

$$\frac{\partial u}{\partial t} = \alpha^2 \frac{\partial^2 u}{\partial x^2}$$

for the conduction of heat along a rod without radiation, subject to the following conditions:

(*i*) u is not infinite for $t \to \infty$,

(*ii*) $\dfrac{\partial u}{\partial x} = 0$ for $x = 0$ and $x = l$,

(*iii*) $u = lx - x^2$ for $t = 0$, between $x = 0$ and $x = l$. (*I.E.T.E. June-2004*)

Ans. $u = \dfrac{l^2}{6} - \dfrac{l^2}{\pi^2}\sum\dfrac{1}{n^2} e^{-4n^2\pi^2\alpha^2 t/l^2} \cos\dfrac{2\pi nx}{l}.$

6. The ends A and B of a rod 30 cm long have their temperatures kept at 20°C and 80°C until steady state conditions prevail. The temperature of the end B is suddenly reduced to 60°C and kept so while the end A is raised to 40°C. Find the temperature distribution in the rod after time t. (*I.E.T.E. W-2002*)

Ans. $u(x, t) = 40 + \dfrac{2x}{3} - \dfrac{40}{\pi}\sum_{n=1}^{\infty}\dfrac{1}{n}\sin\dfrac{n\pi x}{15}\exp\left(-n^2\pi^2 c^2 t/225\right).$

7. The ends A and B of a rod l cm long have the temperature 0°C and 100°C respectively until steady state prevails. The temperature of A is then suddenly raised to 25°C while that of B is reduced to 75°C. Find the temperature distribution in the rod at time t.

Ans. $u = 25 + \dfrac{50x}{l} - \dfrac{100}{\pi}\left[\dfrac{1}{2}e^{-4bt}\sin\dfrac{2\pi x}{l} + \dfrac{1}{2}e^{-16bt}\sin\dfrac{4\pi x}{l} + \cdots\right],$

where $b = (\pi c/l)^2.$

8. The equation for the conduction of heat along a bar of length l is $\dfrac{\partial\theta}{\partial t} = k\dfrac{\partial^2\theta}{\partial x^2},$

neglecting radiation. Find an expression for θ if the ends of the bar are maintained at zero temperature initially and if initially the temperature is T at the centre of the bar and falls uniformly to zero at the ends.

$$\textbf{Ans. } \theta(x, t) = \frac{8T}{\pi^2} \sum_{n=1}^{\infty} \frac{1}{n^2} \sin\frac{n\pi}{2} \sin\frac{n\pi x}{l} \; e^{\frac{-kn^2\pi^2 t}{l^2}}$$

[**Hint:** The initial temperature distribution is given by

$$\theta(x, 0) = \frac{2T}{l} x \text{ for } 0 \le x \le \frac{l}{2} = \frac{2T}{l}(l-x) \text{ for } \frac{l}{2} \le x \le l].$$

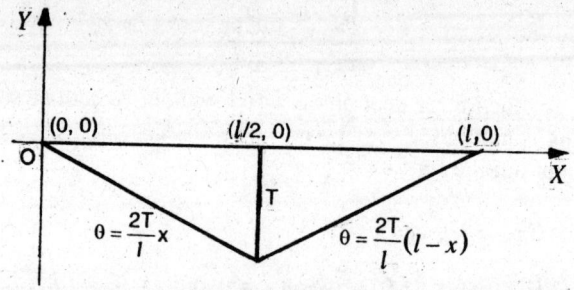

Fig. 20.5.

9. A homogeneous rod of conducting material of length 100 cm has its ends kept at zero temperature and the temperature initially is

$$u(x, 0) = x \left.\right] 0 \le x \le 50$$
$$= 100 - x \left.\right] 50 \le x \le 100.$$

Find the temperature $u(x, t)$ at any time. *(A.M.I.E., W-2003)*

$$\textbf{Ans. } u(x, t) = \frac{400}{\pi^2} \sum_{n=0}^{\infty} \frac{(-1)^n}{(2n+1)^2} e^{\frac{-c^2(2n+1)^2\pi^2 t}{100^2}} \sin\frac{(2n+1)\pi x}{100}$$

10. A uniform rod of length l, whose surface is thermally insulated, is initially at temperature θ_0. At time $t = 0$, one end is suddenly cooled to temperature $0°C$ and subsequentlly maintained to this temperature and at the same time the other end is thermally insulated. Find the temperature at end $x = l$ at any time t.

$$\textbf{Ans. } u(x, t) = \frac{40\theta_0}{\pi} \sum_{n=1}^{\infty} \frac{1}{2n-1} \sin\frac{(2n-1)\pi x}{2l} \; e^{\frac{-c^2(2n-1)^2\pi^2 t}{4l^2}},$$

$$u(l, t) = \frac{40\theta_0}{\pi} \sum_{n=1}^{\infty} \frac{(-1)^{n-1}}{2n-1} \; e^{\frac{-c^2(2n-1)^2\pi^2 t}{4l^2}}.$$

11. A rod of length l has its ends A and B which are kept at 0°C and 75°C, until steady-state conditions prevail. If the temperature of A is suddenly raised to 75°C and that of B to 175°C and maintained thereafter, find the subsequent temperature distribution of the rod.

Ans. $u(x,t) = 75 + \dfrac{100}{l}x - \dfrac{300}{\pi}\displaystyle\sum_{n=1}^{\infty}\dfrac{1}{2n-1}\sin\dfrac{(2n-1)\pi x}{l} \cdot e^{\frac{-c^2(2n-1)^2\pi^2 t}{l^2}}$.

12. The temperatures at the ends $x = 0$ and $x = 50$ cm in length of a rod are held at 0°C and 50°C respectively until steady-state conditions prevail. The two ends of the rod are suddenly insulated. Find the temperature distribution of the rod, assuming that the surface of the rod is impervious to heat flow.

Ans. $u(x,t) = 25 - \dfrac{200}{\pi^2}\displaystyle\sum_{n=1}^{\infty}\dfrac{1}{(2n-1)^2}\cos\dfrac{(2n-1)\pi x}{50} \cdot e^{\frac{-c^2(2n-1)^2\pi^2 t}{2500}}$.

13. A rod of length l has its ends A and B are maintained at 20°C and 40°C respectively until steady-state conditions prevail. The temperature at A is suddently raised to 50°C while that of B is lowered to 10°C and maintained thereafter. Find the subsequent temperature distribution of the rod.

Ans. $u(x,t) = 50 - \dfrac{40x}{l} - \dfrac{60}{\pi}\displaystyle\sum_{n=1}^{\infty}\dfrac{1}{n}\sin\dfrac{2n\pi x}{l} \cdot e^{\frac{-4c^2 n^2\pi^2 t}{l^2}}$.

14. A rod of length l has its lateral surface insulated and is so thin that heat flow in the rod can be regarded as one dimensional. Initially the rod is at the temperature 100°C throughout. At $t = 0$ the temperature at the left end of the rod is suddenly reduced to 50°C and maintained thereafter at this value, while the right end is maintained at 100°C. Let $u(x, t)$ be the temperature at point x in the rod at any subsequent time t.

 (i) Write down the appropriate partial differential equation for $u(x, t)$ with initial and boundary conditions.

 (ii) Solve the differential equation in (i) above using method of separation of variables and show that

$$u(x, t) = 50 + \frac{50x}{l} + \frac{100}{\pi}\sum_{n=1}^{\infty}\frac{1}{n}\sin\frac{n\pi x}{l}\exp\frac{-n^2\pi^2 t}{a^2 l^2},$$

where a^2 is the constant involved in the partial differential equation.

 (I.E.T.E., Dec. 2004)

[Hint: Here we solve the one-dimensional heat equation $\dfrac{\partial^2 u}{\partial x^2} = a^2\dfrac{\partial u}{\partial t}$

subject to the boundary that is, end conditions $u(0, t) = 50$, $u(l, t) = 100$ and the initial condition $u(x, 0) = 100$].

20.9 Two Dimensional Heat Flow

$$\frac{\partial^2 u}{\partial x^2} + \frac{\partial^2 u}{\partial y^2} = a^2 \frac{\partial u}{\partial t}$$

If the temperature at any point of a metal plate in the *XY*-plane depends upon the *x* and *y* co-ordinates and time *t* so that there is no heat-flow normal to the *XY*-plane, then heat flow is called two dimensional. Proceeding as in Art. 20.7 we can prove that the temperature *u* at any point of the plate at any time *t* is governed by the differential equation

$$\frac{\partial^2 u}{\partial x^2} + \frac{\partial^2 u}{\partial y^2} = a^2 \frac{\partial u}{\partial t}. \qquad \qquad ...(20.9)$$

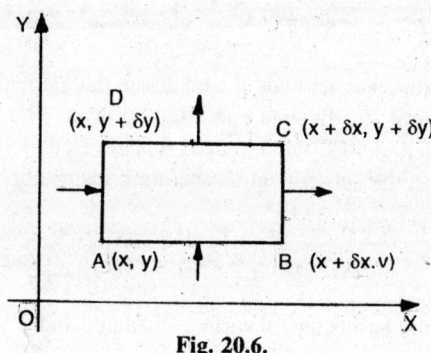

Fig. 20.6.

In the steady state, that is, when temperature *u* does not change with time, $\frac{\partial u}{\partial t} = 0$ and equation (20.9) reduces to the form

$$\frac{\partial^2 u}{\partial x^2} + \frac{\partial^2 u}{\partial y^2} = 0. \qquad \qquad ...(20.10)$$

The above equation is known as **Laplace equation** *in two dimensions.*

This equation is called the Laplace equation in *cartesian co-ordinates* or two dimensional heat flow in steady-state conditions.

Another form of this equation is **Polar form**. This equation is obtained by substituting $x = r \cos \theta$, $y = r \sin \theta$ in the above equation and in the (r, θ) coordinates the Laplace equation is written as

$$r^2 \frac{\partial^2 u}{\partial r^2} + r \frac{\partial u}{\partial r} + \frac{\partial^2 u}{\partial \theta^2} = 0. \qquad \qquad ...(20.11)$$

The polar form of the Laplace equation is useful in solving problems where the configuration is circular or semi-circular or a sector of a circle.

Method of Solving Laplace Equation (Cartesian Form)

The equation is given by

$$\frac{\partial^2 u}{\partial x^2} + \frac{\partial^2 u}{\partial y^2} = 0 \qquad \qquad ...(1)$$

Let $u = XY$ be the solution of (1). Substituting for u in (1) and using the method of separation of variables, we get.

$$Y\frac{d^2 X}{dx^2} + X\frac{d^2 Y}{dy^2} = 0 \text{ or } YX'' + XY'' = 0.$$

This gives

$$\frac{X''}{X} + \frac{Y''}{Y} = 0 \text{ or } \frac{X''}{X} = -\frac{Y''}{Y}. \qquad \qquad ...(2)$$

Since $\dfrac{X''}{X}$ is a function of x alone and $\dfrac{Y''}{Y}$ is function of y alone, and because x and y are independent of each other, relation (2) will hold if each side is constant (otherwise there will be a functional relation between x and y).

The constant can be taken either positive or negative depending upon the *non-zero* boundary conditions.

  (*i*) If the non-zero condition is in terms of x, we assume the constant as negative and write this as $-k^2$.
  (*ii*) If the non-zero condition is in terms of y, we assume the constant as positive and write this as $+k^2$.

Note: If the plate is infinite i.e., it is too long in comparison to its width, then the temperature at infinity is always taken as zero. For example
  (*i*) If the plate is infinite along Y-direction, we write $u(x, \infty) = 0$ for all x and
  (*ii*) If the plate is infinite in X-direction, we write $u(\infty, y) = 0$ for all y.
The procedure will be clear from the following problem.

20.10 Illustrative Examples

Example 20.13 A rectangular plate with insulated surfaces is 8 cm wide and so long compared to its width that it may be considered infinite in length. If the temperature along one short edge y = 0 is given by u(x, 0) = 100 sin (πx/8), 0 < x < 8, while the two long edges, x = 0 and x = 8, as well as the other short edge are kept at 0°C, find the steady state temperature u(x, y).

Solution. In the steady state, the temperature u satisfies the equation

$$\frac{\partial^2 u}{\partial x^2} + \frac{\partial^2 u}{\partial y^2} = 0.$$

Here the *boundary conditions* are

$$u(0, y) = 0 \qquad \qquad ...(1)$$
$$u(8, y) = 0 \qquad \qquad ...(2)$$

$$u(x, 0) = 100 \sin \frac{\pi x}{8} \qquad \qquad ...(3)$$

$$\lim_{y \to \infty} u(x, y) = 0 \text{ that is, } u(x, \infty) = 0 \cdot \qquad \qquad ...(4)$$

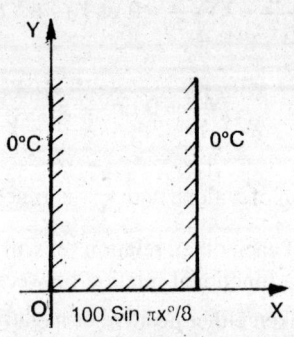

Y

0°C   0°C

O| 100 Sin πx°/8   X

Fig. 20.7

Let $u = XY$ be the solution of the given equation. Substituting this in the given equation, using method of separation of variables, we get

$$X''Y + Y''X = 0 \text{ or } \frac{X''}{X} = -\frac{Y''}{Y} = -k^2 \qquad \text{(because of non-zero condition).}$$

This gives $X'' = -k^2 X$ and $Y'' = K^2 Y$.

The solutions of these equations are

$$X = c_1 \cos kx + c_2 \sin kx,$$
$$Y = c_3 e^{ky} + c_4 e^{-ky}.$$

$$\therefore \quad u = XY = (c_1 \cos kx + c_2 \sin kx)(c_3 e^{ky} + c_4 e^{-ky}). \qquad \qquad ...(5)$$

Now applying the condition (1) in (5) that is, $u(0, y) = 0$, we get

$$0 = c_1(c_3 e^{ky} + c_4 e^{-ky}). \text{ This gives } c_1 = 0.$$

Relation (5) with $c_1 = 0$, gives

$$u = \sin kx(c_5 e^{ky} + c_6 e^{-ky}); \ (c_5 = c_2 c_3; \ c_6 = c_2 c_4). \qquad \qquad ...(6)$$

Again using condition (4) in (6) that is, $u(x, \infty) = 0$, we get $c_5 = 0$.

$\therefore$ Equation (6) with $c_5 = 0$ can be written as

$$u(x, y) = c_6 \sin kx \ e^{-ky}. \qquad \qquad ...(7)$$

Using condition (2) in (7) that is, $u(8, y) = 0$, we obtain

$$0 = c_6 \sin 8 k e^{-ky}. \qquad \qquad ...(8)$$

Now either c_6 is zero or $\sin 8k$ is zero ($\because$ condition is on $x = 8$).

But c_6 which is equal to $c_2 c_4$ from (6) above cannot be zero.

$\therefore$ sin 8 $k = 0$ (If c_2 is zero, the whole X solution vanishes).

Now sin $8k = 0$ gives $8k = n\pi$ $\therefore$ $k = \dfrac{n\pi}{8}$.

Substituting for $k = \dfrac{n\pi}{8}$ in (7), we get

$$u = \sum_{n=1}^{\infty} B_n \sin\frac{n\pi x}{8} e^{-\frac{n\pi y}{8}}. \qquad \ldots(9)$$

Using the third boundary condition, we get from (9)

$$100 \sin\frac{\pi x}{8} = \sum_{n=1}^{\infty} B_n \sin\frac{n\pi x}{8}.$$

Equating the coefficients, $B_1 = 100$, $B_2 = B_3 = \ldots\ldots = 0$; $n = 1$.

$\therefore$ From (9) the solution is given by

$$u(x, y) = 100 \sin\frac{\pi x}{8} e^{\frac{-\pi y}{8}}.$$

Example 20.14 An infinitely long plane uniform plate is bounded by two parallel edges and an end at right angles to them. The breadth is π; this end is maintained at a temperature u_0 at all points and other edges are at zero temperature. Determine the temperature at any point of the plate in steady state. *(I.E.T.E. Dec. 2009, Dec. 2005)*

Solution. In the steady state, the temperature $u(x, y)$ at any point $P(x, y)$ satisfies the differential equation

$$\frac{\partial^2 u}{\partial x^2} + \frac{\partial^2 u}{\partial y^2} = 0. \qquad \ldots(A)$$

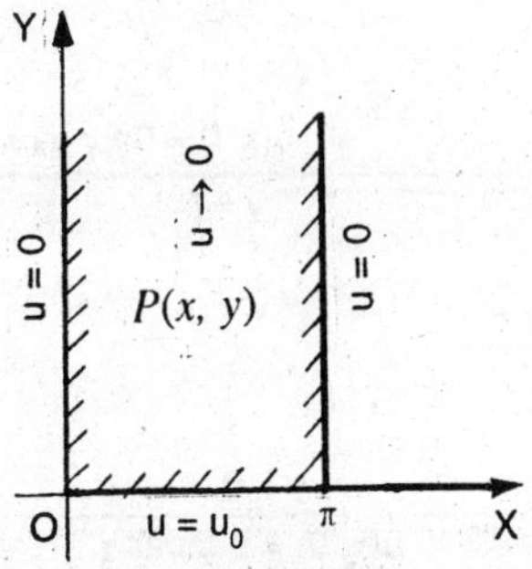

Fig. 20.8

Here the *boundary conditions* are

$$u(0, y) = 0 \qquad \text{...(1)}$$
$$u(\pi, y) = 0 \qquad \text{...(2)}$$
$$u(x, 0) = u_0 \qquad \text{...(3)}$$
$$\lim_{y \to \infty} u(x, y) = 0. \qquad \text{...(4)}$$

By example 20.13, the solution of (A) which satisfies the boundary conditions (1), (2) and (4) is

$$u(x, y) = \sum_{n=1}^{\infty} B_n (\sin nx) e^{-ny} \quad \text{...(5)} \qquad \text{(because here width is } \pi \text{ in place of 8)}$$

$$\therefore \quad u(x, 0) = \sum_{n=1}^{\infty} B_n \sin nx. \quad \text{But } u(x, 0) = u_0 \qquad \text{[boundary condition (3)].}$$

$$\therefore \qquad\qquad u_0 = \sum_{n=1}^{\infty} B_n \sin nx. \qquad \text{...(6)}$$

This is half-range sine series in $(0, \pi)$,

$$\therefore \quad B_n = \frac{2}{\pi} \int_0^{\pi} u_0 \sin nx \, dx = \begin{cases} 0 \ (n \text{ even}) \\ \dfrac{4u_0}{n\pi} (n \text{ odd}) \end{cases}$$

$$\therefore \quad u(x, y) = \frac{4u_0}{\pi} \left[e^{-y} \sin x + \frac{1}{3} e^{-3y} \sin 3x + \frac{1}{5} e^{-5y} \sin 5x + \dots \right].$$

Example 20.15 **An infinitely long metal plate in the form of area is enclosed between $y = 0$ and $y = l$ for positive values of x. The temperature is zero along the edges $y = 0$ and $y = l$ and at ∞. If edge $x = 0$ is kept at a temperature u_0, find the steady state temperature distribution in the plate.**

Solution. In the steady-state, the temperature u satisfies the equation

$$\frac{\partial^2 u}{\partial x^2} + \frac{\partial^2 u}{\partial y^2} = 0. \qquad \text{...(1)}$$

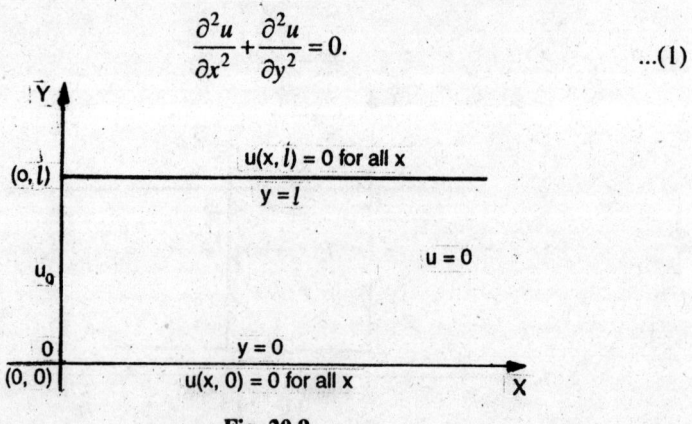

Fig. 20.9

Here the *boundary conditions* are

$$u(x, 0) = 0 \qquad \ldots(2)$$
$$u(x, l) = 0 \qquad \ldots(3)$$
$$u(\infty, y) = 0 \qquad \ldots(4)$$
$$u(0, y) = u_0. \qquad \ldots(5)$$

Let $u = XY$ be the solution of the given equation. Substituting this in the given equation and separating the variables, we get

$$\frac{X''}{X} = -\frac{Y''}{Y} = k^2 \qquad \text{(Non zero condition in } y).$$

Solving the two ordinary equations viz $X'' = k^2 X$ and $Y'' = -k^2 Y$, we get the solution

$$u = XY = (c_1 e^{kx} + c_2 e^{-kx})(c_3 \cos ky + c_4 \sin ky). \qquad \ldots(6)$$

Applying boundary condition (2) in (6), we get $c_3 = 0$.

This give $\qquad u = \left(c_1' e^{kx} + c_2' e^{-kx}\right) \sin ky \; ; \; \left(c_1' = c_1 c_4; \; c_2' = c_2 c_4\right). \qquad \ldots(7)$

Applying boundary condition, (4) in (7), we get $c_1' = 0$. Hence

$$u = c_2' e^{-kx} \sin ky \qquad \ldots(8)$$

Using boundary condition (3) in (8), we obtain

$$0 = c_2' e^{-kx} \sin kl$$

Since $c_2' \neq 0$, $\sin kl = 0$, giving $k = \dfrac{n\pi}{l}$.

$\therefore$ Equation (8) for all values of n can be written as

$$u = \sum c_n \sin \frac{n\pi y}{l} e^{\frac{-n\pi x}{l}}. \qquad \ldots(9)$$

To evalute c_n, we use the boundary condition (5). This gives

$$c_n = \frac{2}{l} \int_0^l u_0 \sin \frac{n\pi y}{l} dy = \frac{2u_0}{n\pi}\left(1 - (-1)^n\right).$$

$$\therefore u(x, y) = \frac{4u_0}{\pi} \sum_{n=1}^{\infty} \frac{1}{2n-1} \sin \frac{(2n-1)\pi y}{l} e^{\frac{-(2n-1)\pi x}{l}}$$

by replacing n by $(2n - 1)$ in (9).

Example 20.16 **Evaluate the steady temperature in a rectangular plate of length 'a' and width b, the sides of which are kept at temperature zero, the lower end is kept at temperature f(x) and the upper edge is kept insulated.**

Solution. In steady state temperature is governed by

$$\frac{\partial^2 u}{\partial x^2} + \frac{\partial^2 u}{\partial y^2} = 0. \qquad \ldots(A)$$

We have to solve the above equation subject to the *conditions*:

$$u(0, y) = 0, \quad \text{and} \quad u(a, y) = 0) \qquad \ldots(1)$$

$$\left(\frac{\partial u}{\partial y}\right)_{y=b} = 0 \quad \text{and } u(x, 0) = f(x). \qquad \qquad ...(2)$$

[We should note the condition $\left(\dfrac{\partial u}{\partial y}\right)_{y=b} = 0$ on the upper edge $y = b$ which is

kept insulated]

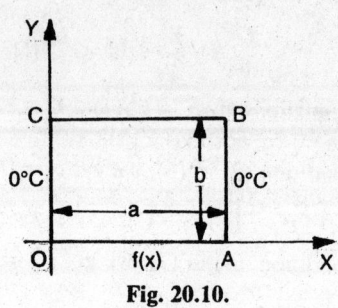

Fig. 20.10.

The solution of (A) satisfying first two conditions is

$$u(x, y) = \sin\frac{n\pi x}{a}\left(b_1 e^{n\pi y/a} + b_2 e^{-n\pi y/a}\right)$$

$$= \sin\frac{n\pi x}{a}\left(c_1 \cosh\frac{n\pi y}{a} + c_2 \sinh\frac{n\pi y}{a}\right)^*.$$

$$\frac{\partial u}{\partial y} = \sin\frac{n\pi x}{a} \cdot \frac{n\pi}{a}\left[c_1 \sinh\frac{n\pi y}{a} + c_2 \cosh\frac{n\pi y}{a}\right].$$

$$\left(\frac{\partial u}{\partial y}\right)_{y=b} = 0 \text{ gives } c_2 = \frac{-c_1 \sinh\dfrac{n\pi b}{a}}{\cosh\dfrac{n\pi b}{a}}.$$

$$\therefore \quad u(x,y) = \sum c_n \sin\frac{n\pi x}{a}\left[\cosh\frac{n\pi y}{a}\cosh\frac{n\pi b}{a} - \sinh\frac{n\pi y}{a}\right.$$

$$\left. \sinh\frac{n\pi b}{a}\right]\text{sech}\frac{n\pi b}{a}.$$

---

* Note: While solving problems on plates which are bounded, we prefer to use hyperbolic function instead of exponential functions. This is mainly due to the reason that hyperbolic functions are convenient to handle while applying boundary conditions in such problems.

$$e^{ky} = \cosh ky + \sinh ky; \quad e^{-ky} = \cosh ky - \sinh ky]$$

Thus $\quad u(x,y) = \sum c_n \sin \dfrac{n\pi x}{a} \cosh \dfrac{n\pi}{a}(y-b)\operatorname{sech}\dfrac{n\pi b}{a}.$ ...(3)

$[\because \cosh(\theta - \phi) = \cosh\theta\cosh\phi - \sinh\theta\sinh\phi]$

Now $\qquad\qquad u(x,0) = \sum c_n \sin\dfrac{n\pi x}{a}\ \left(\because \cosh\dfrac{n\pi b}{a}\operatorname{sech}\dfrac{n\pi b}{a} = 1\right)$

But $\qquad\qquad\qquad u(x,\,0) = f(x).$

$\therefore\quad f(x) = \sum c_n \sin\dfrac{n\pi x}{a},$ where $c_n = \dfrac{2}{a}\displaystyle\int_0^a f(x)\sin\dfrac{n\pi x}{a}dx.$ ...(4)

Hence $u(x,\,y)$ is given by (3) when c_n is given by (4).

Note. *While studying steady state temperature distribution in rectangular sheets it is very convenient and natural that we use Laplace equation in rectangular co-*

ordinates, that is, $\dfrac{\partial^2 u}{\partial x^2} + \dfrac{\partial^2 u}{\partial y^2} = 0.$ *However, in some problems polar co-ordinates*

are more useful. In such problems we require Laplace equation in polar form with r, θ *as independent variables in place of x, y. The polar form of Laplace equation*

$\dfrac{\partial^2 u}{\partial x^2} + \dfrac{\partial^2 u}{\partial y^2} = 0$ *is*

$\dfrac{\partial^2 u}{\partial x^2} + \dfrac{1}{r}\dfrac{\partial u}{\partial r} + \dfrac{1}{r^2}\dfrac{\partial^2 u}{\partial \theta^2} = 0 \qquad$ *or* $\quad r^2\dfrac{\partial^2 u}{\partial x^2} + r\dfrac{\partial u}{\partial r} + \dfrac{\partial^2 u}{\partial \theta^2} = 0.$ ...(20.12)

Example 20.17 The bounding diameter of a semi-circular plate of radius 10 cm is kept at $0°C$ and the temperature along the semi-circular boundary at u_0 until steady state conditions prevail. Find the temperature distribution in the plate.

Solution. Taking the centre O of the circle as pole and diameter as initial line, let the temperature at $P(r, \theta)$ be $u(r, \theta)$.

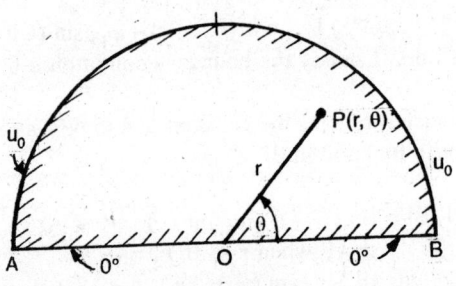

Fig. 20.11.

Now $u(r, \theta)$ satisfies the Laplace equation

$$r^2 \frac{\partial^2 u}{\partial r^2} + r \frac{\partial u}{\partial r} + \frac{\partial^2 u}{\partial \theta^2} = 0. \qquad ...(1)$$

We have to solve (1) subject to the *boundary conditions*
$$u(r, 0) = 0, \ u(r, \pi) = 0, \ u(0, \theta) = 0, \ u(10, \theta) = u_0.$$

Let
$$u = R(r) \ T(\theta).$$

Substituting this in (1) we get
$$r^2 R'' \ T + rR' \ T + RT'' = 0$$

or
$$(r^2 R'' + rR')T + RT'' = 0$$

or
$$\frac{r^2 R'' + rR'}{R} = -\frac{T''}{T} = h \ (\text{say}).$$

$\therefore \qquad r^2 R'' + rR' - hR = 0 \quad$ [Cauchy's Homogeneous equation] $\qquad ...(2)$

and
$$T'' + hT = 0. \qquad ...(3)$$

Putting $r = e^z$ in (2) we get

$$\frac{d^2 R}{dz^2} - hR = 0. \qquad ...(4)$$

Case I. $(h > 0)$. Let $h = k^2$.

Then the solutions of (3) and (4) are
$$T = c_1 \cos k\theta + c_2 \sin k\theta$$

and
$$R = c_3 e^{kz} + c_4 e^{-kz} = c_3 r^k + c_4 r^{-k}.$$

Case II. $h = 0$.

Then
$$T = c_5 \theta + c_6.$$
$$R = c_7 z + c_8 = c_7 \log r + c_8.$$

Case III $(h < 0)$. Let
$$h = -k^2.$$

Then the solutions are

$$T = c_9 e^{k\theta} + c_{10} e^{-k\theta}.$$
$$R = c_{11} \cos kz + c_{12} \sin kz = c_{11} \cos(k \log r) + c_{12} \sin (k \log r).$$

Thus the possible solutions are
$$u(r, \theta) = (c_1 \cos k\theta + c_2 \sin k\theta)(c_3 r^k + c_4 r^{-k}) \qquad ...(A)$$
$$u(r, \theta) = (c_5 \theta + c_6) \ (c_7 \log r + c_8) \qquad ...(B)$$

and $u(r, \theta) = (c_9 e^{k\theta} + c_{10} e^{-k\theta}) \ [c_{11} \cos (k \log r) + c_{12} \sin (k \log r)]. \qquad ...(C)$

Since (B) and (C) don't satisfy the boundary condition $u(0, \theta) = 0$, these are rejected.

[In dealing with such problems the constant h is to be chosen positive as this will give trigonometric function in θ]

Hence

$$u(r, \theta) = (c_1 \cos k\theta + c_2 \sin k\theta)(c_3 r^k + c_4 r^{-k}). \qquad ...(5)$$

As
$$u = 0 \text{ when } r \to 0, c_4 = 0.$$

$\therefore \qquad u = c_3 r^k (c_1 \cos k\theta + c_2 \sin k\theta) = r^k (b_1 \cos k\theta + b_2 \sin k\theta).$

Again
$$u(r, 0) = 0, \text{ we get } b_1 = 0.$$

Hence $\qquad\qquad u = b_2 r^k \sin k\theta.$

Using $\qquad\qquad u(r, \pi) = 0$, we get $0 = b_2\, r^k \sin k\pi.$

$\therefore \qquad\qquad k = n$ (an integer).

Thus $\qquad\qquad u(r, \theta) = b_n\, r^n \sin n\theta$ (Replacing b_2 by b_n).

By superposition principle

$$u(r, \theta) = \Sigma\, b_n\, r^n \sin n\theta \qquad\qquad\text{...(6)}$$

Using the last boundary condition, *i.e.*, $u(10, \theta) = u_0$, we get

$$u_0 = \sum_{n=1}^{\infty} b_n\, 10^n \sin n\theta.$$

Since the right hand side is half-range sine series representing u_0 in $(0, \pi)$

$$b_n = \frac{2}{10^n\, \pi} \int_0^{\pi} u_0 \sin n\theta\, d\theta = \begin{cases} \dfrac{4u_0}{n\pi}\cdot\dfrac{1}{10^n}\,(n \text{ odd}). \\[2mm] 0\ (n \text{ even}). \end{cases}$$

Substituting the value of b_n in (6), the required temperature distribution is given by

$$u(r,\theta) = \frac{4u_0}{\pi}\left[\frac{r}{10}\sin\theta + \frac{1}{3}\left(\frac{r}{10}\right)^3 \sin 3\theta + ...\right].$$

Example 20.18 The bounding diameter of a semi-circular plate of radius 'a' cm is kept at 0°C and the temperature along the semi-circular boundary is given by

$$u(a, \theta) = \begin{cases} 50\,\theta \text{ when } 0 < \theta \le \dfrac{\pi}{2} \\[3mm] 50(\pi - \theta) \text{ when } \dfrac{\pi}{2} < \theta < \pi. \end{cases}$$

Find the steady state temperature function $u(r, \theta)$ $\qquad$ *(Madras, 1997)*

Solution. Here we have to solve

$$r^2 \frac{\partial^2 u}{\partial r^2} + r\frac{\partial u}{\partial r} + \frac{\partial^2 u}{\partial \theta^2} = 0 \qquad\qquad\text{...(1)}$$

subject to the *boundary conditions*

$$u(r, 0) = 0 \qquad\qquad\text{...(A)}$$
$$u(r, \pi) = 0 \qquad\qquad\text{...(B)}$$

$$u(a, \theta) = \begin{cases} 50\,\theta & 0 < \theta \le \dfrac{\pi}{2} \\[3mm] 50(\pi - \theta) & \dfrac{\pi}{2} < \theta < \pi. \end{cases} \qquad\qquad\text{...(C)}$$

The solution of (1) satisfying boundary conditions (A) and (B) by example 20.17, is

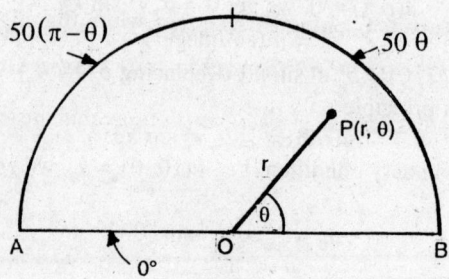

Fig. 20.12.

$$u(r, \theta) = \sum_{n=1}^{\infty} b_n r^n \sin n\theta. \qquad \ldots(2)$$

Using the boundary condition (C), we get

$$u(a, \theta) = \sum_{n=1}^{\infty} b_n a^n \sin n\theta \qquad \ldots(3)$$

where
$$u(a, \theta) = 50\,\theta \quad \text{when } 0 < \theta \leq \pi/2$$
$$= 50(\pi - \theta) \quad \text{when } \pi/2 < \theta < \pi.$$

But (3) is the half-range Fourier sine series for $u(a, \theta)$ in $(0, \pi)$.

$$\therefore \qquad b_n a^n = \frac{2}{\pi} \int_0^\pi u(a, \theta) \sin n\theta \, d\theta$$

$$= \frac{2}{\pi} \left[\int_0^{\pi/2} 50\,\theta \sin n\theta \, d\theta + \int_{\pi/2}^\pi 50(\pi - \theta) \sin n\theta \, d\theta \right] \quad [\text{by}\,(C)]$$

$$= \frac{200}{\pi n^2} \sin \frac{n\pi}{2} \qquad \text{or} \qquad b_n = \frac{200}{\pi a^n n^2} \sin \frac{n\pi}{2}.$$

Putting the value of b_n in (2), the required temperature distribution is

$$u(r, \theta) = \frac{200}{\pi} \sum_{n=1, 3, 5 \ldots} \left[\frac{1}{a^n n^2} \sin \frac{n\pi}{2} r^n \sin n\theta \right]$$

$$= \frac{200}{\pi} \left[\frac{1}{1^2} \left(\frac{r}{a}\right) \sin\theta - \frac{1}{3^2} \left(\frac{r}{a}\right)^3 \sin 3\theta + \frac{1}{5^2} \left(\frac{r}{a}\right)^5 \sin 5\theta - \ldots \right]$$

$$= \frac{200}{\pi} \sum \frac{1}{a^n n^2} \sin \frac{n\pi}{2} r^n \sin n\theta.$$

Example 20.19 A plate in the shape of a truncated quadrant of a circle is bounded by $r = a$, $r = b$, $\theta = 0$ and $\theta = \pi/2$. Its faces are insulated and maintained at temperature $0°C$ along three edges while the edge $r = a$ is kept at temperature $\theta\big((\pi/2) - \theta\big)$. Find the steady-state temperature distribution.

Solution. In the polar form the steady state temperature distribution is given by

$$r^2 \frac{\partial^2 u}{\partial r^2} + r \frac{\partial u}{\partial r} + \frac{\partial^2 u}{\partial \theta^2} = 0. \qquad \text{...(1)}$$

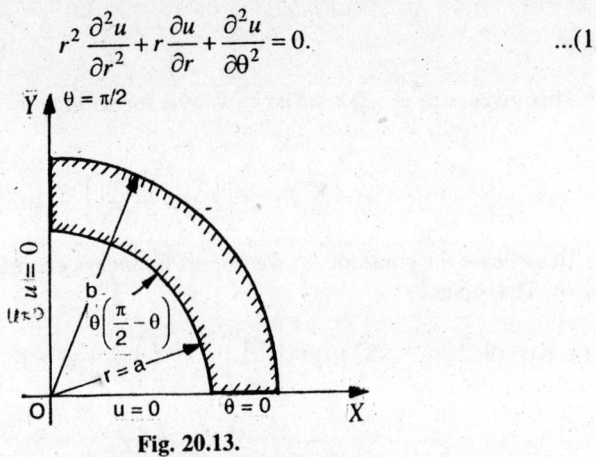

Fig. 20.13.

We have to solve (1) subject to the *boundary conditions*

$$
\left.
\begin{array}{lll}
(i)\ u(r, 0) = 0 & a \le r \le b \\
(ii)\ u(r, \pi/2) = 0 & a \le r \le b \\
(iii)\ u(b, \theta) = 0 & 0 \le \theta \le \pi/2 \\
(iv)\ u(a, \theta) = \theta(\theta/2 - \theta) & 0 \le \theta \le \pi/2
\end{array}
\right\} \qquad \text{...(2)}
$$

The solution of (1) as discussed earlier in example (20.17) is given by

$$u(r, \theta) = (c_1 r^k + c_2 r^{-k})(c_3 \cos k\theta + c_4 \sin k\theta). \qquad \text{...(3)}$$

Using boundary condition 2(*i*), we get

$$0 = (c_1 r^k + c_2 r^{-k})\, c_3. \text{ Therefore, } c_3 = 0.$$

$$u = \left(c_1' r^k + c_2' r^{-k}\right) \sin k\theta. \qquad \left(c_1' = c_1 c_4 ; c_2' = c_2 c_4\right) \qquad \text{...(4)}$$

Applying boundary condition 2(*iii*), we get

$$0 = \left(c_1' b^k + c_2' b^{-k}\right) \sin k\theta.$$

This gives $\quad c_1' b^k + c_2' b^{-k} = 0 \text{ or } c_2' = -c_1' b^k / b^{-k} = -c_1' b^{2k}.$

With this value of c_2', (4) can be written as

$$u(r, \theta) = c_1'\left(r^k - b^{2k} r^{-k}\right) \sin k\theta$$

$$= c_1' b^k \left(\left(\frac{r}{b}\right)^k - \left(\frac{r}{b}\right)^{-k} \right) \sin k\theta. \qquad \qquad ...(5)$$

Again using boundary condition 2(*ii*) i.e., at $\theta = \pi/2$ in (5) above, we get

$$0 = c_1' b^k \left(\left(\frac{r}{b}\right)^k - \left(\frac{r}{b}\right)^{-k} \right) \sin \frac{k\pi}{2}.$$

This gives $\sin \dfrac{k\pi}{2} = 0 = \sin n\pi$ or $k = 2n$ for all n.

$$\therefore \qquad u(r, \theta) = \sum_{n=1}^{\infty} B_n b^{2n} \left(\left(\frac{r}{b}\right)^{2n} - \left(\frac{r}{b}\right)^{-2n} \right) \sin 2n\theta. \qquad ...(6)$$

To evaluate the constant B_n, we use the boundary condition 2 (*iv*) i.e., at $r = a$ in (6). This gives

$$u(a, \theta) = \theta\left(\frac{\pi}{2} - \theta\right) = \sum_{n=1}^{\infty} B_n b^{2n} \left[\left(\frac{a}{b}\right)^{2n} - \left(\frac{a}{b}\right)^{-2n} \right] \sin 2n\theta = \sum_{r=1}^{\infty} B_n' \sin 2n\theta \quad ...(7)$$

where

$$B_n' = B_n b^{2n} \left[\left(\frac{a}{b}\right)^{2n} - \left(\frac{a}{b}\right)^{-2n} \right].$$

Relation (7) is a Fourier half range sine series in $(0, \pi/2)$.

$$\therefore \qquad B_n' = \frac{2}{\pi/2} \int_0^{\pi/2} \theta(\pi/2 - \theta) \sin 2n\theta \, d\theta$$

$$= \frac{4}{\pi} \left[\left(\frac{\pi\theta}{2} - \theta^2\right)\left(-\frac{\cos 2n\theta}{2n}\right) - \left(\frac{\pi}{2} - 2\theta\right)\left(-\frac{\sin 2n\theta}{4n^2}\right) + (-2)\frac{\cos 2n\theta}{8n^3} \right]_0^{\pi/2}$$

$$= \frac{1}{\pi n^3}\left(1 - (-1)^n\right) = \begin{cases} 0 \text{ for } n \text{ even} \\ 2/\pi n^3 \text{ for } n \text{ odd.} \end{cases}$$

Substituating for B_n', B_n is given by

$$B_n b^{2n} = \frac{2}{\pi n^3} \frac{1}{\left(\dfrac{a}{b}\right)^{2n} - \left(\dfrac{a}{b}\right)^{-2n}}$$

$$\therefore \qquad u(r, \theta) = \sum \frac{2}{\pi n^3} \frac{\left[(r/b)^{2n} - (r/b)^{-2n} \right]}{\left[(a/b)^{2n} - (a/b)^{-2n} \right]} \sin 2n\theta, \qquad \qquad n = 1, 3, 5, ...$$

$$= \sum \frac{2}{\pi n^3}\left(\frac{a}{r}\right)^{2n}\left(\frac{r^{4n}-b^{4n}}{a^{4n}-b^{4n}}\right)\sin 2n\theta$$

Putting $n = 2m - 1$, where $m = 1, 2, 3, \ldots\ldots$, we get

$$u(r,\theta) = \sum_{m=1}^{\infty}\frac{2}{\pi}\frac{1}{(2m-1)^3}\left(\frac{a}{r}\right)^{4m-2}\left(\frac{r^{8m-4}-b^{8m-4}}{a^{8m-4}-b^{8m-4}}\right)\sin(4m-2)\theta.$$

Example 20.20 A plate in the form of a sector of a circle is bounded by lines $\theta = 0$, $\theta = \pi/3$ and $r = a$. Its surfaces are insulated and temperature along the boundaries are; u(r, 0) = 0; u(r, $\pi/3$) = 0 and u(a, θ) = $\lambda\theta(\pi/3 - \theta)$. Find the steady state temperature distribution.

Solution. Let $u(r, \theta)$ denote the temperature distribution in the plate. The equation governing the heat flow in polar coordinates is given by

$$r^2\frac{\partial^2 u}{\partial r^2} + r\frac{\partial u}{\partial r} + \frac{\partial^2 u}{\partial \theta^2} = 0 \qquad \ldots(1)$$

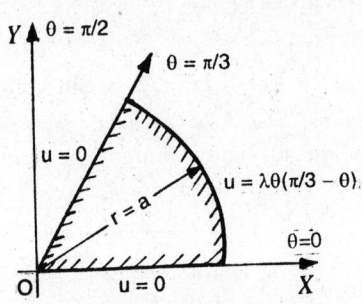

Fig. 20.14.

We seek the solution of (1) subject to the *boundary conditions.*

(i) $u(r, 0) = 0$ for all r \qquad (ii) $u(0, \theta) = 0$, \qquad $0 < \theta < \pi/3$

(iii) $u(r, \pi/3) = 0$ for all r \qquad (iv) $u(a, \theta) = \lambda\theta(\pi/3 - \theta)$, for $0 < \theta < \pi/3$.

As discuss in the previous example, the solution of (1) is given by

$$u(r, \theta) = (c_1 r^k + c_2 r^{-k})(c_3 \cos k\theta + c_4 \sin k\theta). \qquad \ldots(2)$$

Using boundary condition (i) *i.e.,* $u = 0$ at $\theta = 0$, we get

$$0 = (c_1 r^k + c_2 r^{-k})c_3 \text{ which gives } c_3 = 0.$$

$$\therefore \qquad u = \left(c_1' r^k + c_2' r^{-k}\right)\sin k\theta \cdot \quad \left(c_1' = c_1 c_4;\ c_2' = c_2 c_4\right) \qquad \ldots(3)$$

Using boundary conditions (ii) *i.e.,* $u = 0$ at $r = 0$, we get $c_2' = 0$. Hence (3) is written as

$$u = c_1' r^k \sin k\theta. \qquad \ldots(4)$$

Again using condition (*iii*) *i.e.*, $u = 0$ at $\theta = \pi/3$, we get

$$0 = c_1' r^k \sin\left(\frac{\pi k}{3}\right).$$

Since $c_1' \neq 0$, $\sin\left(\frac{\pi k}{3}\right) = 0 = \sin n\pi$. This gives $k = 3n$.

With $k = 3n$, the solution is given by

$$u = \sum_{n=1}^{\infty} c_n' r^{3n} \sin 3n\theta. \qquad \ldots(5)$$

Using condition (*iv*) that is, $u = \lambda\theta(\pi/3 - \theta)$ when $r = a$, we get

$$\lambda\theta\left(\frac{\pi}{3} - \theta\right) = \sum_{n=1}^{\infty} c_n' a^{3n} \sin 3n\theta. \qquad \ldots(6)$$

Relation (6) is a Fourier half range sine series in $(0, \pi/3)$.

The constant $c_n' a^{3n}$ is given by

$$c_n' a^{3n} = \frac{2}{\pi/3} \int_0^{\pi/3} \lambda\theta(\pi/3 - \theta)\sin 3n\theta \, d\theta$$

Integrating the above integral and substituting the boundary conditions, we get

$$c_n' a^{3n} = \frac{6\lambda}{\pi}\left[\frac{-2(-1)^n}{27n^3} + \frac{2}{27n^3}\right]$$

$$= \begin{cases} \dfrac{12\lambda}{27\pi n^3}\left(1 - (-1)^n\right) = 0, & \text{when } n \text{ is even} \\ \dfrac{24\lambda}{27\pi n^3}, & \text{when } n \text{ is odd.} \end{cases} \qquad \therefore \quad c_n' = \frac{8\lambda}{9\pi a^{3n} n^3} \text{ when } n \text{ is odd.}$$

Putting $n = 2m - 1$ and substituting for c_n' in (5), the solution is given by

$$u(r, \theta) = \frac{8\lambda}{9\pi} \sum_{m=1}^{\infty} \left(\frac{r}{a}\right)^{6m-3} \frac{\sin(6m-3)\theta}{(2m-1)^3}.$$

Exercise 20.3

1. A rectangular plate with insulated surfaces is 10 cm wide and so long compared to its width that it may be considered infinite in length without introducing any appreciable error. If the temperature along the short edge $y = 0$ is given by

$$u(x, 0) = 20x, \ 0 \le x \le 5; \ u(x, 0) = 20 (10 - x), \ 5 < x < 10$$

while the two long edges $x = 0$ and $x = 10$ as well as the other short edge are kept at 0°C. Find the steady state temperature at any point (x, y) of the plate.

(Madras, 1997S)

Ans. $u = \dfrac{800}{\pi^2}\left(e^{-\pi y/10} \sin\dfrac{\pi x}{10} - \dfrac{1}{9}e^{-3\pi y/10} \sin\dfrac{3\pi x}{10} + ...\right).$

2. Solve $\dfrac{\partial^2 u}{\partial x^2} + \dfrac{\partial^2 u}{\partial y^2} = 0$, subject to the conditions

$$u(0, y) = u(l, y) = u(x, 0) = 0,$$

and $u(x, a) = \sin(n\pi x)/l.$ *(Mangalore 1997; Osmania 1995)*

[Hint: By Example 20.13, Art 20.10, the solution is of the form

$$u = (a_1 \cos kx + a_2 \sin kx)(a_3 e^{ky} + a_4 e^{-ky}).$$

Now $u(0, y) = 0$ gives $a_1 = 0$, then $u = \sin kx(b_1 e^{ky} + b_2 e^{-ky}).$

$u(l, y) = 0$ gives $k = p\pi/l$ (p being an integer)

$$\therefore \quad u = \sin\frac{p\pi x}{l}\left(b_1 e^{p\pi y/l} + b_2 e^{-p\pi y/l}\right).$$

Again $u(x, 0) = 0$ gives $0 = \sin(p\pi x/l)[b_1 + b_2]$

$$\therefore \qquad b_1 + b_2 = 0 \text{ or } b_2 = -b_1.$$

Thus $u = b_1 \sin\dfrac{p\pi x}{l}\left(e^{p\pi y/l} - e^{-p\pi y/l}\right) = c_p \sin\dfrac{p\pi x}{l} \sinh\dfrac{p\pi y}{l}$, where $2b_1 = c_p.$

Hence the general solution is

$$u = \sum_{p=1}^{\infty} c_p \sin\frac{p\pi x}{l} \sinh\frac{p\pi y}{l}.$$

Lastly $u(x, a) = \sin\dfrac{n\pi x}{l}.$ $\quad \therefore \quad \sin\dfrac{n\pi x}{l} = \sum_{p=1}^{\infty} c_p \sin\dfrac{p\pi x}{l} \sinh\dfrac{p\pi a}{l}.$

Equating coefficients,

$$c_n \sinh\frac{n\pi a}{l} = 1 \text{ or } c_n = 1\bigg/ \sinh\frac{n\pi a}{l},$$

while other c's are zero.

$$\therefore \qquad u(x, y) = \sin\frac{n\pi x}{l}\left[\sinh\frac{n\pi y}{l}\bigg/\sinh\frac{n\pi a}{l}\right].$$

3. A rectangular plate is bounded by the lines $x = 0$, $y = 0$, $x = a$, $y = b$. Its surfaces are insulated and the temperature along the upper horizontal edge

is 100°C while the other three edges are kept at 0°C. Find the steady state temperature function $u(x, y)$.

$$\textbf{Ans. } u = \frac{400}{\pi} \sum_{n=1}^{\infty} \frac{\sin\{(2n-1)\pi x/a\}\sinh\{(2n-1)\pi y/a\}}{(2n-1)\sinh\{(2n-1)\pi b/a\}}.$$

4. Find the steady state temperature at any point of a square plate if two adjacent edges are kept at 0°C and the other two at 100°C.

$$\textbf{Ans. } u(x, y) = \frac{400}{\pi} \sum_{1}^{\infty} \frac{1}{(2n-1)\sinh(2n-1)\pi} \times$$

$$\left[\sin\frac{(2n-1)\pi x}{l}\sinh\frac{(2n-1)\pi y}{l} + \sin(2n-1)\frac{\pi y}{l}\sinh(2n-1)\frac{\pi x}{l}\right],$$

where l is the length of one side.

[**Hint:** Assume $u(x, y) = u_1(x, y) + u_2(x, y)$ where $u_1(x, y)$ and $u_2(x, y)$ are

solutions of $\dfrac{\partial^2 u}{\partial x^2} + \dfrac{\partial^2 u}{\partial y^2} = 0$ and further $u_1(x, y)$ is the temperature at $P(x, y)$

with edge BC kept at 100°C and the other three edges at 0°C. While $u_2(x, y)$ is the temperature at $P(x, y)$ with edge AB kept at 100°C and the other three edges at 0° C].

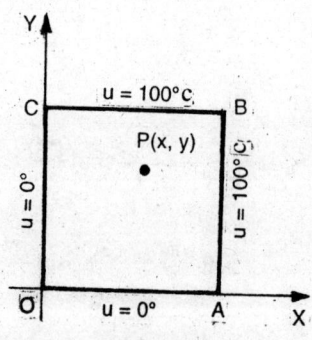

Fig. 20.15.

5. A rectangular metal plate has sides of lengths 'a' and 'b'. The sides $x = 0$, $x = a$, and $y = b$ are insulated while the edge $y = 0$ is kept at temperature $u_0 \cos(\pi x/a)$. Find the temperature 'u' at (x, y) in steady-state.

$$\textbf{Ans: } u = u_0 \left[\cos\frac{\pi x}{a}\,\text{sech}\,\frac{\pi b}{a}\cosh\left\{\frac{\pi(y-b)}{a}\right\}\right].$$

[**Note:** $\cosh a \cosh b \pm \sinh a \sinh b = \cosh(a \pm b)$].

6. A thin semi-circular plate of radius '*a*' has its bounding diameter kept at temperature zero and its circumference at k. Find the temperature distribution in the steady state.

$$\text{Ans.} \quad u(r, \theta) = \frac{4k}{\pi} \sum_{n=1,3,\ldots}^{\infty} \left[\left(\frac{r}{a} \right)^n \frac{1}{n} \sin n\theta \right]$$

7. Obtain the steady state temperature distribution in a semi-circular plate with radius a whose bounding diameter is kept at 0°C while the circumference is kept at 50°C.

$$\text{Ans.} \quad u = \frac{200}{\pi} \sum_{n=1}^{\infty} \frac{(r/a)^{2n-1}}{2n-1} \sin(2n-1)\theta$$

8. A semi-circular plate of radius 10 cm has insulated faces and heat flows in plane curves. The bounding diameter is kept at 0°C and on the circumference the temperature distribution is maintained at

$u(10,\theta) = 400(\pi\theta - \theta^2)/\pi, 0 \le \theta \le \pi.$

Determine the temperature distribution $u(r, \theta)$.

$$\text{Ans.} \quad u(r, \theta) = \frac{3200}{\pi^2} \sum_{n=1}^{\infty} \left[\frac{1}{(2n-1)^3} \left(\frac{r}{10} \right)^{2n-1} \sin(2n-1)\theta \right]$$

9. A semi-circular plate of radius '*a*' cm has insulated faces and heat flows in plane curves. The bounding diameter is kept at 0° C and the semi-circumference is maintained at a temperature given by

$$\frac{k\theta}{\pi} \quad \text{when } 0 \le \theta \le \frac{\pi}{2}$$

and $\qquad k(\pi-\theta)/\pi \quad$ when $\pi/2 \le \theta \le \pi.$

Find the temperature distribution in the steady state.

$$\text{Ans.} \quad u(r, \theta) = \frac{4k}{\pi^2} \sum_{n=1}^{\infty} \frac{(-1)^{n-1}}{(2n-1)^2} \left(\frac{r}{a} \right)^{2n-1} \sin(2n-1)\theta$$

20.11 Solution of Transmission line equations by the Method of Separation of Variables

Example 20.21 Using the method of separation of variables find the current i and voltatge V in a transmission line of l length, t seconds after the ends are suddenly grounded, given that $l(x, 0) = l_0$, $V(x, 0) = \sin\left(\frac{\pi x}{l} \right)$ and that R and G are negligible.

Solution. Method of Separation of Variables. Let z be the dependent variable and x and y two independent variables. In this method we assume the solution to be product of two functions, one of them a function of x alone and the other a function of y alone. In this way the solution is reduced to the solution of ordinary equations.

As R and G are to be neglected we use the equations (Radio equations)

$$\frac{\partial V}{\partial x} = -L\frac{\partial i}{\partial t} \text{ and } \frac{\partial i}{\partial x} = -C\frac{\partial V}{\partial t}.$$

$\therefore$
$$\frac{\partial^2 V}{\partial x^2} = -L\frac{\partial^2 i}{\partial x \partial t} \text{ and } \frac{\partial^2 i}{\partial x \partial t} = -C\frac{\partial^2 V}{\partial t^2}.$$

Eliminating i between these equations, we get

$$\frac{\partial^2 V}{\partial x^2} = LC\frac{\partial^2 V}{\partial t^2}. \qquad \text{...(1)}$$

Since the ends are suddenly grounded, we have
$$V(0, t) = 0, \ V(l, t) = 0. \qquad \text{...(2)}$$

Also the *initial conditions* are $V(x, 0) = V_0 \sin\frac{\pi x}{l}, \quad i(x, 0) = i_0 \qquad \text{...(3)}$

$\therefore$
$$\frac{\partial i(x,0)}{\partial x} = 0 \text{ gives } \frac{\partial V(x,0)}{\partial t} = 0 \quad \left(\because \frac{\partial i}{\partial x} = -C\frac{\partial V}{\partial t}\right) \qquad \text{...(4)}$$

Let $V = X(x)\, T(t)$ be the solution of (1).
Substituting in (1) we get

$$X'' T = LCXT''.$$

$\therefore$
$$\frac{X''}{X} = LC\frac{T''}{T} = -k^2 \ (\text{say}).$$

$\therefore \qquad X'' + k^2 X = 0 \text{ and } LCT'' + K^2 T = 0.$

Solving these equations, we get

$$V = \left(c_1 \cos kx + c_2 \sin kx\right)\left(c_3 \cos\frac{k}{\sqrt{LC}}t + c_4 \sin\frac{k}{\sqrt{LC}}t\right).$$

Using the boundary conditions (2) we get $c_1 = 0$ and $k = n\pi/l$.

$\therefore$
$$V = \sin\frac{n\pi x}{l}\left(A_n \cos\frac{n\pi}{l\sqrt{LC}}t + B_n \sin\frac{n\pi}{l\sqrt{LC}}t\right).$$

Using the initial conditions (4) we get $B_n = 0$.

$\therefore$
$$V = A_n \sin\frac{n\pi x}{l}\cos\frac{n\pi t}{l\sqrt{LC}}.$$

By the initial condition (3) $V_0 \sin \dfrac{\pi x}{l} = A_n \sin \dfrac{n\pi x}{l}$. Therefore, $A_n = V_0$, $n = 1$.

Hence $V(x,t) = V_0 \sin \dfrac{\pi x}{l} \cos \dfrac{\pi t}{l\sqrt{LC}}$ is the required voltage.

To get i we use the equation (1)

From the first, we get $\dfrac{\partial i}{\partial t} = -\dfrac{1}{L}\dfrac{\partial V}{\partial x} = -\dfrac{V_0 \pi}{Ll} \cos \dfrac{\pi x}{l} \cos \dfrac{\pi t}{l\sqrt{LC}}$. ...(5)

From the second $\dfrac{\partial i}{\partial x} = -C \dfrac{\partial V}{dt} = \dfrac{CV_0 \pi}{l\sqrt{LC}} \sin \dfrac{\pi x}{l} \sin \dfrac{\pi t}{l\sqrt{LC}}$. ...(6)

Integrating (6) w.r.t. 't' we get

$$i(x,\, t) = -V_0 \sqrt{\dfrac{C}{L}} \cos \dfrac{\pi x}{l} \sin \dfrac{\pi t}{l\sqrt{LC}} + \phi(x),$$

where $\phi(x)$ is an arbitrary function of x.

This value satisfies (6) if $\phi'(x) = 0$ i.e. $\phi(x)$ is constant. From (3) it is i_0.

$\therefore\ \ i(x,\, t) = i_0 - V_0 \sqrt{\dfrac{C}{L}} \cos \dfrac{\pi x}{l} \sin \dfrac{\pi t}{l\sqrt{LC}}$, which is the required current.

Example 20.22 Neglecting R and G, find the e.m.f. V(x, t) in a line of length *l*, t seconds after the ends were suddenly gorunded, given that

$$i(x,\, 0) = i_0 \text{ and } V(x,\, 0) = E_1 \sin \dfrac{\pi x}{l} + E_5 \sin \dfrac{5\pi x}{l} \qquad (S.V.T.U.,\ 2007)$$

Solution. (*i*) Since R and G are negligible, we use the

Radio equation $\dfrac{\partial^2 V}{\partial x^2} = LC \dfrac{\partial^2 V}{\partial t^2}$. ...(i)

Since the ends are suddenly grounded, we have the boundary conditions *viz.*
$V(0,\, t) = 0$, $V(l,\, t) = 0$ ...(ii)

Also the initial conditions are $i(x,\, 0) = i_0$,

and $V(x,\, 0) = E_1 \sin(\pi x/l) + E_5 \sin(5\pi x/l)$ ...(iii)

Now $\dfrac{\partial i}{\partial x}(x,\, 0) = -C \dfrac{\partial V}{\partial t}$ gives $\dfrac{\partial V}{\partial t}(x,0) = 0$ ...(iv)

Let $V = X(x)\, T(t)$ be the solution of (*i*).

Therefore, (*i*) gives $X''T = LCXT''$

or $\dfrac{X''}{X} = LC \dfrac{T''}{T} = -k^2 \text{ (say)}.$

Therefore $X'' + k^2 X = 0$ and $T'' + \left(k^2 / LC\, T = 0 \right)$

Solving these equations, we get

$$V = \left(c_1 \cos kx + c_2 \sin kx \right) \left(c_3 \cos \dfrac{k}{\sqrt{LC}} \cdot t + c_4 \sin \dfrac{k}{\sqrt{LC}} t \right).$$

Using the boundary conditions (*ii*), we get $c_1 = 0$ and $k = n\pi/l$.

Therefore, $\qquad V = \sin \dfrac{n\pi x}{l} \left(a_n \cos \dfrac{n\pi}{l\sqrt{LC}} t + b_n \sin \dfrac{n\pi}{l\sqrt{LC}} t \right).$

Using the initial condition $\dfrac{\partial V}{\partial t}(x, 0) = 0$ gives $b_n = 0.$

Therefore, $V = a_n \sin \dfrac{n\pi x}{l} \cos \dfrac{n\pi t}{l\sqrt{LC}}$ is a solution.

Thus the most general solution of (*i*) is

$$V = \sum_{n=1}^{\infty} a_n \sin \dfrac{n\pi x}{l} \cos \dfrac{n\pi t}{l\sqrt{LC}}.$$

Finally by the initial condition (*iii*), we have

$$E_1 \sin \dfrac{\pi x}{l} + E_5 \sin \dfrac{5\pi x}{l} = \sum a_n \sin \dfrac{n\pi x}{l}.$$

Therefore, $\qquad a_1 = E_1, a_5 = E_5,$ while all other a's are zero.

Hence $V(x, t) = E_1 \sin \dfrac{\pi x}{l} \cos \dfrac{\pi t}{l\sqrt{LC}} + E_5 \sin \dfrac{5\pi x}{l} \cos \dfrac{5\pi t}{l\sqrt{LC}},$ is the required

solution.

Example 20.23 A transmission line 1000 km long is initially under steady-state conditions with potential 1300 volts at the sending end (x = 0) and 1200 volts at the receiving end (x = 1000). The terminal end of the line is suddenly grounded, but the potential at the source is kept at 1300 volts. Assuming the inductance and leakance to be negligible, find the potential V(x, t).

(Andhra, 2000)

Solution. We use the telegraph equation

$$\dfrac{\partial^2 V}{\partial x^2} = RC \dfrac{\partial V}{\partial t} \quad \text{or} \quad \dfrac{\partial V}{\partial t} = \dfrac{1}{RC} \cdot \dfrac{\partial^2 V}{\partial x^2}. \qquad \text{...(i)}$$

V_s = initial steady voltage satisfying $\dfrac{\partial^2 V}{\partial x^2} = 0 = 1300 - \dfrac{1300 - 1200}{1000} x$

$$= 1300 - 0.1x = V(x, 0). \qquad \ldots (ii)$$

V'_s = Steady voltage (after grounding the terminal end) when steady conditions are ultimately reached $= 1300 - 1.3x$.

$\therefore V(x, t) = V'_s(x) + V_t(x, t)$, where $V_t(x, t)$ is the transient part

$$= (1300 - 1.3x) + \sum_{n=1}^{\infty} b_n \sin \frac{n\pi x}{l} \exp\left\{\frac{-n^2 \pi^2 t}{l^2 RC}\right\}. \qquad \ldots (iii)$$

Putting $t = 0$ in (iii), we have

or $\qquad 1300 - 0.1x = V(x, 0) = 1300 - 1.3x + \displaystyle\sum_{n=1}^{\infty} b_n \sin \frac{n\pi x}{l}$

or $\quad 1.2x = \displaystyle\sum_{n=1}^{\infty} b_n \sin \frac{n\pi x}{l}$ where $b_n = \dfrac{2}{l} \displaystyle\int_0^l 1.2 \sin \frac{n\pi x}{l} dx = \dfrac{2400}{\pi} \dfrac{(-1)^{n+1}}{n}$.

Hence $V(x, t) = 1300 - 1.3x + \dfrac{2400}{\pi} \displaystyle\sum_{n=1}^{\infty} \dfrac{(-1)^{n+1}}{n} \exp\left\{\dfrac{-n^2 \pi^2 t}{1000^2 RC}\right\} \sin \dfrac{n\pi x}{1000}$.

Exercise 20.4

1. In a telegraph wire, the sending end of the line is at a potential e_0, the far end being earthed until steady state conditions prevail. The sending end is suddenly earthed. Show that the potential at a point distant x from the sending end is given by

$$e(x, t) = \frac{2e_0}{\pi} \sum_{n=1}^{\infty} \frac{1}{n} \exp\left\{-\frac{n^2 \pi^2 t}{CRl^2}\right\} \sin \frac{n\pi x}{l},$$

where l is the length of the wire and C, R have their usual meanings.
[**Hint:** Writing e in place of V in the formula of Art. 20.1-4(b).]

we have to solve $\dfrac{\partial^2 e}{\partial x^2} = RC \dfrac{\partial e}{\partial t}$ or $\dfrac{\partial e}{\partial t} = \dfrac{1}{RC} \dfrac{\partial^2 e}{\partial x^2} = c^2 \dfrac{\partial^2 e}{\partial x^2}$ (say).

In steady state $e = \dfrac{e_0}{l}(l - x)$. Hence the *boundary conditions* under which

$\dfrac{\partial e}{\partial t} = c^2 \dfrac{\partial^2 e}{\partial x^2}$ is to be solved are $e(0\, t) = e(l, t) = 0$ $(t \geq 0)$

and $e(x, 0) = e_0(l - x)/l$ for all x in $(0, l)$.

Just like the solution of one dimensional heat equation,

we have $\qquad e(x, t) = (A\cos kx + B\sin kx)e^{-k^2c^2t}$.

Using the first two boundary conditions the general solution is

$$e(x, t) = \sum b_n \sin\frac{n\pi x}{l}\exp\left(-\frac{n^2\pi^2c^2t}{l^2}\right).$$

Last boundary condition gives

$$\frac{e_0}{l}(l-x) = \sum b_n \sin\frac{n\pi x}{l} \quad \text{where } b_n = 2e_0/(n\pi).$$

$$\therefore e(x, t) = \sum_{n=1}^{\infty} \frac{2e_0}{n\pi}\sin\frac{n\pi x}{l}\exp\left(-\frac{n^2\pi^2c^2t}{l^2}\right) = \frac{2e_0}{n\pi}\sum_{n=1}^{\infty}\frac{1}{n}\sin\frac{n\pi x}{l}\exp\left\{\frac{-n^2\pi^2t}{RCl^2}\right\}.]$$

2. A telegraph cable is of length l. Initially the line is uncharged so that $V(x, 0) = 0$. If at $t = 0$ the end $x = l$ is connected to a constant e.m.f. E, find $V(x, t)$ and $i(x, t)$. In particular, show that the current at the receiving end $x = 0$ is given by

$$i(0, t) = \frac{-E}{lR} + \frac{2E}{lR}\sum_{n=1}^{\infty}\left\{(-1)^{n+1}\exp\left(\frac{-n^2\pi^2t}{l^2RC}\right)\right\}.$$

Ans: $V(x, t) = \dfrac{Ex}{l} + 2E\sum_{n=1}^{\infty}\dfrac{(-1)^n}{n\pi}\sin\dfrac{n\pi x}{l}\exp\left\{-\dfrac{n^2\pi^2t}{l^2RC}\right\};$

$$i(x, t) = -\frac{E}{lR} + \frac{2E}{lR}\sum_{n=1}^{\infty}(-1)^{n+1}\cos\frac{n\pi x}{l}\exp\left\{-\frac{n^2\pi^2t}{l^2RC}\right\}.$$

[**Hint:** Here we have to solve the telegraph equation

$$\frac{\partial^2V}{\partial x^2} = RC\frac{\partial V}{\partial t}. \qquad \qquad ...(1)$$

Initially $V(x, 0) = 0$ and finally when the steady state conditions prevail, $V_s = Ex/l$.

To get V in the intermediate period, reckoning time from the instant when the end is connected to e.m.f. E, we assume

$V(x, t) = V_t(x, t) + V_s(x)$

where V_s is the steady state solution at the end and V_t is the transient part of the solution which satisfies (1) and tends to zero as $t \to \infty$ and which satisfies the conditions

$V_t(0, t) = 0$ and $V_t(l, t) = E$].

3. A telephone wire of length l is maintained at a steady state voltage distribution of 20 volts at one end and 12 volts at the other end. The latter end is suddenly grounded at time $t = 0$. Neglect leakage and inductance find the voltage and the current in the wire t seconds later.

Ans. $V(x, t) = \dfrac{20(l-x)}{l} + \dfrac{24}{\pi} \displaystyle\sum_{n=1}^{\infty} \dfrac{(-1)^{n+1}}{n} \sin \dfrac{n\pi x}{l} \cdot e^{-n^2\pi^2 t/RCl^2}$

$i(x, t) = \dfrac{20}{lR} + \dfrac{24}{lR} \displaystyle\sum_{n=1}^{\infty} (-1)^n \cos \dfrac{n\pi x}{l} e^{-n^2\pi^2 t/RCl^2}$.

4. Obtain the solution of the *radio equation,* $\dfrac{\partial^2 V}{\partial x^2} = LC \dfrac{\partial^2 V}{\partial t^2}$

appropriate to the case when a periodic e.m.f. $V_0 \cos pt$ is applied at the end $x = 0$ of the line.

Ans. $V(x, t) = V_0 \cos\left(pt - p\sqrt{LC}\, x\right)$.

Chapter 21

Laplace Transformation

21.1 Introduction

Laplace transformation, also known as operational calculus, originated from the operational methods used by the English Electrical Engineer Heaviside (1850-1925) in solving a variety of physical problems. Heaviside's methods were unsystematic and lacked rigour. It was Bromwich who along with other mathematicians gave rigorous proofs of Heaviside's methods and placed them on firm mathematical foundation.

The Laplace transform methods are widely used in various problems occurring in science and engineering and are particularly useful and effective in solving differential equations arising in initial value problems. By Laplace transformation certain partial differential equations are reduced to ordinary differential equations and ordinary differential equations are reduced to algebraic equations which are easier to solve. The solution of the initial value problem is directly found without first finding the general solution.

21.2 Definition of the Laplace transform

Let $f(t)$ be a function defined for $t \geq 0$. Then the integral $\int_0^\infty e^{-st} f(t) dt$, if it exists, is a function of s, say, $F(s)$. The parameter s may be a real or complex number.

The function $F(s)$ is called the **Laplace transform** of $f(t)$ and is denoted by $\mathcal{L}\{f(t)\}$.

Thus
$$\mathcal{L}\{f(t)\} = F(s) = \int_0^\infty e^{-st} f(t) dt. \qquad \qquad ...(21.1)$$

To get the Laplace transform of $f(t)$ we multiply it by e^{-st} and integrate with respect to t from 0 to ∞. This operation is called **Laplace transformation.**

21.2.1 Notation

In general we shall use a lower case letter to denote the function being transformed

and their corresponding capital letter to denote its Laplace transform; for example,
$$\mathcal{L}\{f(t)\} = F(s), \quad \mathcal{L}\{g(t)\} = G(s), \quad \mathcal{L}\{y(t)\} = Y(s).$$

The symbol $\mathcal{L}$ which transforms $f(t)$ into $F(s)$ is often called the **Laplace transformation operator**.

21.2.2 Linearity of Laplace transformation

Let $f(t)$ and $g(t)$ be any two functions whose Laplace transforms exist. Then, for any two constants α, β, we have
$$\mathcal{L}\{\alpha f(t) + \beta g(t)\} = \alpha \mathcal{L}\{f(t)\} + \beta \mathcal{L}\{g(t)\} = \alpha F(s) + \beta G(s).$$

The above result is known as **Linearity property** of Laplace transformation and follows from its definition.

For
$$\mathcal{L}\{\alpha f(t) + \beta g(t)\} = \int_0^\infty e^{-st}\{\alpha f(t) + \beta g(t)\}\, dt$$

$$= \alpha \int_0^\infty e^{-st} f(t)dt + \beta \int_0^\infty e^{-st} g(t)dt$$

$$= \alpha\mathcal{L}\{f(t)\} + \beta\mathcal{L}\{g(t)\} = \alpha F(s) + \beta G(s). \qquad ...(21.2)$$

21.3 Laplace transform of some elementary functions

Example 21.1 Using the definition, find the Laplace transform of the following functions.

(a) $f(t) = 1$,   (b) $f(t) = t$,   (c) $f(t) = t^2$,   (d) $f(t) = t^n$.

Solution. From the definition, we have

(a) $\mathcal{L}\{f(t)\} = \mathcal{L}(1) = \int_0^\infty e^{-st} \cdot 1\, dt = \left[-\dfrac{1}{s}e^{-st}\right]_0^\infty = \dfrac{1}{s} \qquad s > 0.$

(b) $\mathcal{L}(t) = \int_0^\infty \underset{\text{II}}{e^{-st}} \cdot \underset{\text{I}}{t}\, dt$

$= \left[t\dfrac{e^{-st}}{-s}\right]_0^\infty + \dfrac{1}{s}\int_0^\infty e^{-st} dt$, Integrating by parts, $= 0 + \dfrac{1}{s}\cdot\dfrac{1}{s} = \dfrac{1}{s^2}, s > 0.$

$\left(\because \lim_{t\to\infty} te^{-st} = \lim_{t\to\infty}\dfrac{t}{e^{st}} = \lim_{t\to\infty}\dfrac{1}{se^{st}} = 0 \text{ by L' Hospital's rule.}\right)$

(c) $\mathcal{L}(t^2) = \int_0^\infty \underset{\text{II}}{e^{-st}}\, \underset{\text{I}}{t^2}\, dt = \left[t^2\left(\dfrac{e^{-st}}{-s}\right) - (2t)\left(\dfrac{e^{-st}}{s^2}\right) + (2)\left(\dfrac{e^{-st}}{-s^3}\right)\right]_0^\infty$

(by generalised rule of integration by parts),

$$= \left[\frac{-t^2}{se^{st}} - \frac{2t}{s^2e^{st}} - \frac{2}{s^3e^{st}} \right]_0^\infty = \frac{2}{s^3}, s > 0.$$

(d) $\mathcal{L}\{t^n\} = \int_0^\infty e^{-st}t^n dt$. Putting $st = y$ and $dt = \dfrac{dy}{s}$, we have

$$\mathcal{L}\{t^n\} = \frac{1}{s^{n+1}} \int_0^\infty e^{-y} y^n dy = \frac{\Gamma(n+1)}{s^{n+1}}, \text{ if } s > 0.$$

If n is an integer $\mathcal{L}\{t^n\} = \dfrac{(n)!}{s^{n+1}}$.

Obviously the Laplace transforms of t, t^2, t^3, etc., can also be deduced from $\mathcal{L}\{t^n\}$.

Example 21.2 Find the Laplace transform of the folloing functions.

(a) e^{at}, (b) $\sin at$, (c) $\cos at$, (d) $\sinh \omega t$, (e) $\cosh \omega t$, (f) $e^{-2t + 5}$.

Solution. We have (a) $\mathcal{L}\{e^{at}\} = \int_0^\infty e^{-st} e^{at} dt = \int_0^\infty e^{-(s-a)t} dt$

$$= \left[-\frac{e^{-(s-a)t}}{s-a} \right]_0^\infty = \frac{1}{s-a}, \quad s > a.$$

Note: The Laplace transform of $e^{-3t} = \dfrac{1}{s+3}, s > -3$.

(b) $\mathcal{L}\{\sin at\} = \int_0^\infty e^{-st} \sin at\, dt = \left[\frac{e^{-st}}{s^2 + a^2}(-s\sin at - a\cos at) \right]_0^\infty = \frac{a}{s^2 + a^2}, s > 0.$

$$\left[\because \int e^{bx} \sin cx\, dx = \frac{e^{bx}}{b^2 + c^2}(b\sin cx - c\cos cx) \right]$$

(c) $\mathcal{L}\{\cos at\} = \int_0^\infty e^{-st} \cos at\, dt = \left[\frac{e^{-st}}{s^2 + a^2}(-s\cos at + a\sin at) \right]_0^\infty$

$$\left[\because \int e^{bx} \cos cx\, dx = \frac{e^{bx}}{b^2 + c^2}(b\cos cx + c\sin cx) \right]$$

$$= \frac{s}{s^2 + a^2}, s > 0.$$

Alternative We have $\mathcal{L}\{e^{iat}\} = \dfrac{1}{s-ia}$, by part (a)

or $\qquad \mathcal{L}\{\cos at + i \sin at\} = \dfrac{s+ia}{(s-ia)(s+ia)} = \dfrac{s+ia}{s^2+a^2}$

or $\qquad \mathcal{L}\{\cos at\} + i\,\mathcal{L}\{\sin at\} = \dfrac{s+ia}{s^2+a^2}.$ $\qquad$ [by linearity property]

Equating real and imaginary parts, we obtain

$$\mathcal{L}\{\cos at\} = \frac{s}{s^2+a^2}, \quad \text{and} \quad \mathcal{L}\{\sin at\} = \frac{a}{s^2+a^2}.$$

(d) $\mathcal{L}\{\sinh \omega t\} = \mathcal{L}\left\{\dfrac{e^{\omega t} - e^{-\omega t}}{2}\right\} = \dfrac{1}{2}\left[\mathcal{L}\{e^{\omega t}\} - \mathcal{L}\{e^{-\omega t}\}\right]$ (by linearity property)

$$= \frac{1}{2}\left(\frac{1}{s-\omega} - \frac{1}{s+\omega}\right) = \frac{\omega}{s^2-\omega^2}, s > \omega \geq 0.$$

(e) $\mathcal{L}\{\cosh \omega t\} = \mathcal{L}\left\{\dfrac{1}{2}\left(e^{\omega t} + e^{-\omega t}\right)\right\} = \dfrac{1}{2}\mathcal{L}\{e^{\omega t}\} + \dfrac{1}{2}\mathcal{L}\{e^{-\omega t}\}$

$$= \frac{1}{2}\left(\frac{1}{s-\omega} + \frac{1}{s+\omega}\right) = \frac{s}{s^2-\omega^2}, s > \omega \geq 0.$$

(f) $\mathcal{L}\{e^{-2t+5}\} = \mathcal{L}\{e^{-2t} \cdot e^5\} = e^5\,\mathcal{L}\{e^{-2t}\} = e^5/(s+2), s > -2.$

Table 21.1. Laplace Transform of elementary functions.

Sr. No.	$f(t)$	$\mathcal{L}\{f(t)\} = F(s)$	Sr. No.	$f(t)$	$\mathcal{L}\{f(t)\} = F(s)$
1	1	$\dfrac{1}{s}, s > 0$	6	e^{at}	$\dfrac{1}{s-a}, s > a$
2	t	$\dfrac{1}{s^2}, s > 0$	7	$\sin at$	$\dfrac{a}{s^2+a^2}, s > 0$
3	t^2	$\dfrac{(2)!}{s^3}, s > 0$	8	$\cos at$	$\dfrac{s}{s^2+a^2}, s > 0$
4	t^n $(n = 0,1,2,\ldots)$	$\dfrac{(n!)}{s^{n+1}}, s > 0$	9	$\sinh \omega t$	$\dfrac{\omega}{s^2-\omega^2}, s > \omega$
5	t^α $(\alpha \text{ positive})$	$\dfrac{\Gamma(\alpha+1)}{s^{\alpha+1}},$	10	$\cosh \omega t$	$\dfrac{s}{s^2-\omega^2}, s > \omega$

Remark

In entry 5, $\Gamma(\alpha + 1)$ is the *gamma function* defined by

$$\Gamma(\alpha+1) = \int_0^\infty t^\alpha e^{-t}\,dt, \alpha > -1.$$

A *recursion* or *recurrence formula* for the gamma function is
$\Gamma(\alpha + 1) = \alpha\Gamma(\alpha)$. For our present purposes, however, we need only the following properties:

$$\Gamma(1/2) = \sqrt{\pi}, \Gamma(1) = 1.$$ If α is any positive integer n, then $\Gamma(n + 1) = n!$, thus explaining the relationship of entries 4 and 5 of the table 21.1.

21.4 Sufficient conditions for the existence of Laplace transforms

If $f(t)$ is a *piecewise continuous* function on the interval $[0, \infty)$ and is of **exponential order α for $t \geq 0$, then Laplace transform, $\mathcal{L}\{f(t)\}$ exits for $s > \alpha$.
It is important to note that the above conditions are sufficient but not necessary

---

*Piecewise Continuity
A function $f(t)$ is said to be *piecewise continuous* in an interval if (*i*) the interval can be divided into a finite number of subintervals in each of which $f(t)$ is continuous and (*ii*) the limits of $f(t)$ as t approaches the end points of each subinterval are finite. An example of a piecewise continuous function on a finite interval $a \leq t \leq b$ is shown in Fig. 21.1.

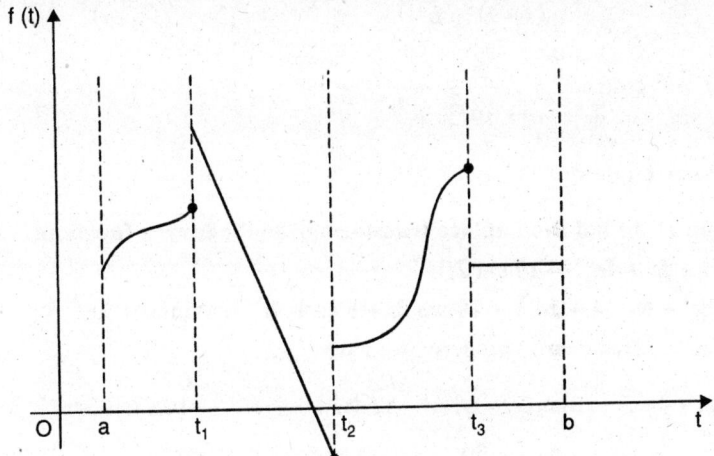

Fig. 21.1. Piecewise continuous function $f(t)$

**Exponential order. A function $f(t)$ is said to be of *exponential order* α, if there exists constants α *and* $M > 0$ such that $|f(t)| \leq Me^{\alpha t}$, $t \geq 0$. Geometrically, this condtion implies that the graph of $f(t)$, $t > 0$ does not grow faster than the graph of the exponential function $\phi(t) = Me^{\alpha t}$, $\alpha > 0$.

for the existence of the Laplace transform. For example, consider the function $f(t) = t^{-1/2}$. This function is not piecewise continuous on the interval $[0, \infty)$, but its Laplace transform exists.

$$\mathcal{L}\{t^{-1/2}\} = \frac{\Gamma(1/2)}{s^{(1/2)}} = \frac{\sqrt{\pi}}{\sqrt{s}} = \sqrt{\pi/s}, \, s > 0.$$

21.5 First translation theorem or first shifting theorem: Replacement of s by $s - a$ in the transform

If $\mathcal{L}\{f(t)\} = F(s)$ and 'a' is any real number, then $\mathcal{L}\{e^{at} f(t)\} = F(s - a)$.
Proof:

Since $F(s) = \int_0^\infty e^{-st} f(t) \, dt$.

$$\therefore F(s-a) = \int_0^\infty e^{-(s-a)t} f(t) \, dt = \int_0^\infty e^{-st} \left[e^{at} f(t) \right] dt = \mathcal{L}\{e^{at} f(t)\}. \quad \text{...(21.3)}$$

The following results can be easily established by virtue of this property:

1. $\mathcal{L}\{e^{at} t^n\} = \dfrac{n!}{(s-a)^{n+1}}$ 2. $\mathcal{L}\{e^{-at} \sin bt\} = \dfrac{b}{(s+a)^2 + b^2}$

3. $\mathcal{L}\{e^{at} \cos bt\} = \dfrac{s-a}{(s-a)^2 + b^2}$ 4. $\mathcal{L}\{e^{at} \sinh \omega t\} = \dfrac{\omega}{(s-a)^2 - \omega^2}$

5. $\mathcal{L}\{e^{at} \cosh \omega t\} = \dfrac{s-a}{(s-a)^2 - \omega^2}$

Illustrative Examples

Example 21.3 Find the Laplace transform of the following functions.
(a) $\{1 + 5t + 4e^{-3t} - 10 \sin 2t\}$,
(b) $\{8t^3 + 6t - 4 + 2e^{-5t} + 10 \sin 3t + 5 \cosh 2t\}$, (c) $\sin(3t + 2)$.

Solution. (a) By linearity property, we have

$$\mathcal{L}\{1 + 5t + 4e^{-3t} - 10 \sin 2t\} = \mathcal{L}\{1\} + 5\mathcal{L}\{t\} + 4\mathcal{L}\{e^{-3t}\} - 10\mathcal{L}\{\sin 2t\}$$

$$= \frac{1}{s} + \frac{5}{s^2} + \frac{4}{s+3} - \frac{20}{s^2 + 4}.$$

(b) By linearity property, we have

$$\mathcal{L}\{8t^3 + 6t - 4 + 2e^{-5t} + 10 \sin 3t + 5 \cosh 2t\}$$

$$= 8\mathcal{L}\{t^3\} + 6\mathcal{L}\{t\} - 4\mathcal{L}\{1\} + 2\mathcal{L}\{e^{-5t}\} + 10\mathcal{L}\{\sin 3t\} + 5\mathcal{L}\{\cosh 2t\}$$

$$= 8 \cdot \frac{3!}{s^4} + 6 \cdot \frac{1}{s^2} - 4 \cdot \frac{1}{s} + 2 \cdot \frac{1}{s+5} + 10 \cdot \frac{3}{s^2+9} + 5 \cdot \frac{s}{s^2-4}$$

$$= \frac{48}{s^4} + \frac{6}{s^2} - \frac{4}{s} + \frac{2}{s+5} + \frac{30}{s^2+9} + \frac{5s}{s^2-4}.$$

(c) $\mathcal{L} \{\sin(3t + 2)\} = \mathcal{L} \{(\sin 3t \cos 2 + \cos 3t \sin 2)\}$

$$= \mathcal{L} \{(\sin 3t \cos 2)\} + \mathcal{L} \{(\cos 3t \sin 2)\}$$

$$= \cos 2 \, \mathcal{L} \{\sin 3t\} + \sin 2 \, \mathcal{L} \{\cos 3t\}$$

$$= \cos 2 \left(\frac{3}{s^2+9} \right) + \sin 2 \left(\frac{s}{s^2+9} \right) = (3 \cos 2 + s \sin 2)/(s^2 + 9).$$

Example 21.4 Find the Laplace transform of each of the following functions.

(a) $(2t^2 - 1)^2$, (b) $4 \cos^2 2t$, (c) $\cos^3 t$, (d) $\sinh^2 2t$, (e) $\sin 3t \cos 4t$.

Solution. (a) $\mathcal{L} \{2t^2 - 1\}^2 = \mathcal{L} \{4t^4 - 4t^2 + 1\} = 4 \mathcal{L} \{t^4\} - 4 \, \mathcal{L} \{t^2\} + \mathcal{L} \{1\}$

$$= 4 \cdot \frac{4!}{s^5} - 4 \cdot \frac{(2)!}{s^3} + \frac{1}{s} = \frac{1}{s^5} (96 - 8s^2 + s^4).$$

(b) $\mathcal{L} \{4 \cos^2 2t\} = 2 \mathcal{L} \{1 + \cos 4t\} = 2 \mathcal{L} \{1\} + 2 \mathcal{L} \{\cos 4t\}$

$$= \frac{2}{s} + \frac{2s}{s^2+16}, s > 0$$

(c) $\mathcal{L} \{\cos^3 t\} = \mathcal{L} \left\{ \frac{\cos 3t + 3 \cos t}{4} \right\} = \frac{1}{4} \mathcal{L} \{\cos 3t\} + \frac{3}{4} L \{\cos t\}$

$$= \frac{1}{4} \cdot \frac{s}{s^2+9} + \frac{3}{4} \cdot \frac{s}{s^2+1} = \frac{s^2 + 7s}{(s^2+9)(s^2+1)}.$$

(d) $\mathcal{L} \{\sinh^2 2t\} = \mathcal{L} \left(\frac{\cosh 4t - 1}{2} \right) = \frac{1}{2} [\mathcal{L} \{\cosh 4t\} - \mathcal{L} \{1\}]$

$$= \frac{1}{2} \left[\frac{s}{s^2-16} - \frac{1}{s} \right] = \frac{8}{s(s^2-16)}.$$

(e) $\mathcal{L} \{\sin 3t \cos 4t\} = \mathcal{L} \left\{ \frac{\sin 7t - \sin t}{2} \right\} = \frac{1}{2} [\mathcal{L} \{\sin 7t\} - \mathcal{L} \{\sin t\}]$

$$= \frac{1}{2} \left(\frac{7}{s^2+49} - \frac{1}{s^2+1} \right) = \frac{3s^2 - 21}{(s^2+1)(s^2+49)}.$$

Example 21.5 Find the Laplace transform of each of the following functions.

(a) $-3/\sqrt{t}$, (b) $5t^4 - 2t^{3/2} + 8$, (c) $3 \sqrt[3]{t} + 4e^{2t}$.

Solution. (a) $L\left\{\dfrac{-3}{\sqrt{t}}\right\} = -3L\{t^{-1/2}\} = \dfrac{-3\Gamma(1/2)}{s^{1/2}} = \dfrac{-3\sqrt{\pi}}{\sqrt{s}} = -3\sqrt{\pi/s}, s > 0.$

(b) $L\{5t^4 - 2t^{3/2} + 8\} = 5L\{t^4\} - 2L\{t^{3/2}\} + 8L\{1\} = \dfrac{5(4!)}{s^5} - \dfrac{2\cdot\Gamma(5/2)}{s^{5/2}} + \dfrac{8}{s}$

$$= \dfrac{5.4!}{s^5} - \dfrac{2\cdot(3/2)(1/2)\Gamma(1/2)}{s^{5/2}} + \dfrac{8}{s} = \dfrac{120}{s^5} - \dfrac{3\sqrt{\pi}}{2s^{5/2}} + \dfrac{8}{s}.$$

(c) $L\{3\sqrt[3]{t} + 4e^{2t}\} = 3L\{t^{1/3}\} + 4L\{e^{2t}\} = \dfrac{3\Gamma(4/3)}{s^{4/3}} + \dfrac{4}{s-2}$

$$= \dfrac{3(1/3)\Gamma(1/3)}{s^{4/3}} + \dfrac{4}{s-2} = \dfrac{\Gamma(1/3)}{s^{4/3}} + \dfrac{4}{s-2}.$$

Example 21.6 Using the first shifting theorem, find the Laplace transform of the following functions.

(a) $t^n e^{-at}$,   (b) $e^{-3t} \sin 4t$,   (c) $\cosh at \cos at$,   (d) $e^{4t} \cosh 5t$.

Solution. (a) We have $L\{t^n\} = \dfrac{n!}{s^{n+1}}$ $(n = 0, 1, 2, 3,...) = F(s)$.

Hence, by the translation theorem, we obtain $L\{e^{-at} t^n\} = \dfrac{n!}{(s+a)^{n+1}}$.

(b) We have $L\{\sin 4t\} = \dfrac{4}{s^2 + 16} = F(s)$. Hence, by the first translation theorem

$$L\{e^{-3t} \sin 4t\} = \dfrac{4}{(s+3)^2 + 16} = \dfrac{4}{s^2 + 6s + 25}.$$

(c) $L\{\cos at\} = \dfrac{s}{s^2 + a^2} = F(s)$.

$$\therefore L\{\cosh at \cos at\} = L\left\{\left(\dfrac{e^{at} + e^{-at}}{2}\right)\cos at\right\}$$

$$= \dfrac{1}{2}\left[L\{e^{at} \cos at\} + L\{e^{-at} \cos at\}\right]$$

$$= \dfrac{1}{2}\left[\dfrac{s-a}{(s-a)^2 + a^2} + \dfrac{s+a}{(s+a)^2 + a^2}\right] \text{(by the translation theorem)} = \dfrac{s^3}{s^4 + 4a^4}.$$

(d) We know $\mathcal{L}\{\cosh 5t\} = \dfrac{s}{s^2 - 25}$.

Hence $\mathcal{L}\{e^{4t} \cosh 5t\} = \dfrac{s-4}{(s-4)^2 - 25} = \dfrac{s-4}{s^2 - 8s - 9}$.

Alternative $\mathcal{L}\{e^{4t} \cosh 5t\} = \mathcal{L}\left\{ e^{4t} \left(\dfrac{e^{5t} + e^{-5t}}{2} \right) \right\} = \dfrac{1}{2} \mathcal{L}\{e^{9t} + e^{-t}\}$

$$= \dfrac{1}{2}\left(\dfrac{1}{s-9} + \dfrac{1}{s+1} \right) = \dfrac{s-4}{s^2 - 8s - 9}.$$

Example 21.7 Using the definition, find the Laplace transform of the function.

$$f(t) = \begin{cases} \cos t, & 0 \le t < \pi, \\ 0, & t \ge \pi. \end{cases}$$

Solution. By definition $\mathcal{L}\{f(t)\} = \int_0^\infty e^{-st} f(t)\, dt = \int_0^\pi e^{-st} f(t)\, dt + \int_\pi^\infty e^{-st} f(t)\, dt$

$$= \int_0^\pi e^{-st} \cos t\, dt + \int_\pi^\infty e^{-st} 0\, dt = \left[\dfrac{e^{-st}}{1+s^2}(-s\cos t + \sin t) \right]_0^\pi + 0 = \dfrac{s(1 + e^{-\pi s})}{1 + s^2}$$

Example 21.8 If $\mathcal{L}\{f(t)\} = F(s)$, show that $\mathcal{L}\{f(at)\} = (1/a)\, F(s/a)$.

Solution. $\mathcal{L}\{f(at)\} = \int_0^\infty e^{-st} f(at)\, dt = \int_0^\infty e^{-s(u/a)} f(u)\, d(u/a)$

$$= \dfrac{1}{a} \int_0^\infty e^{-su/a} f(u)\, du = \dfrac{1}{a} F\left(\dfrac{s}{a} \right), \text{ using the transformation } t = u/a.$$

21.6 Laplace Transform of the Derivatives of f(t)

Theorem. If $f(t)$, $t \ge 0$ is a continuous function and is of exponential order on $[0, \infty)$, and if $f'(t)$ is piecewise continuous on $[0, \infty)$, then $\mathcal{L}\{f'(t)\} = sF(s) - f(0)$.

Proof. Using integration by parts, we have $\mathcal{L}\{f'(t)\} = \int_0^\infty e^{-st} f'(t)\, dt$

$$= \left[e^{-st} f(t) \right]_0^\infty + s \int_0^\infty e^{-st} f(t)\, dt$$

$$= -f(0) + s\, \mathcal{L}\{f(t)\} = -f(0) + sF(s)$$

$$\therefore \quad \mathcal{L}\{f'(t)\} = sF(s) - f(0) \qquad \qquad \text{...(21.4)}$$

Similarly $\qquad \mathcal{L}\{f''(t)\} = s^2 F(s) - sf(0) - f'(0).$

Proof. Since
$$\mathcal{L}\{f'(t)\} = s\mathcal{L}\{f(t)\} - f(0)$$
$$\therefore\ \mathcal{L}\{f''(t)\} = s\mathcal{L}\{f'(t)\} - f'(0)$$
$$= s[s\mathcal{L}\{f(t)\} - f(0)] - f'(0)$$
$$= s^2\mathcal{L}\{f(t)\} - sf(0) - f'(0),$$

That is
$$\mathcal{L}\{f''(t)\} = s^2 F(s) - sf(0) - f'(0). \qquad ...(21.5)$$

In like manner, it can be shown that
$$\mathcal{L}\{f'''(t)\} = s^3 F(s) - s^2 f(0) - sf'(0) - f''(0) \qquad ...(21.6)$$

Laplace transform of the derivative of any order n

$$\mathcal{L}\{f^n(t)\} = s^n F(s) - s^{n-1} f(0) - s^{n-2} f'(0) - ... - sf^{(n-2)}(0) - f^{(n-1)}(0), \quad ...(21.7)$$

where
$$F(s) = \mathcal{L}\{f(t)\}.$$

Example 21.9 Using the formulas of Laplace transforms of derivatives, find the Laplace transforms of the following functions.

(a) $t \sin \omega t$ (b) $\sin^2 t$,   (c) t^3.

Solution. (a) We have $f(t) = t \sin \omega t$. Differentiating, we obtain
$f'(t) = \sin \omega t + \omega t \cos \omega t$,
$$f''(t) = \omega \cos \omega t + \omega \cos \omega t - \omega^2 t \sin \omega t = 2\omega \cos \omega t - \omega^2 t \sin \omega t$$
and $f(0) = 0, f'(t) = 0$. Hence, from the result
$$\mathcal{L}\{f''(t)\} = s^2 \mathcal{L}\{f(t)\} - sf(0) - f'(0).$$

We have $\mathcal{L}\{(2 \omega \cos \omega t - \omega^2 t \sin \omega t)\} = s^2\ \mathcal{L}\{t \sin \omega t\}$

or $2\omega\mathcal{L}(\cos \omega t) - \omega^2\mathcal{L}\{t \sin \omega t\} = s^2\mathcal{L}\{t \sin \omega t\}$

or $(s^2 + \omega^2)\mathcal{L}\{t \sin \omega t\} = 2\omega \cdot \dfrac{s}{s^2 + \omega^2}$ or $\mathcal{L}\{t \sin \omega t\} = \dfrac{2\omega s}{\left(s^2 + \omega^2\right)^2}.$

(b) We have $f(t) = \sin^2 t$ and $f(0) = 0$. Differentiating, we obtain
$f'(t) = 2 \sin t \cos t = \sin 2t$. Using the formula
$\mathcal{L}\{f'(t)\} = s\mathcal{L}\{f(t)\} - f(0)$, we obtain

$$\mathcal{L}\{\sin 2t\} = s\mathcal{L}\{f(t)\},\ \text{ or }\ \frac{2}{s^2 + 4} = s\mathcal{L}\{\sin^2 t\}\ ,\ \text{ or }\ \mathcal{L}\{\sin^2 t\} = \frac{2}{s(s^2 + 4)}.$$

(c) We have $f(t) = t^3, f'(t) = 3t^2, f''(t) = 6t, f'''(t) = 6$,
and $f(0) = f'(0) = f''(0) = 0$. Hence
$$\mathcal{L}\{f'''(t)\} = s^3\ \mathcal{L}\{f(t)\} - s^2 f(0) - sf'(0) - f''(0)$$
or $\mathcal{L}\{6\} = s^3\ \mathcal{L}\{t^3\}$, or $6/s = s^3\ \mathcal{L}\{t^3\}$ or $\mathcal{L}\{t^3\} = 6/s^4$.

21.7 Laplace Transform of the Integral of a Function

Theorem. If $\mathcal{L}\{f(t)\} = F(s)$, then $\mathcal{L}\left\{\displaystyle\int_0^t f(u)\, du\right\} = \dfrac{1}{s}F(s).$

Proof. Let $g(t) = \int_0^t f(u)\,du$. Then $g'(t) = f(t)$ and $g(0) = 0$.

Taking the Laplace transform of both sides, we have

$$\mathcal{L}\{g'(t)\} = s\mathcal{L}\{g(t)\} - g(0) = s\mathcal{L}\{g(t)\} = \mathcal{L}\{f(t)\} = F(s)$$

Thus $\qquad \mathcal{L}\{g(t)\} = \dfrac{F(s)}{s} \quad$ or $\quad \mathcal{L}\left\{\int_0^t f(u)\,du\right\} = \dfrac{F(s)}{s}. \qquad$ ...(21.8)

21.8 Multiplying a function by t^n.

The Laplace transform of the product of a function $f(t)$ with t can be found by differentiating the Laplace transform of $f(t)$. If $F(s) = \mathcal{L}\{f(t)\}$ and $n = 1, 2, 3,...$ and if we assume that interchanging of differentiation and integration is possible,

then $\mathcal{L}\{t^n\, f(t)\} = (-1)^n \dfrac{d^n}{ds^n} F(s)$.

Proof: By definition $\qquad F(s) = \int_0^\infty e^{-st} f(t)\,dt$.

Then by Leibnitz's rule for differentiating under the integral sign

$$\frac{d}{ds} F(s) = F'(s) = \frac{d}{ds} \int_0^\infty e^{-st} f(t)\,dt = \int_0^\infty \frac{\partial}{\partial s}\left\{e^{-st} f(t)\right\} dt$$

$$= -\int_0^\infty e^{-st}\left\{t\, f(t)\right\} dt = -\mathcal{L}\{t\, f(t)\},$$

that is, $\qquad \mathcal{L}\{t\, f(t)\} = -\dfrac{d}{ds}\mathcal{L}\{f(t)\}. \qquad$ ...(21.9)

Similarly $\qquad \mathcal{L}\{t^2 f(t)\} = \mathcal{L}\{t \cdot t\, f(t)\} = -\dfrac{d}{ds}\mathcal{L}\{t\, f(t)\}$.

$$= -\frac{d}{ds}\left(-\frac{d}{ds}\mathcal{L}\{f(t)\}\right) = \frac{d^2}{ds^2}\mathcal{L}\{f(t)\}. \qquad ...(21.10)$$

The preceding two cases suggest the general result for $\mathcal{L}\{t^n f(t)\}$.
If $F(s) = \mathcal{L}\{f(t)\}$ and $n = 1, 2, 3, ...,$ then

$$\mathcal{L}\{t^n f(t)\} = (-1)^n \frac{d^n}{ds^n} F(s) = (-1)^n F^{(n)}(s). \qquad ...(21.11)$$

21.9 Division by t

Let $f(t)$ be piecewise continuous on $[0, \infty)$ and be of exponential order. Then

$$\mathcal{L}\left\{\frac{f(t)}{t}\right\} = \int_s^\infty F(s)\,ds, \text{ provided } \lim_{t\to 0}\left[f(t)/t\right] \text{ exists, where } \mathcal{L}\{f(t)\} = F(s).$$

Proof. From the definition, we have

$$F(s) = \int_0^\infty e^{-st} f(t) dt.$$

Integrating both sides w.r.t. s from s to ∞, we have

$$\int_s^\infty F(s) ds = \int_s^\infty \left[\int_0^\infty e^{-st} f(t) dt \right] ds.$$

As s and t are independent, the order of integration on the right hand side can be changed.

$$\therefore \int_s^\infty F(s) ds = \int_0^\infty \left[\int_s^\infty e^{-st} ds \right] f(t) dt = \int_0^\infty \left[\frac{e^{-st}}{-t} \right]_s^\infty f(t) dt$$

$$= \int_0^\infty e^{-st} \cdot \left[\frac{f(t)}{t} \right] dt = \mathcal{L}\left\{ \frac{f(t)}{t} \right\}.$$

Hence
$$\mathcal{L}\left\{ \frac{f(t)}{t} \right\} = \int_s^\infty F(s) ds. \qquad \qquad ...(21.12)$$

Example 21.10 Find the Laplace transforms of the following functions.

(a) $t \cos 3t$,   (b) $t^2 \sin bt$,   (c) $t^2 e^{3t}$,   (d) $t^2 e^t \sin 4t$.

Solution. (a) Since $\mathcal{L}\{\cos 3t\} = \dfrac{s}{(s^2 + 9)}$, we obtain

$$\mathcal{L}\{t \cos 3t\} = -\frac{d}{ds}\left(\frac{s}{s^2 + 9} \right) = -\frac{(s^2 + 9) - s(2s)}{(s^2 + 9)^2} = \frac{s^2 - 9}{(s^2 + 9)^2}.$$

(b) Since $\mathcal{L}\{\sin bt\} = \dfrac{b}{s^2 + b^2}$, we obtain

$$\mathcal{L}\{t \sin bt\} = -\frac{d}{ds}\left(\frac{b}{s^2 + b^2} \right) = \frac{2bs}{(s^2 + b^2)^2}$$

and
$$\mathcal{L}\{t^2 \sin t\} = -\frac{d}{ds}\left\{ \frac{2bs}{(s^2 + b^2)^2} \right\} = \frac{-2b(s^2 + b^2)^2 + 8bs^2(s^2 + b^2)}{(s^2 + b^2)^4}$$

$$= \frac{2b(3s^2 - b^2)}{(s^2 + b^2)^3}.$$

(c) We have $\mathcal{L}\{e^{3t}\} = \dfrac{1}{(s-3)}$. Hence, $\mathcal{L}\{t^2 e^{3t}\} = (-1)^2 \dfrac{d^2}{ds^2}\left(\dfrac{1}{s-3} \right) = \dfrac{2}{(s-3)^3}$.

(d) Since $\mathcal{L}\{t^2\} = \dfrac{2}{s^3}$. We have

$$\mathcal{L}\{t^2 e^{i4t}\} = \frac{2}{(s-4i)^3} = \frac{2(s+4i)^3}{(s-4i)^3(s+4i)^3} = \frac{2(s+4i)^3}{\left(s^2+16\right)^3}.$$

or

$$\mathcal{L}\{t^2(\cos 4t + i\sin 4t)\} = \frac{2\left(s^3-48s\right)+8i\left(3s^2-16\right)}{\left(s^2+16\right)^3}.$$

Equating imaginary parts on both sides, we have $\mathcal{L}\{t^2 \sin 4t\} = \dfrac{8\left(3s^2-16\right)}{\left(s^2+16\right)^3}.$

Hence, by the first translation theorem, we obtain

$$\mathcal{L}\{e^t t^2 \sin 4t\} = \frac{8\left[3(s-1)^2-16\right]}{\left[(s-1)^2+16\right]^3} = \frac{8\left(3s^2-6s-13\right)}{\left(s^2-2s+17\right)^3}.$$

Example 21.11 Find the Laplace transform of the following functions.

(a) $\dfrac{\sin at}{t}$,　　(b) $\dfrac{e^{-at}-e^{-bt}}{t}$,　　(c) $\dfrac{1-\cos t}{t^2}$.

Solution. (a) By Article 21.9, we obtain

$$\mathcal{L}\left\{\frac{\sin at}{t}\right\} = \int_s^\infty \mathcal{L}\{\sin at\}\, ds = \int_s^\infty \frac{a}{s^2+a^2}\, ds = \left[\tan^{-1}\left(\frac{s}{a}\right)\right]_s^\infty$$

$$= \frac{\pi}{2} - \tan^{-1}\left(\frac{s}{a}\right) = \cot^{-1}\left(\frac{s}{a}\right).$$

(b) Since $\mathcal{L}\{e^{-at}-e^{-bt}\} = \dfrac{1}{s+a} - \dfrac{1}{s+b}$　We have

$$\mathcal{L}\left\{\frac{e^{-at}-e^{-bt}}{t}\right\} = \int_s^\infty \left(\frac{1}{s+a} - \frac{1}{s+b}\right) ds = \left[\log\left(\frac{s+a}{s+b}\right)\right]_s^\infty$$

$$= \left[\log(s+a) - \log(s+b)\right]_s^\infty = \left[\log\left(\frac{1+a/s}{1+b/s}\right)\right]_s^\infty = -\log\frac{s+a}{s+b} = \log\frac{s+b}{s+a}.$$

(c) Since $\mathcal{L}\{1-\cos t\} = \dfrac{1}{s} - \dfrac{s}{s^2+1}$, we have

$$\mathcal{L}\left\{\frac{1-\cos t}{t}\right\} = \int_s^\infty \left(\frac{1}{s} - \frac{s}{s^2+1}\right) ds = \left[\log s - \frac{1}{2}\log(s^2+1)\right]_s^\infty$$

$$= \frac{1}{2}\left[\log\left(\frac{s^2}{s^2+1}\right)\right]_s^\infty = \frac{1}{2}\left[\log\left(\frac{1}{1+\left(\frac{1}{s^2}\right)}\right)\right]_s^\infty = \frac{1}{2}\left[\log 1 - \log\left(\frac{1}{1+\left(\frac{1}{s^2}\right)}\right)\right]$$

$$= -\frac{1}{2}\log\frac{s^2}{s^2+1} = \frac{1}{2}\log\frac{s^2+1}{s^2}.$$

Applying Article 21.9 again, we have

$$\mathcal{L}\left\{\frac{1-\cos t}{t^2}\right\} = \int_s^\infty \frac{1}{2}\log\left(\frac{s^2+1}{s^2}\right) ds = \frac{1}{2}\left[\int_s^\infty \underset{I}{\{\log(s^2+1)} - 2\log s\}\cdot \underset{II}{1}\, ds\right]$$

$$= \frac{1}{2}\left[\{\log(s^2+1)-2\log s\}\cdot s\right]_s^\infty - \frac{1}{2}\int_s^\infty\left(\frac{2s}{s^2+1}-\frac{2}{s}\right)\cdot s\, ds$$

$$= \left[\frac{s}{2}\log\left(\frac{s^2+1}{s^2}\right)\right]_s^\infty + \int_s^\infty \frac{ds}{s^2+1} = \left[\frac{s}{2}\log\left(1+\frac{1}{s^2}\right)\right]_s^\infty + [\tan^{-1} s]_s^\infty$$

$$= -\frac{s}{2}\log\left(1+\frac{1}{s^2}\right)+\left(\frac{\pi}{2}-\tan^{-1} s\right) = \cot^{-1} s - \frac{s}{2}\log\left(1+\frac{1}{s^2}\right).$$

Example 21.12. Prove $\mathcal{L}\left\{\int_0^t \frac{\sin u}{u}\, du\right\} = \frac{1}{s}\tan^{-1}\frac{1}{s}.$

Solution. Let $f(t) = \int_0^t \frac{\sin u}{u}\, du.$ Then $f(0) = 0$ and $f'(t) = \frac{\sin t}{t}$ or $tf'(t) = \sin t.$

Taking the Laplace transform,

$$\mathcal{L}\{t\, f'(t)\} = \mathcal{L}\{\sin t\} \text{ or } -\frac{d}{ds}\{sF(s)-f(0)\} = \frac{1}{s^2+1}$$

that is, $\frac{d}{ds}\{sF(s)\} = -\frac{1}{s^2+1}$. Integrating, $sF(s) = -\tan^{-1} s + c.$

By the initial value theorem, $\lim_{s\to\infty} sF(s) = \lim_{t\to 0} f(t) = f(0) = 0$ so that

$c = \pi/2$. Then $sF(s) = \left(\frac{\pi}{2}\right) - \tan^{-1} s = \tan^{-1}\left(\frac{1}{s}\right)$ or $F(s) = \left(\frac{1}{s}\right)\tan^{-1}\left(\frac{1}{s}\right).$

Alternative. Use the following theorem.

If $\mathcal{L}\{f(t)\} = F(s)$, then $\mathcal{L}\left\{\dfrac{f(t)}{t}\right\} = \displaystyle\int_s^\infty F(u)du$, provided $\lim\limits_{t\to 0} f(t)/t$ exists.

Since $\mathcal{L}\{\sin t\} = \dfrac{1}{s^2+1}$ and $\lim\limits_{t\to 0}\dfrac{\sin t}{t} = 1$, we have

$$\mathcal{L}\left\{\frac{\sin t}{t}\right\} = \int_s^\infty \frac{du}{u^2+1} = \tan^{-1}(1/s).$$

Now applying theorem, $\mathcal{L}\left\{\displaystyle\int_0^t f(u)du\right\} = \dfrac{F(s)}{s}$, we obtain

$$\mathcal{L}\left\{\int_o^t \frac{\sin u}{u}du\right\} = \frac{1}{s}\tan^{-1}\left(\frac{1}{s}\right).$$

Alternative. Using infinite series, we have

$$\int_0^t \frac{\sin u}{u}du = \int_0^t \frac{1}{u}\left(u - \frac{u^3}{3!} + \frac{u^5}{5!} - \frac{u^7}{7!} + \cdots\right)du$$

$$= t - \frac{t^3}{3.3!} + \frac{t^5}{5.5!} - \frac{t^7}{7.7!} + \cdots$$

Then $\mathcal{L}\left\{\displaystyle\int_0^t \frac{\sin u}{u}\right\} = \mathcal{L}\left\{t - \dfrac{t^3}{3.3!} + \dfrac{t^5}{5.5!} - \dfrac{t^7}{7.7!} + \cdots\right\}$

$$= \frac{1}{s^2} - \frac{1}{3.3!}\cdot\frac{3!}{s^4} + \frac{1}{5.5!}\cdot\frac{5!}{s^6} - \frac{1}{7.7!}\cdot\frac{7!}{s^8} + \cdots$$

$$= \frac{1}{s^2} - \frac{1}{3s^4} + \frac{1}{5s^6} - \frac{1}{7s^8} + \cdots$$

$$= \frac{1}{s}\left\{\frac{1/s}{1} - \frac{(1/s)^3}{3} + \frac{(1/s)^5}{5} - \frac{(1/s)^7}{7} + \cdots\right\} = \frac{1}{s}\tan^{-1}\frac{1}{s}.$$

[Using the series, $\tan^{-1}x = x - x^3/3 + x^5/5 - x^7/7 + \cdots$, $|x| < 1$.]

Alternative. Letting $u = tv$, $\displaystyle\int_0^t \frac{\sin u}{u}du = \int_0^t \frac{\sin tv}{v}dv$.

Then $\mathcal{L}\left\{\displaystyle\int_0^t \frac{\sin u}{u}du\right\} = \mathcal{L}\left\{\displaystyle\int_0^t \frac{\sin tv}{v}dv\right\}$

$$= \int_0^\infty e^{-st}\left\{\int_0^t \frac{\sin tv}{v}dv\right\}dt = \int_0^1 \frac{1}{v}\left\{\int_0^\infty e^{-st}\sin tv\, dt\right\}dv$$

$$= \int_0^1 \frac{\mathcal{L}\{\sin tv\}}{v}dv = \int_0^1 \frac{dv}{s^2+v^2} = \frac{1}{s}\tan^{-1}\frac{v}{s}\Big|_0^1 = \frac{1}{s}\tan^{-1}\frac{1}{s}.$$

21.10 Evaluation of Integrals by Laplace transforms

Example 21.13 Evaluate (a) $\int_0^\infty te^{-3t}\sin t\, dt$, $\qquad$ (A.M.I.E., S-2008)

(b) $\int_0^\infty \dfrac{\cos 6t - \cos 4t}{t} dt$, (c) $\int_0^\infty \dfrac{e^{-t} - e^{-3t}}{t} dt$.

Solution. (a) $\mathcal{L}\{t\sin t\} = \int_0^\infty te^{-st}\sin t\, dt = -\dfrac{d}{ds}\mathcal{L}\{\sin t\}$

$= -\dfrac{d}{ds}\left(\dfrac{1}{s^2+1}\right) = \dfrac{2s}{(s^2+1)^2}$.

Then letting $s = 3$, we find $\int_0^\infty te^{-3t}\sin t\, dt = \dfrac{3}{50}$.

(b) $\mathcal{L}\left\{\dfrac{\cos 6t}{t} - \dfrac{\cos 4t}{t}\right\} = \int_0^\infty e^{-st}\dfrac{(\cos 6t - \cos 4t)}{t} dt$

$= \int_s^\infty \dfrac{s}{s^2+36} ds - \int_s^\infty \dfrac{s}{s^2+16} ds = \dfrac{1}{2}\left|\log(s^2+36)\right|_s^\infty - \dfrac{1}{2}\left|\log(s^2+16)\right|_0^\infty$

$= -\dfrac{1}{2}\log(s^2+36) + \dfrac{1}{2}\log(s^2+16) = \dfrac{1}{2}\log\dfrac{s^2+16}{s^2+36}$.

Then letting $s = 0$, we obtain

$\int_0^\infty \dfrac{\cos 6t - \cos 4t}{t} dt = \dfrac{1}{2}\log\dfrac{16}{36} = \log\dfrac{2}{3}$.

(c) $\mathcal{L}\left\{\dfrac{e^{-t}}{t} - \dfrac{e^{-3t}}{t}\right\} = \int_0^\infty e^{-st}\left(\dfrac{e^{-t} - e^{-3t}}{t}\right) dt = \int_s^\infty \left(\dfrac{1}{s+1} - \dfrac{1}{s+3}\right) ds$

$= \left|\log(s+1) - \log(s+3)\right|_s^\infty = \log\left(\dfrac{s+3}{s+1}\right)$.

Then letting $s = 0$, we find $\int_0^\infty \dfrac{e^{-t} - e^{-3t}}{t} dt = \log 3$.

Exercise 21.1

Find the Laplace transform of each of the following functions.

1. $2t^2 + e^{-t} + t^{1/2}$ $\qquad$ **Ans.** $\left(\dfrac{4}{s^3}\right) + \left(\dfrac{1}{s+1}\right) + \dfrac{\sqrt{\pi}}{2s^{3/2}}$.

2. $(3 + 4t)^2$

Ans. $\left(\dfrac{9}{s}\right) + \left(\dfrac{24}{s^2}\right) + \left(\dfrac{32}{s^3}\right).$

3. $4 e^{5t} + 6t^3 - 3 \sin 4t + 2 \cos 2t.$ Ans. $\left(\dfrac{4}{s-5}\right) + \left(\dfrac{36}{s^4}\right) - \left(\dfrac{12}{s^2+16}\right) + \left(\dfrac{2s}{s^2+4}\right).$

4. $(\cos t + \sin t)^2.$

Ans. $(1/s) + \left(2/(s^2 + 4)\right).$

5. $\cos 3t \cos 2t.$

Ans. $(s^3 + 13s)/(s^2 + 1)(s^2 + 25).$

6. $\cos^2 at.$

Ans. $\dfrac{(s^2 + 2a^2)}{s(s^2 + 4a^2)}.$

7. $\sin^3 t.$

Ans. $\dfrac{6}{(s^2 + 1)(s^2 + 9)}.$

8. $\sin(at + b).$

Ans. $\dfrac{(a\cos b + s\sin b)}{(s^2 + a^2)}.$

Find the Laplace transform of the follwing piecewise continuous function

9. (*i*) $f(t) = \begin{cases} 0, & 0 \le t < 2 \\ k, & t \ge 2 \end{cases}$ k constant

Ans. $\dfrac{ke^{-2s}}{s}, s > 0.$

(*ii*) $f(t) = \begin{cases} t, & 0 \le t < 1, \\ 0, & t \ge 1. \end{cases}$

Ans. $\dfrac{e^{-s} - 2e^{-2s} + e^{-3s}}{s^2}.$

(*iii*) $f(t) = \begin{cases} \sin t, & 0 \le t < \pi, \\ 0, & t \ge \pi. \end{cases}$ (*Madras, 2000S*)

Ans. $\dfrac{1 + e^{-\pi s}}{s^2 + 1}.$

10. $f(t) = \begin{cases} \cos t, & 0 \le t < 2\pi, \\ 0, & t \ge 2\pi. \end{cases}$

Ans. $\dfrac{s(1 - e^{-2\pi s})}{(1 + s^2)}.$

Evaluate each of the following:

11. L $\{t^3 e^{-2t}\}.$ (*A.M.I.E., W-2010*)

Ans. $\dfrac{6}{(s+2)^4}.$

12. $\mathcal{L}\{e^{-3t}(2 \cos 5t - 3 \sin 5 t)\}.$ (*J.N.U, 1999*)

Ans. $\dfrac{2s - 9}{s^2 + 6s + 34}.$

13. $\mathcal{L}\{e^{4t} \cosh 5t\}.$

Ans. $\dfrac{s - 4}{s^2 - 8s - 9}.$

14. $\mathcal{L}\{\sinh^3 2t\}$.

    Ans. $\dfrac{48}{(s^2-36)(s^2-4)}$.

15. $\mathcal{L}\{\sinh 2t \cos 2\,t\}$.

    Ans. $\dfrac{2(s^2-8)}{(s^4+64)}$.

16. $\mathcal{L}\{\cosh at \sin at\}$.

    Ans. $\dfrac{a(s^2+2a^2)}{s^4+4a^4}$.

17. $\mathcal{L}\{3t^2 + e^{-t} + t^{1/2}\}$.

    Ans. $\dfrac{6}{s^3} + \dfrac{1}{s+1} + \dfrac{\sqrt{\pi}}{2s^{3/2}}$.

18. $\mathcal{L}\{e^{-3t}\, t^{-1/2}\}$.

    Ans. $\sqrt{\pi/(s+3)}$.

19. $\mathcal{L}\{t \sin at\}$.

    Ans. $\dfrac{2as}{(s^2+a^2)^2}$.

20. $\mathcal{L}\{t \sin^2 3t\}$.

    Ans. $\dfrac{54(s^2+12)}{s^2(s^2+36)^2}$.

21. $\mathcal{L}\{t^2 \cos at\}$.

    Ans. $\dfrac{2s^3-6a^2 s}{(s^2+a^2)^3}$.

22. (i) $\mathcal{L}\{te^{2t} \cos 5t\}$; (ii) $\mathcal{L}\{te^{at} \sin at\}$

    Ans. (i) $\dfrac{(s-2)^2-25}{\left[(s-2)^2+25\right]^2}$; (ii) $\dfrac{2a(s-a)}{(s^2-2as+2a^2)^2}$.

23. $\mathcal{L}\left\{\dfrac{\cos at - \cos bt}{t}\right\}$.

    Ans. $\dfrac{1}{2}\log\dfrac{s^2+b^2}{s^2+a^2}$.

24. $\mathcal{L}\{(1 - e^t)/t\}$. *(Madras 2000 PT, Mangalore 1999)*   **Ans.** $\log\left[\dfrac{(s-1)}{s}\right]$.

25. $\mathcal{L}\left\{\dfrac{\sin^2 t}{t}\right\}$.

    Ans. $\dfrac{1}{4}\log\dfrac{s^2+4}{s^2}$.

26. $\mathcal{L}\left\{\dfrac{\sinh t}{t}\right\}$.

    Ans. $\dfrac{1}{2}\log\left(\dfrac{s+1}{s-1}\right)$.

27. $\mathcal{L}\left\{\dfrac{\sin kt}{t}\right\}$.

    Ans. $\cot^{-1}\left(\dfrac{s}{k}\right)$.

28. Find $\mathcal{L}\{\sin\sqrt{t}\}$ Hence evaluate $\mathcal{L}\left\{\dfrac{\cos\sqrt{t}}{\sqrt{t}}\right\}$. **Ans.** $\dfrac{\sqrt{\pi}}{2s^{3/2}}e^{-1/4s}$, $\dfrac{\sqrt{\pi}}{s^{1/2}}\cdot e^{-1/4s}$

[**Hint.** $\sin\sqrt{t} = \sqrt{t} - \dfrac{(\sqrt{t})^3}{3!} + \dfrac{(\sqrt{t})^5}{5!} - \dfrac{(\sqrt{t})^7}{7!} + \cdots$

$= t^{1/2} - \dfrac{t^{3/2}}{3!} + \dfrac{t^{5/2}}{5!} - \dfrac{t^{7/2}}{7!} + \cdots$ Then the Laplace transform is

$$\mathcal{L}\{\sin\sqrt{t}\} = \frac{\Gamma(3/2)}{s^{3/2}} - \frac{\Gamma(5/2)}{3!s^{5/2}} + \frac{\Gamma(7/2)}{5!s^{7/2}} - \frac{\Gamma(9/2)}{7!s^{9/2}} + \cdots$$

$$= \frac{\sqrt{\pi}}{2s^{3/2}}\left\{1 - (1/2^2 s) + \frac{(1/2^2 s)^2}{2!} - \frac{(1/2^2 s)^3}{3!} + \cdots\right\}$$

$$= \frac{\sqrt{\pi}}{2s^{3/2}}e^{-1/2^2 s} = \frac{\sqrt{\pi}}{2s^{3/2}}e^{-1/4s}. \text{ Thus } \mathcal{L}\left\{\frac{\cos\sqrt{t}}{\sqrt{t}}\right\} = \frac{\sqrt{\pi}}{s^{1/2}}e^{-1/4s}\Big].$$

Evaluate the following integrals

29. (a) $\displaystyle\int_0^\infty t\,e^{-2t}\cos t\,dt$, (b) $\displaystyle\int_0^\infty t^2 e^{-t}\sin t\,dt$, (c) $\displaystyle\int_0^\infty e^{-2t}\sin^3 t\,dt$,

(d) $\displaystyle\int_0^\infty \frac{e^{-3t}-e^{-6t}}{t}\,dt$, (e) $\displaystyle\int_0^\infty e^{-3t}\frac{\sinh t}{t}\,dt$, (f) $\displaystyle\int_0^\infty \frac{\sin t - \sin 2t}{t}e^{-3t}\,dt$,

(g) $\displaystyle\int_0^\infty \frac{1-\cos t}{t}e^{-2t}\,dt$, (h) $\displaystyle\int_0^\infty t^3 e^{-t}\sin t\,dt$, (i) $\displaystyle\int_0^\infty \frac{e^{-t}\sin t}{t}\,dt$.

Ans. (a) 3/25, (b) 1/2, (c) 6/65, (d) log 2, (e) 2^{-1} log 2, (f) $\tan^{-1}(3/2) - \tan^{-1}3$,

(g) $\log(\sqrt{5}/2)$, (h) 0, (i) $\pi/4$.

30. Find Laplace transform of $\dfrac{\sin t}{t}$ and hence show that $\displaystyle\int_0^\infty \frac{\sin t}{t}\,dt = \frac{\pi}{2}$.

[**Hint.** $\displaystyle\int_0^\infty e^{-st}\frac{\sin t}{t}\,dt = \mathcal{L}\left\{\frac{\sin t}{t}\right\} = \int_s^\infty \frac{1}{s^2+1}\,ds$

$= \left|\tan^{-1}s\right|_s^\infty = (\pi/2) - \tan^{-1}s = \cot^{-1}s.$

Putting $s = 0$, we obtain $\displaystyle\int_0^\infty \frac{\sin t}{t}\,dt = \cot^{-1}(0) = \frac{\pi}{2}].$

21.11 Inverse Laplace transforms

Definition: If $\mathcal{L}\{f(t)\} = F(s)$, then $f(t)$ is called the inverse Laplace transform of $F(s)$ and is denoted by $f(t) = \mathcal{L}^{-1}\{F(s)\}$. ...(21.13)

Thus to find $\mathcal{L}^{-1}\{F(s)\}$ we have to find that function of t whose Laplace transform is $F(s)$. For example

$$\mathcal{L}^{-1}\left\{\frac{3}{s^2}\right\} = 3t, \text{ because } \mathcal{L}\{3t\} = \frac{3}{s^2} \text{ and } \mathcal{L}^{-1}\left\{\frac{1}{s+5}\right\} = e^{-5t} \text{ for } \mathcal{L}\{e^{-5t}\} = \frac{1}{s+5}.$$

The inverse Laplace transforms given below follow at once from the results given in the table 21.1 and results given at the end of Art. 21.4

1. $\mathcal{L}^{-1}\left\{\dfrac{1}{s}\right\} = 1..$　　　　　　　　**2.** $\mathcal{L}^{-1}\left\{\dfrac{1}{s-a}\right\} = e^{at}$

3. $\mathcal{L}^{-1}\left\{\dfrac{1}{s^n+1}\right\} = \dfrac{t^n}{n!}$ $(n = 0, 1, 2, 3, ...)$　　**4.** $\mathcal{L}^{-1}\left\{\dfrac{1}{(s-a)^n}\right\} = e^{at}\dfrac{t^{n-1}}{(n-1)!}.$

5. $\mathcal{L}^{-1}\left\{\dfrac{1}{s^2+a^2}\right\} = \dfrac{1}{a}\sin at.$　　　**6.** $\mathcal{L}^{-1}\left\{\dfrac{s}{s^2+a^2}\right\} = \cos at.$

7. $\mathcal{L}^{-1}\left\{\dfrac{1}{s^2-\omega^2}\right\} = \dfrac{1}{\omega}\sinh \omega t.$　　**8.** $\mathcal{L}^{-1}\left\{\dfrac{s}{s^2-\omega^2}\right\} = \cosh \omega t.$

9. $\mathcal{L}^{-1}\left\{\dfrac{1}{(s-a)^2+b^2}\right\} = \dfrac{1}{b}e^{at}\sin bt.$　**10.** $\mathcal{L}^{-1}\left\{\dfrac{s-a}{(s-a)^2+b^2}\right\} = e^{at}\cos bt.$

While determining the inverse Laplace transforms of given function of s, we frequently require the above results. It is, therefore, essential that we are well familiar with them.

There are various other methods available for finding inverse transforms of given functions. Of these, the most important for our purpose is the **method of partial fractions** aplicable when the function to be inverted is a rational algebraic fraction. In such cases the function is resolved into partial fractions and then the inverse transforms can be easily written with the help of *linearity property of inverse Laplace transforms.

The procedure will be clear from some of the examples solved below:

21.12 Illustrative Examples

Example 21.14 Find each of following inverse Laplace transforms.

---

* According to linearity property of inverse Laplace transforms

$\mathcal{L}^{-1}(aF_1(s) + bF_2(s) + ...] = a\mathcal{L}^{-1}F_1(s) + b\mathcal{L}^{-1}F_2(s)...$ where $a, b, ...$ are constants.

(a) $\mathcal{L}^{-1}\left\{\dfrac{4}{s-2}\right\}$, (b) $\mathcal{L}^{-1}\left\{\dfrac{1}{s^5}\right\}$, (c) $\mathcal{L}^{-1}\left\{\dfrac{8s}{s^2+2}\right\}$, (d) $\mathcal{L}^{-1}\left\{\dfrac{1}{2s-5}\right\}$,

(e) $\mathcal{L}^{-1}\left\{\dfrac{3s-12}{s^2+8}\right\}$, (f) $\mathcal{L}^{-1}\left\{\dfrac{12}{4-3s}\right\}$, (g) $\mathcal{L}^{-1}\left\{\dfrac{6}{s^2+4}\right\}$, (h) $\mathcal{L}^{-1}\left\{\dfrac{2s-5}{s^2-9}\right\}$,

(i) $\mathcal{L}^{-1}\left\{\dfrac{3s-8}{s^2+4}-\dfrac{4s-24}{s^2-16}\right\}$.

Solution. (a) $\mathcal{L}^{-1}\left\{\dfrac{4}{s-2}\right\}=4\mathcal{L}^{-1}\left\{\dfrac{1}{s-2}\right\}=4e^{2t}$.

(b) $\mathcal{L}^{-1}\left\{\dfrac{1}{s^5}\right\}=\dfrac{t^4}{4!}=\dfrac{t^4}{24}$.

(c) $\mathcal{L}^{-1}\left\{\dfrac{8s}{s^2+2}\right\}=8\mathcal{L}^{-1}\left\{\dfrac{s}{s^2+\left(\sqrt{2}\right)^2}\right\}=8\cos\sqrt{2}\,t$.

(d) $\mathcal{L}^{-1}\left\{\dfrac{1}{2s-5}\right\}=\dfrac{1}{2}\mathcal{L}^{-1}\left\{\dfrac{1}{s-\dfrac{5}{2}}\right\}=\dfrac{1}{2}\cdot e^{5t/2}$.

(e) $\mathcal{L}^{-1}\left\{\dfrac{3s-12}{s^2+8}\right\}=3\mathcal{L}^{-1}\left\{\dfrac{s}{s^2+\left(2\sqrt{2}\right)^2}\right\}-\dfrac{12}{2\sqrt{2}}\mathcal{L}^{-1}\left\{\dfrac{2\sqrt{2}}{s^2+8}\right\}$

$$=3\cos 2\sqrt{2}\,t-3\sqrt{2}\sin 2\sqrt{2}\,t.$$

(f) $\mathcal{L}^{-1}\left\{\dfrac{12}{4-3s}\right\}=\dfrac{-12}{3}\mathcal{L}^{-1}\left\{\dfrac{1}{s-(4/3)}\right\}=-4e^{4t/3}$.

(g) $\mathcal{L}^{-1}\left\{\dfrac{6}{s^2+4}\right\}=\dfrac{6}{2}\mathcal{L}^{-1}\left\{\dfrac{2}{s^2+(2)^2}\right\}=3\sin 2t$.

(h) $\mathcal{L}^{-1}\left\{\dfrac{2s-5}{s^2-9}\right\}=2\mathcal{L}^{-1}\left\{\dfrac{s}{s^2-9}\right\}-\dfrac{5}{3}\mathcal{L}^{-1}\left\{\dfrac{3}{s^2-9}\right\}$

$$=2\cosh 3t-(5/3)\sinh 3t.$$

(i) $\mathcal{L}^{-1}\left\{\dfrac{3s-8}{s^2+4}-\dfrac{4s-24}{s^2-16}\right\}=3\mathcal{L}^{-1}\left\{\dfrac{s}{s^2+4}\right\}-8\mathcal{L}^{-1}\left\{\dfrac{1}{s^2+4}\right\}$

$$-4\mathcal{L}^{-1}\left\{\dfrac{s}{s^2-16}\right\}+24\mathcal{L}^{-1}\left\{\dfrac{1}{s^2-16}\right\}$$

$$= 3\mathcal{L}^{-1}\left\{\frac{s}{s^2+(2)^2}\right\} - \frac{8}{2}\mathcal{L}^{-1}\left\{\frac{2}{s^2+(2)^2}\right\} - 4\mathcal{L}^{-1}\left\{\frac{s}{s^2-(4)^2}\right\}$$

$$+\frac{24}{4}\mathcal{L}^{-1}\left\{\frac{4}{s^2-(4)^2}\right\}.$$

$$= 3\cos 2t - 4\sin 2t - 4\cosh 2t + 6\sinh 4t.$$

Example 21.15 Find the inverse Laplace transforms of the following functions.

(a) $\dfrac{1}{s^2+6s+15}$, (b) $\dfrac{1}{s^2-4s+20}$, (c) $\dfrac{3s-1}{(s-2)^3}$, (d) $\dfrac{16+3s}{s^2-8s+20}$.

Solution. (a) We write $\dfrac{1}{s^2+6s+15} = \dfrac{1}{(s+3)^2+6} = \dfrac{1}{(s+3)^2+\left(\sqrt{6}\right)^2}$

Hence, $\mathcal{L}^{-1}\left\{\dfrac{1}{s^2+6s+15}\right\} = \mathcal{L}^{-1}\left\{\dfrac{1}{(s+3)^{2*}+\left(\sqrt{6}\right)^2}\right\} = \dfrac{1}{\sqrt{6}}\mathcal{L}^{-1}\left\{\dfrac{\sqrt{6}}{(s+3)^2+\left(\sqrt{6}\right)^2}\right\}$

$$= \left(e^{-3t}\sin\sqrt{6}t\right)/\sqrt{6}.$$

(b) We write $\dfrac{1}{s^2-4s+20} = \dfrac{1}{(s-2)^2+(4)^2}$.

Hence, $\mathcal{L}^{-1}\left\{\dfrac{1}{s^2-4s+20}\right\} = \dfrac{1}{4}\mathcal{L}^{-1}\left\{\dfrac{4}{(s-2)^2+(4)^2}\right\} = \left(e^{2t}\sin 4t\right)/4$.

(c) We write $\dfrac{3s-1}{(s-2)^2} = \dfrac{3(s-2)+5}{(s-2)^2} = \dfrac{3}{(s-2)} + \dfrac{5}{(s-2)^2}$.

Hence, $\mathcal{L}^{-1}\left\{\dfrac{3s-1}{(s-2)^2}\right\} = 3\mathcal{L}^{-1}\left\{\dfrac{1}{s-2}\right\} + 5\mathcal{L}^{-1}\left\{\dfrac{1}{(s-2)^2}\right\} = 3e^{2t} + 5t\,e^{2t}$.

(d) We write $\dfrac{16+3s}{s^2-8s+20} = \dfrac{16+3s}{(s-4)^2+4} = \dfrac{3(s-4)+28}{(s-4)^2+(2)^2}$.

$$= 3\dfrac{(s-4)}{(s-4)^2+(2)^2} + 28\dfrac{1}{(s-4)^2+(2)^2}.$$

---

* **First translation or shifting property.**

Theorem If $\mathcal{L}^{-1}\{F(s)\} = f(t)$, then $\mathcal{L}^{-1}\{F(s-a)\} = e^{at}f(t)$.

$$\therefore \mathcal{L}^{-1}\left\{\frac{16+3s}{s^2-8s+20}\right\} = 3\mathcal{L}^{-1}\left\{\frac{(s-4)}{(s-4)^2+(2)^2}\right\} + \frac{28}{2}\mathcal{L}^{-1}\left\{\frac{2}{(s-4)^2+(2)^2}\right\}$$

$$= 3e^{4t}\cos 2t + 14e^{4t}\sin 2t = (3\cos 2t + 14\sin 2t)e^{4t}.$$

Example 21.16 Find the inverse Laplace transforms of

(a) $\dfrac{2s-11}{s^2-4s+8}$, (b) $\dfrac{3s+7}{s^2+6s+9}$.

Solution. (a) $\mathcal{L}^{-1}\left\{\dfrac{2s-11}{s^2-4s+8}\right\} = \mathcal{L}^{-1}\left\{\dfrac{2s-11}{(s-2)^2+4}\right\} = \mathcal{L}^{-1}\left\{\dfrac{2(s-2)-7}{(s-2)^2+4}\right\}$

$$= 2\mathcal{L}^{-1}\left\{\frac{s-2}{(s-2)^2+2^2}\right\} - \frac{7}{2}\mathcal{L}^{-1}\left\{\frac{2}{(s-2)^2+2^2}\right\}$$

$$= 2e^{2t}\cos 2t - (7/2)e^{2t}\sin 2t = e^{2t}(4\cos 2t - 7\sin 2t)/2.$$

(b) $\mathcal{L}^{-1}\left\{\dfrac{3s+7}{s^2+6s+9}\right\} = \mathcal{L}^{-1}\left\{\dfrac{3s+7}{(s+3)^2}\right\} = \mathcal{L}^{-1}\left\{\dfrac{3(s+3)-2}{(s+3)^2}\right\}$

$$= \mathcal{L}^{-1}\left\{\frac{3}{s+3}\right\} - 2\mathcal{L}^{-1}\left\{\frac{1}{(s+3)^2}\right\} = 3e^{-3t} - 2e^{-3t}\cdot t = e^{-3t}(3-2t).$$

Example 21.17 Find the inverse Laplace transforms of

(a) $\dfrac{3s+7}{s^2-2s-3}$, (b) $\dfrac{s}{(s+1)^2(s^2+1)}$.

Solution. (a) Let $\dfrac{3s+7}{s^2-2s-3} = \dfrac{3s+7}{(s-3)(s+1)} = \dfrac{A}{s-3} + \dfrac{B}{s+1}$.

Multiplying both sides by $(s-3)(s+1)$, we obtain

$3s + 7 = A(s+1) + B(s-3)$.

Putting $s = -1$, we obtain $4 = -4B$ or $B = -1$.

Putting $s = 3$, we obtain $16 = 4A$ or $A = 4$.

Thus $\dfrac{3s+7}{s^2-2s-3} = \dfrac{4}{s-3} - \dfrac{1}{s+1}$.

$$\therefore \mathcal{L}^{-1}\left\{\frac{3s+7}{s^2-2s-3}\right\} = 4\mathcal{L}^{-1}\left\{\frac{1}{s-3}\right\} - \mathcal{L}^{-1}\left\{\frac{1}{s+1}\right\} = 4e^{3t} - e^{-t}.$$

(b) Let $\dfrac{s}{(s+1)^2(s^2+1)} = \dfrac{A}{s+1} + \dfrac{B}{(s+1)^2} + \dfrac{Cs+D}{s^2+1}$.

Multiplying both sides by $(s + 1)^2 (s^2 + 1)$, we obtain

$s = A(s + 1)(s^2 + 1) + B(s^2 + 1) + (Cs + D) (s + 1)^2$.

Putting $s = -1$, $\quad\quad\quad\quad\quad -1 = 2B$ or $B = -1/2$.

Equating coefficient of s^3, $0 = A + C$. ...(i)

Equating coefficients of s^2, $0 = A + B + 2C + D$. ...(ii)

Equating coefficient of s, $1 = A + C + 2D$. ...(iii)

Puttting $s = 0$, $\quad\quad\quad 0 = A + B + D$. ...(iv)

From (i) and (iii), we obtain $1 = 2D$ or $D = 1/2$.

Putting values of B and D in (iv) we have $A = 0$.

Putting values of A in (i), $C = 0$.

Thus $\dfrac{s}{(s+1)^2(s^2+1)} = -\dfrac{1}{2}\cdot\dfrac{1}{(s+1)^2} + \dfrac{1}{2}\cdot\dfrac{1}{s^2+1}$

$\therefore\ \mathcal{L}^{-1}\left\{\dfrac{s}{(s+1)^2(s^2+1)}\right\} = -\dfrac{1}{2}\mathcal{L}^{-1}\left\{\dfrac{1}{(s+1)^2}\right\} + \dfrac{1}{2}\mathcal{L}^{-1}\left\{\dfrac{1}{s^2+1}\right\}$

$= \left(-t\,e^{-t}\right)/2 + \left(\sin t\right)/2 = \left(\sin t - te^{-t}\right)/2.$

Example 21.18 Find (a) $\mathcal{L}^{-1}\left\{\dfrac{2s^2-1}{(s^2+1)(s^2+4)}\right\}$, (b) $\mathcal{L}^{-1}\left\{\dfrac{s}{s^4+s^2+1}\right\}$.

Solution. (a) As all the terms involve s^2, we put $s^2 = p$ and the given fraction

becomes $\dfrac{2p-1}{(p+1)(p+4)}$ which has only linear factors in the denominator.

Let $\dfrac{2p-1}{(p+1)(p+4)} = \dfrac{A}{p+1} + \dfrac{B}{p+4}$. ...(i)

Multiplying both sides by $(p + 1)(p + 4)$, we obtain

$2p - 1 = A(p + 4) + B(p + 1)$.

Putting $p = -1$ and $p = -4$ in turn, we obtain $A = -1$, $B = 3$.

Putting values of A and B in (i) and replacing p by s^2, we have

$$\dfrac{2s^2-1}{(s^2+1)(s^2+4)} = -\dfrac{1}{s^2+1} + \dfrac{3}{s^2+4}.$$

$$\therefore \mathcal{L}^{-1}\left\{\frac{2s^2-1}{(s^2+1)(s^2+4)}\right\} = -\mathcal{L}^{-1}\left\{\frac{1}{s^2+1}\right\} + \frac{3}{2}\mathcal{L}^{-1}\left\{\frac{2}{s^2+4}\right\} = -\sin t + \frac{3}{2}\sin 2t.$$

(b) $\dfrac{s}{s^4+s^2+1} = \dfrac{s}{(s^2+s+1)(s^2-s+1)} = \dfrac{1}{2}\left[\dfrac{1}{s^2-s+1} - \dfrac{1}{s^2+s+1}\right].$

$$\therefore \mathcal{L}^{-1}\left\{\frac{s}{s^4+s^2+1}\right\} = -\frac{1}{2}\mathcal{L}^{-1}\left\{\frac{1}{s^2+s+1}\right\} + \frac{1}{2}\mathcal{L}^{-1}\left\{\frac{1}{s^2-s+1}\right\}$$

$$= -\frac{1}{2}\mathcal{L}^{-1}\left\{1 \div \left[(s+(1/2))^2 + (\sqrt{3}/2)^2\right]\right\}$$

$$+ \frac{1}{2}\mathcal{L}^{-1}\left\{1 \div \left[(s-(1/2))^2 + (\sqrt{3}/2)^2\right]\right\}$$

$$= -\frac{1}{\sqrt{3}}\mathcal{L}^{-1}\left\{\frac{(\sqrt{3}/2)}{(s+(1/2))^2 + (\sqrt{3}/2)^2}\right\} + \frac{1}{\sqrt{3}}\mathcal{L}^{-1}\left\{\frac{(\sqrt{3}/2)}{(s-(1/2))^2 + (\sqrt{3}/2)^2}\right\}$$

$$= -\frac{1}{\sqrt{3}}e^{-t/2}\sin\frac{\sqrt{3}}{2}t + \frac{1}{\sqrt{3}}e^{t/2}\sin\frac{\sqrt{3}}{2}t$$

$$= \frac{2}{\sqrt{3}}\sin\frac{\sqrt{3}}{2}t\left[\frac{e^{t/2}-e^{-t/2}}{2}\right] = \frac{2}{\sqrt{3}}\sin\frac{\sqrt{3}}{2}t \ \sinh\left(\frac{t}{2}\right).$$

Example 21.19 Obtain $f(t)$ when $F(s) = \dfrac{s}{s^4+4a^4}$. *(Andhra, 1999)*

Solution. We have

$$F(s) = \frac{s}{(s^2+2a^2)^2 - (2sa)^2} = \frac{s}{(s^2+2a^2+2sa)(s^2+2a^2-2sa)}$$

$$= \frac{1}{4a}\left[\frac{1}{s^2+2a^2-2sa} - \frac{1}{s^2+2a^2+2sa}\right] \quad \text{(Resolving into partial fractions)}$$

$$= \frac{1}{4a}\left[\frac{1}{(s-a)^2+a^2} - \frac{1}{(s+a)^2+a^2}\right]$$

$$= \frac{1}{4a^2}\left[\frac{a}{(s-a)^2+a^2} - \frac{a}{(s+a)^2+a^2}\right].$$

$$\therefore f(t) = \mathcal{L}^{-1}\{F(s)\} = \frac{1}{4a^2}\left[\mathcal{L}^{-1}\left\{\frac{a}{(s-a)^2+a^2}\right\} - \mathcal{L}^{-1}\left\{\frac{a}{(s+a)^2+a^2}\right\}\right]$$

$$= \frac{1}{4a^2}\left(e^{at}\sin at - e^{-at}\sin at\right)$$

$$= \frac{1}{4a^2}\sin at\left(e^{at} - e^{-at}\right) = \frac{1}{2a^2}\sin at \sinh at.$$

Example 21.20 Find the function of t whose Laplace transform is

$$1/\left(s^2 + 2as + \omega_0^2\right) \text{ when } \omega_0^2 > a^2.$$

Solution. Let $F(s) = \dfrac{1}{(s+a)^2 + \left(\omega_0^2 - a^2\right)}$ $\left(\omega_0^2 > a^2\right)$

$$= \frac{1}{\sqrt{\omega_0^2 - a^2}}\frac{\sqrt{\omega_0^2 - a^2}}{(s+a)^2 + \left(\omega_0^2 - a^2\right)}.$$

$$\therefore f(t) = \mathcal{L}^{-1}\{F(s)\} = \frac{1}{\sqrt{\omega_0^2 - a^2}}\sin\left(\sqrt{\omega_0^2 - a^2}\, t\right)e^{-at}.$$

21.13 Other Inversion Formulae

(A) Inverse Laplace transform of derivatives

If $\mathcal{L}^{-1}\{F(s)\} = f(t)$, then $\mathcal{L}^{-1}\left\{\dfrac{d^n}{ds^n}F(s)\right\} = (-1)^n\, t^n f(t).$ ...(21.14)

(B) Inverse Laplace transform of integrals

If $\mathcal{L}^{-1}\{F(s)\} = f(t)$, then $\mathcal{L}^{-1}\left\{\displaystyle\int_s^\infty F(u)\,du\right\} = \dfrac{f(t)}{t}.$ ...(21.15)

Remark: The above formula is useful in finding $f(t)$ when $F(s)$ is given, provided the inverse transform of the left hand side can be conveniently evaluated.

(C) Multiplication by s^n.

We know $L\{f'(t)\} = sF(s) - f(0)$ where $F(s) = L\{f(t)\}$.

If $\mathcal{L}^{-1}\{F(s)\} = f(t)$, then $\mathcal{L}^{-1}\{s\,F(s)\} = f'(t)$ provided $f(0) = 0$. ...(21.16)

Thus multiplication by s has the effect of *differentiating* $f(t)$.

Generalizations to $\mathcal{L}^{-1}\{s^n F(s)\}$, $n = 2, 3, ...,$ are possible.

(D) Division by s

If $\mathcal{L}^{-1}\{F(s)\} = f(t)$, then $\mathcal{L}^{-1}\left\{\dfrac{F(s)}{s}\right\} = \int_0^t f(u)\,du$ ...(21.17)

Thus division of s (or multiplication by $1/s$) has the effect of *integrating* $f(t)$ from 0 to t.

Generalizations to $\mathcal{L}^{-1}\{F(s)/s^n\}$, $n = 2, 3, ...$ are possible.

(E) Convolution Theorem

Statement: If $\mathcal{L}^{-1}\{F(s)\} = f(t)$ and $\mathcal{L}^{-1}\{G(s)\} = g(t)$,

then $\mathcal{L}^{-1}\{F(s)G(s)\} = \int_0^t f(u)g(t-u)\,du = f * g$.

We call $f * g$ the convolution or faltung of f and g.

Proof: The required result is established if we prove that

$$\mathcal{L}\left[\int_0^t f(u)g(t-u)\,du\right] = F(s)G(s), \quad ...(i) \qquad ...(21.18)$$

where $F(s) = \mathcal{L}\{f(t)\}$ and $G(s) = \mathcal{L}\{g(t)\}$.

Now L.H.S. of (21.18)

$$= \int_0^\infty e^{-st} \int_0^t f(u)g(t-u)\,du\, dt = \int_0^\infty \int_0^t e^{-st} f(u)g(t-u)\,du\, dt .$$

By changing the order of integration, this integral becomes

$$= \int_0^\infty \int_u^\infty e^{-st} f(u)g(t-u)\,dt\, du$$

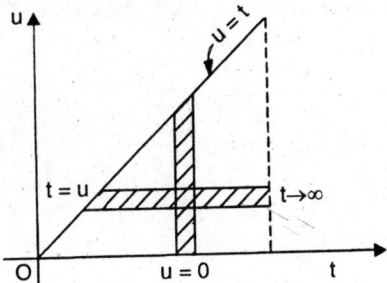

$$= \int_0^\infty e^{-su} f(u) \int_u^\infty e^{-s(t-u)} g(t-u)\,dt\, du$$

$$= \int_0^\infty e^{-su} f(u) \int_u^\infty e^{-sv} g(v)\,dv\, du$$

(by putting $t - u = v$)

$$= \int_0^\infty e^{-su} f(u)\,du \cdot G(s) = F(s) \cdot G(s)$$

   ...(21.19)

Thus the theorem is proved.

Fig. 21.2. Region of integration in the t u-plane.

21.14 Illustrative Examples

Example 21.21 Calculate the inverse transform of

(a) $\log \dfrac{s+1}{s-1}$ (*A.M.I.E.; S-2007*),   (b) $\log\left(1+\dfrac{1}{s^2}\right)$,

(c) $\dfrac{s}{\left(s^2+a^2\right)^2}$,   (*Andhra, 2000*).

Solution. (a) Let $\mathcal{L}^{-1}\left\{\log\dfrac{s+1}{s-1}\right\} = f(t)$. Then by formula 'A', Article 21.13

$$\mathcal{L}^{-1}\left\{\frac{d}{ds}\log\left(\frac{s+1}{s-1}\right)\right\} = -t\, f(t).$$

As $\dfrac{d}{ds}\log\dfrac{s+1}{s-1} = \dfrac{d}{ds}\left[\log(s+1)-\log(s-1)\right] = \dfrac{1}{s+1} - \dfrac{1}{s-1} = \dfrac{-2}{s^2-1}$.

$\therefore \mathcal{L}^{-1}\left(\dfrac{-2}{s^2-1}\right) = -t\,f(t)$ or $-2\sinh t = -t\,f(t)$ or $f(t) = \dfrac{2\sinh t}{t}$.

(b) Let $F(s) = \log\left(1+\dfrac{1}{s^2}\right) = \mathcal{L}\{f(t)\}$.

Then $F'(s) = \dfrac{-2}{s(s^2+1)} = -2\left\{\dfrac{1}{s} - \dfrac{s}{s^2+1}\right\}$.

Thus since $\mathcal{L}^{-1}\{F'(s)\} = -2(1-\cos t) = -t\,f(t)$, or $f(t) = 2(1-\cos t)/t$.

(c) Let $F(s) = \dfrac{1}{s^2+a^2}$. Therefore $f(t) = \mathcal{L}^{-1}\{F(s)\} = \mathcal{L}^{-1}\left\{\dfrac{1}{s^2+a^2}\right\} = \dfrac{1}{a}\sin at$.

We have $F'(s) = \dfrac{d}{ds}\left(\dfrac{1}{s^2+a^2}\right) = \dfrac{-2s}{\left(s^2+a^2\right)^2}$.

By Art. 21.13, Formula A, $\mathcal{L}^{-1}\left\{\dfrac{d}{ds}F(s)\right\} = -t\,f(t)$.

$\therefore \mathcal{L}^{-1}\left\{\dfrac{-2s}{\left(s^2+a^2\right)^2}\right\} = -\dfrac{t\sin at}{a}$,   or $\mathcal{L}^{-1}\left\{\dfrac{s}{\left(s^2+a^2\right)^2}\right\} = \dfrac{t\sin at}{2a}$.

Example 21.22 Calculate the inverse Laplace transform of the following functions.

(a) $\dfrac{s}{\left(s^2+a^2\right)^2}$ (*Andhra, 2000*),

(b) $\dfrac{s^2}{\left(s^2+a^2\right)^2}$. (*Madras 2000 PT*)

Solution. (*a*) If $f(t)$ is the required inverse Laplace transform, then by formula `B` Art. 21.13, we have

$$\frac{f(t)}{t}=\mathcal{L}^{-1}\left\{\int_s^\infty \frac{s}{\left(s^2+a^2\right)^2}\,ds\right\}=\mathcal{L}^{-1}\left\{\frac{1}{2}\int_s^\infty (2s)\left(s^2+a^2\right)^{-2}ds\right\}$$

$$=\mathcal{L}^{-1}\left[-\frac{1}{2}\left(\frac{1}{s^2+a^2}\right)\right]_s^\infty=\frac{1}{2}\mathcal{L}^{-1}\left\{\frac{1}{s^2+a^2}\right\}=\frac{1}{2a}\sin at.$$

Hence, $f(t)=\dfrac{1}{2a}t\sin at.$

(*b*) By part (*a*) $\mathcal{L}^{-1}\left\{\dfrac{s}{\left(s^2+a^2\right)^2}\right\}=\dfrac{1}{2a}t\sin at=f(t)$ say,

Since $f(0)=0$. By formula `C`, Art. 21.13, we obtain

$$\mathcal{L}^{-1}\left\{\frac{s^2}{\left(s^2+a^2\right)^2}\right\}=\mathcal{L}^{-1}\left\{s\cdot\frac{s}{\left(s^2+a^2\right)^2}\right\}=\frac{d}{dt}f(t)$$

$$=\frac{d}{dt}\left(\frac{1}{2a}t\sin at\right)=\frac{1}{2a}(\sin at+at\cos at).$$

Example 21.23 Find the inverse Laplace transform of

(a) $\dfrac{1}{\left(s^2+a^2\right)^2}$, (b) $\dfrac{1}{s^2\left(s^2+a^2\right)}$, (*J.N.T.U., 1995S*), (c) $\dfrac{1}{s^3\left(s^2+1\right)}$.

Solution. (*a*) We have proved in Example 21.22 part (*a*), that

$$\mathcal{L}^{-1}\left\{\frac{s}{\left(s^2+a^2\right)^2}\right\}=\frac{1}{2a}(t\sin at)=f(t)\text{ say}$$

∴ By formula *D* Article 21.13, we have

$$\mathcal{L}^{-1}\left\{\frac{1}{\left(s^2+a^2\right)^2}\right\}=\mathcal{L}^{-1}\left\{\left(\frac{1}{s}\cdot\frac{s}{\left(s^2+a^2\right)^2}\right)\right\}=\int_0^t f(t)\,dt=\int_0^t \frac{1}{2a}u\sin au\,du$$

$$= \frac{1}{2a}\left[u\left(\frac{-\cos au}{a}\right) - 1 \cdot \left(\frac{-\sin au}{a^2}\right)\right]_0^t \qquad \text{(Rule of Integration by parts)}$$

$$= \frac{1}{2a}\left(\frac{-t\cos at}{a} + \frac{\sin at}{a^2}\right) = \frac{1}{2a^3}(\sin at - at\cos at).$$

(b) Since $\mathcal{L}^{-1}\left\{\dfrac{1}{s^2+a^2}\right\} = \dfrac{1}{a}\sin at$. We have by repeated application of formula 'D' Article 21.13.

$$\mathcal{L}^{-1}\left\{\frac{1}{s(s^2+a^2)}\right\} = \int_0^t \frac{1}{a}\sin au \, du = \frac{1}{a^2}(1-\cos at)$$

$$\mathcal{L}^{-1}\left\{\frac{1}{s^2(s^2+a^2)}\right\} = \int_0^t \frac{1}{a^2}(1-\cos au)du = \frac{1}{a^3}(at - \sin at).$$

Alternative Since $\dfrac{1}{s^2(s^2+a^2)} = \dfrac{1}{a^2}\left(\dfrac{1}{s^2} - \dfrac{1}{s^2+a^2}\right)$ (by resolving into partial

fractions)

$$\therefore \mathcal{L}^{-1}\left\{\frac{1}{s^2(s^2+a^2)}\right\} = \frac{1}{a^2}\left[\mathcal{L}^{-1}\left\{\frac{1}{s^2}\right\} - \mathcal{L}^{-1}\left\{\frac{1}{s^2+a^2}\right\}\right]$$

$$= \frac{1}{a^2}\left(t - \frac{1}{a}\sin at\right) = \frac{1}{a^3}(at - \sin at).$$

(c) Since $\mathcal{L}^{-1}\left\{\dfrac{1}{s^2+1}\right\} = \sin t$. We have by repeated application of formula 'D' Article 21.13.

$$\mathcal{L}^{-1}\left\{\frac{1}{s(s^2+1)}\right\} = \int_0^t \sin u \, du = 1 - \cos t,$$

$$\mathcal{L}^{-1}\left\{\frac{1}{s^2(s^2+1)}\right\} = \int_0^t (1-\cos u)du = t - \sin t,$$

$$\mathcal{L}^{-1}\left\{\frac{1}{s^3(s^2+1)}\right\} = \int_0^t (u-\sin u)du = \frac{t^2}{2} - 1 + \cos t.$$

Example 21.24. Apply convolution theorem to evaluate

(a) $\mathcal{L}^{-1}\left\{\dfrac{s}{(s^2+a^2)^2}\right\}$, *(J.N.T.U., 1999, Andhra, 2000)*

(b) $\mathcal{L}^{-1}\left\{\dfrac{s^2}{(s^2+a^2)(s^2+b^2)}\right\}$.　　　　　(*V.T.U.*, *2000*; *Madras*, *1999*)

Solution. (a) Since $\mathcal{L}^{-1}\left\{\dfrac{s}{s^2+a^2}\right\}=\cos at$ and $\mathcal{L}^{-1}\left\{\dfrac{1}{s^2+a^2}\right\}=\dfrac{1}{a}\sin at$.

Therefore, by convolution theorem, we obtain

$$\mathcal{L}^{-1}\left\{\dfrac{s}{(s^2+a^2)^2}\right\}=\mathcal{L}^{-1}\left\{\dfrac{s}{s^2+a^2}\cdot\dfrac{1}{s^2+a^2}\right\}=\int_0^t \cos a(t-u)\dfrac{\sin au}{a}\,du$$

$$=\dfrac{1}{2a}\int_0^t[\sin at+\sin(2au-at)]\,du$$

$$=\dfrac{1}{2a}\left[u\sin at-\dfrac{1}{2a}\cos(2au-at)\right]_0^t=\dfrac{1}{2a}t\sin at.$$

(b) Since $\mathcal{L}^{-1}\left\{\dfrac{s}{s^2+a^2}\right\}=\cos at$ and $\mathcal{L}^{-1}\left\{\dfrac{s}{s^2+b^2}\right\}=\cos bt$.

Therefore, by convolution theorem, we obtain

$$\mathcal{L}^{-1}\left\{\dfrac{s^2}{(s^2+a^2)(s^2+b^2)}\right\}=\mathcal{L}^{-1}\left\{\dfrac{s}{s^2+a^2}\cdot\dfrac{s}{s^2+b^2}\right\}$$

$$=\int_0^t \cos a(t-u)\cos bu\,du$$

$$=\dfrac{1}{2}\int_0^t[\cos\{at-(a-b)u\}+\cos\{at-(a+b)u\}]\,du$$

$$=\dfrac{1}{2}\left[-\dfrac{1}{a-b}\sin\{at-(a-b)u\}-\dfrac{1}{a+b}\sin\{at-(a+b)u\}\right]_0^t$$

$$=-\dfrac{1}{2}\left[\dfrac{\sin bt-\sin at}{a-b}-\dfrac{\sin at+\sin bt}{a+b}\right]=\dfrac{a\sin at-b\sin bt}{a^2-b^2}.$$

Exercise 21.2

Find the inverse Laplace transform of the following functions.

1. $\dfrac{3s-8}{4s^2+25}$.　　　　　**Ans.** $\dfrac{3}{4}\cos\dfrac{5t}{2}-\dfrac{4}{5}\sin\dfrac{5t}{2}$

2. $\dfrac{5s+10}{9s^2-16}$.

   Ans. $\dfrac{5}{9}\cosh\dfrac{4t}{3}+\dfrac{5}{6}\sinh\dfrac{4t}{3}$.

3. $\dfrac{(s+8)}{\left(s^2+4s+5\right)}$.

   Ans. $e^{-2t}(\cos t+6\sin t)$.

4. $\dfrac{2s^2-6s+5}{s^3-6s^2+11s-6}$. *(Andhra, 1999)*

   Ans. $(e^t-2e^{2t}+5e^{3t})/2$.

5. $\dfrac{4}{s^2-s+2}$.

   Ans. $\dfrac{8}{\sqrt{7}}e^{t/2}\sin\left(\sqrt{7}\,t/2\right)$.

6. $\dfrac{(s+2)}{\left(s^2-4s+13\right)}$.

   Ans. $e^{2t}(3\cos 3t+4\sin 3t)/3$.

7. $\dfrac{(s+6)}{\left(s^2+6s+13\right)}$.

   Ans. $e^{-3t}(2\cos 2t+3\sin 2t)/2$.

8. $\dfrac{s+1}{s^2+s+1}$.

   Ans. $\dfrac{e^{-t/2}}{\sqrt{3}}\left(\sqrt{3}\cos\dfrac{\sqrt{3}t}{2}+\sin\dfrac{\sqrt{3}t}{2}\right)$.

9. $\dfrac{(1+2s)}{(s+2)^2(s-1)^2}$.

   Ans. $t(e^t-e^{-2t})/3$.

10. $\dfrac{5s+3}{(s-1)\left(s^2+2s+5\right)}$.

    Ans. $(2e^t-2\,e^{-t}\cos 2t+3\,e^{-t}\sin 2t)/2$.

11. $\dfrac{2s^3+2s^2+4s+1}{\left(s^2+1\right)\left(s^2+s+1\right)}$.

    Ans. $\cos t+2\sin t+e^{-t/2}\left(\cos\dfrac{\sqrt{3}t}{2}-\sqrt{3}\sin\dfrac{\sqrt{3}t}{2}\right)$.

12. $\dfrac{s^3}{\left(s^4+64\right)}$.

    Ans. $\cos 2t\cosh 2t$.

13. $\dfrac{s^2}{\left(s^2+1\right)\left(s^2+4\right)}$.

    Ans. $(2\sin 2t-\sin t)/3$.

14. $\dfrac{a\left(s^2-2a^2\right)}{s^4+4a^4}$.

    Ans. $\cos at\sinh at$.

15. $\dfrac{1}{\left(s^2+2s+5\right)^2}$.

    Ans. $e^{-t}(\sin 2t-2t\cos 2t)/16$.

16. $\dfrac{3s+1}{(s-1)(s^2+1)}$. **Ans.** $2e^t - 2\cos t + \sin t$.

17. $\dfrac{s}{(s^2+9)^2}$. **Ans.** $(t\sin 3t)/6$.

18. $\log(1 + s^{-1})$. *(AMIE., S-2009; Andhra, 1999)* **Ans.** $(1 - e^{-t})$.

19. $\log\left(\dfrac{s+a}{s+b}\right)$. *(Andhra, 2000)* **Ans.** $(e^{-bt} - e^{-at})/t$.

20. $\log\left(1 + \dfrac{a^2}{s^2}\right)$. **Ans.** $2(1 - \cos at)/t$.

21. $\dfrac{1}{s(s+2)^3}$. **Ans.** $\{1 - e^{-2t}(2t^2 + 2t + 1)\}/8$

22. $\dfrac{s+2}{(s^2+4s+5)^2}$. *(Madras, 2000S)* **Ans.** $\dfrac{1}{2}te^{-2t}\sin t$.

23. $\dfrac{1}{s(s+2)^3}$. **Ans.** $\left(1 - e^{-2t} - 2te^{-2t} - 2t^2e^{-2t}\right)/8$.

24. $\dfrac{s^2+6}{(s^2+1)(s^2+4)}$. **Ans.** $\dfrac{1}{3}(5\sin t - \sin 2t)$.

25. $\dfrac{s^3}{(s^4-a^4)}$. **Ans.** $(\cos at + \cosh at)/2$.

26. $\tan^{-1}(2/s)$. *(Andhra, 1999)* **Ans.** $(\sin 2t)/t$.
27. $\cot^{-1}(s+1)$. **Ans.** $(e^{-t}\sin t)/t$.

Apply convolution theorem to evaluate the following function.

28. *(a)* $\mathcal{L}^{-1}\left\{\dfrac{1}{(s-a)(s-b)}\right\}$, *(Madras, 1999)* **Ans.** $(e^{-bt} - e^{-at})/(a - b)$

(b) $\mathcal{L}^{-1}\left\{\dfrac{s^2}{(s^2+4)^2}\right\}$, **Ans.** $\dfrac{1}{4}\sin 2t + \dfrac{t\cos 2t}{2}$

(c) $\mathcal{L}^{-1}\left\{\dfrac{1}{s(s^2+4)}\right\}$, **Ans:** $\dfrac{1}{4}(1 - \cos 2t)$

(d) $\mathcal{L}^{-1}\left\{\dfrac{1}{s^2(s+1)^2}\right\}$. *(Mysore, 1997)* **Ans.** $t(e^{-t} + 1) + 2(e^{-t} - 1)$.

21.15 Applications to Differential Equations

Laplace transformation can be used to solve ordinary as well as partial differential equations. Here we shall apply this method to solve, only linear ordinary differential equations with constant coefficients and simultaneous ordinary differential equations. The advantage of this method over the conventional one, as already pointed out, is that it reduces the problem of solving a differential equation to an algebraic problem. Moreover this method yields the particular solution directly without the necessity of first finding the complementary function and particular integral and then evaluating the arbitrary constants.

Suppose, for example we have to solve the differential equation

$$a\frac{d^2y}{dt^2} + b\frac{dy}{dt} + cy = f(t), \text{ where } a, b, c \text{ constants subject to the conditions}$$

$y = e_1$ and $\dfrac{dy}{dt} = e_2$ when $t = 0$.

This equation can be written in the form
$ay'' + by' + cy = f(t)$ where $y(0) = e_1$ and $y'(0) = e_2$...(i).

Taking the Laplace transform on both sides and denoting

$\mathcal{L}\{y(t)\} = Y(s)$, and $\mathcal{L}\{f(t)\} = F(s)$, we obtain

$a\mathcal{L}\{y''\} + b\mathcal{L}\{y'\} + c\mathcal{L}\{y\} = \mathcal{L}\{f(t)\}$

or $a[s^2Y(s) - sy(0) - y'(0)] + b[sY(s) - y(0)] + cY(s) = F(s)$

or $(as^2 + bs + c) Y(s) = F(s) + a\{sy(0) + y'(0)\} + by(0)$

As $y(0) = e_1$ and $y'(0) = e_2$.

$\therefore (as^2 + bs + c)Y(s) = F(s) + a\{se_1 + e_2\} + be_1$...(ii)

The algebraic equation (ii) is called the **subsidiary equation** of the differential equation (i).

From (ii), we obtain

$$Y(s) = \frac{F(s) + a\left[se_1 + e_2\right] + be_1}{as^2 + bs + c}$$

Taking the inverse Laplace transform, we obtain

$$y(t) = \mathcal{L}^{-1}\{Y(s)\} = \mathcal{L}^{-1}\left[\frac{F(s) + a\{se_1 + e_2\} + be_1}{as^2 + bs + c}\right] \qquad \text{...(21.20)}$$

The procedure followed above is general and is applicable to all linear differential equations with constant coefficients and may be summed up thus.

Procedure to solve Differential Equations

Step 1. *Take Laplace transforms of both sides of the given differential equation using initial conditions. This gives an algebraic equation.*

Step 2. *Solve the algebraic equation and get Y(s) in terms of s and let the solution be Y(s) = F(s).*

Step 3. *Take inverse Laplace transform of both sides. This gives*

$$\mathcal{L}^{-1}\{Y(s)\} = \mathcal{L}^{-1}\{F(s)\} \text{ or } y(t) = f(t), \text{ the required solution.}$$

21.16 Illustrative Examples

Example 21.25 Solve the differential equation

$$\frac{d^2y}{dt^2} - 2\frac{dy}{dt} + 2y = 0, \, y(0) = 1, \, y'(0) = 1.$$

Solution. The given equation can be written in the form $y'' - 2y' + 2y = 0$.

Taking Laplace transform of both sides, we have

$$\mathcal{L}\{y''\} - 2\mathcal{L}\{y'\} + 2\mathcal{L}\{y\} = 0$$

or $[s^2Y(s) - sy(0) - y'(0)] - 2[sY(s) - y(0)] + 2Y(s) = 0.$

Using initial conditions this becomes

$s^2Y(s) - s - 1 - 2sY(s) + 2 + 2Y(s) = 0$

or $\quad (s^2 - 2s + 2) Y(s) = s - 1$

or $\quad y(s) = \dfrac{s-1}{s^2 - 2s + 2} = \dfrac{s-1}{(s-1)^2 + 1}.$

Taking inverse Laplace transform of both sides, we obtain

$$\mathcal{L}^{-1}\{y(s)\} = \mathcal{L}^{-1}\left\{\frac{s-1}{(s-1)^2 + 1}\right\} \text{ or } y(t) = e^t \cos t.$$

Example 21.26 Find the solution of the differential equation

$$y'' - 3y' + 2y = 4 \text{ where } y(0) = 2, \, y'(0) = 3.$$

Solution. Taking Laplace transform of the differential equation, we have

$$\mathcal{L}\{y''\} - 3\mathcal{L}\{y'\} + 2\mathcal{L}\{y\} = \mathcal{L}\{4\}$$

or $s^2Y(s) - sy(0) - y'(0) - 3[sY(s) - y(0)] + 2Y(s) = \dfrac{4}{s}.$

Using the initial conditions, the above equation reduces to

$$s^2Y(s) - 2s - 3 - 3sY(s) + 6 + 2Y(s) = \frac{4}{s}$$

or $(s^2 - 3s + 2)Y(s) = \left(\dfrac{4}{s}\right) + 2s - 3 = (2s^2 - 3s + 4)/s$

or $\qquad Y(s) = \dfrac{2s^2 - 3s + 4}{s(s^2 - 3s + 2)}$

$$= \frac{2}{s} - \frac{3}{s-1} + \frac{3}{s-2}. \quad \text{(Resolving into partial fractions)}$$

Finally, taking inverse Laplace transform of both sides, we have

$$\mathcal{L}^{-1}\{Y(s)\} = \mathcal{L}^{-1}\left\{\frac{2}{s} - \frac{3}{s-1} + \frac{3}{s-2}\right\}$$

or $\qquad y(t) = 2 - 3e^t + 3e^{2t}.$

Example 21.27 Solve the differential equation

$$\frac{d^2y}{dt^2} + \frac{dy}{dy} - 2y = 3\cos 3t - 11\sin 3t,$$

given that $y(0) = 0$ and $y'(0) = 6$.

Solution. Taking Laplace transform of both sides, we have

$$\mathcal{L}\{y''\} + \mathcal{L}\{y'\} - 2\mathcal{L}\{y\} = 3\mathcal{L}\{\cos 3t\} - 11\mathcal{L}\{\sin 3t\}$$

or $\left[s^2Y(s) - sy(0) - y'(0)\right] + \left[sY(s) - y(0)\right] - 2Y(s) = 3 \cdot \dfrac{s}{s^2+9} - 11 \cdot \dfrac{3}{s^2+9}.$

Using initial conditions, we find

$$s^2Y(s) - 6 + sY(s) - 2Y(s) = \frac{(3s-33)}{s^2+9}$$

or $\quad (s^2 + s - 2)Y(s) = \dfrac{3s-33}{s^2+9} + 6 = \dfrac{6s^2+3s+21}{s^2+9},$

or $\quad Y(s) = \dfrac{3(2s^2+s+7)}{(s^2+9)(s^2+s-2)} = \dfrac{3(2s^2+s+7)}{(s^2+9)(s+2)(s-1)}$

$$= \frac{1}{s-1} - \frac{1}{s+2} + \frac{3}{s^2+9}. \quad \text{(Breaking into partial fractions).}$$

Taking inverse Laplace transform of both sides, we obtain

$$y(t) = e^t - e^{-2t} + \sin 3t.$$

Example 21.28 Solve the equation

$$\frac{d^2x}{dt^2} - 3\frac{dx}{dt} + 2x = 1 - e^{2t}, \; x(0) = 1, \; x'(0) = 0.$$

Solution. Taking the Laplace transform of both sides of the differential equation, we have

$$\mathcal{L}\{x''\} - 3\mathcal{L}\{x'\} + 2\mathcal{L}\{x\} = \mathcal{L}\{1\} - \mathcal{L}\{e^{2t}\}$$

or $\qquad \left[s^2X(s) - sx(0) - x'(0)\right] - 3\left[sX(s) - x(0)\right] + 2X(s) = \dfrac{1}{s} - \dfrac{1}{s-2}.$

Using initial conditions this becomes

$$s^2 X(s) - s - 3sX(s) + 3 + 2X(s) = \frac{-2}{s(s-2)}$$

or $(s^2 - 3s + 2)X(s) = s - 3 - \dfrac{2}{s(s-2)} = \dfrac{s^3 - 5s^2 + 6s - 2}{s(s-2)}$

or $X(s) = \dfrac{s^3 - 5s^2 + 6s - 2}{s(s-1)(s-2)^2} = \dfrac{s^2 - 4s + 2}{s(s-2)^2}$

(Cancelling $(s-1)$ from Num. & Den.)

$$= \frac{1}{2}\cdot\frac{1}{s} + \frac{1}{2}\cdot\frac{1}{s-2} - \frac{1}{(s-2)^2}. \quad \text{[(Breaking into partial fractions)]}$$

$$\therefore \mathcal{L}^{-1}\{X(s)\} = \mathcal{L}^{-1}\left\{\frac{1}{2}\cdot\frac{1}{s} + \frac{1}{2}\cdot\frac{1}{s-2} - \frac{1}{(s-2)^2}\right\} \text{ or } x(t) = \frac{1}{2} + \frac{1}{2}e^{2t} - te^{2t}.$$

Example 21.29 Solve $\dfrac{d^2 y}{dt^2} + 2\dfrac{dy}{dt} + 5y = e^{-t}\sin t$, $y(0) = 0, y'(0) = 1$.

Solution. The given equation may be written in the form
$$y'' + 2y' + 5y = e^{-t}\sin t.$$
On Laplace transformation it becomes

$$s^2 Y(s) - sy(0) - y'(0) + 2[sY(s) - y(0)] + 5Y(s) = \frac{1}{(s+1)^2 + 1}.$$

Using initial conditions and simplifying, we obtain

$$(s^2 + 2s + 5)Y(s) - 1 = \frac{1}{s^2 + 2s + 2} \text{ or } (s^2 + 2s + 5)Y(s) = \frac{s^2 + 2s + 3}{s^2 + 2s + 2}$$

or $Y(s) = \dfrac{s^2 + 2s + 3}{(s^2 + 2s + 2)(s^2 + 2s + 5)} = \dfrac{1}{3}\left[\dfrac{1}{s^2 + 2s + 2} + \dfrac{2}{s^2 + 2s + 5}\right].$

(Resolving into partial fraction)

$$\therefore \mathcal{L}^{-1}\{Y(s)\} = \frac{1}{3}\mathcal{L}^{-1}\left\{\frac{1}{s^2 + 2s + 2} + \frac{2}{s^2 + 2s + 5}\right\}$$

$$= \frac{1}{3}\mathcal{L}^{-1}\left\{\frac{1}{(s+1)^2 + 1} + \frac{2}{(s+1)^2 + 2^2}\right\}.$$

or $y(t) = \dfrac{1}{3}\{e^{-t}\sin t + e^{-t}\sin 2t\} = \dfrac{1}{3}e^{-t}(\sin t + \sin 2t)$, the required solution.

Example 21.30 Solve $(D^2 + n^2)x = a \sin(nt + \alpha)$, $x = Dx = 0$, at $t = 0$.

Solution. The given equation is $(D^2 + n^2) x = a(\sin nt \cos \alpha + \cos nt \sin \alpha)$
On Laplace transformation the equation becomes

$$\left[s^2 X(s) - sx(0) - x'(0)\right] + n^2 X(s) = a \cos\alpha \cdot \frac{n}{s^2 + n^2} + a \sin\alpha \frac{s}{s^2 + n^2}.$$

Using initial conditions and simplifying, the above equation becomes

$$X(s) = an \cos\alpha \cdot \frac{1}{\left(s^2 + n^2\right)^2} + a \sin\alpha \cdot \frac{s}{\left(s^2 + n^2\right)^2}.$$

Taking inverse Laplace transform

$$x(t) = an \cos\alpha \cdot \left\{ \frac{1}{2n^3} (\sin nt - nt \cos nt) \right\} + a \sin\alpha \cdot \left(\frac{t}{2n} \sin nt \right)$$

$$= \frac{a}{2n^2} \left[\sin nt \cos\alpha - nt \cos(nt + \alpha) \right].$$

Exercise 21.3

Solve each of the following differential equation by using Laplace transformation method.

1. $(D^2 + 3D + 2)x = 0$, $x(0) = 1$ and $x'(0) = 2$.   **Ans.** $x(t) = 4e^{-t} - 3e^{-2t}$.

2. $y''' + 2y'' - y' - 2y = 0$ given $y(0) = y'(0) = 0$ and $y''(0) = 6$.
   Ans. $y(t) = e^t - 3e^{-t} + 2e^{-2t}$.

3. $y'' + y' - 2y = t$, $y(0) = 1$, $y'(0) = 0$.   **Ans.** $y(t) = (4e^t + e^{-2t} - 2t - 1)/4$.

4. $\dfrac{d^2 y}{dt^2} + 4 \dfrac{dy}{dt} = \sin t$, given that $y(0) = y'(0) = 0$.

   Ans. $y(t) = (17 - e^{-4t} - 16\cos t - 4 \sin t)/68$.

5. $\dfrac{d^2 x}{dt^2} - 3\dfrac{dx}{dt} + 2x = 4e^{2t}$, given that $x(0) = -3$ and $x'(0) = 5$.

   (AMIE, S-2009) **Ans.** $x(t) = 4e^{2t} - 7e^t + 4te^{2t}$.

6. $y'' - 3y' + 2y = 4t + 12e^{-t}$, $y(0) = 6$, $y'(0) = -1$.
   Ans. $y(t) = 3e^t - 2e^{2t} + 2t + 3 + 2e^{-t}$.

7. $y'' + 4y' + 6y = 1 + e^{-t}$, $y(0) = y'(0) = 0$.

   Ans. $y(t) = \dfrac{1}{6} + \dfrac{1}{3}e^{-t} - \dfrac{1}{2}e^{-2t} \cos\sqrt{2}t - \dfrac{\sqrt{2}}{3}e^{-2t} \sin\sqrt{2}t$.

8. $y'' + 4y' + 13y = e^{-t}$, $y(0) = 0$, $y'(0) = 2$.

   Ans. $y(t) = \dfrac{1}{10}\left(e^{-t} - e^{-2t}\cos 3t + \dfrac{19}{3}e^{-2t} \sin 3t \right)$.

9. $\dfrac{d^2x}{dt^2} + x = t \cos 2t$, given that $x(0) = x'(0) = 0$.

> **Ans.** $x(t) = (4 \sin 2t - 5 \sin t - 3t \cos 2t)/9$.

10. $\dfrac{d^2x}{dt^2} - \dfrac{dx}{dt} - 2x = 20 \sin 2t$ when $x(0) = -1$, $x'(0) = 2$.

> **Ans.** $x(t) = 2e^{2t} - 4e^{-t} + \cos 2t - 3 \sin 2t$.

11. $(D^3 + D) x = 2$, $x(0) = 3$, $x'(0) = 1$, $x''(0) = -2$.

> **Ans.** $x(t) = 1 + 2t + 2 \cos t - \sin t$.

12. $(D^2 + 2D + 5)y = e^t \sin t$, where $y(0) = 0$ and $y'(0) = 1$.

> *(Assam, 1999, Mangalore, 1997)*   **Ans.** $y(t) = 11e^{-t}(\sin t + \sin 2t)/3$.

13. $y'' + 9y = \cos 2t$, if $y(0) = 1$, $y(\pi/2) = 1$.

> **Ans.** $y(t) = (4/5)\cos 3t + (4/5)\sin 3t + (1/5)\cos 2t$.

14. $y''' - 3y'' + 3y' - y = t^2 e^t$, given that $y(0) = 1$, $y'(0) = 0$, $y''(0) = -2$.

> **Ans.** $y(t) = e^t - te^t - (t^2 e^t)/2 + (t^5 e^t)/60$. ··

15. $y'' + a^2 y = f(t)$, $y(0) = 1$, $y'(0) = -2$.

> **Ans.** $y(t) = \cos at - \dfrac{2 \sin at}{a} + \dfrac{1}{a}\int_0^t f(u) \sin a(t-u)du$.

16. Find the solution of the initial value problem

$$y' + 3y + 2\int_0^t y(t)dt = t, \; y(0) = 0 . \;\; \textbf{[Hint.} \int_0^t y(t)dt = \frac{Y(s)}{s} \,].$$

> **Ans.** $y(t) = \dfrac{1}{2}e^{-2t} - e^{-t} + \dfrac{1}{2}$.

17. Use Laplace transforms to solve $L\dfrac{di}{dt} + Ri = E$,

where L, R, E are constants and $i = 0$ at $t = 0$.   **Ans.** $i = E(1 - e^{-Rt/L})/R$.

18. A voltage Ee^{-at} is applied at $t = 0$ to a circuit of inductance L and resistance R. Show (by the transform method) that the current at time t is

$$E\big(e^{-at} - e^{-Rt/L}\big)/(R - aL). \hspace{2cm} \text{(V.T.U., 2000)}$$

[Hint. Here we have to solve $L\dfrac{di}{dt} + Ri = Ee^{-at}$, $i(0) = 0$]

21.17 Simultaneous Linear Differential Equations

Laplace transformation can be used to solve simultaneous linear differential equations also. In fact this method reduces a system of simultaneous linear differential equations to a system of linear algebraic equations. The transforms of dependent variables can be found as functions of s by solving these algebraic equations. Then inversion gives the required solution.

Example 21.31 Use Laplace transform to solve

$$(D^2 - D)y + z = 0, \quad (D - 1)y + Dz = 0, \text{ given that for } t = 0,$$
$$y = 0, z = 1 \text{ and } Dy = 0.$$

Solution. Let $\mathcal{L}(y) = Y(s)$ and $\mathcal{L}(z) = Z(s)$.

Taking Laplace transform, we obtain,

$$s^2 Y(s) - sy(0) - y'(0) - sY(s) + y(0) + Z(s) = 0,$$

and $\quad sY(s) - y(0) - Y(s) + sZ(s) - z(0) = 0.$

Using initial conditions, the above equations can be written as

$$(s^2 - s)Y(s) + Z(s) = 0, \qquad \qquad ...(i)$$

and $\qquad (s - 1)Y(s) + sZ(s) = 1. \qquad \qquad ...(ii)$

Solving (i) and (ii) simultaneously, we obtain

$$Y(s) = \begin{vmatrix} 0 & 1 \\ 1 & s \end{vmatrix} \div \begin{vmatrix} s^2 - s & 1 \\ s - 1 & s \end{vmatrix} = -\frac{1}{(s-1)^2(s+1)} = \frac{1}{4} \cdot \frac{1}{s-1} - \frac{1}{2} \frac{1}{(s-1)^2} - \frac{1}{4} \cdot \frac{1}{s+1},$$

and $Z(s) = \begin{vmatrix} s^2 - s & 0 \\ s - 1 & 1 \end{vmatrix} \div \begin{vmatrix} s^2 - s & 1 \\ s - 1 & s \end{vmatrix} = \frac{s}{(s-1)(s+1)} = \frac{1}{2}\left(\frac{1}{s-1} + \frac{1}{s+1}\right).$

Finally $\qquad y(t) = \mathcal{L}^{-1}\{Y(s)\} = \frac{1}{4}\left(e^t - 2t\,e^t - e^{-t}\right),$

and $\qquad z(t) = \mathcal{L}^{-1}\{Z(s)\} = \frac{1}{2}\left(e^t + e^{-t}\right).$

Example 21.32 A mechanical system with two degrees of freedom satisfies the equations

$$2\frac{d^2 x}{dt^2} + 3\frac{dy}{dt} = 4, \qquad 2\frac{d^2 y}{dt^2} - 3\frac{dx}{dt} = 0.$$

Use Laplace transforms to determine x and y at any instant, given that $x, y, \dfrac{dx}{dt}, \dfrac{dy}{dt}$ all vanish at $t = 0$.

Solution. Taking Laplace transforms of the equations, we obtain

$$2\{s^2 X(s) - sx(0) - x'(0)\} + 3\{sY(s) - y(0)\} = \frac{4}{s}, \qquad ...(i)$$

and $\quad 2\{s^2 Y(s) - sy(0) - y'(0)\} - 3\{sX(s) - x(0)\} = 0. \qquad ...(ii)$

Since $x(0) = 0 = y(0) = x'(0) = y'(0)$ therefore, equations (i) and (ii) become

$$2s^2 X(s) + 3sY(s) = \frac{4}{s}, \text{ and } 2s^2 Y(s) - 3sX(s) = 0$$

or $\quad 2sX(s) + 3Y(s) = \dfrac{4}{s^2}, \quad$ ...(iii) $\quad$ and $\quad 3X(s) - 2sY(s) = 0.$ $\qquad$ ...(iv)

Solving (iii) and (iv), we obtain

$$X(s) = \frac{8}{s(4s^2 + 9)} = \frac{8}{9} \cdot \frac{1}{s} - \frac{32}{9} \cdot \frac{s}{4s^2 + 9},$$

and $\qquad Y(s) = \dfrac{12}{s^2(4s^2 + 9)} = \dfrac{4}{3} \cdot \dfrac{1}{s^2} - \dfrac{16}{3} \cdot \dfrac{1}{4s^2 + 9}.$

On inversion we have

$$x(t) = \frac{8}{9}\left(1 - \cos\frac{3}{2}t\right), \quad \text{and} \quad y(t) = \frac{8}{9}\left(\frac{3}{2}t - \sin\frac{3}{2}t\right).$$

Example 21.33 Solve the following set of simultaneous equations with the help of Laplace transforms:

$$\frac{dx}{dt} + 2x - 3y = t, \quad \frac{dy}{dt} - 3x + 2y = e^{2t}, t \ge 0 \text{ where } x = y = 0 \text{ at } t = 0.$$

(AMIE., W-2008)

Solution. Taking Laplace transforms we have

$$sX(s) - x(0) + 2X(s) - 3Y(s) = \frac{1}{s^2}, \quad \text{and} \quad sY(s) - y(0) - 3X(s) + 2Y(s) = \frac{1}{(s-2)}.$$

Using initial conditions and simplifying, we obtain

$$(s+2)X(s) - 3Y(s) = \frac{1}{s^2}, \qquad \text{...(i)}$$

and $\qquad -3X(s) + (s+2)Y(s) = \dfrac{1}{(s-2)}.$ $\qquad$ ...(ii)

Solving (i) and (ii) simultaneously, we obtain

$$X(s) = \begin{vmatrix} \frac{1}{s^2} & -3 \\ \frac{1}{(s-2)} & s+2 \end{vmatrix} \div \begin{vmatrix} s+2 & -3 \\ -3 & s+2 \end{vmatrix}, \text{ and } Y(s) = \begin{vmatrix} s+2 & \frac{1}{s^2} \\ -3 & \frac{1}{(s-2)} \end{vmatrix} \div \begin{vmatrix} s+2 & -3 \\ -3 & s+2 \end{vmatrix}.$$

Thus $\quad X(s) = \dfrac{4(s+1)}{(s+5)(s-2)s^2} = \dfrac{3}{7} \cdot \dfrac{1}{s-2} + \dfrac{16}{175} \cdot \dfrac{1}{s+5} - \dfrac{13}{25} \cdot \dfrac{1}{s} - \dfrac{2}{5} \cdot \dfrac{1}{s^2},$

and
$$Y(s) = \frac{s^2 + 3s + 6}{(s-2)(s+5)s^2} = \frac{4}{7} \cdot \frac{1}{s-2} - \frac{16}{175} \cdot \frac{1}{s+5} - \frac{12}{25} \cdot \frac{1}{s} - \frac{3}{5} \cdot \frac{1}{s^2}.$$

Inversion gives

$$x(t) = \mathcal{L}^{-1}\{X(s)\} = \frac{3}{7}e^{2t} + \frac{16}{175}e^{-5t} - \frac{13}{25} - \frac{2}{5}t,$$

$$y(t) = \mathcal{L}^{-1}\{Y(s)\} = \frac{4}{7}e^{2t} - \frac{16}{175}e^{-5t} - \frac{12}{25} - \frac{3}{5}t.$$

Exercise 21.4

Solve by Laplace transforms the following system of equations

1. $\dfrac{dx}{dt} - 2x + 3y = 0$, $\dfrac{dy}{dt} + 2x - y = 0$, given that $x(0) = 8$ and $y(0) = 3$.

    Ans. $x = 5e^{-t} + 3e^{4t}$, $y = 5e^{-t} - 2e^{4t}$.

2. $\dfrac{dx}{dt} + \dfrac{dy}{dt} + x = -e^{-t}$, $\dfrac{dx}{dt} + \dfrac{2dy}{dt} + 2x + 2y = 0$, subject to the conditions
    $x(0) = -1$, $y(0) = 1$.   **Ans.** $x(t) = -e^{-t}(\cos t + \sin t)$, $y(t) = e^{-t}(1 + \sin t)$.

3. $\dfrac{dx}{dt} - \dfrac{dy}{dt} + 2y = \cos 2t$, $\dfrac{dx}{dt} + \dfrac{dy}{dt} - 2x = \sin 2t$,
    subject to conditions $x(0) = 0$, $y(0) = -1$.

    Ans. $x = \{e^t(\cos t + \sin t) - \cos 2t\}/2$, $y = -e^t(\cos t - \sin t) - \sin 2t$.

4. $3\dfrac{dx_1}{dt} + \dfrac{dx_2}{dt} + 2x_1 = 1$, $\dfrac{dx_1}{dt} + 4\dfrac{dx_2}{dt} + 3x_2 = 0$ $t \geq 0$.
    with initial conditions $x_1 = x_2 = 0$ at $t = 0$.

    Ans. $x_1 = (5 - 2e^{-t} - 3e^{-6t/11})/10$, $x_2 = (e^{-t} - e^{-6t/11})/5$.

5. $\dfrac{dx}{dt} + x + y = e^t$, $2\dfrac{dx}{dt} + \dfrac{dy}{dt} + y = \cos t$ if $x(0) = y(0) = 2$.

    Ans. $x = \cos t - 4 \sin t + e^t - t \sin (t/2)$,
    $y = 3 \cos t + (11 \sin t)/2 - e^t + t(\cos t + \sin t)/2$.

6. $\dfrac{dx}{dt} + y = \sin t$, $\dfrac{dy}{dt} + x = \cos t$; given that $x = 2$ and $y = 0$ when $t = 0$.

    (Madras, 2000)   **Ans.** $x = e^t + e^{-t}$, $y = e^{-t} - e^t + \sin t$.
7. $D^2x + 3x - 2y = 0$, $D^2(x + y) - 3x + 5y = 0$
    subject to $x(0) = y(0) = 0$ and $x'(0) = 3$, $y'(0) = 2$

    Ans: $x = (33 \sin t + \sin 3t)/12$, $y = (11 \sin t - \sin 3t)/4$.

8. $(D^2 - 3D + 2) x + (D - 1)y = 0$, $(D^2 - 5D + 4) y - (D - 1)x = 0$,
   given that $x = Dx = Dy = 0$ at $y = 1$ when $t = 0$.

   Ans: $x = (e^t - e^{3t} + 2t\, e^{3t})/4$, $y = (5e^t - e^{3t} - 2t\, e^{3t})/4$.

9. $D^2x + y = -5 \cos 2t$, $D^2y + x = 5 \cos 2t$,
   where $x(0) = x'(0) = y'(0) = 1$ and $y(0) = -1$.

   Ans. $x = \sin t + \cos 2t$, $y = \sin t - \cos 2t$.

10. Currents x and y in the coupled circuits are given by

$$L\frac{dx}{dt} + Rx + R(x-y) = E, \quad L\frac{dy}{dt} + Ry - R(x-y) = 0.$$

Find x and y in terms of t, given that $x = 0 = y$ when $t = 0$.

Ans: $x = E(4 - 3e^{-Rt/L} - e^{-3Rt/L})/6R$, $y = E(2 - 3e^{-Rt/L} + e^{-3Rt/L})/6R$.

21.18 Laplace Transform of Periodic Functions

Theorem. Let $f(t)$ be piecewise continuous on $[0, \infty)$, be of exponential order and periodic with period $T > 0$.

Then $\mathcal{L}\{f(t)\} = \dfrac{1}{1-e^{-sT}} \displaystyle\int_0^T e^{-st} f(t)dt$, $s > 0$.

Proof: We write the Laplace transform as the sum of the following integrals.

$$\mathcal{L}\{f(t)\} = \int_0^\infty e^{-st} f(t)dt = \int_0^T e^{-st} f(t)dt + \int_T^{2T} e^{-st} f(t)dt$$

$$+ \int_{2T}^{3T} e^{-st} f(t)dt + \ldots\infty \qquad \ldots(21.21)$$

Put $t = u + T$ in the second integral, $t = u + 2T$ in the third integral, ...on the right hand side of the equation (21.21), we have

$$\mathcal{L}\{f(t)\} = \int_0^T e^{-st} f(t)dt + \int_0^T e^{-s(u+T)} f(u+T)du + \int_0^T e^{-s(u+2T)} f(u+2T)du + \cdots$$

Since $f(t)$ is periodic with period T, we obtain

$$\mathcal{L}\{f(t)\} = \int_0^T e^{-su} f(u)du + e^{-sT} \int_0^T e^{-su} f(u)du + e^{-2sT} \int_0^T e^{-su} f(u)du + \cdots$$

$$= \left(1 + e^{-sT} + e^{-2sT} + e^{-3sT+\cdots} \ldots\infty\right) \int_0^T e^{-su} f(u)du$$

$$= \frac{1}{1-e^{-sT}} \int_0^T e^{-st} f(t)dt. \ldots(21.22) \qquad \left(\because \int_0^T e^{-su} f(u)du = \int_0^T e^{-st} f(t)dt\right)$$

Example 21.34 Find the Laplace transform of the function

$$f(t) = \begin{cases} \sin \omega t, & \text{when } 0 < t < \pi/\omega \\ 0, & \text{when } \pi/\omega < t < 2\pi/\omega. \end{cases} \quad \textit{(Coimbatore 1999, Madras 1997)}$$

Solution. Since $f(t)$ is a periodic function with period $2\pi/\omega$.

$$\therefore \quad \mathcal{L}\{f(t)\} = \frac{1}{1-e^{-2\pi s/\omega}} \int_0^{2\pi/\omega} e^{-st} f(t) dt$$

$$= \frac{1}{1-e^{-2\pi s/\omega}} \left[\int_0^{\pi/\omega} e^{-st} \sin \omega t \, dt + \int_{\pi/\omega}^{2\pi/\omega} e^{-st} . 0 \, dt \right]$$

$$= \frac{1}{1-e^{-2\pi s/\omega}} \left[\frac{e^{-st}(-s\sin\omega t - \omega\cos\omega t)}{s^2+\omega^2} \right]_0^{\pi/\omega} = \frac{\omega + \omega e^{-\pi s/\omega}}{(s^2+\omega^2)(1-e^{-2\pi s/\omega})}$$

$$= \frac{\omega}{(1-e^{-\pi s/\omega})(s^2+\omega^2)}.$$

Example 21.35 Find the Laplace transform of the periodic function defined by the triangular wave

$$f(t) = \begin{cases} t, & 0 < t < c \\ 2c - t, & c < t < 2c \end{cases} \text{ and } f(t+2c) = f(t).$$

(Madras, 2000PT; Mangalore, 1997)

Solution. We have $T = 2c$. Therefore

$$\mathcal{L}\{f(t)\} = \frac{1}{1-e^{-2cs}} \left[\int_0^c t \cdot e^{-st} dt + \int_c^{2c} (2c-t) e^{-st} dt \right]$$

$$= \frac{1}{1-e^{-2cs}} \left[-\left\{ \left(\frac{t}{s} + \frac{1}{s^2} \right) e^{-st} \right\}_0^c + \left\{ \left(\frac{-2c}{s} + \left(\frac{t}{s} + \frac{1}{s^2} \right) \right) e^{-st} \right\}_c^{2c} \right]$$

$$= \frac{1}{1-e^{-2cs}} \left[\frac{1}{s^2}(1-2e^{-cs}+e^{-2cs}) \right] = \frac{(1-e^{-cs})^2}{s^2(1-e^{-2cs})}$$

$$= \frac{1-e^{-cs}}{s^2(1+e^{-cs})} = \frac{e^{cs/2}-e^{-cs/2}}{s^2(e^{cs/2}+e^{-cs/2})} = \frac{1}{s^2} \tanh\left(\frac{cs}{2} \right), s > 0.$$

21.19 Unit Step Function or Heaviside Function

The *unit step function* is the function $u(t)$ defined by

$$u(t) = \begin{cases} 0, & if \quad t < 0 \\ 1, & if \quad t \geq 0. \end{cases} \qquad ...(21.23a)$$

The unit step function is also called a *Heaviside function* and is denoted by $H(t)$. If the jump discontinuity is at the point $t = a$, then we define

$$u(t-a) = H(t-a) = \begin{cases} 0, & if \quad t < a, \\ 1, & if \quad t \geq a. \end{cases} \qquad ...(21.23b)$$

The jump is of magnitude 1.
The unit step function $u(t - a)$ is also denoted by $u_a(t)$, that is, we write
$u(t - a) = H(t - a) = u_a(t)$.

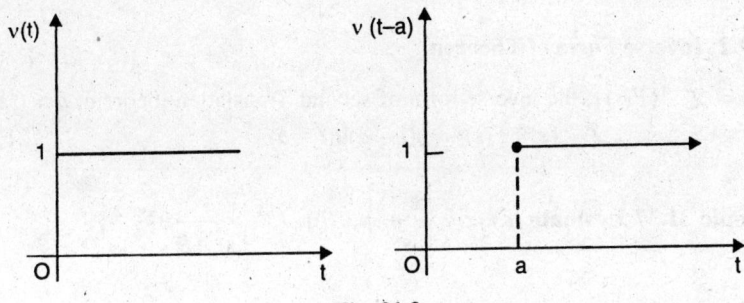

Fig. 21.3

(a) Unit step function $u_0(t)$
   or Heaviside function $H(t)$

(b) Unit step function $u_a(t)$ or
   Heaviside function $H(t - a)$.

Example 21.36 Prove that $\mathcal{L}\{u(t - a)\} = e^{-as}/s$, where $u(t - a)$ is a Heaviside unit step function.

Solution. We have $u(t - a) = \begin{cases} 0, & t < a \\ 1, & t \geq a \end{cases}$ Then

$$\mathcal{L}\{u(t-a)\} = \int_0^a e^{-st}(0)dt + \int_a^\infty e^{-st}(1)dt = \left[\frac{e^{-st}}{-s}\right]_a^\infty = \frac{e^{-as}}{s}.$$

21.19.1 Second Shifting Theorem or Translation on the *t*-axis:
Replacing t by $t - a$ in $f(t)$

If $F(s) = \mathcal{L}\{f(t)\}$ and $a > 0$, then $\mathcal{L}\{f(t - a)u(t - a)\} = e^{-as} F(s)$.

Proof. $\int_0^\infty e^{-st} f(t-a)u(t-a)dt$ can be written as two integrals:

$$\mathcal{L}\{f(t-a)u(t-a)\} = \int_0^a e^{-st} f(t-a) \underbrace{u(t-a)dt}_{\text{zero for } 0 \leq t < a}$$

$$+ \int_a^\infty e^{-st} f(t-a) \underbrace{u(t-a)dt}_{\text{one for } t \leq a}$$

$$= \int_a^\infty e^{-st} f(t-a)dt.$$

Now if we let $v = t - a$, $dv = dt$ in the above integral, then

$$\mathcal{L}\{f(t-a)u(t-a)\} = \int_0^\infty e^{-s(v+a)} f(v)dv = e^{-as} \int_0^\infty e^{-sv} f(v)dv = e^{-as} \mathcal{L}\{f(t)\}$$

$$= e^{-as} F(s). \qquad \qquad ...(21.24)$$

21.19.2 Inverse Form of Theorem

If $f(t) = \mathcal{L}^{-1}\{F(s)\}$, the inverse form of second Translation theorem, $a > 0$ is
$$\mathcal{L}^{-1}\{e^{-as} F(s)\} = f(t-a)u(t-a). \qquad ...(21.25)$$

Example 21.37 Evaluate (a) $\mathcal{L}^{-1}\left\{\dfrac{e^{-2s}}{s^3}\right\}$, (b) $\mathcal{L}^{-1}\left\{\dfrac{s}{s^2+9} e^{-\pi s/2}\right\}$,

(c) $\mathcal{L}^{-1}\left\{\dfrac{e^{-s}}{s(s+1)}\right\}$.

Solution. (a) We have $a = 2$, $F(s) = 1/s^3$, $\mathcal{L}^{-1}\{F(s)\} = t^2/2$.

$$\therefore \mathcal{L}^{-1}\left\{\dfrac{e^{-2s}}{s^3}\right\} = \dfrac{1}{2}(t-2)^2 u(t-2).$$

(b) With $a = \pi/2$, $F(s) = s/(s^2 + 9)$, $\mathcal{L}^{-1}\{F(s)\} = \cos 3t$.

$$\therefore \mathcal{L}^{-1}\left\{\dfrac{s}{s^2+9} e^{-\pi s/2}\right\} = \cos 3\left(t - \dfrac{\pi}{2}\right) u\left(t - \dfrac{\pi}{2}\right).$$

(c) With $a = 1$, $F(s) = \dfrac{1}{s(s+1)} = \left(\dfrac{1}{s} - \dfrac{1}{s+1}\right)$, $\mathcal{L}^{-1}\{F(s)\} = 1 - e^{-t}$

$$\therefore \mathcal{L}^{-1}\left\{\dfrac{e^{-s}}{s(s+1)}\right\} = \left(1 - e^{-(t-1)}\right) u(t-1) = u(t-1) - e^{-(t-1)} u(t-1).$$

Remark: Very often, we may encounter the problem of finding the Laplace transform of a function of the form $f(t)u(t-a)$, where the function $f(t)$ lacks the shifted form $f(t-a)$. To find the Laplace transform of such a function, we use the following procedure. We write

$$f(t)u(t-a) = f(t-a+a) \, u(t-a) = F(t-a) \, u(t-a),$$

where $\qquad F(t-a) = f(t)$ or $F(t) = f(t+a)$.

Using the second shifting theorem, we obtain

$$\mathcal{L}\{F(t-a)u(t-a)\} = e^{-as} \mathcal{L}\{F(t)\} = e^{-as} \mathcal{L}\{f(t+a)\}$$

or $\quad \mathcal{L}\{f(t) \, u(t-a)\} = e^{-as} \mathcal{L}\{f(t+a)\}.$

Example 21.38 Evaluate the following:

(a) $\mathcal{L}\{tu(t-2)\}$,   (b) $\mathcal{L}\{\cos 2t\, u(t-\pi)\}$

Solution. (a) We have $f(t) = t$, $a = 2$, than $f(t+2) = t+2$.

Hence $\mathcal{L}\{tu(t-2)\} = e^{-2s}\mathcal{L}\{t+2\} = e^{-2s}\left(\dfrac{1}{s^2} + \dfrac{2}{s}\right) = \dfrac{e^{-2s}}{s^2} + 2\dfrac{e^{-2s}}{s}$.

(b) We have $f(t) = \cos 2t$, $a = \pi$, then $f(t+\pi) = \cos 2(t+\pi) = \cos(2t+2\pi)$
$= \cos 2t$.

$\therefore \quad \mathcal{L}\{\cos 2t\, u(t-\pi)\} = e^{-\pi s}\mathcal{L}\{\cos 2t\} = \dfrac{s}{s^2+4}e^{-\pi s}$.

21.20 Illustrative Examples

Example 21.39 Express the following functions in terms of unit step function.

(i) $f(t) = \begin{bmatrix} 8 & t < 2 \\ 6 & t \geq 2 \end{bmatrix}$,   (ii) $g(t) = \begin{bmatrix} 4t, & 0 \leq t < 2 \\ 0, & t \geq 2 \end{bmatrix}$,

(iii) $f(t) = \begin{bmatrix} t-1, \\ 3-t \end{bmatrix}$, when $\begin{matrix} 1 < t < 2 \\ 2 < t < 3 \end{matrix}$ (iv) $f(t) = \begin{cases} g(t) & 0 \leq t < a \\ h(t) & t \geq a \end{cases}$

Solution. (i) $f(t) = \begin{bmatrix} 8, & t < 2 \\ 6, & t > 2 \end{bmatrix} = \begin{bmatrix} 8+0 & t < 2 \\ 8-2 & t > 2 \end{bmatrix} = 8 + \begin{bmatrix} 0, & t < 2 \\ -2, & t > 2 \end{bmatrix}$.

$= 8 - 2\begin{bmatrix} 0 & (t < 2) \\ 1 & (t > 2) \end{bmatrix} = 8 - 2u(t-2)$.

Alternative: $f(t) = 8\{u(0) - u(t-2)\} + 6u(t-2) = 8 - 2u(t-2)$.

(ii) $g(t) = \begin{bmatrix} 4t & 0 \leq t < 2 \\ 0 & t \geq 2 \end{bmatrix}$

$= 4t - 4t\,u(t-2) = 4t\left[1 - u(t-2)\right] = 4t\left[u_0(t) - u_2(t)\right]$

(iii) $f(t) = \begin{bmatrix} t-1 & (1 < t < 2) \\ 3-t & (2 < t < 3) \end{bmatrix}$

$= (t-1)\,[u(t-1) - u(t-2)] + (3-t)[u(t-2) - u(t-3)]$
$= (t-1)u(t-1) - 2(t-2)u(t-2) + (t-3)u(t-3)$.

(iv) $f(t) = g(t) - g(t)u(t-a) + h(t)u(t-a)$

(v) $f(t) = f_1(t) - f_1(t)u(t-a) + f_2(t)u(t-a)$.

Example 21.40 Express the function $f(t) = \begin{cases} t-1, & (1 < t < 2) \\ 3-t, & (2 < t < 3), \end{cases}$

in terms of unit step function and obtain its Laplace transform.

Solution. By Example 21.39 (*iii*)

$$f(t) = (t-1)\,u\,(t-1) - 2(t-2)\,u\,(t-2) + (t-3)\,u\,(t-3).$$

Since $\mathcal{L}\{f(t-a)u(t-a)\} = e^{-as}\,F(s)$ where $F(s) = \mathcal{L}\{f(t)\}$.

$$\therefore \mathcal{L}\{f(t)\} = e^{-s} \cdot \frac{1}{s^2} - 2e^{-2s}\frac{1}{s^2} + e^{-3s} \cdot \frac{1}{s^2} = (e^{-s} - 2e^{-2s} + e^{-3s})/s^2.$$

Example 21.41 Find the Laplace transform of the function

(*i*) $f(t) = t^2 u(t-3)$, (*ii*) $f(t) = \sin 2t\,u(t-\pi)$.

Solution. (*i*) We have $f(t) = t^2 u(t-3) = [(t-3)^2 - 9 + 6t]\,u\,(t-3)$
$= [(t-3)^2 + 6\,(t-3) + 9]u(t-3)$.

$$\therefore \mathcal{L}\{t^2 u(t-3)\} = \mathcal{L}[\{(t-3)^2 + 6(t-3) + 9\}u(t-3)]$$

$$= e^{-3s}\left\{ \frac{2}{s^3} + \frac{6}{s^2} + \frac{9}{s} \right\} = \frac{e^{-3s}}{s^3}(2 + 6s + 9s^2).$$

Alternative: We have $f(t) = t^2$ $a = 3$, then $f(t+3) = (t+3)^2$.

$$\therefore \mathcal{L}\{t^2 u(t-3)\} = e^{-3s}\mathcal{L}(t^2 + 6t + 9) = e^{-3s}\left(\frac{2}{s^3} + \frac{6}{s^2} + \frac{9}{s} \right) = e^{-3s}\frac{(2 + 6s + 9s^2)}{s^3}$$

(*ii*) We have $\sin 2t\,u(t-\pi) = \sin(2t - 2\pi + 2\pi)\,u(t-\pi) = \sin 2(t-\pi)u(t-\pi)$

$$\therefore \mathcal{L}\{\sin 2t\,u(t-\pi)\} = e^{-\pi s} \cdot \frac{2}{s^2 + 4}.$$

Alternative: We have $f(t) = \sin 2t$, $a = \pi$, then $f(t+\pi) = \sin 2\,(t+\pi) = \sin 2t$

$$\therefore \mathcal{L}\{\sin 2t\,u(t-\pi)\} = e^{-\pi s} \cdot \frac{2}{s^2 + 4}.$$

Example 21.42 Find the inverse Laplace transform of

(i) $\dfrac{e^{-2s}}{(s-3)}$,   (ii) $\dfrac{(se^{-s/2} + \pi e^{-s})}{s^2 + \pi^2}$,   (iii) $\dfrac{e^{-cs}}{s^2(s+a)}(c > 0)$.   (*V.T.U.*, 2000)

Solution. We know that if

(*i*) $\mathcal{L}^{-1}\{F(s)\} = f(t)$, then $\mathcal{L}^{-1}\{e^{-as}\,F(s)\} = f(t-a)u\,(t-a)$.

With $a = 2$, $F(s) = 1/(s-3)$, $\mathcal{L}^{-1}\{F(s)\} = e^{3t}$

Therefore, $\mathcal{L}^{-1}\left\{ \dfrac{e^{-2s}}{s-3} \right\} = e^{3(t-2)}u(t-2)$.

(ii) $\mathcal{L}^{-1}\left\{\dfrac{se^{-s/2}+\pi e^{-s}}{s^2+\pi^2}\right\} = \mathcal{L}^{-1}\left\{e^{-s/2}\dfrac{s}{s^2+\pi^2}\right\} + \mathcal{L}^{-1}\left\{e^{-s}\cdot\dfrac{\pi}{s^2+\pi^2}\right\}$

$$= \cos\pi\left(t-\frac{1}{2}\right)u\left(t-\frac{1}{2}\right) + \sin\pi(t-1)u(t-1)$$

$$= \sin\pi t\, u\left(t-\frac{1}{2}\right) - \sin\pi t\, u(t-1) = \left\{u\left(t-\frac{1}{2}\right) - u(t-1)\right\}\sin\pi t.$$

(iii) $\mathcal{L}^{-1}\left\{\dfrac{e^{-cs}}{s^2(s+a)}\right\} = \mathcal{L}^{-1}\left\{e^{-cs}\left(-\dfrac{1}{a^2}\cdot\dfrac{1}{s}+\dfrac{1}{a}\dfrac{1}{s^2}+\dfrac{1}{a^2}\cdot\dfrac{1}{s+a}\right)\right\}$

$$= \left[-\frac{1}{a^2}+\frac{1}{a}(t-c)+\frac{1}{a^2}e^{-a(t-c)}\right]u(t-c).$$

Example 21.43 Solve $L\dfrac{di}{dt}+Ri = E(t)$

where $E(t) = \begin{bmatrix} E & (0 < t < a) \\ 0 & (t > a) \end{bmatrix}$ **given that $i = 0$ when $t = 0$.**

Solution. The given differential equation can be written as

$$L\frac{di}{dt}+Ri = E\left[u(t)-u(t-a)\right].$$

Taking Laplace transform of both sides,

$$\mathcal{L}\left[sI(s)-i(0)\right]+RI(s) = E\left[\frac{1}{s}-\frac{e^{-as}}{s}\right]$$

or $(Ls+R)I(s) = E\left(\dfrac{1}{s}-\dfrac{e^{-as}}{s}\right),$ $(\because i = 0, \text{ when } t = 0)$

or $I(s) = \dfrac{E}{s(Ls+R)} - \dfrac{Ee^{-as}}{s(Ls+R)} = \dfrac{E}{R}\left[\dfrac{1}{s}-\dfrac{1}{s+(R/L)} - e^{-as}\left(\dfrac{1}{s}-\dfrac{1}{s+(R/L)}\right)\right]$

$$\therefore\ i(t) = \frac{E}{R}\left[1-e^{-Rt/L}-\left(1-e^{-R(t-a)/L}\right)u(t-a)\right].$$

21.21 Impulse Function

In mechanics we come across problems where very large forces act on bodies for very smal time. Similarly in the study of bending of beams we have point loads which is equivalent to large pressure acting over a very small area. In dealing with such problems the concept of impulse function (also known as **unit impluse function** or **Dirac's delta function**) is introduced below.

Definition. *The function* $\delta(t-a) = \begin{bmatrix} \infty, & if & t = a \\ 0, & if & t \neq a \end{bmatrix}$

such that $\int_{-\infty}^{\infty} \delta(t-a)dt = 1$, is called *impulse function.*

Though there can be no proper function satisfying these conditions, we can consider $\delta(t-a)$ as a limiting case (when $\varepsilon \to 0$) of the function

$$\delta_\varepsilon(t-a) = \begin{cases} \dfrac{1}{\varepsilon}, & a \leq t \leq a+\varepsilon \\ 0, & \text{elsewhere.} \end{cases} \qquad \text{...(21.26)}$$

The Graph of $\delta_\varepsilon(t-a)$ is shown in the figure 21.4.

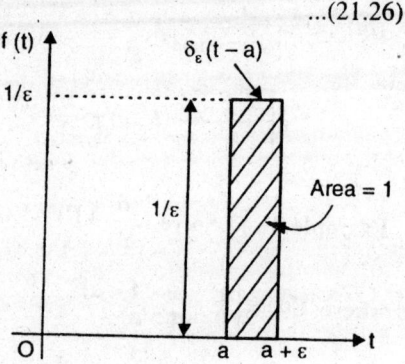

Fig. 21.4

21.21.1 Theorem

If f(t) is a continuous function of t at t = a, then

$$\int_0^\infty f(t)\delta(t-a)\,dt = f(a)$$

Proof. $\int_0^\infty f(t)\delta_\varepsilon(t-a)dt = \dfrac{1}{\varepsilon}\int_a^{a+\varepsilon} f(t)dt$

$= (a+\varepsilon-a)f(\theta)\cdot\dfrac{1}{\varepsilon}$ where θ is some value of t lying between a and $a+\varepsilon$.

$= f(\theta)$ (Mean Value Theorem for integrals).

Taking the limit as $\varepsilon \to 0$, we get

$$\int_0^\infty f(t)\delta(t-a)\,dt = f(a). \qquad \text{...(21.27)}$$

Important Deductions.

(i) $\mathcal{L}\{\delta(t-a)\} = \int_0^\infty e^{-st}\delta(t-a)dt = e^{-as}$. (ii) $\mathcal{L}\{\delta(t)\} = e^{-s \times 0} = e^0 = 1.$

(iii) $\mathcal{L}\{f(t)\delta(t-a)\} = e^{-as}f(a).$

21.21.2 Relation between unit step function and Dirac-delta function

Since $\delta(t-a) = \lim_{\varepsilon \to 0} \delta_\varepsilon(t-a) = \lim_{\varepsilon \to 0} \dfrac{u(t-a)-u(t-a-\varepsilon)}{\varepsilon}$

$$= \frac{d}{dt} u(t-a) = u'(t-a) \qquad \qquad ...(21.28)$$

i.e. $\delta(t-a) = u'(t-a)$ *is a relation between unit step function and Dirac-delta function.*

Example 21.44 The deflection of a beam of length L, clamped, horizontally at both ends and loaded at $x = \dfrac{L}{4}$ at by a weight W is given by

$$EI \frac{d^4 y}{dx^4} = W\delta\left(x - \frac{L}{4}\right). \qquad \qquad ...(i)$$

Find the deflection curve, given that $y = \dfrac{dy}{dx} = 0$ when $x = 0$ and $x = L$.

Solution. Taking Laplace transform of both sides of (*i*), we have

$$EI[s^4 Y(s) - s^3 y(0) - s^2 y'(0) - s y''(0) - y'''(0)] = We^{-Ls/4} - y'''(0)] = We^{-Ls/4}$$

or $EI[s^4 Y(s) - s y''(0) - y'''(0)] = We^{-Ls/4}$ or $EI \, s^4 Y(s) = We^{-Ls/4} + s y_2 + y_3$

where $y_2 = EI \, y''(0)$ and $y_3 = EI y'''(0)$, both constants.

$$\therefore \; EI \, Y(s) = \frac{We^{-Ls/4}}{s^4} + \frac{y_2}{s^3} + \frac{y_3}{s^4}.$$

Inversion gives, $EI \, y(x) = \dfrac{W}{6}\left(x - \dfrac{L}{4}\right)^3 u\left(x - \dfrac{L}{4}\right) + \dfrac{1}{2} y_2 x^2 + \dfrac{y_3 x^3}{6}.$ \qquad ...(ii)

When $x > L/4$ \qquad $EI \, y(x) = \dfrac{W}{6}\left(x - \dfrac{L}{4}\right)^3 + \dfrac{y_2}{2} x^2 + \dfrac{y_3}{6} x^3$

and \qquad $EI \, y' = \dfrac{W}{2}\left(x - \dfrac{L}{4}\right)^2 + y_2 x + \dfrac{y_3}{2} x^2.$

Since $y = y' = 0$ at $x = L$, we have

$$0 = \frac{W}{6}\left(\frac{3L}{4}\right)^3 + \frac{y_2}{2} L^2 + \frac{y_3}{6} L^3 \quad \text{and} \quad 0 = \frac{W}{2}\left(\frac{3L}{4}\right)^2 + y_2 L + \frac{y_3}{2} L^2.$$

Solving the above equations, we get

$$y_2 = 9 \, WL/64 \text{ and } y_3 = -27 \, W/32.$$

Putting these values in (*ii*) we get

$$EI \, y(x) = \frac{W}{6}\left(x - \frac{L}{4}\right)^3 u\left(x - \frac{L}{4}\right) + \frac{9}{128} WL \, x^2 - \frac{9}{64} W \, x^3,$$

which is the required equation.

Exercise 21.5

Find the Laplace transforms of the following periodic functions defined by the wave

1. $f(t) = t$ where $f(t + 1) = f(t)$, $0 \le t < a$, $f(t + a) = f(t)$.

$$\text{Ans. } \frac{1}{s^2} - \frac{ae^{-as}}{s(1-e^{as})}, s > 0.$$

2. $f(t) = \begin{cases} 3t, & 0 \le t < 2 \\ 6, & 2 \le t < 4 \end{cases}$, where $f(t + 4) = f(t)$.

$$\text{Ans. } 3(1 - e^{-2s} - 2se^{-4s})/[s^2(1 - e^{-4s})].$$

3. $f(t) = \sin(\pi t/a)$ for $0 < t < a$.

$$\text{Ans. } \pi a \coth \frac{as}{2} \Big/ (a^2 s^2 + \pi^2).$$

4. Saw-toothed wave function of period T defined by

$f(t) = t/T$ for $0 < t < T$.

$$\text{Ans. } (1/s^2 T) - e^{-sT}/[s(1 - e^{-sT})].$$

5. Express the following functions in terms of unit step functions.

(i) $f(t) = \begin{cases} 1, & 0 \le t < a \\ 2, & a \le t < 2a \\ 3, & 2a \le t < 3a. \end{cases}$ **Ans.** $u(t) + u(t - a) + u(t - 2a) - 3u(t - 3a)$

(ii) $f(t) = \begin{cases} t^2, & 0 < t < 2 \\ 4t, & t > 2. \end{cases}$ **Ans.** $t^2 + (4t - t^2)u(t - 2)$

(iii) $f(t) = \begin{cases} \sin t, & 0 < t < \pi \\ \sin 2t & \pi < t < 2\pi \\ \sin 3t & \quad t > 2\pi. \end{cases}$

$$\text{Ans. } \sin t + (\sin 2t - \sin t)u(t - \pi) + (\sin 3t - \sin 2t)u(t - 2\pi)$$

6. Express $f(t) = \begin{bmatrix} e^{-t}, & 0 < t < 3 \\ 0, & t > 3 \end{bmatrix}$

in terms of unit step function and hence find its Laplace transform.

$$\text{Ans. } e^{-t}\{1 - u(t - 3)\}; \ [1 - e^{-3(s+1)}/(s+1)].$$

7. Find the inverse Laplace transform of

(i) $\dfrac{e^{-as}}{s^2}$, (ii) $\dfrac{e^{-\pi s}}{s^2 + 1}$, (iii) $\dfrac{se^{-as}}{s^2 - \omega^2}$, $a > 0$.

Ans. (i) $(t - a) u(t - a)$, (ii) $-\sin t \, u(t - \pi)$, (iii) $\cosh \omega(t - a) \, u(t - a)$.

8. Find the Laplace transform of the function.

$f(t) = (1 + 2t - 3t^2 + 4t^3) \, u(t - 2)$. $\text{Ans. } e^{-2s}\left[\dfrac{25}{s} + \dfrac{38}{s^2} + \dfrac{42}{s^3} + \dfrac{24}{s^4}\right].$

[**Hint.** By Taylor's Theorem

$$f(t) = (1 + 2t - 3t^2 + 4t^3) = 25 + 38(t-2) + \frac{42}{2!}(t-2)^2 + \frac{24}{3!}(t-2)^3 \Bigg]$$

9. Evaluate $\mathcal{L}\{e^{5t}u(t-1)\}$. **Ans.** $e^{5-s}/(s-5)$.

10. Evaluate $\mathcal{L}\{t\,u(t-1) + t^2\delta(t-1)\}$.

[**Hint.:** Laplace transform $= -\dfrac{d}{ds}\left(\dfrac{e^{-s}}{s}\right) + \dfrac{d^2}{ds^2}(e^{-s})$

$$= e^{-s}(1 + s + s^2)/s^2].$$

11. The deflecttion y of a beam, with stiffness EI and simply supported at its ends $x = 0$ and $x = l$ satisfies the equation

$$EI\frac{d^4y}{dx^4} = w(x), \text{ where } w(x) \text{ is load per unit length and given by}$$

$$w(x) = \begin{bmatrix} w, & l/4 < x < 3l/4, \\ 0, & elsewhere. \end{bmatrix}$$

Find the equation of the deflection curve, given that $y = 0 = y_2$ at $x = 0$ and $x = l$.

Ans. $EIy = \dfrac{1}{24}w\left[\left(x-\dfrac{l}{4}\right)^4 u\left(x-\dfrac{l}{4}\right) - \left(x-\dfrac{3}{4}l\right)^4 \times\right.$

$$\left. u\left(x-\dfrac{3}{4}l\right)\right] + y_1 x + \dfrac{y_3}{6}x^3, \text{ where } y_1 = \dfrac{11}{384}wl^3, y_3 = -\dfrac{l}{4}w.$$

12. A beam is simply supported at its end $x = 0$ and is clampled at the other end $x = l$. It carries a load w at $x = l/4$. Find the resulting deflection at any point, given that deflection y is given by

$$\frac{d^4y}{dx^4} = \frac{w}{EI}\delta\left(x-\frac{l}{4}\right). \quad \textbf{Ans.}\ y = c_1 x + c_2 \frac{x^3}{3!} + \frac{w}{EI}\frac{\left(x-\dfrac{l}{4}\right)^3}{3!}u\left(x-\dfrac{l}{4}\right),$$

$$\text{where } c_1 = \frac{9wl^2}{256EI} \text{ and } c_2 = \frac{-81w}{128EI}.$$

Chapter 22

Finite Differences and Interpolation

22.1 Finite Difference operators and Finite Differences

Let the tabular points or nodal points, $x_0, x_1, ..., x_n$ be equispaced with step length h, that is $x_r = x_0 + rh$, $r = 1, 2,..., n$, or $x_{r+1} - x_r = h$ for all r. The values of x are called the *arguments* and those of y are termed as entries. For equispaced data, we define the following operators.

Forward difference operator Δ

The operator Δ is defined by $\Delta y_r = y_{r+1} - y_r$. Thus the quantities $y_1 - y_0, y_2 - y_1, y_3 - y_2, ...$ are called the *first forward differences* of y. Denoting these differences by $\Delta y_0, \Delta y_1, \Delta y_2 ...$ respectively, we have

$$\Delta y_0 = y_1 - y_0 = f(x_0 + h) - f(x_0)$$
$$\Delta y_1 = y_2 - y_1 = f(x_0 + 2h) - f(x_0 + h)$$

....

....

$$\Delta y_r = y_{r+1} - y_r = f(x_0 + \overline{r+1}\, h) - f(x_0 + rh).$$

Thus we can think of Δ as an operator which when offered y_r as an input produces $y_{r+1} - y_r$ as an output for all r values under consideration.

$$\boxed{y_r} \longrightarrow \boxed{\Delta} \longrightarrow \boxed{y_{r+1} - y_r}$$

The differences of these first differences are called *second differences*. Denoting them by $\Delta^2 y_0, \Delta^2 y_1$, etc., we have

$$\Delta^2 y_0 = \Delta(\Delta y_0) = \Delta y_1 - \Delta y_0 = y_2 - 2y_1 + y_0,$$
$$\Delta^2 y_1 = \Delta(\Delta y_1) = \Delta y_2 - \Delta y_1 = y_3 - 2y_2 + y_1$$

and so on.

Example 22.1 Show that $\Delta^3 y_0 = y_3 - 3y_2 + 3y_1 - y_0$.

Solution. By definition, $\Delta^3 y_0 = \Delta(\Delta^2 y_0) = \Delta[\Delta \cdot (\Delta y_0)] = \Delta(\Delta y_1 - \Delta y_0)$

$$= \Delta(y_2 - y_1) - \Delta(y_1 - y_0) = \Delta y_2 - 2\Delta y_1 + \Delta y_0$$
$$= (y_3 - y_2) - 2(y_2 - y_1) + (y_1 - y_0) = y_3 - 3y_2 + 3y_1 - y_0.$$

In general the nth forward difference is the difference of $(n-1)$th difference and is denoted by

$$\Delta^n y_r = \Delta^{n-1}(\Delta y_r) = \Delta^{n-1}(y_{r+1} - y_r) = \Delta^{n-1} y_{r+1} - \Delta^{n-1} y_r = \Delta[\Delta^{n-1} y_r].$$

The forward difference table 22.1 shows the manner in which the various forward differences can be formed. The first entry is called the leading term and $\Delta y_0, \Delta^2 y_0, \Delta^3 y_0$ etc., are called the leading differences. The forward differences with respect to any point, say x_0, slope in the downward direction $(y_0, \Delta y_0, \Delta^2 y_0, ...)$.

Table 22.1. Forward differences

Argument x	Entry y	First diff. Δy	Second diff. $\Delta^2 y$	Third diff. $\Delta^3 y$
x_0	y_0			
		$\Delta y_0 = y_1 - y_0$		
$x_0 + h$	y_1		$\Delta^2 y_0 = \Delta y_1 - \Delta y_0$	
		$\Delta y_1 = y_2 - y_1$		$\Delta^3 y_0 = \Delta^2 y_1 - \Delta^2 y_0$
$x_0 + 2h$	y_2		$\Delta^2 y_1 = \Delta y_2 - \Delta y_1$	
		$\Delta y_2 = y_3 - y_2$		
$x_0 + 3h$	y_3			

Remark. The nth differences of a polynomial of the nth degree are constant and all higher order differences are zero, when the value of the independent variables are at equal intervals.

Example 22.2 Build upto the constant third difference a difference table of the polynomial $f(x) = 2x^3 - 3x^2 + 4x + 5$ for $x = 0, 1, 2, 3, 4, 5$.

Solution. We have $y = f(x) = 2x^3 - 3x^2 + 4x + 5$. Putting $x = 0, 1, 2, 3, 4, 5$ in the polynomial $f(x)$, we obtain $f(0) = 5$, $f(1) = 8$, $f(2) = 17$, $f(3) = 44$, $f(4) = 101$, $f(5) = 200$.

Difference table

x	$y = f(x)$	Δy	$\Delta^2 y$	$\Delta^3 y$
0	5			
		3		
1	8		6	
		9		12
2	17		18	
		27		12
3	44		30	
		57		12
4	101		42	
		99		
5	200			

Shift Operator E:

The shift (displacement) operator E is defined as the operator that increases the argument of a function by one tabular interval. Thus, by definition,

$Ef_r = Ef(x_r) = f(x_r + h) = f(x_{r+1}) = f_{r+1}$.

Here the input to the operator is f_r. The output is f_{r+1} or if the input to the operator is y_r. The output is y_{r+1}.

$$\boxed{f_r} \longrightarrow \boxed{E} \longrightarrow \boxed{f_{r+1}}$$

The effect of E is to shift or increase the functional value y_r or $f(x_r)$ by one interval to the next higher value y_{r+1}. The higher order operations of E can be similarly defined by

$$E^2 f(x_r) = E(Ef(x_r)) = Ef(x_r + h) = f(x_r + 2h) = y_{r+2}$$
$$E^3 f(x_r) = E^3 y_r = f(x_r + 3h) = y_{r+3}$$

....

....

$$E^n f(x_r) = E^n y_r = f(x_r + nh) = y_{r+n} = E^r y_n.$$

In particular $E^{1/2} y_r = y_{r+1/2}$, $E^{-1/2} y_r = y_{r-1/2}$ and $E^r y_0 = y_r$.

Example 22.3 Evaluate the following

(a) $\Delta\left[\dfrac{1}{f_i}\right]$ (*I.E.T.E., June 2004*), (b) $\Delta^2\left[\dfrac{5x+12}{x^2+5x+6}\right]$, (c) $\Delta^n(x^n)$,

(d) $\Delta^n(1/x)$, (e) $\Delta^n(ab^{cx})$, (f) $\Delta^n(e^{2x+5})$, (*A.M.I.E., W-2000*)

the interval of differencing being unity.

Solution. (a) $\Delta\left(\dfrac{1}{f_i}\right) = \dfrac{1}{f_{i+1}} - \dfrac{1}{f_i} = -\dfrac{f_{i+1} - f_i}{f_i f_{i+1}} = -\dfrac{\Delta f_i}{f_i f_{i+1}}.$

(b) $\Delta^2\left(\dfrac{5x+12}{x^2+5x+6}\right) = \Delta^2\left\{\dfrac{5x+12}{(x+2)(x+3)}\right\} = \Delta^2\left\{\dfrac{2}{x+2} + \dfrac{3}{x+3}\right\}$

$= \Delta\left\{\Delta\left(\dfrac{2}{x+2}\right) + \Delta\left(\dfrac{3}{x+3}\right)\right\} = \Delta\left\{2\left(\dfrac{1}{x+3} - \dfrac{1}{x+2}\right) + 3\left(\dfrac{1}{x+4} - \dfrac{1}{x+3}\right)\right\}$

$= -2\Delta\left\{\dfrac{1}{(x+2)(x+3)}\right\} - 3\Delta\left\{\dfrac{1}{(x+3)(x+4)}\right\}$

$= -2\left\{\dfrac{1}{(x+3)(x+4)} - \dfrac{1}{(x+2)(x+3)}\right\} - 3\left\{\dfrac{1}{(x+4)(x+5)} - \dfrac{1}{(x+3)(x+4)}\right\}$

$$= \frac{4}{(x+2)(x+3)(x+4)} + \frac{6}{(x+3)(x+4)(x+5)} = \frac{2(5x+16)}{(x+2)(x+3)(x+4)(x+5)}.$$

(c) $\Delta^n(x^n) = \Delta^{n-1}\{\Delta x^n\} = \Delta^{n-1}\{(x+1)^n - x^n\} = \Delta^{n-1}\{^nc_1 x^{n-1}\}$

[(On binomial expansion), other terms vanish, having powers $< (n-1)$]

$$= n\Delta^{n-2}\{\Delta x^{n-1}\} = n\Delta^{n-2}\{(x+1)^{n-1} - x^{n-1}\} = n \cdot {}^{n-1}c_1 \Delta^{n-2}(x^{n-2})$$

$$= n(n-1)\Delta^{n-2}(x^{n-2}) = n(n-1)(n-2)\ldots\ldots 3 \cdot 2\Delta x$$

$$= n(n-1)(n-2)\ldots\ldots 3 \cdot 2 \cdot 1 = n! \quad [\text{As } \Delta x = (x+1) - x = 1]$$

(d) $\Delta^n\left(\dfrac{1}{x}\right) = \Delta^{n-1}\left\{\Delta\left(\dfrac{1}{x}\right)\right\} = \Delta^{n-1}\left\{\dfrac{1}{x+1} - \dfrac{1}{x}\right\} = \Delta^{n-1}\left\{\dfrac{-1}{x(x+1)}\right\}$

$$= \Delta^{n-2}\left[\Delta\left\{\frac{-1}{x(x+1)}\right\}\right] = \Delta^{n-2}(-1)\left\{\frac{1}{(x+1)(x+2)} - \frac{1}{x(x+1)}\right\}$$

$$= \Delta^{n-2}(-1)\left\{\frac{-2}{x(x+1)(x+2)}\right\} = \Delta^{n-2}\frac{(-1)^2 2}{x(x+1)(x+2)}$$

$$= \frac{(-1)^n n!}{x(x+1)(x+2)\ldots\ldots(x+n)}.$$

(e) $\Delta^n(ab^{cx}) = \Delta^{n-1}[\Delta(ab^{cx})] = \Delta^{n-1}[ab^{c(x+1)} - ab^{cx}] = (b^c - 1)\,\Delta^{n-1}(ab^{cx})$
$= (b^c - 1)\,\Delta^{n-2}[\Delta(ab^{cx})] = (b^c - 1)\,\Delta^{n-2}\,[ab^{c(x+1)} - ab^{cx}]$
$= (b^c - 1)^2\,\Delta^{n-2}(ab^{cx})$

....
....

$= (b^c - 1)^{n-1}\,\Delta(ab^{cx}) = (b^c - 1)^{n-1}\,(ab^{c(x+1)} - ab^{cx}) = (b^c - 1)^n(ab^{cx}).$

(f) $\Delta^n(e^{2x+5}) = \Delta^{n-1}[\Delta e^{2x+5}] = \Delta^{n-1}[e^{2(x+1)+5} - e^{2x+5}]$
$= (e^2 - 1)\,\Delta^{n-1}(e^{2x+5}) = (e^2 - 1) \cdot \Delta^{n-2}\,[\Delta e^{2x+5}]$
$= (e^2 - 1)^2\,\Delta^{n-2}(e^{2x+5})$

....
....

$= (e^2 - 1)^{n-1}\,\Delta e^{2x+5} = (e^2 - 1)^n\,e^{2x+5}.$

Remark: If the interval of differencing in the value of x is h, then $\Delta^n\,e^{2x+5}$
$= (e^{2h} - 1)^n\,e^{2x+5}.$

Example 22.4 Evaluate the following

(i) $\Delta^3[(1-x)(1-2x)(1-3x)]$,

(A.M.I.E.; S-2007)

(ii) $\Delta^{10}\left[(1-ax)(1-bx^2)(1-cx^3)(1-dx^4)\right].$ *(Madras, 1996S)*

Solution. *(i)*$\Delta^3[(1-x)(1-2x)(1-3x)] = \Delta^3(-6x^3)$, other terms vanishing as they
have powers < 3

$$= -6\Delta^3x^3 = -6\Delta^2(\Delta x^3) = -6\Delta^2\{(x+1)^3 - x^3\}$$
$$= -6\Delta^2(3x^2 + 3x + 1) = -18\Delta^2x^2, \text{ other terms vanishing, having powers} < 2$$
$$= -18\Delta(\Delta x^2) = -18\ \Delta\{(x+1)^2 - x^2\} = -18\ \Delta(2x+1)$$
$$= -36\Delta x, \text{ other term vanishing}$$
$$= -36\{(x+1) - x\} = -36 = -6.3! \qquad\qquad \textbf{[Note: } \Delta^3x^3 = 3!\textbf{]}$$

(ii) $\Delta^{10}[(1-ax)(1-bx^2)(1-cx^3)(1-dx^4)] = \Delta^{10}(abcd\ x^{10})$, other terms vanish
having powers < 10
$$= abcd\ \Delta^{10}(x^{10}) = abcd\ (10!) \qquad\qquad \textbf{[Note: } \Delta^{10}(x^n) = 0 \text{ for } n < 10\textbf{]}$$

Example 22.5 Prove the following relations

(i) $\Delta(f_i g_i) = f_i\ \Delta g_i + g_{i+1}\Delta f_i,$

(ii) $\Delta\left(f_i/g_i\right) = \dfrac{g_i\Delta f_i - f_i\Delta g_i}{g_i g_{i+1}},$ *(I.E.T.E.; June 2004)*

(iii) $\Delta\left(f_i^2\right) = \left(f_i + f_{i+1}\right)\Delta f_i$ *(I.E.T.E.; June 2005)*

Solution. *(i)* $\Delta(f_i \cdot g_i) = f_{i+1}\ g_{i+1} - f_i g_i$
$$= f_{i+1}\ g_{i+1} - f_i\ g_i + f_i\ g_{i+1} - f_i\ g_{i+1}$$
$$= f_i(g_{i+1} - g_i) + g_{i+1}(f_{i+1} - f_i) = f_i\ \Delta g_i + g_{i+1}\ \Delta f_i.$$

(ii) $\Delta\left(\dfrac{f_i}{g_i}\right) = \dfrac{f_{i+1}}{g_{i+1}} - \dfrac{f_i}{g_i} = \dfrac{g_i f_{i+1} - f_i g_{i+1}}{g_i g_{i+1}}$

$$= \dfrac{g_i\left(f_{i+1} - f_i\right) - f_i\left(g_{i+1} - g_i\right)}{g_i g_{i+1}} = \dfrac{g_i\Delta f_i - f_i\Delta g_i}{g_i g_{i+1}}.$$

(iii) $\Delta\left(f_i^2\right) = f_{i+1}^2 - f_i^2 = \left(f_{i+1} + f_i\right)\left(f_{i+1} - f_i\right) = \left(f_{i+1} + f_i\right)\Delta f_i.$

Example 22.6 Evaluate the following, the interval of differencing being h

(i) $\Delta\cot 2^x,$ *(ii)* $\Delta^2\cos 2x.$ *(A.M.I.E., W-1999)*

Solution. *(i)* $\Delta\cot 2^x = \cot 2^{(x+h)} - \cot 2^x = \dfrac{\cos 2^{x+h}}{\sin 2^{x+h}} - \dfrac{\cos 2^x}{\sin 2^x}$

$$= \dfrac{\sin 2^x \cdot \cos 2^{x+h} - \cos 2^x \cdot \sin 2^{x+h}}{\sin 2^x \cdot \sin 2^{x+h}} = \dfrac{\sin\left(2^x - 2^{x+h}\right)}{\sin 2^x \sin 2^{x+h}} = \dfrac{\sin\left\{2^x\left(1-2^h\right)\right\}}{\sin 2^x \sin 2^{x+h}}.$$

(ii) $\Delta^2\cos 2x = \Delta(\Delta\cos 2x) = \Delta[\cos 2(x+h) - \cos 2x]$
$$= \Delta\ \cos 2(x+h) - \Delta\ \cos 2x$$

$$= [\cos 2(x + 2h) - \cos 2(x + h)] - [\cos 2(x + h) - \cos 2x]$$
$$= -2 \sin(2x + 3h) \sin h + 2 \sin (2x + h) \sin h$$
$$= -2 \sin h[\sin (2x + 3h) - \sin(2x + h)]$$
$$= -2 \sin h [(2 \cos(2x + 2h) \sin h] = -4 \sin^2 h \cos(2x + 2h).$$

Backward difference operator ∇

The backward difference operator ∇ is defined by
$$\nabla y_r = y_r - y_{r-1} \text{ or } \nabla f(x_r) = f(x_r) - f(x_r - h).$$
Putting $r = 1, 2, ..., n$. We obtain
$$\nabla y_1 = y_1 - y_0, \qquad \nabla y_2 = y_2 - y_1 \text{ etc.}$$
Similarly, the second and higher order backward differences are defined as
$$\nabla^2 y_r = \nabla(\nabla y_r) = \nabla y_r - \nabla y_{r-1} = y_r - 2y_{r-1} + y_{r-2}$$
$$\nabla^3 y_r = \nabla(\nabla^2 y_r) = \nabla y_r - 2\nabla y_{r-1} + \nabla y_{r-2}$$
$$= (y_r - y_{r-1}) - 2(y_{r-1} - y_{r-2}) + (y_{r-2} - y_{r-3})$$
$$= y_r - 3y_{r-1} + 3y_{r-2} - y_{r-3}, \text{ etc.}$$
and in general, the nth backward difference is the difference of $(n-1)^{th}$ difference, that is
$$\nabla^n y_r = \nabla^{n-1} (\nabla y_r) = \nabla^{n-1}[y_r - y_{r-1}] = \nabla^{n-1} y_r - \nabla^{n-1}y_{r-1} = \nabla[\nabla^{n-1}y_r].$$
The successive backward differences are shown in a Table 22.2

Table 22.2. Backward differences

Argument x	Entry y	First difference ∇y	Second difference $\nabla^2 y$	Third difference $\nabla^3 y$
x_0	y_0			
		$\nabla y_1 = y_1 - y_0$ $= \Delta y_0$		
x_1	y_1		$\nabla^2 y_2 = \nabla y_2 - \nabla y_1$ $= \Delta^2 y_0$	
		$\nabla y_2 = y_2 - y_1$ $= \Delta y_1$		$\nabla^3 y_3 = \nabla^2 y_3 - \nabla^2 y_2$ $= \Delta^3 y_0$
x_2	y_2		$\nabla^2 y_3 = \nabla y_3 - \nabla y_2$ $= \Delta^2 y_1$	
		$\nabla y_3 = y_3 - y_2$ $= \Delta y_2$		
x_3	y_3			

From the Table 22.2, we note that $\Delta^n y_r = \nabla^n y_{r+n}$. The backward differences with respect to any point, say x_n, slope in the upward direction ($y_n, \nabla y_n, \nabla^2 y_n, ...$).

Central difference operator δ

While the forward and backward differences are appropriately applicable for interpolating values near the end points of the tabulated data, the central differences are gainfully used for interpolating values at some interval points of the given data. The central difference operator δ is defined by

$$\delta y_r = y_{r+1/2} - y_{r-1/2} \quad \text{or} \quad \delta f(x_r) = f\left(x_r + \frac{h}{2}\right) - f\left(x_r - \frac{h}{2}\right)$$

or $\delta y_{r+1/2} = y_{r+1} - y_r$. Setting $r = 0, 1, 2, 3, \ldots$, we get
$\delta y_{1/2} = y_1 - y_0$, $\delta y_{3/2} = y_2 - y_1$, $\delta y_{5/2} = y_3 - y_2$ and so on.

In a similar fashion the higher order central differences can be defined by
$\delta^2 y_r = \delta(\delta y_r) = \delta y_{r+1/2} - \delta y_{r-1/2} = y_{r+1} - 2y_r + y_{r-1}$.
$\delta^3 y_r = \delta(\delta^2 y_r) = \delta(y_{r+1} - 2y_r + y_{r-1}) = y_{r+3/2} - 3y_{r+1/2} + 3y_{r-1/2} - y_{r-3/2}$.

We define for any positive integer n
$\delta^n y_r = \delta^{n-1}[\delta y_r] = \delta^{n-1}[y_{r+1/2} - y_{r-1/2}] = \delta^{n-1} y_{r+1/2} - \delta^{n-1} y_{r-1/2} = \delta[\delta^{n-1} y_r]$.

The central differences are exhibited in Table 22.3.

Table 22.3. Central differences

x	y	δy	$\delta^2 y$	$\delta^3 y$
x_0	y_0			
		$\delta y_{1/2} = y_1 - y_0$		
x_1	y_1		$\delta^2 y_1 = \delta y_{3/2} - \delta y_{1/2}$	
		$\delta y_{3/2} = y_2 - y_1$		$\delta^3 y_{3/2} = \delta^2 y_2 - \delta^2 y_1$
x_2	y_2		$\delta^2 y_2 = \delta y_{5/2} - \delta y_{3/2}$	
		$\delta y_{5/2} = y_3 - y_2$		
x_3	y_3			

We note from the above table that the differences of odd orders are manifested by half values and those of even orders by integral values of suffixes.

We note that $\Delta^n y_r = \nabla^n y_{r+n} = \delta^n y_{r+(n/2)}$

Alternative, we can write the central differences centered about x_0 as given in Table 22.4

Table 22.4. Central differences

Argument x	Entry y	1st diff. δy	2nd diff. $\delta^2 y$	3rd diff. $\delta^3 y$	4th diff. $\delta^4 y$
x_{-2}	y_{-2}				
		$\delta y_{-3/2}$			
x_{-1}	y_{-1}		$\delta^2 y_{-1}$		
		$\delta y_{-1/2}$		$\delta^3 y_{-1/2}$	
x_0	y_0		$\delta^2 y_0$		$\delta^4 y_0$
		$\delta y_{1/2}$		$\delta^3 y_{1/2}$	
x_1	y_1		$\delta^2 y_1$		
		$\delta y_{3/2}$			
x_2	y_2				

Averaging (or Mean) Operator μ

The averaging operator μ is defined by the relation

$$\mu y_r = \frac{1}{2}\left(y_{r+1/2} + y_{r-1/2}\right) \quad \text{or} \quad \mu f(x_r) = \frac{1}{2}\left[f\left(x_r + \frac{h}{2}\right) + f\left(x_r - \frac{h}{2}\right)\right]$$

and $\mu\delta y_r = \frac{1}{2}\left(\delta y_{r+1/2} + \delta y_{r-1/2}\right) = \frac{1}{2}\left(y_{r+1} - y_r + y_r - y_{r-1}\right) = \frac{1}{2}\left(y_{r+1} - y_{r-1}\right).$

The averaging operator μ denotes the averaged value at r of the two adjoining values at $r + 1/2$ and $r - 1/2$.

22.2 Relations between the Operators

We know from the definition of finite differences that

$$y_1 - y_0 = \Delta y_0 = \nabla y_1 = \delta y_{1/2} \qquad \qquad ...(22.1)$$

It is obvious from the above relations that it is not the difference $y_1 - y_0$ that changes but the change is effected only in the notations.

(A) Using the forward difference operator, we have

$$\Delta f(x) = f(x + h) - f(x) = Ef(x) - f(x) = (E - 1) f(x).$$

This gives $\qquad \Delta = E - 1$ or $E = 1 + \Delta$ $\qquad \qquad ...(22.2)$

(B) Again for the backward differences

$$\nabla f(x) = f(x) - f(x - h) = f(x) - E^{-1} f(x) = (1 - E^{-1}) f(x).$$

This gives $\nabla = 1 - E^{-1}$ or $E = (1 - \nabla)^{-1}$ $\qquad \qquad ...(22.3)$

From (22.2) and (22.3), we obtain $\nabla - \Delta = -\Delta\nabla$ *(AMIE-W 2003)* $\qquad ...(22.4)$

(C) Also for the Central differences

$$\delta y_r = y_{r+1/2} - y_{r-1/2} \text{ or } \delta f(x) = f\left(x + \frac{h}{2}\right) - f\left(x - \frac{h}{2}\right)$$

$$= E^{1/2}f(x) - E^{-1/2}f(x) = (E^{1/2} - E^{-1/2})f(x)$$

This gives $\qquad\qquad \delta = E^{1/2} - E^{-1/2}$ $\qquad\qquad ...(22.5)$

From (22.5) $\quad \delta = E^{-1/2}(E - 1) = \Delta E^{-1/2} = \Delta(1 + \Delta)^{-1/2},$ *(D.U. 1995)* ...(22.6)

$\delta = E^{1/2}(1 - E^{-1}) = \nabla E^{1/2} = \nabla(1 - \nabla)^{-1/2},$ *(I.E.T.E., W-2006)* $\qquad ...(22.7)$

From (22.5) $\qquad \delta E^{1/2} = E - 1 = \Delta,$ *(A.M.I.T., W-2007)* $\qquad\qquad ...(22.8)$

(D) If D denotes the differential operator d/dx, we have

$$Df(x) = \frac{d}{dx} f(x) = f'(x).$$

Using Taylor series expansion of $f(x + h)$ about x, we obtain

$$f(x+h) = f(x) + hf'(x) + \frac{h^2}{2!} f''(x) + \frac{h^3}{3!} f'''(x) + ...$$

$$= f(x) + hDf(x) + \frac{h^2D^2}{2!}f(x) + \frac{h^3D^3}{3!}f(x) + ...$$

$$= \left(1 + hD + \frac{h^2D^2}{2!} + \frac{h^3D^3}{3!} + ...\right)f(x) = e^{hD}f(x) \qquad ...(22.9)$$

Since $f(x + h) = Ef(x)$, we may write (22.9)
$f(x + h) = Ef(x) = e^{hD} f(x)$. This gives
$$E = e^{hD} \qquad ...(22.10)$$
From relations (22.2), (22.3) and (22.10), we see that the operators D, Δ, ∇ and E are related by
$$E = e^{hD} = 1 + \Delta = (1 - \nabla)^{-1}. \qquad ...(22.11)$$
Taking logarithms on both sides, we obtain
$$hD = \log_e(1 + \Delta) = \log_e(1 - \nabla)^{-1} = -\log_e(1 - \nabla) \qquad ...(22.12)$$

Expanding, we get $\qquad hD = \Delta - \frac{1}{2}\Delta^2 + \frac{1}{3}\Delta^3 - ...... \qquad ...(22.13)$

$$= \nabla + \frac{1}{2}\nabla^2 + \frac{1}{3}\nabla^3 + ... \qquad ...(22.14)$$

Refer relation (22.5), $\delta = E^{1/2} - E^{-1/2} = e^{hD/2} - e^{-hD/2} = 2\sinh(hD/2)$
or $\quad hD = 2\sinh^{-1}(\delta/2)$ $\qquad$ *(I.E.T.E., W-2006)* $\quad ...(22.15)$
(E) The averaging (or mean) operator μ is defined by

$$\mu y_r = \frac{1}{2}\left(y_{r+1/2} + y_{r-1/2}\right) = \frac{1}{2}\left(E^{1/2} + E^{-1/2}\right)y_r. \text{ This gives}$$

$$\mu = \frac{1}{2}\left(E^{1/2} + E^{-1/2}\right) \qquad ...(22.16)$$

$$= \frac{1}{2}\left(e^{hD/2} + e^{-hD/2}\right) = \cosh\left(\frac{hD}{2}\right) \qquad ...(22.17)$$

and $\mu^2 = \frac{1}{4}\left(E^{1/2} + E^{-1/2}\right)^2 = \frac{1}{4}\left[\left(E^{1/2} - E^{-1/2}\right)^2 + 4\right] = \frac{1}{4}(\delta^2 + 4)$, from (22.5)

This gives $\mu = \sqrt{1 + \frac{\delta^2}{4}}$. *(A.M.I.E., W-2008; S-2006)* $\qquad ...(22.18)$

Again $\mu\delta y_r = \frac{1}{2}\left(\delta y_{r+1/2} + \delta y_{r-1/2}\right) = \frac{1}{2}\left(y_{r+1} - y_r + y_r - y_{r-1}\right)$

$$= \frac{1}{2}\left(y_{r+1} - y_{r-1}\right) = \frac{1}{2}\left(E - E^{-1}\right)y_r$$

Hence $\qquad \mu\delta = \dfrac{1}{2}\left(E - E^{-1}\right)$ $\qquad\qquad$ ...(22.19)

$$= \dfrac{1}{2}\left(e^{hD} - e^{-hD}\right) = \sinh(hD) \qquad\qquad \text{...(22.20)}$$

or $\qquad\qquad hD = \sinh^{-1}(\mu\delta).$ $\qquad$ *(I.E.T.E., June 2006)*

Example 22.7 Evaluate the following

(i) $\left(\dfrac{\Delta^2}{E}\right)x^3,$ *(AMIE., W-2004)* $\qquad$ (ii) $\dfrac{\Delta^2 x^3}{Ex^3},$

(iii) $\Delta \log f(x),$ *(AMIE., S-2003)* $\qquad$ (iv) $\left(\dfrac{\Delta^2}{E}\right)e^x \cdot \dfrac{Ee^x}{\Delta^2 e^x}.$ *(AMIE., W-2004)*

(v) $E\Delta f(x)$ and $E\nabla f(x)$ for $x = 1$ with $h = 0.2$. Given $f(x) = xe^x$. The interval of differencing being h. $\qquad$ *(A.M.I.E., W-2004)*

Solution.

(i) $\left(\dfrac{\Delta^2}{E}\right)x^3 = \left\{\dfrac{(E-1)^2}{E}\right\}x^3 = \left(E - 2 + E^{-1}\right)x^3 = Ex^3 - 2x^3 + E^{-1}x^3$

$\qquad\qquad = (x+h)^3 - 2x^3 + (x-h)^3 = 6h^2 x.$

(ii) $\dfrac{\Delta^2 x^3}{Ex^3} = \dfrac{(E-1)^2 x^3}{Ex^3} = \dfrac{\left(E^2 - 2E + 1\right)x^3}{Ex^3} = \dfrac{(x+2h)^3 - 2(x+h)^3 + x^3}{(x+h)^3}$

$\qquad\qquad = \dfrac{6xh^2 + 6h^3}{(x+h)^3} = \dfrac{6h^2}{(x+h)^2}.$

(iii) $\Delta \log f(x) = \log f(x+h) - \log f(x)$

$\qquad\qquad = \log\left[\dfrac{f(x+h)}{f(x)}\right] = \log\left[\dfrac{Ef(x)}{f(x)}\right] = \log\left[\dfrac{(1+\Delta)f(x)}{f(x)}\right]$

$\qquad\qquad = \log\left[\dfrac{f(x) + \Delta f(x)}{f(x)}\right] = \log\left[1 + \dfrac{\Delta f(x)}{f(x)}\right]$

(iv) $\left(\dfrac{\Delta^2}{E}\right)e^x \cdot \dfrac{Ee^x}{\Delta^2 e^x} = \left(\Delta^2 E^{-1}\right)e^x \cdot \dfrac{Ee^x}{\Delta^2 e^x} = \Delta^2 e^{x-h} \cdot \dfrac{Ee^x}{\Delta^2 e^x}$

$\qquad\qquad = e^{-h}\dfrac{\Delta^2 e^x}{\Delta^2 e^x} \cdot Ee^x = e^{-h} \cdot e^{x+h} = e^x.$

(v) $E\Delta f(x) = E\{f(x + h) - f(x)\} = Ef(x + h) - Ef(x) = f(x + 2h) - f(x + h)$.

$\therefore E\Delta(xe^x) = (x + 2h) e^{x+2h} - (x + h)e^{x+h}$. Substituting $x = 1$, $h = 0.2$, we obtain

$E\Delta(xe^x)$ at $(x = 1, h = 0.2)$ is equal to $1.4e^{1.4} - 1.2e^{1.2} = (1.4) (4.0552) -$
(1.2) (3.32012) = 5.67728 - 3.98414 = 1.69314,

and $\qquad E\nabla f(x) = E\{f(x) - f(x - h)\} = f(x + h) - f(x)$.

$\therefore E\nabla(xe^x) = (x + h)e^{x+h} - xe^x$. Substituting the given values, we obtain

$E\nabla(xe^x)$ at $(x = 1, h = 0.2)$ is equal to $1.2e^{1.2} - e = 1.2(3.32012) - 2.71828$
= 1.26586.

Example 22.8 Show that $y_{-3} = y_0 - 3 \nabla y_0 + 3 \nabla^2 y_0 - \nabla^3 y_0$.

Solution. We know that $\nabla = 1 - E^{-1}$ or $E = (1 - \nabla)^{-1}$

$\therefore E^{-3} y_0 = \{(1 - \nabla)^{-1}\}^{-3} y_0 = (1 - \nabla)^3 y_0 = (1 - 3\nabla + 3\nabla^2 - \nabla^3) y_0$
or $y_{-3} = y_0 - 3\nabla y_0 + 3 \nabla^2 y_0 - \nabla^3 y_0$.

Example 22.9 Prove that $\delta^2 y_5 = y_6 - 2y_5 + y_4$.

Solution. We know that $\delta = (E^{1/2} - E^{-1/2})$ or $\delta^2 = (E^{1/2} - E^{-1/2})^2 = (E - 2 + E^{-1})$

$\therefore \delta^2 y_5 = (E - 2 + E^{-1})y_5 = y_6 - 2y_5 + y_4$.

Example 22.10 Prove the following relations

(i) $\Delta + \nabla = \dfrac{\Delta}{\nabla} - \dfrac{\nabla}{\Delta}$, $\qquad\qquad$ *(Madurai, 1996S; I.E.T.E., June 2003)*

(ii) $\Delta^3 y_2 = \nabla^3 y_5$ $\qquad\qquad\qquad$ *(Madras 1999; I.E.T.E., June 2006)*

(iii) $\Delta = \dfrac{1}{2}\delta^2 + \delta\sqrt{1 + \dfrac{1}{4}\delta^2}$

$\qquad\qquad$ *(I.E.T.E., June 2004; Madras, 2000; D.U. 1997; Madurai, 1996)*

(iv) $\mu\delta = \dfrac{1}{2}\Delta + \dfrac{1}{2}\Delta E^{-1}$.

(v) $\sqrt{1 + \mu^2\delta^2} = 1 + \dfrac{1}{2}\delta^2$ $\qquad\qquad$ *(Madras, 1998; I.E.T.E. June 2005)*

Solution. (i) Using $\Delta = E - 1$, $\nabla = 1 - E^{-1}$, we obtain

L.H.S. $= \Delta + \nabla = E - 1 + 1 - E^{-1} = E - E^{-1}$.

R.H.S. $= \dfrac{\Delta}{\nabla} - \dfrac{\nabla}{\Delta} = \dfrac{E-1}{1-E^{-1}} - \dfrac{1-E^{-1}}{E-1} = \dfrac{E(1-E^{-1})}{1-E^{-1}} - \dfrac{E^{-1}(E-1)}{E-1}$

$= E - E^{-1} = $ L.H.S. Hence the result.

(ii) $\Delta^3 y_2 = (E-1)^3 y_2 = (E^3 - 3E^2 + 3E - 1) y_2 = y_5 - 3y_4 + 3y_3 - y_2$ $\qquad$ ...(i)

$\nabla^3 y_5 = (1 - E^{-1})^3 y_5 = (1 - 3E^{-1} + 3E^{-2} - E^{-3}) y_5 = y_5 - 3y_4 + 3y_3 - y_2$ $\qquad$ ...(ii)

From (*i*) and (*ii*), $\Delta^3 y_2 = \nabla^3 y_5$.

(*iii*) We have $\dfrac{1}{2}\delta^2 + \delta\sqrt{1 + \dfrac{1}{4}\delta^2} = \dfrac{1}{2}\left(E^{1/2} - E^{-1/2}\right)^2 + \left(E^{1/2} - E^{-1/2}\right)$

$$\times \sqrt{1 + \dfrac{1}{4}\left(E^{1/2} - E^{-1/2}\right)^2}$$

$$= \dfrac{1}{2}\left(E - 2 + E^{-1}\right) + \left(E^{1/2} - E^{-1/2}\right)\sqrt{\dfrac{E + 2 + E^{-1}}{4}}$$

$$= \dfrac{1}{2}\left(E - 2 + E^{-1}\right) + \dfrac{1}{2}\left(E^{1/2} - E^{-1/2}\right)\left(E^{1/2} + E^{-1/2}\right)$$

$$= \dfrac{1}{2}\left(E - 2 + E^{-1}\right) + \dfrac{1}{2}\left(E - E^{-1}\right) = \dfrac{1}{2}(2E - 2) = E - 1 = \Delta.$$

(*iv*) L.H.S. $= \mu\delta = \dfrac{1}{2}\left(E^{1/2} + E^{-1/2}\right)\left(E^{1/2} - E^{-1/2}\right) = \dfrac{1}{2}\left(E - E^{-1}\right) = \dfrac{1}{2}\left(1 + \Delta - \dfrac{1}{E}\right)$

$$= \dfrac{1}{2}\Delta + \dfrac{1}{2}\left(\dfrac{E - 1}{E}\right) = \dfrac{1}{2}\Delta + \dfrac{1}{2}\Delta E^{-1} = \text{R.H.S.}$$

(*v*) We know that $\delta = E^{1/2} - E^{-1/2}$ and $\mu = \dfrac{1}{2}\left(E^{1/2} + E^{-1/2}\right)$

$$\therefore \ \mu\delta = \dfrac{1}{2}\left(E^{1/2} + E^{-1/2}\right)\left(E^{1/2} - E^{-1/2}\right) = \dfrac{1}{2}\left(E - E^{-1}\right),$$

and $\sqrt{1 + \mu^2 \delta^2} = \sqrt{1 + \dfrac{1}{4}\left(E^2 - 2 + E^{-2}\right)} = \dfrac{1}{2}\left(E + E^{-1}\right).$ ...(*iii*)

Now $1 + \dfrac{1}{2}\delta^2 = 1 + \dfrac{1}{2}\left(E^{1/2} - E^{-1/2}\right)^2 = 1 + \dfrac{1}{2}\left(E - 2 + E^{-1}\right) = \dfrac{1}{2}\left(E + E^{-1}\right)$ ...(*iv*)

From (*iii*) and (*iv*) $\sqrt{1 + \mu^2 \delta^2} = 1 + \dfrac{1}{2}\delta^2.$

Example 22.11 **Find the missing values in the following table**

x	−1	0	1	2	3	4
$y = f(x)$	10	–	8	10	–	50

(Kerala, 1995S)

Solution. Let the missing values be y_1 and y_4. We have the following difference table.

Difference table

x	y	Δy	$\Delta^2 y$	$\Delta^3 y$	$\Delta^4 y$
-1	$y_0 = 10$				
		$y_1 - 10$			
0	y_1		$18 - 2y_1$		
		$8 - y_1$		$3y_1 - 24$	
1	$y_2 = 8$		$y_1 - 6$		$y_4 - 4y_1 + 18$
		2		$y_4 - y_1 - 6$	
2	$y_3 = 10$		$y_4 - 12$		$78 - 4y_4 + y_1$
		$y_4 - 10$		$72 - 3y_4$	
3	y_4		$60 - 2y_4$		
		$50 - y_4$			
4	$y_5 = 50$				

As only four entries y_0, y_2, y_3, y_5 are given, the function y can be represented by a third degree polynomial.

$\therefore$ $\qquad\qquad \Delta^4 y_0 = 0$ and $\Delta^4 y_1 = 0$

i.e., $\qquad\qquad y_4 - 4y_1 + 18 = 0;$ $\qquad 78 - 4y_4 + y_1 = 0$

Solving these, we obtain $y_1 = 10$, $\qquad y_4 = 22$.

Alternative:

$\qquad\qquad \Delta^4 y_0 = 0$ and $\qquad \Delta^4 y_1 = 0$

i.e. $(E - 1)^4 \, y_0 = 0$ and $(E - 1)^4 \, y_1 = 0$

i.e. $(E^4 - 4E^3 + 6E^2 - 4E + 1) \, y_0 = 0$ and $(E^4 - 4E^3 + 6E^2 - 4E + 1) \, y_1 = 0$

i.e., $y_4 - 4y_3 + 6y_2 - 4y_1 + y_0 = 0$ and $y_5 - 4y_4 + 6y_3 - 4y_2 + y_1 = 0$

or $y_4 - 4y_1 + 18 = 0, 78 - 4y_4 + y_1 = 0.$ (on substituting the given values)

Solving these, we get $y_1 = 10$, $\qquad\qquad y_4 = 22$.

22.3 Factorial Polynomials

The continued product of n factors where x is the first factor and successive factors decrease by a constant (we shall take the constant 1 on the assumption that the interval of differencing is unity) is called **factorial polynomial** of nth degree and denoted by $x^{(n)}$.

Thus $\qquad\qquad *x^{(n)} = x \, (x - 1)(x - 2)...(x - n + 1)$.

In particular $x^{(3)} = x(x - 1) \, (x - 2)$, $x^{(4)} = x(x - 1)(x - 2) \, (x - 3)$ and obviously

$x^{(4)} = x(x - 1) \, (x - 2)^{(2)}$. In general $x^{(n)} = x(x-1)^{(n-1)}$.

The importance of these polynomials lies in the fact that they play the same role in calculus of finite differences that the power function x^n plays in ordinary calculus. This is evident from the result (A) and (B) below:

$*x^{(n)}$ is also called as "factorial n^{th} power of x".

(A) $\Delta x^{(n)} = nx^{(n-1)}$

Proof. $\Delta x^{(n)} = (x + 1)^{(n)} - x^{(n)}$

$$= (x + 1) x (x - 1)...[(x + 1) - (n - 1)] - x(x - 1) ...[x - (n - 1)]$$

$$= (x + 1) x (x - 1)... (x - n + 2) - x(x - 1)... (x - n + 1)$$

$$= x(x - 1) ...(x - n + 2) [(x + 1) - (x - n + 1)]$$

$$= x(x - 1) ... ...[x - (n - 2)]. n = n\, x^{(n-1)}.$$

Thus $\Delta x^{(n)} = nx^{(n-1)}.$ ...(22.21)

Notes. (*i*) This result is analogous to $Dx^n = nx^{n-1}$, where $D = \dfrac{d}{dx}$.

(*ii*) In calculus $\left(\dfrac{1}{D}\right) x^n$ or $D^{-1} x^n$ stands for the function whose derivative is x^n.

$\therefore \quad \dfrac{1}{D} x^n = \dfrac{x^{n+1}}{n+1}.$ $\qquad$ $D^{-1} x^n$ is called anti derivative of x^n.

(*iii*) By analogy with the above terminology of calculus

$\dfrac{1}{\Delta} x^{(n)}$ or $\Delta^{-1} x^{(n)}$ is called anti-difference of $x^{(n)}$.

It stands for the function whose first difference is $x^{(n)}$.
From the result (A) it is clear that

(B) $\dfrac{1}{\Delta} x^{(n)} = \dfrac{x^{(n+1)}}{n+1}.$ ...(22.22)

22.3.1 Methods of Expressing any Polynomial in Factorial Notation

From the results (A) and (B) of Art 22.3 it is clear that once a function has been expressed as a series of factorial polynomials, it is very simple to obtain its various differences and anti-differences.

We give below two methods for this purpose by taking some examples.

Example 22.12 Express $2x^3 - 3x^2 + 3x + 15$ in terms of factorial polynomials.

First Method (*Direct Method*)

Solution. Let $2x^3 - 3x^2 + 3x + 15 = A\, x^{(3)} + B\, x^{(2)} + C\, x^{(1)} + D$

$$= Ax(x - 1)(x - 2) + Bx(x - 1) + Cx + D \qquad ...(1)$$

where A, B, C, D are constants to be determined.
Putting $x = 0$ in both sides of (1) we obtain $D = 15$.
Putting $x = 1$, we have

$$2 - 3 + 3 + 15 = C + D = C + 15. \text{ therefore } C = 2$$

Again putting $x = 2$ in (1) we obtain

$$16 - 12 + 6 + 15 = 2B + 2C + D = 2B + 4 + 15$$

or $\qquad 6 = 2B$ i.e., $B = 3$.

Equating Coeff. of x^3 on both sides of (1), we have $A = 2$.

Hence $\qquad 2x^3 - 3x^2 + 3x + 15 = 2x^{(3)} + 3x^{(2)} + 2x^{(1)} + 15.$

Second Method (*Method of synthetic division*)

Let $\qquad 2x^3 - 3x^2 + 3x + 15 = Ax^{(3)} + Bx^{(2)} + Cx^{(1)} + D$

$$= Ax(x - 1)(x - 2) + Bx(x - 1) + (Cx + D) \qquad ...(2)$$

From (2) it is clear that when $2x^3 - 3x^2 + 3x + 15$ is divided by x, the remainder is D and quotient is

$$A(x - 1)(x - 2) + B(x - 1) + C.$$

When this quotient is divided by $(x - 1)$, the remainder is C and quotient is $A(x - 2) + B$.

Dividing the last quotient by $(x - 2)$ the remainder is B and quotient is A.

Now dividing $2x^3 - 3x^2 + 3x + 15$ by x, the remainder is 15 and quotient is $2x^2 - 3x + 3$. Therefore, $D = 15$.

Dividing $2x^2 - 3x + 3$ by $(x - 1)$, the remainder is 2 and quotient is $(2x - 1)$.

$$
\begin{array}{r}
2x-1 \\
x-1\overline{\smash{\big)}\,2x^2-3x+3} \\
\underline{2x^2-2x} \\
-x+3 \\
\underline{-x+1} \\
2
\end{array}
$$
$\qquad \therefore\ C = 2$

Dividing $2x - 1$ by $x - 2$, the remainder is 3 and quotient is 2.

$$
\begin{array}{r}
2 \\
x-2\overline{\smash{\big)}\,2x-1} \\
\underline{2x-4} \\
3
\end{array}
$$
$\qquad \therefore\ B = 3,\ A = 2.$

This explanation was necessary to understand the procedure. In actual practice we write the procedure in the simple and compact form of synthetic division as follows:

$$
\begin{array}{r|rrr|l}
1 & 2 & -3 & 3 & 15 = D \\
  & 0 & 2 & -1 & \\
\hline
2 & 2 & -1 & 2 = C & \\
  & 0 & 4 & & \\
\hline
  & 2 = A & 3 = B & &
\end{array}
$$

$\therefore\ 2x^3 - 3x^2 + 3x + 15 = 2x^{(3)} + 3x^{(2)} + 2x^{(1)} + 15.$

Example 22.13 Express $y = 3x^3 + x^2 + x + 1$ in factorial notation and hence show that $\Delta^3 y = 18$.

Solution. By the method of synthetic division we have

```
1 | 3   1   1 | 1
  | 0   3   4
2 | 3   4 | 5
  | 0   6
  | 3 |10
```

$\therefore$ $\qquad\qquad y = 3x^{(3)} + 10x^{(2)} + 5x^{(1)} + 1$

Now $\qquad\qquad \Delta y = 9x^{(2)} + 20x^{(1)} + 5 \qquad\qquad [\because \Delta x^{(n)} = n\, x^{(n-1)}]$

$\qquad\qquad \Delta^2 y = 18x^{(1)} + 20,\ \Delta^3 y = 18.$

Remark. While using the method of synthetic division, the coeff. of powers of x should be arranged in descending order, writing zero for coeff. of a missing term.

Example 22.14 Represent the function $f(x) = x^4 - 5x^3 + 3x + 4$ and its successive differences in factorial notation. Hence show that $\Delta^5 f(x) = 0$.

Solution. By synthetic division, we obtain

```
1 | 1   -5    0    3 | 4
  | 0    1   -4   -4
2 | 1   -4   -4 | -1
  | 0    2   -4
3 | 1   -2 | -8
  | 0    3
  | 1 | 1
```

$\therefore f(x) = x^{(4)} + x^{(3)} - 8x^{(2)} - x^{(1)} + 4.$

$\Delta f(x) = 4x^{(3)} + 3x^{(2)} - 16x^{(1)} - 1,$

$\Delta^2 f(x) = 12x^{(2)} + 6x^{(1)} - 16,$

$\Delta^3 f(x) = 24x^{(1)} + 6,$

$\Delta^4 f(x) = 24,\ \Delta^5 f(x) = 0.$

Example 22.15 What is the anti-difference of the polynomial

$$f(x) = x^4 - 5x^3 + 3x + 4?$$

Solution. Here we have to find the polynomial whose first difference is the polynomial $f(x)$ or in other words, we have to determine $\dfrac{1}{\Delta} \cdot f(x)$ or $\Delta^{-1} f(x)$.

By Example 22.14, $f(x) = x^{(4)} + x^{(3)} - 8x^{(2)} - x^{(1)} + 4.$

$\therefore \dfrac{1}{\Delta} f(x) = \dfrac{x^{(5)}}{5} + \dfrac{x^{(4)}}{4} - 8\dfrac{x^{(3)}}{3} - \dfrac{x^{(2)}}{2} + 4x^{(1)} + c,$ $\qquad \left[\because \dfrac{1}{\Delta} x^{(n)} = \dfrac{x^{(n+1)}}{n+1}\right]$

where c is an arbitrary constant.

Example 22.16 Obtain the function whose first difference is

$$2x^3 + 3x^2 - 5x + 4.$$

Solution. We first express polynomial $f(x)$ in factorial notation. By the method of synthetic division, we have

$$
\begin{array}{c|ccc|c}
1 & 2 & 3 & -5 & 4 \\
& 0 & 2 & 5 & \\
2 & 2 & 5 & 0 & \\
& 0 & 4 & & \\
\cline{1-3}
\boxed{2} & 9 & & &
\end{array}
\qquad \therefore f(x) = 2x^{(3)} + 9x^{(2)} + 4.
$$

Now, let $F(x)$ be the function whose first difference is $f(x)$. Then $\Delta F(x) = f(x)$

or $F(x) = \dfrac{1}{\Delta} f(x) = \dfrac{1}{\Delta}\left[2x^{(3)} + 9x^{(2)} + 4\right] = \dfrac{2x^{(4)}}{4} + \dfrac{9x^{(3)}}{3} + 4x + c$

$= \dfrac{1}{2}x(x-1)(x-2)(x-3) + 3x(x-1)(x-2) + 4x + c$

$= \dfrac{1}{2}(x^2 - x)(x^2 - 5x + 6) + 3x(x^2 - 3x + 2) + 4x + c = \dfrac{1}{2}(x^4 - 7x^2 + 14x) + c.$

Exercise 22.1

1. Evaluate the following:

   (a) $\Delta\left(\dfrac{1}{x^2 + 5x + 6}\right)$, (b) $\Delta(x!)$, (c) $\Delta^n(e^{ax})$, *(IETE June 2009)* (d) $\Delta^2(ab^{cx})$

   (e) $\Delta^4[x^3(x-1)]$, *(A.M.I.E., W-2002)*, (f) $\Delta[2^x/(x+1)!]$, *(A.M.I.E., W-2003)*
   interval of differencing being unity.

   Ans. (a) $-2/(x+2)(x+3)(x+4)$, (b) $x(x!)$, (c) $(e^a - 1)^n e^{ax}$,

   (d) $(b^c - 1)^2 ab^{cx}$, (e) $4!$, (f) $-\dfrac{(x \cdot 2^x)}{(x+2)!}$.

2. Evaluate the following, interval of differencing being h

   (i) $\Delta \tan^{-1}x$, (ii) $\Delta\left[\dfrac{x^2}{\cos 2x}\right]$, (iii) $\Delta(x^2 + \sin x)$, (iv) $\Delta^2(ab^x)$.

   Ans. (i) $\tan^{-1}\left\{\dfrac{h}{1 + hx + x^2}\right\}$; (ii) $\dfrac{(2hx + h^2)\cos 2x + 2x^2 \sin(h)\sin(2x + h)}{\cos 2(x + h)\cos 2x}$;

   (iii) $h(h + 2x) + 2\sin\dfrac{h}{2}\cos\left(x + \dfrac{h}{2}\right)$; (iv) $a(b^h - 1)^2 \cdot b^x$

3. Prove the following operator relations
   (i) $E\Delta = \Delta E$, (ii) $\Delta^2 = E^2 - 2E + 1$, and $\nabla^2 = E^{-2} - 2E^{-1} + 1$,
   (iii) $\nabla E = E\nabla = \Delta = E - 1$,

(*iv*) $(1 + \Delta)(1 - \nabla) = 1$, (*A.M.I.E., W-2001*) (*v*) $\nabla = \Delta E^{-1}$,

(*vi*) $(\nabla + \Delta)^2 (x^2 + x) = 8$,

(*vii*) $(E^{1/2} + E^{-1/2})(1 + \Delta)^{1/2} = 2 + \Delta$,

(*viii*) $\Delta \nabla = \Delta - \nabla = \nabla \Delta = \delta^2$,   (*I.E.T.E., June 2006; Madras, 1996*)

(*ix*) $\delta^3 y_{1/2} = y_2 - 3y_1 + 3y_0 - y_{-1}$,

(*x*) $\delta = \Delta(1 + \Delta)^{-1/2} = \nabla(1 - \nabla)^{-1/2}$,

(*xi*) $\Delta^3 = \delta^3 E^{3/2}$,   (*A.M.I.E., S-2005*)

(*xii*) $E^{1/2} = \left(1 + \dfrac{\delta^2}{4}\right)^{1/2} + \dfrac{\delta}{2}$,

(*xiii*) $E^{1/2} = \mu + \dfrac{1}{2}\delta$ and $E^{-1/2} = \mu - \dfrac{1}{2}\delta$,

(*xiv*) $\nabla = 1 - E^{-1} = 1 - (1 + \Delta)^{-1} = \delta E^{-1/2}$ (*xv*) $\mu\delta = \dfrac{1}{2}(\Delta + \nabla)$,

(*xvi*) $\nabla^2 = h^2 D^2 - h^3 D^3 + \dfrac{7}{12} h^4 D^4 - \ldots\ldots$

4. Show that

(*i*) $\dfrac{\Delta}{E} x^3$, $(h = 1) = 3x^2 - 3x + 1$ (*A.M.I.E., S-1998*)

(*ii*) $\Delta^4\{x^3(x - 1)\} = 24$ (*A.M.I.E., W-2002*)

(*iii*) $\nabla^2 y_2 = 2$, if $y_0 = 1$, $y_1 = 2$ and $y_2 = 5$ (*A.M.I.E., S-2004*)

(*iv*) $\delta^3 Y_{1/2} = Y_2 - 3Y_1 + 3Y_0 - Y_{-1}$ (*I.E.T.E., Dec. 2004*)

5. Given $u_0 = 3$, $u_1 = 12$, $u_2 = 81$, $u_3 = 200$, $u_4 = 100$, $u_5 = 8$. Find $\Delta^5 u_0$.

                                                       Ans. 755.

6. If $f(x) = e^{ax}$, evaluate $\nabla^n f(x)$, where ∇ is the backward difference operator.

                         (*I.E.T.E., June 2006*) **Ans.** $(1 - e^{-a})^n e^{ax}$

7. Given that $f(1) + f(2) + f(3) = 25$, $f(4) = 29$ and $f(5) + f(6) = 113$, find the value of $f(10)$.   **Ans. 191.**

8. (*a*) Given $\log_{10} 100 = 2$, $\log_{10} 101 = 2.0043$, $\log_{10} 103 = 2.0128$, $\log_{10} 104$ = 2.0170, find $\log_{10} 102$.   (*A.M.I.E., S-1998*) **Ans. 2.0086 approx.**

(*b*) Find the missing values in the following table

x	45	50	55	60	65
$f(x)$	3.0	–	2.0	–	–2.4

                          (*V.T.U., 2001*) **Ans.** $f(50) = 2.925$, $f(60) = 0.225$

9. Find $u_{1/2}$, given $u_{-1} = 202$, $u_0 = 175$, $u_1 = 82$ and $u_2 = 55$.   **Ans. 128.50**

10. Express the following polynomials in terms of factorial polynomials
    (i) $x^3 - x + 1$, (ii) $x^4 - 2x^3 - x$,
    (iii) $3x^3 - 7x^2 + 8x - 1$, (iv) $2x^3 - 3x^2 + 3x - 10$.
    Ans. (i) $x^{(3)} + 3x^{(2)} + 1$, (ii) $x^{(4)} + 4x^{(3)} + x^{(2)} - 2x^{(1)}$
    (iii) $3x^{(3)} + 2x^{(2)} + 4x^{(1)} - 1$, (iv) $2x^{(3)} + 3x^{(2)} + 2x^{(1)} - 10$

11. Represent the function $f(x) = x^4 - 12x^3 + 24x^2 - 30x + 9$
    and its successive differences in factorial notation. Hence show that
    $\Delta^5 f(x) = 0$.   **Ans.** $f(x) = x^{(4)} - 6x^{(3)} - 5x^{(2)} - 17x^{(1)} + 9$,
    $\Delta f(x) = 4x^{(3)} - 18x^{(2)} - 10x^{(1)} - 17$, $\Delta^2 f(x) = 12x^{(2)} - 36x^{(1)} - 10$,
    $\Delta^3 f(x) = 24x^{(1)} - 36$, $\Delta^4 f(x) = 24$, $\Delta^5 f(x) = 0$.

12. Express x^3 in terms of factorial function and hence evaluate $\Delta^2 x^3$.
    Ans. $x^{(3)} + 3x^{(2)} + x^{(1)}$; $6(x + 1)$.

13. Obtain a function whose first difference is
    (i) e^x, (ii) $9x^2 + 11x + 5$, (iii) $x^3 + 3x^2 + 5x + 12$,

    Ans. (i) $\left(\dfrac{e^x}{e-1}\right) + c$, (ii) $3x^3 + x^2 + x + c$, (iii) $\dfrac{1}{4}\left(x^4 + 2x^3 + 5x^2 + 40x\right) + c$

14. Prove that $\left(\dfrac{\Delta^2}{E}\right)u_x \ne \dfrac{\Delta^2 u_x}{Eu_x}$.   (*A.M.I.E., S-2009*)

22.14 Interpolation and Extrapolation or Prediction

Interpolation requires estimating the values of a function $y(x)$ for arguments between x_0, ..., x_n at which the values y_0, ..., y_n are known.

Extrapolation or Prediction involves estimating values of $y(x)$ outside the interval in which the data arguments x_0, ..., x_n fall.

22.5 Newton's Interpolation Formulas

(A) Newton's Forward Difference Interpolation: Let the function $y = f(x)$ take the values y_0, y_1, y_2 ... corresponding to the values x_0, $x_0 + h$, $x_0 + 2h$,... of x. Suppose it is required to evaluate $f(x)$ for $x = x_0 + rh$, where r is any real number.
Using symbolic relations, we have
$E^r f(x) = f(x + rh)$. Since $f(x)$ is approximated by polynomial $P_n(x)$
$\therefore f(x) \approx P_n(x_0 + rh) = E^r f(x_0) = (1 + \Delta)^r y_0$

$$= \left[1 + r\Delta + \frac{r(r-1)}{2!}\Delta^2 + \frac{r(r-1)(r-2)}{3!}\Delta^3\right.$$

$$\left. + \ldots\ldots + \frac{r(r-1)(r-2)\ldots(r-n+1)}{n!}\Delta^n\right] y_0 \quad \text{[using Binomial theorem]}$$

or $$P_n(x_0 + rh) = y_0 + r\Delta y_0 + \frac{r(r-1)}{2!}\Delta^2 y_0 + \frac{r(r-1)(r-2)}{3!}\Delta^3 y_0$$

$$+ \ldots\ldots + \frac{r(r-1)(r-2)\ldots(r-n+1)}{n!}\Delta^n y_0 \qquad \ldots(22.23)$$

where $r = (x - x_0)/h$. The coefficients of $\Delta y_0, \Delta^2 y_0, \ldots\ldots \Delta^n y_0$ are binomial coefficients. The formula (22.23) may also be written as

$$P_n(x_0 + rh) = f_0 + r\Delta f_0 + \frac{r(r-1)}{2!}\Delta^2 f_0 + \ldots\ldots + \frac{r(r-1)\ldots(r-n+1)}{n!}\Delta^n f_0$$

The relation (22.23) is known as **Newton's forward interpolation formula or Newton's formula for equal intervals.**

This formula is used for interpolating the values of y near the top of the table (since all the forward differences are available) and for extrapolating values of y a little backward (*i.e.* to the left) of y_0.

(B) Newton's Backward Difference Interpolation

Using backward difference notation, we have
$$f(x) \approx P_n(x_n + rh) = E^r f(x_n) = (1 - \nabla)^{-r} y_n \qquad (\because E^{-1} = 1 - \nabla)$$

$$= \left[1 + r\nabla + \frac{r(r+1)}{2!}\nabla^2 + \frac{r(r+1)(r+2)}{3!}\nabla^3 + \ldots\ldots \right.$$

$$\left. \ldots\ldots + \frac{r(r+1)(r+2)\ldots(r+n-1)}{n!}\nabla^n \right] y_n \qquad \text{[using Binomial theorem]}$$

or $$P_n(x_n + rh) = y_n + r\nabla y_n + \frac{r(r+1)}{2!}\nabla^2 y_n + \frac{r(r+1)(r+2)}{3!}\nabla^3 y_n + \ldots\ldots$$

$$\ldots\ldots + \frac{r(r+1)(r+2)\ldots\ldots(r+n-1)}{n!}\nabla^n y_n, \qquad \ldots(22.24) \quad \text{where } r = (x - x_n)/h.$$

This may also be written as

$$P_n(x_n + rh) = f_n + r\nabla f_n + \frac{r(r+1)}{2!}\nabla^2 f_n +$$

$$\ldots\ldots + \frac{r(r+1)(r+2)\ldots\ldots(r+n-1)}{n!}\nabla^n f_n.$$

The result (22.24) is known as Newton's backward interpolation formula and this formula is used for interpolating the values of y near the bottom of the table (since all the backward differences are available) and also for extrapolating values of y a little distance ahead (*i.e.* to the right) of y_n.

Error in Newton's Interpolation Formulae

Since $f(x)$ is approximated by $P_n(x)$, the computed results contain errors. The difference $f(x) - P_n(x)$ is called the *error of interpolation* or *truncation error*. In Newton's forward difference interpolation formula, the error of interpolation

$$E_n(f,r) = f(x) - P_n(x) = \frac{r(r-1)......(r-n)}{(n+1)!} h^{n+1} f^{(n+1)}(\xi), \quad ...(22.25)$$

where ξ is some value of x between 0 and n.

This form of the error is practically of no use in computations, as generally we donot know anything about $f^{(n+1)}(\xi)$. However, we can replace the derivative $f^{(n+1)}(\xi)$ as follows:

$$f^{(n+1)}(\xi) \approx \frac{1}{h^{n+1}} \Delta^{n+1} f_0 \qquad ...(22.26)$$

Using (22.26) in (22.25), we obtain

$$Error = E_n(f,r) = \frac{r(r-1).....(r-n)}{(n+1)!} h^{n+1} \cdot \frac{1}{h^{n+1}} \Delta^{n+1} f_0$$

$$\text{or } E_n(f,r) = \frac{r(r-1)......(r-n)}{(n+1)!} \Delta^{n+1} f_0 \qquad ...(22.27)$$

In a similar manner we can express the error in Newton's backward difference interpolation formula in the form

$$E_n(f,r) = \frac{r(r+1)(r+2)....(r+n)}{(n+1)!} \nabla^{n+1} f_n. \qquad ...(22.28)$$

Example 22.17 For the data

x	-3	-2	-1	0	1	2
$f(x)$	-2	-4	-4	-2	2	8

construct the forward and backward difference tables. Using the corresponding interpolation, show that the interpolating polynomial is same.

Solution. The step length is $h = 1$ and $x_0 = -3$, $x_1 = -2$, $x_2 = -1$, $x_3 = 0$, $x_4 = 1$, $x_5 = 2$. We have the following difference table.

Difference table

x	$f(x)$	$\Delta f/\nabla f$	$\Delta^2 f/\nabla^2 f$	$\Delta^3 f/\nabla^3 f$
-3	-2			
		-2		
-2	-4		2	
		0		0
-1	-4		2	
		2		0
0	-2		2	
		4		0
1	2		2	
		6		
2	8			

(a) Using the Newton's forward difference interpolating polynomial and

$$r = \frac{x - x_0}{h} = \frac{x - (-3)}{1} = x + 3, \text{ we obtain}$$

$$P_2(x) = -2 + (x+3)(-2) + \frac{(x+3)(x+3-1)}{2!} \quad (2)$$

$$= -2 - 2x - 6 + x^2 + 5x + 6 = x^2 + 3x - 2.$$

(b) Using the Newton's backward difference interpolating polynomial and

$$r = \frac{x - x_n}{h} = \frac{x - 2}{1} = x - 2, \text{ we obtain}$$

$$P_2(x) = 8 + 6(x-2) + \frac{(x-2)(x-2+1)}{2!}(2)$$

$$= 8 + 6x - 12 + x^2 - 3x + 2 = x^2 + 3x - 2.$$

Example 22.18 For the data

x	-4	-2	0	2	4	6
$f(x)$	-67	-9	1	11	69	223

construct the forward difference table and determine the corresponding interpolating polynomial.

Solution. The step length is $h = 2$, and $x_0 = -4$, $x_1 = -2$, $x_2 = 0$ $x_3 = 2$, $x_4 = 4$, $x_5 = 6$. The difference table is as under:

Forward difference table

x	$f(x)$	Δf	$\Delta^2 f$	$\Delta^3 f$	$\Delta^4 f$
-4	-67				
		58			
-2	-9		-48		
		10		48	
0	1		0		0
		10		48	
2	11		48		0
		58		48	
4	69		96		
		154			
6	223				

We have $r = \dfrac{x - x_0}{h} = \dfrac{x+4}{2}$. The data represents a cubic polynomial.

Using the Newton's forward difference interpolating polynomial,

$$P_3(x) = f_0 + r\Delta f_0 + \frac{r(r-1)}{2!}\Delta^2 f_0 + \ldots + \frac{r(r-1)\ldots(r-n+1)}{n!}\Delta^n f_0,$$

we obtain $P_3(x) = -67 + \left(\dfrac{x+4}{2}\right)(58) + \dfrac{1}{2!}\left(\dfrac{x+4}{2}\right)\left(\dfrac{x+4}{2} - 1\right)(-48)$

$$+ \frac{1}{3!}\left(\frac{x+4}{2}\right)\left(\frac{x+4}{2} - 1\right)\left(\frac{x+4}{2} - 2\right)(48)$$

$$= -67 + \left(\frac{x+4}{2}\right)(58) + \frac{1}{2}\frac{(x+4)(x+2)}{(4)}(-48) + \frac{1}{6}\frac{(x+4)(x+2)x}{(8)}(48)$$

$$= -67 + 29(x+4) - 6(x+2)(x+4) + x(x+2)(x+4) = x^3 + x + 1.$$

Example 22.19 For the data

x	-4	-2	0	2	4	6
$f(x)$	261	19	1	15	253	1291

construct the backward difference table and determine the corresponding interpolating polynomial.

Solution. The step length is $h = 2$ and $x_0 = -4$, $x_1 = -2$, $x_2 = 0$, $x_3 = 2$, $x_4 = 4$, $x_5 = 6$. We have the following difference table.

Backward difference table

x	$f(x)$	∇f	$\nabla^2 f$	$\nabla^3 f$	$\nabla^4 f$	$\nabla^5 f$
-4	261					
		-242				
-2	19		224			
		-18		-192		
0	1		32		384	
		14		192		0
2	15		224		384	
		238		576		
4	253		800			
		1038				
6	1291					

We have $r = \dfrac{x - x_n}{2} = \dfrac{x - 6}{2}$. The data represents a bi-quadratic polynomial.

Using the Newton's backward difference interpolating polynomial,

$$P_4(x) = f_n + r\nabla f_n + \frac{r(r+1)}{2!}\nabla^2 f_n + \frac{1}{3!}r(r+1)(r+2)\nabla^3 f_n + \dots$$

$$\dots + \frac{r(r+1)\cdots\cdots(r+n-1)}{n!}\nabla^n f_n, \text{ we obtain}$$

$$P_4(x) = 1291 + \frac{x-6}{2}(1038) + \left(\frac{x-6}{2}\right)\left(\frac{x-6}{2}+1\right)\frac{800}{2!}$$

$$+\left(\frac{x-6}{2}\right)\left(\frac{x-6}{2}+1\right)\left(\frac{x-6}{2}+2\right)\frac{576}{3!}$$

$$+\left(\frac{x-6}{2}\right)\left(\frac{x-6}{2}+1\right)\left(\frac{x-6}{2}+2\right)\left(\frac{x-6}{2}+3\right)\frac{384}{4!}$$

$$= 1291 + (x-6)519 + (x-4)(x-6)100 + (x-2)(x-4)(x-6)\,12$$

$$+ x(x-2)(x-4)(x-6) = x^4 - x + 1.$$

Example 22.20 For the data

x	0	0.5	1.0	1.5	2.0
$f(x)$	-0.5	1.125	3.5	7.375	13.5

find an approximation to $f(0.25)$ and $f(0.75)$.

Solution. The points 0.25 and 0.75 are near the top of the table of values, we shall use the Newton's forward difference formula.

The data represents a cubic polynomial. Employing Newton's forward difference interpolating polynomial. Here $h = 0.5$ and $r = (x-0)/0.5 = 2x$.

Forward Difference table

x	$f(x)$	Δf	$\Delta^2 f$	$\Delta^3 f$	$\Delta^4 f$
0	−0.500				
		1.625			
0.5	1.125		0.75		
		2.375		0.75	
1.0	3.500		1.50		0
		3.875		0.75	
1.5	7.375		2.25		
		6.125			
2.0	13.500				

$$f(x) \approx P_3(x) = -0.5 + 2x(1.625) + \frac{2x(2x-1)}{2!}(0.75)$$

$$+\frac{2x(2x-1)(2x-2)(0.75)}{3!} = -0.5 + 3.25x + x(2x-1)(0.75)$$

$$+ x(2x-1)(2x-2)(0.25).$$

$$\therefore \quad f(0.25) \approx P_3(0.25) = -0.5 + 3.25(0.25) + 0.25(-0.5)(0.75)$$
$$+ 0.25(-0.5)(-1.5)(0.25)$$
$$= -0.5 + 0.8125 - 0.09375 + 0.046875 = 0.2656, \text{ and}$$

$$f(0.75) \approx P_3(0.75) = -0.5 + 1.5(1.625) + 0.75(0.5)(0.75)$$
$$+ 0.75(0.5)(-0.5)(0.25)$$
$$= -0.5 + 2.4375 + 0.28125 - 0.046875 = 2.1719.$$

Example 22.21 For the data

x	0.0	0.2	0.4	0.6	0.8
$f(x)$	0.500	1.108	1.764	2.516	3.412

find an approximation of $f(0.7)$ and $f(0.5)$.

Solution. The points 0.5 and 0.7 are near the bottom of the table of values, with $h = 0.2$. We shall use the Newton's backward difference formula. We form the following backward difference table.

Backward Difference table

x	$f(x)$	∇f	$\nabla^2 f$	$\nabla^3 f$	$\nabla^4 f$
0	0.500				
		0.608			
0.2	1.108		0.048		
		0.656		0.048	
0.4	1.764		0.096		0
		0.752		0.048	
0.6	2.516		0.144		
		0.896			
0.8	3.412				

The data represents a cubic polynomial. We shall use the following formula with $h = 0.2$.

$$f(x) \approx P_3(x) = f_n + r\nabla f_n + \frac{1}{2}r(r+1)\nabla^2 f_n + \frac{1}{6}r(r+1)(r+2)\nabla^3 f_n .$$

For $x = 0.7$, $r = \dfrac{x - x_n}{h} = \dfrac{0.7 - 0.8}{0.2} = -0.5.$ We obtain

$$f(0.7) \approx P_3(0.7) = 3.412 - 0.5(0.896) + \frac{1}{2}(-0.5)(0.5)(0.144)$$

$$+ \frac{1}{6}(-0.5)(0.5)(1.5)(0.048) = 3.412 - 0.448 - 0.018 - 0.003 = 2.943$$

For $x = 0.5$, $r = \dfrac{x - x_n}{h} = \dfrac{0.5 - 0.8}{0.2} = -1.5,$ we obtain

$$f(0.5) \approx 3.412 - 1.5(0.896) + \frac{1}{2}(-1.5)(-0.5)(0.144)$$

$$+ \frac{1}{6}(-1.5)(-0.5)(0.5)(0.048)$$

$$= 3.412 - 1.344 + 0.054 + 0.003 = 2.125$$

Example 22.22 From the following table, estimate the number of students who have obtained marks between 40 and 45.

Marks	30-40	40-50	50-60	60-70	70-80
No. of students	31	42	51	35	31

(Mangalore, 1999)

Solution. First we prepare the cumulative frequency table

Marks less than (x)	40	50	60	70	80
No of students (y)	31	73	124	159	190

The difference table is as under:

x	y	Δy	$\Delta^2 y$	$\Delta^3 y$	$\Delta^4 y$
40	31				
		42			
50	73		9		
		51		−25	
60	124		−16		37
		35		12	
70	159		−4		
		31			
80	190				

Now we shall find y_{45} that is, number of students with marks less than 45.

Taking $x_0 = 40$, $x = 45$, $h = 10$, we have $r = \dfrac{x - x_0}{h} = \dfrac{45 - 40}{10} = 0.5$. Using Newton's forward interpolation formula, we get

$$y_{45} = y_{40} + r\Delta y_{40} + \frac{r(r-1)}{2!}\Delta^2 y_{40} + \frac{r(r-1)(r-2)}{3!}\Delta^3 y_{40}$$

$$+ \frac{r(r-1)(r-2)(r-3)}{4!}\Delta^4 y_{40}$$

$$= 31 + 0.5(42) + \frac{0.5(-0.5)}{2}(9) + \frac{(0.5)(-0.5)(-1.5)}{6}(-25)$$

$$+ \frac{0.5(-0.5)(-1.5)(-2.5)}{24}(37)$$

$$= 31 + 21 - 1.125 - 1.5625 - 1.4453 \approx 47.87$$

∴ The number of students with marks less than 45 is 48. The number of students with marks less than 40 is 31.

∴ The number of students getting marks between 40 and 45 = 48 − 31 = 17.

Example 22.23 The following table gives the values of $\sin x$ from $x = 45°$ to $60°$ in steps of $5°$. Using Newton's forward interpolation formula find $\sin 51°$ and estimate the error.

x	45°	50°	55°	60°
$f(x) = \sin x$	0.70711	0.76604	0.81915	0.86603

Solution. The point 51° is near the beginning of the table of values, with $h = 5°$. We shall use the Newton's forward difference formula. The forward difference table is shown below:

Forward difference table

x	$f(x)$	Δf	$\Delta^2 f$	$\Delta^3 f$
45°	0.70711			
		0.05893		
50°	0.76604		−0.00582	
		0.05311		−0.00041
55°	0.81915		−0.00623	
		0.04688		
60°	0.86603			

We shall now use Newton's forward difference interpolation formula.

$$f(x) \approx P_3(x) = f_0 + r\Delta f_0 + \frac{1}{2}r(r-1)\Delta^2 f_0 + \frac{1}{6}r(r-1)(r-2)\Delta^3 f_0.$$

For $x = 51°$, $h = 5°$, $r = \dfrac{x - x_0}{h} = \dfrac{51 - 45}{5} = 1.2$. We have

$$\sin 51° \approx 0.70711 + 1.2(0.05893) + \frac{(1.2)(0.2)}{2}(-0.00582)$$

$$+ \frac{(1.2)(0.2)(-0.8)}{6}(-0.00041)$$

$$= 0.70711 + 0.070716 - 0.0006984 + 0.00001312 = 0.777141$$

The error of interpolation $E_n(f, r) = f(x) - P_n(x) = \dfrac{r(r-1)\ldots\ldots(r-n)}{(n+1)!}\Delta^{n+1} f_0$

$$= \frac{(1.2)(1.2 - 1)(1.2 - 2)}{3!}\Delta^3 f_0, \text{ (taking } n = 2)$$

$$= \frac{1.2(0.2)(-0.8)}{6}(-0.00041) = 0.0000131.$$

22.6 Interpolation with Unevenly Spaced Points

The interpolation formulae for the function $y = f(x)$ studied so far, are applicable only when the values of x are given at equidistant intervals. We, therefore, develop two interpolation formulae for unequally spaced values of x.

22.6.1 Lagrange's Interpolation Formula

Let $y = f(x)$ be a function which is equal to $y_0, y_1, y_2, ...,y_n$ when x assumes the values $x_0, x_1, x_2, ..., x_n$. If the values of x are not at equal intervals, Newton's interpolation formula cannot be used. In such cases we find $f(x)$ that is, y for any other value of x by the formula

$$y = \frac{(x-x_1)(x-x_2)...(x-x_n)}{(x_0-x_1)(x_0-x_2)...(x_0-x_n)} y_0 + \frac{(x-x_0)(x-x_2)...(x-x_n)}{(x_1-x_0)(x_1-x_2)...(x_1-x_n)} y_1 +$$

$$......+ \frac{(x-x_0)(x-x_1)...(x-x_{n-1})}{(x_n-x_0)(x_n-x_1)...(x_n-x_{n-1})} y_n.$$

This is known as *Lagrange's interpolation formula* and is used to find the value of a function when values of the independent variable are not equispaced.

Proof. We shall now proceed to prove the above formula.

Let $f(x) = A_1(x-x_1)(x-x_2)... (x-x_n) + A_2(x-x_0)(x-x_2)...(x-x_n) +...$
$+ A_n(x-x_0)(x-x_1)...(x-x_{n-1})$ ...(22.29)

where A's are constants.

Putting $\qquad x = x_0, x_1, ..., x_n$ successively we get

Let $\qquad f(x_0) = A_1 (x_0-x_1)(x_0-x_2)...(x_0-x_n)$

$\qquad\qquad f(x_1) = A_2 (x_1-x_0)(x_1-x_2)...(x_1-x_n)$

$\qquad\qquad ... \qquad\qquad ... \qquad\qquad ... \qquad\qquad ...$

$\qquad\qquad f(x_n) = A_n (x_n-x_0)(x_n-x_1)...(x_n-x_{n-1})$

Putting the values of $A_1, A_2, ..., A_n$ in (22.29), we get

$$f(x) = \frac{(x-x_1)(x-x_2)...(x-x_n)}{(x_0-x_1)(x_0-x_2)...(x_0-x_n)} y_0 + \frac{(x-x_0)(x-x_2)...(x-x_n)}{(x_1-x_0)(x_1-x_2)...(x_1-x_n)} y_1$$

$$+......+ \frac{(x-x_0)(x-x_1)...(x-x_{n-1})}{(x_n-x_0)(x_n-x_1)...(x_n-x_{n-1})} y_n \qquad\qquad ...(22.30)$$

because $\qquad f(x_0) = y_0, \; f(x_1) = y_1$, etc.

The formula (22.30) is known as *Lagrange's interpolation formula.*

The Lagrange formula can also be written in the form

$$y = f(x) = \sum_{r=0}^{n} \frac{\phi(x)}{(x-x_r)\phi'(x_r)} f(x_r), r = 0, 1, ...n, \qquad\qquad ...(22.31)$$

where $\phi(x) = (x-x_0)(x-x_1)...(x-x_n), \qquad \phi'(x_r) = \frac{d}{dx}[\phi(x)]_{x=x_r}$

Since Lagrange's formula is merely a relation between two variables, either of which may be taken as the independent variable, it is evident that by considering y as the independent variable we can write a formula giving x as a function of y. Hence on interchanging x and y in (22.30), we obtain

$$x = \frac{(y-y_1)(y-y_2)....(y-y_n)}{(y_0-y_1)(y_0-y_2)....(y_0-y_n)} x_0 + \frac{(y-y_0)(y-y_2)...(y-y_n)}{(y_1-y_0)(y_1-y_2)...(y_1-y_n)} x_1$$

$$+ .... + \frac{(y-y_0)(y-y_1).....(y-y_{n-1})}{(y_n-y_0)(y_n-y_1)....(y_n-y_{n-1})} x_n. \qquad ...(22.32)$$

Lagrange's formula is used in two ways—(*i*) to find any value of a function when the given values of the independent variable are not equidistant, and (*ii*) to find the value of the independent variable corresponding to a given value of the function. The second type of problems are solved by means of formula (22.32).

Example 22.24 Find the Lagrange interpolating polynomial that fits the following data values.

(i)

x	2	2.5
$f(x)$	4	5.5

Interpolate at $x = 2.2$

(ii)

x	−1	−2	2
$f(x)$	−1	−9	11

Compute $f(0)$.

(A.M.I.E., W-2004)

(iii)

x	−2	1	0	2
$f(x)$	3	−3	1	−1

Interpolate at $x = 1.5$.

Solution. (*i*) We have two data values. Hence, we can fit a linear Lagrange interpolating polynomial. We have

$$P_1(x) = \frac{x-2.5}{2-2.5}(4) + \frac{x-2}{2.5-2}(5.5) = 3x - 2.$$

At $x = 2.2$, we get $f(2.2) \approx P_1(2.2) = 4.6$

(*ii*) We have three data values. We can fit an interpolating polynomial of degree ≤ 2. We have

$$P_2(x) = \frac{(x+2)(x-2)}{(-1+2)(-1-2)}(-1) + \frac{(x+1)(x-2)}{(-2+1)(-2-2)}(-9) + \frac{(x+1)(x+2)}{(2+1)(2+2)}(11)$$

$$= \frac{x^2-4}{3} - \frac{9}{4}(x^2-x-2) + \frac{11}{12}(x^2+3x+2) = -x^2+5x+5.$$

At $x = 0$, we get $f(0) \approx P_2(0) = 5$.

(iii) We have four nodal points. We can fit a polynomial of degree ≤ 3. We have

$$P_3(x) = \frac{(x-1)(x-0)(x-2)}{(-2-1)(-2)(-2-2)}(3) + \frac{(x+2)(x-0)(x-2)}{(1+2)(1-0)(1-2)}(-3)$$

$$+ \frac{(x+2)(x-1)(x-2)}{(0+2)(0-1)(0-2)}(1) + \frac{(x+2)(x-1)(x-0)}{(2+2)(2-1)(2-0)}(-1)$$

$$= -\frac{1}{8}(x^3-3x^2+2x) + (x^3-4x) + \frac{1}{4}(x^3-x^2-4x+4)$$

$$-\frac{1}{8}(x^3+x^2-2x) = x^3-5x+1.$$

At $x = 1.5$, we get $f(1.5) \approx P_3(1.5) = (1.5)^3 - 5(1.5) + 1 = 3.375 - 7.5 + 1 = -3.125$.

Example 22.25 Given $\log_{10} 654 = 2.8156$, $\log_{10} 658 = 2.8182$,

$$\log_{10} 659 = 2.8189, \log_{10} 661 = 2.8202, \text{ find } \log_{10} 656.$$

Solution. We have

x:	654	658	659	661
$\log_{10}x$:	2.8156	2.8182	2.8189	2.8202

Here $x = 656$, $x_0 = 654$, $x_1 = 658$, $x_2 = 659$, $x_3 = 661$.

Since the expression for the polynomial is not required, we can find $\log_{10} 656$, by substituting the above values in the Lagrange formula, we obtain

$$\log_{10} 656 = \frac{(656-658)(656-659)(656-661)}{(654-658)(654-659)(654-661)}(2.8156)$$

$$+ \frac{(656-654)(656-659)(656-661)}{(658-654)(658-659)(658-661)}(2.8182)$$

$$+ \frac{(656-654)(656-658)(656-661)}{(659-654)(659-658)(659-661)}(2.8189)$$

$$+ \frac{(656-654)(656-658)(656-659)}{(661-654)(661-658)(661-659)}(2.8202)$$

$$= \frac{(-2)(-3)(-5)}{(-4)(-5)(-7)}(2.8156) + \frac{(2)(-3)(-5)}{(4)(-1)(-3)}(2.8182)$$

$$+\frac{(2)(-2)(-5)}{(5)(1)(-2)}(2.8189)+\frac{(2)(-2)(-3)}{(7)(3)(2)}(2.8202)$$

$$= 0.6033+7.0455-5.6378+0.8058 = 2.8168.$$

Example 22.26 The following date represents e^{-x}.

x	1	1.2	1.5
$f(x)$	0.36788	0.30119	0.22313

Using Lagrange interpolation, obtain an approximate value of $e^{-1.3}$.
Solution. We have three nodal points. We can fit an interpolating polynomial of degree ≤ 2. We write

$$P_2(x) = \frac{(x-1.2)(x-1.5)}{(1-1.2)(1-1.5)}(0.36788) + \frac{(x-1)(x-1.5)}{(1.2-1)(1.2-1.5)}(0.30119)$$

$$+\frac{(x-1)(x-1.2)}{(1.5-1)(1.5-1.2)}(0.22313)$$

$$= (x^2-2.7x+1.8)\,3.6788 - (x^2-2.5x+1.5)(5.0198) + (x^2-2.2x+1.2)\,1.4875$$

$$= 0.1465x^2 - 0.6557x + 0.8771.$$

Hence, $e^{-1.3} \approx P_2(1.3) = 0.1465\,(1.3)^2 - 0.6557\,(1.3) + 0.8771 = 0.27227.$

22.6.2 Newton's Divided Difference Interpolation

Divided differences

Let the data (x_i, f_i), $i = 0, 1, 2, ..., n$ be given. Then, we define the divided differences as follows.

First divided difference: The first divided difference of any two consecutive data values is defined as

$$f[x_0, x_1] = \frac{f(x_1)-f(x_0)}{x_1-x_0}, \quad .....,$$

$$f[x_i, x_{i+1}] = \frac{f(x_{i+1})-f(x_i)}{x_{i+1}-x_i}, \quad i = 0, 1, ..., n-1 \qquad ...(22.33)$$

Second divided difference: The second divided difference using three consecutive data values is defined as

$$f[x_0, x_1, x_2] = \frac{f[x_1, x_2]-f[x_0, x_1]}{x_2-x_0}$$

Simplifying, we get

$$f\left[x_0, x_1, x_2\right] = \frac{1}{\left(x_2 - x_0\right)} \left[\frac{f_2 - f_1}{x_2 - x_1} - \frac{f_1 - f_0}{x_1 - x_0}\right]$$

$$= \frac{f_0}{\left(x_0 - x_1\right)\left(x_0 - x_2\right)} - \frac{f_1}{\left(x_2 - x_0\right)}\left[\frac{1}{x_2 - x_1} + \frac{1}{x_1 - x_0}\right] + \frac{f_2}{\left(x_2 - x_0\right)\left(x_2 - x_1\right)}$$

$$= \frac{f_0}{\left(x_0 - x_1\right)\left(x_0 - x_2\right)} + \frac{f_1}{\left(x_1 - x_0\right)\left(x_1 - x_2\right)} + \frac{f_2}{\left(x_2 - x_0\right)\left(x_2 - x_1\right)}$$

The nth *divided difference*, is defined as

$$f\left[x_0, x_1, ..., x_n\right] = \frac{f\left[x_1, ..., x_n\right] - f\left[x_0, ..., x_{n-1}\right]}{x_n - x_0} \qquad ...(22.34)$$

which can also be expressed as a linear combination of $f_0, f_1, ..., f_n$.

The divided differences are given in the following table for the data values (x_i, f_i), $i = 0, 1, 2, 3, 4$. *The divided differences are symmetric with respect to their arguments,* that is $f[x_0, x_1] = f[x_1, x_0]$, $f[x_0, x_1, x_2) = f[x_2, x_1, x_0]$ etc.

Table 22.5 Divided differences (d.d) table

x	$f(x)$	First d.d	Second d.d	Third d.d.	Fourth d.d
x_0	f_0				
		$f[x_0, x_1]$			
x_1	f_1		$f[x_0, x_1, x_2]$		
		$f[x_1, x_2]$		$f[x_0, x_1, x_2, x_3]$	
x_2	f_2		$f[x_1, x_2, x_3]$		$f[x_0, x_1, x_2, x_3, x_4]$
		$f[x_2, x_3]$		$f[x_1, x_2, x_3, x_4]$	
x_3	f_3		$f[x_2, x_3, x_4]$		
		$f[x_3, x_4]$			
x_4	f_4				

Newton's divided difference interpolating polynomial is
$$P_n(x) = f_0 + (x - x_0)\, f[x_0, x_1] + (x - x_0)\,(x - x_1)\, f[x_0, x_1, x_2]$$
$$+ ......+ (x - x_0)\,(x - x_1)... (x - x_{n-1})\, f[x_0, x_1, ..., x_n]. \qquad ...(22.35)$$

Example 22.27 Find the third divided difference of $f(x) = 1/x$, using the points x_0, x_1, x_2, x_3.

Solution. We have $f\left[x_0, x_1\right] = \dfrac{f_1 - f_0}{x_1 - x_0} = \dfrac{1}{\left(x_1 - x_0\right)}\left[\dfrac{1}{x_1} - \dfrac{1}{x_0}\right]$

$$= \frac{x_0 - x_1}{x_0 x_1 \left(x_1 - x_0\right)} = -\frac{1}{x_0 x_1}.$$

$$f\left[x_1, x_2\right] = \frac{f_2 - f_1}{x_2 - x_1} = -\frac{1}{x_1 x_2}.$$

Now

$$f\left[x_0, x_1, x_2\right] = \frac{f\left[x_1, x_2\right] - f\left[x_0, x_1\right]}{x_2 - x_0}$$

$$= \frac{1}{\left(x_2 - x_0\right)}\left[-\frac{1}{x_1 x_2} + \frac{1}{x_0 x_1}\right] = \frac{x_2 - x_0}{x_0 x_1 x_2 \left(x_2 - x_0\right)} = \frac{1}{x_0 x_1 x_2}$$

Finally

$$f\left[x_0, x_1, x_2, x_3\right] = \frac{f\left[x_1, x_2, x_3\right] - f\left[x_0, x_1, x_2\right]}{x_3 - x_0}$$

$$= \frac{\dfrac{1}{x_1 x_2 x_3} - \dfrac{1}{x_0 x_1 x_2}}{x_3 - x_0} = \frac{x_0 - x_3}{x_0 x_1 x_2 x_3 \left(x_3 - x_0\right)} = \frac{-1}{x_0 x_1 x_2 x_3}.$$

Example 22.28 For the data

x	-3	-1	0	2	3
$f(x)$	-9	5	3	11	33

construct the divided difference table. Hence, obtain the interpolating polynomial that fits the given data.

Solution. The divided difference table is as follow:

x	$f(x)$	First d.d	Second d.d	Third d.d	Fourth d.d
-3	-9				
		$\dfrac{5+9}{-1+3}=7$			
-1	5		$\dfrac{-2-7}{0+3}=-3$		
		$\dfrac{3-5}{0+1}=-2$		$\dfrac{2+3}{2+3}=1$	
0	3		$\dfrac{4+2}{2+1}=2$		0
		$\dfrac{11-3}{2-0}=4$		$\dfrac{6-2}{3+1}=1$	
2	11		$\dfrac{22-4}{3-0}=6$		
		$\dfrac{33-11}{3-2}=22$			
3	33				

Since the fourth order divided difference is zero, the data represents a third degree polynomial. Using Newton's divided difference interpolating polynomial, we obtain

$$P_3(x) = f_0 + (x - x_0) f[x_0, x_1] + (x - x_0)(x - x_1) f[x_0, x_1, x_2]$$
$$+ (x - x_0)(x - x_1)(x - x_2) f[x_0, x_1, x_2, x_3]$$
$$= -9 + (x + 3)\, 7 + (x + 3)(x + 1)(-3) + (x + 3)(x + 1)(x)(1)$$
$$= -9 + 7x + 21 - 3x^2 - 12x - 9 + x^3 + 4x^2 + 3x = x^3 + x^2 - 2x + 3.$$

Example 22.29 Using Newton's divided difference interpolation, obtain the polynomial which fits the data.

x	-3	-2	-1	1	2	3
$f(x)$	18	12	8	6	8	12

Also obtain the approximate value of $f(0)$ and $f(5)$.   (*I.E.T.E., Dec. 2006*)

Solution. We have $x_0 = -3, f_0 = 18$; $x_1 = -2, f_1 = 12$; $x_2 = -1, f_2 = 8$; $x_3 = 1, f_3 = 6$; $x_4 = 2, f_4 = 8$; $x_5 = 3, f_5 = 12$. The divided differences are given by

$$f[x_0, x_1] = \frac{f_1 - f_0}{x_1 - x_0} = \frac{12 - 18}{-2 + 3} = -6, \quad f[x_1, x_2] = \frac{f_2 - f_1}{x_2 - x_1} = \frac{8 - 12}{-1 - (-2)} = -4$$

$$f[x_2, x_3] = \frac{6 - 8}{1 - (-1)} = -1, \quad f[x_3, x_4] = \frac{8 - 6}{2 - 1} = 2, \quad f[x_4, x_5] = \frac{12 - 8}{3 - 2} = 4$$

$$f[x_0, x_1, x_2] = \frac{f[x_1, x_2] - f[x_0, x_1]}{x_2 - x_0} = \frac{-4 - (-6)}{-1 - (-3)} = 1$$

$$f[x_1, x_2, x_3] = \frac{-1 - (-4)}{1 - (-2)} = 1, \quad f[x_2, x_3, x_4] = \frac{2 + 1}{2 + 1} = 1,$$

$$f[x_3, x_4, x_5] = \frac{4 - 2}{3 - 1} = 1.$$ Since all the second order divided differences are same so third order divided differences are zero.

Hence, we have the following divided difference table.

Table of divided differences

x	$f(x)$	First d.d	Second d.d	Third d.d.
-3	18			
		-6		
-2	12		1	
		-4		0

(*Contd.*)

x	f(x)	First d.d	Second d.d	Third d.d.
-1	8		1	
		-1		0
1	6		1	
		2		0
2	8		1	
		4		
3	12			

The data represents a second degree polynomial. Using Newton's divided difference interpolating polynomial, we obtain

$P_2(x) = f_0 + (x - x_0) f[x_0, x_1] + (x - x_0)(x - x_1) f[x_0, x_1, x_2]$
$= 18 + (x + 3)(-6) + (x + 3)(x + 2)(1) = x^2 - x + 6.$

At $x = 0$, we obtain $f(0) \approx P_2(0) = 6$, and $f(5) \approx P_2(5) = 26$.

Example 22.30 Calculate a fourth divided difference for the following data.

x	0	1	2	4	5
f(x)	0	16	48	88	0

Apply Newton's formula to find the polynomial that fits the given data. What value does this polynomial take at $x = 3$? Compare your results with the Lagrange method.

Solution. We construct the divided difference table for the given data.

x	f(x)	First d.d	Second d.d	Third d.d	Fourth d.d
0	0				
		$\dfrac{16-0}{1-0} = 16$			
1	16		$\dfrac{32-16}{2-0} = 8$		
		$\dfrac{48-16}{2-1} = 32$		$\dfrac{-4-8}{4-0} = -3$	
2	48		$\dfrac{20-32}{4-1} = -4$		$\dfrac{-8+3}{5-0} = -1$
		$\dfrac{88-48}{4-2} = 20$		$\dfrac{-36+4}{5-1} = -8$	
4	88		$\dfrac{-88-20}{5-2} = -36$		
		$\dfrac{-88}{5-4} = -88$			
5	0				

Using Newton's divided difference interpolating polynomial we obtain

$P_4(x) = 0 + 16x + x(x-1)8 + x(x-1)(x-2)(-3) + x(x-1)(x-2)(x-4)(-1)$

$= -x^4 + 4x^3 + 3x^2 + 10x$. Hence, $f(3) \approx P_4(3) = -81 + 108 + 27 + 30 = 84$.

By Lagrange formula

$$P_4(3) = 0 + \frac{3(3-2)(3-4)(3-5)}{1(1-2)(1-4)(1-5)}(16)$$

$$+\frac{3(3-1)(3-4)(3-5)}{2(2-1)(2-4)(2-5)}(48) + \frac{3(3-1)(3-2)(3-5)}{4(4-1)(4-2)(4-5)}(88) + 0$$

$$= -\frac{6}{12}(16) + \frac{12}{12}(48) + \frac{12}{24}(88) = 84.$$

Exercise 22.2

1. For the data

x:	0	1	2	3	4
$f(x)$:	3	6	11	18	27

construct the forward and backward difference tables. Using the corresponding interpolation, show that the interpolating polynomial is same.

(AMIE W-2005) **Ans.** $x^2 + 2x + 3$

2. Using Newton's forward interpolation formula, find $f(x)$ at $x = 8$ from the following table.

x:	0	5	10	15	20	25
$f(x)$:	7	11	14	18	24	32

Ans. 12.814

3. (*a*) The following data are taken from the steam table:

Temp.°C	140	150	160	170	180
pressure kgf/cm²	3.685	4.854	6.302	8.076	10.225

Using Newton's formula, find the pressure of steam for temperature of 142°C and 175°C. [*AMIE, S-2009; IETE S-2009*] **Ans.** 3.899; 9.10 kgf/cm².

(*b*) Construct the difference table from the following values of y. Compute $f(x)$ at (*i*) 1.05, and (*ii*) 1.45.

x	1.0	1.1	1.2	1.3	1.4	1.5
y	0.24197	0.21875	0.19419	0.17137	0.14973	0.12952

(AMIE., W-2005) **Ans.** (*i*) 0.23079, (*ii*) 0.13927.

4. Given $\sin 45° = 0.7071$, $\sin 50° = 0.7660$, $\sin 55° = 0.8192$, $\sin 60° = 0.8660$. Find $\sin 52°$, using Newton's forward interpolation formula. **Ans.** 0.788

5. Given that

x:	1	2	3	4	5
y:	2	5	10	17	26

find the value of $\nabla^2 y_5$. *(AMIE W-2005)* **Ans.** 2

6. Using Newton's backward interpolation formula, find $f(3.8)$ from the following table.

x:	0	1	2	3	4
f(x):	1.00	1.50	2.20	3.10	4.60

Ans. 4.36

7. Apply Newton's backward formula to find a polynomial of degree three which includes the following data.

x:	3	4	5	6
y = f(x):	6	24	60	120

Ans. $x^3 - 3x^2 + 2x$

8. (*a*) The following data gives the melting point of an alloy of lead and zinc, where t is the temperature in °C and p is the percentage of lead in the alloy.

p(%):	60	70	80	90
t:	226	250	276	304

Using Newton's interpolation method, find the melting point of the alloy containing 84% of lead. **Ans.** 286.96°C
(*b*) The table below gives the values of tan x for $0.1 \le x \le 0.30$.

x	0.10	0.15	0.20	0.25	0.30
y = tan x	0.1003	0.1511	0.2027	0.2553	0.3093

Use Newton's formulae for interpolation to obtain the values of (*i*) tan 0.12, and (*ii*) tan 0.5. *(A.M.I.E., W-2008)* **Ans.** (*i*) 0.1205, (*ii*) 0.5543.

9. The following table gives the population of a town during the last six censuses. Estimate, using Newton's interpolation formula, the increase in the population during the period 1986 to 1988.

Year:	1951	1961	1971	1981	1991	2001
Population: (in thousands):	12	15	20	27	39	52

Ans. 2.42 thousands

10. Values of $f(x) = \sqrt{x}$ are listed in following table. Use the data of this table to calculate $\sqrt{155}$.

x	150	152	154	156
$f(x) = \sqrt{x}$	12.247	12.329	12.410	12.490

(*AMIE, W-2006*) **Ans.** 12.453

[**Hint:** Here Newton's backward interpolation formula is convenient.]

11. Use Newton's backward difference polynomial to interpolate at $x = 1.5$, from the following data.

x:	-2	-1	0	1	2
f(x):	20	12	6	2	0

(*IETE, June 2006*) **Ans.** $x^2 - 5x + 6$, 0.75.

12. Using Newton's forward interpolation formula, find the cubic polynomial which takes on the following values.

x:	0	1	2	3
f(x):	1	2	1	10

Also evaluate $f(4)$ by using Newton's backward interpolation formula.

(*AMIE, W-2001*) **Ans.** $2x^3 - 7x^2 + 6x + 1$; 41.

13. The table gives the distances in Kilometers of the visible horizon for the given heights in metres above the earth surface

x (= height):	100	150	200	250	300	350	400
y (= distance):	10.63	13.03	15.04	16.81	18.42	19.90	21.27

find the values of y when $x = 218$ metres and 410 metres.

Ans. 15.70 km; 21.53 km.

14. The data

x:	1	2	3	4	5
f(x):	5	15	35	77	141

represents a third degree polynomial. It is known that $f(3)$ is in error. Find the correct value of $f(3)$. *(IETE, June-2006)* **Ans. 37**

15. Find the polynomial that fits the data.

x:	0.1	0.2	0.3	0.4
f(x):	1.11	1.24	1.39	1.56

(IETE, June 2006) **Ans.** $f(x) = x^2 + x + 1$

16. The data

x:	0	1	2	3	4
f(x):	1	2	10	28	65

represents a polynomial of degree ≤ 3. It is known that $f(2)$ is in error. Find the correct value of $f(2)$. *(IETE, Dec. 2002)* **Ans. 9**

17. Fit a polynomial of minimum degree to the following data.

x:	−1	1	2	4	5
f:(x)	13	15	13	33	67

Ans. $x^3 - 3x^2 + 17$

18. (*a*) Use Lagrange interpolation to find the value of $f(x)$ at $x = 1.5$ from the following table.

x:	−2	−1	1	2
f(x):	−5	1	1	7

Ans. 2 7/8

(*b*) Using Lagrange's interpolation formula, find the missing term in the table.

x:	2	5	8	14
y:	94.8	87.9		68.7

Ans. 81.25

19. Apply Lagrange method to find the value of x when $f(x) = 15$ from the following data.

x:	5	6	9	11
$f(x)$:	12	3	14	16

(*Madras, 2000*) **Ans.** 11.5

20. (*a*) Using Lagrange interpolation formula, find the value of $f(x)$ when $x = 10$, from the following table.

x:	5	6	9	11
$f(x)$:	12	13	14	16

Ans. 14 2/3

(*b*) The following data are taken from the steam tables.

Temp°C (t)	140	160	170	190
Pressure kgf/cm² (P)	3.685	6.302	8.076	12.575

Use Lagrange's interpolation formula to compute the pressure at temperature 165°C.   (*A.M.I.E., S-2007*) **Ans.** 7.14975

21. Given $u_1 = 22$, $u_2 = 30$, $u_4 = 82$, $u_7 = 106$, $u_{12} = 206$, find u_8 using Lagrange interpolation formula.   **Ans.** 120.44

22. Use Lagrange's formula to find the form of $f(x)$, from the following table.

x:	0	2	3	6
$f(x)$:	648	704	729	792

(*Kottayam, 1999*) **Ans.** $(648 + 30x - x^2)$

23. Use Lagrange's interpolation formula to find x corresponding to $f(x) = 7$ from the following table.

x :	1	3	4
$f(x)$:	4	12	19

Ans. 1.86

[**Hint:** Interchange the roles of x and y in the Lagrange interpolation formula].

24. Find the second divided difference of $f(x) = 1/x$, using the points x_0, x_1, x_2.   **Ans.** $1/x_0 \, x_1 \, x_2$.

25. Find the third divided difference of $f(x) = x^4 + x^2 + 1$, using the points 2, 4, 9, 10.   **Ans.** 25

26. (a) Find the third divided difference of the polynomial $x^3 + 3x^2 + 1$ based on the points x_0, x_1, x_2 and x_3. (*IETE, Dec. 2005*) **Ans. 1**
    (b) Given that

x:	0	1	2	4
f(x):	1	1	2	5

find the third order divided difference. (*A.M.I.E., W-2006*) **Ans. (−0.083)**

27. Construct the divided difference table for the following data.

x:	0	1	2	4	5	6
f(x):	1	14	15	5	6	19

Hence, obtain the interpolating polynomial that fits the given data.

Ans. $x^3 - 9x^2 + 21x + 1$

28. Using the following table, find interpolating polynomial that fits the given data.

x:	−1	0	3	6	7
f(x):	3	−6	39	822	1611

Ans. $x^4 - 3x^3 + 5x^2 - 6$

29. Construct the divided difference table for the given data.

x:	0	1	2	5
f(x):	2	3	12	147

Using divided differences interpolation, obtain the interpolating polynomial which fits this data.

Ans. $x^3 + x^2 - x + 2$.

30. The following table gives certain corresponding values of x and $\log_{10} x$.

x:	321.0	322.8	324.2	325.0
$\log_{10} x$:	2.50651	2.50893	2.51081	2.51188

Compute the value of $\log_{10} 323.5$. **Ans. 2.50987**

31. Certain corresponding values of x and $\log_{10} x$ are given below.

x:	300	304	305	307
$\log_{10} x$:	2.4771	2.4829	2.4843	2.4871

Evaluate $\log_{10} 301$, using (*i*) Lagrange's formula, (*ii*) Newton divided difference formulae. *(AMIE, S-2006)* **Ans.** 2.4786

32. Find the polynomial of degree three which takes the values prescribed below.

x:	0	1	2	4
y = f(x):	1	1	2	5

Ans. $\left(-x^3 + 9x^2 - 8x + 12\right)/12$

33. (*a*) Find the first term of the series whose second and subsequent terms are 8, 3, 0, −1, 0. *(Nagpur, 1997)* **Ans.** 15
    (*b*) In the table below the values of y are consecutive terms of a series of which the number 21.6 is the sixth terms. Find the first and the tenth terms of the series.

x:	3	4	5	6	7	8	9
y:	2.7	6.4	12.5	21.6	34.3	51.2	72.9

(AMIE, S-1999) **Ans.** 0.1, 100

34. Using Newton's divided difference interpolation formula, find from the following table the value of y for x = 5.

x:	4.0	4.5	5.5	6.0	7.0
y:	1.58740	1.65096	1.76517	1.81712	1.91293

35. Using Lagrange's interpolation formula, find a polynomial which passes through the points (0, −12), (1, 0), (3, 6), (4, 12).
    Ans. $P_3(x) = x^3 - 7x^2 + 18x - 12$. *(A.M.I.E., S-2010, W-2007)*

36. From the following table, find the value of f(x) at x = 4,

x:	1.5	3	6
f(x):	−0.25	2	20

(A.M.I.E., S-2008) **Ans.** $P_2(x) = x^2 - 3x + 2$; 6.

37. The number of members of a Civil Engineering Society are given below

x:	1987	1988	1989	1991
y:	150	192	241	374

Make the best estimate you can of the numbers of members in 1990.
(A.M.I.E., S-1999) **Ans.** 300.

Example 22.31. Given the values:

x	5	7	11	13	17
$f(x)$	150	392	1452	2366	5202

Evaluate $f(9)$ using Lagrange's formula. *(A.M.I.E., W-2010)*

Solution.

Here $x_0 = 5, x_1 = 7, x_2 = 11, x_3 = 13, x_4 = 17$ and $y_0 = 150, y_1 = 392, y_2 = 1452,$ $y_3 = 2366, y_4 = 5202$. Putting $x = 9$ and substituting the above values in Lagrange's formula, we obtain:

$$f(9) = \frac{(9-7)(9-11)(9-13)(9-17)}{(5-7)(5-11)(5-13)(5-17)}(150)$$

$$+ \frac{(9-5)(9-11)(9-13)(9-17)}{(7-5)(7-11)(7-13)(7-17)}(392)$$

$$+ \frac{(9-5)(9-7)(9-13)(9-17)}{(11-5)(11-7)(11-13)(11-17)}(1452)$$

$$+ \frac{(9-5)(9-7)(9-11)(9-17)}{(13-5)(13-7)(13-11)(13-17)}(2366)$$

$$+ \frac{(9-5)(9-7)(9-11)(9-13)}{(17-5)(17-7)(17-11)(17-13)}(5202)$$

$$= \frac{(2)(-2)(-4)(-8)}{(-2)(-6)(-8)(-12)}(150) + \frac{(4)(-2)(-4)(-8)}{2(-4)(-6)(-10)}(392)$$

$$+ \frac{4(2)(-4)(-8)}{6(4)(-2)(-6)}(1452) + \frac{(4)(2)(-2)(-8)}{(8)(6)(2)(-4)}(2366)$$

$$+ \frac{4(2)(-2)(-4)}{12(10)(6)(4)}(5202)$$

$$= -\frac{50}{3} + \frac{3136}{15} + \frac{3872}{3} - \frac{2366}{3} + \frac{578}{5} = 810.$$

Chapter 23

Numerical Differentiation and Integration

23.1 Numerical Differentiation

Numerical differentiation is the process of calculating the derivatives of a function by means of a set of given values of that function. The problem may be solved by representing the function by an interpolation formula and then differentiating this formula as many times as desired.

If the function is given by a table of values for equal intervals of the independent variable, it should be represented by Newton's interpolation formula. But if given values of the function are at unequal intervals of the independent variable, we represent the function by Lagrange's interpolation formula.

The value of derivatives in terms of differences may then be found by means of those interpolation formulae.

(A) Numerical Differentiation Using Forward Difference Formula

We use the Newton's forward difference interpolating polynomial

$$f(x) = f(x_0 + rh) \approx P_n(x) = f_0 + r\Delta f_0 + \frac{r(r-1)}{2!}\Delta^2 f_0 + \frac{r(r-1)(r-2)}{3!}\Delta^3 f_0$$

$$+ \frac{r(r-1)(r-2)(r-3)}{4!}\Delta^4 f_0 + \cdots\cdots$$

Differentiating with respect to r, we obtain

$$hf'(x_0 + rh) = \Delta f_0 + \frac{2r-1}{2}\Delta^2 f_0 + \frac{3r^2 - 6r + 2}{6}\Delta^3 f_0$$

$$+ \frac{2r^3 - 9r^2 + 11r - 3}{12}\Delta^4 f_0 + \cdots \qquad \text{...(23.1)}$$

Differentiating again,

$$h^2 f''(x_0 + rh) = \Delta^2 f_0 + (r-1)\Delta^3 f_0 + \frac{6r^2 - 18r + 11}{12}\Delta^4 f_0 + \cdots \qquad \text{...(23.2)}$$

$$h^3 f'''(x_0 + rh) = \Delta^3 f_0 + \frac{2r-3}{2}\Delta^4 f_0 + \dots\dots \qquad \dots(23.3)$$

$$h^4 f^{iv}(x_0 + rh) = \Delta^4 f_0 + \dots\dots \qquad \dots(23.4)$$

For the point $x = x_0$ we have $r = 0$. Hence on substituting this value of r in the above formulas, we find the successive derivatives at the tabular point x_0.

$$\mathbf{f'(x_0)} \approx \frac{1}{h}\left(\Delta\mathbf{f_0} - \frac{1}{2}\Delta^2\mathbf{f_0} + \frac{1}{3}\Delta^3\mathbf{f_0} - \frac{1}{4}\Delta^4\mathbf{f_0} + \dots\right) \qquad \dots(23.5)$$

$$\mathbf{f''(x_0)} \approx \frac{1}{h^2}\left(\Delta^2\mathbf{f_0} - \Delta^3\mathbf{f_0} + \frac{11}{12}\Delta^4\mathbf{f_0} - \dots\right) \qquad \dots(23.6)$$

$$\mathbf{f'''(x_0)} \approx \frac{1}{h^3}\left(\Delta^3\mathbf{f_0} - \frac{3}{2}\Delta^4\mathbf{f_0} + \dots\right) \qquad \dots(23.7)$$

$$\mathbf{f^{iv}(x_0)} \approx \frac{1}{h^4}\left(\Delta^4\mathbf{f_0} - \dots\right) \qquad \dots(23.8)$$

Evidently we can find the derivatives in exactly the same way by differentiating Lagrange's formula.

To find the maximum or minimum value of a tabulated function, we compute the necessary differences from the given table, substitute them in the appropriate interpolation formula, put the first derivative of this formula equal to zero, and solve for r. Then x is found from the relation $x = x_0 + rh$.

We can also find the maximum or minimum value of a function by equating to zero the first derivative of Lagrange's formula.

(B) Numerical Differentiation using Backward Difference Formula

Using the Newton's backward difference interpolating polynomial

$$f(x) = f(x_n + rh) \approx P_n(x) = f_n + r\nabla f_n + \frac{r(r+1)}{2!}\nabla^2 f_n$$

$$+ \frac{r(r+1)(r+2)}{3!}\nabla^3 f_n + \frac{r(r+1)(r+2)(r+3)}{4!}\nabla^4 f_n + \dots\dots$$

Differentiating w.r.t., r, we obtain

$$hf'(x_n + rh) = \nabla f_n + \frac{2r+1}{2}\nabla^2 f_n + \frac{3r^2+6r+2}{6}\nabla^3 f_n$$

$$+ \frac{2r^3+9r^2+11r+3}{12}\nabla^4 f_n + \dots\dots \qquad \dots(23.9)$$

$$h^2 f''(x_n + rh) = \nabla^2 f_n + (r+1)\nabla^3 f_n + \frac{6r^2+18r+11}{12}\nabla^4 f_n + \dots \qquad \dots(23.10)$$

$$h^3 f'''\left(x_n + rh\right) = \nabla^3 f_n + \frac{2r+3}{2} \nabla^4 f_n + ... \qquad ...(23.11)$$

$$h^4 f^{iv}\left(x_n + rh\right) = \nabla^4 f_n + ... \qquad ...(23.12)$$

and, at the point $x = x_n$, we have $r = 0$.

Therefore, $f'\left(x_n\right) = \dfrac{1}{h}\left(\nabla f_n + \dfrac{1}{2}\nabla^2 f_n + \dfrac{1}{3}\nabla^3 f_n + \dfrac{1}{4}\nabla^4 f_n + ...\right).$ $\qquad ...(23.13)$

$$f''\left(x_n\right) = \frac{1}{h^2}\left(\nabla^2 f_n + \nabla^3 f_n + \frac{11}{12}\nabla^4 f_n + ...\right). \qquad ...(23.14)$$

$$f'''\left(x_n\right) = \frac{1}{h^3}\left(\nabla^3 f_n + \frac{3}{2}\nabla^4 f_n + ...\right). \qquad ...(23.15)$$

$$f^{iv}\left(x_n\right) = \frac{1}{h^4}\left(\nabla^4 f_n + ...\right). \qquad ...(23.16)$$

23.2 Illustrative Examples

Example 23.1 Find the first, second and third derivatives of the function tabulated below, at the point $x = 1.5$.

x	1.5	2.0	2.5	3.0	3.5	4.0
$f(x)$	3.375	7.000	13.625	24.000	38.875	59.000

(AMIE, W-1998)

Solution. The table of differences is as follows

x	$f(x)$	$\Delta f(x)$	$\Delta^2 f(x)$	$\Delta^3 f(x)$	$\Delta^4 f(x)$
1.5	3.375				
		3.625			
2.0	7.000		3.00		
		6.625		0.75	
2.5	13.625		3.75		0
		10.375		0.75	
3.0	24.000		4.50		0
		14.875		0.75	
3.5	38.875		5.25		
		20.125			
4.0	59.000				

Here $x = x_0 = 1.5$, $h = 0.5$. Therefore, $r = \left(x - x_0\right)/h = 0$.

Using Newton's forward interpolation formula at the point $x = x_0$, we have

$$f'(1.5) = \frac{1}{0.5}\left[3.625 - \frac{1}{2}(3.0) + \frac{1}{3}(0.75)\right] = 4.75,$$

$$f''(1.5) = \frac{1}{(0.5)^2}(3.0 - 0.75) = 9.0, \quad f'''(1.5) = \frac{1}{(0.5)^3}(0.75) = 6.0$$

Example 23.2 From the following table, calculate f' (x) and f"(x) at x = 1.35.

x	1.1	1.2	1.3	1.4	1.5	1.6
f(x)	−1.62628	0.15584	2.45256	5.39168	9.12500	13.83072

Solution. The difference table is as under

x	f(x)	Δf	$\Delta^2 f$	$\Delta^3 f$	$\Delta^4 f$	$\Delta^5 f$
1.1	−1.62628					
		1.78212				
1.2	0.15584		0.5146			
		2.29672		0.1278		
1.3	2.45256		0.6424		0.0240	
		2.93912		0.1518		0.0024
1.4	5.39168		0.7942		0.0264	
		3.73332		0.1782		
1.5	9.12500		0.9724			
		4.70572				
1.6	13.83072					

Here $x = 1.35$, $x_0 = 1.1$, $h = 0.1$. Therefore,

$$r = \frac{(x - x_0)}{h} = \frac{(1.35 - 1.10)}{0.1} = 2.5.$$

Thus

$$f'(x)_{\text{at } x=1.35} = \frac{1}{h}\left[\Delta f_0 + \frac{2r-1}{2}\Delta^2 f_0\right.$$

$$\left. + \frac{3r^2 - 6r + 2}{6}\Delta^3 f_0 + \frac{2r^3 - 9r^2 + 11r - 3}{12}\Delta^4 f_0 + ...\right]$$

$$= \frac{1}{0.1}\left[1.78212 + 2(0.5146) + \frac{3(2.5)^2 - 15 + 2}{6}(0.1278)\right.$$

$$\left. + \frac{2(2.5)^3 - 9(2.5)^2 + 11(2.5) - 3}{12}(0.024) + ...\right]$$

$$= 10(1.78212 + 1.0292 + 0.122475 - 0.001) = 29.32795$$

and $f''(x)$ at $x = 1.35 = \dfrac{1}{h^2}\left[\Delta^2 f_0 + (r-1)\Delta^3 f_0 + \dfrac{6r^2 - 18r + 11}{12}\Delta^4 f_0 - \cdots\right]$

$$= 100\left[0.5146 + 1.5(0.1278) + \frac{6(6.25) - 18(2.5) + 11}{12}(0.024) + \cdots\right]$$

$$= 100(0.5146 + 0.1917 + 0.007 + \cdots) = 71.33.$$

Example 23.3 A rod is rotating in a plane. The following table gives the angle θ (radians) through which the rod has turned for various values of the time t seconds. Calculate the angular velocity and angular acceleration of the rod, when t = 0.7 second and 0.6 second.

t seconds	0.0	0.2	0.4	0.6	0.8	1.0	1.2
θ radians	0.0	0.12	0.49	1.12	2.02	3.20	4.67

Solution. (a) The table of differences is given below

t	θ	$\nabla\theta$	$\nabla^2\theta$	$\nabla^3\theta$	$\nabla^4\theta$
0.0	0.0				
		0.12			
0.2	0.12		0.25		
		0.37		0.01	
0.4	0.49		0.26		0
		0.63		0.01	
0.6	1.12		0.27		0
		0.90		0.01	
0.8	2.02		0.28		**0**
		1.18		**0.01**	
1.0	3.20		**0.29**		
		1.47			
1.2	4.67				

Here $t = 0.7$, $t_n = 1.2$, $h = 0.2$.

Therefore, $\qquad r = \dfrac{(t - t_n)}{h} = \dfrac{(0.7 - 1.2)}{0.2} = -2.5$

From the Newton's backward difference interpolation formula, we have

$$\left(\frac{d\theta}{dt}\right)_{\text{at } t=0.7} = \frac{1}{h}\left[\nabla\theta_n + \frac{2r+1}{2}\nabla^2\theta_n + \frac{3r^2 + 6r + 2}{6}\nabla^3\theta_n\right]$$

$$= \frac{1}{0.2}\left[1.47+(-2)(0.29)+\frac{3(6.25)-15+2}{6}(0.01)\right]$$

$$= 5\left[1.47-0.58-0.0095\right] = 4.4015 \ \text{radian/sec.},$$

and $\left(\dfrac{d^2\theta}{dt^2}\right)_{\text{at } t=0.7} = \dfrac{1}{h^2}\left[\nabla^2\theta_n +(r+1)\nabla^3\theta_n\right]$

$$= \frac{1}{(0.2)^2}[0.29-1.5\times0.01] = 25[0.275] = 6.875 \ \text{radian/sec}^2.$$

(b) At $t = 0.6$, $r = 0$, Substituting these values in the above formulae, we have

$$\therefore \qquad \left(\frac{d\theta}{dt}\right)_{\text{at } t=0.6} = \frac{1}{h}\left[\nabla\theta_n +\frac{1}{2}\nabla^2\theta_n +\frac{1}{3}\nabla^3\theta_n\right]$$

$$= \frac{1}{0.2}\left[0.63+\frac{0.26}{2}+\frac{0.01}{3}\right] = 5[0.7633] = 3.8166 \ \text{radian/sec},$$

and $\left(\dfrac{d^2\theta}{dt^2}\right)_{\text{at } t=0.6} = \dfrac{1}{(0.2)^2}[0.26+0.01] = 6.75 \ \text{radian/sec}^2.$

Example 23.4 Estimate the annual rate of cloth sales of 2000 from the following data:

Year	Sale of cloth (in lakhs of metres)
1985	250
1990	285
1995	328
2005	444

Solution. Let us assume that the annual rate of cloth sale of 2000 be y_1. The table of differences is as under

x	$f(x)$	∇f	$\nabla^2 f$	$\nabla^3 f$	$\nabla^4 f$
1985	250				
		35			
1990	285		8		
		43		y_1-379	
1995	328		y_1-371		$1522-4y_1$
		y_1-328		$1143-3y_1$	
2000	y_1		$772-2y_1$		
		$444-y_1$			
2005	444				

Since fourth differences must be zero by assumptions we have
$$4y_1 = 1522 \text{ or } y_1 = 380.5$$
Rewriting the above table by putting the value of y_1 as 380.5

x	$f(x)$	∇f	$\nabla^2 f$	$\nabla^3 f$
1985	250			
		35		
1990	285		8	
		43		1.5
1995	328		9.5	
		52.5		**1.5**
2000	380.5		**11.0**	
		63.5		
2005	444			

Here $x = 2000$, $x_n = 2005$, and $h = 5$.

Therefore,
$$r = \frac{(x - x_n)}{h} = \frac{(2000 - 2005)}{5} = -1.$$

Using Newton's backward interpolation formula

$$f'(x)_{\text{at } x=2000} = \frac{1}{h}\left[\nabla f_n + \frac{2r+1}{2}\nabla^2 f_n + \frac{3r^2 + 6r + 2}{6}\nabla^3 f_n \right]$$

$$= \frac{1}{5}\left[63.5 - \frac{1}{2}(11.0) - \frac{1}{6}(1.5) \right] = \frac{1}{5}(63.5 - 5.5 - 0.25) = 11.55.$$

Example 23.5 For the set of points $(0, 2)$, $(2, -2)$, $(3, -1)$ evaluate
$$\left(\frac{dy}{dx}\right)_{\text{at } 2}$$

Solution. The difference table is given below

x	y	Δy	$\Delta^2 y$
0	2		
		-4	
2	-2		5
		1	
3	-1		

Since the points are unevenly spaced, we have by Lagrange's formula, the polynomial of degree 2 as

$$P_2(x) = y = \frac{(x-2)(x-3)}{(0-2)(0-3)}(2) + \frac{(x-0)(x-3)}{(2-0)(2-3)}(-2) + \frac{(x-0)(x-2)}{(3-0)(3-2)}(-1)$$

$$= \frac{1}{3}(x-2)(x-3)+x(x-3)-\frac{1}{3}x(x-2).$$

$$\therefore \quad \frac{dy}{dx} = \frac{1}{3}(2x-5)+(2x-3)-\frac{1}{3}(2x-2) = \frac{1}{3}(6x-12) = 2x-4,$$

and thus $\left(\frac{dy}{dx}\right)_{\text{at } x=2} = 2(2)-4 = 0.$

Example 23.6 From the following table, find x, correct to two decimal places, for which y is maximum and find this value of y.

x	1.2	1.3	1.4	1.5	1.6	
y	0.9320	0.9636	0.9855	0.9975	0.9996	*(Madras, 1998)*

Solution. The table of differences is as under

x	y	Δy	$\Delta^2 y$
1.2	0.9320		
		0.0316	
1.3	0.9636		−0.0097
		0.0219	
1.4	0.9855		−0.0099
		0.0120	
1.5	0.9975		−0.0099
		0.0021	
1.6	0.9996		

Let $x_0 = 1.2$. For maximum values of y, $\dfrac{dy}{dr} = 0.$

Differentiating, Newton's forward difference formula with respect to r, terminating after second differences and on putting $\dfrac{dy}{dr} = 0$, we obtain

$0.0316 + \{(2r-1)(-0.0097)/2\} = 0$. Solving for r, we get $r = 3.8$.

Hence $x = x_0 + rh = 1.2 + 3.8(0.1) = 1.58.$

Since 1.58 is closer to $x_n = 1.6$, we use Newton's backward difference formula. This gives

$$y(1.58) = 0.9996 - 0.2(0.0021) + \{-0.2(-0.2+1)(-0.0099)\}/2$$

$$= 0.9996 - 0.0004 + 0.0008 = 1.0.$$

Example 23.7 For a gas being manufactured and stored in a tank, the pressure gauge gave the following readings

Time (minutes)	10	15	20	25
Pressure (atmospheres)	62.21	61.55	61.38	61.70

Find the minimum pressure and the time of its occurrence.

Solution. The difference table is as under

t	p	Δp	$\Delta^2 p$	$\Delta^3 p$
10	62.21			
		−0.66		
15	61.55		0.49	
		−0.17		0
20	61.38		0.49	
		0.32		
25	61.70			

Using Newton's forward difference interpolating polynomial, we obtain,

$$p \approx 62.21 + r(-0.66) + \{r(r-1)(0.49)/2\}.$$

Here p is a function of r. For minimum p, $\dfrac{dp}{dr}$ is zero, that is,

$$\frac{dp}{dr} = (-0.66) + \frac{(2r-1)(0.49)}{2} = 0, \quad \text{or} \quad r = (0.66/0.49) + 0.5 = 1.847.$$

The minimum pressure is

$$p_{min} = 62.21 + 1.847(-0.66) + \{(1.847)(0.847)(0.49)/2\}$$

$$= 62.21 - 1.219 + 0.383 = 61.37 \text{ atmosphere.}$$

The minimum time at which it occurs is

$$t_{min} = 10 + (1.847)(5) = 19.235 \text{ minutes.}$$

Example 23.8 A function $f(x)$ representing the following data values has a minimum in the interval (0.5, 0.8). Find this point of minimum and the minimum value.

x	0.5	0.6	0.7	0.8
$f(x)$	2.3256	1.4632	0.9842	1.3282

Solution. We construct the forward difference interpolating polynomial and differentiate it to find the point of minimum

x	$f(x)$	Δf	$\Delta^2 f$	$\Delta^3 f$
0.5	2.3256			
		−0.8624		
0.6	1.4632		0.3834	
		−0.4790		0.4396
0.7	0.9842		0.8230	
		0.3440		
0.8	1.3282			

We have the Newton's interpolating polynomial as

$$f(x) = 2.3256 + (x-0.5)\left(\frac{-0.8624}{0.1}\right) + (x-0.5)(x-0.6)\left(\frac{0.3834}{0.02}\right)$$

$$+ (x-0.5)(x-0.6)(x-0.7)\left(\frac{0.4396}{0.006}\right).$$

Differentiating and setting it to zero, we obtain

$$f'(x) = 0 = -8.624 + (2x-1.1)19.17 + \left(3x^2 - 3.6x + 1.07\right)\left(\frac{1099}{15}\right)$$

$$= 3297x^2 - 3381.3x + 730.265.$$

The roots of this equation are $x = 0.7164, 0.3092$ of which the first root lies in the interval (0.5, 0.8). The point of minimum is $x \approx 0.7164$ and the minimum value is

$$f(0.7164) = 2.3256 - (0.2164)(8.624) + (0.2164)(0.1164)(19.17)$$
$$+ (0.2164)(0.1164)(0.0164)(73.2667)$$
$$\approx 2.3256 - 1.86623 + 0.48287 + 0.03027 \approx 0.9725$$

[Note that $f''(x) = 6594x - 3381.3$

∴ $f''(0.7164) = 4723.94 - 3381.3 = 1342.64 > 0$]

Exercise 23.1

1. Find the first derivative of $f(x)$ at $x = 0.04$ from the following table

x	0.01	0.02	0.03	0.04	0.05	0.06
$f(x)$	0.1023	0.1047	0.1071	0.1096	0.1122	0.1148

Ans. 4.25583

2. From the following table, find $f'(1.4)$.

x	1.2	1.3	1.4	1.5	1.6
$f(x)$	1.5095	1.6984	1.9043	2.1293	2.3756

Ans. 1.81075

3. Find the numerical value of the first derivative at $x = 0.4$ of the function $f(x)$ defined as under

x	0.1	0.2	0.3	0.4
$f(x)$	1.10517	1.22140	1.34986	1.49182

(AMIE, Summer 1998) **Ans.** 1.49133

4. The following values of x and y are given. Find the value of dy/dx and also of d^2y/dx^2 when $x = 6.5$.

x	5.0	5.5	6.0	6.5	7.0	7.5	8.0
y	3.2188	3.4096	3.5836	3.7436	3.8918	4.0298	4.1588

Ans. 0.3076; −0.04698

5. From the data in the following table, compute $f'(x)$ at $x = 0$.

x	0.00	0.05	0.10	0.15	0.20	0.25
y	0.00000	0.10017	0.20134	0.30452	0.41075	0.52110

Ans. 2.0034

6. Find the first and second derivatives of $\sqrt{x}$ at $x = 2.5$ from the table

x	2.50	2.55	2.60	2.65	2.70	2.75
$\sqrt{x}$	1.58114	1.59687	1.61245	1.62788	1.64317	1.65831

Ans. 0.3160; −0.0600. The correct values are 0.3162, −0.0632.

7. Find the first and second derivatives of the function $y = f(x)$ tabulated below at the point $x = 1.2$

x	1.0	1.2	1.4	1.6	1.8	2.00	2.2
$f(x)$	2.7183	3.3201	4.0552	4.9530	6.0496	7.3891	9.0250

Ans. 3.3205, 3.32

8. From the following table, obtain the value of d^2y/dx^2 at the point $x = 0.96$.

x:	0.96	0.98	1.00	1.02	1.04
y:	0.7825	0.7739	0.7651	0.7563	0.7473

Ans. −1.91675.

9. The population of a certain town (as obtained from census data) is shown in the following table

Year	1966	1976	1986	1996	2006
Population (in thousand)	19.96	39.65	58.81	77.21	94.61

Estimate the rate of growth of population in 1996. **Ans.** 1.80 thous./year

10. From the following table of values of x and y, find dy/dx and d^2y/dx^2 for $x = 1.05$

x	1.00	1.05	1.10	1.15	1.20	1.25	1.30
y	1.00000	1.02470	1.04881	1.07238	1.09544	1.11803	1.14017

(AMIE, W-2000) **Ans.** 0.48763; −0.2144

11. A function $f(x)$ representing the following data has a minimum in the interval (0.5, 0.8). Find this point of minimum and the minimum value.

x	0.5	0.6	0.7	0.8
$f(x)$	1.3254	1.1532	0.9432	1.0514

Ans. 0.7297; 0.9260

23.3 Numerical Integration

It is sometimes difficult or even impossible to evaluate a definite integral $\int_a^b f(x)\,dx$ by method of direct integration. In other cases the integral cannot be expressed in terms of elementary functions or the integrand is defined by means of numerical values given in a tabular form. Then we may obtain approximate numerical values of the definite integral by a numerical integration method.

There are many formulas for numerical integration but only the trapezoidal rule and Simpson's 1/3 rule or simply simpson's rule of integration will be described in this section.

Errors in Computation

Numerical methods give approximate values of quantities that is, they are not exact and contains a number of errors. Errors arise from rounding off, inaccuracy of measured values, truncation and so on. These errors are briefly described as under:

Round-off Errors

Round-off errors result from rounding that is, computation is carried out to only with a specified/fixed number of digits and any excess digits produced in multiplications or divisions are lost.

Truncation Error

Truncation error arises due to the use of approximate formulas which are generally obtained by truncating an infinite series, that is taking only a finite number of terms in an infinite series.

23.4 Trapezoidal Rule

We subdivide the interval of integration $a \le x \le b$ into n subintervals of equal length $h = (b-a)/n$ and in each subinterval approximate f by a broken line of segments (chords) with end points $[a, f(a)], [x_1, f(x_1)], \ldots, [b, f(b)]$ on the curve of f (Fig. 23.1).

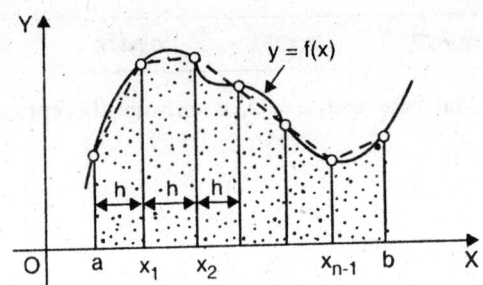

Fig. 23.1. Trapezoidal approximation

Then the area under the curve of f between a and b is approximated by n trapezoids of areas

$$\frac{1}{2}\big[f(a)+f(x_1)\big]h, \ \frac{1}{2}\big[f(x_1)+f(x_2)\big]h, \ldots, \ \frac{1}{2}\big[f(x_{n-1})+f(b)\big]h.$$

By taking their sum we obtain the **trapezoidal rule**

$$\int_a^b f(x)\,dx \approx \frac{h}{2}\Big[f(a)+2\big\{f(x_1)+f(x_2)+\ldots+f(x_{n-1})\big\}+f(b)\Big] \qquad \ldots(23.17)$$

where $h = (b - a)/n$. The $x_1, x_2, \ldots, x_{n-1}$ and a and b are called **nodes**.

The formula (23.17) is called the *composite trapezoidal rule*.

It is clear that the greater the number of intervals (that is, the smaller h is), the closer will be sum of the areas of the trapezoids approach the area under the curve.

Error Bounds and Estimate for the Trapezoidal Rule

The error estimate expression is as follows

$$\left| Error, E_T \right| \leq \frac{(b-a)h^2}{12} M_2, \qquad \qquad ...(23.18)$$

where $M_2 = \max\limits_{a \leq x \leq b} \left| f''(x) \right|$.

If an error tolerance ε is prescribed, it is possible of find n, the number of sub-intervals required to achieve that accuracy.

Since $h = (b-a)/n$, we get from (23.18)

$$\frac{(b-a)^3}{12n^2} M_2 \leq \varepsilon \text{ or } n^2 \geq \frac{(b-a)^3 M_2}{12\varepsilon}. \qquad \qquad ...(23.19)$$

23.5 Illustrative Examples

Example 23.9 Using the following data

x	1	2	3	4
$f(x)$	0.3679	0.1353	0.0498	0.0183

and the trapezoidal rule with $n = 3$, determine the approximate value of $\int_1^4 f(x)dx$.

Solution. We have $a = 1$, $b = 4$. For $n = 3$, we get $h = \dfrac{(4-1)}{3} = 1$. The abscissas are $x_0 = 1$, $x_1 = 2$, $x_2 = 3$, and $x_3 = 4$.

Hence

$$\int_1^4 f(x)dx \approx \frac{h}{2}\left[f(1) + 2\{f(2) + f(3)\} + f(4) \right]$$

$$= \frac{1}{2}\left[0.3679 + 2(0.1353 + 0.0498) + 0.0183 \right] = 0.3782.$$

Example 23.10 Obtain the approximate value of $\int_{-1}^{1} \sqrt{1 - x^2} \cos x\, dx$, using the trapezoidal rule with $n = 8$.

Solution. We have $a = -1$, $b = 1$ and $f(x) = \sqrt{1 - x^2} \cos x$. For $n = 8$, we get $h = (b-a)/n = \{1 - (-1)\}/8 = 0.25$. The abscissas are $x_0 = -1$, $x_1 = -0.75$, $x_2 = -0.50$, $x_3 = -0.25$, $x_4 = 0$, $x_5 = 0.25$, $x_6 = 0.50$, $x_7 = 0.75$, and $x_8 = 1$. The following table is prepared with increasing value of x by 0.25.

x	-1	-0.75	-0.50	-0.25	0	0.25	0.50	0.75	1
$f(x)$	0	0.48401^*	0.76000	0.93818	1	0.93818	0.76000	0.48401	0

By trapezoidal rule $\int_{-1}^{1}\sqrt{1-x^2}\cos x\,dx \approx \dfrac{0.25}{2}\left[0+2(0.48401+0.76+0.93818+1\right.$

$$\left.+0.93818+0.76+0.48401)+0\right] = 0.125(10.72876) = 1.3411$$

Example 23.11 Evaluate $\int_{1}^{2}\cos x\,dx$ using the trapezoidal rule with $n = 2$.

Compare with the exact solution. Find the maximum error.

Solution. We have $a = 1$, $b = 2$, and $f(x) = \cos x$. For $n = 2$, we get

$h = \dfrac{(b-a)}{n} = \dfrac{(2-1)}{2} = 0.5$. The abscissas are $x_0 = 1$, $x_1 = 1.5$ and $x_2 = 2$.

The following table is prepared with increasing value of x by 0.5.

x	1	1.5	2.0
$f(x)$	0.54024^{**}	0.07063	-0.41628

Hence, $\int_{1}^{2}\cos x\,dx = \dfrac{h}{2}\left[f(1)+2f(1.5)+f(2.0)\right]$

$$= \frac{1}{4}[0.54024+0.14126-0.41628] = 0.0663.$$

The exact solution is $[\sin x]_1^2 = \sin 2 - \sin 1 = 0.90924 - 0.84151 = 0.0677$.

The actual error in magnitude is $(0.0677 - 0.0663) = 0.0014$.

We have $f(x) = \cos x, f'(x) = -\sin x, f''(x) = -\cos x$.

Hence $M_2 \le \max_{(1,2)}|\cos x|$ that is, $M_2 \le 0.54024$.

Therefore, $|\text{Error}| \le \dfrac{(b-a)h^2}{12}M_2 = \dfrac{1}{12}(0.54024)h^2$.

---

* For $x = -0.75$, $f(x) = \sqrt{1-x^2}\cos x = \sqrt{1-(-0.75)^2}\cos(-0.75)$

$= \sqrt{1-0.5625}\cos(0.75) = 0.66144\cos 42.972°$ ($\because$ 1radian $= 57.296°$)
$= 0.66144\cos 42°58' = 0.66144 \times 0.73175 = 0.48401$.

** $f(1) = \cos 1 = \cos 57.296° = \cos 57°18' = 0.54024$

For $\qquad h=\dfrac{1}{2}$, $\left|\text{Error}, \varepsilon\right| \le \dfrac{1}{12}(0.54024)\times\dfrac{1}{4}$ that is, $\left|E\right| \le 0.0113$.

Example 23.12 Evaluate the following integrals using trapezoidal rule with $n=4$. Compare with the exact solution.

(i) $\displaystyle\int_0^1 \dfrac{dx}{3+2x}$, $\qquad$ (ii) $\displaystyle\int_0^2 \dfrac{dx}{x^2+2x+10}$.

Find the bound on the error. Also, find the number of sub-intervals required if the error is to be less than 5×10^{-4}.

Solution. (*i*) We have $a=0$, $b=1$ and $f(x)=\dfrac{1}{(3+2x)}$. For $n=4$, we get

$h=\dfrac{(b-a)}{n}=\dfrac{(1-0)}{4}=\dfrac{1}{4}$. The abscissas are $x_0=0$, $x_1=1/4$, $x_2=1/2$, $x_3=3/4$, and $x_4=1.0$. The following table in prepared with increasing value of x by $1/4$.

x	0	1/4	1/2	3/4	1
$f(x)$	1/3	2/7	1/4	2/9	1/5

Hence, $\displaystyle\int_0^1 \dfrac{dx}{3+2x} \approx \dfrac{h}{2}\left[f(0)+2\left\{f\left(\dfrac{1}{4}\right)+f\left(\dfrac{1}{2}\right)+f\left(\dfrac{3}{4}\right)\right\}+f(1)\right]$

$$=\dfrac{1}{8}\left[\dfrac{1}{3}+2\left\{\dfrac{2}{7}+\dfrac{1}{4}+\dfrac{2}{9}\right\}+\dfrac{1}{5}\right]=0.25615.$$

The exact solution is $\dfrac{1}{2}\left[\log_e(3+2x)\right]_0^1 = 0.5\log_e\left(\dfrac{5}{3}\right)$

$$=0.5(0.69897-0.47712)2.30258=0.25541.$$

The actual error in magnitude is 0.00074.

We have $f(x)=\dfrac{1}{3+2x}$, $f'(x)=-\dfrac{2}{(3+2x)^2}$, $f''(x)=\dfrac{8}{(3+2x)^3}$ and

$$M_2=\max_{[0,1]}\left[\dfrac{8}{(3+2x)^3}\right]=\dfrac{8}{27}.$$

Hence, $\qquad |\text{Error}| \le \dfrac{(b-a)h^2}{12}M_2 \le \dfrac{1}{12}\left(\dfrac{8}{27}\right)\left(\dfrac{1}{16}\right) \le 0.00154.$

The actual error is smaller than the maximum magnitude error.

If $\varepsilon = 5 \times 10^{-4}$, than the number of sub-intervals required is given by

$$n^2 \geq \frac{(b-a)^3 M_2}{12\varepsilon} = \frac{1^3 \times 8 \times 10^4}{27 \times 12 \times 5} = 49.38 \text{ or } n \geq 7.03.$$

Since n is an integer, we require $n = 8$.

(*ii*) We have $a = 0$, $b = 2$ and $f(x) = \dfrac{1}{(x^2 + 2x + 10)}$. For $n = 4$, we get

$h = \dfrac{(2-0)}{4} = \dfrac{1}{2}$. The abscissas are $x_0 = 0$, $x_1 = 1/2$, $x_2 = 1$, $x_3 = 3/2$, and $x_4 = 2$.
The following table is prepare with increasing value of x by 1/2.

x	0	1/2	1	3/2	2
$f(x)$	0.1	0.08889	0.07692	0.06557	0.05556

Hence, $\displaystyle\int_0^2 \frac{dx}{x^2 + 2x + 10} \approx \frac{h}{2}\left[f(0) + 2\left\{f\left(\frac{1}{2}\right) + f(1) + f\left(\frac{3}{2}\right)\right\} + f(2)\right]$

$$= \frac{1}{4}\left[0.1 + 2(0.08889 + 0.07692 + 0.06557) + 0.05556\right] = 0.15458.$$

The exact solution is as under:

$$\int_0^2 \frac{dx}{x^2 + 2x + 10} = \int_0^2 \frac{dx}{(x+1)^2 + (3)^2} = \frac{1}{3}\left|\tan^{-1}\left(\frac{x+1}{3}\right)\right|_0^2$$

$$= \frac{1}{3}\left[\frac{\pi}{4} - \tan^{-1}\left(\frac{1}{3}\right)\right] = 0.15455.$$

The actual error in magnitude is 0.00003. We have

$$f(x) = \frac{1}{x^2 + 2x + 10}, \quad f'(x) = -\frac{2(x+1)}{(x^2 + 2x + 10)^2},$$

$$f''(x) = -\frac{\left[2(x^2 + 2x + 10)^2 - 8(x+1)^2(x^2 + 2x + 10)\right]}{(x^2 + 2x + 10)^4} = \frac{6(x^2 + 2x - 2)}{(x^2 + 2x + 10)^3}.$$

$\displaystyle\max_{(0,2)}(x^2 + 2x - 2) = 6$ and $\displaystyle\min_{(0,2)}(x^2 + 2x + 10)^3 = 1000$. Hence, $M_2 \leq \dfrac{6 \times 6}{1000}$.

Therefore, $|\text{Error}| \leq \dfrac{(b-a)h^2}{12} M_2 = \dfrac{2}{12} \times \dfrac{36}{1000} h^2 = \dfrac{3}{500} h^2.$

For $h = 1/2$, $\qquad\qquad$ $|\text{Error}| \leq 0.0015$.

If $\varepsilon = 5 \times 10^{-4}$, then the number of sub-intervals required is given by

$$n^2 \geq \frac{(b-a)^3 M_2}{12\varepsilon} = \frac{8 \times 36 \times 10^4}{1000 \times 12 \times 5} = 48, \text{ or } n = 7.$$

23.6 Simpson's Rule

Simpson's rule for approximating $\int_a^b f(x)\,dx$ is based on approximating the graph

of f with quadratic polynomials that is, parabolic arcs instead of linear polynomials that is, line segments. We approximate the graph of f with parabolic arcs instead of line segments. In simpson's rule, we divide the interval of integration $a \leq x \leq b$ into an **even number** of equal sub-intervals say into $n = 2m$ sub-intervals of length $h = (b-a)/2m$, with end points $x_0 (= a)$, x_1, x_2, ..., x_{2m-1}, x_{2m} $(= b)$; where m is an integer. We have odd number of ordinates. See fig. 23.2.

We now take the first two subintervals, the area of the shaded region under the parabola is

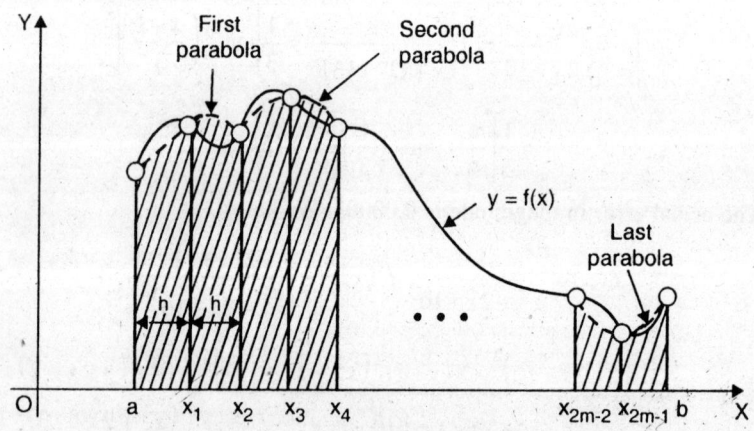

Fig. 23.2. Simpson's rule.

Area $= \dfrac{h}{3}\left[f(a) + 4f(x_1) + f(x_2)\right]$. A similar formula holds for the next two

sub-intervals from x_2 to x_4. Area $= \dfrac{h}{3}\left[f(x_2) + 4f(x_3) + f(x_4)\right]$, and so on. By summing all these m formulas we obtain **Simpson's rule.**

$$\int_a^b f(x)\,dx \approx \frac{h}{3}\Big[f(a)+4\{f(x_1)+f(x_3)+\ldots+f(x_{2m-1})\}$$

$$+2\{f(x_2)+f(x_4)+\ldots+f(x_{2m-2})\}+f(b)\Big]$$

$$=\frac{(b-a)}{3(2m)}\Big[f(a)+4\{f(x_1)+f(x_3)+\ldots+f(x_{2m-1})\}$$

$$+2\{f(x_2)+f(x_4)+\ldots+f(x_{2m-2})\}+f(b)\Big] \qquad \ldots(23.20)$$

This is known as composite *Simpson's 1/3 Rule.*

As in the case of the trapezoidal rule, the greater the number of parts into which the range of integration is divided, the closer will the result be the area under the curve.

Error of Simpson's rule. If the fourth derivative $f^{(iv)}(x)$ exists and is continuous on $a \le x \le b$, the *error* of (23.20) is

$$\left|\text{Error, E}_s\right| \le \frac{(b-a)}{180}h^4 M^4 \qquad \ldots(23.21)$$

where $M_4 = \max\limits_{(a,b)}\left|f^{(iv)}(x)\right|$. If the error tolerance ε is prescribed, it is possible to find $n = 2m$, the number of sub-intervals required to achieve that accuracy. Since $h = (b-a)/2m$, where m is an integer, we get from (23.21)

$$\frac{(b-a)^5 M_4}{2880m^4} \le \varepsilon \text{ or } m^4 \ge \frac{(b-a)^5 M_4}{2880\varepsilon} \qquad \ldots(23.22)$$

Remarks:

1. Simpson's rule is usually much more accurate than the midpoint or trapezoidal rule.

   Application of Simpson's rule

2. The Simpson's rule is not only applicable to evaluate definite integrals representing areas but to evaluate all definite integrals whether they represent volume, surface, mass, moment of inertia, centre of gravity etc.

   (a) *Evaluation of Volumes.* If the various ordinates in formula 23.20 represent equispaced cross-sectional areas, then Simpson's rule gives the volume of the solid. Let $A_0, A_1, A_2, \ldots, A_{2m}$ are the areas of the sections of a solid by parallel planes at equal distances h apart. The volume of the solid

$$=\int_a^b A\,dx \approx \frac{h}{3}\Big[A_0+4(A_1+A_3+\ldots+A_{2m-1})+2(A_2+A_4+\ldots+A_{2m-2})+A_{2m}\Big]$$

$$\ldots(23.23)$$

(b) If the ordinates denote velocities at equal intervals of time, the Simpson's rule gives the distance travelled.

(c) *Evaluation of volume of solid of revolution.* Let the area under the curve $y = f(x)$, above the x-axis, between the lines $x = a$ and $x = b$ revolves about x-axis, the volume of the solid thus formed

$$= \int_a^b \pi [f(x)]^2 \, dx \approx \frac{\pi h}{3}\left[\{f(a)\}^2 + 4\left\{ (f(x_1))^2 + (f(x_3))^2 + ... + (f(x_{2m-1}))^2 \right\} \right.$$

$$\left. +2\left[\{(f(x_2))^2 + (f(x_4))^2 + ... + (f(x_{2m}))^2 \} + (f(b))^2 \right] \right. \qquad ...(23.24)$$

23.7 Illustrative Examples

Example 23.13 (a) From the following data, estimate $\int_{1.1}^{1.9} f(x)\,dx$ as best as you can.

x	1.1	1.2	1.3	1.4	1.5	1.6	1.7	1.8	1.9
$f(x)$	3.95254	4.06587	4.18162	4.29982	4.42051	4.54372	4.66949	4.80135	5.02365

(b) The cross-section of a tree is A sq. cm at a distance x cm from one end. Corresponding values of A and x are given

x	10	30	50	70	90	110	130	150	170
A	120	123	129	131	131	135	142	156	177

Find by Simpson's rule, the volume of the tree in cubic cm between $x = 10$ and $x = 170$.

Solution. (i) Here $n = 8$ and $h = 0.1$. Employing Simpson's 1/3 rule or simply Simpson's rule of integration.

$$\int_{1.1}^{1.9} f(x)\,dx \approx \frac{h}{3}\left[f(x_0) + 4\{f(x_1) + f(x_3) + f(x_5) + f(x_7)\} \right.$$

$$\left. +2\{f(x_2) + f(x_4) + f(x_6)\} + f(x_8) \right]$$

$$= \frac{0.1}{3}\left[3.95254 + 4(4.06587 + 4.29982 + 4.54372 + 4.80135) \right.$$

$$\left. +2(4.18162 + 4.42051 + 4.66949) + 5.02365 \right]$$

$$= \frac{0.1}{3}\left[8.97619+4(17.71076)+2(13.27162)\right] = \frac{1}{30}(106.36247) = 3.54541.$$

(b) To approximate $\int_{10}^{170} f(x)\,dx$ that is, $\int_{10}^{170} A\,dx$, use Simpson's 1/3 rule, with $n = 8$ and $h = 20$.

$$\text{Required Volume} = \int_{10}^{170} A\,dx \approx \frac{20}{3}\left[120+4(123+131+135+156)\right.$$

$$\left.+2(129+131+142)+177\right]$$

$$= \frac{20}{3}(297+2180+804) = \frac{20}{3}(3281) = 21873.33 \text{ cm}^3.$$

Example 23.14 (a) The velocity v (km/min) of a moped which starts from rest, is given at fixed intervals of time t(min) as follows

t	2	4	6	8	10	12	14	16	18	20
v	10	18	25	29	32	20	11	5	2	0

Estimate approximately the total distance covered in 20 minutes.

(b) The velocity v(m/sec) of particle at a distance s(metre) from a point on its path is given by the table

s	0	10	20	30	40	50	60
v	47	58	64	65	61	52	38

Estimate the time taken to travel 60 metre.

(c) A variable force F kgf acts in the direction of motion of a moving body. The following table gives values of F at distances x metres along the path of the body.

x	2	3	4	5	6	7	8	9	10	11	12
F	122	159	188	210	231	246	259	268	277	282	287

Find the work done by the force on the body as its moves from $x = 2$ to $x = 12$.

Solution. (a) If s (km) be the distance covered in time t (min), then $v = ds/dt$ or $ds = v.dt$. It follows on integration that $s = \int v \cdot dt$. Distance covered in 20 minutes is, $s = \int_0^{20} v \, dt$. Since the train starts from rest, $v = 0$, when $t = 0$. Thus

there are 11 values of v. Use Simpson's 1/3 rule, with $h = 2$.

$$\therefore \quad \int_0^{20} v \, dt \approx \frac{2}{3}\left[0 + 4(10 + 25 + 32 + 11 + 2) + 2(18 + 29 + 20 + 5) + 0\right]$$

$$= \frac{2[320 + 144]}{3} = \frac{2(464)}{3} = 309.33 \text{ km.}$$

(b) The time taken to travel 60 metre is $t = \int_0^{60} \frac{1}{v} ds$ (sec). We form a table of $1/v$

against s.

s	0	10	20	30	40	50	60
$1/v$	0.02127	0.01724	0.01563	0.01538	0.0164 0	0.01923	0.02631

Hence, $t = \int_0^{60} \frac{1}{v} ds \approx \frac{10}{3}\left[0.02127 + 4(0.01724 + 0.01538 + 0.01923)\right.$

$$\left. + 2(0.01563 + 0.01640) + 0.02631\right]$$

$$= \frac{10}{3}[0.04758 + 0.20740 + 0.06406] = \frac{10}{3}(0.31904) = 1.06 \text{ seconds.}$$

(c) Work done by the force, $= \int_2^{12} F \cdot dx$. Use Simpson's 1/3 rule, with $h = 1$.

Hence, $\quad \int_2^{12} F \cdot dx \approx \frac{1}{3}\left[122 + 4(159 + 210 + 246 + 268 + 282)\right.$

$$\left. + 2(188 + 231 + 259 + 277) + 287\right]$$

$$= \frac{1}{3}\left[409 + 4(1165) + 2(955)\right] = \frac{1}{3}(6979) = 2326.33 \text{ kgf-m.}$$

Example 23.15 (a) Use Simpson's rule to determine (i) the area under the curve (ii) the volume generated by revolving the area about the x-axis determined by the given data

x	1	2	3	4	5
y	1.8	4.2	7.8	9.2	12.3

(b) A body is in the form of a solid of revolution and is 15 cm long. The following table gives the diameter D in cm of the section at distance x cm from one end. Find the volume of the solid generated.

x	0	2.5	5.0	7.5	10.0	12.5	15.0
D	5	5.5	6.0	6.75	6.25	5.5	5.0

Solution. (*a*) Here $h = 1$.

(*i*) Area under the curve $= \int\limits_{1}^{5} y\,dx \approx \dfrac{1}{3}\Big[1.8 + 4(4.2 + 9.2) + 2(7.8) + 12.3\Big]$

$$= \frac{1}{3}\Big[14.1 + 4(13.4) + 2(7.81)\Big] = 27.767 \text{ sq. units.}$$

(*ii*) Volume of solid generated

$$\approx \frac{\pi \times 1}{3}\Big[(1.8)^2 + 4\{(4.2)^2 + (9.2)^2\} + 2(7.8)^2 + (12.3)^2\Big]$$

$$= \frac{\pi}{3}\Big[3.24 + 4(17.64 + 84.64) + 2(60.84) + 151.29\Big]$$

$$= \frac{\pi}{3}[154.53 + 409.12 + 121.68] = 228.44\,\pi \text{ cubic units.}$$

(*b*) Here $h = 2.5$. By Simpson's rule, volume of the solid generated $= \int\limits_{0}^{15} \dfrac{\pi}{4} D^2\,dx$

$$\approx \frac{\pi}{4}\frac{(2.5)}{3}\Big[(5)^2 + 4\{(5.5)^2 + (6.75)^2 + (5.5)^2\} + 2\{(6.0)^2 + (6.25)^2\} + (5)^2\Big]$$

$$= \frac{2.5\pi}{12}(50 + 424.25 + 150.125) = \frac{2.5\pi}{12} \times 624.375 = 408.817 \text{ cubic cms.}$$

Example 23.16 Using Simpson's rule, evaluate the following

(*i*) $\int\limits_{0}^{\pi/2} e^{\sin x}\, dx$, with n = 8, (*ii*) $\int\limits_{0}^{\pi} \dfrac{\sin x}{x}\,dx$, with seven ordinates,

(*iii*) $\int\limits_{0}^{\pi/2} \sqrt{\sin \theta}\; d\theta$ with n = 6.

Solution: (*i*) We have $a = 0$. $b = \pi/2$ and $f(x) = e^{\sin x}$. For $n = 8$, we get $h = (\pi/2 - 0)/8 = \pi/16$. The following table is prepared with increasing value of x by $\pi/16$.

x	0	$\pi/16$	$\pi/8$	$3\pi/16$	$\pi/4$	$5\pi/16$	$3\pi/8$	$7\pi/16$	$\pi/2$
$\sin x$	0	0.1951	0.3827	0.5555	0.7071	0.8315	0.9239	0.9808	1
$e^{\sin x}$	1	1.215	1.467	1.743	2.028	2.297	2.519	2.667	2.718

Hence, $\displaystyle\int_0^{\pi/2} e^{\sin x}\,dx \approx \frac{\pi}{16 \times 3}[1 + 4(1.215 + 1.743 + 2.297 + 2.667)$

$$+ 2(1.467 + 2.028 + 2.519) + 2.718]$$

$$= \frac{\pi}{48}\left[3.718 + 4(7.922) + 2(6.014)\right] = \frac{\pi}{48}[3.718 + 31.688 + 12.028]$$

$$= \frac{(\pi \times 47.434)}{48} = 3.1058.$$

(*ii*) We have $a = 0$, $b = \pi$ and $f(x) = (\sin x)/x$. For $n = 6$, we get

$h = \dfrac{(\pi - 0)}{6} = \dfrac{\pi}{6}$. We obtain the following table

x	0	$\pi/6 = 30°$	$\pi/3 = 60°$	$\pi/2 = 90°$	$2\pi/3 = 120°$	$5\pi/6 = 150°$	$\pi = 180°$
$\sin x$	0.00	0.50	0.86603	1.00	0.86603	0.50	0.00
$\dfrac{(\sin x)}{x}$	1	$\dfrac{3}{\pi}$	$\dfrac{(2.59809)}{\pi}$	$\dfrac{2}{\pi}$	$\dfrac{(1.29904)}{\pi}$	$\dfrac{(0.6)}{\pi}$	0

Hence, $\displaystyle\int_0^{\pi} \frac{\sin x}{x}\,dx \approx \frac{\pi}{18}\left[1 + 4\left(\frac{3}{\pi} + \frac{2}{\pi} + \frac{0.6}{\pi}\right) + 2\left(\frac{2.59809}{\pi} + \frac{1.29904}{\pi}\right) + 0\right]$

$$= \frac{\pi}{18}\left(1 + \frac{22.4}{\pi} + \frac{7.79426}{\pi}\right) = \frac{33.3371}{18} = 1.8521.$$

(*iii*) Here $h = \dfrac{(\pi/2 - 0)}{6} = \dfrac{\pi}{12}$.

The following table is prepared with increasing value of θ by $\pi/12$.

θ	0	$\dfrac{\pi}{12}=15°$	$\dfrac{\pi}{6}=30°$	$\dfrac{\pi}{4}=45°$	$\dfrac{\pi}{3}=60°$	$\dfrac{5\pi}{12}=75°$	$\dfrac{\pi}{2}=90°$
$\sin\theta$	0.0	0.2588	0.500	0.7071	0.8660	0.9659	1.000
$\sqrt{\sin\theta}$	0	0.5087	0.7071	0.8409	0.9306	0.9828	1.000

Hence, $\displaystyle\int_{0}^{\pi/2} \sqrt{\sin\theta}\,d\theta \approx \frac{\pi}{36}\Big[0+4\,(0.5087+0.8409+0.9828)$

$$+2\,(0.7071+0.9306)+1\Big]=\frac{(\pi\times13.605)}{36}=1.1877.$$

Example 23.17 (a) By Simpson's rule, find the value of $\log_e (3/2)$.

(b) Using Simpson's rule calculate the approximate value of π from

$$\int_{0}^{1}\frac{dx}{1+x^{2}}.$$ *(AMIE, Winter 2005)*

(c) Calculate the value of $\log_e 2$ by finding $\displaystyle\int_{0}^{1}\frac{x^2\,dx}{1+x^3}$, using Simpson's rule

with $n = 4$.   *(AMIE, Summer 1998)*

Solution. (a) $\log_e \dfrac{3}{2}$ may be regarded as $\displaystyle\int_{2}^{3}\frac{1}{x}dx$. Dividing the range into 10 equal

intervals so that $h=\dfrac{(3-2)}{10}=0.1$. We obtain the following table:

x	2.0	2.1	2.2	2.3	2.4	2.5	2.6	2.7	2.8	2.9	3.0
$y=1/x$	0.5000	0.4762	0.4545	0.4348	0.4167	0.4000	0.3846	0.3704	0.3571	0.3448	0.3333

By Simpson's rule,

$$\int_{2}^{3}\frac{dx}{x}\approx\frac{0.1}{3}\Big[0.50+4\,(0.4762+0.4348+0.4000+0.3704+0.3448)$$

$$+2\,(0.4545+0.4167+0.3846+0.3571)+0.3333\Big]$$

$$= \frac{0.1}{3}\left[0.8333+4(2.0262)+2(1.6129)\right]=0.4055$$

(b) Dividing the range into 10 equal intervals, so that $h=\frac{(1-0)}{10}=0.1$. The following table is prepared with increasing value of x by 0.1.

x	0	0.1	0.2	0.3	0.4	0.5	0.6	0.7	0.8	0.9	1.0
$y=\left(1+x^2\right)^{-1}$	1.0000	0.9901	0.9615	0.9174	0.8621	0.8000	0.7353	0.6711	0.6097	0.5525	0.5000

By Simpson's Rule,

$$\int_0^1 \frac{dx}{1+x^2} \approx \frac{0.1}{3}\left[1.00+4(0.9901+0.9174+0.8+0.6711+0.5525)\right.$$

$$+2(0.9615+0.8621+0.7353+0.6097)+0.50\Big]$$

$$=\frac{0.1}{3}\left[1.5+4(3.9311)+2(3.1686)\right]=\frac{1}{30}(1.5+15.7244+6.3372)=0.78538.$$

We also know that

$$\int_0^1 \left(1+x^2\right)^{-1} dx = \left|\tan^{-1} x\right|_0^1 = \tan^{-1} 1 = \frac{\pi}{4}.$$

Therefore, $\frac{1}{4}\pi \approx 0.78538$ or $\pi \approx 3.1415$.

(c) First we shall determine the approximate value of $\int_0^1 \frac{x^2 dx}{1+x^3}$ by employing Simpson's 1/3 rule. We have $a = 0$, $b = 1$, $f(x) = x^2/(1+x^3)$. For $n = 4$, we get $h = \frac{(1-0)}{4} = 0.25$. The following table is prepared with increasing value of x by 0.25.

x	0	0.25	0.50	0.75	1.00
$y=\dfrac{x^2}{\left(1+x^3\right)}$	0	0.06154	0.22222	0.39560	0.5000

Hence, $\displaystyle\int_0^1 \frac{x^2 dx}{1+x^3} \approx \frac{0.25}{3}\left[0+4(0.06154+0.39560)+2(0.22222)+0.50\right]$

$$= 0.25[0.5+1.82856+0.44444]/3 = 0.231083.$$

Also by definite integral, $\displaystyle\int_0^1 \frac{x^2}{1+x^3}\,dx = \frac{1}{3}\int_0^1 \frac{3x^2}{1+x^3}\,dx = \frac{1}{3}\left[\log_e\left(1+x^3\right)\right]_0^1$

$$= \frac{1}{3}\left(\log_e 2 - \log_e 1\right) = \frac{1}{3}\log_e 2 \,.$$

Therefore, $\qquad \log_e 2 = 3\displaystyle\int_0^1 \frac{x^2 dx}{1+x^3} \approx 3(0.231083) = 0.693249.$

Example 23.18 A tank is discharging water through an orifice at a depth x metre below the surface of water whose area is A sq. metres. The following are the values of x for the corresponding values of A

A	1.257	1.39	1.52	1.65	1.809	1.962	2.123	2.295	2.462	2.650	2.827
x	1.50	1.65	1.80	1.95	2.10	2.25	2.40	2.55	2.70	2.85	3.00

Using the formula $(0.018)T = \displaystyle\int_{1.5}^{3.0} \frac{A}{\sqrt{x}}\,dx$, calculate T, the time in seconds for the level of water to drop from 3.0 m to 1.5 m above an orifice.

Solution. Here $h = 0.15$ m. The following table is prepared.

x	1.50	1.65	1.80	1.95	2.10	2.25	2.40	2.55	2.70	2.85	3.00
A	1.257	1.39	1.52	1.65	1.809	1.962	2.123	2.295	2.462	2.650	2.827
$y = \dfrac{A}{\sqrt{x}}$	1.026	1.082	1.133	1.182	1.248	1.308	1.370	1.437	1.498	1.570	1.632

By Simpson's rule,

$$\int_{1.5}^{3.0} \frac{A}{\sqrt{x}}\,dx \approx \frac{0.15}{3}\left[1.026+4(1.082+1.182+1.308+1.437+1.570)\right.$$

$$\left. +2(1.133+1.248+1.370+1.498)+1.632\right]$$

$$= \frac{0.15}{3}[2.658 + 26.316 + 10.498] = \frac{0.15}{3} \times 39.472 = 1.9736.$$

Therefore, $(0.018)T = 1.9736$ that is, $T = \dfrac{(1.9736)}{0.018} = 110$ sec. (nearly).

23.19 Evaluate the following integrals using Simpson's rule with $n = 4$. Compare with the exact solutions

(i) $\displaystyle\int_0^1 \frac{dx}{3+2x}$,   (ii) $\displaystyle\int_0^2 \frac{dx}{x^2+2x+10}$.

Solution. (i) We have $a = 0$, $b = 1$ and $f(x) = \dfrac{1}{(3+2x)}$. For $n = 4$, we obtain

$h = \dfrac{(b-a)}{n} = \dfrac{1}{4}$. The abscissas are $x_0 = 0, x_1 = 1/4, x_2 = 1/2, x_3 = 3/4$ and

$x_4 = 1$. The following table is prepared with increasing values of x by $1/4$.

x	0	1/4	1/2	3/4	1
$f(x)$	1/3	2/7	1/4	2/9	1/5

Hence, $\displaystyle\int_0^1 \frac{dx}{3+2x} \approx \frac{h}{3}\Big[f(0) + 4\{f(1/4) + f(3/4)\} + 2f(1/2) + f(1)\Big]$

$$= \frac{1}{12}\left[\frac{1}{3} + 4\left(\frac{2}{7} + \frac{2}{9}\right) + 2\left(\frac{1}{4}\right) + \frac{1}{5}\right] = 0.25542.$$

The exact solution is $0.5 \log(5/3) = 0.25541$. The error in magnitude is 0.00001.

(ii) We have $a = 0$, $b = 2$ and $f(x) = 1/(x^2 + 2x + 10)$. For $n = 4$, we obtain

$h = \dfrac{(2-0)}{4} = \dfrac{1}{2}$. The abscissas are $x_0 = 0, x_1 = 1/2, x_2 = 1, x_3 = 3/2$ and $x_4 = 2$.
The following table is prepared with increasing value of x by $1/2$.

x	0	1/2	1	3/2	2
$f(x)$	1/10	4/45	1/13	4/61	1/18

Hence, $\displaystyle\int_0^2 \frac{1}{x^2+2x+10}\,dx = \frac{h}{3}\Big[f(0) + 4\{f(1/2) + f(3/2)\} + 2f(1) + f(2)\Big]$

$$= \frac{1}{6}\left[\frac{1}{10} + 4\left(\frac{4}{45} + \frac{4}{61}\right) + 2\left(\frac{1}{13}\right) + \frac{1}{18}\right]$$

$$= \frac{1}{6}\left[0.1 + 4(0.0889 + 0.06557) + 2(0.07692) + 0.05556\right] = \frac{0.92728}{6} = 0.15454.$$

The exact solution is 0.15455. The error in magnitude is 0.00001.

Example 23.20 Evaluate the integral $\int_{-1}^{1} x^2 e^{-x} dx$, by composite Simpson's one-third rule with spacing h = 0.25. You may use any other known numerical method to evaluate the given integral. *(A.M.I.E., S-2008)*

Solution. We have a = -1, b = 1, h = 0.25 and $f(x) = x^2 e^{-x}$. The abscissas are

$x_0 = -1$, $x_1 = -0.75$, $x_2 = -0.50$, $x_3 = -0.25$, $x_4 = 0$, $x_5 = 0.25$, $x_6 = 0.50$,

$x_7 = 0.75$ and $x_8 = 1$. The following table is prepared with increasing values of x by 0.25.

x	-1.0	-0.75	-0.50	-0.25	0	0.25	0.50	0.75	1.0
$f(x)=x^2 e^{-x}$	2.71828	1.19082	0.41218	0.08025	0	0.04868	0.15163	0.26571	0.36788

Hence, $\int_{-1}^{1} x^2 e^{-x} dx \approx \frac{h}{3}\left[f(-1) + 4\{f(-0.75) + f(-0.25) + f(0.25) + f(0.75)\}\right.$

$$\left. + 2\{f(-0.50) + f(0) + f(0.50)\} + f(1.0)\right]$$

$$\approx \frac{0.25}{3}\left[2.71828 + 4(1.19082 + 0.08025 + 0.04868 + 0.26571)\right.$$

$$\left. + 2(0.41218 + 0 + 0.15163) + 0.36788\right]$$

$$= \frac{1}{12}(3.08616 + 6.34184 + 1.12762) = 0.87964.$$

Example 23.21 Apply the trapezoidal and Simpson's 1/3 rule to compute the integral of $\sqrt{x}$ between the arguments 1.00 and 1.30. Use the data of following table. Compare with the correct value of the integral. Estimate the truncation error of the trapezoidal rule and Simpson's rule.

x	1.00	1.05	1.10	1.15	1.20	1.25	1.30
$y = f(x) = \sqrt{x}$	1.00000	1.02470	1.04881	1.07238	1.09544	1.11803	1.14017

Solution. Here $h = 0.05$. Apply the trapezoidal rule, we obtain

$$\int_{1.00}^{1.30} \sqrt{x}\,dx \approx \frac{0.05}{2}[1.00000 + 2(1.02470 + 1.04881 + 1.07238$$

$$+ 1.095544 + 1.11803 + 1.14017] = \frac{0.05}{2}(12.85889) = 0.32147.$$

The correct value is $\frac{2}{3}\left[(1.3)^{3/2} - 1\right] = 0.32149$ to five places, making the actual error 0.00002.

By Simpson's rule, $\int_{1.00}^{1.30} \sqrt{x}\,dx \approx \frac{0.05}{3}[1 + 4(1.02470 + 1.07238 + 1.11803)$

$$+ 2(1.04881 + 1.09544) + 1.14017]$$

$$= \frac{0.05}{3}[2.14017 + 12.86044 + 4.2885] = \frac{0.05}{3}(19.28911) = 0.32149.$$

Truncation error of the trapezoidal rule

We have $\quad f(x) = \sqrt{x},\ f'(x) = x^{-1/2}/2,\ f''(x) = -x^{-3/2}/4.$

$$M_2 = \max_{[1,\,1.30]}\left|\frac{1}{4}x^{-3/2}\right| = \frac{1}{4}.$$

Hence, $\quad \left|\text{Error},\ \varepsilon_T\right| \le \frac{(b-a)h^2}{12} M_2 = 0.00625h^2.$

For $h = 0.05, \left|\text{Error},\ \varepsilon_T\right| \le 0.000016$ which is slightly less than the actual error of 0.00002.

Truncation error of Simpson's rule

$$f'''(x) = \frac{3}{8}x^{-5/2},\ f^{iv}(x) = -\frac{15}{16}x^{-7/2}. \qquad M_4 = \max_{[1,\,1.30]}\left|\frac{15}{16}x^{-7/2}\right| = \frac{15}{16}.$$

Hence, $\left|\text{Error},\ \varepsilon_s\right| \le \frac{(b-a)}{180}h^4 M_4 = \frac{0.30}{180}(0.05)^4\left(\frac{15}{16}\right) \sim 0.00000001$ which is minute.

Example 23.22 A curve passes through the points

(1, 2), (1.5, 2.4), (2, 2.7), (2.5, 2.8), (3, 3), (3.5, 2.6), (4, 2.1) and (4.5, 1.9). Use Simpson's one third rule to obtain the volume of the solid of revolution got by revolving this area about x-axis.

Solution. Here the number of strips h are 7, that is odd and number of values of y that is, ordinates are even. Simpson rule cannot be applied directly. Refer Fig. 23.3. The volume of region $AA'BJ$ when revolved about x-axis is computed by using Pappus theorem on volume of revolution. Simpson's 1/3 rule is applied on the remaining portion.

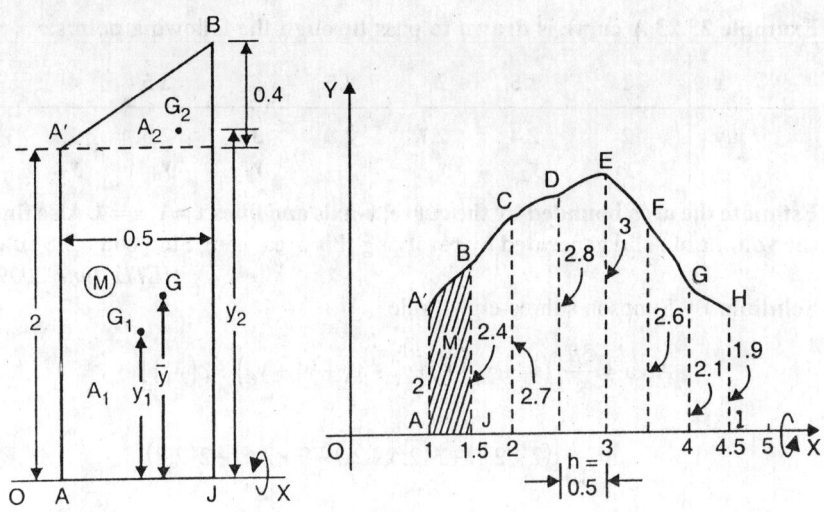

| Fig. 23.3. | Fig. 23.4. |

Refer Fig. 23.3

$A_1 = 2 \times 0.5 = 1$, $A_2 = 1/2 \times 0.5 \times 0.4 = 0.1$, $y_1 = 1$ unit from OX,

$y_2 = 2 + \dfrac{1}{3}(0.4) = \dfrac{32}{15}$. Let $\bar{y}$ be the CG of the whole region (Marked M).

Taking moment about OX.

$$\left(A_1 + A_2\right)\bar{y} = A_1 y_1 + A_2 y_2 \Rightarrow 1 \cdot 1\bar{y} = 1 \times 1 + 0.1 \times \frac{32}{15} \Rightarrow \bar{y} = 1.103.$$

Volume of whole region $M = (1.1) \ 2\pi\bar{y} = (1.1)2\pi(1.103) = 7.626$ cubic units.

Now prepared the following table for computing volume of solid of revolution about X-axis, the portion JBCDEFGHIJ. (Fig. 23.4).

x	1.5	2.0	2.5	3.0	3.5	4.0	4.5
y	2.4	2.7	2.8	3	2.6	2.1	1.9
y^2	5.76	7.29	7.84	9	6.76	4.41	3.61

$\therefore$ Required volume of the solid of revolution

$$= 7.626 + \frac{\pi(0.5)}{3}\left[(5.76+3.61)+4(7.29+9+4.41)+2(7.84+6.76)\right]$$

$$= 7.626 + \frac{\pi}{6}(9.37+82.80+29.20) = 7.626+63.574 = 71.2 \text{ cubic units.}$$

Example 23.23 A curve is drawn to pass through the following points:

x	1	1.5	2	2.5	3	3.5	4
y	2	2.4	2.7	2.8	3	2.6	2.1
	y_1	y_2	y_3	y_4	y_5	y_6	y_7

Estimate the area bounded by the curve, x-axis and lines x = 1, x = 4. Also find the volume of solid generated by revolving this area using Simpson's 3/8 rule.

(IETE, June 2009)

Solution. By Simpson's three-eighth rule

$$\text{Area} = \frac{3h}{8}\left[(y_1+y_7)+3(y_2+y_3+y_5+y_6)+2(y_4)\right]$$

$$= \frac{3}{16}\left[(2+2.1)+3(2.4+2.7+3+2.6)+2(2.8)\right]$$

$$= \frac{3}{16}[4.1+32.1+5.6] = 7.838 \text{ square units.}$$

Required volume of the solid generated

$$= \int_1^4 \pi y^2 dx = \pi \cdot \frac{3h}{8}\left[(y_1^2+y_7^2)+3(y_2^2+y_3^2+y_5^2+y_6^2)+2(y_4^2)\right]$$

$$= \frac{3\pi}{16}\left[(2^2+2.1^2)+3(2.4^2+2.7^2+3^2+2.6^2)+2(2.8)^2\right]$$

$$= \frac{3\pi}{16}\left[(4+4.41)+3(5.76+7.29+9+6.76)+2(7.84)\right]$$

$$= \frac{3\pi}{16}(8.41+86.43+15.68) = \frac{3\pi}{16}\times110.52 = 65.13 \text{ cubic units.}$$

Exercise 23.2

1. Evaluate the following by the trapezoidal rule of integration.

(i) $\displaystyle\int_2^6 \frac{dx}{x}$ with $n = 4$, (ii) $\displaystyle\int_1^3 \log x\, dx$ with $n = 5$,

(iii) $\int_0^1 \sqrt{1+x^3}\, dx$ with $n = 5$,   (iv) $\int_1^5 \sqrt{35+x}\, dx$ with $n = 4$,

(v) $\int_0^6 \dfrac{dx}{1+x^2}$ with $n = 6$,   (vi) $\int_1^5 \log_{10} x\, dx$ with $n = 8$.

    (*AMIE.*, *W-2010*)

                                  (*AMIE.*, *W-2007*)

Ans. (*i*) 1.117, (*ii*) 1.2870, (*iii*) 1.115, (*iv*) 24.654, (*v*) 1.4108.

2. (*a*) Evaluate $\int_0^1 \sqrt{1-x^3}\, dx$ by trapezoidal rule, taking six equal intervals

correct to three decimal places.   (*AMIE, W-2004*)

(*b*) Consider the function $f(x)$ given by

x	0.1	0.2	0.3	0.4	0.5	0.6
$f(x)$	0.425	0.475	0.400	0.450	0.575	0.675

Use composite trapezoidal rule to evaluate the integral $\int_{0.1}^{0.6} \left[f(x) \right] dx.$.

                              (*AMIE, S-2008*) **Ans.** 0.245

3. Use Simpson's 1/3 rule to approximate the integral $\int_0^2 x^3 dx$ with $n = 4$.

                                                  Ans. 4

4. Evaluate by Simpson's 1/3rd rule, $\int_0^1 \dfrac{dx}{1+x}$ taking four number of sub-

intervals and compare the result with the exact value.

         (*AMIE., W-2008; W-2006*) **Ans.** 0.69323; exact value = 0.693147.

5. Given that $e^0 = 1$, $e^1 = 2.72$, $e^2 = 7.39$, $e^3 = 20.09$, $e^4 = 54.60$. Find an

approximate value of $\int_0^4 e^x\, dx$ by (*i*) the trapezoidal rule (*ii*) Simpson's 1/3

rule. Compare it with the actual value.

           Ans. (*i*) 58.00 (*ii*) 53.8733; Actual value = 53.5982

6. Using the following data and the Simpson's 1/3 rule determine the approximate value of $\int_0^6 y\, dx$.

x	0	1	2	3	4	5	6
y	0.146	0.161	0.176	0.190	0.204	0.217	0.230

Ans. 1.13

7. Using Simpson's 1/3 rule, calculate the value of the integral $\int_1^9 y\, dx$, for two
   four and eight intervals.

x	1	2	3	4	5	6	7	8	9
y	2.061	2.312	2.819	3.106	3.670	4.721	6.103	7.95	9.942

Ans. 35.577; 36.687; 36.51

8. (a) Approximate $\int_{-4}^{4} e^{-x^2}\, dx$ by Simpson's 1/3 rule, using intervals with
   $h = 0.5$. *(AMIE, W-2001*

   (b) Calculate by Simpson's 1/3 rule an approximate value of $\int_{-3}^{3} x^4\, dx$ by
   taking seven equidistant ordinates. Compare it with exact value.
   Ans. (a) 1.77241, (b) 98; 97.2

9. Using Simpson's 1/3 rule with four sub-intervals, calculate the value of the
   integral $\int_{1.2}^{1.6}(x+x^{-1})\, dx$. *(AMIE, W-2002)* **Ans. 0.8478**

10. (a) Using Simpson's one-third rule, estimate the area bounded by the curv
    $y = f(x)$, x-axis and ordinate $x = 0$ and $x = 5$ from the following data.

x_i	0	0.5	1	1.5	2	2.5	3	3.5	4	4.5	5
$f(x_i)$	1	2.1	3.2	4	5.1	6.2	7.2	8.3	9	10.2	12

(AMIE, W-2005) **Ans. 30.86**

(b) The following table gives the values of a function at equal intervals.

x	0.0	0.5	1.0	1.5	2.0
$f(x)$	0.3989	0.3521	0.2420	0.1295	0.0540

Evaluate $f(1.8), f'(1.5)$ and $\int\limits_{0}^{2} f(x)\,dx$ stating the formula used.

Ans. 0.0782, –0.1868, 0.4772.

11. Evaluate $\int\limits_{0}^{1} \dfrac{dx}{1+x^2}$ approximately, using (a) the trapezoidal rule, (b)

Simpson's 1/3 rule with five ordinates. *(IETE, June 2009)*

Ans. 0.7828; 0.7854.

12. (a) Evaluate the value of $\log_e 2$ by finding $\int\limits_{0}^{1} \dfrac{2x}{1+x^2}\,dx$, using Simpson's

1/3 rule, by taking $n = 4$. **Ans.** 0.6933

(b) Find an approximate value of $\log_e 5$ by evaluating the integral $\int\limits_{0}^{5} \dfrac{dx}{4x+5}$,

by Simpson's one-third rule, dividing the range into ten equal parts.
(AMIE, S-2007; Madras 2000) **Ans.** 1.61.

13. Find an approximate value of $\int\limits_{0}^{\pi/2} \sqrt{\cos\theta}\,d\theta$, using Simpson's 1/3 rule with

six intervals. **Ans.** 1.1872

14. Using Simpson's 1/3 rule, evaluate the following:

(i) $\int\limits_{0}^{4} \sqrt{x^3 + 3x}\,dx$, with 4 intervals,   (ii) $\int\limits_{0}^{\pi/2} \sqrt{2 + \sin x}\,dx$, with 7 ordinates,

(iii) $\int\limits_{0}^{\pi} \dfrac{\sin^2\theta}{5+4\cos\theta}\,d\theta$, with 5 ordinates,   (iv) $\int\limits_{200}^{1000} \dfrac{dx}{\log_{10} x}$, taking $n = 8$,

(v) $\int\limits_{4}^{5.2} \log_e x\,dx$. **Ans.** (i) 12.07, (ii) 2.545, (iii) 0.388, (iv) 293.4, (v) 1.82785.

15. A river is 80 unit wide. The depth d at a distance x unit from one bank is given by the following table

x	0	10	20	30	40	50	60	70	80
d	0	4	9	7	12	15	14	8	3

Find approximately the area of the cross-section of the river by Simpson's 1/3 rule.   **Ans.** 696.67 sq. units.

16. The ordinates of a curve $y = f(x)$ for given values of x are tabulated below

x	1	2	3	4	5	6	7
y	2.105	2.808	3.614	4.604	5.857	7.451	9.467

Using Simpson's 1/3 rule evaluate the area bounded by the curve, the ordinates at $x = 1$ and $x = 7$ and the x-axis.   **Ans.** 29.99 sq. units

17. The following table gives the velocity v of a particle at time t

t(seconds):	0	2	4	6	8	10	12
v(m/sec):	4	6	16	34	60	94	136

Find the distance moved by the particle in 12 seconds and also the acceleration at $t = 2$ sec.   (*Madras, 1997S*) **Ans.** 552 m; 3 m/sec^2

18. A solid of revolution is formed by rotating about the x-axis the area between the x-axis, the lines $x = 0$ and $x = 1$, and a curve through the points with the following coordinates

x	0.00	0.25	0.50	0.75	1.0
y	1.000	0.9896	0.9589	0.9089	0.8415

Using Simpson's 1/3 rule, estimate the volume of the solid formed, giving the answer to three decimal places.   (*A.M.I.E., S-2006*) **Ans.** 2.819

19. A curve is given by the following pairs of rectangular coordinates. Using Simpson's 1/3 rule, find the volume generated by the revolution of the area under the curve from $x = 1$ to $x = 9$ about the x-axis.

x	1	2	3	4	5	6	7	8	9
y	0	0.6	0.9	1.2	1.5	1.4	1.7	1.8	2

Ans. 14.83 π cubic units

20. A curve is given by the table

x	0	1	2	3	4	5	6
y	0	2	2.5	2.3	2	1.7	1.5

Find by Simpson's 1/3 rule, volume of solid of revolution about x-axis.

Ans. 74.873 units^3.

21. Evaluate $\int_{0.2}^{1.4}(e^x+\sin x)dx$, using Simpson's one third rule of integration.

Taking $n = 12$. *(AMIE, W-1999)* **Ans. 3.64389**

22. Compute the value of the integral

$$I = \int_{0.2}^{1.4}\left(\sin x - \log_e x + e^x\right)dx$$

by (a) the trapezoidal rule (b) Simpson's one-third rule. Taking $n = 12$

Ans. 4.05617; 4.05106; True value = 4.05095

23. Using the data of the following table, compute the integrals

(a) $\int_{0.5}^{1.1}xy\,dx$, (b) $\int_{0.5}^{1.1}y^2dx$, (c) $\int_{0.5}^{1.1}x^2y\,dx$, (d) $\int_{0.5}^{1.1}y^3dx$.

by Simpson's 1/3 rule.

x	0.5	0.6	0.7	0.8	0.9	1.0	1.1
y	0.4804	0.5669	0.6490	0.7262	0.7985	0.8658	0.9281

Ans. (a) 0.3585, (b) 0.3201, (c) 0.3104, (d) 0.2444

24. The value of $\log_e 2$ can be calculated from the integral $\log_e 2 = \int_1^2\frac{1}{x}dx$.

How many subintervals (steps) should be used in the trapezoidal rule to approxi-mate the integral with an error of magnitude less than 10^{-4}?

Ans. $n = 41$.

Example 23.24. Given that

x	1.1	1.2	1.3	1.4	1.5	1.6
y	8.403	8.781	9.129	9.451	9.750	10.031

Find dy/dx at $x = 1.1$. *(A.M.I.E., W-2010)*

Solution. The difference table is

x	y	Δy	$\Delta^2 y$	$\Delta^3 y$	$\Delta^4 y$	$\Delta^5 y$
1.1	8.403					
1.2	8.781	0.378				
1.3	9.129	0.348	-0.030	0.004		
1.4	9.451	0.322	-0.026	0.003	-0.001	0.003
1.5	9.750	0.299	-0.023	0.005	0.002	
1.6	10.031	0.281	-0.018			

We have

$$\left(\frac{dy}{dx}\right)_{x_0} \approx \frac{1}{h}\left[\Delta y_0 - \frac{1}{2}\Delta^2 y_0 + \frac{1}{3}\Delta^3 y_0 - \frac{1}{4}\Delta^4 y_0 + \frac{1}{5}\Delta^5 y_0 - ...\right] \qquad ...(i)$$

Here $h = 0.1$, $x_0 = 1.1$, $\Delta y_0 = 0.378$, $\Delta^2 y_0 = -0.03$ etc.
Substituting these values in (i), we obtain

$$\left(\frac{dy}{dx}\right)_{1.1} \approx \frac{1}{0.1}\left[0.378 - \frac{1}{2}(-0.03) + \frac{1}{3}(0.004) - \frac{1}{4}(-0.001) + \frac{1}{5}(0.003)\right]$$

$$= 10(0.378 + 0.015 + 0.0013 + 0.00025 + 0.0006) = 3.9515.$$

Chapter 24

Statistics and Probability

24.1 *Def. Statistics is the science which deals with the principles and methods used in the collection, presentation, analysis and interpretation of numerical data.* The raw material of statistics is thus the numerical data collected in the beginning of a statistical enquiry. These numerical values are the measurements or observations of a variable quantity or a variable which is affected by a number of causes. For example the data may be the measurements of the dimensions of a mass produced article in certain industry, collected for the purpose of quality control, or it may be the marks of a number of students collected to determine their performance.

When the data consists of a few figures, it can be easily presented and understood. But when the number of figures is very large, a proper classification is essential for analysis and deriving valid inferences.

24.2 Frequency Distribution

One such method of classification is in the form of frequency distribution. The frequency of a value is the number of times a value is repeated. When the number of observations for a variable is small and we have repetitions of the values, we can arrange them in the form of a table according to their magnitudes with corresponding frequencies. Such a table is called *frequency table*.

For example if the variable values are
18, 12, 9, 11, 12, 11, 6, 18, 19, 11, 12, 9, 3, 11, 8, 9, 11, 9, 8, 12, the frequency table is obtained as follows:

Variable value (x_i)	frequency (f_i)
3	1
6	1
8	2
9	4
11	5
12	4
18	2
19	1

When the number of observations is large and the difference between the greatest and smallest values is large, we have classification according to class intervals. For example, let the marks of 50 students of a class be

46, 58, 54, 52, 55, 59, 52, 62, 65, 67, 64, 63, 77, 78, 92, 6, 7, 12, 18, 16, 3, 23, 25, 25, 27, 81, 88, 24, 29, 22, 34, 33, 30, 37, 36, 42, 48, 28, 22, 28, 17, 13, 70, 37, 32, 36, 41, 40, 43, 44.

We can arrange then as follows.

Marks	No. of students getting these marks (Tally Marks)	Frequency
0-10	III	3
10-20	NN	5
20-30	NN NN	10
30-40	NN III	8
40-50	NN II	7
50-60	NN I	6
60-70	NN	5
70-80	III	3
80-90	II	2
90-100	I	1
		Total = 50

Notes:

(1) Border line cases are taken in the higher class. For example a student getting 30 marks belongs to the class 30-40 and not 20-30.

(2) To find the frequency of a class, we consider each observation and put it in the appropriate class by drawing a vertical line. After every four lines, we cross them for the 5th entry.

From the above example it is clear that 50 observations have been divided into 10 groups. These groups are called **classes**. The boundary values 10, 20 etc. are called **class limits**. The higher value of the two limits of a class is called **upper limit** of the class and lower value is known as **lower limit**. The difference between upper limit and lower limit of a class is called its **magnitude** or **class interval**. The number of observations falling within a class is called its **frequency**. The **mid-value** of a class is the value half way between lower and upper limit of that class. For the class 20-30, 25 is the mid value. Mid value is also known as **class mark** or **central value**.

24.3 Measures of Central Tendency

We have seen above how a complex data can be studied after the summarisation process of classification in the form of frequency distribution. For example from the frequency Table of Art. 24.2, we find that 11 students get marks above 60 and

24 students get marks above 40. But these give only a rough idea about its characteristics.

In order to have more accurate idea of the data under study and to compare two series, it is desirable to represent them by numbers. These numbers should neither have the lowest value in the series nor the highest but a value somewhere between the two limits where most of the values are situated. Such representative numbers are called *measures of central tendency.* Some important measures of central tendency are:

(*a*) Arithmetic average or mean (AM)

(*b*) Median (Med.)

(*c*) Mode (Mo)

(*d*) Geomatric mean (GM)

(*e*) Harmonic mean (HM)

We shall briefly discuss them here one by one.

24.4 Arithmetic Mean (AM)

We are all familiar with this measure. It is so popular that whenever we talk of average we mean arithmetic average or arithmetic mean. For example if the daily wages of 5 labourers are 8, 9, 10, 11, 12 rupees respectively then the average wages = (8 + 9 + 10 + 11 + 12)/5 = 10 rupees per head.

Arithmetic mean is a quotient obtain by dividing the sum of observations $\left(\sum x_i \right)$ by the number of observations (n) that is, $\bar{x}$ (read as x bar) $= \sum x_i / n$.

The symbol x_i denotes the observation corresponding to the i-th individual in the sample. If there are n individuals, we write $i = 1, 2, ... n$.

24.4.1 Definition of Arithmetic Mean

(*i*) *For ungrouped data*

If the variate x take the values of $x_1, x_2, ..., x_n$, then the arithmetic mean denoted

by $\bar{x}$ is defined by $\bar{x} = \dfrac{x_1 + x_2 + \cdots + x_n}{n} = \dfrac{1}{n} \sum_{i=1}^{n} x_i.$ ...(24.1)

(*ii*) *For grouped data*

When the values of the variable are given in the form of a frequency table then we multiply each value by the corresponding frequency and find their sum. This sum divided by the total of frequencies is the arithmetic mean (AM or $\bar{x}$)

i.e., $\qquad \bar{x} = \dfrac{f_1 x_1 + f_2 x_2 + ... + f_n x_n}{f_1 + f_2 + ... + f_n}$ ...(24.2)

We can thus write (24.2) in the compact form $\bar{x} = \sum f_i x_i / \sum f_i$ or $\bar{x} = \sum f_i x_i / n$, where $n = \sum f_i$.

In the case of grouped frequency table *i.e.*, when the values are given in the form of class intervals, x is the mid-value of a class and f the corresponding frequency.

Example 24.1 Find the average earning per labourer from the following data:

Daily earning (Rs.)	No. of labourers
8	9
10	12
11	7
12	8
15	1

Solution: Denoting the daily wages by x_i and the corresponding number of labourers by f_i, we construct the following table:

x_i	f_i	$f_i x_i$
8	9	72
10	12	120
11	7	77
12	8	96
15	1	15
Total	$n = \sum f_i = 37$	$\sum f_i x_i = 380$

$\therefore$ Arithmetic mean $= \dfrac{\sum f_i x_i}{\sum f_i} = \dfrac{380}{37} = 10.27$ Rupees.

Example 24.2 Find the arithmetic mean for the following frequency distribution:

Class interval	Frequency
0-4	6
4-8	11
8-12	20
12-16	7
16-20	4

Solution. We make the table as follows:

Classes	x_i	f_i	$f_i x_i$
0-4	2	6	12
4-8	6	11	66
8-12	10	20	200
12-16	14	7	98
16-20	18	4	72
Total		$\Sigma f_i = 48$	$\Sigma f_i x_i = 448$

$$\bar{x} = \Sigma f_i x_i / \Sigma f_i = 448/48 \approx 9.33.$$

24.5 Short methods for Computing the mean

When the frequencies and the values of the variables in the series are given to be large quantities, their products necessarily come out to be large and consequently mathematical labour in the computation of the arithmetic average becomes enormous. To minimize this labour, the following short cut methods prove very handy.

(a) Assumed mean method.

$$\bar{x} = A + \frac{\Sigma f_i d_x}{n} \qquad \qquad ...(24.3)$$

A is called the assumed mean or provisional mean. It is usually a value of the variable x which lies somewhere in the middle of the highest and the lowest value in the given data, preferably the value having the maximum frequency. The symbol d_x is called the deviation of x from A i.e., $d_x = x_i - A$.

(b) Step deviation method

When the values of the variables are given at equal spacing or when the length of the class intervals are equal, a further reduction in labour is achieved by using the step deviation method. This is well expressed mathematically by the formula.

$$\bar{x} = A + \left(\frac{\Sigma f_i u_x}{n} \right) h, \text{ where } u_x = \frac{x_i - A}{h} = \frac{d_x}{h} \qquad ...(24.4)$$

is the step deviation and h, is the length of the class interval.

Example 24.3 Compute the arithmetic mean for the following distribution:

Class Interval	1-4	4-9	9-16	16-27
Frequency	6	12	26	20

Solution. We prepare the following table and apply formula (24.3). Taking assume mean, $A = 12.5$

Size of items	Mid values x_i	f_i	$d_x = \overset{.}{x_i} - A$	$f_i d_x$
1-4	2.5	6	−10	−60
4-9	6.5	12	−6	−72
9-16	12.5	26	0	0
16-27	21.5	20	9	180
Total		$n = \Sigma f_i = 64$		$\Sigma f_i d_x = 48$

The required AM is given by,

$$\overline{x} = A + \frac{\Sigma f_i d_x}{n} = 12.5 + \frac{48}{64} = 12.5 + 0.75 = 13.25.$$

Example 24.4 The marks (out of 100) obtained by different students in an examination are shown in the following frequency table. Compute the mean.

Marks	17.5 to 22.5	22.5 to 27.5	27.5 to 32.5	32.5 to 37.5	37.5 to 42.5	42.5 to 47.5	47.5 to 52.5	52.5 to 57.5	57.5 to 62.5	62.5 to 67.5	67.5 to 72.5
No. of students	2	8	33	80	170	243	213	145	67	35	4

Solution. We shall solve this problem by step deviation method taking $A = 45$ and $h = 5$ (the size of class intervals).

x_i (mid values)	f_i	$d_x = x_i - 45$	$u_x = d_x/5$	$f_i u_x$
20	2	−25	−5	−10
25	8	−20	−4	−32
30	33	−15	−3	−99
35	80	−10	−2	−160
40	170	−5	−1	−170
A→ 45	243	0	0	0
50	213	5	1	213
55	145	10	2	290
60	67	15	3	201
65	35	20	4	140
70	4	25	5	20
Total	$n = \Sigma f_i = 1000$			$\Sigma f_i u_x = 393$

$\therefore$ Mean $= A + \left(\dfrac{\Sigma f_i u_x}{n}\right) \times h$, where $n = \Sigma f_i$

$$= 45 + \frac{393}{1000} \times 5 = 45 + 1.965 \approx 46.965$$

Example 24.5 Find the mean from the following data

Marks	No. of students	Marks	No. of students
Below 10	5	Below 60	60
Below 20	9	Below 70	70
Below 30	17	Below 80	78
Below 40	29	Below 90	83
Below 50	45	Below100	85

Solution. Converting the given comulative frequency distribution into continuous frequency distribution and using step deviation method. Taking $A = 55$ and $h = 10$

Classes	Mid value x_i	f_i	$d_x = x_i - 55$	$u_x = d_x/10$	$f_i u_x$
0-10	5	5	−50	−5	−25
10-20	15	4	−40	−4	−16
20-30	25	8	−30	−3	−24
30-40	35	12	−20	−2	−24
40-50	45	16	−10	−1	−16
50-60	A→55	15	0	0	0
60-70	65	10	10	1	10
70-80	75	8	20	2	16
80-90	85	5	30	3	15
90-100	95	2	40	4	8
Total		$n = \Sigma f_i = 85$			$\Sigma f_i u_x = -56$

$$\therefore \quad \text{Mean} = A + \left(\frac{\Sigma f_i u_x}{n} \right) \times h = 55 + \left(\frac{-56}{85} \right) \times 10 = 48.41$$

24.6 Two important properties of arithmetic mean

1. *The algebraic sum of the deviations of a set of observations from their arithmetic mean is zero.*

Proof. Let $\bar{x}$ be the A.M. and n the number of variates. Then $d = x_i - \bar{x}$.

$$\therefore \quad \Sigma d = \Sigma(x_i - \bar{x}) = \Sigma x_i - \Sigma \bar{x} = n\bar{x} - n\bar{x} = 0. \qquad \left(\because \bar{x} = \frac{\Sigma x_i}{n} \text{ and } \Sigma x_i = n\bar{x} \right)$$

2. *If $\bar{x}$ is the mean of a set $x_1, x_2 ... x_n$ of n observations and $\bar{y}$ is the mean of*

another set $y_1, y_2, ..., y_m$ of m observations, then the arithmetic mean of combined set is given by

$$A.M = \frac{n\bar{x} + m\bar{y}}{n+m}.$$

Proof. $\bar{x} = \frac{\sum x_i}{n}, i = 1, 2,...n$ and $\bar{y} = \frac{\sum y_j}{m}, j = 1, 2,...m.$

$\therefore$ *AM of combined set* $= \frac{\sum x_i + \sum y_j}{m+n} = \frac{n\bar{x} + m\bar{y}}{m+n}.$

Example 24.6 There are 250 observations in a group. The average of first 100 observations is 5 and the average of remaining 150 is $8\frac{1}{3}$. Find the average of the whole group.

Solution. We have $\frac{x_1 + x_2 + ... + x_{100}}{100} = 5$. Therefore $\sum_{i=1}^{100} x_i = 500$ ...(i)

Also $\frac{x_{101} + x_{102} + ... + x_{250}}{150} = \frac{25}{3}$. Therefore $\sum_{j=101}^{250} x_j = 1250.$ ...(ii)

Adding (i) and (ii), we obtain $x_1 + x_2 + ...x_{100} + x_{101} + ... + x_{250} = 1750.$

Thus, A.M. of whole group $= \frac{1750}{250} = 7.$

24.7 Median (Med).

The median of a series of data is defined as that value which divides the whole series into two equal parts.

24.8 Computation of the Median

For various forms of series, the following method can be used for the calculation of the median.

(a) **Raw or ungrouped data.** For finding the median of raw or ungrouped data, we go through the following steps:

   (i) Arrange the data in ascending or descending order of magnitude.

   (ii) If the number of observations n in the data is odd, then $\left(\frac{(n+1)}{2}\right)$th observation is the *median*. If n is even, then the average of $\frac{n}{2}$th and $\left(\frac{n}{2}+1\right)$th observation is the *median*.

Example 24.7 Find the median of the following data:
(i) 35, 31, 12, 16, 29, 74, 45, 40, 49, 57, 62
(ii) 29, 34, 11, 17, 36, 21, 46, 23, 39, 41
Solution. (i) Arranging the data in ascending order, we obtain
12, 16, 29, 31, 35, 40, 45, 49, 57, 62, 74.

Here $n = 11$ (odd). Therefore the median is $\dfrac{(11+1)}{2}$ th i.e. 6th observation

which is 40 and so median = 40.

(ii) Arranging the observations in ascending order, we obtain 11, 17, 21, 23, 29, 34, 36, 39, 41, 46.

Here $n = 10$ (even). Therefore the median is the average of $\left(\dfrac{10}{2}\right)$ th and

$\left(\dfrac{10}{2}+1\right)$ th observations, i.e., 5th and 6th observations, which is $\dfrac{29+34}{2} = 31.5$.

Therefore median = 31.5.

(b) **Calculation of median for Grouped data.** There are two cases:
Case I. When the data is discrete. To determine the median, we follow the following steps.
(a) Arrange the data in ascending or descending order of magnitude along with their frequencies.
(b) Append a column of cumulative frequencies.

(c) (i) If the total frequency n is odd, then $\dfrac{(n+1)}{2}$ th observertion is the *median*.

(ii) If the total frequency n is even, then the mean of $\dfrac{n}{2}$ th and $\left(\dfrac{n}{2}+1\right)$ th obser-

vations is the *median*.

Example 24.8 Find the median for the following data:

Marks (out of 20)	5	9	10	12	13	16	18	20
No. of students	4	5	6	12	11	6	4	2

Solution. We form the table as given below:

Marks (x_i)	No. of Students (f_i)	Comulative Frequency (c_f)
5	4	4
9	5	9
10	6	15
12	12	27
13	11	38

(Contd.)

16	6	44
18	4	48
20	2	50
Total	$\sum f_i = 50$	

Here $n = 50$ (even. Therefore, the median is the mean of $\left(\dfrac{50}{2}\right)th$ and $\left(\dfrac{50}{2}+1\right)th$

observations, *i.e.*, 25th and 26th observations, which are 12 and 12. Thus median

$$= \frac{12+12}{2} = 12.$$

Case II. When the data is continuous and in the form of a frequency distribution. The median, is given by the following formula:

$$Median = l_1 + \left(\frac{n/2 - f_c}{f}\right) \times h \qquad \qquad ...(24.5)$$

where l_1 is the lower limit of the class interval to which group median belongs, f is the frequency of the class-interval to which group median belongs h is the length of class-interval to which group median belongs, n is the total of frequency and f_c is the cumulative frequency of the class preceding the median class.

Remark: For a grouped frequency distribution, first the class limits are made continuous (if not already continuous) and then the above formula is used.

Example 24.9 Find the median for the following data:

Class intervals:	110-119	120-129	130-139	140-149	150-159	160-169	170-179
Frequency:	5	25	40	60	40	25	5

Solution. To compute the median, we have to convert the given distribution to a continuous one. We prepare a table appending a column of cumulative frequency.

Classes	Frequency	Cumulative Frequency
109.5-119.5	5	5
119.5-129.5	25	30
129.5-139.5	40	f 70
139.5-149.5	f: 60	130 ← Median Class
149.5-159.5	40	170
159.5-169.5	25	195
169.5-179.5	5	200
	$n = \sum f = 200$	

We have $\dfrac{n}{2} = 100$. Therefore, median class = 139.5 – 149.5.

$l_1 = 139.5, f_c = 70, f = 60, h = 10$.

$\therefore$ Median $= 139.5 + \left(\dfrac{100 - 70}{60}\right) \times 10 = 144.5$

Example. 24.10 An incomplete frequency function values distribution is given as follows:

Variable	10-20	20-30	30-40	40-50	50-60	60-70	70-80	Total
Frequency	12	30	?	65	?	25	18	229

Given that the median value is 46, determine the missing frequencies using the median formula. *(AMIE, S-2006)*

Solution. Let the frequency of the class 30-40 be f_1 and that of 50-60 be f_2. It is given that $n = 229$.

Making the table for the given data, we have

Class intervals	Frequency	Cumulative Frequency
10-20	12	12
20-30	30	42
30-40	f_1	$42 + f_1$
l_1 40-50 l_2	65	$107 + f_1$
50-60	f_2	$107 + f_1 + f_2$
60-70	25	$132 + f_1 + f_2$
70-80	18	$150 + f_1 + f_2$

$\therefore$ $150 + f_1 + f_2 = 229$, which gives, $f_1 + f_2 = 79$. The median, 46 lies in the class 40-50.

$\therefore$ $l_1 = 40, h = 10, f = 65, f_c = 42 + f_1$ and $n = 229$.

Employing the formula, median $= l_1 + \left(\dfrac{n/2 - f_c}{f}\right) \times h$, we obtain

$46 = 40 + \left(\dfrac{(229/2) - (42 + f_1)}{65}\right) \times 10$ or $46 = 40 + \dfrac{145 - 2f_1}{13}$ which gives $f_1 = 34$.

Therefore, $f_2 = 79 - f_1 = 79 - 34 = 45$.

24.9 Mode

The mode or the modal value is *defined as that value of a variate that occurs most often.* It is also characterised as the point having maximum frequency or greatest density. The mode is of the greatest practical utility to large scale manufacturer of consumer goods.

24.10 Computation of Mode

The following methods prove helpful in the computation of the mode.

(a) *If the measure of each item in a series is known, then the mode is the size of the item which occurs most frequently.*

(b) *For a frequency distribution with arbitrary class intervals, the apparent mode is the class interval which has the maximum frequency.*

(c) *For a frequency distribution with equal class intervals, the mode is computed from the following formula:*

$$\text{Mode} = l_1 + \left(\frac{f_1 - f_0}{2f_1 - f_0 - f_2} \right) h, \qquad \qquad ...(24.6)$$

where l_1 is the lower class-limit of the modal class interval, h is the length of class-interval (or class size), f_1 is the frequecy of the modal class, f_0 and f_2 are respectively, the frequencies of a class just preceding and following the modal class.

Illustration 1. The set 7, 4, 10, 9, 15, 12, 7, 9, 7 has mode 7,

Illustration 2. The set 85, 76, 93, 82 and 96 has no mode.

Illustration 3. The set 2, 3, 4, 4, 4, 5, 5 7, 7, 7, 9 has two modes, 4 and 7, and is called *bimodal*.

Remark. A distribution having only one mode is called *unimodal*.

Example 24.11 Calculate the mean and mode for the following frequency distribution:

Class interval:	0-8	8-16	16-24	24-32	32-40	40-48
Frequency:	8	7	16	24	15	7

Solution. The computation of the mean is performed in the following table:

Class-interval	mid values x_i	$u_x = (x_i - 28)/8$	f_i	$f_i u_x$
0-8	4	−3	8	−24
8-16	12	−2	7	−14
16-24	20	−1	16	−16
l_1 24-32 l_2	A→28	0	24	0
32-40	36	1	15	15
40-48	44	2	7	14
Total			$n = \Sigma f = 77$	$\Sigma f_i u_x = -25$

(i) $\text{Mean} = 28 + \left(\dfrac{-25}{77} \right) \times 8 = 28 - 2.6 = 25.4$

(*ii*) Since the maximum frequency is 24, the mode lies in class 24-32. Employing

the 24.6 formula Mode $= l_1 + \left(\dfrac{f_1 - f_0}{2f_1 - f_0 - f_2}\right) h$. . We have

$l_1 = 24, f_1 = 24, f_0 = 16$ and $f_2 = 15$ and, $h = 8$.

$\therefore$ Mode $= 24 + \left(\dfrac{24 - 16}{2 \times 24 - 16 - 15}\right) \times 8 = 24 + \dfrac{8}{17} \times 8 = 27.76$

Example 24.12 The following table gives the length of life of 150 lamps.

Life (100 hours)	0-4	4-8	8-12	12-16	16-20	20-24	24-28	28-32
No. of lamps:	4	12	40	41	27	13	9	4

Calculate the median and modal life of the lamp.

Solution. We prepare a table as given below:

Class-interval	mid-values x_i	f_i	Cumulative frequency f_c
0-4	2	4	4
4-8	6	12	16
8-12	10	40	56
l_1 12-16 l_2	14	41	97 ← Median Class
16-20	18	27	124
20-24	22	13	137
24-28	26	9	146
28-32	30	4	150
Total		$n = \Sigma f = 150$	

$$\text{Median} = l_1 + \left(\dfrac{n/2 - f_c}{f}\right) \times h \qquad \text{...(1)}$$

Here $n/2 = 75$, 12-16 is the median class. Thus $l_1 = 12$, $f_c = 56$, $f = 41$ and $h = 4$. Using the formula (1), and substituting values, we obtain

$\therefore$ Median $= 12 + \left(\dfrac{75 - 56}{41}\right) \times 4 = 13.8537$

Mode $= l_1 + \left(\dfrac{f_1 - f_0}{2f_1 - f_0 - f_2}\right) \times h = 12 + \left(\dfrac{41 - 40}{2 \times 41 - 40 - 27}\right) \times 4 = 12 + \dfrac{4}{15} = 12.2667$

Since units in class intervals is 100 hours.

$\therefore$ Median life of the lamp $= 1385.37$ hours, and modal life of the lamp $= 1226.67$ hours.

24.11 Relationship between mean, median and mode

When it is difficult to compute mode from a grouped frequency distribution, we may consider the following empirical relationship between mean, median and mode.

$$\text{Mode} = 3 \text{ Median} - 2\text{Mean} \qquad ...(24.7)$$

In Figs. 24.1 and 24.2 below are shown the relative positions of the mean, median and mode for frequency curves which are skewed to the right and left respectively. For symmetrical curves the mean, mode and median all coincides.

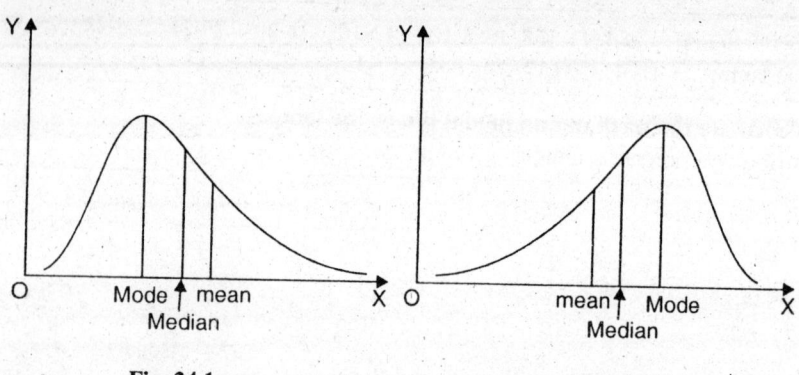

Fig. 24.1   Fig. 24.2

24.12 The Geometric mean G and Harmonic mean H

The geometric mean G of a set of n positive observations $x_1, x_2, x_3, ..., x_n$ is defined as the nth root of the product of the observations.

$$G = \sqrt[n]{x_1 x_2 x_3 \cdots\cdots x_n} \qquad ...(24.8)$$

when the number of items is three or more the task of multiplying the numbers and of extracting the root becomes excessively difficult. To simplify calculations logarithms are used. Geometric mean, G is calculated as follows:

$$\log G = \frac{1}{n}\left[\log x_1 + \log x_2 + \log x_3 + \cdots + \log x_n\right]$$

$$= \frac{1}{n}\sum_{i=1}^{n}\log x_i. \qquad \therefore G = \text{Antilog}\left\{\frac{\sum(\log x_i)}{n}\right\}$$

Illustration 4. Find the geometric mean of the numbers 3, 5, 6, 6, 7, 10, 12.

Solution. Geometric mean, $G = \sqrt[7]{(3)(5)(6)(6)(7)(10)(12)}$.

Taking logarithms on both sides, we obtain

$$\log G = \frac{1}{7}\left(\log 3 + \log 5 + 2\log 6 + \log 7 + \log 10 + \log 12\right)$$

$$= \frac{1}{7}(0.4771 + 0.6990 + 2 \times 0.7782 + 0.8451 + 1.0000 + 1.0792) = 0.8081$$

$\therefore G = $ Antilog $(0.8081) = 6.43$.

24.12.1 Calculation of Geometric mean—Discrete Series

$$G = \text{Antilog}\left[\frac{f_i(\log x_i)}{n}\right] \qquad \qquad ...(24.9)$$

Find the logarithms of the variable x_i, multiply these logarithms with the respective frequencies and obtain the total *i.e.* $\Sigma f_i(\log x_i)$. Divide $\Sigma f_i(\log x_i)$ by the total frequency and take the antilog of the value so obtained.

24.12.2 Calculation of Geometric Mean—Continuous Series

$$G = \text{Antilog}\left[\frac{\Sigma(f \log X)}{N}\right], \qquad \qquad (24.10)$$

where X is the mid value of each class.
Geometric mean in always smaller than arithmetic mean.

24.12.3 The Harmonic Mean *H*

For a given set of non-zero observations, harmonic mean is defined as the reciprocal of the AM of the reciprocals of the observation. So, if a variable x assumes n non-zero values $x_1, x_2, x_3, \ldots, x_n$ then the HM of x is given by

$$H = \frac{n}{\Sigma 1/x_i}. \qquad \qquad ...(24.11)$$

Harmonic mean is always less than the geometric mean.

Illustration 5. Find the harmonic mean H of the numbers 6, 14, 21, 30.

Solution. Harmonic mean or $H = $ Reciprocal $\left(\dfrac{\text{Reci } x_1 + \text{Reci } x_2 + \ldots + \text{Reci } x_n}{n}\right)$

or

$$H = \left(\frac{n}{\text{Reci } x_1 + \text{Reci } x_2 + \ldots + \text{Reci } x_n}\right)$$

$$= \frac{4}{\dfrac{1}{6} + \dfrac{1}{14} + \dfrac{1}{21} + \dfrac{1}{30}} = \frac{4 \times 210}{67} = 12.54.$$

24.12.4 Calculation of Harmonic Mean—Discrete Series

In discrete series, harmonic mean H is computed by applying the following formula:

$$H = \frac{n}{\displaystyle\sum_{i=1}^{n}(f_i/x_i)} \qquad \text{...(24.12)}$$

24.12.5 Calculation of Harmonic mean—Continuous Series

For calculating harmonic mean in continuous series the procedure is the same as applied to discrete series. The only difference is that here we take the reciprocal of the mid-points. The formula is as under:

$$H = \frac{n}{\sum(f_i/X)} \quad \text{or} \quad H = \frac{\sum f_i}{\sum(f_i/X)}, \qquad \text{...(24.13)}$$

where X is the mid value of each class.

Example 24.13 The marks obtained by 25 students in Civil Engineering are as given below. Find their Geometric and Harmonic means.

Marks (x)	11	12	13	14	15
No. of students (f)	3	7	8	5	2

Solution. We prepare the following table for calculation of G.M. and H.M.

x_i	f_i	$1/x_i$	f_i/x_i	$\log x_i$	$f_i \log x_i$
11	3	0.0909	0.2727	1.0414	3.1242
12	7	0.0833	0.5831	1.0792	7.5544
13	8	0.0769	0.6152	1.1139	8.9112
14	5	0.0714	0.3570	1.1461	5.7305
15	2	0.0666	0.1332	1.1761	2.3522
Total	$n = \Sigma f = 25$		$\Sigma f_i/x_i = 1.9612$		$\Sigma f_i \log x_i = 27.6725$

$$GM = \text{Anti}\log\left[\frac{\Sigma(f_i \log x_i)}{n}\right] = \text{Antilog}\left(\frac{27.6725}{25}\right)$$
$$= \text{Antilog}(1.1069) \approx 12.8.$$

$$HM = \frac{n}{\Sigma(f_i/x_i)} = \frac{25}{1.9612} = 12.747.$$

24.12.6 Relationship between Averages

For any set of positive observations, we have the following inequality:

$$AM \geq GM \geq HM \qquad \text{...(24.14)}$$

The equality sign occurs, when all the observations are equal.

24.13 Measures of Dispersion

In the preceeding sections, we have discussed different measures of central tendency and with their aid, it was possible to represent a given distribution by a single number *e.g.* mean, mode, median etc. Let us now consider two distributions representing the marks scored by two groups of 6 students each, viz.

			Students				
	1	2	3	4	5	6	
Marks →							Total Marks
Groups							
↓ A	20	22	23	27	28	30	150
B	15	18	21	29	32	35	150

It is obvious from the given data that mean of each series is 25 and there is no difference between the two series so for as their means are concerned. However, if we examine the items of the two distributions, we find that the marks in one group vary between 20 and 30 while in the second the variation is between 15 and 35. This certainly indicates a difference between the two distributions. The difference arises due to the fact that the divergence from the mean is different in the two groups. In other words the scatter or dispersion from the mean is less in the first group than in the second. This scatter or lack of uniformity in the sizes of the items of a group is known as *dispersion*.

The following are the measures of dispersion which are most commonly used.

(*a*) Range (*b*) Quartile deviation (*c*) Mean deviation (*d*) Standard deviation.
We shall briefly discuss them one by one.

24.14 Range

It is the simplest measure of dispersion. It is given by the difference between the highest and the lowest values of the variable appearing in the distribution. It is a very crude measure of dispersion as it depends only on two extreme items. It is particularly used in quoting interest rates and security prices. For a continuous distribution, *range* is defined as the difference between the lower limit of the smallest class and the upper limit of the largest class.

Symbolically, Range = L–S, (where L is the longest value and S is the smallest value)

$$\text{Relative range} = \frac{\text{Absolute range}}{\text{Sum of the two extremes}}.$$

24.15 Quartile Deviation or Semi-interquartile range

Quartiles are the magnitudes of the items in a series which divide it into four equal parts. The difference between the upper and the lower quartiles *i.e.*, $Q_3 - Q_1$ is called the quartile range and half of this range is defined as quartile deviation.

Mathematically the quartile deviation or $Q \cdot D$ is given by

$$Q \cdot D = \frac{1}{2}(Q_3 - Q_1), \qquad \ldots(24.15)$$

where Q_3 and Q_1 are given by

$$Q_3 = l_1 + \left(\frac{\frac{3}{4}n - f_c}{f}\right) \times h, \qquad \ldots(24.16)$$

where Q_3 is the upper quartile, l_1 is the lower limit of the quartile class, f is the frequency of the quartile class, h is its class intervals, f_c is the cumulative frequency before the quartic class, and n is the total of all the frequencies. Q_1, the lower quartile, is calculated by a similar formula except that in place of $3n/4$ we write $n/4$.

24.16 Mean Deviation

It is defined as the arithmetic mean of all the absolute values of the deviations taken from the mean, median or any other value of a set of observations.

(A) Mean deviation from the mean

(*i*) **For Raw data**, the set of n observation.

$$M.D.(\bar{x}) = \frac{\sum_{i=1}^{n} |x_i - \bar{x}|}{n}, \text{ where } \bar{x} = \sum x_i / n \qquad \ldots(24.17)$$

(*ii*) **For Grouped Data**

$$M.D.(\bar{x}) = \frac{\sum_{i=1}^{k} f_i |x_i - \bar{x}|}{\sum_{i=1}^{k} f_i}, \text{ where } \bar{x} = \frac{\sum f_i x_i}{\sum f_i} \qquad \ldots(24.18)$$

(B) Mean derivation from the median

(*i*) **For Raw data**

$$M.D.(Med.) = \frac{\sum |x_i - Median|}{n}, \qquad \ldots(24.19)$$

(*ii*) **For Grouped Data**

$$M.D.(med.) = \frac{\sum f_i |x_i - Median|}{\sum f_i} \qquad \ldots(24.20)$$

Similarly, mean deviation about mode $= \frac{1}{n} \sum |x_i - Mode| \qquad \ldots(24.21)$

Coefficient of mean deviation about its mean $= \dfrac{M.D.(\bar{x})}{\bar{x}} \times 100$ ...(24.22)

Similarly, coefficient of mean deviation based on median and mode can be found out.

Mean deviation is a better measure than the Range and the Quartile deviation.

Example 24.14 Calculate the mean deviation from the mean and also from the median of the following data:

Height in cm of 7 soldiers: 168, 164, 172, 169, 178, 173, 173.

Solution. (*a*) Mean,

$$\bar{x} = \frac{\sum x_i}{n} = \frac{164+168+169+172+173+173+178}{7} = \frac{1197}{7} = 171 \text{ cm.}$$

The respective $x_i - \bar{x}$ are $-7, -3, -2, 1, 2, 2, 7$.

Therefore, $\sum_{i=1}^{7} |x_i - \bar{x}| = 7+3+2+1+2+2+7 = 24$ and

$$M.D.(\bar{x}) = 24/7 = 3.43 \text{ cm.}$$

(*b*) Here the number of observations is odd ($n = 7$). Arranging the data in ascending order, we have 164, 168, 169, 172, 173, 173, 178. Now Median is $\dfrac{(7+1)}{2}$ th

or 4 th observation. Therefore median = 172 cm.

Hence M.D. (Med) $= \dfrac{\sum |x_i - Median|}{n}$

$$= \frac{\begin{array}{c}|164-172|+|168-172|+|169-172|+|172-172| \\ +|173-172|+|173-172|+|178-172|\end{array}}{7}$$

$$= \frac{8+4+3+0+1+1+6}{7} = \frac{23}{7} = 3.29 \text{ cm.}$$

Example 24.15 Calculate the mean deviation from the mean for the following data:

Score:	x_i	9	11	3	5	13	7
Frequency	f_i	9	5	2	7	1	10

Solution. We construct the following table taking the scores in ascending order.

x_i	f_i	$f_i x_i$	$\lvert x_i - \bar{x} \rvert$	$f_i \lvert x_i - \bar{x} \rvert$
3	2	6	4.65	9.30
5	7	35	2.65	18.55
7	10	70	0.65	6.50
9	9	81	1.35	12.15
11	5	55	3.35	16.75
13	1	13	5.35	5.35
Total	$\Sigma f_i = 34$	$\Sigma f_i x_i = 260$		$\Sigma f_i \lvert x_i - \bar{x} \rvert = 68.60$

Therefore, mean, $\bar{x} = \dfrac{\Sigma f_i x_i}{\Sigma f_i} = \dfrac{260}{34} = 7.65$ (approx.)

and $M \cdot D. (\bar{x}) = \dfrac{\Sigma f_i \lvert x - \bar{x} \rvert}{\Sigma f_i} = \dfrac{68.60}{34} = 2.02$ (approx.)

Example 24.16 Find the mean deviation from the median for the following data:

x_i	3	6	9	12	13	15	21	22
f_i	3	4	5	2	4	5	4	3
Cf_i	3	7	12	14	18	23	27	30

Solution. Here $n = \Sigma f_i = 30$.

Median is the mean of 15th and 16th observations. Since 15th and 16th observations are both 13, the median is $\dfrac{13 + 13}{2} = 13$.

$\therefore \lvert x_i - \text{Median} \rvert$ has respective values 10, 7, 4, 1, 0, 2, 8, 9.

Therefore $\sum f_i \lvert x_i - \text{Median} \rvert = 30 + 28 + 20 + 2 + 0 + 10 + 32 + 27 = 149$.

and $M.D.(\text{Median}) = \dfrac{\sum f_i \lvert x_i - \text{Median} \rvert}{\sum f_i} = \dfrac{149}{30} = 4.97$.

Example 24.17 Find the mean deviation from the mean for the following data:

Classes: 95-105 105-115 115-125 125-135 135-145 145-155

Frequencies: 9   13   26   30   12   10

Solution. We make the table as follows. Let the assumed mean of the data be 120.

Classes	Mid. value x_i	$u_x = \dfrac{x_i - 120}{10}$	f_i	$f_i u_x$	$|x_i - \bar{x}|$	$f_i|x_i - \bar{x}|$
95-105	100	-2	9	-18	25.30	227.70
105-115	110	-1	13	-13	15.30	198.90
115-125	A →120	0	26	0	5.30	137.80
125-135	130	1	30	30	4.70	141.00
135-145	140	2	12	24	14.70	176.40
145-155	150	3	10	30	24.70	247.00
Total			$\Sigma f_i = 100$	$\Sigma f_i u_x = 53$		$\Sigma f_i|x_i - \bar{x}|$ $= 1128.80$

Therefore, $\bar{x} = 120 + \dfrac{\Sigma f_i u_x}{\Sigma f_i} \times h = 120 + \dfrac{53}{100} \times 10 = 125.30.$

$M.D.(\bar{x}) = 1128.80/100 = 11.288$

Example 24.18 Compute the coefficient of mean deviation about median for the following distribution:

Wt. in kgs.:	40-50	50-60	60-70	70-80
No. of persons:	8	12	20	10

Solution. First we compute the median weight.

Computation of median weight

Wt. in kg (Class interval)	No. of persons (frequency, (f_i))	Cummulative frequency (f_c)
40-50	8	8
50-60	12	f_c: 20
l_1 60-70 l_2	20	40
70-80	10	50
Total	$n = \Sigma f_i = 50$	

We find from the table, $n/2 = 25$ lies between the two cumulative frequencies 20 and 40 *i.e.*, $20 < 25 < 40$ and the corresponding class is 60-70. Thus, we have $l_1 = 60, f_c = 20, f = 20$.

$$\text{Median} = l_1 + \dfrac{\dfrac{n}{2} - f_c}{f_i} \times h = \left(60 + \dfrac{25 - 20}{20} \times 10\right) = 62.5 \text{ kg}.$$

Computation of mean deviation of weight from the median.

Weight (kgs)	mid-value (x_i) kgs.	No. of pearsons f_i	$\lvert x_i - Me\rvert$ kgs	$f_1\lvert x_i - Me\rvert$
40-50	45	8	17.50	140
50-60	55	12	7.50	90
60-70	65	20	2.50	50
70-80	75	10	12.50	125
Total		$\sum f_i = 50$		$\sum f_i\lvert x_i - Me\rvert = 405$

Thus mean deviation from median $= \dfrac{\sum f_i\lvert x_i - M_e\rvert}{\sum f_i} = \dfrac{405}{50} = 8.10$ kg.

Hence coefficient of mean deviation based on median

$$= \frac{M.D.(Med.)}{Med.} \times 100 = \frac{8.10}{62.5} \times 100 = 12.96 .$$

24.17 Standard Deviation

It is defined as the square root of the mean of squared deviations, all deviations taken from the actual mean. It is denoted by the Greek letter σ (sigma).

(*i*) **Standard deviation in case of ungrouped data** *i.e.* data having n observations $x_1, x_2, ..., x_n$ is given by

$$\sigma = \sqrt{\frac{\sum_{i=1}^{n}(x_i - \bar{x})^2}{n}}, \text{ where } \bar{x} = \frac{x_1 + x_2 + ... + x_n}{n} \qquad ...(24.23)$$

(*ii*) **Standard deviation in case of ungrouped data without using arithmetic mean:**

(*a*) **Direct method:** The formula in this case, can be expressed as

$$S.D. = \sqrt{\frac{\sum x^2}{n} - \left(\frac{\sum x}{n}\right)^2} \qquad ...(24.24)$$

where $\sum x^2, \sum x$ denotes respectively the sum of the squares and that of the terms of the given data.

(*b*) **Deviation method:** By assuming an arbitrary constant A, we can find from the formula

$$S.D = \sqrt{\frac{\sum_{j=1}^{n} d_j^2}{n} - \left(\frac{\sum_{j=1}^{n} d_j}{n}\right)^2}, \qquad ...(24.25)$$

where $d_j = x_j - A$, $\sum d_j^2$, $\sum d_j$ and n denote the sum of squares of deviations, the sum of the deviations and the number of items respectively.

(iii) **Standard deviation in case of grouped data:**

$$S.D = \sqrt{\frac{\sum_{i}^{m} f_i (x_i - \bar{x})^2}{\sum_{i}^{m} f_i}} \text{ or } \sqrt{\frac{\sum_{i=1}^{m} f_i (x_i - \bar{x})^2}{n}}, \qquad ...(24.26)$$

where x is the mid point of the class interval, $\bar{x} = \sum_{i=1}^{m} f_i x_i \div \sum_{i=1}^{m} f_i$.

(iv) **Standard deviation in case of grouped data without using mean**

(a) **Direct method:** Find the mid-point of the class interval, square it, multiply it with the frequencies and find the sum. Similarly, find the sum of product of mid-point of the class with frequencies. The formula is

$$S.D = \sqrt{\frac{\sum fx^2}{\sum f} - \left(\frac{\sum fx}{\sum f}\right)^2}, \qquad ...(24.27)$$

where x is the mid-point of the class-interval and f are the respective frequencies.

(b) **Deviation Method**

This method can be used only when the class-intervals are of the same size. In this method, we choose a point as assumed mean say A. We have

$$\sigma = h\sqrt{\frac{\sum fu_x^2}{n} - \left(\frac{\sum fu_x}{n}\right)^2}, \text{ where } u_x = \frac{x - A}{h} \qquad ..(24.28)$$

The square of standard deviation *i.e.* σ^2 (Sigma square) is called **variance**.
Example 23.19 Find the standard deviation, from the following **ungrouped data by using mean as well as not using mean.**
12, 6, 7, 3, 15, 10, 18, 5
Solution. *Using mean.* The mean of the given set of observations is given by

$$\bar{x} = \frac{\sum x_i}{n} = \frac{12+6+7+3+15+10+18+5}{8} = \frac{76}{8} = 9.5.$$

The respective $(x_i - \bar{x})^2$ are $2.5^2, 3.5^2, 2.5^2, 6.5^2, 5.5^2, 0.5^2, 8.5^2$ and 4.5^2 so that

$$\sum (x_i - \bar{x})^2 = 6.25 + 12.25 + 6.25 + 42.25 + 30.25 + 0.25 + 72.25 + 20.25 = 190$$

Standard Deviation $\sigma = \sqrt{\dfrac{\Sigma(x_i - \bar{x})^2}{n}} = \sqrt{\dfrac{190}{8}} = \sqrt{23.75} = 4.873.$

(ii) Finding standard deviation without using Arithmetic mean.

(a) *Direct method:*

								Total
x_i:	12	6	7	3	15	10	18	5	76
x_i^2:	144	36	49	9	225	100	324	25	912

$S.D = \sqrt{\dfrac{\Sigma x_i^2}{n} - \left(\dfrac{\Sigma x_i}{n}\right)^2} = \sqrt{\dfrac{912}{8} - \left(\dfrac{76}{8}\right)^2} = \sqrt{114 - 90.25} = 4.873$

(b) *Deviation method:*

Take Assume Mean, A = 10

x_i	$d_x = x_i - A$	d_x^2
12	2	4
6	−4	16
7	−3	9
3	−7	49
15	5	25
A: 10	0	0
18	8	64
5	−5	25
Total	$\Sigma d_x = -4$	$\Sigma d_x^2 = 192$

$S.D. = \sqrt{\dfrac{\Sigma d_x^2}{n} - \left(\dfrac{\Sigma d_x}{n}\right)^2}$

$= \sqrt{\dfrac{192}{8} - \left(\dfrac{-4}{8}\right)^2} = \sqrt{24 - 0.25}$

$= 4.873$

Example 24.20 Calculate the mean, standard deviation and variance for the following data:

Size of item:	6	7	8	9	10	11	12
Frequency:	3	6	9	13	8	5	4

(AMIE, W-2006; V.T.U., 2001)

Solution. Denoting the size of item by x_i, the calculations for finding mean and standard deviation are shown in the following table.

x_i	f_i	$f_i x_i$	$x_i - \bar{x}$	$(x_i - \bar{x})^2$	$f_i(x_i - \bar{x})^2$
6	3	18	−3	9	27
7	6	42	−2	4	24

(Contd.)

8	9	72	−1	1	9
9	13	117	0	0	0
10	8	80	1	1	8
11	5	55	2	4	20
12	4	48	3	9	36
Total	48	432	0	28	124

Here $n = \Sigma f_i = 48$, $\Sigma f_i x_i = 432$, $\Sigma f_i (x_i - \bar{x})^2 = 124$ $\therefore$ $\bar{x} = \dfrac{\Sigma f_i x_i}{n} = \dfrac{432}{48} = 9.$

Variance $(\sigma^2) = \dfrac{\Sigma f_i (x_i - \bar{x})^2}{n} = \dfrac{124}{48} = 2.583.$

and Standard deviation $(\sigma) = \sqrt{2.583} = 1.607.$

Example 24.21 Calculate the standard deviation, variance for the following distribution:

Class interval:	30-40	40-50	50-60	60-70	70-80	80-90	90-100
Frequencies:	3	7	12	15	8	3	2

Solution: Using mean: $d_x = x_i - \bar{x}$.

Class interval	Mid-point x_i	Frequency f_i	$f_i x_i$	$d_x = (x_i - \bar{x})$	d_x^2	$f_i d_x^2$
30-40	35	3	105	−27	729	2187
40-50	45	7	315	−17	289	2023
50-60	55	12	660	−7	49	588
60-70	65	15	975	3	9	135
70-80	75	8	600	13	169	1352
80-90	85	3	255	23	529	1587
90-100	95	2	190	33	1089	2178
Total		$n = \Sigma f_i = 50$	3100			10050

Mean, $\bar{x} = \dfrac{\Sigma f_i x_i}{\Sigma f_i} = \dfrac{3100}{50} = 62.$

$S.D, \sigma = \sqrt{\dfrac{\Sigma f_i d_x^2}{n}} = \sqrt{\dfrac{10050}{50}} = \sqrt{201} = 14.17.$

Variance, $\sigma^2 = 201$

Example 24.22 From the following frequency distribution, calculate th standard deviation using the formula for grouped data:

Class interval	frequency	Class interval	frequency
0-5	2	20-25	21
5-10	5	25-30	16
10-15	7	30-35	8
15-20	13	35-40	3

Solution. Using Deviation method. Denoting the mid value by x_i and takin; $h=5$, A (assumed mean) $= 22.5$, $u_x = (x_i - A)/h$.

Class interval	Mid-point x_i	frequency f_i	$u_x = (x_i - A)/h$	u_x^2	$f_i u_x$	$f_i u_x^2$
0-5	2.5	2	-4	16	-8	32
5-10	7.5	5	-3	9	-15	45
10-15	12.5	7	-2	4	-14	28
15-10	17.5	13	-1	1	-13	13
20-25	A→ 22.5	21	0	0	0	0
25-30	27.5	16	1	1	16	16
30-35	32.5	8	2	4	16	32
35-40	37.5	3	3	9	9	27
Total		n = 75			-9	193

$$\sigma = \sqrt{\frac{\sum f_i u_x^2}{n} - \left(\frac{\sum f_i u_x}{n}\right)^2} \times h = \sqrt{\frac{193}{75} - \left(\frac{-9}{75}\right)^2} \times 5$$

$$= 5\sqrt{\frac{193}{75} - \frac{81}{5625}} \approx 8.0 \cdot$$

24.18 Coefficient of variation

This is another relative measure of dispersion where the coefficient of respective absolute measure is multiplied by 100 to convert the figures into percentages o in relation to 100. When dispersions are measured by the median and quartile: the respective co-efficients of variations are:

(i) Co-efficient of Mean Variation or $C.V = \dfrac{\text{Mean deviation}}{\text{Mean of Median}} \times 100$ ...(24.29

(ii) Co-efficient of Quartile variation or C.V

= (Co-efficient of Quartile Deviation) × 100

$$= \frac{Q_3 - Q_1}{Q_3 + Q_1} \times 100 \qquad \qquad ...(24.30)$$

(iii) Co-efficient of Variation (Karl Pearson's Measure)

$$V = \frac{S.D}{\bar{x}} \times 100 \qquad \qquad ...(24.31)$$

Of the above three measures of co-efficients of variation, the one given by Karl Pearson is most commonly used.

Example 24.23 Fluctuations in the rates of two articles A and B are given below. Find out which of the two shows greater variability.

A.	618	619	616	623	620	624	622	625	622	625	626	625
B.	2152.5	2132.5	2134.25	2132.5	2145	2142.5	2146.25	2130	2146.25	2142.2	2150	2135
	2152.5											

Solution. Arranging the A series in tabular form, we get

x	f	$d_x = x - 622$	d_x^2	fd_x	fd_x^2
616	1	−6	36	−6	36
618	1	−4	16	−4	16
619	1	−3	9	−3	9
620	1	−2	4	−2	4
A→ 622	2	0	0	0	0
623	1	1	1	1	1
624	1	2	4	2	4
625	3	3	9	9	27
626	1	4	16	4	16
Total	12			1	113

Mean, $\bar{x} = 622 + (1/12) = 622.08$

$$\sigma = \sqrt{\frac{113}{12} - \left(\frac{1}{12}\right)^2} = \sqrt{\frac{1355}{144}} = \sqrt{9.41} = 3.06$$

Coefficient of variation of $A = \frac{\sigma}{\bar{x}} \times 100 = \frac{3.06}{622.08} \times 100 = 0.4918$

Arranging B series in tabular form, we get

x_i	f_i	$d_x = x_i - 2141.5$	d_x^2	$f_i d_x$	$f_i d_x^2$
2130.0	1	-11.5	132.25	-11.5	132.25
2132.5	2	-9	81	-18	162
2134.25	1	-7.25	52.5625	-7.25	52.5625
2135	1	-6.5	42.25	-6.5	42.25
A → 2142.5	2	1	1	2	2
2145	1	3.5	12.25	3.5	12.25
2146.25	2	4.75	22.5625	9.5	45.125
2150	1	8.5	72.25	8.5	72.25
2152.5	2	11.0	121.00	22	242
Total	13			2.25	762.6875

Mean, $\bar{x} = 2141.5 + \dfrac{2.25}{13} = 2141.67$

$$\sigma = \sqrt{\frac{762.6875}{13} - \left(\frac{2.25}{13}\right)^2} = \sqrt{58.6682 - 0.0299} = \sqrt{58.6383} = 7.66$$

Coefficient of variation $= \dfrac{\sigma}{\bar{x}} \times 100 = \dfrac{7.66}{2141.67} \times 100 = 0.358$

Hence *A* shows greater variability.

24.19 Skewness

Skewness is defined as lack of symmetry. In an asymmetrical distribution the mean, mode and the median do not coincide and therefore, the range is usually greater on one side of the mode than on the other. The curve has a longer tail on one side. This behaviour of the distribution exhibits skewness.

Since skewness is the measure of the shape of the curve and not of its size, it is advisable to use a ratio as a coefficient of the measure of skewness. This coefficient is given by the following formulae:

(a) Skewness $= \dfrac{\text{mean} - \text{mode}}{\text{Standard deviation}} = \dfrac{\bar{x} - mode}{\sigma}$ ...(24.32)

To avoid use of the mode, we can employ the empirical formula

mean−mode = 3 (mean − median) and define

Skewness $= \dfrac{3(\text{mean} - \text{median})}{\text{Standard deviation}} = \dfrac{3(\bar{x} - \text{median})}{\sigma}$ ...(24.33)

(b) Quartile coefficient of skewness $= \dfrac{Q_3 + Q_1 - 2Med}{Q_3 - Q_1}$ ...(24.34)

Formula (24.32) is generally the one which is used for measuring skewness, However, when the mode is not defined clearly, formula (24.34) can be used.

Skewness can be positive as well as negative. If the mean is greater than either the mode or the median, the curve will have the longer tail on the right and skewness will be positive. On the other hand, if the curve has a longer tail on the left, the mean will be less than the mode and median and the skewness will be negative as is exhibited by the following figures.

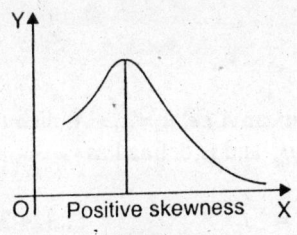

Positive skewness	Negative skewness
Fig. 24.3.	**Fig. 24.4.**

Example 24.24 Find out standard deviation and the coefficient of skewness for the following distribution:

Variable	0–5	5–10	10–15	15–20	20–25	25–30	30–35	35–40
Frequency	2	5	7	13	21	16	8	3

Solution. For finding the mean and standard deviation, we perform the calculations in the following table.

Classes	Variable x_i	f_i	$u_x = \dfrac{x-22.5}{5}$	u_x^2	$f_i u_x$	$f_i u_x^2$
0-5	2.5	2	−4	16	−8	32
5-10	7.5	5	−3	9	−15	45
10-15	12.5	7	−2	4	−14	28
15-20	17.5	13	−1	1	−13	13
20-25	A→22.5	21	0	0	0	0
25-30	27.5	16	1	1	16	16
30-35	32.5	8	2	4	16	32
35-40	37.5	3	3	9	9	27
Total		75			−9	193

$$\text{Mean} = 22.5 + \left(\frac{\Sigma f_i u_x}{\Sigma f_i}\right) \times 5 = 22.5 - \frac{9 \times 5}{75} = 21.9$$

and $\sigma^2 = \left[\dfrac{193}{75} - \left(\dfrac{-9}{75}\right)^2\right] \times 25 = 63.973$. Therefore, Standard deviation, $\sigma \approx 8.0$

The mode is given by

$$\text{Mode} = 20 + \frac{(21-13)}{(42-13-16)} \times 5 = 20 + 3.076 = 23.076$$

$\therefore$ Skewness $= \dfrac{(21.9 - 23.076)}{8} = 0.147$

24.20 Moments

If u denotes the deviation of x from any assumed mean A *i.e.* $u = x_i - A$, then the nth moment about the point $x = A$ is denoted by μ_n' and is defined as

$$\mu_n' = \frac{\sum f_i u^n}{\sum f_i} = \frac{\sum f_i u^n}{n} \text{ where } n = \sum f_i \qquad ...(24.35)$$

The nth moment about the actual mean $\bar{x}$ is denoted by μ_n and is defined as

$$\mu_n = \frac{\sum f(x - \bar{x})^n}{\sum f} \qquad ...(24.36)$$

Putting $n = 0$ in (24.35) and (24.36), we get

$$\mu_0 = \mu_0' = \frac{\sum f}{n} = 1 \qquad ...(24.37)$$

Similarly

$$\mu_1 = \frac{\sum f(x - \bar{x})}{\sum f} = \frac{\sum fx}{\sum f} - \bar{x} = 0 \qquad ...(24.38)$$

$$\mu_1' = \frac{\sum f(x - A)}{\sum f} = \frac{\sum fx}{\sum f} - A = \bar{x} - A = d(say) \qquad ...(24.39)$$

From (24.39), $\bar{x} = A + \mu_1'$. In particular, if we take $A = 0$, we obtain $\bar{x} = \mu_1'$. Hence, mean = first moment about the origin.

$$\mu_2 = \frac{\sum f(x - \bar{x})^2}{\sum f} = \sigma^2 = \text{variance} \qquad ...(24.40)$$

$$\mu_2' = \frac{\sum f(x - A)^2}{\sum f} = \frac{\sum f\{x - \bar{x} + (\bar{x} - A)\}^2}{\sum f} = \frac{\sum f(x - \bar{x} + d)^2}{\sum f}$$

$$= \frac{\Sigma f(x-\bar{x})^2}{\Sigma f} + 2d \frac{\Sigma f(x-\bar{x})}{\Sigma f} + d^2 \frac{\Sigma f}{\Sigma f}$$

$$= \sigma^2 + d^2 \text{ from (24.38) and (24.40)} \qquad \text{...(24.41)}$$

The above relations show that the first moment about the origin is the mean $\bar{x}$ and the second moment about the mean is the *variance*.

24.21 Relation Between the Moments about the Mean and the Moments about any Point

(a) *Moments about the mean in terms of moments about any point.*

Taking $\quad x - \bar{x} = x - A - (\bar{x} - A) = (x - A) - d$, we have

$$\mu_n = \frac{\Sigma f(x-\bar{x})^n}{\Sigma f} = \frac{\Sigma f\{(x-A)-d\}^n}{\Sigma f}$$

$$= \frac{1}{n} \Sigma f\{(x-A)-d\}^n \qquad \text{...(24.42)}$$

Now expanding the right side of (24.42) by Binomial theorem, we get

$$\mu_n = \frac{1}{n}\left[\Sigma f\{(x-A)^n - {}^nC_1 d(x-A)^{n-1} + \cdots + (-1)^n d^n\}\right]$$

$$= \frac{1}{n}\Sigma f(x-A)^n - \frac{1}{n}{}^nC_1 d \Sigma f(x-A)^{n-1} + \cdots + \frac{1}{n}(-1)^n d^n \Sigma f$$

$$= \mu_n' - {}^nC_1 d\,\mu_{n-1}' + {}^nC_2 d^2\mu_{n-2}' + \cdots + (-1)^n d^n \qquad \text{...(24.43)}$$

Putting $n = 2, 3, 4$, etc in (24.43), we get μ_2, μ_3, μ_4 etc in terms of μ_2', μ_3', μ_4' etc.

$$\mu_2 = \mu_2' - 2d\mu_1' + d^2 = \mu_2' - d^2 = \sigma^2 \qquad \text{...(24.44)}$$

$$\mu_3 = \mu_3' - 3d\mu_2' + 3d^2\mu_1' - d^3 = \mu_3' - 3d\mu_2' + 2d^3 \qquad \text{...(24.45)}$$

$$\mu_4 = \mu_4' - 4d\mu_3' + 6d^2\mu_2' - 4d^3\mu_1' + d^4$$

$$= \mu_4' - 4d\mu_3' + 6d^2\mu_2' - 3d^4 \qquad \text{...(24.46)}$$

(b) *Moments about any point in terms of moments about the mean*

We have $x - A = x - \bar{x} + (\bar{x} - A) = (x - \bar{x}) + d$

$$\mu_n' = \frac{\Sigma f(x-A)^n}{\Sigma f} = \frac{1}{n}\Sigma f\{(x-\bar{x})+d\}^n \qquad \text{...(24.47)}$$

Expanding (24.47) by Binomial theorem we get,

$$\mu_n' = \mu_n + {}^nC_1 d\mu_{n-1} + {}^nC_2 d^2\mu_{n-2} + \cdots + d^n \qquad \text{...(24.48)}$$

In particular

$$\mu_2' = \mu_2 + 2d\mu_1 + d^2 = \mu_2 + d^2 = \sigma^2 + d^2 \qquad \text{...(24.49)}$$

$$\mu_3' = \mu_3 + 3d\mu_2 + 3d^2\mu_1 + d^3 = \mu_3 + 3d\mu_2 + d^3 \qquad \ldots(24.50)$$

$$\mu_4' = \mu_4 + 4d\mu_3 + 6d^2\mu_2 + 4d^3\mu_1 + d^4$$

$$= \mu_4 + 4d\mu_3 + 6d^2\mu_2 + d^4 \qquad \ldots(24.51)$$

24.22 Pearson's β and γ coefficients

These coefficients are defined by the following relations.

$$\left. \begin{aligned} \beta_1 &= \frac{\mu_3^2}{\mu_2^3} = \frac{\mu_3^2}{\sigma^6}; \ldots(24.52) \\[2mm] \beta_2 &= \frac{\mu_3}{\mu_2^2} = \frac{\mu_3}{\sigma^4}; \ldots(24.53) \end{aligned} \right\} \text{(β-coefficients)}$$

$$\left. \begin{aligned} \gamma_1 &= +\sqrt{\beta_1} = \frac{\mu_3}{\sigma^3}; \ \ldots(24.54) \\[2mm] \gamma_2 &= \beta_2 - 3 = \frac{\mu_4 - 3\sigma^4}{\sigma^4} \ldots(24.55) \end{aligned} \right\} \text{(γ-coefficients)}$$

These coefficients are all pure numbers because they are the ratios of absolute numbers in which μ_n has the dimension n.

The importance of these numbers lies in the fact that they supply information about the shape of the curve obtained from the frequency distribution. In the case of a symmetrical distribution the moments of odd order about the mean are zero. This gives $\mu_3 = 0$ and therefore, $\beta_1 = 0$. *Hence β_1 defines the departure from symmetry and gives the measure of skewness.*

A frequency distribution may have a flat mode or a sharply peaked mode. The flatness of the mode is called *kurtosis* and this is generally denoted by β_2.

In the case of normal curves, the value of the kurtosis is 3 or $\gamma_2 = 0$, Hence $\gamma_2 = \beta_2 - 3$, defines the excess of kurtosis. The curves for which $\beta_2 < 3$ are called *Platykurtic* and the curves with values of $\beta_2 > 3$ are called *Leptokurtic*.

Platykurtic curves are flatter then the normal curves while Leptokurtic curves are more sharply peaked as shown in the figure below. The normal distribution, fig. (c), which is not very peaked or very flat-topped is called *mesokurtic*.

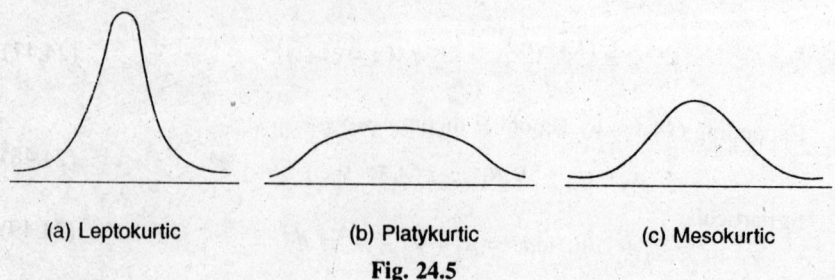

(a) Leptokurtic   (b) Platykurtic   (c) Mesokurtic

Fig. 24.5

It may be remarked that it is only the first four moments which are of practical importance in the study of frequency distribution. As remarked earlier (Art 24.20), the mean is the first moment about origin and the variance the second moment about the mean. The third moment about the mean is the measure of skewness while the fourth moment defines kurtosis *i.e.*, the flatness of the curve.

24.23 Correlation

So far our data for statistical analysis consisted of observations concerning single variable. This data could be heights of students in a particular class, yield of grain per acre in a state etc. There are instances in which two variables are such that an increase or decrease in the value of one variable is accompanied by increase or decrease in the value of the other. Two such variables are said to be correlated.

The correlation is called **direct** or **positive** if two variables move in the same direction *i.e.* an increase (or decrease) on the part of one variable introduces an increase (or decrease) on the part of the other variable. As for example heights and weights of students or yield of food grains per acre, and amount of fertiliser per acre, profit and investment etc are known to be positively correlated.

On the other hand, if the two variables move in the opposite directions *i.e.*, an increase (or a decreases) on the part of one variable result a decrease (or an increase) on the part of the other variable, then the two variables are known to have a **negative** or **inverse** correlation. For example, the price and demand of an items, the profit of insurance co. and the number of claims it has to meet etc. are examples of variables having a negative correlation. If there is no relationship indicated between the variables, we say that they are uncorrelated. As for examples, shoe size and intelligence are uncorrelated. On the basis of what has been stated above, correlation may be defined as *"the amount of similarity, in degree and direction, of variates in corresponding pairs of observations of the two"*.

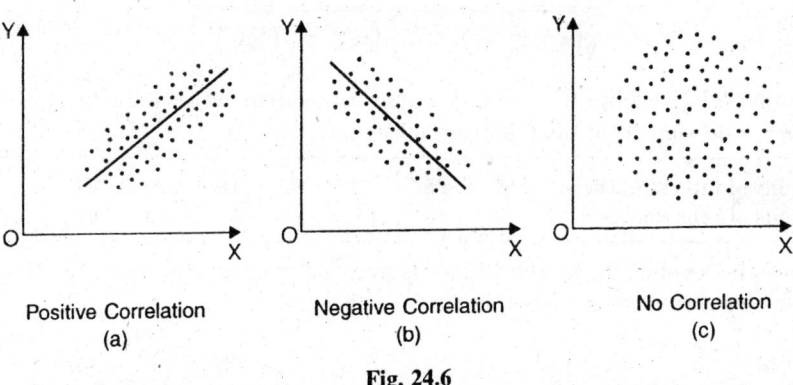

| Positive Correlation | Negative Correlation | No Correlation |
| (a) | (b) | (c) |

Fig. 24.6

24.24 Coefficient of Correlation

For measuring the degree of closeness of linear relationship between the two variables when there is reasonable ground to believe they are correlated, Karl Pearson has developed a formula, called coefficient of correlation. This formula gives the numerical measure of the amount of correlation. This numerical measure is denoted by r. The following are some of the properties of correlation coefficient:

1. The coefficient of correlation is independent of the units measurement.
2. The coefficient of correlation is independent of the choice of both origin and the scale of observations.
3. It lies between -1 and $+1$.

If the variables are x and y, then coefficient of correlation r is given by

$$r = \frac{\Sigma(x_i - \bar{x})(y_i - \bar{y})}{n\,\sigma_x\,\sigma_y}, \qquad \dots(24.56)$$

where symbols have their usual meanings.

If we put $x_i - \bar{x} = X$ and $y_i - \bar{y} = Y$, then, $r = \dfrac{\Sigma XY}{n\,\sigma_x\,\sigma_y}$. $\qquad \dots(24.57)$

Since $\sigma_x = \sqrt{\Sigma X^2 / n}$, $\sigma_y = \sqrt{\Sigma Y^2 / n}$, (24.57) can be written in the form

$$r = \frac{\Sigma XY}{\sqrt{\Sigma X^2}\,\sqrt{\Sigma Y^2}} = \frac{\Sigma XY}{\sqrt{\Sigma X^2 \cdot \Sigma Y^2}}. \qquad \dots(24.58)$$

This formula, which automatically gives the proper sign of r, is called the *product-moment formula*.

Another form of the formula which is quite handy for calculation is

$$r = \frac{n\Sigma x_i y_i - \Sigma x_i \Sigma y_i}{\sqrt{\left[n\Sigma x_i^2 - (\Sigma x_i)^2\right]\cdot\left[n\Sigma y_i^2 - (\Sigma y_i)^2\right]}} \qquad \dots(24.59)$$

Example 24.25 Compute the coefficient of correlation between the heights of fathers and sons from the following table:

Heights of fathers (inches):	65	66	67	68	69	70	71
Heights of sons (inches):	67	68	66	69	72	72	69

Solution: Denoting the heights of fathers by x and those of sons by y, the calculations for correlation are given in the following table:

x_i	y_i	Deviations from the Average $(X = x_i - \bar{x})$	$(Y = y_i - \bar{y})$	XY	X^2	Y^2
65	67	-3	-2	6	9	4
66	68	-2	-1	2	4	1
67	66	-1	-3	3	1	9
68	69	0	0	0	0	0
69	72	1	3	3	1	9
70	72	2	3	6	4	9
71	69	3	0	0	9	0
Total: 476	483	0	0	20	28	32

From this table, $\bar{x}$ = the mean height of fathers $= \dfrac{\Sigma x_i}{n} = \dfrac{476}{7} = 68$ inches.

$\bar{y}$ = the mean height of sons $= \dfrac{\Sigma y}{n} = \dfrac{483}{7} = 69$ inches.

From the above table $\Sigma XY = 20$. $\Sigma X^2 = 28$, $\Sigma Y^2 = 32$.

As per Karl Pearson, coefficient of correlation,

$$r = \frac{\Sigma XY}{\sqrt{\Sigma X^2 \, \Sigma Y^2}} = \frac{20}{\sqrt{28 \times 32}} \approx +0.67. \text{ The correlation is positive.}$$

Example 24.26 The following table gives marks obtained by two students in 10 tests during the year. Find the correlation coefficient.

Test No.	Marks in Math.	Marks in English	Test No.	Marks in Math.	Marks in English
1	77	35	6	35	40
2	54	58	7	90	35
3	27	60	8	25	56
4	52	40	9	56	34
5	14	50	10	60	42

Solution. Let x denote marks in Maths and y marks in English. The calculations for r are as follows:

x_i	y_i	$X = x_i - \bar{x}$	$Y = y_i - \bar{y}$	XY	X^2	Y^2
77	35	28	-10	-280	784	100
54	58	5	13	65	25	169

(Contd.)

x_i	y_i	$X = x_i - \bar{x}$	$Y = y_i - \bar{y}$	XY	X^2	Y^2
27	60	-22	15	-330	484	225
52	40	3	-5	-15	9	25
14	50	-35	5	-175	1225	25
35	40	-14	-5	70	196	25
90	35	41	-10	-410	1681	100
25	56	-24	11	-264	576	121
56	34	7	-11	-77	49	121
60	42	11	-3	-33	121	9
Total: 490	450	0	0	-1449	5150	920

$$\bar{x} = \frac{490}{10} = 49, \ \bar{y} = \frac{450}{10} = 45.$$

Also $\quad \sum XY = -1449, \ \sum X^2 = 5150, \ \sum Y^2 = 920$

$$\therefore \quad r = \frac{\sum XY}{\sqrt{\sum X^2 \sum Y^2}} = \frac{-1449}{\sqrt{5150 \times 920}} = \frac{-1449}{2176.6947} = -0.66.$$

Example 24.27 Compute the correlation coefficient between x and y from the following data:

$n = 10, \ \sum x_i y_i = 220, \ \sum x_i^2 = 200, \ \sum y_i^2 = 262, \ \sum x_i = 40, \ $ and $ \ \sum y_i = 50.$

Solution. Applying formula (24.57), we have

$$r = \frac{n \sum x_i y_i - \sum x_i \sum y_i}{\sqrt{n \sum x_i^2 - \left(\sum x_i\right)^2} \times \sqrt{n \sum y_i^2 - \left(\sum y_i\right)^2}}$$

$$= \frac{10 \times 220 - 40 \times 50}{\sqrt{10 \times 200 - (40)^2} \times \sqrt{10 \times 262 - (50)^2}} = \frac{2200 - 2000}{\sqrt{2000 - 1600} \times \sqrt{2620 - 2500}}$$

$$= \frac{200}{20 \times 10.9545} = 0.913.$$ Thus, there is a good amount of positive correlation between the two variables x and y.

24.25 Short-Computational Formulae

(a) **Assumed mean method.** When the arithmetic means are not whole numbers, it is more convenient to compute the value of r by taking assumed mean A say, for x and B for y. d_x is deviation from assumed average A i.e. $d_x = x_i - A$ and d_y is deviation from assumed average B i.e. $d_y = y_i - B$. When deviations are taken from the assumed average and not the actual average, the formula is as follows:

$$r = \frac{n\Sigma d_x d_y - (\Sigma d_x)(\Sigma d_y)}{\sqrt{\{n\Sigma d_x^2 - (\Sigma d_x)^2\}\{n\Sigma d_y^2 - (\Sigma d_y)^2\}}} \qquad ...(24.60)$$

(b) **Change of origin and change of scale.** Sometimes, *change of origins* and *change of scales* reduces the computational labour to a great extent. When the deviations d_x and d_y have common factors say h and k respectively. Then putting $d_x/h = u_x$ and $d_y/k = v_y$. The correlated coefficient formula is as follows:

$$r = \frac{n\Sigma u_x v_y - (\Sigma u_x)(\Sigma v_y)}{\sqrt{\{n\Sigma u_x^2 - (\Sigma u_x)^2\}\{n\Sigma v_y^2 - (\Sigma v_y)^2\}}} \qquad ...(24.61)$$

24.26 Illustrative Examples

Example 24.28 Calculate the value of r between x and y.

x:	1	3	5	7	8	10
y:	8	12	15	17	18	20

Solution. Here the actual average ($\bar{x}$) of the variable data x_i is 34/6 or 5.66 which is not an integral. To reduce the computational labour to a great extent, we make use of an assumed average. Refer the following table, d_x is deviation from assumed average 7 and d_y is deviation from assumed average 15.

x_i	y_i	d_x $= x_i - 7$	d_y $= y_i - 15$	$d_x d_y$	d_x^2	d_y^2
1	8	-6	-7	42	36	49
3	12	-4	-3	12	16	9
5	→ 15	-2	0	0	4	0
→ 7	17	0	2	0	0	4
8	18	1	3	3	1	9
10	20	3	5	15	9	25

$\Sigma x_i = 34$ $\Sigma y_i = 90$ $\Sigma d_x = -8$ $\Sigma d_y = 0$ $\Sigma d_x d_y = 72$ $\Sigma d_x^2 = 66$ $\Sigma d_y^2 = 96$
$n = 6$

Correlation coefficient is not affected by shift of origin and change of scale. We use the following formula.

$$\therefore \qquad r = \frac{n\Sigma d_x d_y - (\Sigma d_x)(\Sigma d_y)}{\{\sqrt{n\Sigma d_x^2 - (\Sigma d_x)^2}\}\{\sqrt{n\Sigma d_y^2 - (\Sigma d_y)^2}\}}$$

$$= \frac{6(72)-(-8)(0)}{\sqrt{6(66)-(-8)^2}\sqrt{6(96)-0}}$$

$$= \frac{432}{\sqrt{332}\sqrt{576}} \approx 0.988$$

Note. In the above example $\bar{x} = \Sigma x_i/n = 34/6 = 5.66$ and deviations from $\bar{x}$ involve fractions which make computations for r (using the formula $r = \Sigma XY/\sqrt{\Sigma X^2 \Sigma Y^2}$) quite inconvenient. This shows the advantage of using assumed average and not the actual average.

Example 24.29 Calculate r for the following paired data

x:	10	20	30	40	50	60	70	80
y:	32	20	24	36	40	28	48	44

Solution. The correlation coefficient is not affected by change of origin and change of scale. To reduce the computational labour to a great extent we take origin of x and y as 40 and 36 and 10 and 4 are the respective change of scales. We have calculations as follows:

x_i	y_i	$u_x = \dfrac{x_i-40}{10}$	$v_y = \dfrac{y_i-36}{4}$	$u_x v_y$	u_x^2	v_y^2
10	32	−3	−1	3	9	1
20	20	−2	−4	8	4	16
30	24	−1	−3	3	1	9
40	36	0	0	0	0	0
50	40	1	1	1	1	1
60	28	2	−2	−4	4	4
70	48	3	3	9	9	9
80	44	4	2	8	16	4

Total 360 272 $\Sigma u_x = 4$ $\Sigma v_y = -4$ $\Sigma u_x v_y = 28$ $\Sigma u_x^2 = 44$ $\Sigma v_y^2 = 44$

Using the formula:

$$r = \frac{n\Sigma u_x v_y - (\Sigma u_x)(\Sigma v_y)}{\sqrt{n\Sigma u_x^2 - (\Sigma u_x)^2}\sqrt{n\Sigma v_y^2 - (\Sigma v_y)^2}}, \text{ we obtain}$$

Therefore, $r = \dfrac{8(28)-(4)(-4)}{\{\sqrt{8(44)-4^2}\}\{\sqrt{8(44)-(-4)^2}\}} = \dfrac{224+16}{\sqrt{336}\sqrt{336}} \approx 0.714.$

Example 24.30 Calculate correlation coefficient from the following results:

$n = 10$; $\Sigma x = 140$; $\Sigma y = 150$; $\Sigma(x-10)^2 = 180$; $\Sigma(y-15)^2 = 215$;

$\Sigma(x-10)(y-15) = 60$.

Solution: Since the correlation coefficient r is not affected by change of origin of reference, $r_{xy} = r_{uv}$, where $u = x-10$ and $v = y-15$. From the given data, we find

$$\Sigma u = \Sigma(x-10) = \Sigma x - \Sigma 10 = \Sigma x - 10n = 140 - 10(10) = 40,$$

$$\Sigma v = \Sigma(y-15) = \Sigma y - \Sigma 15 = 150 - 15 \times 10 = 0,$$

$$\Sigma u^2 = \Sigma(x-10)^2 = 180, \Sigma v^2 = \Sigma(y-15)^2 = 215,$$

and $\Sigma uv = \Sigma(x-10)(y-15) = 60$.

Now
$$r_{uv} = \frac{n\Sigma uv - (\Sigma u)(\Sigma v)}{\sqrt{\left[n\Sigma u^2 - (\Sigma u)^2\right]\left[n\Sigma v^2 - (\Sigma v)^2\right]}}$$

$$= \frac{10 \times 60 - 40 \times 0}{\sqrt{\left[10 \times 180 - (40)^2\right]\left[10 \times 215 - 0\right]}}$$

$$= \frac{600 - 0}{\sqrt{(1800 - 1600)(2150)}} = \frac{600}{\sqrt{200 \times 2150}} = \frac{6}{\sqrt{2 \times 21.5}} = \frac{6}{\sqrt{43}} \approx 0.915$$

Example 24.31 While calculating correlation coefficient between two variables x and y from 25 pairs of observations, the following results were obtained:

$n = 25$, $\Sigma x = 125$, $\Sigma y = 100$, $\Sigma x^2 = 650$, $\Sigma y^2 = 460$, $\Sigma xy = 508$.

Later it was discovered at the time of checking that the pairs of values

x	y
8	12
6	8

were copied down as

x	y
6	14
8	6

. Obtain the correct value of correlation coefficient.

Solution. To get the correct results, we subtract the incorrect values and add the corresponding correct values.

∴ The correct result would be

$$\Sigma x = 125 - 6 - 8 + 8 + 6 = 125, \qquad \Sigma y = 100 - 14 - 6 + 12 + 8 = 100$$

$$\Sigma x^2 = 650 - 6^2 - 8^2 + 8^2 + 6^2 = 650, \qquad \Sigma y^2 = 460 - 14^2 - 6^2 + 12^2 + 8^2 = 436,$$

$$\Sigma xy = 508 - 6(14) - 8(6) + 8(12) + 6(8) = 520.$$

$$\therefore \text{ Corrected } r(x, y) = \frac{n\sum xy - (\sum x)(\sum y)}{\sqrt{\left[n\sum x^2 - (\sum x)^2\right]}\sqrt{\left[n\sum y^2 - (\sum y)^2\right]}}$$

$$= \frac{25(520) - 125(100)}{\sqrt{\left[25(650) - (125)^2\right]}\sqrt{\left[25(436) - (100)^2\right]}} = \frac{20}{\sqrt{25 \times 36}} \approx 0.667.$$

Example 24.32 Coefficient of correlation between x and y for 50 observations is 0.3. The arithmetic means and standard deviations of x and y are known to be 10 and 6 and 3 and 2 respectively. However, on subsequent verification it was found that one value of x (= 10) and one value of y (= 6) were inaccurate and hence weeded out. With the remaining 49 pairs of values, how is the original value of correlation coefficient affected?

Solution. We know that

$$r = \frac{\sum(x - \bar{x})(y - \bar{y})}{n\sigma_x\sigma_y} \text{ or } \sum(x - \bar{x})(y - \bar{y}) = nr\sigma_x\sigma_y = 50 \times 0.3 \times 3 \times 2 = 90$$

or
$$\sum xy - \bar{x}\sum y - \bar{y}\sum x + \sum\bar{x}\,\bar{y} = 90$$

or
$$\sum xy = 90 + n\bar{x}\,\bar{y} \quad (\because \sum y = n\bar{y}, \sum x = n\bar{x} \text{ and } \sum \bar{x}\,\bar{y} = n\bar{x}\,\bar{y})$$

$$= 90 + 50 \times 10 \times 6 = 3090.$$

We also know that $\sigma_x^2 = \dfrac{\sum x^2}{n} - \left(\dfrac{\sum x}{n}\right)^2 = \dfrac{1}{n}\sum x^2 - \bar{x}^2$

or $\sum x^2 = n\left(\sigma_x^2 + \bar{x}^2\right) = 50(9 + 100) = 5450.$

Similarly, $\sum y^2 = n\left(\sigma_y^2 + \bar{y}^2\right) = 50(4 + 36) = 2000.$

Corrected value

$\sum xy = 3090 - 10 \times 6 = 3030, \sum x = n\bar{x} - 10 = 50 \times 10 - 10 = 490$

$\sum y = n\bar{y} - 6 = 50 \times 6 - 6 = 294, \sum x^2 = 5450 - (10)^2 = 5350,$

and $\sum y^2 = 2000 - (6)^2 = 1964.$

With the actual values of the variables x, y, the correlation coefficient is found from

$$r = \frac{n\sum xy - \sum x\sum y}{\sqrt{\left[n\sum x^2 - (\sum x)^2\right]\left[n\sum y^2 - (\sum y)^2\right]}}.$$

Corrected value of the correlation coefficient with the remaining 49 pairs of values is therefore given by

$$r = \frac{49 \times 3030 - 490 \times 294}{\sqrt{\left[49 \times 5350 - (490)^2\right]\left[49 \times 1964 - (294)^2\right]}}$$

$$= \frac{49 \times 90}{\sqrt{49 \times 450 \times 49 \times 200}} = \frac{49 \times 90}{49 \times 300} = 0.3.$$

Therefore, original value of r remains unaffected.

24.27 Correlation Table

When the number of pairs of values is large they are grouped into a table of double frequency known as *correlation table*. An example of this type of table is given below:

Age of wives	Age of husbands					Total
	20-30	30-40	40-50	50-60	60-70	
15-25	5	9	3	0	0	17
25-35	0	10	**25**	2	0	37
35-45	0	1	12	2	0	15
45-55	0	0	4	16	5	25
55-65	0	0	0	4	2	6
Total	5	20	44	24	7	100

The numbers in the body of the table indicate frequencies. For example the number 25 shown bold in the body of the table indicates that there are 25 couples with ages of husbands between 40-50 and ages of wives between 25-35. Such frequency distribution is known as *bivariate frequency distribution*.

The formula for r for such frequency distribution is

$$r = \frac{n\sum fd_x d_y - (\sum fd_x)n(\sum fd_y)n}{\sqrt{\left[n(\sum fd_x^2) - (\sum fd_x)^2\right]}\sqrt{\left[n(\sum fd_y^2) - (\sum fd_y)^2\right]}} \qquad \text{...(24.62)}$$

where f represents frequencies.

In case deviations for x and y i.e., d_x, d_y have common factors then

$$r = \frac{n\sum fuv - (\sum f_x u)(\sum f_y v)}{\sqrt{n(\sum f_x u^2) - (\sum f_x u)^2}\sqrt{n(\sum f_y v^2) - (\sum f_y v)^2}} \qquad \text{...(24.63)}$$

where $u = d_x/h$ and $v = d_y/k$, h and k having their usual meanings. In case of grouped frequency distribution h and k are magnitudes of classes. The evaluation of r in this case will be clear from the following example.

Example 24.33 Calculate the coefficient of correlation for the correlation table given in article 24.27.

Solution. Let x denote the central values of ages of husbands and y those of wives. The calculations for the value of r are given in the following table.

	$x \rightarrow$	25	35	A → 45	55	65	f_y	$V = \dfrac{y-40}{10}$	$f_y V$	$f_y V^2$
	$y \downarrow$									
	20	5 20	9 18	3 0	0	0	17	−2	−34	68
	30	0	10 10	25 0	2 −2	0	37	−1	−37	37
	B → 40	0	1 0	12 0	2 0	0	15	0	0	0
	50	0	0	4 0	16 16	5 10	25	1	25	25
	60	0	0	0	4 8	2 8	6	2	12	24
	f_x	5	20	44	24	7	$\Sigma f_x = \Sigma f_y$ $= n = 100$	Total	−34 $\Sigma f_y V$	154 $\Sigma f_y V^2$
	$u = \dfrac{x-45}{10}$	−2	−1	0	1	2	Total			
	$f_x u$	−10	−20	0	24	14	$8 = \Sigma f_x u$			
	$f_x u^2$	20	20	0	24	28	$92 = \Sigma f_x u^2$			

The overall table is titled **Age of husbands in years** (columns) and **Age of wives in years** (rows).

As is clear from the table, we have taken $A = 45$ as assumed mean for x and $B = 40$, as assumed mean for y.

Here $\Sigma f_x u = 8$, $\Sigma f_x u^2 = 92$, $\Sigma f_y v = -34$, $\Sigma f_y v^2 = 154$ and $n = 100$.

Thick figures in small squares represent the product fuv

$\therefore \quad \Sigma fuv = 20 + 18 + 10 - 2 + 16 + 10 + 8 + 8 = 88.$

Thus $r = \dfrac{n \Sigma fuv - (\Sigma f_x u)(\Sigma f_y v)}{\sqrt{\left[n\left(\Sigma f_x u^2\right) - \left(\Sigma f_x u\right)^2\right]\left[n\left(\Sigma f_y v^2\right) - \left(\Sigma f_y v\right)^2\right]}}$

$= \dfrac{(100)88 - (8)(-34)}{\sqrt{\{(100)92 - 8^2\}\{100(154) - (-34)^2\}}} = \dfrac{8800 + 272}{\sqrt{9136 \times 14244}}$

$$= \frac{9072}{11407.6} = 0.795 \text{ approx.}$$

Example 24.34 From the following table of bivariate frequency distribution, calculate the coefficient of correlation between heights and weights of children:

Weight (in pounds)	Height (in inches) 40-44	44-48	48-52	52-56	56-60	60-64	Total
35-55	4	40	60	–	–	–	104
55-75	–	–	24	88	12	–	124
75-95	–	–	–	8	32	8	48
95-115	–	–	–	–	4	8	12
115-135	–	–	–	4	–	–	4
135-155	–	–	–	–	4	4	8
Total	4	40	84	100	52	20	300

Solution. Let x denote the central values of height and y-those of weight of children. The calculations for the value of r are given in the following table.

Height in inches

$x \rightarrow$ $y \downarrow$	42	46	A → 50	54	58	62	f_y	$v = \frac{y-85}{20}$	$f_y v$	$f_y v^2$
45	4 16	40 80	60 0	0	0	0	104	-2	-208	416
65	0	0	24 0	88 -88	12 -24	0	124	-1	-124	124
B → 85	0	0	0	8 0	32 0	8 0	48	0	0	0
105	0	0	0	0	4 8	8 24	12	1	12	12
125	0	0	0	4 8	0	0	4	2	8	16
145	0	0	0	0	4 24	4 36	8	3	24	72
f_x	4	40	84	100	52	20	$\Sigma f_x = \Sigma f_y$ = n = 300	Total	-288	640
$u = \frac{x-50}{40}$	-2	-1	0	1	2	3	Total		$= \Sigma f_y v$	$= \Sigma_y v^2$
$f_x u$	-8	-40	0	100	104	60	216 =	$\Sigma f_x u$		
$f_x u^2$	16	40	0	100	208	180	544 =	$\Sigma f_x u^2$		

Weight in pounds

We have taken $A = 50$ as assumed mean for x and $B = 85$ as assumed mean for y.

Here $\Sigma f_x u = 216$, $\Sigma f_x u^2 = 544$, $\Sigma f_y v = -288$, $\Sigma f_y v^2 = 640$ and $n = 300$.

Thick figures in small squares represent the product fuv.

$\therefore \quad \Sigma fuv = 16 + 80 - 88 - 24 + 8 + 24 + 8 + 24 + 36 = 84.$

Thus $r = \dfrac{n \Sigma fuv - (\Sigma f_x u)(\Sigma f_y v)}{\sqrt{\left[n(\Sigma f_x u^2) - (\Sigma f_x u)^2\right]\left[n(\Sigma f_y v^2) - (\Sigma f_y v)^2\right]}}$

$= \dfrac{300(84) - (216)(-288)}{\sqrt{\left[300(544) - (216^2)\right]\left[300(640) - (-288)^2\right]}}$

$= \dfrac{25200 + 62208}{\sqrt{(163200 - 46656) \times (192000 - 82944)}}$

$= \dfrac{87408}{\sqrt{116544 \times 109056}} = \dfrac{87408}{341.385 \times 330.236} \approx 0.775.$

24.28 Rank Correlation

The coefficient of correlation (Pearson's coefficient of correlation) between the variables as have already been discussed is based on the values of the variables. Sometimes these values are not available or it may not be possible to find the magnitudes of the variables. For example we cannot measure intelligence or beauty quantitatively. In such cases it is possible to rank the individuals in some order on the basis of these characteristics. If the ranks assigned are 1 to n, then correlation coefficient between two series of ranks is called *rank correlation coefficient* and is given by

$$R = 1 - \frac{6 \Sigma d^2}{n(n^2 - 1)} \qquad \qquad ...(24.64)$$

where d is the difference between corresponding ranks of the two series and n is the number of items in each series.

Formula (24.64) is known as *Edward Spearman's Formula for rank correlation coefficient.*

If all the d's are zero, then correlation is perfect and $R = 1$ otherwise $-1 \le R < +1$.

Example 24.35 Ten students get the following ranks in two subjects A and B.

A:	3	5	8	4	7	10	2	1	6	9
B:	6	4	9	8	1	2	3	10	5	7

Calculate the rank correlation coefficient.

Solution. Calculation of Rank correlation coefficient.

Rank in A (x)	Rank in B (y)	Rank Difference d = x − y	d²
3	6	−3	9
5	4	1	1
8	9	−1	1
4	8	−4	16
7	1	6	36
10	2	8	64
2	3	−1	1
1	10	−9	81
6	5	1	1
9	7	2	4
Total:			$\Sigma d^2 = 214$

Here $n = 10$, $\Sigma d^2 = 214$.

∴   Rank correlation coefficient, $R = 1 - \dfrac{6 \Sigma d^2}{n(n^2 - 1)} = 1 - \dfrac{6(214)}{10(10^2 - 1)}$

$$= 1 - \frac{1284}{990} = -0.3 \text{ approx.}$$

Example 24.36 Calculate the coefficient of correlation from the data given below by the method of differences:

Series A:	78	89	97	69	59	79	68	57
Series B:	125	137	156	112	107	136	123	108

Solution. Computation of Rank correlation is given in the following table:

A	B	Rank in A (x)	Rank in B (y)	Rank Diff. d = x − y	d²
78	125	4	4	0	0
89	137	2	2	0	0
97	156	1	1	0	0
69	112	5	6	−1	1
59	107	7	8	−1	1
79	136	3	3	0	0
68	123	6	5	1	1
57	108	8	7	1	1
Total					$\Sigma d^2 = 4$

$$\therefore \quad R = 1 - \frac{6\sum d^2}{n(n^2 - 1)} = 1 - \frac{6(4)}{8(64 - 1)} = 1 - \frac{24}{8 \times 63} = \frac{20}{21} = 0.95$$

24.29 Spearman's Modified Formula for Tied Ranks

When the values of the variables are known, we assign rank 1 to the largest value. The next largest is assigned rank 2 and so on. But if two values are equal then there is a tie. In such a situation, these values are assigned the average of the ranks. For example if two items are tied for rank 4, then each is ranked $(4 + 5)/2 = 4.5$ and so on.

Edward spearman's formula for tied rank, the correlation coeffient takes the following form:

$$R = 1 - \frac{6\left\{\sum d^2 + \dfrac{\sum(t^3 - t)}{12}\right\}}{n(n^2 - 1)}, \qquad \qquad ...(24.65)$$

where t is the number of scores or items involved in a tie. The following example will make the procedure clear.

Example 24.37 Find the rank correlation coefficient of the following data:

Series A:	114	108	110	88	97	119	97	100	97	118
Series B:	74	72	83	70	75	82	64	72	68	80

Solution. In series A, the highest score is 119 and, therefore, its rank is 1. The next highest score is 118 and its rank is 2. Proceeding in this way we note that rank of 97 is 7. Since it occur 3 times, there is a tie for 7th place. We find the average rank $(7 + 8 + 9)/3 = 8$ and assign this rank to each of 97. The score next to 97 is 88 and its rank is 10. Similarly the ranks for the second series are determined. There 72 occurs twice and there is a tie for 6th place. The average rank is $(6 + 7)/2 = 6.5$ and this rank is assigned to 72 etc.

Computation of rank correlation coefficient

Series A		Series B		Rank Difference	d^2
Score	Rank (x)	Score	Rank (y)	$d = x - y$	
114	3	74	5	−2	4
108	5	72	6.5	−1.5	2.25
110	4	83	1	3	9
88	10	70	8	2	4

(Contd.)

Series A		Series B		Rank Difference	d^2
Score	Rank (x)	Score	Rank (y)	$d = x - y$	
97	8	75	4	4	16
119	1	82	2	-1	1
97	8	64	10	-2	4
100	6	72	6.5	-0.5	0.25
97	8	68	9	-1	1
118	2	80	3	-1	1
				0	$\Sigma d^2 = 42.50$

By Spearman's modified formula for tied rank, the correlation coefficient is given by

$$R = 1 - \frac{6\left[\Sigma d^2 + \frac{\Sigma(t^3 - t)}{12}\right]}{n(n^2 - 1)}. \text{ Here } \Sigma d^2 = 42.50, n = 10, t = 3, 2$$

$$\therefore \qquad \frac{\Sigma(t^3 - t)}{12} = \frac{3^3 - 3}{12} + \frac{2^3 - 2}{12} = 2 + 0.5 = 2.5$$

Hence $\qquad R = 1 - \dfrac{6\{42.50 + 2.5\}}{10(10^2 - 1)} = 1 - \dfrac{6 \times 45}{990} = 1 - \dfrac{3}{11} \approx 0.73.$

Example 24.38 Compute the coefficient of rank correlation between Commerce marks and Mathematics marks as given below:

Commerce Marks:	80	56	50	48	50	62	60
Mathematics Marks:	90	75	75	65	65	50	65

Solution. This is a case of tied ranks as more than one student share the same mark both for Commerce and Mathematics. For Commerce subject the student receiving 80 marks gets rank 1, one getting 62 marks receives rank 2, the student with 60 marks receives rank 3, student with 56 marks gets rank 4 and since there are two students, each getting 50 marks, each would be receiving a common rank, the average of the next two ranks 5 and 6 i.e. $(5 + 6)/2$ i.e. 5.5 and lastly the last rank 7 goes to the student getting the lowest Commerce marks. In a similar manner, we award ranks to the students with Maths marks.

Table
Computation of Rank correlation between Commerce marks and Maths marks with tied marks.

Commerce mark	Maths. mark	Rank for Com. (x)	Rank for Maths (y)	Difference d = x − y	d²
80	90	1	1	0	0
56	75	4	2.50	1.50	2.25
50	75	5.5	2.50	3	9
48	65	7	5	2	4
50	65	5.5	5	0.50	0.25
62	50	2	7	−5	25
60	65	3	5	−2	4
Total				0	$\sum d^2 = 44.5$

For Commerce mark there is one tie of length 2 and for Maths mark, there are two ties of lengths 2 and 3 respectively, $n = 7$.

Thus
$$\frac{\sum (t^3 - t)}{12} = \frac{(2^3 - 2) + (2^3 - 2) + (3^3 - 3)}{12} = 3.$$

∴
$$R = 1 - \frac{6\left\{\sum d^2 + \dfrac{\sum(t^3 - t)}{12}\right\}}{n(n^2 - 1)} = 1 - \frac{6(44.5 + 3)}{7(7^2 - 1)}$$

$$= 1 - \frac{6 \times 47.5}{7 \times 48} = 1 - 0.848 \approx 0.15.$$

24.30 Curve fitting

Suppose x and y are two related variables and corresponding to a set of values of x, we have measured a set of values of y. Let the pairs of values thus obtained be (x_1, y_1), (x_2, y_2), ..., (x_n, y_n). If these points are plotted on a graph paper, we get a set of points called *scatter diagram*. From this diagram we can find whether the points have the trend of a curve (this can be done by visualising a smooth curve about which these points cluster).

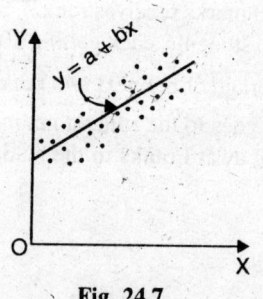

Fig. 24.7.

The general problem of finding approximating curves which fit a given data, is called *curve fitting*.

The scatter diagram of Fig. 24.7 indicates a linear trend and the data can be approximated by a straight line. Obviously no straight line passes exactly through all the points although a number of lines can be drawn which nearly do so.

In the next section we give the procedure to find the line of best fit for a given data. It is based on the method of least squares.

24.31 The lines of best fit

The line of best fit depends upon whether x is independent variable and y the dependent variable; or y independent and x dependent variable. There are two lines of best fit, one corresponding to each of these cases.

(a) x independent and y dependent variable

Procedure to get line of best fit

(i) *Assume that the line approximating the set of points (x_1, y_1) (x_2, y_2), ..., (x_n, y_n) has the equation $y = a + bx$.*   ...(24.66)

(ii) *Form the equations* $\left.\begin{array}{l}\Sigma y = na + b\Sigma x \\ \Sigma xy = a\,\Sigma x + b\Sigma x^2\end{array}\right\}$ ...(24.67)

These equations are called normal equations for this line.

(iii) *Solve the normal equations (24.67) simultaneously for a and b.*

(iv) *Substitute these values of a and b in (24.66). The line thus obtained is the line required.*

(b) y independent and x dependent variable

Procedure to get line of best fit

(i) *Assume the equation to the line to be $x = a + by$*   ...(24.68)

(ii) *Form the normal equations.*

$$\left.\begin{array}{l}\Sigma x = na + b\Sigma y \\ \Sigma xy = a\,\Sigma y + b\Sigma y^2\end{array}\right\}$$ ...(24.69)

(iii) *Solve the equations (24.69) for a and b.*

(iv) *Put these values of a and b in (24.68).*

The resulting line is the line required.

The following solved examples will make the procedure clear.

Example 24.39 Fit a straight line to the following data regarding y as dependent variable.

x:	1	2	3	4	6	8
y:	2.4	3	3.6	4	5	6

Solution. Let the equation to the line be $y = a + bx$

The normal equations to determine a and b are

$$\left.\begin{array}{l}\Sigma y = na + b\Sigma x \\ \Sigma xy = a\,\Sigma x + b\Sigma x^2\end{array}\right\}$$ ...(1)

x	y	xy	x^2
1	2.4	2.4	1
2	3	6	4
3	3.6	10.8	9
4	4	16	16
6	5	30	36
8	6	48	64

Total: $\sum x = 24$ $\sum y = 24$ $\sum xy = 113.2$ $\sum x^2 = 130$

Here $n = 6$, $\sum x = \sum y = 24$, $\sum xy = 113.2$ and $\sum x^2 = 130$.
Putting these values in (1), the normal equations are
$$24 = 6a + 24b, \text{ and } 113.2 = 24a + 130b$$
or $a + 4b = 4$, and $12a + 65b = 56.6$.
On solving, we get $a = 2$, $b = 0.5$.

Hence the required line of best fit is $y = 2 + 0.5x$.

Example 24.40 Fit a straight line to the following data regarding y as independent variable:

x:	1	1.8	3.3	4.5	6.3
y:	0	1	2	3	4

Solution. Let the equation to the line be $x = a + by$

The normal equations to determine a and b are
$$\left.\begin{array}{l} \sum x = na + b\sum y \\ \sum xy = a \sum y + b \sum y^2 \end{array}\right\} \qquad ...(1)$$

x	y	xy	y^2
1	0	0	0
1.8	1	1.8	1
3.3	2	6.6	4
4.5	3	13.5	9
6.3	4	25.2	16

Total: 16.9   10   47.1   30

Here $n = 5$. $\sum x = 16.9$, $\sum y = 10$, $\sum xy = 47.1$, $\sum y^2 = 30$.
Putting these values in (1) we get

$16.9 = 5a + 10b$,   and $47.1 = 10a + 30b$.
On solving, we obtain $a = 0.72$, $b = 1.33$.

$\therefore$ The required line is $x = 0.72 + 1.33y$

Note: When values of variables are large we can reduce simplification work by using suitable transformations (which implies shifting the origin to suitable points). The advantage of this will be clear from the following example.

Example 24.41 Fit a straight line to the following data showing production of a commodity in different years in certain state and find the trend value for 2001.

Year	Production ('000 tons)
2001	10
2002	12
2003	8
2004	10
2005	14

Solution. Let the year be denoted by x and production by y. Since the trend value for a given year is to be found, x is independent and y dependent variable. Now if we keep the values of x as they are, the computation work will be quite laborious. To avoid this we put $X = x - 2003$. Now X is the independent variable.

Let the line of best fit be $y = a + bX$. The normal equations are

$$\Sigma y = na + b \Sigma X$$
$$\Sigma Xy = a\Sigma X + b \Sigma X^2$$

X	y	Xy	X^2
-2	10	-20	4
-1	12	-12	1
0	8	0	0
1	10	10	1
2	14	28	4
Total = 0	54	6	10

Here $n = 5$, $\Sigma X = 0$, $\Sigma y = 54$, $\Sigma Xy = 6$, $\Sigma X^2 = 10$.

$\therefore$ The normal equations are

$54 = 5a + b(0)$ and $6 = a(0) + 10b$ or $5a = 54$ and $10b = 6$.

$\therefore$ $a = 10.8$, $b = 0.6$

Hence the line is $y = 10.8 + 0.6X$. Putting $X = x - 2003$ it becomes

$$y = 10.8 + 0.6(x - 2003)$$

which is the line of best fit required.

The trend value for $x = 2001$ *i.e.,* the value of y when $x = 2001$ is

$y = 10.8 + 0.6(2001 - 2003)$ or $y = 10.8 + (0.6)(-2) = 9.6$.

24.32 Lines of Regression

The lines of best fit of Art 24.31 are also known as *lines of regression*.

(*i*) *When x is treated as independent variable and y as dependent then the line is called* **the line of regression of y on x.**

(*ii*) *When y is treated as independent variable and x as dependent variable, the line is known as* **the line of regression of x on y.**

From the solved examples of Art. 24.31 we know how to find the line of regression of y on x and also the line of regression of x on y.

In example 24.39, we found the line of regression of y on x as $y = 2 + 0.5x$.

In example 24.40, the equation of the line of regression of x on y is

$$x = 0.72 + 1.33\, y.$$

24.33 Correlation and Regression

We have seen that the line of regression of y on x (which is the line of best fit when x is treated as independent variable and y as dependent variable) is

$$y = a + bx \qquad \qquad …(24.70)$$

where a and b are given by the normal equations

$$\Sigma y = na + b\Sigma x \qquad \qquad …(24.71)$$
$$\Sigma xy = a\Sigma x + b\Sigma x^2 \qquad \qquad …(24.72)$$

and n is the number of pairs of values of x and y.

Equation (24.71) can be written in the form

$$\frac{\Sigma y}{n} = a + b\frac{\Sigma x}{n} \quad \text{or} \quad \bar{y} = a + b\bar{x} \qquad …(24.73)$$

Equation (24.73) shows that the point $(\bar{x}, \bar{y})$ lies on (24.70).

Thus the line of regression passes through $(\bar{x}, \bar{y})$, where $\bar{x}$ is the mean of x's and $\bar{y}$ the mean of y's.

Shifting the origin to the point $(\bar{x}, \bar{y})$ we find that (24.70) becomes

$$y - \bar{y} = a + b(x - \bar{x}) \qquad \qquad …(24.74)$$

and (24.71) reduces to the form $\Sigma(y - \bar{y}) = na + b\Sigma(x - \bar{x})$.

Since $\Sigma(y - \bar{y}) = 0$ and $\Sigma(x - \bar{x}) = 0$ (Art. 24.6) Therefore, $a = 0$.

Also (24.72) becomes (as a result of shifting the origin)

$$\Sigma(x - \bar{x})(y - \bar{y}) = b\Sigma(x - \bar{x})^2 \qquad (\because a = 0)$$

or $$\Sigma XY = b\Sigma X^2,$$

where X, Y are deviations of x and y from their means.

$$\therefore \qquad b = \frac{\Sigma XY}{\Sigma X^2} = \frac{\Sigma XY}{n\sigma_X^2} = r\frac{\sigma_Y}{\sigma_X}. \qquad \left(\because r = \frac{\Sigma XY}{n\sigma_X\sigma_Y}\right)$$

Putting the values of a and b in (24.74), the line of regression of y on x beomes

$$y - \bar{y} = r\frac{\sigma_y}{\sigma_x}(x - \bar{x}).$$

$r\dfrac{\sigma_Y}{\sigma_X}\left(= b_{YX}\right)$ is called the **regression coefficient of y on x.** It is, obviously, the slope of this line.

Interchanging x and y we find that the line of regression of x on y is

$$x - \bar{x} = r\frac{\sigma_X}{\sigma_Y}(y - \bar{y}). \qquad\qquad ...(24.75)$$

Here $r\dfrac{\sigma_x}{\sigma_y}\left(= b_{xy}\right)$ is the **regression coefficient of x on y.**

From (24.75) it is clear that this line also passes through $\left(\bar{x}, \bar{y}\right)$.

Summary of important results concerning regression lines.

(i) *When two variables are related, there are two regression lines;*
  (a) *line of regression of y on x,*
  (b) *line of regression of x on y.*
(ii) *In the case of first line, x is independent variable and in the second case y is the independent variable.*
(iii) *The two lines pass through* $\left(\bar{x}, \bar{y}\right)$ *i.e., the means lie on both the lines. Hence, when the equations of these lines are given,* $\bar{x}, \bar{y}$ *can be found by solving them.*

(iv) *When the line of regression of y on x is written in the form* $y = a + bx$, *then*

$b = r\dfrac{\sigma_Y}{\sigma_X}$ *which is called the coefficient of regression of y on x. Obviously it is the slope of this line.*

Similarly, when the line of regression of x on y is written in the form

$x = c + dy$ *then* $d = r\dfrac{\sigma_X}{\sigma_Y}$ *which is called the regression coefficient of x on y. It is not the slope of this line.*

(v) $r^2 = \left(r\dfrac{\sigma_Y}{\sigma_X}\right)\left(r\dfrac{\sigma_X}{\sigma_Y}\right)$ *i.e., product of regression coefficient* $= k^2$ *(say).*

$\therefore$ $r = \pm k.$ **We retain + sign when both regression coeff. are+ve. Otherwise r is negative.**

24.34 Illustrative Examples

Example 24.42 Using the following data, obtain the two regression equations

x:	14	19	24	21	26	22	15	20	19
y	31	36	48	37	50	45	33	41	39

Solution. Presenting the data in tabular form, we get

x	y	$X = x - \bar{x}$	$Y = y - \bar{y}$	XY	X^2	Y^2
14	31	-6	-9	54	36	81
19	36	-1	-4	4	1	16
24	48	4	8	32	16	64
21	37	1	-3	-3	1	9
26	50	6	10	60	36	100
22	45	2	5	10	4	25
15	33	-5	-7	35	25	49
20	41	0	1	0	0	1
19	39	-1	-1	1	1	1
Total 180	360	0	0	193	120	346

$$\bar{x} = \frac{180}{9} = 20, \bar{y} = \frac{360}{9} = 40,$$

$$b_{XY} = \frac{\Sigma XY}{\Sigma Y^2} = \frac{193}{346} = 0.558 \text{ and } b_{YX} = \frac{\Sigma XY}{\Sigma X^2} = \frac{193}{120} = 1.608.$$

Regression equation of x on y is given by

$$x - \bar{x} = b_{xy}(y - \bar{y}) \text{ or, } x - 20 = 0.558(y - 40) \text{ i.e. } x = 0.558y - 2.32$$

Regression equation of y on x is given by

$$y - \bar{y} = b_{YX}(x - \bar{x}) \text{ or } y - 40 = 1.608(x - 20) \text{ that is, } y = 1.608x + 7.84$$

Example 24.43 From the following data find the regression equations, and estimate the likely value of y when x = 100.

x:	78	89	97	69	59	79	68	61
y:	125	137	156	112	107	136	123	108

Solution. We construct the following table:

x	y	$d_x = x - 78$	$d_y = y - 125$	$d_x d_y$	d_X^2	d_Y^2
78	125	0	0	0	0	0
89	137	11	12	132	121	144
97	156	19	31	589	361	961
69	112	-9	-13	117	81	169
59	107	-19	-18	342	361	324
79	136	1	11	11	1	121
68	123	-10	-2	20	100	4
61	108	-17	-17	289	289	289
Total $\Sigma x =$ 600	$\Sigma y =$ 1004	$\Sigma d_x =$ -24	$\Sigma d_y =$ 4	$\Sigma d_x d_y =$ 1500	$\Sigma d_X^2 =$ 1314	$\Sigma d_{Y}^2 =$ 2012

$$\bar{x} = \frac{600}{8} = 75 \text{ and } \bar{y} = \frac{1004}{8} = 125.5.$$

$$b_{xy} = \frac{n\Sigma d_X d_Y - \Sigma d_X \Sigma d_Y}{n\Sigma d_Y^2 - (\Sigma d_y)^2} = \frac{8(1500) - (-24)(4)}{8(2012) - (4)^2} = \frac{12096}{16080} = 0.752.$$

$$b_{yx} = \frac{n\Sigma d_X d_Y - \Sigma d_X \Sigma d_Y}{n\Sigma d_X^2 - (\Sigma d_x)^2} = \frac{8(1500) - (-24)(4)}{8(1314) - (-24)^2} = \frac{12096}{9936} = 1.217.$$

Regression equation of x on y is given by

$x - \bar{x} = b_{xy}(y - \bar{y})$ or $x - 75 = 0.752(y - 125.5)$ i.e., $x = 0.752\,y - 19.376.$

Regression equation of y on x is given by

$y - \bar{y} = b_{yx}(x - \bar{x})$ or $y - 125.5 = 1.217(x - 75)$

that is, $y = 1.217x + 34.225.$

when $x = 100$, $y = 1.217 \times 100 + 34.225 = 155.93$.

Example 24.44 Find the regression equation from the following data;
$\Sigma x = 15$, $\Sigma y = 25$, $\Sigma xy = 83$, $\Sigma x^2 = 55$, $\Sigma y^2 = 135$, $n = 5$. Estimate the values of x and y, if $y = 12$, $x = 8$.

Solution. $\bar{x} = \Sigma x / n = 15/5 = 3$, $\bar{y} = \Sigma y / n = 25/5 = 5$.

$$b_{XY} = r \cdot \frac{\sigma_X}{\sigma_Y} = \frac{n\Sigma xy - \Sigma x \Sigma y}{n\Sigma y^2 - (\Sigma y)^2} = \frac{5(83) - (15)(25)}{5(135) - (25)^2} = \frac{83 - 75}{135 - 125} = \frac{8}{10} = 0.8.$$

$$b_{YX} = r \cdot \frac{\sigma_Y}{\sigma_X} = \frac{n\sum xy - \sum x \sum y}{n\sum x^2 - (\sum x)^2} = \frac{5(83) - (15)(25)}{5(55) - (15)^2} = \frac{83 - 75}{55 - 45} = \frac{8}{10} = 0.8$$

Regression equation of x on y is given by

$$x - \bar{x} = b_{XY}(y - \bar{y}) \quad \text{or} \quad x - 3 = 0.8(y - 5) \text{ that is, } x = 0.8y - 1 \; ...(i)$$

Regression equation of y on x is given by $y - \bar{y} = b_{YX}(x - \bar{x})$

or $y - 5 = 0.8(x - 3)$ that is, $y = 0.8x + 2.6$.   ...(ii)

The line of regression of x on y gives the best estimated value of x for given y. On putting $y = 12$, in (i), we obtain $x = 8.6$.

The line of regression of y on x gives the best estimated value of y on x. We put $x = 8$ in (ii), and obtain $y = 9$.

Example 24.45 **Regression equations of two variables x and y are as follows: 100 x + 87 y − 1913 = 0, and 50x + 100y − 1164 = 0. Find (i) the means, (ii) the regression coefficients, (iii) the coefficient of correlation between x and y.** *(Mysore, 1997S)*

Solution. (i) Since regression equations meet at $(\bar{x}, \bar{y})$, solution to the set of regression equations gives the respective means.

Solving the given equations, we find $x = 15.94$ and $y = 3.67$.

Therefore, $\bar{x} = 15.94$ and $\bar{y} = 3.67$.

(ii) Let us assume that $100x + 87y - 1913 = 0$ is the regression equation of

x on y. This may be written as $x = 19.13 - 0.87y$.

$\therefore \quad b_{XY} = r \cdot \dfrac{\sigma_x}{\sigma_y} = -0.87.$

The other equation $50x + 100y - 1164 = 0$ will then be the regression equation of y on x, which may be written as $y = 11.64 - 0.50x$.

$\therefore \quad b_{YX} = r \cdot \dfrac{\sigma_Y}{\sigma_X} = -0.50.$

Since $b_{XY} \cdot b_{YX} = (-0.87), (-0.50) < 1$, therefore our supposition is correct.

(iii) $r = \sqrt{b_{XY} \cdot b_{YX}} = \sqrt{-0.87 \times (-0.50)} = \pm 0.66$

As the regression coefficients have negative signs, we have to consider negative value for r. Thus $r = -0.66$.

Example 24.46 **In a partially destroyed laboratory record of an analysis of correlation data, the following results only are legible:**

Variance of X = 9, Regression equations 8X − 10Y + 66 = 0, 40X − 18Y = 214. Find on the basis of the above information:

(*i*) The mean values of X and Y, (*ii*) coefficient of correlation between X and Y, and (*iii*) standard deviation of Y.

Solution. (*i*) *Calculation of Mean:* Since the lines of regression pass through the mean value *i.e.* through $(\overline{X}, \overline{Y})$, we have $8\overline{X} - 10\overline{Y} + 66 = 0$ and $40\overline{X} - 18\overline{Y} - 214 = 0$. On solving these equations, we get $\overline{X} = 13, \overline{Y} = 17$.

(*ii*) For finding out the correlation coefficient, we will have to find out the regression coefficients. Since we donot know which of the two regression equations is the equation of X on Y, we make an assumption. Let us take equation $8X - 10Y + 66 = 0$ as the regression equation of X on Y.

This may be written as $X = \dfrac{5}{4}Y - \dfrac{33}{4}$. $\quad \therefore \quad b_{xy} = r \cdot \dfrac{\sigma_x}{\sigma_y} = \dfrac{5}{4}$.

The other equation $40X - 18Y - 214 = 0$ will then be the regression equation of Y on X, which may be written as

$$Y = \dfrac{20}{9}X - \dfrac{107}{9}. \qquad \therefore \quad b_{yx} = r \cdot \dfrac{\sigma_y}{\sigma_x} = \dfrac{20}{9}.$$

Since both the regression coefficients are exceeding 1, our assumption is wrong. Hence $40X - 18Y - 214 = 0$ is the regression equation of X on Y and $8X - 10Y + 66 = 0$ is the regression equation of Y on X.

$$\therefore \quad b_{xy} = \dfrac{9}{20} \text{ and } b_{yx} = \dfrac{4}{5}.$$

Thus $r = \sqrt{b_{xy} \cdot b_{yx}} = \sqrt{\left(\dfrac{9}{20}\right) \times \left(\dfrac{4}{5}\right)} = \pm 0.6$. As the regression coefficients have positive signs, we have to-consider positive value of r. Hence $r = 0.6$.

(*iii*) $\sigma_x = \sqrt{9} = 3, r \cdot \dfrac{\sigma_x}{\sigma_y} = b_{xy} = \dfrac{9}{20}$. Therefore $\sigma_y = 20(0.6)(3)/9 = 4$.

Example 24.47 You are given the following data:

Series	X	Y
Arithmetic mean	36	85
Standard deviation	11	8

Correlation coefficient between X and Y = 0.66. Calculate the regression equations and estimate the value of X when Y = 75.

Solution. We have $\overline{X} = 36, \overline{Y} = 85, \sigma_X = 11, \sigma_Y = 8$ and $r = 0.66$. Regression equation of X on Y is given by

$$X - \overline{X} = r \cdot \dfrac{\sigma_x}{\sigma_y}(Y - \overline{Y}).$$ Substituting the given values, we obtain

$$X - 36 = 0.66 \times \frac{11}{8}(Y - 85) = 0.9075(Y - 85) = 0.9075Y - 77.1375$$

or $X = 0.9075Y - 41.1375$.

Substituting $Y = 75$ in the above equation, we get

$$X = 0.9075(75) - 41.1375 = 68.0625 - 41.1375 = 26.925.$$

Therefore, when $Y = 75$, $X = 26.93$.

Regression equation of Y on X is given by

$$Y - \bar{Y} = r \cdot \frac{\sigma_y}{\sigma_x}(X - \bar{X}).$$ Substituting the given values, we obtain

$$Y - 85 = 0.66 \times \frac{8}{11}(X - 36) = 0.48(X - 36) = 0.48X - 17.28$$

or $Y = 0.48X + 67.72$.

Example 24.48 If θ is the acute angle between the two regression lines in case of two variables x and y show that

$$\tan \theta = \frac{1 - r^2}{r} \cdot \frac{\sigma_x \sigma_y}{\sigma_x^2 + \sigma_y^2}, \text{ where } r, \sigma_x, \sigma_y, \text{ have their usual meanings.}$$

Explain the significance when r = 0 and r = ±1.

(V.T.U., 2001, Andhra, 2000)

Solution. The equations to the line of regression of y on x and x on y are

$$y - \bar{y} = r\frac{\sigma_y}{\sigma_x}(x - \bar{x}) \text{ and } x - \bar{x} = r\frac{\sigma_x}{\sigma_y}(y - \bar{y}) \text{ respectively.}$$

The slope of first line, $m_1 = r(\sigma_y / \sigma_x)$ and slope of second line,

$m_2 = \sigma_y / r(\sigma_x)$.

Since θ is the acute angle between the lines. Then

$$\tan \theta = \frac{m_2 - m_1}{1 + m_1 m_2} = \frac{\dfrac{\sigma_y}{r\sigma_x} - r\dfrac{\sigma_y}{\sigma_x}}{1 + \left(\dfrac{\sigma_y}{r\sigma_x}\right)\left(\dfrac{r\sigma_y}{\sigma_x}\right)} = \frac{\dfrac{\sigma_y}{\sigma_x}\left(\dfrac{1}{r} - r\right)}{1 + \dfrac{\sigma_y^2}{\sigma_x^2}} \text{ or } \tan \theta = \frac{1 - r^2}{r} \cdot \frac{\sigma_x \sigma_y}{\sigma_x^2 + \sigma_y^2}.$$

(i) When $r = 0$, that is, there is no correlation between the variables, then $\tan \theta = \infty \Rightarrow \theta = \pi/2$. So, the two lines of regression are perpendicular to each other.

(ii) When $r = \pm 1$, $\tan \theta = 0$ or $\theta = 0$ or π. The two lines of regression coincide. Hence when $r = \pm 1$ that is, there is perfect positive or negative correlation between x and y, the lines of regression coincide.

Example 24.49 Establish the formula

$$\sigma^2_{x-y} = \sigma^2_x + \sigma^2_y - 2r\,\sigma_x\,\sigma_y,$$

where **r** is the correlation coefficient between **x** and **y**.

Solution. Let $u = x - y$ so that $\bar{u} = \bar{x} - \bar{y}$.

$$\therefore \quad u - \bar{u} = (x - \bar{x}) - (y - \bar{y})$$

or $\quad (u - \bar{u})^2 = (x - \bar{x})^2 + (y - \bar{y})^2 - 2(x - \bar{x})(y - \bar{y}).$

Summing up for n terms, we have

$$\Sigma(u - \bar{u})^2 = \Sigma(x - \bar{x})^2 + \Sigma(y - \bar{y})^2 - 2\Sigma(x - \bar{x})(y - \bar{y})$$

or $\quad \dfrac{\Sigma(u - \bar{u})^2}{n} = \dfrac{\Sigma(x - \bar{x})^2}{n} + \dfrac{\Sigma(y - \bar{y})^2}{n} - 2\dfrac{\Sigma(x - \bar{x})(y - \bar{y})}{n}$

or $\quad \sigma^2_u = \sigma^2_x + \sigma^2_y - 2r\sigma_x\sigma_y$
$\qquad\qquad\qquad\qquad\qquad \left[\because r = \dfrac{\Sigma(x - \bar{x})(y - \bar{y})}{n\sigma_x\sigma_y} \right]$

i.e., $\sigma^2_{x-y} = \sigma^2_x + \sigma^2_y - 2r\sigma_x\sigma_y$, the required result.

Example 24.50 Show that the r, the correlation coefficient always lies between -1 and $+1$.

Solution. Let there be n pairs of observation $(x_1, y_1)(x_2, y_2)......(x_n, y_n)$.

Then $r(X, Y) = \dfrac{\sum_{i=1}^{n}(x_i - \bar{X})(y_i - \bar{Y})}{\sqrt{\sum_{i=1}^{n}(x_i - \bar{X})^2}\,\sqrt{\sum_{i=1}^{n}(y_i - \bar{Y})^2}}.$

Let $x_i - \bar{X} = A_i$ and $y_i - \bar{Y} = B_i$, $i = 1, 2, ...n.$ Then

$r(X, Y) = \dfrac{\sum_{i=1}^{n} A_i B_i}{\sqrt{\sum_{i=1}^{n} A_i^2}\,\sqrt{\sum_{i=1}^{n} B_i^2}}.$ By Cauchy-Schwartz's Inequality, we have

$$\left(\sum_{i=1}^{n} A_i B_i\right)^2 \le \left(\sum_{i=1}^{n} A_i^2\right)\left(\sum_{i=1}^{n} B_i^2\right) \Rightarrow \dfrac{\left(\sum_{i=1}^{n} A_i B_i\right)^2}{\left(\sum_{i=1}^{n} A_i^2\right)\left(\sum_{i=1}^{n} B_i^2\right)} \le 1$$

$$\Rightarrow \left[\dfrac{\sum_{i=1}^{n} A_i B_i}{\sqrt{\sum_{i=1}^{n} A_i^2}\,\sqrt{\sum_{i=1}^{n} B_i^2}}\right]^2 \le 1 \Rightarrow \left[r(X, Y)\right]^2 \le 1 \Rightarrow -1 \le r(X, Y) \le 1.$$

Exercise 24.1

1. Compute the arithmetic mean from the following frequency table:

Height in inches:	58	60	62	64	66	68
No. of persons:	12	14	20	13	8	5

Ans. 67.12 inches approx.

2. Find the mean for the following frequency distribution:

Class interval:	0–8	8–16	16–24	24–32	32–40	40–48
Frequency:	8	7	16	24	15	7

Ans. 25.4

3. The mean of 200 items was 50. Later on, it was discovered that two items were misread as 92 and 8 instead of 192 and 88. Find the correct mean.

Ans. 50.9

[**Hint:** $\bar{x} = \dfrac{\sum x_i}{n}$. Therefore $\sum x_i = n \cdot x = 200 \times 50 = 10,000$.

Correct total: $10,000 - (92 + 8) + (192 + 88) = 10180$ and

Correct mean = $10180/200 = 50.9$].

4. (a) Find the missing frequency from the following data:

Marks:	0–10	10–20	20–30	30–40	40–50	50–60
No. of students:	5	15	20	–	20	10

The arithmetic mean is 34 marks. **Ans.** 180

(b) Given that the mean height of a group of students is 67.45 inches. Find the missing frequencies for the following incomplete distribution of height of 100 students.

Height in inches:	60–62	63–65	66–68	69–71	72–74
No. of persons:	5	18	–	–	8

Ans. 42, 27.

5. (a) Calculate the median for the following data:

Marks obtained	No. of students	Marks obtained	No. of students
Less than 20	0	Less than 70	66
Less than 30	4	Less than 80	82
Less than 40	16	Less than 90	92
Less than 50	30	Less than 100	100
Less than 60	46		

Ans. 62

(b) Compute median from the following data:

Mid-value	Frequency	Mid-value	Frequency
115	6	165	60
125	25	175	38
135	48	185	22
145	72	195	3
155	116		

Ans. 153.8

6. (a) Find the mode for the following distribution:

Class interval	0-10	10-20	20-30	30-40	40-50	50-60	60-70	70-80
Frequency	5	8	7	12	28	20	10	10

(*A.M.I.E., S-2006*) **Ans. 47.67**

(b) The following are the marks obtained by students in a class test. Find the modal marks.

Marks	Students	Marks	Students
32–35	10	48–51	51
36–39	37	52–55	35
40–43	65	56–59	18
44–47	80	60–63	4

[Hint: Take $l_1 = 43.5, h = 4$] **Ans. 44.9**

7. Calculate the mode and the median for the following frequency distribution:

Marks	1-5	6-10	11-15	16-20	21-25	26-30	31-35	36-40	41-45
No. of students	7	10	16	32	24	18	10	5	1

Ans. 18.4, 19.56

8. Find the mean deviation from the mean for the following data:

x_i	5	7	9	10	12	15
f_i	8	6	2	2	2	6

Ans. 3.39

9. (a) Find the mean deviation from the mean for the following data:

Classes	0-10	10-20	20-30	30-40	40-50	50-60
Frequencies	6	8	14	16	14	2

Ans. 10.24

(b) Find the mean deviation about the mean for the following data.

Marks obtained	10-20	20-30	30-40	40-50	50-60	60-70	70-80
No. of students	2	3	8	14	8	3	2

Ans. 10

10. Find the mean deviation from the median for the following data:

x_i	15	21	27	30	35
f_i	3	5	6	7	8

Ans. 5.1

11. The crushing strength of 8 cement concrete experimental blocks, in metric tonnes per sq. cm was 4.8, 4.2, 5.1, 3.8, 4.4, 4.7, 4.1, 4.5.
Find the mean crushing strength and the standard deviation.

Ans. 4.45, 0.36

12. Compute the standard deviation from the following distribution of marks obtained by 90 students:

Marks	20-29	30-39	40-49	50-59	60-69	70-79	80-89	90-99
No. of students	5	12	15	20	18	10	6	4

Ans. 17.65 approximate

13. From the following frequency distribution, compute the standard deviation of 100 students:

Mass in kg	60–62	63–65	66–68	69–71	72–74
No. of students	5	18	42	27	8

Ans. 2.92

14. (a) Calculate the arithmetic average and the standard deviation of the following series:

Expenditure	No. of Students
Below Rs. 5	6
Below Rs. 10	16
Below Rs. 15	28
Below Rs. 20	38
Below Rs. 25	46

Ans. Rs. 6.5 approx.

(b) The mean and standard deviation of 200 items are found to be 60 and 20 respectively. While calculating these measures, two items were wrongly taken as 3 and 67 instead of 13 and 17, find the correct mean and standard deviation. Also determine the correct coefficient of variation.

[**Hint.** We have $\bar{x} = 60$, $\sigma = 20$, $n = 200$. Therefore, $\Sigma x_i = n\bar{x} = 200 \times 60 = 12,000$. Correct $\Sigma x_i = 12000 - (3 + 67) + (13 + 17) = 11960$.

$\therefore$ Correct $\bar{x} = 11960/200 = 59.8$.

Employing, $\sigma = \sqrt{\dfrac{\Sigma x_i^2}{n} - (\bar{x})^2}$ to get $\Sigma x_i^2 = \left\{20^2 + (60)^2\right\}200 = 8,00,000$

Correct $\Sigma x_i^2 = 8,00,000 - (3)^2 - (67)^2 + (13)^2 + (17)^2 = 795960$

$\therefore$ Correct $\sigma = \sqrt{\dfrac{795960}{200} - (59.8)^2} = 20.094$.

Correct C.V. $= \dfrac{\sigma}{\bar{x}} \times 100 = \dfrac{20.094}{59.8} \times 100 = 33.6\%$].

15. Use the frequency distribution of heights in the following table to find the mean height and standard deviation of the 100 male students at *xyz* university.

Height in inch	60–62	63–65	66–68	69–71	72–74
Frequency	5	18	42	27	8

Ans. 67.45; 2.92 approx.

16. The following table shows the marks obtained by 100 candidates in an examination. Calculate the mean, median, mode and standard deviation.

Marks obtained:	1–10	11–20	21–30	31–40	41–50	51–60
No. of candidates:	3	16	26	31	16	8

Ans. 32, 32.3, 33.0, 12.37

17. Calculate the mean, variance and standard deviation for the following frequency distribution:

Classes:	30–40	40–50	50–60	60–70	70–80	80–90	90–100
Frequencies:	3	7	12	15	8	3	2

Ans. 62; 201; 14.17

18. Following runs are scored by two batsmen A and B in a series of innings.

A:	5	26	97	76	112	89	6	108	24	16
B:	51	47	36	60	58	39	44	42	71	50

Find their mean scores and standard deviations. Who is better run scorer and who is more consistent?

Ans. 55.9, 42.0, 49.8, 10.16,: A is better scorer because mean of his scores is greater; B is more consistent because S.D. of his scores is less.

19. Calculate the coefficient of correlation from the following data:

x:	9	8	7	6	5	4	3	2	1
y:	15	16	14	13	11	12	10	8	9

Ans. 0.95

20. Calculate the value of r between X and Y for the values given below:

X:	2	5	7	9	19	17
Y:	25	27	26	29	34	35

Ans. 0.97

21. Find the coefficient of correlation for the following table:

x:	10	14	18	22	26	30
y:	18	12	24	6	30	35

Ans. 0.6

22. Calculate the coefficient of correlation for the following table:

x→ y↓	0–4	4–8	8–12	12–16	Total
0-5	7	0	0	0	7
5-10	6	8	0	0	14
10-15	0	5	3	0	8
15-20	0	7	2	0	9
20-25	0	0	0	9	9
Total	13	20	5	9	47

Ans. 0.8

23. From the following data, compute the coefficient of correlation between x and y.

	x-series	y-series
No. of items:	15	15
Arithmetic mean:	25	18
Sum of squares of deviations from mean	136	138

Summation of product of deviations of x and y series from the respective arithmetic means = 122.

Ans. 0.891

24. From the following data, obtain the value of the correlation coefficient:

$n = 10, \Sigma x = 55, \Sigma y = 40, \Sigma x^2 = 385, \Sigma y^2 = 192$ and $\Sigma(x+y)^2 = 947$.

Ans. −0.6812

25. Find the coefficient of correlation from the following data:

x:	65	63	61	64	68	62	70	66
y:	68	66	68	65	69	66	68	65

Ans. 0.359

26. Examine whether there is any correlation between age and blindness on the basis of the following data:

Age in yrs:	0-10	10-20	20-30	30-40	40-50	50-60	60-70	70-80
No. of persons (in thousands):	90	120	140	100	80	60	40	20
No. of blind persons:	10	15	18	20	15	12	10	6

[**Hint:** Let us denote the mid-value of age in years as x and the no. of blind persons per lakh as y.

Age in yrs	Mid-value x	No. of Persons, P (in thousands)	No. of blind B	No. of blind per lakh $y = (B/P) \times 100$	xy	x^2	y^2
0-10	5	90	10	11	55	25	121
10-20	15	120	15	12	180	225	144
20-30	25	140	18	13	325	625	169
30-40	35	100	20	20	700	1225	400
40-50	45	80	15	19	855	2025	361
50-60	55	60	12	20	1100	3025	400
60-70	65	40	10	25	1625	4225	625
70-80	75	20	6	30	2250	5625	900
Total	320	–	–	150	7090	17000	3120

Substituting the values in the formula

$$r = \frac{n\sum xy - (\sum x)(\sum y)}{\sqrt{\left[n\sum x^2 - (\sum x)^2\right]\left[n\sum y^2 - (\sum y)^2\right]}}$$

$$= \frac{8 \times 7090 - 320 \times 150}{\sqrt{\left[8 \times 17000 - (320)^2\right]\left[8 \times 3120 - (150)^2\right]}}$$

$$= \frac{8720}{183.303 \times 49.5984} = 0.96,$$ which exhibits a very high degree of positive correlation between age and blindness.]

27. In two sets of variables x and y with 20 observations each, the following data were observed:

$\bar{x} = 12, \bar{y} = 15$, S.D.of $x = 3$, S.D.of $y = 4$.

Coefficient of correlation between x and y is 0.4. Later on, it was found that the pair (20, 15) was wrongly taken as (15, 20). Find the correct value of the correlation coefficient. **Ans.** 0.314.

28. Two judges in a beauty contest rank the ten competitors in the following order:

6	4	3	1	2	7	9	8	10	5
4	1	6	7	5	8	10	9	3	2

Do the judges appear to agree in their standard.

Ans. $R = 0.224$. Since R is +ve, the judges appear to agree. The agreement is complete only if $R = 1$.

29. Marks of 12 students in Arithmetic and Algebra are given below:

Arithmetic:	60	34	40	50	45	40	22	43	42	66	64	46
Algebra:	75	32	33	40	45	33	12	30	34	72	41	57

Calculate the rank correlation coefficient. **Ans.** 0.84

30. Find the rank correlation between marks obtained in Mathematics and statistics.

Students:	A	B	C	D	E	F	G	H	I	J
Marks in Maths:	30	20	40	50	30	20	30	50	10	0
Marks in Statistics :	15	40	40	45	20	30	15	50	20	10

Ans. 0.63

31. Find the equations of regression lines for the following values of x and y.

x:	1	2	3	4	5
y:	2	5	3	8	7

Hence find the value of r. **Ans.** $x = 0.5y + 0.5$, $y = 1.3x + 1.1$, $r = 0.81$

32. (a) The observed values of a function are respectively 168, 120, 72 and 63 at the four positions 3, 7, 9 and 10 of the independent variable. What is the best estimate you can give for the value of the function at the position 6 of the independent variable?

[**Hint:** The line of regression of y on x is $y = 217.91 - 15.47x$. When $x = 6$, $y = 125.09$, the required best estimate].

33. Heights of fathers and sons are given in inches:

Height of father:	65	66	67	67	68	69	71	73
Height of son:	67	68	64	68	72	70	69	70

Find the two lines of regression and find the expected average height of son when the height of father is 67.5 inches.

Ans. $y = 0.421x + 39.77$, $x = 0.52y + 32.63$; 68.19

34. The following regression equations were obtained from a correlation table:
$$y = 0.516x + 33.73, \quad x = 0.512y + 33.52.$$
Find the value of (i) r (ii) the mean of x's (iii) the mean of y's:

Ans. $r = 0.514$, $\bar{x} = 69.0268$, $\bar{y} = 69.3476$.

35. From a partially destroyed laboratory only following records could be available: $x = 4y + 5$ and $y = kx + 4$ are the regression lines of x on y and y on x respectively. Show that $0 \le k \le 1/4$ and if $k = 1/16$, find the means of two variables and the coefficient of correlation. Also if var. $x = 9$, find var. y when $k = 1/16$. **Ans.** 28, 23/4; 1/2; $\sigma_y = 3/8$.

36. The following data are given for marks in English and Mathematics in an examination:

Mean marks in English   = 39.5,
Mean marks in Maths   = 47.6
r between marks in English and Maths = 0.42
S.D. of marks in English   = 10.8,
S.D. of marks in Maths   = 16.9.
Form the two lines of regression, calculate the expected average marks in Maths of a candidate who secured 50 marks in English.

Ans. $y - 47.6 = 0.657 (x - 39.5)$, $x - 39.5 = 0.268 (y - 47.6)$, $y = 54.5$.
where x denotes marks in English and y those in Maths.

37. In a partially destroyed laboratory record of an analysis of correlation data, the following results are legible.
Variance of $x = 9$.
Regression equations: $4x - 5y + 33 = 0$ and $20x - 9y = 107$.
What were (a) the mean values of x and y (b) the standard deviation of y (c) the coeff. of correlation between x and y? **(Madras: 1999)**

Ans. $\bar{x} = 13$, $\bar{y} = 17$, $\sigma_y = 4$, $r = 0.6$

38. Establish the formula
$$\sigma_{x+y}^2 = \sigma_x^2 + \sigma_y^2 + 2r\,\sigma_x\sigma_y,$$ where r is the correlation coeff. between x and y.

39. If the regression coefficient of x on y is $-1/6$ and that of y on x is $-3/2$, what is the correlation coefficient between x and y? **Ans.** -0.5

40. Following are the marks in Hindi & English in an annual examination:

	Hindi (x)	English (y)	
Mean	40	50	
S.D.	10	16	$r = +0.5$

Estimate the score of English when score in Hindi is 50 and score in Hindi when score in English is 30.   **Ans.** 58; 34

41. The equation of two lines of regression obtained in a correlation analysis are the followings:

    $2x + 3y - 8 = 0$ and $x + 2y - 5 = 0$.

    Obtain the value of the correlation coefficient and the variance of y, given that the variance of x is 12.   **Ans.** −0.866; 2

42. If $x - (a-2)y + 9 = 0$ and $ax - 3y + 9 = 0$ are the lines of regression of x on y and y on x respectively. Show that $-1 \le a \le 3$.

24.35 Probability

Introduction

The word *"probable"* means "likely", or "most likely to be true", or "likely, though not certain to occur". In other words, we have an uncertain situation where we cannot state with certainty what outcome will really be.

Suppose, we conduct a certain experiment like throwing a die, a card is drawn from a pack of 52 playing cards etc. If a die is tossed in the air, then it is certain that the die will come down, but it is not certain that, say, a 5 will appear. Now consider another experiment that a card is drawn from a well-shuffled pack of playing cards. When a card is drawn from a well-shuffled pack of playing cards, we know that there are 52 cards in a pack of playing cards and the card drawn may be any one of these 52 cards. so there are 52 possible outcomes of this experiment. But we cannot say with certainty which of these 52 outcomes will materialise. In fact, the above statement involves an element of uncertainty. Probability theory is designed to estimate the degree of uncertainty regarding the happening of a given phenomenon.

Before defining probability we shall explain the following basic, widely used concepts and terms.

24.36 Random Experiments and Events

Random Experiment. Any operation that results in two or more outcomes is called an experiment. When all the outcomes of an experiment can be enumerated but which particular outcome will result is not known, the experiment is called a random experiment. For example if a fair coin is tossed, then it has two possible outcomes–"Head up or Tail up" but which face will actually turn up is not known. Similarly, result of rolling a fair die, result of drawing a card from a pack of 52 playing cards–are random events or random experiments. In all these experiments, there are more than one possible outcomes but we are not sure which one of these outcomes actually occurs.

Events: One or more possible outcomes of an experiment are said to form an event, that is any subset of the sample space is an event. For example when two fair coins are tossed, the events are two tails appear, one tail appears, no tail appear, when a card is drawn from a well shuffled pack of playing cards, the following events may be considered: The card is spade. The card is an ace. The card is heart or diamond. An event is said to be *simple* if it corresponds to a single possible outcome of an experiment, whereas the joint occurrence of two or more simple events is called a *compound* event.

For example, tossing a coin once provides us two simple events namely head and tail. On the other hand, getting a head when a coin is tossed twice is an example of composite event as it can be split into the events HT and TH which are both elementary events.

Sample space: All the possible outcomes of a random experiment forms the "Sample space" A sample space is denoted by 'S'. A particular outcome, *i.e.*, an element in *S*, is called a *sample point*.

Sure event: The sample space S of an experiment is the set of all possible outcomes. The event represented by S is called a *sure event*. In the case of throw of a die, the event $S = \{1, 2, 3, 4, 5, 6\}$ is a sure event because one of the outcomes from *S* must occur.

Impossible event: An empty set ϕ is always a subset of a set S. Hence, the emptyset ϕ can always be considered as representing an event of an experiment. But, there is no outcome of the experiment which can belong to ϕ. Hence, the event corresponding to the null set ϕ is known as an *impossible event*. In the case of a throw of a die 'getting a number more than 6', getting a number less than 1 or in the case of a throw of two dice together, getting a sum more than 12 etc., are examples of impossible events.

Equally likely events or Equi-probable events. Two are more outcomes of an experiment are said to be equally likely if any one of them cannot be expected to occur in preference to the others. Thus when a fair coin is tossed the coming up of head or tail is equally likely.

Mutually exclusive events. Two or more events are said to be *mutually exclusive events* if the occurrence of any one of them excludes the occurrence of the other events. For example if a coin is tossed, either the head or tail comes up. Both can't occur simultaneously. Such events are called mutually exclusive events. In terms of sets, two events A and B are said to be mutually exclusive events if A and B are disjoint sets that is if $A \cap B = \phi$.

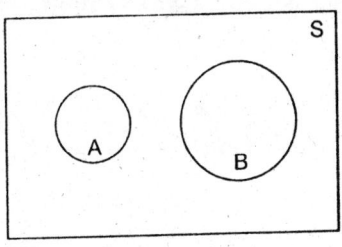

Fig. 24.8. Disjoint events *A* and *B*

Exhausitive set of events. A set of events is said to be *exhaustive* if it includes all the possible events. If we consider the tossing of a coin, head and tail are exhaustive events that is, there is no other third possibility.

Now it is clear to the readers that if we draw a card from a well shuffled pack of playing card, we may get any one of the 52 cards. The 52 cases are mutually exclusive, equally likely and exhaustive. Similarly if a cubical uniform die is thrown, the turning up of each of six faces is equally likely, mutually exclusive and these 6 cases are exhaustive.

Combination of events. This can be done by using the operations "or", "and", "not".

Union of two events. The union of two events A and B denoted by $A \cup B$ is defined as a set of events containing all the sample points of *event A or event B or both* the events. This is shown in Fig. 24.9.

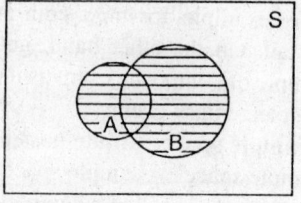

Fig. 24.9. $A \cup B$ is shaded.

Intersection of two events. The *intersection* of two events A and B denoted by $A \cap B$ may be defined as the set containing all the sample points that are common to both the events A and B. This is shown in Fig. 24.10.

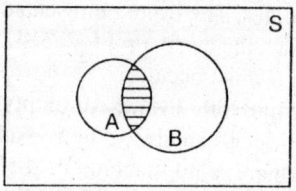

Fig. 24.10. $A \cap B$ is shaded.

Complement of an event. The complement of an event A with respect to the sample space S is the set of all elements of S which are **not** in A. We denote the complement of A by A^c or A' or $\bar{A}$. Thus $\bar{A} = S - A$. Obviously, $A \cup \bar{A} = S$ and $A \cap \bar{A} = \phi$. In Fig. 24.11, the shaded part of the sample space S denotes $\bar{A}$.

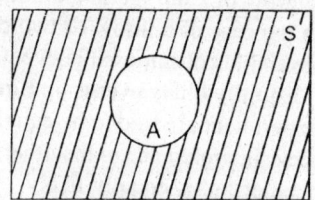

Fig. 24.11. $\bar{A}$ is shaded.

The difference of two events A and B, to be denoted by $A - B$, may be defined as the set of sample points present in set A but **not** in B, that is $A - B = A \cap \bar{B}$. (Refer Fig. 24.12). $B \cap \bar{A}$ is shown in fig. 24.13.

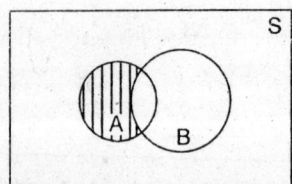

Fig. 24.12. $(A - B)$ is shaded.

Note that $A - B = A \cap \bar{B}$.

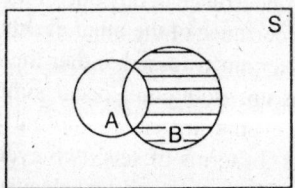

Fig. 24.13. $(B - A)$ is shaded.

Note that $B - A = B \cap \bar{A}$.

24.37 Definition of Probability

If an experiment can result in n exhaustive, mutualy exclusive and equally likely cases and *m* of those are favourable to the event A, then the probability of occurrence of the event A, written as $P(A)$, is defined as the ratio *m/n*.

$\therefore$ The probability of occurrence of $A = P(A)$

$$= \frac{\text{The number of outcomes favourable to event } A}{\text{The total number of exhaustive mutually exclusive and equally likely cases which the experiment can result}} = m/n.$$

Using set notation, we can write $P(A) = n(A)/n(S)$, where $n(S)$ denotes the number of elements (sample points) in the sample space S and likewise $n(A)$ denotes the number of sample points in A. Event A is a subset of S.

The probability of occurrence of an event is a number lies between 0 and 1. If the event cannot occur, its probability is 0. If it must occur, that is, its occurrence is certain, its probability is 1.

The probability of non-occurrence of the event (called its *failure*) is denoted by

$$P(\overline{A}) = (n - m)/n = 1 - (m/n) = 1 - P(A).$$

Remarks: 1. A Fundamental theorem: If an event can happen in any one of m ways and if when this has occurred another event can happen in any one of n ways, then the number of ways in which both events can happen in the specified order is mn.

2. If an event A can happen in m ways and fail in n ways, all these (that is, $m + n$) ways being equally likely, mutually exclusive, and exhaustive then the probability of the happening of A is $m/(m+n)$ and that of its failing is $n/(m+n)$.

In place of saying that probability of happening of an event is $m/(m+n)$, we sometimes say that *the odds are m to n in favour of the event or n to m against the event.*

3. By total number of ways' or possible number of ways' we shall mean "ways which are equally likely, mutually exclusive and exhaustive".

Example 24.51 Find the probability of throwing (*a*) **5** (*b*) **an even number, with an ordinary six faced die,**

Solution. (*a*) In this case there are 6 possible ways out of which only one (that is, appearing of 5) is favourable.

$$\therefore \qquad p = \frac{\text{number of favourable ways}}{\text{total number of ways}} = \frac{1}{6}.$$

(*b*) Here number of favourable ways (that is, appearing of faces with numbers 2, 4, 6) = 3. Therefore $p = 3/6 = 1/2$.

Example 24.52 If a card is drawn from a well shuffled pack, find the probability that the card is (*i*) a queen of spades (*ii*) a king (*iii*) a diamonds.

Solution. (*i*) There are 52 cards in a pack and there is only one queen of spades. Therefore, $p = 1/52$.

(*ii*) There are four Kings in a pack. Hence $p = 4/52 = 1/13$.

(*iii*) There are 13 cards of diamonds. Hence $p = 13/52 = 1/4$.

Example 24.53 A card is drawn from a pack of 52 cards and a gambler bets that it is a spade or an ace. What are the odds against his winning the bet?

Solution. Number of favourable ways *i.e.*, the card is a spade or an ace = 13 + 3 = 16.

The probability that the gambler wins = 16/52 = 4/13 = 4/(4+9).

Hence odds against his winning are 9 to 4.

Example 24.54 Four cards are drawn at random from a pack of well shuffled cards. Find the probability that they are all cards of spades.

Solution. Four cards can be choosen from 13 cards of spades in

$$^{13}C_4 = \frac{13 \cdot 12 \cdot 11 \cdot 10}{4!} \text{ ways.}$$

Total number of possibilities = $^{52}C_4 = \frac{52 \cdot 51 \cdot 50 \cdot 49}{4!}$.

Hence, required probability $= \frac{^{13}C_4}{^{52}C_4} = \frac{13 \cdot 12 \cdot 11 \cdot 10}{52 \cdot 51 \cdot 50 \cdot 49} = \frac{11}{4165} = 0.0026$.

Example 24.55 Two balls are drawn from a bag containing 5 white and 7 red balls. Find the probability that both are red.

Solution. Two red balls can be drawn from 7 red balls in $^7C_2 = \frac{7.6}{2!} = 21$ ways.

Also there are 12 balls in the bag and 2 can be drawn in $^{12}C_2 = \frac{12 \times 11}{2!} = 66$ ways.

Hence, required probability $= {}^7c_2 \div {}^{12}c_2 = 21/66 = 7/22$.

Example 24.56 Five balls are drawn from a bag containing 6 white and 4 black balls. What is the probability that 3 are white and 2 black?

Solution. There are in all 10 balls in the bag and total number of ways of drawing 5 from them = $^{10}C_5$.

Now 3 white balls can be drawn from 6 in $^6C_3$ ways and two black can be drawn from 4 in $^4C_2$ ways and any group of 3 white balls can be associated with 2 black balls in $^6C_3 \times {}^4C_2$ ways (Fundamental Theorm)

Hence   Prob. $= \frac{^6C_3 \times {}^4C_2}{^{10}C_5} = \frac{6 \times 5 \times 4}{3!} \times \frac{4 \times 3}{2!} \times \frac{5!}{10 \cdot 9 \cdot 8 \cdot 7 \cdot 6} = \frac{10}{21}$.

Example 24.57 In a single throw of two dice, what is the probability of getting a sum of 7?

Solution. The sample space consisting of $6 \times 6 = 36$ sample point.

Favourable ways for getting a sum of 7 are: (1, 6), (6, 1), (2, 5), (5, 2), (3, 4), (4, 3) *i.e.* 6 ways.

$$\therefore \qquad \text{Prob.} = \frac{6}{36} = \frac{1}{6}.$$

24.38 Mutually Exclusive Events

We know that two events are called mutually exclusive if both cannot happen simultaneously. For example if a coin is tossed once and head appears then tail cannot appear.

Theorems on Probability

Theorem 1. (Addition rule for probabilities of mutually exclusive events)

Statement. *If $P(A)$ and $P(B)$ are the probabilities for success of two mutually exclusive events A and B then the probability that one or the other event happening is given by $P(A \cup B) = P(A) + P(B)$.*

Proof. Out of n different possible ways, let m_1 be the number of ways favourable to the first event A and m_2, the number of ways favourable to the event B. Thus the number of ways favourable to either A or B is $m_1 + m_2$. The probability P, that either A or B happening is given by

$$P(A \cup B) = \frac{m_1 + m_2}{n} = \frac{m_1}{n} + \frac{m_2}{n} = P(A) + P(B) \qquad \text{...(24.76)}$$

Remarks. 1. If $A_1, A_2, \dots A_n$ are n mutually exclusive events in a sample space S, then

$$P(A_1 \cup A_2 \cup \dots \cup A_n) = P(A_1) + P(A_2) + \dots + P(A_n) \qquad \text{...(24.77)}$$

2. If $A_1, A_2, \dots A_n$ are n mutually exclusive and exhaustive events then

$$P(A_1) + P(A_2) + \dots + P(A_n) = 1.$$

Theorem 2. (Addition rule of probabilities)
If A and B are any two events in a sample space, then the probability that atleast one of the events A or B will occur is given by

$$P(A \cup B) = P(A) + P(B) - P(A \cap B) \qquad \text{...(24.78)}$$

Proof. Consider the venn diagram–Portion I, II, III in Figure 24.14 make up $A \cup B$ and are mutually exclusive. Hence by Theorem I, we get

$$P(A \cup B) = \{P(I) + P(II)\} + P(III)$$

$$= P(A) + [P(B) - P(II)] = P(A) + P(B) - P(A \cap B) \qquad \text{...(24.79)}$$

Note that for mutually exclusive events A and B, that is, if the events A and B donot occur simultaneously we have $A \cap B = \phi$. Hence $P(A \cap B) = P(\phi) = 0.$

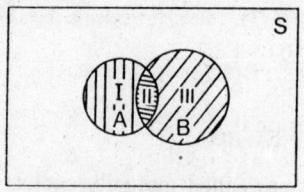

Fig. 24.14. Venn diagram

Then (24.79) reduces to $P(A \cup B) = P(A) + P(B)$ which is same as (24.76).

Remark. If A, B, C are any three events, then

$$P(A \cup B \cup C) = P(A) + P(B) + P(C) - P(A \cap B)$$

$$- P(B \cap C) - P(C \cap A) + P(A \cap B \cap C).$$

Theorem 3. If $\bar{A}$ is the complement of an event A, then $P(\bar{A}) = 1 - P(A)$.

Proof. The sample space S can be decomposed into mutually exclusive events A and $\bar{A}$, that is $S = A \cup \bar{A}$ and $A \cap \bar{A} = \phi$.

Hence $P(A \cup \bar{A}) = P(A) + P(\bar{A}) = P(S) = 1$, thus $P(\bar{A}) = 1 - P(A)$.

Theorem 4. If A and B are two events such that $A \subset B$, then $P(A) \le P(B)$.

Proof. We are given that $A \subset B$. See the venn diagram, Fig. 24.15, we write

$$B = A + (B - A).$$

The events A and $B - A$ are mutually exclusive. Hence

$$P(B) = P[A + (B - A)] = P(A) + P(B - A).$$

Since, $P(B - A) \ge 0$, we obtain $P(B) \ge P(A) \Rightarrow P(A) \le P(B)$.

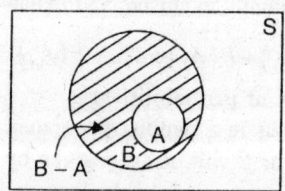

Fig. 24.15. Venn diagram.

24.39 Illustrative Examples

Example 24.58 Two dice are tossed once. Find the probability of getting an even number on the first dice or a total of 8.

Solution. Let the event that getting an even number on the first dice be denoted by A and that getting a total of 8 be denoted by B. We note that $S = 36$.

A: {(2, 1), (2, 2)...... (2, 6), (4, 1), (4, 2)...... (4. 6), (6, 1), (6, 2)...... (6, 6)}

B: {(2, 6), (3, 5), (4, 4), (5, 3), (6, 2)}, and $A \cap B : \{(2, 6), (4, 4), (6, 2)\}$.

Hence $P(A) = 18/36 = 1/2, P(B) = 5/36,$ and $P(A \cap B) = 3/36 = 1/12$.

Therefore P(even number on the first dice or a total of 8) = $P(A \cup B)$

$$= P(A) + P(B) - P(A \cap B) = \frac{1}{2} + \frac{5}{36} - \frac{1}{12} = \frac{5}{9}.$$

Example 24.59 A and B are two mutually exclusive events. If $P(A) = 0.25$, $P(B) = 0.40$ and $P(A \cup B) = 0.50$, find the values of $P(A \cap B)$ and $P(A \cap \bar{B})$.

Solution. We have $P(A \cup B) = P(A) + P(B) - P(A \cap B)$

$\Rightarrow 0.50 = 0.25 + 0.40 - P(A \cap B)$ or $P(A \cap B) = 0.15$.

Now $P(A \cap \bar{B}) = P(A) - P(A \cap B) = 0.25 - 0.15 = 0.10$.

Example 24.60 A number is selected at random from the first 1000 natural numbers. What is the probability that it would be a multiple of 5 or 9?

Solution. We have $S = \{1, 2, 3, \dots, 1000\}$. Let A denote the event the number is divisible by 5. Then A consists of 200 sample points 5, 10, 15,......, 995, 1000 and $P(A) = 200/1000$. Let B denotes the event that the number is divisible by 9. Then B consists of 111 sample points 9, 18, ..., 990, 999 and $P(B) = 111/1000$.

Then event $A \cap B$ consists of sample points 45, 90,...., 945, 990 and $P(A \cap B) = 22/1000$.

Hence the probability that the selected number would be a multiple of 5 or 9 is given by

$$P(A \cup B) = P(A) + P(B) - P(A \cap B)$$

$$= \frac{200}{1000} + \frac{111}{1000} - \frac{22}{1000} = 0.29.$$

Example 24.61 Find the chance of getting a sum of 9 or more in a single throw with two dice.

Solution. The equally likely sample space consisting of 36 sample point. Let A denote the event of throwing 9 in a single throw with two dice. Therefore, A consists of 4 sample points: (3, 6), (4, 5), (5, 4), (6, 3) and hence P(A) =4/36. Let B denote the event of throwing 10 in single throw of two dice. Therefore, B consists of 3 sample points: (4, 6), (5, 5), (6, 4) and P(B) = 3/36. Let C denote the event of throwing 11 in single throw of two dice. Therefore, C consists of sample points: (5, 6), (6, 5), and P (C) = 2/36. Let D denote the event of throwing 12 in a single throw of two dice. Therefore, D consists of only one sample point: (6, 6) and P(D) = 1/36.

Now events A, B, C, D are mutually exclusive and hence we get

$$P(A \cup B \cup C \cup D) = P(A) + P(B) + P(C) + P(D)$$
$$= 4/36 + 3/36 + 2/36 + 1/36 = 10/36 = 5/18.$$

Example 24.62 If A, B, C are mutually exclusive and exhaustive events associated with a random experiment and P(B) = 1.5 P(A) and P(C) = 0.5 P(B), then find P(A).

Solution. Since, A, B, C are mutually exclusive and exhaustive events, we have

$$P(A) + P(B) + P(C) = 1,$$

or $P(A) + 1.5P(A) + (0.5)(1.5)P(A) = 1$ or $P(A) = 4/13.$

Example 24.63 The probability that atleast one of the events A and B occurs is 0.75 and the probability that both the events occur simultaneously is 0.30. Find the probability $P(\overline{A}) + P(\overline{B})$.

Solution. We are given that $P(A \cup B) = 0.75$ and $P(A \cap B) = 0.30.$

Since, $P(A \cup B) = P(A) + P(B) - P(A \cap B).$

We have $P(A) + P(B) = P(A \cup B) + P(A \cap B) = 0.75 + 0.30 = 1.05.$ ...(i)

Also, $P(A) + P(B) = 1 - P(\overline{A}) + 1 - P(\overline{B}) = 2 - P(\overline{A}) - P(\overline{B})$

or $P(\overline{A}) + P(\overline{B}) = 2 - \{P(A) + P(B)\} = 2 - (1.05) = 0.95,$ from (i).

Example 24.64 If $P(A - B) = 0.20$, $P(A) = 0.33$ and $P(B) = 0.50$, what is the probability that out of the two events A and B, only B would occur?

Solution. Refer Fig. 24.16 $P(A - B) = P(A \cap \overline{B}) = P(A) - P(A \cap B)$ ...(i)

and $P(B - A) = P(B \cap \overline{A}) = P(B) - P(A \cap B)$ ...(ii)

Equations (i) and (ii) describe the probabilities of occurrence of the event only A and only B respectively.

We have from (i), $P(A \cap B) = P(A) - P(A - B) = 0.33 - 0.20 = 0.13 \cdot$

The probability that the event B only would occur = $P(B - A)$

$= P(B) - P(A \cap B) = 0.50 - 0.13 = 0.37$

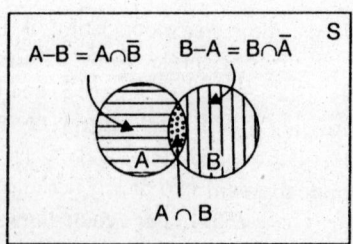

Fig. 24.16. Showing $(A - B)$ and $(B - A)$

Example 24.65 Given that $P(A \cap \bar{B}) = 0.25$, $P(\bar{A} \cap B) = 0.33$ and $P(A \cup B) = 0.75$, find (i) P(B), (ii) P(A).

Solution. (i) We know $P(A \cup B) = P(A) + P(B) - P(A \cap B)$

$$= P(A) - P(A \cap B) + P(B)$$

$$= P(A \cap \bar{B}) + P(B) \qquad \left[\because P(A \cap \bar{B}) = P(A) - P(A \cap B) \right]$$

or $P(B) = P(A \cup B) - P(A \cap \bar{B})$. Substituting the given values, we get

$P(B) = 0.75 - 0.25 = 0.50$.

(ii) $P(A \cup B) = P(A) + P(B) - P(A \cap B)$

$$= P(A) + [P(B) - P(A \cap B)] = P(A) + P(\bar{A} \cap B)$$

$\therefore \quad P(A) = P(A \cup B) - P(\bar{A} \cap B) = 0.75 - 0.33 = 0.42$.

Example 24.66 From a pack of 52 cards three are drawn at random. Find the chance that they are a king, a queen and a knave.

Solution. From a pack of 52 cards, 3 can be drawn in $^{52}C_3$ ways. Now there are 4 kings, 4 queens and 4 knaves. A king can be drawn in $^4C_1$ ways, a queen in $^4C_1$ ways and a knave also is $^4C_1$ ways. Since each can be associated with each of the other, a king, a queen, and a knave can be drawn in

$^4C_1 \times {}^4C_1 \times {}^4C_1 = 4 \times 4 \times 4 = 64$ ways.

$\therefore$ Required probability $= 64 \div {}^{52}C_3 = \dfrac{64 \times 3!}{52 \times 51 \times 50} = \dfrac{16}{5525}$.

Example 24.67 If from a lottery of 30 tickets marked 1, 2, 3,..., 30, four are drawn, find the chance that those marked 1 and 2 are among them.

Solution. Any four tickets can be drawn in $^{30}C_4$ ways.

When two tickets marked 1 and 2 are always there, the remaining two can be drawn from remaining 28 in $^{28}C_2$ ways.

Hence the required chance $= {}^{28}C_2 / {}^{30}C_4 = 2/145$.

Example 24.68 Four persons are chosen at randam from a group containing 3 men, 2 women and 4 children. Show that the probability that exactly two of them are children is 10/21.

Solution. Now 4 persons can be chosen out of 9 in $^9C_4$ ways. Two children can be selected out of 4 in $^4C_2$ ways. The other two can be selected from the remaining (3 men and 2 women) 5 persons in $^5C_2$ ways.

Thus 4 persons such that 2 of them are children can be selected in $^4C_2 \times {}^5C_2$ ways.

Hence the required probability $= \dfrac{\text{No. of favourable cases}}{\text{Total possible cases}} = \dfrac{^4C_2 \times {}^5C_2}{^9C_4} = \dfrac{10}{21}$.

Example 24.69 **One bag contains 4 white and 2 black beads and another contains 3 of each colour. A bead is drawn from each bag. What is the probability that one is white and one is black.**

Solution. One white and one black bead can be drawn in the following two mutually exclusive ways:

(*i*) white from first bag and black from second
(*ii*) black from first and white from second.

p_1, the probability of the first event $= 4/6 \times 3/6 = 1/3$.

p_2, the probability of the second event $= 2/6 \times 3/6 = 1/6$.

∴ Required probability $= 1/3 + 1/6 = 1/2$

Example 24.70 A bag contains 4 white and 2 black balls, and a second bag contains 3 of each colour. A bag is drawn at random and a ball is then selected at random from the bag chosen. What is the probability that the ball selected is white.

Solution. We define the following events:
A: the ball is drawn from the first bag,
B: the ball is drawn from the second bag.

Since, both the bags are equally likely to be selected, we have $P(A) = P(B) = 1/2$.

Now, probability of drawing a white ball from first bag is $p_1 = 4/6 = 2/3$, .

and the probability of drawing a white ball from second bag is $p_2 = 3/6 = 1/2$.

Hence the required probability $= P(A)p_1 + P(B)p_2 = \dfrac{1}{2} \cdot \dfrac{2}{3} + \dfrac{1}{2} \cdot \dfrac{1}{2} = \dfrac{7}{12}$.

Example 24.71 There are 3 boxes with the following composition:

> **Box I: 8 Red + 5 white + 3 Blue balls**
> **Box II: 6 Red + 5 white + 3 Blue balls**
> **Box III: 4 Red + 3 white + 2 Blue balls.**

One of the boxes is selected at random and a ball is drawn from it. What is the probability that the drawn ball is red?

Solution. Let A denote the event that the drawn ball is red. Since any of the 3 boxes may be drawn, we have P(Box I) = P(Box II) = P(Box III) = 1/3.

Also $P(R_1)$ from the first box = probability of drawing a red ball from the first box $= 8/16 = 1/2$.

$P(R_2)$ from Box II $= 6/14 = 3/7$ and $P(R_3)$ from Box III $= 4/9$.

Thus we have

$$P(A) = P(R_1) \times P(Box\ I) + P(R_2) \times P(Box\ II) + P(R_3) \times P(Box\ III)$$

$$= \frac{1}{2} \times \frac{1}{3} + \frac{3}{7} \times \frac{1}{3} + \frac{4}{9} \times \frac{1}{3} = \frac{1}{6} + \frac{1}{7} + \frac{4}{27} = \frac{173}{378}.$$

Example 24.72 Three groups of children contain respectively 3 girls and 1 boy; 2 girls and 2 boys; 1 girl and 3 boys. One child is selected at random from each group. Find the chance that the three selected consist of 1 girl and 2 boys.

Solution. One girl and two boys can be selected in the following mutually exclusive ways.

(*i*) Girl from first group, boy from second and boy from third. p_1 the probability

of this $= \frac{3}{4} \times \frac{2}{4} \times \frac{3}{4} = \frac{9}{32}$.

(*ii*) Boy from first, girl from second and boy from third group. p_2 its probability

$$= \frac{1}{4} \times \frac{2}{4} \times \frac{3}{4} = \frac{3}{32}.$$

(*iii*) Boy from first group, boy from second and girl from third, p_3 the probability

of this $= \frac{1}{4} \times \frac{2}{4} \times \frac{1}{4} = \frac{1}{32}$.

Since these are mutually exclusive ways, the required probability

$$= \frac{9}{32} + \frac{3}{32} + \frac{1}{32} = \frac{13}{32}.$$

Example 24.73 An urn A contains 2 white and 4 black balls. Another urn B contains 5 white and 7 black balls. A ball is transferred from the urn A to B and then a ball is drawn from the urn B. Find the probability that it is white.

(Cochin, 1999)

Solution. There are two mutually exclusive cases: (*i*) when a white ball is transferred from A to B (*ii*) when a black ball is transferred from A to B.

Case 1: The probability of drawing white ball from A is $p_1 = \frac{2}{6} = \frac{1}{3}$. When a white ball is transferred to B, it contains 6 white and 7 black balls. The probability of drawing white ball from B is $p_2 = \frac{6}{13}$.

∴ Probability of transferring white ball to B and then drawing a white ball from B that is, these two events occuring simultaneously is $p_3 = \frac{1}{3} \cdot \frac{6}{13} = \frac{2}{13}$.

Case 2: The probability of drawing a black ball from A is $p_4 = 4/6 = 2/3$ and then the probability of drawing a white ball from B (which now contains 5 white and 8 black balls) is $p_5 = 5/13$.

Probability of these two events occuring simultaneously is $p_6 = \dfrac{2}{3} \cdot \dfrac{5}{13} = \dfrac{10}{39}$.

∴ The required probability is $p = p_3 + p_6 = \dfrac{2}{13} + \dfrac{10}{39} = \dfrac{16}{39}$.

Example 24.74 A batch of 100 lamps is known to have 5% defectives. If 10 lamps are chosen at random, what is the probability that (a) none is defective (b) two are defective.
Solution. (a) There are 95 good and 5 defective lamps. Therefore, the probability p, that all the lamps are good

$$= \frac{^{95}C_{10}}{^{100}C_{10}} = \frac{95 \times 94 \times \cdots \times 86}{100 \times 99 \times \cdots \times 91} = 0.584 .$$

(b) Probability that 2 are defective and 8 good $= \dfrac{^{95}C_8 \times ^5C_2}{^{100}C_{10}} = 0.07 .$

Example 24.75 Three light bulbs are chosen at random from fifteen bulbs of which five are defective. Find the probability that (i) none is defective (ii) exactly one is defective (iii) at least one is defective.
Solution. There are 5 defective and 10 non-defective or good bulbs.
(i) The probability that all three bulbs chosen are good

$$= \frac{^{10}C_3}{^{15}C_3} = \frac{10 \times 9 \times 8}{15 \times 14 \times 13} = \frac{24}{91} .$$

(ii) The probability that of 3 bulbs chosen, 1 is defective and 2 non-defective

$$= \frac{^5C_1 \times ^{10}C_2}{^{15}C_3} = \frac{5 \times 10 \times 9}{2!} \times \frac{3!}{15 \times 14 \times 13} = \frac{45}{91} .$$

(iii) Here the mutually exclusive cases are
(a) one defective, two non-defective, (b) two defective, one non-defective, (c) all three defective.

In case (a) probability $p_1 = \dfrac{^5C_1 \times ^{10}C_2}{^{15}C_3} = \dfrac{45}{91} .$

In case (b) probability $p_2 = \dfrac{^5C_2 \times ^{10}C_1}{^{15}C_3} = \dfrac{10 \times 10}{455} = \dfrac{20}{91} .$

In case (c) probability $p_3 = \dfrac{^5C_3}{^{15}C_3} = \dfrac{10}{455} = \dfrac{2}{91}$.

$\therefore$ p, the probability of at least one defective $= \dfrac{45}{91} + \dfrac{20}{91} + \dfrac{2}{91} = \dfrac{67}{91}$.

(iii) **Alternative.** It is easier to find the probability of none defective which is 24/91.

$\therefore$ Probability of at least one defective

$$= 1 - (\text{Probability of none defective}) = 1 - \dfrac{24}{91} = \dfrac{67}{91}.$$

24.40 Conditional Probability

If A and B are two events, the probability that B occurs under the condition that an event A has occurred is called the *conditional probability* of B given that A has occurred and is denoted by $P(B/A)$. In this case A serve as a new (reduced) sample space, and that probability is the fraction of $P(A)$ which corresponds to $A \cap B$.

Thus $P(B/A) = P(A \cap B)/P(A)$, where $P(A) \neq 0$. ...(24.80)

Similarly, the *conditional probability* of A given that B has occurred is

$$P(A/B) = P(A \cap B)/P(B), \text{ where } P(B) \neq 0. \qquad \qquad ...(24.81)$$

From (24.80) and (24.81), we have the following result.

Theorem 5. (*Multiplication rules of probability*). If A and B are events in a sample space S and $P(A) \neq 0$, $P(B) \neq 0$. Let $P(A/B)$ denote the conditional probability of A when B has occurred. Then,

$$P(A \cap B) = P(A) P(B/A) = P(B) P(A/B). \qquad \qquad ...(24.82)$$

These formula are called multiplication Rules of Probability.

Remark. For three events A, B, C, we have $P(ABC) = P(A) P(B/A) P(C/AB)$. That is, the probability of occurrence of A, B, C is equal to the probability of A times the probability of B given that A has Occurred, times the probability of C given that both A and B have occurred.

Example 24.76 For two events A and B, P(A) = 3/8, P(B) = 5/8 and $P(A \cup B) = 3/4$. Find (i) P(A/B), (ii) P(B/A), (iii) $P(\bar{A}/\bar{B})$, (iv) $P(\bar{A}/B)$, (v) $P(A/\bar{B})$.

Solution. We have $P(A \cup B) = P(A) + P(B) - P(A \cap B)$.

Therefore, $P(A \cap B) = P(A) + P(B) - P(A \cup B) = 3/8 + 5/8 - 3/4 = 1/4$.

Hence (i) $P(A/B) = \dfrac{P(A \cap B)}{P(B)} = \dfrac{1/4}{5/8} = \dfrac{2}{5}$.

(ii) $P(B/A) = \dfrac{P(A \cap B)}{P(A)} = \dfrac{1/4}{3/8} = \dfrac{2}{3}$.

(iii) First compute $P(\overline{B})$ and $P(\overline{A} \cap \overline{B})$.

Now $P(\overline{B}) = 1 - P(B) = 1 - (5/8) = 3/8$.

By De Morgan's law, $\overline{A \cup B} = \overline{A} \cap \overline{B}$; hence

$$P(\overline{A} \cap \overline{B}) = P(\overline{A \cup B}) = 1 - P(A \cup B) = 1 - (3/4) = 1/4.$$

Thus $P(\overline{A}/\overline{B}) = \dfrac{P(\overline{A} \cap \overline{B})}{P(\overline{B})} = \dfrac{1/4}{3/8} = \dfrac{2}{3}$.

(iv) $P(\overline{A}/B) = \dfrac{P(\overline{A} \cap B)}{P(B)} = \dfrac{P(B) - P(A \cap B)}{P(B)} = \dfrac{(5/8) - (1/4)}{(5/8)} = \dfrac{3/8}{5/8} = 3/5$.

(v) $P(A/\overline{B}) = \dfrac{P(A \cap \overline{B})}{P(\overline{B})} = \dfrac{P(A) - P(A \cap B)}{1 - P(B)} = \dfrac{(3/8) - (1/4)}{1 - (5/8)} = \dfrac{1/8}{3/8} = \dfrac{1}{3}$.

Example 24.77 What is the probability that in a single throw of two dice of getting a total of 7 or 11?
Solution. Define the events A: getting a total of 7, B: getting a total of 11.

Now the first die may appear in 6 different ways and according to any one way in which the first appears, the second can appear in 6 ways. The two dice, therefore, may appear in $6 \times 6 = 36$ ways.
The sum of 7 may appear in 6 different ways.

    A: (1, 6), (2, 5), (3, 4), (4, 3), (5, 2), (6, 1).
$\therefore P(A)$ (the probability of getting a sum of 7) = 6/36 = 1/6.
The sum of 11 may appear in two different ways.

    B: (5, 6), (6, 5).
$\therefore P(B)$ (the probability of getting a sum of 11) = 2/36 = 1/18.
$A \cap B = \phi$. Therefore $P(A \cap B) = 0$.
Therefore, P (probability of getting a total of 7 or 11)=$P(A \cup B) = P(A) + P(B)$

$= (1/6) + 1/18 = 2/9$.

Example 24.78 Two dice are thrown twice. What is the probability that the first throw is 7 and the second throw is 11?
Solution. The probability of the first throw of two dice to 7 has been shown (Refer Example 24.77) to be 1/6. The probability for 11 is 1/18.
Hence the probability for first throw to be 7 and second to be 11 is

$(1/6)(1/18) = 1/108$.

Example 24.79 **Two cards are drawn from a pack of 52 cards. Find the probability that they are both aces.**

Also find this probability if the first card is replaced before the second is drawn.

Solution. (*a*) The probability for drawing an ace in the first draw = 4/52 =1/13. Now, when the first ace has been drawn, only 3 aces are left in the remaining 51 cards. Therefore the probability of drawing an ace in the second draw = 3/51=1/17.

Hence the probability of drawing two aces = (1/13)(1/17) = 1/221 = 0.0045.

(*b*) In case the first card is replaced before the second draw, the probability of getting an ace in the second draw is also 4/52 = 1/13.

Hence the probability of drawing two aces = (1/13)(1/13) = 1/169 = .0059.

Example 24.80 A die is thrown twice and the sum of numbers appearing is noted to be 8. What is the conditional probability that the number 2 has appeared atleast once?

Solution. Let the event A define: the number 2 has appeared at least once. The event B define: the sum of the numbers is 8. Then A consists of 11 sample points:

$$A = \{(2,1),(2,2),(2,3),(2,4),(2,5),(2,6),(1,2),(3,2),(4,2),(5,2),(6,2)\}$$

B consists of 5 sample points. Sum of the numbers is 8.

$$B = \{(2,6),(3,5),(4,4),(5,3),(6,2), \text{ and } A \cap B = \{(2,6),(6,2)\}.$$

The total outcomes of throwing two die is 6 × 6 = 36.

Hence, $P(A) = 11/36, P(B) = 5/36, P(A \cap B) = 2/36.$

$$\therefore \quad P(A/B) = \frac{P(A \cap B)}{P(B)} = \frac{2/36}{5/36} = \frac{2}{5}.$$

Example 24.81 Given $P(\bar{A}) = 0.2, P(B) = 0.5$ **and** $P(A \cap \bar{B}) = 0.4$; **find** $P(B/A \cup \bar{B}).$ (*A.M.I.E., S-2003*)

Solution. We have $P(\bar{A}) = 0.2,$ therefore $P(A) = 1 - P(\bar{A}) = 0.8,$ $P(B) = 0.5.$

Therefore $P(\bar{B}) = 1 - 05 = 0.5,$ $P(A \cap \bar{B}) = 0.4.$

We know $P(A \cap \bar{B}) = P(A) - P(A \cap B),$ therefore

$P(A \cap B) = P(A) - P(A \cap \bar{B}) = 0.8 - 0.4 = 0.4.$

Now $B \cap (A \cup \bar{B}) = A \cap B.$ $\therefore P(B \cap (A \cup \bar{B})) = P(A \cap B) = 0.4$

We also know $P(A \cup \bar{B}) = P(A) + P(\bar{B}) - P(A \cap \bar{B}) = 0.8 + 0.5 - 0.4 = 0.9.$

$$\therefore P(B \backslash A \cup \bar{B}) = \frac{P(B \cap (A \cup \bar{B}))}{P(A \cup \bar{B})} = \frac{0.4}{0.9} = 0.44$$

Example 24.82 There are two bags. The first bag contains 10 white and 3 black balls, and second bag contains 3 white and 5 black balls. Two balls are drawn at random from the first bag and are put into the second bag and then a ball is drawn at random from the latter. What is the probability that it is a white ball?

Solution. The two balls drawn from the first bag may be (*i*) both white, (*ii*) one white and one black, (*iii*) both black. Denote these possibilities as events B_1, B_2, B_3 respectively. We have $P(B_1) = {}^{10}C_2 \div {}^{13}C_2 = 15/26$; $P(B_2) = \left({}^{10}C_2 \times {}^3C_1 \right) \div {}^{13}C_2 = 5/13$, $P(B_3) = {}^3C_2 \div {}^{13}C_2 = 1/26$.

When two balls are transferred from first bag to second bag, the second bag will contain:

(*i*) 5 white and 5 black balls, (*ii*) 4 white and 6 black balls, (*iii*) 3 white and 7 black balls.

Denote E: event of drawing a white ball from the second bag after transfer.

Now $P(E/B_1) = 5/10, P(E/B_2) = 4/10, P(E/B_3) = 3/10$.

$$\therefore \quad P(E) = P(B_1) \cdot P(E/B_1) + P(B_2) \cdot P(E/B_2) + P(B_3) \cdot P(E/B_3)$$

$$= \frac{15}{26} \cdot \frac{5}{10} + \frac{5}{13} \cdot \frac{4}{10} + \frac{1}{26} \cdot \frac{3}{10} = \frac{15}{52} + \frac{2}{13} + \frac{3}{260} = \frac{59}{130} = 0.45.$$

24.41 Independent Events

Two events A and B are said to be independent, if the occurrence (or non-occurrence) of A does not affect the probability of occurrence of the B event. This means that the probability of event A does not depend on the occurrence (or non-occurance) of event B and conversely. For example if two coins are tossed, the probability for head on one coin is not affected by the result of the toss on the second coin.

Hence, in this case the conditional probability $P(A/B)$ is same as $P(A)$, that is

$P(A/B) = P(A)$ and $P(B/A) = P(B)$. Substituting in (24.82), we obtain

$$P(A \cap B) = P(A/B)P(B) = P(A) \cdot P(B) \qquad \qquad \dots(24.83)$$

Thus, if two events A and B are such that $P(A \cap B) = P(A)P(B)$, they are called independent events otherwise they are dependent. This is known as *Multiplication Rule of the probability* for independent events.

Remarks. 1. Three events A, B and C are known as independent if the following conditions hold:

$$P(A \cap B) = P(A) \times P(B)$$
$$P(B \cap C) = P(B) \times P(C)$$

$$P(C \cap A) = P(C) \times P(A)$$

$$P(A \cap B \cap C) = P(A) \times P(B) \times P(C)$$

2. If two events A and B are independent, then the following pairs of events are also independent:

(*i*) A and $\bar{B}$, (*ii*) $\bar{A}$ and B, (*iii*) $\bar{A}$ and $\bar{B}$.

Example 24.83 **Prove if the events A and B are independent, then $\bar{A}$ and $\bar{B}$ are also independent.**

Solution. We have $P(A \cap B) = P(A) P(B)$.

$$P(\bar{A} \cap \bar{B}) = P(\overline{A \cup B}) = 1 - P(A \cup B)$$

$$= 1 - [P(A) + P(B) - P(A \cap B)] = 1 - P(A) - P(B) + P(A \cap B)$$

$$= 1 - P(A) - P(B) + P(A) P(B) = [1 - P(A)] - P(B)[1 - P(A)]$$

$$= [1 - P(A)][1 - P(B)] = P(\bar{A}) P(\bar{B}).$$

Hence the events $\bar{A}$ and $\bar{B}$ are also independent.

Example 24.84 **Prove if events A and B are independent, then $\bar{A}$ and B are also independent.**

Solution. Since the events A and B are independent, we have

$$P(A \cap B) = P(A) P(B).$$

The events $A \cap B$ and $\bar{A} \cap B$ are mutually exclusive and $(A \cap B) \cup (\bar{A} \cap B) = B$. (Refer Fig 24.17)

From addition theorem, we obtain

$$P(B) = P(A \cap B) + P(\bar{A} \cap B)$$

or $P(\bar{A} \cap B) = P(B) - P(A \cap B) = P(B) - P(A) P(B)$

$$= P(B)[1 - P(A)] = P(B) P(\bar{A}).$$

Therefore, the events $\bar{A}$ and B are independent.

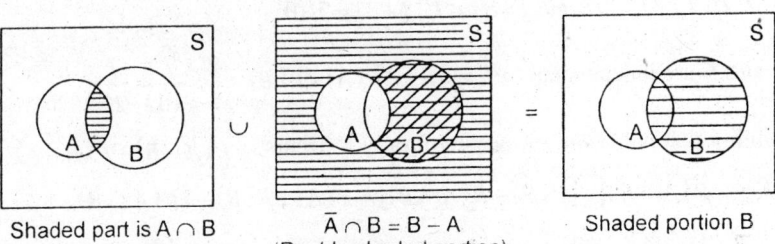

Shaded part is A ∩ B

$\bar{A} \cap B = B - A$
(Double shaded portion)

Shaded portion B

Fig. 24.17.

Example 24.85 A and B are two events such that

$P(A) = 2/5$ and $P(A \cup \bar{B}) = 7/10$. If A and B are independent events, then find P(B).

Solution. Since, A and B are indepencent, A and $\bar{B}$ are also independent. We have

$$P(A \cup \bar{B}) = P(A) + P(\bar{B}) - P(A \cap \bar{B}) = P(A) + P(\bar{B}) - P(A)P(\bar{B}).$$

Substituting the values, we obtain

$$7/10 = (2/5) + P(\bar{B})[1 - 2/5] \Rightarrow P(\bar{B}) = [(7/10) - (2/5)] \div 3/5 = 1/2.$$

Hence $P(B) = 1 - 1/2 = 1/2$.

Example 24.86 A and B are two independent events such that $P(A) = 0.6$ and $P(A \cup B) = 0.8$. Find P(B).

Solution. Since A and B are independent, we have $P(A \cap B) = P(A)P(B)$. Hence

$$P(A \cup B) = P(A) + P(B) - P(A)P(B) = 0.8. \text{ or } P(B)[1 - P(A)] = 0.8 - P(A)$$

or $P(B)[1 - 0.6] = 0.8 - 0.6 = 0.2$ or $P(B) = 0.2/0.4 = 0.5$

Example 24.87 A student A can solve 85% of the problems in a question paper and student B can solve 65% of the problems from the same question paper. If a problem from the question paper is selected at random, what is the probability that atleast one of them will solve the problem?

Solution. The probability that A can solve the problems is $P(A) = 17/20$, and the probability that B can solve the problem is $P(B) = 13/20$. Since the events are independent. The probability that atleast one of them will solve the problem is given by

$$P(A \cup B) = P(A) + P(B) - P(A \cap B) = P(A) + P(B) - P(A)P(B)$$

$$= \frac{17}{20} + \frac{13}{20} - \left(\frac{17}{20}\right)\left(\frac{13}{20}\right) = \frac{379}{400} = 0.948.$$

Example 24.88 The odds in favour of an event is 2:3 and the odds against another event is 3:7. Final the probability that only one of the two events occurs.

Solution. We denote the two event by A and B repectively. We have

$$P(A) = 2/(2+3) = 2/5 \text{ and } P(B) = 7/(7+3) = 7/10.$$

As A and B are independent, $P(A \cap B) = P(A) \times P(B) = \frac{2}{5} \times \frac{7}{10} = \frac{7}{25}$.

Probability that either only A occurs or only B occurs $= P(A - B) + P(B - A)$

$$= [P(A) - P(A \cap B)] + [P(B) - P(A \cap B)] = P(A) + P(B) - 2P(A \cap B)$$

$$= \frac{2}{5} + \frac{7}{10} - 2 \cdot \left(\frac{7}{25}\right) = \frac{27}{50}.$$

Example 24.89 A problem in mechanics is given to three students A, B, C in a class to be solved. The probabilities of their solving the problem are 1/2, 1/3, 1/4 respectively. What is the probability that the problem will be solved?

(V.T.U., 2000)

Solution. The probability that A can solve the problem is 1/2. The probability that A cannot solve the problem is $1-(1/2)=1/2$. Similarly the probabilities that B and C cannot solve the problem are $1-(1/3)$ and $1-(1/4)$ that is, 2/3 and 3/4 respectively.

∴ The probability that A, B and C cannot solve the problem is

$(1/2)(2/3)(3/4)=1/4$.

Hence the probability that the problem will be solved that is, at least one student will solve it $= 1-(1/4)=3/4$.

Example 24.90 An article manufactured by a company consists of two parts A and B. In the process of manufacture of part A, 9 out of 100 are likely to be defective. Similarly 5 out of 100 are likely to be defective in the manufacture of part B. Calculate the probability that the assembled parts will not be defective.

Solution. p_1, the probability of part A being defective $= 9/100$.

p_2, the probability of part B defective $= 5/100$.

∴ $\bar{p}_1$, the probability of A not defective $= 1-p_1 = 1-(9/100) = 91/100$,

and $\bar{p}_2$, the probability of B not defective $= 1-p_2 = 1-(5/100) = 95/100$. Since the events are independent, $p = (91/100)(95/100) = 0.8645$.

Example 24.91 The odds against a certain event are 5 to 2 and the odds in favour of another event, independent of the former are 6 to 5. Find the probability that at least one of the events will occur.

Solution. The probability that first event does not happen $= 5/(5+2) = 5/7$.

The probability that second event does not happen $= 5/(6+5) = 5/11$.

The probability that none of the events occurs $= (5/7)(5/11) = 25/77$.

∴ Probability that at least one occurs $= 1-(25/77) = 52/77$.

Example 24.92 The odds that a Ph.D thesis will be favourably reviewed by three independent examiners are 5 to 2, 4 to 3 and 3 to 4 respectively. What is the probability that a majority approve the thesis?   (I.E.T.E., June 2004)

Solution. The probabilities that the book shall be favourably reviewed by first second and third examiners are 5/7, 4/7, 3/7 respectively. A majority approve the thesis if *atleast* two examiners are favourable.

$\therefore$ Probability that the first two examiners approve the thesis and third examiner reject the thesis $= \dfrac{5}{7} \times \dfrac{4}{7} \times \left(1 - \dfrac{3}{7}\right) = \dfrac{80}{343}$.

Probability that the first and third examiners approve the thesis and second examiner reject the thesis $= \dfrac{5}{7} \times \dfrac{3}{7} \times \left(1 - \dfrac{4}{7}\right) = \dfrac{45}{343}$.

Probability that the second and third examiners approve the thesis and the first reject the thesis $= \dfrac{4}{7} \times \dfrac{3}{7} \times \left(1 - \dfrac{5}{7}\right) = \dfrac{24}{343}$.

Finally, probability that all the three examiners approve the thesis

$$= \dfrac{5}{7} \times \dfrac{4}{7} \times \dfrac{3}{7} = \dfrac{60}{343}.$$

Hence the probability that a majority approve the thesis

$$= \dfrac{80}{343} + \dfrac{45}{343} + \dfrac{24}{343} + \dfrac{60}{343} = \dfrac{209}{343}.$$

Exercise 24.2

1. A bag contains 5 green and 7 red balls. Two balls are drawn. What is the probability that one is green and other is red? **Ans. 35/66**

2. From a set of 17 cards, numbered 1, 2, 3, ..., 17, one is drawn at random. Show that the chance that its number is divisible by 3 or 7 is 7/17.

3. Out of 50 rare books, 3 of which are especially valuable, 5 are stolen at random by a thief. What is the probability that
   (a) none of the three is included? (b) two of the three are included?
   Ans. (a) $47C_5/50C_5$, (b) $\left(3C_2 \times 47C_3\right)/50C_5$.

4. A bag contains 10 white and 15 black balls. Two balls are drawn in succession. What is the probability that
   (i) first is white and second is black? (ii) one is white and one is black?
   Ans. (i) 1/4, (ii) 1/2.

5. If 20% of the bulbs produced by a factory are defective, find the probability that out of 4 bulbs chosen at random atmost 2 bulbs will be defective.
   Ans. 0.9728.

6. An urn contains 10 black and 10 white balls. Find the probability of drawing two balls of the same colour. $\left[\textbf{Hint}: (a) \dfrac{^{10}C_2}{^{20}C_2} + \dfrac{^{10}C_2}{^{20}C_2} = \dfrac{9}{19}\right]$

7. Find the probability of a 4 turning up at least once in two tosses of a fair die.
   Ans. 11/36.

8. A box contains 100 transistors, 20 of which are defective. 10 are selected for inspections. Indicate what is the probability that

(*i*) All 10 are defective, (*ii*) All 10 are good, (*iii*) At least one is defective.

Ans. (*a*) $\dfrac{^{20}C_{10}}{^{100}C_{10}}$, (*ii*) $\dfrac{^{80}C_{10}}{^{100}C_{10}}$, (*iii*) $\dfrac{1 - {}^{80}C_{10}}{^{100}C_{10}}$.

9. A box contains 40 good and 10 defective fuses. If 10 fuses are selected at random, what is the probability that they will all be good.

Ans. $^{40}C_{10} \div {}^{50}C_{10} = 0.0825$.

10. A purse contains 2 silver and 4 copper coins and a second purse contains 4 silver and 4 copper coins. If a coin is selected at randon from one of the two purses, what is the probability that it is a silver coin?

[**Hint.** The chance of choosing 'first purse is 1/2 and if the first purse is chosen, the chance of drawing a silver coin is 2/6 = 1/3. Hence the chance of drawing a silver coin from first purse $= (1/2)(1/3) = 1/6$.

Similarly the chance of drawing a silver coin from the second purse $= (1/2)(4/8) = 1/4$.

Since the events are mutualy exclusive, the probability required, by addition law $= (1/6) + (1/4) = 5/12$].

11. A husband and a wife appear in an interview for two vacancies in the same post. The probability of husband's selection is 1/7 and that of wife's selection is 1/5. What is the probability that

(*i*) both of them will be selected, (*ii*) only one of them will be selected, (*iii*) None of them will be selected? **Ans.** 1/35; 2/7; 24/35.

12. What is the chance that a leap year selected at random will contain 53 Sundays? (*Assam, 1999*) **Ans.** 2/7.

13. If the probabilities of happening of some events are $p_1, p_2, ..., p_n$, then find the probability of happening of at least one of them. (*AMIE., S-2009*) Use this result to find the chance of getting at least one six in a throw of 4 dice.

[**Hint.** The probability that none of the events happens
$= (1 - p_1)(1 - p_2) ...(1 - p_n)$.
Hence the probability that at least one event happens
$= 1 - (1 - p_1)(1 - p_2) ...(1 - p_n)$.
The probability of getting six in a throw of one die = 1/6.
Probability of not getting six $= 5/6$.

Probability of not getting six in a throw of 4 dice $= (5/6)^4$

Probability of getting at least one six $= 1 - (5/6)^4$].

14. A candidate is selected for interview for three posts. For the first post there are 3 candidates, for the second there are 4, and for the third there are 2. What is the chance of his getting at least one post? (*A.M.I.E., S-2001*) **Ans. 3/4.**

15. A manufacturer supplies cheap quarter horse-power motors in lots of 25. A buyer, before taking a lot, tests at random sample of 5 motors and accepts the lot if they are all good; otherwise he rejects the lot. Find the probability that (*a*) he will accept a lot containing 5 defective motors (*b*) he will reject a lot containing only one defective motor.

    Ans. (*a*) $^{20}C_5 / ^{25}C_5 = 0.292$, (*b*) $^{24}C_4 / ^{25}C_5 = 0.20$.

16. A can hit a target 4 times in 5 shots, B 3 times in 4 shots, C twice in 3 shots. Find the probability that (*i*) the target will not be hit (*ii*) the target will be hit by at least one of them.

    [**Hint.** Prob. that target will not be hit $= \left(1 - \dfrac{4}{5}\right)\left(1 - \dfrac{3}{4}\right)\left(1 - \dfrac{2}{3}\right) = \dfrac{1}{60}$

    Probability that target will be hit (*i.e.*, by at least one of them)

    $$= 1 - (1/60) = 59/60].$$

17. If it is 9 to 7 against a person A who is now 35 years of age living till he is 65 and 3 to 2 against a person B now 45 living till he is 75; find the chance that one at least of these person will be alive 30 years hence.

    [**Hint.** Probability that A will die within 30 years $= 9/16$.

    Probability that B will die within 30 years $= 3/5$

    The chance that both will die $= (9/16) \times (3/5) = 27/80$.

    ∴ Probability that both will not be dead *i.e.*, at least one will be alive

    $= 1 - (27/80) = 53/80]$

18. The probability that a 50 year old man will be alive at 60 is 0.83 and the probability that a 45 year old woman will be alive at 55 in 0.87. What is the probability that a man who is 50 and his wife who is 45 will both be alive 10 years hence? Also find the probability that atleast one of them will be alive 10 years hence. **Ans. 0.722, 0.978.**

    $\Big[$ **Hint.** $P(A \cap B) = P(A)P(B); P(A \cup B) = P(A) + P(B) - P(A)P(B) \Big]$

19. 'A' speaks truth in 75 per cent cases and 'B' in 80 per cent cases. In what percentage of cases are they likely to contradict each other in stating the same fact? **Ans. 46%**

    [**Hint.** They contradict if any one of them speaks the truth and other not. Therefore, probability that A speaks the truth but B does not

    $= (75/100)\big[1 - (80/100)\big] = 3/20$. Probability that B speaks the truth but A does not $= (80/100)\big[1 - (75/100)\big] = 1/5$. The two events being mutually exclusive. The required probability $= (3/20) + (1/5) = 7/20]$

20. Of the cigarette smoking population, 70% are men and 30% women. 10% of these men and 20% of these women smoke "WILLS". What is the probability that a person seen smoking "WILLS" will be a man.   [**Hint.** In smoking population: Men smokers = 70%,women smokers = 30%.

    Out of 100, men WILLS smokers $= 70 \times (1/10) = 7 \cdot$

    Out of 100, women WILLS smokers $= 30 \times (1/5) = 6.$

    Probability that a person smoking WILLS is a man $= \dfrac{7}{7+6} = \dfrac{7}{13} \Big]$.

21. An urn contains nine balls, two of which are red, three blue and four black. Three balls are drawn from the urn at random. What is the probability that (*i*) the three balls are of different colours? (*ii*) the three balls are of the same colour?   (*A.M.I.E., S-2000*) **Ans.** (*i*) 2/7, (*ii*) 5/84

22. A card is drawn from a well-shuffled deck of 52 cards. What is the probability of this card being a heart or a king.   **Ans. 4/13.**

    [**Hint.** $P(A) = 13/52$, $P(B) = 4/52$, $P(A \cup B) = 1/52$, $P(A \cap B) = 4/13.$]

23. If A and B be events with $P(A \cup B) = 7/8, P(A \cap B) = 1/4$ and $P(\bar{A}) = 5/8$, find $P(A), P(B)$ and $P(A \cap \bar{B})$. **Ans.** 3/8, 3/4, 1/8.

    $\Big[$ **Hint.** $P(A \cap \bar{B}) = P(A - B) = P(A) - P(A \cap B) = (3/8) - (1/4) = 1/8 \Big]$.

24. The probability that a boy will pass an examination is 3/5 and the probability that the girl pass in the same examination is 2/5. Find the probability that either boy or girl pass in the examination.
    [**Hint.** Event A: Boy pass in the examination. Event B: girl pass in the examination. Both these events are independent, we have

    $$P(A \text{ or } B) = P(A \cup B) = P(A) + P(B) - P(A \cap B)$$

    $$= P(A) + P(B) - P(A) \cdot P(B) = (3/5) + (2/5) - (3/5)(2/5) = 19/25].$$

25. Let A and B be events with $P(A) = 1/3, P(B) = 1/4$ and $P(A \cup B) = 1/2$. Find

    (*i*) $P(A/B)$, (*ii*) $P(B/A)$, (*iii*) $P(A \cap \bar{B})$, (*iv*) $P(A/\bar{B})$.

    Ans. (*i*) 1/3, (*ii*) 1/4, (*iii*) 1/4, (*iv*) 1/3.

    [**Hint:** (*iii*) $P(\bar{B}) = 1 - P(B) = 3/4$. Since A and B are independent, A and $\bar{B}$ are also independent. We obtain $P(A \cap \bar{B}) = P(A) P(\bar{B}) = (1/3)(3/4) = 1/4.$

    (*iv*) $P(A/\bar{B}) = P(A \cap \bar{B})/P(\bar{B}) = 1/4 \div 3/4 = 1/3.$]

26. Let A and B be two events with $P(A) = 1/2, P(B) = 1/3$ and $P(A \cap B) = 1/4$. Find $P(A/B), P(A \cup B), P(\bar{A}/\bar{B})$. (*Mangalore, 1997*). **Ans.** 3/4, 7/12, 3/8.

27. If A and B are two events such that $P(A) = 3/8$, $P(B) = 5/8$ and $P(A \cup B) = 3/4$, find $P(A/B)$ and $P(B/A)$. Also find $P(A \cap \bar{B})$ and $P(A/\bar{B})$.

Ans. 2/5, 2/3, 1/8, 1/3.

24.42 Inverse Probability—Baye's Theorem

If an event has occurred through one of many mutually exclusive and exhaustive events then the conditional probability that it has occurred due to a particular event, is called its *inverse probability* (also known as posteriori probability). A theorem (or rule) named after the English mathematician Thomas Bayes (1702-1761) is used to determine inverse probability.

Bayes' Theorem: *If an event A of S with $P(A) \neq 0$ can occur in conjunction with one of the n mutually exclusive and exhaustive events B_1, B_2,... B_n of the sample space S with $P(B_i) \neq 0$, and if A has actually occurred then the probability that it was preceded by a particular event B_i (i = 1, 2, ..., n) is given by*

$$P(B_i/A) = \frac{P(B_i)P(A/B_i)}{\sum_{i=1}^{n} P(B_i)P(A/B_i)}, i = 1, 2, \ldots, n \qquad \ldots(24.84)$$

Remark. 1. The denominator in equation (24.84) gives the total probability.

2. $P(B_i)$ is known before the experiment and is known as *priori* probability.

3. $P(B_i/A)$ is called *posteriori* probability. It gives the probability of occurrence of the event B_i when the event A has already occurred.

4. Baye's theorem is also called the theorem of *probability of causes*.

Example 24.93 Machines I and II produce 30% and 70% respectively of a factory's output. Machine I produces 3% defectives and machine II produces 4% defectives.

(a) **If an item is defective what is the chance it was produced by machine I?**

(b) **If an item is not defective, what is the chance it was produced by machine I?**

Solution. Denote the events as

B_1: the item was produced by machine I

B_2: the item was produced by machine II

Let the defective item be denoted by M.

Now $P(B_1) = 30/100 = 0.3$; $P(B_2) = 70/100 = 0.7$.

$$P(M/B_1) = 3/100 = 0.03, \ P(M/B_2) = 4/100 = 0.04$$

$$P(\bar{M}/B_1) = 0.97; \ P(\bar{M}/B_2) = 0.96.$$

Applying Bayes' Theorem

(a) $P(B_1/M) = \dfrac{P(B_1)P(M/B_1)}{P(B_1)P(M/B_1)+P(B_2)P(M/B_2)}$

$= \dfrac{(0.3)(0.03)}{(0.3)(0.03)+(0.7)(0.04)} \approx 0.24.$

(b) $P(B_1/\bar{M}) = \dfrac{P(B_1)P(\bar{M}/B_1)}{P(B_1)P(\bar{M}/B_1)+P(B_2)P(\bar{M}/B_2)}$

$= \dfrac{(0.3)(0.97)}{(0.3)(0.97)+(0.7)(0.96)} \approx 0.30 \cdot$

Example 24.94 The chances that doctor A will diagnose a disease X correctly is 60%. The chance that a patient will die by his treatment after correct diagnosis is 40% and the chance of death by wrong diagnosis is 70%. A patient of doctor A who had disease X died. What is the chance that his disease was diagnosed correctly.

Solution. Denote the events as

A_1: The disease is diagnosed correctly.

A_2: The disease is diagnosed incorrectly. E: The patient dies

Then $P(A_1) = 60/100 = 0.6$; $P(A_2) = 0.4$

$P(E/A_1) = 40/100 = 0.4$; $P(E/A_2) = 70/100 = 0.7.$

By Bayes' theorem

$P(A_1/E) = \dfrac{P(A_1)P(E/A_1)}{P(A_1)P(E/A_1)+P(A_2)P(E/A_2)} = \dfrac{(0.6)(0.4)}{(0.6)(0.4)+(0.4)(0.7)} = \dfrac{6}{13}.$

Example 24.95 In a factory which manufactures bolts, machines A, B and C manufacture respectively, 25%, 35% and 40% of the bolts. Of their outputs 5, 4, and 2 per cent are respectively defective bolts. A bolt is drawn at random from the product and is found to be defective. What is the probability that it is manufactured by the machine A, B or C? (*Madras 2000, Mangalore, 1997*)

Solution. Denote the event as

B_1: the bolt is manufactured by machine A,

B_2: the bolt is manufactured by machine B,

B_3: the bolt is manufactured by machine C.

Note that B_1, B_2, B_3 are mutually exclusive and exhaustive events.

Let E = event, 'the bolt is defective'.

The event E occurs with B_1 or with B_2 or with B_3. Given that

$P(B_1) = \dfrac{25}{100} = 0.25, \quad P(B_2) = \dfrac{35}{100} = 0.35, \quad \text{and} \quad P(B_3) = \dfrac{40}{100} = 0.40.$

Again $P(E/B_1)$ = Probability that the bolt drawn is defective given the condition that it is manufactured by machine $A = \dfrac{5}{100} = 0.05$.

Similarly, $P(E/B_2) = 0.04$, $P(E/B_3) = 0.02$.

Hence, by Bayes' Theorem, we have

$$P(B_1/E) = \frac{P(B_1)P(E/B_1)}{P(B_1)P(E/B_1) + P(B_2)P(E/B_2) + P(B_3)P(E/B_3)}$$

$$= \frac{0.25 \times 0.05}{0.25 \times 0.05 + 0.35 \times 0.04 + 0.40 \times 0.02}$$

$$= \frac{0.0125}{0.0125 + 0.014 + 0.008} = \frac{0.0125}{0.0345} = 0.36.$$

Similarly $P(B_2/E) = \dfrac{0.35 \times 0.04}{0.25 \times 0.05 + 0.35 \times 0.04 + 0.40 \times 0.02} = \dfrac{0.014}{0.0345} = 0.41,$

and $\quad P(B_3/E) = \dfrac{0.40 \times 0.02}{0.0345} = \dfrac{0.008}{0.0345} = 0.23.$

Example 24.96 The contents of three urns are as follows:
   urn I: 1 white, 2 black and 3 red balls,
   urn II: 2 white, 1 black and 1 red balls,
   urn III: 4 white, 5 black and 3 red balls.

One urn is chosen at random and two balls are drawn. These happen to be white and red. What is the probability that they come from urn I?

Solution. Denoting the following events.

A: two balls drawn are white and red. Y_1, Y_2 and Y_3 be the events that the balls drawn are from the urn I, urn II, urn III respectively. Then

$$P(Y_1) = P(Y_2) = P(Y_3) = 1/3.$$

Also $P(A/Y_1)$ = Probability of drawing a white ball and a red ball from the urn I

$$= \left({}^1C_1 \times {}^3C_1\right) \Big/ {}^6C_2 = \frac{1 \times 3}{15} = \frac{1}{5}.$$

$P(A/Y_2)$ = Probability of drawing a white ball and a red ball from the urn I

$$= \left({}^2C_1 \times {}^1C_1\right) \Big/ {}^4C_2 = \frac{2 \times 1}{6} = \frac{1}{3}.$$

Similarly $P(A/Y_3) = \dfrac{{}^4C_1 \times {}^3C_1}{{}^{12}C_2} = \dfrac{4 \times 3}{66} = \dfrac{2}{11}.$

Hence, by Bayes' Theorem, we have

$P(Y_1/A)$ = Probability that the balls drawn are from the urn I

$$= \frac{P(Y_1)P(A/Y_1)}{P(Y_1)P(A/Y_1)+P(Y_2)P(A/Y_2)+P(Y_3)P(A/Y_3)}$$

$$= \frac{(1/3)(1/5)}{(1/3)(1/5)+(1/3)(1/3)+(1/3)(2/11)} = \frac{1\times495}{15\times118} = \frac{33}{118}.$$

Thus, the required probability is 33/118.

Example 24.97 In a class of 75 students, 15 were considered to be intelligent, 45 as medium and the rest below average. The probability that a very intelligent student fails in a viva-voce examination is 0.005, the medium student failing has a probability of 0.05, and the correspoding probability for a below average student is 0.15. If a student is known to have passed the viva-voce examination, what is the probability that he is below average?

(*Delhi university, 1997*)

Solution. Denoting A_1, A_2 and A_3 the events that a student is very intelligent, medium and below average respectively and E be the event that a student passes in a viva-voce examination. Given that $P(A_1) = 15/75 = 0.20$,

$P(A_2) = 45/75 = 0.6$, $P(A_3) = 15/75 = 0.20$.
$P(E/A_1) = 1 - 0.005 = 0.995$, $P(E/A_2) = 1 - 0.05 = 0.95$,
$P(E/A_3) = 1 - 0.15 = 0.85$.

The probability that a student is below average has passed the viva-voce examination = $P(A_3/E)$

$$= \frac{P(A_3)P(E/A_3)}{P(A_1)P(E/A_1)+P(A_2)P(E/A_2)+P(A_3)P(E/A_3)}$$

$$= \frac{0.20\times0.85}{0.20\times0.995+0.6\times0.95+0.20\times0.85} = \frac{0.17}{0.199+0.57+0.17} = \frac{0.17}{0.939} = 0.181.$$

Exercise 24.3

1. There are two bags I and II. Bag I contains 2 white and 3 red balls, bag II contains 4 white and 5 red balls. One ball is drawn at random from one of the bags and is found to be red. Find the probability that it was drawn from bag II.
   Ans. 25/32

2. There are three boxes A, B, C containing screws, some of which are defective. The proportion of defective screws in the three boxes are 1/4, 1/8, 1/3 respectively. A box is selected at random and a screw drawn from it at random is found to be defective. What is the probability that the box A was selected?
   Ans. 6/17.

3. An insurance company insured 4000 scooter drivers, 2000 car drivers and 6000 truck drivers. The probabilities of an accident involving a scooter, a car

and a truck are respectively 0.01, 0.03 and 0.15. One of the insured persons meets an accident. What is the probability that he is a car driver?

Ans. 3/50.

4. In the coming year, there will be three candidates for the position of Principal—Dr. Sinha, Mr. Veenu Gopal and Dr. Chatterji–whose chances of getting the appointment are in the proportion 4:2:3 respectively. The probability that Dr. Sinha if selected will abolish co-education in the college is 30%. The probability of Mr. Veenu Gopal and Dr. Chatterji doing the same are respectively 50% and 80%. What is the probability that co-education will be abolished in the college next year. **Ans. 23/45.**

5. A man is known to speak truth 3 out of 4 times. He throws a die and reports that it is a six. Find the probability that it is actually a six. **Ans. 3/8.**

24.43 Random Variables

In many experiments we are interested in *numbers* associated with the outcomes of the experiment. For example, when *n* coins are tossed simultaneously, one may be interested in knowing the number of heads. When a pair of dice is rolled and we ask for the sum of the points on the upturned faces. Whenever we associate a real number with each outcome of an experiment we are considering a *function* whose domain is the set of possible outcomes and whose range is the set of real numbers in question. Such a function is called a *random variable*.

Definition of a Random Variable (RV). Let *S* denote a sample space corresponding to a random experiment *E*. A *random variable X* on a sample space *S* is a function from *S* into the set *R* of real numbers such that the pre-image of every interval of *R* is an event of *S*.

Notations. Capital letters such as *X*, *Y*, *Z* are ordinarily used to denote one-dimensional random variables. A typical outcome of the experiment is usually denoted by the Greek letter ω (omega). Thus $X(\omega)$ denotes that real number which the random variable *X* associates with the outcome ω.

The following are some simple examples of random variables.

Example 1 If the random experiment E consists of two tosses of a coin, the random variable X defining the number of heads will assume the values 0, 1, 2 represented by

Outcome (ω)	TT	TH	HT	HH
Value of X (ω)	0	1	1	2

Thus to each outcome say ω of *E*, there corresponds a real number $X(\omega)$. The experiment *E* contains four outcomes TT, TH, HT and HH. These are also called sample points and they constitute the entire sample space *S* of *E*. Further, since the sample points correspond to outcomes, it is obvious that the random variable $X(\omega)$ is defined for each ω belonging to *S* and thus a random variable corresponds to a real function on *S*.

Example 2 An experiment consists of rolling a die and reading the number of points on the upturned face. The most natural random variable X to consider is

$X(\omega) = \omega$ for $\omega = 1, 2, 3, 4, 5, 6$.

If we are interested in whether the number of points is even or odd, then we can consider a random variable Y, which is defined as follows:

$$Y(\omega) = \begin{cases} 0, \text{if } \omega \text{ is even,} \\ 1, \text{if } \omega \text{ is odd.} \end{cases}$$

The values 0 and 1 are not essential—any two distinct real numbers could be used instead. However, 0 and 1 suggest "even" and "odd" respectively, because they represent the remainder obtained when the outcome ω is divided by 2.

There are two types of random variables *discrete* and *continuous*.

Discrete random variable

A random variable X defined on a discrete sample space is known as a *discrete random variable* if X assume either only a finite number or a countably infinite number of values $x_1, x_2, x_3, \ldots$ called the possible values of X, with positive probabilities. Examples of discrete random variables are marks obtained in a test, number of accidents per month, deaths by cancer, tosses until a die shows the first six etc.

24.44 Probability function of a discrete random variable

Let S be a discrete sample space associated with an experiment and X be discrete random variable which takes the values $x_1, x_2, x_3, \ldots, x_n$. Let each x_i be associated with a number p_i called the probability of x_i and denoted by $p_i = P(X = x_i) = p(x_i)$. The number $p(x_i)$ must satisfy the following conditions:

(*i*) $p(x_i) \geq 0$ for all values of i, and   (*ii*) $\sum\limits_{i=1}^{n} p(x_i) = 1$.

The set of values x_i with their probabilities p_i constitutes the *probability distribution* or the *discrete probability distribution* of the discrete random variable X. The probability function is also called *probability mass function* or *point probability function*.

Example 24.98 A random variable X has the following probability function:

x	0	1	2	3	4	5	6	7	8
$p(x)$	k	$3k$	$5k$	$7k$	$9k$	$11k$	$13k$	$15k$	$17k$

(*i*) Determine the value of k,
(*ii*) Find $p(X < 4)$, $p(X \geq 6)$, $p(3 < X \leq 5)$.
(*iii*) Find the minimum value of λ such that $p(X \leq \lambda) > 1/2$.

Solution. (*i*) If X is a random variable, then $\sum\limits_{i=0}^{8} p(x_i) = 1$.

This gives $k + 3k + 5k + 7k + 9k + 11k + 13k + 15k + 17k = 1$ or $k = 1/81$.

(ii) $p(X < 4) = p(X = 0) + p(X = 1) + p(X = 2) + p(X = 3)$

$\qquad = k + 3k + 5k + 7k = 16k = 16/81$.

$p(X \geq 6) = p(X = 6) + p(X = 7) + p(X = 8)$

$\qquad = 13k + 15k + 17k = 45k = 45/81 = 5/9$.

$p(3 < X \leq 5) = p(X = 4) + p(X = 5) = 9k + 11k = 20k = 20/81$.

(iii) $p(X \leq 1) = 4k = \dfrac{4}{81} < \dfrac{1}{2}; \quad p(x \leq 2) = k + 3k + 5k = 9k = \dfrac{9}{81} < \dfrac{1}{2}$

$p(X \leq 3) = k + 3k + 5k + 7k = 16k = \dfrac{16}{81} < \dfrac{1}{2};$

$p(X \leq 4) = k + 3k + 5k + 7k + 9k = 25k = \dfrac{25}{81} < \dfrac{1}{2}.$

$p(X \leq 5) = k + 3k + 5k + 7k + 9k + 11k = 36k = \dfrac{36}{81} < \dfrac{1}{2}.$

$p(X \leq 6) = k + 3k + 5k + 7k + 9k + 11k + 13k = 49k = \dfrac{49}{81} > \dfrac{1}{2}.$

$\therefore$ The minimum value of λ so that $p(X \leq \lambda) > 1/2$ is 6.

Example 24.99 **From a lot of 10 items containing 3 defectives items, a sample of 4 items are drawn at random without replacement. Let a random variable X denote the number of defective items in the sample. Find the probability distribution of X. Calculate P(X ≤ 1), P(X < 1), P(0 < X < 2).**

Solution. The lot containing 7 non-defective and 3 defective items. Since X denotes the number of defective item. x can take the values 0, 1, 2, 3 such that

For $x = 0$, $p(x) =$ Prob. of getting no defective item $= \dfrac{^7C_4}{^{10}C_4} = \dfrac{1}{6}$.

For $x = 1$, $p(x) =$ Prob. of getting one defective item $= \dfrac{^7C_3 \times {}^3C_1}{^{10}C_4} = \dfrac{1}{2}$.

For $x = 2$, $p(x) =$ Prob. of getting two defective item $= \dfrac{^7C_2 \times {}^3C_2}{^{10}C_4} = \dfrac{3}{10}$.

For $x = 3$, $p(x) =$ Prob. of getting three defective item $= \dfrac{^7C_1 \times {}^3C_3}{^{10}C_4} = \dfrac{1}{30}$.

We have the following probability distribution.

x	0	1	2	3
$p(x)$	1/6	1/2	3/10	1/30

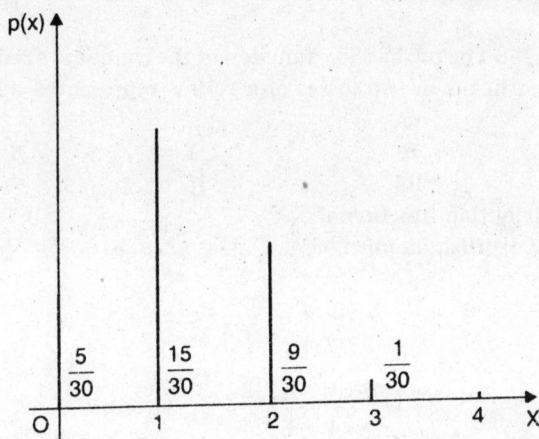

Fig. 24.18 shows the probability function $P(x)$ of the discrete random variable X.

Now $\quad P(X \le 1) = P(X = 0) + P(X = 1) = (1/6) + (1/2) = 2/3.$

$$P(X < 1) = P(X = 0) = 1/6.$$

$$P(0 < X < 2) = P(X = 1) = 1/2.$$

24.45 Distribution function

The distribution function $F(x)$ of random variable X is defined by

$F(x) = P(X \le x)$. It has the following properties.

(i) $0 \le F(x) \le 1$, $\quad$ (ii) If $x_1 < x_2$, then $F(x_1) \le F(x_2)$,

(iii) $P(a \le x \le b) = F(b) - F(a)$.

In the case of a discrete RV, the discrete distribution function is defined by

$$F(x) = \sum_{x_i \le x} p(x_i).$$

If X takes only a finite set of values $x_1, x_2, \ldots \ldots x_n$ then the distribution function

is given by $F(x) = \begin{cases} 0 & -\infty < x < x_1 \\ p(x_1) & x_1 \le x < x_2 \\ p(x_1) + p(x_2) & x_2 \le x < x_3 \\ \vdots & \\ p(x_1) + p(x_2) + \cdots + p(x_n) & \cdots x_n \le x < \infty \end{cases}$

From this definition, we have

$$p(x_i) = P(X = x_i) = F(x_i) - F(x_{i-1}).$$

The distribution function $F(x)$ is also called the *cumulative distribution function* (*cdf*) of X.

Example 24. 100 The probability function of the random variable X denoting the number of heads in two tosses of a coin is represented by the following table:

X	0	1	2
p (x)	1/4	1/2	1/4

Find the distribution function of X.

Solution. The distribution function of $F(x)$ is given by

$$F(x) = \begin{cases} 0 & -\infty < x < 0 \\ 1/4 & 0 \le x < 1 \\ \dfrac{1}{4} + \dfrac{1}{2} = \dfrac{3}{4} & 1 \le x < 2 \\ \dfrac{3}{4} + \dfrac{1}{4} = 1 & 2 \le x < \infty \end{cases}$$

24.46 Mean and variance of a discrete random variable

If X denotes a discrete random variable which can assume the values $x_1, x_2,..., x_n$ with respective probabilities $p_1, p_2, ..., p_n$ where $p_1 + p_2 + ..., p_n = 1$, then the *mean* or *expectation* or *expected value* of X, denoted by $E(X)$ or μ_x or simply E or μ is defined by $E(X) = \mu_x = p_1 x_1 + p_2 x_2 + ... + p_n x_n = \displaystyle\sum_{i=1}^{n} p_i x_i$ ...(24.85)

The *variance* of the random variable X, denoted by Var (X), is defined by

$$\text{Var } (X) = \sigma_X^2 = E\left[\left(X - \mu_x\right)^2\right] = \sum_{i=1}^{n}\left(x_i - \mu_x\right)^2 p_i$$

$$= \sum_{i=1}^{n}\left(x_i^2 - 2\mu_x x_i + \mu_x^2\right)p_i = \sum_{i=1}^{n} x_i^2 p_i - 2\mu_x \sum_{i=1}^{n} x_i p_i + \mu_x^2 \sum_{i=1}^{n} p_i$$

$$= \sum_{i=1}^{n} x_i^2 p_i - 2\mu_x^2 + \mu_x^2$$

$$\left(\text{Using } \sum_{i}^{n} x_i p_i = \mu_x \text{ and } \sum_{i}^{n} p_i = 1\right)$$

$$= \sum_{i=1}^{n} x_i^2 p_i - \mu_x^2 = E\left(X^2\right) - \mu_x^2 = E\left(X^2\right) - \left[E(X)\right]^2$$ ...(24.86)

That is, the variance is equal to the difference between the expectation of the square of the random variable and the square of the expectation of the random variable.

The positive square root of variance is known as standard deviation and is denoted by σ.

Example 24.101 A random variable X has the following probability disribution.

x	1	2	3	4	5
p(x)	k	k	3k	$k^2 + k$	$6k^2$

Find the value of k. Evaluate $P(X < 3)$ and $P(1 < X < 4)$. Determine the distribution function of X. Find the mean, variance of X and standard deviation.

Solution. Since $\sum\limits_{x=1}^{5} p(x) = 1$, we obtain

$k + k + 3k + (k^2 + k) + 6k^2 = 1$, or $7k^2 + 6k - 1 = 0$

or $(7k - 1)(k + 1) = 0$ or $k = 1/7, -1$.

Since $p(x) \geq 0$, the possible value of $k = 1/7$.

Now, $P(X < 3) = P(X = 1) + P(X = 2) = k + k = 2k = 2/7$.

$P(1 < X < 4) = P(X = 2) + P(X = 3) = k + 3k = 4k = 4/7$.

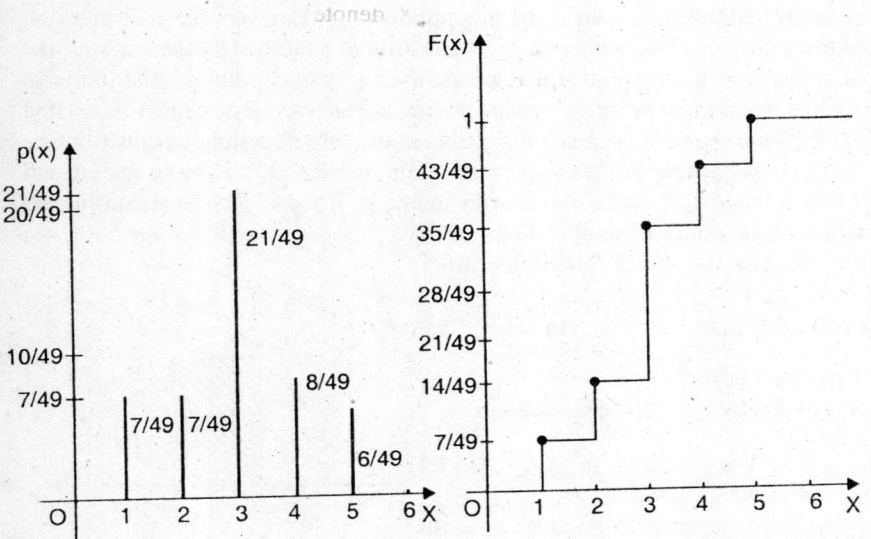

Fig. 24.19 shows the probability function $p(x)$ and distribution function $F(x)$ of the discrete random variable X having the values x given in the table.

TABLE

The results for the probability distribution and distribution function are given below.

x	1	2	3	4	5
$p(x)$	1/7	1/7	3/7	8/49	6/49
$F(x)$	1/7	2/7	5/7	43/49	1

Mean $= \mu_x = \sum x_i p_i = (1)(1/7) + (2)(1/7) + 3(3/7) + 4(8/49) + 5(6/49)$

$= 1/7 + 2/7 + 9/7 + 32/49 + 30/49 = 146/49.$

Variance $= \sigma_x^2 = E(x^2) - [E(x)]^2$

$$= \left[1\left(\frac{1}{7}\right) + 4\left(\frac{1}{7}\right) + 9\left(\frac{3}{7}\right) + 16\left(\frac{8}{49}\right) + 25\left(\frac{6}{49}\right) \right] - \left(\frac{146}{49}\right)^2$$

$$= \frac{502}{49} - \left(\frac{146}{49}\right)^2 = 10.245 - 8.878 = 1.367.$$

Standard deviation $= \sigma_x = \sqrt{1.367} = 1.169.$

24.47 Continuous Random Variables and Distributions

A random variable X is said to be a continuous random variable if it takes all possible values in a given interval I_x. The distribution defined by the variates like ages, heights, weights and temperature are examples of continuous distributions.

Unlike the discrete probability distribution in which a variate can be associated with a definite probability; it is not possible in the case of continuous distribution to assign a set probability to each variate. In such cases it is more useful and practicable to assign probabilities with intervals. Thus if $f(x)$ be a continuous function of the random variable X, the probability that the variable X falls in the infinitesimal small interval $x - \dfrac{dx}{2}$ to $x + \dfrac{dx}{2}$ can be expressed as $f(x)\,dx$.

Symbolically it can be expressed as

$$P\left(x - \frac{dx}{2} \leq X \leq x + \frac{dx}{2} \right) = f(x)\,dx. \quad ...(24.87)$$

Now, $f(x)dx$ represents the area under the curve $y = f(x)$, x-axis and the ordinates $x - (dx/2)$ and $x + (dx/2)$ (Refer Fig. 24.20)

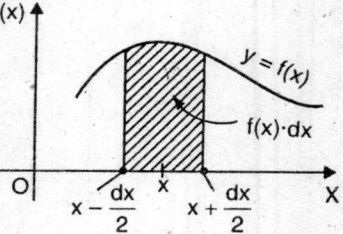

Fig. 24.20. Probability curve.

The function $f(x)$ is called the *probability density function* (*pdf*) of X or simply the *density function*. The expression $f(x)\,dx$, usually written as $dF(x)$ is known as the *probablity differential*. The continuous curve represented by $y = f(x)$ is called the *probability density curve* or briefly the *probability curve*.

The interval of the variate may be finite or infinite but even when the interval is finite it is advisable to consider this as infinite by simple assumption that the density function is zero outside the specified range. Thus if $f(x) = \phi(x)$ is the density function defined for the variate values of X in the interval $a \le x \le b$, it can

be put in the form $\;f(x) = \begin{cases} 0 & \text{if} \quad -\infty \le x < a \quad \text{or} \quad x < a, \\ \phi(x) & \text{if} \quad a \le x \le b, \\ 0 & \text{if} \quad b < x \le \infty \quad \text{or} \quad x > b. \end{cases}$

The Probability density function $f(x)$ possesses the following properties.

(i) $f(x) \ge 0$ for all $x \in I_x$, (ii) $\int_{I_x} f(x)\,dx = 1.$

The property (i) expresses the fact that probabilities can never be negative. Property (ii) stipulates the requirement that the probability of an event that is certain to happen must always be unity.

For a probability density function $f(x)$ the probability that the variate X lies in the interval (a, b) is given by

$$P(a \le X \le b) = \int_a^b f(x)\,dx. \qquad \ldots (24.88)$$

The probability defined by the above integral geometrically denotes the area bounded by the part of the curve $y = f(x)$, the two ordinates $x = a$ and $x = b$ and the x-axis

while $\int_{-\infty}^{\infty} f(x)\,dx = 1 = P(-\infty < X < \infty)$ signifies that the total area under the curve is unity.

24.48 Cumulative distribution function (cdf)

If X is a continuous random variable, then $F(x) = P(X \le x)$ is called the cdf or simply the distribution function that is,

$$F(x) = P(X \le x) = P(-\infty < X \le x) = \int_{-\infty}^{x} f(x)\,dx. \qquad \ldots (24.89)$$

The probability distribution of a continuous random variable is known if either its *pdf* or *cdf* is given.

The cumulative distribution function $F(x)$ has the following important properties (i) $0 \le F(x) \le 1, -\infty < x < \infty.$

(ii) $F(x)$ is a non-decreasing function, that is, if $x_1 < x_2$, then $F(x_1) \le F(x_2)$

(iii) $F(-\infty) = 0$ and $F(\infty) = 1$

(*iv*) $F(x)$ is continuous function of x on the right side.

(*v*) $f(x) = F'(x)$ at all point where $F(x)$ is differentiable.

Since $f(x)$ is always positive, property (*i*) gives that $F(x)$ is a non-decreasing function. Also $f(x)$ can be obtained by differentiating $F(x)$.

Mean and Variance

If X is a continuous random variable and $f(x)$ is the *pdf* of X, then we define

$$\text{Mean} = \mu_x = \int_R x f(x) dx,$$

$$\text{Variance} = \sigma_x^2 = \int_R (x - \mu_x)^2 f(x) dx.$$

Since $\int_R (x - \mu_x)^2 f(x) dx = \int_R (x^2 - 2\mu_x x + \mu_x^2) f(x) dx$

$$= \int_R x^2 f(x) dx - 2\mu_x \int_R x f(x) dx + \mu_x^2 \int_R f(x) dx$$

$$= \int_R x^2 f(x) dx - 2\mu_x \mu_x + \mu_x^2 \cdot 1 = \int_R x^2 f(x) dx - \mu_x^2$$

The mean and variance of the distribution defined on $R = [a, b]$, are defined as

$$\text{Mean} = \mu = \int_a^b x f(x) dx. \qquad \qquad ...(24.90)$$

$$\text{Variance} = \sigma^2 = \int_a^b x^2 f(x) dx - \mu^2. \qquad ....(24.91)$$

$$\Rightarrow \sigma^2 + \mu^2 = \int_a^b x^2 f(x) dx.$$

Example 24.102 Find the constant d so that the function $f(x)$ defined as below may be a density function:

$$f(x) = \begin{cases} 1/d & \text{if} \quad \alpha \le x \le \beta \\ 0 & \text{elsewhere.} \end{cases}$$

Find also the cumulative distribution of the random variable X when d satisfies the requirements for $f(x)$ to be a density function.

Solution. If $d > 0$, $f(x) \ge 0$ for every x. Hence for $f(x)$ to be a density function, we have

$$\int_\alpha^\beta f(x) dx = 1 \quad \text{or} \quad \int_\alpha^\beta \frac{1}{d} dx = 1 \quad \text{or} \quad \frac{1}{d}(\beta - \alpha) = 1 \text{ which gives } d = \beta - \alpha.$$

For the distribution function, we have

$$F(x) = P(X \le x) = P(-\infty < X \le x) = \int_{-\infty}^x f(x) dx$$

If $x < \alpha$, then $F(x) = 0$.

If $\alpha \le x < \beta$, then $F(x) = \int_\alpha^x \frac{1}{d}\,dx = \frac{1}{\beta - \alpha}(x - \alpha)$ and if $x \ge \beta$, then

$$F(x) = \int_\alpha^\beta f(x)\,dx + \int_\beta^x f(x)\,dx = \frac{1}{\beta - \alpha}\int_\alpha^\beta dx + \int_\beta^x 0.\,dx = 1.$$

Thus the required distribution function is $F(x) = \begin{cases} 0, & x < \alpha \\ \dfrac{x - \alpha}{\beta - \alpha}, & \alpha \le x < \beta \\ 1. & x \ge \beta. \end{cases}$

24.49 Illustrative Examples

Example 24.103 Let X be a continuous random variable with distribution

$$f(X) = \begin{cases} \dfrac{1}{6}X + k, & \text{if} \quad 0 \le X \le 3, \\ 0 & \text{elsewhere.} \end{cases}$$

Evaluate (a) the value of k. (b) $P\{1 \le X \le 2\}$.

Solution. (a) The graph of f is shown in Fig. 24.21. Since f is a continuous probability function, the shaded region A must have area equal to 1. Trapezoidal area A [having parallel sides k, $k + (1/2)$ and perpendicular distance between

them 3] $= \dfrac{1}{2}\left(k + k + \dfrac{1}{2}\right)\cdot 3 = 1$ or $k + \dfrac{1}{4} = \dfrac{1}{3}$ or $k = \dfrac{1}{12}$.

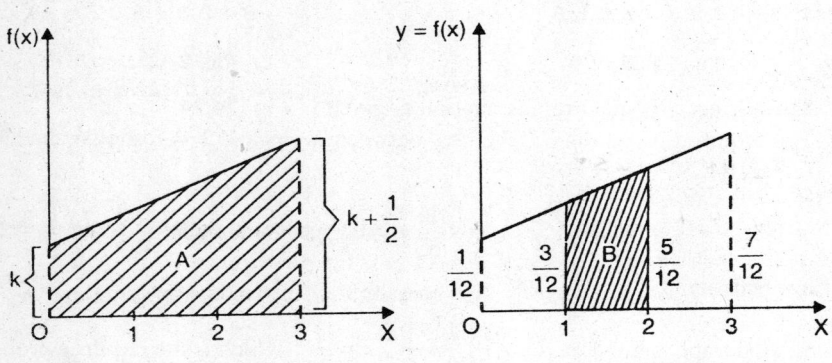

Fig. 24.21.   **Fig. 24.22.**

(b) $P\{1 \le X \le 2\}$ is equal to the area of portion B which is under the graph of f and between ordinates $X = 1$ and $X = 2$ as shown in Fig. 24.22. We have

$$f(X) = \frac{1}{6}X + \frac{1}{12}. \ \therefore \ f(1) = \frac{1}{6} + \frac{1}{12} = \frac{3}{12} \ \text{and} \ f(2) = \frac{2}{6} + \frac{1}{12} = \frac{5}{12}.$$

Trapezoidal area $B = \frac{1}{2}\left(\frac{3}{12} + \frac{5}{12}\right) \cdot 1 = \frac{1}{3}$, which is the required probability.

Example 24.104 A continuous random variable X which can assume only values between $X = 2$ and 8 inclusive, has a density function given by $k(X + 3)$ where 'k' is a constant. Calculate (a) the value of k (b) $P\{3 < x < 5\}$, (c) $P\{X \geq 4\}$, (d) $P\{|X - 5| < 0.5\}$.

Solution. The graph of $p(X) = k(X + 3)$ is a straight line shown in Fig. 24.23. To find k, we proceed as under:

The total area under the line between $X = 2$ and $X = 8$ and above the X-axis must be 1.

At $X = 2$, $p(X) = 5k$; at $X = 8$, $p(x) = 11k$.

Trapezoidal area = $\frac{1}{2}$ (sum of parallel sides) × perpendicular distance between them

$$= \{(5k + 11k) \cdot (6)/2\} = 48k = 1 \text{ or } k = 1/48.$$

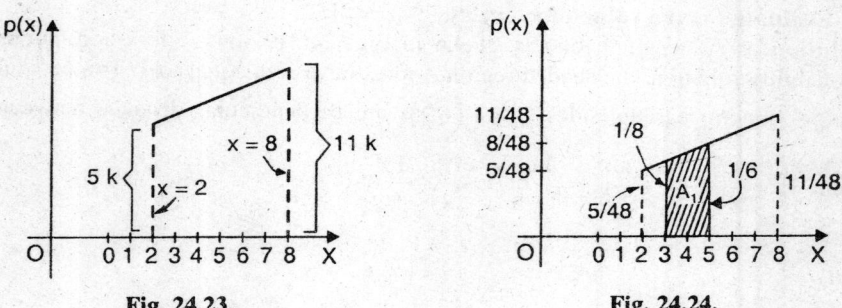

Fig. 24.23.   Fig. 24.24.

The correct graphical representation is given in Fig. 24.24.

(b) The required probability is the area between $X = 3$ and $X = 5$ shown shaded in Fig. 24.24.

$p(3) = \frac{6}{48} = \frac{1}{8}$ and $p(5) = \frac{8}{48} = \frac{1}{6}$ are the ordinates at $X = 3$ and $X = 5$ respectively.

Shaded trapezoidal area $A_1 = \frac{1}{2}\left(\frac{1}{8} + \frac{1}{6}\right) \cdot 2 = \frac{7}{24}$, which is the required probability.

(c) Refer Fig. 24.25. Now $p(4) = 7/48$, $p(8) = 11/48$.

∴ Trapezoidal area $B = \frac{1}{2}\left(\frac{7}{48} + \frac{11}{48}\right) \cdot 4 = \frac{3}{4}$, which is the required probability.

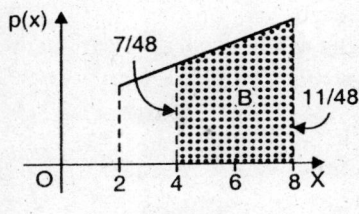

Fig. 24.25.

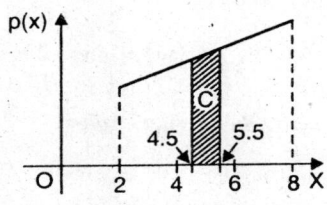

Fig. 24.26.

(d) $|X-5| < 0.5$ may be written as $X - 5 < 0.5$ or $-X + 5 < 0.5$. That is, $X < 5.5$ or $X > 4.5$.

Refer Fig 24.26. Trapezoidal area, 'C' $= \dfrac{1}{2}\left(\dfrac{7.5}{48}+\dfrac{8.5}{48}\right)(1) = \dfrac{1}{6}$, which is the required probability.

Example 24.105 The density function, f, is defined by

$$f(x) = \begin{cases} 0 & \text{if} & x < 0 \\ kx(1-x) & \text{if} & 0 \le x \le 1 \\ 0 & \text{if} & x > 1. \end{cases}$$

Find (i) the value of k; (ii) the distribution function F. (iii) the density and distribution functions, g and G, of the random variable W where $w = x^2$. sketch the graphs of f and F.

Solution. (i) Since $\int_{-\infty}^{+\infty} f(x)dx = 1$,

we have $\int_0^1 kx(1-x)dx = 1$ or $k\left[\dfrac{x^2}{2} - \dfrac{x^3}{3}\right]_0^1 = 1$ or $k = 6$.

(ii) If $x < 0$, $F(x) = 0$, If $x > 1$, $F(x) = 1$.

If $0 \le x \le 1$, $F(x) = \int_0^x 6x(1-x)dx = \left[3x^2 - 2x^3\right]_0^x = x^2(3-2x)$.

Thus F is the function F: Range, $R \to \{y \in R: 0 \le y \le 1\}$ defined by

$$F(x) = \begin{cases} 0, & \text{if} & x < 0, \\ x^2(3-2x), & \text{if} & 0 \le x \le 1, \\ 1. & \text{if} & x > 1. \end{cases}$$

(iii) If $w < 0$, $G(w) = 0$; if $w > 1$, $G(w) = 1$.

If $0 \le w \le 1$, $G(w) = P\{z \in R : 0 \le z \le w\}$

$$= w(3 - 2\sqrt{w}) \quad \text{(Since } x \text{ is positive, } x = \sqrt{w}\text{)}.$$

Thus

$$G(w) = \begin{cases} 0 & \text{if } w < 0 \\ 3w - 2w^{3/2} & \text{if } 0 \le w \le 1 \\ 1 & \text{if } w > 1. \end{cases}$$

Also, $g = dG/dw$, giving

$$g(w) = \begin{cases} 0 & \text{if } w < 0 \\ 3\left(1 - \sqrt{w}\right) & \text{if } 0 \le w \le 1 \\ 0 & \text{if } w > 1. \end{cases}$$

The graph of f, F are shown in Figs. 24.27 and 24.28.

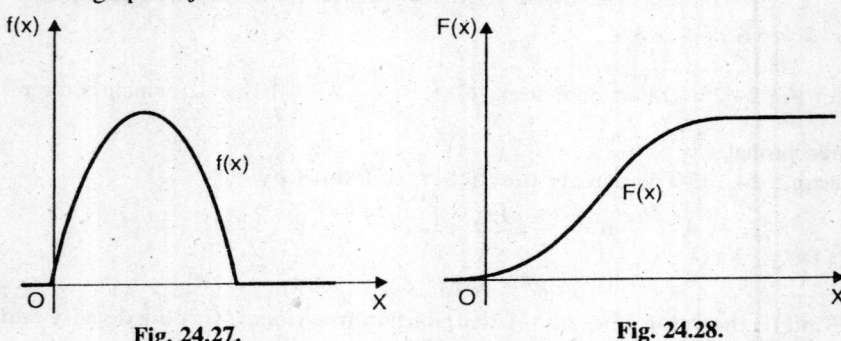

Fig. 24.27. Fig. 24.28.

Example 24.106 If X be a random variate with the following probability distribution.

X	8	12	16	20	24
p(x)	1/8	1/6	3/8	1/4	1/12

Find (a) $E(X)$, (b) $E(X^2)$, (c) $E(2X + 1)^2$, and (d) $E\left[(X - \bar{X})^2\right]$.

Solution. (a) $E(X) = \sum X \cdot p(x) = 8(1/8) + 12(1/6) + 16(3/8) + 20(1/4) + 24(1/12)$
$= 1 + 2 + 6 + 5 + 2 = 16$. This represents the *mean* of the distribution.

(b) $E(X^2) = \sum X^2 \cdot p(x) = (8)^2(1/8) + (12)^2(1/6) + (16)^2(3/8) +$
$(20)^2(1/4) + (24)^2(1/12)$
$= 8 + 24 + 96 + 100 + 48 = 276$. This represents the *second moment* about the origin (zero).

(c) $E(2X + 1)^2 = E(4X^2 + 4X + 1) = 4E(X^2) + 4E(X) + 1$
$= 4(276) + 4(16) + 1 = 1169.$

(d) $E\left[(X - \bar{X})^2\right] = \sum (X - \bar{X})^2 \cdot p(X)$, $\bar{X} = \sum X \div n = 80/5 = 16$
$= (8 - 16)^2(1/8) + (12 - 16)^2(1/6) + (16 - 16)^2(3/8)$
$+ (20 - 16)^2(1/4) + (24 - 16)^2(1/12) = 8 + 8/3 + 4 + 16/3 = 20$

This represents the *variance* of the distribution.

Example 24.107 Find the mean and variance of the random variable **X** whose density function, **f** is defined by

$$f(x) = \begin{cases} 0 & \text{if} \quad x < 0 \\ 4x(1-x^2) & \text{if} \quad 0 \le x \le 1 \\ 0 & \text{if} \quad x > 1. \end{cases}$$

Solution. We have $\mu = E(X) = \int_{-\infty}^{+\infty} x f(x) dx$

$$= \int_0^1 4x^2 (1-x^2) dx \quad [\text{Since } f(x) = 0 \text{ outside the interval } 0 \le x \le 1),$$

$$= \left[4\frac{x^3}{3} - \frac{4x^5}{5} \right]_0^1 = 4\left(\frac{1}{3} - \frac{1}{5} \right) = \frac{8}{15}.$$

The variance of $X = \sigma_x^2 = E\left[(x-\mu)^2 \right] = \int_{-\infty}^{+\infty} (x-\mu)^2 f(x) dx$

$$= \int_{-\infty}^{+\infty} (x^2 - 2\mu x + \mu^2) f(x) dx = \int_{-\infty}^{+\infty} x^2 f(x) dx$$

$$-2\mu \int_{-\infty}^{+\infty} x f(x) dx + \mu^2 \int_{-\infty}^{+\infty} f(x) dx$$

$$= \int_{-\infty}^{+\infty} x^2 f(x) dx - 2\mu \cdot \mu + \mu^2 \cdot 1 \text{ or } \sigma_x^2 + \mu^2 = \int_{-\infty}^{+\infty} x^2 f(x) dx$$

Thus, $\sigma_x^2 + \dfrac{64}{225} = \int_0^1 4x^3 (1-x^2) dx = \left[x^4 - \dfrac{2x^6}{3} \right]_0^1 = \dfrac{1}{3},$

or $\qquad \sigma_x^2 = \dfrac{1}{3} - \dfrac{64}{225} = \dfrac{11}{225}.$

Example 24.108 Suppose the life in hours of a certain kind of radio tube has the probability density function:

$$f(x) = \begin{cases} 100/x^2, & \text{for} \quad x \ge 100, \\ 0, & \text{for} \quad x < 100. \end{cases}$$

Find the distribution function. What is the probability that none of the 3 tubes in a given radio set will have to be replaced during the first 150 hours of operation? What is the probability that all three of the original tubes will be replaced during the first 150 hours? *(IETE., Dec, 2007)*

Solution. Let $F(x)$ be the distribution function, then

$$F(x) = \int_{-\infty}^x f(x) dx = \int_{-\infty}^{100} f(x) dx + \int_{100}^x f(x) dx$$

$$= 0 + \int_{100}^x \frac{100}{x^2} dx = 100 \left[-\frac{1}{x} \right]_{100}^x = 1 - \frac{100}{x}.$$

∴ Probability that a tube will last for 150 hours,

$$F(150) = P(X \le 150) = 1 - \left(\frac{100}{150}\right) = \frac{1}{3}.$$

Hence the probability that all the three tubes will last for 150 hours each *i.e.*, none of the three tubes shall have to be replaced = $(1/3)^3$ = 1/27.

Since the probability that a tube will have to be replaced within 150 hours of its life $= 1 - (1/3) = 2/3$.

Hence the probability that all the three tubes will have to be replaced within 150 hours = $(2/3)^3$ = 8/27.

Example 24.109 **X is a continuous random variable with probability density function given by**

$$f(x) = \begin{cases} 2x^3, & 0 \le x \le 1 \\ 2(2-x)^3, & 1 \le x \le 2 \\ 0, & \text{otherwise} \end{cases}$$

find the standard deviation and the mean deviation about the mean for the random variable X. (*IETE, Dec. 2008*)

Solution. We have mean $= \mu_1' = \int_{-\infty}^{\infty} x f(x)\,dx = \int_0^1 x(2x^3)\,dx + \int_1^2 2x(2-x)^3\,dx$

$$= \left[\frac{2x^5}{5}\right]_0^1 + \left[\frac{-2x^5}{5} + 3x^4 - 8x^3 + 8x^2\right]_1^2 = 1.$$

$$\mu_2' = \int_{-\infty}^{\infty} x^2 f(x)\,dx = \int_0^1 x^2(2x^3) + \int_1^2 2x^2(2-x)^3\,dx$$

$$= \left[\frac{x^6}{3}\right]_0^1 + \left[\frac{-x^6}{3} + 12\frac{x^5}{5} - 6x^4 + 16\frac{x^3}{3}\right]_1^2 = \frac{16}{15}.$$

∴ Variance $= \mu_2' - \mu_1'^2 = \frac{16}{15} - 1 = \frac{1}{15} \Rightarrow S.D. = \sqrt{1/15} = 0.258$

Mean Deviation $= \int_{-\infty}^{\infty} |x - \bar{x}| f(x)\,dx$

$$= \int_0^1 |x-1| 2x^3\,dx + \int_1^2 |x-1| \{2(2-x)^3\}\,dx$$

$$= 2\int_0^1 (x^4 - x^3)\,dx + 2\int_1^2 x(2-x)^3\,dx - 2\int_1^2 (2-x)^3\,dx$$

$$= 2\left[\frac{x^5}{5} - \frac{x^4}{4}\right]_0^1 + 2\left[\frac{-x^5}{5} + \frac{3}{2}x^4 - 4x^3 + 4x^2\right]_1^2 - 2\left[\frac{-x^4}{4} + 2x^3 - 6x^2 + 8x\right]_1^2$$

$$= -\frac{1}{10} + 2\left[\frac{-x^5}{5} + \frac{7x^4}{4} - 6x^3 + 10x^2 - 8x\right]_1^2 = -\frac{1}{10} + \frac{1}{10} = 0$$

Hence mean deviation and standard deviation are 0 and 0.258 respectively.

Exercise 24.4

1. 3 rotten eggs are accidentally mixed with 7 good ones. Find the probability distribution of the number of rotten eggs, if 3 eggs are drawn at random.

Ans.

X	0	1	2	3
P(X)	35/120	63/120	21/120	1/120

2. A random variable X has the following probability function:

X :	0	1	2	3	4	5	6	7
p(X) :	0	k	2k	2k	3k	k^2	$2k^2$	$7k^2 + k$

(*i*) Find k, (*ii*) Evaluate $p(X < 6)$, $p(X \geq 6)$, $p(3 < X \leq 6)$ and $p(0 < X < 5)$

(*iii*) Find the minimum value of α so that $p(X \leq \alpha) > 1/2$.

Ans. 1/10, 81/100; 19/100, 33/100, 4/5; 4.

3. Plot the graph of the cummulative distribution function F of the random variable X with distribution.

x_i	−3	2	6
$p(x_i)$	1/4	1/2	1/4

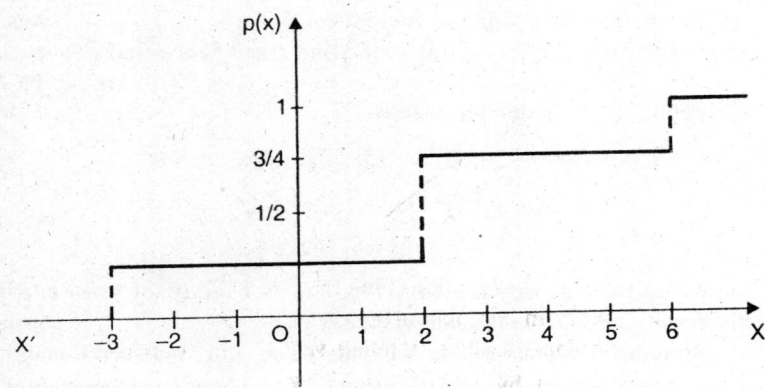

Fig. 24.29.

4. (a) Find (i) $E(X)$, (ii) $E(X^2)$, (iii) $E\left[(X-\bar{X})^2\right]$, and (iv) $E(X^3)$ for the following probability distribution.

X	−10	−20	30
P(X)	1/5	3/10	1/2

Ans. (i) 7, (ii) 590, (iii) 541, (iv) 10900.

(b) A variate X has p.d.f.

x	−3	6	9
P(x)	1/6	1/2	1/3

Find $E(X), E(X^2)$ and $E(2X+1)^2$.   *(IETE-Dec. 2007)*

Ans. 11/2; 93/2; 209.

5. Find the expectation μ, Variance σ^2 and standard deviation σ of the following distribution:

X	1	2	3	4	5	6
p(X)	1/36	3/36	5/36	7/36	9/36	11/36

Ans. 4.47, 1.99, 1.4

6. A function $f(x)$ is defined by
$$f(x) = ce^{-x}, \ 0 \le x \le \infty.$$
Find the value of c which changes $f(x)$ to a probability density function.

Ans. $c = 1$.

7. A function $f(x)$ is defind as follows:
$$f(x) = \begin{cases} kx^{-2} & \text{if} \quad x \ge 1 \\ 0 & \text{if} \quad x < 1. \end{cases}$$
Find the value of k and hence find $P(x < 2)$.   **Ans.**1, 1/2

8. $f(x) = ke^{-2x}$ if $x \ge 0, f(x) = 0$ if $x < 0$; find k and hence find $P(1 < x < 2)$.

Ans. 2, 117/1000.

9. A function $f(x)$ is defined as follows:
$$f(x) = \begin{cases} 0 & \text{if} \quad x < 2 \\ \dfrac{1}{18}(2x+3) & \text{if} \quad 2 \le x \le 4 \\ 0 & \text{if} \quad x > 4. \end{cases}$$

Show that it is a density function. Find the probability that a variate having this density will fall in the interval $2 \le x \le 3$.   **Ans.** 4/9.

10. A continuous random variable X having values only between 0 and 4 has a density function given by $P(X) = (1/2) - kX$, where k is a constant. Calculate (a) the value of k (b) $P\{1 < x < 2\}$.   **Ans.** 1/8; 5/16.

11. State the condition under which a function is said to be a probability density function. Examine whether the function f defined by

$$f(x) = \begin{cases} \dfrac{3}{8}(4x - 2x^2), & 0 \le x \le 2 \\ = 0, & \text{otherwise} \end{cases}$$

is a probability density function.

12. The probability distribution function of a continuous random variable is given by $f(x) = k(3x - x^2 - 2)$, $1 \le x \le 2$. Calculate

(i) the value of k, (ii) $p(x > 3/2)$, (iii) $p(5/4 \le x \le 3/2)$.

Ans. (*i*) $k = 6$, (*ii*) 1/2, (*iii*) 1/4.

$$\left[\textbf{Hint.}(ii)p(x > 3/2) = 1 - p(x \le 3/2) = 1 - \int_{1}^{3/2} 6(3x - x^2 - 2)dx \right].$$

13. Is the following a probability distribution function?

$$F(x) = \begin{cases} 2x, & \text{when} & 0 \le x \le 1 \\ 3 - 2x, & \text{when} & 1 < x \le 3 \\ 0, & \text{elsewhere.} \end{cases}$$

[**Hint.** A probability distribution function must satisfy the following conditions:

(i) $f(x) \ge 0$ and (ii) $\int_{-\infty}^{+\infty} f(x)dx = 1$.] **Ans. No.**

14. Let X be a random variable, a possible choice of $f(x)$ would be

$$f(x) = \begin{cases} 0 & \text{if} \quad x < 0 \\ kx & \text{if} \quad 0 \le x < 50 \\ k(100 - x) & \text{if} \quad 50 \le x < 100 \\ 0 & \text{if} \quad x \ge 100, \text{ where } k \text{ is some number.} \end{cases}$$

The graph of $y = f(x)$ is shown in Fig. 24.30. Calculate (i) k, (ii) probability that P lies between 30 and 60 from A.

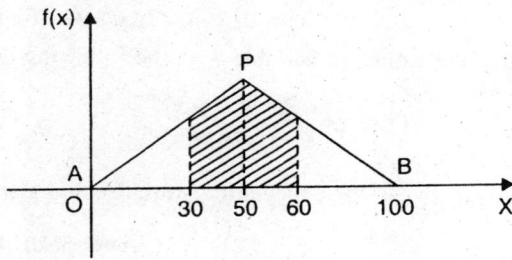

Fig. 24.30.

$$\left[\textbf{Hint. } (i) \int_0^{50} kx\,dx + \int_{50}^{100} k(100-x)\,dx = 1; \right.$$

$$\left. (ii)\ P = \int_{30}^{50} \frac{1}{2500} x\,dx + \int_{50}^{60} \frac{1}{2500}(100-x)\,dx \right]. \qquad \textbf{Ans. } (i)\ 1/2500;\ (ii)\ 0.5.$$

15. A continuous random variable X has a *pdf* $f(x) = 3x^2, 0 \le x \le 1$. Find a and b when $(i)\ P(X \le a) = P(X > a)$ and $(ii) P(X > b) = 0.05$.

    [**Hint.** $P(X \le a) = 0.5 = P(X > a)$.] **Ans.** $(1/2)^{1/3}, (19/20)^{1/3}$.

16. Verify that the following is a distribution function.

    $F(x) = 0$, for $x < 0$,

    $= x/3$, for $0 \le x < 1$,

    $= 1/3$, for $1 \le x < 2$,

    $= x/6$, for $2 \le x < 6$,

    $= 1$, for $x \ge 6$.

17. Define distribution function. Verify that the following is a distribution function.

    $$F(x) = \begin{cases} 0 & \text{when} & x < -a \\ \dfrac{1}{2}\left(\dfrac{x}{a}+1\right) & \text{when} & -a \le x \le a \\ 1 & \text{when} & x > a \end{cases}$$

18. If the density function f, of a distribution is given by

    $$f(x) = \begin{cases} 0 & \text{if} & x < 0 \\ x/2 & \text{if} & 0 \le x \le 1 \\ 1/2 & \text{if} & 1 < x \le 2 \\ \dfrac{1}{2}(3-x) & \text{if} & 2 < x \le 3 \\ 0 & \text{if} & x > 3, \end{cases}$$

    find the distribution function F.

    Ans. $F(x) = 0, x^2/4, (2x-1)/4, (6x-2x^2-5)/4, 1$.

19. Let X be a continuous random variable with the following distribution:

    $$f(x) = \begin{cases} x/2, & \text{if } 0 \le x \le 2 \\ 0, & \text{elsewhere.} \end{cases}$$

    Calculate $p(1 \le x \le 3/2)$; the expectation, variance and standard deviation of X. **Ans.** $5/16; 4/3; 2/9; \sqrt{2}/3$.

20. If X be a continuous random variate with *pdf*

$$f(x) = ax, \qquad 0 \le x < 1$$
$$= a, \qquad 1 \le x \le 2$$
$$= -ax + 3a, \qquad 2 \le x \le 3$$
$$= 0, \text{ elsewhere.}$$

(*i*) Find the value of a, (*ii*) Find the *cdf* of X, (*iii*) Find $P(X \le 1.5)$.

[**Hint.** (*i*) $\int_{-\infty}^{+\infty} f(x)dx = 1$, that is $\int_{-\infty}^{0} f(x)dx + \int_{0}^{1} f(x)dx + \int_{1}^{2} f(x)dx$

$+ \int_{2}^{3} f(x)dx + \int_{3}^{\infty} f(x)dx = 1$, that is $\int_{0}^{1} axdx + \int_{1}^{2} adx + \int_{2}^{3} (3-x)adx = 1$.

This gives $a = 1/2$.

(*ii*) We have $F(x) = P(-\infty < x \le x) = \int_{-\infty}^{x} f(x)dx$

$$x < 0 : F(x) = 0. \quad 0 \le x < 1 : F(x) = \int_{0}^{x} \frac{1}{2}xdx = \frac{x^2}{4}$$

$$1 \le x \le 2 : F(x) = \int_{0}^{1} \frac{x}{2}dx + \int_{1}^{x} \frac{1}{2}dx = \frac{1}{4} + \frac{1}{2}(x-1) = \frac{1}{2}x - \frac{1}{4}$$

$$2 \le x \le 3 : F(x) = \int_{0}^{1} \frac{x}{2}dx + \int_{1}^{2} \frac{1}{2}dx + \int_{2}^{x} \frac{1}{2}(3-x)dx$$

$$= \frac{1}{4} + \frac{1}{2} + \frac{3}{2}(x-2) - \frac{1}{4}(x^2 - 4) = \frac{3}{2}x - \frac{x^2}{4} - \frac{5}{4}$$

$x > 4 : F(x) = 1.$

(*iii*) Now $P(X \le 1.5) = \int_{-\infty}^{1.5} f(x)dx = \int_{-\infty}^{0} f(x)dx + \int_{0}^{1} f(x)dx + \int_{1}^{1.5} f(x)dx$

$$= 0 + \int_{0}^{1} \frac{1}{2}xdx + \int_{1}^{1.5} \frac{1}{2}dx = \frac{1}{4}(1-0) + \frac{1}{2}(1.5-1) = \frac{1}{2}.]$$

21. A continuous probability distribution has density function *f*, where $f(x) = a + bx + cx^2$ for $0 \le x \le 1$ and $f(x) = 0$ outside this range. The mean is 2/3 and the variance is 4/45. Find the values of *a*, *b* and *c*. **Ans.** 1, −4, 6.

22. A random variable X is distributed at random between the values 0 and 1 so that its probability density function is: $f(x) = kx^2(1 - x^3)$, where k is a constant. Find the value of k. Using this value of k, find its mean and standard deviation.

[**Hint:** Since $\int_{-\infty}^{\infty} f(x)dx = 1$, $k \int_0^1 (x^2 - x^5)dx = 1 \Rightarrow k = 6$.

$$\text{Mean} = \mu_1' = \int_{-\infty}^{\infty} xf(x)dx = 6\int_0^1 (x^3 - x^6)dx = \frac{9}{14}.$$

$$\mu_2' = \int_{-\infty}^{\infty} x^2 f(x)dx = 6\int_0^1 (x^4 - x^7)dx = \frac{9}{20}.$$

$$\text{Variance} = \mu_2 = \mu_2' - \mu_1'^2 = \left(\frac{9}{20}\right) - \left(\frac{9}{14}\right)^2 = \frac{9}{245}.$$

$$\text{Therefore S.D.} = \sqrt{\text{Variance}} = \frac{3}{\sqrt{245}}].$$

23. X is a continuous random variable with probability density function given by

$$f(x) = \begin{cases} x^3, & 0 \le x \le 1, \\ (2-x)^3 & 1 \le x < 2. \end{cases}$$

Find the standard deviation and also the mean deviation about the mean.

(*IETE., June 2009*) **Ans.** 1/5, 1/5.

24. A continuous random variable X has a *pdf* $f(x) = kx^2 e^{-x}$, $x \ge 0$ find k. Also find mean and variance of this random variable. (*IETE, June 2009*)

Ans. $k = 1/3$; mean = 3, variance = 3.

24.50 Theoretical Probability Distributions

The probability distributions which are not obtained by actual experiments or observations but are mathematically deduced on certain assumptions are called theoretical probability distributions. Here total probability of 1 is distributed over different values of the random variable X according to some definite law or pattern. We shall discuss the following probability distributions:

    (*i*) Binomial distribution, (*ii*) Poisson distribution, (*iii*) Normal distribution.
    (*iv*) Beta distribution and (*v*) Exponential distribution.

24.51 *Binomial Distribution

The Binomial distribution is a discrete probability distribution. Let the probability

---

* Binomial distribution is also known as *Bernoulli distribution* after James Bernoulli who discovered it at the end of 17th century.

of the success of an event in a single trial be p and probability of failure be $q(= 1 - p)$. It is assumed that p is the same from trial to trial.

We first find the probability that the event succeeds twice in three trials. The different possible ways of 2 successes in 3 trials are $^3C_2$, *i.e.,* 3 and may be shown as follows:

SSF; SFS; FSS, where S stands for success and F for failure.

Now the probability that it succeeds in first two trials and fails in third, (*i.e.,* SSF) is $pp\, q = p^2\, q$.

Similarly the probability of SFS and FSS is also $p^2\, q$.

($\because$ in each case, number of p's is equal to number of times S appears and number of q's is equal to number of F's).

Since $^3C_2$ ways are mutually exclusive, the probability of 2 successes in 3 trials is by addition law

$$^3C_2 p^2 q = {}^3C_2 p^2 q^{3-2}.$$

In general the probability of r successes in n trial is

$$P(X = r) = P(r) = {}^nC_r p^r q^{n-r}, \quad r = 0, 1, 2, ..., n. \quad ...(24.92)$$

where $q = 1 - p$. X is called a binomial random variable with parameters n and p. Denote the binomial distribution by $X \sim B(n, p)$.

For $r = 0$, the probability $= {}^nC_0 p^0 q^n = q^n$ which is the probability that the event fails in all trials.

The probability distribution (24.92) is called *Binomial distribution* because for $r = 0, 1, 2..., n$ it gives probabilities of $0, 1, 2, ..., n$ successes which correspond to successive terms in the binomial expansion

$$(q + p)^n = q^n + {}^nC_1 pq^{n-1} + {}^nC_2 p^2 q^{n-2} + ... + p^n.$$

While using the formula 24.92 of the binomial distribution in solving any problem. We should examine whether all the conditions of Bernoulli trials are satisfied or not. These conditions are given below:

(*i*) There should be a finite number of trials.

(*ii*) All the trials should be independent.

(*iii*) Each trial has exactly two outcomes called success (occurrence of the event) and failure (non-occurrence of the event).

(*iv*) The probability of an outcome remains the same in each trial.

24.51.1 Recurrence formula for the Binomial Distribution

We have $P(r) = {}^nC_r p^r q^{n-r}$ and $P(r+1) = {}^nC_{r+1} p^{r+1} q^{n-r-1}$

Therefore $\dfrac{P(r+1)}{P(r)} = \dfrac{{}^nC_{r+1} p^{r+1} q^{n-r-1}}{{}^nC_r p^r q^{n-r}} = \dfrac{n!\, p^{r+1} q^{n-r-1}}{(r+1)!(n-r-1)!} \cdot \dfrac{r!(n-r)!}{n!\, p^r q^{n-r}}$

$$= \frac{n-r}{r+1} \cdot \frac{p}{q}. \text{ Thus } P(r+1) = \left(\frac{n-r}{r+1}\right) \cdot \frac{p}{q} P(r); r = 0, 1, 2, ..., (n-1) \quad ...(24.93)$$

This is the required *Recurrence* formula. Applying this formula successively, we can find the probabilities $P(1)$, $P(2)$, $P(3)$, etc. if $P(0)$ is known.

Thus, when $r = 0$, we have from (24.93), $P(1) = n(p/q)P(0)$.

But $P(0) = p^o q^n = q^n$. Therefore $P(1) = npq^{n-1}$.

Again, putting $r = 1$ in (24.93), we get

$$P(2) = \frac{n-1}{2} \frac{p}{q} P(1) = \frac{n-1}{2} \cdot \frac{p}{q} \cdot npq^{n-1} = \frac{n(n-1)}{2} p^2 q^{n-2} \text{ and so on.}$$

24.51.2 Mean (or Expectation) and Variance of Binomial Distribution

(A.M.I.E., W-2002) (I.E.T.E. Dec. 2006)

Before we proceed to find the mean and S.D., we must bear in mind that in binomial distribution the variate values are 0, 1, 2,..., n and corresponding to frequencies, we have the probabilities of these values given by $^nC_r p^r q^{n-r}$. Also sum of all these probabilities $= (p + q)^n = 1^n = 1$.

Now, probabilities of getting 0, 1, 2, ..., n successes in n trials are

$$^nC_0 p^0 q^n, \, ^nC_1 p^1 q^{n-1}, \cdots, \, ^nC_n p^n q^0$$

or $\quad q^n, \, ^nC_1 pq^{n-1}, \cdots, p^n \left(\because \, ^nC_0 = \, ^nC_n = 1, p^0 = q^0 = 1 \right)$

$\therefore$ The mean, $\mu = E(X) = \sum_{r=0}^{n} r \cdot P(r) = \sum_{r=0}^{n} {}^nC_r p^r q^{n-r}$

$$= 0(q^n) + 1 \cdot {}^nC_1 pq^{n-1} + 2 \cdot {}^nC_2 p^2 q^{n-2} + \cdots + np^n$$

$$= np\left[q^{n-1} + {}^{n-1}C_1 pq^{n-2} + \cdots + p^{n-1} \right] = np(q+p)^{n-1} = \mathbf{np} \quad ...(24.94)$$

$$\left(\text{as } p+q = 1, \, {}^nC_1 = n, \quad 2 \cdot {}^nC_2 = 2 \cdot \frac{n(n-1)}{2} = n(n-1) = n \cdot {}^{n-1}C_1, \right.$$

$$\left. 3 \cdot {}^nC_3 = 3 \cdot \frac{n(n-1)(n-2)}{3 \cdot 2 \cdot 1} = n \cdot {}^{n-1}C_2. \text{ etc.} \right)$$

The Variance, σ^2. For the binomial distribution, variance σ^2 is given by

$$Var(X) = \sigma^2 = \sum_{r=0}^{n} r^2 P(r) - (\text{mean})^2$$

Now $\sum_{r=0}^{n} r^2 P(r) = 0^2 \cdot q^n + 1^2 \cdot {}^nC_1 pq^{n-1} + 2^2 \cdot {}^nC_2 p^2 q^{n-2} + \cdots + n^2 p^n$

$$= npq^{n-1} + 2n(n-1)p^2q^{n-2} + 3\frac{n(n-1)(n-2)}{2!}p^3q^{n-3} + \cdots + n^2p^n$$

$$= np\left[q^{n-1} + 2(n-1)q^{n-2}p + 3\cdot\frac{(n-1)(n-2)}{2!}q^{n-3}p^2 + \cdots + np^{n-1}\right]$$

$$= np\left[q^{n-1} + (n-1)q^{n-2}p + \frac{(n-1)(n-2)}{2!}q^{n-3}p^2 + \cdots + p^{n-1}\right]$$

$$+ np\left[(n-1)q^{n-2}p + 2\cdot\frac{(n-1)(n-2)}{2!}q^{n-3}p^2 + \ldots + (n-1)p^{n-1}\right]$$

$$= np(q+p)^{n-1} + np(n-1)p\left[q^{n-2} + {}^{n-2}C_1q^{n-3}p + \cdots + p^{n-2}\right]$$

$$= np + n(n-1)p^2(q+p)^{n-2} = np + n(n-1)p^2.$$

Now variance, $\sigma^2 = \left[np + n(n-1)p^2\right] - (\text{mean})^2$

$$= np + n(n-1)p^2 - n^2p^2 = np - np^2 = np(1-p) = \mathbf{npq}$$

Standard deviation, $\sigma = \sqrt{\mathbf{npq}}.$ \hfill ...(24.95)

24.51.3 Moment generating function of binomial distribution

Let X be a binomial variate. Then, the moment generating function (*MGF*) of X is defined by

$$MX(t) = E(e^{tx}) = \sum_{r=0}^{n} e^{tr}\ {}^nC_r p^r q^{n-r}$$

$$= \sum_{r=0}^{n} {}^nC_r\left(e^t p\right)^r q^{n-r} = \left(q + pe^t\right)^n = \left[q + p\left(1 + t + \frac{t^2}{2!} + \frac{t^3}{3!} + \ldots\right)\right]^n$$

$$= \left[q + p + pt + \frac{pt^2}{2!} + \frac{pt^3}{3!} + \ldots\right]^n = \left(1 + pt + \frac{pt^2}{2!} + \frac{pt^3}{3!} + \ldots\right)^n \text{ as } p+q=1.$$
\hfill ...(24.96)

24.52 Illustrative Examples

Example 24.110 If on an average 1 ship in every 10 is wrecked, find the probability that out of 5 ships expected to arrive, 4 atleast will arrive safely.
\hfill (*PTU. 2006*)

Solution. The probability p that a ship is wrecked $= 1/10$.

∴ Probability of its arriving safely $= 1 - (1/10) = 9/10$.

The chance of atleast 4 arriving safely

= chance of 4 arriving safely + chance of 5 arriving safely

$$= ^5C_4 \left(\frac{9}{10}\right)^4 \left(\frac{1}{10}\right) + ^5C_5 \left(\frac{9}{10}\right)^5 = 5 \frac{9^4}{10^5} + \frac{9^5}{10^5} = \frac{9^4}{10^5}(5+9) = \frac{45927}{50000}.$$

Example 24.111 **The probability that a man aged 60 will live to be 70 is 0.65. What is the probability that out of 10 men, now 60, atleast 7 will live to be 70?**   *(IETE, June 2009, June 2008)*

Solution. The probability of success, (*i.e.*, a man living upto 70) is 0.65 and probability of failure $= 1 - 0.65 = 0.35$.

∴ Probability of atleast 7 successes in 10 trials

$$= ^{10}C_7 (0.65)^7 (0.35)^3 + ^{10}C_8 (0.65)^8 (0.35)^2 + ^{10}C_9 (0.65)^9 (0.35) + (0.65)^{10}$$

$$= 0.2523 + 0.1756 + 0.0725 + 0.0135 = 0.514$$

Example 24.112 **If 10% bolts produced by a machine are defective, determine the probability that out of 10 bolts chosen at random (*i*) 1, (*ii*) none, (*iii*) at least 1, (*iv*) atmost two bolts, will be defective.**

Solution. (*i*) Here the probability p, of a bolt being defective

$$= \frac{10}{100} = 0.1.$$ Therefore, $q = 1 - p = 1 - 0.1 = 0.9.$

By Binomial distribution, the probability of 1 defective bolt out of 10

$$= ^{10}C_1 (0.1)^1 (0.9)^9 = 0.03874.$$

(*ii*) The probability that none is defective

$$= ^{10}C_0 (0.1)^0 (0.9)^{10} = (0.9)^{10} = 0.3487$$

(*iii*) Probability of at least one defective

= 1 − (Probability of none defective) = 1 − 0.3487 = 0.6513.

(*iv*) Probability of 2 defective $= ^{10}C_2 (0.1)^2 (0.9)^8 = 0.1937.$

∴ Probability of at most 2 defective

= Probability of none defective + Probability of 1 defective

+ Probability of two defective

= 0.3487 + 0.03874 + 0.1937 = 0.58114.

Example 24.113 **The incidence of occupational disease in an industry is such that the workmen have a 20% chance of suffering from it. What is the probability that out of seven workmen 5 or more will contact the disease?**

Solution. Let p denotes chance of suffering and q chance of not suffering.

∴ $p = 20\% = \dfrac{1}{5}$ and $q = 1 - \left(\dfrac{1}{5}\right) = \dfrac{4}{5}.$ Here $n = 7$.

The binomial expression is

$(q + p)^7 = q^7 + 7q^6p + 21q^5p^2 + 35q^4p^3 + 35q^3p^4 + 21q^2p^5 + 7qp^6 + p^7$.

The probability of 5 or more (*i.e.*, 5, 6, 7) successes is

$$= 21q^2p^5 + 7qp^6 + p^7. \qquad \qquad ...(1)$$

Substituting the values of p and q in (1) we obtain

$$\text{Regd. prob.} = 21\left(\frac{4}{5}\right)^2\left(\frac{1}{5}\right)^5 + 7\left(\frac{4}{5}\right)\left(\frac{1}{5}\right)^6 + \left(\frac{1}{5}\right)^7$$

$$= \frac{336}{78125} + \frac{28}{78125} + \frac{1}{78125} = \frac{365}{78125}.$$

Example 24.114 Out of 320 families with 5 children each how many families would be expected to have (*i*) 2 boys and 3 girls (*ii*) atleast one boy, (*iii*) no girl and (*iv*) at most 2 boys? Assume equal probability for boys and girls.

Solution. Probability of a boy $= p = 1/2$, Probability of a girl $= q = 1/2$.
$N = 320$, $n = 5$.

(*i*) Probability of 2 boys and 3 girls:

$r = 2$, $n - r = 5 - 2 = 3$. Employing the formula: $P(r) = {}^nC_r q^{n-r} p^r$

$$P(r = 2) = {}^5C_2\left(\frac{1}{2}\right)^3\left(\frac{1}{2}\right)^2 = \frac{5.4}{2}\cdot\frac{1}{8}\cdot\frac{1}{4} = \frac{5}{16}.$$

Expected number of families with 2 boys and 3 girls $= 320 \times (5/16) = 100$.

(*ii*) Probability of at least one boy $= 1 -$ probability of no boy

$$= 1 - P(r = 0) = 1 - {}^5C_0\left(\frac{1}{2}\right)^5\left(\frac{1}{2}\right)^0 = 1 - \frac{1}{32} = \frac{31}{32}.$$

$\therefore$ Expected number of families with atleast one boy $= 320 \times \dfrac{31}{32} = 310$.

(*iii*) Prob. (no girl) = Prob. (all 5 boys) $= {}^5C_5\left(\frac{1}{2}\right)^0\left(\frac{1}{2}\right)^5 = \frac{1}{32}$

$\therefore$ Number of families with no girl $= 320 \times \dfrac{1}{32} = 10$.

(*iv*) Prob. (at most 2 boys) = Prob. (no boy) + Prob. (one boy) + Prob. (two boy)

$$= {}^5C_0\left(\frac{1}{2}\right)^5\left(\frac{1}{2}\right)^0 + {}^5C_1\left(\frac{1}{2}\right)^4\left(\frac{1}{2}\right) + {}^5C_2\left(\frac{1}{2}\right)^3\left(\frac{1}{2}\right)^2$$

$$= \frac{1}{32} + 5\cdot\frac{1}{32} + \frac{5.4}{2}\cdot\frac{1}{32} = \frac{1}{2}$$

$\therefore$ Number of families with at most 2 boys $= 320 \times 1/2 = 160$.

Example 24.115 Fit a binomial distribution to the following frequency distribution:

x:	0	1	2	3	4	5	6
f:	13	25	52	58	32	16	4

(Madras; 2000S)

Solution. Here $n = 6$ and $N = \Sigma f = 200$.

$$\therefore \text{ Mean } = \frac{\Sigma f_i x_i}{\Sigma f_i} = \frac{25 + 104 + 174 + 128 + 80 + 24}{200} = \frac{535}{200} = 2.675.$$

Now the mean of the Binomial distribution $= np$

i.e., $np = 6p = 2.675$. $\therefore p = \dfrac{2.675}{6} = 0.446, q = 1 - p = 0.554.$

Hence the binomial distribution to be fitted in $N(q + p)^n$

$$= 200(0.554 + 0.446)^6$$

$$= 200\Big[(0.554)^6 + {}^6C_1(0.554)^5(0.446) + {}^6C_2(0.554)^4(0.446)^2$$

$$+ {}^6C_3(0.554)^3(0.446)^3 + {}^6C_4(0.554)^2(0.446)^4$$

$$+ {}^6C_5(0.554)(0.446)^5 + (0.446)^6\Big]$$

$$= 200[0.0289 + 0.1396 + 0.2810 + 0.3016 + 0.1821 + 0.05866 + 0.00787]$$

$$= 5.78 + 27.92 + 56.2 + 60.32 + 36.42 + 11.73 + 1.57.$$

Therefore, the successive terms in the expansion give the expected or theoretical frequencies (to the nearest integers) which are

x:	0	1	2	3	4	5	6
f:	6	28	56	60	36	12	2

Example 24.116 A target is to be destroyed in a bombing attack. There is 75% chance that any one bomb will strike the target. Assume that two direct hits are required to destroy the target completely. How many bombs must be dropped to give 99% or better chance of completely destroying the target?
Solution. Probability, p the bombs hits the target $= 3/4$, and failure not to hit the target, $q = 1 - p = 1/4$. Let n numbers of bombs be dropped to ensure 99% chance or better to completely destroy the target, X = random variable representing the number of bombs to be used. Hence

$$P(X = r) = {}^nC_r(3/4)^r(1/4)^{n-r}, r = 0, 1, 2, ..., n.$$

We require $P(X \geq 2) \geq 0.99$, or $1 - P(X \leq 1) \geq 0.99$.

or $\quad 1-\left[P(X=0)+P(X=1)\right] \geq 0.99$ or $1-\left(1/4\right)^{n}-n\left(3/4\right)\left(1/4\right)^{n-1} \geq 0.99$

or $\quad \left(1/4\right)^{n}+n\left(3/4\right)\left(1/4\right)^{n-1} \leq 0.01$ or $1+3n \leq (0.01)4^{n}$ or $100+300n \leq 4^{n}$.

By trial, n = 6. $\left(\text{since } 1900 \leq 4^{6}.\right)$

24.53 Poisson Distribution

Poisson distribution is a discrete probability distribution related to the probabilities of events which are extremely rare, but which have a large number of independent opporturnities for occurrence. The number of persons born blind per year in a large city and the number of deaths by horse kick in an army corps, are some of the phenomenae, in which this law is followed.
The probability distribution

$$p(X=r)=\frac{e^{-m}m^{r}}{r!}, r=0,1,2,...,m>0 \qquad \qquad ...(24.97)$$

where $e^{-m}m^{r}/r!$ gives the probability of *r* successes, is called **Poisson distribution**, and X is called a Poisson variate.

Poisson distribution is a limiting case of the Binomial distribution $^{n}C_{r}p^{r}q^{n-r}$ *when p, (the probability of success) is very small, that is,* $p \rightarrow 0$. *n, the number of trials is very large, that is* $n \rightarrow \infty$ *but np = m, (where m is a positive real number) is finite.* When p is small, the event is called a rare event. In practice we call an event rare if the number of trials is at least $50(n \geq 50)$ while np, *i.e.,* m is less than 5.

From (24.97) the probability of 0, 1, 2, ...r,... successes is

$$e^{-m}, e^{-m}m, e^{-m}\frac{m^{2}}{2!},...,e^{-m}\cdot\frac{m^{r}}{r!},... \qquad \qquad ...(24.98)$$

It can be easily seen that sum to infinity of probabilities (24.98) is 1.

$$\left[\because \sum_{r=0}^{\infty} P(r)=e^{-m}\left(1+\frac{m}{1!}+\frac{m^{2}}{2!}+\frac{m^{3}}{3!}+...\text{to } \infty\right)=e^{-m}\cdot e^{m}=1 \right]$$

Poisson distribution as a limiting case of Binomial distribution

We shall deduce the Poisson distribution as a limiting case of Binomial distribution as $n \rightarrow \infty$, $p \rightarrow 0$ but np remains constant, say m.

Let the random variable X be binomially distributed with parameters *n* and *p*.
As $n \rightarrow \infty$, $p \rightarrow 0$ and np = m, we obtain $P(X=r)= ^{n}C_{r}p^{r}q^{n-r}$.

$$= \frac{n(n-1)(n-2)...(n-r+1)}{r!} \cdot \left(\frac{m}{n}\right)^r \left(1-\frac{m}{n}\right)^{n-r}$$

$$= \frac{m^r}{r!}\left[1\left(1-\frac{1}{n}\right)\left(1-\frac{2}{n}\right)...\left(1-\frac{r-1}{n}\right)\right]\left(1-\frac{m}{n}\right)^n \cdot \left(1-\frac{m}{n}\right)^{-r} \qquad ...(24.99)$$

Now $\lim\limits_{n\to\infty}\left(1-\frac{m}{n}\right)^n = e^{-m}, \lim\limits_{n\to\infty}\left(1-\frac{m}{n}\right)^{-r} = 1, \lim\limits_{n\to\infty}\left(1-\frac{a}{n}\right) = 1,$ where 'a' is a

finite number. Therefore, taking limits as $n \to \infty$ in (24.99), we obtain

$$\lim_{\substack{n\to\infty \\ np=m=\text{finite}}} P(X=r) = \frac{m^r}{r!}(1)\cdot e^{-m} = \frac{e^{-m}m^r}{r!}.$$

24.53.1 Mean and Variance of the Poisson distribution   (*A.M.I.E., S-2000*)

The *mean* of a probability distribution is given by

$$E(X) = \mu = \sum_{r=0}^{\infty} r\cdot P(r) = \sum_{r=0}^{\infty} r\, e^{-m}\frac{m^r}{r!}$$

$$= e^{-m}m\sum_{r=1}^{\infty}\frac{m^{r-1}}{(r-1)!} = me^{-m}e^m = \mathbf{m} \qquad ...(24.100)$$

The *variance of a probability distribution* is given by

$$\sigma^2 = \sum_{r=0}^{\infty} r^2 P(r) - \mu^2$$

Now $\displaystyle\sum_{r=0}^{\infty} r^2 P(r) = \sum_{r=0}^{\infty}\left[r(r-1)+r\right]p(r) = \sum_{r=0}^{\infty} r(r-1)P(r) + \sum_{r=0}^{\infty} rP(r)$

$$= \sum_{r=0}^{\infty} r(r-1)P(r) + \mu = \sum_{r=0}^{\infty} r(r-1)\frac{e^{-m}m^r}{r!} + \mu$$

$$= e^{-m}m^2\sum_{2}^{\infty}\frac{m^{r-2}}{(r-2)!} + \mu = e^{-m}m^2\left[1 + \frac{m}{1!} + \frac{m^2}{2!} + ...\right] + \mu$$

$$= e^{-m}\cdot m^2\cdot e^m + \mu = m^2 + \mu.$$

$\therefore$ $\qquad\qquad \sigma^2 = m^2 + \mu - \mu^2 = m^2 + m - m^2 = m \qquad ...(24.101)$

Thus, for Poission distribution **mean = variance = m**

24.53.2 Recurrence formula for the probabilities of Poisson distribution

In poisson distribution, the probability of r successes is given by

$$P(r) = \frac{e^{-m} \cdot m^r}{r!}, \quad r = 0, 1, 2,\dots$$

Setting $r = r + 1$, we obtain

$$P(r+1) = e^{-m} \cdot \frac{m^{r+1}}{(r+1)!} = \frac{m}{r+1} P(r), \, r = 0, 1, 2, 3, \dots \qquad \dots(24.102)$$

This formula is called the *recurrence formula* for the poission distribution. If we know $P(0) = e^{-m}$, then with the formula (24.102), we can calculate all the other probabilities, $P(1)$, $P(2)$, $P(3)$ etc.

24.53.3 Moment generating function (MGF) of Poisson distribution

Let X be a Poisson variate. Then, the moment generating function of X is defined by

$$M_X(t) = E(e^{tX}) = \sum_{r=0}^{\infty} e^{tr} \cdot \left(\frac{e^{-m} m^r}{r!} \right) = e^{-m} \sum_{r=0}^{\infty} \frac{(me^t)^r}{r!}$$

$$= e^{-m} \left[1 + me^t + \frac{(me^t)^2}{2!} + \dots \right] = e^{-m} \cdot e^{me^t} = e^{m(e^t - 1)}. \qquad \dots(24.103)$$

24.54 Illustrative Examples

Example 24.117. Suppose that on an average 1 house in 1000 houses get fire in a year in a district. If there are 2000 houses in that district, find the probability that exactly 5 houses will have fire during the year. (given $e^{-2} = 0.1352$). *(IETE., Dec. 2006)*

Solution. Applying the poisson distribution. $P(X = r) = e^{-m} \cdot \dfrac{m^r}{r!}$

Mean, $m = np = (2000)(1/1000) = 2, r = 5$.

$$\therefore \quad P(X = 5) = e^{-2} \cdot \frac{2^5}{5!} = 0.1352 \times \frac{32}{120} = 0.036.$$

Example 24.118. Ten per cent of tools produced in a certain manufacturing process turn out to be defective. Find the probability that in a sample of 10 tools chosen at random. (*i*) exactly two will be defective, (*ii*) more than one will be the defective.

Solution. Probability to defective tools, $p = 10/100 = 0.1$.

$$m = np = 10(0.1) = 1$$

(*i*) Here $r = 2$, $m = 1$.

Probability of 2 defective tools $= \dfrac{e^{-1} \cdot (1)^2}{2!} = \dfrac{e^{-1}}{2} = \dfrac{1}{2e} = 0.18$ approx,

(*ii*) P(more than 1 defective) $= 1 - P$ (zero defective) $- P$ (one defective)

$$= 1 - e^{-1} - e^{-1} = 1 - \dfrac{2}{e} = \dfrac{e-2}{e} = 0.26 \text{, (approx).}$$

Example 24.119. The mortality rate for a certain disease is 7 per 1000. What is the probability for just 5 deaths from this disease in a group of 400?

Solution. Here $m = np = 400 \times (7/1000) = 2.8$ and $r = 5$

$\therefore$ Probability required $= \dfrac{e^{-2.8} (2.8)^5}{5!} = 0.087$.

Example 24.120. If the probability that an individual suffers a bad reaction from an injection of a given serum is 0.001; determine the probability that out of 2000 individuals (*a*) exactly 3, (*b*) more than 2 individuals will suffer a bad reaction. It is given that $e^{-2} = 0.13534$.

(IETE June 2007; AMIE., W-2004, D.U., 1997)

Solution. Here $p = 0.001$, $n = 2000$. $\therefore m = np = 2000 \times 0.001 = 2$.

(*a*) Probability that 3 individuals suffer from bad reaction $= \dfrac{e^{-2} 2^3}{3!} = \dfrac{4}{3e^2} = 0.18$.

(*b*) Probability that none suffer $= \dfrac{e^{-2} 2^0}{0!} = \dfrac{1}{e^2}$.

Probability that 1 suffer $= \dfrac{e^{-2} 2^1}{1!} = \dfrac{2}{e^2}$.

Probability that 2 suffers $= \dfrac{e^{-2} 2^2}{2!} = \dfrac{2}{e^2}$.

$\therefore$ Prob. of more than 2 suffer from bad reaction

$= 1 - [\text{Probability } (r = 0) + \text{Probability } (r = 1) + \text{Probability } (r = 2)]$

$$= 1 - \left(\dfrac{1}{e^2} + \dfrac{2}{e^2} + \dfrac{2}{e^2} \right) = 1 - \dfrac{5}{e^2} = 0.323$$

Example 24.121. The frequency of accidents per shift in a factory is as shown in the following table:

Accident per shift:	0	1	2	3	4
Frequency (f):	192	100	24	3	1

Calculate the mean number of accidents per shift and the corresponding Poisson distribution and compare with actual observations.

Solution. The mean of Poisson distribution is

$$m = \frac{\Sigma f_i x_i}{\Sigma f_i} = \frac{0 \times 192 + 1 \times 100 + 2 \times 24 + 3 \times 3 + 4 \times 1}{192 + 100 + 24 + 3 + 1} = \frac{161}{320} = 0.503 .$$

Hence the theoretical frequency for r successes is

$$\frac{Ne^{-m}m^r}{r!} = \frac{320e^{-0.503}(0.503)^r}{r!} , \text{ where } r = 0, 1, 2, 3, 4.$$

∴ The theoretical frequencies are

x:	0	1	2	3	4
f:	194	97	24	4	1

Example 24.122. In a certain factory producing cycle tyres there is a small chance of 1 in 500 tyres to be defective. The tyres are supplied in lots of 20. Using Poisson distribution calculate the approximate number of lots containing no defective, one defective and two defective tyres respectively in a consignment of 20,000 tyres. Given $e^{-0.04} = 0.9608$.

Solution. Here $n = 20$, $p = \dfrac{1}{500}$. Therefore, $m = np = \dfrac{20}{500} = 0.04$.

Also $e^{-m} = e^{-0.04} = 0.9608$.

Now $N = 20{,}000/20 = 1000$ lots of 20 each.

The number of lots containing no defective, one defective, two defective tyres are given by

$$Ne^{-m}, Ne^{-m} \cdot m, Ne^{-m}\left(m^2/2\right) \text{ that is,}$$

$$1000(0.9608), 1000(0.9608) \times 0.04, 1000(0.9608) \times \frac{(0.04)^2}{2} \text{ that is,}$$

961, 38, 1 respectively.

Example 24.123. Out of 1000 balls 50 are red and the rest white. If 60 balls are picked at random, what is the probability of picking up (*i*) 3 red balls (*ii*) not more than 3 red balls in the sample? Assume Poisson distribution for the number of red balls picked up in the sample. Given $e^{-3} = 0.0498$.

Solution. Since out of 1000 balls, 50 are red, the probability of drawing a red ball is given by $p = 50/1000 = 1/20$.

(*a*) As $n = 60$, $m = np = 60 \times 1/20 = 3$. ∴ $P(X = 3)$ *i.e.*, probability of drawing 3 red balls

$$= \frac{3^3 e^{-3}}{3!} = \frac{9}{2}(0.0498) = 0.2241$$

(b) P (X not more than 3 red balls)

$$= p(0 \text{ red}) + p(1 \text{ red}) + p(2 \text{ red}) + p(3 \text{ red})$$

$$= \frac{3^0 e^{-3}}{0!} + \frac{3^1 e^{-3}}{1!} + \frac{3^2 e^{-3}}{2!} + \frac{3^3 e^{-3}}{3!}$$

$$= e^{-3}\left(1 + 3 + \frac{3^2}{2!} + \frac{3^3}{3!}\right) = 0.0498(1 + 3 + 4.5 + 4.5) = 0.6474$$

Example 24.124. **A car hire firm has two cars which it hires out day by day. The number of demands for a car on each day is distributed as a Poisson distribution with mean 1.5. Calculate the proportion of days on which (i) car is not used and, (ii) the proportion of days on which some demand is refused.**

(given that $e^{-1.5} = 0.2231$)

(*IETE, June 2009, Dec. 2008S; AMIE-S 2006; JNTU-2003*)

Solution. (i) Here the random variable X, which denotes the number of demands for a car on any day follows Poisson distribution with mean, m = 1.5. The proportion of days on which there are r demands for a car is given by:

$$P(X = r) = \frac{e^{-1.5}(1.5)^r}{r!}; r = 0, 1, 2.....$$

$$\therefore P(X = 0) = \frac{e^{-1.5}(1.5)^0}{0!} = e^{-1.5} = 0.2231 \text{ which gives proportion of days when}$$

no car is used.

(ii) Proportion of days when some demand is refused = Prob. for the number of demands to be more than 2 i.e., $P(X > 2) = 1 - P(X \le 2)$

$$= 1 - [P(X = 0) + P(X = 1) + P(X = 2)] = 1 - e^{-1.5}\left[1 + 1.5 + \frac{(1.5)^2}{2!}\right]$$

$$= 1 - 0.2231 (3.625) = 0.1913.$$

Example 24.125. **Show that in a Poisson distribution with mean 1, then its mean deviation about mean is (2/e) times the standard deviation.**

(*I.E.T.E., Dec. 2005*)

Solution. If p(X = r) denotes probability of r successes, then

$$p(X = r) = \frac{e^{-m}m^r}{r!} = \frac{e^{-1}}{r!} \quad (\because m = 1, \text{ given}); r = 0, 1, 2,...$$

Mean deviation about mean 1 in a Poisson distribution.

$$= \sum_{r=0}^{\infty} p(r)|r - 1| = \sum_{r=0}^{\infty} |r - 1| \frac{e^{-1}}{r!} = e^{-1} \sum_{r=0}^{\infty} \frac{|r - 1|}{r!}$$

$$= e^{-1}\left[\frac{|0-1|}{0!}+\frac{|1-1|}{1!}+\frac{|2-1|}{2!}+\frac{|3-1|}{3!}+\cdots\infty\right]$$

$$= e^{-1}\left[1+0+\frac{2-1}{2!}+\frac{3-1}{3!}+\cdots\cdots\right]$$

$$= e^{-1}\left[1+\left(1-\frac{1}{2!}\right)+\left(\frac{1}{2!}-\frac{1}{3!}\right)+\cdots\infty\right]$$

$$= e^{-1}\left[\left(1+1+\frac{1}{2!}+\frac{1}{3!}+\cdots\infty\right)-\left(\frac{1}{2!}+\frac{1}{3!}+\cdots\infty\right)\right]$$

$$= e^{-1}[e-(e-2)] = 2/e = (2/e) \text{ times the S.D.}$$

Since $\qquad$ S.D. $= \sqrt{m} = 1,$ for $m = 1$

Thus, Mean deviation about mean $= (2/e)$ times the standard deviation.

24.55 Normal Distribution

The normal distribution can be regarded as the limiting form of the Binomial distribution when n is very large and p and q are not very small. It is one of the most important continuous probability distributions because a large number of frequency distributions approximate this distribution. Normal distribution is also called Gaussian distribution.

A continuous random variable X is said to have a normal distribution with parameters μ (mean) and σ^2 (variance) if its probability density function is given by

$$y = f(x) = \frac{1}{\sigma\sqrt{2\pi}}e^{-(x-\mu)^2/(2\sigma^2)} \quad -\infty < x < \infty, \ -\infty < \mu < \infty, \ \sigma > 0 \qquad \text{...(24.104)}$$

The graph of $y = f(x)$ is a bell shaped curve called the normal probability curve. This is shown in the Fig. 24.31. The curve is symmetric about the mean. The maximum occurs at $x = \mu$ and the maximum value is $1/(\sigma\sqrt{2\pi})$. The ordinate $f(x)$ decreases rapidly as $x \to \pm\infty$ and x-axis is an asymptote of the curve. If σ is large, the curve tends to be flat and if σ is small, the curve tends to be more peaked at $x = \mu$. We denote the above normal distribution by

$$X \sim N\left(\mu, \sigma^2\right).$$

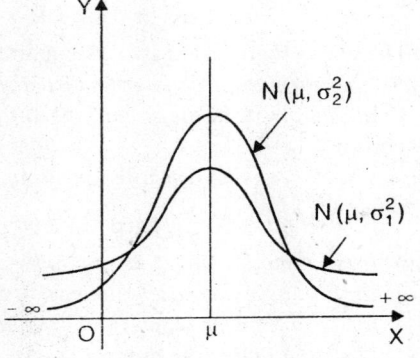

Normal probability curve, $\sigma_1 > \sigma_2$

Fig. 24.31.

The total area bounded by the curve (24.104) and the x-axis is 1. Hence the area bounded by this curve and the two ordinates $x = a$ and $x = b$ $(a < b)$ gives the probability that x lies between a and b. In brief the probability between a and b may be denoted by $P(a < X < b)$.

By the transformation $z = \dfrac{x - \mu}{\sigma}$, (24.104) reduces to the form

$$\phi(z) = \frac{1}{\sqrt{2\pi}} e^{-z^2/2} \qquad -\infty < z < \infty. \qquad \ldots(24.105)$$

This form is called *standard form or standardised normal curve*.

In this z is normally distributed with mean, $\mu = 0$ and variance, $\sigma^2 = 1$. We call z as the standard normal variate and we write $z \sim N(0,1)$.

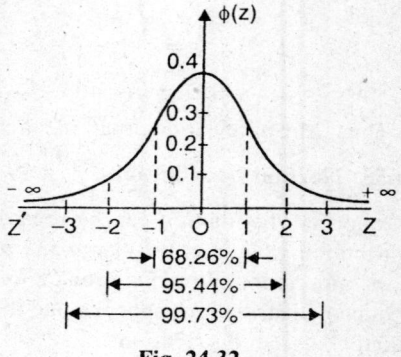

The graph of the standard normal curve given by (24.105) is shown in fig. 24.32.

The area under the normal standardised curve is 1 and area between $z = -1$ and $z = +1$ is 0.6826 (or 68.26% of the total area which is 1).

Area between $z = -2$ and $z = +2$ is 0.9544 (or 95.44% the total area)

Area between $z = -3$ and $z = +3$ is 0.9973 (or 99.73% of the total area)

Fig. 24.32.

Remarks

(*i*) *The normal probability distribution is defined by the equation*

$$y = \frac{1}{\sigma\sqrt{2\pi}} e^{-(x-\mu)^2/(2\sigma^2)}.$$

The curve represented by this equation is called normal curve.

(*ii*) *The normal curve is symmetrical about a vertical line through mean i.e., about $x = \mu$. In the standardized form it is symmetrical about the ordinate at $z = 0$.*

(*iii*) *The curve is known when mean and standard deviation are given. We can verify that if $X \sim N(\mu, \sigma^2)$, then $E(X) = \mu$ and var.$(X) = \sigma^2$.*

(*iv*) *It is unimodal (i.e., has one mode at x = mean). The value of y decreases rapidly as x increases numerically. The two tails of the curve extend infinitely and never touch the horizontal axis. The x-axis is its asymptote.*

(*v*) *The area under the normal curve is unity. The probability for the variable to lie in any interval (a, b) in the range of the variable is given by the area under the normal curve, the ordinates $x = a$ and $x = b$.*

(*vi*) *About 99.73% area lies in the interval* $\mu \pm 3\sigma$.

(*vii*) $\mu \pm \sigma$ *are the points of inflexion.*

(*viii*) *When a variable is normally distributed, the number of its values below mean is equal to the number of values above the mean. Thus mean coincides with the median.*

24.56 The definite integral $\int_0^{z_1} \phi(z)\,dz$ is called the normal probability integral,

where $\phi(z) = \dfrac{e^{-z^2/2}}{\sqrt{2\pi}}$ is the probability density function of the standard normal

variate. The area under the standard normal curve $y = \phi(Z) = \dfrac{1}{\sqrt{2\pi}} e^{-z^2/2}$,

bounded by $Z = 0$ and any positive value of $Z = z_1$ (say) give us the probability $P(0 < Z < z_1)$ and is given in Appendix4–Table I at the end of the text book. The symmetry of the curve about $Z = 0$ permits us to obtain the area between any two values of Z. Therefore,

$$P(0 < Z < z_1) = \frac{1}{\sqrt{2\pi}} \int_0^{z_1} e^{-z^2/2}\,dz = \int_0^{z_1} \phi(z)\,dz.$$

Example 24.126. (a) If $\phi(z) = \dfrac{1}{\sqrt{2\pi}} e^{-z^2/2}$ **is the probability density function of**

the standard normal variate, evaluate normal probability integral (i) to (vi).

(*i*) $\displaystyle\int_0^{0.68} \phi(z)\,dz,$ (*ii*) $\displaystyle\int_{-1.2}^0 \phi(z)\,dz,$ (*iii*) $\displaystyle\int_{1.23}^{1.87} \phi(z)\,dz,$

(*iv*) $\displaystyle\int_{-2.35}^{-0.50} \phi(z)\,dz,$ (*v*) $\displaystyle\int_{-0.80}^{1.53} \phi(z)\,dz,$ (*vi*) $\displaystyle\int_1^{\infty} \phi(z)\,dx.$

(b) Find the area under the normal curve to the right of z = 2.05 and to the left of z = −1.44.

Solution. (*a*) (*i*) This definite integral represents the area under the normal standardized curve between $z = 0$ and $z = 0.68$. (Refer Fig. 24.33)

In the table I in Appendix 4, we proceed downward in the column under z (*i.e.*, the first column) till we reach 0.6. Then we move horizontally to the column marked 8. The result 0.2518 is the required value.

Note. 0.2518 is the probability that z lies between 0 and 0.68 and is denoted by $P(0 \le z \le 0.68) = 0.2518$.

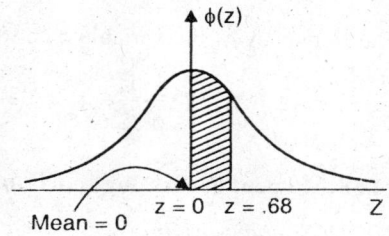

Fig. 24.33.

(*ii*) By symmetry, the area under the normal curve between $z = -1.2$ and $z = 0$ is same as between $z = 0$ and $z = 1.2$.

To get this area we move down the column marked z till we reach 1.2. Then we move right to the column marked 0. The result 0.3849 is the required area.

Note. $P(-1.2 \leq z \leq 0) = 0.3849$.

(*iii*) The required value = Area between $z = 1.23$ and $z = 1.87$ (Refer Fig. 24.34)

= (Area between $z = 0$ and $z = 1.87$) − (Area between $z = 0$ and $z = 1.23$)

= 0.4693 − 0.3907 = 0.0786.

(*iv*) By symmetry the area required =
Area between $z = 0.5$ and $z = 2.35$

= (Area between $z = 0$ and $z = 2.35$) −
(Area between $z = 0$ and $z = 0.5$)

= 0.4906 − 0.1915 = 0.2991.

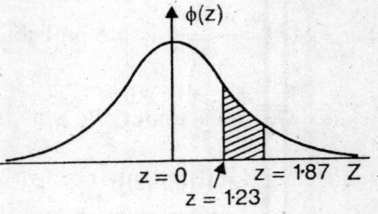

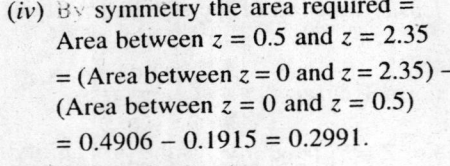

Fig. 24.34.

(*v*) The required value

= (Area between $z = -0.80$ and $z = 0$)

+ (Area between $z = 0$ and $z = 1.53$)

= (Area between $z = 0$ and $z = 0.80$)

+ (Area between $z = 0$ and $z = 1.53$)

= 0.2881 + 0.4370 = 0.7251.

(*vi*) The required value = Area to the right
of $z = 1$ (Refer Fig. 24.35)

= (Area to the right of $z = 0$) −
(Area between $z = 0$ and $z = 1$)

= 0.5 − 0.3413 = 0.1587.

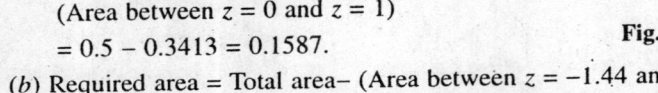

Fig. 24.35.

(*b*) Required area = Total area− (Area between $z = -1.44$ and $z = 0$)

− (Area between $z = 0$ and $z = 2.05$) = 1− 0.4251 − 0.4798 = 0.0951

24.57 The normal curve can be fitted to any given distribution. If μ and σ are the mean and standard deviation of the frequency distribution and N, the total frequency, then the normal frequency distribution is given by

$$\frac{N}{\sigma\sqrt{2\pi}} \, e^{-(x-\mu)^2/(2\sigma^2)}. \qquad \qquad ...(24.106)$$

The frequency of the variable x between a and b is given by

$$N \cdot \frac{1}{\sigma\sqrt{2\pi}} \int_a^b e^{-(x-\mu)^2/(2\sigma^2)} dx.$$

24.58 Moment generating function (MGF) of the distribution

(*i*) *MGF about origin*

The MGF about origin is given by

$$M_X(t) = \int_{-\infty}^{\infty} e^{tx} \cdot f(x)\,dx = \frac{1}{\sigma\sqrt{2\pi}} \int_{-\infty}^{\infty} e^{tx} \cdot e^{-(x-\mu)^2/(2\sigma^2)}\,dx$$

Putting $\dfrac{x-\mu}{\sigma} = z$ that is, $dx = \sigma dz$, we have

$$M_X(t) = \frac{1}{\sqrt{2\pi}} \int_{-\infty}^{\infty} e^{t(\mu+\sigma z)} \cdot e^{-z^2/2}\,dz = \frac{e^{\mu t}}{\sqrt{2\pi}} \int_{-\infty}^{\infty} e^{-(z^2 - 2\sigma t z)/2}\,dz$$

$$= \frac{e^{\mu t}}{\sqrt{2\pi}} \int_{-\infty}^{\infty} e^{-[(z-\sigma t)^2 - \sigma^2 t^2]/2}\,dz = \frac{e^{[\mu t + (\sigma^2 t^2/2)]}}{\sqrt{2\pi}} \int_{-\infty}^{\infty} e^{-(z-\sigma t)^2/2}\,dz$$

$$= \frac{e^{[\mu t + (\sigma^2 t^2/2)]}}{\sqrt{\pi}} \int_{-\infty}^{\infty} e^{-u^2}\,du \qquad\qquad \left[\text{Set } \frac{z - \sigma t}{\sqrt{2}} = u \right]$$

$$= \frac{e^{[\mu t + (\sigma^2 t^2/2)]}}{\sqrt{\pi}} \cdot \sqrt{\pi} = e^{[\mu t + (\sigma^2 t^2/2)]}. \qquad\qquad\qquad ...(24.107)$$

(ii) *MGF of the standard normal distribution* $N(0, 1)$.

The standard normal variate is defined by $z = \dfrac{x-\mu}{\sigma}$ and

$$M_z(t) = \frac{1}{\sigma\sqrt{2\pi}} \int_{-\infty}^{\infty} e^{[(x-\mu)t/\sigma]} \cdot e^{-(x-\mu)^2/(2\sigma^2)}\,dx$$

$$= e^{-\mu t/\sigma} \left[\frac{1}{\sigma\sqrt{2\pi}} \int_{-\infty}^{\infty} e^{xt/\sigma} e^{-(x-\mu)^2/(2\sigma^2)}\,dx \right]$$

$$= e^{-\mu t/\sigma} M_X\left(\frac{t}{\sigma}\right) = e^{-\mu t/\sigma} \cdot e^{\left[\frac{\mu}{\sigma} + \frac{\sigma^2}{2}\left(\frac{t^2}{\sigma^2}\right)\right]} = e^{t^2/2}. \qquad\qquad ...(24.108)$$

Remarks

1. Let X and Y be two independent normal variates with means μ_1, μ_2 and variances σ_1^2, σ_2^2 respectively *i.e.*, $X \sim N(\mu_1, \sigma_1^2)$, $Y \sim N(\mu_2, \sigma_2^2)$. Then the sum and the differences of two independent normal variates is also a normal variate *i.e.*,

$$X + Y \sim N(\mu_1 + \mu_2, \sigma_1^2 + \sigma_2^2),$$

$$X - Y \sim N(\mu_1 - \mu_2, \sigma_1^2 + \sigma_2^2).$$

2. Let $X \sim N(2, 4)$ and $Y \sim N(5, 9)$ are independent variates, the distribution of $3X - Y$ is also normal variates and $3X - Y \sim N(3(2) - 5, (3)^2(4) + 9) = N(1, 45)$.

24.59 Normal distribution as a limiting form of Binomial distribution

Standard normal distribution can be considered as a limiting form of the standard binomial distribution under the following conditions
(*i*) n is very large, $n \to \infty$, (*ii*) neither p nor q is close to zero.

Let $X \sim B(n, p)$. Then, the standard binomial variate is defined as

$$Z = \frac{X - np}{\sqrt{npq}}, X = 0, 1, 2, 3, \dots n.$$

Now, when $X = 0$, $Z = \frac{-np}{\sqrt{npq}} = \sqrt{\frac{np}{q}}$, and

when $X = n$, $Z = \frac{n - np}{\sqrt{npq}} = \frac{nq}{\sqrt{npq}} = \sqrt{\frac{nq}{p}}.$ $(\because p + q = 1)$

Therefore, as $n \to \infty$, Z varies from $-\infty$ to $+\infty$ with increment of $1/\sqrt{npq}$

Central Limit Theorem

Let $X_1, X_2, \dots, X_n$ be a sequence of independent random variables that have the same distribution function and therefore the same mean μ and the same variance σ^2. Then the random variable distribution of

$$Z_n = \frac{X_1 + X_2 + \cdots + X_n - n\mu}{\sigma\sqrt{n}}$$

approaches uniformly the normal distribution with mean 0 and variance 1 as $n \to \infty$. Roughly speaking, the central limit theorem says that in any sequence of repeated trials the standardized sample mean approaches the standard normal curve as the number of trials increase.

24.60 Illustrative Examples

Example 24.127. Fit a normal curve to the following data:

Length of line in cm, x:	8.60	8.59	8.58	8.57	8.56	8.55	8.54	8.53	8.52
Frequency, f:	2	3	4	9	10	8	4	1	1

(PTU 2005)

Solution. Let x denote the length of line and f, the corresponding frequency. We can easily find that

$\mu = 8.563$, $\sigma_x = 0.0173$. and N(total frequency) $= 42$.

∴ The equation of the normal curve fitted to this data is

$$y = \frac{N}{\sigma\sqrt{2\pi}} \, e^{-(x-\mu)^2/(2\sigma^2)}$$

$$= \frac{42}{0.0173\sqrt{2\pi}} \, e^{-(x-8.563)^2/0.0006} \qquad -\infty \le x \le \infty$$

Example 24.128. Assuming that the diameters of 1000 brass plugs taken consecutively from a machine, form a normal distribution with mean 1.9088 cm and standard deviation 0.0050 cm, how many of the plugs are likely to be rejected if the approved diameter is 1.910 ± 0.010 cm?

Solution. The approved range of diameters is 1.90 to 1.92. The corresponding values of the normal variates are

$$z_1 = \frac{1.90 - 1.9088}{0.005} = -1.76, \ z_2 = \frac{1.92 - 1.9088}{0.005} = 2.24.$$

∴ The number of plugs with approved diameters

$$= 1000\left[\int_{-1.76}^{2.24} \frac{1}{\sqrt{2\pi}} e^{-z^2/2} dz \right] = 1000\left[\frac{1}{\sqrt{2\pi}} \int_0^{1.76} e^{-z^2/2} dz + \frac{1}{\sqrt{2\pi}} \int_0^{2.24} e^{-z^2/2} dz \right]$$

$$= 1000\,[0.4608 + 0.4875] = 1000[0.9483] = 948 \text{ approx.}$$

∴ Number of plugs likely to be rejected $= 1000 - 948 = 52$.

Example 24.129. The marks of 500 candidates appearing in an English paper have a mean 41.6 and standard deviation 11.2. Assumming a normal distribution, find how many students obtain first class marks (for which a minimum of 60 marks is necessary).

Solution. The number of students getting first class marks

$$= \frac{500}{11.2\sqrt{2\pi}} \int_{60}^{\infty} e^{-(x-41.6)^2/[2(11.2)^2]} dx \qquad [i.e., 500 \times P(x \ge 60)].$$

Putting $\dfrac{x - 41.6}{11.2} = z$, so that when $x = 60$, $z = 1.643$, the above integral

$$= \frac{500}{\sqrt{2\pi}} \int_{1.643}^{\infty} e^{-z^2/2} dz = 500\left[0.5 - \frac{1}{\sqrt{2\pi}} \int_0^{1.643} e^{-z^2/2} dz \right]$$

$$= 500[0.5 - 0.45] = 500[0.05] = 25 \text{ nearly.} \quad \text{(using Table I in Appendix 4)}$$

Example 24.130. If (i) $X \sim N$ (75, 25), find P_r $(X > 80/X > 77)$.
(ii) $X \sim N$ (10, 4), find $P_r(|X| \le 5)$.

Solution. (i) Since $z_1 = \dfrac{80 - 75}{5} = 1$. Therefore, $P_r(X > 80) = \displaystyle\int_1^{\infty} \phi(z)\,dz$, where $\phi(z) = \dfrac{1}{\sqrt{2\pi}}e^{-z^2}/2 = \int_0^\infty \phi(z)\,dz - \int_0^1 \phi(z)\,dz = 0.5 - 0.3413 = 0.1587$

As $z_2 = \dfrac{77-75}{5} = 0.4$, $P_r(X > 77) = \int_{0.4}^\infty \phi(z)\,dz = \int_0^\infty \phi(z)\,dz - \int_0^{0.4}\phi(z)\,dz$

$$= 0.5 - 0.1554 = 0.3446.$$

Now, $P_r(X > 80/X > 77)$ means probability that $X > 80$ relative to the hypothesis that $X > 77$.

$\therefore$ $P_r(X > 80/X > 77) = P_r(X > 80)/P_r(X > 77) = (0.1587/0.3446) = 0.461.$

[**Note.** $X \sim N(75, 25)$ means that the random variable X is normally distributed with mean = 75 and variance = 25.]

In this case the p.d.f. of X is

$$f(x) = \frac{1}{5\sqrt{2\pi}}\exp\cdot\left(-\frac{(x-75)^2}{2(25)}\right)\!\right].$$

(*ii*) Here $z_1 = \dfrac{5-10}{2} = -2.5$ and $z_2 = \dfrac{-5-10}{2} = -7.5$.

$\therefore$ $P_r(|X| \le 5) = \int_{z_2}^{z_1}\phi(z)\,dz = \int_{-7.5}^{-2.5}\phi(z)\,dz$

$$= \int_{2.5}^{7.5}\phi(z)\,dz = \int_0^{7.5}\phi(z)\,dz - \int_0^{2.5}\phi(z)\,dz = 0.5 - 0.4938 = 0.0062.$$

Example 24.131. (*a*) If $\log_e x$ is normally distributed with mean 1 and variance 4, find $P_r\left(\dfrac{1}{2} < x < 2\right)$ given that $\log_e 2 = 0.693$.

(*b*) In a normal distribution, 31% of the items are under 45 and 8% are over 64. Find the mean and the standard deviation of the distribution. [Given: For a normal distribution the area between z = 0 and z = 0.5 is 0.19 and that between z = 0 and z = 1.4 is 0.42]

(*AMIE, W-2010; PTU 2006, JNTU 2003; IETE June 2003; Madras, 2000S; V.T.U., 2000*)

Solution. (*a*) Let $L = \log_e x$, Here $\mu = 1$, $\sigma^2 = 4$.

$z = \dfrac{L-1}{2} = \dfrac{\log_e x - 1}{2}$. When $x = \dfrac{1}{2}, z = \dfrac{\log(1/2)-1}{2} = \dfrac{-1-\log 2}{2} = -0.8465.$

When $x = 2, z = \dfrac{\log 2 - 1}{2} = \dfrac{0.693-1}{2} = -0.1535.$

$\therefore$ $P_r\left(\dfrac{1}{2} < x < 2\right) = \int_{-0.8465}^{-0.1535}\dfrac{1}{\sqrt{2\pi}}e^{-z^2/2}\,dz = \int_{0.1535}^{0.8465}\dfrac{1}{\sqrt{2\pi}}e^{-z^2/2}\,dz$

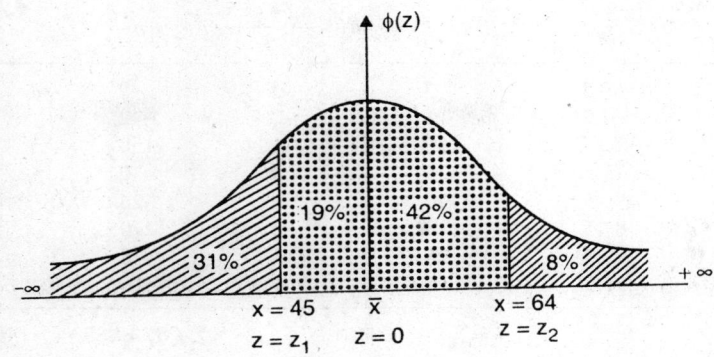

Fig. 24.36.

= (Area between $z = 0$ and $z = 0.8465$) – (Area between $z = 0$ and $z = 0.1535$)

= $0.2996 - 0.0596 = 0.24$ approx.

(b) Let μ be the mean and σ the standard deviation of the distribution. Since the area lying to the left of the ordinate at $x = 45$ is 0.31, the area lying to the right of the ordinate upto the mean is $0.5 - 0.31 = 0.19$. From the table the value of z correponding to the area 0.19 is 0.5 approx.

Hence $\qquad \dfrac{45-\mu}{\sigma} = -0.5$ or $\dfrac{\mu-45}{\sigma} = 0.5.$ ...(i)

Area to the left of the ordinate at $x = 64$ is $0.5 - 0.08 = 0.42$ and the value of z corresponding to this area is 1.4 approx.

Hence $\qquad \dfrac{64-\mu}{\sigma} = 1.4$ ...(ii)

Solving (i) and (ii), we obtain

$\qquad \mu = 50$ and $\sigma = 10.$

Thus mean of the distribution is 50 and the standard deviation is 10.

Example 24.132 Fit a normal curve to the data

Class	60-65	65-70	70-75	75-80	80-85	85-90	90-95	95-100
Frequency	3	21	150	335	326	135	26	4

Compute the expected normal frequencies.

Solution. First, we compute the mean and standard deviation of the distribution.

Class	Mid-value x_i	f_i	$d_x = \dfrac{x_i - 77.5}{5}$	$f_i d_x$	$f_i d_x^2$
60-65	62.5	3	-3	-9	27
65-70	67.5	21	-2	-42	84
70-75	72.5	150	-1	-150	150
75-80 A→	77.5	335	0	0	0
80-85	82.5	326	1	326	326
85-90	87.5	135	2	270	540
90-95	92.5	26	3	78	234
95-100	97.5	4	4	16	64
		$N = \Sigma f_i = 1000$		$\Sigma f_i d_x = 489$	$\Sigma f_i d_x^2 = 1425$

$$\therefore \quad \bar{x} = A + h\frac{\Sigma f_i d_x}{N} = 77.5 + \frac{(489)5}{1000} = 79.945 (= \mu).$$

$$\sigma^2 = h^2 \left[\frac{1}{N} \Sigma f_i d_x^2 - \left(\frac{1}{N} \Sigma f_i d_x \right)^2 \right] = 25 \left[\frac{1425}{1000} - \left(\frac{489}{1000} \right)^2 \right] = 29.647.$$

$\sigma \text{(standard deviation)} = \sqrt{29.647} = 5.445.$

The normal curve is given by

$$f(x) = \frac{1}{5.445\sqrt{2\pi}} \exp\left\{ -\frac{1}{2} \left(\frac{x - 79.945}{5.445} \right)^2 \right\}$$

We construct the following table of values to compute expected normal frequences.

Class	Lower class boundary	$Z = (X - \mu)/\sigma$	$\phi(z) = 1/\sqrt{2\pi}$ $\times \int_{-\infty}^{z} e^{-\mu^2/2} du$	$\Delta\phi(z)$ $= \phi_{z+1} - \phi_z$	Expected frequency $= N\Delta\phi(z)$
Below 60	$-\infty$	$-\infty$	0		
60-65	60	-3.663	0.0001	0.0001	0.1
65-70	65	-2.745	0.0031	0.0030	3.0 } = 3.0
70-75	70	-1.826	0.0341	0.0310	31 = 31
75-80	75	-0.908	0.1821	0.1480	148 = 148
80-85	80	0.0101	0.5040	0.3219	321.9 = 322
85-90	85	0.928	0.8232	0.3192	319.2 = 319
90-95	90	1.847	0.9676	0.1444	144.40 = 144
95-100	95	2.765	0.9972	0.0296	29.60 = 30
Over 100	100	3.683	0.9999	0.0027	2.7 = 3

Example 24.133 It is given that X and Y are independent normal variates with mean 6, 7 and standard deviations 3, 4 respectively. Find the value of K such that

$$P(2X+Y \le K) = P(4x-3Y \ge 4K).$$

Solution. We have $X \sim N(6, 9)$, $Y \sim N(7, 16)$ and X, Y are independent. Define $U = 2X + Y$ and $V = 4X - 3Y$. Then U and V are also normal variates and

$$U = 2X+Y \sim N(2(6)+7, 4(9)+16) = N(19, 52).$$

$$V = 4X-3Y \sim N(4(6)-3(7), 16(9)+9(16)) = N(3, 288).$$

Define, $Z = (U-19)/\sqrt{52}$. Now $Z \sim N(0, 1)$ and

for $\quad U = K, Z = \left[(K-19)\sqrt{52}\right] = z_1$.

Define, $Z = (V-3)/\sqrt{288}$. Again $Z \sim N(0, 1)$ and for $V = 4K$,

$$Z = (4K-3)/\sqrt{288} = z_2. \text{ We require}$$

$$P(2X+Y \le K) = P(4X-3Y \ge 4K)$$

or $\qquad P(Z \le z_1) = P(Z \ge z_2)$

Since the normal probability curve is symmetric about $Z = 0$, this implies that $z_1 = -z_2$. Hence,

$$\frac{K-19}{\sqrt{52}} = -\left[\frac{4K-3}{\sqrt{288}}\right] \Rightarrow \frac{K-19}{\sqrt{13}} = \frac{3-4K}{6\sqrt{2}}$$

or $\qquad K = \dfrac{3\sqrt{13}+114\sqrt{2}}{6\sqrt{2}+4\sqrt{13}} \approx 7.51.$

24.61 Some other distributions: Beta distribution, Exponential distribution

To understand **Beta distribution**, the readers should be familiar with the following definitions.

(a) **Gamma Function. Def.** The gamma function denoted by $\Gamma(n)$ is defined by the improper integral

$$\Gamma(n) = \int_0^\infty x^{n-1} e^{-x} dx, n > 0$$

We note the following values of the Gamma function.

   (i) $\Gamma(n + 1) = n\Gamma(n)$,   (ii) $\Gamma(1) = 1$

(iii) $\Gamma(n + 1) = n!$ for any positive integer m,

   (iv) $\Gamma(1/2) = \sqrt{\pi}$,   (v) $\Gamma(n) = \dfrac{\Gamma(n+1)}{n}(0 > n \ne -1, -2, -3, \cdots)$

(b) **The Beta Function. Def.** The Beta function denoted by $\beta(m, n)$ or $\beta(n, m)$ is defined by the integral.

$$\beta(m,n) = \int_0^1 x^{m-1}(1-x)^{n-1}\, dx \qquad\qquad (m > 0, n > 0)$$

We know that $\beta(m, n) = \beta(n, m)$.

In terms of the Γ function, $\beta(m,n) = \dfrac{\Gamma(m)\Gamma(n)}{\Gamma(m+n)}$ $\quad m, n > 0.$ \qquad ...(24.109)

24.62 Beta Distribution

A random variable X has a **beta distribution** or is a **beta random variable** if X has the interval $(0,1)$ as range and there are positive numbers m and n for which the probability density function f satisfies.

$$f(x) = \begin{cases} \dfrac{1}{\beta(m,n)} x^{m-1}(1-x)^{n-1}, 0 \le x \le 1, & n, m > 0 \\ 0 \text{ elsewhere}, \end{cases}$$

$$= \frac{\Gamma(m+n)}{\Gamma(m)\Gamma(n)} x^{m-1}(1-x)^{n-1},$$

where Γ is the *gamma function* and β is the *beta function*.

24.62.1 Mean and Standard deviation of Beta Distribution (*A.M.I.E., S-1998*)

(a) Mean $= \displaystyle\int_0^1 xf(x)\, dx = \frac{1}{\beta(m, n)}\int_0^1 x \cdot x^{m-1}(1-x)^{n-1}\, dx$

$= \dfrac{1}{\beta(m, n)} \displaystyle\int_0^1 x^m (1-x)^{n-1} dx = \dfrac{\beta(m+1, n)}{\beta(m, n)}$

$= \dfrac{\Gamma(m+1)\Gamma(n)}{\Gamma(m+n+1)} \div \dfrac{\Gamma(m)\Gamma(n)}{\Gamma(m+n)} = \dfrac{m}{m+n}.$ \quad ...(24.110) \quad $\left[\because \Gamma(m+1) = m\Gamma(m)\right]$

(b) Now $\mu_2' = \dfrac{1}{\beta(m, n)} \displaystyle\int_0^1 x^2 \cdot x^{m-1}(1-x)^{n-1}\, dx$

$= \dfrac{1}{\beta(m, n)} \displaystyle\int_0^1 x^{m+1}(1-x)^{n-1}\, dx = \dfrac{1}{\beta(m, n)}\beta(m+2, n)$

$= \dfrac{\Gamma(m+2)\Gamma(n)}{\Gamma(m+n+2)} \div \dfrac{\Gamma(m)\Gamma(n)}{\Gamma(m+n)} = \dfrac{m(m+1)}{(m+n+1)(m+n)}.$

Now Variance, $\sigma^2 = \mu_2' - \mu^2 = \mu_2' - (\text{mean})^2$

$$= \frac{m(m+1)}{(m+n+1)(m+n)} - \left(\frac{m}{m+n}\right)^2 = \frac{mn}{(m+n)^2(m+n+1)}.$$

Standard deviation, $\sigma = \sqrt{\dfrac{mn}{(m+n)^2(m+n+1)}} = \dfrac{1}{m+n}\sqrt{\dfrac{mn}{m+n+1}}$ . ...(24.111)

24.63 Exponential Distribution

Def. The exponential distribution is defined by the density function f

$$f(x) = \begin{cases} 0 & \text{if } x < 0 \\ \lambda e^{-\lambda x} & \text{if } x \geq 0, \end{cases}$$ where λ is a positive constant.

Exponential distributions have a characteristic property which suggests their use in certain problems involving radioactive decay, traffic accidents, and failure of electronic equipment such as vacuum tubes.

24.63.1 Mean and Standard deviation of exponential distribution

$$\text{Mean} = \lambda \int_0^\infty x e^{-\lambda x} dx = \lambda \left\{ \left[x \frac{e^{-\lambda x}}{-\lambda} \right]_0^\infty + \frac{1}{\lambda} \int_0^\infty e^{-\lambda x} dx \right\} = \lambda \left[0 + \frac{1}{\lambda} \left[\frac{e^{-\lambda x}}{-\lambda} \right]_0^\infty \right] = \frac{1}{\lambda}.$$

Now $\mu_2' = \lambda \int_0^\infty x^2 e^{-\lambda x} dx$

$$= \lambda \left[x^2 \left(\frac{e^{-\lambda x}}{-\lambda} \right) - (2x) \left(\frac{e^{-\lambda x}}{\lambda^2} \right) + 2 \left(\frac{e^{-\lambda x}}{-\lambda^2} \right) \right]_0^\infty = \frac{2}{\lambda^2}.$$

Variance, $\sigma^2 = \mu_2' - \mu_1'^2 = \mu_2' - (\text{mean})^2 = \dfrac{2}{\lambda^2} - \dfrac{1}{\lambda^2} = \dfrac{1}{\lambda^2}.$ $\quad (\because \mu_1' = \text{mean})$

$\therefore$ Standard deviation, $\sigma = 1/\lambda.$ [AMIE W-2005; MDU 2004; PTU-2003]

Example 24.134 The p.d.f. of X is given by $f(x) = \lambda e^{-\lambda x}, x \geq 0, \lambda > 0.$ Calculate $P_r[X > E(X)]$.

Solution. $E(X) = \lambda \int_0^\infty x \lambda e^{-\lambda x} dx = \lambda \left[x \left(\frac{e^{-\lambda x}}{-\lambda} \right) - \left(\frac{e^{-\lambda x}}{\lambda^2} \right) \right]_0^\infty = \lambda \left[\frac{1}{\lambda^2} \right] = \frac{1}{\lambda}.$

$\therefore P_r[X > E(X)] = \int_{1/\lambda}^\infty \lambda e^{-\lambda x} dx = \lambda \left| \dfrac{e^{-\lambda x}}{-\lambda} \right|_{1/\lambda}^\infty = -(0 - e^{-1}) = \dfrac{1}{e}.$

Exercise 24.5

1. Incidence of occupational disease in an industry is such that the workers have a 20% chance of suffering from it. What is probability that out of 6 workers chosen at random 4 or more will suffer from the disease?   **Ans. $53/3125$.**

2. The probability that a pen manufactured by a company will be defective is 0.1. If 12 such pens are manufactured, find the probability that
   (a) exactly two will be defective, (b) at least two will be defective, (c) none will be defective. (*V.T.U., 2003, PTU 2002*)

   Ans. 0.2301; 0.3412; 0.2833.

3. A group of 20 aeroplanes are sent on an operational flight. The chance that an aeroplane fails to return from the flight is 5%. Determine the probability that (*i*) one plane does not return (*ii*) at the most 5 planes do not return.

   $$\textbf{Ans. } (i) \ ^{20}C_1 (0.05)(0.95)^{19} \quad (ii) \ \sum_{r=0}^{5} {}^{20}C_r (0.05)^r (0.95)^{20-r}.$$

4. A factory finds that on the average 20% of the bolts produced by a given machine will be defective for certain specified requirements. If 10 bolts are selected at random from the day production of this machine, find the probability that (a) exactly 2 will be defective (b) 2 or more will be defective (c) more than 5 will be defective?

   $$\textbf{Ans. } (a) \ ^{20}C_2 (0.2)^2 (0.8)^8 = 0.302 \quad (b) \ 1 - (0.8)^{10} - 10(0.2)(0.8)^9 = 0.6242$$

   $$(c) \ \sum_{r=6}^{10} {}^{10}C_r (0.2)^r (0.8)^{10-r} = 0.00637.$$

5. (a) The mean and variance of the binomial distribution are 4 and 4/3 respectively. Find $P(X \ge 1)$.   **Ans. 0.9986 (approx.)**

   [**Hint:** $np = 4$, $npq = 4/3$, $\therefore \ q = 1/3$, $p = 1 - q = 2/3$. $\therefore \ n = 4/p = 6$.

   $P(X \ge 1) = 1 - P(X = 0) = 1 - q^n = 1 - (1/3)^6 = 0.9986.$]

   (b) With the usual notations, find the probability of a success for a binomial variate X, if $n = 6$ and $9P(X = 4) = P(X = 2)$.   **Ans. 0.25.**

   (c) Find the binomial distribution for which mean and standard deviation are 6 and 2 respectively.   **Ans.** $p(X = r) = {}^{18}C_r (1/3)^r (2/3)^{18-r} \ r = 0, 1, 2, ..., 18.$

6. A man takes a step forward with probability 0.4 and backward with probability 0.6. Find the probability that at the end of eleven steps, he is just one step away from the starting point.   (*IETE, Dec. 2003*)   **Ans. 3679/10000.**

7. Assuming that half the population of a town consume chocolates and that 100 investigators each take 10 individuals to see whether they are consumers, how many investigators would you expect to report that three people or less were consumers?   (*A.M.I.E., W-1999*)   **Ans. 17.**

8. Out of 800 families with 4 children each, how many familes would you expect to have (*i*) 2 boys and 2 girls, (*ii*) at least one boy, (*iii*) no girl, (*iv*) at most two girls assuming equal probabilities for boys and girls?

$\qquad\qquad$ (*VTU, 2004;IETE June 2002*)   **Ans.** 300; 750; 50; 550.

9. Six dice are thrown together 729 times. How many times do you expect at least three dice to show a five or six. $\qquad\qquad$ (*PTU 1998*) **Ans.** 233.

10. Two players A and B play tennis games. Their chances of winning a game are in the ratio 3:2 respectively. Find A's chance of winning at least two games out of four games played. $\qquad\qquad\qquad\qquad$ **Ans.** 513/625.

$$\left[\textbf{Hint. } P(X \geq 2) = 1 - \left(P(X = 0) + P(X = 1)\right)\right.$$

$$\left. = 1 - {}^4C_0\left(\frac{3}{5}\right)^0\left(\frac{2}{5}\right)^4 - {}^4C_1\left(\frac{3}{5}\right)^1\left(\frac{2}{5}\right)^3 = 1 - \frac{1}{5^4}\left[16 + 4(3)(8)\right] = \frac{513}{625}.\right]$$

11. The probability that a bomb dropped from a plane will strike the target is 20%. If six bombs are dropped, find the probability that (i) exactly 2 will strike the target, (ii) at least 2 will strike the target. $\qquad$ (*MDU 2007*)

$\qquad\qquad\qquad\qquad\qquad\qquad\qquad\qquad\qquad\qquad$ **Ans.** 0.246; 0.345.

12. Fit a binomial distribution for the following data and compare the theoretical frequencies with the actual one:

x:	0	1	2	3	4	5
f:	2	14	20	34	22	8

$\qquad\qquad\qquad\qquad$ (*Madras 1996S*)   **Ans.** $100(0.432 + 0.568)^5$

13. Find the probability that at most 5 defective fuses will be found in a box of 200 fuses if experience shows that 2% of such fuses are defective. ($e^{-4} = 0.0183$). $\qquad\qquad\qquad\qquad\qquad\qquad\qquad$ **Ans.** 0.0183

14. Using Poisson distribution, find the probability that the ace of spades will be drawn from a pack of well shuffled cards at least once in 104 consecutive trials. [**Hint.** $p = 1/52$, $n = 104$, $\therefore m = np = (104)(1/52) = 2$.

$$\text{Reqd. prob} = 1 - (\text{Prob. of not drawing an ace}) = 1 - \frac{e^{-2} 2^0}{0!} = 1 - \frac{1}{e^2} = 0.864]$$

15. If 2% of electric bulbs manufactured by a company are defective, find the probability that in a sample of 100 bulbs (*i*) 0 (*ii*) 1 (*iii*) 2 (*iv*) 3 bulbs will be defective (use Poisson distribution).

$\qquad\qquad\qquad\qquad$ **Ans.** (*i*) 0.135 (*ii*) 0.271 (*iii*) 0.271 (*iv*) 0.180

16. Data was collected over a period of 10 years, showing number of deaths from horse kicks in each of the 200 army corps. The distribution of deaths was as following:

No. of deaths:	0	1	2	3	4
Frequency:	109	65	22	3	1

Fit a Poisson distribution to the above data. *(P.U. 2005)*

[**Hint.** $m = 0.61$, Poisson distribution $= \dfrac{e^{-0.61}(0.61)^r}{r!}$.

Theoretical frequencies are given by $\dfrac{200e^{-0.61}(0.61)^r}{r!}$ for $r = 0, 1, 2, 3, 4$.

From the table $e^{-0.61} = 0.5434$ (approx.). The theoretical frequencies are 108.7, 66.3, 20.2, 4.1 and 0.7 *i.e.*, 109, 66, 20, 4 and 1 respectively.]

17. In a certain factory turning out razor blades, there is a small chance of 0.002 for any blade to be defective. The blades are supplied in packets of 10. Use Poisson distribution to calculate the approximate number of packets containing no defective, one defective and two defective blades respectively in a consignment of 10,000 packets. Use $e^{-0.02} = 0.9802$.

(VTU 2004; UPTU 2002, D.U., 2000)

[**Hint.** Here $N = 10,000$, $n = 10$ and $p = 0.002$, $m = np = 0.02$.

Required frequencies are given by $Ne^{-m}, Ne^{-m} \cdot m, Ne^{-m} \cdot m^2/2!$ that is

10000(0.9802), 10000(0.9802)(0.02), 10000(0.9802)$(0.02^2)/2$ that is, 9802, 196, 2 packets respectively].

18. Wireless sets are manufactured with 25 soldered joints each on the average 1 joint in 500 is defective. How many sets can be expected to be free from defective joints in a consignment of 10,000 sets. **Ans. 9512.**

19. The number of telephone calls arriving on an internal switch board of an office is 90 per hour. Find the probability that at the most 1 to 3 calls in a minutes on the board arrive. (use $e^{-1.5} = 0.2231$) *(P.U. 1998)*. **Ans. 0.711.**

20. A book of 585 pages contains 43 typographical errors. If these errors are randomly distributed throughout the book, what is the probability that 10 pages selected at random, will be free from errors? (Use $e^{-0.735} = 0.4795$).

[**Hint.** Here $p = 43/585 = 0.0735$, $n = 10$, $m = np = 0.735$.

$$P(X = r) = \frac{e^{-m} \cdot m^r}{r!} = \frac{e^{-0.735} \cdot (0.735)^r}{r!}.$$

$\therefore P(\text{no error}) = P(X = 0) = e^{-0.735} = 0.4795)$]

21. (a) Using Poisson's distribution, find the probability that the aces of spades will be drawn from a pack of well-suffled cards at least once in 104 times consecutive trials (Use $e^{-2} = 0.1353$).

(*b*) Six coins are tossed 6400 times. Using the Poisson distribution determine the approximate probability of getting six heads *x* times.

Ans. $\left[e^{-100}(100)^x\right]\Big/x!$

[**Hint.** (*a*) $p(r) = e^{-m}\dfrac{m^r}{r!}$. Here $p = \dfrac{1}{52}$, $n = 104$. $\therefore$ $m = np = 2$.

Probability of drawing ace of spades at least once

$= 1 - p$ (drawing no ace of spades) $= 1 - \dfrac{e^{-2} \cdot 2^0}{0!} = 1 - 0.1353 = 0.8647$]

22. Fit a poisson distribution to the following data and calculate theoretical frequencies:

Deaths	0	1	2	3	4
Frequency	122	60	15	2	1

(*PTU 2006; MDU 2006; V TU 2004; UPTU 2003*)

Ans. Theoretical frequencies are: 121, 61, 15, 2, 1.

23. The distribution of typing mistakes per page committed by a typist is given below. Assuming a Poisson model, find out the expected frequencies:

No. of mistakes /page	(*x*)	0	1	2	3	4	5
No. of pages	(*f*)	142	156	69	27	5	1

Ans. Expected frequencies of Poisson distribution are:

x:	0	1	2	3	4	5
f:	147	147	74	25	6	1

24. If the variance of the Poisson distribution is 2, find the prbabilities for $r = 1$, 2, 3, 4 from the recurrence relation of the poisson distribution. Also find $P(X \ge 4)$. (Use $e^{-2} = 0.1353$)

(*UPTU 2001*)

Ans. 0.2706, 0.2706, 0.1804, 0.0902, 0.1431

25. (*a*) Let *X* be a Poisson random variable such that $2P(X = 0) = P(X = 2)$. Find standard deviation of *X*. **Ans.** $\sqrt{2}$

(*b*) If a random variable *X* has a Poisson distribution such that $p(X = 1) = P(X = 2)$, then find $P(X = 4)$. (*A.M.I.E., S-2009*) **Ans.** $2/3e^2$.

26. If *X* is a poisson variate such that $P(X = 2) = 9P(X = 4) + 90P(X = 6)$. Calculate (*i*) mean (*ii*) standard deviation. **Ans.** 1, 1

27. A manufacturer known from experience that the resistance of resistors he produces is normal with mean = 100Ω and standard deviation = 2Ω. What percentage of resistors will have resistance between 98 and 102Ω?
    (IETE W-2002) **Ans.** 68.26%

28. The average height of soliders of a country is given as 68.22 inches with variance 10.8 sq inch. How many soliders out of 1000 would you expect to be over 72 inches tall? Given that the area under the standard normal curve between $z = 0$ to $z = 0.35$ is 0.1368 and $z = 1.15$ is 0.3749.
    (UPTU 2002) **Ans.** 125.

29. The mean inside diameters of a sample of 200 washers produced by a machine is 5.02 mm and the standard deviation is 0.05 mm. The purpose for which these washers are intended allows a maximum tolerance in the diameter of 4.96 to 5.08 mm; otherwise, the washers are considered defective. Determine percentage of defective washers produced by the machine assuming the diameters are normally distributed. [Area under the standard normal curve between $z = 0$ and $z = 1.2$ is 0.3849] *(A.M.I.E., W-2005)*
    Ans. 23%; Number of defective washers = 47.

30. The length of an item manufactured on an automatic machine tool is a normally distributed random variable with parameters: mean = 10 and variance = 1/200. Find the probability of defective production, if the tolerance is 10 ± 0.05. *(A.M.I.E., W-2001)* **Ans.** 0.48.

31. In sampling a large number of parts manufactured by a machine, the mean number of defectives in a sample of 20 is 2. Out of 1000 such samples, how many would be expected to contain at least 3 defective parts? **Ans.** 324

32. In a test on 2000 electric bulbs, it was found that the life of a particular make, was normally distributed with an average life of 2400 hours and standard deviation of 60 hours. Estimate the number of bulbs likely to burn for (a) more than 2150 hours, (b) less than 1950 hours, and (c) more than 1920 hours but less than 2160 hours. *(VTU, 2005; Assam 1999; PTU 1998)*
    Ans. (a) 68, (b) 184, (c) 1908.

33. A sample of 100 dry battery cells tested to find the length of life produced the following result:
    Mean = 12 hours, standard deviation = 3 hours. Assuming the data to be normally distributed, what percentage of battery cells are expected to have life (a) more than 15 hours, (b) less than 6 hours (c) between 10 and 14 hours. *(UPTU 2004; Madurai 1995)* **Ans.** (a) 15.87% (b) 2.28% (c) 49.74%.

34. A manufacturer produces air mail envelopes whose weight is normally distributed with mean 1.95 gms and standard deviation 0.05 gm. The envelops are sold in lots of 1000. How many envelops in a lot will be weighing between 1.95 gm to 2.00gm each. Given that $\dfrac{1}{\sqrt{2\pi}} \displaystyle\int_{-\infty}^{1} e^{-z^2/2}\, dz = 0.8413$.
    (A.M.I.E., W-2000) **Ans.** 341

35. If $\log_{10}X$ is normally distributed with mean 4 and variance 4, find the probability of $1.202 < X < 8318 \times 10^4$, given that $\log_{10} 1202 = 3.08$ and $\log_{10} 8318 = 3.92$. **Ans. 0.95**

36. In a normal distribution, 7% of the items are under 35 and 89% are under 63. Determine mean and standard deviations of the distribution. [*IETE Dec. 2001*] **Ans. 50.3; 10.36.**

37. The mean height of 500 student's is 151 cm and the standard deviation is 15 cm. Assuming that the heights are normally distributed, find how many students heights lie between 120 and 155 cm. **Ans. 294.**

38. The following table gives a set of observations x of frequency f:

x:	8.53	8.54	8.55	8.56	8.57	8.58	8.59	8.60
f:	1	4	8	10	9	4	3	1

Fit a normal curve to this distribution. **Ans.** $\dfrac{40}{\sqrt{2\pi}}\exp\left[\dfrac{(x-\mu)^2}{2\sigma^2}\right]$,

where $\mu = 8.5628$, and $\sigma = 0.0156$.

39. Fit a normal distribution to the following data:

Variable:	60-62	63-65	66-68	69-71	72-74
Frequency:	5	18	42	27	8

Ans. $y = \dfrac{1}{2.92\sqrt{2\pi}} e^{-(x-67.45)^2/17.06}$

40. Five thousand candidates appeared in a certain examination carrying a maximum of 100 marks. It was found that the marks were normally distributed with mean $\mu = 39.5$ and with standard deviation $\sigma = 12.5$. Determine approximately the number of candidates who secured a first class for which a minimum of 60 marks is necessary. You may use the table given below (d_x denotes the deviation from the mean).

The proportion A of the whole area of the normal curve lying to the left of the ordinate at the deviation d_x/σ is

d_x/σ	1.5	1.6	1.7	1.8
A	0.93319	0.94520	0.95543	0.96407

[**Hint.** We have $\dfrac{d_x}{\sigma} = \dfrac{X-\mu}{\sigma} = \dfrac{60-39.5}{12.5} = 1.64$ Interpolation of the value of A for $d_x/\sigma = 1.64$ can be carried out as below:

d_x/σ	A	ΔA	$\Delta^2 A$	$\Delta^3 A$
1.5	0.93319			
		0.01201		
1.6	0.94520		−0.00178	
		0.01023		0.00019
1.7	0.95543		−0.00159	
		0.00864		
1.8	0.96407			

Here $r = (1.64 - 1.5)/(0.1) = 1.4$. Use Newton's interpolation formula: we have

$$\text{A for } d_x/\sigma \, (= 1.64) = 0.93319 + 1.4 \times 0.01201 + \frac{1.4(1.4-1)}{2!}(-0.00178)$$

$$+ \frac{1.4(0.4)(-0.6)}{3!} \times 0.00019 + \cdots\cdots$$

= 0.93319 + 0.016814 − 0.0004984 − ... = 0.95 (approx).
Hence number of students getting 60 or more than 60 marks is 5%.
∴ Required number of students = 5000 × 0.05 = 250.]

41. The probability density function for the life of a motor car tyre is given by

$f(x) = \lambda e^{-\lambda x}$ for $x \geq 0$, where $\lambda = 0.04$ and x is measured in units of 1500 km. What is the probability that a single tyre will last more than 30 units?

Ans. 0.301

42. Find the points of inflexion of the normal curve

$$f(x) = \frac{1}{4\sqrt{2\pi}} e^{-(x-10)^2/32} \text{ for } -\infty < x < \infty$$ **Ans.** 6 and 14.

[**Hint.** The points of inflexion are given by $\mu - \sigma$ and $\mu + \sigma$].

43. It is given that X and Y are independent normal variates with means 2, 5 and variances 4, 9 respectively. Find the value of K such that

$$P(X+Y \leq K) = P(3X - Y \geq 3K).$$ **Ans.** $K = 2.885$.

44. It is given that X and Y are independent normal variates and $X \sim N(1,4)$,

$Y \sim N(3,16)$. Find the value of K such that

$$P(2X+Y \leq K) = P(4X - Y \geq 2K).$$ **Ans.** $K = 2.487$

24.64 Tests of Significance for Small Samples

In this section, we shall discuss the following tests.

(i) "student's" t-test, (ii) Chi-square (χ^2) test and (iii) F-test.

Before we explain the procedure for using the test, we shall define a few terms:

(a) **Population or Universe.** Population (or Universe) is the totality of all actual or conceivable objects of a class under consideration. In fact it consists of numerical values associated with these objects. For example we speak of population of heights, births, thicknesses of washers, lengths of life of electric bulbs, etc. Population may be finite or infinite, real or imaginary.

(b) **Sample.** It is often impracticable to analyse the whole population for some investigations. For example if we want to find the average heights of an adult Indian, we must have the record of heights of all adults which is difficult to obtain. To overcome this difficulty we use the technique of sampling. The idea of examining a sample is so familiar to us. *Analysis of a handful of grains from the entire stock meant for sale is the study of sample from the population of grains in the stock, to draw some conclusions about the population.*

A sample is a finite number of objects selected from a population.

If the sample is chosen in such a way that the chances of selection of any member of the population are the same it is called a *random sample*.

A sample is called large when the number of observations is ≥ 30, otherwise it is called small sample. Here we shall be concerned with small samples.

(c) **Parameters and statistics.** We shall explain these terms by taking an example. The heights of all adult Indians form a population. The mean, S.D. etc. of the population are called its *parameters*. If we choose a sample from this population, then the mean, S.D. of this sample are called *statistics* while each is a *statistic*.

(d) **Tests of significance.** Let θ be a parameter of the population and θ_0 be the corresponding statistic obtained from a random sample. We start with the assumption that there is no significant difference between the sample statistics and the corresponding population parameter or between two sample statistics. This is termed a *null hypothesis* and denoted by H_0. A hypothesis that is different from the null hypothesis is called an alternative hypothesis and is denoted by H_1.

We defined null hypothesis as $H_0 : \theta = \theta_0$ and alternative hypothesis as

$$H_1 : \theta \neq \theta_0 \text{ (two tailed alternative)}.$$

or $H_1 : \theta > \theta_0$ (right tailed alternative). $\left.\rule{0pt}{18pt}\right\}$ Both are single tailed tests.
or $H_1 : \theta < \theta_0$ (left tailed alternative).

In 5 per cent level of significance, we will be accurate in 95 per cent of the cases and inaccurate in only 5 per cent of the cases while in 1 per cent level of significance, we will be accurate in 99 per cent of the cases and we are likely to be inaccurate in only 1 per cent of the cases. Level of significant indicates the extent of precision of accuracy. The greater level of significance the less will be the precision.

(e) **Significance level:** The significance level stands for the confidence with which the experimenter rejects or retains the null hypothesis. Significance level is usually expressed as a percentage or by the value of α, that is 5% or $\alpha = 0.05$.

(f) **Degrees of freedom:** The number of degrees of freedom of a statistic is the number of independent variates used to compute the statistic. For example, if n is the number of observations in the small sample and m is the number of constraints on them (or m values are already available), that the number of degrees of freedom is given by $v = n - m$. For calculating the mean $\bar{x}$, we use all the observed values $x_1, x_2, .., x_n$. Therefore the mean $\bar{x}$ has n degrees of freedom. Since the S.D. of the sample depends on the mean the standard deviation has $n - 1$ degrees of freedom.

24.65 "Student's" *t*-test

It is a very useful test of significance for small sample size (less than 30) and the population standard deviation σ is unknown.

If $x_1, x_2, ..., x_n$ be the members of a random sample from a normal population $N\left(\mu, \sigma^2\right)$ with mean μ, then $\bar{x}$ (sample mean) $= \sum\limits_{i=1}^{n} x_i \Big/ n$, the sample variance

$$S^2 = \frac{\sum\limits_{i=1}^{n}\left(x_i - \bar{x}\right)^2}{(n-1)}. \text{ The "}t\text{-statistic" is defined as: } t = \frac{\left(\bar{x} - \mu\right)}{S}\sqrt{n}, \quad ... (24.112)$$

with $(n - 1)$ donotes the degrees of freedom. If we calculate t for each sample, we obtain the sampling distribution for t.

This distribution is known as *"Student's" t-distribution* or *t-distribution*.

Assumptions for using t-distributions

(i) The sample size is small (that is, $n \leq 30$).
(ii) Samples are drawn from the normal population.
(iii) Samples are random samples in the sense that the observations are independent.

The t-table. The values of $t_v(\alpha)$ [two tailed test at a few levels of significance and v degrees of freedom] is given in Table II-Appendix IV. The *t*-distribution has a different values for each degrees of freedom.

Applications of *t*-Distribution

The following are some of the uses of the *t*-distribution:
1. To test the significance of the difference between the mean of a small sample and the mean of the population.
2. To test the significance of difference between the means of two small random samples drawn from the same normal population (Independent samples).

3. To test the significance of difference between the means of two small random samples (Dependent samples or paired observations).

4. To test the significance of the correlation coefficient in a small sample.

Case I. *t*-test for significance of the difference between mean of a small sample and mean of population.

To test whether the mean of a random sample drawn from a normal population deviates significantly from a stated value (the hypothetical value of the population mean), when variance of the population is unknown. We calculate the statistic

(i) $t = \dfrac{(\bar{x} - \mu)\sqrt{n}}{S}$, where $\bar{x}$ = the mean of the sample, μ = the mean of the

population, n = the sample size, S = the standard deviation of the sample

$$= \sqrt{\frac{\sum_{i=1}^{n}(x_i - \bar{x})^2}{n-1}} = \sqrt{\frac{\Sigma d^2 - n(\bar{d})^2}{n-1}}, \qquad \ldots(24.113)$$

where d = deviation from the assumed mean and $\bar{d} = \left(\sum d_i\right)/n.$

(*ii*) Compare this calculated value of $|t|$ to the value of t with $(n-1)$ degrees of freedom at level of significance α (given in the table).

(*iii*) If the computed value of $|t|$ exceeds $t_v(\alpha)$ table given in the Appendix 4, we say that the difference between $\bar{X}$ and μ is significant at given level of significance α. We reject the hypothesis and say that the sample is not from a population with mean μ. If $|t| < t_v(\alpha)$, then the null hypothesis H_0 is accepted at α level of significance. We conclude that the difference between $\bar{x}$ and μ is not significant and hence the sample might have been drawn from a population with mean = μ.

Confidence or Fiducial limits of population mean, μ. Assuming that the sample is a random sample from a normal population on unknown mean the 95% confidence limits (level of significance 5%) of the population mean μ are:

$$\bar{x} \pm \frac{S}{\sqrt{n}} t_{n-1}(0.05), \text{ where } t_{n-1}(0.05) \text{ is the 5\% critical value of } t \text{ for } n-1$$

degrees of freedom for a two tailed test.

Similarly, 99% confidence limits (level of significance 1%) for μ are:

$$\bar{x} \pm \frac{S}{\sqrt{n}} t_{n-1}(0.01).$$

The following examples will illustrate this test.

Example 24.135 Compute the student's 't' for the following variable values in

a sample of eight: −4, −2, −2, 0, 2, 2, 3, 3, taking the mean of the universe to be zero.

(A.M.I.E., S-1995)

Solution. The calculation of $\bar{x}$ and S is given in the following table:

x_i	$x_i - \bar{x}$	$\left(x_i - \bar{x}\right)^2$
−4	−4.25	18.0625
−2	−2.25	5.0625
−2	−2.25	5.0625
0	−0.25	0.0625
2	1.75	3.0625
2	1.75	3.0625
3	2.75	7.5625
3	2.75	7.5625
$\Sigma x_i = 2$		$\Sigma\left(x_i - \bar{x}\right)^2 = 49.5$

We find $\bar{x} = \dfrac{2}{8} = 0.25$, $S = \sqrt{\dfrac{\Sigma\left(x_i - \bar{x}\right)^2}{n-1}} = \sqrt{\dfrac{49.5}{7}} = 2.659$. Since $\mu = 0$

The t-statistic is given by $t = \dfrac{\bar{x} - \mu}{S}\sqrt{n} = \dfrac{0.25 - 0}{2.659}\sqrt{8} = 0.266$.

Example 24.136 Ten individuals are choosen at random from a population and their heights are found to be in inches as: 63, 63, 64, 65, 66, 69, 69, 70, 70, 71. Is the assumption of a mean of 65 inches for the normal population reasonable? Test under 5% level of significance. Given that for 9 degrees of freedom the value of student's 't' at 5% level of significance is 2.262.

Solution. We have the mean and standard deviation of the small sample as

$$\bar{x} = \frac{1}{n}\sum_{i=1}^{n} x_i = \frac{1}{10}\left[(2 \times 63) + 64 + 65 + 66 + (2 \times 69) + (2 \times 70) + 71\right] = \frac{670}{10} = 67.$$

$$\Sigma\left(x_i - \bar{x}\right)^2 = 16 + 16 + 9 + 4 + 1 + 4 + 4 + 9 + 9 + 16 = 88,$$

and $S = \sqrt{\dfrac{\Sigma\left(x_i - \bar{x}\right)^2}{(n-1)}} = \sqrt{\dfrac{88}{9}} = 3.13$.

Define

Null hypothesis $H_0 : \mu = 65$ inches. (there is no significant difference in the height)

Alternative hypothesis $H_1 : \mu \neq 65$ inches.

The test statistic is given by $t = \dfrac{\bar{x} - \mu}{S} \sqrt{n} = \dfrac{67 - 65}{3.13} \times \sqrt{10} = 2.02$.

The number of degrees of freedom, $v = 10 - 1 = 9$.

The table value of t for 9 degrees of freedom at 5% level of significance is 2.262. Since the computed value of t is less than the table value, we can say that difference is not significant at 5% level and we conclude that mean height is 65 inches. H_0 is accepted at 5% level of significance. At this level of significance, the data is consistent with the assumption that the population has mean 65 inches.

Example 24.137 **A fertiliser mixing machine is set to give 12 kg of nitrate for every quintal bag of fertiliser. The 100 kg bags are examined. The percentage of nitrate is given below:**

11, 14, 13, 12, 13, 12, 13, 14, 11, 12. Is there reason to believe that the machine is defective? Value of t for 9 degrees of freedom is 2.262.

Solution. Let us take the hypothesis that there is no significant difference between the sample value and the value set for the machine. Applying t-test:

$$t = \frac{(\bar{x} - \mu)}{S} \sqrt{n} \qquad \text{...(i)}$$

Mean and standard deviation are calculating as under:

x_i	$(x_i - A) = d_i$	d_i^2
11	-1	1
14	2	4
13	1	1
$A \rightarrow$ 12	0	0
13	1	1
12	0	0
13	1	1
14	2	4
11	-1	1
12	0	0
$\Sigma x_i = 125$	$\Sigma d_i = 5$	$\Sigma d_i^2 = 13$

We find $\bar{x} = \dfrac{\Sigma x_i}{n} = \dfrac{125}{10} = 12.5, \bar{d} = \dfrac{\Sigma d_i}{n} = \dfrac{5}{10} = 0.5,$

$$S = \sqrt{\frac{\Sigma d_i^2 - n(\bar{d})^2}{n-1}} = \sqrt{\frac{13 - 10(0.5)^2}{10 - 1}} = 1.08$$

Substituting the values in (i), we obtain

$$t = \frac{12.5 - 12}{1.08}\sqrt{10} = \frac{0.5 \times 3.162}{1.08} = 1.464 .$$

We have for 9 degrees of freedom, $t_{0.05} = 2.262$. The calculated value is less than the table value. Hence the hypothesis is accepted. There is no reason to believe that the machine is defective.

Example 24.138 A random sample of size 16 has 53 as mean. The sum of the squares of the deviations taken from mean is 135. Can this sample be regarded as taken from the population having 56 as mean? Obtain 95% and 99% confidence limits of the mean of the population. [For $v = 15, t_{0.01} = 2.947$ and $t_{0.05} = 2.131$]. *(A.M.I.E., S-2003)*

Solution. We have $n = 16$. $\overline{x} = 53$, and $\Sigma(x_i - \overline{x})^2 = 135$. $S^2 = \Sigma(x_i - \overline{x})/n - 1$

$= 135/15 = 9$. We have the test statistic as $t = \dfrac{(\overline{x} - \mu)\sqrt{n}}{S}$. Therefore,

$$t = \frac{53 - 56}{3} \times \sqrt{16} = -4.$$

Define

Null hypothesis H_0 : $\mu = 56$. Alternative hypothesis H_1 : $\mu \neq 56$.

The tabulated value of t at 5% level of significance for 15 degrees of freedom is $t_{0.05} = 2.131$.

As the calculated value of $|t|$ is higher than the table value therefore, H_0 is rejected at 5% level of significance. Hence the result of the experiment does not support the hypothesis that the sample is drawn from the population having 56 as mean.

The confidence limits at 5% level of significance are

$$\overline{x} \pm \left(\frac{S}{\sqrt{n}}\right) t_{0.05} = 53 \pm \frac{3}{4} \times 2.131 = 53 \pm 1.598 \text{ or } (51.40, 54.60).$$

The Confidence limits at 1% level of significance are

$$\overline{x} \pm \left(\frac{S}{\sqrt{n}}\right) t_{0.01} = 53 \pm \frac{3}{4} \times 2.947 = 53 \pm 2.21 \text{ or } (50.79, 55.21).$$

Case 2. Test for significance of the difference between the means of two small random samples drawn from the same normal population (Independent samples)

It is given that two independent random samples of size n_1 and n_2 with means $\overline{x}_1$ and $\overline{x}_2$ and standard deviations S_1 and S_2. We may be interested in testing the hypothesis that the samples come from the same normal population. To carry out the test, we proceed as under.

We calculate the test statistic *t:*

$$t = \frac{\bar{x}_1 - \bar{x}_2}{S} \times \sqrt{\frac{n_1 n_2}{n_1 + n_2}}, \qquad \text{...(24.108)}$$

where S = combined standard deviation

The value of S is calculated by the following formula:

$$S = \sqrt{\frac{\Sigma(x_1 - \bar{x}_1)^2 + \Sigma(x_2 - \bar{x}_2)^2}{(n_1 - 1) + (n_2 - 1)}}. \qquad \text{...(24.114)}$$

If the calculated value of t be $> t_{0.05}$ or $(t_{0.01})$, the difference between the sample means is said to be significant at 5% or (1%) level of significance; otherwise the data are said to be consistent with the hypothesis.

Remarks: 1. In formula (24.109), when the actual means are in fractions the deviations should be taken from assumed means. In such a case the combined standard deviation is obtained by applying the following formula:

$$S = \sqrt{\frac{\Sigma(x_1 - A_1)^2 + \Sigma(x_2 - A_2)^2 - n_1(\bar{x}_1 - A_1)^2 - n_2(\bar{x}_2 - A_2)^2}{n_1 + n_2 - 2}} \qquad \text{...(24.115)}$$

where
$\quad A_1$ = Assumed mean of the first sample,

$\quad A_2$ = Assumed mean of the second sample,

$\quad \bar{x}_1$ = Actual mean of the first sample

$\quad \bar{x}_2$ = Actual mean of the second sample

The degrees of freedom, $v = (n_1 + n_2 - 2)$.

2. When we are given the number of observations n_1 and n_2 standard deviation S_1 and S_2 respectively of the two samples, the pooled estimate of standard deviation can be obtained as follows:

$$S = \sqrt{\frac{(n_1 - 1)S_1^2 + (n_2 - 1)S_2^2}{n_1 + n_2 - 2}} \qquad \text{...(24.116)}$$

The following examples will illustrate this test:

Example 24.139 Two different-types of drugs A and B were tried on certain patients for increasing weight. 8 persons were given drug A and 10 persons were given drug B. The increase in weight in kgm is given below:

Drug A:	2	3	6	8	10	1	2	8		
Drug B:	5	6	7	1	11	4	3	8	6	9

Do the two drugs differ significantly with regard to their effect in increasing weight (the table value of t for $v = 16$ at 5% level is 2.120).

Solution. We have $n_1 = 8$, $n_2 = 10$. We first compute the means of two independent samples and combined standard deviation of samples.

$$\bar{x}_1 = \frac{\sum x_1}{n_1} = \frac{40}{8} = 5; \bar{x}_2 = \frac{\sum x_2}{n_2} = \frac{60}{10} = 6.$$

Define H_0: $\bar{x}_1 = \bar{x}_2$ and H_1: $\bar{x}_1 \neq \bar{x}_2$.

We have, degrees of freedom = $n_1 + n_2 - 2 = 16$. Given $t_{16}(0.05) = 2.120$

The combined standard deviation is approximated as

$$S = \sqrt{\frac{\sum(x_1 - \bar{x}_1)^2 + \sum(x_2 - \bar{x}_2)^2}{n_1 + n_2 - 2}} = \sqrt{\frac{82 + 78}{10 + 8 - 2}} = \sqrt{\frac{160}{16}} = 3.162$$

The test statistic is given by

$$t = \frac{\bar{x}_1 - \bar{x}_2}{S\sqrt{\frac{1}{n_1} + \frac{1}{n_2}}} = \frac{5 - 6}{3.162}\sqrt{\frac{8 \times 10}{8 + 10}} = -\frac{1}{3.162} \times 2.108 = -0.666.$$

	Drug A			Drug B		
x_1	$(x_1 - \bar{x}_1)$ $= (x_1 - 5)$	$(x_1 - \bar{x}_1)^2$	x_2	$(x_2 - \bar{x}_2)$ $= (x_2 - 6)$	$(x_2 - \bar{x}_2)^2$
2	–3	9	5	–1	1
3	–2	4	6	0	0
6	1	1	7	1	1
8	3	9	1	–5	25
10	5	25	11	5	25
1	–4	16	4	–2	4
2	–3	9	3	–3	9
8	3	9	8	2	4
			6	0	0
			9	3	9
$\sum x_1 = 40$		$\sum(x_1 - \bar{x}_1)^2$ $= 82$	$\sum x_2 = 60$		$\sum(x_2 - \bar{x}_2)^2$ $= 78$

Since $|t| = 0.666 < 2.12$. We accept the null hypothesis that there is no significant difference and we may conclude that the drugs 'A' and 'B' donot differ significantly as regard their effect on increase in weight.

Example 24.140 For a random sample of 10 pigs, fed on diet A, the increase in weight in pounds in a certain period were:

10, 6, 16, 17, 13, 12, 8, 14, 15, 9.

For another random sample of 12 pigs, fed on diet *B*, the increase in the same period were:

$$7, 13, 22, 15, 12, 14, 18, 8, 21, 23, 10, 17.$$

Test whether the diets *A* and *B* differ significantly as regards their effect on increase in weight? (The table value of *t* for $v = 20$ at 5% level is 2.09).

Solution. Let us take the null hypothesis that diets *A* and *B* donot differ significantly with regard to their effect on increase in weight. Applying *t*-test:

$$t = \frac{\overline{x}_1 - \overline{x}_2}{S}\sqrt{\frac{n_1 n_2}{n_1 + n_2}}, \text{ where } S = \sqrt{\frac{\Sigma(x_1 - \overline{x}_1)^2 + \Sigma(x_2 - \overline{x}_2)^2}{(n_1 - 1) + (n_2 - 1)}}.$$

We have $n_1 = 10$, $n_2 = 12$. We first compute the means of two independent samples and combined standard deviation of samples.

Pigs fed on diet A			Pigs fed on diet B		
Increase in wt x_1	*Deviations from mean value* $\overline{x}_1 = 12$ $(x_1 - \overline{x}_1)$	$(x_1 - \overline{x}_1)^2$	*Increase in wt* x_2	*Deviations from mean value* $\overline{x}_2 = 15$ $(x_2 - \overline{x}_2)$	$(x_2 - \overline{x}_2)^2$
10	−2	4	7	−8	64
6	−6	36	13	−2	4
16	4	16	22	7	49
17	5	25	15	0	0
13	1	1	12	−3	9
12	0	0	14	−1	1
8	−4	16	18	3	9
14	2	4	8	−7	49
15	3	9	21	6	36
9	−3	9	23	8	64
			10	−5	25
			17	2	4
$\Sigma x_1 = 120$		$\Sigma(x_1 - \overline{x}_1)^2$ $= 120$	$\Sigma x_2 = 180$		$\Sigma(x_2 - \overline{x}_2)^2$ $= 314$

$$\overline{x}_1 = \frac{\Sigma x_1}{n_1} = \frac{120}{10} = 12 \text{ pounds.} \qquad \overline{x}_2 = \frac{\Sigma x_2}{n_2} = \frac{180}{12} = 15 \text{ pounds.}$$

$$S = \sqrt{\frac{\Sigma(x_1 - \overline{x}_1)^2 + \Sigma(x_2 - \overline{x}_2)^2}{(n_1 - 1) + (n_2 - 1)}} = \sqrt{\frac{120 + 314}{10 + 12 - 2}} = \sqrt{\frac{434}{20}} = 4.66.$$

we have the test statistic as

$$t = \frac{\bar{x}_1 - \bar{x}_2}{S} \cdot \sqrt{\frac{n_1 n_2}{n_1 + n_2}} = \frac{12 - 15}{4.66} \times \sqrt{\frac{10 \times 12}{10 + 12}} = \frac{-3}{4.66} \times 2.34 = -1.51,$$

dof, $v = n_1 + n_2 - 2 = 10 + 12 - 2 = 20$.

For $v = 20$, the table value of t at 5 per cent level is 2.09. The calculated value of $|t|$ is less than the table value and hence the experiment provides no evidence against the hypothesis. We, therefore, conclude that diets A and B donot differ significantly as regards their effect on increase in weight.

Example 24.141 Intelligence test given to two groups of boys and girls gave the following results:

	Size	Mean marks	S.D.
Girls	12	84	10
Boys	8	81	12

Examine if the difference of mean scores significant.

Solution. Let us take the hypothesis that there is no significant difference in the mean scores obtained by girls and boys. Applying t-test of the difference of means,

$$t = \frac{\bar{x}_1 - \bar{x}_2}{S} \sqrt{\frac{n_1 n_2}{n_1 + n_2}}. \text{ Here } n_1 = 12, \bar{x}_1 = 84, S_1 = 10, n_2 = 8, \bar{x}_2 = 81, S_2 = 12.$$

$$S = \sqrt{\frac{(n_1 - 1)S_1^2 + (n_2 - 1)S_2^2}{n_1 + n_2 - 2}} = \sqrt{\frac{11(10)^2 + 7(12)^2}{12 + 8 - 2}} = \sqrt{\frac{1100 + 1008}{18}}$$

$$= \sqrt{\frac{2108}{18}} = 10.82.$$

Substituting the values in the above formula, we obtain

$$t = \frac{84 - 81}{10.82} \sqrt{\frac{12 \times 8}{12 + 8}} = \frac{3}{10.82} \times 2.19 = 0.607.$$

For *dof,* $v = 18$, $t_{0.05} = 2.10$.

The calculated value of t is less than the table value. Hence, the hypothesis is accepted. We therefore conclude that there is no significant difference in the mean score obtained by girls and boys.

Case 3. t-Test for Difference of means of two samples (Dependent samples or paired observations)

In the previous test it was assumed that the two samples were independent that is, the values of observations in one sample donot depend on the other. However, there are many situations in which this condition does not hold true. Two samples are said to be dependent when the elements in one sample are related to those in the other in any significant or meaningful manner. This implies that the pairs of observation (x_i, y_i) belong to same sample unit. In this case, we consider the d_i (difference between each pair of observations) $= x_i - y_i$, $\bar{d} = \bar{x} - \bar{y}$ and define the null hypothesis as $H_0 : \bar{d} = 0$ and alternate hypothesis as $H_1 : \bar{d} \neq 0$. When samples are dependent they comprise the same number of elementary units. The t-test based on paired observations is defined as

$$t = \frac{\bar{d}\sqrt{n}}{S}, \quad ...(24.117)$$

with $n-1$ degrees of freedom and $S =$ the standard deviation of the differences. The value of S is calculated as follows:

$$S = \sqrt{\frac{\Sigma(d_i - \bar{d})^2}{(n-1)}} \quad \text{or} \quad \sqrt{\frac{\Sigma d_i^2 - n(\bar{d})^2}{(n-1)}}.$$

The following examples will illustrate the test.

Example 24.142 Eleven persons were appointed in a clerical position in an office. Their performance was noted by giving a test and the marks recorded out of 30. They were given 4 month's training and again they were given a test and marks were recorded out of 30.

Employees	A	B	C	D	E	F	G	H	I	J	K
Marks obtd. before training	23	20	19	21	18	20	18	17	23	16	19
Marks obtd. after training	24	19	22	18	20	22	20	20	23	20	17

By applying the t-test can it be concluded that the employees have benefited by the training? Test at 5% level of significance.

Solution. The data pertains to the marks obtained by the same set of candidates. If x_i, y_i denote the marks obtained in the two tests, we obtain the values of $d_i = x_i - y_i$, as shown in the following table.

Employees	Marks obtd. in Ist test x_i	Marks obtd. in IInd Test y_i	Increments $d_i = x_i - y_i$	d_i^2
A	23	24	−1	1
B	20	19	1	1
C	19	22	−3	9
D	21	18	3	9
E	18	20	−2	4

(Contd.)

F	20	22	−2	4
G	18	20	−2	4
H	17	20	−3	9
I	23	23	0	0
J	16	20	−4	16
K	19	17	2	4

Total = 11		$\Sigma d_i = -11$	$\sum d_i^2 = 61$

We find $\bar{d} = \dfrac{\Sigma d_i}{n} = -\dfrac{11}{11} = -1$, $S = \sqrt{\dfrac{\Sigma d_i^2 - n(\bar{d})^2}{n-1}} = \sqrt{\dfrac{61 - 11(-1)^2}{11-1}} = \sqrt{5} = 2.236$.

We define

Null hypothesis. $H_0 : \bar{d} = 0$ (the employees have not benefitted from training)

Alternate hypothesis. $H_i : \bar{d} < 0$ (the employees have benefitted from training).

We shall use the one tailed test. Now t (0.05) for one tailed test = t (0.1) for two tailed test, with the degrees of freedom, $v = n - 1 = 10$. The value of t for $P = 0.05$ and $v = 10$ is 1.812. The test statistic is given by

$$t = \frac{\bar{d}\sqrt{n}}{S} = \frac{-1 \times \sqrt{11}}{2.236} = \frac{-3.317}{2.236} = -1.483.$$

We find $|t| = 1.483 < 1.812$. Hence, we accept the null hypothesis that the employees have not benefitted from the training.

Example 24.143 Nine persons to whom a drink was given, registered the following increments in blood pressure:

$$7, 3, -1, 4, -3, 5, 6, -4, -1.$$

Show that the data donot indicate that the drink is responsible for these increments.

Solution. Here we are given the increments in blood pressure, that is $d_i = (x_i - y_i)$.

d_i	d_i^2
7	49
3	9
−1	1
4	16
−3	9
5	25

(Contd.)

6	36
-4	16
-1	1

$$\Sigma d_i = 16 \qquad \Sigma d_i^2 = 162$$

We find $\overline{d} = \Sigma d_i / n = \dfrac{16}{9} = 1.778$, $S = \sqrt{\dfrac{\Sigma d_i^2 - n(\overline{d})^2}{n-1}} = \sqrt{\dfrac{162 - 9(1.778)^2}{8}}$

$$= \sqrt{\dfrac{162 - 28.452}{8}} = 4.086.$$

We define

Null hypothesis. $H_0 : \overline{d} = 0$. (there is no difference in the blood pressure readings of the persons).

Alternate hypothesis. $H_i : \overline{d} < 0$ (the drink results in an increase in blood pressure).

We shall use the one tailed test. Now $t (0.05)$ for one tailed test $= t (0.01)$ for two tailed test, with the degrees of freedom $= 8$. The value of t for $P = 0.05$ and $v = 8$ is 1.86. The test statistic is given by $t = \dfrac{\overline{d}\sqrt{n}}{S}$.

Substituting the values in the test formula, we obtain

$t = \dfrac{1.778\sqrt{9}}{4.086} = 1.305$. We find $t = 1.305 < 1.86$. Hence we accept the null hypothesis that the drink in general is not responsible by an increase in blood pressure.

24.66 The Chi-Square (χ^2) Test (Non-Parametric Test)

When a fair coin is tossed 100 times, we expect from theoretical consideration that heads will appear 50 times and tails 50 times. But in practice these results are rarely achieved *i.e.* the results obtained in an experiment donot agree exactly with the theoretical results. The *magnitude of discrepency between observation and theory is given by the quantity* χ^2 (*read chi-square*). If $\chi^2 = 0$, the observed and theoretical frequencies completely agree, while if $\chi^2 > 0$, they do not agree exactly. The larger the value of χ^2, the greater is the discrepancy between observed and expected frequencies.

Definition. Let a distribution be given. Suppose that $O_1, O_2, ..., O_n$ be a set of observed (experimental) frequencies and $e_1, e_2, ..., e_n$ be the corresponding set of expected (or theoretical) frequencies, then χ^2 is given by

$$\chi^2 = \frac{(O_1 - e_1)^2}{e_1} + \frac{(O_2 - e_2)^2}{e_2} + \ldots + \frac{(O_n - e_n)^2}{e_n}$$

$$= \sum_{i=1}^{n} \frac{(O_i - e_i)^2}{e_i} \text{ with } \sum_{i=1}^{n} O_i = \sum_{i=1}^{n} e_i. \qquad \ldots(24.118)$$

Simplifying (24.118), we get

$$\chi^2 = \sum_{i=1}^{n} \frac{1}{e_i}\left(O_i^2 + e_i^2 - 2O_i e_i\right) = \sum_{i=1}^{n}\left(\frac{O_i^2}{e_i} + e_i - 2 O_i\right)$$

$$= \sum_{i=1}^{n}\left(\frac{O_i^2}{e_i}\right) + \sum_{i=1}^{n} e_i - 2\sum_{i=1}^{n} O_i = \sum_{i=1}^{n}\left(\frac{O_i^2}{e_i}\right) - N \qquad \ldots(24.119)$$

Since $\displaystyle\sum_{i=1}^{n} O_i = \sum_{i=1}^{n} e_i = N$ is the total frequency.

We can use either of the expressions in (24.118) or (24.119). The χ^2-statistic defined in (24.118) depends only on the observed and expected frequencies and the degrees of freedom. It is independent of the population parameters. Hence, the test is also known as *non-parametric test* or distribution free test.

The number of degrees of freedom, (v or *dof*) are obtained by the rule $v = n - k$, where k refers to the number of independent constraints. If the data is given in a series of n numbers, then $v = n - 1$. In general, when we fit a binomial distribution the number of degrees of freedom is one less than the number of classes; when we fit a poisson distribution, the degrees of freedom are 2 less than the number of classes, because we use the total frequency and the arithmetic mean to get the parameter of the poisson distribution. When we fit a normal curve, the number of degrees of freedom are 3 less than the number of classes, because in this fitting we use the total frequency, mean and standard deviation.

Goodness of fit. The value of χ^2 may be used to determine how well a set of observations fits a given distribution, χ^2 therefore, provides a test of goodness of fit. It is a measure to examine the agreement or disagreement between the observed frequencies and expected frequencies. It may be used to examine the validity of some hypothesis about an observed frequency distribution. Individual theoretical frequencies of the classes should not be small. Ideally, $O_i \geq 10$. If a theoretical frequency is less than 10, then it is combined with the preceeding or the succeeding frequencies, so that the combined frequency is ≥ 10. This procedure is called the *method of pooling*. (Refer Q. 18, Exercise 24.6) The degrees of freedom is adjusted accordingly.

Significance Tests

Let $\chi_n^2(\alpha)$ denote the value of χ^2-distribution with n degrees of freedom at α level of significance. A few of these values are given in Table-III (Appendix 4). If the computed $\chi^2 < \chi_n^2(\alpha)$, then the null hypothesis H_0 is accepted, that is the difference between the observed and expected (values) frequencies is not significant at α level of significance. We conclude that the given sample is drawn from the hypothetical population. If the computed $\chi^2 > \chi_n^2(\alpha)$, then the null hypothesis is rejected and we conclude that the random sample is not from the hypothetical population.

Example 24.144 The following table shows the observed and expected frequencies in tossing a die 120 times. Test the hypothesis that the die is fair, using 5% level of significance.

Face	1	2	3	4	5	6
Observed frequency	25	17	15	23	24	16
Expected frequency	20	20	20	20	20	20

Given for $v = 5$, χ^2 at 5% level is 11.07.

Solution. The χ^2-statistic is given by $\chi^2 = \Sigma_i \left[\dfrac{(O_i - e_i)^2}{e_i} \right]$

$$= \frac{(25-20)^2}{20} + \frac{(17-20)^2}{20} + \frac{(15-20)^2}{20} + \frac{(23-20)^2}{20} + \frac{(24-20)^2}{20} + \frac{(16-20)^2}{20}$$

$$= \frac{25}{20} + \frac{9}{20} + \frac{25}{20} + \frac{9}{20} + \frac{16}{20} + \frac{16}{20} = 5.00.$$

The degrees of freedom, $v = 6 - 1 = 5$.
Tabulated value of χ^2 at 5% level of significance for 5 degrees of freedom is 11.07.

Since the calculated value of $\chi^2 = 5.00 < 11.07$, we accept the null hypothesis H_0. We conclude that we cannot reject the hypothesis that the die is fair.

Example 24.145 The theory predicts the proportion of beans in the four groups A, B, C and D should be in the ratio 9 : 3 : 3 : 1. In an experiment among 1600 beans, the number in the four groups were 882, 313, 287 and 118. Does the experimental result support the theory? *(AMIE., W-2006)*

Solution. The total number of beans = 882 + 313 + 287 + 118 = 1600. Since the expected numbers are in poportion 9 : 3 : 3 : 1 (and 9 + 3 + 3 + 1 = 16), we would expect.

$\frac{9}{16}(1600), \frac{3}{16}(1600), \frac{3}{16}(1600), \frac{1}{16}(1600)$ beans *i.e.*, 900, 300, 300, 100 beans.

Then $\chi^2 = \frac{(882-900)^2}{900} + \frac{(313-300)^2}{300} + \frac{(287-300)^2}{300} + \frac{(118-100)^2}{100}$

$= \frac{324}{900} + \frac{169}{300} + \frac{169}{300} + \frac{324}{100} = \frac{4254}{900} = 4.73$ and *dof*, $\nu = 4 - 1 = 3$.

For $\nu = 3$, we have χ^2 at 5% level = 7.82.

Since the calculated value of χ^2 is less than that of the tabulated value, we accept the null hypotesis H_0. We conclude that the theory and experiment are in agreement.

Example 24.146 A sample survey of 320 families with 5 children each revealed the following distribution:

Numbers of families	18	56	110	88	40	8
Numbers of Boys	5 boys	4 boys	3 boys	2 boys	1 boy	0 boy
Numbers of Girls	0 girl	1 girl	2 girls	3 girls	4 girls	5 girls

Is the data consistent with the hypothesis that male and female births are equally possible? Test at 5% level of significance.

Solution. Let p = probability of a male birth, and $q = 1 - p$ = probability of female birth. We define the null hupothesis as H_o : male and female births are equally possible. $\therefore p = q = 1/2$. The probabilities of (5 boys), (4 boys and 1 girl), (3 boys, 2 girls), ..., (5 girls) are given by the terms in the binomial expansion

$(p+q)^5 = p^5 + {}^5C_1 p^4 q + {}^5C_2 p^3 q^2 + {}^5C_3 p^2 q^3 + {}^5C_4 pq^4 + q^5$

$= p^5 + 5p^4 q + 10p^3 q^2 + 10p^2 q^3 + 5pq^4 + q^5$

$\therefore$ $P(5$ boys and 0 girl$) = (1/2)^5 = 1/32$.

$P(4$ boys and 1 girl$) = 5(1/2)^4(1/2) = 5/32$.

$P(3$ boys and 2 girls$) = 10(1/2)^3 (1/2)^2 = 10/32$.

$P(2$ boys and 3 girls$) = 10(1/2)^2(1/2)^3 = 10/32$.

$P(1$ boy and 4 girls$) = 5(1/2) (1/2)^4 = 5/32$.

$P(0$ boy and 5 girls$) = (1/2)^5 = 1/32$.

Then the expected number of families with 5, 4, 3, 2, 1 and 0 boys are obtained respectively by multiplying the above probabilities by 320, and the results are 10, 50, 100, 100, 50, 10.

Hence $\chi^2 = \frac{(18-10)^2}{10} + \frac{(56-50)^2}{50} + \frac{(110-100)^2}{100} + \frac{(88-100)^2}{100}$

$+ \frac{(40-50)^2}{50} + \frac{(8-10)^2}{10}$

$$= \frac{64}{10} + \frac{36}{50} + \frac{100}{100} + \frac{144}{100} + \frac{100}{50} + \frac{4}{10} = 11.96.$$

We note that the values of p and q are not obtained from the given data.

The degrees of freedom, $v = 6 - 1 = 5$. At 5% level for 5 degrees of freedom the table value of χ^2 is 11.07. The calculated value of χ^2 is 11.96. Since calculated value of χ^2 is greater than tabulated value, it is significant at 5% level of significance. Hence, we reject the null hypothesis H_0. We conclude that male and female births are not equally possible.

Example 24.147 Fit a Poisson distribution for the following data and test the goodness of fit at 5% level of significance.

x:	0	1	2	3	4
Frequency:	419	352	154	56	19

(*V.T.U.*, *2001*)

Solution. The Poisson distribution is given by

$$P(X = r) = \frac{e^{-m}m^r}{r!}, r = 0, 1, 2.....$$

The mean of the Poisson distribution is given by

$$m = \frac{\Sigma fx}{\Sigma f} = \frac{352 + 308 + 168 + 76}{1000} = \frac{904}{1000} = 0.904. \quad e^{-m} = e^{-0.904} = 0.4049.$$

N = the total frequency = 1000. The expected frequencies are given by

$$e_i = N\frac{e^{-m}m^r}{r!}, \quad r = 0, 1, 2, 3, 4, \text{ we have the frequencies}$$

O_i	419	352	154	56	19	$\Sigma O_i = 1000$
e_i	405	366	166	50	13	$\Sigma e_i = 1000$

In order that the total observed and expected frequencies may agree, e_i are adjusted to make them whole numbers such that $\Sigma e_i = 1000$. (In case, if a theoretical frequency is less than 10, then it is combined with the preceeding or the succeeding frequencies such that the combined frequency is ≥ 10. Here of course, none of the frequencies <10).

The chi-square statistic is given by

$$\chi^2 = \frac{(419-405)^2}{405} + \frac{(352-366)^2}{366} + \frac{(154-166)^2}{166} + \frac{(56-50)^2}{50} + \frac{(19-13)^2}{13}$$

$$= 0.484 + 0.536 + 0.867 + 0.72 + 2.77 = 5.38.$$

Since the mean of the theoretical distribution has been estimated from the given data so the degrees of freedom, $v = 5 - 2 = 3$.

For $v = 3$, the tabulated value of χ^2 at 5% level is 7.82. Since the calculated value of $\chi^2 < \chi^2_{0.05}$, the agreement between the fact and theory is good and hence the Poisson distribution provides a good fit to the data.

Example 24.148 **Fit a normal curve to the data:**

Class	30-35	35-40	40-45	45-50	50-55	55-60	60-65
Frequency	3	32	125	230	120	35	5

Compute the expected normal frequencies and test the goodness of fit at 5% level of significance.

Solution. First, we compute the mean and standard deviation of the distribution.

Class	Mid-value x_i	f_i	$d_x = \dfrac{x_i - 47.5}{5}$	$f_i d_x$	$f_i d_x^2$
30-35	32.5	3	-3	-9	27
35-40	37.5	32	-2	-64	128
40-45	42.5	125	-1	-125	125
45-50	A→ 47.5	230	0	0	0
50-55	52.5	120	1	120	120
55-60	57.5	35	2	70	140
60-65	62.5	5	3	15	45

$$N = \sum f_i = 550 \qquad \sum f_i d_x = 7 \qquad \sum f_i d_x^2 = 585$$

$$\bar{x} = A + h \frac{\sum f_i d_x}{N} = 47.5 + \frac{7 \times 5}{550} = 47.5636 \,(= \mu)$$

$$\sigma^2 = h^2 \left[\frac{1}{N} \sum f_i d_x^2 - \left(\frac{1}{N} \sum f_i d_x \right)^2 \right] = 25 \left[\frac{585}{550} - \left(\frac{7}{550} \right)^2 \right] = \begin{array}{l} 25 \times 1.06344 \\ = 26.586. \end{array}$$

σ (Standard deviation) $= \sqrt{26.586} = 5.1562$.

The normal curve is given by

$$f(x) = \frac{1}{5.1562 \sqrt{2\pi}} \exp - \left\{ \frac{1}{2} \left(\frac{x - 47.5636}{5.1562} \right)^2 \right\}.$$

We construct the following table of values to compute expected normal frequencies.

Class	Lower class boundary	$Z = (x-\mu)/\sigma$	$\phi(z) = \dfrac{1}{\sqrt{2\pi}} \times \displaystyle\int_{-\infty}^{z} e^{-u^2/2}\,du$	$\Delta\phi(z)$ $= \phi_{z+1} - \phi_z$	Expected frequency $= N\Delta\phi(z)$
Below 30	$-\infty$	$-\infty$	0		
				0.0003	$\left.\begin{array}{l}0.165\\3.905\end{array}\right\} = 4$
30-35	30	-3.4063	0.0003		
				0.0071	
35-40	35	-2.4366	0.0074		$35.09 = 35$
				0.0638	
40-45	40	-1.4669	0.0712		$131.06 = 131$
				0.2383	
45-50	45	-0.4972	0.3095		$204.71 = 205$
				0.3722	
50-55	50	0.4725	0.6817		$134.03 = 134$
				0.2437	
55-60	55	1.4422	0.9254		$36.63 = 37$
				0.0666	
60-65	60	2.4119	0.9920		$4.18 = 4$
				0.0076	
Over 65	65	3.3816	0.9996		

Grouping the classes so that each class frequency is ≥ 10, we obtain

O_i	35	125	230	120	40
e_i	39	131	205	134	41
$O_i - e_i$	-4	-6	25	-14	-1

The χ^2-statistic is given by

$$\chi^2 = \Sigma_i \frac{\left[(o_i - e_i)^2\right]}{e_i} = \frac{16}{39} + \frac{36}{131} + \frac{625}{205} + \frac{196}{134} + \frac{1}{41} = 5.221$$

The given data has been used to find μ, σ and Σe_i.

Hence degree of freedom, $v = 5-3 = 2$. Now $\chi_2^2(0.05) = 5.991$.

Since $\chi^2 = 5.221 < 5.991$. We accept the null hypothesis. We conclude that the fitting of normal distribution to the given data is satisfactory and can be accepted.

24.67 F-test

The object of the F-test is to find whether the two independent estimates of population variance differ significantly, or whether the two samples may be regarded as drawn from the normal population having the same variance.

Given two independent random samples $x_i\,(i = 1, 2,..,n_1)$ and $y_j\,(j = 1, 2,..,n_2)$ of size n_1 and n_2 respectively with means $\bar{x}$ and $\bar{y}$ and standard deviations S_x and S_y. For carrying out the test of significance, we calculate the F-statistic.

We defined F-statistic as:

$$F = \frac{S_x^2}{S_y^2}, \text{ with } v_1 \text{ and } v_2 \text{ degrees of freedom} \qquad ...(24.120)$$

$$s_x^2 = \frac{\Sigma(\bar{x}_i - \bar{x})^2}{(n_1 - 1)} \text{ with } v_1 = n_1 - 1 \text{ degrees of freedom and } s_y^2 = \frac{\Sigma(y_j - \bar{y})^2}{(n_2 - 1)}, \text{ with}$$

$v_2 = n_2 - 1$ degrees of freedom.

It should be noted that the *numerator is always the greater variance.*

If the calculated value of F exceeds $F_{0.05}$ for $[(n_1 - 1), (n_2 - 1)]$ degrees of freedom, we say that the ratio is significant at 5% level. If $F < F_{0.05}$ we say that the same could have come from two normal populations with the same variance.

Refer Table IV (Appendix 4) given at the end of the text. This table provide the critical values of F for different values of v_1 and v_2 at 5% and 1% level of significance respectively.

Remark. To test, if two small random samples have been drawn from the same normal population, we must apply both the F-test and t-test. Using F-test, we test the equality of the population variances. If equality of variances is shown, then we apply *t*-test-for the significance of the differences of the two sample means.

Example 24.149 Two random samples of sizes 8 and 10 gave the sum of squares of deviations from their respective means as 84.4 and 102.6 respectively. Can they be regarded as drawn from normal population with the same variance? You are given that at 5% level of significance, critical value of F for (7, 9) degrees of freedom is 3.29 and for (8, 10) degrees of freedom, its value is 3.07.

(Madurai, 1995)

Solution. We are given:

$$n_1 = 8, \ \Sigma(x_i - \bar{x})^2 = 84.4 \ ; \ n_2 = 10, \ \Sigma(y_j - \bar{y})^2 = 102.6.$$

$$\therefore \quad s_x^2 = \frac{\Sigma(x_i - \bar{x}_1)^2}{(n_1 - 1)} = \frac{84.4}{7} = 12.06; \qquad s_y^2 = \frac{\Sigma(y_j - \bar{y})^2}{(n_2 - 1)} = \frac{102.6}{9} = 11.4.$$

Now $s_x^2 > s_y^2$. Hence we take $v_1 = n_1 - 1 = 7$ and $v_2 = n_2 - 1 = 9$. We define

Null hypothesis $H_0 : s_x^2 = s_y^2$

Alternate hypothesis $H_1 : s_x^2 \neq s_y^2$

Now, the *F*-statistic is given by $F = \dfrac{s_x^2}{s_y^2} = \dfrac{12.06}{11.4} = 1.058.$

Tabulated $F_{0.05}$ (7, 9) = 3.29 (Given)

The calculated value of F is less than the table value *i.e.*, $1.058 < 3.29$. Hence, we accept the null hypothesis H_0. The two random samples might have come from two normal populations with the same variance.

Example 24.150 Two random samples of sizes 10 and 12 gave the following values of the variable.

Sample I (x):	20	16	26	27	23	22	18	24	25	19		
Sample II (y):	27	33	42	35	32	34	38	28	41	43	30	37

Test the difference of the estimates of the population variances at 5% level of significance.

Solution. Let us take the null hypothesis H_0: that the samples have been drawn from normal populations with same variance.

Computations for s_x^2 and s_y^2.

	Sample I			Sample II		
x_i	$x_i - \bar{x}$	$(x_i - \bar{x})^2$	y_i	$y_j - \bar{y}$	$(y_j - \bar{y})^2$
20	-2	4	27	-8	64
16	-6	36	33	-2	4
26	4	16	42	7	49
27	5	25	→ 35	0	0
23	1	1	32	-3	9
→ 22	0	0	34	-1	1
18	-4	16	38	3	9
24	2	4	28	-7	49
25	3	9	41	6	36
19	-3	9	43	8	64
			30	-5	25
			37	2	4
$\Sigma x_1 = 220$		$\Sigma(x_i - \bar{x})^2$ $= 120$	$\Sigma y_j = 420$		$\Sigma(y_j - \bar{y})^2$ $= 314$

$$\bar{x} = \frac{220}{10} = 22, \; n_1 = 10, \; \Sigma(x_i - \bar{x})^2 = 120.$$

$$\bar{y} = \frac{420}{12} = 35, \; n_2 = 12, \; \Sigma(y_j - \bar{y})^2 = 314.$$

$$s_x^2 = \frac{\Sigma(x_i - \bar{x})^2}{(n_1 - 1)} = \frac{120}{9} = 13.333. \quad s_y^2 = \frac{\Sigma(y_j - \bar{y})^2}{(n_2 - 1)} = \frac{314}{11} = 28.545.$$

Note that $S_y^2 > S_x^2$. We take $v_1 = 11$, $v_2 = 9$.

The F-statistic is given by $F = 28.545/13.333 = 2.141$.

At 5% level of significance, we have from table $F_{0.05}(11, 9) = 3.105$. Since the calculated value of F is less than the table value. We accept the null hypothesis.

We conclude that there is no significant difference between the population variances.

Example 24.151 The values in two random samples are given below.

Sample 1	31	22	28	29	27	35
Sample 2	27	21	25	23	26	

Can we conclude that the two samples are drawn from the same population? Test at 5% level of significance.

Solution. We shall use both the F-test and t-test to draw a conclusion.

F-test *Null hypothesis* $H_0 : \sigma_x^2 = \sigma_y^2 = \sigma^2$ *i.e.* the population variance donot differ significantly.

Alternative hypothesis $H_1 : \sigma_x^2 \neq \sigma_y^2$.

F-statistic is given by $F = \dfrac{s_x^2}{s_y^2}, \left(\text{if } s_x^2 > s_y^2\right)$

Computations for s_x^2 and s_y^2.

Sample I			Sample II		
x_i	$x_i - \bar{x}$	$\left(x_i - \bar{x}_1\right)^2$	y_i	$y_j - \bar{y}$	$\left(y_j - \bar{y}\right)^2$
31	2.33	5.43			
22	−6.67	44.49	27	2.60	6.76
28	−0.67	0.45	21	−3.40	11.56
29	0.33	0.11	25	0.60	0.36
27	−1.67	2.79	23	−1.40	1.96
35	6.33	40.07	26	1.60	2.56
Σx_i		$\Sigma\left(x_i - \bar{x}\right)^2$	Σy_j		$\Sigma\left(y_j - \bar{y}\right)^2$
= 172		= 93.34	= 122		= 23.2

We find $\bar{x} = 172/6 = 28.67$; $n_1 = 6$, $\Sigma\left(x_1 - \bar{x}\right)^2 = 93.34$.

$\bar{y} = 122/5 = 24.40$; $n_2 = 5$, $\Sigma\left(y_j - \bar{y}\right)^2 = 23.2$

$$s_x^2 = \frac{\Sigma(x_i - \overline{x})^2}{n_1 - 1} = \frac{93.34}{5} = 18.67; \; s_y^2 = \frac{\Sigma(y_j - \overline{y})^2}{n_2 - 1} = \frac{23.2}{4} = 5.8.$$

The *F*-statistic is given by

$$F = \frac{s_x^2}{s_y^2} = \frac{18.67}{5.8} = 3.218.$$

At 5% level of significance, $F_{0.05}$ (5, 4) = 6.26. Since the calculated value of *F* is less than the table value, we accept the null hypothesis. We conclude that the difference between the estimates of population variances is not significant.

Now, we use the **t-test**. Define, $H_0' : \mu_x = \mu_y$ and $H_1' : \mu_x \ne \mu_y$.

The *t*-statistic is given by $t = \dfrac{\overline{x} - \overline{y}}{S} \sqrt{\dfrac{n_1 n_2}{n_1 + n_2}}$,

where $S = \sqrt{\dfrac{\Sigma(x_i - \overline{x})^2 + \Sigma(y_i - \overline{y})^2}{(n_1 - 1) + (n_2 - 1)}} = \sqrt{\dfrac{93.34 + 23.2}{6 + 5 - 2}} = 3.598..$

$$\therefore t = \frac{28.67 - 24.40}{3.598} \times \sqrt{\frac{30}{11}} = 1.186 \times 1.651 = 1.958,$$

with $v = n_1 + n_2 - 2 = 9$ degree of freedom. At 5% level of significance, we get from table II (Appendix 4) the value of *t* as 2.262. Since, *t* = 1.958 < 2.262, we accept the null hypothesis that there is no significant difference between the means of the two samples.

Exercise 24.6

1. A random sample of six steel beams has mean compressive strength of 58392 psi with standard deviation of 648 psi. Test the null hypothesis $H_0 : \mu = 58000$ psi against the alternative hypothesis $H_1 : \mu > 58000$ psi at 5%, level of significance (value of |*t*| at 5 degrees of freedom and 5% significance level is 2.015). Here μ denotes the population mean.　　　　(*A.M.I.E., S-2000*)

　　　　　　　　　　　　　　　　　　Ans.: Hypothesis H_0 is accepted.

2. A random sample of 16 values from a normal population showed a mean of 41.5 cm and a sum of squares of deviations from this mean equals to 135 cm². Show that the assumption of a 43.5 cm for the population mean is not reasonable. Test at 5 per cent level of significance.

3. Samples of two types of electric light tubes were tested for length of life and the following data were obtained:

	Type I	Type II
Sample size	8	7
Sample means	1234 hours	1036 hours
Sample standard deviation	36 hours	40 hours

Is the difference in the means sufficient to warrant that type I is superior to type II regarding the length of life?

Ans.: Calculated value of $t = 10.091$. Significant.

4. For the following data examine if the means of two samples differ significantly:

Sample I:	$n_1 = 12$,	$\bar{X}_1 = 57.2$	$s_1 = 3.41$
Sample II:	$n_2 = 7$,	$\bar{X}_2 = 52.3$	$s_2 = 3.62$

Ans.: No significant difference between means.

5. The sales data of an item in six shops before and after a special promotional compaign are as under:

Shops:	A	B	C	D	E	F
Before campaign:	53	28	31	48	50	42
After campaign:	58	29	30	55	56	45

Can the campaign be judged to be a success? Test at 5% level of significance.

Ans.: Promotional campaign has been successful.

6. A certain stimulus administered to each of 12 patients resulted in the following increases of blood pressure: 5, 2, 8, −1, 3, 0, −2, 1, 5, 0, 4, 6. Can it be concluded that the stimulus will in general be accompanied by an increase in blood pressure? (*Madras 1996*)

Ans.: Stimulus will result in increase of blood pressure.

7. The following table gives the number of accidents that took place in an industry during various days of the week. Test if accidents are uniformly distributed over the week.

Day	Mon.	Tue.	Wed.	Thus.	Fri.	Sat.
No. of accidents	14	18	12	11	15	14

Ans.: $\chi^2 = 2.1428$; table value $\chi_5^2(0.05) = 11.07$. H_0 is accepted.

8. Find χ^2 for the following table:

Class	A	B	C	D	E
Obs. frequency	8	29	44	15	4
Expec. or theoretical frequency	7	24	38	24	7

[**Hint:** Incase, the expected frequencies are less than 10, we group together such classes. Here some frequencies being less than 10, we regroup the data as:

Class	A and B	C	D and E
Obs. frequency	8 + 29 = 37	44	15 + 4 = 19
Theoretical frequency	7 + 24 = 31	38	24 + 7 = 31

$$\chi^2 = \sum_{i=1}^{3} \frac{(O_i - e_i)^2}{e_i} = \frac{(37-31)^2}{31} + \frac{(44-38)^2}{38} + \frac{(19-31)^2}{31} = 6.75 \text{ approx.}].$$

9. In his experiments with peas, Gregor Mendel observed that 315 were round and yellow, 108 were round and green, 101 were wrinkled and yellow, and 32 were wrinkled and green. According to his theory of heredity, the numbers should be in the proportion 9:3:3:1. Is there any evidence to doubt his theory at the (*i*) 0.01 and (*ii*) 0.05 significance levels?

$$\left(\text{Given } \chi^2_{0.01} = 11.35 \text{ and } \chi^2_{0.05} = 7.82 \text{ for 3 } dof\right) \qquad (A.M.I.E., W\text{-}2005)$$

Ans.: The theory and experiment are in agreement.

10. Fit a Poisson distribution for the following data and test the goodness of fit at 5% level of significance.

x:	0	1	2	3	4	5
f:	110	170	130	60	23	7

[**Hint:** Mean, $m = \Sigma fx / \Sigma f = 737/500 = 1.474$. The theoretical frequencies are:

$$N \cdot \frac{e^{-m} m^r}{r!} \text{ i.e., } \frac{500 e^{-1.474} (1.474)^r}{r!}, \, r = 0, 1, 2, 3, 4, 5.$$

The expected frequencies are: 115, 169, 124, 61, 22, 9.
The values of e_i are adjusted to make them whole numbers such that $\Sigma e_i = 500$. We have the frequencies.

O_i	110	170	130	60	23	7	$\Sigma O_i = 500$
e_i	115	169	124	61	22	9	$\Sigma e_i = 500$

Grouping the classes, so that each class frequency is ≥ 10, we obtain

O_i	110	170	130	60	(23 + 7)	$\Sigma O_i = 500$
e_i	115	169	124	61	(22 + 9)	$\Sigma e_i = 500$

$$\chi^2 = \sum_i^n \left[\frac{(O_i - e_i)^2}{e_i} \right] = \frac{25}{115} + \frac{1}{169} + \frac{36}{124} + \frac{1}{61} + \frac{1}{31} = 0.560$$

Tabulated value of χ^2 at 5% level for $5 - 2 = 3$ $dof = 7.815$. Hence Poisson distribution is satisfactory].

11. Two random samples gave the following results:

Sample	Size	Sample mean	Sum of squares of deviations from mean
I	10	15	90
II	12	14	108

Test whether the samples are drawn from the same population.

[Given: $F_{0.05}(9,11) = 2.90$, $F_{0.05}(11, 9) = 3.10$ and $t_{0.05}(20) = 2.086$, $t_{0.05}(22) = 2.074$]

> **Ans:** $H_0 : \sigma_1^2 = \sigma_2^2$, $F = 1.018 < 2.90$; not significant.
> $H_0' : \mu_1 = \mu_2$, $t = 0.742 < t_{0.05} \, 20$; not significant; Both the hypothesis accepted. Given samples drawn from the same normal population.

12. Two random samples of sizes 9 and 7 gave the sum of squares of deviations from their respective means as 175 and 95 respectively. Can they be regarded as drawn from normal populations with the same variance? Given: At 5% level of significance, $F_{0.05}(8, 6) = 4.15$.

Ans. $F = 1.381 < 4.15$. The random samples are from the same population.

13. Two independent sample of sizes 7 and 6 had the following values:

Sample A	28	30	32	33	31	29	34
Sample B	29	30	30	24	27	28	

Test at 5% level of significance whether the samples have been drawn from normal populations having the same variance. (*Madras, 1997*)

Ans. The random samples are from the same population.

14. Two random sample have the following values.

Sample 1	15	25	16	20	22	24	21	17	19	23		
Sample 2	35	31	25	38	26	29	32	34	33	27	29	31

Test at 5% level of significance whether the two samples could have come from the same population.

Ans. The random samples are not from the same population.

APPENDICES

Appendices

APPENDIX 1

Some Constants

$e = 2.71828$

$\pi = 3.14159$

$\sqrt{2} = 1.41421$

$\sqrt{3} = 1.73205$

$\sqrt{10} = 3.16228$

$1° = 60' = 3600'' = 0.01745$ rad

1 radian $= 57° 17' 44.81''$

$\qquad = 57{\cdot}29578°$

$\log_e 2 = 0.69315$

$\log_e 3 = 1.09861$

$\log_e 10 = 2.30259$

$\log_{10} e = 0.43429$

$\log_e \pi = 1.14473$

$\log_{10} \pi = 0.49715$

π radians $= 180$ degrees

Some Important Conversion Factors

1 inch = 2.54000 cm

100 sq. metres = 1 are

100 hectares = 1 sq. kilometre

1 Quintal = 100 kgm

1 H.P = 178.298 cal/sec = 0.74570 KW

°F = °C (1.8) + 32

1 acre = 4840 yd^2 = 4046. 8564 m^2

100 ares = 1 hectare

1 lb = 453.59 gm

1 Kilogm. · weights = 2.20462 lb.

1 gallon of water = 4.5460 litres

1 metric Ton (Tonne) = 1000 kgm

1 KW = 1000 walts = 238.662 cal/sec.

Greek Alphabet

α	Alpha	ι	Iota	ρ	Rho
β	Beta	κ	Kappa	σ, Σ	Sigma
γ, Γ	Gamma	λ, Λ	Lambda	τ	Tau
δ, Δ	Delta	μ	Mu	υ, Y	Upsilon
$\varepsilon, \in$	Epsilon	ν	Nu	$\varphi, \phi\ \Phi$	Phi
ζ	Zeta	ξ	Xi	χ	Chi
η	Eta	o	Omicron	ψ, Ψ	Psi
$\theta, \vartheta, \Theta$	Theta	π	Pi	ω, Ω	Omega

APPENDIX A.2

Revision of essential formulae

ALGEBRAIC

1. Laws of Exponents

$$a^m.a^n = a^{m+n}, (ab)^m = a^m b^m, (a^m)^n = a^{mn}, a^{m/n} = \sqrt[n]{a^m}$$

If $a \neq 0, a^m/a^n = a^{m-n}, a^0 = 1, a^{-m} = 1/a^m.$

2. Cross-multiplication

If $a_1 x + b_1 y + c_1 = 0$ and $a_2 x + b_2 y + c_2 = 0$,

then $$\frac{x}{b_1 c_2 - b_2 c_1} = \frac{y}{c_1 a_2 - c_2 a_1} = \frac{1}{a_1 b_2 - a_2 b_1}$$

$$\therefore \quad x = \frac{b_1 c_2 - b_2 c_1}{a_1 b_2 - a_2 b_1}, y = \frac{c_1 a_2 - c_2 a_1}{a_1 b_2 - a_2 b_1}.$$

3. Quadratic equations

The solutions of the quadratic equation $ax^2 + bx + c = 0$ are $x = \dfrac{-b \pm \sqrt{b^2 - 4ac}}{2a}$.

If α abd β are the roots of this quardratic equation with $\alpha > \beta$, then $\alpha + \beta = -b/a$, $\alpha\beta = c/a$. The roots are (*i*) real and different, (*ii*) equal (*iii*) imaginary according as the discriminant $b^2 - 4ac$ is (*i*) positive, (*ii*) zero (*iii*) negative.

4. The Cubic equation

If α, β, γ are the roots of cubic equation $ax^3 + bx^2 + cx + d = 0$, then $\alpha + \beta + \gamma = -b/a$, $\alpha\beta + \beta\gamma + \gamma\alpha = c/a$, $\alpha\beta\gamma = -d/a$.

5. The Factor therorem

If for $x = a, f(x) = 0$, then $(x - a)$ is a factor of $f(x)$ or $x = a$ is a root of $f(x) = 0$.

6. Series

If the series in arithmetical progression (A.P.) is $a + (a + d) + (a + 2d) + ...,$ *i.e.* first term a and the common difference d. Then nth term of A.P. is $a + (n - 1) d$.

S_n = Sum to n terms of the A.P. $= \dfrac{n}{2}(a + l)$, where l is the last term

$$= \frac{n}{2}[2a + (n-1)d].$$

If the series in geometrical progression (G.P.) is $a + ar + ar^2 + ...$, a is called the first term and the constant number $r \neq 0$ is called the common ratio of the G.P. Then nth term of G.P. $= ar^{n-1}$.

$$S_n = \text{Sum to } n \text{ terms of G.P.} = \frac{a(1-r^n)}{1-r}, \text{ if } |r| < 1, = \frac{a(r^n -1)}{r-1}, \text{ if } |r| > 1.$$

If $|r| < 1$, $r^n \to 0$ as $n \to \infty$. Therefore S_n or S, sum to infinity $= \dfrac{a}{1-r}$.

Sum of first n natural numbers, *i.e.* $1 + 2 + 3 + ... + n$ is $S_1 = \dfrac{n(n+1)}{2}$.

Sum of squares of the first n natural numbers *i.e.*, $1^2 + 2^2 + 3^2 + ... + n^2$ is

$$S_2 = \frac{n(n+1)(2n+1)}{6}.$$

Sum of cubes of the first n natural numbers *i.e.*, $1^3 + 2^3 + 3^3 + ... + n^3$ is

$$S_3 = \left[\frac{n(n+1)}{2}\right]^2.$$

7. Permutations and Combinations

$$n! = n(n-1)(n-2)...4 \cdot 3 \cdot 2 \cdot 1. \qquad 0! = 1, \quad {}^nP_r = \frac{n!}{(n-r)!}$$

$${}^nC_r = \frac{n!}{r!(n-r)!}, 0 \leq r \leq n \text{ and } n \text{ is a non-negative integer } = \frac{{}^nP_r}{r!}.$$

$${}^nC_0 = 1 = {}^nC_n, \quad {}^nC_{n-r} = {}^nC_r, \quad {}^nC_r + {}^nC_{r-1} = {}^{n+1}C_r.$$

8. Binomial Theorem

(*i*) For any positive integer n,

$$(a+b)^n = a^n + na^{n-1}b + \frac{n(n-1)}{1.2}a^{n-2}b^2 + \frac{n(n-1)(n-2)}{1.2.3}a^{n-3}b^3 +$$

$$\cdots + nab^{n-1} + b^n.$$

(*ii*) Certain particular cases of Binomial Formula, where we assume $|x| < 1$

$$(1+x)^{-1} = 1 - x + x^2 - x^3 + \cdots + (-1)^r x^r + \cdots$$

$$(1+x)^{-2} = 1 - 2x + 3x^2 - \cdots + (-1)^r (r+1)x^r + \cdots$$

$$(1+x)^{-3} = 1 - 3x + 6x^2 - \ldots + (-1)^r \frac{(r+1)(r+2)}{2} x^r + \ldots$$

$$(1-x)^{-1} = 1 + x + x^2 + x^3 + \cdots + x^r + \cdots$$

$$(1-x)^{-2} = 1 + 2x + 3x^2 + 4x^3 + \cdots + (r+1)x^r + \cdots$$

$$(1-x)^{-3} = 1 + 3x + 6x^2 + \cdots + \frac{(r+1)(r+2)}{2} x^r + \ldots$$

9. Determinants

(*i*) Det. of order 2: $\begin{vmatrix} a_1 & b_1 \\ a_2 & b_2 \end{vmatrix} = a_1 b_2 - a_2 b_1 .$

(*ii*) Det. of order 3: $\begin{vmatrix} a_1 & b_1 & c_1 \\ a_2 & b_2 & c_2 \\ a_3 & b_3 & c_3 \end{vmatrix} = a_1 \begin{vmatrix} b_2 & c_2 \\ b_3 & c_3 \end{vmatrix} - b_1 \begin{vmatrix} a_2 & c_2 \\ a_3 & c_3 \end{vmatrix} + c_1 \begin{vmatrix} a_2 & b_2 \\ a_3 & b_3 \end{vmatrix} .$

LOGARITHMS

Laws of logarithms

(*i*) $\log_a (mn) = \log_a m + \log_a n$ (*ii*) $\log_a \left(\dfrac{m}{n} \right) = \log_a m - \log_a n$

(*iii*) $\log_a m^n = n \log_a m$ (*iv*) $\log_b r = \dfrac{\log_a r}{\log_a b}$ (Base-changing formula)

(Note that we can use any base in place of (a))

Remark: $\log_a 1 = 0$, $\log_a a = 1$, $\log_a a^x = x$, $e^{\log x} = x$.

MENSURATION

(A) Formulas from plane Geometry:

Notations

a, b and c are sides of a triangle; $s = (a+b+c)/2$ = semi-perimeter; h = altitude; p = perimeter; $\angle A$, $\angle B$ and $\angle C$ are angles of a triangle, r = radius of a circle, C = circumference, R = radius of a sphere.

Triangle:

1. Area of a triangle $= = \dfrac{1}{2} bh.$

$$= \sqrt{s(s-a)(s-b)(s-c)} \text{ (Hero's formula)}$$

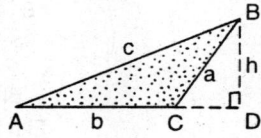

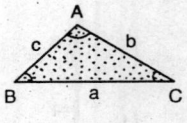

$$= \frac{1}{2}ab\sin C = \frac{1}{2}ca\sin B = \frac{1}{2}bc\sin A$$

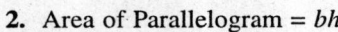

$$= a^2 \cdot \frac{\sqrt{3}}{4} \quad \text{(for an equilateral triangle)}$$

$$= \frac{a^2 \sin B \sin C}{2\sin A} = \frac{b^2 \sin C \sin A}{2\sin B} = \frac{c^2 \sin A \sin B}{2\sin C}$$

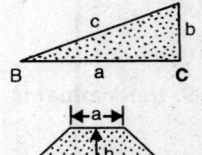

$$= \frac{1}{2}ab \quad \text{(right angled triangle, hypotenuse } c\text{)}$$

2. Area of Parallelogram $= bh$

3. Area of Trapezoid $= \frac{1}{2}(a+b)h$

4. Area of circle $= \pi r^2, C = 2\pi r$

5. Area of sector $= r^2\theta/2$, Arc length $= r\theta$, (θ in radians)

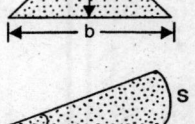
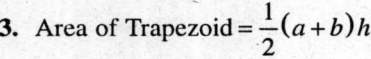
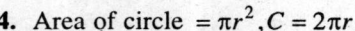

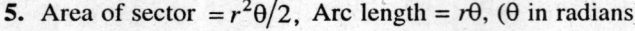
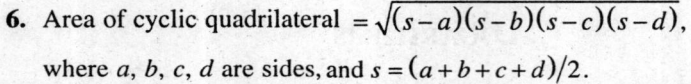

6. Area of cyclic quadrilateral $= \sqrt{(s-a)(s-b)(s-c)(s-d)}$,

where a, b, c, d are sides, and $s = (a+b+c+d)/2$.

7. Area of an ellipse $= \pi\,ab$; (semi-major axis a and semi-minor axis b)

(B) Formulas from Solid Geometry:

1. Right Circular cylinder:

Volume $= \pi r^2 h$; Curved surface, $S = 2\pi rh$.

2. Right circular Cone:

Volume $= \pi r^2 h/3$; Curved surface, $S = \pi rl = \pi r\sqrt{r^2 + h^2}$.

3. Any cylinder or Prism with parallel Bases

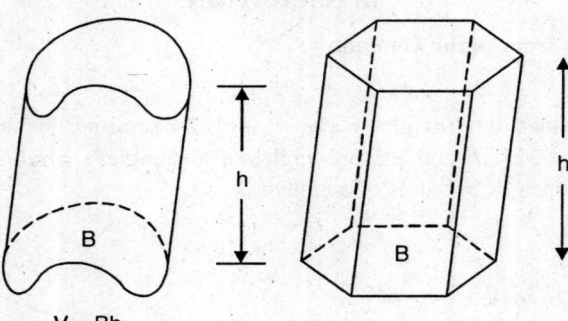

V = Bh

Fig. Volume = (Base area) × height = (B)(h)

4. Frustum of Pyramids and Cones:

$$\text{Volume} = \frac{h}{3}\left[A_1 + A_2 + \sqrt{A_1 A_2}\right]; \text{ where } h = \text{height},$$

A_1 and A_2 are the areas of ends.

5. Frustum of right Circular Cone

$$V = \frac{\pi}{3}\left(r_1^2 + r_1 r_2 + r_2^2\right)h.$$

6. Sphere:

Volume $= 4\pi R^3/3$, Surface, $S = 4\pi R^2$.

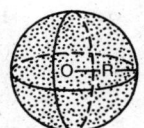

TRIGONOMETRIC

1. Radian: An angle subtended at the centre of a circle by an arc equal in length to the radius of a circle is said to have a measure of 1 radian.

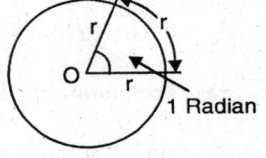

1 Radian

$\theta = l/r$, where l is the length of arc which subtends angle θ (in radians) at the centre of the circle with radius r.

π radians $= 180°$.

2. Pythagorean Identities

$\sin^2\theta + \cos^2\theta = 1,$ $\qquad \sec^2\theta = 1 + \tan^2\theta,$ $\qquad \csc^2\theta = 1 + \cot^2\theta.$

3. Sign Identities

$\sin(-\theta) = -\sin\theta,$ $\qquad \cos(-\theta) = \cos\theta,$ $\qquad \tan(-\theta) = -\tan\theta.$

4. Complement Identities

$$\sin\left(\frac{\pi}{2}-\theta\right) = \cos\theta, \quad \cos\left(\frac{\pi}{2}-\theta\right) = \sin\theta, \quad \tan\left(\frac{\pi}{2}-\theta\right) = \cot\theta.$$

sin	all
tan	cos

$$\sin\left(\theta-\frac{\pi}{2}\right) = -\cos\theta, \quad \cos\left(\theta-\frac{\pi}{2}\right) = \sin\theta, \quad \sin\left(\frac{\pi}{2}+\theta\right) = \cos\theta,$$

$$\cos\left(\frac{\pi}{2}+\theta\right) = -\sin\theta.$$

5. Supplement Identities

$$\sin(\pi-\theta) = \sin\theta, \quad \cos(\pi-\theta) = -\cos\theta, \quad \tan(\pi-\theta) = -\tan\theta.$$

$$\sin(\pi+\theta) = -\sin\theta, \quad \cos(\pi+\theta) = -\cos\theta, \quad \tan(\pi+\theta) = \tan\theta.$$

6. Values of trigonometric functions

Angle	0°	π/6	π/4	π/3	π/2	π	3π/2	2π
sin	0	1/2	$1/\sqrt{2}$	$\sqrt{3}/2$	1	0	−1	0
cos	1	$\sqrt{3}/2$	$1/\sqrt{2}$	1/2	0	−1	0	1
tan	0	$1/\sqrt{3}$	1	$\sqrt{3}$	not defined	0	not defined	0

7. Trigonometric Functions of Sum and Difference

$\sin(A \pm B) = \sin A \cos B \pm \cos A \sin B.$

$\cos(A \pm B) = \cos A \cos B \mp \sin A \sin B.$

$$\tan(A \pm B) = \frac{\tan A \pm \tan B}{1 \mp \tan A \tan B}.$$

8. Trigonometric Functions of Multiples and sub multiples Angles

$$\sin 2A = 2\sin A \cos A = \frac{2\tan A}{1 + \tan^2 A}.$$

$$\cos 2A = \cos^2 A - \sin^2 A, = 2\cos^2 A - 1, = 1 - 2\sin^2 A = \frac{1 - \tan^2 A}{1 + \tan^2 A}.$$

$$\sin^2 \frac{A}{2} = \frac{1 - \cos A}{2}, \quad \cos^2 \frac{A}{2} = \frac{1 + \cos A}{2}.$$

$$\tan 2A = \frac{2\tan A}{1 - \tan^2 A}, \quad \sin 3A = 3\sin A - 4\sin^3 A, \quad \cos 3A = 4\cos^3 A - 3\cos A,$$

$$\tan 3A = \frac{3\tan A - \tan^3 A}{1 - 3\tan^2 A}.$$

9. Transformation of a Product into a sum and vice versa

$$\sin A \cos B = \frac{1}{2}\sin(A + B) + \frac{1}{2}\sin(A - B),$$

$$\cos A \sin B = \frac{1}{2}\sin(A + B) - \frac{1}{2}\sin(A - B),$$

$$\cos A \cos B = \frac{1}{2}\cos(A + B) + \frac{1}{2}\cos(A - B),$$

$$\sin A \sin B = \frac{1}{2}\cos(A - B) - \frac{1}{2}\cos(A + B).$$

$$\sin C + \sin D = 2\sin\frac{C+D}{2}\cos\frac{C-D}{2}, \quad \sin C - \sin D = 2\cos\frac{C+D}{2}\sin\frac{C-D}{2},$$

$$\cos C + \cos D = 2\cos\frac{C+D}{2}\cos\frac{C-D}{2}, \quad \cos C - \cos D = -2\sin\frac{C+D}{2}\sin\frac{C-D}{2}.$$

10. Law of sines or the Sine Formula

If A, B and C are the angles of a triangle and if a, b and c are the lengths of the sides opposite to A, B and C, respectively, then

$$\frac{\sin A}{a} = \frac{\sin B}{b} = \frac{\sin C}{c}.$$

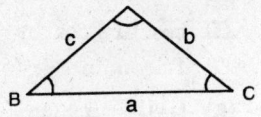

11. Law of Cosines or the Cosine Formulae

$$a^2 = b^2 + c^2 - 2bc\cos A,$$
$$b^2 = c^2 + a^2 - 2ca\cos B,$$
$$c^2 = a^2 + b^2 - 2ab\cos B.$$

12. Projection Formulae

$$a = b\cos C + c\cos B, \; b = c\cos A + a\cos C, \; c = a\cos B + b\cos A.$$

13. Circumradius of the triangle, when sides are known:

$$R = \frac{a}{2\sin A} = \frac{b}{2\sin B} = \frac{c}{2\sin C} = \frac{abc}{4\Delta}.$$

14. Inradius of the triangle

$$r = \frac{\Delta}{s} = \frac{\text{area of the triangle}}{\text{semi-perimeter}}.$$

15. Inverse Circular Functions

(a) (i) $\sin^{-1}\dfrac{1}{x} = \operatorname{cosec}^{-1}x, \, x \geq 1$, (ii) $\cos^{-1}\dfrac{1}{x} = \sec^{-1}x, \, x \geq 1$.

(iii) $\tan^{-1}\dfrac{1}{x} = \cot^{-1}x, \, x > 0$.

(b) (i) $\sin^{-1}(-x) = -\sin^{-1}x; \, x \in [-1,1]$ (ii) $\tan^{-1}(-x) = -\tan^{-1}x, x \in R$.

(iii) $\operatorname{cosec}^{-1}(-x) = -\operatorname{cosec}^{-1}x, |x| \geq 1$.

(c) (i) $\cos^{-1}(-x) = \pi - \cos^{-1}x, x \in [-1,1]$, (ii) $\sec^{-1}(-x) = \pi - \sec^{-1}x, |x| \geq 1$.

(iii) $\cot^{-1}(-x) = \pi - \cot^{-1}x, x \in R$.

(d) *(i)* $\sin^{-1} x + \cos^{-1} x = \dfrac{\pi}{2}, x \in [-1,1],$ *(ii)* $\tan^{-1} x + \cot^{-1} x = \dfrac{\pi}{2}, x \in R,$

     (iii) $\operatorname{cosec}^{-1} x + \sec^{-1} x = \dfrac{\pi}{2}, |x| \geq 1.$

(e) $\sin^{-1} x + \sin^{-1} y = \sin^{-1}\left(x\sqrt{1-y^2} + y\sqrt{1-x^2}\right).$

(f) $\cos^{-1} x + \cos^{-1} y = \cos^{-1}\left(xy - \sqrt{1-x^2}\sqrt{1-y^2}\right).$

(g) *(i)* $\tan^{-1} x + \tan^{-1} y = \tan^{-1}\dfrac{x+y}{1-xy},$ if $xy < 1$

     (ii) $\tan^{-1} x - \tan^{-1} y = \tan^{-1}\dfrac{x-y}{1+xy},$ if $xy > -1.$

(h) *(i)* $2\tan^{-1} x = \sin^{-1}\dfrac{2x}{1+x^2}, |x| \leq 1,$ *(ii)* $2\tan^{-1} x = \cos^{-1}\dfrac{1-x^2}{1+x^2}, x \geq 0.$

     (iii) $2\tan^{-1} x = \tan^{-1}\dfrac{2x}{1-x^2}, -1 < x < 1.$

COMPLEX NUMBERS AND FUNCTIONS

1. Every complex number $x + iy$ can always be expressed in the form $r(\cos\theta + i\sin\theta)$, where $r = \sqrt{x^2 + y^2}$ and $\theta = \tan^{-1}(y/x)$.

2. De'Moivre's Theorem

Let $z = r(\cos\theta + i\sin\theta)$. Then $z^n = r^n\left[\cos n\theta + i\sin n\theta\right]$, where n is any positive or negative integer or a real rational number.

$$z^{-n} = r^{-n}\left[\cos n\theta - i\sin n\theta\right] = \dfrac{1}{r^n\left[\cos n\theta + i\sin n\theta\right]} = \dfrac{1}{z^n}.$$

3. Exponential, circular functions of a complex variable, z

(i) e^z or $\exp(z) = e^{x+iy} = e^x \cdot e^{iy} = e^x(\cos y + i\sin y),$

(ii) Exponential form of $z(= x + iy) = re^{i\theta}$ *(iii)* $\sin z = \dfrac{e^{iz} - e^{-iz}}{2i},$

$(iv)\ \cos z = \dfrac{e^{iz} + e^{-iz}}{2},\quad (v)\ \tan z = \dfrac{\sin z}{\cos z} = \dfrac{e^{iz} - e^{-iz}}{i\left(e^{iz} + e^{-iz}\right)}.$

4. Euler's Theorem

$e^{i\theta} = \cos\theta + i\sin\theta$, where θ is real or complex.

5. Hyperbolic Functions, If x be real or complex,

(a) $\sinh x = \dfrac{e^x - e^{-x}}{2}$, $\cosh x = \dfrac{e^x + e^{-x}}{2}$, $\tanh x = \dfrac{\sinh x}{\cosh x} = \dfrac{e^x - e^{-x}}{e^x + e^{-x}}$,

$\coth x = 1/\tanh x = \cosh x/\sinh x$, $\operatorname{sech} x = 1/\cosh x$, $\operatorname{cosech} x = 1/\sinh x$.

(b) *Fundamental formulae*

$\cosh^2 x - \sinh^2 x = 1$, $\operatorname{sech}^2 x + \tanh^2 x = 1$, $\coth^2 x - \operatorname{cosech}^2 x = 1$.

6. Relations between hyperbolic and circular functions

$\sin ix = i\sinh x;\quad \cos ix = \cosh x;\quad \tan ix = i\tanh x.$

7. Inverse Hyperbolic function

$\sinh^{-1} x = \log_e\left(x + \sqrt{x^2 + 1}\right);$ $\qquad\qquad \cosh^{-1} x = \log_e\left(x + \sqrt{x^2 - 1}\right)$

$\tanh^{-1} x = \dfrac{1}{2}\log\dfrac{1 + x}{1 - x}.$

EXPONENTIAL AND LOGARITHMIC SERIES

1. Exponential Series

$$e^x = 1 + \frac{x}{1!} + \frac{x^2}{2!} + \frac{x^3}{3!} + \cdots + \frac{x^n}{n!} + \cdots \infty,\quad e = 1 + \frac{1}{1!} + \frac{1}{2!} + \frac{1}{3!} + \cdots \infty$$

2. Logarithmic Series: If $|x| < 1$, then $\log(1 + x) = x - \dfrac{x^2}{2} + \dfrac{x^3}{3} - \dfrac{x^4}{4} + \cdots$

Some Important Series

$\sin x = x - \dfrac{x^3}{3!} + \dfrac{x^5}{5!} - \dfrac{x^7}{7!} + $ $\qquad \log(1 - x) = -\left(x + \dfrac{x^2}{2} + \dfrac{x^3}{3} + \cdots \infty\right)$

$\cos x = 1 - \dfrac{x^2}{2!} + \dfrac{x^4}{4!} - \dfrac{x^6}{6!} + \cdots$ $\qquad \sinh x = x + \dfrac{x^3}{3!} + \dfrac{x^5}{5!} + \dfrac{x^7}{7!} + \cdots \infty$

$a^x = e^{x\log a} = 1 + \dfrac{\log a}{1!}x + \dfrac{(\log a)^2}{2!}x^2 + \dfrac{(\log a)^3}{3!}x^3 + \cdots \infty$

Polar Co-ordinates

$$x = r\cos\theta,\ y = r\sin\theta,\quad r = \sqrt{x^2 + y^2},\ \theta = \tan^{-1}(y/x),\quad dxdy = x\,dr\,d\theta.$$

APPENDIX 3

Some Derivative and Integral Formulas

■ **Derivative Formulas** u and v are functions of x, c_1, and c_2 are constants.

1. $\dfrac{d}{dx}(c_1 u + c_2 v) = c_1 u' + c_1 v'$ (Sum Rule)

2. $\dfrac{d}{dx}(uv) = uv' + vu'$ (Product Rule),

3. $\dfrac{d}{dx}\left(\dfrac{u}{v}\right) = \dfrac{vu' - uv'}{v^2}$ (Quotient Rule),

4. $\dfrac{dy}{dx} = \dfrac{dy}{du}\cdot\dfrac{du}{dx}$ (Chain Rule),   5. $\dfrac{d}{dx}(x^n) = nx^{n-1}$,

6. $\dfrac{d}{dx}(\log_e x) = \dfrac{1}{x}$,   7. $\dfrac{d}{dx}(e^{ax}) = ae^{ax}$,

8. $\dfrac{d}{dx}(a^x) = a^x \log_e a$,   9. $\dfrac{d}{dx}(\log_a x) = \dfrac{\log_a e}{x} = \dfrac{1}{x\log_e a}$,

10. $\dfrac{d}{dx}(\sin x) = \cos x$,   11. $\dfrac{d}{dx}(\cos x) = -\sin x$,

12. $\dfrac{d}{dx}(\tan x) = \sec^2 x$,   13. $\dfrac{d}{dx}(\cot x) = -\text{cosec}^2 x$,

14. $\dfrac{d}{dx}(\sec x) = \sec x \tan x$,   15. $\dfrac{d}{dx}(\text{cosec } x) = -\text{cosec } x \cot x$,

16. $\dfrac{d}{dx}(\sinh x) = \cosh x$.   17. $\dfrac{d}{dx}(\cosh x) = \sinh x$,

18. $\dfrac{d}{dx}(\tanh x) = \sec h^2 x$,   19. $\dfrac{d}{dx}(\coth x) = -\text{cosech}^2 x$,

20. $\dfrac{d}{dx}(\text{sech } x) = -\text{sech } x \tanh x$,   21. $\dfrac{d}{dx}\text{cosech } x = -\text{cosech } x \coth x$,

22. $\dfrac{d}{dx}(\sin^{-1} x) = \dfrac{1}{\sqrt{1-x^2}}$,   23. $\dfrac{d}{dx}(\cos^{-1} x) = \dfrac{-1}{\sqrt{1-x^2}}$

24. $\dfrac{d}{dx}\left(\tan^{-1} x\right) = \dfrac{1}{1+x^2}$ **25.** $\dfrac{d}{dx}\cot^{-1} x = \dfrac{-1}{1+x^2}$

26. $\dfrac{d}{dx}\sec^{-1} x = \dfrac{1}{x\sqrt{x^2-1}}, x > 1$ **27.** $\dfrac{d}{dx}\csc^{-1} x = \dfrac{-1}{x\sqrt{x^2-1}}, x > 1$

28. $\dfrac{d}{dx}\sinh^{-1} x = \dfrac{1}{\sqrt{x^2+1}},$ **29.** $\dfrac{d}{dx}\cosh^{-1} x = \dfrac{\pm 1}{\sqrt{x^2-1}}, \left(x^2 > 1\right),$

30. $\dfrac{d}{dx}\tanh^{-1} x = \dfrac{1}{1-x^2}, \left(x^2 < 1\right),$ **31.** $\dfrac{d}{dx}\coth^{-1} x = \dfrac{-1}{x^2-1}, \left(x^2 > 1\right),$

32. $\dfrac{d}{dx}\operatorname{sech}^{-1} x = \dfrac{\pm 1}{x\sqrt{1-x^2}}, \left(x^2 < 1\right),$ **33.** $\dfrac{d}{dx}\operatorname{cosech}^{-1} x = \dfrac{\pm 1}{x\sqrt{x^2+1}}.$

■ **Integral Formulas** *u* and *v* are functions of *x*; c_1, and c_2 are constants.

1. $\displaystyle\int \left(c_1 u + c_2 v\right) dx = c_1 \int u\,dx + c_2 \int v\,dx + c,$

2. $\displaystyle\int uv'\,dx = uv - \int u'v\,dx + c, \left(\begin{array}{c}\text{integration}\\ \text{by parts}\end{array}\right)$ **3.** $\displaystyle\int x^n dx = \dfrac{x^{n+1}}{n+1} + c \;\; (n \ne -1),$

4. $\displaystyle\int \dfrac{1}{x} dx = \log_e |x| + c,$ **5.** $\displaystyle\int e^{ax} dx = \dfrac{1}{a} e^{ax} + c,$

6. $\displaystyle\int a^x dx = \dfrac{a^x}{\log_e a} + c,$ **7.** $\displaystyle\int \sin ax\, dx = -\dfrac{1}{a}\cos ax + c,$

8. $\displaystyle\int \cos ax\, dx = \dfrac{1}{a}\sin ax + c,$ **9.** $\displaystyle\int \sec^2 x\, dx = \tan x + c,$

10. $\displaystyle\int \operatorname{cosec}^2 x\, dx = -\cot x + c,$ **11.** $\displaystyle\int \sec x \tan x\, dx = \sec x + c$

12. $\displaystyle\int \operatorname{cosec} x \cot x\, dx = -\operatorname{cosec} x + c,$

13. $\displaystyle\int \tan ax\, dx = -\dfrac{1}{a}\log|\cos ax| + c = \dfrac{1}{a}\log|\sec ax| + c,$

14. $\displaystyle\int \cot ax\, dx = \dfrac{1}{a}\log|\sin ax| + c,$

15. $\displaystyle\int \sec ax \, dx = \begin{cases} \dfrac{1}{a}\log|\sec ax + \tan ax| + c \\ \log|\tan(\pi/4 + ax/2)| + c \end{cases}$,

16. $\displaystyle\int \csc x \, dx = \begin{cases} \log|\csc x - \cot x| + c \\ \log\left|\tan\dfrac{x}{2}\right| + c \end{cases}$,

17. $\displaystyle\int \dfrac{dx}{\sqrt{a^2 - x^2}} = \sin^{-1}\dfrac{x}{a} + c$,   **18.** $\displaystyle\int \dfrac{dx}{x^2 + a^2} = \dfrac{1}{a}\tan^{-1}\dfrac{x}{a} + c$,

19. $\displaystyle\int \sinh x \, dx = \cosh x + c$,   **20.** $\displaystyle\int \cosh x \, dx = \sinh x + c$,

21. $\displaystyle\int \dfrac{dx}{\sqrt{x^2 + a^2}} = \begin{cases} \sinh^{-1}\dfrac{x}{a} + c \\ \log_e\left(x + \sqrt{x^2 + a^2}\right) + c, \end{cases}$

22. $\displaystyle\int \dfrac{dx}{\sqrt{x^2 - a^2}} = \begin{cases} \cosh^{-1}\dfrac{x}{a} + c \\ \log_e\left|x + \sqrt{x^2 - a^2}\right| + c \end{cases}$,

23. $\displaystyle\int \dfrac{1}{a^2 - x^2} \, dx = \dfrac{1}{2a}\log\left|\dfrac{x+a}{x-a}\right| + c$,   **24.** $\displaystyle\int \dfrac{dx}{x^2 - a^2} = \dfrac{1}{2a}\log\left|\dfrac{x-a}{x+a}\right| + c$,

25. $\displaystyle\int \sqrt{a^2 - x^2} \, dx = \dfrac{x}{2}\sqrt{a^2 - x^2} + \dfrac{a^2}{2}\sin^{-1}\dfrac{x}{a} + c$,

26. $\displaystyle\int \sqrt{x^2 + a^2} \, dx = \dfrac{x}{2}\sqrt{x^2 + a^2} + \dfrac{a^2}{2}\log\left|x + \sqrt{x^2 + a^2}\right| + c$,

27. $\displaystyle\int \sqrt{x^2 - a^2} \, dx = \dfrac{x}{2}\sqrt{x^2 - a^2} - \dfrac{a^2}{2}\log\left|x + \sqrt{x^2 - a^2}\right| + c$,

28. $\displaystyle\int \log x \, dx = x\log x - x + c$,

29. $\displaystyle\int e^{ax} \sin bx \, dx = \dfrac{e^{ax}}{a^2 + b^2}(a\sin bx - b\cos bx) + c$

30. $\displaystyle\int e^{ax} \cos bx \, dx = \dfrac{e^{ax}}{a^2 + b^2}(a\cos bx + b\sin bx) + c$

Appendix 4
Table I: NORMAL DISTRIBUTION
Areas under the Standard Normal Curve
from 0 to Z

φ (z)

z	0	1	2	3	4	5	6	7	8	9
0.0	0.0000	0.0040	0.0080	0.0120	0.0160	0.0199	0.0239	0.0279	0.0319	0.0359
0.1	0.0398	0.0438	0.0478	0.0517	0.0557	0.0596	0.0636	0.0675	0.0714	0.0754
0.2	0.0793	0.0832	0.0871	0.0910	0.0948	0.0987	0.1026	0.1064	0.1103	0.1141
0.3	0.1179	0.1217	0.1255	0.1293	0.1331	0.1368	0.1406	0.1443	0.1480	0.1517
0.4	0.1554	0.1591	0.1628	0.1664	0.1700	0.1736	0.1772	0.1808	0.1844	0.1879
0.5	0.1915	0.1950	0.1985	0.2019	0.2054	0.2088	0.2123	0.2157	0.2190	0.2224
0.6	0.2258	0.2291	0.2324	0.2357	0.2389	0.2422	0.2454	0.2486	0.2518	0.2549
0.7	0.2580	0.2612	0.2642	0.2673	0.2704	0.2734	0.2764	0.2794	0.2823	0.2852
0.8	0.2881	0.2910	0.2939	0.2967	0.2995	0.3023	0.3051	0.3078	0.3106	0.3133
0.9	0.3159	0.3186	0.3212	0.3238	0.3264	0.3289	0.3315	0.3340	0.3365	0.3389
1.0	0.3413	0.3438	0.3461	0.3485	0.3508	0.3531	0.3554	0.3577	0.3599	0.3621
1.1	0.3643	0.3665	0.3686	0.3708	0.3729	0.3749	0.3770	0.3790	0.3810	0.3830
1.2	0.3849	0.3869	0.3888	0.3907	0.3925	0.3944	0.3962	0.3980	0.3997	0.4015
1.3	0.4032	0.4049	0.4066	0.4082	0.4099	0.4115	0.4131	0.4147	0.4162	0.4177
1.4	0.4192	0.4207	0.4222	0.4236	0.4251	0.4265	0.4279	0.4292	0.4306	0.4319
1.5	0.4332	0.4345	0.4357	0.4370	0.4382	0.4394	0.4406	0.4418	0.4429	0.4441
1.6	0.4452	0.4463	0.4474	0.4484	0.4495	0.4505	0.4515	0.4525	0.4535	0.4545
1.7	0.4554	0.4564	0.4573	0.4582	0.4591	0.4599	0.4608	0.4616	0.4625	0.4633
1.8	0.4641	0.4649	0.4656	0.4664	0.4671	0.4678	0.4686	0.4693	0.4699	0.4706
1.9	0.4713	0.4719	0.4726	0.4732	0.4738	0.4744	0.4750	0.4756	0.4761	0.4767
2.0	0.4772	0.4778	0.4783	0.4788	0.4793	0.4798	0.4803	0.4808	0.4812	0.4817
2.1	0.4821	0.4826	0.4830	0.4834	0.4838	0.4842	0.4846	0.4850	0.4854	0.4857
2.2	0.4861	0.4864	0.4868	0.4871	0.4875	0.4878	0.4881	0.4884	0.4887	0.4890
2.3	0.4893	0.4896	0.4898	0.4901	0.4904	0.4906	0.4909	0.4911	0.4913	0.4916
2.4	0.4918	0.4920	0.4922	0.4925	0.4927	0.4929	0.4913	0.4932	0.4934	0.4936
2.5	0.4938	0.4940	0.4941	0.4943	0.4945	0.4946	0.4948	0.4949	0.4951	0.4952
2.6	0.4953	0.4955	0.4956	0.4957	0.4959	0.4960	0.4961	0.4962	0.4963	0.4964
2.7	0.4965	0.4966	0.4967	0.4968	0.4969	0.4970	0.4971	0.4972	0.4973	0.4974
2.8	0.4974	0.4975	0.4976	0.4977	0.4977	0.4978	0.4979	0.4979	0.4980	0.4981
2.9	0.4981	0.4982	0.4982	0.4983	0.4984	0.4984	0.4985	0.4985	0.4986	0.4986
3.0	0.4987	0.4987	0.4987	0.4988	0.4988	0.4989	0.4989	0.4989	0.4990	0.4990
3.1	0.4990	0.4991	0.4991	0.4991	0.4992	0.4992	0.4992	0.4992	0.4993	0.4993
3.2	0.4993	0.4993	0.4994	0.4994	0.4994	0.4994	0.4994	0.4995	0.4995	0.4995
3.3	0.4995	0.4995	0.4995	0.4996	0.4996	0.4996	0.4996	0.4996	0.4996	0.4997
3.4	0.4997	0.4997	0.4997	0.4997	0.4997	0.4997	0.4997	0.4997	0.4997	0.4998
3.5	0.4998	0.4998	0.4998	0.4998	0.4998	0.4998	0.4998	0.4998	0.4998	0.4998
3.6	0.4998	0.4998	0.4999	0.4999	0.4999	0.4999	0.4999	0.4999	0.4999	0.4999
3.7	0.4999	0.4999	0.4999	0.4999	0.4999	0.4999	0.4999	0.4999	0.4999	0.4999
3.8	0.4999	0.4999	0.4999	0.4999	0.4999	0.4999	0.4999	0.4999	0.4999	0.4999
3.9	0.5000	0.5000	0.5000	0.5000	0.5000	0.5000	0.5000	0.5000	0.5000	0.5000

Appendix 4
Table II: t-DISTRIBUTION
Values of |t| with probability P and degrees of freedom v

P → ↓ v	0.9	0.10	0.05	0.02	0.01
1	0.158	6.314	12.706	31.821	63.657
2	0.142	2.920	4.303	6.965	9.925
3	0.137	2.353	3.182	4.541	5.841
4	0.134	2.132	2.776	3.747	4.604
5	0.132	2.015	2.571	3.365	4.032
6	0.131	1.943	2.447	3.143	3.707
7	0.130	1.895	2.365	2.998	3.496
8	0.130	1.860	2.306	2.896	3.355
9	0.129	1.833	2.262	2.821	3.250
10	0.129	1.812	2.228	2.764	3.169
11	0.129	1.796	2.201	2.718	3.106
12	0.128	1.782	2.179	2.681	3.055
13	0.128	1.771	2.160	2.650	3.012
14	0.128	1.761	2.145	2.624	2.977
15	0.128	1.753	2.131	2.602	2.947
16	0.128	1.746	2.120	2.583	2.921
17	0.128	1.740	2.110	2.567	2.898
18	0.127	1.734	2.101	2.552	2.878
19	0.127	1.729	2.093	2.539	2.861
20	0.127	1.725	2.086	2.528	2.845
21	0.127	1.721	2.080	2.518	2.831
22	0.127	1.717	2.074	2.508	2.819
23	0.127	1.714	2.069	2.500	2.807
24	0.127	1.711	2.064	2.492	2.797
25	0.127	1.708	2.060	2.485	2.787
30	0.127	1.697	2.042	2.457	2.750
40	0.126	1.684	2.021	2.423	2.704
60	0.126	1.671	2.000	2.390	2.660
120	0.126	1.658	1.980	2.358	2.617
∞	0.126	1.645	1.960	2.326	2.576

Appendix 4
Table III: CHI-SQUARE DISTRIBUTION
Values of χ^2 with probability P and dof-ν

$P \rightarrow$ $\downarrow \nu$	0.99	0.95	0.10	0.05	0.02	0.01
1	0.0002	0.004	2.706	3.841	5.412	6.635
2	0.0201	0.103	4.605	5.991	7.824	9.210
3	0.115	0.352	6.251	7.815	9.837	11.345
4	0.297	0.711	7.779	9.488	11.668	13.277
5	0.554	1.145	9.236	11.070	13.388	15.086
6	0.872	1.635	10.645	12.592	15.033	16.812
7	1.238	2.167	12.017	14.067	16.622	18.475
8	1.646	2.733	13.362	15.507	18.168	20.090
9	2.088	3.325	14.684	16.919	19.670	21.666
10	2.558	3.940	15.987	18.307	21.161	23.209
11	3.053	4.575	17.275	19.675	22.618	24.725
12	3.571	5.226	18.549	21.026	24.054	26.217
13	4.107	5.892	19.812	22.362	25.472	27.688
14	4.660	6.571	21.064	23.685	26.873	29.141
15	5.229	7.261	22.307	24.996	28.259	30.578
16	5.812	7.962	23.542	26.296	29.633	32.000
17	6.408	8.672	24.768	27.587	30.995	33.409
18	7.015	9.390	25.989	28.869	32.346	34.805
19	7.633	10.117	27.204	30.114	33.687	36.191
20	8.260	10.851	28.412	31.410	35.020	37.566
21	8.897	11.581	29.615	32.671	36.343	38.932
22	9.542	12.338	30.813	33.924	37.659	40.289
23	10.196	13.091	32.007	35.172	38.968	41.638
24	10.856	13.848	33.196	36.415	40.270	42.980
25	11.524	14.611	34.382	37.652	41.566	44.314
26	12.198	15.379	35.563	38.885	42.856	45.642
27	12.879	16.151	36.741	40.113	44.140	46.963
28	13.565	16.928	37.916	41.337	45.419	48.278
29	14.256	17.708	39.087	42.557	46.693	49.558
30	14.953	18.493	40.256	43.773	47.962	50.892

Appendix 4
Table IV: F-DISTRIBUTION WITH (v_1, v_2)
DEGREES OF FREEDOM
F-Table: 5% and 1% level of significance

$v_1 \rightarrow$ $\downarrow v_2$	1	2	3	4	5	6	8	12	24	∞
1	161.4	199.5	215.7	224.6	230.2	234.0	238.9	243.9	249.0	253.4
	4052	4999	5403	5625	5764	5859	5981	6106	6234	6366
2	18.51	19.00	19.16	19.25	19.30	19.33	19.87	19.41	19.45	19.50
	98.49	99.01	99.17	99.25	99.30	99.33	99.36	99.42	99.46	99.50
3	10.13	9.55	9.28	9.12	9.01	8.94	8.84	8.74	8.64	8.53
	34.12	30.81	29.46	28.71	28.24	27.91	27.49	27.05	26.60	26.12
4	7.71	6.94	6.59	6.39	6.26	6.16	6.04	5.91	5.77	5.63
	21.20	18.00	16.69	15.98	15.52	15.21	14.80	14.37	13.93	13.46
5	6.61	5.79	5.41	5.19	5.05	4.95	4.82	4.68	4.53	4.36
	16.26	13.27	12.06	11.39	10.97	10.67	10.27	9.89	9.47	9.02
6	5.99	5.14	4.76	4.53	4.39	4.28	4.15	4.00	3.84	3.67
	13.74	10.92	9.78	9.15	8.75	8.47	8.10	7.72	7.31	6.88
7	5.59	4.74	4.35	4.12	3.97	3.87	3.73	3.57	3.41	3.23
	12.25	9.55	8.45	7.85	7.46	7.19	6.84	6.47	6.07	5.65
8	5.32	4.46	4.07	3.84	3.69	3.58	3.44	3.28	3.12	2.93
	11.26	8.65	7.59	7.01	6.63	6.37	6.03	5.67	5.28	4.86
9	5.12	4.26	3.86	3.63	3.48	3.37	3.23	3.07	2.90	2.71
	10.56	8.02	6.99	6.42	6.06	5.80	5.47	5.11	4.73	4.31
10	4.96	4.10	3.71	3.48	3.33	3.22	3.07	2.91	2.74	2.54
	10.04	7.56	6.55	5.99	5.64	5.39	5.06	4.71	4.33	3.91
12	4.75	3.88	3.49	3.26	3.11	3.00	2.85	2.69	2.50	2.30
	9.33	6.93	5.95	5.41	5.06	4.82	4.50	4.16	3.78	3.36
14	4.60	3.74	3.34	3.11	2.96	2.85	2.70	2.53	2.35	2.13
	8.86	6.51	5.56	5.03	4.69	4.46	4.14	3.80	3.43	3.00
16	4.49	3.63	3.24	3.01	2.85	2.74	2.59	2.42	2.24	2.01
	8.53	6.23	5.29	4.77	4.44	4.20	3.89	3.55	3.18	2.75
18	4.41	3.55	3.16	2.93	2.77	2.66	2.51	2.34	2.15	1.92
	8.28	6.01	5.09	4.58	4.25	4.01	3.71	3.37	3.01	2.57
20	4.35	3.49	3.10	2.87	2.71	2.60	2.45	2.28	2.08	1.84
	8.10	5.85	4.94	4.43	4.10	3.87	3.56	3.23	2.86	2.42
25	4.24	3.38	2.99	2.76	2.60	2.49	2.34	2.16	1.96	1.71
	7.77	5.57	4.68	4.18	3.86	3.63	3.32	2.99	2.62	2.17
30	4.17	3.32	2.92	2.69	2.53	2.42	2.27	2.09	1.89	1.62
	7.56	5.39	4.51	4.02	3.70	3.47	3.17	2.84	2.47	2.01
40	4.08	3.23	2.84	2.61	2.45	2.34	2.18	2.00	1.79	1.51
	7.31	5.18	4.31	3.83	3.51	3.29	2.99	2.66	2.29	181
60	4.00	3.15	2.76	2.52	2.37	2.25	2.10	1.92	1.70	1.39
	7.08	4.98	4.13	3.65	3.34	3.12	2.82	2.50	2.12	1.60
120	3.92	3.07	2.68	2.45	2.29	2.17	2.02	1.83	1.61	1.25
	6.85	4.79	3.95	3.48	3.17	2.96	2.66	2.34	1.95	1.38
∞	3.84	2.99	2.60	2.37	2.21	2.09	1.94	1.75	1.52	1.00
	6.64	4.60	3.78	3.32	3.02	2.80	2.51	2.18	1.79	1.00

APPENDIX 5

WINTER 2010; AN 209

ENGINEERING MATHEMATICS

1. (a) Find the eigenvalues and corresponding eigenvectors of the matrix

$$A = \begin{pmatrix} 1 & 0 & 0 \\ 0 & 2 & 1 \\ 2 & 0 & 3 \end{pmatrix}$$

Refer Example 16.8 on page 16.9

Ans. 1, 2, 3; $[-1, -1, 1]^T$, $[0, 1, 0]^T$, $[0, -1, 1]^T$.

(b) Evaluate

$$\int_C [(x + y)\, dx - x^2 dy + (y + z)\, dz]$$

where C is $x^2 = 4y$, $z = x$, $0 \le x \le 2$.

See Example 14.49 on page 14.57.

(c) Find the surface area of the solid generated by revolving the curve $x^2 + (y - b)^2 = a^2$ about x-axis.

See Example 9.33 on page 9.31.

2. (a) Find the normal vector and the equation of tangent plane to the surface $z = \sqrt{x^2 + y^2}$ at the point $(3, 4, 5)$. Also, find the divergence of the vector field $(x^2y^2 - z^3)\, i + (2xyz)\, j + (e^{xyz})\, k$.

Solution

(i) Let $f(x, y, z) = z - \sqrt{x^2 + y^2} = 0$ be the surface. Then, the normal vector is given by

$$\nabla f = \left(i\frac{\partial}{\partial x} + j\frac{\partial}{\partial y} + k\frac{\partial}{\partial z} \right)(z - \sqrt{x^2 + y^2})$$

$$= -\frac{x}{\sqrt{x^2 + y^2}}\, i - \frac{y}{\sqrt{x^2 + y^2}}\, j + k = -\frac{x}{z}\, i - \frac{y}{z}\, j + k.$$

At $(3, 4, 5)$, the normal vector is given by

$$\nabla f(3, 4, 5) = -\frac{3}{5}\, i - \frac{4}{5}\, j + k.$$

The tangent plane at the point $(3, 4, 5)$ is given by

$$-\frac{3}{5}(x - 3) - \frac{4}{5}(y - 4) + (z - 5) = 0 \text{ or } 3x + 4y - 5z = 0.$$

(ii) Let $v = (x^2 y^2 - z^3) i + (2xyz) j + (e^{xyz}) k$

$$\therefore \text{ div } v = \frac{\partial}{\partial x}(x^2 y^2 - z^3) + \frac{\partial}{\partial y}(2xyz) + \frac{\partial}{\partial z}(e^{xyz})$$

$$= 2xy^2 + 2xz + xye^{xyz}.$$

(b) Verify whether the vectors $x_1 = (1 \ \ 3 \ \ 4 \ \ 2)$, $x_2 = (3 \ \ -5 \ \ 2 \ \ 2)$ and $x_3 = (2 \ -1 \ \ 3 \ \ 2)$ are linearly-dependent. If so, express one of these vectors as a linear combination of others.

Refer Example 15.22(a) on page 15.63. **Ans.** $x_1 = 2x_3 - x_2$.

(c) Find the total length of the curve $r = a \sin^3(\theta/3)$.

See Example 7.9(c) on page 7.12.

3. (a) Discuss the convergence of the following series:

(i) $1 + \dfrac{2!}{2^2} + \dfrac{3!}{3^3} + \dfrac{4!}{4^4} + \dots.$

Solution. Given series is $\Sigma u_n = \displaystyle\sum_{n=1}^{\infty} \frac{n!}{n^n}$.

Here $\dfrac{u_n}{u_{n+1}} = \dfrac{n!}{n^n} \cdot \dfrac{(n+1)^{n+1}}{(n+1)!} = \dfrac{n!(n+1)^n.(n+1)}{n^n.(n+1)\,n!}$

$$= \frac{(n+1)^n}{n^n} = \left(1 + \frac{1}{n}\right)^n.$$

Therefore, $\displaystyle\lim_{n \to \infty} \frac{u_n}{u_{n+1}} = \lim_{n \to \infty} \left(1 + \frac{1}{n}\right)^n = e$, which is > 1.

Hence the given series is convergent.

(ii) $\displaystyle\sum_{n=1}^{\infty} \frac{1}{\sqrt{n} + \sqrt{n+1}}$

Ans. Divergent.

(b) Apply Green's theorem to evaluate

$$\int_C [(2x^2 - y^2)\, dx + (x^2 + y^2)dy]$$

where C is the boundary of the area enclosed by the x-axis and upper-half of the circle.

$$x^2 + y^2 = a^2.$$

Solution. Here $P(x, y) = 2x^2 - y^2, Q(x, y) = x^2 + y^2$.

By Green's theorem

$$\oint_C (2x^2 - y^2)dx + (x^2 + y^2)\,dy = \oint_C P(x, y)dx + Q(x, y)dy$$

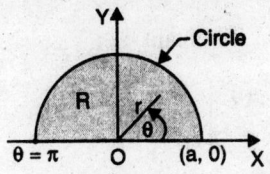

Fig. 1.

$$= \iint_R \left(\frac{\partial Q}{\partial x} - \frac{\partial P}{\partial y} \right) dxdy = \iint_R \left[\frac{\partial}{\partial x}(x^2 + y^2) - \frac{\partial}{\partial y}(2x^2 - y^2) \right] dxdy$$

$$= 2 \iint_R (x + y) \, dxdy, \text{ where } R \text{ is the region shown in Fig. 1.}$$

$$= 2 \int_{r=0}^{a} \int_{\theta=0}^{\pi} r(\cos \theta + \sin \theta).r \, d\theta \, dr$$

[Changing to polar coordinates (r, θ), r varies from 0 to a and θ varies from 0 to π]

$$= 2 \int_{r=0}^{a} \int_{\theta=0}^{\pi} r^2 (\cos \theta + \sin \theta) \, d\theta dr = 2 \int_{r=0}^{a} r^2 [\sin \theta - \cos \theta]_0^\pi \, dr$$

$$= 4 \left[\frac{r^3}{3} \right]_0^a = \frac{4}{3} a^3 .$$

(c) Find the value(s) of λ for which the equations

$(\lambda - 1) x + (3\lambda + 1) y + 2\lambda z = 0$

$(\lambda - 1) x + 2 (2\lambda - 1) y + (\lambda + 3) z = 0$

$2x + (3\lambda + 1) y + 3(\lambda - 1) z = 0$

have non-trivial solution(s). Find the ratios $x : y : z$ when λ has the smallest of these values. What happens when λ has the greatest of these values?

Solution. For the given system of equations to have a non-trivial solution, the determinant of the coefficient matrix should be zero.

That is, $\begin{vmatrix} \lambda - 1 & 3\lambda + 1 & 2\lambda \\ \lambda - 1 & 4\lambda - 2 & \lambda + 3 \\ 2 & 3\lambda + 1 & 3(\lambda - 1) \end{vmatrix} = 0$ (operate $R_2 \to R_2 - R_1$)

or, $\begin{vmatrix} \lambda - 1 & 3\lambda + 1 & 2\lambda \\ 0 & \lambda - 3 & 3 - \lambda \\ 2 & 3\lambda + 1 & 3(\lambda - 1) \end{vmatrix} = 0$ (operate $C_3 \to C_3 + C_2$)

or, $\begin{vmatrix} \lambda - 1 & 3\lambda + 1 & 5\lambda + 1 \\ 0 & \lambda - 3 & 0 \\ 2 & 3\lambda + 1 & 2(3\lambda - 1) \end{vmatrix} = 0$

Expanding along R_2, we obtain

$$(\lambda - 3)\begin{vmatrix} \lambda - 1 & 5\lambda + 1 \\ 2 & 2(3\lambda - 1) \end{vmatrix} = 0, \text{ or } 2(\lambda - 3)[(\lambda - 1)(3\lambda - 1) - (5\lambda + 1)] = 0$$

or $6\lambda(\lambda - 3)^2 = 0$ or $\lambda = 0, 3$.

(i) When $\lambda = 0$, equations becomes

$-x + y = 0, -x - 2y + 3z = 0, 2x + y - 3z = 0$. On solving these equations, we get $x = y = z$. Thus $x : y : z = 1 : 1 : 1$.

(ii) When $\lambda = 3$, equations becomes identical.

4. (a) Find the shortest distance between the line $2x + y = 10$ and the ellipse $9x^2 + 4y^2 = 36$.

See Example 12.59 on page 12.69.

(b) Evaluate the integral

$$\int\int_R e^{x^2} \, dx \, dy$$

where the region R is given by $R : 2y \le x \le 2$ and $0 \le y \le 1$.

See Q. 17 on page 13.23.

5. (a) Solve the following differential equations:

(i) $(1 + y^2) \, dx = (\tan^{-1} y - x) \, dy$

See Example 17.20 on page 17.24.

(ii) $\dfrac{d^2 y}{dx^2} - 2\dfrac{dy}{dx} + y = xe^x \sin x$

See Example 18.17(iv) on page 18.28.

(b) Use difference operator technique to get missing values in the following table:

x	45	50	55	60	65
y	30	–	2.0	–	-2.4

Refer Example 22.11 on page 22.12.

Ans. $f(50) = 2.925, f(60) = 0.225$.

6. (a) Solve

$$\frac{dy}{dx} + \frac{y \cos x + \sin y + y}{\sin x + x \cos y + x} = 0, \, y = 0 \text{ when } x = 0.$$

See Example 17.52 on page 17.54.

(b) Solve the equation

$$y'' + 3y' + 2y = 2e^x$$

by using the method of variation of parameters.

Ans. $y(x) = c_1 e^{-x} + c_2 e^{-2x} + (e^x/3)$.

(c) Find the Fourier series expansion of the periodic function
$$f(x) = x, \quad -\pi \le x \le \pi, \quad f(x + 2\pi) = f(x).$$

Ans. Solution: Here the function $f(x)$ is an odd function on $[-\pi, \pi]$. The Fourier coefficients are obtained as follows:

$$a_0 = \frac{1}{\pi} \int_{-\pi}^{\pi} x \, dx = 0.$$

$$a_n = \frac{1}{\pi} \int_{-\pi}^{\pi} x \cos nx \, dx = 0. \quad (x \cos nx \text{ is an odd function on } [-\pi, \pi].$$

$$b_n = \frac{1}{\pi} \int_{-\pi}^{\pi} x \sin nx \, dx = \frac{2}{\pi} \int_{0}^{\pi} x \sin nx \, dx,$$

$$(x \sin nx \text{ is an even function on } [-\pi, \pi].$$

$$= \frac{2}{\pi} \left[-x \left(\frac{\cos nx}{n} \right) + \left(\frac{\sin nx}{n^2} \right) \right]_{0}^{\pi}$$

$$= \frac{2}{\pi} \left[\frac{-\pi \cos n\pi}{n} \right] = \frac{2}{n}(-\cos n\pi) = \frac{2}{n}(-1)^{n+1}.$$

Therefore, the Fourier expansion of the given function on

$[-\pi, \pi]$ is given by $f(x) = \sum_{n=1}^{\infty} b_n \sin nx$

That is, $x = 2 \left[\sin x - \frac{1}{2} \sin 2x + \frac{1}{3} \sin 3x - \frac{1}{4} \sin 4x + ... \right].$

7. (a) Find the cubic polynomial which takes the following values:

x	0	1	2	3
$f(x)$	1	2	1	10

Hence or otherwise, evaluate $f(4)$.

Refer Example 22.18 on page 22.22.

Ans. $P_3(x) = 2x^3 - 7x^2 + 6x + 1, f(4) = 41.$

(b) In a normal distribution, 31% of the items are under 45 and 8% are over 64. Find the mean and standard deviation of the distribution.

[Given: $P(z \ge 0.5) = 0.31$, $P(z \le 1.4) = 0.92$, where z is standard normal variate.]

See Example 24.131(b) on page 24.136.

(c) Given that

x	1.1	1.2	1.3	1.4	1.5	1.6
y	8.403	8.781	9.129	9.451	9.750	10.031

Find dy/dx at $x = 1.1$.

See Example 23.24 on page 23.37.

8. (a) The data given below show the test scores made by 10 salesmen on an intelligence test and their weekly sales:

Salesman	1	2	3	4	5	6	7	8	9	10
Test Score	40	70	50	60	80	50	90	40	60	60
Sales (in 'ooo)	2.5	6.0	4.5	5.0	4.5	2.0	5.5	3.0	4.5	3.0

Calculate the regression line of sales on test scores and estimate the most probable weekly sales if a salesman makes a score of 70.

Solution: First we find the mean of test scores, x, and mean of sales, y, by "Assumed mean method". We prepare the following table:

Test scores x	Sales y	$d_x = x_i - A$	$d_y = y_i - B$	$(d_x)(d_y)$	d_x^2	d_y^2
40	2.5	−20	−2.0	40	400	4
70	6.0	10	1.5	15	100	2.25
50	B→4.5	−10	0	0	100	0
A→60	5.0	0	0.5	0	0	0.25
80	4.5	20	0	0	400	0
50	2.0	−10	−2.5	25	100	6.25
90	5.5	30	1.0	30	900	1
40	3.0	−20	−1.5	30	400	2.25
60	4.5	0	0	0	0	0
60	3.0	0	−1.5	0	0	2.25

$\Sigma d_x = 0$ $\Sigma d_y = -4.5$ $\Sigma d_x d_y = 140$ $\Sigma d_x^2 = 2400$ $\Sigma d_y^2 = 18.25$

Therefore

$$\bar{x} = A = \frac{\Sigma d_x}{n} = 60 + (0/10) = 60, \quad \bar{y} = B + \frac{\Sigma d_y}{n} = 4.5 + (-4.5/10) = 4.05.$$

Regression line of sales (y) on scores (x) is given by

$$y - \bar{y} = r\left(\frac{\sigma_y}{\sigma_x}\right)(x - \bar{x}) = b_{yx}(x - \bar{x}), \quad \text{where } b_{yx} \text{ is the regression coefficient of y on x. It is obviously, the slope of this line.}$$

or, $$y - \bar{y} = \frac{n\Sigma d_x d_y - \Sigma d_x \Sigma d_y}{n\Sigma d_x^2 - (\Sigma d_x)^2}(x - \bar{x}).$$

Substituting the values, we obtain

$$y - 4.05 = \frac{10(140) - 0}{10(2400) - 0}(x - 60)$$

= 0.06 $(x - 60)$ or $y = 0.06x + 0.45$, the required regression line.

For $x = 70$, $y = 0.06(70) + 0.45 = 4.65$.

(b) Given the values:

x	5	7	11	13	17
$f(x)$	150	392	1452	2366	5202

Evaluate $f(9)$ using Lagrange's formula.

See Example 22.31 on page 22.44.

(c) Evaluate

$$\int_0^6 dx / (1 + x^2)$$

by using Trapezoidal rule. Take $h = 1$.

Refer Example 23.10 on page 23.14. **Ans.** 1.4108.

9. Answer the following:

(i) A binomial distribution has mean 20 and standard deviation 4. Find the parameters of the distribution.

Solution: Let X be a binomial variate with parameters n and p. We know

Mean $= np$ and standard deviation $= \sqrt{npq}$.

Therefore, $\dfrac{npq}{np} = \dfrac{16}{20}, \Rightarrow q = 0.8. \ p = 1 - q = 1 - 0.8 = 0.2$ and $n = 20/p$

$= 20/0.2 = 100$. Thus parameters n and p are 100 and 1/5 respectively.

(ii) Verify whether the vector field

$$v = xyz \ (yz\mathbf{i} + xz\mathbf{j} + xy\mathbf{k})$$

is conservative.

Refer Example 14.55 on page 14.62. **Ans.** Yes.

(iii) Find the value(s) of k for which the vectors $(1, k, 5)$, $(1, -3, 2)$, $(2, -1, 1)$ will form a basis in R^3.

See Example 15.79 on page 15.106.

(iv) Verify whether the matrix $\begin{pmatrix} 2 & 1 \\ 1 & 0 \end{pmatrix}$ is diagonalizable.

See Q. 31 on page 16.40. **Ans.** Yes.

(v) Consider the function $f(x, y) = \tan^{-1} (y/x)$. Find the value of $x\dfrac{\partial f}{\partial x} + y\dfrac{\partial f}{\partial y}$.

Ans. Zero.

(vi) Evaluate $\lim_{x \to 0}$ $[\log_e (1 + x)]/\sin x$.

Solution. $\lim\limits_{x \to 0} \dfrac{[\log_e (1 + x)]}{\sin x}$ $\left(\dfrac{0}{0} \text{ form}\right) = \lim\limits_{x \to 0} [1/(1 + x)]/\cos x = 1.$

(vii) Find the coefficient of x^3 in the expansion of $e^{\sin x}$.

See Q. 15 on page 2.47. **Ans.** Zero.

(viii) Find the Laplace transform of the function $e^{-2t}.t^3$.

Ans. By the translation theorem, we obtain

$$L\{e^{-2t}.t^3\} = \frac{3!}{(s + 2)^{3+1}} = \frac{6}{(s + 2)^4}.$$

(ix) Obtain the function whose first difference is $9x^2 + 11x + 5$.

Refer Example 22.16 on page 22.16. **Ans.** $3x^3 + x^2 + x + c.$

(x) Find complementary function of the differential equation

$$\frac{d^2y}{dx^2} - 2\frac{dy}{dx} + y = \sec 2x.$$

Ans. $y_c(x) = (c_1 + c_2 x) e^x.$